Gene and Cell Therapy

Therapeutic Mechanisms and Strategies

Third Edition

Gene and Cell Therapy

Therapeutic Mechanisms and Strategies

Third Edition

Edited by **Nancy Smyth Templeton**

CRC Press
Taylor & Francis Group
Boca Raton   London   New York

CRC Press is an imprint of the
Taylor & Francis Group, an **informa** business

CRC Press
Taylor & Francis Group
6000 Broken Sound Parkway NW, Suite 300
Boca Raton, FL 33487-2742

© 2009 by Taylor & Francis Group, LLC
CRC Press is an imprint of Taylor & Francis Group, an Informa business

No claim to original U.S. Government works
Printed in the United States of America on acid-free paper
10 9 8 7 6 5 4 3 2 1

International Standard Book Number-13: 978-0-8493-8768-5 (Hardcover)

Library of Congress Cataloging-in-Publication Data

Gene and cell therapy : therapeutic mechanisms and strategies / editor, Nancy Smyth Templeton. -- 3rd ed.
    p. ; cm.
  Includes bibliographical references and index.
  ISBN 978-0-8493-8768-5 (alk. paper)
  1. Gene therapy. 2. Cellular therapy. I. Templeton, Nancy Smyth, 1952- II. Title.
  [DNLM: 1. Gene Therapy--methods. 2. Cell Transplantation--methods. QZ 52 G3252 2009]

RB155.8.G468 2009
615.8'95--dc22
                                               2008011642

Visit the Taylor & Francis Web site at
http://www.taylorandfrancis.com

and the CRC Press Web site at
http://www.crcpress.com

Contents

PART I   Delivery Systems and Therapeutic Strategies

PART II   *Other Therapeutic Strategies*

PART III   Gene Expression and Detection

PART IV   Disease Targets and Therapeutic Strategies

PART V Clinical Trials and Regulatory Issues

Foreword

BIRTH AND INFANCY OF GENE THERAPY

Major and daring new developments in biomedicine and particularly in clinical studies are often circuitous and beset with the need to understand mechanisms of the inevitable setbacks to refine concepts, invent new tools, and set new directions. In the modern era of medical science, this was certainly true of indispensable modern therapeutic methods such as bone marrow transplantation, many aspects of cancer chemotherapy, the application of monoclonal antibodies to improved diagnosis and therapy, and many other fields. As these breakthroughs evolved from laboratory curiosities to effective therapies, there were long periods of setbacks and clinical failures, harm to patients so deep that at times paralyzed scientific investigation and fueled public doubts about the validity of the approaches. For instance, human clinical applications of bone marrow transplantation first appeared late in the 1950s but it was not until well into the mid-1970s that the survival rate of transplant patients with a variety of hematological diseases rose to above 1%–2%. Chemotherapy for the uniformly fatal childhood lymphocytic leukemia began in 1948 and has now achieved a cure rate of well over 80% or better. But such successes did not happen overnight, and it took more than 15 years or so before the cure rate climbed over 8%–10%. Diagnostic and therapeutic applications of monoclonal antibody methodology had a promising and even a splashy birth in the early 1970s but then virtually vanished for more than a decade because of failures and setbacks. It is only now assuming a prominent place in clinical application.

Although the concepts of human gene therapy emerged first early in the 1970s, it was not possible to apply these ideas readily to experimental testing let alone to human clinical use because of the paucity of well-developed efficient and safe methods of gene transfer. With the development of the retrovirus vectors in the early 1980s, the impetus for rapid delivery to human clinical needs became irresistible and human clinical studies began in 1989–1990. Over the next few years, to great fanfare and soaring expectations, many additional clinical studies were undertaken, many new vectors systems were developed, successes were often thought to be imminent, an entire new biotechnology industry was born, and claims were even made for clinical success.

We are all much wiser now and we realize the errors of those early days during which gene therapy was simplistically imagined to be straightforward. It was not to be, and we have learned and continue to learn the many ways in which the current tools available to us in the clinical setting for genetic attack on disease are still immature and in need of refining and a great deal of troubleshooting. But one cannot troubleshoot unless there is trouble, and of course there have been a number of very prominent adverse events in gene therapy—induction of childhood leukemia in the course of successful treatment for childhood immunodeficiency, patient deaths in some studies, and abandoned clinical studies in the face of troubling setbacks with other disease studies. Based on the experience with transplantation methods for bone marrow and other organs, chemotherapy, monoclonal antibody technology, and many other aspects of modern medicine, we now should realize that such a history is to be expected and the expected evolution of difficult new technologies should not be interpreted to be a "failure," as happens disturbingly often in the public media, some very influential members of which seem to have little understanding of the ways in which biomedicine progresses and apparently less of an appetite for learning and illuminating this history for the public.

Gene therapy must now be seen in historical terms to have taken a route from concept to unequivocal clinical success that is typical of new biomedicine. And there has been clinical success even with the admittedly imperfect tools so far available. Children with X-linked severe combined immunodeficiency (X-SCID) and adenosine deaminase deficiency-severe combined immunodeficiency (ADA-SCID) have received real therapy and most of these treated children, but sadly not all, are now leading normal childhood lives [1,2]. Therapy does not mean uncomplicated success in every case—no therapy promises or delivers this. Nevertheless, there is little doubt that additional clinical successes are imminent—forms of cancer, genetic and degenerative diseases of the eye, and others.

What is needed for the next phase for making gene therapy an established and broadly utilized method of treatment, not just occasional proof-of-principle clinical studies, but rather for far broader application for many more common human diseases? How can the still sparse clinical successes become treatments of choice for more of the human health scourges rather than occasional curiosities? One might begin to argue that the field has indeed achieved such a status, at least for diseases such as ADA-SCID and possibly X-SCID. But the needs are still enormous for most other target diseases. As is so thoroughly outlined in this book, a prime need is the development of ever more efficient, safe, and targeted methods of delivery of genes and other elements designed to modify expression of resident genes. We often pay lip service to the notion that there will be no magic bullet and universally useful vectors: there will be a need for many kinds of viral and nonviral vectors tailor-fit for specific clinical needs. But at the same time, we all can occasionally become so enamored with some systems even to the point of shifting developmental resources from important systems. For instance, there has recently been a palpable embrace of clinical applications of the adeno-associated virus (AAV) vector system as the most current promising one for many clinical applications. As important as that system is, resources for production and clinical application of other important delivery systems have been reduced far too much to the point of endangering important development and early clinical trials with other vector systems. This lesson may have come all the more pertinent with the recent death of a patient in the United States who was suffering from severe but not immediately life-threatening rheumatoid arthritis and who was enrolled in a study using direct joint delivery of an AAV vector expressing an inhibitor of tumor necrosis factor alpha (TNF-α). The causes of death in this case [3] have not yet been clarified but the bulk of the current evidence points to the likelihood that the patient was already immunocompromised by other earlier and concomitant treatments, that she was thereby susceptible to the overwhelming opportunistic infection that was the immediate cause of her death, and that the AAV vector system or the transgene played little if any role in the adverse event. Although these are not yet proven, the tragic event does serve to remind us that we must not become complacent about the potential for disruption of important homeostatic systems from any gene transfer manipulation in human patients and experimental subjects and that even this vector delivery system, generally thought to be very safe, must be understood in greater detail than we presently do. And so we need better vectors.

Wide-scale gene therapy also would be greatly enhanced by a major effort to develop clinical applications for the powerful new methods for modifying endogenous gene expression with RNA interference methods, and for genetic sequence correction through zinc-finger or other site-specific recombination approaches rather than relying entirely on genome augmentation with foreign transgenes to complement aberrant functions. The recent discovery of the powerful and ubiquitous role of RNAs in regulating normal endogenous gene expression will enable many therapeutic approaches to disease treatment. These methods receive appropriately prominent discussion in this book.

Of course, possibly the most important element in making gene therapy more widely effective is the need to achieve a far deeper understanding of disease and its pathogenesis. In many cases, even for monogenic disease, we understand too little to devise the most powerful genetic means to intervene.

This book is an attempt to put together many of the technical advances in gene delivery, in new approaches to the regulation and modification of gene expression, and in its clinical application. The underlying understanding of this volume is that the new intellectual achievement represented by gene therapy requires years and that the road toward that goal will not always be smooth. New technology is and will continue to be needed, if only it were otherwise, as in the case of Athena, the Greek goddess of so many areas of human intellect and work—wisdom, war, justice, the arts, industriousness, weaving, patroness of craftsmanship and diplomacy, inventor of the flute, and more. She had so very much to do and, amazingly, she was fully armed and prepared for all of her tasks even as she sprang at birth from the head of her father Zeus. She needed no period of learning; her tasks and her tools must have all been so clear and straightforward for her. Unfortunately, most human accomplishments are not as clearly defined and we usually require long periods of learning, errors, missteps, and relearning. Gene therapy is one such human endeavor and I am sure that Athena would have understood this and has been patient with us. Even so, we can all take great pleasure and pride in being present and even taking part in the birth and early rearing of this nascent field of medicine. Its future development will be exciting and important.

Theodore Friedmann
Department of Pediatrics
School of Medicine, University of California, San Diego

REFERENCES

1. Thrasher AJ, Gaspar HB, Baum C, Modlich U, Schambach A, Candotti F, Otsu M, Sorrentino B, Scobie L, Cameron E, Blyth K, Neil J, Abina SH, Cavazzana-Calvo M, Fischer A. Gene therapy: X-SCID transgene leukaemogenicity. *Nature* 443: E5–E7.
2. Gaspar HB, Bjorkegren E, Parsley K, Gilmour KC, King D, Sinclair J, Zhang F, Giannakopoulos A, Adams S, Fairbanks LD, Gaspar J, Henderson L, Xu-Bayford JH, Davies EG, Veys PA, Kinnon C, Thrasher AJ. Successful reconstitution of immunity in ADA-SCID by stem cell gene therapy following cessation of PEG-ADA and use of mild preconditioning. *Mol. Ther.* 14:505–513 (2006).
3. Friedmann, T. Paying attention to the trees, not the forest. Hastings Forum, August 1, 2007.

Preface

Since the publication of the first edition of this book in 2000, gene therapy and cell therapy clinical trials have yielded some remarkable successes and some disappointing failures. In many research laboratories, valuable insights have been gained that will enable us to overcome many of the difficulties encountered to date and to move these therapies toward routine use in the clinic. As with any new therapy, such as heart transplantation, many failures must be expected before treatment becomes routine. Long-term success requires that we rationally assess each failure and determine what additional knowledge is required to solve each problem we encounter. Once again, the authors of this third edition have provided thoughtful insights into the challenges that must be met for specific applications of gene and cell therapies.

The ongoing development of these therapies draws from many disciplines including cell biology, virology, molecular biology, medicine, genetics, immunology, biochemistry, physiology, chemistry, biophysics, molecular imaging, microbiology, pharmacology, toxicology, and others. This book includes topics in each of these areas and has been planned to facilitate its ongoing use as a textbook for classes in gene and cell therapies and gene transfer. The chapters are written so that readers from all backgrounds including students, physicians, scientists, and other interested persons can understand the current status of gene and cell therapies and how these diverse disciplines contribute to these fields. Several contributors have broadened their introductions to facilitate use of this edition as an educational textbook and have contributed information to provide context for their focused research efforts. Hopefully, this volume will provide readers with a broad knowledge of the aspects and tools available in the evolving fields of gene therapy and cell therapy.

Nancy Smyth Templeton

Editor

Dr. Nancy Smyth Templeton obtained a PhD in molecular biology and biochemistry and had her postdoctoral stint at the National Institutes of Health, Bethesda, Maryland. Currently, she is a tenure-track assistant professor in the Department of Molecular and Cellular Biology, a member of the MCB graduate faculty, and a principal investigator. She also has a secondary appointment in the Department of Molecular Physiology and Biophysics and at the Center for Cell and Gene Therapy.

Dr. Templeton has created and patented novel liposomes for increased systemic delivery of nucleic acids at the NIH, a technology licensed by several companies, for research and for therapeutics. She has served on numerous review panels, study sections, professional committees, and also as an ad hoc reviewer for several peer-reviewed journals. She has edited two books and has 48 publications to her credit. She is a member of many professional societies and has presented over 137 seminars as an invited speaker or moderator including national and international meetings. She is a consultant for several biotech and pharmaceutical companies and also has worked in biotech companies. She has worked through sponsored research agreements for several companies including her Baylor start-up company and has two issued patents and several patent applications submitted. Several investigators at the Baylor College of Medicine, the M.D. Anderson Cancer Center, and other institutions have been and will use Dr. Templeton's nonviral therapeutics in upcoming Phase I/Phase II clinical trials for the treatment of lung, breast, head and neck, and pancreatic cancers, and hepatitis B and C infections.

Dr. Templeton has also mentored numerous students, medical students, and postdoctoral fellows, and has taught classes for first-year medical students. She has also received two Fulbright and Jaworski Faculty Excellence awards in 2005 and in 2007 for creating enduring educational materials and for teaching and evaluation, respectively.

Contributors

Kari J. Airenne
University of Kuopio
Kuopio, Finland

Nikiforos Ballian
The University of Wisconsin Hospital
  and Clinics
Madison, Wisconsin

David L. Bartlett
University of Pittsburgh
Pittsburgh, Pennsylvania

Christopher Baum
Hannover Medical School
Hannover, Germany

C. Frank Bennett
ISIS Pharmaceuticals
Carlsbad, California

Jean Bennett
University of Pennsylvania
Philadelphia, Pennsylvania

Kristian Berg
Institute for Cancer Research
Oslo, Norway

Lilia Bi
Food and Drug Administration
Rockville, Maryland

Sandip Biswal
Stanford School of Medicine
Stanford, California

Catherine M. Bollard
Baylor College of Medicine
Houston, Texas

Anette Bonsted
Institute for Cancer Research
Oslo, Norway

William J. Bowers
University of Rochester School
  of Medicine and Dentistry
Rochester, New York

Karsten Brand
Institute of General Pathology
University Hospital Heidelberg
Heidelberg, Germany

Malcolm K. Brenner
Baylor College of Medicine
Houston, Texas

Nicola Brunetti-Pierri
Baylor College of Medicine
Houston, Texas

F. Charles Brunicardi
Baylor College of Medicine
Houston, Texas

William Buitrago
Baylor College of Medicine
Houston, Texas

Haim Burstein
Targeted Genetics
  Corporation
Seattle, Washington

Michele P. Calos
Stanford School of Medicine
Stanford, California

Matthew C. Canver
University of Pennsylvania
Philadelphia, Pennsylvania

Barrie J. Carter
Targeted Genetics
  Corporation
Seattle, Washington

Lawrence Chan
Baylor College of Medicine
Houston, Texas

Ana S. Coroadinha
ITQB/IBET
Oeiras, Portugal

Pedro E. Cruz
ITQB/IBET
Oeiras, Portugal

Ronald G. Crystal
Weill Medical College of Cornell
  University
New York, New York

David T. Curiel
University of Alabama
  at Birmingham
Birmingham, Alabama

Ming-Shen Dai
Queen Mary's School of Medicine
  and Dentistry
London, United Kingdom

Mu-Shui Dai
Indiana University School
  of Medicine
Indianapolis, Indiana

Antonin R. de Fougerolles
Alnylam Pharmaceuticals, Inc.
Cambridge, Massachusetts

Giuseppe De Rosa
Università degli Studi di Napoli
  Federico
Naples, Italy

Jason DeRouchey
National Institutes of Health
Bethesda, Maryland

Suresh de Silva
University of Rochester School
  of Medicine and Dentistry
Rochester, New York

Gianpietro Dotti
Baylor College of Medicine
Houston, Texas

Ruxandra Draghia-Akli
VGX Pharmaceuticals
The Woodlands, Texas

Christine Dufès
University of Strathclyde
Glasgow, United Kingdom

Dwaine F. Emerich
InCytu, Inc.
Providence, Rhode Island

Nicholas D. Evans
Imperial College
London, United Kingdom

Ruth S. Everett
The University of Arkansas
  for Medical Sciences
Little Rock, Arizona

Elias Fattal
School of Pharmacy
Châtenay-Malabry, France

Howard J. Federoff
Georgetown University
  Medical Center
Washington, DC

John M. Felder III
Baylor College of Medicine
Houston, Texas

D. J. Fink
University of Michigan and VA Ann
  Arbor Healthcare System
Ann Arbor, Michigan

A. R. Frampton, Jr.
University of Pittsburgh School
  of Medicine
Pittsburgh, Pennsylvania

William A. Frazier
Washington University School
  of Medicine
St. Louis, Missouri

Jayme R. Gallegos
Oregon Health and Science University
Portland, Oregon

Sanjiv Sam Gambhir
Stanford University
Stanford, California

Richard Geary
ISIS Pharmaceuticals
Carlsbad, California

Jean-Pierre Gillet
National Institutes of Health
Bethesda, Maryland

Joel N. Glasgow
University of Alabama at Birmingham
Birmingham, Alabama

J. C. Glorioso
University of Pittsburgh School
  of Medicine
Pittsburgh, Pennsylvania

W. F. Goins
University of Pittsburgh School
  of Medicine
Pittsburgh, Pennsylvania

Margaret A. Goodell
Baylor College of Medicine
Houston, Texas

Michael M. Gottesman
National Cancer Institute
National Institutes of Health
Bethesda, Maryland

Neil R. Hackett
Weill Medical College of Cornell
  University
New York, New York

Peter Hafkemeyer
Kreiskrankenhaus Emmendingen
Emmendingen, Germany

Greg Hardee
ISIS Pharmaceuticals
Carlsbad, California

Cary O. Harding
Oregon Health and Science University
Portland, Oregon

Daniel Harries
The Hebrew University
Jerusalem, Israel

Hansjörg Hauser
HZI
Braunschweig, Germany

Akseli Hemminki
University of Helsinki and Helsinki
  University Central Hospital
Helsinki, Finland

Scott Henry
ISIS Pharmaceuticals
Carlsbad, California

Helen E. Heslop
Baylor College of Medicine
Houston, Texas

Anders Høgset
PCI Biotech AS
Oslo, Norway

Mitchell E. Horwitz
Duke University
Durham, North Carolina

Christine A. Hrycyna
Purdue University
West Lafayette, Indiana

Jeff S. Isenberg
National Institutes of Health
Bethesda, Maryland

Larry G. Johnson
The University of Arkansas
  for Medicine Sciences
Little Rock, Arizona

Pei Lee Kan
University of London School
  of Pharmacy
London, United Kingdom

Elizabeth M. Kang
National Institutes of Health
Bethesda, Maryland

Mark A.F. Kendall
The University of Queensland
Brisbane, Queensland, Australia

Amir S. Khan
VGX Pharmaceuticals
The Woodlands, Texas

Chava Kimchi-Sarfaty
Food and Drug Administration
Bethesda, Maryland

David Kirn
Jennerex Biotherapeutics
San Francisco, California

Olli H. Laitinen
Vactech Oy
Tampere, Finland

Caroline Lee
National University of Singapore
Singapore

Thomas Licht
Schlossberg Clinic
Oberstaufen, Germany

Kuan-Yin Karen Lin
Baylor College
 of Medicine
Houston, Texas

Douglas W. Losordo
Northwestern University Feinberg
 School of Medicine
Chicago, Illinois

Hua Lu
Indiana University School
 of Medicine
Indianapolis, Indiana

Tobias Maetzig
Hannover Medical School
Hannover, Germany

Anssi J. Mähönen
University of Kuopio
Kuopio, Finland

Harry L. Malech
National Institutes of Health
Bethesda, Maryland

Devika Soundara Manickam
Wayne State University
Detroit, Michigan

Marina Mata
University of Michigan
 and VA Ann
Arbor Healthcare System
Ann Arbor, Michigan

J. Andrea McCart
University of Toronto
Toronto, Ontario, Canada

Philip Ng
Baylor College of Medicine
Houston, Texas

Kazuhiro Oka
Baylor College of Medicine
Houston, Texas

Fatma Okur
Baylor College of Medicine
Houston, Texas

Bert W. O'Malley Jr.
University of Pennsylvania
Philadelphia, Pennsylvania

David Oupický
Wayne State University
Detroit, Michigan

V. Adrian Parsegian
National Institutes of Health
Bethesda, Maryland

Ira Pastan
National Institutes of Health
Bethesda, Maryland

Richard W. Peluso
Targeted Genetics
 Corporation
Seattle, Washington

Zhaohui Peng
SiBiono GeneTech Co., Ltd.
Shenzhen, China

Alexander Philipp
Ludwig-Maximilians University
Munich, Germany

Christian Plank
Technische Universität München
Munich, Germany

Rudolf Podgornik
University of Ljubljana
Ljubljana, Slovenia and National
 Institutes of Health
Bethesda, Maryland

Julia M. Polak
Imperial College
London, United Kingdom

Maria Rajecki
University of Helsinki
Helsinki, Finland

Mari Raki
University of Helsinki
Helsinki, Finland

Sunetra Ray
David Geffen School of Medicine,
 University of California,
 Los Angeles
Los Angeles, California

David D. Roberts
National Institutes of Health
Bethesda, Maryland

Teresa Rodrigues
ITQB/IBET
Oeiras, Portugal

Jerome G. Roncalli
Northwestern University
Chicago, Illinois

Dennis R. Roop
Regenerative Medicine and Stem Cell
 Biology
Aurora, Colorado

John J. Rossi
City of Hope Beckman Research
 Institute
Duarte, California

Raphaël F. Rousseau
Baylor College of Medicine
Houston, Texas

Susan Leanne Samson
Baylor College of Medicine
Houston, Texas

Barbara Savoldo
Baylor College of Medicine
Houston, Texas

Axel Schambach
Hannover Medical School
Hannover, Germany

Andreas G. Schätzlein
University of London School
 of Pharmacy
London, United Kingdom

Franz Scherer
Technische Universität München
Munich, Germany

Lisa J. Scherer
City of Hope Beckman Research
 Institute
Duarte, California

Shiri Shinar
National Institutes of Health
Bethesda, Maryland

Stephanie L. Simek
Food and Drug Administration
Rockville, Maryland

Olga Sirin
Baylor College of Medicine
Houston, Texas

Hemult H. Strey
State University of New York
Stony Brook, New York

Richard E. Sutton
Baylor College of Medicine
Houston, Texas

Eric Swayze
ISIS Pharmaceuticals
Carlsbad, California

Nancy Smyth Templeton
Baylor College of Medicine
Houston, Texas

Siok K. Tey
Baylor College of Medicine
Houston, Texas

Christopher G. Thanos
Brown University
Providence, Rhode Island

Lloyd Tillman
ISIS Pharmaceuticals
Carlsbad, California

Jörn Tongers
Northwestern University
Chicago, Illinois

Ijeoma F. Uchegbu
University of London School
of Pharmacy
London, United Kingdom

Georges Vassaux
Université de Nantes, Nantes
Atlantique Universités
Nantes, France

Nuria Vilaboa
Hospital Universitario La Paz
and CIBER-BBN
Madrid, Spain

Richard Voellmy
HSF Pharmaceuticals S.A.
Pully, Switzerland

Ernst Wagner
Ludwig-Maximilians University
Munich, Germany

David B. Weiner
University of Pennsylvania
Philadelphia, Pennsylvania

D. Wolfe
Diamyd Inc.
Pittsburgh, Pennsylvania

Jon A. Wolff
University of Wisconsin–Madison
Madison, Wisconsin

Lauren E. Woodard
Stanford University School
of Medicine
Stanford, California

Jian Yan
University of Pennsylvania
Philadelphia, Pennsylvania

Vijay Yechoor
Baylor College of Medicine
Houston, Texas

Seppo Ylä-Herttuala
University of Kuopio
Kuopio, Finland

Ye-Zi You
University of Science
and Technology
Hefei, China

Qing Yu
SiBiono GeneTech Co., Ltd.
Shenzhen, China

Jingya Zhu
SiBiono GeneTech Co., Ltd.
Shenzhen, China

Thomas P. Zwaka
Baylor College of Medicine
Houston, Texas

Part I

Delivery Systems and Therapeutic Strategies

1 Retroviral Vectors for Cell and Gene Therapy

Axel Schambach, Tobias Maetzig, and Christopher Baum

CONTENTS

1.1 INTRODUCTION

Gene therapy is defined as the insertion of genes into an individual's cells and tissues, undertaken to treat or prevent a disease. Basically, gene therapy aims to reach two important goals:

1. Substitution for a defective or missing factor (e.g., factor IX or adenosine deaminase [ADA]) by introducing an intact copy of the faulty gene or ideally repairing it [1]. Instead of supplementing the missing factor (protein), gene therapy directly targets the molecular cause.
2. Introduction of a gene to modify the way a cell functions. For example, for the treatment of cancer, the therapeutic gene product can encode a potentially suicidal protein, which mediates sensitivity to a precursor substance (cytotoxic prodrug).

The simple treatment of cells with naked DNA is often not sufficient to reach efficient gene transfer. If the DNA can be introduced into a cell via a "gene shuttle," gene transfer is by far more efficient. Viruses represent naturally evolved shuttle systems for nucleic acids.

Gene transfer vehicles have been developed on the basis of retrovirus, adenovirus, adeno-associated virus, herpes virus, and many others [2]. Furthermore, various physical gene transfer systems have been established, e.g., the complexation of DNA with condensing and membrane traversing agents (liposomes and cationic polymers) [3].

However, most of these approaches mediate only temporary episomal gene maintenance and expression. The advent of episomally replicating DNA circles still has to demonstrate its potential for persisting therapeutic genetic modification of somatic cells with a high replicative potential [4,5]. In contrast retroviral vectors have the ability for stable integration into the host cell genome and allow long-term expression, so that theoretically a single administration could have a sustained, potentially even, lifelong curative effect [6].

The first clinical gene therapy study using a retroviral vector was initiated in 1990, undertaken to introduce the ADA gene into T-lymphocytes from patients with ADA deficiency [7–9]. Now in the year 2007, over 1300 clinical protocols have been approved worldwide. In close to one quarter of these studies retroviral vectors have been used (overview in http://www.wiley.co.uk/genetherapy/ clinical/).

First trials targeting hematopoietic stem cells (HSCs) have been successfully conducted and proven clinically benefit for the patients as shown for X-linked severe combined immunodeficiency (SCID-X1), ADA-dependent SCID (SCID-ADA) [6,8,10], and chronic granulomatous disease (CGD) [11]. Promising approaches in adoptively transferred T cells are also emerging [12,13]. At the same time, evidence for dose-limiting side effects of retroviral gene delivery into HSCs has been obtained in animal models and clinical trials [11,14–18]. Previously anticipated to be a finite risk of minor importance [19], cancer induction by random insertion of vectors in the vicinity of cellular proto-oncogenes, resulting

3

in their upregulation, now represents a significant concern associated with retroviral gene delivery. In this somewhat heated situation, the present chapter explores basic principles of gammaretroviral vector design and avenues for future research. Gammaretroviral vector production and lentiviral vectors are covered in different chapters of this book.

1.2 RETROVIRUSES AND THEIR LIFE CYCLE

Retroviruses are lipid-enveloped viruses. Each particle's capsid harbors two copies of a linear, plus-stranded RNA genome in the size of 7–11 kb. The family *Retroviridae* contains various viruses that have shown potential utility for gene therapy, such as the mammalian and avian type C retroviruses (gammaretroviruses, e.g., murine leukemia virus [MLV]), lentiviruses (e.g., human immunodeficiency virus type 1 [HIV-1]), and spumaviruses (e.g., foamy virus) [20].

Retroviruses share significant similarities in their virion structure, genomic organization, and method of replication. For the development of retroviral gene transfer vectors, the understanding of retroviral replication was a fundamental prerequisite. Section 1.3 explains the replication of MLV (Figure 1.1).

Infection with a retrovirus starts with the recognition of a surface receptor (e.g., for ecotropic MLV this is mCAT1, a cationic amino acid transporter) allowing fusion with the cell membrane or receptor-mediated endocytosis. Entry then proceeds with the uncoating of the viral particle, a still mysterious process whose exact mechanism remains to be defined. A ribonucleoprotein (RNP) complex composed of viral and cellular proteins with yet unknown structure is released into the cytoplasm. At this or a later stage in the cytoplasmic journey, retroviral RNA may be released and subject to immediate translation. While this early release and translation of retroviral mRNA clearly represents a byproduct of the retroviral life cycle, it may nevertheless be of interest as a tool for transient cell modification [21].

The paradigmatic use of retroviral vectors, however, results from their ability to deliver integration-competent DNA into cells. This occurs as follows: Still associated with the capsid, the viral RNA is reverse transcribed into linear double-stranded DNA by the retroviral enzyme reverse transcriptase (RT). For reverse transcription to be initiated and completed, several motifs are required on the retroviral RNA. The most important ones are the primer binding site (PBS), where a cell-derived tRNA serves as the primer for the RT-polymerase,

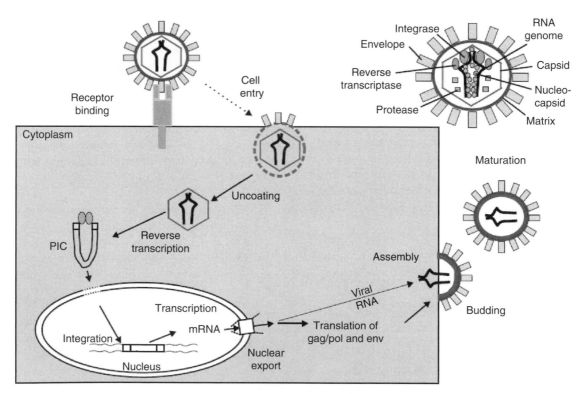

FIGURE 1.1 Retroviral life cycle as shown for a simple gammaretrovirus. After receptor binding the virus fuses with the target cell (or may by endocytosed), and the RNP complex is released into the cytoplasm. The RT performs the synthesis of double-stranded DNA, and the arising preintegration complex (PIC) gains access to the nucleus (most efficient during mitosis). The viral integrase (IN) catalyzes the integration into the host cell genome. The integrated viral DNA is called provirus. After transcription the viral mRNA is exported to the cytoplasm. Unspliced RNAs encode for the structural proteins (gag), replication enzymes (pol), and spliced RNAs for the envelope (env) proteins. Genomic viral mRNA and the translated gag, pol, and env proteins are transported to the plasma membrane, where virus assembly and budding are initiated. The fully infectious virion is created by proteolytic rearrangements. In the right upper corner an (enlarged) mature virus with its RNA genome, nucleocapsid, RT, IN, protease, spherical capsid, matrix, and envelope proteins is shown.

and the polypurine tract (PPT) which primes synthesis of the second DNA strand by forming an RNA hairpin that is resistant to the RNAseH activity of RT. The complex steps of reverse transcription are further supported by the sequence homology of the R-region that builds both start and end (if we ignore the polyA tail) of the retroviral genomic RNA, and may involve intra- and intermolecular jumps of RT. The two strands of RNA present in the same virion may thus support recombination events. Furthermore, a mutation rate of about 1 in 10 kb processed bases is a typical outcome of reverse transcription.

The product of reverse transcription remains associated with some viral proteins (most notably the IN) and various cellular proteins to form the so-called PIC. In the various steps between entry and formation of the PIC, several host factors may block retroviral replication as part of an innate immunity against retroviruses. Factors targeting capsid sequences encoded by gag (such as TRIM5α) may be responsible for major species-specific differences in retroviral infectivity [22]. The NB-tropic MLV gag sequences are unrestricted in both rodent and human cells and used in all major MLV-based gene vector systems, allowing relatively straightforward dosing studies in the murine model.

Intracellular transport of the PIC follows the cytoskeleton along the highways of the cells, the microtubules, and is dependent on host proteins [23]. Gammaretroviral PICs cannot actively cross nuclear membranes, possibly by missing a linker molecule to the nuclear import machinery. Only breakdown of the nuclear membrane during mitosis exposes the chromosomes to the PIC [24]. Having encountered host DNA, the viral IN incorporates the linear DNA into the host cell genome. Prior to integration, circular forms of the retroviral DNA may be formed to result in so-called 2-LTR-circles and 1-LTR-circle. Having lost the ability for IN-mediated integration, these circles typically represent dead ends of the retroviral life cycle. In the context of lentiviral vectors that are able to transduce resting cells, these unintegrated circular forms become increasingly popular for sustained cell modification [25,26]. In replicating cells, the long-terminal repeat (LTR) circles are expected to be lost during mitosis because they lack the capacity for chromosomal attachment.

Integration of the retroviral DNA into cellular chromosomes is highly efficient, occurs without major structural modifications of the neighboring cellular sequences, and does not form transgene concatemers, but is untargeted. Somewhat dependent on its location in the genome and the associated chromatin context, an integrated provirus may serve as an efficient gene expression cassette. The transcriptional control elements are largely contained in the LTR. Both LTRs contain enhancer sequences that may interact with neighboring promoters. Although the LTRs are sequence identical, their role in gene expression is different. The 5′LTR typically serves as the promoter for the polymerase II complex while the 3′LTR provides transcriptional termination and polyadenylation signals. The exact mechanisms regulating the differential utilization of the promoter and termination signals remain to be elucidated. The most likely explanation is read-through suppression of the 3′ promoter by the transcript originating in the 5′LTR, and suppression of the 5′LTR-located polyadenylation signal by its proximity to the transcriptional start site (TSS) [27].

All replication-competent retroviruses have open reading frames (ORFs) that harbor the structural proteins (gag), the replication enzymes (pol), and the viral glycoproteins (env). The simple gammaretroviruses have only one primary transcript, which serves as genomic and messenger RNA (mRNA) for the viral gag and pol proteins. The viral glycoproteins are synthesized from a spliced mRNA, formed by a splice donor (SD) located shortly behind the PBS and a splice acceptor (SA) that is found in the downstream sequences of pol. The recognition of these sequences is incomplete, allowing the coexpression of unspliced and spliced RNA. Suboptimal consensus sequences and, in the case of MLV, SD suppression by a complex RNA secondary structure have been identified as the mechanisms underlying partial splicing [28,29]. The spliced intron corresponds to the gag/pol coding sequence. In gammaretroviral vectors, the intron can be reduced to the size of the packaging motif [28,30].

Gag and pol are synthesized as polyproteins from the cytosolic ribosomes. The viral env glycoproteins are translated at the rough endoplasmic reticulum, and transported and processed via the vesicular transport pathway. Gag self-assembles in close proximity to the cellular membrane. In the assembly process preferentially those mRNAs are packaged that carry the packaging signal ψ, mediated by specific interaction with the nucleocapsid domains of gag. Another nucleic acid to be incorporated to sustain the life cycle is a tRNA that matches the PBS. As mentioned above, this tRNA is a gift of the host cell, not encoded by the viral genome.

Freshly synthesized env represents a protein precursor for a bipartite transmembrane protein. The env precursor is incorporated during budding of the viral particle as part of the membrane coat. After virus release has taken place at the cell membrane, extracellular conversion to the mature, infectious virion follows. Virion maturation thus only occurs after budding and involves cleavage of the viral polyproteins into functional subunits by the viral protease (for further reading see [31]). Targeted proteolysis thus generates four distinct proteins out of gag (matrix, p12, capsid, nucleocapsid), three out of pol (protease, RT, and IN), and two out of env (surface and transmembrane part). The necessity for particle maturation in the extracellular phase of the life cycle is one principle by which self-superinfection of a retrovirally infected cell is avoided. The second principle is the saturation of the env receptor by de novo synthesized env proteins. Blocking self-infection is an important goal in the design of retroviral packaging cells, to avoid the accumulation of potentially recombined proviral vector copies [32].

1.3 RETROVIRAL VECTORS AND SPLIT PACKAGING DESIGN

To take advantage of the retroviral life cycle for vector development, one would need to include a foreign gene, if possible

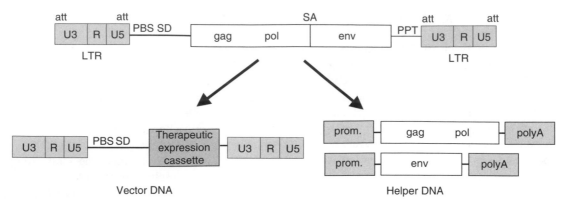

FIGURE 1.2 Split-packaging design. A strategy for engineering a virus into a vector. The general genome organization of a simple retrovirus is shown above with the LTRs (LTR, with U3, R, and U5 regions), the PBS, splice signals (SD, SA), the packaging signal ψ, the PPT, and the attachment signals of the IN (att). ORFs for the structural proteins (gag), replication enzymes (pol), and the envelope glycoproteins (env) are indicated. To construct a (replication-deficient) gene transfer tool, the cis-acting elements (LTR and leader region) and the ORFs for gag, pol, and env are divided on separate plasmids (vector DNA and helper DNA). Gag, pol, and env are encoded by helper plasmids (prom., promoter; polyA, polyA signal) that can be delivered as a single molecule or split into different DNA molecules for safety reasons. The helper DNA lacks the packaging signal (ψ) and thus cannot be packaged into a viral particle. The vector DNA contains the therapeutic expression cassette, flanked by the LTR. It harbors a packaging signal (ψ) for efficient packaging into the viral particle.

along with its regulatory sequences, into the virus genome and, at the same time, prevent viral gene expression and disable viral replication. To achieve this goal, the retroviral coding sequences have to be removed, which creates at least 6 kb space for the transgene of interest. If the cis-acting sequences required for nuclear export of unspliced RNA, assembly, reverse transcription, and integration are all preserved, this operation may result in a very efficient replication-deficient vector (Figure 1.2). These cis-acting sequences comprise the R-region stem loop (RSL), the packaging signal (ψ), the PBS, the PPT, and those parts of the LTR that are essential for reverse transcription and integration (R, U5, and small "attachment" sequences located in the beginning of U3 and the end of U5) (Figure 1.3). Further enhancing vector

safety, designer PBS sequences may be incorporated that match only artificial tRNAs, which need to be expressed from transgenic sequences in packaging cells [33].

The viral structural proteins (gag) and replication enzymes (pol) as well as the envelope (env) glycoprotein are encoded on separate helper plasmids which should avoid any sequence overlap with the transfer vector. Typically gag–pol is expressed from one plasmid and env from another. In the case of MLV, this is easy to achieve as the minimal cis-elements necessary for vector construction do not overlap with the coding sequences [28,30].

This so-called split-packaging design makes recombination back to a replication-competent retrovirus a very unlikely event and thus contributes to biosafety. However, still used in

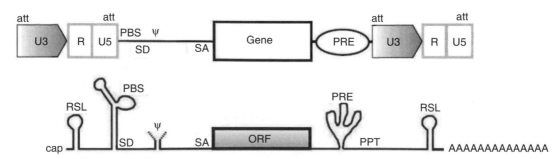

FIGURE 1.3 LTR-driven retroviral vectors. The schematic view of an LTR vector (compare Figure 1.2) and its corresponding RNA (below) is given. Several cis-acting elements and stable RNA secondary structures are depicted. Within the R-region of the LTR a stable stem-loop structure was mapped (RSL; [110]), which is thought to increase cytoplasmic RNA levels. The intron, located in the leader region (SD, SA) and surrounding the packaging signal ψ may further enhance cytoplasmic RNA levels. The PBS is embedded in another stem-loop structure originating in the U5. Furthermore, the woodchuck hepatitis virus posttranscriptional regulatory element (PRE), which is located in the 3'UTR enhances retroviral RNA levels by increasing mRNA stability, export and polyadenylation. ORF, open reading frame.

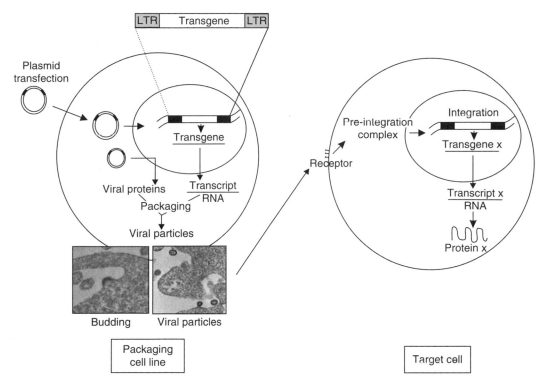

FIGURE 1.4 Production of viral particles and transduction of target cells. Production of infectious retroviral particles from a packaging cell line, transfected with a plasmid encoding the retroviral vector using the calcium phosphate method or lipofection. The packaging cell line (often derived from 293 T cells) provides the viral proteins which are either introduced via stable transfection (e.g., Phoenix cells, Nolan Lab, Stanford) or transiently cotransfected together with the retroviral vector. The vector encoded packaging signal ψ is recognized by the nucleocapsid proteins and allows the specific packaging of the retroviral RNA (including the transgene sequence). Infectious particles are released from the cell membrane (i.e., budding, left electron microscopy picture; full particles right; pictures kindly provided by Dr. G. Rutter, Hamburg). The cell culture supernatant—containing the viral particles—is harvested, filtered, and transferred onto target cells. Retroviral particles enter the target cells through interaction with a specific membrane receptor. After uncoating, the viral RNA is reverse transcribed into DNA. The DNA reaches the nucleus and is integrated into the host cell genome. Because of the lack of other viral structural proteins, no new infectious particles can be generated. Transgene expression can be monitored by standard techniques, including northern and western blot, real-time PCR, ELISA, FACS, and immunofluorescence.

many laboratories are early generations of gammaretroviral vectors that do not provide a strict separation of vector and helper sequences. When using such constructs, the occasional generation of replication-competent recombinants is an expected outcome.

Replication-incompetent viral particles can only perform a "single-round" infection of the target cell (Figure 1.4). The viral structural proteins recognize the ψ-containing vector RNA, but not the helper (ψ-deficient) nucleic acid, resulting in the packaging of the vector genome into an infectious particle. After entry of the particle into the target cell, only the vector nucleic acid is reverse transcribed and can integrate into the genome of the target cell. Since neither structural proteins nor replication enzymes are encoded by the target cell, the generation of replication-competent virus is prevented. A potential advantage of gammaretroviral vectors is that no exogenous human retrovirus has been described so far that is able to mobilize vector mRNA for infection of other cells or organisms.

Production of gammaretroviral vector particles can be achieved after transient cotransfection of the helper plasmids with a plasmid encoding the transfer vector, or using stable packaging lines in which the transfer vector's DNA is introduced in a transient or stable form (Figure 1.3). For more details regarding the gammaretroviral packaging technology, refer to Chapter 2 by Cruz et al.

1.4 SELF-INACTIVATING VECTOR DESIGN

Self-inactivating (SIN) retroviral vectors lack enhancer–promoter sequences in the U3 region of their LTRs and rather make use of an internal promoter to initiate transcription of the transgene cassette (Figure 1.5). This design has been first described for gammaretroviral vectors in 1986 [34] and was subsequently adopted for lentiviral and spumaviral vectors [35,36]. The SIN design has some important advantages compared to the conventional LTR design: It reduces the risk of insertional upregulation of neighboring genes by decreasing long-distance enhancer interactions or promoter activity from the 3′LTR [37,38]; it further lowers the risk of recombination to create a replication-competent retrovirus; if a cell expressing the SIN transgene is superinfected by a replicating retrovirus,

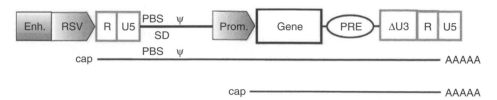

FIGURE 1.5 SIN retroviral vectors. SIN vectors are devoid of promoter/enhancer elements in the 3′U3 region and are therefore dependent on an internal promoter. Moreover, this internal promoter can be chosen according to the tissue specificity and expression strength needed for the vector application. To overcome transcriptional interference and to produce sufficient amounts of packageable genomic RNA (upper), a strong promoter should be taken to drive the genomic message (e.g., Rous sarcoma virus [RSV] promoter plus SV40 enhancer).

the SIN-vector RNA (due to the lack of genomic packageable RNA) is unlikely to be mobilized; and finally, deleting the strong retroviral enhancer–promoter sequences avoids potential interference with more cell-type specific or inducible transcriptional control systems that are introduced to express the gene(s) of interest.

A few rules have to be considered to construct efficient SIN vectors. Reverse transcription copies the 3′U3 region into the 5′LTR, so that the deletion of enhancer–promoter elements in the 3′U3 of the vector plasmid will affect both LTRs. For the same reason, modifications of the plasmid's 5′U3 region, as recommended above to achieve high titers, do not alter the nature of the vector's genomic sequences packaged in the virions, provided that the modified 5′U3 region still uses the same TSS as the original retrovirus.

For construction of SIN vectors, all the enhancer–promoter sequences, which are located behind the IN attachment site and terminate in the TATA box, should be removed from the 3′U3 region. Some earlier generations of SIN vectors still included the TATA box, which was removed later to further improve safety [39,40].

Unlike HIV-derived lentiviral SIN vectors, the first generation gammaretroviral SIN vectors were compromised by a severe drop in infectious titers. Inclusion of the PRE from woodchuck hepatitis virus (wPRE) increased infectious titers above 1×10^6 transducing units per milliliter [39]. Nevertheless, infectious titers remained substantially reduced when compared to conventional LTR-driven retroviral vectors. Of note, SIN vectors produce two types of RNA in the packaging cell line. The first is the genomic RNA harboring the packaging ψ needed for efficient packaging, the second the shorter RNA initiated by the internal promoter (Figure 1.5). Since the availability of packageable, genomic RNA is a major determinant for a productive virus titer, efforts have been initiated to enhance the production of genomic RNA from SIN vector plasmids. Incorporating promoters with a high processivity in the 5′LTR of the SIN plasmid may substantially increase the titer of SIN vectors, under optimal circumstances to those achieved with the archetypal LTR-driven vectors. Such promoters can either be constitutively active (e.g., RSV, cytomegalovirus, or combinations with additional enhancers) or inducible versions (e.g., tetracycline responsive promoters) [35,40,41].

Having deleted the enhancer–promoter sequences from the 3′U3 region of the vector plasmid creates space for the incorporation of new elements that may further improve vector performance. The U3 region typically tolerates insertions of ~500 bp of foreign sequences without a significant loss of titer. This only applies if these sequences follow the basic requirements of foreign sequences for retroviral vector genomes. Introduced sequences should be devoid of

- Recombinogenic motifs (e.g., those formed by direct or inverted sequence repeats)
- Transcriptional terminators (to avoid premature termination of the transcript in packaging cells)
- Strong splice sites
- Complicated secondary structures (that are poorly resolved during reverse transcription)
- Promoters forming significant amounts of antisense message (to avoid RNA interference during the production process)

If suitable sequences are introduced in the residual U3 region (Figure 1.6) after reverse transcription these will be present in two copies, i.e., at both ends of the vector's genome. This feature is particularly useful for the introduction of insulator elements and transcriptional termination sequences. Small expression cassettes, e.g., shRNA genes driven by polymerase III promoters [42], may also be introduced into the U3 region, resulting in "double copy vectors." Even the R-region is amenable for such "double copy" modifications [43]. Although the latter architecture has not found many friends, it still underscores the flexibility of the gammaretroviral genome.

However, gammaretroviral sequence flexibility is not unlimited. For reasons that remain to be identified, certain expression cassettes are significantly more stable in a lentiviral vector context. The most prominent example is the β-globin gene cassette that combines large cis-regulatory sequences and introns. Due to the presence of strong cellular introns, this cassette needs to be introduced in antisense orientation to the vector's genomic mRNA. While gammaretroviral vectors fail to give good titers with this insert, lentiviral vectors could be produced to titers that allow a convincing correction of disease in

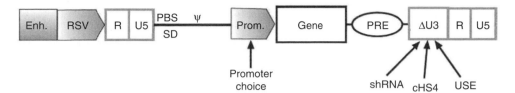

FIGURE 1.6 SIN vectors within the 3'UTR and the 3'ΔU3 region. The deletion of promoter/elements of the 3'U3 (ΔU3) creates space for the incorporation of useful features. Interestingly, these elements will be present (due to the mode of reverse transcription) in its integrated form. For RNAi applications short hairpin RNA (shRNA) expression cassettes, consisting of a Pol. I promoter and a shRNA RNA, can be introduced. In addition, chromatin insulators (e.g., cHS4) and polyadenylation enhancers (upstream polyadenylation enhancers, USE) can be incorporated here (for details see text).

preclinical animal models [44–48]. If the mechanisms for the differential performance of lentiviral and gammaretroviral globin vectors will be identified, similarly complex cassettes could potentially be adopted for gammaretroviral vectors.

1.5 TERMINATION ENHANCERS AND INSULATORS

As the retroviral termination signal is present in both LTRs, the element in the 5'LTR has to be suppressed to allow for efficient transcription of the retroviral mRNA species. Gammaretroviral and lentiviral termination signals differ in their configuration and mechanism of suppression. While in lentiviruses cross-talk with the SD is involved in suppression of the 5'termination sequence, the exact mechanism remains to be defined for gammaretroviruses [27]. Vectors derived from either retrovirus may cause significant amounts of transcriptional read-through into downstream cellular sequences. Read-through is further enhanced in the SIN design, suggesting that the wild-type U3 sequences contribute to transcriptional termination [49,50]. Cellular termination signals cannot be simply incorporated into retroviral vectors as they may trigger premature termination of the transcript in packaging cells, thus causing a severe loss of titer. Two elements have been shown to tighten termination of retroviral vector sequences while even increasing the yields of infectious particles. The first is the PRE derived from wPRE. This rather long sequence (>500 bases) forms extended secondary structures and may act at several levels of posttranscriptional gene regulation [51–55]. Evidence for a role in enhancing transcriptional termination has been obtained by us and others [54,55]. Significantly shorter sequences that may be even more potent termination enhancers than the wPRE are upstream polyadenylation enhancer elements (USE) derived from other viruses or cellular genes. These U-rich motifs facilitate the binding of crucial cellular termination factors to accomplish more efficient utilization of the retroviral polyadenylation site [56]. Using a panel of reporter assays, including assays detecting transcripts of integrated vectors, we showed that USEs can largely suppress

residual read-through of gammaretroviral and lentiviral SIN vectors [54].

Insulators are cis-elements that inhibit enhancer or silencer interactions between the integrated transgene and the neighboring chromatin [44,57–63]. These sequences are of interest to reduce the notorious position-dependent variegation effects, to increase the transcriptional autonomy of the transgene and avoid silencing, thus allowing for a more homogenous transcription level in polyclonal situations. The second aim for the incorporation of insulators is to reduce long-distance effects of the integrated transgene on neighboring cellular genes (insertional mutagenesis). Several groups have provided evidence that insulators may be helpful to reach these goals, although they may not seem to act in a very dominant manner if the vector's enhancer is strong [44,57–63]. Defining more powerful insulators and analyzing their potential context-dependence are important activities of ongoing research.

1.6 COEXPRESSION STRATEGIES

Gene therapy trials performed today often make use of monocistronic vectors which allow the correction of monogenic diseases. For more complex applications such as the additional use of selection markers, different coexpression strategies (Figure 1.7) can be applied which have shown to be functional in cell and animal models, although most of them have different (dis)advantages and are yet to be tested in human trials.

In general, coexpression of at least two transgenes can be achieved either by (post)transcriptional (splicing, multiple promoters) or by translational mechanisms (IRES, 2A cleavage, fusion proteins). One of the most effective coexpression systems has been evolved by retroviruses. Every retrovirus depends on the generation of packageable genomic RNA. By controlling splicing of the genomic message a balanced ratio of subgenomic gag–pol and env molecules is obtained which is crucial for the survival of the virus. The principle of balanced splicing can also be employed for the coexpression of transgenes from retroviral vectors but it is problematic, especially in the context of the SIN design. Splicing cannot be suppressed

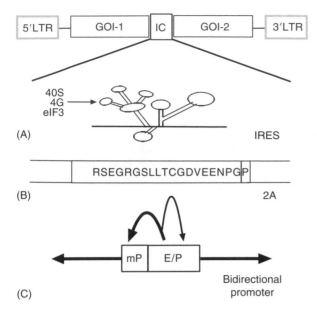

FIGURE 1.7 Coexpression strategies. Bicistronic expression can be achieved via several strategies. For efficient coexpression the internal cassette (IC) can be formed by internal ribosome entry sites, 2A "self-cleavage" sites and single or bidirectional promoters. (A) Internal ribosome entry sites recruit parts of the translation machinery (e.g., eukaryotic initiation factor 4G and eIF3 plus 40S ribosome). (B) Self-cleavage indicated for the Thosea asigna virus 2A-cleavage site (T2A). The peptide consists of 20 amino acids. Cleavage occurs via a co-translational ribosome skipping mechanism between the C-terminal Glycin and Prolin residues, leaving the indicated amino acids attached to the end of the 5′ protein and the start of the 3′ protein. (C) Bidirectional promoters consist of two antisense oriented (minimal = m) promoters (P) sharing the same enhancer (E) but driving transgene expression in opposite directions.

in the packaging cell, which ultimately leads to a drop in titer and lower transduction efficiencies. In SIN vectors, the intron might be deleted prior to packaging of the genomic RNA, unless it contains the packaging signal [30,39,52].

A simpler approach can be taken by generating fusion proteins (e.g., formed by a therapeutic and a marker protein) although their biological function has to be empirically tested, because the fusion might induce misfolding or interfere with interaction partners. The most elaborate fusion partners described are GFP (and its derivatives) and luciferase which can be fused to many genes of interest at the C- or N-terminus.

Most cellular mRNAs carry a 5′ m7G cap which protects the RNA against degradation and serves as the ribosome binding site. Translation is then initiated at the first AUG start codon detected by the scanning ribosome (cap-dependent translation). In contrast to cap-dependent translation, the IRES sequence (i.e., internal ribosomal entry site) forms a secondary structure recruiting ribosomes and all translation initiation factors (except the cap binding protein eIF4E) directly to the initiation codon of a downstream ORF. In order to achieve bicistronic gene expression, the IRES has to be located between both genes. Translation is then cap-dependent upstream but cap independent downstream of the IRES.

Several IRES sequences have been found in viral and eukaryotic genes allowing for bicistronic transgene expression. Their size ranges between 10 to 600 nt [64]. The most widely used IRES in gene therapy vectors today is derived from encephalomyocarditis virus and is ~600 nt long. Although viral vectors are routinely equipped with IRES sequences, they are not the optimal choice when equimolar amounts of both transgenes are needed since the suboptimal recognition of the IRES typically leads to a reduced expression of its 3′gene.

In contrast to IRES, almost equal expression levels can be achieved by using effective 2A "self-cleavage" sites [65,66]. Here, both transgenes are encoded as fusion proteins, with the 2A peptide interspaced. This peptide consists of ~20 amino acids with a C-terminal prolyl–glycyl–prolyl (PG|P) sequence which will cotranslationally be cleaved by a ribosomal skipping mechanism between the glycyl–prolyl residues. Thus, the beginning of the 2A peptide will remain attached to the C-terminus of the first and an additional prolyl residue to the N-terminus of the second protein. Noteworthily, never reported to impair protein functions nor to be immunogenic, antibodies have been successfully raised against the FMDV 2A tag fused to the N-terminal protein [67]. Advantages of the 2A-sequences are not only their small size (~60 nt) but also the potential to incorporate multiple 2A-sequences between multiple transgenes. In doing so, Vignali and colleagues could give rise to all four subunits of the CD3 receptor complex [68].

Instead of employing translation dependent coexpression mechanisms it is also possible to rely on elements regulating transcription, namely promoters, which can be arranged in different settings. The simplest setting consists of an LTR-driven vector with an additional internal promoter, in this way controlling two ORFs. This can also be achieved by two internal promoters in a SIN-vector configuration. Since the message initiated from the first promoters runs over the transcription factor binding sites of the downstream promoter, it might be silenced by transcriptional suppression. Additionally, promoter interference might hamper efficient gene expression which makes empirical tests for all promoter–transgene combinations essential. A more elegant approach using a bidirectional promoter was recently published by Amendola and colleagues [69]. This approach uses two antisense oriented internal core promoters sharing the same enhancer, as previously introduced by Bujard et al. for the simultaneous regulation of two genes from the same tetracycline-inducible promoter [70]. Although bidirectional promoters are a sophisticated coexpression alternative, these constructs generally suffer from reduced titers, since the generated antisense message interferes with packageable genomic RNA production.

1.7 TARGETING INTEGRATION

IN has been identified as the most important determinant of the semi-random integration pattern of retroviruses [71]. HIV-1 and MLV differ in their propensity to integrate in the 5 kb neighborhood of a gene's TSS (twofold higher risk for MLV), next to DNAse1 hypersensitive sites (10-fold elevated for

MLV) and within transcribed sequences (twofold elevated for HIV) [71]. Other retroviral INs are less biased in their preference for actively transcribed genes. However, even avian sarcoma leukosis virus which has the least bias for active genes and no TSS preference [72] is still a potent insertional mutagen *in vivo* [73], possibly because of its strong enhancer/promoter sequences within the U3 [74]. Irrespective of the IN used, it thus seems wise to restrict the number of insertion events when modifying cell types that have an increased risk of malignant transformation, such as HSCs.

The strictest approach to overcome the issue of insertional mutagenesis would be the generation of a retroviral IN that acts in a site-specific manner. As IN is organized in distinct subdomains, the DNA-recognition site has been altered to adopt site-specific modules. Proof-of-principle for modified target specificity has been achieved *in vitro*, but so far no highly specific system has been described for targeted gene insertion in mammalian cells that works with similar efficiency as unmodified retroviral vectors [75]. To reach this important goal, the field follows various approaches. One of the most promising is the cotransduction of cells with IN-deficient lentiviral vectors that provide LTR circles as a template for homologous recombination, triggered by site-specific DNA double-strand breaks that can be introduced using transiently expressed endonucleases [76,77]. Determining residual off-target effects and increasing the efficiency in primary cells represent challenges for future research.

1.8 PSEUDOTYPING

Gammaretroviral vectors are pretty flexible in their choice of potential envelope proteins. A variety of foreign glycoproteins has been successfully incorporated into vector particles. The most widely used envelope proteins are the murine ecotropic envelope (which restricts infection to rodent cells) [78], the amphotropic envelope (which transduces both human and rodent cells) [79], the envelope derived from gibbon ape leukemia virus [80], or the one from feline retrovirus RD114 [81] (both of which transduce human cells much more efficiently than mouse cells). Examples for virtually ubiquitously infectious glycoproteins are those derived from vesicular stomatitis virus (VSV) [82] or the lymphocytic choriomeningitis virus [83]. Approaches to obtain retargeted designer envelope proteins are reviewed elsewhere [84]. One of the challenges in envelope retargeting is to preserve the ability for fusion of the viral particle with the target cell following recognition of the surface structure.

The choice of the envelope protein will primarily depend upon the receptor status of the target cell and the ability to concentrate vector stocks. Very stable preparations that allow repeated concentration by ultracentrifugation can be obtained when using the VSV glycoprotein. However, cytotoxic effects and contamination of vector preparation with cellular microvesicles limit the enthusiasm for this pseudotype [85].

1.9 APPLICATIONS

For a long time, key applications of gammaretroviral vectors have been the expression of oncogenes in cancer research, and the creation of potentially therapeutic vectors for the genetic engineering of HSCs and T-lymphocytes. For the study of oncogenes or other sequences that may have an impact on cell survival, researchers typically use LTR-driven gammaretroviral vectors that coexpress a fluorescent protein as part of a bicistronic message, the fluorescent protein typically being under control of an internal ribosomal entry site. If such vectors trigger malignant transformation of cells, it may be a good idea to investigate the insertion site for potential evidence of collaborating cellular sequences that may have been selected based on the survival advantage mediated by insertional mutagenesis in the given experimental system [86–89].

For the therapeutic use of gammaretroviral LTR-driven vectors, there is increasing evidence for insertional mutagenesis as a dose-limiting toxicity. This is particularly relevant for the genetic modification of HSCs [11,14–18]. Future attempts to use gammaretroviral vectors in this setting should be based upon carefully redesigned vectors (if possible a SIN design without strong enhancers that are active in HSCs) and a proactive validation of residual genotoxicity in emerging preclinical models [37,38,62]. The use of earlier generations of LTR-driven vectors may have to be restricted to conditions for which an appropriate safety profile is established or to a situation of an imminent medical need when alternative vectors are not yet available. Importantly, insertional side effects are highly context-dependent and vary for yet unknown reasons even in different clinical studies targeting HSCs [90–92]. In contrast to the situation encountered in HSCs, no insertional side effects have been detected to date in recipients of gammaretrovirally transduced human T cells, although infused cell numbers are usually much higher than in the trials targeting HSCs [93].

A still hot topic in basic research is the propensity of gammaretroviral LTR vectors to be epigenetically silenced in embryonic stem cells (ESCs). This feature, most prominent for Moloney MLV, has actually been very welcome in the literally revolutionary creation of "induced pluripotent cells" by retroviral transduction of differentiated somatic cells with a cocktail of transcription factor sequences [94,95]. In this context, robust silencing of the transduced sequences, some of which are clearly oncogenic, may have represented a major advantage of the vectors used.

Modifications of the gammaretroviral LTR enhancers and the PBS—which overlaps with the binding site for a still ill-defined repressor—have resulted in vectors that increase the chance of transgene expression in ESCs [96]. Similarly designed vectors also showed better long-term expression in HSCs, where the repressor targeting the PBS sequences also operates [97–100]. However, if transduced ESCs were induced to differentiate, these modified LTR-driven vectors were still silenced with a very high probability [101], in contrast to lentiviral vectors with a SIN design and a cellular promoter [102].

Initial reports suggested that lentiviral vectors are generally more resistant to silencing than gammaretroviral vectors. However, carefully designed studies have found no clear evidence for this hypothesis [103]. Vector architecture and content thus seem to play a major role in determining the susceptibility of retrovirally transduced sequences to silencing in stem cells and their differentiated progeny [103,104]. For modification of ESCs and the generation of transgenic animals using viral vectors, lentiviral SIN vectors with internal cellular promoters currently represent the state of the art [102,105]. The side-by-side evaluation of gammaretroviral and lentiviral vectors will play an important role in the elucidation of the underlying mechanisms.

1.10 OUTLOOK

In the era of lentiviral vectors that are able to transduce quiescent cells and appear to be more robust than gammaretroviral vectors in the incorporation of complex transgene cassettes, one may wonder which interesting features remain to be offered by gammaretroviral vectors.

The first nice feature is their simplicity. The gammaretroviral technology can easily be adapted in every laboratory that has experience with cell culture, plasmid cloning, and transfection. Unless very complex cassettes are introduced, infectious titers are robust and most target cells that replicate in culture are readily transduced when using an appropriate envelope protein. Of note, the generation of lentiviral vectors can be similarly efficient and simple. If wanted, both lentiviral and gammaretroviral vectors can be restricted to biosafety level 1, when using the ecotropic envelope for pseudotyping [106,107]. The second feature and currently a real advantage that is of interest for clinical use is the availability of stable gammaretroviral packaging conditions that are suitable for upscaling to clinical grade vector production. Lentiviral packaging systems are still struggling to reach this important goal, although promising advances have been reported [108,109]. The third point of interest is the relatively low concern of vector genome mobilization and the absence of preexisting immunity to gammaretroviral particle-dependent antigens. This may be of particular relevance in gene therapy for patients suffering from HIV infection. The fourth potential advantage is the advent of SIN vectors with reduced long-distance enhancer activity. Using such vectors, the insertion profile of gammaretroviral vectors may not be as problematic as it is in the context of either LTR-driven vectors or SIN vectors containing a strong internal enhancer [37,38]. Envisaging the availability of a fully insulated SIN vector, the gammaretroviral insertion pattern may even be superior to the lentiviral one, as it may reduce the residual risk of gene disruption. The fifth argument for the continued use of gammaretroviral vectors is their established safety record in applications other than HSCs. Finally, as the diversity of human gene therapy increases and patients may need to be treated more than once with viral vectors or vector-modified cells, adaptive immunity may recognize vector-associated antigens that can be present for several days in transduced cells. It will thus be important to have access to a variety of alternative vector systems, such as lentiviral, foamyviral, and gammaretroviral when considering systems for stable genetic modification. These points may suffice to illustrate that gammaretroviral technology, as many other vector systems, is very potent when applied to a suitable cell type but under a high pressure for innovations to further increase efficiency and biosafety. Considering the above, the flexibility and simplicity of the gammaretroviral life cycle creates a fertile ground for basic and translational vectorology.

ACKNOWLEDGMENTS

This work was supported by grants from the Deutsche Forschungsgemeinschaft (DFG SPP1230, to Z.L. and C.B.), the European Union (CONSERT and CLINIGENE to C.B.), the Bundesministerium für Bildung und Forschung (BMBF TreatID, to C.B.), the National Cancer Institute (R01-CA 107492-01A2, to C.B.), the Else-Kröner Foundation (to A.S.), and German National Merit Foundation (Studienstiftung; to T.M.).

REFERENCES

1. Mulligan, R. C. 1993. The basic science of gene therapy. *Science* 260: 926–932.
2. Thomas, C. E., Ehrhardt, A., and Kay, M. A. 2003. Progress and problems with the use of viral vectors for gene therapy. *Nature Rev Genet* 4: 346–358.
3. Louise, C. 2006. Nonviral vectors. *Methods Mol Biol* 333: 201–226.
4. Glover, D. J., Lipps, H. J., and Jans, D. A. 2005. Towards safe, non-viral therapeutic gene expression in humans. *Nat Rev Genet* 6: 299–310.
5. Papapetrou, E. P., Ziros, P. G., Micheva, I. D., Zoumbos, N. C., and Athanassiadou, A. 2006. Gene transfer into human hematopoietic progenitor cells with an episomal vector carrying an S/MAR element. *Gene Ther* 13: 40–51.
6. Cavazzana-Calvo, M., et al. 2000. Gene therapy of human severe combined immunodeficiency (SCID)-X1 disease. *Science* 288: 669–672.
7. Blaese, R. M. and Culver, K. W. 1992. Gene therapy for primary immunodeficiency disease. *Immunodefic Rev* 3: 329–349.
8. Aiuti, A., et al. 2002. Correction of ADA-SCID by stem cell gene therapy combined with nonmyeloablative conditioning. *Science* 296: 2410–2413.
9. Ferrari, G., et al. 1991. An in vivo model of somatic cell gene therapy for human severe combined immunodeficiency. *Science* 251: 1363–1366.
10. Gaspar, H. B., et al. 2004. Gene therapy of X-linked severe combined immunodeficiency by use of a pseudotyped gammaretroviral vector. *Lancet* 364: 2181–2187.
11. Ott, M. G., et al. 2006. Correction of X-linked chronic granulomatous disease by gene therapy, augmented by insertional activation of MDS1-EVI1, PRDM16 or SETBP1. *Nat Med* 12: 401–409.
12. Bonini, C., et al. 1997. HSV-TK gene transfer into donor lymphocytes for control of allogeneic graft-versus-leukemia. *Science* 276: 1719–1724.
13. Morgan, R. A., et al. 2006. Cancer regression in patients after transfer of genetically engineered lymphocytes. *Science* 314: 126–129.

14. Li, Z., et al. 2002. Murine leukemia induced by retroviral gene marking. *Science* 296: 497.

15. Hacein-Bey-Abina, S., et al. 2003. LMO2-associated clonal T cell proliferation in two patients after gene therapy for SCID-X1. *Science* 302: 415–419.

16. Kustikova, O. S., et al. 2005. Clonal dominance of hematopoietic stem cells triggered by retroviral gene marking. *Science* 308: 1171–1174.

17. Seggewiss, R., et al. 2006. Acute myeloid leukemia associated with retroviral gene transfer to hematopoietic progenitor cells of a rhesus macaque. *Blood* 107: 3865–3867.

18. Cavazzana-Calvo, M. and Fischer, A. 2007. Gene therapy for severe combined immunodeficiency: Are we there yet? *J Clin Invest* 117: 1456–1465.

19. Moolten, F. L. and Cupples, L. A. 1992. A model for predicting the risk of cancer consequent to retroviral gene therapy. *Hum Gene Ther* 3: 479–486.

20. Baum, C., Schambach, A., Bohne, J., and Galla, M. 2006. Retrovirus vectors: Toward the plentivirus? *Mol Ther* 13: 1050–1063.

21. Galla, M., Will, E., Kraunus, J., Chen, L., and Baum, C. 2004. Retroviral pseudotransduction for targeted cell manipulation. *Mol Cell* 16: 309–315.

22. Goff, S. P. 2004. Retrovirus restriction factors. *Mol Cell* 16: 849–859.

23. Anderson, J. L. and Hope, T. J. 2005. Intracellular trafficking of retroviral vectors: Obstacles and advances. *Gene Ther* 12: 1667–1678.

24. Suzuki, Y. and Craigie, R. 2007. The road to chromatin-nuclear entry of retroviruses. *Nat Rev Microbiol* 5: 187–196.

25. Yanez-Munoz, R. J., et al. 2006. Effective gene therapy with nonintegrating lentiviral vectors. *Nat Med* 12: 348–353.

26. Philpott, N. J. and Thrasher, A. J. 2007. Use of nonintegrating lentiviral vectors for gene therapy. *Hum Gene Ther* 18: 483–489.

27. Furger, A., Monks, J., and Proudfoot, N. J. 2001. The retroviruses human immunodeficiency virus type 1 and Moloney murine leukemia virus adopt radically different strategies to regulate promoter-proximal polyadenylation. *J Virol* 75: 11735–11746.

28. Kim, S. H., Yu, S. S., Park, J. S., Robbins, P. D., An, C. S., and Kim, S. 1998. Construction of retroviral vectors with improved safety, gene expression, and versatility. *J Virol* 72: 994–1004.

29. Kraunus, J., Zychlinski, D., Heise, T., Galla, M., Bohne, J., and Baum, C. 2006. Murine leukemia virus regulates alternative splicing through sequences upstream of the 5′ splice site. *J Biol Chem* 281: 37381–37390.

30. Hildinger, M., Abel, K. L., Ostertag, W., and Baum, C. 1999. Design of 5′ untranslated sequences in retroviral vectors developed for medical use. *J Virol* 73: 4083–4089.

31. Coffin, J., Hughes, S. H., Varmus, H. E., and Miller, A. D. (Eds.). 2000. *Retroviruses*, Cold Spring Harbor Laboratory Press, Planview.

32. Vogt, B., et al. 2001. Lack of superinfection interference in retroviral vector producer cells. *Hum Gene Ther* 12: 359–365.

33. Lund, A. H., Duch, M., Lovmand, J., Jorgensen, P., and Pedersen, F. S. 1997. Complementation of a primer binding site-impaired murine leukemia virus-derived retroviral vector by a genetically engineered tRNA-like primer. *J Virol* 71: 1191–1195.

34. Yu, S. F., et al. 1986. Self-inactivating retroviral vectors designed for transfer of whole genes into mammalian cells. *Proc Natl Acad Sci U S A* 83: 3194–3198.

35. Dull, T., et al. 1998. A third-generation lentivirus vector with a conditional packaging system. *J Virol* 72: 8463–8471.

36. Trobridge, G., Josephson, N., Vassilopoulos, G., Mac, J., and Russell, D. W. 2002. Improved foamy virus vectors with minimal viral sequences. *Mol Ther* 6: 321–328.

37. Modlich, U., et al. 2006. Cell-culture assays reveal the importance of retroviral vector design for insertional genotoxicity. *Blood* 108: 2545–2553.

38. Montini, E., et al. 2006. Hematopoietic stem cell gene transfer in a tumor-prone mouse model uncovers low genotoxicity of lentiviral vector integration. *Nat Biotechnol* 24: 687–696.

39. Kraunus, J., et al. 2004. Self-inactivating retroviral vectors with improved RNA processing. *Gene Ther* 11: 1568–1578.

40. Schambach, A., et al. 2006. Overcoming promoter competition in packaging cells improves production of self-inactivating retroviral vectors. *Gene Ther* 13: 1524–1533.

41. Xu, K., Ma, H., McCown, T. J., Verma, I. M., and Kafri, T. 2001. Generation of a stable cell line producing high-titer self-inactivating lentiviral vectors. *Mol Ther* 3: 97–104.

42. Scherr, M., Battmer, K., Ganser, A., and Eder, M. 2003. Modulation of gene expression by lentiviral-mediated delivery of small interfering RNA. *Cell Cycle* 2: 251–257.

43. Adam, M. A., Osborne, W. R., and Miller, A. D. 1995. R-region cDNA inserts in retroviral vectors are compatible with virus replication and high-level protein synthesis from the insert. *Hum Gene Ther* 6: 1169–1176.

44. Puthenveetil, G., et al. 2004. Successful correction of the human beta-thalassemia major phenotype using a lentiviral vector. *Blood* 104: 3445–3453.

45. Imren, S., et al. 2004. High-level beta-globin expression and preferred intragenic integration after lentiviral transduction of human cord blood stem cells. *J Clin Invest* 114: 953–962.

46. May, C., Rivella, S., Chadburn, A., and Sadelain, M. 2002. Successful treatment of murine beta-thalassemia intermedia by transfer of the human beta-globin gene. *Blood* 99: 1902–1908.

47. Pawliuk, R., et al. 2001. Correction of sickle cell disease in transgenic mouse models by gene therapy. *Science* 294: 2368–2371.

48. Persons, D. A., et al. 2003. Successful treatment of murine beta-thalassemia using in vivo selection of genetically modified, drug-resistant hematopoietic stem cells. *Blood* 102: 506–513.

49. Zaiss, A. K., Son, S., and Chang, L. J. 2002. RNA 3′ readthrough of oncoretrovirus and lentivirus: Implications for vector safety and efficacy. *J Virol* 76: 7209–7219.

50. Yang, Q., Lucas, A., Son, S., and Chang, L. J. 2007. Overlapping enhancer/promoter and transcriptional termination signals in the lentiviral long terminal repeat. *Retrovirology* 4: 4.

51. Zufferey, R., Donello, J. E., Trono, D., and Hope, Z. J. 1999. Woodchuck hepatitis virus posttranscriptional regulatory element enhances expression of transgenes delivered by retroviral vectors. *J Virol* 73: 2886–2892.

52. Schambach, A., Wodrich, H., Hildinger, M., Bohne, J., Krausslich, H. G., and Baum, C. 2000. Context dependence of different modules for posttranscriptional enhancement of gene expression from retroviral vectors. *Mol Ther* 2: 435–445.

53. Hope, T. 2002. Improving the post-transcriptional aspects of lentiviral vectors. *Curr Top Microbiol Immunol* 261: 179–189.

54. Schambach, A., Galla, M., Maetzig, T., Loew, R., and Baum, C. 2007. Improving transcriptional termination of self-inactivating gamma-retroviral and lentiviral vectors. *Mol Ther* 15: 1167–1173.

55. Higashimoto, T., et al. 2007. The woodchuck hepatitis virus post-transcriptional regulatory element reduces readthrough transcription from retroviral vectors. *Gene Ther* 14: 1298–1304.

56. Gilmartin, G. M., Fleming, E. S., Oetjen, J., and Graveley, B. R. 1995. CPSF recognition of an HIV-1 mRNA 3′-processing enhancer: Multiple sequence contacts involved in poly(A) site definition. *Genes Dev* 9: 72–83.

57. Bell, A. C., West, A. G., and Felsenfeld, G. 1999. The protein CTCF is required for the enhancer blocking activity of vertebrate insulators. *Cell* 98: 387–396.

58. Emery, D. W., Yannaki, E., Tubb, J., and Stamatoyannopoulos, G. 2000. A chromatin insulator protects retrovirus vectors from chromosomal position effects. *Proc Natl Acad Sci U S A* 97: 9150–9155.

59. Rivella, S., Callegari, J. A., May, C., Tan, C. W., and Sadelain, M. 2000. The cHS4 insulator increases the probability of retroviral expression at random chromosomal integration sites. *J Virol* 74: 4679–4687.

60. Ramezani, A., Hawley, T. S., and Hawley, R. G. 2003. Performance- and safety-enhanced lentiviral vectors containing the human interferon-beta scaffold attachment region and the chicken beta-globin insulator. *Blood* 101: 4717–4724.

61. Aker, M., et al. 2007. Extended core sequences from the cHS4 insulator are necessary for protecting retroviral vectors from silencing position effects. *Hum Gene Ther* 18: 333–343.

62. Evans-Galea, M. V., Wielgosz, M. M., Hanawa, H., Srivastava, D. K., and Nienhuis, A. W. 2007. Suppression of clonal dominance in cultured human lymphoid cells by addition of the cHS4 insulator to a lentiviral vector. *Mol Ther* 15: 801–809.

63. Robert-Richard, E., Richard, E., Malik, P., Ged, C., de Verneuil, H., and Moreau-Gaudry, F. 2007. Murine retroviral but not human cellular promoters induce in vivo erythroid-specific deregulation that can be partially prevented by insulators. *Mol Ther* 15: 173–182.

64. Li, T. and Zhang, J. 2004. Stable expression of three genes from a tricistronic retroviral vector containing a picornavirus and 9-nt cellular internal ribosome entry site elements. *J Virol Methods* 115: 137–144.

65. Klump, H., Schiedlmeier, B., Vogt, B., Ryan, M., Ostertag, W., and Baum, C. 2001. Retroviral vector-mediated expression of HoxB4 in hematopoietic cells using a novel coexpression strategy. *Gene Ther* 8: 811–817.

66. Szymczak, A. L. and Vignali, D. A. 2005. Development of 2A peptide-based strategies in the design of multicistronic vectors. *Expert Opin Biol Ther* 5: 627–638.

67. Ryan, M. D. and Drew, J. 1994. Foot-and-mouth disease virus 2A oligopeptide mediated cleavage of an artificial polyprotein. *EMBO J* 13: 928–933.

68. Szymczak, A. L., et al. 2004. Correction of multi-gene deficiency in vivo using a single "self-cleaving" 2A peptide-based retroviral vector. *Nat Biotechnol* 22: 589–594.

69. Amendola, M., Venneri, M. A., Biffi, A., Vigna, E., and Naldini, L. 2005. Coordinate dual-gene transgenesis by lentiviral vectors carrying synthetic bidirectional promoters. *Nat Biotechnol* 23: 108–116.

70. Baron, U., Freundlieb, S., Gossen, M., and Bujard, H. 1995. Co-regulation of two gene activities by tetracycline via a bidirectional promoter. *Nucleic Acids Res* 23: 3605–3606.

71. Lewinski, M. K., et al. 2006. Retroviral DNA integration: Viral and cellular determinants of target-site selection. *PLoS Pathog* 2: e60.

72. Mitchell, R. S., et al. 2004. Retroviral DNA integration: ASLV, HIV, and MLV show distinct target site preferences. *PLoS Biol* 2: E234.

73. Hayward, W. S., Neel, B. G., and Astrin, S. M. 1981. Activation of a cellular onc gene by promoter insertion in ALV-induced lymphoid leukosis. *Nature* 290: 475–480.

74. Derse, D., et al. 2007. Human T-cell leukemia virus type 1 integration target sites in the human genome: Comparison with those of other retroviruses. *J Virol* 81: 6731–6741.

75. Bushman, F. D. 2003. Targeting survival: Integration site selection by retroviruses and LTR-retrotransposons. *Cell* 115: 135–138.

76. Urnov, F. D., et al. 2005. Highly efficient endogenous human gene correction using designed zinc-finger nucleases. *Nature* 435: 646–651.

77. Cathomen, T. and Weitzman, M. D. 2005. Gene repair: Pointing the finger at genetic disease. *Gene Ther* 12: 1415–1416.

78. Morita, S., Kojima, T., and Kitamura, T. 2000. Plat-E: An efficient and stable system for transient packaging of retroviruses. *Gene Ther* 7: 1063–1070.

79. Cone, R. D. and Mulligan, R. C. 1984. High-efficiency gene transfer into mammalian cells: Generation of helper-free recombinant retrovirus with broad mammalian host range. *Proc Natl Acad Sci U S A* 81: 6349–6353.

80. Miller, A. D., Garcia, J. V., von Suhr, N., Lynch, C. M., Wilson, C., and Eiden, M. V. 1991. Construction and properties of retrovirus packaging cells based on gibbon ape leukemia virus. *J Virol* 65: 2220–2224.

81. Takeuchi, Y., Cosset, F. L., Lachmann, P. J., Okada, H., Weiss, R. A., and Collins, M. K. 1994. Type C retrovirus inactivation by human complement is determined by both the viral genome and the producer cell. *J Virol* 68: 8001–8007.

82. Yang, Y., et al. 1995. Inducible, high-level production of infectious murine leukemia retroviral vector particles pseudotyped with vesicular stomatitis virus G envelope protein. *Hum Gene Ther* 6: 1203–1213.

83. Beyer, W. R., Westphal, M., Ostertag, W., and von Laer, D. 2002. Oncoretrovirus and lentivirus vectors pseudotyped with lymphocytic choriomeningitis virus glycoprotein: Generation, concentration, and broad host range. *J Virol* 76: 1488–1495.

84. Sandrin, V., Russell, S. J., and Cosset, F. L. 2003. Targeting retroviral and lentiviral vectors. *Curr Top Microbiol Immunol* 281: 137–178.

85. Pichlmair, A., et al. 2007. Tubulovesicular structures within vesicular stomatitis virus G protein-pseudotyped lentiviral vector preparations carry DNA and stimulate antiviral responses via Toll-like receptor 9. *J Virol* 81: 539–547.

86. Du, Y., Spence, S. E., Jenkins, N. A., and Copeland, N. G. 2005. Cooperating cancer-gene identification through oncogenic-retrovirus-induced insertional mutagenesis. *Blood* 106: 2498–2505.

87. Kustikova, O. S., et al. 2007. Retroviral vector insertion sites associated with dominant hematopoietic clones mark "stemness" pathways. *Blood* 109: 1897–1907.

88. Li, Z., et al. 2007. Insertional mutagenesis by replication-defective retroviral vectors encoding the large T oncogene. *Ann NY Acad Sci* 1106: 95–113.

89. Miething, C., et al. 2007. Retroviral insertional mutagenesis identifies RUNX genes involved in chronic myeloid leukemia disease persistence under imatinib treatment. *Proc Natl Acad Sci U S A* 104: 4594–4599.

90. Deichmann, A., et al. 2007. Vector integration is nonrandom and clustered and influences the fate of lymphopoiesis in SCID-X1 gene therapy. *J Clin Invest* 117: 2225–2232.

91. Schwarzwaelder, K., et al. 2007. Gammaretrovirus-mediated correction of SCID-X1 is associated with skewed vector integration site distribution in vivo. *J Clin Invest* 117: 2241–2249.

92. Aiuti, A., et al. 2007. Multilineage hematopoietic reconstitution without clonal selection in ADA-SCID patients treated with stem cell gene therapy. *J Clin Invest* 117: 2233–2240.

93. Recchia, A., et al. 2006. Retroviral vector integration deregulates gene expression but has no consequence on the biology and function of transplanted T cells. *Proc Natl Acad Sci U S A* 103: 1457–1462.

94. Takahashi, K. and Yamanaka, S. 2006. Induction of pluripotent stem cells from mouse embryonic and adult fibroblast cultures by defined factors. *Cell* 126: 663–676.

95. Meissner, A., Wernig, M., and Jaenisch, R. 2007. Direct reprogramming of genetically unmodified fibroblasts into pluripotent stem cells. *Nat Biotechnol* 25: 1177–1181.

96. Grez, M., Akgun, E., Hilberg, F., and Ostertag, W. 1990. Embryonic stem cell virus, a recombinant murine retrovirus with expression in embryonic stem cells. *Proc Natl Acad Sci U S A* 87: 9202–9206.

97. Hawley, R. G., Lieu, F. H., Fong, A. Z., and Hawley, T. S. 1994. Versatile retroviral vectors for potential use in gene therapy. *Gene Ther* 1: 136–138.

98. Baum, C., Hegewisch-Becker, S., Eckert, H. G., Stocking, C., and Ostertag, W. 1995. Novel retroviral vectors for efficient expression of the multidrug-resistance (mdr-1) gene in early hemopoietic cells. *J Virol* 69: 7541–7547.

99. Challita, P. M., Skelton, D., el-Khoueiry, A., Yu, X. J., Weinberg, K., and Kohn, D. B. 1995. Multiple modifications in cis elements of the long terminal repeat of retroviral vectors lead to increased expression and decreased DNA methylation in embryonic carcinoma cells. *J Virol* 69: 748–755.

100. Haas, D. L., et al. 2003. The Moloney murine leukemia virus repressor binding site represses expression in murine and human hematopoietic stem cells. *J Virol* 77: 9439–9450.

101. Laker, C., et al. 1998. Host cis-mediated extinction of a retrovirus permissive for expression in embryonal stem cells during differentiation. *J Virol* 72: 339–348.

102. Hamaguchi, I., et al. 2000. Lentivirus vector gene expression during ES cell-derived hematopoietic development in vitro. *J Virol* 74: 10778–10784.

103. Ellis, J. 2005. Silencing and variegation of gammaretrovirus and lentivirus vectors. *Hum Gene Ther* 16: 1241–1246.

104. Hong, S., et al. 2007. Functional analysis of various promoters in lentiviral vectors at different stages of in vitro differentiation of mouse embryonic stem cells. *Mol Ther* 15: 1630–1639.

105. Lois, C., Hong, E. J., Pease, S., Brown, E. J., and Baltimore, D. 2002. Germline transmission and tissue-specific expression of transgenes delivered by lentiviral vectors. *Science* 295: 868–872.

106. Hanawa, H., et al. 2002. Comparison of various envelope proteins for their ability to pseudotype lentiviral vectors and transduce primitive hematopoietic cells from human blood. *Mol Ther* 5: 242–251.

107. Schambach, A., et al. 2006. Lentiviral vectors pseudotyped with murine ecotropic envelope: Increased biosafety and convenience in preclinical research. *Exp Hematol* 34: 588–592.

108. Strang, B. L., Ikeda, Y., Cosset, F. L., Collins, M. K., and Takeuchi, Y. 2004. Characterization of HIV-1 vectors with gammaretrovirus envelope glycoproteins produced from stable packaging cells. *Gene Ther* 11: 591–598.

109. Relander, T., et al. 2005. Gene transfer to repopulating human CD34 + cells using amphotropic-, GALV-, or RD114-pseudo-typed HIV-1-based vectors from stable producer cells. *Mol Ther* 11: 452–459.

110. Trubetskoy, A. M., Okenquist, S. A., and Lenz, J. 1999. R region sequences in the long terminal repeat of a murine retrovirus specifically increase expression of unspliced RNAs. *J Virol* 73: 3477–3483.

2 Production of Retroviral Vectors: From the Producer Cell to the Final Product

Pedro E. Cruz, Ana S. Coroadinha, Teresa Rodrigues, and Hansjörg Hauser

CONTENTS

2.1 INTRODUCTION

Retroviral vectors have become an important tool for gene transfer, murine leukemia virus (MLV)-based retroviral vectors have been used over 25 years in gene therapy clinical trials. Application of these vectors leads to integration of the transgene into the host-cell chromosome resulting in prolonged transgene expression; thus, they are particularly suited for diseases where long-term expression of the therapeutic gene is desirable. Traditionally, retroviruses have been the

TABLE 2.1

Examples of *ex Vivo* Clinical Trials Where Onco-Retroviral Vectors Were Used as Gene Therapy Vectors for the Treatment of Several Immunodeficiencies

Therapeutic Gene	Target Disease	Administration	Dose per Patient	Transduction Efficiency (%)	Producer Cells	References
Gp91phox	X-CGD	Transduced CD34$^+$ cells	3.6–5.1×10^6 transduced cells/kg	40–45	PG13	(Ott et al., 2006)
γ_c chain	SCID-X1		5.7–6.1×10^6 transduced cells/kg	20–40	ψ-CRIP	(Cavazzana-Calvo et al., 2000)
ADA	ADA-SCID	Transduced T-cells	0.2–2.2×10^6 transduced cells/kg	21–25	Gp + Am12	(Aiuti et al., 2002)
			1–10×10^9 transduced cells	1–10	PA317	(Blaese et al., 1995)

vectors of choice for *ex vivo* transduction of hematopoietic stem cells (Table 2.1). The first clinical trial where retroviral vectors were applied occurred in 1990 to correct adenosine deaminase (ADA) deficiency (Blaese et al., 1995). White blood cells isolated from two children with severe combined immunodeficiency (ADA–SCID) were transduced *ex vivo* with an MLV-based vector expressing ADA and a neomycin marker. Neomycin resistant cells were isolated and introduced into the patients. The treatment improved the physical conditions of the patients and the ADA-containing provirus was stable in the blood for several years. It was concluded that gene therapy could be an effective addition to the treatment for some patients with this severe immunodeficiency disease. Since these first studies, gene therapy has become the focus of a continuously expanding research activity. The demonstration of the full correction of the SCID-X1 phenotype in infants and the recent treatment of CGD further demonstrated the efficacy of retroviral vectors (Cavazzana-Calvo et al., 2000; Ott et al., 2006). Retroviral vectors have been used for the treatment of other monogenic disorders *ex vivo*, the Gaucher disease, by transferring the glucocerebrosidase gene into hematopoietic cells (Dunbar et al., 1998) and hypercholesterolemia, by transferring the low-density lipoprotein receptor into hepatocytes (Grossman et al., 1994).

Retroviruses have also shown some promising results in cancer therapy and bone marrow transplantation. In tumor therapy, retroviral vectors have been mainly applied *ex vivo* to transduce immunomodulatory molecules (IL-2 and IFN-γ) into neoplastic cells and tumor infiltrating lymphocytes (Tan et al., 1996; Nemunaitis et al., 1998; Palmer et al., 1999). In other cases, retroviral vectors were applied *in vivo* for the treatment of melanoma, glioblastoma, ovarian, breast, prostate, and lung cancers (Roth and Cristiano, 1997). The introduction of the *thymidine kinase* suicide gene and the administration of ganciclovir showed that the treatment of graft-versus-host disease was efficient (Bonini et al., 1997).

Clinical trials to treat HIV-1 infection with therapeutic retroviral vectors were also performed. One of the strategies was the *ex vivo* modification of syngenic lymphocytes using retroviral vectors to suppress the expression of HIV genes (Nabel et al., 1994; Morgan and Walker, 1996; Wong-Staal et al., 1998). A second strategy involves the modification of autologous fibroblasts to express a part of the HIV envelope gene so that a host immune response could be elicited (Nabel et al., 1994; Morgan and Walker, 1996; Wong-Staal et al., 1998).

2.2 BIOLOGY OF RETROVIRUSES

2.2.1 VIRAL PARTICLE

Retroviruses are complex enveloped particles with a lipid bilayer surrounding a nucleocapsid (NC) containing the genetic material and the enzymes necessary for replication (Coffin et al., 1997). They belong to the RNA family of enveloped viruses with the ability to "reverse transcribe" their genome from RNA to DNA (Weiss, 1998). Virions measure around 120 nm in diameter and contain two identical copies of positive strand RNA genome, forming a dimer, complexed with nucleocapsid proteins. The enzymes reverse transcriptase (RT), integrase (IN), and protease (PR) are also contained in the nucleocapsid, the inner portion of the virus (Palu et al., 2000). A protein shell, formed by capsid proteins (CA), encloses the NC, the inner portion of the viral core (Jones and Morikawa, 1998). The matrix (MA) proteins form a layer outside the core and interact with the envelope, this consists of a lipid bilayer derived from the cellular membrane, which surrounds the viral core particle (Coffin et al., 1997). Anchored on the membrane envelope the viral envelope glycoproteins are responsible for virus interaction with specific receptors on the host cell. The envelope is a protein complex formed by two protein subunits: transmembrane (TM) and surface (SU) connected by disulfide linkage, the latter binding to the cellular receptor and the former participating in the fusion of the viral and cellular membranes (Opstelten et al., 1998; Lavillette et al., 2000) (Figure 2.1).

2.2.2 RETROVIRUS GENOME AND VIRAL REPLICATION

On the basis of their genome structure, retroviruses can be classified into simple (e.g., MLV) or complex retroviruses (e.g., HIV) (Coffin et al., 1997). Both encode four gene

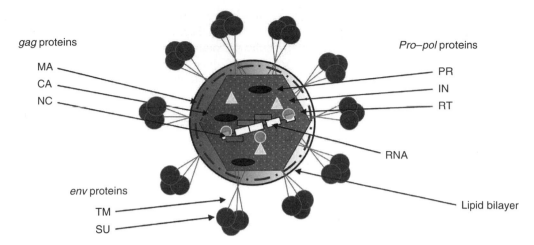

FIGURE 2.1   Structure of a retrovirus particle.

families, *gag* (group specific antigen), *pro* (PR), *pol* (polymerase), and *env* (envelope) (Figure 2.2). The *gag* sequence codes for the three main structural proteins: MA, CA, and NC (additionally also codes for p12 which assists in the virion assembly). The *pro* gene, allocated between *gag* and *pol*, encodes PRs responsible for Gag and Gag–Pol cleavage and viral particle maturation during budding. The *pol* sequences encode the enzymes RT and IN, the former being responsible for transcription of viral RNA to DNA during the infection process and the latter catalyzing the integration of the proviral DNA into the host genome. The *env* sequence encodes the two subunits of the envelope glycoprotein, SU and TM (Figures 2.1 and 2.2).

Efficient gene transduction and integration depends on a number of *cis*-acting sequences present in the retroviral genome: two long-terminal repeats (LTRs), a viral packaging signal (ψ), signals for reverse transcription such as the primer binding site (PBS), and the polypurine tract (PPT). The LTRs contain elements required to drive expression, reverse transcription and integration into the host chromosome (Hu and Pathak, 2000). The packaging signal (ψ) is the sequence that interacts with the viral proteins allowing the specific packaging of the viral RNA. PBS and PPT are required for reverse transcription.

The full-length RNA transcript produced by the integrated provirus encodes the precursor proteins for *gag*, *pro*, and *pol*

but in the unspliced form serves as the viral genome. It is assembled into progeny virions at the cell membrane. A portion of the full-length RNA is also spliced to form *env* mRNA which encodes a precursor for the virion membrane protein complex. Thus, the full-length RNA serves three essential functions: as genomic RNA, as mRNA for *gag*, *pro*, and *pol* proteins, and as pre-RNA for *env* mRNA.

The envelope proteins suffer posttranslation modifications in the Golgi, the addition of N-linked carbohydrate side chains. The cleavage of the precursors occurs also in the Golgi apparatus into the SU and TM peptides. They are transported, associated to each other, to the cell membrane where assembly of the virus occurs (Hunter and Swanstrom, 1990; Hansen et al., 1993).

Hence, the retrovirus replication cycle can be summarized in three main steps: (1) interaction and fusion of the viral particles with the cell surface, (2) uncoating of the virus, reverse transcription and integration of the viral genome, and (3) synthesis, assembly, and budding of newly formed virus. In step (1) the viral envelope glycoproteins bind to specific receptor complexes on the host-cell surface, leading to membrane fusion of the virus to the cell, and to subsequent internalization of the virus. Subsequently in step (2) the virus core is released into the cytoplasm where it is partially degraded to form a large nucleoprotein (pre-integration complex) that is transported to the nucleus (Palu et al., 2000). For MLV the entry in the nucleus is mitose-dependent (Roe et al., 1993; Lewis and Emerman, 1994). Oncoretroviruses are not able to pass through the pores of the nuclear membrane and only gain access to it when the nuclear membrane is disassembled during mitosis. The double-stranded proviral DNA synthesized by RT in the cytoplasm is integrated in the host's genome by viral enzymes. In step (3) the integrated provirus is transcribed into new viral genomic RNA and into viral proteins, all of this making use of cellular enzymes and machineries. Subsequently assembling of virions occurs, forming new viruses that bud out of the host cell (Coffin et al., 1997).

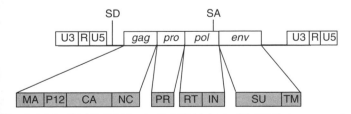

FIGURE 2.2   Proviral structure of murine leukaemia virus with a genome size of 8.8 kb showing the different gene families and the viral proteins each codifies.

2.3 CONVERTING A VIRUS INTO A VECTOR: REPLICATION-DEFECTIVE RETROVIRAL VECTORS

In most gene therapy applications it is not desirable to deliver replication-competent retroviruses (RCRs) because they may spread beyond the target tissue and cause adverse effects. Therefore, most recombinant retroviral systems designed for gene delivery are divided into a vector, expressing the therapeutic gene and a helper construct that encodes the viral proteins, but limiting the ability of the virus to replicate (Miller et al., 1986). The helper or packaging functions are provided by a helper virus, as co-transfected plasmids, but mostly in a packaging cell format. The latter has the advantage of generating higher virus titers, reproducible batches, and higher safety. Helper or packaging cells are therefore more suitable for clinical applications and thus the following chapters will only focus on this application.

2.3.1 HELPER OR PACKAGING CELLS

Several packaging cell lines producing retroviral vectors have been generated; in these, the sequences expressing the viral genes, *gag*, *pro*, *pol*, and *env* (packaging functions) are supplied in *trans* using molecular constructs which cannot be packaged into retroviral particles. The therapeutic gene is supplied into a genetic vector that mimics the structure of the viral genome by containing minimal *cis*-acting sequences, which allow its efficient incorporation into viral particles, its reverse transcription, integration, and expression in the target cells (Palu et al., 2000).

Retroviral vectors have been derived from several viruses, namely avian, simian, or feline retroviruses. However, the most common and widely used is derived from a murine retrovirus, the MLV. In the design of retroviral packaging cell lines, the problem of spontaneous formation of RCRs has been addressed and this led to different types of helper cell lines that can be divided into "generations." The

first generation of packaging cells was produced by providing the packaging functions with a retroviral genome where the packaging signal was deleted from the 5′ untranslated region, thus preventing the incorporation of the packaging functions into the viral particles (Cone and Mulligan, 1984). However, one recombination event between the vector and the packaging genome could easily lead to the creation of recombinant helper free viruses (Miller et al., 1986). Further modifications were introduced in the second generation packaging cells where the 3′LTR and the second strand initiation site were replaced with the polyadenylation site of SV40. Additionally, the 5′ end of the 5′LTR was deleted. Two recombination events would now be required to restore a replication-competent virus. The well-known PA317 cell line was constructed in this way (Table 2.2). The third generation further separates the constructs that express, *gag–pol* and *env*. In the case of ψ-CRIP and ψ-CRE cells, *gag–pro–pol* and *env* were cloned in separate constructs (Figure 2.3) (Danos and Mulligan, 1988). The resulting system requires three independent recombination events to generate replication-competent virus, which is statistically very improbable. Nevertheless, replication-competent viruses can still occur in third generation cell lines (Chong and Vile, 1996; Chong et al., 1998). This was reduced by decreasing the homology in the vector construct, using different LTR species to those used in the packaging functions (Cosset et al., 1995) or alternatively, using heterologous promoters such as CMV (Soneoka et al., 1995; Rigg et al., 1996). The majority of the cell lines developed for stable production of retroviral vectors are derived from murine NIH/3T3 cells but have been slowly substituted by human cell lines such as HT1080 or HEK 293 (Rigg et al., 1996) (see Section 4.4).

2.3.2 ENVELOPE PROTEINS AND VECTOR TROPISM

The separation of the three retroviral components, *gag–pol*, viral genome, and envelope, in different constructs in the packaging cell line allowed recombinant retroviral vectors to be

TABLE 2.2
Retroviral Vector Producer Cell Lines

Producer Cell Name	Cell Line	Envelope	Maximal Titers (IP/mL)
PA317	Murine NIH/3T3	Amphotropic	3×10^6
PG13	Murine NIH/3T3	GaLV	5×10^6
ψ-CRIP	Murine NIH/3T3	Amphotropic	6×10^6
Gp + *env*Am12	NIH 3T3	Amphotropic	1×10^6
FLY A4	Human HT1080	Amphotropic	2×10^7
FLY RD18	Human HT1080	RD114	1.2×10^5
Te Fly Ga18	Human Te671	GaLV	5×10^6
Phoenix	Human 293	Amphotropic	1×10^5
293-SPA	Human 293	Amphotropic	6×10^6
PUZIkat2	Human 293	Amphotropic	7×10^6
Flp293A	Human 293	Amphotropic	2×10^7
293 FLEX	Human 293	GaLV	3×10^6
CEM FLY	Human CEM	Amphotropic	2.2×10^6

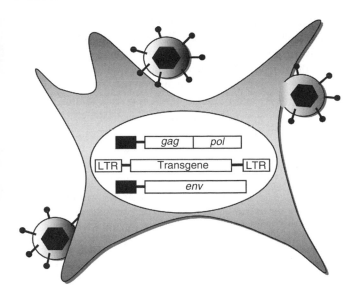

FIGURE 2.3 Structure of a retroviral producer cell: the helper functions (*gag, pro, pol,* and *env*) are supplied in two different constructs and the vector expressing the transgene or therapeutic gene is supplied in a third construct (only this is packaged in the retroviral particles).

pseudotyped with SU proteins from different viruses. This technique can be useful to determine the tropism of a virus that can be either expanded to an ensemble of target cells or restricted to a specific cell type. A wide range of packaging cell lines with different pseudotypes is currently available (Table 2.2).

Four subgroups of naturally occurring MLVs were identified based on their receptor recognition in mouse cells. Accordingly, their envelope proteins are commonly referred to: ecotropic, amphotropic, polytropic, and xenotropic. For example, the most commonly used envelope to pseudotype retroviral vectors are the amphotropic MLV envelopes 4070A or 10A1, which can transduce nearly all mammalian cells (Miller and Chen, 1996). Other retroviral helper cells produce pseudotyped viruses that are derived from surface glycoproteins from different retrovirus species or of different virus families. Pseudotyping with the glycoprotein from GaLV (gibbon ape leukemia virus) has come into practice as it transduces hematopoietic cells more efficiently than the same retroviral vectors with amphotropic pseudotype (Miller et al., 1991). The glycoprotein from the cat endogenous virus RD114 allows efficient transduction of human cells and, in addition, confers partial resistance to human serum (Takeuchi et al., 1994). The HIV-1 envelope targets specific CD4 positive cells and can be incorporated in MLV particles (Schnierle et al., 1996; Lodge et al., 1998). SU proteins of less related virus have also been successfully incorporated into retroviral vectors. This is the case with VSV-G (vesicular stomatitis virus G protein) that allows efficient infection of a broad range of target cells. VSV-G pseudotyped viral vectors have the advantage of resisting ultracentrifugation due to the monomeric structure of VSV-G. Unfortunately, this protein can be toxic to cells which constrains its constitutive expression in stable packaging cell lines (Burns et al., 1993; Yee et al., 1994a,b).

Genetically or chemically modified envelope glycoproteins have also been developed for cell or tissue specific targeting. Retroviral envelope proteins can tolerate a variety of genetically encoded modifications. Polypeptides can be fused to N-terminally truncated envelope glycoproteins, thereby changing the SU portion of the envelope to interact with other cell surface molecules. This is often achieved by deletion of a part of the coding region of SU and replacing it with regions of other proteins such as erythropoietin, heregulin, insulin-like growth factor I, and single chain variable fragment antibodies against various proteins (Chu et al., 1994; Kasahara et al., 1994; Chu and Dornburg, 1995; Han et al., 1995; Somia et al., 1995; Jiang et al., 1998; Konishi et al., 1998; Chadwick et al., 1999; Jiang and Dornburg, 1999).

Another pseudotyping strategy makes use of bifunctional adaptor molecules to redirect retrovirus attachment. The adaptor works as a molecular bridge between the viral envelope and a receptor molecule on the cell surface. Multivalent antibody linkers or biotin streptavidin bridges are examples of such. However, efficient transduction was generally impaired by the lack of proper envelope–receptor interaction (Goud et al., 1988; Etienne-Julan et al., 1992). The earliest example of a chemical modification of the binding characteristics was the modification of ecotropic MLV particles by the addition of lactose to their surface in order to target the asialoglycoprotein receptor on human hepatocytes (Neda et al., 1991).

2.3.3 DESIGN OF VIRAL GENOME AND THERAPEUTIC TRANSGENE

The simplest approach to generate a viral vector is to use the retroviral LTR promoter to control the expression of the desired cDNA or protein of interest. The minimal *cis*-acting sequences needed are the following regions initially identified as essential for the correct function of the vector: two LTRs, the PBS, the packaging signal extending into the 5′ of the *gag* sequence and, the downstream untranslated region containing the PPT.

The design of the vector genome can be modified in order to include several genetic elements that render it more efficient in specific treatments and also in order to increase the vector safety. Alternative internal promoters and IRES (internal ribosomal entry site) elements can be used to drive the expression of one or more proteins. Selectable markers (e.g., neomycin, thymidine kinase, hypoxanthine–guanine phosphoribosyl transferase) can be used to facilitate selection of transduced cells.

To increase vector safety, the regulatory sequences in the LTRs can be deleted as in self-inactivating or self-deleting vectors (Yu et al., 1986). The viral vector genome design and safety will be thoroughly discussed in another chapter.

2.3.4 RCRs: ACCIDENTAL FORMATION OF NEWLY RECOMBINED VIRUSES

Depending on the construction of helper cell lines retrovirus-mediated gene transfer can give rise to newly recombined viruses (Bosselman et al., 1987; Scarpa et al., 1991; Otto et al., 1994). As a consequence, infected cells obtain undesired properties or even become producer cells themselves, a process which renders the transfer uncontrollable.

There are two predominant mechanisms that are responsible for alteration of the properties of infectants: recombination and retrofection. Homologous recombination is a frequent process. The efficiency depends on the degree of homology between the retroviral vector and the helper function. As a result of a recombination event immobilized helper functions regain their capacity to replicate autonomously and thus RCRs spread through the culture. The RCRs represent the main safety problem accompanying the retrovirus-mediated gene transfer; amphotropic RCRs obtained from recombination events in MLV-based packaging cells were shown to replicate in nonhuman primates and to induce lymphoma (Donahue et al., 1992; Vanin et al., 1994).

Apart from the frequent recombination during transfection of cells (extrachromosomal DNA recombination), during cultivation of retroviral helper cells two additional recombination pathways have to be considered. Generally, recombination may occur within transfected packaging cells at the DNA level or during reverse transcription at the RNA level. While DNA recombination is a purely cellular process performed by the cellular repair machinery, RNA recombination is coupled with the retroviral life cycle and mediated by a viral enzyme, RT.

A prerequisite for RNA-based recombination is the co-packaging of the vector and the RNA into the same virus particle. Although the RNA lacks the ψ packaging signal it may be packaged at a low rate. It was shown that the deletion of the ψ packaging signal may reduce the rate of packaging only about 10^3–10^4 fold (Mann and Baltimore, 1985; Stuhlmann et al., 1990). Therefore, packaging cell lines producing 10^6 virus particles per milliliter release a significant number of virions carrying the helper RNA. These helper RNAs may recombine and thereby give rise to the formation of RCRs.

To increase the level of safety much effort was spent to decrease the risk of recombination in packaging cells. The subdivision of the helper genome in two independent transcription units resulted in helper cells in which more than one recombination event is required for the formation of RCRs. In addition, the extent of homology between vector and helper genes was significantly reduced. In these packaging cells (e.g., ΩE cells [Morgenstern and Land, 1990]) the frequency of RCR formation is significantly reduced.

Further investigations showed that even cellular RNAs without any retroviral elements may be packaged to a low extent, a process which was termed retrofection (Linial, 1987; Linial and Miller, 1990). In MLV-based packaging cells the rate of retrofection is reduced 10^6–10^8 fold if compared to functional retroviral vectors (Stuhlmann et al., 1990). The risk arising from the packaging of cellular genes is probably low as their RNAs may not compensate for any viral function and thus could not contribute to the formation of RCRs. However, it cannot be excluded that certain cellular RNAs may form a structure that is similar to the retroviral ψ sequence

and are preferentially transmitted. A safety problem arises if these RNAs are able to change the properties of infected cells, as for example the oncogenes do.

Another potential problem using the retrovirus-mediated gene transfer arises from endogenous retroviral elements that are naturally present in the cells, such as endogenous retroviruses with the vector helper functions. In particular, rodent cells carry a huge amount of endogenous retroviral elements. Usually, they are not able to form infectious viruses themselves; however, they are potential candidates for recombination events both at DNA and RNA level and thus may contribute to the formation and release of new, autonomously replicating retroviruses. Initially the majority of the cell lines developed for stable production of retroviral vectors are derived from NIH/3T3 cells (Danos and Mulligan, 1988; Morgenstern and Land, 1990; Miller et al., 1991) which were found to express endogenous MLV sequences (Irving et al., 1993). These can be packaged and participate in recombination to form replication-competent viruses (Scadden et al., 1990; Cosset et al., 1993; Vanin et al., 1994). Thus, non-rodent cells (e.g., human cells) were introduced and are nowadays preferred. Recently constructed packaging cell lines are derived from human cell lines 293 that do not have MLV-related endogenous sequences (Rigg et al., 1996; Patience et al., 1998).

2.4 STATE OF THE ART STABLE RETROVIRAL VECTOR PRODUCER CELL LINES

One of the major disadvantages of retroviral vectors concerns the difficulty in manufacturing. The retroviral vector productions yield relatively low titers as a result of low producer cell line productivity and short vector half-life (Wikstrom et al., 2004). However, gene transfer clinical protocols often require large amounts of biologically active recombinant retrovirus (Table 2.1). Thus, the successful implementation of retroviral vectors in such trials depends to a great extent on the existence of high titer stable producer cells.

2.4.1 CLASSICAL RETROVIRAL PRODUCER CELL LINES

Historically, most packaging cell lines used to produce recombinant retrovirus have been derived from murine cell lines, e.g., NIH/3T3. These are being replaced by human cell lines. This is not only due to the possibility of packaging endogenous retroviral sequences and recombination of these with the vector as mentioned above, but also because of the glycosylation pattern of the murine cell lines that render the virus sensitive to human complement. It is generally considered that the presence of galactosyl(α1–3)galactose carbohydrate moieties produced by murine cells in the retroviral vector envelope is rapidly detected by antibodies and inactivated by the human complement system (Cosset et al., 1995; Pensiero et al., 1996; Takeuchi et al., 1996; Takeuchi et al., 1997), although other translational modifications might also play a role (Mason et al., 1999). The murine packaging cell lines also have the disadvantage of producing lower retroviral infectious titers

(Naviaux et al., 1996; Davis et al., 1997). Producer cell lines based on human cell lines were therefore developed being able to produce retroviral vectors resistant to complement inactivation (Cosset et al., 1995). The human embryonic kidney cell line HEK 293 has been shown to produce virus that yields higher transduction efficiencies (Finer et al., 1994; Rigg et al., 1996) as compared to murine ψ-CRIP-based producer cells (Davis et al., 1997). Table 2.2 is a partial list of some of the available retroviral producer cell lines.

2.4.2 Novel Retroviral Vector Producer Cell Lines

The viral titer produced by packaging cells limits the use of retroviral vectors in gene therapy. The titer depends on a number of items and their interplay, including the nature of the cell line, the expression of the vector and the helper functions, the balance of helper function expression, and on the cultivation conditions. It seems that a maximal expression of the viral vector construct genome, which expresses the transgene, is one of the limiting conditions. The need for screening for appropriate expression levels of the viral vector indicates that this parameter is critical for virus production. This is supported by previous work indicating that optimal virus production depends on a high retroviral vector transgene transcription and balanced expression of the retroviral helper genes encoding *gag–pol* and *env* (Yap et al., 2000).

In addition, the expression of the helper functions must be sufficient to allow packaging of the transcribed viral RNA. Thus, in the development of helper cell lines, the maximization of helper function expression was an important issue. Improvement was achieved by different means. Extensive screenings have to be undertaken in order to identify clones that lead to a high-titer virus production with stability for the time of production.

The inserting of selectable markers in the packaging constructs downstream of the viral genes, so they are translated from the same transcript after ribosomal reinitiation, assuring maximal expression of the viral genes revealed to be an elegant and efficient method (Cosset et al., 1995). Therefore, to facilitate screening for efficient vector expression many viral vectors contain a (selectable) marker gene. This marker gene allows screening for stable integration and high level, long-term expression. This approach is often needed in settings where the therapeutic transgene itself does not support high throughput screenings. However, such marker genes can impose significant problems in therapeutic settings. One concern is that an immune response against the foreign gene(s) is provoked. This can result in the elimination of transduced cells, thereby limiting or excluding the therapeutic success. Another concern is raised since insertional leukemogenesis was observed in a mouse model due to the transduction of a marker gene that was considered before to be biologically inert (Li et al., 2002; Baum et al., 2003). Thus, it is now state of the art to exclude any coding sequences apart from the therapeutic gene. Several approaches have addressed this point. One strategy is to express the resistance marker from an independent expression cassette outside the retroviral

vector cassette (Xu et al., 2004). Alternatively, a loxP flanked marker gene is integrated into the retroviral vector to be excised upon Cre-mediated recombination (Wildner et al., 1998; Loew et al., 2004). While the latter approach allows efficient selection of producer clones it is not compatible with large size therapeutic vectors.

Once having isolated a cell clone with a favorable titer, it has to be optimized with respect to large-scale culture conditions, including the adaptation to production media (Merten, 2004). This is time consuming and this work has to be performed for each producer cell clone independently. For clinical applications, production under GMP conditions has to be established. This requires a detailed safety testing including an accurate and precise characterization of the producer cell and the stability of recombinant virus production. In total, the process starting from transfection of a packaging cell line to a suitable clone usually takes 6 months or more (Sinn et al., 2005).

The use of helper cell lines described so far requires the establishment of the vector into the chromosomal DNA of the packaging cell. This is accompanied by random integration. The properties of cell clones generated in this way and in particular the titer is largely determined by the nature of the integration site and the influence of neighboring chromosomal elements. Based on the work of Karreman et al. (1996), Schucht et al. (2006) and Coroadinha et al. (2006a) developed a new strategy that provides well-defined high-titer retroviral producer cells at high speed. This strategy excludes random integration of the vector DNA into the host's chromosomal DNA. Instead, it relies on the reuse of prescreened favorable retroviral vector integration sites. These integration loci that are tagged with Flp-recombinase recognition sites confer stable, high-level expression. These integration loci were identified by extensive screening for high and stable expression of the retroviral vector transgene. In subsequent steps, the expression constructs for *gag–pol* and *env* were introduced and cell clones were screened for maximal expression of these genes.

Different retroviral vectors were specifically integrated into the preselected loci via Flp-mediated recombination (Figure 2.4). All isolated clones showed correct integration and identical titers for each of the vectors. Up to 2.5×10^7 infectious particles (IPs) per 10^6 cells in 24 h were obtained within 3 weeks. Also, high-titer producer cells for a 8.9 kb collagen VII therapeutic vector were obtained (Schucht et al., 2006). These data provide evidence that the precise integration of viral vectors into defined chromosomal loci leads to highly predictable virus production (Schucht et al., 2006; Coroadinha et al., 2006a). The method speeds up the time to clinical or research application and supports marker free viral vector production. It further provides a tool for evaluation of different retroviral vector designs under identical conditions.

Site specific integration of retroviral vectors into predefined loci is further characterized by (1) high speed, (2) safety due to the defined integration of the vector within the packaging cell line, and (3) favorable production conditions by preadaptation of the packaging cell line to culture conditions and media.

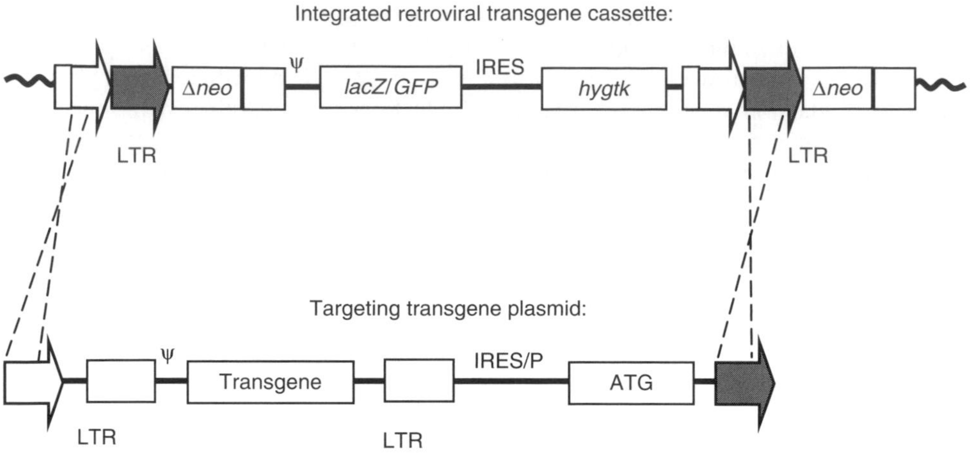

FIGURE 2.4 Tag and target methodology for the cassette exchange by Flp-mediated recombination in Flp293A and 293 Flex retroviral producer cell lines. The tagging retroviral plasmid contains a reporter gene, *lacZ* or *GFP*, *hygtk* positive and negative selection markers, two FRT sites, and a defective *neo* gene. The targeting plasmid contains the two FRT sequences flanking the gene of interest and an ATG sequence that will restore the *neo* gene in the tagged clone after Flp-mediated recombination.

2.5 PRODUCTION OF RETROVIRAL VECTORS

The production of clinical grade retroviral vectors has to be compliant with the stringent regulatory requirements imposed by the FDA and the EMEA. Biological safety and compliance with the guidelines for good manufacturing practices (cGMP) must be assured whenever clinical trials are envisaged. To avoid potential immunological problems during clinical trials all production activities have to be performed in an aseptic environment and all potentially unsafe contaminants have to be removed. Raw materials used in the production will have to be free from adventitious viruses, as these will not be removed during purification (specially enveloped viruses) due do their similarity with retroviral vectors. For this purpose several actions must be taken from the beginning, including the selection of producer cell, production system, and culture conditions.

After the construction of the producer cell line it is necessary to establish a master cell bank (MCB) consisting of frozen cell stocks derived from a single-cloned cell, a working cell bank (WCB) derived from the MCB and maximum end of production cells (MEPC) comprising cells having undergone a number of population doublings beyond the end of the production process. The creation of the MCB involves a number of steps starting with the choice of the best producer clone. Table 2.3 below provides a timetable for MCB generation and vector production and certification.

To initiate a production run typically one to two vials of cells from the MCB are thawed and cells are expanded in tissue culture flasks. The cell culture is then scaled-up by increasing the number of flasks until sufficient cells are finally obtained to inoculate the appropriate number of cell factories or roller bottles to produce the clinical batch. The inoculum density ranges from 1 to 2×10^4 cells/cm^2 for cell factories and the cell cube in reported GMP production runs (Eckert et al., 2000; Wikstrom et al., 2004).

2.5.1 CURRENT PRODUCTION SYSTEMS

The ongoing landmark clinical trials presented in Table 2.1 suggest the use of 10^8–10^9 transduced cells per patient. With an average transduction efficiency of 25% and for a multiplicity of infection of 5, one should expect a need of 10^9–10^{10} retroviral vectors per patient. At the current production titer of 10^6 IPs per milliliter this represents a need for 1–10 L per patient.

Since only four patients are recruited for initial studies, the production systems used to date are considered for small

---

TABLE 2.3
Timeline for MCB Generation and Vector Production

Step	Time
Transfection of packaging cells	1 week
Population supernatant screening for titer	1 week
Clone isolation	3 weeks
Clone screening for high titer producers	2 weeks
Candidate clones expansion	2 weeks
Evaluation of candidate clones for titer and function	4 weeks minimum
Preparation of the MCB (100–200 vials)	3 weeks
Evaluation of the MCB for titer and function	3 weeks minimum
Production run–generation of 20 L per run	4 weeks
Confirmation of titer and function	2 weeks minimum
Certification of the MCB and clinical supernatant	14 weeks
Preparation of product release records to include QA audit	2 weeks

Source: Adapted from Cornetta, K., Matheson, L., and Ballas, C., *Gene. Ther.,* 12, S28, 2005.

scale, preferably disposable systems. These include cell factories, large T-flasks, and roller bottles as the first choice.

Although these systems can be considered as not being the state of the art for mass production, they fit perfectly the current needs for the production of clinical batches of retroviral vectors. In fact, the use of these systems is widespread being the choice of the National Gene Vector Laboratory (Indianapolis, IN) in over 30 production runs since 1995 (Cornetta et al., 2005). Even today, roller bottles, the cell cube, and cell factories are being used for the production of clinical grade material (Eckert et al., 2000; Wikstrom et al., 2004; Przybylowski et al., 2006). In these systems, the vector production under GMP conditions ranges from 10 to 40 L, again in line with the needs for clinical trials. For example, both the cell cube (25 stack) and the 20 roller bottle system are able to provide total yields in the order of 10^{10}–10^{11} IPs using Phoenix Frape-1 cells (Wikstrom et al., 2004).

This somewhat conservative approach to retroviral vector production derives from a number of facts. First, when going from a small flask to a number of larger flasks, the scale-up is linear, with roller bottles being the typical example. This reduces uncertainty as the medium volume and cell numbers are increased proportionally to the growth area, similar titers are normally obtained. This makes validation of the production process relatively easy and a preferred option remains for sporadic low supernatant volume demands. A second argument relates to the fact that all large-scale systems utilize at least some non-disposable parts that need to be cleaned and sterilized before reuse, thus leading to the performance of extensive validation studies.

Finally, for most phase I or II applications full optimization of the production process is often not financially justifiable. The focus is thus put on to the consistency of the harvested material, the compliance with the quality specifications, and the attainment of a titer that is acceptable for the intended application.

There are, nonetheless, a number of alternative culture systems for retroviral vector production (Merten, 2004). These include a number of systems of greater scalability such as fixed bed bioreactors and stirred tanks that allow improving the titers by 10 to 100-fold reaching more than 10^7 IPs per milliliter.

In spite of the relatively large number of cells that grow attached to surfaces (including PA317, ψ-CRIP, and PG13 cells, among others), some cells can be grown as suspension cultures after adaptation. These include the recent HEK 293-based packaging cells. In this case, suspension cultures may be the only option. Thus, stirred tanks would be more appropriate and all the advantages referred above for these bioreactors can be used to increase both cell concentration and retrovirus titer (Garnier et al., 1994). Nevertheless, even in the cases where retroviruses may represent a viable solution for other diseases with higher incidence such as cancer (Lenz et al., 2002; Braybrooke et al., 2005), given the validation issues the trend will probably continue to be the use of disposable bioreactors such as the stirred tanks from Hyclone/Thermo and the Wave bioreactor.

2.5.2 Culture Conditions

As with all other animal cells, mammalian cells used for the production of retrovirus are also affected by nutrient limitation (oxygen and glucose) and accumulation of byproducts such as ammonia and lactate. As a consequence, producer cell metabolism has been studied in order to better understand the factors influencing both vector productivity and vector stability (Coroadinha et al., 2006b). In addition, both osmotic pressure and the carbon source affect the specific retrovirus productivity and vector stability (Coroadinha et al., 2006c; Coroadinha et al., 2006d), thus creating a challenge for the definition of new medium formulations.

Culture temperature has been shown to affect the final retrovirus titer in the large majority of the reports. In fact, culture temperature assumes a particularly important role as it influences both the rates of retrovirus production and degradation (Cruz et al., 2000). According to Le Doux et al. (1999) an optimal temperature exists because the rate of retrovirus degradation has a higher increase with temperature than the rate of retrovirus production in the range between 28°C and 37°C. In fact, there are several retroviral vector clinical batches currently being produced at 32°C (Lamers et al., 2006a; Lamers et al., 2006b).

2.5.3 Operation Mode

Although the production of retroviral vectors is usually performed as a batch culture, due to the low stability of these vectors at 37°C (Le Doux et al., 1999; Pizzato et al., 2001; Beer et al., 2003; Carmo et al., 2006), several harvests should be performed. The use of this stepwise perfusion mode has the advantage of recovering retroviruses from the culture broth at shorter residence times with visible benefits upon retrovirus quality, derived from reduced degradation and increased final IP number per batch. The collection of culture supernatants in cell factories can be performed every 24 h up to 4 days (Eckert et al., 2000). The same time interval can be used in continuous perfusion mode in the cell cube with equivalent gains in total IP yields per batch (Wikstrom et al., 2004). The several harvests are then stored at −80°C after clarification before they are thawed and pooled for quality control and release.

Medium exchange is not common in large scale production of recombinant proteins and monoclonal antibodies in suspension cultures. However, when using adherent cells in cell factories, roller bottles, cell cube, or fixed bed bioreactors at the 10–40 L scale, this does not pose relevant challenges. In fact, since the current production protocols use medium containing 10% serum during cell expansion phase and serum-free medium during the production phase, the ability to perform medium exchange is practically a prerequisite to the system (Eckert et al., 2000).

2.5.4 Culture Medium

The production of retroviral vectors strongly affects vector downstream processing particularly in what concerns the producer cells and culture medium used. Producer cell–contaminant proteins and DNA as well as some media additives, like serum, need to be removed to meet regulatory agencies quality standards. The development of production processes that use serum-free media is an important step towards prevention of immunological responses to viral preparations and simplification of the downstream strategy. Retroviral vectors are usually produced by adherent cell lines that secrete variable quantities of extracellular MA proteins, partially consisting of proteoglycans (Merten, 2004). These macromolecules are negatively charged and influence the transduction efficiency of retroviral vectors (Le Doux et al., 1996; Le Doux et al., 1998). Cell lines grown in suspension are known to produce less extracellular MA (Merten, 2004) and, combined with higher process scalability potential, they offer attractive advantages in larger scale retroviral vectors production.

2.6 PURIFICATION OF RETROVIRAL VECTORS

Purification of retroviral vectors from cell culture derived contaminants (e.g., producer cells, DNA, proteins from culture media or released by the producer cells) is mandatory to prevent toxicity, inflammation, or immune response in the individuals undergoing therapy (Tuschong et al., 2002). Till recently, downstream processing (i.e., the process by which the retroviral vectors are separated from contaminants) of retroviral vectors for phase I clinical trials has been only based on the separation of producer cells and debris which is insufficient regarding the increasingly stringent quality standards set by the regulatory agencies (FDA and EMEA). The removal of the contaminants requires the use of separation technology, usually based on chromatography and membrane processes (Segura et al., 2006a; Rodrigues et al., 2007a). Nevertheless, the special properties of these complex products require technological improvement of the existing purification processes and development of specific technology to increase productivity and throughput, while maintaining biological activity of the final product.

2.6.1 TECHNOLOGICAL CHALLENGES

The purification of retroviral vectors, like other macromolecular assemblies, still remains a challenge due to their large particle size, low diffusion rates, and complex molecular surfaces (Lozinsky et al., 2003). Most of the problems are associated with the low-titer production systems (usually between 10^5 and 10^7 IP per milliliter) (Merten, 2004), the inherent instability of retroviral vectors (Layne et al., 1992), the presence of transduction inhibitors (Le Doux et al., 1996), and intrinsic contaminants such as noninfectious vectors. Downstream processing of retroviral vectors needs to be fast, with few purification steps, and efficient in the removal of the contaminants as well as in the maintenance of viral infectivity. Different applications for the retroviral vectors may also require different levels of quality and biological potency of the therapeutic preparations. Thus, the definition of the purification process is also dependent on the final application; retroviral

vectors purified for *in vivo* therapy will need to fulfill more strict quality demands.

2.6.2 PROPERTIES OF RETROVIRAL VECTORS CRITICAL IN DOWNSTREAM PROCESSING

The development of a purification process demands a deep knowledge on the physical, chemical, and biological properties of retroviral vectors. Retroviral vectors are enveloped viruses with a complex macromolecular structure of 80–150 nm diameter (Salmeen et al., 1976; Coffin et al., 1997) (Figure 2.1). Due to their large size they have low diffusivity (10^{-8} cm^2/s) and their density is 1.15–1.16 g/cm^3 (Coffin et al., 1997). Mature and infectious retroviral vectors have a morphology with a round or sometimes slightly angular core, centered in the middle of the particles (Yeager et al., 1998). They are composed by 60%–70% protein, 30%–40% lipid (derived from the plasma membrane of the producer cell line), 2%–4% carbohydrate, and 1% RNA (Andreadis et al., 1999). The lipid membrane of retroviral vectors is studded with glycoprotein projections, between 100 and 300 per particle, the envelope proteins (Andreadis et al., 1999). Approximately three quarters of the total retroviral vectors protein corresponds to the Gag proteins; the rest includes the viral envelope glycoproteins and, to a lesser extent, the viral enzymes RT, IN, and PR (Coffin et al., 1997). Some cytoplasmic and membrane cellular proteins are also known to be incorporated during vector assembly and budding (Ott, 1997). The lipid composition of the retroviral particle is derived from the producer cell plasma membrane during budding, but is enriched in sphingomyelin and cholesterol (Quigley et al., 1971; Aloia et al., 1993; Beer et al., 2003).

The isoelectric point of Moloney MLV occurs at very low pH values thus, viral particles are negatively charged at physiological pH (Rimai et al., 1975).

Retroviral vectors are known to be sensitive to environmental conditions and to lose infectivity relatively fast; thus the knowledge of optimal conditions for maintenance of retroviral stability is critical in downstream processing. During production retroviral vectors bud from the producer cell line and start to loose infectivity due to their labile nature at 37°C, with a half-life of about 8 h at this temperature (Higashikawa and Chang, 2001; Beer et al., 2003). As a consequence, few of the particles in a preparation are infectious, the ratio of total to IPs being typically above 100:1 (Higashikawa and Chang, 2001). Retroviral vectors are more stable at low temperatures, with half-lives higher than 100 h at 4°C (Le Doux et al., 1999; Beer et al., 2003). Besides temperature, retroviral vectors are also sensitive to acidic or basic pH, having a bell shaped pH profile with an optimum pH between 5.5 and 8.0 (Ye et al., 2004). High salt concentrations have also been reported to affect retroviral stability (Segura et al., 2005).

As mentioned above, the stability of retroviral vectors also depends on a number of producer cell and production process related variables such as the producer cell line type, viral membrane cholesterol levels, pseudotyping, production medium osmolarity, and production temperature (Beer et al., 2003; Carmo et al., 2006; Coroadinha et al., 2006a; Coroadinha et al., 2006d). Thus, each type of vector will also have different tolerance to environmental factors.

2.6.3 Target Contaminants

2.6.3.1 Proteins

Protein impurities are the most abundant contaminants in retroviral vectors supernatants. They arise from the serum usually used to supplement cell growth media or are secreted by the producer cells. Serum albumin is the most abundant protein contaminant that comes from serum, followed by immunoglobulins (Anderson and Anderson, 2002). Evaluation of protein contamination is usually performed by SDS-PAGE with silver staining analysis and Western-blot directed to protein contaminants or vector proteins, usually aids gel interpretation (Figure 2.5) (Segura et al., 2006). Other techniques to determine and quantify protein contamination include HPLC analysis (Transfiguracion et al., 2003) and ELISA quantitation of BSA (Richieri et al., 1998; Schonely et al., 2003).

2.6.3.2 DNA

The major source of contaminating DNA are the host producer cells but, if transient transfection is used in the production of retroviral vectors, plasmid contamination can also occur (Chen et al., 2001; Sastry et al., 2004). Concern about nucleic acid contamination arises from the possibility of cellular transformation events in the patient. Contaminant DNA limits are usually dependent on product and application and set for approval to the regulatory authorities. FDA recommends a limit of 10 ng DNA/dose (O'Keeffe et al., 1999).

2.6.3.3 Transduction Inhibitors

Transduction inhibitors interfere with the ability of the retroviral vectors to infect target cells. Thus, they decrease the efficacy of the vector preparations and increase the necessary dose to achieve desired potency. Several types of transduction inhibitors have been identified, including noninfectious particles (Seppen et al., 2000; Transfiguracion et al., 2003), free envelope proteins (Landazuri and Le Doux, 2006), and negatively charged proteoglycans (Le Doux et al., 1996).

2.6.4 Purification of Retroviral Vectors

The first protocols used for the purification of retroviral vectors were similar to those performed for the native viruses and based on their size and density, using methods such as ultracentrifugation and density gradients (Faff et al., 1993; Andreadis et al., 1999; Reiser, 2000). According to these protocols, retroviral vectors are first separated from cell debris by low-speed centrifugation or filtration through 0.45 μm membranes, and afterwards pelleted by ultracentrifugation and separated from the extracellular medium. The resulting pellet is resuspended in buffer and the retroviral vectors sedimented to equilibrium by ultracentrifugation, usually in a sucrose gradient. Recovery of infectious vectors is usually very low after ultracentrifugation (<1%) due to the degradation of the viral vectors (Burns et al., 1993; Powell et al., 2000). Moreover, these methods are time-consuming, difficult to scale-up and allow for coprecipitation of impurities that have a similar density to the virus, like membrane vesicles from host cells (Bess et al., 1997; Seppen et al., 2000). Therefore, they are not suitable for clinical applications. Other purification processes, well established for protein therapeutics, are wiser options.

2.6.4.1 Membrane Separation Processes

Membrane filtration is defined as the separation of two or more components from a fluid suspension stream using membranes, based primarily on size differences. The membrane acts as a selective barrier as it permits the passage of components smaller than the pore size and retains larger components.

As retroviral vectors are nanoscale particles, membrane separation can be applied for clarification (cells and cellular debris are retained by the large pore membranes while the viral particles are allowed to pass), concentration, and partial purification of retroviral supernatants (the smaller pore membranes retain retroviral vectors while small proteins cross the membrane barrier). Strategies for clarification and concentration

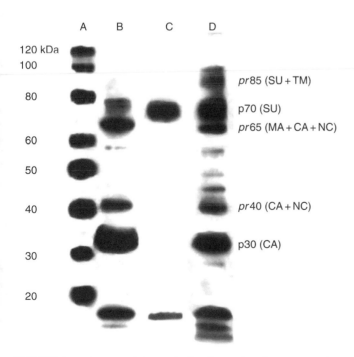

FIGURE 2.5 Western blot analysis of proteins from an amphotropic retroviral vector using different types of primary antibodies: A, molecular weight markers; B, rat monoclonal antibody against p30 Gag protein of MLV (produced by hybridoma cell line R187, ATCC no CRL-1912); C, rat monoclonal antibody against p69/70 envelope proteins (83A25) (Evans et al., 1990); D, swine polyclonal anti-amphoteric MLV antibody (Cat# 77S000445, Viromed, USA).

of viral stocks using membrane technology include microfiltration and ultrafiltration (UF), correspondingly.

Traditionally, separation of cells from produced stocks is performed using microfiltration membranes with 0.45 μm pores. However, blockage of the membrane pores with cells and debris often hinders the passage of retroviral vectors resulting in lower recoveries (Reeves and Cornetta, 2000; Rodrigues et al., 2007). To circumvent this problem, step filtration with decreasing pore size can be used allowing a significant increase in retroviral vectors recovery (Reeves and Cornetta, 2000; Rodrigues et al., 2007). Other factors might also impact the recovery after microfiltration, including the membrane characteristics and retroviral vectors stability.

UF is an attractive process with a number of important advantages: it is fast, robust, flexible, and useful for concentrating and washing feed streams (change the pH, salt concentration, and solute conditions of the retained product) prior to chromatography. UF has been applied with success in the concentration and partial purification of retroviral vectors, mostly at laboratory scale, using centrifugation dependent modules (Saha et al., 1994; Miller et al., 1996; Parente and Wolfe, 1996; Reiser, 2000; Yamada et al., 2003) or stirred vessels (Miller et al., 1996; Cruz et al., 2000; Transfiguracion et al., 2003). Retroviral supernatants can be ultra/dialfiltered using various pore sizes, ranging from conventional 20 to 500 kDa molecular weight cut-off membranes (Table 2.4) up to 35 and 75 nm pore microfiltration membranes (Makino et al., 1994). The use of UF membranes with high cut-off pores is advantageous as it should, in principle, result in higher permeate fluxes (decreased process time) and higher purification factors due to removal of both large and low molecular weight proteins. Nevertheless, lower recoveries might

TABLE 2.4
Comparison of Purification Processes for Retroviral Vectors

Vector	Process Steps	Recovery (%)	Concentration Factor (CF)	Final Titer (IP/mL)	Reference
VSV-G, MLV derived	Microfiltration				
	Ultra/diafiltration				
	Heparin affinity Chrom.	38.3	6	5.71×10^6	(Segura et al., 2005)
	SEC				
VSV-G, HIV-1 derived	Step microfiltration				
	Ultra/diafiltration				
	Benzonase treatment	~30	10	3.29×10^8	
	SEC (Sephacryl S500)				
	Sterile filtration				
	Step microfiltration				(Slepushkin et al., 2003)
	AEX Membrane Chrom.				
	Ultra/diafiltration				
	Benzonase treatment	~30	20	3.29×10^8	
	Diafiltration				
	Sterile filtration				
VSV-G, MLV derived	Ultracentrifugation				
	SEC	19	24	4.8×10^7	(Transfiguracion et al., 2003)
	Ultra/diafiltration				
Ampho-MLV	Microfiltration				
	Ultra/diafiltration				
	AEXc	26	85	3.2×10^8	(Rodrigues et al., 2007b)
	Ultra/diafiltration				

Source: Adapted from Rodrigues, T., et al., *J. Gene Med.* 9, 233, 2007b.

Note: AEXc, anion-exchange chromatography; Chrom., chromatography; CF, concentration factor; IP, infectious particles; SEC, size exclusion chromatography.

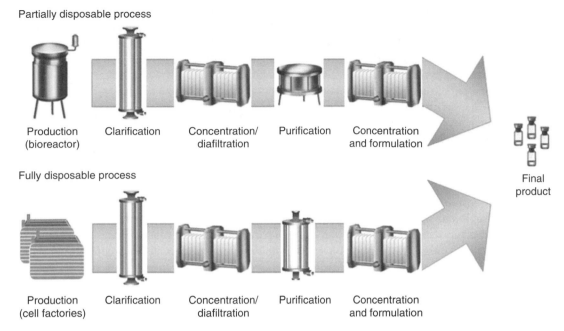

FIGURE 2.6 Downstream processing of retroviral vectors from partially to fully disposable processes. (Adapted from images kindly provided by Sartorius AG.)

occur if the vector and membranes pore size are within the same order of magnitude (Makino et al., 1994).

Several membrane geometries use tangential flow filtration (TFF), which is easily scalable. This geometry minimizes build-up of solids on the membrane thus, decreased fouling. The most often reported TFF geometries for retroviral vectors concentration and clarification are flat sheet cassettes or hollow-fibers (Segura et al., 2006; Rodrigues et al., 2007). Together with some developments in the field of disposable downstream technology, TFF constitutes an interesting option for the production of retroviral vectors for clinical use (Table 2.4, Figure 2.6).

Although competing with ultracentrifugation in terms of achievable concentration and volume reduction factors, the advantages of UF in terms of throughput, scalability, and maintenance of vector biological activity and integrity are indisputable.

2.6.4.2 Chromatographic Methods

Chromatography is the predominant technology used for purification of biomolecules and can often be operated both as fixed or fluidized/expanded bed and also in membrane configurations. Due to the labile nature of retroviral vectors, conditions for chromatographic purification are quite restricted, limiting the range of chromatographic media and buffers that can be used. An initial screening of conditions is advisable (Rodrigues et al., 2006). Besides having high binding capacity for adsorption and elution of retroviral vectors in nondenaturing conditions, the adsorbents need also to separate retroviral vectors from contaminants derived from host-cell

material and cell culture medium. Several properties of retroviral vectors have been explored for selective adsorption/desorption to chromatographic media, including charge, size, and specific interaction with SU protein domains or unique ligand binding sites (Segura et al., 2006; Rodrigues et al., 2007).

Several affinity chromatography strategies have been evaluated for the purification of retroviral vectors, namely heparin affinity (Segura et al., 2005), streptavidin–biotin affinity (Hughes et al., 2001; Williams et al., 2005), and immobilized metal affinity (IMAC) (Ye et al., 2004). Retroviral vectors purification through interaction with heparin matrices results in high IP recoveries, approximately 60% (Segura et al., 2005). Nevertheless, many cellular and serum contaminant proteins have also affinity to heparin; thus, another purification step is necessary to remove such contaminants (Segura et al., 2005). Although the implementation of purification strategies including streptavidin–biotin affinity and IMAC is possible, the use of technologies that require modification of the vectors will pose obstacles for regulatory approval. Additionally, the desorption reagents used to elute the retroviral vectors impair their biological activity (Rodrigues et al., 2007).

Anion-exchange chromatography (AEXc) involves the adsorption of negatively charged retroviral vectors to positively charged chromatographic resins (Richieri et al., 1998; Kuiper et al., 2002; Rodrigues et al., 2006). Retroviral vectors bind strongly to AEX matrices thus, desorption is usually accomplished at salt concentrations around 1 M NaCl. AEXc has been successfully used for the purification of both retroviruses and lentiviruses allowing IP recoveries of up to 70%

(Rodrigues et al., 2007). Diethylaminoethyl ligands were considered the best anion-exchangers for capture/elution of retroviral vectors and in combination with tentacle MA technology they offer the highest recoveries of IPs after elution (Rodrigues et al., 2006). AEX chromatography offers milder conditions and also higher purification factors without the need for modification of the vector itself (Rodrigues et al., 2007). Moreover, it has been used in the preparation of other types of viruses for conventional vaccine and gene therapy trials (Konz et al., 2005; Tancevski et al., 2006), hence presents itself as a good candidate for retroviral vector purification.

Retroviral vectors can also be efficiently separated from cellular and medium contaminants by size exclusion chromatography (SEC) usually with recoveries above 70% (McGrath et al., 1978; Slepushkin et al., 2003; Transfiguracion et al., 2003). SEC is a powerful separation tool but it is difficult to scale-up, has very low throughput and works with low linear flow rates, thus increasing process time. Also it does not offer advantages in terms of final purity when combined with AEXc, thus it is not advantageous in terms of clinical application (Rodrigues et al., 2007).

The introduction of disposable membrane chromatography, already available with several chemistries in the development of purification processes for retroviral vectors (Figure 2.6) will be very beneficial in terms of clinical application, due to higher process flexibility and lower validation costs (Ghosh, 2002; Thommes and Etzel, 2007).

2.6.5 STORAGE OF RETROVIRAL VECTORS

The efficacy of retroviruses as gene delivery agents depends on the assurance of efficient and stable transfer of the transgene into the target tissue. Therefore, the maintenance of viral vectors potency until administration is of germane importance for the success of the clinical application.

Downstream processing, final product formulation, and storage temperature are determinants of the half-life of retroviral vectors. Due to this fact, reported vector half-lives range from 19 days to 18 months for retroviral vectors stored at −80°C (Wikstrom et al., 2004; Carmo et al., 2006). Historically, long-term storage of viral preparations requires cryopreservation where small volume aliquots of viral stocks are stored at −80°C until use or, alternatively, kept in a lyophilized form at 4°C (Harris, 1954).

Although little is known regarding the degradation and stabilization mechanisms of retroviral vectors, it has been shown that the addition of cryoprotectants (e.g., sucrose and trehalose) decreases the loss of infectivity of viral vectors and enhances their stability during long-term storage (Bieganski et al., 1998; Cruz et al., 2006). The need for such formulations creates an important challenge as purified vectors seem to be more sensitive than those from clarified supernatants that constitute today's common practice (Carmo et al., 2006).

Furthermore, the relevance of this point is strengthened by the fact that the vector production under GMP conditions, the necessary safety and potency testing and the subsequent final product release for clinical application may take 6 to 12 months. Along with the time necessary to obtain the approval of the clinical protocol by the regulatory agencies, the maintenance of the potency has to be assured for at least 1 to 2 years.

2.7 QUALITY ASSESSMENT

Several orthogonal methods are necessary to assess quality of retroviral preparations due to the complexity of the product and also due to its final clinical application. Thus, the quality of a purified retroviral vectors preparation is described by appropriate dose, potency, purity, and safety standards (safety of retroviral vectors preparations is addressed in a different chapter). These quality tools aid the downstream process development and provide reassurance that preliminary *in vitro* studies are conducted with material that will be representative of that to be used in clinical trials.

2.7.1 DOSE AND POTENCY

Dose and potency are important characteristics of a retroviral preparation and are closely related. Potency can be defined as a measure of the capacity of the target cells, transduced by retroviral vectors preparations, to express the transgene product. For potency either *in vitro* or *in vivo* tests, or both, have been specifically designed for the transgene and target cells in question (Slepushkin et al., 2003). This is an important parameter to measure during clinical application. Parameters that usually define the dose of a purified retroviral vectors preparation to be used in clinical trials are the biological activity, which constitutes a measure of the IP content, the total particles, and the total protein content (Table 2.5). There is no accurate method to measure infectious and total particles due to the complexity of the target. Nevertheless, combinations of the orthogonal methods presented in Table 2.5 provide satisfactory and quantitative answers that aid process development.

2.7.2 PURITY

Being complex macromolecular assemblies of proteins, lipids, and RNA evaluation of retroviral vectors purity is cumbersome. Most of the difficulty arises from the unavailability of a well characterized standard of biologically active and pure retroviral vectors. Moreover, retroviral vectors incorporate components of the producer cells, mostly proteins, during the budding process, both within the lipid bilayer and inside the viral particle (Ott, 1997). All these characteristics greatly increase the difficulty in the determination of which sample components come from the vector and which are indeed contaminants. Due to the complexity of retroviral vector products, demonstration of product purity needs to be performed using wide range of methods, including physicochemical and immunological techniques (Table 2.5).

TABLE 2.5
Summary of the More Relevant Quality Assays Used for Retroviral Vectors

Test	Target	Assays	References
Dose	Particle number	RT-PCR	Carmo et al., 2004
		HPLC	Transfiguracion et al., 2004
		Electron microscopy	Higashikawa and Chang, 2001
		RT activity	Kwon et al., 2003
	Infectious titer	Transduction assays	Rill et al., 1992; Hughes et al., 2001
		Proviral insertion	Sastry et al., 2002
	Total protein	BCA, Bradford, Lowry	EuropeanPharmacopeia, 2005
Potency	Transgene	Quantitation of transgene expression	Slepushkin et al., 2003
Purity	Viral protein profile	Silver stained SDS-PAGE	Schonely et al., 2003
		Western blot	Figure 2.5
	Contaminants:		
	DNA	Total DNA quantitation	CBER, 1998
		HPLC	
		PCR	
	Proteins	Silver stained SDS-PAGE	O'Keeffe et al., 1999
		HPLC	
		ELISA	Schonely et al., 2003
Safety	Mycoplasma		EuropeanPharmacopeia, 2005
	Sterility		
	Pyrogen/endotoxins	LAL assay	ICH, 1999
		U.S.P. Rabbit pyrogen test	CBER, 2000
	Adventitious viruses	Viral antigen and antibody	
		Immunoassay	Sastry et al., 2005
		PCR	
	RCR	Amplification in a suitable cell line followed by: PG-4 S$^+$ L$^-$ assay, PCR, PERT	Printz et al., 1995; Forestell et al., 1996; Martineau et al., 1997

2.7.3 SAFETY

Contamination with RCRs constitutes a primary safety concern regarding clinical application. Being structurally similar to non-replicative vectors, their separation during downstream processing is virtually impossible. Thus, procedures for testing the master, WCBs, and vector containing supernatants are critically important during the entire manufacture process development (CBER, 2000; Anson, 2004). Sensitive assays for the screening of RCR are required as these recombination events occur at low frequency and high levels of replication incompetent particles can interfere with the detection methods (Forestell et al., 1996). Most of the methods currently used to detect RCR in preparations for use in clinical trials are based on a combination of biologic assays and molecular detection (Table 2.5).

The elimination of adventitious viruses from vector preparations is another safety concern. Usually strategies are implemented at the end of the downstream process to eliminate such contaminants. However, they result in the inactivation of the retroviral vectors which is undesirable. Therefore, probable contamination sources (e.g., producer cell line and reagents from animal sources) should be shown free of adventitious virus contamination (EMEA, 1997). Finally the end-product must be tested for the presence of other microorganisms, namely mycoplasma, bacteria, and fungi and also for the presence of endotoxins and pyrogens (Table 2.5).

2.8 CONCLUDING REMARKS

Retroviral vectors continue to be one of the preferential vehicles for gene therapy as they offer several advantages relative to other vectors. Current *ex vivo* gene therapy applications use clinical grade retroviral culture supernatants produced in suboptimal conditions regarding maximum viral infectivity, purification, and maintenance of stability. In clinical settings, phase I and II trials are slowly moving to phase III, as safety concerns are being solved. This drives research to a more technological level since larger retroviral preparation volumes will require better producer cells, more sophisticated production methods, and postharvest processing. Future *in vivo* gene therapy applications of retroviral vectors will require concentrated and purified vector preparations. The generation of such purified retrovirus clinical batches with assured long-term storage is still an important challenge. The loss of viral infectivity during downstream processing represents a limitation and is still an issue. Consequently, further developments in retroviral vector production directed to increase productivity, stability, and recovery after purification and storage are mandatory to ensure the vector titers and quality required for clinical applications.

REFERENCES

Aiuti, A., S. Slavin, M. Aker, F. Ficara, S. Deola, A. Mortellaro, S. Morecki, G. Andolfi, A. Tabucchi, F. Carlucci, E. Marinello, F. Cattaneo, S. Vai, P. Servida, R. Miniero, M. G. Roncarolo, and C. Bordignon 2002. Correction of ADA-SCID by stem cell gene therapy combined with nonmyeloablative conditioning. *Science* 296(5577): 2410–2413.

Aloia, R. C., H. Tian, and F. C. Jensen 1993. Lipid composition and fluidity of the human immunodeficiency virus envelope and host cell plasma membranes. *Proc Natl Acad Sci USA* 90(11): 5181–5185.

Anderson, N. L. and N. G. Anderson 2002. The human plasma proteome: History, character, and diagnostic prospects. *Mol Cell Proteomics* 1(11): 845–867.

Andreadis, S. T., C. M. Roth, J. M. Le Doux, J. R. Morgan, and M. L. Yarmush 1999. Large-scale processing of recombinant retroviruses for gene therapy. *Biotechnol Prog* 15(1): 1–11.

Anson, D. S. 2004. The use of retroviral vectors for gene therapy-what are the risks? A review of retroviral pathogenesis and its relevance to retroviral vector-mediated gene delivery. PG - 9. *Genet Vaccines Ther* 2(1): 9.

Baum, C., J. Dullmann, Z. Li, B. Fehse, J. Meyer, D. A. Williams, and C. von Kalle 2003. Side effects of retroviral gene transfer into hematopoietic stem cells. *Blood* 101(6): 2099–2114.

Beer, C., A. Meyer, K. Muller, and M. Wirth 2003. The temperature stability of mouse retroviruses depends on the cholesterol levels of viral lipid shell and cellular plasma membrane. *Virology* 308(1): 137–146.

Bess, J. W., Jr., R. J. Gorelick, W. J. Bosche, L. E. Henderson, and L. O. Arthur 1997. Microvesicles are a source of contaminating cellular proteins found in purified HIV-1 preparations. *Virology* 230(1): 134–144.

Bieganski, R. M., A. Fowler, J. R. Morgan, and M. Toner 1998. Stabilization of active recombinant retroviruses in an amorphous dry state with trehalose. *Biotechnol Prog* 14(4): 615–620.

Blaese, R. M., K. W. Culver, A. D. Miller, C. S. Carter, T. Fleisher, M. Clerici, G. Shearer, L. Chang, Y. Chiang, P. Tolstoshev, J. J. Greenblatt, S. A. Rosenberg, H. Klein, M. Berger, C. A. Mullen, W. J. Ramsey, L. Muul, R. A. Morgan, and W. F. Anderson 1995. T lymphocyte-directed gene therapy for ADA-SCID: Initial trial results after 4 years. *Science* 270(5235): 475–480.

Bonini, C., G. Ferrari, S. Verzeletti, P. Servida, E. Zappone, L. Ruggieri, M. Ponzoni, S. Rossini, F. Mavilio, C. Traversari, and C. Bordignon 1997. HSV-TK gene transfer into donor lymphocytes for control of allogeneic graft-versus-leukemia. *Science* 276(5319): 1719–1724.

Bosselman, R. A., R. Y. Hsu, J. Bruszewski, S. Hu, F. Martin, and M. Nicolson 1987. Replication-defective chimeric helper proviruses and factors affecting generation of competent virus: Expression of Moloney murine leukemia virus structural genes via the metallothionein promoter. *Mol Cell Biol* 7(5): 1797–1806.

Braybrooke, J. P., A. Slade, G. Deplanque, R. Harrop, S. Madhusudan, M. D. Forster, R. Gibson, A. Makris, D. C. Talbot, J. Steiner, L. White, O. Kan, S. Naylor, M. W. Carroll, S. M. Kingsman, and A. L. Harris 2005. Phase I study of MetXia-P450 gene therapy and oral cyclophosphamide for patients with advanced breast cancer or melanoma. *Clin Cancer Res* 11(4): 1512–1520.

Burns, J. C., T. Friedmann, W. Driever, M. Burrascano, and J. K. Yee 1993. Vesicular stomatitis virus G glycoprotein pseudotyped retroviral vectors: Concentration to very high titer and efficient gene transfer into mammalian and nonmammalian cells. *Proc Natl Acad Sci U S A* 90(17): 8033–8037.

Carmo, M., T. Q. Faria, H. Falk, A. S. Coroadinha, M. Teixeira, O.-W. Merten, C. Gény-Fiamma, P. M. Alves, O. Danos, A. Panet, M. J. T. Carrondo, and P. E. Cruz 2006. Relationship between retroviral vector membrane and vector stability. *J Gen Virol* 87: 1349–1356.

Carmo, M., C. Peixoto, A. S. Coroadinha, P. M. Alves, P. E. Cruz, and M. J. T. Carrondo 2004. Quantitation of MLV-based retroviral vectors using real-time RT-PCR. *J Virol Methods* 119(2): 115–119.

Cavazzana-Calvo, M., S. Hacein-Bey, G. de Saint Basile, F. Gross, E. Yvon, P. Nusbaum, F. Selz, C. Hue, S. Certain, J. L. Casanova, P. Bousso, F. L. Deist, and A. Fischer 2000. Gene therapy of human severe combined immunodeficiency (SCID)-X1 disease. *Science* 288(5466): 669–672.

CBER 1998. Guidance for Industry: Guidance for Human Somatic Cell Therapy and Gene Therapy, FDA.

CBER 2000. Supplemental Guidance on Testing for Replication Competent Retrovirus in Retroviral Vector Based Gene Therapy Products and During Follow-up of Patients in Clinical Trials Using Retroviral Vectors, FDA.

Chadwick, M. P., F. J. Morling, F. L. Cosset, and S. J. Russell 1999. Modification of retroviral tropism by display of IGF-I. *J Mol Biol* 285(2): 485–494.

Chen, J., L. Reeves, N. Sanburn, J. Croop, D. A. Williams, and K. Cornetta 2001. Packaging cell line DNA contamination of vector supernatants: Implication for laboratory and clinical research. *Virology* 282(1): 186–197.

Chong, H., W. Starkey, and R. G. Vile 1998. A replication-competent retrovirus arising from a split-function packaging cell line was generated by recombination events between the vector, one of the packaging constructs, and endogenous retroviral sequences. *J Virol* 72(4): 2663–2670.

Chong, H. and R. G. Vile 1996. Replication-competent retrovirus produced by a 'split-function' third generation amphotropic packaging cell line. *Gene Ther* 3(7): 624–629.

Chu, T. H. and R. Dornburg 1995. Retroviral vector particles displaying the antigen-binding site of an antibody enable cell-type-specific gene transfer. *J Virol* 69(4): 2659–2663.

Chu, T. H., I. Martinez, W. C. Sheay, and R. Dornburg 1994. Cell targeting with retroviral vector particles containing antibody-envelope fusion proteins. *Gene Ther* 1(5): 292–299.

Coffin, J. M., S. H. Hughes, and H. E. Varmus 1997. Historical introduction to the general properties of retroviruses. *Retroviruses.* CSHL Press, NY.

Cone, R. D. and R. C. Mulligan 1984. High-efficiency gene transfer into mammalian cells: Generation of helper-free recombinant retrovirus with broad mammalian host range. *Proc Natl Acad Sci U S A* 81(20): 6349–6353.

Cornetta, K., L. Matheson, and C. Ballas 2005. Retroviral vector production in the National Gene Vector Laboratory at Indiana University. *Gene Ther* 12 Suppl 1: S28–S35.

Coroadinha, A. S., R. Schucht, L. Gama-Norton, D. Wirth, H. Hauser, and M. J. Carrondo 2006a. The use of recombinase mediated cassette exchange in retroviral vector producer cell lines: Predictability and efficiency by transgene exchange. *J Biotechnol* 124(2): 457–468.

Coroadinha, A. S., P. M. Alves, S. S. Santos, P. E. Cruz, O. W. Merten, and M. J. Carrondo 2006b. Retrovirus producer cell line metabolism: Implications on viral productivity. *Appl Microbiol Biotechnol* 72(6): 1125–1135.

Coroadinha, A. S., J. Ribeiro, A. Roldao, P. E. Cruz, P. M. Alves, O. W. Merten, and M. J. Carrondo 2006c. Effect of medium sugar source on the production of retroviral vectors for gene therapy. *Biotechnol Bioeng* 94(1): 24–36.

Coroadinha, A. S., A. C. Silva, E. Pires, A. Coelho, P. M. Alves, and M. J. Carrondo 2006d. Effect of osmotic pressure on the production of retroviral vectors: Enhancement in vector stability. *Biotechnol Bioeng* 94(2): 322–329.

Cosset, F. L., A. Girod, F. Flamant, A. Drynda, C. Ronfort, S. Valsesia, R. M. Molina, C. Faure, V. M. Nigon, and G. Verdier 1993. Use of helper cells with two host ranges to generate high-titer retroviral vectors. *Virology* 193(1): 385–395.

Cosset, F. L., Y. Takeuchi, J. L. Battini, R. A. Weiss, and M. K. Collins 1995. High-titer packaging cells producing recombinant retroviruses resistant to human serum. *J Virol* 69(12): 7430–7436.

Cruz, P. E., D. Goncalves, J. Almeida, J. L. Moreira, and M. J. Carrondo 2000. Modeling retrovirus production for gene therapy. 2. Integrated optimization of bioreaction and downstream processing. *Biotechnol Prog* 16(3): 350–357.

Cruz, P. E., A. C. Silva, A. Roldão, M. Carmo, M. J. T. Carrondo, and P. M. Alves 2006. Screening of novel excipients for improving the stability of retroviral and adenoviral vectors. *Biotechnol Prog* 22(2): 568–576.

Danos, O. and R. C. Mulligan 1988. Safe and efficient generation of recombinant retroviruses with amphotropic and ecotropic host ranges. *Proc Natl Acad Sci U S A* 85(17): 6460–6464.

Davis, J. L., R. M. Witt, P. R. Gross, C. A. Hokanson, S. Jungles, L. K. Cohen, O. Danos, and S. K. Spratt 1997. Retroviral particles produced from a stable human-derived packaging cell line transduce target cells with very high efficiencies. *Hum Gene Ther* 8(12): 1459–1467.

Donahue, R. E., S. W. Kessler, D. Bodine, K. McDonagh, C. Dunbar, S. Goodman, B. Agricola, E. Byrne, M. Raffeld, R. Moen, J. Bacher, K. M. Zsebo, and A. W. Nienhuis 1992. Helper virus induced T cell lymphoma in nonhuman primates after retroviral mediated gene transfer. *J Exp Med* 176(4): 1125–1135.

Dunbar, C. E., D. B. Kohn, R. Schiffmann, N. W. Barton, J. A. Nolta, J. A. Esplin, J. Pensiero, Z. Long, C. Lockey, R. V. Emmons, S. Csik, S. Leitman, C. B. Krebs, C. Carter, R. O. Brady, and S. Karlsson 1998. Retroviral transfer of the glucocerebrosidase gene into CD34 + cells from patients with Gaucher disease: In vivo detection of transduced cells without myeloablation. *Hum Gene Ther* 9(17): 2629–2640.

Eckert, H. G., K. Kuhlcke, A. J. Schilz, C. Lindemann, N. Basara, A. A. Fauser, and C. Baum 2000. Clinical scale production of an improved retroviral vector expressing the human multidrug resistance 1 gene (MDR1). *Bone Marrow Transplant* 25 Suppl 2: S114–S117.

EMEA 1997. Quality of Biotechnological Products: Viral Safety Evaluation Of Biotechnology Products Derived From Cell Lines Of Human Or Animal Origin, ICH Q5A, European Agency for the Evaluation of Medicinal Products.

Etienne-Julan, M., P. Roux, S. Carillo, P. Jeanteur, and M. Piechaczyk 1992. The efficiency of cell targeting by recombinant retroviruses depends on the nature of the receptor and the composition of the artificial cell-virus linker. *J Gen Virol* 73 (Pt 12): 3251–3255.

EuropeanPharmacopeia 2005. European Pharmacopeia.

Evans, L. H., R. P. Morrison, F. G. Malik, J. Portis, and W. J. Britt 1990. A neutralizable epitope common to the envelope glycoproteins of ecotropic, polytropic, xenotropic, and amphotropic murine leukemia viruses. *J Virol* 64(12): 6176–6183.

Faff, O., B. A. Murray, V. Erfle, and R. Hehlmann 1993. Large scale production and purification of human retrovirus-like particles related to the mouse mammary tumor virus. *FEMS Microbiol Lett* 109(2–3): 289–296.

Finer, M. H., T. J. Dull, L. Qin, D. Farson, and M. R. Roberts 1994. Kat: A high-efficiency retroviral transduction system for primary human T lymphocytes. *Blood* 83(1): 43–50.

Forestell, S. P., J. S. Dando, E. Bohnlein, and R. J. Rigg 1996. Improved detection of replication-competent retrovirus. *J Virol Methods* 60(2): 171–178.

Garnier, A., J. Cote, I. Nadeau, A. Kamen, and B. Massie 1994. Scale-up of the adenovirus expression system for the production of recombinant protein in human 293S cells. *Cytotechnology* 15(1–3): 145–155.

Ghosh, R. 2002. Protein separation using membrane chromatography: Opportunities and challenges. *J Chromatogr* A 952(1–2): 13–27.

Goud, B., P. Legrain, and G. Buttin 1988. Antibody-mediated binding of a murine ecotropic Moloney retroviral vector to human cells allows internalization but not the establishment of the proviral state. *Virology* 163(1): 251–254.

Grossman, M., S. E. Raper, K. Kozarsky, E. A. Stein, J. F. Engelhardt, D. Muller, P. J. Lupien, and J. M. Wilson 1994. Successful ex vivo gene therapy directed to liver in a patient with familial hypercholesterolaemia. *Nat Genet* 6(4): 335–341.

Han, X., N. Kasahara, and Y. W. Kan 1995. Ligand-directed retroviral targeting of human breast cancer cells. *Proc Natl Acad Sci U S A* 92(21): 9747–9751.

Hansen, M., L. Jelinek, R. S. Jones, J. Stegeman-Olsen, and E. Barklis 1993. Assembly and composition of intracellular particles formed by Moloney murine leukemia virus. *J Virol* 67(9): 5163–5174.

Harris, R. J. C. (Ed.) 1954. The preservation of viruses. *Biological Applications of Freezing and Drying*. New York, Academic Press, pp. 201–214.

Higashikawa, F. and L. Chang 2001. Kinetic analyses of stability of simple and complex retroviral vectors. *Virology* 280(1): 124–131.

Hu, W. S. and V. K. Pathak 2000. Design of retroviral vectors and helper cells for gene therapy. *Pharmacol Rev* 52(4): 493–511.

Hughes, C., J. Galea-Lauri, F. Farzaneh, and D. Darling 2001. Streptavidin paramagnetic particles provide a choice of three affinity-based capture and magnetic concentration strategies for retroviral vectors. *Mol Ther* 3(4): 623–630.

Hunter, E. and R. Swanstrom 1990. Retrovirus envelope glycoproteins. *Curr Top Microbiol Immunol* 157: 187–253.

ICH 1999. Viral safety evaluation of Biotechnology products derived from cell lines of human origin, Harmonized Tripartite Guideline. Q5A (R1).

Irving, J. M., L. W. Chang, and F. J. Castillo 1993. A reverse transcriptase-polymerase chain reaction assay for the detection and quantitation of murine retroviruses. *Biotechnology* (N Y) 11(9): 1042–1046.

Jiang, A., T. H. Chu, F. Nocken, K. Cichutek, and R. Dornburg 1998. Cell-type-specific gene transfer into human cells with retroviral vectors that display single-chain antibodies. *J Virol* 72(12): 10148–10156.

Jiang, A. and R. Dornburg 1999. In vivo cell type-specific gene delivery with retroviral vectors that display single chain antibodies. *Gene Ther* 6(12): 1982–1987.

Jones, I. M. and Y. Morikawa 1998. The molecular basis of HIV capsid assembly. *Rev Med Virol* 8(2): 87–95.

Karreman, S., H. Hauser, and C. Karreman 1996. On the use of double FLP recognition targets (FRTs) in the LTR of retroviruses for the construction of high producer cell lines. *Nucleic Acids Res* 24(9): 1616–1624.

Kasahara, N., A. M. Dozy, and Y. W. Kan 1994. Tissue-specific targeting of retroviral vectors through ligand-receptor interactions. *Science* 266(5189): 1373–1376.

Konishi, H., T. Ochiya, K. A. Chester, R. H. Begent, T. Muto, T. Sugimura, and M. Terada 1998. Targeting strategy for gene delivery to carcinoembryonic antigen-producing cancer cells by retrovirus displaying a single-chain variable fragment antibody. *Hum Gene Ther* 9(2): 235–248.

Konz, J. O., A. L. Lee, J. A. Lewis, and S. L. Sagar 2005. Development of a purification process for adenovirus: Controlling virus aggregation to improve the clearance of host cell DNA. *Biotechnol Prog* 21(2): 466–472.

Kuiper, M., R. M. Sanches, J. A. Walford, and N. K. Slater 2002. Purification of a functional gene therapy vector derived from Moloney murine leukaemia virus using membrane filtration and ceramic hydroxyapatite chromatography. *Biotechnol Bioeng* 80(4): 445–453.

Kwon, Y. J., G. Hung, W. F. Anderson, C. A. Peng, and H. Yu 2003. Determination of infectious retrovirus concentration from colony-forming assay with quantitative analysis. *J Virol* 77(10): 5712–5720.

Lamers, C. H., P. van Elzakker, S. C. Langeveld, S. Sleijfer, and J. W. Gratama (2006a). Process validation and clinical evaluation of a protocol to generate gene-modified T lymphocytes for immunogene therapy for metastatic renal cell carcinoma: GMP-controlled transduction and expansion of patient's T lymphocytes using a carboxy anhydrase IX-specific scFv transgene. *Cytotherapy* 8(6): 542–553.

Lamers, C. H., R. A. Willemsen, P. van Elzakker, B. A. van Krimpen, J. W. Gratama, and R. Debets (2006b). Phoenix-ampho outperforms PG13 as retroviral packaging cells to transduce human T cells with tumor-specific receptors: Implications for clinical immunogene therapy of cancer. *Cancer Gene Ther* 13(5): 503–509.

Landazuri, N. and J. M. Le Doux 2006. Complexation with chondroitin sulfate C and polybrene rapidly purifies retrovirus from inhibitors of transduction and substantially enhances gene transfer. *Biotechnol Bioeng* 93(1): 146–158.

Lavillette, D., A. Ruggieri, S. J. Russell, and F. L. Cosset 2000. Activation of a cell entry pathway common to type C mammalian retroviruses by soluble envelope fragments. *J Virol* 74(1): 295–304.

Layne, S. P., M. J. Merges, M. Dembo, J. L. Spouge, S. R. Conley, J. P. Moore, J. L. Raina, H. Renz, H. R. Gelderblom, and P. L. Nara 1992. Factors underlying spontaneous inactivation and susceptibility to neutralization of human immunodeficiency virus. *Virology* 189(2): 695–714.

Le Doux, J. M., H. E. Davis, J. R. Morgan, and M. L. Yarmush 1999. Kinetics of retrovirus production and decay. *Biotechnol Bioeng* 63(6): 654–662.

Le Doux, J. M., J. R. Morgan, R. G. Snow, and M. L. Yarmush 1996. Proteoglycans secreted by packaging cell lines inhibit retrovirus infection. *J Virol* 70(9): 6468–6473.

Le Doux, J. M., J. R. Morgan, and M. L. Yarmush 1998. Removal of proteoglycans increases efficiency of retroviral gene transfer. *Biotechnol Bioeng* 58(1): 23–34.

Lenz, H. J., W. F. Anderson, F. L. Hall, and E. M. Gordon 2002. Clinical protocol. Tumor site specific phase I evaluation of safety and efficacy of hepatic arterial infusion of a matrix-targeted retroviral vector bearing a dominant negative cyclin G1 construct as intervention for colorectal carcinoma metastatic to liver. *Hum Gene Ther* 13(12): 1515–1537.

Lewis, P. F. and M. Emerman 1994. Passage through mitosis is required for oncoretroviruses but not for the human immunodeficiency virus. *J Virol* 68(1): 510–516.

Li, Z., J. Dullmann, B. Schiedlmeier, M. Schmidt, C. von Kalle, J. Meyer, M. Forster, C. Stocking, A. Wahlers, O. Frank, W. Ostertag, K. Kuhlcke, H. G. Eckert, B. Fehse, and C. Baum 2002. Murine leukemia induced by retroviral gene marking. *Science* 296(5567): 497.

Linial, M. 1987. Creation of a processed pseudogene by retroviral infection. *Cell* 49(1): 93–102.

Linial, M. L. and A. D. Miller 1990. Retroviral RNA packaging: sequence requirements and implications. *Retroviruses: Strategies of Replication*. R. Swanstrom and P. K. Vogt (Eds.). Berlin, Springer, pp. 125–152.

Lodge, R., R. A. Subbramanian, J. Forget, G. Lemay, and E. A. Cohen 1998. MuLV-based vectors pseudotyped with truncated HIV glycoproteins mediate specific gene transfer in CD4 + peripheral blood lymphocytes. *Gene Ther* 5(5): 655–664.

Loew, R., N. Selevsek, B. Fehse, D. von Laer, C. Baum, A. Fauser, and K. Kuehlcke 2004. Simplified generation of high-titer retrovirus producer cells for clinically relevant retroviral vectors by reversible inclusion of a lox-P-flanked marker gene. *Mol Ther* 9(5): 738–746.

Lozinsky, V. I., I. Y. Galaev, F. M. Plieva, I. N. Savina, H. Jungvid, and B. Mattiasson 2003. Polymeric cryogels as promising materials of biotechnological interest. *Trends Biotechnol* 21(10): 445–451.

Makino, M., G. Ishikawa, K. Yamaguchi, Y. Okada, K. Watanabe, Y. Sasaki-Iwaki, S. Manabe, M. Honda, and K. Komuro 1994. Concentration of live retrovirus with a regenerated cellulose hollow fiber, BMM. *Arch Virol* 139(1–2): 87–96.

Mann, R. and D. Baltimore 1985. Varying the position of a retrovirus packaging sequence results in the encapsidation of both unspliced and spliced RNAs. *J Virol* 54(2): 401–407.

Martineau, D., W. M. Klump, J. E. McCormack, N. J. DePolo, E. Kamantigue, M. Petrowski, J. Hanlon, D. J. Jolly, S. J. Mento, and N. Sajjadi 1997. Evaluation of PCR and ELISA assays for screening clinical trial subjects for replication-competent retrovirus. *Hum Gene Ther* 8(10): 1231–1241.

Mason, J. M., D. E. Guzowski, L. O. Goodwin, D. Porti, K. C. Cronin, S. Teichberg, and R. G. Pergolizzi 1999. Human serum-resistant retroviral vector particles from galactosyl (alpha1–3) galactosyl containing nonprimate cell lines. *Gene Ther* 6(8): 1397–1405.

McGrath, M., O. Witte, T. Pincus, and I. L. Weissman 1978. Retrovirus purification: method that conserves envelope glycoprotein and maximizes infectivity. *J Virol* 25(3): 923–927.

Merten, O. W. 2004. State-of-the-art of the production of retroviral vectors. *J Gene Med* 6 Suppl 1: S105–S124.

Miller, A. D. and C. Buttimore 1986. Redesign of retrovirus packaging cell lines to avoid recombination leading to helper virus production. *Mol Cell Biol* 6(8): 2895–2902.

Miller, A. D. and F. Chen 1996. Retrovirus packaging cells based on 10A1 murine leukemia virus for production of vectors that use multiple receptors for cell entry. *J Virol* 70(8): 5564–5571.

Miller, A. D., J. V. Garcia, N. von Suhr, C. M. Lynch, C. Wilson, and M. V. Eiden 1991. Construction and properties of retrovirus packaging cells based on gibbon ape leukemia virus. *J Virol* 65(5): 2220–2224.

Miller, A. D., D. R. Trauber, and C. Buttimore 1986. Factors involved in production of helper virus-free retrovirus vectors. *Somat Cell Mol Genet* 12(2): 175–183.

Miller, D. L., P. J. Meikle, and D. S. Anson 1996. A rapid and efficient method for concentration of small volumes of retroviral supernatant. *Nucleic Acids Res* 24(8): 1576–1577.

Morgan, R. A. and R. Walker 1996. Gene therapy for AIDS using retroviral mediated gene transfer to deliver HIV-1 antisense TAR and transdominant Rev protein genes to syngeneic lymphocytes in HIV-1 infected identical twins. *Hum Gene Ther* 7(10): 1281–1306.

Morgenstern, J. P. and H. Land 1990. Advanced mammalian gene transfer: High titre retroviral vectors with multiple drug selection markers and a complementary helper-free packaging cell line. *Nucleic Acids Res* 18(12): 3587–3596.

Nabel, G. J., B. A. Fox, L. Post, C. B. Thompson, and C. Woffendin 1994. A molecular genetic intervention for AIDS: Effects of a transdominant negative form of Rev. *Hum Gene Ther* 5(1): 79–92.

Naviaux, R. K., E. Costanzi, M. Haas, and I. M. Verma 1996. The pCL vector system: rapid production of helper-free, high-titer, recombinant retroviruses. *J Virol* 70(8): 5701–5705.

Neda, H., C. H. Wu, and G. Y. Wu 1991. Chemical modification of an ecotropic murine leukemia virus results in redirection of its target cell specificity. *J Biol Chem* 266(22): 14143–14146.

Nemunaitis, J., C. Bohart, T. Fong, W. Meyer, G. Edelman, R. S. Paulson, D. Orr, V. Jain, J. O'Brien, J. Kuhn, K. J. Kowal, S. Burkeholder, J. Bruce, N. Ognoskie, D. Wynne, D. Martineau, and D. Ando 1998. Phase I trial of retroviral vector-mediated interferon (IFN)-gamma gene transfer into autologous tumor cells in patients with metastatic melanoma. *Cancer Gene Ther* 5(5): 292–300.

O'Keeffe, R. S., M. D. Johnston, and N. K. Slater 1999. The affinity adsorptive recovery of an infectious herpes simplex virus vaccine. *Biotechnol Bioeng* 62(5): 537–545.

Opstelten, D. J., M. Wallin, and H. Garoff 1998. Moloney murine leukemia virus envelope protein subunits, gp70 and Pr15E, form a stable disulfide-linked complex. *J Virol* 72(8): 6537–6545.

Ott, D. E. 1997. Cellular proteins in HIV virions. *Rev Med Virol* 7(3): 167–180.

Ott, M. G., M. Schmidt, K. Schwarzwaelder, S. Stein, U. Siler, U. Koehl, H. Glimm, K. Kuhlcke, A. Schilz, H. Kunkel, S. Naundorf, A. Brinkmann, A. Deichmann, M. Fischer, C. Ball, I. Pilz, C. Dunbar, Y. Du, N. A. Jenkins, N. G. Copeland, U. Luthi, M. Hassan, A. J. Thrasher, D. Hoelzer, C. von Kalle, R. Seger, and M. Grez 2006. Correction of X-linked chronic granulomatous disease by gene therapy, augmented by insertional activation of MDS1-EVI1, PRDM16 or SETBP1. *Nat Med* 12(4): 401–409.

Otto, E., A. Jones-Trower, E. F. Vanin, K. Stambaugh, S. N. Mueller, W. F. Anderson, and G. J. McGarrity 1994. Characterization of a replication-competent retrovirus resulting from recombination of packaging and vector sequences. *Hum Gene Ther* 5(5): 567–575.

Palmer, K., J. Moore, M. Everard, J. D. Harris, S. Rodgers, R. C. Rees, A. K. Murray, R. Mascari, J. Kirkwood, P. G. Riches, C. Fisher, J. M. Thomas, M. Harries, S. R. Johnston, M. K. Collins, and M. E. Gore 1999. Gene therapy with autologous, interleukin 2-secreting tumor cells in patients with malignant melanoma. *Hum Gene Ther* 10(8): 1261–1268.

Palu, G., C. Parolin, Y. Takeuchi, and M. Pizzato 2000. Progress with retroviral gene vectors. *Rev Med Virol* 10(3): 185–202.

Parente, M. K. and J. H. Wolfe 1996. Production of increased titer retrovirus vectors from stable producer cell lines by superinfection and concentration. *Gene Ther* 3(9): 756–760.

Patience, C., Y. Takeuchi, F. L. Cosset, and R. A. Weiss 1998. Packaging of endogenous retroviral sequences in retroviral vectors produced by murine and human packaging cells. *J Virol* 72(4): 2671–2676.

Pensiero, M. N., C. A. Wysocki, K. Nader, and G. E. Kikuchi 1996. Development of amphotropic murine retrovirus vectors resistant to inactivation by human serum. *Hum Gene Ther* 7(9): 1095–1101.

Pizzato, M., O. W. Merten, E. D. Blair, and Y. Takeuchi 2001. Development of a suspension packaging cell line for production of high titre, serum-resistant murine leukemia virus vectors. *Gene Ther* 8(10): 737–745.

Powell, S. K., M. A. Kaloss, A. Pinkstaff, R. McKee, I. Burimski, M. Pensiero, E. Otto, W. P. Stemmer, and N. W. Soong 2000. Breeding of retroviruses by DNA shuffling for improved stability and processing yields. *Nat Biotechnol* 18(12): 1279–1282.

Printz, M., J. Reynolds, S. J. Mento, D. Jolly, K. Kowal, and N. Sajjadi 1995. Recombinant retroviral vector interferes with the detection of amphotropic replication competent retrovirus in standard culture assays. *Gene Ther* 2(2): 143–150.

Przybylowski, M., A. Hakakha, J. Stefanski, J. Hodges, M. Sadelain, and I. Riviere 2006. Production scale-up and validation of packaging cell clearance of clinical-grade retroviral vector stocks produced in cell factories. *Gene Ther* 13(1): 95–100.

Quigley, J. P., D. B. Rifkin, and E. Reich 1971. Phospholipid composition of Rous sarcoma virus, host cell membranes and other enveloped RNA viruses. *Virology* 46(1): 106–116.

Reeves, L. and K. Cornetta 2000. Clinical retroviral vector production: step filtration using clinically approved filters improves titers. *Gene Ther* 7(23): 1993–1998.

Reiser, J. 2000. Production and concentration of pseudotyped HIV-1-based gene transfer vectors. *Gene Ther* 7(11): 910–913.

Richieri, S. P., R. Bartholomew, R. C. Aloia, J. Savary, R. Gore, J. Holt, F. Ferre, R. Musil, H. R. Tian, R. Trauger, P. Lowry, F. Jensen, D. J. Carlo, R. Z. Maigetter, and C. P. Prior 1998. Characterization of highly purified, inactivated HIV-1 particles isolated by anion exchange chromatography. *Vaccine* 16(2–3): 119–129.

Rigg, R. J., J. Chen, J. S. Dando, S. P. Forestell, I. Plavec, and E. Bohnlein 1996. A novel human amphotropic packaging cell line: High titer, complement resistance, and improved safety. *Virology* 218(1): 290–295.

Rill, D. R., R. C. Moen, M. Buschle, C. Bartholomew, N. K. Foreman, J. Mirro, Jr., R. A. Krance, J. N. Ihle, and M. K. Brenner 1992. An approach for the analysis of relapse and marrow reconstitution after autologous marrow transplantation using retrovirus-mediated gene transfer. *Blood* 79(10): 2694–2700.

Rimai, L., I. Salmeen, D. Hart, L. Liebes, M. A. Rich, and J. J. McCormick 1975. Electrophoretic mobilities of RNA tumor viruses. Studies by Doppler-shifted light scattering spectroscopy. *Biochemistry* 14(21): 4621–4627.

Rodrigues, T., M. J. Carrondo, P. M. Alves, and P. E. Cruz 2007a. Purification of retroviral vectors for clinical application: Biological implications and technological challenges. *J Biotechnol* 127(3): 520–541.

Rodrigues, T., A. Carvalho, M. Carmo, M. J. Carrondo, P. M. Alves, and P. E. Cruz 2007b. Scaleable purification process for gene therapy retroviral vectors. *J Gene Med* 9(4): 233–243.

Rodrigues, T., A. Carvalho, A. Roldão, M. J. T. Carrondo, P. M. Alves, and P. E. Cruz 2006. Screening anion-exchange chromatographic matrices for isolation of onco-retroviral vectors. *J Chromatogr B Analyt Technol Biomed Life Sci* 837(1–2): 59–68.

Roe, T., T. C. Reynolds, G. Yu, and P. O. Brown 1993. Integration of murine leukemia virus DNA depends on mitosis. *Embo J* 12(5): 2099–2108.

Roth, J. A. and R. J. Cristiano 1997. Gene therapy for cancer: What have we done and where are we going? *J Natl Cancer Inst* 89(1): 21–39.

Saha, K., Y. C. Lin, and P. K. Wong 1994. A simple method for obtaining highly viable virus from culture supernatant. *J Virol Methods* 46(3): 349–352.

Salmeen, I., L. Rimai, R. B. Luftig, L. Libes, E. Retzel, M. Rich, and J. J. McCormick 1976. Hydrodynamic diameters of murine mammary, Rous sarcoma, and feline leukemia RNA tumor viruses: Studies by laser beat frequency light-scattering spectroscopy and electron microscopy. *J Virol* 17(2): 584–596.

Sastry, L., T. Johnson, M. J. Hobson, B. Smucker, and K. Cornetta 2002. Titering lentiviral vectors: Comparison of DNA, RNA and marker expression methods. *Gene Ther* 9(17): 1155–1162.

Sastry, L., Y. Xu, R. Cooper, K. Pollok, and K. Cornetta 2004. Evaluation of plasmid DNA removal from lentiviral vectors by benzonase treatment. *Hum Gene Ther* 15(2): 221–226.

Sastry, L., Y. Xu, L. Duffy, S. Koop, A. Jasti, H. Roehl, D. Jolly, and K. Cornetta 2005. Product-enhanced reverse transcriptase assay for replication-competent retrovirus and lentivirus detection. *Hum Gene Ther* 16(10): 1227–1236.

Scadden, D. T., B. Fuller, and J. M. Cunningham 1990. Human cells infected with retrovirus vectors acquire an endogenous murine provirus. *J Virol* 64(1): 424–427.

Scarpa, M., D. Cournoyer, D. M. Muzny, K. A. Moore, J. W. Belmont, and C. T. Caskey 1991. Characterization of recombinant helper retroviruses from Moloney-based vectors in ecotropic and amphotropic packaging cell lines. *Virology* 180(2): 849–852.

Schnierle, B. S., D. Moritz, M. Jeschke, and B. Groner 1996. Expression of chimeric envelope proteins in helper cell lines and integration into Moloney murine leukemia virus particles. *Gene Ther* 3(4): 334–342.

Schonely, K., C. Afable, V. Slepushkin, X. Lu, K. Andre, J. Boehmer, K. Bengston, M. Doub, R. Cohen, D. Berlinger, T. Slepushkina, Z. Chen, Y. Li, G. Binder, B. Davis, L. Humeau, and B. Dropulic 2003. QC release testing of an HIV-1 based lentiviral vector lot and transduced cellular product. *Bioprocessing* 2: 39–47.

Schucht, R., A. S. Coroadinha, M. A. Zanta-Boussif, E. Verhoeyen, M. J. Carrondo, H. Hauser, and D. Wirth 2006. A new generation of retroviral producer cells: Predictable and stable virus production by Flp-mediated site-specific integration of retroviral vectors. *Mol Ther* 14(2): 285–292.

Segura, M., A. Kamen, and A. Garnier 2006a. Downstream processing of oncoretroviral and lentiviral gene therapy vectors. *Biotechnol Adv* 24(3): 321–337.

Segura, M. D., A. Garnier, and A. Kamen 2006b. Purification and characterization of retrovirus vector particles by rate zonal ultracentrifugation. *J Virol Methods* 133(1): 82–91.

Segura, M. M., A. Kamen, P. Trudel, and A. Garnier 2005. A novel purification strategy for retrovirus gene therapy vectors using heparin affinity chromatography. *Biotechnol Bioeng* 90(4): 391–404.

Seppen, J., S. Barry, G. M. Lam, N. Ramesh, and W. R. Osborne 2000. Retroviral preparations derived from PA317 packaging cells contain inhibitors that copurify with viral particles and are devoid of viral vector RNA. *Hum Gene Ther* 11(5): 771–775.

Sinn, P. L., S. L. Sauter, and P. B. McCray, Jr. 2005. Gene therapy progress and prospects: Development of improved lentiviral and retroviral vectors—design, biosafety, and production. *Gene Ther* 12(14): 1089–1098.

Slepushkin, V., N. Chang, R. Cohen, Y. Gan, B. Jiang, E. Deausen, D. Berlinger, G. Binder, K. Andre, L. Humeau, and B. Dropulic 2003. Large-scale purification of a lentiviral vector by size exclusion chromatography or mustang Q ion exchange capsule. *BioProcessing J* 2: 89–95.

Somia, N. V., M. Zoppe, and I. M. Verma 1995. Generation of targeted retroviral vectors by using single-chain variable fragment: An approach to in vivo gene delivery. *Proc Natl Acad Sci U S A* 92(16): 7570–7574.

Soneoka, Y., P. M. Cannon, E. E. Ramsdale, J. C. Griffiths, G. Romano, S. M. Kingsman, and A. J. Kingsman 1995. A transient three-plasmid expression system for the production of high titer retroviral vectors. *Nucleic Acids Res* 23(4): 628–633.

Stuhlmann, H., M. Dieckmann, and P. Berg 1990. Transduction of cellular neo mRNA by retrovirus-mediated recombination. *J Virol* 64(12): 5783–5796.

Takeuchi, Y., F. L. Cosset, P. J. Lachmann, H. Okada, R. A. Weiss, and M. K. Collins 1994. Type C retrovirus inactivation by human complement is determined by both the viral genome and the producer cell. *J Virol* 68(12): 8001–8007.

Takeuchi, Y., S. H. Liong, P. D. Bieniasz, U. Jager, C. D. Porter, T. Friedman, M. O. McClure, and R. A. Weiss 1997. Sensitization of rhabdo-, lenti-, and spumaviruses to human serum by galactosyl(alpha1–3)galactosylation. *J Virol* 71(8): 6174–6178.

Takeuchi, Y., C. D. Porter, K. M. Strahan, A. F. Preece, K. Gustafsson, F. L. Cosset, R. A. Weiss, and M. K. Collins 1996. Sensitization of cells and retroviruses to human serum by (alpha 1–3) galactosyltransferase. *Nature* 379(6560): 85–88.

Tan, Y., M. Xu, W. Wang, F. Zhang, D. Li, X. Xu, J. Gu, and R. M. Hoffman 1996. IL-2 gene therapy of advanced lung cancer patients. *Anticancer Res* 16(4A): 1993–1998.

Tancevski, I., A. Wehinger, J. R. Patsch, and A. Ritsch 2006. In vivo application of adenoviral vectors purified by a Taqman Real Time PCR-supported chromatographic protocol. *Int J Biol Macromol* 39(1–3): 77–82.

Thommes, J. and M. Etzel 2007. Alternatives to chromatographic separations. *Biotechnol Prog* 23(1): 42–45.

Transfiguracion, J., H. Coelho, and A. Kamen 2004. High-performance liquid chromatographic total particles quantification of retroviral vectors pseudotyped with vesicular stomatitis virus-G glycoprotein. *J Chromatogr B Analyt Technol Biomed Life Sci* 813(1–2): 167–173.

Transfiguracion, J., D. E. Jaalouk, K. Ghani, J. Galipeau, and A. Kamen 2003. Size-exclusion chromatography purification of high-titer vesicular stomatitis virus G glycoprotein-pseudotyped retrovectors for cell and gene therapy applications. *Hum Gene Ther* 14(12): 1139–1153.

Tuschong, L., S. L. Soenen, R. M. Blaese, F. Candotti, and L. M. Muul 2002. Immune response to fetal calf serum by two adenosine deaminase-deficient patients after T cell gene therapy. *Hum Gene Ther* 13(13): 1605–1610.

Vanin, E. F., M. Kaloss, C. Broscius, and A. W. Nienhuis 1994. Characterization of replication-competent retroviruses from nonhuman primates with virus-induced T-cell lymphomas and observations regarding the mechanism of oncogenesis. *J Virol* 68(7): 4241–4250.

Weiss, R. 1998. Viral RNA-dependent DNA polymerase RNA-dependent DNA polymerase in virions of Rous sarcoma virus. *Rev Med Virol* 8(1): 3–11.

Wikstrom, K., P. Blomberg, and K. B. Islam 2004. Clinical grade vector production: analysis of yield, stability, and storage of gmp-produced retroviral vectors for gene therapy. *Biotechnol Prog* 20(4): 1198–1203.

Wildner, O., F. Candotti, E. G. Krecko, K. G. Xanthopoulos, W. J. Ramsey, and R. M. Blaese 1998. Generation of a conditionally neo(r)-containing retroviral producer cell line: effects of neo(r) on retroviral titer and transgene expression. *Gene Ther* 5(5): 684–691.

Williams, S. L., M. E. Eccleston, and N. K. Slater 2005. Affinity capture of a biotinylated retrovirus on macroporous monolithic adsorbents: Towards a rapid single-step purification process. *Biotechnol Bioeng* 89(7): 783–787.

Wong-Staal, F., E. M. Poeschla, and D. J. Looney 1998. A controlled, Phase 1 clinical trial to evaluate the safety and effects in HIV-1 infected humans of autologous lymphocytes transduced with a ribozyme that cleaves HIV-1 RNA. *Hum Gene Ther* 9(16): 2407–2425.

Xu, L., K. Tsuji, H. Mostowski, M. Otsu, F. Candotti, and A. S. Rosenberg 2004. A convenient method for positive selection of retroviral producing cells generating vectors devoid of selectable markers. *J Virol Methods* 118(1): 61–67.

Yamada, K., D. M. McCarty, V. J. Madden, and C. E. Walsh 2003. Lentivirus vector purification using anion exchange HPLC leads to improved gene transfer. *Biotechniques* 34(5): 1074–1078, 1080.

Yap, M. W., S. M. Kingsman, and A. J. Kingsman 2000. Effects of stoichiometry of retroviral components on virus production. *J Gen Virol* 81(Pt 9): 2195–2202.

Ye, K., S. Jin, M. M. Ataai, J. S. Schultz, and J. Ibeh 2004. Tagging retrovirus vectors with a metal binding peptide and one-step purification by immobilized metal affinity chromatography. *J Virol* 78(18): 9820–9827.

Yeager, M., E. M. Wilson-Kubalek, S. G. Weiner, P. O. Brown, and A. Rein 1998. Supramolecular organization of immature and mature murine leukemia virus revealed by electron cryomicroscopy: Implications for retroviral assembly mechanisms. *Proc Natl Acad Sci U S A* 95(13): 7299–7304.

Yee, J. K., T. Friedmann, and J. C. Burns 1994a. Generation of high-titer pseudotyped retroviral vectors with very broad host range. *Methods Cell Biol* 43 Pt A: 99–112.

Yee, J. K., A. Miyanohara, P. LaPorte, K. Bouic, J. C. Burns, and T. Friedmann 1994b. A general method for the generation of high-titer, pantropic retroviral vectors: Highly efficient infection of primary hepatocytes. *Proc Natl Acad Sci U S A* 91(20): 9564–9568.

Yu, S. F., T. von Ruden, P. W. Kantoff, C. Garber, M. Seiberg, U. Ruther, W. F. Anderson, E. F. Wagner, and E. Gilboa 1986. Self-inactivating retroviral vectors designed for transfer of whole genes into mammalian cells. *Proc Natl Acad Sci U S A* 83(10): 3194–3198.

3 Adenovirus Vectors for Gene Therapy

Neil R. Hackett and Ronald G. Crystal

CONTENTS

3.1 INTRODUCTION

The emergence of recombinant DNA as a tool to study medicine quickly promulgated the concept of cloned genes as therapeutics. As originally conceived, the concept of gene therapy was simply to introduce a wild-type copy of a deficient gene into cells to restore function *in trans* [1–4]. Viewed in this way, the technical challenge was to efficiently deliver the gene to the appropriate cell and have it expressed for sufficient time, or readminister as often as needed, for the therapeutic application. Adenovirus (Ad) gene transfer vectors offer one strategy to achieve this. The focus on Ad for gene transfer was based on basic research establishing the biology of Ad, and the knowledge that Ad efficiently delivers the viral genome to the target cells. Importantly, Ad is not oncogenic in humans, the genomes of common Ad are completely defined, the Ad genome can be easily modified, and recombinant Ad can be readily produced in large quantities and highly concentrated without modifying the ability of the virus to infect cells.

In retrospect, the original goal of using Ad as simple delivery systems to permanently complement genetic defects seems naive. Whereas Ad gene transfer vectors can achieve robust expression of the transgene in many target organs, expression of the transgene is limited in time, resulting from a complex combination of innate and adaptive immune host defenses against the virus [5,6]. In this context, Ad vectors in their present form are most useful in applications where transient (days to weeks) expression is sufficient to have the desired therapeutic effect. For applications where persistent expression is required to achieve a therapeutic goal, there are still many challenges before and if Ad vectors will be successful. In this chapter we summarize the biology of Ad, the construction and use of first generation Ad vectors, the current status of advanced forms of Ad vectors, clinical applications of

Ad vectors, and the future prospects of using Ad in gene transfer applications.

3.2 BIOLOGY OF Ads

Adenoviruses are a group of double-stranded DNA viruses which infect a variety of vertebrate hosts including rodents, chickens, and primates. Human Ad has been isolated from several sources including the upper respiratory tracts of military recruits with respiratory infections, adenoids, conjunctiva, and the stool of infants with diarrhea [7]. As with other viruses, there is an immune response to Ad infection which includes the production of neutralizing antibodies, defined as antibodies which prevent Ad infection *in vitro*. Neutralizing sera have been used to distinguish >50 different Ad serotypes which are divided into subgroups A through E [8]. The presence of antibodies against one serotype generally protects against reinfection by the same serotype. In the context of gene therapy, serotypes 5 and 2 of the subgroup C have been used most because their structure and biology is well described and there are convenient biologic reagents available to produce recombinant subgroup C Ad gene transfer vectors in large quantities. Ads of subgroups C cause various respiratory infections either as outbreaks in confined groups (such as military recruits) or in children, sometimes associated with conjunctival infections [9]. The known predilection of subgroup C in the respiratory tract led to the initial uses of Ad gene therapy for the treatment of cystic fibrosis (CF), and thus the focus on serotypes 2 and 5.

3.2.1 STRUCTURE

Adenovirus consists of an icosahedral protein capsid of approximately 70–100 nm diameter and, within that capsid, a single copy of a double-stranded DNA molecule of length approximately 36,000 bp (Figure 3.1, reviewed in Ref. [10]). In the context of gene therapy, the fiber, penton base, and hexon are the most important capsid proteins. The 20 triangular faces of the viral capsid are built from hexon, the major capsid protein. The 240 hexon capsomeres in the capsid are each trimers comprising three copies of the 105 kDa hexon subunit with each trimer interacting with six others in a pseudo-equivalent fashion. The three-dimensional structure of hexon shows that the homotrimer has loops which project out from the capsid surface [11]. Capsid proteins VI, VIII, and IX are associated with hexon and their role is to stabilize the capsid structure. The 12 capsid vertices are made up of the penton capsomere, a complex of five copies of the penton base and three copies of fiber. Each penton capsomere interacts with five hexon capsomeres, one from each of the five faces that converge at the vertex. The fiber protein projects outward from the penton base. The DNA is wrapped in the histone-like core protein VII and there is a terminal protein attached to the 5′ end of each strand of the DNA.

Neutralizing antibodies are directed primarily against epitopes located on the loops of the hexon. This is expected, as the loops project from the surface of the virus where they

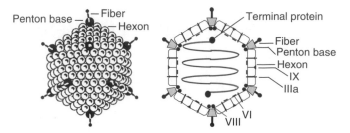

FIGURE 3.1 Structure of the Ad capsid. Shown (left) is a three-dimensional representation and (right) a simplified cross-section of the capsid showing the deployment of the capsid proteins and Ad genome. The capsid is an icosahedron with 20 faces and 12 vertices. The faces are composed of hexons, each comprised of trimers of the hexon protein. The hexons are trapezoid shaped, with three loops on top, extending from the face of the capsid. The loops represent the variable regions that differ among serotypes and are the major epitopes for neutralizing antibodies. Proteins IX and VIII are associated with the hexon and are thought to stabilize the capsid. The vertices are composed of a fiber and penton base. The fiber has three domains, the base which interacts with penton, the shaft, and the knob. The knob interacts with a high affinity receptor on the target cell, and the shaft holds the virus away from the surface of the cell, depending on the length of the shaft. The penton base interacts with the hexon and the fiber and contains epitopes that interact with integrins on the cell surface. The 36 kb double-stranded DNA genome is wrapped around capsid core protein VII and the terminal protein is attached to the two 5′ ends of the Ad genome. (Adapted from Shenk T., *Fields Virology*, Lippincott-Raven Publishers, Inc., Philadelphia, 1996.)

are accessible to antibodies. When the primary structures of the capsids of different serotypes are compared, related Ad differs most in these loops, suggesting the selective pressures applied by the immune system result in the emergence of mutations in the external hexon loops [12].

The fiber protein is a trimer consisting of three domains, base, shaft, and knob. The N-terminal base domain interacts with the penton base. The shaft includes an extended domain consisting of variable numbers of a 15 amino acid pseudo-repeats. The number of repeats, and therefore the length of the shaft, varies between 23 copies for the subgroup A viruses and 6 copies for the subgroup B viruses. The distal C-terminal domain of the fiber protein, referred to as the "knob," interacts with the high affinity receptor on the surface of the target cell. The high affinity receptor for Ads except those of subgroup B is referred to as "CAR" (coxsackie and adenovirus receptor), reflecting the fact that the coxsackie B viruses and most serotypes of Ad share the same receptor [13,14]. CAR is a single membrane-spanning protein with two extracellular immunoglobulin-like domains. Apart from acting as a virus receptor, the function of CAR is unknown.

A sequence motif on the penton base is involved in internalization of the virus after high affinity CAR–fiber interaction. In serotypes 2 and 5 the amino acid motif arginine–glycine–aspartate (RGD) interacts with $\alpha_v\beta_3$ and $\alpha_v\beta_5$ integrins of the cell surface and this interaction is essential for efficient internalization [15].

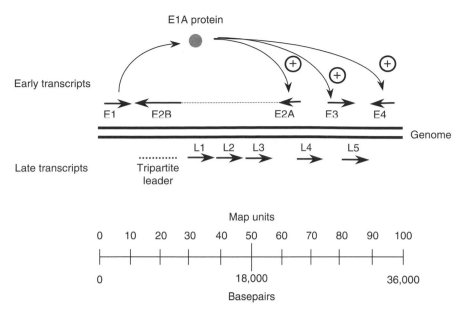

FIGURE 3.2 Structure and transcription of the major genes of the Ad type 5 genome. Schematic summary of the transcription of Ad during lytic infection. The genome is represented as two parallel lines and is divided by the scale shown underneath into 100 map units (1 map unit = 360 bp). There are nine major complex transcription units divided into early (above the genome) and late transcripts (below). The four early transcripts are produced before the commencement of DNA replication and specify regulatory proteins and proteins required for DNA replication. Upon initial infection of a cell, the E1A protein is produced from transcripts in the E1 region. E1A is a major regulatory factor required for transcription of E1B, E2, E3, and E4. In replication-deficient Ad vectors, the E1 region is deleted. Proteins coded by the E2 and E4 regions are required for late gene transcription. The E3 region codes for proteins that help the virus evade host defenses. All late transcripts rightwards originate at the same point and are produced by alternate splicing. The tripartite leader sequence is present at the 5′ end of all late transcripts. The L3 region specifies hexon, the L5 region specifies fiber, and the L2 region specifies penton.

For Ad type 5, the most commonly used Ad for gene transfer vectors, the complete 35,935 bp DNA sequence is known. For convenient reference, the genome is divided into 100 equally sized map units. A detailed transcription map at various time points postinfection is used to divide the genome into interspersed early (E) and late (L) regions (Figure 3.2, reviewed in Ref. [10]). There is considerable transcriptional overlap among the genes, making manipulation of some areas of the genome difficult. Each of the five early genes is comprised of a complex transcription unit with alternative sites for transcription initiation, termination, and splicing. The E1A and E1B genes are transcribed rightwards at the left hand end of the genome close to the DNA replication origin and DNA packaging signal. The E4 region is transcribed leftward at the right hand end of the genome. Distal to the E1 and E4 regions are the termini of the DNA which are inverted copies of the same sequence. Replication of the ends of the DNA is achieved by the attachment of terminal protein to the 5′ end of the DNA which acts as a primer to initiate unidirectional replication. This terminal protein is one of the components of the E2 transcriptional unit which is transcribed leftward commencing at map unit 75. The remaining early transcription unit is the E3 gene which is transcribed rightward commencing at map unit 77.

The five late genes are expressed after the beginning of DNA replication and encode the viral structural proteins. These late transcripts are all transcribed rightwards originating from map unit 17 and contain the same three part leader sequence before alternate splicing generates different mature mRNAs.

3.2.2 VIRAL REPLICATION

The Ad viral life cycle is understood best for subgroup C, which is another factor in the choice of Ad5 and Ad2 as gene therapy vectors (Figure 3.3). The knob of the fiber protein binds to the CAR receptor [13] followed by an interaction of the RGD sequence in the penton base with cell surface $\alpha_V\beta_3$ or $\alpha_V\beta_5$ integrins [15]. Excess soluble integrins inhibit Ad internalization but not binding, suggesting that penton base–integrin interaction is instrumental in internalization [16]. Ad modified with deletion of the RGD motif replicates effectively *in vitro* so the penton base–integrin interaction is likely related to efficiency of Ad infection, but is not essential [17]. The Ad enters the cell by endocytosis into clathrin coated pits, a process that can be blocked by dynamin inhibitors [18]. After endocytosis, Ad is rapidly released into the cytoplasm prior to extensive endosome fusion. The virus proceeds rapidly to the nucleus, probably actively transported on microtubules using the dynein motor [19], and then binds to the surface of the nucleus near the nuclear pore [20]. Using fluorescent viruses, this process has been shown to be efficient and rapid, with >90% of Ad5 delivered to the nucleus within 1 h [21]. At the nuclear membrane, the DNA and terminal protein are internalized by an unknown mechanism and are assembled into the nuclear scaffold for active transcription.

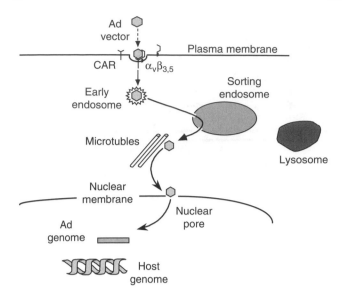

FIGURE 3.3 Trafficking of Ad from membrane to nucleus. The initial contact between the virus and cell is mediated by the knob of fiber and the CAR. This allows the secondary interaction between the penton and $\alpha_v\beta_3$ or $\alpha_v\beta_5$ integrins which is required for internalization. The initial internalization is via by coated pits which give rise to coated vesicles. After a very short interval, prior to fusion of early endosomes into sorting endosomes, a conformational change in the viral capsid allows escape of the virus into the cytoplasm. Microtubules carry the virus towards the nucleus. The whole capsid attaches to the outside of the nucleus but only the DNA and terminal protein are inserted into the nucleus itself, where they are assembled onto the nuclear matrix to allow transcription. (Courtesy of P. Leopold, Weill Medical College of Cornell University. With permission.)

With wild-type Ad, the viral E1A gene is transcribed immediately after infection [10]. After alternate splicing, the E1 mRNAs are translated into the two E1A proteins essential for transcription of other early viral mRNAs. E1A proteins promote the expression of cellular genes needed for DNA replication by interacting with the retinoblastoma susceptibility protein (Rb) which normally suppress entry into S phase of the cell cycle by complexing with the host transcription factor E2F. E1A also interacts with a number of cellular transcriptional factors to promote the assembly of complexes that promote transcription of other early adenoviral genes. Among the important downstream products induced by E1A is the product of the E1B gene, which blocks the apoptotic pathway through interaction with p53 long enough for a productive viral infection. The E1B 55 kDa protein also complexes with the ORF6 protein from the E4 region to modulate expression of the viral late genes which begin to be expressed around 6 h postinfection. At that time, DNA replication begins and the transcription of late genes commences, providing the capsid components that assemble into mature virions [10]. The new virions are assembled in the nucleus, necessitating transport of capsid proteins into the nucleus. As the viral infection proceeds, the integrity and viability of the cells decreases, but the mechanism of viral release from the cell is not understood.

In the context of gene therapy, the E3 region is important as it encodes immunosuppressive functions that work through two mechanisms [22]. The E3 gp19 kDa protein prevents major histocompatibility complex class I-mediated antigen presentation on the cell surface, thereby inhibiting the differentiation of cytotoxic T lymphocytes directed against viral antigens [22]. The E3 14.7 and E3 10.4 kDa proteins inhibit apoptosis of infected cells initiated by fas/fas ligand or tumor necrosis factor (TNF) [23]. The promoter for the E3 region requires E1 products, and thus in E1$^-$ deleted Ad vectors, the presence or absence of the E3 region is not relevant (see Section 3.4.3).

Transcripts from the E2 region specify the three nonhost proteins directly involved in DNA replication: The DNA polymerase, the single-stranded DNA binding protein (ssDBP), and the preterminal protein [10]. Like other viruses, Ad has developed a specific strategy for the faithful replication of the DNA ends. The last 103 nt at both ends of the genome consist of inverted copies of the identical sequence. The terminal protein binds covalently to the 5' end and acts as a primer for DNA synthesis by the adenoviral DNA polymerase of the leading strand starting at either end. DNA polymerase proceeds by a strand displacement mechanism creating a duplex and a displaced strand, which is sequestered by the ssDBP and has terminal protein attached to one end. Base pairing of the ends of the single strand creates a panhandle structure with ends identical to those of the duplex. Reformation of duplex from the single stranded form occurs by the same mechanism with the Ad polymerase initiating at the terminal protein and displacing the ssDBP. Interestingly, the viral genomes undergoing replication are at a different location from those being transcribed [24]. DNA is packaged into capsids as directed by a DNA sequence close to the left hand end of the virus. The efficiency of packaging depends on the length of DNA; genomes greater than 105% or less than 95% of the normal length propagate much less efficiently [25,26].

The E4 region plays important roles in the viral life cycle by promoting the selective expression of viral genes at the expense of cellular genes. For example, the E4-ORF3 and ORF6 proteins inhibit the transport of transcripts of cellular genes from nucleus to cytoplasm while promoting the transport of late viral transcripts [10]. The E4 region is therefore essential for viral gene expression and subsequent viral replication.

3.3 CONSTRUCTION AND USE OF FIRST-GENERATION Ad VECTORS

3.3.1 CONSTRUCTION

Although the pathology associated with wild-type Ad infections is generally mild, there is a potential risk of using fully replication-competent Ad (RCA) for gene transfer in that the inflammatory host responses to Ad infection may alter organ function [9]. There is also the possibility of overwhelming infection if Ad replication is allowed to progress when there are deficiencies in the host defense system. Since the E1A products are essential for expression of other early and late genes and DNA replication, the most direct approach to eliminating replication is to delete the E1A genes. To produce E1⁻ Ad vectors, the classic approach is to transfect the recombinant E1⁻ Ad vector genome into the human embryonic kidney cell line 293, a cell line originally established by transforming primary cells with Ad5 [27,28]. The 293 cells contain approximately 11 map units of the Ad5 genome, originating at the left hand end [29].

One example of a so-called first-generation Ad vector expresses the human CF transmembrane conductance regulator (CFTR) cDNA under control of the constitutively highly active cytomegalovirus immediate/early promoter (CMV, Figure 3.4, [30]). A polyadenylation site is located following the cDNA and the whole expression cassette in a left to right orientation replaces the E1A and part of the E1B genes. Since the expression cassette is 5601 bp in length while the E1 deletion is 3062 bp, it is necessary to delete part of the E3 region

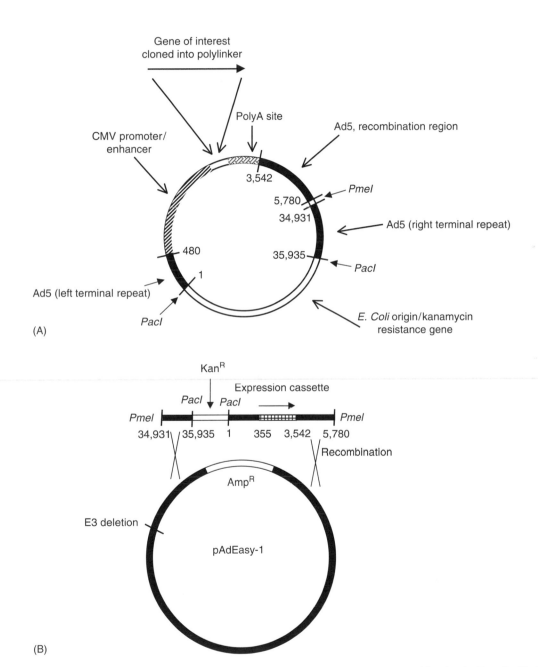

(A)

(B)

FIGURE 3.4 Production of the genome of a replication-deficient Ad vector by homologous recombination in *E. coli*. (A) A new cDNA is cloned into the kanamycin resistance plasmid pShuttleCMV between the CMV promoter and the polyadenylation site of SV40. Analogous plasmids are available which allow other promoters to be used [31]. The expression cassette lies between two portions of the Ad genome, nt 1–480 and 3542–5780. (B) Homologous recombination between Pmel linearized pShuttleCMV containing the gene of interest and supercoiled pEasy-1 results in a kanamycin resistant plasmid with a full length Ad genome bordered by Pacl sites. Plasmid pEasy-1 is used to make E1⁻E3⁻E4⁺ vectors or pEasy-2 to make E1⁻E3⁻E4⁻ vectors. The recombinant prepared in large amounts for transfection into E1 complementing cell lines where it will give rise to the desired vector.

in order to construct the vector. Since the E3 region is nonessential *in vitro*, this deletion does not affect propagation of the replication-deficient virus in 293 cells. However, for some therapeutic genes, if the extra space is not necessary, the E3 region can be retained.

The genomes of first generation (E1⁻, E3⁻) Ad are now typically constructed and amplified as *E. coli* plasmids (Figure 3.4) followed by transfection into 293 cells for the production of vector. In one widely used system [31], the Ad genome is created by homologous recombination between a shuttle plasmid and a backbone plasmid. The shuttle plasmid consists of the extreme right and left hand ends (map units 1–16 and 97–100) of the Ad genome with the expression cassette for the transgene in the deleted E1 region (map units 2–10 deleted). The backbone plasmid contains mostly the Ad genome (map units 10–100) except for the extreme left hand end and a deletion of E3. Recombinants between the shuttle and backbone plasmids containing the whole vector genome are selected by appropriate antibiotic resistance markers.

Once made, a new vector is plaque purified repeatedly in 293 cells (to remove any contaminating wild-type virus) and is then propagated to produce the required amounts of the vector. Under standard laboratory conditions, it is possible to produce up to 2×10^{13} viral particles from fifty 150 mm cell culture plates (about 10^9 293 cells). The recombinant Ad is easily purified from cell lysates on equilibrium cesium chloride density gradients. After purification, the vector is assayed for infectivity by plaquing efficiency on 293 cells, for the presence of contaminating RCA by the plaquing efficiency on A549 cells (an E1⁻ cell line) [32], and for the activity of the transgene (using whatever assay is relevant). Titer on 293 cells gives the titer in plaque-forming units (pfu) per mL. This has historically been the activity unit used to standardize doses for experimental animals and patients. However, it has become evident that the pfu is an arbitrary, poorly reproducible measurement, and thus most laboratories now use particle units (pu) as the dosing unit, based on the premise that highly purified viruses made by a standard protocol represent a uniform population of potentially infectious units. The particle count is calculated from the absorbance at 260 nm using the formula: $1A_{260} = 1.25 \times 10^{12}$ particles/mL, and is typically 10 to 100 times the titer in pfu [33].

3.3.2 In Vitro Studies

The methods outlined above have been used to make a large number of first generation E1⁻E3⁻ adenoviral vectors. Occasionally, there appears to be an inhibitory effect of transgene expression on virus production. Conditional systems to suppress transgene expression during production have been devised to overcome this limitation [34,35]. Among the most widely used Ad vectors are those which express readily monitored reporter genes such as β-galactosidase, luciferase, chloramphenicol acetyl transferase (CAT), and green fluorescent protein (GFP) [36,37]. As a control, viruses with the same promoter driving expression of no transgene (AdNull) are used.

Using the reporter gene Ad vectors, there are many studies examining the ease of gene transfer to different primary cells and cell lines. Some primary epithelial cells are easily infected by wild-type Ad type 5 and, as expected, are easily transfected by adenoviral vectors. By contrast, macrophages [38] and lymphocytes [39,40] are more difficult to infect and only very high multiplicities of infection in concentrated cell suspensions are effective. The discovery that CAR is the adenoviral receptor partially accounts for the relative ease of infection. There are several studies in which the overexpression of CAR was shown to be sufficient to make an otherwise refractory cell line susceptible to gene transfer by Ad [38,41]. But integrins and post-internalization factors must also affect the efficiency of gene transfer.

A large number of cancer cell lines have been shown to be susceptible to Ad-mediated gene transfer including cells derived from hepatoma [42,43], glioblastoma [44], myeloma [45,46], melanoma [47], prostate [48], and ovarian cancer [49,50]. On the other hand, lymphoma cell lines are resistant to infection [46]. Studies in which cells are infected *in vitro* are instructive in indicating what cell types and therefore diseases might be candidates for adenoviral gene therapy. It is difficult to evaluate if studies with reporter genes show that therapeutic levels of proteins are achievable in any cell type due to the use of Ad vectors with different promoters, reporter genes, multiplicities of infection, and times of exposure.

In cells infected *in vitro* with E1 deleted, replication-deficient Ad vectors, a low level of transcription of early and late genes [51,52] as well as a small amount of DNA synthesis [53] can be detected. The reason for this is not entirely understood, but is hypothesized to result from E1-like activities in the target cell which support expression of Ad genes. In dividing cells there would also be a high level of E2F which would support adenoviral transcription. However, measurements of viral load in cultures suggest that this does not translate into the production of infectious viral particles in the absence of contaminating wild-type Ad. While cells may continue to divide after infection, the absolute level of vector does not increase or decrease even if the amount of vector per cell does decrease to the point when a small minority of cells are infected. On the basis of this evidence, the Ad genome is likely to remain episomal and is not integrated into the cellular genome, although this is difficult to prove, as it is very hard to detect integrated DNA at a very low frequency.

3.3.3 In Vivo Studies and Tissue Specificity of Gene Transfer

The feasibility of Ad vector-mediated gene transfer *in vitro* posed the question of the efficiency of this vector system *in vivo*. Since Ad gene transfer vectors are made from human Ads, there was no a priori reason to believe they would infect rodents or other model animals. Some early studies used cotton rats since this species had been shown previously to be permissive for replication of human Ads [54–56]. For example, intratracheal administration of a replication-deficient virus expressing the reporter gene β-galactosidase to cotton rats resulted in expression of β-galactosidase in the airway epithelium [57]. Numerous other animals have been used to demonstrate efficient Ad vector-mediated gene transfer, including rats, mice, pigs, rabbits, and nonhuman primates. From these studies, a number of general conclusions can be drawn. Importantly, many tissues can be infected based on the route of administration. As expected from the tropism of Ad5, the transgene delivered to the respiratory epithelium is readily expressed after intranasal or intratracheal administration. But intravenous injection into rodents results primarily in transgene expression in the liver and spleen [58–60]. It is not known if the hepatocytes or hepatic endothelium account for this tropism since surprisingly, the preference for liver does not correspond to the distribution of the CAR receptor among organs [13]. Direct injection into the peritoneum [36], kidney [61], pancreas [62], cerebral spinal fluid [63], skeletal muscle [64], brain [65], cardiac muscle [66], and into the coronary artery [67], and many other tissues result in local expression of the transgene. But the absolute efficiency of gene transfer and expression and leakage to other organs has seldom been calculated and it is often unclear if therapeutic levels of transgene expression can be achieved.

3.3.4 HOST RESPONSES

Another important concept that emerged from *in vivo* studies in experimental animals was the short duration of transgene expression mediated by Ad vectors. Typically, transgene expression levels peak in 1–7 days and decline rapidly to undetectable levels by 2–4 weeks (Figure 3.5). This is true for most routes of administration with the exception of direct

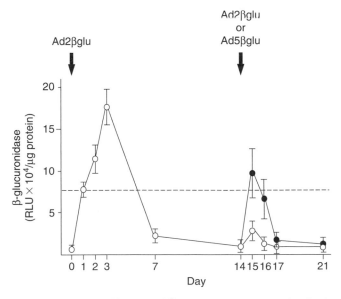

FIGURE 3.5 Quantification of β-glucuronidase expression in the lung over time following repeat administration of the same serotype vector or a vector from an alternate serotype. β-glucuronidase expression in the lung after initial intratracheal administration of Ad2βglu (10^{11} particles) followed 14 days later by intratracheal administration of either the same vector (Ad2βglu, 10^9 pfu, ○), or a vector of the alternate serotype (Ad5βglu, 10^9 pfu, ● [73]).

injection into a few immunoprivileged tissues such as brain [68–70]. Immediately upon administration, the innate immune system serves to eliminate a large amount of vector. Using viral DNA levels as a means to monitor viral clearance from the liver, approximately 90% of an intravenous bolus is cleared after 24 h [71]. Inhibitors of the reticulo-endothelium system reduce this early loss of vector, suggesting that macrophages are responsible for early vector clearance [72].

It is known that infection by wild-type human Ads results in a strong immune response in experimental animals and humans. It was not clear whether replication-deficient gene therapy vectors would have the same effect since the net expression of viral genes would be so much lower and the tissues involved would be different from those involved in natural infections. In practice, both cellular and humoral anti-vector immune responses are observed in rodents after intratracheal [73–75], intravenous [76,77], and intraperitoneal [36] administration. Antibodies against various adenoviral proteins including hexon are induced which can be detected by Western analysis and neutralizing assays. Anti-vector cytotoxic T lymphocyte (CTL) responses are also observed. The CTL are assumed, to eliminate cells infected by the vector *in vivo*.

When a transgene is used which is foreign to the host, CTL and antibodies are usually, but not always, detected against the expression product of transgene [52,78,79]. The basis for this is not fully understood. In classical antigen presentation pathways, a cellular immune response would result from antigens expressed in the antigen presenting cell and presented in the context of MHC class I. This would imply that Ad infects antigen presenting cells directly and there is good evidence both *in vivo* and *in vitro* that this occurs [80]. But *in vivo* there is also a strong humoral anti-transgene response to many proteins which would not be expected to be taken up by antigen presenting cells and presented by the class II dependent pathway. It is possible that apoptosis of Ad infected cells followed by uptake of apoptotic cells by antigen presenting cells is critical (cross presentation). The strong anti-transgene product response has proven useful in using Ad as a carrier for genetic vaccination in situations where both cellular humoral immunity is desired as described below.

These observations lead to the hypothesis that the immune response is essential for the elimination of adenoviral vectors. While the initial rapid phase of Ad clearance attributed to the innate immune system was equally rapid in immunodeficient mice and immunocompetent mice [81], overall duration of gene expression has been shown to be much longer in immunodeficient mice [81–83]. Practically, this suggests ways in which a partial, transient deficiency in the immune system might be exploited to prolong the expression of a therapeutic gene. Conventional immunosuppressants such as corticosteroids [84] and cyclosporin [85–87] have been used to reduce anti-Ad immune responses and increase the duration of transgene expression in experimental animals. The danger of applying this to human subjects is that opportunistic infections or infection from contaminating wild-type Ad may result. But greater persistence of Ad vectors can also be achieved via simultaneous systemic administration of molecules such as antibodies against CD40 ligand [88,89] and CTLA4Ig [90,91] which block interaction between T cells and antigen presenting cells. These immunomodulators could be used locally, possibly co-expressed on the same Ad as the therapeutic gene, to give more specific immunosupression.

The humoral and cellular responses also evoke immunological memory which prevents effective gene expression following subsequent administration of the same vector (Figure 3.5). Neutralizing antibodies sequester the readministered vector before it infects cells and cause its immune clearance. Thus, the barriers to readministration of a second vector should be serotype-specific, a concept that has been proven in experimental studies (Figure 3.5) [73]. Cellular immune system memory can also eliminate any readministered vectors that escape neutralization and infect host cells. The determinants of the cellular immune response are more conserved between serotypes and a second vector of different serotype is eliminated faster from immune animals than from naive animals (Figure 3.5).

3.4   IMPROVED Ad VECTORS

Two salient points emerge from the data discussed above. The first is that only some cells and tissues can be efficiently infected by Ad vectors. The second is that there is a strong immune response against adenoviral vectors which results in elimination of cells infected by the vector and the inability to achieve effective gene transfer and expression following readministration of the same vector. A number of approaches are being developed that might mitigate these problems.

3.4.1   Elimination of Replication-Competent Ad

Adenoviral vectors are produced in the 293 cell line which provides *in trans* the E1 functions that render them conditionally replication competent, permitting vector growth. The difficulty with this approach to propagate vectors is that there is the possibility of homologous recombination between the replication-deficient vector and the chromosomal copy of the Ad5 genome (Figure 3.6A) [32]. This inevitably occurs at a low frequency resulting in the production of E1+, E3− RCA.

First-generation Ad vector

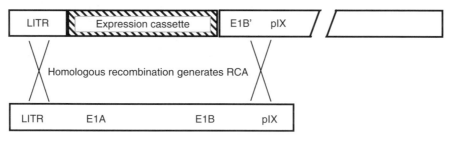

Adenoviral genome in 293 cells

(A)

Ad vector with complete E1 deletion

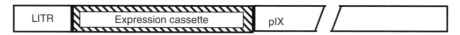

No shared sequences for homologous recombination

Adenoviral genome in perC6 cells

(B)

FIGURE 3.6   Production of replication-competent Ad by homologous recombination between Ad vector and genome of 293 cells. The 293 cell line (open rectangle) contains nucleotides 1–4344 from Ad type 5 including the left inverted terminal repeat (LITR), the E1A and E1B genes, and the adjacent protein IX gene (pIX). The E1 deletion in most first-generation vectors (top rectangle) stretches from nucleotides 355–3328 which is replaced by the expression cassette for the therapeutic gene. Therefore, two homologous recombination events (crossed line) can occur which restore the E1 region and give a replication-competent (albeit E3−) virus. In the second example, the extent of the E1 deletion in the vector has been extended to encompass all of the E1A and E1B genes. At the same time, the E1A and E1B genes in the complementing cell line (i.e., perC6 cells [163]) have no flanking sequence and expression is driven by the phosphoglycerol kinase promoter. As a result there is no homology at either end and only two illegitimate recombination events can result in the production of RCA. As a result the frequency is very low.

TABLE 3.1

Ad Deletions and Complementing Cell Lines

Cell Line (Genotype)	Deletions Complemented	Comments	References
293 (E1A/B⁺)	E1⁻	Human embryonic kidney cells transformed by nucleotides 1–4344 of Ad5	[28]
911 (E1A/B⁺)	E1⁻	Human embryonic retinoblast cells containing nucleotides 79–5789 of Ad5. Enhanced plaquing efficiency over 293	[165]
perC6 (E1A/B⁺)	E1⁻	Nucleotides 459–3510 of Ad5 driven by phosphoglycerate kinase promoter. Reduced production of RCA compared to 911 and 293	[163]
293-E4 (E1A/B⁺, E4)	E1⁻, E4⁻	293 derivative with E4 gene expression driven by mouse α-inhibin promoter	[166]
293-ORF6 (E1A/B⁺, E4(ORF6)⁺)	E1⁻, E4⁻	293 derivative expressing E4-ORF6 from metallothionein promoter	[95]
IGRP2 (E1A/B⁺, E4(ORF6/7)⁺)	E1⁻, E4⁻	293 derivative with ORF6 and 7 of E4 driven by MMTV-LTR promoter	[167]
VK2–20 & VK10–9 (E1A/B⁺, pIX⁺ E4⁺)	E1⁻, E4⁻, pIX⁻	293 derivative expressing E4 and pIX from MMTV-LTR or metallothionein promoters allows larger genes to be inserted in E1 region	[168]
293 (E1A/B⁺, pTP⁺)	E1⁻, TP⁻	293 derivative expressing terminal protein inducible by tetracycline	[169]
293 (E1A/B⁺, pTP⁺, pol⁺)	E1⁻, TP⁻, pol⁻	293 derivative expressing both terminal protein and DNA polymerase	[170]
293-C2 (E1⁺, E2A⁺)	E1⁻, E2A⁻	293 derivative with 5.9 kb fragment of Ad5 containing E2a region	[171]
AE1–2a (E1A/B⁺, E2A⁺)	E1⁻, E2A⁻	Lung epidermal carcinoma line A549 derivative transformed by E1 and E2A gene under glucocorticoid responsive promoters	[53]
AE1–2a (E1A/B⁺, E2A⁺)	E1⁻, E2A⁻	293 derivative with tetracycline repressible E2a gene	[172]
293-Cre (E1A/B⁺, Cre⁺)	All viral genes (with lox containing helper)	293 derivative expressing Cre recombinase which excises packaging signal from helper virus	[173]

Once formed, RCA will outgrow the replication-deficient gene therapy vector *in vitro*. To minimize the production of RCA, Ad is plaque purified several times on 293 cells and exhaustively tested for the presence of RCA which is readily detected on the basis of its ability to form plaques on E1 negative cell lines such as the human lung epithelial cell line A549 [32]. For clinical studies, adenoviral vectors should be uncontaminated by RCA (level <1 RCA per dose). However, the preparation of Ad vector of this quality is difficult and most *in vivo* animal and *in vitro* studies are done with preparations with uncharacterized levels of RCA that are probably >1 RCA per 10⁸ particles.

To reduce RCA production, two approaches can be used: reduce the size of the trans-complementing E1 region in the cell line or increase the size of the E1 deletion in the vector. Several cell lines have been developed that have less of the Ad genome (compared to 293 cells), while retaining the ability to supply the E1A and E1B functions *in trans* and the high productivity of the 293 cell line (Table 3.1). The E1 deletion in the first generation clinical vectors was smaller than optimal, retaining 31% of the 3′ end of the E1B gene. Deleting this sequence in conjunction with cell lines that express E1A/B from non-adenoviral promoter allows production of Ads in circumstances where there is no overlap between the cellular sequence and the vector (Figure 3.6B). In these cell lines, RCA is virtually eliminated since it can only arise through two illegitimate recombination events, an occurrence that is very rare.

3.4.2 Vectors with Additional Early Gene Deletions

First-generation Ad vectors permit limited Ad gene expression and DNA replication which probably contributes to the immune response against the vector. In addition, the possibility of making RCA during propagation is a potentially dangerous feature. By making additional mutations or deletions in the Ad genome, both of these problems can be avoided. One implementation is to make an E1⁻ vector with an E2A mutation that renders the vector replication incompetent at 37°C [92]. Such temperature sensitive vectors can be propagated at 32°C, but cannot replicate in the mammalian host at 37°C even if E1-like activities were present. Further, if homologous recombination in 293 cells during production results in an E1⁺ viral genome, the ability to replicate in a mammalian host is not restored. *In vivo* studies show that this defect reduces the inflammatory response following vector administration and permits larger transgene expression [53,93,94]. However, the temperature sensitive mutation is unstable and partially replication competent at 37°C.

Other mutations in early genes have been utilized in Ad vectors, including partial and complete E2 and E4 deletions. Both of these genes are essential for viral replication and therefore necessitate the production of cell lines that complement both the E2 or E4 deletion as well as the E1 deletion. In general, this has been achieved using 293 cells transfected with the appropriate E2 or E4 gene driven by an inducible promoter (Table 3.1). In a typical example, E4 deleted vectors were constructed in a cell line which expressed the E4 OFR6 behind an inducible metallothionine promoter [95]. Cell lines of this type are more difficult to work with than 293 cells, and the efficiency of vector production is often lower. There is inconsistent data on whether additional genomic deletions result in a blunted immune response and whether this translates into longer persistence of transgene expression. Some studies are complicated by the immune response to the foreign transgene as well as to the vector and by the tendency of the commonly used CMV promoter to be inactivated over time without vector elimination. For example, one report [96] indicates that a complete E4 deletion has no effect on the time course of gene expression in immunocompetent animals after administration to lung or liver. This is at variance with other reports [52,97,98] showing that E4 deletion results in a reduced immune response and longer transgene expression. The details of vector construction, route of administration, promoter used [99], dose, and genotype of the recipient are critical to the efficacy of E4 (and other) deletion(s). It is likely that studies in humans will be necessary to determine if additional genomic deletions have an impact of the duration of expression of the therapeutic gene and whether this translates into a significant clinical impact.

3.4.3 E3 Restored Vectors

The E3 region encodes genes that repress host response to infection through both reducing antigen presentation on the cell surface and protecting infected cells against TNF-α and fas/fas L-mediated apoptosis. E3 is deleted in most Ad vectors to make room for the transgene within the length constraints of packaging. Arguing that E3 expression might increase persistence of vectors, several groups have sought to restore one or more E3 function. In mice, expression of the E3gp19K protein has been shown to reduce MHC class I expression in vitro [100], and reduce CTL levels in vivo [101], but there are contradictory results on whether this translates into prolonged persistence [100,102]. On the other hand, two studies have shown that the whole E3 region does in fact prolong vector persistence in vivo. This is true whether the E3 region is expressed from an exogenous promoter in the E1 region [103] or expressed from its own promoter in the normal position [104]. Since E1 function is required for E3 expression, the latter result is surprising and may indicate that a low level of E3 expression is sufficient to prolong persistence. As with other modifications of the viral backbone, the critical question is if the expression of E3+ vectors would be longer in humans when administered in the route intended for therapy. This has not yet been answered.

3.4.4 Helper-Dependent Vectors

On the premise that any adenoviral gene expression would cause an immune response, some investigators have developed methods to eliminate all the Ad genes from the vector. In fact, the size constraints imposed by some large genes such

as dystrophin require that most of the Ad genome be deleted simply to make space for the therapeutic gene [105]. An additional benefit of the higher capacity is the possibility of using the endogenous promoter and control elements such as matrix attachment regions to exactly regulate the expression level of the therapeutic gene [106]. But, deletion of all Ad genes requires that those functions be provided in trans for vector production. This is achieved by using helper viruses that are deficient in packaging. In an early implementation, a helper virus with defective packaging signals was used which is packaged into virion with much lower efficiency than the therapeutic virus which has two intact packaging signals. By this method, a mixed lysate is formed with two viruses that differ in size and therefore can be separated on cesium chloride equilibrium density gradients [107].

A refinement of this technology utilizes the lox/Cre system to negatively select for helper virus in a coinfection of helper and helper-dependent virus vectors [108]. In the lox/Cre system, the DNA recombinase Cre from bacteriophage lambda efficiently mediates recombination between lox sites; thus sequences between two lox sites in the same orientation are deleted. A helper virus called psi5 has been engineered that has lox sites flanking the packaging signal so that Cre recombinase excises the packaging signal and prevents packaging of the genome. The psi5 vector propagates normally in 293 cells but packages inefficiently in 293 derivatives expressing the Cre recombinase. Coinfection of psi5 and helper-dependent vector results primarily in packaging of the helper-dependent vector using proteins specified by the psi5 genome.

Helper-dependent vectors are beginning to attain widespread use and give long-term transgene expression in models where the recipients are immunologically tolerant to the expressed transgene. A helper-dependent vector expressing the human α1-antitrypsin genomic configuration from the α1-antitrypsin promoter can be administered to mice [109] and baboons [110] resulting in prolonged high-level gene expression with little decline over 10 months. Similar data has been reported for factor IX [111], erythropoietin, and dystrophin. In the case of apoE deficient mice, the genetic defect which results in high serum cholesterol has been corrected for over 2 years by administration of a helper-dependent vector expressing the wild-type apoE [112]. It is clear that the toxicology is for helper-dependent vectors is less than for first generation vectors in some situations but the acute response to vector, particularly in recipients which are immune to Ad type 5 is not affected by use of helper-dependent vectors.

3.4.5 Sero-Switch Vectors

The induction of neutralizing antibodies is presumably one of the barriers against successful readministration of the same Ad gene transfer vector. In experimental animals, this has been observed for intravenous [113], intratracheal [73,114], and intraperitoneal [36] administration where the vector is exposed to antibodies prior to contact with the tissue. Effective readministration of the same vector has also been demonstrated to be possible using direct tissue injection [115].

Since prior infection by one Ad does not protect against infection by a different serotype [73], using different serotypes of gene therapy vector should allow readministration of the same transgene. Extensive testing of this concept is difficult because existing vectors are only of serotypes 2 and 5 since the E1 functions of these two viruses can be efficiently complemented by the 293 cell line. For example (Figure 3.5), when rats were administered intratracheally with Ad of serotype 2 expressing β-glucuronidase, there was expression which peaked at 3 days and declined to undetectable levels by 14 days. Readministration of the same vector resulted in a very low level of gene expression due to the immune response to the first vector administration. But a second administration of a vector expressing the same transgene but of serotype 5 resulted in a level of gene expression at day one comparable to that seen in a naive rat. The decline to baseline was faster than in naive animals, probably due to the elimination of infected cells by the cellular immune system using epitopes conserved between serotypes 2 and 5 [73].

These data posed the question as to which epitopes are responsible for preventing readministration of the same serotype. Vectors have been constructed with the capsid of serotype 5 but with the fiber gene of serotype 7a. Fortunately, the fiber protein from serotype 7a interacts with the penton base of serotype 5 allowing the assembly of serotype 7a/5 chimeric capsids. *In vivo* experiments show that the fiber switch from 5 to 7a does not facilitate readministration suggesting that the immune response to fiber is not the barrier that prevents readministration of the same serotype [116]. This is consistent with the concept that the primary humoral immune response is directed to the external loops of the hexon protein. Vectors have been constructed in which the hexon gene of Ad5 has been replaced by that of Ad serotype 2[117]. Even though the level of serological cross-reaction between the pure Ad5 capsid and the variant with hexon from Ad2 was low, the hexon switch did not allow successful readministration *in vivo*. This illustrates the importance of other arms of the immune system and the diversity of epitopes which are involved in immune response to gene therapy vectors.

It is not clear if sero-switching is a viable strategy for long-term gene therapy. First, the difficulty of making efficient complementing cell lines for viruses of different subgroups is unknown. Second, the delivery of a therapeutic gene to the target tissue and its subsequent expression would not necessarily be the same among serotypes. Finally, even if this could be achieved, the therapeutic advantage of expressing the transgene once for a few days rather than several times for a few days might not be all that great in terms of genetic disease where persistent gene therapy is needed.

3.4.6 VECTORS WITH MODIFIED TROPISM

The specificity of adenoviral infections *in vitro* is dictated by the presence of the CAR receptor and integrins. The role of these receptors *in vivo* is less certain since the cell types to which a vector is exposed becomes a critical issue. Since a majority of vector administered intravenously in rodents is found in the liver, gene therapy for other tissues requires both detargeting from the normal trafficking route to the liver as well as a delivery system to the target [118,119]. For lung epithelium, intratracheal administration is feasible. Direct injection is possible in some other applications, for example, into the myocardium or directly into a tumor. But vector that is inadvertently injected into capillaries or drains into the circulation through the lymphatic system will find its way to the liver. Alternatively, there are cell types such as lymphocytes [40] that are very difficult to infect with type 5 Ad vectors. These considerations raise questions about whether vectors can be retargetted to cells or tissues of interest, or at least if the expression of the transgene could be limited to that tissue.

Many vectors use the CMV immediate/early promoter/enhancer which was chosen on the basis that it directs a high level of transgene expression and is expressed in most tissues studied. But many times, expression in a specific tissue or cell type is more desirable and expression in other tissues might be toxic. Therefore, the promoters of genes specific to a cell type have sometimes been used for specific applications [120–122]. For example, carcinoembryonic antigen (CEA) and α-fetoprotein (AFP) are tumor specific antigens that are not expressed by normal cells. When a therapeutic gene is expressed from an Ad vector with an AFP promoter, expression should be confined to specific tumor cells expressing AFP and not normal cells. Kaneko et al. [120] have shown this theory to be correct and have further demonstrated that the expected selectivity is maintained *in vivo*. In this context, the vector Av1AFPTK1 [expressing thymidine kinase from herpes simplex virus (HSV-TK) from the AFP promoter] can prevent tumor growth in gancyclovir-treated nude mice implanted with an AFP-expressing tumor cell line but not in identical mice implanted with a control (non-AFP expressing) tumor cell line. By contrast, the vector Av1TK1 which expresses HSV-TK from the Rous sarcoma virus promoter protects gancyclovir-treated nude mice regardless which cell line is used to transduce the tumor.

Alternatively, modifications of the fiber/high affinity receptor interaction or the penton–integrin interaction might be used to modify tissue tropism. In this context, the sero-switch vectors described above [116] as well as capsid chimeras with part of the Ad3 fiber [123,124] or the fiber gene from Ad17 [125] might be more effective in certain tissues since, a priori, different serotypes of wild-type Ads with different known pathologies should target different tissues. In an extensive survey of the tropism of Ad5-derived vectors but with fibers derived from different serotypes, the fiber genes of Ad16 was found to be better at targeting fibroblasts and chondrocytes, that of Ad35 better at targeting dendritic cells and melanocytes and Ad50 better at targeting myobalsts and hematopoietic stem cells [126].

Some groups have taken the approach of directly screening for serotypes that replicate preferentially in brain or lung epithelium. In both cases, wild-type strains were screened for efficient replication and certain subgroup D viruses, including serotype 17, were identified to replicate more efficiently. On this basis, a serotype 2 virus with the fiber from Ad17 was constructed and used to study infection of various cell types *in vitro*. This hybrid is much more efficient at gene transfer to human umbilical vein endothelial cells, neurons, glioma cell lines, and lung epithelial cells than the pure Ad2 gene transfer vector [125]. However, it is sufficiently proficient in infecting 293 cells that it can still be propagated and titered for production purposes. It is not known if this translates into better gene transfer efficiency *in vivo.*

A number of other approaches to modifying tropism have been identified (Table 3.2). For example, the fiber protein can tolerate some manipulation without impairing virus production. The determination of the three-dimensional structure of fiber in conjunction with mutagenesis studies [127] assists in the identification of amino acids essential for the interaction with CAR and identification of domains where insertions might be tolerated without grossly affecting structure. An early modification to fiber was the addition of an oligolysine motif to the C-terminal of the fiber protein, giving the virus an affinity for polyanions such as heparin sulfate [128]. This profoundly affects the cell types that can be infected *in vitro,* allowing cells lacking CAR, such as vascular smooth muscle cells and B-cells, to be infected. It has also been shown that this oligolysine addition allows for more efficient gene transfer to smooth muscle cells *in vivo.* Additional manipulations to either the C-terminus or in the HI loop of fiber (summarized in Table 3.2) have been described to modify the tropism of Ad vectors. The most widely used is the addition of an additional RGD integrin binding motif to the fiber knob domain thereby increasing the efficiency of infection of some important cell types including ovarian cancer cells, fibroblasts, and dendritic cells [129–131]. In a study that may point to future developments, phage display technology was used to identify peptide motifs that preferentially target human umbilical vein endo-

thelial cells. When this peptide sequence was incorporated into the HI loop of fiber, Ads were generated with a high preference for endothelium [129,130]. A similar approach was used to target neural precursor cells in the brain [132].

Bispecific antibodies have been used as a reagent to direct Ad towards particular cell types [40,133,134]. For example, using a bispecific antibody conjugate with one arm binding the Ad fiber and the other binding the epidermal growth factor (EGF) receptor [133], it was possible to increase the specificity of Ad vectors towards glioma cell lines with low levels of CAR but high levels of the EGF receptor.

The route by which Ad reaches the tissue and cell type is complex, especially after intravenous injection. The half-life of Ad in blood is about 2 min and the interactions with cellular protein components of blood are unknown [135]. In addition the degree to which Ad penetrate endothelium in various tissues is unknown meaning that *in vivo* tissue tropism is not a simple outcome of the abundance of CAR primary receptor and integrin secondary receptor. Moreover, in humans there will generally be antibodies against Ad so the formation of complexes with antibody will further complicate tropism. If retargeting is to be successful after an intravenous injection, both detargeting from the normal pathways [119] as well as retargeting to the novel pathways must be achieved.

A number of groups have eliminated the amino acids in the knob of fiber that is essential for interaction with CAR. This makes the resultant vectors difficult to propagate in 293 cells and therefore substitute pseudoreceptor-ligand systems have to be devised. For example, the addition of an HA epitope from the influenzae hemagglutinin gene permits a 293 derivative with an anti-HA single-chain antibody to be used

TABLE 3.2
Examples of Retargeting of Ad Gene Transfer Vectors

Modification	Target	Rationale	References
Ad5(fiber 7a)	Cells expressing Ad7 high affinity receptors	Replace whole of Ad5 fiber by that of Ad7a (subgroup B)	[116]
Penton LDV	Lymphocytes, monocytes	Replace RGD $\alpha_v\beta_5$-binding motif in penton base by LDV which interacts with $\alpha_4\beta1$ integrin	[17]
Conjugations to anti-Ad Fabs	Tumor cells	Conjugate ligands (e.g., FGF, folate) to Fab fragments specific to fiber; bind to vector before delivery to cells bearing cognate receptor	[174,175]
Bifunctional antibody	Tumor cells	Retarget to EGFR[a] expressing cells; bind virus with bispecific antibody against EGFR and Ad fiber	[133]
Bifunctional antibody	Smooth muscle cells, endothelium	Create vector with penton modified to express defined epitope (AdFLAG); use bispecific antibody versus FLAG and α_v integrin (or E selectin) to target cells expressing α_v integrin (or E selectin)	[176]
Bispecific antibody (anti-CD3)	T cells	Use AdFLAG vector and bispecific antibody (Anti-FLAG, anti-CD3) to target CD3[+] T cells	[40]
Oligolysine	Cells with surface heparin sulfate	Add seven lysine residues to C-terminus of fiber	[128,177]
Fiber RGD	Cells expressing $\alpha_v\beta_3$ and $\alpha_v\beta_5$ integrins	Add RGD integrin binding motif to knob of fiber	[41,129]
Ad5(fiber 17)	Endothelium, lung epithelium, brain	Replace fiber gene of Ad2 vector with that of Ad17 (subgroup D)	[125]
Ad5/9 (short shaft)	Melanoma, glioma, smooth muscle cells	Place knob of Ad9 on shaft of Ad5 after eight repeats	[178]

[a] EGFR, epidermal growth factor receptor.

for propagation of vectors unable to interact with CAR or integrins [119]. The interaction of integrin and penton base has been modified both by elimination of the RGD motif in penton base and by its replacement by the LDV motif which should promote interaction with $\alpha_4\beta_1$ integrins, characteristic of lymphocytes and monocytes [17].

3.5 HUMAN APPLICATIONS

Before commencing clinical studies, batches of vector must be made under FDA current good manufacturing practice (GMP) and satisfy rigorous lot release criteria. As with all drugs, the identity (now including full DNA sequencing), purity, and potency of the vector must be verified in validated assays. For safety reasons, the absence of RCA is a major concern and vector destined for human usage must contain <1 RCA per dose. As dose increases, the challenge of making vector free of RCA in the 293 cell line becomes more problematic. Vectors must also be verified for the absence of not only human viruses but also animal viruses including porcine and bovine viruses that may be carried through from reagents used in cell culture. Taken together, these studies impose a high hurdle of cost and time before having a vector suitable for human administration on hand. Most production is now done in dedicated facilities with high overhead to maintain the required trained personnel, facility integrity, and quality control units.

Human clinical studies also require prior toxicology studies in experimental animals assessing the effects both vector and transgene expression. There is now considerable human data on the effects of administration of first generation Ad5-based vectors, so the effects of the transgene are usually the focus. Depending on the novelty of the application and the phase to which clinical studies have progressed, toxicology studies may need to conform to FDA good laboratory practice and may require nonhuman primates in addition to rodents. The information derived from toxicology studies is necessarily limited due to biological differences between experimental animals and humans, especially as it relates to the innate and acquired immune response. Experimental animals are naive to human Ads so the immune response to vector is primary while most humans have preexisting immunity to Ad5 and therefore the immune response is secondary. There are differences in antigen processing and presentation between inbred animals and humans and the effects of some E3 genes are limited to human MHC proteins. In addition, the potential effects of RCA are different since human Ads do not replicate in rodent cells but some human Ad are actually oncogenic in rodents but not in humans.

The final step of commencing a clinical study is obtaining regulatory approval which is also complex. Gene therapy, in general, and Ad, in particular, are generally perceived as being particularly hazardous and therefore multiple levels of review are required. The local institutional review board (IRB) and institutional biosafety committee (IBC) must approve as well as the recombinant DNA advisory committee of the National Institutes of Health. The investigational new drug application represents the last phase of approval. Each group has slightly different concerns and the final result is a protocol that is conservative yet likely to be safe, ethical, and informative. All revisions to the protocol must be approved by all regulatory groups and therefore keeping the protocol current needs well-organized management. The actual clinical study must be performed under good clinical practice (GCP) with timely reporting of all adverse events to the FDA, RAC, IRB, and IBC. In addition, many institutions have added a data safety monitoring board to further ensure the safety of gene therapy studies.

Data on models of disease in experimental animals provided sufficient basis to proceed to Ad vector-mediated gene transfer in humans. The first human studies were commenced in 1993 and a total of 228 studies are listed in the current update of the database of the RAC. These have generally been small phase I or phase II studies (Table 3.3). While the first studies were directed at CF, the currently active studies are focused on cancer and cardiovascular disease. In total, 75% of the listed studies involve cancer with the same vector being studied in a number of different tumors or different patient subsets or concurrently with different concurrent therapies. In general, capsid modified vectors and second generation vectors have not been used in humans and it is unknown if the advantages these new vectors sometimes offer in experimental animals apply in humans.

In general, gene transfer of Ad vectors has been safe and well tolerated [136,137]. The notable exception is that intravenous administration results in dose-dependent toxicity that has resulted in one death [138–141]. This reaction is rapid and is believed to be related to the innate immune system [142]. In addition to showing safety, several studies have shown that there is also effective delivery of the vector to the patient and subsequent expression of the therapeutic gene. Demonstrations of actual therapeutic benefit will require critical placebo controlled testing. If an Ad vector does prove to be effective in a large study, there remain substantial production and support issues that will require a significant investment of time and money to provide the reliable supply of a marketed drug.

3.5.1 GENETIC DISEASES

When gene therapy is conceived as the addition of a good copy of a defective gene, it is natural that the initial focus was on genetic diseases such as CF. As described above, the feasibility of intratracheal administration to the lung was demonstrated in animals and the first clinical trial of an Ad vector was to the airway epithelium of CF patients [143]. The initial study was a dose-escalating safety study in which vector was administered to the bronchi in 20 mL of fluid. It became clear that this volume was not well tolerated and so subsequent studies used smaller volumes, or a spray of aerosolized vector into the bronchi. The relative accessibility of the site of administration allows that samples of respiratory epithelium can be recovered by bronchial brushing and the presence of vector and therapeutic gene expression can be assessed repetitively. By sensitive quantitative PCR methods, the expression of the CFTR gene delivered by the vector is seen at the site of

TABLE 3.3
Clinical Studies of Adenoviral Gene Therapy[a]

Application/Approach	Transgene	Indication	Routes	References
Genetic disease	CFTR	CF	Intranasal/respiratory tract	[143,144,179–182]
	OTC	OTC deficiency	Intravenous	[138]
	p53	Li-Fraumeni syndrome	Intratumor	[183]
Cancer: Minimize side effects	Aquaporin	Radiation treatment of cancer	Salivary gland	[184]
Cancer: Reporter gene	LacZ	Lung cancer	Intratumor	[185]
Cancer: Growth suppresser genes	bclxs	Breast cancer	*Ex vivo* purge, stem cells	[186]
	mda7	Breast cancer, melanoma	Intratumor	[187–189]
	p16	Prostate cancer	Intratumor	[190]
	p53	Bladder cancer	Intravesicle	[192,193]
	p53	Head/neck squamous cell carcinoma, hepatic metastasis, NSCLC, oral carcinoma, pre-malignant oral cancer, prostate cancer, thyroid cancer, breast cancer, glioma	Intratumor	[191,194–201,201–209]
	p53	Hepatic metastasis, advanced cancer	Intrahepatic artery, intravenous	[191,210,211]
	p53	Ovarian cancer	Intraperitoneal	[212–214]
	RTVP-1	Prostate cancer	Intratumor	[215]
	Domininat negative EGF receptor	Glioma	Intratumor	[216]
Cancer: Immunotherapy	AFP + GM-CSF	Hepatocellular carcinoma	Intravenous	[217]
	B7.1	Metastatic cancer (melanoma/breast)	Intratumor	[218]
	B7.1	Renal carcinoma	*Ex vivo*, autologous tumor cells	[219]
	CD154	CLL	*Ex vivo*, autologous tumor or B cells Intranodal	[220–222]
	CD40L	Lung cancer, espohogeal cancer	Intratumor	[223]
	CD40L/IL-2	CLL, acute leukemia	*Ex vivo*, autologous tumor cells, bone marrow, or fibroblasts	[224,225]
	fVII/Fc	Melanoma	Intratumor	[226]
	GA733–2 (EpCAM)	Colon cancer	Intradermal	[227]
	GM-CSF	Melanoma, NSCLC, breast cancer, ovarian, AML	*Ex vivo*, autologous tumor cells	[228–230]
	CCL-21	NSCLC	*Ex vivo*, dendritic cells	[231]
	IFN-α	Bladder cancer	Intra-vesicle	[232]
	IFN-β	Glioma, ovarian cancer	Intratumor	[233,234]
	IFN-γ	Melanoma, lymphoma	Intratumor	[235,236]
	IL-12	Primary or metastatic liver cancer, GI carcinomas	Intratumor *ex vivo* (dendritic cells)	[237–240]
	IL-2	Neuorblastoma, prostate cancer, metastatic breast cancer, metastatic melanoma	*Ex vivo*, autologous tumor cells	[241–247]
	IL-2	NSCLC	Intratumor	[247]
	TNF-α	Metastatic/recurrent tumor, sarcoma, pancreatic cancer, head and neck cancer, melanoma, rectal cancer, esophageal cancer	Intratumor	[248–253]
	IL-7	NSCLC	*Ex vivo*, Autologous DC	[254]
Cancer: Immunotherapy + tumor antigen	MART-1	Melanoma	Various *in vivo* and *ex vivo*	[255–257]
	PSA	Prostate cancer	Intradermal	[258]
	L523S	NSCLC	Intramuscular	[259]
	P501	Prostate cancer	Intratumor	[258]
	p53	SCLC	*Ex vivo* infection of dendritic cells	[260]

TABLE 3.3 (continued)
Clinical Studies of Adenoviral Gene Therapy[a]

Application/Approach	Transgene	Indication	Routes	References
	LMP2A + LMP1	Nasopharyngeal carcinoma	*Ex vivo* transduced antigen presenting cells	[261]
	MUC1 + CD40L	Prostate cancer	Subcutaneous	[262]
Cancer: Oncolytic virus	E1A	Glioma, solid tumors, metastatic colon cancer, pancreatic cancer, prostate cancer, ovarian cancer	Intratumor Intravenous Intraperitoneal	[263–275]
	Prostate-specific oncolytic virus	Prostate cancer	Intratumor	[268,269,276]
Cancer: Oncolytic virus + prodrug activation	CD/HSV-TK	Pancreatic cancer, prostate cancer, penile cancer	Intratumor	[277–279]
Cancer: Oncolytic virus + marker gene	hNIS	Prostate	Intratumor	[280]
Cancer: Prodrug activation	Cytosine deaminase (CD)/HSV-TK	Prostate cancer	Intratumor	[281]
	CD	Hepatic metastasis	Intratumor	[282]
	HSV-TK	Brain tumor, glioma, head/neck squamous cell carcinoma, hepatic metastasis, melanoma, pancreatic cancer, NSCLC, ovarian cancer, prostate, retinoblastoma, mesothelioma	Intratumor	[240,283–298]
	HSV-TK	Mesothelioma	Intrapleural	[299]
	HSV-TK	Ovarian cancer	Intraperitoneal	[300,301]
	Nitroreductase	Liver cancer	Intratumor	[302]
	hNIS	Head and neck cancer, prostate cancer		[303,304]
Cancer: Prodrug activation + immunotherapy	HSV-TK/IL2	Hepatic metastasis	Intratumor	[305]
Cardiovascular: Angiogenesis	VEGF	Coronary artery disease (CAD)	Intramyocardium	[146,147,306–308]
	FGF-4	CAD	Intracoronary	[309–311]
	PDGF-B	Peripheral vascular disease (PVD)	Intradermal/ulcer	[312] (Selective genetics)
	VEGF	PVD	Intramuscular	[313]
	HIF-1/VP16	PVD, CAD	Intramuscular Intramyocardial	[314]
	PDGF-B	Diabetic ulcers	Collagen gel on ulcer	[315]
Cardiovascular: Anti-restenosis	VEGF-D	AV anastomosis	Collagen collar in graft	[316]
	iNOS	AV anastomosis	Intra-graft	[317]
	Adenylate cyclase 6	Congestive heart failure	Intracoronary	[318]
Cardiovascular: Anti-angiogenesis	PEDF	Macular degeneration	Intravitreal	[319]
Glaucoma	p21	Glaucoma	Subconjunctival	[320]
Infectious disease	Engineered zinc finger nuclease	HIV-suppress receptor levels	Intravenous	(Sangamo Biosceinces)
	env + rev, gag + proteasc, gag, pol, and nef	HIV	Oral or intranasal	[321] (Merck/NIH consortium)
		Adenoviral infection	*Ex vivo* stimulation of CTL	[322]
Normals	CD	None	Intradermal, aerosol to airway	[145,148]

Note: AFP, α-fetoprotein; CD, cytosine deaminase; CD40L, CD40 ligand; CEA, carcinoembryonic antigen; CFTR, cystic fibrosis transmembrane regulator; FGF, fibroblast growth factor; GMCSF, granulocyte macrophage colony stimulating factor; HER2, c-erb B2/neu protein; HIF, hypoxia inducible factor; hNIS, human sodium iodide symporter; HSVTK, herpes simplex virus thymidine kinase; IFN, interferon; IL, interleukin; OTC, ornithine transcarboxylase; PDGF, platelet-derived growth factor; PEDF, pigment epithelium derived factor; PSA, prostate-specific antigen; RTVP-1, related to testes-specific, vespid, and pathogenesis protein-1; TNF, tumor necrosis factor; VEGF, vascular endothelial growth factor.

[a] Compiled from the 228 protocols in the RAC database as of August 2007 and database searches. Protocols are organized first by general area, transgene, indication, and then route of administration/approach. Where possible the references are for clinical data and if not then supporting animal efficacy and/or toxicology.

administration at vector doses of 5×10^8 pfu and greater. The level of vector-derived CFTR mRNA is approximately 5% of the level of expression of the endogenous CFTR gene and this is believed to be sufficient for therapeutic effect. However, this level of expression is only achieved for a period of a few days and expression rapidly declines to baseline by 30 days [144]. Interestingly, the administration of the vector to the airway does not lead to a significant immune response against Ad reflected in either neutralizing antibodies [145] nor Ad-specific T-cell proliferation.

Since the initial safety of vectors expressing CFTR was demonstrated, a study with repetitive administration has been completed. The important result of this study is that expression is reduced or eliminated in subsequent administrations as expected from the data from experimental animals, presumably from the immune response to the first dose [144].

To date, only one study of adenoviral vectors for any metabolic disease has been initiated. This is not surprising since all animal data suggests only a short time of expression of genes delivered by Ad vectors to most tissues, and the rationale for a human study was not firmly established. Ornithine transcarbamylase (OTC) deficiency is a recessive metabolic disorder of nitrogen metabolism. A E1$^-$, E4$^-$ deleted Ad vector expressing the cDNA for OTC was constructed and administered by the intrahepatic route to adults with partial OTC deficiency and safety parameters and the efficiency of gene transfer are currently being assessed [138]. During this study it became apparent that a large dose of intravenous vector could be fatal. The use of a helper-dependent vector for the treatment of hemophilia by intravenous administration was stopped when toxicity was observed.

3.5.2 ONCOLOGY

Due to the unknown safety profile of Ad vectors, it is generally easier to design the early human trials for life-threatening disease. Of the protocols listed by the RAC, 75% involve cancer. Four basic approaches can be identified: local prodrug activation, tumor suppressor genes, immunotherapy, and oncolytic viruses.

One of the first strategies of human gene therapy for cancer was to locally deliver novel enzymes that metabolize prodrugs into the active chemotherapy agent. The general concept of these studies is that local activation of the prodrug in the tumor will concentrate the active agent in the tumor, thus limiting the systemic toxicity from the active drug. Two genes have been used in human clinical trials: the HSV-TK and the *E. coli* CD gene. The HSV-TK protein activates the prodrug ganciclovir to ganciclovir monophosphate, an inhibitor of DNA polymerase. For CD, the prodrug is 5-fluorocytosine, which is activated by CD into the active chemotherapeutic agent 5-fluorouracil. For both agents and activating enzymes, a theoretical benefit is the bystander effect in which active drug would be excreted from the vector-infected cells to kill the neighboring cells of the tumor. Thus, it is not essential to infect every cell of the tumor with the Ad vector.

Currently active protocols apply the prodrug strategy to many types of cancer including prostate cancer, CNS malignancies, ovarian cancer, mesothelioma, hepatic metastases of colon cancer, and squamous cell carcinoma of the head and neck (Table 3.3). Most studies involve phase I/II studies with intratumoral injection of escalating doses of vector prior to chemotherapy with the prodrug and subsequent scheduled surgery. Tumor removal provides samples for analysis of vector levels, for expression of the therapeutic gene, and for analysis of the activation of prodrug and histological studies for cell death and inflammation. The primary end point of these studies is safety, which has been established in some cases.

Tumor suppresser genes have also been used in human clinical studies: p53 (for ovarian cancer, prostate cancer, squamous cancer of the head and neck, breast cancer, non-small cell lung cancer, hepatic carcinoma, and hepatic metastases), retinoblastoma susceptibility gene (for bladder cancer), mda7 gene for melanoma, and p16 gene for prostate cancer. The concept is that tumor cells have defective tumor suppresser genes that cannot limit cell division, but restoration of the wild-type gene will limit cell division. The theoretical limitation of using anti-proliferative genes for tumor therapy is that they will only inhibit proliferation of the cell they infect and have no *cis* effect on neighboring cells. The trial designs are generally similar to those for the prodrug strategy.

A third general approach to Ad gene therapy for cancers has been immunostimulatory genes (Table 3.3). Several different genes have been used in human studies, including CD40 ligand for chronic lymphocytic leukemia, granulocyte macrophage colony stimulating factor for melanoma and non-small cell lung cancer, interleukin-2 for neuroblastoma, MART-1 (a melanoma-specific antigen) and B7 (CD80) for melanoma. The concept of immunostimulatory gene therapy is to promote the natural immune surveillance and elimination of tumors, which express abnormal antigens by giving a general boost to the cellular immune system (e.g., with IL-2) or with a tumor-specific antigen (e.g., MART-1) and anti-erbB-2 single-chain antibodies (for ovarian cancer) (Table 3.3).

A novel anti-proliferative approach has been used in human studies using conditionally replication-competent viruses. As described above, the E1B gene is essential to protect Ad infected cells from apoptosis and its mode of action is through interaction with p53. It follows that E1B function would only be effective in p53 positive cells, but not in p53 deficient tumor cells, i.e., E1B positive viruses would induce apoptosis only in p53 negative tumor cells, while normal cells should be protected by the p53 gene. In this context, E1A negative, E1B positive viruses have been demonstrated in animal models to show selective cytolytic effects against tumors. The same viruses have been used in phase I and phase II studies of human ovarian cancers, pancreatic cancer, and head and neck cancer with direct intratumoral injection in conjunction with chemotherapy. These studies are now being extended to phase III testing.

The simplest approach to therapy is to directly administer the vector to the tumor. But for some special applications an *ex vivo* approach is possible. This is complex due to the need to protect the cells cultured and infected *ex vivo* from adventitious agents while they are outside the patient. Three approaches have been used. The most common is to infect the tumor cells with an immunostimulatory vector (e.g., Ad expressing a cytokine) expecting to evoke an anti-tumor immune response upon returning the cells to the patient. Alternatively, dendritic cells, a potent antigen presenting cell, have been infected *ex vivo* with Ad expressing a known tumor antigen to evoke an immune response upon returning to the donor. Finally, bone marrow derived stem cells have been purged of possible contaminants by infection by an antiproliferative vector prior to returning to a donor.

3.5.3 CARDIOVASCULAR

With the observation that there is only short-term gene expression from Ad vectors, the question arose as to which medical applications might benefit from transient expression of a therapeutic gene. The general area of tissue repair and engineering emerged as a good candidate where secreted growth factors would initiate the desired cascade of tissue remodeling which, once initiated, would not require the continuous presence of the therapeutic gene. For example, expression of vascular endothelial growth factor (VEGF) after injection of an Ad vector expressing VEGF into rat retroperitoneal fat pad is brief, reverting to baseline after 10 days (Figure 3.7). By contrast, the VEGF protein induces an angiogenic response that persists long after the stimulus has disappeared.

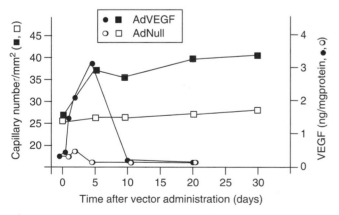

FIGURE 3.7 Time course of gene expression and anatomical response after administration of Ad expressing VEGF. The retroperitoneal fat pad of rats was injected with 5×10^8 pfu of either AdVEGF (a first-generation E1-E3- vector expressing the 165 amino acid form of human VEGF, solid symbols) or the control vector expressing no transgene (AdNull, open symbols). At intervals, animals were anesthetized, a laporotomy performed, and the fat pad photographed. The number of vessels crossing a circle of 1 cm diameter centered on the injection site was measured (left axis: □, ■). The fat pad was also homogenized and the level of VEGF determined by ELISA (right axis: ○, ● [164]).

On the basis of extensive preclinical testing in animal models of angiogenesis, a number of groups performed clinical studies of adenoviral gene transfer to ischemic heart or ischemic limb in an effort to induce the growth of new vessels and to increase blood flow. While there is no rigorous efficacy data, there is preliminary evidence of improved cardiac function in a number of patients receiving either Ad expressing FGF-4 in the coronary arteries or Ad expressing VEGF121 in the myocardium [146,147]. Moreover, when injected into muscles of patients with peripheral vascular disease Ad expressing VEGF resulted in marginal improvements in some parameters of blood flow to that limb.

3.5.4 NORMALS

The early Ad gene therapy trials demonstrated that, while effective gene transfer could be achieved, persistence of expression is clearly a problem for Ad vectors. Although there was clearly an immune response to the vector and possibly the transgene itself, the biology of that response is not well understood. Animal models, particularly those involving inbred mice, have limited utility in predicting the immune response in humans. To assess the human–host response to Ad vectors, two trials with normal subjects have been performed using intradermal [148] or intratracheal administration of an Ad vector expressing the *E. coli* CD gene. The intent of these trials is to describe the immune response in humans to an E1-, E3- Ad vector to provide a background to assess more advanced vectors on a rational basis.

3.6 FUTURE PROSPECTS

3.6.1 DECREASING VECTOR ELIMINATION

A number of approaches have been developed which should reduce the immune response to Ad vectors, prolong transgene expression, and enhance the efficiency of readministration. The basic hypothesis is that by reducing adenoviral gene expression, there should be a decrease in the host response to the vector and an increase in persistence. This is observed in some experimental animal models but not others. The basic problem posed by these data is whether prolonged persistence and reduced host response will be observed in humans with an administration route compatible with treatment. The only way to answer this question will be to perform the appropriate clinical studies in humans.

The limitations of Ad have prompted some investigators to make hybrids that exploit the wide range of cell types infected by Ad but allow persistence using features of other viruses such as retroviruses. For example, retroviruses can be produced in situ by coinfection of cells by two Ad vectors [149,150]. The first Ad contains the expression cassette for the therapeutic gene flanked by the retrovirus terminal repeats with the necessary *cis* sequences for packaging, all of this being transcribed from a CMV promoter. The second Ad expresses the *trans* factors (pol/gag/env) required for assembly of infectious retroviral particles which will then infect the neighboring cells and result in long-term gene therapy. The use of such hybrids has been demonstrated both *in vitro* and *in vivo* but does not overcome the need of retroviral vectors for dividing cells, a limitation that might be overcome by making Ad/lentivirus hybrids or Ad/adenoassociated virus hybrids [151,152].

Phage display technology has provided an approach to select a peptide sequence with desirable binding properties. This has been exploited in selecting phages that target to different tissues after intravenous injection, presumably through interacting with the endothelium of that tissue. It is likely that these peptide motifs can be incorporated into the knob of the Ad fiber to facilitate targeting of Ad vectors to a desired tissue. This requires that the knob-modified vector be able to propagate in 293 cells, but strategies have been developed to overcome this production hurdle [119].

3.6.2   APPLICATIONS FOR TRANSIENT GENE THERAPY

The technical innovations described above are at best laboratory proofs that will require extensive animal studies before clinical testing. But the clinical data to date suggests that success with currently available Ad vectors is possible in applications where transient expression might be sufficient. For example, studies of therapeutic angiogenesis for coronary artery disease described above are a prototype of this type of application. Medical indications like cancer, infectious disease, tissue remodeling (angiogenesis, recovery from surgery, stroke, or injury) are areas where development might be most appropriate. On the other hand, metabolic and genetic disease, autoimmune disease, and other chronic conditions would seem to need substantial advances in adenoviral vector design or more likely some kind of hybrid vector before they become treatable on a persistent basis. Importantly, the knowledge of the cellular and host response to Ad infection in humans is still quite rudimentary and will need to be described in much greater detail before more rational approaches to prolonging expression can be devised.

3.6.3   AD VACCINE CARRIERS

The strong humoral and cellular immune response to the expressed transgene suggests that Ad vectors may be effective as vaccines for infectious diseases and cancer. It is possible that there is an inflammatory response to Ad capsids or to residual Ad gene expression that enhances the immune response over that obtained by expression from a plasmid vector. Alternatively, the ability of Ad to directly infect dendritic cells *in vivo* [80,153] and express the antigen gene may result in antigen presentation by the class I pathway. Examples of exploiting the cellular anti-transgene response include eliciting a cellular immune response to tumor antigens as discussed earlier. This property may also be useful in elimination of virus infected cells [154–156] or malaria [157]. It is clear that antigens expressed by Ad also elicit a strong humoral immunity that can be used to block initial infection by viruses [158,159]. In the context that most humans have been infected by wild-type Ads and have preexisting antibodies [160], it is

possible that use of Ad derived from other mammals may provide more effective vaccines [161,162].

REFERENCES

1. Brody SL and Crystal RG. Adenovirus-mediated in vivo gene transfer. *Ann N Y Acad Sci* 1994; 716: 90–101.
2. Bramson JL, Graham FL, and Gauldie J. The use of adenoviral vectors for gene therapy and gene transfer in vivo. *Curr Opin Biotechnol* 1995; 6: 590–595.
3. Trapnell BC and Gorziglia M. Gene therapy using adenoviral vectors. *Curr Opin Biotechnol* 1994; 5: 617–625.
4. Kozarsky KF and Wilson JM. Gene therapy: Adenovirus vectors. *Curr Opin Genet Dev* 1993; 3: 499–503.
5. Hackett NR, Kaminsky SM, Sondhi D, and Crystal RG. Antivector and antitransgene host responses in gene therapy. *Curr Opin Mol Ther* 2000; 2: 376–382.
6. Crystal RG. Transfer of genes to humans: Early lessons and obstacles to success. *Science* 1995; 270: 404–410.
7. Horwitz MS, Adenoviruses. In: BN Fields, DM Knipe and PM Howley, (Eds.), *Fields Virology*. Lippincott-Raven Publishers, Inc., Philadelphia, 1996, pp. 2149–2171.
8. Bailey A and Mautner V. Phylogenetic relationships among adenovirus serotypes. *Virology* 1994; 205: 438–452.
9. Ginsberg HS and Prince GA. The molecular basis of adenovirus pathogenesis. *Infect Agents Dis* 1994; 3: 1–8.
10. Shenk T, Adenoviridae: The viruses and their replication. In: BN Fields, DM Knipe and PM Howley, (Eds.), *Fields Virology*. Lippincott-Raven Publishers, Inc., Philadelphia, 1996, pp. 2111–2148.
11. Athappilly FK, Murali R, Rux JJ, Cai Z, and Burnett RM. The refined crystal structure of hexon, the major coat protein of adenovirus type 2, at 2.9 A resolution. *J Mol Biol* 1994; 242: 430–455.
12. Crawford-Miksza L and Schnurr DP. Analysis of 15 adenovirus hexon proteins reveals the location and structure of seven hypervariable regions containing serotype-specific residues. *J Virol* 1996; 70: 1836–1844.
13. Bergelson JM, Cunningham JA, Droguett G, Kurt-Jones EA, Krithivas A, Hong JS, Horwitz MS, Crowell RL, and Finberg RW. Isolation of a common receptor for Coxsackie B viruses and adenoviruses 2 and 5. *Science* 1997; 275: 1320–1323.
14. Roelvink PW, Lizonova A, Lee JG, Li Y, Bergelson JM, Finberg RW, Brough DE, Kovesdi I, and Wickham TJ. The coxsackievirus-adenovirus receptor protein can function as a cellular attachment protein for adenovirus serotypes from subgroups A, C, D, E, and F. *J Virol* 1998; 72: 7909–7915.
15. Wickham TJ, Filardo EJ, Cheresh DA, and Nemerow GR. Integrin alpha v beta 5 selectively promotes adenovirus mediated cell membrane permeabilization. *J Cell Biol* 1994; 127: 257–264.
16. Mathias P, Wickham T, Moore M, and Nemerow G. Multiple adenovirus serotypes use alpha v integrins for infection. *J Virol* 1994; 68: 6811–6814.
17. Wickham TJ, Carrion ME, and Kovesdi I. Targeting of adenovirus penton base to new receptors through replacement of its RGD motif with other receptor-specific peptide motifs. *Gene Ther* 1995; 2: 750–756.
18. Wang K, Huang S, Kapoor-Munshi A, and Nemerow G. Adenovirus internalization and infection require dynamin. *J Virol* 1998; 72: 3455–3458.

19. Leopold PL, Kreitzer G, Miyazawa N, Rempel S, Pfister KK, Rodriguez-Boulan E, and Crystal RG. Dynein- and microtubule-mediated translocation of adenovirus serotype 5 occurs after endosomal lysis. *Hum Gene Ther* 2000; 11: 151–165.

20. Trotman LC, Mosberger N, Fornerod M, Stidwill RP, and Greber UF. Import of adenovirus DNA involves the nuclear pore complex receptor CAN/Nup214 and histone H1. *Nat Cell Biol* 2001; 3: 1092–1100.

21. Leopold PL, Ferris B, Grinberg I, Worgall S, Hackett NR, and Crystal RG. Fluorescent virions: Dynamic tracking of the pathway of adenoviral gene transfer vectors in living cells. *Hum Gene Ther* 1998; 9: 367–378.

22. Wold WS, Tollefson AE, and Hermiston TW. E3 transcription unit of adenovirus. *Curr Top Microbiol Immunol* 1995; 199: 237–274.

23. Shisler J, Yang C, Walter B, Ware CF, and Gooding LR. The adenovirus E3–10.4K/14.5K complex mediates loss of cell surface Fas (CD95) and resistance to Fas-induced apoptosis. *J Virol* 1997; 71: 8299–8306.

24. Puvion-Dutilleul F and Puvion E. Sites of transcription of adenovirus type 5 genomes in relation to early viral DNA replication in infected HeLa cells. A high resolution in situ hybridization and autoradiographical study. *Biol Cell* 1991; 71: 135–147.

25. Alemany R, Dai Y, Lou YC, Sethi E, Prokopenko E, Josephs SF, and Zhang WW. Complementation of helper-dependent adenoviral vectors: Size effects and titer fluctuations. *J Virol Methods* 1997; 68: 147–159.

26. Bett AJ, Prevec L, and Graham FL. Packaging capacity and stability of human adenovirus type 5 vectors. *J Virol* 1993; 67: 5911–5921.

27. Graham FL and Prevec L. Methods for construction of adenovirus vectors. *Mol Biotechnol* 1995; Jun; 3: 207–220.

28. Graham FL, Smiley J, Russell WC, and Nairn R. Characteristics of a human cell line transformed by DNA from human adeno-virus type 5. *J Gen Virol* 1977; 36: 59–74.

29. Louis N, Evelegh C, and Graham FL. Cloning and sequencing of the cellular-viral junctions from the human adenovirus type 5 transformed 293 cell line. *Virology* 1997; 233: 423–429.

30. Rosenfeld MA, Yoshimura K, Trapnell BC, Yoneyama K, Rosenthal ER, Dalemans W, Fukayama M, Bargon J, Stier LE, Stratford-Perricaudet L, Perricaudet M, Guggino WB, Pavirani A, Lecocq J-P, and Crystal RG. In vivo transfer of the human cystic fibrosis transmembrane conductance regulator gene to the airway epithelium. *Cell* 1992; 68: 143–155.

31. He TC, Zhou S, da Costa LT, Yu J, Kinzler KW, and Vogelstein B. A simplified system for generating recombinant adenoviruses. *Proc Natl Acad Sci U S A* 1998; 95: 2509–2514.

32. Hehir KM, Armentano D, Cardoza LM, Choquette TL, Berthelette PB, White GA, Couture LA, Everton MB, Keegan J, Martin JM, Pratt DA, Smith MP, Smith AE, and Wadsworth SC. Molecular characterization of replication-competent variants of adenovirus vectors and genome modifications to prevent their occurrence. *J Virol* 1996; 70: 8459–8467.

33. Mittereder N, March KL, and Trapnell BC. Evaluation of the concentration and bioactivity of adenovirus vectors for gene therapy. *J Virol* 1996; 70: 7498–7509.

34. Bruder JT, Appiah A, Kirkman WM, III, Chen P, Tian J, Reddy D, Brough DE, Lizonova A, and Kovesdi I. Improved production of adenovirus vectors expressing apoptotic transgenes. *Hum Gene Ther* 2000; 11: 139–149.

35. Gall JG, Lizonova A, EttyReddy D, McVey D, Zuber M, Kovesdi I, Aughtman B, King CR, and Brough DE. Rescue and production of vaccine and therapeutic adenovirus vectors expressing inhibitory transgenes. *Mol Biotechnol* 2007; 35: 263–273.

36. Setoguchi Y, Jaffe HA, Chu CS, and Crystal RG. Intraperitoneal in vivo gene therapy to deliver alpha 1-antitrypsin to the systemic circulation. *Am J Respir Cell Mol Biol* 1994; 10: 369–377.

37. Mittal SK, Bett AJ, Prevec L, and Graham FL. Foreign gene expression by human adenovirus type 5-based vectors studied using firefly luciferase and bacterial beta-galactosidase genes as reporters. *Virology* 1995; 210: 226–230.

38. Kaner RJ, Worgall S, Leopold PL, Stolze E, Milano E, Hidaka C, Ramalingam R, Hackett NR, Singh R, Bergelson J, Finberg R, Falck-Pedersen E, and Crystal RG. Modification of the genetic program of human alveolar macrophages by adenovirus vectors in vitro is feasible but inefficient, limited in part by the low level of expression of the coxsackie/adenovirus receptor. *Am J Respir Cell Mol Biol* 1999; 20: 361–370.

39. Leon RP, Hedlund T, Meech SJ, Li S, Schaack J, Hunger SP, Duke RC, and De Gregori J. Adenoviral-mediated gene transfer in lymphocytes. *Proc Natl Acad Sci U S A* 1998; 95: 13159–13164.

40. Wickham TJ, Lee GM, Titus JA, Sconocchia G, Bakacs T, Kovesdi I, and Segal DM. Targeted adenovirus-mediated gene delivery to T cells via CD3. *J Virol* 1997; 71: 7663–7669.

41. Hidaka C, Milano E, Leopold PL, Bergelson JM, Hackett NR, Finberg RW, Wickham TJ, Kovesdi I, Roelvink P, and Crystal RG. CAR-dependent and CAR-independent pathways of adenovirus vector-mediated gene transfer and expression in human fibroblasts. *J Clin Invest* 1999; 103: 579–587.

42. Arbuthnot PB, Bralet MP, Le Jossic C, Dedieu JF, Perricaudet M, Brechot C, and Ferry N. In vitro and in vivo hepatoma cell-specific expression of a gene transferred with an adenoviral vector. *Hum Gene Ther* 1996; 7: 1503–1514.

43. Huang H, Chen SH, Kosai K, Finegold MJ, and Woo SL. Gene therapy for hepatocellular carcinoma: Long-term remission of primary and metastatic tumors in mice by interleukin-2 gene therapy in vivo. *Gene Ther* 1996; 3: 980–987.

44. Kock H, Harris MP, Anderson SC, Machemer T, Hancock W, Sutjipto S, Wills KN, Gregory RJ, Shepard HM, Westphal M, and Maneval DC. Adenovirus-mediated p53 gene transfer suppresses growth of human glioblastoma cells in vitro and in vivo. *Int J Cancer* 1996; 67: 808–815.

45. Prince HM, Dessureault S, Gallinger S, Krajden M, Sutherland DR, Addison C, Zhang Y, Graham FL, and Stewart AK. Efficient adenovirus-mediated gene expression in malignant human plasma cells: Relative lymphoid cell resistance. *Exp Hematol* 1998; 26: 27–36.

46. Wattel E, Vanrumbeke M, Abina MA, Cambier N, Preudhomme C, Haddada H, and Fenaux P. Differential efficacy of adenoviral mediated gene transfer into cells from hematological cell lines and fresh hematological malignancies. *Leukemia* 1996; 10: 171–174.

47. Bonnekoh B, Greenhalgh DA, Bundman DS, Eckhardt JN, Longley MA, Chen SH, Woo SL, and Roop DR. Inhibition of melanoma growth by adenoviral-mediated HSV thymidine kinase gene transfer in vivo. *J Invest Dermatol* 1995; 104: 313–317.

48. Hall SJ, Mutchnik SE, Chen SH, Woo SL, and Thompson TC. Adenovirus-mediated herpes simplex virus thymidine kinase gene and ganciclovir therapy leads to systemic activity against spontaneous and induced metastasis in an orthotopic mouse model of prostate cancer. *Int J Cancer* 1997; 70: 183–187.

49. Behbakht K, Benjamin I, Chiu HC, Eck SL, Van Deerlin PG, Rubin SC, and Boyd J. Adenovirus-mediated gene therapy of ovarian cancer in a mouse model. *Am J Obstet Gynecol* 1996; 175: 1260–1265.

50. Tong XW, Block A, Chen SH, Contant CF, Agoulnik I, Blankenburg K, Kaufman RH, Woo SL, and Kieback DG. In vivo gene therapy of ovarian cancer by adenovirus-mediated thymidine kinase gene transduction and ganciclovir administration. *Gynecol Oncol* 1996; 61: 175–179.

51. Christ M, Lusky M, Stoeckel F, Dreyer D, Dieterle A, Michou AI, Pavirani A, and Mehtali M. Gene therapy with recombinant adenovirus vectors: Evaluation of the host immune response. *Immunol Lett* 1997; 57: 19–25.

52. Gao GP, Yang Y, and Wilson JM. Biology of adenovirus vectors with E1 and E4 deletions for liver-directed gene therapy. *J Virol* 1996; 70: 8934–8943.

53. Gorziglia MI, Kadan MJ, Yei S, Lim J, Lee GM, Luthra R, and Trapnell BC. Elimination of both E1 and E2 from adenovirus vectors further improves prospects for in vivo human gene therapy. *J Virol* 1996; 70: 4173–4178.

54. Rosenfeld MA, Siegfried W, Yoshimura K, Yoneyama K, Fukayama M, Stier LE, Paakko PK, Gilardi P, Stratford-Perricaudet LD, Perricaudet M, Jallat S, Pavirani A, Lecocq J-P, and Crystal RG. Adenovirus-mediated transfer of a recombinant alpha 1-antitrypsin gene to the lung epithelium in vivo. *Science* 1991; 252: 431–434.

55. Zabner J, Petersen DM, Puga AP, Graham SM, Couture LA, Keyes LD, Lukason MJ, St George JA, Gregory RJ, and Smith AE. Safety and efficacy of repetitive adenovirus-mediated transfer of CFTR cDNA to airway epithelia of primates and cotton rats. *Nat Genet* 1994; 6: 75–83.

56. Yei S, Mittereder N, Wert S, Whitsett JA, Wilmott RW, and Trapnell BC. In vivo evaluation of the safety of adenovirus-mediated transfer of the human cystic fibrosis transmembrane conductance regulator cDNA to the lung. *Hum Gene Ther* 1994; 5: 731–744.

57. Mastrangeli A, Danel C, Rosenfeld MA, Stratford-Perricaudet L, Perricaudet M, Pavirani A, Lecocq JP, and Crystal RG. Diversity of airway epithelial cell targets for in vivo recombinant adenovirus-mediated gene transfer. *J Clin Invest* 1993; 91: 225–234.

58. Peeters MJ, Patijn GA, Lieber A, Meuse L, and Kay MA. Adenovirus-mediated hepatic gene transfer in mice: Comparison of intravascular and biliary administration. *Hum Gene Ther* 1996; 7: 1693–1699.

59. Hackett NR, El Sawy T, Lee LY, Silva I, O'Leary J, Rosengart TK, and Crystal RG. Use of quantitative TaqMan real-time PCR to track the time-dependent distribution of gene transfer vectors in vivo. *Mol Ther* 2000; 2: 649–656.

60. Huard J, Lochmuller H, Acsadi G, Jani A, Massie B, and Karpati G. The route of administration is a major determinant of the transduction efficiency of rat tissues by adenoviral recombinants. *Gene Ther* 1995; 2: 107–115.

61. Zhu G, Nicolson AG, Zheng XX, Strom TB, and Sukhatme VP. Adenovirus-mediated beta-galactosidase gene delivery to the liver leads to protein deposition in kidney glomeruli. *Kidney Int* 1997; 52: 992–999.

62. Raper SE and De Matteo RP. Adenovirus-mediated in vivo gene transfer and expression in normal rat pancreas. *Pancreas* 1996; 12: 401–410.

63. Bajocchi G, Feldman SH, Crystal RG, and Mastrangeli A. Direct in vivo gene transfer to ependymal cells in the central nervous system using recombinant adenovirus vectors. *Nat Genet* 1993; 3: 229–234.

64. Xing Z, Ohkawara Y, Jordana M, Graham FL, and Gauldie J. Adenoviral vector-mediated interleukin-10 expression in vivo: Intramuscular gene transfer inhibits cytokine responses in endotoxemia. *Gene Ther* 1997; 4: 140–149.

65. Barkats M, Bilang-Bleuel A, Buc-Caron MH, Castel-Barthe MN, Corti O, Finiels F, Horellou P, Revah F, Sabate O, and Mallet J. Adenovirus in the brain: Recent advances of gene therapy for neurodegenerative diseases. *Prog Neurobiol* 1998; 55: 333–341.

66. Mack CA, Patel SR, Schwarz EA, Zanzonico P, Hahn RT, Ilercil A, Devereux RB, Goldsmith SJ, Christian TF, Sanborn TA, Kovesdi I, Hackett N, Isom OW, Crystal RG, and Rosengart TK. Biologic bypass with the use of adenovirus-mediated gene transfer of the complementary deoxyribonucleic acid for vascular endothelial growth factor 121 improves myocardial perfusion and function in the ischemic porcine heart. *J Thorac Cardiovasc Surg* 1998; 115: 168–176.

67. French BA, Mazur W, Ali NM, Geske RS, Finnigan JP, Rodgers GP, Roberts R, and Raizner AE. Percutaneous transluminal in vivo gene transfer by recombinant adenovirus in normal porcine coronary arteries, atherosclerotic arteries, and two models of coronary restenosis. *Circulation* 1994; 90: 2402–2413.

68. Parr MJ, Wen PY, Schaub M, Khoury SJ, Sayegh MH, and Fine HA. Immune parameters affecting adenoviral vector gene therapy in the brain. *J Neurovirol* 1998; 4: 194–203.

69. Di Polo A, Aigner LJ, Dunn RJ, Bray GM, and Aguayo AJ. Prolonged delivery of brain-derived neurotrophic factor by adenovirus-infected Muller cells temporarily rescues injured retinal ganglion cells. *Proc Natl Acad Sci U S A* 1998; 95: 3978–3983.

70. Byrnes AP, MacLaren RE, and Charlton HM. Immunological instability of persistent adenovirus vectors in the brain: Peripheral exposure to vector leads to renewed inflammation, reduced gene expression, and demyelination. *J Neurosci* 1996; 16: 3045–3055.

71. Worgall S, Leopold PL, Wolff G, Ferris B, Van Roijen N, and Crystal RG. Role of alveolar macrophages in rapid elimination of adenovirus vectors administered to the epithelial surface of the respiratory tract. *Hum Gene Ther* 1997; 8: 1675–1684.

72. Wolff G, Worgall S, van Rooijen N, Song WR, Harvey BG, and Crystal RG. Enhancement of in vivo adenovirus-mediated gene transfer and expression by prior depletion of tissue macrophages in the target organ. *J Virol* 1997; 71: 624–629.

73. Mack CA, Song WR, Carpenter H, Wickham TJ, Kovesdi I, Harvey BG, Magovern CJ, Isom OW, Rosengart T, Falck-Pedersen E, Hackett NR, Crystal RG, and Mastrangeli A. Circumvention of anti-adenovirus neutralizing immunity by administration of an adenoviral vector of an alternate serotype. *Hum Gene Ther* 1997; 8: 99–109.

74. Chirmule N, Hughes JV, Gao GP, Raper SE, and Wilson JM. Role of E4 in eliciting CD4 T-cell and B-cell responses to adenovirus vectors delivered to murine and nonhuman primate lungs. *J Virol* 1998; 72: 6138–6145.

75. Yang Y, Su Q, and Wilson JM. Role of viral antigens in destructive cellular immune responses to adenovirus vector-transduced cells in mouse lungs. *J Virol* 1996; 70: 7209–7212.

76. Gahery-Segard H, Juillard V, Gaston J, Lengagne R, Pavirani A, Boulanger P, and Guillet JG. Humoral immune response to the capsid components of recombinant adenoviruses: Routes of immunization modulate virus-induced Ig subclass shifts. *Eur J Immunol* 1997; 27: 653–659.

77. Ilan Y, Jona VK, Sengupta K, Davidson A, Horwitz MS, Roy-Chowdhury N, and Roy-Chowdhury J. Transient immunosuppression with FK506 permits long-term expression of therapeutic genes introduced into the liver using recombinant adenoviruses in the rat. *Hepatology* 1997; 26: 949–956.

78. Song W, Kong HL, Traktman P, and Crystal RG. Cytotoxic T lymphocyte responses to proteins encoded by heterologous transgenes transferred in vivo by adenoviral vectors. *Hum Gene Ther* 1997; 8: 1207–1217.

79. Tripathy SK, Black HB, Goldwasser E, and Leiden JM. Immune responses to transgene-encoded proteins limit the stability of gene expression after injection of replication-defective adeno-virus vectors. *Nat Med* 1996; 2: 545–550.

80. Zhang Y, Chirmule N, Gao GP, Qian R, Croyle M, Joshi B, Tazelaar J, and Wilson JM. Acute cytokine response to systemic adenoviral vectors in mice is mediated by dendritic cells and macrophages. *Mol Ther* 2001; 3: 697–707.

81. Worgall S, Wolff G, Falck-Pedersen E, and Crystal RG. Innate immune mechanisms dominate elimination of adenoviral vectors following in vivo administration. *Hum Gene Ther* 1997; 8: 37–44.

82. Dai Y, Schwarz EM, Gu D, Zhang WW, Sarvetnick N, and Verma IM. Cellular and humoral immune responses to adenoviral vectors containing factor IX gene: Tolerization of factor IX and vector antigens allows for long-term expression. *Proc Natl Acad Sci U S A* 1995; 92: 1401–1405.

83. Michou AI, Santoro L, Christ M, Julliard V, Pavirani A, and Mehtali M. Adenovirus-mediated gene transfer: Influence of transgene, mouse strain and type of immune response on persistence of transgene expression. *Gene Ther* 1997; 4: 473–482.

84. Sullivan DE, Dash S, Du H, Hiramatsu N, Aydin F, Kolls J, Blanchard J, Baskin G, and Gerber MA. Liver-directed gene transfer in non-human primates. *Hum Gene Ther* 1997; 8: 1195–1206.

85. Geddes BJ, Harding TC, Hughes DS, Byrnes AP, Lightman SL, Conde G, and Uney JB. Persistent transgene expression in the hypothalamus following stereotaxic delivery of a recombinant adenovirus: Suppression of the immune response with cyclosporin. *Endocrinology* 1996; 137: 5166–5169.

86. Elshami AA, Kucharczuk JC, Sterman DH, Smythe WR, Hwang HC, Amin KM, Litzky LA, Albelda SM, and Kaiser LR. The role of immunosuppression in the efficacy of cancer gene therapy using adenovirus transfer of the herpes simplex thymidine kinase gene. *Ann Surg* 1995; 222: 298–307.

87. Cassivi SD, Liu M, Boehler A, Tanswell AK, Slutsky AS, Keshavjee S, and Todd STRJ. Transgene expression after adenovirus-mediated retransfection of rat lungs is increased and prolonged by transplant immunosuppression. *J Thorac Cardiovasc Surg* 1999; 117: 1–7.

88. Scaria A, St George JA, Gregory RJ, Noelle RJ, Wadsworth SC, Smith AE, and Kaplan JM. Antibody to CD40 ligand inhibits both humoral and cellular immune responses to adenoviral vectors and facilitates repeated administration to mouse airway. *Gene Ther* 1997; 4: 611–617.

89. Yang Y, Su Q, Grewal IS, Schilz R, Flavell RA, and Wilson JM. Transient subversion of CD40 ligand function diminishes immune responses to adenovirus vectors in mouse liver and lung tissues. *J Virol* 1996; 70: 6370–6377.

90. Jooss K, Turka LA, and Wilson JM. Blunting of immune responses to adenoviral vectors in mouse liver and lung with CTLA4Ig. *Gene Ther* 1998; 5: 309–319.

91. Guibinga GH, Lochmuller H, Massie B, Nalbantoglu J, Karpati G, and Petrof BJ. Combinatorial blockade of calcineurin and CD28 signaling facilitates primary and secondary therapeutic gene transfer by adenovirus vectors in dystrophic (mdx) mouse muscles. *J Virol* 1998; 72: 4601–4609.

92. Yang Y, Nunes FA, Berencsi K, Gonczol E, Engelhardt JF, and Wilson JM. Inactivation of E2a in recombinant adenoviruses improves the prospect for gene therapy in cystic fibrosis. *Nat Genet* 1994; 7: 362–369.

93. Engelhardt JF, Litzky L, and Wilson JM. Prolonged transgene expression in cotton rat lung with recombinant adenoviruses defective in E2a. *Hum Gene Ther* 1994; 5: 1217–1229.

94. Engelhardt JF, Ye X, Doranz B, and Wilson JM. Ablation of E2A in recombinant adenoviruses improves transgene persistence and decreases inflammatory response in mouse liver. *Proc Natl Acad Sci U S A* 1994; 91: 6196–6200.

95. Brough DE, Lizonova A, Hsu C, Kulesa VA, and Kovesdi I. A gene transfer vector-cell line system for complete functional complementation of adenovirus early regions E1 and E4. *J Virol* 1996; 70: 6497–6501.

96. Brough DE, Hsu C, Kulesa VA, Lee GM, Cantolupo LJ, Lizonova A, and Kovesdi I. Activation of transgene expression by early region 4 is responsible for a high level of persistent transgene expression from adenovirus vectors in vivo. *J Virol* 1997; 71: 9206–9213.

97. Dedieu JF, Vigne E, Torrent C, Jullien C, Mahfouz I, Caillaud JM, Aubailly N, Orsini C, Guillaume JM, Opolon P, Delaere P, Perricaudet M, and Yeh P. Long-term gene delivery into the livers of immunocompetent mice with E1/E4-defective adenoviruses. *J Virol* 1997; 71: 4626–4637.

98. Wang Q, Greenburg G, Bunch D, Farson D, and Finer MH. Persistent transgene expression in mouse liver following in vivo gene transfer with a delta E1/delta E4 adenovirus vector. *Gene Ther* 1997; 4: 393–400.

99. Armentano D, Zabner J, Sacks C, Sookdeo CC, Smith MP, St George JA, Wadsworth SC, Smith AE, and Gregory RJ. Effect of the E4 region on the persistence of transgene expression from adenovirus vectors. *J Virol* 1997; 71: 2408–2416.

100. Schowalter DB, Tubb JC, Liu M, Wilson CB, and Kay MA. Heterologous expression of adenovirus E3-gp19K in an E1a-deleted adenovirus vector inhibits MHC I expression in vitro, but does not prolong transgene expression in vivo. *Gene Ther* 1997; 4: 351–360.

101. Lee MG, Abina MA, Haddada H, and Perricaudet M. The constitutive expression of the immunomodulatory gp19k protein in E1−, E3− adenoviral vectors strongly reduces the host cytotoxic T cell response against the vector. *Gene Ther* 1995; 2: 256–262.

102. Bruder JT, Jie T, McVey DL, and Kovesdi I. Expression of gp19K increases the persistence of transgene expression from an adenovirus vector in the mouse lung and liver. *J Virol* 1997; 71: 7623–7628.

103. Ilan Y, Droguett G, Chowdhury NR, Li Y, Sengupta K, Thummala NR, Davidson A, Chowdhury JR, and Horwitz MS. Insertion of the adenoviral E3 region into a recombinant viral vector prevents antiviral humoral and cellular immune responses and permits long-term gene expression. *Proc Natl Acad Sci U S A* 1997; 94: 2587–2592.

104. Poller W, Schneider-Rasp S, Liebert U, Merklein F, Thalheimer P, Haack A, Schwaab R, Schmitt C, and Brackmann HH. Stabilization of transgene expression by incorporation of E3 region genes into an adenoviral factor IX vector and by transient anti-CD4 treatment of the host. *Gene Ther* 1996; 3: 521–530.

105. Gilbert R, Nalbantoglu J, Howell JM, Davies L, Fletcher S, Amalfitano A, Petrof BJ, Kamen A, Massie B, and Karpati G. Dystrophin expression in muscle following gene transfer with a fully deleted ("gutted") adenovirus is markedly improved by trans-acting adenoviral gene products. *Hum Gene Ther* 2001; 12: 1741–1755.

106. Pastore L, Morral N, Zhou H, Garcia R, Parks RJ, Kochanek S, Graham FL, Lee B, and Beaudet AL. Use of a liver-specific promoter reduces immune response to the transgene in adenoviral vectors. *Hum Gene Ther* 1999; 10: 1773–1781.

107. Mitani K, Graham FL, Caskey CT, and Kochanek S. Rescue, propagation, and partial purification of a helper virus-dependent adenovirus vector. *Proc Natl Acad Sci U S A* 1995; 92: 3854–3858.

108. Parks RJ, Chen L, Anton M, Sankar U, Rudnicki MA, and Graham FL. A helper-dependent adenovirus vector system: Removal of helper virus by Cre-mediated excision of the viral packaging signal. *Proc Natl Acad Sci U S A* 1996; 93: 13565–13570.

109. Morral N, Parks RJ, Zhou H, Langston C, Schiedner G, Quinones J, Graham FL, Kochanek S, and Beaudet AL. High doses of a helper-dependent adenoviral vector yield supraphysiological levels of alpha1-antitrypsin with negligible toxicity. *Hum Gene Ther* 1998; 9: 2709–2716.

110. Morral N, O'Neal WK, Rice K, Leland MM, Piedra PA, Aguilar-Cordova E, Carey KD, Beaudet AL, and Langston C. Lethal toxicity, severe endothelial injury, and a threshold effect with high doses of an adenoviral vector in baboons. *Hum Gene Ther* 2002; 13: 143–154.

111. Ehrhardt A and Kay MA. A new adenoviral helper-dependent vector results in long-term therapeutic levels of human coagulation factor IX at low doses in vivo. *Blood* 2002; 99: 3923–3930.

112. Kim IH, Jozkowicz A, Piedra PA, Oka K, and Chan L. Lifetime correction of genetic deficiency in mice with a single injection of helper-dependent adenoviral vector. *Proc Natl Acad Sci U S A* 2001; 98: 13282–13287.

113. Smith TA, White BD, Gardner JM, Kaleko M, and McClelland A. Transient immunosuppression permits successful repetitive intravenous administration of an adenovirus vector. *Gene Ther* 1996; 3: 496–502.

114. Dong JY, Wang D, Van Ginkel FW, Pascual DW, and Frizzell RA. Systematic analysis of repeated gene delivery into animal lungs with a recombinant adenovirus vector. *Hum Gene Ther* 1996; 7: 319–331.

115. Kagami H, Atkinson JC, Michalek SM, Handelman B, Yu S, Baum BJ, and O'Connell B. Repetitive adenovirus administration to the parotid gland: Role of immunological barriers and induction of oral tolerance. *Hum Gene Ther* 1998; 9: 305–313.

116. Gall J, Kass-Eisler A, Leinwand L, and Falck-Pedersen E. Adenovirus type 5 and 7 capsid chimera: Fiber replacement alters receptor tropism without affecting primary immune neutralization epitopes. *J Virol* 1996; 70: 2116–2123.

117. Gall JG, Crystal RG, and Falck-Pedersen E. Construction and characterization of hexon-chimeric adenoviruses: Specification of adenovirus serotype. *J Virol* 1998; 72: 10260–10264.

118. Wickham TJ. Targeting adenovirus. *Gene Ther* 2000; 7: 110–114.

119. Einfeld DA, Brough DE, Roelvink PW, Kovesdi I, and Wickham TJ. Construction of a pseudoreceptor that mediates transduction by adenoviruses expressing a ligand in fiber or penton base. *J Virol* 1999; 73: 9130–9136.

120. Kaneko S, Hallenbeck P, Kotani T, Nakabayashi H, McGarrity G, Tamaoki T, Anderson WF, and Chiang YL. Adenovirus-mediated gene therapy of hepatocellular carcinoma using cancer-specific gene expression. *Cancer Res* 1995; 55: 5283–5287.

121. Rothmann T, Katus HA, Hartong R, Perricaudet M, and Franz WM. Heart muscle-specific gene expression using replication defective recombinant adenovirus. *Gene Ther* 1996; 3: 919–926.

122. Dematteo RP, McClane SJ, Fisher K, Yeh H, Chu G, Burke C, and Raper SE. Engineering tissue-specific expression of a recombinant adenovirus: Selective transgene transcription in the pancreas using the amylase promoter. *J Surg Res* 1997; 72: 155–161.

123. Stevenson SC, Rollence M, Marshall-Neff J, and McClelland A. Selective targeting of human cells by a chimeric adenovirus vector containing a modified fiber protein. *J Virol* 1997; 71: 4782–4790.

124. Krasnykh VN, Mikheeva GV, Douglas JT, and Curiel DT. Generation of recombinant adenovirus vectors with modified fibers for altering viral tropism. *J Virol* 1996; 70: 6839–6846.

125. Chillon M, Bosch A, Zabner J, Law L, Armentano D, Welsh MJ, and Davidson BL. Group D adenoviruses infect primary central nervous system cells more efficiently than those from group C. *J Virol* 1999; 73: 2537–2540.

126. Havenga MJ, Lemckert AA, Ophorst OJ, van Meijer M, Germeraad WT, Grimbergen J, van Den Doel MA, Vogels R, van Deutekom J, Janson AA, de Bruijn JD, Uytdehaag F, Quax PH, Logtenberg T, Mehtali M, and Bout A. Exploiting the natural diversity in adenovirus tropism for therapy and prevention of disease. *J Virol* 2002; 76: 4612–4620.

127. Roelvink PW, Mi LG, Einfeld DA, Kovesdi I, and Wickham TJ. Identification of a conserved receptor-binding site on the fiber proteins of CAR-recognizing adenoviridae. *Science* 1999; 286: 1568–1571.

128. Bouri K, Feero WG, Myerburg MM, Wickham TJ, Kovesdi I, Hoffman EP, and Clemens PR. Polylysine modification of adenoviral fiber protein enhances muscle cell transduction. *Hum Gene Ther* 1999; 10: 1633–1640.

129. Dmitriev I, Krasnykh V, Miller CR, Wang M, Kashentseva E, Mikheeva G, Belousova N, and Curiel DT. An adenovirus vector with genetically modified fibers demonstrates expanded tropism via utilization of a coxsackievirus and adenovirus receptor-independent cell entry mechanism. *J Virol* 1998; 72: 9706–9713.

130. Kibbe MR, Murdock A, Wickham T, Lizonova A, Kovesdi I, Nie S, Shears L, Billiar TR, and Tzeng E. Optimizing cardiovascular gene therapy: Increased vascular gene transfer with modified adenoviral vectors. *Arch Surg* 2000; 135: 191–197.

131. Wickham TJ, Tzeng E, Shears LL, Roelvink PW, Li Y, Lee GM, Brough DE, Lizonova A, and Kovesdi I. Increased in vitro and in vivo gene transfer by adenovirus vectors containing chimeric fiber proteins. *J Virol* 1997; 71: 8221–8229.

132. Schmidt A, Haas SJ, Hildebrandt S, Scheibe J, Eckhoff B, Racek T, Kempermann G, Wree A, and Putzer BM. Selective targeting of adenoviral vectors to neural precursor cells in the hippocampus of adult mice: New prospects for in situ gene therapy. *Stem Cells* 2007; 25: 2910–2918.

133. Miller CR, Buchsbaum DJ, Reynolds PN, Douglas JT, Gillespie GY, Mayo MS, Raben D, and Curiel DT. Differential susceptibility of primary and established human glioma cells to adenovirus infection: Targeting via the epidermal growth factor receptor achieves fiber receptor-independent gene transfer. *Cancer Res* 1998; 58: 5738–5748.

134. Wickham TJ, Haskard D, Segal D, and Kovesdi I. Targeting endothelium for gene therapy via receptors up-regulated during angiogenesis and inflammation. *Cancer Immunol Immunother* 1997; 45: 149–151.

135. Alemany R, Suzuki K, and Curiel DT. Blood clearance rates of adenovirus type 5 in mice. *J Gen Virol* 2000; 81: 2605–2609.

136. Crystal RG, Harvey BG, Wisnivesky JP, O'Donoghue KA, Chu KW, Maroni J, Muscat JC, Pippo AL, Wright CE, Kaner RJ, Leopold PL, Kessler PD, Rasmussen HS, Rosengart TK, and Hollmann C. Analysis of risk factors for local delivery of low- and intermediate-dose adenovirus gene transfer vectors to individuals with a spectrum of comorbid conditions. *Hum Gene Ther* 2002; 13: 65–100.

137. Harvey BG, Maroni J, O'Donoghue KA, Chu KW, Muscat JC, Pippo AL, Wright CE, Hollmann C, Wisnivesky JP, Kessler PD, Rasmussen HS, Rosengart TK, and Crystal RG. Safety

of local delivery of low- and intermediate-dose adenovirus gene transfer vectors to individuals with a spectrum of morbid conditions. *Hum Gene Ther* 2002; 13: 15–63.

138. Raper SE, Yudkoff M, Chirmule N, Gao GP, Nunes F, Haskal ZJ, Furth EE, Propert KJ, Robinson MB, Magosin S, Simoes H, Speicher L, Hughes J, Tazelaar J, Wivel NA, Wilson JM, and Batshaw ML. A pilot study of in vivo liver-directed gene transfer with an adenoviral vector in partial ornithine transcarbamylase deficiency. *Hum Gene Ther* 2002; 13: 163–175.

139. Lieber A, He CY, Meuse L, Schowalter D, Kirillova I, Winther B, and Kay MA. The role of Kupffer cell activation and viral gene expression in early liver toxicity after infusion of recombinant adenovirus vectors. *J Virol* 1997; 71: 8798–8807.

140. Morrissey RE, Horvath C, Snyder EA, Patrick J, and MacDonald JS. Rodent nonclinical safety evaluation studies of SCH 58500, an adenoviral vector for the p53 gene. *Toxicol Sci* 2002; 65: 266–275.

141. Gallo-Penn AM, Shirley PS, Andrews JL, Tinlin S, Webster S, Cameron C, Hough C, Notley C, Lillicrap D, Kaleko M, and Connelly S. Systemic delivery of an adenoviral vector encoding canine factor VIII results in short-term phenotypic correction, inhibitor development, and biphasic liver toxicity in hemophilia A dogs. *Blood* 2001; 97: 107–113.

142. Muruve DA. The innate immune response to adenovirus vectors. *Hum Gene Ther* 2004; 15: 1157–1166.

143. Crystal RG, McElvaney NG, Rosenfeld MA, Chu CS, Mastrangeli A, Hay JG, Brody SL, Jaffe HA, Eissa NT, and Danel C. Administration of an adenovirus containing the human CFTR cDNA to the respiratory tract of individuals with cystic fibrosis. *Nat Genet* 1994; 8: 42–51.

144. Harvey BG, Leopold PL, Hackett NR, Grasso TM, Williams PM, Tucker AL, Kaner RJ, Ferris B, Gonda I, Sweeney TD, Ramalingam R, Kovesdi I, Shak S, and Crystal RG. Airway epithelial CFTR mRNA expression in cystic fibrosis patients after repetitive administration of a recombinant adenovirus. *J Clin Invest* 1999; 104: 1245–1255.

145. Harvey BG, Hackett NR, El Sawy T, Rosengart TK, Hirschowitz EA, Lieberman MD, Lesser ML, and Crystal RG. Variability of human systemic humoral immune responses to adenovirus gene transfer vectors administered to different organs. *J Virol* 1999; 73: 6729–6742.

146. Rosengart TK, Lee LY, Patel SR, Sanborn TA, Parikh M, Bergman GW, Hachamovitch R, Szulc M, Kligfield PD, Okin PM, Hahn RT, Devereux RB, Post MR, Hackett NR, Foster T, Grasso TM, Lesser ML, Isom OW, and Crystal RG. Angiogenesis gene therapy: Phase I assessment of direct intramyocardial administration of an adenovirus vector expressing VEGF121 cDNA to individuals with clinically significant severe coronary artery disease. *Circulation* 1999; 100: 468–474.

147. Rosengart TK, Lee LY, Patel SR, Kligfield PD, Okin PM, Hackett NR, Isom OW, and Crystal RG. Six-month assessment of a phase I trial of angiogenic gene therapy for the treatment of coronary artery disease using direct intramyocardial administration of an adenovirus vector expressing the VEGF121 cDNA. *Ann Surg* 1999; 230: 466–470.

148. Harvey BG, Worgall S, Ely S, Leopold PL, and Crystal RG. Cellular immune responses of healthy individuals to intradermal administration of an E1⁻E3⁻ adenovirus gene transfer vector. *Hum Gene Ther* 1999; 10: 2823–2837.

149. Bilbao G, Feng M, Rancourt C, Jackson WHJ, and Curiel DT. Adenoviral/retroviral vector chimeras: A novel strategy to achieve high-efficiency stable transduction in vivo. *FASEB J* 1997; 11: 624–634.

150. Ramsey WJ, Caplen NJ, Li Q, Higginbotham JN, Shah M, and Blaese RM. Adenovirus vectors as transcomplementing

templates for the production of replication defective retroviral vectors. *Biochem Biophys Res Commun* 1998; 246: 912–919.

151. Fisher KJ, Kelley WM, Burda JF, and Wilson JM. A novel adenovirus-adeno-associated virus hybrid vector that displays efficient rescue and delivery of the AAV genome. *Hum Gene Ther* 1996; 7: 2079–2087.

152. Gao GP, Qu G, Faust LZ, Engdahl RK, Xiao W, Hughes JV, Zoltick PW, and Wilson JM. High-titer adeno-associated viral vectors from a Rep/Cap cell line and hybrid shuttle virus. *Hum Gene Ther* 1998; 9: 2353–2362.

153. Labow D, Lee S, Ginsberg RJ, Crystal RG, and Korst RJ. Adenovirus vector-mediated gene transfer to regional lymph nodes. *Hum Gene Ther* 2000; 11: 759–769.

154. He Z, Wlazlo AP, Kowalczyk DW, Cheng J, Xiang ZQ, Giles-Davis W, and Ertl HC. Viral recombinant vaccines to the E6 and E7 antigens of HPV-16. *Virology* 2000; 270: 146–161.

155. Schadeck EB, Partidos CD, Fooks AR, Obeid OE, Wilkinson GW, Stephenson JR, and Steward MW. CTL epitopes identified with a defective recombinant adenovirus expressing measles virus nucleoprotein and evaluation of their protective capacity in mice. *Virus Res* 1999; 65: 75–86.

156. Zolla-Pazner S, Lubeck M, Xu S, Burda S, Natuk RJ, Sinangil F, Steimer K, Gallo RC, Eichberg JW, Matthews T, and Robert-Guroff M. Induction of neutralizing antibodies to T-cell line-adapted and primary human immunodeficiency virus type 1 isolates with a prime-boost vaccine regimen in chimpanzees. *J Virol* 1998; 72: 1052–1059.

157. Rodrigues EG, Zavala F, Nussenzweig RS, Wilson JM, and Tsuji M. Efficient induction of protective anti-malaria immunity by recombinant adenovirus. *Vaccine* 1998; 16: 1812–1817.

158. Makimura M, Miyake S, Akino N, Takamori K, Matsuura Y, Miyamura T, and Saito I. Induction of antibodies against structural proteins of hepatitis C virus in mice using recombinant adenovirus. *Vaccine* 1996; 14: 28–36.

159. Xiang ZQ, Yang Y, Wilson JM, and Ertl HC. A replication-defective human adenovirus recombinant serves as a highly efficacious vaccine carrier. *Virology* 1996; 219: 220–227.

160. Cohen CJ, Xiang ZQ, Gao GP, Ertl HC, Wilson JM, and Bergelson JM. Chimpanzee adenovirus CV-68 adapted as a gene delivery vector interacts with the coxsackievirus and adenovirus receptor. *J Gen Virol* 2002; 83: 151–155.

161. Xiang Z, Gao G, Reyes-Sandoval A, Cohen CJ, Li Y, Bergelson JM, Wilson JM, and Ertl HC. Novel, chimpanzee serotype 68-based adenoviral vaccine carrier for induction of antibodies to a transgene product. *J Virol* 2002; 76: 2667–2675.

162. Mittal SK, Prevec L, Graham FL, and Babiuk LA. Development of a bovine adenovirus type 3-based expression vector. *J Gen Virol* 1995; 76: 93–102.

163. Fallaux FJ, Bout A, van dV I, van den Wollenberg DJ, Hehir KM, Keegan J, Auger C, Cramer SJ, van Ormondt H, van der Eb AJ, Valerio D, and Hoeben RC. New helper cells and matched early region 1-deleted adenovirus vectors prevent generation of replication-competent adenoviruses. *Hum Gene Ther* 1998; 9: 1909–1917.

164. Magovern CJ, Mack CA, Zhang J, Rosengart TK, Isom OW, and Crystal RG. Regional angiogenesis induced in nonischemic tissue by an adenoviral vector expressing vascular endothelial growth factor. *Hum Gene Ther* 1997; 8: 215–227.

165. Fallaux FJ, Kranenburg O, Cramer SJ, Houweling A, van Ormondt H, Hoeben RC, and van der Eb AJ. Characterization of 911: A new helper cell line for the titration and propagation of early region 1-deleted adenoviral vectors. *Hum Gene Ther* 1996; 7: 215–222.

166. Wang Q, Jia XC, and Finer MH. A packaging cell line for propagation of recombinant adenovirus vectors containing two lethal gene-region deletions. *Gene Ther* 1995; 2: 775–783.

167. Yeh P, Dedieu JF, Orsini C, Vigne E, Denefle P, and Perricaudet M. Efficient dual transcomplementation of adenovirus E1 and E4 regions from a 293-derived cell line expressing a minimal E4 functional unit. *J Virol* 1996; 70: 559–565.

168. Krougliak V and Graham FL. Development of cell lines capable of complementing E1, E4, and protein IX defective adenovirus type 5 mutants. *Hum Gene Ther* 1995; 6: 1575–1586.

169. Langer SJ and Schaack J. 293 cell lines that inducibly express high levels of adenovirus type 5 precursor terminal protein. *Virology* 1996; 221: 172–179.

170. Amalfitano A and Chamberlain JS. Isolation and characterization of packaging cell lines that coexpress the adenovirus E1, DNA polymerase, and preterminal proteins: Implications for gene therapy. *Gene Ther* 1997; 4: 258–263.

171. Zhou H, O'Neal W, Morral N, and Beaudet AL. Development of a complementing cell line and a system for construction of adenovirus vectors with E1 and E2a deleted. *J Virol* 1996; 70: 7030–7038.

172. Zhou H and Beaudet AL. A new vector system with inducible E2a cell line for production of higher titer and safer adenoviral vectors. *Virology* 2000; 275: 348–357.

173. Hardy S, Kitamura M, Harris-Stansil T, Dai Y, and Phipps ML. Construction of adenovirus vectors through Cre-lox recombination. *J Virol* 1997; 71: 1842–1849.

174. Rogers BE, Douglas JT, Ahlem C, Buchsbaum DJ, Frincke J, and Curiel DT. Use of a novel cross-linking method to modify adenovirus tropism. *Gene Ther* 1997; 4: 1387–1392.

175. Goldman CK, Rogers BE, Douglas JT, Sosnowski BA, Ying W, Siegal GP, Baird A, Campain JA, and Curiel DT. Targeted gene delivery to Kaposi's sarcoma cells via the fibroblast growth factor receptor. *Cancer Res* 1997; 57: 1447–1451.

176. Wickham TJ, Segal DM, Roelvink PW, Carrion ME, Lizonova A, Lee GM, and Kovesdi I. Targeted adenovirus gene transfer to endothelial and smooth muscle cells by using bispecific antibodies. *J Virol* 1996; 70: 6831–6838.

177. Wickham TJ, Roelvink PW, Brough DE, and Kovesdi I. Adenovirus targeted to heparan-containing receptors increases its gene delivery efficiency to multiple cell types. *Nat Biotechnol* 1996; 14: 1570–1573.

178. Roelvink PW, Kovesdi I, and Wickham TJ. Comparative analysis of adenovirus fiber-cell interaction: Adenovirus type 2 (Ad2) and Ad9 utilize the same cellular fiber receptor but use different binding strategies for attachment. *J Virol* 1996; 70: 7614–7621.

179. Perricone MA, Morris JE, Pavelka K, Plog MS, O'Sullivan BP, Joseph PM, Dorkin H, Lapey A, Balfour R, Meeker DP, Smith AE, Wadsworth SC, and St GJ. Aerosol and lobar administration of a recombinant adenovirus to individuals with cystic fibrosis. II. Transfection efficiency in airway epithelium. *Hum Gene Ther* 2001; 12: 1383–1394.

180. Knowles MR, Hohneker KW, Zhou Z, Olsen JC, Noah TL, Hu PC, Leigh MW, Engelhardt JF, Edwards LJ, and Jones KR. A controlled study of adenoviral-vector-mediated gene transfer in the nasal epithelium of patients with cystic fibrosis. *N Engl J Med* 1995; 333: 823–831.

181. Zuckerman JB, Robinson CB, McCoy KS, Shell R, Sferra TJ, Chirmule N, Magosin SA, Propert KJ, Brown-Parr EC, Hughes JV, Tazelaar J, Baker C, Goldman MJ, and Wilson JM. A phase I study of adenovirus-mediated transfer of the human cystic fibrosis transmembrane conductance regulator gene to a lung segment of individuals with cystic fibrosis. *Hum Gene Ther* 1999; 10: 2973–2985.

182. Zabner J, Ramsey BW, Meeker DP, Aitken ML, Balfour RP, Gibson RL, Launspach J, Moscicki RA, Richards SM, and Standaert TA. Repeat administration of an adenovirus vector encoding cystic fibrosis transmembrane conductance regulator to the nasal epithelium of patients with cystic fibrosis. *J Clin Invest* 1996; 97: 1504–1511.

183. Senzer N, Nemunaitis J, Nemunaitis M, Lamont J, Gore M, Gabra H, Eeles R, Sodha N, Lynch FJ, Zumstein LA, Menander KB, Sobol RE, and Chada S. p53 therapy in a patient with Li-Fraumeni syndrome. *Mol Cancer Ther* 2007; 6: 1478–1482.

184. Zheng C, Goldsmith CM, Mineshiba F, Chiorini JA, Kerr A, Wenk ML, Vallant M, Irwin RD, and Baum BJ. Toxicity and biodistribution of a first-generation recombinant adenoviral vector, encoding aquaporin-1, after retroductal delivery to a single rat submandibular gland. *Hum Gene Ther* 2006; 17: 1122–1133.

185. Tursz T, Cesne AL, Baldeyrou P, Gautier E, Opolon P, Schatz C, Pavirani A, Courtney M, Lamy D, Ragot T, Saulnier P, Andremont A, Monier R, Perricaudet M, and Le CT. Phase I study of a recombinant adenovirus-mediated gene transfer in lung cancer patients. *J Natl Cancer Inst* 1996; 88: 1857–1863.

186. Han JS, Nunez G, Wicha MS, and Clarke MF. Targeting cancer cell death with a bcl-XS adenovirus. *Springer Semin Immunopathol* 1998; 19: 279–288.

187. Cunningham CC, Chada S, Merritt JA, Tong A, Senzer N, Zhang Y, Mhashilkar A, Parker K, Vukelja S, Richards D, Hood J, Coffee K, and Nemunaitis J. Clinical and local biological effects of an intratumoral injection of mda-7 (IL24; INGN 241) in patients with advanced carcinoma: A phase I study. *Mol Ther* 2005; 11: 149–159.

188. Tong AW, Nemunaitis J, Su D, Zhang Y, Cunningham C, Senzer N, Netto G, Rich D, Mhashilkar A, Parker K, Coffee K, Ramesh R, Ekmekcioglu S, Grimm EA, Van Wart HJ, Merritt J, and Chada S. Intratumoral injection of INGN 241, a nonreplicating adenovector expressing the melanoma-differentiation associated gene-7 (mda-7/IL24): Biologic outcome in advanced cancer patients. *Mol Ther* 2005; 11: 160–172.

189. Fisher PB, Sarkar D, Lebedeva IV, Emdad L, Gupta P, Sauane M, Su ZZ, Grant S, Dent P, Curiel DT, Senzer N, and Nemunaitis J. Melanoma differentiation associated gene-7/interleukin-24 (mda-7/IL-24): Novel gene therapeutic for metastatic melanoma. *Toxicol Appl Pharmacol* 2006; 224: 300–307.

190. Allay JA, Steiner MS, Zhang Y, Reed CP, Cockroft J, and Lu Y. Adenovirus p16 gene therapy for prostate cancer. *World J Urol* 2000; 18: 111–120.

191. Tolcher AW, Hao D, de BJ, Miller A, Patnaik A, Hammond LA, Smetzer L, Van Wart HJ, Merritt J, Rowinsky EK, Takimoto C, Von HD, and Eckhardt SG. Phase I, pharmacokinetic, and pharmacodynamic study of intravenously administered Ad5CMV-p53, an adenoviral vector containing the wild-type p53 gene, in patients with advanced cancer. *J Clin Oncol* 2006; 24: 2052–2058.

192. Pagliaro LC, Keyhani A, Williams D, Woods D, Liu B, Perrotte P, Slaton JW, Merritt JA, Grossman HB, and Dinney CP. Repeated intravesical instillations of an adenoviral vector in patients with locally advanced bladder cancer: A phase I study of p53 gene therapy. *J Clin Oncol* 2003; 21: 2247–2253.

193. Kuball J, Wen SF, Leissner J, Atkins D, Meinhardt P, Quijano E, Engler H, Hutchins B, Maneval DC, Grace MJ, Fritz MA, Storkel S, Thuroff JW, Huber C, and Schuler M. Successful adenovirus-mediated wild-type p53 gene transfer in patients with bladder cancer by intravesical vector instillation. *J Clin Oncol* 2002; 20: 957–965.

194. Clayman GL, Frank DK, Bruso PA, and Goepfert H. Adenovirus-mediated wild-type p53 gene transfer as a surgical adjuvant in advanced head and neck cancers. *Clin Cancer Res* 1999; 5: 1715–1722.

195. Breau RL and Clayman GL. Gene therapy for head and neck cancer. *Curr Opin Oncol* 1996; 8: 227–231.

196. Merritt JA, Roth JA, and Logothetis CJ. Clinical evaluation of adenoviral-mediated p53 gene transfer: Review of INGN 201 studies. *Semin Oncol* 2001; 28: 105–114.

197. Pisters LL, Pettaway CA, Troncoso P, McDonnell TJ, Stephens LC, Wood CG, Do KA, Brisbay SM, Wang X, Hossan EA, Evans RB, Soto C, Jacobson MG, Parker K, Merritt JA, Steiner MS, and Logothetis CJ. Evidence that transfer of functional p53 protein results in increased apoptosis in prostate cancer. *Clin Cancer Res* 2004; 10: 2587–2593.

198. Swisher SG, Roth JA, Komaki R, Gu J, Lee JJ, Hicks M, Ro JY, Hong WK, Merritt JA, Ahrar K, Atkinson NE, Correa AM, Dolormente M, Dreiling L, El-Naggar AK, Fossella F, Francisco R, Glisson B, Grammer S, Herbst R, Huaringa A, Kemp B, Khuri FR, Kurie JM, Liao Z, McDonnell TJ, Morice R, Morello F, Munden R, Papadimitrakopoulou V, Pisters KM, Putnam JB, Jr., Sarabia AJ, Shelton T, Stevens C, Shin DM, Smythe WR, Vaporciyan AA, Walsh GL, and Yin M. Induction of p53-regulated genes and tumor regression in lung cancer patients after intratumoral delivery of adenoviral p53 (INGN 201) and radiation therapy. *Clin Cancer Res* 2003; 9: 93–101.

199. Swisher SG, Roth JA, Nemunaitis J, Lawrence DD, Kemp BL, Carrasco CH, Connors DG, El-Naggar AK, Fossella F, Glisson BS, Hong WK, Khuri FR, Kurie JM, Lee JJ, Lee JS, Mack M, Merritt JA, Nguyen DM, Nesbitt JC, Perez-Soler R, Pisters KM, Putnam JB, Jr., Richli WR, Savin M, Schrump DS, Shin DM, Shulkin A, Walsh GL, Wait J, Weill D, and Waugh MK. Adenovirus-mediated p53 gene transfer in advanced non-small-cell lung cancer. *J Natl Cancer Inst* 1999; 91: 763–771.

200. Schuler M, Herrmann R, De Greve JL, Stewart AK, Gatzemeier U, Stewart DJ, Laufman L, Gralla R, Kuball J, Buhl R, Heussel CP, Kommoss F, Perruchoud AP, Shepherd FA, Fritz MA, Horowitz JA, Huber C, and Rochlitz C. Adenovirus-mediated wild-type p53 gene transfer in patients receiving chemotherapy for advanced non-small-cell lung cancer: Results of a multicenter phase II study. *J Clin Oncol* 2001; 19: 1750–1758.

201. Cristofanilli M, Krishnamurthy S, Guerra L, Broglio K, Arun B, Booser DJ, Menander K, Van Wart HJ, Valero V, and Hortobagyi GN. A nonreplicating adenoviral vector that contains the wild-type p53 transgene combined with chemotherapy for primary breast cancer: Safety, efficacy, and biologic activity of a novel gene-therapy approach. *Cancer* 2006; 107: 935–944.

202. Clayman GL, El-Naggar AK, Lippman SM, Henderson YC, Frederick M, Merritt JA, Zumstein LA, Timmons TM, Liu TJ, Ginsberg L, Roth JA, Hong WK, Bruso P, and Goepfert H. Adenovirus-mediated p53 gene transfer in patients with advanced recurrent head and neck squamous cell carcinoma. *J Clin Oncol* 1998; 16: 2221–2232.

203. Shimada H, Matsubara H, Shiratori T, Shimizu T, Miyazaki S, Okazumi S, Nabeya Y, Shuto K, Hayashi H, Tanizawa T, Nakatani Y, Nakasa H, Kitada M, and Ochiai T. Phase I/II adenoviral p53 gene therapy for chemoradiation resistant advanced esophageal squamous cell carcinoma. *Cancer Sci* 2006; 97: 554–561.

204. Dummer R, Bergh J, Karlsson Y, Horowitz JA, Mulder NH, Huinink DTB, Burg G, Hofbauer G, and Osanto S. Biological activity and safety of adenoviral vector-expressed wild-type p53 after intratumoral injection in melanoma and breast cancer patients with p53-overexpressing tumors. *Cancer Gene Ther* 2000; 7: 1069–1076.

205. Lang FF, Bruner JM, Fuller GN, Aldape K, Prados MD, Chang S, Berger MS, McDermott MW, Kunwar SM, Junck LR, Chandler W, Zwiebel JA, Kaplan RS, and Yung WK. Phase I trial of adenovirus-mediated p53 gene therapy for recurrent glioma: Biological and clinical results. *J Clin Oncol* 2003; 21: 2508–2518.

206. Zhang S, Xu G, Liu C, Xiao S, Sun Y, Su X, Cai Y, Li D, and Xu B. Clinical study of recombinant adenovirus-p53 (Adp53) combined with hyperthermia in advanced cancer (a report of 15 cases). *Int J Hyperthermia* 2005; 21: 631–636.

207. Nemunaitis J, Swisher SG, Timmons T, Connors D, Mack M, Doerksen L, Weill D, Wait J, Lawrence DD, Kemp BL, Fossella F, Glisson BS, Hong WK, Khuri FR, Kurie JM, Lee JJ, Lee JS, Nguyen DM, Nesbitt JC, Perez-Soler R, Pisters KM, Putnam JB, Richli WR, Shin DM, Walsh GL, Merritt J, and Roth J. Adenovirus-mediated p53 gene transfer in sequence with cisplatin to tumors of patients with non-small-cell lung cancer. *J Clin Oncol* 2000; 18: 609–622.

208. Weill D, Mack M, Roth J, Swisher S, Proksch S, Merritt J, and Nemunaitis J. Adenoviral-mediated p53 gene transfer to non-small cell lung cancer through endobronchial injection. *Chest* 2000; 118: 966–970.

209. Fujiwara T, Tanaka N, Kanazawa S, Ohtani S, Saijo Y, Nukiwa T, Yoshimura K, Sato T, Eto Y, Chada S, Nakamura H, and Kato H. Multicenter phase I study of repeated intratumoral delivery of adenoviral p53 in patients with advanced non-small-cell lung cancer. *J Clin Oncol* 2006; 24: 1689–1699.

210. Atencio IA, Grace M, Bordens R, Fritz M, Horowitz JA, Hutchins B, Indelicato S, Jacobs S, Kolz K, Maneval D, Musco ML, Shinoda J, Venook A, Wen S, and Warren R. Biological activities of a recombinant adenovirus p53 (SCH 58500) administered by hepatic arterial infusion in a phase 1 colorectal cancer trial. *Cancer Gene Ther* 2006; 13: 169–181.

211. Habib NA, Hodgson HJ, Lemoine N, and Pignatelli M. A phase I/II study of hepatic artery infusion with wtp53-CMV-Ad in metastatic malignant liver tumours. *Hum Gene Ther* 1999; 10: 2019–2034.

212. Buller RE, Runnebaum IB, Karlan BY, Horowitz JA, Shahin M, Buekers T, Petrauskas S, Kreienberg R, Slamon D, and Pegram M. A phase I/II trial of rAd/p53 (SCH 58500) gene replacement in recurrent ovarian cancer. *Cancer Gene Ther* 2002; 9: 553–566.

213. Buller RE, Shahin MS, Horowitz JA, Runnebaum IB, Mahavni V, Petrauskas S, Kreienberg R, Karlan B, Slamon D, and Pegram M. Long term follow-up of patients with recurrent ovarian cancer after Ad p53 gene replacement with SCH 58500. *Cancer Gene Ther* 2002; 9: 567–572.

214. Wolf JK, Bodurka DC, Gano JB, Deavers M, Ramondetta L, Ramirez PT, Levenback C, and Gershenson DM. A phase I study of Adp53 (INGN 201; ADVEXIN) for patients with platinum- and paclitaxel-resistant epithelial ovarian cancer. *Gynecol Oncol* 2004; 94: 442–448.

215. Naruishi K, Timme TL, Kusaka N, Fujita T, Yang G, Goltsov A, Satoh T, Ji X, Tian W, Abdelfattah E, Men T, Watanabe M, Tabata K, and Thompson TC. Adenoviral vector-mediated RTVP-1 gene-modified tumor cell-based vaccine suppresses the development of experimental prostate cancer. *Cancer Gene Ther* 2006; 13: 658–663.

216. Chung TD and Broaddus WC. Molecular targeting in radiotherapy: Epidermal growth factor receptor. *Mol Interv* 2005; 5: 15–19.

217. Meng WS, Butterfield LH, Ribas A, Dissette VB, Heller JB, Miranda GA, Glaspy JA, McBride WH, and Economou JS. Alpha-Fetoprotein-specific tumor immunity induced by plasmid prime-adenovirus boost genetic vaccination. *Cancer Res* 2001; 61: 8782–8786.

218. Boxhorn HK, Smith JG, Chang YJ, Guerry D, Lee WM, Rodeck U, Turka LA, and Eck SL. Adenoviral transduction of melanoma cells with B7–1: Antitumor immunity and immunosuppressive factors. *Cancer Immunol Immunother* 1998; 46: 283–292.

219. Antonia SJ and Seigne JD. B7–1 gene-modified autologous tumor-cell vaccines for renal-cell carcinoma. *World J Urol* 2000; 18: 157–163.

220. Messmer D and Kipps TJ. CD154 gene therapy for human B-cell malignancies. *Ann N Y Acad Sci* 2005; 1062: 51–60.

221. Dicker F, Kater AP, Prada CE, Fukuda T, Castro JE, Sun G, Wang JY, and Kipps TJ. CD154 induces p73 to overcome the resistance to apoptosis of chronic lymphocytic leukemia cells lacking functional p53. *Blood* 2006; 108: 3450–3457.

222. Battle TE, Wierda WG, Rassenti LZ, Zahrieh D, Neuberg D, Kipps TJ, and Frank DA. In vivo activation of signal transducer and activator of transcription 1 after CD154 gene therapy for chronic lymphocytic leukemia is associated with clinical and immunologic response. *Clin Cancer Res* 2003; 9: 2166–2172.

223. Kikuchi T and Crystal RG. Anti-tumor immunity induced by in vivo adenovirus vector-mediated expression of CD40 ligand in tumor cells. *Hum Gene Ther* 1999; 10: 1375–1387.

224. Rousseau RF, Biagi E, Dutour A, Yvon ES, Brown MP, Lin T, Mei Z, Grilley B, Popek E, Heslop HE, Gee AP, Krance RA, Popat U, Carrum G, Margolin JF, and Brenner MK. Immunotherapy of high-risk acute leukemia with a recipient (autologous) vaccine expressing transgenic human CD40L and IL-2 after chemotherapy and allogeneic stem cell transplantation. *Blood* 2006; 107: 1332–1341.

225. Biagi E, Rousseau R, Yvon E, Schwartz M, Dotti G, Foster A, Havlik-Cooper D, Grilley B, Gee A, Baker K, Carrum G, Rice L, Andreeff M, Popat U, and Brenner M. Responses to human CD40 ligand/human interleukin-2 autologous cell vaccine in patients with B-cell chronic lymphocytic leukemia. *Clin Cancer Res* 2005; 11: 6916–6923.

226. Tang Y, Borgstrom P, Maynard J, Koziol J, Hu Z, Garen A, and Deisseroth A. Mapping of angiogenic markers for targeting of vectors to tumor vascular endothelial cells. *Cancer Gene Ther* 2007; 14: 346–353.

227. Gutzmer R, Li W, Sutterwala S, Lemos MP, Elizalde JI, Urtishak SL, Behrens EM, Rivers PM, Schlienger K, Laufer TM, Eck SL, and Marks MS. A tumor-associated glycoprotein that blocks MHC class II-dependent antigen presentation by dendritic cells. *J Immunol* 2004; 173: 1023–1032.

228. Kusumoto M, Umeda S, Ikubo A, Aoki Y, Tawfik O, Oben R, Williamson S, Jewell W, and Suzuki T. Phase 1 clinical trial of irradiated autologous melanoma cells adenovirally transduced with human GM-CSF gene. *Cancer Immunol Immunother* 2001; 50: 373–381.

229. Nemunaitis J. GVAX (GMCSF gene modified tumor vaccine) in advanced stage non small cell lung cancer. *J Control Release* 2003; 91: 225–231.

230. Soiffer R, Hodi FS, Haluska F, Jung K, Gillessen S, Singer S, Tanabe K, Duda R, Mentzer S, Jaklitsch M, Bueno R, Clift S, Hardy S, Neuberg D, Mulligan R, Webb I, Mihm M, and Dranoff G. Vaccination with irradiated, autologous melanoma cells engineered to secrete granulocyte-macrophage colony-stimulating factor by adenoviral-mediated gene transfer augments antitumor immunity in patients with metastatic melanoma. *J Clin Oncol* 2003; 21: 3343–3350.

231. Yang SC, Hillinger S, Riedl K, Zhang L, Zhu L, Huang M, Atianzar K, Kuo BY, Gardner B, Batra RK, Strieter RM, Dubinett SM, and Sharma S. Intratumoral administration of dendritic cells overexpressing CCL21 generates systemic antitumor responses and confers tumor immunity. *Clin Cancer Res* 2004; 10: 2891–2901.

232. Adam L, Black PC, Kassouf W, Eve B, McConkey D, Munsell MF, Benedict WF, and Dinney CP. Adenoviral mediated interferon-alpha 2b gene therapy suppresses the pro-angiogenic effect of vascular endothelial growth factor in superficial bladder cancer. *J Urol* 2007; 177: 1900–1906.

233. Eck SL, Alavi JB, Judy K, Phillips P, Alavi A, Hackney D, Cross P, Hughes J, Gao G, Wilson JM, and Propert K. Treatment of recurrent or progressive malignant glioma with a recombinant adenovirus expressing human interferon-beta (H5.010CMVhIFN-beta): A phase I trial. *Hum Gene Ther* 2001; 12: 97–113.

234. Sterman DH, Gillespie CT, Carroll RG, Coughlin CM, Lord EM, Sun J, Haas A, Recio A, Kaiser LR, Coukos G, June CH, Albelda SM, and Vonderheide RH. Interferon beta adenoviral gene therapy in a patient with ovarian cancer. *Nat Clin Pract Oncol* 2006; 3: 633–639.

235. Khorana AA, Rosenblatt JD, Sahasrabudhe DM, Evans T, Ladrigan M, Marquis D, Rosell K, Whiteside T, Phillippe S, Acres B, Slos P, Squiban P, Ross M, and Kendra K. A phase I trial of immunotherapy with intratumoral adenovirus-interferon-gamma (TG1041) in patients with malignant melanoma. *Cancer Gene Ther* 2003; 10: 251–259.

236. Dummer R, Hassel JC, Fellenberg F, Eichmuller S, Maier T, Slos P, Acres B, Bleuzen P, Bataille V, Squiban P, Burg G, and Urosevic M. Adenovirus-mediated intralesional interferon-gamma gene transfer induces tumor regressions in cutaneous lymphomas. *Blood* 2004; 104: 1631–1638.

237. Sangro B, Mazzolini G, Ruiz J, Herraiz M, Quiroga J, Herrero I, Benito A, Larrache J, Pueyo J, Subtil JC, Olague C, Sola J, Sadaba B, Lacasa C, Melero I, Qian C, and Prieto J. Phase I trial of intratumoral injection of an adenovirus encoding interleukin-12 for advanced digestive tumors. *J Clin Oncol* 2004; 22: 1389–1397.

238. Mazzolini G, Alfaro C, Sangro B, Feijoo E, Ruiz J, Benito A, Tirapu I, Arina A, Sola J, Herraiz M, Lucena F, Olague C, Subtil J, Quiroga J, Herrero I, Sadaba B, Bendandi M, Qian C, Prieto J, and Melero I. Intratumoral injection of dendritic cells engineered to secrete interleukin-12 by recombinant adenovirus in patients with metastatic gastrointestinal carcinomas. *J Clin Oncol* 2005; 23: 999–1010.

239. Sung MW, Chen SH, Thung SN, Zhang DY, Huang TG, Mandeli JP, and Woo SL. Intratumoral delivery of adenovirus-mediated interleukin-12 gene in mice with metastatic cancer in the liver. *Hum Gene Ther* 2002; 13: 731–743.

240. Sung MW, Yeh HC, Thung SN, Schwartz ME, Mandeli JP, Chen SH, and Woo SL. Intratumoral adenovirus-mediated suicide gene transfer for hepatic metastases from colorectal adenocarcinoma: Results of a phase I clinical trial. *Mol Ther* 2001; 4: 182–191.

241. Bowman L, Grossmann M, Rill D, Brown M, Zhong WY, Alexander B, Leimig T, Coustan-Smith E, Campana D, Jenkins J, Woods D, Kitchingman G, Vanin E, and Brenner M. IL-2 adenovector-transduced autologous tumor cells induce antitumor immune responses in patients with neuroblastoma. *Blood* 1998; 92: 1941–1949.

242. Russell HV, Strother D, Mei Z, Rill D, Popek E, Biagi E, Yvon E, Brenner M, and Rousseau R. Phase I trial of vaccination with autologous neuroblastoma tumor cells genetically modified to secrete IL-2 and lymphotactin. *J Immunother* 2007; 30: 227–233.

243. Trudel S, Trachtenberg J, Toi A, Sweet J, Li ZH, Jewett M, Tshilias J, Zhuang LH, Hitt M, Wan Y, Gauldie J, Graham FL, Dancey J, and Stewart AK. A phase I trial of adenovector-mediated delivery of interleukin-2 (AdIL-2) in high-risk localized prostate cancer. *Cancer Gene Ther* 2003; 10: 755–763.

244. Belldegrun A, Tso CL, Zisman A, Naitoh J, Said J, Pantuck AJ, Hinkel A, deKernion J, and Figlin R. Interleukin 2 gene therapy for prostate cancer: Phase I clinical trial and basic biology. *Hum Gene Ther* 2001; 12: 883–892.

245. Stewart AK, Lassam NJ, Quirt IC, Bailey DJ, Rotstein LE, Krajden M, Dessureault S, Gallinger S, Cappe D, Wan Y, Addison CL, Moen RC, Gauldie J, and Graham FL. Adenovector-mediated gene delivery of interleukin-2 in metastatic breast cancer and melanoma: Results of a phase 1 clinical trial. *Gene Ther* 1999; 6: 350–363.

246. Trudel S, Li Z, Dodgson C, Nanji S, Wan Y, Voralia M, Hitt M, Gauldie J, Graham FL, and Stewart AK. Adenovector engineered interleukin-2 expressing autologous plasma cell vaccination after high-dose chemotherapy for multiple myeloma—a phase 1 study. *Leukemia* 2001; 15: 846–854.

247. Griscelli F, Opolon P, Saulnier P, Mami-Chouaib F, Gautier E, Echchakir H, Angevin E, Le CT, Bataille V, Squiban P, Tursz T, and Escudier B. Recombinant adenovirus shedding after intratumoral gene transfer in lung cancer patients. *Gene Ther* 2003; 10: 386–395.

248. Macgill RS, Davis TA, Macko J, Mauceri HJ, Weichselbaum RR, and King CR. Local gene delivery of tumor necrosis factor alpha can impact primary tumor growth and metastases through a host-mediated response. *Clin Exp Metastasis* 2007; 24: 521–531.

249. McLoughlin JM, McCarty TM, Cunningham C, Clark V, Senzer N, Nemunaitis J, and Kuhn JA. TNFerade, an adenovector carrying the transgene for human tumor necrosis factor alpha, for patients with advanced solid tumors: Surgical experience and long-term follow-up. *Ann Surg Oncol* 2005; 12: 825–830.

250. Senzer N, Mani S, Rosemurgy A, Nemunaitis J, Cunningham C, Guha C, Bayol N, Gillen M, Chu K, Rasmussen C, Rasmussen H, Kufe D, Weichselbaum R, and Hanna N. TNFerade biologic, an adenovector with a radiation-inducible promoter, carrying the human tumor necrosis factor alpha gene: A phase I study in patients with solid tumors. *J Clin Oncol* 2004; 22: 592–601.

251. Sharma A, Mani S, Hanna N, Guha C, Vikram B, Weichselbaum RR, Sparano J, Sood B, Lee D, Regine W, Muhodin M, Valentino J, Herman J, Desimone P, Arnold S, Carrico J, Rockich AK, Warner-Carpenter J, and Barton-Baxter M. Clinical protocol. An open-label, phase I, dose-escalation study of tumor necrosis factor-alpha (TNFerade Biologic) gene transfer with radiation therapy for locally advanced, recurrent, or metastatic solid tumors. *Hum Gene Ther* 2001; 12: 1109–1131.

252. Mundt AJ, Vijayakumar S, Nemunaitis J, Sandler A, Schwartz H, Hanna N, Peabody T, Senzer N, Chu K, Rasmussen CS, Kessler PD, Rasmussen HS, Warso M, Kufe DW, Gupta TD, and Weichselbaum RR. A phase I trial of TNFerade biologic in patients with soft tissue sarcoma in the extremities. *Clin Cancer Res* 2004; 10: 5747–5753.

253. Rasmussen H, Rasmussen C, Lempicki M, Durham R, Brough D, King CR, and Weichselbaum R. TNFerade biologic: Preclinical toxicology of a novel adenovector with a radiation-inducible promoter, carrying the human tumor necrosis factor alpha gene. *Cancer Gene Ther* 2002; 9: 951–957.

254. Miller PW, Sharma S, Stolina M, Butterfield LH, Luo J, Lin Y, Dohadwala M, Batra RK, Wu L, Economou JS, and Dubinett SM. Intratumoral administration of adenoviral interleukin 7 gene-modified dendritic cells augments specific antitumor immunity and achieves tumor eradication. *Hum Gene Ther* 2000; 11: 53–65.

255. Rosenberg SA, Zhai Y, Yang JC, Schwartzentruber DJ, Hwu P, Marincola FM, Topalian SL, Restifo NP, Seipp CA, Einhorn JH, Roberts B, and White DE. Immunizing patients with metastatic melanoma using recombinant adenoviruses encoding MART-1 or gp100 melanoma antigens. *J Natl Cancer Inst* 1998; 90: 1894–1900.

256. Tsao H, Millman P, Linette GP, Hodi FS, Sober AJ, Goldberg MA, and Haluska FG. Hypopigmentation associated with an adenovirus-mediated gp100/MART-1-transduced dendritic cell vaccine for metastatic melanoma. *Arch Dermatol* 2002; 138: 799–802.

257. Schumacher L, Ribas A, Dissette VB, McBride WH, Mukherji B, Economou JS, and Butterfield LH. Human dendritic cell maturation by adenovirus transduction enhances tumor antigen-specific T-cell responses. *J Immunother (1997)* 2004; 27: 191–200.

258. Lubaroff DM, Konety B, Link BK, Ratliff TL, Madsen T, Shannon M, Ecklund D, and Williams RD. Clinical protocol: Phase I study of an adenovirus/prostate-specific antigen vaccine in men with metastatic prostate cancer. *Hum Gene Ther* 2006; 17: 220–229.

259. Nemunaitis J, Meyers T, Senzer N, Cunningham C, West H, Vallieres E, Anthony S, Vukelja S, Berman B, Tully H, Pappen B, Sarmiento S, Arzaga R, Duniho S, Engardt S, Meagher M, and Cheever MA. Phase I trial of sequential administration of recombinant DNA and adenovirus expressing L523S protein in early stage non-small-cell lung cancer. *Mol Ther* 2006; 13: 1185–1191.

260. Antonia SJ, Mirza N, Fricke I, Chiappori A, Thompson P, Williams N, Bepler G, Simon G, Janssen W, Lee JH, Menander K, Chada S, and Gabrilovich DI. Combination of p53 cancer vaccine with chemotherapy in patients with extensive stage small cell lung cancer. *Clin Cancer Res* 2006; 12: 878–887.

261. Bollard CM, Gottschalk S, Leen AM, Weiss H, Straathof KC, Carrum G, Khalil M, Wu MF, Huls MH, Chang CC, Gresik MV, Gee AP, Brenner MK, Rooney CM, and Heslop HE. Complete responses of relapsed lymphoma following genetic modification of tumor-antigen presenting cells and T-lymphocyte transfer. *Blood* 2007; 110: 2838–2845.

262. Tang Y, Zhang L, Yuan J, Akbulut H, Maynard J, Linton PJ, and Deisseroth A. Multistep process through which adenoviral vector vaccine overcomes anergy to tumor-associated antigens. *Blood* 2004; 104: 2704–2713.

263. Chiocca EA, Abbed KM, Tatter S, Louis DN, Hochberg FH, Barker F, Kracher J, Grossman SA, Fisher JD, Carson K, Rosenblum M, Mikkelsen T, Olson J, Markert J, Rosenfeld S, Nabors LB, Brem S, Phuphanich S, Freeman S, Kaplan R, and Zwiebel J. A phase I open-label, dose-escalation, multi-institutional trial of injection with an E1B-Attenuated adenovirus, ONYX-015, into the peritumoral region of recurrent malignant gliomas, in the adjuvant setting. *Mol Ther* 2004; 10: 958–966.

264. Nemunaitis J, Senzer N, Sarmiento S, Zhang YA, Arzaga R, Sands B, Maples P, and Tong AW. A phase I trial of intravenous infusion of ONYX-015 and enbrel in solid tumor patients. *Cancer Gene Ther* 2007; 14: 885–893.

265. Nemunaitis J, Cunningham C, Buchanan A, Blackburn A, Edelman G, Maples P, Netto G, Tong A, Randlev B, Olson S, and Kirn D. Intravenous infusion of a replication-selective adenovirus (ONYX-015) in cancer patients: Safety, feasibility and biological activity. *Gene Ther* 2001; 8: 746–759.

266. Reid TR, Freeman S, Post L, McCormick F, and Sze DY. Effects of Onyx-015 among metastatic colorectal cancer patients that have failed prior treatment with 5-FU/leucovorin. *Cancer Gene Ther* 2005; 12: 673–681.

267. Mulvihill S, Warren R, Venook A, Adler A, Randlev B, Heise C, and Kirn D. Safety and feasibility of injection with an E1B-55 kDa gene-deleted, replication-selective adenovirus

66 Gene and Cell Therapy: Therapeutic Mechanisms and Strategies

(ONYX-015) into primary carcinomas of the pancreas: A phase I trial. *Gene Ther* 2001; 8: 308–315.

268. Small EJ, Carducci MA, Burke JM, Rodriguez R, Fong L, van UL, Yu DC, Aimi J, Ando D, Working P, Kirn D, and Wilding G. A phase I trial of intravenous CG7870, a replication-selective, prostate-specific antigen-targeted oncolytic adenovirus, for the treatment of hormone-refractory, metastatic prostate cancer. *Mol Ther* 2006; 14: 107–117.

269. DeWeese TL, van der PH, Li S, Mikhak B, Drew R, Goemann M, Hamper U, DeJong R, Detorie N, Rodriguez R, Haulk T, DeMarzo AM, Piantadosi S, Yu DC, Chen Y, Henderson DR, Carducci MA, Nelson WG, and Simons JW. A phase I trial of CV706, a replication-competent, PSA selective oncolytic adenovirus, for the treatment of locally recurrent prostate cancer following radiation therapy. *Cancer Res* 2001; 61: 7464–7472.

270. Vasey PA, Shulman LN, Campos S, Davis J, Gore M, Johnston S, Kirn DH, O'Neill V, Siddiqui N, Seiden MV, and Kaye SB. Phase I trial of intraperitoneal injection of the E1B-55-kd-gene-deleted adenovirus ONYX-015 (dl1520) given on days 1 through 5 every 3 weeks in patients with recurrent/refractory epithelial ovarian cancer. *J Clin Oncol* 2002; 20: 1562–1569.

271. Page JG, Tian B, Schweikart K, Tomaszewski J, Harris R, Broadt T, Polley-Nelson J, Noker PE, Wang M, Makhija S, Aurigemma R, Curiel DT, and Alvarez RD. Identifying the safety profile of a novel infectivity-enhanced conditionally replicative adenovirus, Ad5-delta24-RGD, in anticipation of a phase I trial for recurrent ovarian cancer. *Am J Obstet Gynecol* 2007; 196: 389–389.

272. Nemunaitis J, Khuri F, Ganly I, Arseneau J, Posner M, Vokes E, Kuhn J, McCarty T, Landers S, Blackburn A, Romel L, Randlev B, Kaye S, and Kirn D. Phase II trial of intratumoral administration of ONYX-015, a replication-selective adenovirus, in patients with refractory head and neck cancer. *J Clin Oncol* 2001; 19: 289–298.

273. Rudin CM, Cohen EE, Papadimitrakopoulou VA, Silverman S Jr., Recant W, El-Naggar AK, Stenson K, Lippman SM, Hong WK, and Vokes EE. An attenuated adenovirus, ONYX-015, as mouthwash therapy for premalignant oral dysplasia. *J Clin Oncol* 2003; 21: 4546–4552.

274. Hecht JR, Bedford R, Abbruzzese JL, Lahoti S, Reid TR, Soetikno RM, Kirn DH, and Freeman SM. A phase I/II trial of intratumoral endoscopic ultrasound injection of ONYX-015 with intravenous gemcitabine in unresectable pancreatic carcinoma. *Clin Cancer Res* 2003; 9: 555–561.

275. Reid T, Galanis E, Abbruzzese J, Sze D, Wein LM, Andrews J, Randlev B, Heise C, Uprichard M, Hatfield M, Rome L, Rubin J, and Kirn D. Hepatic arterial infusion of a replication-selective oncolytic adenovirus (dl1520): Phase II viral, immunologic, and clinical endpoints. *Cancer Res* 2002; 62: 6070–6079.

276. Dilley J, Reddy S, Ko D, Nguyen N, Rojas G, Working P, and Yu DC. Oncolytic adenovirus CG7870 in combination with radiation demonstrates synergistic enhancements of antitumor efficacy without loss of specificity. *Cancer Gene Ther* 2005; 12: 715–722.

277. Freytag SO, Barton KN, Brown SL, Narra V, Zhang Y, Tyson D, Nall C, Lu M, Ajlouni M, Movsas B, and Kim JH. Replication-competent adenovirus-mediated suicide gene therapy with radiation in a preclinical model of pancreatic cancer. *Mol Ther* 2007; 15: 1600–1605.

278. Freytag SO, Khil M, Stricker H, Peabody J, Menon M, Peralta-Venturina M, Nafziger D, Pegg J, Paielli D, Brown S, Barton K, Lu M, guilar-Cordova E, and Kim JH. Phase I study of replication-competent adenovirus-mediated double suicide gene

therapy for the treatment of locally recurrent prostate cancer. *Cancer Res* 2002; 62: 4968–4976.

279. Freytag SO, Stricker H, Peabody J, Pegg J, Paielli D, Movsas B, Barton KN, Brown SL, Lu M, and Kim JH. Five-year follow-up of trial of replication-competent adenovirus-mediated suicide gene therapy for treatment of prostate cancer. *Mol Ther* 2007; 15: 636–642.

280. Barton KN, Freytag SO, Nurushev T, Yoo S, Lu M, Yin FF, Li S, Movsas B, Kim JH, and Brown SL. A model for optimizing adenoviral delivery in human cancer gene therapy trials. *Hum Gene Ther* 2007; 18: 562–572.

281. Freytag SO, Rogulski KR, Paielli DL, Gilbert JD, and Kim JH. A novel three-pronged approach to kill cancer cells selectively: Concomitant viral, double suicide gene, and radiotherapy. *Hum Gene Ther* 1998; 9: 1323–1333.

282. Ohwada A, Hirschowitz EA, and Crystal RG. Regional delivery of an adenovirus vector containing the *Escherichia coli* cytosine deaminase gene to provide local activation of 5-fluorocytosine to suppress the growth of colon carcinoma metastatic to liver. *Hum Gene Ther* 1996; 7: 1567–1576.

283. Eck SL, Alavi JB, Alavi A, Davis A, Hackney D, Judy K, Mollman J, Phillips PC, Wheeldon EB, and Wilson JM. Treatment of advanced CNS malignancies with the recombinant adenovirus H5.010RSVTK: A phase I trial. *Hum Gene Ther* 1996; 7: 1465–1482.

284. Germano IM, Fable J, Gultekin SH, and Silvers A. Adenovirus/herpes simplex-thymidine kinase/ganciclovir complex: Preliminary results of a phase I trial in patients with recurrent malignant gliomas. *J Neurooncol* 2003; 65: 279–289.

285. Morris JC, Ramsey WJ, Wildner O, Muslow HA, guilar-Cordova E, and Blaese RM. A phase I study of intralesional administration of an adenovirus vector expressing the HSV-1 thymidine kinase gene (AdV.RSV-TK) in combination with escalating doses of ganciclovir in patients with cutaneous metastatic malignant melanoma. *Hum Gene Ther* 2000; 11: 487–503.

286. Chevez-Barrios P, Chintagumpala M, Mieler W, Paysse E, Boniuk M, Kozinetz C, Hurwitz MY, and Hurwitz RL. Response of retinoblastoma with vitreous tumor seeding to adenovirus-mediated delivery of thymidine kinase followed by ganciclovir. *J Clin Oncol* 2005; 23: 7927–7935.

287. Herman JR, Adler HL, guilar-Cordova E, Rojas-Martinez A, Woo S, Timme TL, Wheeler TM, Thompson TC, and Scardino PT. In situ gene therapy for adenocarcinoma of the prostate: A phase I clinical trial. *Hum Gene Ther* 1999; 10: 1239–1249.

288. van der Linden RR, Haagmans BL, Mongiat-Artus P, van Doornum GJ, Kraaij R, Kadmon D, guilar-Cordova E, Osterhaus AD, van der Kwast TH, and Bangma CH. Virus specific immune responses after human neoadjuvant adenovirus-mediated suicide gene therapy for prostate cancer. *Eur Urol* 2005; 48: 153–161.

289. Malaeb BS, Gardner TA, Margulis V, Yang L, Gillenwater JY, Chung LW, Macik G, and Koeneman KS. Elevated activated partial thromboplastin time during administration of first-generation adenoviral vectors for gene therapy for prostate cancer: Identification of lupus anticoagulants. *Urology* 2005; 66: 830–834.

290. Shalev M, Kadmon D, Teh BS, Butler EB, guilar-Cordova E, Thompson TC, Herman JR, Adler HL, Scardino PT, and Miles BJ. Suicide gene therapy toxicity after multiple and repeat injections in patients with localized prostate cancer. *J Urol* 2000; 163: 1747–1750.

291. Trask TW, Trask RP, guilar-Cordova E, Shine HD, Wyde PR, Goodman JC, Hamilton WJ, Rojas-Martinez A, Chen SH,

Woo SL, and Grossman RG. Phase I study of adenoviral delivery of the HSV-tk gene and ganciclovir administration in patients with current malignant brain tumors. *Mol Ther* 2000; 1: 195–203.

292. Sterman DH, Treat J, Litzky LA, Amin KM, Coonrod L, Molnar-Kimber K, Recio A, Knox L, Wilson JM, Albelda SM, and Kaiser LR. Adenovirus-mediated herpes simplex virus thymidine kinase/ganciclovir gene therapy in patients with localized malignancy: Results of a phase I clinical trial in malignant mesothelioma. *Hum Gene Ther* 1998; 9: 1083–1092.

293. Fujita T, Teh BS, Timme TL, Mai WY, Satoh T, Kusaka N, Naruishi K, Fattah EA, guilar-Cordova E, Butler EB, and Thompson TC. Sustained long-term immune responses after in situ gene therapy combined with radiotherapy and hormonal therapy in prostate cancer patients. *Int J Radiat Oncol Biol Phys* 2006; 65: 84–90.

294. Nasu Y, Saika T, Ebara S, Kusaka N, Kaku H, Abarzua F, Manabe D, Thompson TC, and Kumon H. Suicide gene therapy with adenoviral delivery of HSV-tK gene for patients with local recurrence of prostate cancer after hormonal therapy. *Mol Ther* 2007; 15: 834–840.

295. Teh BS, guilar-Cordova E, Kernen K, Chou CC, Shalev M, Vlachaki MT, Miles B, Kadmon D, Mai WY, Caillouet J, Davis M, Ayala G, Wheeler T, Brady J, Carpenter LS, Lu HH, Chiu JK, Woo SY, Thompson T, Butler and EB. Phase I/II trial evaluating combined radiotherapy and in situ gene therapy with or without hormonal therapy in the treatment of prostate cancer—a preliminary report. *Int J Radiat Oncol Biol Phys* 2001; 51: 605–613.

296. Teh BS, Ayala G, Aguilar L, Mai WY, Timme TL, Vlachaki MT, Miles B, Kadmon D, Wheeler T, Caillouet J, Davis M, Carpenter LS, Lu HH, Chiu JK, Woo SY, Thompson T, guilar-Cordova E, and Butler EB. Phase I-II trial evaluating combined intensity-modulated radiotherapy and in situ gene therapy with or without hormonal therapy in treatment of prostate cancer-interim report on PSA response and biopsy data. *Int J Radiat Oncol Biol Phys* 2004; 58: 1520–1529.

297. Kubo H, Gardner TA, Wada Y, Koeneman KS, Gotoh A, Yang L, Kao C, Lim SD, Amin MB, Yang H, Black ME, Matsubara S, Nakagawa M, Gillenwater JY, Zhau HE, and Chung LW. Phase I dose escalation clinical trial of adenovirus vector carrying osteocalcin promoter-driven herpes simplex virus thymidine kinase in localized and metastatic hormone-refractory prostate cancer. *Hum Gene Ther* 2003; 14: 227–241.

298. Ayala G, Satoh T, Li R, Shalev M, Gdor Y, guilar-Cordova E, Frolov A, Wheeler TM, Miles BJ, Rauen K, Teh BS, Butler EB, Thompson TC, and Kadmon D. Biological response determinants in HSV-tk + ganciclovir gene therapy for prostate cancer. *Mol Ther* 2006; 13: 716–728.

299. Sterman DH, Recio A, Vachani A, Sun J, Cheung L, DeLong P, Amin KM, Litzky LA, Wilson JM, Kaiser LR, and Albelda SM. Long-term follow-up of patients with malignant pleural mesothelioma receiving high-dose adenovirus herpes simplex thymidine kinase/ganciclovir suicide gene therapy. *Clin Cancer Res* 2005; 11: 7444–7453.

300. Hasenburg A, Fischer DC, Tong XW, Rojas-Martinez A, Nyberg-Hoffman C, Orlowska-Volk M, Kohlberger P, Kaufman RH, Ramzy I, Guilar-Cordova E, and Kieback DG. Histologic and immunohistochemical analysis of tissue response to adenovirus-mediated herpes simplex thymidine kinase gene therapy of ovarian cancer. *Int J Gynecol Cancer* 2002; 12: 66–73.

301. Hasenburg A, Tong XW, Fischer DC, Rojas-Martinez A, Nyberg-Hoffman C, Kaplan AL, Kaufman RH, Ramzy I, Guilar-Cordova E, and Kieback DG. Adenovirus-mediated thymidine kinase gene therapy in combination with topotecan for patients with recurrent ovarian cancer: 2.5-year follow-up. *Gynecol Oncol* 2001; 83: 549–554.

302. Palmer DH, Mautner V, Mirza D, Oliff S, Gerritsen W, van dS, Jr., Hubscher S, Reynolds G, Bonney S, Rajaratnam R, Hull D, Horne M, Ellis J, Mountain A, Hill S, Harris PA, Searle PF, Young LS, James ND, and Kerr DJ. Virus-directed enzyme prodrug therapy: Intratumoral administration of a replication-deficient adenovirus encoding nitroreductase to patients with resectable liver cancer. *J Clin Oncol* 2004; 22: 1546–1552.

303. Gaut AW, Niu G, Krager KJ, Graham MM, Trask DK, and Domann FE. Genetically targeted radiotherapy of head and neck squamous cell carcinoma using the sodium-iodide symporter (NIS). *Head Neck* 2004; 26: 265–271.

304. Dwyer RM, Schatz SM, Bergert ER, Myers RM, Harvey ME, Classic KL, Blanco MC, Frisk CS, Marler RJ, Davis BJ, O'Connor MK, Russell SJ, and Morris JC. A preclinical large animal model of adenovirus-mediated expression of the sodium-iodide symporter for radioiodide imaging and therapy of locally recurrent prostate cancer. *Mol Ther* 2005; 12: 835–841.

305. Nasu Y, Bangma CH, Hull GW, Yang G, Wang J, Shimura S, McCurdy MA, Ebara S, Lee HM, Timme TL, and Thompson TC. Combination gene therapy with adenoviral vector-mediated HSV-tk + GCV and IL-12 in an orthotopic mouse model for prostate cancer. *Prostate Cancer Prostatic Dis* 2001; 4: 44–55.

306. Fuchs S, Dib N, Cohen BM, Okubagzi P, Diethrich EB, Campbell A, Macko J, Kessler PD, Rasmussen HS, Epstein SE, and Kornowski R. A randomized, double-blind, placebo-controlled, multicenter, pilot study of the safety and feasibility of catheter-based intramyocardial injection of AdVEGF121 in patients with refractory advanced coronary artery disease. *Catheter Cardiovasc Interv* 2006; 68: 372–378.

307. Hedman M, Hartikainen J, Syvanne M, Stjernvall J, Hedman A, Kivela A, Vanninen E, Mussalo H, Kauppila E, Simula S, Narvanen O, Rantala A, Peuhkurinen K, Nieminen MS, Laakso M, and Yla-Herttuala S. Safety and feasibility of catheter-based local intracoronary vascular endothelial growth factor gene transfer in the prevention of postangioplasty and in-stent restenosis and in the treatment of chronic myocardial ischemia: Phase II results of the Kuopio Angiogenesis Trial (KAT). *Circulation* 2003; 107: 2677–2683.

308. Stewart DJ, Hilton JD, Arnold JM, Gregoire J, Rivard A, Archer SL, Charbonneau F, Cohen E, Curtis M, Buller CE, Mendelsohn FO, Dib N, Page P, Ducas J, Plante S, Sullivan J, Macko J, Rasmussen C, Kessler PD, and Rasmussen HS. Angiogenic gene therapy in patients with nonrevascularizable ischemic heart disease: A phase 2 randomized, controlled trial of AdVEGF(121) (AdVEGF121) versus maximum medical treatment. *Gene Ther* 2006; 13: 1503–1511.

309. Grines C, Rubanyi GM, Kleiman NS, Marrott P, and Watkins MW. Angiogenic gene therapy with adenovirus 5 fibroblast growth factor-4 (Ad5FGF-4): A new option for the treatment of coronary artery disease. *Am J Cardiol* 2003; 92: 24N–31N.

310. Grines CL, Watkins MW, Helmer G, Penny W, Brinker J, Marmur JD, West A, Rade JJ, Marrott P, Hammond HK, and Engler RL. Angiogenic Gene Therapy (AGENT) trial in patients with stable angina pectoris. *Circulation* 2002; 105: 1291–1297.

311. Grines CL, Watkins MW, Mahmarian JJ, Iskandrian AE, Rade JJ, Marrott P, Pratt C, and Kleiman N. A randomized, double-blind, placebo-controlled trial of Ad5FGF-4 gene therapy and its effect on myocardial perfusion in patients with stable angina. *J Am Coll Cardiol* 2003; 42: 1339–1347.

312. Margolis DJ, Crombleholme T, and Herlyn M. Clinical protocol: Phase I trial to evaluate the safety of H5.020CMV.PDGF-B for the treatment of a diabetic insensate foot ulcer. *Wound Repair Regen* 2000; 8: 480–493.

313. Rajagopalan S, Mohler E, III, Lederman RJ, Saucedo J, Mendelsohn FO, Olin J, Blebea J, Goldman C, Trachtenberg JD, Pressler M, Rasmussen H, Annex BH, and Hirsch AT. Regional angiogenesis with vascular endothelial growth factor (VEGF) in peripheral arterial disease: Design of the RAVE trial. *Am Heart J* 2003; 145: 1114–1118.

314. Rajagopalan S, Olin J, Deitcher S, Pieczek A, Laird J, Grossman PM, Goldman CK, McEllin K, Kelly R, and Chronos N. Use of a constitutively active hypoxia-inducible factor-1alpha transgene as a therapeutic strategy in no-option critical limb ischemia patients: Phase I dose-escalation experience. *Circulation* 2007; 115: 1234–1243.

315. Margolis DJ, Cromblehome T, Herlyn M, Cross P, Weinberg L, Filip J, and Propert K. Clinical protocol. Phase I trial to evaluate the safety of H5.020CMV.PDGF-b and limb compression bandage for the treatment of venous leg ulcer: Trial A. *Hum Gene Ther* 2004; 15: 1003–1019.

316. Fuster V, Charlton P, and Boyd A. Clinical protocol. A phase IIb, randomized, multicenter, double-blind study of the efficacy and safety of Trinam (EG004) in stenosis prevention at the graft-vein anastomosis site in dialysis patients. *Hum Gene Ther* 2001; 12: 2025–2027.

317. Barbato JE and Tzeng E. iNOS gene transfer for graft disease. *Trends Cardiovasc Med* 2004; 14: 267–272.

318. Hammond HK. Adenylyl cyclase gene transfer in heart failure. *Ann N Y Acad Sci* 2006; 1080: 426–436.

319. Campochiaro PA, Nguyen QD, Shah SM, Klein ML, Holz E, Frank RN, Saperstein DA, Gupta A, Stout JT, Macko J, DiBartolomeo R, and Wei LL. Adenoviral vector-delivered pigment epithelium-derived factor for neovascular age-related macular degeneration: Results of a phase I clinical trial. *Hum Gene Ther* 2006; 17: 167–176.

320. Atencio IA, Chen Z, Nguyen QH, Faha B, and Maneval DC. p21WAF-1/Cip-1 gene therapy as an adjunct to glaucoma filtration surgery. *Curr Opin Mol Ther* 2004; 6: 624–628.

321. Catanzaro AT, Koup RA, Roederer M, Bailer RT, Enama ME, Moodie Z, Gu L, Martin JE, Novik L, Chakrabarti BK, Butman BT, Gall JG, King CR, Andrews CA, Sheets R, Gomez PL, Mascola JR, Nabel GJ, and Graham BS. Phase 1 safety and immunogenicity evaluation of a multiclade HIV-1 candidate vaccine delivered by a replication-defective recombinant adenovirus vector. *J Infect Dis* 2006; 194: 1638–1649.

322. Myers GD, Bollard CM, Wu MF, Weiss H, Rooney CM, Heslop HE, and Leen AM. Reconstitution of adenovirus-specific cell-mediated immunity in pediatric patients after hematopoietic stem cell transplantation. *Bone Marrow Transplant* 2007; 39: 677–686.

4 Modified Adenoviruses for Gene Therapy

Joel N. Glasgow, Akseli Hemminki, and David T. Curiel

CONTENTS

4.1   INTRODUCTION

While still in its infancy, gene-based therapy has emerged as a potentially powerful therapeutic platform. The concept of gene therapy as a rational molecular intervention is simple, that is, to correct or eradicate defective tissues via delivery of nucleic acids. However, practical implementation of safe, highly effective gene-based interventions has been difficult due to the stringent and complex requirements of this paradigm. Indeed, the introduction of functional foreign nucleic acids into mammalian target cells without oncogenicity or significant toxicity or immune response represents a daunting task unique in biomedicine. This realization has spawned the field of vector design, which has for 20 years sought to engineer gene delivery vector systems compatible with the stringent mandates of human clinical use.

Vectors based on human adenovirus (Ad) serotypes 2 and 5 continue to show utility as gene therapy vectors, particularly in the context of cancer gene therapy, owing to several innate biological characteristics: Replication-deficient Ad vectors display *in vivo* stability and superior gene transfer efficiency to numerous dividing and nondividing cell targets *in vivo* and rarely cause severe disease in humans. Further, production parameters for clinical grade Ad vectors are well established. In 2007, Ad vectors were employed in one-fourth of clinical trials worldwide (326 of 1309) with two-thirds of all trials being for cancer (871 of 1309) [1].

Nonetheless, clinical trial results of Ad vectors have clearly exposed the need to advance Ad vector technology. Relatively lackluster clinical performance of Ad-based agents to date has provided clear direction for vector modifications designed to improve efficacy and safety. In this regard, two distinct approaches have been employed: (1) transductional targeting, which limits the entry of agents to target cells and (2) transcriptional targeting, which restricts expression of transgenes (or Ad replication, in some cases) to target tissues by using tissue- or tumor-specific promoters (TSPs). The following is a discussion of Ad biology, barriers to transductional targeting, and a review of the vector modifications applied toward transforming the human Ad into a clinically effective therapeutic agent.

4.2   Ad STRUCTURE

The family Adenoviridae contains over 100 serotypes, including 52 human serotypes that are divided among 7 species (A–G) based on genome homology and organization, oncogenicity, and hemagglutination properties [2–6]. The human Ad is a nonenveloped icosahedral particle that encapsulates up to a 36-kilobase double-stranded DNA genome. The Ad capsid is comprised of several minor and three major capsid proteins: hexon is the most abundant structural component with 720 copies per virion and constitutes the bulk of the protein shell; 5 penton monomers form the penton base platform at each of the 12 capsid vertices to which the 12 fiber homotrimers attach (Figure 4.1). The distal tip of each fiber is composed of a globular knob domain, which serves as the major viral attachment site for cellular receptors. Hexon appears to play only a structural role as a coating protein, while the penton base and the fiber are responsible for distinct virion–cell interactions that constitute Ad tropism. Detailed structures of hexon [7–11], penton base [12], and fiber [13,14] have been determined by crystallography; the high-resolution structure of the entire virion has been determined by various methods [15,16].

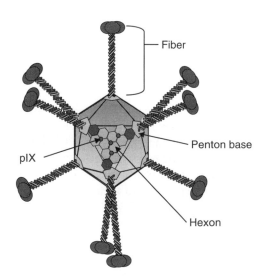

FIGURE 4.1 (See color insert following blank page 206.)   Generalized Ad structure depicting major structural components of a wild-type Ad capsid. Hexon is depicted as a hexagon with 12 homotrimers per capsid face (only 12 of 240 trimers are shown). The penton base, comprised of five subunits at each vertex, and trimeric fiber and pIX structures are shown. Ad serotype-2 and -5 capsids contain a 36-kilobase double-stranded DNA genome (not shown).

4.3   Ad TRANSDUCTION PATHWAY

Intense research efforts into Ad biology have revealed crucial steps involved in gene transfer. The tropism of Ad is determined by two distinct virus–cell interactions: attachment to a primary receptor molecule at the cell surface, followed by interaction with molecules responsible for virion internalization. Initial high-affinity binding of the virion occurs via direct binding of the fiber knob domain to its cognate primary cellular receptor, which is the 346 amino acid coxsackie and adenovirus receptor (CAR) glycoprotein for most serotypes including Ad2 and Ad5, which are widely used in gene therapy approaches [17,18]. Other receptors have been described for Ad5, although the nature of their interactions with the Ad5 virion is unclear and their roles appear limited. These receptors include heparin sulfate glycosaminoglycans [19,20], class I major histocompatability complex [21], and vascular cell adhesion molecule-1 [22]. Receptor binding is followed by receptor-mediated endocytosis of the virion via interaction of penton base Arg–Gly–Asp (RGD) motifs with cellular integrins including $\alpha v \beta 3$ and $\alpha v \beta 5$ [23], $\alpha v \beta 1$ [24], and $\alpha_3 \beta 1$ and $\alpha_5 \beta 1$ [25]. The virus enters the cell in clathrin-coated vesicles [26] and is transported to endosomes. Acidification of the endosome results in stepwise virion disassembly and subsequent release of the virus core into the cytosol, where it docks at the nuclear membrane. Further capsid disassembly and subsequent nuclear import of the Ad genome are then achieved via interaction of genome-associated Ad core proteins with transportin and importin components of the nuclear pore complex [27–29].

4.4   TRANSDUCTIONAL TARGETING OF Ad

Targeted gene delivery is ultimately predicated on the ability of the vector to discriminate between target and nontarget cells via interaction with unique cell- or disease-specific surface markers. However, the CAR dependence of Ad5 transduction results in a scenario wherein nontarget but high-CAR cells can be infected, whereas target tissues, if low in CAR, remain poorly infected. This may be of particular relevance in Ad-based cancer gene therapy, as increased CAR expression appears to have a growth inhibitory effect on some cancer cell lines, while loss of CAR expression correlates with tumor progression and advanced disease. Indeed, a lack or downregulation of CAR has been reported for various tumor types, such as ovarian, prostate, lung, breast and colorectal cancer, melanoma, glioma, and rhabdomyosarcoma [30–39] (reviewed in [40,41]). Further, this may be a general phenomenon associated with the carcinogenesis of various tumor types, as inverse correlation with tumor grade has been suggested [37,38]. CAR is a 346 amino acid glycoprotein containing a single membrane-spanning domain and is localized to tight junctions [42] where it mediates homotypic cell adhesion [43,44]. The functions of CAR are not fully understood, but it plays role in cell adhesion and perhaps cell cycle regulation, as it has a role in suppression of tumor growth [38]. Interestingly, a preliminary report has suggested inverse correlation between activity of the tumor-associated Raf–MEK–ERK pathway and CAR expression [45]. For the treatment of cancer with Ad-based vectors, these associations are problematic since CAR-dependent approaches may not be useful against advanced disease if CAR is variably expressed in these tumors. Thus, targeting of Ad to tumor cells should be useful for increasing the clinical efficacy and safety of approaches. Further, considering the widespread expression of CAR, targeting approaches may be advantageous for increasing the specificity of any clinical Ad gene therapy application.

The biodistribution of Ad-based vectors *in vivo*, however, is not determined solely by receptor biodistribution [46]. Intravenous administration of Ad results in accumulation in the liver, spleen, heart, lung, and kidneys of mice, although these tissues may not necessarily be the highest in CAR expression [47,48]. Instead, the degree of blood flow and the structure of the vasculature in each organ probably contribute to the biodistribution. This is true with regard to the liver in particular, which retains the majority of systemically administered Ad particles via hepatic macrophage (Kupffer cell) uptake [49,50] and hepatocyte transduction [51], leading to Ad-mediated inflammation and liver toxicity [52–55]. Thus, the nature of Ad–host interactions that determine the fate of systemically applied Ad has come under considerable scrutiny.

To this end, initial attempts to "de-target" the liver were based on the supposition that CAR- and integrin-based interactions were required for liver transduction *in vivo*. Strategies to inhibit hepatocyte or liver Kupffer cell uptake by ablating CAR- or integrin-binding motifs in the Ad capsid have been largely unsuccessful, however, indicating that native Ad tropism determinants contribute little to vector hepatotropism *in vivo* [56–60]. These data notwithstanding, work by several groups has implicated the fiber protein as a major structural determinant of liver tropism *in vivo* (reviewed by Nicklin et al. [61]). For example, shortening of the native fiber shaft domain of the Ad 5 fiber [62] or replacement of the Ad5 shaft with the short Ad3 shaft domain [63] was shown to attenuate liver uptake *in vivo* following intravenous delivery. In related work, Smith and coworkers examined the role of a putative heparan sulfate proteoglycan (HSPG)-binding motif, KKTK, in the third repeat of the native fiber shaft [64]. Replacement of this motif with an irrelevant peptide sequence reduced reporter gene expression in the liver by 90%. This was also the first suggestion of the importance of HSPG as Ad receptors *in vivo*.

Critical work by Shayakhmetov and colleagues shed light on these apparently contradictory lines of evidence by uncovering a major role for coagulation factor IX (fIX) and complement component C4-binding protein (C4BP) in hepatocyte and Kupffer cell uptake of intravenous Ad [65]. A key finding was that these blood factors mediated *in vivo* tropism by cross-linking Ad particles to hepatocellular HSPG and the LDL receptor-related protein (reviewed in [66]). Kupffer cell sequestration of Ad particles was likewise heavily dependent on Ad association with fIX and C4BP. Importantly, Ad5 vectors containing fibers genetically modified to ablate fIX and C4BP binding provided 50-fold lower liver transduction with

reduced inflammation and hepatotoxicity. Extending this work, this group also showed that Ad5 virions bind to circulating platelets *in vivo*, resulting in their aggregation, entrapment in liver sinusoids, and eventual clearance by Kupffer cells [49]. Further, Ad sequestration in organs was reduced by platelet depletion before systemic vector injection.

These efforts serve to highlight the complexity of vector–host interplay, and have identified genetic modifications that have important practical implications for designing safer and more effective Ad-based vectors for clinical applications. In the absence of a clinically defined upper limit for ectopic liver transduction in humans, it is clear that the concepts of "de-targeting" and "re-targeting" must be simultaneously employed to allow for maximum Ad vector efficacy at the lowest possible dose. Two distinct approaches have been employed to transductionally target Ad-based therapeutic vectors: (1) adapter molecule-based targeting and (2) genetic targeting via structural manipulation of the Ad capsid.

4.4.1 Adapter-Based Ad Targeting

The formation of a molecular bridge between the Ad vector and a cell surface receptor constitutes the adapter-based concept of Ad targeting (Figure 4.2). Adapter function is performed by so called bispecific molecules that cross-link the Ad vector to alternative cell surface receptors, bypassing the native CAR-based tropism. This approach is predicated by the aforementioned two-step entry mechanism of the Ad virion wherein attachment is distinct from internalization. In this way, alternative means of cellular attachment do not impede Ad cell entry. The majority of current adapter-based Ad targeting approaches incorporate the two mandates of delivery targeting, that of ablation of native CAR-dependent Ad tropism and formation of a novel tropism to previously identified cellular receptors. Bispecific adapter molecules include, but are not limited to bispecific antibodies, chemical conjugates

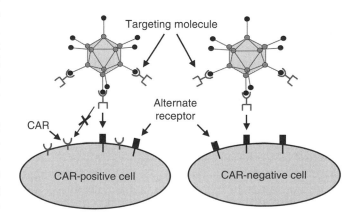

FIGURE 4.2 Adapter molecules for receptor-specific Ad targeting. A generalized adapter molecule ablates native CAR-based tropism and targets Ad to an alternate cellular receptor molecule. The dual specificity of the adapter molecule for both the Ad and the alternative receptor provides novel, CAR-independent cell binding.

between antibody fragments (Fab), and cell-selective ligands such as folate, Fab–antibody conjugates using antibodies against target cell receptors, Fab–peptide ligand conjugates, and multidomain recombinant fusion proteins comprised of soluble CAR and peptide ligands or recombinant Fab.

The first *in vitro* demonstration of Ad targeting via the adapter method resulted in CAR-independent, folate receptor-mediated cellular uptake of the virion by cancer cells overexpressing this receptor [67]. This was accomplished using a bispecific conjugate consisting of an anti-knob neutralizing Fab chemically linked to folate. A similar targeting adapter comprised of the same anti-knob Fab as above fused to basic fibroblast growth factor (FGF2) was utilized to target Ad vectors to FGF receptor-positive Kaposi's sarcoma and ovarian cancer substrates [68,69]. Upon intraperitoneal injection of Ad-Fab-FGF2 coding for herpes simplex virus type I thymidine kinase (HSV-TK) into mice bearing human ovarian cancer, survival was prolonged [70]. Importantly, this targeting system also reduced hepatic toxicity and resulted in increased survival in melanoma [71] and ovarian cancer [72] xenograft mouse models. Other Fab–ligand conjugates targeted against Ep-CAM, Tag-72, EGF receptor, CD40, and other cell markers have been employed in a similar manner with promising results [31,34,73–77].

Dmitriev and colleagues developed a more elegant alternative to chemical conjugates by creating a single recombinant fusion molecule formed by a truncated, soluble form of CAR (sCAR) fused to either CD40 [78] or epidermal growth factor (EGF) [79]. Using the latter, a ninefold increase in reporter gene expression was achieved in several EGFR-overexpressing cancer cell lines compared with untargeted Ad or EGFR-negative cells *in vitro*. EGF-directed targeting to EGFR-positive cells was shown to be dependent on cell surface EGFR density, an additional confirmation of Ad targeting specificity. Addressed also was the issue of virion–adapter complex stability, a critical issue if targeting adapters are to be employed *in vivo*. In this regard, preformed Ad/sCAR–EGF complexes subjected to gel filtration purification showed the same targeting profile as those not purified, indicating adequate Ad–adapter complex stability. To further increase Ad/sCAR–ligand complex stability, Kashentseva and colleagues developed a trimeric sCAR-anti-c-*erb*B2 single chain antibody adapter molecule. The trimeric sCAR-c-*erb*B2 adapter displayed increased affinity for the Ad fiber knob while augmenting gene transfer up to 17-fold in six c-*erb*B2-positive breast and ovarian cancer cell lines *in vitro* [80]. In a similar work, Itoh and colleagues demonstrated improved efficiency of sCAR-based fusion molecule adapters [81]. Kim et al. showed that adapter trimerization yielded a 100-fold increase in infection of CAR-deficient human diploid fibroblasts compared to the monomeric sCAR adapter [82]. Importantly, *in vivo* employment of a nontargeted trimeric sCAR adapter attenuated liver transduction in mice following intravenous administration, indicating excellent *in vivo* stability of this Ad–trimeric adapter complex.

The above proof-of-principle studies and others have rationalized the further testing of targeting adapters *in vivo*. In this regard, Reynolds et al. employed a novel bispecific adapter composed of an anti-knob Fab chemically conjugated to a monoclonal antibody (9B9) raised against angiotensin-converting enzyme (ACE), a surface molecule expressed preferentially on pulmonary capillary endothelium and upregulated in various disease states of the lung [75]. Following peripheral intravenous injection of the Ad/Fab-9B9 complex, reporter transgene expression and viral DNA in the lung was increased 20-fold over untargeted Ad and reporter gene expression in the liver, a nontarget, high-CAR organ, was reduced by 83%. Further, ACE-targeted gene delivery of bone morphogenetic protein receptor type 2 to the pulmonary vascular endothelium of rats reduced hypoxia-induced pulmonary hypertension, as reflected by reductions in pulmonary artery and right ventricular pressures, right ventricular hypertrophy, and muscularization of distal pulmonary arterioles [83].

In another adapter-based *in vivo* targeting model, Everts and coworkers used a bifunctional adapter molecule comprised of sCAR fused to a single-chain antibody (MFE-23) directed against carcinoembryonic antigen (CEA) [84]. Systemic administration of the Ad/sCAR-MFE-23 adapter complex increased gene expression in CEA-positive murine lung by 10-fold and reduced liver transduction, resulting in an improved lung-to-liver ratio of gene expression compared with untargeted Ad. In another study, the sCAR-MFE adapter was used to target Ad vectors to CEA-expressing cells *in vitro*, CEA-positive subcutaneous tumor grafts, and hepatic tumor grafts following systemic administration of the Ad/sCAR-MFE complex [85]. Of note, Ad/sCAR-MFE liver transduction was reduced by over 95% and CEA-negative tumors *in vivo* were not transduced, suggesting remarkable safety and specificity profiles for this approach.

Overall, adapter-based Ad targeting studies provide compelling evidence that Ad tropism modification can be achieved by targeting alternate cellular receptors and that this modality augments gene delivery to CAR-deficient target cells *in vitro*. Adapter-targeted vectors have also performed well *in vivo*, although data so far are limited. While single-component systems have been favored for employment in human gene therapy trials, rigorous analysis of the pharmacodynamics and pharmacokinetics, and systemic stability of vector–adapter complexes may provide the rationale for clinical translation.

4.4.2 AD TARGETING VIA GENETIC MODIFICATION: FIBER

Genetic manipulation of capsid proteins has yielded increasingly promising data in terms of Ad targeting. Direct capsid Ad modification may be advantageous over adapter molecules in that all virions harbor an identical modification and lack of stability is less likely. Redirection of Ad tropism via genetic capsid modification is conceptually elegant, but genetic targeting efforts must work within narrow structural constraints. The success of this approach depends upon modulation of the complex protein structure–function relationships that result in Ad tropism modification, without disrupting the innate molecular interactions required for proper biological function. Specifically,

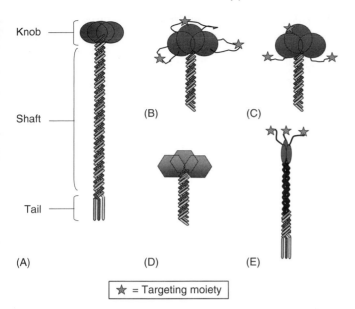

Knob

Shaft

Tail

(A)   (B)   (C)   (D)   (E)

⭐ = Targeting moiety

FIGURE 4.3 (See color insert following blank page 206.) Genetically modified fibers for Ad targeting. (See text for specific examples of each genetic targeting approach and targeting moiety used.) (A) Trimeric wild-type Ad fiber; tail, shaft, and knob regions are shown. (B) Ad5 fiber knob with a constrained targeting peptide in the flexible H-I loop. (C) Ad5 fiber knob containing a C-terminal targeting ligand. (D) Pseudotyped chimeric fiber bearing an alternate human or animal serotype knob domain. (E) "De-knobbed" fiber containing a heterologous trimerization sequence and a C-terminal targeting moiety.

genetic modification can affect fiber trimerization, production, stability, and ultimately packaging of infectious virions. On the basis of a clear understanding of Ad infection biology, development of genetically targeted vectors has rationally focused on the fiber, the primary capsid determinant of Ad tropism. In general, there have been three basic strategies for genetic tropism modification via structural modification of the Ad fiber: (1) so-called fiber pseudotyping, (2) ligand incorporation into the fiber knob domain, and (3) de-knobbing of the fiber coupled with ligand addition (Figure 4.3).

4.4.2.1   Fiber Pseudotyping

As previously mentioned, clinically relevant tissues are often refractory to Ad5 infection, including several cancer cell types, owing to negligible CAR levels. Ad fiber pseudotyping, the genetic replacement of either the entire fiber or knob domain with its structural counterpart from another human Ad serotype that recognizes a cellular receptor other than CAR, was first accomplished by Krasnykh et al. to circumvent CAR deficiency [86]. This vector expresses a chimeric fiber containing the Ad serotype 3 knob domain (Ad5/3) and demonstrates the same CAR-independent cell recognition as does Ad3. The use of another Ad5/3 vector selectively targeted low-CAR lymphoid cell lines *in vitro*, whereas these cells were refractory to Ad5 infection [87]. Ad5/3 has been useful for retargeting Ad5 to low-CAR primary ovarian carcinoma cells and cell lines [88,89] and primary glioma

[90–92] *in vitro* and *in vivo*. Other fiber pseudotyped Ad vectors display CAR-independent tropism by virtue of the natural diversity in receptor recognition found in human species B and D fibers [93]. In this regard, primary receptors for species B Ads have been recently identified, including the complement regulatory protein CD46 [94–97], CD80, and CD86 [98], although an additional unknown receptor is postulated [95]. Species D Ad receptors include CD46 and α(2–3)-linked sialic acid, a common element of glycolipids [99–101]. Fiber pseudotyping has identified chimeric vectors with superior infectivity to Ad5 in several clinically relevant cell types including primary ovarian carcinoma cells [88,89,102], vascular endothelial cells (ECs) [103], dendritic cells [104], B-cells [87], CD34 + hematopoietic cells [105], synovial tissue [106], human cardiovascular tissue [107], and others [108,109]. Interestingly, this strategy has been extended to exploit fiber elements from nonhuman Ads [110,111] and the fiber-like σ1 reovirus attachment protein, which targets cells expressing junctional adhesion molecule [112,113].

4.4.2.2   Genetic Ligand Incorporation

To circumvent variable expression of CAR, targeting ligands have been incorporated into the Ad knob domain without ablating native CAR binding. This has resulted in Ad vectors with expanded, rather than restricted, cell recognition. These efforts are based on rigorous structural analysis of the knob domain and have exploited two separate locations within the knob that tolerate genetic manipulation without loss of fiber function, the C-terminus and the HI-loop. Since the C-terminus of the Ad knob is solvent exposed, extension of the knob peptide to include a targeting peptide moiety is conceptually simple. Ads with C-terminal integrin-binding RGD motifs and polylysine ligands have yielded some promising results *in vitro* and *in vivo*, but other peptide ligands were rendered ineffective in the C-terminus structural context [114], presumably due to steric or other inhibition. On this basis, Krasnykh et al. inserted a FLAG peptide sequence into an exposed loop structure that connects β-sheets H and I (HI-loop) within the Ad5 knob, showing that this locale is structurally permissive to modification [115]. Indeed, the Ad5 HI-loop tolerates peptide insertions up to 100 amino acids with minimal negative effects on virion integrity, thus suggesting considerable potential for ligand incorporation at this site [116]. On this basis, Magnusson and colleagues developed a targeting approach via incorporation of complex ligands in the HI-loop. This group genetically inserted tandem HER2/neu reactive Affibody molecules in the HI-loop of a CAR binding-ablated fiber, resulting in CAR-independent, HER2-mediated cell infection *in vitro* [117]. Dmitriev and coworkers introduced an integrin-binding RGD peptide sequence into the HI-loop. The resulting vector, Ad5lucRGD, used the RGD–cellular integrin interaction to enhance gene delivery to ovarian cancer cell lines and primary tumors versus unmodified Ad5 [33,118,119]. The expanded tropism of this vector has been useful in several other cancer contexts including carcinomas of the ovary, pancreas, colon cancer, and head and neck carcinomas, all of which frequently display highly variable CAR levels (reviewed in [120]). Wu et al. demonstrated that Ad vectors with a double-fiber modification consisting of a C-terminal polylysine stretch, which interacts with heparan sulfates, and the HI-loop RGD provided increased infectivity in several CAR-deficient cell lines [121], as well as human pancreatic islet cells [122], ovarian carcinoma [123], and cervical cancer cells *in vivo* [124]. Other targeting peptides have functioned in the HI-loop locale, including a vascular EC-binding motif SIGYLPLP [125]. This fiber modification also provided cancer cell-selective infection [126].

Korokhov et al., Volpers et al., and others [127] have developed similar targeting approaches that embody elements of both genetic fiber modification and adapter-based targeting by incorporating the immunoglobulin (Ig)-binding domain of *Staphylococcus aureus* protein A into the fiber C-terminus or HI-loop [128,129]. As a result, these fiber-modified vectors form stable complexes with a wide variety of targeting molecules containing the Fc region of Ig. This provides the opportunity to screen numerous targeting molecules directed against a host of cell-surface elements. This approach was used to target and activate dendritic cells via an Fc-single chain antibody directed against CD40 [130]. This system was also used to target ovarian cancer cells via an antibody directed against mesothelin [131], as well as the pulmonary endothelium in a rat model *in vivo* [132].

4.4.2.3 Knob-Deleted Fibers with Complex Ligands

The structural conflicts emerging from knob modifications and the observation that fiber-deleted Ad vectors could be produced [133,134] provided the conceptual basis for replacing the native fiber with knobless fibers. Virions containing a knobless fiber would be ablated for CAR binding, a hallmark of targeted Ad vectors. Simultaneous addition of a targeting ligand to the knobless fiber would result in a more specifically targeted Ad. The technical barrier to this approach is the innate trimerization function of the fiber knob domain, required for proper fiber function and capsid assembly. A solution was devised by addition of a foreign trimerization motif to replace the native fiber or knob or both [135]. Krasnykh et al. replaced the fiber and knob domains with bacteriophage T4 fibritin containing a C-terminal 6-His motif [136]. This novel Ad variant lacks the ability to interact with CAR and demonstrated up to a 100-fold increase in reporter gene expression to cells presenting an artificial 6-His binding receptor. A similar de-knobbing strategy was employed by Magnussen and colleagues, wherein an integrin-binding RGD motif was utilized, resulting in selective infection of integrin-expressing cell lines *in vitro* [137], as well as human glandular cells [138]. On the basis of feasibility of knob replacement with foreign trimerization motifs, an elegant system was devised wherein the trimeric CD40 ligand was fused to the C-terminus of a knobless fiber [139]. Notably, this vector provided CD40-specific gene delivery *in vivo* following systemic delivery [140]. Further, this vector accomplished CD40-mediated infection of human monocyte-derived dendritic

cells, suggesting possible utility for cancer vaccine "antigen loading" approaches. In addition, Ad vectors simultaneously incorporating multiple fiber types with distinct receptor specificities have been proposed [141,142].

Modification of Ad tropism via antibody-based adapter molecules has demonstrated considerable target specificity, as discussed above. On this basis, the development of single-component Ad vectors with genetically incorporated recombinant antibodies, antibody-derived moieties, or other multidomain ligands has been a long-standing field milestone. However, genetic incorporation of any moiety into the Ad capsid requires that the peptide be compatible with the nonreducing environment within the cytosol and nucleus, wherein Ad capsid proteins are translated and assembled. Capsid incorporation of several classes of complex targeting ligands, including single-chain antibodies (scFv) and growth factors, has been severely hampered by the innate biosynthetic incompatibilities between the ligand and Ad capsid proteins, resulting in unstable or insoluble ligands and reduced Ad replication [143]. On this basis, rational development of complex ligands with cytoplasmic solubility and stability will be required for their application to Ad vectors. Exemplifying this concept, Hedley and colleagues developed a single-component antibody targeted Ad vector by incorporating a cytoplasmically stable scFv into a de-knobbed fiber [144]. This vector targeted its cognate epitope expressed at the cell surface, suggesting that cytoplasmic stability of the targeting molecule, per se, allows retention of antigen recognition in the context of genetic capsid incorporation. In another landmark report, Ulasov et al. created an Ad vector targeted to a glioma-associated receptor, the α2 chain of the interleukin-13 receptor (IL-13Rα2), by fusing the full length human IL-13 cytokine was fused to the C-terminus of a de-knobbed fiber containing the T4 fibritin trimerization motif [145]. This targeted vector provided IL-13Rα2-specific gene delivery to passaged and primary human glioma cells *in vitro* and in a xenograft model *in vivo*.

4.4.3 AD TARGETING VIA GENETIC MODIFICATION: OTHER CAPSID LOCALES

The field-wide appreciation of the difficulty of incorporating complex ligands into the Ad fiber has prompted the identification of other capsid proteins amenable to genetic ligand incorporation. These approaches have the potential, through increased capsid valency and unique capsid microenvironments, to incorporate an increased number of ligands per virion. To date, the hexon capsid protein as well as minor capsid protein pIX have been used as platforms for incorporation of heterologous peptides.

4.4.3.1 Hexon

Hexon is the largest (952 amino acids) and most abundant capsid protein, and as such is an attractive locale for peptide ligand incorporation due to both its surface exposure and high valency (240 hexon homotrimers per virion). The primary sequence of the hexon monomer is highly conserved among human serotypes with exception of nine nonconserved hypervariable regions (HVRs) of unknown function found mainly within solvent-exposed loops at the surface [8,146,147]. In this regard, Vigne and coworkers exploited hexon hypervariable region 5 (HVR5), a loop structure in hexon, as a site for incorporation of an integrin-binding RGD motif. Notably, virion stability was unaffected by the addition of the foreign peptide, while providing enhanced, fiber-independent transduction to low-CAR vascular smooth muscle cells [148]. Extending this work, Wu et al. identified HVRs 2, 3, and 5–7 as hexon sites tolerating 6-histidine (6-His) motifs without adverse affects to virion formation or stability [149]. Six-His motifs in HVRs 2 and 5 mediated virion binding to anti-6-His antibodies; however, 6-His-mediated viral infection of cells was not observed, in contrast to the findings of Vigne et al. as mentioned earlier, highlighting the importance of the nature and the length of the incorporated peptides. While not related to Ad vector targeting per se, it bears mentioning that genetic alterations of hexon HVRs have been used as a method for evading host neutralizing antibodies (NAbs) [150], as most NAbs are directed against hexon [151].

4.4.3.2 Protein IX

Protein IX (pIX) has emerged as a versatile capsid locale well suited for display of ligands, with utility for both targeting and imaging modalities (reviewed in Ref. [152]). The 14.3 kDa pIX is the smallest of the minor capsid proteins, a subset of capsid proteins that generally function to stabilize the capsid shell. In the mature Ad virion, 80 pIX homotrimers [153] stabilize hexon–hexon interactions during capsid assembly, and it is therefore termed a "cement" protein. Indeed, virions deleted for pIX have decreased thermostability and a DNA capacity that is approximately 2 kb less than the normal length [154–156]. Interest in employing pIX as a capsid site for incorporation of peptide ligands stemmed from the observation that the C-terminus of pIX is located at the capsid surface [157,158], which prompted several groups to explore the fusion of several polypeptides to this terminus.

Dmitriev and colleagues first reported the incorporation of functional targeting peptides at the pIX C-terminus by inserting polylysine or FLAG motifs, resulting in augmented, CAR-independent gene transfer via binding to cellular heparan sulfate moieties [159]. Similarly, Vellinga and colleagues fused an integrin-binding RGD peptide to pIX, and used α-helical spacers (up to 7.5 nm in length and 113 amino acids) to extend the RGD motif away from the virion surface [160].

Increased gene transfer to CAR-deficient endothelioma cells was observed with increased spacer length, giving support to the notion that the pIX C-terminus may reside in a cavity formed by surrounding hexons. In a proof-of-principle study, this same group fused a hyperstable scFv directed against β-galactosidase onto a 7.5 nm pIX C-terminal extension [161]. Of note, the scFv was functional in this pIX structural context, as evidenced by binding of Ad particles to β-galactosidase protein in vitro.

To evaluate the utility of multiple ligand types at the pIX capsid locale, Campos and coworkers fused to the pIX-C-terminus to a 71-amino acid fragment of the *Propionibacterium shermanii* 1.3S transcarboxylase protein, which functions as a biotin acceptor peptide (BAP) [162]. During virus propagation the BAP is metabolically biotinylated, rendering this virus compatible with a host of avidin-tagged ligands including peptides, antibodies, and carbohydrates. Importantly, it was noted in this study that coupling transferrin to virions via pIX-BAP resulted in specific transferrin receptor-mediated infection of C2C12 cells, but the use of an antibody directed against the transferrin receptor (CD71) did not. This dichotomy was not observed when these two ligands successfully redirected Ad tropism when incorporated into fiber. The authors speculate the difference was not due to a lack of receptor recognition by the pIX-anti-CD71 complex, but rather a difference between the dissociation of targeted fiber and pIX from the Ad particle in endosomes, resulting in trapping of the pIX-anti-CD71 variant, but not the fiber-anti-CD71 in the endosome. If this notion is fully validated, it will represent a key finding showing that the nature of pIX-incorporated ligands may influence successful redirection of Ad infection.

pIX has also been used for the display of relatively large imaging molecules such as C-terminal fusions. While Ad vector imaging is beyond the scope of this chapter, the successful incorporation of the 240 amino acid enhanced green fluorescent protein (eGFP) into pIX bears mentioning because of the size and complexity of this fusion. Of note, the presence of the pIX–eGPF fusion in purified Ad virions did not appreciably decrease virus viability or capsid stability, and has allowed monitoring of Ad localization in vitro and in vivo [163,164]. Other Ad vectors have been reported that harbor complex imaging polypeptides at the pIX C-terminus, such as herpes simplex virus type 1 thymidine kinase [165], a firefly luciferase–thymidine kinase fusion [166], and monomeric red fluorescent protein [167]. As a whole, these studies have established pIX as a highly relevant capsid locale marked by the highest structural compatibility with diverse targeting and imaging ligands observed to date. On the basis of its surface capsid position, pIIIa has been proposed as a platform for ligand display for modification of Ad tropism [168,169]. The minor capsid protein pIIIa is a 67 kDa monomer that is cleaved at the C-terminus during maturation of the virion, giving rise to the final 63.5 kDa form. Initial high-resolution imaging studies originally indicated that pIIIa is an elongated protein penetrating the capsid and is located along the icosahedral edges of the virion [15]. However, later structural studies performed at higher resolution place pIIIa below the outer surface of the virion, likely precluding its utility for foreign peptide display [170].

In the aggregate, these results highlight that genetic manipulation of a variety of Ad capsid proteins is currently feasible, and has brought to fruition several novel targeting and imaging paradigms. These successes confirm a level of capsid flexibility that was largely unexpected. There remains, however, an ongoing struggle to identify true targeting ligands that are structurally and biosynthetically compatible with Ad capsid formation and stability.

4.5 TRANSCRIPTIONAL TARGETING: USE OF TISSUE-SPECIFIC PROMOTERS

Transductional targeting approaches attempt to restrict Ad vector entry into target cells. In contrast, transcriptional targeting does not change vector tropism but instead restricts gene expression to target tissue by placing virally encoded transgenes under the control of cellular promoters that are specifically active or overactive in the target tissue. The ideal TSP element would exhibit the widest differential between "target on/liver off" expression profiles, key to ablation of liver toxicity from any ectopically localized vector. One of the first tissue-specific promoters (TSPs) explored for cancer was the CEA promoter, expressed in most gastric, pancreatic, and lung cancers [171]. For hepatomas, the promoter of the α-fetoprotein (AFP) has been investigated [172,173], while the L-plastin promoter (LP-P) was used in ovarian and breast cancer cell lines [174,175]. Other promoters tested for ovarian cancer include DF3 [176], Cox2 and midkine [177], mesothelin [131], and the secretory leukoprotease inhibitor (SLPI) promoter [178,179] as well as the ovarian-specific promoter-1 [180]. The Cox-2 promoter has also been explored in the context of gastric carcinomas [181]. The CXCR4 promoter was shown to have general transcriptional selectivity for cancer cells [182], breast cancer cell lines, and primary cells [183] as well as renal cancer cell lines [184]. The survivin promoter has emerged as a potentially powerful element for transcriptional control in glioma and melanoma, and exhibits a "liver-off" activity profile [185,186].

Osteocalcin (OC) is a bone protein expressed in osteotropic tumors and differentiated osteoblasts, as well as numerous solid tumors, including osteosarcoma and prostate cancer [187,188]. An Ad utilizing the OC promoter to drive *HSV-TK* expression in prostate cancer cells resulted in destruction of tumor cells *in vitro* and in subcutaneous or bone tumor xenografts. Interestingly, tissue-specific toxicity was seen in bone metastases. OC is expressed in osteoblastic lesions, thus offering the possibility for cotargeting of regenerating bone and tumor. Further, many types of cancers metastatic to the bone could be amenable to treatment.

The SLPI gene is expressed in several different carcinomas, including ovarian cancer. Its expression in normal organs, such as the liver, is low. Therefore, the SLPI promoter was utilized to drive transgene expression in ovarian cancer cell lines and primary tumor cells isolated from patient samples [178]. The promoter was activated in both cell lines and primary tumor cells in an Ad context *in vitro*. A murine orthotopic model of peritoneally disseminated ovarian cancer was used to demonstrate high tumor gene expression versus low liver expression with the SLPI promoter, and that Ad-delivered *HSV-TK* under the control of the SLPI promoter was able to increase survival.

4.5.1 VASCULATURE-SPECIFIC PROMOTERS

The vascular endothelium is an attractive therapeutic target for cancer, vascular, and cardiopulmonary diseases, and is easily accessible to intravascular vector administration. The vascular endothelial growth factor receptor type-1 (flt-1) promoter has been used to restrict transgene expression to the endothelium in rodents while reducing liver gene expression [83,84,189]. E-selectin expression is minimal in normal blood vessels but high in the capillaries of tumors, and the promoter was used for driving gene expression in an Ad. Upon infection, EC lines expressed high levels of reporter gene expression, while non-EC lines showed low expression. The addition of TNF-α, an inducer of the promoter, further increased E-selectin's activity [190]. The murine pre-proendothelin-1 (PPE-1) promoter was also used as a TSP for adenoviral-mediated delivery to EC cells. Systemic administration to lung tumor-bearing mice resulted in gene expression in the new vasculature of primary tumors [191]. The Roundabout-4 receptor (Robo4) is a transmembrane receptor with endothelial-specific expression in embryonic vasculature as well as in the adult [192,193]. On this basis, the Robo4 promoter has been employed for transcriptional targeting of Ad gene expression to the endothelium [194].

4.5.2 TREATMENT-RESPONSIVE PROMOTERS

Another strategy for cancer gene therapy involves regulating gene expression with another form of treatment, such as radiation (reviewed in [195]). For example, the early growth response gene-1 (EGR-1) promoter, which is radiation inducible, has been used as a TSP for the specific expression of *lacZ* and *HSV-TK* in glioma and hepatocellular carcinoma cells. Radiation-induced transcription of EGR-1 in these cells was accomplished with relatively low doses [196]. Another approach for dynamically controlling promoter expression involves chemically inducible promoters. For example, a tetracycline-activated promoter can be used to regulate gene expression and subsequent protein production by giving oral tetracycline [197]. Withdrawal of the drug rapidly abrogates gene expression. Similarly, mifepristone- (RU486) and tamoxifen-inducible systems for controlled transgene expression have been developed [198,199].

4.6 DOUBLE-TARGETED Ad VECTORS

Transductional and transcriptional targeting can be combined to create double targeted viruses. Conceivably, this approach could be synergistic with regard to safety and efficacy. Initial proof-of-concept was achieved by using the vasculature-specific flt-1 promoter and a lung endothelium-targeted adapter strategy [189]. Impressively, the tumor-to-liver ratio of gene expression was increased 300,000-fold when both targeting modalities were utilized. Also, double targeting for ovarian cancer has been achieved *in vitro* and *in vivo* [179]. Transductional targeting with a sCAR-fibritin-antiErbB2-sFv adapter was able to increase gene transfer to target cells while reducing transduction of nontarget cells. When combined with transcriptional targeting with the SLPI promoter, an increase in selectivity was seen. Also, the transductional targeting increased the level of SLPI-mediated transgene expression in target cells, thereby compensating for the lower gene expression typically seen with TSPs.

4.7 TRANSCRIPTIONAL AND TRANSDUCTIONAL TARGETING OF CONDITIONALLY REPLICATING Ad

Although nonreplicating first generation Ad vectors have provided high *in vitro* and *in vivo* transduction rates and good safety data, clinical cancer trials have suggested that the single agent antitumor effect may not be sufficient for all treatment approaches [40]. As early as the 1950s, the use of viruses that replicate and spread specifically inside tumors has been suggested as a way to improve tumor penetration with an additional benefit of local amplification of effect [200].

To this end, conditionally replicative adenovirus (CRAd) agents have been explored. These viruses are genetically modified to take advantage of tumor-specific changes that allow preferential replication of the virus in target cells [41,201,202]. The Ad replication cycle causes oncolysis of the cell, resulting in the release of newly generated virions and subsequent infection of neighboring cells. Thus, the antitumor effect is not delivered with a transgene but by replication of the virus per se. In theory, the oncolytic process continues as long as target cells for the virus persist. There are two main ways to control viral replication. One method is the control of replication regulators, such as the viral early genes, with tumor-specific promoter elements. The other method involves introduction of deletions in the viral genome that requires specific cellular factors to compensate the effects of these deletions. Further, both approaches can be combined with the potential for increased specificity.

Numerous tumor-associated promoters have been used to control viral replication (reviewed in Refs. [120,203]). Typically, the promoter element is placed to control expression of *E1A*, the crucial regulator of Ad replication, sometimes combined with other genes such as *E1B* or *E4*. An interesting concept is targeting CRAds to tumor vasculature [204]. However, this strategy is more challenging to study preclinically, as animal models are unavailable; ECs derive from the host in xenograft systems and murine cells do not support replication of human Ads. To further increase the oncolytic effect, transgenes for cytokines or prodrug-activating enzymes have been included in CRAds [205,206]. This approach may also allow noninvasive imaging and abrogation of virus replication in case of toxicity.

Heretofore, two approaches have been utilized for creation of deletion-type CRAds. The first was ONYX-015 (initially reported as *dl*1520), which has two mutations in the gene coding for the E1B 55-kDa protein [207,208]. The purpose of this protein is binding and inactivation of p53 in infected cells, for induction of S-phase, which is required for virus replication. Thus, this virus should only replicate in cells with an aberrant p53-p14ARF pathway, a common

feature in human tumors [209]. While this is still a subject of debate, initial studies suggested that this agent replicates more effectively in tumor than in normal cells [210–213]. Unfortunately, the function of E1B55 kD is not limited to p53 binding, which causes inefficient replication compared to wild-type Ad [207,208,214].

The second group of deletion mutant CRAds have a 24 bp deletion in the constant region (CR) 2 of *E1A* [215,216]. This domain of the E1A protein is responsible for binding the retinoblastoma tumor suppressor/cell cycle regulator protein (Rb), thereby allowing Ad to induce S-phase entry. Therefore, viruses with this type of deletion have reduced ability to overcome the G1–S checkpoint and replicate efficiently only in cells where this interaction is not necessary, e.g., tumor cells defective in the *Rb-p16* pathway. Appropriately, this pathway seems to be inactive in most human tumors [217]. It has been shown that replication of CR2-deleted viruses is attenuated in nonproliferating normal cells [215,216]. Importantly, abrogation of replication was also demonstrated when Rb was reintroduced into otherwise permissive cells [215]. Ads with mutations in both CR1 and CR2 of *E1A* have also been found to replicate selectively in tumor cells although increases in comparison to just CR2-deleted CRAds have not yet been demonstrated [218–221].

As Ad5-based vectors, CRAds have native CAR-dependent tropism. Unfortunately, CAR expression is low or variable in many types of clinical cancers [30–32,37–39,91]. Nevertheless, even CRAds with wild-type tropism have shown evidence of clinical utility [222–224]. These initial successes suggested that if efficiency of infection and specificity of replication of these agents could be enhanced, significant improvements in clinical efficacy could be gained. This was corroborated by a clear demonstration of the relationship between infectivity and oncolytic potency [225–227]. Consequently, genetic capsid modifications that allow CAR-independent or enhanced infectivity in nonreplicative Ads have been applied to CRAds, resulting in impressive gains in preclinical efficacy. For example, Ad5-Δ24RGD features the integrin-binding RGD-4C peptide displayed from the HI-loop of the Ad5 fiber [228], and displays similar oncolytic potency to wild-type virus in ovarian cancer cells. Further, this virus is able to replicate in ovarian cancer primary cell spheroids and results in significantly prolonged survival in an orthotopic model of ovarian [229] and lung cancer [230]. In addition, Ad5-Δ24RGD has also been evaluated against osteosarcoma substrates [231,232].

On the basis of superior infectivity of Ad vectors containing human serotype 3 fiber knob domains in ovarian substrates [89,233], a Δ24-based CRAd featuring the serotype 3 knob (Ad5/3-Δ24) was created [234]. This agent demonstrated dramatic antitumor efficacy in ovarian cancer cell lines, primary tumor specimens, and in orthotopic animal models of ovarian cancer. The first tumor-associated promoter-controlled infectivity-enhanced CRAd has been constructed and evaluated on ovarian, pancreatic, and gastric cancer substrates [235–237].

A major problem in assessing preclinical CRAd efficacy and safety is a severely limited number of appropriate animal models. Human Ads (including CRAds) do not replicate productively in commonly used animal models. Therefore, meaningful safety data are difficult to obtain, and efficacy data is likely skewed because of deficient immune responses in xenograft models. Further, evaluation of host–virus interactions and their modulation has not been possible. To address this field-wide deficiency, Hemminki and colleagues have proposed the development of syngeneic CRAds for use in animal models of cancer [238]. To allow the evaluation of Ad5-based CRAds, the cotton rat has been put forth as a model [239,240]. Since this rodent is semipermissive for human Ad replication, this immunocompetent tumor model allows investigation of host immune system effects on the CRAd–tumor interaction as well as the effects on normal host cells *in vivo*. Another promising model system for CRAd evaluation is the Syrian hamster. Human Ads replicate well in Syrian hamster cell lines *in vitro*, and demonstrate significant antitumor efficacy following injection into Syrian hamster tumors *in vivo* [241].

4.8   CLINICAL TRIALS WITH MODIFIED Ad

No clinical trials have yet been completed with targeted Ads. Nevertheless, some trials are in progress. For example, the first transductionally targeted CRAd trial with Ad5-Δ24RGD has received funding from the National Cancer Institute Rapid Access to Intervention Development program for vector production costs and has undergone full preclinical toxicological analysis in cotton rats [242]. As of this writing, a Phase I clinical trial is under way at the University of Alabama at Birmingham for recurrent epithelial ovarian cancer (Ronald D. Alvarez, M.D., personal communication).

4.9   FUTURE DIRECTIONS

Despite the advances in Ad-based gene therapy vectors described in this chapter, several obstacles remain. The vascular endothelial wall is a significant physical barrier prohibiting access of systemically administered vectors to tumors and other target tissues. To overcome this obstacle, strategies need to be developed to route Ad vectors via transcytosis pathways through the endothelium. As an example, Zhu et al. redirected Ad vectors to the transcytosing transferrin receptor pathway, using the bifunctional adapter molecule [243]. The transcytosed Ad virions retained the ability to infect cells, establishing the feasibility of this approach. However, efficiency of Ad trafficking via this pathway is poor, and current efforts are directed toward exploring other transcytosing pathways such as the melanotransferrin pathway [244], the poly immunoglobulin A receptor pathway [245], or caveolae-mediated transcytosis pathways [246]. One can envision the development of mosaic Ad vectors incorporating both targeting ligands directed to such transcytosis pathways as well as ligands mediating subsequent targeting and infection of target cells present beyond the vascular wall.

4.10   CONCLUSIONS

Ad-based vectors are the most widely used platform for gene delivery because of their efficacy in gene expression in dividing and nondividing cells. They are of particular utility for cancer gene therapy applications, where temporary gene expression is acceptable or even beneficial. The history of Ad-based gene therapy studies clearly illustrates and confirms the critical linkage between improved vector design and improvement in therapeutic potential. Indeed, clinical breakthroughs have been dependent on advances in vector development. With regard to Ad-mediated cancer treatment, high-level tumor transduction remains a key developmental hurdle. To this end, both Ad and CRAd vectors possessing infectivity enhancement and targeting capabilities should be evaluated in the most stringent model systems possible. Advanced Ad-based vectors with imaging, targeting, and therapeutic capabilities have yet to be fully employed; however, the feasibilities leading to this accomplishment are now being established.

ACKNOWLEDGMENTS

This work was supported by a research grant from Actelion Pharmaceuticals (J.N.G.); by the EU FP6 THERADPOX and APOTHERAPY, HUCH research funds (EVO), by the Sigrid Juselius Foundation, Academy of Finland, the Emil Aaltonen Foundation, and the Finnish Cancer Society (A.H.); by the U.S. National Institutes of Health grant 5P01 CA104177, and grant W81XWH-05-1-0035 from the U.S. Department of Defense (D.T.C.).

REFERENCES

1. Clinicals Trials Database. The Journal of Gene Medicine Clinical Trials Database. http://www.wiley.co.uk/genmed/clinical, 2007.
2. Benkö M, Harrach B, Both G et al. Family *Adenoviridae*. In: C Fauquet; MA Mayo; J Maniloff; U Desselberger; L Ball, editors. *Virus Taxonomy. VIIIth Report of the International Committee on Taxonomy of Viruses*. New York: Elsevier; 2005; pp. 213–228.
3. Davison AJ, Benko M, and Harrach B. Genetic content and evolution of adenoviruses. *J Gen Virol* 2003; 84: 2895–2908.
4. Mei YF and Wadell G. Epitopes and hemagglutination binding domain on subgenus B:2 adenovirus fibers. *J Virol* 1996; 70: 3688–3697.
5. Shenk T. Adenoviridae: and their replication. In: B Fields; P Howley; D Knipe, editors. *Virology*. New York: Raven Press; 1996; pp. 2111–2148.
6. Jones MS, 2nd, Harrach B, Ganac RD et al. New adenovirus species found in a patient presenting with gastroenteritis. *J Virol* 2007; 81: 5978–5984.
7. Stewart PL, Burnett RM. Adenovirus structure by X-ray crystallography and electron microscopy. *Curr Top Microbiol Immunol* 1995; 199 (Pt 1): 25–38.
8. Rux JJ, Kuser PR, and Burnett RM. Structural and phylogenetic analysis of adenovirus hexons by use of high-resolution x-ray crystallographic, molecular modeling, and sequence-based methods. *J Virol* 2003; 77: 9553–9566.
9. Roberts MM, White JL, Grutter MG et al. Three-dimensional structure of the adenovirus major coat protein hexon. *Science* 1986; 232: 1148–1151.

10. Pichla-Gollon SL, Drinker M, Zhou X et al. Structure-based identification of a major neutralizing site in an adenovirus hexon. *J Virol* 2007; 81: 1680–1689.

11. Rux JJ and Burnett RM. Large-scale purification and crystallization of adenovirus hexon. *Methods Mol Med* 2007; 131: 231–250.

12. Zubieta C, Schoehn G, Chroboczek J et al. The structure of the human adenovirus 2 penton. *Mol Cell* 2005; 17: 121–135.

13. van Raaij MJ, Louis N, Chroboczek J et al. Structure of the human adenovirus serotype 2 fiber head domain at 1.5 A resolution. *Virology* 1999; 262: 333–343.

14. van Raaij MJ, Mitraki A, Lavigne G et al. A triple beta-spiral in the adenovirus fibre shaft reveals a new structural motif for a fibrous protein. *Nature* 1999; 401: 935–938.

15. Stewart PL, Fuller SD, and Burnett RM. Difference imaging of adenovirus: bridging the resolution gap between X-ray crystallography and electron microscopy. *Embo J* 1993; 12: 2589–2599.

16. Fabry CM, Rosa-Calatrava M, Conway JF et al. A quasi-atomic model of human adenovirus type 5 capsid. *Embo J* 2005; 24: 1645–1654.

17. Tomko RP, Xu R, and Philipson L. HCAR and MCAR: the human and mouse cellular receptors for subgroup C adenoviruses and group B coxsackieviruses. *Proc Natl Acad Sci U S A* 1997; 94: 3352–3356.

18. Bergelson JM, Cunningham JA, Droguett G et al. Isolation of a common receptor for Coxsackie B viruses and adenoviruses 2 and 5. *Science* 1997; 275: 1320–1323.

19. Dechecchi MC, Melotti P, Bonizzato A et al. Heparan sulfate glycosaminoglycans are receptors sufficient to mediate the initial binding of adenovirus types 2 and 5. *J Virol* 2001; 75: 8772–8780.

20. Dechecchi MC, Tamanini A, Bonizzato A et al. Heparan sulfate glycosaminoglycans are involved in adenovirus type 5 and 2-host cell interactions. *Virology* 2000; 268: 382–390.

21. Hong SS, Karayan L, Tournier J et al. Adenovirus type 5 fiber knob binds to MHC class I alpha2 domain at the surface of human epithelial and B lymphoblastoid cells. *Embo J* 1997; 16: 2294–2306.

22. Chu Y, Heistad D, Cybulsky MI et al. Vascular cell adhesion molecule-1 augments adenovirus-mediated gene transfer. *Arterioscler Thromb Vasc Biol* 2001; 21: 238–242.

23. Wickham TJ, Mathias P, Cheresh DA et al. Integrins alpha v beta 3 and alpha v beta 5 promote adenovirus internalization but not virus attachment. *Cell* 1993; 73: 309–319.

24. Li E, Brown SL, Stupack DG et al. Integrin alpha(v)beta1 is an adenovirus coreceptor. *J Virol* 2001; 75: 5405–5409.

25. Davison E, Diaz RM, Hart IR et al. Integrin alpha5beta1-mediated adenovirus infection is enhanced by the integrin-activating antibody TS2/16. *J Virol* 1997; 71: 6204–6207.

26. Meier O, Boucke K, Hammer SV et al. Adenovirus triggers macropinocytosis and endosomal leakage together with its clathrin-mediated uptake. *J Cell Biol* 2002; 158: 1119–1131.

27. Trotman LC, Mosberger N, Fornerod M et al. Import of adenovirus DNA involves the nuclear pore complex receptor CAN/Nup214 and histone H1. *Nat Cell Biol* 2001; 3: 1092–1100.

28. Saphire AC, Guan T, Schirmer EC et al. Nuclear import of adenovirus DNA in vitro involves the nuclear protein import pathway and hsc70. *J Biol Chem* 2000; 275: 4298–4304.

29. Hindley CE, Lawrence FJ, and Matthews DA. A role for transportin in the nuclear import of adenovirus core proteins and DNA. *Traffic* 2007; 8: 1313–1322.

30. Li Y, Pong RC, Bergelson JM et al. Loss of adenoviral receptor expression in human bladder cancer cells: a potential impact on the efficacy of gene therapy. *Cancer Res* 1999; 59: 325–330.

31. Miller CR, Buchsbaum DJ, Reynolds PN et al. Differential susceptibility of primary and established human glioma cells to adenovirus infection: targeting via the epidermal growth factor receptor achieves fiber receptor-independent gene transfer. *Cancer Res* 1998; 58: 5738–5748.

32. Cripe TP, Dunphy EJ, Holub AD et al. Fiber knob modifications overcome low, heterogeneous expression of the coxsackievirus-adenovirus receptor that limits adenovirus gene transfer and oncolysis for human rhabdomyosarcoma cells. *Cancer Res* 2001; 61: 2953–2960.

33. Dmitriev I, Krasnykh V, Miller CR et al. An adenovirus vector with genetically modified fibers demonstrates expanded tropism via utilization of a coxsackievirus and adenovirus receptor-independent cell entry mechanism. *J Virol* 1998; 72: 9706–9713.

34. Kelly FJ, Miller CR, Buchsbaum DJ et al. Selectivity of TAG-72-targeted adenovirus gene transfer to primary ovarian carcinoma cells versus autologous mesothelial cells in vitro. *Clin Cancer Res* 2000; 6: 4323–4333.

35. Vanderkwaak TJ, Wang M, Gomez-Navarro J et al. An advanced generation of adenoviral vectors selectively enhances gene transfer for ovarian cancer gene therapy approaches. *Gynecol Oncol* 1999; 74: 227–234.

36. Kasono K, Blackwell JL, Douglas JT et al. Selective gene delivery to head and neck cancer cells via an integrin targeted adenoviral vector. *Clin Cancer Res* 1999; 5: 2571–2579.

37. Okegawa T, Li Y, Pong RC et al. The dual impact of coxsackie and adenovirus receptor expression on human prostate cancer gene therapy. *Cancer Res* 2000; 60: 5031–5036.

38. Okegawa T, Pong RC, Li Y et al. The mechanism of the growth-inhibitory effect of coxsackie and adenovirus receptor (CAR) on human bladder cancer: a functional analysis of CAR protein structure. *Cancer Res* 2001; 61: 6592–6600.

39. Shayakhmetov DM, Li ZY, Ni S et al. Targeting of adenovirus vectors to tumor cells does not enable efficient transduction of breast cancer metastases. *Cancer Res* 2002; 62: 1063–1068.

40. Hemminki A and Alvarez RD. Adenoviruses in oncology: A viable option? *BioDrugs* 2002; 16: 77–87.

41. Kanerva A and Hemminki A. Adenoviruses for treatment of cancer. *Ann Med* 2005; 37: 33–43.

42. Cohen CJ, Shieh JT, Pickles RJ et al. The coxsackievirus and adenovirus receptor is a transmembrane component of the tight junction. *Proc Natl Acad Sci U S A* 2001; 98: 15191–15196.

43. Walters RW, Freimuth P, Moninger TO et al. Adenovirus fiber disrupts CAR-mediated intercellular adhesion allowing virus escape. *Cell* 2002; 110: 789–799.

44. Honda T, Saitoh H, Masuko M et al. The coxsackievirus-adenovirus receptor protein as a cell adhesion molecule in the developing mouse brain. *Brain Res Mol Brain Res* 2000; 77: 19–28.

45. Anders M, Christian C, McMahon M et al. Inhibition of the Raf/MEK/ERK pathway up-regulates expression of the coxsackievirus and adenovirus receptor in cancer cells. *Cancer Res* 2003; 63: 2088–2095.

46. Fechner H, Haack A, Wang H et al. Expression of coxsackie adenovirus receptor and alphav-integrin does not correlate with adenovector targeting in vivo indicating anatomical vector barriers. *Gene Ther* 1999; 6: 1520–1535.

47. Wood M, Perrotte P, Onishi E et al. Biodistribution of an adenoviral vector carrying the luciferase reporter gene following intravesical or intravenous administration to a mouse. *Cancer Gene Ther* 1999; 6: 367–372.

48. Reynolds P, Dmitriev I, and Curiel D. Insertion of an RGD motif into the HI loop of adenovirus fiber protein alters the distribution of transgene expression of the systemically administered vector. *Gene Ther* 1999; 6: 1336–1339.

49. Stone D, Liu Y, Shayakhmetov D et al. Adenovirus-platelet interaction in blood causes virus sequestration to the reticuloendothelial system of the liver. *J Virol* 2007; 81: 4866–4871.

50. Tao N, Gao GP, Parr M et al. Sequestration of adenoviral vector by Kupffer cells leads to a nonlinear dose response of transduction in liver. *Mol Ther* 2001; 3: 28–35.

51. Connelly S. Adenoviral vectors for liver-directed gene therapy. *Curr Opin Mol Ther* 1999; 1: 565–572.

52. Lieber A, He CY, Meuse L et al. The role of Kupffer cell activation and viral gene expression in early liver toxicity after infusion of recombinant adenovirus vectors. *J Virol* 1997; 71: 8798–8807.

53. Peeters MJ, Patijn GA, Lieber A et al. Adenovirus-mediated hepatic gene transfer in mice: comparison of intravascular and biliary administration. *Hum Gene Ther* 1996; 7: 1693–1699.

54. Alemany R, Suzuki K, and Curiel DT. Blood clearance rates of adenovirus type 5 in mice. *J Gen Virol* 2000a; 81: 2605–2609.

55. Worgall S, Wolff G, Falck-Pedersen E et al. Innate immune mechanisms dominate elimination of adenoviral vectors following in vivo administration. *Hum Gene Ther* 1997; 8: 37–44.

56. Alemany R and Curiel DT. CAR-binding ablation does not change biodistribution and toxicity of adenoviral vectors. *Gene Ther* 2001; 8: 1347–1353.

57. Smith TA, Idamakanti N, Marshall-Neff J et al. Receptor interactions involved in adenoviral-mediated gene delivery after systemic administration in non-human primates. *Hum Gene Ther* 2003; 14: 1595–1604.

58. Smith T, Idamakanti N, Kylefjord H et al. In vivo hepatic adenoviral gene delivery occurs independently of the coxsackievirus-adenovirus receptor. *Mol Ther* 2002; 5: 770–779.

59. Martin K, Brie A, Saulnier P et al. Simultaneous CAR- and alpha V integrin-binding ablation fails to reduce Ad5 liver tropism. *Mol Ther* 2003; 8: 485–494.

60. Mizuguchi H, Koizumi N, Hosono T et al. CAR- or alphav integrin-binding ablated adenovirus vectors, but not fiber-modified vectors containing RGD peptide, do not change the systemic gene transfer properties in mice. *Gene Ther* 2002; 9: 769–776.

61. Nicklin SA, Wu E, Nemerow GR et al. The influence of adenovirus fiber structure and function on vector development for gene therapy. *Mol Ther* 2005; 12: 384–393.

62. Vigne E, Dedieu JF, Brie A et al. Genetic manipulations of adenovirus type 5 fiber resulting in liver tropism attenuation. *Gene Ther* 2003; 10: 153–162.

63. Breidenbach M, Rein DT, Wang M et al. Genetic replacement of the adenovirus shaft fiber reduces liver tropism in ovarian cancer gene therapy. *Hum Gene Ther* 2004; 15: 509–518.

64. Smith TA, Idamakanti N, Rollence ML et al. Adenovirus serotype 5 fiber shaft influences in vivo gene transfer in mice. *Hum Gene Ther* 2003; 14: 777–787.

65. Shayakhmetov DM, Li ZY, Ni S et al. Analysis of adenovirus sequestration in the liver, transduction of hepatic cells, and innate toxicity after injection of fiber-modified vectors. *J Virol* 2004; 78: 5368–5381.

66. Baker AH, McVey JH, Waddington SN et al. The influence of blood on in vivo adenovirus bio-distribution and transduction. *Mol Ther* 2007; 15: 1410–1416.

67. Douglas JT, Rogers BE, Rosenfeld ME et al. Targeted gene delivery by tropism-modified adenoviral vectors. *Nat Biotechnol* 1996; 14: 1574–1578.

68. Goldman CK, Rogers BE, Douglas JT et al. Targeted gene delivery to Kaposi's sarcoma cells via the fibroblast growth factor receptor. *Cancer Res* 1997; 57: 1447–1451.

69. Rogers BE, Douglas JT, Ahlem C et al. Use of a novel cross-linking method to modify adenovirus tropism. *Gene Ther* 1997; 4: 1387–1392.

70. Rancourt C, Rogers BE, Sosnowski BA et al. Basic fibroblast growth factor enhancement of adenovirus-mediated delivery of the herpes simplex virus thymidine kinase gene results in augmented therapeutic benefit in a murine model of ovarian cancer. *Clin Cancer Res* 1998; 4: 2455–2461.

71. Gu DL, Gonzalez AM, Printz MA et al. Fibroblast growth factor 2 retargeted adenovirus has redirected cellular tropism: evidence for reduced toxicity and enhanced antitumor activity in mice. *Cancer Res* 1999; 59: 2608–2614.

72. Printz MA, Gonzalez AM, Cunningham M et al. Fibroblast growth factor 2-retargeted adenoviral vectors exhibit a modified biolocalization pattern and display reduced toxicity relative to native adenoviral vectors. *Hum Gene Ther* 2000; 11: 191–204.

73. Haisma HJ, Pinedo HM, Rijswijk A et al. Tumor-specific gene transfer via an adenoviral vector targeted to the pan-carcinoma antigen EpCAM. *Gene Ther* 1999; 6: 1469–1474.

74. Heideman DA, Snijders PJ, Craanen ME et al. Selective gene delivery toward gastric and esophageal adenocarcinoma cells via EpCAM-targeted adenoviral vectors. *Cancer Gene Ther* 2001; 8: 342–351.

75. Reynolds PN, Zinn KR, Gavrilyuk VD et al. A targetable, injectable adenoviral vector for selective gene delivery to pulmonary endothelium in vivo. *Mol Ther* 2000; 2: 562–578.

76. Tillman BW, de Gruijl TD, Luykx-de Bakker SA et al. Maturation of dendritic cells accompanies high-efficiency gene transfer by a CD40-targeted adenoviral vector. *J Immunol* 1999; 162: 6378–6383.

77. Hakkarainen T, Hemminki A, Pereboev AV et al. CD40 is expressed on ovarian cancer cells and can be utilized for targeting adenoviruses. *Clin Cancer Res* 2003; 9: 619–624.

78. Pereboev AV, Asiedu CK, Kawakami Y et al. Coxsackievirus-adenovirus receptor genetically fused to anti-human CD40 scFv enhances adenoviral transduction of dendritic cells. *Gene Ther* 2002; 9: 1189–1193.

79. Dmitriev I, Kashentseva E, Rogers BE et al. Ectodomain of coxsackievirus and adenovirus receptor genetically fused to epidermal growth factor mediates adenovirus targeting to epidermal growth factor receptor-positive cells. *J Virol* 2000; 74: 6875–6884.

80. Kashentseva EA, Seki T, Curiel DT et al. Adenovirus targeting to c-erbB-2 oncoprotein by single-chain antibody fused to trimeric form of adenovirus receptor ectodomain. *Cancer Res* 2002; 62: 609–616.

81. Itoh A, Okada T, Mizuguchi H et al. A soluble CAR-SCF fusion protein improves adenoviral vector-mediated gene transfer to c-Kit-positive hematopoietic cells. *J Gene Med* 2003; 5: 929–940.

82. Kim J, Smith T, Idamakanti N et al. Targeting adenoviral vectors by using the extracellular domain of the coxsackie-adenovirus receptor: improved potency via trimerization. *J Virol* 2002; 76: 1892–1903.

83. Reynolds AM, Xia W, Holmes MD et al. Bone morphogenetic protein type 2 receptor gene therapy attenuates hypoxic pulmonary hypertension. *Am J Physiol Lung Cell Mol Physiol* 2007; 292: L1182–L1192.

84. Everts M, Kim-Park SA, Preuss MA et al. Selective induction of tumor-associated antigens in murine pulmonary vasculature using double-targeted adenoviral vectors. *Gene Ther* 2005; 12: 1042–1048.

85. Li HJ, Everts M, Pereboeva L et al. Adenovirus tumor targeting and hepatic untargeting by a coxsackie/adenovirus receptor ectodomain anti-carcinoembryonic antigen bispecific adapter. *Cancer Res* 2007; 67: 5354–5361.

86. Krasnykh VN, Mikheeva GV, Douglas JT et al. Generation of recombinant adenovirus vectors with modified fibers for altering viral tropism. *J Virol* 1996; 70: 6839–6846.

87. Von Seggern DJ, Huang S, Fleck SK et al. Adenovirus vector pseudotyping in fiber-expressing cell lines: improved transduction of Epstein–Barr virus-transformed B cells. *J Virol* 2000; 74: 354–362.

88. Kanerva A, Wang M, Bauerschmitz GJ et al. Gene transfer to ovarian cancer versus normal tissues with fiber-modified adenoviruses. *Mol Ther* 2002b; 5: 695–704.

89. Kanerva A, Mikheeva GV, Krasnykh V et al. Targeting adenovirus to the serotype 3 receptor increases gene transfer efficiency to ovarian cancer cells. *Clin Cancer Res* 2002; 8: 275–280.

90. Zheng S, Ulasov IV, Han Y et al. Fiber-knob modifications enhance adenoviral tropism and gene transfer in malignant glioma. *J Gene Med* 2007; 9: 151–160.

91. Ulasov IV, Tyler MA, Zheng S et al. CD46 represents a target for adenoviral gene therapy of malignant glioma. *Hum Gene Ther* 2006; 17: 556–564.

92. Ulasov IV, Rivera AA, Han Y et al. Targeting adenovirus to CD80 and CD86 receptors increases gene transfer efficiency to malignant glioma cells. *J Neurosurg* 2007; 107: 617–627.

93. Havenga MJ, Lemckert AA, Ophorst OJ et al. Exploiting the natural diversity in adenovirus tropism for therapy and prevention of disease. *J Virol* 2002; 76: 4612–4620.

94. Gaggar A, Shayakhmetov DM, and Lieber A. CD46 is a cellular receptor for group B adenoviruses. *Nat Med* 2003; 9: 1408–1412.

95. Segerman A, Arnberg N, Erikson A et al. There are two different species B adenovirus receptors: sBAR, common to species B1 and B2 adenoviruses, and sB2AR, exclusively used by species B2 adenoviruses. *J Virol* 2003; 77: 1157–1162.

96. Segerman A, Atkinson JP, Marttila M et al. Adenovirus type 11 uses CD46 as a cellular receptor. *J Virol* 2003; 77: 9183–9191.

97. Sirena D, Lilienfeld B, Eisenhut M et al. The human membrane cofactor CD46 is a receptor for species B adenovirus serotype 3. *J Virol* 2004; 78: 4454–4462.

98. Short JJ, Pereboev AV, Kawakami Y et al. Adenovirus serotype 3 utilizes CD80 (B7.1) and CD86 (B7.2) as cellular attachment receptors. *Virology* 2004; 322: 349–359.

99. Arnberg N, Edlund K, Kidd AH et al. Adenovirus type 37 uses sialic acid as a cellular receptor. *J Virol* 2000; 74: 42–48.

100. Arnberg N, Kidd AH, Edlund K et al. Adenovirus type 37 binds to cell surface sialic acid through a charge-dependent interaction. *Virology* 2002; 302: 33–43.

101. Wu E, Trauger SA, Pache L et al. Membrane cofactor protein is a receptor for adenoviruses associated with epidemic keratoconjunctivitis. *J Virol* 2004; 78: 3897–3905.

102. Kanerva A, Zinn KR, Chaudhuri TR et al. Enhanced therapeutic efficacy for ovarian cancer with a serotype 3 receptor targeted oncolytic virus. *Mol Ther* 2003; 8: 449–458.

103. Zabner J, Chillon M, Grunst T et al. A chimeric type 2 adenovirus vector with a type 17 fiber enhances gene transfer to human airway epithelia. *J Virol* 1999; 73: 8689–8695.

104. Rea D, Havenga MJ, van Den Assem M et al. Highly efficient transduction of human monocyte-derived dendritic cells with subgroup B fiber-modified adenovirus vectors enhances transgene-encoded antigen presentation to cytotoxic T cells. *J Immunol* 2001; 166: 5236–5244.

105. Shayakhmetov DM, Papayannopoulou T, Stamatoyannopoulos G et al. Efficient gene transfer into human CD34(+) cells by a retargeted adenovirus vector. *J Virol* 2000; 74: 2567–2583.

106. Goossens PH, Havenga MJ, Pieterman E et al. Infection efficiency of type 5 adenoviral vectors in synovial tissue can be enhanced with a type 16 fiber. *Arthritis Rheum* 2001; 44: 570–577.

107. Havenga MJ, Lemckert AA, Grimbergen JM et al. Improved adenovirus vectors for infection of cardiovascular tissues. *J Virol* 2001; 75: 3335–3342.

108. Gall J, Kass-Eisler A, Leinwand L et al. Adenovirus type 5 and 7 capsid chimera: fiber replacement alters receptor tropism without affecting primary immune neutralization epitopes. *J Virol* 1996; 70: 2116–2123.

109. Chillon M, Bosch A, Zabner J et al. Group D adenoviruses infect primary central nervous system cells more efficiently than those from group C. *J Virol* 1999; 73: 2537–2540.

110. Stoff-Khalili MA, Rivera AA, Glasgow JN et al. A human adenoviral vector with a chimeric fiber from canine adenovirus type 1 results in novel expanded tropism for cancer gene therapy. *Gene Ther* 2005; 12: 1696–1706.

111. Glasgow JN, Kremer EJ, Hemminki A et al. An adenovirus vector with a chimeric fiber derived from canine adenovirus type 2 displays novel tropism. *Virology* 2004; 324: 103–116.

112. Tsuruta Y, Pereboeva L, Glasgow JN et al. Reovirus sigma1 fiber incorporated into adenovirus serotype 5 enhances infectivity via a CAR-independent pathway. *Biochem Biophys Res Commun* 2005; 335: 205–214.

113. Mercier GT, Campbell JA, Chappell JD et al. A chimeric adenovirus vector encoding reovirus attachment protein sigma1 targets cells expressing junctional adhesion molecule 1. *Proc Natl Acad Sci U S A* 2004; 101: 6188–6193.

114. Wickham TJ, Tzeng E, Shears LL, 2nd et al. Increased in vitro and in vivo gene transfer by adenovirus vectors containing chimeric fiber proteins. *J Virol* 1997; 71: 8221–8229.

115. Krasnykh V, Dmitriev I, Mikheeva G et al. Characterization of an adenovirus vector containing a heterologous peptide epitope in the HI loop of the fiber knob. *J Virol* 1998; 72: 1844–1852.

116. Belousova N, Krendelchtchikova V, Curiel DT et al. Modulation of adenovirus vector tropism via incorporation of polypeptide ligands into the fiber protein. *J Virol* 2002; 76: 8621–8631.

117. Magnusson MK, Henning P, Myhre S et al. Adenovirus 5 vector genetically re-targeted by an Affibody molecule with specificity for tumor antigen HER2/neu. *Cancer Gene Ther* 2007; 14: 468–479.

118. Hemminki A, Wang M, Desmond RA et al. Serum and ascites neutralizing antibodies in ovarian cancer patients treated with intraperitoneal adenoviral gene therapy. *Hum Gene Ther* 2002; 13: 1505–1514.

119. Hemminki A, Belousova N, Zinn KR et al. An adenovirus with enhanced infectivity mediates molecular chemotherapy of ovarian cancer cells and allows imaging of gene expression. *Mol Ther* 2001; 4: 223–231.

120. Bauerschmitz GJ, Barker SD, and Hemminki A. Adenoviral gene therapy for cancer: From vectors to targeted and replication competent agents (Review). *Int J Oncol* 2002; 21: 1161–1174.

121. Wu H, Seki T, Dmitriev I et al. Double modification of adenovirus fiber with RGD and polylysine motifs improves coxsackievirus-adenovirus receptor-independent gene transfer efficiency. *Hum Gene Ther* 2002; 13: 1647–1653.

122. Contreras JL, Wu H, Smyth CA et al. Double genetic modification of adenovirus fiber with RGD polylysine motifs significantly enhances gene transfer to isolated human pancreatic islets. *Transplantation* 2003; 76: 252–261.

123. Wu H, Han T, Lam JT et al. Preclinical evaluation of a class of infectivity-enhanced adenoviral vectors in ovarian cancer gene therapy. *Gene Ther* 2004; 11: 874–878.

124. Rein DT, Breidenbach M, Wu H et al. Gene transfer to cervical cancer with fiber-modified adenoviruses. *Int J Cancer* 2004; 111: 698–704.

125. Nicklin SA, White SJ, Watkins SJ et al. Selective targeting of gene transfer to vascular endothelial cells by use of peptides isolated by phage display. *Circulation* 2000; 102: 231–237.

126. Nicklin SA, Dishart KL, Buening H et al. Transductional and transcriptional targeting of cancer cells using genetically engineered viral vectors. *Cancer Lett* 2003; 201: 165–173.

127. Henning P, Andersson KM, Frykholm K et al. Tumor cell targeted gene delivery by adenovirus 5 vectors carrying knobless fibers with antibody-binding domains. *Gene Ther* 2005; 12: 211–224.

128. Volpers C, Thirion C, Biermann V et al. Antibody-mediated targeting of an adenovirus vector modified to contain a synthetic immunoglobulin g-binding domain in the capsid. *J Virol* 2003; 77: 2093–2104.

129. Korokhov N, Mikheeva G, Krendelshchikov A et al. Targeting of adenovirus via genetic modification of the viral capsid combined with a protein bridge. *J Virol* 2003; 77: 12931–12940.

130. Korokhov N, de Gruijl TD, Aldrich WA et al. High efficiency transduction of dendritic cells by adenoviral vectors targeted to DC-SIGN. *Cancer Biol Ther* 2005; 4: 289–294.

131. Breidenbach M, Rein DT, Everts M et al. Mesothelin-mediated targeting of adenoviral vectors for ovarian cancer gene therapy. *Gene Ther* 2005; 12: 187–193.

132. Balyasnikova IV, Metzger R, Visintine DJ et al. Selective rat lung endothelial targeting with a new set of monoclonal antibodies to angiotensin I-converting enzyme. *Pulm Pharmacol Ther* 2005; 18: 251–267.

133. Von Seggern DJ, Chiu CY, Fleck SK et al. A helper-independent adenovirus vector with E1, E3, and fiber deleted: structure and infectivity of fiberless particles. *J Virol* 1999; 73: 1601–1608.

134. Falgout B and Ketner G. Characterization of adenovirus particles made by deletion mutants lacking the fiber gene. *J Virol* 1988; 62: 622–625.

135. Papanikolopoulou K, Forge V, Goeltz P et al. Formation of highly stable chimeric trimers by fusion of an adenovirus fiber shaft fragment with the foldon domain of bacteriophage t4 fibritin. *J Biol Chem* 2004; 279: 8991–8998.

136. Krasnykh V, Belousova N, Korokhov N et al. Genetic targeting of an adenovirus vector via replacement of the fiber protein with the phage T4 fibritin. *J Virol* 2001; 75: 4176–4183.

137. Magnusson MK, Hong SS, Boulanger P et al. Genetic retargeting of adenovirus: novel strategy employing "deknobbing" of the fiber. *J Virol* 2001; 75: 7280–7289.

138. Gaden F, Franqueville L, Magnusson MK et al. Gene transduction and cell entry pathway of fiber-modified adenovirus type 5 vectors carrying novel endocytic peptide ligands selected on human tracheal glandular cells. *J Virol* 2004; 78: 7227–7247.

139. Belousova N, Korokhov N, Krendelshchikova V et al. Genetically targeted adenovirus vector directed to CD40-expressing cells. *J Virol* 2003; 77: 11367–11377.

140. Izumi M, Kawakami Y, Glasgow JN et al. In vivo analysis of a genetically modified adenoviral vector targeted to human CD40 using a novel transient transgenic model. *J Gene Med* 2005; 7: 1517–1525.

141. Pereboeva L, Komarova S, Mahasreshti PJ et al. Fiber-mosaic adenovirus as a novel approach to design genetically modified adenoviral vectors. *Virus Res* 2004; 105: 35–46.

142. Takayama K, Reynolds PN, Short JJ et al. A mosaic adenovirus possessing serotype Ad5 and serotype Ad3 knobs exhibits expanded tropism. *Virology* 2003; 309: 282–293.

143. Magnusson MK, Hong SS, Henning P et al. Genetic retargeting of adenovirus vectors: functionality of targeting ligands and their influence on virus viability. *J Gene Med* 2002; 4: 356–370.

144. Hedley SJ, Auf der Maur A, Hohn S et al. An adenovirus vector with a chimeric fiber incorporating stabilized single chain antibody achieves targeted gene delivery. *Gene Ther* 2006; 13: 88–94.

145. Ulasov IV, Tyler MA, Han Y et al. Novel recombinant adenoviral vector that targets the interleukin-13 receptor alpha2 chain permits effective gene transfer to malignant glioma. *Hum Gene Ther* 2007; 18: 118–129.

146. Athappilly FK, Murali R, Rux JJ et al. The refined crystal structure of hexon, the major coat protein of adenovirus type 2, at 2.9 A resolution. *J Mol Biol* 1994; 242: 430–455.

147. Crawford-Miksza L and Schnurr DP. Analysis of 15 adenovirus hexon proteins reveals the location and structure of seven hypervariable regions containing serotype-specific residues. *J Virol* 1996; 70: 1836–1844.

148. Vigne E, Mahfouz I, Dedieu JF et al. RGD inclusion in the hexon monomer provides adenovirus type 5-based vectors with a fiber knob-independent pathway for infection. *J Virol* 1999; 73: 5156–5161.

149. Wu H, Han T, Belousova N et al. Identification of sites in adenovirus hexon for foreign peptide incorporation. *J Virol* 2005; 79: 3382–3390.

150. Roberts DM, Nanda A, Havenga MJ et al. Hexon-chimaeric adenovirus serotype 5 vectors circumvent pre-existing anti-vector immunity. *Nature* 2006; 441: 239–243.

151. Sumida SM, Truitt DM, Lemckert AA et al. Neutralizing antibodies to adenovirus serotype 5 vaccine vectors are directed primarily against the adenovirus hexon protein. *J Immunol* 2005; 174: 7179–7185.

152. Parks RJ. Adenovirus protein IX: a new look at an old protein. *Mol Ther* 2005; 11: 19–25.

153. Vellinga J, van den Wollenberg DJ, van der Heijdt S et al. The coiled-coil domain of the adenovirus type 5 protein IX is dispensable for capsid incorporation and thermostability. *J Virol* 2005; 79: 3206–3210.

154. Ghosh-Choudhury G, Haj-Ahmad Y, and Graham FL. Protein IX, a minor component of the human adenovirus capsid, is essential for the packaging of full length genomes. *Embo J* 1987; 6: 1733–1739.

155. Colby WW and Shenk T. Adenovirus type 5 virions can be assembled in vivo in the absence of detectable polypeptide IX. *J Virol* 1981; 39: 977–980.

156. Boulanger P, Lemay P, Blair GE et al. Characterization of adenovirus protein IX. *J Gen Virol* 1979; 44: 783–800.

157. Akalu A, Liebermann H, Bauer U et al. The subgenus-specific C-terminal region of protein IX is located on the surface of the adenovirus capsid. *J Virol* 1999; 73: 6182–6187.

158. Rosa-Calatrava M, Grave L, Puvion-Dutilleul F et al. Functional analysis of adenovirus protein IX identifies domains involved in capsid stability, transcriptional activity, and nuclear reorganization. *J Virol* 2001; 75: 7131–7141.

159. Dmitriev IP, Kashentseva EA, and Curiel DT. Engineering of adenovirus vectors containing heterologous peptide sequences in the C terminus of capsid protein IX. *J Virol* 2002; 76: 6893–6899.

160. Vellinga J, Rabelink MJ, Cramer SJ et al. Spacers increase the accessibility of peptide ligands linked to the carboxyl terminus of adenovirus minor capsid protein IX. *J Virol* 2004; 78: 3470–3479.

161. Vellinga J, de Vrij J, Myhre S et al. Efficient incorporation of a functional hyper-stable single-chain antibody fragment protein-IX fusion in the adenovirus capsid. *Gene Ther* 2007; 14: 664–670.

162. Campos SK, Parrott MB, and Barry MA. Avidin-based targeting and purification of a protein IX-modified, metabolically biotinylated adenoviral vector. *Mol Ther* 2004; 9: 942–954.

163. Meulenbroek RA, Sargent KL, Lunde J et al. Use of adenovirus protein IX (pIX) to display large polypeptides on the virion–generation of fluorescent virus through the incorporation of pIX-GFP. *Mol Ther* 2004; 9: 617–624.

164. Le LP, Everts M, Dmitriev IP et al. Fluorescently labeled adenovirus with pIX-EGFP for vector detection. *Mol Imaging* 2004; 3: 105–116.

165. Li J, Le LP, Sibley DA et al. Genetic incorporation of HSV-1 thymidine kinase into the adenovirus protein IX for functional display on the virion. *Virology* 2005; 338: 247–258.

166. Matthews QL, Sibley DA, Wu H et al. Genetic incorporation of a herpes simplex virus type 1 thymidine kinase and firefly luciferase fusion into the adenovirus protein IX for functional display on the virion. *Mol Imaging* 2006; 5: 510–519.

167. Le LP, Le HN, Dmitriev IP et al. Dynamic monitoring of oncolytic adenovirus in vivo by genetic capsid labeling. *J Natl Cancer Inst* 2006; 98: 203–214.

168. Vellinga J, Van der Heijdt S, and Hoeben RC. The adenovirus capsid: major progress in minor proteins. *J Gen Virol* 2005; 86: 1581–1588.

169. Dmitriev I, Kashentseva EA, Seki T et al. Utilization of minor capsid polypeptides IX and IIIa for adenovirus targeting. *Mol Ther* 2001; 3: S167.

170. Saban SD, Silvestry M, Nemerow GR et al. Visualization of alpha-helices in a 6-angstrom resolution cryoelectron microscopy structure of adenovirus allows refinement of capsid protein assignments. *J Virol* 2006; 80: 12049–12059.

171. Tanaka T, Kanai F, Lan KH et al. Adenovirus-mediated gene therapy of gastric carcinoma using cancer-specific gene expression in vivo. *Biochem Biophys Res Commun* 1997; 231: 775–779.

172. Kaneko S, Hallenbeck P, Kotani T et al. Adenovirus-mediated gene therapy of hepatocellular carcinoma using cancer-specific gene expression. *Cancer Res* 1995; 55: 5283–5287.

173. Arbuthnot PB, Bralet MP, Le Jossic C et al. In vitro and in vivo hepatoma cell-specific expression of a gene transferred with an adenoviral vector. *Hum Gene Ther* 1996; 7: 1503 1514.

174. Chung I, Schwartz PE, Crystal RG et al. Use of L-plastin promoter to develop an adenoviral system that confers transgene expression in ovarian cancer cells but not in normal mesothelial cells. *Cancer Gene Ther* 1999; 6: 99–106.

175. Peng XY, Won JH, Rutherford T et al. The use of the L-plastin promoter for adenoviral-mediated, tumor-specific gene expression in ovarian and bladder cancer cell lines. *Cancer Res* 2001; 61: 4405–4413.

176. Tai YT, Strobel T, Kufe D et al. In vivo cytotoxicity of ovarian cancer cells through tumor-selective expression of the BAX gene. *Cancer Res* 1999; 59: 2121–2126.

177. Casado E, Gomez-Navarro J, Yamamoto M et al. Strategies to accomplish targeted expression of transgenes in ovarian cancer for molecular therapeutic applications. *Clin Cancer Res* 2001; 7: 2496–2504.

178. Barker SD, Coolidge CJ, Kanerva A et al. The secretory leukoprotease inhibitor (SLPI) promoter for ovarian cancer gene therapy. *J Gene Med* 2003; 5: 300–310.

179. Barker SD, Dmitriev IP, Nettelbeck DM et al. Combined transcriptional and transductional targeting improves the specificity and efficacy of adenoviral gene delivery to ovarian carcinoma. *Gene Ther* 2003; 10: 1198–1204.

180. Bao R, Selvakumaran M, and Hamilton TC. Targeted gene therapy of ovarian cancer using an ovarian-specific promoter. *Gynecol Oncol* 2002; 84: 228–234.

181. Yamamoto M, Alemany R, Adachi Y et al. Characterization of the cyclooxygenase-2 promoter in an adenoviral vector and its application for the mitigation of toxicity in suicide gene therapy of gastrointestinal cancers. *Mol Ther* 2001; 3: 385–394.

182. Zhu ZB, Makhija SK, Lu B et al. Transcriptional targeting of adenoviral vector through the CXCR4 tumor-specific promoter. *Gene Ther* 2004; 11: 645–648.

183. Stoff-Khalili MA, Stoff A, Rivera AA et al. Preclinical evaluation of transcriptional targeting strategies for carcinoma of the breast in a tissue slice model system. *Breast Cancer Res* 2005; 7: R1141–1152.

184. Haviv YS, van Houdt WJ, Lu B et al. Transcriptional targeting in renal cancer cell lines via the human CXCR4 promoter. *Mol Cancer Ther* 2004; 3: 687–691.

185. Lu B, Makhija SK, Nettelbeck DM et al. Evaluation of tumor-specific promoter activities in melanoma. *Gene Ther* 2005; 12: 330–338.

186. Zhu ZB, Makhija SK, Lu B et al. Transcriptional targeting of tumors with a novel tumor-specific survivin promoter. *Cancer Gene Ther* 2004; 11: 256–262.

187. Ko SC, Cheon J, Kao C et al. Osteocalcin promoter-based toxic gene therapy for the treatment of osteosarcoma in experimental models. *Cancer Res* 1996; 56: 4614–4619.

188. Koeneman KS, Kao C, Ko SC et al. Osteocalcin-directed gene therapy for prostate-cancer bone metastasis. *World J Urol* 2000; 18: 102–110.

189. Reynolds PN, Nicklin SA, Kaliberova L et al. Combined transductional and transcriptional targeting improves the specificity of transgene expression in vivo. *Nat Biotechnol* 2001; 19: 838–842.

190. Walton T, Wang JL, Ribas A et al. Endothelium-specific expression of an E-selectin promoter recombinant adenoviral vector. *Anticancer Res* 1998; 18: 1357–1360.

191. Varda-Bloom N, Shaish A, Gonen A et al. Tissue-specific gene therapy directed to tumor angiogenesis. *Gene Ther* 2001; 8: 819–827.

192. Park KW, Morrison CM, Sorensen LK et al. Robo4 is a vascular-specific receptor that inhibits endothelial migration. *Dev Biol* 2003; 261: 251–267.

193. Huminiecki L, Gorn M, Suchting S et al. Magic roundabout is a new member of the roundabout receptor family that is endothelial specific and expressed at sites of active angiogenesis. *Genomics* 2002; 79: 547–552.

194. Preuss MP, Barnes JA, Glasgow JN et al. Transcriptional targeting of gene expression to the endothelium using the roundabout-4 receptor promoter. *Mol Ther* 2007; 15: S49.

195. Han Z, Wang H, and Hallahan DE. Radiation-guided gene therapy of cancer. *Technol Cancer Res Treat* 2006; 5: 437–444.

196. Manome Y, Kunieda T, Wen PY et al. Transgene expression in malignant glioma using a replication-defective adenoviral vector containing the Egr-1 promoter: activation by ionizing radiation or uptake of radioactive iododeoxyuridine. *Hum Gene Ther* 1998; 9: 1409–1417.

197. Pitzer C, Schindowski K, Pomer S et al. In vivo manipulation of interleukin-2 expression by a retroviral tetracycline (tet)-regulated system. *Cancer Gene Ther* 1999; 6: 139–146.

198. Wang L, Hernandez-Alcoceba R, Shankar V et al. Prolonged and inducible transgene expression in the liver using gutless adenovirus: a potential therapy for liver cancer. *Gastroenterology* 2004; 126: 278–289.

199. Zerby D, Sakhuja K, Reddy PS et al. In vivo ligand-inducible regulation of gene expression in a gutless adenoviral vector system. *Hum Gene Ther* 2003; 14: 749–761.

200. Southam CM, Noyes WF, and Mellors R. Virus in human cancer cells in vivo. *Virology* 1958; 5: 395.

201. Oosterhoff D and van Beusechem VW. Conditionally replicating adenoviruses as anticancer agents and ways to improve their efficacy. *J Exp Ther Oncol* 2004; 4: 37–57.

202. Nettelbeck DM. Virotherapeutics: conditionally replicative adenoviruses for viral oncolysis. *Anticancer Drugs* 2003; 14: 577–584.

203. Everts B and van der Poel HG. Replication-selective oncolytic viruses in the treatment of cancer. *Cancer Gene Ther* 2005; 12: 141–161.

204. Savontaus MJ, Sauter BV, Huang TG et al. Transcriptional targeting of conditionally replicating adenovirus to dividing endothelial cells. *Gene Ther* 2002; 9: 972–979.

205. Freytag SO, Rogulski KR, Paielli DL et al. A novel three-pronged approach to kill cancer cells selectively: concomitant viral, double suicide gene, and radiotherapy. *Hum Gene Ther* 1998; 9: 1323–1333.

206. Wildner O, Blaese RM, and Morris JC. Therapy of colon cancer with oncolytic adenovirus is enhanced by the addition of herpes simplex virus-thymidine kinase. *Cancer Res* 1999; 59: 410–413.

207. Barker DD and Berk AJ. Adenovirus proteins from both E1B reading frames are required for transformation of rodent cells by viral infection and DNA transfection. *Virology* 1987; 156: 107–121.

208. Bischoff JR, Kirn DH, Williams A et al. An adenovirus mutant that replicates selectively in p53-deficient human tumor cells. *Science* 1996; 274: 373–376.

209. Ries SJ, Brandts CH, Chung AS et al. Loss of p14ARF in tumor cells facilitates replication of the adenovirus mutant dl1520 (ONYX-015). *Nat Med* 2000; 6: 1128–1133.

210. Hay JG, Shapiro N, Sauthoff H et al. Targeting the replication of adenoviral gene therapy vectors to lung cancer cells: the importance of the adenoviral E1b-55kD gene. *Hum Gene Ther* 1999; 10: 579–590.

211. Heise C, Ganly I, Kim YT et al. Efficacy of a replication-selective adenovirus against ovarian carcinomatosis is dependent on tumor burden, viral replication and p53 status. *Gene Ther* 2000; 7: 1925–1929.

212. Heise C, Sampson-Johannes A, Williams A et al. ONYX-015, an E1B gene-attenuated adenovirus, causes tumor-specific cytolysis and antitumoral efficacy that can be augmented by standard chemotherapeutic agents. *Nat Med* 1997; 3: 639–645.

213. Rothmann T, Hengstermann A, Whitaker NJ et al. Replication of ONYX-015, a potential anticancer adenovirus, is independent of p53 status in tumor cells. *J Virol* 1998; 72: 9470–9478.

214. Dix BR, Edwards SJ, and Braithwaite AW. Does the antitumor adenovirus ONYX-015/dl1520 selectively target cells defective in the p53 pathway? *J Virol* 2001; 75: 5443–5447.

215. Fueyo J, Gomez-Manzano C, Alemany R et al. A mutant oncolytic adenovirus targeting the Rb pathway produces anti-glioma effect in vivo. *Oncogene* 2000; 19: 2–12.

216. Heise C, Hermiston T, Johnson L et al. An adenovirus E1A mutant that demonstrates potent and selective systemic anti-tumoral efficacy. *Nat Med* 2000; 6: 1134–1139.

217. Sherr CJ. Cancer cell cycles. *Science* 1996; 274: 1672–1677.

218. Doronin K, Toth K, Kuppuswamy M et al. Tumor-specific, replication-competent adenovirus vectors overexpressing the adenovirus death protein. *J Virol* 2000; 74: 6147–6155.

219. Doronin K, Kuppuswamy M, Toth K et al. Tissue-specific, tumor-selective, replication-competent adenovirus vector for cancer gene therapy. *J Virol* 2001; 75: 3314–3324.

220. Balague C, Noya F, Alemany R et al. Human papillomavirus E6E7-mediated adenovirus cell killing: selectivity of mutant adenovirus replication in organotypic cultures of human keratinocytes. *J Virol* 2001; 75: 7602–7611.

221. Nettelbeck DM, Rivera AA, Balague C et al. Novel oncolytic adenoviruses targeted to melanoma: specific viral replication and cytolysis by expression of E1A mutants from the tyrosinase enhancer/promoter. *Cancer Res* 2002; 62: 4663–4670.

222. Yu W and Fang H. Clinical trials with oncolytic adenovirus in China. *Curr Cancer Drug Targets* 2007; 7: 141–148.

223. Nemunaitis J, Ganly I, Khuri F et al. Selective replication and oncolysis in p53 mutant tumors with ONYX-015, an E1B-55kD gene-deleted adenovirus, in patients with advanced head and neck cancer: a phase II trial. *Cancer Res* 2000; 60: 6359–6366.

224. Khuri FR, Nemunaitis J, Ganly I et al. A controlled trial of intratumoral ONYX-015, a selectively-replicating adenovirus, in combination with cisplatin and 5-fluorouracil in patients with recurrent head and neck cancer. *Nat Med* 2000; 6: 879–885.

225. Shinoura N, Yoshida Y, Tsunoda R et al. Highly augmented cytopathic effect of a fiber-mutant E1B-defective adenovirus for gene therapy of gliomas. *Cancer Res* 1999; 59: 3411–3416.

226. Douglas JT, Kim M, Sumerel LA et al. Efficient oncolysis by a replicating adenovirus (ad) in vivo is critically dependent on tumor expression of primary ad receptors. *Cancer Res* 2001; 61: 813–817.

227. Hemminki A, Dmitriev I, Liu B et al. Targeting oncolytic adenoviral agents to the epidermal growth factor pathway with a secretory fusion molecule. *Cancer Res* 2001; 61: 6377–6381.

228. Suzuki K, Fueyo J, Krasnykh V et al. A conditionally replicative adenovirus with enhanced infectivity shows improved oncolytic potency. *Clin Cancer Res* 2001; 7: 120–126.

229. Bauerschmitz GJ, Lam JT, Kanerva A et al. Treatment of ovarian cancer with a tropism modified oncolytic adenovirus. *Cancer Res* 2002; 62: 1266–1270.

230. Sarkioja M, Kanerva A, Salo J et al. Noninvasive imaging for evaluation of the systemic delivery of capsid-modified adenoviruses in an orthotopic model of advanced lung cancer. *Cancer* 2006; 107: 1578–1588.

231. Witlox AM, Van Beusechem VW, Molenaar B et al. Conditionally replicative adenovirus with tropism expanded towards integrins inhibits osteosarcoma tumor growth in vitro and in vivo. *Clin Cancer Res* 2004; 10: 61–67.

232. Graat HC, Witlox MA, Schagen FH et al. Different susceptibility of osteosarcoma cell lines and primary cells to treatment with oncolytic adenovirus and doxorubicin or cisplatin. *Br J Cancer* 2006; 94: 1837–1844.

233. Kanerva A, Wang M, Bauerschmitz GJ et al. Gene transfer to ovarian cancer versus normal tissues with fiber-modified adenoviruses. *Mol Ther* 2002; 5: 695–704.

234. Kanerva A, Zinn KR, Chaudhuri TR et al. Enhanced therapeutic efficacy for ovarian cancer with a serotype 3 receptor-targeted oncolytic adenovirus. *Mol Ther* 2003; 8: 449–458.

235. Kanerva A, Lam J, Yamamoto M et al. A cyclooxygenase-2 promoter based conditionally replicating adenovirus with enhanced infectivity for treatment of ovarian carcinoma. *Mol Ther* 2002; 5: S414.

236. Ono HA, Davydova JG, Adachi Y et al. Promoter-controlled infectivity-enhanced conditionally replicative adenoviral vectors for the treatment of gastric cancer. *J Gastroenterol* 2005; 40: 31–42.

237. Yamamoto M, Davydova J, Wang M et al. Infectivity enhanced, cyclooxygenase-2 promoter-based conditionally replicative adenovirus for pancreatic cancer. *Gastroenterology* 2003; 125: 1203–1218.

238. Hemminki A, Kanerva A, Kremer EJ et al. A canine conditionally replicating adenovirus for evaluating oncolytic virotherapy in a syngeneic animal model. *Mol Ther* 2003; 7: 163–173.

239. Toth K, Spencer JF, Tollefson AE et al. Cotton rat tumor model for the evaluation of oncolytic adenoviruses. *Hum Gene Ther* 2005; 16: 139–146.

240. Toth K, Spencer JF, and Wold WS. Immunocompetent, semi-permissive cotton rat tumor model for the evaluation of oncolytic adenoviruses. *Methods Mol Med* 2007; 130: 157–168.

241. Thomas MA, Spencer JF, and Wold WS. Use of the Syrian hamster as an animal model for oncolytic adenovirus vectors. *Methods Mol Med* 2007; 130: 169–183.

242. Page JG, Tian B, Schweikart K et al. Identifying the safety profile of a novel infectivity-enhanced conditionally replicative adenovirus, Ad5-delta24-RGD, in anticipation of a phase I trial for recurrent ovarian cancer. *Am J Obstet Gynecol* 2007; 196: 389 e1–9; discussion e9–10.

243. Zhu ZB, Makhija SK, Lu B et al. Transport across a polarized monolayer of Caco-2 cells by transferrin receptor-mediated adenovirus transcytosis. *Virology* 2004; 325: 116–128.

244. Moroo I, Ujiie M, Walker BL et al. Identification of a novel route of iron transcytosis across the mammalian blood–brain barrier. *Microcirculation* 2003; 10: 457–462.

245. Mostov KE. Transepithelial transport of immunoglobulins. *Annu Rev Immunol* 1994; 12: 63–84.

246. McIntosh DP, Tan XY, Oh P et al. Targeting endothelium and its dynamic caveolae for tissue-specific transcytosis in vivo: a pathway to overcome cell barriers to drug and gene delivery. *Proc Natl Acad Sci U S A* 2002; 99: 1996–2001.

5 Helper-Dependent Adenoviral Vectors for Gene Therapy

Nicola Brunetti-Pierri and Philip Ng

CONTENTS

5.1 INTRODUCTION

Gene therapy is the amelioration of disease through the use of nucleic acid. This broad definition reflects the large spectrum of diseases that can be potentially considered for treatment, as well as the many methods of introducing the nucleic acid into target cells, which may or may not be directly affected by the disease. This chapter focuses on one very specific area of gene therapy research: the development and application of helper-dependent adenoviral vectors (HDAds) (also referred to as gutless, gutted, mini, fully deleted, high-capacity, Δ, pseudo, encapsidated adenovirus [Ad] mini-chromosome) for gene therapy. Successful gene therapy requires a gene transfer vector that is able to efficiently transduce the target cells *in vivo* and provide high level and long-term transgene expression with minimal toxicity. How HDAds have measured up to these expectations are reviewed in this chapter.

5.2 ADENOVIRUS

The Ad has a non-enveloped icosahedral capsid containing a linear double-stranded DNA genome of ~30–40 kb. Of the ~50 serotypes of human Ad, the most extensively characterized are serotypes 2 (Ad2) and 5 (Ad5) of subgroup C (reviewed in [1]). The 36 kb genomes of Ad2 and Ad5 are flanked by inverted terminal repeats (ITRs) which are the only sequences required in *cis* for viral DNA replication. A *cis*-acting packaging signal, required for encapsidation of the genome, is located near the left ITR (relative to the conventional map of Ad). The Ad genome can be roughly divided into two sets of genes (Figure 5.1): the early region genes, E1A, E1B, E2, E3, and E4, are expressed before DNA replication and the late region genes, L1 to L5, are expressed to high levels after initiation of DNA replication. The E1A transcription unit is the first early region to be expressed during

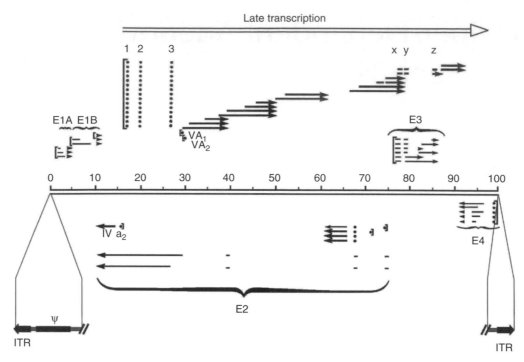

FIGURE 5.1   Transcription map of human Ad serotype 5. The 100 map unit (~36 kb) genome is divided into four early region transcription units, E1–E4, and five families of late mRNA, L1–L5, which are alternative splice products of a common late transcript expressed from the MLP located at 16 map units. Four smaller transcripts, pIX, IVa, and VA RNA's I and II, are also produced. The 103 bp ITRs are located at the termini of the genome and are involved in viral DNA replication, and the packaging signal (ψ) located from nucleotides 190 to 380 at the left end is involved in packaging of the genome into virion capsids.

viral infection and it encodes two major E1A proteins that are involved in transcriptional regulation of the virus and stimulation of the host cell to enter an S phase-like state. The two major E1B proteins are necessary for blocking host mRNA transport, stimulating viral mRNA transport, and blocking E1A-induced apoptosis. The E2 region encodes proteins required for viral DNA replication and can be divided into two subregions; E2a encodes the 72 kD DNA-binding protein and E2b encodes the viral DNA polymerase and terminal protein precursor. The E3 region, which is dispensable for virus growth in cell culture, encodes at least seven proteins most of which are involved in host immune evasion. The E4 region encodes at least six proteins, some functioning to facilitate DNA replication, enhance late gene expression, and decrease host protein synthesis. The late region genes are expressed from a common major late promoter (MLP) and are generated by alternative splicing of a single transcript. Most of the late mRNAs encode virion structural proteins. In addition to the early and late region genes, four other small transcripts are also produced. The gene encoding protein IX (pIX) is colinear with E1B but uses a different promoter and is expressed at an intermediate time, as is the pIVa2 gene. Other late transcripts include the RNA polymerase III transcribed VA RNA I and II.

Virus infection is initiated through the Ad fiber protein binding to the primary coxsackie-adenovirus receptors (CAR) on the cell surface [2–4] followed by a secondary interaction between the virion penton base and $\alpha_v\beta_3$ and $\alpha_v\beta_5$ integrins

[5]. The efficiency with which Ad binds and enters the cell is directly related to the level of primary and secondary receptors present on the cell surface [6,7]. Penton–integrin interaction triggers Ad internalization by endocytosis following which the virion escapes from the early endosome into the cytosol prior to lysosome formation [8,9]. The virion is sequentially disassembled during translocation along the microtubule network towards the nucleus where the viral DNA is released into the nucleus [10]. Once in the nucleus, viral DNA replication, beginning 6–8 h postinfection, and assembly of progeny virions occur. The entire life cycle takes about 24–36 h, generating about 10^4 virions per infected cell. Ads have never been implicated as a cause of malignant disease in their natural host and in immunocompetent humans most infections caused by the most common serotypes are relatively mild and self-limiting. The reader is referred to the excellent review by Shenk [1] for a more comprehensive discussion of Ads.

5.3   FIRST-GENERATION ADENOVIRAL VECTORS

Ads are excellent mammalian gene transfer vectors due to their ability to efficiently infect a variety of quiescent and proliferating cell types from various species to direct high-level transgene expression. Consequently, Ad vectors are extensively used as potential recombinant viral vaccines, for high-level protein production in cultured cells, and for gene therapy (for reviews see [11–15]). First-generation Ad vectors (FGAds) typically have foreign DNA inserted in place of early region 1 (E1). E1-deleted vectors are replication deficient and are propagated in E1 complementing cells such as 293 [16]. Typically, FGAds also have a deletion in the nonessential E3 region to maximize cloning capacity. The fundamental principle underlying all current methods of constructing FGAds is based on the discovery that up to 10% of Ad viral DNA molecules become circularized following infection of mammalian cells [17]. This permitted the cloning of the entire Ad genome as an infectious bacterial plasmid which could be manipulated with relative ease by standard molecular biology techniques. This Ad genomic plasmid could be stably propagated in *E. coli* and was capable of generating infectious virus following transfection into permissive mammalian cells at efficiencies comparable to purified virion DNA. An excellent review by Danthinne and Imperiale [18] covers the plethora of methods for constructing FGAd. While FGAds remain very useful for many applications, it has become clear that transgene expression *in vivo* is only transient. Several factors contribute to this, including strong innate and inflammatory responses to the vector [19,20], acute and chronic toxicity due to low level viral gene expression from the vector backbone [21], and generation of anti-Ad cytotoxic T-lymphocytes due to de novo viral gene expression [22–25] or processing of virion proteins [26]. While high-level transient transgene expression afforded by FGAd may be adequate, or even desirable, for many gene transfer and gene therapy applications, the toxicity and transient nature of expression kinetics renders these vectors unsuitable in cases where prolonged, stable expression is required.

5.4 MULTIPLY DELETED ADENOVIRAL VECTORS

While the deletion of E1 in FGAds theoretically results in a replication-defective vector following transduction of cells devoid of trans-complementing E1 functions, it has become clear that this is not the case and that the E2, E3, and E4 promoters are active resulting in viral DNA replication and expression of the late viral genes, especially following high multiplicities of infection [27–30]. Therefore, in an attempt to further attenuate Ad, additional deletions of essential viral genes have been pursued. Examples of these include deletion or mutation of the E2 or E4 regions in addition to E1. These multiply deleted Ad vectors (also referred to as second or third generation Ad vectors) are helper virus independent for propagation but require the generation of new producer cell lines that trans-complement the additional deletion [30]. Despite the potential offered by these multiply deleted Ad vectors, viral coding sequences still remain and therefore so does the potential for their expression. The advantages of multiply deleted Ad over FGAd remain controversial as some studies show them to be superior in terms of toxicity and longevity of transgene expression [31–37] while others do not [21,38–41]. Detailed discussion of multiply deleted Ad vectors is beyond the scope of this chapter and the reader is referred to an excellent and comprehensive review covering this subject by Parks and Amalfitano [42].

5.5 HELPER-DEPENDENT ADENOVIRAL VECTORS

5.5.1 HISTORICAL PERSPECTIVE

In theory, the easiest way to completely eliminate the toxicity associated with viral gene expression is to delete all of the viral coding sequences from the vector. Ads with large deletions of viral sequences were among the very first vectors reported. For example, in 1970, Lewis and Rowe [43] described a recombinant Ad in which essential viral genes were replaced with SV40 DNA. This replacement rendered the hybrids replication defective and they could only be propagated in the presence of a coinfecting wild-type helper Ad which provided trans-complementation. A recombinant Ad was also described in 1981 by Deuring et al. [44] following repeated high multiplicity of infection passage of wild-type Ad serotype 12 through human KB cells. These hybrids were found to contain symmetrically duplicated human chromosomal DNA flanked by approximately 1 kb of DNA from the left end of the Ad genome. As in the case of the Ad-SV40 hybrids, these symmetrical recombinants (SYRECs) were replication defective and could only be propagated in the presence of a coinfecting helper Ad12 by trans-complementation. The SYRECs were maintained for years in this manner and could be partially purified from the helper virus by CsCl density equilibrium centrifugation owing to differences in their genome sizes which bestowed different buoyant densities.

The defective helper virus-dependent Ad-SV40 and SYRECs hybrids demonstrated the possibility of generating vectors completely devoid of Ad coding sequences. In principle, only ~500 bp of cis-acting Ad sequences necessary for DNA replication (ITR) and encapsidation (ψ) are required for propagation of these HDAds in the presence of a coinfecting helper virus. The advantages of HDAds are considerable: like FG and multiply deleted Ads, HDAds would retain the ability to very efficiently transduce a wide variety of cell types from numerous species in a cell cycle independent manner. However, unlike FGAd and multiply deleted Ads, deletion of all viral coding sequences would drastically reduce vector-mediated toxicity and significantly prolong the duration of transgene expression as described in detail below.

5.5.2 EARLY SYSTEMS FOR GENERATING HDADS

Although, as mentioned above, the very first Ad vectors were defective, helper-dependent viruses, with the development of E1 complementing 293 cells, vector systems focused on FGAd (E1-deleted) vectors because these were much easier to isolate and propagate than HDAds. The possibility of using HDAds for gene transfer was reexamined in studies reported by Mitani et al. [45]. These investigators used a β-galactosidase-neomycin fusion gene to replace 7.3 kb of essential Ad sequences in an Ad genomic plasmid. While this modification did not remove all of the viral coding sequences, it did render the recombinant defective and helper dependent. This vector was rescued by co-transfection of 293 cells with purified Ad2

DNA which provided helper functions. Following X-gal staining, 1%–5% of the resulting plaques turned blue indicating rescue of the recombinant virus. Blue plaque isolates were expanded by serial propagation on 293 cells and finally purified by CsCl ultracentrifugation. Fractionation through the gradient resulted in partial purification of the vector from the helper virus due to the difference in their buoyant densities. Significantly, the ability of the vector to transduce cells *in vitro* and express the reporter gene was demonstrated. However, the yield of vector was quite low. Furthermore, helper virus contamination remained rather high, at about 200- to 500-fold greater excess over the vector. In addition, the genome of the vector had undergone rearrangement.

Using a different strategy, Fisher et al. [46] constructed a plasmid bearing a 5.5 kb HDAd genome containing the Ad 5′ ITR and packaging signal, LacZ reporter gene and 3′ ITR. In this system, 293 cells were infected with an FGAd to serve as a helper virus and subsequently transfected with the HDAd plasmid DNA. The HDAd was amplified by serial coinfections and finally purified by CsCl ultracentrifugation. Using this method, partial purification of the vector could be achieved, but with the helper virus still 10- to 100-fold excess. Additionally, considerable genomic rearrangements, in the form of concatemerization of the vector, were observed, which were likely due to the small size of the vector genome [47]. Nevertheless, the vector was demonstrated to be capable of transducing cells *in vitro*. Using this strategy, a vector bearing CFTR [46] and dystrophin [48] expression cassettes was generated and shown to transduce cells *in vitro*.

Another strategy for generating HDAds was reported by Kumar-Singh and Chamberlain [49]. In this system, 293 cells were co-transfected with a plasmid bearing the HDAd genome and purified Ad DNA to provide helper functions. The HDAd contained a LacZ reporter gene and dystrophin cDNA while the helper virus contained the alkaline phosphatase (AP) reporter gene. The vector was amplified by serial propagation and purified by CsCl ultracentrifugation. This resulted in a final vector preparation with 4% helper virus contamination as determined by AP:LacZ ratio. Importantly, this vector was capable of transducing myogenic cell cultures and expressing dystrophin in myotubes.

These early systems showed that HDAds could indeed be generated and could transduce a variety of target cells *in vitro* to direct transgene expression. However, they also emphasized the need to further improve production strategies as the high levels of helper virus contamination, low vector yield and, in many cases, vector genome rearrangement were clearly obstacles, which needed to be addressed before the full potential of HDAd could be realized. In particular, relying solely on physical separation between the vector and the helper by CsCl ultracentrifugation was clearly inadequate to achieve the desired vector purities. Therefore, strategies to preferentially inhibit helper virus propagation while not affecting its ability to trans-complement the HDAd were required.

One system designed to specifically address preferential inhibition of helper virus propagation was reported in 1996 by Kochanek et al. [50]. This strategy was based on early studies

in Ad packaging by Gräble and Hearing [51], which showed that deletion of 91 bp from the packaging signal severely reduced, but did not abolish, encapsidation of the Ad genome into virions while not affecting viral DNA replication. More importantly, Gräble and Hearing observed a competition for packaging in cells co-infected with two Ads, one containing a wild-type packaging signal and the other having a mutant packaging signal: the former was packaged preferentially, over the latter. Taking advantage of these observations, Kochanek et al. deleted this 91 bp deletion from the Ad packaging signal of the helper viral genome thus impairing its ability to be packaged but not affecting its ability to replicate and thus trans-complement the HDAd genome [50]. In addition, because the HDAd genome contained the wild-type Ad packaging signal, it would be preferentially packaged over the helper viral genome. The combination of this strategy in conjunction with CsCl ultracentrifugation resulted in a final vector preparation with ~1% helper virus contamination. A similar strategy, but using a different packaging signal mutation, was employed by Alemany et al. [52] to generate HDAds that were used in the only human trial to date (see Section 5.6.1.1). These strategies represented a significant improvement over the previous methods in terms of lower levels of helper virus contamination. However, helper virus contamination remained relatively high. Furthermore, the 91 bp packaging signal deletion resulted in 90-fold reduction in yield of helper virus. From a practical standpoint, this would render production of helper virus stocks problematic since large amounts would be needed to produce the considerable quantities of HDAd needed for clinical applications.

5.5.3 Cre/loxP System for Generating HDAds

The first efficient method for generating HDAd was the Cre/loxP system developed by Graham and coworkers in 1996 [53] (Figure 5.2). In this system the HDAd genome is first constructed in a bacterial plasmid. Minimally, the HDAd genome contains the expression cassette of interest and ~500 bp of cis-acting Ad sequences necessary for vector DNA replication (ITRs) and packaging (ψ). As described in detail in Section 5.5.4, inclusion of stuffer DNA in the HDAd genome is often required for efficient packaging. To rescue the HDAd (that is, to convert the "plasmid form" of the HDAd genome into the "viral form"), the plasmid is first digested with the appropriate restriction enzyme to liberate the HDAd genome from the bacterial plasmid sequences. In this method, 293 cells expressing Cre are then transfected with the linearized HDAd genome and subsequently infected with the helper virus. The helper virus bears a packaging signal flanked by loxP sites and thus following infection of 293 Cre cells, the packaging signal is excised from the helper viral genome by Cre-mediated site-specific recombination between the loxP sites. This renders the helper viral genome unpackagable but still able to undergo DNA replication and thus trans-complement the replication and encapsidation of the HDAd genome. The titer of the HDAd is increased by serial coinfection of 293 Cre cells with the HDAd and the helper virus. Similar systems based on

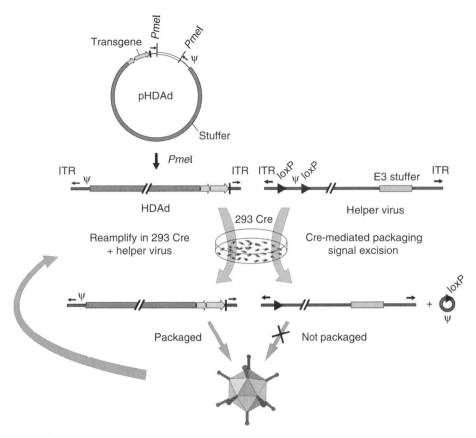

FIGURE 5.2 (See color insert following blank page 206.) Cre/loxP system for generating HD vectors. The HDAd contains only ~500 bp of cis-acting Ad sequences required for DNA replication (ITRs) and packaging (ψ), the remainder of the genome consists of the desired transgene (yellow) and non-Ad "stuffer" sequences (green). The HDAd genome is constructed as a bacterial plasmid (pHDAd) and is liberated by restriction enzyme digestion (e.g., *Pme*I). To rescue the HDAd, the liberated genome is transfected into 293 cells expressing Cre and infected with a helper virus bearing a packaging signal (ψ) flanked by loxP sites. The helper viral genome also contains a stuffer sequence in E3 to prevent the formation of RCA in 293-derived cells. Cre-mediated excision of ψ renders the helper virus genome unpackagable, but still able to replicate and provide all of the necessary *trans*-acting factors for propagation of the HDAd. The titer of the HD vector is increased by serial coinfections of 293 Cre cells with the HDAd and the helper virus.

using the yeast site specific recombinase FLP to catalyze recombination between 34 bp frt sites were subsequently developed [54,55].

The Cre/loxP method of producing HDAd described above is a complex three-component system. Consequently efficient production of large quantities of high quality vector has been difficult, permitting production of only modest amounts of HDAd only suitable for small animal experiments or low-dose large animal studies. While results from these studies have been encouraging, improved methods to permit efficient and reliable production of large quantities of HDAd were needed to fully evaluate the safety and efficacy of these vectors in preclinical studies with larger animal models and ultimately for clinical applications. The production problem was addressed by the development of an improved system comprised of a suspension adapted producer cell line that expresses very high levels of Cre, a helper virus resistant to mutation and refined protocols [56]. With these improvements, $>1 \times 10^{13}$ viral particles (vp) can be easily produced from 3 L of cells within 2 weeks of initial vector rescue, with

specific yields of $>10,000$ vp/cell and with exceedingly low helper virus contamination of 0.4%–0.1% without relying on CsCl purification and 0.02%–0.01% following CsCl purification as determined by DNA-based assays. This technological advancement has significantly improved our ability to assess this promising gene therapy technology, especially in large animal models and, perhaps ultimately, for clinical applications. Detailed methodologies for producing HDAd are described elsewhere [57].

5.5.4 CHARACTERISTICS OF THE HELPER-DEPENDENT VECTOR

In addition to the minimal cis-acting Ad sequences required for DNA replication and encapsidation, Sandig et al. [58] showed that inclusion of a small segment of noncoding Ad sequence from the E4 region adjacent to the right ITR increases vector yields, possibly by enhancing packaging of the HDAd DNA. Early studies of Ad have established a maximum packaging capacity of 105% of the wild-type genome

(~37.8 kb) [59]. Subsequently, using the Cre/loxP system for generating HDAds, Parks and Graham [47] established that the minimum genome size for efficient packaging into Ad virions was 27 kb. Vector genome sizes above the maximum packaging capacity were not efficiently packaged, if at all. Vector genome sizes below 27 kb were inefficiently packaged and frequently underwent DNA rearrangements to produce genomes closer to that of wild-type viruses (approximately 36 kb) [46,47,60]. Therefore, the size of the vector is an important consideration: for efficient packaging and stability, the vector genome should be between ~75% (>27 kb) [47] and 105% (<37.8 kb) [59] of wild-type. Since the minimal Ad cis-acting sequences and the transgene of interest are usually well below the minimal size required for efficient packaging the vector must often include stuffer DNA. The choice of stuffer DNA is important with regard to vector stability, replication efficiency, and *in vivo* performance [58,61,62]. While it remains unclear what constitutes a good stuffer, in general, noncoding eukaryotic DNA has been preferred while repetitive elements and unnecessary homology with the helper virus should be avoided to ensure vector stability and prolonged transgene expression.

In addition to the absolute size of the HDAd, its relative size compared to the helper virus has an impact on the level of helper virus contamination. This is because Cre-mediated selection against the helper virus, while efficient, is not absolute [56,63]; (see Section 5.5.6). If the genome size of the HDAd is sufficiently different from that of the helper virus then the two species can be physically separated by CsCl ultracentrifugation due to their different buoyant densities. HDAds with genome size between 28 and 31 kb are ideal because they are efficiently packaged, replicated, and easily separated from residual helper viruses (35 to 37 kb) following CsCl ultracentrifugation.

It is also becoming clear that the nature of the transgene has a significant influence on the degree and duration of transgene expression. Specifically, transgenes in their native genomic context have been demonstrated to be superior to their cDNA counterparts with respect to level and duration of expression [62,64,65]. This is likely due to a more physiological regulation of gene expression. In this regard, HDAds, because of the large cloning capacity of 37 kb, offer the advantage of potentially transferring many genes in their genomic context whereas other vectors (e.g., FGAd and multiply deleted Ads, retroviral, lentiviral, or adeno-associated viral vectors) cannot, due to their limited cloning capacity. In the case of cDNA, expression from a tissue-specific promoter was found to be superior to ubiquitous promoters in terms of toxicity and duration of expression [62,66]. This has been attributed to a reduction of transgene expression in transduced antigen presenting cells (APCs) following systemic delivery of vectors that carry expression cassettes under the control of tissue-specific promoters [66]. Expression from tissue-specific promoters has been demonstrated to be more "tissue-specific" within the context of a HDAd than an FGAd [67]. This may be due to the influence of Ad sequences from the FGAd backbone on transgene expression specificity which does not occur

with HDAds [67]. In addition, use of tissue-specific promoters may result in more robust amplification of the HDAd since high-level transgene expression in the producer cells may negatively impact viral replication during serial coinfections, especially if the transgene product is toxic to the producer cells.

Accurate and reliable methods for measuring the vector dose are crucial for preclinical and clinical investigation. There are two methods of expressing the concentration of Ad-based vectors: (1) Physical titer is the total concentration of vp obtained by measuring the amount of DNA in a vector preparation and is expressed as vp per milliliter. (2) Infectious titer is the concentration of vector particles capable of infecting permissive cells to cause cytopathic effect following an *in vitro* infection assay and is expressed as infectious units (IU) per milliliter. It should be noted that not all vp in a preparation are infectious. Therefore, a comparison of the vp:IU ratio provides a measure of the infectivity of the vector preparation. Determining the infectious titer of FGAd and multiply deleted Ads is straightforward because they cause cytopathic effect following infection of permissive cells. However, determining the infectious titer of HDAd is not straightforward because they do not cause cytopathic effect in transduced cells due to the absence of viral genes. If the HDAd contains a reporter transgene then the infectious titer can be determined. However, this is not the case with the majority of therapeutic HDAds thus rendering the infectivity of HDAd preparations difficult to ascertain and, as described below, accurate determination of vector infectivity is critical.

Following the tragic death of a partial OTC-deficient patient given systemic E1,E4-deleted Ad [68], the National Institutes of Health (NIH) Office of Biotechnology Activities established a Recombinant DNA Advisory Committee (RAC) working group on Ad safety and toxicity (AdSAT) to conduct a comprehensive review and analysis of the scientific, safety, and ethical issues associated with Ad-based human gene transfer [69]. One important deficiency cited by the AdSAT was a lack of standardization in determining the physical and infectious vector titers and they recommended the development of a fully characterized Ad reference standard (ARM) to address this problem [69,70]. The ARM is intended to be used as a reference standard in determining the physical and infectious titers of clinical grade Ad vectors. Furthermore, the FDA has recently recommended that clinical grade Ad vectors should have a vp:IU ratio of no more than 30:1 [69]. Demonstrating this requirement for HDAd has been problematic because traditional infectious titer assays are not possible as described above. Although a number of methods have been described to determine the infectious titer of HDAd [71–73], one study is particularly relevant as it is the only one to directly compare the infectivity of the HDAd to the ARM [74]. This assay is based on coinfection of 293 cells with the ARM and the HDAd which permits direct comparison of the amounts of ARM and HDAd virions that gain cellular entry. Because the assay relies on coinfection rather than comparisons between two or more independent single-infections (as is the case with all other methods described to date), the test article (the vector) and the reference standard (the ARM) experience identical conditions thereby eliminating the numerous variables associated with time-to-time, person-to-person, laboratory-to-laboratory, and protocol-to-protocol variations. Moreover, it eliminates the inherent variability of biological assays that are significant hurdles in the accurate determination of physical and infectious Ad titers. This universal coinfection assay permits reliable comparisons within and between studies thus satisfying important criteria for clinical studies.

5.5.5 Characteristics of the Helper Virus

The most commonly used helper virus used for HDAd production is a serotype 5, FGAd (E1-deleted) with its packaging signal flanked by loxP sites (Figure 5.2). As with all FGAds propagated in 293 or 293-derived cells, the potential exists for the generation of replication-competent Ad (RCA; E1+) as a consequence of homologous recombination between the helper virus and the Ad sequences present in 293 cells. To prevent the formation of RCA, a stuffer sequence was inserted into the E3 region to render any E1+ recombinants too large to be packaged [53]. The length of the E3 stuffer is such that the total size of the helper virus genome is <105% of wild-type but >105% following homologous recombination with Ad sequences in the 293 cells. As of this writing, the emergence of RCA has yet to be reported using helper viruses with an E3 stuffer. In contrast, RCA is readily detected using helper viruses without an E3 stuffer when propagated in 293 and 293-derived cells [41,53]. The choice of sequence used as stuffer in the E3 region may be important as it has been observed that some E3 inserts result in poor virus propagation, perhaps due to interference with fiber expression (F. L. Graham, unpublished data). While it remains unclear what sequences constitute a good E3 stuffer, in general, noncoding sequences with no homology to the HDAd would be preferred. A number of sequences have been found to be stable and not to adversely affect virus propagation, including bacterial plasmid sequences, bacteriophage λ sequences, and human DNA [53–55,58,75]. The use of other E1-complementing cell lines engineered to preclude the formation of RCA is another option [76,77].

Since the HDAd genome does not integrate into the host chromosomes, but rather presumably remains episomal, it is likely that transgene expression will not be permanent. Should transgene expression fade over time, it would be desirable to simply readminister the vector. Unfortunately, this simplistic approach is not possible because the initial administration elicits immunity in the form of neutralizing anti-Ad antibody that renders subsequent readministrations ineffective. One strategy, known as "serotype switching," may help to overcome this problem. In addition to the Ad serotype 5-based helper virus, Parks et al. have generated a serotype 2 helper virus [75]. Helper viruses based on serotypes 1 and 6 have also been generated (F. L. Graham, unpublished results). Therefore, genetically identical HDAds of different serotypes can be generated simply by changing the serotype of the helper virus used for vector amplification. Parks et al. demonstrated that mice injected with a serotype

2 HDAd produced serotype 2 neutralizing antibodies which prevented successful transduction with the same serotype HDAd [75] but did not prevent successful transduction following administration of a serotype 5 HDAd. Successful readministration of HDAd of alternative serotypes has also been demonstrated by Kim et al. [64] (see Section 5.6.1). As discussed in Section 5.2, there are ~50 human serotypes of Ad. Therefore, it may be possible to create a panel of different serotype helper viruses and use these to generate different serotype but genetically identical HDAds. These HDAds could then be given sequentially when transgene expression wanes from the previous vector administration. Because of the large number of Ad serotypes, this could theoretically continue for the lifetime of the patient, although development of the many helper viruses needed for this approach might prove challenging.

5.5.6 Helper Virus Contamination

HDAd produced using the Cre/loxP systems is invariably contaminated with helper virus. Currently, the final purity of HDAd preparations is dependent on two enrichment steps. The first is Cre-mediated packaging signal excision during vector amplification. It has been shown that the level of helper virus contamination is related to the amount of Cre expressed by the producer cells; higher amounts of intracellular Cre result in lower levels of helper virus contamination as a result of higher efficiency packaging signal excision [56,78]. Using producer cells with very high levels of Cre expression, this enrichment step alone results in about 0.4%–0.1% helper virus contamination as determined by DNA-based methods of quantification (Southern blot hybridization and real-time PCR) [56]. Provided that the genome sizes of the HDAd and the helper virus are sufficiently different, further enrichment can be obtained by CsCl ultracentrifugation which can further reduce contamination levels to 0.02%–0.01% [56]. Although current systems cannot produce HDAd free of helper virus, it should be noted that the helper virus is essentially an E1-deleted FGAd at low contaminating amounts which are far below the much higher quantities of FGAd that have been given to numerous patients in clinical trials without adverse affect. It should also be noted that in mouse models, intravenous [41] or intramuscular [79] injection of HDAd with up to 10% helper virus contamination did not reduce the duration of transgene expression or result in significantly higher toxicity compared to preparations with only 0.1%–0.5% contamination. In summary, the low helper virus contamination levels achievable with optimal systems should not compromise safety and efficacy. Nevertheless, an accurate determination of helper virus contamination levels is important since high levels of helper virus contamination may affect experimental results and compromise safety. Knowing the level of helper virus contamination also allows for meaningful comparisons between different studies. A variety of methods for determining helper virus contamination have been reported, all of which can be divided into two basic categories. The first is based on determining the infectious titer of the contaminating helper virus in a HDAd preparation. These methods included plaque assay on 293 cells (or 293-derived cells) or infectious unit assay based on the presence of a reporter transgene in the helper virus. Using these infectious titer assays, the level of helper virus contamination is often reported as a percent or a ratio of infectious helper virus in the vector preparation. However, these methods suffer from a number of shortcomings. First the units of measurement for the HDAd and helper virus are not the same and thus not directly comparable making it difficult to accurately determine the level of helper virus contamination. Second, it has been shown that infectious assays are 10- to 50-fold less sensitive than DNA-based assays described below, thus grossly underestimating the true level of contamination and may not be capable of detecting very low levels of helper virus contamination [41,72]. Alternative methods of determining helper virus contamination use direct measurement of the amount of HDAd and helper viral DNA in a vector preparation. This has been accomplished by standard Southern blot hybridization analysis [54,56,72] or quantitative real-time PCR assays [41,55,56,58]. Not only are these methods more sensitive and reliable, but also the unit of measurement is the same for both the HDAd and the helper virus therefore allowing the level of helper virus contamination to be expressed simply as a percentage of the total DNA.

5.6   *IN VIVO* STUDIES

As of this writing, numerous examples of *in vivo* HDAd-mediated gene transfer through different route of administration (intravenous, intramuscular, brain, subretinal, and intratumoral injection, airway administration) in various small and large animal disease models have been reported

(Table 5.1). The purpose of this chapter is not to provide a comprehensive review of all of these studies. Instead, examples of particular significance or interest are described.

5.6.1   Liver-Directed Gene Therapy

The liver is very attractive target for gene therapy because it is the affected organ in many genetic diseases, the fenestrated structure of its endothelium permits exposure of the parenchymal cells to intravenously delivered vector, and permits secretion of vector encoded therapeutic proteins into the circulation for systemic delivery. To date numerous examples of *in vivo* liver-directed gene therapy for disease models using HDAd have been reported (summarized in Table 5.1). In general, all these studies have demonstrated that HDAd can lead to long-term phenotypic correction in the absence of chronic toxicity supporting the potential of HDAd for clinical applications. The correction of hypercholesterolemia in apolipoprotein E (apoE)-deficient mice is a paradigmatic example of this great potential [64]. In this study the efficacy of FGAd encoding mouse apoE cDNA (FG-Ad5-cE) was compared to a HDAd encoding the mouse apoE cDNA (HD-Ad5-cE), or a HDAd bearing the mouse genomic apoE locus (HD-Ad5-gE). Intravenous injection of ApoE deficient mice with 5×10^{12} vp/kg of FG-Ad5-cE resulted in an immediate fall in plasma cholesterol levels to within normal range (Figure 5.3A). However, this effect was transient and plasma cholesterol levels increased after 28 days, returning to pretreatment levels by 112 days. Correlative with the plasma cholesterol levels, the levels of plasma apoE immediately increased shortly after injection but rapidly declined to pretreatment levels by day 28 (Figure 5.3B). Similarly, intravenous injection of 7.5×10^{12}

TABLE 5.1

Examples of *in Vivo* Studies Using HDAd Vectors in Different Disease Models

Disease		Target Organ	Species	References
Hemophilias	Hemophilia A	Liver	Mouse/dog/human	[41,87,187–190]
	Hemophilia B	Liver	Mouse/dog	[71,86,191]
Inborn errors of metabolism	OTC deficiency	Liver	Mouse	[192]
	GSD 1a	Liver	Mouse	[193]
	Crigler-Najjar syndrome	Liver	Rat	[81]
	ApoE deficiency	Liver	Mouse	[64]
	Familial hypercholesterolemia	Liver	Mouse	[194–196]
	Pompe disease	Liver	Mouse	[197]
Lung	Cystic fibrosis	Lung	Mouse	[146]
Muscular diseases	Muscular dystrophy	Muscle	Mouse	[160–165,167–169,171]
Neurodegenerative diseases	Huntington disease	Brain	Mouse	[180]
Inherited diseases of the eye	Retinitis pigmentosa	Eye	Mouse	[185]
Other monogenic diseases	Leptin deficiency	Liver	Mouse	[60]
Multifactorial disorders	Diabetes mellitus	Liver	Mouse	[82]
	Systemic hypertension	Liver	Mouse	[198]
	Erectile dysfunction	Penile corpora cavernosa	Rat	[199]
Cancer	Liver cancer	Liver tumor	Mouse	[200]
	Breast cancer	Breast tumor	Mouse	[201]

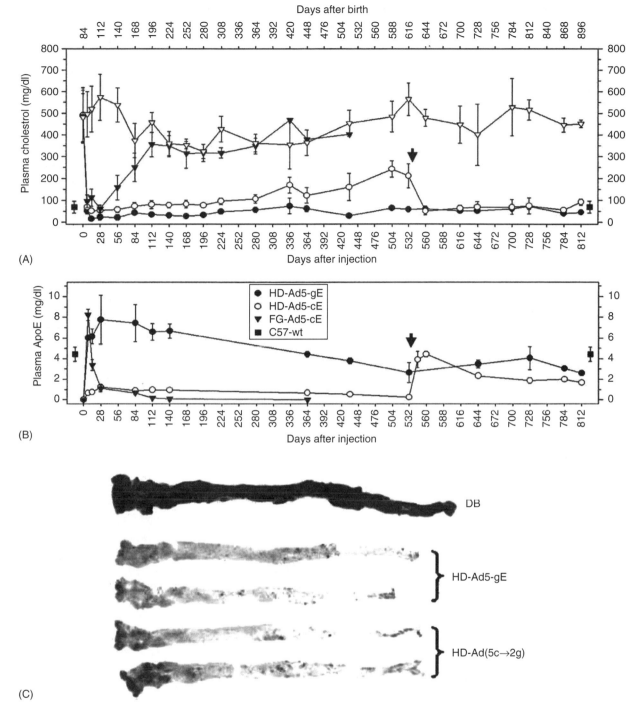

FIGURE 5.3 HDAd-mediated phenotypic correction of atherosclerosis in ApoE-deficient mice. (A) Plasma cholesterol and (B) plasma ApoE levels in ApoE-deficient mice injected with dialysis buffer (white triangle), FG-Ad5-cE (black triangle), HD-Ad5-gE (black circle) or HD-Ad5-cE followed (indicated by the bold arrow) by HD-Ad2-gE (white circle). (C) Aortas from HDAd-treated and control dialysis buffer (DB) treated mice stained with Oil Red at 2.3 years postinjection. Atherosclerotic lesion areas, stained red, determined by quantitative morphometry were 91.45 mm^2 for DB-treated animals; 0.81 and 0.31 mm^2 for HD-Ad5-gE-treated animals; and 5.89 and 1.74 mm^2 for HD-Ad5-cE followed by HD-Ad2-gE-treated animals. (Modified from Kim, I.H., Jozkowicz, A., Piedra, P.A., Oka, K., and Chan, L., *Proc. Natl. Acad. Sci. U S A*, 98, 13282, 2001. With permission.)

particles/kg of HD-Ad5-cE produced a complete and immediate lowering of plasma cholesterol to normal levels, but in contrast to FG-Ad5-cE, the reduced levels lasted about a year before gradually increasing (Figure 5.3A). ApoE appeared in plasma within a week and remained at a level ~25% of wild-

type (<10% of normal plasma levels of apoE is sufficient to maintain normal plasma cholesterol) but slowly declined to <10% of wild-type after about 1.5 years (Figure 5.3B), at which time plasma cholesterol levels rose to ~50% of untreated mice (Figure 5.3A). Intravenous injection of 7.5×10^{12}

particles/kg of HD-Ad5-gE also resulted in a complete and immediate lowering of plasma cholesterol to subnormal levels for about 9 months, with levels subsequently staying within the normal range for the rest of the natural lifespan of the animal (2.5 years) (Figure 5.3A). In this case, plasma apoE reached ~200% wild-type levels within 4 weeks and remained at supraphysiological levels for >4 months at which time it slowly declined to about wild-type levels at 1 year and remained at 60%–90% physiological concentrations for the lifetime of the animals (2.5 years) (Figure 5.3B). As the duration of ApoE expression from HD-Ad5-gE was superior to HD-Ad5-cE it would appear that genomic based transgenes may be more effective than cDNA based transgenes. Simple re-administration of the vector when transgene expression wanes is not possible due to the potent neutralizing anti-Ad antibody response that is elicited by the first administration. Indeed, mice previously treated with HD-Ad5-cE could not be successfully retreated again with the same vector. One solution to overcome this problem is to administer a vector of a different serotype. To evaluate this strategy, Kim et al. generated a serotype 2 version of the genomic ApoE vector (HD-Ad2-gE) [64] and showed that it could be successfully administered to mice previously treated with the serotype 5 HD-Ad5-cE to lower plasma cholesterol levels and raise plasma ApoE levels for the remainder of the animal lives (Figure 5.3A and B). Aortas in all mice, examined at 2.3 years after treatment with HDAds, were essentially free of atherosclerotic lesions as determined by quantitative morphometry (Figure 5.3C) demonstrating that a single injection of HDAd encoding ApoE could confer lifetime protection against aortic atherosclerosis. Kim et al. also investigated the associated toxicities and found that whereas injection of FG-Ad5-cE resulted in significant hepatotoxicity as indicated by significant elevation of AST and ALT (>10- to 20-fold), no such evidence of damage was observed following injection of any of the various HDAd constructs, even after a second administration with the serotype 2 HDAd [64].

In summary, this study demonstrated that a single intravenous injection of HDAd resulted in life-long expression of the therapeutic transgene and permanent phenotypic correction of a genetic disease. Second, negligible hepatotoxicity was associated with HDAd administrations. Third, this study showed the large cloning capacity of the vector which permits delivery of transgenes in their native chromosomal context results in superior kinetics and duration of expression. Indeed, should transgene expression diminish over time, administration of an alternative serotype HDAd is effective at circumventing the humoral immune response generated by the initial treatment.

Encouraging results have also been reported in the Crigler-Najjar syndrome animal model. Crigler-Najjar syndrome type I is a recessively inherited disorder caused by a deficiency of uridine diphospho-glucuronosyl transferase 1 A1 (UTG1A1) and characterized by severe unconjugated hyperbilirubinemia resulting in jaundice and increased risk of kernicterus. Current therapy consists of cumbersome and inconvenient life-long phototherapy to prevent kernicterus or liver transplantation

[80]. Toietta et al. [81] showed that a single systemic injection of a HDAd expressing UTG1A1 at doses of 1×10^{13} vp/kg or 3×10^{12} vp/kg resulted in life-long expression of UTG1A1 and permanent phenotypic correction of hyperbilirubinemia in the Gunn rats, the Crigler-Najjar syndrome animal model. However, at a lower dose of 6×10^{11} vp/kg, only partial, life-long correction was observed.

An interesting application of HDAd for the treatment of diabetes mellitus has also been reported. In this study, two HDAds, one expressing *Neurod*1 (a transcription factor expressed in developing and adult β-cells of the pancreas), and the other expressing *betacellulin* (a β-cell growth factor), were co-injected into diabetic mice at the dose of 3×10^{11} vp and 1×10^{11} vp, respectively [82]. Following systemic vector administration clusters of cells with immunohistochemical and ultrastructural properties similar to pancreatic islet were surprisingly noted in the liver of the treated mice. Remarkably, the diabetic mice also showed a normalization of glucose levels for the duration of the experiment of at least 120 days. These results showed the presence of endocrine pancreatic precursors in the liver and the possibility to manipulate their development to induce pancreatic islet formation by HDAd-mediated gene transfer.

Importantly, long-term expression by HDAd has also recapitulated in a clinically relevant large animal model [83–85]. In the first study addressing the utility of HDAds in a large animal model, three baboons were intravenously injected with $3.3–3.9 \times 10^{11}$ vp/kg of HDAd expressing hAAT [85]. The hAAT expression persisted for more than 1 year in two of the three animals (Figure 5.4). Maximum levels of serum hAAT of 3–4 mg/mL were reached 3–4 weeks postinjection in these two baboons and slowly declined to 8% and 19% of the highest levels after 24 and 16 months, respectively. The slow decline in hAAT expression was attributed to the fact that the baboons were young (7.5 and 9 months old) when injected, and that the decrease in hAAT concentrations was correlative to the growth of the animals. The third baboon injected with vector had significantly lower levels of serum hAAT which rapidly declined to undetectable levels after 2 months due to antibodies formation against the human AAT protein (Figure 5.4). It was significant that no abnormalities in blood cell counts and liver enzymes were observed in these three baboons at any time, starting 3 days postinjection. These results were in sharp contrast with the findings observed in the baboons injected with the FGAd expressing hAAT, in which the transgene levels were measurable for only 3–5 months after the injection (Figure 5.4). In the case of FGAd, this was shown not to be due to the generation of anti-hAAT antibodies but was attributed to the generation of a cellular immune response against viral proteins expressed from the vector backbone resulting in the elimination of vector transduced hepatocytes. These early experiments convincingly demonstrated that HDAd was superior to FGAd with respect to duration of transgene expression and hepatotoxicity in mice and, significantly, in a nonhuman primate. In another large animal study, the intravenous injection of 3×10^{12} vp/kg of a

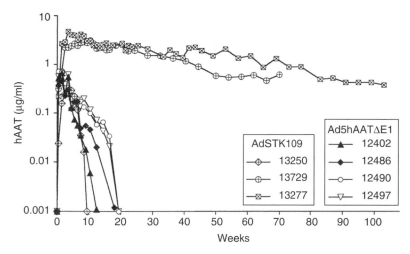

FIGURE 5.4   Serum levels of hAAT in baboons following intravenous administration of the HDAd AdSTK109 or the FGAd AdhAATΔE1. Baboons 12402 and 12486 were injected with 6.2×10^{11} particles/kg of AdhAATΔE1. Baboons 12490 and 12497 were injected with 1.4×10^{12} particles/kg of AdhAATΔE1. Baboons 13250, 13250, and 13277 were injected with 3.3×10^{11} particles/kg, 3.9×10^{11} particles/kg, and 3.6×10^{11} particles/kg, respectively, of AdSTK109. (From Morral, N., O'Neal, W., Rice, K., Leland, M., Kaplan, J., Piedra, P.A., Zhou, H., Parks, R.J., Velji, R., Aguilar-Cordova, E., Wadsworth, S., Graham, F.L., Kochanek, S., Carey, K.D., and Beaudet, A.L., *Proc. Natl. Acad. Sci. U S A*, 96, 12816, 1999. With permission.)

HDAd expressing the canine coagulation factor IX in two hemophilia B dogs resulted in sustained phenotypic improvement of the bleeding diathesis for the duration of the experiment of at least 604 and 446 days [86].

5.6.1.1   Human Clinical Trail

To date, a single human clinical trial has been conducted using HDAd for liver-directed gene therapy. GenStar Therapeutics Corporation sponsored a hemophilia A gene therapy trial in 2001 using HDAd at a dose of 4.3×10^{11} vp/kg delivered by peripheral intravenous injection into one FVIII-deficient patient. That study was never published and little information is known about the outcomes of the trial. For example, the level of helper virus contamination was never disclosed, but the early system of relying on preferential packaging of the HDAd over a helper virus bearing a mutant packaging signal was notorious for high helper virus contamination levels [52,87]. According to the RAC [88]:

> Approximately 5 hours after intravenous infusion of investigational agent (Mini-AdFVIII) the subject developed fever, chills, achiness, back pain, and headache. The fever peaked at 102.6°F, approximately 8 hours after investigational agent infusion, and resolved by about 12 hours. The subject, who has a history of multiple spontaneous bleeds, experienced a spontaneous hemarthrosis of the knee on Day 1 post investigational agent infusion, treated in the usual manner with recombinant factor VIII (r-AHF) with resolution of the bleeding event. The subject also experienced elevations in liver enzyme values with ALT peaking on post-infusion Day 7 and AST values peaking on post-infusion Day 7 and transient declines in factor VII levels and platelet counts. All laboratory values returned to baseline by Day 19 and were not considered clinically serious. Event attributions were not addressed in this submission.

In addition, no measurable increase in FVIII was detected in this single patient, and because the FDA required a log lower dose for the next patient, the trial was subsequently closed by the sponsor [87]. However, because the data was never formally published many important questions regarding this trial remain unanswered and it is impossible to critically interpret what little information is available. This, however, has not prevented unsubstantiated opinions and speculations regarding this failed trial. Although the outcome of this trial was disappointing, the knowledge and technological advancements acquired since 2001 along with the recent improvements in vector production, purity, and characterizations outlined in this chapter may yet resurrect HDAd as an important vector for liver-directed gene therapy, especially considering the plethora of encouraging and compelling preclinical studies obtained since this trial.

5.6.1.2   Threshold Effect to Hepatocytes Transduction

Several preclinical studies in small and large animal models clearly have demonstrated the tremendous potential of HDAd for liver-directed gene therapy (see Section 5.6.1). However, relatively high vector doses were required to achieve the results presented in those studies because such high doses are required to achieve efficient hepatic transduction following systemic intravascular delivery. It is now clear that there is a nonlinear dose response to hepatic transduction, with low doses yielding very low to undetectable levels of transgene expression, but with higher doses resulting in disproportionately high levels of transgene expression. Kupffer cells of the liver appear to play a significant role in the nonlinear dose response by sequestering intravenous Ad [89,90]. Antibodies, both specific and nonspecific for Ad, may also play a significant role in the nonlinear dose response. Studies have shown that the threshold effect to hepatic transduction by Ad is less in antibody-deficient *Rag*-1 and μMT mice [89,90]. Perhaps opsonization of the virion by antibodies may enhance the efficiency of Fc-receptor mediated vector uptake by macrophages such as Kupffer cells. Furthermore, systemic administration of Ad vectors likely results in widespread transduction of a large number of various extrahepatic cell types (e.g., blood cells, endothelium, spleen, lung, etc.) which may also be important components of the barrier to efficient hepatocyte transduction. For example, although the liver takes up more vector than any other individual organ following systemic injection, the total amount of vector that is distributed throughout the body in mice [91], nonhuman primates [92,93], and a human patient [68] is significant on a vector genome copy number per microgram DNA basis.

It has also become clear that the liver microarchitecture plays a critical role in the efficiency of hepatocyte transduction by Ad vectors. The liver fenestrations normally form a structural barrier preventing large circulating macromolecules in the hepatic blood from accessing the hepatocytes. Several studies suggest that the size of the endothelial fenestrations of the liver (≤100 nm) plays a key role in the efficiency of Ad-mediated hepatocyte transduction (Ad virion ≥100 nm) (see Section 5.6.1.4) [94–96]. Specifically, it was demonstrated that there is a positive correlation between the size of the fenestrations and the efficiency of hepatic transduction following systemic administration of Ad [94,95].

5.6.1.3 Acute Toxicity

FGAd and multiply deleted Ad-mediated toxicity appear biphasic; not only does transduction by these early generation Ads, but not HDAd, cause chronic toxicity as described above due to viral gene expression from the vector backbone (late phase), but also results in acute toxicity (early phase). The acute response occurs immediately following vector administration, its severity is dose-dependent and is characterized by elevations in serum pro-inflammatory cytokines consistent with activation of the innate inflammatory immune response [91,92,97,98]. Indeed, the death of a partial OTC-deficient patient, whose clinical course was marked by systemic inflammatory response syndrome (SIRS), disseminated intravascular coagulation and multi-organ failure, was attributed to the acute toxicity from the administration of a second generation (E1 and E4-deleted) Ad vector [68]. Unlike FGAd-mediated chronic toxicity, several lines of evidence suggest that this acute toxicity is not dependent on viral gene expression from the vector backbone. First, this response is observed very shortly after vector injection, likely prior to viral gene expression [91,92,97]. Second, the response is observed following injection of vector rendered transcriptionally inactive by exposure to psoralen and UV in both mice [91,97] and nonhuman primates [92]. These observations have led to the hypothesis that the viral capsid is responsible for triggering the acute inflammatory response and therefore HDAd would also provoke an identical response since the viral capsid is presumably identical for both classes of vectors. Indeed, this was confirmed in nonhuman primates in which dose-dependent

acute toxicity, consistent with activation of the innate inflammatory immune response, was observed following systemic administration of HDAd [99]. In this study, two baboons were systemically injected, via peripheral vein, with 5.6×10^{12} vp/kg or 1.1×10^{13} vp/kg. Approximately 50% hepatocyte transduction was achieved at a dose of 5.6×10^{12} vp/kg (Figure 5.5A). This animal suffered mild and transient acute toxicity consistent with activation of the innate inflammatory response as evident by elevations of serum cytokine IL-6 (Figure 5.5E) but completely recovered by 24 h postinjection. Hepatocyte transduction of 100% was achieved in the animal injected with 1.1×10^{13} vp/kg (Figure 5.5B). However, this animal suffered severe acute toxicity with a stronger inflammatory response as evident by the dramatic increase in serum IL-6 (Figure 5.5F). This animal became moribund and was euthanized 8.5 h postinjection due to poor condition. Transduction of other organs and tissues was also observed in these animals and was significantly more extensive in the high dose animal. For example, substantial transduction of the spleen was noted at 5.6×10^{12} vp/kg (Figure 5.5C) but was even more extensive at 1.1×10^{13} vp/kg (Figure 5.5D). This suggested that a large proportion of the vector dose was not taken up by the liver but that widespread vector dissemination occurred following systemic injection which, as discussed below, may play a role in contributing to the severity of the inflammatory response.

The role of the viral capsid in causing the acute toxicity was further confirmed by Muruve et al. [100] showing that intravenous injection of either FGAd or HDAd into mice induced an acute expression of inflammatory cytokine and chemokine genes in the liver, including interferon-inducible protein 10, macrophage inflammatory protein 2, and TNF-α. Furthermore, like FGAd, HDAd induced the recruitment of CD11b-positive leukocytes to the transduced liver within hours of administration. However, FGAd, but not HDAd, also induced a second phase of liver inflammation, consisting of inflammatory gene expression and CD3-positive lymphocytic infiltrates 7 days postinjection. In contrast, beyond 24 h no infiltrates or expression of inflammatory genes was detected in the livers of mice injected with HDAd. These results demonstrate that HDAd induced intact innate but attenuated adaptive immune responses *in vivo*.

The mechanism(s) responsible for Ad-mediated activation of the acute inflammatory response is not known, although several have been postulated. Vector uptake and activation to secrete proinflammatory cytokines by macrophages and dendritic cells in the spleen have been implicated [91,92]. Indeed, the spleen appears to be a major contributor to IL-12 elevation following systemic Ad in mice [91]. Vector transduction of endothelial cells [101,102], peripheral mononuclear cells [103], and Ad-mediated complement activation [104–106] have all been implicated to play a role in acute toxicity. Several studies also suggest that antibodies may play a role. Both neutralizing anti-Ad antibodies [93,104] and non-neutralizing or naturally occurring (nonspecific cross-reacting) antibodies [89,90,104] may contribute to acute toxicity, perhaps by opsonizing the vp and rendering them more susceptible to

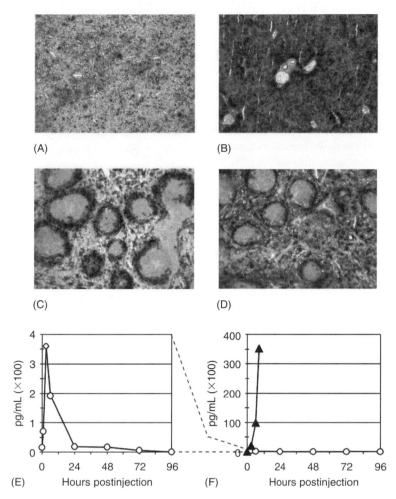

FIGURE 5.5 (See color insert following blank page 206.)   High dose systemic injection of HDAd into nonhuman primates. Baboon injected with (A, C, and E) 5.6×10^{12} vp/kg or (B, D, and F) 1.1×10^{13} vp/kg of HDAd-LacZ. X-gal stained (A and B) liver and (C and D) spleen and (E and F) serum IL-6 from injected baboons. Circles, 5.6×10^{12} vp/kg; triangles, 1.1×10^{13} vp/kg. Note different scales in E and F. (Adapted from Brunetti-Pierri, N., Palmer, D.J., Beaudet, A.L., Carey, K.D., Finegold, M., and Ng, P., *Hum. Gene Ther.*, 15, 35, 2004.)

Fc-mediated uptake by macrophages which in turn may become activated to secrete proinflammatory cytokines. For example, systemic vector administration into nonhuman primates resulted in significantly higher IL-6 levels in animals with neutralizing anti-Ad antibodies compared to naïve animals [93]. Toll-like receptors (TLRs), crucial components in pathogen recognition processes, are recently emerging as another important players in Ad-mediated acute toxicity [107–113].

As discussed above (see Section 5.6.1.2), Kupffer cells of the liver are primarily responsible for the nonlinear dose response by preferentially sequestering systemic Ad, at least in mice, and their activation following uptake of Ad has been implicated to play a role in the acute innate inflammatory response [114]. However, their precise role in acute toxicity remains to be fully elucidated and warrants further discussion because of their potential role in HDAd-mediated, liver-direct gene therapy. Most studies in mice utilizing selective depletion of Kupffer cells prior to Ad administration did not investigate markers of innate inflammatory response such as serum IL-6

[19,115,116]. Of the studies in which this was addressed, depletion of Kupffer cells in mice by gadolinium chloride nearly eliminated TNF but resulted in a more robust IL-6 increase and did not affect NF-κB following systemic Ad compared to mice bearing Kupffer cells [117]. In another study, depletion of Kupffer cells by clodronate liposomes in mice prior to systemic Ad appeared to have no effect on the serum levels of IL-2, IL-4, IL-5, IL-6, TNF-α, and INF-γ [114]. Likewise, Brunetti-Pierri et al. showed that the amount of Kupffer cell transduction did not correlate with serum IL-6 and IL-12 levels [96] and Manickan et al. showed that Kupffer cell death as a consequence of Ad uptake did not result in elevated serum IL-6 [118]. It is also interesting to note that systemic injection of low vector doses into nonhuman primates and mice resulting in low efficiency hepatocyte transduction almost certainly results in substantial transduction of Kupffer cells since they are the barrier responsible for the threshold effect to efficient hepatocyte transduction [89,90]. Yet, these animals exhibit little, if any, manifestations of acute toxicity. Taken together, these studies suggest that vector uptake by Kupffer cells, at least alone, may not necessarily provoke a potent inflammatory response. Indeed it has been suggested that Kupffer cells may, in fact, play a protective role [119]. Clearly, a better understanding of the role of Kupffer cells in Ad-mediated acute toxicity is needed. Another contributing factors may be the variations in responses to vector due to inter-individual variations or pre-existing pathological conditions. Smith et al., for example, have shown that the biodistribution of adenoviral vectors is altered in cirrhotic rats due to the presence of pulmonary intravascular macrophages, which cause a shift in vector uptake from the liver to the lungs. Interestingly, the cirrhotic rats exhibited a potent activation of proinflammatory cytokines as compared to control non-cirrhotic rats [120]. These may have been a contributing factor in the death of one partial OTC-deficient patient given recombinant Ad [68,121].

Ultimately, the precise mechanism responsible for activation of the acute inflammatory response by systemic Ad is multifactorial and complex and remains to be fully elucidated. However, regardless of the precise mechanism, it is likely that a threshold of innate immune activation must first be attained, as a consequence of high dose and systemic exposure of the vector to many cell types and blood borne components, before severe and lethal acute toxicity manifests. Evidence of robust activation of the acute inflammatory response is observed in both rodents and nonhuman primates given comparable systemic high dose Ad (on a per kg basis). However, it is important to emphasize that unlike primates, lethal SIRS does not develop in rodents and in fact, such high doses are well tolerated. This may reflect species to species differences in the quality of the innate immune response or sensitivities of the end organs to pathologic sequelae [91,92]. This likely accounts for the plethora of studies reporting negligible toxicity in mice given high dose HDAd and underscores the importance of safety and toxicity evaluations in larger animals.

5.6.1.4 Overcoming the Threshold to Hepatic Transduction and Acute Toxicity

Given the tremendous potential of HDAd for liver-directed gene therapy, several groups have investigated different strategies to overcome the threshold to hepatocyte transduction and the obstacle of the acute toxicity. Because the severity of the acute response is dose-dependent, one of these approaches aimed at preferential targeting of the vector to the liver thereby allowing the use of lower vector doses. For example, injection of HDAd directly into the surgically isolated liver in nonhuman primates was shown to achieve higher efficiency hepatic transduction with reduced systemic vector dissemination resulting in stable, high level, long-term transgene expression for the duration of the observation period of at least 665 days in one animal and at least 560 days in two other animals (Figure 5.6) [84]. Interestingly, this stability in transgene expression contrasts an earlier study in which transgene expression levels steadily declined to ~10% of peak levels over 1–2 years (Figure 5.4) [85]. The apparent difference in stability of transgene expression in these two studies may be attributed, in part, to differences in the age of the animals at the time of vector administration (juvenile baboons in the study by Morral et al. [85] and adult baboons in the study by Brunetti-Pierri et al. [84]) and, in part, to differences in the backbone sequences, cis-acting elements, and transgene [58,61,62]). Importantly, no chronic toxicity was evident during the entire observation period in either study. However, although the surgical approach has generated encouraging results, the method is not clinically attractive due to its invasiveness.

One outstanding question pertaining to the relevance of liver-directed HDAd-mediated gene therapy is whether high efficiency hepatic transduction necessarily results in severe acute toxicity. This was investigated using the technique of hydrodynamic tail vein injection to delivery HDAd into mice [96]. Hydrodynamic injection entails the rapid injection of a large volume (100 mL/kg injected within 7 s). Interestingly, hydrodynamic injection of HDAd resulted in significantly improved hepatic transduction efficiency, reduced systemic vector dissemination, and, importantly, appeared to reduce the severity of the innate inflammatory response. These results suggested that high efficiency hepatic transduction does not, at least alone, necessarily provoke a potent inflammatory

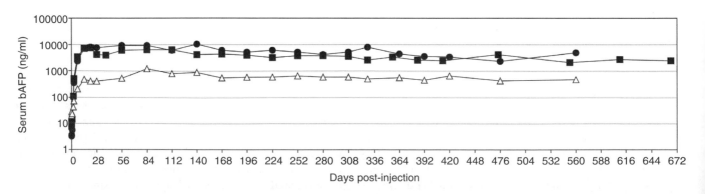

FIGURE 5.6 Stable, high level, long-term transgene expression in nonhuman primates following injection of HDAd directly into the isolated liver. Duration of transgene expression following administration of a HDAd expressing bAFP in three baboons. Black symbols represent 1×10^{12} vp/kg and white symbols represent 1×10^{11} vp/kg. (From Brunetti-Pierri, N., Ng, T., Iannitti, D.A., Palmer, D.J., Beaudet, A.L., Finegold, M.J., Carey, K.D., Cioffi, W.G., and Ng, P., *Hum. Gene Ther.*, 17, 391, 2006.)

response and that systemic vector dissemination may play a major role in the severity of the inflammatory response [96]. As discussed in Section 5.6.1.2, the size of the hepatic fenestrations plays a key role in the efficiency of hepatocytes transduction by Ad vectors and it is believed that the rapid injection of the large volume increases the intrahepatic pressure thereby enlarging the ≤100 nm fenestrations to permit extravasation of the ≥100 HDAd virion increasing access to the hepatocyte [96]. Importantly, this study also suggested that the therapeutic index of HDAd may be improved by preferential delivery into the liver. Unfortunately, this method of delivery is not feasible in large animals because of the rapid large volumes injected. However, considering its tremendous potential, a minimally invasive method to mimic hydrodynamic injection using balloon occlusion catheters was developed to achieve preferential hepatic delivery of the HDAd (Figure 5.7) [83]. This approach was developed to mimic the effect of hydrodynamic injection in larger animals but without rapid, large volume injection. In this method, hepatic venous outflow is occluded using two balloon occlusion catheters percutaneously placed in the inferior vena cava (IVC), above and below the hepatic veins (HV) (Figure 5.7). Because blood entering the liver from the hepatic artery (HA) and portal vein (PV) remains unobstructed, an increase in intrahepatic pressure is achieved, mimicking the high pressures achieved by systemic hydrodynamic injection in the mice. This pseudo-hydrodynamic injection method resulted in high efficiency hepatic transduction with minimal toxicity and stable, long-term transgene expression of at least 413 days [83].

An alternative potential approach to increase targeting to the liver may entail the use of drugs to enlarge liver fenestration [94]. Indeed, *N*-acetylcysteine combined with transient liver ischemia and Na-decanoate have been shown to increase the size of sinusoidal fenestrae and augment Ad-mediated

hepatocyte transduction [95]. However, further studies are needed to determine the real clinical potential of these pharmacological approaches of increasing hepatocyte transduction by Ad for human applications.

Alternative strategies of "masking" the viral capsid have also been reported to attenuate the severity of the innate inflammatory response. Yotnda et al. [122] demonstrated that systemic injection of Ad encapsidated within bilamellar cationic liposomes resulted in a 70%–80% decrease in serum IL-6 compared to unencapsidated virions without compromising hepatic transduction efficiency. Likewise, two independent groups were able to demonstrate that systemic administration of PEGylated Ad into mice resulted in a 50%–70% reduction in serum IL-6 compared to unPEGylated vector without compromising hepatic transduction efficiency [123,124]. Although the precise mechanism by which these masking strategies operate to attenuate the acute inflammatory response is not fully understood, they nevertheless offer an intriguing and potentially important avenue for increasing the therapeutic index of HDAd.

5.6.2 Lung Directed Gene Therapy

FGAds have been the most extensively used vector for pulmonary gene transfer with the goal of treating cystic fibrosis (CF). Although once thought to be an ideal vector for CF gene therapy, more than a decade of research has revealed a number of serious shortcomings and the enthusiasm for FGAd has greatly diminished. First, pulmonary delivery of FGAd in small animals, large animals, and humans is inefficient [125–129]. It was discovered that the cellular receptor for Ad (and other viral vectors) resided on the basolateral surface of the airway epithelial cells and that the tight junctions prevented vector–receptor interactions required for transduction [130]. However, the transient disruption of the tight junctions can significantly increase the efficiency of transduction thus dramatically decreasing the vector dose required to achieve therapeutic levels of transduction. Various strategies have been proposed to relax the tight junctions and improve adenoviral entry into airway epithelia and they include: calcium phosphate co-precipitates [131], EGTA [132], EDTA [133], polycations [134], sodium caprate [135], L-α-lysophosphatidylcholine [136,137], and other agents. Second, pulmonary delivery of FGAd resulted in dose-dependent inflammation and pneumonia [125,138–141] beginning about 3–4 days post-administration and became progressively more severe before eventually resolving. This has been attributed to expression of the viral genes present in the vector backbone of FGAds which is directly cytotoxic and also provokes an adaptive cellular immune response against the transduced cells consequently resulting in transient transgene expression and long-term, chronic toxicity [21–25,39]. This dose-dependent toxicity was partially addressed by using second generation Ad vectors deleted/mutated of other early viral genes, such as E2 or E4, in addition to E1 which further diminished, but did not eliminate pulmonary inflammation and pneumonia, likely as a result of continued leaky expression of the viral late genes [23,32,35,142].

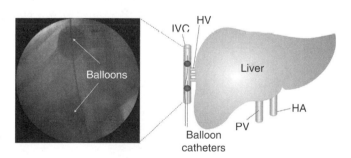

FIGURE 5.7 Pseudo-hydrodynamic injection of HDAd into nonhuman primates. Two balloon occlusion catheters are percutaneously positioned in the IVC above and below the HV. Inflation of the balloons results in hepatic venous outflow occlusion from the HV while blood inflow from the PV and HA increases the intrahepatic pressure. Following 30 min of occlusion, the balloon occlusion catheters are removed from the animal and HDAd is administered by systemic peripheral intravenous injection. (From Brunetti-Pierri, N., Stapleton, G.E., Palmer, D.J., Zuo, Y., Mane, V.P., Finegold, M.J., Beaudet, A.L., Leland, M.M., Mullins, C.E., and Ng, P., *Mol. Ther.*, 15, 732, 2007. With permission.)

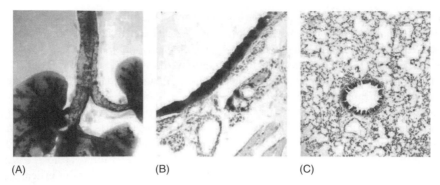

(A)   (B)   (C)

FIGURE 5.8 (See color insert following blank page 206.)   Epithelia transduction of the (A) proximal and distal airway, (B) trachea, and (C) bronchiole of mice 3 days post-intranasal administration of 2×10^{10} vp of HDAd-K18LacZ. Blue areas represent HDAd transduced cells. (From Toietta, G., Koehler, D.R., Finegold, M.J., Lee, B., Hu, J., and Beaudet, A.L., *Mol. Ther.*, 7, 649, 2003. With permission.)

Toietta et al. [143] compared the safety and efficacy of HDAd versus FGAd in mice for pulmonary transduction. In that study, a HDAd (called HDAd-K18LacZ) which expresses a nuclear-localized bacterial β-galactosidase reporter protein from the control elements of the human cytokeratin 18 (K18) gene was constructed. The expression pattern of K18 is similar to that of Cftr [144]; in the lungs, K18, like Cftr, is expressed in the epithelium of large airways and bronchioles and in submucosal glands, but with little expression in the alveoli. In contrast to commonly used viral promoters, the K18 promoter is less likely to suffer host "shut-off" and could reduce immune stimulation resulting in inappropriate expression in APCs [66]. The large cloning capacity of HDAds makes this vector ideal to accommodate the relatively large K18 control elements (4.1 kb) and reporter or therapeutic cDNAs. To assess the utility of this vector, mice were first given a 25 μL intranasal administration of 0.4 M EGTA to disrupt tight junctions. Thirty minutes later, 2×10^{10} vp of HDAd-K18LacZ in 50 μL was administered intranasally and their lungs were analyzed 3 days later. Extensive transduction was observed in the proximal and distal airways (Figure 5.8A). Trachea sections revealed extensive transduction of the epithelium and submucosal glands (Figure 5.8B). Extensive transduction was also

achieved in the bronchiolar epithelium with little transduction of the alveolar cells (Figure 5.8C). To directly compare the toxicity of HDAd versus FGAd, EGTA and then 4×10^{10} vp of either HDAd or FGAd were intranasally administered to mice and their lungs were examined 3 days later. As expected, administration of FGAd resulted in pulmonary inflammation with focal peribronchial lymphocytic infiltrates and focal alveolar macrophages (Figure 5.9A). In contrast, the lungs of mice given HDAd were free of inflammation (Figure 5.9B) and indistinguishable from saline treated animals. The duration of HDAd-mediated pulmonary transgene expression was also investigated using HDAd-K18hAFP, a HDAd expressing the human α-fetoprotein (hAFP) as a reporter gene [145] from the K18 promoter. In these studies, mice were first pretreated with EGTA and, 30 min later, 2×10^{10} vp of HDAd-K18hAFP in 50 μL was intranasally administered. Serum and BAL fluid were collected from these mice at various times post-administration and revealed the presence of hAFP. Importantly, serum levels peaked at 50 IU/mL at 1 week and then decreased over time but remained detectable for at least 15 weeks post-administration. AFP immunohistochemistry performed on the lungs revealed the presence of hAFP positive airway epithelia cells, consistent with the interpretation that the serum

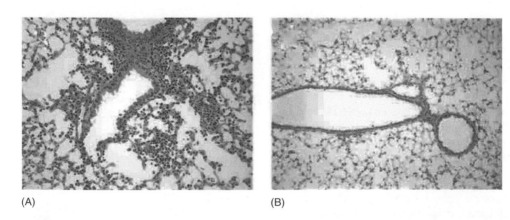

(A)   (B)

FIGURE 5.9 (See color insert following blank page 206.)   Histology of representative mouse lung 3 days post-intranasal administration of a total of 4×10^{10} vp of an (A) FGAd or (B) HDAd. (From Toietta, G., Koehler, D.R., Finegold, M.J., Lee, B., Hu, J., and Beaudet, A.L., *Mol. Ther.*, 7, 649, 2003. With permission.)

and BAL hAFP had originated from these transduced cells. Taken together, the results of this study indicated that intranasal administration of HDAd following disruption of tight junctions resulted in high efficiency pulmonary transduction and use of the K18 promoter greatly restricts expression to the desired target cell types relevant to CF gene therapy. Additionally, HDAd, unlike FGAd, do not cause pulmonary inflammation (presumably due to the absence of viral gene expression), at least in mice, and at least when examined 3 days post-administration. Moreover, HDAd-mediated pulmonary transgene expression could persist for at least 15 weeks.

To test the therapeutic potential of HDAd for CF gene therapy, Koehler et al. constructed a vector called HDAd-K18CFTR bearing the human CFTR cDNA expressed from the K18 control elements [146]. This vector was found to express properly localized CFTR in cultured cells and in the apical airway epithelia of mice following intranasal administration of 2×10^{11} vp (preceded by EGTA pretreatment). Moreover, CFTR RNA and protein were present in whole lung and bronchioles for the duration of the observation period of at least 28 days. Importantly, the authors also demonstrated that HDAd-K18CFTR could improve resistance to acute lung infection in cftr knockout mice [146]. In these experiments, cftr knockout mice or their wild-type litter mates were given either saline or 2×10^{10} vp of HDAd-K18CFTR by intranasal administration following pretreatment with EGTA. These mice were then challenged 7 days later with a clinical isolate of *Burkholderia cepacia* (Bcc). Five days post-Bcc, the lungs were harvested for histopathology. Wild-type mice given either saline (Figure 5.10A) or HDAd (Figure 5.10B) prior to Bcc had minimal to moderate patchy inflammation in the form of bronchiolitis and scattered foci of mild to moderate pneumonia. As expected, cftr knockout mice given saline prior to Bcc had marked inflammation in the form of severe pneumonia and bronchiolitis, inflammatory cell infiltration, exudation, consolidation, respiratory cell hyperplasia, necrosis, and high lung bacterial counts (Figure 5.10C). In sharp contrast, cftr knockout mice given HDAd-K18CFTR prior to Bcc had significantly less severe histopathology similar to what was observed in the wild-type littermates given Bcc and, importantly, their lungs harbored no live Bcc (Figure 5.10D). Taken together, these results indicate that HDAd can express properly localized CFTR in the appropriate target cell types for CF gene therapy *in vivo* as well as suggest that this vector could benefit CF patients by reducing susceptibility to opportunistic pathogens.

Despite the encouraging results in mice described above the requirement for two separate administrations—one to deliver EGTA to disrupt the tight junction and then a second, 30 min later, to deliver the vector—is suboptimal in terms of safety and efficacy because this increases the procedure time

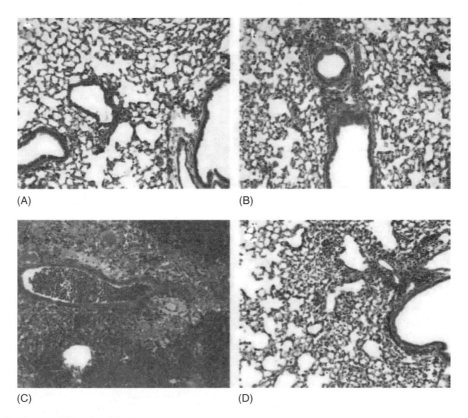

(A) (B)

(C) (D)

FIGURE 5.10 (See color insert following blank page 206.) Lung histopathology following Bcc challenge. Representative H&E stained sections from wild-type littermates (A and B) and CFTR knockout mice (C and D) receiving saline (A and C) or HDAd-K18hCFTR (B and D) prior to bacterial challenge. (From Koehler, D.R., Sajjan, U., Chow, Y.H., Martin, B., Kent, G., Tanswell, A.K., McKerlie, C., Forstner, J.F., and Hu, J., *Proc. Natl. Acad. Sci. U S A*, 100, 15364, 2003. With permission.)

and compromises transduction efficiency because both the EGTA and the vector must be applied to the same location, which is not guaranteed in the case of separate, independent administrations. This obstacle was subsequently addressed by Koehler et al. [137] who demonstrated that high efficiency pulmonary transduction by HDAd can be obtained by formulating the vector in 0.1% L-α-lysophosphatidylcholine (LPC) thus permitting a single administration containing the vector and the tight junction opening agent. The intranasal delivery as performed in mice (spontaneous liquid inhalation) is not applicable to larger animals and a clinically relevant method of vector delivery was therefore developed by Koehler et al. [137]. In this study, an intracorporeal nebulizing catheter called the AeroProbe (Trudell Medical International) designed to aerosolize material directly into the trachea and lungs was utilized. Using the AeroProbe, Koehler et al. [137] aerosolized 1.25 mL of 0.1% LPC containing 5×10^{11} vp HDAd-K18LacZ into the lungs of 2.8 kg rabbits. The X-gal staining revealed extensive transduction of the trachea (Figure 5.11A). Although the intratracheal aerosolization was intended to deliver vector to the entire lung, X-gal staining revealed high interlobular variation. Nevertheless, in those lobes that received vector (Figure 5.11B through D), exceedingly high and unprecedented levels of transduction were achieved in all cell types of the proximal and distal airway epithelium (Figure 5.11E), from the trachea to terminal bronchioles. All rabbits, including those given LPC only as controls, exhibited a transient decrease in dynamic lung compliance immediately following aerosol delivery. Fever and mild-to-moderate patchy pneumonia without edema were also observed leading the investigators to speculate that LPC may have been a contributing factor and that these may be eliminated/minimized by further optimizing the dose of LPC or vector. Nevertheless, this study is significant because it is the first to demonstrate high efficiency transduction of the airway epithelium in a large animal which had been a major obstacle to CF gene therapy. The above described strategy of delivering HDAd in rabbits has recently been applied to nonhuman primates and has yielded similar encouraging preliminary results [147].

Despite the promising and compelling studies described above, several outstanding issues still remain to be addressed. First, the efficiency of transduction will likely be greatly reduced in severely diseased CF lungs possibly rendering it necessary to treat less affected lungs that may be more amenable to gene transfer. Second, the requirement to transiently open the tight junctions in order to achieve efficient transduction raises safety concerns since this may allow bacteria to penetrate the epithelium, especially from CF patient airway epithelium colonized with several microbial pathogens [148]. However, this potential concern should be weighed against potential benefits. Risk would depend upon the patient particular colonizing strain(s) and amounts, as well as their serum level of antibodies against the bacteria. Intravenous and inhaled antibiotics routinely prescribed to CF patients could be administered prophylactically around the time of gene therapy. Third, the duration of pulmonary transgene expression from HDAd remains unknown. In the case of liver-directed

gene therapy, HDAd has been shown to mediate transgene expression for 1 year or more in small and large animal models (see Section 5.6.1). However, the rate of turnover of airway epithelial cells is greater than hepatocytes and may likely be

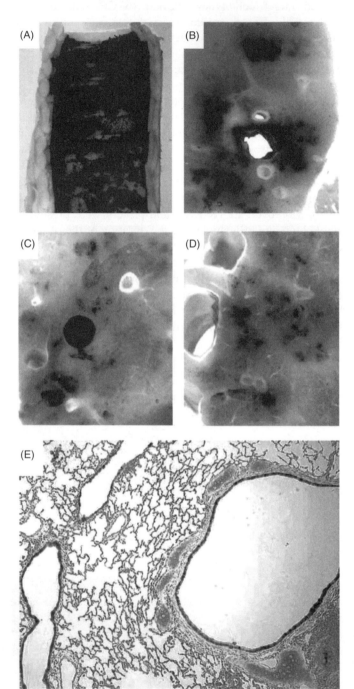

FIGURE 5.11 (See color insert following blank page 206.) Pulmonary transduction in 2.8 kg rabbits following AeroProbe-mediated intratracheal aerosolization of 1.25 mL 0.1% LPC containing 5×10^{11} vp of HDAd-K18LacZ. X-gal stained (A) trachea, (B) right upper lobe, (C) left lower lobe, (D) right lower lobe, and (E) bronchus and bronchioles. Blue areas represent HDAd transduced cells. (From Koehler, D.R., Frndova, H., Leung, K., Louca, E., Palmer, D., Ng, P., McKerlie, C., Cox, P., Coates, A.L., and Hu, J., *J. Gene Med.*, 7, 1409, 2005. With permission.)

even greater in the CF lung. Fourth, vector re-administration may be challenging due to an adaptive humoral immune response against the HDAd capsid. Strategies such as transient immunosuppression at the time of vector administration [34,149–153], masking the vector with PEG [154,155], or switching to a different serotype vector have been successful for intrapulmonary re-administration [156,157] and may need to be considered. It should be noted that the vector genotype may influence re-administration. For example, Wilson and coworkers have demonstrated that, unlike first-generation Ad vectors (E1-deleted), second generation Ad vectors (E1 and E4-deleted) failed to generate a neutralizing antibody response against the vector in the BAL fluid in both mice and rhesus monkeys following intrapulmonary delivery [142]. The authors hypothesized that the absence of viral E4 gene expression from the second generation Ad vector resulted in an attenuated humoral response due to reduced inflammation and further implied that re-administration would therefore be possible although this was not directly demonstrated. Because HDAd are deleted of all viral genes, including E4, an attenuated adaptive humoral response may result but this remains to be determined as is its affect, if any, on re-administration. Fifth, uniform delivery of vector to all lung lobes will likely be important and this has yet to be achieved. With respect to the AeroProbe, further optimization will be needed and one possibility is to use the AeroProbe in conjunction with a bronchoscope to permit targeted lobar aerosolization [158].

5.6.3 MUSCLE DIRECTED GENE THERAPY

Duchenne muscular dystrophy (DMD) is a lethal X-linked degenerative muscle disease caused by mutations in the dystrophin gene, with a frequency of 1 in 3500 male births. Dystrophin is an essential structural component of the skeletal muscle cell membrane, linking intracellular actin filaments with the dystrophin-associated proteins in the sarcolemma. Dystrophin deficiency results in instability of the muscle cell membrane causing muscle fiber degeneration. Because of lack of effective treatments, DMD has been considered a prime candidate for gene therapy, the goal of which is to transfer normal copies of the dystrophin gene into the muscle fibers of affected patients. Various gene therapy vectors have been investigated and FGAd has actually been the first vector to successfully deliver a human dystrophin cDNA by intramuscular administration into *mdx* mice, a genetic and physiopathological model for DMD [159]. The length of the dystrophin cDNA (14 kb) precluded its inclusion into most gene therapy viral vectors and HDAds with their large cloning capacity can accommodate not only one full length dystrophin cDNA but also two copies of the gene [160,161]. Indeed, the first *in vivo* application of HDAds was transduction of skeletal muscle for DMD gene therapy [50]. HDAd vector expressing full-length dystrophin has been shown to restore in the skeletal muscle the full dystrophin–glycoprotein complex, resulting in a reduced level of muscle degeneration and amelioration of the physiological and pathological indices of muscle disease

[160–164]. Compared to *mdx* mice treated as neonates, higher doses of vector are needed to treat adult *mdx* mice because of inefficient infection of mature muscle by Ad [163]. Moreover, direct intramuscular injection of HDAd encoding full-length murine dystrophin resulted in only transient expression in adult-injected mice [164] because of a humoral immune response against the dystrophin protein. Importantly, such a response has not been observed in immunodeficient SCID mice suggesting that sustained expression could be achieved in the absence of an immune response to the transgene product [164,165]. A similar outcome could be predicted to occur in humans as well, since many DMD patients have large dystrophin gene deletions preventing the expression of epitopes encoded by deleted exons. In these cases the dystrophin produced by HDAd-transduced cells could be perceived as a neoantigen thus preventing long-term expression [166]. To overcome this problem, co-delivery of immunomodulatory molecules such as CTL4AIg have been successfully employed in immunocompetent mice to prolonged dystrophin expression [167,168].

HDAd vectors have also been explored for in utero gene therapy of DMD. The immaturity of the immune system of the fetus coupled with the survival advantage of dystrophin-expressing muscle fibers over dystrophin-deficient muscle fibers may offer a significant advantage for this promising therapeutic strategy. In this approach HDAds were found to be less toxic than the FGAd and able to drive stable transgene expression and restoration of the disarcoglycan complex [169].

Respiratory insufficiency due to progressive weakness of the diaphragm and other respiratory muscles is the main cause of death of DMD patients when they are in their late teens or early twenties [170]. Therefore, diaphragm-directed gene therapy has been investigated in the *mdx* mouse using HDAd injected directly into the diaphragm exposed through a laparotomy [171]. This approach has lead to reversal of functional abnormalities of dystrophic diaphragms but at the same time it also points to the difficulties of DMD gene therapy which, in order to be effective, requires the transduction of a critical number of muscle fibers throughout the body.

The muscle remains an attractive tissue for transduction because, like the liver, it may serve a more general purpose in gene therapy: (1) muscle constitutes as much as 40% of the total body mass and much of it is readily accessible, (2) skeletal myocytes can be transduced *in vivo*, (3) skeletal myofibers have a relatively long half-life and therefore represent a stable platform for transgene expression, (4) muscle is highly vascularized and skeletal muscle can efficiently secrete recombinant proteins into the circulation for systemic delivery, and (5) high seroprevalence in the adult human population of preexisting anti-Ad neutralizing antibodies, an obstacle for intravenous vector delivery, may be minimized through localized delivery into the muscle. This strategy was investigated in a study by Maione et al. [79] in which mice were intramuscularly injected with a HDAd expressing mouse erythropoietin as a marker gene. All mice intramuscularly injected with a dose of 1×10^6 IU or 3×10^5 IU retained transgene expression for at least 4 months. At a dose of 3×10^6 IU, 30% of the mice slowly lost transgene expression. In contrast, all mice injected intramuscularly with FGAd lost transgene expression by day 21. The effect of preexisting anti-Ad antibody on the effectiveness of intramuscular administration was also investigated. The authors found that successful intramuscular gene transfer could be accomplished in pre-immunized mice although a 30- to 100-fold higher dose was required to achieve 87% and 100%, respectively, the levels of transgene expression in naïve mice. This is in sharp contrast to intravenous gene transfer in which 60% of pre-immunized mice were completely refractory to transduction while the remaining 40% were transduced but exhibited transgene expression levels comparable to naïve mice intravenously injected with 1000-fold lower dose.

5.6.4 BRAIN- AND EYE-DIRECTED GENE THERAPY

The use of adenoviral vectors to deliver genes to the nervous system holds great promise for therapeutic applications. Stable, long-term transgene expression, in the absence of toxicity, is critical if gene therapy is to be successful in treating a variety of central nervous system (CNS) disorders ranging from simple monogenic disorders (such as spinal muscular atrophy, Lesch–Nyhan syndrome, leukodystrophies, lysosomal storage diseases, amyotrophic lateral sclerosis, Huntington's disease) to complex diseases, including common neurodegenerative disorders such as Parkinson's disease and Alzheimer's disease. Because of their ability to infect postmitotic cells, including cells of the CNS [172], and to mediate long-term transgene expression, Ad-based vectors are particularly attractive for these applications. Unlike the rapid decline observed in transgene expression in peripheral organs following intravenous administration, FGAd-mediated transduction of adult brain cells leads to stable transgene expression [173,174]. Brain intraparenchymal injection of FGAd vectors in fact elicits a minimal, transient local inflammation which does not compromise sustained vector-driven expression. It is thought that FGAd-mediated long-term transgene expression occurs because the brain is relatively protected from the effects of the immune response, and in fact, Ad injection into the brain results in an ineffective T cell response against brain-transduced cells [175]. However, the immune system can and does respond to antigenic stimuli in the brain [176], and if a peripheral immune response against Ad is elicited after natural infection or vector readministration, loss of transgene expression and chronic inflammation are observed [177]. Interestingly, these detrimental effects are not seen when HDAd is used [177,178]. Moreover, when directly compared with FGAd, HDAd can mediate significantly higher transgene expression levels and induce a substantially reduced inflammatory and immune response as shown by a reduction in the increases seen in proinflammatory cytokines (IL-1β and TNF-α) [179].

Encouraging preliminary results, including a significant inhibition of huntingtin protein aggregation, have been recently reported in a Huntington's disease animal model following stereotaxic injection into the striatum of a HDAd vector expressing a short hairpin RNA silencing the Huntington disease gene [180]. However, this type of approach is far from being ready for clinical applications since the delivered vector had limited brain distribution not extending beyond a few millimeters from the needle track and because demonstration of efficacy using clinically relevant endpoints is still lacking. Being a complex organ with discrete and intricate interconnections between various types of cells, the brain is a complicated target for vector delivery aiming to correct a disease with diffuse involvement. Diseases requiring localized gene therapy to discrete set of neurons such as Parkinson's disease or brain tumors are more amenable for treatment. For brain cancer applications FGAd vectors have been used successfully in clinical trial for glioblastoma multiforme and intratumoral injection has been associated with increased survival in two different trials [181,182]. Given the high risk that FGAd treatment of glioblastoma multiforme can be compromised by exposure to natural Ad infection, HDAd vectors encoding a regulatable therapeutic genes could offer a significantly safer and more effective treatment for patients with this type as well as other forms of brain cancer [183,184].

There have been a limited number of studies investigating HDAd vectors for eye gene therapy. In one of these studies, subretinal injections of HDAd vectors were shown to transduce and rescue cells from the neurosensory retina in a mouse model of retinal degeneration following [185]. Moreover, HDAd was shown to be able to mediate long-term therapeutic genes in the retinal pigment epithelium (RPE) following subretinal injection without evidence of adverse immune reactions or significant toxicity [186].

5.7 CONCLUDING REMARKS

HDAd possesses many characteristics that make them attractive vectors for gene therapy of a wide variety of genetic and acquired diseases. However, acute toxicity provoked by the viral capsid is the most significant obstacle currently hindering the clinical application of this otherwise promising technology. Indeed, the tragic death of Jesse Gelsinger has justifiably raised serious concerns over the use of all Ad-based vectors, especially for intravascular delivery and liver-directed gene therapy. However, the potential of using HDAd, including intravascular delivery and liver-directed gene therapy, should not be dismissed but should instead proceed with caution considering the wealth of encouraging and compelling studies amassed to date. Studies to elucidate the mechanism(s) of Ad-mediated activation of the innate inflammatory response are clearly needed. Perhaps with a better understanding of this phenomenon, strategies can be developed to minimize, if not eliminate, this serious but ultimately surmountable problem of acute toxic response. When this is accomplished, then HDAd should be able to provide sustained, high level transgene expression with no further toxicity.

ACKNOWLEDGMENTS

PN is supported by the National Institutes of Health (P50 HL59314, R01 DK067324, and P51 RR13986), the Texas

Affiliate of the American Heart Association (0465102Y), and the Cystic Fibrosis Foundation (NG0530 and NG05G0). NB-P is supported by the National Institutes of Health (K99 DK077447) and the Texas Affiliate of the American Heart Association (0765032Y).

REFERENCES

1. Shenk, T., *Adenoviridae: The viruses and Their Replication, in Fields Viology*, Fields, B. N., Knipe, D. M., Howely, P. M., (Eds.), Philadelphia, PA, Lipponcott-Raven Publishers, 1996, pp. 2111–2148.
2. Bergelson, J. M., Cunningham, J. A., Droguett, G., Kurt-Jones, E. A., Krithivas, A., Hong, J. S., Horwitz, M. S., Crowell, R. L., and Finberg, R. W., Isolation of a common receptor for Coxsackie B viruses and adenoviruses 2 and 5, *Science*, 275, 1320–1323, 1997.
3. Hong, S. S., Karayan, L., Tournier, J., Curiel, D. T., and Boulanger, P. A., Adenovirus type 5 fiber knob binds to MHC class I alpha2 domain at the surface of human epithelial and B lymphoblastoid cells, *Embo. J.*, 16, 2294–2306, 1997.
4. Tomko, R. P., Xu, R., and Philipson, L., HCAR and MCAR: The human and mouse cellular receptors for subgroup C adenoviruses and group B coxsackieviruses, *Proc. Natl. Acad. Sci. USA*, 94, 3352–3356, 1997.
5. Wickham, T. J., Mathias, P., Cheresh, D. A., and Nemerow, G. R., Integrins alpha v beta 3 and alpha v beta 5 promote adenovirus internalization but not virus attachment, *Cell*, 73, 309–319, 1993.
6. Wickham, T. J., Segal, D. M., Roelvink, P. W., Carrion, M. E., Lizonova, A., Lee, G. M., and Kovesdi, I., Targeted adenovirus gene transfer to endothelial and smooth muscle cells by using bispecific antibodies, *J. Virol.*, 70, 6831–6838, 1996.
7. Goldman, M., Su, Q., and Wilson, J. M., Gradient of RGD-dependent entry of adenoviral vector in nasal and intrapulmonary epithelia: Implications for gene therapy of cystic fibrosis, *Gene Ther.*, 3, 811–818, 1996.
8. Mellman, I., The importance of being acid: The role of acidification in intracellular membrane traffic, *J. Exp. Biol.*, 172, 39–45, 1992.
9. Leopold, P. L., Ferris, B., Grinberg, I., Worgall, S., Hackett, N. R., and Crystal, R. G., Fluorescent virions: Dynamic tracking of the pathway of adenoviral gene transfer vectors in living cells, *Hum. Gene Ther.*, 9, 367–378, 1998.
10. Greber, U. F., Willetts, M., Webster, P., and Helenius, A., Step-wise dismantling of adenovirus 2 during entry into cells, *Cell*, 75, 477–486, 1993.
11. Berkner, K. L., Development of adenovirus vectors for the expression of heterologous genes, *Biotechniques*, 6, 616–629, 1988.
12. Graham, F. L. and Prevec, L., Vaccines: New approaches to immunological problems. In *Adenovirus-based Expression Vectors and Recombinant Vaccines*, Ellis, R. W., (Ed.), Boston, MA, Butterworth-Heinemann, 1992, pp. 363–389.
13. Hitt, M. M., Addison, C. L., and Graham, F. L., Human adenovirus vectors for gene transfer into mammalian cells, *Adv. Pharmacol.*, 40, 137–206, 1997.
14. Hitt, M. M., Parks, R., and Graham, F. L., The development of human gene therapy. In *Structure and Genetic Organization of Adenovirus Vectors*, Friedman, T., (Ed.), Cold Spring Harbor, NY, Cold Spring Harbor Laboratory Press, 1999, pp. 61–86.
15. Breyer, B., Jiang, W., Cheng, H., Zhou, L., Paul, R., Feng, T., and He, T. C., Adenoviral vector-mediated gene transfer for human gene therapy, *Curr. Gene Ther.*, 1, 149–162, 2001.
16. Graham, F. L., Smiley, J., Russell, W. C., and Nairn, R., Characteristics of a human cell line transformed by DNA from human adenovirus type 5, *J. Gen. Virol.*, 36, 59–74, 1977.
17. Ruben, M., Bacchetti, S., and Graham, F., Covalently closed circles of adenovirus 5 DNA, *Nature*, 301, 172–174, 1983.
18. Danthinne, X. and Imperiale, M. J., Production of first generation adenovirus vectors: A review, *Gene Ther.*, 7, 1707–1714, 2000.
19. Wolff, G., Worgall, S., van Rooijen, N., Song, W. R., Harvey, B. G., and Crystal, R. G., Enhancement of in vivo adenovirus-mediated gene transfer and expression by prior depletion of tissue macrophages in the target organ, *J. Virol.*, 71, 624–629, 1997.
20. Worgall, S., Wolff, G., Falck-Pedersen, E., and Crystal, R. G., Innate immune mechanisms dominate elimination of adenoviral vectors following in vivo administration, *Hum. Gene Ther.*, 8, 37–44, 1997.
21. Morral, N., O'Neal, W., Zhou, H., Langston, C., and Beaudet, A., Immune responses to reporter proteins and high viral dose limit duration of expression with adenoviral vectors: Comparison of E2a wild type and E2a deleted vectors, *Hum. Gene Ther.*, 8, 1275–1286, 1997.
22. Dai, Y., Schwarz, E. M., Gu, D., Zhang, W. W., Sarvetnick, N., and Verma, I. M., Cellular and humoral immune responses to adenoviral vectors containing factor IX gene: Tolerization of factor IX and vector antigens allows for long-term expression, *Proc. Natl. Acad. Sci. USA*, 92, 1401–1405, 1995.
23. Yang, Y., Nunes, F. A., Berencsi, K., Furth, E. E., Gonczol, E., and Wilson, J. M., Cellular immunity to viral antigens limits E1-deleted adenoviruses for gene therapy, *Proc. Natl. Acad. Sci. USA*, 91, 4407–4411, 1994.
24. Yang, Y., Li, Q., Ertl, H. C., and Wilson, J. M., Cellular and humoral immune responses to viral antigens create barriers to lung-directed gene therapy with recombinant adenoviruses, *J. Virol.*, 69, 2004–2015, 1995.
25. Yang, Y., Xiang, Z., Ertl, H. C., and Wilson, J. M., Upregulation of class I major histocompatibility complex antigens by interferon gamma is necessary for T-cell-mediated elimination of recombinant adenovirus-infected hepatocytes in vivo, *Proc. Natl. Acad. Sci. USA*, 92, 7257–7261, 1995.
26. Kafri, T., Morgan, D., Krahl, T., Sarvetnick, N., Sherman, L., and Verma, I., Cellular immune response to adenoviral vector infected cells does not require de novo viral gene expression: Implications for gene therapy, *Proc. Natl. Acad. Sci. USA*, 95, 11377–11382, 1998.
27. Nevins, J. R., Mechanism of activation of early viral transcription by the adenovirus E1A gene product, *Cell*, 26, 213–220, 1981.
28. Imperiale, M. J., Kao, H. T., Feldman, L. T., Nevins, J. R., and Strickland, S., Common control of the heat shock gene and early adenovirus genes: Evidence for a cellular E1A-like activity, *Mol. Cell. Biol.*, 4, 867–874, 1984.
29. Mittereder, N., Yei, S., Bachurski, C., Cuppoletti, J., Whitsett, J. A., Tolstoshev, P., and Trapnell, B. C., Evaluation of the efficacy and safety of in vitro, adenovirus-mediated transfer of the human cystic fibrosis transmembrane conductance regulator cDNA, *Hum. Gene Ther.*, 5, 717–729, 1994.
30. Amalfitano, A., Hauser, M. A., Hu, H., Serra, D., Begy, C. R., and Chamberlain, J. S., Production and characterization of improved adenovirus vectors with the E1, E2b, and E3 genes deleted, *J. Virol.*, 72, 926–933, 1998.
31. Engelhardt, J. F., Ye, X., Doranz, B., and Wilson, J. M., Ablation of E2A in recombinant adenoviruses improves transgene persistence and decreases inflammatory response in mouse liver, *Proc. Natl. Acad. Sci. USA*, 91, 6196–6200, 1994.

32. Engelhardt, J. F., Litzky, L., and Wilson, J. M., Prolonged trans-gene expression in cotton rat lung with recombinant adenoviruses defective in E2a, *Hum. Gene Ther.*, 5, 1217–1229, 1994.

33. Gao, G. P., Yang, Y., and Wilson, J. M., Biology of adenovirus vectors with E1 and E4 deletions for liver-directed gene therapy, *J. Virol.*, 70, 8934–8943, 1996.

34. Yang, Y., Nunes, F. A., Berencsi, K., Gonczol, E., Engelhardt, J. F., and Wilson, J. M., Inactivation of E2a in recombinant adenoviruses improves the prospect for gene therapy in cystic fibrosis, *Nat. Genet.*, 7, 362–369, 1994.

35. Goldman, M. J., Litzky, L. A., Engelhardt, J. F., and Wilson, J. M., Transfer of the CFTR gene to the lung of nonhuman primates with E1-deleted, E2a-defective recombinant adenoviruses: A preclinical toxicology study, *Hum. Gene Ther.*, 6, 839–851, 1995.

36. Dedieu, J. F., Vigne, E., Torrent, C., Jullien, C., Mahfouz, I., Caillaud, J. M., Aubailly, N., Orsini, C., Guillaume, J. M., Opolon, P., Delaere, P., Perricaudet, M., and Yeh, P., Long-term gene delivery into the livers of immunocompetent mice with E1/E4-defective adenoviruses, *J. Virol.*, 71, 4626–4637, 1997.

37. Wang, Q., Greenburg, G., Bunch, D., Farson, D., and Finer, M. H., Persistent transgene expression in mouse liver following in vivo gene transfer with a delta E1/delta E4 adenovirus vector, *Gene Ther.*, 4, 393–400, 1997.

38. Fang, B., Wang, H., Gordon, G., Bellinger, D. A., Read, M. S., Brinkhous, K. M., Woo, S. L., and Eisensmith, R. C., Lack of persistence of E1- recombinant adenoviral vectors containing a temperature-sensitive E2A mutation in immunocompetent mice and hemophilia B dogs, *Gene Ther.*, 3, 217–222, 1996.

39. O'Neal, W. K., Zhou, H., Morral, N., Aguilar-Cordova, E., Pestaner, J., Langston, C., Mull, B., Wang, Y., Beaudet, A. L., and Lee, B., Toxicological comparison of E2a-deleted and first-generation adenoviral vectors expressing alpha1-antitrypsin after systemic delivery, *Hum. Gene Ther.*, 9, 1587–1598, 1998.

40. Lusky, M., Christ, M., Rittner, K., Dieterle, A., Dreyer, D., Mourot, B., Schultz, H., Stoeckel, F., Pavirani, A., and Mehtali, M., In vitro and in vivo biology of recombinant adenovirus vectors with E1, E1/E2A, or E1/E4 deleted, *J. Virol.*, 72, 2022–2032, 1998.

41. Reddy, P. S., Sakhuja, K., Ganesh, S., Yang, L., Kayda, D., Brann, T., Pattison, S., Golightly, D., Idamakanti, N., Pink-staff, A., Kaloss, M., Barjot, C., Chamberlain, J. S., Kaleko, M., and Connelly, S., Sustained human factor VIII expression in hemophilia A mice following systemic delivery of a gutless adenoviral vector, *Mol. Ther.*, 5, 63–73, 2002.

42. Amalfitano, A. and Parks, R. J., Separating fact from fiction: Assessing the potential of modified adenovirus vectors for use in human gene therapy, *Curr. Gene Ther.*, 2, 111–133, 2002.

43. Lewis, A. M., Jr. and Rowe, W. P., Isolation of two plaque variants from the adenovirus type 2-simian virus 40 hybrid population which differ in their efficiency in yielding simian virus 40, *J. Virol.*, 5, 413–420, 1970.

44. Deuring, R., Klotz, G., and Doerfler, W., An unusual symmetric recombinant between adenovirus type 12 DNA and human cell DNA, *Proc. Natl. Acad. Sci. U S A*, 78, 3142–3146, 1981.

45. Mitani, K., Graham, F. L., Caskey, C. T., and Kochanek, S., Rescue, propagation, and partial purification of a helper virus-dependent adenoviral vector, *Proc. Natl. Acad. Sci. U S A*, 92, 3854–3858, 1995.

46. Fisher, K. J., Choi, H., Burda, J., Chen, S. J., and Wilson, J. M., Recombinant adenovirus deleted of all viral genes for gene therapy of cystic fibrosis, *Virology*, 217, 11–22, 1996.

47. Parks, R. J. and Graham, F. L., A helper-dependent system for adenovirus vector production helps define a lower limit for efficient DNA packaging, *J. Virol.*, 71, 3293–3298, 1997.

48. Haecker, S. E., Stedman, H. H., Balice-Gordon, R. J., Smith, D. B., Greelish, J. P., Mitchell, M. A., Wells, A., Sweeney, H. L., and Wilson, J. M., In vivo expression of full-length human dystrophin from adenoviral vectors deleted of all viral genes, *Hum. Gene Ther.*, 7, 1907–1914, 1996.

49. Kumar-Singh, R. and Chamberlain, J. S., Encapsidated adenovirus minichromosomes allow delivery and expression of a 14 kb dystrophin cDNA to muscle cells, *Hum. Mol. Genet.*, 5, 913–921, 1996.

50. Kochanek, S., Clemens, P. R., Mitani, K., Chen, H. H., Chan, S., and Caskey, C. T., A new adenoviral vector: Replacement of all viral coding sequences with 28 kb of DNA independently expressing both full-length dystrophin and beta-galactosidase, *Proc. Natl. Acad. Sci. U S A*, 93, 5731–5736, 1996.

51. Gräble, M. and Hearing, P., Adenovirus type 5 packaging domain is composed of a repeated element that is functionally redundant, *J. Virol.*, 64, 2047–2056, 1990.

52. Alemany, R., Dai, Y., Lou, Y. C., Sethi, E., Prokopenko, E., Josephs, S. F., and Zhang, W. W., Complementation of helper-dependent adenoviral vectors: Size effects and titer fluctuations, *J. Virol. Methods*, 68, 147–159, 1997.

53. Parks, R. J., Chen, L., Anton, M., Sankar, U., Rudnicki, M. A., and Graham, F. L., A helper-dependent adenovirus vector system: Removal of helper virus by Cre-mediated excision of the viral packaging signal, *Proc. Natl. Acad. Sci. U S A*, 93, 13565–13570, 1996.

54. Ng, P., Beauchamp, C., Evelegh, C., Parks, R., and Graham, F. L., Development of a FLP/frt system for generating helper-dependent adenoviral vectors, *Mol. Ther.*, 3, 809–815, 2001.

55. Umana, P., Gerdes, C. A., Stone, D., Davis, J. R., Ward, D., Castro, M. G., and Lowenstein, P. R., Efficient FLPe recombinase enables scalable production of helper-dependent adenoviral vectors with negligible helper-virus contamination, *Nat. Biotechnol.*, 19, 582–585, 2001.

56. Palmer, D. and Ng, P., Improved system for helper-dependent adenoviral vector production, *Mol. Ther.*, 8, 846–852, 2003.

57. Palmer, D.J. and Ng, P., *Methods for the Production and Characterization of Helper-dependent Adenoviral Vectors*, Cold Spring Harbor Press, 2007.

58. Sandig, V., Youil, R., Bett, A. J., Franlin, L. L., Oshima, M., Maione, D., Wang, F., Metzker, M. L., Savino, R., and Caskey, C. T., Optimization of the helper-dependent adenovirus system for production and potency in vivo, *Proc. Natl. Acad. Sci. U S A*, 97, 1002–1007, 2000.

59. Bett, A. J., Prevec, L., and Graham, F. L., Packaging capacity and stability of human adenovirus type 5 vectors, *J. Virol.*, 67, 5911–5921, 1993.

60. Morsy, M. A., Gu, M., Motzel, S., Zhao, J., Lin, J., Su, Q., Allen, H., Franlin, L., Parks, R. J., Graham, F. L., Kochanek, S., Bett, A. J., and Caskey, C. T., An adenoviral vector deleted for all viral coding sequences results in enhanced safety and extended expression of a leptin transgene, *Proc. Natl. Acad. Sci. U S A*, 95, 7866–7871, 1998.

61. Parks, R. J., Bramson, J. L., Wan, Y., Addison, C. L., and Graham, F. L., Effects of stuffer DNA on transgene expression from helper-dependent adenovirus vectors, *J. Virol.*, 73, 8027–8034, 1999.

62. Schiedner, G., Hertel, S., Johnston, M., Biermann, V., Dries, V., and Kochanek, S., Variables affecting in vivo performance of high-capacity adenovirus vectors, *J. Virol.*, 76, 1600–1609, 2002.

63. Ng, P., Evelegh, C., Cummings, D., and Graham, F. L., Cre levels limit packaging signal excision efficiency in the Cre/loxP helper-dependent adenoviral vector system, *J. Virol.*, 76, 4181–4189, 2002.

64. Kim, I. H., Jozkowicz, A., Piedra, P. A., Oka, K., and Chan, L., Lifetime correction of genetic deficiency in mice with a single injection of helper-dependent adenoviral vector, *Proc. Natl. Acad. Sci. U S A*, 98, 13282–13287, 2001.

65. Schiedner, G., Morral, N., Parks, R. J., Wu, Y., Koopmans, S. C., Langston, C., Graham, F. L., Beaudet, A. L., and Kochanek, S., Genomic DNA transfer with a high-capacity adenovirus vector results in improved in vivo gene expression and decreased toxicity, *Nat. Genet.*, 18, 180–183, 1998.

66. Pastore, L., Morral, N., Zhou, H., Garcia, R., Parks, R. J., Kochanek, S., Graham, F. L., Lee, B., and Beaudet, A. L., Use of a liver-specific promoter reduces immune response to the transgene in adenoviral vectors, *Hum. Gene Ther.*, 10, 1773–1781, 1999.

67. Shi, C. X., Hitt, M., Ng, P., and Graham, F. L., Superior tissue-specific expression from tyrosinase and prostate-specific antigen promoters/enhancers in helper-dependent compared with first-generation adenoviral vectors, *Hum. Gene Ther.*, 13, 211–224, 2002.

68. Raper, S. E., Chirmule, N., Lee, F. S., Wivel, N. A., Bagg, A., Gao, G. P., Wilson, J. M., and Batshaw, M. L., Fatal systemic inflammatory response syndrome in a ornithine transcarbamylase deficient patient following adenoviral gene transfer, *Mol. Genet. Metab.*, 80, 148–158, 2003.

69. Assessment of adenoviral vector safety and toxicity: Report of the National Institutes of Health Recombinant DNA Advisory Committee, *Hum. Gene Ther.*, 13, 3–13, 2002.

70. Hutchins, B., Sajjadi, N., Seaver, S., Shepherd, A., Bauer, S. R., Simek, S., Carson, K., and Aguilar-Cordova, E., Working toward an adenoviral vector testing standard, *Mol. Ther.*, 2, 532–534, 2000.

71. Ehrhardt, A. and Kay, M. A., A new adenoviral helper-dependent vector results in long-term therapeutic levels of human coagulation factor IX at low doses in vivo, *Blood*, 99, 3923–3930, 2002.

72. Kreppel, F., Biermann, V., Kochanek, S., and Schiedner, G., A DNA-based method to assay total and infectious particle contents and helper virus contamination in high-capacity adenoviral vector preparations, *Hum. Gene Ther.*, 13, 1151–1156, 2002.

73. Puntel, M., Curtin, J. F., Zirger, J. M., Muhammad, A. K., Xiong, W., Liu, C., Hu, J., Kroeger, K. M., Czer, P., Sciascia, S., Mondkar, S., Lowenstein, P. R., and Castro, M. G., Quantification of high-capacity helper-dependent adenoviral vector genomes in vitro and in vivo, using quantitative Taq-Man real-time polymerase chain reaction, *Hum. Gene Ther.*, 17, 531–544, 2006.

74. Palmer, D. J. and Ng, P., Physical and infectious titers of helper-dependent adenoviral vectors: A method of direct comparison to the adenovirus reference material, *Mol. Ther.*, 10, 792–798, 2004.

75. Parks, R., Evelegh, C., and Graham, F., Use of helper-dependent adenoviral vectors of alternative serotypes permits repeat vector administration, *Gene Ther.*, 6, 1565–1573, 1999.

76. Fallaux, F. J., Bout, A., van der Velde, I., van den Wollenberg, D. J., Hehir, K. M., Keegan, J., Auger, C., Cramer, S. J., van Ormondt, H., van der Eb, A. J., Valerio, D., and Hoeben, R. C., New helper cells and matched early region 1-deleted adenovirus vectors prevent generation of replication-competent adenoviruses, *Hum. Gene Ther.*, 9, 1909–1917, 1998.

77. Schiedner, G., Hertel, S., and Kochanek, S., Efficient transformation of primary human amniocytes by E1 functions of Ad5: Generation of new cell lines for adenoviral vector production, *Hum. Gene Ther.*, 11, 2105–2116, 2000.

78. Ng, P., Parks, R. J., and Graham, F. L., Preparation of helper-dependent adenoviral vectors, *Methods Mol. Med.*, 69, 371–388, 2002.

79. Maione, D., Della Rocca, C., Giannetti, P., D'Arrigo, R., Liberatoscioli, L., Franlin, L. L., Sandig, V., Ciliberto, G., La Monica, N., and Savino, R., An improved helper-dependent adenoviral vector allows persistent gene expression after intramuscular delivery and overcomes preexisting immunity to adenovirus, *Proc. Natl. Acad. Sci. U S A*, 98, 5986–5991, 2001.

80. Strauss, K. A., Robinson, D. L., Vreman, H. J., Puffenberger, E. G., Hart, G., and Morton, D. H., Management of hyperbilirubinemia and prevention of kernicterus in 20 patients with Crigler-Najjar disease, *Eur. J. Pediatr.*, 165, 306–319, 2006.

81. Toietta, G., Mane, V. P., Norona, W. S., Finegold, M. J., Ng, P., McDonagh, A. F., Beaudet, A. L., and Lee, B., Lifelong elimination of hyperbilirubinemia in the Gunn rat with a single injection of helper-dependent adenoviral vector, *Proc. Natl. Acad. Sci. U S A*, 102, 3930–3935, 2005.

82. Kojima, H., Fujimiya, M., Matsumura, K., Younan, P., Imaeda, H., Maeda, M., and Chan, L., NeuroD-betacellulin gene therapy induces islet neogenesis in the liver and reverses diabetes in mice, *Nat. Med.*, 9, 596–603, 2003.

83. Brunetti-Pierri, N., Stapleton, G. E., Palmer, D. J., Zuo, Y., Mane, V. P., Finegold, M. J., Beaudet, A. L., Leland, M. M., Mullins, C. E., and Ng, P., Pseudo-hydrodynamic delivery of helper-dependent adenoviral vectors into non-human primates for liver-directed gene therapy, *Mol. Ther.*, 15, 732–740, 2007.

84. Brunetti-Pierri, N., Ng, T., Iannitti, D. A., Palmer, D. J., Beaudet, A. L., Finegold, M. J., Carey, K. D., Cioffi, W. G., and Ng, P., Improved hepatic transduction, reduced systemic vector dissemination, and long-term transgene expression by delivering helper-dependent adenoviral vectors into the surgically isolated liver of nonhuman primates, *Hum. Gene Ther.*, 17, 391–404, 2006.

85. Morral, N., O'Neal, W., Rice, K., Leland, M., Kaplan, J., Piedra, P. A., Zhou, H., Parks, R. J., Velji, R., Aguilar-Cordova, E., Wadsworth, S., Graham, F. L., Kochanek, S., Carey, K. D., and Beaudet, A. L., Administration of helper-dependent adenoviral vectors and sequential delivery of different vector serotype for long term liver directed gene transfer in baboons, *Proc. Natl. Acad. Sci. U S A*, 96, 12816–12821, 1999.

86. Brunetti-Pierri, N., Nichols, T. C., McCorquodale, S., Merricks, E., Palmer, D. J., Beaudet, A. L., and Ng, P., Sustained phenotypic correction of canine hemophilia B after systemic administration of helper-dependent adenoviral vector, *Hum. Gene Ther.*, 16, 811–820, 2005.

87. White, G. I. and Monahan, P.E., *Gene Therapy for Hemophilia A.*, Blackwell, 2005.

88. Recombinant DNA Advisory Committee, Serious, Possibly Associated and Unexpected adverse events reported for Human Gene Transfer Protocols. Reporting Period: 05/01/01–08/01/01. http://www4.od.nih.gov/oba/rac/SAE_rpts/Mod0901s/Sep01_MODs.htm.

89. Tao, N., Gao, G. P., Parr, M., Johnston, J., Baradet, T., Wilson, J. M., Barsoum, J., and Fawell, S. E., Sequestration of adenoviral vector by Kupffer cells leads to a nonlinear dose response of transduction in liver, *Mol. Ther.*, 3, 28–35, 2001.

90. Schiedner, G., Hertel, S., Johnston, M., Dries, V., van Rooijen, N., and Kochanek, S., Selective depletion or blockade of Kupffer cells leads to enhanced and prolonged hepatic transgene expression using high-capacity adenoviral vectors, *Mol. Ther.*, 7, 35–43, 2003.

91. Zhang, Y., Chirmule, N., Gao, G. P., Qian, R., Croyle, M., Joshi, B., Tazelaar, J., and Wilson, J. M., Acute cytokine response to systemic adenoviral vectors in mice is mediated by dendritic cells and macrophages, *Mol. Ther.*, 3, 697–707, 2001.

92. Schnell, M. A., Zhang, Y., Tazelaar, J., Gao, G. P., Yu, Q. C., Qian, R., Chen, S. J., Varnavski, A. N., LeClair, C., Raper, S. E.,

and Wilson, J. M., Activation of innate immunity in nonhuman primates following intraportal administration of adenoviral vectors, *Mol. Ther.*, 3, 708–722, 2001.

93. Varnavski, A. N., Zhang, Y., Schnell, M., Tazelaar, J., Louboutin, J. P., Yu, Q. C., Bagg, A., Gao, G. P., and Wilson, J. M., Preexisting immunity to adenovirus in rhesus monkeys fails to prevent vector-induced toxicity, *J. Virol.*, 76, 5711–5719, 2002.

94. Lievens, J., Snoeys, J., Vekemans, K., Van Linthout, S., de Zanger, R., Collen, D., Wisse, E., and De Geest, B., The size of sinusoidal fenestrae is a critical determinant of hepatocyte transduction after adenoviral gene transfer, *Gene Ther.*, 11, 1523–1531, 2004.

95. Snoeys, J., Lievens, J., Wisse, E., Jacobs, F., Duimel, H., Collen, D., Frederik, P., and De Geest, B., Species differences in transgene DNA uptake in hepatocytes after adenoviral transfer correlate with the size of endothelial fenestrae, *Gene Ther.*, 14, 604–612, 2007.

96. Brunetti-Pierri, N., Palmer, D. J., Mane, V., Finegold, M., Beaudet, A. L., and Ng, P., Increased hepatic transduction with reduced systemic dissemination and proinflammatory cytokines following hydrodynamic injection of helper-dependent adenoviral vectors, *Mol. Ther.*, 12, 99–106, 2005.

97. Muruve, D. A., Barnes, M. J., Stillman, I. E., and Libermann, T. A., Adenoviral gene therapy leads to rapid induction of multiple chemokines and acute neutrophil-dependent hepatic injury in vivo, *Hum. Gene Ther.*, 10, 965–976, 1999.

98. Liu, Q. and Muruve, D. A., Molecular basis of the inflammatory response to adenovirus vectors, *Gene Ther.*, 10, 935–940, 2003.

99. Brunetti-Pierri, N., Palmer, D. J., Beaudet, A. L., Carey, K. D., Finegold, M., and Ng, P., Acute toxicity after high-dose systemic injection of helper-dependent adenoviral vectors into nonhuman primates, *Hum. Gene Ther.*, 15, 35–46, 2004.

100. Muruve, D. A., Cotter, M. J., Zaiss, A. K., White, L. R., Liu, Q., Chan, T., Clark, S. A., Ross, P. J., Meulenbroek, R. A., Maelandsmo, G. M., and Parks, R. J., Helper-dependent adenovirus vectors elicit intact innate but attenuated adaptive host immune responses in vivo, *J. Virol.*, 78, 5966–5972, 2004.

101. Ramalingam, R., Rafii, S., Worgall, S., Hackett, N. R., and Crystal, R. G., Induction of endogenous genes following infection of human endothelial cells with an E1(−) E4(+) adenovirus gene transfer vector, *J. Virol.*, 73, 10183–10190, 1999.

102. Morral, N., O'Neal, W. K., Rice, K., Leland, M. M., Piedra, P. A., Aguilar-Cordova, E., Carey, K. D., Beaudet, A. L., and Langston, C., Lethal toxicity, severe endothelial injury, and a threshold effect with high doses of an adenoviral vector in baboons, *Hum. Gene Ther.*, 13, 143–154, 2002.

103. Higginbotham, J. N., Seth, P., Blaese, R. M., and Ramsey, W. J., The release of inflammatory cytokines from human peripheral blood mononuclear cells in vitro following exposure to adenovirus variants and capsid, *Hum. Gene Ther.*, 13, 129–141, 2002.

104. Cichon, G., Boeckh-Herwig, S., Schmidt, H. H., Wehnes, E., Muller, T., Pring-Akerblom, P., and Burger, R., Complement activation by recombinant adenoviruses, *Gene Ther.*, 8, 1794–1800, 2001.

105. Jiang, H., Wang, Z., Serra, D., Frank, M. M., and Amalfitano, A., Recombinant adenovirus vectors activate the alternative complement pathway, leading to the binding of human complement protein C3 independent of anti-ad antibodies, *Mol. Ther.*, 10, 1140–1142, 2004.

106. Kiang, A., Hartman, Z. C., Everett, R. S., Serra, D., Jiang, H., Frank, M. M., and Amalfitano, A., Multiple innate inflammatory responses induced after systemic adenovirus vector delivery depend on a functional complement system, *Mol. Ther.*, 14, 588–598, 2006.

107. Basner-Tschakarjan, E., Gaffal, E., O'Keeffe, M., Tormo, D., Limmer, A., Wagner, H., Hochrein, H., and Tuting, T., Adenovirus efficiently transduces plasmacytoid dendritic cells resulting in TLR9-dependent maturation and IFN-alpha production, *J. Gene Med.*, 8, 1300–1306, 2006.

108. Iacobelli-Martinez, M. and Nemerow, G. R., Preferential activation of Toll-like receptor nine by CD46-utilizing adenoviruses, *J. Virol.*, 81, 1305–1312, 2007.

109. Hartman, Z. C., Black, E. P., and Amalfitano, A., Adenoviral infection induces a multi-faceted innate cellular immune response that is mediated by the toll-like receptor pathway in A549 cells, *Virology*, 358, 357–372, 2007.

110. Hartman, Z. C., Kiang, A., Everett, R. S., Serra, D., Yang, X. Y., Clay, T. M., and Amalfitano, A., Adenovirus infection triggers a rapid, MyD88-regulated transcriptome response critical to acute-phase and adaptive immune responses in vivo, *J. Virol.*, 81, 1796–1812, 2007.

111. Cerullo, V., Seiler, M. P., Mane, V., Brunetti-Pierri, N., Clarke, C., Bertin, T. K., Rodgers, J. R., and Lee, B., Toll-like receptor 9 triggers an innate immune response to helper-dependent adenoviral vectors, *Mol. Ther.*, 15, 378–385, 2007.

112. Hensley, S. E. and Amalfitano, A., Toll-like receptors impact on safety and efficacy of gene transfer vectors, *Mol. Ther.*, published online June 5, 2007.

113. Zhu, J., Huang, X., and Yang, Y., Innate immune response to adenoviral vectors is mediated by both Toll-like receptor-dependent and -independent pathways, *J. Virol.*, 81, 3170–3180, 2007.

114. Schiedner, G., Bloch, W., Hertel, S., Johnston, M., Molojavyi, A., Dries, V., Varga, G., Van Rooijen, N., and Kochanek, S., A hemodynamic response to intravenous adenovirus vector particles is caused by systemic Kupffer cell-mediated activation of endothelial cells, *Hum. Gene Ther.*, 14, 1631–1641, 2003.

115. Kuzmin, A. I., Finegold, M. J., and Eisensmith, R. C., Macrophage depletion increases the safety, efficacy and persistence of adenovirus-mediated gene transfer in vivo, *Gene Ther.*, 4, 309–316, 1997.

116. Kuzmin, A. I., Galenko, O., and Eisensmith, R. C., An immunomodulatory procedure that stabilizes transgene expression and permits readministration of E1-deleted adenovirus vectors, *Mol. Ther.*, 3, 293–301, 2001.

117. Lieber, A., He, C. Y., Meuse, L., Schowalter, D., Kirillova, I., Winther, B., and Kay, M. A., The role of Kupffer cell activation and viral gene expression in early liver toxicity after infusion of recombinant adenovirus vectors, *J. Virol.*, 71, 8798–8807, 1997.

118. Manickan, E., Smith, J. S., Tian, J., Eggerman, T. L., Lozier, J. N., Muller, J., and Byrnes, A. P., Rapid Kupffer cell death after intravenous injection of adenovirus vectors, *Mol. Ther.*, 13, 108–117, 2006.

119. Jooss, K. and Chirmule, N., Immunity to adenovirus and adeno-associated viral vectors: Implications for gene therapy, *Gene Ther.*, 10, 955–963, 2003.

120. Smith, J. S., Tian, J., Lozier, J. N., and Byrnes, A. P., Severe pulmonary pathology after intravenous administration of vectors in cirrhotic rats, *Mol. Ther.*, 9, 932–941, 2004.

121. Raper, S. E., Yudkoff, M., Chirmule, N., Gao, G. P., Nunes, F., Haskal, Z. J., Furth, E. E., Propert, K. J., Robinson, M. B., Magosin, S., Simoes, H., Speicher, L., Hughes, J., Tazelaar, J., Wivel, N. A., Wilson, J. M., and Batshaw, M. L., A pilot study of in vivo liver-directed gene transfer with an adenoviral vector in partial ornithine transcarbamylase deficiency, *Hum. Gene Ther.*, 13, 163–175, 2002.

122. Yotnda, P., Chen, D. H., Chiu, W., Piedra, P. A., Davis, A., Templeton, N. S., and Brenner, M. K., Bilamellar cationic liposomes protect adenovectors from preexisting humoral immune responses, *Mol. Ther.*, 5, 233–241, 2002.

123. Mok, H., Palmer, D. J., Ng, P., and Barry, M. A., Evaluation of polyethylene glycol modification of first-generation and helper-dependent adenoviral vectors to reduce innate immune responses, *Mol. Ther.*, 11, 66–79, 2005.

124. Croyle, M. A., Le, H. T., Linse, K. D., Cerullo, V., Toietta, G., Beaudet, A., and Pastore, L., PEGylated helper-dependent adenoviral vectors: Highly efficient vectors with an enhanced safety profile, *Gene Ther.*, 12, 579–587, 2005.

125. Joseph, P. M., O'Sullivan, B. P., Lapey, A., Dorkin, H., Oren, J., Balfour, R., Perricone, M. A., Rosenberg, M., Wadsworth, S. C., Smith, A. E., St George, J. A., and Meeker, D. P., Aerosol and lobar administration of a recombinant adenovirus to individuals with cystic fibrosis. I. Methods, safety, and clinical implications, *Hum. Gene Ther.*, 12, 1369–1382, 2001.

126. Grubb, B. R., Pickles, R. J., Ye, H., Yankaskas, J. R., Vick, R. N., Engelhardt, J. F., Wilson, J. M., Johnson, L. G., and Boucher, R. C., Inefficient gene transfer by adenovirus vector to cystic fibrosis airway epithelia of mice and humans, *Nature*, 371, 802–806, 1994.

127. Harvey, B. G., Leopold, P. L., Hackett, N. R., Grasso, T. M., Williams, P. M., Tucker, A. L., Kaner, R. J., Ferris, B., Gonda, I., Sweeney, T. D., Ramalingam, R., Kovesdi, I., Shak, S., and Crystal, R. G., Airway epithelial CFTR mRNA expression in cystic fibrosis patients after repetitive administration of a recombinant adenovirus, *J. Clin. Invest.*, 104, 1245–1255, 1999.

128. Zuckerman, J. B., Robinson, C. B., McCoy, K. S., Shell, R., Sferra, T. J., Chirmule, N., Magosin, S. A., Propert, K. J., Brown-Parr, E. C., Hughes, J. V., Tazelaar, J., Baker, C., Goldman, M. J., and Wilson, J. M., A phase I study of adenovirus-mediated transfer of the human cystic fibrosis transmembrane conductance regulator gene to a lung segment of individuals with cystic fibrosis, *Hum. Gene Ther.*, 10, 2973–2985, 1999.

129. Perricone, M. A., Morris, J. E., Pavelka, K., Plog, M. S., O'Sullivan, B. P., Joseph, P. M., Dorkin, H., Lapey, A., Balfour, R., Meeker, D. P., Smith, A. E., Wadsworth, S. C., and St George, J. A., Aerosol and lobar administration of a recombinant adenovirus to individuals with cystic fibrosis. II. Transfection efficiency in airway epithelium, *Hum. Gene Ther.*, 12, 1383–1394, 2001.

130. Pickles, R. J., Fahrner, J. A., Petrella, J. M., Boucher, R. C., and Bergelson, J. M., Retargeting the coxsackievirus and adenovirus receptor to the apical surface of polarized epithelial cells reveals the glycocalyx as a barrier to adenovirus-mediated gene transfer, *J. Virol.*, 74, 6050–6057, 2000.

131. Lee, J. H., Zabner, J., and Welsh, M. J., Delivery of an adenovirus vector in a calcium phosphate coprecipitate enhances the therapeutic index of gene transfer to airway epithelia, *Hum. Gene Ther.*, 10, 603–613, 1999.

132. Chu, Q., St George, J. A., Lukason, M., Cheng, S. H., Scheule, R. K., and Eastman, S. J., EGTA enhancement of adenovirus-mediated gene transfer to mouse tracheal epithelium in vivo, *Hum. Gene Ther.*, 12, 455–467, 2001.

133. Wang, G., Zabner, J., Deering, C., Launspach, J., Shao, J., Bodner, M., Jolly, D. J., Davidson, B. L., and McCray, P. B. Jr., Increasing epithelial junction permeability enhances gene transfer to airway epithelia in vivo, *Am. J. Respir. Cell Mol. Biol.*, 22, 129–138, 2000.

134. Kaplan, J. M., Pennington, S. E., St George, J. A., Woodworth, L. A., Fasbender, A., Marshall, J., Cheng, S. H., Wadsworth, S. C.,

Gregory, R. J., and Smith, A. E., Potentiation of gene transfer to the mouse lung by complexes of adenovirus vector and polycations improves therapeutic potential, *Hum. Gene Ther.*, 9, 1469–1479, 1998.

135. Johnson, L. G., Vanhook, M. K., Coyne, C. B., Haykal-Coates, N., and Gavett, S. H., Safety and efficiency of modulating paracellular permeability to enhance airway epithelial gene transfer in vivo, *Hum. Gene Ther.*, 14, 729–747, 2003.

136. Limberis, M., Anson, D. S., Fuller, M., and Parsons, D. W., Recovery of airway cystic fibrosis transmembrane conductance regulator function in mice with cystic fibrosis after single-dose lentivirus-mediated gene transfer, *Hum. Gene Ther.*, 13, 1961–1970, 2002.

137. Koehler, D. R., Frndova, H., Leung, K., Louca, E., Palmer, D., Ng, P., McKerlie, C., Cox, P., Coates, A. L., and Hu, J., Aerosol delivery of an enhanced helper-dependent adenovirus formulation to rabbit lung using an intratracheal catheter, *J. Gene Med.*, 7, 1409–1420, 2005.

138. Simon, R. H., Engelhardt, J. F., Yang, Y., Zepeda, M., Weber-Pendleton, S., Grossman, M., and Wilson, J. M., Adenovirus-mediated transfer of the CFTR gene to lung of nonhuman primates: Toxicity study, *Hum. Gene Ther.*, 4, 771–780, 1993.

139. Yei, S., Mittereder, N., Wert, S., Whitsett, J. A., Wilmott, R. W., and Trapnell, B. C., In vivo evaluation of the safety of adenovirus-mediated transfer of the human cystic fibrosis transmembrane conductance regulator cDNA to the lung, *Hum. Gene Ther.*, 5, 731–744, 1994.

140. Wilmott, R. W., Amin, R. S., Perez, C. R., Wert, S. E., Keller, G., Boivin, G. P., Hirsch, R., De Inocencio, J., Lu, P., Reising, S. F., Yei, S., Whitsett, J. A., and Trapnell, B. C., Safety of adenovirus-mediated transfer of the human cystic fibrosis transmembrane conductance regulator cDNA to the lungs of nonhuman primates, *Hum. Gene Ther.*, 7, 301–318, 1996.

141. Harvey, B. G., Maroni, J., O'Donoghue, K. A., Chu, K. W., Muscat, J. C., Pippo, A. L., Wright, C. E., Hollmann, C., Wisnivesky, J. P., Kessler, P. D., Rasmussen, H. S., Rosengart, T. K., and Crystal, R. G., Safety of local delivery of low- and intermediate-dose adenovirus gene transfer vectors to individuals with a spectrum of morbid conditions, *Hum. Gene Ther.*, 13, 15–63, 2002.

142. Chirmule, N., Hughes, J. V., Gao, G. P., Raper, S. E., and Wilson, J. M., Role of E4 in eliciting CD4 T-cell and B-cell responses to adenovirus vectors delivered to murine and nonhuman primate lungs, *J. Virol.*, 72, 6138–6145, 1998.

143. Toietta, G., Koehler, D. R., Finegold, M. J., Lee, B., Hu, J., and Beaudet, A. L., Reduced inflammation and improved airway expression using helper-dependent adenoviral vectors with a K18 promoter, *Mol. Ther.*, 7, 649–658, 2003.

144. Koehler, D. R., Hannam, V., Belcastro, R., Steer, B., Wen, Y., Post, M., Downey, G., Tanswell, A. K., and Hu, J., Targeting transgene expression for cystic fibrosis gene therapy, *Mol. Ther.*, 4, 58–65, 2001.

145. O'Neal, W. K., Rose, E., Zhou, H., Langston, C., Rice, K., Carey, D., and Beaudet, A. L., Multiple advantages of alpha-fetoprotein as a marker for in vivo gene transfer, *Mol. Ther.*, 2, 640–648, 2000.

146. Koehler, D. R., Sajjan, U., Chow, Y. H., Martin, B., Kent, G., Tanswell, A. K., McKerlie, C., Forstner, J. F., and Hu, J., Protection of Cftr knockout mice from acute lung infection by a helper-dependent adenoviral vector expressing Cftr in airway epithelia, *Proc. Natl. Acad. Sci. U S A*, 100, 15364–15369, 2003.

147. Hiatt, P., Brunetti-Pierri, N., Koehler, D., Ruth McConnell, R., Katkin, J., Palmer, D. J., Dimmock, D., Hu, J., Finegold, M., Beaudet, A. L., Carey, K. D., Rice, K., and Ng, P., Aerosol

delivery of helper-dependent adenoviral vector into nonhuman primate lungs results in high efficiency pulmonary transduction with minimal toxicity, *Mol. Ther.*, 11, 317, 2005.

148. Seidner, S. R., Jobe, A. H., Ikegami, M., Pettenazzo, A., Priestley, A., and Ruffini, L., Lysophosphatidylcholine uptake and metabolism in the adult rabbit lung, *Biochim. Biophys. Acta*, 961, 328–336, 1988.

149. Yang, Y., Su, Q., Grewal, I. S., Schilz, R., Flavell, R. A., and Wilson, J. M., Transient subversion of CD40 ligand function diminishes immune responses to adenovirus vectors in mouse liver and lung tissues, *J. Virol.*, 70, 6370–6377, 1996.

150. Lei, D., Lehmann, M., Shellito, J. E., Nelson, S., Siegling, A., Volk, H. D., and Kolls, J. K., Nondepleting anti-CD4 antibody treatment prolongs lung-directed E1-deleted adenovirus-mediated gene expression in rats, *Hum. Gene Ther.*, 7, 2273–2279, 1996.

151. Jooss, K., Yang, Y., and Wilson, J. M., Cyclophosphamide diminishes inflammation and prolongs transgene expression following delivery of adenoviral vectors to mouse liver and lung, *Hum. Gene Ther.*, 7, 1555–1566, 1996.

152. Kaplan, J. M. and Smith, A. E., Transient immunosuppression with deoxyspergualin improves longevity of transgene expression and ability to readminister adenoviral vector to the mouse lung, *Hum. Gene Ther.*, 8, 1095–1104, 1997.

153. Chirmule, N., Truneh, A., Haecker, S. E., Tazelaar, J., Gao, G., Raper, S. E., Hughes, J. V., and Wilson, J. M., Repeated administration of adenoviral vectors in lungs of human CD4 transgenic mice treated with a nondepleting CD4 antibody, *J. Immunol.*, 163, 448–455, 1999.

154. O'Riordan, C. R., Lachapelle, A., Delgado, C., Parkes, V., Wadsworth, S. C., Smith, A. E., and Francis, G. E., PEGylation of adenovirus with retention of infectivity and protection from neutralizing antibody in vitro and in vivo, *Hum. Gene Ther.*, 10, 1349–1358, 1999.

155. Croyle, M. A., Chirmule, N., Zhang, Y., and Wilson, J. M., "Stealth" adenoviruses blunt cell-mediated and humoral immune responses against the virus and allow for significant gene expression upon readministration in the lung, *J. Virol.*, 75, 4792–4801, 2001.

156. Mastrangeli, A., Harvey, B. G., Yao, J., Wolff, G., Kovesdi, I., Crystal, R. G., and Falck-Pedersen, E., "Sero-switch" adenovirus-mediated in vivo gene transfer: Circumvention of anti-adenovirus humoral immune defenses against repeat adenovirus vector administration by changing the adenovirus serotype, *Hum. Gene Ther.*, 7, 79–87, 1996.

157. Mack, C. A., Song, W. R., Carpenter, H., Wickham, T. J., Kovesdi, I., Harvey, B. G., Magovern, C. J., Isom, O. W., Rosengart, T., Falck-Pedersen, E., Hackett, N. R., Crystal, R. G., and Mastrangeli, A., Circumvention of anti-adenovirus neutralizing immunity by administration of an adenoviral vector of an alternate serotype, *Hum. Gene Ther.*, 8, 99–109, 1997.

158. Hiatt, P., Brunetti-Pierri, N., McConnell, R., Palmer, D. J., Zuo, Y., Finegold, M., Beaudet, A., and Ng, P., Bronchoscope-guided, targeted lobar aerosolization of HDAd into nonhuman primate lungs results in uniform, high level pulmonary transduction, long term transgene expression and negligible toxicity, *Mol. Ther.*, 15, S161, 2007.

159. Ragot, T., Vincent, N., Chafey, P., Vigne, E., Gilgenkrantz, H., Couton, D., Cartaud, J., Briand, P., Kaplan, J. C., Perricaudet, M., and Kahn, A., Efficient adenovirus-mediated transfer of a human minidystrophin gene to skeletal muscle of mdx mice, *Nature*, 361, 647–650, 1993.

160. Dudley, R. W., Lu, Y., Gilbert, R., Matecki, S., Nalbantoglu, J., Petrof, B. J., and Karpati, G., Sustained improvement of muscle function one year after full-length dystrophin gene

transfer into mdx mice by a gutted helper-dependent adenoviral vector, *Hum. Gene Ther.*, 15, 145–156, 2004.

161. Gilbert, R., Dudley, R. W., Liu, A. B., Petrof, B. J., Nalbantoglu, J., and Karpati, G., Prolonged dystrophin expression and functional correction of mdx mouse muscle following gene transfer with a helper-dependent (gutted) adenovirus-encoding murine dystrophin, *Hum. Mol. Genet.*, 12, 1287–1299, 2003.

162. Clemens, P. R., Kochanek, S., Sunada, Y., Chan, S., Chen, H. H., Campbell, K. P., and Caskey, C. T., In vivo muscle gene transfer of full-length dystrophin with an adenoviral vector that lacks all viral genes, *Gene Ther.*, 3, 965–972, 1996.

163. DelloRusso, C., Scott, J. M., Hartigan-O'Connor, D., Salvatori, G., Barjot, C., Robinson, A. S., Crawford, R. W., Brooks, S. V., and Chamberlain, J. S., Functional correction of adult mdx mouse muscle using gutted adenoviral vectors expressing full-length dystrophin, *Proc. Natl. Acad. Sci. USA*, 99, 12979–12984, 2002.

164. Gilchrist, S. C., Ontell, M. P., Kochanek, S., and Clemens, P. R., Immune response to full-length dystrophin delivered to Dmd muscle by a high-capacity adenoviral vector, *Mol. Ther.*, 6, 359–368, 2002.

165. Gilbert, R., Liu, A., Petrof, B., Nalbantoglu, J., and Karpati, G., Improved performance of a fully gutted adenovirus vector containing two full-length dystrophin cDNAs regulated by a strong promoter, *Mol. Ther.*, 6, 501–509, 2002.

166. Ohtsuka, Y., Udaka, K., Yamashiro, Y., Yagita, H., and Okumura, K., Dystrophin acts as a transplantation rejection antigen in dystrophin-deficient mice: Implication for gene therapy, *J. Immunol.*, 160, 4635–4640, 1998.

167. Jiang, Z., Schiedner, G., van Rooijen, N., Liu, C. C., Kochanek, S., and Clemens, P. R., Sustained muscle expression of dystrophin from a high-capacity adenoviral vector with systemic gene transfer of T cell costimulatory blockade, *Mol. Ther.*, 10, 688–696, 2004.

168. Jiang, Z., Schiedner, G., Gilchrist, S. C., Kochanek, S., and Clemens, P. R., CTLA4Ig delivered by high-capacity adenoviral vector induces stable expression of dystrophin in mdx mouse muscle, *Gene Ther.*, 11, 1453–1461, 2004.

169. Bilbao, R., Reay, D. P., Wu, E., Zheng, H., Biermann, V., Kochanek, S., and Clemens, P. R., Comparison of high-capacity and first-generation adenoviral vector gene delivery to murine muscle in utero, *Gene Ther.*, 12, 39–47, 2005.

170. Smith, P. E., Calverley, P. M., Edwards, R. H., Evans, G. A., and Campbell, E. J., Practical problems in the respiratory care of patients with muscular dystrophy, *N. Engl. J. Med.*, 316, 1197–1205, 1987.

171. Matecki, S., Dudley, R. W., Divangahi, M., Gilbert, R., Nalbantoglu, J., Karpati, G., and Petrof, B. J., Therapeutic gene transfer to dystrophic diaphragm by an adenoviral vector deleted of all viral genes, *Am. J. Physiol. Lung Cell. Mol. Physiol.*, 287, L569–L576, 2004.

172. Persson, A., Fan, X., Widegren, B., and Englund, E., Cell type- and region-dependent coxsackie adenovirus receptor expression in the central nervous system, *J. Neurooncol.*, 78, 1–6, 2006.

173. Le Gal La Salle, G., Robert, J. J., Berrard, S., Ridoux, V., Stratford-Perricaudet, L. D., Perricaudet, M., and Mallet, J., An adenovirus vector for gene transfer into neurons and glia in the brain, *Science*, 259, 988–990, 1993.

174. Davidson, B. L., Allen, E. D., Kozarsky, K. F., Wilson, J. M., and Roessler, B. J., A model system for in vivo gene transfer into the central nervous system using an adenoviral vector, *Nat. Genet.*, 3, 219–223, 1993.

175. Byrnes, A. P., Wood, M. J., and Charlton, H. M., Role of T cells in inflammation caused by adenovirus vectors in the brain, *Gene Ther.*, 3, 644–651, 1996.

176. Perry, V. H., Andersson, P. B., and Gordon, S., Macrophages and inflammation in the central nervous system, *Trends Neurosci.*, 16, 268–273, 1993.

177. Thomas, C. E., Schiedner, G., Kochanek, S., Castro, M. G., and Lowenstein, P. R., Peripheral infection with adenovirus causes unexpected long-term brain inflammation in animals injected intracranially with first-generation, but not with high-capacity, adenovirus vectors: Toward realistic long-term neurological gene therapy for chronic diseases, *Proc. Natl. Acad. Sci. U S A*, 97, 7482–7487, 2000.

178. Xiong, W., Goverdhana, S., Sciascia, S. A., Candolfi, M., Zirger, J. M., Barcia, C., Curtin, J. F., King, G. D., Jaita, G., Liu, C., Kroeger, K., Agadjanian, H., Medina-Kauwe, L., Palmer, D., Ng, P., Lowenstein, P. R., and Castro, M. G., Regulatable gutless adenovirus vectors sustain inducible transgene expression in the brain in the presence of an immune response against adenoviruses, *J. Virol.*, 80, 27–37, 2006.

179. Zou, L., Yuan, X., Zhou, H., Lu, H., and Yang, K., Helper-dependent adenoviral vector-mediated gene transfer in aged rat brain, *Hum. Gene Ther.*, 12, 181–191, 2001.

180. Huang, B., Schiefer, J., Sass, C., Landwehrmeyer, G. B., Kosinski, C. M., and Kochanek, S., High-capacity adenoviral vector-mediated reduction of huntingtin aggregate load in vitro and in vivo, *Hum. Gene Ther.*, 18, 303–311, 2007.

181. Germano, I. M., Fable, J., Gultekin, S. H., and Silvers, A., Adenovirus/herpes simplex-thymidine kinase/ganciclovir complex: Preliminary results of a phase I trial in patients with recurrent malignant gliomas, *J. Neurooncol.*, 65, 279–289, 2003.

182. Immonen, A., Vapalahti, M., Tyynela, K., Hurskainen, H., Sandmair, A., Vanninen, R., Langford, G., Murray, N., and Yla-Herttuala, S., AdvHSV-tk gene therapy with intravenous ganciclovir improves survival in human malignant glioma: A randomised, controlled study, *Mol. Ther.*, 10, 967–972, 2004.

183. Candolfi, M., Curtin, J. F., Xiong, W. D., Kroeger, K. M., Liu, C., Rentsendorj, A., Agadjanian, H., Medina-Kauwe, L., Palmer, D., Ng, P., Lowenstein, P. R., and Castro, M. G., Effective high-capacity gutless adenoviral vectors mediate transgene expression in human glioma cells, *Mol. Ther.*, 14, 371–381, 2006.

184. Candolfi, M., Pluhar, G. E., Kroeger, K., Puntel, M., Curtin, J., Barcia, C., Muhammad, A. K., Xiong, W., Liu, C., Mondkar, S., Kuoy, W., Kang, T., McNeil, E. A., Freese, A. B., Ohlfest, J. R., Moore, P., Palmer, D., Ng, P., Young, J. D., Lowenstein, P. R., and Castro, M. G., Optimization of adenoviral vector-mediated transgene expression in the canine brain in vivo, and in canine glioma cells in vitro, *Neuro-Oncol*, 9(3), 245–258, 2007.

185. Kumar-Singh, R. and Farber, D. B., Encapsidated adenovirus mini-chromosome-mediated delivery of genes to the retina: Application to the rescue of photoreceptor degeneration, *Hum. Mol. Genet.*, 7, 1893–1900, 1998.

186. Kreppel, F., Luther, T. T., Semkova, I., Schraermeyer, U., and Kochanek, S., Long-term transgene expression in the RPE after gene transfer with a high-capacity adenoviral vector, *Invest. Ophthalmol. Vis. Sci.*, 43, 1965–1970, 2002.

187. Balague, C., Zhou, J., Dai, Y., Alemany, R., Josephs, S. F., Andreason, G., Hariharan, M., Sethi, E., Prokopenko, E., Jan, H. Y., Lou, Y. C., Hubert-Leslie, D., Ruiz, L., and Zhang, W. W., Sustained high-level expression of full-length human factor VIII and restoration of clotting activity in hemophilic mice using a minimal adenovirus vector, *Blood*, 95, 820–828, 2000.

188. Chuah, M. K., Schiedner, G., Thorrez, L., Brown, B., Johnston, M., Gillijns, V., Hertel, S., Van Rooijen, N., Lillicrap, D., Collen, D., VandenDriessche, T., and Kochanek, S., Therapeutic factor VIII levels and negligible toxicity in mouse and dog models of hemophilia A following gene therapy with high-capacity adenoviral vectors, *Blood*, 101, 1734–1743, 2003.

189. Brown, B. D., Shi, C. X., Powell, S., Hurlbut, D., Graham, F. L., and Lillicrap, D., Helper-dependent adenoviral vectors mediate therapeutic factor VIII expression for several months with minimal accompanying toxicity in a canine model of severe hemophilia A, *Blood*, 103, 804–810, 2004.

190. McCormack, W. M., Jr., Seiler, M. P., Bertin, T. K., Ubhayakar, K., Palmer, D. J., Ng, P., Nichols, T. C., and Lee, B., Helper-dependent adenoviral gene therapy mediates long-term correction of the clotting defect in the canine hemophilia A model, *J. Thromb. Haemost.*, 4, 1218–1225, 2006.

191. Ehrhardt, A., Xu, H., Dillow, A. M., Bellinger, D. A., Nichols, T. C., and Kay, M. A., A gene-deleted adenoviral vector results in phenotypic correction of canine hemophilia B without liver toxicity or thrombocytopenia, *Blood*, 102, 2403–2411, 2003.

192. Mian, A., McCormack, W. M., Jr., Mane, V., Kleppe, S., Ng, P., Finegold, M., O'Brien, W. E., Rodgers, J. R., Beaudet, A. L., and Lee, B., Long-term correction of ornithine transcarbamylase deficiency by WPRE-mediated overexpression using a helper-dependent adenovirus, *Mol. Ther.*, 10, 492–499, 2004.

193. Koeberl, D. D., Sun, B., Bird, A., Chen, Y., Oka, K., and Chan, L., Efficacy of helper-dependent adenovirus vector-mediated gene therapy in murine glycogen storage disease type Ia, *Mol. Ther.*, 15, 1253–1258, 2007.

194. Oka, K., Pastore, L., Kim, I. H., Merched, A., Nomura, S., Lee, H. J., Merched-Sauvage, M., Arden-Riley, C., Lee, B., Finegold, M., Beaudet, A., and Chan, L., Long-term stable correction of low-density lipoprotein receptor-deficient mice with a helper-dependent adenoviral vector expressing the very low-density lipoprotein receptor, *Circulation*, 103, 1274–1281, 2001.

195. Belalcazar, L. M., Merched, A., Carr, B., Oka, K., Chen, K. H., Pastore, L., Beaudet, A., and Chan, L., Long-term stable expression of human apolipoprotein A-I mediated by helper-dependent adenovirus gene transfer inhibits atherosclerosis progression and remodels atherosclerotic plaques in a mouse model of familial hypercholesterolemia, *Circulation*, 107, 2726–2732, 2003.

196. Nomura, S., Merched, A., Nour, E., Dieker, C., Oka, K., and Chan, L., Low-density lipoprotein receptor gene therapy using helper-dependent adenovirus produces long-term protection against atherosclerosis in a mouse model of familial hypercholesterolemia, *Gene Ther.*, 11, 1540–1548, 2004.

197. Kiang, A., Hartman, Z. C., Liao, S., Xu, F., Serra, D., Palmer, D. J., Ng, P., and Amalfitano, A., Fully deleted adenovirus persistently expressing GAA accomplishes long-term skeletal muscle glycogen correction in tolerant and nontolerant GSD-II mice, *Mol. Ther.*, 13, 127–134, 2006.

198. Schillinger, K. J., Tsai, S. Y., Taffet, G. E., Reddy, A. K., Marian, A. J., Entman, M. L., Oka, K., Chan, L., and O'Malley, B. W., Regulatable atrial natriuretic peptide gene therapy for hypertension, *Proc. Natl. Acad. Sci. U S A*, 102, 13789–13794, 2005.

199. Magee, T. R., Ferrini, M., Garban, H. J., Vernet, D., Mitani, K., Rajfer, J., and Gonzalez-Cadavid, N. F., Gene therapy of erectile dysfunction in the rat with penile neuronal nitric oxide synthase, *Biol. Reprod.*, 67, 1033–1041, 2002.

200. Wang, L., Hernandez-Alcoceba, R., Shankar, V., Zabala, M., Kochanek, S., Sangro, B., Kramer, M. G., Prieto, J., and Qian, C., Prolonged and inducible transgene expression in the liver using gutless adenovirus: A potential therapy for liver cancer, *Gastroenterology*, 126, 278–289, 2004.

201. Shi, C. X., Long, M. A., Liu, L., Graham, F. L., Gauldie, J., and Hitt, M. M., The human SCGB2A2 (mammaglobin-1) promoter/enhancer in a helper-dependent adenovirus vector directs high levels of transgene expression in mammary carcinoma cells but not in normal nonmammary cells, *Mol. Ther.*, 10, 758–767, 2004.

6 Adeno-Associated Virus and AAV Vectors for Gene Delivery

Barrie J. Carter, Haim Burstein, and Richard W. Peluso

CONTENTS

6.1 INTRODUCTION

In this chapter we summarize the state of development of adeno-associated virus (AAV) vectors and provide an overview of AAV as well as some historical comment on early seminal studies. We discuss the key advances in the recent years including improvements in vector production, the expanded understanding of AAV and AAV vector biology, studies on the applications for persistent gene expression, and update the extensive series of AAV vector clinical trials. We have not attempted to provide an exhaustive collection of references on development of AAV vectors. Other sources provide general reviews of AAV [1,2] and summaries on early development of AAV vectors [3–7].

AAV vectors have a number of advantageous properties as gene delivery vehicles. The parental virus has never been shown to cause disease. AAV vectors are the smallest and most chemically defined particulate gene delivery system and potentially could be classified as well characterized biologics for therapeutic applications. AAV vectors contain no viral genes that could elicit undesirable cellular immune responses and appear not to induce inflammatory responses. The primary host response that might impact use of AAV vectors is a neutralizing antibody response. The vectors readily transduce dividing or nondividing cells and can persist essentially for the lifetime of the cell. Thus, AAV vectors can mediate impressive long-term gene expression when administered *in vivo*. Consequently, these vectors may be well suited for applications where the vector is delivered infrequently and where any potential host antibody response to the AAV capsid protein may be less inhibitory. One limitation for AAV vectors is the limited DNA payload capacity of about 4.5 kb per particle.

The lack of good production systems that could generate high titer vectors was an early obstacle to the development of AAV vectors, but this has been overcome through significant advances in both upstream production and downstream purification of AAV vectors. Clinical development of AAV vectors has progressed significantly and studies of an AAV vector in cystic fibrosis (CF) patients [8–13] have been extended to phase II trials. Other AAV vectors then entered clinical trials for hemophilia B [14] and to date at least 20 AAV vectors have been introduced into clinical trials for a wide variety of therapeutic and prophylactic applications [15,16] (Table 6.1).

Since the second edition of this book [17] there have been further advances in application of AAV vectors in many animal models and analysis of AAV vector safety profiles and host cell responses. Additional advances in understanding the structure and biology of AAV vectors including uptake into cells, trafficking to the cell nucleus, and the mechanism of genome persistence suggest possible ways to modify the biological targeting of AAV vectors, to enhance transduction efficiency and to overcome the packaging limitation. Also, there have been substantial advances in manufacture of AAV vectors [18,19].

Most of the early studies on AAV used AAV serotype 2, but genomes of serotypes 1 through 6 were cloned and sequenced [20]. Many additional AAV serotypes and sequences have been isolated [21], particularly from human and in nonhuman primate tissues and an extensive clade structure is now emerging [22–24]. The biological properties of individual serotypes include differences in the interactions with cellular receptors [25]. Other studies are now providing information on the structure of the AAV capsids and how their interactions with the cell may be modified. In addition, the resolution of the crystal structure of AAV2 [26] provided the foundation for structural studies of many other AAV serotypes [27]. Thus, together with additional studies on cellular trafficking pathways, it may be possible to modify the targeting of AAV vectors as well as to enhance their transduction efficiency [28,29].

Studies on the mechanism of persistence of vector genomes in transduced cells indicate that this involves formation of polymeric DNA structures or concatemers. Concatemers also can be formed between two different vector genomes introduced into the same cell. This provides a way to partly circumvent the packaging limit of AAV by dividing a gene expression cassette between two AAV vectors (dual vectors) and allowing recombination in the cell to generate the intact expression cassette.

6.2 ADENO-ASSOCIATED VIRUS

6.2.1 AAV Discovery

AAV is a small, DNA-containing virus which belongs to the family Parvoviridae within the genus *Dependovirus*. AAV originally was observed as a contaminant of laboratory preparations of adenovirus, then was recognized as a virus that was different from adenovirus but was dependent upon adenovirus for its replication [30,31]. Soon after the discovery of AAV in laboratory stocks of adenoviruses, it was isolated from humans [32]. AAV has not been associated with any disease but has

TABLE 6.1

Current and Completed Clinical Trials with AAV Vectors

Disease Indication[a]	Protein Expressed	Capsid	Route of Administration	Phase	Subject Number Vector (Placebo)[b]	Status[c]
CF	CFTR	AAV2	Lung (bronchoscope), nose	I	25	Completed
CF	CFTR	AAV2	Maxillary sinus	I	10	Completed
CF	CFTR	AAV2	Maxillary Sinus	II	23	Completed
CF	CFTR	AAV2	Lung via aerosol	I	12	Completed
CF	CFTR	AAV2	Lung via aerosol	II	20 [18]	Completed
CF	CFTR	AAV2	Lung via aerosol	II	50 [50]	Completed
Hemophilia	FIX	AAV2	Intramuscular	I	9	Completed
Hemophilia	FIX	AAV2	Hepatic artery	I	6	Ended
Arthritis	TNFR:Fc	AAV2	Intra-articular	I	11 [4]	Completed
Arthritis	TNFR:Fc	AAV2	Intra-articular	I/II	90 [30]	Ongoing
HIV vaccine	HIV1 gag-pro-rt	AAV2	Intramuscular	I	64 [16]	Completed
HIV vaccine	HIV1 gag-pro-rt	AAV2	Intramuscular	II	70 [21]	Ongoing
Muscular dystrophy	Sarcoglycan	AAV2	Intramuscular	I	1	Ended
Muscular dystrophy	Mini-dystrophin	AAV2.5	Intramuscular	I	6	Ongoing
Muscular dystrophy	Mini-dystrophin	AAV1	Intramuscular	I	6	Ongoing
Hereditary emphysema	AAT	AAV2	Intramuscular	I	12	Completed
Hereditary emphysema	AAT	AAV1	Intramuscular	I	12	Ongoing
Parkinson's	GAD65,GAD67	AAV2	Intracranial	I	12	Completed
Canavan's	AAC	AAV2	Intracranial	I	21	Ongoing
Batten's	CLN2	AAV2	Intracranial	I	10	Ongoing
Alzheimer's	NGF	AAV2	Intracranial	I	6	Ongoing
Parkinson's	AADC	AAV2	Intracranial	I	15	Ongoing
Parkinson's	NTN (neurturin)	AAV2	Intracranial	I	12	Completed
Parkinson's	NTN (neurturin)	AAV2	Intracranial	II	34 [17]	Ongoing
Dyslipidemia	LPL	AAV1	Intramuscular	I	8	Completed
Dyslipidemia	LPL	AAV1	Intramuscular	I/II	8	Ongoing
Retinal degeneration	RPE65	AAV2	Subretinal	I	12	Ongoing
Chronic heart failure	SERCA2	AAV1	Intracoronary	I	12	Ongoing
Prostate cancer	GM-CSF	AAV2	Ex vivo, intradermal	I	9	Complete
Prostate cancer	GM-CSF	AAV2	Ex vivo, intradermal	I/II	80	Complete
Prostate cancer	GM-CSF	AAV2	Ex vivo, intradermal	III	300 [294]	Ongoing
Prostate cancer	GM-CSF	AAV2	Ex vivo, intradermal	III	300 [294]	Ongoing
Malignant melanoma	B7–2, IL-7	AAV2	Ex vivo, intradermal	I	?	Ended

Source: Modified from Carter, B.J., *Hum. Gene. Ther.*, 16, 541, 2005.

[a] All of these trials involve *in vivo* delivery of AAV vectors except for the cancer indications in which allogenic cells transduced *ex vivo* with an AAV vector are administered as a therapeutic cancer vaccine.

[b] For ongoing trials, the number of subjects to be administered vector is indicated but does not necessarily reflect the enrollment to date. For placebo-controlled trials, the planned number of placebo subjects is indicated in parentheses. Projected enrollment numbers are not available for some trials.

[c] For trials which are finished, "completed" indicates that subject enrollment was achieved essentially as originally planned, whereas "ended" indicates that enrollment was stopped short of the original goal. Ongoing indicates that the trial is under way as of mid-2007.

been isolated from humans, initially in association with infections by adenovirus [32]. More recently, endogenous AAV DNA sequences have been isolated from a variety of human tissue samples including adenoids, tonsils, leukocytes, lymph nodes, liver, muscle, and cervix [21–23,33–36]. Thus it is suggested [24] that AAV generally may enter humans via the otopharyngeal route in association with adenovirus and then spread to other sites perhaps as a result of replication.

Adenoviruses are efficient helpers for AAV replication in *in vitro* cell culture. The epidemiological evidence (Section 6.2.2) suggests that adenovirus is the normal helper in natural infections in humans. Also, adenovirus has been shown to be an efficient helper for AAV replication in animals such as rhesus macaques [37]. Herpesviruses can also function as helpers for AAV replication *in vitro* but they are much less efficient. In contrast, there is no evidence that herpesviruses can act as a helper in humans or any animal species. In addition to not being associated with causing any diseases, it is not apparent that AAV plays any role in altering or affecting the course of *in vivo* infection with either type of helper, adenovirus or herpesvirus [24].

Multiple serotypes of AAV now have been distinguished. AAV1, AAV2, AAV3, and AAV4 have extensive DNA homology and significant serologic overlap [38–41], but AAV5 is somewhat less related [42,43]. AAV2 and AAV3 are the most frequently isolated from humans [32] whereas AAV5 has been isolated from humans only once [44]. AAV4 is a simian isolate that does not infect humans and AAV1 originally may have been isolated from a simian source [45]. More recent isolates, AAV6 [46], AAV7, and AAV8 [21], are discussed below. Other AAVs have been identified in a variety of other animal species [47] but less characterization of these AAV isolates has been reported [1]. However, it has been noted that the entire genus *Dependovirus* appears to be derived from avian AAVs [23].

6.2.2 EPIDEMIOLOGY

In the United States, a significant proportion of the population over age 10 may be seropositive for AAV2 and AAV3 [32]. AAV2 and AAV3 appear to be transmitted primarily in nursery populations in conjunction with the helper adenovirus and thus appear to be replication defective also in the natural human host. A signal epidemiological study of AAV was carried out in a population of children in an orphanage in Washington DC in whom seroconversion to AAV was observed during the course of an adenovirus infection [32]. In primary infected individuals, the virus may be shed in body fluids including sputum and stool. It is noteworthy that the early epidemiological studies also analyzed neutralizing antibody responses to AAV in humans. The presence of serum neutralizing antibody against AAV2 or AAV3 did not prevent reinfection of humans but did prevent shedding of the virus [32]. This observation is relevant for the use of AAV as a gene delivery vector because it suggests that repeat delivery may be feasible [48] and that shedding of vector may be less likely.

6.2.3 BIOLOGY OF AAV LIFE CYCLE

AAV is a defective parvovirus that replicates *in vitro* only in cells in which certain functions are provided by a coinfecting helper virus which is generally an adenovirus but may be a herpesvirus [1]. AAV has both a broad host range and wide cell and tissue specificity and replicates in many cell lines of human, simian or rodent origin provided an appropriate helper virus is present. There may be some limitations to AAV tissue specificity *in vivo* or at least some significant differences in efficiency of transduction of different tissues and organs. These limitations may reflect the receptor and co-receptors apparently utilized by different AAV serotypes for entry into cells as well as cellular trafficking of AAV. This aspect of AAV biology is becoming of increasing importance for development of AAV vectors. A second set of parameters that may

impact AAV tissue and organ specificity and its replication reflect the nature of the helper function provided by helper viruses.

An additional event required by AAV to function efficiently as a gene delivery vehicle is the need to convert the incoming single-stranded DNA genome to a double-stranded molecule to permit transcription and gene expression. This process is termed single-strand (SS) conversion or metabolic activation [49] and the rate at which it occurs may depend in part on the physiological state of the host cell, but the process may be accelerated by treatment of the cell with genotoxic agents or by certain helper virus functions.

Infection of certain cell lines by AAV in the absence of helper functions results in its persistence as a latent provirus integrated into the host cell genome [50,51]. In such cell lines, the integrated AAV genome may be rescued and replicated to yield a burst of infectious progeny AAV particles if the cells are superinfected with a helper virus such as adenovirus. Importantly, in cultured cells, AAV exhibits a high preference for integration at a specific region, the AAVS1 site, on human chromosome 19 [52,53]. The efficiency and specificity of this process is mediated by the AAV *rep* gene [54–56]. Rep-deleted AAV vectors do not retain specificity for integration into this chromosome 19 region [57] and generally do not integrate efficiently but remain as episomes (Section 6.9).

6.2.4 MODE OF CELL ENTRY AND HOST TROPISM

AAV appears to have a broad host range and different AAV serotypes replicate *in vitro* in many human cells and a variety of simian and rodent cell lines if a helper virus with the appropriate host range is also present. AAV also infects various animal species and human isolates of AAV will grow in mice or monkeys if the appropriate mouse or monkey adenovirus is also present. This indicated that cellular receptors for AAV were likely to be relatively common on many cell types.

Initial experiments demonstrated [58] that AAV2 particles can use heparin sulfate proteoglycans (HSPG) as a receptor and some cell lines that do not produce HSPG are impaired for AAV binding and infection. Further studies showed that AAV2 also utilizes a co-receptor for efficient internalization and two possible co-receptors, $\alpha_v\beta_5$ integrin [59] and human fibroblast growth factor receptor 1 (FGFR1) [60] were identified. It is of interest that the $\alpha_v\beta_5$ integrin co-receptor, which is used for a similar purpose by adenovirus type 2 and 5, is preferentially located on airway epithelial cells in the more distal areas of the conducting airway [61]. This may be important for use of an AAV2 gene therapy vector for CF since the distal airway is the region of the lung most impacted by the disease. It is also noteworthy that FGFR is expressed in most tissues but is of highest abundance in skeletal muscle and neuroblasts and glioblasts in the brain and these two organs appear to be good targets for AAV2 transduction.

The existence of more than one co-receptor suggests that AAV may have multiple mechanisms for cell entry and there is already some evidence to support this concept [62]. Also, real-time imaging of entry into HeLa cells by individual AAV2 particles labeled with the dye Cy5 [63] showed that endocytosis was rapid and that some particles could reach the nucleus within 15 min of first contacting the cell. However, there was evidence of free diffusion of both endosomes and AAV particles and also evidence for movement of each of these entities being driven by cellular motor proteins. Furthermore, cellular trafficking events following endocytosis of AAV, that may involve the ubiquitin-proteosome pathway, appear to play a significant role [64–67].

Cell entry may also be impacted by the route of delivery and AAV vectors can transduce airway cells when delivered directly to the lung [68,69], or brain cells [69] or myocytes [70,71] when delivered directly to these organs. However, when delivered intravenously by tail vein injection in mice [72,73], the vector preferentially accumulated in the liver and this may reflect both the presence of a much more porous vasculature in the liver and also the small size of the AAV particles. The small size of the AAV particle also may be of advantage in passing through the basal lamina pores in muscle and thus accessing a large number of myoblasts and myotubes.

The initial observations on AAV cellular entry and tropism are being greatly expanded upon in the context of AAV vectors. Recent development of AAV vectors based on different capsids has increased interest and provided many more detailed insights into the multiple receptors and co-receptors utilized by various AAV serotypes as well as generating much more information on cellular trafficking of AAV particles. This is discussed in more detail in Sections 6.7 and 6.8 below.

6.3 AAV MOLECULAR BIOLOGY

6.3.1 Particle Structure

AAV is a nonenveloped particle about 20 nm in diameter with icosahedral symmetry which is stable to heat, mild proteolytic digestion, and nonionic detergents. The AAV particle is comprised of a protein coat, containing the three capsid proteins, VP1, VP2, and VP3, which encloses a linear single-stranded DNA genome having a molecular weight of 1.5×10^6. The VP1, VP2, and VP3 proteins are present in the viral capsid in the ratio of 1:1:8. The DNA represents 25% by mass of the particle that therefore exhibits a high buoyant density (1.41 g/cm^3) in cesium chloride. The relative stability of the AAV particle is an important property since it can withstand robust purification procedures and this facilitates scaled-up production of AAV vectors.

The crystal structure of AAV2 was determined at 3.0 Å resolution by x-ray crystallography [26] and this revealed several interesting features that have greatly facilitated efforts aimed modifying the AAV capsid to alter targeting specificity. The structure (Figure 6.1) shows that each capsid comprises 60 protein subunits arranged in $T = 1$ icosahedral symmetry. All of the amino acids, except the 14 amino-terminal residues of VP1, could be localized in the structure. The surface of the capsid shows a distinctive topology with three peaks clustering around each threefold icosahedral axis (Figure 6.1). Each

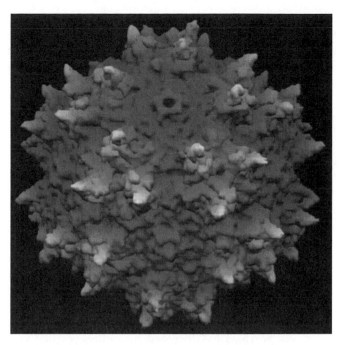

FIGURE 6.1 Structure of AAV serotype 2. The structure was determined by x-ray crystallography at a resolution of 3.0 Å. The surface topology is shown drawn to scale. The view is down a twofold axis (center of the virus) with threefold left and right of center, and fivefold above and below. Overall, the outside surface is positively charged with a prominent ring of symmetry-related positive patches in a depression surrounding the fivefold axis. (Reprinted from Xie, Q., Bu, W., Bhatia, S., Hare, H.J., Somasundaram, T., Azzi, A., and Chapman, M.S., *Proc. Natl. Acad. Sci. U S A*, 99, 10405, 2002. With permission.)

of these threefold proximal peaks is formed from two interacting protein subunits and the sides of these peaks appear to be the regions that mediate the receptor binding interactions with heparin sulfate. Additional crystal structures of other AAV serotypes including AAV4 and AAV8 have also been determined. Most recently the structure of the AAV8 capsid was determined at 2.6 Å resolution [74]. The overall topology of the common viral capsid protein of AAV8 is similar to that of AAV2 and AAV4 with an eight-strand β-barrel and long loops between the β-strands but there are significant differences at the surface protrusions around the two-, three-, and fivefold axes. At least for AAV2, these latter structures have important roles in transduction efficiency and antibody recognition. The surface of AAV2 has a basic charge at the region associated with heparin binding but at the analogous region AAV8 has a reduced distribution of basic charge. This is consistent with the lack of heparin binding by AAV8.

All other parvoviruses contain a phospholipase A2 activity in the capsid protein and AAV has a similar activity that is located in the unique amino-terminal region of the VP1 protein [75]. This activity appears to be required to mediate exit from endosomes and is important for infectivity [76,77]. No posttranslational modifications of the AAV capsid have been identified. Up to eight potential *O*- or *N*-glycosylation motifs, of which three are on the surface near the protrusions at the threefold axes, can be identified in the AAV2 capsid. However, a recent analysis of purified AAV2 capsids using peptide mapping and high-resolution mass spectrometry showed that these sites are not glycosylated [78].

A novel feature of AAV is that although each particle contains only one single-stranded genome, strands of either complementary sense, "plus" or "minus" strands, are packaged into individual particles. Equal numbers of AAV particles contain either a plus or minus strand. Both strands are equally infectious and AAV displays single-hit kinetics for infectivity [79]. When DNA is extracted from AAV particles, the plus and minus strands anneal to generate duplex molecules of 3.0×10^6 molecular weight. However, Crawford and his colleagues [80] showed on the basis of a careful physical characterization of AAV particles that each particle appeared to contain DNA with a molecular weight of 1.5×10^6. They suggested that the only way to reconcile this conundrum was to propose that individual plus and minus strands must be packaged into individual particles. An elegant proof of this conundrum was provided by Rose and his colleagues [81] who made two preparations of AAV particles, in which one preparation had thymidine substituted by bromodeoxyuridine (BudR) and the other was un-substituted. The two preparations of particles were mixed prior to extraction of DNA. Analysis of the duplex DNA obtained upon extraction showed components with intermediate density formed by individual strands from substituted or un-substituted particles that had annealed during extraction. This constituted formal proof of the novel DNA strand segregation exhibited by AAV during packaging of its DNA. BudR substitution of AAV DNA also permits separation of the plus and minus strands and this was used along with 5′ end labeling and restriction endonuclease cleavage to determine the strand polarity of the AAV genome [82].

6.3.2 AAV Genome Structure

Most of the fundamental studies of AAV genome structure and function were derived from AAV2. More recently, additional serotypes have been examined and most show a relatively similar genome arrangement and expression pattern although there are some differences particularly with respect to transcription as described below.

The AAV2 DNA genome [38,83] is 4681 nucleotides long and includes one copy of the 145 nucleotide long inverted terminal repeat (ITR) at each end and a unique sequence region of 4391 nucleotides long that contains two main open reading frames for the *rep* and *cap* genes (Figure 6.2). The unique region contains three transcription promoters p_5, p_{19}, and p_{40} that are used to express the *rep* and *cap* genes. Transcripts from all three promoters are terminated at a polyA site at the right hand end of the genome. The ITR sequences are required in *cis* to provide functional origins of replication (ORI) as well as signals for encapsidation, integration into the cell genome, and rescue from either host cell chromosomes or

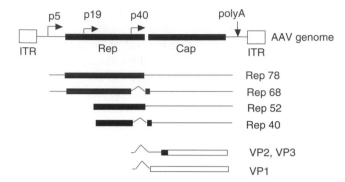

FIGURE 6.2 Structure of the AAV2 genome. The AAV2 genome is shown schematically at the top. Open boxes indicate the ITRs (replication origins) and the arrows indicate the three transcription promoters, p5, p19, and p40. The polyA site is indicated by the downward arrow. The six RNA transcripts are shown below with the introns indicated by the caret marks. The coding regions for the four rep proteins (Rep78, Rep68, Rep52, and Rep40) are shown by the solid rectangular boxes and for the viral capsid proteins (VP1, VP2, and VP3) by the open boxes. The three capsid proteins VP1, VP2, and VP3 all comprise the same common VP3 amino acid. The VP2 contains an additional N-terminal sequence (shown by the black box). VP1 comprises all of the VP2 sequence plus an additional N-terminal sequence.

recombinant plasmids. The genomes of other AAV serotypes have been sequenced and appear to have a structure similar to AAV2 [40–42,46].

The *rep* gene is transcribed from two promoters, p_5 and p_{19}, to generate two families of transcripts and two families of rep proteins (Figure 6.1). In addition, splicing of these mRNAs yields two different carboxyl terminal regions in the rep proteins. The capsid gene is expressed from transcripts from the p40 promoter that accumulate as two 2.3 kb mRNAs that are alternately spliced. The majority 2.3 kb transcript codes for the VP3 protein initiated from a consensus AUG initiation codon. However, at about a 10-fold lower frequency, translation of this transcript also occurs slightly upstream at a nonconsensus ACG initiation codon to yield VP2. The minority 2.3 kb mRNA is spliced to an alternate 3′ donor site 30 nucleotides upstream and this retains an AUG codon that is used to initiate translation of VP1. Thus, VP1 and VP2 have the same polypeptide sequence as VP3 but have additional aminoterminal sequences. This elegant arrangement results in generation of VP1, VP2, and VP3 in ratios of about 1:1:8 that are the same as the ratio of these proteins in the viral particle.

During construction of AAV vectors it was observed that the AAV2 ITR can function directly as a transcription promoter at a low level [84,85]. Recent studies [86] have shown that the AAV5 ITR is a very efficient transcription promoter. AAV5 transcription also has some other differences in that the two transcripts from the rep gene are terminated at a polyA site in the middle of the genome. Also, in AAV5 the rep protein does not appear to regulate expression of the *cap* gene as occurs in AAV2 [86].

6.3.3 REPLICATION

In a productive infection in the presence of a helper such as adenovirus [4,87], the infecting parental AAV SS genome is converted to a parental duplex replicating form (RF) by a self-priming mechanism that takes advantage of the ability of the ITR to form a hairpin structure (Figure 6.3). The parental RF molecule is then amplified to form a large pool of progeny RF molecules in a process that requires both the helper functions and the AAV *rep* gene products, Rep78 and Rep68. AAV RF genomes are a mixture of head-to-head or tail-to-tail multimers or concatemers and are precursors to progeny SS DNA genomes that are packaged into preformed empty AAV capsids [87,88]. Rep52 and Rep40 interact with the preformed capsid apparently to provide a DNA helicase function for DNA packaging [89].

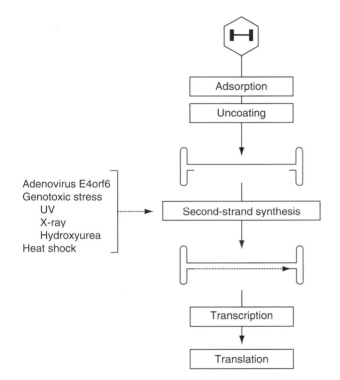

Adenovirus E4orf6
Genotoxic stress
UV
X-ray
Hydroxyurea
Heat shock

FIGURE 6.3 Metabolic pathway of AAV genomes in cells. Binding of AAV to cells is independent of helper virus functions. Trafficking of AAV to the nucleus may be enhanced by agents that interact with the ubiquitin pathway and proteosome processing. Conversion of the infecting SS genome to a duplex structure (or parental RF) through the process of metabolic activation (second-strand synthesis) can occur independently of helper virus. This process may be enhanced by infection with helper adenovirus genes such as E4orf6 or by other metabolic insults including genotoxic stress or heat shock. Treatments that enhance metabolic activation may enhance gene expression from the vector template. The single strand and duplex strands are drawn to show the ITR in the base-paired hairpin conformation that allows self-priming of replication to form a duplex template using cellular DNA polymerases. For further details see text. (Reprinted from Ferrari, F.K., Samulski, T., Shenk, T., and Samulski, R.J., *J. Virol.*, 70, 3227, 1996. With permission.)

The kinetics of AAV replication and assembly has been investigated [87,90]. In human HeLa or 293 cells simultaneously infected with AAV and adenovirus, there are three phases of the growth cycle. In the first 8–10h, the cell becomes permissive for AAV replication as a result of expression of a subset of adenovirus genes including E1, E2A, E4, and the VA RNA [82]. During this period, the infecting AAV genome is converted to the initial parental duplex RF DNA by self-priming from the terminal base-paired 3′ hydroxyl group provided by the ability of the ITR to form a self-paired hairpin. This initial generation of a duplex genome also provides a template for transcription and expression of AAV proteins. In a second phase, from about 10–20h after infection, the bulk of the AAV rep and cap proteins are synthesized, and there is a large amplification of monomeric and concatemeric duplex AAV RF genomes to a constant level [90]. During the third phase of AAV growth, between 16 and 30 h, SS progeny molecules are synthesized by a strand-displacement replication mechanism and packaged into preformed capsids followed by accumulation of mature, infectious AAV particles [88,90].

The rep proteins perform important biochemical functions [4]. Rep68 and Rep78 bind to the ITR and are site-specific, strand-specific endonucleases that cleave the hairpin in an RF molecule at the site that is the 5′ terminus of the mature strand. In addition, these proteins contain an ATP-binding site which is important for the enzymatic activity but not for binding to the ITR. Further, Rep78 and Rep 68 have both DNA and RNA helicase activity. These rep proteins also regulate transcription [4,6,7]. Rep78 is a negative auto-regulator of the p5 promoter, i.e., of its own synthesis, but is an activator of the p40 promoter to enhance capsid protein production. Rep52 and Rep40 do not bind to the ITR but provide a helicase in assembly of mature particles. Also, the smaller rep proteins are anti-repressors and block the negative auto-regulation of p_5 by Rep78 [87,91].

The AAV replication cycle is highly coordinated with respect to expression of rep and cap proteins and the relationship between replication and assembly [90,92]. Any vector production process that provides the rep and cap functions by complementation may decrease the efficiency of this highly regulated process. Nonetheless, AAV has one of the largest burst sizes of any virus and following infection of cells with AAV and adenovirus as helper, this may be well in excess of 100,000 particles per cell [1]. These considerations imply that a high yield of AAV vector particles per cell theoretically is attainable. Attaining high specific productivity is of crucial importance in developing scaled-up vector production because the ability to obtain maximum yields ideally requires high specific productivity (yield of particles per cell) or large biomass (total number of cells). Maximizing the specific productivity may avoid unnecessary increases in biomass.

6.3.4 GENETICS

The cloning of infectious AAV genomes in bacterial plasmids facilitated a molecular genetic analysis of AAV [93,94]. These studies showed that the *rep* and *cap* genes are required in *trans* to provide functions for replication and encapsidation of viral genomes, respectively, and that the ITR is required in *cis* [4,6,83]. Mutations in the ITR have an Ori phenotype and cannot be complemented in *trans*.

Mutations that affect the Rep78 and Rep68 proteins have a Rep phenotype and are deficient for both the bulk replication and amplification of duplex RF molecules and for accumulation of SS, progeny genomes. A mutation that affected only the Rep52 and Rep40 proteins showed an Ssd phenotype in which duplex RF replication occurred normally but no SS progeny DNA accumulated.

The *cap* gene encodes the proteins VP1, VP2, and VP3 that share a common overlapping sequence but VP1 and VP2 contain additional amino terminal sequence. All three proteins are required for capsid production. Mutations that affect VP2 or VP3 have a cap phenotype, and block capsid assembly and prevent any accumulation of SS DNA. This indicates that VP3 and VP2 are primarily responsible for forming the capsid and that SS DNA does not accumulate unless it can be packaged into capsids. Mutations that affect only the amino terminus of the VP1 protein do not prevent accumulation of capsids or SS DNA but no infectious AAV particles accumulate. This phenotype has been described as either Inf or Lip (low infectivity particles).

These genetic studies together with additional biochemical studies show that Rep68 and Rep78 are required for replication, that VP2 and VP3 are required to form the capsid and that Rep52 and Rep40 appear to act in concert together with VP1 to encapsidate the DNA and stabilize the particles [4,5,89]. An additional role of VP1 appears to be to provide the phospholipase activity that is required for infectivity [75]. Thus, the Inf and Lip phenotypes may reflect the impact of various mutations on the lipase activity for endosomal release or on the packaging interaction between the capsid and rep proteins and the genome [95] (see also Section 6.8.1).

The rep proteins exhibit several pleiotropic regulatory activities including positive and negative regulation of AAV genes and expression from some other viral or cellular promoters, as well as inhibitory effects on the host cell. Because of the inhibitory effects of expression of *rep* gene products on cell growth, expression of rep proteins in stable cell lines was difficult to achieve, and this delayed development of AAV packaging cell lines [19]. For this reason, various approaches to AAV vector production have employed transient transfection of cells with AAV vector plasmids and complementing *rep–cap* plasmids. However, even in these transfection systems, the closely coordinated regulation of *rep* and *cap* gene expression and the interactions of the three AAV promoters [92] are important considerations in optimizing vector production.

6.4 AAV LATENCY AND PERSISTENCE

6.4.1 Latency

A U.S. Government screening program to assess human cell lines for vaccine production led to the observation that infection of primary cultures of human embryonic kidney cells with adenovirus resulted in rescue of infectious AAV. The hypothesis that some cultures may have carried a latent form of AAV was tested directly [50] by infecting a human cell line, Detroit 6, with AAV at a high multiplicity of infection and passaging the cell cultures until no infectious AAV genomes were present which required at least 10 cell passages. Following this, superinfection of the cultures with adenovirus resulted in rescue of infectious AAV. This provided an important demonstration of a way in which AAV may survive in a cell if conditions are not permissive for replication.

6.4.2 Integration

Analysis of a human cell line, Detroit 6, carrying latent AAV showed that the cells contained a relatively low number of AAV genomes that were integrated into the host cell chromosome mostly as tandem repeats. Early studies of cell lines stably transduced with AAV vectors expressing selectable markers, also showed that most stable copies in the cell existed as tandem repeats with a head-to-tail conformation [51]. Analysis of chromosomal flanking sequences showed that, for wild-type AAV, 50%–70% of these integration events occurred in a defined region [52,53]. When wild-type AAV infects human cell lines in culture, up to 50%–70% of these integration events occur at a region known as the AAVS1 site on chromosome 19 at 19q13ter. Both the specificity and efficiency of AAV integration are mediated by the AAV rep protein that binds to the ITR and to a site in the AAVS1 site on human chromosome 19 [54]. These studies were performed on cells in culture using high multiplicities of infection and it is noteworthy that naturally occurring, latent, integrated AAV genomes had not been well characterized in humans or any other animal species until recently. One study of wtAAV infection of rhesus macaques in the presence and absence of adenovirus infection detected by PCR amplification a single wtAAV–AAVS1 host cell DNA junction in only one of nine animals, suggesting that integration of AAV even under favorable conditions is very rare [96]. More recently [24], analysis of naturally occurring AAV detected wtAAV in only 9 of 175 human tissues samples that included tonsil, spleen, and lung. Furthermore all of the genomes appeared to be unintegrated episomal copies with no AAV–AAVs1 junctions being detected and the only possible cell–AAV junction that was detected appeared to be in a highly repeated element in chromosome 1. Consequently, the available evidence indicates that integration of wtAAV *in vivo* does not reflect the experimental *in vitro* observations, but rather appears to be a rare event and mostly AAV genomes appear to persist as episomes, as has also been demonstrated for AAV vectors (Section 6.4.3).

6.4.3 Vector Persistence

Some early studies with AAV vectors expressing selectable markers suggested that these vectors integrated also at AAVS1 [52,97]. But the vectors used in these studies also were contaminated by wild-type AAV particles and may represent rare integration events mediated by *rep* and enriched by the selectable marker. In contrast, AAV vectors that contain no AAV *rep* coding sequences and no selectable marker have reduced efficiency and specificity for integration at the chromosome 19 AAVS1 site [56,57], and more usually persist in an episomal state. Initial evidence for this episomal persistence came from fluorescent in situ hybridization (FISH) analysis of cell lines transduced with AAV vectors which carried a low number of copies of an AAV vector as determined by Southern hybridization. FISH analysis of interphase nuclei compared to FISH analysis of metaphase chromosomes showed a reduced proportion of the cells carrying all the copies at a metaphase chromosomal site [57].

Studies performed *in vivo* in a variety of animal models now indicate that, in the absence of selective pressure, AAV vectors generally persist as episomal genomes. A number of studies have now shown that AAV can persist for extended periods when administered *in vivo* [68–72] and that the predominant form of the persisting vector genomes appears to be multimeric structures which are head-to-tail concatemers [98–101] that are circular [102–104]. How these head-to-tail multimers are formed is unknown but it cannot involve the normal AAV replication process because that requires rep protein and gives only head-to-head or tail-to-tail concatemers. Whether the circular concatemers are integration intermediates, as has been suggested for AAV integration [55,105], is also unknown. However, available evidence indicates that the majority of these head-to-tail concatemers are episomal and that integrated copies of vector in organs such as liver or muscle are very rare at least as judged from studies of AAV2 and AAV1 in rodents, rabbits, and nonhuman primates [96,106–108].

6.4.4 TARGETED INTEGRATION

Although AAV vectors that do not contain the rep gene do not integrate at the AAVS1 site at any significant frequency, they can be directed to integrate at the chromosome 19 site by supplying the rep gene in *trans*. For instance, a plasmid containing an AAV vector comprised of a reporter gene between the AAV ITRs and having a rep gene also in the plasmid, but outside the ITRs, resulted in integration of the AAV ITR vector cassette into the chromosomal site at 19q in human 293 cells *in vitro* [109]. Also, a baculovirus vector that contained an AAV reporter gene vector and separately contained an AAV rep gene was able to direct integration of the AAV vector into the AAVS1 site in human 293 cells [110]. Similarly, co-transduction of an epithelial tumor cell line (HeLa) or a hepatoma cell line (HepG2) with two hybrid Ad/AAV viruses, one carrying an AAV vector and the other expressing the rep gene resulted in integration of the AAV vector into the AAVS1 site [111].

In another study, the ability of AAV vectors to carry sequences that could mediate homologous recombination with chromosomal sequences was examined in cultured human cell lines. Homologous recombination at specific chromosomal sites could be attained [97] and additional studies showed that careful design of the interacting homologous sequences could

raise the efficiency of the process to about 1% [112]. However, this still requires a selective marker so its application to gene therapy may be limited.

6.5 AAV PERMISSIVITY

6.5.1 HELPER FUNCTIONS PROVIDED BY OTHER VIRUSES

The precise mechanism of the helper function provided by adenovirus or other helper viruses has not been clearly defined. These helper functions may be complex but relatively indirect and probably affect cellular physiology rather than providing viral proteins with specific functions in the AAV replication system. Studies with adenovirus [91] have clearly defined that only a limited set of adenovirus genes are required and these comprise the early genes E1A, E1B, E2A, E4orf6, and the VA RNA. The primary role of E1A is to transcriptionally activate the other adenovirus genes, but it may also transcriptionally activate the AAV p5 promoter. E1B and the E4orf6 protein of adenovirus interact to form a complex and these two genes provide the minimal function required to permit AAV DNA replication. The E2A gene of adenovirus has a complex function because it is a SS DNA binding protein that is directly involved in adenovirus replication but it also has an important role in regulating adenovirus gene expression. The role of E2A for AAV appears not to be a DNA replication function but to involve enhancement of AAV gene expression and particularly expression of AAV capsid protein. The VA RNA is also important in maximizing the level of AAV gene expression.

6.5.2 ALTERNATE PATHWAYS TO PERMISSIVITY

The concept that the helper virus renders the cell permissive by enhancing AAV replication in an indirect way is consistent with the evidence that helper virus genes do not appear to provide enzymatic functions required for AAV DNA replication and that these functions are provided by AAV rep protein and the cellular DNA replication apparatus [91]. This is also consistent with the observations that in certain cells lines, particularly if they are transformed with an oncogene, helper-independent replication of AAV DNA can occur if the cells are also treated with genotoxic agents such as UV or x-irradiation or with hydroxyurea [113,114]. In these circumstances, a small proportion of the cells could be rendered permissive for AAV replication but the level of replication and production of infectious AAV was very low.

6.5.3 REPLICATION OR PERSISTENCE

Two distinguishable phases of the AAV life cycle occur in permissive or nonpermissive cells. In either case the infecting single-stranded genome is converted to a duplex structure. There is evidence that the single-stranded genome may be converted to either linear duplexes or circular duplex molecules [102,103]. However, in permissive cells, in the presence of helper virus, this duplex genome then appears to follow the pathway of bulk replication using the self-priming property of the ITR to yield a large pool of head-to-head and tail-to-tail RF molecules and ultimately a large burst of progeny particles. In nonpermissive cells, in the absence of helper, these genomes follow a pathway that leads to generation of head-to-tail concatemers that persist as episomes or become integrated into the host cell chromosome. In this nonpermissive state there are two important parameters that may have different consequences for AAV or AAV vectors. In either case there is no helper functions provided by another virus. However, for wild-type AAV, the rep gene is present and therefore may be expressed. This may explain why AAV integrates efficiently into the chromosome 19. For an AAV vector, the rep gene is not present and thus vectors may progress through the integration pathway more slowly, or not at all, and rather follow the pathway of persistence as a circular episome [106,107].

6.6 AAV VECTORS

6.6.1 Design of AAV Vectors

The ability to generate AAV vectors was facilitated by the observation that molecular cloning of double-strand (DS) AAV DNA into bacterial plasmids followed by transfection into helper virus-infected mammalian cells resulted in rescue and replication of the AAV genome free of any plasmid sequence to yield a burst of infectious AAV particles [5]. This rescue may occur by a mechanism analogous to that used in reactivation of a latent provirus after superinfection of cells with adenovirus [115]. The general principles of AAV vector construction [4,5,7] are based upon modifying the molecular clones by substituting the AAV coding sequence with foreign DNA to generate a vector plasmid. In the vector, only the *cis* acting ITR sequences must be retained. The vector plasmid is introduced into producer cells that are also rendered permissive by an appropriate helper virus such as adenovirus. In order to achieve replication and encapsidation of the vector genome into AAV particles, the vector plasmid must be complemented for the *trans*-acting AAV *rep* and *cap* functions that were deleted in construction of the vector plasmid. AAV vector particles can be purified and concentrated from lysates of such producer cells.

The AAV capsid has three important effects for AAV vectors. There is a limit of about 5 kb of DNA that can be packaged in an AAV vector particle. This places constraints on inclusion of very large cDNAs and may also limit the ability to include extensive regulatory control sequences in the vector. The capsid also interacts with the AAV receptor and co-receptors on host cells and thus mediates cell entry. The capsid may also induce immune responses that could limit delivery of AAV vectors for some applications [116,117]. Except for the limitation on packaging size and the requirements for ITRs, there are no obvious limitations on the design of gene cassettes in AAV vectors. The ITR can function as a transcription promoter [84–86] but does not interfere with other promoters. Tissue-specific promoters appear to retain specificity [118,119] and a number of other regulated expression systems have now been used successfully in AAV

vectors [120–123]. Introns function in AAV vectors and may enhance expression, and more than one promoter and gene cassette can be inserted in the same vector. Importantly, transcription from AAV does not seem to be susceptible to *in vivo* silencing as shown by expression for over 6 years after delivery via multiple routes to multiple species, including humans [70,71,124–128].

6.6.2 Production of AAV Vectors

AAV has one of the largest burst sizes of any virus as noted above and this implies that a high yield of AAV vector particles per cell theoretically is attainable [1,7]. Maximizing the specific productivity may avoid unnecessary increases in biomass during vector production. However, the cytostatic properties of the AAV rep protein presented an obstacle to generation of stable packaging cell lines for producing AAV vectors [129]. Consequently, AAV vector production initially was based on transient transfection of a vector plasmid and a second plasmid, to provide complementing rep and cap functions, into adenovirus-infected cells, usually the transformed human 293 cell line. The original vector production systems yielded a mixture of AAV vector and adenovirus particles and exhibited relatively low specific productivity [4,5]. Furthermore, recombination between the vector plasmid and complementing plasmids generated pseudo wild-type or replication-competent AAV (rcAAV) or other recombinant AAV species [130,131].

Upstream production of AAV vectors can now be accomplished by transfection-based methods in which adenovirus infection is replaced by DNA transfection with the relevant adenovirus genes, with engineered mammalian cell-based AAV production systems that do not require DNA transfection, recombinant herpes simplex viruses carrying AAV genes and genomes, or recombinant baculovirus (Section 6.6.3). Each of the vector production methods can give a specific productivity in excess of 10^4 vector particles per cell, some much higher. However, the DNA transfection systems may be limited in terms of scale-up for potential commercial vector manufacturing. Downstream processing and purification are now usually accomplished by chromatographic procedures that give a higher degree of purity than the earlier purification procedures using ultracentrifugation.

6.6.3 Complementation Systems

Multiple approaches have been taken with respect to upstream production of AAV vectors. First, in DNA transfection-based procedures, various modifications have been made to the complementing *rep–cap* cassette in an attempt to enhance specific productivity and to decrease production of rcAAV. One group demonstrated that expression of rep and cap proteins may be limiting [132] but two other studies [133,134] suggested that cap proteins were limiting due to downregulation of cap by increased production of rep. A packaging plasmid which has the Rep78/68 expression downregulated by changing the initiation codon AUG to ACG was reported to give higher cap expression and higher yields of vector particles [134].

The only adenovirus genes required for full helper function are E1, E2A, E4, and VA, and transfection of the latter three genes into cells which contain the E1 genes, such as human 293 cells, can provide full permissivity for AAV [91]. The infectious adenovirus can be replaced as the helper with a plasmid containing only the adenovirus E2A, E4, and VA genes [135,136] that, together with the E1A genes supplied by 293 cells, provide a complete helper function in the absence of adenovirus production. Another group [137] used a plasmid containing nearly all of the adenovirus genome except the E1 region, but this yielded infectious adenovirus, probably by recombination with the E1 region in the cell. All of these systems require transfection with three plasmids, for vector, rep–cap, and adenovirus helper function, respectively. In contrast, Grimm et al. [138] combined all of the three adenovirus genes and the rep–cap genes into a single plasmid. In general, all of these approaches increased vector productivity compared to earlier systems such as pAAV/Ad [139], and productivities of at least 10^4 particles per cell have been reported. Nevertheless, these approaches still require DNA transfection and may be unwieldy for production scale-up.

An alternate approach to AAV vector production is to generate stable cell lines that contain the rep and cap complementing genes (packaging cells) or the vector genome or both (producer cells). In order to avoid DNA transfection, the cells must still be infected by a helper virus, adenovirus, but this can be removed readily as a result of advances in downstream purification processes (Section 6.6.5). Rescue of vector from a cell line having the vector stably integrated was demonstrated by transfecting the cells with a rep–cap helper plasmid and infecting with adenovirus [140]. Stable cell lines containing a rep gene, capable of generating functional rep protein, were constructed by Yang et al. [129] who replaced the p5 promoter with a heterologous promoter. Clark et al. [141] generated cell lines containing the rep and cap gene cassettes but deleted for AAV ITRs. Furthermore, the vector plasmid could be stably incorporated into the packaging cells to yield AAV vector producer cell lines [141–143]. Producer cell lines provide a scalable AAV vector production system that does not require manufacturing of DNA and may reduce generation of rcAAV. However, a new producer cell line must be generated for each individual AAV vector and this may be laborious.

A modification of the packaging cell line method is to use a cell line containing a rep–cap gene cassette that is then infected with an Ad/AAV hybrid virus. The Ad/AAV hybrid is an E1 gene-deleted adenovirus containing the AAV-ITR vector cassette [144,145]. After infection of cells containing the rep–cap genes, the AAV-ITR cassette is excised from the Ad/AAV, amplified and then packaged into AAV particles. This allows the same packaging cell line to be used for production of different AAV vectors simply by changing the Ad/AAV hybrid virus but it requires coinfection with adenovirus to provide the E1 gene function. The Ad/AAV hybrid viruses can also be used as delivery vehicles for AAV vectors [146–148] but this might suffer some disadvantages such as

induction of innate immune responses characteristic of the adenovirus capsid interaction with cells. Another packaging cell system was described [149] in which the packaging cell contains both a *rep–cap* cassette and the AAV ITR vector cassette, and both cassettes are attached to an SV40 replication origin. Also in the cells is a SV40 T-antigen gene that is under control of the *tet*-regulated system such that addition of doxycycline induces T-antigen that in turn results in amplification of the *rep–cap* and the vector cassettes. Subsequent infection of the cells with adenovirus renders the cells permissive for vector production.

Herpes simplex virus can be used in production of AAV vectors by generating HSV/AAV hybrid viruses. One approach [150,151] utilizes an HSV/AAV hybrid virus in which the AAV rep–cap genes, under control of their native promoters or altered promoters, were inserted into the HSV genome. This HSV/AAV *rep–cap* virus could generate AAV vector when infected into cells lines along with a transfected AAV vector plasmid or into cell lines carrying an AAV vector provirus. Alternatively, an HSV/AAV hybrid virus was constructed by inserting an AAV ITR vector cassette between HSV genome replication origins and then packaging this construct into an HSV particle [152–156]. Cells can be coinfected with each of the recombinant herpes simplex recombinants, generating AAV vectors [157].

A recent approach to scalable AAV vector manufacture utilizes insect cells. In this system, baculovirus vectors containing rep–cap gene cassettes or the AAV-ITR vector cassette are used to infect insect cells and AAV vectors can be generated [158]. Since the original report, several groups have worked to develop this method of vector manufacture. Problems with instability of the recombinant baculoviruses and low infectivity of the vector particles were identified, and solutions were offered. Progress has been made in scaling this method, and it may become a viable approach to large-scale vector production, once all technical issues are identified and resolved [159–161].

6.6.4 REPLICATION COMPETENT OR WILD-TYPE AAV

Wild-type AAV is not a human pathogen but generation of pseudo wild-type or rcAAV during vector production needs to be avoided for several reasons. The presence of pseudo wild-type AAV in vector preparations may increase the likelihood of vector mobilization following a helper virus infection in the patient. This in turn could increase the likelihood of cellular immune responses to AAV proteins as well as causing significant alterations in the biology of the vector because of the pleiotropic effects of the rep proteins. The earliest AAV vectors [4,5] were produced by co-transfection with helper plasmids that had overlapping homology with the vector and this generated vector particles contaminated with pseudo wild-type AAV due to homologous recombination. Reduction of the overlapping AAV sequence homology between the vector and helper plasmids reduced, but did not eliminate, generation of pseudo wild-type AAV [130,139,140].

A combination of vector plasmid and packaging plasmid in which the AAV region containing the p_5 promoter was not present in either plasmid prevented generation of wtAAV but some pseudo wild-type AAV (rcAAV) was generated at very low frequency by recombination [130,140]. This recombination was decreased to undetectable levels in a packaging system (split-gene packaging) carrying rep and cap genes in separate cassettes [130] so that three or four recombination events would be required to generate rcAAV. An alternate approach to decreasing pseudo wild-type or recombinant AAV is to insert a large intron within the rep gene in the helper plasmid so that any recombinants would tend to be too large to package in AAV particles [162].

It is likely that all vector production systems may have a propensity to generate pseudo wild-type AAV or other recombinants of AAV at some low frequency because it is not possible to eliminate nonhomologous recombination in DNA. Recent analysis of pseudo wild-type AAV recombinants has revealed that they appear to form through a mismatch repair pathway involving short regions of imperfect sequence identity of a size (~8–20 nt) that is small enough to occur at random in any stretch of DNA (B. Thorne, personal communication). Standardized assays to analyze and evaluate such rcAAV or recombinant species in AAV vector preparations have not yet been developed. However, to detect replication competent species, an assay that employs two cycles of amplification in non-*rep/cap* expressing cells and then a sensitive readout such as hybridization or PCR (rather than rep or cap immunoassays) is likely to be required [144].

6.6.5 Purification

Historically, AAV was purified by proteolytic digestion of cell lysates in the presence of detergents followed by banding in CsCl gradients to concentrate and purify the particles and separate adenovirus particles. Significant progress has been made in the downstream processing of AAV vectors and this has led to much higher quality and purity. This is critically important for preclinical studies and clinical trials. The reliance upon the original CsCl centrifugation techniques is being abandoned because it is a cumbersome procedure that does not provide high purity, it may inactivate some AAV vector and it is difficult to envision its use for commercial production. Several groups have employed nonionic iodixonal gradients as an initial bulk-recovery method [163,164]. A variety of chromatographic methods using standard ion-exchange resins; hydrophobic interaction resins; and antibody, heparin, or sialic acid affinity resins in both conventional filter and HPLC formats have been employed [138,161,165–184]. These results demonstrate the feasibility of developing scalable and well-controlled purification methods for AAV vectors that could lead to commercialization.

6.6.6 Assay of AAV Vectors

Characterization and testing of AAV vectors are critical elements for clinical development. While there is variability in assay methods and properties measured, the minimal testing should include a measure of vector strength by determining the concentration of capsids containing vector genomes, a measure of total capsids, a measure of "infectivity," an assessment of vector purity, and a measure of the vector potency.

Early in the development of AAV vectors, the concentration was measured using dot-blot hybridization of vector genomes and comparing the signal intensity to a standard curve derived from a plasmid containing the vector genome [185]. Recently, more precise and quantitative assays have largely supplanted the dot-blot assay. Measurement of vector concentration by real-time PCR amplification of the vector genome has resulted in increased precision and accuracy [166]. These assays measure the number of vector genomes that are encapsidated and thus protected from digestion by DNAse. Consequently, vector concentration is expressed in units of DNAse-resistant particles (DRP). This type of assay can be used to facilitate reliable quantitation of vectors during process development for production of multiple AAV serotypes [151].

Infectivity measurements of vector preparations are important mainly to establish a consistency of vector manufacturing rather than being an indication of vector potency *in vivo*. By their nature, AAV vectors are replication defective, so *in vitro* measurements of infectivity are artificial and generally require complementation with *rep* and *cap* as well as helper virus functions. Initially, vectors were measured by an infectious center assay in which wtAAV as a source of rep and cap and adenovirus were added along with dilutions of vectors to infect cells [186]. Cells that were replicating the vector genome were then scored by DNA probe hybridization to individual cells collected on a filter, hence the term infectious center. The development of cell lines containing rep and cap [187] has allowed infectivity measurements of vectors to be performed in a 96-well format assay that yields data with higher precision than the infectious center assay, and is amenable to a high throughput format that can be used for purification development [188]. In conjunction with the real-time PCR assay, this type of infectivity assay yields data on particle to infectivity (P:I) ratios of vectors with high precision thus aiding process development. It is important to recognize that the apparent particle to infectivity ratio of vectors based on different serotypes of AAV may be quite different due to the natural variation that exists in receptor usage and potentially in intracellular trafficking among the serotypes. A cell line containing rep and cap may be more efficiently transduced by one serotype as compared to a second serotype, yielding differences in P:I ratios that may not be reflective of differences in vector potency *in vivo* but rather a difference in the ability of a given serotype to efficiently infect the *rep/cap*-expressing cell line used in the assay.

It is imperative to establish an assay to measure expression of the transgene. For a phase I clinical trial it may be acceptable to evaluate transgene expression in cultured cells following transduction by the vector as a function of increasing multiplicity of infection using a Western blot of cell extracts. However, for later-stage clinical development, a more relevant potency assay will need to be established and applied to all vector lots during testing and release. The nature of the potency assay depends on the properties of the expressed transgene product. For example, if the product is a cytokine, a potency assay could evaluate the binding of the cytokine to its receptor, or a downstream signal generated in the cell as a result of the cytokine binding.

Other aspects of vector preparations are also important to measure. The ratio of empty to full capsids can be determined using a capsid ELISA [189], negative-stain electron microscopy analysis [171,172], optical measurements [173], velocity sedimentation analysis, or a measure of total protein compared to concentration of genome-containing vectors. Vector purity is usually assessed by acrylamide gel electrophoresis. For clinical development it is important to measure the residual contaminants including host cell protein and DNA, as well as serum components. In cases where the vector is manufactured from a cell line of tumor origin, such as HeLa cells or A549 cells, a determination of the residual amount and size of host cell DNA is an important part of the release testing. If the vector is of high purity, its aggregation state can be assessed by using dynamic laser light scattering to measure the hydrodynamic radius. This provides a size distribution of particles in solution and the relative amount of particles in each size class [174]. This measure is useful for vectors at high concentration and can aid in the development of formulations that maintain vectors in a monomeric, nonaggregated state.

6.7 GENERAL PROPERTIES OF AAV VECTORS

6.7.1 SEROTYPES OF AAV

The growing appreciation of natural variation among AAV isolates and application of this knowledge to vector production and usage has added to the attractiveness of rAAV as a gene therapy modality. There are 11 serotypes of AAV reported, and more than 100 capsid sequence variants have been characterized from a number of species, both human and nonhuman [175]. Some of these new variant capsids have been used to make vectors, thus expanding the range of capsid sequences that can be employed with the goals of increasing vector transduction efficiency, decreasing vector dose, improving targeting using inherent tropisms, and avoidance of a preexisting host neutralizing antibody response.

Complete coding sequence information is available for serotypes AAV1 to AAV6, AAV10, and AAV11 [38–43,46,47] as well as capsid gene sequences for AAV7, AAV8, and for over 100 capsid gene sequences isolated from multiple animal and human species [21,23]. AAV6 is a recombinant between AAV1 and AAV2. In addition, there are two isolates and sequences of AAV3 that differ from each other in a number of amino acids in both rep and cap [46]. Receptor usage by these viruses varies and is reflective of their sequence diversity [58–60,190–198]. Sialic acid and heparin are the most commonly used primary receptors for cell binding. A number of secondary receptors used for cellular entry have been identified. Some serotypes are able to use multiple secondary receptors, such as AAV2 with at least four. This variation in receptor

usage, at least in part, is responsible for the increasingly appreciated efficiency of vectors of different serotypes to transduce specific cell types and tissues *in vivo*. The serotypes of AAV are highly related to each other with the notable exception of AAV5 [42,43]. AAV5 is the most distantly related of the serotypes, and also displays a serotype-specific terminal resolution site (TRS) in its ITR [195]. Even though rep proteins from other serotypes will bind the AAV5 ITR, they do not efficiently cleave at the TRS.

Most early studies with AAV vectors utilized serotype 2. A number of animal studies demonstrated transgene expression that persisted for months to years in muscle, liver, CNS tissues, lung, and retina [68–71,98–101,109,118,124]. As other serotypes of AAV have been used in vector production, it has become apparent that there are distinct and often advantageous differences among vector serotypes in their efficiency of transgene expression in various tissues and cell types, at least in small animal models [175]. It is difficult to list a hierarchy of transduction efficiencies of target organs relative to individual serotypes because there are conflicting data in the literature both with regard to organ tropism and the threshold of expression differences as reported from different laboratories. These differences may reflect differences among strains of animals used in the studies, or differences in vectors produced by different investigators. It is worth noting that the dramatically increased efficiency of transgene expression by several of the newer serotypes and isolates of AAV capsids relative to the classic AAV2-based vectors may not translate when tested in larger animals, so caution should be exercised when evaluating data from only mice and rats, for example [176].

Cellular receptors play an important role in AAV vector tropism. For instance, the increased transduction efficiency for AAV5-based vectors in mouse airway may be due in part to the demonstrated increase in binding to the apical surface of polarized airway epithelial cells [62,199–202]. However, cellular trafficking of vectors also plays a role in transduction efficiency [67]. Approximately 5% of hepatocytes are transduced as measured by transgene expression, following delivery of AAV2-based vectors to mice in spite of the fact that every cell contains vector DNA one day following administration [203]. This shows that events postbinding and internalization are important for successful transduction (Section 6.8.2). Interestingly, when vectors based on AAV5 or AAV8 were administered to mice, increased numbers of hepatocytes were transduced, up to 100% and increased levels of transgene expression were observed [21,204,205].

In the murine CNS, AAV4 vectors transduce ependymal cells almost exclusively, while AAV5 vectors transduce both neurons and astrocytes [206]. In retina, a number of studies have demonstrated large differences among serotypes in the ability to transduce photoreceptor cells and the retinal pigmented epithelium [175,190,207,208].

Vectors based on AAV6 and AAV8 have been shown to efficiently escape from the blood stream and transduce most muscle groups in small animals when administered systemically [209–212]. Thus AAV6 and AAV8 may have potential for treating diseases such as muscular dystrophy in which multiple muscle groups are affected. A recent report demonstrates nearly complete transduction of a perfused skeletal muscle in both mice and monkeys with AAV6 and AAV8 vectors, but not AAV1 vectors [213]. This result makes it likely that a successful treatment for a number of muscle diseases could be developed with these serotypes.

While it is clear that there is great variability among different AAV serotypes and capsid sequences in the ability to transduce cells, tissues, or organs efficiently depending upon the route of administration and the capsid sequence, further work is needed to understand the relative importance of the factors contributing to differences in transduction efficiencies with different serotypes of AAV vectors.

6.8 VECTOR METABOLISM

Early studies of AAV vectors led to the suggestion that the major factor limiting the efficiency of transduction was the process of converting the SS genome to a duplex molecule. More recent studies are beginning to reveal a complex pathway of events that impact upon the function of AAV vectors. Thus, delivery of the AAV vector genome to the cell nucleus may be influenced by the availability of specific receptors and co-receptors for AAV binding and entry and by cellular trafficking including a potential diversion of AAV particles into an ubiquitin-mediated proteosome degradation pathway. After the genome is successfully delivered into the nucleus, a succession of events converts the vector into larger concatameric molecules. Most of these concatameric molecules appear to be maintained for prolonged periods, perhaps for the lifetime of the cell, within the nucleus as episomal molecules, and very few if any integrate into the host genome [24,107,108].

6.8.1 CELLULAR BINDING AND TRAFFICKING OF AAV VECTORS

The ability of a vector particle to interact with a specific receptor molecule on a target cell is critical for successful transduction. The primary binding receptors identified, heparin sulfate and sialic acid, are commonly found on many cells and are also the receptor molecules utilized by a large number of viruses besides AAV. This suggests that additional receptors that lend more specificity to attachment and penetration of cells might exist and several such co-receptors have been identified (Section 6.7.1).

While cellular attachment is the critical first event, it does not necessarily imply that the vector will be able to efficiently transduce the cell. This has become increasingly apparent from a more detailed understanding of the trafficking and uncoating of AAV vectors [62–67]. For example, polarized human airway epithelial cells are transduced with varying efficiencies by AAV2-based vectors depending on the route of delivery and entry from the basolateral surface results in about a 200-fold increase in gene expression in the cells

compared to vector administered from the apical surface [62,214]. Surprisingly, the difference in vector attachment to the two cell surfaces is only about fivefold. This finding led to the discovery that the vectors traffic differently in these cells depending on the side of the polarized airway to which they bind [62,214]. Vectors administered from the apical surface are modified by ubiquitination. The addition of proteosome or ubiquitin-ligase inhibitors led to an increase in transduction following apical administration of vectors. These comparisons of differences in binding and differences in transduction efficiency have led to an appreciation that cellular trafficking of vectors is important for successful transduction, and that it may differ between cell types and even within individual polarized cells between the apical and basal surfaces.

After binding to the cellular receptor, AAV2 is internalized by an endocytosis mechanism that for heparin binding appears to be mainly via clathrin-coated pits [215,216] although some clathrin-independent uptake may also occur [217,218]. Additional studies using fluorescent Cy3-conjugated AAV2 vector particles showed that endocytosis can be mediated by an $\alpha_v\beta_5$ integrin/Rac1 dependent mechanism and that subsequent trafficking to the nucleus requires activation of PI3-kinase pathways as well as functional microtubules and microfilaments [217,218]. Evidence for the involvement of early endosomes in AAV trafficking, as well as the involvement of microtubules and microfilaments, exists mainly from the use of inhibitors [64,65]. Recent studies indicate that AAV can traffic through late and recycling endosomal compartments in a dose-dependent manner [66]. Escape of vector particles from endosomal compartments into the cytoplasm appears to be mediated by a latent phospholipase activity resident in the N-terminus of VP1 [75,219] that becomes externalized and active in the endosome. Transcytosis of AAV vector particles has been reported to allow trafficking through endothelial and epithelial barriers that would normally restrict their spread [220].

There is consensus that, in the absence of helper virus, there is a block at the step of import of viral vector genomes into the nucleus of the cell [216,221,222]. However, there is no clear understanding as to how or where the viral genome is uncoated or how the vector enters the nucleus, although it does not appear to use the nuclear pore complex for this step [222,223]. Evidence has been presented that, once the capsid enters the nucleus, the rate of disassembly may vary among serotypes and this may influence transgene expression levels in a tissue or target cell type [224,225]. Recent work from Akache et al. [226] provides strong evidence that cellular proteases, cathepsins, act on the incoming capsid in the endosome, perhaps leading to efficient uncoating once the cathepsin-cleaved capsids enter the nucleus of the cell. Nuclear entry and uncoating are areas of AAV biology that remain incompletely understood.

6.8.2 VECTOR DNA METABOLISM

In order for rAAV vectors to transduce a cell, the SS DNA genome must be converted into a double-stranded form [49] as shown in Figure 6.3. Several reports, based upon transduction of cells *in vitro*, indicated that this single strand to double strand conversion step might be rate limiting in the absence of helper virus coinfection [227–229]. The adenovirus E4 or F6 protein has been shown to increase the level of transgene expression and to increase second-strand DNA synthesis [230]. More recently a cellular protein that bound to the single-stranded d-sequence of the AAV ITR was identified as the well-known FK506-binding protein (FKBP-52). This protein has been implicated in controlling the conversion of SS to DS DNA in vector-infected cells and phosphorylation of FKBP-52 influences its ability to bind the d-sequence [231–234]. When phosphorylated at tyrosine residues (by the epidermal growth factor receptor protein tyrosine kinase), FKBP-52 binds to the SS d-sequence region of the ITR that is present in infecting vector genomes, and second-strand DNA synthesis is impaired. The efficiency of transduction in a number of cell types *in vitro* and *in vivo* correlates with the phosphorylation state of FKBP-52. In HeLa cells, overexpression of a cellular phosphatase (TC-PTP) that can use the FKBP-52 protein as a substrate led to dephosphorylation of the FKBP-52, an increase in AAV second-strand DNA synthesis, and an increase in transgene expression. Transgenic mice expressing either the wt or a catalytically mutant form of the phosphatase were created. Hematopoietic stem cells from transgenic mice expressing the wt TC-PTP phosphatase were transduced by an AAV2 vector, but those from mice expressing the phosphatase-negative mutant were not. These results suggest that the block to second-strand DNA synthesis is due to binding of FKBP-52 to the d-sequence of infecting vector genomic DNA and that this binding is regulated by phosphorylation.

A second mechanism for conversion of SS DNA to DS DNA has been proposed. In this model, annealing of negative and positive sense SS DNA genomes, both of which are packaged with equal efficiency into individual AAV particles, occurs in cells to form ds DNA in the absence of second-strand DNA synthesis. There is some experimental evidence for this self-annealing model both *in vitro* and *in vivo* [235] although it is not consistent with observations that AAV infection displays single-hit kinetics [79]. However, it is possible that double-stranded DNA may be formed via either pathway in cells transduced with rAAV vectors.

Further support for the idea that conversion of single-stranded vector genomes into transcriptionally active double-stranded forms is crucial for transduction and is a rate-limiting step comes from a study demonstrating that vector genomes smaller than half the size of the AAV genome are packaged in multiple ways [236]. Particles contain either a single vector genome, two copies of the small vector genome, or a covalently linked double-stranded hairpin molecule equivalent to a replicative intermediate formed during vector genome replication. In a separate study, vectors containing these small self-complementary genomes (scAAV) were shown to be relatively insensitive to the enhancing effects of adenovirus on transduction, resistant to the effects of DNA synthesis inhibitors, and displayed altered kinetics of transgene

expression *in vivo* [237]. When these vectors were administered to mice, an increase in transduction from the usual 5% to more than 50% of hepatocytes was observed [205,238,239]. These results strengthen the model that second-strand DNA synthesis is a rate-limiting step for transduction, at least in hepatocytes. These self-complementary vector constructs are attracting attention since they seem to overcome the barriers to efficient transduction seen with many single-stranded AAV vectors [240].

Once the vector genome is converted into a double-stranded DNA molecule, a number of fates have been reported. One group of studies convincingly demonstrates that the viral genome circularizes, and that these circular monomers recombine at the AAV ITRs to form larger circular concatemers of head-to-head, head-to-tail, and tail-to-tail arrangements both *in vitro* and *in vivo* [102,103,241,242]. These genomic concatemeric molecules, containing a "double-D" ITR structure (an ITR bracketed on each side by the d-sequence and presumably formed by an ITR-ITR recombination event), persist extra-chromosomally as episomes and are responsible for the long-term persistent transgene expression seen with rAAV vectors. SCID mice that are deficient in DNA-dependent protein kinase activity (DNA-PK) lack the ability to convert rAAV genomes into circular concatemers [243]. Rather the concatemers formed in these mice appear to be linear molecules, suggesting that DNA-PK activity is involved in the formation of circular episomes. In both normal and SCID mice, transgene expression persisted for at least one year at similar levels. A second group of reports [235,244,245] suggest that vector genome concatamerization occurs by recombination of linear monomeric genomes. It is possible that both mechanisms of concatamerization are operative and perhaps there are tissue-specific differences, such as muscle or liver, in which one or the other pathway is more likely to occur. Despite the differences reported on the substrate for concatemer formation, it is clear that long-term transgene expression is mainly mediated by episomal concatemers of viral genomes rather than integrated molecules [106,107]. A recent report provides evidence that there are multiple fates for the double-stranded DNA once it is formed, and that the rate-limiting step for successful transduction of a cell leading to long-term transgene expression may be stabilization of the double-stranded vector DNA [246].

Evidence that DNA repair and recombination are directly involved in circularization or concatamerization of AAV vector genomes is supported by recent insights into possible biochemical mechanisms of their formation [243,245]. In fibroblasts from a patient with ataxia telangiectasia (ATM) there is greatly enhanced formation of AAV vector circular forms and enhanced integration of the head-to-tail concatemers as proviral genomes [247]. The ATM gene is a PI-3 kinase that regulates the p53-dependent cell cycle checkpoint and apoptotic pathways and in these ATM cells the DNA double-strand break (DSB) repair systems that normally can be activated by UV irradiation appear to be already activated maximally. Consequently, AAV vectors in these cells yield a high level of transduction and this is not activated further by

UV irradiation [247], in contrast to the observations in normal cells [227–229]. Additional evidence that DSB repair pathways are involved in regulating AAV transduction comes from observations that the proteins Ku86 and Rad52, which are known to recognize DNA hairpin structures and DNA termini and to promote repair of DSB, could associate with the AAV DNA ITR [248].

6.8.3 DUAL VECTORS

The ability of AAV vectors to form concatemers and the evidence that intermolecular recombination is directly involved in their generation has been used to extend the capacity of AAV vectors beyond the 4.5 kb payload limit in the dual-vector system. It is now possible to divide a gene expression cassette that is up to 9 kb in size between two AAV vectors and, following coinfection with both vectors, take advantage of the intermolecular recombination to generate the intact expression cassette. This process has been demonstrated both *in vitro* and *in vivo* and shows remarkable efficiency [244,249–251].

Both *cis*-activation [249] and *trans*-splicing [250] modes of dual vectors have been described. In *cis*-activation, one vector carries a high efficiency enhancer of transcription and the other vector carries the transcription promoter and the gene sequence to be expressed. Recombination after infection places the enhancer in *cis* with the expression cassette and increases transcription. In the *trans*-splicing mode, one vector carries the transcription promoter and the 5' part of a gene, and the other vector carries the 3' part of the gene. By judicious arrangement of splice donor and acceptor sites, the appropriate mRNA can be derived from the read-through transcript using the heteroconcatemeric template. Interestingly, the intervening ITR sequence does not appear to inhibit read-through transcription or RNA splicing [249,250]. It does appear, however, that in these dual-vector constructs, some transcription may be seen directly from the ITR either because of its own promoter activity or because of enhancement from the *cis*-acting elements provided by the second vector [249,250].

Cis-activation was demonstrated using a vector containing super-enhancer elements comprising parts of the SV40 and CMV enhancers together with a second vector containing the luciferase gene. This dual enhancer combination gave robust expression of luciferase in mouse skeletal muscle at levels that compared well with that from administration of a single vector expressing the same elements [249]. In another study, two vectors containing a LacZ reporter cDNA or an enhancer promoter cassette from human elongation factor EIFα, respectively, were shown to yield reporter gene expression in livers of mice after dual administration by portal vein injection [244].

Trans-splicing was demonstrated using one vector that contained a transcription promoter and the 5' part of the erythropoietin (EPO) genome locus and a second vector that contained the 3' region of the EPO locus [250]. Dual injection of the two vectors into mouse skeletal muscle provided therapeutic levels of EPO sufficient to protect the mice from adenine

induced anemia. Similarly, another group used two vectors, one containing the CMV promoter with the 5' half of the LacZ gene and the second containing the 3' half of the LacZ gene [251]. These vectors also gave robust expression of intact LacZ after dual injection into mouse skeletal muscle. Improvements in the design of these *trans*-splicing has led to greatly increased efficiency of transgene expression [252,253], with a recent report demonstrating equivalent transgene expression from a set of *trans*-splicing vectors compared to a single AAV vector encoding the same final mRNA.

The dual-vector approach to packaging and expressing large genes in rAAV vectors was extended to an investigation of the use of homologous recombination between two vectors, each containing an overlapping DNA sequence to drive recombination [254,255]. The data demonstrated that non-ITR-mediated recombination between two genomes can occur and result in expression of a large transgene. However, the efficiency of this homologous recombination process may vary depending on the tissue type. In mice, the efficiency in muscle was low and yielded less than 1% of the level of expression seen from an intact vector, whereas in lung the efficiency approached that seen with an intact vector construct.

The dual-vector approaches to packaging and expressing larger transgenes in AAV particles offer a way to almost double the payload capacity of AAV vectors that should extend the utility of this vector system to larger genes. Careful analysis will be required to determine additional safety issues or risks this approach might present for clinical use. For instance, for some applications such as hemophilia A, it may be deleterious to express a partial FVIII protein. Nonetheless, as these systems are better characterized they may well prove to have clinical utility.

6.8.4 CAPSID MODIFICATIONS AND TARGETING OF VECTORS

There is an interest in modifying the capsid structure of AAV vectors to target transgene expression to a specific cell type [28,29]. Multiple approaches have been taken. One approach involves attaching ligands. Thus, retargeting was obtained by the use of a bi-specific antibody having one arm specific for AAV capsids and the second arm specific for a receptor on a cell [256] while not involving capsid modifications, this method did achieve vector targeting to the specificity of the antibody used. Another group conjugated growth factors to the capsids [257] and achieved increased transduction in cells expressing receptors for the growth factors. A third group engineered an immunoglobulin-binding domain into the capsid of AAV and then used this modified particle to bind various targeting antibodies [258].

A second approach to vector targeting involves alterations or additions directly to the amino acid sequence of the capsid proteins. The overall goal of this approach is to replace the natural receptor interactions of the capsid with a receptor interaction that would be engineered into the capsids as an amino acid sequence or receptor ligand. There are a number of studies that describe such attempts at vector retargeting, and some reports describe a systematic analysis of insertions into the capsid and the effect on vector yield and tropism [258–261]. It is possible to generate vector particles with modified tropism as a result of amino acid substitution or addition to the capsid and further work will reveal the usefulness of this approach for clinical gene therapy applications. Peptide ligands that are specific for various receptors have been successfully engineered into AAV2 capsid proteins VP1, 2, and 3 with the resultant vector target specificity being achieved [258,260,262–264]. Capsid retargeting can also be achieved by screening a library of capsid protein mutants containing random insertions of short amino acid sequences on a target cell type of interest [265–267].

Several groups have produced vectors containing a mosaic of capsid proteins from two different serotypes [268,269]. The resulting particles retained properties from each serotype, but some combinations led to properties unique to the mosaic. Work is under way in several labs to identify domains of one serotype that lend it a unique property, and then combine these domains into chimeric capsids [270,271]. Lastly, Warrington et al. [272] demonstrated that capsids of AAV could be formed without VP2, and then constructed AAV particles containing assorted peptides fused to the N-terminus of VP2, including green fluorescent protein. This approach may allow the generation of specifically targeted vectors but this objective has yet to be achieved.

6.9 PRECLINICAL STUDIES, HOST RESPONSES, AND TOXICITY

AAV vectors will most likely be applied to target nondividing cells with infrequent delivery in view of the episomal persistence of the vector. A substantial body of studies have now been carried out to analyze the safety and biodistribution of AAVs vector to support a number of clinical trials. These studies have generally shown a good safety profile and provided information on host immune responses, which for AAV vectors appear to be relatively modest. In addition, the biodistribution studies have allowed a determination of the integration frequency of AAV vector genomes and support the concept that generally AAV vectors persist as episomes. Nevertheless, it is important to assess the likelihood of integration and the possibility of resulting deleterious effects, and to assess host responses particularly if they would limit application of AAV vectors.

There have been few reports or indications of toxicity mediated by AAV vectors. The first safety studies of AAV vectors were conducted to support the clinical trials of an AAV-CFTR vector. Toxicity was examined extensively for a vector expressing the CF transmembrane regulator (CFTR) protein following delivery of these AAV-CFTR vector particles directly to the lung in rabbits and nonhuman primates. In rabbits, the vector persisted and expressed for at least 6 months but no short- or long-term toxicity was observed and there was no indication of T-cell infiltration or inflammatory responses [68]. Similarly, in rhesus macaques, AAV-CFTR vector particles were delivered directly to one lobe of a lung and also persisted and expressed for at least 6 months [273]. Furthermore, no toxicities were observed by pulmonary function testing, by radiological examination, analysis of blood gases and cell counts and differential in bronchoalveolar lavage or by gross morphological examination or histopathological examination or organ tissues [273]. Importantly, preclinical studies of the biodistribution of the AAV-CFTR vector following pulmonary delivery in rabbits and macaques showed that there was minimal spread of vector to organs outside of the lung, no vector in gonads, and no toxicity in any organ. However, use of different delivery routes may lead to more extensive biodistribution.

Studies in rhesus macaques were also performed to determine if the AAV-CFTR vector could be shed or mobilized from a treated individual [37]. AAV-CFTR particles were delivered to the lower right lobe of the lung and high doses of both wild-type AAV particles and an adenovirus capable of replicating in macaques were administered to the nose of the animals. These studies indicated that the vector was not readily mobilized and that the probability of vector shedding and transmission to other individuals was likely to be low. The favorable safety profile of AAV in these preclinical studies has been predictive of a similar safety profile observed in clinical trials of the AAV-CFTR vector in CF patients [9–13]. These clinical studies showed that there was only a very low level of vector shedding in sputum after administration to the lung and in a study in conjunction with an aerosol delivery trial, it was demonstrated that there was no transmission of vector to others such as health care workers [274].

An AAV–FIX vector was studied in preclinical rodent and canine models in support of clinical trials in hemophilia B patients via intramuscular injection or intravenous delivery via the hepatic portal vein. The vector showed a good safety profile when administered by intramuscular injection to hemophilia B dogs that were reflected in the intramuscular injection clinical trial [14] (Section 6.10). However, in a second clinical trial, following intravenous hepatic administration of the AAV–FIX vector, patient semen was positive for the vector genome for a few weeks. This indicated that vector had apparently been distributed to gonads but further investigation indicated that the vector was not present in motile sperm [275]. A more extensive study of AAV–FIX vector delivery by both intramuscular and intravenous routes, using a sensitive DNA-PCR assay, showed that there was a dose-dependent detection of vector genomes sequences in the gonads of males of several animal species including mice, rats, rabbits, and dogs [275]. However, although testis tissue of these species was positive for vector for a short period after delivery, in both rabbits and dogs, semen and sperm were negative for vector sequences, suggesting that the risk of inadvertent germline transmission of vector sequences after intramuscular or intravenous delivery is extremely low [276,277]. Additional studies of AAV2 vectors expressing FIX [278] or α-1-antitrypsin (AAT) [279] genes, after intramuscular delivery into rhesus macaques, again showed excellent safety profiles and did not detect transmission of vector sequences to gonads. On the other hand, an AAV1 α-1-antitrypsin vector delivered to mice and rabbits by intramuscular injection did show appearance of vector sequences in rabbit semen but this was dose-dependent and declined over a period of 60 days. Additional safety and biodistribution studies have been reported for an AAV1-calpain vector delivered intravenously to mice [280]. Several safety and biodistribution studies for AAV2 or AAV4 vectors delivered subretinally in rodents, dogs, and cynomolgus monkeys have been reported and all show minimal toxicity of the vector and almost no spread of vector beyond the retina [281–284]. Although one study of subretinal delivery of an AAV2 vector did show ocular inflammation, it was subsequently demonstrated to be due to impurities in the vector preparation [282,283]. Even if biodistribution studies show transfer of vector genome sequences to gonads, provided that the signal is dose dependent and disappears over time and does not appear in motile sperm, then vertical transmission is considered highly unlikely.

As discussed in Section 6.4.3, a variety of studies indicate that AAV vectors persist for prolonged periods as episomal structures and that integrated copies of AAV vector genomes are very rare, including in such tissues as liver and muscle [96,106–108]. Some of the more extensive biodistribution studies with AAV2 and AAV1 vectors in rodents and rabbits [106–108] using several hundred animals facilitated an estimation of the maximum integration frequency that is likely. These studies [106] lead to the estimation that the integration frequency of AAV vectors is several orders of magnitude lower than the spontaneous rate of mutation for human genomes so that the likelihood of insertional mutagenesis by AAV vectors is very low. Indeed, using similar methods of analysis the integration frequency of AAV vectors in animals is as rare as that following transduction with adenovirus or plasmid DNA vectors. After discussion of this information at a public meeting [285], the U.S. FDA issued guidelines for long-term follow-up of gene transfer research subjects in clinical trials which indicate that AAV is considered as a nonintegrating vector.

Consistent with its low integration frequency, AAV has never been associated with any disease or shown to promote tumorigenesis. Indeed, several studies show that AAV or AAV ITRs can inhibit tumorigenesis [286,287]. Also, studies in rhesus macaques infected with wild-type AAV2 by intramuscular, intravenous, and intranasal routes showed that although there were some antibody responses against the AAV capsid, there was no cellular immune response to AAV components and there was no indication of any tumorigenesis [96]. However, there was an anecdotal observation of tumors occurring in mice having a homozygous mutation in the β-glucuronidase gene, that were treated as neonates via head vein injection with an AAV vector expressing β-glucuronidase (BGUS). These mice exhibit the lysosomal storage disease mucopolysaccharidosis VII (Sly syndrome) that is characteristic of the human disease and, if left untreated, the animals die in several months. The mice treated with the AAV-BGUS vector survived up to 18 months and showed a remarkable biochemical and physiological correction of the disease [288]. However, at 12–18 months some of these mice developed hepatic tumors but vector genomes could not be found in all of the tumors and not at any higher frequency than in nontumor tissue from the same animals [289]. Thus, these tumors did not appear to be caused by insertional mutagenesis and clonal expansion. A more recent study by the same group [290] in the same mouse model did report integrated vector in the tumors but there is no explanation as to the disparity with the previous result. Furthermore, most other studies of AAV vectors, including intravenous delivery in the same animal model and a variety of other animal disease models, have never shown any tumor formation [289,291,292]. The only other study that has reported tumors, after transduction of OTC deficient mice with an AAV-LacZ vector, concluded that the phenomenon was apparently related to expression of the transgene [293]. The tumors in the MPS VII mice remain as an unexplained observation since in the first study there were no sham-treated control mice that survived as long as the treated mice [288]. In the second study a promoterless vector was used but it still retained the transcriptionally active ITR so it may be an inadequate control [289]. It is possible that the tumor formation is specific for this disease, or the particular animal model, or for ectopic production of BGUS, or to some other unknown cause and does not have any direct relation to the use of an AAV vector. However, some caution needs to be exercised, and MPS VII disease possibly may not be an attractive candidate for gene therapy with an AAV vector.

Host immune responses, including innate immune responses, cellular immunity, or humoral antibody responses, may hinder the use of some gene therapy vectors. However, administration of AAV vectors does not induce pro-inflammatory cytokines, unlike adenovirus vectors [48]. Cellular binding and biodistribution of AAV vectors is controlled by the viral capsid. A recent examination of AAV infection of human diploid fibroblasts using gene array chips showed very little change in gene expression profiles upon binding of AAV capsids with only a nonpathogenic response and no induction of immune response or stress response genes [294]. In contrast adenovirus binding induced a wide array of genes associated with pathogenic responses, including immune and stress response genes. Also, AAV vectors are replication defective and contain no viral genes, so cellular immune responses against the viral components should not be evoked readily. In all of the *in vivo* studies of AAV vectors in rodents, rabbits, or rhesus macaques, there has been little evidence of cellular immune responses to viral components.

AAV vectors may be used mainly for clinical applications requiring only infrequent delivery, but potential humoral immune responses against the viral capsid, either preexisting in the human population or induced by vector administration, must be considered [3,28,29]. Reinfection of humans by AAV is not prevented by serum neutralizing antibodies [3,31,32]. However, more extensive studies will be required to assess whether induction of neutralizing antibody responses will pose any limitations to AAV vectors.

Induction of anti-AAV capsid antibody responses after vector administration may reduce the efficiency of transduction upon re-administration [70,295,296]. This depends upon the route of administration [96,297,298] and also may depend upon the quality of the vector preparations. In one study in which two AAV vectors expressing the reporter genes bacterial β-galactosidase or human alkaline phosphatase were successively administered to lungs of rabbits, expression from the second administered vector was impaired and this was ascribed to a neutralizing antibody response [299]. Similar studies in mice also implied that neutralizing antibodies impaired re-administration of AAV vectors but that this could be partially or completely overcome by transient immunosuppression with anti-CD40 ligand antibodies or soluble CTLA4-immunoglobulin at the time of the initial vector administration [296,300]. However, interpretation of such studies may be complicated by the expression of the foreign reporter proteins that could represent confounding variables. Furthermore, other studies showed that immune responses to AAV capsid were greatly reduced following airway administration of AAV [96] and that vector transduction could be seen after at least three repeated administrations to the lung of rabbits [48]. Thus, up to three successive administrations of AAV vectors to rabbit lungs over a 20-week period did not prevent gene expression from the third delivery of vector [48] in spite of a dose-dependent rise in serum antibody against the AAV capsid. This was reflected in clinical trials of an AAV-CFTR vector administered to the airway of CF patients. Subjects who were not serum antibody positive for AAV capsid antibody prior to vector administration did seroconvert for serum AAV antibody in a dose-dependent fashion [9–13].

It remains to be determined how neutralizing antibody responses to AAV vector capsids might impact applications of AAV vectors. This will most likely require studies in humans to determine if the various animal models such as rodents or rabbits are predictive for the immune response to AAV vectors in humans and whether such immune responses will pose any limitations to their therapeutic application. As discussed in Section 6.10, in clinical trials so far, preexisting anticapsid antibody in serum has not appeared to impact delivery of an AAV-CFTR vector by aerosol to the lung or an AAV–FIX vector to muscle, but intravenous delivery of an AAV–FIX vector to the liver may have been impacted. For relatively infrequent administration of AAV vectors, transient immune blockade [295,296] has been suggested but may not be an attractive option for therapeutic use of AAV or any other gene delivery vectors. It has been suggested also that vectors with capsids of different serotypes could be used for subsequent administrations [202]. However, although in rodents and rabbits the several serotypes of AAV do not generate cross-reacting antibodies this may not be predictive for humans. Recent examination of various human sera indicates substantial cross-reaction between, for instance, AAV1 and AAV2 anticapsid antibodies which is not seen in smaller animals (R. Peluso, unpublished data).

Immune responses to the transgene expressed by an AAV vector vary and may depend upon the route of delivery. Both MHC class II-restricted antibody responses and MHC class I cytotoxic T-lymphocytes have been reported but this may vary with the route of administration (297). In some studies, such as intramuscular delivery in mice, there was no immune response to an expressed foreign reporter gene such as bacterial β-galactosidase and it was suggested that AAV may be a poor adjuvant or may not readily infect professional antigen presenting cells in muscle [301,302]. However, an AAV vector expressing the herpes simplex virus type2 gB protein was delivered intramuscularly into mice and elicited both MHC Class I-restricted CTL responses against the gB protein and anti-gB antibodies [303]. Following intramuscular delivery of an AAV human FIX vector [273,304,305] there was an antibody response, but not a CTL response, against the FIX protein. The rules governing immune responses to foreign transgenes following AAV vector delivery remain to be elucidated more fully but AAV may have utility as a viral vaccination vector (Section 6.10.8).

A recent observation in a clinical trial in hemophiliacs of a potential AAV vector capsid specific CD8+ CTL response [306] led to an additional report that healthy subjects carry AAV capsid-specific CD8+ T-cells and that AAV-mediated gene transfer can stimulate their expansion [117]. However, the mechanism of this is not understood [117] and it has not yet been possible to model this phenomenon in animals [307–309]. Interestingly, AAV2-induced human T-cells [117] also proliferated upon exposure to alternate AAV serotypes, indicating that other serotypes are unlikely to evade capsid-specific immune responses. Surprisingly, this has not been the case in animal studies, since no such expansion occurred in mice after AAV-mediated gene transfer. Moreover, when mice were immunized with adenoviral vectors expressing AAV capsid, they developed lytic CD8+ T-cells in blood, lymph tissues and liver, but these were unable to eliminate AAV-transduced hepatocytes following infusion with an AAV vector expressing human factor IX under a hepatocyte-specific promoter, and FIX expression was sustained and comparable in AAV-preimmune and naïve animals [307]. Furthermore, in AAV2-pretreated hemophilia B dogs, canine FIX expression increased from less than 1%–16% of normal levels when treated with an AAV2/8 vector, and a high level of expression has lasted for more than 2 years. Additionally, no significant liver toxicity or canine FIX-specific antibodies have been detected in these animals [308]. A more recent analysis suggested that the apparent CTL response observed by Manno et al. [306] could actually be directed at an epitope expressed from a +1 reading frame shift in the factor IX transgene ([310], R.J. Samulski, personal communication).

If the above observations are substantiated, they would imply that in the human subject, hepatocytes displaying AAV capsid sequences complexed to major histocompatability complex class I molecules on their surface of transduced cells were destroyed by capsid-specific CD8+ cells. However, the AAV capsid is not encoded in the vector and is introduced to hepatocytes temporally during vector infusion, so one potential solution to mitigate the host T-cell response in humans and to allow long-term expression of FIX may be to induce transient immunosuppression until AAV capsid is cleared from the transduced cells. The safety of co-administration of immunosuppressive agents mycophenolate mofetil and tacro-limus, during AAV8-FIX vector infusion to the hepatic artery of rhesus macaques, was tested recently and resulted in successful transduction of the liver and expression of FIX without toxicity [311]. However, these animals did not develop elevated liver enzymes or T-cell response to AAV8 capsid also in the absence of immunosuppression. Therefore, it is not yet known if these immunomodulators can really mitigate the anti-AAV capsid T-cell response observed in the human subjects during the clinical trial.

With the exception of the potential CTL response in a hemophiliac, the preclinical studies for AAV2 vectors, and to a more limited extent for AAV1, have been quite predictive of the benign safety profile observed thus far in clinical trials (Section 6.10). The predictive accuracy of efficacy in various animal models with respect to clinical benefits awaits evaluation in more advanced clinical trials. There appears to be a reasonable predictability for scaling of delivery efficiency, at least as judged for AAV2 vectors, when comparing small animals such as rodent to larger animals including rabbits, dogs, and monkeys. As noted in Section 6.7.1, there are many additional AAV serotypes being evaluated as vectors, but the predictably for scaling from rodents to larger animals has yet to be established. Indeed a recent evaluation of AAV8 and AAV9 vectors in mouse and dog models of hemophilia A showed that neither vector scaled predictably to the larger animals [177,311].

6.10 AAV VECTOR APPLICATIONS: CLINICAL TRIALS

AAV vectors are now being developed for therapeutic applications. The first AAV vector to undergo clinical trials was an AAV2 vector expressing the CFTR cDNA that advanced to phase II clinical trials in CF patients [8–13]. Although testing of this CFTR vector has been discontinued more than 20 additional AAV vectors have been introduced into clinical trials for a wide variety of therapeutic and prophylactic applications (Table 6.1). In this section we describe a series of applications that are being developed with AAV vectors that either are now in clinical trials or are expected to enter clinical testing in the next several years. Most of the examples that we will discuss are chronic diseases that affect various organ systems or vaccine applications, and require different routes of delivery. These examples serve to illustrate the potential wide applicability of AAV vectors particularly for persistent gene expression in nondividing or slowly dividing cells. Most of the clinical trials have used vectors having AAV2 capsids but several more recent trials have begun testing of vectors with capsids of other serotypes such as AAV1 (Table 6.1).

6.10.1 Cystic Fibrosis

CF appeared to be an attractive target for gene therapy particularly in view of the relative accessibility of the lung for installation of delivery vectors. CF is a lethal autosomal recessive disease that is caused by mutations in the gene coding for the CFTR protein. The defect in the CFTR protein, which is a chloride ion channel expressed in epithelial cells, leads to complex biochemical changes in several organs including lung and often the exocrine pancreas. In the lung, there is a decreased mucociliary clearance, increased bacterial colonization, and a chronic neutrophil-dominated inflammatory response that leads to progressive destruction of tissue in the conducting airways. The usual cause of morbidity and eventual mortality in CF patients is a progressive loss of lung function. Thus, the goal of a gene therapy for CF is to deliver the CFTR cDNA into the epithelial cells in the conducting airways of the lung. One challenge in constructing an AAV vector expressing the CFTR protein is that the coding region requires a minimum cDNA size of 4.4 kb thus leaving very little space for a transcription promoter. In order to package this cDNA into a single AAV vector, advantage was taken of the discovery of the transcription promoter properties of the AAV ITR [84]. Thus an AAV vector expressing the CFTR cDNA was introduced into clinical trials in CF patients by delivery of the vector to the lung and the nasal epithelium [8,11–13] in addition into the maxillary sinus [9,10].

The lung is the primary target for CF gene therapy but delivery to the maxillary sinus was undertaken as a novel attempt to obtain an early indication of the potential of the AAV CFTR vector. The maxillary sinuses of CF patients exhibit chronic inflammation and bacterial colonization that is reflective of some aspects of the CF disease in the lung and in some CF patients who have undergone surgical bilateral antrostomy, the maxillary sinus is accessible to instillation of vectors and for sampling and biopsy. In an initial trial, 15 sinuses of CF patients were treated with increasing doses of the AAV-CFTR vector [9]. DNA PCR analysis of sinus biopsies showed dose-dependent delivery of the vector genome that persisted in the sinus for at least 70 days after instillation. In epithelial cell surfaces of CF patients, the transmembrane potential is hyperpolarized compared to normal patients because of the absence of a functional CFTR chloride channel. In the treated sinuses at the higher doses of vector, there was some reversal of the electrophysiologic defect, thus providing suggestive evidence for expression of the CFTR protein from the delivered vector.

A follow-up, double-blinded, randomized and placebo-controlled phase II study was performed in 23 CF patients in whom vector was administered to one maxillary sinus, while the contralateral sinus received a placebo treatment [10]. This study confirmed the safety of the AAV-CFTR vector delivered to the sinus. Additionally, the anti-inflammatory cytokine interleukin-10 showed a significant difference between vector- and placebo-treated sinuses over a 90 day period. This suggested that modulation of cytokine levels might provide a useful surrogate marker for additional trials. Because several patients had participated in the previous phase I trial, the phase II study further suggested that the vector remained safe after multiple administrations to the sinus without induction of serum neutralizing antibodies [10].

The first clinical test of administration to the lung [8] was a single dose administration, dose escalation, open label trial using a bronchoscope to install the vector into one lobe. In the same trial, a second dose of vector was administered to the nasal epithelium in a blinded, placebo-controlled manner. Administration of the vector was well tolerated but measurement of DNA and RNA molecular endpoints from biopsy samples to assess delivery and expression and *in vivo* electrophysiologic assessment of *trans* membrane epithelial potential difference (TEPD) in the nasal epithelium was difficult. Interestingly, when biopsy samples of nasal epithelium taken several days after *in vivo* administration of the vector were cultured *in vitro*, vector DNA and expressed CFTR mRNA were detected and there was some change in the TEPD [230]. This at least showed that vector administered *in vivo* could provide CFTR function when cells were subsequently tested *in vitro*.

A phase I, single administration, dose escalation trial designed to assess safety and delivery of AAV-CFTR by inhaled aerosol to the lung was carried out in 12 adult CF patients exhibiting mild lung disease. The vector was well tolerated and no apparent safety concerns were demonstrated in the study. Vector administration at the highest dose of 10^{13} genome-containing particles resulted in significant gene transfer to the airway. A clear dose–response relationship was observed in vector gene transfer over 30 days but the level of vector DNA declined over 90 days [11]. Following this study, a multidose, double-blinded, placebo-controlled, and randomized phase II trial was conducted in 37 CF patients with mild CF disease. Patients received three doses (10^{13} DRP per dose) at monthly intervals, administered by inhaled aerosol. This trial also was notable for enrolling patients down to age 12, which is important because the eventual target population to treat CF is likely to be younger patients. In this trial, the multidose vector administration was safe and well tolerated and interestingly, there appeared to be an increase in pulmonary function measured by spirometry at 30 days and a decrease of the pro-inflammatory cytokine IL-8 in the sputum [12]. A second phase IIb trial conducted in 100 CF subjects failed to confirm these effects on lung function or IL-8 [13] and highlighted an emerging general problem in many trials in CF patients. In such trials the behavior of placebo arms is unpredictable and coupled with the difficulty of evaluating gene expression in small airways this makes development of any gene therapy for CF a daunting task.

For CF clinical trials, molecular endpoint measures to evaluate gene expression have proven particularly challenging. First there is no good assay for CFTR protein. DNA and RNA PCR can be conducted on airways cells that are obtained via a brush inserted through a bronchoscope. This has several disadvantages including poor quality of the samples that often contain very few cells. Consequently, it has been possible to obtain data from DNA PCR but it has proven extremely difficult to obtain reliable RNA samples. In the clinical trials of the AAV CFTR vector in lungs of CF patients, gene transfer has been demonstrated but data on RNA expression has not been obtained. Other problematic issues are that the invasive bronchoscopy procedure disrupts other measures of pulmonary function and inflammation. Furthermore, the bronchoscopic procedure can only provide samples from the upper airways whereas the vector needs to function in the lower airways which are the primary site of the disease. Consequently, the difficulties with molecular endpoints and the variable and unpredictable behavior of placebo arms in clinical trials currently makes clinical development of a gene therapy for CF a challenging task, with any vector, not just AAV [312,313]. Even for AAV vectors, it is therefore difficult to assess the likelihood of success for other approaches that have been suggested such as use of alternative AAV capsid serotypes [314], or a truncated CFTR cDNA thus allowing space for a larger, perhaps more efficient promoter [315].

6.10.2 Hemophilia B

Hemophilia is a severe X-linked recessive disease that results from mutations in the gene for either blood coagulation factor VIII (FVIII) in hemophilia A or IX (FIX) in hemophilia B. The absence of functional FVIII or FIX leads to severe bleeding diathesis. Expression of these clotting factors at less than 1% of normal levels results in severe disease whereas levels over 5% of the normal range appear to be sufficient for normal function. Levels between 1% and 5% lead to much milder disease and prophylactic delivery of these clotting factors at these levels of 1% of the normal levels (2–10 and 50–250 ng/mL for FVIII and FIX, respectively) can decrease the risk of spontaneous bleeding into joints and soft tissues and lower the risk of fatal intracranial bleeding. However, the FVIII and FIX proteins have a short half-life and this has stimulated interest in developing gene therapy approaches in which the clotting factors may be produced more persistently. Several groups have provided evidence that FIX protein can be expressed for prolonged periods, in both murine and canine models, after delivery of AAV–FIX vectors either by portal vein injection to target the liver or by intramuscular injection [99,100,316,317]. Generally, the levels of expression, as measured by the whole blood clotting time (WBCT) assay, have indicated that at least a partial correction of the defect in both the murine and canine disease models can be achieved and suggest that accumulation of therapeutic amounts of FIX may be achievable in humans.

AAV vector delivery of the human factor IX gene into either immunodeficient or immunocompetent mice by portal-vein injection into liver resulted in prolonged expression of factor IX for up to 36 weeks at serum levels of 250 to 2000 ng/mL which is equivalent to about one fifth of the normal human level [99]. Similar portal-vein delivery of human factor IX in hemophiliac dogs resulted in expression of factor IX at about 1% of normal canine levels, an absence of inhibitors and sustained partial correction of WBCT for at least 8 months [316]. In another study by Herzog et al. [99], intramuscular injection of an AAV human FIX vector into immunodeficient mice led to prolonged expression at about 350 ng/mL in serum for at least 6 months but in immunocompetent mice there was generation of inhibitory antibody. These same investigators [317] subsequently showed that in hemophiliac dogs, intramuscular injection of high doses of the AAV-canine FIX vector achieved expression for over 17 months and demonstrated a stable, dose-dependent partial correction of the WBCT. Moreover, at the highest dose there was a partial correction of the activated partial thromboplastin time (APTT) which may be a more reliable measure than WBCT.

Development of AAV vectors to deliver FVIII for hemophilia A faces an extra challenge in that the FVIII protein cDNA is over 7 kb which is too large to package into a single AAV vector. However, the FVIII protein is processed by excision of an internal B-domain such that the secreted protein consists of heavy and light chains. Removal of the B-domain DNA sequence reduces the size of the cDNA to about 4.2 kb which can be packaged into a single AAV vector with space to include only small regulatory elements. Because of this size restraint AAV vectors have not entered clinical development for hemophilia A.

For hemophilia clinical trials, in contrast to the CF trials, a molecular endpoint is readily available because the level of the FIX or FVIII in serum can be directly measured and the level of protein will likely be related to degree of clinical benefit. In contrast, DNA or RNA measures again would require invasive biopsy especially following portal vein delivery to the liver.

A clinical study of intramuscular injection of an AAV vector expressing human FIX in adults with severe hemophilia B was recently concluded [318,319]. This was a phase I open label dose escalation trial in a total of eight adult men with severe hemophilia B due to a missense mutation in FIX and with serum levels of FIX below 1% of normal. Three dose levels of vector, 2×10^{11} vg/kg ($n = 3$), 6×10^{11} vg/kg ($n = 3$), and 1.8×10^{12} vg/kg ($n = 2$), were administered, under ultrasound guidance to avoid large blood vessels, by injection at multiple sites in muscles of one or two legs. The treatment was generally well tolerated and there were essentially no vector-related toxicities in any of the three dose cohorts. Analysis of biodistribution in body fluids showed very little extramuscular distribution of vector and importantly no vector was detected in any semen sample over 24 weeks of the study. Analysis of muscle biopsies detected both the vector DNA and vector-specific mRNA as well as accumulation of FIX at the injection sites. Although all treated patients showed elevation of anti-capsid antibody levels in serum, the levels of gene transfer and expression in muscle did not correlate with either pre- or posttreatment levels of AAV2 capsid antibody. Importantly, there was no formation of inhibitory antibodies to FIX. However, the levels of circulating FIX in serum were disappointing and there was no dose response with only four of eight patients showing any detectable level from 1% to 1.4% at any time point over the 24 week period. There were no significant changes in clinical endpoints. Remarkably a muscle biopsy obtained from one subject 3.7 years following treatment with the highest vector dose, 1.8×10^{12} vg/kg, revealed a low level of local FIX vector DNA and FIX protein expression. This suggests that improved muscle delivery and transduction particularly of myocytes may be needed to achieve an effective

treatment [125]. Indeed, a recent study demonstrated that intravascular delivery of vector to skeletal muscle provides higher levels of circulating FIX and long-term cures in hemophilic dogs at vector doses comparable to those previously used in humans, due to widespread transduction of skeletal muscle [320]. Other strategies, such as the use of pseudotyped AAV2/1 vector and a myocyte-specific synthetic promoter to control FIX expression, have resulted in improved efficacy of AAV-mediated gene transfer to skeletal muscle in murine and canine hemophilia B models [321–323]. However, the increase in the levels of FIX expression was also accompanied by the formation of inhibitory antibodies and further studies are needed to define vector doses and treatment regimens that result in tolerance rather than immunity to FIX.

A second phase I clinical trial with the AAV2–FIX vector was initiated in patients with severe hemophilia B patients, but in this trial the vector was administered via hepatic artery infusion to target the liver which may be the natural source of FIX production. This trial was based on preclinical studies in hemophiliac dogs in which therapeutic levels of FIX were sustained over a prolonged period with a dose of vector at 10^{12} vg/kg [316–322,324]. In this open-label trial seven subjects with severe hemophilia B were treated at three vector dose levels of 1×10^{11} vg/kg, 10^{12} vg/kg, and 2×10^{12} vg/kg and there were several significant observations. Following hepatic infusion of the AAV2–FIX vector at the lower doses, patient semen samples were positive for the vector genome for several weeks [275]. However, as noted in Section 6.9, this did not represent vector in motile sperm, suggesting that the risk of inadvertent germline transmission of vector sequences after intramuscular or intravenous delivery is extremely low [276]. Two additional patients then were treated at the intermediate dose. Administration of the two lower doses of vector was safe and well tolerated and no toxicities were noted but there was also no significant therapeutic level of FIX protein produced.

Administration of the highest dose of 2×10^{12} vg/kg to two subjects resulted in several observations with regard to possible immune responses that differed from the preclinical studies [244]. Both subjects showed detectable levels of FIX expression, but expression was transient and peaked at 3% in one and 11% in the other at 4 and 10 weeks postadministration, respectively. In the latter subject, the loss of expression was accompanied by a transient rise in serum transaminase levels and in a third subject treated subsequently with slightly lower vector dose, transaminitis was again documented. Importantly, the transient therapeutic levels of FIX were achieved in subjects having low preexisting neutralizing antibody (NAB) titers to AAV2, but not in those with higher AAV2 NAB [306]. Thus a high level of preexisting NAB to AAV2 in some proportion of the human population might interfere with efficient in vivo gene delivery in a clinical setting [306]. The sudden loss of a level of expression of 11% accompanied with a contemporaneous transaminitis was suggestive of a CTL response and an AAV capsid-specific CD8+ T-cell clone was expanded from this subject and showed cytolytic activity against cells that displayed capsid antigen [306].

However, the mechanism of this is not understood [117] and it has not yet been possible to model this phenomenon in animals as discussed in Section 6.9.

6.10.3 MUSCULAR DYSTROPHY

The muscular dystrophies are a clinically and genetically heterogeneous group of disorders that show myofiber degeneration and regeneration and are characterized by progressive muscle wasting and weakness of variable distribution and severity. They are associated with mutations in genes encoding several classes of proteins ranging from extracellular matrix and integral membrane proteins to cytoskeletal proteins, but also include a heterogeneous group of proteins including proteases, nuclear proteins, and signaling molecules [325].

The most common myopathy in children, Duchenne muscular dystrophy (DMD), is a severe X-linked neuromuscular disease that affects approximately 1 of every 3500 males born, and is caused by recessive mutations in the gene for the muscle protein dystrophin. This lack of a functional dystrophin protein in DMD results in loss of muscle fiber integrity by disrupting the physical linkage between the actin cytoskeleton within the muscle fiber and the extracellular matrix. Affected boys begin manifesting signs of disease early in life, cease walking at the beginning of the second decade, and often die due to cardiac arrest or respiratory insufficiency by age 20 years [326].

An effective gene therapy for muscular dystrophy would address an unmet medical need but several major challenges must be overcome. The first challenge is that the dystrophin gene is the largest known gene with a full-length cDNA that is 14 kb. However, dystrophin can retain significant function even when missing large portions of its sequence. Large, inframe deletions in the central-rod domain often lead to the milder Becker muscular dystrophy (BMD) [327]. The *mdx* mouse is a naturally occurring murine model that has a premature stop codon generated by a point mutation in exon 23 of the dystrophin gene [328]. Functional analysis of dystrophin structural domains in transgenic dystrophin-deficient *mdx* mice revealed multiple regions of the protein that can be deleted in various combinations to generate potentially highly functional mini-dystrophin genes [329]. Several groups have generated functional miniature versions of the human dystrophin gene that can be readily packaged into AAV vectors [330–335]. When injected into the muscle of *mdx* mice, efficient and stable expression was noted in a majority of myofibers and the missing dystrophin and dystrophin-associated protein complexes were restored onto the plasma membrane. This treatment ameliorated dystrophic pathology in the *mdx* muscle and led to normal myofiber morphology, histology, and cell membrane integrity.

Another particularly challenging problem is that widespread expression in the skeletal and cardiac muscle is a prerequisite for successful gene therapy of muscular dystrophies. But systemic gene transfer to striated muscle is hampered by the vascular endothelium that represents a barrier to distribution of vector via the circulation. One approach uses intravenous administration of a pseudotyped AAV2/6 vector together with vascular endothelial growth factor (VEGF) to permeabilize blood vessels. In adult mice this resulted in efficient whole body muscle-specific transduction of skeletal and cardiac muscle [209]. In *mdx* mice with the microdystrophin gene expressed from a muscle-specific promoter, widespread dystrophin expression was observed throughout the skeletal muscles and resulted in partial phenotypic correction of the limb muscles that were less susceptible to contraction-induced injury [210]. A follow-up study in dystrophin–utrophin double-knockout mice resulted in restoration of dystrophin protein expression in the respiratory, cardiac, and limb musculature of these mice, considerably reducing skeletal muscle pathology and extending their lifespan [211]. Similarly, extensive cardiac transduction, with microdystrophin correctly localized at the periphery of the cardiac myocytes and functionally associated with the sarcolemmal membrane was obtained following a single systemic injection of AAV2/6 carrying the microdystrophin transgene in the presence of VEGF [211]. Significantly, in this study, microdystrophin gene transfer corrected the baseline end-diastolic volume defect in the *mdx* mouse heart and prevented cardiac pump failure.

Despite the significant progress made in transduction of the mouse musculature using alternative AAV serotypes and VGEF, key safety issues to be resolved include the use of VEGF and the resulting extended biodistribution of the vector genomes in many other tissues beside the skeletal muscle, including the brain and testes. Interestingly, vectors based on the AAV8 capsid, at least in neonatal mice, can cross the blood vessels barrier and disseminate efficiently in skeletal and cardiac muscle in the absence of VEGF, and lead to prolonged transduction in these tissues [212].

Other strategies have evaluated ways to express larger dystrophins. Transcript rescue is based on the observation that exon skipping naturally occurs at low frequency during dystrophin mRNA processing and can restore the disrupted reading frame and lead to the production of rescued protein [336,337]. Persistent exon skipping that removed the mutated exon on the dystrophin mRNA of the *mdx* mouse was achieved following intramuscular injection or intra-arterial perfusion of the lower limb with a pseudotyped AAV2/1 vector expressing antisense sequences linked to a modified U7 small nuclear RNA [338]. This led to sustained production of functional dystrophin at physiological levels in entire groups of muscle and the correction of the muscular dystrophy in the absence of an immune response against the rescued dystrophin [338]. With this strategy, body-wide rescue of dystrophin synthesis and recovery of muscle strength was obtained following systemic (tail vein) delivery of a pseudotyped AAV2/1 vector-mediated antisense-U1 small nuclear RNA targeting exon 51 of dystrophin gene in *mdx* mice [339].

Larger dystrophins may also be expressed using the *trans*-splicing dual AAV vector approach (discussed in Section 6.8.3) [244,249–251]. Thus, a 6 kb mini-dystrophin gene was efficiently delivered and functionally expressed in the skeletal muscle of *mdx* mice by splitting the dystrophin coding sequence between two *trans*-splicing pseudotyped AAV2/6 vectors [340]. Furthermore, the dystrophic pathology was ameliorated and muscle strength improved in both adult, and to a lesser degree in aged mice, but in the absence of cellular immune response.

In contrast to the relative success of AAV-mediated microdystrophin gene transfer and restoration of essential muscle functions in inbred murine models of muscular dystrophy, intramuscular administration of AAV2 or AAV2/6 vectors to random-bred wild-type dogs elicited an apparent robust cellular immune responses against the AAV capsid proteins that were sufficient to largely eliminate transgene expression [341]. However, a brief course of immunosuppression shortly prior and following AAV administration to the muscle was sufficient to permit long term and robust expression of a canine microdystrophin gene in the skeletal muscle of a dog model for DMD and its expression restored localization of components of the dystrophin-associated protein complex at the muscle membrane [342].

A phase I clinical trial for the delivery of mini-dystrophin by AAV to boys with DMD is currently under way at Columbus (Ohio) Children's Hospital. In this trial, six boys with DMD will undergo injections of the AAV-mini-dystrophin gene into the biceps muscle of one arm and a placebo in the other arm in a blinded manner. The AAV serotype used in this trial is AAV2.5 which differs from AAV2 capsid at five amino acids, resulting in increased transduction efficiency. After 6 weeks, an analysis of the injected muscle tissue's microscopic appearance, as well as extensive assessment of health and strength of the trial participants, may reveal whether the treatment is likely to be safe and result in persistent production of the essential microdystrophin protein in muscle cells.

The limb girdle muscular dystrophies (LGMD) are a heterogeneous group of inherited autosomal recessive neuromuscular diseases characterized by proximal muscular weakness and variable progression of symptoms. LGMD disease is caused by mutations in a number of genes including one of the four small (cDNA < 2 kb) muscle sarcoglycan genes (α, β, χ, δ) expressed predominantly in striated muscle. These transmembrane glycoproteins associate with each other in equal stoichiometry to form the sarcoglycan complex, and a deficiency of one component typically leads to partial or complete absence of all the other sarcoglycan proteins on the sarcolemma. The Bio 14.6 cardiomyopathic hamster is a naturally occurring LGMD model due to a deletion in the δ-sarcoglycan gene. Administration of AAV δ-sarcoglycan vectors to these animals either by intramuscular or intravascular administration led to genetic, biochemical, histological, and functional rescue of relatively large regions of muscle [343–345]. Mice that are null mutants for χ-sarcoglycan exhibit severe muscle pathologies that can be partly corrected if treated at less than 3 weeks of age by intramuscular injections of an AAV vector expressing the χ-sarcoglycan gene from a muscle-specific promoter [345]. Also, AAV α- or β-sarcoglycan vectors can rescue an α- or β-sarcoglycan defect in the corresponding knockout mouse model. Interestingly, while the β-sarcoglycan vector showed long-term sustained expression for more than 21 months and led to widespread biochemical and histological

rescue of the dystrophic muscle, transduction of myofibers by the α-sarcoglycan vector was transient and correlated with induction of significant immune response [346]. The transience of the latter vector was attributed to cytotoxicity resulting from over expression, by more than 100-fold over normal levels of the protein rather than due to an immune response to the transgene. In a more recent study, substitution of the very strong ubiquitous promoter with a muscle-specific promoter resulted in efficient and sustained expression of α-sarcoglycan with correct sarcolemmal localization and without evidence for toxicity following intra-arterial injection of AAV2/1-$\alpha\beta$-sarcoglycan into the limbs of an LGMD type 2D mice [347]. Transgene expression resulted in restoration of the sarcoglycan complex, histological improvement, membrane stabilization, and correction of pseudohypertrophy. More importantly, α-sarcoglycan expression produced full rescue of the contractile force deficits and stretch sensibility and led to an increase of the global activity of the animals when both posterior limbs were injected. A phase I clinical trial of an AAV2 to deliver sarcoglycan genes to subjects with LGMD was initiated but enrolled and treated only one patient [348]. However, this is now being revisited using a pseudotyped AAV2/1 vector [346].

6.10.4 Central Nervous System Disease

The brain is an attractive organ for gene therapy, because production of biologically active molecules within the brain might circumvent poor penetration of compounds that are delivered systemically due to a tight vascular blood–brain barrier. Treatment of most of these diseases is not easily approachable by a protein therapy because of the difficulty associated with repeated delivery of short-lived proteins to the brain. Local gene expression might also focus therapy in specific brain regions, thereby avoiding exposure of other areas to agents that might cause undesirable effects.

6.10.4.1 Parkinson's Disease

Parkinson's disease (PD) is a common progressive neurodegenerative movement disorder that affects 5% of the population over 65 years of age. In this disease there is loss of dopaminergic neurons, mainly in the substantia nigra, which leads to deficiency of the neurotransmitter dopamine (DA) in the striatum. The clinical symptoms typically appear after extensive loss of 60%–80% of the dopaminergic neurons, which is correlated with the DA deficiency. DA is synthesized from tyrosine, first through conversion to L-dihydroxyphenylalanine (L-DOPA) by tyrosine hydroxylase (TH), and then to DA by aromatic amino acid decarboxylase (AADC). L-DOPA therapy is very efficacious but response declines as the disease progresses and is complicated by adverse side effects. Thus, the different enzymes involved in DA synthesis have been targeted for gene transfer replacement therapy in order to restore dopaminergic stimulation of the striatum.

In the 6-hydroxydopamine (OHDA)-lesioned rat model of PD, a single intrastriatal infusion of an AAV2-AADC vector led to an enhanced conversion of L-DOPA to restore DA to 50% of normal levels 12 weeks after vector administration [349] and persistence of transgene expression for at least 1 year [350]. Co-expression of TH and AADC, using two separate AAV vectors in this Parkinsonian rat model resulted in more effective DA production and more remarkable behavioral recovery, compared with the expression of TH alone [351]. Convection-enhanced delivery of the AAV2-AADC vector into six sites in the striatum of neurotoxin 1-methyl-4-phenyl-1,2,3,6-tetrahydropyridine (MPTP)-treated monkeys restored AADC activity to levels that exceeded the normal range [352] and expression was observed for 6 years [353]. A clinical trial of an AAV2-AADC vector has been initiated in PD patients [354].

Clinical trials with AAV vectors to test a novel nondopaminergic approach to treatment of PD are under way [354,355] based on a hypothesis that re-establishment of normal brain activity within motor circuits might reverse motor deficits of PD. The strategy is to administer a mixture of two different AAV2 vectors each expressing respectively the gene for one of the two isoforms of the enzyme glutamic acid decarboxylase (GAD-65 and GAD-67) by unilateral injection into the subthalamic nucleus (STN) region of the brain. GAD catalyzes synthesis of GABA, the major inhibitory neurotransmitter in the brain. The STN has a central role in the region of the brain that is responsible for regulation of movement and becomes disinhibited in PD. The concept is that expression of GAD will cause the STN to produce γ-aminobutyric acid (GABA) and convert excitatory STN projections into inhibitory projections. This inhibitory of STN activity in turn may palliate the motor symptoms of PD. Previous studies in humans have demonstrated that reduction of subthalamic nucleus activity by electrical stimulation, lesioning, or GABA infusion could ameliorate signs of advanced PD [355], whereas animal studies indicate that AAV-GAD appears to improve brain function and signs of the disease without toxic effects [356–359]. The phase I clinical study was aimed to assess the safety and tolerability of AAV-GAD vectors in 12 PD patients over a period of 1 year, using a single-arm, open label, dose-escalation design. Patients were also assessed clinically both off and on medication at baseline and at different intervals following gene therapy. After 1 year of follow-up of all patients, there were no adverse events related to gene therapy. Importantly, significant improvements in motor scores, based on the Unified Parkinson's Disease Rating Scale (UPDRS), were noted predominantly on the side of the body that was contralateral to surgery as early as 3 months following surgery and persisted up to 12 months. Additionally, PET scans demonstrated a substantial reduction in thalamic metabolism that was restricted to the treated hemisphere, and a correlation between clinical motor scores and brain metabolism in the supplementary motor area [360]. However, caution must be exerted in interpretation of these results because there was no sham-operated control group and the placebo effect in PD clinical trials can be significant [361], especially after surgical intervention [362]. Consequently, no firm conclusions can be drawn about efficacy, despite objective improvement in motor function and functional imaging.

Another strategy for gene therapy of PD involves the use of neuronal-specific growth factors to regenerate or halt ongoing degeneration of dopaminergic neurons of the substantia nigra. Glial cell line-derived neurotrophic factor (GDNF) is a potent neurotrophic molecule for nigral dopaminergic neurons both *in vitro* [363] and *in vivo* [364]. In the 6-OHDA-lesioned rat model of PD, peri-nigral injection of an AAV vector encoding the rat GDNF, three weeks prior to a striatal 6-OHDA treatment, resulted in stable transgene expression for 10 weeks and significant protection of neurons from degeneration in the substantia nigra. However, there was no recovery of striatal TH-containing fibers and no recovery of motor function, suggesting that striatal dopaminergic recovery is necessary for functional improvement [365]. In another study, both nigral and striatal long-term transduction of up to 6 months provided significant protection of nigral DA neurons against 6-OHDA-induced degeneration. However, only the rats receiving AAV-GDNF in the striatum displayed behavioral recovery, accompanied by significant re-innervation of the lesioned striatum [366]. In both studies [365,366], AAV-GDNF was administered before or shortly after the injection of the neurotoxin 6-OHDA.

Studies of AAV-GDNF vectors have not advanced to clinical trials but clinical trials with an analogous neurotropic growth factor, neurturin (NTN), have begun [354]. NTN is a member of the GDNF family, and is known to repair damaged and dying midbrain dopaminergic neurons. An AAV2-NTN vector exhibited both long-term efficacy and safety in preclinical rat and monkey models of PD [367]. In a phase I open-label study, conducted in 12 patients with advanced PD, an AAV2-NTN vector encoding the neurotrophic factor NTN was stereotactically delivered to the putamen, a region in the midbrain affected by the degeneration of dopaminergic neurons, at two different doses [354]. Patients receiving the low dose demonstrated approximately 36% improvement in UPDRS motor "off" scores by 9 months and patients receiving a fourfold higher vector dose showed similar effects 3 months earlier. This effect was maintained for at least 12 months, the final follow-up time point in the study. In addition, patients also demonstrated a 50% reduction in off time and a doubling of good quality "on" time without dyskinesias. Similarly to the AAV-GAD vector approach, these results with AAV-NTN are encouraging but require careful placebo-controlled studies to be confirmed. Consequently, a follow-up on phase II double-blind, controlled clinical trial is currently enrolling 51 patients with advanced PD. Patients will be administered AAV2-NTN to the putamen via stereotactic neurosurgery and will be followed-up for 12 months for safety and efficacy.

6.10.4.2 Canavan's Disease

A clinical trial to treat Canavan's disease, a childhood leukodystrophy, is being conducted [250]. This disease results from an autosomal recessive mutation in the gene for aspartoacyclase (ASPA) that causes a toxic accumulation of the metabolite *N*-acetyl-aspartate (NAA). In this phase I safety trial up to 10 patients aged between 3 months and 6 years received intracranial infusions via six cranial burr holes of AAV2-ASPA vector to the parenchyma of the brain at a maximum dose of 1×10^{12} vector genomes per subject. The treatment was well tolerated and only low level of immune response to AAV2 was detected in a few subjects [368].

6.10.4.3 Batten's Disease

One form of the autosomal recessive disorders known as Batten disease is late infantile neuronal ceroid lipofuscinosis (LINCL) which is caused by a mutation in the gene *CLN2* that encodes a proteolytic enzyme tripeptidyl peptidase. Absence of the enzyme leads to a lysosomal storage disorder with accumulation of lipofuscin and consequent neurodegeneration. Disease onset occurs at about age 2–4 years with progressive neurologic decline, cognitive, impairment, blindness, seizures, ataxia, and death at age 8–10. Other than symptomatic relief there is no treatment for this disease and again a protein therapy is unlikely. An AAV2 vector expressing the human CLN2 cDNA from the CB promoter is now being tested in a clinical trial in which the vector will be administered to the whole brain via fine catheters placed at six different subcortical sites in order to obtain a wide distribution [369]. The total dose per patient is 3.6×10^{12} particles and up to ten patients, four with severe disease and then six with more moderate disease, will be treated. The primary endpoint measure is neurological assessment using the LINCL clinical rating scale and secondary endpoints involve brain imaging.

6.10.4.4 Alzheimer's Disease

Alzheimer's disease leads to progressive decline in memory and broad cognitive functions and involves multiple brain systems but the most critical for decline in memory and cognition appears to be loss of cholinergic neurons. There are several million patients in the United States and there is no treatment to prevent disease progression. There is some evidence that β-nerve growth factor (NGF) may induce a trophic response in cholinergic neurons. Consequently, a phase I/II dose escalation, randomized controlled study was recently initiated [370] to assess primarily the safety and tolerability of an AAV2-NGF vector delivered to the basal forebrain region by stereotactic injection. Additional measures will include cognitive scores and imaging analyses. The phase I portion of the trial which is ongoing will examine the safety of the vector at two different doses before proceeding to the phase II trial.

6.10.5 OPHTHALMIC DISEASE

AAV vectors may be well suited for efficient long-term treatment of ocular diseases because they efficiently and stably transduce retinal pigment epithelium (RPE) and photoreceptor cells following subretinal injection [119,371–374]. Retinitis pigmentosa (RP) comprises a group of inherited retinal degenerative diseases that lead to progressive reduction in visual field extent and impairment of visual acuity. The disease is triggered by mutations in various genes that cause degeneration and death of photoreceptors by apoptotic pathways [375].

Modulation of photoreceptor apoptosis may offer an effective therapeutic approach to RP. AAV vectors carrying various genes including those encoding for ciliary neurotrophic factor (CNTF), fibroblast growth factors FGF-2, FGF-5, and FGF-18, and glial cell line-derived neurotrophic factor (GDNF) were evaluated in rodent models of RP subretinal injections. In general, these studies demonstrated long-term expression of the transgene, delayed photoreceptor degeneration, increased rod photoreceptor survival, and functional improvement [376–380].

Other therapeutic approaches to RP are aimed at specific mutant genes. Twelve percent of Americans with the blinding disease, autosomal retinitis pigmentosa (ADRP), carry an autosomal dominant P23H mutation in their rhodopsin gene and a similar transgenic rat model of this disease is available. Delivery with an AAV vector of a ribozyme targeted at this mutation in the rodent model, protected photoreceptors from death and resulted in significantly slowing the degenerative disease for 3 months [381]. The gene *Prph2* encodes the photoreceptor-specific membrane glycoprotein peripherin-2 and mutations in this gene cause several photoreceptor dystrophies, including autosomal dominant retinitis pigmentosa and macular dystrophy. A common feature of these diseases is the loss of photoreceptor function, also seen in the retinal degeneration slow ($Prph2^{Rd2/Rd2}$) mouse. Subretinal injection of AAV-*Prph2* in these mice resulted in stable restoration of photoreceptor ultrastructure and electrophysiological function correction [382].

Control of angiogenesis in the retina is essential to the preservation of vision. Ocular neovascularization (NV) is a major threat to vision and a complicating feature of many eye diseases including proliferative diabetic retinopathy (PDR), age-related macular degeneration (AMD), and retinopathy of prematurity (ROP). Regulation of vascularization in the mature retina involves a balance between endogenous positive growth factors, such as VEGF, and inhibition of angiogenesis, such as pigment epithelium-derived factor (PEDF). Several studies examined ocular administration of AAV vectors to mice exhibiting ischemia-induced retinal NV. Expression of anti-angiogenic proteins, either PEDF or the Kringle domains 1–3 of angiostatin (K1K3), gave sustained therapeutic levels of PEDF and K1K3 in the mouse eye and significantly reduced the level of retinal NV [383,384]. Expression of a soluble VEGF receptor also led to significant reduction in the number of neovascular endothelial cells and inhibited retinal NV [385,386].

Leber congenital amaurosis (LCA) is a severe, early-onset form of retinal degeneration involving both rods and cones. LCA is caused by mutations in one of at least eight different genes, three of which are RPE-specific, the remainder being expressed in photoreceptors [387]. The largest amount of preclinical data assembled to date has been on the form of LCA due to mutations in the RPE65 gene. The RPE65 gene encodes a 65 kDa RPE-specific protein that is required for the conversion of vitamin A to 11-cis retinal by the RPE and thus for the regeneration of the rod visual pigment. Mutations in the RPE65 gene are responsible for approximately 10% of Leber autosomal recessive severe, infantile-onset retinal dystrophies [388].

Most patients with such RPE65 mutations present in infancy with bilateral visual impairment and severe night blindness [389,390]. Peripheral visual fields may be preserved in early childhood but later become severely constricted. Visual acuity varies from 20/60 to 20/200 in early childhood but there is gradual deterioration in vision so that by late teens or early adult life there is little residual vision.

RPE65–/– knockout mice and Briard dogs do not generate adequate levels of visual pigment, have very poor vision and severely depressed light- and dark-adapted ERG responses, and thus are good models of human LCA with RPE65 mutations [391]. Several groups have demonstrated proof-of-principle for gene therapy in the *RPE65–/–* knockout mice and Briard dogs following subretinal injection of AAV2-RPE65 vector which resulted in significant improvement of visual function using electrophysiology and behavioral assessments [386,392]. Stable improvement in visual function has been maintained in many animals for longer than 3 years [282]. The demonstration of long-term functional improvement following gene-replacement of RPE65 in preclinical models has supported proposals for clinical trials of recombinant AAV-mediated gene therapy for patients with LCA. Indeed, a phase I/II clinical trial has recently began in the United Kingdom in young adults and children who have LCA (R. Ali, personal communication). In this clinical study, the RPE65 gene under the control of its own promoter is being delivered via subretinal injection of an AAV2 vector. The purpose of this clinical study is to evaluate the safety and efficacy of this approach in patients. It is anticipated that the youngest patients would be more likely to benefit the most from this treatment. Two additional clinical trials of AAV2 vectors expressing RPE65 for LCA have also obtained regulatory approval in the United States [283,284,393].

6.10.6 Rheumatoid Arthritis

Rheumatoid arthritis (RA), the most common inflammatory joint disease, is a chronic autoimmune disorder that affects approximately 1% of the population and causes significant disability [394]. The etiology of RA is largely unknown, although current evidence suggests contributions from both environmental and genetic components [395]. The chronic inflammation in the arthritic joint is characterized by recruitment of immune cells, including lymphocytes, macrophages, and plasma cells, leading to massive thickening of the synovium accompanied by release of inflammatory mediators, ultimately leading to invasion and destruction of articular cartilage and bone. At the molecular level, chronic inflammatory arthritis is characterized by diminution of T-cell factors and an abundance of cytokines and growth factors such as interleukin-6 (IL-6), tumor necrosis factor α (TNF-α), and interleukin-1β (IL-1β) that are produced by macrophages and synovial fibroblasts and play a major role in the progression of joint destruction [396]. IL-1β, in particular, is a key cytokine that induces cartilage degradation, whereas TNF-α is a major cytokine involved in the joint inflammation [397].

Conventional treatment to manage the symptoms of arthritis uses general anti-inflammatory agents, including both steroidal and nonsteroidal drugs, and disease-modifying drugs such as methotrexate. However, none of these pharmacologic agents have yet proven effective in halting the progression of disease. Recent introduced biologic agents are more effective in ameliorating arthritis symptoms and slowing the disease progression. In particular, inhibitors of TNF-α and IL-1β have proven effective in preclinical studies and in human clinical trials. Four biologic therapeutics based upon use of soluble TNF-α receptor immunoglobulin Fc (TNFR:Fc, etanercept), anti-TNF-α monoclonal antibodies (Humira, Remicade), and IL-1 receptor antagonist (IL-1Ra, Kineret) are now approved for the treatment of RA.

The use of biologics for arthritis therapy nevertheless presents several challenges. The relatively short half-life of these proteins dictates frequent, daily or weekly dosing and effective levels of the therapeutic protein may not be maintained for extended periods. In addition, therapeutic proteins are administered systemically and may have reduced bioavailability in some affected joints. Gene transfer may be an efficient means of delivery of these biological agents. Persistent transgene expression could circumvent the need for frequent repeat dosing and facilitate attainment of steady levels of the product. Furthermore, localized gene transfer might increase bioavailability in nonresponding joints.

AAV vectors encoding genes that selectively target key mediators, or that interfere with key biological processes, involved in the pathogenesis of RA have been evaluated in rodent models of inflammatory arthritis. In mice with collagen-induced arthritis (CIA), an AAV vector encoding interleukin-4 (IL-4) demonstrated protection, for up to 7 months, from articular cartilage destruction and amelioration of disease severity [398,399]. Also, an AAV vector expressing the viral interleukin-10 (vIL-10) under the transcriptional control of the regulated TetON system was evaluated following intramuscular administration to mice with CIA. Expression of vIL-10, specifically induced by doxycycline, persisted for at least 4 months and reduced significantly the incidence and severity of arthritis [399]. In TNF-α transgenic mice, intra-articular injection of AAV encoding soluble TNF receptor type I (TNFR I) significantly decreased synovial hyperplasia, and cartilage and bone destruction [400]. An AAV vector encoding IL-1Ra cDNA was evaluated in a lipopolysaccharide (LPS)-induced arthritis model in rats using therapeutic, recurrence, and preventative protocols [401]. IL-1Ra expression was upregulated by LPS-induced joint inflammation and proved efficacious in all protocols. Importantly, the IL-1Ra transgene persisted for more than 3 months and could be induced to express therapeutic levels of soluble IL-1Ra upon LPS administration. This resulted in suppression of inflammation and IL-1β production in the treated knee joints.

In the streptococcal cell wall (SCW)-induced rat arthritis model, administration of an AAV vector encoding the rat TNFR:Fc fusion gene, either systemically (intramuscular), or locally (intra-articular), resulted in profound suppression of arthritis [402]. This was reflected in decreased inflammatory cell infiltration, pannus formation, cartilage and bone destruction, and mRNA expression of joint proinflammatory cytokines. Moreover, administration of the vector to one joint suppressed arthritis in the contralateral joint. These studies formed the basis of a phase I dose escalation trial, to assess the safety of intra-articular delivery of AAV2 containing the human TNFR: Fc fusion gene (tgAAC94) in subjects with inflammatory arthritis who were not on concomitant systemic TNF-α antagonist therapy [403]. Eleven of the fifteen subjects enrolled in the trial were randomized to receive either one of two escalating dose levels of tgAAC94, while the other four received a placebo. The trial contained a placebo arm primarily to assess whether any adverse events were attributable to an intra-articular injection itself, as opposed to an intra-articular injection of tgAAC94. The conclusions from the analysis of trial results were that tgAAC94 was well-tolerated at doses up to 1×10^{11} DRP/mL of joint volume and that no drug-related serious adverse events were observed. The study was not powered to show efficacy but subjects administered a single dose of tgAAC94, showed some continued improvement in signs and symptoms in treated subjects ($n = 11$). Some improvement was also noted in mean tenderness and swelling scores in subjects receiving placebo.

Currently, the tgAAC94 vector is being tested in a double-blinded, placebo-controlled phase I/II clinical trial in subjects with inflammatory arthritis. In this study of 120 subjects, three dose levels of tgAAC94 are being evaluated in subjects with RA, psoriatic arthritis, or ankylosing spondylitis, who may or may not be receiving concomitant treatments of anti-TNF-α therapy, but continue to experience inflammation in one or more joints (404). In the first segment of this study, subjects are receiving a single intra-articular injection of tgAAC94 or placebo in the affected joint and are monitored until swelling in the target joint reaches predetermined criteria for reinjection. At that time, those who were injected with tgAAC94 as well as those subjects initially injected with placebo will receive an injection of tgAAC94 in the affected joint as part of the open-label segment of the study. The primary endpoint of the study is to establish the safety of higher doses and of repeat administration of tgAAC94 into the joints of subjects with and without concomitant TNF-α inhibitor therapy. Secondary endpoints include evaluation of pain and swelling, duration of response, overall disease activity following intra-articular administration of tgAAC94 into affected joints, patient and functional assessment of target joint, and joint inflammation and damage as assessed by magnetic resonance imaging.

6.10.7　Other Diseases

AAV vectors have also entered into clinical trials for a number of other indications (Table 6.1). Hereditary emphysema is caused by a mutation in the gene for AAT which is normally synthesized in the liver and transported via serum to the lung where it inhibits the activity of the neutrophil elastase. The absence of AAT may lead to progressive destruction of lung tissue and premature mortality. The disease

can be treated by weekly intravenous injection of the protein AAT, but this is costly and the protein is in short supply. The general steady state level of AAT in blood is about 1 mg/mL, so relatively high expression is required from any gene therapy approach. Preclinical studies in rodents and larger animals suggest that a level of AAT protein in serum between 500 and 1000 ug/mL can be achieved and maintained for over 1 year [124,405].

On the basis of these results, a phase I clinical trial was conducted [406]. In this trial, an AAV2-AAT vector was administered by intramuscular injection to subjects with hereditary emphysema who all had the common PI*Z mutation in the AAT gene. This mutation leads to a defective salt bridge in the AAT protein. Three dose cohorts were administered with vector via intramuscular injection at doses up to 6.9×10^{13}. Most of the subjects were on systemic AAT protein therapy, and even though this systemic protein therapy was discontinued before administration of the vector, the wash-out period may not have been long enough and the serum level of the protein expressed from the vector could not be determined. A second trial, using an AAV1 is currently under way, in the hope that this serotype vector will give much higher expression in the muscle [407].

Two other clinical trials are currently being conducted with AAV1 vectors for treatment, in one case for hereditary hypertriglyceridemia caused by mutations in the gene for lipoprotein lipase (LPL) and in another case for a chronic disease, congestive heart failure (Table 6.1). For LPL deficiency one-phase I/II trial was conducted in the Netherlands in eight subjects who were administered the AAV1-LPL vector intramuscularly. Some decreases in serum levels of triglycerides were observed but there was no evidence of expression of the LPL transgene. A second phase II trial is currently being conducted in a population of LPL deficient subjects in Quebec, Canada. One characteristic of chronic heart failure is a progressive loss of contractility of the heart muscle and thus decreased ejection volume. Cycling of calcium plays a critical role in the cardiac contraction–relaxation cycle. In particular, the sarcoplasmic reticulum calcium ATPase pump (SERCA2a) is required in cardiac myocardium to recycle calcium back into the sarcoplasmic reticulum during each diastolic period. Subjects with chronic heart failure exhibit a progressive loss of activity of SERCA2a in their myocardium. Consequently, a clinical trial is currently under way to deliver the gene for this enzyme using an AAV1 vector, which is particularly efficient in cardiomyocytes, via infusion through the coronary artery.

6.10.8　Vaccines

Early reports on administration to animals of AAV vectors expressing some neo-antigenic transgenes demonstrated the long-term durability of transgene expression in the muscle and showed that the use of AAV vectors did not result necessarily in an immune response against the vector-encoded transgene [70,71,316,408]. However, other studies in mice showed that under certain conditions both humoral and antigen-specific T-cell responses to the transgene protein could be observed. Production of these immune responses probably depends upon several parameters that include the nature of the neo-antigen and its level of expression and localization within transduced cells, the mouse strain used, and the vector dose. Transgene-specific immune responses have been observed in mice following administration of AAV vectors expressing genes encoding *Escherichia coli* β-galactosidase [301,302,409], ovalbumin [298], HSV-2 glycoproteins B (gB) and D (gD) [303], influenza virus hemagglutinin (HA) and HIV-1 Env [409–411], human α-1-antitrypsin (hAAT) [124], and clotting factor IX [305].

The ability to elicit immune responses implies that there is no general mechanism by which AAV blunts the immune system. While the mechanism of antigen presentation following administration of recombinant AAV vectors to the muscle is not clear, one report suggests that activation of T-cells in the draining lymph nodes occurs exclusively through cross-presentation by antigen-presenting cells rather than by direct transduction of dendritic cells (DC) [301]. Whether cross-presentation is a phenomenon that occurs for all transgenes in the context of recombinant AAV vectors is unknown but it suggests a possible role for these vectors as vehicles for prophylactic or therapeutic vaccines. For example, one study demonstrated that a recombinant AAV vector expressing a secreted HSV-2 gB led to the activation of gB-specific CTL, which were most likely activated via cross-presentation of the secreted protein by DC [264]. In contrast, mice injected intramuscularly with an AAV vector expressing ovalbumin developed a robust humoral response to the transgene product but only a minimal ovalbumin-specific CTL response [298].

The use of AAV to stimulate an anti-human immunodeficiency virus (HIV-1) response has been tested in mice. A single, intramuscular injection of an AAV vector encoding the HIV-1 *env*, *tat*, and *rev* genes (AAV–HIV) induced robust, long-term production of HIV-1-specific serum IgG and MHC class I-restricted CTL activity [411]. HIV-specific cell-mediated immunity was enhanced strongly by co-administration with an AAV vector encoding interleukin-12, while boosting with AAV–HIV resulted in rapid and strong HIV-1-specific humoral responses. When AAV–HIV was administered orally, a strong systemic and regional HIV-specific humoral immunity and MHC class I-restricted CTL response was induced which significantly reduced viral load after intrarectal challenge with a recombinant vaccinia virus expressing the HIV-1 *env* gene [412].

In rhesus macaques a single dose, intramuscular administration of an AAV vector expressing the simian immunodeficiency virus (SIV) major structural genes resulted in long-term CD8+, antigen-specific CTL responses against multiple SIV protein epitopes that were similar to responses observed in monkeys directly infected with pathogenic SIV. Neutralizing antibody responses were also robust and persisted for more than 1 year. More recent studies to examine the efficacy of AAV-SIV vaccines in an macaque-SIV challenge model showed that immunized macaques were able to significantly lower replication of a live, virulent SIV after intravenous challenge

at peak (2 weeks) and set-point (10 weeks) compared with mock-vaccinated control animals [413,414].

An HIV vaccine candidate, tgAAC09, that utilizes an AAV2 vector to deliver genes encoding the HIV proteins gag, protease and reverse transcriptase, is currently being tested in clinical trials. In a phase I, double-blind, placebo-controlled, dose-escalation safety study of tgAAC09, 80 healthy volunteers in Europe and India received a single intramuscular injection of tgAAC09 at different doses. Additionally, 21 of the 50 European volunteers received a booster vaccination of either tgAAC09 at the highest dose tested, or placebo. The vaccination with tgAAC09 was safe and well tolerated and stimulated a modest immune response against gag. HIV gag-specific T-cell responses were observed in 20% of participants receiving the highest dose of tgAAC09 tested. A phase II clinical study of tgAAC09 currently under way at multiple sites in South Africa, Zambia, and Uganda is being conducted to evaluate the impact of a higher dose of tgAAC09 and boost vaccination on the strength and duration of immune responses.

6.11 SUMMARY

The work summarized in this chapter illustrates the increasingly rapid progress that is now being made in developing therapeutic applications of AAV vectors. The early clinical testing of AAV vectors for the treatment of CF was extremely important in initiating the regulatory environment for AAV vectors and demonstrating the inherent good safety profile of AAV vectors. As more groups have extended investigations to additional *in vivo* models, the potential utility of AAV vectors as therapeutic gene delivery vehicles has gained more widespread interest. The level of clinical activity in testing many additional AAV vectors has been accelerated in recent years. The development of more sophisticated production systems for AAV vectors has enhanced both the quantity and quality of vectors that can be produced. Additional work to modify the transduction efficiency by judicious choice of serotype for the capsid or by modification of the capsid structure may expand the use and potency of AAV vectors. In addition, further understanding of the intracellular metabolism of AAV vectors also increase the sophistication with which these vectors can be deployed.

REFERENCES

1. BJ Carter. The growth cycle of adeno-associated virus. In: P Tjissen (Ed.), *Handbook of Parvoviruses*, Vol I, CRC Press, Boca Raton, Florida, 1989, pp. 55–168.
2. TR Flotte and KI Berns. Adeno-associated virus: A ubiquitous commensal of mammals. *Hum Gene Ther* 16:401–407, 2005.
3. BJ Carter. Parvoviruses as vectors. In: P Tjissen (Ed.), *Handbook of Parvoviruses*, Vol II, CRC Press, Boca Raton, Florida, 1989, pp. 155–168.
4. N Muzyczka. Use of adeno-associated virus as a generalized transduction vector in mammalian cells. *Curr Top Microbiol Immunol* 158:97–129, 1992.
5. BJ Carter. Adeno-associated virus vectors. *Curr Opin Biotechnol* 3:533–539, 1992.
6. TR Flotte and BJ Carter. Adeno-associated virus vectors for gene therapy. *Gene Ther* 2:357–362, 1995.

7. BJ Carter. Adeno-associated virus and the development of adeno-associated virus vectors: A historical perspective. *Mol Ther* 10:981–989, 2004.

8. TR Flotte, PL Zeitlin, TC Reynolds, AE Heald, P Pedersen, S Beck, CK Conrad, L Baras-Ernst, M Humphries, K Sullivan, R Wetzel, G Taylor, BJ Carter, and WB Guggino. Phase I trial of intranasal and endobronchial administration of a recombinant adeno-associated virus serotype 2 (rAAV2)-CFTR vector in adult cystic fibrosis patients: A two part clinical study. *Hum Gene Ther* 14:1079–1088, 2003.

9. JA Wagner, AH Messner, ML Moran, R Daifuku, K Kouyama, JK Desch, S Manley, AM Norbash, CK Conrad, S Friborg, T Reynolds, WB Guggino, RB Moss, BJ Carter, JJ Wine, TR Flotte, and P Gardner. Safety and biological efficacy of an adeno-associated virus vector-cystic fibrosis transmembrane regulator (AAV-CFTR) in the cystic fibrosis maxillary sinus. *Laryngoscope* 109:266–274, 1999.

10. JA Wagner, IB Nepomuceno, AH Messner, ML Moran, EP Batson, S DiMiceli, BW Brown, JK Desch, AM Norbash, CK Conrad, WB Guggino, TR Flotte, JJ Wine, BJ Carter, RB Moss, and P Gardner. A phase II, double-blind, randomized, placebo-controlled clinical trial of tgAAVCF using maxillary sinus delivery in CF patients with antrostomies. *Hum Gene Ther* 13:1349–1359, 2002.

11. ML Aitken, RB Moss, DA Waltz, ME Dovey, MR Tonelli, SC McNamara, RL Gibson, BW Ramsey, BJ Carter, and TC Reynolds. A phase I study of aerosolized administration of tgAAVCF to CF subjects with mild lung disease. *Hum Gene Ther* 12:1907–1916, 2001.

12. RB Moss, D Rodman, LT Spencer, ML Aitken, PL Zeitlin, D Waltz, C Milla, AS Brody, JP Clancy, B Ramsey, N Hamblett, and AE Heald. Repeated adeno-associated virus serotype 2 aerosol-mediated cystic fibrosis transmembrane regulator gene transfer to the lungs of patients with cystic fibrosis: A multicenter, double-blind, placebo-controlled trial. *Chest* 125:509–521, 2004.

13. RB Moss, C Milla, J Colombo, F Accurso, PL Zeitlin, JP Clancy, LT Spencer, J Pilewski, DA Waltz, HL Dorkin, T Ferkol, M Plan, B Ramsey, BJ Carter, DB martin, and AE Heald. Repeated aerosolized AAV-CFTR for treatment of cystic fibrosis: A randomized placebo-controlled phase 2B trial. *Hum Gene Ther* 18:726–732, 2007.

14. MA Kay, CS Manno, MV Rogni, PJ Larson, LB Couto, A McClelland, B Glader, AJ Chen, SJ Tai, RW Herzog, V Arruda, F Johnson, C Scallen, E Skarsgard, AW Flake, and KA High. Evidence for gene transfer and expression of factor IX in hemophilia B patients treated with an AAV vector. *Nat Genet* 24:257–261, 2000.

15. BJ Carter. Adeno-associated virus vectors in clinical trials. *Hum Gene Ther* 16:541–550, 2005.

16. BJ Carter. Clinical development with adeno-associated virus vectors. In: JR Kerr, SF Cotmore, ME Bloom, RM Linden, and CR Parrish (Eds.), *Parvoviruses*, Hodder Arnold, London, UK, 2006, pp. 499–510.

17. BJ Carter, HB Burstein, and RW Peluso. Adeno-associated virus and adeno-associated virus vectors for gene delivery. In: NS Templeton (Ed.), *Gene and Cell Therapy: Therapeutic Mechanisms and Strategies*, 2nd Edition, Marcel Dekker, New York, 2000, pp. 41–49.

18. S Zolotukhin. Production of recombinant adeno-associated virus vectors. *Hum Gene Ther* 16:551–557, 2005.

19. RW Peluso. The manufacture of adeno-associated vectors. In: B Dropulic and BJ Carter (Eds.), *Concepts in Genetic Medicine*, John Wiley & Sons, New York, 2007, pp. 245–252.

20. JE Rabinowitz and R Samulski. Adeno-associated virus expression systems for gene transfer. *Curr Opin Biotechnol* 9:470–475, 1998.

21. GP Gao, MR Alvira, L Wang, R Calcedo, J Johnston, and JM Wilson. Novel adeno-associated viruses from rhesus monkeys as vectors for human gene therapy. *Proc Natl Acad Sci U S A* 99:11854–11859, 2002.

22. GP Gao, MR Alvira, S Somanathan, Y Lu, LH Vandenberghe, JJ Rux, R Calcedo, J Sanmiguel, Z Abbas, and JM Wilson. Adeno-associated viruses undergo substantial evolution in primates during natural infections. *Proc Natl Acad Sci U S A* 100:6081–6086, 2003.

23. GP Gao, LH Vandennerghe, MR Alvira, Y Lu, R Calcedo, X Zhou, and JM Wilson. Clades od adeno-associated viruses are widely disseminated in human tissues. *J Virol* 78:6381–6388, 2004.

24. CL Chen, RL Jensen, BC Schnepp, MJ Connell, R Shell, TJ Sferra, JS Bartlett, KR Clark, and PR Johnson. Molecular characterization of adeno-associated viruses infecting children. *J Virol* 79:14781–14792, 2005.

25. C Summerford and RJ Samulski. Adeno-associated viral vectors for gene therapy. *Biogenic Amines* 4:451–475, 1998.

26. Q Xie, W Bu, S Bhatia, HJ Hare, T Somasundaram, A Azzi, and MS Chapman. The atomic structure of adeno-associated virus (AAV-2), a vector for gene therapy. *Proc Natl Acad Sci U S A* 99:10405–10410, 2002.

27. M Agbandje and MS Chapman. Correlating structure with function in the viral capsid. In: JR Kerr, SF Cotmore, ME Bloom, RM Linden, and CR Parrish (Eds.), *Parvoviruses*, Hodder Arnold, London, UK, 2006, pp. 125–139.

28. JE Rabinowitz and RJ Samulski. Building a better vector: Manipulation of AAV virions. *Virology* 278:301–308, 2000.

29. N Muzyczka and KH Warrington, Jr. Custom adeno-associated virus capsids: The next generation of recombinant vectors with novel tropism. *Hum Gene Ther* 16:408–416, 2005.

30. RW Atchison, BC Casto, and McD Hammon. Adenovirus-associated defective virus particles. *Science* 49:754–756, 1965.

31. MD Hoggan, NR Blacklow, and WP Rowe. Studies of small DNA viruses found in various adenovirus preparations: Physical, biological and immunological characteristics. *Proc Natl Acad Sci U S A* 55:1467–1472, 1966.

32. NR Blacklow. Adeno-associated viruses of humans. In: JR Pattison (Ed.), *Parvoviruses and Human Disease*. CRC Press, Boca Raton, Florida, 1988, pp. 165–174.

33. Z Grossman, E Mendelson, F Brok-Simoni, F Mileguir, Y Litner, G Rechavi, and B Ramot. Detection of adeno-associated virus type 2 in human peripheral blood cells. *J Gen Virol* 73:961–966, 1992.

34. L Han, TH Parmley, S Keith, KJ Kozlowski, LJ Smith, and PL Hermonat. High prevalence of adeno-associated virus (AAV) type rep DNA in cervical materials: AAV may be sexually transmitted. *Virus Genes* 12:47–52, 1973.

35. CM Waltz, TR Amisi, JR Schlehofer, L Gissman, A Schneider, and M Muller. Detection of infectious adeno-associated virus particles in human cervical biopsies. *Virology* 247:97–105, 1998.

36. M Friedman-Einat, Z Grossman, F Mileguir, Z Smetana, M Ashkenazi, G Barkai, N Varsano, E Glick, and E Mendelson. Detection of adeno-associated virus type 2 sequences in the human genital tract. *J Clin Microbiol* 35:71–78, 1997.

37. SA Afione, CK Conrad, WG Kearns, S Chunduru, R Adams, TC Reynolds, WB Guggino, GR Cutting, BJ Carter, and TR Flotte. In vivo model of adeno-associated virus vector persistence and rescue. *J Virol* 70:3235–3241, 1996.

38. A Srivastava, EW Lusby, and KI Berns. Nucleotide sequence and organization of adeno-associated virus 2 genome. *J Virol* 45:555–564, 1983.

39. S Miramatsu, H Mizukami, N Young, and KE Brown. Nucleotide sequencing and generation of an infectious clone of adeno-associated virus 3. *Virology* 221:208–217, 1996.

40. JA Chiorini, L Yang, Y Liu, B Safer, and RM Kotin. Cloning of adeno-associated virus type 4 (AAV4) and generation of recombinant AAV4 particles. *J Virol* 71:6823–6833, 1967.

41. W Xiao, N Chirmule, SC Berta, B McCullough, G Gao, and JM Wilson. Gene therapy vectors based on adeno-associated virus type 1. *J Virol* 73:3994–4003, 1999.

42. J Chiorini, L Yang, Y Liu, B Safer, and R Kotin. Cloning and characterization of adeno-associated virus type 5. *J Virol* 73:1309–1319, 1999.

43. U Bantel-Schaal, H Delius, R Schmidt, and H zur Hausen. Human adeno-associated virus type 5 is only distally related to other known primate helper-dependent Parvoviruses. *J Virol* 73:939–947, 1999.

44. B Georg-Fries, S Biederlack, L Wolf, and H zur Hausen. Analysis of proteins, helper-dependence and seroepidemiology of a new human parvovirus. *Virology* 134:64–71, 1984.

45. WP Parks, JL Melnick, R Rongey, and HD Mayor. Physical assays and growth cycle studies of a defective adeno satellite virus. *J Virol* 1:171–176, 1967.

46. EA Rutledge, CL Halbert, and DW Russell. Infectious clones of vectors derived from adeno-associated virus (AAV) serotypes other than AAV type 2. *J Virol* 72:309–319, 1998.

47. S Mori, L Wang, T Takeuchi, and T Kanda. Two novel adeno-associated viruses from cynomolgus monkeys: Pseudotyping characterization of capsid protein. *Virology* 330:375–383, 2004.

48. SE Beck, LA Jones, K Chesnut, SM Walsh, TC Reynolds, BJ Carter, FB Askin, TR Flotte, and WB Guggino. Repeated delivery of adeno-associated virus vectors to the rabbit airway. *J Virol* 73:9446–9455, 1999.

49. BJ Carter. The promise of adeno-associated virus vectors. *Nat Biotechnol* 14:1725–1726, 1996.

50. MD Hoggan, GF Thomas, and FB Johnson. Continuous carriage of adenovirus associated virus genome in cell culture in the absence of helper adenovirus. 1973 Proceedings of the Fourth Lepetit Colloquium, pp. 41–47, North Holland, Amsterdam, 1973.

51. SK McLauglin, P Collis, PL Hermonat, and N Muzyczka. Adeno-associated virus general transduction vectors: Analysis of proviral structure. *J Virol* 62:1963–1973, 1998.

52. RM Kotin, M Siniscalco, RJ Samulski, XD Zhu, L Hunter, CA Laughlin, S McLaughlin, N Muzyczka, M Rocchi, and KI Berns. Site-specific integration by adeno-associated virus. *Proc Natl Acad Sci U S A* 87:2211–2215, 1990.

53. RJ Samulski, X Zhu, S Xiao, JD Brook, DE Housman, N Epstein, and LA Hunter. Targeted integration of adeno-associated virus (AAV) into human chromosome 19. *EMBO J* 10:3941–3950, 1991.

54. MD Weitzman, SRM Kyostio, RM Kotin, and RA Owens. Adeno-associated virus (AAV) Rep proteins mediate complex formation between AAV DNA and its integration site in human DNA. *Proc Natl Acad Sci U S A* 91:5808–5812, 1994.

55. RM Linden, P Ward, C Giraud, E Winocour, and KI Bern. Site-specific integration by adeno-associated virus. *Proc Natl Acad Sci U S A* 93:11288–11294, 1996.

56. SM Young, DM McCarty, N Degtyareva, and RJ Samulski. Role of adeno-associated virus rep protein and human chromosome 19 in site-specific recombination. *J Virol* 74:3953–3966, 2000.

57. WG Kearns, SA Afione, SB Fulmer, MG Pang, D Erikson, L Egan, MJ Landrum, TR Flotte, and GR Cutting. Recombinant adeno-associated virus (AAV-CFTR) vectors do not integrate in a site-specific fashion in an immortalized epithelial cell line. *Gene Ther* 3:748–755, 1996.

58. C Summerford and RJ Samulski. Membrane-associated heparan sulfate proteoglycan is a receptor for adeno-associated virus type 2 virions. *J Virol* 72:1438–1445, 1998.

59. C Summerford, JS Bartlett, and RJ Samulski. $\alpha_v\beta_5$ integrin: A co-receptor for adeno-associated virus type 2 infection. *Nat Med* 5:78–82, 1999.

60. K Qing, C Mah, J Hansen, S Zhou, V Dwarki, and A Srivastava. Human fibroblast growth factor receptor 1 is a co-receptor for infection by adeno-associated virus 2. *Nat Med* 5:71–77, 1999.

61. M Goldman, Q Su, and JM Wilson. Gradient of RGD-dependent entry of adenoviral vector in nasal and intrapulmonary epithelia: Implications for gene therapy of cystic fibrosis. *Gene Ther* 3:811–818, 1996.

62. D Duan, Y Yue, Y Yan, PB McCray, and JF Engelhardt. Polarity influences the efficiency of recombinant adeno-associated virus infection in differentiated airway epithelia. *Hum Gene Ther* 9:2761–2776, 1998.

63. G Seisenberger, MU Reid, T Endress, H Buning, M Hallek, and C Brauchle. Real-time single-molecule imaging of the infection pathway of an adeno-associated virus. *Science* 294:1029–1932, 2001.

64. J Hansen, K Qing, and A Srivastava. Adeno-associated virus type 2 mediated gene transfer: Altered endocytotic processing enhances transduction efficiency. *J Virol* 75:4080–4090, 2001.

65. U Bantel-Schaal, B Hub, and J Kartenbeck. Endocytosis of adeno-associated virus type 5 leads to accumulation of virus particles in the Golgi compartment. *J Virol* 76:2340–2349, 2002.

66. W Ding, LN Zhang, C Yeaman, and JF Engelhardt. rAAV2 traffics through both the late and the recycling endosomes in a dose-dependant fashion. *Mol Ther* 13:671–682, 2006.

67. Y Yan, R Zak, GWG Luxton, TC Ritchie, U Bantel-Schaal, and JF Engelhardt. Ubiquitination of both adeno-associated virus type 2 and 5 capsid proteins affects the transduction efficiency of recombinant vectors. *J Virol* 76:2043–2053, 2002.

68. TR Flotte, SA Afione, R Solow, SA McGrath, C Conrad, PL Zeitlin, WB Guggino, and BJ Carter. In vivo delivery of adeno-associated vectors expressing the cystic fibrosis transmembrane conductance regulator to the airway epithelium. *Proc Natl Acad Sci U S A* 93:10163–10617, 1993.

69. MG Kaplitt, P Leone, RJ Samulski, X Xiao, D Pfaff, KL O'Malley, and M During. Long-term gene expression and phenotypic correction using adeno-associated virus vectors in the mammalian brain. *Nat Genet* 8:148–154, 1994.

70. X Xiao, J Li, and RJ Samulski. Efficient long-term gene transfer into muscle tissue of immunocompetent mice by adeno-associated virus vector. *J Virol* 70:8098–8108, 1996.

71. PD Kessler, GM Podsakoff, X Chen, SA McQuiston, PC Colosi, LA Matelis, GJ Kurtzman, and B Byrne. Gene delivery to skeletal muscle results in sustained expression and systemic delivery of a therapeutic protein. *Proc Natl Acad Sci U S A* 93:14082–14087, 1996.

72. DD Koerberl, IE Alexander, CL Halbert, DW Russell, and AD Miller. Persistent expression of human clotting factor IX from mouse liver after intravenous injection of AAV vectors. *Proc Natl Acad Sci U S A* 94:1426–1431, 1997.

73. S Ponnazhagan, P Mukherjee, MC Yoder, X-S Wang, SZ Zhou, J Kaplan, S Wadsworth, and A Srivastava. Adeno-associated virus type 2-mediated gene transfer in vivo: Organ-tropism and expression of transduced sequences in mice. *Gene* 190:203–210, 1997.

74. HJ Nam, MD Lane, E Padron, B Gurda, R McKenna, E Kohlbrenner, G Aslandi, B Byrne, N Muzyczka, S Zolotukhin, and M Agbandje-Mckenna. Structure of adeno-associated virus serotype 8, a gene therapy vector. *J Virol* 81:12260–12271, 2007.

75. Z Zadori, J Szelei, M-C Lacoste, Y Li, S Gariepy, P Raymond, M Allaire, IR Nabi, and P Tjissen. A viral phospholipase A2 is required for parvovirus infectivity. *Dev Cell* 1:291–302, 2001.

76. M Vihinen, S Suikkanen, and CR Parrish. Pathways of cell infection by parvoviruses and adeno-associated viruses. *J Virol* 78:6709–6714, 2004.

77. A Girod, CE Wobus, Z Zadori, M Reid, K Leike, P Tijssen, JA Kleinscmidt, and M Hallek. The VP1 capsid protein of adeno-associated virus type 2 is carrying a phospholipase A2 domain required for virus infectivity. *J Gen Virol* 83:973–978, 2002.

78. S Murray, CL Nilsson, JT Hare, MR Emmett, A Korostelev, H Ongley, AG Marshall, and MS Chapman. Characterization of the capsid protein glycosylation of adeno-associated virus type 2 by high-resolution mass spectrometry. *J Virol* 80:6171–6176.

79. NR Blacklow, MD Hoggan, and WP Rowe. Immunofluorescent studies on the potentiation of adenovirus-associated virus by adenovirus 7. *J Exp Med* 125:755–762, 1967.

80. LV Crawford, EAC Follett, MG Burdon, and DJ McGeoch. The DNA of a minute virus of mice. *J Gen Virol* 4:37–48, 1969.

81. JA Rose, KI Berns, MD Hoggan, and FJ Koczot. Evidence for a single-stranded adenovirus-associated virus genome: Formation of a DNA density hybrid upon release of viral DNA. *Proc Natl Acad Sci U S A* 64:863–869, 1969.

82. BJ Carter, G Khoury, and DT Denhardt. Physical map and strand polarity of specific fragments of adeno-associated virus DNA produced by endonuclease R.EcoRI. *J Virol* 16:559–568, 1975.

83. JW Smuda and BJ Carter. Adeno-associated viruses having nonsense mutations in the capsid gene: Growth in mammalian cells having an inducible amber suppressor. *Virology* 84:310–318, 1991.

84. TR Flotte, PL Zeitlin, R Solow, S Afione, RA Owens, D Markakis, M Drum, WB Guggino, and BJ Carter. Expression of the cystic fibrosis transmembrane conductance regulator from a novel adeno-associated virus promoter. *J Biol Chem* 268:3781–3790, 1993.

85. RP Haberman, TJ McCown, and RJ Samulski. Novel transcriptional regulatory signals in the adeno-associated virus terminal repeat A/D junction element. *J Virol* 74:8732–8739, 2000.

86. J Qui, R Nayak, GE Tullis, and DJ Pintel. Characterization of the transcription profile of adeno-associated virus type 5 reveals a number of unique features compared to previously characterized adeno-associated viruses. *J Virol* 76:12435–12447, 2002.

87. BJ Carter, E Mendelson, and JP Trempe. AAV DNA replication, integration and genetics. In: P. Tjissen (Ed.), *Handbook of Parvoviruses*, Vol I, CRC Press, Boca Raton, Florida, 1989, pp. 169–226.

88. MW Myers and BJ Carter. 1980. Adeno-associated virus assembly. *Virology* 102:71–82, 1980.

89. J King, D Dubielzig, D Grimm, and J Kleinschmidt. DNA helicase-mediated packaging of adeno-associated virus type 2 genomes into pre-formed capsids. *EMBO J* 20:3282–3291, 2001.

90. B Redemann, E Mendelson, and BJ Carter. Adeno-associated virus rep protein synthesis during productive infection. *J Virol* 63:873–88, 1998.

91. BJ Carter. Adeno-associated virus helper functions. In: P Tjissen (Ed.), *Handbook of Parvoviruses*, Vol I, CRC Press, Boca Raton, Florida, 1989, pp. 255–282.

92. DJ Perriera, DM McCarty, and N Muzyczka. The adeno-associated virus (AAV) rep protein acts as both a repressor and an activator to regulate AAV transcription during a productive infection. *J Virol* 71:1079–1088, 1997.

93. RJ Samulski, KI Berns, N Tan, and N Muzyczka. Cloning of infectious adeno-associated virus into pBR322: Rescue of intact virus from the recombinant plasmid in human cells. *Proc Natl Acad Sci U S A* 79:2077–2081, 1982.

94. CA Laughlin, JD Tratschin, H Coon, and BJ Carter. Cloning of infectious adeno-associated virus genomes in bacterial plasmids. *Gene* 23:65–73, 1983.

95. S Bleker, M Pawlita, and JA Kleinschmidt. Impact of capsid conformation and rep–cap interactions on adeno-associated type 2 genome packaging. *J Virol* 80:810–820, 2006.

96. YJ Hernandez, J Wang, WG Kearns, S Wiler, A Poirer, and TR Flotte. Latent adeno-associated virus infection elicits humoral but not cell-mediated immune responses in a non-human primate model. *J Virol* 73:8549–8558, 1999.

97. DW Russell and RK Hirata. Human gene targeting by viral vectors. *Nat Genet* 18:325–330, 1998.

98. KJ Fisher, K Jooss, J Alston, Y Yang, SH Haecker, K High, R Pathak, S Raper, and LM Wilson. Recombinant adeno-associated virus vectors for muscle directed gene therapy. *Nat Med* 3:306–312, 1997.

99. RW Herzog, JN Hagstrom, S-H Kung, SJ Tai, JM Wilson, KJ Fisher, and KA High. Stable gene transfer and expression of human blood coagulation factor IX after intramuscular injection of recombinant adeno-associated virus. *Proc Natl Acad Sci U S A* 94:5804–5809, 1997.

100. RO Snyder, SK Spratt, C Lagarde, C Bohl, B Kaspar, B Sloan, LK Cohen, and O Danos. Efficient and stable adeno-associated virus-mediated transduction of the skeletal muscle of adult immunocompetent mice. *Hum Gene Ther* 8:1891–1900, 1997.

101. CH Miao, RO Snyder, DB Schowalter, GA Patijn, B Donahue, B Winther, and MA Kay. The kinetics of rAAV integration into the liver. *Nat Genet* 19:13–14, 1998.

102. D Duan, P Sharma, J Yang, Y Yue, L Dudus, Y Zhang, KJ Fisher, and JF Engelhardt. Circular intermediates of recombinant adeno-associated virus have defined structural characteristics responsible for long-term episomal persistence in muscle tissue. *J Virol* 72:8568–8577, 1998.

103. D Duan, P Sharma, L Dudus, Y Zhang, S Sanlioglu, Z Yan, Y Yue, Y Ye, R Lester, J Yang, KJ Fisher, and JF Engelhardt. Formation of adeno-associated virus circular genomes is differentially regulated by adenovirus E4orf6 and E2A gene expression. *J Virol* 73:161–169, 1999.

104. N Vincent-Lacaze, RO Snyder, R Gluzman, D Bohl, C Lagarde, and O Danos. Structure of adeno-associated virus vector DNA following transduction of the skeletal muscle. *J Virol* 73:1949–1955, 1999.

105. C McKeon and RJ Samulski. NIDDK workshop on AAV vectors: Gene transfer into quiescent cells. *Hum Gene Ther* 7:1615–1619, 1996.

106. H Nakai, SR Yant, TA Storm, S Fuess, L Meuse, and MA Kay. Extra-chromosomal recombinant adeno-associated virus vector genomes are primarily responsible for stable liver transduction in vivo. *J Virol* 75:6969–6976, 2001.

107. BC Schnepp, KR Clark, DL Klemanski, CA Pacak, and PR Johnson. Genetic fate of recombinant adeno-associated virus vector genomes in muscle. *J Virol* 77:3495–3505, 2003.

108. BC Schnepp, MC Soult, T Allen, P Anklesaria, PR Johnson, and K Munson. Biodistribution and integration assessment of AAV1gag-PR-RT(ΔRNaseH) after intramuscular administration in rabbits. *Mol Ther* 13:S190, 2006.

109. RT Surosky, M Urabe, SG Godwin, SA McQuiston, GJ Kurtzman, K Ozawa, and G Natsoulis. Adeno-associated

virus rep proteins target DNA sequences to a unique locus in the human genome. *J Virol* 71:7951–7959, 1997.

110. F Palombo, A Monciotti, A Recchia, R Cortese, G Ciliberto, and N La Monica. Site-specific integration in mammalian cells mediated by a new hybrid baculovirus-adeno-associated virus vector. *J Virol* 72:5025–5034, 1998.

111. A Recchia, RJ Parks, S Lamartina, C Toniatti, L Pieroni, F Palombo, G Ciliberto, FL Graham, R Cortese, and N La Monica. Site-specific integration mediated by a hybrid adenovirus/adeno-associated virus vector. *Proc Natl Acad Sci U S A* 96:2615–2620, 1999.

112. R Hirata, J Chamberlain, R Dong, and DW Russell. Targeted transgene insertion into human chromosomes by adeno-associated virus vectors. *Nat Biotechnol* 20:735–737, 2002.

113. B Yakobson, TA Hrynko, MJ Peak, and E Winocour. Replication of adeno-associated virus in cells irradiated with UV light at 254 nm. *J Virol* 63:1023–1030, 1988.

114. U Bantel-Schaal and H zur Hausen. Adeno-associated viruses inhibit SV40 DNA amplification and replication of Herpes simplex virus in SV40-transformed hamster cells. *Virology* 164:64–74, 1988.

115. P Ward, P Elisa, and RM Linden. Rescue of the adeno-associated virus genome from a plasmid vector: Evidence for rescue by replication. *J Virol* 77:11480–11490, 2003.

116. CD Scallan, H Jiang, T Liu, S Patarroyo-White, JM Sommer, S Zhou, LB Couto, and GF Pierce. Human immunoglobulin inhibits liver transduction by AAV vectors at low AAV2 neutralizing titers in SCID mice. *Blood* 107:1810–1817, 2006.

117. F Mingozzi, MV Maus, DJ Hui, DE Sabatino, SL Murphy, JE Rasko, MV Ragni, CS Manno, J Somme, H Jiang, GF Pierce, HC Ertl, and KA High. CD8(+) T-cell responses to adeno-associated virus capsid in humans. *Nat Med* 13:419–422, 2007.

118. AL Peel, S Zolotukhin, GW Schrimsher, N Muzyczka, and PJ Reier. Efficient transduction of green fluorescent protein in spinal cord neurons using adeno-associated virus vectors containing cell type-specific promoters. *Gene Ther* 4.16–24, 1997.

119. JG Flannery, S Zolotukhin, MI Vaquero, MM LaVail, N Muzyczka, and WW Hauswirth. Efficient photoreceptor-targeted gene expression in vivo by recombinant adeno-associated virus. *Proc Natl Acad Sci U S A* 94:6916–6921, 1997.

120. KC Rendahl, D Quiroz, M Ladner, M Coyne, J Seltzer, WC manning, and LA Escobedo. Tightly regulated long-term erythropoietin expression in vivo using tet-inducible recombinant adeno-associated viral vectors. *Hum Gene Ther* 13: 335–342, 2002.

121. D Favre, V Blouin, R Provost, R Spisek, F Porrot, D Bohl, F Marme, Y Cherel, A Salvetti, B Hurtrel, J-M Heard, Y Riviere, and P Mouiller. Lack of an immune response against the tetracycline-dependent transactivator correlates with long-term doxycycline-regulated transgene expression in nonhuman primates after intramuscular injection of recombinant adeno-associated virus. *J Virol* 76:11605–11611, 2002.

122. A Aurrichio, VM Rivera, T Clackson, EE O'Connor, AM Magure, MJ Tolentino, J Bennett, and JM Wilson. Pharmacological regulation of protein expression from adeno-associated viral vectors in the eye. *Mol Ther* 6:238–242, 2002.

123. L Jiang, S Rampalli, D George, C Press, EG Bremer, MR O'Gorman, and MC Bohn. Tight regulation from a single tet off rAAV vector as demonstrated by flow cytometry and real time PCR. *Gene Ther* 11:1057–1067, 2004.

124. S Song, M Morgan, T Ellis, A Poirier, K Chesnut, J Wang, M Brantly, N Muzyczka, BJ Byrne, M Atkinson, and TR Flotte.

Sustained secretion of human alpha-1-antitrypsin from murine muscle transduced with adeno-associated virus vectors. *Proc Natl Acad Sci U S A* 95:14384–14388, 1998.

125. H Jiang, GF Pierce, MC Ozelo, EV DePaul, JA Vargas, P Smith, J Sommer, A Luk, CS Manno, KA High, and VR Arruda. Evidence of multiyear factor IX expression by AAV-mediated gene transfer to skeletal muscle in an individual with severe hemophilia B. *Mol Ther* 14:452–455, 2006.

126. C Lebherz, A Auricchio, AM Maguire, VM Rivera, W Tang, RL Grant, T Clackson, J Bennett, and JM Wilson. Long-term inducible gene expression in the eye via adeno-associated virus gene transfer in nonhuman primates. *Hum Gene Ther* 16:178–186, 2005.

127. VM Rivera, GP Gao, RL Grant, MA Schnell, PW Zoltick, LW Rozamus, T Clackson, and JM Wilson. Long-term pharmacologically regulated expression of erythropoietin in primates following AAV-mediated gene transfer. *Blood* 105:1424–1430, 2005.

128. S Sandalon, EM Bruckheimer, KH Lustig, and Burstein H. Long-term suppression of experimental arthritis following intramuscular administration of a pseudotyped AAV2/1-TNFR:Fc vector. *Mol Ther* 15:264–269, 2007.

129. Q Yang, F Chen, and JP Trempe. Characterization of cell lines that inducibly express the adeno-associated virus Rep proteins. *J Virol* 68:4847–4856, 1994.

130. A Allen, DJ Debelak, TC Reynolds, and AD Miller. Identification and elimination of replication-competent adeno-associated virus (AAV) that can arise by non-homologous recombination during AAV vector production. *J Virol* 71:6816–6822, 1997.

131. XS Wang, B Khuntirat, K Qing, S Ponnazhagan, DM Kube, S Zhou, VJ Dwarki, and A Srivastava. Characterization of wild-type adeno-associated virus type 2-like particles generated during recombinant viral vector production and strategies for their elimination. *J Virol* 72:5472–5480, 1998.

132. P-D Fan and J-Y Dong. Replication of rep–cap genes is essential for the high-efficiency production of recombinant AAV. *Hum Gene Ther* 8:87–98, 1997.

133. K Vincent, ST Piraino, and SC Wadsworth. Analysis of recombinant adeno-associated virus packaging and requirements for rep and cap gene products. *J Virol* 71:1897–1905, 1997.

134. J Li, RJ Samulski, and X Xiao. Role for highly regulated rep gene expression in adeno-associated virus vector production. *J Virol* 71:5236–5243, 1997.

135. X Xiao, J Li, and RJ Samulski. Production of high-titer recombinant adeno-associated virus vectors in the absence of helper adenovirus. *J Virol* 72:2224–2232, 1998.

136. T Matushita, S Elliger, C Elliger, G Podskaoff, K Villareal, GJ Kurtzman, Y Iwaki, and P Colosi. Adeno-associated virus vectors can be efficiently produced without helper virus. *Gene Ther* 5:938–945, 1998.

137. AS Salvetti, S Orev, G Chadeuf, D Favre, Y Cherel, P Champion-Arnaud, J David-Ameline, and P Mouillier. Factors influencing recombinant adeno-associated virus production. *Hum Gene Ther* 9:695–706, 1998.

138. D Grimm, A Kern, K Rittnet, and AJ Kleinschmidt. Novel tools for production and purification of recombinant adeno-associated virus vectors. *Hum Gene Ther* 9:2745–2760, 1998.

139. RJ Samulski, LS Chang, and TE Shenk. Helper-free stocks of recombinant adeno-associated viruses: Normal integration does not require viral gene expression. *J Virol* 63:3822–3828, 1989.

140. TR Flotte, X Barrazza-Ortiz, R Solow, SA Afione, BJ Carter, and WB Guggino. An improved system for packaging recombinant adeno-associated virus vectors capable of in vivo transduction. *Gene Ther* 2:39–47, 1995.

141. KR Clark, F Voulgaropoulou, DM Fraley, and PR Johnson. Cell lines for the production of recombinant adeno-associated virus. *Hum Gene Ther* 6:1329–1341, 1995.

142. X Liu, F Voulgaropoulou, R Chen, PR Johnson, and KR Clark. Selective rep–cap gene amplifications as a mechanism for high-titer recombinant AAV production from stable cell lines. *Mol Ther* 2:394–403, 2000.

143. D Farson, TC Harding, L Tao, J Liu, S Powell, V Vimal, S Yendluri, K Koprivnikar, K Ho, C Twitty, P Husak, A Lin, RO Snyder, and BA Donahue. Development and characterization of a cell line for large-scale, serum-free production of recombinant adeno-associated viral vectors. *J Gene Med* 6:1369–1381, 2004.

144. G-P Gao, G Qu, LZ Faust, RK Engdahl, W Xiao, JV Hughes, PW Zoltick, and JM Wilson. High-titer adeno-associated viral vectors from a rep/cap cell line and hybrid shuttle virus. *Hum Gene Ther* 9:2353–2362, 1998.

145. XL Liu, KR Clark, and PR Johnson. Production of recombinant adeno-associated virus vectors using a packaging cell line and a hybrid recombinant virus. *Gene Ther* 6:293–299, 1999.

146. KJ Fisher, WM Kelley, JF Burda, and JM Wilson. A novel adenovirus-adeno-associated virus hybrid vector that displays efficient rescue and delivery of the AAV genome. *Hum Gene Ther* 7:2079–2087, 1996.

147. A Lieber, DS Steinwarder, CA Carlson, and MA Kay. Integrating adenovirus-adeno-associated virus hybrid vectors devoid of all genes. *J Virol* 73:9314–9324, 1999.

148. Z Sandolon, DM Gnatenko, WF Bahou, and P Hearing. Adeno-associated (AAV) rep protein enhances the generation of a recombinant mini-adenovirus utilizing an Ad/AAV hybrid virus. *J Virol* 74:10381–10389, 2000.

149. I Inoue and DW Russell. Packaging cells based on inducible gene amplification for the production of adeno-associated virus vectors. *J Virol* 72:7024–7031, 1998.

150. JE Conway, S Zolotukhin, N Muzyczka, GS Hayward, and BJ Byrne. Recombinant adeno-associated virus type 2 replication and packaging is entirely supported by a herpes simplex virus type 1 amplicon expressing rep and cap. *J Virol* 71:8780–8789, 1997.

151. JT Wustner, S Arnold, M Lock, JC Richardson, VB Himes, G Kurtzman, and RW Peluso. Production of recombinant adeno-associated type 5 (rAAV5) vectors using recombinant herpes simplex viruses containing *rep* and *cap*. *Mol Ther* 6:510–518, 2002.

152. KM Johnston, D Jacoby, PA Pechan, C Fraefel, P Borghesani, D Schuback, RJ Dunn, FI Smith, and XO Breakfield. HSV/AAV hybrid amplicon vectors extend transgene expression in human glioma cells. *Hum Gene Ther* 8:359–370, 1997.

153. X Zhang, M de Alwis, SL Hart, FW Fitzke, SC Inglis, MEG Boursnell, RL Levinsky, C Kinnon, RR Ali, and AJ Thrasher. High-titer recombinant adeno-associated virus production from replicating amplicons and herpes virus deleted for glycoprotein H. *Human Gene Ther* 10:2527–2537, 1999.

154. JM Booth, A Mistry, X Li, A Thrasher, and RS Coffin. Transfection-free and scalable recombinant AAV vector production using HSV/AAV hybrids. *Gene Ther* 11:829–837, 2004.

155. E Toublanc, A Benraiss, D Bonnin, V Blouin, N Brument, N Cartier, AL Epstein, A., P Moullier, and A Salvetti. Identification of a replication-defective herpes simplex virus for recombinant adeno-associated virus type 2 (rAAV2) particle assembly using stable producer cell lines. *J Gene Med* 6:555–564, 2004.

156. KK Hwang, T Mandell, H Kintner, S Zolotukhin, R Snyder, and BJ Byrne. High titer recombinant adeno-associated virus production using replication deficient herpes simplex viruses type 1. *Mol Ther* 7:S14, 2003.

157. DR Knop and H Harrell. Bioreactor production of recombinant herpes simplex virus vectors. *Biotechnol Prog* 23:715–721, 2007.

158. M Urabe, C Ding, and RM Kotin. Insect cells as a factory to produce adeno-associated virus type 2 vectors. *Hum Gene Ther* 13:1935–1943, 2002.

159. MG Aucoin, M Perrier, and AA Kamen. Production of adeno-associated viral vectors in insect cells using triple infection: Optimization of baculovirus concentration ratios. *Biotechnol Bioeng* 95:1081–1092, 2006.

160. A Negrete and RM Kotin. Production of recombinant adeno-associated viral vectors using two bioreactor confirmations at different scales. *J Virol Methods* 145:155–161, 2007.

161. M Urabe, T Nakakura, K-Q Xin, Y Obara, H Mizukami, A Kume, RM Kotin, and K Ozawa. Scalable generation of high-titer recombinant adeno-associated virus type 5 in insect cells. *J Virol* 80:1874–1885, 2006.

162. L Cao, Y Liu, M During, and W Xiao. High-titer wild-type free recombinant adeno-associated virus vector production using intron-containing helper plasmids. *J Virol* 74:11456–11463, 2000.

163. S Zolotukhin, BJ Byrne, E Mason, I Zolotukhin, M Potter, K Chesnut, C Summerford, RJ Samulski, and N Muzyczka. Recombinant adeno-associated virus purification using novel methods improves infectious titer and yield. *Gene Ther* 6:973–985, 1999.

164. WT Hermans, O ter Brakke, A Dijkhuisen, MA Sonnemans, D Grimm, JA Kleinschmidt, and J Verhaagen. Purification of recombinant adeno-associated virus by iodixonal gradient ultracentrifugation allows rapid and reproducible preparation of vector stocks for gene transfer in the nervous system. *Hum Gene Ther* 10:1885–1891, 1999.

165. K Tamayose, Y Hirai, and T Shimada. A new strategy for large-scale preparation of high titer recombinant adeno-associated virus vectors by using packaging cell lines and sulfonated cellulose column chromatography. *Hum Gene Ther* 7:507–513, 1996.

166. RW Clark, X Liu, JP McGrath, and PR Johnson. Highly purified recombinant adeno-associated virus vectors are biologically active and free of detectable helper and wild type viruses. *Hum Gene Ther* 10:1031–1039, 1999.

167. G Gao, G Gu, MS Burnhan, J Huang, N Chirmule, B Joshi, Q-C Yu, IA Marsh, CM Conceicao, and JM Wilson. Purification of recombinant adeno-associated virus vectors by column chromatography and its performance in vivo. *Hum Gene Ther* 11:2079–2091, 2000.

168. DJ Debelak, J Fisher, S Iuliano, D Sesholtz, DL Sloane, and EM Atkinson. Cation exchange high-performance liquid chromatography of recombinant adeno-associated virus type 2. *J Chromatogr* 740:195–202, 2000.

169. CR O'Riordan, AL Lachappelle, KA Vincent, and SC Wadsworth. Scaleable chromatographic purification process for recombinant adeno-associated virus (rAAV). *J Gene Med* 2:444–454, 2000.

170. L Drittanti, C Jenny, K Poulard, A Samba, P Manceau, N Soria, N Vincent, O Danos, and M Vega. Optimized helper virus-free production of high quality adeno-associated virus vectors. *J Gene Med* 3:59–71, 2001.

171. N Kaludov, B Handelman, and JA Chiorini. Scalable purification of adeno-associated virus type 2, 4, or 5 using ion-exchange chromatography. *Hum Gene Ther* 13:1235–1243, 2002.

172. N Brument, R Morenweisser, V Blouin, E Toublanc, I Raimbaud, Y Cherel, S Folliot, F Gaden, P Boulanger, G Kroner-Lux, and

P Moullier. A versatile and scalable two-step ion-exchange chromatography process for the purification of recombinant adeno-associated virus serotypes 2 and 5. *Mol Ther* 6: 678–688, 2002.

173. JM Sommer, PH Smith, S Parthasarathy, J Isaacs, S Vijay, J Kieran, SK Powell, A McClelland, and JF Wright. Quantification of adeno-associated virus particles and empty capsids by optical density measurement. *Mol Ther* 7:122–128, 2003.

174. JF Wright, T Le, J Prado, J Bahr-Davidson, PH Smith, Z Zhen, JM Sommers, GF Pierce, and G Qu. Identification of factors that contribute to recombinant AAV2 particle aggregation and methods to prevent its occurrence during vector purification and formulation. *Mol Ther* 12:171–178, 2005.

175. Z Wu, A Asokan, and RJ Samulski. Adeno-associated virus serotypes: Vector toolkit for human gene therapy. *Mol Ther* 14:316–327, 2006.

176. H Jiang, D Lillicrap, S Patarroyo-White, T Liu, X Qian, CD Scallan, S Powell, T Keller, M McMurray, A Labelle, D Nagy, JA Vargas, S Zhou, LB Couto, and GF Pierce. Multiyear therapeutic benefit of AAV serotypes 2, 6 and 8 delivering Factor VIII to hemophilia A mice and dogs. *Blood* 108:107–115, 2006.

177. R Sarkar, M Mucci, S Addya, R Tretreault, DA Bellinger, TC Nichols, and HH Kazazian. Long-term efficacy of adeno-associated virus serotypes 8 and 9 in hemophilia A dogs and mice. *Hum Gene Ther* 17:427–439, 2006.

178. A Aurrichio, M Hildinger, E O'Connor, G-P Gao, and JM Wilson. Isolation of highly infectious and pure adeno-associated virus type 2 vectors with a single-step gravity-flow column. *Hum Gene Ther* 12:71–76, 2001.

179. A Aurrichio, E O'Connor, M Hildinger, and LM Wilson. A single-step affinity column for purification of serotype-5 based adeno-associated viral vectors. *Mol Ther* 4:372–379, 2001.

180. RH Smith, C Ding, and RM Kotin. Serum-free production and column purification of adenoassociated virus type 5. *J Virol Methods* 114:115–124, 2003.

181. AM Davidoff, CTC Ng, S Sleep, J Gray, S Azam, Y Zhao, JH McIntosh, M Karimipoor, and AC Nathwani. Purification of recombinant adeno-associated virus type 8 vectors by ion exchange chromatography generates clinical grade vector stocks. *J Virol Methods* 121:209–215, 2004.

182. MJ Blankinship, P Gregorevic, JM Allen, SQ Harper, H Harper, CL Halbert, DA Miller, and JS Chamberlin. Efficient transduction of skeletal muscle using vectors based on adeno-associated virus type 6. *Mol Ther* 10:671–678, 2004.

183. G Qu, J Bahr-Davidson, J Prado, A Tai, F Cataniag, J McDonnell, J Zhou, B Hauck, J Luna, JM Sommer, P Smith, S Zhou, P Colosi, KA High, GF Pierce, and JF Wright. Separation of adeno-associated virus type 2 empty particles from genome containing vectors by anion-exchange column chromatography. *J Virol Methods* 140:183–192, 2007.

184. BA Thorne, P Quigley, G Nicols, C Moore, E Pastor, D Price, JW Ament, RK Takeya, and RW Peluso. Characterizing clearance of helper adenovirus by a clinical rAAV1 manufacturing process. *Biologicals* 36:7–18, 2008.

185. A Salvetti, A Oreve, G Cahdeuf, D Favre, Y Cherel, P Champion-Arnaud, J David-Ameline, and P Moullier. Factors influencing recombinant adeno-associated virus production. *Hum Gene Ther* 9:695–706, 1998.

186. JS Bartlett and RJ Samulski. Methods for construction and propagation of recombinant adeno-associated virus vectors. In: *Methods in Molecular Medicine, Gene Therapy Protocols*. Humana Press, Inc., Totowa, NJ, 2000, pp. 25–40.

187. KR Clark, F Voulgaropoulou, and PR Johnson. A stable cell line carrying adenovirus inducible rep and cap genes allows fore infectivity titration of adeno-associated virus vectors. *Gene Ther* 3:1124–1132, 1996.

188. EM Atkinson, DJ Debelak, LA Hart, and TC Reynolds. A high-throughput hybridization method for titer determination of viruses ands gene therapy vectors. *Nucleic Acids Res* 26: 2821–2823, 1998.

189. DR Grimm, A Kern, M Pawlita, F Ferrari, R Samulski, and JA Kleinschmdit. Titration of AAV-2 particles via a novel capsid ELISA: Packaging of genomes limits production of recombinant AAV-2. *Gene Ther* 6:1322–1330, 1999.

190. RW Walters, SM Yi, S Keshavjee, KE brown, MJ Welsh, JA Chiorini, and J Zabner. Binding of adeno-associated virus type 5 to 2,3-linked sialic acid is required for gene transfer. *J Biol Chem* 276:20610–20616, 2001.

191. Y Kashiwakura, K Tamayose, K Iwabuchi, Y Hirai, T Shimada, K Matsumoto, T Nakamura, M Watanabe, K Oshimi, and H Daida. Hepatocyte growth factor receptor is a coreceptor for adeno-associated virus type 2 infection. *J Virol* 79:609–614, 2005.

192. Z Wu, E Miller, M Agbandje-McKenna, and RJ Samulski. Alpha 2,3 and alpha 2,6 N-linked sialic acids facilitate efficient binding and transduction by adeno-associated viruses types 1 and 6. *J Virol* 80:9093–9103, 2006.

193. B Akache, D Grimm, K Pandey, SR Yant, H Xu, and MA Kay. The 36/67-kilodalton laminin receptor is a receptor for adeno-associated virus serotypes 8, 2, 3, and 9. *J Virol* 80:9831–9839, 2006.

194. G Di Pasquale, D Scuderio, A Monks, and JA Chiorini. Identification of PDFGR as a receptor for AAV5 transduction by comparative gene expression analysis using cDNA microarrays. *Nat Med* 9:1306–1312, 2003.

195. JA Chorini, SA Afione, and RM Kotin. Adeno-associated virus (AAV) type 5 rep protein cleaves a unique terminal resolution site compared to other AAV serotypes. *J Virol* 73:4293–4298, 1999.

196. H Chao, Y Liu, J Rabinowitz, C Li, RJ Samulski, and CE Walsh. Several log increase in therapeutic transgene delivery by distinct adeno-associated viral serotype vectors. *Mol Ther* 2:619–623, 2000.

197. J Rabinowitz, F Rolling, C Li, H Conrath, W Xiao, X Xiao, and RJ Samulski. Cross-packaging of a single adeno-associated virus (AAV) type 2 vector genome into multiple AAV serotypes enables transduction with broad specificity. *J Virol* 76:791–801, 2002.

198. M Hildinger, A Aurricchio, G Gao, L Wang, N Chirmule, and JM Wilson. Hybrid vectors based on adeno-associated virus serotypes 2 and 5 for muscle-directed gene transfer. *J Virol* 75:6199–6203, 2001.

199. J Zabner, M Seiler, R Walters, RM Kotin, W Fulgeras, BL Davidson, and AJ Chiorini. Adeno-associated virus type 5 (AAV5) but not AAV2 binds to the apical surfaces of airway epithelial cells and facilitates gene transfer. *J Virol* 74:3652–3658, 2000.

200. A Aurichio, E O'Connor, D Weiner, G-P Gao, M Hildinger, L Wang, R Calcedo, and JM Wilson. Noninvasive gene transfer to the lung for systemic delivery of therapeutic proteins. *J Clin Inv* 110:499–504, 2002.

201. CL Halbert, JM Allen, and AD Miller. Adeno-associated virus type 6 (AAV6) vectors mediate efficient transduction of airway epithelium in mouse lungs compared to that of AAV2 vector. *J Virol* 75:6615–6624, 2001.

202. CL Halbert, EA Rutledge, JM Allen, DW Russell, and AD Miller. Repeat transduction in the mouse lung by using

adeno-associated virus vectors with different serotypes. *J Virol* 74:1524–1552, 2000.

203. CH Maio, H Nakai, AR Thompson, TA storm, W Chiu, RO Snyder, and MA Kay. Nonrandom transduction of recombinant adeno-associated virus vectors in mouse hepatocytes in vivo: Cell cycling does not influence hepatocyte transduction. *J Virol* 74:3793–3803, 2000.

204. F Mingozzi, J Schuttrumpf, VR Arruda, Y liu, Y-L Liu, KA High, W Xiao, and RW Herzog. Improved hepatic gene transfer by using an adeno-associated virus serotype 5 vector. *J Virol* 76:10497–10502, 2002.

205. D Grimm, KL Streetz, CJ Loping, TA Storm, K Pandey, CR Davis, P Marion, F Salazar, and MA Kay. Fatality in mice due to oversaturation of cellular microRNA/short hairpin RNA pathways. *Nature* (London) 441:537–541, 2006.

206. B Davidson, C Stein, J Heth, L Martins, R Kotin, T Derksen, J Zabner, A Rhodes, and J Chiorini. Recombinant adeno-associated virus type 2, 4 and 5 vectors: Transduction of variant cell types and regions in mammalian central nervous system. *Proc Natl Acad Sci U S A* 97:3428–3432, 2000.

207. A Aurricchio, A Kobinger, G Anand, M Hildinger, E O'Connor, AM McGuire, JM Wilson, and J Bennett. Exchange of surface proteins impacts on viral vector cellular specificity and transduction characteristics: The retina as a model. *Hum Mol Gen* 10:3075–3081, 2001.

208. GS Yang, M Schmidt, Z Yan, JD Lindbloom, TC Harding, BA Donohue, JF Engelhardt, R Kotin, and BL Davidson. Virus-mediated transduction of murine retina with adeno-associated virus: Effects of viral capsid and genome size. *J Virol* 76:7651–7660, 2002.

209. P Gregorevic, MJ Blankinship, JM Allen, RW Crawford, L Meuse, DG Miller, DW Russell, and JS Chamberlain. Systemic delivery of genes to striated muscles using adeno-associated viral vectors. *Nat Med* 10:828–834, 2004.

210. P Gregorevic, JM Allen, E Minami, MJ Blankinship, M Haraguchi, L Meuse, E Finn, ME Adams, SC Froehner, CE Murry, and JS Chamberlain. rAAV-microdystrophin preserves muscle function and extends lifespan in severely dystrophic mice. *Nat Med* 12:787–789, 2006.

211. D Townsend, MJ Blankinship, JM Allen, P Gregorevic, JS Chamberlain, and JM Metzger. Systemic administration of micro-dystrophin restores cardiac geometry and prevents dobutamine-induced cardia pump failure. *Mol Ther* 15:1086–1092, 2007.

212. Z Wang, T Zhu, C Qiao, L Zhou, B Wang, J Zhang, C Chen, J Li, and X Xiao. Adeno-associated virus serotype 8 efficiently delivers genes to muscle and heart. *Nat Biotechnol* 23:321–328, 2005.

213. LR Rodino-Klapac, PML Jansses, CL Montgomery, BD Coley, LG Chicoine, KR Clark, and JR Mendell. A translational approach for limb vascular delivery of the micro-dystrophin gene without high volume or high pressure for treatment of Duchenne muscular dystrophy. *J Translational Med* 5:45, 2007.

214. D Duan, Y Yue, Z Yan, J Yang, and JF Engelhardt. Endosomal processing limits gene transfer to polarized airway epithelia by adeno-associated virus. *J Clin Inv* 105:1573–1587, 2000.

215. D Duan, Q Li, AW Kao, Y Yue, JE Pessin, and JF Engelhardt. Dynamin is required for recombinant adeno-associated virus type 2 infection. *J Virol* 73:10371–10376, 1999.

216. JS Bartlett, R Wilcher, and RJ Samulski. Infectious entry pathway of adeno-associated and adeno-associated virus vectors. *J Virol* 74:2777–2785, 2000.

217. AM Douar, K Poulard, D Stockholm, and O Danos. Intracellular trafficking of adeno-associated virus vectors: Routing to the late endosomal compartment and proteosome degradation. *J Virol* 75:1824–1833, 2000.

218. S Sanlioglu, PK Benson, J Yang, EM Atkinson, T Reynolds, and JF Engelhardt. Endocytosis and nuclear trafficking of adeno-associated virus type 2 are controlled by Rac1 and phosphatidylinositol-3 kinase activation. *Virology* 74:9184–9196, 2000.

219. F Sonntag, S Bleker, B Leuchs, R Fischer, and JA Kleinschmidt. Adeno-associated virus type 2 capsids with externalized VP1/VP2 trafficking domains are generated prior to passage through the cytoplasm and are maintained until uncoating occurs in the nucleus. *J Virol* 80:10040–10054, 2006.

220. G Di Pasquale and JA Chiorini. AAV transcytosis through barrier epithelia and endothelium. *Mol Ther* 13:506–516, 2006.

221. J Hansen, K Qing, HJ Kwon, C Mah, and A Srivastava. Impaired intracellular trafficking of adeno-associated virus type 2 vectors limits efficient transduction of murine fibroblasts. *J Virol* 74: 992–996, 2000.

222. W Xiao, KH Warrington, P Hearing, J Hughes, and N Muzyczka. Adenovirus-facilitated nuclear translocation of adeno-associated virus type 2. *J Virol* 76:11505–11517, 2002.

223. J Hansen, K Qing, and A. Srivastava. Infection of purified nuclei by adeno-associated virus 2. *Mol Ther* 4:289–296, 2001.

224. CE Thomas, TA Storm, Z Huang, and MA Kay. Rapid uncoating of vector genomes is the key to efficient liver transduction with pseudotyped adeno-associated virus vectors. *J Virol* 78:3110–3122, 2004.

225. I Sipo, H Felchner, S Pinkert, L Suckau, X Wang, S Weger, and W Poller. Differential internalization and nuclear uncoating of self-complementary adeno-associated virus pseudotype vectors as determinants of cardiac cell transduction. *Gene Ther* 14:1319–1329, 2007.

226. B Akache, D Grimm, X Shen, S Fuess, SR Yant, DS Glazer, and MA Kay. A two-hybrid screen identifies cathepsins B and L as uncoating factors for adeno-associated virus 2 and 8. *Mol Ther* 15:330–339, 2007.

227. FK Ferrari, T Samulski, T Shenk, and RJ Samulski. Second-strand synthesis is a rate limiting step for efficient transduction by recombinant adeno-associated virus vectors. *J Virol* 70:3227–3234, 1996.

228. KJ Fisher, GP Gao, MD Weitzman, R DeMatteo, JF Burda, and JM Wilson. Transduction with recombinant adeno-associated virus vectors for gene therapy is limited by leading strand synthesis. *J Virol* 70:520–532, 1996.

229. DW Russell, IE Alexander, and AD Miller. DNA synthesis and topoisomerase inhibitors increase transduction by adeno-associated virus vectors. *J Virol* 92:5719–5723, 1996.

230. S Sanioglu, D Duan, and JF Engelhardt. Two independent molecular pathways for recombinant adeno-associated virus genomes conversion occur after UV-C and E4orf6 augmentation of transduction. *Hum Gene Ther* 10:591–602, 1999.

231. K Qing, J Hansen, KA Weigel-Kelley, M Tan, S Zhou, and A Srivastava. Adeno-associated virus type 2-mediated gene transfer: Role of cellular FKBP52 protein in transgene expression. *J Virol* 75:8968–8976, 2001.

232. K Qing, W Li, L Zhong, M Tan, J Hansen, KA Weigel-Kelley, L Chen, MC Yoder, and A Srivastava. Adeno-associated virus 2-mediated gene transfer: Role of cellular T cell protein tyrosine phosphatase in transgene expression and established cell line in vitro and transgenic mice in vivo. *J Virol* 77:2741–2746, 2003.

233. L Zhong, W Li, Z Yang, Y Li, K Qing, KA Weigel-Kelley, MC Yoder, W Shou, and A Srivastava. Improved transduction

of primary murine hepatocytes by recombinant adeno-associated virus 2 vectors in vivo. *Gene Ther* 11:1165–1169, 2004.

234. L Zhong, W Zhao, J Wu, B Li, S Zolotukhin, L Govindasamy, M Agbandje-McKenna, and A Srivastava. A dual role of EGFR protein tyrosine kinase signaling in ubiquitination of AAV2 capsids and viral second-strand DNA synthesis. *Mol Ther* 15:1323–1330, 2007.

235. H Nakai, TA Storm, and MA Kay. Recruitment of single-stranded recombinant adeno-associated vector genomes and intermolecular recombination are responsible for stable transduction of liver in vivo. *J Virol* 74:9451–9463, 2000.

236. RK Hirata and DW Russell. Design and packaging of adeno-associated virus gene targeting vectors. *J Virol* 74:4612–4620, 2000.

237. DM McCarty, PE Monahan, and RJ Samulski. Self-complementary recombinant adeno-associated virus (scAAV) vectors promote efficient transduction independently of DNA synthesis. *Gene Ther* 8:1248–1254, 2001.

238. DM McCarty, H Fu, and J Samulski. Self-complementary rAAV (scAAV) vectors overcome barriers to efficient transduction in mouse liver cells. *Mol Ther* 5:S3, 2002.

239. Z Whang, H Ma, L Sun, J Zhiang, J Li, and X Xiao. Double stranded AAV offers far more than doubled gene transfer. *Mol Ther* 5:S185, 2002.

240. H Fu, J Muenzer, RJ Samulski, G Breese, J Sifford, X Zeng, and DM McCarty. Self-complementary adeno-associated virus scrotype 2 vector: Global distribution and broad dispersion of AAV-mediated transgene expression in mouse brain. *Mol Ther* 8:911–917, 2003.

241. D Duan, Z Yan, Y Yue, and JF Engelhardt. Structural analysis of adeno-associated virus transduction circular intermediates. *Virology* 261:8–14, 1999.

242. J Yang, W Zhou, Y Zhang, T Zidon, T Ritchie, and JE Engelhardt. Concatemerization of adeno-associated virus circular genomes occurs through intermolecular recombination. *J Virol* 73:9468–9477, 1999.

243. S Song, P Laipis, KI Berns, and T Flotte. Effect of DNA-dependent protein kinase on the molecular fate of the rAAV2 genome in skeletal muscle. *Proc Natl Acad Sci U S A* 98:4084–4088, 2000.

244. H Nakai, TA Storm, and MA Kay. Increasing the size of rAAV-mediated expression cassettes in vivo by intermolecular joining of two complementary vectors. *Nat Biotechnol* 18:527–532, 2000.

245. Z-Y Chen, SR Yant, C-Y He, L Meuse, S Shen, and MA Kay. Linear DNAs concatemerize in vivo and result in sustained transgene expression in mouse liver. *Mol Ther* 3:403–409, 2001.

246. J Wang, J Xie, H Lu, L Chen, B Hauck, RJ Samulski, and W Xiao. Existence of transient functional double-stranded DNA intermediates during recombinant AAV transduction. *Proc Natl Acad Sci U S A* 104:13104–13109, 2007.

247. S Sanlioglu, P Benson, and JE Engelhardt. Loss of ATM function enhances recombinant adeno-associated virus transduction and integration through pathways similar to UV irradiation. *Virology* 268:68–78, 2000.

248. L Zentilin, A Marcello, and M Giacca. Involvement of cellular double-stranded DNA break binding proteins in processing of the recombinant adeno-associated virus genome. *J Virol* 75:12279–12287, 2001.

249. D Duan, Y Yue, Z Yan, and JE Engelhardt. A new dual vector approach to enhance recombinant adeno-associated virus-mediated gene expression through intermolecular *cis* activation. *Nat Med* 6:595–598, 2000.

250. Z Yan, Y Zhang, D Duan, and JE Engelhardt. Trans-splicing vectors expand the utility of adeno-associated virus for gene therapy. *Proc Natl Acad Sci U S A* 97:6716–6721, 2000.

251. L Sun, J Li, and X Xiao. Overcoming adeno-associated virus size limitation through viral DNA heterodimerization. *Nat Med* 6:599–602, 2000.

252. Z Yan, DCM Lei-Butters, Y Zhang, R Zak, and JF Engelhardt. Hybrid adeno-associated virus bearing nonhomologous inverted terminal repeats enhances dual-vector reconstruction of minigenes in vivo. *Hum Gene Ther* 18:81–87, 2007.

253. A Ghosh, Y Yue, C Long, B Bostick, and D Duan. Efficient whole-body transduction with *trans*-splicing adeno-associated viral vectors. *Mol Ther* 15:750–755, 2007.

254. CL Halbert, JM Allen, and AD Miller. Efficient mouse airway transduction following recombination between AAV vectors carrying parts of a larger gene. *Nat Biotechnol* 20:697–701, 2002.

255. D Duan, Y Yue, and JF Engelhardt. Expanding AAV packaging capacity with *trans*-splicing or overlapping vectors: A quantitative comparison. *Mol Ther* 4:383–391, 2001.

256. JS Bartlett, J Kleinschmidt, RC Boucher, and RJ Samulski. Targeted adeno-associated virus vector transduction of non-permissive cells mediated by a bi-specific F($\alpha\beta$'Y$_2$) antibody. *Nat Biotechnol* 17:181–185, 1999.

257. S Ponnazhagen, G Mahendra, S Kumar, JA Thompson, and M Castillas. Conjugate-based targeting of recombinant adeno-associated type 2 vectors by using avidin-linked ligands. *J Virol* 76:12900–12907, 2002.

258. A Girod, M Ried, M Wobus, H Lahm, K Leike, J Kleinschmidt, G Delage, and M Hallek. Genetic capsid modifications allow efficient re-targeting of adeno-associated virus type 2. *Nat Med* 5:1052–1056, 1999.

259. M Moskalenko, L Chen, M van Roey, BA Donahue, RO Snyder, JG McArthur, and SD Patel. Epitope mapping of human adeno-associated virus type 2 neutralizing anti-antibodies: Implications for gene therapy and structure. *J Virol* 74:1761–1766, 2000.

260. P Wu, W Xiao, T Conlon, J Hughes, M Agbandje-McKenna, T Ferkol, T Flotte, and N Muzyczka. Analysis of the adeno-associated type 2 (AAV2) capsid gene and construction of AAV vectors with altered tropism. *J Virol* 74:8635–8647, 2000.

261. JE Rabinowitz, W Xiao, and RJ Samulski. Insertional mutagenesis of AAV2 capsid and the production of recombinant virus. *Virology* 265:274–285, 1999.

262. MU Reid, A Gorod, K Leike, H Buning, and M Hallek. Adeno-associated virus capsids displaying immunoglobulin-binding domains permit antibody-mediated vector retargeting to specific cell surface receptors. *J Virol* 76:4559–4566, 2002.

263. M Grifman, M Trepel, P Speece, L Baetriz-Gilbert, W Arap, R Pasquilini, and MD Weitzman. Incorporation of tumor targeting peptides into recombinant adeno-associated virus capsids. *Mol Ther* 3:964–970, 2001.

264. W Shi, GS Arnold, and JS Bartlett. Insertional mutagenesis of the adeno-associated virus type 2 (AAV2) capsid gene and generation of AAV 2 vectors targeted to alternative cell-surface receptors. *Hum Gene Ther* 12:1697–1711, 2001.

265. L Perabo, J Endell, S King, K Lux, D Goldnau, M Hallek, and H Buning. Combinatorial engineering of a gene therapy vector: Directed evolution of adeno-associated virus. *J Gene Med* 8:155–162, 2006.

266. DA Watercamp, OJ Muller, Y Ying, M Trepel, and JA Kleinschmidt. Isolation of targeted AAV2 vectors from novel virus display libraries. *J Gene Med* 8:1307–1318, 2006.

267. N Maheshri, JT Hoerber, BK Kaspar, and DV Schaffer. Directed evolution of adeno-associated virus yields enhanced gene delivery vectors. *Nat Biotechnol* 24:198–204, 2006.

268. B Hauck, L Chen, and W Xiao. Generation and characterization of chimeric AAV vectors. *Mol Ther* 7:419–425, 2003.

269. JE Rabinowitz, DE Bowles, SM Faust, JG Ledford, SE Cunningham, and RJ Samulski. Cross-dressing the virion: The transcapsidation of adeno-associated virus serotypes functionally defines subgroups. *J Virol* 78:4421–4432, 2004.

270. B Hauk and W Xiao. Characterization of tissue tropism determinants of adeno-associated virus type 1. *J Virol* 77:2768–2774, 2003.

271. X Shen, T Storm, and MA Kay. Characterization of the relationship of AAV capsid domain swapping to liver transduction efficiency. *Mol Ther* 15:1955–1962, 2007.

272. KH Warrington, OS Gorbatyuk, JK Harrison, SR Opie, S Zolotukhin, and N Muzyczka. Adeno-associated virus type 2 VP2 capsid protein is nonessential and can tolerate large peptide insertions at its N terminus. *J Virol* 78:6595–6609, 2004.

273. CK Conrad, SS Allen, SA Afione, TC Reynolds, SE Beck, M Fee-Maki, X Barrazza-Ortiz, R Adams, FB Askin, BJ Carter, WB Guggino, and TR Flotte. Safety of single-dose administration of an adeno-associated virus (AAV-CFTR) vector in the primate lung. *Gene Ther* 3:658–668, 1996.

274. GA Croteau, DB Martin, J camp, M Yost, C Conrad, PL Zeitlin, and AE Heald. Evaluation of exposure and health care worker response to nebulized administration of tgAAVCF to patients with cystic fibrosis. *Ann Occup Hyg* 48:673–681, 2004.

275. MA Kay and KA High. US department of health and human services national institutes of health recombinant DNA advisory committee. Minutes of meeting. *Hum Gene Ther* 13:1663–1673, 2002.

276. VR Arruda, PA Fields, R Milner, L Wainwright, MP de Miguel, PJ Donovan, RW Herzog, TC Nicholls, JA Biegel, M Razavi, M Dake, D huff, AW Flake, L Couto, MA Kay, and KA High. Lack of germline transmission of vector sequences following systemic administration of recombinant AAV2 vector in males. *Mol Ther* 4:586–596, 2001.

277. J Schuettrumpt, JH Li, LB Couto, K Addya, DGB Leonard, Z Zhen, J Summer, and VR Arruda. Inadvertent germline transmission of AAV2 vector: Findings in a rabbit model correlate with those in a human clinical trial. *Mol Ther* 13:1064–1073, 2006.

278. D Favre, N Provost, V Blouin, G Blancho, Y Cherel, A Salvetti, and P Moullier. Immediate and long-term safety of recombinant adeno-associated virus injection into nonhuman primate muscle. *Mol Ther* 4:559–568, 2001.

279. S Song, M Scott-Jorgensen, J Wang, A Poirer, J Crawford, M Campbell-Thompson, and TR Flotte. Intramuscular administration of recombinant adeno-associated virus 2 α-1 antitrypsin (rAAV-SERPINA1) vectors in a non-human primate model: Safety and immunologic aspects. *Mol Ther* 6:329–336, 2002.

280. M Bartoli, C Roudaut, S Martin, F Fougerousse, L Suel, J Poupiot, E Gicquel, F Noulet, O Danos, and I Richard. Safety and efficacy of AAV-mediated calpain 3 gene transfer in a mouse model of limb-girdle muscular dystrophy. *Mol Ther* 13:250–259, 2005.

281. G Le Meur, K Stieger, AJ Smith, M Weber, JY Deschamps, D Nivard, A Mendes-Madeira, N Provost, Y Pereon, Y Cherel, RR Ali, C Hamel, P Moullier, and F Rolling. Restoration of vision in RPE65-deficient Briard dogs using an AAV serotype 4 vector that specifically targets the retinal pigmented epithelium. *Gene Ther* 14:292–303, 2007.

282. GM Acland, GD Aguirre, J Bennett, TS Aleman, AV Cideciyan, J Bennicelli, NS Dejneka, SE Pearce-Kelling, AM Maguire, K Palczewski, WW Hauwirth, and SG Jacobson. Long-term restoration of rod and cone vision by single dose rAAV-mediated gene transfer to the retina in a canine model of childhood blindness. *Mol Ther* 12:1072–1081. 2005.

283. SG Jacobson, GM ACland, GD Aguirre, TS Aleman, SB Schwarz, AV Cideciyan, CJ Zeiss, AM Komaromy, S Kaushal, AJ Roman, EAM Windsor, A Sumuaroka, SE Pierce-kelling, TJ Conlon, VA Chiodo, AL Boye, TR Flotte, AM Maguire, J Bennett, and WH Hauswirth. Safety of recombinant adeno-associated virus type 2-RPE65 vector delivered by ocular subretinal injection. *Mol Ther* 13:1074–1084, 2006.

284. SG Jacobson, SL Boye, TS Aleman, TJ Conlon, CJ Zeiss, AJ Roman, AV Cideciyan, SB Schwarz, AM Kamoromy, M Doobrajh, AY Cheung, A Sumaroka, SE Pierce-Kelling, GD Aguirre, A Kaushal, AE Maguire, TR Flotte, and WH Hauswirth. Safety in non-human primates of ocular AAV2-RPE65 a candidate treatment form blindness in Leber congenital amaurosis. *Hum Gene Ther* 17:1–14, 2006.

285. K Nyberg, BJ Carter, T Chen, TR Flotte, S Rose, D Rosenblum, SL Simek, and C Wilson. Long-term follow-up of participants in human gene transfer research. *Mol Ther* 10:976–980, 2004.

286. LM dela Maza and BJ Carter. Inhibition of adenovirus oncogenicity in hamsters by adeno-associated virus DNA. *J Natl Cancer Inst* 67:1323–1326, 1981.

287. K Raj, P Ogston, and P Beard. Virus-mediated killing of cells that lack p53 activity. *Nature* (London) 412:914–917, 2001.

288. TM Daly, KK Ohlemiller, MS Roberts, CA Vogler, and MS Sands. Prevention of systemic clinical disease in MPS VII mice following AAV-mediated neonatal gene transfer. *Gene Ther* 8:1291–1298, 2001.

289. A Donsante, C Vogler, N Muzyczka, JM Crawford, J Barker, T Flotte, M Campbell-Thompson, T Daly, and MS Sands. Observed incidence of tumorigenesis in long-term rodent studies of rAAV vectors. *Gene Ther* 8:1343–1346, 2001.

290. A Donsante, DG Miller, Y Li, C Vogler, EM Brunt, DW Russell, and MS Sands. AAV vector integration sites in mouse hepatocellular carcinoma. *Science* 317:477, 2007.

291. S Song, J Embury, PJ Laipis, KI Berns, JM Crawford, and TR Flotte. Stable therapeutic serum levels of alpha-1 antitrypsin (AAT) after portal vein injection of recombinant adeno-associated virus (AAV) vectors. *Gene Ther* 8:1299–1306, 2001.

292. P Bell, L Wang, C Lebherz, DB Fleider, MS Bove, D Wu, GP Gao, JM Wilson, and NA Wivel. No evidence for tumorigenesis of AAV vectors in a large-scale study in mice. *Mol Ther* 12:299–306, 2005.

293. D Moscioni, H Morizono, RJ McCarter, A Stern, L Cabrera-Luque, A Hoang, J Sanmiguel, D Wu, P Pell, GP Gao, SE Raper, JM Wilson, and ML Batshaw. Long-term correction of ammonia metabolism and prolonged survival in orthine transcarbamylase-deficient mice following liver-directed treatment with adeno-associated viral vectors. *Mol Ther* 14:25–44, 2006.

294. JL Stillwell and RJ Samulski. Role of viral vectors and virion shells in cellular gene expression. *Mol Ther* 9:337–346, 2004.

295. N Chirmule, W Xiao, A Truneh, M Schnell, JV Hughes, P Zoltich, and LM Wilson. Humoral immunity to adeno-associated virus type 2 vectors following administration to murine and non-human primate muscle. *J Virol* 74:2420–2425, 2000.

296. CL Halbert, TA Standaert, CB Wilson, and AD Miller. Successful readministration of adeno-associated virus vectors to

the mouse lung requires transient immunosuppression during initial exposure. *J Virol* 72:9795–9805, 1998.

297. W Xiao, N Chirmule, MA Schnell, J Tazelaar, JV Hughes, and JM Wilson. Route of administration determines induction of T-cell independent humoral responses to adeno-associated virus vectors. *Mol Ther* 1:323–329, 2000.

298. G Brockstedt, GM Podsakoff, L Fong, G Kurtzman, W Mueller-Ruchholtz, and E Engelmann. Induction of immunity to antigens expressed by recombinant adeno-associated virus depends on the route of administration. *Clin Immunol* 2:67–75, 1999.

299. CL Halbert, TA Standaert, ML Aitken, IE Alexander, DW Russell, and AD Miller. Transduction by adeno-associated virus vectors in the rabbit airway: Efficiency, persistence and readministration. *J Virol* 71:5932–5941, 1997.

300. WC Manning, S Zhou, MP Bland, JA Escobedo, and V Dwarki. Transient immunosuppression allows transgene expression following readministration of adeno-associated virus vectors. *Hum Gene Ther* 9:477–485, 1998.

301. Y Zhang, N Chirmule, G-P Gao, and JM Wilson. CD40-ligand dependent activation of cytotoxic T lymphocytes by adeno-associated virus vectors in vivo: Role of immature dendritic cells. *J Virol* 74:8003–8010, 2000.

302. K Jooss, Y Yang, KJ Fisher, and JM Wilson. Transduction of dendritic cells by DNA viral vectors directs the immune response to transgene products in muscle fibers. *J Virol* 72:4212–4223, 1998.

303. WC Manning, X Paliard, S Zhou, MP Bland, Ay Lee, K Hong, CM Walker, JA Escobedo, and V Dwarki. Genetic immunization with adeno-associated virus vectors expressing herpes simplex type 2 glycoproteins B and D. *J Virol* 71:7960–7962, 1997.

304. PA Fields, DW Kowalczyk, VR Arruda, E Armstrong, ML McClelland, JW Hogstrom, EJ Pasi, HCJ Erth, RW Herzog, and KA High. Role of vector in activation of T cell subsets in immune responses against the secreted transgene product Factor IX. *Mol Ther* 1:225–231, 2000.

305. RW Herzog, PA Fields, VR Arruda, JO Brubaker, E Armstrong, D McClintock, DA Bellinger, LB Couto, TC Nichols, and KA High. Influence of vector dose on Factor IX-specific T and B cell responses in muscle-directed gene therapy. *Hum Gene Ther* 13:1281–1291, 2002.

306. CS Manno, VR Arruda, GF Pierce, B Glader, M Ragni, JJE Rasko, MC Ozelo, K Hoots, P Blatt, B Konkle, M Dake, R Kaye, M Razavi, A Zaiko, J Zehnder, PK Rustagi, H Nakai, A Chew, D Leonard, JF Wright, RR Lessard, JM Sommer, M Tigges, D Sabatino, A Luk, H Jiang, F Mingozzi, L Couto, HC Ertl, KA High, and MA Kay. Successful transduction of liver in hemophilia by AAV-factor IX and limitations imposed by the host immune response. *Nat Med* 12:342–347, 2006.

307. H Li, SL Murphy, W Giles-Davis, S Edmonson, Z Xiang, Y Li, MO Lasaro, KA High, and HCJ Ertl. Pre-existing AAV capsid-specific CD8 + T cells are unable to eliminate AAV-transduced hepatocytes. *Mol Ther* 15:792–800, 2007.

308. L Wang, R Calcedo, TC Nichols, DA Bellinger, A Dillow, IM Verma, and JM Wilson. Sustained correction of disease in naïve and AAV2-pretreated hemophilia B dogs: AAV2/8-mediated, liver-directed gene therapy. *Blood* 105:3079–3086, 2005.

309. C Li, M Hirsch, A Asokan, B Zeithami, H Ma, T Kafri, and RJ Samulski. Adeno-associated virus type 2 (AAV2) capsid-specific cytotoxic T lymphocytes eliminate only vector-transduced cells coexpressing the AAV2 capsid protein. *J Virol* 81:7540–7547, 2007.

310. H Jiang, LB Couto, S Pattarroy-White, T Liu, D Nagy, JA Vargas, S Zhou, CD Scallan, J Sommer, S Vijay, F Mingozzi, KA High, and GF Pierce. Effects of transient immunosuppression on adenoassociated, virus-mediated, liver-directed gene transfer in rhesus macaques and implications for human gene therapy. *Blood* 108:3321–3328, 2006.

311. TR Flotte, F Schweibert, PL Zeitlin, BJ Carter, and WB Guggino. Correlation between DNA transfer and cystic fibrosis airway epithelial cell correction after recombinant adeno-associated virus serotype 2 gene therapy. *Hum Gene Ther* 16:921–928, 2005.

312. TR Flotte, P Ng, DE Dylla, PB McCray, G Wang, JK Kolls, and J Hu. Viral vector-mediated and cell-based therapies for treatment of cystic fibrosis. *Mol Ther* 15:229–241, 2007.

313. CL Halbert, SH Lam, and AD Miller. High-efficiency promoter-dependent transduction by adeno-associated virus type 6 vectors in mouse lung. *Hum Gene Ther* 18:344–354, 2007.

314. AC Fischer, CI Smith, L Cebotaru, X Zhang, FB Askin, J Wright, SE Guggino, RJ Adams, T Flotte, and WB Guggino. Expression of a truncated cystic fibrosis transmembrane regulator with an AAV5-pseudotyped vector in primates. *Mol Ther* 15:756–763, 2007.

315. RO Snyder, C Miao, L Meuse, J Tubb, BA Donahue, F Lin, DW Stafford, S Patel, AR Thompson, T Nichols, MS Read, DA Bellinger, KM Brinkhous, and MA Kay. Correction of hemophilia B in canine and murine models using recombinant adeno-associated vectors. *Nat Med* 5:64–69, 1999.

316. RW Herzog, EY Yang, LB Couto, JN Hagstrom, D Elwell, PA Fields, M Burton, BNellinger DA, MS Read, KM Brinkhous, GM Podsakoff, TC Nichols, GJ Kurtzman, and KA High. Long-term correction of canine hemophilia by gene transfer of blood coagulation factor IX mediated by adeno-associated viral vector. *Nat Med* 5:56–63, 1999.

317. MA Kay, CS Manno, MV Ragni, PJ Larson, LB Couto, A McClelland, B Glader, AJ Chew, SJ Tai, RW Herzog, VR Arruda, F Johnson, C Scallan, E Skarsgard, AW Flake, and KA High. Evidence for gene transfer and expression of factor IX in hemophilia B patients treated with an AAV vector. *Nat Med* 24:201–202, 2000.

318. CS Manno, AJ Chew, S Hutchinson, PJ Larson, RW Herzog, VR Arruda, SJ Tai, MV Ragni, A Thompson, M Ozelo, LB Couto, DGB Leonard, FA Johnson, A McClelland, C Scallan, E Skarsgard, AW Flake, MA Kay, KA High, and B Glader. AAV-mediated factor IX gene transfer to skeletal muscle in patients with severe hemophilia B. *Blood* 101:2963–2972, 2003.

319. VR Arruda, HH Stedman, TC Nichols, ME Haskins, M Nicholson, RW Herzog, LB Couto, and KA High. Regional intravascular delivery of AAV-2-F.IX to skeletal muscle achieves long-term correction of hemophilia B in a large animal model. *Blood* 105:3458–3646, 2005.

320. YL Liu, F Mingozzi, SM Rodriguez-Colon, S Joseph, E Dobrzynski, T Suzuki, KA High, and RW Herzog. Therapeutic levels of factor IX expression using a muscle-specific promoter and adeno-associated virus serotype 1 vector. *Hum Gene Ther* 15:783–792, 2004.

321. VR Arruda, J Schuettrumpf, RW Herzog, TC Nichols, N Robinson, Y Lofti, F Mingozzi, W Xiao, LB Couto, and KA High. Safety and efficacy of factor IX gene transfer to skeletal muscle in murine and canine hemophilia B models by adeno-associated viral vector serotype 1. *Blood* 103:85–92, 2004.

322. L Wang, TC Nichols, MS Read, DA Bellinger, and IM Verma. Sustained expression of therapeutic level of factor IX in hemophilia B dogs by AAV-mediated gene therapy in liver. *Mol Ther* 1:154–158, 2000.

323. JD Mount, RW Herzog, DM Tilson, SA Goodman, N Robinson, ML McCleland, D Bellinger, TC Nichols, VR Arruda, CD Jr. Lothrop, and KA High. Sustained phenotypic correction of hemophilia B dogs with a factor IX null mutation by liver-directed gene therapy. *Blood* 99:2670–2676, 2002.

324. AE Emery. The muscular dystrophies. *Lancet* 359:687–695, 2002.

325. O Ibraghimov-Beskrovnaya, JM Ervasti, CJ Leveille, CA Slaughter, SW Sernett, and KP Campbell. Primary structure of dystrophin-associated glycoproteins linking dystrophin to the extracellular matrix. *Nature* (London) 355:696–702, 1999.

326. SB England, LV Nicholson, MA Johnson, S Forrest, DR Love, EE Zubrzycka-Gaarn, DE Bulman, JB Harris, and KE Davies. Very mild muscular dystrophy associated with the deletion of 46% of dystrophin. *Nature* (London) 343:180–182, 1990.

327. P Sicinski, Y Geng, AS Ryder-Cook, EA Barnard, MG Darliso, and PJ Barnad. The molecular basis of muscular dystrophy in the mdx mouse: A point mutation. *Science* 244:1578–1580, 1989.

328. SF Phelps, MA Hauser, NM Cole, JA Rafael, RT Hinkle, JA Faulkner, and JS Chamberlain. Expression of full-length and truncated dystrophin mini-genes in transgenic mdx mice. *Hum Mol Genet* 4:1251–1258, 1995.

329. B Wang, J Li, and X Xiao. Adeno-associated virus vector carrying human minidystrophin genes effectively ameliorates muscular dystrophy in *mdx* mouse model. *Proc Natl Acad Sci U S A* 7:13714–13719, 2000.

330. SA Fabb, DJ Wells, P Serpente, and G Dickson. Adeno-associated virus vector gene transfer and sarcolemmal expression of a 144 kDa micro-dystrophin effectively restores the dystrophin-associated protein complex and inhibits myofiber degeneration in nude/mdx mice. *Hum Mol Genet* 11:733–741, 2002.

331. SQ Harper, MA Hauser, C DelloRusso, D Duan, RW Crawford, SF Phelps, HA Harper, AS Robinson, JF Engelhardt, SV Brooks, and JS Chamberlain. Modular flexibility of dystrophin: Implications for gene therapy of Duchenne muscular dystrophy. *Nat Med* 8:253–261, 2002.

332. D Dressman, YP Tsao, A Sakamoto, EP Hoffman, and X Xiao. rAAV vector-mediated sarcoglycan gene transfer in a hamster model for limb girdle muscular dystrophy. *Gene Ther* 6:74–82, 1999.

333. M Yoshimura, M Sakamoto, M Ikemoto, Y Mochizuki, K Yussa, Y Miyagoe-Suzuki, and S Takeda. AAV vector-mediated microdystrophin expression in a relatively small percentage of mdx myofibers improved the mdx phenotype. *Mol Ther* 10:821828, 2004.

334. M Liu, Y Yue, SQ Harper, RW Grange, JS Chamberlain, and D Duan. Adeno-associated virus-mediated microdystrophin expression protects young mdx muscle from contraction-induced injury. *Mol Ther* 11:245–256, 2005.

335. LV Nicholson. The "rescue" of dystrophin synthesis in boys with Duchenne muscular dystrophy. *Neuromuscul Disord* 3:525–31, 1993.

336. QL Lu, GE Morris, SD Wilton, T Ly, OV Artemyeva, P Strong, and TA Partridge. Massive idiosyncratic exon skipping corrects the nonsense mutation in dystrophic mouse muscle and produces functional revertant fibers by clonal expansion. *J Cell Biol* 148:985–996, 2000.

337. A Goyenvalle, A Vulin, F Fougerousse, F Leturcq, JC Kaplan, L Garcia, and O Danos. Rescue of dystrophic muscle through U7 snRNA-mediated exon skipping. *Science* 306:1796–1799, 2004.

338. MA Denti, A Rosa, G D'Antona, O Sthandier, FG De Angelis, C Nicoletti, M Allocca, O Pansarasa, V Parente, A Musaro, A Auricchio, R Bottinelli, and I Bozzoni. Body-wide gene therapy of Duchenne muscular dystrophy in the *mdx* mouse model. *Proc Natl Acad Sci U S A* 103:3758–3763, 2006.

339. Y Lai, Y Yue, A Ghosh, JF Engelhardt, JS Chamberlain, and D Duan. Efficient in vivo gene expression by *trans*-splicing adeno-associated viral vectors. *Nat Biotechnol* 23:1435–1439, 2005.

340. Z Wang, JM Allen, SR Riddle, P Gregorevic, R Strob, SJ Tapscott, JS Chamberlain, and CS Kuhr. Immunity to adeno-associated virus-mediated gene transfer in a random-bred canine model of Duchenne muscular dystrophy. *Hum Gene Ther* 18:18–26, 2007.

341. Z Wang, CS Kuhr, JM Allen, M Blankinship, P Gregorevic, JS Chamberlain, SJ Tapscott, and R Storb. Sustained AAV-mediated dystrophin expression in a canine model of Duchenne muscular dystrophy with a brief course of immunosuppression. *Mol Ther* 15:1160–1166, 2007.

342. JP Greelish, LT Su, EB Lankford, JM Burkman, H Chen, SK Konig, IM Mercier, PR Desjardins, MA Mitchell, XG Zheng, L Leferovich, GP Gao, RJ Balice-Gordon, JM Wilson, and HL Stedman. Stable restoration of the sarcoglycan complex in dystrophic muscle perfused with histamine and a recombinant adeno-associated viral vector. *Nat Med* 5:439–443, 1999.

343. X Xiao, J Li, YP Tsao, D Dressman, EP Hoffman, and JF Watchko. Full functional rescue of a complete muscle (TA) in dystrophic hamsters by adeno-associated virus vector-directed gene therapy. *J Virol* 74:1436–1442, 2000.

344. L Cordier, AA Hack, MO Scott, ER Barton-Davis, G Gao, JM Wilson, EM McNally, and HL Sweeney. Rescue of skeletal muscles of γ-sarcoglycan-deficient mice with adeno-associated virus-mediated gene transfer. *Mol Ther* 1:119–129, 2000.

345. D Dressman, K Araishi, M Imamura, T Sasaoka, LA Liu, E Engvall, and EP Hoffman. Delivery of α- and β-sarcoglycan by recombinant adeno-associated virus: Efficient rescue of muscle, but differential toxicity. *Hum Gene Ther* 13:1631–1646, 2002.

346. F Fougerousse, M Bartoli, J Poupiot, L Arandel, M Durand, N Guerchet, E Gicquel, O Danos, and I Richard. Phenotypic correction of α-sarcoglycan deficiency by intra-arterial injection of a muscle-specific serotype 1 rAAV vector. *Mol Ther* 15:53–61, 2007.

347. H Stedman, J Mendell, JM Wilson, R Finke, and A-L Kleckner. Phase I clinical trial utilizing gene therapy for limb girdle muscular dystrophy with α-, β-, γ-, or δ-sarcoglycan gene delivered with intramuscular instillations of adeno-associated vectors. *Hum Gene Ther* 11:777–790, 2000.

348. R Sanchez-Pernaute, J Harvey-White, J Cunningham, and KS Bankiewicz. Functional effect of adeno-associated virus mediated gene transfer of aromatic L-amino acid decarboxylase into the striatum of 6-OHDA-lesioned rats. *Mol Ther* 4:324–330, 2001.

349. E Leff, SK Spratt, RO Snyder, and RJ Mandel. Long-term restoration of striatal L-aromatic amino acid decarboxylase activity using recombinant adeno-associated viral vector gene transfer in a rodent model of Parkinson's disease. *Neuroscience* 92:185–196, 1999.

350. DS Fan, M Ogawa, KI Fujimoto, K Ikeguchi, Y Ogasawara, M Urabe, M Nishizawa, I Nakano, M Yoshida, I Nagatsu, H Ichinose, T Nagatsu, GJ Kurtzman, and K Ozawa. Behavioral recovery in 6-hydroxydopamine-lesioned rats by cotransduction of striatum with tyrosine hydroxylase and aromatic L-amino acid decarboxylase genes using two separate adeno-associated virus vectors. *Hum Gene Ther* 9:2527–2535, 1998.

351. JR Forsayeth, JL Eberling, LM Sanfter, Z Zhen, P Pivirotto, J Bringas, J Cunningham, and KS Bankiewicz. A dose-ranging study of AAV-hAADC therapy in Parkinsonian monkeys. *Mol Ther* 14:571–578, 2006.

352. KS Bankiewicz, J Forsayeth, JL Eberling, R Sanchez-Pernaute, P Pivirotto, J Bringas, P Herscovitch, RE Carson, W Eckleman, B Reutter, and J Cunningham. Long-term clinical improvement in MTMP-lesioned primates after gene therapy with AAV-hAADC. *Mol Ther* 14:564–570.

353. M Fiandaca, J Forsayeth, and K Banckiewicz. Current status of gene therapy trials for Parkinson's disease. *Exp Neurol* 209:51–57, 2008.

354. C Hamani, JA Saint-Cyr, J Fraser, M Kaplitt, and AM Lozano. The subthalamic nucleus in the context of movement disorders. *Brain* 127:4–20, 2004.

355. ME Emborg, M Carbon, JE Holden, MJ During, Y Ma, C Tang, J Moirano, H Fitzsimons, BZ Roitberg, E Tuccar, A Roberts, MG Kaplitt, and D Eidelberg. Subthalamic glutamic acid decarboxylase gene therapy: Changes in motor function and cortical metabolism. *J Cereb Blood Flow Metab* 27:501–509, 2007.

356. MG Kaplitt, P Leone, RJ Samulski, X Xiao, DW Pfaff, KL O'Malley, and JM During. Long term gene expression and phenotypic correction using adeno-associated virus vectors in the mammalian brain. *Nat Genet* 8:148–154, 1994.

357. B Lee, H Lee, YR Nam, JH Oh, YH Cho, and JW Chang. Enhanced expression of glutamate decarboxylase 65 improves symptoms of rat parkinsonian models. *Gene Ther* 12:1215–1222, 2005.

358. J Luo, MG Kaplitt, HL Fitzsimons, DS Zuzga, Y Liy, ML Oshinsky, and MJ During. Subthalamic GAD gene therapy in a Parkinson's disease rat model. *Science* 298:425–429, 2002.

359. MG Kaplitt, A Feigin, C Tang, HL Fitzsimons, P Mattis, PA Lawlor, RJ Bland, D Young, K Strybing, D Eidelberg, and MJ During. Safety and tolerability of gene therapy with an adeno-associated virus (AAV) borne GAD gene for Parkinson's disease: An open label, phase I trial. *The Lancet* 369:2097–2105, 2007.

360. R dela Fuente-Fernandez, M Schulzer, and AJ Stoessl. Placebo mechanisms and reward circuity: Clues from Parkinson's disease. *Biol Psychiatry* 56:61–71, 2004.

361. C McRac, E Cherin, TG Yamazaki, G Diem, AH Vo, D Russell, JH Ellgring, S Fahn, P Greene, S Dillon, H Winfield, KB Bjugstad, and CR Freed. Effects of perceived treatment on quality of life and medical outcomes in a double-blind placebo surgery trial. *Arch Gen Psychiatry* 61:412–420, 2004.

362. G Hou, LF Lin, and C Mytilineou. Glial cell line-derived neurotrophic factor exerts neurotrophic effects on dopaminergic neurons in vitro and promotes their survival and regrowth after damage by 1-methyl-4-phenylpyridinium. *J Neurochem* 66:74–82, 1996.

363. M Kearns and DM Gash. GDNF protects nigral dopamine neurons against 6-hydroxydopamine in vivo. *Brain Res* 672:104–111, 1996.

364. RJ Mandel, SK Spratt, RO Snyder, and SE Leff. Midbrain injection of recombinant adeno-associated virus encoding rat glial cell line-derived neurotrophic factor protects nigral neurons in a progressive 6-hydroxydopamine-induced degeneration model of Parkinson's disease in rats. *Proc Natl Acad Sci U S A* 94:14083–14088, 1997.

365. D Kirik, C Rosenblad, A Bjorklund, and RJ Mandel. Long-term rAAV-mediated gene transfer of GDNF in the rat Parkinson's model: Intrastriatal but not intranigral transduction promotes functional regeneration in the lesioned nigrostriatal system. *J Neurosci* 20:4686–4700, 2000.

366. M Gasmi, CD Herzog, EP Brandon, JJ Cunningham, GA Ramirez, ET Ketchum, and RT Bartus. AAV2 mediated delivery of human neuturin to the rat nigrastriatal system: Long-term efficacy and tolerability of CERE-120 for Parkinson's disease. *Mol Ther* 15:62–68, 2007.

367. SWJ McPhee, CG Janson, RJ Samulski, AS Camp, J Francis, D Shera, L Lioutermann, M Feely, A Freese, and P Leone. Immune responses to AAV in a phase I study for Canavan disease. *J Gen Med* 8:577–588, 2006.

368. RG Crystal, D Sondhi, NR Hackett, SM Kaminsky, S Worgal, P Stieg, M Souweidane, S Hosain, L Heier, D Ballon, M Kaplitt, BM Greenwald, JD Howell, K Strybing, J Dyke, and H Voss. Administration of a replication-deficient adeno-associated virus gene transfer vector expressing the human CLN2 cDNA to the brain of children with late infantile neuronal ceroid lipofuscinosis: Clinical protocol. *Hum Gene Ther* 15:1131–1154, 2004.

369. DA Bennett. Minutes of NIH recombinant DNA Advisory Committee March, 2004. Available at URL www4.od.nih.gov/oba/rac/meeting.html.

370. L Dudus, V Anand, GM Acland, SJ Chen, JM Wilson, KJ Fisher, AM Maguire, and J Bennett. Persistent transgene product in retina, optic nerve and brain after intraocular injection of rAAV. *Vision Res* 39:2545–2553, 1999.

371. C Jomary, KA Vincent, J Grist, MJ Neal, and SE Jones. Rescue of photoreceptor function by AAV-mediated gene transfer in a mouse model of inherited retinal degeneration. *Gene Ther* 4:683–690, 1997.

372. J Guy, X Qi, and WW Hauswirth. Adeno-associated viral-mediated catalase expression suppresses optic neuritis in experimental allergic encephalomyelitis. *Proc Natl Acad Sci U S A* 95:13847–13852, 1998.

373. RR Ali, MB Reichel, M De Alwis, N Kanuga, C Kinnon, RJ Levinsky, DM Hunt, SS Bhattacharya, and AJ Thrasher. Adeno-associated virus gene transfer to mouse retina. *Hum Gene Ther* 9:81–86, 1998.

374. JK Phelan and D Bok. A brief review of retinitis pigmentosa and the identified retinitis pigmentosa genes. *Mol Vision* 6:116–124, 2000.

375. FQ Liang, NS Dejneka, DR Cohen, NV Krasnoperova, J Lem, AM Maguire, L Dudus, KJ Fisher, and J Bennett. AAV-mediated delivery of ciliary neurotrophic factor prolongs photoreceptor survival in the rhodopsin knockout mouse. *Mol Ther* 3:241–248, 2001.

376. D Lau, LH McGee, S Zhou, KG Rendahl, WC Manning, JA Escobedo, and JG Flannery. Retinal degeneration is slowed in transgenic rats by AAV-mediated delivery of FGF-2. *Invest Ophthalmol Vis Sci* 41:3622–3633, 2000.

377. ES Green, KG Rendahl, S Zhou, M Ladner, M Coyne, R Srivastava, WC Manning, and JG Flanery. Two animal models of retinal degeneration are rescued by recombinant adeno-associated virus-mediated production of FGF-5 and FGF-18. *Mol Ther* 3:507–515, 2001.

378. LH McGee, H Abel, WW Hauswirth, and JG Flannery. Glial cell line derived neurotrophic factor delays photoreceptor degeneration in a transgenic rat model of retinitis pigmentosa. *Mol Ther* 4:622–629, 2001.

379. D Bok, D Yasumura, MT Matthes, A Ruiz, JL Duncan, AV Chappelow, S Zolutukhin, W Hauswirth, and MM LaVail. Effects of adeno-associated virus-vectored ciliary neurotrophic factor on retinal structure and function in mice with a P216L rds/peripherin mutation. *Exp Eye Res* 74:719–735, 2002.

380. AS Lewin, KA Drenser, WW Hauswirth, S Nishikawa, D Yasumura, JG Flannery, and LaVailMM: Ribozyme rescue

of photoreceptor cells in a transgenic model of autosomal dominant retinitis pigmentosa. *Nat Med* 4:967–971, 1998.

381. Ali, GM Sarra, C Stephens, MD Alwis, JW Bainbridge, PM Munro, S Fauser, MB Reichel, C Kinnon, DM Hunt, SS Bhattacharya, and AJ Thrasher. Restoration of photoreceptor ultrastructure and function in retinal degeneration slow mice by gene therapy. *Nature Genet* 25:306–310, 2000.

382. BJ Raisler, KI Berns, MB Grant, D Beliaev, and WW Hauswirth. Adeno-associated virus type-2 expression of pigmented epithelium-derived factor or Kringles 1–3 of angiostatin reduce retinal neovascularization. *Proc Natl Acad Sci U S A* 99:8909–8914, 2002.

383. K Mori, P Gehlbach, S Yamamoto, E Duh, DJ Zack, Q Li, KI Berns, BJ Raisler, WW Hauswirth, and PA Campochiaro. AAV-mediated gene transfer of pigment epithelium-derived factor inhibits choroidal neovascularization. *Invest Ophthalmol Vis Sci* 43:1994–2000, 2002.

384. JW Bainbridge, A Mistry, M De Alwis, E Paleolog, A Baker, AJ Thrasher, and RR Ali. Inhibition of retinal neovascularisation by gene transfer of soluble VEGF receptor sFlt-1. *Gene Ther* 9:320–326, 2002.

385. YK Lai, WY Shen, M Brankov, CM Lai, IJ Constable, and PE Rakoczy. Potential long-term inhibition of ocular neovascularisation by recombinant adeno-associated virus-mediated secretion gene therapy. *Gene Ther* 9:804–813, 2002.

386. R Allikmets. Leber congenital amaurosis: A genetic paradigm. *Ophthalmic Genet* 25:67–79, 2004.

387. FP Cremers, JA van den Hurk, and AI den Hollander. Molecular genetics of Leber congenital amaurosis. *Hum Mol Genet* 11:1169–1176, 2002.

388. CP Hamel, JM Griffoin, L Lasquellec, C Bazalgette, B Arnaud, and B Retinal dystrophies caused by mutations in RPE65: Assessment of visual functions. *Br J Ophthalmol* 85:424–427. 2001.

389. S Yzer, LI van den Born, J Schuil, HY Kroes, MM van Genderen, FN Boonstra, HB van den, HG Brunner, RK Koenekoop, and FP Cremers. A Tyr368His RPE65 founder mutation is associated with variable expression and progression of early onset retinal dystrophy in 10 families of a genetically isolated population. *J Med Genet* 40:709–713, 2003.

390. K Narfstrom, A Wrigstad, B Ekesten, and SEG Nilsson. Hereditary retinal dystrophy in the briard dog: Clinical and hereditary characteristics. *Prog Vet Comp Ophthalmol* 4:85–92, 1994.

391. GM Acland, GD Aguirre, J Ray, Q Zhang, TS Aleman, AV Cideciyan, SE Pearce-Kelling, V Anand, Y Zeng, AM Maguire, SG Jacobson, WW Hauswirth, and J Bennett. Gene therapy restores vision in a canine model of childhood blindness. *Nat Genet* 28:92–95, 2001.

392. J Wilson. Humility in clinical trials. *Mol Ther* 15:1571–1572, 2007.

393. J Bennett. Commentary. An aye for eye gene therapy. *Hum Gene Ther* 17:177–179, 2006.

394. EJ Harris. Rheumatoid arthritis. Pathophysiology and implications for therapy. *New Engl J Med* 322:1277–1289, 1990.

395. M Brennan and M Feldman. Cytokines in autoimmunity. *Curr Opin Immunol* 4:754–759, 1992.

396. WP Arend and JM Dayer. Inhibition of the production and effects of interleukin-1 and tumor necrosis factor α in rheumatoid arthritis. *Arthritis Rheum* 38:151–160, 1995.

397. S Watanabe, T Imagawa, GP Boivin, G Gao, JM Wilson, and R Hirsch. Adeno-associated virus mediates long-term gene transfer and delivery of chondroprotective IL-4 to murine synovium. *Mol Ther* 2:147–152, 2002.

398. V Cottard, D Mulleman, P Bouille, M Mezzina, MC Boissier, and N Bessis. Adeno-associated virus-mediated delivery of IL-4 prevents collagen-induced arthritis. *Gene Ther* 7:1930–1939, 2000.

399. F Apparailly, V Millet, D Noel, C Jacquet, J Sany, and C Jorgensen. Tetracycline-inducible interleukin-10 gene transfer mediated by an adeno-associated virus: Application to experimental arthritis. *Hum Gene Ther* 13:1179–1188, 2002.

400. HG Zhang, J Xie, P Yang, Y Wang, L Xu, D Liu, HC Hsu, T Zhou, CK Edwards, and JD Mountz. Adeno-associated virus production of tumor necrosis factor receptor neutralizes tumor necrosis factor alpha and reduces arthritis. *Hum Gene Ther* 11:2431–2442, 2000.

401. RY Pan, SL Chen, X Xiao, DW Liu, HJ Peng, and YP Tsao. Therapy and prevention of arthritis by recombinant adeno-associated virus vector with delivery of interleukin-1 receptor antagonist. *Arthritis Rheum* 43:289–297, 2000.

402. JMK Chan, WW Jin, T Stepan, H Burstein, and SM Wahl. Suppression of SCW-induced arthritis with recombinant TNFR:Fc gene transfer. *Mol Ther* 6:1–10, 2002.

403. A Heald and G Pate, 13E04/13G01 Study teams, P Anklesaria. Clinical studies of intra-articular administration of a recombinant adeno-associated vector containing a TNF-α antagonist gene in inflammatory arthritis. *Mol Ther* 13:S419, 2006.

404. PJ Mease, AE Heald, and P Anklesaria, 13G01 study group. A phase I/II clinical study of intra-articular administration of a recombinant adeno-associated vector containing a TNF-α antagonist gene in inflammatory arthritis. *Mol Ther* 15:S401, 2007.

405. TR Flotte, TJ Conlon, A Poirier, M Campbell-Thompson, and BJ Byrne. Preclinical characterization of a recombinant adeno-associated virus type 1 pseudotyped vector demonstrates dose-dependent injection site inflammation and dissemination of vector genomes to distant sites. *Hum Gene Ther* 18:245–256, 2007.

406. ML Brantly, LT Spencer, M Humphries, TJ Conlon, CT Spencer, A Poirier, W Garlington, D Baker, S Song, KI Berns, N Muzyczka, RO Snyder, BJ Byrne, and TR Flotte. Phase I trial of intramuscular injection of a recombinant adeno-associated serotype 2 AAV vector in AAT-deficient adults. *Hum Gene Ther* 17:1–10, 2006.

407. TR Flotte, ML Brantly, LT Spencer, M Humphries, TJ Conlon, CT Spencer, A Poirer, W Garlington, D Baker, S Song, KI Brens, N Muzyczka, RO Snyder, S Washer, and BJ Byrne. Phase I clinical trials of intramuscular injection of rAAV2 and rAAV1-pseudotyped versions of an alpha-1-antitrypsin (AAT) vector in AAT-deficient adults. *Mol Ther* 15:S402, 2007.

408. KR Clark, TJ Sferra, and PR Johnson. Recombinant adeno-associated viral vectors mediate long-term transgene expression in muscle. *Hum Gene Ther* 8:659–669, 1997.

409. A Sarukhan, C Soudais, O Danos, and K Jooss. Factors influencing cross-presentation of non-self antigens expressed from recombinant adeno-associated virus vectors. *J Gene Med* 3:260–270, 2001.

410. A Sarukhan, S Camugli, B Gjata, H von Boehmer, O Danos, and K Jooss. Successful interference with cellular immune responses to immunogenic proteins encoded by recombinant viral vectors. *J Virol* 75:269–277, 2001.

411. KQ Xin, M Urabe, J Yang, K Nomiyama, H Mizukami, K Hamajima, H Nomiyama, T Saito, M Imai, J Monahan, K Okuda, K Ozawa, and K Okuda. A novel recombinant adeno-associated virus vaccine induces a long-term humoral immune response to human immunodeficiency virus. *Hum Gene Ther* 12:1047–1061, 2001.

412. KQ Xin, T Ooki, H Mizukami, K Hamajima, K Okudela, K Hashimoto, Y Kojima, N Jounai, Y Kumamoto, S Sasaki, D Klinman, K Ozawa, and K Okuda. Oral administration of recombinant adeno-associated virus elicits human immuno-deficiency virus-specific immune responses. *Hum Gene Ther* 13:1571–1581, 2002.

413. PR Johnson, BC Schnepp, DC Montefiori, NL Letvin, and KR Clark. Novel SIV/HIV vaccines based on recombinant adeno-associated virus vectors. *Mol Ther* 23:S15–S16, 2001.

414. PR Johnson, BC Schnepp, MJ Connell, D Rohne, S Robinson, GR Krivulka, CI Lord, R Zinn, DC Montefiori, NL Letvin, and KR Clark. Novel adeno-associated virus vector vaccine restricts replication of simian immunodeficiency virus in macaques. *J Virol* 79:955–965, 2005.

7 Therapeutic Gene Transfer with Replication-Defective Herpes Simplex Viral Vectors

D. Wolfe, W.F. Goins, A.R. Frampton, Jr., Marina Mata, D.J. Fink, and J.C. Glorioso

CONTENTS

7.1 INTRODUCTION

7.1.1 GENERAL OVERVIEW

Successful gene therapy requires not only the identification of therapeutic genes but also the design and construction of suitable vehicles for delivery and expression of these genes to target tissues *in vivo*. Identification of genes with therapeutic potential and engineering effective gene transfer vectors are rapidly proceeding, suggesting that gene therapy is likely an imminent clinical reality. Following inevitable setbacks in any innovative approach, gene therapy researchers have spent considerable time and effort in the engineering of safe and effective gene transfer vectors. Combining the natural traits of the desired vector system with the disease specific attributes will significantly increase the likelihood of efficacious gene transfer. For example, use of a neurotrophic vector

system for gene transfer to the nervous system likely increases the odds for success. Similarly, use of a gene product naturally involved in pain relief, such as preproenkephalin that directs production of opioid peptides, also increases the likelihood of success. Combining the natural biology of the disease with the natural biology of the gene transfer vector is not only intuitive, but we have found it is also effective.

With considerable progress in vector design, the number of gene therapy clinical trials has been steadily increasing. Several successful gene therapy clinical trials have recently re-invigorated gene transfer research and have begun fulfilling the promise for creating a new age of molecular medicine in which genomics and proteomics merge with genetic diagnostics and therapeutic gene transfer. We speculate that the age of efficacious gene therapy will soon be upon us.

In order to overcome technical hurdles, the development of strategies for vector targeting, modifications to increase transgene stability, regulation of gene expression, an circumvention of undesirable immune responses and effects of vector integration remain at the forefront of gene therapy research. Delivery of transgenes into target cells can be accomplished using either viral or nonviral vectors, with viral strategies remaining the most prevalent in human clinical trials. Virus-mediated gene delivery requires efficient methods for vector construction, vector production, and target cell infection. In this chapter, the relative merits and potential applications of replication-defective genomic herpes simplex virus type 1 (HSV-1) vectors are discussed. In the context of vector development, the natural history of HSV infection in the host is reviewed, highlighting unique features of the virus biology. Recent advances in therapeutic gene transfer to the nervous system using HSV-based vectors are also discussed.

7.1.2 Advantages of HSV Vectors

Successful therapeutic gene transfer will require construction of vectors tailored to specific applications, often utilizing the natural attributes of the base virus. The human herpesviruses naturally reside within the peripheral nervous system (PNS), an attribute we take advantage of by targeting PNS disease. Replication-defective HSV-based vectors represent promising candidate vectors for several types of gene therapy applications that include neuropathological disorders, cancer, pain control, autoimmune syndromes, and metabolic diseases.

Herpesviruses, large DNA viruses with the potential to accommodate multiple transgene cassettes, have evolved mechanisms that allow lifelong persistence in a nonintegrated latent state without causing disease in an immune-competent host. Among the herpesviruses, HSV-1 is an attractive vehicle for gene transfer to the nervous system because natural infection of humans results in a usually benign, lifelong persistence of viral genomes in neurons. This latent state is characterized by the absence of lytic viral protein expression and the presence of these latent genomes does not alter nerve cell function or survival. The HSV-1 genome contains a unique, neuron-specific promoter complex that remains active during latency, the latency active promoter (LAP). This promoter can be adapted to express therapeutic proteins without compromising the latent state or stimulating immune rejection of transduced cells. The establishment of latency does not appear to require the expression of viral lytic functions. Essential genes required for expression of the viral lytic functions can therefore be deleted to create completely replication-defective vectors that nonetheless effectively establish a latent state, but cannot cause disease or reactivate from latency. Experimental HSV infection is not limited to neurons; the virus is capable of infecting most mammalian cell types, and does not require cell division for infection and gene expression. Accordingly, HSV may be generally useful for gene transfer to a variety of non-neuronal tissues, particularly where short-term transgene expression is required to achieve a therapeutic effect.

Considerable technical progress has been made in developing HSV-1 into a practical gene transfer vector, including the development of efficient methods for vector construction and high-titer vector production. Recombinant HSV vectors expressing a multitude of gene products have been applied to models of human disease with a great degree of efficacy and without overt safety concerns. The obstacles requiring satisfactory resolution in order to realize the full potential of these vectors include (1) elimination of residual vector toxicity, (2) design of promoter cassettes that provide for sufficient level and duration of transgene expression, and (3) targeting of transgene expression to specific cell populations through the use of tissue-specific promoters, or by altering the virus host range through modifying receptor utilization for attachment and entry. This chapter concentrates on the design, production, and utilization of replication-deficient genomic HSV vectors.

7.2 VECTOR DESIGN STRATEGIES

7.2.1 Biology of the Viral Lytic Cycle

HSV-1 is a double-stranded DNA virus whose capsid is surrounded by a dense layer of proteins, the tegument, which is contained within a lipid bilayer envelope (Figure 7.1A). Glycoproteins embedded in the viral envelope mediate infection of the host cell, which takes place in two identifiable stages: (1) attachment to the cell surface and (2) fusion with the cell membrane, resulting in virus penetration. The envelope of HSV-1 contains at least 10 glycoproteins (gB, gC, gD, gE, gG, gH, gI, gJ, gL, and gM) and 4 nonglycosylated integral membrane proteins (products of the UL20, UL34, UL45, and UL49.5 genes). Of the 10 glycoproteins, gB, gD, gH, and gL are essential for viral infection [1–4], whereas gC, gE, gG, gI, gJ, and gM are dispensable for infection *in vitro* [5–7] yet can contribute to the replication and spread of the virus *in vivo* in the nervous system.

Attachment of the viral particle is mediated by several glycoproteins [5,6,8]. The sequential attachment steps in infection result in fusion of the viral envelope with the cell surface membrane and entry of the viral capsid into the cell cytoplasm. Even though the molecular events of penetration are not well understood, it is clear that multiple viral glycoproteins are required (e.g., gB, gD, gH/gL) [2,4,9–11]. In addition, following new virion assembly, viral glycoproteins are also involved in a less well-defined process of egress and release of mature particles from the infected cell membrane. Recent work has now shown a link between the viral envelope proteins and the capsid via several of the tegument proteins that explains the formation of active centers of virus assembly. In particular, the VP1/2 and UL37 inner tegument proteins have been shown to associate with intranuclear capsids [12] and that the outer tegument layer proteins VP16, VP22, and UL11 can bind to the cytoplasmic tails of gD, gE, and gH [13–15], thereby providing a framework within the infected cell where particle assembly occurs. The redundancy in these interactions explains previous data where deletion of single glycoprotein or tegument protein genes did not result in decreased virus release and overall titer, however, the deletion

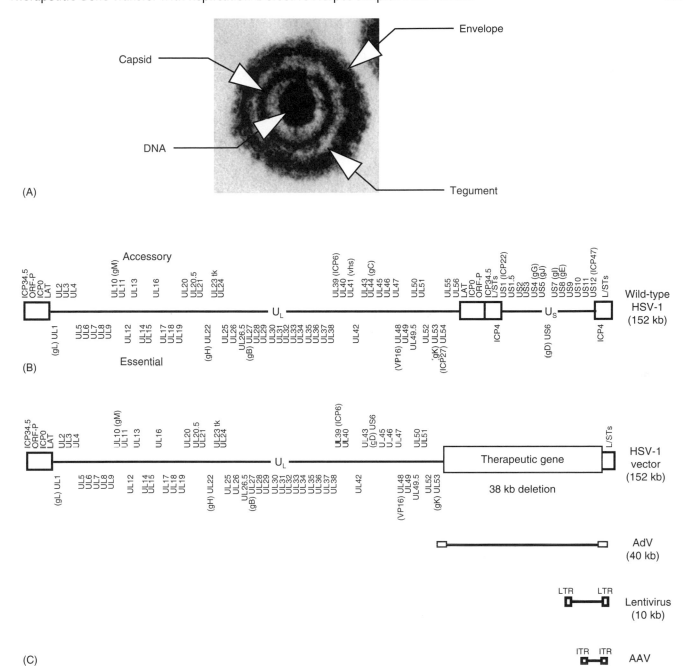

FIGURE 7.1 HSV-1 virion structure, genome organization, and comparison of viral vector payload capacities. (A) Electron micrograph of the HSV particle showing the capsid, tegument, and glycoprotein-containing lipid envelope. (B) Schematic representation of the HSV genome showing the unique long (UL) and unique short (US) segments, each bounded by inverted repeat (IR) elements. The location of the essential genes, which are required for viral replication *in vitro*, and the nonessential or accessory genes, which may be deleted without affecting replication *in vitro*, are indicated. (C) Schematic diagram of various viral vector genomes currently in use for gene transfer and therapy studies, including the overall size of the entire vector genome. The HSV-1 vector, which contains a 38 kb deletion of sequences comprising the joint region and the entire US segment of the viral genome (ICP4−, ICP22−, ICP27−), can accommodate foreign transgene sequences that are larger than lentivirus or AAV vectors and equivalent in size to the complete adenoviral (AdV) genome.

of multiple functions has greatly impaired virion assembly [16]. Viral particles are also capable of spreading from cell to cell across cell junctions, a process requiring the functional activities of several glycoproteins that are not required for initial infection (e.g., gI/gE) [17,18] and also the function of at least some tegument proteins [19,20].

The genome structure of HSV can be divided into viral genes that are essential or accessory for replication in cell culture (Figure 7.1B). The accessory functions may be deleted without significantly hampering virus growth in culture. However, removal of essential genes necessitates the use of complementing cell lines that express the essential products in order to propagate these viral recombinants. In human infections, HSV binds to and enters epidermal cells following direct contact with an infected individual that is shedding virus or has an active lesion. Following virus attachment, the viral capsid penetrates the surface membranes of epithelial cells of the skin or mucosa and is transported to the nuclear membrane where viral DNA is injected through a nuclear pore (Figure 7.2A). Once inside the nucleus, the viral DNA is circularized and transported to nuclear domain 10 (ND10) structures

[21,22], where the immediate early (IE) genes are expressed as part of the sequential cascade of lytic gene synthesis [23] (Figure 7.2B). Transcription of the five IE genes (ICP0, ICP4, ICP22, ICP27, and ICP47) does not require de novo viral protein synthesis. Expression of the IE genes is controlled by promoters that contain one or more copies of an enhancer element responsive to the viral tegument protein VP16 (aka Vmw 65 or TIF), a transactivator that is transported into the nucleus along with viral DNA [24–26]. The essential IE genes ICP4 and ICP27 encode products required for expression of the early (E) and late (L) genes [27–30]; the former (E) gene class specifying primarily enzyme functions required for viral DNA synthesis and the latter (L) comprising primarily virion structural

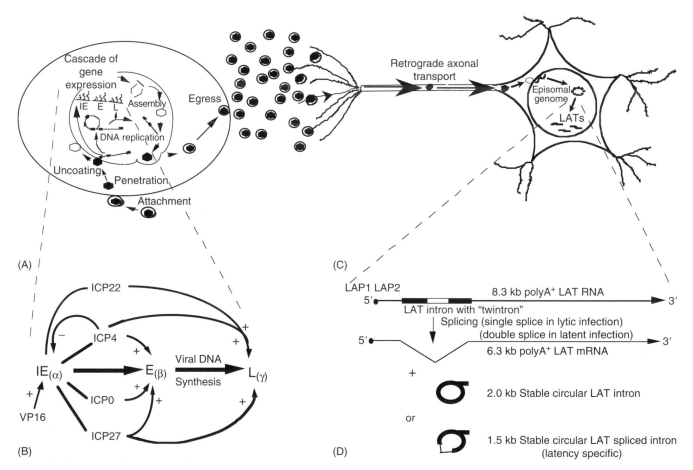

FIGURE 7.2 HSV-1 life cycle in the host. (A) Lytic infection. Primary lytic infection of epithelial or mucosal cells results from the attachment and penetration of HSV particles to host cells, a complex process involving many HSV surface glycoproteins. Following transport of the capsid to the nuclear membrane and injection of linear dsDNA into the nucleus, the genome circularizes and begins to express the lytic HSV gene functions in a highly regulated sequential cascade, yielding the expression of proteins involved in viral DNA synthesis and virion structural components. Following assembly of newly synthesized particles within the nucleus, virion maturation results in the egress of these virions from the infected cell. (B) Schematic diagram of the sequential cascade of lytic gene expression. The five IE or α genes are expressed immediately upon infection through transactivation by the VP16 tegument protein. The ICP4, ICP27, and ICP0 IE gene functions are responsible for the activation of the early or β genes that are primarily involved in viral DNA synthesis. In addition, ICP4 acts to shut-off expression of the IE genes. Following viral DNA replication, ICP4, ICP22, and ICP27 participate in the activation of true viral late or γ genes, which mainly encode virion structural components. (C) Latent infection. When virion particles encounter and bind to axonal termini that innervate the site of primary infection, viral capsids are transported in a retrograde manner to the nerve cell body. At this point the circular viral genome can persist as an episomal molecule in a latent state within the neuron, wherein viral lytic gene expression is silenced and a series of latency-associated transcripts (LATs) are produced. (D) Gene expression during latency. The major 2.0 kb LAT arises from the large 8.3 kb polyA+ through a splicing event that yields an unstable 6.3 kb LAT and a circular LAT lariat of 2.0 kb. The location of the LAP regions LAP1 and LAP2 relative to the LATs is depicted.

components. ICP4 regulates viral promoter function [31], whereas ICP27 affects the processing and transport of viral RNA [32–34]. The IE gene products ICP0 and ICP22 contribute to viral gene transcription but are not essential to virus replication in cultured cells [35–38]. ICP0 is a promiscuous transactivator that exerts its effect prior to the transcription initiation event; it is not a DNA-binding protein [39]. ICP22 has been found to regulate the level of ICP0 expression [40]. ICP47 does not affect transcription but rather has been reported to interfere with the transporter of antigen presentation (TAP) that is responsible for loading major histocompatibility (MHC) class I molecules with antigenic peptides [41–43]. Expression of late genes is dependent on both viral DNA synthesis and IE gene functions [29,31,44,45]. Following translation of the late gene products, which become viral structural components of the capsid, tegument, and envelope, genome-length copies of viral DNA are packaged into the newly assembled capsids. Tegument proteins accumulate around the capsid and the immature particle buds through the inner nuclear membrane followed by progressing through the transgolgi network where the viral glycoproteins are localized. Double-membrane-enveloped virus containing virus-encoded glycoproteins modified by the Golgi apparatus enzymes fuse with the cell membrane forming a mature, extracellular virus particle with a single membrane bilayer [46]. The infectious particles can infect neighboring cells by cell-to-cell transmission or can be released for infection of distal cells. With the exception of sensory neurons, cell lysis accompanies productive viral infection.

7.2.2 COMPLEMENTATION OF ESSENTIAL GENES AND ELIMINATION OF CYTOTOXICITY

Because HSV genes are expressed in a sequential cascade during lytic infection [23], removal of the single essential IE gene ICP4 severely inhibits the expression of later E and L genes [27], resulting in a defective vector incapable of producing virus particles. In addition, the IE gene products, with the exception of ICP47, are individually toxic to most cell types when expressed at high levels by transfection [47]. The elimination of multiple IE genes reduces the cytotoxicity of HSV-based vectors for cell lines and primary neuronal cell cultures [48–52] (Figure 7.3A and B). Cytotoxicity of a genomic HSV vector deleted for ICP4, ICP22, and ICP27 in cultured Vero cells is reduced compared with a virus-only deficient for ICP4 [52]. Such mutants are also less toxic to primary neurons [49] or undifferentiated cells such as umbilical cord blood-derived stem cells [53] which may involve the ability of these cells to regulate the activity of ICP0 or the downstream effects of ICP0, as the protein is less stable in primary neurons than other cell types [54], or may relate to the antiviral state of the target cell.

Interestingly, ICP0 appears to be the key regulatory protein encoded by the virus that determines vector-associated toxicity and controls the level of transgene expression from the viral genome. The role of ICP0 in infection pertaining to gene expression and toxicity can be evaluated in mutant viruses lacking the various IE genes. A virus defective for the expression of all five IE genes was developed and demonstrated to be completely nontoxic at all multiplicities of infection tested [55]. However, transgene expression from this virus was very low in most cells, suggesting that expression of ICP0 is essential for high levels of transgene expression from such depleted viruses. It was also determined that the expression of ICP0 in the absence of the other IE proteins has profound effects on infected non-neuronal cells [52]. However, infected cells did not divide or synthesize DNA. Curiously, and unlike all previously generated HSV mutants, the infected cells did not round up, but remained metabolically active for several days in a state conducive to high-level transgene expression. Additional published studies demonstrated that both the prolonged high-level transgene expression and the induction of cell cycle arrest were functions of ICP0 [55,56]. Subsequently, published studies have demonstrated that the expression of ICP0 from defective viruses results in a cellular gene expression profile consistent with cell cycle arrest at G1/S [57–59] or G2/M [60,61]. Therefore, residual toxicity in non-neuronal cells of a vector lacking all IE genes except ICP0 is due to the expression of ICP0. ICP0 has been shown to be a potent transactivator of viral and cellular promoters in transient assays [39,62] and provides for efficient viral gene expression and growth *in vitro* and *in vivo* [63,64]. Several activities have been described for ICP0 but the functional relevance of these remains to be clarified. ICP0 localizes to and disperses complex nuclear structures called ND10s/PODs/NBs/Kr, which are thought to be involved in the proliferative state of the cell [21,65] and are involved in acting as a storing house for cellular factors that are stimulated in response to a variety of cellular insults such as DNA damage due to irradiation or chemicals, as well as participating in the IFN response to virus infection. Two E3 ubiquitin ligase domains have been identified in ICP0 [66–68] which may explain the ability of ICP0 to induce the proteosomal degradation of many of the ND10 components such as PML and Sp100 [67,69,70], CENP-A [61], CENP-C [71], CENP-B [72], DNA-PK [73,74], and p53 [66]. It has also been shown to associate with a 135-kDa ubiquitin-specific protease USP7 [75,76], and other ubiquitin components such as cdc34 [68], and also UbcH5a and UbcH6a [66,67,74] where it is believed that its association with USP7 blocks its ability to ubiquitinate itself thereby leading to its own degradation by the proteosomal machinery [77]. In addition, it associates with the BMAL-1 transcription factor [78] and the p60 protein [79]. It also interacts with molecules involved in translation [80] and cell cycle regulation such as CyclinD3 [81] and binds to HDACs [82,83] and plays a role in dissociating HDACs from the HDAC1/2:REST:Co-REST complex [84], which is thought to play a major role in activating gene expression. ICP0 also modulates the innate response to the virus by altering the IFN response to HSV through IRF3, CBP, and p300 [85,86] that may also enable the cell to repress expression from the viral genome. Recent evidence using PML, IFNαR, and STAT1 knockout mice has shown that STAT1 and the IFNs play a role in repression while PML does not [87]. Therefore, ICP0 has the potential to alter many aspects of host cell metabolism. However, neurons regulate and process ICP0 differently and

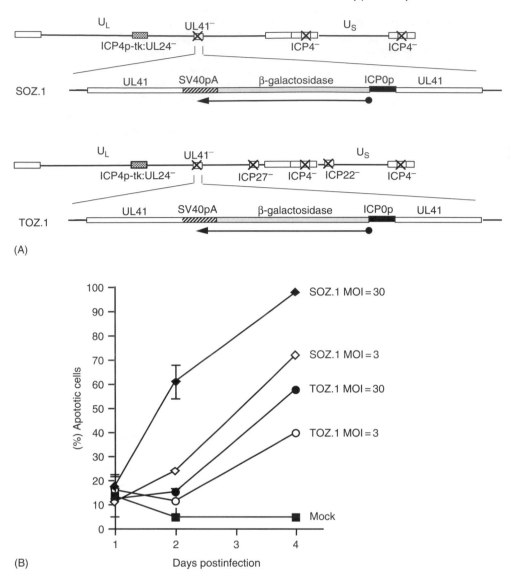

FIGURE 7.3 Reduced toxicity following infection of primary dorsal root ganglia neurons. (A) Schematic representation of the first-generation SOZ.1 (ICP4⁻, UL41⁻:ICP0p-lacZ:UL24⁻:ICP4p-tk) and the third-generation TOZ.1 (ICP4⁻, ICP22⁻, ICP27⁻, UL41⁻: ICP0p-lacZ, UL24⁻:ICP4p-tk) replication-defective vectors displaying the lacZ transgene inserted into the UL41 locus under the control of the ICP0 promoter using the SV40 polyadenylation signal. (B) Number of primary dorsal root ganglion (DRG) neurons undergoing apoptosis was determined at various times following infection with either SOZ.1 or TOZ.1. Even at an MOI of 30, TOZ.1 was less toxic than the first-generation vector (SOZ.1) at the lower MOI (3.0).

are not dramatically affected by ICP0 expression [54] since they lack ND10 structures. Moreover, proteosome inhibitors affect ICP0 degradation in neurons and other cells that tie with ICP0s function as an E3-Ubiquitin ligase [66–68]. The decreased half-life of ICP0 in neurons also contributes to the overall decreased toxicity in these cells.

We engineered numerous mutant background vectors for gene therapy applications that require short-term, high-level expression by deleting multiple viral functions, including the IE genes ICP4, ICP22, ICP27, and ICP47. We chose to remove the ICP47 gene in some vectors to avoid interference with antigen presentation in applications intended for induction of specific immunity, but we did not delete the ICP0 gene because this gene product improves transgene expression and permits

efficient construction of recombinant vectors [56,88]. In these mutant backgrounds, we eliminated the virion host shut-off function (vhs) encoded by UL41 because this virus tegument component indiscriminately interferes with translation of mRNA in infected cells [89–91]. A similar vector background has been used to express transgenes for up to 21 days in cultured primary neurons without causing neuronal cell death [49]. For applications involving infection of cancer cells *in vivo*, the transient arrest of cell division offered by ICP0 and subsequent recovery of cell growth at high multiplicity of infection should prove advantageous because transduced cells produce high levels of transgene product prior to induction of cell death. For other applications in non-neuronal targets, it may be necessary to delete ICP0 or employ a modified form of the gene that enables sufficient therapeutic gene expression from the viral backbone in the absence of observable vector-related cytotoxicity. We are currently investigating numerous approaches in a matrix deleting these gene products in combinations.

7.2.3 VECTOR TRANSGENE CAPACITY

The treatment of monogenic diseases typically requires only limited vector capacity, but complex applications may require the delivery of large or multiple independent genetic sequences. A comparison of the genome structure and capacity of several current vector systems is shown in Figure 7.1C. The overall size of the HSV-1 genome (152 kb) represents an attractive feature for employing the vector for the transfer of large amounts of exogenous genetic sequences. Approximately one-half of the HSV-1 coding sequences are nonessential for virus replication in cell culture and therefore may be deleted to increase transgene capacity without blocking viral replication (Figure 7.1B). The latency region of the virus genome represents approximately 8 kb of sequence that can be removed and the joint region of the virus is composed of 15 kb of redundant sequence that can be eliminated without compromising virus replication [92]. In one line of experiments we removed in excess of 26 kb of the HSV-1 genome (Wolfe, D., unpublished data, 2007). At least 44 kb of HSV sequence can potentially be removed and vectors propagated in cells engineered to complement just three viral functions (ICP4, ICP27, and gD). Transgene expression cassettes can also be inserted into deleted essential gene loci to avoid transfer of foreign sequences to wild-type virus by recombination that could potentially occur between the vector and wild-type genomes *in vivo*. We have observed that some nonessential genes (e.g., IE genes ICP0 and ICP22) are toxic to some cell types, yet the products of these genes are required for high-titer vector production. The toxicity of these products makes it difficult to produce a complementing cell line carrying these genes. However, it is possible to engineer the promoters for these genes in a manner to make their function dependent on viral IE genes present only in complementing cells [52]. By the judicious selection of viral gene deletions and promoter alterations, high-titer vectors can be produced with minimal complementation.

We have developed a panel of novel HSV-1 vectors with a background suitable for expression of multiple transgenes using a rapid gene insertion procedure [93]. To take advantage of the reduced cytotoxicity resulting from the deletion of ICP4, ICP22, and ICP27 genes [49,51], we designed a single vector in which nine viral genes were deleted, removing a total of 11.6 kb of viral DNA that was replaced with multiple transgenes under control of different promoters. These HSV multigene vectors were constructed with either four or five independent transgenes at distinct loci [94] with all the transgenes simultaneously expressed for up to 7 days. These multigene vectors demonstrate the potential for using HSV-1 vectors for the expression of complex sets of transgenes that have coordinated or complementary functions.

Recent work inserting bacterial artificial chromosome (BAC) elements into large viruses including HSV promises to increase the speed of recombinant construction as well as allow the creation of extensively modified vectors without the necessity of creating complementing cell lines. We have introduced the BAC elements into multiply defective HSV vectors and propagated them in *Escherichia coli* (Wolfe, D., 2007, unpublished data). Transfection of HSV:BAC DNA purified from bacteria into complementing cells is much more efficient than using viral DNA purified from mammalian cells. Infectious virus produced can be recovered back into *E. coli* for further modification. Several methods have been used to modify the viral genome in bacteria with high frequencies of recombinant production in a greatly shortened time frame. We have further modified vector genomes and have constructed recombinant replication-defective BAC vectors that have been used to incorporate cellular cDNA libraries using recombineering (Wolfe, D., 2007, unpublished data) as well as engineering additional methods for recombinant engineering in bacteria.

7.2.4 AMPLICONS AS AN ALTERNATIVE VECTOR SYSTEM

Often referred to as "defective" HSV-1 vectors, amplicons are plasmids engineered to contain both an HSV origin of replication and packaging signals as well as a bacterial origin of replication [95]. Amplicons are propagated in bacteria and then cotransfected with a defective HSV "helper" virus to create a mixed population of HSV particles containing either the defective HSV helper genome or concatemers of the plasmid packaged within an HSV capsid. In concept, the production of virion-packaged amplicons uses transient complementation of the entire HSV genome to provide the needed replication machinery and viral structural components. A comparison of replication-defective genomic HSV vectors and helper virus-free amplicons is shown in Figure 7.4A and B. Amplicons have been used to express reporter genes [96–102] or biologically active peptides [103–112] transiently in tissue culture systems.

In vivo, a prolonged expression of both a *lacZ* reporter gene [100,113–117] and of the TH gene [113] following amplicon injection into brain has been reported. However, the production of amplicons requires repeated passaging of the amplicon/helper virus preparation, which results in the emergence of recombinant wild-type virus that, although estimated to occur at the low frequency of 10^{-5} [113,114], results in the death of 10% of infected animals in experiments *in vivo* [113]. The production of true helper virus-free amplicon preparations using multiple restriction fragments of the helper virus genome that lack packaging signals has recently been reported [118]. However, the maximal yield obtained with that method has remained low (<10^7 pfu/mL), and expression *in vivo* has not been fully tested [118]. The presence of cytotoxic helper virus and the generation of replication-competent contaminants represent technical hurdles to the effective production and use of amplicons in human patients. Helper-free amplicons will likely require the development of new helper systems to make their use practical enough for human applications.

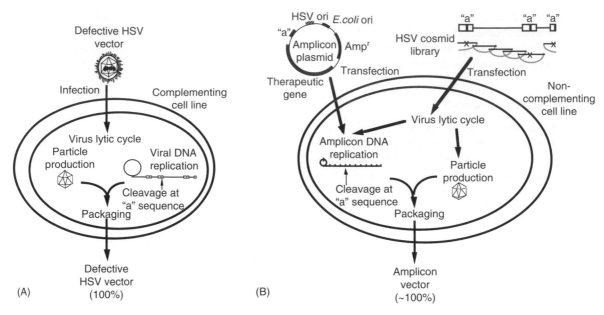

FIGURE 7.4   Strategies for HSV-1 vector design. (A) The production of defective full-length genomic HSV vectors is carried out in cell lines engineered to provide the deleted essential genes in *trans*. These vectors can be produced in high titers, are capable of long-term persistence in neurons *in vivo*, can accommodate large or multiple transgenes, and are incapable of replicating in neurons or other cells because of the missing essential genes. (B) Helper-free amplicons can be readily propagated in bacteria using the bacterial origin of replication (*E. coli* ORI), and then transfected into a noncomplementing cell line along with five cosmids that encompass the entire HSV genome. Unlike the standard amplicon system in which the final preparation consists either of a mixture of amplicon concatemers and defective HSV particles, only amplicon concatemers get packaged into new virus particles because the overlapping cosmids lack the HSV packaging sequence ("a" sequence). The helper-free amplicon preparations suffer from low titer yields, decreased stability of the amplicon DNA, and decreased transgene payload.

Improvements in the amplicon-packaging system using Cre–lox [119] among other methods have resulted in preparations that are less immunogenic when injected into the CNS of rodents [120] that represents a drastic improvement in the use of the amplicon system. With these modifications, amplicons still continue to be used for *in vivo* delivery to the CNS to deliver neurotrophin-3 (NT-3) or the NMDA receptor [121] or ATM [122] or to treat cancer [123,124]. Recently, they have been employed as a vaccine vector with some success [125–128] as their lower overall titers and immunogenic nature compared to the genomic replication-defective HSV vectors prove to be an advantage for vaccine approaches.

Hybrid amplicon vectors have been produced with either AAV [129] or EBV [130] that allow rep-mediated site-specific integration of the transgene cassette (AAV) [131,132] or enable long-term plasmid DNA maintenance in cells of non-neuronal origin (EBV). Finally, as mentioned previously, HSV:BAC constructs [133,134] have now been employed to incorporate either large transgene cassettes containing larger natural promoters and full-length eukaryotic genes [135] or for the introduction of cDNA libraries (Wolfe, D., 2007, unpublished data).

7.3   REDIRECTING VECTOR INFECTION

One method to achieve cell-specific expression of the therapeutic gene is targeting vector recognition and infection of cells to unique receptors present principally on target cells.

Targeted viral infection requires (1) the identification of cell-specific surface receptor(s) to which viral binding/entry can be directed and (2) the modification of viral glycoproteins to recognize novel receptors while eliminating the binding of these viral ligands to native receptors, a process that ideally should be accomplished without compromising infectivity.

Initial binding of HSV particles to cell surface heparan sulfate (HS) and other glycosaminoglycans (GAGs) [9,143–149] is mediated by exposed domains of glycoproteins C [150–154] and B [145,151], as evidenced by (1) diminished cell attachment of gC and gB mutants, (2) interference with cell attachment by pretreatment with anti-gB or anti-gC antibodies, (3) reduced attachment of virus after enzymatic removal of cell surface GAGs, and (4) competition studies using heparin. The role of gB is complex as this glycoprotein functions both in the initial binding of virus to the cell surface (nonessential for cell entry) and in later stages in the virion-cell fusion program (essential for cell entry). Molecular dissection of the domains of gB that are responsible for each distinct process has allowed for the generation of a gB mutant that exhibits a reduced capacity for binding to GAGs without a loss of its fusogenic activity [151]. This mutant, generated by deletion of a positively charged lysine-rich region within gB (designated gBpK⁻), is an important resource in the generation of viral vectors with targeted cell binding and preserved entry (Figure 7.5). The initial gB/gC-HS/GAG-mediated cell attachment greatly enhances but is not essential for subsequent events in the cell entry cascade; thus, cell attachment represents a reasonable target for strategies aimed at effecting restricted cell entry.

Attachment of HSV-1 to cell surface HS is followed by gD-mediated attachment to a second cell receptor. Work by several labs has identified gD cognate receptors utilized for

both virus attachment and penetration. The first herpes virus entry mediator (HVEM or HveA), a member of the tumor necrosis factor-α (TNF-α)/nerve growth factor (NGF) receptor family, was identified by screening a cDNA expression library in HSV-1 resistant CHO cells for clones that enabled virus infection [155]. Domains of gD that contribute to HveA binding have been identified in virus infection inhibition studies using monoclonal antibodies that recognize residues 11–19

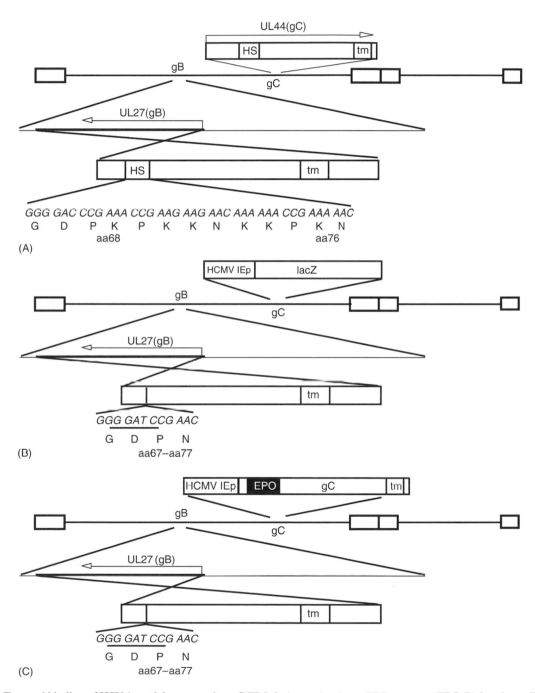

FIGURE 7.5   Targeted binding of HSV-1 particles expressing gC:EPO fusion molecules to EPO receptor (EPO-R)-bearing cells. (A) Diagram of the KOS wild-type HSV-1 genome depicting the location of the two HSV glycoproteins (gB and gC) involved in HS binding. The HS-binding domain of gB consisting of a series of polylysine (pK) residues is shown in greater detail. (B) The KgBpK⁻gC⁻ recombinant virus deleted for binding to HS was constructed by deleting the pK region from the essential gB gene and deletion of the nonessential gC gene by insertion of an HCMV IEp-lacZ expression cassette into the gC locus. (C) The gC:EPO2 recombinant (KgBpK⁻gCEPO2) was constructed by introducing EPO into the gC gene, replacing aa# 1–162 in the KgBpK⁻gC⁻ recombinant virus, which could readily be purified by X-gal staining.

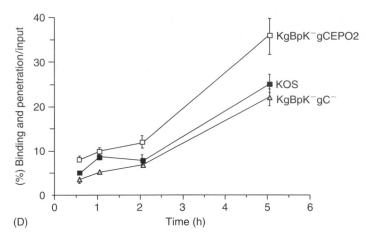

FIGURE 7.5 (continued) (D) The percentage of radiolabeled wild-type HSV-1 (KOS), the recombinant deleted for HS-binding (KgBpK⁻gC⁻), and the EPO-expressing (KgBpK⁻gCEPO2) viruses that bound to K566 cells bearing the EPO-R was determined and is expressed compared with input virus. These data demonstrate that the KgBpK⁻gCEPO2 recombinant virus binds to the EPO-R and this binding conferred increased infectivity of the KgBpK⁻gCEPO2 virus for cells bearing the EPO-R (K566).

and 222–252 [156]. Further insight into the interacting domains of gD and HveA was acquired upon the crystallization of gD bound to HveA. Data acquired from these studies identified critical amino acids primarily within the amino termini of HveA and gD that are critical for gD-HveA complex formation [157–160]. In addition to the HveA receptor, a number of other gD entry mediators have been identified, including HveB (nectin-2) [161], 3-O-sulfated HS [162], and HveC (nectin-1) [163,164]. HveC is ubiquitously expressed on most cells, including epithelial and neuronal cells, and is generally considered to be the main HSV-1 entry receptor. HveC is a member of the immunoglobulin superfamily and has no structural relation to HveA [164]. A type I integral membrane protein, HveC contains a variable (V) domain and two constant (C) domains, but attachment of HSV-1 via gD to only the V domain is necessary and sufficient to initiate fusion of the virus with the cell [165]. Recently, we and other laboratories discovered the first mutants of gD that do not bind to HveC, but can still mediate virus entry through an interaction with HveA [158,166–168].

The consequence of the sequential attachment steps in infection is fusion of the virus envelope with a cellular membrane and subsequent virus entry [5,6]. A role for gD in virus penetration is supported by evidence that attached virus can be neutralized by anti-gD antibody and virus mutants deleted for gD attach to cells but do not penetrate [4,10,11]. Mutants deleted for gH/gL [169,170] or gB [1,171] are also blocked in virus penetration but are not defective in attachment. Both gB and gD are capable of inducing syncytia if expressed on the cell surface at low pH, supporting a possible role for both molecules in fusion [172]. Fuller and Lee proposed that entry involves a cascade of events in which gD initiates the fusion process [9]. According to their hypothesis, a fusion bridge is most likely formed by the action of gB followed by extension of the bridge and virus release requiring the fusogenic activities

of gH/gL. Recent data suggest that, while gB, gH, and gL are all required for fusion, the order of glycoprotein activation is gD-gH-gL, which form a hemifusion intermediate before gB completes the fusion process [173,174]. In addition to external entry, HSV-1 is also capable of infecting neighboring cells by moving transcellularly across cell membranes. This process, referred to as cell to cell or lateral spread, can be distinguished from infection by exocytosed particles through its ability to occur in the presence of virus neutralizing antibodies as well as by deletion of certain accessory envelope glycoproteins (e.g., gI/gE) which prevents lateral spread but not initial infection [17,18].

Entry of HSV-1 can occur through fusion of the viral envelope with the plasma membrane [5,6,175] or an internal endosomal/phagosomal membrane depending upon the cell type [176–179]. Regardless of the mode of entry, after fusion of the viral envelope with a cell membrane, HSV-1 virions are deposited into the cell cytoplasm and the viral DNA is ultimately delivered to the nucleus where transcription and replication of the viral genome ensues.

In early studies designed to alter the cell attachment properties of HSV, we focused on the elimination of GAG-binding by removing the GAG-binding domains from the virus envelope, while leaving intact the viral determinants that mediate cell entry. We hypothesized that vector targeting may be achieved by replacement of the HS-binding domains with ligands that specifically recognize and bind alternate cell receptors. A double-mutant virus, KgBpK⁻ gC⁻, was derived from wild-type KOS strain (Figure 7.5A and B). In this mutant, the coding sequence for the nonessential gC gene was removed from the viral genome and the wild-type gB sequence replaced by the gB:pK⁻ mutation [180] (Figure 7.5B). The resulting virus demonstrated an 80% reduction in binding to Vero cells compared with wild-type virus. By replacing the HS-binding domain of gC with the coding sequence of erythropoietin (EPO) in the background of the

KgBpK⁻gC⁻ mutant virus (Figure 7.5C), we demonstrated that the gC:EPO fusion protein was incorporated into the budding virion and that recombinant virus was specifically retained on a soluble EPO-receptor column [180]. The gC:EPO virus demonstrated a twofold increase in infection of K562 cells (Figure 7.5D), which express the EPO receptor. The EPO-expressing particle also stimulated the proliferation of FD-EPO cells [180], an EPO-dependent cell line, indicating that virus binding to the EPO-receptor had occurred.

In a similar fashion, HSV-1 was also engineered to bind to hepatocytes through the expression of preS1 and a preS1 peptide from hepatitis B virus [181]. Expression of these molecules from an HSV-1 recombinant increased entry into hepatocytes compared to the parental virus that was impaired in binding to GAGs. However, while this recombinant displayed increased entry kinetics on hepatocytes, wild-type HSV-1 entry was even more efficient than the recombinant indicating that the preferred mode of entry in these cells was mediated by an interaction of gD with one or more of the natural receptors. Grandi et al. replaced the HS-binding domain of gC with a His-tag sequence and showed that this vector exhibited a fourfold increase in binding to 293T cells expressing a pseudo-His-tag receptor [182,183]. Collectively, these data showed that HSV-1 vectors can be generated that bind to cells expressing an alternate receptor without disrupting the fusogenic properties of HSV-1.

The next step in the development of redirected HSV vectors was the replacement of HveA- and HveC-binding domains within gD with targeting molecules that could effectively redirect the tropism of HSV vectors. The first data showing that this strategy was a feasible one was reported by Zhou et al. who retargeted an HSV-1 recombinant to IL-13Rα2-expressing cells by insertion of an IL-13 ligand into gC and gD [184]. Virus binding to GAGs was significantly reduced through elimination of the HS-binding domains of gB and gC, and the insertion of IL-13 into the N-terminus of gD destroyed the ability of this recombinant to bind the HveA receptor, but binding to HveC remained intact. This HSV recombinant was shown to efficiently enter and replicate in cells that only express the IL-13Rα2 receptor. These results showed that, in addition to elimination of most of the HS binding, the HveA-binding domain of gD could be altered without affecting virus entry. However, this recombinant was not impaired for entry on cells bearing the HveC receptor. In a subsequent study, the same group inserted the urokinase plasminogen activator (uPA) into the same position of gD and showed that this recombinant could enter cells that express the uPA receptor [185]. Taken together, these data showed that different ligands could be inserted into the amino terminus of gD and effectively retarget the virus. In a recent report, these authors showed that a mutation of amino acid 34 of gD (V34S) in the IL-13Rα2-retargeted virus enabled this recombinant to exclusively enter cells bearing the IL-13Rα2 receptor [186].

Menotti et al. showed that, in addition to ligand insertion, single-chain antibodies against alternate receptors could also be used to redirect HSV tropism. In their study, a single-chain antibody that recognizes the Her2-neu receptor, which is highly expressed on breast tumor cells, was inserted into gD at the same location as the IL-13 and uPA ligands [184–186]. Recombinant viruses harboring the scFv were able to enter cells bearing only the Her2-neu receptor, but could also infect cells expressing nectin-1. These data showed that scFvs, which are relatively large constructs, could be accommodated by HSV and used to target HSV to alternate receptors.

An alternative approach to introducing novel ligands into viral glycoproteins for targeted HSV entry involves the development of soluble adapter molecules designed to interact simultaneously with a cell-specific receptor and virion gD. In our model study [187], the gD-binding V-domain of HveC was fused to a single-chain antibody (scFv) to the EGF receptor in an effort to target EGFR-bearing cells. The adapter molecule enabled HSV-1 entry into naturally non-permissive CHO cells expressing the human EGFR [187], but not into CHO cells lacking the receptor, and entry was not observed when the antibody portion of the adapter was replaced with that of a different specificity. The degree of EGFR-scFv-adapter-mediated entry increased with viral dose, reaching approximately 65% of the cells at an MOI of three compared to 75% infection of HveC bearing CHO cells at the same MOI in the absence of adapter. Adapter-mediated virus entry into CHO-EGFR cells was inhibited in a dose-dependent manner by antibodies to gD, EGFR, and the HveC V domain as well as by soluble gD and soluble V domain, establishing its dependence on accessible EGFR, adapter V domain, and viral gD. Moreover, efficient adapter-mediated entry required the presence of cellular GAGs. Together, these results showed that a bi-specific adapter molecule can efficiently redirect HSV infection using a process that does not require the generation of a new recombinant HSV vector for each retargeting attempt.

7.4 VECTOR TRANSGENE EXPRESSION

7.4.1 Lytic Gene Promoters

We have utilized many divergent viral and cellular promoters in the background of replication-defective viral vectors [49,53,94,188–190]. Typically, lytic viral gene promoters display transient activity in both neuronal and non-neuronal cells. Thus, these viral IE promoters are effective for applications that require only transient transgene expression. Alternatively, the use of transient promoters can allow the design and application of these vectors in clinical trials in order to minimize unforeseen effects of long-term gene expression. Following success in initial clinical trials, long-term promoters may be substituted to allow extended therapeutic benefit from a single-vector application. The human cytomegalovirus (HCMV) IE gene promoter produced vigorous transgene expression for up to 21 days postinfection *in vivo* and in neuronal cell cultures in the background of a vector deleted for ICP4, ICP27, and ICP22 [49]. Studies in rabbits and primates demonstrated transgene expression under control of the HCMV promoter for at least 1 year following infection of rabbit joints, although this result is not typical and may have been influenced by the biological activity of the NGF transgene [191]. Other promoters, such as those from SV40 and various retroviral LTRs, are also transiently active [192,193] following infection of brain. Cellular promoters such as the muscle-specific muscle creatinine kinase (MCK) enhancer support muscle-specific expression in myotubes in culture [188]. Several reports have described long-term transgene expression from similar promoters in neurons [194–197], but expression declined significantly after the first week and may represent activity of aborted reactivation events. However, many therapeutic applications will require prolonged transgene expression from latent genomes. For this purpose, we and other laboratories have studied the native latency gene promoters in considerable detail.

7.4.2 BIOLOGY OF LATENCY AND THE LAP

Following infection of epithelial cells of skin or mucous membrane, viral particles come into contact with sensory neuron axon terminals in which the particle is transported along microtubules to the nerve cell body where the viral DNA enters the nucleus [198,199] (Figure 7.2C). Although the virus can express lytic functions in sensory neurons [200–202], lytic gene expression is curtailed through a set of largely undefined molecular events and the virus enters a latent state. Latency is typified by expression of a series of LATs from the repeat regions flanking the long unique segment (U_L) of the viral genome partially antisense to and overlapping the 3′ end of the ICP0 mRNA [203–205] (Figure 7.2D). Two colinear, non-polyadenylated (poly A⁻) LAT RNA species of 2.0 and 1.5 kb [206,207], which accumulate in the nuclei of latently infected neurons [208], are stable nonlinear intron lariats derived from a large, unstable 8.3 kb polyadenylated primary transcript [209–213]. It has recently been demonstrated that this region of the genome remains transcriptionally active while transcription from the remainder of the genome is repressed due to histone deacetylation [214,215] and that a CCTC element within LAP2 may be responsible for this difference in chromatin structure [216,217] as we had previously suggested [218,219]. The virus can remain latent for the life of the individual, although sporadically viral genomes may reactivate in response to a variety of stimuli including immune suppression, ultraviolet light, fever, and stress [199,220].

The promoter/regulatory region that controls LAT expression is of interest to gene therapy applications because that promoter remains active during latency when all other viral promoters are silenced. LAT expression is differentially directed by two independent LAPs, LAP1 [221–224] and LAP2 [219,222,225]. LAP1, predominantly responsible for LAT expression during latency [222,223], is located 5′ proximal to the unstable 8.3 kb LAT. LAP2 is primarily responsible for LAT expression during lytic infection, but is also capable of driving low-level expression of LAT in the absence of LAP1 during latency [222] and is located immediately upstream to the 2.0 kb LAT intron [219,222,225]. Deletion of both LAP1 and LAP2 completely eliminates expression of detectable levels of LAT during latency *in vivo* in animals [222], demonstrating that both promoters contribute to LAT expression during latency in neurons *in vivo*.

The continuous expression of the LAT region of the HSV genome during neuronal latency suggests that it should be possible to exploit the LAT promoter to express therapeutic genes from latent viral genomes in neurons. Both LAP1 and LAP2 have been employed to achieve long-term transgene expression from the HSV vector genome during latency. For example, a LAP1-β-globin recombinant produced transgene expression in murine peripheral neurons during latency [223], but the level of product decreased over time [226]. Other examples include recombinants with LAP1 driving expression of β-glucuronidase [227], NGF [226], β-galactosidase [225,226,228], or murine α-interferon [192] that either displayed a similar expression pattern or were not active in latently infected animals [192,196,226]. These data suggest that LAP1 may lack the *cis*-elements required for long-term transgene expression in the context of the HSV viral genome. However, long-term transgene expression in PNS neurons or neurons of the spinal cord was achieved when LAP1 was juxtaposed to the Moloney murine leukemia virus (MoMLV) promoter [229–232], unlike recombinants employing either LAP1 or the LTR alone, suggesting that the elements responsible for extended expression lie elsewhere within the LAT promoter/regulatory region. These *cis*-acting elements may be complemented by elements within the MoMLV promoter, thus allowing transgene expression to continue during latency. When LAP2 was added to the LAP1-reporter cassette in the ectopic site within the genome, long-term expression within PNS neurons was restored although the transcription start site was not determined [231]. The LAP1–LAP2 complex was capable of driving long-term transgene expression when a lacZ reporter gene cassette was introduced into the LAT intron in the native LAT locus [233] or when a LAP2-lacZ expression cassette was present in an ectopic locus within the viral genome [219]. We have also shown that a LAP2-NGF cassette present either in the tk (Figure 7.6A and B) or Us3 loci of the vector expressed this gene product in latently infected rodent neurons both in culture and *in vivo* [234–236], and that expression of the therapeutic gene was sufficient to block toxic insult *in vitro* or in diabetic animals. Recently, we have been able to use LAP2 alone to direct expression of NGF [237] or in conjunction with the HCMV IE promoter to drive expression of NT-3 [237,238] to achieve long-term neuroprotection in the PNS of rodents that correlated with functional recovery. While LAP2 activity in CNS neurons appears somewhat weaker than in PNS neurons, we are easily able to achieve therapeutic levels of GDNF expression that result in a change in rotational behavior of LAP2-GDNF vector-injected 6-hydroxydopamine (6-OHDA) treated animals even when the vector was injected 6 months prior to treatment of the rats with 6-OHDA, whereas animals treated with the HCMV IEp-GDNF vector were only capable of restoring behavior when injected 1 week prior to toxic insult [239]. This demonstrates the overall utility of LAP2 for therapeutic gene expression in neurons of the PNS, CNS, and spinal cord. Of interest, we have shown that both the HCMV IE gene promoter and LAP2 are capable of long-term NGF expression in rabbits following intra-articular injection with a replication-defective HSV

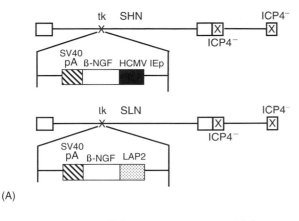

(A)

	3 days		14 days	
	Cell viability	Protective enzymes induced*	Cell viability	Protective enzymes induced*
SLN	−	−	+	+
SHN	+	+	−	−
Control	−	−	−	−
NGF protein	+	+	+	+

(B)

FIGURE 7.6 HSV vector-mediated expression of NGF protects neurons from peroxide toxicity. (A) Vectors SHN and SLN are replication-defective due to a deletion of the essential IE gene ICP4. SHN expresses NGF from the HCMV IE promoter from within the HSV-TK locus while SLN utilizes the latency-active promoter, LAP2, to drive NGF expression. (B) Transduction of primary DRG neurons with SHN induces cellular enzymes that protect neurons from peroxide toxicity at three days postinfection but not at 14 days. Conversely, SLN expression of NGF takes several days to accumulate until protection is evident at 14 days postinfection. Using these divergent promoter systems we are able to direct expression of neuroactive gene products at high levels for a short term (HCMV) or at somewhat lower levels long term (up to one year).

vector [191] and in rabbit fat tissue explants and subcutaneous adipose tissue following direct injection of vector into the rabbit fat-pad [240]. These results have uncovered two important findings: First, that the HCMV IE promoter can remain active long term although we do not fully understand the mechanism behind this activity; and second, that LAP2 is active in cells other than neurons such as adipose cells and cells of the ligaments. These findings will be important for designing treatment regimens where continuous therapeutic protein production will be required to achieve a therapeutic outcome in tissues other than those of the nervous system.

7.4.3 DRUG-REGULATED PROMOTERS

For many applications, it may be desirable to regulate expression of a therapeutic transgene *in vivo*, either to limit

toxic effects from high-level expression or more closely resemble physiological expression profiles. We created a viral vector with regulatable transgene expression using an autoregulatory loop that consisted of a promoter with 5 tandem copies of the 17-bp Gal4 DNA recognition element that could be transactivated by vector-encoded chimeric Gal4/VP16 protein (Figure 7.7A), based on the ability of the chimeric protein to transactivate promoters containing this site [241–243] despite the repressive presence of nucleosomes [244,245]. The constitutive Gal4/VP16 transactivator was able to induce transgene expression from a Gal4-sensitive minimal promoter in the background of the virus [189]. Regulation was achieved by replacing the constitutive transactivator with a chimeric molecule consisting of the hormone-binding domain of the mutated progesterone receptor fused to the transactivation domain of VP16 and DNA-binding domain of Gal4 [246,247]. In the presence of the progesterone analog RU486, the inactive chimeric transactivator assumes a conformation allowing it to bind to and transactivate the Gal4 recognition site-containing promoter driving transgene expression. Compared with control (Figure 7.7B), completion of the autoregulatory loop following administration of RU486 (Figure 7.7C) resulted in substantial enhancement of expression of the transgene in the CNS [189] (Figure 7.7D). Following infection of the rat hippocampus with the regulatable virus, levels of viral vector-derived transgene expression are stimulated by administration of the inducing agent RU486 [248].

7.5 APPLICATIONS

7.5.1 HSV GENE TRANSFER FOR NEUROPATHY AND PAIN

The PNS, specifically the sensory neurons with cell bodies in the DRG and peripheral axonal terminals in the skin, represents a natural target for the therapeutic use of HSV-mediated gene transfer. Recombinant replication-defective HSV-based vectors, like the parental wild-type virus, target with high efficiency to DRG neurons from skin inoculation, in part because sensory nerve terminals expresses high amounts of the HveC HSV receptor [249]. Following fusion of the viral envelope with the cell membrane, specific interactions between viral capsid proteins and the cytoplasmic motor ATPase dynein along with its multisubunit cofactor dynactin ensure efficient retrograde axonal transport along microtubules to the nucleus [250,251] where the viral DNA is injected through a nuclear pore. Because our recombinant vectors are incapable of replication, the recombinant viral genome is forced into a pseudo-latent state as an intranuclear but extrachromosomal element that may persist for the life of the host.

There are two major conditions of peripheral nerve that are particularly well suited for treatment by HSV-mediated gene transfer. The first is peripheral polyneuropathy, a condition resulting from degeneration of peripheral sensory neurons.

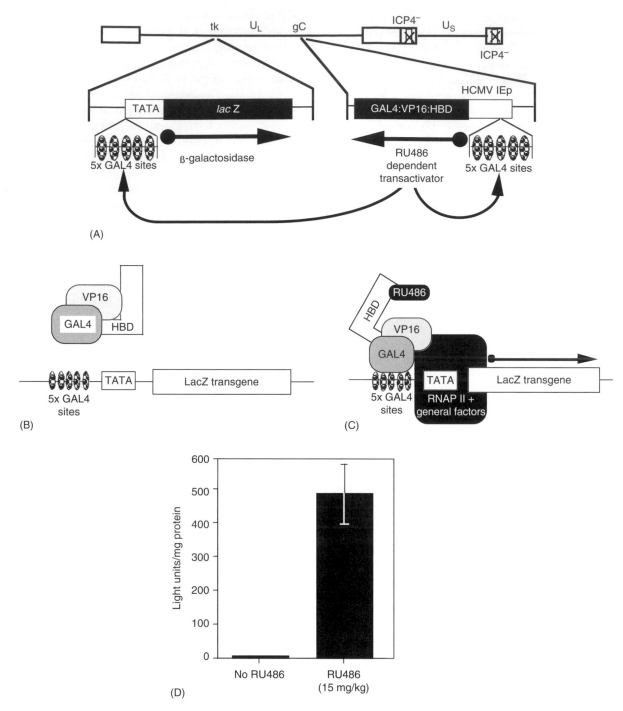

FIGURE 7.7 *In vivo* regulated transgene activation from HSV vectors by drug-inducible recombinant transactivator (RTA). (A) GLVP RTA vector contains the inducible chimeric transactivator GAL4:VP16:HBD, composed of the yeast GAM DNA-binding domain fused to the HSV VP16 transactivation domain fused in frame to a mutant form of the progesterone receptor hormone-binding domain, which can be activated by RU486. In addition, the vector contains a promoter–reporter cassette (GAL4TATA-lacZ) that is responsive to activation by the transactivator (RTA). (B) In the absence of the drug, the GAL4:VP16:HBD RTA cannot bind to the GAL4 sites in the minimal TATA box promoter, resulting in no expression of the ß-galactosidase transgene. (C) Following administration of RU486, the tripartite transactivator undergoes a conformational change that allows the RTA to bind to the GAL4-binding sites to yield ß-galactosidase expression. (D) Quantitation of ß-galactosidase transgene expression in the presence of RU486 displays activation of the transgene promoter by the inducible transactivator following injection of the vector into rat CNS.

The most common cause of neuropathy in the developed world is diabetes; diabetic polyneuropathy is a very common condition 100 times more prevalent than Parkinson's disease, Alzheimer's disease, and multiple sclerosis combined and a cause of substantial morbidity. The second condition that may be approached using HSV-mediated gene transfer is chronic pain. In both polyneuropathy and chronic pain, therapeutic peptides of proven efficacy have been identified. For example, polyneuropathy can be prevented in animal models by treatment with neurotrophic factors; pain can be substantially ameliorated by the delivery of opioid peptides. However, it is difficult to deliver these short-lived peptide factors in adequate doses to achieve therapeutic effects without causing intolerable side effects that result from the widespread expression of the receptors for these peptides throughout the nervous system and in non-nervous structures in the body. Theoretically, targeted gene delivery to the DRG using HSV vectors can be used to express a neurotrophic factor to prevent neurodegeneration, or to express a pain-relieving peptide to modify nociceptive neurotransmission at the spinal level, while avoiding the off-target effects that are engendered by systemic administration of these same agents.

7.5.1.1 Neuropathy

The initial proof-of-principle studies of HSV-mediated gene transfer to treat neuropathy examined the effect of vector-mediated expression of NT-3 in neuropathy caused by pyridoxine (vitamin B6). Pyridoxine overdose causes a selective degeneration of large DRG neurons that are known to express the Trk C high affinity NT-3 receptor, and previous work had demonstrated that daily subcutaneous co-administration of NT-3 (5–20 mg/kg) with pyridoxine prevented the development of neuropathy [252]. Using an identical paradigm of pyridoxine intoxication we demonstrated that subcutaneous inoculation of a replication-incompetent HSV vector expressing NT-3 prevented the degeneration of large myelinated fibers characteristic of neuropathy as measured by electrophysiologic, morphologic, and behavioral outcomes [253]. In this model, neuropathy develops subacutely over the course of 2 weeks. In a subsequent study, we demonstrated that with appropriate choice of promoter, a prolonged neuroprotective effect can be achieved. Rats inoculated subcutaneously with a nonreplicating HSV vector containing the NT-3 coding sequence under the control of the HSV LAP2 element [219] protected against the neurotoxic effects of pyridoxine administered five and a half months after vector inoculation [237], using the same measures examined in the subacute intoxication paradigm.

We subsequently examined the effect of HSV-mediated expression of NGF or NT-3 in the clinically relevant model of neuropathy caused by administration of a chemotherapeutic drug. Peripheral sensory polyneuropathy is a dose-related predictable off-target effect of many chemotherapeutic agents like cisplatin [254] that are used to treat cancer, and the effect on peripheral nerve either limits the use of these agents or results in toxicity that impairs quality of life. Subcutaneous

inoculation of nonreplicating HSV vectors expressing either NGF or NT-3 three days prior to the onset of treatment with cisplatin prevented the development of neuropathy assessed by standard electrophysiologic, behavioral, and histologic parameters [255].

HSV vectors have similar neuroprotective effects in protecting against the development of diabetic polyneuropathy. Administration of streptozotocin (STZ) to mice results in degeneration of beta cells, and the resultant hypoinsulinemic state serves as a model of type 1 diabetes. These animals develop a mild predominantly sensory neuropathy characterized by reductions in sensory nerve amplitude, and distal axonal degeneration that can be characterized by the loss of sensory fibers in the skin peripherally and a loss of neuropeptide containing afferent terminals in the spinal cord centrally. In initial studies we demonstrated that subcutaneous inoculation of an NGF-expressing HSV vector 2 weeks after the onset of diabetes protected against the development of neuropathy measured 4 weeks later [256]. We subsequently demonstrated that similar neuroprotective effects could be achieved using an HSV vector engineered to express the vascular endothelial growth factor (VEGF) in DRG neurons *in vivo* [257]. More recently, we have shown that subcutaneous inoculation 2 weeks after the onset of diabetes of an HSV vector expressing NT-3 under the control of the LAP2 element is effective in preserving sensory nerve function measured by electrophysiologic and histologic parameters [238]. Interestingly, in these experiments we found that HSV-mediated expression of NT-3 protected small-diameter fibers expressing substance P (SP) and calcitonin gene-related peptide (CGRP) despite the fact that these fibers are conventionally thought of as expressing predominantly the high affinity NGF receptor TrkA. This observation is congruent with our previous observation that HSV-mediated expression of NGF protects large diameter (predominantly TrkC-expressing) DRG fibers from the toxic effects of pyridoxine overdose [253], and suggests that local continuous expression of these trophic factors may provide more potent effects than would be anticipated from studies of bolus peptide administration.

In each of the experimental models described above, the HSV vector was inoculated prior to the onset of neuropathy. We have not examined whether the vectors might be used to enhance recovery from neuropathy, but achieving prevention alone would be a major advance in the treatment of neuropathy. Most sensory neuropathies develop gradually over a period of months to years; patients with diabetes, for example, characteristically present with loss of sensation in their toes, but there is no available therapy that will prevent progression of neuropathy. Gene transfer at that stage, if it prevented the development of a severe sensory neuropathy, would represent a substantial advance in our therapeutic armamentarium. Neuropathy caused by chemotherapeutic drugs represents one model in which the onset of neuropathy is temporally defined and pretreatment is possible, so that this category of neuropathy may serve as a convenient model for the first human trials of HSV gene therapy for neuropathy.

7.5.1.2 Pain

Using HSV vectors that express either inhibitory neurotransmitters or anti-inflammatory peptides, an analogous approach using HSV-mediated gene transfer can be employed to modify nociceptive neurotransmission. Wilson and colleagues first reported that a replication-competent HSV-based vector engineered to express preproenkephalin could be injected subcutaneously into the dorsum of the foot to transduce DRG and produce an antihyperalgesic effect as demonstrated by an increased latency to withdraw the foot from noxious heat after sensitization of C-fibers by application of capsaicin, or sensitization of Aδ-fibers by application of dimethyl sulfoxide [258]. These investigators subsequently demonstrated similar effects on nociceptive neurotransmission in primates [259].

We and others extended these results to models of chronic pain using a nonreplicating HSV vector engineered to express enkephalin. We demonstrated that subcutaneous inoculation of the vector reduced spontaneous pain behaviors in the delayed phase of the formalin test of inflammatory pain [260], and in the selective spinal nerve ligation (SNL) model of chronic neuropathic pain [261]. The effect of vector-mediated enkephalin production was continuous over time, and was reversed by opiate antagonists naloxone or naltrexone. Transduction of DRG with the enkephalin-expressing vector enhanced the effect of morphine. Uninoculated animals with neuropathic pain responded to intraperitoneal morphine with an ED_{50} of 1.8 mg/kg but animals inoculated with the enkephalin-expressing vector demonstrated a reduction in the ED_{50} of morphine to 0.15 mg/kg [261]. In addition, expression of the delta opioid receptor agonist enkephalin from the vector continued to produce analgesic effects in animals that had been rendered tolerant to the analgesic effect of morphine [261].

Pain resulting from cancer metastatic to bone has features of both inflammatory and neuropathic pain. To evaluate the potential therapeutic effect of HSV-mediated gene transfer and expression of preproenkephalin (PE) in pain due to cancer, we tested the vector in the osteogenic sarcoma model in the mouse. Tumor-injected mice demonstrated spontaneous pain, increasing to 2 weeks after tumor inoculation and remaining at the same level up to 4 weeks postinoculation. Subcutaneous inoculation of the enkephalin-expressing vector 1 week after tumor implantation resulted in a substantial and significant reduction in spontaneous behavior recorded 2 and 3 weeks after tumor implantation (1 and 2 weeks after vector inoculation), an effect that was reversed by intrathecal naltrexone.

Taken together, the results of studies demonstrate that a nonreplicating genomic HSV vector expressing PE is antinociceptive in models of neuropathic pain, inflammatory pain, and pain resulting from cancer in rodent models. A proposal for the first human trial employing these vectors in the treatment of intractable pain resulting from cancer metastatic to a vertebral body was presented to the Recombinant DNA Advisory Committee at the National Institutes of Health in June 2002. In the near future, we should be able to determine if this approach will be as successful in treating the human disease as it has been in the animal models.

In addition to expression of preproenkephalin in various animal models of pain, we have successfully constructed and utilized numerous other HSV-based vectors expressing a variety of genes in several different pain models, two of which will be briefly reviewed below. We engineered a vector to express glutamic acid decarboxylase (GAD) in DRG following peripheral footpad inoculation (Figure 7.8A). In the SNL model of neuropathic pain, the HSV-GAD vector effectively reversed both the resultant thermal hyperalgesia as well as mechanical allodynia (Figure 7.8B). These effects, which also suppressed c-Fos induction in the spinal cord, lasted at least 6 weeks. Importantly, these antinociceptive effects were effectively re-established by reinoculation of the vector without loss of effect or development of tolerance (Figure 7.8B). This ability to effectively re-administer vector and re-establish the therapeutic effect is a very important attribute and is in contrast with most other gene transfer vector systems.

Another use of the GAD vector is to address the large unmet need for therapies focusing on pain due to spinal cord injury. Spinal cord injury pain is, in general, not effectively treated with standard opioid drugs leading to the large unmet need for this indication. Delivery of HSV-GAD vector by peripheral footpad inoculation leads to constitutive γ-amino butyric acid (GABA) release in the spinal cord. GABA, a major inhibitory neurotransmitter, is involved in regulation of neuronal activity and mimicked by Baclofen that has shown some positive effects in clinical trials for spinal cord injury pain. Both thermal hyperalgesia and mechanical allodynia associated with a T13 hemisection were alleviated with HSV-GAD. Spinal CGRP levels were attenuated by GAD treatment and these antinociceptive effects were antagonized by bicuculline or phaclofen. We are currently evaluating specific indications and market potential for a GAD-based therapeutic prior to human clinical trials.

Processed enkephalin binds with high affinity to the ∂-opioid receptor while the widely used acute pain drug morphine, as well as most other common pain medications, binds to the μ-opioid receptor. Recently, endomorphins have been identified from brain homogenates and may represent endogenous μ-receptor ligands [262,263]. The endomorphins are carboxyamidated tetrapeptides that have been found to be effective in several animal models of pain. However, while the peptide is present in nervous tissues from vertebrates, the corresponding gene has not been identified in the genome nor has any transcript been identified. In order to improve potential endomorphin approaches to pain management, we engineered a tripartite gene product to direct the synthesis, processing, amidation, and regulated secretion of endomorphin-2 (EM-2) with the amino acid sequence of YPFF-NH$_2$ [264]. This synthetic EM-2 gene was recombined into a replication-defective HSV vector and EM-2 expression confirmed by HPLC-mass spectroscopy and radioimmunoassay following *in vitro* and *in vivo* transduction, respectively. The EM-2 vector was found to effectively

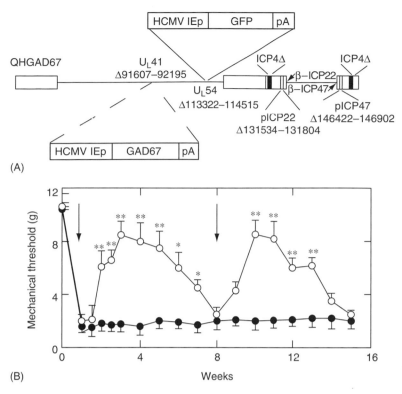

FIGURE 7.8 HSV-mediated GAD delivery for neuropathic pain therapy. (A) Schematic diagram of the replication-defective HSV vector engineered to express GAD67 as well as the marker gene GFP. The GAD expression vector is defective for both essential IE genes ICP4 and ICP27 (U_L54). In addition, deletion of the TATGARAT element within the promoters for ICP22 and ICP47 restricts their expression to ICP4 complementing cell lines. The GAD67 cDNA is driven by the HCMV IE promoter from within the U_L41 locus while GFP is expressed from the U_L54 (ICP27) locus. (B) Ligation of the L5 spinal nerve in rats caused a significant decrease in the threshold to tactile stimulation, which persisted for more than 4 months and was unaffected by inoculation of control vector (arrow; closed circles). Subcutaneous inoculation of the GAD expression vector (arrow; open circles) produced an antiallodynic evidenced by an increase in the mechanical threshold required to stimulate a pain response. Seven weeks after vector administration the antinociceptive effect diminished. Upon reinoculation (second arrow) the antiallodynic effect was re-established that lasted for an additional 7 weeks.

reverse thermal hyperalgesia and mechanical allodynia, hallmarks of SNL induced neuropathic pain [264]. This effect was reversed by intrathecal CTOP administration, a μ-opioid receptor specific antagonist.

In summary, we have utilized the innate aspects of our replication-defective HSV vector platform and the natural biology of pain transmission to develop several avenues toward a novel pain therapeutic. Using our platform, we are able to achieve relatively long-term effects (several weeks/dose) with the HCMV promoter and extended (months/years) using our LAP2 system. Importantly, with this platform, we are able to re-establish the therapeutic effect by reinoculation. Moreover, with the genetic flexibility of our HSV platform, we are able to include several distinct approaches, for example, GABA receptor agonism with the GAD gene and the ∂-opioid receptor agonism with the enkephalin gene, into a single vector; an approach we are currently evaluating.

7.5.2 CANCER

Cancer gene therapy may offer a treatment modality to patients who have exhausted all other standard treatment

regimens such as surgery, chemotherapy, and radiation therapy. There are a number of considerations in applying gene therapy to the treatment of cancer that include the selection of the appropriate therapeutic gene(s), the specific effect or mechanism, target tissue, and method of gene delivery. The overriding problem is that cancer is generally a systemic disease, and thus, even if gene transfer is effective in destroying a tumor locally, metastases will promote continued disease.

Strategies to treat cancer by gene therapy can be considered in three categories: (1) tumor cell destruction using conditionally replicating viruses that selectively replicate in and kill tumor cells [265,266] compared with the surrounding normal tissue, (2) tumor cell destruction by expression of transgenes whose products induce cell death, or sensitize the cells to chemo-[267] or radiation therapy [268], and (3) tumor vaccination through expression of transgenes whose products recruit, activate, or costimulate immunity or provide tumor antigens. The latter approach is more likely to be effective in treating metastatic disease. Because these strategies are complementary, it has also been suggested that they can be used in combination. Examples of these various approaches include the use of (1) prodrug-activating genes such as thymidine kinase (TK) or cytosine deaminase (CD) [269]; (2) cytokines such as TNF-α, γIFN, and various interleukins [270]; (3) MHC products such as costimulatory molecules (B7.1) [271–273]; (4) allotypic class I or class II molecules [271–274]; and (5) tumor antigens [275,276], which together may assist in the recruitment and activation of nonspecific inflammatory responses [277] or the induction of tumor-specific immunity.

The first strategy involves the use of vectors that are replication competent, but depend on attributes unique to the tumor cell to support viral growth. For example, E1b-deficient AdV vectors can replicate in tumor cells mutant for p53 but generally not in normal cells [265,278–280]. Thus, the intent of this strategy is to provide a mechanism for virus spread locally in the tumor to increase the number of infected tumor cells. However, this treatment is limited to p53-defective tumors. Similarly, HSV vectors have been engineered that replicate in dividing cells, such as tumor cells, but not in normal neurons. The use of conditional replication-competent viruses could in theory allow for spread in tumor tissue without damaging normal brain, thereby increasing the specificity and effectiveness compared with nonreplicating vectors that express transgenes that augment tumor cell killing. HSV vectors of this type include mutants lacking the γ 34.5 gene, which is required for growth specifically in neurons [281–283]. Deletion of this gene alone (1716), or in combination with the UL39 ribonucleotide reductase (RR) large subunit (G207), creates viruses that are highly compromised for their ability to replicate in and kill neuronal cells, yet that retain the ability to replicate in and kill dividing tumor cells.

Although these vectors were originally used to treat animal models of malignant glioma [284–288] they have now been employed to treat breast [289,290], lung [291,292], head and neck [293], melanoma [294,295], colorectal [296–298], prostate [299–302], ovarian [303,304], peritoneal [305–307], bladder [300,308], renal [309], cervical [310], and gallbladder [311] tumors in various animal models, demonstrating their utility. In addition, they have also been used in conjunction with suicide gene therapy (SGT) [312], low-dose ionizing radiotherapy [298,310], and chemotherapeutic agents like cisplatinum [292,313].

The second approach to treat cancer involves SGT for the treatment of cancer in experimental animals and in phase I human clinical protocols [267,314–318]. This strategy uses the bystander destruction of tumor cells mediated by a variety of mechanisms other than virus spread, including the recruitment of natural killer (NK) cells by expressing the appropriate cytokines, the activation of anticancer drugs at the tumor site that kill multiple tumor cells in addition to those transduced by the vector, and the use of antigens and cytokine-expressing genes to elicit specific antitumor immunity. Transfer of the HSV gene TK into tumor cells results in tumor cell death when combined with the antiviral drug ganciclovir (GCV). TK has been shown to convert the prodrug into a toxic nucleoside analog that, upon incorporation into nascent DNA, results in the interruption of DNA replication by chain termination. A uniquely powerful characteristic of the TK-GCV approach is that only a small fraction of the tumor cells need to be transduced with the suicide gene to result in significant antitumor activity, an activity known as the "bystander effect" [315,317–320]. It has been demonstrated that cell-to-cell transfer of activated GCV via gap junctions between transduced tumor cells and untransduced neighboring cells is a major mechanism of the bystander effect [321–324]. A variation on this strategy is to introduce cell lines into the tumor site that produce TK-expressing retroviruses that then infect tumor cells locally. Although setting up virus production factories in this manner is logical, the practice of this strategy has been disappointing because the xenogeneic producer cell lines induce rapid inflammatory processes that lead to brain swelling and little detectable gene transfer [325]. Although these TK-GCV SGT strategies suffer from limitations, approaches of this nature are under evaluation for efficacy in phase I–II clinical trials for patients with brain tumors; however, they have met with limited success.

In our initial experiments using replication-defective HSV vectors expressing HSV-TK for SGT, we employed both vectors deleted for single (ICP4⁻) and multiple (ICP4⁻/ICP27⁻/ICP22⁻) essential IE genes that expressed the TK gene from an HSV IE promoter to ensure its expression from the replication-defective virus backbone (Figure 7.9A). We tested the ability of these TK overexpressing replication-defective HSV vectors to act as a treatment for established tumors in rodent glioma models and demonstrated significant increases in survival following administration of the HSV-TK vector and GCV [326,327]. However, the magnitude of the bystander effect was inversely proportional to the overall toxicity of the vector. Thus, more cytotoxic vectors like the ICP4 deletion mutant resulted in the cytotoxic death of the transduced tumor cells before they were able to produce and release significant levels of modified GCV, thereby dampening the overall bystander-mediated killing [326]. We have now seen similar results using vectors that express CD alone or in combination with HSV-TK [328], suggesting that further modifications will be required to achieve more effective tumor cell killing.

To augment the cell killing seen in SGT we have taken two approaches. In the first approach, we attempted to augment the bystander effect by altering the make up of tumor cell gap junction complexes. Connexins are the major components of gap junctions [329] that play a major role in intercellular communication to control homeostasis and cell proliferation. Reduced intercellular communication through gap junctions has been regularly observed in transformed cells [330–332] and may be due to reduced connexin expression. Retrovirus-mediated introduction of the connexin 43 (Cx43) gene limited the growth of transformed cells that shared the characteristic of reduced gap junctional activity [333], and other studies have reported similar findings supporting the suggestion that connexins alone have tumor-suppressor activity [334–337]. Gap junctions also play a critical role in the HSV-TK/GCV bystander effect by enabling the transfer of activated GCV from TK-positive to neighboring TK-negative cells [321,338–341]. Tumor cells having reduced gap junction formation are less susceptible to the bystander effect [341], suggesting that transfer of connexin genes into

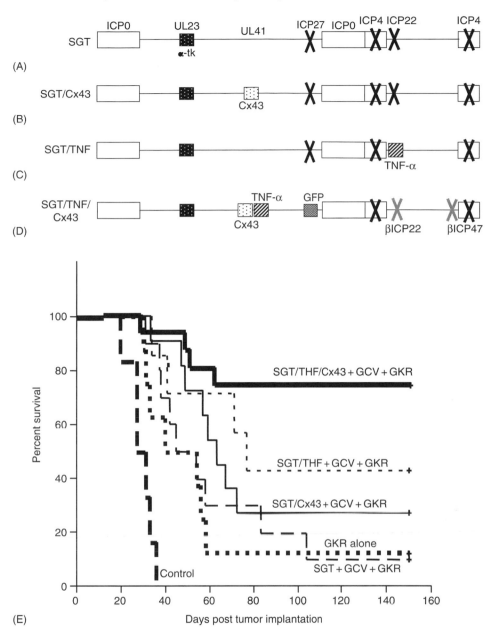

FIGURE 7.9 HSV-vector mediated tumor cell killing. Diagrams of replication-defective HSV-1 expression vectors for expressing (A) HSV-TK (SGT), (B) HSV-TK and connexin 43 (SGT/Cx43), (C) HSV-TK and TNF-α (SGT/TNF), and (D) HSV-TK, connexin 43, and TNF-α (SGT/TNF/Cx43). All vectors express ICP0 from both copies of the IR element flanking the UL segment of the HSV genome, and TK as an IE gene from a copy of the ICP4 IE promoter replacing the native TK promoter in the UL23 gene locus. All vectors have the ICP4 and ICP27 genes deleted, and all but the SGT/TNF/Cx43 vector have inactivating deletions in ICP22. This vector has the IE genes ICP22 and ICP47 turned into early genes by deletion of the TAATGARAT promoters element converting these to early (β) genes. The SGT/Cx43 vector contains the Cx43 gene driven by the ICP0 IE promoter inserted into the U_L41 locus. The SGT/TNF vector has TNF inserted into the ICP22 gene locus under control of the HCMV IE promoter. The SGT/TNF/Cx43 vector has both the ICP0p-Cx43 and HCMV IEp-TNF expression cassettes inserted into the U_L41 locus. The ability of these vectors to destroy tumors and effect animal survival in the immuno-competent rat 9L tumor model was evaluated by intratumoral vector injection 3 days following implantation of the 9L tumor cells into the frontal lobe of Fisher rats. Animals received i.p. injections of GCV at the time of vector injections for 10 consecutive days and gamma-knife radiosurgery (GKR) at 2 days postvector injection.

these cells to restore or augment intercellular communication would improve the effectiveness of HSV-TK/GCV therapy. Indeed, it has been demonstrated both *in vitro* and *in vivo* that the bystander effect can be potentiated by the expression of connexin [340,342,343]. The bystander effect requires connexin expression not only by the TK-positive GCV-activating cell, but also by the TK-negative recipient cell [341]. This suggests that gene therapy approaches aimed at codelivery of connexin with HSV-TK to mediate an enhanced bystander effect may not be extremely effective against tumors that express no connexin at all as a study of 17 cell lines measuring bystander effects and gap junctional activity [339] demonstrated. Furthermore, overexpression of connexin in the transduced cells will increase transfer of activated GCV, providing even limited connexin expression in neighboring cells.

We tested the potential benefit of coexpressing connexin with HSV-TK for the treatment of glioblastoma [344]. Human U-87MG tumor cells express detectable amounts of Cx43 and are sensitive to bystander killing *in vitro*, but this effect is not sufficient to control tumor formation with TK-transduced cells and GCV treatment alone [330]. Replication-defective vectors (ICP4$^-$/ICP27$^-$/ICP22$^-$) that we engineered to express Cx43 (Figure 7.9B) were able to enhance TK-GCV tumor cell killing in cultures of U-87MG tumor cells that express Cx43 [344]. Thus, although U-87MG cells already showed a good bystander effect, the expression of additional Cx43 in a fraction of the population further enhanced the effectiveness of GCV treatment. We also were able to demonstrate a pronounced bystander effect with L929 cells [344] that did not show a bystander effect when infected with a TK-expressing vector, in agreement with recent findings by others [341]. This indicated that vector-directed connexin expression enabled bystander killing among these otherwise bystander-resistant cells. Together, these results suggested that vector-directed Cx43 expression should be beneficial regardless of whether the target cells express significant levels of connexin.

These *in vitro* assays were extended to animals using an *ex vivo* flank tumor model and *in vivo* using animals bearing U-87MG tumors in the CNS [344]. Dramatically, all animals implanted with Cx43 vector-infected cells and treated with GCV were tumor free 1 week after cessation of GCV administration, whereas no other animal was tumor free at this time, including animals treated with Cx43 vector alone or the TK-expressing vector plus GCV. Moreover, all animals in this treatment group were alive at the end of the observation period (72d), whereas no animals in any of the other groups survived past day 41. Encouraged by these promising results, experiments were initiated to test the effectiveness of combined Cx43/TK gene delivery *in vivo* because the *ex vivo* approach enables one to infect every tumor cell that does not readily mimic the actual situation in human patients with glioblastoma. In the *in vivo* experiments, virus was injected directly into the tumor mass 3 days following tumor cell implantation and GCV was administered for 10 days after vector inoculation. In these studies, all animals in every treatment group died by day 50, whereas 50% of the Cx43-expression vector-treated animals survived past 50 days and one-third were still alive at the end of the study (70d), indicating a beneficial effect of connexin/TK gene codelivery *in vivo*.

In the second approach to augment SGT, we created vectors that express cytokines in the hope of stimulating a host response to the tumor. There has been a considerable amount of recent interest in using cytokine genes, costimulatory molecules, tumor antigens, and recruitment molecules to enhance the immune response to the tumor. Antitumor immunity should prove effective in treatment of metastatic cancer. The

development of antitumor immunity could circumvent the need for replication-competent vectors because tumor-specific cytotoxic T lymphocytes (CTLs) constantly move through the brain parenchyma searching for target cells. A growing body of literature suggests that local expression of cytokines can enhance CTL activation at least in animal model systems and these bear testing in human brain cancer. HSV offers the potential for combinational gene therapy in this regard because multiple immunomodulatory genes can be recombined into the virus and comparatively tested [94].

TNF-α has been demonstrated to possess an array of antitumor activities, including potent cytotoxicity exerted directly on tumor cells [345], enhancement of the expression of HLA antigens [346] and ICAM-1 [347] on tumor cell surfaces, enhancement of interleukin-2 receptors on lymphocytes [348], and stimulation of such effector cells as NK cells, lymphokine-activated killer cells, and CTLs [348–351]. However, despite this promising antitumor profile, the clinical use of TNF-α has been constrained by the toxicity of systemic TNF-α delivery [349,351]. This problem could be minimized by local production of TNF-α at the site of tumor growth, which may allow for effective use of this cytokine as an antitumor agent. Furthermore, TNF-α has a radiosensitizing ability that could optimize its antitumor effects. In an effort to augment the effectiveness of HSV-TK-mediated SGT, we created the replication-defective, triple IE gene-deleted HSV vector (ICP4$^-$, ICP27$^-$, ICP22$^-$) that expresses HSV-TK and TNF-α (Figure 7.9C).

In vitro studies demonstrated that high levels of TNF-α could be detected in the media of vector-infected cells during the first 24h period, but this was followed by a precipitous decline in production on day 2 and subsequent days with the protein being no longer detectable on day 7 [327]. The bioactivity of TNF-α produced in this experiment was tested by exposure of cultured TNF-α-sensitive L929 fibrosarcoma cells to medium collected after the first day of infection. A dramatic reduction in cell viability was observed for cells treated with medium from TNF vector-infected cells, and this reduction was essentially identical over time to that seen with unconditioned medium supplemented with 10 ng recombinant TNF-α protein [327]. These results demonstrated that TNF-α produced by the TNF vector was biologically active and comparable in specific activity to recombinant TNF-α. We then determined that intracellular production of TNF-α could enhance HSV-TK/GCV-mediated cell killing of both the TNF-α-sensitive L929 cell line, as well as the TNF-α-resistant U-87MG cell line. Although the mechanism is unclear, this enhancement indicated that the combination of vector-directed TNF-α expression and HSV-TK expression with GCV treatment could be beneficial not only against TNF-α-sensitive tumor cells such as L929, but also against TNF-α-resistant tumors.

To determine if the increased effectiveness of combination gene treatment evident *in vitro* could also be observed *in vivo*, we first tested the effect of intratumoral vector injection followed by GCV treatment on established L929 tumors (TNF-α sensitive) in the flanks of immune-competent mice

[327]. Tumor treatment with the TNF vector plus GCV resulted in significantly greater growth inhibition and extended animal survival compared with all other treatments. These results demonstrated the promise of combination TNF/TK gene therapy for the treatment of TNF-sensitive tumors and added an incentive to test the same treatment against TNF-resistant tumors. However, the results in the U-87MG TNF-α-resistant intracerebral tumor model in immunodeficient mice were not as significant as those observed with the L929 model, although 2 of 14 animals in the TNF/GCV treatment group survived past 80 days [327]. This reduced response in the U-87MG model may be due to the fact that the tumor cell killing mediated by TNF is only the result of the cytotoxic effects of TNF expression intracellularly and is not augmented by its immunomodulatory ability. Together, these results suggest that combined HSV-TK/TNF therapy is beneficial, but that a more effective strategy may be required.

Fractionated radiotherapy has been shown to confer a small but significant survival benefit to patients with glioblastoma. Unfortunately, the dose of radiotherapy that may be tolerated by the brain (about 60 Gy) is inadequate for tumor eradication. To circumvent inherent toxicity problems, techniques have been developed that allow radiation focused to the tumor bed, allowing a higher dose to be delivered (radiosurgery). This enables eradication of the central portion of the tumor, but does not allow delivery of an augmented radiation dose to the tumor periphery. GKR allows for precise delivery of a single high dose of radiation to brain tumors without opening the skull. In this technique, tumors are targeted by the application of a tightly focused high-energy radiation field, which results in minimal collateral damage to the surrounding normal tissue. Radiosurgery has been used for boost irradiation of patients with malignant glial tumors, in addition to conventional wide-margin fractionated radiotherapy [352,353]. Unfortunately, glioma cells are often seen invading the normal tissue surrounding the tumor, often migrating along normal white matter tracts [354,355]. This feature of glioma is largely responsible for the inability to achieve a surgical cure by resection and the correspondingly poor prognosis. We have therefore examined ways in which the response to radiotherapy may be enhanced by gene delivery. One such approach involves selectively sensitizing tumor cells to radiotherapy. This would confer a major advantage, in that the sensitized cells could be effectively killed by a low dose of radiotherapy that is not toxic to surrounding brain tissue. In this strategy, we combine GKR with the injection of replication-defective HSV vectors expressing HSV-TK, Cx43, and TNF.

In the first series of experiments, we used GKR in conjunction with vector-mediated TNF expression because TNF has previously been shown to have a synergistic effect with ionizing radiation when delivered as a recombinant protein [356–362], by plasmid-based delivery [363–366] or using an AdV vector [367–371]. Moreover, the TNF-α approach has proven safe and shown some efficacy in human phase I clinical trials [371,372]. Experiments were carried out to determine whether HSV-TK/TNF gene transfer along with GCV treatment was more effective in the presence of low-dose

gamma-knife radiation. Both the TNF/TK vector and GKR alone were effective in protecting a proportion of animals from tumor growth and animal death when used in the U-87MG model of glioblastoma in nude mice [373]. The results using the TNF/TK vector with SGT and GKR demonstrated that the combination of TNF, GCV, and GKR was superior to other treatments, such that 89% of the animals in that treatment group surviving for the length of the study (75 days) and 67% of the animals were found to be tumor free at 75 days [373].

We then extended these results to an immunocompetent tumor model that may more closely mimic the human disease by carrying out survival studies comparing the efficiency of HSV-TK, TNF, and Cx43 in combination with GCV, with and without radiosurgery using the 9L gliosarcoma model in immunocompetent Fisher 344 rats [374]. In this study, GKR was found to enhance the survival of 9L intracranial tumor-bearing rats compared with the control (Figure 7.9E); however, treatment resulted in an overall survival of 15%, similar to what has been reported in human patients [352,375,376]. Combining the TNF and Cx43 genes in an HSV-TK/ICP0 vector (Figure 7.9D) further improved animal survival (Figure 7.9E). Eleven of fifteen animals treated with the HSV-TK/TNF/Cx43 vector, GCV, and GKR survived for more than 150 days compared with 7 of 15 treated with TNF, GCV, and GKR. These results demonstrate that our most effective current strategy for the treatment of animal brain tumors involves the multigene vector, which simultaneously expresses TNF-α Cx43, HSV-TK, and HSV-ICP0, combined with radiosurgery and GCV treatment. This combination approach employs genes whose products that are tumoricidal (HSV-TK, TNF), augment this process by increasing the bystander effect (Cx43), sensitize tumor cells to radiation (TNF), and stimulate the host immune response to the tumor (TNF). This combination approach may also prove to be effective against metastatic disease, but remains to be tested in these animal models.

7.6 SUMMARY AND FUTURE DIRECTIONS

This overview of HSV biology and gene transfer has focused on the use of highly defective HSV genomic vectors that are blocked very early in the virus lytic cycle. These vectors express few viral functions and are highly reduced in vector toxicity, even for primary neurons in culture that are readily killed by less-defective HSV vectors. Moreover, these vector backgrounds are suitable for expression of multiple transgenes or single large genes (e.g., dystrophin) in applications where expression of single- or multiple-gene products are required to achieve a therapeutic outcome (e.g., tumor cell killing, vaccination). Expression of these transgenes can be coordinated, even sequentially, using strategies similar to those employed by the virus to regulate its own genes. Expression can also be controlled by drug-sensitive transactivators, which may prove to be important for regulating the timing and duration of transgene expression. HSV vectors may be most suited for expression of genes in the nervous system where the virus has evolved to remain lifelong in a latent state. The highly defective viruses deleted for multiple IE genes are able to efficiently establish residence in neurons and serve as a platform for long-term gene expression using the LAP2 system. These mutants are unable to reactivate from latency and cannot spread to other nerves or tissues following cell transduction. Delivery of these vectors requires direct inoculation of tissue to achieve direct contact with neurons. Ideally, HSV vectors would be most effective if infection could be targeted to specific cell types using enveloped particles defective for their normal receptor recognition ligands, but modified to contain novel attachment and entry functions. This area of research is still very early in development, and it remains to be determined to what extent this will be feasible. Finally, it should be emphasized that current viral delivery systems may each become reduced to highly defective transfer vectors, retaining only those elements required for vector DNA maintenance and transgene expression. Fortunately, the natural biology of many persistent viruses, including HSV-1, indicates that long-term vector maintenance will be possible and we continue to learn from the highly evolved biology of persistent and latent viruses in order to mimic their strategies for gene transfer and therapy.

REFERENCES

1. Cai, W., B. Gu, and S. Person. 1988. Role of glycoprotein B of herpes simplex virus type 1 in viral entry and cell fusion. *J. Virol.* 62:2596–2604.
2. Desai, P., P. Schaffer, and A. Minson. 1988. Excretion of non-infectious virus particles lacking glycoprotein H by a temperature-sensitive mutant of herpes-simplex virus type 1: Evidence that gH is essential for virion infectivity. *J. Gen. Virol.* 69:1147–1156.
3. Hutchinson, L., K. Goldsmith, D. Snoddy, H. Ghosh, F. L. Graham, and D. C. Johnson. 1992. Identification and characterization of a novel herpes simplex virus glycoprotein, gK, involved in cell fusion. *J. Virol.* 66:5603–5609.
4. Ligas, M. and D. Johnson. 1988. A herpes simplex virus mutant in which glycoprotein D sequences are replaced by β-galactosidase sequences binds to but is unable to penetrate into cells. *J. Virol.* 62:1486–1494.
5. Spear, P. 1993. Entry of alpha-herpes viruses into cells. *Sem. Virol.* 4:167–180.
6. Spear, P. 1993. Membrane fusion induced by herpes simplex virus. In: J. Bentz (Ed.), *Viral Fusion Mechanisms.* CRC Press, Boca Raton, Florida, pp. 201–232.
7. Steven, A. and P. Spear. 1997. Herpesvirus capsid assembly and envelopment. In: W. Chiu, R. Burnett, and R. Garcea (Eds.), *Structural Biology of Viruses.* Oxford University Press, New York, pp. 512–533.
8. Mettenleiter, T. C. 1994. Initiation and spread of α-herpesvirus infections. *Trends Micro.* 2:2–3.
9. Fuller, A. and W. Lee. 1992. Herpes simplex virus type 1 entry through a cascade of virus-cell interactions requires different roles of gD and gH in penetration. *J. Virol.* 66:5002–5012.
10. Fuller, A. and P. Spear. 1985. Specificities of monoclonal and polyclonal antibodies that inhibit adsorption of herpes simplex virus to cells and lack of inhibition by potent neutralizing antibodies. *J. Virol.* 55:475–482.
11. Highlander, S., S. Sutherland, P. Gage, D. Johnson, M. Levine, and J. Glorioso. 1987. Neutralizing monoclonal antibodies specific for herpes simplex virus glycoprotein D inhibit virus penetration. *J. Virol.* 61:3356–3364.
12. Bucks, M. A., K. J. O'Regan, M. A. Murphy, J. W. Wills, and R. J. Courtney. 2007. Herpes simplex virus type 1 tegument proteins VP1/2 and UL37 are associated with intranuclear capsids. *Virology* 361:316–324.
13. Chi, J. H., C. A. Harley, A. Mukhopadhyay, and D. W. Wilson. 2005. The cytoplasmic tail of herpes simplex virus envelope glycoprotein D binds to the tegument protein VP22 and to capsids. *J. Gen. Virol.* 86:253–261.
14. Farnsworth, A., T. W. Wisner, and D. C. Johnson. 2007. Cytoplasmic residues of herpes simplex virus glycoprotein gE required for secondary envelopment and binding of tegument proteins VP22 and UL11 to gE and gD. *J. Virol.* 81:319–331.
15. O'Regan, K. J., M. A. Bucks, M. A. Murphy, J. W. Wills, and R. J. Courtney. 2007. A conserved region of the herpes simplex virus type 1 tegument protein VP22 facilitates interaction with the cytoplasmic tail of glycoprotein E (gE). *Virology* 358:192–200.
16. Farnsworth, A., K. Goldsmith, and D. C. Johnson. 2003. Herpes simplex virus glycoproteins gD and gE/gI serve essential but redundant functions during acquisition of the virion envelope in the cytoplasm. *J. Virol.* 77:8481–8494.
17. Dingwell, K., C. Brunetti, R. Hendricks, Q. Tang, M. Tang, A. Rainbow, and D. Johnson. 1994. Herpes simplex virus glycoproteins E and I facilitate cell-to-cell spread in vivo and across junctions of cultured cells. *J. Virol.* 68:834–845.
18. Dingwell, K. S., L. C. Doering, and D. C. Johnson. 1995. Glycoproteins E and I facilitate neuron-to-neuron spread of herpes simplex virus. *J. Virol.* 69:7087–7098.
19. Duffy, C., J. H. Lavail, A. N. Tauscher, E. G. Wills, J. A. Blaho, and J. D. Baines. 2006. Characterization of a UL49-null mutant: VP22 of herpes simplex virus type 1 facilitates viral spread in cultured cells and the mouse cornea. *J. Virol.* 80:8664–8675.
20. Saksena, M. M., H. Wakisaka, B. Tijono, R. A. Boadle, F. Rixon, H. Takahashi, and A. L. Cunningham. 2006. Herpes simplex virus type 1 accumulation, envelopment, and exit in growth cones and varicosities in mid-distal regions of axons. *J. Virol.* 80:3592–3606.
21. Maul, G. G. and R. D. Everett. 1994. The nuclear location of PML, a cellular member of the C3HC4 zinc-binding domain protein family, is rearranged during herpes simplex virus infection by the C3HC4 viral protein ICP0. *J. Gen. Virol.* 75 (Pt 6): 1223–1233.
22. Mullen, M.-A., S. Gerstberger, D. M. Ciufo, J. D. Mosca, and G. S. Hayward. 1995. Evaluation of colocalization interactions between the IE110, IE175, and IE63 transactivator proteins of herpes simplex virus within subcellular punctate structures. *J. Virol.* 69:476–491.
23. Honess, R. and B. Roizman. 1974. Regulation of herpes simplex virus macromolecular synthesis. I. Cascade regulation of the synthesis of three groups of viral proteins. *J. Virol.* 14:8–19.
24. Campbell, M. E. M., J. W. Palfeyman, and C. M. Preston. 1984. Identification of herpes simplex virus DNA sequences which encode a trans-acting polypeptide responsible for stimulation of immediate early transcription. *J. Mol. Biol.* 180:1–19.
25. Mackem, S. and B. Roizman. 1982. Differentiation between alpha promoter and regulatory regions of herpes simplex virus type 1: The functional domains and sequence of a movable alpha regulator. *Proc. Natl. Acad. Sci. U S A* 79:4917–4921.
26. Preston, C., M. Frame, and M. Campbell. 1988. A complex formed between cell components and an HSV structural polypeptide binds to a viral immediate early gene regulatory DNA sequence. *Cell* 52:425–434.

27. DeLuca, N. A. and P. A. Schaffer. 1985. Activation of immediate-early, early, and late promoters by temperature-sensitive and wild-type forms of herpes simplex virus type 1 protein ICP4. *Mol. Cell. Biol.* 5:1997–2008.

28. Dixon, R. A. F. and P. A. Schaffer. 1980. Fine-structure mapping and functional analysis of temperature-sensitive mutants in the gene encoding the herpes simplex virus type 1 immediate early protein VP175. *J. Virol.* 36:189–203.

29. McCarthy, A., L. McMahan, and P. Schaffer. 1989. Herpes simplex virus type 1 ICP27 deletion mutants exhibit altered patterns of transcription and are DNA deficient. *J. Virol.* 63:18–27.

30. Sacks, W., C. Greene, D. Aschman, and P. Schaffer. 1985. Herpes simplex virus type 1 ICP27 is essential regulatory protein. *J. Virol.* 55:796–805.

31. DeLuca, N., A. McCarthy, and P. Schaffer. 1985. Isolation and characterization of deletion mutants of Herpes simplex virus type 1 in the gene encoding immediate-early regulatory protein ICP4. *J. Virol.* 56:558–570.

32. Sandri-Goldin, R. and M. Hibbard. 1996. The herpes simplex virus type 1 regulatory protein ICP27 coimmunoprecipitates with anti-sm antiserum, and the C terminus appears to be required for this interaction. *J. Virol.* 70:108–118.

33. Sandri-Goldin, R., M. Hibbard, and M. Hardwicke. 1995. The C-terminal repressor region of herpes simplex virus type 1 ICP27 is required for the redistribution of small nuclear ribonucleoprotein particles and splicing factor SC25; however, these alterations are not sufficient to inhibit host cell splicing. *J. Virol.* 69:6063–6076.

34. Sandri-Goldin, R. M. 1998. Interactions between a herpes simplex virus regulatory protein and cellular mRNA processing pathways. *Methods* 16:95–104.

35. Prod'hon, C., I. Machuca, H. Berthomme, A. Epstein, and B. Jacquemont. 1996. Characterization of regulatory functions of the HSV-1 immediate-early protein ICP22. *Virology* 226:393–402.

36. Rice, S., M. Long, V. Lam, and C. Spencer. 1994. RNA polymerase II is aberrantly phosphorylated and localized to viral replication compartments following herpes simplex virus infection. *J. Virol.* 68:988–1001.

37. Sacks, W. R. and P. A. Schaffer. 1987. Deletion mutants in the gene encoding the herpes simplex virus type 1 immediate-early protein ICP0 exhibit impaired growth in cell culture. *J. Virol.* 61:829–839.

38. Stow, N. and E. Stow. 1986. Isolation and characterization of a herpes simplex virus type 1 mutant containing a deletion within the gene encoding the immediate early polypeptide Vmw 110. *J. Gen. Virol.* 67:2571–2585.

39. Everett, R. D. 1984. Transactivation of transcription by herpes simplex virus products: Requirements for two herpes simplex virus type 1 immediate early polypeptides for maximum activity. *EMBO J.* 3:3135–3141.

40. Carter, K. and B. Roizman. 1996. The promoter and transcriptional unit of a novel herpes simplex virus 1 alpha gene are contained in, and encode a protein in frame with, the open reading frame of the alpha22 gene. *J. Virol.* 70:172–178.

41. Fruh, K., K. Ahn, H. Djaballah, P. Sempe, P. M. van Endert, R. Tampe, P. A. Petersen, and Y. Yang. 1995. A viral inhibitor of peptide transporters for antigen presentation. *Nature* 375:415–418.

42. Hill, A., P. Jugovic, I. York, G. Russ, J. Bennink, J. Yewdell, H. Ploegh, and D. Johnson. 1995. Herpes simplex virus turns off the TAP to evade host immunity. *Nature* 375:411–415.

43. York, I., C. Roop, D. Andrews, S. Riddell, F. Graham, and D. Johnson. 1994. A cytosolic herpes simplex virus protein inhibits antigen presentation to CD8+ T lymphocytes. *Cell* 77:525–535.

44. Holland, L. E., K. P. Anderson, C. Shipman, and E. K. Wagner. 1980. Viral DNA synthesis is required for efficient expression of specific herpes simplex virus type 1 mRNA. *Virology* 101:10–24.

45. Mavromara-Nazos, P., and B. Roizman. 1987. Activation of herpes simplex virus 1 γ2 genes by viral DNA replication. *Virology* 161:593–598.

46. Roizman, B. and A. Sears. 1990. Herpes simplex viruses and their replication. In: B. Fields, D. Knipe, R. Chanock, M. Hirsch, J. Melnick, T. Monath, and B. Roizman (Eds.), *Virology*, 2nd edn. Raven Press, Ltd., New York, pp. 1795–1841.

47. Johnson, P. A., A. Miyanohara, F. Levine, T. Cahill, and T. Friedmann. 1992. Cytotoxicity of a replication-defective mutant herpes simplex virus type 1. *J. Virol.* 66:2952–2965.

48. Johnson, P., M. Wang, and T. Friedmann. 1994. Improved cell survival by the reduction of immediate-early gene expression in replication-defective mutants of herpes simplex virus type 1 but not by mutation of the viron host shutoff function. *J. Virol.* 68:6347–6362.

49. Krisky, D., D. Wolfe, W. Goins, P. Marconi, R. Ramakrishnan, M. Mata, R. Rouse, D. Fink, and J. Glorioso. 1998. Deletion of multiple immediate early genes from herpes simplex virus reduces cytotoxicity and permits long-term gene expression in neurons. *Gene Ther.* 5:1593–1603.

50. Marconi, P., D. Krisky, T. Oligino, P. L. Poliani, R. Ramakrishnan, W. F. Goins, D. J. Fink, and J. C. Glorioso. 1996. Replication-defective herpes simplex virus vectors for gene transfer in vivo. *Proc. Natl. Acad. Sci. U S A* 93:11319–11320.

51. Samaniego, L., A. Webb, and N. DeLuca. 1995. Functional interaction between herpes simplex virus immediate-early proteins during infection: Gene expression as a consequence of ICP27 and different domains of ICP4. *J. Virol.* 69:5705–5715.

52. Wu, N., S. Watkins, P. Schaffer, and N. DeLuca. 1996. Prolonged gene expression and cell survival after infection by a herpes simplex virus mutant defective in the immediate-early genes encoding ICP4, ICP27, and ICP22. *J. Virol.* 70:6358–6368.

53. Wolfe, D., J. B. Wechuck, D. M. Krisky, J. P. Goff, W. F. Goins, A. Ozuer, M. E. Epperly, J. S. Greenberger, D. J. Fink, and J. C. Glorioso. 2004. Delivery of herpes simplex virus-based vectors to stem cells. *Methods Mol. Biol.* 246:339–352.

54. Chen, X., J. Li, M. Mata, J. Goss, D. Wolfe, J. Glorioso, and D. Fink. 2000. Herpes simplex virus type 1 ICP0 protein does not accumulate in the nucleus of primary neurons in culture. *J. Virol.* 74:10132–10141.

55. Samaniego, L. A., L. Neiderhiser, and N. A. DeLuca. 1998. Persistence and expression of the herpes simplex virus genome in the absence of immediate-early proteins. *J. Virol.* 72:3307–3320.

56. Samaniego, L., N. Wu, and N. DeLuca. 1997. The herpes simplex virus immediate-early protein ICP0 affects transcription from the viral genome and infected-cell survival in the absence of ICP4 and ICP27. *J. Virol.* 71:4614–4625.

57. Hobbs, W., and N. DeLuca. 1999. Perturbation of cell cycle progression and cellular gene expression as a function of herpes simplex virus ICP0. *J. Virol.* 73:8245–8255.

58. Lomonte, P. and R. D. Everett. 1999. Herpes simplex virus type 1 immediate-early protein Vmw110 inhibits progression of cells through mitosis and from G(1) into S phase of the cell cycle. *J. Virol.* 73:9456–9467.

59. Song, B., K. C. Yeh, J. Liu, and D. M. Knipe. 2001. Herpes simplex virus gene products required for viral inhibition of expression of G1-phase functions. *Virology* 290:320–328.

60. Everett, R. D., P. Lomonte, T. Sternsdorf, R. van Driel, and A. Orr. 1999. Cell cycle regulation of PML modification and ND10 composition. *J. Cell Sci.* 112(Pt 24):4581–4588.

61. Lomonte, P., K. F. Sullivan, and R. D. Everett. 2001. Degradation of nucleosome-associated centromeric histone H3-like protein CENP-A induced by herpes simplex virus type 1 protein ICP0. *J. Biol. Chem.* 276:5829–5835.

62. Quinlan, M. P. and D. M. Knipe. 1985. Stimulation of expression of a herpes simplex virus DNA-binding protein by two viral factors. *Mol. Cell. Biol.* 5:957–963.

63. Cai, W. and P. A. Schaffer. 1989. Herpes simplex virus type 1 ICP0 plays a critical role in the de novo synthesis of infectious virus following transfection of viral DNA. *J. Virol.* 63:4579–4589.

64. Cai, W. and P. A. Schaffer. 1992. Herpes simplex virus type 1 ICP0 regulates expression of immediate-early, early, and late genes in productively infected cells. *J. Virol.* 66:2904–2915.

65. Everett, R. and G. Maul. 1994. HSV-1 IE protein Vmw110 causes redistribution of PML. *EMBO J.* 13:5062–5069.

66. Boutell, C. and R. D. Everett. 2003. The herpes simplex virus type 1 (HSV-1) regulatory protein ICP0 interacts with and Ubiquitinates p53. *J. Biol. Chem.* 278:36596–36602.

67. Gu, H. and B. Roizman. 2003. The degradation of promyelocytic leukemia and Sp100 proteins by herpes simplex virus 1 is mediated by the ubiquitin-conjugating enzyme UbcH5a. *Proc. Natl. Acad. Sci. U S A* 100:8963–8968.

68. Hagglund, R. and B. Roizman. 2002. Characterization of the novel E3 ubiquitin ligase encoded in exon 3 of herpes simplex virus-1-infected cell protein 0. *Proc. Natl. Acad. Sci. U S A* 99:7889–7894.

69. Muller, S. and A. Dejean. 1999. Viral immediate-early proteins abrogate the modification by SUMO-1 of PML and Sp100 proteins, correlating with nuclear body disruption. *J. Virol.* 73:5137–5143.

70. Parkinson, J. and R. D. Everett. 2000. Alphaherpesvirus proteins related to herpes simplex virus type 1 ICP0 affect cellular structures and proteins. *J. Virol.* 74:10006–10017.

71. Everett, R. D. 2000. ICP0 induces the accumulation of colocalizing conjugated ubiquitin. *J. Virol.* 74:9994–10005.

72. Lomonte, P. and E. Morency. 2007. Centromeric protein CENP-B proteasomal degradation induced by the viral protein ICP0. *FEBS Lett.* 581:658–662.

73. Lees-Miller, S. P., M. C. Long, M. A. Kilvert, V. Lam, S. A. Rice, and C. A. Spencer. 1996. Attenuation of DNA-dependent protein kinase activity and its catalytic subunit by the herpes simplex virus type 1 transactivator ICP0. *J. Virol.* 70:7471–7477.

74. Parkinson, J., S. P. Lees-Miller, and R. D. Everett. 1999. Herpes simplex virus type 1 immediate-early protein vmw110 induces the proteasome-dependent degradation of the catalytic subunit of DNA-dependent protein kinase. *J. Virol.* 73:650–657.

75. Everett, R. D., M. Meredith, A. Orr, A. Cross, M. Kathoria, and J. Parkinson. 1997. A novel ubiquitin-specific protease is dynamically associated with the PML nuclear domain and binds to a herpesvirus regulatory protein. *EMBO J.* 16:566–577.

76. Meredith, M., A. Orr, and R. Everett. 1994. Herpes simplex virus type 1 immediate-early protein Vmw 110 binds strongly and specifically to a 135 kDa cellular protein. *Virology* 200:457–469.

77. Geoffroy, M. C., G. Chadeuf, A. Orr, A. Salvetti, and R. D. Everett. 2006. Impact of the interaction between herpes simplex virus type 1 regulatory protein ICP0 and ubiquitin-specific protease USP7 on activation of adeno-associated virus type 2 rep gene expression. *J. Virol.* 80:3650–3654.

78. Kawaguchi, Y., R. Bruni, and B. Roizman. 1997. Interaction of herpes simplex virus 1 alpha regulatory protein ICP0 with elongation factor 1delta: ICP0 affects translational machinery. *J. Virol.* 71:1019–1024.

79. Bruni, R., B. Fineschi, W. O. Ogle, and B. Roizman. 1999. A novel cellular protein, p60, interacting with both herpes simplex virus 1 regulatory proteins ICP22 and ICP0 is modified in a cell-type-specific manner and is recruited to the nucleus after infection. *J. Virol.* 73:3810–3817.

80. Kawaguchi, Y. C., R. Bruni, and B. Roizman. 1997. Interaction of the herpes simplex virus 1 regulatory protein ICP0 with elongation factor 1d: ICP0 affects translational machinery. *J. Virol.* 71:1019–1024.

81. Kawaguchi, Y., C. Van Sant, and B. Roizman. 1997. Herpes simplex virus 1 alpha regulatory protein ICP0 interacts with and stabilizes the cell cycle regulator cyclin D3. *J. Virol.* 71:7328–7336.

82. Lomonte, P., J. Thomas, P. Texier, C. Caron, S. Khochbin, and A. L. Epstein. 2004. Functional interaction between class II histone deacetylases and ICP0 of herpes simplex virus type 1. *J. Virol.* 78:6744–6757.

83. Poon, A. P., H. Gu, and B. Roizman. 2006. ICP0 and the US3 protein kinase of herpes simplex virus 1 independently block histone deacetylation to enable gene expression. *Proc. Natl. Acad. Sci. U S A* 103:9993–9998.

84. Gu, H., Y. Liang, G. Mandel, and B. Roizman. 2005. Components of the REST/CoREST/histone deacetylase repressor complex are disrupted, modified, and translocated in HSV-1-infected cells. *Proc. Natl. Acad. Sci. U S A* 102:7571–7576.

85. Eidson, K. M., W. E. Hobbs, B. J. Manning, P. Carlson, and N. A. DeLuca. 2002. Expression of herpes simplex virus ICP0 inhibits the induction of interferon-stimulated genes by viral infection. *J. Virol.* 76:2180–2191.

86. Melroe, G. T., L. Silva, P. A. Schaffer, and D. M. Knipe. 2007. Recruitment of activated IRF-3 and CBP/p300 to herpes simplex virus ICP0 nuclear foci: Potential role in blocking IFN-beta induction. *Virology* 360:305–321.

87. Halford, W. P., C. Weisend, J. Grace, M. Soboleski, D. J. Carr, J. W. Balliet, Y. Imai, T. P. Margolis, and B. M. Gebhardt. 2006. ICP0 antagonizes Stat 1-dependent repression of herpes simplex virus: Implications for the regulation of viral latency. *Virol. J.* 3:44.

88. Krisky, D. M., P. C. Marconi, T. J. Oligino, R. J. Rouse, D. J. Fink, J. B. Cohen, S. C. Watkins, and J. C. Glorioso. 1998. Development of herpes simplex virus replication-defective multigene vectors for combination gene therapy applications. *Gene Ther.* 5:1517–1530.

89. Kwong, A. D. and N. Frenkel. 1987. Herpes simplex virus-infected cells contain a function(s) that destabilizes both host and viral mRNAs. *Proc. Natl. Acad. Sci. U S A* 84:1926–1930.

90. Oroskar, A. and G. Read. 1989. Control of mRNA stability by the virion host shutoff function of herpes simplex virus. *J. Virol.* 63:1897–1906.

91. Read, G. S. and N. Frenkel. 1983. Herpes simplex virus mutants defective in the virion-associated shutoff of host polypeptide synthesis and exhibiting abnormal synthesis of α (immediate early) viral polypeptides. *J. Virol.* 46:498–512.

92. Jenkins, F. J., M. J. Casadaban, and B. Roizman. 1985. Application of the mini-Mu-phage for target-sequence-specific insertional mutagenesis of the herpes simplex virus genome. *Proc. Natl. Acad. Sci. U S A* 82:4773–4777.

93. Krisky, D., P. Marconi, T. Oligino, R. Rouse, D. Fink, and J. Glorioso. 1997. Rapid method for construction of recombinant HSV gene transfer vectors. *Gene Ther.* 4:1120–1125.

94. Krisky, D., P. Marconi, T. Oligino, R. Rouse, D. Fink, J. Cohen, S. Watkins, and J. Glorioso. 1998. Development of herpes

simplex virus replication-defective multigene vectors for combination gene therapy applications. *Gene Ther.* 5:1517–1530.

95. Spaete, R. and N. Frenkel. 1982. The herpes simplex virus amplicon: A new eucaryotic defective-virus cloning amplifying vector. *Cell* 30:295–304.

96. Carroll, N. M., E. A. Chiocca, K. Takahashi, and K. K. Tanabe. 1996. Enhancement of gene therapy specificity for diffuse colon carcinoma liver metastases with recombinant herpes simplex virus. *Ann. Surg.* 224:323–329; discussion 329–330.

97. Freese, A. and A. Geller. 1991. Infection of cultured striatal neurons with a defective HSV-1 vector: Implications for gene therapy. *Nucleic Acids Res.* 19:7219–7223.

98. Geller, A. and X. Breakefield. 1988. A defective HSV-1 vector expresses *Escherichia coli* ß-galactosidase in cultured peripheral neurons. *Science* 241:1667–1669.

99. Geller, A. and A. Freese. 1990a. Infection of cultured central nervous system neurons with a defective herpes simplex virus 1 vector results in stable expression of *Escherichia coli* ß-galactosidase. *Proc. Natl. Acad. Sci. U S A* 87: 1149–1153.

100. Ho, D. Y., J. R. McLaughlin, and R. M. Sapolsky. 1996. Inducible gene expression from defective herpes simplex virus vectors using the tetracycline-responsive promoter system. *Brain Res. Mol. Brain Res.* 41:200–209.

101. Pechan, P. A., M. Fotaki, R. L. Thompson, R. Dunn, M. Chase, E. A. Chiocca, and X. O. Breakefield. 1996. A novel "piggyback" packaging system for herpes simplex virus amplicon vectors. *Hum. Gene Ther.* 7:2003–2013.

102. Shering, A. F., D. Bain, K. Stewart, A. L. Epstein, M. G. Castro, G. W. Wilkinson, and P. R. Lowenstein. 1997. Cell type-specific expression in brain cell cultures from a short human cytomegalovirus major immediate early promoter depends on whether it is inserted into herpesvirus or adenovirus vectors. *J. Gen. Virol.* 78:445–459.

103. Battleman, D., A. Geller, and M. Chao. 1993. HSV-1 vector-mediated gene transfer of the human nerve growth factor receptor p75hNGFR defines high-affinity NGF binding. *J. Neurosci.* 13.941–951.

104. Bergold, P. J., P. Casaccia-Bonnefil, X. L. Zeng, and H. J. Federoff. 1993. Transsynaptic neuronal loss induced in hippocampal slice cultures by a herpes simplex virus vector expressing the GluR6 subunit of the kainate receptor. *Proc. Natl. Acad. Sci. U S A* 90:6165–6169.

105. Casaccia-Bonnefil, P., E. Benedikz, H. Shen, A. Stelzer, D. Edelstein, M. Geschwind, M. Brownlee, H. J. Federoff, and P. J. Bergold. 1993. Localized gene transfer into organotypic hippocampal slice cultures and acute hippocampal slices. *J. Neurosci. Methods* 50:341–351.

106. Geller, A., M. During, J. Haycock, A. Freese, and R. Neve. 1993. Long-term increases in neurotransmitter release from neuronal cells expressing a constitutively active adenylate cyclase from a herpes simplex virus type 1 vector. *Proc. Natl. Acad. Sci. U S A* 90:7603–7607.

107. Geschwind, M., J. Kessler, A. Geller, and H. Federoff. 1994. Transfer of the nerve growth factor gene into cell lines and cultured neurons using a defective herpes simplex virus vector. Transfer to the NGF gene into cells by a HSV-1 vector. *Brain Res.* 24:327–335.

108. Ho, D., E. Mocarski, and R. Sapolsky. 1993. Altering central nervous system physiology with a defective herpes simplex virus vector expressing the glucose transporter gene. *Proc. Natl. Acad. Sci. U S A* 90:3655–3659.

109. Jia, W. W., Y. Wang, D. Qiang, F. Tufaro, R. Remington, and M. Cynader. 1996. A bcl-2 expressing viral vector protects cortical neurons from excitotoxicity even when administered several hours after the toxic insult. *Brain Res. Mol. Brain Res.* 42:350–353.

110. Lawrence, M. S., D. Y. Ho, R. Dash, and R. M. Sapolsky. 1995. Herpes simplex virus vectors overexpressing the glucose transporter gene protect against seizure-induced neuron loss. *Proc. Natl. Acad. Sci. U S A* 92:7247–7251.

111. Miyatake, S., A. Iyer, R. L. Martuza, and S. D. Rabkin. 1997. Transcriptional targeting of herpes simplex virus for cell-specific replication. *J. Virol.* 71:5124–5132.

112. Miyatake, S., R. L. Martuza, and S. D. Rabkin. 1997. Defective herpes simplex virus vectors expressing thymidine kinase for the treatment of malignant glioma. *Cancer Gene Ther.* 4:222–228.

113. During, M., J. Naegele, K. O'Malley, and A. Geller. 1994a. Long-term behavioral recovery in parkinsonian rats by an HSV vector expressing tyrosine hydroxylase. *Science* 266:1399–1403.

114. Geller, A., K. Keyomarsi, J. Bryan, and A. Pardee. 1990b. An efficient deletion mutant packaging system for defective herpes simplex virus vectors: Potential applications to human gene therapy and neuronal physiology. *Proc. Natl. Acad. Sci. U S A* 87:8950–8954.

115. Geller, A. I., L. Yu, Y. Wang, and C. Fraefel. 1997. Helper virus-free herpes simplex virus-1 plasmid vectors for gene therapy of Parkinson's disease and other neurological disorders. *Exp. Neurol.* 144:98–102.

116. Jin, B., M. Belloni, B. Conti, H. Federoff, R. Starr, J. Son, H. Baker, and T. Joh. 1996. Prolonged in vivo gene expression driven by a tyrosine hydroxylase promoter in a defective herpes simplex virus amplicon vector. *Hum. Gene Ther.* 7:2015–2024.

117. New, K. C. and S. D. Rabkin. 1996. Co-expression of two gene products in the CNS using double-cassette defective herpes simplex virus vectors. *Brain Res. Mol. Brain Res.* 37: 317–323.

118. Fraefel, C., S. Song, F. Lim, P. Lang, L. Yu, Y. Wang, P. Wild, and A. I. Geller. 1996. Helper virus-free transfer of herpes simplex virus type 1 plasmid vectors into neural cells. *J. Virol.* 70:7190–7197.

119. Zaupa, C., V. Revol-Guyot, and A. L. Epstein. 2003. Improved packaging system for generation of high-level noncytotoxic HSV-1 amplicon vectors using Cre-loxP site-specific recombination to delete the packaging signals of defective helper genomes. *Hum. Gene Ther.* 14:1049–1063.

120. Olschowka, J. A., W. J. Bowers, S. D. Hurley, M. A. Mastrangelo, and H. J. Federoff. 2003. Helper-free HSV-1 amplicons elicit a markedly less robust innate immune response in the CNS. *Mol. Ther.* 7:218–227.

121. Arvanian, V. L., W. J. Bowers, A. Anderson, P. J. Horner, H. J. Federoff, and L. M. Mendell. 2006. Combined delivery of neurotrophin-3 and NMDA receptors 2D subunit strengthens synaptic transmission in contused and staggered double hemisected spinal cord of neonatal rat. *Exp. Neurol.* 197: 347–352.

122. Cortes, M. L., C. J. Bakkenist, M. V. Di Maria, M. B. Kastan, and X. O. Breakefield. 2003. HSV-1 amplicon vector-mediated expression of ATM cDNA and correction of the ataxia-telangiectasia cellular phenotype. *Gene Ther.* 10:1321–1327.

123. Reinblatt, M., R. H. Pin, W. J. Bowers, H. J. Federoff, and Y. Fong. 2005. Herpes simplex virus amplicon delivery of a hypoxia-inducible soluble vascular endothelial growth factor receptor (sFlk-1) inhibits angiogenesis and tumor growth in pancreatic adenocarcinoma. *Ann. Surg. Oncol.* 12:1025–1036.

124. Saydam, O., D. L. Glauser, I. Heid, G. Turkeri, M. Hilbe, A. H. Jacobs, M. Ackermann, and C. Fraefel. 2005. Herpes simplex

virus 1 amplicon vector-mediated siRNA targeting epidermal growth factor receptor inhibits growth of human glioma cells in vivo. *Mol. Ther.* 12:803–812.

125. Bowers, W. J., M. A. Mastrangelo, H. A. Stanley, A. E. Casey, L. J. Milo, Jr., and H. J. Federoff. 2005. HSV amplicon-mediated Abeta vaccination in Tg2576 mice: Differential antigen-specific immune responses. *Neurobiol. Aging* 26:393–407.

126. Delman, K. A., J. S. Zager, J. J. Bennett, S. Malhotra, M. I. Ebright, P. F. McAuliffe, M. W. Halterman, H. J. Federoff, and Y. Fong. 2002. Efficacy of multiagent herpes simplex virus amplicon-mediated immunotherapy as adjuvant treatment for experimental hepatic cancer. *Ann. Surg.* 236:337–342; discussion 342–343.

127. Gorantla, S., K. Santos, V. Meyer, S. Dewhurst, W. J. Bowers, H. J. Federoff, H. E. Gendelman, and L. Poluektova. 2005. Human dendritic cells transduced with herpes simplex virus amplicons encoding human immunodeficiency virus type 1 (HIV-1) gp120 elicit adaptive immune responses from human cells engrafted into NOD/SCID mice and confer partial protection against HIV-1 challenge. *J. Virol.* 79:2124–2132.

128. Hocknell, P. K., R. D. Wiley, X. Wang, T. G. Evans, W. J. Bowers, T. Hanke, H. J. Federoff, and S. Dewhurst. 2002. Expression of human immunodeficiency virus type 1 gp120 from herpes simplex virus type 1-derived amplicons results in potent, specific, and durable cellular and humoral immune responses. *J. Virol.* 76:5565–5580.

129. Lam, P., K. M. Hui, Y. Wang, P. D. Allen, D. N. Louis, C. J. Yuan, and X. O. Breakefield. 2002. Dynamics of transgene expression in human glioblastoma cells mediated by herpes simplex virus/adeno-associated virus amplicon vectors. *Hum. Gene Ther.* 13:2147–2159.

130. Sena-Esteves, M., Y. Saeki, S. M. Camp, E. A. Chiocca, and X. O. Breakefield. 1999. Single-step conversion of cells to retrovirus vector producers with herpes simplex virus-Epstein–Barr virus hybrid amplicons. *J. Virol.* 73:10426–10439.

131. Bakowska, J. C., M. V. Di Maria, S. M. Camp, Y. Wang, P. D. Allen, and X. O. Breakefield. 2003. Targeted transgene integration into transgenic mouse fibroblasts carrying the full-length human AAVS1 locus mediated by HSV/AAV rep(+) hybrid amplicon vector. *Gene Ther.* 10:1691–1702.

132. Liu, Q., C. F. Perez, and Y. Wang. 2006. Efficient site-specific integration of large transgenes by an enhanced herpes simplex virus/adeno-associated virus hybrid amplicon vector. *J. Virol.* 80:1672–1679.

133. Stavropoulos, T. A. and C. A. Strathdee. 1998. An enhanced packaging system for helper-dependent herpes simplex virus vectors. *J. Virol.* 72:7137–7143.

134. Strathdee, C. A. and M. R. McLeod. 2000. A modular set of helper-dependent herpes simplex virus expression vectors. *Mol. Ther.* 1:479–485.

135. Wade-Martins, R., E. R. Smith, E. Tyminski, E. A. Chiocca, and Y. Saeki. 2001. An infectious transfer and expression system for genomic DNA loci in human and mouse cells. *Nat. Biotechnol.* 19:1067–1070.

136. Rogers, B. E., J. T. Douglas, C. Ahlem, D. J. Buchsbaum, J. Frincke, and D. T. Curiel. 1997. Use of a novel cross-linking method to modify adenovirus tropism. *Gene Ther.* 4:1387–1392.

137. Wickham, T. J., D. Haskard, D. Segal, and I. Kovesdi. 1997. Targeting endothelium for gene therapy via receptors upregulated during angiogenesis and inflammation. *Cancer Immunol. Immunother.* 45:149–151.

138. Wickham, T. J., E. Tzeng, L. L. Shears II, P. W. Roelvink, Y. Li, G. M. Lee, D. E. Brough, A. Lizonova, and I. Koveski. 1997.

Increased in vitro and in vivo gene transfer by adenovirus vectors containing chimeric fiber proteins. *J. Virol.* 71:8221–8229.

139. Kasahara, N., M. Dozy, and Y. W. Kan. 1994. Tissue-specific targeting of retroviral vectors through ligand-receptor interactions. *Science* 255:1373–1376.

140. Marin, M., D. Noel, S. Valsesia-Wittman, F. Brockly, M. Etienne-Julan, S. Russel, F. L. Cosset, and M. Piechaczyk. 1996. Targeted infection of human cells via major histocompatibility complex class I molecules by moloney murine leukemia virus-derived viruses displaying single-chain antibody fragment-envelope fusion proteins. *J. Virol.* 70:2957–2962.

141. Russell, S. J., R. E. Hawkins, and G. Winter. 1993. Retroviral vectors displaying functional antibody fragments. *Nucleic Acids Res.* 21:1081–1085.

142. Somia, N. V., M. Zoppe, and I. M. Verma. 1995. Generation of targeted retroviral vectors by using single-chain variable fragment: An approach to in vivo gene delivery. *Proc. Natl. Acad. Sci. U S A* 92:7570–7574.

143. Banfield, B., Y. Leduc, L. Esford, K. Schubert, and F. Tufaro. 1995. Sequential isolation of proteoglycan synthesis mutants by using herpes simplex virus as a selective agent: Evidence for a proteoglycan-independent virus entry pathway. *J. Virol.* 69:3290–3298.

144. Gruenheid, S., L. Gatzke, H. Meadows, and F. Tufaro. 1993. Herpes simplex virus infection and propagation in a mouse L cell mutant lacking heparan sulfate proteoglycans. *J. Virol.* 67:93–100.

145. Herold, B., R. Visalli, N. Susmarski, C. Brandt, and P. Spear. 1994. Glycoprotein C-independent binding of herpes simplex virus to cells requires cell surface heparan sulfate and glycoprotein B. *J. Gen. Virol.* 75:1211–1222.

146. Shieh, M., D. WuDunn, R. Montgomery, J. Esko, and P. Spear. 1992. Cell surface receptors for herpes simplex virus are heparan sulfate proteoglycans. *J. Cell Biol.* 116:1273–1281.

147. Spear, P., M. Shieh, B. Herold, D. WuDunn, and T. Koshy. 1992. Heparan sulfate glycosaminoglycans as primary cell surface receptors for herpes simplex virus. *Adv. Exp. Med. Biol.* 313:341–353.

148. Williams, R. and S. Straus. 1997. Specificity and affinity of binding of herpes simplex virus type 2 glycoprotein B to glycosaminoglycans. *J. Virol.* 71:1375–1380.

149. WuDunn, D. and P. Spear. 1989. Initial interaction of herpes simplex virus with cells is binding to heparan sulfate. *J. Virol.* 63:52–58.

150. Herold, B. C., S. I. Gerber, T. Polonsky, B. J. Belval, P. N. Shaklee, and K. Holme. 1995. Identification of structural features of heparin required for inhibition of herpes simplex virus type 1 binding. *Virology* 206:1108–1116.

151. Laquerre, S., R. Argnani, D. Anderson, S. Zucchini, R. Manservigi, and J. Glorioso. 1998. Heparan sulfate proteoglycan binding by herpes simplex virus type 1 glycoproteins B and C which differ in their contribution to virus attachment, penetration, and cell-to-cell spread. *J. Virol.* 72:6119–6130.

152. Tal-Singer, R., C. Peng, M. Ponce de Leon, W. R. Abrams, B. W. Banfield, F. Tufaro, G. H. Cohen, and R. J. Eisenberg. 1995. Interaction of herpes simplex virus glycoprotein gC with mammalian cell surface molecules. *J. Virol.* 69:4471–4483.

153. Trybala, E., T. Berghtrom, B. Svennerholm, S. Jeansson, J. C. Glorioso, and S. Olufsson. 1994. Localization of the functional site in herpes simplex virus type 1 glycoproteins C involved in binding to cell surface heparan sulfate. *J. Gen. Virol.* 75:743–752.

154. Trybala, E., J. Liljeqvist, B. Svennerholm, and T. Bergstrom. 2000. Herpes simplex virus types 1 and 2 differ in their interaction with heparan sulfate. *J. Virol.* 74:9106–9114.

155. Montgomery, R., M. Warner, B. Lum, and P. Spear. 1996. Herpes simplex virus 1 entry into cells mediated by a novel member of the TNF/NGF receptor family. *Cell* 87:427–436.

156. Nicola, A., M. Ponce de Leon, R. Xu, W. Hou, J. Whitbeck, C. Krummenacher, R. Montgomery, P. Spear, R. Eisenberg, and G. Cohen. 1998. Monoclonal antibodies to distinct sites on herpes simplex virus (HSV) glycoprotein D block HSV binding to HVEM. *J. Virol.* 72:3595–3601.

157. Carfi, A., S. H. Willis, J. C. Whitbeck, C. Krummenacher, G. H. Cohen, R. J. Eisenberg, and D. C. Wiley. 2001. Herpes simplex virus glycoprotein D bound to the human receptor HveA. *Mol. Cell* 8:169–179.

158. Connolly, S. A., D. J. Landsburg, A. Carfi, J. C. Whitbeck, Y. Zuo, D. C. Wiley, G. H. Cohen, and R. J. Eisenberg. 2005. Potential nectin-1 binding site on herpes simplex virus glycoprotein D. *J. Virol.* 79:1282–1295.

159. Connolly, S. A., D. J. Landsburg, A. Carfi, D. C. Wiley, G. H. Cohen, and R. J. Eisenberg. 2003. Structure-based mutagenesis of herpes simplex virus glycoprotein D defines three critical regions at the gD-HveA/HVEM binding interface. *J. Virol.* 77:8127–8140.

160. Connolly, S. A., D. J. Landsburg, A. Carfi, D. C. Wiley, R. J. Eisenberg, and G. H. Cohen. 2002. Structure-based analysis of the herpes simplex virus glycoprotein D binding site present on herpesvirus entry mediator HveA (HVEM). *J. Virol.* 76:10894–10904.

161. Warner, M., R. Geraghty, W. Martinez, R. Montgomery, J. Whitbeck, R. Xu, R. Eisenberg, G. Cohen, and P. Spear. 1998. A cell surface protein with herpesvirus entry activity (HveB) confers susceptibility to infection by mutants of herpes simplex virus type 1, herpes simplex virus type 2, and pseudorabies virus. *Virology* 246:179–189.

162. Shukla, D., J. Liu, P. Blaiklock, N. W. Shworak, X. Bai, J. D. Esko, G. H. Cohen, R. J. Eisenberg, R. D. Rosenberg, and P. G. Spear. 1999. A novel role for 3-O-sulfated heparan sulfate in herpes simplex virus 1 entry. *Cell* 99:13–22.

163. Geraghty, R., A. Fridberg, C. Krummenacher, G. Cohen, R. Eisenberg, and P. Spear. 2001. Use of chimeric nectin-1(HveC)-related receptors to demonstrate that ability to bind alphaherpesvirus gD is not necessarily sufficient for viral entry. *Virology* 285:366–375.

164. Geraghty, R., C. Krummenacher, G. Cohen, R. Eisenberg, and P. Spear. 1998. Entry of alphaherpesviruses mediated by poliovirus receptor-related protein 1 and poliovirus receptor. *Science* 280:1618–1620.

165. Krummenacher, C., A. Rux, J. Whitbeck, M. Ponce-de-Leon, H. Lou, I. Baribaud, W. Hou, C. Zou, R. Geraghty, P. Spear, R. Eisenberg, and G. Cohen. 1999. The first immunoglobulin-like domain of HveC is sufficient to bind herpes simplex virus gD with full affinity, while the third domain is involved in oligomerization of HveC. *J. Virol.* 73: 8127–8137.

166. Manoj, S., C. R. Jogger, D. Myscofski, M. Yoon, and P. G. Spear. 2004. Mutations in herpes simplex virus glycoprotein D that prevent cell entry via nectins and alter cell tropism. *Proc. Natl. Acad. Sci. U S A* 101:12414–12421.

167. Tsvitov, M., A. R. Frampton, Jr., W. A. Shah, S. K. Wendell, A. Ozuer, Z. Kapacee, W. F. Goins, J. B. Cohen, and J. C. Glorioso. 2007. Characterization of soluble glycoprotein D-mediated herpes simplex virus type 1 infection. *Virology* 360:477–491.

168. Yoon, M. and P. G. Spear. 2004. Random mutagenesis of the gene encoding a viral ligand for multiple cell entry receptors to obtain viral mutants altered for receptor usage. *Proc. Natl. Acad. Sci. U S A* 101:17252–17257.

169. Forrester, A., H. Farrell, G. Wilkinson, J. Kaye, N. Davis-Poynter, and T. Minson. 1992. Construction and properties of a mutant of herpes simplex virus type 1 with glycoprotein H coding sequences deleted. *J. Virol.* 66:341–348.

170. Roop, C., L. Hutchinson, and D. Johnson. 1993. A mutant herpes simplex virus type 1 unable to express glycoprotein L cannot enter cells, and its particles lack glycoprotein H. *J. Virol.* 67:2285–2297.

171. Cai, W., S. Person, S. Warner, J. Zhou, and J. Glorioso. 1987. Linker-insertion nonsense and restriction-site deletion mutations of the gB glycoprotein gene of herpes simplex virus type 1. *J. Virol.* 61:714–721.

172. Butcher, M., K. Raviprakash, and H. P. Ghosh. 1990. Acid pH-induced fusion of cells by herpes simplex virus glycoproteins gB and gD. *J. Biol. Chem.* 265:5862–5868.

173. Subramanian, R. P., J. E. Dunn, and R. J. Geraghty. 2005. The nectin-1alpha transmembrane domain, but not the cytoplasmic tail, influences cell fusion induced by HSV-1 glycoproteins. *Virology* 339:176–191.

174. Subramanian, R. P. and R. J. Geraghty. 2007. Herpes simplex virus type 1 mediates fusion through a hemifusion intermediate by sequential activity of glycoproteins D, H, L, and B. *Proc. Natl. Acad. Sci. U S A* 104:2903–2908.

175. Roizman, B. and A. Sears. 1996. Herpes simplex viruses and their replication. In: B. Fields, D. Knipe, P. Howley, R. Chanock, M. Hirsch, J. Melnick, T. Monath, and B. Roizman (Eds.), *Fields Virology*, 3rd edn., Lippincott-Raven, Philadelphia, Pennsylvania, pp. 2231–2295.

176. Clement, C., V. Tiwari, P. M. Scanlan, T. Valyi-Nagy, B. Y. Yue, and D. Shukla. 2006. A novel role for phagocytosis-like uptake in herpes simplex virus entry. *J. Cell Biol.* 174:1009–1021.

177. Milne, R. S., A. V. Nicola, J. C. Whitbeck, R. J. Eisenberg, and G. H. Cohen. 2005. Glycoprotein D receptor-dependent, low-pH-independent endocytic entry of herpes simplex virus type 1. *J. Virol.* 79:6655–6663.

178. Nicola, A. V., A. M. McEvoy, and S. E. Straus. 2003. Roles for endocytosis and low pH in herpes simplex virus entry into HeLa and Chinese hamster ovary cells. *J. Virol.* 77:5324–5332.

179. Nicola, A. V. and S. E. Straus. 2004. Cellular and viral requirements for rapid endocytic entry of herpes simplex virus. *J. Virol.* 78:7508–7517.

180. Laquerre, S., D. Anderson, D. Stolze, and J. Glorioso. 1998. Recombinant herpes simplex virus type 1 engineered for targeted binding to erythropoietin receptor bearing cells. *J. Virol.* 72:9683–9697.

181. Argnani, R., L. Boccafogli, P. C. Marconi, and R. Manservigi. 2004. Specific targeted binding of herpes simplex virus type 1 to hepatocytes via the human hepatitis B virus preS1 peptide. *Gene Ther.* 11:1087–1098.

182. Grandi, P., M. Spear, X. O. Breakefield, and S. Wang. 2004. Targeting HSV amplicon vectors. *Methods* 33:179–186.

183. Grandi, P., S. Wang, D. Schuback, V. Krasnykh, M. Spear, D. T. Curiel, R. Manservigi, and X. O. Breakefield. 2004. HSV-1 virions engineered for specific binding to cell surface receptors. *Mol. Ther.* 9:419–427.

184. Zhou, G., G.-J. Ye, W. Debinski, and B. Roizman. 2002. Engineered herpes simplex virus 1 is dependent on IL13Rα2 receptor for cell entry and independent of glycoprotein D receptor interaction. *Proc. Natl. Acad. Sci. U S A* 99:15124–15129.

185. Kamiyama, H., G. Zhou, and B. Roizman. 2006. Herpes simplex virus 1 recombinant virions exhibiting the amino terminal fragment of urokinase-type plasminogen activator can enter cells via the cognate receptor. *Gene Ther.* 13:621–629.

186. Zhou, G. and B. Roizman. 2005. Characterization of a recombinant herpes simplex virus 1 designed to enter cells via the

IL13Ralpha2 receptor of malignant glioma cells. *J. Virol.* 79:5272–5277.

187. Nakano, K., R. Asano, K. Tsumoto, H. Kwon, W. F. Goins, I. Kumagai, J. B. Cohen, and J. C. Glorioso. 2005. Herpes simplex virus targeting to the EGF receptor by a gD-specific soluble bridging molecule. *Mol. Ther.* 11:617–626.

188. Akkaraju, G. R., J. Huard, E. P. Hoffman, W. F. Goins, R. Pruchnic, S. C. Watkins, J. B. Cohen, and J. C. Glorioso. 1999. Herpes simplex virus vector-mediated dystrophin gene transfer and expression in MDX mouse skeletal muscle. *J. Gene Med.* 1:280–289.

189. Oligino, T., P. L. Poliani, P. Marconi, M. A. Bender, M. C. Schmidt, D. J. Fink, and J. C. Glorioso. 1996. In vivo transgene activation from an HSV-based gene vector by GAL4:VP16. *Gene Ther.* 3:892–899.

190. Wolfe, D., A. Niranjan, A. Trichel, C. Wiley, A. Ozuer, E. Kanal, D. Kondziolka, D. Krisky, J. Goss, N. DeLuca, M. Murphey-Corb, and J. C. Glorioso. 2004. Safety and biodistribution studies of an HSV multigene vector following intracranial delivery to non-human primates. *Gene Ther.* 11:1675–1684.

191. Wolfe, D., W. Goins, T. Kaplan, S. Capuano, J. Fradette, M. Murphey-Corb, J. Cohen, P. Robbins, and J. Glorioso. 2001. Systemic accumulation of biologically active nerve growth factor following intra-articular herpesvirus gene transfer. *Mol. Ther.* 3:61–69.

192. Mester, J. C., P. Pitha, and J. C. Glorioso. 1995. Anti-viral activity of herpes simplex virus vectors expressing alpha-interferon. *Gene Ther.* 3:187–196.

193. Rasty, S., P. Thatikunta, J. Gordon, K. Khalili, S. Amini, and J. Glorioso. 1996. Human immunodeficiency virus tat gene transfer to the murine central nervous system using a replication-defective herpes simplex virus vector stimulates transforming growth factor beta 1 gene expression. *Proc. Natl. Acad. Sci. U S A* 93:6073–6078.

194. Bloom, D. C. and R. G. Jarman. 1998. Generation and use of recombinant reporter viruses for study of herpes simplex virus infections in vivo. *Methods* 16:117–125.

195. Carpenter, D. E. and J. G. Stevens. 1996. Long-term expression of a foreign gene from a unique position in the latent herpes simplex virus genome. *Hum. Gene Ther.* 7:1447–1454.

196. Lokensgard, J. R., D. C. Bloom, A. T. Dobson, and L. T. Feldman. 1994. Long-term promoter activity during herpes simplex virus latency. *J. Virol.* 68:7148–7158.

197. Preston, C. M., R. Mabbs, and M. J. Nicholl. 1997. Construction and characterization of herpes simplex virus type 1 mutants with conditional defects in immediate early gene expression. *Virology* 229:228–239.

198. Cook, M. L. and J. G. Stevens. 1973. Pathogenesis of herpetic neuritis and ganglionitis in mice: Evidence of intra-axonal transport of infection. *Infect. Immun.* 7:272–288.

199. Stevens, J. G. 1989. Human herpesviruses: A consideration of the latent state. *Microbiol. Rev.* 53:318–332.

200. Rodahl, E. and J. Stevens. 1992. Differential accumulation of herpes simplex virus type 1 latency-associated transcripts in sensory and autonomic ganglia. *Virology* 189:385–388.

201. Speck, P. and A. Simmons. 1992. Synchronous appearance of antigen-positive and latently infected neurons in spinal ganglia of mice infected with a virulent strain or herpes simplex virus. *J. Gen. Virol.* 73:1281–1285.

202. Speck, P. G. and A. Simmons. 1991. Divergent molecular pathways of productive and latent infection with a virulent strain of herpes simplex virus type 1. *J. Virol.* 65:4004–4005.

203. Croen, K. D., J. M. Ostrove, L. J. Dragovic, J. E. Smialek, and S. E. Straus. 1987. Latent herpes simplex virus in human trigeminal ganglia. Detection of an immediate early gene "anti-sense" transcript by in situ hybridization. *New Engl. J. Med.* 317:1427–1432.

204. Deatly, A. M., J. G. Spivack, E. Lavi, and N. W. Fraser. 1987. RNA from an immediate early region of the HSV-1 genome is present in the trigeminal ganglia of latently infected mice. *Proc. Natl. Acad. Sci. U S A* 84:3204–3208.

205. Stevens, J. G., E. K. Wagner, G. B. Devi-Rao, M. L. Cook, and L. T. Feldman. 1987. RNA complementary to a herpesviruses α gene mRNA is prominent in latently infected neurons. *Science* 255:1056–1059.

206. Deatly, A. M., J. G. Spivack, E. Lavi, D. O'Boyle, and N. W. Fraser. 1988. Latent herpes simplex virus type 1 transcripts in peripheral and central nervous systems tissues of mice map to similar regions of the viral genome. *J. Virol.* 62:749–756.

207. Wagner, E. K., W. M. Flanagan, G. B. Devi-Rao, Y. F. Zhang, J. M. Hill, K. P. Anderson, and J. G. Stevens. 1988b. The herpes simplex virus latency-associated transcript is spliced during the latent phase of infection. *J. Virol.* 62:4577–4585.

208. Devi-Rao, G., J. Aguilar, M. Rice, H. J. Garza, D. Bloom, J. Hill, and E. Wagner. 1997. Herpes simplex virus genome replication and transcription during induced reactivation in the rabbit eye. *J. Virol.* 71:7039–7047.

209. Alvira, M. R., J. B. Cohen, W. F. Goins, and J. C. Glorioso. 1999. Genetic studies exposing the splicing events involved in HSV-1 latency associated transcript (LAT) production during lytic and latent infection. *J. Virol.* 73:3866–3876.

210. Farrell, M. J., A. T. Dobson, and L. T. Feldman. 1991. Herpes simplex virus latency-associated transcript is a stable intron. *Proc. Natl. Acad. Sci. U S A* 88:790–794.

211. Krummenacher, C., J. Zabolotny, and N. Fraser. 1997. Selection of a nonconsensus branch point is influenced by an RNA stem-loop structure and is important to confer stability to the herpes simplex virus 2-kilobase latency-associated transcript. *J. Virol.* 71:5849–5860.

212. Rodahl, E. and L. Haarr. 1997. Analysis of the 2-kilobase latency-associated transcript expressed in PC12 cells productively infected with herpes simplex virus type 1: Evidence for a stable, nonlinear structure. *J. Virol.* 71:1703–1707.

213. Zabolotny, J., C. Krummenacher, and N. W. Fraser. 1997. The herpes simplex virus type 1 2.0-kilobase latency-associated transcript is a stable intron which branches at a guanosine. *J. Virol.* 71:4199–4208.

214. Amelio, A. L., N. V. Giordani, N. J. Kubat, J. E. O'Neil, and D. C. Bloom. 2006. Deacetylation of the herpes simplex virus type 1 latency-associated transcript (LAT) enhancer and a decrease in LAT abundance precede an increase in ICP0 transcriptional permissiveness at early times postexplant. *J. Virol.* 80:2063–2068.

215. Kubat, N. J., A. L. Amelio, N. V. Giordani, and D. C. Bloom. 2004. The herpes simplex virus type 1 latency-associated transcript (LAT) enhancer/rcr is hyperacetylated during latency independently of LAT transcription. *J. Virol.* 78:12508–12518.

216. Amelio, A. L., P. K. McAnany, and D. C. Bloom. 2006. A chromatin insulator-like element in the herpes simplex virus type 1 latency-associated transcript region binds CCCTC-binding factor and displays enhancer-blocking and silencing activities. *J. Virol.* 80:2358–2368.

217. Chen, Q., L. Lin, S. Smith, J. Huang, S. L. Berger, and J. Zhou. 2007. CTCF-dependent chromatin boundary element between the latency-associated transcript and ICP0 promoters in the herpes simplex virus type 1 genome. *J. Virol.* 81:5192–5201.

218. French, S. W., M. C. Schmidt, and J. C. Glorioso. 1996. Involvement of an HMG protein in the transcriptional activity

of the herpes simplex virus latency active promoter 2. *Mol. Cell. Biol.* 16:5393–5399.

219. Goins, W. F., L. R. Sternberg, K. D. Croen, P. R. Krause, R. L. Hendricks, D. J. Fink, S. E. Straus, M. Levine, and J. C. Glorioso. 1994. A novel latency-active promoter is contained within the herpes simplex virus type 1 U$_L$ flanking repeats. *J. Virol.* 68:2239–2252.

220. Blondeau, J. M., F. Y. Aoki, and G. B. Glavin. 1993. Stress-induced reactivation of latent herpes simplex virus infection in rat lumbar dorsal root ganglia. *J. Psychosomatic Res.* 37: 843–849.

221. Batchelor, A. H. and P. O. O'Hare. 1990. Regulation and cell-type-specific activity of a promoter located upstream of the latency-associated transcript of herpes simplex virus type 1. *J. Virol.* 64:3269–3279.

222. Chen, X., M. C. Schmidt, W. F. Goins, and J. C. Glorioso. 1995. Two herpes simplex virus type-1 latency active promoters differ in their contribution to latency-associated transcript expression during lytic and latent infection. *J. Virol.* 69:7899–7908.

223. Dobson, A. T., F. Sederati, G. Devi-Rao, W. M. Flanagan, M. J. Farrell, J. G. Stevens, E. K. Wagner, and L. T. Feldman. 1989. Identification of the latency-associated transcript promoter by expression of rabbit β-globin mRNA in mouse sensory nerve ganglia latently infected with a recombinant herpes simplex virus. *J. Virol.* 63:3844–3851.

224. Zwaagstra, J., H. Ghiasi, A. B. Nesburn, and S. L. Wechsler. 1989. In vitro promoter activity associated with the latency-associated transcript gene of herpes simplex virus type 1. *J. Gen. Virol.* 70:2163–2169.

225. Nicosia, M., S. L. Deshmane, J. M. Zabolotny, T. Valyi-Nagy, and N. W. Fraser. 1993. Herpes simplex virus type 1 Latency-Associated Transcript (LAT) promoter deletion mutants can express a 2-kilobase transcript mapping to the LAT region. *J. Virol.* 67:7276–7283.

226. Margolis, T. P., D. C. Bloom, A. T. Dobson, L. T. Feldman, and J. G. Stevens. 1993. Decreased reporter gene expression during latent infection with HSV LAT promoter constructs. *Virology* 197:585–592.

227. Wolfe, J. H., S. L. Deshmane, and N. W. Fraser. 1992. Herpes-virus vector gene transfer and expression of β-glucuronidase in the central nervous system of MPS VII mice. *Nat. Genet.* 1:379–384.

228. Lokensgard, J. R., L. T. Feldman, and H. Berthomme. 1997. The latency-associated promoter of herpes simplex virus type 1 requires a region downstream of the transcription start site for long-term expression during latency. *J. Virol.* 71:6714–6719.

229. Lilley, C. E., R. H. Branston, and R. S. Coffin. 2001. Herpes simplex virus vectors for the nervous system. *Curr. Gene Ther.* 1:339–358.

230. Lilley, C. E., F. Groutsi, Z. Han, J. A. Palmer, P. N. Anderson, D. S. Latchman, and R. S. Coffin. 2001. Multiple immediate-early gene-deficient herpes simplex virus vectors allowing efficient gene delivery to neurons in culture and widespread gene delivery to the central nervous system in vivo. *J. Virol.* 75:4343–4356.

231. Palmer, J. A., R. H. Branston, C. E. Lilley, M. J. Robinson, F. Groutsi, J. Smith, D. S. Latchman, and R. S. Coffin. 2000. Development and optimization of herpes simplex virus vectors for multiple long-term gene delivery to the peripheral nervous system. *J. Virol.* 74:5604–5618.

232. Perez, M. C., S. P. Hunt, R. S. Coffin, and J. A. Palmer. 2004. Comparative analysis of genomic HSV vectors for gene delivery to motor neurons following peripheral inoculation in vivo. *Gene Ther.* 11:1023–1032.

233. Ho, D. Y. and E. S. Mocarski. 1989. Herpes simplex virus latent RNA (LAT) is not required for latent infection in the mouse. *Proc. Natl. Acad. Sci. U S A* 86:7596–7600.

234. Goins, W. F., K. A. Lee, J. D. Cavalcoli, M. E. O'Malley, S. T. DeKosky, D. J. Fink, and J. C. Glorioso. 1999. Herpes simplex virus type 1 vector-mediated expression of nerve growth factor protects dorsal root ganglia neurons from peroxide toxicity. *J. Virol.* 73:519–532.

235. Goins, W. F., N. Yoshimura, M. W. Phelan, T. Yokoyama, M. O. Fraser, H. Ozawa, N. J. Bennett, W. C. de Groat, J. C. Glorioso, and M. B. Chancellor. 2001. Herpes simplex virus mediated nerve growth factor expression in bladder and afferent neurons: Potential treatment for diabetic bladder dysfunction. *J. Urol.* 165:1748–1754.

236. Sasaki, K., M. B. Chancellor, W. F. Goins, M. W. Phelan, J. C. Glorioso, W. C. de Groat, and N. Yoshimura. 2004. Gene therapy using replication-defective herpes simplex virus vectors expressing nerve growth factor in a rat model of diabetic cystopathy. *Diabetes* 53:2723–2730.

237. Chattopadhyay, M., D. Wolfe, M. Mata, S. Huang, J. C. Glorioso, and D. J. Fink. 2005. Long-term neuroprotection achieved with latency-associated promoter-driven herpes simplex virus gene transfer to the peripheral nervous system. *Mol. Ther.* 12:307–313.

238. Chattopadhyay, M., M. Mata, J. Goss, D. Wolfe, S. Huang, J. C. Glorioso, and D. J. Fink. 2007. Prolonged preservation of nerve function in diabetic neuropathy in mice by herpes simplex virus-mediated gene transfer. *Diabetologia* 50: 1550–1558.

239. Puskovic, V., D. Wolfe, J. Goss, S. Huang, M. Mata, J. C. Glorioso, and D. J. Fink. 2004. Prolonged biologically active transgene expression driven by HSV LAP2 in brain in vivo. *Mol. Ther.* 10:67–75.

240. Fradette, J., D. Wolfe, W. F. Goins, S. Huang, R. M. Flanigan, and J. C. Glorioso. 2005. HSV vector-mediated transduction and GDNF secretion from adipose cells. *Gene Ther.* 12:48–58.

241. Carey, M., J. Leatherwood, and M. Ptashne. 1990. A potent GAL4 derivative activates transcription at a distance in vitro. *Science* 247:710–712.

242. Chasman, D. I., M. Leatherwood, M. Carey, M. Ptashne, and R. D. Kornberg. 1989. Activation of yeast polymerase II transcription by herpesvirus VP16 and GAL4 derivative in vitro. *Mol. Cell. Biol.* 9:4746–4749.

243. Sadowski, I., J. Ma, S. Triezenberg, and M. Ptashne. 1988. GAL4/VP16 is an unusually potent transcriptional activator. *Nature* 335:563–564.

244. Axelrod, J. D., M. S. Reagan, and J. Majors. 1993. GAL4 disrupts a repressing nucleosome during activation of GAL 1 transcription in vivo. *Genes Devel.* 7:857–869.

245. Xu, L., W. Schaffner, and D. Rungger. 1993a. Transcription activation by recombinant GAL4/VP16 in the *Xenopus* oocyte. *Nucleic Acids Res.* 21:2775.

246. Vegeto, E., G. F. Allan, W. T. Schrader, M.-J. Tsai, D. P. McDonnell, and B. W. O'Malley. 1992. The mechanism of RU486 antagonism is dependent on the conformation of the carboxy-terminal tail of the human progesterone receptor. *Cell* 69:703–713.

247. Wang, Y., B. O'Malley, Jr, S. Tsai, and B. O'Malley. 1994. A novel regulatory system for gene transfer. *Proc. Natl. Acad. Sci. U S A* 91:8180–8184.

248. Oligino, T., P. L. Poliani, Y. Wang, S. Y. Tsai, B. W. O'Malley, and J. C. Glorioso. 1998. Drug inducible transgene expression in brain using a herpes simplex virus vector. *Gene Ther.* 5:491–496.

249. Mata, M., M. Zhang, X. Hu, and D. Fink. 2001. HveC (nectin-1) is expressed at high levels in sensory neurons, but not in motor neurons of the rat peripheral nervous system. *J. Neuro-Virol.* 7:1–5.

250. Dohner, K., K. Radtke, S. Schmidt, and B. Sodeik. 2006. Eclipse phase of herpes simplex virus type 1 infection: Efficient dynein-mediated capsid transport without the small capsid protein VP26. *J. Virol.* 80:8211–8224.

251. Mabit, H., M. Y. Nakano, U. Prank, B. Saam, K. Dohner, B. Sodeik, and U. F. Greber. 2002. Intact microtubules support adenovirus and herpes simplex virus infections. *J. Virol.* 76:9962–9971.

252. Helgren, M. E., K. D. Cliffer, K. Torrento, C. Cavnor, R. Curtis, P. S. DiStefano, S. J. Wiegand, and L. RM. 1997. Neurotrophin-3 administration attenuates deficits of pyridoxine-induced large-fiber sensory neuropathy. *J. Neurosci.* 17:372–382.

253. Chattopadhyay, M., D. Wolfe, S. Huang, J. Goss, J. C. Glorioso, M. Mata, and D. J. Fink. 2002. In vivo gene therapy for pyridoxine-induced neuropathy by herpes simplex virus-mediated gene transfer of neurotrophin-3. *Ann. Neurol.* 51:19–27.

254. Krarup-Hansen, A., S. Helweg-Larsen, H. Schmalbruch, M. Rorth, and C. Krarup. 2007. Neuronal involvement in cisplatin neuropathy: Prospective clinical and neurophysiological studies. *Brain* 130:1076–1088.

255. Chattopadhyay, M., J. Goss, D. Wolfe, W. C. Goins, S. Huang, J. C. Glorioso, M. Mata, and D. J. Fink. 2004. Protective effect of herpes simplex virus-mediated neurotrophin gene transfer in cisplatin neuropathy. *Brain* 127:929–939.

256. Goss, J. R., W. F. Goins, D. Lacomis, M. Mata, J. C. Glorioso, and D. J. Fink. 2002. Herpes simplex-mediated gene transfer of nerve growth factor protects against peripheral nerruropathy in streptozotocin-induced diabetes in the mouse. *Diabetes* 51:2227–2232.

257. Chattopadhyay, M., D. Krisky, D. Wolfe, J. C. Glorioso, M. Mata, and D. J. Fink. 2005. HSV-mediated gene transfer of vascular endothelial growth factor to dorsal root ganglia prevents diabetic neuropathy. *Gene Ther.* 12:1377–1384.

258. Wilson, S. P., D. C. Yeomans, M. A. Bender, Y. Lu, W. F. Goins, and J. C. Glorioso. 1999. Antihyperalgesic effects of infection with a preproenkephalin-encoding herpes virus. *Proc. Natl. Acad. Sci. U S A* 96:3211–3216.

259. Yeomans, D. C., Y. Lu, C. E. Laurito, M. C. Peters, G. Vota-Vellis, S. P. Wilson, and G. D. Pappas. 2006. Recombinant herpes vector-mediated analgesia in a primate model of hyperalgesia. *Mol. Ther.* 13:589–597.

260. Goss, J. R., M. Mata, W. F. Goins, H. H. Wu, J. C. Glorioso, and D. J. Fink. 2001. Antinociceptive effect of a genomic herpes simplex virus-based vector expressing human proenkephalin in rat dorsal root ganglion. *Gene Ther.* 8:551–556.

261. Hao, S., M. Mata, W. Goins, J. C. Glorioso, and D. J. Fink. 2003. Transgene-mediated enkephalin release enhances the effect of morphine and evades tolerance to produce a sustained antiallodynic effect in neuropathic pain. *Pain* 102:135–142.

262. Zadina, J. E. 2002. Isolation and distribution of endomorphins in the central nervous system. *Jpn. J. Pharmacol.* 89:203–208.

263. Zadina, J. E., L. Hackler, L. J. Ge, and A. J. Kastin. 1997. A potent and selective endogenous agonist for the mu-opiate receptor. *Nature* 386:499–502.

264. Wolfe, D., S. Hao, J. Hu, R. Srinivasan, J. Goss, M. Mata, D. J. Fink, and J. C. Glorioso. 2007. Engineering an endomorphin-2 gene for use in neuropathic pain therapy. *Pain.* 133:29–38.

265. Bischoff, J. R., D. H. Kirn, A. Williams, C. Heise, S. Horn, M. Muna, L. Ng, J. Nye, A. Sampson-Johannes, A. Fattaey, and F. McCormick. 1996. An adenovirus mutant that replicates selectively in p53-deficient human tumor cells. *Science* 274:373–376.

266. Markert, J., M. Medlock, S. Rabkin, G. Gillespie, T. Todo, W. Hunter, C. Palmer, F. Feigenbaum, C. Tornatore, F. Tufaro, and R. Martuza. 2000. Conditionally replicating herpes simplex virus mutant, G207 for the treatment of malignant glioma: Results of a phase I trial. *Gene Ther.* 7:867–874.

267. Moolten, F. L. 1986. Tumor chemosensitivity conferred by inserted herpes thymidine kinase genes: Paradigm for a prospective cancer control strategy. *Cancer Res.* 46:5276–5281.

268. Hanna, N. N., H. J. Mauceri, J. D. Wayne, D. E. Hallahan, D. W. Kufe, and R. R. Weichselbaum. 1997. Virally directed cytosine deaminase/5-fluorocytosine gene therapy enhances radiation response in human cancer xenografts. *Cancer Res.* 57:4205–4209.

269. Freeman, S. M., K. A. Whartenby, J. L. Freeman, C. N. Abboud, and A. J. Marrogi. 1996. In situ use of suicide genes for cancer therapy. *Sem. Oncol.* 23:31–45.

270. Finke, S., B. Trojaneck, P. Moller, D. Schadendorf, A. Neubauer, D. Huhn, and I. G. Schmidt-Wolf. 1997. Increase of cytotoxic sensitivity of primary human melanoma cells transfected with the interleukin-7 gene to autologous and allogeneic immunologic effector cells. *Cancer Gene Ther.* 4:260–268.

271. Chen, L., S. Ashe, W. Brady, I. Hellstrom, K. E. Hellstrom, J. A. Ledbetter, P. McGowan, and P. S. Linsley. 1992. Costimulation of antitumor immunity by the B7 counterreceptor for the T lymphocyte molecules CD28 and CTLA-4. *Cell* 71:1093–1102.

272. Galea-Lauri, J., F. Farzaneh, and J. Gaken. 1996. Novel costimulators in the immune gene therapy of cancer. *Cancer Gene Ther.* 3:202–214.

273. Katsanis, E., Z. Xu, M. A. Bausero, B. B. Dancisak, K. B. Gorden, G. Davis, G. S. Gray, P. J. Orchard, and B. R. Blazar. 1995. B7-1 expression decreases tumorigenicity and induces partial systemic immunity to murine neuroblastoma deficient in major histocompatibility complex and costimulatory molecules. *Cancer Gene Ther.* 2:39–46.

274. Schmidt, W., P. Steinlein, M. Buschle, T. Schweighoffer, E. Herbst, K. Mechtler, H. Kirlappos, and M. L. Birnstiel. 1996. Transloading of tumor cells with foreign major histocompatibility complex class I peptide ligand: A novel general strategy for the generation of potent cancer vaccines. *Proc. Natl. Acad. Sci. U S A* 93:9759–9763.

275. Henderson, R. A. and O. J. Finn. 1996. Human tumor antigens are ready to fly. *Adv. Immunol.* 62:217–256.

276. Pecher, G. and O. J. Finn. 1996. Induction of cellular immunity in chimpanzees to human tumor-associated antigen mucin by vaccination with MUC-1 cDNA-transfected Epstein–Barr virus-immortalized autologous B cells. *Proc. Natl. Acad. Sci. U S A* 93:1699–1704.

277. Herrlinger, U., C. M. Kramm, K. M. Johnston, D. N. Louis, D. Finkelstein, G. Reznikoff, G. Dranoff, X. O. Breakefield, and J. S. Yu. 1997. Vaccination for experimental gliomas using GM-CSF-transduced glioma cells. *Cancer Gene Ther.* 4:345–352.

278. Goodrum, F. D. and D. A. Ornelles. 1997. The early region 1B 55 kD oncoprotein of adenovirus relieves growth restrictions imposed on viral replication by the cell cycle. *J. Virol.* 71:548–561.

279. Hall, A. R., B. R. Dix, S. J. O'Carroll, and A. W. Braithwaite. 1998. p53-dependent cell death/apoptosis is required for a productive adenovirus infection. *Nat. Med.* 4:1068–1072.

280. Ridgway, P. J., A. R. Hall, C. J. Myers, and A. W. Braithwaite. 1997. p53/E1b58 kDa complex regulates adenovirus replication. *Virology* 237:404–413.

281. Chambers, R., G. Y. Gillespie, L. Soroceanu, S. Andreansky, S. Chatterjee, J. Chou, B. Roizman, and R. Whitley. 1995. Comparison of genetically engineered herpes simplex viruses for the treatment of brain tumors in a scid mouse model of human malignant glioma. *Proc. Natl. Acad. Sci. U S A* 92:1411–1415.

282. MacLean, A., M. ul-Fareed, L. Robertson, J. Harland, and S. Brown. 1991. Herpes simplex virus type 1 deletion variants 1714 and 1716 pinpoint neurovirulence-related sequences in Glasgow strain 17+ between immediate early gene 1 and the "a" sequence. *J. Gen. Virol.* 72:631–639.

283. Whitley, R., E. Kern, S. Chatterjee, J. Chou, and B. Roizman. 1993. Replication establishment of latency, and induced reactivation of herpes simplex virus γ_i 34.5 deletion mutants in rodent models. *J. Clin. Invest.* 91:2837–2843.

284. Kesari, S., B. Randazzo, T. Valyi-Nagy, Q. Huang, S. Brown, A. MacLean, V.-Y. Lee, J. Trojanowski, and N. Fraser. 1995. Therapy of experimental human brain tumors using a neuroattenuated herpes simplex virus mutant. *Lab. Invest.* 73:636–648.

285. Lasner, T., S. Kesari, S. Brown, V. Lee, N. Fraser, and J. Trojanowski. 1996. Therapy of a murine model of pediatric brain tumors using a herpes simplex virus type-1 ICP34.5 mutant and demonstration of viral replication within the CNS. *J. Neuropathol. Exp. Neurol.* 55:1259–1269.

286. McKie, E., A. MacLean, A. Lewis, G. Cruickshank, R. Rampling, S. Barnett, P. Kennedy, and S. Brown. 1996. Selective in vitro replication of herpes simplex virus type 1 (HSV-1) ICP34.5 null mutants in primary human CNS tumours—evaluation of a potentially effective clinical therapy. *Brit. J. Cancer* 74:745–752.

287. Mineta, T., S. Rabkin, and R. Martuza. 1994. Treatment of malignant gliomas using ganciclovir-hypersensitive, ribonucleotide reductase-deficient herpes simplex viral mutant. *Cancer Res.* 54:3963–3966.

288. Yazaki, T., H. Manz, S. Rabkin, and R. Martuza. 1995. Treatment of human malignant meningiomas by G207, a replication-competent multimutated herpes simplex virus 1. *Cancer Res.* 55:4752–4756.

289. Toda, M., S. Rabkin, and R. Martuza. 1998. Treatment of human breast cancer in a brain metastatic model by G207, a replication-competent multimutated herpes simplex virus 1. *Hum. Gene Ther.* 9:2177–2185.

290. Wu, A., A. Mazumder, R. Martuza, X. Liu, M. Thein, K. Meehan, and S. Rabkin. 2001. Biological purging of breast cancer cells using an attenuated replication-competent herpes simplex virus in human hematopoietic stem cell transplantation. *Cancer Res.* 61:3009–3015.

291. Lambright, E., D. Caparrelli, A. Abbas, T. Toyoizumi, G. Coukos, K. Molnar-Kimber, and L. Kaiser. 1999. Oncolytic therapy using a mutant type-1 herpes simplex virus and the role of the immune system. *Ann. Thorac. Surg.* 68:1756–1762.

292. Toyoizumi, T., R. Mick, A. Abbas, E. Kang, L. Kaiser, and K. Molnar-Kimber. 1999. Combined therapy with chemotherapeutic agents and herpes simplex virus type 1 ICP34.5 mutant (HSV-1716) in human non-small cell lung cancer. *Hum. Gene Ther.* 10:3013–3029.

293. Carew, J., D. Kooby, M. Halterman, H. Federoff, and Y. Fong. 1999. Selective infection and cytolysis of human head and neck squamous cell carcinoma with sparing of normal mucosa by a cytotoxic herpes simplex virus type 1 (G207). *Hum. Gene Ther.* 10:1599–1606.

294. Randazzo, B., M. Bhat, S. Kesari, N. Fraser, and S. Brown. 1997. Treatment of experimental subcutaneous human melanoma with a replication-restricted herpes simplex virus mutant. *J. Invest. Dermatol.* 108:933–937.

295. Randazzo, B., S. Kesari, R. Gesser, D. Alsop, J. Ford, S. Brown, A. Maclean, and N. Fraser. 1995. Treatment of experimental intracranial murine melanoma with a neuroattenuated herpes simplex virus 1 mutant. *Virology* 211:94–101.

296. Endo, T., M. Toda, M. Watanabe, Y. Iizuka, T. Kubota, M. Kitajima, and Y. Kawakami. 2002. *In situ* cancer vaccination with a replication-conditional HSV for the treatment of liver metastasis of colon cancer. *Cancer Gene Ther.* 9:142–148.

297. Kooby, D., J. Carew, M. Halterman, J. Mack, J. Bertino, L. Blumgart, H. Federoff, and Y. Fong. 1999. Oncolytic viral therapy for human colorectal cancer and liver metastases using a multi-mutated herpes simplex virus type-1 (G207). *FASEB J.* 13:1325–1334.

298. Stanziale, S., H. Petrowsky, J. Joe, G. Roberts, J. Zager, N. Gusani, L. Ben-Porat, M. Gonen, and Y. Fong. 2002. Ionizing radiation potentiates the antitumor efficacy of oncoloytic herpes simplex virus G207 by upregulating ribonucleotide reductase. *Surgery* 132:353–359.

299. Cozzi, P., P. Burke, A. Bhargav, W. Heston, B. Huryk, P. Scardino, and Y. Fong. 2002. Oncolytic viral gene therapy for prostate cancer using two attenuated, replication-competent, genetically engineered herpes simplex viruses. *Prostate* 53: 95–100.

300. Cozzi, P., S. Malhotra, P. McAuliffe, D. Kooby, H. Federoff, B. Huryk, P. Johnson, P. Scardino, W. Heston, and Y. Fong. 2001. Intravesical oncolytic viral therapy using attenuated, replication-competent herpes simplex viruses G207 and Nv1020 is effective in the treatment of bladder cancer in an orthotopic syngeneic model. *FASEB J.* 15:1306–1308.

301. Oyama, M., T. Ohigashi, M. Hoshi, M. Murai, K. Uyemura, and T. Yazaki. 2000. Oncolytic viral therapy for human prostate cancer by conditionally replicating herpes simplex virus 1 vector G207. *Jpn. J. Cancer Res.* 91:1339–1344.

302. Walker, J., K. McGeagh, P. Sundaresan, T. Jorgensen, S. Rabkin, and R. Martuza. 1999. Local and systemic therapy of human prostate adenocarcinoma with the conditionally replicating herpes simplex virus vector G207. *Hum. Gene Ther.* 10:2237–2243.

303. Coukos, G., A. Makrigiannakis, E. Kang, D. Caparelli, I. Benjamin, L. Kaiser, S. Rubin, S. Albelda, and K. Molnar-Kimber. 1999. Use of carrier cells to deliver a replication-selective herpes simplex virus-1 mutant for intraperitoneal therapy of epithelial ovarian cancer. *Clin. Cancer Res.* 5:1523–1537.

304. Coukos, G., A. Makrigiannakis, S. Montas, L. Kaiser, T. Toyozumi, I. Benjamin, S. Albelda, S. Rubin, and K. Molnar-Kimber. 2000. Multi-attenuated herpes simplex virus-1 mutant G207 exerts cytotoxicity against epithelia ovarian cancer but not normal mesothelium and is suitable for intraperitoneal oncolytic therapy. *Cancer Gene Ther.* 7:275–283.

305. Bennett, J., K. Delman, B. Burt, A. Mariotti, S. Malhotra, J. Zager, H. Petrowsky, S. Mastorides, H. Federoff, and Y. Fong. 2002. Comparison of safety, delivery, and efficacy of two oncolytic herpes viruses (G207 and NV1020) for peritoneal cancer. *Cancer Gene Ther.* 9:935–945.

306. Bennett, J., D. Kooby, K. Delman, P. McAuliffe, M. Halterman, H. Federoff, and Y. Fong. 2000. Antitumor efficacy of regional oncolytic viral therapy for peritoneally disseminated cancer. *J. Mol. Med.* 78:166–174.

307. Lambright, E., E. Kang, S. Force, M. Lanuti, D. Caparrelli, L. Kaiser, S. Albelda, and K. Molnar-Kimber. 2000. Effect of preexisting anti-herpes immunity on the efficacy of herpes simplex viral therapy in a murine intraperitoneal tumor model. *Mol. Ther.* 2:387–393.

308. Oyama, M., T. Ohigashi, M. Hoshi, J. Nakashima, M. Tachibana, M. Murai, K. Uyemura, and T. Yazaki. 2000.

Intravesical and intravenous therapy of human bladder cancer by the herpes vector G207. *Hum. Gene Ther.* 11:1683–1693.

309. Oyama, M., T. Ohigashi, M. Hoshi, M. Murai, K. Uyemura, and T. Yazaki. 2001. Treatment of human renal cell carcinoma by a conditionally replicating herpes vector G207. *J. Urol.* 165:1274–1278.

310. Blank, S., S. Rubin, G. Coukos, K. Amin, S. Albelda, and K. Molnar-Kimber. 2002. Replication-selective herpes simplex virus type 1 mutant therapy of cervical cancer is enhanced by low-dose radiation. *Hum. Gene Ther.* 13:627–639.

311. Nakano, K., T. Todo, K. Chijiiwa, and M. Tanaka. 2001. Therapeutic efficacy of G207, a conditionally replicating herpes simplex virus type 1 mutant, for gallbladder carcinoma in immunocompetent hamsters. *Mol. Ther.* 3:431–437.

312. Todo, T., S. Rabkin, and R. Martuza. 2000. Evaluation of ganciclovir-mediated enhancement of the antitumoral effect in oncolytic, multimutated herpes simplex virus type 1 (G207) therapy of brain tumors. *Cancer Gene Ther.* 7:939–946.

313. Chahlavi, A., S. Rabkin, T. Todo, P. Sundaresan, and R. Martuza. 1999. Effect of prior exposure to herpes simplex virus 1 on viral vector-mediated tumor therapy in immunocompetent mice. *Gene Ther.* 6:1751–1758.

314. Barba, D., J. Hardin, J. Ray, and F. H. Gage. 1993. Thymidine kinase-mediated killing of rat brain tumors. *J. Neurosurg.* 79:729–735.

315. Caruso, M., Y. Panis, S. Gagandeep, D. Houssin, J. L. Salzmann, and D. Klatzmann. 1993. Regression of established macroscopic liver metastases after in situ transduction of a suicide gene. *Proc. Natl. Acad. Sci. U S A* 90:7024–7028.

316. Culver, K., Z. Ram, S. Walbridge, H. Ishii, E. Oldfield, and R. Blaese. 1992. In vivo gene transfer with retroviral vector-producer cells for treatment of experimental brain tumors. *Science* 256:1550–1552.

317. Ezzeddine, Z. D., R. L. Martuza, D. Platika, M. P. Short, A. Malick, B. Choi, and X. O. Breakefield. 1991. Selective killing of glioma cells in culture and in vivo by retrovirus transfer of the herpes simplex virus thymidine kinase gene. *New Biol.* 3:608–614.

318. Ram, Z., K. W. Culver, S. Walbridge, R. M. Blaese, and E. H. Oldfield. 1986. In situ retroviral-mediated gene transfer for the treatment of brain tumors in rats. *Cancer Res.* 46:5276–5281.

319. Burger, P. and B. Scheithauer. 1994. Tumors of the central nervous system, *Atlas of Tumor Pathology*, vol. Third Series, Fascicle 10. Armed Forces Institute of Pathology, Washington, DC, p. 59.

320. Short, M. P., B. C. Choi, J. K. Lee, A. Malick, X. O. Breakefield, and R. L. Martuza. 1990. Gene delivery to glioma cells in rat brain by grafting of a retrovirus packaging cell line. *J. Neurosci. Res.* 27:427–439.

321. Bi, W. L., L. M. Parysek, R. Warnick, and P. J. Stambrook. 1993. In vitro evidence that metabolic cooperation is responsible for the bystander effect observed with HCV Tk retroviral gene therapy. *Hum. Gene Ther.* 4:725–731.

322. Freeman, S., C. Abboud, K. Whartenby, C. Packman, D. Koeplin, F. Moolten, and G. Abraham. 1993. The "Bystander Effect": Tumor regression when a fraction of the tumor mass is genetically modified. *Cancer Res.* 53:5274–5283.

323. Kato, K., J. Yoshida, M. Mizuno, K. Sugita, and N. Emi. 1994. Retroviral transfer of herpes simplex thymidine kinase into glioma cells targeting of gancyclovir cytotoxic effect. *Neurol. Med. Chir. (Tokyo)* 34:339–344.

324. Wu, J. K., W. G. Cano, S. A. Meylaerts, P. Qi, F. Vrionis, and V. Cherington. 1994. Bystander tumoricidal effect in the treatment of experimental brain tumors. *Neurosurgery* 35:1094–1102.

325. Oshiro, E. M., J. J. Viola, E. H. Oldfield, S. Walbridge, J. Bacher, J. A. Frank, R. M. Blaese, and Z. Ram. 1995. Toxicity studies and distribution dynamics of retroviral vectors following intrathecal administration of retroviral vector-producer cells. *Cancer Gene Ther.* 2:87–95.

326. Moriuchi, S., D. Krisky, P. Marconi, M. Tamura, K. Shimizu, T. Yoshimine, J. Cohen, and J. Glorioso. 2000. HSV vector cytotoxicity is inversely correlated with effective TK/GCV suicide gene therapy of rat gliosarcoma. *Gene Ther.* 7:1483–1490.

327. Moriuchi, S., T. Oligino, D. Krisky, P. Marconi, D. Fink, J. Cohen, and J. Glorioso. 1998. Enhanced tumor-cell killing in the presence of ganciclovir by HSV-1 vector-directed co-expression of human TNF-α and HSV thymidine kinase. *Cancer Res.* 58:5731–5737.

328. Moriuchi, S., D. Wolfe, M. Tamura, T. Yoshimine, F. Miura, J. B. Cohen, and J. C. Glorioso. 2002. Double suicide gene therapy using a replication defective herpes simplex virus vector reveals reciprocal interference in a malignant glioma model. *Gene Ther.* 9:584–591.

329. Bennett, M. and V. Verselis. 1992. Biophysics of gap junctions. *Semin. Cell Biol.* 3:29–47.

330. Colombo, B. M., S. Benedetti, S. Ottolenghi, M. Mora, B. Pollo, G. Poli, and G. Finocchiaro. 1995. The "bystander effect": Association of U-87 cell death with ganciclovir-mediated apoptosis of nearby cells and lack of effect in athymic mice. *Hum. Gene Ther.* 6:763–772.

331. Naus, C. C., J. F. Bechberger, and D. L. Paul. 1991. Gap junction gene expression in human seizure disorder. *Exp. Neurol.* 111:198–203.

332. Yamasaki, H. 1996. Role of disrupted gap junctional intercellular communication in detection and characterization of carcinogens. *Mutat. Res.* 365:91–105.

333. Mehta, P. P., A. Hotz-Wagenblatt, B. Rose, D. Shalloway, and W. R. Loewenstein. 1991. Incorporation of the gene for a cell-cell channel protein into transformed cells leads to normalization of growth. *J. Membr. Biol.* 124:207–225.

334. Loewenstein, W. R. and Y. Kanno. 1966. Intercellular communication and the control of tissue growth: Lack of communication between cancer cells. *Nature* 209:1248–1249.

335. Mehta, P. P., J. S. Bertram, and W. R. Loewenstein. 1986. Growth inhibition of transformed cells correlates with their junctional communication with normal cells. *Cell* 44:187–196.

336. Shinoura, N., L. Chen, M. A. Wani, Y. G. Kim, J. J. Larson, R. E. Warnick, M. Simon, A. G. Menon, W. L. Bi, and P. J. Stambrook. 1996. Protein and messenger RNA expression of connexin43 in astrocytomas: Implications in brain tumor gene therapy. *J. Neurosurg.* 84:839–845.

337. Zhu, D., S. Caveney, G. M. Kidder, and C. C. Naus. 1991. Transfection of C6 glioma cells with connexin 43 cDNA: Analysis of expression, intercellular coupling, and cell proliferation. *Proc. Natl. Acad. Sci. U S A* 88:1883–1887.

338. Elshami, A. A., A. Saavedra, H. Zhang, J. C. Kucharczuk, D. C. Spray, G. I. Fishman, K. M. Amin, L. R. Kaiser, and S. M. Albelda. 1996. Gap junctions play a role in the "bystander effect" of the herpes simplex virus thymidine kinase/ganciclovir system in vitro. *Gene Ther.* 3:85–92.

339. Fick, J., F. G. n. Barker, P. Dazin, E. M. Westphale, E. C. Beyer, and M. A. Israel. 1995. The extent of heterocellular communication mediated by gap junctions is predictive of bystander tumor cytotoxicity in vitro. *Proc. Natl. Acad. Sci. U S A* 92:11071–11075.

340. Mesnil, M., C. Piccoli, G. Tiraby, K. Willecke, and H. Yamasaki. 1996. Bystander killing of cancer cells by herpes simplex virus thymidine kinase gene is mediated by connexins. *Proc. Natl. Acad. Sci. U S A* 93:1831–1835.

341. Touraine, R. L., N. Vahanian, W. J. Ramsey, and R. M. Blaese. 1998. Enhancement of the herpes simplex virus thymidine kinase/ganciclovir bystander effect and its antitumor efficacy in vivo by pharmacologic manipulation of gap junctions. *Hum. Gene Ther.* 9:2385–2391.

342. Dilber, M. S., M. R. Abedi, B. Christensson, B. Bjorkstrand, G. M. Kidder, C. C. Naus, G. Gahrton, and C. I. Smith. 1997. Gap junctions promote the bystander effect of herpes simplex virus thymidine kinase in vivo. *Cancer Res.* 57:1523–1528.

343. Vrionis, F. D., J. K. Wu, P. Qi, M. Waltzman, V. Cherington, and D. C. Spray. 1997. The bystander effect exerted by tumor cells expressing the herpes simplex virus thymidine kinase (HSVtk) gene is dependent on connexin expression and cell communication via gap junctions. *Gene Ther.* 4:577–585.

344. Marconi, P., M. Tamura, S. Moriuchi, D. Krisky, W. Goins, J. Cohen, and J. Glorioso. 2000. Connexin43-enhanced suicide gene therapy using herpes viral vectors. *Mol. Ther.* 1:71–81.

345. Han, S. K., S. L. Brody, and R. G. Crystal. 1994. Suppression of in vivo tumorigenicity of human lung cancer cells by retrovirus-mediated transfer of the human tumor necrosis factor-alpha cDNA. *Am. J. Resp. Cell. Mol. Biol.* 11:270–278.

346. Pfizenmaier, K., K. Pfizenmaier, P. Scheurich, C. Schluter, and M. Kronke. 1987. Tumor necrosis factor enhances HLA-A, B, C and HLA-DR gene expression in human tumor cells. *J. Immunol.* 138:975–980.

347. Watanabe, Y., K. Kuribayashi, S. Miyatake, K. Nishihara, E. Nakayama, T. Taniyama, and T. Sakata. 1989. Exogenous expression of mouse interferon cDNA in mouse neuroblastoma C1300 cells results in reduced tumorigenicity by augmented antitumor immunity. *Proc. Nat. Acad. Sci. U S A* 86:9456–9460.

348. Ostensen, M. E., D. L. Thiele, and P. E. Lipsky. 1987. Enhancement of human natural killer cell function by the combined effects of tumor necrosis factor alpha or interleukin-1 and interferon-alpha or interleukin-2. *J. Biol. Res. Mod.* 8:53–61.

349. Owen-Schaub, L. B., J. U. Gutterman, and E. A. Grimm. 1988. Synergy of tumor necrosis factor and interleukin 2 in the activation of human cytotoxic lymphocytes: Effect of tumor necrosis factor alpha and interleukin 2 in the generation of human lymphokine-activated killer cell cytotoxicity. *Cancer Res.* 48:788–792.

350. Plaetinck, G., W. Declercq, J. Tavernier, M. Jabholz, and W. Fiers. 1987. Recombinant tumor necrosis factor can induce interleukin 2 receptor expression and cytolytic activity in a rat x mouse T cell hybrid. *Eur. J. Immunol.* 17:1835–1838.

351. Ranges, G. E., I. S. Figari, T. Espevik, and M. A. Palladino Jr. 1987. Inhibition of cytotoxic T cell development by transforming growth factor beta and reversal by recombinant tumor necrosis factor alpha. *J. Exp. Med.* 166:991–998.

352. Kondziolka, D., J. Flickinger, D. Bissonette, M. Bozik, and D. Lunsford. 1997. Survival benefit of stereotactic radiosurgery for patients with malignant glial neoplasms. *Neurosurgery* 41:776–785.

353. Sarkaria, J. N., M. P. Mehta, J. S. Loeffler, J. M. Buatti, R. J. Chappell, A. B. Levin, E. Alexander, 3rd, W. A. Friedman, and T. J. Kinsella. 1994. Stereotactic radiosurgery improves survival in malignant gliomas compared with the RTOG recursive partitioning analysis. *Int. J. Radiat. Oncol. Biol. Phys.* 30:164.

354. Kelly, K. A., J. M. Kirkwood, and D. S. Kapp. 1984. Glioblastoma multiforms: Pathology, natural history and treatment. *Cancer Treat. Rev.* 11:1–26.

355. McComb, R. D. and D. D. Bigner. 1984. The biology of malignant gliomas—a comprehensive survey. *Clin. Neuropathol.* 3:93–106.

356. Gridley, D., J. Archambeau, M. Andres, X. Mao, K. Wright, and J. Slater. 1997. Tumor necrosis factor-alpha enhances antitumor effects of radiation against glioma xenografts. *Oncol. Res.* 9:217–227.

357. Gridley, D., W. Glisson, and J. Uhm. 1994. Interaction of tumour necrosis factor-alpha and radiation against human colon tumour cells. *Ther. Immunol.* 1:25–31.

358. Gridley, D., S. Hammond, and B. Liwnicz. 1994. Tumor necrosis factor-alpha augments radiation effects against human colon tumor xenografts. *Anticancer Res.* 14:1107–1112.

359. Huang, P., A. Allam, L. Perez, A. Taghian, J. Freeman, and H. Suit. 1995. The effect of combining recombinant human tumor necrosis factor-alpha with local radiation on tumor control probability of a human glioblastoma multiforme xenograft in nude mice. *Int. J. Radiat. Oncol. Biol. Phys.* 32:93–98.

360. Kimura, K., C. Bowen, S. Spiegel, and E. Gelmann. 1999. Tumor necrosis factor-α sensitizes prostate cancer cells to γ-irradiation-induced apoptosis. *Cancer Res.* 59:1606–1614.

361. Leonard, M., R. Jeffs, J. Gearhart, and D. Coffey. 1992. Recombinant human tumor necrosis factor enhances radiosensitivity and improves animal survival in murine neuroblastoma. *J. Urol.* 148:743–746.

362. Sersa, G., V. Willingham, and L. Milas. 1988. Anti-tumor effects of tumor necrosis factor alone or combined with radiotherapy. *Int. J. Cancer* 42:129–134.

363. Baher, A., M. Andres, J. Folz-Holbeck, J. Cao, and D. Gridley. 1999. A model using radiation and plasmid-mediated tumor necrosis factor-alpha gene therapy for treatment of glioblastomas. *Anticancer Res.* 19:2917–2924.

364. Gridley, D., J. Baer, J. Cao, G. Miller, D. Kim, T. Timiryasova, I. Fodor, and J. Slater. 2002. TNF-alpha gene and proton radiotherapy in an orthotopic brain tumor model. *Int. J. Oncol.* 21:251–259.

365. Kim, D., M. Andres, J. Li, E. Kajioka, G. Miller, A. Seynhaeve, L. Ten Hagen, and D. Gridley. 2001. Liposome-encapsulated tumor necrosis factor-alpha enhances the effects of radiation against human colon tumor xenografts. *J. Interferon Cytokine Res.* 21:885–897.

366. Weichselbaum, R., D. Hallahan, M. Beckett, H. Mauceri, H.-M. Lee, V. Sukhatme, and D. Kufe. 1994. Gene therapy targeted by radiation preferentially radiosensitizes tumor cells. *Cancer Res.* 54:4266–4269.

367. Chung, T., H. Mauceri, D. Hallahan, J. Yu, S. Chung, W. Grdina, S. Yajnik, and D. Kufe. 1998. Tumor necrosis factor-alpha-based gene therapy enhances radiation cytotoxicity in human prostate cancer. *Cancer Gene Ther.* 5:344–349.

368. Gupta, V., J. Park, N. Jaskowiak, H. Mauceri, S. Seetharam, R. Weichselbaum, and M. Posner. 2002. Combined gene therapy and ionizing radiation is a novel approach to treat human esophageal adenocarcinoma. *Ann. Surg. Oncol.* 9:500–504.

369. Mauceri, H., N. Hanna, J. Wayne, D. Hallahan, S. Hellman, and R. Weichselbaum. 1996. Tumor necrosis factor alphpa (TNF-alpha) gene therapy targeted by ionizing radiation selectively damages tumor vasculature. *Cancer Res.* 56:4311–4314.

370. Staba, M., H. Mauceri, D. Kufe, D. Hallahan, and R. Weichselbaum. 1998. Adenoviral TNF-alpha gene therapy and radiation damage tumor vasculature in a human malignant glioma xenograft. *Gene. Ther.* 5:293–300.

371. Weicheselbaum, R., D. Kufe, S. Hellman, H. Rasmussen, C. Richter King, P. Fischer, and H. Mauceri. 2002. Radiation-induced tumour necrosis factor-α expression: Clinical application of transcriptional and physical targeting of gene therapy. *Lancet Onc.* 3:665–671.

372. Sharma, A., S. Mani, N. Hanna, C. Guha, B. Vikram, R. Weichselbaum, J. Sparano, B. Sood, D. Lee, W. Regine, M. Muhodin, J. Valentino, J. Herman, P. Desimone, S. Arnold, J. Carrico, A. Rockich, J. Warner-Carpenter, and M. Barton-Baxter. 2001. Clinical Protocol: An open-label, phase-1, dose-escalation study of tumor necrosis factor-α (TNF-α Biologic) gene

transfer with radiation therapy for locally advanced, recurrent, or metastatic solid tumors. *Hum. Gene Ther.* 12:1109–1131.

373. Niranjan, A., S. Moriuchi, L. Lunsford, D. Kondziolka, J. Flickinger, W. Fellows, S. Rajendiran, M. Tamura, J. Cohen, and J. Glorioso. 2000. Effective treatment of experimental glioblastoma by HSV vector-mediated TNF-α and HSV-tk gene transfer in combination with radiosurgery and ganciclovir administration. *Mol. Ther.* 2:114–120.

374. Niranjan, A., D. Wolfe, M. Tamura, M. K. Soares, D. M. Krisky, L. D. Lunsford, S. Li, W. Fellows-Mayle, N. A. DeLuca, J. B. Cohen, and J. C. Glorioso. 2003. Treatment of rat gliosarcoma brain tumors by HSV-based multigene therapy combined with radiosurgery. *Mol. Ther.* 8:530–542.

375. Kondziolka, D., A. Patel, L. Lunsford, A. Kassam, and J. Flickinger. 1999. Stereotactic radiosurgery plus whole brain radiotherapy versus radiotherapy alone for patients with multiple brain metastases. *Int. J. Radiat. Oncol. Biol. Phys.* 45:427–434.

376. Larson, D., P. Gutin, M. McDermott, K. Lamborn, P. Sneed, W. Wara, J. Flickinger, D. Kondziolka, L. Lunsford, W. Hudgins, G. Friehs, K. Haselsberger, K. Leber, G. Pendl, S. Chung, R. Coffey, R. Dinapoli, E. Shaw, S. Vermeulen, R. Young, M. Hirato, H. Inoue, C. Ohye, and T. Shibazaki. 1996. Gamma knife for glioma: Selection factors and survival. *Int. J. Radiat. Oncol. Biol. Phys.* 36:1045–1053.

8 Oncolytic Viruses for Treatment of Cancer

Mari Raki, Maria Rajecki, David Kirn, and Akseli Hemminki

CONTENTS

8.1 INTRODUCTION

Although the treatment of most cancers has improved steadily, metastatic solid tumors remain incurable. Despite frequent responses, refractory clones often emerge rapidly and the disease becomes refractory to available treatment modalities. Although chemotherapeutic agents and radiation therapy target various cellular structures and pathways, the majority of them kill cancer cells through induction of apoptosis and selectivity is based mostly on more rapid replication of tumor cells in comparison to normal cells. As malignant cells are characterized by an ability to adapt to the environment, apoptosis-resistant clones frequently develop following standard treatment. Furthermore, during subsequent treatment regimens, resistance factors are shared and therefore loss of response typically occurs more rapidly, and there is a tendency for cross-resistance between agents. Therefore, new agents with novel mechanisms of action and lacking cross-resistance to currently available approaches are needed.

Due to safety concerns, gene therapy approaches have usually been based on viruses that are unable to replicate in infected cells. Although replication-deficient viruses expressing therapeutic transgenes have provided high *in vitro* and *in vivo* preclinical efficacy and good clinical safety data, trials have demonstrated that the utility of these agents may be limited when faced with advanced and bulky disease [1]. Viruses that replicate and spread specifically inside the tumor have been suggested as a way to improve penetration of and dissemination within solid tumor masses [2–4]. Oncolytic viruses used in various cancer gene therapy approaches take advantage of tumor-specific changes that allow preferential replication of the virus in target cells (Figure 8.1). The viral replication cycle causes oncolysis of the cell, resulting in the release of newly generated virions and subsequent infection of neighboring cells. Normal tissue is spared due to lack of replication. Thus, the antitumor effect is delivered by the actual replication of the virus. Therefore, a transgene is not necessarily required, but can be used for additional efficacy. Importantly, the viral replication cycle allows dramatic local amplification of the input dose, and in theory, the oncolytic process continues as long as target cells persist, and vascular dissemination can also occur to distant sites.

8.2 HISTORY OF ONCOLYTIC VIROTHERAPY

Oncolytic virotherapy originates from the beginning of the last century and is based on observations of tumor regressions in cancer patients contracting a viral infection. In 1904, a patient with chronic myelogenous leukemia had a dramatic decrease in white blood cells during a "flu-like" illness. During the 1950s–1980s, further examples were reported: varicella-induced remission of acute lymphoblastic leukemia [5] and measles-induced regressions of leukemias [6], Burkitt's

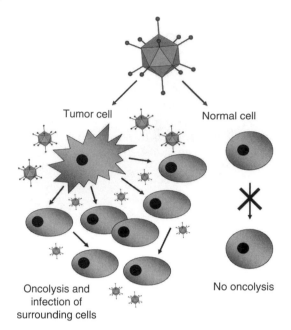

FIGURE 8.1 Principle of oncolysis viruses. Replication kills tumor cells while features engineered into the virus genome prevent replication in normal cells.

lymphoma [7] and Hodgkin's lymphoma [8]. A chicken farmer's gastric cancer responded following occupational exposure to Newcastle disease virus [9].

Similar responses occurred following purposeful vaccination of cancer patients. In 1912, a cervical carcinoma responded following inoculation of an attenuated rabies vaccine [10]. This led to an experiment in 1920 by Levaditi and Nicolau, who documented vaccinia oncolysis of murine tumors [11]. Subsequently, a human trial was performed in 1940 by using an attenuated rabies virus against melanoma [12]. By 1950s, Southam et al. were treating patients suffering from various types of cancers and frequent remissions were reported [13,14]. A long-lasting benefit was reported in a patient with chronic lymphocytic leukemia, who sustained a complete response lasting over 3 years following systemic vaccinia virus dissemination from a vaccination site in the

skin [15]. In the 1980s, measles virus vaccination was associated with responses in two patients with Hodgkin's lymphoma [16,17]. An overview of historical trials with oncolytic viruses has been listed on Table 8.1.

8.3 ONCOLYTIC VIRUSES

An oncolytic virus considered for treatment of cancer should have several important properties. Virus must be able to infect, replicate in, and kill human tumor cells including nondividing cancer cells [2–4]. From a safety standpoint, the parental virus should cause only mild, well characterized human disease. Further, a nonintegrating virus would decrease the risk of acquiring unwanted changes to the host genome. Viral replication in normal tissues might be preventable by genetic engineering or by utilizing agents that are capable of inactivating virus if necessary [18]. Also, a genetically stable virus that can be easily purified to high titers according to current good manufacturing practices (cGMP) is desirable.

There are various approaches for achieving selective viral replication in tumor cells. As the cellular changes induced by viral infection are often similar to changes acquired during tumor development, it is not surprising that some viruses (measles, vesicular stomatitis, and Newcastle disease viruses), have inherent selectivity for cancer cells. However, many popular oncolytic viruses (adenovirus, herpes simplex, and vaccinia viruses) are genetically modified to minimize toxicity to normal tissues and to maximize the tolerated dose.

One way to restrict replication to malignant tissues is to use tumor-specific promoters (TSPs). Another strategy involves engineering deletions in viral genes critical for efficient replication and cytotoxicity in normal but not in cancer cells. This may seem complex but is facilitated by the intriguing similarities between carcinogenesis and cellular takeover by DNA viruses. This may pertain to general cellular physiology: There are a handful of central growth control pathways that need to be overcome, regardless of whether the purpose is immortal growth or utilization of the cell for virus production. As an alternative, or better yet, to complement

TABLE 8.1
Historical Oncolytic Virus Trials

Type of Virus	Patients	Responses	Safety	Year	References
Sera or tissue extract containing Hepatitis B virus	22	4 reductions in tumor size, 7 clinical improvements	14 got hepatitis, 1 died of hepatitis	1949	[142]
Glandular fever sera	5	3 brief remissions	Good	1952	[143]
Egypt 101 (West Nile)	34	4 brief tumor regressions	2 encephalitis, fever and nausea in other patients	1952	[13]
Adenovirus	30	65% of injections caused local tumor necrosis	Good, vaginal hemorrhages from necrotic tumors	1956	[144]
Mumps	90	37 good responses (>50% reduction in tumor size), 42 PR (<50% reduction in tumor size)	Good, in 7 patients bleeding and fever	1974	[145]

transcriptional targeting such as described above, transductional modification of viral coat proteins can be genetically or chemically performed to alter virus tropism for more effective gene delivery to tumor cells and less transduction of normal cells.

Here, some of the most popular oncolytic virus platforms will be reviewed. Emphasis is placed on agents that have clinical data available and central pieces of that data will also be presented.

8.3.1 Adenovirus

Adenoviruses are nonenveloped, double-stranded DNA viruses whose native pathogenesis typically involves mild upper respiratory tract, ocular, and gastrointestinal infections. The icosahedral virus capsid encapsulates an up to 38 kb linear DNA genome whose main structural components include hexon, fiber, and penton [19]. Penton units are located at each of the 12 vertices of the capsid, and the receptor binding fibers protrude from the penton bases. Infection by serotype 5 (Ad5), which is most commonly used for gene therapy, starts by binding of the fiber knob domain to the primary receptor, the coxsackie-adenovirus receptor (CAR) [20]. Binding is followed by internalization of the virus, mediated by the interaction of a penton base Arg–Gly–Asp (RGD) motif and cellular $\alpha_v\beta$ integrins, which triggers endocytosis of the virion via clathrin-coated pits [21]. Thereafter, viral DNA is transported to the nucleus through a rather poorly understood microtubule-mediated process, and viral genes are expressed. These processes are best understood *in vitro*. *In vivo*, it is conceivable that the fiber may have important interactions with blood factors, with implications on biodistribution, at least in mice [22,23].

Adenovirus is an attractive vector for gene therapy due to its high efficiency of infection of both dividing and quiescent cells, ease of high titer production, and stability *in vivo* [3,4]. Adenoviral DNA does not integrate into the host genome, resulting in a low risk of mutagenesis in humans. Additionally, the molecular biology of adenovirus has been well characterized and its safety has been demonstrated in a large number of clinical trials [4]. The limited duration of viral gene expression may render adenoviruses less desirable for the treatment of hereditary diseases, where long-term expression is needed, but is adequate for cancer gene therapy, where the purpose is to kill the target cells.

The first cancer trials with replicating adenoviruses were done already in the 1950s [24]. Ten different serotypes of wild-type adenoviruses were applied intratumorally, intra-arterially, by both routes, or intravenously into patients with cervical carcinoma. No significant toxicity was reported. A "marked to moderate local tumor response" was reported in over half of the patients. However, systemic responses were not detected and all patients eventually had tumor progression. Although these clinical studies were not performed according to current clinical research standards and therefore should be interpreted with caution, they suggest that replication competent or even wild-type adenoviruses can be safely administered to humans and that viruses can replicate in tumors for therapeutic effect.

8.3.1.1 Type I Oncolytic Adenoviruses

With type I oncolytic adenoviruses, tumor-specific replication is achieved by introduction of loss-of-function mutations to the virus genome that requires specific cellular factors to compensate for the effects of these mutations. Most approaches are based on deletions in immediately-early (*E1A*) or early (*E1B*) adenoviral genes resulting in mutant E1 proteins unable to bind cellular proteins necessary for viral replication in normal cells, but not in cancer cells, because of transcomplementation in the latter [4].

The first and most widely studied oncolytic adenovirus *dl*1520 (ONYX-015) carries two mutations in the gene coding for the E1B-55 kDa protein [25]. One purpose of this protein is binding and inactivation of p53 in infected cells, for induction of S-phase, which is required for effective virus replication. Thus, this virus may have selectivity for cells with an aberrant p53–p14^ARF pathway, a common feature of human tumors. Initial studies suggested that this agent replicates selectively in cancer cells lacking functional p53 [25–28]. However, contradicting studies have suggested that cells with functional p53 also support *dl*1520 replication [29,30]. It has become clear that factors independent of p53 play critical roles in determining the sensitivity of cells to *dl*1520. Loss of function of p14^ARF, which normally stabilizes p53, can result in inactivation of p53 [31]. Adenoviral proteins other than E1B-55 kDa, including E4 or F6, E1B-19 kDa, and E1A, have effects on p53 function [32]. Further, the functions of E1B-55 kDa are not limited to p53 binding, but include also mRNA transport [33], and therefore the virus replicates inefficiently compared with the wild-type adenovirus. Nevertheless, taken together, these data seem to suggest more effective replication of *dl*1520 in tumor cells in comparison to normal cells. Importantly, this has later been confirmed in humans and one reason may be the relative protection from the immune system offered by the tumor environment [34].

Ad5-Δ24 and *dl*922–947 are closely related viruses that carry a 24 bp deletion in the constant region 2 (CR2) of the E1A gene [35,36]. This domain of the E1A protein is responsible for binding of the retinoblastoma tumor suppressor/cell cycle regulator protein (Rb) for induction of S-phase entry. Therefore, viruses with this type of deletion have reduced ability to overcome the G1-S checkpoint and replicate selectively in cells deficient in the Rb-p16 pathway, including most if not all cancer cells [37]. These agents are promising anticancer agents as they are not attenuated in comparison to wild-type viruses and some reports suggest that they may be even more oncolytic than wild type in tumor cells [36]. It should be noted, however, that Rb is phosphorylated in rapidly replicating normal cells which could allow a degree of replication of Δ24 generation agents. Further, their safety has not yet been demonstrated in human trials, although they are expected to begin soon [4,38].

8.3.1.2 Clinical Experience with *dl*1520 and Closely Related Viruses

The safety and antitumor efficacy of *dl*1520 type viruses has been tested in numerous phase I/II/III clinical trials with various tumor types and several routes of administration, with and without concomitant conventional treatments (Table 8.2). The goal has been to sequentially increase systemic exposure to the virus after safety with localized delivery has been shown. Following demonstration of safety and biological activity by the intratumoral route, trials were sequentially initiated to study intracavitary, intra-arterial, and eventually intravenous administration of *dl*1520 [34].

Intratumoral delivery of *dl*1520 has been most extensively studied in the context of head and neck cancer [39–43]. Treatment has been well tolerated without dose-limiting toxicity and virus has been shown to replicate in tumor cells but not in peritumoral tissue. Antitumor activity has been demonstrated by shrinkage of tumors. In a phase II study of intratumorally administered *dl*1520 to 40 patients, 3 complete and 2 partial responses were reported [42]. However, the most encouraging results were seen in a phase II combination trial

---

TABLE 8.2
Trials with *dl*1520 and H101

Cancer Type	Phase	Patients	Safety	Responses	References
Recurrent head and neck cancer	I	22	Good	No objective responses, 5 MRI detected necrosis at injection site	[39]
	II	37 (30 evaluated)	5 patients experienced grade 4 adverse effects	8 CR, 11 PR lasting over 5 months	[40]
	II	14 (9 evaluated)	No SAE, 2 grade 3 mucositis	3 CR, 1 MR, 3 PR, and 2 SD	[146]
	II	37	Good	2 CR, 3 PR, 3 MR, 9 SD, and 7 PD	[41]
	II	40[a]	Good, serious adverse effects probably related to treatment in 5 patients	Standard therapy: 2 CR, 2 PR, 2 MR, 10 SD, and 13 PD Hyperfractioned therapy: 1 CR, 1 MR, 3 SD, and 2 PD	[42]
Refractory primary and secondary liver tumors	I	9	Good	6 SD and 1 PD	[147]
Pancreatic cancer	I	23	Good	No objective responses	[45]
Solid tumors with metastasis	I	10	Good	1 MR, 9 SD	[55]
Liver metastasis of colorectal carcinoma	I	11	Good	1 PR	[52]
Recurrent epithelial ovarian cancer	I	16	No SAE, 1 grade 3 abdominal pain and diarrhea	No objective responses	[49]
Hepatocellular carcinoma	I/II	5	Good	1 PR, 4 PD	[47]
Gastrointestinal carcinomas metastatic to liver	II	27	No SAE, grade 3/4 hyperbilirubinemia in 2 patients, grade 4 systemic inflammatory response	3 PR, 4 MR, 9 SD, and 11 PD	[53]
Pancreatic carcinoma	I/II	21	Good	2 PR, 2 MR, 6 SD, and 11 PD	[46]
Hepatobiliary carcinoma	II	20 (16 evaluated)	Grade 3 adverse effects including 3 hepatic toxicity, anemia, infection and atrial fibrillation	1 PR, 1 SD and 8 patients had >50% decrease in tumor markers	[148]
Metastatic colorectal cancer	II	18	Good	No responses	[56]
Premalignant oral dysplasia	I/II	22 (19 evaluated)	Good	7 histologic resolutions, 1 improvement in dysplasia grade	[51]
Late stage cancers	II	50	No SAE, one grade 4 hepatic dysfunction and hematologic toxicity	3 CR and 11 PR	[149]
Recurrent malignant glioma	I	24	Good	1 SD, 23 PD	[50]
Advanced sarcomas	I/II	6	Good	1 PR	[150]
Refractory metastatic colorectal carcinoma	II	24	Good	2 PR, 11 SD, treatment of 8 patients was stopped due to progression	[54]

Note: CR, complete response; MRI, magnetic resonance imagining; MR, minor response; PD, progressive disease; PR, partial response; SD, stable disease; SAE, serious adverse effects.

[a] 30 patients received 1 injection daily (standard therapy) and 10 patients received 2 injections daily (hyperfractioned therapy).

of intratumorally delivered *dl*1520 with 5-fluorouracil (5-FU) and cisplatin to 37 patients with recurrent head and neck cancer [40]. Treatment resulted in 27% and 36% complete and partial responses, respectively.

Recently, intratumoral injection of H101, an oncolytic adenovirus quite similar to *dl*1520, was evaluated in a phase III trial against squamous cell cancer of head and neck or esophagus [43,44]. Combination treatment with 5-FU and cisplatin resulted in a response rate of 78.8% (41/52) versus 39.6% (21/53) with chemotherapy alone ($P < .0001$). Virus injected patients had more mild flu-like symptoms and injection site reactions but severe toxicity was not increased. This is a landmark trial in several regards. It is the first randomized demonstration of safety and efficacy of oncolytic viruses in humans. Also, a doubling of the response rate represents a magnitude of improvement rarely seen in oncology, which allowed a highly significant result in a relatively small trial. The oncology and gene therapy communities eagerly await confirmation of the results in further trials performed also outside China. Nevertheless, these data and response rates are well in accord with an earlier independent phase II trial performed in the United States [40], which supports the validity of the H101 trial.

Intratumoral administration of *dl*1520 has been utilized in nonrandomized trials with several other cancer types including pancreatic cancer [45,46], hepatocellular carcinoma [47], and oral carcinoma [48]. However, results from these trials have been less promising as responses to treatment have been rare.

Intracavitary administration of *dl*1520 has been tested in patients with recurrent ovarian cancer [49], recurrent malignant glioma [50], and premalignant oral dysplasia [51]. A phase I ovarian cancer trial with intraperitoneal delivery of *dl*1520 resulted in grade 3 abdominal pain and diarrhea in one patient but the maximum tolerated dose was not reached, and the "maximum affordable dose" was 10^{11} pfu [49]. Disappointingly, there were no clinical or radiological responses in any patients.

After the safety and preliminary efficacy of intratumoral *dl*1520 had been established, phase I/II hepatic artery infusion trials were carried out among patients with colorectal liver metastases who had failed first-line chemotherapy. Although the treatment was well tolerated, no objective responses were seen after single-agent treatment [52,53]. However, more promising results were obtained when *dl*1520 was given via hepatic artery in combination with intravenous 5-FU and leucovorin to patients who failed to respond to *dl*1520 alone. One patient demonstrated a partial response and 10 had stable disease. A recent study demonstrated that this combination resulted in prolonged stable disease in nearly half (48%) of the patients [54]. The first intravenous trial with *dl*1520 resulted in no evidence of antitumor activity in patients with advanced carcinoma metastatic to the lung [55]. In another intravenous study, stable disease was seen in some patients with metastatic colorectal cancer but only for a few months [56]. In both reports, *dl*1520 was rapidly cleared from the circulation.

In summary, *dl*1520 has been well tolerated even at the highest doses that could be administered (2×10^{12}–2×10^{13} pfu) by any route of administration. The lack of clinically significant toxicity in the liver or other organs has been remarkable. Flu-like symptoms have been the most common toxicities and have been more frequent in patients receiving intravascular treatment. Viral replication has been documented in head/neck and colorectal tumors following intratumoral or intra-arterial administration. Heretofore, single-agent antitumoral activity has been limited to head and neck cancers. However, favorable and potentially synergistic interaction with chemotherapy has been discovered in multiple tumor types and by multiple routes of administration [4,57,58].

8.3.1.3 Type II Oncolytic Adenoviruses

With type II oncolytic adenoviruses, tumor-specific promoters replace endogenous viral promoters to restrict viral replication to target tissues expressing the TSP. Usually, a TSP is placed to control *E1A*, but alternatively or in addition, other early genes can also be regulated. Various promoters have been used to limit viral replication to desired tissues, including α-fetoprotein [59,60], DF3/MUC1 [61], midkine [62], cyclooxygenase-2 [63,64], human telomerase reverse transcriptase [65,66], and secretory leukoprotease inhibitor [67–69].

Human prostate specific antigen (PSA) and rat probasin promoters have been utilized for prostate cancer-specific replication. CG7060 has the PSA promoter and enhancer controlling *E1A* expression [70]. In a phase I trial of locally recurrent prostate cancer, CG7060 was well tolerated following intraprostatic injection to 20 patients and resulted in dose-dependent reductions in PSA in some patients [71]. Similar results were obtained in a phase I/II dose-escalation trial with intraprostatic CG7870 [72], which has the rat probasin promoter controlling *E1A* and the PSA promoter and enhancer driving *E1B* [73]. In contrast to CG7060, this virus has an intact E3 region. CG7870 was recently delivered intravenously to 23 patients with hormone-refractory metastatic prostate cancer [74]. Although PSA levels decreased in some patients, neither partial nor complete formal PSA responses were observed.

Oncolytic adenoviruses combining both type I and type II approaches have been constructed [75,76]. Such double control of viral replication seems advantageous over a single control approach, because it seems to be possible to gain specificity without loss off efficacy.

8.3.2 Herpes Simplex Virus

Herpes simplex virus type 1 (HSV-1) is an enveloped, double-stranded DNA virus that frequently causes recurrent oropharyngeal or genital cold sores [77]. HSV-1 is an attractive vector for gene therapy as up to 30 kb of the 152 kb viral genome can be replaced with transgene cassettes. HSV-1 has natural tropism for neuronal tissue and therefore clinical applications have mainly focused on treating brain tumors and neurodegenerative diseases. Pathogenicity of HSV-1 can be reduced by mutating one or more of the crucial virulence genes resulting in selective replication competence in dividing cells.

G207 is an HSV-1 mutant that carries deletions in both loci of neurovirulence gene $\gamma34.5$ and an inactivating insertion in the *UL39* gene [78]. Although originally designed for brain tumor therapy, the virus has shown efficacy against various solid tumors *in vitro* and *in vivo* [77]. Intratumoral G207 was evaluated in a phase I dose-escalation study for recurrent malignant glioma [79]. No shedding of G207 was detectable and no toxicity or encephalitis was reported even with the highest dose 3×10^9 pfu. Phase II trials with G207 are now ongoing.

HSV1716 is another $\gamma34.5$ deleted oncolytic HSV-1 that has been evaluated in phase I clinical trials against malignant glioma [80–82] and melanoma [83]. The virus has been successfully delivered at relatively low doses (1×10^3–1×10^5 pfu) without evidence of adverse events. In addition to glioma and melanoma, HSV1716 is currently in clinical trials for treatment of head and neck cancer and mesothelioma.

NV1020 is a genetically engineered HSV-1 containing a deletion of one copy of $\gamma34.5$, *UL24*, and *UL56*. The deleted region has been replaced with a genome fragment derived from HSV-2. NV1020 was initially developed as a herpes vaccine but it has also been used as an oncolytic agent against various tumors [77]. In a phase I trial, NV1020 was administered via the hepatic artery to patients with colorectal carcinoma liver metastases [84,85]. The treatment was well tolerated and the virus is currently in phase II trials.

OncoVex^{GM-CSF}, a $\gamma34.5$ deleted HSV-1 strain expressing a human immunostimulatory molecule GM-CSF, has shown promising antitumor effects both *in vitro* and *in vivo* [86], and has been tested in a phase I trial against cutaneous or subcutaneous metastasis of various cancer types. Side effects related to treatment were mostly mild, and shrinkage or no progression of injected lesions were seen [87].

8.3.3 VACCINIA VIRUS

Vaccinia virus is a member of the poxvirus family [88]. It is related to smallpox, and has been used as a smallpox vaccine in millions of humans. Vaccinia is typically a mild pathogen in humans, and it may cause rash, fever, and body aches. It is an enveloped, double-stranded DNA virus with a genome of approximately 200 kb. The large genome allows insertion of large transgenes for virotherapeutic purposes.

Intratumoral treatment with vaccine strains of vaccinia has resulted in significant and reproducible antitumor efficacy in numerous clinical trials. Vaccinia-induced responses have often been durable, complete, and distant from the site of injection [89–92]. Also systemic efficacy following intravenous delivery has been suggested [91,93]. Objective tumor responses occurred in the lungs of two patients with widely metastatic adenocarcinoma. A third patient with multiple myeloma had an objective serologic response.

The clinical use of vaccinia as an anticancer agent has traditionally focused on the induction of an antitumor immune response, since the virus is highly immunogenic [92,94–97]. However, conditionally replicating deletion mutants have been recently developed [88]. Most genetically modified vaccinia viruses have the thymidine kinase (TK) gene deleted, which might help to give selectivity for dividing cells. This deletion makes the virus dependent on host cell nucleotides, high amounts of which are available in dividing cells. TK-deficient vaccinia virus mutants were shown to replicate selectively in tumor cells [98]. To further enhance tumor-specificity, a combined TK and vaccinia growth factor (VGF) deleted virus was generated [99]. In murine tumor models, the TK/VGF double-deletion mutant displayed higher tumor-targeting capacity and potent antitumor activity with reduced viral pathogenicity. Also other viral genes such as SP-1 and SP-2, have been successfully deleted for improved tumor selectivity [100]. These genes encode serine protease inhibitors homologous to proteins known to be upregulated in cancer cells. Therefore, they may be dispensable for replication in tumor cells but are required in normal cells. Additional tumor selectivity may result from the size of the virus. Leaky tumor vasculature may allow preferential extravasation in comparison to intact normal vessels.

8.3.4 MEASLES VIRUS

Wild-type measles virus is a negative-strand RNA virus causing rash, fever, cough, and conjunctivitis. Although the disease is usually mild, it causes almost 1 million deaths of children each year worldwide [101]. Wild-type measles does not propagate in cells derived from human neoplasms. However, certain members of the Edmonston vaccine lineage (MV-Edm) are selectively oncolytic against various types of human malignancies, probably due to mutations enhancing the ability of virus to interact with CD46, a receptor highly expressed by tumor cells. Thus, due to altered receptor-specificity, attenuated pathogenicity, and ability to selectively replicate on cancer cells, MV-Edm is a promising agent for cancer gene therapy.

Attenuated MV-Edm has shown potent antitumor activity against a variety of human malignancies *in vivo* [101]. Initially, there was no convenient way to evaluate the replication and persistence of the virus *in vivo*, and the oncolytic potency of MV-Edm in different human tumor xenograft models was deemed highly variable. Therefore, MV-Edm-derived oncolytic measles viruses expressing either soluble marker peptide carcinoembryonic antigen (CEA) (MV-CEA [102]) or thyroidal sodium iodide symporter (NIS) (MV-NIS [103]) were developed. Currently, intraperitoneal delivery of MV-CEA and systemic administration of MV-NIS are under evaluation in phase I clinical trials against recurrent human ovarian cancer and myeloma, respectively. Progress in the former trial has been presented in an intermediate report [104]. Out of the first 12 patients, 5 have displayed stable disease for 2–8 months, while significant CA125 decreases have been seen in 3 patients. No serious side effects have been seen and dose escalation continues.

8.3.5 Newcastle Disease Virus

Newcastle disease virus (NDV) is a negative-strand RNA virus that causes severe respiratory and central nervous system diseases to fowl [101]. However, to humans it is usually a mild pathogen causing conjunctivitis. Tissue culture-adapted strains of the virus show reduced virulence towards normal cells but oncolytic activity in human cancer cells, possibly due to cancer-specific defects in the interferon signaling pathway.

The first promising report of efficacy with NDV was published in 1964 [105]. A patient with myelogenous leukemia responded to intravenous NDV treatment. Subsequently, various naturally attenuated NDV strains have been reported useful in the treatment of neoplastic diseases. The oncolytic potency of NDV strain 73-T has been demonstrated in several human cancer xenografts and in clinical trials [101]. A follow-up of stage III melanoma patients treated with 73-T oncolysate in 1975 reported a 55% overall 15 year survival [106].

MTH-68/H is an *in vitro* passaged NDV strain administered initially by inhalational to patients with advanced solid tumors [107]. Objective responses were reported in 55% of patients. Intravenous MTH-68/H was given to 15 glioblastoma patients [108]. Daily infusions were associated with tumor shrinkage in seven patients, although interpretation is complicated by concomitant treatments.

Oncolytic NDV strain PV 701 has been evaluated in a phase I clinical trial in patients with advanced solid cancers [109]. Seventy-nine patients were treated with single or multiple intravenous doses of PV 701. The maximum tolerated dose was 1.2×10^{10} pfu/m^2 with the most common adverse effects being flu-like symptoms that occurred mainly after the first dose and were decreased in number and severity with each subsequent dose. Desensitization with a lower dose enabled the utilization of a higher dose 1.2×10^{11} pfu/m^2. Objective responses occurred at higher dose levels, and progression-free survival ranged from 4 to 31 months. A recent phase I study utilized a two-step desensitization approach (two low doses followed by higher dose) to successfully improve patient tolerability to PV 701 treatment [110]. Another phase I study sought to improve patient tolerability to PV 701 treatment by slowing the intravenous infusion rate. This regimen diminished adverse effects and tumor responses were reported [111].

Systemic delivery of lentogenic NDV-HUJ strain was recently tested in a phase I/II trial in patients with recurrent glioblastoma multiforme [112]. The maximum tolerated dose was not reached but a complete response was seen in one patient.

8.4 FUTURE DIRECTIONS

In summary, although preclinical studies utilizing oncolytic viruses for treating neoplastic diseases have been mostly promising, data from clinical trials is less convincing (Tables 8.2 and 8.3). Overall, the safety of these approaches has been very good, but objective antitumor responses following intravascular delivery have been rare thus far. Nevertheless, transduction of tumor cells or secondary viral burst has been demonstrated in many cases. Many reports confirm that if intratumoral injection can be performed, responses can be seen regularly. Furthermore, many viruses (including adenovirus, vaccinia, reovirus, and NDV) can achieve tumor transduction following systemic delivery. Nevertheless, the available clinical data implies that transduction efficacy continues to be the limiting factor for clinical benefit. Several strategies are currently being explored to improve transduction of tumor cells and effective penetration of solid tumors (Figure 8.2). In particular, spreading within the tumor can be limited by physical barriers such as stromal cells and matrix, necrotic, hypoxic, or hyperbaric regions. Also, methods for increasing bioavailability following intravenous delivery seem appealing.

In the context of oncolytic adenoviruses, all published trials rely on CAR for entry into cells. It has been demonstrated that a major determinant of the oncolytic potency of replicating agents is their capability of infecting target cells [113]. Unfortunately, concurrent studies have suggested that expression of CAR is frequently dysregulated in many types of advanced cancers [3]. Various strategies have been evaluated to genetically modify adenovirus tropism in order to circumvent CAR deficiency, for increased transduction of tumor cells and reduced normal tissue tropism.

Ad5-Δ24RGD is Rb-p16 pathway selective oncolytic adenovirus featuring an integrin binding RGD-4C modification in the HI-loop of the fiber knob [114]. Fiber pseudotyping has also been evaluated. Substitution of the knob domain of Ad5 with the corresponding domain of serotype 3 (Ad3) allows binding and entry through the Ad3 receptor, which is expressed to high degree on ovarian cancer cells [115,116]. Ad5/3-Δ24 is another Rb-p16 pathway selective oncolytic adenovirus that contains a chimeric fiber featuring a serotype 3 knob [117]. Recent studies have demonstrated that these both agents deliver a powerful antitumor effect to many types of cancer cells *in vitro*, to clinical cancer specimens fresh from patients, and in orthotopic models of ovarian cancer, and both viruses are now proceeding towards clinical testing on ovarian cancer patients [117–119].

Although pretargeting with adaptors requiring conjugation prior to infection is perhaps not optimal for replicating agents, secretory adapter molecules (produced in conjunction with replication) could be more useful for allowing targeting of viral progeny. Coinfection of a replication-deficient virus coding sCAR-EGF with Ad5-Δ24 resulted in enhanced oncolysis [120]. sCAR is a secretory variant of the adenovirus receptor, while the other end of the fusion molecule binds to the epidermal growth factor receptor (EGFR), which is often highly expressed on advanced tumors. However, when this approach was tested in a single-component oncolytic virus, D24sCAR-EGF, efficacy was not increased [121]. This suggested that the expression of biologically active adapter proteins can interfere with virus production and oncolysis. In contrast, van Beusechem et al. introduced a bispecific single chain (csFv) antibody 425-S11, which recognizes EGFR and

TABLE 8.3

Modern Trials with Other Oncolytic Viruses

Name	Virus Type	Tumor Selectivity	Phase	Patients	Safety	Results	References
CV706	Adeno	Prostate	I	20	Good	5 PSA reductions of ≥50%	[71]
Ad5-CD/TKrep	Adeno	—	I	15	No SAE, 2 patients had grade 3 lymphopenia	Mean PSA half-life was significantly shorter in patients who were administered >1 week of prodrug therapy than in patients who received prodrugs for 1 week ($P < .02$).	[125]
CG7870	Adeno	Prostate	I	23	One serious grade 3 fatigue, 2 grade 3 fevers	No response	[74]
NDV-HUJ	NDV	—	I/II	14	Good	1 CR	[112]
PV 701	NDV	—	I	79	SAE included fever, diarrhea, leucopenia, thrombocytopenia, fatigue, dehydration, and were mostly related to the first cycle of treatment	1 CR, 1 PR, and 7 MR	[109]
PV 701	NDV	—	I	18	No SAE, transient grade 4 rise in liver enzymes and pain at tumor site	1 CR, 3 PR, and 2 MR	[111]
MTH-68/H	NDV	—	—	14	Good	4 CR, 4 patients still alive (2004) after 5–9 years from the beginning of the therapy	[108]
MV-EZ	Measles	—	I	5	Good	1 PR, 2 SD, and 2 PD	[151]
G207	HSV-1	—	I	20	No SAE, hemorrhages at injection site and weakness	14 patients had a decrease in the enhanced mass measured by MRI	[79]
HSV1716	HSV-1	—	I/II	12	Good	2 SD, 10 died for progressive disease	[82]
OncoVex-GM-CSF	HSV-1	—	I	30	No SAE, more grade 3/4 side effects in single-dose group	3 SD, no progression, or shrinkage of injected lesions were observed	[87]
NV1020	HSV-1	—	I	12	No SAE, grade 3 diarrhea, leukocytosis and liver enzyme elevation in 3 patients	2 MR, 7 SD, 3 PD	[85]
Vaccinia/ GM-CSF RV	Vaccinia	—	I	7	Good	1 CR, 1PR	[92]
rV-PSA	Vaccinia	—	I	33	Good	14 SD for 6 months, 9 SD for 11–25 months, 6 SD at the completion. Responses were measured as PSA levels in 3 successive monthly determinations.	[95]

Note: CR, complete response; HSV-1, herpes simplex virus 1; MRI, magnetic resonance imaging; MR, minor response; NDV, new castle disease virus; PD, progressive disease; PR, partial response; PSA, prostate specific antigen; SAE, serious adverse effects; SD, stable disease.

fiber knob, into the *E3* region of Ad5-Δ24. This secretory retargeting moiety increased oncolytic potency on CAR-deficient glioma cells [122].

For further potentiation, replication competent viruses can be armed with therapeutic transgenes such as cytokines, suicide genes, fusogenic, proteolytic, or antiangiogenic moieties [123]. For example, Ad5-CD/TKrep contains a fusion suicide gene expressing both *Escherichia coli* cytosine deaminase (CD) and herpes simplex virus thymidine kinase (TK) which can convert systemically administered prodrugs 5-fluorocytosine and ganciclovir into toxic agents. The activated drugs can also spread into surrounding cells delivering a powerful bystander effect. In a phase I study with recurrent prostate cancer patients, Ad5-CD/TKrep treatment appeared safe and also resulted in frequent decreases in PSA levels [124,125].

A powerful approach for increasing efficacy is utilization of oncolytic viruses in combination with conventional anticancer therapies in a multimodal antitumor approach. Oncolytic tumor killing differs from conventional anticancer therapies, providing a possibility for additive or synergistic interactions. Further, the toxicity profiles may be different,

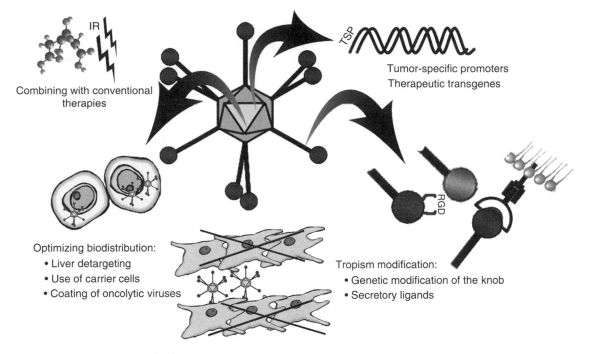

FIGURE 8.2 Ways to improve oncolytic viruses.

which could result in enhanced efficacy without significantly increased adverse effects. There are several studies suggesting enhanced cell killing activity when oncolytic viruses have been combined with chemotherapeutic agents [26,40,60, 126–132] or radiation [71,72,133–137].

Finally, the biodistribution of many viruses may not be optimal for systemic delivery. For example, adenovirus has strong liver tropism, at least in mice. Intriguing approaches for improving the bioavailability of oncolytic viruses include ablation of hepatic Kupffer cells [138], using carrier cells [139,140] or coating of viruses with synthetic molecules such as polyethylene glycol [141]. Further, preliminary evidence suggests that certain blood factors such as platelets or members of the coagulation cascade may be of important determinants of the fate of intravascular adenovirus [23]. These aspects may be relevant for other oncolytic viruses as well, but much work needs to be done to before those viruses are understood as well as adenovirus.

8.5 CONCLUSIONS

In summary, oncolytic viruses have emerged, or in fact reemerged, as promising developmental agents for treatment of cancer refractory to more conventional treatments. Only one phase III trial has been completed thus far, demonstrating that field is still in its infancy. Nevertheless, the only completed trial was a positive one with regard to safety and efficacy. Uniquely, oncolytic viruses face much more complex regulatory, production and intellectual property issues than other novel, but more conventional therapeutics such as small molecular inhibitors and monoclonal antibodies. Overlapping patents, lay concerns over gene delivery, and standardization

of biological production systems may be more challenging obstacles than any of the scientific ones mentioned above. Nevertheless, the complexity and capacity for adaptation exhibited by advanced cancers suggests that all new cancer drugs will eventually have an opportunity for demonstrating their potency. The single most impressive feature of oncolytic viruses may be their outstanding capacity for rapid killing of tumor cells. In a world where most new cancer drugs are cytostatic or stabilizing by nature, the ability to kill cells and concurrently mount a protective immune reaction may well prove valuable for avoiding resistance and relapse.

ACKNOWLEDGMENTS

This work was supported by Helsinki Biomedical Graduate School, EU FP6 THERADPOX & APOTHERAPY, HUCH Research Funds (EVO), Sigrid Juselius Foundation, Academy of Finland, Emil Aaltonen Foundation, Finnish Cancer Society, and University of Helsinki.

REFERENCES

1. Puumalainen, A.M. et al., Beta-galactosidase gene transfer to human malignant glioma in vivo using replication-deficient retroviruses and adenoviruses, *Hum. Gene Ther.*, 9, 1769, 1998.
2. Kirn, D., Martuza, R.L., and Zwiebel, J., Replication-selective virotherapy for cancer: Biological principles, risk management and future directions, *Nat. Med.*, 7, 781, 2001.
3. Bauerschmitz, G.J., Barker, S.D., and Hemminki, A., Adenoviral gene therapy for cancer: From vectors to targeted and replication competent agents (review), *Int. J. Oncol.*, 21, 1161, 2002.
4. Kanerva, A. and Hemminki, A., Adenoviruses for treatment of cancer, *Ann. Med.*, 37, 33, 2005.

5. Bierman, H.R. et al., Remissions in leukemia of childhood following acute infectious disease: Staphylococcus and streptococcus, varicella, and feline panleukopenia, *Cancer*, 6, 591, 1953.

6. Pasquinucci, G., Possible effect of measles on leukaemia, *Lancet*, 1, 136, 1971.

7. Bluming, A.Z. and Ziegler, J.L., Regression of Burkitt's lymphoma in association with measles infection, *Lancet*, 2, 105, 1971.

8. Taqi, A.M. et al., Regression of Hodgkin's disease after measles, *Lancet*, 1, 1112, 1981.

9. Csatary, L.K., Viruses in the treatment of cancer, *Lancet*, 2, 825, 1971.

10. DePace, N., Sulla scomparsa di un enorme cancro vegetante del collo dell'utero senza cura chirurgia, *Ginecologia*, 9, 82, 1912.

11. Levaditi, C. and Nicolau, S., Sur le culture du virus vaccinal dans les neoplasmes epithelieux, *CR Soc Biol*, 86, 988, 1922.

12. Pack, G.T., Note on the experimental use of rabies vaccine for melanomatosis, *Arch. Dermatol.*, 62, 694, 1950.

13. Southam, C.M. and Moore, A.E., Clinical studies of viruses as antineoplastic agents with particular reference to Egypt 101 virus, *Cancer*, 5, 1025, 1952.

14. Southam, C.M., Noyes, W.F., and Mellors, R., Virus in human cancer cells in vivo, *Virology*, 5, 395, 1958.

15. Hansen, R.M. and Libnoch, J.A., Remission of chronic lymphocytic leukemia after smallpox vaccination, *Arch. Intern. Med.*, 138, 1137, 1978.

16. Greentree, L.B., Hodgkin's disease: Therapeutic role of measles vaccine, *Am. J. Med.*, 75, 928, 1983.

17. Schattner, A., Therapeutic role of measles vaccine in Hodgkin's disease, *Lancet*, 1, 171, 1984.

18. Kanerva, A. et al., Chlorpromazine and apigenin reduce adenovirus replication and decrease replication associated toxicity, *J. Gene Med.*, 9, 3, 2007.

19. Russell, W.C., Update on adenovirus and its vectors, *J. Gen. Virol.*, 81, 2573, 2000.

20. Bergelson, J.M. et al., Isolation of a common receptor for Coxsackie B viruses and adenoviruses 2 and 5, *Science*, 275, 1320, 1997.

21. Wickham, T.J. et al., Integrins alpha v beta 3 and alpha v beta 5 promote adenovirus internalization but not virus attachment, *Cell*, 73, 309, 1993.

22. Shayakhmetov, D.M. et al., Adenovirus binding to blood factors results in liver cell infection and hepatotoxicity, *J. Virol.*, 79, 7478, 2005.

23. Parker, A.L. et al., Multiple vitamin K-dependent coagulation zymogens promote adenovirus-mediated gene delivery to hepatocytes, *Blood*, 108, 2554, 2006.

24. Huebner, R.J. et al., Studies on the use of viruses in the treatment of carcinoma of the cervix, *Cancer*, 9, 1211, 1956.

25. Bischoff, J.R. et al., An adenovirus mutant that replicates selectively in p53-deficient human tumor cells, *Science*, 274, 373, 1996.

26. Heise, C. et al., ONYX-015, an E1B gene-attenuated adenovirus, causes tumor-specific cytolysis and antitumoral efficacy that can be augmented by standard chemotherapeutic agents, *Nat. Med.*, 3, 639, 1997.

27. Heise, C.C. et al., Intravenous administration of ONYX-015, a selectively replicating adenovirus, induces antitumoral efficacy, *Cancer Res.*, 59, 2623, 1999.

28. Rogulski, K.R. et al., In vivo antitumor activity of ONYX-015 is influenced by p53 status and is augmented by radiotherapy, *Cancer Res.*, 60, 1193, 2000.

29. Goodrum, F.D. and Ornelles, D.A., p53 status does not determine outcome of E1B 55-kilodalton mutant adenovirus lytic infection, *J. Virol.*, 72, 9479, 1998.

30. Rothmann, T. et al., Replication of ONYX-015, a potential anticancer adenovirus, is independent of p53 status in tumor cells, *J. Virol.*, 72, 9470, 1998.

31. Ries, S.J. et al., Loss of p14ARF in tumor cells facilitates replication of the adenovirus mutant dl1520 (ONYX-015), *Nat. Med.*, 6, 1128, 2000.

32. Dobner, T. et al., Blockage by adenovirus E4orf6 of transcriptional activation by the p53 tumor suppressor, *Science*, 272, 1470, 1996.

33. Yew, P.R., Liu, X., and Berk, A.J., Adenovirus E1B oncoprotein tethers a transcriptional repression domain to p53, *Genes Dev.*, 8, 190, 1994.

34. Kirn, D., Clinical research results with dl1520 (Onyx-015), a replication-selective adenovirus for the treatment of cancer: What have we learned? *Gene Ther.*, 8, 89, 2001.

35. Fueyo, J. et al., A mutant oncolytic adenovirus targeting the Rb pathway produces anti-glioma effect in vivo, *Oncogene*, 19, 2, 2000.

36. Heise, C. et al., An adenovirus E1A mutant that demonstrates potent and selective systemic anti-tumoral efficacy, *Nat. Med.*, 6, 1134, 2000.

37. Sherr, C.J., Cancer cell cycles, *Science*, 274, 1672, 1996.

38. Page, J.G. et al., Identifying the safety profile of a novel infectivity-enhanced conditionally replicative adenovirus, Ad5-delta24-RGD, in anticipation of a phase I trial for recurrent ovarian cancer, *Am. J. Obstet. Gynecol.*, 196, 389.e1, 2007.

39. Ganly, I. et al., A phase I study of Onyx-015, an E1B attenuated adenovirus, administered intratumorally to patients with recurrent head and neck cancer, *Clin. Cancer Res.*, 6, 798, 2000.

40. Khuri, F.R. et al., A controlled trial of intratumoral ONYX-015, a selectively-replicating adenovirus, in combination with cisplatin and 5-fluorouracil in patients with recurrent head and neck cancer, *Nat. Med.*, 6, 879, 2000.

41. Nemunaitis, J. et al., Selective replication and oncolysis in p53 mutant tumors with ONYX-015, an E1B-55 kD gene-deleted adenovirus, in patients with advanced head and neck cancer: A phase II trial, *Cancer Res.*, 60, 6359, 2000.

42. Nemunaitis, J. et al., Phase II trial of intratumoral administration of ONYX-015, a replication-selective adenovirus, in patients with refractory head and neck cancer, *J. Clin. Oncol.*, 19, 289, 2001.

43. Yu, W. and Fang, H., Clinical trials with oncolytic adenovirus in China, *Curr. Cancer. Drug Targets*, 7, 141, 2007.

44. Xia, Z.J. et al., Phase III randomized clinical trial of intratumoral injection of E1B gene-deleted adenovirus (H101) combined with cisplatin-based chemotherapy in treating squamous cell cancer of head and neck or esophagus. *Ai Zheng*, 23, 1666, 2004.

45. Mulvihill, S. et al., Safety and feasibility of injection with an E1B-55 kDa gene-deleted, replication-selective adenovirus (ONYX-015) into primary carcinomas of the pancreas: A phase I trial, *Gene Ther.*, 8, 308, 2001.

46. Hecht, J.R. et al., A phase I/II trial of intratumoral endoscopic ultrasound injection of ONYX-015 with intravenous gemcitabine in unresectable pancreatic carcinoma, *Clin. Cancer Res.*, 9, 555, 2003.

47. Habib, N. et al., Clinical trial of E1B-deleted adenovirus (dl1520) gene therapy for hepatocellular carcinoma, *Cancer Gene Ther.*, 9, 254, 2002.

48. Morley, S. et al., The dl1520 virus is found preferentially in tumor tissue after direct intratumoral injection in oral carcinoma, *Clin. Cancer Res.*, 10, 4357, 2004.

49. Vasey, P.A. et al., Phase I trial of intraperitoneal injection of the E1B-55-kd-gene-deleted adenovirus ONYX-015 (dl1520) given on days 1 through 5 every 3 weeks in patients with

recurrent/refractory epithelial ovarian cancer, *J. Clin. Oncol.*, 20, 1562, 2002.

50. Chiocca, E.A. et al., A phase I open-label, dose-escalation, multi-institutional trial of injection with an E1B-Attenuated adenovirus, ONYX-015, into the peritumoral region of recurrent malignant gliomas, in the adjuvant setting, *Mol. Ther.*, 10, 958, 2004.

51. Rudin, C.M. et al., An attenuated adenovirus, ONYX-015, as mouthwash therapy for premalignant oral dysplasia, *J. Clin. Oncol.*, 21, 4546, 2003.

52. Reid, T. et al., Intra-arterial administration of a replication-selective adenovirus (dl1520) in patients with colorectal carcinoma metastatic to the liver: A phase I trial, *Gene Ther.*, 8, 1618, 2001.

53. Reid, T. et al., Hepatic arterial infusion of a replication-selective oncolytic adenovirus (dl1520): Phase II viral, immunologic, and clinical endpoints, *Cancer Res.*, 62, 6070, 2002.

54. Reid, T.R. et al., Effects of Onyx-015 among metastatic colorectal cancer patients that have failed prior treatment with 5-FU/leucovorin, *Cancer Gene Ther.*, 12, 673, 2005.

55. Nemunaitis, J. et al., Intravenous infusion of a replication-selective adenovirus (ONYX-015) in cancer patients: Safety, feasibility and biological activity, *Gene Ther.*, 8, 746, 2001.

56. Hamid, O. et al., Phase II trial of intravenous CI-1042 in patients with metastatic colorectal cancer, *J. Clin. Oncol.*, 21, 1498, 2003.

57. Raki, M. et al., Gene transfer approaches for gynecological diseases, *Mol. Ther.*, 14, 154, 2006.

58. Crompton, A.M. and Kirn, D.H., From ONYX-015 to armed vaccinia viruses: The education and evolution of oncolytic virus development, *Curr. Cancer. Drug Targets*, 7, 133, 2007.

59. Hallenbeck, P.L. et al., A novel tumor-specific replication-restricted adenoviral vector for gene therapy of hepatocellular carcinoma, *Hum. Gene Ther.*, 10, 1721, 1999.

60. Li, Y. et al., A hepatocellular carcinoma-specific adenovirus variant, CV890, eliminates distant human liver tumors in combination with doxorubicin, *Cancer Res.*, 61, 6428, 2001.

61. Kurihara, T. et al., Selectivity of a replication-competent adenovirus for human breast carcinoma cells expressing the MUC1 antigen, *J. Clin. Invest.*, 106, 763, 2000.

62. Adachi, Y. et al., A midkine promoter-based conditionally replicative adenovirus for treatment of pediatric solid tumors and bone marrow tumor purging, *Cancer Res.*, 61, 7882, 2001.

63. Yamamoto, M. et al., Infectivity enhanced, cyclooxygenase-2 promoter-based conditionally replicative adenovirus for pancreatic cancer, *Gastroenterology*, 125, 1203, 2003.

64. Kanerva, A. et al., A cyclooxygenase-2 promoter-based conditionally replicating adenovirus with enhanced infectivity for treatment of ovarian adenocarcinoma, *Gene Ther.*, 11, 552, 2004.

65. Wirth, T. et al., A telomerase-dependent conditionally replicating adenovirus for selective treatment of cancer, *Cancer Res.*, 63, 3181, 2003.

66. Irving, J. et al., Conditionally replicative adenovirus driven by the human telomerase promoter provides broad-spectrum antitumor activity without liver toxicity, *Cancer Gene Ther.*, 11, 174, 2004.

67. Barker, S.D. et al., The secretory leukoprotease inhibitor (SLPI) promoter for ovarian cancer gene therapy, *J. Gene Med.*, 5, 300, 2003.

68. Barker, S.D. et al., Combined transcriptional and transductional targeting improves the specificity and efficacy of adenoviral gene delivery to ovarian carcinoma, *Gene Ther.*, 10, 1198, 2003.

69. Rein, D.T. et al., A fiber-modified, secretory leukoprotease inhibitor promoter-based conditionally replicating adenovirus for treatment of ovarian cancer, *Clin. Cancer Res.*, 11, 1327, 2005.

70. Rodriguez, R. et al., Prostate attenuated replication competent adenovirus (ARCA) CN706: A selective cytotoxic for prostate-specific antigen-positive prostate cancer cells, *Cancer Res.*, 57, 2559, 1997.

71. DeWeese, T.L. et al., A phase I trial of CV706, a replication-competent, PSA selective oncolytic adenovirus, for the treatment of locally recurrent prostate cancer following radiation therapy, *Cancer Res.*, 61, 7464, 2001.

72. DeWeese, T., Anterbery, E., and Michalski, J., A phase I/II dose escalation trial of the intraprostatic injection of CG7870, a prostate specific antigen-dependent oncolytic adenovirus, in patients with locally recurred prostate cancer following definitive radiotherapy, *Mol. Ther.*, 7, S446, 2003.

73. Yu, D.C. et al., The addition of adenovirus type 5 region E3 enables calydon virus 787 to eliminate distant prostate tumor xenografts, *Cancer Res.*, 59, 4200, 1999.

74. Small, E.J. et al., A phase I trial of intravenous CG7870, a replication-selective, prostate-specific antigen-targeted oncolytic adenovirus, for the treatment of hormone-refractory, metastatic prostate cancer, *Mol. Ther.*, 14, 107, 2006.

75. Nettelbeck, D.M. et al., Novel oncolytic adenoviruses targeted to melanoma: Specific viral replication and cytolysis by expression of E1A mutants from the tyrosinase enhancer/promoter, *Cancer Res.*, 62, 4663, 2002.

76. Bauerschmitz, G.J. et al., Triple-targeted oncolytic adenoviruses featuring the Cox2 promoter, E1A transcomplementation, and serotype chimerism for enhanced selectivity for ovarian cancer cells, *Mol. Ther.*, 14, 164, 2006.

77. Varghese, S. and Rabkin, S.D., Oncolytic herpes simplex virus vectors for cancer virotherapy, *Cancer Gene Ther.*, 9, 967, 2002.

78. Mineta, T. et al., Attenuated multi-mutated herpes simplex virus-1 for the treatment of malignant gliomas, *Nat. Med.*, 1, 938, 1995.

79. Markert, J.M. et al., Conditionally replicating herpes simplex virus mutant, G207 for the treatment of malignant glioma: Results of a phase I trial, *Gene Ther.*, 7, 867, 2000.

80. Rampling, R. et al., Toxicity evaluation of replication-competent herpes simplex virus (ICP 34.5 null mutant 1716) in patients with recurrent malignant glioma, *Gene Ther.*, 7, 859, 2000.

81. Papanastassiou, V. et al., The potential for efficacy of the modified (ICP 34.5(-)) herpes simplex virus HSV1716 following intratumoural injection into human malignant glioma: A proof of principle study, *Gene Ther.*, 9, 398, 2002.

82. Harrow, S. et al., HSV1716 injection into the brain adjacent to tumour following surgical resection of high-grade glioma: Safety data and long-term survival, *Gene Ther.*, 11, 1648, 2004.

83. MacKie, R.M., Stewart, B., and Brown, S.M., Intralesional injection of herpes simplex virus 1716 in metastatic melanoma, *Lancet*, 357, 525, 2001.

84. Fong, Y. et al., Phase I study of a replication-competent herpes simplex oncolytic virus for treatment of hepatic colorectal metastases. *ASCO Annual Meeting*, 2002.

85. Kemeny, N. et al., Phase I, open-label, dose-escalating study of a genetically engineered herpes simplex virus, NV1020, in subjects with metastatic colorectal carcinoma to the liver, *Hum. Gene Ther.*, 17, 1214, 2006.

86. Liu, B.L. et al., ICP34.5 deleted herpes simplex virus with enhanced oncolytic, immune stimulating, and anti-tumour properties, *Gene Ther.*, 10, 292, 2003.

87. Hu, J.C. et al., A phase I study of OncoVEXGM-CSF, a second-generation oncolytic herpes simplex virus expressing granulocyte macrophage colony-stimulating factor, *Clin. Cancer Res.*, 12, 6737, 2006.

88. Thorne, S.H., Hwang, T.H., and Kirn, D.H., Vaccinia virus and oncolytic virotherapy of cancer, *Curr. Opin. Mol. Ther.*, 7, 359, 2005.

89. Milton, G.W. and Brown, M.M., The limited role of attenuated smallpox virus in the management of advanced malignant melanoma, *Aust. N. Z. J. Surg.*, 35, 286, 1966.

90. Hunter-Craig, I. et al., Use of vaccinia virus in the treatment of metastatic malignant melanoma, *Br. Med. J.*, 2, 512, 1970.

91. Arakawa, S., Jr. et al., Clinical trial of attenuated vaccinia virus AS strain in the treatment of advanced adenocarcinoma. Report on two cases, *J. Cancer Res. Clin. Oncol.*, 113, 95, 1987.

92. Mastrangelo, M.J. et al., Intratumoral recombinant GM-CSF-encoding virus as gene therapy in patients with cutaneous melanoma, *Cancer Gene Ther.*, 6, 409, 1999.

93. Roenigk, H.H., Jr. et al., Immunotherapy of malignant melanoma with vaccinia virus, *Arch. Dermatol.*, 109, 668, 1974.

94. Livingston, P.O. et al., Serological response of melanoma patients to vaccines prepared from VSV lysates of autologous and allogeneic cultured melanoma cells, *Cancer*, 55, 713, 1985.

95. Eder, J.P. et al., A phase I trial of a recombinant vaccinia virus expressing prostate-specific antigen in advanced prostate cancer, *Clin. Cancer Res.*, 6, 1632, 2000.

96. Marshall, J.L. et al., Phase I study in advanced cancer patients of a diversified prime-and-boost vaccination protocol using recombinant vaccinia virus and recombinant nonreplicating avipox virus to elicit anti-carcinoembryonic antigen immune responses, *J. Clin. Oncol.*, 18, 3964, 2000.

97. Gulley, J. et al., Phase I study of a vaccine using recombinant vaccinia virus expressing PSA (rV-PSA) in patients with metastatic androgen-independent prostate cancer, *Prostate*, 53, 109, 2002.

98. Puhlmann, M. et al., Vaccinia as a vector for tumor-directed gene therapy: Biodistribution of a thymidine kinase-deleted mutant, *Cancer Gene Ther.*, 7, 66, 2000.

99. McCart, J.A. et al., Systemic cancer therapy with a tumor-selective vaccinia virus mutant lacking thymidine kinase and vaccinia growth factor genes, *Cancer Res.*, 61, 8751, 2001.

100. Guo, Z.S. et al., The enhanced tumor selectivity of an oncolytic vaccinia lacking the host range and antiapoptosis genes SPI-1 and SPI-2, *Cancer Res.*, 65, 9991, 2005.

101. Russell, S.J., RNA viruses as virotherapy agents, *Cancer Gene Ther.*, 9, 961, 2002.

102. Peng, K.W. et al., Non-invasive in vivo monitoring of trackable viruses expressing soluble marker peptides, *Nat. Med.*, 8, 527, 2002.

103. Dingli, D. et al., Image-guided radiovirotherapy for multiple myeloma using a recombinant measles virus expressing the thyroidal sodium iodide symporter, *Blood*, 103, 1641, 2004.

104. Galanis, E. et al., Phase I trial of intraperitoneal (ip) administration of a measles virus (MV) strain expressing the human carcinoembryonic antigen (CEA) in ovarian cancer patients, *Mol. Ther.*, 13, S281, 2006.

105. Wheelock, E.F. and Dingle, J.H., Observations on the repeated administration of viruses to a patient with acute leukaemia. A preliminary report, *N. Engl. J. Med.*, 271, 645, 1964.

106. Batliwalla, F.M. et al., A 15-year follow-up of AJCC stage III malignant melanoma patients treated postsurgically with Newcastle disease virus (NDV) oncolysate and determination of alterations in the CD8 T cell repertoire, *Mol. Med.*, 4, 783, 1998.

107. Csatary, L.K. et al., Attenuated veterinary virus vaccine for the treatment of cancer, *Cancer Detect. Prev.*, 17, 619, 1993.

108. Csatary, L.K. et al., MTH-68/H oncolytic viral treatment in human high-grade gliomas, *J. Neurooncol.*, 67, 83, 2004.

109. Pecora, A.L. et al., Phase I trial of intravenous administration of PV701, an oncolytic virus, in patients with advanced solid cancers, *J. Clin. Oncol.*, 20, 2251, 2002.

110. Laurie, S.A. et al., A phase 1 clinical study of intravenous administration of PV701, an oncolytic virus, using two-step desensitization, *Clin. Cancer Res.*, 12, 2555, 2006.

111. Hotte, S.J. et al., An optimized clinical regimen for the oncolytic virus PV701, *Clin. Cancer Res.*, 13, 977, 2007.

112. Freeman, A.I. et al., Phase I/II trial of intravenous NDV-HUJ oncolytic virus in recurrent glioblastoma multiforme, *Mol. Ther.*, 13, 221, 2006.

113. Douglas, J.T. et al., Efficient oncolysis by a replicating adenovirus (ad) in vivo is critically dependent on tumor expression of primary ad receptors, *Cancer Res.*, 61, 813, 2001.

114. Suzuki, K. et al., A conditionally replicative adenovirus with enhanced infectivity shows improved oncolytic potency, *Clin. Cancer Res.*, 7, 120, 2001.

115. Kanerva, A. et al., Gene transfer to ovarian cancer versus normal tissues with fiber-modified adenoviruses, *Mol. Ther.*, 5, 695, 2002.

116. Kanerva, A. et al., Targeting adenovirus to the serotype 3 receptor increases gene transfer efficiency to ovarian cancer cells, *Clin. Cancer Res.*, 8, 275, 2002.

117. Kanerva, A. et al., Enhanced therapeutic efficacy for ovarian cancer with a serotype 3 receptor-targeted oncolytic adenovirus, *Mol. Ther.*, 8, 449, 2003.

118. Bauerschmitz, G.J. et al., Treatment of ovarian cancer with a tropism modified oncolytic adenovirus, *Cancer Res.*, 62, 1266, 2002.

119. Lam, J.T. et al., Replication of an integrin targeted conditionally replicating adenovirus on primary ovarian cancer spheroids, *Cancer Gene Ther.*, 10, 377, 2003.

120. Hemminki, A. et al., Targeting oncolytic adenoviral agents to the epidermal growth factor pathway with a secretory fusion molecule, *Cancer Res.*, 61, 6377, 2001.

121. Hemminki, A. et al., Production of an EGFR targeting molecule from a conditionally replicating adenovirus impairs its oncolytic potential, *Cancer Gene Ther.*, 10, 583, 2003.

122. van Beusechem, V.W. et al., Conditionally replicative adenovirus expressing a targeting adapter molecule exhibits enhanced oncolytic potency on CAR-deficient tumors, *Gene Ther.*, 10, 1982, 2003.

123. Hermiston, T.W. and Kuhn, I., Armed therapeutic viruses: Strategies and challenges to arming oncolytic viruses with therapeutic genes, *Cancer Gene Ther.*, 9, 1022, 2002.

124. Freytag, S.O. et al., Phase I study of replication-competent adenovirus-mediated double suicide gene therapy for the treatment of locally recurrent prostate cancer, *Cancer Res.*, 62, 4968, 2002.

125. Freytag, S.O. et al., Phase I study of replication-competent adenovirus-mediated double-suicide gene therapy in combination with conventional-dose three-dimensional conformal radiation therapy for the treatment of newly diagnosed, intermediate- to high-risk prostate cancer, *Cancer Res.*, 63, 7497, 2003.

126. Chahlavi, A. et al., Replication-competent herpes simplex virus vector G207 and cisplatin combination therapy for head and neck squamous cell carcinoma, *Neoplasia*, 1, 162, 1999.

127. Heise, C., Lemmon, M., and Kirn, D., Efficacy with a replication-selective adenovirus plus cisplatin-based chemotherapy:

Dependence on sequencing but not p53 functional status or route of administration, *Clin. Cancer Res.*, 6, 4908, 2000.

128. You, L., Yang, C.T., and Jablons, D.M., ONYX-015 works synergistically with chemotherapy in lung cancer cell lines and primary cultures freshly made from lung cancer patients, *Cancer Res.*, 60, 1009, 2000.

129. Yu, D.C. et al., Antitumor synergy of CV787, a prostate cancer-specific adenovirus, and paclitaxel and docetaxel, *Cancer Res.*, 61, 517, 2001.

130. Cinatl, J., Jr. et al., Potent oncolytic activity of multimutated herpes simplex virus G207 in combination with vincristine against human rhabdomyosarcoma, *Cancer Res.*, 63, 1508, 2003.

131. Raki, M. et al., Combination of gemcitabine and Ad5/3-Δ24, a tropism modified conditionally replicating adenovirus, for the treatment of ovarian cancer, *Gene Ther.*, 12, 1198, 2005.

132. Dilley, J. et al., Oncolytic adenovirus CG7870 in combination with radiation demonstrates synergistic enhancements of antitumor efficacy without loss of specificity, *Cancer Gene Ther.*, 12, 715, 2005.

133. Advani, S.J. et al., Enhancement of replication of genetically engineered herpes simplex viruses by ionizing radiation: A new paradigm for destruction of therapeutically intractable tumors, *Gene Ther.*, 5, 160, 1998.

134. Bradley, J.D. et al., Ionizing radiation improves survival in mice bearing intracranial high-grade gliomas injected with genetically modified herpes simplex virus, *Clin. Cancer Res.*, 5, 1517, 1999.

135. Chen, Y. et al., CV706, a prostate cancer-specific adenovirus variant, in combination with radiotherapy produces synergistic antitumor efficacy without increasing toxicity, *Cancer Res.*, 61, 5453, 2001.

136. Chung, S.M. et al., The use of a genetically engineered herpes simplex virus (R7020) with ionizing radiation for experimental hepatoma, *Gene Ther.*, 9, 75, 2002.

137. Blank, S.V. et al., Replication-selective herpes simplex virus type 1 mutant therapy of cervical cancer is enhanced by low-dose radiation, *Hum. Gene Ther.*, 13, 627, 2002.

138. Schiedner, G. et al., Selective depletion or blockade of Kupffer cells leads to enhanced and prolonged hepatic transgene expression using high-capacity adenoviral vectors, *Mol. Ther.*, 7, 35, 2003.

139. Komarova, S. et al., Mesenchymal progenitor cells as cellular vehicles for delivery of oncolytic adenoviruses, *Mol. Cancer. Ther.*, 5, 755, 2006.

140. Hakkarainen, T. et al., Human mesenchymal stem cells lack tumor tropism but enhance the antitumor activity of oncolytic adenoviruses in orthotopic lung and breast tumors, *Hum. Gene Ther.*, 2007.

141. O'Riordan, C.R. et al., PEGylation of adenovirus with retention of infectivity and protection from neutralizing antibody in vitro and in vivo, *Hum. Gene Ther.*, 10, 1349, 1999.

142. Hoster, H.A., Zanes, R.P., and von Haam, E., The association of "viral" hepatitis and Hodgkin's disease (a preliminary report), *Cancer Res.*, 9, 473, 1949.

143. Taylor, A.W., Effects of glandular fever infection in acute leukaemia, *Br. Med. J.*, 1, 589, 1953.

144. Georgiades, J. et al., Research on the oncolytic effect of APC viruses in cancer of the cervix uteri; preliminary report, *Biul. Inst. Med. Morsk. Gdansk.*, 10, 49, 1959.

145. Asada, T., Treatment of human cancer with mumps virus, *Cancer*, 34, 1907, 1974.

146. Lamont, J.P. et al., A prospective phase II trial of ONYX-015 adenovirus and chemotherapy in recurrent squamous cell carcinoma of the head and neck (the Baylor experience), *Ann. Surg. Oncol.*, 7, 588, 2000.

147. Habib, N.A. et al., E1B-deleted adenovirus (dl1520) gene therapy for patients with primary and secondary liver tumors, *Hum. Gene Ther.*, 12, 219, 2001.

148. Makower, D. et al., Phase II clinical trial of intralesional administration of the oncolytic adenovirus ONYX-015 in patients with hepatobiliary tumors with correlative p53 studies, *Clin. Cancer Res.*, 9, 693, 2003.

149. Lu, W. et al., Intra-tumor injection of H101, a recombinant adenovirus, in combination with chemotherapy in patients with advanced cancers: A pilot phase II clinical trial, *World J. Gastroenterol.*, 10, 3634, 2004.

150. Galanis, E. et al., Phase I-II trial of ONYX-015 in combination with MAP chemotherapy in patients with advanced sarcomas, *Gene Ther.*, 12, 437, 2005.

151. Heinzerling, L. et al., Oncolytic measles virus in cutaneous T-cell lymphomas mounts antitumor immune responses in vivo and targets interferon-resistant tumor cells, *Blood*, 106, 2287, 2005.

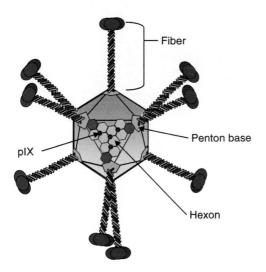

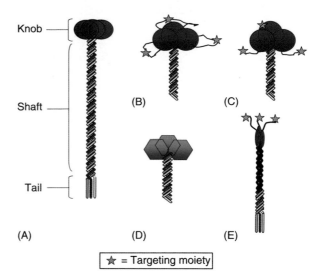

FIGURE 4.1 Generalized Ad structure depicting major structural components of a wild-type Ad capsid. Hexon is depicted as a hexagon with 12 homotrimers per capsid face (only 12 of 240 trimers are shown). The penton base, comprised of five subunits at each vertex, and trimeric fiber and pIX structures are shown. Ad serotype-2 and -5 capsids contain a 36-kilobase double-stranded DNA genome (not shown).

FIGURE 4.3 Genetically modified fibers for Ad targeting. (See text for specific examples of each genetic targeting approach and targeting moiety used.) (A) Trimeric wild-type Ad fiber; tail, shaft, and knob regions are shown. (B) Ad5 fiber knob with a constrained targeting peptide in the flexible H-I loop. (C) Ad5 fiber knob containing a C-terminal targeting ligand. (D) Pseudotyped chimeric fiber bearing an alternate human or animal serotype knob domain. (E) "De-knobbed" fiber containing a heterologous trimerization sequence and a C-terminal targeting moiety.

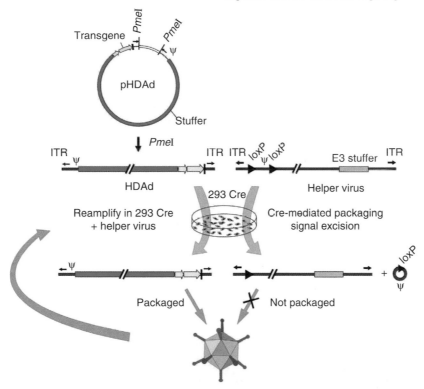

FIGURE 5.2 Cre/loxP system for generating HD vectors. The HDAd contains only ~500 bp of cis-acting Ad sequences required for DNA replication (ITRs) and packaging (ψ), the remainder of the genome consists of the desired transgene (yellow) and non-Ad "stuffer" sequences (green). The HDAd genome is constructed as a bacterial plasmid (pHDAd) and is liberated by restriction enzyme digestion (e.g., *Pme*I). To rescue the HDAd, the liberated genome is transfected into 293 cells expressing Cre and infected with a helper virus bearing a packaging signal (ψ) flanked by loxP sites. The helper viral genome also contains a stuffer sequence in E3 to prevent the formation of RCA in 293-derived cells. Cre-mediated excision of ψ renders the helper virus genome unpackagable, but still able to replicate and provide all of the necessary *trans*-acting factors for propagation of the HDAd. The titer of the HD vector is increased by serial coinfections of 293 Cre cells with the HDAd and the helper virus.

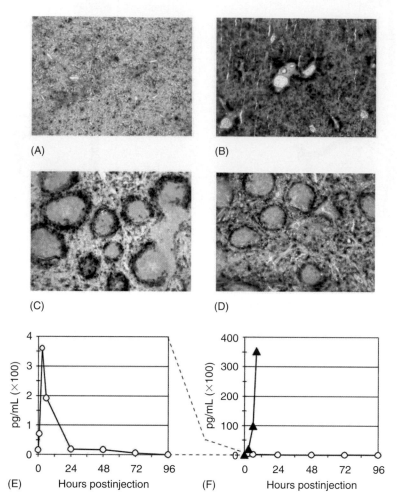

FIGURE 5.5 High dose systemic injection of HDAd into nonhuman primates. Baboon injected with (A, C, and E) 5.6×10^{12} vp/kg or (B, D, and F) 1.1×10^{13} vp/kg of HDAd-LacZ. X-gal stained (A and B) liver and (C and D) spleen and (E and F) serum IL-6 from injected baboons. Circles, 5.6×10^{12} vp/kg; triangles, 1.1×10^{13}vp/kg. Note different scales in E and F. (Adapted from Brunetti-Pierri, N., Palmer, D.J., Beaudet, A.L., Carey, K.D., Finegold, M., and Ng, P., *Hum. Gene Ther.*, 15, 35, 2004.)

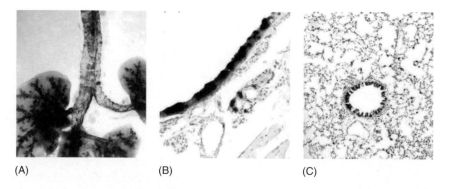

FIGURE 5.8 Epithelia transduction of the (A) proximal and distal airway, (B) trachea, and (C) bronchiole of mice 3 days post-intranasal administration of 2×10^{10} vp of HDAd-K18LacZ. Blue areas represent HDAd transduced cells. (From Toietta, G., Koehler, D.R., Finegold, M.J., Lee, B., Hu, J., and Beaudet, A.L., *Mol. Ther.*, 7, 649, 2003. With permission.)

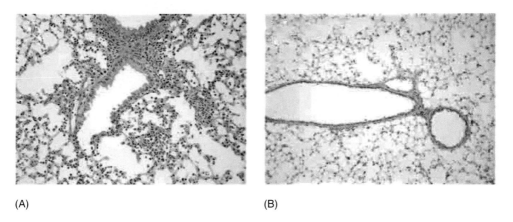

(A)   (B)

FIGURE 5.9   Histology of representative mouse lung 3 days post-intranasal administration of a total of 4×10^{10} vp of an (A) FGAd or (B) HDAd. (From Toietta, G., Koehler, D.R., Finegold, M.J., Lee, B., Hu, J., and Beaudet, A.L., *Mol. Ther.*, 7, 649, 2003. With permission.)

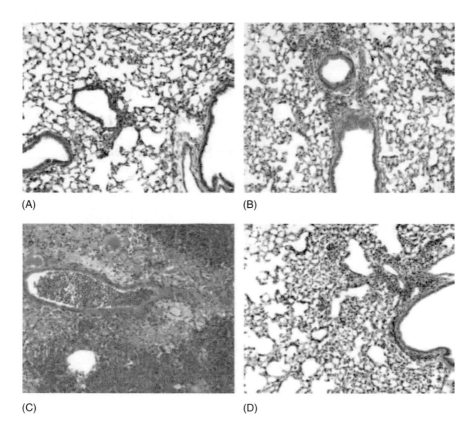

FIGURE 5.10   Lung histopathology following Bcc challenge. Representative H&E stained sections from wild-type littermates (A and B) and CFTR knockout mice (C and D) receiving saline (A and C) or HDAd-K18hCFTR (B and D) prior to bacterial challenge. (From Koehler, D.R., Sajjan, U., Chow, Y.H., Martin, B., Kent, G., Tanswell, A.K., McKerlie, C., Forstner, J.F., and Hu, J., *Proc. Natl. Acad. Sci. U S A*, 100, 15364, 2003. With permission.)

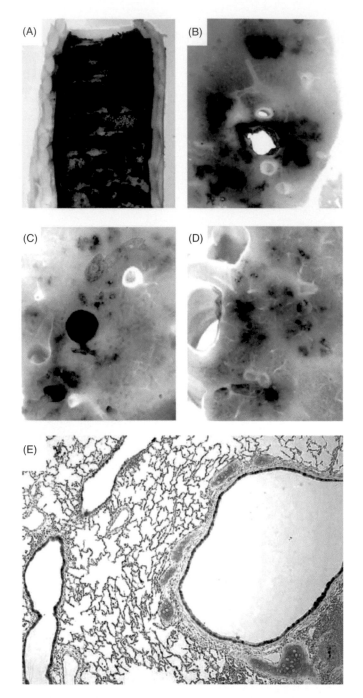

FIGURE 5.11 Pulmonary transduction in 2.8 kg rabbits following AeroProbe-mediated intratracheal aerosolization of 1.25 mL 0.1% LPC containing 5×10^{11} vp of HDAd-K18LacZ. X-gal stained (A) trachea, (B) right upper lobe, (C) left lower lobe, (D) right lower lobe, and (E) bronchus and bronchioles. Blue areas represent HDAd transduced cells. (From Koehler, D.R., Frndova, H., Leung, K., Louca, E., Palmer, D., Ng, P., McKerlie, C., Cox, P., Coates, A.L., and Hu, J., *J. Gene Med.*, 7, 1409, 2005. With permission.)

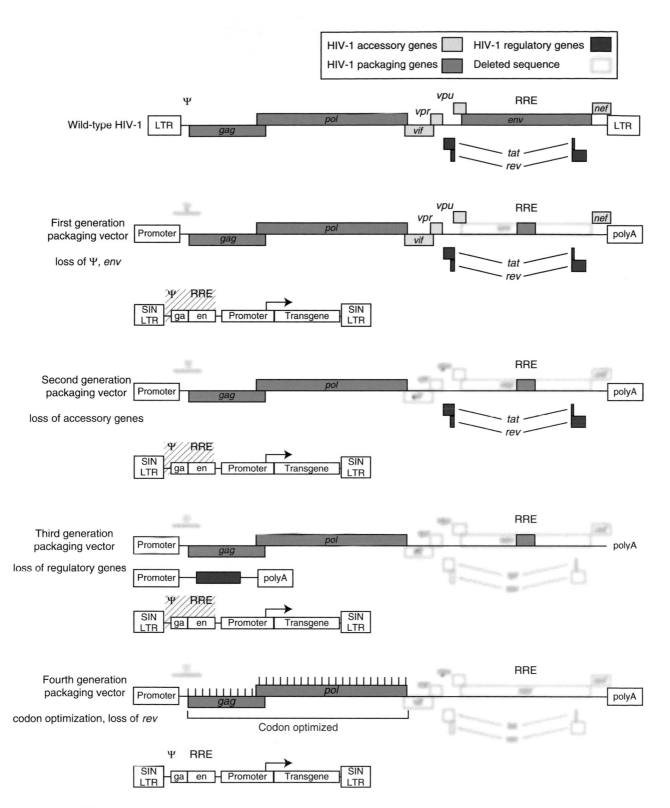

FIGURE 9.7 Wild-type HIV-1 and the various generations of lentiviral packaging vectors for comparison. A minimal transfer vector is shown alongside each packaging vector and areas of sequence homology between the two are indicated by shading.

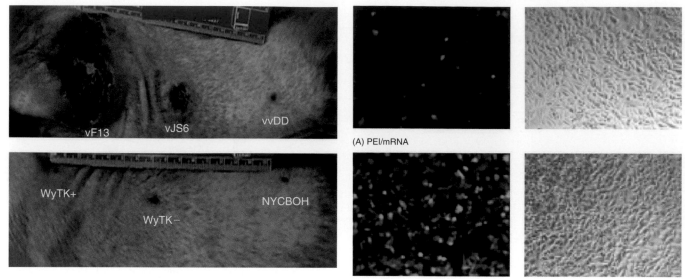

(A) PEI/mRNA

(B) HIS6 RPC/mRNA

FIGURE 10.4 Cutaneous lesions 8 days after intradermal injection of 106 pfu of the various vaccinia viruses: vF13 (wild-type WR strain), vJS6 (TK– WR strain), vvDD (TK–/VGF– WR strain), WyTK+ (wild-type Wyeth strain), WyTK–(TK– Wyeth strain), NYCBOH (wild-type New York City Board of Health strain). (From Naik, A.M., et al., *Hum. Gene Ther.*, 17, 31, 2006. With permission.)

FIGURE 13.5 Histidine-rich RPCs mediate efficient delivery of mRNA. Fluorescent images of PC-3 cells expressing green fluorescent protein (GFP) following transfection with 0.5 μg cap-*GFP*-A64 mRNA condensed with (i) 25 kDa PEI or (ii) reducible polycation (RPC). Phase-contrast images of transfected PC-3 cells are shown in the corresponding right-hand column. (From Read, M. L. et al., *Nucleic Acids Res.*, 33, e86, 2005. With permission.)

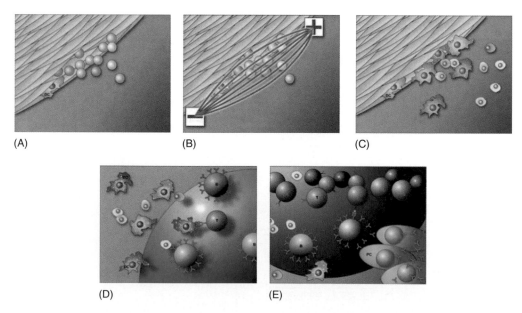

(A) (B) (C)

(D) (E)

FIGURE 16.1 Graphic representation of the delivery of macromolecules by direct injection followed by electroporation to the skeletal muscle for vaccination purposes. First, (A) the DNA vaccine is administered into skeletal muscle; (B) application of an electric field (electroporation); (C) contact of the expressed antigens with dendritic cells (DC) and subsequent activation; (D) activated DCs presenting antigens to B cells (B) and antigen fragments to T cells (T); (E) activated T-cells presenting antigens to B cells, activation of principal components (PC) for the generation of an immune response to the expressed antigen.

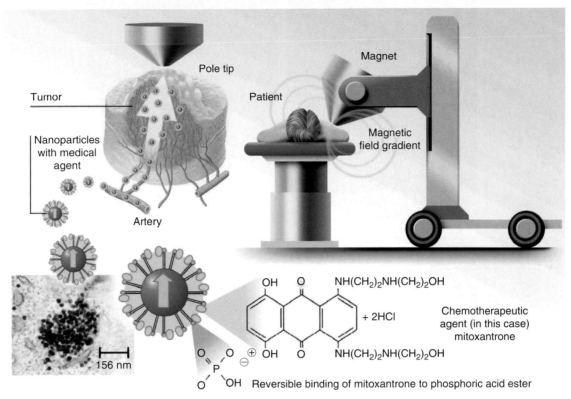

Image at lower left shows tumor tissue imbued with magnetic nanoparticles

FIGURE 17.1 Magnetic drug targeting. An anticancer drug (e.g., mitoxantrone) bound to magnetic particles is injected into a blood vessel (here tumor-feeding blood vessel) of the patient and is concentrated in the target tissue (e.g., tumor) by an external magnetic field. (From Siemens AG, Pictures of the Future 01/2007, page 64, www.siemens.com/pof. With permission.)

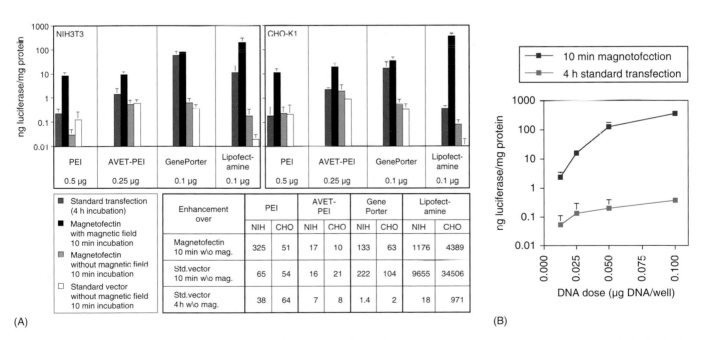

FIGURE 17.6 Enhancements achieved with nonviral magnetofection. (A) Comparison of magnetofections (with magnetic field 10 min incubation) with their corresponding standard transfections (4 h incubation) in NIH 3T3 and CHO-K1 cells. In the table, enhancements of magnetofection over transfections without (w/o) magnet are specified. AVET, adenovirus enhanced transfection. (B) Comparison of Lipofectamine magnetofection and Lipofectamine standard transfection in CHO–K1 cells with regard to their DNA dose–response profile. (From Scherer, F., Anton, M., Schillinger, U., Henke, J., Bergemann, C., Krüger, A., Gänsbacher, B., and Plank, C., *Gene Ther.*, 9, 102, 2002. With permission.)

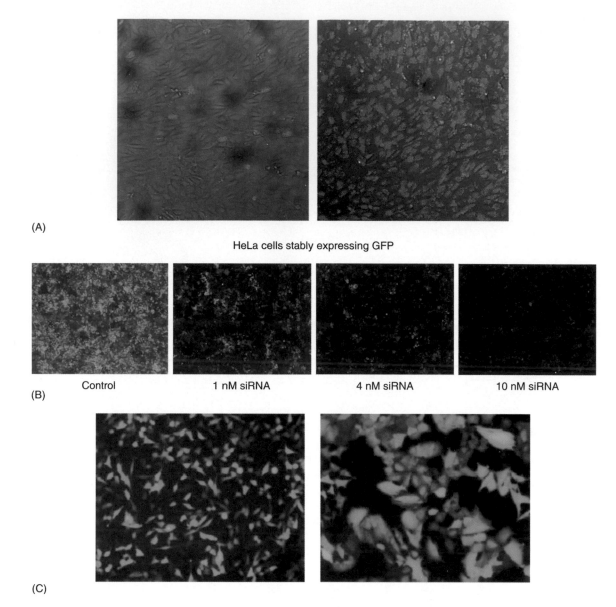

FIGURE 17.7 Some examples for the applicability of magnetofection. (A) Magnetofection of primary HUVEC-C cells with fluorescence-labeled phosphorothioate oligonucleotides (ODN). Right panel: In the presence of a magnetic field, up to 90% of the cells had oligonucleotide bound or internalized within 15 min of incubation with complexes of ODN with magnetic nanoparticles. Left panel: In the absence of a magnet, only very little binding/uptake was observed within this short period. (From Krotz, F., Wit, C., Sohn, H.Y., Zahler, S., Gloe, T., Pohl, U., and Plank, C., *Mol. Ther.*, 7, 700, 2003. With permission.) (B) Magnetofection of HeLa cells stably expressing GFP with siRNA to knock down GFP expression. From left to right: untreated cells (control), 1 nM siRNA directed against GFP, 4 nM siRNA, 10 nM siRNA, respectively, associated with magnetic nanoparticles and incubated on the cells in the presence of a magnetic field. This experiment was carried out in a 96-well plate. During transfection, the cell culture plate was positioned on a magnetic plate, which is shown in Figure 17.5A. The magnetic nanoparticles used here are a commercially available magnetofection reagents. (From the homepage of Oz Biosciences (www.ozbiosciences.com).) (C) Magnetofection of HeLa (left panel) and A549 (human non-small cell lung carcinoma) (right panel) cells with a pseudo-type HI virus carrying a GFP reporter gene. (From the homepage of OZ Biosciences (www.ozbiosciences.com).)

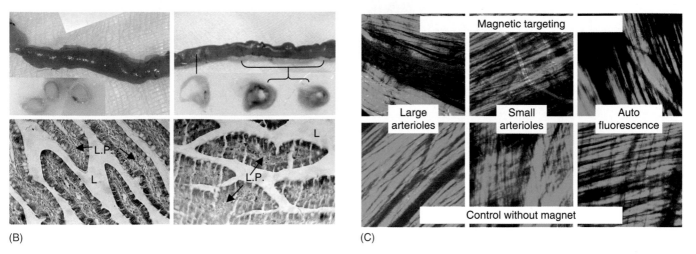

(B)

(C)

FIGURE 17.12 (B) X-gal staining performed 48 h after magnetic particle/DNA complexes were applied to the ilea of rats in the absence (left panels) and under the influence of a magnetic field for 20 min (right panels). Blue staining reveals efficient gene delivery only in the presence of magnet (right panels). Upper panels: intestinal tubes after X-gal stain. Insets: cross-sections of tubes. Lower panels: Paraffin sections counterstained with eosin, 400× magnification. X-gal staining is found in the lamina propria. L, lumen; LP, lamina propria. (From Scherer, F., Anton, M., Schillinger, U., Henke, J., Bergemann, C., Krüger, A., Gänsbacher, B., and Plank, C., *Gene Ther.*, 9, 102, 2002. With permission.) (C) Injection of fluorescence-labeled ODN/magnetic particle associates into the femoral artery caused site-specific accumulation in the vessels of the ipsilateral mouse cremaster at which a magnet was placed (upper panel, left: large arterioles, middle: small arterioles, right: autofluorescence before injection of magnetic complexes) but not into the contralateral cremaster, which was not in contact with a magnetic field (lower panel). (From Krotz, F., Wit, C., Sohn, H.Y., Zahler, S., Gloe, T., Pohl, U., and Plank, C., *Mol. Ther.*, 7, 700, 2003. With permission.)

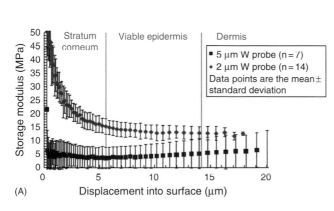

(A)

FIGURE 18.2 Mechanical properties (mean ± standard deviation) as a function of displacement obtained with microprobes indented into murine ears. (A) Storage modulus with 5 and 2 μm microprobes. (B) Stress with a 2 μm microprobe. (Adapted from Kendall, M.A.F., Chong, Y., and Cock, A., *Biomaterials*, 28, 4968, 2007. With permission.)

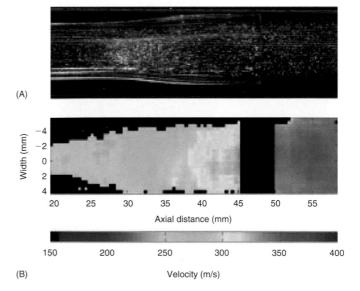

(A)

(B)

FIGURE 18.5 (A) Raw image and (B) derived PIV velocity map of the instantaneous particle flowfield of a CST prototype, taken 225 μs after diaphragm rupture. The payload was 2.2 mg of 39 μm diameter polystyrene spheres. (Adapted from Kendall, M.A.F., *Shock Waves J.*, 12, 22, 2002. With permission.)

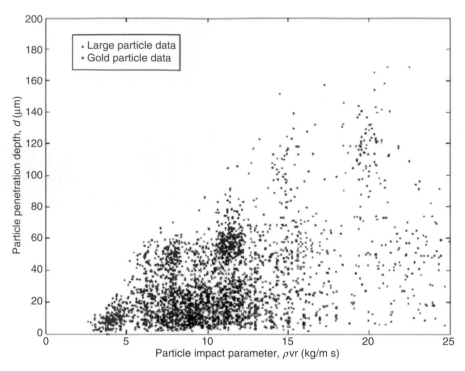

FIGURE 18.8 Raw gold and larger particle penetration into excised human skin as a function of the particle radius, density, and impact velocity. (Adapted from Kendall, M.A.F., Mitchell, T.J., and Wrighton-Smith, P., *J. Biomech.*, 37, 1733, 2004. With permission.)

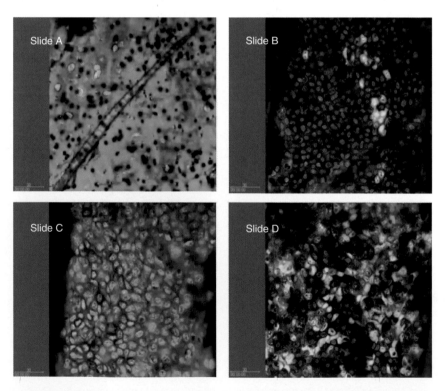

FIGURE 18.11 Murine tissue stained for live–dead cell discrimination with a cocktail of acridine orange/ethidium bromide and imaged with near-infrared two-photon excitation. Cells emitting red fluorescence are dead whereas cells emitting green fluorescence are alive. In slide A, the dark spots are perforations in the SC caused by microparticle bombardment. Slide B is the corresponding viable epidermis 11.3 μm below the SC. Slides A and B represent an image set of "high particle density" shown in Figure 18.10. Slide C is the viable epidermis of the control (9.7 μm below the SC) where particle delivery to the tissue was not made. Slice D is representative of the viable epidermis of tissue with intermediate particle densities (10.5 μm below the SC). Images were collected with LaserSharp software (Carl Zeiss, Hertfordshire, U.K.). (From Raju, P.A., Truong, N.K., and Kendall, M.A.F., *Vaccine*, 24, 4644, 2006. With permission.)

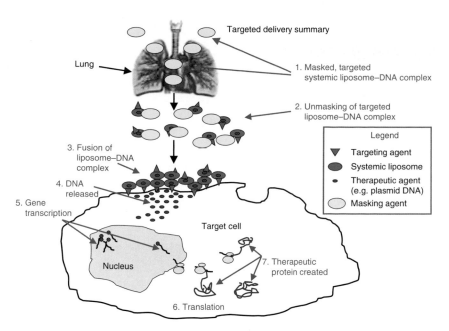

FIGURE 19.7 Optimized strategy for delivery and gene expression in the target cell. Optimization of many steps is required to achieve targeted delivery, shielding from nonspecific uptake in nontarget organs and tissues, deshielding, fusion with the cell membrane, entry of nucleic acids into the cell and to the nucleus, and production of gene expression of a cDNA cloned in a plasmid.

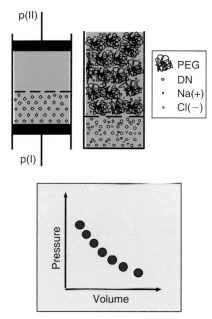

FIGURE 21.4 Osmotic stress method. (From Parsegian, V.A., Rand, R.P., Fuller, N.L., and Rau, D.C., *Methods Enzymol.*, 1986, 127, 400.) DNA liquid crystals are equilibrated against solutions of a neutral polymer (such as PEG or PVP, depicted as disordered coils). These solutions are of known osmotic pressure, pH, temperature, and ionic composition. (From Rau, D.C., Lee, B.K., and Parsegian, V.A., *Proc. Natl. Acad. Sci. USA*, 1984, 81, 2621.) Equilibration of DNA under the osmotic stress of external polymer solution is effectively the same as exerting mechanical pressure on the DNA subphase with a piston that passes water and small solutes but not DNA. After equilibration under this known stress, DNA separation is measured either by x-ray scattering, if the DNA subphase is sufficiently ordered, or by densitometry. (From Strey, H.H., Parsegian, V.A., and Podgornik, R., *Phys. Rev. Lett.,* 1997, 78, 895.) DNA density and osmotic stress thus determined immediately provide an equation of state (osmotic pressure as a function of the density of the DNA subphase) to be codified in analytic form over an entire phase diagram.

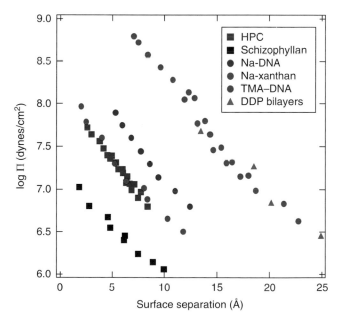

FIGURE 21.5 Interactions between biological macromolecules show striking universality at close surface-to-surface separations (or equivalently at very large densities). Hydroxypropyl cellulose, schizophyllan, different DNA salts, xanthan, and DDP bilayers at small intermolecular separations (given in terms of the separation between effective molecular surfaces of the interacting molecules) all show a strong repulsive interactions decaying with about the same characteristic decay length. The log-linear plot is thus more or less a straight line. (Composite data, courtesy D.C. Rau.)

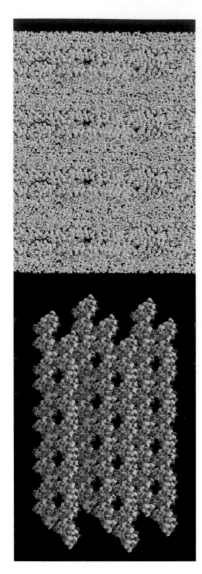

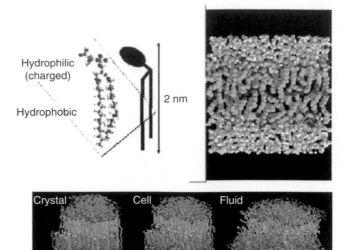

Lipid	Bending regidity (10^{-19} J)	Area compression (mN/m)
DMPC	1.15	145
SOPC	0.9	200
EggPC	0.4–2	
Plasma membrane	0.2–2	700
Red blood cell	0.13–0.3	450

FIGURE 21.20 The lipid bilayer. A lipid molecule has a hydrophilic and a hydrophobic part (here shown is the phosphatidylserine molecule that has a charged headgroup). At high enough densities, lipid molecules assemble into a lipid bilayer. Together with membrane proteins as its most important component, the lipid bilayer is the underlying structural component of biological membranes. The degree of order of the lipids in a bilayer depends drastically on temperature and goes through a sequence of phases (see main text): Crystalline, gel, and fluid, as depicted in the middle drawing. The box at the bottom gives sample values of bilayer bending rigidity and area compressibility for some biologically relevant lipids and one well-studied cell membrane. (Graphics courtesy of B.B Brooks, and M. Hodoscek, NIH).

FIGURE 21.11 Thermally excited conformational fluctuations in a multilamellar membrane array (small molecules are waters and long chain molecules are phospholipids) or in a tightly packed polyelectrolyte chain array (the figure represents a hexagonally packed DNA array) leading to collisions between membranes or polyelectrolyte chains. These collisions contribute an additional repulsive contribution to the total osmotic pressure in the array, a repulsion that depends on the average spacing between the fluctuating objects. (Graphics courtesy of B.B. Brooks and M. Hodoscek, NIH)

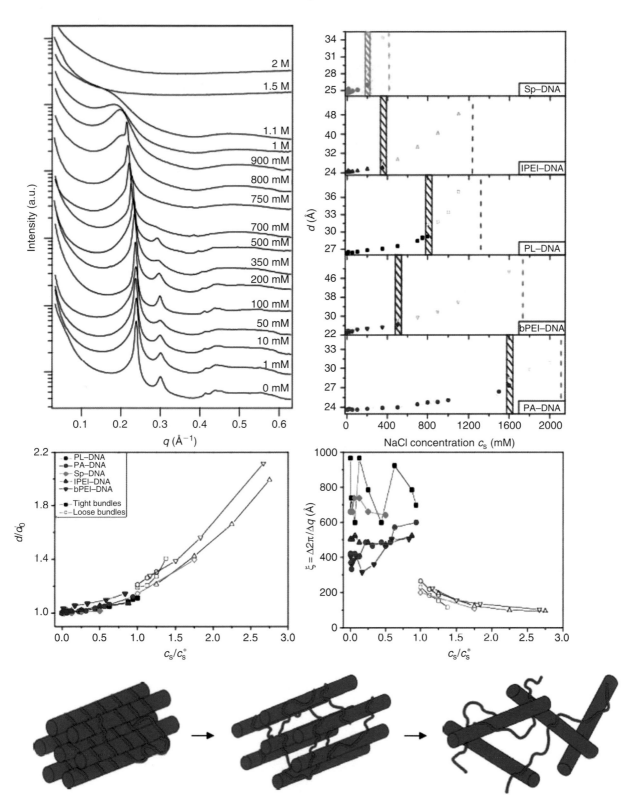

FIGURE 21.30 Phase behavior of polyplexes with increasing monovalent salt concentration. Universal phase behavior is observed for all systems with an initial swelling of a tightly packed hexagonal array of DNA rods. As a critical salt concentration, c_s^*, dependent on polycation, the onset of a coexistence regime is observed between tight and loose bundles. In-plane correlation lengths, ξ, representative of long-range order within the arrays show a sharp decrease upon crossing c_s^*. When scaled with respect to c_s^*, both swelling ratio (d/d_o) and ξ collapse onto universal behavior. With increasing salt, the loose bundles lose both positional and orientational order with an increase in a network structure at the expense of the loose bundle regime. At sufficiently high salt concentration, or dilute polymer concentration, all Coulombic interactions are screened and the samples disassociate completely into the dilute phase.

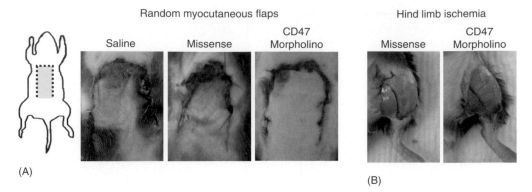

FIGURE 22.4   Improved survival of fixed ischemic injury following CD47 morpholino treatment of dorsal skin flaps and ischemic hind limbs in C57BL/6 mice. Wild-type C57BL/6 age- and sex-matched mice underwent 1 × 2 cm random myocutaneous flaps. Flaps were treated at the time of surgery with saline, a mismatched morpholino control or a target CD47 morpholino, and tissue survival determined (A). Wild-type C57BL/6 age- and sex-matched mice underwent ligation of the femoral artery at the inguinal ligament. Limbs were treated at the time of surgery by direct intramuscular injection of vehicle (saline), a mismatched control morpholino sequence, or a CD47 morpholino.

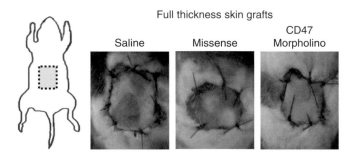

FIGURE 22.5   Improved survival of full thickness skin grafts following CD47 morpholino treatment in C57BL/6 mice. Wild-type C57BL/6 age- and sex-matched mice underwent 1 × 1 cm full thickness skin grafts. At the time of surgery, grafts and wound beds were treated by direct injection with saline (vehicle), a missense control morpholino, or a CD47 morpholino (10 μM), recognizing the murine sequence of the protein. Tissue survival was assessed on postoperative day 7.

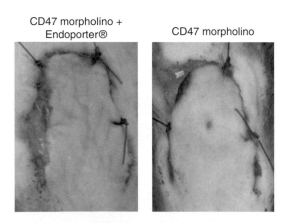

FIGURE 22.6   CPP does not enhance the efficacy of a CD47 morpholino for ischemic injury. Wild-type C57BL/6 age- and sex-matched mice underwent 1 × 2 cm random myocutaneous flaps. Flaps were treated at the time of surgery with a targeted CD47 morpholino plus the manufacture's recommended cell uptake agent (Endoporter®) in saline or a CD47 morpholino in saline alone and tissue survival determined.

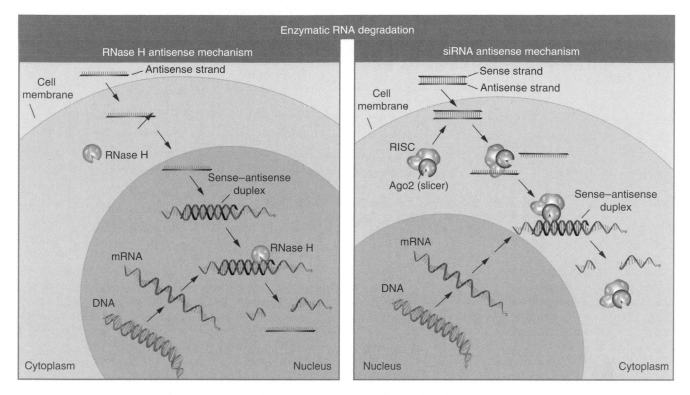

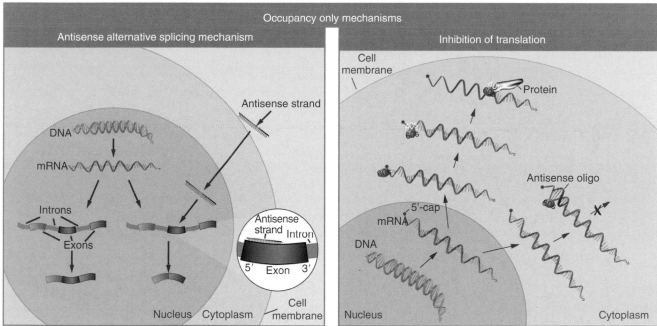

FIGURE 23.2 Antisense mechanisms of action. Cartoon depicting three different mechanisms by which an antisense oligonucleotide can inhibit expression of a targeted gene product by hybridization to the mRNA or pre-mRNA, which codes for the gene product.

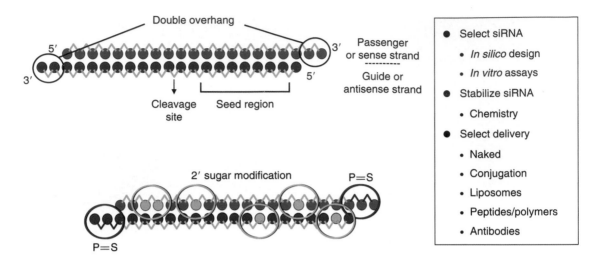

FIGURE 24.2 Developing RNAi therapeutics. Steps involved in developing an RNAi therapeutic. This three-step process begins with *in silico* design and *in vitro* screening of target siRNA, is followed by incorporating stabilizing chemical modifications on lead siRNA as required, and ends with selection and *in vivo* evaluation of delivery technologies appropriate for the target cell type/organ and the disease setting. A schematic illustration of some of the important features of siRNA structure (two base pair overhangs, seed region, and mRNA cleavage site) is shown. An example of an optimized siRNA molecule, which incorporates chemical modifications to increase nuclease stability and to minimize off-targeting, is shown below. The phosphorothioate and 2′ base modifications shown are illustrative of exo-nuclease and endo-nuclease stabilizing chemistries, respectively; other chemistries also exist that confer similar properties on an siRNA duplex. (From Bumcrot, D. et al. *Nat. Chem. Biol.*, 2006, 2, 711; de Fougerolles, A.R. et al., *Nat. Rev. Drug Discovery*, 6, 443, 2007.)

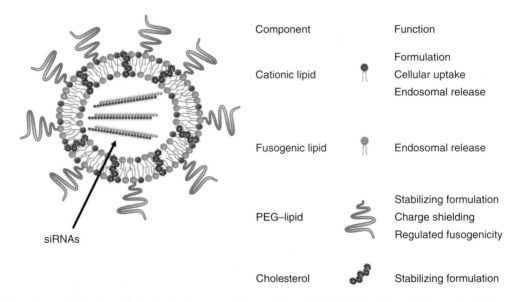

FIGURE 24.4 Structure and function of a cationic liposome. Composition of a SNALP is shown. Other cationic liposome formulations also exist, which may or may not contain a fusogenic or PEG–lipid. Ratios of different components also can vary between formulations.

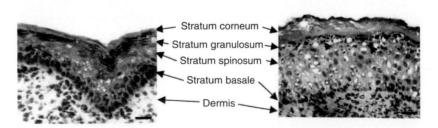

FIGURE 27.2 Hematoxylin/eosin staining of sections of 17.5-day-old mouse embryo skin (left) and of tissue formed from ESCs grown on a matrix secreted by a skin cell line (right). (From Coraux, C., Hilmi, C., Rouleau, M., Spadafora, A., Hinnrasky, J., Ortonne, J.P., et al. *Curr. Biol.*, 13, 849, 2003. With permission.)

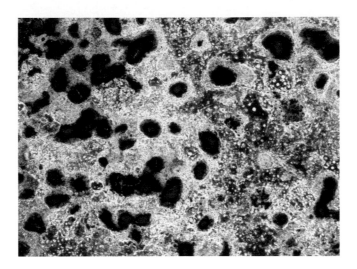

FIGURE 27.3 Phase contrast and fluorescence composite image of murine ESCs grown for 21 days in the presence of osteogenic supplements (β-glycerophosphate, ascorbate, and dexamethasone). Bone nodules are indicated by areas that stain positively for alizarin red S, which fluoresces when bound to calcified ECM.

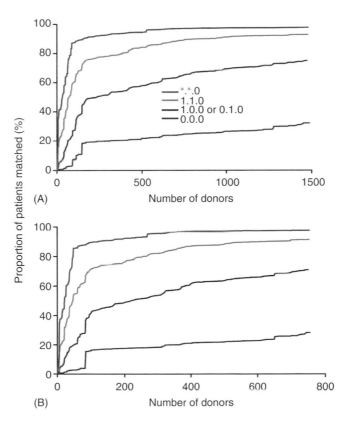

FIGURE 27.4 Graphs of estimates of how many ESC donors are required to provide different degrees of HLA matching in potential recipients. (0.0.0) indicates a zero HLA-A, HLA-B, and HLA-DR mismatch; (1.0.0 or 0.1.0) indicates a zero HLA-DR mismatch with no more than a single HLA-A or HLA-B mismatch; (1.1.0) indicates a zero HLA-DR mismatch with no more than a single HLA-A and a single HLA-B mismatch; and (*.*. 0) indicates a zero HLA-DR mismatch. (A) was obtained using data from 1500 consecutive cadaveric organ donors and (B) 760 consecutive blood-group-O cadaveric organ donors. (From Zhan, X., Dravid, G., Ye, Z., Hammond, H., Shamblott, M., Gearhart, J., et al., *Lancet*, 364, 163, 2004. With permission.)

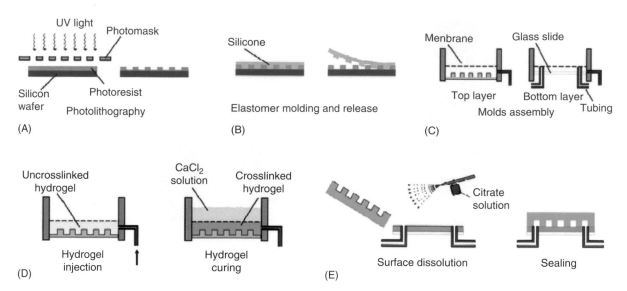

FIGURE 27.5 Process by which patterned hydrogel scaffolds can be constructed. Silicon photoresists are first patterned by exposure to UV light (A). A curable gel, such as silicone can then be molded over the photoresist (B) and used as a mold to create structure in a biocompatible hydrogel such as alginate that may contain cells (C and D). Channels in the hydrogel can then be perfused with a liquid such as growth medium (E). (From Cabodi, M., Choi, N.W., Gleghorn, J.P., Lee, C.S.D., Bonassar, L.J., and Stroock, A.D., *J. Am. Chem. Soc.*, 127, 13788, 2005. With permission.)

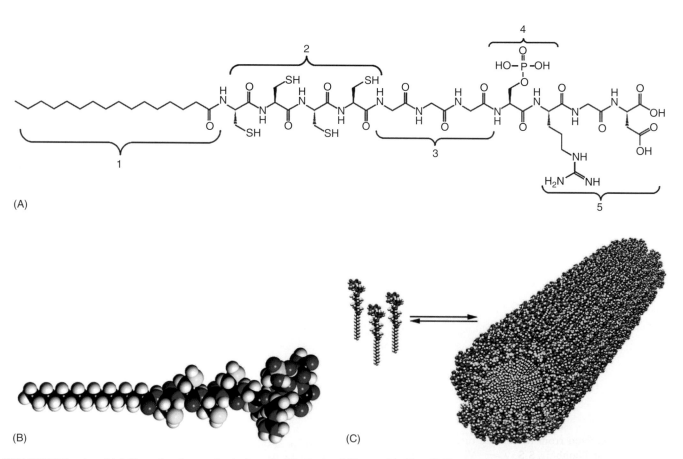

FIGURE 27.7 Amphiphilic molecules can be designed with a hydrophilic peptide "head" (5) and a hydrophobic tail (1) (A). These peptides self-assemble to form thread-like nanofibres (B and C). (From Hartgerink, J.D., Beniash, E., and Stupp, S.I., *Science*, 294, 1684, 2001. With permission.)

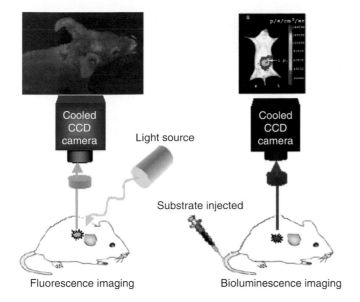

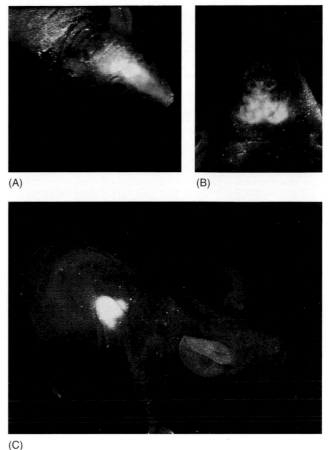

(A)   (B)

(C)

FIGURE 30.1 Basic principles of optical CCD imaging (fluorescence/bioluminescence). There are fundamentally two different types of optically based imaging systems: fluorescence imaging, which uses emitters such as GFP, wavelength-shifted GFP mutants, RFP, smart probes and near-infrared fluorescent (NIRF) probes, and bioluminescence imaging, which utilize systems such as firefly luciferase/D-Luciferin or Renilla luciferase/coelenterazine. Emission of light from fluorescent markers requires external light excitation while bioluminescent systems generate light de novo when the appropriate substrates/cofactors are made available. In both cases, light emitted from either system can be detected with a thermoelectrically cooled CCD camera since they emit light in the visible light range (400–700 nm) to NIR range (~800 nm). Cooled to −120°C to −150°C, these cameras can detect weakly luminescent sources within a light-tight chamber. Being exquisitely sensitive to light, these desktop camera systems allow for quantitative analysis of the data. Image shown above the "fluorescence imaging" schematic is a representative image obtained from a glioma model, which expresseses RFP (image used with permission from Anticancer, Inc.). The method of imaging bioluminescent sources living subjects with a CCD camera is relatively straightforward: The animal is anesthetized, subsequently injected with the substrate, and immediately placed in the light-tight chamber. A light photographic image of the animal is obtained which is followed by a bioluminescence image captured by the cooled CCD camera positioned above the subject within the confines of the dark chamber. The two images are subsequently superimposed on one another by a computer, and relative location of luciferase activity is inferred from the composite image. An adjacent color scale confers relative concentration of luciferase activity. Sample image above the "Bioluminescence imaging" schematic is a typical image obtained with this technology. In this specific example, image was obtained after intravenous injection of coelenterazine into a mouse containing intraperitoneal Renilla luciferase-expressing tumor cells. Significant bioluminescence is detected from the region of the xenograft. (From Bhaumik, S. and Gambhir, S.S., *Proc. Natl. Acad. Sci. USA*, 99, 377, 2002. With permission.)

FIGURE 30.5 Optical (fluorescence) imaging of transgene expression. The adenoviral (vAd) vector AdCMV5GFPAE1/AE3 [vAd-GFP] (Quantum, Montreal, Canada) constitutively expresses an EGFP, which is driven by a cytomegalo virus (CMV) promoter. The vector was delivered to the brain after an upper midline scalp incision and creation of a parietal skull window. Twenty microliters containing 8×10^{10} plaque-forming units (pfu)/ml vAd-GFP per mouse was injected into the brain. Twenty four hours later, fluorescence imaging of the entire animal (lower magnification) was carried out in a light box illuminated by blue light fiber optics, which provided the external excitation wavelength and imaged by using the cooled color CCD camera. Emitted fluorescence was collected through a long-pass filter GG475 on a three-chip thermoelectrically cooled, color CCD camera. Images of $1,024 \times 724$ pixels are captured directly on a personal computer or continuously through video output on a high-resolution video recorder. Images are subsequently processed for contrast and brightness and analyzed with imaging software. Higher magnification images (not shown here) can be accomplished by using a fluorescence stereomicroscope equipped with a 50 W mercury lamp. In this scenario, selective excitation of GFP is produced through a D425/60 bandpass filter and 470 DCXR dichroic mirror. Emitted fluorescence are captured and processed as described above. Images (A) and (B) demonstrate GFP transgene expression following adenoviral delivery to the brain. Image (C) demonstrates Ad-CMV-GFP delivery to the livery via portal vein cannulation. (Courtesy of Anticancer, Inc. and from Yang, M., Baranov, E., Moossa, A.R., Penman, S., and Hoffman, R.M., *Proc. Natl. Acad. Sci. USA*, 97, 12278, 2000. With permission.)

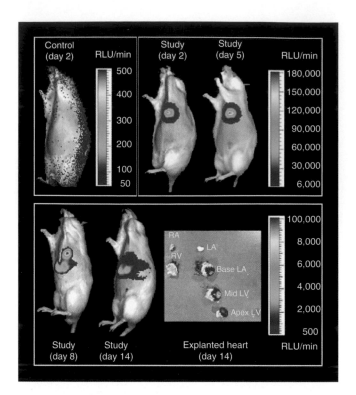

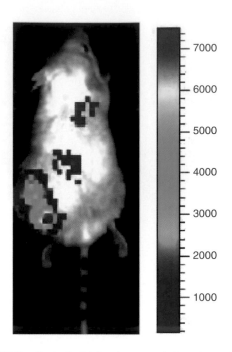

FIGURE 30.6 Optical (bioluminescence) imaging of cardiac reporter gene expression. Replication-defective adenovirus carrying firefly luciferase (*Fluc*) driven by a constitutive cytomegalovirus (CMV) promoter (Ad-CMV-*Fluc*, 1×10^9 pfu) was injected directly into the myocardium (anterolateral wall) of a rat. Images obtained 2, 5, 8, and 14 days later from a cooled CCD camera demonstrate significant cardiac emissions from FL activity ($P < 0.05$ vs. control). By day 8, luciferase activity is seen in the liver, which is probably from spillover of adenoviral vector into the systemic circulation and subsequent hepatic transfection of the virus via coxsackie-adenovirus receptors on hepatocytes. On day 14, the heart of the same rat was explanted after whole-body imaging was performed and sliced into three sections (bottom right). Firefly luciferase activity is localized at anterolateral wall of left ventricle along the site of virus injection. Control rats, which received an intracardiac injection of Ad-CMV-HSV1-sr39*tk* (1×10^9 pfu), demonstrate no significant firefly luciferase activity 2 days after the injection (upper left). Please note the bioluminescent scales are different for control rat, study rat days 2 to 5, and study rat days 8 to 14 to account for the wide range of cardiac firefly luciferase activity observed. Scales are a quantitative indicator of light photons detected (relative light units [RLU]/minute [min]). RA indicates right atrium; RV, right ventricle; and LV, left ventricle. (Reproduced from Wu, J.C., Inubushi, M., Sundaresan, G., Schelbert, H.R., and Gambhir, S.S., *Circulation*, 105, 1631, 2002. With permission.)

FIGURE 30.7 Optical (bioluminescence) imaging of targeted transgene expression using a tissue-specific promoter. One way to target gene therapy is through the use of tissue-specific promoters. However, most tissue-specific promoters yield low levels of transcription. In this example, certain key regulatory elements of the promoter and enhancer of prostate-specific antigen (PSA) have been multimerized to yield a construct, PSE-BC, which is 20-fold more active than the native PSA promoter/enhancer. Following incorporation into an adenovirus vector (AdPSE-BC-luc) and subsequent intratumoral injection into a human prostate cancer xenograft model (LAPC series), firefly luciferase expression can be seen in the main tumor xenograft (left flank) as well as other extratumoral sites, such as the back and chest, in this male SCID mice 21 days after vector delivery. Detailed histologic analysis of the xenograft and extratumoral sites demonstrates that firefly luciferase expression is restricted to the prostate tumor and prostate metastases, respectively. The metastases, in this case, are located in the spine and lung. By comparison, CCD imaging and histologic analysis of xenografts injected with AdCMV-luc show markedly diminished expression of firefly luciferase in the xenograft and increased nonspecific expression in the liver at 21 days postinjection (figure not shown). Results from this study indicate that tissue-specific transgene expression is possible and that CCD imaging can be used to track firefly luciferase-marked tumor cells. Scale indicates the number of photons detected (RLU/min). (Reproduced from Adams, J.Y., Johnson, M., Sato, M. et al., *Nat. Med.*, 8, 891, 2002. With permission.)

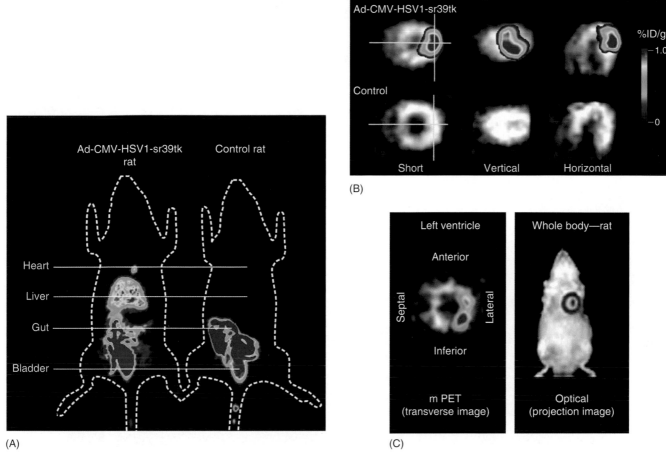

FIGURE 30.8 Micropet and optical (bioluminescence) imaging of cardiac reporter gene delivery. (A) Imaging cardiac gene expression using adenoviral-mediated mutant thymidine kinase (HSV1-sr39*tk*) as PET reporter gene and [18F]FHBG as PET reporter probe. Trapping of tracer occurs only in cells expressing the reporter gene. At day 4, whole-body microPET image of a rat shows focal cardiac [18F]FHBG activity at the site of intramyocardial Ad-CMV-HSV1-sr39*tk* injection. Liver [18F]FHBG activity is also seen because of systemic adenoviral leakage with transduction of hepatocytes. Control rat injected with Ad-CMV-*Fluc* shows no [18F]FHBG activity in either the cardiac of hepatic regions. Radiolabeled probe is always "visible" with radionuclide imaging regardless of whether it has localized to its target or not. As a result, radionuclide-based images will exhibit a certain degree of nonspecific tracer localization since "unbound" reporter probe is metabolized through either the enterohepatic or urinary system or both. In this example, nonspecific reporter probe activity gut and bladder activities are seen for both study and control rats because of route of [18F]FHBG clearance. (B) Tomographic views of cardiac microPET images. The [13N]NH₃ (gray scale) images of perfusion are superimposed on [18F]FHBG images (color scale), demonstrating HSV1-sr39*tk* reporter gene expression. [18F]FHBG activity is seen in the anterolateral wall for experimental rat compared with background signal in control rat. Perpendicular lines represent the axis for vertical and horizontal cuts. Color scale is expressed as % ID/g. (C) Comparison of typical images obtained with PET (left) and optical imaging (right). The optical method is more sensitive (at limited depths), easier to perform, and demonstrates minimal background noise. With PET, we can see that the transgene was delivered to the anterolateral aspect of the left ventricle. Such spatial resolution is not afforded by *in vivo* optical imaging at this time. (Reproduced from Wu, J.C., Inubushi, M., Sundaresan, G., Schelbert, H.R., and Gambhir, S.S., *Circulation*, 105, 1631, 2002; Wu, J.C., Inubushi, M., Sundaresan, G., Schelbert, H.R., and Gambhir, S.S., *Circulation*, 106, 180, 2002. With permission.)

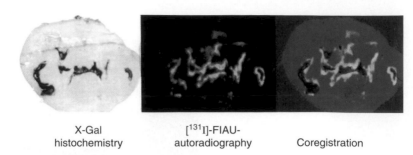

X-Gal
histochemistry

[^{131}I]-FIAU-
autoradiography

Coregistration

FIGURE 30.13 Tracking HSV infection with [^{131}I]FIAU using autoradiography. Tracking wild-type HSV-1 infection with radionuclide-based techniques can be accomplished using the virus' native TK gene and a reporter probe such as radiolabeled [^{131}I]FIAU. To help corroborate imaging findings with histochemistry findings, a replication conditional, oncolytic recombinant HSV-1 virus vector, hrR3, containing a *lacZ* insertional mutation within the *RR* gene locus, has been prepared. Following injection of the vector into rat gliosarcoma xenografts, tumors were processed for tissue-sectioning, autoradiography, and β-galactosidase-stained histology. Image coregistration of tumor histology, HSV-1-*tk* related radioactivity (assessed by [^{131}I]FIAU autoradiography), and *lacZ* gene expression (assessed by β-galactosidase staining) demonstrated a characteristic pattern of gene expression around the injection sites. A narrow band of *lacZ* gene expression immediately adjacent to necrotic tumor areas is observed, and this zone is surrounded by a rim of HSV-1-*tk*-related radioactivity, primarily in viable-appearing tumor tissue. PET images (not shown) of injected tumors in the intact animal have also been performed using [^{124}I]FIAU as a reporter probe; the areas of PET-labeled probe uptake correlate well with the β-galactosidase-stained photomicrographs. (Reproduced from Jacobs, A., Tjuvajev, J.G., Dubrovin, M. et al., *Cancer Res.*, 61, 2983, 2001. With permission.)

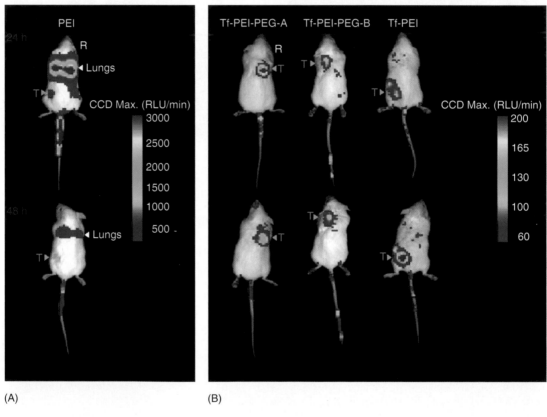

(A)   (B)

FIGURE 30.14 Tracking Tf targeted polyethylenimine (PEI)-mediated gene delivery using optical bioluminescence imaging. Delivery of the bioluminescence reporter gene, firefly luciferase (*Fluc*), by CMV-*Fluc* DNA/PEI polyplexes and subsequent *Fluc* expression can be imaged in living mice using a cooled CCD camera. Additionally, the biodistribution of modified PEI polycation complexes, altered with molecules such as Tf or polyethylene glycol (PEG), can be studied in this manner. Tf targeting has been shown to improve the transfection efficiency in certain tumor cell lines and PEG modification has been shown to improve circulation times of DNA/PEI complexes and prevent their nonspecific uptake by the reticuloendothelial system. All CCD images are of living mice carrying N2A xenograft 24 or 48 h after intravenous injection of various DNA/PEI polyplexes. Site of tumor is indicated (T). (A) PEI (positive control) treated animals show relatively high *Fluc* expression (using 1× D-Luciferin) in the lungs as compared with the tumor. The activity on the left-hind limb is from the N2A cell tumor (T). Nonspecific tail activity occurs at the DNA/PEI polyplex injection site. All *Fluc* expression decreases at 48 h. (B) Tf-PEI-PEG-A, Tf-PEI-PEG-B, and Tf-PEI treated mice show *Fluc* expression (using 2× D-luciferin) in the tumor (T) and tail regions, but no detectable signal in the lungs. For each formulation, expression in the tumor varied over 24 to 48 h. All images are quantitated as indicated by the two scales (RLU/min). (Reproduced from Hildebrandt, I.J., Iyer, M., Wagner, E., and Gambhir, S.S., *Gene Ther.*, in press. With permission.)

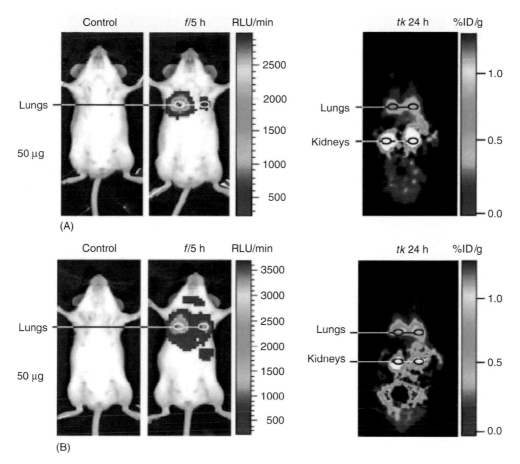

FIGURE 30.15 Tracking cationic lipid-mediated reporter gene delivery using optical (bioluminescence) and PET imaging. Cationic lipids associate with negatively charged DNA to form complexes that bind to cell surfaces by way of electrostatic interaction, thereby allowing a nonviral means of gene transfer. Distribution of systemic administration of DNA–lipid complexes in mice is demonstrated by delivering prepared DNA–lipid complexes that carry optical and PET reporter genes. CMV-*fl* plasmid DNA (CMV promoter driving expression of firefly luciferase [*fl*] gene) was mixed with cationic lipid, 1,2-dioleoyl-3-trimethylammonium-propoane (DOTAP) and cholesterol, to form *fl* DNA–lipid complexes. Similar procedure was used to prepare HSV1-sr39*tk* DNA–lipid complex (*tk* DNA–lipid complex). (A, B) show images following administration of 50 and 75 μg of each *fl* and *tk* DNA–lipid complexes via tail vein injection into CD-1 mice, respectively. Bioluminescent images (left images) were obtained 5 h after injection of the vector and 5 min after intraperitoneal injection of D-Luciferin. MicroPET images (right images) were obtained 24 h after vector delivery and 1 h after [18F]FHBG injection. Control mice (left) optical images obtained prior to administration of D-luciferin. Optical and PET images demonstrate that lungs are primary organs for transgene expression. Increased dose of DNA–lipid complex results in greater pulmonary transgene expression. Activity seen in the kidneys in the microPET images is the result of excreted, unsequestered reporter probe, [18F]FHBG. (Reproduced from Iyer, M., Berenji, M., Templeton, N.S., and Gambhir, S.S., *Mol. Ther.*, 6, 555, 2002. With permission.)

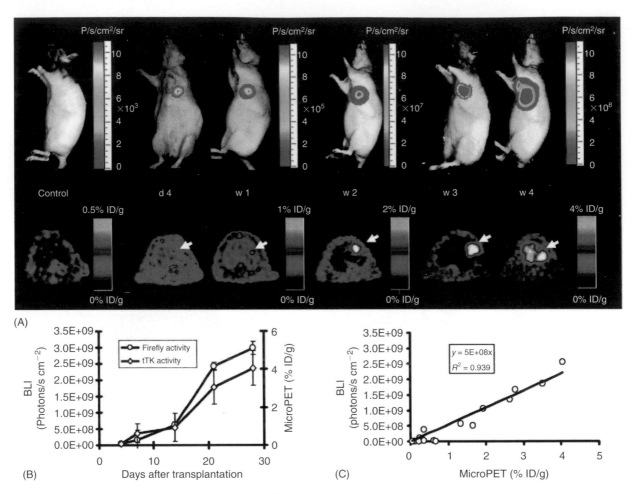

FIGURE 30.16 Molecular imaging of transplanted ES cells with bioluminescence and PET imaging. (A) Embryonic stem cells stably transduced with a triple fusion reporter gene mrfp-fluc-ttk were transplanted in the heart of athymic rats. Animals were subsequently imaged for 4 weeks to assess cell survival. A representative study animal is shown in the figure. Significant bioluminescence (top) and PET (bottom) signals were observed at day 4, week 1, week 2, week 3, and week 4 following transplantation. In contrast, control animals had background activities only. (B) Quantification of imaging signals showed a drastic increase of fluc and ttk activities from week 2 to week 4. Extracardiac signals were observed during subsequent weeks. (C) Quantification of cell signals showed a robust *in vivo* correlation between bioluminescence and PET imaging ($r^2 = 0.92$). BLI indicates bioluminescence. (Reproduced from Cao, F., Lin, S., Xie, X. et al., *Circulation*, 113, 1005, 2006. With permission.)

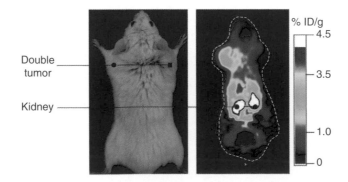

FIGURE 30.17 Noninvasive imaging of T-cell trafficking following adoptive transfer. Tumors were implanted subcutaneously on the shoulders of the mouse in the figure: A Moloney murine sarcoma virus–Moloney murine leukemia virus (M-MSV/MMuLV) tumor on its left shoulder and a control P815 tumor on its right shoulder. T-cells from an animal carrying an M-MSV/M-MuLV tumor were transfected with a retrovirus expressing both the HSV1sr39Tk and GFP reporter proteins. Transfected cells were first sorted by FACS and then injected intraperitoneally into the mouse bearing the M-MSV/M-MuLV and P815 tumor, and the animals were imaged by micoPET following [18F]FHBG injection. Homing of the cells to the M-MSV/M-MuLV tumor, but not the P815 tumor, was apparent within day 13 of injection. (Reproduced from Dubey, P., Su, H., Adonai, N. et al., *Proc. Natl. Acad. Sci. USA*, 100, 1232, 2003. With permission.)

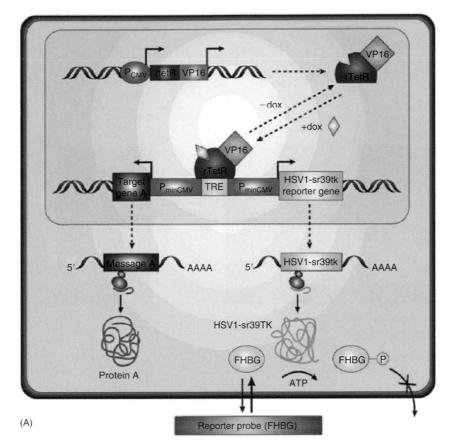

FIGURE 30.18 Indirect PET imaging using a bi-directional transcriptional approach. (A) Target (therapeutic) gene expression can be measured indirectly by imaging reporter gene expression if expression of the two genes is linked. Both genes can be simultaneously expressed from two minimal CMV promoters that are regulated by a single bidirectional TRE. The rTetR–VP16 fusion protein is produced constitutively from a CMV promoter. When the rTetR–VP16 fusion protein binds doxycycline, this complex binds to the TRE regulatory sequence and substantially enhances expression from the two minimal CMV promoters. The target gene A in one coding region and a reporter gene (for example, a reporter kinase such as HSV1-*sr39tk*) in the alternative coding region are transcribed simultaneously into two mRNA molecules. Translation of the two mRNA molecules yields two distinct proteins in amounts that are directly correlated with each other.

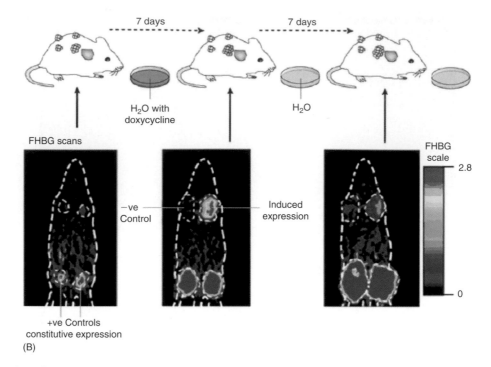

FIGURE 30.18 (continued) (B) Quantitative imaging of the locations and magnitude of PET reporter gene expression by trapping of a PET tracer inside the cell (for example, by phosphorylation of [^{18}F]FHBG by the HSV1-sr39TK reporter protein) provides an indirect measure of target gene expression. Sequential microPET imaging studies of a nude mouse carrying four tumors. Four tumor cell lines, two positive controls (constitutive reporter gene expression), one negative control, and one inducible line (reporter gene expression) induced by doxycycline, were injected subcutaneously into four separate sites in a single mouse. When tumors reached a size of at least 5 mm, the mouse was imaged with 9-(4-[^{18}F] fluoro-3-hydroxymethylbutyl)guanine ([^{18}F]FHBG). Doxycycline was then added to the water supply for 7 days. The mouse was then scanned again with [^{18}F]FHBG. Doxycycline was removed from the water supply for the next 7 days, and the mouse was again scanned with [^{18}F]FHBG. The locations of the four tumors and the mouse outline are shown by the dotted regions of interest. All images are 1–2 mm coronal sections through the four tumors. The % ID/g (% injected dose per gram tissue) scale for [^{18}F]FHBG is shown on the right. The negative control tumors show no gene expression and the positive control tumors show increased expression over the time course. The tumor on the top right, with inducible gene expression, initially does not accumulate [^{18}F]FHBG, then at 7 days after addition of doxycycline, induction of reporter gene expression traps [^{18}F]FHBG. Seven days after withdrawal of doxycycline, there is decreased induction and minimal trapping of [^{18}F]FHBG. The [^{18}F]FHBG image signal correlates well with target gene expression (not shown). (Reproduced from Gambhir, S.S., *Nat. Rev. Cancer*, 2, 683, 2002. With permission.)

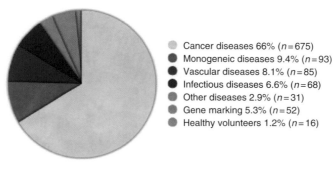

The Journal of Gene Medicine, © 2005 John Wiley & Sons Ltd   www.wiley.co.uk/genmed/clinical

FIGURE 34.1 Percentage of diseases targeted by clinical gene therapy trials. (http://www.wiley.co.uk/genmed/clinical/)

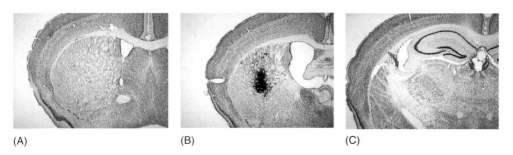

(A)   (B)   (C)

FIGURE 39.1 Stereotactic delivery of an HSV amplicon vector expressing β-galactosidase into the mouse striatum. Mice were injected with 1 × 10⁵ transduction units of HSVlac using a microprocessor-controlled pump. Animals were sacrificed and perfused 4 days post-transduction and X-gal histochemistry was performed on 40 μm sections. Sections representative of the injection site (Panel A), a site anterior of the injection (Panel B), and a site posterior of the injection (Panel C). All sections were counterstained with thionin and acquired at a magnification of 2.5×. The photomicrographs indicate that focal delivery of a viral vector can be achieved within the brain. (From Brooks, A.I., et al., *J. Neurosci. Methods*, 80(2), 137, 1998. © 1998 Elsevier Science B.V. With permission.)

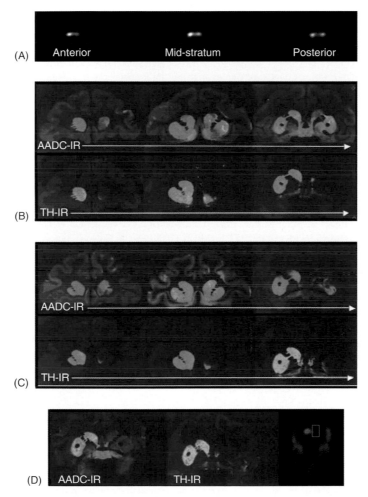

FIGURE 39.3 Immunohistochemistry of AADC in hemilesioned monkeys before and after AAV-hAADC treatment. (A) Three [¹⁸F]FMT PET coronal images of an AAV-hAADC-treated monkey striatum at 18 months show the strongest AADC transgene expression in targeted area (midstriatum) and correlate with AADC immunostaining. (B–D) In (B) and (C), color-coded immunostaining against TH and AADC in progressive anterior to posterior coronal slices in two representative AAV-hAADC monkeys is shown at 3 years after AAV delivery. (D) shows TH and AADC staining in a representative AAV-LacZ monkey at 2 years after AAV delivery. The pattern of TH and AADC immunoreactivity, normally colocalized in intact dopaminergic pathways, is very different in these MPTP-lesioned animals. Note the dramatic reduction of anti-TH staining in the right stratum of all animals compared to the intact left side, confirming profound loss of dopaminergic fibers and terminals on the lesioned side (B–D). Anti-AADC staining shows endogenous AADC in the nonlesioned, left striatum and almost complete restoration of the enzyme within the right striatum of AAV-hAADC animals, demonstrating widespread distribution of AADC transgene expression (B and C). No such AADC staining or [¹⁸F]FMT PET signal is seen on right side of control animals (D). The box in the PET image in (D) indicates the region of infusion of AAV-LacZ. (From Bankiewicz, K.S., et al., *Mol. Ther.*, 14(4), 564, 2006. © 2006 The American Society of Gene Therapy. With permission.)

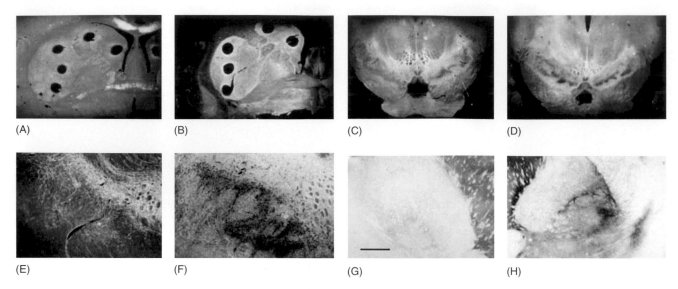

FIGURE 39.4 Delivery of a lentiviral vector expressing the GDNF gene in Parkinsonian nonhuman primates. Rhesus monkeys received the dopaminergic toxin MPTP unilaterally (right side) to establish a parkinsonian state. One week later, the animals received ipsilateral infusions of lentiviral vector expressing GDNF (lenti-GDNF) or one expressing the reporter protein β-galactosidase (lenti-βGal). Three months following treatment the animals were sacrificed and immunohistochemistry was performed. Panels A and B depict low-power dark-field photomicrographs through the right striatum of TH-immunostained sections of MPTP-treated monkeys treated with lenti-βGal (Panel A) or lenti-GDNF (Panel B). There appeared to be a comprehensive diminution of TH immunoreactivity in the caudate and putamen of lenti-βGal-treated animals, while a nearly normal level of TH immunoreactivity was seen in the animals receiving lenti-GDNF. Low power (Panels C and D) and medium power (Panels E and F) photomicrographs are shown of a TH-immunostained section through the substantia nigra of animals treated with lenti-βGal (Panels C and E) and lenti-GDNF (Panels D and F). Note the loss of TH-immunoreactive neurons in the lenti-βGal-treated animals on the side of the MPTP infusion. TH-immunoreactive sprouting fibers, as well as an above normal number of TH-positive nigral perikarya, are observed in lenti-GDNF-treated animals on the side of the MPTP injection. Panels G and H depict bright-field low-power photomicrographs of a TH-immunostained section from a lenti-GDNF-treated monkey. Note the normal TH-immunoreactive fiber density through the globus pallidus on the intact side that was not treated with lenti-GDNF (Panel G). In contrast, an enhanced network of TH-immunoreactive fibers is seen on the side treated with both MPTP and lenti-GDNF. Scale bar in (G) represents the following magnifications: Panels A, B, C, and D at 3500 μm; Panels E, F, G, and H at 1150 μm. (From Kordower, J.H., et al., *Science*, 290(5492), 767, 2000. © 2000 AAAS. With permission.)

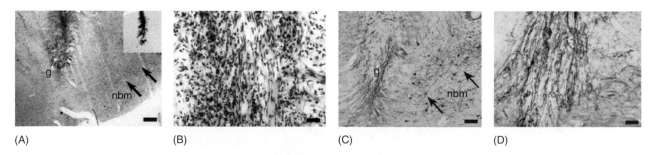

FIGURE 39.5 Trophic response to NGF in the human brain. (A,B) Nissl stain of autologous, NGF-secreting cell implant in brain of individual with Alzheimer's disease 5 weeks after treatment. Graft (g) adjacent to nucleus basalis of Meynert (nbm; arrows). Inset, robust mRNA encoding NGF by in situ hybridization within graft. Scale bar in (A) 247 μm; in (B) 24 μm. Note proximity of graft to nbm seen in similar perspective in c at higher magnification. (C,D) Immunocytochemistry for cholinergic neurons (p75) shows graft implant on left (g) and adjacent neurons of nbm (arrows). Higher magnification (d) shows dense penetration of cholinergic axons into graft. Scale bar in (C) 82 μm; in (D) 11 μm. (From Tuszynski, M.H., et al., *Nat. Med.*, 11(5), 551, 2005. © 2005 Nature Publishing Group. With permission.)

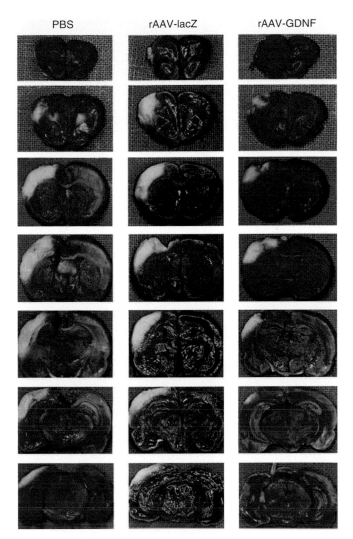

FIGURE 39.10 Injection of rAAV-GDNF markedly reduces cortical infarction induced by middle cerebral arterial ligation in rats. The right middle cerbral artery and bilateral common carotid arterial were occluded for 90 min. Animals received a PBS, rAAV-lacZ (10^{10} viral particles), or rAAV-GDNF (10^{10} viral particles) unilateral infusion during arterial occlusion, were sacrificed 72 h later, and their brains were coronally sectioned (2 mm thickness) for TTC staining. White areas represent infarcted zones in the cerebral cortex. Rats receiving rAAV-GDNF exhibited a marked reduction in infarct size as compared to animals receiving the rAAV-lacZ or PBS control. (From Tsai, T.H., et al., *ExNeurol.*, 166(2), 266, 2000. © 2000 Elsevier Science B.V. With permission.)

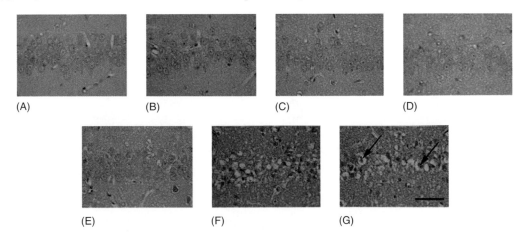

FIGURE 39.11 Representative photographs of pyramidal neurons in the hippocampal CA1 regions after treatment (4 and 6 h after ischemia) with SeV vector following ischemic injury. Sham-operated gerbils (A), SeV/GDNF (B,E), SeV/NGF (C,F), or SeV/GFP (D,G) were administered intraventricularly 4 h (B,C,D) or 6 h (E,F,G) after ischemic insult. The sections were stained with hematoxylin and eosin. Scale bar = 50 μm. (From Shirakura, M., et al, *Gene Ther.*, 11(9), 784, 2004. © 2004 Nature Publishing Group. With permission.)

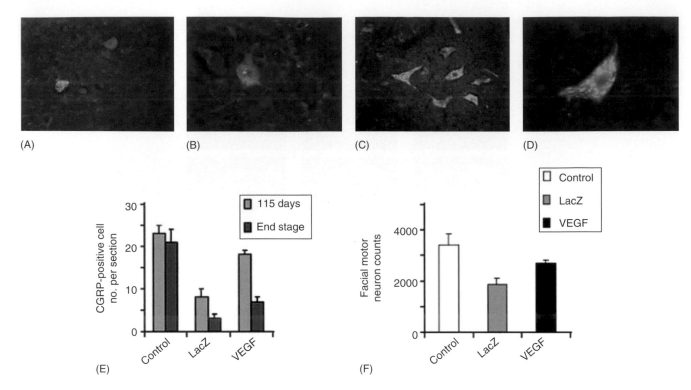

FIGURE 39.13 VEGF gene therapy protects spinal and brainstem motor neurons in SOD1^{G93A} transgenic mice. (A)–(D) Immunohisto-chemistry showing CGRP-positive neurons in lumbar spinal cord of EIAV-LacZ (A,B) and EIAV-VEGF injected animals (C,D). (E) Cell counts of surviving lumbar spinal cord motor neurons in control (wild-type), EIAV-LacZ, and EIAV-VEGF–treated SOD1^{G93A} mice at 115 days of age (blue) and at the end stage of disease (red). (F) Quantification of facial nucleus motor neurons in animals injected with EIAV-LacZ, EIAV-VEGF, and control animals at the end stage of disease. (From Azzouz, M., et al., *ExNeurol.*, 141(2), 225, 1996. © 2004 Nature Publishing Group. With permission.)

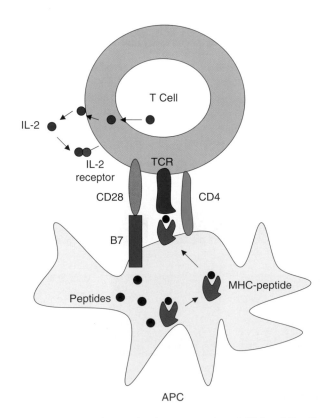

FIGURE 44.3 Effective T cell activation by APC. The interaction between antigen-MHC and T cell receptor leads to the expression of IL-2 receptor. This T cell proliferates when the second signal is provided from APC's costimulatory proteins. CD28–CD80/CD86 ligation initiates the production of IL-2 production and leads to the proliferation of activated T cell.

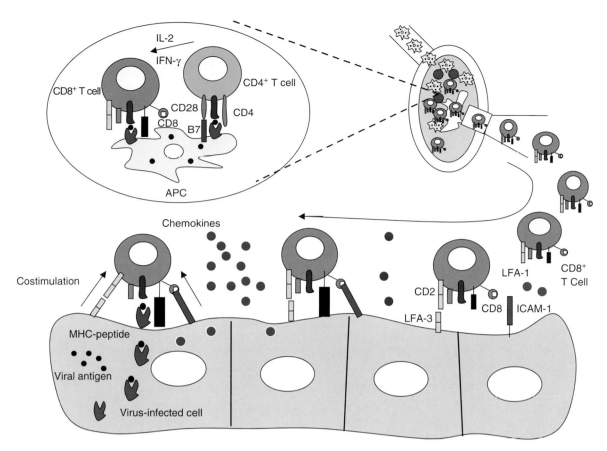

FIGURE 44.8 Regulation of CD8+ T cell expansion by adhesion molecules and chemokines in the periphery. Specific adhesion molecules and chemokines provide modulatory signals to CD8+ T cells in effector stage. This network of cytokine, chemokine, costimulatory molecules, and adhesion molecules represents a coordinated regulation and maintenance of effector T cells in the periphery.

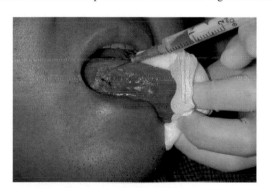
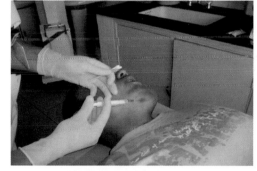

Intratumoral injection of Gendicine

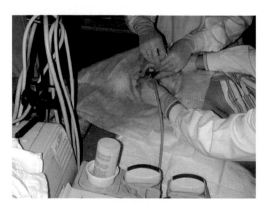
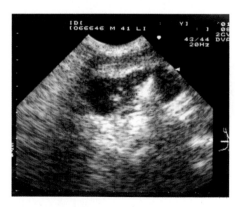

(A)   Injection of Gendicine
       under ultrasonic guidance

(B)   Showing a strong echo
       after Gendicine injection

FIGURE 49.3 Injection of Gendicine.

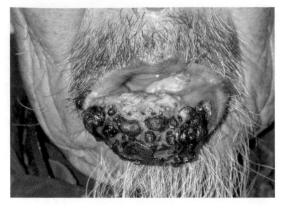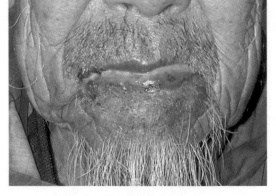

Before Gendicine administration   2 Months after Gendicine treatment

Male, 67 years, suffering from SCC of lower lip, T4N2aM0, was enrolled into the group
and treated with Gendicine in combination with chemotherapy (CP, BLM and MTX).
Ten injections of Gendicine (4×10^{12} VP/injection, one injection every 5 days)
and two courses of chemotherapy. The tumor showed complete clinical recovery.

FIGURE 49.5   Efficacy of Gendicine in combination with chemotherapy for treatment of advanced squamous cell carcinoma.

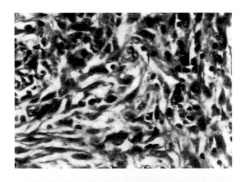

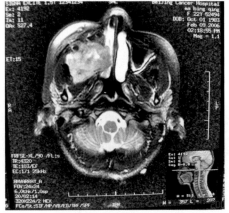

Eiomyosarcoma

13 Months after 6 injections
of Gendicine in combination
with a 50 Gy radiation
therapy

Tumor showing CR

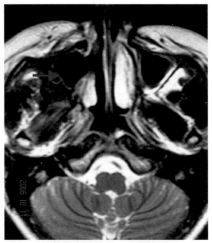

After Gendicine treatment

Before Gendicine treatment

FIGURE 49.6   Efficacy of Gendicine in combination with radiation therapy for treatment of eiomyosarcoma.

9 Lentiviral Vectors

John M. Felder III and Richard E. Sutton

CONTENTS

9.1   INTRODUCTION

Lentiviruses are a genus of the family *Retroviridae* (retroviruses) [1] that have been modified to be used as gene transfer agents. "Lenti-," from the Latin for "slow," refers to the well-known properties of these viruses as human and animal pathogens that are characterized by long incubation periods and persistent infection. Members of the lentivirus genus have garnered much attention from both scientists and the public at large since the early 1980s, when the then-phenomenon and now-pandemic of AIDS introduced the world to the most well-known and feared lentivirus: human immunodeficiency virus type 1 (HIV).

Although HIV is still the most well publicized and widely studied of the lentiviruses, there are other members of this group. All are pathogenic, but during the last decade these viruses have held specific interest for researchers not only in their capacity as infectious agents but now as promising tools (vectors) for the delivery of genetic material. Lentiviruses possess several features, such as capacity for a large (up to 9 kbp DNA) genetic payload and a minimal immune response, that make them attractive as gene transfer agents. The singular attribute, however, that most distinguishes this class of virus from other viral vectors (particularly from conventional retroviral vectors) is their ability to transduce and permanently modify nondividing cells. This property was initially demonstrated in cell culture but has since been reproduced in many experiments *in vivo*. The excitement generated by these results has fostered a large amount of research, some of which has recently lead to the first use of these vectors in human trials; an encouraging step towards the ultimate goal of utilizing the unique properties of lentiviral vectors in the clinical therapeutic setting.

This chapter is written for readers with at least a basic understanding of molecular biology and its methods. It aims to provide an overview of the historical development, design features, and production of these vectors as well as the opportunities and obstacles surrounding their use. The references provide examples and leads for further information, and are not meant to be exhaustive.

9.1.1 BIOLOGY OF THE LENTIVIRUSES: LIFE CYCLE AND GENETIC GEOGRAPHY

A good portion of this chapter is dedicated to understanding the construction and manipulation of lentiviral vectors. It is difficult to consider manipulation of a vector, however, without first having a comprehension of its components. There, it is necessary to first discuss the life cycle and genomic architecture of the native lentiviruses that are the template for all of the lentiviral vectors.

Lentiviruses, like all retroviruses, are enveloped RNA viruses that fuse with the membrane of a target cell and deposit a viral core within. Once inside, viral core particles reverse transcribe their RNA genome into duplex DNA and then integrate it into the host cell genome (see Figure 9.1B for details of life cycle). Following integration, long-terminal repeat (LTR)-driven RNA transcription occurs, such that viral proteins are translated and progeny virions are assembled, budded, and released to infect other cells [1,2]. This life cycle makes lentiviruses attractive vectors for gene therapy because it means that they are able to carry therapeutic sequences within their genomes and then integrate the DNA (including the therapeutic sequence) into that of the host cell. This ability to alter the host cell genome by inserting genetic sequences of interest is one essential aim of gene therapy and these modifications can be tailored by manipulation of the vector genome.

The proviral genome of HIV is shown in Figure 9.2, as are those of several other lentiviruses (feline immunodeficiency virus [FIV], equine infectious anemia virus [EIAV], simian immunodeficiency virus [SIV], and bovine immunodeficiency virus [BIV]) that have been used to construct vectors. The features of HIV and HIV-based vectors will be considered more extensively here than other lentiviral vectors, since efforts at developing vectors based upon HIV have thus far dominated the field.

A working knowledge of the HIV genome is vital for an understanding of both normal viral gene function and vector design. When approaching the relatively complex genome of HIV, it may be helpful to first functionally divide it into the two broad categories of *cis* and *trans* components, which can then be individually considered. The *cis* functions of HIV are carried out by segments of RNA or DNA that serve important functions, without being translated into protein. Important *cis*-acting elements of HIV include the LTRs, donor and acceptor splice sites, Rev responsive element (RRE), and packaging

signal (Ψ) [4]. The LTR is a regulatory sequence of DNA found at both the 5′ and 3′ ends of the provirus that acts as a promoter of transcription for the viral genome. The RRE is a structured RNA element within the *env* gene (see the following paragraph) which, when bound by the HIV protein Rev, allows an RNA transcript to be exported unspliced from the nucleus. Ψ consists of two short sequences on either side of the major splice donor site that form structured RNA elements that allow recognition and selective packaging of viral genomic RNA into the viral capsid (CA) before viral assembly and membrane budding [5,6].

The *trans*-acting components of the HIV genome are composed of nine open reading frames (Figure 9.2). Three of these can be considered to make up the "backbone" of the virus, since they are essential for replication and are shared among all retroviruses. These three open reading frames contain the conserved genes encoding the polyproteins Gag, Pol, and Env (shown shaded in diagonal lines, Figure 9.2). These three polyproteins are posttranslationally cleaved into the structural and enzymatic proteins that perform the vital functions necessary to the life cycle of any retrovirus: Env is cleaved into two structural proteins, SU (surface or gp120) and TM (transmembrane or gp41) which together make up the outer membrane envelope protein gp160. Gag is proteolytically cleaved into matrix (MA), CA, NC (nucleocapsid), and p6. Collectively, these four structural components aggregate to form the proteinacious core which houses the virus' RNA genome. Pol is also cleaved into the three essential viral enzymes PR (protease), RT (reverse transcriptase), and IN (integrase) which are responsible for proteolytic cleavage of most viral polyprotein precursors, reverse transcription of the viral RNA genome into DNA, and integration of viral DNA in the host cell genome, respectively [7].

The remaining six open reading frames contain encode genes that are not found in simple oncoretroviruses and that distinguish HIV as a lentivirus. Two of these genes, *tat* and *rev* (shown shaded with horizontal lines, Figure 9.2), code for regulatory proteins. Tat plays a critical role in producing large quantities of mRNA transcripts from the integrated proviral genome by acting as a transactivator of transcriptional elongation. Specifically, Tat forms a complex with the cellular proteins cyclin T1 and CDK9 which hyperphosphorylates the carboxyterminal domain of RNA polymerase II [8–10]. This modification increases the processivity of RNAPII, resulting in a 100- to 500-fold enhancement in the transcription of the HIV genome [11].

Rev, the other regulatory protein of HIV, is responsible for allowing unspliced or singly spliced species of messenger RNA to exit the nucleus intact. This function is crucial to viral replication, since the cell's default pathway is to splice all nascent mRNA transcripts as part of the nuclear processing required before export of mature mRNAs through nuclear pores. If all transcripts of the HIV genome were treated in this manner, there would be no full-length mRNA transcripts of the genome to be packaged into progeny virions. Unspliced and singly spliced mRNAs are also necessary for the proper translation of Gag/Pol and some HIV accessory proteins. Rev is

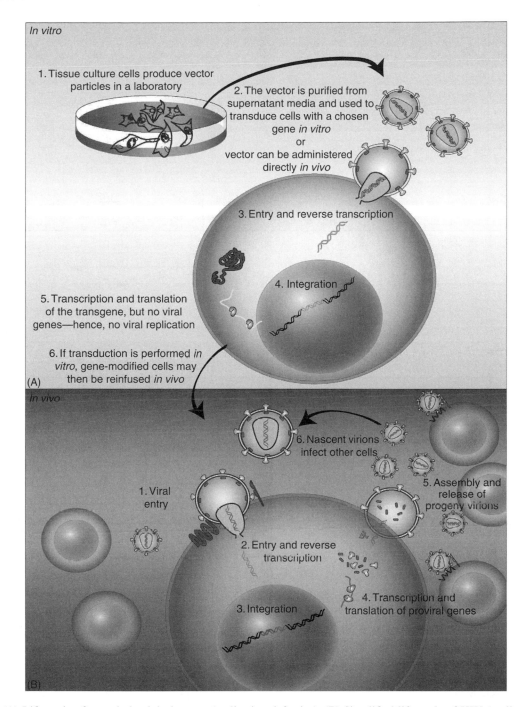

FIGURE 9.1 (A) Life cycle of generic lentiviral vector (replication-defective). (B) Simplified life cycle of HIV (replication-competent). Released virus is infectious and capable of infecting other cells.

able to overcome the cell's default splicing mechanism by multimerizing along the RRE, a structured RNA element within the intron shared by Tat and Rev. Cellular factors bind to a nuclear export signal within Rev and these complexes (containing the unspliced RNA) are exported from the nucleus [12].

The final four genes code for the so-called accessory proteins, which are also unique to lentiviruses: Vif, Nef, Vpr, and Vpu (shown in speckled shading, Figure 9.2). These proteins are not strictly necessary for viral replication *in vitro*, but all

play one or more roles in disease progression or pathogenesis in man. Vif has recently been found to play an important role in rendering cells susceptible to HIV infection by causing the proteasomal degradation of APOBEC3 proteins, which normally restrict retroviral replication by mutating or degrading viral cDNA [13–15]. The *in vivo* importance and exact function of the other accessory proteins are less clear, although it has been established that they modulate multiple host cell processes [7,16]. For example, Nef is a protein produced from a fully spliced mRNA that augments viral infectivity and rates

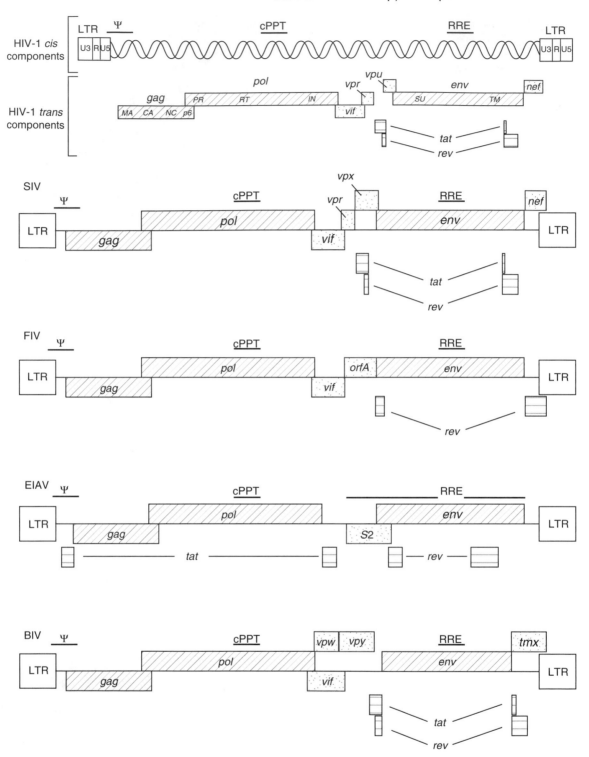

FIGURE 9.2 Genomes of lentiviruses that have been used as the bases for vector construction. Genes coding for structural proteins are shaded with diagonal lines, regulatory genes with horizontal lines, and accessory genes are speckled. Ψ, packaging signal; RRE, rev responsive element; cPPT, central polypurine tract; LTR, long terminal repeat.

of viral DNA synthesis following infection [17,18]. It is known to exert pleotropic effects, including interference with cellular signal transduction pathways, alteration of normal cellular trafficking patterns, and downregulation of CD4 and class I major histocompatibility complexes (MHCs) [19,20] from the cell surface. Vpu acts as an adapter protein to aid in the endocytosis and degradation of surface CD4 molecules, and is important for optimal virus release in certain cell types [21–25]. Finally, Vpr [26,27] is known to cause cell-cycle arrest in the G(2)/M phase [28–30] and to modulate gene transcription [31].

Vpr has also been implicated in the ability of HIV to infect nondividing cells. Vpr has a noncanonical nuclear localization signal and it has been demonstrated to directly interact with cellular nucleoporins and cause physical disruptions in the cell's nuclear envelope [32–35].

Several nonhuman lentiviruses [36], shown in Figure 9.2, have also been used to construct vectors. The standard *cis* and *trans* retroviral features are conserved among all of them, but accessory proteins vary somewhat in number and function. All of the viruses have a Rev/RRE system that allows export of unspliced or partially spliced viral RNA from the nucleus. The Tat and Vif proteins and functions also appear to be conserved with the exception that EIAV has no Vif protein and FIV has no recognizable Tat, although some Tat-like functions are performed by ORFA [37]. In addition, the non-primate lentiviruses encode other genes of unknown function, such as EIAV's S2 gene and dUTPase in the Pol polypeptide. The variety of accessory proteins among different lentiviruses and the uncertainty regarding the nature of some of their functions presents another level of complexity. However, while both human and nonhuman lentiviral accessory proteins all doubtlessly play important roles in viral pathogenesis *in vivo*, it is generally the case that they are not needed for replication *in vitro*. As we will discuss shortly, they are also not needed for effective *in vitro* gene delivery, and so for the purposes of gene therapy and vector construction, can largely be removed and ignored.

9.1.2 NUCLEAR IMPORT AND INTEGRATION IN NONDIVIDING CELLS

As mentioned in the introduction, the ability to infect nondividing cells has been a key issue driving the development of vectors based on lentiviruses. This is in part due to the fact that most oncoretroviruses (such as murine leukemia virus), whose use and prominence in gene therapy predates that of lentiviruses, do not share this property and must instead target actively dividing cells, whose nuclear membranes are periodically disassembled, in order to gain access to chromatin for integration [38,39]. The discovery that lentiviruses could infect nondividing cells [40–42] provided a vector whose shared heritage with oncoretroviruses meant that it maintained the desirable properties of such vectors but was able to transduce a range of target tissues (such as neurons) that were previously inaccessible via retroviral vector technology. This is an important ability for viruses such as HIV, whose initial targets are likely to be terminally differentiated tissue macrophages found near mucosal surfaces and whose later targets may include quiescent helper T cells, CNS microglia, and tissue macrophages.

After fusing with the cellular membrane and being deposited in a target cell, HIV-1 viral cores shed their CA protein and begin the process of reverse transcribing double-stranded cDNA, forming what is know as a reverse transcription complex (RTC). When reverse transcription is completed and viral cDNA is ready for import into the nucleus, the complex is termed a preintegration complex (PIC) and contains viral

and cellular proteins along with the cDNA [43]. The PIC traverses an intact nuclear membrane via an energy-dependent mechanism [44], likely utilizing nuclear pore complexes. Exactly how the PIC (whose diameter is more than twice that of the pores themselves) is able to cross the intact nuclear envelope remains controversial. Traditionally, nuclear import of HIV PICs has been attributed to three seemingly redundant nuclear localization signals (NLSs) [45]: a canonical NLS within the structural protein MA [46,47], a bipartite, canonical NLS within the enzyme IN [48], and a noncanonical sequence within Vpr [32–35]. However, no sole determinant of import has been reproducibly identified and instead the current picture suggests a multitude of contributing factors govern import [43]. Recent evidence points to several previously unconsidered factors that may underlie PIC import, including the viral central DNA "flap," structure formed during reverse transcription [49], the viral CA protein [50,51], and the cellular proteins importin 7 and nucleoporin 98 [43].

Integration into host chromatin by IN is another important event in the viral life cycle to be considered, since it allows permanent, irreversible modification of the host cell genome. At this point, relatively little is known about that factors that determine the choice of HIV integration sites within the host cell chromatin. Several genomic analyses have observed that the pattern of integration occurs preferentially within genes [52], but importantly throughout the transcriptional unit of each gene [53]. This contrasts to the 5′ bias of murine leukemia viral vector integration sites, and it may improve the safety profile of HIV-based vectors compared to those of MLV. Recently LEDGF/p75 [54] has been shown to modulate the interaction between HIV PICs and host cell chromatin via a tethering-like mechanism and may be an essential host factor for integration.

9.2 CREATION AND FEATURES OF INITIAL LENTIVIRAL VECTORS

The purpose, life cycle, and structure of lentiviral vectors differ considerably from that of their parent viruses. With the understanding of lentivirus biology gained in Section 9.1, we can now contrast these differences in detail as we explore the vectors themselves.

9.2.1 HOW ARE LENTIVIRAL VECTORS DIFFERENT THAN LENTIVIRUSES?

The "purpose" of the lentiviral life cycle is to deliver genes needed for viral replication to a cell and then replicate those genes for delivery to another cell. In contrast, the purpose of a lentiviral vector is to deliver a "particular" gene or set of genes, chosen by its designer, into a cell in order to effect a chosen function (beyond mere replication) within the cell. Unlike their parent viruses, a lentiviral vector typically does not replicate its components once it integrates into the target cell.

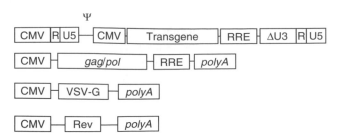

FIGURE 9.3 A typical lentiviral transfer vector (top), along with plasmids encoding requisite trans-acting genes. CMV, cytomegalovirus promoter; R, repeat region of LTR; U5, 5′ untranslated region; ΔU3, deleted 3′ untranslated region; polyA, polyadenylation sequence. Only the relevant portions of plasmids are shown.

The life cycles of lentiviruses and vector (see Figure 9.1) are also different, although they can both be conceptually divided into two stages: (1) viral entry, reverse transcription, transport into the nucleus, and integration correspond to the use of the vector to carry heterologous genes into target cells and (2) the subsequent provirus transcription, synthesis of viral proteins, packaging of the viral genome, maturation of the viral particle, and export from the cell correspond to the production of the vector particle [55]. Methods for producing vector particles are discussed in detail in later sections, but it is important to note now that vector production is markedly different than virus production. Whereas viruses autonomously reproduce and spread themselves within and among cells, vectors are nearly always initially generated in a laboratory using tissue culture cells to assemble specific viral and transgene components that are chosen in advance. Following generation, vector particles are then collected and used to transduce a target cell of choice.

Genomes are another point of difference between lentiviruses and vectors. While the genomes of lentiviruses such as HIV-1 are notoriously susceptible to small mutations, they are nonetheless relatively fixed structurally, and one genome can be put forth for each species of virus that accurately portrays the organization and gene products for that virus (Figure 9.2). Vector genomes, in contrast, may vary widely, depending upon the design of the vector (see Figures 9.3 through 9.7). It would be difficult to draw a prototypical vector, since there are so many markedly different permutations of this concept (nonetheless, see Figure 9.3 for the sake of comparison between viral and vector genomes). Vector genomes may

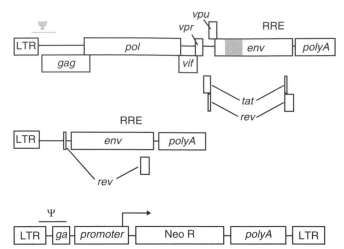

FIGURE 9.5 The three-component system introduced by Poznansky et al. Gray shaded regions indicate deletions. "ga" is a truncated *gag* ORF.

include some, all, or very few of the cis and trans elements that are found in their parent viruses, while they invariably contain some genes that are not. In addition, vector genomes are often divided into several pieces whereas lentiviral genomes are always entirely contained on a single, continuous strand of RNA or DNA. Certain factors, such as the intended application and included safety measures that determine the design and organization of a vector genome, will be discussed in detail below.

9.2.2 Turning Virus into Vector

Given the essential difference in purpose between a virus and a vector defined above, the hypothetical design of the simplest

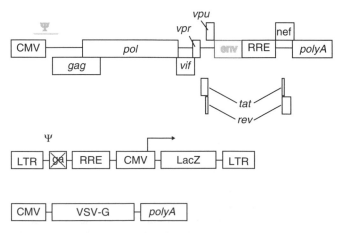

FIGURE 9.6 Vector system introduced by Naldini et al. At top is first-generation packaging vector, middle is transfer vector, and bottom is VSV-G expression construct. Gray shaded regions indicate deletions. "ga" is a truncated *gag* ORF that has been frame-shifted.

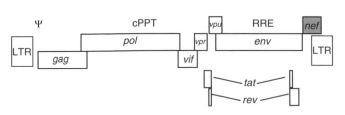

FIGURE 9.4 pHXB-CAT1. The shaded area was replaced with the chloramphenicol acetyltransferase gene; otherwise, this simple vector is very similar to wild-type HIV-1.

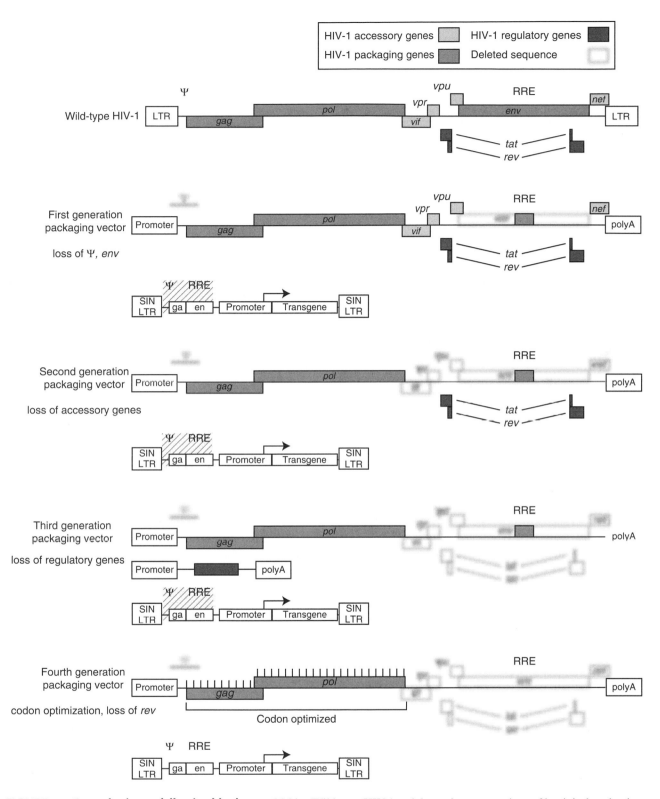

FIGURE 9.7 (See color insert following blank page 206.) Wild-type HIV-1 and the various generations of lentiviral packaging vectors for comparison. A minimal transfer vector is shown alongside each packaging vector and areas of sequence homology between the two are indicated by shading.

possible lentiviral vector would be nothing more than a parent virus that had been modified to include an exogenous gene chosen by the designer. In fact, the historical record of vector development matches these requirements: The first lentiviral vectors were constructed by Sodroski and coworkers in 1989 and were nothing more than a T-tropic HIV-1 plasmid provirus (HXBc2) that had been altered to include a chloramphenicol acetyltransferase (CAT) reporter gene in place of nonessential *nef*, near the 3′ end of the virus (termed pHXB-CAT1) (Figure 9.4). When this plasmid was transfected* into Jurkat T cells, it was found to produce infectious virus that expressed CAT activity in proportion to the level of infection of a cell culture [56]. The replicating entity used in this experiment meets our criterion of "vector" since it transferred the CAT gene; however, this earliest vector was actually replication competent, used all of the viral gene products (except Nef) in their native forms[†], and was produced in a manner similar to the "wild-type" virus that served as the experimental control.

Shortly thereafter the same group of investigators introduced the next level of complexity in vector architecture when they designed an HIV-based research tool to quickly assay the effects of different HIV envelope mutations on viral entry into cells. The assay relied upon pHXB-CAT1 as described above. The plasmid was further altered to include a 580 bp deletion in the *env* gene (termed pHXBΔenvCAT) such that envelope was defective and the virus could not replicate. Full-length envelope was provided *in trans* by co-transfecting cells (COS-1 or Jurkat) with both pHXBΔenvCAT and a separate gp160 expression plasmid. Transfected cells proceeded to produce infectious but replication-defective virus. The presence of the packaging signal, Ψ, on transcripts of pHXBΔenvCAT allowed its RNA to be incorporated into nascent viral core particles, while the gp160 RNA was not. The harvested virions could then be used to transduce target cells with only a single cycle of replication [57].

This laboratory "tool" intended to test different gp160s was an important step in the evolution towards modern vectors. Instead of a self-replicating entity, the gene transfer agent now served to permanently modify target cells with a gene of interest and then remain in that cell, without the ability to spread from cell to cell. This method makes available the desirable ability of lentiviruses to infect and alter cells while separating it from replication. It also demonstrates that producer cells can be transfected with a variety of different plasmids, and RNAs are incorporated into infectious viral particles based on the presence of the packaging signal [5]. Already one can see how this sort of technology would be very applicable to gene therapy. Creation of replication-defective vectors in order to carry a gene of interest and provision of requisite *trans*-acting factors is a mainstay of lentiviral and retroviral vector technology.

Contemporaneous with the above innovations, Page et al. constructed an HIV vector that similarly included a deletion in *env*, making it replication defective, but also introduced several other important innovations. First, by replacing the region of deleted envelope with the drug resistance gene *hypoxanthine–guanine phosphoribosyl-transferase* (*gpt*), this vector allowed the first quantitative measurements of titer (number of infectious units or IU per mL) [58]. This type of

measurement cannot be obtained by similar vectors encoding CAT. Co-transfection of cells with the *gpt*-containing vector and a gp160 expression cassette resulted in a cell-free virus titer of 825–3000 IU/mL. Ultrastructural features of these replication-defective viral particles were comparable to those of wild-type HIV.

The other important innovation in vector design that was introduced by this group was the demonstration that defective HIV viral cores could be packaged not only by HIV's native gp160, but also by the heterologous envelope proteins of other viruses. Moloney murine leukemia virus amphotrophic envelope 4070A (A-MoMLV) was used for this purpose [58], and later human T-cell leukemia virus type I (HTLV-I) envelope [59]. This process, known as pseudotyping, had been pioneered previously (like many of the "innovations" in lentiviral vector technology) with oncoretroviral vectors, but was an important step in the evolution of lentiviral vectors since it extended their host cell range, broadening the potential applications in which they might be useful. As will be discussed shortly, continued experimentation in the arena of pseudotyping would eventually yield an enormous boon to the utility of these vectors.

9.2.3 MAKING LENTIVIRAL VECTORS SAFER AND MORE USEFUL

Although the constructs described above had features that distinguished them as vectors rather than viruses, they were still considered to be molecular tools for the study of HIV and other viruses and not for gene transfer. Others first suggested using lentiviral vectors for gene therapy when they constructed a Tat-inducible vector and argued that such a vector could be used to transduce HIV-infected cells, where a therapeutic gene would preferentially be expressed in the presence of Tat [60]. A vector encoding a Tat-inducible trans-dominant interfering mutant version of *env* selectively inhibited viral replication in cells that had been infected by HIV and produced Tat [61].

As with all vectors, the issue of safety is paramount. Safety concerns regarding the use of lentiviral vectors are largely engendered both by the fact that HIV is a well-known and potentially lethal human pathogen, and by the possibility of regenerating some form of replicating virus (by DNA or even RNA recombination events in the producer cell). The fear of unwittingly generating replication-competent lentivirus (RCL) has its roots in a well-known precedent wherein replication-competent retrovirus was generated in MLV vector production systems and was linked to the development of lymphoma in monkeys [62,63]. Additionally, most lentiviruses slowly but eventually destroy the immune system and are transmissible.

As the design of lentiviral vectors has progressed, steps have been sequentially taken to minimize the possibility of producing RCL. The first of these measures appeared early in the historical development of these vectors when a three-component vector system was created in 1991 (Figure 9.5). Investigators, who had recently identified the packaging signal Ψ near the 5′ LTR of HIV [5], realized that any RNA sequence containing it could be efficiently packaged into an HIV core if

---

* See Section 9.4 for a better description of the process of transfection.
[†] There was a small deletion in the 3′ end of the *env* gene that was found not to affect the kinetics of replication.

the necessary trans-acting HIV elements (e.g., Gag, Pol, Env) were present. Likewise, any RNA lacking Ψ would not be packaged. Thus, the transfer vector did not need to contain any coding sequences but instead simply the requisite cis-acting elements. A transfer vector that consisted of only the necessary cis-acting sequences flanking a neomycin phosphotransferase gene (an antibiotic selection marker) driven by an internal promoter was constructed [64]. All of the proteins required to package the transfer vector were provided in *trans* by co-transfection with either a full-length provirus with a small deletion in the Ψ (so it would not be packaged) or with both an HIV envelope expression construct and a separate provirus that had a deletion in Ψ and envelope. When all *trans*-acting functions were provided by two separate expression plasmids (i.e., utilizing the full three-component system), the generation of replication-competent viruses was reduced to an undetectable level. This work of course paralleled that of retroviral vectors in terms of splitting components to improve safety. Later vectors also went on to incorporate additional, improved safety features that will be discussed shortly.

With evidence that the risk of generating RCL could be minimized, the final significant obstacle to the realistic use of lentiviral vectors in gene therapy was the efficient production of vector particles at high-enough titers to be useful. A major breakthrough occurred in the mid-1990s when, extending the technique of pseudotyping and borrowing a trick from the retroviral vector field, Naldini et al. demonstrated that high-titer, pseudotyped HIV particles could be generated using vesicular stomatitis virus (VSV)-G (glycoprotein) as an envelope [65]. Using a vector system with extensive deletions of viral genes from the packaging vector and employing the heterologous VSV-G envelope, titers of 1×10^5 IU/mL (as measured using a LacZ transgene) after transient transfection of 293T cells were achieved (Figure 9.6). Because of increased particle stability conferred by VSV-G pseudotyping, vector supernatants resulting from such transfections could be concentrated up to 1000-fold by ultracentrifugation without signficant loss of IU, resulting in titers in the range of 1×10^8 IU/mL. Additional refinements and improvements led to titers of 10^9 IU/mL, which is quite useful for a number of *in vivo* applications [55,66]. Additionally, VSV-G is apparently "pan-tropic," and thus confers a very wide range of potential cellular target types when used as the pseudotyping glycoprotein.

The discovery of high-titer VSV-G pseudotyping was enough to abolish the final obstacle to realistic consideration of the use of lentiviral vectors for gene therapy purposes. However, in the same landmark study, Naldini et al. also provided a remarkable incentive to do so when they demonstrated that HIV-based vectors (like the intact virus) could efficiently and persistently transduce rat neurons, human macrophages, and other nondividing cells. This cemented the importance of lentiviral vectors as a potential tool with true clinical utility and created a large amount of excitement in the field since, as mentioned previously, oncoretroviral vectors were the predominant tool being used for long-term gene transfer at the time but they lacked the ability to transduce nondividing cells. The adaptation of VSV-G pseudotyping to safer, multicomponent

HIV-based vectors and proof of their ability to transduce nondividing cells could be considered the start of the development of modern lentiviral vectors [67]. Thus, systems such as the one used by Naldini et al. are often referred to as first-generation lentiviral vectors.

9.3 DESIGNS AND PROPERTIES OF MODERN LENTIVIRAL VECTORS

9.3.1 FOUR GENERATIONS OF MODERN LENTIVIRAL VECTORS

The steps taken by Naldini et al. in 1996 ushered in an era of intense interest in the use of lentiviral vectors for gene therapy which drew many new researchers to the field. Many of those seeking to improve and expand upon the exciting new results began by using the vector system that was successfully employed by Naldini et al. This system, pictured in Figure 9.6, employed three separate plasmids for production to minimize the generation of RCL: (1) a transfer vector containing the transgene of interest along with the minimum viral *cis*-acting sequences necessary for packaging, reverse transcription, and integration, (2) a packaging vector encoding all necessary trans-acting genes minus envelope but also with its 5′ and 3′ LTRs replaced by an heterologous CMV promoter and an SV40 poly(A) site, respectively, and (3) a envelope construct expressing typically VSV-G (or amphotrophic MLV envelope). As mentioned previously, the impressive success of this system in achieving high titer and transducing nondividing cells led to its widespread adoption and hence it became known as a first-generation lentiviral vector. However, even in a multicomponent system such as the first-generation vector, it is theoretically possible to generate RCL by homologous recombination of the viral sequences that are shared by the packaging construct, the transfer vector, and possibly by endogenous retroviral or other DNA sequences within the producer cell. Therefore, as research in the field continued, so did efforts to reduce sequence homology between components of the vector system by identifying the minimal *cis* and *trans*-acting sequences required for effective vector production.

The ideas behind this process are simple: (1) remove as many viral genes as possible and (2) minimize sequence overlap or identity between transfer and packaging vector. In practice, however, changes along these lines proceeded via a series of discrete modifications, with the advent of each giving rise to another distinct "generation" of vector. It is useful to trace the history of these developments since each one represents another important aspect of modern vector design (Figure 9.7).

First-generation vectors contained a number of undesirable sequence homologies (see Figure 9.7) that are seemingly unavoidable at first glance, since they contain genes essential for proper packaging of the transfer vector. Therefore, if the goal is to reduce sequence homology, a logical place to begin is by deleting those genes which are not known to be essential for packaging. This is in fact what took place in the creation of the second-generation HIV-based vectors when the remaining viral accessory genes (*nef*, *vif*, *vpr*, and *vpu*) were deleted from the packaging construct and the resulting vector was still capable of transducing noncycling cells and sustaining transgene expression just as effectively as first-generation vectors [68,69]. By removing many of the genes thought necessary for HIV pathogenesis but not essential for HIV to function as a vector, the generation of RCL became less probable, and any RCL generated would be unlikely to be pathogenic [70].

Second-generation vectors lacking any HIV accessory genes made the recombinant generation of a pathogenic RCL very unlikely. However, since even the smallest safety risk might preclude the clinical use of such vectors, third-generation lentiviral vectors focused on improvements that would further increase their theoretical biosafety. Namely, they attempted to decrease any remaining sequence homology and cripple any potential RCL even further by reducing the reliance on HIV regulatory genes (*tat* and *rev*) in the vector system. As described earlier, Tat serves an indispensable role in the transcription of genes from viral LTRs. Thus, in order to achieve independence from Tat, all constructs within the vector system would have to rely on promoters other than HIV LTRs. The packaging and pseudotyping constructs had been using heterologous (usually CMV) promoters since the first generation of vectors, so it was only the transfer vector whose promoter required modification. Since some parts of the HIV LTRs are needed for reverse transcription and integration of the vector, the entire LTR could not be replaced by a heterologous promoter. Instead, a portion of it, the 5′ "U3" sequence, was selectively replaced by a portion of the LTR from an avian retrovirus, resulting in a chimeric promoter that allowed Tat-independent transcription in producer cell lines [71].

rev has proven to be more difficult to eliminate than *tat*, since, as discussed in Section 9.1.1, it is needed for the nuclear export of unspliced RNAs, such as those of the packaging construct and transfer vector. In addition, it is well known that HIV coding sequences contain *cis*-repressive elements, also called inhibitory sequences (INS), that produce instability in transcripts and cause mRNAs to be targeted for destruction in the absence of Rev. For both of these reasons, the *rev* gene was still included in third-generation vector systems, but was moved to its own separate plasmid. Doing so would theoretically decrease the chances of RCL generation and thus improve the margin of safety. Separation of viral sequences usually means that a greater number of discrete recombination events need to take place in order to reconstruct an RCL, and of course the statistical probability of this happening decreases markedly as the number of events increases. So, because expression of viral genes of the packaging construct require Rev, relegating Rev to a separate plasmid makes the generation of RCL dependent upon one more recombination event than would have been necessary in second-generation vectors, thus decreasing its likelihood even further.

By the third generation of lentiviral vectors in 1998, the contribution of HIV to the vector had been reduced to a fraction of *cis*-acting sequences and to only three genes, *gag*, *pol*, and *rev*, in the separate packaging constructs, compared with the nine genes necessary for the *in vivo* replication and pathogenesis of wild-type HIV [71]. Because of their excellent transduction efficiency and very good theoretical biosafety profile, derivatives of these vectors are still what is predominantly in use today. Nonetheless, efforts have continued at optimizing the vector genome, leading to a more loosely defined "fourth generation" of lentivectors that share several features aimed at the further reduction of sequence homology and a decreased reliance on viral genes. The primary trend shared among this latest showing of vector designs is expression independent of Rev. One method of achieving this has been to replace the RRE with heterologous viral sequences known to enhance the export and stability of unspliced RNAs by interacting with cellular factors [72]. The Mason-Pfizer monkey virus constitutive transport element (CTE) is the classical representative of this group but more recently, functional analogs of it such as the posttranscriptional control element (PCE) of the spleen necrosis virus and even the human nuclear protein Sam68 have been suggested as alternatives [73–77]. Although the inclusion of such *cis*-acting sequences has been shown to provide some success in replacing the functions of Rev, all such mechanisms share the same set of problems: All show a decrease in titer compared with systems that use Rev/RRE, and all have an arguably decreased theoretical biosafety profile compared with third-generation vectors because of the elimination of dependency on a separate Rev plasmid for expression of viral genes in the packaging construct. Finally, such systems do nothing to address the sequence homology remaining between Ψ in the transfer vector (which necessarily includes several hundred base pairs of *gag* for proper encapsidation—see Figure 9.7) and *gag* in the packaging vector.

In order to address these problems, another, very different, approach termed "codon optimization" was introduced in 2000. Codon optimization consists of introducing a mutation in the wobble codon position into the majority of codons in the *gag–pol* packaging construct, preserving the primary amino acid sequence but effectively eliminating any areas of homology with naturally occurring HIV-1 *gag–pol* sequences [78]. This heavily modified sequence eliminated the problem of DNA homology between portions of *gag* in the packaging construct and transfer vector.* It also had the added benefit of inactivating the INS within *gag–pol*, which made expression of the packaging construct completely independent of Rev. Kotsopoulou et al. were also able to show that deletion of several hundred nucleotides of the remaining *gag* sequence surrounding Ψ in the transfer vector removed enough INS to allow efficient export independent of the RRE while keeping enough of Ψ intact to allow efficient packaging. The end result was a true fourth-generation vector genome that eliminated all significant remaining areas of sequence homology and whose titers were only fivefold lower than third-generation systems based upon wild-type HIV sequence that utilizes Rev and the RRE. Most investigtors, especially those that are

---

* The transfer vector retained w.t. *gag* coding sequence.

laboratory-based, are sufficiently comfortable with the safety and efficacy of third-generation vectors and thus do not use those of the fourth generation.

9.3.2 Optimization of the Transfer Vector

Until the introduction of third-generation lentiviral vectors, most improvements in vector systems had focused on the packaging construct. From about 1998 onwards, optimization of the transfer vector itself also became an important priority beginning, as usual, with a focus on biosafety issues.

Unlike packaging constructs, alterations of the transfer vector always concern *cis*-acting elements, since the no *trans*-acting genes are present. Initial attempts to improve the safety of transfer vectors were carried out by deleting nonnecessary viral sequences or replacing them with exogenous *cis*-acting genetic elements [72]. First-generation transfer vectors were already rather minimal, consisting of only the viral LTRs, tRNA primer binding site, packaging signal (including several hundred base pair of *gag*), donor and acceptor splice sites, polypurine tracts, and RRE (part of the *env* ORF). Attempts have been made since then to define exactly the extent to which each of these sequences is necessary. One of the best examples of this is the packaging signal, whose exact boundaries are important because they unavoidably overlap with the *gag* ORF (providing an unfortunate sequence homology with the packaging vector) [6]. When possible, remaining viral sequences were attenuated, a measure that is exemplified by the frameshifting of residual *gag* sequence in the first-generation transfer vectors used by Naldini et al. [79]. Elimination of viral sequences from the transgene served the important purpose of reducing sequence homology, but also created space that could be filled with larger transgene "payloads." In a minimal transfer vector, the packaging capacity of lentiviral vectors is generally considered to be 5–8 kb; however, unlike AAV or adenovirus where the limit is strict, the packaging efficiency is inversely proportional to the insert size [80].

Drawing from the long experience of the retroviral vector field, it was obvious to those pondering the safety features of lentiviral transfer vectors that even after minimizing viral *cis* sequences, those viral sequences that did remain would have some sort of biological activity after integration into target cells and could still be conceivably mobilized, especially by a virus that provided the requisite trans-acting sequences. The most ideal situation would of course be to selectively introduce the gene of interest into a target cell (along with an active promoter to ensure that it is transcribed) without any other sequences. The prospect of how to do this posed yet another engineering question whose answer would have a significant impact on vector design. Luckily, vectors created in response to this problem had already existed for many years in the closely related retroviral vector field [81,82]: The so-called self-inactivating vectors (SIN) cleverly take advantage of the viral replicative cycle (Figure 9.8). The U3 region is known to contain the TATA box, transcription factor binding sites, and other sequences important for the promotion of transcription. This is why the U3 region of the 5′ LTR was replaced by

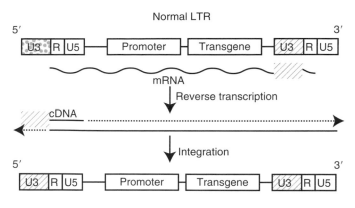

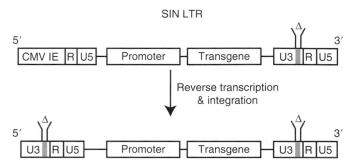

FIGURE 9.8 Safety modifications of the transfer vector. Because the 5′ U3 region acts as a promoter, it is not part of the mRNA transcript. Therefore, during normal reverse transcription a 5′ U3 must be regenerated from the 3′ U3. Thus, both U3 regions of the integrated vector will reflect the 3′ U3 of the original transfer vector. A SIN vector is created by introducing a deletion in the promotor/enhancer region of U3 (Δ) which is duplicated during reverse transcription to the 5′ LTR, thereby inactivating it.

heterologous promoters beginning with third-generation vectors, as discussed above. SIN vectors are created by deleting the 3′ U3 region of the viral LTR (except for a small region termed ATT needed for integration [83]). During the normal process of reverse transcription, this deletion is duplicated and transferred to the 5′ LTR so that upon integration of the viral cDNA, both LTRs are inactivated. This also means that the integrated form of the transfer vector lacks the heterologous promoter that is found in the 5′ LTR of its plasmid and RNA precursors.

In the retroviral vector field, SIN vectors had been plagued by poor titers compared to conventional vectors. When adopted by the lentiviral vector field in 1998 [71,84–86], SIN vectors seemed to be of equivalent if not greater utility compared to conventional vectors. One of the initial constructs had no significant decrease in titer and was able to transduce more cell types in greater numbers than a comparable non-SIN vector when injected into the subretinal space of rats [85]. This observation has led to the recognition that in addition to providing increased biosafety, SIN vectors have the added benefit of reducing promoter interference by inactivating the LTR. Less promoter interference meant that transgene expression could be driven more efficiently by the heterologous promoter included in the transfer vector, thus reducing the variability of expression between tissues caused by the activity of the LTR in some cell types but not others.

The adaptation and successful use of basic SIN constructs have been by far the most important improvement to the biosafety of transfer vectors. However, some of their putative ability to improve biosafety has recently come under question. Whereas it was once thought that these vectors are virtually impossible to mobilize once introduced into a target cell, even if the target is infected with wild-typed, replication-competent HIV [87], more recent work suggests that full-length transcripts competent for encapsidation, transduction, and integration are made at low levels even from an SIN LTR [88]. This may be due to the presence of transcription factor binding sites within the 5' untranslated "leader" region between the LTR and the truncated *gag* ORF of the transfer vector. Additionally, although it was initially thought that the risk of triggering cellular oncogenes by the enhancer activity of the LTR is diminished in SIN vectors [85], more recent studies have found that by interfering with normal polyA signals, SIN LTRs allow considerably greater 3' transcriptional readthrough than wild-type LTRs. In fact, the level of read-through is similar to that of MLV-based vectors [89,90]. Whether that would promote expression of oncogenes found near the vector integration site is not known.

Of course, there are likely to be solutions for all of these problems. It has already been shown that the inclusion of a heterologous poly(A) signal in the 3' U5 region of SIN LTRs can terminate transcription efficiently and prevent 3' read-through [86,89] and that selective mutation of certain sites in the 5' leader region significantly reduces the production of full-length transcripts from the SIN LTR [88]. Further modifications of the transfer vector, particularly to reduce the likelihood of vector mobilization, are also foreseeable. For example, it has been shown that MLV has a propensity to delete large direct sequence repeats during the process of reverse transcription, and that if a transfer vector is constructed with direct repeats flanking its packaging signal, the packaging signal is efficiently deleted following transduction. This property has been exploited in MLV to create the so-called Ψ⁻ vectors wherein nearly 90% of proviruses show a deleted packaging signal and mobilization of the vector in the presence of a helper virus is subsequently reduced 28,000-fold [91]. It has also been shown that the frequency of direct repeat deletions in HIV is very similar to that of MLV [92], so it seems likely that such strategies will eventually be employed in the creation of even safer lentiviral transfer vectors. In the meantime, however, the theoretical biosafety of this platform has progressed to such an apparently high degree that newer developments in the evolution of transfer vectors have focused less on safety concerns and more on improving performance, flexibility, and vector capabilities.

As with biosafety modifications, optimizations of the performance and capabilities of transfer vectors have proceeded through the modification of *cis*-acting sequences (Figure 9.9). Generally, *cis*-acting sequences have been added to the transfer vector in response to challenges posed by specific gene therapy-related issues. A good early example of one of these problems is the difficulty that was experienced in achieving high levels of transgene expression even from tissues that had

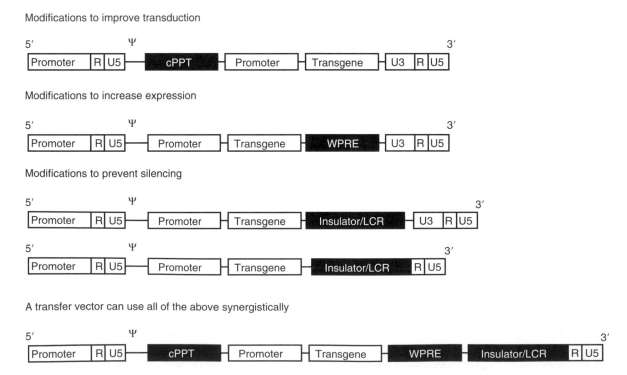

FIGURE 9.9 Modifications that enhance transfer vector performance. cPPT, central polypurine tract; WPRE, woodchuck hepatitis virus posttranscriptional regulatory element; LCR, locus-control region; "insulator," an example would be the HS4 core from the β-globin LCR.

been successfully transduced. Since it was well known that many steps, both transcriptional and posttranscriptional, are involved in the regulation of gene expression, investigators reasoned that the simplest point at which to intercede may therefore be at the posttranscriptional step. This line of reasoning led to the introduction and subsequent widespread inclusion into vector design of the woodchuck hepatitis virus posttranscriptional regulatory element (WPRE), a *cis*-acting RNA element shown to increase the level of transgene expression by two- to fivefold over non-WPRE-containing vectors [93]. It has been postulated to accomplish this by several mechanisms including mRNA stabilization, an interaction with the nuclear export machinery, and enhanced efficiency of transcriptional termination or polyA [94]. In the past several years, concerns about possible tumorigenic effects of this hepadnavirus-derived element [95] have led to the development of an optimized version with truncated viral ORFs and deleted promoter regions [94]. Moreover, experimentation with *cis*-acting elements such as the spleen necrosis virus PCE, cytomegalovirus immediate-early enhancer (CMV IE), murine stem cell leukemia virus LTR, and others continues to yield novel combinations that, when incorporated into the transfer vector, produce even greater yields of transgene expression [77,96].

In the field of retroviral vectors, it is well-known that even achieving efficient transduction and high levels of transgene expression are not always enough—often transgenes are eventually subject to the phenomenon of transcriptional silencing. This problem also exists in the lentiviral field, although to a lesser degree because, perhaps, of a lower CpG dinucleotide content in the HIV genome [97]. The widespread use of SIN vectors has largely abrogated the issue of silencing by deleting those elements within the LTR thought to be involved in mediating these effects. Nonetheless, when silencing remains a particular problem it can often be modulated by the addition of certain positive regulatory elements to the transfer vector [72]. Among these are the so-called DNAse hypersensitive regions which act in *cis* to prevent nucleosome formation around themselves. Thus, inclusion of such a region in the transfer vector can cause chromatin surrounding the transgene to be more "open" and accessible to transcription factors, and less prone to histone-dependent silencing. The classic example of this is incorporation of portions of the β-globin locus control region (LCR) to prevent silencing of the β-globin transgene gene in cells of the hematopoietic lineage [98]. Conceptually similar to but mechanistically different from the LCR are "chromatin insulators," which protect gene expression from neighboring enhancers or silencers [99]. The best characterized of these is the HS4 core from the β-globin LCR, which has been shown to facilitate persistant transgene expression in human hematopoietic cells [100]. S/MARs (scaffold/MA attachment regions) are DNA sequences that mediate attachment of chromatin loops to proteinacious MA or scaffold proteins of the nuclear envelope. It is thought that by doing so, they may make nearby chromatin more accessible and therefore act as insulator elements if present in the vicinity of transgenes. Adding S/MARs to lentivectors has so

far resulted in increased liver and hepatocyte transduction efficiency [101,102] and an increase in the duration time of transgene expression in human HSCs [100].

Other approaches to improving the effectiveness of transfer vectors, instead of attempting to improve transgene expression, focus on the earlier step of transduction. After it was realized that a DNA flap structure formed during reverse transcription was important for nuclear translocation [49], some investigators reasoned that this could be a rate-limiting step in lentiviral transduction. They therefore attempted to rescue the efficiency of this process by the inclusion of a *cis*-acting element, termed the central polypurine tract (cPPT). The cPPT is a portion of the HIV-1 *pol* ORF that coordinates in *cis* a central strand displacement event that occurs during reverse transcription and results in a 99-nucleotide plus-strand overlap known as the central DNA flap (for a detailed review of this process, see [103]). This central, three-strand flap mediates a late step involved in the nuclear import of HIV's genome (although it is not essential). During the early development of HIV-based vectors, this homologous sequence was not included in the transfer vector because its importance was not appreciated. Recent reports, however, have found that if incorporated in any of a variety of positions in the transfer vector [104], it can confer markedly improved transduction efficiency in certain cell types, notably HSCs [105–107]. Both the WPRE and cPPT have been shown to be quite potent in enhancing titer or transduction effiency in some situations, but are ineffective in others [45,55,108]. Regardless, the combination of these two elements is often present in modern transfer vectors (especially since their inclusion is not detrimental) and, when used in conjunction with a third- or fourth-generation packaging construct, has been referred to as an "advanced generation" vector [109].

9.3.2.1 Multicistronic Expression

The features discussed above and others have created such a multiplicity of design options that engineering a lentiviral transfer vector with efficient transduction and transgene expression is now quite common and feasible in most situations. Thus, attention has been turned towards other practical aspects of transfer vector design, including the ability to express more than one transgene. Vectors designed for this purpose are deemed "multicistronic," and such designs have improved the flexibility and capabilities of the vectors considerably. For instance, a bicistronic vector may be used to selectively enrich for a population of transduced cells *in vitro* by encoding both a transgene of interest and a selectable resistance marker, such as *hygro*. When a population of cells is then exposed to hygromycin, only transduced cells actively producing the transgene will survive, allowing them to be studied in isolation. The most common method of achieving this type of bicistronic expression is by inserting an internal ribosome entry site (IRES) between transgenes. Transcription of both genes will be initiated by the heterologous promoter included in the expression cassette. At the stage of protein synthesis, however, the 5′ mRNA cap will drive translation of the upstream transgene, while translation of the downstream

ORF will begin after ribosome entry at the IRES. More than two transgenes may be expressed by using multiple, distinct IRESs [110]. In a striking example of this principle, three distinct catecholaminergic enzymes from one expression cassette were encoded on a single vector to successfully intervene in a rat model of Parkinson's disease [111].

Designs utilizing an IRES between transgenes are by far the most commonly used for multicistronic expression. However, it is an unfortunate truth that transgenes downstream from the IRES may suffer from decreased expression relative to upstream transgenes [112]. In response to this, a recent approach to multicistronic expression vectors has been engineered that involves the use of a synthetic, bidirectional promoter situated between two transgenes in opposite orientations to drive the transcription of both. In this `design, an efficient promoter such as that of the phosphoglycerate kinase gene (PGK) initially recruits RNAP to the transcription site, and then flanking minimal core promoters, such as minCMV, direct RNAP either upstream or downstream to achieve expression of both genes [113,114] (Figure 9.10).

Following nature's cues, splice sites have also been used to express more than one transgene. In one construct two transgenes are separated by the RRE; the upstream transgene is expressed in unspliced mRNAs and the downstream transgene is expressed when mRNAs are spliced. Variable deletions of the RRE can alter the balance between expression of the two genes by fostering either more (longer RRE) or less (shorter RRE) nuclear export of unspliced mRNA [115].

Finally, "pseudo-multicistronic" expression can be realized by expressing a hybrid fusion protein to carry out more than one function [116–118]. Many of these multi-transgene expression constructs have also been employed in retroviral vector design.

9.3.3 SPECIALIZED ASPECTS OF VECTOR DESIGN

9.3.3.1 Pseudotyping

As previously discussed, a major step towards clinical use of lentiviral vectors came when they were first pseudotyped with VSV-G [41]. VSV-G imparted vector particles with increased stability (allowing concentration by centrifugation and some freeze-thawing), and a very wide target cell tropism (possibly pan-tropism). Because of these properties, VSV-G has been used in making the vast majority of vector particles to date. However, VSV-G is not the "perfect" pseudotype for every application. To begin with, the VSV-G protein is toxic to producer cells (by causing membrane fusion) and therefore sustained production of VSV-G-pseudotyped particles poses a challenge (discussed later). Moreover, the broad tropism of VSV-G, though valuable, can be an impediment to progress when targeted transduction of specific tissues is required.

In response to these obstacles, a number of other pseudotyping options have been explored, and HIV-based vectors have proven amenable to pseudotyping with a remarkably wide variety of heterologous viral glycoproteins and envelopes while still retaining their basic functionality. In general, new pseudotypes are created either to provide a specific tissue tropism or delivery method, look for a nontoxic alternative to VSV-G, provide higher titers, or simply explore what novel properties such pseudotypes might possess. Some of the more intriguing examples include Ebola-glycoprotein pseudotyped

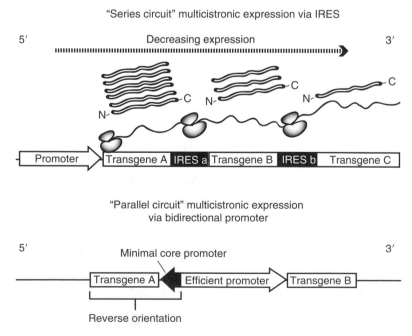

FIGURE 9.10 Methods for achieving multiple transgene expression from a single transfer vector. mRNA levels (and consequently amount of protein) of a transgene may decrease after each IRES. A bidirectional promoter may provide more equal levels of protein expression for two transgenes.

vectors for delivery to airway epithelium, and rabies glycoprotein-pseudotyped vectors for retrograde transport of the vector genome from the axons of neurons to the nucleus [119]. This latter example raises the possibility of introducing genes into the easily accessible muscle of a clinical subject and transducing the nuclei of neurons in the not-easily-accessible spinal cord, with prospects for treating motor neuron diseases [55,120].

Every glycoprotein confers a slightly different set of qualities or properties on the vector system and so each pseudotype may have its own potential niche, either in research or therapeutics. Table 9.1 below summarizes some of the glycoproteins that have been employed to date, along with their titers, tropisms, and potential uses.

9.3.3.2   Vector Targeting

There are a variety of circumstances in which it is easy to appreciate the value of a vector that is targeted to certain cell types. The most prominent of these is, of course, *in vivo* gene therapy, where a simplified, idealized scenario would involve a single, parenteral administration of a dose of therapeutic vector to permanently modify a disease state involving a specific tissue, organ, or structure. In such an ideal scenario, vector targeting would be necessary both to avoid depletion of limited numbers of viral particles and to specifically transduce certain cell types. But how to best accomplish this valuable goal?

One obvious approach to this problem is to directly inject the vector into the tissue of interest. This primitive strategy has been demonstrated to be effective in certain animal models, notably by stereotactic injection of vectors into the substantia nigra [111] and subretinal space [149] of rats, and may prove to be useful for certain localized *in vivo* gene therapy applications in man. However, true vector containment using this approach is uncertain and this tactic also falls short of the idealized scenario described above, wherein any condition ought to be treatable by a limited number of injections using a nonspecific site, such as an easily accessible vein. More elegant methods of vector targeting involve modification of the vector itself rather than modification of the delivery method, and are generally aimed at creating vectors that show gene expression limited to specific, selectable tissues under the conditions described above.

There are two general paradigms for achieving this type of vector targeting, and the first of these to be attempted is also the most intuitive: altering the surface of virion particles so that they specifically interact with certain cell types but not others. The goal of this type of modification is best exemplified by certain naturally occurring virus–host interactions, with HIV-1 itself being a prime example. The envelope glycoprotein of HIV-1, gp160, is known to initially and specifically interact with the cellular surface protein CD4, such that only cells bearing surface CD4 molecules (and co-receptor) can be infected by the virus. This type of specificity is exactly what is desired for gene therapy applications and so from very early on, vector makers have turned to the use of pseudotyping as a mechanism to confer particular tropsims on vector particles that would be suitable for *in vivo* targeting.

Pseudotyping confers tissue tropism via interactions between components of the chosen viral envelope and their counterpart elements on cell surfaces. The recognition of this feature in HIV-1, as mentioned above, prompted several studies to suggest the potential use of gp160 "pseudotyped" particles as CD4+ cell-targeted vectors for anti-HIV therapy [150,151]. Since lentiviral vectors have shown themselves to be permissive to pseudotyping with a wide range of viral glycoproteins, it was naturally hoped that different, heterologous pseudotypes would provide the same kind of specificity for other desirable tissues that gp160 was known to provide for CD4 positive cells. Unfortunately, the type of specificity seen with HIV gp160 seems generally to be the exception rather than the rule. Pseudotyping lentiviral vector particles with heterologous envelopes nearly always produces an altered range of particle tropism, with range being the operative word. It is true that some pseudotypes are vastly superior for transducing certain cells types: The Ebola Zaire (EboZ) glycoprotein has shown clear superiority to other glycoproteins in transducing apical airway epithelia, for instance [138]. However, it is almost always the case that those heterologous envelopes that afford high titers and stable vector particles (necessary features for therapeutic utility) do not confer any meaningful level of specific tissue tropism in the context of our idealized conditions. This holds true for EboZ, which has been shown to also transduce liver, heart, and muscle tissues [152]. The ultimate example is VSV-G, whose consistently high titer and particle stability have helped to make it the standard bearer against which most other pseudotypes are compared but whose extraordinarily wide tropism has made it very non-useful in targeted transductions *in vivo*.

For all of these reasons, straightforward heterologous pseudotyping has failed to provide a uniform targeting solution that would work within the context of our ideal scenario. This is not to say that pseudotyping as a targeting mechanism will not be useful in the future of gene therapy; however, it may be that the ideal use for many pseudotypes will be in conjunction with specialized delivery methods to a localized space where selection among only a few cell types is necessary. An example of this would be a localized injection into the subretinal space for the treatment of retinitis pigmentosa: If VSV-G is used as the pseudotype, both photoreceptors and cells of the retinal pigment epithelium (RPE) are transduced, whereas if Mokola virus glycoprotein is used, transduction is limited to the RPE [153]. Clearly, neither pseudotype would offer an effective targeting solution if administered via peripheral vein. However, if injected in the subretinal space where only two cell types are directly available, Mokola pseudotypes can be used to selectively treat retinitis pigmentosa due to defects in the RPE, whereas VSV-G might be better for retinitis pigmentosa due to defects in the photoreceptors themselves.

More recently, some investigators are continuing to work within the same paradigm, but have shifted their efforts away from simple pseudotyping and towards engineering viral envelope glycoproteins with modifications designed to enhance their targeting abilities. Initial efforts towards this

TABLE 9.1

Properties of Lentiviral Vectors Pseudotyped with Various Heterologous Viral Glycoproteins

Family	Genus	Species/Envelope	Vector	Comments/Titer (IU/mL)	References
Rhabdoviridae	Vesiculovirus	VSV-G	HIV-1 HIV-2 FIV EIAV SIV BIV JDV CAEV	Very wide tropism. Particles durable against high-speed centrifugation. Causes syncytium/cytotoxicity ito producer cells. Susceptible to complement-mediated degradation which can be minimized by PEGylation Titer: 10^5–10^7	[41,125–132]
	Lyssavirus	Rabies	HIV-1	Rabies confers retrograde transport in neuronal axons	[122,123,133,134]
		Mokola	EIAV	Mokola selectively transduces RPE upon subretinal injection	
			SIV	Titer: 2×10^6	
Arenaviridae	Arenavirus	Lymphocytic choriomeningitis virus (LCMV)	HIV-1 FIV SIV	Low toxicity, may be best for pancreatic islet cells	[134–136]
Togaviridae	Alphavirus	Ross River virus	HIV-1 FIV	Transduces hepatocytes upon systemic administration, transduces glia > neurons on administration to brain. Titers [after concentration]: 1.5×10^8 (FIV), 6×10^7 (HIV-1)	[137,138,139]
		Sindbis virus	HIV-1	pH-dependent endosomal entry. Useful for vector targeting	[140]
		Venezuelan equine encephalitis virus	HIV-1	Broad tropism, pH-dependent endosomal entry, immunity distinct from other alphaviruses Titer: 10^6	[141]
Filoviridae	Filovirus	Ebola Reston (EboR)	HIV-1	Efficiently transduces airway epithelium (EboZ > EboR)	[142,143]
		EboZ	HIV-1		
		Marburg	FIV		
Retroviridae	Betaretrovirus	Jaagsiekte sheep virus (JSRV)	FIV	May be more specific to type II cells than other airway epithelial cell types	[144,145]
			HIV-1	Titer: $>10^8$ (after concentration)	
	Gammaretrovirus	Moloney murine leukemia virus 4070 envelope	HIV-1 SIV	Able to transduce most cells Titer: 1×10^5	[41,59,134,146]
		Feline endogenous retrovirus (RD114)	HIV-1 SIV	More efficient and less toxic than VSV-G in cells of the hematopoietic system	[134,147]
Coronaviridae	Coronavirus	SARS-CoV	HIV-1	Transduces airway epithelium	[148]
Orthomyxoviridae	Influenzavirus D	Influenza virus hemagglutinin	HIV-1	Transduces airway epithelium Titer: 10^6	[149]
Baculoviridae	Nucleopolyhedrovirus	Autographa californica multiple nucleopolyhedro virus (AcMNPV)–GP64	HIV-1	Poor transducer of hematopoietic cells. Efficiently transduces airway epithelium	[149–151]
			FIV	Titers 10^7–10^9 (FIV)	
Bunyaviridae	Hantavirus	Hantavirus (HTNV)	HIV-1	Enhanced transduction of vascular cells	[152]

end involved the modification of glycoproteins to include a heterologous polypeptide or ligand, which could then bind to an appropriate target on cell surfaces, even if that cell did not express the original receptor for the particular glycoprotein used. One obvious possibility for targeting with this type of system is to alter a surface glycoprotein, for example that of amphotropic MLV, to include a single-chain antibody against a specified target such as a tumor cell antigen. All vector particles bearing the altered surface glycoprotein would then have a targeting specificity for tumor cells, because of the antibody now displayed on the virion surface.

This intuitive concept was initially tested on retroviral and lentiviral vectors using retroviral glycoproteins. However, efforts to this end met with little success in part because the fusogenic activity of retroviral envelope glycoproteins, which is crucial for cellular entry of the virion after it has bound its target, appears to be largely dependent upon intrinsic properties of the specific retroviral receptor, which is bypassed by using a heterologous peptide or ligand for targeting instead of the unmodified retroviral glycoprotein (for an in-depth review of these topics, see Ref. [154]). Thus, more recently attention has shifted away from using retroviral envelope glycoproteins and towards other heterologous glycoproteins that are known to initiate virion-cellular fusion through a mechanism independent of interaction with the virion receptor. The two best characterized such glycoproteins are the HA (hemagglutinin) glycoprotein of influenza virus and the E1/E2 glycoproteins of Sindbis virus (E2 is responsible for receptor binding and E1 mediates cell fusion). Both of these glycoproteins have a binding domain that binds to a non cell-type specific receptor, such as sialic acid, which is not involved in cell fusion. Instead, after binding to the cell surface the virion is taken up by the cell in an endocytic vesicle. Once activated by acidic pH inside the endosome, a second component of the viral glycoprotein, termed the *fusogen*, acts to mediate virion–vesicular fusion and entry of the viral contents into the cytoplasm [154].

This separation of the functions of binding and fusion (in contrast to retroviral glycoproteins where both cell binding and fusion depend upon specific interactions between the envelope glycoprotein and the cellular receptor) has recently allowed the creation of a number of vector designs that have shown successful targeting *in vivo*. Examples of this include the use of Sindbis-pseudotyped lentiviral vectors whose E2 glycoproteins have been modified to encode single-chain monoclonal antibodies. When delivered via the tail vein of a mouse, the vectors have been shown to specifically target and transduce melanoma xenograft cells that display the antigen [155] or to cells expressing the chemokine receptor CCR5 [156]. At the time of this writing, the most recent group to take advantage of this sort of design has opted to include a monoclonal antibody (in this case, αCD20) in the producer cell transfection mix, not within the fusogenic glycoproteins themselves. This is possible since many cell surface proteins are incorporated into viral particles. By completely separating the binding and fusing proteins, the activity of the fusogen was not disturbed, resulting in a greater titer. Using lentiviral vectors based on this system and administered again through the mouse tail vein, Yang and coworkers were able to show specific transduction of xenografted human B cells bearing CD20 [136]. Systems based on such pH-dependent fusogenic pseudotypes combined with a heterologous targeting moiety seem to show great promise for *in vivo* gene transfer, with the only provision being that the targeted cellular receptor must be efficiently endocytosed after virion binding.

In addition to modifying the surface of virion particles, other important advances that have been made in vector targeting may be largely described under a second, distinct paradigm: These efforts are aimed not at controlling which cells are transduced by vector, but instead controlling which transduced cells express the vector's transgene (Figure 9.11).

The classic method for exerting this type of control has been to include tissue specific promoters and enhancers in the transfer vector to drive expression of the transgene. Thus, whereas a standard transfer vector might use the constitutively active CMV promoter to drive transgene expression, a vector optimized to target only liver cells would use a promoter such as the albumin gene promoter (ALB), which is normally only transcriptionally active in hepatocytes that produce albumin. Using this approach, many groups have shown tissue-specific expression even when many different

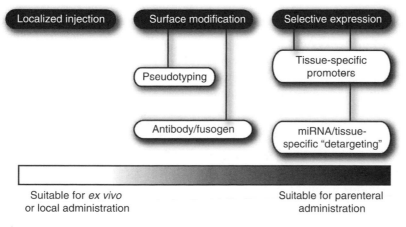

FIGURE 9.11 Vector targeting.

tissue types are known to have been transduced. In fact, the ALB promoter driving a GFP or human factor IX transgene was used to show that transgene expression was essentially limited to hepatocytes after systemic administration via mouse tail vein [157]. One obvious benefit of this type of targeting is high-level, tissue-specific expression of the transgene. However, another important benefit of limiting expression to the target tissue using tissue-specific promoters is that cells of the immune system, even if transduced, will not express the transgene. This is important because direct transgene expression within antigen presenting cells (such as dendritic cells or macrophages) may result in more efficient antigen processing and presentation to the immune system than selective expression in parenchymal cells [158,159]. The consequences of this type of efficient immune recognition of the transgene are often a swift immune response aimed at eliminating transgene expression, thus negating any potential therapeutic effect achieved by gene transfer in the first place. After systemic administration into the tail vein of immunocompetent mice of a vector with a liver-specific promoter, expression of Factor IX was limited to hepatocytes and continued for a prolonged period without a significant immune response. When the same experiment was performed using a vector with a CMV promoter, expression occurred in many tissue types including those of the immune system, and a transgene-specific immune response quickly reduced expression of the foreign antigen [160].

Unfortunately, even when tissue-specific promoters are used, neutralizing antibodies against the transgene and immune-mediated vector clearance may still be observed, albeit at a reduced incidence and extent [160,161]. To further counteract this detrimental immune response, Brown and colleagues have capitalized on the recent proliferation of research in the area of miRNAs to create a completely novel lentiviral vector system capable of further downregulating the immune response by specifically "de-targeting" cells of hematopoietic lineage for transgene expression [161]. The mechanism is based on the recognition that miRNAs of the family miR-142 are expressed only in hematopoietic lineages [162] and exploits this fact by incorporating four tandem copies of short (23 bp) sequences (termed miR-T) complementary to members of the miR-142 family into the 3′ UTR of the GFP transgene. Thus, whenever an mRNA transcript of the transgene is made in a hematopoietic cell, miR-142 miRNAs (which are constitutively present in these cells) will bind to the tandem complementary miR-T sequences in the nascent transcript, forming a short segment of double-stranded RNA, which then reduces translation of the mRNA in those cells. In nonhematopoietic cells, by contrast, miR-142 miRNAs are not expressed so that binding to the miR-T sequences never occurs, double-stranded RNA is not formed and transcripts are translated at normal levels. miRNA-mediated restriction of transgene expression in this manner is very robust, vastly reducing transgene expression compared to controls even when greater than 100 copies of the vector genome were present within a target cell. Following intravenous

administration of the vector, many different organs and cell types were transduced but it was always the case that within these organs, no immune cells or other cells of hematopoietic lineage expressed the transgene. This hematopoietic de-targeting prevented immune-mediated vector clearance better than tissue-specific promoters, resulting in more stable, long-term gene expression [161].

Clearly, this new approach offers a powerful mechanism to regulate a transgene and it is not difficult to envision vector systems that utilize both miRNA-mediated immune cell de-targeting along with tissue-specific promoter-based targeting to maximize transgene expression in the tissue of interest and minimize it in immune cells. Indeed, it is conceivable that future vector systems will utilize these elements along with features of the previously discussed envelope modification targeting paradigm to produce vectors that restrict transgene distribution both by selective transduction and selective expression.

9.3.3.3 Regulatable Vectors

Vector targeting systems as described above provide a means of control over which cells express a transgene. However, it is obvious that a prerequisite for almost any case of intervention into the genetic homeostasis of a human being will be control over not only where a transgene is expressed, but also if, when, and to what degree it is expressed [163]. This need to create gene therapy vectors whose products behave more like traditional pharmaceuticals has not gone unnoticed by investigators in the lentiviral vector field, and to date several systems, deemed regulatable or inducible vectors, have been designed and tested that allow conditional expression of a transgene (Figure 9.12).

Although HIV-1 is known to form pools of latently infected cells *in vivo*, it nonetheless is lacking any easily manipulable or well-understood intrinsic mechanism for regulating its gene expression. Therefore, most regulatory systems employed to date in lentiviral vector design have their origins in other organisms, and have been refined through experimentation down to a series of cis-acting DNA elements which have been adapted to function in eukaryotic cells and can be easily transplanted by cloning into any vector. The prototypical example of these regulatory systems is derived from the tetracycline-resistance operon of *Escherichia coli*, in which gene transcription is negatively regulated by the tetracycline repressor (*tet*R) protein. In the presence of tetracycline, *tet*R does not bind to its operators located within the promoter

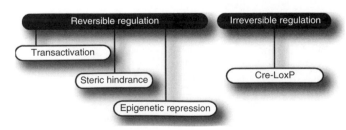

FIGURE 9.12 Regulatable vectors.

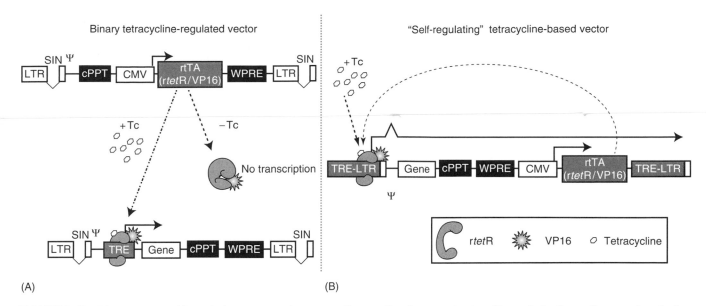

FIGURE 9.13 Binary versus self-regulating vectors using tetracycline-mediated transactivation. Shown is the "tet on" system, but the "tet off" system would look similar. rtTA, reverse tetracycline-controlled transactivator of transcription; rtetR, reverse tetracycline repressor; VP16, transcription activation domain from HSV; TRE, tetracycline-responsive element (a series of repeated tetracycline operator [tetO] sequences). Part A shows a binary system that requires two transfer vectors; B shows a self-regulating, single transfer vector.

region of the operon and thus allows transcription [164]. This operon was first modified to control gene activity in eukaryotic systems by engineering the "Tet off" system by fusing tetR to the herpes simplex virus (HSV)-VP16 transcription activation domain to create tTA, a tetracycline-controlled transactivator that strongly promotes transcription at modified tetracycline operator (tetO) sequences in the absence of tetracycline (Tc) [164]. These tetO seqeunces could be placed upstream of genes of interest, acting as a promoter for their transcription in the presence of tTA. The same group later created a "Tet-on" version of the system by mutagenizing tetR to make it dependent upon Tc for DNA binding (a reversal of its wild-type binding properties), and then fusing this mutant protein to VP16 once again to generate a reverse Tc-controlled transactivator, dubbed "rtTA" [165] (see Figure 9.13 for a schematic representation of this mechanism).

Tetracycline-regulatable promoters rapidly became a well-characterized and widespread tool in molecular biology and, following its successful use in adenoviral and AAV vectors, were incorporated into the lentiviral vector platform by several groups in 2000 [166,167]. By using a single vector that encoded tTA and a tet-responsive element (TRE)* just upstream of the GFP transgene, Kafri et al. were able to show a more than 500-fold increase in transgene expression upon Tc withdrawal *in vitro* and a similarly dramatic difference in rat brains *in vivo*. Importantly, the vectors were produced at titers equivalent to conventional vectors (>10⁹ IU/mL after concentration) [166].

The vector introduced by Kafri et al. was the entry point of small-molecule regulation into the lentiviral vector field and, although it was sound as a proof-of-concept, nonetheless

contained undesirable limitations. Notably, although Tc withdrawal increased transgene expression >500-fold, background expression of transgene was observed even in the presence of Tc. This was thought to be due to the activity of enhancer elements in the U3 region of the HIV LTR. Other investigators have employed a variety of strategies to reduce this activity, including replacing U3 with an alternative promoter [168], inserting an insulator element [169], or using SIN vectors [170], all with some success. Unwanted interactions between the Tc-responsive fusion protein and Tc-responsive promoter element were also found to play a role in driving background transgene expression. These concerns have been largely addressed by the engineering of alternative, synthetic versions of both tTA and rtTA that display improved *in vivo* stability, greater Tc sensitivity, and practically nonexistent background expression of transgene [171]. Concurrently, "second generation" synthetic TREs have been engineered to have optimal promoter activity and better fidelity with tetracycline-responsive fusion proteins. When tested in the setting of a lentiviral vector, these optimized chimeric transactivators and TREs have been shown to provide lower background expression while maintaining reasonable levels of induction [169].

With the inclusion of all of the above refinements, the most modern tetracycline-regulatable vectors based on a Tc-responsive fusion protein can achieve inducible transgene expression repeatedly and reproducibly, with low background, over a range spanning 2 orders of magnitude. Binary vector systems, in which one transfer vector encodes a version of tTA or rtTA and the other encodes a TRE and transgene, appear to achieve the tightest control, lowest background, and greatest inducible range. However, for gene therapy purposes, it is thought that a single vector approach would be advantageous,

---

* Seven copies of the Tet operon binding site flanked downstream by a minimal CMV promoter.

since proper regulation in the two vector system mandates that every target cell be transduced by at least one copy of each vector. Transfer vectors that incorporate both the transgene and all of the elements needed for Tc regulation (e.g., the TRE and a chimeric transactivator) have been called "self-regulating" and, while they are more convenient therapeutically, generally suffer from internal promoter interference and have a lower maximum level of transgene induction (Figure 9.13). Further challenges thus lie ahead in the design of a self-regulating vector that has both undetectable background and acceptable induction levels, although recent work has shown that optimal arrangement of the internal elements of a self-regulating transfer vector can achieve expression levels equivalent to those of a traditional CMV-driven noninducible vector [168].

The use of tTA or rtTA along with a TRE is the prototypical and most common method of exercising Tc regulation over vector transgene expression. However, it is not the only scheme for doing so. Effective regulation in a lentiviral vector has been achieved using tetR (not fused to a transactivator) and tetracycline operator sites placed between the TATA box and the transcription start site of the hCMV major immediate-early promoter [172]. This construct more closely resembles the structure of the original prokaryotic *tet* operon in that transcription is blocked by bound tetR until Tc is added, whereupon the promoter becomes open to the activity of the cell's normal transcriptional machinery.

More recently, Szulc et al. have described a novel twist on the idea of a Tc-regulated fusion protein for transcriptional control of the transfer vector [173]. Instead of using a protein that transactivates transcription from the TRE (a frequent source of "leaky" backgrounds expression), a fusion protein was constructed that consisted of the tetracycline repressor (tetR) and KRAB, a protein domain known to mediate epigenetic silencing and heterochromatin formation. The resultant protein, dubbed tTRKRAB, epigenetically silences 2–3 kb of surrounding sequence when recruited to a tetracycline operator site, thus drastically reducing the possibility of transcription in the off state. Importantly, transgene expression was under tight control in multiple target tissues *in vitro* and *in vivo* (in rat brain), even after several on/off cycles. This system was also shown to accommodate both RNA polymerase (Pol) II promoter-driven transgenes and Pol III-controlled sequences encoding small inhibitory hairpin RNAs (shRNAs). Using such a vector, the expression of an endogenous gene was reduced using an appropriate shRNA, while simultaneously producing a transgene in transduced cells—a scenario with important implications in gene therapy, where it may be necessary to knockdown a mutant endogenous gene product while replacing it with a functional wild-type version.

To date, the majority of regulatable lentiviral vectors have used elements of the *E. coli tet* resistance operon and are structured to respond to the administration of tetracycline or one of its derivatives (commonly doxycyline). Tetracycline-based regulation has probably become predominant because of its extensive characterization and track record of successful use, acceptable range of inducible expression, the relatively small sequence size of its regulatory elements, and the availability, safety, and favorable tissue penetration properties of the tetracycline class of drugs. It is not, however, the only inducible system that has been applied to gene regulation in lentiviral vectors. Others include regulatory systems responsive to the *Drosophila* hormone ecdysone [174], the human hormone progesterone (or its analog mifepristone) [175], and gaseous acetaldehyde [176]. An elegantly designed regulatable vector based on Cre–LoxP recombination has also been demonstrated to be effective [177], but lacks the reversibility of other systems. Each of these alternatives have proven their feasibility, but some are hindered by certain unalterable aspects of their nature. For instance, the genes required for a single self-regulating vector that responds to ecdysone are large enough (around 9 kb simply to express GFP) to raise concern over the packaging efficiency of the vector [174]. Equally, Cre–LoxP-based inducible vectors, while very effective, are not reversible and require a second transduction step by a vector expressing Cre in order to achieve recombination [177]. Finally, none of these alternative systems has yet matched the extensive use that has been enjoyed by Tc-based designs on account of their relative simplicity, ease of construction, and inducibility characteristics.

9.3.3.4 Control over Integration

Retroviruses and lentiviruses have gained popularity as agents of gene transfer largely because features of their natural life cycle correspond very favorably to the aims of gene therapy. These viruses are able to enter cells and permanently modify the target genome via the integration of new genetic material. However, therapeutic use of these viruses has required a series of modifications intended to retain their desirable properties while nullifying their pathogenicity. So far, this has been achieved by providing essentially wild-type proteins (from the packaging vector) to carry out the desirable infection and integration functions of the viral life cycle, while modifying the contents of the viral genome (in the transfer vector) to control expression from the integrated vector. Custom tailoring of the *cis* and *trans* elements of the transfer vector has allowed exclusion of most viral sequence as well as some control over not only what is expressed from the provirus, but also when and where it is expressed, thus markedly increasing biosafety. Host genome integration, however, is an unavoidable step that can reduce biosafety.

Many mechanisms exist to provide control over safety issues stemming from what is inserted, but virtually none over dangers that stem from where it is inserted. The most pressing of these dangers is posed by the very process of integration that is such an attractive feature of the lentiviral technology. Retroviruses, of which the lentiviruses are a subgroup, were first identified by their ability to cause tumours in animals [2]. For those animal retroviruses that do not carry an oncogene, the cause of the cancer is the same in each case: proviral integration that has transcriptionally activated a cellular proto-oncogene. After proto-oncogene activation, uncontrolled cellular proliferation may result. This type of insertional oncogenesis was always recognized as a potential threat of retrovirally mediated gene therapy, but has been the focus of especially intense interest since 2003, when leukemia developed in several of the children with X-linked SCID following *ex vivo* gene transfer of the common γ-chain cDNA to CD34+ hematopoietic stem cells using an MuLV vector [178]. Although insertional oncogenesis has been demonstrated to be much less likely with lentiviral than retroviral vectors [179] (and in the natural setting no lentivirus has been shown to cause malignancy directly), this is a risk that is understandably still taken very seriously in the context of human gene therapy trials. It is also theoretically possible that vector integration will disrupt a tumor suppressor gene, which may lead to cancer in the setting of haploinsufficiency.

To improve control over the process of vector integration and reduce the chances of such a catastrophe occurring, investigators have pursued two general lines of research. The first and most investigated of these consists of various attempts to direct the HIV PIC towards a single or limited number of sites in the host genome that is favorable or non-detrimental to the host (i.e., unlikely to cause insertional oncogenesis or tumor suppressor inactivation). In the absence of any modification, HIV appears to integrate in a quasi-random manner along the entire transcriptional unit, with the only discernable preference being for active genes [53,180]. In order to direct integration to distinct sites, fusions of IN with proteins that have strong, site-specific DNA recognition domains have been used, with limited success. The earliest effort to this end was utilizing a fusion of HIV-1 IN and the λ repressor protein. When the purified λR-IN fusion protein was incubated *in vitro* along with substrate lentiviral DNA and phage λ target DNA, integration was found to occur preferentially in the regions surrounding λ operator sites, suggesting that tethering IN to a target could control site selection [181]. Subsequent studies obtained similar *in vitro* results with fusions of IN and the *E. coli* protein LexA [182], or the DNA binding domain of the zinc-finger protein zif268 [183]. These early fusion protein experiments were as much an effort to understand the biology of lentiviral integration as they were an attempt to control it. Because of this, they were generally performed as *in vitro* chemical reactions between purified proteins and nucleic acids. Additionally, all of these experiments relied upon proteins whose DNA-binding domains recognized fixed sequences that were not necessarily unique or even present within the human genome, potentially limiting their utility. Many of these efforts were doomed because IN has nonspecific DNA-binding activity coupled with the fact that the human genome is a vast target.

The results of the XL-SCID trial, however, have prompted extensions of this work. Specifically, Tan et al. have recently developed fusions of IN and a polydactyl protein (E2C) which contains six zinc-finger domains and recognizes an 18 bp stretch that is unique within the human genome. In addition, each of the six zinc-finger domains can potentially be developed as a modular building block for specific recognition of each of the 64 possible 5'-NNN-3' sequences, thus making any prespecified 18 bp sequence of DNA a potential target for

directed integration [184]. More recently, the same group has incorporated several variants of the IN/E2C fusion into vector particles and used them to infect HeLa cells. They were able to show directed integration near DNA flanking the E2C recognition sequence at a relative specificity that was perhaps 10-fold greater than wild-type IN. Demonstrating some efficacy in living cells was an encouraging step for this technology, however, the vast majority of integration events still took place in a seemingly undirected fashion and packaging of the fusion protein relied upon using a Vpr fusion that resulted in very low titers compared to controls [185].

Another strategy to achieve directed integration takes advantage of growing knowledge regarding the interaction between lentiviral PICs and eukaryotic nuclear proteins. Recent investigations have shown that a number of endogenous nuclear proteins contribute to or are essential for proper HIV integration. One of these proteins, LEDGF/p75, has been shown to interact with both IN and with chromosomal DNA and may play a part in directing HIV integration preferentially towards transcriptional units [54,186]. By fusing LEDGF/p75 to the λ repressor protein, Ciuffi et al. have shown that an unmodified IN can be recruited to integrate in positions closely flanking the λ operator site via what seems to be a protein–protein "tethering" mechanism [187]. This method has yet to be tested in cells, and it achieved directed integration at only very low levels. By relying on the alteration of cofactor nuclear proteins rather than IN itself, it offers a different potential approach for exerting control over integration site selection.

While the ultimate goal of safe and site-specific integration will undoubtedly continue to draw further attention, methods for directing integration to a particular site with high efficiency and specificity are currently in their infancy. It will require a tremendous technical leap forward to fully direct integration in cells. As an alternative to such endeavors, another means to control integration has been explored: foregoing it altogether (Figure 9.14).

By introducing a variety of mutations into the coding sequence of HIV IN, several vectors have been constructed that can still transduce nondividing cells, but are unable to integrate their cDNA into host cell chromatin [188–191]. Typically, IN catalytic activity is reduced by four orders-of-magnitude by these mutants. A certain percentage of transfer vector cDNA becomes circularized and persists as a nuclear episome, termed "extrachromosomal DNA" (E-DNA), which is transcriptionally active, albeit at lower levels than integrated proviral DNA [188]. The E-DNA is stable within a cell's nucleus but since it lacks an origin of replication, will quickly be "diluted" in a population of dividing cells. It is also subject to endonuclease degradation. Hence, nonintegrating vectors seem very well suited to provide either transient expression in a dividing cell population, or relatively long-term expression in quiescent, terminally differentiated cells. As an encouraging proof-of-concept, Yanez-Munoz et al. have used a nonintegrating lentiviral vector to show substantial long-term rescue of several clinically relevant rodent models of retinal degeneration [190]. When equipped with an SV40 origin of replication,

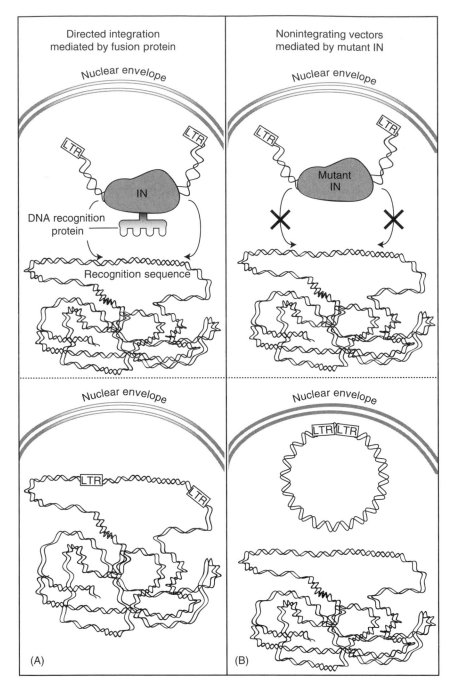

FIGURE 9.14 (A) Site-specific integration mediated by IN fused to a protein with DNA recognition domains. (B) Lentiviral vectors can generate persistent, nonintegrating episomal DNA if packaged with a defective IN.

circular E-DNA persisted and provided long-term expression in dividing cell lines (such as 293T) that also expressed the SV40 large T antigen. Such a vector could be useful in an approach that allows conditional replication of the E-DNA only in cells that provide a required viral or cellular *trans*-acting factor, such as in virally infected cells, or cancer cells expressing transforming proteins [188].

Nonintegrating lentiviral vectors retain most of the promising features of their traditional, integrating counterparts: a large transgene capacity, transduction of nondividing cells, flexibilty in pseudotyping, and a paucity of immune response.

By preserving these features but removing the major safety drawback of insertional oncogenesis, they signify a major step forward. The advantages of nonintegrating vectors extend beyond the obvious, such as diminishing the magnitude of negative positional effects, silencing or extinction of expression, and making it unlikely that any RCL generated would be replication competent [189]. Even a more modest viewpoint would have to concede that while IN-defective vectors may not become standard for every application, they markedly increase the versatility of the lentiviral vector platform while quelling fears related to insertional oncogenesis. And, unlike with schemes for directing integration, excitement surrounding the potential of nonintegrating lentiviral vectors seems all the more well founded by recent demonstrations of their effectiveness *in vivo* [189,190]. It should be emphasized, however, that for uncertain reasons levels of transgene expression are reduced compared to that of integrating vectors.

9.4 PRODUCTION, PURIFICATION, AND TESTING

Section 9.3 was devoted to an in-depth exploration of the various components and design features of modern lentiviral vectors. Section 9.4 discusses the very practical aspects of

how these intangible designs are translated into actual, physical vector particles and how these particles are then purified and prepared for use in either the laboratory or clinical settings.

9.4.1 PRODUCTION OF LENTIVIRAL VECTORS

Vectors based on lentiviruses, like most other viral vectors, are generally produced in one of two ways: (1) by transient transfection of highly transfectable cells (generally HEK 293T) or (2) from stable vector-producing cell lines (VCLs) (Figure 9.15).

Transient transfection is the most commonly used method of lentiviral vector production because it can be performed

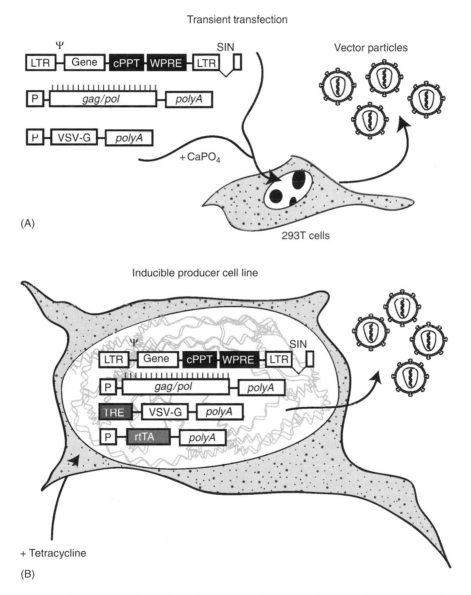

FIGURE 9.15 Two common methods for producing lentiviral vectors. (A) In transient transfection, plasmids coding for each element (transfer vector, packaging construct, envelope) of the vector particle are introduced into naive 293T cells in the presence of a transfection reagent such as calcium phosphate. The cells produce vector for a few days before succumbing to the cytotoxicity of vector components such as PR and VSV-G. (B) In a VCL, cells already contain stable copies of the various vector genes. Production is initiated according to the administration or withdrawal of tetracycline. Although producer cells will die, uninduced cells can be maintained/expanded in tissue culture for future inductions. Only the relevant portions of plasmids are shown. Elements of the tetracycline-based regulation system are shaded in gray. Cell shown in B is larger only for the sake of convenience.

rapidly, inexpensively, and it provides suitable titers for most laboratory applications. In this process, separate plasmids encoding the transfer vector (with its transgene), the packaging construct (containing viral structural proteins and enzymes), an envelope protein (most commonly VSV-G), and any necessary regulatory proteins (usually just Rev) are mixed together and introduced to naive 293T cells in the presence of a reagent capable of mediating transfection. Once inside the cells, vector RNA and protein products are produced and assembled by the producer cells, as directed by the genetic information contained in the DNA plasmids. The process is termed "transient," since it is only expected to produce vector for 2–3 days before the cells begin to suffer from the toxicity of viral proteins.

VCLs, in contrast, are engineered to already contain some or all of the components necessary to produce vector and cells of this lineage will produce vector particles throughout their lifetimes (Figure 9.15). They can be created either via stable transfection or transduction with the necessary packaging, transfer, and envelope sequences. The creation of stable VCLs is an important step in bringing a viral vector-based gene therapy into clinical trials because they offer a nonvariable, scalable source of vector production that can be characterized in detail and scrutinized for safety concerns before their products are used in human trials. In addition, the use of stable producer cell lines eliminates the problem of carrying over plasmid DNA into vector batches as well as the risk of homologous recombination between transfected plasmids (although recombination at other levels to generate RCL is still possible), both of which are concerns with transient transfection.

When lentiviral vectors arrived as a feasible alternative for gene therapy, long experience garnered previously in the retroviral vector field allowed work on development of a lentiviral VCL to begin immediately, and several groups have since reported the creation of stable, well-characterized packaging cell lines for the production of lentiviral vectors [192,193–197]. However, despite the existence of these systems, there has been a significant barrier to their widespread adoption, and transient transfection currently remains the predominant method by which these vectors are produced. This is probably the case for a number of reasons, but the foremost of these are related to the cytotoxicity caused by various viral proteins (notably PR [198]) and the fusogenic VSV-G, which is very commonly used as an envelope protein. To get around this problem, most lentiviral VCLs employ inducible expression cassettes to limit expression of VSV-G and viral proteins to periodic intervals [194,196,199], however, these are generally accompanied by titer loss and unregulated basal activity. Even so, toxicity to producer cells has reduced the life span of most producer cell lines to 4–7 days at maximum, which is hardly longer than the 3 or so days of production obtained with transient transfection and contrasts unfavorably with MLV-based packaging cells that use nontoxic envelopes and can produce vector for 2 weeks or more with one run. Some of the difficulties of sustained lentiviral production appear to be offset if the vector is introduced into the packaging cell line by transduction rather than transfection and one group has

used this technique to create a cell line, "STAR," capable of generating high titer HIV-based lentiviral vectors (>10^7 TU/mL) for at least 3 months [193,200]. However, this cell line uses a gammaretroviral pseudotype, and so does not solve the significant issue of VSV-G toxicity to VCLs.

The relative importance of each of these production methods to the future of the field remains unclear: Traditionally, VCLs have been seen as a prerequisite for the production of clinical-grade vectors, however, the first clinical trial using lentiviral vectors has proceeded using transient transfection as its production method [201,202]. Both methods continue to be refined, with new VCLs frequently being introduced [192,197,203] and transient transfection being honed through various optimizations in reagents and reaction protocols [204–206].

9.4.2 Processing and Purification

After production, a variety of protocols can be implemented to purify the resultant vector in a concentrated form. The most common method by far is ultracentrifugation and resuspension [66,206]. This process generally consists of collecting supernatant from producer cells, clarifying it by an initial filtration step to remove cellular debris, ultracentrifugation to produce a viral pellet, resuspension in an isotonic buffer over several hours, and occasionally further purification (such as by centrifugation through a sucrose cushion) [206]. These maneuvers typically concentrate supernatant 100–1000-fold, taking an unpurified preparation of producer cell supernatant at a titer of 10^6 TU/mL to a final titer of 10^8–10^9 TU/mL in the purified preparation [55,206]. Such a preparation is generally of sufficient titer and purity for most laboratory purposes, including cell culture work and administration to animals. In fact, this ease of preparation has been one of the driving factors behind the widespread adoption of these vectors in laboratory science.

Ultracentrifugation and resuspension, however, is not an ideal technique for every application. To begin with, although VSV-G pseudotyped vector particles are stable under these conditions, most other viral glycoproteins cannot tolerate such high g forces. Additionally, since preparative ultracentrifuge rotors such as Beckman's SW28 only accommodate up to 240 mL, production scale-up to 100 L for clinical use is problematic. Third, ultracentrifugation also leads to pelleting of cellular membranes, aggregated proteins, and other debris, which may be problematic in terms of immune response. For example, expression of VSV-G leads to the formation of tubovesicular structures (TVS) carrying VSV-G along with a variety of plasmid and cellular DNAs. If these TVS are not removed, the DNA may be recognized by toll-like receptors to initiate an innate immune response [207].

For all of these reasons, alternative protocols are very desirable and a multitude of different options are available. Some of these focus intensely on specific areas of processing, such as concentration. To this end, a method has been developed to rapidly concentrate vector particles from large volumes of supernatant by coprecipitation with poly-L-lysine [208], and others have described similar methods [209].

Similar to ultracentrifugation, however, these methods tend to cause precipitation of cellular debris or other impurities that can lead to undesirable or unpredictable immune responses [210]. Therefore, while convenient for concentration, such methods may mandate additional steps for purification [211].

A very different approach to vector concentration and manipulation rests on modification of the vector particles themselves. Copackaging of vector particles with biotinylated proteins from producer cell membranes has been described, allowing vectors to be subsequently manipulated via streptavidin-equipped paramagnetic particles [212,213]. Similarly, using a novel peptide insertion library, Yu and Schaffer have created a variant of the VSV-G protein that retains its functionality yet includes a His_6 tag, allowing facile purification with a Ni-nitrilotriacetic acid affinity chromatography column [214]. Though exciting, such methods are relatively new and have yet to be extensively adopted, thus their current impact is unclear.

Concerns over purity have led to other innovative methods intended to simplify or improve purification and reduce the likelihood of unwanted antigens in the vector preparation. Among these are sucrose-gradient fractionation of concentrated vector, and the use of serum-free media in producer cells [210], along with the technique of tangential flow ultrafiltration [215]. Because of a need to produce large amounts of highly purified vector, protocols intended for clinical use often employ different methods. The first approved clinical trial with a lentiviral vector used a combination of ultrafiltration, benzonase digestion (to remove contaminant DNA) [216], and size-exclusion chromatography [217]. Anion exchange and other column chromatography methods, in which vector is initially adsorbed and then eluted, have also been described [217–219]. These offer hope for further improving scalability by allowing more rapid throughput of unpurified supernatant than can be achieved compared to other methods.

9.4.3 TITER DETERMINATION

As lentiviral vectors are relied upon more and more heavily as laboratory tools and as they begin to make their way into clinical use, it is becoming absolutely essential to have a reliable means of determining the titer or potency of a given preparation of vector. A multitude of different methods to evaluate vector titers have been described and are commonly used [220]; however none of these has the universality required to become the much-needed standard measure of vector potency.

Methods for titering lentiviral vectors can be roughly divided into functional and nonfunctional assays [221]. Nonfunctional assays are generally aimed at characterizing and somehow quantifying the contents of a vector preparation by measuring either (1) p24 (Capsid) antigen ELISA, (2) vector genomic RNA content (typically by quantitative RT-PCR), or (3) endogenous RT activity. While all of these methods provide quantitative information, the values often only poorly correlate with vector transduction efficiency or gene expression. This is probably due to multiple reasons,

but some obvious examples for each case include the facts that (1) Capsid ELISA results will include quantitation of free and aggregated p24 that is not associated with transducing vector particles, (2) vector genomic RNA titers will include partial RNA transcripts or full-length RNA incorporated into defective particles, and (3) RT may be associated with bald or noninfectious particles [221]. In general, these nonfunctional assays tend to overestimate the functional vector titer. As an example, based upon CA amounts, it has been estimated that only about 1% of viral particles present in vector preparations are infectious [222] and that genomic RNA values overestimate functional eGFP titers (TU/mL) by anywhere from 200- to 10,000-fold [222,223]. In addition, nonfunctional assays do not take into account the nature of the vector construct (such as the inclusion of additional cis-acting sequences or cell-specific promoter) or cell type to be transduced, both of which can have a substantial impact on functional titer.

Although poor at predicting gene transfer efficiency, measurements of CA and genomic RNA content may be useful for experimentally normalizing vector preparations [221] such that two batches of vector produced under the same conditions and with a very similar Capsid ELISA value are likely to provide similar functional titers if used under identical transduction conditions. A Capsid ELISA, however, cannot substitute for a functional assay.

By focusing evaluation on steps that follow successful transduction, the so-called functional titering assays may give a more meaningful interpretation of the number of infectious vector particles within a preparation and, in some cases, the capability of said particles to express transgene at a particular level in a particular cell type. Of these, the most commonly used by far is the addition of vector to target cells in serial dilutions, followed by assessment of transgene or marker expression either by quantifying colony formation (in the case of antibiotic resistance reporter genes), staining (for LacZ and other enzymatic reporters), or fluorescence-activated cell sorting (for autofluorescence or cell surface proteins). Titer is measured in the linear dilution range. End-point dilution is not suitable for every vector since it relies upon the presence of a reporter gene (which may not be present). Additionally, target cell type and precise transduction conditions may complicate functional titering [224].

Functional titer posttransduction may also be measured by determining the total number of integrated vector DNA copies by quantitative PCR (qPCR), normalizing to the number of target cells by qPCR of an endogenous cellular gene such as β-actin. However, because of positional variation in transgene expression, vector DNA qPCR assays tend to overestimate functional eGFP titers anywhere from 6- to 60-fold [220,223,225–227].

In efforts to make a quantitative assessment of functional titer in situations where a reporter gene is not included in the transfer vector, several groups have found that transgene mRNA expression correlates very closely to reporter gene expression [221,225]. Thus, posttransduction quantitative RT-PCR (qRT-PCR) of transgene mRNA (often with a primer directed against the WPRE or other nucleotide sequence included in the transcript) seems well suited to measuring the functional titer of transgenes whose protein products are not readily quantifiable, as may often be the case. Such a method does require a normalization control for interpretative purposes.

9.4.4 RCL and Other Safety Issues in Manufacturing

The primary safety concern in the production of lentiviral vectors is the possibility of reconstitution of an RCL either by homologous or nonhomologous recombination between the distributed transfer, packaging, and envelope vector components introduced into producer cells (either at the level of DNA plasmids or RNA), or possibly by nonhomologous recombination of vector components with human endogenous retroviral elements in the producer cell genome. The separation of cis and trans elements of the vector into separate transfer and packaging vector plasmids, the reliance on a separate pseudotyping plasmid for the provision of a functional envelope, the removal of all nonnecessary viral genes, and various other measures such as codon-optimization and truncation of remaining viral ORFs that were discussed in earlier sections make it exceedingly unlikely that a recombination event between plasmids could regenerate a replication-competent entity. Generation of RCL has not yet been reported in the published literature. Nonetheless, careful scrutiny for the presence of any potential RCL is warranted by both the precedent of replication-competent retroviruses being generated in MLV-based production systems and the link to lymphoma in monkeys [63], and the fact that HIV is a well-known and potentially lethal human pathogen. An RCL could have highly unpredictable properties and, depending upon its structure, could be transmitted both horizontally and vertically.

For these reasons, screens for RCL are a mandatory step in the testing of vector preparations to be used in clinical trials. It is generally agreed that any such assay should begin by amplification of potential RCL in a permissive cell line in tissue culture. The main difficulty thus far in settling upon an acceptable test for the presence of RCL has been that since an unintentionally generated RCL is still only a theoretical possibility, its properties are unknown and hence it is difficult to design a detection assay with a suitable positive control. To get around this, a variety of different test methods have been proposed, with most of these focusing on generic aspects of lentiviral biology that would presumably have to be included in any possible variant of an RCL. Among these, perhaps the most likely to emerge as universally acceptable indicators of RCL activity are PCR-based protocols such as the very sensitive product-enhanced reverse transcription assay (PERT) and its derivations [228–230]. The PERT assay uses a defined RNA template (usually phage RNA) as a substrate for RT activity. If RT is present, the predefined RNA template will be reverse-transcribed into cDNA which can then be amplified by quantitative PCR, giving a measure of the amount of RT present in the sample.

Thus, PERT and the similar fluorescence PERT (F-PERT) are only successful in the presence of exogenously provided RT such as would be present in a preparation contaminated with RCL and ought to be capable of detecting any imaginable configuration of RCL since all of these, by definition, would have to contain functional RT. Furthermore, PERT and its derivations have been evaluated by several groups in the context of an assay for RCL [231–233] and shown by some to be at least as sensitive as other available methods for the detection of RCL after amplification in a permissive cell line, even in the presence of competing vector. PERT can be carried out as a single-tube, one-step assay and various improvements to the technique have been described that facilitate ease of handling and reproducibility while reducing background and maintaining sensitivity (<10 virions) [234,235].

Several other assays for RCL exist and are used in conjunction with or in place of the PERT assay. Standard or quantitative PCR or RT-PCR assays to detect VSV-G nucleic acid sequences and recombination between *gag* and the transfer vector (in the region of Ψ) have been described [236,237], as have ELISAs that measure CA amounts during blind serial passaging of tissue culture supernatant, with the expectation being that an RCL will have the structural HIV genes intact [238]. These assays share the common problem that they depend upon the presence of a predicted element within the postulated RCL and may thus lack the requisite broad sensitivity to detect unusual forms that may be produced by recombination of vector elements with genomic retroviral elements endogenous in the producer cells. However, conventional assays can discern recombinants arising solely from components of the vector system, as demonstrated by the detection of as little as one man-made RCL in a background of 2.5×10^8 IU of vector using Capsid ELISA, after a period of biological amplification [238]. With regards to lentiviral vector clinical trials, none of these methods are CLIA-certified or approved by the FDA, and at the moment the FDA is examining the assay methodologies of the involved investigators on a case-by-case basis.

For a summary of the various steps in vector production, processing, and safety testing, see Figure 9.16.

9.5 NON-HIV-BASED LENTIVIRAL VECTORS

HIV-based vectors offer great promise as a platform for gene therapy, but it was recognized very early on that safety concerns would be the primary factor hampering or inhibiting their eventual entry into the clinical arena. Many investigators have therefore sought to retain the desirable traits of HIV-1-based vectors while diminishing their perceived safety risks (namely the generation of an RCL) by turning to other members of the lentivirus family that are closely related but have decreased or absent pathogenicity in man. A number of less-familiar lentiviruses have been engineered into vectors for the transduction of nondividing cells. Most nonhuman lentiviral vectors have been found to behave very similarly to HIV: Generally all of them are capable of VSV-G-pseudotyping and transduction of nondividing cells; however, they differ somewhat in titer and the types of cells they can transduce efficiently.

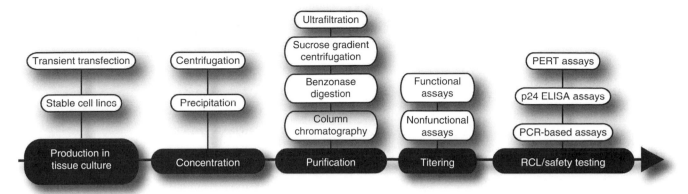

FIGURE 9.16 Generic sequence of processing events that take place after vector production. Some of the more common protocols are shown as variations.

HIV-2, a relative of HIV-1 that induces a milder course of disease and a slower progression to AIDS [239], was among the first of the lentiviruses to be modified as a vector, following in the footsteps of HIV [121,240,241]. At roughly the same time, a proliferation of animal lentiviruses that are not considered to be pathogenic to man were all introduced as vectors within a few years of each another. Vectors based upon FIV [122], EIAV [123], SIV [115,124,242,243], BIV [125], jembrana disease virus (JDV—another bovine lentivirus) [126], and caprine arthritis encephalitis virus (CAEV) [244] all serve as potential alternatives to HIV for gene delivery. All of them have been demonstrated to transduce mitotically inactive cells, and most of them have advanced from a first generation (helper genome with an envelope deletion) through a second generation (elimination of unnecessary accessory proteins with an attempt to precisely define the packaging signal), to a third generation in which all of the accessory and regulatory proteins except for Rev have been eliminated [55,73,127,245–247]. In the case of EIAV and BIV, even codon-optimized vectors have been generated [233,248], and beneficial cis-acting sequences, such as the cPPT/CTS and WPRE, have been included.

Because they were among the first to be introduced and have reasonable VSV-G-pseudotype titers (unlike CAEV [244]), FIV, EIAV, and SIV have become the most frequently employed animal lentiviral vectors. This extensive use has led to the availability of relatively fine-tuned vector designs and their properties, including integration site preferences [249–251], have been increasingly characterized. Each of them has played relatively prominent roles: FIV vectors were one of the first to be used in demonstrating that RNAi-induced transcriptional gene silencing is operational in human cells [252], and an EIAV vector may be the initial one to be used clinically to treat Parkinson's disease (see http://www.oxfordbiomedica.co.uk/news/2006-ob-24.htm). As the best characterized of the animal lentiviral vectors, FIV, EIAV, and SIV all appear to have applications similar to those of HIV-based vectors, with notable caveats. Some reports suggest EIAV vectors may be more subject to silencing

in human cells than are HIV vectors or have lower transgene expression levels, despite equivalent transduction efficiencies [253–255], and FIV vectors may be inferior for transducing human hematopoietic cells [256]. Comparative performance interpretations are complicated by the fact that the best of each system are usually not used, although it makes sense that a well-adapted human pathogen (such as HIV) is likely to be superior in terms of infectability, transgene levels, and persistence in human cells.

Many arguments have been put forth for the further development of vectors based on non-primate lentiviruses, with most of these based on the purported safety advantages of these vectors over those based upon HIV. While it is true that animal lentiviruses such as FIV and EIAV lack the capability to replicate efficiently in human cells, it is not clear whether or not their corresponding vectors are truly advantageous over those based upon HIV. Some argue instead in the favor of HIV vectors, citing the very low probably of generating an RCL with modern vector systems, and the preferability of generating an RCL based on a known pathogen with effective treatments, versus an RCL based on an animal virus whose properties could be very unpredictable if given the ability to spread in human tissues [257]. It remains to be seen which lentivirus, if any, will be the superior template for human gene therapy, but in the meantime animal lentiviral vectors have established themselves as flexible and useful tools for many *in vitro* applications and are not likely to disappear in the near future.

9.6 LENTIVIRAL VECTOR UTILITY

The intrinsic capability to transduce and permanently modify targets at any stage of the cell cycle as well as the wide tropism afforded them by pseudotyping has rapidly made lentiviral vectors a tool of leading importance in an astounding number of different applications, both at the bench and bedside. The list of possible uses of these vectors seems likely to grow as novel technologies like RNAi are discovered and increase in importance. That being said, an exhaustive description of every potential use for these vectors is beyond the scope of this chapter and so preference will be given to those with likely applications in gene therapy, although a few illustrative examples of their importance in basic research will be discussed first.

9.6.1 Lentiviral Vectors in Basic Research

As mentioned earlier, lentiviral vectors were originally conceived of in the late 1980s by Sodroski and collaborators for the purpose of studying viral biology, with a focus on cell binding and entry. Although they have since become widely regarded as promising therapeutic tools, these vectors continue to make important contributions to an increased understanding of lentiviral replication.

It has long been known that the innate tropisms of lentiviruses and their derivative vectors prevents them from efficiently infecting or transducing cells of certain species, with the one of the best known examples being the failure of HIV-based vectors to transduce some species of monkey cells. Recently, Stremlau et al. uncovered one mechanism behind this "restriction" by introducing a cDNA library derived from rhesus macaques into a human cell line [258]. The human cells were then screened for clones that were resistant to transduction with HIV-based vectors encoding eGFP but still permissive to transduction of a similar SIV vector. When such a clone was isolated, it was found to contain a cDNA encoding TRIM5α, and this gene was later confirmed to be a key host cell factor mediating the type of species–species restriction that explains not only the curious inability of HIV vectors to transduce certain monkey cells, but also the broader biological phenomenon as to why some retroviruses are restricted to specific species.

The discovery of TRIM5α demonstrated that lentiviral vectors may have much to contribute towards uncovering host cell factors involved in viral replication. More recently, however, these vectors have begun to be applied widely to the study of mammalian biology, with powerful implications. While these vectors have long been used to permanently modify cells (either to produce a desired alteration or explore novel properties), they are now being combined with the power of genomics and RNAi to do so on a previously unattainable scale. cDNA and shRNA libraries cloned into lentiviral vectors are making it possible to perform genome-wide gain-of-function or loss-of-function screens in essentially any cell type *in vitro* or perhaps even *in vivo*. Furthermore, this technology can be combined with regulatable vector technology to allow for the development of conditional phenotypic screens, especially if the gain or loss-of-function is cytotoxic or lethal [259–262].

Finally, because HIV vectors are resistant to transcriptional silencing, transduction of either oocytes or early embryos offers a truly efficient and relatively simple method for producing transgenic animals, not necessarily limited to rodent species. The applications for this technology range from generation of disease models to the use of animals for production of human therapeutic proteins. Specialized vectors used in transgenesis can produce either permanent loss-of-function models (via mating with Cre-expressing animals) or tunable RNAi knockdown or transgene expression (using

regulatable vectors). This should allow a high degree of temporal manipulation over the transgenic phenotype, which should be useful not only in the study of disease progression but also in aspects of animal physiology [173,263].

9.6.2 Target Tissues and Therapeutic Models (Preclinical Models and Clinical Applications of Lentiviral Vectors)

As with any gene delivery approach, lentiviral vectors are not ideal for all gene transfer applications. However, one of the exciting aspects about them has been the identification of tissues and diseases where the technology seems to have great potential and there are no other current significant alternative approaches. Table 9.2 shows several candidate diseases in which success has been achieved in preclinical animal models; note that it is focused on illnesses that affect the central nervous system (CNS) and the hematopoietic system. The low immunogenicity and persistent expression of lentiviral vectors make them well suited for neuronal cell transduction in the context of chronic diseases, and their ability to transduce quiescent hematopoietic stem cells opens the door to effective, lasting treatment for a variety of genetic and infectious diseases (e.g., sickle cell anemia and AIDS, respectively) that involve multiple lineages of the hematopoietic system. Predictably, future gene therapeutics will likely focus on these promising areas (e.g., www.oxfordbiomedica.co.uk/ and http://www.genethon.com/index.php?id=1&L=1). Because of the characteristics of the vector, there has been comparatively little work in targeting cancerous tumors [155,264–266]. However, there is a significant interest in using lentiviral vectors to "vaccinate" dendritic cells or T cells with tumor antigens, invoking a potentially effective immune response against various cancers [211,267–269]. At this point, a large number and variety of primary cells and tissues have been efficiently transduced by lentiviral vectors, and it would be no more than educated supposition to say which disease and transgene will be targeted next for phase I clinical trials. Below, however, we briefly discuss clinical trials involving these vectors.

9.6.3 Lentiviral Vectors in Clinical Trials

The latest generation of lentiviral vectors contains no open reading frames from the parental viruses and, when integrated into the host genome, should only express the transgene of interest. These advanced vectors take advantage of a number of features such as split-coding and codon optimization to improve safety and increase structural/enzymatic gene expression. Given that HIV is a known human pathogen, there will likely always be safety concerns regarding vectors derived from it. Patient well-being and assessment of risk to benefit is paramount in clinical trials. Recruiting patients already infected with wild-type HIV to evaluate potential gene therapeutic strategies offers a favorable risk-to-benefit ratio for the first clinical trials with these vectors [282], with the caveat that there could be recombination between wild-type virus

TABLE 9.2

Animal Models Showing Disease Improvement after Therapeutic Transduction with Lentiviral Vectors

Disease	Gene	Animal	Method	Correction	Comments	References
Parkinson's disease	GDNF	Rhesus monkey, Rx MPTP	Direct CNS injection	Reversed functional defects, prevented NS degeneration	8 month F/U	[270]
Parkinson's disease	Tyrosine hydroxylase (TH), aromatic amino acid dopa decarboxylase (AADC), GTP cyclohydrolase 1 (CH1)	Rat, Rx 6-OH DA Rhesus monkey PD model	Stereotactic injection to striatum	Consistent, complete reversal of phenotype in chronic macaque model	Three genes for DA synthesis delivered by a single vector. 5 month F/U	[111,271]
Retinitis pigmentosa	cGMP PDE β	Mouse rd	Subretinal injection	Expression confirmed in photoreceptors	6 month F/U	[272]
Metachromatic leukodystrophy	Arylsulfatase A	Mouse KO	Direct CNS injection	Decreased neuropathology and hippocampal learning impairments	Hippocampal neuronal degeneration central to disease	[273]
Mucopoly-saccharidosis type VII	β-glucuronidase	Mouse KO	Direct CNS injection	Reversal of histopathology	Regression in entire brain	[274]
Sickle cell disease	Anti-sickling β-globin	Mouse with sickle globin	Transduction of progenitors	Improved anemia and pathology	Vector included LCR	[275]
Thalassemia	β-globin	Mouse KO	Transduction of progenitors	Improved anemia and RBC morphology	Vector included LCR	[3]
ALS	VEGF	Superoxide dismut-ase (SOD1) (G93A) mouse	Injection into muscle with retrograde transport to spinal cord	Delayed onset, slowed progression of dz. Increased life expectancy ~30%	Effective use of Rabies G pseudotyped retrograde transport	[120]
ALS	RNAi against SOD1	SOD1(G93A) mouse	Injection into muscle with retrograde transport to spinal cord	100% delay in onset of symptoms, 80% extension in survival	Effective use of Rabies G pseudotyped retrograde transport, RNAi	[276]
Alzheimer's disease	siRNA targeting β-secretase (BACE1)	Amyloid precursor protein (APP) transgenic mouse	Direct CNS injection	Reduced amyloid production and behavioral deficits		[277]
Leber congenital amaurosis	RPE65	Rpe65^rd12/rd12 mouse	Subretinal injection	Significant functional improvement on ERG	Nonintegrating vector used	[190]
Cyclic neutropenia	G-CSF	Gray collie	Intramuscular injection	Substantially increased neutrophil count, correction of immunodeficiency		[278]
SCID	ADA	ADA (–/–) mouse	Ex vivo transduction of bone marrow, irradiation, reimplantation	Restored normal immune function, including antibody production	F/U 6 months	[279]
Wiskott–Aldrich syndrome (WAS)	WAS protein (WASP)	WAS (–/–) mouse	Ex vivo transduction of HSCs, irradiation, reimplantation	Significant engraftment of WASP expressing hematopoietic cells		[280]
Exudative age-related macular degeneration (ARMD)	Endostatin or angiostatin	Mouse with laser-induced choroidal neovascularization (CNV)	Subretinal injection	>50% reduction in CNV area. Also reduction in vascular hyperpermeability	12 month F/U	[281]

and vector components. If VSV-G is encoded within such a recombinant expanded host range and germline transmission is possible.

A variety of strategies have been proposed to utilize HIV-based vectors to combat HIV. These are aimed primarily at HSC and their resultant progeny or at T cells directly and are intended to protect them from HIV using approaches such as inducible suicide genes [283], antisense RNA [284], siRNA/shRNA [285], and a multitude of others [286]. In January 2003, the first clinical trial using lentiviral vectors was initiated in the United States with VRX496, an HIV-based vector engineered to conditionally express an antisense RNA against envelope glycoprotein in the presence of regulatory proteins provided by wild-type virus. After establishing clinical-grade production and quality assurance methods [217,237], autologous peripheral CD4+ T cells were bead-activated, transduced *ex vivo* with high-titer vector stocks, and then reinfused into HIV+ patients failing conventional anti-retroviral therapy. The results of the recently published phase I safety and tolerability data indicate that the procedure was well tolerated and follow-up studies over 2 years have not detected any adverse clinical events [202]. Although vector-virus recombinants have not yet been excluded and the vector appears to be safe thus far, it is difficult to draw meaningful conclusions regarding efficacy from such a limited phase I trial.

Importantly, analysis of the patient blood samples after the T cell reinfusion revealed wild-type viral particles housing RNA transcripts of the vector. This process, known as vector "mobilization," may be beneficial here since it presumably allows spread of the vector to other CD4+ T cells, where it may have a protective effect. However, in many other circumstances, vector mobilization would be undesirable and for reasons outlined above vector mobilization or recombination after superinfection with HIV is considered one of the greatest safety concerns surrounding the use of these vectors. For example, the generation in man of an RCL with unpredictable or unwanted properties would pose complex biosafety and ethical issues that may be difficult to address [282].

Other safety issues regarding these vectors trace their origins to the development of leukemia resulting from insertional oncogenesis in four patients treated with γ-retroviral vectors in a gene therapy trial for X-linked SCID (see Refs. [90,178] and [http://www.esgct.org/upload/4th_CaseofLeukemia1.pdf]). Vectors based upon lentiviruses are thought to be less likely to pose this sort of threat because (1) they exhibit less of a propensity than retroviruses to integrate near the 5′ end of genes, where promoters are usually located, (2) lentiviral LTRs exhibit less read-through of the 3′ LTR than MLV-based vectors (although SIN lentiviral vectors do exhibit MLV-like read-through unless a heterologous polyA addition signal is included at their 3′ end) [89], and (3) initial studies comparing the potential for oncogenesis among retro- and lentiviral vectors suggest that lentiviruses do not have the same oncogenic potential as their retroviral cousins [179], although further study is needed.

Overall, the clinical safety issues surrounding lentiviral vectors are real, but very manageable. The excellent safety

and tolerability results of the first lentiviral gene therapy trial will likely open the door for these vectors to be used in applications outside of HIV infection. Indeed, as this chapter goes to press, seven new human gene therapy trials utilizing lentiviral vectors are ongoing or just beginning. The majority of these are for the treatment of late-stage HIV+ patients, but one is for the totally unrelated condition of mucopolysaccharidosis type VII and the other two are aimed towards treating melanoma and lymphoma (http://www.gemcris.od.nih.gov/). Several commercial entities may enter the fray soon with lentiviral vector treatments for Parkinson's disease, WAS, various retinopathies, and perhaps other conditions.

9.7　SUMMARY

Lentiviral vectors are at an exciting stage of development. The realization of their unique properties as a gene transfer agent has led to an explosion in investigation and innovation in this field, with thousands of published reports being generated in the span of roughly a decade. All of this interest has resulted in a large number of increasingly tailored and sophisticated vector designs being available and an ever-increasing knowledge of how effectively these vectors might one day be put to use for very well-defined therapeutic applications. Additionally, the general pervasiveness of these vectors in the last few years has lead to their adoption for powerful new basic science techniques with the potential to accelerate biological research in other fields. Although these vectors may be years away from FDA-licensure, encouraging preliminary results coupled with the growing number of clinical entrants suggests that true clinical applicability may be less than a decade away.

REFERENCES

1. Goff, S., Retroviridae: The retroviruses and their replication, in *Fields Virology*, D. Knipe and P. Howley, Editors. 2001, Lippincott, Williams and Wilkins: Philadelphia. pp. 1871–1939.
2. Vogt, P.K., Historical introduction to the general properties of retroviruses, in *Retroviruses*, J.M. Coffin, S.H. Hughes, and H.E. Varmus, Editors. 1997, CSHL Press: Cold Spring Harbor, NY. pp. 1–26.
3. Moreau-Gaudry, F., et al., High-level erythroid-specific gene expression in primary human and murine hematopoietic cells with self-inactivating lentiviral vectors. *Blood*, 2001. 98(9): 2664–2672.
4. Vogt, V.M., Retroviral virions and genomes, in *Retroviruses*, J.M. Coffin, S.H. Hughes, and H.E. Varmus, Editors. 1997, CSHL Press: Cold Spring Harbor, NY. pp. 27–69.
5. Lever, A., et al., Identification of a sequence required for efficient packaging of human immunodeficiency virus type 1 RNA into virions. *J Virol*, 1989. 63(9): 4085–4087.
6. McBride, M.S. and A.T. Panganiban, The human immunodeficiency virus type 1 encapsidation site is a multipartite RNA element composed of functional hairpin structures. *J Virol*, 1996. 70(5): 2963–2973.
7. Frankel, A.D. and J.A. Young, HIV-1: Fifteen proteins and an RNA. *Annu Rev Biochem*, 1998. 67: 1–25.
8. Herrmann, C.H. and A.P. Rice, Lentivirus Tat proteins specifically associate with a cellular protein kinase, TAK, that

hyperphosphorylates the carboxyl-terminal domain of the large subunit of RNA polymerase II: Candidate for a Tat cofactor. *J Virol*, 1995. 69(3): 1612–1620.

9. Peruzzi, F., The multiple functions of HIV-1 Tat: Proliferation versus apoptosis. *Front Biosci*, 2006. 11: 708–717.

10. Yang, X., et al., TAK, an HIV Tat-associated kinase, is a member of the cyclin-dependent family of protein kinases and is induced by activation of peripheral blood lymphocytes and differentiation of promonocytic cell lines. *Proc Natl Acad Sci U S A*, 1997. 94(23): 12331–12336.

11. Jones, K.A. and B.M. Peterlin, Control of RNA initiation and elongation at the HIV-1 promoter. *Annu Rev Biochem*, 1994. 63: 717–743.

12. Pollard, V.W. and M.H. Malim, The HIV-1 Rev protein. *Annu Rev Microbiol*, 1998. 52: 491–532.

13. Marin, M., et al., HIV-1 Vif protein binds the editing enzyme APOBEC3G and induces its degradation. *Nat Med*, 2003. 9(11): 1398–1403.

14. Harris, R.S. and M.T. Liddament, Retroviral restriction by APOBEC proteins. *Nat Rev Immunol*, 2004. 4(11): 868–877.

15. Sheehy, A.M., N.C. Gaddis, and M.H. Malim, The antiretroviral enzyme APOBEC3G is degraded by the proteasome in response to HIV-1 Vif. *Nat Med*, 2003. 9(11): 1404–1407.

16. Steffens, C.M. and T.J. Hope, Recent advances in the understanding of HIV accessory protein function. *Aids*, 2001. 15 Suppl 5: S21–S26.

17. Kestler, H.W., III, et al., Importance of the nef gene for maintenance of high virus loads and for development of AIDS. *Cell*, 1991. 65(4): 651–662.

18. Chowers, M.Y., et al., Optimal infectivity in vitro of human immunodeficiency virus type 1 requires an intact nef gene. *J Virol*, 1994. 68(5): 2906–2914.

19. Roeth, J.F. and K.L. Collins, Human immunodeficiency virus type 1 Nef: Adapting to intracellular trafficking pathways. *Microbiol Mol Biol Rev*, 2006. 70(2): 548–563.

20. Das, S.R. and S. Jameel, Biology of the HIV Nef protein. *Indian J Med Res*, 2005. 121(4): 315–332.

21. Willey, R.L., et al., Human immunodeficiency virus type 1 Vpu protein induces rapid degradation of CD4. *J Virol*, 1992. 66(12): 7193–7200.

22. Schubert, U., et al., The two biological activities of human immunodeficiency virus type 1 Vpu protein involve two separable structural domains. *J Virol*, 1996. 70(2): 809–819.

23. Hout, D.R., et al., Vpu: A multifunctional protein that enhances the pathogenesis of human immunodeficiency virus type 1. *Curr HIV Res*, 2004. 2(3): 255–270.

24. Neil, S.J., et al., HIV-1 Vpu promotes release and prevents endocytosis of nascent retrovirus particles from the plasma membrane. *PLoS Pathog*, 2006. 2(5): e39.

25. Rucker, E., et al., Vpr and Vpu are important for efficient human immunodeficiency virus type 1 replication and CD4 + T-cell depletion in human lymphoid tissue ex vivo. *J Virol*, 2004. 78(22): 12689–12693.

26. Andersen, J.L. and V. Planelles, The role of Vpr in HIV-1 pathogenesis. *Curr HIV Res*, 2005. 3(1): 43–51.

27. Le Rouzic, E. and S. Benichou, The Vpr protein from HIV-1: Distinct roles along the viral life cycle. *Retrovirology*, 2005. 2: 11.

28. Jowett, J.B.M., et al., The human immunodeficiency virus type 1 vpr gene arrests infected T cells in the $G_2 + M$ phase of the cell cycle. *J Virol*, 1995. 69: 6304–6313.

29. Goh, W.C., et al., HIV-1 Vpr increases viral expression by manipulation of the cell cycle: A mechanism for selection of Vpr in vivo. *Nat Med*, 1998. 4(1): 65–71.

30. Yoshizuka, N., et al., Human immunodeficiency virus type 1 Vpr-dependent cell cycle arrest through a mitogen-activated protein kinase signal transduction pathway. *J Virol*, 2005. 79(17): 11366–11381.

31. Cui, J., et al., The role of Vpr in the regulation of HIV-1 gene expression. *Cell Cycle*, 2006. 5(22): 2626–2638.

32. de Noronha, C.M., et al., Dynamic disruptions in nuclear envelope architecture and integrity induced by HIV-1 Vpr. *Science*, 2001. 294(5544): 1105–1108.

33. Fouchier, R.A., et al., Interaction of the human immunodeficiency virus type 1 Vpr protein with the nuclear pore complex. *J Virol*, 1998. 72(7): 6004–6013.

34. Sherman, M.P., et al., Nucleocytoplasmic shuttling by human immunodeficiency virus type 1 Vpr. *J Virol*, 2001. 75(3): 1522–1532.

35. Sherman, M.P., et al., Insights into the biology of HIV-1 viral protein R. *DNA Cell Biol*, 2002. 21(9): 679–688.

36. Desrosiers, R., Nonhuman lentiviruses, in *Fields Virology*, D. Knipe and P. Howley, Editors. 2001, Lipincott William and Wilkins: Philadelphia. pp. 2095–2121.

37. Chatterji, U., A. de Parseval, and J.H. Elder, Feline immunodeficiency virus OrfA is distinct from other lentivirus transactivators. *J Virol*, 2002. 76(19): 9624–9634.

38. Lewis, P.F. and M. Emerman, Passage through mitosis is required for oncoretroviruses but not for the human immunodeficiency virus. *J Virol*, 1994. 68(1): 510–516.

39. Roe, T., et al., Integration of murine leukemia virus DNA depends on mitosis. *Embo J*, 1993. 12(5): 2099–2108.

40. Lewis, P., M. Hensel, and M. Emerman, Human immunodeficiency virus infection of cells arrested in the cell cycle. *Embo J*, 1992. 11(8): 3053–3058.

41. Naldini, L., et al., In vivo delivery and stable transduction of nondividing cells by a lentiviral vector. *Science*, 1996. 272: 263–267.

42. Freed, T. and A. Martin, HIVs and their replication, in *Fields Virology*, D. Knipe and P. Howley, Editors. 2001, Lippincott, Williams and Wilkin: Philadelphia, PA. pp. 1971–2041.

43. Anderson, J.L. and T.J. Hope, Intracellular trafficking of retroviral vectors: Obstacles and advances. *Gene Ther*, 2005. 12(23): 1667–1678.

44. Bukrinsky, M.I., et al., Active nuclear import of human immunodeficiency virus type 1 preintegration complexes. *Proc Natl Acad Sci U S A*, 1992. 89(14): 6580–6584.

45. Dvorin, J.D. and M.H. Malim, Intracellular trafficking of HIV-1 cores: Journey to the center of the cell. *Curr Top Microbiol Immunol*, 2003. 281: 179–208.

46. von Schwedler, U., R.S. Kornbluth, and D. Trono, The nuclear localization signal of the matrix protein of human immunodeficiency virus type 1 allows the establishment of infection in macrophages and quiescent T lymphocytes. *Proc Natl Acad Sci U S A*, 1994. 91: 6992–6996.

47. Bukrinsky, M.I., et al., A nuclear localization signal within HIV-1 matrix protein that governs infection of non-dividing cells. *Nature*, 1993. 365(6447): 666–669.

48. Gallay, P., et al., HIV-1 infection of nondividing cells through the recognition of integrase by the importin/karyopherin pathway. *Proc Natl Acad Sci U S A*, 1997. 94: 9825–9830.

49. Zennou, V., et al., HIV-1 genome nuclear import is mediated by a central DNA flap. *Cell*, 2000. 101(2): 173–185.

50. Yamashita, M. and M. Emerman, Capsid is a dominant determinant of retrovirus infectivity in nondividing cells. *J Virol*, 2004. 78(11): 5670–5678.

51. Dismuke, D.J. and C. Aiken, Evidence for a functional link between uncoating of the human immunodeficiency virus type

1 core and nuclear import of the viral preintegration complex. *J Virol*, 2006. 80(8): 3712–3720.

52. Liu, H., et al., Integration of human immunodeficiency virus type 1 in untreated infection occurs preferentially within genes. *J Virol*, 2006. 80(15): 7765–7768.

53. Schroder, A.R., et al., HIV-1 integration in the human genome favors active genes and local hotspots. *Cell*, 2002. 110(4): 521–529.

54. Ciuffi, A., et al., A role for LEDGF/p75 in targeting HIV DNA integration. *Nat Med*, 2005. 11(12): 1287–1289.

55. Jolly, D., Lentiviral vectors, in *Gene and Cell Therapy: Therapeutic Mechanisms and Strategies*, N.S. Templeton, Editor. 2004, Marcel Dekker: NY. pp. 131–145.

56. Terwilliger, E.F., et al., Construction and use of a replication-competent human immunodeficiency virus (HIV-1) that expresses the chloramphenicol acetyltransferase enzyme. *Proc Natl Acad Sci U S A*, 1989. 86: 3857–3861.

57. Helseth, E., et al., Rapid complementation assays measuring replicative potential of human immunodeficiency virus type 1 envelope glycoprotein mutants. *J Virol*, 1990. 64(5): 2416–2420.

58. Page, K.A., N.R. Landau, and D.R. Littman, Construction and use of a human immunodeficiency virus vector for analysis of virus infectivity. *J Virol*, 1990. 64(11): 5270–5276.

59. Landau, N.R., K.A. Page, and D.R. Littman, Pseudotyping with human T-cell leukemia virus type I broadens the human immunodeficiency virus host range. *J Virol*, 1991. 65(1): 162–169.

60. Buchschacher, G.L., Jr. and A.T. Panganiban, Human immunodeficiency virus vectors for inducible expression of foreign genes. *J Virol*, 1992. 66(5): 2731–2739.

61. Buchschacher, G.L., Jr., E.O. Freed, and A.T. Panganiban, Cells induced to express a human immunodeficiency virus type 1 envelope gene mutant inhibit the spread of wild-type virus. *Hum Gene Ther*, 1992. 3(4): 391–397.

62. Otto, E., et al., Characterization of a replication-competent retrovirus resulting from recombination of packaging and vector sequences. *Hum Gene Ther*, 1994. 5(5): 567–575.

63. Vanin, E.F., et al., Characterization of replication-competent retroviruses from nonhuman primates with virus-induced T-cell lymphomas and observations regarding the mechanism of oncogenesis. *J Virol*, 1994. 68(7): 4241–4250.

64. Poznansky, M., et al., Gene transfer into human lymphocytes by a defective human immunodeficiency virus type 1 vector. *J Virol*, 1991. 65(1): 532–536.

65. Naldini, L., et al., In vivo gene delivery and stable transduction of nondividing cells by a lentiviral vector. *Science*, 1996. 272(5259): 263–267.

66. Burns, J.C., et al., Vesicular stomatitis virus G glycoprotein pseudotyped retroviral vectors: Concentration to very high titer and efficient gene transfer into mammalian and nonmammalian cells. *Proc Natl Acad Sci U S A*, 1993. 90(17): 8033–8037.

67. Sutton, R.E., Cellular transduction by lentiviral vectors, in *Viral Vectors: Basic Science and Gene Therapy*, A. Cid-Arregui and A. Garcia-Carranca, Editors. 2000, Eaton Publishing: Natick, MA. pp. 485–499.

68. Zufferey, R., et al., Multiply attenuated lentiviral vector achieves efficient gene delivery in vivo. *Nat Biotechnol*, 1997. 15(9): 871–875.

69. Kim, V.N., et al., Minimal requirements for a lentivirus vector based on human immunodeficiency virus type 1. *J Virol*, 1998. 72: 811–816.

70. Naldini, L. and I.M. Verma, Lentiviral vectors. *Adv Virus Res*, 2000. 55: 599–609.

71. Dull, T., et al., A third-generation lentivirus vector with a conditional packaging system. *J Virol*, 1998. 72(11): 8463–8471.

72. Delenda, C., Lentiviral vectors: Optimization of packaging, transduction and gene expression. *J Gene Med*, 2004. 6 Suppl 1: S125–S138.

73. Pandya, S., et al., Development of an Rev-independent, minimal simian immunodeficiency virus-derived vector system. *Hum Gene Ther*, 2001. 12(7): 847–857.

74. Bray, M., et al., A small element from the Mason-Pfizer monkey virus genome makes human immunodeficiency virus type 1 expression and replication Rev-independent. *Proc Natl Acad Sci U S A*, 1994. 91(4): 1256–1260.

75. Reddy, T.R., et al., Inhibition of HIV replication by dominant negative mutants of Sam68, a functional homolog of HIV-1 Rev. *Nat Med*, 1999. 5(6): 635–642.

76. Roberts, T.M. and K. Boris-Lawrie, The 5′ RNA terminus of spleen necrosis virus stimulates translation of nonviral mRNA. *J Virol*, 2000. 74(17): 8111–8118.

77. Yilmaz, A., et al., Coordinate enhancement of transgene transcription and translation in a lentiviral vector. *Retrovirology*, 2006. 3: 13.

78. Kotsopoulou, E., et al., A Rev-independent human immunodeficiency virus type 1 (HIV-1)-based vector that exploits a codon-optimized HIV-1 gag–pol gene. *J Virol*, 2000. 74(10): 4839–4852.

79. Naldini, L., et al., In vivo gene delivery and stable transduction of nondividing cells by a lentiviral vector [see comments]. *Science*, 1996. 272(5259): 263–267.

80. Sinn, P.L., S.L. Sauter, and P.B. McCray, Jr., Gene therapy progress and prospects: Development of improved lentiviral and retroviral vectors—design, biosafety, and production. *Gene Ther*, 2005. 12(14): 1089–1098.

81. Yu, S.F., et al., Self-inactivating retroviral vectors designed for transfer of whole genes into mammalian cells. *Proc Natl Acad Sci U S A*, 1986. 83: 3194–3198.

82. Hwang, J.J., L. Li, and W.F. Anderson, A conditional self-inactivating retrovirus vector that uses a tetracycline-responsive expression system. *J Virol*, 1997. 71(9): 7128–7131.

83. Reicin, A.S., et al., Sequences in the human immunodeficiency virus type 1 U3 region required for in vivo and in vitro integration. *J Virol*, 1995. 69(9): 5904–5907.

84. Zufferey, R., et al., Self-inactivating lentivirus vector for safe and efficient In vivo gene delivery [In Process Citation]. *J Virol*, 1998. 72(12): 9873–9880.

85. Miyoshi, H., et al., Development of a self-inactivating lentivirus vector. *J Virol*, 1998. 72(10): 8150–8157.

86. Iwakuma, T., Y. Cui, and L.J. Chang, Self-inactivating lentiviral vectors with U3 and U5 modifications. *Virology*, 1999. 261(1): 120–132.

87. Bukovsky, A.A., J.P. Song, and L. Naldini, Interaction of human immunodeficiency virus-derived vectors with wild-type virus in transduced cells. *J Virol*, 1999. 73(8): 7087–7092.

88. Logan, A.C., et al., Integrated self-inactivating lentiviral vectors produce full-length genomic transcripts competent for encapsidation and integration. *J Virol*, 2004. 78(16): 8421–8436.

89. Zaiss, A.K., S. Son, and L.J. Chang, RNA 3′ readthrough of oncoretrovirus and lentivirus: Implications for vector safety and efficacy. *J Virol*, 2002. 76(14): 7209–7219.

90. Hacein-Bey-Abina, S., et al., LMO2-associated clonal T cell proliferation in two patients after gene therapy for SCID-X1. *Science*, 2003. 302(5644): 415–419.

91. Delviks, K.A., W.S. Hu, and V.K. Pathak, Psi-vectors: Murine leukemia virus-based self-inactivating and self-activating retroviral vectors. *J Virol*, 1997. 71(8): 6218–6224.

92. An, W. and A. Telesnitsky, Frequency of direct repeat deletion in a human immunodeficiency virus type 1 vector during

reverse transcription in human cells. *Virology*, 2001. 286(2): 475–482.

93. Zufferey, R., et al., Woodchuck hepatitis virus posttranscriptional regulatory element enhances expression of transgenes delivered by retroviral vectors. *J Virol*, 1999. 73(4): 2886–2892.

94. Schambach, A., et al., Woodchuck hepatitis virus posttranscriptional regulatory element deletcd from X protein and promoter sequences enhances retroviral vector titer and expression. *Gene Ther*, 2006. 13(7): 641–645.

95. Kingsman, S.M., K. Mitrophanous, and J.C. Olsen, Potential oncogene activity of the woodchuck hepatitis posttranscriptional regulatory element (WPRE). *Gene Ther*, 2005. 12(1): 3–4.

96. Ramezani, A., T.S. Hawley, and R.G. Hawley, Lentiviral vectors for enhanced gene expression in human hematopoietic cells. *Mol Ther*, 2000. 2(5): 458–469.

97. Berkhout, B. and F.J. van Hemert, The unusual nucleotide content of the HIV RNA genome results in a biased amino acid composition of HIV proteins. *Nucleic Acids Res*, 1994. 22(9): 1705–1711.

98. May, C., et al., Therapeutic haemoglobin synthesis in beta-thalassaemic mice expressing lentivirus-encoded human beta-globin. *Nature*, 2000. 406(6791): 82–86.

99. Burgess-Beusse, B., et al., The insulation of genes from external enhancers and silencing chromatin. *Proc Natl Acad Sci U S A*, 2002. 99 Suppl 4: 16433–16437.

100. Ramezani, A., T.S. Hawley, and R.G. Hawley, Performance- and safety-enhanced lentiviral vectors containing the human interferon-beta scaffold attachment region and the chicken beta-globin insulator. *Blood*, 2003. 101(12): 4717–4727.

101. Park, F. and M.A. Kay, Modified HIV-1 based lentiviral vectors have an effect on viral transduction efficiency and gene expression in vitro and in vivo. *Mol Ther*, 2001. 4(3): 164–173.

102. Park, F., K. Ohashi, and M.A. Kay, The effect of age on hepatic gene transfer with self-inactivating lentiviral vectors in vivo. *Mol Ther*, 2003. 8(2): 314–323.

103. De Rijck, J., et al., Lentiviral nuclear import: A complex interplay between virus and host. *Bioessays*, 2007. 29(5): 441–451.

104. De Rijck, J., B. Van Maele, and Z. Debyser, Positional effects of the central DNA flap in HIV-1-derived lentiviral vectors. *Biochem Biophys Res Commun*, 2005. 328(4): 987–994.

105. Follenzi, A., et al., Gene transfer by lentiviral vectors is limited by nuclear translocation and rescued by HIV-1 pol sequences. *Nat Genet*, 2000. 25(2): 217–222.

106. Sirven, A., et al., Enhanced transgene expression in cord blood CD34(+)-derived hematopoietic cells, including developing T cells and NOD/SCID mouse repopulating cells, following transduction with modified trip lentiviral vectors. *Mol Ther*, 2001. 3(4): 438–448.

107. Sirven, A., et al., The human immunodeficiency virus type-1 central DNA flap is a crucial determinant for lentiviral vector nuclear import and gene transduction of human hematopoietic stem cells. *Blood*, 2000. 96(13): 4103–4110.

108. Dvorin, J.D., et al., Reassessment of the roles of integrase and the central DNA flap in human immunodeficiency virus type 1 nuclear import. *J Virol*, 2002. 76(23): 12087–12096.

109. Bonci, D., et al., "Advanced" generation lentiviruses as efficient vectors for cardiomyocyte gene transduction in vitro and in vivo. *Gene Ther*, 2003. 10(8): 630–636.

110. Mitta, B., et al., Advanced modular self-inactivating lentiviral expression vectors for multigene interventions in mammalian cells and in vivo transduction. *Nucleic Acids Res*, 2002. 30(21): e113.

111. Azzouz, M., et al., Multicistronic lentiviral vector-mediated striatal gene transfer of aromatic L-amino acid decarboxylase, tyrosine hydroxylase, and GTP cyclohydrolase I induces sustained transgene expression, dopamine production, and functional improvement in a rat model of Parkinson's disease. *J Neurosci*, 2002. 22(23): 10302–10312.

112. Mizuguchi, H., et al., IRES-dependent second gene expression is significantly lower than cap-dependent first gene expression in a bicistronic vector. *Mol Ther*, 2000. 1(4): 376–382.

113. Amendola, M., et al., Coordinate dual-gene transgenesis by lentiviral vectors carrying synthetic bidirectional promoters. *Nat Biotechnol*, 2005. 23(1): 108–116.

114. Ben-Dor, I., et al., Lentiviral vectors harboring a dual-gene system allow high and homogeneous transgene expression in selected polyclonal human embryonic stem cells. *Mol Ther*, 2006. 14(2): 255–267.

115. Nakajima, T., et al., Development of novel simian immunodeficiency virus vectors carrying a dual gene expression system. *Hum Gene Ther*, 2000. 11(13): 1863–1874.

116. Chan, L., et al., IL-2/B7.1 (CD80) fusagene transduction of AML blasts by a self-inactivating lentiviral vector stimulates T cell responses in vitro: A strategy to generate whole cell vaccines for AML. *Mol Ther*, 2005. 11(1): 120–131.

117. Murthy, R.C., et al., Corneal transduction to inhibit angiogenesis and graft failurc. *Invest Ophthalmol Vis Sci*, 2003. 44(5): 1837–1842.

118. Lai, Z. and R.O. Brady, Gene transfer into the central nervous system in vivo using a recombinanat lentivirus vector. *J Neurosci Res*, 2002. 67(3): 363–371.

119. Mazarakis, N.D., et al., Rabies virus glycoprotein pseudotyping of lentiviral vectors enables retrograde axonal transport and access to the nervous system after peripheral delivery. *Hum Mol Genet*, 2001. 10(19): 2109–2121.

120. Azzouz, M., et al., VEGF delivery with retrogradely transported lentivector prolongs survival in a mouse ALS model. *Nature*, 2004. 429(6990): 413–417.

121. Poeschla, F., et al., Identification of a human immunodcficiency virus type 2 (HIV-2) encapsidation determinant and transduction of nondividing human cells by HIV-2-bascd lentivirus vectors. *J Virol*, 1998. 72(8): 6527–6536.

122. Poeschla, E.M., F. Wong-Staal, and D.J. Looney, Efficient transduction of nondividing human cells by felinc immunodeficiency virus lentiviral vectors. *Nat Med*, 1998. 4(3): 354–357.

123. Olsen, J.C., Gene transfer vectors derived from equine infectious anemia virus. *Gene Ther*, 1998. 5(11): 1481–1487.

124. Mangeot, P.E., et al., Development of minimal lentivirus vectors derived from simian immunodeficiency virus (SIVmac251) and their use for gene transfer into human dendritic cells. *J Virol*, 2000. 74(18): 8307–8315.

125. Berkowitz, R., et al., Construction and molecular analysis of gene transfer systems derived from bovine immunodeficiency virus. *J Virol*, 2001. 75(7): 3371–3382.

126. Metharom, P., et al., Development of disablcd, rcplication-defective gene transfer vectors from the Jembrana disease virus, a new infectious agent of cattle. *Vet Microbiol*, 2001. 80(1): 9–22.

127. Mselli-Lakhal, L., et al., Gene transfer system derived from the caprine arthritis-encephalitis lentivirus. *J Virol Methods*, 2006. 136(1–2): 177–184.

128. Croyle, M.A., et al., PEGylation of a vesicular stomatitis virus G pseudotyped lentivirus vector prevents inactivation in serum. *J Virol*, 2004. 78(2): 912–921.

129. Mochizuki, H., et al., High-titer human immunodeficiency virus type 1-based vector systems for gene delivery into nondividing cells. *J Virol*, 1998. 72(11): 8873–8883.

130. Duisit, G., et al., Five recombinant simian immunodeficiency virus pseudotypes lead to exclusive transduction of retinal pigmented epithelium in rat. *Mol Ther*, 2002. 6(4): 446–454.

131. Kobinger, G.P., et al., Transduction of human islets with pseudotyped lentiviral vectors. *Hum Gene Ther*, 2004. 15(2): 211–219.

132. Stein, C.S., I. Martins, and B.L. Davidson, The lymphocytic choriomeningitis virus envelope glycoprotein targets lentiviral gene transfer vector to neural progenitors in the murine brain. *Mol Ther*, 2005. 11(3): 382–389.

133. Kahl, C.A., et al., Human immunodeficiency virus type 1-derived lentivirus vectors pseudotyped with envelope glycoproteins derived from Ross River virus and Semliki Forest virus. *J Virol*, 2004. 78(3): 1421–1430.

134. Jakobsson, J., et al., Efficient transduction of neurons using Ross River glycoprotein-pseudotyped lentiviral vectors. *Gene Ther*, 2006. 13(12): 966–973.

135. Kang, Y., et al., In vivo gene transfer using a nonprimate lentiviral vector pseudotyped with Ross River virus glycoproteins. *J Virol*, 2002. 76(18): 9378–9388.

136. Yang, L., et al., Targeting lentiviral vectors to specific cell types in vivo. *Proc Natl Acad Sci U S A*, 2006. 103(31): 11479–11484.

137. Kolokoltsov, A.A., S.C. Weaver, and R.A. Davey, Efficient functional pseudotyping of oncoretroviral and lentiviral vectors by Venezuelan equine encephalitis virus envelope proteins. *J Virol*, 2005. 79(2): 756–763.

138. Kobinger, G.P., et al., Filovirus-pseudotyped lentiviral vector can efficiently and stably transduce airway epithelia in vivo. *Nat Biotechnol*, 2001. 19(3): 225–230.

139. Sinn, P.L., et al., Lentivirus vectors pseudotyped with filoviral envelope glycoproteins transduce airway epithelia from the apical surface independently of folate receptor alpha. *J Virol*, 2003. 77(10): 5902–5910.

140. Sinn, P.L., et al., Gene transfer to respiratory epithelia with lentivirus pseudotyped with Jaagsiekte sheep retrovirus envelope glycoprotein. *Hum Gene Ther*, 2005. 16(4): 479–488.

141. Sinn, P.L., et al., Inclusion of Jaagsiekte sheep retrovirus proviral elements markedly increases lentivirus vector pseudotyping efficiency. *Mol Ther*, 2005. 11(3): 460–469.

142. Stitz, J., et al., Lentiviral vectors pseudotyped with envelope glycoproteins derived from gibbon ape leukemia virus and murine leukemia virus 10A1. *Virology*, 2000. 273(1): 16–20.

143. Zhang, X.Y., V.F. La Russa, and J. Reiser, Transduction of bone-marrow-derived mesenchymal stem cells by using lentivirus vectors pseudotyped with modified RD114 envelope glycoproteins. *J Virol*, 2004. 78(3): 1219–1229.

144. Kobinger, G.P., et al., Human immunodeficiency viral vector pseudotyped with the spike envelope of severe acute respiratory syndrome coronavirus transduces human airway epithelial cells and dendritic cells. *Hum Gene Ther*, 2007. 18(5): 413–422.

145. Sinn, P.L., et al., Persistent gene expression in mouse nasal epithelia following feline immunodeficiency virus-based vector gene transfer. *J Virol*, 2005. 79(20): 12818–12827.

146. Schauber, C.A., et al., Lentiviral vectors pseudotyped with baculovirus gp64 efficiently transduce mouse cells in vivo and show tropism restriction against hematopoietic cell types in vitro. *Gene Ther*, 2004. 11(3): 266–275.

147. Kang, Y., et al., Persistent expression of factor VIII in vivo following nonprimate lentiviral gene transfer. *Blood*, 2005. 106(5): 1552–1558.

148. Qian, Z., et al., Targeting vascular injury using Hantavirus-pseudotyped lentiviral vectors. *Mol Ther*, 2006. 13(4): 694–704.

149. Miyoshi, H., et al., Stable and efficient gene transfer into the retina using an HIV-based lentiviral vector. *Proc Natl Acad Sci U S A*, 1997. 94(19): 10319–10323.

150. Furuta, R.A., et al., Use of a human immunodeficiency virus type 1 Rev mutant without nucleolar dysfunction as a candidate for potential AIDS therapy. *J Virol*, 1995. 69(3): 1591–1599.

151. Corbeau, P. and F. Wong-Staal, Anti-HIV effects of HIV vectors. *Virology*, 1998. 243(2): 268–274.

152. MacKenzie, T.C., et al., Efficient transduction of liver and muscle after in utero injection of lentiviral vectors with different pseudotypes. *Mol Ther*, 2002. 6(3): 349–358.

153. Auricchio, A., et al., Exchange of surface proteins impacts on viral vector cellular specificity and transduction characteristics: The retina as a model. *Hum Mol Genet*, 2001. 10(26): 3075–3081.

154. Bartosch, B. and F.L. Cosset, Strategies for retargeted gene delivery using vectors derived from lentiviruses. *Curr Gene Ther*, 2004. 4(4): 427–443.

155. Morizono, K., et al., Lentiviral vector retargeting to P-glycoprotein on metastatic melanoma through intravenous injection. *Nat Med*, 2005. 11(3): 346–352.

156. Aires da Silva, F., et al., Cell type-specific targeting with sindbis pseudotyped lentiviral vectors displaying anti-CCR5 single-chain antibodies. *Hum Gene Ther*, 2005. 16(2): 223–234.

157. Follenzi, A., et al., Efficient gene delivery and targeted expression to hepatocytes in vivo by improved lentiviral vectors. *Hum Gene Ther*, 2002. 13(2): 243–260.

158. Esslinger, C., P. Romero, and H.R. MacDonald, Efficient transduction of dendritic cells and induction of a T-cell response by third-generation lentivectors. *Hum Gene Ther*, 2002. 13(9): 1091–1100.

159. Jooss, K., et al., Transduction of dendritic cells by DNA viral vectors directs the immune response to transgene products in muscle fibers. *J Virol*, 1998. 72(5): 4212–4223.

160. Follenzi, A., et al., Targeting lentiviral vector expression to hepatocytes limits transgene-specific immune response and establishes long-term expression of human antihemophilic factor IX in mice. *Blood*, 2004. 103(10): 3700–3709.

161. Brown, B.D., et al., Endogenous microRNA regulation suppresses transgene expression in hematopoietic lineages and enables stable gene transfer. *Nat Med*, 2006. 12(5): 585–591.

162. Chen, C.Z., et al., MicroRNAs modulate hematopoietic lineage differentiation. *Science*, 2004. 303(5654): 83–86.

163. Gossen, M., Conditional gene expression: Intelligent designs. *Gene Ther*, 2006. 13(17): 1251–1252.

164. Gossen, M. and H. Bujard, Tight control of gene expression in mammalian cells by tetracycline-responsive promoters. *Proc Natl Acad Sci U S A*, 1992. 89(12): 5547–5551.

165. Gossen, M., et al., Transcriptional activation by tetracyclines in mammalian cells. *Science*, 1995. 268(5218): 1766–1769.

166. Kafri, T., et al., Lentiviral vectors: Regulated gene expression. *Mol Ther*, 2000. 1(6): 516–521.

167. Reiser, J., et al., Development of multigene and regulated lentivirus vectors. *J Virol*, 2000. 74(22): 10589–10599.

168. Vigna, E., et al., Efficient Tet-dependent expression of human factor IX in vivo by a new self-regulating lentiviral vector. *Mol Ther*, 2005. 11(5): 763–775.

169. Pluta, K., et al., Tight control of transgene expression by lentivirus vectors containing second-generation tetracycline-responsive promoters. *J Gene Med*, 2005. 7(6): 803–817.

170. Vigna, E., et al., Robust and efficient regulation of transgene expression in vivo by improved tetracycline-dependent lentiviral vectors. *Mol Ther*, 2002. 5(3): 252–261.

171. Urlinger, S., et al., Exploring the sequence space for tetracy-cline-dependent transcriptional activators: Novel mutations yield expanded range and sensitivity. *Proc Natl Acad Sci U S A*, 2000. 97(14): 7963–7968.

172. Ogueta, S.B., F. Yao, and W.A. Marasco, Design and in vitro characterization of a single regulatory module for efficient control of gene expression in both plasmid DNA and a self-inactivating lentiviral vector. *Mol Med*, 2001. 7(8): 569–579.

173. Szulc, J., et al., A versatile tool for conditional gene expression and knockdown. *Nat Methods*, 2006. 3(2): 109–116.

174. Galimi, F., et al., Development of ecdysone-regulated lentiviral vectors. *Mol Ther*, 2005. 11(1): 142–148.

175. Sirin, O. and F. Park, Regulating gene expression using self-inactivating lentiviral vectors containing the mifepristone-inducible system. *Gene*, 2003. 323: 67–77.

176. Hartenbach, S. and M. Fussenegger, Autoregulated, bidirectional and multicistronic gas-inducible mammalian as well as lentiviral expression vectors. *J Biotechnol*, 2005. 120(1): 83–98.

177. Ventura, A., et al., Cre-lox-regulated conditional RNA interference from transgenes. *Proc Natl Acad Sci U S A*, 2004. 101(28): 10380–10385.

178. Hacein-Bey-Abina, S., et al., A serious adverse event after successful gene therapy for X-linked severe combined immunodeficiency. *N Engl J Med*, 2003. 348(3): 255–256.

179. Montini, E., et al., Hematopoietic stem cell gene transfer in a tumor-prone mouse model uncovers low genotoxicity of lentiviral vector integration. *Nat Biotechnol*, 2006. 24(6): 687–696.

180. Mitchell, R.S., et al., Retroviral DNA integration: ASLV, HIV, and MLV show distinct target site preferences. *PLoS Biol*, 2004. 2(8): E234.

181. Bushman, F.D., Tethering human immunodeficiency virus 1 integrase to a DNA site directs integration to nearby sequences. *Proc Natl Acad Sci U S A*, 1994. 91(20): 9233–9237.

182. Goulaouic, H. and S.A. Chow, Directed integration of viral DNA mediated by fusion proteins consisting of human immunodeficiency virus type 1 integrase and *Escherichia coli* LexA protein. *J Virol*, 1996. 70(1): 37–46.

183. Bushman, F.D. and M.D. Miller, Tethering human immunodeficiency virus type 1 preintegration complexes to target DNA promotes integration at nearby sites. *J Virol*, 1997. 71(1): 458–464.

184. Tan, W., et al., Fusion proteins consisting of human immunodeficiency virus type 1 integrase and the designed polydactyl zinc finger protein E2C direct integration of viral DNA into specific sites. *J Virol*, 2004. 78(3): 1301–1313.

185. Tan, W., et al., Human immunodeficiency virus type 1 incorporated with fusion proteins consisting of integrase and the designed polydactyl zinc finger protein E2C can bias integration of viral DNA into a predetermined chromosomal region in human cells. *J Virol*, 2006. 80(4): 1939–1948.

186. Ciuffi, A. and F.D. Bushman, Retroviral DNA integration: HIV and the role of LEDGF/p75. *Trends Genet*, 2006. 22(7): 388–395.

187. Ciuffi, A., et al., Modulating target site selection during human immunodeficiency virus DNA integration in vitro with an engineered tethering factor. *Hum Gene Ther*, 2006. 17(9): 960–967.

188. Vargas, J., Jr., et al., Novel integrase-defective lentiviral episomal vectors for gene transfer. *Hum Gene Ther*, 2004. 15(4): 361–372.

189. Philippe, S., et al., Lentiviral vectors with a defective integrase allow efficient and sustained transgene expression in vitro and in vivo. *Proc Natl Acad Sci U S A*, 2006. 103(47): 17684–17689.

190. Yanez-Munoz, R.J., et al., Effective gene therapy with nonintegrating lentiviral vectors. *Nat Med*, 2006. 12(3): 348–353.

191. Nightingale, S.J., et al., Transient gene expression by nonintegrating lentiviral vectors. *Mol Ther*, 2006. 13(6): 1121–1132.

192. Strang, B.L., et al., Human immunodeficiency virus type 1 vectors with alphavirus envelope glycoproteins produced from stable packaging cells. *J Virol*, 2005. 79(3): 1765–1771.

193. Ikeda, Y., et al., Continuous high-titer HIV-1 vector production. *Nat Biotechnol*, 2003. 21(5): 569–572.

194. Xu, K., et al., Generation of a stable cell line producing high-titer self-inactivating lentiviral vectors. *Mol Ther*, 2001. 3(1): 97–104.

195. Kafri, T., et al., A packaging cell line for lentivirus vectors. *J Virol*, 1999. 73(1): 576–584.

196. Klages, N., R. Zufferey, and D. Trono, A stable system for the high-titer production of multiply attenuated lentiviral vectors. *Mol Ther*, 2000. 2(2): 170–176.

197. Cockrell, A.S., et al., A trans-lentiviral packaging cell line for high-titer conditional self-inactivating HIV-1 vectors. *Mol Ther*, 2006. 14(2): 276–284.

198. Kaplan, A.H. and R. Swanstrom, Human immunodeficiency virus type 1 Gag proteins are processed in two cellular compartments. *Proc Natl Acad Sci U S A*, 1991. 88(10): 4528–45232.

199. Sparacio, S., et al., Generation of a flexible cell line with regulatable, high-level expression of HIV Gag/Pol particles capable of packaging HIV-derived vectors. *Mol Ther*, 2001. 3(4): 602–612.

200. Strang, B.L., et al., Characterization of HIV-1 vectors with gammaretrovirus envelope glycoproteins produced from stable packaging cells. *Gene Ther*, 2004. 11(7): 591–598.

201. Lu, X., et al., Safe two-plasmid production for the first clinical lentivirus vector that achieves >99% transduction in primary cells using a one-step protocol. *J Gene Med*, 2004. 6(9): 963–973.

202. Levine, B.L., et al., Gene transfer in humans using a conditionally replicating lentiviral vector. *Proc Natl Acad Sci U S A*, 2006. 103(46): 17372–17377.

203. Ni, Y., et al., Generation of a packaging cell line for prolonged large-scale production of high-titer HIV-1-based lentiviral vector. *J Gene Med*, 2005. 7(6): 818–834.

204. Mitta, B., M. Rimann, and M. Fussenegger, Detailed design and comparative analysis of protocols for optimized production of high-performance HIV-1-derived lentiviral particles. *Metab Eng*, 2005. 7(5–6): 426–436.

205. Koldej, R., et al., Optimisation of a multipartite human immunodeficiency virus based vector system; control of virus infectivity and large-scale production. *J Gene Med*, 2005. 7(11): 1390–1399.

206. Tiscornia, G., O. Singer, and I.M. Verma, Production and purification of lentiviral vectors. *Nat Protoc*, 2006. 1(1): 241–245.

207. Pichlmair, A., et al., Tubulovesicular structures within vesicular stomatitis virus G protein-pseudotyped lentiviral vector preparations carry DNA and stimulate antiviral responses via Toll-like receptor 9. *J Virol*, 2007. 81(2): 539–547.

208. Zhang, B., et al., A highly efficient and consistent method for harvesting large volumes of high-titre lentiviral vectors. *Gene Ther*, 2001. 8(22): 1745–1751.

209. Pham, L., et al., Concentration of viral vectors by co-precipitation with calcium phosphate. *J Gene Med*, 2001. 3(2): 188–194.

210. Baekelandt, V., et al., Optimized lentiviral vector production and purification procedure prevents immune response

after transduction of mouse brain. *Gene Ther*, 2003. 10(23): 1933–1940.

211. Breckpot, K., J.L. Aerts, and K. Thielemans, Lentiviral vectors for cancer immunotherapy: Transforming infectious particles into therapeutics. *Gene Ther*, 2007. 14(11): 847–862.

212. Nesbeth, D., et al., Metabolic biotinylation of lentiviral pseudotypes for scalable paramagnetic microparticle-dependent manipulation. *Mol Ther*, 2006. 13(4): 814–822.

213. Chan, L., et al., Conjugation of lentivirus to paramagnetic particles via nonviral proteins allows efficient concentration and infection of primary acute myeloid leukemia cells. *J Virol*, 2005. 79(20): 13190–13194.

214. Yu, J.H. and D.V. Schaffer, Selection of novel vesicular stomatitis virus glycoprotein variants from a peptide insertion library for enhanced purification of retroviral and lentiviral vectors. *J Virol*, 2006. 80(7): 3285–3292.

215. Geraerts, M., et al., Upscaling of lentiviral vector production by tangential flow filtration. *J Gene Med*, 2005. 7(10): 1299–1310.

216. Sastry, L., et al., Evaluation of plasmid DNA removal from lentiviral vectors by benzonase treatment. *Hum Gene Ther*, 2004. 15(2): 221–226.

217. Slepushkin, V., et al., Large-scale purification of a lentiviral vector by size exclusion chromatography or Mustang Q ion exchange capsule. *Bioprocessing Journal*, 2003. 2: 89–95.

218. Yamada, K., et al., Lentivirus vector purification using anion exchange HPLC leads to improved gene transfer. *Biotechniques*, 2003. 34(5): 1074–1078, 1080.

219. Scherr, M., et al., Efficient gene transfer into the CNS by lentiviral vectors purified by anion exchange chromatography. *Gene Ther*, 2002. 9(24): 1708–1714.

220. Delenda, C. and C. Gaillard, Real-time quantitative PCR for the design of lentiviral vector analytical assays. *Gene Ther*, 2005. 12 Suppl 1: S36–S50.

221. Geraerts, M., et al., Comparison of lentiviral vector titration methods. *BMC Biotechnol*, 2006. 6: 34.

222. Scherr, M., et al., Quantitative determination of lentiviral vector particle numbers by real-time PCR. *Biotechniques*, 2001. 31(3): 520, 522, 524, passim.

223. Sastry, L., et al., Titering lentiviral vectors: Comparison of DNA, RNA and marker expression methods. *Gene Ther*, 2002. 9(17): 1155–1162.

224. Zhang, B., et al., The significance of controlled conditions in lentiviral vector titration and in the use of multiplicity of infection (MOI) for predicting gene transfer events. *Genet Vaccines Ther*, 2004. 2(1): 6.

225. Lizee, G., et al., Real-time quantitative reverse transcriptase-polymerase chain reaction as a method for determining lentiviral vector titers and measuring transgene expression. *Hum Gene Ther*, 2003. 14(6): 497–507.

226. Butler, S.L., M.S. Hansen, and F.D. Bushman, A quantitative assay for HIV DNA integration in vivo. *Nat Med*, 2001. 7(5): 631–634.

227. Martin-Rendon, E., et al., New methods to titrate EIAV-based lentiviral vectors. *Mol Ther*, 2002. 5(5 Pt 1): 566–570.

228. Silver, J., et al., An RT-PCR assay for the enzyme activity of reverse transcriptase capable of detecting single virions. *Nucleic Acids Res*, 1993. 21(15): 3593–3594.

229. Pyra, H., J. Boni, and J. Schupbach, Ultrasensitive retrovirus detection by a reverse transcriptase assay based on product enhancement. *Proc Natl Acad Sci U S A*, 1994. 91(4): 1544–1548.

230. Lovatt, A., et al., High throughput detection of retrovirus-associated reverse transcriptase using an improved fluorescent product enhanced reverse transcriptase assay and its compari-

son to conventional detection methods. *J Virol Methods*, 1999. 82(2): 185–200.

231. Rohll, J.B., et al., Design, production, safety, evaluation, and clinical applications of nonprimate lentiviral vectors. *Methods Enzymol*, 2002. 346: 466–500.

232. Sastry, L., et al., Product-enhanced reverse transcriptase assay for replication-competent retrovirus and lentivirus detection. *Hum Gene Ther*, 2005. 16(10): 1227–1236.

233. Miskin, J., et al., A replication competent lentivirus (RCL) assay for equine infectious anaemia virus (EIAV)-based lentiviral vectors. *Gene Ther*, 2006. 13(3): 196–205.

234. Sears, J.F. and A.S. Khan, Single-tube fluorescent product-enhanced reverse transcriptase assay with Ampliwax (STF-PERT) for retrovirus quantitation. *J Virol Methods*, 2003. 108(1): 139–142.

235. Fan, X.Y., et al., A modified single-tube one-step product-enhanced reverse transcriptase (mSTOS-PERT) assay with heparin as DNA polymerase inhibitor for specific detection of RTase activity. *J Clin Virol*, 2006. 37(4): 305–312.

236. Sastry, L., et al., Certification assays for HIV-1-based vectors: Frequent passage of gag sequences without evidence of replication-competent viruses. *Mol Ther*, 2003. 8(5): 830–839.

237. Schonely, K., et al., QC release testing of an HIV-1 based lentiviral vector lot and transduced cellular product. *Bioprocessing Journal*, 2003. 2: 39–47.

238. Escarpe, P., et al., Development of a sensitive assay for detection of replication-competent recombinant lentivirus in large-scale HIV-based vector preparations. *Mol Ther*, 2003. 8(2): 332–341.

239. Bock, P.J. and D.M. Markovitz, Infection with HIV-2. *Aids*, 2001. 15 Suppl 5: S35–S45.

240. Corbeau, P., G. Kraus, and F. Wong-Staal, Transduction of human macrophages using a stable HIV-1/HIV-2-derived gene delivery system. *Gene Ther.*, 1998. 5: 99–104.

241. Arya, S.K., M. Zamani, and P. Kundra, Human immunodeficiency virus type 2 lentivirus vectors for gene transfer: Expression and potential for helper virus-free packaging. *Hum Gene Ther*, 1998. 9(9): 1371–1380.

242. Negre, D., et al., Characterization of novel safe lentiviral vectors derived from simian immunodeficiency virus (SIV-mac251) that efficiently transduce mature human dendritic cells. *Gene Ther*, 2000. 7(19): 1613–1623.

243. Hofmann, W., et al., Species-specific, postentry barriers to primate immunodeficiency virus infection. *J Virol*, 1999. 73(12): 10020–10028.

244. Mselli-Lakhal, L., et al., Defective RNA packaging is responsible for low transduction efficiency of CAEV-based vectors. *Arch Virol*, 1998. 143(4): 681–695.

245. Lin, Y.L., et al., Feline immunodeficiency virus vectors for efficient transduction of primary human synoviocytes: Application to an original model of rheumatoid arthritis. *Hum Gene Ther*, 2004. 15(6): 588–596.

246. Mukherjee, S., et al., A HIV-2-based self-inactivating vector for enhanced gene transduction. *J Biotechnol*, 2007. 127(4): 745–757.

247. Mustafa, F., et al., Sequences intervening between the core packaging determinants are dispensable for maintaining the packaging potential and propagation of feline immunodeficiency virus transfer vector RNAs. *J Virol*, 2005. 79(21): 13817–13821.

248. Molina, R.P., et al., A synthetic Rev-independent bovine immunodeficiency virus-based packaging construct. *Hum Gene Ther*, 2004. 15(9): 865–877.

249. Hacker, C.V., et al., The integration profile of EIAV-based vectors. *Mol Ther*, 2006. 14(4): 536–545.

250. Kang, Y., et al., Integration site choice of a feline immunodeficiency virus vector. *J Virol*, 2006. 80(17): 8820–8823.

251. Monse, H., et al., Viral determinants of integration site preferences of simian immunodeficiency virus-based vectors. *J Virol*, 2006. 80(16): 8145–8150.

252. Morris, K.V., et al., Small interfering RNA-induced transcriptional gene silencing in human cells. *Science*, 2004. 305(5688): 1289–1292.

253. O'Rourke, J.P., et al., Comparison of gene transfer efficiencies and gene expression levels achieved with equine infectious anemia virus- and human immunodeficiency virus type 1-derived lentivirus vectors. *J Virol*, 2002. 76(3): 1510–1515.

254. Ikeda, Y., et al., Gene transduction efficiency in cells of different species by HIV and EIAV vectors. *Gene Ther*, 2002. 9(14): 932–938.

255. Siapati, E.K., et al., Comparison of HIV- and EIAV-based vectors on their efficiency in transducing murine and human hematopoietic repopulating cells. *Mol Ther*, 2005. 12(3): 537–546.

256. Price, M.A., et al., Expression from second-generation feline immunodeficiency virus vectors is impaired in human hematopoietic cells. *Mol Ther*, 2002. 6(5): 645–652.

257. Quinonez, R. and R.E. Sutton, Lentiviral vectors for gene delivery into cells. *DNA Cell Biol*, 2002. 21(12): 937–951.

258. Stremlau, M., et al., The cytoplasmic body component TRIM5alpha restricts HIV-1 infection in old world monkeys. *Nature*, 2004. 427(6977): 848–853.

259. Root, D.E., et al., Genome-scale loss-of-function screening with a lentiviral RNAi library. *Nat Methods*, 2006. 3(9): 715–719.

260. Moffat, J., et al., A lentiviral RNAi library for human and mouse genes applied to an arrayed viral high-content screen. *Cell*, 2006. 124(6): 1283–1298.

261. Bailey, S.N., et al., Microarrays of lentiviruses for gene function screens in immortalized and primary cells. *Nat Methods*, 2006. 3(2): 117–122.

262. Wiznerowicz, M., J. Szulc, and D. Trono, Tuning silence: Conditional systems for RNA interference. *Nat Methods*, 2006. 3(9): 682–688.

263. Pfeifer, A., Lentiviral transgenesis—a versatile tool for basic research and gene therapy. *Curr Gene Ther*, 2006. 6(4): 535–542.

264. De Palma, M., M.A. Venneri, and L. Naldini, In vivo targeting of tumor endothelial cells by systemic delivery of lentiviral vectors. *Hum Gene Ther*, 2003. 14(12): 1193–1206.

265. Bao, L., et al., Stable transgene expression in tumors and metastases after transduction with lentiviral vectors based on human immunodeficiency virus type 1. *Hum Gene Ther*, 2004. 15(5): 445–456.

266. Miletic, H., et al., Selective transduction of malignant glioma by lentiviral vectors pseudotyped with lymphocytic choriomeningitis virus glycoproteins. *Hum Gene Ther*, 2004. 15(11): 1091–1100.

267. Lizee, G., M.I. Gonzales, and S.L. Topalian, Lentivirus vector-mediated expression of tumor-associated epitopes by human antigen presenting cells. *Hum Gene Ther*, 2004. 15(4): 393–404.

268. Dullaers, M. and K. Thielemans, From pathogen to medicine: HIV-1-derived lentiviral vectors as vehicles for dendritic cell based cancer immunotherapy. *J Gene Med*, 2006. 8(1): 3–17.

269. Dullaers, M., et al., Induction of effective therapeutic antitumor immunity by direct in vivo administration of lentiviral vectors. *Gene Ther*, 2006. 13(7): 630–640.

270. Kordower, J.H., et al., Neurodegeneration prevented by lentiviral vector delivery of GDNF in primate models of Parkinson's disease. *Science*, 2000. 290(5492): 767–773.

271. Wong, L.F., et al., Lentivirus-mediated gene transfer to the central nervous system: Therapeutic and research applications. *Hum Gene Ther*, 2006. 17(1): 1–9.

272. Takahashi, M., et al., Rescue from photoreceptor degeneration in the rd mouse by human immunodeficiency virus vector-mediated gene transfer. *J Virol*, 1999. 73(9): 7812–7816.

273. Consiglio, A., et al., In vivo gene therapy of metachromatic leukodystrophy by lentiviral vectors: Correction of neuropathology and protection against learning impairments in affected mice. *Nat Med*, 2001. 7(3): 310–316.

274. Brooks, A.I., et al., Functional correction of established central nervous system deficits in an animal model of lysosomal storage disease with feline immunodeficiency virus-based vectors. *Proc Natl Acad Sci U S A*, 2002. 99(9): 6216–6221.

275. Pawliuk, R., et al., Correction of sickle cell disease in transgenic mouse models by gene therapy. *Science*, 2001. 294(5550): 2368–2371.

276. Ralph, G.S., et al., Silencing mutant SOD1 using RNAi protects against neurodegeneration and extends survival in an ALS model. *Nat Med*, 2005. 11(4): 429–433.

277. Singer, O., et al., Targeting BACE1 with siRNAs ameliorates Alzheimer disease neuropathology in a transgenic model. *Nat Neurosci*, 2005. 8(10): 1343–1349.

278. Yanay, O., et al., An adult dog with cyclic neutropenia treated by lentivirus-mediated delivery of granulocyte colony-stimulating factor. *Hum Gene Ther*, 2006. 17(4): 464–469.

279. Mortellaro, A., et al., Ex vivo gene therapy with lentiviral vectors rescues adenosine deaminase (ADA)-deficient mice and corrects their immune and metabolic defects. *Blood*, 2006. 108(9): 2979–2988.

280. Dupre, L., et al., Efficacy of gene therapy for Wiskott–Aldrich syndrome using a WAS promoter/cDNA-containing lentiviral vector and nonlethal irradiation. *Hum Gene Ther*, 2006. 17(3): 303–313.

281. Balaggan, K.S., et al., EIAV vector-mediated delivery of endostatin or angiostatin inhibits angiogenesis and vascular hyperpermeability in experimental CNV. *Gene Ther*, 2006. 13(15): 1153–1165.

282. Kohn, D.B., Lentiviral vectors ready for prime-time. *Nat Biotechnol*, 2007. 25(1): 65–66.

283. Dropulic, B., M. Hermankova, and P.M. Pitha, A conditionally replicating HIV-1 vector interferes with wild-type HIV-1 replication and spread. *Proc Natl Acad Sci U S A*, 1996. 93(20): 11103–11108.

284. Lu, X., et al., Antisense-mediated inhibition of human immunodeficiency virus (HIV) replication by use of an HIV type 1-based vector results in severely attenuated mutants incapable of developing resistance. *J Virol*, 2004. 78(13): 7079–7088.

285. Novina, C.D., et al., siRNA-directed inhibition of HIV-1 infection. *Nat Med*, 2002. 8(7): 681–686.

286. Poluri, A., M. van Maanen, and R.E. Sutton, Genetic therapy for HIV/AIDS. *Expert Opin Biol Ther*, 2003. 3(6): 951–963.

10 Vaccinia Viral Vectors

J. Andrea McCart and David L. Bartlett

CONTENTS

10.1 INTRODUCTION

Vaccinia virus has been used clinically as a vaccine for smallpox for over 150 years, and thus is associated with a rich history and extensive clinical experience [1,2]. In 1798 Edward Jenner demonstrated protection from smallpox by vaccination with the cowpox virus obtained from infected milkmaids. In the early nineteenth century this evolved into the use of vaccinia virus [3]. The precise origin of vaccinia virus is difficult to identify, as the virus has no known natural host, leading some to suggest that it may have arisen from mutations in the cowpox or smallpox viruses [4]. However, it is more likely that vaccinia represents a distinct strain that is extinct in its natural host or so rare that it is difficult to identify naturally [5].

Vaccinia virus was widely used as a vaccine for the eradication of smallpox until 1978. Since that time it has been used experimentally as an *in vitro* gene expression vector and to express foreign genes as a vaccine for infectious agents and cancer [6]. Vaccinia has many advantages over other viruses as a vector for gene delivery. Vaccinia has a wide host range. Its genome has been completely sequenced facilitating the creation of recombinant vectors. It can hold up to 25 kb of foreign DNA without a need for viral deletions [7]. Recombinant vectors are easily produced in high titers for use *in vivo* [8].

Vaccinia virus has previously been modified to carry various antigens, cytokines, and immunostimulatory molecules [9]. It has also become evident that several properties make it useful as an oncolytic virus for cancer therapy [10]. Any cell infected with the virus is rapidly killed, and cell-to-cell spread

is efficient. Also, a natural tumor tropism exists in animal models. Applications requiring long-term gene expression, however, are not feasible with vaccinia. This chapter will review the relevant aspects of vaccinia biology necessary for its use as a gene transfer vector, and review the preclinical and clinical development of vaccinia virus for gene therapy and oncolytic virotherapy.

10.2 BIOLOGY OF VACCINIA

Poxviruses are classified into two subfamilies, chordopoxvirinae (vertebrate poxviruses) and entomopoxvirinae (insect pox viruses), and at least 46 species [11]. The classification scheme is based on host range, sequence homology, and antigenicity. Vaccinia virus is a member of the orthopoxvirus genus. It is genetically distinct from both cowpox virus and variola virus (smallpox). All members of the orthopoxvirus genus have immune cross reactivity and are genetically stable. This allowed for the complete eradication of variola virus in 1977 (last case of endemic smallpox).

Multiple strains of vaccinia viruses exist (Table 10.1). As vaccination became widespread throughout the world, numerous centers produced and maintained the vaccine in different ways, resulting in numerous strains, which differ in characteristics, pathogenicity, and host range. The New York City Board of Health (NYCBH) strain was obtained from England in 1856 and was originally used for small pox vaccination in the United States [5]. The western reserve (WR) strain is a laboratory derivative of this strain and appears to be one of the more virulent

TABLE 10.1
Vaccinia Strains Reported in the Literature

NYCBH	Tashkent	Ikeda
Wyeth	USSR	IHD
WR	Evans	DIs
Copenhagen	Praha	LC16
Lister	LIVP	EM63
MVA	Tian Tan	IC
		AS

strains in laboratory animals and nonhuman primates. It has not been utilized in patients to date. A recent comparison of vaccinia strains has suggested that WR has superior tumor lytic effects [12]. Another derivative, the Wyeth strain, was produced by Wyeth as a smallpox vaccine and is the backbone commonly used for experimental vaccines in clinical trials. The modified vaccinia Ankara (MVA) strain was developed through multiple rounds of infection in avian cells. This strain is highly attenuated and does not replicate in human cells [13].

As with all poxviruses, the vaccinia virus is a double-stranded DNA virus whose entire life cycle exists within the cytoplasm of eukaryotic cells. The virus contains an outer envelope as well as an internal membrane and it carries the enzymes required for initiation of transcription. The genome of the Copenhagen strain of vaccinia virus was completely sequenced and reported in 1990 [14] (other strains have been sequenced subsequently [15]). The genome consists of double-stranded DNA with inverted terminal repeats and a terminal hairpin loop, which mimics a large, circular, single-stranded DNA. The genome consists of 191,636 base pairs encoding approximately 2063 proteins of 65 or more amino acids in length. It is among the largest viruses in size, averaging 270 × 350 nm in the shape of a brick [16].

Vaccinia has two infectious forms, the mature virus (MV) and the enveloped virus (EV) [17]. MV is the form recovered during viral purification *in vitro*, as it is released on cellular disruption. Purification of EV is quite difficult as the envelope is too fragile to withstand the purification process. While attachment to the cell surface is different for MV and EV [18], recent studies have shown that cell entry by vaccinia occurs by fusion with the host cell membrane [19]. This is mediated by an entry-fusion complex consisting of eight viral proteins: A16, A21, A28, G3, G9, H2, J5, and L5 [20]. This complex is also required for cell-to-cell spread of EV suggesting that the outer envelop is lost prior to fusion. Recently a low pH, endosomal entry pathway was identified [21]. The MV contains several other proteins on its outer envelope including A17L, A27L, and D8L [22–28] which likely play a role in viral attachment. D8L was one of the first membrane proteins identified in the MV. It is nonessential in the viral life cycle but may mediate MV binding to cell surface chondroitin sulfate [26,27]. A27L mediates vaccinia interaction with cell surface heparan sulphate [25,29]. Virus infection was inhibited by up to 60% in the presence of soluble heparin [25] depending on the cell line used. EV is believed to be responsible for cell-to-cell spread and long-range transmission

of vaccinia virus *in vivo* [30]. Six proteins (encoded by A33R, A34R, A36R, A56R, B5R, and F13L) are EV-specific [11]. A56R has no effect on infectivity or spread if mutated [31].

It may be possible to circumvent normal receptor requirements by engineering vaccinia virus to bind to alternative cell surface molecules. Consistent with this, expression of an ScFv to erbB2 on the surface of the EV (created as a fusion with A56R) was shown to bind erbB2 by ELISA [31]. Creation of a fusion protein between an ScFv and A56R is technically feasible and may direct binding of the EV to a specific antigen or cell type. Fusions of other surface proteins, including B5R have been reported [32].

The life cycle of vaccinia is illustrated in Figure 10.1. Vaccinia virus (as with all poxviruses) spends its entire life cycle in the cytoplasm of the host cell and has never been shown to integrate into the host genome. Although originally it was thought that vaccinia virus has very few interactions with host-cellular proteins, host proteins are important. The CCR5 chemokine receptor has been shown to mediate a cascade of downstream signaling events required for late gene transcription in some cell lines [33]. The transcription factor YYI is utilized for initiation and termination of intermediate and late gene transcription [34–36]. The virus induces a profound cytopathic effect very soon after viral entry, as early viral enzymes completely shut down host cell functions. By 4–6 h after infection there is almost complete inhibition of host protein synthesis. This allows for very efficient expression of viral genes and viral replication. In fact, approximately 10,000 copies of the viral genome are made within 12 h of infection [37]. Half of these are incorporated into mature virions and released.

After vaccinia enters the cell transcription begins. Three stages of transcription—early, intermediate, and late—have been described, each with its own specific promoters and transcription factors [38]. The enzymes and other proteins required for transcription are contained within the viral core along with the viral genomic DNA [36]. Proteins needed for viral replication are synthesized at the early (pre-replicative) stage of infection. A DNA-dependent RNA polymerase is contained within the viral core leading to the synthesis of early messenger RNA (mRNA). Early mRNAs appear within minutes of viral infection [36]. Translation of this RNA forms early proteins, which are involved in uncoating of the viral DNA, DNA replication, and intermediate transactivation for transcription of intermediate mRNA. Intermediate mRNA is then expressed which encodes for late transactivators leading to late mRNA synthesis. Late proteins include structural proteins for membrane formation and early transcription factors to be incorporated into the new virus particle. Only a relatively small number of proteins are required for DNA synthesis, making the system simple and largely autonomous.

The vaccinia double-stranded DNA genome replicates in the cytoplasm forming multiple concatamers of the genome. These concatamers are then resolved into individual genomes which are encapsulated along with the early transcription factors into Golgi-derived membranes. The first stage in the formation of infectious particles is the development of viral crescents composed of lipid and viral protein. To date, the

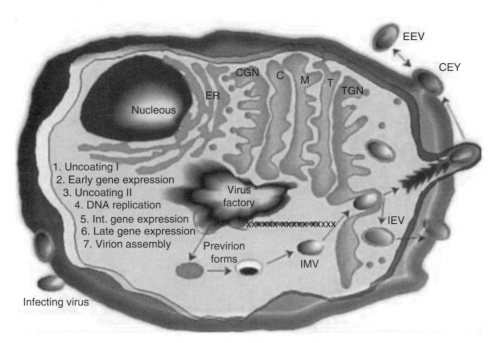

FIGURE 10.1 Vaccinia virus replication cycle. A diagram of the infected cell is shown with an exaggerated view of the endoplasmic reticulum (ER), *cis-*, *medial-*, *trans-*Golgi and the *trans-*Golgi network (C, M, T, and TGN, respectively). The major stages of the virus life cycle are listed. Following late gene expression, previrion forms assemble to form intracellular mature virus (MV). The MV is targeted to the TGN and following envelopment, intracellular enveloped virus (IEV) is formed. IEV are propelled to the cell surface by the polymerization of actin filaments. Once released the virus may remain attached to the membrane as cell-associated enveloped virus (CEV) or be released into the medium as extracellular enveloped virus (EV). (From Grosenbach, D.W. and Hruby, D.E., *Front. Biosci.*, 3, 354, 1998. With permission.)

origin of these crescents is disputed. Currently, the crescent is thought to be composed of a single-lipid bilayer without continuity to cellular membranes [39]. These crescents then coalesce into immature virus that lack infectivity. Immature virus becomes MV by condensation of the core and processing of core proteins (Figure 10.2). MV is transported to sites at which it becomes wrapped with two additional membranes. These membranes are derived from trans-Golgi network membranes that have been modified by the inclusion of virus-encoded proteins and ultimately become part of the EV outer envelope. These wrapped, intracellular enveloped viruses move along microtubules to the cell surface where the outer membrane fuses with the plasma membrane, exposing the viruses on the cell surface. If the virus is retained or reattached it is called the cell-associated EV (CEV), but if released becomes EV. Upon reaching the plasma membrane vaccinia switches from microtubule-dependent transport to the formation of actin tails needed for cell-to-cell spread of virus [40]. This is dependent on phosphorylation of an EV protein A36R which recently has been shown to be mediated by multiple families of tyrosine kinases [40–42] and likely contributes to wide host range and rapid spread of vaccinia virus. The A34R gene product plays a role in holding the virus to the cell surface. The WR strain exists almost exclusively as a cell-associated virus [43]; however, mutations in A34R can lead to increased EV release and decreased CEV [44].

This entire life cycle occurs very rapidly. Initial RNA transcripts are detectable within 20 min of infection and DNA replication begins 1–2 h after infection. Within several hours after infection the majority of mRNA within the cytoplasm is from vaccinia-encoded genes. The entire replication cycle occurs in approximately 12 h [6].

10.3 CONSTRUCTION OF RECOMBINANT VECTORS

Homologous recombination occurs naturally during the replication of vaccinia virus, thus lending itself towards efficient insertion of foreign DNA. The creation of recombinant vaccinia vectors is relatively simple. The issues to be considered when creating a recombinant vaccinia vector include choosing a site for proposed recombination, choosing a selection method(s), and choosing a promoter for the foreign gene.

A shuttle plasmid is first created where a foreign gene expressed off a vaccinia promoter is flanked by vaccinia DNA sequences. Care must be taken that the foreign gene does not contain vaccinia transcription termination signals for early promoters (TTTTTNT) [45]. The most common site of recombination has been the vaccinia thymidine kinase (TK) gene. Insertion of genes into the TK locus eliminates functional viral TK, leading to attenuation of the virus *in vivo* [46]. Recombinations into numerous other loci have been performed, including intergenic segments such that no functional deletion occurs [47,48]. The functional analysis of many vaccinia genes has been defined through insertional deletion.

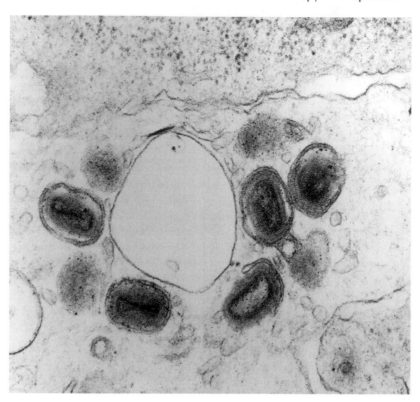

FIGURE 10.2 Electron micrograph of intracytoplasmic MV form of vaccinia. The virions have a characteristic brick shape with a biconcave central core. (Courtesy of Maria Tsokos, and Mones Abu-Asab, Laboratory of Pathology, NCI. With permission.)

A wide range of vaccinia promoters are available for expression of transgenes. It is necessary to use vaccinia promoters for creation of the recombinant vectors, as these are specific for vaccinia polymerase. Eukaryotic promoters will not function in vaccinia infection as the host cell polymerase is not present in the cytoplasm where vaccinia transcription occurs. Several natural and synthetic early and late promoters have been described with various levels of activity [49,50]. The native vaccinia promoters are generally very strong and compare favorably to other viral promoters used in other viral vectors. The synthetic early/late promoter described by Chakrabarti et al. [50] has led to consistent, reliable high levels of gene expression in numerous systems tested.

Several methods for selection of recombinant viruses are available. Growth in the presence of the thymidine analogue BdUr can be used to select for a TK negative phenotype in select cells after recombination into the TK locus [51]. Others have commonly used the selection gene xanthine–guanine phosphoribosyltransferase (XGRPT) which allows for selective growth in media containing mycophenolic acid [52]. Positive selection through replacement of an essential gene previously deleted from a backbone virus grown on permissive cell lines is also available [53]. β-Galactosidase and green fluorescent protein can aid in selection of recombinants. Once the shuttle plasmid is constructed, it can be transfected into a cell that has been infected with vaccinia. Homologous recombination leads to the insertion of the foreign gene into 0.1% of progeny virus

genomes [6]. The use of at least 3–5 rounds of selection ensures that there is no contaminating parental virus.

In order to improve the efficiency of creating recombinant vaccinia viruses, Domi et al. [54,55] have developed vaccinia-bacterial artificial chromosome (VAC–BAC) technology to allow one-step insertion of a gene of interest into the VAC–BAC plasmid, selection in *Escherichia coli,* and then transfection of the DNA into fowlpox-infected helper cells to generate infectious vaccinia virions. This avoids the need for multiple time-consuming plaque purifications.

10.4 *IN VITRO* GENE TRANSFER VECTOR

It has long been recognized that vaccinia is a valuable tool for expression of foreign genes *in vitro* [56,57]. It is relatively easy to make recombinant viruses and to grow and purify the virus. Large inserts can be accepted in the genome, and strong synthetic promoters lead to high levels of protein expression from infected cells. The virus has a broad tropism and will infect and replicate in most mammalian cells. The high efficiency of expression with vaccinia obviated the need for other vectors which permanently integrated into the genome and the cloning of high-expressing cells. Infection with recombinant vaccinia leads to high levels of expression of foreign genes that are processed in the appropriate way such that their function can be studied. Vaccinia infection of cells at an multiplicity of infection (MOI) of 1.0 leads to greater than 99% of cells expressing the gene of interest in most cell lines. Rather than make recombinants, simple plasmid transfection in a virally infected cell leads to efficient expression of foreign genes. The gene must be placed under control of a vaccinia promoter. The backdrop of vaccinia infection allows for cytoplasmic transcription avoiding the inefficient process of trafficking into the nucleus. This leads to greater than 90% of cells expressing foreign genes after simple plasmid transfection [58]. This efficient transient expression system has been utilized for the functional analysis of innumerable foreign proteins over the years. Inactivation of the virus with psoralen and UV light can result in an efficient expression vector that does not cause a cytopathic effect [59]. This may be important for the functional analysis of some proteins.

10.5 HOST IMMUNE RESPONSE TO VACCINIA

The immune response to vaccinia viral vectors serves as our paradoxical friend and foe in the attempt to develop vaccinia as an effective vector for gene therapy. On the one hand, the vigorous immune response is desirable because we believe that it enhances its potential as a vaccine. On the other hand, the vigorous immune response leads to premature clearance of the virus before adequate levels of replication have occurred, thus decreasing the level of transgene expression and possibly the overall efficacy.

Most viruses which infect human cells are also endemic in the population, and therefore many patients will have circulating antibodies against the viruses and preformed cellular precursors. Vaccinia is unique in that it is not endemic to humans, and since wide scale smallpox immunizations terminated in the 1970s, young patients will not have been exposed to the virus. Most cancer patients, however, are older and have been exposed to vaccinia. As with other viral vectors, reinfection is possible after prior exposure. Laboratory workers and military personnel who undergo revaccination for smallpox usually form pox vesicles in the skin, despite prior vaccination. This has also been demonstrated in tumor vaccine trials in patients previously immunized [60]. Workers at vaccine production plants in the past suffered from repeated skin infections with vaccinia and pox lesions in the skin. Some viruses can avoid circulating antibodies by mutating their coat proteins and changing serotype. This is not seen with vaccinia virus. However, vaccinia has evolved expression of immunosuppressive proteins [61–63]. Viral surface proteins act to inhibit complement, and the extracellular envelope is known to be almost completely resistant to antibody neutralization [64].

Both cellular immunity and neutralizing antibodies play a role in protection from vaccinia infection. The T-cell response to vaccinia seems to be quite potent and is probably more important than antibodies in the primary host resistance to the virus. Progressive vaccinia infection correlates with a defect in cell-mediated immunity [65]. In murine models, lacking a functional T-cell population, vaccinia is able to replicate and express genes within tumor cells at high levels for greater than 30 days [66]. On the other hand, in an immunocompetent host the window of gene expression only lasts for about 8 days with high levels of gene activity lasting approximately 4 days [67].

The success of vaccinia as a gene therapy vector relies on its efficiency *in vivo*, thus vaccinia virus possesses a wide range of immune evasion strategies in order to survive (Table 10.2). Understanding and manipulating these factors may optimize the vector for clinical use. If one examines

---

TABLE 10.2
Vaccinia Gene Products Which Inhibit the Immune Response

Vaccinia Gene	Known/Putative Function	References
A44L	Steroid synthesis	[162]
A46R	Putative IL-1 antagonist, TLR inhibitor	[83,163]
A52R	Putative inhibitor of TLR signaling	[83,164]
A53R	Soluble TNF receptor	[90]
B5R	Inhibits complement	[165]
B8R	IFN-γ soluble receptor	[61,62]
B13R (SPI-2)	Inhibits IL-1β converting enzyme	[139,145]
B15R	IL-1β soluble receptor	[61]
B18R	IFN-α/β soluble receptor	[61,62]
B22R (SPI-1)	Binds cathepsin G	[139,145]
B29R/CKBP	Soluble chemokine-binding protein	[61]
C3L/VCP	Complement binding protein	[62,89]
C12L	Binds and inhibits IL-18	[63,81]
E3L	Binds dsRNA to block PKR activation	[61,62,166,167]
F1L	Inhibits cytochrome C	[168]
K1L	Inhibits NF-κB activation	[169]
K3L	Prevents phosphorylation of eIF2α	[61,62,167]
N1L	Inhibits NF-κB	[170,171]

these factors closely, it is clear that the majority of them encode for proteins that are able to actively suppress both innate immunity and the development of a T helper 1 (Th1) immune response. For example, vaccinia virus has adopted at least three different genes whose product can block the function of the type 1 interferon family members IFN-α and β [68–70]. These factors are secreted by a variety of cells in response to innate danger signals. They can induce an antiviral state and upregulate adaptive immune functions. Vaccinia also carries genes for multiple inhibitors of chemokines, some of the earliest substances produced during the initiation of an immune response [71–73].

Vaccinia encodes for at least three factors that can directly block the function of IFN-γ, one of the most potent Th1 cytokines [70,74–76]. In addition, vaccinia encodes for the recently described IL-18 binding protein (IL-18BP) [77–81]. IL-18BP is a naturally produced soluble factor that blocks the binding of IL-18 to its cognate receptor. IL-18BP has been shown to be one of the most potent inhibitors to the development of a Th1-biased immune response [79]. Vaccinia virus also encodes for several other immunosuppressive factors including factors to block complement activation, IL-1β soluble receptor, and soluble TNF-receptor antagonist [72,82–90]. These observations suggest that subverting the early innate immune response and slowing development of Th1 responses is important for the efficacy of vaccinia therapy.

Other studies have confirmed the critical role of the Th1 response to clearance of vaccinia viral infection. Van den Broek et al. examined the effect of Th1 (IFN-γ, IL-12) and Th2 (IL-4, IL-10) balance in the clearance of vaccinia virus in mice using cytokine knockouts [91]. Vaccinia viral replication was enhanced in IL-12 and IFN-γ knockout mice, with IL-12–/– demonstrating greater susceptibility to infection than IFN-γ deficient mice. Interestingly, development of anti-vaccinia CTL was completely abrogated in IL-12–/– mice but remained normal in IFN-γ–/–. In contrast, IL-4 and IL-10 deficient mice showed marked enhancement of vaccinia viral clearance, suggesting that these cytokines naturally suppress the host response to vaccinia. IL-10–/– mice exhibited greater inhibition of viral replication than IL-4 deficient mice. When the effects of each of these cytokines on vaccinia infection was examined in recombinant viral constructs, local expression of IL-4 showed a much greater inhibition of host responses. In fact, while the absence of IL-10 resulted in improved clearance of vaccinia virus which was mediated by increased levels of IL-6 and IL-1, the local expression of IL-10 had little to no effect on viral clearance. Similarly, Deonarain et al. have shown that IFN-α/β knockout mice demonstrate markedly enhanced susceptibility to vaccinia-viral infection [92].

There are several strategies that have been investigated to circumvent this problem of premature immune clearance. First, one could create a virus that is less recognizable by the immune system. This could be accomplished by mutating the viral coat of the vaccinia virus to make it less cross-reactive with antibodies. However, the Poxviridae and in particular vaccinia virus are antigenically very complex and it is unlikely that one or two mutations in viral envelope genes could significantly alter

antibody recognition. Further any mutations in the viral envelope may decrease the infectivity of the virus. Another strategy would be to develop other poxviruses that are able to selectively infect and lyse human tumor cells but do not cross-react immunologically with vaccinia virus. Viruses from the Yatapox genus infect monkeys and secondarily have infected monkey caretakers [93]. These viruses do not cross-react with vaccinia, yet they cause human disease and replicate in human cells. The yaba-like disease (YLD) virus is under investigation as another replicating poxvirus for tumor-directed gene therapy [18]. Avian poxviruses also do not cross-react with vaccinia virus and have become popular expression vectors [94]. They do not replicate in human cells and are less efficient vectors overall.

Another approach to circumventing premature clearance of a vaccinia vector is to create a viral recombinant that actively suppresses host cellular immune responses. Several groups have reported that insertion of Th2 like cytokines such as IL-4 or IL-10 into vaccinia virus increases *in vivo* viral replication and slows host clearance of infection [91,95]. However, creation of a virus that is not recognized by the immune system obviously creates serious safety concerns for the population as a whole as unforeseen events could lead to a pathogenic virus which is not immunologically cleared.

A third approach to improving *in vivo* viral replication involves reversible, transient host immunosuppression. Because of the growth of knowledge in solid organ transplantation we now have available multiple immunosuppressive agents that can very precisely target specific pathways of the host immune response. This knowledge combined with our growing understanding of the immune response to vaccinia virus should allow us to reversibly slow the immune response theoretically allowing for more efficient *in vivo* viral replication in the tumor, higher transgene expression and greater oncolysis. This will be a feasible approach with vaccinia mutants exhibiting tumor selectivity.

10.6 *IN VIVO* PATHOGENICITY AND BIODISTRIBUTION

Vaccinia virus has been used as a live vaccine in the smallpox eradication program, and more recently as a vaccine against cancer [96]. It has not been widely accepted as a potential tumor-directed gene therapy vector, however, due to concerns regarding the safety of a systemically administered, replicating virus. With the recent heightened interest in the smallpox vaccine due to threats of bioterrorism more information has emerged regarding the safety of the smallpox (vaccinia) directly and secondary spread to healthcare workers, immunocompromised patients, and family members [97]. Although it is generally considered to be a relatively safe vector for vaccinations, a defined risk exists for generalized vaccinia, vaccinia-associated encephalitis, vaccinia necrosum, and eczema vaccinatum in infants and the immunosuppressed population, specifically those with deficits in cellular immunity [98–103]. Vaccinia-associated encephalitis is a recognized complication of smallpox vaccination that can lead to death, and vaccinia can be recovered from the central nervous

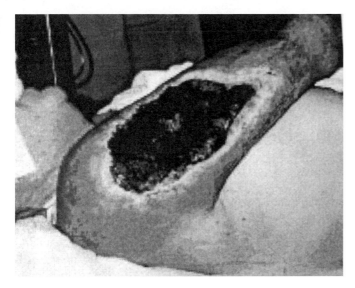

FIGURE 10.3 Example of vaccinia necrosum in a 66-year-old male 50 days after vaccination with a vaccinia melanoma cell lysate. The man had chronic lymphocytic leukemia. (From Wills V.L., et al., *Aust. N. Z. J. Surg.*, 70, 149, 2000. With permission.)

system [102]. Classically, vaccinia infection in immunosuppressed patients leads to a progressive necrotic ulcer known as vaccinia necrosum. This ulcer can progress to destroy significant amounts of tissue, leaving exposed bone, requiring tissue grafts or amputation [104] (Figure 10.3). Often this dramatic local infection does not lead to systemic viral spread. Patients with eczema, however, can get vaccinia infection of eczematous skin throughout the body (eczema vaccinatum). A large viral load such as this leads to fever and malaise and can lead to death from a "septic syndrome." The risk of secondary spread of vaccinia has been estimated to be in the range of 10% and of secondary cases the mortality rate was 11% [97]. As with the vaccine, risks of secondary infection are most common in children, patients with dermatological conditions such as eczema, and immunocompromised individuals [97].

The exact cause of vaccinia pathogenicity and lethality in animal models is difficult to determine. Mice moribund from wild-type vaccinia infection have demonstrated high levels of circulating inflammatory cytokines, and it is suspected that the systemic inflammatory response syndrome (SIRS) plays a significant role in viral pathogenicity as opposed to specific organ dysfunction (Naiak, A. et al., unpublished data).

The biodistribution of systemic injection of wild-type and TK-deleted vaccinia virus has been studied extensively in many animal models, including mice, rats, rabbits, and primates [10,67,105]. The highest titers of recoverable virus are always from the ovary. Many-fold less virus is recoverable from the liver, spleen, lung, and brain. Cutaneous pox lesions can form which have high titers of vaccinia. Mutations (such as TK-deletion) which attenuate the virus by making it less efficient for replication in nondividing cells are less recoverable from normal organs after systemic injection [106].

10.7 SAFETY CONSIDERATIONS AS A GENE THERAPY VECTOR

Safety considerations for cancer gene therapy vectors include direct pathogenicity of the virus, toxicity of the therapeutic gene product, genome insertion with risk for malignant transformation and germ line mutations, teratogenesis, and the ability to recombine with endemic virus or spontaneously mutate to form a more virulent pathogen. Because vaccinia is a cytoplasmic virus, the viral DNA does not transport to the nucleus and therefore integration into the genome is very unlikely. In addition, there is no known latent infection with vaccinia virus and all cells infected by the virus will be killed by the virus. In addition, since poxviruses are not endemic in the population, it is extremely unlikely for recombinations to occur in patients between attenuated strains and wild-type strains which would result in a more virulent virus with world health implications [107]. The stability of the virus has already been proven during vaccination as part of the small pox eradication program, so it is unlikely for spontaneous mutations to occur which would change the pathogenicity.

On the other hand, the properties that make it a useful virus for tumor-directed gene therapy also make it potentially dangerous. It replicates efficiently in human cells, and its pathogenicity as a systemically delivered virus in humans is unknown. The scarification of the skin during vaccination for small pox results in viral replication in the dermis, pox formation over 5–7 days, followed by an aggressive immune response against the virus which eliminates the virus and prevents systemic spread. A permanent scar in the skin results from the infection. It is not difficult to imagine that if such an infection occurred in an organ such as the brain that this could result in a poor outcome. During vaccination for smallpox, some patients with T-cell deficient immune systems suffered progressive systemic infection and death from vaccinia [65]. In vaccine trials for HIV patients, deaths have been attributed to systemic viremia in the setting of an immunocompromised host [108]. While intradermal delivery is quite safe for the vaccine strains, more virulent strains such as WR delivered systemically may be more pathogenic. These viruses need to be carefully examined in preclinical toxicology studies prior to human trials.

Vaccinia and other poxviruses have been identified, designed, or treated such that they no longer replicate in human cells, but still efficiently express genes. These include the MVA strain (attenuated by serial passage in chick embryo fibroblasts, until it no longer replicated in human cells), NYVAC (life cycle blocked prior to DNA replication in non-avian cells), fowl poxvirus, and entomopox viruses [109–111]. Vaccinia can also be reliably inactivated using UV light and psoralen such that early genes are still expressed, but no cytopathic effect or replication occurs [59]. Also, viral mutants can be constructed with deletions in essential genes preventing replication except in cell lines where the gene is compensated for by stable integration into the genome. While nonreplicating viruses improve the safety profile, they would not be expected to be efficient for the purpose of tumor-directed gene therapy. Any mutations which result in improved tumor specificity and decreased systemic virulence should be considered. An efficient strain of vaccinia virus such as WR may be mutated to inhibit replication in nondividing cells, but maintain efficient replication in tumor cells. This would significantly decrease viral pathogenicity but maintain the efficiency of vaccinia as a vector. An enzyme/prodrug approach may inhibit viral replication and provide a switch for turning off infection prior to host toxicity. Treatment with 5-fluorocytosine (5-FC) prolonged survival in a model where mice were administered a lethal dose of vaccinia expressing the cytosine deaminase gene [66]. This was the original intention of "suicide genes" and needs to be explored further in vaccinia [112].

Other strategies towards improving the safety of this vector have been described [46,47,96,113,114]. One strategy to attenuate the virus has been the deletion of genes required to evade the host immune response [61]. As discussed above, vaccinia expresses several proteins that interfere with the host response to viral infection. These include inhibitors of apoptosis such as the serpins SPI-1(B22R) and SPI-2(B13R), inhibitors of cytokines such as interferon, interleukins, tumour necrosis factor, and mechanisms to evade complement (vaccinia complement control protein). Deletions of many of these proteins have led to attenuation of the virus. Safety can also be improved by actively improving the host immune response against the vector. Vaccinia engineered to express inflammatory cytokines is rapidly cleared by the host, leading to decreased pathogenicity. IL-2 expressing vaccinia is rapidly cleared by NK cells, leading to marked attenuation. IFN-γ expressing vaccinia was also less virulent [115]. Another strategy to attenuate the virus has been the deletion of genes required for viral DNA synthesis. Deletion of the vaccinia TK gene or vaccinia growth factor (VGF) genes renders the virus dependent on the nucleotide pool of the host cell [46,47]. A mutant vaccinia virus with both the TK and VGF genes deleted was highly attenuated in nude mice [116].

10.8 CLINICAL EXPERIENCE

10.8.1 Smallpox Vaccination

Extensive clinical experience exists with vaccinia virus as a vaccine for the eradication of smallpox. The most common commercial preparation used in the United States was the Wyeth Dryvax [117]. It is the only vaccine available today. The virus was produced by infection of live calves by dermal scarification, followed by physical scraping of the skin. Future vaccines will be produced on cell lines, and ongoing trials are comparing strains for safety and efficacy [104]. The vaccine is delivered by scarification of the skin. The lyophilized virus is reconstituted and spread on the skin. A scarification needle is then used to penetrate the dermis through the vaccinia coat in multiple places. Effective vaccination is indicated by the development of pustules 6–10 days after vaccination. The pustules represent replicating vaccinia within the dermis. Live virus can be recovered from the pustules from days

3 through 14 after vaccination. There is a direct relationship between the intensity and extent of virus multiplication in the skin and the magnitude and duration of antibody response. The immune protection seems to last a lifetime, including both circulating antibodies and memory T cells [118]. The current recommendation, however, is to be boosted with vaccinia every 10 years.

Adverse events occurred in about 1250 per million vaccinations, as described above, including vaccinia necrosum, vaccinia-associated encephalitis, and eczema vaccinatum [118]. Aggressive dermal replication occurred almost exclusively in patients who were T-cell immunodeficient. The majority of deaths occurred in infants who suffered postvaccinal encephalitis. The risk of complications increased with the more virulent strains of virus used in Austria and Denmark. Despite worldwide use of this live virus vaccine, no reported adverse events related to mutation of the virus to a more aggressive phenotype was ever reported. No viral-induced tumor formation has been reported. Overall the virus is remarkably safe for use in humans, despite controlled viral replication in the skin of a potentially destructive virus.

10.8.2 Other Vaccines

After proven success as a vaccine responsible for the elimination of endemic smallpox in the world, the obvious leap towards using vaccinia as a vaccine for other indications was made, and vaccinia was engineered to express antigens from other infectious agents. Likewise as tumor antigens were recognized and defined, vaccinia was used as a cancer vaccine. The size of the vector allows for flexibility in engineering, such that immune enhancing genes and antigen genes can be recombined together into the genome. In general, these approaches do not rely on targeting of any specific tissues and may not require viral replication. Rather they are designed to take advantage of the immune stimulatory effects of the complex viral particle and the efficient transcriptional machinery of the virus. For safety considerations, nonreplicating vaccinia mutants were developed. The known inflammatory response to the vector combined with expression of tumor antigens and immunostimulatory molecules holds promise for cancer therapy. The potential seems great, but controversy exists as to whether complex immunogenic viruses like vaccinia may be less effective as vaccine vectors against proteins foreign to the virus. It is possible that weaker epitopes from tumor-associated antigens are effectively hidden by the strongly immunogenetic vaccinia proteins [119].

Vaccinia virus has been utilized in multiple clinical trials as vaccines for treatment of a variety of tumors as well as treatment of infectious diseases such as rabies, malaria, papilloma virus, mycobacterium, and HIV (Table 10.3). Vaccinia virus has been delivered as subcutaneous, intramuscular, intratumoral, and intravesical (bladder) injections in clinical immunotherapy trials without significant vector related toxicity [60,120,121]. Doses of up to 10^9 plaque forming units (pfu) have been delivered safely. Intravenous injection of fowlpox virus has been performed with no significant toxicity; however,

TABLE 10.3

Representative Clinical Trials with Poxviruses

Vector	Results	References
HIVAC-1e (gp160)	HIV immunity in healthy controls	[172]
Vaccinia-CEA	No clinical response	[173]
Vaccinia-GmCSF	Regression of injected lesions	[60]
Vaccinia-CEA	No clinical response	[129]
Vaccinia-IL-2	No clinical response	[138]
NYVAC-JEV	Neutralizing Ab to JEV	[174]
Vaccinia: PSA	Stabilization of PSA levels	[126]
Vaccinia: HPV	Responses in cervical cancer	[175]
Fowlpox-CEA-TRICOM modified DC's	Safe and CEA immune responses	[176]
DNA-Vaccinia: Mel3	Failed immunization	[177]
ALVAC-Mage	3% partial response	[178]
Fowlpox-vaccinia: CEA-TRICOM	1 of 58 complete response	[179]
Vaccinia-fowlpox NY-ESO-1	1 of 8 complete response	[180]
DNA-MVA: TRAP for malaria	Partial protection against *P. falciparum*	[181]
VV-HPV16 and 18	17% clinical response in intraepithelial neoplasia	[182]
Vaccinia/fowlpox tyrosinase	12.5% clinical response rate in melanoma	[183]
Vaccinia-fowlpox: PSA-TRICOM	PSA immune responses	[184]
MVA-5T4 for colorectal	5T4-specific immune responses	[185]
MVA-E2	Regression of flat condyloma	[186]

this species does not replicate in human cells [122]. No systemic injection of a replicating vaccinia virus has been performed in human trials.

Approximately 69 clinical trials have been reported in the English literature using poxviruses (PubMed search); 39 of these are tumor vaccines and 30 are for infectious diseases. Over half use replicating vaccinia virus and about 20% use MVA. Allogeneic cell lysate vaccines incorporating vaccinia virus have been explored the most in clinical trials. In these studies vaccinia was not used as a vector, but as an immunogen, and the virus was not replication competent. A phase III randomized, double blind, multi-institutional trial of an allogeneic vaccinia virus-augmented melanoma cell lysate (VMO) vaccine was performed with 250 patients from 11 centers [123]. A 10% survival advantage to VMO-treated patients was detected; however, this was not statistically significant. Hersey et al. in Australia also reported an improved survival in patients treated with an allogeneic vaccinia melanoma cell lysate (VMCL) vaccine [124,125]. In many of these trials, DTH response and development of antibodies to tumor antigens correlates with disease free survival. No vaccinia pathogenicity was observed.

Eder et al. reported a phase I trial of vaccinia expressing prostate specific antigen (PSA) in prostate cancer patients [126]. The virus was delivered intradermally every 4 weeks for three doses. No significant toxicities were related to the virus, which was a Wyeth strain. A cutaneous reaction consistent with viral replication was seen in all patients treated with 2.65×10^7 pfu vaccinia or greater. After the third dose, 14 of 19 patients continued to demonstrate cutaneous replication. Several patients developed T-cell immune responses associated with prolonged stabilization of their cancer.

Preclinical and clinical data suggest an advantage to a prime-and-boost strategy using two different vectors [127]. This has become popular with poxviruses, given the immunologically diverse poxviruses available as expression vectors. Vaccinia virus has been combined with fowlpox virus, Sindbis virus, peptides, and plasmid DNA. A recent clinical study examined recombinant vaccinia expressing NY-ESO-1 followed by recombinant fowlpox-NY-ESO-1 in a prime boost setting for cancer patients. One of eight patients with melanoma had a complete response [128]. Vaccinia expressing CEA has been studied clinically as a priming vaccine followed by a boost with avipox expressing CEA [129]. This regimen consisted of 1×10^7 pfu Wyeth strain vaccinia injected intradermally, and it was well tolerated. Specific T-cell immune responses were generated without clinical responses.

A good example of the utility of vaccinia as an immune vector is the development of rV-CEA TRICOM by Greiner et al. [130]. This vaccinia expresses a triad of costimulatory molecules: B7.1, ICAM-1, and LFA-3 along with CEA for a vaccine against CEA expressing cancers. A phase I study of Fowlpox-CEA(6D)-TRICOM sequentially with Vaccinia-CEA(6D)-TRICOM with GM-CSF was reported demonstrating 40% stable disease for at least 4 months, and 1 of 58 patients had a complete pathologic response [131]. Poxviruses have also been used to infect dendritic cells and express antigens and costimulatory molecules for vaccination. A phase I trial was reported of dendritic cells infected with fowlpox encoding CEA and costimulatory molecules, demonstrating safety but minimal clinical efficacy [132].

10.8.3 REPLICATION-SELECTIVE VACCINIA: ONCOLYTIC THERAPY

The extent of experience with vaccinia over the years and its proven safety record has lead to acceptance of exploration of this vector in more novel delivery systems for terminal cancer patients. As the field of tumor-directed cancer gene therapy has developed, the study of tumor-selective replicating viruses has become an important endeavor. Because of the known ability of vaccinia to destroy tissue (a complication of smallpox vaccination known as vaccinia necrosum) [133] vaccinia has also been developed as a direct oncolytic virus for cancer therapy. The concept of utilizing replication-competent viruses to selectively destroy tumors is quite appealing. Numerous viruses have been explored as tumor-selective replicating vectors, including adenovirus, herpes simplex virus, reovirus, newcastle disease virus, autonomous parvovirus, measles virus, vesicular stomatitis virus, coxsackievirus A21, and vaccinia virus [134–136]. Advantages and disadvantages exist for each of these vectors, and some limitations are common to all vectors.

Vectors are most limited by inefficient replication *in vivo*, inefficient tumor targeting, and safety concerns. Vaccinia has many characteristics that overcome these limitations: (1) It has a quick, efficient life cycle, forming mature virions in just 6h after infection. (2) It spreads efficiently cell to cell thus increasing the efficacy of an *in vivo* infection. (3) It has a large genome that can accept over 25 kb of inserted DNA without deletions. (4) Vaccinia virus carries its own strong promoters capable of achieving very high levels of transgene expression. (5) It can infect a wide range of human tissues but does not cause any known human disease. (6) Lastly, there is a large body of knowledge about its biology and extensive experience with it clinically as part of the smallpox vaccination program.

Preclinical development has focused on mutating the WR strain of vaccinia virus to make it replication selective for tumor cells [66,116,137–139]. As well as being more tumor lytic *in vitro* [12], the WR strain of vaccinia virus appears to be more efficient *in vivo* than other strains used in vaccination trials. An intradermal injection of 10^6 pfu of a wild-type WR strain of vaccinia in rhesus macaques led to a necrotic ulcer of $108\,cm^2$ in diameter in only 8 days, without systemic spread of the virus (Figure 10.4). This compared to <1 cm for NYCBH and Wyeth strains [140]. This ability to quickly spread, express genes, and destroy tissue to this extent is unique among current vectors in clinical and preclinical development.

Other nontumor sites of dividing cells need to be considered when developing tumor-selective viruses. These sites, known to suffer toxicity from chemotherapy agents, such as bone marrow-derived cells and gastrointestinal mucosa are not affected by systemic vaccinia in murine, rat, rabbit, or primate studies [10]. We and others have studied the wild-type virus and found that after intravenous injection, the highest amounts of virus can be recovered from the tumor and ovary, with minimal to no virus being recovered from other organs [12,67,105,116,141]. The natural tropism of this virus to tumor is surprising and the mechanism of this is not

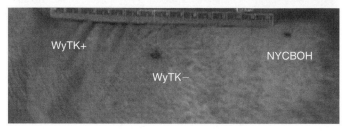

FIGURE 10.4 (See color insert following blank page 206.) Cutaneous lesions 8 days after intradermal injection of 106 pfu of the various vaccinia viruses: vF13 (wild-type WR strain), vJS6 (TK– WR strain), vvDD (TK–/VGF– WR strain), WyTK+ (wild-type Wyeth strain), WyTK–(TK– Wyeth strain), NYCBOH (wild-type New York City Board of Health strain). (From Naik, A.M., et al., *Hum. Gene Ther.*, 17, 31, 2006. With permission.)

established. This tumor tropism was demonstrated in numerous tumor models, including murine colon cancer and melanoma, rat sarcoma, human colon cancer in nude mice, and rabbit kidney cancer. Historically, smallpox virus was noted to have tropism for injured and irritated skin. This is thought to be secondary to histamine release leading to leaky vasculature allowing for transfer of the virus out of the circulation. Vaccinia is a large virus particle (350 nm in diameter) and requires leaky vasculature for extravasation into tissues. This leaky vasculature is lacking in tissues such as the GI mucosa. Despite bone marrow-derived cells having ready access to circulating vaccinia, no bone marrow toxicity is encountered, even in animals succumbing to viral pathogenicity (unpublished observations). Bone marrow-derived cells are known not to infect well by vaccinia for unknown reasons. The one place that replicating virus is recovered is ovarian follicles. Notably tumors and ovarian follicles are both known to be sites of vascular endothelial growth factor (VEGF) production and leaky vasculature [142]. As demonstrated by immunohistochemistry, vaccinia tropism to the ovary is specific for ovarian follicles without infection or spread through normal ovarian parenchyma [116]. In addition, local hyperthermia, which increases the "leakiness" of tumor vessels, was shown to enhance viral delivery [143].

In order to develop more tumor-selective vaccinia viruses, a number of mutations have been introduced. The WR strain of vaccinia was mutated such that the VGF gene and the TK genes were deleted [12,116]. The TK gene is important for nucleotide synthesis and DNA replication, and is near essential for replication in nondividing cells where the host nucleotide pool is low. VGF is a protein that is expressed early by vaccinia virus and is secreted by infected cells. It binds growth factor receptors on surrounding resting cells and stimulates them to proliferate. This increases the available nucleotides in these resting cells, priming them for vaccinia infection. Deleting both the TK and VGF genes leads to near complete abrogation of replication in resting cells, without decreasing the ability of the virus to replicate in the tumor environment as many tumors are known to have upregulated expression of EGF [12]. This mutant virus was found to have markedly enhanced tumor specificity (Figure 10.5). When tested in rhesus macaques it was found to be completely nonpathogenic when delivered intravenously at doses of up to 10^9 pfu [140]. Intradermal inoculation at 10^6 pfu demonstrated no viral replication. Nevertheless, 4 days after intravenous virus delivery, equal titers of wild-type and double-deleted virus could be recovered from subcutaneous tumors in mice [116]. This virus can be given systemically at doses of 10^9 pfu to a nude mouse without pathogenicity, and leads to marked responses in established subcutaneous tumors (Figure 10.6) [116]. The double-deleted WR strain virus has shown remarkable efficacy against the National Cancer Institute (NCI) panel of tumor cell lines (Figure 10.7) [12], and is significantly more tumor lytic than an oncolytic adenovirus Onxy-015 already in clinical trials.

A subsequent mutated WR vaccinia virus was made with deletions in SPI-1 and SPI-2 [139,144]. Viral serpin genes are required to help combat the host response to virus infection and SPI-1 and SPI-2 have been shown to inhibit apoptosis (Table 10.2) [145]. Normal cells and tissues would be expected to undergo apoptosis after infection with this mutant virus limiting infection and spread, yet viral replication should proceed normally in tumors which are known to harbor mutations in the apoptotic pathways [146]. This virus was able to replicate in tumors as well as the wild-type virus but was much less toxic. Systemic delivery led to

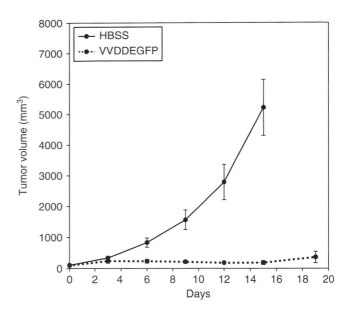

FIGURE 10.6   Antitumor response after systemic injection of 10^9 pfu of the TK/VGF-deleted virus in nude mice bearing 5 mm MC-38 tumors. (From McCart J.A., et al., *Cancer Res.*, 61, 8751, 2001.)

subcutaneous tumor regressions in both immunosuppressed and immunocompetent mice [139].

Other viral strains such as Wyeth, Lister, and LIVP have been similarly mutated to improve the tumor tropism of these viruses [147–149]. Wyeth and Lister are already attenuated vaccine strains of vaccinia virus and have been both further attenuated and rendered selective for dividing cells by insertional mutagenesis of the TK gene [147,148]. The LIVP strain was attenuated by three deletions: A56R encoding hemagglutinin, J2R encoding TK, and F14.5R an uncharacterized but highly conserved ORF. This led to improved survival of nude mice compared to wild-type virus. In contrast to the WR strain where tropism to the ovary is quite marked [116], LIVP is naturally not tropic to ovaries thus no viral recovery from the ovary was noted [149]. High-tumor replication and regression of established tumors in immunosuppressed mice was demonstrated.

In addition to the above mutations, which contribute to the tumor-selectivity of vaccinia virus, the antitumor efficacy has been enhanced by engineering transgenes into the deleted loci. The Wyeth and WR strain viruses were engineered to express GMCSF in order to enhance the antitumor immune response [12,147]. Tumor-bearing, immunocompetent rats and rabbits showed significantly reduced tumor burdens after vaccinia-GMCSF therapy [147]. Suicide genes such as cytosine deaminase and purine nucleotide phosphorylase [66,150,151] have also been used to improve the tumor responses. Interestingly, some prodrugs were also able to inhibit vaccinia replication [66] and may be able to be used as a safety switch if the viral infection becomes uncontrollable. The ability to noninvasively track vaccinia virus biodistribution and activity has been made possible by incorporating imaging transgenes such as the human somatostatin receptor

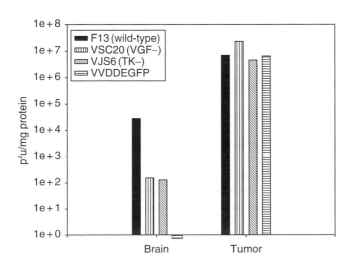

FIGURE 10.5   Differential viral recovery (median viral titers) from brain and tumor after systemic injection of 10^7 pfu wild-type, VGF-deleted, TK-deleted, and TK/VGF-deleted vaccinia virus in MC-38 tumor-bearing mice.

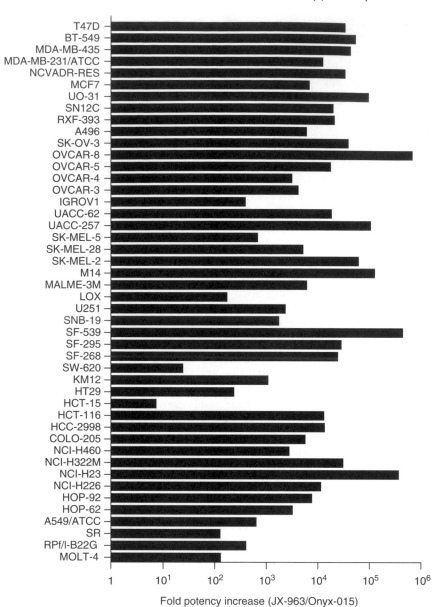

FIGURE 10.7 Cytopathic effect of JX-963 (vvDD expressing GM-CSF) versus Onyx-015 on a panel of human tumor cell lines. EC_{50} values were determined 3 days following infection of tumor cell lines with JX-963 and 6 days after Onyx-015 infection. The ratio of Onxy-015 EC_{50} to the JX-963 EC_{50} was plotted (a value greater than 1.0 indicates that JX-963 was more potent). (From Thorne S.H., et al., *J. Clin. Invest.*, 117, 3350, 2007. With permission.)

(SSTR2) for nuclear imaging (Figure 10.8) [152], green fluorescent protein for fluorescent imaging [116,149,153], and luciferase for bioluminescent imaging [148].

Clinical trials, focused directly on the oncolytic activity of vaccinia virus after intratumoral injections, have recently begun to recruit patients with metastatic melanoma (www. clinicaltrials.gov). Roenigk et al. described in 1974 direct injection of melanoma with vaccinia from standard vialed smallpox vaccine (Wyeth strain) at unknown concentrations in 20 patients at 2 week intervals [154]. Numerous interesting antitumor responses of injected lesions were described. Even older studies from the 1960s describe the

treatment of warts with direct intralesional vaccinia injection with success [155].

Mastrangelo et al. reported their results of intratumoral injection of a NYCBH vaccinia strain expressing GM-CSF into cutaneous melanoma [60,156]. This was a phase I trial of escalating doses up to 2×10^7 pfu per lesion and 8×10^7 pfu per session (multiple lesions injected). Patients were administered twice weekly intratumoral injections over 6 weeks. Systemic toxicity was limited to mild flu-like symptoms that resolved within 24 h and local inflammation at the injection site with doses of greater than or equal to 10^7 pfu per lesion. All patients were vaccinated against vaccinia within weeks

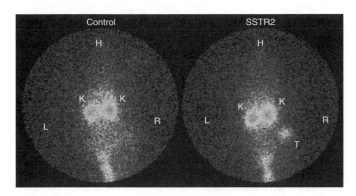

FIGURE 10.8 Posterior whole body images of tumor-bearing athymic mice 1 week after i.p. injection with vaccinia virus and 24 h after i.v. (tail vein) injection with ¹¹¹In-pentetreotide. Tumor (T) is visible on the right flank of vvDD-SSTR2-injected mouse (right) but not in the control vvDD-GFP-injected mouse (left). Prominent visualization of both kidneys (K) is noted in all animals as well as the tail (site of injection). The right (R) and left (L) sides of the mouse as well as the head (H) are indicated. (From McCart J.A., et al., *Mol. Ther.*, 10, 553, 2004. With permission.)

prior to receiving the vaccinia-GM-CSF. Interesting responses were seen in five of the seven patients treated. Three patients had mixed responses with complete regression of treated and untreated dermal metastases, one patient had a partial response with regression of injected and uninjected regional dermal metastases, and one patient with multiple dermal metastases confined to the scalp achieved a complete remission. This group plans to extend their observations with continued clinical trials using this vector.

10.8.4 TREATMENT OF POXVIRUS INFECTION

The treatment of orthopoxvirus infections has become of widespread interest recently due to the terroristic threat of biologic warfare with smallpox virus. As well, clinical trials utilizing vaccinia as a gene delivery vector would benefit greatly from a drug or compound which could turn off viral replication. Vaccinia immunoglobulin (VIG) is the only approved product available for treating complications of vaccinia infection. VIG has been owned by the Department of Defense, with a small amount available through the Centers for Disease Control [103]. No randomized controlled clinical trials have been performed to evaluate therapeutic efficacy, and therefore there is doubt as to its effectiveness in established complications from vaccinia. A randomized trial examining concomitant treatment with vaccinia and VIG demonstrated a significantly lower rate of postvaccinial encephalitis. Recommendations from the Bioterrorism Preparedness and Response Program suggest VIG as first line therapy and cidofovir as second line therapy for severe adverse reactions to vaccinia vaccines [157].

Numerous antiviral drugs have been tested in animal models of orthopoxvirus infections [158] (Table 10.4). Of the licensed antiviral compounds, cidofovir has the greatest potential for protection against and treatment of vaccinia infections [159,160]. Numerous other agents have also been identified that demonstrate efficacy against different poxviruses. The pharmacokinetics and safety profile for most of these agents have not been defined in humans, so their true utility will have to be determined in the future.

Suicide genes engineered into the virus may function to decrease viral replication upon addition of the prodrug. This has been demonstrated experimentally both *in vitro* and *in vivo* using a vaccinia expressing cytosine deaminase followed by the addition of 5-FC [66]. Prolonged survival after inoculation with a lethal dose of vaccinia in a murine model was achieved with addition of prodrug compared to controls. Future studies with more potent enzyme/prodrug systems may enhance this effect. Combinations of antiviral drugs and prodrugs may improve outcome.

The recent finding that the cell-to-cell spread of vaccinia is dependent upon the phosphorylation of src- and abl-family tyrosine kinases has enabled the development of a new family of drugs. Drugs such as ST-571 (Gleevec), an Abl-family kinase inhibitor, improve survival from a lethal vaccinia challenge in mice by inhibiting viral dissemination [41].

TABLE 10.4
Antiviral Compounds Effective *in Vivo* against Vaccinia

Animal Model	Compound	References
Vaccinia keratitis in rabbits	Ribavirin	[187]
Vaccinia tail lesion formation in mice following i.v. injection	Interferon	[188]
Vaccinia tail lesion formation in mice following i.v. injection	Polyacrylic acid	[189]
Vaccinia tail lesion formation in mice following i.v. injection	Ara-C, ribavirin, 5-iodo-dUrd, 5-ethyl-dUrd, 5-thiocyano-dUrd	[189]
Vaccinia tail lesion formation in mice following i.v. injection	C-c³ Ado	[190]
Vaccinia tail lesion formation in mice following i.v. injection	3-Deazaneplanocin A, Ara-A	[191]
Vaccinia tail lesion formation in mice following i.v. injection	(S)-HPMPA	[192]
Vaccinia-related death in SCID mice following i.v. injection	(S)-HPMPC	[193]
Vaccinia-related death in SCID mice following i.v. injection	H961 (diacetate ester prodrug of S2242)	[194]
Vaccinia tail lesion formation in mice following i.v. injection	ST-246	[195]
Vaccinia-related death in C57BL/6 mice following i.n. injection	STI-571(Gleevec)	[41]

The development of tightly regulated inducible gene expression systems would allow for *in vivo* induction of genes, which would be toxic to the virus itself and inhibit viral replication. This has been difficult in vaccinia virus, because of its unique transcription system. Nevertheless, Traktman et al. have reported on a tetracycline-inducible expression system in vaccinia [161]. This system should be explored further to demonstrate the potential for *in vivo* induction of genes that are self-toxic.

10.9 SUMMARY

Vaccinia virus is an interesting gene expression vector, which is worthy of continued exploration as a gene therapy vector. It is an efficient, destructive virus with some element of baseline tumor specificity. Powerful transcription machinery can lead to very high levels of therapeutic gene expression within tumor cells, and its immunogenicity may lead to an improved immunotherapy application. Mutations lead to tumor-specific replication and direct oncolytic applications. Because the vaccinia virus can include multiple genes it would be possible to simultaneously express toxic genes, multiple suicide genes, cytokine genes, costimulatory genes, HLA genes, and tumor antigens.

Compared to other replicating vectors such as herpes virus and adenovirus, the study of vaccinia as a tumor-directed vector is in its infancy. Over time the advantages of this vector may become more apparent, and its applicability may be more significant as the population ages and more cancer patients have not been vaccinated against smallpox. Further understanding of the biology of the virus will improve our ability to manipulate it to our advantage and enhance its potential as a vector for tumor-directed gene therapy.

REFERENCES

1. Niemialtowski M.G. et al. Controlling orthopoxvirus infections—200 years after Jenner's revolutionary immunization, *Arch. Immunol. Ther. Exp. (Warsz.)*, 44, 373, 1996.
2. Mastrangelo M.J. et al. Poxvirus vectors: Orphaned and underappreciated, *J. Clin. Invest.*, 105, 1031, 2000.
3. Esposito J.J. and Fenner F. In: Knipe D.M. and Howley P.M., (Eds.), *Fields Virology*. Philadelphia, Lippincott, Williams and Wilkins, 2001, pp. 2885–2921.
4. Baxby D. In: Quinnan GV, Jr. (Ed.), *Vaccinia Viruses as Vectors for Vaccine Antigens*. New York, Elsevier, 1985, pp. 3–8.
5. Fenner F., Wittek R., and Dumbell K.R. *The Orthopoxviruses*. New York, Academic Press, Inc., 1989, pp. 143–170.
6. Moss B. Vaccinia virus: A tool for research and vaccine development, *Science*, 252, 1662, 1991.
7. Smith G.L. and Moss B. Infectious poxvirus vectors have capacity for at least 25,000 base pairs of foreign DNA, *Gene*, 25, 21, 1983.
8. Moss B. Poxvirus vectors: Cytoplasmic expression of transferred genes, *Curr. Opin Genet. Dev.*, 3, 86, 1993.
9. Bartlett D.L. In: Hernaiz Driever P. and Rabkin S.D. (Eds.), *Replication-Competent Viruses for Cancer Therapy*. Basel, Karger, 2001, pp. 130–159.
10. Zeh H.J. and Bartlett D.L. Development of a replication-selective, oncolytic poxvirus for the treatment of human cancers, *Cancer Gene Ther.*, 9, 1001, 2002.
11. Moss B. In: Knipe D.M. and Howley P.M. (Eds.), *Fields Virology*. Philadelphia, Lippincott, Williams and Wilkins, 2001; 2849–2884.
12. Thorne S.H. et al. Rational strain selection and engineering creates a broad-spectrum, systemically effective oncolytic poxvirus, JX-963, *J. Clin. Invest.*, 117, 3350, 2007.
13. Sutter G. and Moss B. Novel vaccinia vector derived from the host range restricted and highly attenuated MVA strain of vaccinia virus, *Dev. Biol. Stand.*, 84, 195, 1995.
14. Goebel S.J. et al. The complete DNA sequence of vaccinia virus, *Virology*, 179, 247, 1990.
15. Antoine G. et al. The complete genomic sequence of the modified vaccinia Ankara strain: Comparison with other orthopoxviruses, *Virology*, 244, 365, 1998.
16. Dubochet J. et al. Structure of intracellular mature vaccinia virus observed by cryoelectron microscopy, *J. Virol.*, 68, 1935, 1994.
17. Smith G.L. and Vanderplasschen A. Extracellular enveloped vaccinia virus. Entry, egress, and evasion, *Adv. Exp. Med. Biol.*, 440, 395, 1998.
18. Hu Y. et al. Yaba-like disease virus: An alternative replicating poxvirus vector for cancer gene therapy, *J. Virol.*, 75, 10300, 2001.
19. Carter G.C. et al. Entry of the vaccinia virus intracellular mature virion and its interactions with glycosaminoglycans, *J. Gen. Virol.*, 86, 1279, 2005.
20. Senkevich T.G. et al. Poxvirus multiprotein entry-fusion complex, *Proc. Natl. Acad. Sci. U S A*, 102, 18572, 2005.
21. Townsley A.C. et al. Vaccinia virus entry into cells via a low-pH-dependent endosomal pathway, *J. Virol.*, 80, 8899, 2006.
22. Betakova T., Wolffe E.J., and Moss B. Membrane topology of the vaccinia virus A17L envelope protein, *Virology*, 261, 347, 1999.
23. Wolffe E.J. et al. Vaccinia virus A17L open reading frame encodes an essential component of nascent viral membranes that is required to initiate morphogenesis, *J. Virol.*, 70, 2797, 1996.
24. Sanderson C.M., Hollinshead M., and Smith G.L. The vaccinia virus A27L protein is needed for the microtubule-dependent transport of intracellular mature virus particles, *J. Gen. Virol.*, 81, 47, 2000.
25. Chung C.S. et al. A27L protein mediates vaccinia virus interaction with cell surface heparan sulfate, *J. Virol.*, 72, 1577, 1998.
26. Sodeik B. et al. Assembly of vaccinia virus: Incorporation of p14 and p32 into the membrane of the intracellular mature virus, *J. Virol.*, 69, 3560, 1995.
27. Hsiao J.C., Chung C.S., and Chang W. Vaccinia virus envelope D8L protein binds to cell surface chondroitin sulfate and mediates the adsorption of intracellular mature virions to cells, *J. Virol.*, 73, 8750, 1999.
28. Wallengren K. et al. The A17L gene product of vaccinia virus is exposed on the surface of IMV, *Virology*, 290, 143, 2001.
29. Ho Y. et al. The oligomeric structure of vaccinia viral envelope protein A27L is essential for binding to heparin and heparan sulfates on cell surfaces: A structural and functional approach using site-specific mutagenesis, *J. Mol. Biol.*, 349, 1060, 2005.
30. Law M. and Smith G.L. Antibody neutralization of the extracellular enveloped form of vaccinia virus, *Virology*, 280, 132, 2001.
31. Galmiche M.C. et al. Expression of a functional single chain antibody on the surface of extracellular enveloped vaccinia virus as a step towards selective tumour cell targeting, *J. Gen. Virol.*, 78 (Pt 11), 3019, 1997.
32. Katz E., Wolffe E.J., and Moss B. The cytoplasmic and transmembrane domains of the vaccinia virus B5R protein target a

chimeric human immunodeficiency virus type 1 glycoprotein to the outer envelope of nascent vaccinia virions, *J. Virol.*, 71, 3178, 1997.

33. Rahbar R. et al. Vaccinia virus activation of CCR5 invokes tyrosine phosphorylation signaling events that support virus replication, *J. Virol.*, 80, 7245, 2006.

34. Broyles S.S. et al. Transcription factor YY1 is a vaccinia virus late promoter activator, *J. Biol. Chem.*, 274, 35662, 1999.

35. Zhu M., Moore T., and Broyles S.S. A cellular protein binds vaccinia virus late promoters and activates transcription in vitro, *J. Virol.*, 72, 3893, 1998.

36. Broyles S.S. Vaccinia virus transcription, *J. Gen. Virol.*, 84, 2293, 2003.

37. Salzman N.P. The rate of formation of vaccinia deoxyribonucleic acid and vaccinia. virus, *Virology*, 10, 150, 1960.

38. Beaud G. Vaccinia virus DNA replication: A short review, *Biochimie*, 77, 774, 1995.

39. Sodeik B. and Krijnse-Locker J. Assembly of vaccinia virus revisited: De novo membrane synthesis or acquisition from the host? *Trends Microbiol.*, 10, 15, 2002.

40. Newsome T.P., Scaplehorn N., and Way M. SRC mediates a switch from microtubule- to actin-based motility of vaccinia. virus, *Science*, 306, 124, 2004.

41. Reeves P.M. et al. Disabling poxvirus pathogenesis by inhibition of Abl-family tyrosine kinases, *Nat. Med.*, 11, 731, 2005.

42. Newsome T.P. et al. Abl collaborates with Src family kinases to stimulate actin-based motility of vaccinia virus, *Cell Microbiol.*, 8, 233, 2006.

43. Moss B. Poxvirus entry and membrane fusion, *Virology*, 344, 48, 2006.

44. Blasco R., Sisler J.R., and Moss B. Dissociation of progeny vaccinia virus from the cell membrane is regulated by a viral envelope glycoprotein: Effect of a point mutation in the lectin homology domain of the A34R gene, *J. Virol.*, 67, 3319, 1993.

45. Yuen L. and Moss B. Oligonucleotide sequence signaling transcriptional termination of vaccinia virus early genes, *Proc. Natl. Acad. Sci. U S A*, 84, 6417, 1987.

46. Buller R.M. et al. Decreased virulence of recombinant vaccinia virus expression vectors is associated with a thymidine kinase-negative phenotype, *Nature*, 317, 813, 1985.

47. Buller R.M. et al. Deletion of the vaccinia virus growth factor gene reduces virus virulence, *J. Virol.*, 62, 866, 1988.

48. Fathi Z. et al. Intragenic and intergenic recombination between temperature-sensitive mutants of vaccinia virus, *J. Gen. Virol.*, 72 (Pt 11), 2733, 1991.

49. Davison A.J. and Moss B. Structure of vaccinia virus early promoters, *J. Mol. Biol.*, 210, 749, 1989.

50. Chakrabarti S., Sisler J.R., and Moss B., Compact, synthetic, vaccinia virus early/late promoter for protein expression, *Biotechniques*, 23, 1094, 1997.

51. Earl P.L. and Moss B. In: Ausubel F.M., Kinston R., Kingston R.E., Moore D.D., Seidman J.G., Smith J. et al. (Eds.), *Current Protocols in Molecular Biology*. New York, Greene/Wiley Interscience, 1998; 16.15.1–16.18.11.

52. Boyle D.B. and Coupar B.E. A dominant selectable marker for the construction of recombinant poxviruses, *Gene*, 65, 123, 1988.

53. Perkus M.E., Limbach K., and Paoletti E. Cloning and expression of foreign genes in vaccinia virus, using a host range selection system, *J. Virol.*, 63, 3829, 1989.

54. Domi A. and Moss B. Cloning the vaccinia virus genome as a bacterial artificial chromosome in *Escherichia coli* and recovery of infectious virus in mammalian cells, *Proc. Natl. Acad. Sci. U S A*, 99, 12415, 2002.

55. Domi A. and Moss B. Engineering of a vaccinia virus bacterial artificial chromosome in *Escherichia coli* by bacteriophage lambda-based recombination, *Nat. Methods*, 2, 95, 2005.

56. Panicali D. and Paoletti E. Construction of poxviruses as cloning vectors: Insertion of the thymidine kinase gene from herpes simplex virus into the DNA of infectious vaccinia virus, *Proc. Natl. Acad. Sci. U S A*, 79, 4927, 1982.

57. Mackett M., Smith G.L., and Moss B. Vaccinia virus: A selectable eukaryotic cloning and expression vector, *Proc. Natl. Acad. Sci. U S A*, 79, 7415, 1982.

58. Cochran M.A., Mackett M., and Moss B. Eukaryotic transient expression system dependent on transcription factors and regulatory DNA sequences of vaccinia virus, *Proc. Natl. Acad. Sci. U S A*, 82, 19, 1985.

59. Tsung K. et al. Gene expression and cytopathic effect of vaccinia virus inactivated by psoralen and long-wave UV light, *J. Virol.*, 70, 165, 1996.

60. Mastrangelo M.J. et al. Intratumoral recombinant GM-CSF-encoding virus as gene therapy in patients with cutaneous melanoma, *Cancer Gene Ther.*, 6, 409, 1999.

61. Smith G.L. et al. Vaccinia virus immune evasion, *Immunol. Rev.*, 159, 137, 1997.

62. Moss B. and Shisler J.L. Immunology 101 at poxvirus U: Immune evasion genes, *Semin. Immunol.*, 13, 59, 2001.

63. Haga I.R. and Bowie A.G. Evasion of innate immunity by vaccinia virus, *Parasitology*, 130 Suppl, S11–S25, 2005.

64. Smith G.L., Vanderplasschen A., and Law M. The formation and function of extracellular enveloped vaccinia virus, *J. Gen. Virol.*, 83, 2915, 2002.

65. Fenner F., Wittek R., and Dumbell K.R. *The Orthopoxviruses.* New York, Academic Press, Inc., 1989, pp. 85–141.

66. McCart J.A. et al. Complex interactions between the replicating oncolytic effect and the enzyme/prodrug effect of vaccinia-mediated tumor regression, *Gene Ther.*, 7, 1217, 2000.

67. Puhlmann M. et al. Vaccinia as a vector for tumor-directed gene therapy: Biodistribution of a thymidine kinase-deleted mutant, *Cancer Gene Ther.*, 7, 66, 2000.

68. Alcami A., Symons J.A., and Smith G.L. The vaccinia virus soluble alpha/beta interferon (IFN) receptor binds to the cell surface and protects cells from the antiviral effects of IFN, *J. Virol.*, 74, 11230, 2000.

69. Colamonici O.R. et al. Vaccinia virus B18R gene encodes a type I interferon-binding protein that blocks interferon alpha transmembrane signaling, *J. Biol. Chem.*, 270, 15974, 1995.

70. Symons J.A., Alcami A., and Smith G.L. Vaccinia virus encodes a soluble type I interferon receptor of novel structure and broad species specificity, *Cell*, 81, 551, 1995.

71. Seet B.T. and McFadden G. Viral chemokine-binding proteins, *J. Leukoc. Biol.*, 72, 24, 2002.

72. Alcami A. et al. Blockade of chemokine activity by a soluble chemokine binding protein from vaccinia virus, *J. Immunol.*, 160, 624, 1998.

73. Mahalingam S. and Karupiah G. Modulation of chemokines by poxvirus infections, *Curr. Opin. Immunol.*, 12, 409, 2000.

74. Alcami A. and Smith G.L. The vaccinia virus soluble interferon-gamma receptor is a homodimer, *J. Gen. Virol.*, 83, 545, 2002.

75. Najarro P., Traktman P., and Lewis J.A. Vaccinia virus blocks gamma interferon signal transduction: Viral VH1 phosphatase reverses Stat1 activation, *J. Virol.*, 75, 3185, 2001.

76. Alcami A. and Smith G.L. Vaccinia, cowpox, and camelpox viruses encode soluble gamma interferon receptors with novel broad species specificity, *J. Virol.*, 69, 4633, 1995.

77. Smith V.P., Bryant N.A., and Alcami A. Ectromelia, vaccinia and cowpox viruses encode secreted interleukin-18-binding proteins, *J. Gen. Virol.*, 81, 1223, 2000.

78. Calderara S., Xiang Y., and Moss B. Orthopoxvirus IL-18 binding proteins: Affinities and antagonist activities, *Virology*, 279, 22, 2001.

79. Novick D. et al. Interleukin-18 binding protein: A novel modulator of the Th1 cytokine response, *Immunity*, 10, 127, 1999.

80. Born T.L. et al. A poxvirus protein that binds to and inactivates IL-18, and inhibits NK cell response, *J. Immunol.*, 164, 3246, 2000.

81. Symons J.A. et al. The vaccinia virus C12L protein inhibits mouse IL-18 and promotes virus virulence in the murine intranasal model, *J. Gen. Virol.*, 83, 2833, 2002.

82. Engelstad M., Howard S.T., and Smith G.L. A constitutively expressed vaccinia gene encodes a 42-kDa glycoprotein related to complement control factors that forms part of the extracellular virus envelope, *Virology*, 188, 801, 1992.

83. Bowie A. et al. A46R and A52R from vaccinia virus are antagonists of host IL-1 and toll-like receptor signaling, *Proc. Natl. Acad. Sci. U S A*, 97, 10162, 2000.

84. Howard S.T., Chan Y.S., and Smith G.L. Vaccinia virus homologues of the Shope fibroma virus inverted terminal repeat proteins and a discontinuous ORF related to the tumor necrosis factor receptor family, *Virology*, 180, 633, 1991.

85. Smith G.L., Howard S.T., and Chan Y.S. Vaccinia virus encodes a family of genes with homology to serine proteinase inhibitors, *J. Gen. Virol.*, 70 (Pt 9), 2333, 1989.

86. Alcami A. and Smith G.L. A soluble receptor for interleukin-1 beta encoded by vaccinia virus: A novel mechanism of virus modulation of the host response to infection, *Cell*, 71, 153, 1992.

87. Spriggs M.K. et al. Vaccinia and cowpox viruses encode a novel secreted interleukin-1-binding protein, *Cell*, 71, 145, 1992.

88. Kotwal G.J. et al. Inhibition of the complement cascade by the major secretory protein of vaccinia virus, *Science*, 250, 827, 1990.

89. Jha P. and Kotwal G.J. Vaccinia complement control protein: Multi-functional protein and a potential wonder drug, *J. Biosci.*, 28, 265, 2003.

90. Alcami A. et al. Vaccinia virus strains Lister, USSR and Evans express soluble and cell-surface tumour necrosis factor receptors, *J. Gen. Virol.*, 80 (Pt 4), 949, 1999.

91. van Den B.M. et al. IL-4 and IL-10 antagonize IL-12-mediated protection against acute vaccinia virus infection with a limited role of IFN-gamma and nitric oxide synthetase 2, *J. Immunol.*, 164, 371, 2000.

92. Deonarain R. et al. Impaired antiviral response and alpha/beta interferon induction in mice lacking beta interferon, *J. Virol.*, 74, 3404, 2000.

93. Grace J.T. Jr. and Mirand E.A. Yaba virus infection in humans, *Exp. Med. Surg.*, 23, 213, 1965.

94. Paoletti E. Applications of pox virus vectors to vaccination: An update, *Proc. Natl. Acad. Sci. U S A*, 93, 11349, 1996.

95. Sharma D.P. et al. Interleukin-4 mediates down regulation of antiviral cytokine expression and cytotoxic T-lymphocyte responses and exacerbates vaccinia virus infection in vivo, *J. Virol.*, 70, 7103, 1996.

96. Moss B. Genetically engineered poxviruses for recombinant gene expression, vaccination, and safety, *Proc. Natl. Acad. Sci. U S A*, 93, 11341, 1996.

97. Sepkowitz K.A. How contagious is vaccinia? *N. Engl. J. Med.*, 348, 439, 2003.

98. Lane J.M. and Millar J.D. Risks of smallpox vaccination complications in the United States, *Am. J. Epidemiol.*, 93, 238, 1971.

99. Robinson M.J. et al. A fatal case of progressive vaccinia—clinical and pathological studies, *Aust. Paediatr. J*, 13, 125, 1977.

100. Turkel S.B. and Overturf G.D. Vaccinia necrosum complicating immunoblastic sarcoma, *Cancer*, 40, 226, 1977.

101. Keane J.T. et al. Progressive vaccinia associated with combined variable immunodeficiency, *Arch. Dermatol.*, 119, 404, 1983.

102. Gurvich E.B. and Vilesova I.S. Vaccinia virus in postvaccinal encephalitis, *Acta Virol.*, 27, 154, 1983.

103. Enserink M. Public health. Treating vaccine reactions: Two lifelines, but no guarantees, *Science*, 298, 2313, 2002.

104. Enserink M. Bioterrorism. In search of a kinder, gentler vaccine, *Science*, 296, 1594, 2002.

105. Peplinski G.R. et al. In vivo murine tumor gene delivery and expression by systemic recombinant vaccinia virus encoding interleukin-1beta, *Cancer J. Sci. Am.*, 2, 21, 1996.

106. Whitman E.D. et al. In vitro and in vivo kinetics of recombinant vaccinia virus cancer-gene therapy, *Surgery*, 116, 183, 1994.

107. Sandvik T. et al. Naturally occurring orthopoxviruses: Potential for recombination with vaccine vectors, *J. Clin. Microbiol.*, 36, 2542, 1998.

108. Dorozynski A. and Anderson A. Deaths in vaccine trials trigger French inquiry, *Science*, 252, 501, 1991.

109. Wang M. et al. Active immunotherapy of cancer with a non-replicating recombinant fowlpox virus encoding a model tumor-associated antigen, *J. Immunol.*, 154, 4685, 1995.

110. Li Y., Hall R.L., and Moyer R.W. Transient, nonlethal expression of genes in vertebrate cells by recombinant entomopoxviruses, *J. Virol.*, 71, 9557, 1997.

111. Blanchard T.J. et al. Modified vaccinia virus Ankara undergoes limited replication in human cells and lacks several immunomodulatory proteins: Implications for use as a human vaccine, *J. Gen. Virol.*, 79 (Pt 5), 1159, 1998.

112. Plautz G., Nabel E.G., and Nabel G.J. Selective elimination of recombinant genes in vivo with a suicide retroviral vector, *New Biol.*, 3, 709, 1991.

113. Flexner C. et al. Attenuation and immunogenicity in primates of vaccinia virus recombinants expressing human interleukin-2, *Vaccine*, 8, 17, 1990.

114. Ramshaw I.A. et al. Recovery of immunodeficient mice from a vaccinia virus/IL-2 recombinant infection, *Nature*, 329, 545, 1987.

115. Kohonen-Corish M.R. et al. Immunodeficient mice recover from infection with vaccinia virus expressing interferon-gamma, *Eur. J. Immunol.*, 20, 157, 1990.

116. McCart J.A. et al. Systemic cancer therapy with a tumor-selective vaccinia virus mutant lacking thymidine kinase and vaccinia growth factor genes, *Cancer Res.*, 61, 8751, 2001.

117. Baxby D. Smallpox vaccination techniques; from knives and forks to needles and pins, *Vaccine*, 20, 2140, 2002.

118. Demkowicz W.E., Jr. et al. Human cytotoxic T-cell memory: Long-lived responses to vaccinia virus, *J. Virol.*, 70, 2627, 1996.

119. Smith C.L. et al. Immunodominance of poxviral-specific CTL in a human trial of recombinant-modified vaccinia Ankara 2, *J. Immunol.*, 175, 8431, 2005.

120. Lattime E.C. et al. In situ cytokine gene transfection using vaccinia virus vectors, *Semin. Oncol.*, 23, 88, 1996.

121. Rosenberg S.A. et al. Human gene marker/therapy clinical protocols, *Hum. Gene Ther.*, 11, 919, 2000.

122. Rosenberg S.A. et al. Recombinant fowlpox viruses encoding the anchor-modified gp100 melanoma antigen can generate antitumor immune responses in patients with metastatic melanoma 1, *Clin. Cancer Res.*, 9, 2973, 2003.

123. Wallack M.K. et al. Surgical adjuvant active specific immunotherapy for patients with stage III melanoma: The final analysis of data from a phase III, randomized, double-blind, multicenter vaccinia melanoma oncolysate trial, *J. Am. Coll. Surg.*, 187, 69, 1998.

124. Hersey P. et al. Evidence that treatment with vaccinia melanoma cell lysates (VMCL) may improve survival of patients with stage II melanoma. Treatment of stage II melanoma with viral lysates, *Cancer Immunol. Immunother.*, 25, 257, 1987.

125. Hersey P. Evaluation of vaccinia viral lysates as therapeutic vaccines in the treatment of melanoma, *Ann. N. Y. Acad. Sci.*, 690, 167, 1993.

126. Eder J.P. et al. A phase I trial of a recombinant vaccinia virus expressing prostate-specific antigen in advanced prostate cancer, *Clin. Cancer Res.*, 6, 1632, 2000.

127. Irvine K.R. et al. Enhancing efficacy of recombinant anticancer vaccines with prime/boost regimens that use two different vectors 3, *J. Natl. Cancer Inst.*, 89, 1595, 1997.

128. Jager E. et al. Recombinant vaccinia/fowlpox NY-ESO-1 vaccines induce both humoral and cellular NY-ESO-1-specific immune responses in cancer patients 5, *Proc. Natl. Acad. Sci. U S A*, 103, 14453, 2006.

129. Marshall J.L. et al. Phase I study in advanced cancer patients of a diversified prime-and-boost vaccination protocol using recombinant vaccinia virus and recombinant nonreplicating avipox virus to elicit anti-carcinoembryonic antigen immune responses, *J. Clin. Oncol.*, 18, 3964, 2000.

130. Greiner J.W. et al. Vaccine-based therapy directed against carcinoembryonic antigen demonstrates antitumor activity on spontaneous intestinal tumors in the absence of autoimmunity, *Cancer Res.*, 62, 6944, 2002.

131. Marshall J.L. et al. Phase I study of sequential vaccinations with fowlpox-CEA(6D)-TRICOM alone and sequentially with vaccinia-CEA(6D)-TRICOM, with and without granulocyte-macrophage colony-stimulating factor in patients with carcinoembryonic antigen-expressing carcinomas 6, *J. Clin. Oncol.*, 23, 720, 2005.

132. Morse M.A. et al. Phase I study of immunization with dendritic cells modified with fowlpox encoding carcinoembryonic antigen and costimulatory molecules 4, *Clin. Cancer Res.*, 11, 3017, 2005.

133. Wills V.L. et al. Vaccinia necrosum: A forgotten disease, *Aust. N. Z. J. Surg.*, 70, 149, 2000.

134. Hernaiz D.P. and Rabkin S.D. Replication competent viruses for cancer therapy. Hernaiz D.P. and Rabkin S.D. (Eds.), [22], 1–187. 2001. Basel, Karger. Monographs in Virology. Doerr, H. W.

135. Kirn D. Virotherapy for cancer: Current status, hurdles, and future directions, *Cancer Gene Ther.*, 9, 959, 2002.

136. Russell S.J. RNA viruses as virotherapy agents, *Cancer Gene Ther.*, 9, 961, 2002.

137. Gnant M.F. et al. Systemic administration of a recombinant vaccinia virus expressing the cytosine deaminase gene and subsequent treatment with 5-fluorocytosine leads to tumor-specific gene expression and prolongation of survival in mice, *Cancer Res.*, 59, 3396, 1999.

138. Mukherjee S. et al. Replication-restricted vaccinia as a cytokine gene therapy vector in cancer: Persistent transgene expression despite antibody generation, *Cancer Gene Ther.*, 7, 663, 2000.

139. Guo Z.S. et al. The enhanced tumor selectivity of an oncolytic vaccinia lacking the host range and antiapoptosis genes SPI-1 and SPI-2, *Cancer Res.*, 65, 9991, 2005.

140. Naik A.M. et al. Intravenous and isolated limb perfusion delivery of wild type and a tumor-selective replicating mutant vaccinia virus in nonhuman primates, *Hum. Gene Ther.*, 17, 31, 2006.

141. Gnant M.F. et al. Tumor-specific gene delivery using recombinant vaccinia virus in a rabbit model of liver metastases, *J. Natl. Cancer Inst.*, 91, 1744, 1999.

142. Goede V. et al. Analysis of blood vessel maturation processes during cyclic ovarian angiogenesis, *Lab Invest.*, 78, 1385, 1998.

143. Chang E. et al. Targeting vaccinia to solid tumors with local hyperthermia, *Hum. Gene Ther.*, 16, 435, 2005.

144. Yang S. et al. A new recombinant vaccinia with targeted deletion of three viral genes: Its safety and efficacy as an oncolytic virus, *Gene Ther.*, 14, 638, 2007.

145. Shisler J.L. and Moss B. Immunology 102 at poxvirus U: Avoiding apoptosis, *Semin. Immunol.*, 13, 67, 2001.

146. Igney F.H. and Krammer P.H. Death and anti-death: Tumour resistance to apoptosis, *Nat. Rev. Cancer*, 2, 277, 2002.

147. Kim J.H. et al. Systemic armed oncolytic and immunologic therapy for cancer with JX-594, a targeted poxvirus expressing GM-CSF, *Mol. Ther.*, 14, 361, 2006.

148. Hung C.F. et al. Vaccinia virus preferentially infects and controls human and murine ovarian tumors in mice, *Gene Ther.*, 14, 20, 2007.

149. Zhang Q. et al. Eradication of solid human breast tumors in nude mice with an intravenously injected light-emitting oncolytic vaccinia virus, *Cancer Res.*, 67, 10038, 2007.

150. Puhlmann M. et al. Thymidine kinase-deleted vaccinia virus expressing purine nucleoside phosphorylase as a vector for tumor-directed gene therapy, *Hum. Gene Ther.*, 10, 649, 1999.

151. Chalikonda S. et al. Oncolytic virotherapy for ovarian carcinomatosis using a replication-selective vaccinia virus armed with a yeast cytosine deaminase gene, *Cancer Gene Ther.*, 15, 115, 2008.

152. McCart J.A. et al. Oncolytic vaccinia virus expressing the human somatostatin receptor SSTR2: Molecular imaging after systemic delivery using 111In-pentetreotide, *Mol. Ther.*, 10, 553, 2004.

153. Yu Y.A. et al. Visualization of tumors and metastases in live animals with bacteria and vaccinia virus encoding light-emitting proteins, *Nat. Biotechnol.*, 22, 313, 2004.

154. Roenigk H.H., Jr. et al. Immunotherapy of malignant melanoma with vaccinia virus, *Arch. Dermatol.*, 109, 668, 1974.

155. Israel R.M. Treatment of warts by vaccination, *Arch. Dermatol.*, 100, 222, 1969.

156. Mastrangelo M.J., Maguire H.C., and Lattime E.C. Intralesional vaccinia/GM-CSF recombinant virus in the treatment of metastatic melanoma, *Adv. Exp. Med. Biol.*, 465, 391, 2000.

157. Cono J., Casey C.G., and Bell D.M. Smallpox vaccination and adverse reactions. Guidance for clinicians, *MMWR Recomm. Rep.*, 52, 1, 2003.

158. De C.E. Vaccinia virus inhibitors as a paradigm for the chemotherapy of poxvirus infections, *Clin. Microbiol. Rev.*, 14, 382, 2001.

159. De C.E. Cidofovir in the treatment of poxvirus infections, *Antiviral Res.*, 55, 1, 2002.

160. Smee D.F. et al. Effects of cidofovir on the pathogenesis of a lethal vaccinia virus respiratory infection in mice, *Antiviral Res.*, 52, 55, 2001.

161. Traktman P. et al. Elucidating the essential role of the A14 phosphoprotein in vaccinia virus morphogenesis: Construction and characterization of a tetracycline-inducible recombinant, *J. Virol.*, 74, 3682, 2000.

162. Reading P.C., Moore J.B., and Smith G.L. Steroid hormone synthesis by vaccinia virus suppresses the inflammatory response to infection, *J. Exp. Med.*, 197, 1269, 2003.

163. Stack J. et al. Vaccinia virus protein A46R targets multiple Toll-like-interleukin-1 receptor adaptors and contributes to virulence, *J. Exp. Med.*, 201, 1007, 2005.

164. Harte M.T. et al. The poxvirus protein A52R targets Toll-like receptor signaling complexes to suppress host defense, *J. Exp. Med.*, 197, 343, 2003.

165. Vanderplasschen A. et al. Extracellular enveloped vaccinia virus is resistant to complement because of incorporation of host complement control proteins into its envelope, *Proc. Natl. Acad. Sci. U S A*, 95, 7544, 1998.

166. Chang H.W., Watson J.C., and Jacobs B.L. The E3L gene of vaccinia virus encodes an inhibitor of the interferon-induced, double-stranded RNA-dependent protein kinase, *Proc. Natl. Acad. Sci. U S A*, 89, 4825, 1992.

167. Davies M.V. et al. The E3L and K3L vaccinia virus gene products stimulate translation through inhibition of the double-stranded RNA-dependent protein kinase by different mechanisms, *J. Virol.*, 67, 1688, 1993.

168. Stewart T.L., Wasilenko S.T., and Barry M. Vaccinia virus F1L protein is a tail-anchored protein that functions at the mitochondria to inhibit apoptosis, *J. Virol.*, 79, 1084, 2005.

169. Shisler J.L. and Jin X.L. The vaccinia virus K1L gene product inhibits host NF-kappaB activation by preventing IkappaBalpha degradation, *J. Virol.*, 78, 3553, 2004.

170. DiPerna G. et al. Poxvirus protein N1L targets the I-kappaB kinase complex, inhibits signaling to NF-kappaB by the tumor necrosis factor superfamily of receptors, and inhibits NF-kappaB and IRF3 signaling by toll-like receptors, *J. Biol. Chem.*, 279, 36570, 2004.

171. Billings B. et al. Lack of N1L gene expression results in a significant decrease of vaccinia virus replication in mouse brain, *Ann. N. Y. Acad. Sci.*, 1030, 297, 2004.

172. Graham B.S. et al. Determinants of antibody response after recombinant gp160 boosting in vaccinia-naive volunteers primed with gp160-recombinant vaccinia virus. The National Institute of Allergy and Infectious Diseases AIDS Vaccine Clinical Trials Network, *J. Infect. Dis.*, 170, 782, 1994.

173. Tsang K.Y. et al. Generation of human cytotoxic T cells specific for human carcinoembryonic antigen epitopes from patients immunized with recombinant vaccinia-CEA vaccine, *J. Natl. Cancer Inst.*, 87, 982, 1995.

174. Kanesa-thasan N. et al. Safety and immunogenicity of NYVAC-JEV and ALVAC-JEV attenuated recombinant Japanese encephalitis virus—poxvirus vaccines in vaccinia-nonimmune and vaccinia-immune humans, *Vaccine*, 19, 483, 2000.

175. Adams M. et al. Clinical studies of human papilloma vaccines in pre-invasive and invasive cancer, *Vaccine*, 19, 2549, 2001.

176. Morse M.A. et al. Phase I study of immunization with dendritic cells modified with fowlpox encoding carcinoembryonic antigen and costimulatory molecules, *Clin. Cancer Res.*, 11, 3017, 2005.

177. Smith C.L. et al. Immunodominance of poxviral-specific CTL in a human trial of recombinant-modified vaccinia Ankara, *J. Immunol.*, 175, 8431, 2005.

178. van B.N. et al. Tumoral and immunologic response after vaccination of melanoma patients with an ALVAC virus encoding MAGE antigens recognized by T cells, *J. Clin. Oncol.*, 23, 9008, 2005.

179. Marshall J.L. et al. Phase I study of sequential vaccinations with fowlpox-CEA(6D)-TRICOM alone and sequentially with vaccinia-CEA(6D)-TRICOM, with and without granulocyte-macrophage colony-stimulating factor in patients with carcinoembryonic antigen-expressing carcinomas, *J. Clin. Oncol.*, 23, 720, 2005.

180. Jager E. et al. Recombinant vaccinia/fowlpox NY-ESO-1 vaccines induce both humoral and cellular NY-ESO-1-specific immune responses in cancer patients, *Proc. Natl. Acad. Sci. U S A*, 103, 14453, 2006.

181. Dunachie S.J. et al. A DNA prime-modified vaccinia virus ankara boost vaccine encoding thrombospondin-related adhesion protein but not circumsporozoite protein partially protects healthy malaria-naive adults against Plasmodium falciparum sporozoite challenge, *Infect. Immun.*, 74, 5933, 2006.

182. Fiander A.N. et al. Prime-boost vaccination strategy in women with high-grade, noncervical anogenital intraepithelial neoplasia: Clinical results from a multicenter phase II trial, *Int. J. Gynecol. Cancer*, 16, 1075, 2006.

183. Lindsey K.R. et al. Evaluation of prime/boost regimens using recombinant poxvirus/tyrosinase vaccines for the treatment of patients with metastatic melanoma, *Clin. Cancer Res.*, 12, 2526, 2006.

184. Arlen P.M. et al. Clinical safety of a viral vector based prostate cancer vaccine strategy, *J. Urol.*, 178, 1515, 2007.

185. Harrop R. et al. Vaccination of colorectal cancer patients with modified vaccinia ankara encoding the tumor antigen 5T4 (TroVax) given alongside chemotherapy induces potent immune responses, *Clin. Cancer Res.*, 13, 4487, 2007.

186. Albarran Y.C. et al. MVA E2 recombinant vaccine in the treatment of human papillomavirus infection in men presenting intraurethral flat condyloma: A phase I/II study, *BioDrugs*, 21, 47, 2007.

187. Sidwell R.W. et al. Effect of 1-beta-D-ribofuranosyl1–1,2,4-triazole-3-carboxamide (virazole, ICN 1229) on herpes and vaccinia keratitis and encephalitis in laboratory animals, *Antimicrob. Agents Chemother.*, 3, 242, 1973.

188. De C.E. and De S.P. Effect of interferon, polyacrylin acid, and polymethacrylic acid on tail lesions on mice infected with vaccinia virus, *Appl. Microbiol.*, 16, 1314, 1968.

189. De C.E. et al. Effect of cytosine, arabinoside, iododeoxyuridine, ethyldeoxyuridine, thiocyanatodeoxyuridine, and ribavirin on tail lesion formation in mice infected with vaccinia virus, *Proc. Soc. Exp Biol. Med.*, 151, 487, 1976.

190. De C.E. et al. Broad-spectrum antiviral activity of adenosine analogues, *Antiviral Res.*, 4, 119, 1984.

191. Tseng C.K. et al. Synthesis of 3-deazaneplanocin A, a powerful inhibitor of S-adenosylhomocysteine hydrolase with potent and selective in vitro and in vivo antiviral activities, *J. Med. Chem.*, 32, 1442, 1989.

192. De C.E., Holy A., and Rosenberg I. Efficacy of phosphonylmethoxyalkyl derivatives of adenine in experimental herpes simplex virus and vaccinia virus infections in vivo, *Antimicrob. Agents Chemother.*, 33, 185, 1989.

193. Neyts J. and De C.E. Efficacy of (S)-1-(3-hydroxy-2-phosphonylmethoxypropyl)cytosine for the treatment of lethal vaccinia virus infections in severe combined immune deficiency (SCID) mice, *J. Med. Virol.*, 41, 242, 1993.

194. Neyts J. and De C.E. Efficacy of 2-amino-7-(1,3-dihydroxy-2-propoxymethyl)purine for treatment of vaccinia virus (orthopoxvirus) infections in mice, *Antimicrob. Agents Chemother.*, 45, 84, 2001.

195. Yang G. et al. An orally bioavailable antipoxvirus compound (ST-246) inhibits extracellular virus formation and protects mice from lethal orthopoxvirus Challenge, *J. Virol.*, 79, 13139, 2005.

196. Grosenbach D.W. and Hruby D.E. Biology of vaccinia virus acylproteins, *Front. Biosci.*, 3, 354, 1998.

11 Baculovirus-Mediated Gene Transfer: An Emerging Universal Concept

Kari J. Airenne, Anssi J. Mähönen, Olli H. Laitinen, and Seppo Ylä-Herttuala

CONTENTS

11.1 INTRODUCTION

First commercial gene therapy treatments based on an adenovirus serotype 5 were launched recently in China [1–3]. Owing to the tighter regulatory rules, the first gene-based medicine in the Western world is expected to enter markets at the earliest by 2008 [4,5]. The emerging success is, however, still shadowed by an overall limited clinical success and the appearance of fourth cancer case in the French trial of X-SCID, laying clouds especially on the use of integrating vectors in gene delivery [6,7]. Indeed, much still remains to be solved before gene therapy becomes a standard clinical practice and further development of vectors is necessary [8]. In this context, baculoviruses have raised increasing interest because they have a long history as safe insecticides [9] and gene delivery tools in insect cells [10,11], and the long-lived dogma of incompatibility with vertebrate cells has been revised [12–14].

Baculoviruses are common in nature and even in our food, and have been known for hundreds of years [9,15]. Still, no diseases have been linked to baculoviruses in any organism outside the phylum *Arthropoda* [16]. Baculoviruses have been studied since 1920s as insecticides against forestry and agriculture pests [9,17] and, therefore, a lot of data are available about their biology [16] and biosafety [11,18]. The baculovirus expression vector system (BEVS) became a popular choice for recombinant protein production during the late 1980s and 1990s, with a large number of commercially available reagents [10].

It became evident already in early 1980s that baculoviruses can penetrate into nontarget cells nonproductively, including many human cell lines. No viral replication or gene expression could be detected [19,20]. However, in 1985, Carbonell et al. [21] reported a first successful transduction of mammalian cells by a recombinant baculovirus bearing a promoter active in target cells (Rous sarcoma virus [RSV] long terminal repeat promoter) as part of the recombinant baculovirus genome. In 1995, Hofmann et al. [22] confirmed these results. They were also the first to suggest baculovirus-mediated gene therapy. There was a 10-year gap between these two papers probably because Carbonell and Miller [23,24] later claimed that their initial findings of low-level gene expression in mammalian cells were due to pseudotransduction. During the late 1990s, the concept of baculovirus-mediated gene transfer was further verified and the list of suitable target cells is still continuously increasing.

The fact that baculoviruses are efficiently destroyed by complement [25] delayed the first successful *in vivo* applications of baculovirus-mediated gene transfer to the beginning of the new millennium [14], although the feasibility of these viruses in an *ex vivo* perfusion model was reported soon after baculoviruses were suggested as tools for gene therapy [25]. Several reports have also demonstrated the use of baculoviruses *in vivo* applications, especially in immune-privileged tissues such as brain, eye, or testis [26–31]. Indeed, baculoviruses offer many advantages compared with other viral vectors in terms of safety, high capacity for the incorporation of foreign DNA, and easy production (Table 11.1). This chapter provides an overview of the history as well as current status of baculovirus-mediated gene delivery. Future trends are also discussed. For a more detailed background on the biology of baculoviruses, the reader can refer to an excellent book, *The Baculoviruses*, edited by Miller [16] and a laboratory manual by O'Reilly et al. [10] describing recombinant protein production in insect cells by BEVS.

11.2 BACULOVIRUSES

Baculoviruses constitute a family of viruses, the *Baculoviridae*, including more than 600 known members [32–34], which infect permissively only arthropod hosts. The double-stranded circular DNA genome (80–180 kbp) of baculoviruses [34–36] is condensed into a nucleoprotein structure known as a core [37]. The core is located within a flexible rod-shaped capsid, averaging 25–50 nm in diameter and 250–300 nm in length [34,38–40] and can expand relatively freely to accommodate even very large DNA inserts [41]. The core and the capsid are known collectively as a nucleocapsid. Membrane-enveloped nucleocapsids are referred to as virus particles or virions [10].

Traditionally, baculoviruses are divided into two morphologically distinct genera: nuclear polyhedrosis viruses (NPVs) and granulosis viruses (GVs). However, recently an updated classification according to baculovirus genetic relationships has been suggested. Comparisons of 29 baculovirus genomes indicated that baculovirus phylogeny follows the classification of the hosts more closely than morphological traits. The updated classifications include four genera: alphabaculovirus (lepidopteran-specific NPV), betabaculovirus (lepidopteran-specific granuloviruses), gammabaculovirus (hymenopteran-specific NPV), and deltabaculovirus (dipteran-specific NPV) [42]. In the NPV group, virions that obtain an envelope from nuclear membrane are occluded within a paracrystalline protein matrix (occluded virions [OVs]), forming large (1–15 μm) polyhedral inclusion bodies (PIBs) containing multiple virions. NPVs are further distinguished on the basis of whether they contain a single nucleocapsid (SNPV) or multiple nucleocapsids (MNPVs) per envelope in the polyhedrin matrix [43]. In contrast to NPVs, GVs have only a single virion embedded in a very small inclusion body [10].

Baculoviruses exist in two distinct forms that play different roles during natural life cycle of the virus (Figure 11.1). Occlusion-derived virions (ODVs) are responsible for horizontal transmission between insect hosts, whereas systemic spread within the insect and propagation in tissue culture is dependent on budded viruses (BVs) [16,44,45]. Structurally, BVs and ODVs differ by the origin and composition of their envelopes [46]. BVs acquire their envelopes by budding through the plasma membrane whereas envelopes of ODVs are derived from the nuclear membrane of the insect cell. They also differ in the mechanisms by which they enter the host cells. BVs enter the cells by adsorptive endocytosis [47],

TABLE 11.1
Advantages of Baculoviruses in Gene Delivery

Advantages	References
Easy and fast preparation at high titers (>10^{10} pfu/mL)	[10,14,83,97,98,169]
Capability to transduce all type of cells from any origin	[11,12,20,103,105,106,142,143,147]
High capacity for DNA inserts (>100 kb)	[10,207,208]
Safe, no replication in mammalian cells (risk group 1 agents)	[20,238]
Little or low cytotoxicity even at a very high moi	[25,107,114,115,358]
Viruses stable at 4°C protected from the light	[359,360]
FDA approved production system	[264,266]
Automated virus generation possible for a high-throughput applications (drug screening, expression cassette optimization, etc.)	[97,98,254,358]

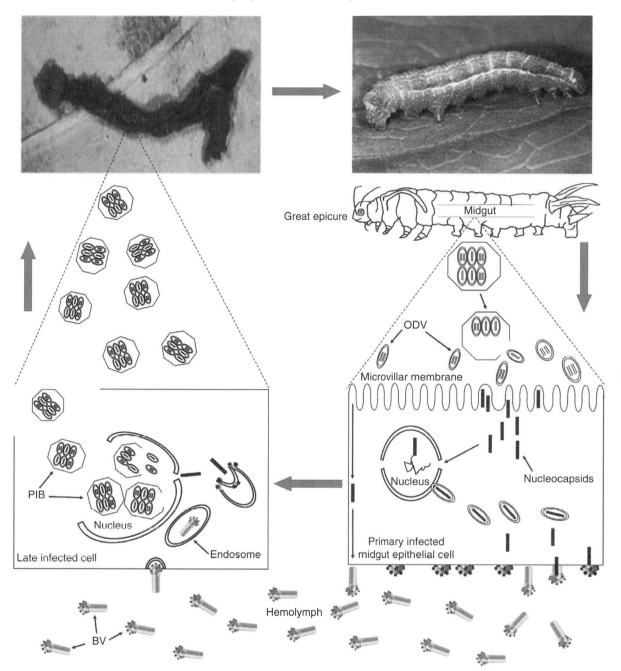

FIGURE 11.1 Life cycle of AcMNPV. Larva ingests virions containing PIBs with their nourishment. In the alkaline environment of the midgut, the PIBs break down and occlusion-derived virions (ODVs) infect primarily epithelia of the gut wall. ODVs fuse directly to the plasmalemma of epithelium and the nucleocapsids are transported into the nucleus. A minor proportion of nucleocapsids may travel through the cells in a process called transcytosis. The infection spreads through hemolymph inside the larva via BVs that contain an essential protein gp64 on their envelopes. Later in the infection, virions are packed into PIBs inside the nucleus of the infected cells. PIBs are released to the environment ready to infect the next hosts (horizontal infection) when cells lyse and larva dies. ⬡, PIB; ⬮, OVs; ▬▬✸, BV; 🦠, gp64 adhesion point in the plasma membrane of infected cells; |, nucleocapsid.

but ODVs enter the midgut epithelial cells via direct membrane fusion at the cell surface [48,49]. PIBs are formed during late phase of natural infection by embedding the virions in the crystalline protein matrix, which is composed mostly of polyhedrin protein [50]. PIBs enable the horizontal infection of larva by contaminating the plant on which the larva feeds [51]. The virions are protected from environmental factors within the PIB, but in the alkaline midgut of the larva, the crystalline polyhedrin matrix is solubilized [52] and the released virions enter the midgut cells by fusion with the membrane of the microvilli [53]. During the lytic cycle of infection, the cells release BVs from the basolateral area of the midgut cells [54]. The spread of infection within the insect occurs from the midgut, trough tracheal cells, to most tissues via the hemolymph [55–58]. Eventually, the larva dies and the PIBs that are produced in the very late phase of infection are released into the environment, and the cycle begins again. Most of the naturally occurring baculoviruses kill their target hosts within 4–7 days [17].

Baculovirus infection can be divided into early, late, and very late phases. Biologically these phases correspond to reprogramming the cell for virus replication (BV and PIB production). In the early phase (the first 6h), the virus prepares infected cell for viral DNA replication. This phase is also known as viral synthesis phase. Virus-specific RNAs can be detected in the cells by 30min postinfection (pi) [59]. The late phase extends from 6h pi to ~20–24h pi. During this viral structural phase, late genes are expressed and the production of BVs starts around 12h pi. Progeny nucleocapsids leave the nucleus and are transported onto the plasma membrane where they acquire their envelope. The occlusion-specific phase begins around 20h pi. Production of infectious BVs decreases and packaging of virus particles into polyhedrin matrix begins, followed by the cell lysis [10]. The packaging into polyhedrin matrix (i.e., production of PIBs) does not usually take place with BEVS, because the polyhedrin gene has been deleted from most baculovirus genomes used in biotechnology procedures.

11.2.1 AUTOGRAPHA CALIFORNICA MULTIPLE NUCLEOPOLYHEDROVIRUS AND BACULOVIRUS EXPRESSION VECTOR SYSTEM

The prototype of the family *Baculoviridae* and the most extensively studied baculovirus is the *Autographa californica* multiple nucleopolyhedrovirus (AcMNPV). Its genome (~134 kbp) has been sequenced and predicted to contain 154 open reading frames [60]. Although the three-dimensional structure remains to be determined, the components of its cigar-shaped loosely enveloped virion are fairly well known [61,62]. The vp39 represent the major capsid protein of AcMNPV (Figure 11.2). It is randomly distributed over the capsid surface. The major envelope glycoprotein of the BV and ODV form is gp64 and p74, respectively. The gp64 is believed to be responsible for the formation of so-called peplomer structure at one end of the virion (Figure 11.2). The nucleocapsid of ODV and BV is similar but their envelopes differ in the composition (Figure 11.2). The differences in lipid and protein composition of BV and OV envelopes reflect different origin and functions during the virus life cycle [46]. Basic DNA-binding protein, p6.9, is shown to be present in the virions of several baculoviruses, including AcMNPV. It may become phosphorylated upon entry into the insect cells, which may result in unpackaging of the viral DNA [63,64].

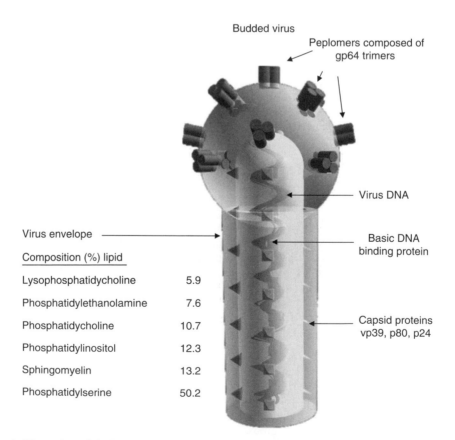

FIGURE 11.2 Schematic illustration of the budded form of AcMNPV. (Modified from Funk, C.J., Braunagel, S.C., and Rohrmann, G.F., *The Baculoviruses*, Plenum Press, New York, 1997.)

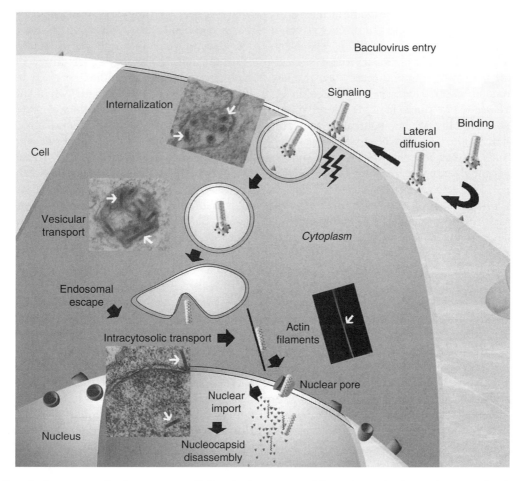

FIGURE 11.3 Baculovirus penetration into a target cell. Baculovirus enters cell by endocytosis via attachment and activation of unknown receptors. Nucleocapsid escapes endosome in a pH-dependent process mediated by baculovirus major envelope glycoprotein gp64. Actin is involved in intracytoplasmic movement and nuclear entry of a large nucleocapsid. Nucleocapsid disassembly takes place in the nucleus, and the double-stranded DNA genome is exposed. Transient transgene expression is promoter dependent. See text for more detailed description. Authentic electron and confocal microscopy pictures represent virions from the corresponding steps of entry. Nucleocapsids are shown by white arrows.

AcMNPV enters insect cells by adsorptive endocytosis [47] (Figure 11.3). The role of the major envelope glycoprotein, gp64, is essential for viral entry. It mediates virus binding to the cell surface and pH-dependent escape of the AcMNPV capsid from the endosomes [65]. Glycoprotein gp64 is necessary and sufficient also for virus preparation, because it is needed for efficient virion budding from the insect cells [66]. The cellular receptors for AcMNPV attachment and entry still remain to be revealed [67–69].

The range of transfer plasmids and parent viruses available for the AcMNPV, and the characteristics (growth, expression level) of the cell lines supporting AcMNPV, have made it a standard for eukaryotic protein production. The most commonly used cell lines for AcMNPV are Sf-9 and Sf-21AE. They both originate from IPLB-Sf-21 cells, which were derived from *Spodoptera frugiperda* pupal ovarian tissue [70]. A cell line derived from *Trichoplusia ni* egg cell homogenates (BTI-TN-5B1-4 = High Five), however, has become a popular alternative to Sf cell lines due to the fact that these cells have been shown to produce up to 28-fold more secreted

proteins than any other insect cell lines [71–73]. These cell lines grow well in both monolayer and suspension cultures, and a serum-free culture medium can be used if desired.

To enable expression of a recombinant protein in insect cells, the gene for the desired protein is usually placed under a strong polyhedrin promoter (*polh*) of AcMNPV [10]. For other cells, a promoter active in the target cells, such as CMV (cytomegalovirus), RSV, or CAG (chicken β-actin promoter) must be used because *polh* is inactive in these cells. The *polh* is normally responsible for the synthesis of polyhedrin, which can constitute up to 50% of the total protein of the infected cell. Fortunately, polyhedrin is not essential to virus replication or infection in the cell culture [74]. However, the use of the *polh* promoter may be restricted in some cases by the fact that it is activated very late in the infection cycle at a point where the host cell machinery for posttranslational modifications is compromised. Problems with the *polh* promoter have been encountered, especially with proteins whose biological activity depends on proper glycosylation [75]. In such cases, the use of an alternative strong viral promoter that initiates transcription earlier in the infection cycle (while the host post-translational modification pathways are still functional) could be useful. Indeed, Chazenbalk and Rapoport [76] were able to produce a more glycosylated and functional form of extracellular domain of the human thyrotropin receptor under a late basic protein promoter (*late*). However, engineered insect cells or virions are needed to mimic mammalian cell glycosylation of expressed proteins. While mammalian cells produce compositionally more complex N-glycans containing terminal sialic acids, insect cells mostly produce simpler N-glycans with terminal mannose residues [13,77]. Other promoters, which will be activated earlier than *polh*, include promoters for the p10 gene (*p10*), the major capsid protein gene (*vp39*), the basic 6.9-kD protein gene (*cor*), and the viral ie1 gene (*ie1*). These promoters are available in a variety of baculovirus plasmids [10,78–80].

11.2.2 Preparation of Recombinant Baculoviruses

Owing to the large size of baculovirus genome (80–180 kbp), a homologous recombination procedure was originally adopted to insert foreign genes into baculovirus genome, instead of conventional plasmid cloning techniques [81]. In practice, the target gene is subcloned into a transfer (donor) vector containing a suitable promoter, flanked by baculovirus DNA derived from a nonessential locus, such as the polyhedrin gene of AcMNPV. The viral DNA and transfer plasmid are then cotransfected into insect cells where the recombination events take place. Typically, 0.1%–1% of the resulting progeny is recombinant, which complicates their identification. If the target gene is inserted into the polyhedrin locus, altered plaque morphology of the recombinant viruses can be used for the identification of recombinant viruses. The cells in which the nuclei do not contain PIBs contain a recombinant virus. However, the detection of the desired PIB-minus plaque phenotype against the background of >99% wild-type parental viruses is difficult. Viral identification may be facilitated by the introduction of a *lacZ* cassette (β-galactosidase) along with the foreign gene, which enables the detection of the recombinant viruses according to a blue color [10]. Drug selection may also be used [82].

Several techniques have been developed to further facilitate and speed up the construction of recombinant baculoviruses [80,83]. By using a unique restriction site (*Bsu*36 I) at the polyhedrin locus, Kitts and coworkers [84] were able to linearize the double-stranded circular genome of AcMNPV. The linearization reduced the background of wild-type viruses and as a result, 10%–25% of the progeny viruses were recombinant. To obtain an even higher proportion of recombinants (85%–99%), Kitts and Possee [85] further modified the AcMNPV genome to enable *Bsu*36 I digestion to also remove an essential gene (ORF 1629) from the AcMNPV genome. Infective viruses will only be reconstituted by recombination with the transfer vector carrying the gene of interest, whereby an intact ORF 1629 will be restored to the genome. The system also enables a blue/white color selection of the recombinant viruses. However, it suffers

from the need of time-consuming plaque assays to purify the recombinant virus. A version of this system has been developed in which the target gene is amplified with specific primers using the polymerase chain reaction. This enables the ligase-free coupling of the linearized transfer vector and amplified gene in the mixture, which then can be used directly to transfect insect cells with cut viral DNA. The avoidance of cloning steps in *Escherichia coli* speeds up the construction of the recombinant viruses and simplifies manipulation of toxic genes [86,87]. To ease the manipulation of ORF 1629–deleted baculovirus genome, Je et al. [88] and Zhao et al. [89] have described a bacmid form of this virus that can be maintained in *E. coli*. These bacmids should also allow background-free virus preparation. However, generation of baculoviruses still relies on the inefficient homologous recombination in insect cells, and extraction and restriction enzyme treatment of the large ORF 1629–deleted viral genome (134 kbp) is troublesome.

Construction of the recombinant baculovirus genomes by direct cloning was first reported by Ernst and coworkers [90]. They introduced the I-*Sce*I homing endonuclease site into the AcMNPV genome by homologous recombination. The new virus genome, called Ac-omega, can be cut with I-*Sce*I meganuclease, and the target gene bearing compatible ends can be ligated straight to the linearized Ac-omega DNA under a polyhedrin promoter. A similar direct cloning system, Homingbac, utilizes I-*Ceu*I homing endonuclease for the direct cloning of transgenes into *UDP-ecdyglycosyltransferase* locus of several baculovirus species. All Homingbac viruses were designed to retain the polyhedra phenotype so that they could be inoculated *per os* to insects [91]. Direct and directional cloning into baculovirus genome by *Bsu*36 I has also been reported [92,93]. Direct cloning is simpler than homologous recombination in insect cells, but owing to the normal background problems encountered with conventional cloning techniques, the need of plaque purification cannot be entirely avoided. In addition, handling of the large AcMNPV genome (134 kbp) makes the method inconvenient. Another *in vitro* technique based on the preparation of recombinant baculoviruses by Cre–*loxP* system has also been described [94]. The advantage of this method is a possibility to recover cloned inserts from baculovirus genome into Cre/*loxP* compatible plasmids. However, only up to 50% of the viral progeny are recombinants.

To avoid laborious and time-consuming plaque purification process, the genetic material can be introduced into the baculovirus genome in yeast or bacterial cells. Patel and coworkers [87] reported a novel method of propagating the viral genome by homologous recombination in the yeast *Saccharomyces cerevisae*, where the appropriate recombinants can be more easily selected. Viruses are then obtained by transfecting insect cells. This method is rapid (pure recombinant virus within 10–12 days), and it ensures that there is no parental virus background. It also eliminates the need for time-consuming plaque assays, and multiple recombinants can be readily isolated. The major disadvantages of this system are the need of experience in yeast culturing and the incompatibility of traditional transfer vectors with the system. An even faster, and currently the most popular, approach for generating recombinant baculoviruses uses site-specific transposition with Tn7 to insert foreign genes into bacmid DNA (virus genome) propagated in *E. coli* cells. The *E. coli* clones containing recombinant bacmids are selected by color (β-galactosidase), and the DNA purified from a single white colony is used to transfect insect cells [95]. The system has the same advantages as the yeast system but is faster (pure recombinant virus within 7–10 days) and easier to work with for those not familiar with the yeast cells. A variant of the transposon-based baculovirus generation system was published in 1995 [96]. The poor selection features of the original system have recently been overcome by a modified donor vector (pBVboost) and an improved selection scheme of the baculovirus bacmids in *E. coli* with a mutated levansucrase gene from *Bacillus amyloliquefaciens*. The new selection schema bypasses the disadvantages associated with the original transposition-based generation of baculovirus genomes in *E. coli* while retaining the simple, rapid, and convenient virus production [97,98]. Baculovirus generation in *E. coli* by genetically integrated cloning strategy (MAGIC) [99] or phage λ red system [100] has also been described.

11.2.3　Transduction of Vertebrate Cells and Other Nontarget Cells

Baculoviruses have extensively been studied in the past for their ability to infect nontarget cells for safety considerations with regard to their use as biological pesticides (see Section 11.2.8). Early work by Volkman and Goldsmith demonstrated that baculoviruses were able to enter certain cell lines derived from vertebrate species [20]. Thirty-five nontarget host cell lines, 23 of human and 12 of nonhuman vertebrate origin, were exposed to AcMNPV. However, no evidence of viral gene expression was obtained. This study was in accordance with the earlier reports showing uptake of AcMNPV by several vertebrate cell lines with no evidence of viral replication [19,101]. Carbonell and coworkers [21] were the first to show in 1985 that by constructing a recombinant baculovirus bearing a suitable promoter, marker gene expression in nontarget cells can be obtained. Because only a low level of marker gene expression was observed in the studied cells, it was claimed by the authors later (1987) that marker protein was carried into the cells with the virus and not actively expressed in the cells [23,24]. Many publications since Carbonell and coworkers' initial findings have, however, confirmed that AcMNPVs containing mammalian expression cassettes can enter vertebrate cells and express reporter genes under the control of suitable promoters. Indeed, a universal concept is currently emerging that by using the latest generation vectors and optimizing procedures (see below), any cell from any source can efficiently be transduced by AcMNPV. Ample lists of tested mammalian cells for the baculovirus-mediated gene delivery can be found from several articles [12,102–107].

11.2.3.1　Mammalian Cells

Hofmann et al. [22,107] were the first to demonstrate in 1995, a decade after the pioneering work by Carbonell and coworkers [21], that a recombinant AcMNPV containing the luciferase gene under a CMV promoter can efficiently drive transgene expression in human hepatocytes (Huh7 and HepG2), as well as in primary hepatocytes of human and rabbit origin. In accordance with these results, Boyce and Bucher [108] showed next year that a virus carrying *lacZ* reporter gene under the control of RSV promoter led to a high-level expression of the marker gene in human hepatocellular carcinoma line HepG2, as well as in primary rat hepatocytes. More recently, primary rat and human hepatic stellate cells (HSCs) and human hepatoma FLC4 cells were shown to be highly susceptible to baculovirus-mediated gene delivery [102,109]. These findings favor the potential of baculovirus vectors for liver-directed gene therapy.

A high-level expression of marker gene can be achieved not only in hepatic but also in a wide range of other cell lines [102,104,106,107,110–113]. Shoji et al. [110] transduced a panel of mammalian cells with a baculovirus vector carrying a marker gene in comparison with a replication-defective adenovirus vector. In this study, high-level luciferase activity was detected not only in human hepatocytes but also in other cell lines such as monkey kidney cells (COS7), porcine kidney cells (CPK), and human cervix carcinoma cells (Hela). Furthermore, the same level of marker gene expression was observed in these cells by both viruses, but as an advantage for the baculovirus, much lower cytotoxicity was associated with it than adenovirus at a high multiplicity of infection (moi). The inert nature of baculovirus in vertebrate cells compared with adenovirus was also demonstrated in cells such as Hela [114], SHSY-5Y (human neuroblastoma), and C3A (human liver) [115]. The use of CAG promoter in the study by Shoji and coworkers [110] resulted in a 10-fold higher level of luciferase gene activity than in the previous study with a CMV promoter [22]. This indicates the importance of the promoter and the expression cassette per se for successful transduction of target cells not only by baculoviruses but also by any other gene delivery vector [28,116–119]. High-level expression of T7 RNA polymerase was also directed by CAG in HepG2 and CPK cells, and lower expression was observed in some other cell lines [111]. Condreay et al. [106] demonstrated that recombinant baculoviruses containing green fluorescent protein (GFP) gene under the CMV promoter can transduce a wide range of mammalian cell types originating from different tissues. Cell lines of hepatic origin were transduced efficiently, as described earlier, but notable gene expression was also detected in cell lines derived from kidney tissue (Cos-7, BHK, CV-1, 293), and other nonhepatic cell lines such as keratinocytes (W12, primary human keratinocytes), bone marrow fibroblasts, and osteosarcoma cells (MG-63). The lowest efficiency of transduction and level of GFP expression was seen in cell lines of hematopoietic origin, such as THP-1, U937, K562, Raw264.7, and P388D1.

Adherent cells such as WI-38 (lung fibroblast), MCF-7 (gastric carcinoma), BGC-223 (breast carcinoma), and DMS-114 (small cell lung cancer) were found to be more permissive for baculovirus-mediated gene delivery than suspension cell lines such as P815 (mouse mastocytoma) or CNE (human nasopharyngeal carcinoma) in a study that included 20 different cell lines of human, mouse, and monkey origins [103]. Some human osteogenic sarcoma cells (U-2OS, Saos-2, and Saos-LM2) [105], prostate cancer cells (PC3) [120], and prostatic stromal cells (CHPSCs) [121,122] have proved to be good targets for baculovirus. Studies aiming to define pleiotropic effects of IE2 protein of human CMV revealed human primary cells (human foreskin fibroblasts [HFF] and human umbilical vein endothelial cells [HUVEC]) as particularly good targets for the baculovirus [123,124]. In line with this, efficient gene transfer has also been obtained in primary human and mouse pancreatic islet cells [125]. Recombinant baculovirus expressed the glycoprotein gB of pseudorabies virus (PrV) under the CAG promoter in various mammalian cell lines and produced specific antibodies in mice against PrV [126]. High levels of expression of PrV gB were observed in many porcine kidney cell lines such as CPK, SK-H, and CPK-NS, and in hamster kidney cells (BHK-21). High efficiency gene transfer into mouse, hamster, monkey, pig, and human kidney cells was reported also by Liang and coworkers [127].

The potential of baculovirus for therapeutic applications was tested by a virus containing a p53 tumor suppressor gene under the control of the CMV promoter together with an anticancer drug, adriamycin [128]. Greater than 95% of Saos-2 cells were killed by the combination of the recombinant virus (moi of 100) and adriamycin (35 ng/mL), suggesting that the combination treatment greatly enhanced apoptosis of the tumor cells. Overexpression of the human homolog of the drosophila tumor suppressor l(2)Tid led to p53 independent inhibition of cell proliferation and induced apoptosis in human osteosarcoma cells [129]. Comparison between baculoviral and adenoviral vectors in cardiovascular cells supported AcMPNV as a potential vector for cardiovascular diseases such as restenosis [130]. To delay and reverse the degeneration of intervertebral disc disease, human Sox9 gene, an essential transcription factor for synthesis of type II collagen was expressed in rabbit vertebral pulp cells [131].

11.2.3.2 Stem and Progenitor Cells

Embryonic and adult stem and progenitor cells hold great therapeutic and research potential because they can, in principle, differentiate into any cell lineage in the body due to their high self-renewal capacity and plasticity [132–135]. Mesenchymal stem cells (MSCs) derived from blood of bone marrow and umbilical cord were efficiently transduced by AcMNPV. The growth of the MSCs was not compromised and the cells retained their plasticity by differentiating into adipogenic, chondrogenic, and osteogenic lineages. A trend similar to adenovirus transduction was observed that the transgene expression decreased as the MCS-derived progenitors differentiated toward the end-stage mature cells

[136–138]. Baculovirus-mediated expression of growth factors in rabbit articular chondrocytes proved the potential also to modulate the differentiation status of progenitor cells by AcMNPV for the cartilage engineering [139]. The efficacy of a baculoviral vector equipped with GFP under a human elongation factor 1-α promoter in human embryonic stem cells (hESCs) was up to 80% [140]. Transduction had no effect on hESCs growth, phenotype, or pluripotency. Extended transgene expression was achieved by a hybrid baculovirus carrying also rep gene and inverted terminal repeat sequences from AAV. However, the percentage of hESCs exhibiting stable EGF expression was only 2% [140].

11.2.3.3 Avian Cells

The suitability of baculovirus for gene delivery into avian cells was observed in virus-treated chicken primary myoblasts and whole-embryonic fibroblasts over 80% of cells expressing GFP after transduction [141]. Another study found the primary cells obtained from embryonic day 9 or day 15 chick embryo, or day 13 duck embryo, permissive for baculovirus transduction. Similar transduction efficacy was detected in all tested cells derived from different tissues of day 15 chick embryo (liver, heart, and lung). Duck cells seemed to be more permissive than chicken cells [142].

11.2.3.4 Fish Cells

Fish are subjected to many diseases and vaccination could be used to prevent infections. Therefore, five fish cell lines derived from tilapia, carp, salmon, and rainbow trout were studied for their ability to permit baculovirus-mediated gene delivery. As a result, a VSV-G pseudotyped baculovirus that contained the lacZ gene under CAG was shown to be capable of transducing all cells except rainbow trout gonad cells (RTG-2) [143]. Transduction of zebrafish embryos was demonstrated by Wagle and Jesuthasan [144]. By injecting virus into specific tissues and using right promoters, both the location and time of gene expression could be controlled. The results favor an idea of using AcMNPV for the functional genomics studies in zebrafish [145].

11.2.3.5 Other Non-Lepidopteran Hosts

Non-lytic and efficient gene delivery (up to 100% cells) was gained in fruit fly cells (Drosophila S2) transduced with a baculovirus containing gene of interest under the control of the promoter for the D. melanogaster heat-shock protein 70 [146]. Similar virus was also used successfully to study developmental biology in Drosophila, Tribolium, and Xenopus laevis embryos [147]. Recently, AcMNPV vector was reported to transfer and express EGFP in the larval and pupal honeybees [148].

11.2.4 TRANSDUCTION OPTIMIZATION

Baculovirus-mediated gene delivery can substantially be enhanced by optimized transduction conditions and vector design [98]. Although AcMNPV can enter any cell from any origin, universal entry does not guarantee efficient gene expression. Correct intracellular traffic and nuclear entry is also vital (see Section 11.2.6).

Cell culture medium has significant impact on the successful cell transduction. It has been observed that, especially, the commonly used Dulbeccos's modified Eagle's medium (DMEM) hinders baculovirus-mediated gene delivery [149,150]. This may also explain, along with possible clonal selection of cells lines, the published and partially contradictory results regarding susceptibility of different cell lines to baculovirus-mediated transduction by independent groups [150,151]. Systematic exploration of medium constituents suggested that HCO_3^-, rather than divalent cations such as Mg^{2+}, Ca^{2+}, or serum, exerts the inhibitory effect in a concentration-dependent manner [149]. Transient substitution of DMEM with PBS in transduction has been suggested as a solution to avoid DMEM associated problems. However, cells can be sensitive to cultivation in PBS [152]. Comparison of different standard culture mediums for transduction revealed that RPMI 1640, in general, gives the best results. Cells readily adapt to this medium without problems [150].

Several medium supplements have been reported to aid baculovirus-mediated gene delivery. Usage of histone deacetylase inhibitors (HDIs) like sodium butyrate or trichostatin A in culture medium usually markedly increases baculovirus-mediated gene expression [14,106]. The effect is supposed to be mainly due to the more exposed expression cassette structure [153]. These agents also enhance adenovirus- and retrovirus-mediated transgene expression *in vivo* and *in vitro* [14,154,155]. Interestingly, it was recently published that the U_S3 or $U_S3.5$ protein kinase of HSV could substitute HDI effect in U2OS cells [156,157]. Co-expression or baculovirus capsid display [151] of these kinases or their analogs might, thus, provide a useful tool to improve baculovirus-mediated gene expression. Microtubule interfering substances, such as nocodazole or vinblastine, augment baculovirus-mediated gene delivery because cytoplasmic transport and nuclear entry of baculovirus nucleocapsid seems to be hindered by intact microtubule network [40,158]. The HDI inhibitors and cytoskeleton interfering substances, however, are typically harmful to cells, which limits their use [159].

To obtain best results, transduction should also be performed in subconfluent cell culture. Established intercellular connections between cells grown to overconfluence disturb baculovirus penetration into cells. Disruption of the cell–cell junctions by a calcium chelator, ethylene glycol-bis(β-amino-ethylether)-N,N,N',N'-tetraacetic acid (EGTA), significantly improved the transduction efficiency [160–162]. Baculovirus entry into cells or tissues with tight cellular junctions may thus require transient breakage of intercellular linkages.

Better results have also been achieved by modification of the standard transduction protocol. Extended incubation time (>4 h) allows improved virus uptake, and transduction at temperatures under 37°C has been suggested to improve results by prolonging half-life of the virus in transduction [20,103,143,152,163]. In addition, since baculoviruses are not toxic to vertebrate cells, repeated exposure of cells with the virus (supertransduction) can extend the otherwise transient transgene expression [164,165]. Although working with purified virus is recommended [27,98,166–169], it is possible to transduce vertebrate cells with a virus in the insect cell medium diluted at least 1:1 to the desired culture medium [103,152,170]. This is useful, especially, for the screening purposes [171].

Baculovirus envelope modification is yet another option to achieve enhanced gene delivery. The major envelope glycoprotein, gp64, has been harnessed for this purpose [172,173]. However, truncated vesicular stomatitis virus G transmembrane glycoprotein (VSV-G) [174] or a transmembrane region of neuraminidase [175] can also be utilized to display peptides or proteins on the baculovirus surface. Several reports of baculoviruses pseudotyped with intact VSV-G provide evidence that VSV-G is also able to enhance transduction efficiency of baculovirus in vertebrate cells *in vitro* and *in vivo* [29,68,112,176]. VSV-G has a broad host range because its entry into the cells seems to not depend on the presence of any specific receptor but a phospholipid component of the plasma membrane [177]. VSV-G mediates pH-dependent membrane fusion in endosomes [178], and has earlier been used to pseudotype and stabilize other enveloped viruses such as murine retroviruses and lentiviruses [179–181]. In baculovirus studies, VSV-G was cloned under the control of the *polh* promoter so that it was expressed in infected insect cells but not in vertebrate cells. The level of marker gene expression in HepG2 cells treated with a pseudotyped virus was 10-fold higher than in the same cells treated with a wild-type virus. Pieroni et al. [176] showed that VSV-G enhances the transduction efficacy of pseudotyped baculovirus also *in vivo*. VSV-G or measles virus receptor CD46 can complement the gp64 and, therefore, allows productive infection, replication, and propagation of the gp64 deleted baculovirus in Sf9 insect cells [182,183]. However, the virus propagation was inefficient and delayed as compared with the wild-type virus, but virions were similar in morphology to the wild-type viruses. A major problem of VSV-G, cytotoxicity [180,184,185], can be avoided by a baculovirus vector displaying in addition to gp64 only a 21-amino acid ectodomain, in conjunction with the TM and CTD domains, of VSV-G (VSV-GED) [186]. In accordance with VSV-G pseudotyping, the VSV-GED virus resulted in several folds higher transduction efficiency in several cell lines as compared to the control virus. Not only the number of transduced cells was increased but also the level of reporter gene expression was higher in the transduced cells. Enhanced gene delivery *in vivo* was also detected. VSV-GED has several advantages compared to the VSV-G pseudotyping and provides a novel tool to improve baculovirus-mediated gene delivery without compromising high viral titers. Other examples of reported beneficial molecules for virus surface modification include the mouse hepatitis virus S protein [68], extra copy of gp64 [68], avian influenza virus hemagglutinin [187], RGD-motif [188], tumor-homing peptides [189], a short peptide motif from gp350/220 of EBV [190], and avidin [191]. Avidin displaying virus, Baavi, provides versatile potential not only to vector targeting but also imaging (see Section 11.3), being compatible with plethora of commercially available biotinylated molecules [192]. Pegylation is yet another possibility to modify baculovirus transduction efficacy [193].

Finally, the choice of promoter is important. CMV promoter is, in general, a good first choice to drive transgene expression using baculovirus vectors [194], but Chicken β-actin promoter has been shown to be more active in some cells [110]. The latter also suffers less from epigenetic silencing [195] and drive well expression in neurons as opposite to CMV [196,197]. Cell-type-specific gene expression has been achieved with tissue and organism-specific promoters [144,146,198–201]. Enhancers can be used to boost promoter activity. Examples include the cytomegalovirus enhancer [199,200,202] and the homologous region sequence (hr1) of AcMNPV [203]. The Woodchuck hepatitis virus posttranscriptional regulatory element (WPRE) was very recently recommended to be included in all baculovirus expression cassettes [150]. The WPRE-bearing virus showed remarkable better transgene expression in all tested cell lines. In some cells, the impact was comparable to earlier-mentioned and well-demonstrated sodium butyrate effect but having an advantage of being nontoxic. Transactivator strategy has also been applied successfully to baculoviruses to drive efficient and cell-specific transgene expression [163,204].

11.2.5 Baculovirus Hybrid Vectors

BEVS has become a popular choice for a virus-like particle (VLP) generation in vaccine production and for a basic research [205]. Indeed, the extraordinary capacity of AcMNPV to carry foreign DNA (no known limit for foreign DNA, but over 100 kbp tolerated) allows construction of recombinant viruses bearing also large expression cassettes [10,206–208]. This is a valuable property for many purposes and not least for a hybrid vector construction, as shown by several authors. Palombo and coworkers [209] constructed a baculovirus-adeno-associated virus (Bac-AAV) hybrid vector to prolong the transient nature of baculovirus-mediated transgene expression. The idea was to use the natural integration capacity of AAV to carry the transgene cassette into a defined region (chromosome 19) of the host cell genome [209]. β-galactosidase and hygromycin resistance gene excision from the baculovirus backbone vector and a subsequent integration into the genome of 293 cells was shown to occur with significant frequency. However, integration into the desired region occurred only in a fraction of the clones, and nonspecific integration and multiple insertions of the marker gene were detected. The Wang group has applied a similar strategy to flank an expression cassette with the AAV inverted terminal repeats to improve and extend gene expression in neural cells [200,210], to treat cancer [211], and to transduce hESCs [140].

A system for the production of gutless adenovirus vectors (FD-AdVs) that does not require helper adenoviruses was described by Cheshenko and coworkers [207]. The helper virus was replaced by the baculovirus–adenovirus hybrid vector (Bac/Ad) containing a Cre recombinase-excisable copy of the packaging-deficient adenovirus genome. 293-Cre cells were transfected with the FD-AdV plasmid containing a transgene cassette, packaging signals, and two copies of the inverted terminal repeats, and were followed by transduction by the Bac/Ad. High titer FD-AdV virus preparations (10^8 pfu/mL) were attained. However, the system has to be further improved to avoid generation of replication-competent viruses during a large-scale production. Yap et al. [111] constructed a recombinant baculovirus carrying a cDNA of the bacteriophage T7 RNA polymerase under the control of the CAG promoter. High-level expression of this enzyme in various mammalian cell lines was observed after the baculovirus transduction. A plasmid bearing the entire poliovirus genome under the T7 promoter yielded a high titer of infectious poliovirus in the HeLa cells after prior transduction with the baculovirus.

To efficiently propagate and study hepatitis B virus (HBV) and hepatitis C virus (HCV) in cultured cells, recombinant baculoviruses carrying the HBV [212–216] and HCV [206,217,218] cDNAs under mammalian promoters have been prepared. The control of the gene expression in both the HCV minigenome and the full-length HCV construct was also investigated [219]. In addition, Tet-off and ecdysome/ponasterone-inducible (pon) systems were compared to control the gene expression. The tetracycline-controlled system gave a low basal activity and was highly inducible in almost 100% of HepG2 cells. Hepatitis–baculovirus hybrid vectors, thus, represent a simple and highly flexible system for studying the effects of antivirals or cytokines on HBV [220] and HCV production, and for the understanding of their replication and pathogenesis at the molecular level.

To investigate biology of the human cytomegalovirus (HCMV), genes encoding immediate early proteins of the HCMV were cloned into baculoviruses under the CAG promoter [221]. These viruses provided a new strategy for efficient isolation of the HCMV viruses with mutations in essential genes. In another study, a hybrid baculovirus was used to carry woodchuck hepatitis virus (WHV) pregenome under CMV promoter into WC-3 woodchuck hepatoma cells to study the effect of adenovirus superinfection on WHV replication [222].

11.2.6 Entry Mechanisms into the Vertebrate Cells

The BV form of AcMNPV enters cells by adsorptive endocytosis in a multistep process including virus attachment, cellular entry (endocytosis), vesicular transport, endosomal escape, intracytosolic transport, nuclear entry, capsid disassembly, and gene expression (Figure 11.3). The exact molecular mechanisms leading to infection of insect cells, or transduction of vertebrate cells, remain still poorly understood. Results from both the insect and vertebrate cells hint involvement of clathrin-mediated endocytosis in the virion uptake [40,47,223]. Very recent observations in vertebrate cells suggested also involvement of macropinocytosis, and caveolae route, in baculovirus entry [224,225]. Indeed, several endocytic routes may be involved depending on cell type and culture conditioning [226].

van Loo et al. [40] showed that baculoviruses can also transduce nondividing cells and that the mechanism of the viral capsid transport into the nucleus before uncoating is apparently identical to that of insect cells (Figure 11.3) [227]. Pig kidney cells (Pk1) were arrested in S phase with aphidicolin, a reversible blocker of DNA polymerase, to show that the arrested cells can be transduced similarly as untreated mitotic cells. Baculovirus capability to transduce dormant cells has been verified by independent studies [228,229].

Some studies have suggested presence of specific binding sites for the baculovirus uptake into insect cells. However, the precise nature of these cell surface receptors is still unknown [47,223]. AcMNPV binding studies revealed existence of 3.1×10^{10} binding sites with avidity of 2.3×10^{10} M^{-1} in Sf21 cells [47] in line with the results obtained in High Five cells [223]. The affinity of AcMNPV against unknown receptors was in a similar range also in other insect cell lines but the amount of available binding sites varied between the studied cells. Attachment seemed to be limited by diffusion and large virion size [223]. In vertebrate cells, both receptor and nonspecific charge-mediated viral uptake have been suggested for the still speculative receptors [22,67,68]. The effect of the titer and virus competition studies in the cell culture suggested that gene transfer into hepatocytes might be due to the asialoglycoprotein receptor [22]. However, study in Pk1 cells, which do not express this receptor, did not support this conclusion [40]. Contrary to this specific receptor hypothesis, baculoviruses were reported to enter several cell lines by absorptive endocytosis, which does not require any interaction with a high-affinity receptor but rather interaction with a heterogeneous cell surface motifs [67]. Electrostatic charges were shown to be important. Negatively charged epitopes at the cell membrane appeared to be critical for baculovirus–cell interactions and the subsequent entry. Heparan sulfate proteoglycan seemed to be an important docking motif. Enzymatic removal of heparan sulfate groups from the cell surface caused a significant reduction in transduction [67]. The mode of baculovirus entry into the target mammalian cells was further studied by constructing recombinant baculoviruses encoding gp64 (extra copy), VSV-G, mouse hepatitis virus S protein, or GFP to compare susceptibility of various cell lines to these recombinant baculoviruses [68]. Increased amounts of gp64 or foreign envelope protein (VSV-G, MHVS) on the virus surface (envelope) caused higher expression than the control virus in several mammalian cell lines. Furthermore, this study indicated that phospholipids, such as phosphatidic acid or phosphatidylinositol, on the cell surface played an important role in the transduction of mammalian cells by baculovirus, whereas heparin and heparan sulfate did not. The published results are thus contradictory, and further studies are needed to resolve the exact nature of baculovirus receptors. However, the receptors may well represent a general cell surface motif such as heparan sulfate, phospholipid, integrin, or lectin because baculovirus can penetrate to plethora of cells. Indeed, as discussed in previous chapters, current data suggest that AcMNPV has a universal entry capacity.

The virion attachment on the cell surface, as well as pH-dependent escape of the AcMNPV nucleocapsid from the endosomes, is mediated by the major envelope glycoprotein, gp64 [65]. The precise vesicular routing beyond virion entry, leading to endosomal escape, is currently not known. Because AcMNPV is able to penetrate dozens of cells, and of different origin [20,102–105,142–144,147], gp64 has raised recently a lot of interest in pseudotyping and stabilizing of other enveloped viruses such as lentiviruses (HIV [230–233] and FIV [234,235]) or human respiratory syncytial virus (HRSV) [236,237]. The noncytotoxic nature of gp64 is a major advantage compared to VSV-G in virus production and transduction [186].

AcMNPV infection induces a series of actin rearrangements in insect cells, and actin plays an important role also in transduction of vertebrate cells although no viral replication takes place in these cells [20,238]. The major capsid protein of AcMNPV, vp39, may have a role on the nucleocapsid intracytoplasmic movement by hastening it [239]. The minor capsid protein, p78/83 [240], may assist on this, and has an important role in mediating nuclear actin accumulation and assembly vital for virus progeny generation [241–243]. Contrary to actin, microtubules are not essential to transduction. Disintegration of microtubules by nocodazole or vinblastine enhanced baculovirus-mediated gene delivery in vertebrate cells, suggesting that microtubules may pose a mechanical barrier for nucleocapsid movement toward and entry into the nucleus [40,158]. Indeed, the intracytoplasmic movement and nuclear entry are those steps in which baculovirus entry into vertebrate cells seems to be mostly restricted [244,245].

Disassembly of the viral nucleocapsid and exposure of nucleoprotein core take place in the nucleus. Phosphorylation of the basic viral protein, p6.9 (VP12), may control the disassembly according to studies performed in insect cells [64,246,247]. The p6.9 functions like a histone in the condensation of the viral DNA. The basic protein is substituted for host cell histones after disassembly and viral genome is thus under epigenetic control of the host cell [141,248,249].

11.2.7 TOWARD THERAPEUTIC APPLICATIONS

11.2.7.1 Gene Therapy

Current evidence suggests that baculoviruses provide also an effective tool for gene delivery *in vivo* (Table 11.2). The first attempts of *in vivo* gene transfer with baculovirus were performed in the liver parenchyma of rats and mice [25]. Several attempts were also undertaken to inject baculoviruses directly into systemic and intraportal circulation [25]. These experiments resulted in undetectable transgene expression, suggesting that the virus was somehow inactivated by serum components. This led to several studies in immune-compromised animals. When Hofmann et al. [107] injected a β-galactosidase expressing baculovirus directly into the liver parenchyma of C5-deficient immunocompromised mice, a few transduced hepatocytes were detected around the injection site. They also injected recombinant baculoviruses into the Huh7-derived

TABLE 11.2

Summary of the Preclinical Data of Baculovirus-Mediated Gene Transfer *in Vivo*

Study	Target Animal	Positive Organ/Tissue	Positive Cell Types	Notes	References
Hoffmann et al., 1998	C-5 deficient mice	Liver	Hepatocytes		[107]
Airenne et al., 2000	Rabbits	Adventitia	Fibroblasts Smooth muscle cells	Delivery by collar devise	[14]
Sarkis et al., 2000	Nude mice, BALB/c mice, and Sprague–Dawley rat	Brain	Striatum: astrocytes, corpus callosum, ependymal layer, and neuronal cells		[28]
Haeseleer et al., 2001	Mice	Retina Corneus	Retinal pigment epithelial cells Retinal inner nuclear layer Corneal endothelial cells Ganglion cell layer Muller cells Lens epithelium Photoreceptor cells		[27]
Hüser et al., 2001	Neonatal Wistar rats	Liver parenchyma	Not determined	The first report of a complement protected BV	[252]
Kircheis et al., 2001	Complement deficient A/J mice	Liver Spleen Kidney Lungs	Not determined	Delivery through tail vein	[250]
Pieroni et al., 2001	DBA/2J, BALB/c, and C57BL/6 mice	Quadriceps femoris muscle	Myofibers	Transgene expression mouse strain desendant	[176]
Lehtolainen et al., 2002	Normal Wistar rat	Hindbrain Forebrain Subarachnoidal space Spleen Heart Lung	Cuboid epithelium of the choroid plexus Endothelial cells of brain microvessels	Very high transduction rate in choroid plexus cells ($76 \pm 14\%$)	[26]
Abe et al., 2003	BALB/c female mice	Intramuscular, intranasal, and intradermal (administration route)	Not determined	Vaccine against influenza virus Intranasal route most efficient in vaccination	[270]
Tani et al., 2003	BALB/c mice	Cerebral cortex Mouse testes Seminiferous tubules	Astrocytes Pyramidal cells Basal and sertoli cells	A synthetic protease inhibitor to protect BV from the complement	[29]
Facciabene et al., 2004	BALB/c female mice	Intramuscular, intranasal, and subcutaneous (administration route)	Not determined	Vaccine against hepatitis C virus. Only intramuscular administration led to clear humoral response	[271]
Li et al., 2004	Adult male Wistar rats	Striatum, cerebral cortex, substantia nigra, retina, visual cortex, lateral geniculate nucleus, and superior colliculus	Neurons	Axonal transport explain virus spreading to distant tissues	[201]
Laitinen et al., 2005	Female Wistar rats	Brain	Cuboid epithelium of choroids plexus	Multipurpose simultaneous transgene vector allowing expression in mammalian, insects, and bacteria	[254]
Hoare et al., 2005	MF-1 mice	Liver	Parenchymal cells	Complement inactivation by soluble complement receptor type 1 prior virus treatment allowed successful portal vein administration	[251]

TABLE 11.2 (continued)
Summary of the Preclinical Data of Baculovirus-Mediated Gene Transfer *in Vivo*

Study	Target Animal	Positive Organ/Tissue	Positive Cell Types	Notes	References
Li et al., 2005	Male Wistar rats	Brain Eye	Neurons	CMV enhancer/platelet-derived growth factor hybrid promoter targeted transgene expression effectively to neuronal cells	[199]
Wang and Wang, 2005	Male Wistar rats	Brain Striatum Cerebral cortex	Astrocytes in striatum Cerebral cortex neurons	AAV inverted terminal repeats/CMV enhancer/glial fibrillary acidic protein hybrid promoter allowed 90 days long astrocyte-specific gene expression	[210]
Räty et al., 2006	Male Wistar rats	Brain	Cuboid epithelium of choroids plexus	The first report of noninvasive magnetic resonance imaging of viral biodistribution using avidin-coated virus and biotinylated USPIO particles	[253]
Kaikkonen et al., 2006	Female BDIX rats	Hindbrain	Rat: cuboid epithelium of choroids plexus, brain microvessels, endothelial cells, epithelial lining of the lateral ventricles and the cerebral aqueduct, and subarachnoidal membrane	VSV-GED pseudotyping enhanced transduction and changed viral biodistribution pattern both in rat and rabbit	[186]
	New Zealand white rabbit	Forebrain Rabbit muscle	Rabbit: muscle cells		
Wang et al., 2006	Rat	Rat C6-Luc glioma xenograft in brain	Xenograft cells, not directly measured but evaluated by A-chain of diphtheria toxin transgene's ability to suppress xenograft growth	Glial fibrillary acidic protein promoter to restrict the transgene expression to cells of glial origin	[211]
Liu et al., 2006	New Zealand white rabbits	Intervertebral disc	Nucleus pulposus cells		[30]
Kim et al., 2006	BALB/c (6 week) mice	Liver Spleen Lung Kidney Brain Heart	Not determined	Pegylation of baculoviruses increased transduction efficiency	[193]
Strauss et al., 2007	BALB/c female mice	Intramuscular (administration route)	Not determined	Vaccine against malaria. Virus both displaying and expressing the antigene was the most efficient	[286]
Räty et al., 2007	Male Wistar rats	Liver Kidney Lung Brain Spleen	Tubular epithelial cells of the kidney	Several administration routes were compared and viral biodistribution was studied by noninvasive SPECT/CT imaging modality	[345]

human hepatocarcinomas generated in nude mice (T-cell deficient) and got low gene transfer efficiency. A systemic baculovirus gene vector delivery into complement-deficient Neuro2a tumor-bearing A/J mice resulted in transgene expression primarily in liver, spleen, and kidney, but significant expression was also found in the tumor [250]. Pegylation of baculovirus was reported to enable brain transduction in mice after intravenous application. GFP expression was also detected in liver, spleen, lung, heart, and kidney [193]. Intraportal application with a complement inactivator led to some hepatic expression of *Lac Z* marker gene in MF-1 mice [251]. In contrast to other related reports, this study also reported toxicity associated to a high dose systemic administration of baculoviruses. Injection of decay acceleration factor (DAF)-displaying complement-resistant baculovirus into the liver parenchyma of complement-sufficient neonatal Wistar rats caused an enhanced expression of the marker gene suggesting that generation of complement-resistant vectors could improve the gene transfer efficiency *in vivo* [252]. However, direct intrahepatic injection of the complement-resistant AcMNPV-DAF-βgal vector in adult rats resulted only in single positive cells distant from the injection site.

Delivery methods that allow gene transfer in the absence of serum, or to the sites where viruses are not exposed to the complement, have led to more successful experiments in immune-competent animals. BALB/c mice, nude mice, and Sprague–Dawley rats were injected with recombinant baculovirus directly into the striatum of brain [26,28]. Marker gene expression was detected in the striatum, the corpus callosum, and the ependymal layer, indicating the ability of baculoviruses to transduce neural cells *in vivo*. Transduced cells were identified mainly as astrocytes with only a few positive neurones. No difference was detected between the three species or between the cobra venom factor (CVF) (an inhibitor of the complement system) treated and untreated groups, suggesting immune-privileged nature of the brain. Transduction efficiency, tropism, and biodistribution of the baculoviruses after local delivery into the brain have also been studied in comparison to adenoviruses in BDIX and Wistar rats [26]. In this study, baculoviruses were found to transduce cuboid epithelium of the choroid plexus cells very efficiently (76% ± 14%). A clear difference was observed with the adenovirus vector when injected into the corpus callosum; adenoviruses did not transduce the choroid plexus cells, whereas ventricular ependymal lining and cells in the corpus callosum were transduced with a high efficacy. Adenovirus injection into the striatum resulted in an effective transduction near the injection site and in the corpus callosum. Both viruses lead to transgene expression in endothelial cells of brain microvessels throughout the forebrain [26]. AcMNPV's high tendency to transduce choroid plexus cells was verified by other studies conducted in rat brain [186,253,254]. Transgene expression could be detected also in the walls of lateral ventricles and in subarachnoid membranes when the vector was pseudotyped with the truncated VSV-G molecule (VSV-GED) [186].

Wang and coworkers demonstrated advantage of using a tissue-specific hybrid promoter in rat brain. The promoter of human platelet-derived growth factor β-chain and synapsin-1, fused to enhancer of CMV immediately early promoter, drove efficient neuron-specific gene expression [199,204]. Astrocyte-specific gene expression was achieved by glial fibrillary acidic protein promoter [200]. Driven by neural-specific promoter, transcriptionally targeted transduction was detectable not only in neurons in injection sites but also in remote target regions. Axonal transport of the baculovirus vector was suggested to explain the results [201]. An expression cassette flanked with the ITRs of AAV boosted and extended transgene expression in rat brain [210]. The strength of a promoter could also be improved by a chimeric transactivator strategy [204]. As a step forward to real therapy, astrocyte-specific baculovirus was recently used to treat malignant glioma. A virus expressing A-chain of diphtheria toxin effectively suppressed tumor development in a rat C6 xenograft model [211]. Efficient transduction of intervertebral disc, an encapsulated and avascular immune-privileged tissue, was reported in rabbit [30]. Gene-delivery efficacy of nucleus pulposus cells compared favorable to reported results with adenovirus and retrovirus vector. In another study, lumbar intrathecal injection into the cerebrospinal fluid was used to transduce rat dorsal root ganglia cells [31].

In a different approach, carotid arteries of New Zealand white rabbits were successfully transduced by recombinant baculoviruses using a collar device. This system allowed gene delivery with minimal exposure to complement [14]. Transient expression in the adventitial cells was observed with an efficacy and duration comparable to adenoviruses. Recombinant baculoviruses have also been tested for direct administration into a mouse eye by subretinal injections. A strong expression of the marker gene in retinal pigment epithelial cells was reported by this study [27]. Intravitreal injection of the virus resulted in the marker gene expression in the corneal endothelium, lens, retinal pigment epithelial cells, and retina. The ocular tissue contains areas where antigens are not subjected to the complement pathway and therefore makes it a good target for baculovirus-mediated gene therapy. High gene delivery efficacy was observed also in immune privileged mice testis [29].

Direct injection of recombinant baculoviruses into the quadriceps femoris muscle of BALB/c and C57BL6 mice resulted in a transient expression of β-galactosidase. Expression levels were 5- to 10-fold higher when VSV-G pseudotyped baculoviruses were used [176]. The authors also used C5-deficient mice where a higher and more sustained gene expression (up to 178 days) was observed than in BALB/c and C57BL6 mice, suggesting a different gene transfer efficiency between the different mouse strains and the importance of the complement activity.

11.2.7.2 RNAi

Gene silencing by RNA interference (RNAi) is a powerful approach for a loss-of-function studies in fundamental biological questions [255,256]. It provides also new therapeutic platform to treat common diseases [257]. The facile and high throughout-compatible preparation of baculoviruses [97,254] as well as their inert nature [114,115] make baculoviruses as attractive alternatives for RNAi delivery in vertebrate cells. Nicholson and coworkers were the first to show that a baculovirus-delivered U6-based short hairpin RNA (shRNA against lamin A/C) performed well in Saos2, Hepg2, Huh7, and primary human HSCs [258]. In another study, a hybrid HI promoter efficiently suppressed target gene expression by up to 95% in cultured human NT2 neural precursor and C6 rat glioma cells, and by 82% in rat brain [202]. Suppression of porcine arterivirus [259], HBV [260], HCV [261], and influenza virus A and B [262] replication by baculovirus-mediated shRNA delivery provides a proof of principle of baculovirus as a potential vehicle for RNAi-mediated antiviral therapy. A related strategy, taking advantage of a baculovirus-delivered ribozyme against HIV-1 U5 gene, was successful to inhibit HIV-1 replication substantially *in vitro* [263].

11.2.7.3 Vaccination

BEVS has shown to be useful in a subunit vaccine production [205]. A cervical cancer vaccine, Cervarix (GlaxoSmithKline PLC), against the two most prevalent cancer-causing types of the human papillomavirus (HPV 16 and 18), was recently approved for a clinical use in woman aged 10–45 years old in Australia and Europe, and it is expected to be launched soon in USA too. This bivalent HPV-16/18 VLP vaccine [264] is the first BEVS-based product for clinical use having an exceptional market potential. Another example of an advanced baculovirus based subunit vaccine is the influenza vaccine, FluBlok (Protein Sciences Corporation), which has already shown HS efficacy in clinical trials [265–267]. FluBlok uses recombinant HA proteins derived from current circulating influenza viruses that are selected by the World Health Organization and the Center for Disease Control on an annual basis [268]. There are several well-recognized disadvantages in the standard use of embryonated eggs for vaccine production such as a slow vaccine generation time (about 6 months), possible disrupt of the supply by avian influenza virus, safety concerns due to live flu virus, and need to adapt all new influenza variants for high-yield growth. BEVS avoids all these and allows moving a flu vaccine into full production in about 6–8 weeks from the time the target strain is identified [269]. FluBlok is expected to reach markets during 2008–2009 flu seasons.

In addition to subunit vaccines containing recombinant proteins produced in insect cells, baculovirus-mediated gene delivery, baculovirus display of desired antigens, or infected insect cells per se may prove to be usable for vaccination too [172,173,205]. Induction of specific antibodies was observed in mice that were inoculated intranasally or intramuscularly with a virus encoding PrV glycoprotein B. Higher levels of antibodies were detected in serum after intramuscular rather than intranasal inoculation [126]. Although intramuscular or intraperitoneal immunization elicited higher antibody titers, protection from a lethal challenge of the influenza virus was only achieved by intranasal application of hemagglutinin encoding baculovirus [270]. Better immune response against E2 glycoprotein of HCV

and carcinoembryonic antigen (CEA) was gained in mice with VSV-G pseudotyped virus [271]. Intramuscular application was also found in this study to give the best response. In accordance, a VSV-G pseudotyped baculovirus expressing murine telomerase reverse transcriptase induced antitumor immunity in a murine glioma model after direct injection into muscle [272].

Lindley and coworkers [273] reported a novel use of the baculovirus gp64-display system [172,173] as an efficient method to produce monoclonal antibodies against desired antigens. Immunization with a whole virus eased and speeded up antibody production by eliminating the need for laborious purification of a target antigen before immunization. Furthermore, the immunostimulatory nature of virions alleviated the need for other adjuvants [274,275]. Same strategy was used to generate monoclonal antibodies against human peroxisome proliferator-activated receptors and the amino-terminal domain of human hepatocyte nuclear factor 4α1/α2/α3 [276,277]. This method might also be useful for direct vaccination with desired displayed antigens, as suggested by Tami and coworkers [278]. In support of this, baculovirus-displayed proteins have been proven to elicit efficiently antigen-specific responses and vaccination potential in mice against bovine herpes virus 1 protein [279], Rinderpest virus and *Peste des petits ruminants* virus [280], malaria [281], and acute respiratory syndrome-associated coronavirus (SARS-CoV) [282–284]. Furthermore, surface-displayed hemagglutinin may provide an alternative method as an influenza vaccination [187,285]. A virus that both displayed and expressed *Plasmodium falciparum* circumsporozoite protein (CSP) was superior in inducing anti-CSP immune responses [286]. This study showed also that baculovirus transduction not only induced efficient antibody responses but also triggered maturation of professional antigen presenting dendritic cells (DCs) leading to specific antibody and CD4(+) and CD8(+) T-cell responses against CSP in mice after i.m. vector injection. DC induction by AcMNPV was detected also in human monocyte-derived DCs [287] and in mouse bone marrow–derived DCs [288].

Vaccination with infected cells is still an interesting additional possibility. Lyophilized insect cells infected with interferon-β eradicated subcutaneous tumors of UV2237m fibrosarcoma and K-1735M2 melanoma in syngenic mice, inhibited tumor growth and hepatic metastasis in syngenic and orthotopic model of the CT-26 murine colon cancer, and was effective in immunotherapy against occult brain metastasis [289,290]. In another study, fixed and sonified insect cells containing Japanese encephalitis virus (JEV) proteins were used to detect T-cell responses of volunteer donors who had experienced subclinical JEV infection [291]. This approach has applications in the preliminary screening of potential vaccine antigens. When mice received virions or infected insect cells bearing the fusion protein gp64-site A of a foot and mouth disease virus (FMDV), a specific and protecting immune response to FMDV was elicited [274]. Whole cell immunization protocol was used also in brucellosis vaccine study [292].

11.2.8 SAFETY

Extensive safety testing of baculoviruses has been conducted, which has led to their classification onto the lowest biohazard level [11]. These studies have revealed that nucleopolyhedroviruses are harmless to and unable to replicate in microorganisms, noninsect invertebrate cell lines, vertebrate cells, vertebrates, plants, and nonarthropod invertebrates. These trials have included long-term carcinogenicity and teratogenicity tests, tests in primates, and tests in humans [18]. Many different mammalian species have been studied, including rats, mice, dogs, guinea pigs, monkeys, and humans. In these tests, baculoviruses were administered by oral, intravenous, intracerebral, intramuscular, and topical routes. No toxicity, allergic responses, or pathogenicity associated with the baculoviruses were detected [293]. The safety of baculoviruses is also underscored by the fact that we are exposed everyday to baculovirus particles present in large numbers in our environment and food. Yet, no diseases have been linked to baculoviruses [15,294].

11.2.8.1 Restricted Replication

The safety of baculoviruses is secured at several levels of restrictions in baculovirus infectivity [295]. The polyhedrin matrix is essential to horizontal transfer of the virions by protecting them from the environment. The alkaline midgut of insects facilitates the dissolution of the polyhedrin matrix, leading to primary infection of the larva (Figure 11.1). Organisms such as birds and mammals that lack such alkaline conditions in their digestive tract or other potential points of entry, such as respiratory tract, are not infected by ODV [296–298]. Tissue or cell-type specificity may create the second level of restriction, although the baculovirus host range does not appear to be limited at the point of entry into the target cells [20,68]. This is particularly true for AcMNPV. The subsequent steps following the virus entry by adsorptive endocytosis may be much more important [151,158,245]. Although escape from the endosomes certainly creates barrier, it was found that nucleocapsids seemed to enter into the cytoplasm even in the cells in which no marker gene expression was detected [151]. The block in transduction of vertebrate cells by AcMNPV seems, thus, to lie in the defective intracytosolic transport or entry of the nucleocapsid into the nucleus [158,244]. This is in agreement with the known ability of gp64 to mediate pH-dependent membrane fusion in endosomes [65]. Finally, if the nucleocapsid reaches the nucleus, strictly guided molecular mechanisms that cover the expression of the baculovirus genome remain in place. Indeed, baculoviruses propagate only in insect cells and are inherently unable to replicate in mammalian cells [104,238,299,300]. This is an advantage when baculoviruses are used for gene therapy, because risks related to the rise of replication-competent viruses during the virus production can be avoided. Replication-competent viruses are the major concerns with most of the current main stream gene delivery vectors based on human pathogens [301].

11.2.8.2 Guided Viral Gene Expression

Although AcMNPV DNA does not replicate or persist in mammalian cells [238], baculoviral early genes are transcribed by host DNA polymerase II [302,303]. Furthermore, a product of early gene (baculovirus transactivator IE1), which is able to activate several early, late, and very late viral promoters, was demonstrated to be functional in mammalian cells [304,305]. Adenoviral proteins were also capable of transactivating a heterologous insect virus promoter in mammalian cells [306]. In addition, the baculoviral early-to-late promoter, which activity is dependent on baculoviral early gene expression, drove gene expression cell type dependently in some mammalian cells [307]. So, it is not surprising that detectable mRNA levels of baculoviral genes have been recently reported in some vertebrate cells [245,308,309]. However, the relevance of the detected viral gene expression in safety terms remains to be studied further because these results were based on indirect or sensitive PCR-based methods. The extent and timing of the viral transcription, known to be important for productive infection in insect cells, and most importantly, evidence of any translation of viral transcripts, remain mainly to be resolved. The few studies concerning these issues this far have been able to show some translation of the viral mRNA (IE2) only in HepG2 cells [229,310].

11.2.8.3 Genome Size

The large size of the baculovirus genome (~134 kbp) raises a theoretical concern of homologous recombination of it (or parts of it) into the target cell genome. In the worst scenario, this might lead to malignancy of the target cell as reported in the case of naturally integrating retrovirus and AAV vectors [7,311–315]. However, integration of AcMNPV genomic fragments into the mammalian genome has been shown to take place *in vitro* only in the presence of selection pressure [106,316]. In these studies, Chinese hamster ovary (CHO) cells, which were stably transduced by selection pressure with recombinant baculoviruses, expressed GFP at least 25 passages [106]. Analysis of the baculovirus-derived DNA indicated that DNA derived from the viral vector had stably integrated into the transduced CHO cells [106]. Integration into the CHO genome occurred as small fragments (5–18 kb) via illegitimate recombination [316]. Two of the clonal cell lines maintained starting levels of the GFP expression over a 5 month period with and without selection. The two remaining clones, however, showed a loss of the marker gene expression. Baculoviruses can, thus, be used for the preparation of stable cell lines. No evidence of integration of AcMNPV genomic fragments into the target cell genome has been found without concomitant antibiotic selection pressure [238,317].

11.2.8.4 Immune Response

One of the biggest challenges in gene therapy is the immune response of the host. The host defense mechanisms function both at the cellular level by generating cytotoxic T-cells and at the humoral level by generating antibodies against foreign antigens. Cellular immunity eliminates the transduced cells, whereas humoral immunity protects against the repeated administration of the vector [176,318,319]. The host may recognize not only the vector and the transgene product but also the foreign DNA, which makes the vector design even more challenging [320]. However, the eye, nervous system, and reproductive organs possess immune-privileged regions [321] and, therefore, provide good targets also for the baculovirus-mediated gene therapy [26,27,29–31].

As discussed earlier, immune responses against baculoviruses were suggested as a reason for no detectable transgene expression when mice and rats were treated with systemic, intraportal, or direct injections into the liver parenchyma [25]. Therefore, the influence of untreated and heat-inactivated serums from different species was tested for baculovirus transduction efficiency. The results indicated that the classical complement (C) cascade inactivates baculoviruses rapidly and the extent of inactivation varies from one species to another [25,322]. Incubation of baculoviruses in the serum of either C3- or C4-deficient guinea pigs did not cause any significant neutralization of the baculovirus, suggesting that C3 and C4 components are essential [25,322]. In line with these findings, the direct injection of recombinant baculoviruses into C5-deficient mice resulted in a higher and longer-lasting expression of the marker gene than that in mice, which were not complement deficient [176]. In *ex vivo* experiments that excluded the C system, detectable levels of the marker gene expression were found in human liver segments perfused by the baculovirus vectors [25,107]. More detailed investigations have demonstrated that the classical pathway of the C system and assembly of the very late C components are essential for the inactivation of the baculovirus in human serum, indicating the presence of IgM or IgG antibodies against baculoviruses [322]. These antibodies are most probably part of the innate self/nonself pattern recognition immune system detecting antigens without a known history of immunization [286,323,324]. Involvement of the alternative C pathway in baculovirus inactivation has also been suggested [251].

A couple of efforts have been taken to protect baculovirus from inactivation by the complement system [325,326]. The treatment of human serum with functional blocking agents (CVF, anti-C5 monoclonal antibodies) against the components of the C-cascade increased vector survival significantly in a dose-dependent manner *in vitro* [322]. Soluble complement receptor type 1 (sCR1), a potent inhibitor of both the classical and alternative C pathways, increased baculovirus survival in human serum in a dose-dependent manner [327] and reduced intraportal application associated toxicity in mice [251]. A complete baculovirus survival was achieved by the soluble complement receptor at concentration of $100\,\mu g/mL$ of serum [327]. Incorporation of the complement-regulatory protein DAF into the viral envelope was also shown to improve gene transfer efficiency in neonatal Wistar rats *in vivo* [252]. Synthetic inhibitor of complement (FUT-175) was also effective [29]. Pegylation [193] and VSV-G pseudotyping [29] may also protect baculoviruses from the complement attack.

Baculoviruses have been shown to stimulate antiviral activity in mammalian cells by promoting cytokine production. Baculovirus exposure resulted in the activation of TNF-α, IL-1α, and IL-1β expression in the primary hepatocyte cultures and production of interferons (IFNs) in some mammalian cells [328,329]. Kupffer cells were present in the primary hepatocyte cultures and were most probably responsible for the cytokine production [328]. The IFN-stimulating activity of the baculoviruses required live virus [329]. Contribution of the toll-like receptors for eliciting a strong innate immune response in mice was first suggested in a study that showed an AcMNPV-induced secretion of TNF-α and IL-6 as well as increased expression of activation ligands (CD69 and mature macrophage Ag receptor) in murine RAW264.7 macrophages [270]. Further studies revealed an important role of the toll-like receptor 9 and MyD88-dependent signaling pathway on activation of the immune cells via baculoviral DNA [330]. Role of viral DNA on promoting humoral and CD8+ T-cell adaptive responses as well as inducing maturation of dendritic cells and the production of inflammatory cytokines (IFN-$\alpha\beta$) against coadministered antigen was further supported by studies of Herva-Stubbs and coworkers [275]. However, other viral components and recognition pathways are also involved on the priming of the immune system [275,330]. The induction of antiviral effects (cytokine production) in vertebrate cells is also cell-type dependent. No cytokine production was detected in HepG2, Huh7, BHK, or L929 cells while IFN synthesis in dendritic, macrophage, and SMMC-7721 cells was evident [331,332]. No correlation existed between AcMNPV transgene expression and induction of INF production [332]. In accordance with these studies, baculovirus-mediated periadventitial gene transfer was found to result in immune responses [14]. Interestingly, however, only a modest microglia response was seen in rat brain after the baculovirus transduction whereas the adenovirus gene transfer led to a strong microglia response [26].

11.3 FUTURE ASPECTS

The genuine advantages of baculovirus vectors (Table 11.1), and the recent advances in vector design as well as in transduction optimization, certainly make AcMNPV a powerful vector for a universal gene delivery into any cell from any origin *in vitro* [97,98,254]. However, despite the profound existing knowledge about the baculovirus safety, biology, and factors important for efficient cell transduction *in vitro*, further demand for vector development and studies concerning AcMNPV behavior, especially *in vivo*, still exists. These issues need to be addressed before clinical trials with baculovirus vectors can be launched. Common challenges shared with all gene therapy vectors such as truly transcriptionally and positionally targeted gene delivery also largely remain to be resolved.

An ideal gene therapy vector should transduce only the desired target cells with a high efficacy. Therefore, a strong interest in developing targeted gene delivery vectors has emerged [333–339]. As a first step toward targeted baculovirus transduction, Ojala and coworkers [340] described baculovirus vectors displaying either a functional single chain antibody fragment (scFv) specific for the carcinoembryonic antigen or the synthetic IgG-binding domains derived from protein A of *Staphylococcus aureus*. Display of the targeting moieties on the viral surface was achieved through fusion to the N-terminus of gp64 (Figure 11.2) [341]. Specific binding of the gp64 fusion viruses to mammalian target cells could be demonstrated by fluorescence and confocal microscopy. However, no enhancement of the viral entry or gene transfer into the mammalian cells could be observed by monitoring GFP expression. Indeed, it is well known that a specific ligand–receptor interaction does not necessarily guarantee efficient transduction of the target cells [333,342]. Correct internalization, escape from endosomes, transport of the genetic material into the nucleus, and efficient transcription and translation are required as well [150,151,158,226,343,344]. The first baculovirus showing also enhanced gene expression beyond specific binding was based on avidin display on virus surface. Displaying of this biotin binding protein on virus provided a flexile tool for virus targeting [191] and biodistribution imaging [253,345]. More recent papers have described other targeting moieties enabling also enhanced gene expression. Examples include SARS-CoV S protein (lung cells A549, NCI-H520, HFL-1, and MRC-5) [282], a peptide motif of EBV (B-cells Raji, HR1, B95–8, BJAB, and DG75) [190], an RGD-motif derived from coxsackievirus A9 and human parechovirus (A549 lung cells) [188] or FMDV [346], RGD tumor homing peptides LyP-1, F3, and CGKRK (human breast carcinoma cells MDA-MB-435 and HepG2 hepatocytes) [189]. The true targeting of baculoviruses, however, is hampered by the extreme tropism of baculovirus, which is mediated by the gp64 envelope glycoprotein. The gp64 is vital for virus budding, attachment, and endosomal escape, and must, thus, also exist as an intact form in the virus envelope to guarantee efficient virus generation [341,346]. A first step forward for a successful pseudotyping of baculovirus without concomitant native (genomic) gp64 expression was published by Kitagawa and coworkers [183]. In this paper, a measles virus receptor (CD46) pseudotyped gp64-null baculovirus was prepared in Sfgp64 cells stably producing gp64 or by transfecting insect cell with a recombinant viral genome (bacmids). However, titers of the pseudotyped viruses were low and revertant viruses that had incorporated the gp64 gene were generated after passaging in Sfgp64 cells. Further method development is, thus, needed for efficient generation of truly targeted baculoviruses.

Further knowledge on molecular mechanisms behind viral cellular entry will create the basis for further engineering of the baculoviruses, allowing better possibilities for cell/tissue-specific entrance also *in vivo*. It is also important to learn more about host response and biodistribution of viral particles after different administration routes *in vivo*. Indeed, a first study using magnetic resonance imaging (MRI) for virus particle biodistribution determination was performed with a baculovirus [253]. This study confirmed the high tropism of baculoviruses against choroids plexus cells in the ventricles of rat brain. Single-photon emission computed tomography (SPECT/CT) was also used recently for baculovirus biodistribution studies [345]. This study revealed new aspects of baculovirus trafficking in an immune competent rat model after i.f., i.p., i.m., and i.c.v. administration. The results suggested virus spread via the lymphatic system and extensive accumulation of virus particles into the kidneys, especially after i.p. administration, was observed. Transduction of kidney was confirmed by reverse transcriptase–polymerase chain reaction and immunohistochemistry. Baculovirus may, thus, be beneficial for the treatment of kidney diseases. These studies show that MRI and SPECT/CT can provide useful and important information of vector behavior in an immune-competent animal in a noninvasive way.

Current baculovirus vectors allow only transient transgene expression, which is desirable in the treatment of cancer [4] or cardiovascular disorders [347,348]. However, more prolonged expression is needed for the treatment of genetic diseases [349]. AAV baculovirus hybrid vectors have shown some success in this regard [140,200,209–211]. A strategy based on EBV episomal replication elements (OriP/EBNA-1) was also recently reported to prolong baculovirus-mediated gene delivery [350]. The use of episomal maintenance for the extension of transgene expression has the advantage of avoidance of possible insertional mutagenesis problems [351–353]. Controlled and inducible gene expression will be highly desirable in vectors capable of sustained gene expression [354–356].

A novel system where the fusion protein (enhanced GFP) is displayed in large quantities on the surface of the baculovirus capsid without compromising the viral titer or functionality is thus far used in virus entry mechanism studies [158,229] but has also potential to open remaining intracellular blocks hindering baculovirus transduction [244]. The system is based on the production of the desired peptides or proteins as either C-terminally or N-terminally linked fusion proteins with the baculovirus major capsid protein, vp39 (Figure 11.2). As the penetration of the baculovirus into mammalian cells and the release of the viral capsid into the cytoplasm seems to be a general phenomenon, the concept of baculovirus-mediated therapy may be further extended, with the possibility of using baculovirus capsid as a shuttle to transport also therapeutic proteins directly into the cells as an alternative to traditional protein transduction strategies [357].

In conclusion, the latest advances in baculovirus technology have further strengthened the concept of a baculovirus-mediated gene delivery with many advantages compared to current gene therapy vectors (Table 11.1). Baculoviruses are especially useful for large expression constructs allowing the use of multigene strategies. Inherent safety, together with the ease and speed of production of these vectors, creates a powerful and universal concept to test different gene delivery constructs also in a high-throughput setup.

ACKNOWLEDGMENTS

We thank Ms. Minna Kaikkonen, Ms. Johanna Laakkonen, and Ms. Emilia Makkonen for comments regarding the manuscript; Dr. Jani Räty for graphics in Figures 11.1 through 11.3, Dr. Susan E. Rice Mahr for the photograph of infected larva, and Dr. Franklin Dlott for the photograph of healthy Alfalfa looper in Figure 11.1. We also thank Dr. Maija Vihinen-Ranta and Ms. Johanna Laakkonen for the EM and confocal pictures in Figure 11.3.

REFERENCES

1. Peng, Z., Current status of gendicine in China: Recombinant human Ad-p53 agent for treatment of cancers, *Hum. Gene Ther.*, 16, 1016, 2005.
2. Wilson, J.M., Gendicine: The first commercial gene therapy product, *Hum. Gene Ther.*, 16, 1014, 2005.
3. Guo, J. and Xin, H., Chinese gene therapy. Splicing out the West? *Science*, 314, 1232, 2006.
4. Immonen, A., Vapalahti, M., Tyynela, K., Hurskainen, H., Sandmair, A., Vanninen, R., Langford, G., Murray, N., and Yla-Herttuala, S., AdvHSV-tk gene therapy with intravenous ganciclovir improves survival in human malignant glioma: A randomised, controlled study, *Mol. Ther.*, 10, 967, 2004.
5. Adis I Limited, INGN 201: Ad-p53, Ad5CMV-p53, Adenoviral p53, p53 Gene Therapy—Introgen, RPR/INGN 201, *Drugs R. D.*, 8, 176, 2007.
6. Cavazzana-Calvo, M. and Fischer, A., Gene therapy for severe combined immunodeficiency: Are we there yet? *J. Clin. Invest.*, 117, 1456, 2007.
7. Frederickson, R.M., Integrating ideas on insertional mutagenesis by gene transfer vectors, *Mol. Ther.*, 15, 1228, 2007.
8. Somia, N. and Verma, I.M., Gene therapy: Trials and tribulations, *Nat. Rev. Genet.*, 1, 91, 2000.
9. Summers, M.D., Milestones leading to the genetic engineering of baculoviruses as expression vector systems and viral pesticides, *Adv. Virus Res.*, 68, 3, 2006.
10. O'Reilly, D.R., Miller, L.K., and Luckov, V.A., *Baculovirus Expression Vectors. A Laboratory Manual*, Oxford University Press, New York, 1994.
11. Kost, T.A. and Condreay, J.P., Innovations-biotechnology: Baculovirus vectors as gene transfer vectors for mammalian cells: Biosafety considerations, *J. Am. Biol. Safety Assoc.*, 7, 167, 2002.
12. Hu, Y.C., Baculovirus vectors for gene therapy, *Adv. Virus Res.*, 68, 287, 2006.
13. Kost, T.A., Condreay, J.P., and Jarvis, D.L., Baculovirus as versatile vectors for protein expression in insect and mammalian cells, *Nat. Biotechnol.*, 23, 567, 2005.
14. Airenne, K.J., Hiltunen, M.O., Turunen, M.P., Turunen, A.M., Laitinen, O.H., Kulomaa, M.S., and Yla-Herttuala, S., Baculovirus-mediated periadventitial gene transfer to rabbit carotid artery, *Gene Ther.*, 7, 1499, 2000.
15. Miller, L.K., Introduction to the baculoviruses, in: *The Baculoviruses*, 1st ed., Miller, L.K., Ed., Plenum Press, New York, 1997, p. 1.
16. Miller, L.K., *The Baculoviruses*, 1st ed., Plenum Press, New York, 1997.
17. Black, B.C., Brennan, L.A., Dierks, P.M., and Gard, I.E., Commercialization of baculoviral insecticides, in: *The Baculoviruses*, 1st ed., Miller, L.K., Ed., Plenum Press, New York, 1997, p. 341.
18. Burges, H.D., Croizier, G., and Huger, J., A review of safety tests on baculoviruses, *Entomaphaga*, 25, 329, 1980.
19. Granados, R.R., Replication phenomena of insect viruses in vivo and in vitro., in: *Safety Aspects of Baculoviruses as Biological Insecticides*, Miltenburger, H.G., Ed., Bundesministerium fur Forschung und Technologie, Bonn, 1978, p. 163.
20. Volkman, L.E. and Goldsmith, P.A., In vitro survey of *Autographa californica* nuclear polyhedrosis virus interaction with nontarget vertebrate host cells, *Appl. Environ. Microbiol.*, 45, 1085, 1983.
21. Carbonell, L.F., Klowden, M.J., and Miller, L.K., Baculovirus-mediated expression of bacterial genes in dipteran and mammalian cells, *J. Virol.*, 56, 153, 1985.
22. Hofmann, C., Sandig, V., Jennings, G., Rudolph, M., Schlag, P., and Strauss, M., Efficient gene transfer into human hepatocytes by baculovirus vectors, *Proc. Natl. Acad. Sci. U. S. A.*, 92, 10099, 1995.
23. Carbonell, L.F. and Miller, L.K., Baculovirus interaction with nontarget organisms: A virus-borne reporter gene is not expressed in two cell lines, *Appl. Environ. Microbiol.*, 53, 1412, 1987.
24. Carbonell, L.F. and Miller, L.K., Genetic engineering of viral pesticides: Expression of foreing genes in nonpermissive cells, *Mol. Strateg. Crop Prot.*, V48, 235, 1987.
25. Sandig, V., Hofmann, C., Steinert, S., Jennings, G., Schlag, P., and Strauss, M., Gene transfer into hepatocytes and human liver tissue by baculovirus vectors, *Hum. Gene Ther.*, 7, 1937, 1996.
26. Lehtolainen, P., Tyynela, K., Kannasto, J., Airenne, K.J., and Yla-Herttuala, S., Baculoviruses exhibit restricted cell type specificity in rat brain: A comparison of baculovirus- and adenovirus-mediated intracerebral gene transfer in vivo, *Gene Ther.*, 9, 1693, 2002.
27. Haeseleer, F., Imanishi, Y., Saperstein, D.A., and Palczewski, K., Gene transfer mediated by recombinant baculovirus into mouse eye, *Invest. Ophthalmol. Vis. Sci.*, 42, 3294, 2001.
28. Sarkis, C., Serguera, C., Petres, S., Buchet, D., Ridet, J.L., Edelman, L., and Mallet, J., Efficient transduction of neural cells in vitro and in vivo by a baculovirus-derived vector, *Proc. Natl. Acad. Sci. U. S. A.*, 97, 14638, 2000.
29. Tani, H., Limn, C.K., Yap, C.C., Onishi, M., Nozaki, M., Nishimune, Y., Okahashi, N., Kitagawa, Y., Watanabe, R., Mochizuki, R., Moriishi, K., and Matsuura, Y., In vitro and in vivo gene delivery by recombinant baculoviruses, *J. Virol.*, 77, 9799, 2003.
30. Liu, X., Li, K., Song, J., Liang, C., Wang, X., and Chen, X., Efficient and stable gene expression in rabbit intervertebral disc cells transduced with a recombinant baculovirus vector, *Spine*, 31, 732, 2006.
31. Wang, X., Wang, C., Zeng, J., Xu, X., Hwang, P.Y., Yee, W.C., Ng, Y.K., and Wang, S., Gene transfer to dorsal root ganglia by intrathecal injection: Effects on regeneration of peripheral nerves, *Mol. Ther.*, 12, 314, 2005.
32. Martignoni, M.E. and Iwai, P.J., *A Catalog of Viral Diseases of Insects, Mites and Ticks*, 4th ed., US Department of Agriculture, Forest Service, Pacific Northwest Research Station, Portland, OR, 1986.
33. Gröner, A., Specificity and safety of baculoviruses, in: *The Biology of Baculoviruses, Vol 1*, Granados, R.R. and Federici, B.A., Eds., CRC Press, Boca Raton, 1986, p. 177.
34. Theilmann, D.A., Blissard, G.W., Bonning, B., Jehle, J., O'Reilly, D.R., Rohrmann, G.F., Thiem, S., and Vlak, J.M., Family Baculoviridae, in: *Virus Taxonomy: Eighth Report of the International Committee on Taxonomy of Viruses*, Fauquet, C.M., Mayo, M.A., Maniloff, J., Desselberger, U., and Ball, L.A., Eds., Elsevier, London, 2005, p. 177.
35. Summers, M.D. and Anderson, D.L., Granulosis virus deoxyribonucleic acid: A closed, double-stranded molecule, *J. Virol.*, 9, 710, 1972.

36. Burgess, S., Molecular weights of lepidopteran baculoviruses DNAs: Derivation by electron microscopy, *J. Gen. Virol.*, 37, 501, 1977.

37. Tweeten, K.A., Bulla, L.A., and Consigli, R.A., Characterization of an extremely basic protein derived from granulosis virus nucleocapsid, *J. Virol.*, 33, 866, 1980.

38. Harrap, K.A., The structure of nuclear polyhedrosis viruses. II. The virus particle, *Virology*, 50, 124, 1972.

39. Williams, G.V. and Faulkner, P., Cytological changes and viral morphogenesis during baculovirus infection, in: *The Baculoviruses*, 1st ed., Miller, L.K., Ed., Plenum Press, New York, 1997, 64.

40. van Loo, N.D., Fortunati, E., Ehlert, E., Rabelink, M., Grosveld, F., and Scholte, B.J., Baculovirus infection of nondividing mammalian cells: Mechanisms of entry and nuclear transport of capsids, *J. Virol.*, 75, 961, 2001.

41. Fraser, M.J., Ultrastructural observations of virion maturation in *Autographa californica* nuclear polyhedrosis virus infected *Spodoptera frugiperda* cell cultures, *J. Ultrastruct. Mol. Struct. Res.*, 95, 189, 1986.

42. Jehle, J.A., Blissard, G.W., Bonning, B.C., Cory, J.S., Herniou, E.A., Rohrmann, G.F., Theilmann, D.A., Thiem, S.M., and Vlak, J.M., On the classification and nomenclature of baculoviruses: A proposal for revision, *Arch. Virol.*, 151, 1257, 2006.

43. Miller, L.K., Baculoviruses as gene expression vectors, *Annu. Rev. Microbiol.*, 42, 177, 1988.

44. Blissard, G.W., Baculovirus–insect cell interactions, *Cytotechnology*, 20, 73, 1996.

45. Rohrmann, G.F., Baculovirus structural proteins, *J. Gen. Virol.*, 73, 749, 1992.

46. Braunagel, S.C. and Summers, M.D., *Autographa californica* nuclear polyhedrosis virus, PDV, and ECV viral envelopes and nucleocapsids: Structural proteins, antigens, lipid and fatty acid profiles, *Virology*, 202, 315, 1994.

47. Wang, P., Hammer, D.A., and Granados, R.R., Binding and fusion of *Autographa californica* nucleopolyhedrovirus to cultured insect cells, *J. Gen. Virol.*, 78, 3081, 1997.

48. Granados, R.R., Early events in the infection of *Hiliothis zea* midgut cells by a baculovirus, *Virology*, 90, 170, 1978.

49. Summers, M.D., Electron microscopic observations on granulosis virus entry, uncoating and replication processes during infection of the midgut cells of *Trichoplusia ni*, *J. Ultrastruct. Res.*, 35, 606, 1971.

50. Harrap, K.A., The structure of nuclear polyhedrosis viruses. I. The inclusion body, *Virology*, 50, 114, 1972.

51. Granados, R.R. and Federici, B.A. *The Biology of Baculoviruses*, CRC Press, Boca Raton, 1986.

52. Harrap, K.A. and Longworth, J.F., An evaluation of purification methods for baculoviruses, *J. Invertebr. Pathol.*, 24, 55, 1974.

53. Granados, R.R. and Williams, K.A., In vivo infection and replication of baculoviruses, in: *The Biology of Baculoviruses*, Granados, R.R. and Federici, B.A., Eds., CRC Press, Boca Raton, 1986, p. 89.

54. Keddie, B.A., Aponte, G.W., and Volkman, L.E., The pathway of infection of *Autographa californica* nuclear polyhedrosis virus in an insect host, *Science*, 243, 1728, 1989.

55. Federici, B.A., Baculovirus pathogenesis, in: *The Baculoviruses*, 1st ed., Miller, L.K., Ed., Plenum Press, New York, 1997, p. 33.

56. Ohkawa, T., Washburn, J.O., Sitapara, R., Sid, E., and Volkman, L.E., Specific binding of *Autographa californica* M nucleopolyhedrovirus occlusion-derived virus to midgut cells of *Heliothis virescens* larvae is mediated by products of pif genes Ac119 and Ac022 but not by Ac115, *J. Virol.*, 79, 15258, 2005.

57. Washburn, J.O., Lyons, E.H., Haas-Stapleton, E.J., and Volkman, L.E., Multiple nucleocapsid packaging of *Autographa californica* nucleopolyhedrovirus accelerates the onset of systemic infection in *Trichoplusia ni*, *J. Virol.*, 73, 411, 1999.

58. Engelhard, E.K., Kam-Morgan, L.N., Washburn, J.O., and Volkman, L.E., The insect tracheal system: A conduit for the systemic spread of *Autographa californica* M nuclear polyhedrosis virus, *Proc. Natl. Acad. Sci. U. S. A.*, 91, 3224, 1994.

59. Chisholm, G.E. and Henner, D.J., Multiple early transcripts and splicing of the *Autographa californica* nuclear polyhedrosis virus IE-1 gene, *J. Virol.*, 62, 3193, 1988.

60. Ayres, M.D., Howard, S.C., Kuzio, J., Lopez-Ferber, M., and Possee, R.D., The complete DNA sequence of *Autographa californica* nuclear polyhedrosis virus, *Virology*, 202, 586, 1994.

61. Funk, C.J., Braunagel, S.C., and Rohrmann, G.F., Baculovirus structure, in: *The Baculoviruses*, 1st ed., Miller, L.K., Ed., Plenum Press, New York, 1997, p. 7.

62. Slack, J. and Arif, B.M., The baculoviruses occlusion-derived virus: Virion structure and function, *Adv. Virus Res.*, 69, 99, 2007.

63. Wilson, M.E. and Consigli, R.A., Characterization of a protein kinase activity associated with purified capsids of the granulosis virus infection *Plodia interpunctella*, *Virology*, 143, 516, 1985.

64. Wilson, M.E. and Consigli, R.A., Functions of a protein kinase activity associated with purified capsids of the granulosis virus infecting *Plodia interpunctella*, *Virology*, 143, 526, 1985.

65. Blissard, G.W. and Wenz, J.R., Baculovirus gp64 envelope glycoprotein is sufficient to mediate pH-dependent membrane fusion, *J. Virol.*, 66, 6829, 1992.

66. Oomens, A.G. and Blissard, G.W., Requirement for GP64 to drive efficient budding of *Autographa californica* multicapsid nucleopolyhedrovirus, *Virology*, 254, 297, 1999.

67. Duisit, G., Saleun, S., Douthe, S., Barsoum, J., Chadeuf, G., and Moullier, P., Baculovirus vector requires electrostatic interactions including heparan sulfate for efficient gene transfer in mammalian cells, *J. Gene Med.*, 1, 93, 1999.

68. Tani, H., Nishijima, M., Ushijima, H., Miyamura, T., and Matsuura, Y., Characterization of cell-surface determinants important for baculovirus infection, *Virology*, 279, 343, 2001.

69. Hynes, R.O., Integrins: Versatility, modulation, and signaling in cell adhesion, *Cell*, 69, 11, 1992.

70. Vaughn, J.L., Goodwin, R.H., Tompkins, G.J., and McCawley, P., The establishment of two cell lines from the insect *Spodoptera frugiperda* (Lepidoptera; Noctuidae), *In Vitro*, 13, 213, 1977.

71. Wickham, T.J., Davis, T., Granados, R.R., Shuler, M.L., and Wood, H.A., Screening of insect cell lines for the production of recombinant proteins and infectious virus in the baculovirus expression system, *Biotechnol. Prog.*, 8, 391, 1992.

72. Davis, T.R., Wickham, T.J., McKenna, K.A., Granados, R.R., Shuler, M.L., and Wood, H.A., Comparative recombinant protein production of eight insect cell lines, *In Vitro Cell Dev. Biol. Anim.*, 29A, 388, 1993.

73. Wickham, T.J., Nemerow, G.R., Wood, H.A., and Shuler, M.L., Comparison of different cell lines for the production of recombinant baculovirus proteins, in: *Baculovirus Expression Protocols*, Richardson, C.D., Ed., Humana Press, Totowa, 1995, p. 385.

74. Smith, G.E., Fraser, M.J., and Summers, M.D., Molecular engineering of the *Autographa californica* nuclear polyhedrosis virus genome: Deletion mutations within the polyhedrin gene, *J. Virol.*, 46, 584, 1983.

75. Rosa, D., Campagnoli, S., Moretto, C., Guenzi, E., Cousens, L., Chin, M., Dong, C., Weiner, A.J., Lau, J.Y., Choo, Q.L., Chien, D., Pileri, P., Houghton, M., and Abrignani, S., A quantitative test to estimate neutralizing antibodies to the hepatitis C virus:

Cytofluorimetric assessment of envelope glycoprotein 2 binding to target cells, *Proc. Natl. Acad. Sci. U. S. A.*, 93, 1759, 1996.

76. Chazenbalk, G.D. and Rapoport, B., Expression of the extracellular domain of the thyrotropin receptor in the baculovirus system using a promoter active earlier than the polyhedrin promoter. Implications for the expression of functional highly glycosylated proteins, *J. Biol. Chem.*, 270, 1543, 1995.

77. Harrison, R.L. and Jarvis, D.L., Protein N-glycosylation in the baculovirus-insect cell expression system and engineering of insect cells to produce "mammalianized" recombinant glycoproteins, *Adv. Virus Res.*, 68, 159, 2006.

78. Miller, L.K., Baculoviruses: High-level expression in insect cells, *Curr. Opin. Genet. Dev.*, 3, 97, 1993.

79. Jarvis, D.L. and Finn, E.E., Modifying the insect cell N-glycosylation pathway with immediate early baculovirus expression vectors, *Nat. Biotechnol.*, 14, 1288, 1996.

80. Jones, I. and Morikawa, Y., Baculovirus vectors for expression in insect cells, *Curr. Opin. Biotechnol.*, 7, 512, 1996.

81. Smith, G.E., Summers, M.D., and Fraser, M.J., Production of human beta interferon in insect cells with a baculovirus expression vector, *Mol. Cell. Biol.*, 3, 2156, 1983.

82. Godeau, F., Saucier, C., and Kourilsky, P., Replication inhibition by nucleoside analogues of a recombinant *Autographa californica* multicapsid nuclear polyhedrosis virus harboring the herpes thymidine kinase gene driven by the IE-1(0) promoter: A new way to select recombinant baculoviruses, *Nucleic Acids Res.*, 20, 6239, 1992.

83. Davies, A.H., Current methods for manipulating baculoviruses, *Biotechnology (N. Y.)*, 12, 47, 1994.

84. Kitts, P.A., Ayres, M.D., and Possee, R.D., Linearization of baculovirus DNA enhances the recovery of recombinant virus expression vectors, *Nucleic Acids Res.*, 18, 5667, 1990.

85. Kitts, P.A. and Possee, R.D., A method for producing recombinant baculovirus expression vectors at high frequency, *Biotechniques*, 14, 810, 1993.

86. Bishop, D.H.L., Novy, R., and Mierendorf, R., The BacVector system: Simplified cloning and protein expression using novel baculovirus vectors, *Innovations*, 4, 1, 1995.

87. Patel, G., Nasmyth, K., and Jones, N., A new method for the isolation of recombinant baculovirus, *Nucleic Acids Res.*, 20, 97, 1991.

88. Je, Y.H., Chang, J.H., Choi, J.H., Roh, J.Y., Jin, B.R., O'Reilly, D.R., and Kang, S.K., A defective viral genome maintained in *Esherichia coli* for the generation of baculovirus expression vectors, *Biotechnol. Lett.*, 23, 575, 2001.

89. Zhao, Y., Chapman, D.A., and Jones, I.M., Improving baculovirus recombination, *Nucleic Acids Res.*, 31, E6, 2003.

90. Ernst, W.J., Grabherr, R.M., and Katinger, H.W., Direct cloning into the *Autographa californica* nuclear polyhedrosis virus for generation of recombinant baculoviruses, *Nucleic Acids Res.*, 22, 2855, 1994.

91. Lihoradova, O.A., Ogay, I.D., Abdukarimov, A.A., Azimova, S., Lynn, D.E., and Slack, J.M., The Homingbac baculovirus cloning system: An alternative way to introduce foreign DNA into baculovirus genomes, *J. Virol. Methods*, 140, 59, 2007.

92. Ma, Q., Zhou, L., Ma, L., and Huo, K., Directional and direct cloning strategy for high-throughput generation of recombinant baculoviruses, *Biotechniques*, 41, 453, 2006.

93. Lu, A. and Miller, L.K., Generation of recombinant baculoviruses by direct cloning, *Biotechniques*, 21, 63, 1996.

94. Peakman, T.C., Harris, R.A., and Gewert, D.R., Highly efficient generation of recombinant baculoviruses by enzymatically medicated site-specific in vitro recombination, *Nucleic Acids Res.*, 20, 495, 1992.

95. Luckow, V.A., Lee, S.C., Barry, G.F., and Olins, P.O., Efficient generation of infectious recombinant baculoviruses by site-specific transposon-mediated insertion of foreign genes into a baculovirus genome propagated in *Escherichia coli*, *J. Virol.*, 67, 4566, 1993.

96. Leusch, M.S., Lee, S.C., and Olins, P.O., A novel host-vector system for direct selection of recombinant baculoviruses (bacmids) in *Escherichia coli*, *Gene*, 160, 191, 1995.

97. Airenne, K.J., Peltomaa, E., Hytonen, V.P., Laitinen, O.H., and Yla-Herttuala, S., Improved generation of recombinant baculovirus genomes in *Escherichia coli*, *Nucleic Acids Res.*, 31, e101, 2003.

98. Airenne, K.J., Laitinen, O.H., Mähönen, A.J., and Ylä-Herttuala, S., Safe, simple, and high-capacity gene delivery into insect and vertebrate cells by recombinant baculoviruses, in: *Gene Transfer: Delivery and Expression of DNA and RNA*, Friedmann, T. and Rossi, J., Eds., Cold Spring Harbor laboratory Press, New York, 2007, p. 313.

99. Yao, L.G., Liu, Z.C., Zhang, X.M., Kan, Y.C., and Zhou, J.J., A high efficient method for generation of recombinant Bombyx mori nuclear polyhedrosis virus Bacmid and large-scale expression of foreign proteins in silkworm larvae, *Biotechnol. Appl. Biochem.*, 48, 45, 2007.

100. Hou, S., Chen, X., Wang, H., Tao, M., and Hu, Z., Efficient method to generate homologous recombinant baculovirus genomes in *E. coli*, *Biotechniques*, 32, 783, 786, 788, 2002.

101. McIntosh, A.H. and Shamy, R., Effects of the nuclear polyhedrosis virus (NPV) of *Autographa californica* on a vertebrate viper cell line, *Ann. N. Y. Acad. Sci.*, 266, 327, 1975.

102. Matsuo, E., Tani, H., Lim, C., Komoda, Y., Okamoto, T., Miyamoto, H., Moriishi, K., Yagi, S., Patel, A.H., Miyamura, T., and Matsuura, Y., Characterization of HCV-like particles produced in a human hepatoma cell line by a recombinant baculovirus, *Biochem. Biophys. Res. Commun.*, 340, 200, 2006.

103. Cheng, T., Xu, C.Y., Wang, Y.B., Chen, M., Wu, T., Zhang, J., and Xia, N.S., A rapid and efficient method to express target genes in mammalian cells by baculovirus, *World J. Gastroenterol.*, 10, 1612, 2004.

104. Kost, T.A. and Condreay, J.P., Recombinant baculoviruses as mammalian cell gene-delivery vectors, *Trends Biotechnol.*, 20, 173, 2002.

105. Song, S.U., Shin, S.H., Kim, S.K., Choi, G.S., Kim, W.C., Lee, M.H., Kim, S.J., Kim, I.H., Choi, M.S., Hong, Y.J., and Lee, K.H., Effective transduction of osteogenic sarcoma cells by a baculovirus vector, *J. Gen. Virol.*, 84, 697, 2003.

106. Condreay, J.P., Witherspoon, S.M., Clay, W.C., and Kost, T.A., Transient and stable gene expression in mammalian cells transduced with a recombinant baculovirus vector, *Proc. Natl. Acad. Sci. U. S. A.*, 96, 127, 1999.

107. Hofmann, C., Wolfgang, L., and Strauss, M., The baculovirus vector system for gene delivery into hepatocytes, *Gene Ther. Mol. Biol.*, 1, 231, 1998.

108. Boyce, F.M. and Bucher, N.L., Baculovirus-mediated gene transfer into mammalian cells, *Proc. Natl. Acad. Sci. U. S. A.*, 93, 2348, 1996.

109. Gao, R., McCormick, C.J., Arthur, M.J., Ruddell, R., Oakley, F., Smart, D.E., Murphy, F.R., Harris, M.P., and Mann, D.A., High efficiency gene transfer into cultured primary rat and human hepatic stellate cells using baculovirus vectors, *Liver*, 22, 15, 2002.

110. Shoji, I., Aizaki, H., Tani, H., Ishii, K., Chiba, T., Saito, I., Miyamura, T., and Matsuura, Y., Efficient gene transfer into various mammalian cells, including non-hepatic cells, by baculovirus vectors, *J. Gen. Virol.*, 78, 2657, 1997.

111. Yap, C.C., Ishii, K., Aoki, Y., Aizaki, H., Tani, H., Shimizu, H., Ueno, Y., Miyamura, T., and Matsuura, Y., A hybrid baculovirus-T7 RNA polymerase system for recovery of an infectious virus from cDNA, *Virology*, 231, 192, 1997.

112. Barsoum, J., Brown, R., McKee, M., and Boyce, F.M., Efficient transduction of mammalian cells by a recombinant baculovirus having the vesicular stomatitis virus G glycoprotein, *Hum. Gene Ther.*, 8, 2011, 1997.

113. Zhu, Y., QI, Y., Liu, Z., and Xiao, G., Baculovirus-mediated gene transfer into mammalian cells, *Sci. China*, 41, 473, 1998.

114. Wang, L., Shan, L., Yin, J., Zhao, M., Su, D., and Zhong, J., The activation of lytic replication of Epstein–Barr virus by baculovirus-mediated gene transduction, *Arch. Virol.*, 151, 2047, 2006.

115. Andersson, M., Warolen, M., Nilsson, J., Selander, M., Sterky, C., Bergdahl, K., Sorving, C., James, S.R., and Doverskog, M., Baculovirus-mediated gene transfer and recombinant protein expression do not interfere with insulin dependent phosphorylation of PKB/Akt in human SHSY-5Y and C3A cells, *BMC. Cell Biol.*, 8, 6, 2007.

116. Palmiter, R.D., Sandgren, E.P., Avarbock, M.R., Allen, D.D., and Brinster, R.L., Heterologous introns can enhance expression of transgenes in mice, *Proc. Natl. Acad. Sci. U. S. A.*, 88, 478, 1991.

117. Lacy-Hulbert, A., Thomas, R., Li, X.P., Lilley, C.E., Coffin, R.S., and Roes, J., Interruption of coding sequences by heterologous introns can enhance the functional expression of recombinant genes, *Gene Ther.*, 8, 649, 2001.

118. Donello, J.E., Loeb, J.E., and Hope, T.J., Woodchuck hepatitis virus contains a tripartite posttranscriptional regulatory element, *J. Virol.*, 72, 5085, 1998.

119. De Geest, B., van Linthout, S., Lox, M., Collen, D., and Holvoet, P., Sustained expression of human apolipoprotein A-I after adenoviral gene transfer in C57BL/6 mice: Role of apolipoprotein A-I promoter, apolipoprotein A-I introns, and human apolipoprotein E enhancer, *Hum. Gene Ther.*, 11, 101, 2000.

120. Stanbridge, L.J., Dussupt, V., and Maitland, N.J., Baculoviruses as vectors for gene therapy against human prostate cancer, *J. Biomed. Biotechnol.*, 2, 79, 2003.

121. Takahashi, R., Nishimura, J., Hirano, K., Naito, S., and Kanaide, H., Functional role of PKC in contraction of cultured human prostatic stromal cells, *J. Cell Biochem.*, 96, 65, 2005.

122. Takahashi, R., Nishimura, J., Seki, N., Yunoki, T., Tomoda, T., Kanaide, H., and Naito, S., RhoA/Rho kinase-mediated Ca(2+) sensitization in the contraction of human prostate, *Neurourol. Urodyn.*, 26, 547, 2007.

123. Kronschnabl, M., Marschall, M., and Stamminger, T., Efficient and tightly regulated expression systems for the human cytomegalovirus major transactivator protein IE2p86 in permissive cells, *Virus Res.*, 83, 89, 2002.

124. Kronschnabl, M. and Stamminger, T., Synergistic induction of intercellular adhesion molecule-1 by the human cytomegalovirus transactivators IE2p86 and pp71 is mediated via an Sp1-binding site, *J. Gen. Virol.*, 84, 61, 2003.

125. Ma, L., Tamarina, N., Wang, Y., Kuznetsov, A., Patel, N., Kending, C., Hering, B.J., and Philipson, L.H., Baculovirus-mediated gene transfer into pancreatic islet cells, *Diabetes*, 49, 1986, 2000.

126. Aoki, H., Sakoda, Y., Jukuroki, K., Takada, A., Kida, H., and Fukusho, A., Induction of antibodies in mice by a recombinant baculovirus expressing pseudorabies virus glycoprotein B in mammalian cells, *Vet. Microbiol.*, 68, 197, 1999.

127. Liang, C.Y., Wang, H.Z., Li, T.X., Hu, Z.H., and Chen, X.W., High efficiency gene transfer into mammalian kidney cells using baculovirus vectors, *Arch. Virol.*, 149, 51, 2004.

128. Song, S.U. and Boyce, F.M., Combination treatment for osteosarcoma with baculoviral vector mediated gene therapy (p53) and chemotherapy (adriamycin), *Exp. Mol. Med.*, 33, 46, 2001.

129. Cheng, H., Cenciarelli, C., Nelkin, G., Tsan, R., Fan, D., Cheng-Mayer, C., and Fidler, I.J., Molecular mechanism of hTid-1, the human homolog of Drosophila tumor suppressor l(2)Tid, in the regulation of NF-kappaB activity and suppression of tumor growth, *Mol. Cell Biol.*, 25, 44, 2005.

130. Grassi, G., Kohn, H., Dapas, B., Farra, R., Platz, J., Engel, S., Cjsareck, S., Kandolf, R., Teutsch, C., Klima, R., Triolo, G., and Kuhn, A., Comparison between recombinant baculo- and adenoviral-vectors as transfer system in cardiovascular cells, *Arch. Virol.*, 151, 255, 2006.

131. Liu, X.Y., Yang, S.H., Liang, C.Y., Song, J.H., Li, K.H., and Chen, X.W., Construction of recombinant baculovirus Ac-CMV-hSox9 for gene therapy of intervertebral disc degeneration, *Chin. J. Traumatol.*, 10, 94, 2007.

132. Mimeault, M. and Batra, S.K., Concise review: Recent advances on the significance of stem cells in tissue regeneration and cancer therapies, *Stem Cells*, 24, 2319, 2006.

133. Alhadlaq, A. and Mao, J.J., Mesenchymal stem cells: Isolation and therapeutics, *Stem Cells Dev.*, 13, 436, 2004.

134. Kashofer, K. and Bonnet, D., Gene therapy progress and prospects: Stem cell plasticity, *Gene Ther.*, 12, 1229, 2005.

135. Yates, F. and Daley, G.Q., Progress and prospects: Gene transfer into embryonic stem cells, *Gene Ther.*, 13, 1431, 2006.

136. Ho, Y.C., Chung, Y.C., Hwang, S.M., Wang, K.C., and Hu, Y.C., Transgene expression and differentiation of baculovirus-transduced human mesenchymal stem cells, *J. Gene Med.*, 7, 860, 2005.

137. Ho, Y.C., Lee, H.P., Hwang, S.M., Lo, W.H., Chen, H.C., Chung, C.K., and Hu, Y.C., Baculovirus transduction of human mesenchymal stem cell-derived progenitor cells: Variation of transgene expression with cellular differentiation states, *Gene Ther.*, 13, 1471, 2006.

138. Lee, H.P., Ho, Y.C., Hwang, S.M., Sung, L.Y., Shen, H.C., Liu, H.J., and Hu, Y.C., Variation of baculovirus-harbored transgene transcription among mesenchymal stem cell-derived progenitors leads to varied expression, *Biotechnol. Bioeng.*, 97, 649, 2007.

139. Sung, L.Y., Lo, W.H., Chiu, H.Y., Chen, H.C., Chung, C.K., Lee, H.P., and Hu, Y.C., Modulation of chondrocyte phenotype via baculovirus-mediated growth factor expression, *Biomaterials*, 28, 3437, 2007.

140. Zeng, J., Du, J., Zhao, Y., Palanisamy, N., and Wang, S., Baculoviral vector-mediated transient and stable transgene expression in human embryonic stem cells, *Stem Cells*, 25, 1055, 2007.

141. Ping, W., Ge, J., Li, S., Zhou, H., Wang, K., Feng, Y., and Lou, Z., Baculovirus-mediated gene expression in chicken primary cells, *Avian Dis.*, 50, 59, 2006.

142. Song, J., Liang, C., and Chen, X., Transduction of avian cells with recombinant baculovirus, *J. Virol. Methods*, 135, 157, 2006.

143. Leisy, D.J., Lewis, T.D., Leong, J.A., and Rohrmann, G.F., Transduction of cultured fish cells with recombinant baculoviruses, *J. Gen. Virol.*, 84, 1173, 2003.

144. Wagle, M. and Jesuthasan, S., Baculovirus-mediated gene expression in zebrafish, *Mar. Biotechnol. (N. Y.)*, 5, 58, 2003.

145. Wagle, M., Grunewald, B., Subburaju, S., Barzaghi, C., Le, G.S., Chan, J., and Jesuthasan, S., EphrinB2a in the zebrafish retinotectal system, *J. Neurobiol.*, 59, 57, 2004.

146. Lee, D.F., Chen, C.C., Hsu, T.A., and Juang, J.L., A baculovirus superinfection system: Efficient vehicle for gene transfer into Drosophila S2 cells, *J. Virol.*, 74, 11873, 2000.

147. Oppenheimer, D.I., MacNicol, A.M., and Patel, N.H., Functional conservation of the wingless-engrailed interaction as shown by a widely applicable baculovirus misexpression system, *Curr. Biol.*, 9, 1288, 1999.

148. Ando, T., Fujiyuki, T., Kawashima, T., Morioka, M., Kubo, T., and Fujiwara, H., In vivo gene transfer into the honeybee using a nucleopolyhedrovirus vector, *Biochem. Biophys. Res. Commun.*, 352, 335, 2007.

149. Shen, H.C., Lee, H.P., Lo, W.H., Yang, D.G., and Hu, Y.C., Baculovirus-mediated gene transfer is attenuated by sodium bicarbonate, *J. Gene Med.*, 9, 470, 2007.

150. Mahonen, A.J., Airenne, K.J., Purola, S., Peltomaa, E., Kaikkonen, M.U., Riekkinen, M., Heikura, T., Kinnunen, K., Roschier, M.M., Wirth, T., and Yla-Herttuala, S., Post-transcriptional regulatory element boosts baculovirus-mediated gene-expression in vertebrate cells, *J. Biotechnol.*, 131, 1, 2007.

151. Kukkonen, S.P., Airenne, K.J., Marjomaki, V., Laitinen, O.H., Lehtolainen, P., Kankaanpaa, P., Mahonen, A.J., Raty, J.K., Nordlund, H.R., Oker-Blom, C., Kulomaa, M.S., and Yla-Herttuala, S., Baculovirus capsid display: A novel tool for transduction imaging, *Mol. Ther.*, 8, 853, 2003.

152. Hsu, C.S., Ho, Y.C., Wang, K.C., and Hu, Y.C., Investigation of optimal transduction conditions for baculovirus-mediated gene delivery into mammalian cells, *Biotechnol. Bioeng.*, 88, 42, 2004.

153. Berger, S.L., Gene regulation. Local or global? *Nature*, 408, 412, 415, 2000.

154. Gaetano, C., Catalano, A., Palumbo, R., Illi, B., Orlando, G., Ventoruzzo, G., Serino, F., and Capogrossi, M.C., Transcriptionally active drugs improve adenovirus vector performance in vitro and in vivo, *Gene Ther.*, 7, 1624, 2000.

155. Tobias, C.A., Kim, D., and Fischer, I., Improved recombinant retroviral titers utilizing trichostatin A, *Biotechniques*, 29, 884, 2000.

156. Poon, A.P., Gu, H., and Roizman, B., ICP0 and the US3 protein kinase of herpes simplex virus 1 independently block histone deacetylation to enable gene expression, *Proc. Natl. Acad. Sci. U. S. A.*, 103, 9993, 2006.

157. Poon, A.P. and Roizman, B., Mapping of key functions of the herpes simplex virus 1 U(S)3 protein kinase: The U(S)3 protein can form functional heteromultimeric structures derived from overlapping truncated polypeptides, *J. Virol.*, 81, 1980, 2007.

158. Salminen, M., Airenne, K.J., Rinnankoski, R., Reimari, J., Valilehto, O., Rinne, J., Suikkanen, S., Kukkonen, S., Yla-Herttuala, S., Kulomaa, M.S., and Vihinen-Ranta, M., Improvement in nuclear entry and transgene expression of baculoviruses by disintegration of microtubules in human hepatocytes, *J. Virol.*, 79, 2720, 2005.

159. Hunt, L., Batard, P., Jordan, M., and Wurm, F.M., Fluorescent proteins in animal cells for process development: Optimization of sodium butyrate treatment as an example, *Biotechnol. Bioeng.*, 77, 528, 2002.

160. Bilello, J.P., Delaney, W.E., Boyce, F.M., and Isom, H.C., Transient disruption of intercellular junctions enables baculovirus entry into nondividing hepatocytes, *J. Virol.*, 75, 9857, 2001.

161. Bilello, J.P., Cable, E.E., and Isom, H.C., Expression of E-cadherin and other paracellular junction genes is decreased in iron-loaded hepatocytes, *Am. J. Pathol.*, 162, 1323, 2003.

162. Bilello, J.P., Cable, E.E., Myers, R.L., and Isom, H.C., Role of paracellular junction complexes in baculovirus-mediated gene transfer to nondividing rat hepatocytes, *Gene Ther.*, 10, 733, 2003.

163. Ramos, L., Kopec, L.A., Sweitzer, S.M., Fornwald, J.A., Zhao, H., Mcallister, P., McNulty, D.E., Trill, J.J., and Kane, J.F., Rapid expression of recombinant proteins in modified CHO cells using the baculovirus system, *Cytotechnology*, 38, 37, 2002.

164. Wang, K.C., Wu, J.C., Chung, Y.C., Ho, Y.C., Chang, M.D., and Hu, Y.C., Baculovirus as a highly efficient gene delivery vector for the expression of hepatitis delta virus antigens in mammalian cells, *Biotechnol. Bioeng.*, 89, 464, 2005.

165. Hu, Y.C., Tsai, C.T., Chang, Y.J., and Huang, J.H., Enhancement and prolongation of baculovirus-mediated expression in mammalian cells: Focuses on strategic infection and feeding, *Biotechnol. Prog.*, 19, 373, 2003.

166. Nakowitsch, S., Kittel, C., Ernst, W., Egorov, A., and Grabherr, R., Optimization of baculovirus transduction on FreeStyle293 cells for the generation of influenza B/Lee/40, *Mol. Biotechnol.*, 34, 157, 2006.

167. Transfiguracion, J., Jorio, H., Meghrous, J., Jacob, D., and Kamen, A., High yield purification of functional baculovirus vectors by size exclusion chromatography, *J. Virol. Methods*, 142, 21, 2007.

168. Barsoum, J., Concentration of recombinant baculovirus by cation-exchange chromatography, *Biotechniques*, 26, 834, 838, 840, 1999.

169. Wu, C., Soh, K.Y., and Wang, S., Ion-exchange membrane chromatography method for rapid and efficient purification of recombinant baculovirus and baculovirus gp64 protein, *Hum. Gene Ther.*, 2007.

170. Lee, H.P. and Hu, Y.C., Expression in mammalian cells using BacMam viruses, in: *Expression Systems: Methods Express*, Dyson, M.R. and Durocher, Y., Eds., Scion Publishing Limited, Oxford, 2007, p. 261.

171. Philipps, B., Forstner, M., and Mayr, L.M., A baculovirus expression vector system for simultaneous protein expression in insect and mammalian cells, *Biotechnol. Prog.*, 21, 708, 2005.

172. Oker-Blom, C., Airenne, K.J., and Grabherr, R., Baculovirus display strategies: Emerging tools for eukaryotic libraries and gene delivery, *Brief. Funct. Genomic. Proteomic.*, 2, 244, 2003.

173. Makela, A.R. and Oker-Blom, C., Baculovirus display: A multifunctional technology for gene delivery and eukaryotic library development, *Adv. Virus Res.*, 68, 91, 2006.

174. Chapple, S.D. and Jones, I.M., Non-polar distribution of green fluorescent protein on the surface of *Autographa californica* nucleopolyhedrovirus using a heterologous membrane anchor, *J. Biotechnol.*, 95, 269, 2002.

175. Borg, J., Nevsten, P., Wallenberg, R., Stenstrom, M., Cardell, S., Falkenberg, C., and Holm, C., Amino-terminal anchored surface display in insect cells and budded baculovirus using the amino-terminal end of neuraminidase, *J. Biotechnol.*, 114, 21, 2004.

176. Pieroni, L., Maione, D., and La Monica, N., In vivo gene transfer in mouse skeletal muscle mediated by baculovirus vectors, *Hum. Gene Ther.*, 12, 871, 2001.

177. Mastromarino, P., Conti, C., Goldoni, P., Hauttecoeur, B., and Orsi, N., Characterization of membrane components of the erythrocyte involved in vesicular stomatitis virus attachment and fusion at acidic pH, *J. Gen. Virol.*, 68, 2359, 1987.

178. Eidelman, O., Schlegel, R., Tralka, T.S., and Blumenthal, R., pH-dependent fusion induced by vesicular stomatitis virus glycoprotein reconstituted into phospholipid vesicles, *J. Biol. Chem.*, 259, 4622, 1984.

179. Naldini, L., Blomer, U., Gallay, P., Ory, D., Mulligan, R., Gage, F.H., Verma, I.M., and Trono, D., In vivo gene delivery and stable transduction of nondividing cells by a lentiviral vector, *Science*, 272, 263, 1996.

180. Burns, J.C., Friedmann, T., Driever, W., Burrascano, M., and Yee, J.K., Vesicular stomatitis virus G glycoprotein pseudotyped retroviral vectors: Concentration to very high titer and efficient gene transfer into mammalian and nonmammalian cells, *Proc. Natl. Acad. Sci. U. S. A.*, 90, 8033, 1993.

181. Poeschla, E.M., Wong-Staal, F., and Looney, D.J., Efficient transduction of nondividing human cells by feline immunodeficiency virus lentiviral vectors, *Nat. Med.*, 4, 354, 1998.

182. Mangor, J.T., Monsma, S.A., Johnson, M.C., and Blissard, G.W., A GP64-null baculovirus pseudotyped with vesicular stomatitis virus G protein, *J. Virol.*, 75, 2544, 2001.

183. Kitagawa, Y., Tani, H., Limn, C.K., Matsunaga, T.M., Moriishi, K., and Matsuura, Y., Ligand-directed gene targeting to mammalian cells by pseudotype baculoviruses, *J. Virol.*, 79, 3639, 2005.

184. Park, F., Ohashi, K., and Kay, M.A., Therapeutic levels of human factor VIII and IX using HIV-1-based lentiviral vectors in mouse liver, *Blood*, 96, 1173, 2000.

185. Baum, C., Schambach, A., Bohne, J., and Galla, M., Retrovirus vectors: Toward the plentivirus? *Mol. Ther.*, 13, 1050, 2006.

186. Kaikkonen, M.U., Raty, J.K., Airenne, K.J., Wirth, T., Heikura, T., and Yla-Herttuala, S., Truncated vesicular stomatitis virus G protein improves baculovirus transduction efficiency in vitro and in vivo, *Gene Ther.*, 13, 304, 2006.

187. Yang, D.G., Chung, Y.C., Lai, Y.K., Lai, C.W., Liu, H.J., and Hu, Y.C., Avian influenza virus hemagglutinin display on baculovirus envelope: Cytoplasmic domain affects virus properties and vaccine potential, *Mol. Ther.*, 15, 989, 2007.

188. Matilainen, H., Makela, A.R., Riikonen, R., Saloniemi, T., Korhonen, E., Hyypia, T., Heino, J., Grabherr, R., and Oker-Blom, C., RGD motifs on the surface of baculovirus enhance transduction of human lung carcinoma cells, *J. Biotechnol.*, 125, 114, 2006.

189. Makela, A.R., Matilainen, H., White, D.J., Ruoslahti, E., and Oker-Blom, C., Enhanced baculovirus-mediated transduction of human cancer cells by tumor-homing peptides, *J. Virol.*, 80, 6603, 2006.

190. Ge, J., Huang, Y., Hu, X., and Zhong, J., A surface-modified baculovirus vector with improved gene delivery to B-lymphocytic cells, *J. Biotechnol.*, 129, 367, 2007.

191. Raty, J.K., Airenne, K.J., Marttila, A.T., Marjomaki, V., Hytonen, V.P., Lehtolainen, P., Laitinen, O.H., Mahonen, A.J., Kulomaa, M.S., and Yla-Herttuala, S., Enhanced gene delivery by avidin-displaying baculovirus, *Mol. Ther.*, 9, 282, 2004.

192. Wilchek, M. and Bayer, E.A., Introduction to avidin-biotin technology, *Methods Enzymol.*, 184, 5, 1990.

193. Kim, Y.K., Park, I.K., Jiang, H.L., Choi, J.Y., Je, Y.H., Jin, H., Kim, H.W., Cho, M.H., and Cho, C.S., Regulation of transduction efficiency by pegylation of baculovirus vector in vitro and in vivo, *J. Biotechnol.*, 125, 104, 2006.

194. Spenger, A., Ernst, W., Condreay, J.P., Kost, T.A., and Grabherr, R., Influence of promoter choice and trichostatin A treatment on expression of baculovirus delivered genes in mammalian cells, *Protein Expr. Purif.*, 38, 17, 2004.

195. Michalkiewicz, M., Michalkiewicz, T., Geurts, A.M., Roman, R.J., Slocum, G.R., Singer, O., Weihrauch, D., Greene, A.S., Kaldunski, M.L., Verma, I.M., Jacob, H.J., and Cowley, A.W., Jr., Efficient transgenic rat production by a lentiviral vector, *Am. J. Physiol. Heart Circ. Physiol.*, 293, H881, 2007.

196. Kasparov, S., Suitability of hCMV for viral gene expression in the brain, *Nat. Methods*, 4, 379, 2007.

197. Liew, C.G., Draper, J.S., Walsh, J., Moore, H., and Andrews, P.W., Transient and stable transgene expression in human embryonic stem cells, *Stem Cells*, 25, 1521, 2007.

198. Park, S.W., Lee, H.K., Kim, T.G., Yoon, S.K., and Paik, S.Y., Hepatocyte-specific gene expression by baculovirus pseudotyped with vesicular stomatitis virus envelope glycoprotein, *Biochem. Biophys. Res. Commun.*, 289, 444, 2001.

199. Li, Y., Yang, Y., and Wang, S., Neuronal gene transfer by baculovirus-derived vectors accommodating a neurone-specific promoter, *Exp. Physiol.*, 90, 39, 2005.

200. Wang, C.Y. and Wang, S., Astrocytic expression of transgene in the rat brain mediated by baculovirus vectors containing an astrocyte-specific promoter, *Gene Ther.*, 13, 1447, 2006.

201. Li, Y., Wang, X., Guo, H., and Wang, S., Axonal transport of recombinant baculovirus vectors, *Mol. Ther.*, 10, 1121, 2004.

202. Ong, S.T., Li, F., Du, J., Tan, Y.W., and Wang, S., Hybrid cytomegalovirus enhancer-h1 promoter-based plasmid and baculovirus vectors mediate effective RNA interference, *Hum. Gene Ther.*, 16, 1404, 2005.

203. Viswanathan, P., Venkaiah, B., Kumar, M.S., Rasheedi, S., Vrati, S., Bashyam, M.D., and Hasnain, S.E., The homologous region sequence (hrl) of *Autographa californica* multinucleocapsid polyhedrosis virus can enhance transcription from non-baculoviral promoters in mammalian cells, *J. Biol. Chem.*, 278, 52564, 2003.

204. Liu, B.H., Yang, Y., Paton, J.F., Li, F., Boulaire, J., Kasparov, S., and Wang, S., GAL4-NF-kappaB fusion protein augments transgene expression from neuronal promoters in the rat brain, *Mol. Ther.*, 14, 872, 2006.

205. van Oers, M.M., Vaccines for viral and parasitic diseases produced with baculovirus vectors, *Adv. Virus Res.*, 68, 193, 2006.

206. Fipaldini, C., Bellei, B., and La Monica, N., Expression of hepatitis C virus cDNA in human hepatoma cell line mediated by a hybrid baculovirus-HCV vector, *Virology*, 255, 302, 1999.

207. Cheshenko, N., Krougliak, N., Eisensmith, R.C., and Krougliak, V.A., A novel system for the production of fully deleted adenovirus vectors that does not require helper adenovirus, *Gene Ther.*, 8, 846, 2001.

208. Hartley, J.L., Cloning technologies for protein expression and purification, *Curr. Opin. Biotechnol.*, 17, 359, 2006.

209. Palombo, F., Monciotti, A., Recchia, A., Cortese, R., Ciliberto, G., and La Monica, N., Site-specific integration in mammalian cells mediated by a new hybrid baculovirus-adeno-associated virus vector, *J. Virol.*, 72, 5025, 1998.

210. Wang, C.Y. and Wang, S., Adeno-associated virus inverted terminal repeats improve neuronal transgene expression mediated by baculoviral vectors in rat brain, *Hum. Gene Ther.*, 16, 1219, 2005.

211. Wang, C.Y., Li, F., Yang, Y., Guo, H.Y., Wu, C.X., and Wang, S., Recombinant baculovirus containing the diphtheria toxin A gene for malignant glioma therapy, *Cancer Res.*, 66, 5798, 2006.

212. Delaney, W.E. and Isom, H.C., Hepatitis B virus replication in human HepG2 cells mediated by hepatitis B virus recombinant baculovirus, *Hepatology*, 28, 1134, 1998.

213. Delaney, W.E., Miller, T.G., and Isom, H.C., Use of the hepatitis B virus recombinant baculovirus-HepG2 system to study the effects of (-)-beta-2′,3′-dideoxy-3′-thiacytidine on replication of hepatitis B virus and accumulation of covalently closed circular DNA, *Antimicrob. Agents Chemother.*, 43, 2017, 1999.

214. Delaney, W.E., Edwards, R., Colledge, D., Shaw, T., Torresi, J., Miller, T.G., Isom, H.C., Bock, C.T., Manns, M.P., Trautwein, C., and Locarnini, S., Cross-resistance testing of antihepadnaviral compounds using novel recombinant baculoviruses which encode drug-resistant strains of hepatitis B virus, *Antimicrob. Agents Chemother.*, 45, 1705, 2001.

215. Abdelhamed, A.M., Kelley, C.M., Miller, T.G., Furman, P.A., and Isom, H.C., Rebound of hepatitis B virus replication in HepG2 cells after cessation of antiviral treatment, *J. Virol.*, 76, 8148, 2002.

216. Heipertz, R.A., Jr., Miller, T.G., Kelley, C.M., Delaney, W.E., Locarnini, S.A., and Isom, H.C., In vitro study of the effects of precore and lamivudine-resistant mutations on hepatitis B virus replication, *J. Virol.*, 81, 3068, 2007.

217. McCormick, C.J., Brown, D., Griffin, S., Challinor, L., Rowlands, D.J., and Harris, M., A link between translation of the hepatitis C virus polyprotein and polymerase function; possible consequences for hyperphosphorylation of NS5A, *J. Gen. Virol.*, 87, 93, 2006.

218. Street, A., Macdonald, A., McCormick, C., and Harris, M., Hepatitis C virus NS5A-mediated activation of phosphoinositide 3-kinase results in stabilization of cellular beta-catenin and stimulation of beta-catenin-responsive transcription, *J. Virol.*, 79, 5006, 2005.

219. McCormick, C.J., Rowlands, D.J., and Harris, M., Efficient delivery and regulable expression of hepatitis C virus full-length and minigenome constructs in hepatocyte-derived cell lines using baculovirus vectors, *J. Gen. Virol.*, 83, 383, 2002.

220. Shaw, T., Bartholomeusz, A., and Locarnini, S., HBV drug resistance: Mechanisms, detection and interpretation, *J. Hepatol.*, 44, 593, 2006.

221. Dwarakanath, R.S., Clark, C.L., McElroy, A.K., and Spector, D.H., The use of recombinant baculoviruses for sustained expression of human cytomegalovirus immediate early proteins in fibroblasts, *Virology*, 284, 297, 2001.

222. Zhu, Y., Cullen, J.M., Aldrich, C.E., Saputelli, J., Miller, D., Seeger, C., Mason, W.S., and Jilbert, A.R., Adenovirus-based gene therapy during clevudine treatment of woodchucks chronically infected with woodchuck hepatitis virus, *Virology*, 327, 26, 2004.

223. Wickham, T.J., Shuler, M.L., Hammer, D.A., Granados, R.R., and Wood, H.A., Equilibrium and kinetic analysis of *Autographa californica* nuclear polyhedrosis virus attachment to different insect cell lines, *J. Gen. Virol.*, 73 (Pt 12), 3185, 1992.

224. Matilainen, H., Rinne, J., Gilbert, L., Marjomaki, V., Reunanen, H., and Oker-Blom, C., Baculovirus entry into human hepatoma cells, *J. Virol.*, 79, 15452, 2005.

225. Long, G., Pan, X., Kormelink, R., and Vlak, J.M., Functional entry of baculovirus into insect and mammalian cells is dependent on clathrin-mediated endocytosis, *J. Virol.*, 80, 8830, 2006.

226. Marsh, M. and Helenius, A., Virus entry: Open sesame, *Cell*, 124, 729, 2006.

227. Granados, R.R. and Lawler K.A., In vivo pathway of *Autographa californica* baculovirus invasion and infection, *Virology*, 108, 297, 1981.

228. Lee, H.P., Chen, Y.L., Shen, H.C., Lo, W.H., and Hu, Y.C., Baculovirus transduction of rat articular chondrocytes: Roles of cell cycle, *J. Gene Med.*, 9, 33, 2007.

229. Laakkonen, J.P., Kaikkonen, M.U., Ronkainen, P.H., Ihalainen, T.O., Niskanen, E.A., Hakkinen, M., Salminen, M., Kulomaa, M.S., Ylä-Herttuala, S., Airenne, K.J., and Vinhinen-Ranta, M., Baculovirus-mediated immediate early gene expression and nuclear reorganization in human cells, *Cell. Microbiol.*, 10, 667, 2007.

230. Kumar, M., Bradow, B.P., and Zimmerberg, J., Large-scale production of pseudotyped lentiviral vectors using baculovirus GP64, *Hum. Gene Ther.*, 14, 67, 2003.

231. Schauber, C.A., Tuerk, M.J., Pacheco, C.D., Escarpe, P.A., and Veres, G., Lentiviral vectors pseudotyped with baculovirus gp64 efficiently transduce mouse cells in vivo and show tropism restriction against hematopoietic cell types in vitro, *Gene Ther.*, 11, 266, 2004.

232. Guibinga, G.H. and Friedmann, T., Baculovirus GP64-pseudotyped HIV-based lentivirus vectors are stabilized against complement inactivation by codisplay of decay accelerating factor (DAF) or of a GP64-DAF fusion protein, *Mol. Ther.*, 11, 645, 2005.

233. Kremer, K.L., Dunning, K.R., Parsons, D.W., and Anson, D.S., Gene delivery to airway epithelial cells in vivo: A direct comparison of apical and basolateral transduction strategies using pseudotyped lentivirus vectors, *J. Gene Med.*, 9, 362, 2007.

234. Kang, Y., Xie, L., Tran, D.T., Stein, C.S., Hickey, M., Davidson, B.L., and McCray, P.B., Jr., Persistent expression of factor VIII in vivo following nonprimate lentiviral gene transfer, *Blood*, 106, 1552, 2005.

235. Sinn, P.L., Burnight, E.R., Hickey, M.A., Blissard, G.W., and McCray, P.B., Jr., Persistent gene expression in mouse nasal epithelia following feline immunodeficiency virus-based vector gene transfer, *J. Virol.*, 79, 12818, 2005.

236. Oomens, A.G. and Wertz, G.W., The baculovirus GP64 protein mediates highly stable infectivity of a human respiratory syncytial virus lacking its homologous transmembrane glycoproteins, *J. Virol.*, 78, 124, 2004.

237. Sastre, P., Oomens, A.G., and Wertz, G.W., The stability of human respiratory syncytial virus is enhanced by incorporation of the baculovirus GP64 protein, *Vaccine*, 25, 5025, 2007.

238. Tjia, S.T., zu Altenschildesche, G.M., and Doerfler, W., *Autographa californica* nuclear polyhedrosis virus (AcNPV) DNA does not persist in mass cultures of mammalian cells, *Virology*, 125, 107, 1983.

239. Charlton, C.A. and Volkman, L.E., Penetration of *Autographa californica* nuclear polyhedrosis virus nucleocapsids into IPLB Sf 21 cells induces actin cable formation, *Virology*, 197, 245, 1993.

240. Lanier, L.M. and Volkman, L.E., Actin binding and nucleation by *Autographa californica* M nucleopolyhedrovirus, *Virology*, 243, 167, 1998.

241. Ohkawa, T., Rowe, A.R., and Volkman, L.E., Identification of six *Autographa californica* multicapsid nucleopolyhedrovirus early genes that mediate nuclear localization of G-actin, *J. Virol.*, 76, 12281, 2002.

242. Goley, E.D., Ohkawa, T., Mancuso, J., Woodruff, J.B., D'Alessio, J.A., Cande, W.Z., Volkman, L.E., and Welch, M.D., Dynamic nuclear actin assembly by Arp2/3 complex and a baculovirus WASP-like protein, *Science*, 314, 464, 2006.

243. Xu, H., Yao, L., Lu, S., and QI, Y., Host filamentous actin is associated with *Heliothis armigera* single nucleopolyhedrosis virus (HaSNPV) nucleocapsid transport to the host nucleus, *Curr. Microbiol.*, 54, 199, 2007.

244. Kukkonen, S.P., Airenne, K.J., Marjomaki, V., Laitinen, O.H., Lehtolainen, P., Kankaanpaa, P., Mahonen, A.J., Raty, J.K., Nordlund, H.R., Oker-Blom, C., Kulomaa, M.S., and Ylä-Herttuala, S., Baculovirus capsid display: A novel tool for transduction imaging, *Mol. Ther.*, 8, 853, 2003.

245. Kitajima, M., Hamazaki, H., Miyano-Kurosaki, N., and Takaku, H., Characterization of baculovirus *Autographa californica* multiple nuclear polyhedrosis virus infection in mammalian cells, *Biochem. Biophys. Res. Commun.*, 343, 378, 2006.

246. Funk, C.J. and Consigli, R.A., Phosphate cycling on the basic protein of *Plodia interpunctella* granulosis virus, *Virology*, 193, 396, 1993.

247. Oppenheimer, D.I. and Volkman, L.E., Proteolysis of p6.9 induced by cytochalasin D in *Autographa californica* M nuclear polyhedrosis virus-infected cells, *Virology*, 207, 1, 1995.

248. Wilson, M.E. and Miller, L.K., Changes in the nucleoprotein complexes of a baculovirus DNA during infection, *Virology*, 151, 315, 1986.

249. Wilson, M.E. and Price, K.H., Association of *Autographa californica* nuclear polyhedrosis virus (AcMNPV) with the nuclear matrix, *Virology*, 167, 233, 1988.

250. Kircheis, R., Wightman, L., Schreiber, A., Robitza, B., Rossler, V., Kursa, M., and Wagner, E., Polyethylenimine/DNA complexes shielded by transferrin target gene expression to tumors after systemic application, *Gene Ther.*, 8, 28, 2001.

251. Hoare, J., Waddington, S., Thomas, H.C., Coutelle, C., and McGarvey, M.J., Complement inhibition rescued mice allowing observation of transgene expression following intraportal delivery of baculovirus in mice, *J. Gene Med.*, 7, 325, 2005.

252. Hüser, A., Rudolph, M., and Hofmann, C., Incorporation of decay-accelerating factor into the baculovirus envelope generates complement-resistant gene transfer vectors, *Nat. Biotechnol.*, 19, 451, 2001.

253. Raty, J.K., Liimatainen, T., Wirth, T., Airenne, K.J., Ihalainen, T.O., Huhtala, T., Hamerlynck, E., Vihinen-Ranta, M., Narvanen, A., Yla-Herttuala, S., and Hakumaki, J.M., Magnetic resonance imaging of viral particle biodistribution in vivo, *Gene Ther.*, 13, 1440, 2006.

254. Laitinen, O.H., Airenne, K.J., Hytonen, V.P., Peltomaa, E., Mahonen, A.J., Wirth, T., Lind, M.M., Makela, K.A., Toivanen, P.I., Schenkwein, D., Heikura, T., Nordlund, H.R., Kulomaa, M.S., and Yla-Herttuala, S., A multipurpose vector system for the screening of libraries in bacteria, insect and mammalian cells and expression in vivo, *Nucleic Acids Res.*, 33, e42, 2005.

255. Fewell, G.D. and Schmitt, K., Vector-based RNAi approaches for stable, inducible and genome-wide screens, *Drug Discov. Today*, 11, 975, 2006.

256. Kim, D.H. and Rossi, J.J., Strategies for silencing human disease using RNA interference, *Nat. Rev. Genet.*, 8, 173, 2007.

257. Mack, G.S., MicroRNA gets down to business, *Nat. Biotechnol.*, 25, 631, 2007.

258. Nicholson, L.J., Philippe, M., Paine, A.J., Mann, D.A., and Dolphin, C.T., RNA interference mediated in human primary cells via recombinant baculoviral vectors, *Mol. Ther.*, 11, 638, 2005.

259. Lu, L., Ho, Y., and Kwang, J., Suppression of porcine arterivirus replication by baculovirus-delivered shRNA targeting nucleoprotein, *Biochem. Biophys. Res. Commun.*, 340, 1178, 2006.

260. Isom, H.C., Starkey, J.L., and Chiari, E., Recombinant baculovirus expressing anti-HBsAg shRNA inhibits HBV cccDNA amplification in HepG2 cells, *FASEB J.*, 21, A1137–A113a, 2007.

261. Suzuki, H., Kaneko, H., Tamai, N., Miyano-Kurosaki, N., Hashimoto, K., Shimotohno, K., and Takaku, H., Suppression of HCV RNA replication by baculovirus-mediated shRNA expression, *Nucleic Acids Symp. Ser. (Oxf)*, 49, 339, 2005.

262. Saitoh, H., Miyano-Kurosaki, N., and Takaku, H., Inhibition of influenza virus A and B production by RNA interference, *Antiviral Res.*, 70, 50, 2006.

263. Kaneko, H., Suzuki, H., Abe, T., Miyano-Kurosaki, N., and Takaku, H., Inhibition of HIV-1 replication by vesicular stomatitis virus envelope glycoprotein pseudotyped baculovirus vector-transduced ribozyme in mammalian cells, *Biochem. Biophys. Res. Commun.*, 349, 1220, 2006.

264. Harper, D.M., Franco, E.L., Wheeler, C., Ferris, D.G., Jenkins, D., Schuind, A., Zahaf, T., Innis, B., Naud, P., De Carvalho, N.S., Roteli-Martins, C.M., Teixeira, J., Blatter, M.M., Korn, A.P., Quint, W., and Dubin, G., Efficacy of a bivalent L1 virus-like particle vaccine in prevention of infection with human papillomavirus types 16 and 18 in young women: A randomised controlled trial, *Lancet*, 364, 1757, 2004.

265. Treanor, J.J., Schiff, G.M., Couch, R.B., Cate, T.R., Brady, R.C., Hay, C.M., Wolff, M., She, D., and Cox, M.M., Dose-related safety and immunogenicity of a trivalent baculovirus-expressed influenza-virus hemagglutinin vaccine in elderly adults, *J. Infect. Dis.*, 193, 1223, 2006.

266. Treanor, J.J., Schiff, G.M., Hayden, F.G., Brady, R.C., Hay, C.M., Meyer, A.L., Holden-Wiltse, J., Liang, H., Gilbert, A., and Cox, M., Safety and immunogenicity of a baculovirus-expressed hemagglutinin influenza vaccine: A randomized controlled trial, *JAMA*, 297, 1577, 2007.

267. Safdar, A., Rodriguez, M.A., Fayad, L.E., Rodriguez, G.H., Pro, B., Wang, M., Romaguera, J.E., Goy, A.H., Hagemeister, F.B., McLaughlin, P., Bodey, G.P., Kwak, L.W., Raad, I.I., and Couch, R.B., Dose-related safety and immunogenicity of baculovirus-expressed trivalent influenza vaccine: A double-blind, controlled trial in adult patients with non-Hodgkin B cell lymphoma, *J. Infect. Dis.*, 194, 1394, 2006.

268. Kelley, G. and Cox, M., Insects join the flu fight, *Pharm. Discov. Dev.*, May/June, 28, 2006.

269. Safdar, A. and Cox, M.M., Baculovirus-expressed influenza vaccine. A novel technology for safe and expeditious vaccine production for human use, *Expert. Opin. Investig. Drugs*, 16, 927, 2007.

270. Abe, T., Takahashi, H., Hamazaki, H., Miyano-Kurosaki, N., Matsuura, Y., and Takaku, H., Baculovirus induces an innate immune response and confers protection from lethal influenza virus infection in mice, *J. Immunol.*, 171, 1133, 2003.

271. Facciabene, A., Aurisicchio, L., and La, M.N., Baculovirus vectors elicit antigen-specific immune responses in mice, *J. Virol.*, 78, 8663, 2004.

272. Kim, C.H., Yoon, J.S., Sohn, H.J., Kim, C.K., Paik, S.Y., Hong, Y.K., and Kim, T.G., Direct vaccination with pseudotype baculovirus expressing murine telomerase induces anti-tumor immunity comparable with RNA-electroporated dendritic cells in a murine glioma model, *Cancer Lett.*, 250, 276, 2007.

273. Lindley, K.M., Su, J.L., Hodges, P.K., Wisely, G.B., Bledsoe, R.K., Condreay, J.P., Winegar, D.A., Hutchins, J.T., and Kost, T.A., Production of monoclonal antibodies using recombinant baculovirus displaying gp64-fusion proteins, *J. Immunol. Methods*, 234, 123, 2000.

274. Tami, C., Peralta, A., Barbieri, R., Berinstein, A., Carrillo, E., and Taboga, O., Immunological properties of FMDV-gP64 fusion proteins expressed on SF9 cell and baculovirus surfaces, *Vaccine*, 23, 840, 2004.

275. Hervas-Stubbs, S., Rueda, P., Lopez, L., and Leclerc, C., Insect baculoviruses strongly potentiate adaptive immune responses by inducing type I IFN, *J. Immunol.*, 178, 2361, 2007.

276. Tanaka, T., Takeno, T., Watanabe, Y., Uchiyama, Y., Murakami, T., Yamashita, H., Suzuki, A., Aoi, R., Iwanari, H., Jiang, S.Y., Naito, M., Tachibana, K., Doi, T., Shulman, A.I., Mangelsdorf, D.J., Reiter, R., Auwerx, J., Hamakubo, T., and Kodama, T., The generation of monoclonal antibodies against human peroxisome proliferator-activated receptors (PPARs), *J. Atheroscler. Thromb.*, 9, 233, 2002.

277. Jiang, S., Tanaka, T., Iwanari, H., Hotta, H., Yamashita, H., Kumakura, J., Watanabe, Y., Uchiyama, Y., Aburatani, H., Hamakubo, T., Kodama, T., and Naito, M., Expression and localization of P1 promoter-driven hepatocyte nuclear factor-4alpha (HNF4alpha) isoforms in human and rats, *Nucl. Recept.*, 1, 5, 2003.

278. Tami, C., Farber, M., Palma, E.L., and Taboga, O., Presentation of antigenic sites from foot-and-mouth disease virus on the surface of baculovirus and in the membrane of infected cells, *Arch. Virol.*, 145, 1815, 2000.

279. Peralta, A., Molinari, P., Conte-Grand, D., Calamante, G., and Taboga, O., A chimeric baculovirus displaying bovine herpesvirus-1 (BHV-1) glycoprotein D on its surface and their immunological properties, *Appl. Microbiol. Biotechnol.*, 75, 407, 2007.

280. Rahman, M.M., Shaila, M.S., and Gopinathan, K.P., Baculovirus display of fusion protein of Peste des petits ruminants virus and hemagglutination protein of rinderpest virus and immunogenicity of the displayed proteins in mouse model, *Virology*, 317, 36, 2003.

281. Yoshida, S., Kondoh, D., Arai, E., Matsuoka, H., Seki, C., Tanaka, T., Okada, M., and Ishii, A., Baculovirus virions displaying Plasmodium berghei circumsporozoite protein protect mice against malaria sporozoite infection, *Virology*, 316, 161, 2003.

282. Chang, Y.J., Liu, C.Y., Chiang, B.L., Chao, Y.C., and Chen, C.C., Induction of IL-8 release in lung cells via activator protein-1 by recombinant baculovirus displaying severe acute respiratory syndrome-coronavirus spike proteins: Identification of two functional regions, *J. Immunol.*, 173, 7602, 2004.

283. Shih, Y.P., Chen, C.Y., Liu, S.J., Chen, K.H., Lee, Y.M., Chao, Y.C., and Chen, Y.M., Identifying epitopes responsible for neutralizing antibody and DC-SIGN binding on the spike glycoprotein of the severe acute respiratory syndrome coronavirus, *J. Virol.*, 80, 10315, 2006.

284. Feng, Q., Liu, Y., Qu, X., Deng, H., Ding, M., Lau, T.L., Yu, A.C., and Chen, J., Baculovirus surface display of SARS coronavirus (SARS-CoV) spike protein and immunogenicity of the displayed protein in mice models, *DNA Cell Biol.*, 25, 668, 2006.

285. Lu, L., Yu, L., and Kwang, J., Baculovirus surface-displayed hemagglutinin of H5N1 influenza virus sustains its authentic cleavage, hemagglutination activity, and antigenicity, *Biochem. Biophys. Res. Commun.*, 358, 404, 2007.

286. Strauss, R., Huser, A., Ni, S., Tuve, S., Kiviat, N., Sow, P.S., Hofmann, C., and Lieber, A., Baculovirus-based vaccination vectors allow for efficient induction of immune responses against *Plasmodium falciparum* circumsporozoite protein, *Mol. Ther.*, 15, 193, 2007.

287. Schutz, A., Scheller, N., Breinig, T., and Meyerhans, A., The *Autographa californica* nuclear polyhedrosis virus AcNPV induces functional maturation of human monocyte-derived dendritic cells, *Vaccine*, 24, 7190, 2006.

288. Hashimoto, K., Suzuki, T., Sakai, R., Miyazawa, Y., Saito, R., Yamamoto, H., Nakayama, T., Miyano-Kurosaki, N., and Takaku, H., Innate immunity activation in mouse dendritic cells infected by Baculovirus, *J. Immunol.*, 178, LB39–LB3c, 2007.

289. Ozawa, S., Lu, W., Bucana, C.D., Kanayama, H.O., Shinohara, H., Fidler, I.J., and Dong, Z., Regression of primary murine colon cancer and occult liver metastasis by intralesional injection of lyophilized preparation of insect cells producing murine interferon-beta, *Int. J. Oncol.*, 22, 977, 2003.

290. Lu, W., Su, J., Kim, L.S., Bucana, C.D., Donawho, C., He, J., Fidler, I.J., and Dong, Z., Active specific immunotherapy against occult brain metastasis, *Cancer Res.*, 63, 1345, 2003.

291. Kumar, P., Uchil, P.D., Sulochana, P., Nirmala, G., Chandrashekar, R., Haridattatreya, M., and Satchidanandam, V., Screening for T cell-eliciting proteins of Japanese encephalitis virus in a healthy JE-endemic human cohort using recombinant baculovirus-infected insect cell preparations, *Arch. Virol.*, 148, 1569, 2003.

292. Bae, J.E., Schurig, G.G., and Toth, T.E., Mice immune responses to *Brucella abortus* heat shock proteins. Use of baculovirus recombinant-expressing whole insect cells, purified *Brucella abortus* recombinant proteins, and a vaccinia virus recombinant as immunogens, *Vet. Microbiol.*, 88, 189, 2002.

293. Ignoffo, C.M. and Heimpel, A.M., The nuclear-polyhedrosis virus of *Heliothis zea* (Boddie) and *Heliothis virescens* (Fabricus). V. Toxicity-pathogenicity of virus to white mice and guinea pigs, *J. Invert. Pathol.*, 7, 329, 1965.

294. Murillo, R., Elvira, S., Munoz, D., Williams, T., and Caballero, P., Genetic and phenotypic variability in *Spodoptera exigua* nucleopolyhedrovirus isolates from greenhouse soils in southern Spain, *Biol. Control*, 38, 157, 2006.

295. Miller, L.K. and Lu, A., The molecular basis of baculovirus host range, in: *The Baculoviruses*, 1st ed., Miller, L.K., Ed., Plenum Press, New York, 1997, p. 217.

296. Morel, G. and Fouillaud, M., Presence of microorganisms and viral inclusion bodies in the nests of the paper wasp *Polistes hebraeus* Fabricus (Hymenoptera, Vespidae), *J. Invertebr. Pathol.*, 60, 210, 1992.

297. Lautenschlager, R.A. and Podgwaite, J.D., Passage rates of nucleopolyhedrosis virus by avian and mammalian predators of the gypsy moth, *Environ. Entomol.*, 8, 210, 1979.

298. Entwistle, P.F., Adams, P.H.W., and Evans, H.F., Epizootiology of a nuclear polyhedrosis virus in European spruce sawly (*Gilpinia hercyniae*): The rate of passage of infective virus through the gut of birds during cage tests, *J. Invertebr. Pathol.*, 15, 173, 1978.

299. Groner, A., Granados, R.R., and Burand, J.P., Interaction of *Autographa californica* nuclear polyhedrosis virus with two nonpermissive cell lines, *Intervirology*, 21, 203, 1984.

300. Boyce, F.M., Non-chemical gene transfer into mammalian cells. A new DNA transduction system based on baculovirus, *Innovations*, 9, 1, 1999.

301. Fallaux, F.J., van der Eb, A.J., and Hoeben, R.C., Who's afraid of replication-competent adenoviruses? *Gene Ther.*, 6, 709, 1996.

302. Morris, T.D. and Miller, L.K., Promoter influence on baculovirus-mediated gene expression in permissive and nonpermissive insect cell lines, *J. Virol.*, 66, 7397, 1992.

303. Hoopes, R.R., Jr. and Rohrmann, G.F., In vitro transcription of baculovirus immediate early genes: Accurate mRNA initiation by nuclear extracts from both insect and human cells, *Proc. Natl. Acad. Sci. U. S. A.*, 88, 4513, 1991.

304. Murges, D., Kremer, A., and Knebel-Morsdorf, D., Baculovirus transactivator IE1 is functional in mammalian cells, *J. Gen. Virol.*, 78 (Pt 6), 1507, 1997.

305. Dai, X., Willis, L.G., Huijskens, I., Palli, S.R., and Theilmann, D.A., The acidic activation domains of the baculovirus transactivators IE1 and IE0 are functional for transcriptional activation in both insect and mammalian cells, *J. Gen. Virol.*, 85, 573, 2004.

306. Knebel, D. and Doerfler, W., Activation of an insect baculovirus promoter in mammalian cells by adenovirus functions, *Virus Res.*, 8, 317, 1987.

307. Liu, Y.K., Chu, C.C., and Wu, T.Y., Baculovirus ETL promoter acts as a shuttle promoter between insect cells and mammalian cells, *Acta Pharmacol. Sin.*, 27, 321, 2006.

308. Fujita, R., Matsuyama, T., Yamagishi, J., Sahara, K., Asano, S., and Bando, H., Expression of *Autographa californica* multiple nucleopolyhedrovirus genes in mammalian cells and upregulation of the host beta-actin gene, *J. Virol.*, 80, 2390, 2006.

309. Kenoutis, C., Efrose, R.C., Swevers, L., Lavdas, A.A., Gaitanou, M., Matsas, R., and Iatrou, K., Baculovirus-mediated gene delivery into mammalian cells does not alter their transcriptional and differentiating potential but is accompanied by early viral gene expression, *J. Virol.*, 80, 4135, 2006.

310. Regev, A., Rivkin, H., Gurevitz, M., and Chejanovsky, N., New measures of insecticidal efficacy and safety obtained with the 39 K promoter of a recombinant baculovirus, *FEBS Lett.*, 580, 6777, 2006.

311. Marshall, E., Clinical research. Gene therapy a suspect in leukemia-like disease, *Science*, 298, 34, 2002.

312. Li, Z., Dullmann, J., Schiedlmeier, B., Schmidt, M., von Kalle, C., Meyer, J., Forster, M., Stocking, C., Wahlers, A., Frank, O., Ostertag, W., Kuhlcke, K., Eckert, H.G., Fehse, B., and Baum, C., Murine leukemia induced by retroviral gene marking, *Science*, 296, 497, 2002.

313. Pearson, H., Liver tumours temper hopes for gene-therapy technique, *Nature*, 413, 9, 2001.

314. Donsante, A., Vogler, C., Muzyczka, N., Crawford, J.M., Barker, J., Flotte, T., Campbell-Thompson, M., Daly, T., and Sands, M.S., Observed incidence of tumorigenesis in long-term rodent studies of rAAV vectors, *Gene Ther.*, 8, 1343, 2001.

315. Woods, N.B., Muessig, A., Schmidt, M., Flygare, J., Olsson, K., Salmon, P., Trono, D., von Kalle, C., and Karlsson, S., Lentiviral vector transduction of NOD/SCID repopulating cells results in multiple vector integrations per transduced cell: Risk of insertional mutagenesis, *Blood*, 2002.

316. Merrihew, R.V., Clay, W.C., Condreay, J.P., Witherspoon, S.M., Dallas, W.S., and Kost, T.A., Chromosomal integration of transduced recombinant baculovirus DNA in mammalian cells, *J. Virol.*, 75, 903, 2001.

317. Brusca, J., Summers, M., Couch, J., and Courtney, L., *Autographa californica* nuclear polyhedrosis virus efficiently enters but does not replicate in poikilothermic vertebrate cells, *Intervirology*, 26, 207, 1986.

318. Halbert, C.L., Rutledge, E.A., Allen, J.M., Russell, D.W., and Miller, A.D., Repeat transduction in the mouse lung by using adeno-associated virus vectors with different serotypes, *J. Virol.*, 74, 1524, 2000.

319. Yang, Y., Li, Q., Ertl, H.C., and Wilson, J.M., Cellular and humoral immune responses to viral antigens create barriers to lung-directed gene therapy with recombinant adenoviruses, *J. Virol.*, 69, 2004, 1995.

320. Krieg, A.M., Minding the Cs and Gs, *Mol. Ther.*, 1, 209, 2000.

321. Ferguson, T.A. and Griffith, T.S., A vision of cell death: Insights into immune privilege, *Immunol. Rev.*, 156, 167, 1997.

322. Hofmann, C. and Strauss, M., Baculovirus-mediated gene transfer in the presence of human serum or blood facilitated by inhibition of the complement system, *Gene Ther.*, 5, 531, 1998.

323. Platt, J.L., Knocking out xenograft rejection, *Nat. Biotechnol.*, 20, 231, 2002.

324. Bianchet, M.A., Odom, E.W., Vasta, G.R., and Amzel, L.M., A novel fucose recognition fold involved in innate immunity, *Nat. Struct. Biol.*, 9, 628, 2002.

325. Kirschfink, M., Targeting complement in therapy, *Immunol. Rev.*, 180, 177, 2001.

326. Barrington, R., Zhang, M., Fischer, M., and Carroll, M.C., The role of complement in inflammation and adaptive immunity, *Immunol. Rev.*, 180, 5, 2001.

327. Hofmann, C., Huser, A., Lehnert, W., and Strauss, M., Protection of baculovirus-vectors against complement-mediated inactivation by recombinant soluble complement receptor type 1, *Biol. Chem.*, 380, 393, 1999.

328. Beck, N.B., Sidhu, J.S., and Omiecinski, C.J., Baculovirus vectors repress phenobarbital-mediated gene induction and stimulate cytokine expression in primary cultures of rat hepatocytes, *Gene Ther.*, 7, 1274, 2000.

329. Gronowski, A.M., Hilbert, D.M., Sheehan, K.C., Garotta, G., and Schreiber, R.D., Baculovirus stimulates antiviral effects in mammalian cells, *J. Virol.*, 73, 9944, 1999.

330. Abe, T., Hemmi, H., Miyamoto, H., Moriishi, K., Tamura, S., Takaku, H., Akira, S., and Matsuura, Y., Involvement of the toll-like receptor 9 signaling pathway in the induction of innate immunity by baculovirus, *J. Virol.*, 79, 2847, 2005.

331. McCormick, C.J., Challinor, L., Macdonald, A., Rowlands, D.J., and Harris, M., Introduction of replication-competent hepatitis C virus transcripts using a tetracycline-regulable baculovirus delivery system, *J. Gen. Virol.*, 85, 429, 2004.

332. Liang, C., Song, J., Hu, Z., and Chen, X., Group I but not group II NPV induces antiviral effects in mammalian cells, *Sci. China C. Life Sci.*, 49, 467, 2006.

333. Verma, I.M. and Weitzman, M.D., Gene therapy: Twenty-first century medicine, *Annu. Rev. Biochem.*, 74, 711, 2005.

334. Robson, T. and Hirst, D.G., Transcriptional targeting in cancer gene therapy, *J. Biomed. Biotechnol.*, 2003, 110, 2003.

335. Brandwijk, R.J., Griffioen, A.W., and Thijssen, V.L., Targeted gene-delivery strategies for angiostatic cancer treatment, *Trends Mol. Med.*, 13, 200, 2007.

336. Noureddini, S.C. and Curiel, D.T., Genetic targeting strategies for adenovirus, *Mol. Pharm.*, 2, 341, 2005.

337. Campos, S.K. and Barry, M.A., Current advances and future challenges in adenoviral vector biology and targeting, *Curr. Gene Ther.*, 7, 189, 2007.

338. Yu, J.H. and Schaffer, D.V., Advanced targeting strategies for murine retroviral and adeno-associated viral vectors, *Adv. Biochem. Eng. Biotechnol.*, 99, 147, 2005.

339. Peng, K.W., Strategies for targeting therapeutic gene delivery, *Mol. Med. Today*, 5, 448, 1999.

340. Ojala, K., Mottershead, D.G., Suokko, A., and Oker-Blom, C., Specific binding of baculoviruses displaying gp64 fusion proteins to mammalian cells, *Biochem. Biophys. Res Commun.*, 284, 777, 2001.

341. Grabherr, R., Ernst, W., Oker-Blom, C., and Jones, I., Developments in the use of baculoviruses for the surface display of complex eukaryotic proteins, *Trends Biotechnol.*, 19, 231, 2001.

342. Russell, S.J. and Cosset, F.L., Modifying the host range properties of retroviral vectors, *J. Gene Med.*, 1, 300, 1999.

343. Anderson, J.L. and Hope, T.J., Intracellular trafficking of retroviral vectors: Obstacles and advances, *Gene Ther.*, 12, 1667, 2005.

344. Meier, O. and Greber, U.F., Adenovirus endocytosis, *J. Gene Med.*, 6(Suppl 1), S152, 2004.

345. Raty, J.K., Liimatainen, T., Huhtala, T., Kaikkonen, M.U., Airenne, K.J., Hakumaki, J.M., Narvanen, A., and Yla-Herttuala, S., SPECT/CT imaging of baculovirus biodistribution in rat, *Gene Ther.*, 14, 930, 2007.

346. Ernst, W., Schinko, T., Spenger, A., Oker-Blom, C., and Grabherr, R., Improving baculovirus transduction of mammalian cells by surface display of a RGD-motif, *J. Biotechnol.*, 126, 237, 2006.

347. Yla-Herttuala, S. and Martin, J.F., Cardiovascular gene therapy, *Lancet*, 355, 213, 2000.

348. Rissanen, T.T. and Yla-Herttuala, S., Current status of cardiovascular gene therapy, *Mol. Ther.*, 15, 1233, 2007.

349. Hacein-Bey-Abina, S., Le Deist, F., Carlier, F., Bouneaud, C., Hue, C., De Villartay, J.P., Thrasher, A.J., Wulffraat, N., Sorensen, R., Dupuis-Girod, S., Fischer, A., Davies, E.G., Kuis, W., Leiva, L., and Cavazzana-Calvo, M., Sustained

correction of X-linked severe combined immunodeficiency by ex vivo gene therapy, *N. Engl. J. Med.*, 346, 1185, 2002.

350. Shan, L., Wang, L., Yin, J., Zhong, P., and Zhong, J., An OriP/EBNA-1-based baculovirus vector with prolonged and enhanced transgene expression, *J. Gene Med.*, 8, 1400, 2006.

351. van Craenenbroeck, K., Vanhoenacker, P., and Haegeman, G., Episomal vectors for gene expression in mammalian cells, *Eur. J. Biochem.*, 267, 5665, 2000.

352. Conese, M., Auriche, C., and Ascenzioni, F., Gene therapy progress and prospects: Episomally maintained self-replicating systems, *Gene Ther.*, 11, 1735, 2004.

353. Jackson, D.A., Juranek, S., and Lipps, H.J., Designing nonviral vectors for efficient gene transfer and long-term gene expression, *Mol. Ther.*, 14, 613, 2006.

354. Lewandoski, M., Conditional control of gene expression in the mouse, *Nat. Rev. Genet.*, 2, 743, 2001.

355. Toniatti, C., Bujard, H., Cortese, R., and Ciliberto, G., Gene therapy progress and prospects: Transcription regulatory systems, *Gene Ther.*, 11, 649, 2004.

356. Weber, W. and Fussenegger, M., Pharmacologic transgene control systems for gene therapy, *J. Gene Med.*, 8, 535, 2006.

357. Ford, K.G., Souberbielle, B.E., Darling, D., and Farzaneh, F., Protein transduction: An alternative to genetic intervention? *Gene Ther.*, 8, 1, 2001.

358. Kost, T.A., Condreay, J.P., Ames, R.S., Rees, S., and Romanos, M.A., Implementation of BacMam virus gene delivery technology in a drug discovery setting, *Drug Discov. Today*, 12, 396, 2007.

359. Jarvis, D.L. and Garcia, A., Jr., Long-term stability of baculoviruses stored under various conditions, *Biotechniques*, 16, 508, 1994.

360. Jorio, H., Tran, R., and Kamen, A., Stability of serum-free and purified baculovirus stocks under various storage conditions, *Biotechnol. Prog.*, 22, 319, 2006.

12 Bacterial Vector in Gene Therapy

Ming-Shen Dai and Georges Vassaux

CONTENTS

Gene therapy is defined as the delivery of genetic material for a therapeutic purpose. Because of their nature, bacteria have essentially been used as vaccination vectors. Current gene therapy, through nucleic acid vaccination in prophylactic or therapeutic purposes, brings the hope to monogenic/polygenic diseases, cancer, and ischemic heart disease [1]. DNA vaccines are eukaryotic expression vectors encoding one or several antigens with/without additional genes encoding for modulatory functions. With improving plasmid purification/dosages, antigen-presenting cells (APCs) targeting, and technologies for modulation of immune responses, substantial progress has been made in the past decade in overcoming the major inadequacies of first-generation gene-transfer vectors.

To induce an efficient immune response, effective antigen proteins or DNA delivery technologies are required. In addition, the vectors must protect the therapeutic DNA vaccine from degradation and clearance by the host immune system. A variety of viral and nonviral delivery systems are available for basic and clinical research. Viral vectors derived from retroviruses, adenoviruses, lentiviruses, poxviruses, parvoviruses, and herpesviruses are the most frequently used [2]. Apart from those, naked plasmid DNA or its combination with compounds increasing the efficiency of cell membrane penetration (cationic lipids, lipoplexes etc.) has also been used [3,4]. Moreover, gene transfer can also occur from bacteria to a very broad range of recipients that include yeast [5] and plants [6], and several laboratories have reported a relatively high frequency of functional gene transfer from bacteria to mammalian cells [7–14].

Because of their nature properties, bacterial vectors are particularly suited for gene transfer in APCs. They will naturally elicit an immune response which can be beneficial as the bacteria will have a built-in adjuvant effect [15]. The key is to match the correct (and safe) dosage, vector, and adjuvant response with the desired immune reaction. Typically, a bacterial infection will innately lead to APC cytokine (TNF-α, IL-1, IL-6) release [16] and attract other APCs and polymorphonuclear leukocytes to the site of infection [17]. Surrounding tissue is at risk during the digestion of the foreign bacteria as APCs and polymorphonuclear leucocytes also release microcidal agents (lysozyme, bactericidal/permeability-increasing protein, acid hydrolases, nitric oxide, and oxygen free radicals) that can damage nearby endothelial cells [18,19] and thus cause potential undesired side effects (resulting from cytokine release). However, tempering these reactions while delivering a specific antigen is the goal of bacterially mediated immunomodulation.

12.1 LIVE OR ATTENUATED BACTERIA

The history of bacterial vaccine development began with the use of bacillus Calmette–Guérin (BCG) vaccine against tuberculosis (TB) by Albert Calmette and Camille Guèrin at the Pasteur Institute (Lille, France) in 1927 [20] and continued with the development of the currently licensed Ty21a vaccine for typhoid [21]. With the progress of modern molecular biology, these bacteria can be processed to retain their adjuvant role in mounting immune response and carry desired antigens, proteins, or plasmid DNA as prophylactic/therapeutic vaccines. Several experimental tools exist to further influence the antigen delivery process in precision and flexibility [7,22]. To date, extensive research has been conducted using different live, live attenuated as well as killed bacterial strains as vectors. Live bacterial vaccines can be regarded as a self-limiting organism stimulating an immune response to one or more expressed antigens. The use of live bacteria to induce an immune response to itself or to a carried vaccine component is an attractive strategy. Advantages of live bacterial vaccines include their mimicry of a natural infection, intrinsic adjuvant properties, and their possibility to be administered orally [23]. However, the utilization of live bacteria to infect the host may potentially be associated with pathologies. Therefore, attenuation of bacteria virulence was developed in order to balance the pathogenicity/reactogenicity and preserve the ability to elicit a protective immune response. Attenuation of the pathogenic bacterial strains can be achieved through abrogation of virulence factors [24,25], metabolic crippling [26], or intracellular replication-deficient [27] by directed mutagenesis using recombinant DNA methods. These include attenuated *Listeria monocytogenes* and commensal *Bifidobacterium longum* representing Gram-positive bacteria, and Gram-negative bacteria such as attenuated *Yersinia* spp., *Shigella* spp., *Salmonella* spp., and apathogenic *E. coli* [7,28,29]. A variety of attenuated mutant strains of *Salmonella* and *Shigella* deleted in genes for cell wall synthesis [9,30] and genes for RNA/DNA precursor synthesis [31,32] have already been used for bactofection. Auxotrophic mutants of *Salmonella* and *Shigella* defective in aromatic acid biosynthesis [33,34] have been studied as vaccine carriers as well. However, attenuations of the live bacterial pathogenicity may influence the overall performance of the strain, i.e., the efficacy of DNA delivery or the immunogenicity of the vaccine [23,35]. Even though the applications of live attenuated bacterial vaccines are extensive and has lead to more than 2000 published papers, only very few of the promising candidates have survived the licensing process and become registered [36] reflecting the difficulty in developing a commercial live vaccine. Furthermore, human clinical trials using *S. typhimurium* (VNP20009) have shown dose-limiting toxicity despite extensive attenuation [37]. As an alternative approach, nonliving bacterial components/lysate or whole bacterial cells have been used for both human and animal mucosal immunization and been shown to provide humoral and cellular responses. In addition, the bacterial ghost (BG) system [38], which is a nonliving bacterial envelope made from protein E-mediated lysis of Gram-negative bacteria in combination with antigen packaging. Since this BG system do not involve host DNA or live organisms, they exhibit improved potency with regard to target antigens compared to conventional approaches, they are versatile with regards to DNA or protein antigen choice and size. Killed but metabolically active (KBMA) vaccines are based on bacterial nucleotide excision repair mutants, and are inactivated by photochemical treatment with a synthetic highly reactive psoralen known as S-59 [39]. The KBMA vaccine approach is based on the hypothesis that the increased sensitivity of *uvr* mutant strains to photochemical inactivation is correlated with randomly distributed, infrequent psoralen crosslinks, leaving intact the ability of a population of inactivated bacteria to express its genes, while preventing productive growth and the ability to cause disease in the immunized host. Another strategy of killed-bacteria is to use formalin to fix the bacterial vectors but retain the ability of delivery antigen or protein [40–42].

12.2 MECHANISMS OF GENE TRANSFER BY BACTERIAL VECTORS

The first step of gene transfer by bacterial vectors lies in the entry of the delivery vehicle uptake by host target cells (Figure 12.1). The bacterial cell is typically confined to spherical or cylindrical shapes between 1 and 5 μm which can innately target the phagocytic cells. When professional phagocytic cells such as macrophages or dendritic cells are targeted, this entry is likely to happen through phagocytosis. This contact will activate these phagocytic cells, i.e., the APCs. Gram-negative cells contain lipopolysaccharide (LPS), whereas Gram-positive cells posses lipoteichoic acid. These and other features (unmethylated cytosine–phosphate–guanine [CpG] dinucleotides [43–45], peptidoglycan [46], flagellin [47], and bacterial lipoproteins [48]), the so-called pathogen-associated molecular patterns (PAMPs), naturally influence bacterial uptake and act as natural adjuvants to improve APCs activation

FIGURE 12.1 Strategies of bacteria vectors in vaccinations. Different strategies are undertaken by the bacterial vectors in delivering the DNA vaccines. (1) After bacterial vectors penetrating into the mammalian cells, the vectors are destructed or undergo lysis in the cytoplasm of target cells. The plasmid carrying therapeutic gene ( ) is released and get into the nucleus and expressed by the eukaryotic transcription and translation machinery ( ). (2) The passenger antigen ( ) is expressed within the bacterial compartment and delivered into the target cells. These nonsecretory antigens can only be released into the cytoplasm of target cell upon the phagosomal lysis of vector. Further antigen processing and presentation will happen essentially in APCs. (3) Bacterial vector do not enter the eukaryotic cell, but express and secret the therapeutical transgene in the intercellular space. (4) The bacterial vector can escape from lysosomal/phagosomal lysis. The transgene is expressed after penetrating into the cytoplasm of target cell by the prokaryotic transcription and translation machinery and the expressed antigenic protein is being secreted into the cytoplasm.

through binding to receptors of the Toll-like receptor (TLR) family [44,49]. In the case of non-phagocytic cells such as those of the intestinal epithelium, there are two major strategies for bacteria to gain entry into a eukaryotic cell [50,51]. For certain genera such as *Salmonella* or *Shigella*, contact between the bacteria and the host results in the secretion by the bacteria of a set of invasion proteins that triggers intracellular signaling events. For example, upon the contact of *S. typhimurium* with host cells results in activation of a specialized protein secretion system (type III), which leads to cytoskeletal rearrangement, membrane ruffling, and bacterial uptake by pinocytosis [50]. For other genera such as *Yersinia* or *Listeria*, binding of a single bacterial protein to a particular ligand on the host cell surface is necessary and sufficient to trigger entry by a zipper-like mechanism. After internalization, the bacteria are localized in the phagosomal vacuoles and are targeted for degradation. Therefore, escaping from the vacuolar compartment is essential to the delivery DNA vaccines.

Shigella spp. and *L. monocytogenes* can disrupt the endosomal/lysosomal membrane, escape into the host cell's cytoplasm, and spread from one infected cell to an adjacent one by exploiting the cell's actin polymerization for intracellular motility [52,53]. This process of phagosomal escape is particularly well understood and exploited in vectors based on bacteria replicating in the cytosol [10,54]. In *L. monocytogenes*, the pore-forming cytolysin listeriolysin O (LLO), which is encoded by *hly*, plays an essential role in the escaping step. Strains with mutations in *hly* that inactivate the pore-forming activity of LLO are unable to escape from the primary phagosome [55,56]. Once spreading to the cytosol, wild-type *L. monocytogenes* will replicate [57]. An attenuated *L. monocytogenes* strain has been engineered to undergo self-destruction in the cell cytosol by production of a phage lysine under the control of the promoter of actA, which is preferentially activated when the bacteria are in the cytosol [10,27]. Another bacterial vector taking advantage of the pore-forming activity of the LLO protein is the invasive *E. coli* [11,40]. In addition to the *hly* locus from *L. monocytogenes* encoding the LLO protein, the *inv* locus encoding the invasin protein from *Yersinia pseudotuberculosis* has been inserted into this bacterium [11]. Invasin binds to β1-integrin expressed at the surface of mammalian cells and this binding is necessary and sufficient for entry of the whole bacterium into the mammalian cell [51]. Therefore, invasin expression restricts the tropism of these bacteria to non-phagocytic cells expressing β1-integrin [58,59], while invasin-negative recombinant *E. coli* can be used to target professional phagocytic cells [41,42]. In the case of bacterial delivery vectors that remain in the lysosome/phagosome, such as *Salmonella*, the mechanism of delivery remains unclear [12].

Another interesting approach might be used in bacterial delivery systems [7]. Bacteria are not used for the gene transfer but the proteins secreted by bacteria persist in the target tissues. More recently, the utilization of double-stranded RNA (dsRNA or RNA interference, the so-called RNAi) to silence target genes has potential therapeutic applications that are widely acknowledged [29,60]. Recent progress in this strategy was made by using nonpathogenic bacteria to induce gene silencing in target cells, both *in vitro* and *in vivo*. Bacteria-mediated induction of RNAi was established by demonstrating target-specific gene silencing after transfer of double-stranded RNA from *E. coli* in the nematode *Caenorhabditis elegans* [61,62]. Another example of bacteria-mediated RNAi transfer for functional genomics is postulated by Xiang et al. who developed an siRNA delivering system using nonpathogenic *E. coli* engineered to transcribe shRNAs from a plasmid containing the invasin gene *inv* and the LLO gene *HlyA*, which encode two bacterial factors needed for successful transfer of the shRNAs into mammalian cells. Upon oral or intravenous administration, *E. coli* encoding shRNA against CTNNB1 (catenin β-1) induce significant gene silencing in the intestinal epithelium and in human colon cancer xenografts in mice [63,64]. Another *in vivo* study carried out by Zhang et al. using attenuated *S. typhimurium* to deliver STAT3-specific RNAi, bacterial vector could preferentially home to tumor tissues, inhibit tumor growth, and extend the survival in experimented mice [65]. In the light of these results obtained previously, the delivery of dsRNA or eukaryotic expression plasmids encoding siRNAs into mammalian cells can be envisaged in the near future.

12.3 APPLICATION OF BACTERIAL VECTORS

12.3.1 INFECTIOUS DISEASES

At the end of the nineteenth century, killed bacteria started to be first used as vaccination against infectious disease. In early twentieth century, killed *Salmonella typhi* was used both in England and Germany by the end of the First World War. However, it is only until 1960 that well controlled field trials of two forms of killed *S. typhi* vaccine provided unequivocal evidence of their considerable but incomplete protective effect [33,66].

Mucosal membranes remain the most frequent portals of entry of pathogenic microorganisms. Mucosal immunization with whole formalin killed *Pseudomonas aeruginosa*, *Branhamella catarrhalis*, nontypable *Haemophilus influenzae*, or *Streptococcus pneumoniae* results in enhanced homologous bacterial clearance from the lung of immune animals challenged with live bacteria [67]. This has prompted studies aimed at the development of intranasal or enteral vaccination protocols in the protection of mucosa. Furthermore, following the attenuation of bacterial pathogenicity and carrying antigen-expressing plasmids, those bacterial vectors exploit more flexible strategies in infectious disease (Table 12.1). For example, attenuated strains of *Yersinia* spp., *S. enterica* serovar *typhi* and *Typhimurium*, *Shigella* spp., and *L. monocytogenes* have been successfully applied as DNA delivery vectors against certain infectious diseases mainly of bacterial and viral origin. Antigens were derived from viruses such as HIV, measles, hepatitis B, hepatitis C, herpes simplex 2, cytomegalovirus, human papillomavirus 16, and pseudorabies. Similarly, a variety of bacterial antigens from *L. monocytogenes*, *Mycobacterium tuberculosis*,

TABLE 12.1

Bacterial Vectors in the Treatment of Infectious Diseases

Bacterial Vector	Encoding Bacterial Antigen	Host Immune Responses			References
		Humoral	Cellular	Clinical	
Shigella flexneri	β-galactosidase (*E. coli*)	+			[9,175]
Salmonella typhimurium	LLO (*Listeria*)	+		+	[34]
Salmonella typhi (CVD 915)	Tetanus toxin (Frag C)	+			[32,69]
Shigella flexneri 2a (CVD 1204)	Tetanus toxin (Frag C)	+			[31]
Yersinia enterocolitica	Bacterioferritin (BFR) and P39 (*Brucella*)	+		+	[68]
Salmonella typhimurium	LLO and ActA (*L. monocytogenes*)	+	+	+	[70,71]
Salmonella enterica serovar Typhi	*Bacillus anthracis* (PA)	+	Cytokine +		[72]
Salmonella typhimurium	*Plasmodium falciparum* (SSP-2)		ELISPOT +		[73]
Salmonella enterica serovar Typhi	*Plasmodium falciparum* (tCSP)		Cytokine +		[74]
Salmonella typhimurium	*H pylori* ureB and IL-2	+		+	[76]
Salmonella typhimurium	PEB1 (*C. jejuni*)	+		–	[176]
Salmonella typhimurium	Hemagglutinin H (measles)	+		+	[77]
Shigella flexneri	MV F, HA, or NP (measles)	+	+	+	[30]
Shigella flexneri	gp120 (HIV-1)	+		+	[79,177]
Shigella flexneri	SF2 gag (HIV-1)		+		[81]
Salmonella typhimurium	Gag (HIV)	+	+		[82]
Salmonella typhimurium	HBsAg (hepatitis B virus)	±	+		[83,84]
Salmonella typhimurium	Nonstructural region 3 (hepatitic C virus-NS3)		+	+	[85]
Salmonella typhimurium	Glycoprotein D (HSV-2)		+	+	[88]
Salmonella typhimurium	HPV16 VLPs	+			[89,90]
Shigella flexneri	Hemagglutinin (influenza)	+	+	+	[87]
Listeria	Nucleoprotein + LLO (influenza)		+		[86]
Listeria monocytogenes	Antigen 85 complex and MPB/MPT51 (*Mycobacterium tuberculosis*)		+		[93]
Salmonella typhimurium	Ag85a (*M. tuberculosis*)		+	+	[94]

Chlamydia trachomatis, *Brucella abortus*, *Pseudomonas aeruginosa*, and *Clostridium tetani* were exploited through bacteria-mediated DNA transfer. Antigens from eukaryotic pathogens such as *Penicillium marneffei* were also delivered using bacterial DNA carriers [29].

Oral administration in mice of an attenuated *Yersinia enterocolitica* carrying a eukaryotic expression plasmid encoding *Brucella* genes elicited humoral and cellular responses. This immune response, in conjunction with potential cross-reacting antibodies against LPS of both *Yersinia* and *Brucella*, induced protection against a *Brucella* challenge [68].

In earlier studies by Pasetti et al., intranasal vaccination of *Salmonella typhi* or *Shigella flexneri* 2a carrying plasmids encoding fragment C (Frag C) of tetanus toxin could elicit higher titers of Frag C antibodies than those induced by intramuscular inoculation of bacteria carrying a prokaryotic expression plasmid in which the expression of the same antigen was inducible *in vivo* [31,32]. Those pioneer studies demonstrated that bacteria can serve as a vehicle for the delivery of foreign antigens to the systemic immune system and intranasal

inoculation is sufficient to induce an immune response against both the live vector and heterologous antigen as oral inoculation [69]. Another set of studies exploited *S. typhimurium* as a gene delivery vector, when β-galactosidase was used as a transgene, specific cytotoxic T-lymphocytes (CTLs) and T-helper (Th) cells, mainly of the Th1 type, as well as specific antibodies could be detected after a single oral vaccination [34]. A very similar immune response was elicited by *S. typhimurium* vectors expressing two virulence factors of *L. monocytogenes* (LLO and ActA) [70]. These vaccines could protect the mice from a lethal challenge with *L. monocytogenes* and partially protective after single dose [71]. β-galactosidase as well as a fusion antigen of the *P. aeruginosa* outer membrane protein with the fimbriae was administered orally or nasally. Multiple nasal administrations were necessary to obtain a T-cell response in the spleen compared with a single oral one [13,71]. Specific IgGs were observed in the gut, saliva, and serum after oral administration, whereas nasal administration gave rise to antibodies in the lungs, saliva, and serum, with hardly any antigen-specific IgA detected [71]. Independently, partial protective responses against *Chlamydia* were obtained in the lungs of mice after oral administration of *Salmonella* encoding the major outer membrane protein of *Chlamydia trachomatis* [13]. To increase the immunogenicity of the expression plasmid, Galen et al. showed that the vaccination of engineering of cytolysin A hemolysin (ClyA) fused to the encoded antigen, domain 4 (D4) moiety of *Bacillus anthracis* protective antigen (PA), lead to a higher titer of anti-PA IgG production [72]. For parasitic infection, Gomez-Duarte et al. demonstrated the immunogenicity of *S. typhi* secreting *Plasmodium falciparum* sporozoite surface protein (SSP-2) and the cytokine responses from nasal immunization in mice [73]. Furthermore, the same group developed a prime-boost strategy consisting of nasal mucosal delivery of *P. falciparum* circumsporozoite surface protein PfCSP exported from a *S. typhi* followed by an intradermal PfCSP DNA boosting. The mice in the prime-boost group developed higher frequencies of IFN-γ-secreting cells [74]. Shen et al. reported the use of *L. monocytogenes* carrying IL-12 cDNA to manipulate host immune responses against *Leishmania* major-infection [75], which is the first demonstration that a gene introduced into a host by *Listeria* works to modulate the murine host immune response against infections *in vivo*. More recently, attenuated *S. typhimurium* vaccine encoding *Helicobater pylori* UreB gene and mouse IL-2 gene was constructed and tested in mice model which showed the protective effect against *H. pylori* infection [76] which may contribute to the development of a human-use *H pylori* DNA vaccine to treat peptic ulcer.

In the past decade, developing a measles vaccine that can be administered in the presence of maternal antibodies has posed a great challenge. A *S. flexneri* and *S. Typhi Ty21a* were employcd as DNA vaccinc encoding measles antigens. Intranasal immunization with the *Shigella* carrier as well as intraperitoneal immunization with the *Salmonella* carrier resulted in strong, boostable measles-specific cellular immune responses albeit weak humoral responses in mice [30,77]. Another priority for the global health issue is to develop an effective vaccine to treat/prevent HIV infection. The overwhelming number of HIV transmissions takes place via mucosal surfaces and hence the mucosal immunity stimulated by HIV vaccine plays a vital role in protection against HIV. Shata et al. demonstrated *S. typhimurium ΔaroA* carrying an HIV-1 envelope (Env) DNA vaccine administered to mice stomach and led to Env-specific CD8+ T cell responses, both in mucosal and systemic lymphoid tissue. By contrast, intramuscular vaccination with the naked Env DNA vaccine induced systemic CD8+ T cells only [78]. Likewise, oral immunization with an HIV glycoprotein (gp120) DNA vaccine via *S. typhimurium ΔaroA* induced strong systemic and mucosal CD8+ T-cell responses [78]. In addition, *S. flexneri ΔaroA ΔiscA* was employed to deliver a DNA vaccine encoding gp120 [79]. Intranasal immunization of mice elicited strong CD8+ T-cell responses comparable to those induced by intramuscular injection of the plasmid or intraperitoneal immunization with a recombinant vaccinia-*env* vector. When *S. enterica* serovar *Typhi* Ty21a, *S. enterica* serovar *Typhimurium ΔaroA* and *S. flexneri Δasd* were compared in

the intranasal mouse model, *Shigella* induced the strongest gp120-specific CD8+ T cell responses, followed by *S. enterica* serovar *Typhimurium* and Ty21a, respectively. In a side-by-study [80], Karpenko et al. compared the efficacy of naked DNA vaccines alone, attenuated bacteria, and virus-like particles (VLPs) as carriers for DNA delivery in inducing immune responses against an HIV polyepitope antigen. For this purpose, mice were immunized intramuscularly with naked DNA plasmid or with encapsulated VLPs containing the eukaryotic HIV-polyepitope CTL-immunogen (TCI) expression plasmid or immunized mucosally with attenuated *Salmonella* carrying the identical plasmid. The artificial VLPs and the recombinant *Salmonella* vaccine both induced HIV-specific immune responses superior to those obtained after vaccination with naked DNA alone. Compared to VLPs as delivery vectors, *Salmonella* vaccination revealed slightly higher humoral and cellular immune responses [80]. Finally, Xu et al. proposed the use of an attenuated *Shigella* defective in LPS-O-antigen synthesis and carrying an HIV gag DNA vaccine as a boost, after priming by intramuscular DNA injection [81]. Gag specific T-cell responses were detected in the spleen and lungs and the prime/boost strategy resulted in a dramatically increased T-cell response in the lungs [81]. Another group in Japan by Tsunetsugu-Yokota et al. also demonstrated that an oral *Salmonella* vaccine carrying HIV gag potentiated the intestinal CTL responses in nasally primed mice with gag p24 and cholera toxin adjuvant [82].

Bacterial vaccines also bring some hope to the treatment of chronic viral hepatitis disease. Oral vaccination with *S. typhimurium* encoding a hepatitis B virus surface antigen (HBsAg) proved to induce stronger CTL responses in Balb/c mice than recombinant HBsAg vaccination [83]. The same research group also showed that a single administration of these bacteria vaccine to transgenic mice expressing HBsAg in the liver led to loss of expression of the antigen in hepatocytes [84]. This loss of HBsAg was accompanied by hepatic flare that subsided after 3 weeks, while the suppression of HBsAg expression continued in the absence of overt liver pathology for the remaining duration of the experiment (12 weeks). This single administration of the recombinant bacteria induced CTLs, Th1 T cells, and HBsAg-specific IgG2 subclass antibodies as further proof that immune tolerance against the viral antigen had been broken. A similar *S. typhimurium* was also used in bacteria-mediated DNA vaccination against the non-structural region 3 (NS3) of hepatitis C virus in HLA-A2.1 transgenic mice [85]. A single oral administration induced A2.1-restricted CTLs, INF-γ-producing T cells, and resistance against a challenge with NS3-expressing vaccinia virus [85].

Other DNA vaccinations mediated by bacteria include vaccination against influenza [86,87] with *Listeria* or a recombinant strain of *S. flexneri*, vaccination against herpes simplex virus-2 by attenuated *S. typhimurium* [88], vaccination against human papillomavirus type16 by *S. typhimurium* [89,90], vaccination against influenza virus by *S. flexneri* [87], and vaccination against pseudo-rabies virus using either nonpathogenic *E. coli* administered intramuscularly [91] or a swine-adapted strain of *S. choleraesuis* [92].

TB is evolving as the leading killer of young adults worldwide among all the single infectious agents nowadays. The recent increase in TB patients especially in developing countries is caused among others, by pandemic HIV, war and political instability, drug resistance, and increasing poverty. The BCG vaccine has been effective in controlling childhood TB but does not confer protection against adult TB of endogenous or exogenous infection. Therefore, efforts are being made to develop alternative weapons to fight TB infection. Miki et al. [93] reported protective cellular immunity against *Mycobacterium tuberculosis* using the *L. monocytogenes* Δ2 strain DNA vaccine encoding *M. tuberculosis* antigens. Parida et al. [94] employed the *S. enterica* serovar *Typhimurium* Δ*aroA* strain as DNA vaccine carrier for the delivery of the *M. tuberculosis* Ag85A antigen. Comparative animal studies in mice were performed using the intranasal as well as the oral route for the application of the *Salmonella* vaccine carrier. To this end, bactofection was superior to vaccination with naked DNA as monitored by the bacterial load in different organs after intravenous challenge with *M. tuberculosis*. The immune response as monitored by IL-2 and IFN-γ levels was higher after administration of the vaccine carrier by the intranasal compared to the oral route.

12.4 CANCER IMMUNOTHERAPY

Cancer is the second leading cause of death in industrialized countries. Since increasing number of patients are suffering from different forms of cancer, traditional cytotoxic chemotherapy is often not curative or durably responsive when tumor cells are disseminated. Cancer immunotherapy is a promising approach due to its potential eradication of disseminated tumor cells. Vaccines against cancer face a difficult challenge as their targets are usually tumor-associated self-antigens that are poorly immunogenic, undergo mutations, and often suppress immune responses at the cellular level. Therefore, targeting specific tumor antigen, inducing *in vivo* effective CTL responses, and breaking self-tolerance should be the main aims when developing cancer vaccines.

Anecdotal case reports from more than 200 years ago describe tumor regression in patients with severe bacterial infections [95]. Application of bacteria in cancer therapy was pioneered independently by Friedrich Fehleisen and William B Coley in the late 1800s and early 1900s [96,97], leading eventually to immunomodulation for the treatment of cancer. Many more recent studies have now shown the potential of genetically engineered bacteria as tumor-targeting vectors in human cancer therapy (Table 12.2). Among the bacterial

TABLE 12.2
Bacterial Vectors in Cancer Treatment

| Bacterial Vector | Encoding Tumor Antigen | Host Immune Responses | | | References |
		Humoral	Cellular	Clinical	
Salmonella typhimurium	β-Galactosidase (model Ag)	+	+	+	[98,99]
Salmonella typhimurium	gp100 (melanoma)	+	+	+	[103]
Salmonella typhimurium	gp100 (melanoma)	+	+	−	[104]
Salmonella typhimurium	CD40L			+	[121]
Salmonella typhimurium	Ubiquintinated gp100 and TRP-2 (melanoma)		+	+	[119]
Salmonella typhimurium	Tyrosine hydroxylase (neuroblastoma)			+	[101,102]
Clostridium sporogenes	Cytosine deaminase			+	[142]
Salmonella typhimurium	Cytosine deaminase			+	[150]
Salmonella typhimurium	Murine multidrug resistance-1 (MDR-1)		+	+	[118]
Salmonella typhimurium	Survivin and CCL21		+	+	[130]
Salmonella typhimurium	CEA		+	+	[105,106]
Salmonella typhimurium	Fos-related Ag 1 co-expressing IL-18		+	+	[128,129]
Salmonella choleraesuis	Thrombospondin-1 (TSP-1) gene			+	[114]
Salmonella typhimurium	Endoglin (CD105)		+	+	[113]
Salmonella typhimurium	NKG2D ligand-pH60 or surviving		+	+	[131]
Bifidobacterium longum	Endostatin gene		+	+	[136,137]
Salmonella typhimurium	Legumain (tumor-associated macrophages)		+	+	[117]
Salmonella typhimurium	Murine PDGF receptor-β		+	+	[115]
Salmonella typhimurium	Murine vascular endothelial growth factor receptor-2 (FLK-1)		+	+	[109,111]

vectors designed for cancer vaccine, *Salmonella* strains have essentially been used to deliver DNA for therapeutic applications in oncology. In the first instance, "model" tumor antigens such as β-galactosidase was encoded in eukaryotic expression vectors carried by strains of *Salmonella* [98,99]. Oral administration of these bacterial strains delivery of DNA vaccines encoding the model tumor antigens protected the mice against challenges with fibrosarcoma [98] or renal carcinoma [99]. In addition, when real murine/human tumor antigen/epitope was expressed in *Salmonella* vector, the tumor protective effect still could be demonstrated in neuroblastoma [100–102], melanoma [103,104], and adenocarcinoma [105,106].

Attacking the tumor's vasculature has been documented as an effective strategy to inhibit tumor growth/metastasis. This antiangiogenic intervention has been anticipated in *Salmonella* expressing vascular endothelial growth factor receptor (FLK-1 or Fra-1) [107–111]. Recently, Endoglin (CD105), a co-receptor in the TGF-β receptor complex that is overexpressed on proliferating endothelial cells in the breast tumor neovasculature, was delivered by attenuated *S. typhimurium*. In a prophylactic setting, a pronounced CD8$^+$ T-cell response was induced which effectively suppressed dissemination of pulmonary metastases in mice [112,113]. Another strategy used the delivery of the endogenous angiogenic inhibitor, thrombospondin-1 (TSP-1), by *S. choleraesuis* demonstrated to be effective for the treatment of primary and metastatic melanomas [114]. Platelet derived growth factor receptor beta (PDGFRβ) expressed by a *Salmonella* vaccine also showed the suppression of angiogenesis *in vivo* and reduction of tumor stromal cell proliferation [115]. Combined application of cyclophosphamide with a DNA vaccine targeting PDGFRβ not only completely inhibited the growth of different tumor types but also led to tumor rejections in mice [116]. Legumain is highly upregulated on macrophages in many tumor tissues and consequently applicable for tumor therapy. Immunization of mice with *Salmonella* vaccine encoding for Legumain induced a robust CD8$^+$ T-cell response against tumor-associated macrophages and in turn led to a suppression of tumor angiogenesis, tumor growth, and metastasis by profoundly altering the tumor microenvironment [117]. Emergence of acquired multidrug resistance (MDR) remains a major challenge in the treatment of cancer following chemotherapeutic drugs and thus MDR gene serves an alternative cancer vaccine target. *Salmonella* vaccine against MDR-1 have also inhibited tumor growth and metastasis in preclinical studies [118]. This also enabled the further combination of targeting MDR-1 with tradition cytotoxic therapy in effective treatments of cancer.

It has been shown that the immune responses mediated from bacteria-delivering DNA vaccines can be dramatically improved by modifying the expression vectors or co-expression of immunostimulatory molecules. Optimization of the bacterial vectors carrying DNA vaccine was shown to be achieved by different means. (1) The expression levels of delivered antigen can be enhanced by adding a posttranscriptional regulatory acting RNA elements. In experimental murine model, the antitumor effect can be further augmented by boosting with antibody–cytokine fusion protein boost that targets IL-2 to the tumor microenvironment [101]. (2) Antigen-fusion to ubiquitin can also increase antigen processing and effective presentation in order to break peripheral T cell tolerance to a self-antigen [119,120]. (3) Co-expression of invariant chain leads to preferential presentation in the context of MHC class II molecules [104]. (4) Expression of CD40L [121] alone or co-expression of CD40L/antigen [120] upregulated the expression of Fas, B7–1, and B7–2 molecules and improved antigen presentation to T cells. (5) Fusion of cytokine gene into DNA vaccine produced immune-modulatory effects, such as IL-2 [122], IL-4/IL-18 [123,124], GM-CSF/IL-12 [125], IL-12/VEGFR2 (or FLK-1) [126], IFN-γ [127], IL-18 [128] (or combined FRA-1) [128,129], CCL21 with apoptosis protein inhibitor (survivin) [130], NKG2D ligands [131,132]. (6) Co-expression of LLO in addition to tumor antigen caused lysosomal escape and the antigenic epitope to be presented onto MHC class I complex [41,42,58]. (7) From the experience in bacterial vaccine against HIV infection [81,82], the use of prime-boost strategy to enhance and prolong CTL responses is also applied in the field of cancer vaccine, such as primed by naked DNA vaccine and boosted by bacterial vaccine or primed by bacterial vaccine boosted by antigen-loaded DCs [99].

Hypoxic or necrotic regions are characteristic of solid tumors in many murine and human tumors, including the majority of primary tumors of the breast and uterine cervix. The limitation of cancer gene therapy approaches in solid tumor treatment is the delivery vectors, which achieve specific high-level expression within tumor tissues or the tumor environment following systemic or parenteral administration. Accordingly, gene therapy for solid tumors that exploits and targets gene expression in hypoxic tumor cells is currently being investigated [133]. It is known that certain species of anaerobic bacteria, including the genera *Clostridium* and *Bifidobacterium*, can selectively germinate and grow in the hypoxic regions of solid tumors after intravenous injection [134,135]. In such cases, bacteria proliferate between tumor cells and thus the therapeutic gene is not introduced into the tumor cell, but rather the therapeutic protein is produced in the tumor. Candidate bacterial vectors for gene introduction include species of *Bifidobacterium*, *Clostridium*, and *Salmonella*. *Bifidobacterium* is a normal bacterial flora in the intestine, and is the nonpathogenic. *B. longum* transformed with plasmid containing the gene for human endostatin, a potent inhibitor of angiogenesis, has been shown to affect liver tumor growth inhibition in Balb/c mice [136,137]. In the study by Theys et al. described that *C. acetobutylicum* expressing murine-TNF-α is a possible tool for cancer therapy [138]. The use of anaerobic bacterial vaccine for the selective delivery of prodrug-activating enzymes also has been proved to be efficient in preclinical studies [139–142].

In these studies, the *E. coli* enzyme cytosine deaminase (CD) [140,142,143] and nitroreductase [139] were expressed in *Clostridium* and were shown to convert the nontoxic prodrugs 5-fluorocytosine and CB1954, respectively, into toxic compounds capable of diffusing in the tumors and killing the cancer cells through a bystander effect. Using a similar principle, a strain of *Clostridia* was engineered in which a radioresponsive promoter drove the expression of TNF-α [144]. Gram-negative *Salmonella* have also been proposed as oncolytic agents. In contrast to obligate anaerobic bacteria such as *Clostridia* and *Bifidobacteria*, *Salmonella* are facultative anaerobic bacteria and have the potential to colonize oxygenated small metastatic lesions as well as large tumors with a hypoxic centre. The anaerobic bacteria per se also could exert oncolytic effect and thus potentiate the cytotoxic chemotherapy. In the study by Lee et al., *Salmonella choleraesuis* in combination with cisplatin acted additively to retard tumor growth and extensively prolong the survival time of the mice bearing hepatomas or lung tumors. This enhanced antitumor immune responses, manifested by increased infiltrating neutrophils, CD8[+] T cells, and apoptotic cells in the tumors, represents a promising strategy for the treatment of primary and metastatic tumors [145]. Dang et al. assessed 26 different strains of anaerobic bacteria systemically for their proliferative capacity to grow within avascular compartments of transplanted tumors, one (*Clostridium novyi*) appeared particularly promising [146]. This research group also demonstrated that intravenous injection of *C. novyi-NT* can potentiate the treatment effect from selected anti-microtubule agents [147] or radiotherapy [148] in animal models without excessive toxicity. Another study carried out by Cheong et al. demonstrated the *C. novyi-NT* plus a single dose of liposomal doxorubicin enhanced the tumor eradication effect [149].

One study applied this bacterial delivery of prodrug-activating enzyme into clinical setting [150]. Nemunaitis et al. used *Salmonella* vector expressing *E. coli* CD in the therapy of chemotherapy-refractory colorectal cancer patients. Although the study was a pilot study with only three participating patients, the results are very promising and point towards the potential of this procedure [150]. An earlier study published by Toso et al. applying *Salmonella* vaccine (VNP20009) in clinical cancer patients with metastatic melanoma demonstrated that *Salmonella* vaccine could be safely administered intravenously to human. The maximum tolerated dose was found to be 3×10^8 cfu/m², with thrombocytopenia, anemia, persistent bacteremia, hyperbilirubinemia, diarrhea, vomiting, nausea, elevated alkaline phosphatase, and hypophosphatemia as the dose-limiting toxicities. Some tumor colonization was observed upon injection of highest tolerated dose. However, no clinical antitumor effects were seen in their study [37]. Altering the kinetics of infusion failed to improve response or colonization in a small number of subsequently treated patients [151]. Another phase I trial of systemic administration of *Salmonella* vaccine to dogs with the spontaneous tumor model showed acceptable toxicity results in detectable bacterial colonization of tumor tissue and significant antitumor activity in tumor-bearing dogs [152].

12.4.1 OTHER GENETIC DEFICIENCIES

Although most studies using bacteria for various gene therapy-related procedures developed for the treatment of infectious or neoplastic diseases, expectations have been put into the usage of bacterial vaccine in other clinical entities like inflammatory diseases, genetic deficiencies, or vascular diseases.

Inflammatory bowel disease (IBD) is a significant health-care problem and the pathogenesis is complex and relies on the interaction between three essential factors: genetic susceptibility, intestinal bacteria, and the gut mucosal immune response. The traditional treatment by immuno-modulation strategies usually lacks organ specificity and results in unpleasant side effects. As an alternative, administration of an engineered *Lactococcus lactis* secreting IL-10 led to local production of the anti-inflammatory cytokine and prevented the development of colitis in different mouse models [153]. This strain was also modified to carry a mutation disabling the thymidylate synthase, resulting in a strain dependent on the availability of thymine or thymidine in the local microenvironment, which was less likely to accumulate in the environment [154]. Another bacterial vector used is the invasive *E. coli* [11], expressing invasin from *Yersinia* and LLO from *Listeria* and carrying a eukaryotic expression plasmid expressing TGF-β1 [155]. Oral administration of these bacteria led to a significant reduction of the severity of experimental colitis in mice, with vector-specific transcripts detected in colonic and extra-colonic tissues such as the lungs, liver, and spleen. Further modification of the CMV promoter replacing by the inflammation inducible IL-8 promoter has shown the improved expression of TGF-β1 in the inflamed colon without extracolonic expression and affecting the therapeutic effects [155].

Bacterial vectors are capable of transferring a functional gene copy into target cells (airway epithelium) in various models of monogenic diseases like cystic fibrosis (CF). Fajac et al. have shown that *E. coli* expressing invasin and LLO is able to deliver GFP gene into normal or CF airway epithelial cells [156]. A major problem of this study was the unsatisfying selectivity of gene transfer. Krusch et al. have used *L. monocytogenes* to deliver CF transmembrane conductance regulator (CFTR) to mammal cells indicating the possibilities of bactofection usage in the treatment of inherited disorders [157]. Invasive strains of *E. coli* are also capable of transferring CFTR into target cells (airway epithelium) in CF [156,158,159]. These results demonstrate that bacteria allow the cloning, propagation, and transfer of large intact and functional genomic DNA fragments and their subsequent direct delivery into cells for functional analysis.

In the last years several studies have indicated the possibilities of VEGF in the therapy of ischemic heart disease. In the study by Celec et al., *E. coli* vectors, which produce angiogenic factors, VEGF, proved to induce neovascularization in experimental animals [160]. This strategy provides a new modality for experimental angiogenesis and may be applicable in ischemic diseases. On the contrary, VEGF may also be served as the target in artherosclerotic diseases. Vaccination against VEGFR2, carried by *S. typhimurium*, led to inhibition of angiogenesis, slowing down the progression of preexisting advanced atherosclerosis, and increased neointima formation. These results indicate the significant role of VEGFR2 (+) cells in cardiovascular diseases and DNA vaccination targeting these cells may contribute to the development of novel therapies against atherosclerosis [161].

12.5 EMERGING TECHNOLOGIES

One of the problems encountered when trying to transfect mammalian cells with large DNA molecules is the possibility of mechanical breakage of these large molecules during the purification process. In that context, the utilization of bacteria to transfer large DNA molecules would simplify the procedure. The delivery of bacterial artificial chromosomes was first demonstrated into HeLa cells using an invasive *E. coli* [162]. Direct DNA transfer of up to around 1 Mb was demonstrated and as the bacterial vector is equipped with an inducible recombination system, modifications of the bacterial artificial chromosome sequences should be possible. More recently, efficient transfer of an α-satellite DNA cloned into a P1-based artificial chromosome was stably delivered into the HT1080 cell line and efficiently generated human artificial chromosomes de novo [159]. In the same report, a 160 kb construct containing CFTR gene was transferred into the same cells, where it was transcribed and correctly spliced [159]. In a study in mice, large DNA molecules carrying the viral genome of the murine cytomegalovirus (MCMV) were transferred using *E. coli* and *S. typhimurium* as delivery vectors [163]. This transfer led to a productive virus infection that resulted in elevated titers of specific anti-MCMV antibodies, protection against lethal MCMV challenge, and strong expression of additional genes introduced into the viral genome. Thus, the reconstitution of infectious virus from live attenuated bacteria presents a novel concept for multivalent virus vaccines launched from bacterial vectors.

The second issue in bacterial vector therapy is to monitor the delivery efficiency after vaccination. Soghomonyan et al. demonstrated that using *Salmonella* (VNP20009) expressing HSV-thymidine kinase reporter gene was able to selectively localize within murine tumor models and to effectively sequester a radiolabeled nucleoside analogue, and thus could be analyzed by quantitative autoradiography and positron emission tomography (PET). The ability to noninvasively detect *Salmonella* vectors by PET imaging has the potential to be conducted in a clinical setting, and could aid in development of these vectors by demonstrating the efficiency and duration of targeting as well as indicating the locations of tumors [164].

It is clear that each vector interacts with the elements of the innate immune system in complicated ways that can impact both the quantity and quality of immune responses elicited by the vector. The immune system can normally respond to bacterial vector and passenger antigen stimulation but mechanisms of immune regulation in hosts will be started to induce self-tolerance. The past 10 years have witnessed a renaissance in the field of active mechanisms of immune regulation, and much of this points to a particular group of T cells, currently described as CD4+ CD25+ Foxp3+ regulatory T (Treg) cells which play a suppressive role in antitumor vaccinations [165]. Several studies have demonstrated the adjuvant effects with affecting Treg cells obtained from additional gene co-expressing in bacterial vectors. Hussain et al. demonstrated that the Treg cell inhibition is responsible for the differences in vaccine-mediated antitumor responses and less Treg cells was found in the mice vaccinated with *Listeria* vector expressing LLO and antigen [166,167]. Furthermore, Nitcheu-Tefit et al. showed that the incorporation of LLO in the *E. coli* vaccine can affect Treg cell inhibitory function and may have important implications for enhancing antitumor vaccination strategies in humans [168]. By the increasing number of studies regarding the role of Treg cells in cancer patients [169,170], there is a clear rationale for developing bacterial vaccines to manipulate these regulatory influences and augment antitumor immunity.

12.6 BIOSAFETY ISSUE

The ideal bacterial vaccine has to meet the balance between the biosafety and immunogenicity. There is no single bacterial vector approved for human utilization in gene therapy despite the extensive research that has been conducted or still is undergoing in the field of DNA vaccination. The development of candidate vaccine focuses mainly on immunological aspects, e.g., to the induction of the desired immune response against the delivered antigen. However, the lack of relevant animal models during the preclinical development phase makes it hard to predict the immune response and hence the outcome of clinical trials with candidate bacterial vectors.

For bacterial gene therapy two major types of safety issues can be distinguished [171]. The first issue involves the safety of the bacterial strain chosen to deliver the genetic material of interest, e.g., the risks associated with the release of the nonrecombinant bacterium. The second involves the risks that are associated with the delivered foreign genetic material. Several guidelines have been developed by the competent regulatory authorities that address these issues resulting in an extensive documentation necessary to characterize the strains and also to characterize the impact on the environment [171].

The main concerns about the safety of bacterial vaccine strains are reversion to virulence, the establishment of systemic infections, horizontal gene transferring, or translocation to organs others than the side of administration of the vaccine [23,172]. A further concern is whether the delivered bacteria could have a negative impact on the indigenous microflora or establish a permanent colonization of the intestine. Also the unintentional transfer of the bacterial cells to other individuals is a point of concern, especially, if they are shed into the environment and are capable of surviving in environments other than the intestines. Additional safety issues that have to be considered are the sequence of the therapeutic gene and the way in which the recombinant strain is constructed. Bacteria-mediated DNA vaccine integration is very unlikely but may potentially cause mutagenesis if the plasmid insertion reduces the activity of a tumor suppressor or increases the activity of an oncogene. Previous integration by Ledwith et al. study has shown that the probability of any DNA molecule integrating randomly into the chromosome, however, is quite low following intramuscular DNA injection [173].

Upon modifying the immune responses of DNA vaccines by co-administration of genes encoding immunostimulatory cytokines, these vaccines need to be tested either in animal species that respond to the encoded human cytokine or to use expression of analogous animal genes to avoid of chronic inflammation or generalized immunosuppression. It has been demonstrated that the vaccination with live bacterial vaccines could induce the formation of auto-reactive anti-DNA antibodies, and several authors have tried to assess the possibility that vaccination has an impact on the generation of autoimmune responses [174]. However, the relation between vaccine and autoimmunity remains obscure and subsequent publications did not support this theory of generating autoimmune reactions. However, this remains an important point, especially when a bacterial vaccine expresses a real tumor antigen, which is usually a self-antigen.

12.7   FUTURE DIRECTION AND CONCLUSION

As more and more antigens are identified for various infectious diseases or cancer, bacterial vectors will offer a shuttle for introducing the newly identified antigens to the immune system. Furthermore, it should be possible to coordinately express more then one antigen at a time (i.e., a panel of disease specific antigens) to mount a complete immune response against a particular disease in one particular administration. Although the experience from LLO and Invasin expressed in bacterial vaccine has greatly improved the antigen delivery and intracellular processing, one obstacle most likely still hindering gene delivery is the intracellular trafficking of gene cargo to the nucleus. Current systems might be able to efficiently enter the cell cytoplasm and release genetic cargo, but the use of designed mechanisms for facilitating nuclear transport has not been investigated thoroughly. Also, the dosage, location of vaccination, or optimal prime/boost strategy in bacterial vaccine therapy still need to be further defined.

In summary, bacterial vectors in gene therapy seems to be a valuable tool with a specific niche of application, mostly in immunotherapy.

REFERENCES

1. Mountain A. Gene therapy: The first decade. *Trends Biotechnol.* 2000;18:119–128.
2. Gardlik R, Palffy R, Hodosy J, Lukacs J, Turna J, and Celec P. Vectors and delivery systems in gene therapy. *Med Sci Monit.* 2005;11:RA110–121.
3. Page DT and Cudmore S. Innovations in oral gene delivery: Challenges and potentials. *Drug Discov Today.* 2001;6:92–101.
4. Glover DJ, Lipps HJ, and Jans DA. Towards safe, non-viral therapeutic gene expression in humans. *Nat Rev Genet.* 2005; 6:299–310.

5. Heinemann JA and Sprague GF, Jr. Bacterial conjugative plasmids mobilize DNA transfer between bacteria and yeast. *Nature.* 1989;340:205–209.
6. Lessl M and Lanka E. Common mechanisms in bacterial conjugation and Ti-mediated T-DNA transfer to plant cells. *Cell.* 1994;77:321–324.
7. Palffy R, Gardlik R, Hodosy J, et al. Bacteria in gene therapy: Bactofection versus alternative gene therapy. *Gene Ther.* 2006;13:101–105.
8. Lewis GK. Live-attenuated *Salmonella* as a prototype vaccine vector for passenger immunogens in humans: Are we there yet? *Expert Rev Vaccines.* 2007;6:431–440.
9. Sizemore DR, Branstrom AA, and Sadoff JC. Attenuated *Shigella* as a DNA delivery vehicle for DNA-mediated immunization. *Science.* 1995;270:299–302.
10. Dietrich G, Bubert A, Gentschev I, et al. Delivery of antigen-encoding plasmid DNA into the cytosol of macrophages by attenuated suicide *Listeria monocytogenes. Nat Biotechnol.* 1998;16:181–185.
11. Grillot-Courvalin C, Goussard S, Huetz F, Ojcius DM, and Courvalin P. Functional gene transfer from intracellular bacteria to mammalian cells. *Nat Biotechnol.* 1998;16:862–866.
12. Grillot-Courvalin C, Goussard S, and Courvalin P. Bacteria as gene delivery vectors for mammalian cells. *Curr Opin Biotechnol.* 1999;10:477–481.
13. Loessner H and Weiss S. Bacteria-mediated DNA transfer in gene therapy and vaccination. *Expert Opin Biol Ther.* 2004; 4:157–168.
14. Schaffner W. Direct transfer of cloned genes from bacteria to mammalian cells. *Proc Natl Acad Sci U S A.* 1980;77: 2163–2167.
15. Steinman RM. Dendritic cells and the control of immunity: Enhancing the efficiency of antigen presentation. *Mt Sinai J Med.* 2001;68:160–166.
16. Luster MI, Germolec DR, Yoshida T, Kayama F, and Thompson M. Endotoxin-induced cytokine gene expression and excretion in the liver. *Hepatology.* 1994;19:480–488.
17. Mulligan MS, Smith CW, Anderson DC, et al. Role of leukocyte adhesion molecules in complement-induced lung injury. *J Immunol.* 1993;150:2401–2406.
18. Lukacs NW, Hogaboam C, Campbell E, and Kunkel SL. Chemokines: Function, regulation and alteration of inflammatory responses. *Chem Immunol.* 1999;72:102–120.
19. Qureshi ST, Gros P, and Malo D. The Lps locus: Genetic regulation of host responses to bacterial lipopolysaccharide. *Inflamm Res.* 1999;48:613–620.
20. Natali F and Ramon P. BCG vaccination. *Rev Pneumol Clin.* 1994;50:268–274.
21. Germanier R and Fuer E. Isolation and characterization of Gal E mutant Ty 21a of *Salmonella typhi*: A candidate strain for a live, oral typhoid vaccine. *J Infect Dis.* 1975;131:553–558.
22. Daudel D, Weidinger G, and Spreng S. Use of attenuated bacteria as delivery vectors for DNA vaccines. *Expert Rev Vaccines.* 2007;6:97–110.
23. Detmer A and Glenting J. Live bacterial vaccines—a review and identification of potential hazards. *Microb Cell Fact.* 2006;5:23.
24. Kaper JB, Lockman H, Baldini MM, and Levine MM. Recombinant nontoxinogenic Vibrio cholerae strains as attenuated cholera vaccine candidates. *Nature.* 1984;308:655–658.
25. Levine MM, Kaper JB, Herrington D, et al. Safety, immunogenicity, and efficacy of recombinant live oral cholera vaccines, CVD 103 and CVD 103-HgR. *Lancet.* 1988;2:467–470.
26. Chatfield SN, Fairweather N, Charles I, et al. Construction of a genetically defined *Salmonella typhi* Ty2 aroA, aroC mutant

for the engineering of a candidate oral typhoid-tetanus vaccine. *Vaccine*. 1992;10:53–60.

27. Bouwer HG, Alberti-Segui C, Montfort MJ, Berkowitz ND, and Higgins DE. Directed antigen delivery as a vaccine strategy for an intracellular bacterial pathogen. *Proc Natl Acad Sci U S A*. 2006;103:5102–5107.

28. Parsa S and Pfeifer B. Engineering bacterial vectors for delivery of genes and proteins to antigen-presenting cells. *Mol Pharm*. 2007;4:4–17.

29. Vassaux G, Nitcheu J, Jezzard S, and Lemoine NR. Bacterial gene therapy strategies. *J Pathol*. 2006;208:290–298.

30. Fennelly GJ, Khan SA, Abadi MA, Wild TF, and Bloom BR. Mucosal DNA vaccine immunization against measles with a highly attenuated *Shigella flexneri* vector. *J Immunol*. 1999;162:1603–1610.

31. Anderson RJ, Pasetti MF, Sztein MB, Levine MM, and Noriega FR. DeltaguaBA attenuated *Shigella flexneri* 2a strain CVD 1204 as a *Shigella* vaccine and as a live mucosal delivery system for fragment C of tetanus toxin. *Vaccine*. 2000;18:2193–2202.

32. Pasetti MF, Anderson RJ, Noriega FR, Levine MM, and Sztein MB. Attenuated deltaguaBA *Salmonella typhi* vaccine strain CVD 915 as a live vector utilizing prokaryotic or eukaryotic expression systems to deliver foreign antigens and elicit immune responses. *Clin Immunol*. 1999;92:76–89.

33. Stocker BA. Aromatic-dependent *Salmonella* as anti-bacterial vaccines and as presenters of heterologous antigens or of DNA encoding them. *J Biotechnol*. 2000;83:45–50.

34. Darji A, Guzman CA, Gerstel B, et al. Oral somatic transgene vaccination using attenuated *S. typhimurium*. *Cell*. 1997;91:765–775.

35. Weiss S and Chakraborty T. Transfer of eukaryotic expression plasmids to mammalian host cells by bacterial carriers. *Curr Opin Biotechnol*. 2001;12:467–472.

36. Dietrich G, Griot-Wenk M, Metcalfe IC, Lang AB, and Viret JF. Experience with registered mucosal vaccines. *Vaccine*. 2003;21:678–683.

37. Toso JF, Gill VJ, Hwu P, et al. Phase I study of the intravenous administration of attenuated *Salmonella typhimurium* to patients with metastatic melanoma. *J Clin Oncol*. 2002;20:142–152.

38. Mayr UB, Walcher P, Azimpour C, Riedmann E, Haller C, and Lubitz W. Bacterial ghosts as antigen delivery vehicles. *Adv Drug Deliv Rev*. 2005;57:1381–1391.

39. Brockstedt DG, Bahjat KS, Giedlin MA, et al. Killed but metabolically active microbes: A new vaccine paradigm for eliciting effector T-cell responses and protective immunity. *Nat Med*. 2005;11:853–860.

40. Higgins DE, Shastri N, and Portnoy DA. Delivery of protein to the cytosol of macrophages using *Escherichia coli* K-12. *Mol Microbiol*. 1999;31:1631–1641.

41. Radford KJ, Higgins DE, Pasquini S, et al. A recombinant *E. coli* vaccine to promote MHC class I-dependent antigen presentation: Application to cancer immunotherapy. *Gene Ther*. 2002;9:1455–1463.

42. Radford KJ, Jackson AM, Wang JH, Vassaux G, and Lemoine NR. Recombinant *E. coli* efficiently delivers antigen and maturation signals to human dendritic cells: Presentation of MART1 to CD8+ T cells. *Int J Cancer*. 2003;105:811–819.

43. Roberts TL, Dunn JA, Terry TD, et al. Differences in macrophage activation by bacterial DNA and CpG-containing oligonucleotides. *J Immunol*. 2005;175:3569–3576.

44. Hemmi H, Takeuchi O, Kawai T, et al. A Toll-like receptor recognizes bacterial DNA. *Nature*. 2000;408:740–745.

45. Medzhitov R. CpG DNA: Security code for host defense. *Nat Immunol*. 2001;2:15–16.

46. Visser L, Jan de Heer H, Boven LA, et al. Proinflammatory bacterial peptidoglycan as a cofactor for the development of central nervous system autoimmune disease. *J Immunol*. 2005;174:808–816.

47. Gewirtz AT, Navas TA, Lyons S, Godowski PJ, and Madara JL. Cutting edge: Bacterial flagellin activates basolaterally expressed TLR5 to induce epithelial proinflammatory gene expression. *J Immunol*. 2001;167:1882–1885.

48. Aliprantis AO, Yang RB, Mark MR, et al. Cell activation and apoptosis by bacterial lipoproteins through toll-like receptor-2. *Science*. 1999;285:736–739.

49. Poltorak A, Ricciardi-Castagnoli P, Citterio S, and Beutler B. Physical contact between lipopolysaccharide and toll-like receptor 4 revealed by genetic complementation. *Proc Natl Acad Sci U S A*. 2000;97:2163–2167.

50. Galan JE and Bliska JB. Cross-talk between bacterial pathogens and their host cells. *Annu Rev Cell Dev Biol*. 1996;12:221–255.

51. Marra A and Isberg RR. Common entry mechanisms. Bacterial pathogenesis. *Curr Biol*. 1996;6:1084–1086.

52. Goebel W and Kuhn M. Bacterial replication in the host cell cytosol. *Curr Opin Microbiol*. 2000;3:49–53.

53. Ogawa M and Sasakawa C. Intracellular survival of *Shigella*. *Cell Microbiol*. 2006;8:177–184.

54. Grillot-Courvalin C, Goussard S, and Courvalin P. Wild-type intracellular bacteria deliver DNA into mammalian cells. *Cell Microbiol*. 2002;4:177–186.

55. Gaillard JL, Berche P, Mounier J, Richard S, and Sansonetti P. In vitro model of penetration and intracellular growth of *Listeria monocytogenes* in the human enterocyte-like cell line Caco-2. *Infect Immun*. 1987;55:2822–2829.

56. Michel E, Reich KA, Favier R, Berche P, and Cossart P. Attenuated mutants of the intracellular bacterium *Listeria monocytogenes* obtained by single amino acid substitutions in listeriolysin O. *Mol Microbiol*. 1990;4:2167–2178.

57. Portnoy DA, Chakraborty T, Goebel W, and Cossart P. Molecular determinants of *Listeria monocytogenes* pathogenesis. *Infect Immun*. 1992;60:1263–1267.

58. Critchley RJ, Jezzard S, Radford KJ, et al. Potential therapeutic applications of recombinant, invasive *E. coli*. *Gene Ther*. 2004;11:1224–1233.

59. Critchley-Thorne RJ, Stagg AJ, and Vassaux G. Recombinant *Escherichia coli* expressing invasin targets the Peyer's patches: The basis for a bacterial formulation for oral vaccination. *Mol Ther*. 2006;14:183–191.

60. Karagiannis TC and El-Osta A. RNA interference and potential therapeutic applications of short interfering RNAs. *Cancer Gene Ther*. 2005;12:787–795.

61. Timmons L, Court DL, and Fire A. Ingestion of bacterially expressed dsRNAs can produce specific and potent genetic interference in *Caenorhabditis elegans*. *Gene*. 2001;263:103–112.

62. Timmons L and Fire A. Specific interference by ingested dsRNA. *Nature*. 1998;395:854.

63. Li CX, Parker A, Menocal E, Xiang S, Borodyansky L, and Fruehauf JH. Delivery of RNA interference. *Cell Cycle*. 2006;5:2103–2109.

64. Xiang S, Fruehauf J, and Li CJ. Short hairpin RNA-expressing bacteria elicit RNA interference in mammals. *Nat Biotechnol*. 2006;24:697–702.

65. Zhang L, Gao L, Zhao L, et al. Intratumoral delivery and suppression of prostate tumor growth by attenuated *Salmonella enterica* serovar *typhimurium* carrying plasmid-based small interfering RNAs. *Cancer Res*. 2007;67:5859–5864.

66. Engels EA, Falagas ME, Lau J, and Bennish ML. Typhoid fever vaccines: A meta-analysis of studies on efficacy and toxicity. *BMJ.* 1998;316:110–116.

67. Kyd JM and Cripps AW. Killed whole bacterial cells, a mucosal delivery system for the induction of immunity in the respiratory tract and middle ear: An overview. *Vaccine.* 1999;17:1775–1781.

68. Al-Mariri A, Tibor A, Lestrate P, Mertens P, De Bolle X, and Letesson JJ. *Yersinia enterocolitica* as a vehicle for a naked DNA vaccine encoding *Brucella abortus* bacterioferritin or P39 antigen. *Infect Immun.* 2002;70:1915–1923.

69. Pickett TE, Pasetti MF, Galen JE, Sztein MB, and Levine MM. In vivo characterization of the murine intranasal model for assessing the immunogenicity of attenuated *Salmonella enterica* serovar Typhi strains as live mucosal vaccines and as live vectors. *Infect Immun.* 2000;68:205–213.

70. Bauer H, Darji A, Chakraborty T, and Weiss S. *Salmonella*-mediated oral DNA vaccination using stabilized eukaryotic expression plasmids. *Gene Ther.* 2005;12:364–372.

71. Darji A, zur Lage S, Garbe AI, Chakraborty T, and Weiss S. Oral delivery of DNA vaccines using attenuated *Salmonella typhimurium* as carrier. *FEMS Immunol Med Microbiol.* 2000;27:341–349.

72. Galen JE, Zhao L, Chinchilla M, et al. Adaptation of the endogenous *Salmonella enterica* serovar Typhi clyA-encoded hemolysin for antigen export enhances the immunogenicity of anthrax protective antigen domain 4 expressed by the attenuated live-vector vaccine strain CVD 908-htrA. *Infect Immun.* 2004;72:7096–7106.

73. Gomez-Duarte OG, Pasetti MF, Santiago A, Sztein MB, Hoffman SL, and Levine MM. Expression, extracellular secretion, and immunogenicity of the *Plasmodium falciparum* sporozoite surface protein 2 in *Salmonella* vaccine strains. *Infect Immun.* 2001;69:1192–1198.

74. Chinchilla M, Pasetti MF, Medina-Moreno S, et al. Enhanced immunity to *Plasmodium falciparum* circumsporozoite protein (PfCSP) by using *Salmonella enterica* serovar Typhi expressing PfCSP and a PfCSP-encoding DNA vaccine in a heterologous prime-boost strategy. *Infect Immun.* 2007; 75:3769–3779.

75. Shen H, Kanoh M, Liu F, Maruyama S, and Asano Y. Modulation of the immune system by *Listeria monocytogenes*-mediated gene transfer into mammalian cells. *Microbiol Immunol.* 2004;48:329–337.

76. Xu C, Li ZS, Du YQ, et al. Construction of recombinant attenuated *Salmonella typhimurium* DNA vaccine expressing *H. pylori* ureB and IL-2. *World J Gastroenterol.* 2007;13: 939–944.

77. Pasetti MF, Barry EM, Losonsky G, et al. Attenuated *Salmonella enterica* serovar Typhi and *Shigella flexneri* 2a strains mucosally deliver DNA vaccines encoding measles virus hemagglutinin, inducing specific immune responses and protection in cotton rats. *J Virol.* 2003;77:5209–5217.

78. Shata MT, Reitz MS, Jr., DeVico AL, Lewis GK, and Hone DM. Mucosal and systemic HIV-1 Env-specific CD8(+) T-cells develop after intragastric vaccination with a *Salmonella* Env DNA vaccine vector. *Vaccine.* 2001;20:623–629.

79. Shata MT and Hone DM. Vaccination with a *Shigella* DNA vaccine vector induces antigen-specific CD8(+) T cells and antiviral protective immunity. *J Virol.* 2001;75:9665–9670.

80. Karpenko LI, Nekrasova NA, Ilyichev AA, et al. Comparative analysis using a mouse model of the immunogenicity of artificial VLP and attenuated *Salmonella* strain carrying a DNA-vaccine encoding HIV-1 polyepitope CTL-immunogen. *Vaccine.* 2004;22:1692–1699.

81. Xu F, Hong M, and Ulmer JB. Immunogenicity of an HIV-1 gag DNA vaccine carried by attenuated *Shigella. Vaccine.* 2003;21:644–648.

82. Tsunetsugu-Yokota Y, Ishige M, and Murakami M. Oral attenuated *Salmonella enterica* serovar Typhimurium vaccine expressing codon-optimized HIV type 1 Gag enhanced intestinal immunity in mice. *AIDS Res Hum Retroviruses.* 2007; 23:278–286.

83. Woo PC, Wong LP, Zheng BJ, and Yuen KY. Unique immunogenicity of hepatitis B virus DNA vaccine presented by live-attenuated *Salmonella typhimurium. Vaccine.* 2001; 19:2945–2954.

84. Zheng BJ, Ng MH, Chan KW, et al. A single dose of oral DNA immunization delivered by attenuated *Salmonella typhimurium* down-regulates transgene expression in HBsAg transgenic mice. *Eur J Immunol.* 2002;32:3294–3304.

85. Wedemeyer H, Gagneten S, Davis A, Bartenschlager R, Feinstone S, and Rehermann B. Oral immunization with HCV-NS3-transformed *Salmonella*: Induction of HCV-specific CTL in a transgenic mouse model. *Gastroenterology.* 2001;121:1158–1166.

86. Ikonomidis G, Paterson Y, Kos FJ, and Portnoy DA. Delivery of a viral antigen to the class I processing and presentation pathway by *Listeria monocytogenes. J Exp Med.* 1994;180: 2209–2218.

87. Vecino WH, Quanquin NM, Martinez-Sobrido L, et al. Mucosal immunization with attenuated *Shigella flexneri* harboring an influenza hemagglutinin DNA vaccine protects mice against a lethal influenza challenge. *Virology.* 2004;325:192–199.

88. Flo J, Tisminetzky S, and Baralle F. Oral transgene vaccination mediated by attenuated Salmonellae is an effective method to prevent *Herpes simplex* virus-2 induced disease in mice. *Vaccine.* 2001;19:1772–1782.

89. Baud D, Benyacoub J, Revaz V, et al. Immunogenicity against human papillomavirus type 16 virus-like particles is strongly enhanced by the PhoPc phenotype in *Salmonella enterica* serovar Typhimurium. *Infect Immun.* 2004;72:750–756.

90. Baud D, Ponci F, Bobst M, De Grandi P, and Nardelli-Haefliger D. Improved efficiency of a *Salmonella*-based vaccine against human papillomavirus type 16 virus-like particles achieved by using a codon-optimized version of L1. *J Virol.* 2004;78:12901–12909.

91. Shiau AL, Chu CY, Su WC, and Wu CL. Vaccination with the glycoprotein D gene of pseudorabies virus delivered by nonpathogenic *Escherichia coli* elicits protective immune responses. *Vaccine.* 2001;19:3277–3284.

92. Shiau AL, Chen YL, Liao CY, Huang YS, and Wu CL. Prothymosin alpha enhances protective immune responses induced by oral DNA vaccination against pseudorabies delivered by *Salmonella choleraesuis. Vaccine.* 2001;19:3947–3956.

93. Miki K, Nagata T, Tanaka T, et al. Induction of protective cellular immunity against Mycobacterium tuberculosis by recombinant attenuated self-destructing *Listeria monocytogenes* strains harboring eukaryotic expression plasmids for antigen 85 complex and MPB/MPT51. *Infect Immun.* 2004;72:2014–2021.

94. Parida SK, Huygen K, Ryffel B, and Chakraborty T. Novel bacterial delivery system with attenuated *Salmonella typhimurium* carrying plasmid encoding Mtb antigen 85A for mucosal immunization: Establishment of proof of principle in TB mouse model. *Ann N Y Acad Sci.* 2005;1056:366–378.

95. Hall S. *A Commotion in the Blood.* New York: Henry Holt and Company. 1998.

96. Coley W. Contribution to the knowledge of sarcoma. *Ann Surgery.* 1891;14:199–220.

97. Coley W. Late results of the treatment of inoperable sarcoma by the mixed toxins of erysipelas and *Bacillus prodigiosus*. *Am J Med Sci*. 1906;131:375–430.

98. Paglia P, Medina E, Arioli I, Guzman CA, and Colombo MP. Gene transfer in dendritic cells, induced by oral DNA vaccination with *Salmonella typhimurium*, results in protective immunity against a murine fibrosarcoma. *Blood*. 1998;92:3172–3176.

99. Zoller M and Christ O. Prophylactic tumor vaccination: Comparison of effector mechanisms initiated by protein versus DNA vaccination. *J Immunol*. 2001;166:3440–3450.

100. Huebener N, Lange B, Lemmel C, et al. Vaccination with minigenes encoding for novel "self" antigens are effective in DNA-vaccination against neuroblastoma. *Cancer Lett*. 2003;197:211–217.

101. Pertl U, Wodrich H, Ruehlmann JM, Gillies SD, Lode HN, and Reisfeld RA. Immunotherapy with a posttranscriptionally modified DNA vaccine induces complete protection against metastatic neuroblastoma. *Blood*. 2003;101:649–654.

102. Lode HN, Pertl U, Xiang R, Gaedicke G, and Reisfeld RA. Tyrosine hydroxylase-based DNA-vaccination is effective against murine neuroblastoma. *Med Pediatr Oncol*. 2000;35:641–646.

103. Cochlovius B, Stassar MJ, Schreurs MW, Benner A, and Adema GJ. Oral DNA vaccination: Antigen uptake and presentation by dendritic cells elicits protective immunity. *Immunol Lett*. 2002;80:89–96.

104. Weth R, Christ O, Stevanovic S, and Zoller M. Gene delivery by attenuated *Salmonella typhimurium*: Comparing the efficacy of helper versus cytotoxic T cell priming in tumor vaccination. *Cancer Gene Ther*. 2001;8:599–611.

105. Niethammer AG, Primus FJ, Xiang R, et al. An oral DNA vaccine against human carcinoembryonic antigen (CEA) prevents growth and dissemination of Lewis lung carcinoma in CEA transgenic mice. *Vaccine*. 2001;20:421–429.

106. Zhou H, Luo Y, Mizutani M, et al. A novel transgenic mouse model for immunological evaluation of carcinoembryonic antigen-based DNA minigene vaccines. *J Clin Invest*. 2004;113:1792–1798.

107. Reisfeld RA, Niethammer AG, Luo Y, and Xiang R. DNA vaccines designed to inhibit tumor growth by suppression of angiogenesis. *Int Arch Allergy Immunol*. 2004;133:295–304.

108. Reisfeld RA, Niethammer AG, Luo Y, and Xiang R. DNA vaccines suppress tumor growth and metastases by the induction of anti-angiogenesis. *Immunol Rev*. 2004;199:181–190.

109. Niethammer AG, Xiang R, Becker JC, et al. A DNA vaccine against VEGF receptor 2 prevents effective angiogenesis and inhibits tumor growth. *Nat Med*. 2002;8:1369–1375.

110. Zhou H, Luo Y, Mizutani M, Mizutani N, Reisfeld RA, and Xiang R. T cell-mediated suppression of angiogenesis results in tumor protective immunity. *Blood*. 2005;106:2026–2032.

111. Luo Y, Markowitz D, Xiang R, Zhou H, and Reisfeld RA. FLK-1-based minigene vaccines induce T cell-mediated suppression of angiogenesis and tumor protective immunity in syngeneic BALB/c mice. *Vaccine*. 2007;25:1409–1415.

112. Needham DJ, Lee JX, and Beilharz MW. Intra-tumoural regulatory T cells: A potential new target in cancer immunotherapy. *Biochem Biophys Res Commun*. 2006;343:684–691.

113. Lee SH, Mizutani N, Mizutani M, et al. Endoglin (CD105) is a target for an oral DNA vaccine against breast cancer. *Cancer Immunol Immunother*. 2006;55:1565–1574.

114. Lee CH, Wu CL, and Shiau AL. Systemic administration of attenuated *Salmonella choleraesuis* carrying thrombospondin-1 gene leads to tumor-specific transgene expression, delayed tumor growth and prolonged survival in the murine melanoma model. *Cancer Gene Ther*. 2005;12:175–184.

115. Kaplan CD, Kruger JA, Zhou H, Luo Y, Xiang R, and Reisfeld RA. A novel DNA vaccine encoding PDGFRbeta suppresses growth and dissemination of murine colon, lung and breast carcinoma. *Vaccine*. 2006;24:6994–7002.

116. Loeffler M, Kruger JA, and Reisfeld RA. Immunostimulatory effects of low-dose cyclophosphamide are controlled by inducible nitric oxide synthase. *Cancer Res*. 2005;65:5027–5030.

117. Luo Y, Zhou H, Krueger J, et al. Targeting tumor-associated macrophages as a novel strategy against breast cancer. *J Clin Invest*. 2006;116:2132–2141.

118. Niethammer AG, Wodrich H, Loeffler M, et al. Multidrug resistance-1 (MDR-1): A new target for T cell-based immunotherapy. *FASEB J*. 2005;19:158–159.

119. Xiang R, Lode HN, Chao TH, et al. An autologous oral DNA vaccine protects against murine melanoma. *Proc Natl Acad Sci U S A*. 2000;97:5492–5497.

120. Xiang R, Primus FJ, Ruehlmann JM, et al. A dual-function DNA vaccine encoding carcinoembryonic antigen and CD40 ligand trimer induces T cell-mediated protective immunity against colon cancer in carcinoembryonic antigen-transgenic mice. *J Immunol*. 2001;167:4560–4565.

121. Urashima M, Suzuki H, Yuza Y, Akiyama M, Ohno N, and Eto Y. An oral CD40 ligand gene therapy against lymphoma using attenuated *Salmonella typhimurium*. *Blood*. 2000;95:1258–1263.

122. Niethammer AG, Xiang R, Ruehlmann JM, et al. Targeted interleukin 2 therapy enhances protective immunity induced by an autologous oral DNA vaccine against murine melanoma. *Cancer Res*. 2001;61:6178–6184.

123. Agorio C, Schreiber F, Sheppard M, et al. Live attenuated *Salmonella* as a vector for oral cytokine gene therapy in melanoma. *J Gene Med*. 2007;9:416–423.

124. Rosenkranz CD, Chiara D, Agorio C, et al. Towards new immunotherapies: Targeting recombinant cytokines to the immune system using live attenuated *Salmonella*. *Vaccine*. 2003;21:798–801.

125. Yuhua L, Kunyuan G, Hui C, et al. Oral cytokine gene therapy against murine tumor using attenuated *Salmonella typhimurium*. *Int J Cancer*. 2001;94:438–443.

126. Feng KK, Zhao HY, Qiu H, Liu JX, and Chen J. Combined therapy with flk1-based DNA vaccine and interleukin-12 results in enhanced antiangiogenic and antitumor effects. *Cancer Lett*. 2005;221:41–47.

127. Paglia P, Terrazzini N, Schulze K, Guzman CA, and Colombo MP. In vivo correction of genetic defects of monocyte/macrophages using attenuated *Salmonella* as oral vectors for targeted gene delivery. *Gene Ther*. 2000;7:1725–1730.

128. Luo Y, Zhou H, Mizutani M, Mizutani N, Reisfeld RA, and Xiang R. Transcription factor Fos-related antigen 1 is an effective target for a breast cancer vaccine. *Proc Natl Acad Sci U S A*. 2003;100:8850–8855.

129. Luo Y, Zhou H, Mizutani M, et al. A DNA vaccine targeting Fos-related antigen 1 enhanced by IL-18 induces long-lived T-cell memory against tumor recurrence. *Cancer Res*. 2005;65:3419–3427.

130. Xiang R, Mizutani N, Luo Y, et al. A DNA vaccine targeting survivin combines apoptosis with suppression of angiogenesis in lung tumor eradication. *Cancer Res*. 2005;65:553–561.

131. Zhou H, Luo Y, Lo JF, et al. DNA-based vaccines activate innate and adaptive antitumor immunity by engaging the NKG2D receptor. *Proc Natl Acad Sci U S A*. 2005;102:10846–10851.

132. Zhou H, Luo Y, Kaplan CD, et al. A DNA-based cancer vaccine enhances lymphocyte cross talk by engaging the NKG2D receptor. *Blood*. 2006;107:3251–3257.

133. Dachs GU, Patterson AV, Firth JD, et al. Targeting gene expression to hypoxic tumor cells. *Nat Med*. 1997;3:515–520.

134. Fujimori M. Genetically engineered bifidobacterium as a drug delivery system for systemic therapy of metastatic breast cancer patients. *Breast Cancer*. 2006;13:27–31.

135. Yazawa K, Fujimori M, Amano J, Kano Y, and Taniguchi S. Bifidobacterium longum as a delivery system for cancer gene therapy: Selective localization and growth in hypoxic tumors. *Cancer Gene Ther*. 2000;7:269–274.

136. Fu GF, Li X, Hou YY, Fan YR, Liu WH, and Xu GX. Bifidobacterium longum as an oral delivery system of endostatin for gene therapy on solid liver cancer. *Cancer Gene Ther*. 2005;12:133–140.

137. Xu YF, Zhu LP, Hu B, et al. A new expression plasmid in Bifidobacterium longum as a delivery system of endostatin for cancer gene therapy. *Cancer Gene Ther*. 2007;14:151–157.

138. Theys J, Nuyts S, Landuyt W, et al. Stable *Escherichia coli–Clostridium acetobutylicum* shuttle vector for secretion of murine tumor necrosis factor alpha. *Appl Environ Microbiol*. 1999;65:4295–4300.

139. Lemmon MJ, van Zijl P, Fox ME, et al. Anaerobic bacteria as a gene delivery system that is controlled by the tumor microenvironment. *Gene Ther*. 1997;4:791–796.

140. Fox ME, Lemmon MJ, Mauchline ML, et al. Anaerobic bacteria as a delivery system for cancer gene therapy: In vitro activation of 5-fluorocytosine by genetically engineered clostridia. *Gene Ther*. 1996;3:173–178.

141. Minton NP. Clostridia in cancer therapy. *Nat Rev Microbiol*. 2003;1:237–242.

142. Liu SC, Minton NP, Giaccia AJ, and Brown JM. Anticancer efficacy of systemically delivered anaerobic bacteria as gene therapy vectors targeting tumor hypoxia/necrosis. *Gene Ther*. 2002;9:291–296.

143. Theys J, Landuyt W, Nuyts S, et al. Specific targeting of cytosine deaminase to solid tumors by engineered *Clostridium acetobutylicum*. *Cancer Gene Ther*. 2001;8:294–297.

144. Nuyts S, Van Mellaert L, Theys J, et al. Radio-responsive recA promoter significantly increases TNFalpha production in recombinant clostridia after 2 Gy irradiation. *Gene Ther*. 2001;8:1197–1201.

145. Lee CH, Wu CL, Tai YS, and Shiau AL. Systemic administration of attenuated *Salmonella choleraesuis* in combination with cisplatin for cancer therapy. *Mol Ther*. 2005;11:707–716.

146. Dang LH, Bettegowda C, Huso DL, Kinzler KW, and Vogelstein B. Combination bacteriolytic therapy for the treatment of experimental tumors. *Proc Natl Acad Sci U S A*. 2001;98:15155–15160.

147. Dang LH, Bettegowda C, Agrawal N, et al. Targeting vascular and avascular compartments of tumors with C. novyi-NT and anti-microtubule agents. *Cancer Biol Ther*. 2004;3:326–337.

148. Bettegowda C, Dang LH, Abrams R, et al. Overcoming the hypoxic barrier to radiation therapy with anaerobic bacteria. *Proc Natl Acad Sci U S A*. 2003;100:15083–15088.

149. Cheong I, Huang X, Bettegowda C, et al. A bacterial protein enhances the release and efficacy of liposomal cancer drugs. *Science*. 2006;314:1308–1311.

150. Nemunaitis J, Cunningham C, Senzer N, et al. Pilot trial of genetically modified, attenuated *Salmonella* expressing the E. coli cytosine deaminase gene in refractory cancer patients. *Cancer Gene Ther*. 2003;10:737–744.

151. Heimann DM and Rosenberg SA. Continuous intravenous administration of live genetically modified *Salmonella Typhimurium* in patients with metastatic melanoma. *J Immunother*. 2003;26:179–180.

152. Thamm DH, Kurzman ID, King I, et al. Systemic administration of an attenuated, tumor-targeting *Salmonella typhimurium* to dogs with spontaneous neoplasia: Phase I evaluation. *Clin Cancer Res*. 2005;11:4827–4834.

153. Steidler L, Hans W, Schotte L, et al. Treatment of murine colitis by Lactococcus lactis secreting interleukin-10. *Science*. 2000;289:1352–1355.

154. Steidler L, Neirynck S, Huyghebaert N, et al. Biological containment of genetically modified Lactococcus lactis for intestinal delivery of human interleukin 10. *Nat Biotechnol*. 2003;21:785–789.

155. Castagliuolo I, Beggiao E, Brun P, et al. Engineered *E. coli* delivers therapeutic genes to the colonic mucosa. *Gene Ther*. 2005;12:1070–1078.

156. Fajac I, Grosse S, Collombet JM, et al. Recombinant *Escherichia coli* as a gene delivery vector into airway epithelial cells. *J Control Release*. 2004;97:371–381.

157. Krusch S, Domann E, Frings M, et al. *Listeria monocytogenes* mediated CFTR transgene transfer to mammalian cells. *J Gene Med*. 2002;4:655–667.

158. Griesenbach U, Geddes DM, and Alton EW. Gene therapy progress and prospects: Cystic fibrosis. *Gene Ther*. 2006;13:1061–1067.

159. Laner A, Goussard S, Ramalho AS, et al. Bacterial transfer of large functional genomic DNA into human cells. *Gene Ther*. 2005;12:1559–1572.

160. Celec P, Gardlik R, Palffy R, et al. The use of transformed *Escherichia coli* for experimental angiogenesis induced by regulated in situ production of vascular endothelial growth factor—an alternative gene therapy. *Med Hypotheses*. 2005;64:505–511.

161. Hauer AD, van Puijvelde GH, Peterse N, et al. Vaccination against VEGFR2 attenuates initiation and progression of atherosclerosis. *Arterioscler Thromb Vasc Biol*. 2007.

162. Narayanan K and Warburton PE. DNA modification and functional delivery into human cells using *Escherichia coli* DH10B. *Nucleic Acids Res*. 2003;31:e51.

163. Cicin-Sain L, Brune W, Bubic I, Jonjic S, and Koszinowski UH. Vaccination of mice with bacteria carrying a cloned herpesvirus genome reconstituted in vivo. *J Virol*. 2003;77:8249–8255.

164. Soghomonyan SA, Doubrovin M, Pike J, et al. Positron emission tomography (PET) imaging of tumor-localized *Salmonella* expressing HSV1-TK. *Cancer Gene Ther*. 2005;12:101–108.

165. Kronenberg M and Rudensky A. Regulation of immunity by self-reactive T cells. *Nature*. 2005;435:598–604.

166. Hussain SF and Paterson Y. CD4+ CD25+ regulatory T cells that secrete TGFbeta and IL-10 are preferentially induced by a vaccine vector. *J Immunother (1997)*. 2004;27:339–346.

167. Hussain SF and Paterson Y. What is needed for effective antitumor immunotherapy? Lessons learned using *Listeria monocytogenes* as a live vector for HPV-associated tumors. *Cancer Immunol Immunother*. 2005;54:577–586.

168. Nitcheu-Tefit J, Dai MS, Critchley-Thorne RJ, et al. Listeriolysin O expressed in a bacterial vaccine suppresses CD4+ CD25 high regulatory T cell function in vivo. *J Immunol*. 2007;179:1532–1541.

169. Betts GJ, Clarke SL, Richards HE, Godkin AJ, and Gallimore AM. Regulating the immune response to tumours. *Adv Drug Deliv Rev*. 2006;58:948–961.

170. Lizee G, Radvanyi LG, Overwijk WW, and Hwu P. Improving antitumor immune responses by circumventing immuno regulatory cells and mechanisms. *Clin Cancer Res*. 2006;12:4794–4803.

171. Favre D and Viret JF. Biosafety evaluation of recombinant live oral bacterial vaccines in the context of European regulation. *Vaccine.* 2006;24:3856–3864.

172. Medina E and Guzman CA. Use of live bacterial vaccine vectors for antigen delivery: Potential and limitations. *Vaccine.* 2001;19:1573–1580.

173. Ledwith BJ, Manam S, Troilo PJ, et al. Plasmid DNA vaccines: Investigation of integration into host cellular DNA following intramuscular injection in mice. *Intervirology.* 2000;43:258–272.

174. Cohen AD and Shoenfeld Y. Vaccine-induced autoimmunity. *J Autoimmun.* 1996;9:699–703.

175. Sizemore DR, Branstrom AA, and Sadoff JC. Attenuated bacteria as a DNA delivery vehicle for DNA-mediated immunization. *Vaccine.* 1997;15:804–807.

176. Sizemore DR, Warner B, Lawrence J, Jones A, and Killeen KP. Live, attenuated *Salmonella typhimurium* vectoring Campylobacter antigens. *Vaccine.* 2006;24:3793–3803.

177. Devico AL, Fouts TR, Shata MT, Kamin-Lewis R, Lewis GK, and Hone DM. Development of an oral prime-boost strategy to elicit broadly neutralizing antibodies against HIV-1. *Vaccine.* 2002;20:1968–1974.

13 Redox-Responsive Polymer-Based Gene Delivery Systems

David Oupický, Ye-Zi You, and Devika Soundara Manickam

CONTENTS

13.1   INTRODUCTION

Polyelectrolyte complexes of nucleic acids with polycations (polyplexes) are investigated as promising delivery vectors for a variety of nucleic acid therapeutics. In particular, polyplexes capable of responding to environmental changes or stimuli by altering their properties and behavior seem to promise a significant improvement of the efficacy of the delivery process. Variety of external and endogenous stimuli can be utilized to improve overall efficiency of polyplexes. While the external stimuli are by nature physical (ultrasound, heat, magnetic field, light), the endogenous ones offer a wider variety ranging from simple chemical stimuli to complex biochemical stimuli. Irrespective of its nature, a given stimulus can improve polyplex efficiency by one of three distinct mechanisms. First, a stimulus can be used to facilitate spatial and temporal control of the release of nucleic acids from the polyplexes. Examples of successfully used stimuli to release nucleic acids from polyplexes include pH gradient and redox potential gradient [1–3]. Second, a stimulus can be used to alter properties of polyplexes and thus allow overcoming selected barriers. This approach is relatively less investigated than the first one and can be based on both external and endogenous stimuli (heat, ultrasound, pH gradient, redox potential gradient). An example of such an approach is to use hyperthermia to improve transfection activity [4–7]. Third, a stimulus can be used to favorably alter physiological properties of target tissues to increase the efficiency of polyplexes. This approach creates potential risks as well as benefits and is limited to external stimuli such as acoustic cavitation [8–11]. Here, we discuss the properties and current state of the art of polyplexes responsive to redox potential gradients.

13.2   REDOX-RESPONSIVE POLYPLEXES

One of the several stimuli, which has been utilized for improving the efficiency of nucleic acid delivery, is the redox potential gradient existing between extracellular and intracellular environments. The existence of a high redox potential gradient between oxidizing extracellular space and the reducing environment of subcellular organelles has been exploited by incorporating disulfide bonds into the structure of the delivery vectors to provide them with the capability to release the therapeutic nucleic acids selectively in the subcellular reducing space. The original interest in gene delivery systems controlled by redox potential gradients was guided by the need to transiently enhance stability of the vectors during the delivery. Although both polyplexes and lipid-based delivery systems [12–17] have been developed that are controlled by redox potential gradients, this chapter focuses exclusively on the polyplex vectors.

13.2.1   BIOLOGICAL RATIONALE

A redox potential gradient exists between extracellular environment and various subcellular organelles in normal as well as pathological states. Disulfide bonds present in the structure of polyplexes are readily reduced in the reducing intracellular environment, while they are generally preserved in the oxidizing extracellular space. The intracellular reduction of disulfide bonds in polyplexes is most likely mediated by thiol/disulfide exchange reactions with small redox molecules like glutathione (GSH) and thioredoxin; either alone or with the help of redox enzymes. GSH (L-γ-glutamyl-L-cysteinylglycine) is the most abundant intracellular thiol present in millimolar concentrations inside the cell but only in micromolar concentrations in the blood plasma [18]. GSH has multiple direct and indirect functions in many critical cellular processes like synthesis of proteins and DNA, amino acid transport, enzyme activity, metabolism, and protection of cells [19]. GSH also serves as a reductant by functioning to destroy free radicals, hydrogen peroxide, and other peroxides. It also functions as a storage form of cysteine. The intracellular GSH concentration is an additive function of both its oxidized (GSSG) and reduced form (GSH). The GSH redox ratio ([GSH]:[GSSG]) is maintained and determined by the activity of GSH reductase, NADPH concentrations, and transhydrogenase activity. The redox state of the GSH/GSSG couple is often used as an indicator of the overall redox environment of the cell. In contrast to other redox systems in the cell, it is important to realize that not only the [GSH]:[GSSG] ratio, but also absolute concentrations of GSH and GSSG are important for the estimation of the redox state of the cell [20]. This is because the Nernst equation for the reduction potential of the GSH/GSSG couple depends on $[GSH]^2/[GSSG]$. The GSH concentration and redox ratio ([GSH]:[GSSG]) are different in various subcellular compartments, supporting different roles of this redox molecule in the different compartments [21]. Because of the different GSH levels in various subcellular organelles, the local GSH concentration and [GSH]:[GSSG] ratio are important for predicting the location and rate of the disulfide reduction.

The majority of GSH is usually found in the cytosol (1–11 mM), which is also the principal site of GSH biosynthesis [22–24]. The most reducing environment in the cell is usually found within the nucleus, where it is required for DNA synthesis and repair and to maintain a number of transcription factors in reduced state [25–27]. The nuclear GSH levels are typically greater than those found in the cytosol and can reach up to 20 mM [21,27,28]. Another major pool of GSH in the cell is found in mitochondria (~5 mM) [29,30]. Both mitochondrial and nuclear GSH pools are at least partially independent of the cytosolic pool. In contrast to the reducing environment found in the nucleus and mitochondria, the endoplasmic reticulum (ER) is more oxidizing than the cytosol [31,32]. The ratio [GSH]:[GSSG] typically ranges from 1:1 to 3:1, which is relatively low compared with the overall ratio in the cell (usually >30:1). Whereas the above concentrations can serve as a general guideline when assessing the influence of redox environment on the polyplexes, one should keep in mind that the redox state of the cell is not static. It depends on a variety of factors including the stage of the cell cycle and biological status of the cell in general [20].

Despite numerous studies, the redox state of endosomes and lysosomes remains a contested issue. Even though some evidence suggests the possibility of endosomal and lysosomal reduction of disulfide bonds [33–37], recent report provides evidence that both the endosomal and lysosomal environments are oxidizing (similar to ER) and disulfide reduction proceeds rather inefficiently [38]. As a way of explaining the previous observations of disulfide reduction in endosomes, it was suggested that such reductions in endocytosed macromolecules could have proceeded in Golgi [39]. As acknowledged by the authors of the study on the oxidizing environment in endosomes and lysosomes, however, one cannot exclude the possibility that a small subset of lysosomes could be reducing or that only some disulfides within certain proteins could be susceptible to endo/lysosomal cleavage. Because redox polyplexes typically contain a large number of disulfides, it is unlikely that the limited reducing potential of endo/lysosomes plays an important role in their subcellular trafficking. The subcellular distribution of reducing capability and current understanding of subcellular trafficking of polyplexes therefore suggests that intracellular reduction of disulfide bonds in redox-sensitive polyplexes proceeds preferentially in the cytoplasm and nucleus (Figure 13.1).

Whereas the reducing nature of the intracellular environment is a well-established fact, the presence of reactive oxygen species and absence of a redox buffer means that the extracellular space is predominantly oxidizing ([GSH] ~1.5 μM; [GSH]:[GSSG] ~7:1) [18]. Despite the oxidizing nature of the extracellular environment, however, the presence of redox-active thiols in numerous proteins on the cellular plasma membrane suggests that at least the microenvironment of the cell surface can support disulfide reductions [39–41]. The redox activity of the plasma membrane is closely correlated with the levels of redox enzymes at the plasma membrane [42–44]. The maintenance of the thiol groups is mediated by the transfer or shuffling of hydrogens and electrons between the cysteine thiols of these surface proteins [45]. The total levels of redox-active thiols on the surface of cells range from ~4 to ~30 nmol per 10^6 cells [45,46]. Whereas the intracellular reducing environment has been widely utilized in gene delivery, the reducing microenvironment of the plasma membrane received considerably less attention. To our knowledge, there has been only one report describing an attempt to utilize the cell membrane thiols to improve gene delivery by increasing cellular uptake of DNA [47] and one report proposing to utilize the cell surface reduction as a trigger to disassemble thin DNA films [48]. Whether the reducing potential of the plasma membrane is sufficient enough to have any significant, negative or positive, effect on the activity of gene delivery systems, however, remains to be investigated [39].

13.2.2 Preparation Methods of Redox-Responsive Polyplexes

Vast majority of the reported redox-responsive polyplexes contains reducible disulfide bonds in the structure of the polycations. There are only a few examples of polyplexes in which

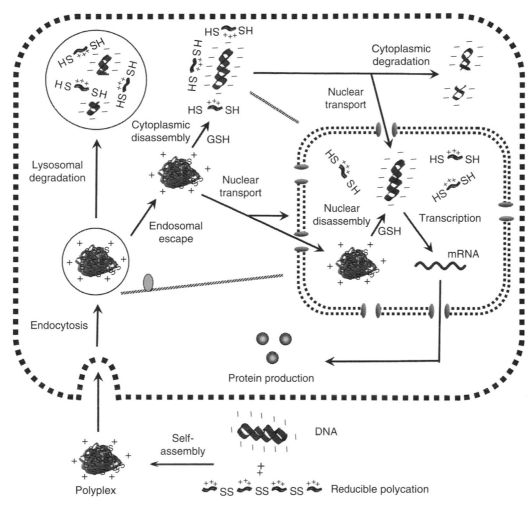

FIGURE 13.1　Schematic representation of subcellular trafficking of redox-responsive DNA polyplexes. (Adapted from Oupicky, D., et al., *Expert Opin. Drug Deliv.*, 2, 653, 2005.)

the disulfides are associated with the nucleic acid [49]. The introduction of disulfide bonds into polyplexes is relatively simple. Three different methods are primarily employed in achieving this task (Figure 13.2) and are reviewed in this section: (1) polycations containing disulfide bonds are synthesized first and used to prepare the polyplexes; (2) oligocations containing two or more thiols are used to prepare the polyplexes, which is followed by DNA-template polymerization within the polyplexes leading to high-molecular weight reducible polymers; (3) disulfide-containing cross-linkers are used to cross-link pre-formed polyplexes. Although each of the three main methods has its own set of advantages and disadvantages, the recent trends seem to favor the first method in which the disulfide-containing polycations are synthesized first and then used to formulate the polyplexes.

13.2.2.1　Synthesis of Disulfide-Containing Polycations

There has been an increased interest in recent years in the synthetic methods of preparing disulfide-containing polycations. Synthesizing polycations containing the disulfide bonds already before polyplex formation (as opposed to creating the

disulfides after the polyplexes are formed) allows not only better characterization of the polymers, but also a more diverse and controlled placement of the disulfide bonds (Figure 13.2A). Historically, probably the first example of a polycation with disulfides in the backbone intended as gene delivery carrier has been described by Balakirev et al. [50]. The authors prepared a cationic derivative of lipoic acid and its polymerization via opening the 1,2-dithiolane ring provided the desired reducible polycation.

Among the most readily available methods for synthesis of reducible polycations is to react oligo- and polyamines with commercially available disulfide-containing cross-linkers. Such approach was applied to the synthesis of reducible polycations based on low-molecular weight poly(ethylenimine) (PEI) and a variety of oligoamines such as spermine or pentaethylene hexamine [51–54]. Because of the multivalent nature of the polycations used and random nature of the cross-linking, the synthesized reducible polycations are usually branched or hyperbranched polymers. While these features to some extent reduce the definition of the product, the simplicity of the reactions offers the possibility to synthesize small polycation libraries and fast screening of numerous polycations [52].

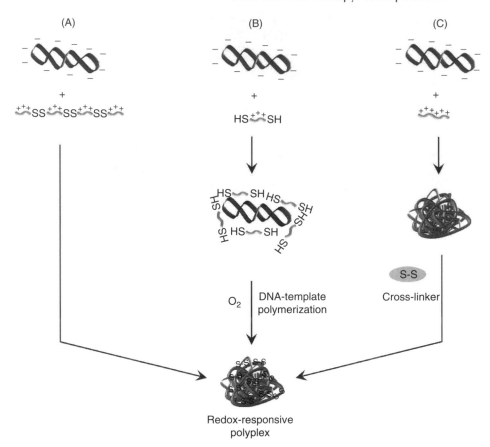

FIGURE 13.2 Methods of preparation of disulfide-containing polyplexes. (A) Polyplexes are prepared using polycations containing disulfide bonds, (B) DNA-template polymerization: Polyplexes are prepared using oligocations containing two or more thiols followed by DNA-template polymerization within the polyplexes, and (C) polyplexes are prepared and then cross-linked using disulfide-containing cross-linkers.

Another versatile synthetic approach to reducible polycations relies on mild oxidation of thiol groups introduced into the structure of oligo- and polycations. Oxidation of thiol-containing polycations offers tremendous versatility as virtually any polycation can be thiolated. In particular, peptides are especially well suited for this approach as thiol groups can be introduced with unprecedented control. Cationic peptides containing cysteine terminal groups have been polymerized into linear polypeptides via oxidation of the thiol groups with a mild oxidizing agent such as dimethylsulfoxide [55–58]. Reducible cationic polypeptides were prepared not only based on a single peptide, but also on a combination of several peptides with different amino acid sequences, thus offering an easy way of preparing polymers containing multiple biological functions [59–61].

While easy availability and potential inherent biological activity made peptide-based polycations among the most attractive for preparation of redox-sensitive polyplexes, recent reports also indicate increased interest in new types of synthetic reducible polycations.

Synthesis of well-defined analogues of reducible linear PEI by oxidation of α,ω-bisthiol-oligoethylenimines has been described by Park and colleagues [62]. The synthesized

reducible PEI shows reduced cytotoxicity and comparable transfection activity when compared with control nonreducible PEI. The synthesized reducible PEI represent better defined polymers in comparison with earlier attempts to prepare reducible PEI by random cross-linking of PEI oligomers [51,52].

In a search for efficient polymer carriers of nucleic acid therapeutics, methacrylate- and acrylate-based polycations have been widely investigated due to convenient synthesis by free-radical polymerization and the possibility to optimize their properties by copolymerization of a number of ionic and nonionic comonomers [63]. Poly(2-dimethylaminoethyl methacrylate) (PDMAEMA) emerged as one of the most promising polycations in this class [63–68]. Substantial cytotoxicity, however, impeded PDMAEMA prospects in gene delivery and prompted the development of biodegradable DMAEMA alternatives [66]. While synthesis of reducible polypeptides by oxidation of terminal thiol groups has been reported by numerous investigators, the traditional way of preparing methacrylate polymers by free-radical polymerization leads to polymers with high polydispersity and with poorly defined end-group chemistries; both of which severely limit the possibility of synthesizing the necessary

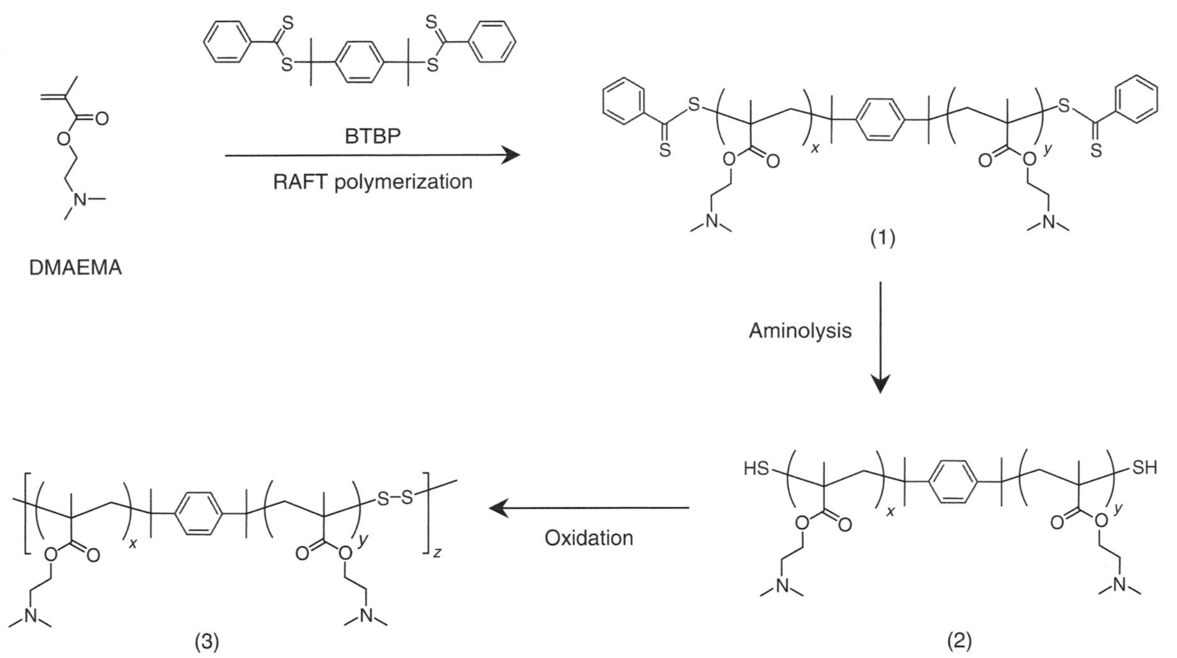

FIGURE 13.3 Synthetic approach to reducible polycations via RAFT polymerization demonstrated on the example of DMAEMA. (From You, Y.Z. et al., *Biomacromolecules*, 8, 2038, 2007; You, Y., et al., *J. Control. Release*, 122, 217, 2007.)

thiol-terminated polymers. Two recent studies report, for the first time, a synthesis of reducible PDMAEMA copolymers using reversible addition-fragmentation chain transfer (RAFT) polymerization and their use in gene delivery [69,70]. RAFT polymerization is an exceptionally versatile tool for the synthesis of α,ω-functionalized polymers of narrow molar mass distributions and it has opened new possibilities for the synthesis of previously unattainable (meth)acrylate copolymers [71–74]. RAFT polymerization is arguably more versatile with respect to monomer choice than other living free-radical polymerization techniques as it can be applied to virtually any type of monomers without the necessity of protecting functional groups. Reducible PDMAEMA was synthesized as shown in Figure 13.3. Difunctional RAFT agent (BTBP) was used in an AIBN-initiated radical polymerization of DMAEMA to obtain α,ω-dithioester-functionalized oligomers (1). The terminal dithioester groups of (1) were converted into thiol groups by aminolysis with hexylamine. Final reducible PDMAEMA (3) was synthesized by oxidation of the terminal thiol groups of (2) (Figure 13.3). Reducibility of the synthesized PDMAEMA was confirmed by treatment with dithiothreitol (DTT), which resulted in the decrease of molar mass to values similar to the original dithiol-DMAEMA oligomer [69,70].

Reducible poly(amidoamine)s (PAAs) emerged as yet another recent example of the increased interest in reducible polycations. PAAs are a family of water-soluble and biodegradable polymers obtained by the polyaddition of primary or secondary aliphatic amines to bisacrylamides (Figure 13.4). The first example of PAAs containing disulfide linkages along their backbones synthesized by the stepwise polyaddition of 2-methylpiperazine to *N,N'*-bis(acryloyl)cystamine

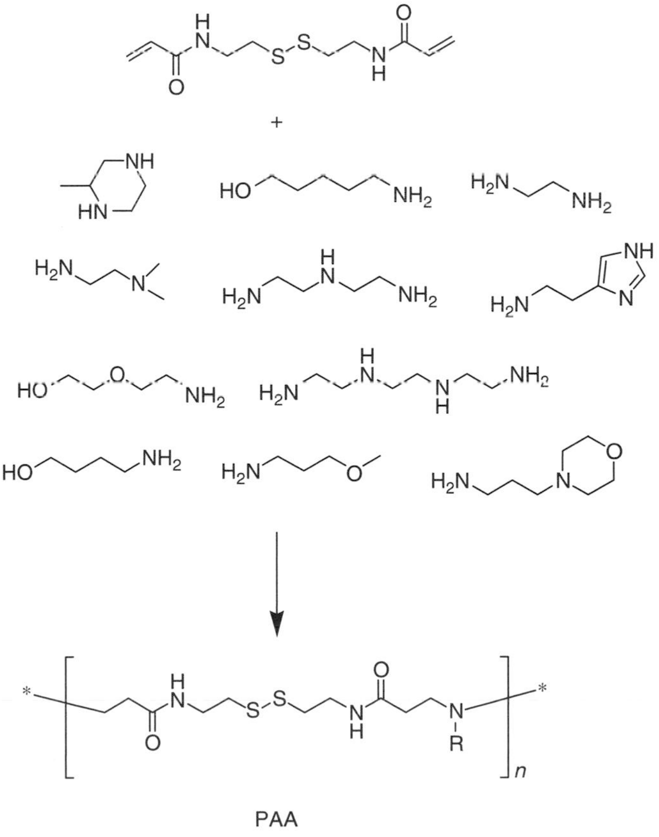

FIGURE 13.4 Examples of synthesized reducible PAAs. (From Emilitri, E., Ranucci, E., and Ferruti, P., *J. Polym. Sci. Pol. Chem.*, 43, 1404, 2005; Lin, C. et al., *Bioconjug. Chem.*, 18, 138, 2007; Christensen, L.V. et al., *Bioconjug. Chem.*, 17, 1233, 2006.)

and *N,N'*-bis(acryloyl)-(L)-cystine was reported by Ferruti et al. [75]. Synthesis of a small library of reducible PAAs suitable for use in gene delivery was reported in several recent publications by Kim and Feijen by Michael addition reactions of different oligoamines with the cystamine bisacrylamide linker [76–78]. All the reducible PAAs were shown to be reduced to low-molecular weight products after disulfide reduction.

Alternative strategy aimed at reducing the affinity of polycations to nucleic acids upon cellular internalization uses polymers containing disulfide bonds in the side chain. For example, quantitative substitution of amino groups of PLL with 3-(2-aminoethyldithio)propionyl residues resulted in the reducible disulfide bonds located in the side chains of the polycation. Release of DNA from such polyplexes was observed upon treatment with DTT and GSH [79]. More recently, biodegradable poly(β-amino ester) with disulfide side chain was synthesized and used as an efficient gene delivery carrier [80]. Finally, an interesting example of redox-responsive polyplexes in which the disulfide reduction leads to a disassembly of necklace-like rotaxane structures was reported by Ooya et al. [81].

13.2.2.2 DNA Template-Assisted Polymerization

The second method of introducing disulfide bonds into the structure of polyplexes relies on the use of cationic molecules containing at least two thiol functionalities and their oxidation in situ (Figure 13.2B). When polyplexes are formed using this method, the local concentration of the thiol groups increases substantially, which leads to an easy aerial oxidation without the need for any additional oxidizing agent. This approach, also called DNA template-assisted polymerization, has been widely explored using short cationic peptides containing multiple cysteine residues by Rice and colleagues. The original studies were performed with peptides consisting of several lysines and different number (1–5) of cysteines [82]. Using peptides with three and more cysteines resulted in the formation of polyplexes stabilized by disulfide cross-links. The general applicability of this approach was confirmed for peptides with various amino acid sequences as well as peptides containing additional nonpeptide functionalities like poly(ethylene glycol) (PEG) and carbohydrates [58,83–87].

Direct comparison was conducted of the properties of DNA polyplexes prepared either by using the DNA-template-assisted polymerization or by using pre-formed reducible polypeptide [58]. A cationic peptide with a sequence derived from the HIV-1 Tat was used in those studies. Even though some differences were reported in the physical properties of the two types of polyplexes, no major effect of the preparation method was observed on the transfection properties of the polyplexes.

Similar approach has been applied by Kataoka et al. to the formation of disulfide-stabilized polymer micelles/polyplexes for delivery of various nucleic acids [88–91]. In this alternative approach, thiol groups are introduced into the cationic blocks using standard chemical methods such as use of Traut's

reagent and *N*-succinimidyl 3-(2-pyridyldithio)-propionate (SPDP), and the cross-linking disulfide bonds are then formed in situ by aerial oxidation as above.

13.2.2.3 Polyplex Cross-Linking Using Low Molecular Weight Reducible Cross-Linkers

Probably the simplest way of providing polyplexes with stabilizing disulfide bonds is to use commercially available disulfide-containing cross-linking agents (Figure 13.2C) [53,92–94]. For example, complexes of DNA with poly(L-lysine) (PLL) were efficiently and reversibly stabilized by cross-linking with dimethyl-3,3′-dithiobispropionimidate (DTBP) via unpaired amino groups of PLL [92,93]. These low-molecular weight cross-linking agents can also be used in the DNA-template copolymerizations with DNA-condensing cationic molecules to form linear reversible polycations [95]. This approach, despite its simplicity, has not found a widespread use because of the need for excess cross-linkers due to competing hydrolysis, poor reproducibility, and difficulties to analyze the extent of the reaction in the obtained polyplexes.

13.2.3 BIOLOGICAL ACTIVITY AND MECHANISM OF ACTION

The studies of the redox-responsive polyplexes focused mostly on their *in vitro* properties with only limited amount of data available on the *in vivo* behavior and properties. Here, we will discuss the properties of redox-sensitive polyplexes that we believe represent the main advantages for their use and the rationale for further development.

13.2.3.1 Physicochemical Properties of Redox-Responsive Polyplexes

Physicochemical properties are known to affect biological properties of polyplexes. Changes in the physicochemical properties caused by the reduction of disulfides in the polyplexes are therefore likely to be reflected in changes of biological activity. The most obvious consequence of the reduction of disulfides in polyplexes is a decrease of the affinity between the nucleic acids (e.g., DNA) and the polycations. This in turn results in a faster and easier polyplex disassembly and DNA release. While polyplexes based on linear reducible polycations show only quantitative changes in the disassembly with disulfide reduction [56,60,77], cross-linked polyplexes typically exhibit full stabilization against disassembly (i.e., DNA release), which is then reversed after reduction [53,93]. Introduction of disulfide bonds into polycations usually increases their hydrophobicity, although no evidence suggests that this change has any influence on activity of the polyplexes [48,75,96]. For the reduction-induced changes in properties to have any effect on biological activity of polyplexes, they must occur on the same timescale as the extracellular and subcellular steps of delivery. Available evidence clearly points out that the disulfide reduction in the polyplexes proceeds with high rates and is typically completed within 1 h (usually faster) at

GSH concentrations found in the cytoplasm [50,62,80]. Conversely, the reduction rates at GSH concentrations typically found in blood plasma are almost negligible. This means that, unlike many polyplexes based on hydrolytically degradable polycations, the redox-responsive polyplexes show a significant spatial selectivity of their responses, and that the rates at which the responses to reducing environment occur can realistically influence the delivery process.

13.2.3.2 *In Vitro* Properties of Redox Polyplexes

The easy intracellular reversibility of disulfide bonds in the redox-sensitive polyplexes proved to be an advantageous feature for efficiently delivering a variety of nucleic acids, including plasmid DNA, mRNA, antisense oligonucleotides, and siRNA. The reduced affinity between the nucleic acids and polycations upon the intracellular reduction of disulfide bonds is believed to enhance availability of the nucleic acids for their biological function. Due to the subcellular compartmentalization of the GSH, it may be even possible to improve the selectivity of cellular delivery by controlling the location of the redox-induced disassembly. Indeed, based on the current knowledge, we predict that most likely RNA-based therapeutics will benefit from a potential to improve the selectivity of delivery based on GSH levels and subcellular distribution.

The intracellular reduction of disulfides within the polyplexes has been clearly demonstrated in numerous studies. The reduction is believed to increase the rates of polyplex disassembly and thus promote better availability of the delivered nucleic acids for transcription/translation machinery (Figure 13.1). It is reasonable to hypothesize that changes in intracellular GSH concentrations will manifest themselves in altered activity of the redox polyplexes. The anticipated relationship between the extent and rates of intracellular degradation and transgene activity then offers a real opportunity to modulate polyplex activity in disease states associated with altered cellular redox state. Available evidence suggests that indeed artificially changing cellular GSH levels leads to changes in polyplex biological activity, albeit the differences are often small. Most of the published data were obtained with DNA polyplexes. In general, increasing the cellular GSH levels by incubating the cells with a membrane-permeable derivative of GSH leads to an increase in transfection activity of redox DNA polyplexes [50,56,77,94]; supporting the notion that increasing the GSH concentration leads to increased disulfide reduction and improved DNA availability. A recent report, however, shows a significant decrease of transfection activity after artificial increase of GSH levels in prostate carcinoma PC-3 cells [57]. These results suggest that an optimum intracellular GSH concentration might exist for each cell line that will result in the highest transgene expression; and that a straightforward relationship between intracellular GSH concentrations and polyplex activity could be an oversimplified view. If increasing the GSH concentration leads to enhanced transfection activity, then, depleting intracellular GSH by inhibiting its de novo synthesis would be expected to have an opposite effect. Most of the available reports confirm this hypothesis by demonstrating a small decrease in DNA transfection of redox polyplexes after incubating cells with buthionine sulfoximine (BSO), which inhibits γ-glutamylcysteine synthetase necessary for cytosolic GSH biosynthesis [50,56,57,77,94]. Similar effect of GSH inhibition on the reduction of activity was confirmed also for siRNA redox polyplexes [97].

As already mentioned, delivery of plasmid DNA has been by far the most investigated application of the redox-sensitive polyplexes. Number of studies in a variety of cell lines showed transfection activities comparable or better than those observed for PEI and commercially available lipid formulations [51,55–57,60,76,77,79,82,83,90]. The benefits of the redox-sensitive polyplexes for efficient delivery have been demonstrated also for mRNA. Unlike DNA, only cationic lipids were successfully used to efficiently deliver mRNA. High intracellular stability of polyplexes based on high-molecular weight polycations was believed to be responsible for the extremely low mRNA expression [98]. In comparison, polyplexes formed using smaller polycations could mediate relatively high levels of gene expression. Recent studies document that using redox-sensitive polyplexes results in a high mRNA activity, even when high-molecular weight polycations are used (Figure 13.5) [56,57]. These results suggest that cytosolic reduction of the disulfide bonds in the polyplexes results in an increased bioavailability of mRNA.

Recently published studies confirm the applicability of redox-sensitive polyplexes also for efficient delivery of small nucleic acids such as siRNA and antisense oligonucleotides [49,57,97,99]. Similar to mRNA, siRNA acts in the cytoplasm and thus its delivery is significantly simplified by not needing to traverse the nuclear membrane to initiate its action. One can hypothesize that the reducing environment leads to environmentally triggered release of siRNA in the cytoplasm. This was clearly demonstrated by an efficient silencing of human vascular endothelial growth factor (VEGF) in PC-3 cells [97]. The redox polyplexes exhibited much higher RNAi activity against VEGF expression than control PEI polyplexes ($\sim$50% vs. $\sim$20% silencing). Further evidence of the advantageous use of redox polyplexes for siRNA delivery was demonstrated by the ability of high-molecular weight, reducible, histidine-rich polycations to deliver siRNA targeted against neutrophin receptor p75^{NTR} in primary cultures of adult rat dorsal root ganglion cells [57]. While the redox polyplexes showed a complete knock-down, PEI polyplexes caused only around 60% knock-down in expression, which was attributed to the restricted availability of the siRNA in the cytoplasm.

Efficiency of a different approach to deliver siRNA using redox-sensitive polyplexes was demonstrated by the ability of disulfide-linked conjugates of siRNA with membrane-penetrating peptides to cause efficient silencing in a wide variety of cell lines [99]. Cytoplasmic delivery of antisense oligodeoxynucleotides (ODN) using redox-sensitive polyplexes also was reported recently [49]. The authors synthesized redox-sensitive polyion complex micelles composed of polycations and a PEG–ODN conjugate bearing a disulfide

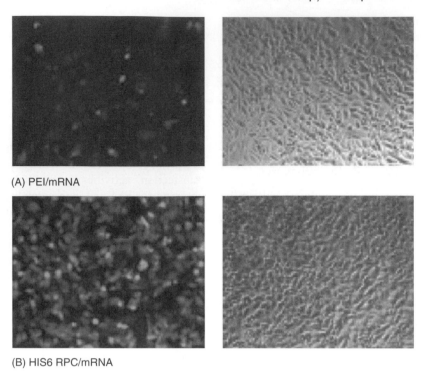

(A) PEI/mRNA

(B) HIS6 RPC/mRNA

FIGURE 13.5 (See color insert following blank page 206.) Histidine-rich RPCs mediate efficient delivery of mRNA. Fluorescent images of PC-3 cells expressing green fluorescent protein (GFP) following transfection with 0.5 μg cap-*GFP*-A64 mRNA condensed with (A) 25 kDa PEI or (B) reducible polycation (RPC). Phase-contrast images of transfected PC-3 cells are shown in the corresponding right-hand column. (From Read, M. L. et al., *Nucleic Acids Res.*, 33, e86, 2005. With permission.)

linkage between the PEG and ODN segments and demonstrated enhanced cytoplasmic delivery of ODN.

13.2.3.3 Decreased Toxicity of Reducible Polycations

The toxicity of polycations such as PEI is a well-known phenomenon [100–102]. Available data on cytotoxicity of the reducible polycations and their DNA polyplexes show that all the reducible polycations exhibit only minimal toxicity when compared with control nonreducible polycations [56–58, 76–78]. In fact, the observed cytotoxicity is often similar to that obtained for low-molecular weight degradation products of the reducible polycations. The observed low cytotoxicity of the disulfide-containing polycations is most likely a direct consequence of the reduced binding affinity for the intracellular membranes and vital proteins and nucleic acids after rapid intracellular reduction and subsequent decrease of molecular weight of these polycations. The available results also suggest that significant contribution to the cytotoxicity of polycations originates from their intracellular action and that direct destabilization of the plasma membrane may not be the major contributor to the polycation toxicity [60].

13.2.3.4 *In Vivo* Properties of Redox Polyplexes

The studies reported so far focused on two aspects of the *in vivo* behavior of redox-sensitive polyplexes, namely on the stability of their disulfide bonds in plasma circulation [61,93,94] and on the hepatic transfection efficiency [85,91].

Although blood plasma environment is generally oxidizing, reduction of disulfide bonds can potentially occur due to the presence of low concentrations of cysteine and GSH. Premature reduction of disulfides in the polyplexes prior to their arrival at the target cells could severely compromise their transfection activity as evidenced by experiments with polyplexes pre-incubated with reducing agents before *in vitro* transfection experiments [56]. The compatibility of redox-sensitive polyplexes with plasma environment was tested with two formulations. First, polyplexes based on reducible polypeptide prepared from Cys(Lys)$_{10}$Cys were stabilized by surface cross-linking with multivalent HPMA copolymers and their extended plasma circulation was confirmed in mice [61]. Second, PLL/DNA polyplexes were cross-linked with disulfide-containing cross-linker and sterically stabilized by PEG [93]. The polyplexes then exhibited significantly extended plasma circulations times, which increased with increasing cross-linking density. This was documented by an increase in the amount of the polyplexes remaining in the circulation 30 min after intravenous injection compared with non-cross-linked controls. Almost 40% of the injected dose (ID) was found in circulation for highly cross-linked polyplexes, 27% for intermediately cross-linked polyplexes, whereas only 5.8% of control non-cross-linked polyplexes remained in circulation 30 min postinjection [93]. These findings were recently confirmed for disulfide cross-linked PEI polyplexes, although the improvement in the plasma circulation was not as substantial due to the absence of PEG in the polyplex formulations [94].

Nevertheless, the area under the plasma concentration–time curve for 0–60 min increased to 2435% ID/mL min for the highest cross-linking degree tested, compared to 766% ID/mL min for the control polyplexes. These observations indicate that the reduction-mediated destabilization of the polyplexes in blood plasma is not significant, if any, within the timeframe observed and that redox-sensitive polyplexes are suitable for systemic application. These results also are supported by previous studies on long-circulating liposomes containing disulfide bonds [103].

In vivo transfection efficiency of redox-sensitive polyplexes was evaluated using hepatocyte-targeted formulations [85]. The polyplexes were prepared using three thiol-containing peptides with different functions: (i) hepatocyte targeting, (ii) blocking nonspecific RES recognition, and (iii) DNA condensation. The structure of the DNA-condensing peptide was varied to modulate the metabolic stability of DNA in the liver. Four different peptides based on L- or D-amino acids and containing either cysteine or penicillamine residues were studied. It was found that positively charged polyplexes exhibited higher metabolic stability in the liver than negatively charged polyplexes. Replacing L-amino acids with D-amino acids also resulted in significantly increased metabolic half-life of DNA in the liver, while using sterically stabilized disulfides by replacing cysteine with penicillamine had no influence on metabolic stability. This suggests that disulfide reduction proceeds readily in the hepatocytes and is not the rate-limiting step of hepatic metabolism of DNA. Finally, the redox-sensitive polyplexes were able to mediate prolonged expression of transgene (secreted alkaline phosphatase) *in vivo*, which peaked at day 7 postinjection and returned to baseline levels by day 12.

Recent report indicates suitability of the redox polyplexes for delivering VEGF gene to ischemic myocardium to achieve tissue regeneration [104]. VEGF gene delivery can reduce myocardial infarct size through cardiomyocyte regeneration, enhancement of cardiomyocyte viability, and neovascular proliferation. Using a rabbit infarct model, the author showed that the redox-responsive polyplexes produce higher VEGF concentrations than controls. The best formulation showed VEGF expression nearly fourfold higher than the luciferase control and twofold higher than a lipid-based positive control at half the transfection amount [104].

REFERENCES

1. Lynn, D.M. and Langer, R., Degradable poly(beta-amino esters): Synthesis, characterization, and self-assembly with plasmid DNA. *J. Am. Chem. Soc.*, 122, 10761, 2000.
2. Kwok, K.Y., Yang, Y.S., and Rice, K.G., Evolution of cross-linked non-viral gene delivery systems. *Curr. Opin. Mol. Ther.*, 3, 142, 2001.
3. Soundara Manickam, D. and Oupicky, D., Polyplex gene delivery modulated by redox potential gradients. *J. Drug Target*, 14, 519, 2006.
4. Zintchenko, A., Ogris, M., and Wagner, E., Temperature dependent gene expression induced by PNIPAM-based copolymers: Potential of hyperthermia in gene transfer. *Bioconjug. Chem.*, 17, 766, 2006.
5. Bisht, H.S. et al. Temperature-controlled properties of DNA complexes with poly(ethylenimine)-graft-poly(*N*-isopropylacrylamide). *Biomacromolecules*, 7, 1169, 2006.
6. Oupicky, D. et al., Temperature controlled behavior of self-assembly gene delivery vectors based on complexes of DNA with poly(L-lysine)-graft-poly(*N*-isopropylacrylamide). *Macromolecules*, 36, 6863, 2003.
7. Lavigne, M.D. et al., Enhanced gene expression through temperature profile-induced variations in molecular architecture of thermoresponsive polymer vectors. *J. Gene Med.*, 9, 44, 2007.
8. Pitt, W.G., Husseini, G.A., and Staples, B.J., Ultrasonic drug delivery—a general review. *Expert Opin. Drug Deliv.*, 1, 37, 2004.
9. Miller, D.L. and Song, J., Tumor growth reduction and DNA transfer by cavitation-enhanced high-intensity focused ultrasound in vivo. *Ultrasound Med. Biol.*, 29, 887, 2003.
10. Unger, E.C. et al., Gene delivery using ultrasound contrast agents. *Echocardiogr. J. Card.*, 18, 355, 2001.
11. Greenleaf, W.J. et al., Artificial cavitation nuclei significantly enhance acoustically induced cell transfection. *Ultrasound Med. Biol.*, 24, 587, 1998.
12. Saito, G., Swanson, J.A., and Lee, K.D., Drug delivery strategy utilizing conjugation via reversible disulfide linkages: Role and site of cellular reducing activities. *Adv. Drug Deliv. Rev.*, 55, 199, 2003.
13. Tang, F.X. and Hughes, J.A., Introduction of a disulfide bond into a cationic lipid enhances transgene expression of plasmid DNA. *Biochem. Biophys. Res. Commun.*, 242, 141, 1998.
14. Tang, F.X. and Hughes, J.A., Use of dithiodiglycolic acid as a tether for cationic lipids decreases the cytotoxicity and increases transgene expression of plasmid DNA in vitro. *Bioconjug. Chem.*, 10, 791, 1999.
15. Byk, G. et al., Reduction-sensitive lipopolyamines as a novel nonviral gene delivery system for modulated release of DNA with improved transgene expression. *J. Med. Chem.*, 43, 4377, 2000.
16. Chittimalla, C. et al., Monomolecular DNA nanoparticles for intravenous delivery of genes. *J. Am. Chem. Soc.*, 127, 11436, 2005.
17. Dauty, E., Behr, J.P., and Remy, J.S., Development of plasmid and oligonucleotide nanometric particles. *Gene Ther.*, 9, 743, 2002.
18. Jones, D.P. et al., Glutathione measurement in human plasma. Evaluation of sample collection, storage and derivatization conditions for analysis of dansyl derivatives by HPLC. *Clin. Chim. Acta*, 275, 175, 1998.
19. Meister, A. and Anderson, M.E., Glutathione. *Annu. Rev. Biochem.*, 52, 711, 1983.
20. Schafer, F.Q. and Buettner, G.R., Redox environment of the cell as viewed through the redox state of the glutathione disulfide/glutathione couple. *Free Radic. Biol. Med.*, 30, 1191, 2001.
21. Smith, C.V. et al., Compartmentation of glutathione: Implications for the study of toxicity and disease. *Toxicol. Appl. Pharmacol.*, 140, 1, 1996.
22. Gilbert, H.F., Molecular and cellular aspects of thiol-disulfide exchange. *Adv. Enzymol.*, 63, 69, 1990.
23. Kosower, N.S. and Kosower, E.M., The glutathione status of cells. *Int. Rev. Cytol.*, 54, 109, 1978.
24. Hwang, C., Lodish, H.F., and Sinskey, A.J., Measurement of glutathione redox state in cytosol and secretory pathway of cultured cells. *Methods Enzymol.*, 251, 212, 1995.
25. Arrigo, A.P., Gene expression and the thiol redox state. *Free Radic. Biol. Med.*, 27, 936, 1999.

26. Wu, X. et al., Physical and functional sensitivity of zinc finger transcription factors to redox change. *Mol. Cell. Biol.*, 16, 1035, 1996.

27. Bellomo, G. et al., Demonstration of nuclear compartmentalization of glutathione in hepatocytes. *Proc. Natl. Acad. Sci. U S A*, 89, 4412, 1992.

28. Soboll, S. et al., The content of glutathione and glutathione S-transferases and the glutathione peroxidase activity in rat liver nuclei determined by a non-aqueous technique of cell fractionation. *Biochem. J.*, 311 (Pt 3), 889, 1995.

29. Wahllander, A. et al., Hepatic mitochondrial and cytosolic glutathione content and the subcellular distribution of GSH-S-transferases. *FEBS Lett.*, 97, 138, 1979.

30. Lash, L.H., Putt, D.A., and Matherly, L.H., Overexpression of a mitochondrial glutathione transporter protects NRK-52E cells from chemically induced apoptosis. *J. Am. Soc. Nephrol.*, 13, 292A, 2002.

31. Braakman, I., Helenius, J., and Helenius, A., Manipulating disulfide bond formation and protein folding in the endoplasmic reticulum. *EMBO J.*, 11, 1717, 1992.

32. Hwang, C., Sinskey, A.J., and Lodish, H.F., Oxidized redox state of glutathione in the endoplasmic reticulum. *Science*, 257, 1496, 1992.

33. Shen, W.C., Ryser, H.J., and LaManna, L., Disulfide spacer between methotrexate and poly(D-lysine). A probe for exploring the reductive process in endocytosis. *J. Biol. Chem.*, 260, 10905, 1985.

34. Fivaz, M. et al., Differential sorting and fate of endocytosed GPI-anchored proteins. *EMBO J.*, 21, 3989, 2002.

35. Phan, U.T., Arunachalam, B., and Cresswell, P., Gamma-interferon-inducible lysosomal thiol reductase (GILT)—Maturation, activity and mechanism of action. *J. Biol. Chem.*, 275, 25907, 2000.

36. Saito, G., Amidon, G.L., and Lee, K.D., Enhanced cytosolic delivery of plasmid DNA by a sulfhydryl-activatable listeriolysin O/protamine conjugate utilizing cellular reducing potential. *Gene Ther.*, 10, 72, 2003.

37. Collins, D.S., Unanue, E.R., and Harding, C.V., Reduction of disulfide bonds within lysosomes is a key step in antigen processing. *J. Immunol.*, 147, 4054, 1991.

38. Austin, C.D. et al., Oxidizing potential of endosomes and lysosomes limits intracellular cleavage of disulfide-based antibody-drug conjugates. *Proc. Natl. Acad. Sci. U S A*, 102, 17987, 2005.

39. Feener, E.P., Shen, W.C., and Ryser, H.J.P., Cleavage of disulfide bonds in endocytosed macromolecules—a processing not associated with lysosomes or endosomes. *J. Biol. Chem.*, 265, 18780, 1990.

40. Ryser, H.J., Mandel, R., and Ghani, F., Cell surface sulfhydryls are required for the cytotoxicity of diphtheria toxin but not of ricin in Chinese hamster ovary cells. *J. Biol. Chem.*, 266, 18439, 1991.

41. Sahaf, B. et al., The extracellular microenvironment plays a key role in regulating the redox status of cell surface proteins in HIV-infected subjects. *Arch. Biochem. Biophys.*, 434, 26, 2005.

42. Donoghue, N. and Hogg, P.J., Characterization of redox-active proteins on cell surface. *Methods Enzymol.*, 348, 76, 2002.

43. Gilbert, H.F., Protein disulfide isomerase and assisted protein folding. *J. Biol. Chem.*, 272, 29399, 1997.

44. Donoghue, N. et al., Presence of closely spaced protein thiols on the surface of mammalian cells. *Protein Sci.*, 9, 2436, 2000.

45. Jiang, X.M. et al., Redox control of exofacial protein thiols/disulfides by protein disulfide isomerase. *J. Biol. Chem.*, 274, 2416, 1999.

46. Laragione, T. et al., Redox regulation of surface protein thiols: Identification of integrin alpha-4 as a molecular target by using redox proteomics. *Proc. Natl. Acad. Sci. U S A*, 100, 14737, 2003.

47. Kichler, A. et al., Efficient gene delivery with neutral complexes of lipospermine and thiol-reactive phospholipids. *Biochem. Biophys. Res. Commun.*, 209, 444, 1995.

48. Blacklock, J. et al., Disassembly of layer-by-layer films of plasmid DNA and reducible TAT polypeptide. *Biomaterials*, 28, 117, 2007.

49. Oishi, M. et al., Supramolecular assemblies for the cytoplasmic delivery of antisense oligodeoxynucleotide: Polyion complex (PIC) micelles based on poly(ethylene glycol)-SS-oligodeoxynucleotide conjugate. *Biomacromolecules*, 6, 2449, 2005.

50. Balakirev, M., Schoehn, G., and Chroboczek, J., Lipoic acid-derived amphiphiles for redox-controlled DNA delivery. *Chem. Biol.*, 7, 813, 2000.

51. Gosselin, M.A., Guo, W.J., and Lee, R.J., Efficient gene transfer using reversibly cross-linked low molecular weight polyethylenimine. *Bioconjug. Chem.*, 12, 989, 2001.

52. Kloeckner, J., Wagner, E., and Ogris, M., Degradable gene carriers based on oligomerized polyamines. *Eur. J. Pharm. Sci.*, 29, 414, 2006.

53. Neu, M. et al., Stabilized nanocarriers for plasmids based upon cross-linked poly(ethylene imine). *Biomacromolecules*, 7, 3428, 2006.

54. Wang, Y.X., Chen, P., and Shen, J.C., The development and characterization of a glutathione-sensitive cross-linked polyethylenimine gene vector. *Biomaterials*, 27, 5292, 2006.

55. Oupicky, D., Parker, A.L., and Seymour, L.W., Laterally stabilized complexes of DNA with linear reducible polycations: Strategy for triggered intracellular activation of DNA delivery vectors. *J. Am. Chem. Soc.*, 124, 8, 2002.

56. Read, M.L. et al., Vectors based on reducible polycations facilitate intracellular release of nucleic acids. *J. Gene. Med.*, 5, 232, 2003.

57. Read, M.L. et al., A versatile reducible polycation-based system for efficient delivery of a broad range of nucleic acids. *Nucleic Acids Res.*, 33, e86, 2005.

58. Soundara Manickam, D. et al., Influence of TAT-peptide polymerization on properties and transfection activity of TAT/DNA polyplexes. *J. Control. Release*, 102, 293, 2005.

59. Chen, C.P. et al., Synthetic PEGylated glycoproteins and their utility in gene delivery. *Bioconjug. Chem.*, 18, 371, 2007.

60. Soundara Manickam, D. and Oupicky, D., Multiblock reducible copolypeptides containing histidine-rich and nuclear localization sequences for gene delivery. *Bioconjug. Chem.*, 17, 1395, 2006.

61. Zhou, Q.H. et al., Ultrasound-enhanced transfection activity of HPMA-stabilized DNA polyplexes with prolonged plasma circulation. *J. Control. Release*, 106, 416, 2005.

62. Lee, Y. et al., Visualization of the degradation of a disulfide polymer, linear poly(ethylenimine sulfide), for gene delivery. *Bioconjug. Chem.*, 18, 13, 2007.

63. Reschel, T. et al., Physical properties and in vitro transfection efficiency of gene delivery vectors based on complexes of DNA with synthetic polycations. *J. Control. Release*, 81, 201, 2002.

64. Verbaan, F.J. et al., The fate of poly(2-dimethyl amino ethyl)methacrylate-based polyplexes after intravenous administration. *Int. J. Pharm.*, 214, 99, 2001.

65. De Smedt, S.C., Demeester, J., and Hennink, W.E., Cationic polymer based gene delivery systems. *Pharm. Res.*, 17, 113, 2000.

66. Funhoff, A.M. et al., Cationic polymethacrylates with covalently linked membrane destabilizing peptides as gene delivery vectors. *J. Control. Release*, 101, 233, 2005.

67. Van De Wetering, P. et al., 2-(dimethylamino)ethyl methacrylate based (co)polymers as gene transfer agents. *J. Control. Release*, 53, 145, 1998.

68. Jones, R.A., Poniris, M.H., and Wilson, M.R., pDMAEMA is internalised by endocytosis but does not physically disrupt endosomes. *J. Control. Release*, 96, 379, 2004.

69. You, Y.Z. et al., A versatile approach to reducible vinyl polymers via oxidation of telechelic polymers prepared by reversible addition fragmentation chain transfer polymerization. *Biomacromolecules*, 8, 2038, 2007.

70. You, Y. et al., Reducible poly(2-dimethylaminoethyl methacrylate): Synthesis, cytotoxicity, and gene delivery activity. *J. Control. Release*, 122, 217, 2007.

71. McCormick, C.L. and Lowe, A.B., Aqueous RAFT polymerization: Recent developments in synthesis of functional water-soluble (Co)polymers with controlled structures. *Acc. Chem. Res.*, 37, 312, 2004.

72. Patton, D.L. et al., A facile synthesis route to thiol-functionalized alpha,w-telechelic polymers via reversible addition fragmentation chain transfer polymerization. *Macromolecules*, 38, 8597, 2005.

73. Goh, Y.K. and Monteiro, M.J., Novel approach to tailoring molecular weight distribution and structure with a difunctional RAFT agent. *Macromolecules*, 39, 4966, 2006.

74. Qiu, X.P. and Winnik, F.M., Facile and efficient one-pot transformation of RAFT polymer end groups via a mild aminolysis/Michael addition sequence. *Macromol. Rapid Commun.*, 27, 1648, 2006.

75. Emilitri, E., Ranucci, E., and Ferruti, P., New poly(amidoamine)s containing disulfide linkages in their main chain. *J. Polym. Sci. Pol. Chem.*, 43, 1404, 2005.

76. Lin, C. et al., Novel bioreducible poly(amido amine)s for highly efficient gene delivery. *Bioconjug. Chem.*, 18, 138, 2007.

77. Christensen, L.V. et al., Reducible poly(amido ethylenimine)s designed for triggered intracellular gene delivery. *Bioconjug. Chem.*, 17, 1233, 2006.

78. Lin, C. et al., Linear poly(amido amine)s with secondary and tertiary amino groups and variable amounts of disulfide linkages: Synthesis and in vitro gene transfer properties. *J. Control. Release*, 116, 130, 2006.

79. Pichon, C. et al., Poly[Lys-(AEDTP)]: A cationic polymer that allows dissociation of pDNA/cationic polymer complexes in a reductive medium and enhances polyfection. *Bioconjug. Chem.*, 13, 76, 2002.

80. Zugates, G.T. et al., Synthesis of poly(beta-amino ester)s with thiol-reactive side chains for DNA delivery. *J. Am. Chem. Soc.*, 128, 12726, 2006.

81. Ooya, T. et al., Biocleavable polyrotaxane—Plasmid DNA polyplex for enhanced gene delivery. *J. Am. Chem. Soc.*, 128, 3852, 2006.

82. McKenzie, D.L., Kwok, K.Y., and Rice, K.G., A potent new class of reductively activated peptide gene delivery agents. *J. Biol. Chem.*, 275, 9970, 2000.

83. McKenzie, D.L., et al., Low molecular weight disulfide cross-linking peptides as nonviral gene delivery carriers. *Bioconjug. Chem.*, 11, 901, 2000.

84. Park, Y. et al., Synthesis of sulfhydryl cross-linking poly(ethylene glycol)-peptides and glycopeptides as carriers for gene delivery. *Bioconjug. Chem.*, 13, 232, 2002.

85. Kwok, K.Y. et al., In vivo gene transfer using sulfhydryl cross-linked PEG-peptide/glycopeptide DNA co-condensates. *J. Pharm. Sci.*, 92, 1174, 2003.

86. Trentin, D. et al., Peptide-matrix-mediated gene transfer of an oxygen-insensitive hypoxia-inducible factor-1alpha variant for local induction of angiogenesis. *Proc. Natl. Acad. Sci. U S A*, 103, 2506, 2006.

87. Trentin, D., Hubbell, J., and Hall, H., Non-viral gene delivery for local and controlled DNA release. *J. Control. Release*, 102, 263, 2005.

88. Kakizawa, Y., Harada, A., and Kataoka, K., Environment-sensitive stabilization of core-shell structured polyion complex micelle by reversible cross-linking of the core through disulfide bond. *J. Am. Chem. Soc.*, 121, 11247, 1999.

89. Kakizawa, Y., Harada, A., and Kataoka, K., Glutathione-sensitive stabilization of block copolymer Micelles composed of antisense DNA and thiolated poly(ethylene glycol)-block-poly(L-lysine): A potential carrier for systemic delivery of antisense DNA. *Biomacromolecules*, 2, 491, 2001.

90. Miyata, K. et al., Block catiomer polyplexes with regulated densities of charge and disulfide cross-linking directed to enhance gene expression. *J. Am. Chem. Soc.*, 126, 2355, 2004.

91. Miyata, K. et al., Freeze-dried formulations for in vivo gene delivery of PEGylated polyplex micelles with disulfide crosslinked cores to the liver. *J. Control. Release*, 109, 15, 2005.

92. Trubetskoy, V.S. et al., Caged DNA does not aggregate in high ionic strength solutions. *Bioconjug. Chem.*, 10, 624, 1999.

93. Oupicky, D., Carlisle, R.C., and Seymour, L.W., Triggered intracellular activation of disulfide crosslinked polyelectrolyte gene delivery complexes with extended systemic circulation in vivo. *Gene Ther.*, 8, 713, 2001.

94. Neu, M. et al., Crosslinked nanocarriers based upon poly(ethylene imine) for systemic plasmid delivery: In vitro characterization and in vivo studies in mice. *J. Control. Release*, 118, 370, 2007.

95. Trubetskoy, V.S. et al., Self-assembly of DNA-polymer complexes using template polymerization. *Nucleic Acids Res.*, 26, 4178, 1998.

96. Lee, Y., Mo, H., Koo, H., Park, J.-Y., Cho, M.Y., Jin, G.-W., and Park, J.-S, Visualization of the degradation of a disulfide polymer, linear poly(ethylenimine sulfide) for gene delivery. *Bioconjug. Chem.*, 18, 13, 2007.

97. Hoon Jeong, J. et al., Reducible poly(amido ethylenimine) directed to enhance RNA interference. *Biomaterials*, 28, 1912, 2007.

98. Bettinger, T. et al., Peptide-mediated RNA delivery: A novel approach for enhanced transfection of primary and post-mitotic cells. *Nucleic Acids Res.*, 29, 3882, 2001.

99. Muratovska, A. and Eccles, M.R., Conjugate for efficient delivery of short interfering RNA (siRNA) into mammalian cells. *FEBS Lett.*, 558, 63, 2004.

100. Godbey, W.T., Wu, K.K., and Mikos, A.G., Poly(ethylenimine)-mediated gene delivery affects endothelial cell function and viability. *Biomaterials*, 22, 471, 2001.

101. Symonds, P. et al., Low and high molecular weight poly(L-lysine)s/poly(L-lysine)-DNA complexes initiate mitochondrial-mediated apoptosis differently. *FEBS Lett.*, 579, 6191, 2005.

102. Moghimi, S.M. et al., A two-stage poly(ethylenimine)-mediated cytotoxicity: Implications for gene transfer/therapy. *Mol. Ther.*, 11, 990, 2005.

103. Zhang, J.X. et al., Pharmaco attributes of dioleoyl phosphatidyl ethanolamine/cholesteryl hemisuccinate liposomes containing different types of cleavable lipopolymers. *Pharmacol. Res.*, 49, 185, 2004.

104. Christensen, L.V. et al., Reducible poly(amido ethylenediamine) for hypoxia-inducible VEGF delivery. *J. Control. Release*, 118, 254, 2007.

14 Polymers and Dendrimers for Gene Delivery in Gene Therapy

Ijeoma F. Uchegbu, Christine Dufès, Pei Lee Kan, and Andreas G. Schätzlein

CONTENTS

14.1   INTRODUCTION

The completion of the Human Genome Project [1] is a step towards understanding the genetic basis of disease and will doubtless provide more information on the interplay between genes and ill health. Genes are nucleic acid materials which serve as repositories for the amino acid sequences of the cell's work horses, the proteins. Proteins in turn control cell physiology and cell biochemistry. Mutated genes can give rise to nonfunctional proteins or pathogenic proteins and ultimately disease. Such disease-causing genes may be used as therapeutic targets in gene therapy. Gene therapy is important not just because it is an alternative means of treating disease but also because it offers the hope of treatments for currently incurable diseases such as cystic fibrosis, sickle cell anemia, and cancer. Such diseases are either hereditary or acquired and can be either monogenetic in origin (due to the mutation of a single gene) or originate from the malfunctioning of more than one gene. Examples of monogenetic diseases are cystic fibrosis [2], sickle cell anemia [3], and severe combined immune deficiency disease [4]. The exact genetic basis of diseases such as cancer is more complex and typically is the result of multiple mutations. In gene therapy, the therapeutic exogenous gene encodes for the replacement copy of the missing or faulty gene (Figure 14.1). Alternatively, as in the treatment of certain cancers, the therapeutic gene may encode for an enzyme capable of specific activation of a pro-drug [5], or encode for a therapeutic protein [6].

Another aspect of gene therapy is the use of genes as vaccines. Genes used in the prevention of infectious diseases are genes encoding for specific antigens which will

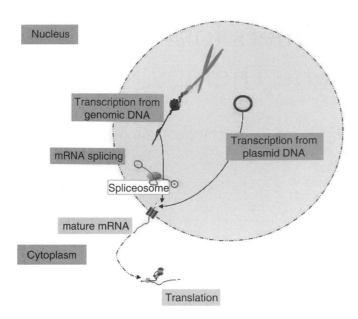

FIGURE 14.1 Replacement gene therapy. A therapeutic plasmid could give rise to a therapeutic protein which either replaces a nonfunctional protein from a mutated gene or has a therapeutic effect.

ultimately produce prophylactic antibodies [7]. Gene therapy is also possible using small interfering RNA sequences (siRNAs) to effect gene silencing. These materials inhibit translation and the disease-causing protein is no longer expressed (Figure 14.2).

Although it is possible to combat diseases with gene therapy, the current paucity of marketed gene therapeutics hints at the fact that reducing this concept to practice is rather problematic. One key difficulty lies in the delivery of the therapeutic gene or siRNA [8–10]. The gene therapeutic, if administered for the treatment of a disease such as cancer or cystic fibrosis must evade degradation by extracellular nucleases, resist deactivation by other extracellular components, and traverse both the plasma and nuclear membranes intact in order to access the transcription locus and produce the therapeutic protein. Each of these stages is fraught with a huge potential for failure and to achieve effective gene therapy these transport barriers must be overcome with the aid of delivery systems. It is in the delivery of genes that polymers and dendrimers are applied. Delivery systems fall into two main classes: viruses and synthetic compounds (nonviral systems). Nonviral gene delivery

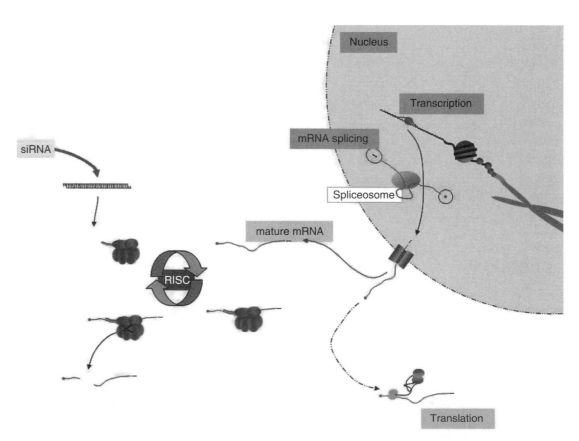

FIGURE 14.2 Exogenous siRNAs are incorporated into the RNA-induced silencing complex (RISC). The RISC is a large protein-siRNA complex which targets RNA transcripts for degradation. Binding of the siRNA is in the double-stranded form but unwinding of the strand yields a single-stranded antisense template to guide recognition of the target mRNA. The RISC precursor complex (~250 kDa) is transformed in the presence of ATP into the active 100 kDa complex that can endonucleolytically cleave the substrate mRNA in regions homologous to the siRNA template. The cleaved mRNA fragments are released and the RISC complex can bind and process the next mRNA molecule.

systems may be prepared from polymers, lipids, or dendrimers [8]. This chapter is concerned with the delivery of genes and other nucleotides using polymers and dendrimers (Figures 14.3 and 14.4).

Polymer- and dendrimer-based gene therapeutics comprise of anionic DNA electrostatically bound via its phosphate groups to amine polymers or dendrimers (Figures 14.5 and 14.6). These amine polymers and dendrimers are protonated at physiological pH and thus possess a positive charge and the electrostatic complexes are termed polyplexes in the case of polymers and dendriplexes in the case of dendrimers (Figures 14.5 and 14.6). Polyplexes or dendriplexes present as particles of 40–1000 nm in size [11,12] and colloidal particles are obtained largely by the manipulation of dendrimer/polymer DNA charge ratio [11,12].

Gene therapy is not a brand new concept as the first gene therapy clinical trial took place in May 1989 [13]. The trial involved cancer patients and retroviral gene transfer into human somatic cells in an *ex vivo* approach. Cells were harvested, infected with the gene of interest, and then reconstituted back into the patients. This clinical trial signaled the start of a long journey and 2 years later the first nonviral gene therapy clinical trial was conducted [14]. Generally viral vectors are thought to have superior transfection efficiencies over that seen with nonviral vectors [15,16] and as such the former have been most frequently investigated in clinical trials [17]. However, serious safety issues have been associated with the use of viral vectors over the last 5 years such as insertional mutagenesis with retroviral vectors [18] and a fatal response to the use of adenoviral gene delivery [19]. These events make the hunt for safe, effective, and preferably nonviral gene transfer systems even more important than was previously thought. This notwithstanding, the commercial launch of the viral gene medicine, Gendicine, in China is highly encouraging [20]. This gene therapeutic comprises an adenovirus carrying the p53 apoptosis causing

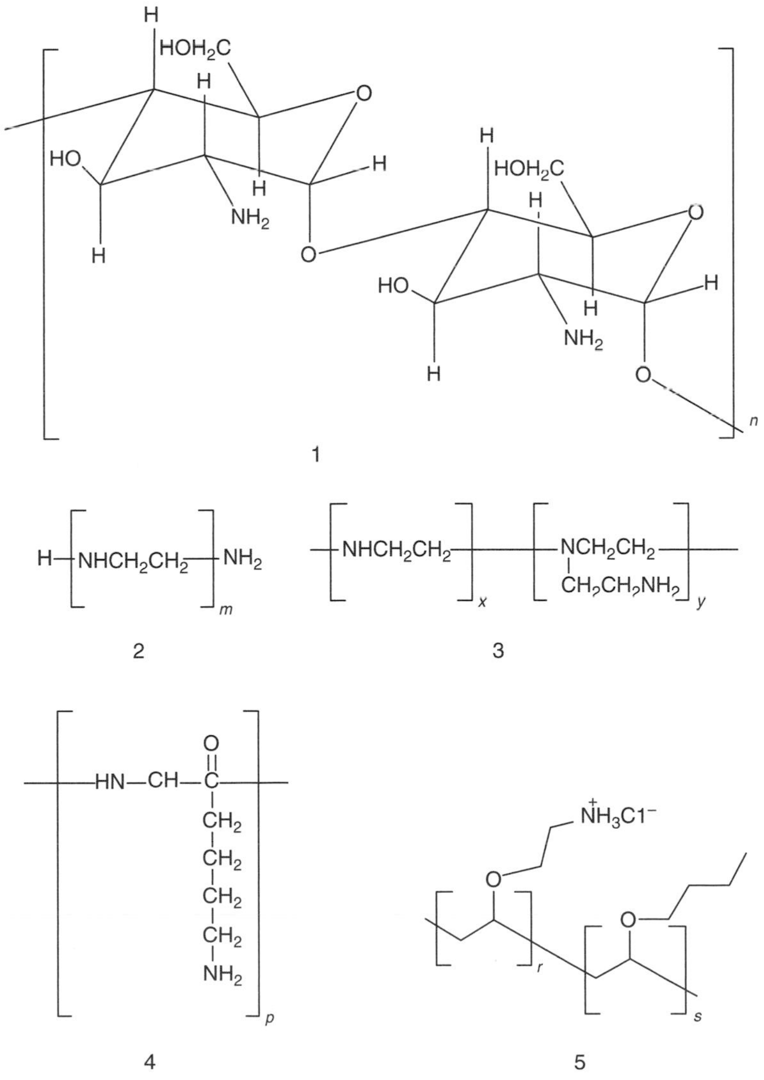

FIGURE 14.3 Examples of some cationic gene (polymers 1–4) and siRNA (polymer 5 only) delivery polymers. 1, chitosan; 2, linear poly(ethylenimine) (PEI); 3, branched PEI; 4, poly(L-lysine); 5, polyvinyl ether. Various derivatives of polymers 1–5 have also been used to deliver other nucleotides for gene therapy.

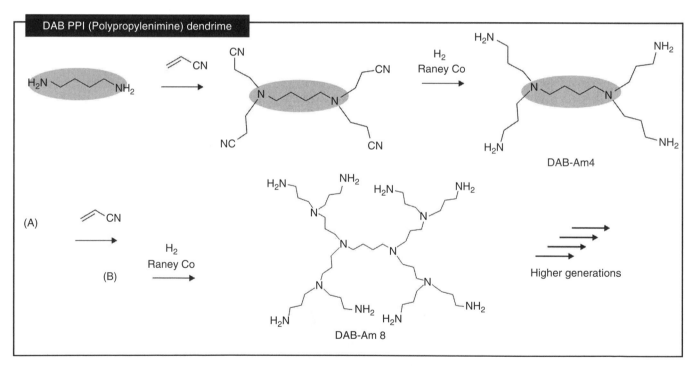

FIGURE 14.4 Convergent synthesis of polyamidoamine and poly(propylenimine) dendrimers; these dendrimers are active as gene transfer agents.

gene and it is injected locally into the tumor for the treatment of head and neck cancers. At the time of writing, this particular form of gene therapy was only licensed for use in China. The fact that viruses are largely not amenable to systemic delivery (e.g., used in intravenous injections) also militates against their widespread use. This limitation is especially important when considering the treatment of cancer, for example, as micro metastases may only be combated with

the use of systemically active therapeutics. Metastasis is the event that leads to a cancer patient's death.

Currently gene therapy research is focused on achieving biodegradable gene delivery options that offer specificity of targeting to the desired cells on systemic administration, transfection efficiencies on par with viruses, and possibly long-term gene expression by sustained release mechanisms; this is the desire although we are not there yet [8,9].

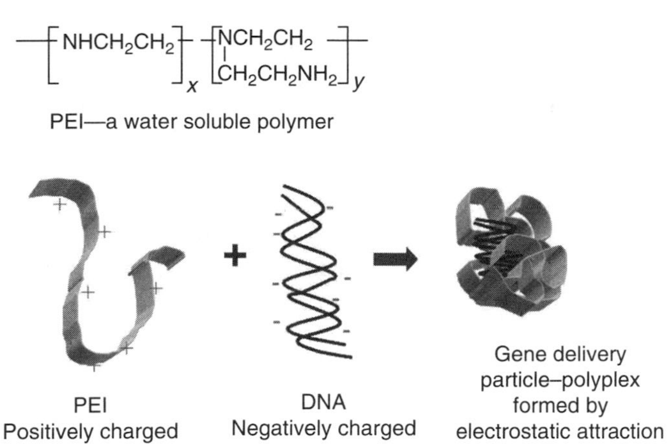

$$-\left[NHCH_2CH_2\right]_x-\left[\begin{array}{c}NCH_2CH_2\\|\\CH_2CH_2NH_2\end{array}\right]_y$$

PEI—a water soluble polymer

PEI
Positively charged

$+$

DNA
Negatively charged

$\rightarrow$

Gene delivery
particle–polyplex
formed by
electrostatic attraction

FIGURE 14.5 Polyplex formation on electrostatic binding of genes and polymers. (Reproduced from Uchegbu, I.F. and Schatzlein, A.G., *Polymers in Drug Delivery*, CRC Press, Boca Raton, 2006. With permission.)

Before considering the various carriers (polymers and dendrimers) in detail, it is thus pertinent to first review the biological barriers to delivery encountered by DNA medicines.

14.2   BIOLOGICAL BARRIERS TO GENE DELIVERY

Nucleic acids, unless significantly chemically modified, tend to be rapidly degraded in tissues or in the systemic circulation [21]. The package of nucleic acids within synthetic particles made from cationic lipids, polymers, or dendrimers [8,9,22,23], which complex DNA or oligonucleotides, via electrostatic interactions, limits the extracellular degradation of nucleic acids. The complexation process leads to a nanoparticle and an excess of the cationic carrier ensures that the nanoparticle carries a positive charge at physiological pH. This cationic nanoparticle encounters a number of biological barriers before locating the nuclei of its target cell types.

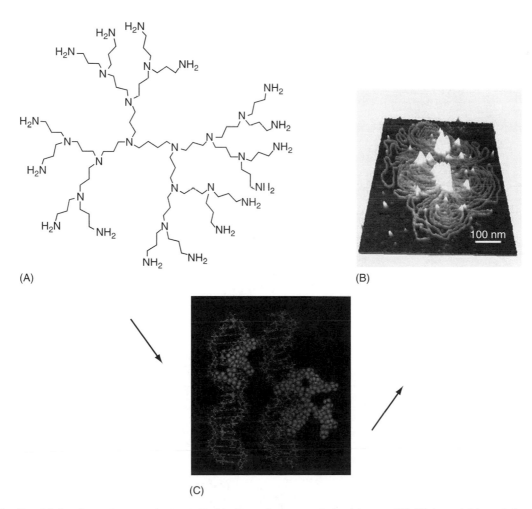

(A)

(B)

(C)

FIGURE 14.6 Dendriplex formation on electrostatic binding of genes and dendrimers. (A) Water soluble poly(propylenimine)G3 dendrimer, (B) atomic force microscopy image of poly(propylenimine)G2 dendrimer–DNA dendriplexes (dendrimer, DNA weight ratio = 5:1). (Adapted from Zinselmeyer, B.H. et al., *Pharm. Res.*, 19, 960, 2002.) (C) molecular modeling simulation of (from left to right) poly(propylenimine)G2 dendrimer–DNA binding and poly(propylenimine)G3 dendrimer–DNA binding. (Adapted from Schatzlein, A.G. et al., *J. Control. Rel.*, 101, 247, 2005.)

Most of the work on the biological barriers to gene transfer has been carried out on polyplexes as opposed to dendriplexes and hence much of the evidence supporting the presence of these biological barriers is drawn from the polyplex literature. In mechanistic terms it appears that dendriplexes and polyplexes behave in a similar manner and differences between the mechanism of action of polyplexes and dendriplexes, at the cellular level at least, appear to be subtle rather than profound. However, some differences in

the biodistribution of gene expression have been noted, for example, when polyplexes are compared to dendriplexes produced from low-molecular weight dendrimers; the latter produce superior tumoricidal results when incorporated in a cancer gene medicine [6].

The biological barriers faced by nucleic acid medicines may be broadly classed as being of an extracellular and intracellular nature (Figure 14.7). Examples of the extracellular barriers to gene delivery include susceptibility to enzymatic

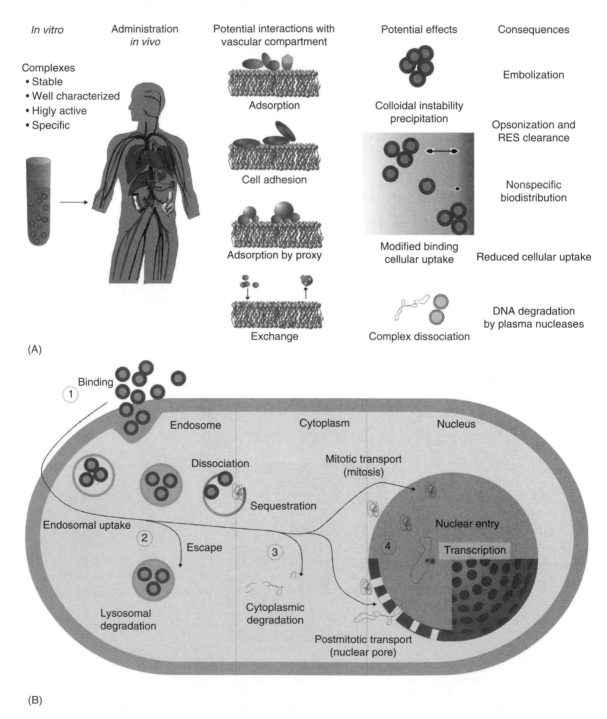

FIGURE 14.7   (A) Extracellular barriers to the delivery of therapeutic genes. (B) Intracellular barriers to the delivery of therapeutic genes. (Reproduced from Uchegbu, I.F. and Schatzlein, A.G., *Polymers in Drug Delivery*, CRC Press, Boca Raton, 2006. With permission.)

degradation by nucleases in the serum and extracellular fluid [24], the presence of a mucus coat on certain cell surfaces [25], such as those required to be breached in the treatment of cystic fibrosis, vulnerability to inactivation by interactions with blood components [26], the possibility of recognition and elimination by immunological defense systems [27], and the possibility of uptake by nontarget organs and tissues [28]. Once inside the cell, the intracellular barriers to gene expression come into play and these include entrapment and degradation within the endosomes and lysosomes [29,30], vulnerability to cytoplasmic nuclease enzymatic attack [31], and the relative impenetrability of the nuclear membrane [32,33]. Once safely inside the nucleus the genes are transcribed to produce mRNA which is in turn transported to the cytoplasm and eventually translated into the therapeutic protein. The biological barriers described above must be overcome in order to achieve a clinically relevant level of gene expression.

14.3 EXTRACELLULAR BARRIERS TO GENE DELIVERY

14.3.1 INTERACTIONS WITH PLASMA AND BLOOD COMPONENTS

Upon entering the systemic circulation, the injected exogenous gene will be enzymatically degraded by the nucleases present in the serum [24]. However, upon incorporation of the gene within polyplexes or dendriplexes [34,35], the gene does acquire some resistance to degradation by serum nucleases [36]. The polyplexes then need to travel to their target site of action. One factor that is crucial in allowing the polyplexes to travel unhindered to their target site is their particle size [37]. A colloidal (<500 nm) particle size is ideal. On intravenous injection of colloidal polyplexes, there is an interaction with serum proteins which leads to polyplex aggregation and an increase in particle size [26,28,38]. This size increase is thought to hinder transport through the fine capillaries and tissues and result in polyplexes being entrapped within the first capillary bed encountered, that of the lung [26,28].

Exogenous gene expression is predominantly observed in the lung following intravenous administration of polyplexes [26,28]. It is clear that if polyplexes are to be targeted to tissues other than the lung, the polyplexes must resist aggregation prior and subsequent to *in vivo* application [26,28,39]. However, most polyplexes tend to aggregate at physiological salt concentrations and are also bound by electrostatic attractions to negatively charged blood components such as serum proteins and blood cells [26,28]. Specifically, when negatively charged serum proteins are bound to the polyplexes, charge neutralization occurs, leading to an increase in polyplex particle size [28]. Such an increase in the size of the polyplex not only results in gene expression occurring predominantly in the lung but also ultimately reduces the level of gene transfer in a target region distal to the lung [26]. One other possible consequence of polyplexes interacting with serum proteins is that the polyplexes may then be cleared from the blood by the macrophages of the reticuloendothelial system (in the liver

spleen and bone marrow) in common with other intravenously injected particulates [27]. Polyplex aggregation may be suppressed by the covalent attachment of poly(ethylene oxide) moieties to the polyplex [26].

Unless they have been administered locally polyplexes and dendriplexes need to travel in the vascular compartment to reach the appropriate organ/site of action. If the target site lies in the parenchyma the particles then need to extravasate and travel through the interstitium to finally reach the target cells. While this is obviously not trivial further challenges exist. When the positively charged synthetic vector systems are administered *in vivo*, the nonspecific nature of electrostatic interactions leads to promiscuous interaction of the particles with biological surfaces, macromolecules, and cells. In the process complexes may be destabilized [40].

14.3.2 UPTAKE BY THE TARGET CELL

Even if the uncomplexed and naked therapeutic gene manages to evade devastation by the serum nucleases it still needs to gain entry to the cell and the first barrier to penetration into the cell is posed by the inherent nature of the gene. DNA faces significant difficulty in entering the cells due to its hydrophilic nature, large size, and polyanionic character. The surface charge of the cell membrane is negative [41] and it is envisaged that this cell surface will repel the approach of the large anionic DNA molecule. Polyplexes can assist in facilitating DNA uptake as they usually carry a positive charge [8]. Via this positive charge, the polyplexes are able to interact with the anionic surface charge of the cells and facilitate endosomal uptake into the cell [42]. Additionally, binding and uptake of the poly(amidoamine) dendrimer Superfect depends on cellular membrane cholesterol [43], as does the binding of lipoplexes [44], and any uncomplexed dendrimer present in the formulation also interacts with membranes although it is not clear whether the ability of the uncomplexed dendrimer to interact with membranes [45–47] is relevant for the uptake and intracellular processing of dendriplexes.

However, the electrostatic binding to the anionic cell surfaces is of a nonspecific nature and can often lead to inefficiency in transfection when complexes are given by an intravenous route due to electrostatic interactions with nontargeted cells. These nonspecific interactions must be suppressed if gene expression with the exogenous gene is to be targeted to particular cells. In order to increase the specificity of uptake into the target cells, homing devices are incorporated in the vector/DNA complex. Examples of such homing devices used to increase the specificity of recognition by the desired target cells include ligands such as galactose [48], *N*-acetyl galactosamine [49], antibodies [50], and transferrin [51]. These ligands will obviously be more efficient if the nonspecific interactions with blood components and cells are reduced. A commonly used strategy to reduce nonspecific interactions is the shielding of cationic surface charges using poly(ethylene oxide) moieties [26,49].

Additionally certain extracellular barriers to gene transfer are disease-specific [52]. In cystic fibrosis patients, barriers such as the thick mucus which coats the cystic fibrosis lung epithelium and the lack of specific receptors that may be exploited for receptor-mediated uptake have been cited as some of the key barriers to transfection with mucus being a formidable barrier for the gene delivery particle to traverse [53]. Generally, although not specifically applied to poly-plexes, strategies to overcome these barriers such as the use of mucolytic agents have had limited success [54].

14.4 INTRACELLULAR BARRIERS TO GENE DELIVERY

Upon arrival in the cell the exogenous gene is once again faced with a number of hurdles which prevent it reaching the nucleus, the site of transcription. The polyplex is usually taken up by endocytosis and DNA must escape from the endosome in order to avoid degradation, traverse the cytoplasm intact, and in turn cross the nuclear membrane to gain entry to the nucleus (Figure 14.7B).

14.4.1 ENDOSOMAL ESCAPE

If the gene fails to escape from the endosome the net result is a reduction in gene expression [37]. Some polyplexes can facili-tate endosomal escape [55,56]. Cationic polyplexes prepared from cationic polymers such as PEI, by virtue of their high level

of amino groups, are believed to act by buffering the acidic con-tents of these vacuoles [55–57]. The accompanying increase in the pH prevents the action of the degradative enzymes and even-tually enables a rupture of the endosome due to an increase in the ionic and latterly water content of the endosome [29,55–59]. The ionic content of the endosome is increased by an energy-driven influx of chloride ions and protons in response to the intraendosomal buffering by the polymer. This hypothesized mechanism of action by polyamines is termed the proton sponge hypothesis. There is now good evidence that poly(amidoamine) and poly(propylenimine) dendrimers also exploit their inherent buffer capacity to disrupt endosomes [59].

One of the newer strategies to facilitate endosomal escape is the use of an amine polymer with the amine shielded by a poly(ethylene oxide) substituent. An acid cleavable linker between the amine function and the poly(ethylene oxide) moiety enables intracellular cleavage of the poly(ethylene oxide), amine bond, releasing the amine function to facilitate endosomal escape (Figure 14.8) [49].

Peptides, although not strictly polymers, are also able to promote endosomal escape of gene transfer systems and a description of their actions is included for completeness. Fusogenic peptides undergo a pH-triggered conformational change on entering the endosomes which leads to membrane destabilization and the release of entrapped DNA [30,60]. Examples of endosomolytic peptides are the viral peptides

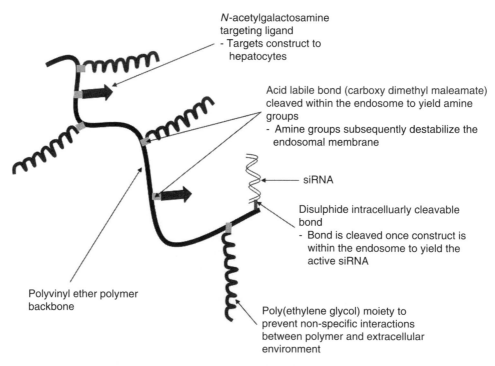

N-acetylgalactosamine targeting ligand
- Targets construct to hepatocytes

Acid labile bond (carboxy dimethyl maleamate) cleaved within the endosome to yield amine groups
- Amine groups subsequently destabilize the endosomal membrane

siRNA

Disulphide intracelluarly cleavable bond
- Bond is cleaved once construct is within the endosome to yield the active siRNA

Polyvinyl ether polymer backbone

Poly(ethylene glycol) moiety to prevent non-specific interactions between polymer and extracellular environment

FIGURE 14.8 Schematic representation of the "dynamic polyconjugates" used to target anti-apolipoprotein B siRNA to hepatocytes. Both the polymer bonds of poly(ethylene oxide) and N-acetylgalactosamine are acid-cleavable carboxy dimethyl maleamate bonds, destined to be cleaved within the endosome. This water soluble conjugate is designed to be taken up, via the galactose units, by the asialoglycoprotein receptor on the hepatocytes; endosomal cleavage of the acid labile maleamate bonds will subsequently release the N-acetylgalactosamine and poly(ethylene oxide) groups to reveal amines. The free amine groups are now available to disrupt the endosome and finally cleavage of the disulphide bonds results in release of the siRNA moiety. (Redrawn from Rozema et al., *Proc. Natl. Acad. Sci. U S A*, 104, 12982, 2007.)

such as the amino terminal domain of the influenza virus hemagglutinin HA2 subunit [30] and the synthetic mimics of amphiphilic anionic viral peptides [60].

14.4.2 STABILITY IN CYTOPLASM

After escaping from the endosomes, DNA may be sequestered in the cytoplasm and subjected to the onslaught of cytoplasmic nucleases [31]. Microinjected plasmid DNA injected into the cytoplasm undergoes rapid degradation by cytoplasmic nucleases with an apparent half-life of 50–90 min [31]. The cytoplasm is a hostile environment for the exogenous gene as was exemplified by one study which reported that transgene expression was observed in 13% of the cells when 10,000 copies of naked DNA were microinjected into the cytoplasm whereas less than 10 copies of naked DNA were needed to induce the same level of gene expression when DNA was microinjected into the nucleus [61]. It is possible that polymeric gene delivery systems which protect DNA from nuclease degradation [62,63] may protect DNA from intracytoplasmic degradation.

14.4.3 NUCLEAR IMPORT

Assuming that DNA remains stable within the cytoplasm, there is still a requirement for DNA to enter the nucleus for transcription to occur and breaching the nuclear barrier is undoubtedly the most formidable challenge of all. The transport of DNA from the cytoplasmic medium into the nucleus is limited by the presence of the nuclear envelope. In dividing eukaryotic cells, nucleocytoplasmic transfer of DNA can occur when the nuclear envelope breaks down during mitosis [15]. Cells in the nondividing phase however are normally resistant to nucleocytoplasmic transfer of plasmid DNA [15]. In nondividing cells the nucleocytoplasmic exchange of molecules occurs through the nuclear pore complexes (NPC) that span the nuclear envelope [64]. Hence, the nuclear envelope acts like a molecular sieve enabling small aqueous molecules of up to 9 nm in diameter (<17 kDa) to diffuse freely through the NPC [64–67]. However, larger molecules of up to 25 nm (>41 kDa) such as plasmid DNA and larger DNA fragments undergo a sequence-specific active transport process involving multiple cellular components [64–67].

Although there is very little direct evidence that polymers actually assist the translocation of DNA or oligonucleotides to the nucleus, PEI has been shown to translocate to the nucleus whether complexed with DNA or in the free form [68] and fluorescence microscopy studies of dendrimer–antisense complexes (Oregon green conjugated poly(amidoamine)$_{G5}$ and carboxytetramethylrhodamine labeled antisense oligonucleotide) suggest that dendrimers have the ability to accumulate to some extent in the nucleus [69]. Also it appears that separation of the nucleotide from the dendrimer is not a prerequisite for activity as poly(amidoamine)$_{G5}$ dendrimer antisense oligonucleotide complexes are active in the nucleus (the nucleus is the site of action of the oligonucleotide in question) although a large proportion of antisense oligonucleotide in the nucleus seemed to be still complexed [69].

Common strategies to encourage nuclear localization include exploiting the nucleocytoplasmic transport machinery by modifying the plasmid DNA with specific sequences so that it can be recognized by cellular factors as a nuclear import substrate [70,71]. A common example of such modification is to attach a DNA nuclear localization signal, for example, the simian virus 40 (SV40) enhancer domain to the plasmid construct, aiding the recognition of these plasmid DNA constructs by transcription factors, with subsequent nuclear import of the resultant complex [72]. The use of an SV40 enhancer sequence improves nuclear delivery *in vivo* and in turn gene transfer [70].

Furthermore, oligopeptide sequences known as nuclear localization signals (NLS), which direct transport to the nucleus, may be linked to plasmid DNA using electrostatic attractions or covalent bonds [73–76]. The site for covalent attachment of NLS to plasmid DNA has to be chosen with care, and the golden rule in this matter is to prevent binding of the NLS to the expression cassette in order to ensure that the transcription ability of the transgene is preserved [77]. These NLS are mainly composed of positively charged oligopeptides made up of sequences of lysine or arginine residues.

14.5 OVERCOMING THE BIOLOGICAL BARRIERS TO GENE DELIVERY

As can be appreciated from the foregoing account, some of the biological barriers faced by naked nucleotides may be overcome by the use of polyplexes and dendriplexes, although new barriers to targeted gene transfer may be introduced by the use of cationic complexing agents. In order to overcome the biological barriers detailed above, a gene therapy vector would include a poly(cationic) sequence for DNA condensation, a stealth type coating to evade detection by the macrophages of the reticuloendothelial system—preferably one that may be shed at the site of cell entry—a colloidal stabilizing entity to prevent colloidal instability in the blood and accumulation in lung capillaries, ligands facilitating cell-specific entry or the site-specific uncovering of a cationic surface to facilitate cell entry, an endosomolytic component which could also be a polycation, and finally nuclear localization signals. Although the ideal system described above does not exist at present, various polymers have been evaluated for their ability to protect and deliver genes across the barriers outlined above and these individual polymers are treated below.

Herman Staudinger's realization that macromolecules consisted not, as previously thought, of aggregates but were in fact long, chain like molecules, started a long journey of discovery which has since led to a wide array of new polymeric materials that have affected every aspect of our lives. Polymeric gene delivery systems are usually positively charged at physiological pH (cationic) [8] and the most commonly used molecules are shown in Figure 14.3. Cationic polymers, by virtue of their possession of protonable groups at physiological pH (amine groups) are able to undergo electrostatic interactions with DNA, the latter of which is anionic at physiological pH [8]. DNA is compacted within the electrostatic complex, a process termed DNA condensation and the colloidal particles which result from this process are known as polyplexes (Figure 14.5). It is these polyplexes that enable the transport of DNA across the various biological barriers to its nuclear destination. In the delivery of siRNA however, a polyplex is not used, but rather a polymer, siRNA conjugate [49]. There is also some evidence that the noncationic poly(ethylene oxide)–poly(propylene oxide)–poly(ethylene oxide) triblock copolymers facilitate gene transfection on direct injection into muscle tissue [78], and despite the fact that cationic polymers are widely used in preclinical studies, the few polymeric gene therapy clinical trials on record have involved the use of poly(vinyl pyrrolidone) [79] and poly(lactic acid)-co-glycolic acid [80], both of which are noncationic at physiological pH.

14.6 POLY(L-LYSINE)

The first cationic polymer-based gene delivery vehicle was poly-L-lysine, the use of which was first reported in 1987 [81]. In this pioneering effort, poly(L-lysine) was conjugated to asialoorosomucoid for targeted gene delivery to liver hepatocytes [81]. On its own, poly(L-lysine), although able to efficiently bind DNA, is not an effective gene transfer agent [82] with adenovirus particles [83], histidyl residues [84], or lysomotropic agents such as chloroquine [85] being required for gene transfer to be observed. Additionally poly(L-lysine)-based polyplexes are associated with their own intrinsic cytotoxicity and although this characteristic may be alleviated somewhat by converting the polymer to an amphiphile [82,86], or by glycosylation [87], unnecessary cytotoxicity associated with an intrinsic poor activity makes this polymer an unlikely first choice material. Recently, L-lysine–L-histidine copolymers have been found to deliver genes, with the L-lysine moieties effecting DNA binding and the L-histidine moieties effecting endosomal escape of the polyplex [88].

A great deal of the early knowledge gained by working on poly(L-lysine) served to elucidate the problems confronting nonviral gene delivery and paved way for the development of the more efficient systems that followed. This first generation polymer can be said to have provided researchers in the field with a valuable learning experience although it is unlikely to feature as a gene delivery system in a marketed product.

14.7 POLY(ETHYLENIMINE)

PEI (Figure 14.3) is one of the most efficient cationic polymer gene transfer systems available, although it has not yet been approved for clinical use. PEI's *in vivo* gene transfer ability has been proven on both local (intraventricular [89], intratracheal [90], and intratumoral [38]) and systemic [91] administration. PEI exists in a number of molecular weight formats (0.42–800 kDa) and transfection efficiency is highest at a molecular weight of between 12 and 70 kDa [92,93] and the most commonly used PEI molecular weight is 22–25 kDa [90,91]. Both linear and branched formats of the polymer exist (Figure 14.1) and the linear molecule is a more efficient gene

transfer agent than the branched material [90,91,94], although the reason for this is unclear at present. Despite the fact that PEI is a successful gene transfer agent for experimental animal studies, there have been reports of unacceptable toxicity with the use of this polymer [26,57,90,95] and even nontoxic doses are associated with the upregulation of numerous genes associated with adverse events [96] in contrast to the use of nontoxic doses of chitosan. Toxicity is undoubtedly modulated by reducing the quantity of protonable amine groups per molecule either through the attachment of poly(ethylene oxide) chains to PEI [97,98] or by the methylation of secondary and tertiary amines to give quaternary ammonium groups [98]. However, an improved biocompatibility with these two methods usually comes at the expense of a reduction in activity. Branched PEI (MW ~ 25 kDa), unlike linear poly-(ethylenimine) (MW ~ 22 kDa), also induces proinflammatory markers of tissue damage (the chemokine receptors CCR5 and CCR3) [99]. It thus appears as if the linear material enjoys a superior biocompatibility profile. Recently low-molecular weight (MW < 2 kDa) cross-linked PEI, incorporating biodegradable cross-links between individual 2 kDa fragments, was shown to result in improved biocompatibility without the loss of gene transfer activity [100], with the high-molecular weight cross-linked material facilitating gene transfer into the cell and the degraded low-molecular weight fragments resulting thereafter being less toxic, due to their low-molecular weight.

Every third atom of the PEI molecule is a nitrogen atom, resulting in a densely charged backbone composed of 25% primary amines, 50% secondary amines, and 25% tertiary amines. These amine groups are important for DNA binding and enabling the therapeutic transgene escape from the endosome on uptake into the cell [58]. The proton sponge hypothesis (Section 14.4) has been put forward to explain the mechanism by which PEI enables escape of the exogenous gene from the endosome subsequent to cellular uptake.

Gene therapeutics may be administered locally or systemically but one of the problems associated with the systemic (intravenous) administration of cationic lipids and cationic polymers such as PEI is the fact that gene transfection occurs predominantly in the lung endothelium [91]. Apart from poly(propylenimine) dendrimers [101], for example, which transfect the liver in mouse models, very few nonviral systems have been found not to predominantly transfect the lung. This passive targeting to the lung is thought to be the result of aggregation of the polyplexes in the blood and their entrapment within the lung capillaries (Section 14.3) [26]. While transfection of the lung endothelial cells may occasionally be welcome, it is sometimes necessary to achieve gene expression in other areas such as tumors located at sites remote from the lung or the site of injection. To reduce the likelihood of passive lung targeting, poly(ethylene oxide) has been grafted on to PEI [97,102]. Conjugation of poly(ethylene oxide) units to PEI reduces the surface charge of the polyplexes and prevents their aggregation and localization within the lung capillaries [26,103,104]. However while poly(ethylene oxide) chains improve the colloidal stability of these particulate formulations by providing a steric hindrance to particle aggregation

[26,97], this strategy is often accompanied by diminished polyplex gene transfer activity, largely due to poor cellular uptake of the poly(ethylene oxide)-covered polyplex [97,103]. In order to counteract these cellular uptake problems, it is thus necessary to apply a ligand at the distal end of the poly(ethylene oxide) groups, which facilitates receptor-mediated uptake.

A variety of other PEI derivatives have been employed in an effort to improve the gene transfer ability of PEI polyplexes. One such modification which has proved successful is the attachment of the nuclear localization signal D-mellitin to PEI [105]. Other modifications, including the use of amphiphilic PEIs [98,106–108] or the conjugation of poly(ethylene oxide) to PEI [98,109], have met with limited success, with improvements in toxicity sometimes being observed but with no simultaneous improvements in gene transfer efficiency.

In summary, it can be concluded that PEI is an efficient gene transfer polymer, however it appears that the cytotoxicity of this molecule has so far yet to be ameliorated to the extent that it may be considered as a suitable excipient candidate in the clinical development of gene therapeutics. Although PEI, by virtue of its tumorostatic activity, when injected intravenously [6], may serve as a lead for the development of anticancer gene medicines. However, what the field requires for noncancer gene therapeutics is gene carrier PEI derivatives that are orders of magnitude less cytotoxic than the current crop of PEI polymers, but which also maintain the *in vivo* gene transfer activity of the current molecules. One promising approach is the use of low-molecular weight PEI fragments cross-linked with biodegradable cross-links [100].

14.8 CHITOSAN

Chitosan (Figure 14.1) is a carbohydrate polymer, derived from chitin; the latter a by-product of the shell fish industry and was first reported as a gene delivery agent in the mid-1990s [110,111]. Chitosan is composed of *N*-acetyl-D-glucosamine and D-glucosamine monomers linked by β(1,4)glycosidic bonds. The presence of amino groups in chitosan gives chitosan the ability to condense DNA. Chitosan is attractive as a gene delivery tool because of its good biocompatibility profile when compared to cationic liposomes [112–115], polyamine polymers [116], and polyamine dendrimers [114]. Chitosan is not as amine dense as PEI or poly(L-lysine) (Figure 14.3) and this property could be responsible for its good biocompatibility and also possibly for its relatively poor gene transfer ability [113,115,117–119]. Chitosan's favorable biocompatibility has prompted researchers to find ways to optimize gene transfer with this agent. Controlling molecular weight appears to be the key and an optimum gene transfer activity lies between a degree of polymerization of 7 and 635 [119–121]. Additionally, increasing the charge density by incorporating a permanent positive charge in the molecule in the form of a trimethyl quaternary ammonium group appears to offer some marginal benefit [115] and the incorporation of targeting groups such as galactose improves targeting to hepatocytes [111,113]. Urocanic acid groups have aided gene transfer, the latter possibly by aiding endosomal escape [116]. However chitosan, while being able to achieve gene transfer to some extent, appears to lack the required level of efficiency that would be needed to allow it to be developed for the clinical delivery of genes.

14.9 OTHER GENE DELIVERY POLYMERS

The cationic polymers discussed above have to date served as key players in the gene delivery field. Other polyamino acids with protonable nitrogen atoms, necessary for DNA binding, have also been studied, such as poly(L-histidine) [122] and poly(L-ornithine) [86] derivatives. In addition various other polymers have been studied as gene transfer systems. Among these are a group of biodegradable polymers, e.g., poly[α-(4-aminobutyl)-L-glycolic acid] [123–125], and poly(β-amino esters) [126], designed to degrade subsequent to having performed their delivery function. The poly(β-amino esters) are possible candidates for microparticle genetic vaccines as the intradermal administration of poly(β-amino ester)-based microparticles encapsulating a gene encoding for an octapeptide epitope which stimulates polyclonal CD8 T cell responses in mice, results in an antigen specific, immune-mediated tumor growth delay [127]. The use of biodegradable gene carriers is aimed at minimizing the cytotoxicity associated with the accumulation of such agents within the cell and hence is aimed at improving the biocompatibility of the gene transfer agent. Other polymers that have been developed are thermosensitive amine polymers designed to form a gel in situ on administration and thus provide a depot system capable of sustained DNA delivery [128]. There is also some evidence that the noncationic poly(ethylene oxide)–poly(propylene oxide)–poly(ethylene oxide) triblock copolymers improve the gene transfection seen on direct injection of genes into muscle tissue [78], and both polyvinyl pyrrolidone [79] and poly(lactic acid-co-glycolic acid) [80] have been used in clinical studies.

14.10 DENDRIMERS

Dendrimers (from the Greek "dendron": tree, and "meros": part) are highly ordered, branched monodisperse macromolecules [129] (Figure 14.4). Such dendritic structures first emerged in a new class of polymers named "cascade molecules," initially reported by Vögtle and his group at the end of the 1970s [130]. Further development by Tomalia's group [131], and Newkome's group [132,133] gave rise to larger dendritic structures. These hyper-branched molecules were called "dendrimers" or "arborols" (from the Latin "arbor" for tree). For a historical view see review [134]. Their unique molecular architecture means that dendrimers have a number of distinctive properties which differentiate them from other polymers; specifically they are not the result of statistical polymerization events but are built up in a stepwise fashion from a core group (a convergent method of synthesis: Figure 14.4) or by the addition of dendrimer arms to a core group (a divergent method of synthesis). The controlled synthetic approach means that dendrimers tend to be monodisperse with a well-defined size and structure. Dendrimers used in gene delivery are given in Figure 14.4.

Dendrimers used for gene delivery contain amine functional groups and are usually protonated at physiological pH (cationic); such amine groups, which when protonated may be electrostatically bound to DNA phosphate groups to give dendriplexes (Figure 14.6) [12,34,101,135–137]. This electrostatic interaction protects DNA from degradation [34,35].

Poly(amidoamine) and poly(propylenimine) dendrimers bind DNA at a 1:1 stoichiometry of primary amine to phosphate, but it is only at higher ratios that more stable and efficient complexes are being formed [12,136]. These dendriplexes are heterogenous [135] and with some systems, e.g., poly(amidoamine)$_{G7}$–DNA complexes (where the subscript G7 refers to the generation, i.e., generation 7, see Figure 14.4), more than 90% of transfection activity lies in just 10%–20% of complexes (found in the low density aqueous soluble fraction). It is apparent that the formation of dendriplex nanoparticles needs to be finely tuned to optimize transfection efficiency.

Poly(amidoamine) and poly(propylenimine) dendrimers efficiently deliver nucleic acids into cells, ranging from small oligonucleotides to plasmids and artificial chromosomes [6,12,69,138–144]. The higher generation poly(amidoamine) dendrimers (e.g., generation 5–generation 10) efficiently transfect cells and are more efficient gene carriers than the lower generation materials [145,146]. For these higher generation dendrimers a near exponential increase in efficiency with generation has been observed [145]. On the other hand toxicity has been shown to be directly related to molecular weight [147,148]. For example, the higher generation poly(amidoamine) dendrimers lead to the formation of transient holes (G7) or the enlargement of existing holes (G5) [45] and a tendency to cause hemolysis and changes in erythrocyte morphology have been linked to the presence of surface primary amine groups [148]. Thus a balance needs to be struck between the transfection enhancements seen with the higher generation poly(amidoamine) dendrimers and the related toxic effects. Degradation of these higher generation dendrimers to reduce the branch density in individual molecules increases the activity of the higher generation poly(amidoamine) dendrimers considerably, enhancing activity by almost 50-fold [146].

The ability of the poly(propylenimine) dendrimers to bind DNA is generation dependent as is their cytotoxicity [12]. Optimum activity without attendant compromising cytotoxicity was found in the lower generation poly(propylenimine) dendrimers, specifically poly(propylenimine)$_{G3}$ (DAB-AM16) [12]. Dendrimer cytotoxicity is reduced by the formation of dendriplexes or the conversion of dendrimer primary and tertiary amines to their quaternary ammonium functions [12]. Additionally the quaternary ammonium dendriplexes were still transfection competent, with transfection occurring mainly in the liver [12]. From the foregoing, reducing primary amine density appears to be generally linked to improvements in biocompatibility. Attempts to bring the higher generation (above generation 3) poly(propylenimine) dendrimers into the gene therapy arena have focused on ameliorating their cytotoxicity and the fourth generation dendrimers have thus been modified by amidation of the terminal primary amines followed by the conversion of the internal tertiary amines into quaternary ammonium groups [149] and in other instances by the addition of guanidine groups to the terminal primary amines [150].

When higher generation poly(amidoamine) dendriplexes [200 µg DNA complexed with 650 µg poly(amidoamine)$_{G9}$] are injected intravenously, gene expression occurs mainly in the lung parenchyma but not in other organs [151]. Lower generation dendrimer conjugate [poly(amidoamine)$_{G3}$–α-cyclodextrin] dendriplexes, however, transfect primarily the spleen [152], while the lower generation poly(propylenimine) (generation 2 and generation 3) dendriplexes [50 µg DNA complexed with 250 µg poly(propylenimine)$_{G3}$] transfect primarily the liver [101].

14.11 SYSTEMIC TARGETING

Genetic therapies often require to be targeted to specific organs, tissues, and cell types. Targeting is either achieved by exploiting the use of targeting ligands or by exploiting a poorly understood propensity for dendriplexes and polyplexes to target specific tissues.

14.11.1 LIVER TARGETING

Liver targeting of gene therapeutics may facilitate the gene therapy of hepatic carcinomas such as hepatocellular carcinoma. Passive targeting to the liver has been achieved using poly(propylenimine)$_{G3}$ dendriplexes formulated at a nitrogen–phosphate ratio of 16 (N:P ratio = 16) [101]. This passive liver targeting is unusual with polymer and dendrimer gene delivery systems, which typically transfect the lung [91,151] on intravenous administration, as do cationic lipoplexes (cationic liposome–DNA electrostatic complexes) [153]. The poly(propylenimine)$_{G3}$ system appears to avoid the lung due to a hypothesized mechanism involving diminished aggregation within the blood, the diminished aggregation itself a result of the lower molecular weight of poly(propylenimine)$_{G3}$ (MW = 1.684 kDa) when compared to PEI polymers (MW ~ 25 kDa) or transfection-competent poly(amidoamine) dendrimers (MW > 28 kDa). However, while gene transfection is found in the liver with these low-molecular weight poly(propylenimine) dendrimers, it is not clear if poly(propylenimine)$_{G3}$ dendriplexes target hepatocytes or whether these dendriplexes target other liver cell types.

Targeting gene expression to hepatocytes has been achieved by conjugating either asialoorosomucoid [81] or galactose [154–156] to gene delivery polymers or dendrimers. The galactose moiety has been used in elegant experiments to target siRNAs to hepatocytes [49]. Rozema's group [49] have demonstrated that delivery of siRNA may be achieved by using a polyvinyl ether conjugated siRNA in which conjugation is achieved via an intracellularly cleavable disulphide bond. This polymer–siRNA conjugate is then further linked to both poly(ethylene oxide) units—to prevent nonspecific extracellular interactions and N-acetyl galactosamine to achieve hepatocyte targeting (Figure 14.8). The polymer–siRNA conjugate effectively silenced the apolipoprotein B gene, which is involved in cholesterol transport and serum cholesterol levels were also decreased [49]. Unfortunately silencing of the apolipoprotein B gene was accompanied by a fatty liver in experimental animals and hence apolipoprotein B gene silencing in this manner will still require additional effort before it becomes a legitimate therapeutic target. Competing technologies employing lipid modified poly(amidoamine) dendrimers, termed interfering nanoparticles [157] or lipid particles termed stable nucleic acid lipid particles, [158] also lead to apolipoprotein B silencing and reduced blood cholesterol levels.

14.11.2 Lung Targeting

On intravenous injection of PEI polyplexes, gene transfection occurs predominantly in the lung endothelium [91], due to a hypothesized mechanism involving aggregation of the polyplexes in the blood and their entrapment within the lung capillaries [26]. For the treatment of lung cancers such as nonsmall cell lung cancer, the commonest form of lung cancer and cystic fibrosis, it will be optimal to transfect the cells lining the airways as opposed to the cells lining the lung capillaries. However, evidence that inhaled therapies may be useful is not yet apparent.

14.11.3 Tumor Targeting

Cancer gene therapy is currently limited by the difficulty of efficiently delivering therapeutic genes to remote tumors and metastases by systemic administration [5,40]. However, poly(propylenimine)$_{G3}$ dendriplexes on intravenous administration induce tumor gene expression [6]. When murine xenografts were treated by intravenous injection of poly(propylenimine)$_{G3}$ dendriplexes containing a tumor necrosis factor (TNF-α) expression plasmid under control of a tumor-specific promoter, regression of established tumors was observed in 100% of the animals (Figure 14.9). The treatment (5 injections over 10 days) also led to an excellent long-term response (80% complete and 20% partial response at 17 weeks). The antitumor activity is the result of a combination of the effects of the tumor-specific expression of TNF-α and an intrinsic anti-proliferative effect of the dendrimer [6]. This novel anti-proliferative effect was also observed with other cationic polymers. The lack of apparent toxicity and significant weight loss compared to untreated controls suggests that the treatment is relatively well tolerated [6]. Furthermore, the administration of poly(propylenimine)$_{G3}$ dendriplexes containing genes expressing either p53, a minimal p53-derived apoptotic peptide, or a small hairpin RNA (shRNA) sequence that targets the P73 apoptosis inhibitor, iASPP, results in tumor regression, independent of p53 status [143]. The gene expressing the shRNA sequence leads to downregulation of iASPP and consequently activation of the p73 apoptosis pathway [143]. These studies conducted by Bell and coworkers were the first to validate p73 as a target in cancer therapy. It is

possible that the good tumor gene expression seen with the poly(propylenimine)$_{G3}$ dendriplexes could be linked to the ability of these dendriplexes to avoid the lung (Section 14.11.1). Others have observed that low-molecular weight (MW ~ 4 kDa) oligoethylenimine pseudodendrimers also direct gene expression away from the lung tissue and to tumor tissue on intravenous administration, when compared to linear (MW ~ 25 kDa) PEI [159].

Exploiting receptor-mediated uptake is another method of targeting gene therapeutics to tumors and as such both folate [160] and transferrin [51,161] ligands have been used for tumor targeting. PEI–TNF-α polyplexes are able to achieve tumor regression, on systemic administration through a combination of gene transfer [162] and an intrinsic anti-proliferative activity of the carrier PEI (Figure 14.9) [6]. Tumor targeting with PEI polyplexes was achieved by using poly(ethylene oxide) groups to prevent nonspecific interactions and a transferrin ligand to facilitate receptor-mediated uptake into tumor cells [162].

14.12 LOCAL DELIVERY OF GENES TO SPECIFIC ORGANS OR TISSUES

Due to the problems surrounding systemic gene therapy, as detailed above, a number of local delivery strategies have been attempted. In actual fact the world's first licensed gene therapeutic (Gendicine) [163] was licensed for intratumoral delivery in the treatment of head and neck cancer and the intratumoral delivery of genes, although not practical in all cases, does yield favorable results. For example, the intratumoral delivery of a poly(amidoamine) suicide gene (HSV-tk) dendriplex [100 μg HSV-tk and 300 μg poly(amidoamine) dendrimer] led to a pronounced growth delay [164]. This plasmid which expresses an enzyme which activates the prodrug ganciclovir contained Epstein–Barr virus (Epstein–Barr virus nuclear antigen gene, EBNA1/OriP) sequences with the ability to replicate and persist in the nucleus of the transfected cells. The animals received up to four weekly cycles (single injection of complex followed by 100 mg kg^{-1} per day of the prodrug Ganciclovir for 6 days) [164]. Plasmids containing the EBNA/OriP system increase gene expression eightfold when compared to normal plasmids and when the EBNA1/OriP system is used in conjunction with a vector expressing *Fas* ligand; the injection of 10 μg plasmid complexed with the dendrimer (Superfect) at a ratio of 10:1 (w/w) also led to a pronounced tumor growth delay [165]. A growth delay was also demonstrated after intratumoral injection of plasmids coding for the anti-angiogenic peptide angiostatin or the tissue inhibitor of metalloproteinase-2 genes in special dendrimer/plasmid/oligonucleotide complexes [166].

With the aim of developing treatments against blindness caused by neovascularization, oligonucleotide delivery to the back of the eye has been successfully attempted by the intravitreal injection of complexes with antisense oligonucleotides and a lipid L-lysine dendrimer [167]. The oligonucleotide therapy inhibits neovascularization of the choroidea by downregulation of VEGF over a period of up to 2 months.

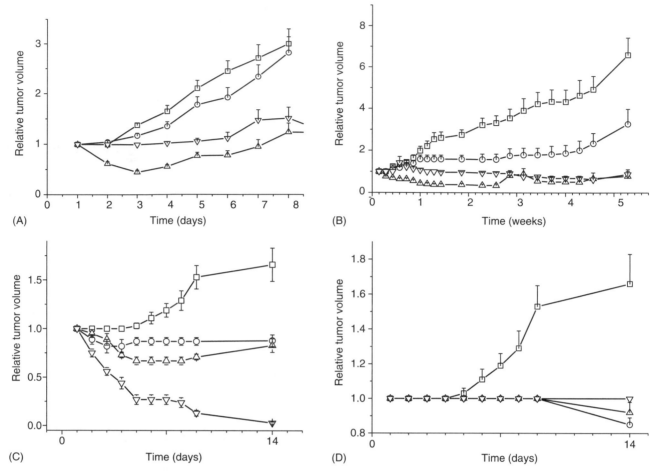

FIGURE 14.9 Experimental systemic tumor gene therapy with cationic polymers. (A) Therapeutic effects of the systemic mTNF-α gene therapy (5 injections given every second day) against established LS174T colorectal adenocarcinoma (ATCC CCL-188). (B) C33a (ATCC HTB31) cervix carcinoma xenografts. DNA-dendrimer nanoparticles (Δ) are superior to the administration of naked mTNF-α DNA (O) and dendrimer alone [poly(propylenimine)G3, ∇] and lead to tumor regression and prolonged tumor growth retardation when compared to tumors in untreated animals (□). (C) Various cationic polymers are used to treat established A431 xenografts either complexed with mTNF-α expression plasmid or (D) as polymers alone. Poly(propylenimine)G3 dendrimers (Δ), linear PEI (O), and fractured poly(amidoamine) dendrimer (∇) and respective complexes were given by single tail-vein injection (treatment day = day 0). (Adapted from Dufes, C. et al., *Cancer Res.*, 65, 8079, 2005.)

Gene therapy targeted at the central nervous system (CNS) is attractive for the treatment of cognitive decline or pain. However, delivery to the CNS has only been achieved using local delivery strategies. Low-molecular weight (MW ~ 600 Da) PEIs have been linked via an activated β-cyclodextrin to produce a gene transfer polymer, which on injection into the spinal cord produced transfection levels similar to that produced by high-molecular weight (MW ~ 25 kDa) PEIs [168]. Additionally, the PEI–β-cyclodextrin conjugates are less cytotoxic, when compared to high-molecular weight (MW ~ 25 kDa)-branched PEI [168]. Gene therapy of the CNS has also been revealed as a mechanism to control neuropathic pain [169], as the intrathecal administration of PEI polyplexes consisting of a plasmid encoding for interleukin 10 (IL-10) leads to prolonged pain control.

Gene therapy of cardiovascular tissue could be useful in cardiac transplantation and evidence of gene delivery to cardiac tissue has been presented. Using direct injection in a murine

cardiac transplant model poly(amidoamine)$_{G5}$ dendriplexes demonstrated more widespread and prolonged expression compared to the naked plasmid and when combined with a viral IL-10 gene were able to prolong graft survival [170]. The efficiency of the procedure was improved at a higher nitrogen to phosphate charge ratio of 20:1 [171], and in combination with electroporation [171].

14.13   CLINICAL TRIALS

Out of the 1340 gene therapy trials currently on public record, 55% utilize viral vectors [17] and to date no clinical studies utilizing polymers have been reported in the peer reviewed literature. However, the National Institutes of Health do record three clinical trial registrations involving the use of polyvinyl pyrrolidone gene medicines [79]. Polyvinyl pyrrolidone has been used to deliver recombinant human insulin-like growth factor-I genes by direct injection into muscles to treat peripheral neuropathy and muscle weakness and has been used to deliver both human interferon α (hIFN-α) and human interleukin-12 (hIL-12) formulations intratumorally into squamous cell carcinomas of the head and neck [79]. Additionally polyvinyl pyrrolidone has been used to deliver hIFN-α and hIL-12 intratumorally to metastatic melanoma patients with unresectable carcinomas [79]. The results of these trials have not been publicized. There are also reports of the use of poly(lactic acid)-co-glycolic acid gene medicines being used for DNA vaccination studies [80], although once again results are not available in the public domain. Despite evidence showing cationic polymers to be superior gene transfer agents than polymers possessing a neutral or anionic charge at physiological pH, clinical investigators have adopted a cautious approach and have opted for excipients that are generally regarded as safe [172], such as polyvinyl pyrrolidone and poly(lactic acid-co-glycolic acid), both of which are nonamine polymers, and hence not cationic at physiological pH.

14.14 CONCLUSION

A number of polymers and dendrimers, by virtue of possessing a cationic charge at physiological pH, have been found to be suitable candidates for the carriage of genes across the various biological barriers outlined above. While the lower generation poly(propylenimine) dendrimers, and the 25 kDa PEIs and polyvinyl ether polymers have demonstrated the ability to bring about a therapeutic response in preclinical models, the search for efficient and safe gene delivery systems is far from over. Researchers have achieved a modicum of success with these first generation agents, and it is likely that some of these agents will move into clinical development. It is feasible that a licensed nonviral gene therapeutic will emerge from the cationic polymer or cationic dendrimer class of agents, however, what has been learnt from these early studies on polymeric and dendrimer systems will continue to assist in the development of more efficient agents no matter from where such agents originate. An ideal gene delivery system has to be able to shuttle the gene safely to the nuclei of its target tissue with the gene enjoying travel that has limited encounters with degradative influences. We know that certain cationic dendrimers or polymers can fulfill these requirements, and providing the cytotoxicity of the carriers are either appropriately channeled to facilitate tumor cell kill or otherwise curtailed, it is likely that the first class of synthetic gene delivery agent that completes the clinical development journey may be a cationic dendrimer or polymer with targeting ability.

REFERENCES

1. McIlwain, C., World leaders heap praise on human genome landmark, *Nature*, 405, 983–984, 2000.
2. Porteous, D.J. and Dorin, J.R., Gene therapy for cystic fibrosis—where and when? *Hum. Mol. Genetics*, 2, 211–212, 1993.
3. Pawliuk, R. et al., Correction of sickle cell disease in transgenic mouse models by gene therapy, *Science*, 294, 2368–2371, 2001.
4. Cavazzana-Calvo, M. et al., Gene therapy of human severe combined immunodeficiency (SCID)-X1 disease, *Science*, 288, 669–672, 2000.
5. Schatzlein, A.G., Non-viral vectors in cancer gene therapy: Principles and progress, *Anti-Cancer Drugs*, 12, 275–304, 2001.
6. Dufes, C. et al., Synthetic anti-cancer gene medicine exploiting intrinsic anti-tumour activity of cationic vector to cure established tumours, *Cancer Res.*, 65, 8079–8084, 2005.
7. Pandha, H.S. et al., Gene therapy: Recent progress in the clinical oncology arena, *Curr. Opin. Mol. Ther.*, 2, 362–375, 2000.
8. Brown, M.D., Schatzlein, A., and Uchegbu, I.F., Gene delivery with synthetic (non viral) carriers, *Int. J. Pharm.*, 229, 1–21, 2001.
9. Pack, D.W. et al., Design and development of polymers for gene delivery, *Nature Rev. Drug Discov.*, 4, 581–593, 2005.
10. Gary, D.J., Puri, N., and Won, Y.Y., Polymer-based siRNA delivery: Perspectives on the fundamental and phenomenological distinctions from polymer-based DNA delivery, *J. Control. Rel.*, 121, 64–73, 2007.
11. Ogris, M. et al., The size of DNA/transferrin-PEI complexes is an important factor for gene expression in cultured cells, *Gene Ther.*, 5, 1425–1433, 1998.
12. Zinselmeyer, B.H. et al., The lower-generation polypropylenimine dendrimers are effective gene-transfer agents, *Pharm. Res.*, 19, 960–967, 2002.
13. Rosenberg, S.A. et al., Gene transfer into humans-immunotherapy of patients with advanced melanoma, using tumor-infiltrating lymphocytes modified by retroviral gene transduction, *New Engl. J. Med.*, 323, 570–578, 1990.
14. Nabel, G.J. et al., Direct gene transfer with DNA-liposome complexes in melanoma: Expression, biologic activity, and lack of toxicity in humans, *Proc. Natl. Acad. Sci. U S A*, 90, 11307–11311, 1993.
15. Brunner, S. et al., Cell cycle dependence of gene transfer by lipoplex polyplex and recombinant adenovirus, *Gene Ther.*, 7, 401–407, 2000.
16. Verma, I.M. and Somia, N., Gene therapy-promises, problems and prospects, *Nature*, 389, 239–242, 1997.
17. Journal of Gene Medicine, Gene Therapy Clinical Trials, *J. Gene. Med.*, http://www.wiley.co.uk/genetherapy/clinical/, 2007.
18. Marshall, E., Second child in French trial is found to have leukemia, *Science*, 299, 320, 2003.
19. Marshall, E., Gene therapy death prompts review of adenovirus vector, *Science*, 286, 2244–2245, 2000.
20. Zhenzhen, L. et al., Health biotechnology in China—reawakening of a giant, *Nat. Biotechnol.*, 22, DC13–DC18, 2004.
21. Niven, R. et al., Biodistribution of radiolabeled lipid-DNA complexes and DNA in mice, *J. Pharm. Sci.*, 87, 1292–1299, 1998.
22. Wagner, E., Strategies to improve DNA polyplexes for in vivo gene transfer: Will "artificial viruses" be the answer? *Pharm. Res.*, 21, 8–14, 2004.
23. Dufes, C., Uchegbu, I.F., and Schatzlein, A.G., Dendrimers in gene delivery, *Adv. Drug Deliv. Rev.*, 57, 2177–2202, 2005.
24. Hashida, M. et al., Pharmacokinetics and targeted delivery of proteins and genes, *J. Control. Rel.*, 41, 91–97, 1996.
25. Alton, E.W. et al., Towards gene therapy for cystic fibrosis: A clinical progress report, *Gene Ther.*, 5, 291–292, 1998.
26. Ogris, M. et al., PEGylated DNA/transferrin-PEI complexes: Reduced interaction with blood components, extended circulation in blood and potential for systemic gene delivery, *Gene Ther.*, 6, 595–605, 1999.

27. Patel, H.M., Serum opsonins and liposomes: Their interactions and opsonophagocytosis, *Crit. Rev. Ther. Drug Carrier Syst.*, 9, 39–90, 1992.

28. Dash, P.R. et al., Factors affecting blood clearance and in vivo distribution of polyelectrolyte complexes for gene delivery, *Gene Ther.*, 6, 643–650, 1999.

29. Behr, J.P., The proton sponge: A means to enter cells viruses never thought of, *Med. Sci.*, 12, 56–58, 1996.

30. Wagner, E. et al., Influenza-virus hemagglutinin-Ha-2 N-terminal fusogenic peptides augment gene-transfer by transferrin polylysine DNA complexes—toward a synthetic virus-like gene-transfer vehicle, *Proc. Natl. Acad. Sci. U S A*, 89, 7934–7938, 1992.

31. Lechardeur, D. et al., Metabolic stability of plasmid DNA in the cytosol: A potential barrier to gene transfer, *Gene Ther.*, 6, 482–497, 1999.

32. Zabner, J. et al., Cellular and molecular barriers to gene-transfer by a cationic lipid, *J. Biol. Chem.*, 270, 18997–19007, 1995.

33. Goldfarb, D.S. et al., Synthetic peptides as nuclear localization signals, *Nature*, 322, 641–644, 1986.

34. Bielinska, A.U., Kukowska-Latallo, J.F., and Baker, J.R., The interaction of plasmid DNA with polyamidoamine dendrimers: Mechanism of complex formation and analysis of alterations induced in nuclease sensitivity and transcriptional activity of the complexed DNA, *Biochim. Biophys. Acta*, 1353, 180–190, 1997.

35. Abdelhady, H.G. et al., Direct real-time molecular scale visualisation of the degradation of condensed DNA complexes exposed to DNase I, *Nucleic Acids Res.*, 31, 4001–4005, 2003.

36. Chiou, H.C. et al., Enhanced resistance to nuclease degradation of nucleic acids complexed to asialoglycoprotein-polylysine carriers, *Nucleic Acids Res.*, 22, 5439–5446, 1994.

37. Nishikawa, M. and Huang, L., Nonviral vectors in the new millennium: Delivery barriers in gene transfer, *Hum. Gene Ther.*, 12, 861–870, 2001.

38. Coll, J.L. et al., In vivo delivery to tumors of DNA complexed with linear polyethylenimine, *Hum. Gene Ther.*, 10, 1659–1666, 1999.

39. Kircheis, R., Wightman, L., and Wagner, E., Design and gene delivery activity of modified polyethylenimines, *Adv. Drug Deliv. Rev.*, 53, 341–358, 2001.

40. Schatzlein, A.G., Targeting of synthetic gene delivery systems, *J. Biomed. Biotechnol.*, 2003, 149–158, 2003.

41. Singh, A.K., Kasinath, B.S., and Lewis, E.J., Interactions of polycations with cell-surface negative charges of epithelial cells, *Biochim. Biophys. Acta*, 1120, 337–342, 1992.

42. Mislick, K.A. and Baldeschwieler, J.D., Evidence for the role of proteoglycans in cation-mediated gene transfer, *Proc. Natl. Acad. Sci. U S A*, 93, 12349–12354, 1996.

43. Manunta, M. et al., Gene delivery by dendrimers operates via a cholesterol dependent pathway, *Nucleic Acids Res.*, 32, 2730–2739, 2004.

44. Zuhorn, I.S., Kalicharan, R., and Hoekstra, D., Lipoplex-mediated transfection of mammalian cells occurs through the cholesterol-dependent clathrin-mediated pathway of endocytosis, *J. Biol. Chem.*, 277, 18021–18028, 2002.

45. Hong, S.P. et al., Interaction of poly(amidoamine) dendrimers with supported lipid bilayers and cells: Hole formation and the relation to transport, *Bioconjug. Chem.*, 15, 774–782, 2004.

46. Zhang, Z.Y. and Smith, B.D., High-generation polycationic dendrimers are unusually effective at disrupting anionic vesicles: Membrane bending model, *Bioconjug. Chem.*, 11, 805–814, 2000.

47. Lai, J.C., Yuan, C.L., and Thomas, J.L., Single-cell measurements of polyamidoamine dendrimer binding, *Ann. Biomed. Eng.*, 30, 409–416, 2002.

48. Plank, C. et al., Gene transfer into hepatocytes using asialoglycoprotein receptor mediated endocytosis of DNA complexed with an artificial tetra-antennary galactose ligand, *Bioconjug. Chem.*, 3, 533–539, 1992.

49. Rozema, D.B. et al., Dynamic polyconjugates for targeted in vivo delivery of siRNA to hepatocytes, *Proc. Natl. Acad. Sci. U S A*, 104, 12982–12987, 2007.

50. Trubetskoy, V.S. et al., Use of n-terminal modified poly(l-lysine) antibody conjugate as a carrier for targeted gene delivery in mouse lung endothelial-cells, *Bioconjug. Chem.*, 3, 323–327, 1992.

51. Wagner, E. et al., Transferrin-polycation conjugates as carriers for DNA uptake into cells, *Proc. Natl. Acad. Sci. U S A*, 87, 3410–3414, 1990.

52. Pilewski, J.M., Gene therapy for airway diseases-continued progress towards identifying and overcoming barriers to efficiency, *Am. J. Respir. Cell Mol. Biol.*, 27, 117–121, 2002.

53. Ferrari, S., Geddes, D.M., and Alton, E., Barriers to and new approaches for gene therapy and gene delivery in cystic fibrosis, *Adv. Drug Deliv. Rev.*, 54, 1373–1393, 2002.

54. Ferrari, S. et al., Mucus altering agents as adjuncts for nonviral gene transfer to airway epithelium, *Gene Ther.*, 8, 1380–1386, 2001.

55. Akinc, A. et al., Exploring polyethylenimine-mediated DNA transfection and the proton sponge hypothesis, *J. Gene. Med.*, 7, 657–663, 2005.

56. Boussif, O., Zanta, M.A., and Behr, J.P., Optimized galenics improve in vitro gene transfer with cationic molecules up to 1000-fold, *Gene Ther.*, 3, 1074–1080, 1996.

57. Boussif, O. et al., A versatile vector for gene and oligonucleotide transfer into cells in culture and in vivo: Polyethylenimine, *Proc. Natl. Acad. Sci. U S A*, 92, 7297–7301, 1995.

58. Remy, J.S. et al., Gene transfer with lipospermines and polyethylenimines, *Adv. Drug Deliv. Rev.*, 30, 85–95, 1998.

59. Sonawane, N.D., Szoka, F.C., and Verkman, A.S., Chloride accumulation and swelling in endosomes enhances DNA transfer by polyamine-DNA polyplexes, *J. Biol. Chem.*, 278, 44826–44831, 2003.

60. Murthy, N. et al., The design and synthesis of polymers for eukaryotic membrane disruption, *J. Control. Rel.*, 61, 137–143, 1999.

61. Pollard, H. et al., Polyethylenimine but not cationic lipids promotes transgene delivery to the nucleus in mammalian cells, *J. Biol. Chem.*, 273, 7507–7511, 1998.

62. Goh, S.L. et al., Cross-linked microparticles as carriers for the delivery of plasmid DNA for vaccine development, *Bioconjug. Chem.*, 15, 467–474, 2004.

63. Brus, C. et al., Efficiency of polyethylenimines and polyethylenimine-graft-poly (ethylene glycol) block copolymers to protect oligonucleotides against enzymatic degradation, *Eur. J. Pharm. Biopharm.*, 57, 427–430, 2004.

64. Ludtke, J.J. et al., A nuclear localization signal can enhance both the nuclear transport and expression of 1 kb DNA, *J. Cell Sci.*, 112, 2033–2041, 1999.

65. Nigg, E.A., Nucleocytoplasmic transport: Signals, mechanisms and regulation, *Nature*, 386, 779–787, 1997.

66. Ohno, M., Fornerod, M., and Mattaj, I.W., Nucleocytoplasmic transport: The last 200 nanometers, *Cell*, 92, 327–336, 1998.

67. Peters, R. et al., Fluorescence microphotolysis to measure nucleocytoplasmic transport in vivo and in vitro, *Biochem. Soc. Trans.*, 14, 821–822, 1986.

68. Godbey, W.T., Wu, K.K., and Mikos, A.G., Tracking the intracellular path of poly(ethylenimine)/DNA complexes for gene delivery, *Proc. Natl. Acad. Sci. U S A*, 96, 5177–5181, 1999.

69. Yoo, H. and Juliano, R.L., Enhanced delivery of antisense oligonucleotides with fluorophore-conjugated PAMAM dendrimers, *Nucleic Acids Res.*, 28, 4225–4231, 2000.

70. Young, J.L., Benoit, J.N., and Dean, D.A., Effect of a DNA nuclear targeting sequence on gene transfer and expression of plasmids in the intact vasculature, *Gene Ther.*, 10, 1465–1470, 2003.

71. Vacik, J. et al., Cell-specific nuclear import of plasmid DNA, *Gene Ther.*, 6, 1006–1014, 1999.

72. Wilson, G.L. et al., Nuclear import of plasmid DNA in digitonin-permeabilized cells requires both cytoplasmic factors and specific DNA sequences, *J. Biol. Chem.*, 274, 22025–22032, 1999.

73. Aronsohn, A.I. and Hughes, J.A., Nuclear localization signal peptides enhance cationic liposome-mediated gene therapy, *J. Drug Target.*, 5, 163–169, 1998.

74. Zanta, M.A., Belguise-Valladier, P., and Behr, J.P., Gene delivery: A single nuclear localization signal peptide is sufficient to carry DNA to the cell nucleus, *Proc. Natl. Acad. Sci. U S A*, 96, 91–96, 1999.

75. Adam, S.A. and Gerace, L., Cytosolic proteins that specifically bind nuclear localization signals are receptors for nuclear import, *Cell*, 66, 837–847, 1991.

76. Robbins, J. et al., Two independent basic domains in nucleoplasmin nuclear targeting sequence: Identification of a class of bipartite nuclear targeting sequence, *Cell*, 64, 615–623, 1991.

77. Cartier, R. and Reszka, R., Utilization of synthetic peptides containing nuclear localization signals for nonviral gene transfer systems, *Gene Ther.*, 9, 157–167, 2002.

78. Bello-Roufai, M., Lambert, O., and Pitard, B., Relationships between the physicochemical properties of an amphiphilic triblock copolymers/DNA complexes and their intramuscular transfection efficiency, *Nucleic Acids Res.*, 35, 728–739, 2007.

79. National Institute of Health, Genetic Modification Clinical Research Information System, http://www.gemcris.od.nih.gov/, 2007.

80. Putnam, D., Polymers for gene delivery across length scales, *Nat. Mater.*, 5, 439–451, 2006.

81. Wu, G.Y. and Wu, C.H., Receptor-mediated in vitro gene transformation by a soluble DNA carrier system, *J. Biol. Chem.*, 262, 4429–4432, 1987.

82. Brown, M.D. et al., Preliminary characterization of novel amino acid based polymeric vesicles as gene and drug delivery agents, *Bioconjug. Chem.*, 11, 880–891, 2000.

83. Curiel, D.T. et al., Adenovirus enhancement of transferrin-polylysine-mediated gene delivery, *Proc. Natl. Acad. Sci. U S A*, 88, 8850–8854, 1991.

84. Midoux, P. and Monsigny, M., Efficient gene transfer by histidylated poly-L-Lysine/pDNA complexes, *Bioconjug. Chem.*, 10, 406–411, 1999.

85. Pouton, C.W. et al., Polycation-DNA complexes for gene delivery: A comparison of the biopharmaceutical properties of cationic polypeptides and cationic lipids, *J. Control. Rel.*, 53, 289–299, 1998.

86. Brown, M.D. et al., In vitro and in vivo gene transfer with poly(amino acid) vesicles, *J. Control. Rel.*, 93, 193–211, 2003.

87. Boussif, O. et al., Synthesis of polyallylamine derivatives and their use as gene transfer vectors in vitro, *Bioconjug. Chem.*, 10, 877–883, 1999.

88. Leng, Q.X. et al., Histidine-lysine peptides as carriers of nucleic acids, *Drug News Perspect.*, 20, 77–86, 2007.

89. Goula, D., Remy, J.S., P. Erbacher, M. Wasowicz, G. Levi, B. Abdallah, and B.A. Demeneix, Size, diffusibility and transfection performance of linear PEI/DNA complexes in the mouse central nervous system, *Gene Ther.*, 5, 712–717, 1998.

90. Ferrari, S. et al., Exgen 500 is an efficient vector for gene delivery to lung epithelial cells in vitro and in vivo, *Gene Ther.*, 4, 1100–1106, 1997.

91. Bragonzi, A. et al., Comparison between cationic polymers and lipids in mediating systemic gene delivery to the lungs, *Gene Ther.*, 6, 1995–2004, 1999.

92. Godbey, W.T., Wu, K.K., and Mikos, A.G., Size matters: Molecular weight affects the efficiency of poly(ethylenimine) as a gene delivery vehicle, *J. Biomed. Mater. Res.*, 45, 268–275, 1999.

93. Fischer, D. et al., A novel non-viral vector for DNA delivery based on low molecular weight, branched polyethylenimine: Effect of molecular weight on transfection efficiency and cytotoxicity, *Pharm. Res.*, 16, 1273–1279, 1999.

94. Wightman, L. et al., Different behavior of branched and linear polyethylenimine for gene delivery in vitro and in vivo, *J. Gene. Med.*, 3, 362–372, 2001.

95. Godbey, W.T., Wu, K.K., and Mikos, A.G., Poly(ethylenimine)-mediated gene delivery affects endothelial cell function and viability, *Biomaterials*, 22, 471–480, 2001.

96. Regnstrom, K. et al., Gene expression profiles in mouse lung tissue after administration of two cationic polymers used for nonviral gene delivery, *Pharm. Res.*, 23, 475–482, 2006.

97. Kichler, A., M. Chillon, C. Leborgne, O. Danos, and B. Frisch, Intranasal gene delivery with a polyethylenimine-PEG conjugate, *J. Control. Rel.*, 81, 379–388, 2002.

98. Brownlie, A., Uchegbu, I.F., and Schatzlein, A.G., PEI-based vesicle-polymer hybrid gene delivery system with improved biocompatibility, *Int. J. Pharm.*, 274, 41–52, 2004.

99. Jeong, G.J. et al., Biodistribution and tissue expression kinetics of plasmid DNA complexed with polyethylenimines of different molecular weight and structure, *J. Control. Rel.*, 118, 118–125, 2007.

100. Thomas, M. et al., Cross-linked small polyethylenimines: While still nontoxic, deliver DNA efficiently to mammalian cells in vitro and in vivo, *Pharm. Res.*, 22, 373–380, 2005.

101. Schatzlein, A.G. et al., Preferential liver gene expression with polypropylenimine dendrimers, *J. Control. Rel.*, 101, 247–258, 2005.

102. Tang, G.P. et al., Polyethylene glycol modified polyethylenimine for improved CNS gene transfer effects of PEGylation extent, *Biomaterials*, 24, 2351–2362, 2003.

103. Nguyen, H.K. et al., Evaluation of polyether-polyethylenimine graft copolymers as gene transfer agents, *Gene Ther.*, 7, 126–138, 2000.

104. Sung, S.J. et al., Effect of polyethylene glycol on gene delivery of polyethylenimine, *Biol. Pharm. Bull.*, 26, 492–500, 2003.

105. Ogris, M. et al., Melittin enables efficient vesicular escape and enhanced nuclear access of nonviral gene delivery vectors, *J. Biol. Chem.*, 276, 47550–47555, 2001.

106. Han, S.O., Mahato, R.I., and Kim, S.W., Water-soluble lipopolymer for gene delivery, *Bioconjug. Chem.*, 12, 337–345, 2001.

107. Thomas, M. and Klibanov, A.M., Enhancing polyethylenimine's delivery of plasmid DNA into mammalian cells, *Proc. Natl. Acad. Sci. U S A*, 99, 14640–14645, 2002.

108. Wang, D. et al., Novel branched poly(ethylenimine)—cholesterol water soluble lipopolymers for gene delivery, *Biomacromolecules*, 3, 1197–1207, 2002.

109. Merdan, T. et al., PEGylation of poly(ethylene imine) affects stability of complexes with plasmid DNA under in vivo conditions in a dose-dependent manner after intravenous injection into mice, *Bioconjug. Chem.*, 16, 785–792, 2005.

110. Mumper, R.J. et al., Novel polymeric condensing carriers for gene delivery, *Proceed. Control. Rel. Soc.*, 178–179, 1995.

111. Murata, J., Ohya, Y., and Ouchi, T., Possibility of application of quaternary chitosan having pendant galactose residues as gene delivery tool, *Carbohydr. Polym.*, 29, 69–74, 1996.

112. Corsi, K. et al., Mesenchymal stem cells, MG63 and HEK293 transfection using chitosan-DNA nanoparticles, *Biomaterials*, 24, 1255–1264, 2003.

113. Gao, S.Y. et al., Galactosylated low molecular weight chitosan as DNA carrier for hepatocyte-targeting, *Int. J. Pharm.*, 255, 57–68, 2003.

114. Li, X.W. et al., Sustained expression in mammalian cells with DNA complexed with chitosan nanoparticles, *Biochim. Biophys. Acta-Gene Struct. Expression*, 1630, 7–18, 2003.

115. Thanou, M. et al., Quaternized chitosan oligomers as novel gene delivery vectors in epithelial cell lines, *Biomaterials*, 23, 153–159, 2002.

116. Kim, T.H. et al., Efficient gene delivery by urocanic acid-modified chitosan, *J. Control. Rel.*, 93, 389–402, 2003.

117. Leong, K.W. et al., DNA-polycation nanospheres as non-viral gene delivery vehicles, *J. Control. Rel.*, 53, 183–193, 1998.

118. Erbacher, P. et al., Chitosan-based vector/DNA complexes for gene delivery: Biophysical characteristics and transfection ability, *Pharm. Res.*, 15, 1332–1339, 1998.

119. MacLaughlin, F.C. et al., Chitosan and depolymerised chitosan oligomers as condensing carriers for in vivo plasmid delivery, *J. Control. Rel.*, 56, 259–272, 1998.

120. Ishii, T., Okahata, Y., and Sato, T., Mechanism of cell transfection with plasmid/chitosan complexes, *Biochim. Biophys. Acta-Biomembr.*, 1514, 51–64, 2001.

121. Uchegbu, I.F. et al., Gene transfer with three amphiphilic glycol chitosans—the degree of polymerisation is the main controller of transfection efficacy, *J. Drug Target.*, 12, 527–539, 2004.

122. Putnam, D. et al., Polyhistidine-PEG: DNA nanocomposites for gene delivery, *Biomaterials*, 24, 4425–4433, 2003.

123. Lim, Y.B. et al., Biodegradable polyester, poly alpha-(4 aminobutyl)-L-glycolic acid, as a non-toxic gene carrier, *Pharm. Res.*, 17, 811–816, 2000.

124. Lee, M. et al., Prevention of autoimmune insulitis by delivery of a chimeric plasmid encoding interleukin-4 and interleukin-10, *J. Control. Rel.*, 88, 333–342, 2003.

125. Maheshwari, A. et al., Biodegradable polymer-based interleukin-12 gene delivery: Role of induced cytokines, tumor infiltrating cells and nitric oxide in anti-tumor activity, *Gene Ther.*, 9, 1075–1084, 2002.

126. Zugates, G.T. et al., Poly(beta-amino ester)s for DNA delivery, *Isr. J. Chem.*, 45, 477–485, 2005.

127. Little, S.R. et al., Poly-beta amino ester-containing microparticles enhance the activity of nonviral genetic vaccines, *Proc. Natl. Acad. Sci. U S A*, 101, 9534–9539, 2004.

128. Hinrichs, W.L. et al., Thermosensitive polymers as carriers for gene delivery, *J. Control. Rel.*, 60, 249–259, 1999.

129. Klajnert, B. and Bryszewska, M., Dendrimers: Properties and applications, *Acta Biochim. Pol.*, 48, 199–208, 2001.

130. Buhleier, E., Wehner, W., and Vogtle, F., Cascade-chain-like and nonskid-chain-like syntheses of molecular cavity topologies, *Synthesis*, 155–158, 1978.

131. Tomalia, D.A. et al., A new class of polymers—Starburst-dendritic macromolecules, *Polymer J.*, 17, 117–132, 1985.

132. Newkome, G.R. et al., Cascade molecules.2. Synthesis and characterization of a benzene[9]3-arborol, *J. Am. Chem. Soc.*, 108, 849–850, 1986.

133. Newkome, G.R. et al., Micelles.1. Cascade molecules—a new approach to Micelles—a [27]-arborol, *J. Org. Chem.*, 50, 2003–2004, 1985.

134. Ottaviani, M.F. et al., Interactions between starburst dendrimers and mixed DMPC/DMPA- Na vesicles studied by the spin label and the spin probe techniques, supported by transmission electron microscopy, *Langmuir*, 18, 2347–2357, 2002.

135. Bielinska, A.U. et al., DNA complexing with polyamidoamine dendrimers: Implications for transfection, *Bioconjug. Chem.*, 10, 843–850, 1999.

136. Tang, M.X. and Szoka, F.C., The influence of polymer structure on the interactions of cationic polymers with DNA and morphology of the resulting complexes, *Gene Ther.*, 4, 823–832, 1997.

137. Chen, W., Turro, N.J., and Tomalia, D.A., Using ethidium bromide to probe the interactions between DNA and dendrimers, *Langmuir*, 16, 15–19, 2000.

138. Bielinska, A. et al., Regulation of in vitro gene expression using antisense oligonucleotides or antisense expression plasmids transfected using starburst PAMAM dendrimers, *Nucleic Acids Res.*, 24, 2176–2182, 1996.

139. Delong, R. et al., Characterization of complexes of oligonucleotides with polyamidoamine starburst dendrimers and effects on intracellular delivery, *J. Pharm. Sci.*, 86, 762–764, 1997.

140. Yoo, H., Sazani, P., and Juliano, R.L., PAMAM dendrimers as delivery agents for antisense oligonucleotides, *Pharm. Res.*, 16, 1799–1804, 1999.

141. Hollins, A.J. et al., Evaluation of generation 2 and 3 poly(propylenimine) dendrimers for the potential cellular delivery of antisense oligonucleotides targeting epidermal growth factor receptor, *Pharm. Res.*, 21, 458–466, 2004.

142. Santhakumran, L.M., Thomas, T., and Thomas, T.J., Enhanced cellular uptake of a triplex-forming oligonucleotide by nanoparticle formation in the presence of polypropylenimine dendrimers, *Nucleic Acid Res.*, 32, 2102–2112, 2004.

143. Bell, H.S. et al., A p53-derived apoptotic peptide derepresses p73 to cause tumor regression in vivo, *J. Clin. Investig.*, 117, 1008–1018, 2007.

144. de Jong, G. et al., Efficient in-vitro transfer of a 60-Mb mammalian artificial chromosome into murine and hamster cells using cationic lipids and dendrimers, *Chromosome Res.*, 9, 475–485, 2001.

145. Kukowska-Latallo, J.F. et al., Efficient transfer of genetic material into mammalian-cells using Starburst polyamidoamine dendrimers, *Proc. Natl. Acad. Sci. U S A*, 93, 4897–4902, 1996.

146. Tang, M.X., Redemann, C.T., and Szoka, F.C., In vitro gene delivery by degraded polyamidoamine dendrimers, *Bioconjug. Chem.*, 7, 703–714, 1996.

147. Haensler, J. and Szoka, F.C., Polyamidoamine cascade polymers mediate efficient transfection of cells in culture, *Bioconjug. Chem.*, 4, 372–379, 1993.

148. Malik, N. et al., Dendrimers: Relationship between structure and biocompatibility in vitro, and preliminary studies on the biodistribution of I-125-labelled polyamidoamine dendrimers in vivo, *J. Control. Rel.*, 65, 133–148, 2000.

149. Tack, F. et al., Modified poly(propylene imine) dendrimers as effective transfection agents for catalytic DNA enzymes (DNAzymes), *J. Drug Target.*, 14, 69–86, 2006.

150. Tziveleka, L.A. et al., Synthesis and characterization of guanidinylated poly(propylene imine) dendrimers as gene transfection agents, *J. Control. Rel.*, 117, 137–146, 2007.

151. Kukowska-Latallo, J.F. et al., Intravascular and endobronchial DNA delivery to murine lung tissue using a novel, nonviral vector, *Hum. Gene Ther.*, 11, 1385–1395, 2000.

152. Kihara, F. et al., In vitro and in vivo gene transfer by an optimised a-cyclodextrin conjugate with polyamidoamine dendrimer, *Bioconjug. Chem.*, 14, 342–350, 2003.

153. Liu, Y. et al., Factors influencing the efficiency of cationic liposome-mediated intravenous gene delivery, *Nature Biotechnol.*, 15, 167–173, 1997.

154. Han, J. and Yeom, Y.I., Specific gene transfer mediated by galactosylated poly-L-lysine into hepatoma cells, *Int. J. Pharm.*, 202, 151–160, 2000.

155. Perales, J.C. et al., Gene transfer in vivo: Sustained expression and regulation of genes introduced into the liver by receptor-targeted uptake, *Proc. Natl. Acad. Sci. U S A*, 91, 4086–4090, 1994.

156. Kim, K.S. et al., Bifunctional compounds for targeted hepatic gene delivery, *Gene Ther.*, 14, 704–708, 2007.

157. Baigude, H. et al., Design and creation of new nanomaterials for therapeutic RNAi, *ACS Chem. Biol.*, 2, 237–241, 2007.

158. Zimmermann, T.S. et al., RNAi-mediated gene silencing in non-human primates, *Nature*, 441, 111–114, 2006.

159. Russ, V. et al., Novel biodegradeable oligoethylenimine acrylate ester-based pseudodendrimers for in vitro and in vivo gene transfer, *Gene Ther.*, doi: 10.11038/sj.gt.3303046, 2007.

160. Mislick, K.A. et al., Transfection of folate-polylysine DNA complexes: Evidence for lysosomal delivery, *Bioconjug. Chem.*, 6, 512–515, 1995.

161. Cotten, M. et al., Transferrin-polycation-mediated introduction of DNA into human leukemic cells: Stimulation by agents that affect the survival of transfected DNA or modulate transferrin receptor levels, *Proc. Natl. Acad. Sci. U S A*, 87, 4033–4037, 1990.

162. Kircheis, R. et al., Tumor-targeted gene delivery: An attractive strategy to use highly active effector molecules in cancer treatment, *Gene Ther.*, 9, 731–735, 2002.

163. Peng, Z., Current status of Gendicine in China: Recombinant Ad-p53 agent for treatment of cancers, *Hum. Gene Ther.*, 16, 1016–1027, 2005.

164. Maruyama-Tabata, H. et al., Effective suicide gene therapy in vivo by EBV-based plasmid vector coupled with polyamidoamine dendrimer, *Gene Ther.*, 7, 53–60, 2000.

165. Kato, N. et al., Efficient gene transfer from innervated muscle into rat peripheral and central nervous systems using a non-viral haemagglutinating virus of Japan (KVJ)-liposome method, *J. Neurochem.*, 85, 810–815, 2003.

166. Vincent, L. et al., Efficacy of dendrimer-mediated angiostatin and TIMP-2 gene delivery on inhibition of tumor growth and angiogenesis: In vitro and in vivo studies, *Int. J. Cancer*, 105, 419–429, 2003.

167. Marano, R.J. et al., Inhibition of in vitro VEGF expression and choroidal neovascularization by synthetic dendrimer peptide mediated delivery of a sense oligonucleotide, *Exp. Eye Res.*, 79, 525–535, 2004.

168. Tang, G.P. et al., Low molecular weight polyethylenimines linked by beta-cyclodextrin for gene transfer into the nervous system, *J. Gene. Med.*, 8, 736–744, 2006.

169. Milligan, E.D. et al., Intrathecal polymer-based interleukin-10*gene delivery for neuropathic pain, *Neuron Glia Biol.*, 2, 293–308, 2006.

170. Qin, L. et al., Efficient transfer of genes into murine cardiac grafts by Starburst polyamidoamine dendrimers, *Hum. Gene Ther.*, 9, 553–560, 1998.

171. Wang, Y.O. et al., DNA/dendrimer complexes mediate gene transfer into murine cardiac transplants ex vivo, *Mol. Ther.*, 2, 602–608, 2000.

172. Rowe, R.C., Sheskey, P.J., and Owen, S.C., *Handbook of Pharmaceutical Excipients*, Pharmaceutical Press, London, 2006.

15 Receptor-Targeted Polyplexes for DNA and siRNA Delivery

Alexander Philipp and Ernst Wagner

CONTENTS

15.1 INTRODUCTION

Natural mechanisms enable the active transport of macromolecules into cells, such as receptor-mediated endocytosis, macropinocytosis, phagocytosis, and other mechanisms used also by various viruses and microorganisms to infect cells. These natural pathways can be used in a new, artificial setting as means of a transfer route to deliver nucleic acid into cells. A series of intra- and extracellular barriers have to be overcome before a delivered gene can exert the required effect. In addition, the host organism has developed defense mechanism to protect the genome against the uptake of exogenous nucleic acids. For example, many processes including the immune response can counteract viral infection processes, which present a significant problem for the development of viral gene vectors. For this reason, nonviral vectors have been developed with useful characteristics such as reduced immunogenicity, and pharmaceutical advantages (simple synthesis and large-scale production). However, they show far lower transfection efficiency compared to viral vectors.

More effective nonviral delivery systems are needed. These can succeed in therapy only if properly directed towards the target site, i.e., the diseased tissue, for example, a tumor. Thus, targeting is one of the major bottlenecks in gene therapy. Once the target cell is reached, intracellular uptake and release of the therapeutic nucleic acid in bioactive form is the second major delivery task. This chapter reviews the development of polymer-based nonviral transfer systems (polyplexes) which exploit receptor-mediated delivery to target and transfer therapeutic nucleic acids into cells.

The first receptor-targeted polyplexes and their *in vivo* use were described already 20 years ago [1,2]. Their further development and current concepts on types of therapeutic nucleic acids, polymer conjugates, polyplex formation, and targeting across extracellular and intracellular barriers are reviewed. Recent designs of bioresponsive polyplexes that during the delivery process change their characteristics to fit best for the various delivery steps, and current therapeutic polyplex concepts are also discussed.

15.2 DEVELOPMENT OF POLYPLEXES

15.2.1 NUCLEIC ACIDS FOR GENE THERAPY

Nucleic acids with therapeutic potential include plasmid DNA (pDNA) or synthetic nucleic acids such as antisense oligonucleotides, ribozymes, small interfering RNA (siRNA), other double-stranded nucleic acids like decoy DNA [3] or double-stranded RNA, polyinosine–cytosine (polyIC) [4], and RNA aptamers [5]. The various types of nucleic acids achieve different effects at the molecular genetic level. Thus, pDNA vectors are mainly used for intranuclear delivery to replace or to substitute a specific genetic function in the target cell resulting in a *gain of gene function*. The transcriptional/translational machinery amplifies the effector molecules to therapeutically relevant levels, which act directly in the producer cell or can be transported to other cells. In contrary, *loss of gene function* is often mediated by intracytoplasmic delivery of synthetic antisense oligonucleotides or siRNA reducing the expression of endogenous genes in a sequence-specific manner [6]. In the siRNA approach the catalytic nature of the formed RNA-induced silencing complex (RISC) in degrading the target mRNA leads to knockdown of the target gene; this effect, however, is limited to the transfected cells. The different molecular and biophysical properties (size, stability) and delivery requirements for pDNA (requires nuclear uptake, optionally persistent expression) and siRNA (cytoplasmic delivery, transient) also influence the therapeutic strategies.

15.2.2 NUCLEIC ACID CARRIER MOLECULES

The efficacy of gene therapy depends on several factors. Direct delivery of "naked" nucleic acids, i.e., in the absence of a carrier, can be only rarely applied with reasonable efficiency, such as in case of naked pDNA in intramuscular vaccination or hydrodynamic delivery [7,8]. In systemic applications, problems like undesired interactions with blood components, degradation, and complement activation occur. For safe and efficient transfer into the cells, nucleic acids need to be stabilized. This can be achieved by using carrier molecules that condense and compact the negatively charged nucleic acids into particles of virus-like dimensions. The carriers can be cationic polymers or lipids which lead to noncovalent complex formation into "lipoplexes" or "polyplexes" [9]. This chapter focuses on polyplexes, i.e., polymer-based systems.

Polyplexes are formed by nucleic acids and polycationic polymers like polylysine (PLL), polyethylenimine (PEI), or polyamidoamine (PAMAM) dendrimers [10–12]. Gene transfer efficiency strongly varies between the different formulations. Within commonly used polymers, PEI and PAMAM dendrimers are the most effective polycations with excellent and consistent transfection efficiency on several cell lines. The buffering capacity of these polymers offers the opportunity to escape from the endosome (proton sponge effect) [13]. Drawbacks of PEI and PAMAM dendrimers are significant toxicity and lack of degradability. PLL, in contrast, is biodegradable but ineffective in transfection efficiency unless endosomal disruptive agents are included. A series of other nucleic acid-binding polycations have also been used: for example, polyarginine [14], protamine [15], nuclear high-mobility group proteins [16], histones [17], and biodegradable polymers as described below.

The cationic residues within the polyplex can enhance binding to the cell and may also mediate the transfer of nucleic acid to the cytoplasm by disruption of the vesicular membranes. Thus, condensing the nucleic acid is essential for efficient transfer into cells; however, not every polymer that binds and condenses nucleic acid results in efficient gene transfer.

In addition, the polymeric carrier should also protect nucleic acids from degradation by serum nucleases and shield them from undesired interactions with the physiological environment due to neutralizing the negative charges. In reality, the carrier is unable to carry out all the tasks meaning that in some cases its primary role is to bind and protect nucleic acids from the environment.

Furthermore the polymeric carrier has to meet different requirements depending on the type of nucleic acid. In case of siRNA a release of siRNA from the carrier molecule is necessary in the cytoplasm to allow forming the RISC complex. In case of pDNA, nuclear entry, vector unpacking, and transcription are the necessary steps. Any of these steps might be rate limiting depending on the nature of the carrier. Carrier molecules should interact with the genetic material in a nondamaging reversible manner which in most cases is provided by noncovalent electrostatic interaction. In few cases, also reversible covalent attachment has been exploited, for example, covalent linkage of siRNA with cholesterol which was found to greatly enhance activity of siRNA [18].

15.2.3 New Cationic Polymeric Carriers

Commonly used nucleic acid carriers like PEI are limited by inherent toxicity and moderate efficiency. Thus, novel biodegradable carriers with improved efficiency, less toxicity, and a better biocompatibility are required [19–24]. Polycationic carriers with high-molecular weight (HMW) often show acute- and long-term toxicity. In contrast, low-molecular weight (LMW) polymers often show reduced toxicity but also low transfection efficiency and *in vivo* stability. Bioreversible cross-linking of LMW polymers into larger molecules, either by reducible disulfides or hydrolyzable linkages, is one encouraging strategy. Several approaches were performed in cross-linking LMW-PEI via ester- or amid-linkages to receive a strongly enhanced gene transfer efficiency *in vitro* and *in vivo* [19,22,25]. Most importantly, these polymers were less toxic than standard PEI. Moreover, ester linkages were also used for synthesis of biodegradable copolymers of PLL and polyethylene glycol (PEG) [26].

Several strategies for biodegradable carriers are based on reductive cleavage of disulfide bonds [27]. Read et al. synthesized reducible polycations based on short histidine and oligolysine residues enhancing gene transfer efficiency *in vitro* for pDNA, mRNA, and siRNA [28]. Intracellular cleavage of disulfide bonds leads to release of the pDNA, and histidine residues mediate endosomal escape. The reducible polycations showed promising transfection efficiency; toxicity was not observed under conditions where PLL was strongly toxic. In combination with PEG, disulfide cross-linked polymers were also applied to facilitate gene transfer *in vivo* [29].

Davis and colleagues designed novel cationic β-cyclodextrin-based polymers (βCDPs) [30], whereby the incorporation of cyclodextrin in the amidine-containing polymers reduced toxicity of the polymer by up to three orders of magnitude. Modification of βCDPs with PEG and transferrin (Tf) was used successfully for *in vivo* tumor targeting of RNA-cleaving DNA enzymes [31]. Adamantane–PEG–Tf conjugates fuse with the polyplex by forming adamantane–cyclodextrin inclusion complexes. Also cyclodextrins were used for grafting PEI which enables coating of polyplexes with adamantane–PEG conjugates [32].

It has to be emphasized that successful pDNA formulations are not always suitable for siRNA delivery. Leng et al. [33] demonstrated that a branched carrier consisting of lysine and histidine residues designed for pDNA was not able to mediate siRNA delivery. Compositions of various related carriers (with different degree of branching, length of terminal arms, changed histidine-lysine-sequences) generated a suitable carrier for siRNA with minimal toxicity and luciferase knockdown activity. Hassani et al. tested different formulations to deliver siRNA into the mouse brain *in vivo* [34]. They found out that linear PEI, a powerful pDNA delivery agent, was not suitable for efficient siRNA delivery at concentrations where cationic lipid formulations were effective. Aigner et al. showed that PEI is able to deliver siRNA into distant subcutaneous tumors [35]. Fractionation of standard PEI generated a low-molecular weight fraction with superior transfection efficiency in pDNA and siRNA delivery [36].

15.3 CELLULAR TARGETING AND ENDOCYTOSIS

15.3.1 Endocytosis of Polyplexes

Endocytosis is the major pathway of entry into cells. However, various endocytic pathways [37,38] operate in eukaryotic cells, i.e., clathrin-dependent and -independent pathways, the latter including phagocytosis, macropinocytosis, and caveolae-mediated internalization. Understanding cellular uptake and intracellular processing of nonviral gene delivery systems is a key aspect in developing more efficient vectors. In one study small particles (<200 nm) were found to internalize via clathrin-mediated endocytosis, whereas large particles (>500 nm) internalized via caveolae-mediated endocytosis [39]. In many cell types, positively charged polyplexes like pDNA/PEI polyplexes internalize through adhesion to negatively charged transmembrane heparan sulfate-proteoglycans (HSPG) [40,41]. Clathrin-dependent receptor-mediated endocytosis (coated pit endocytosis) involves the binding of a ligand to a specific cell surface receptor, resulting in the clustering of the ligand/receptor complexes in clathrin coated pits, invagination into the cell and taking off of the coated pits from the cell surface membrane to form intracellular coated vesicles, and maturation (uncoating, fusion of vesicles) into endosomes. Within these endosomes, ligands and receptors are sorted to their appropriate intracellular destination (e.g., lysosome, Golgi apparatus, nucleus, or cell surface membrane). The clathrin-independent endocytosis [42] resulting in uncoated pits includes phagocytosis, pinocytosis, and potocytosis. Phagocytosis is a mechanism of internalizing large particles and microorganisms (>0.5 μm). This mode of internalization is initiated by receptors on the phagocyte recognizing the particle either directly or indirectly via opsonization. Internalization is primarily mediated via pseudopod action rather than pits (invaginations) on the cell surface. The wrapped particle, initially situated in the early phagosome, is eventually destroyed along the endocytosed pathway. Thus, cell capability to recognize particles via receptors and to form pseudopods seems to be a major characteristic mediating phagocytic internalization. Potocytosis and transcytosis may use caveolae as routes for internalization [38,43]. Many receptors contain motifs in their cytoplasmic domains that act as recognition sequences for initiating enhanced intracellular uptake process of macromolecules. With ligand interaction, the rate of receptor internalization is increased.

Gersdorff et al. [44] evaluated the impact of clathrin- and caveolae/lipid-raft-dependent endocytosis on cell entry and transfection efficiency of PEI polyplexes. Interestingly, the productive endocytotic pathway was varying in different cell types. For example, in COS-7 cells the clathrin-dependent pathway was the main contributor to the transfection process, whereas in HeLa cells the lipid-raft-dependent pathway was more relevant, consistent with other literature [45].

15.3.2 Ligand Conjugates for Targeting Polyplexes

Gene transfer vehicles have to accomplish several major delivery tasks, especially transferring the nucleic acid from the site of administration to the surface of the target cells, and facilitating the internalization into the cells. Incorporation of receptor ligands has been considered to promote these tasks, i.e., targeting and enhanced cellular internalization of the nucleic acid (Figure 15.1). For *in vivo* application, both processes are substantial for successful gene therapy.

Targeting ligands can be proteins like Tf, asialoglycoproteins, epidermal growth factor, or they can be small natural/synthetic molecules, e.g., folic acid, peptides, or sugar derivatives. Examples of ligands that have been used as conjugates with polycationic carriers for targeted gene transfer are shown in Table 15.1 and Refs. [2,15,46–83]. In the selection of a ligand several aspects have to be considered. Some ligands are very tissue-specific (e.g., asialoglycoproteins/hepatocytes), whereas others are not (e.g., Tf/iron supply to many cell types). Some ligands are internalized very efficiently (e.g., Tf, anti-CD3 antibody bound to the T-cell receptor associated surface molecule CD3), whereas some others may be internalized either very slowly or not at all. Some ligands may carry charges and interact with other proteins and surfaces (e.g., FGF with blood), others not or may even act as a shield (e.g., serum protein Tf). Thus, the proper choice of ligand for efficient gene transfer is important. Ligands have been successfully incorporated into polyplexes by conjugation to polycations such as PLL, PEI, protamines, and histones, or also intercalating DNA-binding compounds (Table 15.1).

One commonly used ligand is the serum glycoprotein Tf for targeted gene delivery. The Tf-receptor required for iron uptake into cells is over-expressed in tumor cells due to the higher demand of iron for their growth. Tf as a protein ligand combines both a shielding function of the vectors and a

targeting function towards the tumor site [84,85]. It was shown that incorporation of Tf or anti-Tf receptor single-chain antibody Fv (anti-Tf scFv) fragments into polyplexes and lipoplexes resulted in enhanced gene transfer efficiencies *in vitro* and *in vivo* [86–94]. Conjugation of PEG to pDNA/PEI polyplexes or lipoplexes containing Tf or anti-Tf scFv fragments as targeting ligand further improved applications for *in vivo*.

After systemic administration of Tf–PEG-conjugated vectors gene expression was mainly found at the tumor site [92,93,95]. Tail vein injection of Tf-conjugated pDNA/PEI polyplexes into tumor-bearing mice resulted in 100-fold higher luciferase reporter gene expression in the distant tumors in comparison with the other organs [88,92] and maintains over 3 days upon single injection. Systemic application of Tf polyplexes of pDNA encoding TNF-α into tumor-bearing mice induced tumor necrosis and inhibition of tumor growth in several mouse tumor models [89,95].

A cationic cyclodextrin carrier containing Tf as targeting ligand and PEG for shielding was developed for systemic siRNA delivery [96]. Repeated systemic delivery of siRNA against the Ewing's sarcoma-specific chromosomal translocation t (11:22) inhibited growth of metastatic Ewing's sarcoma in a murine model.

EGF is another largely investigated ligand for tumor targeting [4,19,97–99]. The EGF-receptor (EGFR) is upregulated in several tumors including epithelial tumors, glioblastoma, and hepatocellular carcinoma. Conjugation of EGF to PEGylated PEI polyplexes resulted in strong enhancement of gene expression *in vitro* [98]. Also upon systemic application in hepatocellular carcinoma-bearing mice, a 10-fold enhanced gene expression was found at the tumor site [47]. Shir et al. recently described therapeutic effects of EGF-targeted polyplexes containing poly IC observing the complete regression of established EGFR over-expressing glioblastomas in nude mice [4].

Another tumor target is the folate receptor which shows a narrow expression on healthy tissue whereas it is upregulated in numerous cancer types [100]. Efforts in the development of folate receptor-targeted vectors by using folate itself as a targeting ligand show promising tumor-specific gene transfer [101].

Similar to tumor cells, also tumor vascular endothelial cells over-express certain surface markers that are only upregulated in neoangiogenic tumor blood vessels or exist at very low levels on normal blood vessels. Such surface receptors include, for example, the integrins like αvβ3 and αvβ5. Several approaches to target gene transfer carriers towards the integrins use synthetic peptides with the RGD sequence [102–104]. RGD-targeted PEI polyplexes have shown enhanced gene transfer efficiency compared to PEI. Thereby the receptor-mediated gene transfer depends on the degree of RGD substitution of targeted PEI conjugates [105]. Coupling RGD via a PEG spacer to PEI, targeting was partially reduced, possibly because of hiding the RGD sequence inside the PEG curtain [104,105]. Suh et al. demonstrated the influence of optimal composition of RGD–PEG–PEI conjugates on transfection efficiency showing that a growing degree of PEG-RGD incorporated onto PEI leads to decreased transfection efficiency

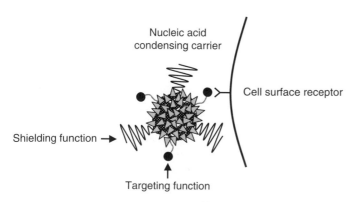

FIGURE 15.1 Assembly of polyplexes. Gene transfer particles are composed of nucleic acid, ligand–polycation conjugates and additional elements, e.g., shielding agents. Surface characteristics of polyplexes (charge density, degree of coating with ligand/shielding agents) can influence cell binding. Unspecific interactions can be reduced by shielding with PEG. Cell targeting is mediated by ligand–receptor interaction. Examples of targeting ligand/receptor systems can be found in Table 15.1.

TABLE 15.1

Ligands Used in Receptor-Targeted Polyplexes

Receptor	Ligand	Polycation Used
Airway cells	Surfactant proteins A and B	PLL
Arterial wall	Artery wall-binding peptide	PEG–PLL
ASGP-R	Asialoglycoproteins	PLL, oligolysine, bisacridine,
	Synthetic galactosylated ligands	polyornithine, PEI
Carbohydrates	Lectins	PLL
CD3	Anti-CD3	PLL, PEI
CD5	Anti-CD5	PLL
CD13	NGR	PEI
CD44	Hyaluronic acid fragments	PPI
CD117	Steel factor	PLL
	Anti-CD117	
EGF receptor	EGF, EGF peptide	PLL, PEI, PLL plus gal4 domain
	Anti-EGF	
	TGF-α	
ErbB2	Anti-ErbB2	PLL plus gal4 domain
Fc receptor	IgG	PLL
FGF2 receptor	Basic FGF	PLL
Folate receptor	Folate	PLL, PEI
Hepatocyte basolateral surface	Malarial circumsporozoite protein	PLL
Her2	Anti-HER2	PLL, PEI
Insulin receptor	Insulin	Cation-modified albumin, PLL
Integrin	RGD peptide	Oligolysine, PEG–PEI
LDL receptor family (hepatocytes)	Receptor-associated protein	PLL, poly(ᴅ)lysine
Mannose receptor (macrophages)	Synthetic ligands, mannosylated	PLL
Nerve growth factor (NGF) receptor TrkA	NGF-served synthetic peptide	Oligolysine
Neuroblastoma	Antibody ChCE7	PEG–PEI
Ovarian carcinoma cell surface antigen OA3	Antibody OV-TL16 Fab' fragment	
PECAM (lung endothelium)	Anti-PECAM antibody	PEI
Poly-immunoglobulin receptor	Antisecretory component Fab' fragment	PLL
Serpin-enzyme receptor	Peptide ligand	PLL
Surface immunoglobulin	Anti-IgG, Anti-idiotype	PLL
Thrombomodulin	Antithrombomodulin	PLL
Tn carbohydrate	Anti-Tn	PLL
Transferrin receptor	Transferrin Anti-TfR scFv	PLL, protamine, ethidium dimer, PEI, PEG PEI

Note: PEG, polyethylene glycol; PEI, polyethylenimine; PLL, polylysine; PPI, polypropylenimine.

[104]. Recently, RGD–PEG–PEI conjugates have been successfully tested for systemic antiangiogenic siRNA therapy [106]. Repeated intravenous administration into neuroblastoma-bearing mice resulted in sequence-specific inhibition of tumor growth.

As a further ligand for targeting angiogenic blood vessels, the NGR peptide (=Asn-Gly-Arg motif) has been investigated [107]. NGR shows highest affinity for aminopeptidase N (APN, also known as CD13) which plays an important role in tumor invasion [108]. Moffatt et al. recently reported promising *in vivo* results using a CD13 targeted PEG–PEI polyplex resulting in enhanced gene expression at the tumor site [56]. In these polyplexes CNGRC (the cyclic form of a pentapeptide containing the NGR motif) conjugated with PEG is attached

to PEI via noncovalent phenyldiboronic acid/salicylhydroxamic acid bridges. *In vivo* application for p53 gene transfer polyplexes showed encouraging therapeutic effects, i.e., significant regression of non-small-cell lung carcinoma. Moreover, this vector targeted tumor tissue and tumor-associated endothelial cells but not normal cells [109].

Other possibilities, e.g., using antibodies to target specific cells, have been investigated successfully. Anti-CD3 antibody, coupled to pDNA polyplexes, has been shown to be very efficient in targeting T cells via binding the CD3 T-cell receptor complex [110]. Also attachment of modified MIP-1 beta to the surface of pDNA–liposome complexes targeting CCR5 displayed on the surface of helper T cells and macrophages increased gene expression in circulating lymphocytes approximately sixfold [111]. Malignant B cells have been successfully targeted by anti-idiotype antibodies [112].

The majority of approaches attaches ligands to polyplexes by conjugation to polycations. An alternative mode to incorporate ligands into polyplexes employs nucleic acid-binding domains derived from transcription factors to attach a protein component to pDNA [113]. For example, pDNA containing GAL4 recognition motifs can interact with GAL4 containing carrier molecules. Cell-specific ligands can be combined with the GAL4 domain to form a chimeric fusion protein to allow gene delivery to specific cells, e.g., GAL4/invasin fusion protein has been shown to transfect target cells in an invasin receptor-dependent manner [114]. However, for complex formation, and hence condensation, an additional condensing agent such as PLL is required.

15.4 INTRACELLULAR TRAFFICKING AND BARRIERS

Once the nucleic acid has been internalized into the target cell, several intracellular barriers (Figure 15.2) have to be overcome for successful gene expression. To tackle the

intracellular barriers, a polyplex needs to escape from the endosome, survive the cytoplasmic environment, traffic the cytoplasmic environment targeting the nucleus with subsequent nuclear entry, and be disassembled in the nucleus so it can be recognized by the cell's transcription machinery.

Overcoming these intracellular barriers can be achieved by certain carrier molecules [115] and cell-targeting ligands, although exploiting other factors, e.g., additional endosomal releasing mechanisms and nuclear localization signals, might strongly improve the capacity of a polyplex to overcome intracellular barriers.

15.4.1 ENDOSOMAL RELEASE

The first major intracellular barrier that greatly compromises the efficiency of efficient gene transfer is the entrapment and degradation of the polyplex within intracellular vesicles after taking off of the coated pits from the plasma membrane (Figure 15.2). Entrapment and degradation can be seen as two separate barriers, because overcoming vesicular degradation would only result in accumulation of the transferred gene in the vesicles limiting further transport to

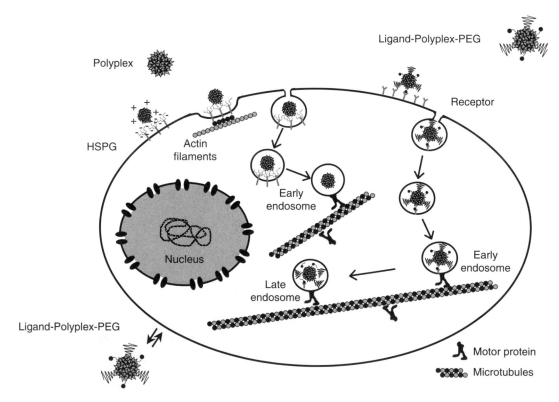

FIGURE 15.2 Intracellular release and trafficking of polyplexes. Polyplexes have to be stabilized and shielded to avoid unspecific interactions in the blood circulation. Incorporation of PEG into the polyplex is one approach to mediate these characteristics. After binding to the cell surface either by specific ligand–target-receptor interaction or unspecific interaction of positively charged polyplexes to negatively charged transmembrane heparan sulfate-proteoglycans (HSPG), polyplexes are internalized into endosomal vesicles. The acidic environment triggers cleavage of bioreversibly linked PEG from the polyplex enhancing release from the vesicle to the cytosol. Polyplexes entrapped in endocytic vesicles can be effectively transported into the perinuclear region. In the initial phase during endocytosis into vesicles, polyplexes are colocalized with the actin cytoskeleton, followed by rapid intracellular migration along microtubules. The nucleus is separated from the cytoplasm by the nuclear envelope, that is stretched by nuclear pore complexes (NPC) mediating active transport of macromolecules from the cytoplasm to the nucleus.

the nucleus. Thus, after cellular internalization, the transferred gene needs to overcome enzymatic degradation during vesicular fusion into lysosomes, and afterwards it needs to be released from the vesicles to traffic the cytoplasm and target the nucleus.

Several strategies have been developed to ensure the protection and release of polyplexes from intracellular vesicles involving incorporation of vesicular destructive elements to polyplexes, which perturb the integrity of vesicle membranes allowing the contents to get into the cytoplasm in a nondamaging manner. Application of endosomal-releasing elements to the transfection agents has shown to enhance gene transfer. Lysosomotropic agents, membrane active peptides and proteins, and also photosynthesizing compounds possess properties that disrupt vesicular membranes promoting the release of polyplexes.

15.4.1.1 Lysosomotropic and Membrane-Disruptive Agents

Lysosomotropic agents are weak-base amines that can specifically inhibit lysosomal function [116,117], e.g., ammonium chloride and the weakly basic alkylamines such as methylamine, propyl amine, or chloroquine. These agents are termed lysosomotropic because they accumulate in the endosomes which is partly due to the initially low lysosomal pH, and partly because of continuous perfusion of protons into the endosome.

In some cell lines, e.g., erythroid cell line K562, gene expression is strongly enhanced by adding chloroquine to the transfection medium [48]. Chloroquine seems to accumulate in the endosomal compartment acting osmotically, vacuolarizing, and eventually disrupting the vesicle. In this case, the transferred gene is protected and released from intracellular vesicles. Bafilomycin or monensin, two other agents that also prevent endosomal acidification but do not accumulate, show no enhanced transfection efficiency. Regarding chloroquine, its effectiveness in enhancing gene transfer is also dependent on cell type, which ensures vesicular accumulation of the lysosomotropic agent. For example, the K562 cells have a defect in their vesicular pump system, determining the enhanced accumulation of chloroquine in endocytic vesicles in comparison to other cells [118]. The use of chloroquine is limited due to toxic properties; analogs are being generated for better performance [119].

Many biological processes such as entry of viruses, bacteria, and toxins into cells, or the activities of antimicrobial peptides are examples how efficiently nature can modulate membrane barriers and promote endosomal escape. The membrane reorganization process is the result of specific actions of certain membrane disruptive elements (peptides and proteins).

In most cases, the membrane active area in such elements are peptide domains with amphipathic sequences. Under appropriate conditions the amphipathic sequences can interact with lipid membranes, perturbing them, which can be specifically used to influence gene transfer investigating various

endosomolytic agents [120]. Membrane active peptides can be derived from viral peptide sequences such as the N-terminus of influenza virus hemagglutinin subunit HA-2 [121,122] or the N-terminus of rhinovirus VP-1, from other natural sources such as melittin from bee venom [123,124], or they may be designed synthetically from the derived peptides by molecular modeling, for example GALA, KALA, EGLA, or JTS1 [125–130] or other artificial sequences [131,132].

The peptides can be incorporated into polyplexes by covalently linking to the polycation. Another possibility to achieve attachment is simple noncovalent ionic interaction between positively charged polyplexes and the negatively charged residues of membrane destabilizing elements. Membrane destabilizing peptides have shown to enhance receptor-mediated gene transfer up to 1000-fold.

Also natural nonviral endosome-destabilizing agents including listeriolysin O [133] were incorporated into polyplexes. Melittin conjugates [123,124,134] have shown to enhance gene transfer efficiency, however toxicity limits their application. Melittin shows pronounced lytic activity also at neutral pH which is responsible for toxic side effects. To overcome this limitation, Rozema et al. modified the lysines of melittin with a dimethylmaleic anhydride derivative masking the lytic activity at neutral pH. Upon endosomal acidification, the masking groups are removed and the lytic activity of melittin is restored [135]. Meyer et al. coupled melittin modified with dimethylmaleic anhydride covalently to PLL resulting in enhancement of gene transfer efficiency and also reducing the toxicity of both melittin and PLL [136]. Other approaches to pH-dependent lytic activity include acidic melittin analogs [137] or the incorporation of melittin into bioreducible copolymers [134].

15.4.1.2 Photochemical Membrane Disruption

Endosomal escape is a major drawback for efficient gene delivery. To overcome this a method called photochemical internalization (PCI) [138,139] was developed. Cells are treated with photosensitizing compounds, such as tetra (4-sulfonatophenyl) porphine or aluminium phthalocyanine, that locate in the membranes of endosomes. These amphiphilic photosensitizers become activated upon illumination and induce formation of reactive oxygen species, destroying endosomal membrane structures which lead to release of polyplexes into the cytoplasm. PCI enhances gene transfer of viral and nonviral carrier like adenovirus, lipoplexes, and polyplexes with or without receptor targeting [99,140–142]. Glucosylated PEI with increased solubility and transfection efficiency in comparison to unmodified PEI was used for polyplex formation injected directly into the tumors. Luciferase reporter gene expression after photosensitization was increased up to 100-fold in comparison to nonirradiated tumors.

Kataoka et al. developed a system assembling the photosensitizer and the gene carrier in one ternary complex [143]. pDNA was complexed with cationic peptides and afterwards anionic photosensitizers, a dendrimeric phthalocyanine, were attached to the polyplex. These ternary polyplexes achieved *in vitro* more than 100-fold increase of luciferase reporter gene expression in comparison to transfections without PCI; *in vivo* gene expression was detectable only at the laser irradiated site.

In vitro, PCI already showed encouraging results in improving gene delivery of polyplexes [144,145]. For lipoplexes the PCI effect was variable and depends on the type of liposomes used [141]. Moreover, receptor-mediated gene delivery in combination with PCI worked very well combining targeting and endosomal release facilities. Kloeckner et al. showed a 2- to 600-fold enhanced transfection efficiency of EGF-targeted polyplexes *in vitro* compared to transfections without PCI [99]. Ndoye et al. recently showed promising therapeutic effects in an HNSCC model *in vivo* with PEI-mediated p53 gene transfer combined with PCI [146].

15.4.1.3 Cationic Proton Sponge Polymers

In contrast to nucleic acid-binding polycations such as PLL, there are other polycations like PEI or PAMAM dendrimers that possess intrinsic endosomolytic activity [147], thus not requiring the presence of additional endosomolytic agents. Most of these agents have buffering capacity below physiological pH. This buffering capacity is due to residues of these agents being not protonated at physiological pH, making them efficient "proton sponges." Upon acidification in the intracellular vesicle further protonation of the polymers occurs which triggers chloride influx, resulting in osmotic endosome swelling, and thus, destabilization and rupture of the intracellular vesicle membrane, resulting in the escape of the included polyplex [148,149]. Intracellular trafficking studies measuring intravesicular pH using a pH-sensitive dye confirmed that PEI polyplexes can buffer the endocytic vesicles while PLL polyplexes do not [150].

Histidinylated PLL was also found to have enhanced transfection properties, presumably mediated by a related vesicular escape mechanism [151]. It can be noted that several polymers effective in gene transfer are proton sponges; however, not every cationic proton sponge is an effective gene transfer agent [152].

15.4.2 CYTOPLASMIC TRAFFICKING, NUCLEAR ENTRY, VECTOR UNPACKING

Besides receptor-mediated endocytosis and release from endosome the delivery systems encounter several tasks such as, cytosolic transport towards the nucleus, release of pDNA/siRNA, and nuclear import. To date the order of these processes is unclear, whether cytoplasmic trafficking proceeds before or also after endosomal escape, whether unpackaging occurs before or after nuclear entry. Different scenarios have been observed, however, it remains unclear which of the observed routes are the productive ones. Investigators have attempted to incorporate nuclear localization signals into polyplexes for improved nuclear targeting, without convincing success. In contrast to pDNA-based therapeutics, nuclear transport is irrelevant for siRNA, where targets are primarily localized in the cytoplasm.

15.4.2.1 Cytoplasmic Trafficking

Free cytosolic pDNA or also pDNA polyplexes are too large to freely diffuse within the cytosol [153]. Polyplexes entrapped in endocytic vesicles, however, can be effectively transported into the perinuclear region. Single particle tracking studies using PEI polyplexes in living cells [154] found colocalization of polyplexes with the actin cytoskeleton in the initial phase during endocytosis into vesicles, followed by rapid intracellular migration along microtubules. In a subsequent study [155], EGF-targeted polyplexes were analyzed in respect to internalization into EGF-receptor expressing cells. Uptake kinetics and internalization dynamics of EGF-targeted polyplexes were faster and more efficient internalization (within 5 min in average) compared with untargeted PEI polyplexes (more than 30 min). Transport phase I was characterized by very slow, actin cytoskeleton-mediated movement of the particles at the cell surface and during internalization. Within the cell during phase II, endosomal particles displayed increased velocities with normal and confined diffusion (due to cytoskeleton-restricted movement) in the cytoplasm. Phase III was characterized by very fast active transport of polyplexes within vesicles along the microtubules [154,155].

15.4.2.2 Nuclear Entry

The nucleus is separated from the cytoplasm by the nuclear envelope, which consists of two chemically distinct membranes, the inner and outer membrane (continuous to the endoplasmic reticulum), and in between the perinuclear cisterna space. NPCs stretch the nuclear envelope, and active transport of macromolecules from the cytoplasm to the nucleus occurs through these complexes.

There are several distinct nuclear import signals which guide the import of proteins into the nucleus. These signals are part of the primary sequence of the protein destined to be targeted into the nucleus. The best characterized ones are the SV40 LTA "classical" NLS (nuclear localization signal) and M9 (an import signal of hnRNP AI protein) import signals. The utilization of such NLS signals for polyplex delivery has been attempted, however, no convincing break-through has been achieved [156].

The nuclear envelope represents an obvious barrier for polyplex-mediated gene transfer. How polyplexes find their way to the nucleus, whether it occurs before or after polyplex dissociation is not fully understood. Rudolph et al. analyzed the dissociation of pDNA from polyplexes after cellular uptake and describe the interaction of gene vectors with t-RNA mediating the dissociation [157].

In dividing cells pDNA may passively enter the nucleus during cell division when the nuclear membrane is broken down [158]. Although rupture of the nuclear envelope is not a necessity for some PEI polyplexes to penetrate the nucleus [115,159], transfection efficiency of polyplexes in general is critically dependent on cell division [160]. However, many cell types are nondividing with the nuclear membrane staying intact, thus nuclear entry and trafficking may represent an additional intracellular barrier for successful *in vivo* gene transfer.

15.4.2.3 Vector Unpacking

The cell nucleus is saturated containing large amounts of DNA, RNA, and proteins. In addition, nuclear processes, e.g., replication, transcription, translation, and DNA repair processes, are constantly active in specific compartments, resulting in a nuclear jam [161,162].

Within all these cellular components and nuclear processes, the polyplex needs to become exposed enabling the nuclear expression machinery to recognize and express it. Early disassembly and release of the nucleic acid from the carrier molecule in the cytoplasm may prevent efficient transfer to the nucleus, hence obliterate expression. Polyplexes most likely disassemble in the nucleus which is a characteristic desirable for successful gene transfer and expression [115].

The release of pDNA complexed with polycations, including linear and branched polyethylenimine (LPEI, BPEI) and PLL, was analyzed with confocal microscopy using fluorescence resonance energy transfer (FRET) between a pair of donor–acceptor fluorescent dyes (fluorescein and Cy3) tagged on a single pDNA molecule. pDNA complexed with LPEI underwent a more rapid unpacking after endosomal release in comparison to pDNA complexed with BPEI. These intracellular characteristics showed a clear correlation to their gene transfer efficiency. The pDNA/LPEI polyplexes revealed a considerably higher and faster gene expression compared with pDNA/BPEI polyplexes. In the pDNA/PLL polyplexes, neither endosome escape nor pDNA disassembly was observed [163].

Another study describes a quantitative comparison of the cellular uptake and subsequent intracellular distribution (e.g., endosome, cytosol, and nucleus) of exogenous pDNA transfected by viral and nonviral vectors *in vitro*. As a model, adenovirus (Ad) and Lipofectamine Plus (LFN) were used for comparison since they are highly potent and widely used viral and nonviral vectors, respectively. The findings indicate that LFN requires three orders of magnitude more intranuclear gene copies to exhibit a gene expression comparable to that of the Ad. This suggests that the difference in transfection efficiency principally arises from differences in nuclear transcription efficiency and not from a difference in intracellular trafficking between Ad and LFN [164].

15.4.3 Persistence of Gene Expression

Several factors threaten the persistence of transferred pDNA within the nucleus: (1) degradation by intranuclear nucleases, (2) pDNA loss, mainly during cell division, although pDNA may also be rapidly lost even when cells are not dividing, (3) loss of transfected cell, because of apoptotic, inflammatory, or immune response, (4) silencing of the introduced gene by transcriptional shut-off, (5) possibly inefficient intranuclear trafficking.

pDNA loss can be prevented by including specific sequences to the transferred gene that ensure either integration of the pDNA into host chromosome, or extrachromosomal replication with equal segregation to daughter cells. These persistence ensuring sequences can be derived from certain viruses or from chromosomes.

Retroviruses and adeno-associated virus (AVV) stably insert their genome into host genome. The integration mechanisms have been characterized (retrovirus: LTR sequences, integrase protein; AAV: ITR sequences, rep protein) and may be exploited by incorporation of the corresponding nucleic acid and protein elements into a virus like particle (polyplex).

The sleeping beauty (SB) transposon system has been recently evaluated providing encouraging results [165,166]. SB transposase catalyses the excision of a transposon from a donor molecule and its integration into a cellular chromosome.

The phase integrase C31 (phiC31) system represents another novel technology for gene therapy. The phiC31 integrase can integrate introduced pDNA into preferred locations in unmodified mammalian genomes, resulting in robust, long-term expression of the integrated transgene [167].

Other viruses such as herpes virus, e.g., EBV, can persist in infected cells without integrating their genome into the host. This persistence is partially due to replication property of viral pDNA via cis-acting origin of replication which is activated by the trans-acting gene product of the viral EBNA-1 and additional nuclear retention mechanisms. The viral persistence mechanisms can be utilized in designing extrachromosomal replicating pDNA constructs (episomal vectors) by integrating the appropriate sequence elements, recognizable in mammalian cells, into the pDNA construct being transferred (e.g., EBV Ori P, EBNA-1). pDNA constructs containing these sequences have the ability to replicate once per cell cycle with nuclear retention without interfering with the host chromosomes. An alternative to using viral origin of replication, human genomic sequences may be used to mediate pDNA construct replication [168].

The inhibition of gene expression by RNA interference hosts a high potential for application in therapy of human diseases. However, while exogenous applications of siRNA efficiently inhibit gene expression, these effects are only transient in mammalian cells. Lipp et al. designed a short hairpin RNA-expression cassette to target the bcr-abl oncogene that was then introduced into the nonviral plasmid vector pEPI-1 [169,170] which replicates episomally in the absence of selection. Forty-two days after transfection of the bcr-abl-positive cell line K562 the bcr-abl-dependent growth rate was found to be drastically reduced in K562 cells. Western analysis revealed a more than 90% reduction in the expression of the fusion protein bcr-abl while the expression of the bcr protein remained unaffected. This and further studies demonstrate that the vector system pEPI-1 can be used for efficient long-term gene expression including specific gene suppression by using a short hairpin RNA transcription unit.

15.5 EXTRACELLULAR BARRIERS TO GENE DELIVERY

Upon *in vivo* application polyplexes face both intracellular and extracellular barriers. A schematic diagram of the potential intracellular and extracellular fate of a polyplex is shown in Figure 15.3. Strategies have to be developed in designing polyplexes that are able to survive blood or other biological fluids, and to escape extracellular physical barriers in order to reach the target cells. The specific strategies must consider the physicochemical properties of polyplexes, e.g., size, shape/flexibility, overall charge, charge density, and nonelectrostatic interactions at the surface. Besides considering these physical properties, one could imagine to use active endogenous cellular transport mechanisms such as transcytosis [171].

In systemic administration, i.e., intravenously application of the vector, the circulatory pathway/environment and various nontarget cells and organs, encountered by the polyplex, are major challenges [172]. Local administration methods such as direct injection of the vectors into the target region or topical administration are not confronted with the circulation problems, but, nevertheless, still confronted with barriers such as extracellular matrix or inflammatory and immune responses.

15.5.1 Physical Restrictions of Transfection Particles

Size seems to be a general critical factor for drug targeting [173]. Because of size restriction, several hundred nanometer large polyplexes are not able to penetrate endothelial and epithelial barriers [174], or extravasate from the vascular to the interstitial space (Figure 15.4). Particle size is also an important factor when considering organ clearance and intraorgan distribution. For example, particles that are too large to pass through the vascular endothelium to the liver parenchyma are engulfed and degraded by liver Kupffer cells, i.e., phagocytic cells located next to the vascular epithelial cells.

The structure of condensed nucleic acid polyplexes has been analyzed in several reports, e.g., [15,175–178]. Polyplexes have been characterized by electron microscopy and atomic force microscopy (shape, size), laser light scattering (size), their electrophoretic mobility (reflects charge and size of complexes), zeta potential measurements (charge), circular dichroism (conformation of pDNA), or centrifugation techniques (molecular weight and condensation). The results give some insight on generating small polyplexes. In addition, the compact polyplexes are more stable against enzymatic or mechanical degradation, which takes place during nucleic acid transport to the target cells/tissue.

However, preparation of effective polyplexes for *in vivo* delivery of nucleic acids remains a major hurdle. The extent of nucleic acid condensation depends on a number of variables including the ratio of positively charged nucleic acid-binding element (cationic carrier) to negatively charged nucleic acid. Also the size and modification of the nucleic acid-binding element, size and sequence of the nucleic acid, and even the procedure of complex formation [49] are strongly influencing the *in vitro* and *in vivo* gene transfer efficiency. The net charge

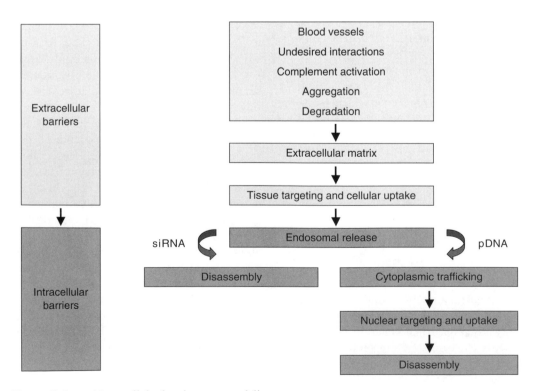

FIGURE 15.3 Extracellular and intracellular barriers to gene delivery.

Systemic application:

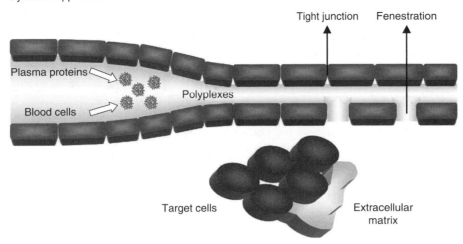

FIGURE 15.4 Vascular barriers to gene delivery. Route of polyplexes across the vascular barrier into the interstitial space, avoiding interactions with the extracellular matrix in order to reach the target cells/tissue.

of the nucleic acid/cationic carrier complex affects its solubility. Polyplexes with either an excess of nucleic acid or positively charged carrier are stabilized in solution by the negative or positive charges. At molar charge ratios (positive charges of carrier nitrogens to negative charges of nucleic acid phosphates) close to electroneutrality, often aggregation of polyplexes is observed unless the carrier contains hydrophilic domains such as PEG for stabilization. The procedure of complex formation still has to be improved in generating homogenous and stable polyplexes capable of overcoming the physical barriers.

For example, Wagner et al. described one approach for polyplex formation [15]. Flash mixing of dilute compounds in physiological phosphate-free buffer results in formation of kinetically controlled polyplexes. At ratios of electroneutrality or higher, donut- and rod-like particles of 80–120 nm in diameter are formed for PLL polyplexes. Polyplexes containing Tf-conjugated PLL have increased solubility compared to unmodified PLL. For this type of polyplex formation a strong influence of parameters such as pDNA concentration or charge ratio, and ionic strength of solution was observed. Mixing pDNA/PEI polyplexes at N/P (PEI nitrogen: DNA phosphate) molar ratios below 6/1 in 150 mM saline results in rapid aggregation, which can be avoided by complex formation at low ionic strength (e.g., glucose balanced buffer) generating particles with an average diameter of approximately 40–50 nm [178]. It was also found that incorporation of hydrophilic PEG residues into polyplexes stabilize and prevent their aggregation [179].

Furthermore, purification of pDNA/PEI polyplexes from free PEI demonstrated low cellular and systemic toxicity, whereas also high transfection efficiency was achieved at high pDNA concentrations. Size exclusion chromatography (SEC) or electrophoresis was used to purify the pDNA/PEI polyplexes, which did not change their size and zeta-potential [180,181].

15.5.2 Undesired Interactions with Plasma, Degradative Enzymes, Matrix, and Nontarget Tissue

Polyplexes, when administered *in vivo*, are surrounded by a variety of compounds present in blood plasma. Salts, lipids, carbohydrates, proteins, or enzymes contribute to changes in the physicochemical properties of the polyplex. Some of these factors (opsonins) may coat the polyplex, causing aggregation, dissociation, or degradation of the polyplex. This may influence the composition of the polyplex as well as the bioavailability. Thus, the polyplexes, even when reaching the target cells/tissue, may no longer exhibit the physical properties necessary for efficient gene transfer into cells.

Previous studies have demonstrated the inactivation of pDNA/PLL polyplexes by blood components [182], whereby one of the factors was identified as the complement system [183]. Upon incubation of polyplexes with human plasma, specific proteins (IgM, fibrinogen, fibronectin, and complement C3) bind to the complexes. By coating pDNA/PEI polyplexes with PEG through covalent coupling to PEI, interaction with blood components was found to be strongly reduced [179].

Once the polyplex has traveled across the vascular barrier into the interstitial space, it has to avoid interactions with the extracellular matrix in order to reach the target cells/tissue (Figure 15.4). The extracellular matrix comprises different combinations of collagens, proteoglycans, hyaluronic acid, fibronectin, and other glycoproteins. These components could act as barriers by binding the polyplexes, e.g., hyaluronic acid binds cations very effectively. This implies that polyplexes resulting in a net positive charge could also interact with the extracellular matrix like binding, dissociation, or aggregation. Thus, optimizing the polyplexes bearing an overall low charge ratio close to neutrality is a possibility to avoid such interactions.

15.5.3 Inflammatory and Immunological Responses

As a result of introducing foreign molecules into the body, individual immune cells are stimulated to produce antibodies (humoral immunity). In addition to this humoral response, specific T cells may also be activated (cellular immunity). These two processes are the specific immune response. There is, however, also the nonspecific immune response, including phagocytosis, inflammation, and other nonspecific host resistance mechanisms such as the complement system [184]. These nonspecific mechanisms are developed immediately against any foreign molecule, even though the host has never encountered. Thus, the nonspecific immune response is a major extracellular barrier for the polyplex.

Inflammatory response is a major problem for any gene delivery system [185]. During an inflammatory response, neutrophil granulocytes and to a lesser extent macrophages migrate out of the capillaries into the surrounding tissue. At the site of inflammation the phagocytes recognize the foreign molecules via receptors on their surface leading to subsequent phagocytosis. Attachment is greatly enhanced and specified upon opsonization of foreign molecules, e.g., by the C3b component of complement. Both neutrophil granulocytes and macrophages have receptors which specifically bind to C3b allowing them to recognize their target.

Positively charged polyplexes have the ability to activate the complement system [183]. The positive charges are responsive to the complement protein C3b. Opsonization of such particles by C3b leads to the initiation of a cascade of events presumably resulting in the clearance of polyplexes by the reticuloendothelial system. Coating the positive charges of the polyplexes with other macromolecules may inhibit interactions with components of the complement system, hence decrease complement activation and clearance of the complexes from the blood circulation [186,187].

15.6 BIORESPONSIVE SYSTEMS: SYNTHETIC VIRUSES

Transport requirements during delivery of nucleic acids can change with time; they differ in the extracellular phase (protection) from the intracellular phase (release and activation). Consistently, the bioresponsiveness of delivery systems is a key issue of prospective developments in nucleic acid delivery [6]. Viruses, the most efficient gene delivery vectors, sense their environment and respond to the biological surrounding in a dynamic manner. Mimicking nature by integrating bioresponsive domains into synthetic carriers (Synthetic viruses) can be advantageous as demonstrated in first experimental approaches.

For example, PEG shielding is beneficial for systemic circulation of polyplexes but could also hinder intracellular release. To overcome this PEG dilemma, Szoka et al. took advantage of differences in the pH of biological compartments.

They prepared several pDNA lipoplexes consisting of a cationic lipid mixture and a pH-sensitive PEG-diorthoester-distearoylglycerol lipid. In vitro experiments showed enhanced luciferase reporter gene expression (up to three orders of magnitude) of pH-sensitive lipoplexes compared to the stable shielded ones [188].

A similar strategy was successful in case of polyplexes. Bioresponsive PEGylation with pH-labile hydrazone or acetal linkages were used for in vitro and in vivo gene transfer [189,190]. PEI polyplexes shielded with these pH-sensitive PEG conjugates lose their PEG shield at low pH, i.e., in endosome. Such polyplexes, when targeted for the Tf or EGF receptor-mediated endocytosis pathway, displayed up to 100-fold higher gene transfer efficiency compared to polyplexes with a stable PEG shield.

A different approach to PEGylate particle-based systems was accomplished by Oishi et al. linking PEG directly to siRNA or antisense oligonucleotides [191,192]. For this purpose they used an acid-labile thioproprionate linker and mixed the grafted oligonucleotide afterwards with PLL to form polyion complex micelles (PIC). Targeted acid-labile PICs showed significant higher inhibition of gene expression in vitro compared to the stable PICs in a human hepatoma model.

Besides the pH gradient, changes in the concentration of specific enzymes can also be used as trigger. Harashima et al. designed a cleavable PEG-lipid containing a peptidic substrate for matrix metalloproteinase occurring in the extracellular space of tumor tissues [193]. The cleavable PEG was formulated with pDNA lipoplexes in comparison to standard noncleavable PEG-lipids mediating an approximately 100-fold enhanced gene expression in vivo.

Murthy et al. masked the endosomolytic property of poly(propylacrylic acid) conjugates via acid-degradable acetal linkages [194]. The triggered removal of masking groups by acidification in the endosome was also responsible for restoring the lytic activity of melittin [135]. Chen et al. took advantage of an intracellular reducing environment for reactivation of the lytic activity of melittin [195]. A lysine-modified, disulfide-linked poly-melittin analog was created loosing its lytic activity after polyplex formation. Intracellular reduction leads to release of lytic melittin which improves endosomal release and gene transfer efficiency.

Listeriolysin O (LLO) is a natural membrane-disruptive protein from Listeria monocytogenes, which contains a single cysteine residue involved in endosomal membrane binding and disruption. Conjugation of LLO by this cysteine via disulfide bond formation to protamine [133] results in the loss of the lytic activity, which is recovered by intracellular reduction. The transfection efficiency was enhanced by three orders of magnitude when the LLO–protamine conjugate was included in pDNA/protamine polyplexes.

The difference in intracellular redox potential has also been used as trigger for intracellular release of the nucleic acid, including pDNA, mRNA, and siRNA, delivered by reducible polycations [28].

15.7 THERAPEUTIC APPLICATIONS OF POLYPLEXES

Although gene transfer via receptor-targeted polyplexes is mainly successful *in vitro*, several reports have demonstrated the ability to deliver nucleic acids efficiently to target cells *in vivo*. This has been enabled by using the various methods to overcome the intra- and extracellular barriers, but also to a large extent by shielding of the polyplexes reducing their positive surface charge (zeta potential), hence preventing interactions with blood components. Shielding methods include those with PEG [179], poly-*N*-(2-hydroxypropyl)methacrylamide (pHPMA) [196], poloxamer [197], or targeting ligand density [85].

Numerous pDNA, siRNA, and related nucleic acid formulations have shown very encouraging anticancer effects *in vivo*. The therapeutic effects aimed at are, i.e., interference with neoangiogenesis, reducing tumor cell proliferation, induction of apoptosis, or activation of the immune system.

For disseminated cancer, pDNA and siRNA have to be applied systematically using particle-based systems protecting their content from degradation and transferring it to the cancer tissue, while encountering many barriers on the *in vivo* delivery route. Improved targeting, efficient intracellular delivery, reduced toxicity, and bioresponsiveness are key elements for additional optimization.

15.7.1 Targeted Delivery of Therapeutic pDNA

15.7.1.1 Targeted Delivery of pDNA To Different Tissues

First encouraging results for *in vivo* gene transfer, via intravenous injection, were obtained by targeting the liver asialoglycoprotein receptor using asialoorosomucoid covalently linked to PLL carrying the CAT marker gene [2,50,198–200]. Gene expression proved to be liver specific as other organs, e.g., kidney, spleen and lungs, did not produce detectable quantities of CAT. Gene expression persistence was improved by applying partial hepatectomy, a procedure for stimulating hepatic regeneration and pDNA synthesis, beginning 12 h after surgery resulting in CAT activity up to 11 weeks postsurgery.

Related approaches have been used to target the liver with polyplexes of different sizes. Less than 20 nm small galactosylated pDNA/PLL polyplexes encoding human factor IX have been shown to target the hepatic asialoglycoprotein receptor, with up to 140 days of detectable protein in serum of transfected rats, achieved without partial hepatectomy [51]. In another approach, the polymeric immunoglobulin receptor has been targeted by 25 nm pDNA polyplexes bearing antigen-binding fragment of an antibody enabling gene transfer to rat pneumocytes following intravenous administration [52,53].

This type of small PLL polyplexes have been further developed into PEGylated PLL polyplexes. These have been recently tested in clinical trials in CF patients; transient correction of the CFTR ion channel in nasal epithelium was demonstrated [201].

Polyplexes based on PLL do not demonstrate efficient gene transfer *in vivo*. However, modifications such as addition of endosomal disruptive agents like adenoviruses [54] resulted in efficient gene expression in airway epithelium after intratracheal application [202], and in tumors after intratumoral application [203,204].

The lung has been targeted via systemic tail vein injection of mice by pDNA/LPEI polyplexes [205,206]. High levels of gene expression were observed in the lungs while lower expression in the other organs (including heart, spleen, kidney, and liver). However, systemic gene delivery with positively charged pDNA/PEI polyplexes has strong side effects resulting in lung embolism and severe toxicity [85,207]. The lung can be more specifically targeted by aerosol delivery of PEI polyplexes. Rudolph et al. showed that aerosolized pDNA/PEI polyplexes yielded transfection levels 15-fold higher than a 140-fold higher dose of the same vector applied directly via intratracheal intubation to the lungs of mice [208,209]. The strategy was also successful in sheep as a large animal model for optimizing cystic fibrosis gene therapy. Dose-related toxicity was reduced by aerosol administration compared to direct instillation [210].

15.7.1.2 Targeted pDNA Polyplexes for Cancer Therapy

Receptor-targeted polyplexes can be used to systemically deliver genes into tumors. Intravenous application of standard positively charged pDNA/PEI-Tf polyplexes through the tail vein into tumor-bearing mice resulted in gene expression in the tail and lung; no expression could be observed in the tumor, but serious toxicity [179,204]. However, surface shielding of the Tf-linked polyplexes with PEG via covalent coupling to PEI leads to stabilizing of polyplexes in size and preventing of interactions with plasma proteins and erythrocyte aggregation [179]. Tail vein injecting these PEGylated polyplexes in mice leads to less toxicity and further enhancement of gene expression [179,204]. Similar results were obtained with other Tf polyplexes with electroneutral surface, e.g., shielded optimized adenovirus-linked PLL-Tf polyplexes [204] or PEI-Tf polyplexes with a higher content of Tf as shielding agent [85].

As subcutaneous tumors are not directly supplied by main blood vessels, delivery is dependent on peripheral blood supply. Thus, the polyplexes probably target such distant tumors in a combinatorial passive targeting (i.e., overcoming extracellular barriers, EPR = enhanced permeability and retention effect as described for tumors [211]) and active targeting mechanisms (i.e., specific receptor targeted endocytosis).

Another report demonstrated that intravenous injection of PEGylated EGF-containing pDNA/PEI polyplexes results in a highly specific expression in human hepatocellular carcinoma

(HCC) tumors. Following intravenous injection into human HCC xenograft-bearing SCID mice, luciferase reporter gene expression was predominantly found in the tumor, with levels up to two logs higher than in the liver [47].

TNF-α is a cytokine with immunostimulatory and cytotoxic properties, particularly affecting the tumor vasculature. pDNA encoding TNF-α has been systemically delivered to tumor-bearing mice resulting in expression in tumor cells near to the feeding vessel. Locally produced TNF-α is assumed to be responsible for the obliteration of the tumor vasculature resulting in tumor necrosis [89]. For this purpose Tf-conjugated PEI polyplexes were applied via tail vein, which can additionally contain a PEG shield. Delivery of a luciferase reporter gene resulted in gene transfer into distant subcutaneous neuroblastoma tumors approximately 100-fold higher than in other organs [95,179] and specificity was confirmed by luciferase live imaging in mice [92]. Repeated systemic application of TNF-α polyplexes into tumor-bearing mice induced tumor necrosis and inhibition of tumor growth in four murine tumor models [89,95]. Due to the fact that TNF-α gene expression was largely localized within the tumor, no significant systemic TNF-α related toxicities could be observed. Therapeutic effects were enhanced by the combination of tumor-targeted TNF gene therapy with DOXIL, a liposomal doxorubicin formulation that passively accumulates within tumors [6].

Intravenous TNF-α gene therapy was also effective in treatment of three other established tumor models: epidermoid carcinoma, cervix carcinoma, and colorectal carcinoma [212]. In this study different polyplex formulations were applied based on polypropylenimine dendrimers, and the TNF-alpha gene expression constructs were transcriptionally targeted by telomerase gene promoters. The cationic dendrimer carrier was found to exhibit pDNA-independent, intrinsic antitumor activity, extending from growth retardation to complete tumor regression. These results let one suggest that such combined effects of targeted nucleic acids with polymer-based antitumor activity may be a good basis for efficient gene therapy.

Polyplexes consisting of pDNA and glucosylated PEI have been used for p53 *in vivo* gene transfer in mice bearing head and neck squamous cell carcinoma xenografts. Combining the treatment with PCI technology resulted in enhanced intracellular release of endocytosed polyplexes by photosensitizer and light [142,144], leading to a 20-fold increased and also sustained gene expression. Furthermore weekly treatment repeated for 7 weeks resulted in tumor growth inhibition in all animals [146].

15.7.2 TUMOR-TARGETED DELIVERY OF THERAPEUTIC siRNA

Application of siRNA or siRNA expressing vectors can trigger specific antitumoral effects [213,214]. Cell-type specific antibody-mediated siRNA delivery was achieved with the fusion antibody fragments and the nucleic acid binding polycation protamine [215]. In preliminary experiments, the HIV envelope protein was used for antibody targeting to an artificial epitope in engineered B16 tumor cells. Intratumoral or intravenous injection of a cocktail of siRNAs for c-myc, MDM2, and VEGF complexed with anti (Env)-protamine reduced the growth of envelope-expressing subcutaneous B16 tumors. An ErbB2 single-chain antibody, protamine fusion delivered siRNAs specifically into ErbB2-expressing cancer cells.

A polymeric cationic cyclodextrin carrier was developed by the incorporation of PEG, for shielding and Tf as targeting ligand, for systemic siRNA delivery [96]. In a murine model of metastatic Ewing's sarcoma, repeated systemic delivery of siRNA against the EWS-FLI1 translocation resulted in significantly reduction of tumor growth. The presence of Tf in the formulation was required for this therapeutic effect.

For antiangiogenic cancer treatment, ideally tumor sites should be in contact with the therapeutic antiangiogenic factors for an extended time period. Specific and extended production of the factors in all tumor sites goes beyond the possibility of pDNA delivery technologies that are currently available. However, the extended localized production of therapeutic factors, which are secreted into the systemic circulation and suppress the tumor neovascularization, has been demonstrated.

Systemic antiangiogenic siRNA therapy has also been attempted using targeted and shielded polyplexes. siRNA targeting VEGF receptor-2 was complexed with RGD peptide–PEG–PEI conjugates. These RGD peptides act as ligands for integrins over-expressed on the tumor neovasculature [106]. Repeated intravenous administration into tumor-bearing mice resulted in siRNA sequence-specific inhibition of tumor growth by more than 90% compared to the control siRNA.

15.7.3 TARGETED DELIVERY OF OTHER THERAPEUTIC NUCLEIC ACIDS

Other therapeutic nucleic acids include antisense oligonucleotides, ribozymes, or aptamers. PEI polyplexes have been applied to stabilize RNA ribozymes against degradation [216]. After intraperitoneal injections in a mouse xenograft model, prolonged blood circulation and intact ribozymes in the distant tumor were obtained. PEI-complexed ribozymes targeted against the pleiotrophin growth factor resulted in intratumoral target gene knock-down and a distinct reduction of tumor growth.

Synthetic antiproliferative poly IC RNA, a strong activator of apoptosis and interferon response, was targeted to glioblastoma by PEI–PEG–EGF polyplexes [4]. Intratumoral delivery of EGF-targeted poly IC induced rapid apoptosis of the target cells and complete regression of intracranial tumors in nude mice, without opposing toxicity on normal brain tissue. To achieve this effect, the incorporation of endosomal releasing agents such as melittin in the polyplexes was required. Expression of several cytokines and "bystander killing" of untransfected tumor cells was found out *in vitro* and *in vivo*. Similarly, delivery of poly IC completely eliminated EGF-receptor over-expressing breast cancer and adenocarcinoma xenografts. The best effects were obtained with formulations using a novel carrier combining all delivery functions in one polymer conjugate, i.e., melittin–PEI–PEG–EGF.

15.8 CONCLUSION

Since the design of the first targeted polyplexes [1] 20 years ago the field has progressed steadily. Lessons have been learnt on the type of targeting ligands and polymers that can be successfully used. Knowledge on the extracellular and intracellular barriers greatly increased. Polyplexes have been developed into products that have been tested in clinical gene therapy trials *ex vivo* [217] and *in vivo* [201]. Systemic targeting of pDNA and siRNA polyplexes has been demonstrated in several animal models. Understanding the barriers has resulted in the design of the next generation of bioresponsive polyplexes (synthetic viruses) which hopefully will combine the effectiveness of natural viruses with the safety of synthetic pharmaceutical products.

REFERENCES

1. Wu GY and Wu CH. Receptor-mediated in vitro gene transformation by a soluble DNA carrier system. *J Biol Chem* 1987, 262:4429–4432.
2. Wu GY and Wu CH. Receptor-mediated gene delivery and expression in vivo. *J Biol Chem* 1988, 262:14621–14624.
3. Tomita N, Azuma H, Kaneda Y, Ogihara T, and Morishita R. Application of decoy oligodeoxynucleotides-based approach to renal diseases. *Curr Drug Targets* 2004, 5:717–733.
4. Shir A, Ogris M, Wagner E, and Levitzki A. EGF receptor-targeted synthetic double-stranded RNA eliminates glioblastoma, breast cancer, and adenocarcinoma tumors in mice. *PLoS Med* 2006, 3:e6.
5. Ulrich H, Trujillo CA, Nery AA, Alves JM, Majumder P, Resende RR, and Martins AH. DNA and RNA aptamers: From tools for basic research towards therapeutic applications. *Comb Chem High Throughput Screen* 2006, 9:619–632.
6. Wagner E, Kircheis R, and Walker GF. Targeted nucleic acid delivery into tumors: New avenues for cancer therapy. *Biomed Pharmacother* 2004, 58:152–161.
7. Zhang G, Budker V, and Wolff JA. High levels of foreign gene expression in hepatocytes after tail vein injections of naked plasmid DNA. *Hum Gene Ther* 1999, 10:1735–1737.
8. Liu F, Song Y, and Liu D. Hydrodynamics-based transfection in animals by systemic administration of plasmid DNA. *Gene Ther* 1999, 6:1258–1266.
9. Felgner PL, Barenholz Y, Behr JP, Cheng SH, Cullis P, Huang L, Jessee JA, Seymour L, Szoka F, Thierry AR, Wagner E, and Wu G. Nomenclature for synthetic gene delivery systems. *Hum Gene Ther* 1997, 8:511–512.
10. De Smedt SC, Demeester J, and Hennink WE. Cationic polymer based gene delivery systems. *Pharm Res* 2000, 17:113–126.
11. Wagner E. Strategies to improve DNA polyplexes for in vivo gene transfer: Will "artificial viruses" be the answer? *Pharm Res* 2004; 21:8–14.
12. Tiera MJ, Winnik FO, and Fernandes JC. Synthetic and natural polycations for gene therapy: State of the art and new perspectives. *Curr Gene Ther* 2006, 6:59–71.
13. Boussif O, Lezoualc'h F, Zanta MA, Mergny MD, Scherman D, Demeneix B, and Behr JP. A versatile vector for gene and oligonucleotide transfer into cells in culture and in vivo: Polyethylenimine. *Proc Natl Acad Sci U S A* 1995, 92:7297–7301.
14. Emi N, Kidoaki S, Yoshikawa K, and Saito H. Gene transfer mediated by polyarginine requires a formation of big carrier-complex of DNA aggregate. *Biochem Biophys Res Commun* 1997, 231:421–424.
15. Wagner E, Cotten M, Foisner R, and Birnstiel ML. Transferrin-polycation-DNA complexes: The effect of polycations on the structure of the complex and DNA delivery to cells. *Proc Natl Acad Sci U S A* 1991, 88:4255–4259.
16. Mistry AR, Falciola L, Monaco L, Tagliabue R, Acerbis G, Knight A, Harbottle RP, Soria M, Bianchi ME, Coutelle C, and Hart SL. Recombinant HMG1 protein produced in *Pichia pastoris*: A nonviral gene delivery agent. *Biotechniques* 1997, 22:718–729.
17. Demirhan I, Hasselmayer O, Chandra A, Ehemann M, and Chandra P. Histone-mediated transfer and expression of the HIV-1 tat gene in Jurkat cells. *J Hum Virol* 1998, 1:430–440.
18. Soutschek J, Akinc A, Bramlage B, Charisse K, Constien R, Donoghue M, Elbashir S, Geick A, Hadwiger P, Harborth J, John M, Kesavan V, Lavine G, Pandey RK, Racie T, Rajeev KG, Rohl I, Toudjarska I, Wang G, Wuschko S, Bumcrot D, Koteliansky V, Limmer S, Manoharan M, and Vornlocher HP. Therapeutic silencing of an endogenous gene by systemic administration of modified siRNAs. *Nature* 2004, 432:173–178.
19. Kloeckner J, Boeckle S, Persson D, Roedl W, Ogris M, Berg K, and Wagner E. DNA polyplexes based on degradable oligoethylenimine-derivatives: Combination with EGF receptor targeting and endosomal release functions. *J Control Release* 2006, 116:115–122.
20. Zhao Z, Wang J, Mao HQ, and Leong KW. Polyphosphoesters in drug and gene delivery. *Adv Drug Deliv Rev* 2003, 55:483–499.
21. Kim YH, Park JH, Lee M, Kim YH, Park TG, and Kim SW. Polyethylenimine with acid-labile linkages as a biodegradable gene carrier. *J Control Release* 2005, 103:209–219.
22. Thomas M, Ge Q, Lu JJ, Chen J, and Klibanov AM. Cross-linked small polyethylenimines: While still nontoxic, deliver DNA efficiently to mammalian cells in vitro and in vivo. *Pharm Res* 2005, 22:373–380.
23. de Wolf HK, Luten J, Snel CJ, Oussoren C, Hennink WE, and Storm G. In vivo tumor transfection mediated by polyplexes based on biodegradable poly(DMAEA)-phosphazene. *J Control Release* 2005;
24. Zhong Z, Song Y, Engbersen JF, Lok MC, Hennink WE, and Feijen J. A versatile family of degradable non-viral gene carriers based on hyperbranched poly(ester amine)s. *J Control Release* 2005, 109:317–329.
25. Forrest ML, Koerber JT, and Pack DW. A degradable polyethylenimine derivative with low toxicity for highly efficient gene delivery. *Bioconjug Chem* 2003, 14:934–940.
26. Ahn CH, Chae SY, Bae YH, and Kim SW. Synthesis of biodegradable multi-block copolymers of poly(L-lysine) and poly(ethylene glycol) as a non-viral gene carrier. *J Control Release* 2004, 97:567–574.
27. Saito G, Swanson JA, and Lee KD. Drug delivery strategy utilizing conjugation via reversible disulfide linkages: Role and site of cellular reducing activities. *Adv Drug Deliv Rev* 2003, 55:199–215.
28. Read ML, Singh S, Ahmed Z, Stevenson M, Briggs SS, Oupicky D, Barrett LB, Spice R, Kendall M, Berry M, Preece JA, Logan A, and Seymour LW. A versatile reducible polycation-based system for efficient delivery of a broad range of nucleic acids. *Nucleic Acids Res* 2005, 33:e86.

29. Kwok KY, Park Y, Yang Y, McKenzie DL, Liu Y, and Rice KG. In vivo gene transfer using sulfhydryl cross-linked PEG-peptide/glycopeptide DNA co-condensates. *J Pharm Sci* 2003, 92:1174–1185.

30. Hwang SJ, Bellocq NC, and Davis ME. Effects of structure of beta-cyclodextrin-containing polymers on gene delivery. *Bioconjug Chem* 2001, 12:280–290.

31. Pun SH, Tack F, Bellocq NC, Cheng J, Grubbs BH, Jensen GS, Davis ME, Brewster M, Janicot M, Janssens B, Floren W, and Bakker A. Targeted delivery of RNA-cleaving DNA enzyme (DNAzyme) to tumor tissue by transferrin-modified, cyclodextrin-based particles. *Cancer Biol Ther* 2004, 3:641–650.

32. Pun SH, Bellocq NC, Liu A, Jensen G, Machemer T, Quijano E, Schluep T, Wen S, Engler H, Heidel J, and Davis ME. Cyclodextrin-modified polyethylenimine polymers for gene delivery. *Bioconjug Chem* 2004, 15:831–840.

33. Leng Q, Scaria P, Zhu J, Ambulos N, Campbell P, and Mixson AJ. Highly branched HK peptides are effective carriers of siRNA. *J Gene Med* 2005, 7:977–986.

34. Hassani Z, Lemkine GF, Erbacher P, Palmier K, Alfama G, Giovannangeli C, Behr JP, and Demeneix BA. Lipid-mediated siRNA delivery down-regulates exogenous gene expression in the mouse brain at picomolar levels. *J Gene Med* 2005, 7:198–207.

35. Urban-Klein B, Werth S, Abuharbeid S, Czubayko F, and Aigner A. RNAi-mediated gene-targeting through systemic application of polyethylenimine (PEI)-complexed siRNA in vivo. *Gene Ther* 2005, 12:461–466.

36. Werth S, Urban-Klein B, Dai L, Hobel S, Grzelinski M, Bakowsky U, Czubayko F, and Aigner A. A low molecular weight fraction of polyethylenimine (PEI) displays increased transfection efficiency of DNA and siRNA in fresh or lyophilized complexes. *J Control Release* 2006, 112:257–270.

37. Mukherjee S, Ghosh RN, and Maxfield FR. Endocytosis. *Physiol Rev* 1997, 77:759–803.

38. Anderson RG, Kamen BA, Rothberg KG, and Lacey SW. Potocytosis: Sequestration and transport of small molecules by caveolae. *Science* 1992, 255:410–411.

39. Rejman J, Oberle V, Zuhorn IS, and Hoekstra D. Size-dependent internalization of particles via the pathways of clathrin- and caveolae-mediated endocytosis. *Biochem J* 2004, 377:159–169.

40. Mislick KA and Baldeschwieler JD. Evidence for the role of proteoglycans in cation-mediated gene transfer. *Proc Natl Acad Sci U S A* 1996, 93:12349–12354.

41. Kopatz I, Remy JS, and Behr JP. A model for non-viral gene delivery: Through syndecan adhesion molecules and powered by actin. *J Gene Med* 2004, 6:769–776.

42. Sandvig K and van Deurs B. Endocytosis without clathrin. *Trends Cell Biol* 1994, 4:275–277.

43. Smart EJ, Mineo C, and Anderson RG. Clustered folate receptors deliver 5-methyltetrahydrofolate to cytoplasm of MA104 cells. *J Cell Biol* 1996, 134:1169–1177.

44. von Gersdorff K, Sanders NN, Vandenbroucke R, De Smedt SC, Wagner E, and Ogris M. The internalization route resulting in successful gene expression depends on both cell line and polyethylenimine polyplex type. *Mol Ther* 2006, 14: 745–753.

45. Rejman J, Bragonzi A, and Conese M. Role of clathrin- and caveolae-mediated endocytosis in gene transfer mediated by lipo- and polyplexes. *Mol Ther* 2005, 12:468–474.

46. Sosnowski BA, Gonzalez AM, Chandler LA, Buechler YJ, Pierce GF, and Baird A. Targeting DNA to cells with basic fibroblast growth factor (FGF2). *J Biol Chem* 1996, 271: 33647–33653.

47. Wolschek MF, Thallinger C, Kursa M, Rossler V, Allen M, Lichtenberger C, Kircheis R, Lucas T, Willheim M, Reinisch W, Gangl A, Wagner E, and Jansen B. Specific systemic nonviral gene delivery to human hepatocellular carcinoma xenografts in SCID mice. *Hepatology* 2002, 36:1106–1114.

48. Cotten M, Langle-Rouault F, Kirlappos H, Wagner E, Mechtler K, Zenke M, Beug H, and Birnstiel ML. Transferrin-polycation-mediated introduction of DNA into human leukemic cells: Stimulation by agents that affect the survival of transfected DNA or modulate transferrin receptor levels. *Proc Natl Acad Sci U S A* 1990, 87:4033–4037.

49. Wagner E, Ogris M, and Zauner W. Polylysine-based transfection systems utilizing receptor-mediated delivery. *Adv Drug Deliv Rev* 1998, 30:97–113.

50. Chowdhury NR, Hays RM, Bommineni VR, Franki N, Chowdhury JR, Wu CH, and Wu GY. Microtubular disruption prolongs the expression of human bilirubin-uridinediphosphoglucuronate-glucuronosyltransferase-1 gene transferred into Gunn rat livers. *J Biol Chem* 1996, 271:2341–2346.

51. Perales JC, Ferkol T, Beegen H, Ratnoff OD, and Hanson RW. Gene transfer in vivo: Sustained expression and regulation of genes introduced into the liver by receptor-targeted uptake. *Proc Natl Acad Sci U S A* 1994, 91:4086–4090.

52. Ferkol T, Perales JC, Eckman E, Kaetzel CS, Hanson RW, and Davis PB. Gene transfer into the airway epithelium of animals by targeting the polymeric immunoglobulin receptor. *J Clin Invest* 1995, 95:493–502.

53. Ferkol T, Pellicena PA, Eckman E, Perales JC, Trzaska T, Tosi M, Redline R, and Davis PB. Immunologic responses to gene transfer into mice via the polymeric immunoglobulin receptor. *Gene Ther* 1996, 3:669–678.

54. Wagner E, Zatloukal K, Cotten M, Kirlappos H, Mechtler K, Curiel DT, and Birnstiel ML. Coupling of adenovirus to transferrin-polylysine/DNA complexes greatly enhances receptor-mediated gene delivery and expression of transfected genes. *Proc Natl Acad Sci U S A* 1992, 89:6099–6103.

55. Moffatt S and Cristiano RJ. Uptake characteristics of NGR-coupled stealth PEI/pDNA nanoparticles loaded with PLGA-PEG-PLGA tri-block copolymer for targeted delivery to human monocyte-derived dendritic cells. *Int J Pharm* 2006, 321:143–154.

56. Moffatt S, Wiehle S, and Cristiano RJ. Tumor-specific gene delivery mediated by a novel peptide-polyethylenimine-DNA polyplex targeting aminopeptidase N/CD13. *Hum Gene Ther* 2005, 16:57–67.

57. Baatz JE, Bruno MD, Ciraolo PJ, Glasser SW, Stripp BR, Smyth KL, and Korfhagen TR. Utilization of modified surfactant-associated protein B for delivery of DNA to airway cells in culture. *Proc Natl Acad Sci U S A* 1994, 91:2547–2551.

58. Ross GF, Morris RE, Ciraolo G, Huelsman K, Bruno M, Whitsett JA, Baatz JE, and Korfhagen TR. Surfactant protein A-polylysine conjugates for delivery of DNA to airway cells in culture. *Hum Gene Ther* 1995, 6:31–40.

59. Plank C, Zatloukal K, Cotten M, Mechtler K, and Wagner E. Gene transfer into hepatocytes using asialoglycoprotein receptor mediated endocytosis of DNA complexed with an artificial tetra-antennary galactose ligand. *Bioconjug Chem* 1992, 3:533–539.

60. Midoux P, Mendes C, Legrand A, Raimond J, Mayer R, Monsigny M, and Roche AC. Specific gene transfer mediated by lactosylated poly-L-lysine into hepatoma cells. *Nucleic Acids Res* 1993, 21:871–878.

61. Merwin JR, Noell GS, Thomas WL, Chiou HC, DeRome ME, McKee TD, Spitalny GL, and Findeis MA. Targeted delivery

of DNA using YEE(GalNAcAH)3, a synthetic glycopeptide ligand for the asialoglycoprotein receptor. *Bioconjug Chem* 1994, 5:612–620.

62. Chiu MH, Tamura T, Wadhwa MS, and Rice KG. In vivo targeting function of N-linked oligosaccharides with terminating galactose and N-acetylgalactosamine residues. *J Biol Chem* 1994, 269:16195–16202.

63. Morimoto K, Nishikawa M, Kawakami S, Nakano T, Hattori Y, Fumoto S, Yamashita F, and Hashida M. Molecular weight-dependent gene transfection activity of unmodified and galactosylated polyethyleneimine on hepatoma cells and mouse liver. *Mol Ther* 2003, 7:254–261.

64. Merwin JR, Carmichael EP, Noell GS, DeRome ME, Thomas WL, Robert N, Spitalny G, and Chiou HC. CD5-mediated specific delivery of DNA to T lymphocytes: Compartmentalization augmented by adenovirus. *J Immunol Methods* 1995, 186:257–266.

65. Liu X, Tian PK, Ju DW, Zhang MH, Yao M, Cao XT, and Gu JR. Systemic genetic transfer of p21WAF-1 and GM-CSF utilizing of a novel oligopeptide-based EGF receptor targeting polyplex. *Cancer Gene Ther* 2003, 10:529–539.

66. Fominaya J, Uherek C, and Wels W. A chimeric fusion protein containing transforming growth factor-alpha mediates gene transfer via binding to the EGF receptor. *Gene Ther* 1998, 5:521–530.

67. Uherek C, Fominaya J, and Wels W. A modular DNA carrier protein based on the structure of diphtheria toxin mediates target cell-specific gene delivery. *J Biol Chem* 1998, 273: 8835–8841.

68. Curiel TJ, Cook DR, Bogedain C, Jilg W, Harrison GS, Cotten M, Curiel DT, and Wagner E. Efficient foreign gene expression in Epstein–Barr virus-transformed human B-cells. *Virology* 1994, 198:577–585.

69. Mislick KA, Baldeschwieler JD, Kayyem JF, and Meade TJ. Transfection of folate-polylysine DNA complexes: Evidence for lysosomal delivery. *Bioconjug Chem* 1995, 6:512–515.

70. Gottschalk S, Cristiano RJ, Smith LC, and Woo SL. Folate receptor mediated DNA delivery into tumor cells: Potosomal. *Gene Ther* 1994, 1:185–191.

71. Ding ZM, Cristiano RJ, Roth JA, Takacs B, and Kuo MT. Malarial circumsporozoite protein is a novel gene delivery vehicle to primary hepatocyte cultures and cultured cells. *J Biol Chem* 1995, 270:3667–3676.

72. Foster BJ and Kern JA. HER2-targeted gene transfer. *Hum Gene Ther* 1997, 8:719–727.

73. Chiu SJ, Ueno NT, and Lee RJ. Tumor-targeted gene delivery via anti-HER2 antibody (trastuzumab, Herceptin) conjugated polyethylenimine. *J Control Release* 2004, 97:357–369.

74. Hart SL, Harbottle RP, Cooper R, Miller A, Williamson R, and Coutelle C. Gene delivery and expression mediated by an integrin-binding peptide. *Gene Ther* 1995, 2:552–554.

75. Kim TG, Kang SY, Kang JH, Cho MY, Kim JI, Kim SH, and Kim JS. Gene transfer into human hepatoma cells by receptor-associated protein/polylysine conjugates. *Bioconjug Chem* 2004, 15:326–332.

76. Erbacher P, Bousser MT, Raimond J, Monsigny M, Midoux P, and Roche AC. Gene transfer by DNA/glycosylated polylysine complexes into human blood monocyte-derived macrophages. *Hum Gene Ther* 1996, 7:721–729.

77. Ferkol T, Perales JC, Mularo F, and Hanson RW. Receptor-mediated gene transfer into macrophages. *Proc Natl Acad Sci U S A* 1996, 93:101–105.

78. Zeng J, Too HP, Ma Y, Luo ES, and Wang S. A synthetic peptide containing loop 4 of nerve growth factor for targeted gene delivery. *J Gene Med* 2004, 6:1247–1256.

79. Coll JL, Wagner E, Combaret V, Metchler K, Amstutz H, Iacono-Di-Cacito I, Simon N, and Favrot MC. In vitro targeting and specific transfection of human neuroblastoma cells by chCE7 antibody-mediated gene transfer. *Gene Ther* 1997, 4:156–161.

80. Merdan T, Callahan J, Petersen H, Kunath K, Bakowsky U, Kopeckova P, Kissel T, and Kopecek J. Pegylated polyethylenimine-fab' antibody fragment conjugates for targeted gene delivery to human ovarian carcinoma cells. *Bioconjug Chem* 2003, 14:989–996.

81. Li S, Tan Y, Viroonchatapan E, Pitt BR, and Huang L. Targeted gene delivery to pulmonary endothelium by anti-PECAM antibody. *Am J Physiol Lung Cell Mol Physiol* 2000, 278:504–511.

82. Ziady AG, Ferkol T, Dawson DV, Perlmutter DH, and Davis PB. Chain length of the polylysine in receptor-targeted gene transfer complexes affects duration of reporter gene expression both in vitro and in vivo. *J Biol Chem* 1999, 274: 4908–4916.

83. Thurnher M, Wagner E, Clausen H, Mechtler K, Rusconi S, Dinter A, Birnstiel ML, Berger EG, and Cotten M. Carbohydrate receptor-mediated gene transfer to human T leukaemic cells. *Glycobiology* 1994, 4:429–435.

84. Xu L, Pirollo KF, Tang WH, Rait A, and Chang EH. Transferrin-liposome-mediated systemic p53 gene therapy in combination with radiation results in regression of human head and neck cancer xenografts. *Hum Gene Ther* 1999, 10:2941–2952.

85. Kircheis R, Wightman L, Schreiber A, Robitza B, Rossler V, Kursa M, and Wagner E. Polyethylenimine/DNA complexes shielded by transferrin target gene expression to tumors after systemic application. *Gene Ther* 2001, 8:28–40.

86. Wagner E, Zenke M, Cotten M, Beug H, and Birnstiel ML. Transferrin-polycation conjugates as carriers for DNA uptake into cells. *Proc Natl Acad Sci U S A* 1990, 87:3410–3414.

87. Kircheis R, Kichler A, Wallner G, Kursa M, Ogris M, Felzmann T, Buchberger M, and Wagner E. Coupling of cell-binding ligands to polyethylenimine for targeted gene delivery. *Gene Ther* 1997, 4:409–418.

88. Kircheis R, Blessing T, Brunner S, Wightman L, and Wagner E. Tumor targeting with surface-shielded ligand–polycation DNA complexes. *J Control Release* 2001, 72:165–170.

89. Kircheis R, Ostermann E, Wolschek MF, Lichtenberger C, Magin-Lachmann C, Wightman L, Kursa M, and Wagner E. Tumor-targeted gene delivery of tumor necrosis factor-alpha induces tumor necrosis and tumor regression without systemic toxicity. *Cancer Gene Ther* 2002, 9:673–680.

90. Xu L, Huang CC, Huang W, Tang WH, Rait A, Yin YZ, Cruz I, Xiang LM, Pirollo KF, and Chang EH. Systemic tumor-targeted gene delivery by anti-transferrin receptor scFv-immunoliposomes. *Mol Cancer Ther* 2002, 1:337–346.

91. Tros d, I, Arangoa MA, Moreno-Aliaga MJ, and Duzgunes N. Enhanced gene delivery in vitro and in vivo by improved transferrin-lipoplexes. *Biochim Biophys Acta* 2002; 1561:209–221.

92. Hildebrandt IJ, Iyer M, Wagner E, and Gambhir SS. Optical imaging of transferrin targeted PEI/DNA complexes in living subjects. *Gene Ther* 2003, 10:758–764.

93. Yu W, Pirollo KF, Rait A, Yu B, Xiang LM, Huang WQ, Zhou Q, Ertem G, and Chang EH. A sterically stabilized immunolipoplex for systemic administration of a therapeutic gene. *Gene Ther* 2004, 11:1434–1440.

94. da Cruz MT, Cardoso AL, de Almeida LP, Simoes S, and de Lima MC. Tf-lipoplex-mediated NGF gene transfer to the CNS: Neuronal protection and recovery in an excitotoxic model of brain injury. *Gene Ther* 2005, 12:1242–1252.

95. Kursa M, Walker GF, Roessler V, Ogris M, Roedl W, Kircheis R, and Wagner E. Novel Shielded Transferrin-Polyethylene Glycol-Polyethylenimine/DNA Complexes for Systemic Tumor-Targeted Gene Transfer. *Bioconjug Chem* 2003, 14:222–231.

96. Hu-Lieskovan S, Heidel JD, Bartlett DW, Davis ME, and Triche TJ. Sequence-specific knockdown of EWS-FLI1 by targeted, nonviral delivery of small interfering RNA inhibits tumor growth in a murine model of metastatic Ewing's sarcoma. *Cancer Res* 2005, 65:8984–8992.

97. Chen J, Gamou S, Takayanagi A, and Shimizu N. A novel gene delivery system using EGF receptor-mediated endocytosis. *FEBS Lett* 1994, 338:167–169.

98. Blessing T, Kursa M, Holzhauser R, Kircheis R, and Wagner E. Different strategies for formation of pegylated EGF-conjugated PEI/DNA complexes for targeted gene delivery. *Bioconjug Chem* 2001, 12:529–537.

99. Kloeckner J, Prasmickaite L, Hogset A, Berg K, and Wagner E. Photochemically enhanced gene delivery of EGF receptor-targeted DNA polyplexes. *J Drug Target* 2004, 12:205–213.

100. Hilgenbrink AR and Low PS. Folate receptor-mediated drug targeting: From therapeutics to diagnostics. *J Pharm Sci* 2005, 94:2135–2146.

101. Guo W and Lee RJ. Receptor-targeted gene delivery via folate-conjugated polyethylenimine. *AAPS Pharmsci* 1999, 1: Article 19.

102. Harbottle RP, Cooper RG, Hart SL, Ladhoff A, McKay T, Knight AM, Wagner E, Miller AD, and Coutelle C. An RGD-oligolysine peptide: A prototype construct for integrin-mediated gene delivery. *Hum Gene Ther* 1998, 9:1037–1047.

103. Erbacher P, Remy JS, and Behr JP. Gene transfer with synthetic virus-like particles via the integrin-mediated endocytosis pathway. *Gene Ther* 1999, 6:138–145.

104. Suh W, Han SO, Yu L, and Kim SW. An angiogenic, endothelial-cell-targeted polymeric gene carrier. *Mol Ther* 2002, 6:664–672.

105. Kunath K, Merdan T, Hegener O, Haberlein H, and Kissel T. Integrin targeting using RGD-PEI conjugates for in vitro gene transfer. *J Gene Med* 2003, 5:588–599.

106. Schiffelers RM, Ansari A, Xu J, Zhou Q, Tang Q, Storm G, Molema G, Lu PY, Scaria PV, and Woodle MC. Cancer siRNA therapy by tumor selective delivery with ligand-targeted sterically stabilized nanoparticle. *Nucleic Acids Res* 2004, 32: e149.

107. Pasqualini R, Koivunen E, Kain R, Lahdenranta J, Sakamoto M, Stryhn A, Ashmun RA, Shapiro LH, Arap W, and Ruoslahti E. Aminopeptidase N is a receptor for tumor-homing peptides and a target for inhibiting angiogenesis. *Cancer Res* 2000, 60:722–727.

108. Fujii H, Nakajima M, Saiki I, Yoneda J, Azuma I, and Tsuruo T. Human melanoma invasion and metastasis enhancement by high expression of aminopeptidase N/CD13. *Clin Exp Metastasis* 1995, 13:337–344.

109. Moffatt S, Wiehle S, and Cristiano RJ. A multifunctional PEI-based cationic polyplex for enhanced systemic p53-mediated gene therapy. *Gene Ther* 2006, 13:1512–1523.

110. Buschle M, Cotten M, Kirlappos H, Mechtler K, Schaffner G, Zauner W, Birnstiel ML, and Wagner E. Receptor-mediated gene transfer into human T lymphocytes via binding of DNA/CD3 antibody particles to the CD3 T cell receptor complex. *Hum Gene Ther* 1995, 6:753–761.

111. Zhang Y, Lu H, LiWang P, Sili U, and Templeton NS. Optimization of gene expression in nonactivated circulating lymphocytes. *Mol Ther* 2003, 8:629–636.

112. Schachtschabel U, Pavlinkova G, Lou D, and Köhler H. Antibody-mediated gene delivery for B-cell lymphoma in vitro. *Cancer Gene Ther* 1996, 3:365–372.

113. Fominaya J and Wels W. Target cell-specific DNA transfer mediated by a chimeric multidomain protein. Novel non-viral gene delivery system. *J Biol Chem* 1996, 271:10560–10568.

114. Paul RW, Weisser KE, Loomis A, Sloane DL, LaFoe D, Atkinson EM, and Overell RW. Gene transfer using a novel fusion protein, GAL4/invasin. *Hum Gene Ther* 1997, 8: 1253–1262.

115. Pollard H, Remy JS, Loussouarn G, Demolombe S, Behr JP, and Escande D. Polyethylenimine but not cationic lipids promotes transgene delivery to the nucleus in mammalian cells. *J Biol Chem* 1998, 273:7507–7511.

116. Seglen PO. Inhibitors of lysosomal function. *Methods Enzymol* 1983, 96:737–764.

117. Pless DD and Wellner RB. In vitro fusion of endocytic vesicles: Effects of reagents that alter endosomal pH. *J Cell Biochem* 1996, 62:27–39.

118. Sipe DM, Jesurum A, and Murphy RF. Absence of Na +, K(+)-ATPase regulation of endosomal acidification in K562 erythroleukemia cells. Analysis via inhibition of transferrin recycling by low temperatures. *J Biol Chem* 1991, 266: 3469–3474.

119. Cheng J, Zeidan R, Mishra S, Liu A, Pun SH, Kulkarni RP, Jensen GS, Bellocq NC, and Davis ME. Structure-function correlation of chloroquine and analogues as transgene expression enhancers in nonviral gene delivery. *J Med Chem* 2006, 49:6522–6531.

120. Plank C, Zauner W, and Wagner E. Application of membrane-active peptides for drug and gene delivery across cellular membranes. *Adv Drug Deliv Rev* 1998, 34:21–35.

121. Wagner E, Plank C, Zatloukal K, Cotten M, and Birnstiel ML. Influenza virus hemagglutinin HA-2 N-terminal fusogenic peptides augment gene transfer by transferrin-polylysine-DNA complexes: Toward a synthetic virus-like gene-transfer vehicle. *Proc Natl Acad Sci U S A* 1992, 89:7934–7938.

122. Mechtler K and Wagner E. Gene transfer mediated by influenza virus peptides: The role of peptide sequence. *New J Chem* 1997, 21:105–111.

123. Ogris M, Carlisle RC, Bettinger T, and Seymour LW. Melittin enables efficient vesicular escape and enhanced nuclear access of nonviral gene delivery vectors. *J Biol Chem* 2001, 276:47550–47555.

124. Boeckle S, Wagner E, and Ogris M.C- versus N-terminally linked melittin-polyethylenimine conjugates: The site of linkage strongly influences activity of DNA polyplexes. *J Gene Med* 2005, 7:1335–1347.

125. Zauner W, Blaas D, Kuechler E, and Wagner E. Rhinovirus-mediated endosomal release of transfection complexes. *J Virol* 1995, 69:1085–1092.

126. Plank C, Oberhauser B, Mechtler K, Koch C, and Wagner E. The influence of endosome-disruptive peptides on gene transfer using synthetic virus-like gene transfer systems. *J Biol Chem* 1994, 269:12918–12924.

127. Gottschalk S, Sparrow JT, Hauer J, Mims MP, Leland FE, Woo SL, and Smith LC. A novel DNA-peptide complex for efficient gene transfer and expression in mammalian cells. *Gene Ther* 1996, 3:48–57.

128. Kichler A, Mechtler K, Behr JP, and Wagner E. Influence of membrane-active peptides on lipospermine/DNA complex mediated gene transfer. *Bioconjug Chem* 1997, 8:213–221.

129. Wyman TB, Nicol F, Zelphati O, Scaria PV, Plank C, and Szoka FC, Jr. Design, synthesis, and characterization of a

cationic peptide that binds to nucleic acids and permeabilizes bilayers. *Biochemistry* 1997, 36:3008–3017.

130. Li W, Nicol F, and Szoka FC, Jr. GALA: A designed synthetic pH-responsive amphipathic peptide with applications in drug and gene delivery. *Adv Drug Deliv Rev* 2004, 56:967–985.

131. Kichler A, Leborgne C, Marz J, Danos O, and Bechinger B. Histidine-rich amphipathic peptide antibiotics promote efficient delivery of DNA into mammalian cells. *Proc Natl Acad Sci U S A* 2003, 100:1564–1568.

132. Yu W, Pirollo KF, Yu B, Rait A, Xiang L, Huang W, Zhou Q, Ertem G, and Chang EH. Enhanced transfection efficiency of a systemically delivered tumor-targeting immunolipoplex by inclusion of a pH-sensitive histidylated oligolysine peptide. *Nucleic Acids Res* 2004, 32:e48.

133. Saito G, Amidon GL, and Lee KD. Enhanced cytosolic delivery of plasmid DNA by a sulfhydryl-activatable listeriolysin O/protamine conjugate utilizing cellular reducing potential. *Gene Ther* 2003, 10:72–83.

134. Chen CP, Kim JS, Steenblock E, Liu D, and Rice KG. Gene transfer with poly-melittin peptides. *Bioconjug Chem* 2006, 17:1057–1062.

135. Rozema DB, Ekena K, Lewis DL, Loomis AG, and Wolff JA. Endosomolysis by masking of a membrane-active agent (EMMA) for cytoplasmic release of macromolecules. *Bioconjug Chem* 2003, 14:51–57.

136. Meyer M, Zintchenko A, Ogris M, and Wagner E. A dimethylmaleic acid-melittin-polylysine conjugate with reduced toxicity, pH-triggered endosomolytic activity and enhanced gene transfer potential. *J Gene Med* 2007, 9:797–805.

137. Boeckle S, Fahrmeir J, Roedl W, Ogris M, and Wagner E. Melittin analogs with high lytic activity at endosomal pH enhance transfection with purified targeted PEI polyplexes. *J Control Release* 2006, 112:240–248.

138. Berg K, Selbo PK, Prasmickaite L, Tjelle TE, Sandvig K, Moan J, Gaudernack G, Fodstad O, Kjolsrud S, Anholt H, Rodal GH, Rodal SK, and Hogset A. Photochemical internalization: A novel technology for delivery of macromolecules into cytosol. *Cancer Res* 1999, 59:1180–1183.

139. Prasmickaite L, Hogset A, Selbo PK, Engesaeter BO, Hellum M, and Berg K. Photochemical disruption of endocytic vesicles before delivery of drugs: A new strategy for cancer therapy. *Br J Cancer* 2002, 86:652–657.

140. Hogset A, Ovstebo EB, Prasmickaite L, Berg K, Fodstad O, and Maelandsmo GM. Light-induced adenovirus gene transfer, an efficient and specific gene delivery technology for cancer gene therapy. *Cancer Gene Ther* 2002, 9:365–371.

141. Hellum M, Hogset A, Engesaeter BO, Prasmickaite L, Stokke T, Wheeler C, and Berg K. Photochemically enhanced gene delivery with cationic lipid formulations. *Photochem Photobiol Sci* 2003, 2:407–411.

142. Ndoye A, Merlin JL, Leroux A, Dolivet G, Erbacher P, Behr JP, Berg K, and Guillemin F. Enhanced gene transfer and cell death following p53 gene transfer using photochemical internalisation of glucosylated PEI-DNA complexes. *J Gene Med* 2004, 6:884–894.

143. Nishiyama N, Iriyama A, Jang WD, Miyata K, Itaka K, Inoue Y, Takahashi H, Yanagi Y, Tamaki Y, Koyama H, and Kataoka K. Light-induced gene transfer from packaged DNA enveloped in a dendrimeric photosensitizer. *Nat Mater* 2005, 4:934–941.

144. Hogset A, Prasmickaite L, Engesaeter BO, Hellum M, Selbo PK, Olsen VM, Maelandsmo GM, and Berg K. Light directed gene transfer by photochemical internalisation. *Curr Gene Ther* 2003, 3:89–112.

145. Prasmickaite L, Hogset A, Tjelle TE, Olsen VM, and Berg K. Role of endosomes in gene transfection mediated by photochemical internalisation (PCI). *J Gene Med* 2000, 2:477–488.

146. Ndoye A, Dolivet G, Hogset A, Leroux A, Fifre A, Erbacher P, Berg K, Behr JP, Guillemin F, and Merlin JL. Eradication of p53-mutated head and neck squamous cell carcinoma xenografts using nonviral p53 gene therapy and photochemical internalization. *Mol Ther* 2006, 13:1156–1162.

147. Akinc A, Thomas M, Klibanov AM, and Langer R. Exploring polyethylenimine-mediated DNA transfection and the proton sponge hypothesis. *J Gene Med* 2005, 7:657–663.

148. Sonawane ND, Szoka FC, Jr., and Verkman AS. Chloride accumulation and swelling in endosomes enhances DNA transfer by polyamine-DNA polyplexes. *J Biol Chem* 2003, 278:44826–44831.

149. Kichler A, Leborgne C, Coeytaux E, and Danos O. Polyethylenimine-mediated gene delivery: A mechanistic study. *J Gene Med* 2001, 3:135–144.

150. Kulkarni RP, Mishra S, Fraser SE, and Davis ME. Single cell kinetics of intracellular, nonviral, nucleic acid delivery vehicle acidification and trafficking. *Bioconjug Chem* 2005, 16:986–994.

151. Pichon C, Guerin B, Refregiers M, Goncalves C, Vigny P, and Midoux P. Zinc improves gene transfer mediated by DNA/cationic polymer complexes. *J Gene Med* 2002, 4:548–559.

152. Funhoff AM, van Nostrum CF, Koning GA, Schuurmans-Nieuwenbroek NM, Crommelin DJ, and Hennink WE. Endosomal escape of polymeric gene delivery complexes is not always enhanced by polymers buffering at low pH. *Biomacromolecules* 2004, 5:32–39.

153. Lukacs GL, Haggie P, Seksek O, Lechardeur D, Freedman N, and Verkman AS. Size-dependent DNA mobility in cytoplasm and nucleus. *J Biol Chem* 2000, 275:1625–1629.

154. Bausinger R, von Gersdorff K, Braeckmans K, Ogris M, Wagner E, Brauchle C, and Zumbusch A. The transport of nanosized gene carriers unraveled by live-cell imaging. *Angew Chem Int Ed Engl* 2006, 45:1568–1572.

155. de Bruin K, Ruthardt N, von Gersdorff K, Bausinger R, Wagner E, Ogris M, and Brauchle C. Cellular dynamics of EGF receptor-targeted synthetic viruses. *Mol Ther* 2007, 15:1297–1305.

156. van der Aa MA, Koning GA, d'Oliveira C, Oosting RS, Wilschut KJ, Hennink WE, and Crommelin DJ. An NLS peptide covalently linked to linear DNA does not enhance transfection efficiency of cationic polymer based gene delivery systems. *J Gene Med* 2005, 7:208–217.

157. Huth S, Hoffmann F, von Gersdorff K, Laner A, Reinhardt D, Rosenecker J, and Rudolph C. Interaction of polyamine gene vectors with RNA leads to the dissociation of plasmid DNA-carrier complexes. *J Gene Med* 2006, 8:1416–1424.

158. Wilke M, Fortunati E, van den BM, Hoogeveen AT, and Scholte BJ. Efficacy of a peptide-based gene delivery system depends on mitotic activity. *Gene Ther* 1996, 3:1133–1142.

159. Brunner S, Furtbauer E, Sauer T, Kursa M, and Wagner E. Overcoming the nuclear barrier: cell cycle independent nonviral gene transfer with linear polyethylenimine or electroporation. *Mol Ther* 2002, 5:80–86.

160. Brunner S, Sauer T, Carotta S, Cotten M, Saltik M, and Wagner E. Cell cycle dependence of gene transfer by lipoplex, polyplex and recombinant adenovirus. *Gene Ther* 2000, 7:401–407.

161. Laemmli UK and Tjian R. A nuclear traffic jam—unraveling multicomponent machines and compartments. *Curr Opin Cell Biol* 1996, 8:299–302.

162. Singer RH and Green MR. Compartmentalization of eukaryotic gene expression: Causes and effects. *Cell* 1997, 91: 291–294.

163. Itaka K, Harada A, Yamasaki Y, Nakamura K, Kawaguchi H, and Kataoka K. In situ single cell observation by fluorescence resonance energy transfer reveals fast intra-cytoplasmic delivery and easy release of plasmid DNA complexed with linear polyethylenimine. *J Gene Med* 2004, 6:76–84.

164. Hama S, Akita H, Ito R, Mizuguchi H, Hayakawa T, and Harashima H. Quantitative comparison of intracellular trafficking and nuclear transcription between adenoviral and lipoplex systems. *Mol Ther* 2006, 13:786–794.

165. Liu G, Aronovich EL, Cui Z, Whitley CB, and Hackett PB. Excision of sleeping beauty transposons: Parameters and applications to gene therapy. *J Gene Med* 2004, 6:574–583.

166. Ohlfest JR, Demorest ZL, Motooka Y, Vengco I, Oh S, Chen E, Scappaticci FA, Saplis RJ, Ekker SC, Low WC, Freese AB, and Largaespada DA. Combinatorial antiangiogenic gene therapy by nonviral gene transfer using the sleeping beauty transposon causes tumor regression and improves survival in mice bearing intracranial human glioblastoma. *Mol Ther* 2005, 12:778–788.

167. Calos MP. The phiC31 integrase system for gene therapy. *Curr Gene Ther* 2006, 6:633–645.

168. Calos MP. The potential of extrachromosomal replicating vectors for gene therapy. *Trends Genet* 1996, 12:463–466.

169. Jenke AC, Stehle IM, Herrmann F, Eisenberger T, Baiker A, Bode J, Fackelmayer FO, and Lipps HJ. Nuclear scaffold/matrix attached region modules linked to a transcription unit are sufficient for replication and maintenance of a mammalian episome. *Proc Natl Acad Sci U S A* 2004, 101: 11322–11327.

170. Jenke AC, Eisenberger T, Baiker A, Stehle IM, Wirth S, and Lipps HJ. The nonviral episomal replicating vector pEPI-1 allows long-term inhibition of bcr-abl expression by shRNA. *Hum Gene Ther* 2005, 16:533–539.

171. Middleton J, Neil S, Wintle J, Clark-Lewis I, Moore H, Lam C, Auer M, Hub E, and Rot A. Transcytosis and surface presentation of IL-8 by venular endothelial cells. *Cell* 1997, 91:385–395.

172. Bijsterbosch MK, Manoharan M, Rump ET, De Vrueh RL, van Veghel R, Tivel KL, Biessen EA, Bennett CF, Cook PD, and van Berkel TJ. In vivo fate of phosphorothioate antisense oligodeoxynucleotides: Predominant uptake by scavenger receptors on endothelial liver cells. *Nucleic Acids Res* 1997, 25:3290–3296.

173. Davis SS. Biomedical applications of nanotechnology–implications for drug targeting and gene therapy. *Trends Biotechnol* 1997, 15:217–224.

174. Lampugnani MG and Dejana E. Interendothelial junctions: Structure, signalling and functional roles. *Curr Opin Cell Biol* 1997, 9:674–682.

175. Dunlap DD, Maggi A, Soria MR, and Monaco L. Nanoscopic structure of DNA condensed for gene delivery. *Nucleic Acids Res* 1997, 25:3095–3101.

176. Labat M, Steffan AM, Brisson C, Perron H, Feugeas O, Furstenberger P, Oberling F, Brambilla E, and Behr JP. An electron microscopy study into the mechanism of gene transfer with lipopolyamines. *Gene Ther* 1996, 3:1010–1017.

177. Wolfert MA and Seymour LW. Atomic force microscopic analysis of the influence of the molecular weight of poly(L)lysine on the size of polyelectrolyte complexes formed with DNA. *Gene Ther* 1996, 3:269–273.

178. Ogris M, Steinlein P, Kursa M, Mechtler K, Kircheis R, and Wagner E. The size of DNA/transferrin-PEI complexes is an important factor for gene expression in cultured cells. *Gene Ther* 1998, 5:1425–1433.

179. Ogris M, Brunner S, Schuller S, Kircheis R, and Wagner E. PEGylated DNA/transferrin-PEI complexes: Reduced interaction with blood components, extended circulation in blood and potential for systemic gene delivery. *Gene Ther* 1999, 6:595–605.

180. Boeckle S, von Gersdorff K, van der Piepen S, Culmsee C, Wagner E, and Ogris M. Purification of polyethylenimine polyplexes highlights the role of free polycations in gene transfer. *J Gene Med* 2004, 6:1102–1111.

181. Fahrmeir J, Gunther M, Tietze N, Wagner E, and Ogris M. Electrophoretic purification of tumor targeted polyethylenimine-based polyplexes reduces toxic side effects in vivo. *J Control Release* 2007, 122:236–245.

182. Wagner E, Curiel D, and Cotten M. Delivery of drugs, proteins and genes into cells using transferrin as a ligand for receptor-mediated endocytosis. *Adv Drug Del Rev* 1994, 14:113–136.

183. Plank C, Mechtler K, Szoka FC, Jr., and Wagner E. Activation of the complement system by synthetic DNA complexes: A potential barrier for intravenous gene delivery. *Hum Gene Ther* 1996, 7:1437–1446.

184. Wolbink GJ, Brouwer MC, Buysmann S, ten Berge IJ, and Hack CE. CRP-mediated activation of complement in vivo: Assessment by measuring circulating complement-C-reactive protein complexes. *J Immunol* 1996, 157:473–479.

185. Rodman DM, San H, Simari R, Stephan D, Tanner F, Yang Z, Nabel GJ, and Nabel EG. In vivo gene delivery to the pulmonary circulation in rats: Transgene distribution and vascular inflammatory response. *Am J Respir Cell Mol Biol* 1997, 16:640–649.

186. Miller N and Whelan J. Progress in transcriptionally targeted and regulatable vectors for genetic therapy. *Hum Gene Ther* 1997, 8:803–815.

187. Wiertz EJ, Mukherjee S, and Ploegh HL. Viruses use stealth technology to escape from the host immune system. *Mol Med Today* 1997, 3:116–123.

188. Li W, Huang Z, MacKay JA, Grube S, and Szoka FC, Jr. Low-pH-sensitive poly(ethylene glycol) (PEG)-stabilized plasmid nanolipoparticles: Effects of PEG chain length, lipid composition and assembly conditions on gene delivery. *J Gene Med* 2005, 7:67–79.

189. Walker GF, Fella C, Pelisek J, Fahrmeir J, Boeckle S, Ogris M, and Wagner E. Toward synthetic viruses: Endosomal pH-triggered deshielding of targeted polyplexes greatly enhances gene transfer in vitro and in vivo. *Mol Ther* 2005, 11:418–425.

190. Knorr V, Allmendinger L, Walker GF, Paintner FF, and Wagner E. An Acetal-Based PEGylation Reagent for pH-Sensitive Shielding of DNA Polyplexes. *Bioconjug Chem* 2007, 2007 May 4; [Epub ahead of print].

191. Oishi M, Nagatsugi F, Sasaki S, Nagasaki Y, and Kataoka K. Smart polyion complex micelles for targeted intracellular delivery of PEGylated antisense oligonucleotides containing acid-labile linkages. *Chembiochem* 2005, 6:718–725.

192. Oishi M, Sasaki S, Nagasaki Y, and Kataoka K. pH-responsive oligodeoxynucleotide (ODN)-poly(ethylene glycol) conjugate through acid-labile beta-thiopropionate linkage: Preparation and polyion complex micelle formation. *Biomacromolecules* 2003, 4:1426–1432.

193. Hatakeyama H, Akita H, Kogure K, Oishi M, Nagasaki Y, Kihara Y, Ueno M, Kobayashi H, Kikuchi H, and Harashima H. Development of a novel systemic gene delivery system for cancer therapy with a tumor-specific cleavable PEG-lipid. *Gene Ther* 2007, 14:68–77.

194. Murthy N, Campbell J, Fausto N, Hoffman AS, and Stayton PS. Design and synthesis of pH-responsive polymeric carriers that target uptake and enhance the intracellular delivery of oligonucleotides. *J Control Release* 2003, 89: 365–374.

195. Xenariou S, Griesenbach U, Ferrari S, Dean P, Scheule RK, Cheng SH, Geddes DM, Plank C, and Alton EW. Using magnetic forces to enhance non-viral gene transfer to airway epithelium in vivo. *Gene Ther* 2006, 13:1545–1552.

196. Dash PR, Read ML, Fisher KD, Howard KA, Wolfert M, Oupicky D, Subr V, Strohalm J, Ulbrich K, and Seymour LW. Decreased binding to proteins and cells of polymeric gene delivery vectors surface modified with a multivalent hydrophilic polymer and retargeting through attachment of transferrin. *J Biol Chem* 2000, 275:3793–3802.

197. Lemieux P, Guerin N, Paradis G, Proulx R, Chistyakova L, Kabanov A, and Alakhov V. A combination of poloxamers increases gene expression of plasmid DNA in skeletal muscle. *Gene Ther* 2000, 7:986–991.

198. Wu CH, Wilson JM, and Wu GY. Targeting genes: Delivery and persistent expression of a foreign gene driven by mammalian regulatory elements in vivo. *J Biol Chem* 1989, 264:16985–16987.

199. Wu GY, Wilson JM, Shalaby F, Grossman M, Shafritz DA, and Wu CH. Receptor-mediated gene delivery in vivo. Partial correction of genetic analbuminemia in Nagase rats. *J Biol Chem* 1991; 266:14338–14342.

200. Chowdhury NR, Wu CH, Wu GY, Yerneni PC, Bommineni VR, and Chowdhury JR. Fate of DNA targeted to the liver by asialoglycoprotein receptor-mediated endocytosis in vivo. Prolonged persistence in cytoplasmic vesicles after partial hepatectomy. *J Biol Chem* 1993, 268:11265–11271.

201. Davis PB and Cooper MJ. Vectors for airway gene delivery. *AAPS J* 2007, 9:E11-E17.

202. Gao L, Wagner E, Cotten M, Agarwal S, Harris C, Romer M, Miller L, Hu PC, and Curiel D. Direct in vivo gene transfer to airway epithelium employing adenovirus-polylysine-DNA complexes. *Hum Gene Ther* 1993, 4:17–24.

203. Nguyen DM, Wiehle SA, Roth JA, and Cristiano RJ. Gene delivery into malignant cells in vivo by a conjugated adenovirus/DNA complex. *Cancer Gene Therapy* 1997, 4:183–190.

204. Kircheis R, Schuller S, Brunner S, Ogris M, Heider KH, Zauner W, and Wagner E. Polycation-based DNA complexes for tumor-targeted gene delivery in vivo. *J Gene Med* 1999, 1:111–120.

205. Goula D, Benoist D, Mantero S, Merlo G, Levi G, and Demeneix BA. Polyethylenimine-based intravenous delivery of transgenes to mouse lung. *Gene Ther* 1998, 5:1291–1295.

206. Zou SM, Erbacher P, Remy JS, and Behr JP. Systemic linear polyethylenimine (L-PEI)-mediated gene delivery in the mouse. *J Gene Med* 2000, 2:128–134.

207. Chollet P, Favrot MC, Hurbin A, and Coll JL. Side-effects of a systemic injection of linear polyethylenimine-DNA complexes. *J Gene Med* 2002, 4:84–91.

208. Rudolph C, Schillinger U, Ortiz A, Plank C, Golas MM, Sander B, Stark H, and Rosenecker J. Aerosolized nanogram quantities of plasmid DNA mediate highly efficient gene delivery to mouse airway epithelium. *Mol Ther* 2005, 12:493–501.

209. Rudolph C, Ortiz A, Schillinger U, Jauernig J, Plank C, and Rosenecker J. Methodological optimization of polyethylenimine (PEI)-based gene delivery to the lungs of mice via aerosol application. *J Gene Med* 2005, 7:59–66.

210. McLachlan G, Baker A, Tennant P, Gordon C, Vrettou C, Renwick L, Blundell R, Cheng SH, Scheule RK, Davies L, Painter H, Coles RL, Lawton AE, Marriott C, Gill DR, Hyde SC, Griesenbach U, Alton EW, Boyd AC, Porteous DJ, and Collie DD. Optimizing aerosol gene delivery and expression in the ovine lung. *Mol Ther* 2007, 15:348–354.

211. Maeda H. The enhanced permeability and retention (EPR) effect in tumor vasculature: The key role of tumor-selective macromolecular drug targeting. *Adv Enzyme Regul* 2001, 41:189–207.

212. Dufes C, Keith WN, Bilsland A, Proutski I, Uchegbu IF, and Schatzlein AG. Synthetic anticancer gene medicine exploits intrinsic antitumor activity of cationic vector to cure established tumors. *Cancer Res* 2005, 65:8079–8084.

213. Behlke MA. Progress towards in vivo use of siRNAs. *Mol Ther* 2006, 13:644–670.

214. Meyer M and Wagner E. Recent developments in the application of plasmid DNA-based vectors and small interfering RNA therapeutics for cancer. *Hum Gene Ther* 2006, 17:1062–1076.

215. Song E, Zhu P, Lee SK, Chowdhury D, Kussman S, Dykxhoorn DM, Feng Y, Palliser D, Weiner DB, Shankar P, Marasco WA, and Lieberman J. Antibody mediated in vivo delivery of small interfering RNAs via cell-surface receptors. *Nat Biotechnol* 2005, 23:709–717.

216. Aigner A, Fischer D, Merdan T, Brus C, Kissel T, and Czubayko F. Delivery of unmodified bioactive ribozymes by an RNA-stabilizing polyethylenimine (LMW-PEI) efficiently down-regulates gene expression. *Gene Ther* 2002, 9:1700–1707.

217. Schreiber S, Kampgen E, Wagner E, et al. Immunotherapy of metastatic malignant melanoma by a vaccine consisting of autologous interleukin 2-transfected cancer cells: outcome of a phase I study. *Hum Gene Ther* 1999, 10:983–993.

16 Electroporation of Plasmid-Based Vaccines and Therapeutics

Ruxandra Draghia-Akli and Amir S. Khan

CONTENTS

16.1 INTRODUCTION

Electroporation (EP), electrotransfer, or electrokinetic delivery is a technique to introduce nucleic acids and other macromolecules into cells both *in vitro* and *in vivo*, using conditions that induce transient cellular membrane permeabilization. Within the last decade, EP has evolved from an experimental technique derived from *in vitro* applications [1] to a robust preclinical and clinical delivery method; this physical method for *in vivo* non-viral gene therapy is currently used in laboratory animals, large mammals, and human clinical trials [2,3].

EP can be applied to deliver a multitude of macromolecules, from drugs with poor systemic bioavailability, such as bleomycin [4,5] or cisplatin [6], to nucleic acid fragments, oligonucleotides [7], siRNA [8], and plasmids encoding hormones, cytokines, enzymes, or antigens to a wide variety of tissues, such as solid tumors, muscle, in particular skeletal muscle, skin, and liver. Indeed, most EP applications involve delivery of plasmids to different organs and tissues (selected studies are summarized in the following Tables). As with the EP technique itself, plasmids evolved from being a substitute for more-or-less successful viral therapies to attractive candidates for development of clinically relevant products. Advantages of plasmids include simplicity of design, structure, rapid molecular biology manipulation of relevant components, and relative ease of manufacturing under good manufacturing practices (GMP) [9]. The classic disadvantage, low expression level *in vivo*, was resolved by the optimization of EP and other physical methods of delivery adapted to individual applications, such as the hydrodynamic or ultrasound methods [10]. In this chapter, we focus on the clinical developments of this technique that occurred within the last 3 years.

16.2 MECHANISTIC STUDIES OF ELECTROPORATION

A number of investigators focused on the mechanisms of DNA entry into cells [11], and in particular of DNA behavior and distribution after EP, both at cellular and tissue level, keeping in mind that as a prerequisite to a successful EP, reversible undamaging target cell permeabilization is mandatory [12]. The general mechanistic hypothesis is that during EP transient pores are formed as a function of the transmembrane voltage; during the period of membrane destabilization, macromolecules present in the immediate extracellular medium around target cells can gain access into the intracellular space [13]. It has been hypothesized that when an electromagnetic potential is applied to the cells, firstly, the cell membrane is charging, followed by the creation of pores, and ultimately by the evolution of pore radii. The transmembrane potential, number of pores, and distribution of pore radii were shown to be a function of time and position on the cell surface [14]. Fundamental studies showed that permeabilization of cells occurs only if the local electrical field at the target site reaches a critical value [15,16], typically higher than 200 mV. After the EP pulses are completed, a slow resealing occurs in seconds to minutes. Nevertheless, some changes in the membrane properties and decrease in tissue impedance may be measured and remain present for hours [17,18]. It is thus no surprise that the degree of permeabilization of cells is dependent upon the electric field intensity, specific impedance of the target tissue, length and number of pulses, shape, size, architecture and type of electrodes, buffer formulation, and cell size [19]. The entry of DNA and other macromolecules into cells is still debated. It was hypothesized as either an electrophoretic facilitation during pulses, with DNA molecules reoriented and drawn toward the anode (similarly to *in vitro* EP) [12], or passive diffusion of DNA through the permeabilized membrane defect [16]. As the authors point out, an interesting implication for this latest mechanism would be the ability to transfer neutral particles, such as PEGylated compounds, by EP.

Despite the large variation in EP conditions used for the target tissue or organ [2], the expression achieved in many of these cases represents efficacious levels of gene products, some of which are within the range that is necessary to treat human diseases or to induce a robust immune response. In studies of the electric field-mediated enhancement of gene and drug delivery, different types of electrodes have been used. These include clamp, caliper [20] or four-plated electrodes [21], tweezers [22],

paddles [23], needle arrays [24], electrode-reservoir devices [25], flexible electrode arrays which may be used with endoscopic and laparoscopic devices [26], as well as bathtub-type electrodes for internal organs, such as kidney [27]. While in rodent experiments, external electrodes are prevalent, in large mammals, including humans, the electrode design must be adapted to the application, organ, and animal species.

16.3   TISSUE-SPECIFIC EXPRESSION

16.3.1   Skeletal Muscle

Since the initial report by Aihara and Miyazaki [28], delivering a plasmid encoding for interleukin-5 (IL-5) to mice, *in vivo* EP of the skeletal muscle for vaccines and therapeutic proteins has become widely used. Table 16.1 summarizes some of the recent proof-of-concept literature reports in which EP were used to enhance plasmid delivery to skeletal muscle using reporter plasmids. Although the devices, conditions, methods, and animal models substantially differ, all studies

TABLE 16.1

Electroporation of Macromolecules into Muscle for Proof of Concept Reporter Gene Studies

Plasmid Product or Drug	Muscle Type	Species	Endpoints	References
Luciferase	Muscle and liver	Wistar rats	Controlled EP pulses yield higher luciferase expression	[12]
GFP (green fluorescent protein)	FDB (flexor digitorum brevis), soleus, TA (tibialis anterior), extensor digitorum muscles	C57Bl/6 mice	Testing of a mammalian protein expression system	[60]
Luciferase	Muscle and skin	Mice	Expression is reduced when reporter plasmid is mixed with Ca²⁺ ions	[61]
SEAP (secreted embryonic alkaline phosphatase), luciferase, mouse IgG, β-gal (β-galactosidase)	Quadriceps and TA muscles	Mice, rats	Optimal EP parameters for long-term expression versus DNA vaccination	[62]
JNK	Muscle	Mice	JNK over-expression affected the phosphorylation state of kinases in skeletal muscle	[63]
GFP, blue fluorescent protein (BFP), β-gal	EDL (extensor digitorum longum)	NMRI mice	Gene expression after co-injection of reporter plasmids	[64]
GFP	Tibialis cranialis (TC) and EDL	C57Bl/6 mice & Wistar rats	Gene expression, muscle performance, and histology of injected muscles	[65]
GFP, luciferase, erythropoietin (EPO)	Tc	C57bl/6 mice	Hemoglobin levels with regulatable promoter	[66]
GFP, luciferase, nNOS	Bladder muscle	Rats	EP conditions for nNOS immunoreactivity and NO release	[67]
SEAP	TA muscle, semimembranosus	Mice, pigs	Serum SEAP levels with various EP conditions	[24]
Luciferase	Muscle	Zebrafish, Indian carp	Promoter strength, EP parameters affect gene expression	[68]
p53	Muscle	C57bl/6 mice	Transfection efficiency of ultrasound versus EP in muscle	[69]
β-gal and myostatin shRNA	TA muscle	Rats	Myostatin mRNA and protein expression and muscle mass	[70]
β-gal, dystrophin, utrophin	Muscle	mdx mice, immune deficient	Transfection efficiency, muscle fiber damage, duration of expression	[31]
Luciferase, GFP, Muscarinic-3 receptor	Bladder	Rats	M3 immunohistochemistry, bladder muscle contractile responses	[71]

TABLE 16.1 (continued)

Electroporation of Macromolecules into Muscle for Proof of Concept Reporter Gene Studies

Plasmid Product or Drug	Muscle Type	Species	Endpoints	References
GFP, luciferase	TA muscle	Balb/c mice, S-D rats	Satellite cell activation, gene expression in muscle cell types	[72]
IL-4, GFP, insulin-like growth factor I (IGF-I)	TA muscle	C57Bl/10 mice	Gene expression, muscle damage following EP, muscle fiber contractility	[73]
β-gal	Muscle	NIH Swiss mice	Effects of intra-arterial versus IM injection, electrode geometry, and DNA vehicle on gene expression	[74]
SEAP or EGFP-expressing	TA muscle or gastrocnemius	C57/Bl6 Mice	Induction of anti-SEAP antibodies, Optimization of EP parameters	[59]
Plasmids	Muscle or Skin		Dependent on target muscle, formulation and current intensity	
Influenza Hemagglutinin, Neuraminidase and SEAP	Muscle	Pigs	SEAP expression, hemagglutination inhibition titers, histopathology on muscle biopsy samples	

conclude that plasmid injection followed by EP substantially enhances plasmid uptake over injection alone by two to three orders of magnitude, and that the method can be successfully used to deliver nucleic acids for preclinical and clinical applications. As myofibers are postmitotic, multinucleated cells, and muscles are physiologically very well vascularized and accessible for treatment, this organ can be used as a bioreactor for the persistent production and secretion of proteins into the blood stream. Similar to nucleic acid injection alone, the relative level of production would depend on the type of muscle (predominately slow [type I], such as soleus, or the fast [type II], such as the gastrocnemius–plantaris complex in mice) [29] that impacts transcription [30] as well as fat and fiber content, age and species [24,31].

The applications of intramuscular EP-mediated gene transfer cover a large spectrum of diseases: vaccination (Table 16.2), and therapeutic applications, including autoimmune diseases, including diabetes, malignancies, renal disease, anemia, endocrine dysfunctions (Table 16.3), prevention of drug toxicity, etc. In the case of plasmid encoding non-immunogenic species-specific transgene products, the duration of gene expression, and clinically relevant levels of therapeutic proteins,

TABLE 16.2

Electroporation of Macromolecules into Muscle for Vaccination Purposes

Plasmid Product or Drug	Muscle Type	Species	Endpoints	References
CKLF1	Muscle	Mice	Antibody production	[75]
APC-specific fusion proteins	Muscle	Mice	Increasing immunogenicity of plasmid DNA vaccines	[76]
Tuberculosis (TB) antigen	Quadriceps muscle	Mice	Increased Th1-directed response to TB antigen using DNA vaccine and EP	[77]
Interleukin-23	Pre-tibial muscle	C57Bl/6 mice	Antitumor immunity	[78]
Human hepatitis B virus antibody cDNA	Muscle	Scid mice	Serum antibody levels following IM-EP and hydrodynamic delivery	[79]
Viral interleukin-10 (IL-10)	Bilateral tibialis anterior (TA) muscle	Mice	Cytokine mRNA levels, macroscopic and histological examination in arthrogen-induced arthritis model	[80]
HBsAg, green fluorescent protein (GFP), luciferase, ovalbumin	TA muscle	Balb/c mice	Immunogenicity following application of EP few days prior to DNA vaccination	[81]
Her-2/neu (EC-TM)	Muscle	Syrian hamsters	Prophylactic vaccination using EP increases immunogenic response to chemically induced tumor challenge	[82]
Her-2/neu (EC-TM)	Muscle	Balb-neut mice	Increased immunogenicity of vaccine to protect against tumor development in transgenic mice	[83]
GFP, Haemonchus contortus antigen	Semi-membranous/ semitendinosus	Sheep	Gene expression, humoral responses and immune memory to vaccine	[84]
β-galactosidase (β-gal), mouse IgG2b-anti-NIP mAb	Glutaeobiceps and/or semitendinosus	Rabbits, sheep	Gene expression localization, humoral response following DNA vaccination, mAb production	[46]
Recombinant mAbs	Quadriceps, gluteus	Mice, sheep	Monoclonal antibody production, expression in mice, and sheep	[85]
Luciferase, IL-12, Japanese encephalitis virus envelope protein	Quadriceps muscle	C3H/hen mice	Antibody titers, protection from lethal viral challenge	[86]
Foot-and-mouth disease viral antigens P1, VP1	Muscle	Mice	Neutralizing antibodies, protection from viral challenge	[87]

(continued)

TABLE 16.2 (continued)
Electroporation of Macromolecules into Muscle for Vaccination Purposes

Plasmid Product or Drug	Muscle Type	Species	Endpoints	References
Plasmids expressing HIV gag and Rhesus macaque IL-15	Muscle	Indian Rhesus macaques	Cellular immune response, antigen-specific CD4+ and CD8+ T cells	[88]
HIV-1 env and gag	Muscle	Cynomolgus macaques	Cellular immune responses, robust anti-envelope antibodies	[89]
Optimized gag of HIV clades A-D and Rhesus IL-12	Muscle	Rhesus macaques	Cellular and humoral responses following DNA vaccination	[90]

TABLE 16.3
Electroporation of Macromolecules into Muscle for Therapeutic Purposes

Plasmid Product or Drug	Muscle Type	Species	Endpoints	References
α-MSH (melanocyte stimulating hormone)	TA muscle	ICR mice	Attenuate liver fibrosis	[91]
IL-10	TA muscle	ICR mice	Attenuation of carbon tetrachloride-induced liver fibrosis	[92]
IGF-I/Shh	Gastroc./soleus muscle	Mice	IGF-I and Shh expression can regulate bone maintenance	[93]
Human BMP-2	Gastroc. muscle	Wistar rats	Bone formation	[94]
IL1-Ra	Gastroc. muscle	DBA/1 mice	Reduction of collagen-induced arthritis	[95]
Angiopoietin-I and VEGF	Hindlimb muscle	Rats	Capillary density and limb necrosis in ischemic vascular disease model	[96]
β-gal and human TSH receptor	Muscle	Balb/c mice	Production of a mouse model of Graves disease	[97]
Soluble p75 TNF-receptor linked to Fc hIgG1	Gastroc. muscle	Balb/c mice	Histopathological examination of inflammation and myocardial fibrosis after viral myocarditis-induced injury	[98]
Luciferase, GFP, GAPD siRNA	TA muscle	GFP TG mice	Ability of siRNA to block gene expression	[99]
VEGF 164	TA muscle	Diabetic mice	Recovery from sensory deficits in a mouse model of diabetic neuropathy	[100]
Murine factor VIII	Gastroc. muscle	FVIII-def. Mice	Survival after hemostatic challenge	[20]
Rapsyn	Muscle	Rats	Compound muscle action potentials, acetylcholine receptors in experimentally induced *Myasthenia gravis*	[101]
hIL-10 (CMV promoter)	TA muscle	F344 rats	Lung transplantation rejection	[102]
hIL-10 (ubiquitin promoter)	TA muscle	F344 rats	Lung transplantation rejection	[103]
hIL-10	TA muscle	F344 rats	Gene expression, cardiac transplantation rejection	[104]
mIL-10	TA muscle	Lewis rats	Reduced mast cell density, cardiac histamine concentration, and mast cell growth in acute myocarditis	[105]
DGAT1	TA muscle	Rats	Triglyceride storage, lipid content in treated muscle	[106]
IGF-I	TA muscle	Rats	Muscle mass, muscle content in glucocorticoid-treated animals	[107]
Flt3L	TA muscle	C57Bl/6 mice	Gene expression, dendritic cell stimulation, antitumor effect	[108]
Chemokine-like factor-1	Muscle	Balb/c mice	Bronchial and bronchiolar wall abnormalities as a model for respiratory syndromes	[109]
α-MSH	TA muscle	ICR mice	Reversal of hepatic fibrosis	[91]
Mutated leptin	Leg muscle, liver	ICR mice	Body weight, fat tissue weight, neuropeptide Y expression in hypothalamus	[110]
Human leptin–human insulin precursor secretion signal peptide fusion protein	Sural muscle	Mice, S-D rats, rabbits	Blood glucose, serum leptin levels, leptin antibody levels, immunohisto-chemistry for expression	[111]
TGF type II receptor	Femoral muscle	C57Bl/6 mice	Expression in serum, anti-TGF-β1 therapy for attenuation of apoptosis, lung injury, and fibrosis	[112]
Calcitonin-gene related peptide (CGRP)	Lower limb muscles	Wistar rats	Neointimal formation reduced after balloon injury, increased iNOS, decreased apoptosis, and proliferation of vascular smooth muscle cells	[113]
Hepatocyte growth factor (HGF)	TA muscle	Hamsters	HGF levels, cardiac function, decreased ventricular fibrosis in cardiomyopathic animals	[114]
GFP, mdx correction	TA muscle	C57Bl/10 mdx mice	GFP expression, mdx mouse dystrophin gene correction	[115]

Note: Gastroc, Gastrocnemius.

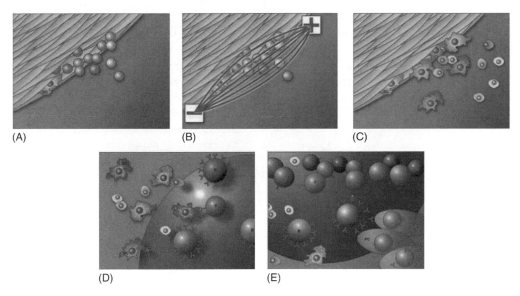

FIGURE 16.1 (See color insert following blank page 206.) Graphic representation of the delivery of macromolecules by direct injection followed by electroporation to the skeletal muscle for vaccination purposes. First, (A) the DNA vaccine is administered into skeletal muscle; (B) application of an electric field (electroporation); (C) contact of the expressed antigens with dendritic cells (DC) and subsequent activation; (D) activated DCs presenting antigens to B cells (B) and antigen fragments to T cells (T); (E) activated T-cells presenting antigens to B cells, activation of principal components (PC) for the generation of an immune response to the expressed antigen.

such as erythropoietin or growth hormone releasing hormone, was reported to be at least 9 to 15 months longer in species such as dogs [32], nonhuman primates [18], or cattle [33] compared to rodents [13]. For vaccination applications, strong humoral and cellular responses have been described after nucleic acid delivery by EP in nonhuman primates [34] and other models, sometimes after one single vaccination [35], for both infectious diseases, and genetic immunization [36]. A schematic representation of EP for vaccination purposes is provided in Figure 16.1. Collectively, these studies provide the evidence that adequate levels of secreted proteins can be achieved using plasmids in a simple, safe, and efficient manner, with significant potential for gene transfer and vaccination for large animals and humans.

16.3.2 LIVER

The liver represents a primary target for gene therapeutic treatment of many metabolic diseases, cancers, hepatitis, and other pathologies. Recombinant viral vectors are by far prevalently used to introduce genes into the liver, despite concerns about their safely and potential adverse effects. Another physical delivery technique, the hydrodynamic method, which involves delivery of plasmids or siRNA [37,38] in relatively large volumes, to the hepatic vasculature, gained popularity lately. Electroporation constitutes now an alternative method to deliver transgenes to the liver, mostly for modeling and basic studies concerning EP (Table 16.4), as well as functional identification of regulatory elements in gene sequences [39].

TABLE 16.4
Electroporation of Macromolecules into Liver

Plasmid Product or Drug	Organ	Species	Endpoints	References
HGF	Liver	Rats	Prevention of radiation-induced liver damage	[116]
N/A	Liver	S-D rats	Minimally invasive tissue ablation	[117]
N/A	Liver	Rats	Measuring electrical impedance and conductivity	[118]
N/A	Liver	Rabbits	Tissue permeabilization by taking into account tissue conductivity change	[119]
Luciferase, SEAP, β-gal, factor VIII	Liver	Balb/c, F-VIII-deficient mice	Vector distribution; phenotypic correction of hemophilic mice	[120]
Luciferase	Liver	Mice	Electroporation of the liver following hydrodynamically injected PEG-particles increases	[121]
Luciferase, β-gal	Liver	ddy mice	Electroporation of the liver following IV administered naked plasmid DNA resulted in 100-fold higher expression	[122]

The technique is also performed in conjunction with chemotherapy for different malignancies, as electrochemotherapy [40]; interestingly, tumor ablation has been achieved by locally delivering high intensity electric fields. These advances in liver gene delivery may provide powerful tools for basic research or potential clinical application studies.

16.3.3 SKIN

In the last few years, DNA delivery to skin has emerged as a major site for EP-mediated gene transfer applications requiring local or systemic distribution of a transgene product, and in particular for DNA vaccination [41,42] (Table 16.5). The intradermic (ID) space can be limited as a target organ because the volume of formulation that can be delivered is small. A small volume adapted for the ID or subcutaneous injections is translated in comparatively small plasmid doses that can be used in IM injections. Nevertheless, as a frontline of the host's defense against pathogens, skin is well equipped for immune surveillance [43], and nucleic acid doses that can elicit an immune response are much lower than doses needed to ensure a long-term production of a therapeutic protein. Electroporation of skin induces a mild and reversible impairment of the barrier function of the skin, a decrease in skin resistance, and a transient decrease in blood flow. Neither severe inflammation nor necroses are generally observed [44]. Because of surface access, skin EP is relatively painless and can be used for prophylactic vaccination [45,46]. EP has been used to enhance the

transdermal delivery of drugs, such as methotrexate [47]. After direct plasmid injection into skin, transfected cells are typically restricted to the epidermis. However, when EP is applied after the injection, larger numbers of adipocytes, fibroblasts, and dendritic-like cells within the dermal and subdermal tissues, as well as lymph nodes draining electropermeabilized sites, are transfected [48]. While mostly used as a target for vaccination purposes, an example of therapeutic use is the delivery of a keratinocyte growth factor expressing plasmid to the skin was shown to improve wound healing in a mouse model [49]. Gene therapy to enhance wound healing is a rapidly evolving field with several techniques being investigated including EP [50]. Further research is ongoing to improve the delivery of DNA using novel surface electrodes [21] to improve transfection efficiency while decreasing tissue damage. Transdermal drug delivery is another potential application for skin EP, if safety, efficacy, cost-effectiveness can be demonstrated [51].

16.3.4 TUMORS

Electroporation continues to be used as an anticancer treatment, though there are both successes and setbacks in this field [2,52]. Electrochemotherapy or enhanced delivery of chemotherapeutic drugs, especially bleomycin or cisplatin [53], to solid tumors has been used successfully for many years. Melanoma and metastatic melanoma, head and neck squamous cell carcinomas, and basal cell carcinoma are among the most studied tumor

TABLE 16.5
Electroporation of Macromolecules into Skin

Plasmid Product or Drug	Organ	Species	Endpoints	References
Sulphorhodamine (fluorescent drug surrogate)	Stratum corneum	Humans	Pain level and feasibility of procedure	[25]
India ink	Skin	Guinea pigs	Localization of India ink particles in stratum corneum and other skin layers	[123]
Mannitol	Skin	Mice	Iontophoresis increased skin permeability	[124]
Human insulin	Skin	Rats	More basic pH improved percutaneous absorption of insulin with iontophoresis	[125]
Salicylic acid	Skin	Rats	Drug time course in extracellular fluid	[126]
K^b-binding OVA-peptide SIINFEKL	Skin	Mice	Enhanced CTL response of transdermally delivered antigenic protein using needle-free vaccination	[127]
Methotrexate (MTX)	Skin	Mice	Serum MTX increased using transdermal delivery and needle-free EP arrays	[128]
Hepatitis B surface antigen	Skin	Rabbits	Humoral and cellular immune responses postvaccination	[42]
Hepatitis B surface antigen	Skin	Rabbits	Safety of skin EP procedure, skin viability post-procedure	[129]
Smallpox antigens	Skin	Mice	Protection from lethal intranasal poxvirus challenge	[130]
Luciferase, GFP	Skin	Wistar rats	Duration of expression, effect of electric pulses on skin	[131]
LacZ	Skin	Rats	Location and duration of expression, tissue damage	[132]
Keratinocyte growth factor	Skin	Mice	Wound healing in diabetic mouse model	[49]
Keratinocyte growth factor-1	Skin	S-D rats	Cutaneous wound healing in sepsis-based model of diabetic wound healing	[133]
bFGF	Skin	Rats	Area of necrosis and neovascularization, reduction of ischemic skin flap	[134]
TGF-β1	Skin	Mice	Re-epithelialization rate, collagen synthesis, and angiogenesis in diabetic mice	[135]
GFP	Skin	Pig	Improved gene expression with high plasmid concentrations	[136]
HIV DNA vaccine, IL-12	Skin	Rhesus macaques	Cellular and humoral responses	

TABLE 16.6
Electroporation of Macromolecules into Tumors: Chemical and Physical Disruption

Drug or Plasmid Product	Organ	Species	Endpoints	References
Bleomycin	Melanoma	Humans	Increased effect of bleomycin on melanoma metastases in skin with EP	[5]
Bleomycin	Melanoma	Humans	Tumor response rate, treatment tolerance	[137]
Bleomycin	Head, neck tumors	Humans	Safety, effectiveness, and side effects of bleomycin on patients	[138]
Bleomycin	Mucosal melanoma	Dogs	Tumor response rate, survival	[139]
Bleomycin	Tumors	Mice	Low-dose bleomycin delivery to tumor xenografts	[26]
Bleomycin	LPB sarcoma	Mice	Local tumor control, response to ionizing radiation treatments	[140]
Bleomycin	Colon carcinoma	Balb/c nude mice	Tumor growth, tumor regression, accumulation of inflammatory cells	[141]
Bleomycin	Lewis lung carcinoma	C57Bl/6 mice	Tumor growth, extent of tumor necrosis, tumor doubling time	[4]
Photosensitizing agents	Hepatoma A22 tumors	Mice	Accumulation of photosensitizers in tumors	[142]
Doxorubicin	Fibro-sarcoma	Swiss mice	Tumor growth kinetics in response to chemotherapy/radiotherapy combinations	[143]
TNF-related apoptosis-inducing ligand (TRAIL), 5-FU	Renal cell carcinoma	SCID mice	Tumor growth suppression and apoptosis, liver function	[144]
Cisplatin	Oral fibro-sarcoma	Hamsters	Tumor size	[145]
Cisplatin, p53	Murine sarcomas	Mice	Tumor growth, tumor regression	[146]
Cisplatin, platinum^{2+} complexes	Mammary carcinoma	Mice	Tumor growth inhibition, tumor toxicity	[147]
N/A	Implanted tumor cells	Mice	Electric potential distribution and resistance in implanted tumors	[148]

models [54] using electrochemotherapy. Ablation of cancer cells using physical properties of EP has also been demonstrated [55]. Clinical trials using this method for the treatment of solid tumors have been conducted in humans and other species. Recently, investigators focused on plasmid delivery to tumors as a means to increase long-term antitumor immunity [56] successfully, to inhibit angiogenesis or to reduce tumor volume. Table 16.6 contains reports pertaining to direct EP of the tumor either as a physical disruption method to induce apoptosis or to increase uptake of chemotherapeutic agents in tumor cells. Studies that use EP to improve efficacy of plasmid vectors and other nucleic acids are depicted in Table 16.7. Delivering these compounds using highly effective EP methods may form the basis for future human applications.

TABLE 16.7
Electroporation of Macromolecules into Tumors: Nucleic Acid Expression

Plasmid Product or Drug	Organ	Species	Endpoints	References
GFP	Tumor	Rats	Transfection efficiency in P22 rat tumors	[149]
GFP	Human solid tumor masses	Mice	EP using flexible electrode system induces tumor regression and decreased tumor growth in tumor xenografts	[150]
Melanin transcription factor siRNA	Implanted tumor cells	Mice	Tumor growth and progression of implanted melanomas	[151]
Viral protein R	Implanted tumor	Mice	Regression and apoptosis of subcutaneously implanted melanomas	[152]
IL-12	Tumors	Mice	T-cell infiltration, CTL activity, angiogenesis inhibition, IL-12 expression	[153]
IL-12	Implanted tumor	Mice	Weight change, tumor response, and serum IL-12 levels in tumor-bearing mice	[154]
B7.1, IL-12	SCCVII or TRAMP tumors	Mice	Tumor regression, CTL activity, and serum levels of B7.1, IL-12	[155]
Murine IL-12	Liver tumor	C3h mice	Treat solid malignant tumors and recurrences of hepatocellular carcinoma	[40]
IL-12, herpes simplex virus thymidine kinase (HSVtk)	CT26 tumors	Balb/c mice	Number of lung metastases and survival time in tumor-challenged mice	[156]
Mouse IL-12	Hepatocellular carcinoma	Mice	Electro-sonoporation elevated serum IL-12 and IFN-γ, survival, inhibition of metastases, reduced microvessels	[157]
Mouse IL-24	Hepatoma cells	Mice	Tumor growth, survival, tumor vascularization, liver metastasis	[158]
IL-15	Melanoma	Mice	Tumor volume/size, tumor regression, and long-term survival in B16 melanoma	[159]

(continued)

TABLE 16.7 (continued)
Electroporation of Macromolecules into Tumors: Nucleic Acid Expression

Plasmid Product or Drug	Organ	Species	Endpoints	References
GM-CSF, B7–1	Tumors	Balb/c mice	Tumor regression, metastases	[160]
IFN-α	Skin T-cell lymphoma	Humans	Response of mycosis fungoides to IFN-α treatment	[161]
Endostatin, HSVtk	Mammary tumors	Mice	Tumor volume, metastasis, intratumor microvessel density	[162]
Diphtheria toxin, GFP	Mammary carcinomas	SCID mice	Tumor volume, GFP expression, tumor necrotic area, DNA synthesis, lung metastasis, tumor microvessel density	[163]
Wild- or mutant-type p27Kip1	Oral cancer cell B88 xenografts	Nude mice	Apoptotic cell death, tumor growth inhibition	[164]

16.3.5 OTHERS: EYE, LUNG, CARDIAC MUSCLE, VASCULAR SMOOTH MUSCLE, AND TESTES

Studies examining transfection of various cell types in the eye using EP are summarized in Table 16.8. Recently, electric pulse-mediated plasmid transfer has been used to deliver transgenes to cornea in an effort to ensure long-lasting expression of a desired protein either for treatment of a local disease or for research purposes. Results of these studies demonstrate that EP is an excellent method for delivering genes to multiple cell layers within the cornea with extremely high levels of gene expression and little, if any, inflammatory response or tissue damage. Retinal ganglion cells also have been targeted. After a first demonstration showing retrograde labeling of up to 41% of the total ganglion cells in the electro-injected area, other studies with BDNF gene transferred by *in vivo* EP showed protection of axotomized retinal ganglion cells against apoptosis.

Investigators have also demonstrated the potential application of plasmid EP in many other tissues including the lung, heart, vascular smooth muscle (as previously reviewed by Dean [57]), and testes (Table 16.9).

16.4 SUMMARY AND FUTURE DIRECTIONS

Electroporation has moved from numerous animal experiments to use in on-going human clinical trials, as previously reviewed [2]. Furthermore, the field of EP is rapidly evolving to overcome the early questions of pain associated with the procedure, feasibility, and efficacy of the procedure in humans to deliver chemotherapeutic drugs, DNA vaccines, siRNA, and therapeutic plasmids to multiple organs of the body. Recent developments in the field of EP involve fundamental and pre-clinical studies using devices and techniques that are capable of controlling the pulse pattern, time between the pulses, their length, duration, electric field intensity and spatial distribution,

TABLE 16.8
Electroporation of Macromolecules into Cell Types in the Eye

Plasmid Product or Drug	Organ	Species	Endpoints	References
IL-10	Cornea	Mice	Duration of expression with long-acting promoter and nuclear localization, corneal cell turnover	[165]
GFP, other reporters	Retin	S-D rats	Controlled expression of reporter genes	[166]
Luciferase	Retinal pigment epithelium	Mice	Transcription factor, gene expression	[167]
Luciferase	Retinal pigment epithelium	Rabbits	Reporter gene expression, tissue damage	[168]
Luciferase	Subretinal space	Mice	Reporter gene expression time course and duration	[169]
Chimeric TNF-α soluble receptor	Ciliary muscle	Rats	Protein expression in aqueous humor, pathology/structural damage to tissue	[170]
Ephrin-A receptor	Retinal ganglion cell	Ferrets	Inducing developmental abnormalities in projections from eye to brain	[171]
GFP, luciferase	Retinal pigment epithelium	Rats	Plasmid expression duration using phiC31 integrase	[172]
Taurine upregulated gene 1 (TUG1)	Retina	S-D rats	Malformations in photoreceptor differentiation using RNAi for TUG1	[173]
β-gal	Retinal pigment epithelium	Balb/c mice	Reporter expression duration and localization	[174]
GDNF, GFP	Retinal ganglion cells	Rats	Reporter gene transfer efficiency, tissue damage, retinal ganglion cell survival	[175]
GFP	Corneal and retinal epithelium	Chicken embryos (in ovo)	GFP expression analysis by immunohistochemistry, lens epithelial cell proliferation	[176]

TABLE 16.9

Electroporation of Macromolecules in Lung, Cardiac, Vascular Muscle, or Testes Tissue

Plasmid Product or Drug	Organ	Species	Endpoints	References
Hepatocyte growth factor	Lung	Rats	Reduced injury/apoptosis after bleomycin-induced pulmonary fibrosis	[177]
Luciferase	Lung	Rats	Gene expression as assessed by imaging and RLU/mg protein	[178]
β1 subunit of Na^+, K^+-ATPase	Lung	Rats	Alveolar fluid clearance	[179]
Luciferase, β-gal, α1 and β1Na^+, K^+-ATPase subunits	Lung	Balb/c mice	Alveolar fluid clearance, respiratory mechanics, protection from LPS-induced lung injury	[180]
Luciferase, GFP	Lung	Mice, sheep	Reporter gene activity and duration	[181]
Luciferase	Lung	Mice	Inflammatory response as measured by proinflammatory cytokines in bronchioalveolar fluid	[182]
Luciferase, β-gal	Liver, skin, kidney, spleen, muscle, IV	Mice	Gene expression localization in various organs and with/without EP, duration of expression	[183]
N/A	Heart	Pigs	Epicardial atrial ablation using irreversible EP for atrial fibrillation	[184]
TNF receptor-associated factor 6	Carotid artery	Rabbits	After stent implantation, cell replication was prevented in both the intima and media	[185]
Yellow fluorescent protein	Testes	Syrian Golden hamster	Expression in epididymal sperm, duration of expression, no adverse effects on testicular integrity, and sperm quality	[186]
Egg activation factor (PLC zeta), yellow fluorescent protein	Testes	Rats	Gene transfer and localization in mature sperm	[187]
GFP, luciferase	Testes	S-D rats	Comparison of promoter activity, localization of expression	[188]
β-gal	Testes	Mice	Effects of expression on spermatogenesis, duration of expression	[189]
β-gal	Testes	Mice	Localization and duration of expression, comparison of administration methods	[190]
Dmc1 (DNA recombinase)	Testes	Mice	RNA interference with spermatocyte function	[191]
Rat EPO	Testes	S-D rats	Testicular weight, numbers of spermatids, and sperm	[192]

and various other parameters. The number, shape, nature, and relative placement of electrodes to the target organ are also intensely studied in numerous animal models [58]. Tissue damage that has been previously associated with EP can be minimized by optimization of conditions, and has been shown to be mild to moderate, and transitory at a series of electric field intensities and plasmid quantities [59]. These optimizations are/will be applied into human clinical trials. Only in the last 30 months, more than 300 articles have been published relating to *in vivo* EP and its applications. This progression indicates the acceptance of EP as a potential clinical technique.

Delivering nucleic acids and macromolecules into cells and tissue by physical techniques has been an impediment for years. The preclinical and clinical development of *in vivo* EP represents a quantum leap in this field. Although its exact mechanism remains elusive, this reliable and consistent method entered into human clinical trials in the last year. Preliminary results from these trials, presented at recent international meetings are demonstrating its usefulness and the potential of EP to transform nonviral therapies into a novel medical field.

REFERENCES

1. Neumann E, Schaefer-Ridder M, Wang Y, and Hofschneider PH 1982 Gene transfer into mouse lyoma cells by electroporation in high electric fields. *EMBO J* 1:841–845.

2. Prud'homme GJ, Glinka Y, Khan AS, and Draghia-Akli R 2006 Electroporation-enhanced nonviral gene transfer for the prevention or treatment of immunological, endocrine and neoplastic diseases. *Curr Gene Ther* 6:243–273.

3. Prud'homme GJ, Draghia-Akli R, and Wang Q 2007 Plasmid-based gene therapy of diabetes mellitus. *Gene Ther* 14:553–564.

4. Satkauskas S, Batiuskaite D, Salomskaite-Davalgiene S, and Venslauskas MS 2005 Effectiveness of tumor electrochemotherapy as a function of electric pulse strength and duration. *Bioelectrochemistry* 65:105–111.

5. Gaudy C, Richard MA, Folchetti G, Bonerandi JJ, and Grob JJ 2006 Randomized controlled study of electrochemotherapy in the local treatment of skin metastases of melanoma. *J Cutan Med Surg* 10:115–121.

6. Rebersek M, Cufer T, Cemazar M, Kranjc S, and Sersa G 2004 Electrochemotherapy with cisplatin of cutaneous tumor lesions in breast cancer. *Anticancer Drugs* 15:593–597.

7. Nunamaker EA, Zhang HY, Shirasawa Y, Benoit JN, and Dean DA 2003 Electroporation-mediated delivery of catalytic oligodeoxynucleotides for manipulation of vascular gene expression. *Am J Physiol Heart Circ Physiol* 285:H2240-H2247.

8. Inoue A, Takahashi KA, Mazda O, Terauchi R, Arai Y, Kishida T, Shin-Ya M, Asada H, Morihara T, Tonomura H, Ohashi S, Kajikawa Y, Kawahito Y, Imanishi J, Kawata M, and Kubo T 2005 Electro-transfer of small interfering RNA ameliorated arthritis in rats. *Biochem Biophys Res Commun* 336: 903–908.

9. Hebel HL, Attra HE, Khan AS, and Draghia-Akli R 2006 Successful parallel development and integration of a plasmid-based biologic, container/closure system and electrokinetic delivery device. *Vaccine* 24:4607–4614.

10. Lavigne MD and Gorecki DC 2006 Emerging vectors and targeting methods for nonviral gene therapy. *Expert Opin Emerg Drugs* 11:541–557.

11. Wolff JA and Budker V 2005 The mechanism of naked DNA uptake and expression. *Adv Genet* 54:3–20.

12. Cukjati D, Batiuskaite D, Andre F, Miklavcic D, and Mir LM 2007 Real time electroporation control for accurate and safe in vivo non-viral gene therapy. *Bioelectrochemistry* 70:501–507.

13. Trollet C, Bloquel C, Scherman D, and Bigey P 2006 Electrotransfer into skeletal muscle for protein expression. *Curr Gene Ther* 6:561–578.

14. Krassowska W and Filev PD 2007 Modeling electroporation in a single cell. *Biophys J* 92:404–417.

15. Favard C, Dean DS, and Rols MP 2007 Electrotransfer as a non viral method of gene delivery. *Curr Gene Ther* 7:67–77.

16. Liu F, Heston S, Shollenberger LM, Sun B, Mickle M, Lovell M, and Huang L 2006 Mechanism of in vivo DNA transport into cells by electroporation: Electrophoresis across the plasma membrane may not be involved. *J Gene Med* 8:353–361.

17. Grafstrom G, Engstrom P, Salford LG, and Persson BR 2006 99mTc-DTPA uptake and electrical impedance measurements in verification of in vivo electropermeabilization efficiency in rat muscle. *Cancer Biother Radiopharm* 21:623–635.

18. Zampaglione I, Arcuri M, Cappelletti M, Ciliberto G, Perretta G, Nicosia A, La MN, and Fattori E 2005 In vivo DNA gene electro-transfer: A systematic analysis of different electrical parameters. *J Gene Med* 7:1475–1481.

19. Becker SM and Kuznetsov AV 2006 Numerical modeling of in vivo plate electroporation thermal dose assessment. *J Biomech Eng* 128:76–84.

20. Long YC, Jaichandran S, Ho LP, Tien SL, Tan SY, and Kon OL 2004 FVIII gene delivery by muscle electroporation corrects murine hemophilia A. *J Gene Med* 7:494–505.

21. Heller LC, Jaroszeski MJ, Coppola D, McCray AN, Hickey J, and Heller R 2007 Optimization of cutaneous electrically mediated plasmid DNA delivery using novel electrode. *Gene Ther* 14:275–280.

22. Sato M 2005 Intraoviductal introduction of plasmid DNA and subsequent electroporation for efficient in vivo gene transfer to murine oviductal epithelium. *Mol Reprod Dev* 71:321–330.

23. Dean DA 2003 Electroporation of the vasculature and the lung. *DNA Cell Biol* 22:797–806.

24. Khan AS, Pope MA, and Draghia-Akli R 2005 Highly efficient constant-current electroporation increases in vivo plasmid expression. *DNA Cell Biol* 24:810–818.

25. Pliquett U and Weaver JC 2007 Feasibility of an electrode-reservoir device for transdermal drug delivery by noninvasive skin electroporation. *IEEE Trans Biomed Eng* 54:536–538.

26. Soden DM, Larkin JO, Collins CG, Tangney M, Aarons S, Piggott J, Morrissey A, Dunne C, and O'Sullivan GC 2006 Successful application of targeted electrochemotherapy using novel flexible electrodes and low dose bleomycin to solid tumours. *Cancer Lett* 232:300–310.

27. Isaka Y, Yamada K, Takabatake Y, Mizui M, Miura-Tsujie M, Ichimaru N, Yazawa K, Utsugi R, Okuyama A, Hori M, Imai E, and Takahara S 2005 Electroporation-mediated HGF gene transfection protected the kidney against graft injury. *Gene Ther* 12:815–820.

28. Aihara H and Miyazaki J 1998 Gene transfer into muscle by electroporation in vivo. *Nat Biotechnol* 16:867–870.

29. Brown PA, Khan AS, Pope MA, and Draghia-Akli R 2007 Muscle characteristics affect plasmid expression following electroporation in both large and small animals., *Mol Ther* 15 (Suppl 1), p S349.

30. Calvo S, Vullhorst D, Venepally P, Cheng J, Karavanova I, and Buonanno A 2001 Molecular dissection of DNA sequences and factors involved in slow muscle-specific transcription. *Mol Cell Biol* 21:8490–8503.

31. Molnar MJ, Gilbert R, Lu Y, Liu AB, Guo A, Larochelle N, Orlopp K, Lochmuller H, Petrof BJ, Nalbantoglu J, and Karpati G 2004 Factors influencing the efficacy, longevity, and safety of electroporation-assisted plasmid-based gene transfer into mouse muscles. *Mol Ther* 10:447–455.

32. Tone CM, Cardoza DM, Carpenter RH, and Draghia-Akli R 2004 Long-term effects of plasmid-mediated growth hormone releasing hormone in dogs. *Cancer Gene Ther* 11:389–396.

33. Brown PA and Draghia-Akli R 2004 Improved clinical outcome in cats and dogs in chronic renal failure after plasmid-mediated GHRH treatment. Society for Neuroscience Abstract 2004.

34. Luckay A, Sidhu MK, Kjeken R, Megati S, Chong SY, Roopchand V, Garcia-Hand D, Abdullah R, Braun R, Montefiori DC, Rosati M, Felber BK, Pavlakis GN, Mathiesen I, Israel ZR, Eldridge JH, and Egan MA 2007 Effect of plasmid DNA vaccine design and in vivo electroporation on the resulting vaccine-specific immune responses in rhesus macaques. *J Virol* 81:5257–5269.

35. Tsang C, Babiuk S, van Drunen Littel-van den Hurk, Babiuk LA, and Griebel P 2007 A single DNA immunization in combination with electroporation prolongs the primary immune response and maintains immune memory for six months. *Vaccine* 25:5485–5494.

36. Aurisicchio L, Mennuni C, Giannetti P, Calvaruso F, Nuzzo M, Cipriani B, Palombo F, Monaci P, Ciliberto G, and La MN 2007 Immunogenicity and safety of a DNA prime/adenovirus boost vaccine against rhesus CEA in nonhuman primates. *Int J Cancer* 120:2290–2300.

37. McAnuff MA, Rettig GR, and Rice KG 2007 Potency of siRNA versus shRNA mediated knockdown in vivo. *J Pharm Sci* 96:2922–2930.

38. Suda T, Gao X, Stolz DB, and Liu D 2007 Structural impact of hydrodynamic injection on mouse liver. *Gene Ther* 14:129–137.

39. Lagor WR, Heller R, de Groh ED, and Ness GC 2007 Functional analysis of the hepatic HMG-CoA reductase promoter by in vivo electroporation. *Exp Biol Med* (Maywood) 232:353–361.

40. Harada N, Shimada M, Okano S, Suehiro T, Soejima Y, Tomita Y, and Maehara Y 2004 IL-12 gene therapy is an effective therapeutic strategy for hepatocellular carcinoma in immunosuppressed mice. *J Immunol* 173:6635–6644.

41. Foldvari M, Babiuk S, and Badea I 2006 DNA delivery for vaccination and therapeutics through the skin. *Curr Drug Deliv* 3:17–28.

42. Medi BM, Hoselton S, Marepalli RB, and Singh J 2005 Skin targeted DNA vaccine delivery using electroporation in rabbits. I: efficacy. *Int J Pharm* 294:53–63.

43. Cui Z, Dierling A, and Foldvari M 2006 Non-invasive immunization on the skin using DNA vaccine. *Curr Drug Deliv* 3:29–35.

44. Dujardin N, Staes E, Kalia Y, Clarys P, Guy R, and Preat V 2002 In vivo assessment of skin electroporation using square wave pulses. *J Control Release* 79:219–227.

45. Frederickson RM, Carter BJ, and Pilaro AM 2003 Nonclinical toxicology in support of licensure of gene therapies. *Mol Ther* 8:8–10.

46. Tjelle TE, Salte R, Mathiesen I, and Kjeken R 2006 A novel electroporation device for gene delivery in large animals and humans. *Vaccine* 24:4667–4670.

47. Wong TW, Zhao YL, Sen A, and Hui SW 2005 Pilot study of topical delivery of methotrexate by electroporation. *Br J Dermatol* 152:524–530.

48. Drabick JJ, Glasspool-Malone J, King A, and Malone RW 2001 Cutaneous transfection and immune responses to intradermal nucleic acid vaccination are significantly enhanced by in vivo electropermeabilization. *Mol Ther* 3:249–255.

49. Marti G, Ferguson M, Wang J, Byrnes C, Dieb R, Qaiser R, Bonde P, Duncan MD, and Harmon JW 2004 Electroporation transfection with KGF-1 DNA improves wound healing in a diabetic mouse model. *Gene Ther* 11:1780–1785.

50. Branski LK, Pereira CT, Herndon DN, and Jeschke MG 2007 Gene therapy in wound healing: present status and future directions. *Gene Ther* 14:1–10.

51. Brown MB, Martin GP, Jones SA, and Akomeah FK 2006 Dermal and transdermal drug delivery systems: current and future prospects. *Drug Deliv* 13:175–187.

52. Cemazar M, Golzio M, Sersa G, Rols MP, and Teissie J 2006 Electrically-assisted nucleic acids delivery to tissues in vivo: Where do we stand? *Curr Pharm Des* 12:3817–3825.

53. Giardino R, Fini M, Bonazzi V, Cadossi R, Nicolini A, and Carpi A 2006 Electrochemotherapy a novel approach to the treatment of metastatic nodules on the skin and subcutaneous tissues. *Biomed Pharmacother* 60:458–462.

54. Byrne CM and Thompson JF 2006 Role of electrochemotherapy in the treatment of metastatic melanoma and other metastatic and primary skin tumors. *Expert Rev Anticancer Ther* 6:671–678.

55. Miller L, Leor J, and Rubinsky B 2005 Cancer cells ablation with irreversible electroporation. *Technol Cancer Res Treat* 4:699–705.

56. Stevenson FK, Ottensmeier CH, Johnson P, Zhu D, Buchan SL, McCann KJ, Roddick JS, King AT, McNicholl F, Savelyeva N, and Rice J 2004 DNA vaccines to attack cancer. *Proc Natl Acad Sci U S A* 101 Suppl 2:14646–52. Epub;%2004 Aug 3:14646–14652.

57. Dean DA 2005 Nonviral gene transfer to skeletal, smooth, and cardiac muscle in living animals. *Am J Physiol Cell Physiol* 289:C233-C245.

58. Draghia-Akli R and Khan AS 2007 In vivo electroporation of gene sequences for therapeutic and vaccination applications. *Recent Patents on DNA & Gene Sequences* 1:207–213.

59. Draghia-Akli R, Khan AS, Brown PA, Pope MA, Wu L, Hirao L, and Weiner DB 2008 Parameters for DNA vaccination using adaptive constant-current electroporation in mouse and pig models. *Vaccine* (2008), doi:10.1016/j.vaccine.2008.03.071.

60. Difranco M, Neco P, Capote J, Meera P, and Vergara JL 2006 Quantitative evaluation of mammalian skeletal muscle as a heterologous protein expression system. *Protein Expr Purif* 47:281–288.

61. Zhao YG, Lu HL, Peng JL, and Xu YH 2006 Inhibitory effect of Ca²⁺ on in vivo gene transfer by electroporation. *Acta Pharmacol Sin* 27:307–310.

62. Gronevik E, von Steyern FV, Kalhovde JM, Tjelle TE, and Mathiesen I 2005 Gene expression and immune response kinetics using electroporation-mediated DNA delivery to muscle. *J Gene Med* 7:218–227.

63. Fujii N, Boppart MD, Dufresne SD, Crowley PF, Jozsi AC, Sakamoto K, Yu H, Aschenbach WG, Kim S, Miyazaki H, Rui L, White MF, Hirshman MF, and Goodyear LJ 2004 Overexpression or ablation of JNK in skeletal muscle has no effect on glycogen synthase activity. *Am J Physiol Cell Physiol* 287:C200–C208.

64. Rana ZA, Ekmark M, and Gundersen K 2004 Coexpression after electroporation of plasmid mixtures into muscle in vivo. *Acta Physiol Scand* 181:233–238.

65. Hojman P, Zibert JR, Gissel H, Eriksen J, and Gehl J 2007 Gene expression profiles in skeletal muscle after gene electrotransfer. *BMC Mol Biol* 8:56.

66. Hojman P, Gissel H, and Gehl J 2007 Sensitive and precise regulation of haemoglobin after gene transfer of erythropoietin to muscle tissue using electroporation. *Gene Ther* 14:950–959.

67. Iwashita H, Yoshida M, Nishi T, Otani M, and Ueda S 2004 In vivo transfer of a neuronal nitric oxide synthase expression vector into the rat bladder by electroporation. *BJU Int* 93:1098–1103.

68. Rambabu KM, Rao SH, and Rao NM 2005 Efficient expression of transgenes in adult zebrafish by electroporation. *BMC Biotechnol* 5:29.

69. Kusumanto YH, Mulder NH, Dam WA, Losen MH, Meijer C, and Hospers GA 2007 Improvement of in vivo transfer of plasmid DNA in muscle: Comparison of electroporation versus ultrasound. *Drug Deliv* 14:273–277.

70. Magee TR, Artaza JN, Ferrini MG, Vernet D, Zuniga FI, Cantini L, Reisz-Porszasz S, Rajfer J, and Gonzalez-Cadavid NF 2006 Myostatin short interfering hairpin RNA gene transfer increases skeletal muscle mass. *J Gene Med* 8:1171–1181.

71. Otani M, Yoshida M, Iwashita H, Kawano Y, Miyamae K, Inadome A, Nishi T, and Ueda S 2004 Electroporation-mediated muscarinic M3 receptor gene transfer into rat urinary bladder. *Int J Urol* 11:1001–1008.

72. Peng B, Zhao Y, Lu H, Pang W, and Xu Y 2005 In vivo plasmid DNA electroporation resulted in transfection of satellite cells and lasting transgene expression in regenerated muscle fibers. *Biochem Biophys Res Commun* 338:1490–1498.

73. Schertzer JD, Plant DR, and Lynch GS 2005 Optimizing plasmid-based gene transfer for investigating skeletal muscle structure and function *Mol Ther* 13:795–803.

74. Taylor J, Babbs CF, Alzghoul MB, Olsen A, Latour M, Pond AL, and Hannon K 2004 Optimization of ectopic gene expression in skeletal muscle through DNA transfer by electroporation. *BMC Biotechnol* 4:11.

75. Chen Y, Zhang T, Li T, Han W, Zhang Y, and Ma D 2005 Preparation and characterization of a monoclonal antibody against CKLF1 using DNA immunization with in vivo electroporation. *Hybridoma* (Larchmt) 24:305–308.

76. Fredriksen AB, Sandlie I, and Bogen B 2006 DNA vaccines increase immunogenicity of idiotypic tumor antigen by targeting novel fusion proteins to antigen-presenting cells. *Mol Ther* 13:776–785.

77. Gronevik E, Mathiesen I, and Lomo T 2005 Early events of electroporation-mediated intramuscular DNA vaccination potentiate Th1-directed immune responses. *J Gene Med* 7:1246–1254.

78. Kaiga T, Sato M, Kaneda H, Iwakura Y, Takayama T, and Tahara H 2007 Systemic administration of IL-23 induces potent antitumor immunity primarily mediated through Th1-type response in association with the endogenously expressed IL-12. *J Immunol* 178:7571–7580.

79. Kitaguchi K, Toda M, Takekoshi M, Maeda F, Muramatsu T, and Murai A 2005 Immune deficiency enhances expression of recombinant human antibody in mice after nonviral in vivo gene transfer. *Int J Mol Med* 16:683–688.

80. Kuroda T, Maruyama H, Shimotori M, Higuchi N, Kameda S, Tahara H, Miyazaki J, and Gejyo F 2006 Effects of viral interleukin 10 introduced by in vivo electroporation on arthrogen-induced arthritis in mice. *J Rheumatol* 33:455–462.

81. Peng B, Zhao Y, Xu L, and Xu Y 2007 Electric pulses applied prior to intramuscular DNA vaccination greatly improve the vaccine immunogenicity. *Vaccine* 25:2064–2073.

82. Berta GN, Mognetti B, Spadaro M, Trione E, Amici A, Forni G, Di CF, and Cavallo F 2005 Anti-HER-2 DNA vaccine protects Syrian hamsters against squamous cell carcinomas. *Br J Cancer* 93:1250–1256.

83. Quaglino E, Mastini C, Iezzi M, Forni G, Musiani P, Klapper LN, Hardy B, and Cavallo F 2005 The adjuvant activity of BAT antibody enables DNA vaccination to inhibit the progression of established autochthonous Her-2/neu carcinomas in BALB/c mice. *Vaccine* 23:3280–3287.

84. Scheerlinck JP, Karlis J, Tjelle TE, Presidente PJ, Mathiesen I, and Newton SE 2004 In vivo electroporation improves immune responses to DNA vaccination in sheep. *Vaccine* 22:1820–1825.

85. Tjelle TE, Corthay A, Lunde E, Sandlie I, Michaelsen TE, Mathiesen I, and Bogen B 2004 Monoclonal antibodies produced by muscle after plasmid injection and electroporation. *Mol Ther* 9:328–336.

86. Wu CJ, Lee SC, Huang HW, and Tao MH 2004 In vivo electroporation of skeletal muscles increases the efficacy of Japanese encephalitis virus DNA vaccine. *Vaccine* 22:1457–1464.

87. Yang NS, Wang JH, Lin KF, Wang CY, Kim SA, Yang YL, Jong MH, Kuo TY, Lai SS, Cheng RH, Chan MT, and Liang SM 2005 Comparative studies of the capsid precursor polypeptide P1 and the capsid protein VP1 cDNA vectors for DNA vaccination against foot-and-mouth disease virus. *J Gene Med* 7:708–717.

88. Rosati M, Valentin A, Jalah R, Patel V, von Gegerfelt AS, Bergamaschi C, Alicea C, Weiss D, Treece J, Pal R, Markham P, Marques E, August J, Khan AS, Draghia-Akli R, Felber B, and Pavlakis G 2007 Increased immune responses in rhesus macaques by DNA vaccination combined with electroporation. *Vaccine* (2008), doi:10.1016/j.vaccine.2008.03.090.

89. Cristillo AD, Weiss D, Hudacik L, Restrepo S, Galmin L, Suschak J, Draghia-Akli R, Markham P, and Pal R 2008 Persistent antibody and T cell responses induced by HIV-1 DNA vaccine delivered by electroporation. *Biochem Biophys Res Commun* 366:29–35.

90. Hirao L, Wu L, Khan AS, Hokey D, Yan J, Dai A, Betts M, Draghia-Akli R, and Weiner DB 2008 Combined effects of IL-12 and electroporation enhances the potency of DNA vaccination in macaques. *Vaccine* (2008), doi:10.1016/j.vaccine. 2008.02.036.

91. Wang CH, Lee TH, Lu CN, Chou WY, Hung KS, Concejero AM, and Jawan B 2006 Electroporative alpha-MSH gene transfer attenuates thioacetamide-induced murine hepatic fibrosis by MMP and TIMP modulation. *Gene Ther* 13:1000–1009.

92. Chou WY, Lu CN, Lee TH, Wu CL, Hung KS, Concejero AM, Jawan B, and Wang CH 2006 Electroporative interleukin-10 gene transfer ameliorates carbon tetrachloride-induced murine liver fibrosis by MMP and TIMP modulation. *Acta Pharmacol Sin* 27:469–476.

93. Alzghoul MB, Gerrard D, Watkins BA, and Hannon K 2004 Ectopic expression of IGF-I and Shh by skeletal muscle inhibits disuse-mediated skeletal muscle atrophy and bone osteopenia in vivo. *FASEB J* 18:221–223.

94. Kawai M, Bessho K, Maruyama H, Miyazaki J, and Yamamoto T 2005 Human BMP-2 gene transfer using transcutaneous in vivo electroporation induced both intramembranous and endochondral ossification. *Anat Rec A Discov Mol Cell Evol Biol* 287:1264–1271.

95. Jeong JG, Kim JM, Ho SH, Hahn W, Yu SS, and Kim S 2004 Electrotransfer of human IL-1Ra into skeletal muscles reduces the incidence of murine collagen-induced arthritis. *J Gene Med* 6:1125–1133.

96. Jiang J, Jiangl N, Gao W, Zhu J, Guo Y, Shen D, Chen G, and Tang J 2006 Augmentation of revascularization and prevention of plasma leakage by angiopoietin-1 and vascular endothelial growth factor co-transfection in rats with experimental limb ischaemia. *Acta Cardiol* 61:145–153.

97. Kaneda T, Honda A, Hakozaki A, Fuse T, Muto A, and Yoshida T 2007 An improved Graves' disease model established by using in vivo electroporation exhibited long-term immunity to hyperthyroidism in BALB/c mice. *Endocrinology* 148:2335–2344.

98. Kim JM, Lim BK, Ho SH, Yun SH, Shin JO, Park EM, Kim DK, Kim S, and Jeon ES 2006 TNFR-Fc fusion protein expressed by in vivo electroporation improves survival rates and myocardial injury in coxsackievirus induced murine myocarditis. *Biochem Biophys Res Commun* 344:765–771.

99. Kishida T, Asada H, Gojo S, Ohashi S, Shin-Ya M, Yasutomi K, Terauchi R, Takahashi KA, Kubo T, Imanishi J, and Mazda O 2004 Sequence-specific gene silencing in murine muscle induced by electroporation-mediated transfer of short interfering RNA. *J Gene Med* 6:105–110.

100. Murakami T, Arai M, Sunada Y, and Nakamura A 2006 VEGF 164 gene transfer by electroporation improves diabetic sensory neuropathy in mice. *J Gene Med* 8:773–781.

101. Losen M, Stassen MH, Martinez-Martinez P, Machiels BM, Duimel H, Frederik P, Veldman H, Wokke JH, Spaans F, Vincent A, and De Baets MH 2005 Increased expression of rapsyn in muscles prevents acetylcholine receptor loss in experimental autoimmune myasthenia gravis. *Brain* 128:2327–2337.

102. Pierog J, Gazdhar A, Stammberger U, Gugger M, Hyde S, Mathiesen I, Grodzki T, and Schmid RA 2005 Synergistic effect of low dose cyclosporine A and human interleukin 10 overexpression on acute rejection in rat lung allotransplantation. *Eur J Cardiothorac Surg* 27:1030–1035.

103. Stammberger U, Bilici M, Gugger M, Gazdhar A, Hamacher J, Hyde SC, and Schmid RA 2006 Prolonged amelioration of acute lung allograft rejection by overexpression of human interleukin-10 under control of a long acting ubiquitin C promoter in rats. *J Heart Lung Transplant* 25:1474–1479.

104. Tavakoli R, Gazdhar A, Pierog J, Bogdanova A, Gugger M, Pringle IA, Gill DR, Hyde SC, Genoni M, and Schmid RA 2006 Electroporation-mediated interleukin-10 overexpression in skeletal muscle reduces acute rejection in rat cardiac allografts. *J Gene Med* 8:242–248.

105. Palaniyandi SS, Watanabe K, Ma M, Tachikawa H, Kodama M, and Aizawa Y 2004 Inhibition of mast cells by interleukin-10 gene transfer contributes to protection against acute myocarditis in rats. *Eur J Immunol* 34:3508–3515.

106. Roorda BD, Hesselink MK, Schaart G, Moonen-Kornips E, Martinez-Martinez P, Losen M, De Baets MH, Mensink RP, and Schrauwen P 2005 DGAT1 overexpression in muscle by in vivo DNA electroporation increases intramyocellular lipid content. *J Lipid Res* 46:230–236.

107. Schakman O, Gilson H, de C, V, Lause P, Verniers J, Havaux X, Ketelslegers JM, and Thissen JP 2005 Insulin-like growth factor-I gene transfer by electroporation prevents skeletal muscle atrophy in glucocorticoid-treated rats. *Endocrinology* 146:1789–1797.

108. Shimao K, Takayama T, Enomoto K, Saito T, Nagai S, Miyazaki J, Ogawa K, and Tahara H 2005 Cancer gene therapy using in vivo electroporation of Flt3-ligand. *Int J Oncol* 27:457–463.

109. Tan YX, Han WL, Chen YY, Ouyang NT, Tang Y, Li F, Ding PG, Ren XL, Zeng GQ, Ding J, Zhu T, Ma DL, and Zhong NS 2004 Chemokine-like factor 1, a novel cytokine, contributes

to airway damage, remodeling and pulmonary fibrosis. *Chin Med J (Engl)* 117:1123–1129.

110. Xiang L, Murai A, and Muramatsu T 2005 Mutated-leptin gene transfer induces increases in body weight by electroporation and hydrodynamics-based gene delivery in mice. *Int J Mol Med* 16:1015–1020.

111. Wang XD, Tang JG, Xie XL, Yang JC, Li S, Ji JG, and Gu J 2005 A comprehensive study of optimal conditions for naked plasmid DNA transfer into skeletal muscle by electroporation. *J Gene Med* 7:1235–1245.

112. Yamada M, Kuwano K, Maeyama T, Yoshimi M, Hamada N, Fukumoto J, Egashira K, Hiasa K, Takayama K, and Nakanishi Y 2006 Gene transfer of soluble transforming growth factor type II receptor by in vivo electroporation attenuates lung injury and fibrosis. *J Clin Pathol* 60:916–920.

113. Wang W, Sun W, and Wang X 2004 Intramuscular gene transfer of CGRP inhibits neointimal hyperplasia after balloon injury in the rat abdominal aorta. *Am J Physiol Heart Circ Physiol* 287:H1582–H1589.

114. Komamura K, Tatsumi R, Miyazaki J, Matsumoto K, Yamato E, Nakamura T, Shimizu Y, Nakatani T, Kitamura S, Tomoike H, Kitakaze M, Kangawa K, and Miyatake K 2004 Treatment of dilated cardiomyopathy with electroporation of hepatocyte growth factor gene into skeletal muscle. *Hypertension* 44:365–371.

115. Wong SH, Lowes KN, Quigley AF, Marotta R, Kita M, Byrne E, Kornberg AJ, Cook MJ, and Kapsa RM 2005 DNA electroporation in vivo targets mature fibers in dystrophic mdx muscle. *Neuromuscul Disord* 15:630–641.

116. Chi CH, Liu IL, Lo WY, Liaw BS, Wang YS, and Chi KH 2005 Hepatocyte growth factor gene therapy prevents radiation-induced liver damage. *World J Gastroenterol* 11:1496–1502.

117. Edd JF, Horowitz L, Davalos RV, Mir LM, and Rubinsky B 2006 In vivo results of a new focal tissue ablation technique: Irreversible electroporation. *IEEE Trans Biomed Eng* 53:1409–1415.

118. Ivorra A, and Rubinsky B 2007 In vivo electrical impedance measurements during and after electroporation of rat liver. *Bioelectrochemistry* 70:287–295.

119. Sel D, Cukjati D, Batiuskaite D, Slivnik T, Mir LM, and Miklavcic D 2005 Sequential finite element model of tissue electropermeabilization. *IEEE Trans Biomed Eng* 52:816–827.

120. Jaichandran S, Yap ST, Khoo AB, Ho LP, Tien SL, and Kon OL 2006 In vivo liver electroporation: Optimization and demonstration of therapeutic efficacy. *Hum Gene Ther* 17:362–375.

121. Kawano T, Yamagata M, Takahashi H, Niidome Y, Yamada S, Katayama Y, and Niidome T 2006 Stabilizing of plasmid DNA in vivo by PEG-modified cationic gold nanoparticles and the gene expression assisted with electrical pulses. *J Control Release* 111:382–389.

122. Sakai M, Nishikawa M, Thanaketpaisarn O, Yamashita F, and Hashida M 2005 Hepatocyte-targeted gene transfer by combination of vascularly delivered plasmid DNA and in vivo electroporation. *Gene Ther* 12:607–616.

123. Ortega VV, Martinez AF, Gascon JY, Sanchez NA, Banos MA, and Rubiales FC 2006 Transdermal transport of India ink by electromagnetic electroporation in Guinea pigs: An ultrastructural study. *Ultrastruct Pathol* 30:65–74.

124. Tokumoto S, Mori K, Higo N, and Sugibayashi K 2005 Effect of electroporation on the electroosmosis across hairless mouse skin in vitro. *J Control Release* 105:296–304.

125. Tokumoto S, Higo N, and Sugibayashi K 2006 Effect of electroporation and pH on the iontophoretic transdermal delivery of human insulin. *Int J Pharm* 326:13–19.

126. Murthy SN, Zhao YL, Hui SW, and Sen A 2005 Electroporation and transcutaneous extraction (ETE) for pharmacokinetic studies of drugs. *J Control Release* 105:132–141.

127. Zhao YL, Murthy SN, Manjili MH, Guan LJ, Sen A, and Hui SW 2006 Induction of cytotoxic T-lymphocytes by electroporation-enhanced needle-free skin immunization. *Vaccine* 24:1282–1290.

128. Wong TW, Chen CH, Huang CC, Lin CD, and Hui SW 2006 Painless electroporation with a new needle-free microelectrode array to enhance transdermal drug delivery. *J Control Release* 110:557–565.

129. Medi BM and Singh J 2006 Skin targeted DNA vaccine delivery using electroporation in rabbits II. Safety. *Int J Pharm* 308:61–68.

130. Hooper JW, Golden JW, Ferro AM, and King AD 2007 Smallpox DNA vaccine delivered by novel skin electroporation device protects mice against intranasal poxvirus challenge. *Vaccine* 25:1814–1823.

131. Pavselj N and Preat V 2005 DNA electrotransfer into the skin using a combination of one high- and one low-voltage pulse. *J Control Release* 106:407–415.

132. Maruyama H, Miyazaki J, and Gejyo F 2005 Epidermis-targeted gene transfer using in vivo electroporation. *Methods Mol Biol* 289:431–436.

133. Lin MP, Marti GP, Dieb R, Wang J, Ferguson M, Qaiser R, Bonde P, Duncan MD, and Harmon JW 2006 Delivery of plasmid DNA expression vector for keratinocyte growth factor-1 using electroporation to improve cutaneous wound healing in a septic rat model. *Wound Repair Regen* 14:618–624.

134. Fujihara Y, Koyama H, Nishiyama N, Eguchi T, and Takato T 2005 Gene transfer of bFGF to recipient bed improves survival of ischemic skin flap. *Br J Plast Surg* 58:511–517.

135. Lee PY, Chesnoy S, and Huang L 2004 Electroporatic delivery of TGF-beta1 gene works synergistically with electric therapy to enhance diabetic wound healing in db/db mice. *J Invest Dermatol* 123:791–798.

136. Hirao LA, Wu L, Khan AS, Satishchandran A, Draghia-Akli R, and Weiner DB 2008 Intradermal/subcutaneous immunization by electroporation improves plasmid vaccine delivery and potency in pigs and rhesus macaques. *Vaccine* 26:440–448.

137. Byrne CM, Thompson JF, Johnston H, Hersey P, Quinn MJ, Michael HT, and McCarthy WH 2005 Treatment of metastatic melanoma using electroporation therapy with bleomycin (electrochemotherapy). *Melanoma Res* 15:45–51.

138. Tijink BM, de BR, Van Dongen GA, and Leemans CR 2006 How we do it: Chemo-electroporation in the head and neck for otherwise untreatable patients. *Clin Otolaryngol* 31:447–451.

139. Spugnini EP, Dragonetti E, Vincenzi B, Onori N, Citro G, and Baldi A 2006 Pulse-mediated chemotherapy enhances local control and survival in a spontaneous canine model of primary mucosal melanoma. *Melanoma Res* 16:23–27.

140. Kranjc S, Cemazar M, Grosel A, Sentjurc M, and Sersa G 2005 Radiosensitising effect of electrochemotherapy with bleomycin in LPB sarcoma cells and tumors in mice. *BMC Cancer* 5:115.

141. Gunji Y, Uesato M, Miyazaki S, Shimada H, Matsubara H, Nabeya Y, Kouda K, Makino H, Kouzu T, and Ochiai T 2005 Generation of antitumor immunity against large colon tumors by repeated runs of electrochemotherapy. *Hepatogastroenterology* 52:770–774.

142. Tamosiunas M, Bagdonas S, Didziapetriene J, and Rotomskis R 2005 Electroporation of transplantable tumour for the enhanced accumulation of photosensitizers. *J Photochem Photobiol B* 81:67–75.

143. Shil P, Kumar A, Vidyasagar PB, and Mishra KP 2006 Electroporation enhances radiation and doxorubicin-induced toxicity in solid tumor in vivo. *J Environ Pathol Toxicol Oncol* 25:625–632.

144. Matsubara H, Mizutani Y, Hongo F, Nakanishi H, Kimura Y, Ushijima S, Kawauchi A, Tamura T, Sakata T, and Miki T 2006 Gene therapy with TRAIL against renal cell carcinoma. *Mol Cancer Ther* 5:2165–2171.

145. Fulimoto T, Maeda H, Kubo K, Sugita Y, Nakashima T, Sato E, Tanaka Y, Madachi M, Aiba M, and Kameyama Y 2005 Enhanced anti-tumour effect of cisplatin with low-voltage electrochemotherapy in hamster oral fibrosarcoma. *J Int Med Res* 33:507–512.

146. Grosel A, Sersa G, Kranjc S, and Cemazar M 2006 Electrogene therapy with p53 of murine sarcomas alone or combined with electrochemotherapy using cisplatin. *DNA Cell Biol* 25:674–683.

147. Cemazar M, Pipan Z, Grabner S, Bukovec N, and Sersa G 2006 Cytotoxicity of different platinum (II) analogues to human tumour cell lines in vitro and murine tumour in vivo alone or combined with electroporation. *Anticancer Res* 26:1997–2002.

148. Mossop BJ, Barr RC, Henshaw JW, Zaharoff DA, and Yuan F 2006 Electric fields in tumors exposed to external voltage sources: Implication for electric field-mediated drug and gene delivery. *Ann Biomed Eng* 34:1564–1572.

149. Cemazar M, Wilson I, Dachs GU, Tozer GM, and Sersa G 2004 Direct visualization of electroporation-assisted in vivo gene delivery to tumors using intravital microscopy—spatial and time dependent distribution. *BMC Cancer* 4:81.

150. Soden D, Larkin J, Collins C, Piggott J, Morrissey A, Norman A, Dunne C, and O'Sullivan GC 2004 The development of novel flexible electrode arrays for the electrochemotherapy of solid tumour tissue. (Potential for endoscopic treatment of inaccessible cancers). *Conf Proc IEEE Eng Med Biol Soc* 5:3547–3550.

151. Nakai N, Kishida T, Shin-Ya M, Imanishi J, Ueda Y, Kishimoto S, and Mazda O 2007 Therapeutic RNA interference of malignant melanoma by electrotransfer of small interfering RNA targeting Mitf. *Gene Ther* 14:357–365.

152. McCray AN, Ugen KE, Muthumani K, Kim JJ, Weiner DB, and Heller R 2006 Complete regression of established subcutaneous B16 murine melanoma tumors after delivery of an HIV-1 Vpr-expressing plasmid by in vivo electroporation. *Mol Ther* 14:647–655.

153. Li S, Zhang L, Torrero M, Cannon M, and Barret R 2005 Administration route- and immune cell activation-dependent tumor eradication by IL12 electrotransfer. *Mol Ther* 12:942–949.

154. Heller L, Merkler K, Westover J, Cruz Y, Coppola D, Benson K, Daud A, and Heller R 2006 Evaluation of toxicity following electrically mediated interleukin-12 gene delivery in a B16 mouse melanoma model. *Clin Cancer Res* 12:3177–3183.

155. Liu J, Xia X, Torrero M, Barrett R, Shillitoe EJ, and Li S 2006 The mechanism of exogenous B7.1-enhanced IL-12-mediated complete regression of tumors by a single electroporation delivery. *Int J Cancer* 119:2113–2118.

156. Goto T, Nishi T, Kobayashi O, Tamura T, Dev SB, Takeshima H, Kochi M, Kuratsu J, Sakata T, and Ushio Y 2004 Combination electro-gene therapy using herpes virus thymidine kinase and interleukin-12 expression plasmids is highly efficient against murine carcinomas in vivo. *Mol Ther* 10:929–937.

157. Yamashita Y, Shimada M, Minagawa R, Tsujita E, Harimoto N, Tanaka S, Shirabe K, Miyazaki J, and Maehara Y 2004 Muscle-targeted interleukin-12 gene therapy of orthotopic hepatocellular carcinoma in mice using in vivo electrosonoporation. *Mol Cancer Ther* 3:1177–1182.

158. Chen WY, Cheng YT, Lei HY, Chang CP, Wang CW, and Chang MS 2005 IL-24 inhibits the growth of hepatoma cells in vivo. *Genes Immun* 6:493–499.

159. Ugen KE, Kutzler MA, Marrero B, Westover J, Coppola D, Weiner DB, and Heller R 2006 Regression of subcutaneous B16 melanoma tumors after intratumoral delivery of an IL-15-expressing plasmid followed by in vivo electroporation. *Cancer Gene Ther* 13:969–974.

160. Collins CG, Tangney M, Larkin JO, Casey G, Whelan MC, Cashman J, Murphy J, Soden D, Vejda S, McKenna S, Kiely B, Collins JK, Barrett J, Aarons S, and O'Sullivan GC 2006 Local gene therapy of solid tumors with GM-CSF and B7-1 eradicates both treated and distal tumors. *Cancer Gene Ther* 13:1061–1071.

161. Peycheva E, Daskalov I, and Tsoneva I 2007 Electrochemotherapy of Mycosis fungoides by interferon-alpha. *Bioelectrochemistry* 70:283–286.

162. Shibata MA, Morimoto J, Doi H, Morishima S, Naka M, and Otsuki Y 2007 Electrogene therapy using endostatin, with or without suicide gene therapy, suppresses murine mammary tumor growth and metastasis. *Cancer Gene Ther* 14:268–278.

163. Shibata MA, Miwa Y, Miyashita M, Morimoto J, Abe H, and Otsuki Y 2005 Electrogene transfer of an Epstein–Barr virus-based plasmid replicon vector containing the diphtheria toxin A gene suppresses mammary carcinoma growth in SCID mice. *Cancer Sci* 96:434–440.

164. Harada K, Supriatno, Kawaguchi S, Onoue T, Kawashima Y, Yoshida H, and Sato M 2005 High antitumor activity using intratumoral injection of plasmid DNA with mutant-type p27Kip1 gene following in vivo electroporation. *Oncol Rep* 13:201–206.

165. Zhou R and Dean DA 2007 Gene transfer of interleukin 10 to the murine cornea using electroporation. *Exp Biol Med* (Maywood) 232:362–369.

166. Matsuda T and Cepko CL 2007 Controlled expression of transgenes introduced by in vivo electroporation. *Proc Natl Acad Sci U S A* 104:1027–1032.

167. Esumi N, Kachi S, Campochiaro PA, and Zack DJ 2007 VMD2 promoter requires two proximal E-box sites for its activity in vivo and is regulated by the MITF-TFE family. *J Biol Chem* 282:1838–1850.

168. Chalberg TW, Vankov A, Molnar FE, Butterwick AF, Huie P, Calos MP, and Palanker DV 2006 Gene transfer to rabbit retina with electron avalanche transfection. *Invest Ophthalmol Vis Sci* 47:4083–4090.

169. Kachi S, Esumi N, Zack DJ, and Campochiaro PA 2006 Sustained expression after nonviral ocular gene transfer using mammalian promoters. *Gene Ther* 13:798–804.

170. Bloquel C, Bejjani R, Bigey P, Bedioui F, Doat M, BenEzra D, Scherman D, and Behar-Cohen F 2006 Plasmid electrotransfer of eye ciliary muscle: Principles and therapeutic efficacy using hTNF-alpha soluble receptor in uveitis. *FASEB J* 20:389–391.

171. Huberman AD, Murray KD, Warland DK, Feldheim DA, and Chapman B 2005 Ephrin-As mediate targeting of eye-specific projections to the lateral geniculate nucleus. *Nat Neurosci* 8:1013–1021.

172. Chalberg TW, Genise HL, Vollrath D, and Calos MP 2005 phiC31 integrase confers genomic integration and long-term transgene expression in rat retina. *Invest Ophthalmol Vis Sci* 46:2140–2146.

173. Young TL, Matsuda T, and Cepko CL 2005 The noncoding RNA taurine upregulated gene 1 is required for differentiation of the murine retina. *Curr Biol* 15:501–512.

174. Kachi S, Oshima Y, Esumi N, Kachi M, Rogers B, Zack DJ, Campochiaro PA 2005 Nonviral ocular gene transfer. *Gene Ther* 12:843–851.

175. Ishikawa H, Takano M, Matsumoto N, Sawada H, Ide C, Mimura O, and Dezawa M 2005 Effect of GDNF gene transfer into axotomized retinal ganglion cells using in vivo electroporation with a contact lens-type electrode. *Gene Ther* 12:289–298.

176. Chen YX, Krull CE, and Reneker LW 2004 Targeted gene expression in the chicken eye by in ovo electroporation. *Mol Vis* 10:874–883.

177. Gazdhar A, Fachinger P, van I.C, Pierog J, Gugger M, Friis R, Schmid RA, and Geiser T 2007 Gene transfer of hepatocyte growth factor by electroporation reduces bleomycin-induced lung fibrosis. *Am J Physiol Lung Cell Mol Physiol* 292:L529–L536.

178. Gazdhar A, Bilici M, Pierog J, Ayuni EL, Gugger M, Wetterwald A, Cecchini M, and Schmid RA 2006 In vivo electroporation and ubiquitin promoter—a protocol for sustained gene expression in the lung. *J Gene Med* 8:910–918.

179. Machado-Aranda D, Adir Y, Young JL, Briva A, Budinger GR, Yeldandi AV, Sznajder JI, and Dean DA 2005 Gene transfer of the Na+, K+-ATPase beta1 subunit using electroporation increases lung liquid clearance. *Am J Respir Crit Care Med* 171:204–211.

180. Mutlu GM, Machado-Aranda D, Norton JE, Bellmeyer A, Urich D, Zhou R, and Dean DA 2007 Electroporation-mediated gene transfer of the Na+, K+-ATPase rescues endotoxin-induced lung injury. *Am J Respir Crit Care Med* 176:582–590.

181. Pringle IA, McLachlan G, Collie DD, Sumner-Jones SG, Lawton AE, Tennant P, Baker A, Gordon C, Blundell R, Varathalingam A, Davies LA, Schmid RA, Cheng SH, Porteous DJ, Gill DR, and Hyde SC 2007 Electroporation enhances reporter gene expression following delivery of naked plasmid DNA to the lung. *J Gene Med* 9:369–380.

182. Zhou R, Norton JE, Zhang N, and Dean DA 2007 Electroporation-mediated transfer of plasmids to the lung results in reduced TLR9 signaling and inflammation. *Gene Ther* 14:775–780.

183. Thanaketpaisarn O, Nishikawa M, Yamashita F, and Hashida M 2005 Tissue-specific characteristics of in vivo electric gene: transfer by tissue and intravenous injection of plasmid DNA. *Pharm Res* 22:883–891.

184. Lavee J, Onik G, Mikus P, and Rubinsky B 2007 A novel non-thermal energy source for surgical epicardial atrial ablation: Irreversible electroporation. *Heart Surg Forum* 10:E162-E167.

185. Miyahara T, Koyama H, Miyata T, Shigematsu H, Inoue J, Takato T, and Nagawa H 2006 Inflammatory responses involving tumor necrosis factor receptor-associated factor 6 contribute to in-stent lesion formation in a stent implantation model of rabbit carotid artery. *J Vasc Surg* 43:592–600.

186. Hibbitt O, Coward K, Kubota H, Prathalingham N, Holt W, Kohri K, and Parrington J 2006 In vivo gene transfer by electroporation allows expression of a fluorescent transgene in Hamster testis and epididymal sperm and has no adverse effects upon testicular integrity or sperm quality. *Biol Reprod* 74:95–101.

187. Coward K, Kubota H, Hibbitt O, McIlhinney J, Kohri K, and Parrington J 2006 Expression of a fluorescent recombinant form of sperm protein phospholipase C zeta in mouse epididymal sperm by in vivo gene transfer into the testis. *Fertil Steril* 85 Suppl 1:1281–1289.

188. Kirby JL, Yang L, Labus JC, Lye RJ, Hsia N, Day R, Cornwall GA, and Hinton BT 2004 Characterization of epididymal epithelial cell-specific gene promoters by in vivo electroporation. *Biol Reprod* 71:613–619.

189. Umemoto Y, Sasaki S, Kojima Y, Kubota H, Kaneko T, Hayashi Y, and Kohri K 2005 Gene transfer to mouse testes by electroporation and its influence on spermatogenesis. *J Androl* 26:264–271.

190. Kubota H, Hayashi Y, Kubota Y, Coward K, and Parrington J 2005 Comparison of two methods of in vivo gene transfer by electroporation. *Fertil Steril* 83 Suppl 1:1310–1318.

191. Shoji M, Chuma S, Yoshida K, Morita T, and Nakatsuji N 2005 RNA interference during spermatogenesis in mice. *Dev Biol* 282:524–534.

192. Dobashi M, Goda K, Maruyama H, and Fujisawa M 2005 Erythropoietin gene transfer into rat testes by + electroporation may reduce the risk of germ cell loss caused by cryptorchidism. *Asian J Androl* 7:369–373.

17 Magnetofection: Using Magnetic Particles and Magnetic Force to Enhance and to Target Nucleic Acid Delivery

Franz Scherer and Christian Plank

CONTENTS

17.1 INTRODUCTION

We define magnetofection as nucleic acid delivery under the influence of a magnetic field acting on nucleic acid vectors, which are associated with magnetic nanoparticles. Such methods have been developed independently by several research groups [1–14]. Since we have first introduced the term magnetofection [15], it has become widely used in the scientific literature.

Nowadays, the introduction of nucleic acids into cells is a well-established and widely used tool in research and has become a highly promising strategy for therapeutic applications. By introducing nucleic acids into cells, a wide variety of cellular processes can be influenced or exploited to some particular purpose by having nucleic acids interact with cellular components or having genetic information encoded in nucleic acids expressed in cells. Almost 40 years of research has yielded highly efficient shuttles for nucleic acid delivery, which are genetically modified viruses and their synthetic counterparts and which are commonly known as viral and nonviral vectors. Yet, further improvements with respect to the efficiency and the specificity of nucleic acid delivery are required and can be achieved. This is mostly evident *in vivo* where the target cells not only are often refractory to transfection or transduction but also the access to the target cells is limited by multiple barriers in addition to those that are prevalent on the cellular level. The latter are relatively well defined. Limiting steps on the cellular level include, for example, overcoming the plasma membrane, escape from internal vesicles, and suitable intracellular localization (e.g., if required, intracellular transport toward the nucleus, nuclear uptake, and suitable localization within the nucleus). Nature itself has optimized viruses in overcoming such limitations. Consequently, nonviral vectors were often designed to mimic essential viral functions that allow them to infect cells and were equipped with modules that mediate binding to cells, uptake via natural transport processes, escape from endosomes, and nuclear uptake. No matter the level of perfection to which the nature has evolved viruses or how sophisticated nonviral vector engineers have been in copying their design, vectors, in the first place, need to find their target cells. Independent of vector type, vector contact with target cells is the primary event in, and a prerequisite for, a successful transfection/transduction process. The frequency and total number of productive cell binding events are dependent on the vector concentration in the culture medium and are limited by diffusion under standard cell culture conditions. This is due to the fact that viral vectors, in general, and nonviral vectors, usually, are nanoparticles and as such can reach target cells in culture only by diffusion unless further measures are taken.* As diffusion is a slow process and many vector types are subject to time-dependent inactivation under cell culture conditions (as well as being toxic in very high concentrations), measures that accelerate vectors toward target cells and, thus, increase the vector concentration at the cell surface can improve transfection/transduction efficiency [16]. In this context, the most convenient measure in cell culture has been exploiting gravitation. For example, vectors have been associated with dense silica particles to overcome the diffusion limitation [17]. Precipitate formation with nucleic acids is, in fact, the oldest nonviral method to achieve nucleic acid delivery in cell culture [18] and also nowadays, gravitation is exploited unwittingly by most researchers performing in vitro transfections with commercially available reagents. Most cationic lipids and polycations form precipitates with nucleic acids in salt-containing solution. Also, centrifugal force [19–22] and convective flow of transfection medium toward target cells [23] enhance nucleic acid delivery by accelerating vectors toward the cells to be transfected. All these methods are physical means of vector targeting. For obvious reasons, some of these methods cannot be applied in vivo. Magnetofection is another physical means of vector targeting, which at least has the potential for efficient in vivo application.

17.2 MAGNETIC DRUG TARGETING AND THE PRINCIPLE OF MAGNETOFECTION

Certain medical indications of localized disease predictably will profit from methods of localized drug delivery. Typical examples include solid tumors before systemic spread where surgical excision is not an option or where conventional means of therapy have been exhausted. Standard therapy with cytotoxic drugs involves "flooding" the patient with medication to achieve a sufficiently high drug concentration at the disease site. Not surprisingly, such treatments are frequently associated with severe side effects that compromise the patient's quality of life, if not worse.

Already in the mid-1960s, researchers attempted the first steps of exploiting magnetic fields to localize carbonyl iron in intracranial aneurisms and in this manner produce magnetically localized thrombi [24–27]. In the 1970s, Widder and colleagues introduced the concept of magnetic drug targeting [28], designed for accumulating drugs mostly in tumors upon administration into the circulation. Active agents are associated with magnetically responsive materials in the nano- to micrometer size range and are accumulated at the site of disease by a suitable magnetic field (Figure 17.1). The magnetic carrier materials are mostly iron oxides of various compositions, which can be of natural or synthetic origin [16,29–31]. In a first step, Widder and coworkers prepared albumin microspheres with entrapped iron oxide and adriamycin, infused them into the caudal artery of rat tails, and demonstrated magnetic retention in a tail segment exposed to a permanent magnet. In another study, they applied magnetic albumin microspheres with entrapped doxorubicin in a Yoshida sarcoma rat model. A 100-fold higher dose of free doxorubicin was required to achieve the same drug level in the tumor as with the magnetically targeted drug [32]. The treatment was therapeutically effective in that it resulted in total tumor remission in a high percentage of experimental animals. In contrast, animals treated with free doxorubicin, placebo microspheres, or nonlocalized doxorubicin microspheres exhibited a significant increase in tumor size with metastases and subsequent death in 90%–100% of the animals [33,34]. Other researchers obtained similar results [35–41]. After extensive preclinical examinations, Lübbe et al. were the first to apply magnetic drug targeting in a phase I clinical trial with 14 cancer patients where other therapeutic options had been exhausted [42–45]. The treatment was well tolerated and retardation of tumor growth and even local remissions were observed [45]. Nevertheless, one conclusion of the study was that improvements of magnetic drug targeting were required to make it more effective, at least in cancer therapy. Alexiou et al. tried to improve the efficacy in a tumor-bearing rabbit model by using the strongest magnetic field (generated through an electromagnet) ever applied in magnetic drug targeting [46,47]. The results were quite favorable. On the other hand, an independent clinical phase II/III trial for the treatment of hepatocellular carcinoma with a magnetic carrier composed of elementary iron [48–50] had to be discontinued as the clinical endpoints could not be met with statistical significance. In summary, at least in animal models it has been clearly demonstrated that (1) magnetic drug targeting is feasible even if the drug administration site is remote from the target site under magnetic field influence [43,51], (2) that magnetic particles can extravasate under the

---

* For lipoplexes and polyplexes, this holds true with the limitation that they often tend to aggregate, depending on their surface characteristics and on the ionic strength of the medium. For their aggregated forms, gravitation is of course the driving force that lets them sediment on cells.

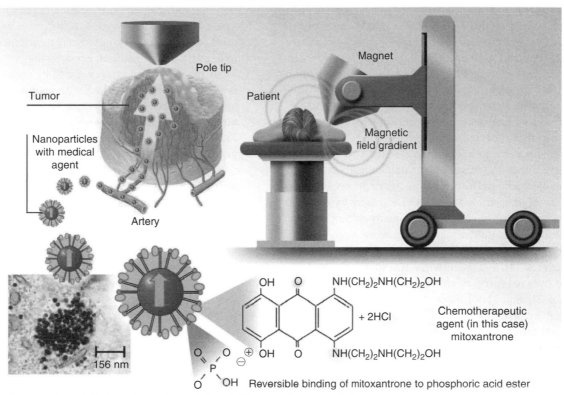

Magnetic fields tackle tumors

Pole tip

Tumor

Nanoparticles with medical agent

Artery

Magnet

Patient

Magnetic field gradient

156 nm

OH O   NH(CH₂)₂NH(CH₂)₂OH

+ 2HCl

OH O   NH(CH₂)₂NH(CH₂)₂OH

Chemotherapeutic agent (in this case) mitoxantrone

Reversible binding of mitoxantrone to phosphoric acid ester

Image at lower left shows tumor tissue imbued with magnetic nanoparticles

FIGURE 17.1 (See color insert following blank page 206.)   Magnetic drug targeting. An anticancer drug (e.g., mitoxantrone) bound to magnetic particles is injected into a blood vessel (here tumor-feeding blood vessel) of the patient and is concentrated in the target tissue (e.g., tumor) by an external magnetic field. (From Siemens AG, Pictures of the Future 01/2007, page 64, www.siemens.com/pof. With permission.)

influence of the magnetic field [37,52,53], and (3) that the magnetic carriers are well tolerated.

Similar motivations that inspire the development of magnetic drug targeting apply to developing nucleic acid or gene therapies. The objective is to achieve therapeutic benefit for patients with ever improved specificity and less side effects, particularly in cases where conventional treatments fail. However, achieving and maintaining therapeutic drug levels at target sites is an even bigger challenge with nucleic acids or vectors than it is with conventional drugs. Therefore, we and others have considered magnetic targeting of nucleic acids or gene vectors to be a promising concept.

The principle of magnetofection, as we and others have reported, is quite simple; although several technical hurdles had to be overcome to make it work efficiently. Naked nucleic acids or gene vectors are bound to magnetic nanoparticles, and in analogy to magnetic drug targeting, a magnetic field is used to accumulate nucleic acids or vectors at target cells (Figure 17.2). *In vitro*, this is straightforward, *in vivo*, several challenges remain as are discussed later.

The first to pursue this idea were Kuehnle and Kuehnle in a patent application filed in 1994 (U.S. patent 5516670 issued in 1996) where they describe torpedo-like magnetic particles

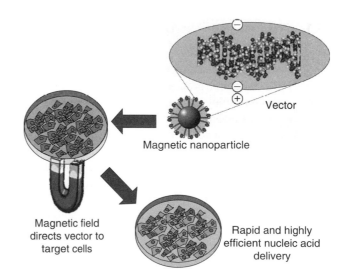

Vector

Magnetic nanoparticle

Magnetic field directs vector to target cells

Rapid and highly efficient nucleic acid delivery

FIGURE 17.2   Magnetofection in cell culture. Nonviral or viral vectors associated with magnetic nanoparticles (magnetofectins) are accelerated onto a cell layer in a culture dish through magnetic force. The result is rapid and efficient transfection, even at lower vector doses. (Modified from Plank, C., Schillinger, U., Scherer, F., Bergemann, C., and Remy, J.S., *Biol. Chem.*, 384, 737, 2003. With permission.)

that may be suitable to deliver DNA as well as an apparatus generating a magnetic field for delivery. Chan describes the use of pulsating magnetic fields for introducing nucleic acids that are associated with magnetic nanoparticles into cells in a patent application filed in 1996 (U.S. patent 5753477 issued in 1998). These patents contain interesting ideas and instructive experimental setups. But unfortunately, significant experimental results are missing. Only in the year 2000, Mah et al. presented the first scientific abstract on associating AAV vectors reversibly with magnetic microspheres of varying diameters, which led to increased vector transduction efficiencies [54]. We have first reported on magnetofection and publicly used this term at a scientific conference in the year 2000 [15]. The work of Hughes and coworkers on magnetically enhanced retroviral nucleic acid delivery was the first to be published in a peer-reviewed journal [1] followed briefly later by our own work [2,55] and Mah's paper on magnetically enhanced AAV vector-mediated gene delivery [3]. In the same year, Pandori and coworkers reported about adenovirus–magnetic microbead conjugates [4] and adenoviral transduction enhanced and guided by a magnetic field. Since then, numerous papers have been published about magnetofection method development or that involve the use of magnetofection methods as research tools (see also Tables 17.1 and 17.2). Also, magnetofection reagents have become commercially available in the meantime. Promising more recent developments include the use of a special type of alternating magnetic fields to enhance gene delivery [56] and a method called nanotube spearing where DNA is immobilized on nickel-embedded nanotubes, which are speared by magnetic force into cells *in vitro* [57].

17.3 WHAT IS NEEDED TO PRACTICE MAGNETOFECTION?

17.3.1 Nucleic Acids and/or Gene Vectors

It has turned out that cellular uptake during magnetofection involves the same mechanisms that also govern conventional transfection/transduction processes [22]. Therefore, the chemical, physical, and biological characteristics of nucleic acids and vectors are as essential in magnetofection as they are in conventional nucleic acid delivery. For example, we have found that endosomolytic agents like the influenza peptide INF7 promote transfection efficiency in magnetofection as they do in conventional transfection [2]. Important characteristics of vectors and their components when to be assembled with magnetic nanoparticles are predominantly charge, size, and the possibility of chemical coupling reactions. Accordingly, strategies of assembling vectors with magnetic particles can be developed and magnetic particle synthesis can be adjusted to the needs of the assembly process.

17.3.2 Magnetic Particles

To be useful in magnetofection, magnetic particles need to fulfill some essential features (text with modifications from Ref. [58]):

1. They need to comprise functionality which allows them to be associated with nucleic acids, gene vectors, and, if necessary, third components (e.g., a polycationic or lipidic transfection reagent).
2. The association with nucleic acids, gene vectors, or third components must not impair their functionality in delivery and intracellular processing.
3. Their magnetic properties need to be such that they or at least their formulations with nucleic acids, vectors, or further components can be attracted toward target cells using "reasonable" magnetic force. Reasonable in this context means that the device generating the magnetic force must be affordable enough and must have dimensions compatible with handling in a research laboratory. Reasonable magnetic force also means that the magnetic field itself does not unduly impair cellular functions.
4. They must be physically and chemically stable enough to be stored over several months in a pharmaceutically acceptable suspension medium.
5. Their magnetic core and their surface coatings have to be biocompatible, so that they can be applied in living cells and organisms.

Magnetic particles that meet these requirements and that have been used in magnetofection consist of a magnetic iron oxide core (e.g., magnetite [Fe_3O_4] or maghemite [Fe_2O_3]) and are coated with different, usually charged polymers (Figure 17.3). Although magnetic nanoparticle synthesis is textbook knowledge [59,60], it is still a science and an art in its own rights [61]. Among the many different synthetic procedures that yield magnetic nanomaterials, essentially two types can be discriminated: one involving high temperature, usually carried out in organic solvent and often starting from organometallic precursor compounds. These procedures yield extraordinarily uniform magnetic nanomaterials whose particle size reportedly can be tuned by adjusting the reaction conditions [62]. The resulting magnetic nanoparticles are usually coated with hydrophobic compounds such as oleic acid and are dispersed in organic solvent. Their use in biological applications requires special procedures to transfer them into aqueous environment while avoiding agglomeration (for example, providing them with a second, amphipathic coating layer). The other type of synthetic procedure is less sophisticated and involves relatively mild aqueous reaction conditions throughout the synthesis. Magnetic particles derived with this latter synthetic procedure have been mostly used in magnetofection applications [63] (a detailed protocol is provided in Ref. [58]). The synthesis starts from aqueous, sterile filtered, and degassed solutions of iron(II) and iron(III) salts. A primary precipitate is formed by the addition of a base such as aqueous ammonia in the presence of a compound intended to serve as a stabilizing layer on the newly formed particle surface. The material is then heated to 90°C and magnetic material forms. After cooling to room temperature, various sonication, dialysis, or washing steps are carried out in order to remove excess stabilizing compound. A large variety of particles can

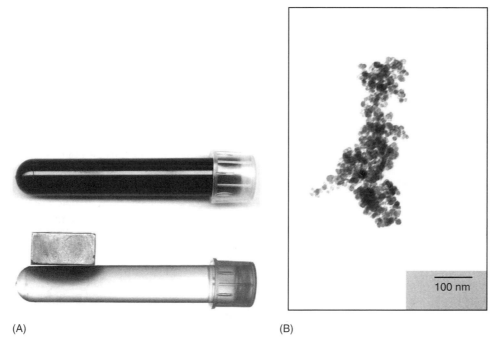

(A)   (B)

FIGURE 17.3 Magnetic particles. (A) Brown iron oxide nanoparticles suspended in water are attracted by an applied permanent magnet. (B) Transmission electron micrographs of magnetic particles coated with PEI. No stain was used and therefore only the electron-dense iron oxide crystals but no PEI molecules are visible. The magnetic particles are organized in multidomain structures of irregular shape. (From Plank, C., Schillinger, U., Scherer, F., Bergemann, C., and Remy, J.S., *Biol. Chem.*, 384, 737, 2003.)

be generated using this basic protocol with different stabilizers such as derivatized dextrans or starch or polycations such as polyethylenimine or polylysine or protamine sulfate, or polyanions such as citrate, polyaspartic acid, phosphorylated starch, polyacrylic acid, arabinic acid, and many other compounds. The hydrodynamic diameters of such particles can range from 20 nm to the micrometer-range [60,64,65]. Recently, Chorny and coworkers have reported on a novel way of preparing polylactide magnetic particles [66]. Their procedure involves an emulsification-solvent evaporation method in the presence of PEI and polylactide applied to oleate-coated magnetic nanoparticles (synthesized similarly as described above). Also, this magnetic carrier for plasmid DNA turned out to be highly useful in magnetically driven gene delivery.

The "responsiveness" of magnetic particles to magnetic fields is dependent, on the one hand, on their material properties and, on the other hand, on the characteristics of the magnetic field. The important material properties are the magnetic susceptibility and the saturation magnetization that determine the magnetic moment. The magnetic force acting on magnetic nanoparticles is dependent on their size, the magnetic field gradient, and the magnetic flux density (below the limit of magnetic saturation of the particles). To our knowledge, so far only iron oxide magnetic nano- and microparticles have been used in magnetofection applications. Compared with elementary iron, iron–cobalt, iron–platinum, or other magnetic nanomaterials, the iron oxides have relatively low saturation magnetization. Nevertheless, their magnetic properties are entirely sufficient at least for magnetofection in cell culture where the target cells can be brought in closest proximity to

the source of a magnetic gradient field. Maximum magnetic responsiveness is of course desirable for intravascular applications *in vivo*. However, in this case, the major material characteristics in favor of iron oxides are their proven biocompatibilty [67]. These materials have been in clinical use as contrast agents in magnetic resonance imaging (MRI) for many years. Iron oxide magnetic nanoparticles are biodegradable and their iron sooner or later is fed into the iron storage pool, which, for example, is of several grams in the human body. In this context, it is interesting to note that chemists were not the first to design iron oxide particles. Magnetite crystals are found in living organisms such as magnetotactic bacteria, in the central nervous systems of fish, and even in the human brain [68–71]. At least for fish, magnetite has been defined as a magnetoreceptor essential for orientation along the earth's magnetic field [70], and there is experimental evidence that similar mechanisms apply in migratory birds [72].

Finally, a few remarks on the term magnetic and on particle size: the literature in magnetic drug targeting and MRI often refers to superparamagnetic iron oxide nanoparticles (SPION). In very simplified terms, superparamagnetic behavior means that particles are magnetized only when they are in an external magnetic field and show no residual magnetization once the external magnetic field is removed [29]. This is due to the so-called Brownian relaxation. With very small particles, magnetization of the particles is immediately destroyed by thermal motion once the external field has been turned off. True superparamagnetism is very difficult to achieve at room temperature and is dependent both on magnetic core size and shell properties (for example, aggregation phenomena can destroy superparamagnetic behavior) [73,74]. Concerning the purpose of magnetofection in cell culture, properties like superparamagnetism are not required. It is even such that there appear to be no special requirements on magnetic particle core size for magnetofection above a certain lower limit and below an unknown upper limit. The lower limit is dictated by basic laws of physics and chemistry and reasonable magnetic force as defined above. We have found that particles with a magnetite crystallite size of 9–11 nm are superior to smaller particles with 3–4 nm crystallite size as components of the magnetic transfection vectors for magnetofection. Concerning the unknown upper size limit, it is remarkable that, for example, in Hughes's work, magnetic microparticles linked by the extremely stable biotin–streptavidin interaction to retroviral vectors were fully compatible with viral infectivity [1]. In summary, for magnetic targeting, particularly *in vivo*, it is certainly desirable to have magnetic particles with maximum magnetic susceptibility, as long as they are biocompatible. However, so far there is no indication that magnetofection in cell culture would benefit from maximizing magnetic force acting on magnetic nucleic acid vectors (passages of the above text are from Ref. [58]).

17.3.3 ASSOCIATING NUCLEIC ACIDS OR VECTORS WITH MAGNETIC PARTICLES

In general, possible ways of association are biological (e.g., receptor–ligand), physical (e.g., electrostatic, hydrophilic, hydrophobic), and chemical (covalent) conjugation.

A very straightforward possibility of linkage is to mix preassembled nucleic acid vectors (e.g., polyplexes, lipoplexes, or viruses) with charged magnetic nanoparticles and to incubate them for several minutes in physiological salt-solution until they associate through physical self-assembly (Figure 17.4A). This strategy is based on the natural tendency of polyelectrolyte nanoparticles to aggregate at elevated ionic strength, also known as salt-induced colloid aggregation [75]. As nucleic acid vectors and the magnetic particles described above are usually charged nanoparticles, they are amenable to this phenomenon. In this way, even similarly charged particles will associate because their repulsive Coulomb forces can be overcome by the attractive van der Waals forces when the particles approach each other closely enough. At elevated ionic strength (e.g., physiological salt concentration), such an approach is possible because the extension of the so-called diffuse layer of counterions surrounding charged particle surfaces is greatly reduced compared with salt-free conditions. Thus, for binding a charged nucleic acid vector, it does not play an important role if the magnetic nanoparticles used are positively or negatively charged. Interestingly, binding via salt-induced aggregation works not only when preassembled vectors are mixed with charged magnetic particles but also when vector components and magnetic particles are mixed step by step. In our experience, the mixing order can play a role, but not a major one. More important are the ratios of the various components (magnetic particles, nucleic acid, or preassembled vector, third components such as polycations or cationic lipids). In binary systems (for example, polycation-coated magnetic nanoparticles plus viral vector particles), the association behavior is usually simple. At a certain ratio of the components, magnetic particles and vectors will be quantitatively associated with one another [2] (Figure 17.4B and C). When either component is in excess, either free magnetic particles or free vector particles will coexist with magnetic particle–vector complexes. In ternary systems, for example, PEI-coated magnetic particles plus nucleic acid plus free PEI, there may be a complex binding behavior. In salt-free aqueous suspension, the nucleic acid is able to bind to the magnetic nanoparticles but will be competed off by free PEI. In contrast, in salt-containing solution, the nucleic acid can be competed off the magnetic nanoparticles by free PEI, but the resulting nucleic acid-PEI nanoparticles can then undergo salt-induced colloid aggregation with the magnetic nanoparticles. At large excess of free PEI, the abundant species can be PEI-nucleic acid particles coexisting with magnetic nanoparticles not associated with the nucleic acid. Therefore, it is essential that for any combination of magnetic nanoparticles with nucleic acids or vectors, the binding behavior is determined. This can be carried out most accurately with radioactive-labeled vectors or vector components as we have described previously [2,16] (Figure 17.4B and C). In general, however, any commercially available transfection reagent can be transformed into a magnetic vector via salt-induced aggregation [16,76]. Apart from actually determining the association behaviors, also serial titrations of vector compositions can be carried out and the optimal vector composition can be determined in transfection experiments. If magnetic field application displays a strong enhancing effect on transfection efficiency, one can bona fide assume that the vector or the nucleic acid was indeed associated with magnetic particles. Magnetic vector compositions can be further refined by carrying out chemistry on any of the components pre- or post-assembly in order to attach effectors of nucleic acid delivery such as targeting moieties, endosomolytic components, or molecules for steric stabilization [77]. Although salt-induced colloid aggregation provides a simple and efficient binding method, it has to be considered, on the one hand, that very large aggregates could lead to embolism when injected into blood vessels. Appropriate aggregate sizes can be chosen by incubation times in salt-containing medium (Figure 17.4D). On the other hand, noncovalently associated magnetic vector preparations may dissociate in the blood stream due to opsonization or mediated by the extracellular matrix in target organs. Therefore, different ways of magnetic vector preparation can be advantageous for *in vivo* applications.

Another approach exploiting electrostatic interactions for vector assembly was chosen by Hirao et al. [5] who prepared magnetic nanoparticle-containing cationic liposomes, which can be associated with DNA. In a further physical strategy, Haim et al. [11] formed complexes between lentiviral vectors and negatively charged magnetite nanoparticles (coated with starch-phosphate) by colloidal clustering which was facilitated by divalent cations. Similarly, for virus concentration,

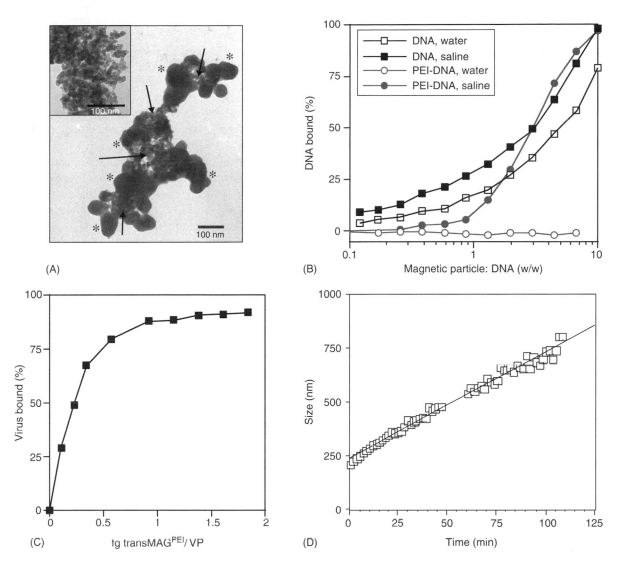

FIGURE 17.4 Binding of nucleic acid (vectors) to iron oxide (magnetite) nanoparticles coated with the polycation PEI. (A) Electron micrograph showing an extended associate of magnetic particles (arrows) and PEI-DNA vectors (asterisks) produced through salt-induced aggregation. For comparison, the inset shows only the iron oxide crystals. (From Plank, C., Anton, M., Rudolph, C., Rosenecker, J., and Krotz, F., *Expert Opin. Biol. Ther.*, 3, 745, 2003. With permission.) (B) Percentage of radio-labeled DNA bound to magnetic particles in dependence of the particle/DNA weight ratio. Only in salt-containing solution (150 mM NaCl) binding of positively charged PEI-DNA vectors is possible. This phenomenon is the result of salt-induced aggregation. (From Scherer, F., et al., *Gene Ther.*, 9, 102, 2002. With permission.) (C) Percentage of radio-labeled adenovirus bound through salt-induced aggregation to magnetic particles in dependence of the particle/virus ratio. (From Plank, C., Schillinger, U., Scherer, F., Bergemann, C., and Remy, J.S., *Biol. Chem.*, 384, 737, 2003. With permission.) (D) Time-dependent growth of salt-induced aggregates (measured by dynamic light scattering) resulting from magnetic particles plus PEI-DNA in 150 mM NaCl. The particles aggregate with approximately linear kinetics starting at 217 ± 2 nm and remaining in the submicrometer range within 2 h. (From Plank, C., Schillinger, U., Scherer, F., Bergemann, C., and Remy, J.S., *Biol. Chem.*, 384, 737, 2003.)

Iwata et al. bound sulfonated magnetic beads to porcine parvovirus and poliovirus via divalent cations like Zn^{2+} or Cu^{2+} [78].

Of course, not only physical binding can be exploited for nucleic acid or vector association with magnetic particles but also biological conjugation. Hughes et al. [1] used three different biological strategies to bind retroviral vectors to magnetic microparticles. They conjugated streptavidinylated magnetic particles to (1) a biotinylated antibody directed against the retroviral vectors, (2) biotinylated lectin, which binds to retroviral vectors, and (3) biotinylated retroviruses. Chan et al. and Nesbeth et al. linked streptavidin magnetic microspheres to biotinylated nonviral producer cell proteins that were incorporated into lentiviral surfaces [10,79]. Further, Pandori et al. coupled biotinylated adenoviral vectors to streptavidin-coated magnetic microparticles [4]; Mah et al. used magnetic avidin-microspheres to bind biotinylated heparan sulfate, which was reversibly bound to adeno-associated viruses [3]; and Raty et al. associated avidin-displaying baculoviruses to biotinylated magnetic particles [9].

An interesting chemical strategy for binding was used by Cao et al. [80]. They conjugated plasmid DNA to dextran iron oxide nanoparticles by an oxidation–reduction reaction [80].

One can assume that biotin–streptavidin bridges between vectors and magnetic particles will be more stable *in vivo* than noncovalent linkages. On the other hand, the noncovalent ways of association may facilitate intracellular processing due to the reversibility of the linkage. These questions require further experimentation. In summary, any of the described methods yielded very favorable results in magnetofection applications *in vitro*.

17.3.4 Magnets

To an average citizen and even to most scientists, if they are not physicists or electrical engineers, magnetic fields still have something mysterious. One cannot see these fields and usually one cannot feel them. Much remains to be elucidated on their impacts on cell physiology. But concerning their actions on magnetic nanoparticles, there is nothing mysterious at all. The magnetic force acting on magnetic nanoparticles is described by the following formula [81]:

$$F_{\mathrm{m}} = (m \cdot \nabla)B = \frac{1}{2} \cdot \frac{\Delta\chi V}{\mu_0} \cdot \nabla B^2$$

where

m is the magnetic moment of the particle
B is the magnetic flux density
$\Delta\chi$ is the magnetic susceptibility difference of between the magnetic particles and the surrounding medium
V is the volume of the particles
μ_0 is the permeability of vacuum
∇B is the magnetic field gradient

This means that the essential parameters for magnetic accumulation are the volume (and thus the third power of the radius) of the particle and the field gradient. The magnetic flux density (measured in tesla [T]) only has an influence below the limit of saturation magnetization of the particles, which is a material property (measured in $A \cdot m^2/kg$). Magnetic nanoparticles will move in a magnetic field only if it is not homogenous, if the particles experience a field gradient (measured in T/m). In other words, they will migrate toward the highest density of magetic field lines. Further implications of this formula are discussed toward the end of the chapter. In magnetofection, the applied magnet has the task to move the magnetic particle–nucleic acid associates to the target site. This means in practice that target sites ought to be subject to a magnetic flux density, which is sufficient to cause saturation magnetization of the magnetic drug carrier and ought to be subject to the highest field gradient possible. This implies that for *in vivo* applications, magnets ought to be tailor-made according to the anatomy of the target site to optimize

magnetic retention. This is all the more important as magnetic flux density and field gradient rapidly decrease with the distance from the source of the magnetic field.

For magnetofection in cell culture, the requirements are not that challenging. As described above, the magnetic force acting on magnetic vectors is proportional to the size of the magnetic particle associated with the vector. Hughes et al. [1], for example, have used micrometer-sized magnetic particles in their work [1]. This is probably the reason why a thin magnetic sheeting (Bisiflex II) was sufficient to attract their vector (a retrovirus) to the target cells (though, during a 24-h incubation period). In most cases, however, rare earth permanent magnets are used (e.g., neodymium–iron–boron magnets), which produce magnetic flux densities of around 1 T at their surfaces. The field gradients they produce are dependent on the magnet geometry. We have, for example, developed a magnetic plate in the 96-well format where cylindrical shaped Nd–Fe–B magnets are inserted in an acrylic glass template in strictly alternating orientation (Figure 17.5A). These plates produce a magnetic flux density between 0.13 and 0.24 T and a magnetic field gradient between 67 and 123 T/m at a distance of ~2 mm from the individual magnets' surfaces. This would be the position where the cells are located when seeded in a 96-well plate and positioned on this magnetic plate. This magnetic induction leads to approximately 80% saturation of particles consisting of almost pure magnetite for which an induction of 0.5–0.6 T is necessary to achieve 100% saturation (Mykhaylyk O., unpublished data, 2005). Although induction and gradient decrease rapidly with increasing distance from the surface of the magnet, for magnetic sedimentation in cell culture this magnetic device is very efficient. In the meantime, such devices are commercially available (www.ozbiosciences.com, www.chemicell.com). Individual magnets can be bought from numerous companies, so anyone can build their own devices for magnetofection [12].

Nd–Fe–B magnets were also used successfully for some *in vivo* applications where the magnet could be applied directly onto the target tissue (e.g., onto the ear vein of pigs, the ear artery of rabbits, the surface of tumors or through surgery onto the ileum of rats, the stomach of mice, and the testicles of mice) [2,82,83]. The magnets used in these proof-of-principle experiments were not optimized in design and shape. Whether magnetic retention against the viscous drag force of the blood flow is feasible or not is dependent on many parameters. On the basis of the physical considerations, it is not possible, for example, to retain individual nonaggregated nanoparticles upon first pass with reasonable magnetic force at high blood flow rates, which are prevalent, for example, in the aorta (20 cm/s). On the other hand, magnetic retention is feasible even upon first pass at more moderate flow rates and is increased if a magnetic formulation passes the target site several times (this is the so-called avalanche effect). Particles that have been deposited upon first pass are a source of strong field gradients themselves in an external magnetic field and contribute to the retention of further particles at second and third pass, etc.

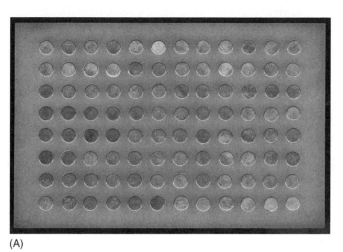

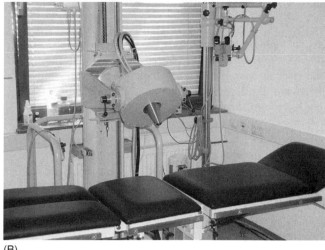

(A)   (B)

FIGURE 17.5 Magnets. (A) Example for an array of permanent Nd–Fe–B magnets on which a 96-well plate can be positioned. The cylindrical magnets are inserted in an acrylic glass template in strictly alternating polarization to permit a relatively homogenous sedimentation of magnetic particles on cells growing in the wells. (B) Electromagnet constructed for magnetic drug targeting by Siemens AG. The magnet is high scalable in field gradient. (From www.siemens.de/pof. With permission.)

As mentioned above, for *in vivo* applications, magnetic devices ought to be tailor-made in the optimal case. Recently, Siemens AG, Germany, has developed a prototype electromagnet for magnetic drug targeting (Figure 17.5B). This magnet produces a field gradient of more than 100 T/m at the pole tip and at a distance of 22 mm of 10 T/m [84]. The magnetic flux density is scalable and is sufficient to induce magnetic saturation of iron oxide nanoparticles in a target volume at some distance of the pole shoe, which would, for example, be suitable for magnetic drug targeting in superficial tumors. Also Gleich and colleagues have reported on the design of electromagnets for magnetic drug targeting [85] and have come up with similar designs as Siemens. Evaluation of these magnets for *in vivo* magnetic drug targeting and magnetofection is in process.

It has been mentioned in Section 17.1 that one of the earliest accounts of magnetofection-type methods can be found in the patent literature. In his patent issued in 1998, Chan describes the use of magnetic field pulses acting on oligonucleotides bound to magnetic nanoparticles (U.S. patent 5753477). The equipment he used produced extremely strong field pulses (micro- to millisecond pulses of 2–50 T). Chan describes that his method worked best if he first induced magnetization of the particles (and at the same time orientation along the field lines) with a permanent magnet and then applied the strong field pulses.

A novel technique using alternating magnetic fields in a magnetofection-type method has been described recently by Kamau and colleagues [56]. The magnetic device they use (the so-called dynamic marker) produces a relatively complex pulsating field with a sinus type wave perpendicular to the cell culture plate overlayed with a field modulation in the plane of the cell culture plate at low frequency. It is speculated that

such alternating fields may cause some oscillation of magnetic particles, which may facilitate cellular uptake when the particles are bound to cell surfaces. What really happens is not understood. In any case, using this technique, Kamau et al. achieved quite substantial improvements in the percentage of transfected cells in a variety of cell lines. Again, a combination of premagnetization with a permanent field followed by application of the dynamic field resulted in synergistic enhancements.

Highly surprising results have been reported by Cai and coworkers who report on a method they call nanotube spearing [57]. They prepared carbon nanotubes grown by plasma-enhanced chemical vapor deposition, which have ferromagnetic nickel in their tips. DNA was bound to such nanotubes. Gene delivery even to very difficult transfect cells such as primary B cells was highly successful when the cells were immobilized on grids or coverslips and were exposed to the carbon nanotube spears in a simple beaker, which was placed on a simple magnetic stirrer operated at 1200 rpm. Obviously, the weak rotating field of a laboratory magnetic stirrer is sufficient to induce rotation or oscillation in the magnetic nanotubes. Also, in this case, exposition to a permanent field combined with the application of the rotating field resulted in synergistic enhancements of gene delivery.

In summary, efficient magnetofection in cell culture can be achieved with very simple magnetic devices even with weak gradient fields and can reportedly be improved with still somewhat exotic nonstatic fields. For the latter, substantial research is required to understand their mechanisms of action. *In vivo* magnetofection can be mediated with simple permanent magnets; however, it is expected that magnetic drug targeting, in general, will greatly profit from tailor-made magnets.

17.4 USE OF MAGNETOFECTION

The previous chapters have given an overview on the historical development of magnetofection techniques and on the tools that are required to perform magnetofection. In this section, we discuss magnetofection techniques in cell culture and *in vivo* and highlight its fundamental features which in summary are the following:

1. Magnetofection appears to be applicable with any known nucleic acid vector and nucleic acid type.
2. Magnetofection can improve the dose–response profile of many vectors when compared with nucleic acid delivery with the same vector under conventional conditions. This improvement applies to the overall expression of a transfected/transduced gene or the overall downregulation of target gene expression (antisense, siRNA) as well as to the percentage of transfected/transduced cells. This improvement allows the user to save valuable vector material.
3. Magnetofection can greatly improve the transfection kinetics of many vector types.
4. Magnetofection can be used to localize nucleic acid delivery *in vitro* and *in vivo*.
5. The transfection/transduction mechanism with magnetofection does not appear to differ fundamentally from conventional transfection/transduction. The major impact of magnetic field influence in cell culture appears to be the rapid sedimentation of the full

applied vector dose on the target cells within a short period. In this manner, transfection/transduction likely proceeds in a more synchronized manner than in conventional transfection/transduction.

17.4.1 MAGNETOFECTION IN CELL CULTURE

17.4.1.1 Comparison to Standard Transfections/ Transductions with Regard to the Nucleic Acid Transfer Efficiency

In our first published reports on magnetofection, we have compared reporter gene expression upon transfection/transduction with a variety of nonviral and viral vectors under conventional and magnetofection conditions [2,55]. For example, two different cell types (NIH 3T3 and CHO-K1) were incubated for 4 h or only for 10 min with four different standard vectors (PEI-DNA, PEI-DNA-inactivated adenovirus, GenePorter-DNA, and Lipofectamine-DNA) under conventional transfection conditions or under magnetofection conditions for 10 min (meaning the same vector associated with magnetic nanoparticles under the influence of a magnetic field) (Figure 17.6A). Depending on vector type and incubation time, the observed enhancements in luciferase reporter gene expression were quite substantial. In this example, magnetofection was always superior to conventional transfection when 10 min incubation times were compared and was usually superior even when 10 min magnetofection was compared with 4 h conventional transfection. The enhancing effect was particularly strong with

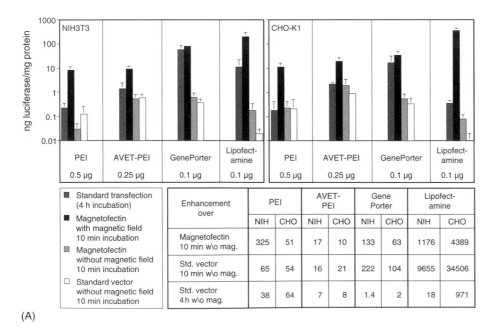

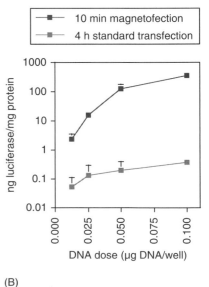

(A)   (B)

FIGURE 17.6 (See color insert following blank page 206.)   Enhancements achieved with nonviral magnetofection. (A) Comparison of magnetofections (with magnetic field 10 min incubation) with their corresponding standard transfections (4 h incubation) in NIH 3T3 and CHO-K1 cells. In the table, enhancements of magnetofection over transfections without (w/o) magnet are specified. AVET, adenovirus enhanced transfection. (B) Comparison of Lipofectamine magnetofection and Lipofectamine standard transfection in CHO–K1 cells with regard to their DNA dose–response profile. (From Scherer, F., Anton, M., Schillinger, U., Henke, J., Bergemann, C., Krüger, A., Gänsbacher, B., and Plank, C., *Gene Ther.*, 9, 102, 2002. With permission.)

Lipofectamine but was less pronounced (10 min magnetofection vs. 10 min conventional) or absent (10 min magnetofection vs. 4 h conventional) with GenePorter. With Lipofectamine, the advantage in the dose–response profile was such pronounced that magnetofection with a low vector dose yielded higher reporter gene expression levels than an eightfold higher dose of the conventional vector (Figure 17.6B) [76]. At that time, we speculated that the observed enhancements were due to a higher number of vector–cell contacts within the incubation time mediated by magnetofection versus conventional transfection. Later, Gersting et al. verified this assumption when they examined adhesion patterns of fluorescently labeled gene transfer complexes on airway epithelial (16HBE) cells by fluorescence microscopy [86].

When establishing a novel method or research tool in life sciences, an important question is whether other researchers can reproduce the findings. Are the protocols that one has generated robust and comprehensible enough to be reproducible, was the interpretation of results unbiased enough? The question is also, whether a novel transfection method is only successful in a few cell types with selected vector types under very special conditions or whether there is universal applicability. We were lucky in that several research groups have reported about the success of magnetically guided nucleic acid delivery, which we called magnetofection, at the same time as we did. Confirmations of the utility of the method keep accumulating.

17.4.1.2 Enhancements in Tranfection/Transduction

Gersting et al. found that in airway epithelial cells (16HBE cell line and primary cells) magnetofection was, with an incubation time of 15 min, more or at least equally efficient in gene transfer as standard PEI-polyfection with a 4 h incubation time. Furthermore, the favorable dose–response relationship has been confirmed [86]. Morishita et al. showed that magnetic field–guided delivery significantly enhances transfections with HVJ-E vectors and that the required HVJ titers can be reduced [87]. In primary human umbilical vein endothelial cells (HUVECs), magnetofection with various cationic lipids and PEI strongly improved gene transfer (up to 360-fold) while cytotoxicity remained at acceptable levels [88]. Improved transfection efficiencies and DNA dose–response profiles through magnetofection were also observed with the lipofection reagent Metafectene in NIH 3T3 cells [16] and with the lipofection reagent DMRIE in CT26 cells [76]. In the latter experiment, it was demonstrated that not only the overall transgene expression but also the percentage of transfected cells can be enhanced by magnetofection. In a further publication of Krötz et al., it was shown that magnetofection with various lipid vectors and PEI does improve the transfection efficiencies and dose–response relationships not only with plasmid DNA but also with antisense oligodeoxynucleotides (ODNs) in HUVEC cells. It turned out that ODN-magnetofection with its shorter incubation time is significantly less toxic than the corresponding standard transfection [82]. But it has to be mentioned that in contrast to magnetofection with FUGENE plus plasmid DNA [88], magnetofection with FUGENE plus antisense-ODNs was less efficient than the standard FUGENE transfection. Among the many comparisons, this is one of the few cases where magnetofection led to a decrease in transfection efficiency. In 293T cells, novel DNA fragments (PCR products of amplified EGFP gene) for transfections were tested by Kamau et al. and they found that with these molecules magnetofection leads to strong enhancements [56]. Earlier work from Isalan and coworkers had already demonstrated the utility of arrays of magnetic beads coated with PCR products [12]. Magnetofection also increased the efficiency of transfections of siRNA. We demonstrated that efficient knock down of stable eGFP expression in HT1080 cells with linear PEI and synthetic siRNA was only achieved through magnetofection [16] and not with linear PEI-siRNA alone. This result also indicates the potential of magnetofection for nucleic acid transfer into cells which are difficult to transfect with standard methods.

17.4.1.3 Applicability of Magnetofection with Different Nucleic Acid Types

Among the examined nucleic acid molecules are the DNA plasmids (Figure 17.6) [2,88,89], antisense ODN (Figure 17.7A) [82,90], PCR products [12,56], siRNAs (Figure 17.7B) [16, 90–92], mRNAs, dsRNAs, and shRNAs (www.ozbiosciences.com). Magnetofection is suitable to achieve not only overexpression of a transfected/transduce gene but also silencing of gene expression.

17.4.1.4 Applicability with Different Vector Types or Transfection Reagents

Experiments showed that magnetofection is useful for both nonviral and viral vectors. Tested nonviral vectors are polyplexes with branched (Figure 17.6A) and linear PEI [16], polylysine (pL), and dendrimer (Superfect from Qiagen, unpublished data). Additionally, endosomolytic agents like a synthetic influenza virus peptide [93] or a chemically inactivated adenovirus [2] could be successfully integrated into polyplex-based magnetic DNA complexes. Examples for lipoplex-based vectors used are Lipofectamine and GenePorter (Figure 17.6A), DOTAP, DMRIE-C, Effectene [76], Metafectene [16], Fugene [88], and the lipopolyamine DreamFect (www.ozbiosciences.com). As mentioned, magnetofection can be combined with the HVJ-E vector (hemagglutinating virus of Japan-envelope) [87]. Viral vectors tested in combination with magnetic particles are adenoviruses [2,4,94], retroviruses (Figure 17.7C) [1,2,10,11,95], measles viruses [13], adeno-associated viruses [3], herpes viruses (HSV-1), alpha viruses (Sindbis virus), polyomaviruses (SV40) [6], baculoviruses [9], rhabdoviruses (VSV) (www.ozbiosciences.com), porcine parvoviruses (PPV), polyoviruses, and monkey cytomegaloviruses [78]. An essential finding with viral vectors is that magnetofection, due to its favorable dose–response relationship can compensate for low viral titers.

Particularly interesting results have been obtained with adenoviruses. These vectors rank among the most efficient

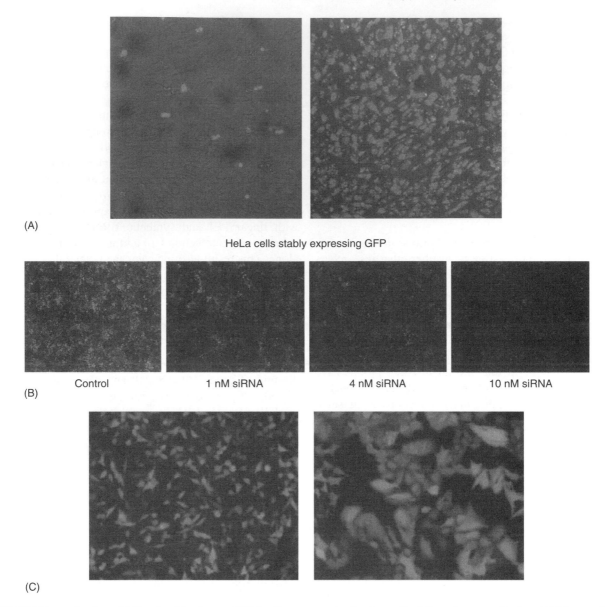

HeLa cells stably expressing GFP

FIGURE 17.7 (See color insert following blank page 206.) Some examples for the applicability of magnetofection. (A) Magnetofection of primary HUVEC-C cells with fluorescence-labeled phosphorothioate oligonucleotides (ODN). Right panel: In the presence of a magnetic field, up to 90% of the cells had oligonucleotide bound or internalized within 15 min of incubation with complexes of ODN with magnetic nanoparticles. Left panel: In the absence of a magnet, only very little binding/uptake was observed within this short period. (From Krotz, F., Wit, C., Sohn, H.Y., Zahler, S., Gloe, T., Pohl, U., and Plank, C., *Mol. Ther.*, 7, 700, 2003. With permission.) (B) Magnetofection of HeLa cells stably expressing GFP with siRNA to knock down GFP expression. From left to right: untreated cells (control), 1 nM siRNA directed against GFP, 4 nM siRNA, 10 nM siRNA, respectively, associated with magnetic nanoparticles and incubated on the cells in the presence of a magnetic field. This experiment was carried out in a 96-well plate. During transfection, the cell culture plate was positioned on a magnetic plate, which is shown in Figure 17.5A. The magnetic nanoparticles used here are a commercially available magnetofection reagents. (From the homepage of Oz Biosciences (www.ozbiosciences.com).) (C) Magnetofection of HeLa (left panel) and A549 (human non-small cell lung carcinoma) (right panel) cells with a pseudo-type HI virus carrying a GFP reporter gene. (From the homepage of OZ Biosciences (www. ozbiosciences.com).)

vehicles for nucleic acid delivery. They can infect a broad range of host cell types, dividing and nondividing cells, and have a large capacity to package foreign gene inserts. However, these vectors are inefficient in transducing cells that do not express the coxsackie and adenovirus receptor (CAR) or are even incapable of doing so. Among such cells are the important targets of gene therapy such as various tumor cell types. Therefore, it was particularly encouraging that magnetofection can mediate adenoviral transduction of cells like NIH 3T3 that express little or no CAR (Figure 17.8) [2]. In our work, this effect has been achieved with PEI-coated magnetic nanoparticles that might mediate cell binding via their positive charge. But Pandori and coworkers [4] got similar results with biotinylated adenovirus bound to streptavidin beads. This would suggest that electrostatic interaction with the plasma membrane is not the driving force of adenoviral magnetofection in cells

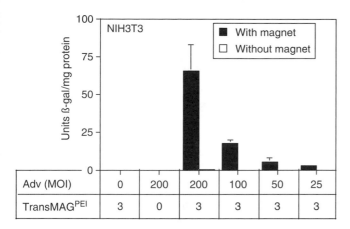

FIGURE 17.8 Infection of NIH 3T3 cells with adenoviruses (Adv) containing the LacZ gene. As NIH 3T3 cells express only little or no CAR, no transduction can be detected at a multiplicity of infection (MOI) of 200 with the virus alone. With magnetic particles (3 μg/ six-well) and magnet (magnetofection) even at lower MOIs gene transfer is obtained. (From Scherer, F., Anton, M., Schillinger, U., Henke, J., Bergemann, C., Krüger, A., Gänsbacher, B., and Plank, C., *Gene Ther.*, 9, 102, 2002. With permission.)

that are nonpermissive to adenoviral infection. In analogy to what has been found with adenoviruses, Kadota and coworkers reported that measles viruses associated with magnetic nanoparticles were to transduce cells lacking the SLAM-receptor under magnetofection conditions [13]. Thus, in summary, magnetofection is a particularly interesting tool in viral transduction as it allows working with low viral titers (and thus saves time and money), allows localizing viral delivery (see further below), and is able to expand the host range of some viruses to otherwise nonpermissive cells.

17.4.1.5 Applicability of Magnetofection in Different Cell Types

There are many examples of successfully magnetofected cell lines described in publications (Table 17.1). Researchers use magnetofection primarily as a tool in basic research rather than continuing establishing or refining the method itself. An updated list of references where magnetofection has been used successfully can be found at the Web page of OZ Biosciences (Marseille, France; www.ozbiosciences.com). The cell types that have been magnetofected include cell lines

TABLE 17.1
References to Magnetofection in Cell Culture

Origin/Description	Cell Line	Type	References
Acute T-cell lymphoma	Jurkat	Human	[96]
Adherent β-lymphoblastoid	B95a	Marmoset	[13]
Adrenocortical	H295R	Human	[97–99]
Aortic smooth muscle cells	A-10	Rat	[66]
Breast carcinoma	MCF-7, MDA-MB-453	Human	[100–102]
Bronchial epithelial	BEAS-2B, 16HBE	Human	[86,22]
Cervical epithelial carcinoma	HeLa, C12S	Human	[3,13,56,83,103]
Colon carcinoma	CT26	Murine	[76]
Chronic myelogenous leukemia	K562	Human	[2]
Embryonic fibroblasts	MEF	Murine	[104,105]
Embryonic fibroblasts	NIH-3T3	Murine	[2,103]
Embryonic kidney	HEK-293, 293T	Human	[56,106,107]
Endothelial cells	SVEC	Murine	[106]
Fibroblast	3Y1	Rat	[108]
Fibrosarcoma	L929, HT1080	Murine, human	[13,16]
Glioblastoma	A172, KS-1, NYGM, T98G, U251, U373, U87, YH-13, YK6–1	Human	[109]
Glioma	D-17P4	Rat	[4]
Head and neck carcinoma	B11, HNSCCs	Human	[105,106,110]
Intestinal endocrine	STC-1	Murine	[111]
Kidney	COS7, Vero, Vero E6, BHK-21	Monkey, hamster	[13,14,56,89,108,112]
Laryngeal epithelium	Hep2	Human	[106]
Lung epithelial carcinoma	H441	Human	[113]
Microvascular endothelium	HMEC-1	Human	[114]
Neuroblastoma	SH-SY5Y, SK-N-BE2	Human	[115,116]
Oral cavity carcinoma	HN12	Human	[106]
Osteosarcoma	HOS, Saos-2	Human	[5,117]

(continued)

TABLE 17.1 (continued)
References to Magnetofection in Cell Culture

Origin/Description	Cell Line	Type	References
Ovary	CHO	Hamster	[2,13,118]
Pancreatic	βTC-tet cells	Murine	[119]
Prostate carcinoma	PC-3	Human	[102]
Renal cortical	M-1	Murine	[113]
T-cell leukemia	MOLT-4	Human	[96]
T lymphocyte	H9	Human	[117]
Tongue squamous carcinoma	Cal27	Human	[106]

Origin/Description	Primary Cells	Type	References
Gastric glands	Adherent gastric cells	Human, mouse	[120,121]
	Aortic smooth muscle	Human	[122]
Articular	Chondrocytes	Rabbit, human	[83,123]
Aorta (PAEC), cord blood, umbilical vein (HUVEC)	Endothelial cells	Porcine, human, bovine, rat	[11,12,14,82,88,106, 107,110,124–126]
Lung	Epithelial	Mouse	[86]
Fetal fibers	Fibroblasts	Mouse	[127,128]
Brain tumor (GBM)	Glioblastoma	Human	[109]
Megakaryocyte erythroid progenitor	MEPs	Mouse	[129]
Skeletal myotubes	Myoblasts	Mouse	[128]
Gastric	Myofibroblasts	Human	[130]
Hippocampal, cortical (embryonic DRG), vagal afferent	Neurons	Rat, mouse	[116,131–134]
PBL, PBMC	Peripheral lymphocytes	Human, macaques	[2,135]
Articular	Synoviocytes	Sheep	[56]

and primary cells many of which are considered difficult to transfect/transduce. From these references, it can be concluded that magnetofection is an efficient transfection/transduction method for a large number of different cells and has become a valuable tool in the life sciences. Figure 17.7 shows results we have obtained in HeLa cells and HUVECs.

In summary, usually magnetofection is significantly more efficient than the corresponding standard transfection or transduction, but there are cases in which magnetofection is only equally or even less efficient. The high efficiency at tolerable toxicity makes magnetofection a useful tool for examinations in hard-to-transfect/transduce and sensitive primary cells. Owing to the often improved nucleic acid dose–response profiles and the reduced incubation times, magnetofection may be particularly useful for high throughput applications.

17.4.1.6 Critical Parameters in Magnetofection

As the magnetofection appears to involve similar mechanisms as conventional transfection/transduction (see below), it is dependent on the potency of the parent vector composition associated with magnetic nanoparticles. For example, components such as endosomolytic substances, which improve nonviral gene delivery, also improve nonviral magnetofection.

As the association of nucleic acids/vectors is a prerequisite for magnetofection to work, the ratio of magnetic particles to nucleic acids/vectors is an essential parameter [13,76]. In nonviral magnetofections, usually a magnetic particle/DNA ratios between 0.5 and 6 (w/w) yield optimum efficiencies (Figure 17.9A). Using an insufficient amount of magnetic nanoparticles may lead to insuffi-

cient vector binding and insufficient magnetic sedimentation. An excessive amount of magnetic nanoparticles may lead to toxicity [87]. Similarly, the overall vector dose is an important parameter and needs to be optimized. The optimal dose is usually lower in magnetofection than in standard transfections [88].

Another important parameter is the incubation time of cells with vectors and the exposure time to a magnetic field. Magnetofection experiments with lipoplexes revealed that only a few minutes of magnetic field influence can be sufficient to yield optimum gene transfer efficiency [16,76]. This turned out to be an advantage when the cells to be transfected were sensitive to longer incubation times as was the case with HUVECs in transfections with antisense oligonucleotides [82], (Figure 17.9B). To achieve efficient ODN uptake with conventional means, long incubation times were required that were toxic to the cells. Magnetofection conditions required only a few minutes of incubation, which was far less toxic to the cells. Also with lentiviral vectors it has been demonstrated that very short incubation times (1 min) are sufficient to achieve optimal transduction with magnetofection procedures [11].

Finally, as mentioned before, further improvements in magnetofection can be achieved with alternating fields [56] (Figure 17.9C), a topic that certainly requires further examination.

17.4.1.7 Mechanism of Magnetofection in Cell Culture

Already our initial experiments had indicated that the uptake mechanism of magnetic complexes, at least with nonviral vectors, is probably the same as for their nonmagnetic counterparts.

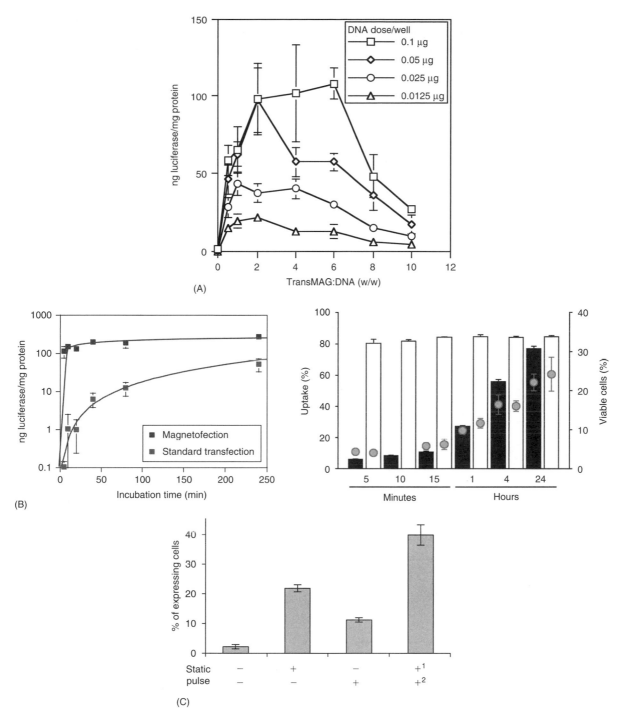

FIGURE 17.9 Optimizing magnetofection parameters. (A) Magnetofection of CHO–K1 cells with GenePorter liposomes at different magnetic particle/DNA ratios and DNA doses. (From Plank, C., Schillinger, U., Scherer, F., Bergemann, C., and Remy, J.S., *Biol. Chem.*, 384, 737, 2003. With permission.) (B) Lipofections with different incubation times for cells with vectors. Left panel: In contrast to the standard GenePorter transfection (grey), in GenePorter magnetofection (black) already 5 min incubation are sufficient to obtain a gene transfer efficiency close to the optimum. (Modified from Plank, C., Anton, M., Rudolph, C., Rosenecker, J., and Krotz, F., *Expert Opin. Biol. Ther.*, 3, 745, 2003.) Right panel: In magnetofection (white bars) with Effectene an incubation time of 5 min leads to nearly maximum ODN uptake while in the corresponding standard transfection (black bars) 24 h are necessary. Additionally, 5 min are significantly less toxic to cells (see circles) than longer incubation times. (From Krotz, F., Wit, C., Sohn, H.Y., Zahler, S., Gloe, T., Pohl, U., and Plank, C., *Mol. Ther.*, 7, 700, 2003. With permission.) (C) 293T cells were transfected with magnetic particle/DNA complexes and placed either on a static magnetic plate for 5 min, or a pulsating magnetic field was applied for 5 min, or the pulsating field was applied after the static field. The combination of premagnetization with a static field followed by the application of a pulsating field yielded significant enhancements in the percentage of transfected cells. (From Kamau, S.W., Hassa, P.O., Steitz, B., Petri-Fink, A., Hofmann, H., Hofmann-Amtenbrink, M., von Rechenberg, B., and Hottiger, M.O. *Nucl. Acid Res.*, 34, e40, 2006. With permission.)

Magnetofection with a binary complex consisting only of naked DNA and PEI-coated magnetic nanoparticles worked far less efficiently than magnetofection with ternary complexes comprising also endosomolytic agents such as the influenza peptide INF7, free PEI, or chemically inactivated adenovirus. This strongly indicated that the uptake mechanism involved endocytosis. This was confirmed by Huth and coworkers [22] who have performed a detailed examination on the uptake mechanism of PEI-DNA complexes in comparison to the same vector associated with magnetic nanoparticles. They carried out transfections in the presence and in the absence of magnetic field influence and various inhibitors of cellular uptake mechanisms. The results clearly showed that the uptake mechanism for this vector type, no matter whether under magnetofection or conventional conditions, is mainly endocytosis (Figure 17.10) including clathrin-dependent pathways as well

as caveolae-mediated endocytic uptake. Whether this similarity of uptake mechanisms applies also to lipoplexes and viral vectors has not been examined so far.

An important question was whether the magnetic field would have any further impact on transfection (as measured by reporter gene expression) beyond the rapid sedimentation of the vector on the target cells. It is well documented that low frequency electromagnetic fields (EMFs) and static magnetic fields can have biological effects on cells and tissues. It is assumed that their primary site of action is the plasma membrane [141,142] and that for example phagocytosis [143] or cellular metabolic activity [144] can be reduced by magnetic fields. Further, it is known that EMFs activate genes under the control of EMF-sensitive promoters [145], that static electromagnets can induce the expression of oncogenes in tumor cell lines [146,147] and that a permanent magnetic field (300 mT)

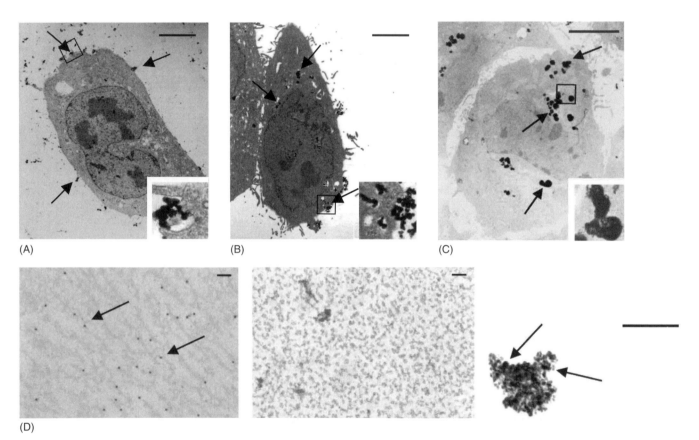

FIGURE 17.10 Electron microscopy of HeLa cells magnetofected with PEI-containing magnetofectins. If not otherwise stated the scale bars indicate 5 μm. The insets show a higher magnification of the labeled area. (A) Cells were exposed to magnetic DNA complexes and a magnetic field for 15 min. Arrows indicate the electron-dense iron oxide particles. The DNA is not electron dense and is not visible in this micrograph. The magnetic particles are concentrated around the cell and their uptake already starts. (B) Cells were exposed to magnetic DNA complexes and a magnetic field for 15 min and subsequently they were incubated for 24 h without magnet in fresh medium. In this micrograph, magnetic particles are found within the cell and the accumulation in endosomal structures suggests endocytosis as uptake mechanism. (C) The only difference to (B) is that the DNA is gold-labeled to make it electron dense. Thus, not only the magnetic particles but also the DNA is visible. This micrograph, especially the inset, reveals that magnetic particles and gold-labeled DNA are cointernalized into the cell. For better recognition, both components and their complexation are illustrated in (D). (D) Electron microscopic presentation of the components of magnetic particle/gold-labeled DNA complexes. In the left panel, gold-labeled DNA (scale bar: 100 nm), in the middle panel iron oxide particles (scale bar 100 nm), and in the right panel complexes (scale bar: 200 nm) are shown. Arrows indicate gold-labeled DNA. (From Huth, S., Lausier, J., Gersting, S.W., Rudolph, C., Plank, C., Welsch, U., and Rosenecker, J., *J. Gene. Med.*, 6, 923, 2004. With permission.)

changes the expression of some genes in *Escherichia coli* [148]. We examined whether the exposition of cells to Nd–Fe–B permanent magnets during conventional transfection (i.e., without magnetic particles) has an impact on reporter gene expression and found no such difference [93]. Huth et al. sedimented PEI-DNA complexes and the same vector associated with magnetic nanoparticles on the target cells by centrifugation and only subsequently applied a magnetic field [22]. There was no enhancement in reporter gene expression upon exposition to the magnetic field even with cells that received magnetic nanoparticle-associated vectors. Thus, it can be concluded that the enhancing effect of a magnetic field is mainly the result of accelerated vector sedimentation and not of changes in cell physiology, enhancement of cellular uptake, or activation of reporter gene expression.

In summary, the following mechanism is proposed for magnetofections. The magnetic nucleic acid vectors are concentrated efficiently by magnetic force on the cell surface within minutes and their uptake starts immediately. The uptake of magnetofectins follows a similar mechanism as the corresponding standard vectors but it occurs earlier and is more synchronized. The high efficiency in nucleic acid transfer with magnetofection is probably mainly a result of efficient sedimentation and, therefore, high availability of nucleic acid vectors at the cell surface for uptake.

17.4.1.8 Targeting (Localization) of Nucleic Acid Delivery with Magnetofection

The major motivation of magnetic drug targeting comes, of course, from *in vivo* applications. Concerning magnetic targetability, experiments in cell culture have served as models. Permanent magnets of different shape were positioned underneath the culture dishes in which magnetofection experiments were carried out. It turned out that gene delivery becomes confined to an area which is determined by the shape of the magnet in use. This was demonstrated with nonviral vectors and adenoviruses (Figure 17.11) [82,2]. Further, e.g., Hughes et al. [1], Pandori et al. [4], Mah et al. [3], and Raty et al. [9] showed magnetic field–guided targeting with retroviruses, adenoviruses, adeno-associated viruses, and baculoviruses. This confinement is the result of the physics of interaction of magnetic nanoparticles with magnetic fields that have already been discussed. The particles are attracted toward the highest density of magnetic field lines, which is always found at edges and tips of magnets. Therefore, by choosing the shape of a permanent magnet, nucleic acid delivery can be targeted (localized) to a subpopulation of cells within a culture dish. In this way, magnetic field–guided nucleic acid vector delivery could offer, e.g., the possibility to evaluate cells transfected with different nucleic acids compared to the untransfected control cells within the same well or it could enable the examination of the influence of secreted transgene-encoded factors on neighboring untransfected cells. A step in this direction has been Isalan's work who used magnetic field–guided delivery of PCR products for multiple, parallel cell transfections on microscope cover-slip arrays [12].

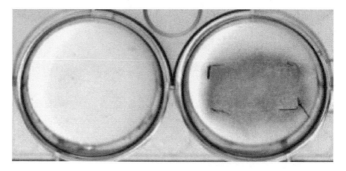

FIGURE 17.11 Targeting (localization) of gene delivery with adenoviral magnetofection. Adenoviral vectors carrying the lacZ gene were associated with magnetic particles and incubated on NIH 3T3 cells in the absence (left), or in the presence (right) of a magnetic field. A rectangular permanent magnet was placed to the underside of the right well during incubation. Efficient gene transfer and reporter gene expression (indicated through the blue X-gal stain) is confined to the area where the magnet had been positioned. Similar experiments are shown in Refs. [16,55]. (From the homepage of OZ Biosciences (www.ozbiosciences.com).)

Thus, as already mentioned, in combination with the shorter incubation times (reduced from hours to minutes) and the improved nucleic acid dose–response relationships, magnetofection may even be a method of choice for automated high throughput screening of genes and of therapeutically useful sequences.

17.4.2 MAGNETOFECTION *IN VIVO*

Although there is plenty of evidence for the utility of magnetofection in cell culture (see Table 17.1), the reports on magnetofection *in vivo* are far less abundant (Table 17.2). We have provided proof-of-principle that magnetic guidance of transfection in blood vessels upon intravascular administration is feasible. PEI-containing magnetic DNA complexes were injected into ear veins of pigs or ear arteries of rabbits and a permanent magnet was placed above the blood vessel downstream of the injection site [16,2]. Reporter gene expression was detected at the injection sites and at the position of the magnet. Upon intra-arterial administration, expression was also found downstream of the magnet position, possibly due to passive retention of the DNA complexes in the capillary beds of the rabbit ear vasculature. Except for the injection sites, no expression was found in the contralateral ear veins or arteries. These findings suggest that magnetofection upon intravascular administration may have potential if the vascular endothelium is the target tissue and if a suitable magnet can be positioned in direct vicinity of the target site. Appropriate medical indications may include the localized delivery of angiogenic growth factor genes in the case of ischemic diseases, or, on the other hand, the delivery of anti-angiogenic factor genes to the tumor vasculature for tumor therapy, just to name two examples.

Krötz et al. [82] have reported that at least in a particular mouse model, magnetic targeting of nucleic acids also works if the magnetically targeted site is not in direct vicinity of the

TABLE 17.2

In Vivo, ex Vivo Citations

Species	Description	References
Rat	Stomach	[2]
Rat	Jejunum	[16]
Feline	Fibrosarcoma	[83]
Hamster	Isolated arteries	[90]
Human	Human acute myeloid leukemia blasts	[10]
Mouse	Abdominal cavity, intravascular	[82]
Mouse	Intramuscular	[3]
Mouse	Airway epithelium	[136]
Mouse	Intravenous administration Transfection in lung, brain, spleen, and kidney	[7]
Mouse	Intravascular, localized delivery in skin chamber model	[137]
Mouse	Direct injection into the liver	[87]
Rabbit	Ear artery	[16]
Sheep	Synovium	[138–140]
Swine	Ear vein	[2]
Swine	Airway epithelium	[86]

injection site. They injected magnetofectins containing fluorescence-labeled antisense ODNs into the femoral arteries of mice, and during injection a permanent magnet was applied to one testicle, which was exposed by surgery. Confocal fluorescence microscopy revealed that specific uptake of ODNs was only observed in cremaster muscle blood vessels of testicles, which were exposed to a magnetic field (Figure 17.12C) and not on the contralateral site in the same animal.

Whether magnetofection upon intravascular administration has any therapeutic potential has still not been demonstrated. In the case of nonviral compositions associated with magnetic nanoparticles via noncovalent interactions, similar limitations apply upon intravascular administration as applicable to nonviral vectors in general. These include strong interactions with blood components, vector inactivation, dissociation of the composition, or rapid removal by the reticulo-endothelial system. Further, current limitations of the method with respect to intravascular administrations are discussed further below.

However, magnetofection also has potential in orthotopic administrations. In a rat model, we administered nonviral magnetic vectors into the ileum lumen after laparatomy in the presence of a magnetic field (Figure 17.12B). Reporter gene expression was found specifically where the permanent magnet had been positioned [2]. A similar result was found when magnetic particle–adenovirus associates were applied into the stomachs (which were exposed after laparatomy) of mice while a magnet was positioned outside the stomach wall [2]. In summary, the results of these experiments revealed that reporter gene transfer was strongly enhanced in the area under influence of a magnetic field whereas without application of a magnet (controls), either no or much poorer transfection/transduction was achieved. Interestingly, effective localized

reporter gene delivery was obtained even in the gut and in the stomach where harsh conditions (degradative enzymes and low pH) prevail.

Unfortunately, not in all applications a magnetic field enhanced the nucleic acid transfer efficiency. For example, Xenariou et al. examined nonviral gene transfer to the murine nasal epithelium and application of a magnetic field did not improve the transfection efficiency [136]. Analogously, in the experiments of Morishita et al. who directly injected maghemite nanoparticles associated with HVJ-E vectors into the liver of mice, a magnet placed onto the liver resulted in no significant local enhancement [87]. Nevertheless, in both studies very valuable results were obtained because both Xenariou et al. and Morishita et al. already observed that magnetic nanoparticles alone can lead to enhanced gene transfer. Concerning the design of magnetic nanoparticles, Morishita and colleagues found out that in contrast to *in vitro* experiments *in vivo*, a coating of magnetic particles with heparin, but not with protamine sulfate, works enhancing. This finding indicates that a magnetic particle-coating that is efficient in cell culture is not necessarily successful *in vivo* and vice versa. However, as already mentioned, up-to-date magnetic field–guided nucleic acid delivery does not work in all animal models and one possible explanation is that for some applications, magnetic fields are not strong enough. Therefore, improvements in magnetic field technology and eventually also in magnetic vector design are desirable. Some ideas are presented in Section 17.5.2.

Important insights about the tolerability of magnetic field–guided gene delivery *in vivo* were gained in a preliminary study in sheep by Galuppo et al. [138]. Intra-articular injection of PEI-coated iron oxide nanoparticles linked to plasmids and application of a static and a pulsating magnetic field resulted in mild-to-moderate inflammatory responses in the majority of synovial membrane samples whereas larger particles (200–250 nm) tended to be associated with more inflammation than smaller ones (50 nm). Nevertheless, intra-articular application of all magnetic nanoparticles was well tolerated. With regard to the tolerability of iron oxide, Weissleder et al. found in earlier studies that iron oxide particles used as contrast agents in MRI are fully biocompatible [67]. After intravenous application in rats, the particles were cleared by macrophages in liver and spleen, the iron oxides were degraded in lysosomes via hydrolytic enzymes, and the resulting elementary iron was integrated into the natural iron metabolism (e.g., incorporation into hemoglobin). Additionally, Weissleder and coworkers showed that in rats and beagle dogs, a relatively high dose of 167 mg iron/kg body mass still had no toxic effects on the liver or other organs and they mentioned that for clinical MRI a dose of approximately 1 mg iron/kg is proposed. Therefore, e.g., the 76.9 μg iron oxide particles/kg applied intravenously in pigs for magnetic nucleic acid targeting (Figure 17.12A) [2] can be assumed to be safe. A further interesting finding about iron oxide nanoparticles was obtained by Xiang et al. who showed that intravenously delivered polylysine-coated iron oxide nanoparticles can even transfer DNA across the blood–brain barrier to glial cells and

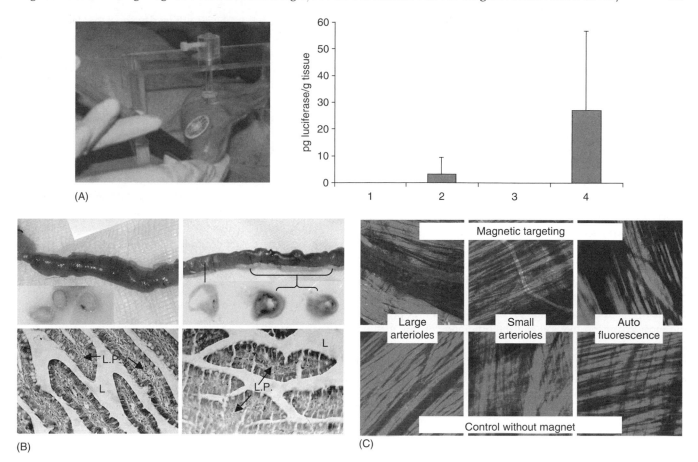

FIGURE 17.12 (See B and C in color insert following blank page 206.) (A) Left panel: Experimental setup: A permanent magnet, attached to a plunger adjustable for height, was placed above the ear vein of a pig without causing any pressure on the blood vessel. Subsequently, magnetic gene vectors (brown suspension in the syringe) were injected via a cannula into the ear vein upstream of the magnet. Right panel: Luciferase gene expression in the ear veins after injection with or without a magnet placed downstream. The graph shows that on the average without magnet (1 and 3) no significant gene transfer was monitored whereas with magnet, maximum values were obtained underneath the magnetic field. (B) X-gal staining performed 48 h after magnetic particle/DNA complexes were applied to the ilea of rats in the absence (left panels) and under the influence of a magnetic field for 20 min (right panels). Blue staining reveals efficient gene delivery only in the presence of magnet (right panels). Upper panels: intestinal tubes after X-gal stain. Insets: cross-sections of tubes. Lower panels: Paraffin sections counterstained with eosin, 400× magnification. X-gal staining is found in the lamina propria. L, lumen; LP, lamina propria. (From Scherer, F., Anton, M., Schillinger, U., Henke, J., Bergemann, C., Krüger, A., Gänsbacher, B., and Plank, C., *Gene Ther.*, 9, 102, 2002. With permission.) (C) Injection of fluorescence-labeled ODN/magnetic particle associates into the femoral artery caused site-specific accumulation in the vessels of the ipsilateral mouse cremaster at which a magnet was placed (upper panel, left: large arterioles, middle: small arterioles, right: autofluorescence before injection of magnetic complexes) but not into the contralateral cremaster, which was not in contact with a magnetic field (lower panel). (From Krotz, F., Wit, C., Sohn, H.Y., Zahler, S., Gloe, T., Pohl, U., and Plank, C., *Mol. Ther.*, 7, 700, 2003. With permission.)

neurons of the brain. Thus iron oxide nanoparticles (eventually in combination with a magnetic field) may have potential for gene therapy of CNS diseases [7] if such findings can be reproduced by other researchers.

Speculating about the future, magnetofection may become a useful *in vivo* research tool for the local examination of gene function similar to as it is now used in cell culture. Of course, we and others also envisage therapeutic applications. A first step toward tumor therapy was undertaken in a veterinary clinical study using immuno gene therapy for feline fibrosarcoma [83]. This is one of the most frequent tumors in cats and has an extremely high recurrence rate upon standard therapy, which is surgical excision. The gene coding for human GM-CSF (granulocyte macrophage colony stimulating factor) in a nonviral magnetic formulation was administered twice in a 1 week interval before surgery into the biologically active margins of the fibrosarcoma, and a permanent magnet was fixed on the tumor adjacent to the injection site during 1 h after vector injection. The immunohistochemistry showed that the GM-CSF gene was expressed in the tumor and some tissue penetration by the vector could be observed. The preliminary clinical outcome after a phase II study with more than 20 patients is a significant increase in tumor-free survival of the cats from only 23% at the 1 year time point in the case of standard therapy (surgery only) to 52% with presurgical magnetofection of the human GM-CSF gene. The long-term follow up will be very interesting as it will reveal the true benefits of this treatment.

17.5 PLACE OF MAGNETOFECTION IN THE FIELD OF NUCLEIC ACID DELIVERY, LIMITATIONS, REQUIRED IMPROVEMENTS, AND POTENTIAL ADVANCES

Now that the benefits and potentials of magnetofection have been highlighted comprehensively, it is time to acknowledge that the method is not a solution for any limitation in nucleic acid delivery. After many years of research, conventional forms of delivery have been advanced successfully to clinical trials. Other forms of targeting and other physical methods of delivery are highly promising. Up to date, side by side comparisons or combinations, for example, with electroporation, sonoporation, hydroporation, aerosolization, ballistic methods, occlusion of the blood outflow from the target organ, or biological targeting are still missing. In the absence of further evidence, any of the named methods needs to be considered as promising as magnetofection is. Electroporation is probably the most advanced method and the efficiency of this method *in vitro* and *in vivo*, particularly with nondividing cells, still needs to be achieved with magnetofection. When compared to all the other physical methods, the major advantage of magnetofection is that it is able to combine simplicity, nonexpensiveness, localization of delivery, enhanced efficiency, and reduction of incubation time and of vector doses.

17.5.1 LIMITATIONS AND REQUIRED IMPROVEMENTS

As has been described above, there are strong indications that the mechanism of magnetofection involves natural uptake mechanisms into cells that are also prevalent for conventional methods. This also means that limitations apply. Once a natural uptake mechanism is saturated, no further improvements can be expected from an increased vector dose at the plasma membrane, which at least *in vitro* can easily be achieved by magnetofection. This explains why, although in principle any vector can be used in magnetofection, not necessarily any vector can be improved by doing so. In certain cases, the association with magnetic particles may even turn out inhibitory.

Similar considerations apply *in vivo*. So far, no magnetic vector formulation has been described that overcomes the limitations of opsonization, vector inactivation, and dissociation that also apply for conventional vectors. The same efforts of vector shielding and providing them with biological targeting functionality dedicated to conventional vector development need to be dedicated to magnetic vector formulation. Only with long-circulating magnetic vectors it can be expected that substantial magnetic vector accumulation at a target site will be feasible, particularly if the target site is not in direct vicinity of the administration site. In addition, substantial physical limitations apply to magnetofection *in vivo*, in particular upon administration into the blood circulation. First of all, not any target structure is accessible to the source of a magnetic field of high enough flux density and field gradient. It has been described above that these quantities decline rapidly with increasing distance from the source of the field. Field gradients

so far cannot be produced arbitrarily in space. Hence, as of yet magnetic drug targeting is limited mostly to superficial target structures. Other essential parameters are the hydrodynamic forces prevalent *in vivo*. Magnetic force needs at least to compensate for blood flow rates in the target structure. In cases where the endothelium does not represent the actual target tissue, extravasation is also required. And although extravasation from the blood circulation by a yet unknown mechanism has been observed in magnetic drug targeting [83,28], this alone will not warrant uptake into target cells. Tissue penetration and uptake into target cells also remain among the major limitations where magnetic vectors are administered directly into a target tissue.

Thus, given all the potential advantages and benefits of magnetic drug targeting and magnetofection that have been discussed here, it becomes clear that major advances are required before the method can be considered superior to any other way of nucleic acid delivery *in vivo*.

17.5.2 POTENTIAL ADVANCES

With the above considerations and from what has been described earlier in this chapter, it becomes clear that potential advances need to come from two sides: improved magnetic formulations and the side of physical forces that are applied. The objective is optimizing retention, tissue penetration, and cellular uptake.

It can be expected that progress will come from a combination of physical targeting with biological targeting. Vector shielding ought to bring about sufficient stability *in vivo* and providing magnetic vectors with targeting ligands may improve target cell specificity and cellular uptake once the vector has been magnetically accumulated in the target tissue.

Advances that can be contributed by tailor-made magnetic field generating devices have already been discussed. The major goal is maximizing magnetic field gradients at target sites. In this respect, an interesting suggestion has been made by Babincova and Babinec [149]. They placed a ferromagnetic wire alongside a tube, which served as a model blood vessel and positioned the whole setup in a strong homogenous magnetic field. Because of its ferromagnetic properties, the wire induces strong local field gradients in the exogenous homogenous field. This effect can be used advantageously to trap magnetic particles against hydrodynamic force. Therefore, Babincova and Babinec suggest that the strong homogenous magnetic fields of MRI equipment may be exploited for magnetic retention, which then could be several magnitudes more efficient than with permanent magnets. The procedure would, however, require a minimal invasive implantation of ferromagnetic material at the target site.

As has been described before, the magnetic force acting on magnetic nanoparticles [81] is

$$F_\mathrm{m} = (m \cdot \nabla) B = \frac{1}{2} \cdot \frac{\Delta\chi V}{\mu_0} \cdot \nabla B^2$$

where

 m is the magnetic moment of the particle

 B is the magnetic flux density

 $\Delta\chi$ is the magnetic susceptibility difference of between the magnetic particles and the surrounding medium

 μ_0 is the permeability of vacuum

 ∇B is the magnetic field gradient

 V is the volume of the particles

The latter is proportional to the third power of the particle radius. Upon intravascular administration, the magnetic force needs to counterbalance the viscous drag force caused by the blood flow, which, as a first approximation, is described by Stoke's law

$$F_D = 6\pi R\eta\Delta\upsilon$$

and which is dependent on the first power of the particle radius. This means, based on a very simplified consideration, one can predict that increasing the particle size will over-proportionally be in favor of the magnetic force. On the other hand, constraints on a suitable particle size for *in vivo* applications are given by the diameter of blood capillaries (around 5 μm). For most envisaged applications, embolisms need to be avoided.

Therefore, magnetic formulations are required that appear to be large diameter magnetic particles to the magnetic field, which, however, have inherent flexibility to travel through the blood capillaries like blood cells do. Liposomes containing an optimized amount of magnetic nanoparticles in their aqueous lumen appear to be suitable for this purpose [150,151]. Our own current approach is to combine sonoporation and magnetofection. Microbubbles are gas bubbles usually of a few micrometers in diameter surrounded by a shell, which can consist of proteins, polymers, lipids, or combinations thereof. Because of their gas content, they are excellent reflectors of ultrasound. Consequently, they are used as contrast agents in medical ultrasound imaging. Both low molecular weight drugs and nucleic acids can be associated with microbubbles in various ways. In addition, microbubbles can be targeted exploiting receptor–ligand type interactions (including antigen–antibody interactions) [152,153]. Drug-loaded microbubbles hold potential as "magic bullet" agents to deliver drugs to precise locations in the body, these precise locations are determined by where the ultrasound energy is focused [153,154]. The interaction of microbubbles and ultrasound in the low MHz range leads to cavitation, microstreaming, and, eventually to, bubble burst and consequent drug release [155]. In addition, cavitation can lead to microvessel rupture leading to increased permeability of the endothelial barrier [156,157]. This effect has been used to deliver nanoparticles and even red blood cells to the interstitium of rat skeletal muscle [158]. Cavitation nuclei formed by microbubbles have also been used to permeabilize the blood–brain barrier [159]. Essential for a combination with magnetofection is that microbubbles have been used very successfully in nucleic acid delivery in a variety of tissues [3,152,153,160–170].

On the basis of this state of the art, we reasoned that micro-bubbles may be ideal carriers to incorporate a high quantity of magnetic nanoparticles as well as active agents such as drugs or nucleic acids. This idea had already been published in 2000 by Soetanto and coworkers [171]. We postulate that the flexibility of this carrier should be sufficient to "squeeze" through blood capillaries and that magnetic accumulation of the carrier against hydrodynamic forces ought to be facilitated compared with "free," physically uncoupled magnetic nanoparticles. This is because in magnetic microbubbles, one carrier object would comprise a plurality of magnetic particles that would be immobilized in the bubble shell in close vicinity and thus Brownian relaxation would be greatly reduced. Active agents, once magnetically accumulated at a target site would be released locally by the application of ultrasound, which might also improve tissue penetration by the active agent. Last but not least, magnetic microbubbles might be highly useful contrast agents both for ultrasound and MRI.

In the meantime, we have succeeded in manufacturing magnetic microbubbles and we have confirmed that indeed the magnetic accumulation of magnetic nanoparticles against hydrodynamic forces is greatly superior in microbubble formulation than in the form of free magnetic nanoparticles. This ongoing work is still unpublished. Promising preliminary results have been presented at scientific conferences [172–174]. Currently, we are examining magnetically targeted and ultrasound-triggered delivery of nucleic acids *in vivo*. In one study, magnetic microbubbles are injected into the circulation and magnetic accumulation and ultrasound-triggered deposition of nucleic acids is observed by intravital microscopy in a mouse dorsal skin chamber model. The other model, exemplary for topical administration, is a ventral skin flap model. Magnetic microbubbles carrying the gene of an angiogenic factor are injected subcutaneously. Later on, a skin flap is excised and repositioned. The encouraging result is that only the combination of magnetic field and ultrasound yields a local deposition of nucleic acids in the circulation model or an improved skin-flap survival in the orthotopic model. One of our goals is to target endothelial cells in the vasculature of tumors with the help of magnetic microbubbles and in this manner deposit anticancer agents there, be it classical pharmaceuticals or nucleic acids.

Through combined efforts in magnetic field physics, in magnetic particle physics and chemistry, in pharmaceutical formulations, and in medical application, we envisage that magnetofection some time may become a therapeutic option for many diseases. The coming years will show whether the existing technical hurdles will be overcome.

17.6 SUMMARY

Magnetofection is a recently developed method that applies magnetic particles in combination with magnetic force to enhance and to target nucleic acid delivery *in vitro* and *in vivo*. It is universally applicable to different types of nucleic acid molecules (e.g., plasmid DNA, oligodesoxynucleotides, PCR products, siRNA, mRNA), vectors (nonviral and viral), and cells (including primary and others that are hard to transfect/transduce). In cell culture, magnetofection improves conventional (standard) nucleic acid transfer methods like polyfection, lipofection, or viral transduction with regard to efficiency, nucleic acid dose–response profile, and reduced incubation time in many cases. In animal experiments, it is demonstrated that magnetofection can also enable targeted (localized) and efficient transfection/transduction *in vivo*. But it also has to be mentioned that for a broader *in vivo* applicability and especially for the development of clinical applications many significant improvements are necessary.

ACKNOWLEDGMENT

This work has been supported by grants from the German government and the European union. Financial support of the German Excellence initiative via the "Nanosystems Initiative Munich (NIM)" is gratefully acknowledged.

REFERENCES

1. Hughes, C. et al., Streptavidin paramagnetic particles provide a choice of three affinity-based capture and magnetic concentration strategies for retroviral vectors, *Mol Ther*, 3, 623–630, 2001.
2. Scherer, F. et al., Magnetofection: Enhancing and targeting gene delivery by magnetic force in vitro and in vivo, *Gene Ther*, 9, 102–109, 2002.
3. Mah, C. et al., Improved method of recombinant AAV2 delivery for systemic targeted gene therapy, *Mol Ther*, 6, 106–112, 2002.
4. Pandori, M., Hobson, D., and Sano, T., Adenovirus-microbead conjugates possess enhanced infectivity: A new strategy for localized gene delivery, *Virology*, 299, 204–212, 2002.
5. Hirao, K. et al., Targeted gene delivery to human osteosarcoma cells with magnetic cationic liposomes under a magnetic field, *Int J Oncol*, 22, 1065–1071, 2003.
6. Satoh, K. et al., Virus concentration using polyethyleneimine-conjugated magnetic beads for improving the sensitivity of nucleic acid amplification tests, *J Virol Methods*, 114, 11–19, 2003.
7. Xiang, J.J. et al., IONP-PLL: A novel non-viral vector for efficient gene delivery, *J Gene Med*, 5, 803–817, 2003.
8. Campos, S.K., Parrott, M.B., and Barry, M.A., Avidin-based targeting and purification of a protein IX-modified, metabolically biotinylated adenoviral vector, *Mol Ther*, 9, 942–954, 2004.
9. Raty, J.K. et al., Enhanced gene delivery by avidin-displaying baculovirus, *Mol Ther*, 9, 282–291, 2004.
10. Chan, L. et al., Conjugation of lentivirus to paramagnetic particles via nonviral proteins allows efficient concentration and infection of primary acute myeloid leukemia cells, *J Virol*, 79, 13190–13194, 2005.
11. Haim, H., Steiner, I., and Panet, A., Synchronized infection of cell cultures by magnetically controlled virus, *J Virol*, 79, 622–625, 2005.
12. Isalan, M. et al., Localized transfection on arrays of magnetic beads coated with PCR products, *Nat Methods*, 2, 113–118, 2005.
13. Kadota, S. et al., Enhancing of measles virus infection by magnetofection, *J Virol Methods*, 128, 61–66, 2005.
14. Morishita, N. et al., Magnetic nanoparticles with surface modification enhanced gene delivery of HVJ-E vector, *Biochem Biophys Res Commun*, 334, 1121–1126, 2005.
15. Plank, C. et al., Magnetofection: Enhancement and localization of gene delivery with magnetic particles under the influence of a magnetic field, *J Gene Med*, 2(Suppl.), S24, 2000.
16. Plank, C. et al., Enhancing and targeting nucleic acid delivery by magnetic force, *Expert Opin Biol Ther*, 3, 745–758, 2003.
17. Luo, D. and Saltzman, W.M., Enhancement of transfection by physical concentration of DNA at the cell surface, *Nat Biotechnol*, 18, 893–895, 2000.
18. Graham, F.L. and van der Eb, A.J., Transformation of rat cells by DNA of human adenovirus 5, *Virology*, 54, 536–539, 1973.
19. Bunnell, B.A. et al., High-efficiency retroviral-mediated gene transfer into human and nonhuman primate peripheral blood lymphocytes, *Proc Natl Acad Sci U S A*, 92, 7739–7743, 1995.
20. Boussif, O., Zanta, M.A., and Behr, J.P., Optimized galenics improve in vitro gene transfer with cationic molecules up to 1000-fold, *Gene Ther*, 3, 1074–1080, 1996.
21. O'Doherty, U., Swiggard, W.J., and Malim, M.H., Human immunodeficiency virus type 1 spinoculation enhances infection through virus binding, *J Virol*, 74, 10074–10080, 2000.
22. Huth, S. et al., Insights into the mechanism of magnetofection using PEI-based magnetofectins for gene transfer, *J Gene Med*, 6, 923–936, 2004.
23. Chuck, A.S., Clarke, M.F., and Palsson, B.O., Retroviral infection is limited by Brownian motion, *Hum Gene Ther*, 7, 1527–1534, 1996.
24. Alksne, J.F. and Fingerhut, A.G., Magnetically controlled metallic thrombosis of intracranial aneurysms. A preliminary report, *Bull Los Angeles Neurol Soc*, 30, 153–155, 1965.
25. Fingerhut, A.G. and Alksne, J.F., Thrombosis of intracranial aneurysms. An experimental approach utilizing magnetically controlled iron particles, *Radiology*, 86, 342–343, 1966.
26. Alksne, J.F., Fingerhut, A.G., and Rand, R.W., Magnetically controlled focal intravascular thrombosis in dogs, *J Neurosurg*, 25, 516–525, 1966.
27. Meyers, P.H. et al., Pathologic studies following magnetic control of metallic iron particles in the lymphatic and vascular system of dogs as a contrast and isotopic agent, *Am J Roentgenol Radium Ther Nucl Med*, 96, 913–921, 1966.
28. Widder, K.J., Senyel, A.E., and Scarpelli, G.D., Magnetic microspheres: A model system of site specific drug delivery in vivo, *Proc Soc Exp Biol Med*, 158, 141–146, 1978.
29. Fahlvik, A.K., Klaveness, J., and Stark, D.D., Iron oxides as MR imaging contrast agents, *J Magn Reson Imaging*, 3, 187–194., 1993.
30. Carlin, R.L., *Magnetochemistry*, Springer, Heidelberg, 1986.
31. Weiss, A. and Witte, H., *Magnetochemie: Grundlagen und Anwendungen*, Wiley/VCH, Weinheim, 1997.
32. Senyei, A.E. et al., In vivo kinetics of magnetically targeted low-dose doxorubicin, *J Pharm Sci*, 70, 389–391, 1981.
33. Widder, K.J. et al., Tumor remission in Yoshida sarcoma-bearing rts by selective targeting of magnetic albumin microspheres containing doxorubicin, *Proc Natl Acad Sci U S A*, 78, 579–581, 1981.
34. Widder, K.J. et al., Selective targeting of magnetic albumin microspheres containing low-dose doxorubicin: Total remission in Yoshida sarcoma-bearing rats, *Eur J Cancer Clin Oncol*, 19, 135–139, 1983.
35. Gupta, P.K. and Hung, C.T., Comparative disposition of adriamycin delivered via magnetic albumin microspheres in presence and absence of magnetic field in rats, *Life Sci*, 46, 471–479, 1990.
36. Gupta, P.K. and Hung, C.T., Effect of carrier dose on the multiple tissue disposition of doxorubicin hydrochloride administered via magnetic albumin microspheres in rats, *J Pharm Sci*, 78, 745–748, 1989.

37. Gupta, P.K., Hung, C.T., and Rao, N.S., Ultrastructural disposition of adriamycin-associated magnetic albumin microspheres in rats, *J Pharm Sci*, 78, 290–294, 1989.

38. Gupta, P.K. and Hung, C.T., Magnetically controlled targeted micro-carrier systems, *Life Sci*, 44, 175–186, 1989.

39. Gupta, P.K. and Hung, C.T., Targeted delivery of low dose doxorubicin hydrochloride administered via magnetic albumin microspheres in rats, *J Microencapsul*, 7, 85–94, 1990.

40. Kato, T. et al., [An approach to magnetically controlled cancer chemotherapy. I. Preparation and properties of ferromagnetic mitomycin C microcapsules (author's transl)], *Nippon Gan Chiryo Gakkai Shi*, 15, 876–880, 1980.

41. Kato, T. et al., Magnetic microcapsules for targeted delivery of anticancer drugs, *Appl Biochem Biotechnol*, 10, 199–211, 1984.

42. Lubbe, A.S. et al., Preclinical experiences with magnetic drug targeting: Tolerance and efficacy, *Cancer Res*, 56, 4694–4701, 1996.

43. Lubbe, A.S. et al., Clinical experiences with magnetic drug targeting: A phase I study with 4'-epidoxorubicin in 14 patients with advanced solid tumors, *Cancer Res*, 56, 4686–4693, 1996.

44. Lübbe, A.S. and Bergemann, C., *Selected Preclinical and First Clinical Experiences with Magnetically Targeted 4-Epidoxorubicin in Patients with Advanced Solid Tumors*, Plenum Press, New York, London, 1997.

45. Lemke, A.J. et al., MRI after magnetic drug targeting in patients with advanced solid malignant tumors, *Eur Radiol*, 14, 1949–1955, 2004.

46. Alexiou, C. et al., Locoregional cancer treatment with magnetic drug targeting, *Cancer Res*, 60, 6641–6648, 2000.

47. Alexiou, C. et al., Targeting cancer cells: Magnetic nanoparticles as drug carriers, *Eur Biophys J*, 35, 446–450, 2006.

48. Johnson, J. et al., The MTC technology: A platform technology for the site-specific delivery of pharmaceutical agents, *Eur Cells Mater*, 3, 12–15, 2002.

49. Rudge, S. et al., Adsorption and desorption of chemotherapeutic drugs from a magnetically targeted carrier (MTC), *J Control Release*, 74, 335–340, 2001.

50. Goodwin, S.C. et al., Single-dose toxicity study of hepatic intra-arterial infusion of doxorubicin coupled to a novel magnetically targeted drug carrier, *Toxicol Sci*, 60, 177–183, 2001.

51. Lubbe, A.S., Alexiou, C., and Bergemann, C., Clinical applications of magnetic drug targeting, *J Surg Res*, 95, 200–206, 2001.

52. Widder, K.J. et al., Selective targeting of magnetic albumin microspheres to the Yoshida sarcoma: Ultrastructural evaluation of microsphere disposition, *Eur J Cancer Clin Oncol*, 19, 141–147, 1983.

53. Goodwin, S.C. et al., Targeting and retention of magnetic targeted carriers (MTCs) enhancing intra-arterial chemotherapy, *J Magn Magn Mater*, 194, 132–139, 1999.

54. Mah, C. et al., Microsphere-mediated delivery of recombinant AAV vectors in vitro and in vivo, *Mol Ther*, 1, S239, 2000.

55. Plank, C. et al., Magnetofection: Enhancing and targeting gene delivery by magnetic force, *Eur Cells Mater*, 3, 79–80, 2002.

56. Kamau, S.W. et al., Enhancement of the efficiency of non-viral gene delivery by application of pulsed magnetic field, *Nucleic Acids Res*, 34, e40, 2006.

57. Cai, D. et al., Highly efficient molecular delivery into mammalian cells using carbon nanotube spearing, *Nat Methods*, 2, 449–454, 2005.

58. Mykhaylyk, O. et al., Generation of magnetic non-viral gene transfer agents and magnetofection in vitro, *Nat Protocols*, Nat. Protocols, 2, 2391–2411, 2007.

59. Cornell, R.M. and Schwertmann, U., *The Iron Oxides*, Wiley-VCh, Weinheim, 2003.

60. Schwertmann, U. and Cornell, R.M., *Iron Oxides in the Laboratory*, Wiley-VCh, Weinheim, 2000.

61. Hyeon, T., Chemical synthesis of magnetic nanoparticles, *Chem Commun (Camb)*, 927–934, 2003.

62. Park, J. et al., Ultra-large-scale syntheses of monodisperse nanocrystals, *Nat Mater*, 3, 891–895, 2004.

63. Mykhaylyk, O. et al., Magnetic nanoparticle formulations for DNA and siRNA delivery, *J Magn Magn Mater*, 311, 275–281, 2007.

64. Bergemann, C., Magnetic particle for transport of diagnostic or therapeutic agent, German Patent appl. DE19624426, 1998.

65. Pilgrimm, H., Superparamagnetic particles, process for producing the same and their use, United States Patent 5,916,539, 1999.

66. Chorny, M. et al., Magnetically driven plasmid DNA delivery with biodegradable polymeric nanoparticles, *FASEB J*, 21, 2510–2519, 2007.

67. Weissleder, R. et al., Superparamagnetic iron oxide: Pharmacokinetics and toxicity, *AJR Am J Roentgenol*, 152, 167–173, 1989.

68. Kirschvink, J.L., Kobayashi-Kirschvink, A., and Woodford, B.J., Magnetite biomineralization in the human brain, *Proc Natl Acad Sci U S A*, 89, 7683–7687, 1992.

69. Mertl, M., Magnetic cells: Stuff of legend? *Science*, 283, 775, 1999.

70. Diebel, C.E. et al., Magnetite defines a vertebrate magnetoreceptor, *Nature*, 406, 299–302, 2000.

71. Lohmann, K.J. and Johnsen, S., The neurobiology of magnetoreception in vertebrate animals, *Trends Neurosci*, 23, 153–159, 2000.

72. Beason, R., Behavioural evidence for the use of magnetic material in magnetoreception by a migratory bird, *J Exp Biol*, 198, 141–146, 1995.

73. Mikhaylova, M. et al., Superparamagnetism of magnetite nanoparticles: Dependence on surface modification, *Langmuir*, 20, 2472–2477, 2004.

74. Prozorov, R. et al., Magnetic irreversibility and relaxation in assembly of ferromagnetic nanoparticles, *Phys Rev B*, 59, 6956 LP–6965, 1999.

75. Hiemenz, P., *Principles of Colloid and Surface Chemistry*, 2nd ed, Marcel Dekker, Inc., New York, 1986.

76. Plank, C. et al., The magnetofection method: Using magnetic force to enhance gene delivery, *Biol Chem*, 384, 737–747, 2003.

77. Ogris, M. et al., PEGylated DNA/transferrin-PEI complexes: Reduced interaction with blood components, extended circulation in blood and potential for systemic gene delivery, *Gene Ther*, 6, 595–605, 1999.

78. Iwata, A. et al., Virus concentration using sulfonated magnetic beads to improve sensitivity in nucleic acid amplification tests, *Biol Pharm Bull*, 26, 1065–1069, 2003.

79. Nesbeth, D. et al., Metabolic biotinylation of lentiviral pseudotypes for scalable paramagnetic microparticle-dependent manipulation, *Mol Ther*, 13, 814–822, 2006.

80. Cao, Z.G. et al., Preparation and feasibility of superparamagnetic dextran iron oxide nanoparticles as gene carrier, *Ai Zheng*, 23, 1105–1109, 2004.

81. Zborowski, M. et al., Continuous cell separation using novel magnetic quadrupole flow sorter, *J Magn Magn Mater*, 194, 224–230, 1999.

82. Krotz, F. et al., Magnetofection—a highly efficient tool for antisense oligonucleotide delivery in vitro and in vivo, *Mol Ther*, 7, 700–710, 2003.

83. Schillinger, U. et al., Advances in magnetofection—magnetically guided nucleic acid delivery, *J Magn Magn Mater*, 293, 501–508, 2005.

84. Alexiou, C. et al., A high field gradient magnet for magnetic drug targeting, *IEEE Trans Appl Superconduct*, 16, 1527–1530, 2006.

85. Gleich, B. et al., Design and evaluation of magnetic fields for nanoparticle drug targeting in cancer, *IEEE Transactions Nanotechnol*, 6, 164–170, 2007.

86. Gersting, S.W. et al., Gene delivery to respiratory epithelial cells by magnetofection, *J Gene Med*, 6, 913–922, 2004.

87. Morishita, N. et al., Magnetic nanoparticles with surface modification enhanced gene delivery of HVJ-E vector, *Biochem Biophys Res Commun*, 334, 1121–1126, 2005.

88. Krotz, F. et al., Magnetofection potentiates gene delivery to cultured endothelial cells, *J Vasc Res*, 40, 425–434, 2003.

89. Takeda, S. et al., Novel drug delivery system by surface modified magnetic nanoparticles, *J Nanosci Nanotechnol*, 6, 3269–3276, 2006.

90. Krotz, F. et al., The tyrosine phosphatase, SHP-1, is a negative regulator of endothelial superoxide formation, *J Am Coll Cardiol*, 45, 1700–1706, 2005.

91. Smith, C., Sharpening the tools of RNA interference, *Nat Methods*, 3, 475–486, 2006.

92. Cao, Z.G. et al., Effects of small interfering RNA magnetic nanoparticles combination with external magnetic fields on survivin gene expression of bladder cancer cells and apoptosis, *Zhonghua Wai Ke Za Zhi*, 44, 1248–1251, 2006.

93. Scherer, F., Establishment of magnetofection—a novel method using superparamagnetic nanoparticles and magnetic force to enhance and to target nucleic acid delivery. Ph.D. thesis, Ludwig–Maximilians Universität, Munich, Germany, 2006.

94. Mok, H. et al., Evaluation of polyethylene glycol modification of first-generation and helper-dependent adenoviral vectors to reduce innate immune responses, *Mol Ther*, 11, 66–79, 2005.

95. Tai, M.F. et al., Generation of magnetic retroviral vectors with magnetic nanoparticles, *Rev Adv Mater Sci*, 5, 319–323, 2003.

96. Minami, R. et al., RCAS1 induced by HIV-Tat is involved in the apoptosis of HIV-1 infected and uninfected CD4+ T cells, *Cell Immunol*, 243, 41–47, 2006.

97. Romero, D.G. et al., Angiotensin II-mediated protein kinase D activation stimulates aldosterone and cortisol secretion in H295R human adrenocortical cells, *Endocrinology*, 147, 6046–6055, 2006.

98. Romero, D.G. et al., Disabled-2 is expressed in adrenal zona glomerulosa and is involved in aldosterone secretion, *Endocrinology*, 148, 2644–2652, 2007.

99. Romero, D.G. et al., Adrenal transcription regulatory genes modulated by angiotensin II and their role in steroidogenesis, *Physiol Genomics*, 30, 26–34, 2007.

100. Wei, W., Xu, C., and Wu, H., Magnetic iron oxide nanoparticles mediated gene therapy for breast cancer—an in vitro study, *J Huazhong Univ Sci Technolog Med Sci*, 26, 728–730, 2006.

101. Wang, J. et al., Lewis X oligosaccharides targeting to DC-SIGN enhanced antigen-specific immune response, *Immunology*, 121, 174–182, 2007.

102. Pasder, O. et al., Downregulation of Fer induces PP1 activation and cell-cycle arrest in malignant cells, *Oncogene*, 25, 4194–4206, 2006.

103. Mykhaylyk, O. et al., Magnetic nanoparticle formulations for DNA and siRNA delivery, *J Magn Magn Mater*, 311, 275–281, 2007.

104. Seki, T. et al., Fused protein of deltaPKC activation loop and PDK1-interacting fragment (deltaAL-PIF) functions as a pseudosubstrate and an inhibitory molecule for PDK1 when expressed in cells, *Genes Cells*, 11, 1051–1070, 2006.

105. Basile, J.R. et al., MT1-MMP controls tumor-induced angiogenesis through the release of semaphorin 4D, *J Biol Chem*, 282, 6899–6905, 2007.

106. Basile, J.R. et al., Semaphorin 4D provides a link between axon guidance processes and tumor-induced angiogenesis, *Proc Natl Acad Sci U S A*, 103, 9017–9022, 2006.

107. Deleuze, V. et al., TAL-1/SCL and its partners E47 and LMO2 up-regulate VE-cadherin expression in endothelial cells, *Mol Cell Biol*, 27, 2687–2697, 2007.

108. Huang, P., Senga, T., and Hamaguchi, M., A novel role of phospho-beta-catenin in microtubule regrowth at centrosome, *Oncogene*, 26, 4357–4371, 2007.

109. Fukushima, T. et al., Silencing of insulin-like growth factor-binding protein-2 in human glioblastoma cells reduces both invasiveness and expression of progression-associated gene CD24, *J Biol Chem*, 282, 18634–18644, 2007.

110. Basile, J.R., Afkhami, T., and Gutkind, J.S., Semaphorin 4D/plexin-B1 induces endothelial cell migration through the activation of PYK2, Src, and the phosphatidylinositol 3-kinase-Akt pathway, *Mol Cell Biol*, 25, 6889–6898, 2005.

111. Kim, E.A. et al., Phosphorylation and transactivation of Pax6 by homeodomain-interacting protein kinase 2, *J Biol Chem*, 281, 7489–7497, 2006.

112. Mizutani, T. et al., Mechanisms of establishment of persistent SARS-CoV-infected cells, *Biochem Biophys Res Commun*, 347, 261–265, 2006.

113. Schmidt, C.M. et al., Efficient downregulation of ENaC activity by synthetic siRNAs, *Mol Ther*, 13, S267, 2006.

114. Sapet, C. et al., Thrombin-induced endothelial microparticle generation: Identification of a novel pathway involving ROCK-II activation by caspase-2, *Blood*, 108, 1868–1876, 2006.

115. Kaneko, M. et al., Activity of a novel PDGF beta-receptor enhancer during the cell cycle and upon differentiation of neuroblastoma, *Exp Cell Res*, 312, 2028–2039, 2006.

116. Baer, K. et al., PICK1 interacts with [alpha]7 neuronal nicotinic acetylcholine receptors and controls their clustering, *Mol Cell Neurosci*, 35, 339–355, 2007.

117. Thomas, J.A., Ott, D.E., and Gorelick, R.J., Efficiency of human immunodeficiency virus type 1 postentry infection processes: Evidence against disproportionate numbers of defective virions, *J Virol*, 81, 4367–4370, 2007.

118. Pinto, M.P. et al., The import competence of a peroxisomal membrane protein is determined by Pex19p before the docking step, *J Biol Chem*, 281, 34492–34502, 2006.

119. Hollander, K., Bar-Chen, M., and Efrat, S., Baculovirus p35 increases pancreatic beta-cell resistance to apoptosis, *Biochem Biophys Res Commun*, 332, 550–556, 2005.

120. Varro, A. et al., Increased gastric expression of MMP-7 in hypergastrinemia and significance for epithelial-mesenchymal signaling, *Am J Physiol Gastrointest Liver Physiol*, 292, G1133–G1140, 2007.

121. Steele, I. et al., Helicobacter and gastrin stimulate Reg1 expression in gastric epithelial cells through distinct promoter elements, *Am J Physiol Gastrointest Liver Physiol*, 293, G347–G354, 2007.

122. Ambasta, R.K. et al., Direct interaction of the novel Nox proteins with p22phox is required for the formation of a functionally active NADPH oxidase, *J Biol Chem*, 279, 45935–45941, 2004.

123. Recklies, A.D. et al., Inflammatory cytokines induce production of CHI3L1 by articular chondrocytes, *J Biol Chem*, 280, 41213–41221, 2005.

124. Doshida, M. et al., Raloxifene increases proliferation and up-regulates telomerase activity in human umbilical vein endothelial cells, *J Biol Chem*, 281, 24270–24278, 2006.

125. Kaur, S. et al., Robo4 signaling in endothelial cells implies attraction guidance mechanisms, *J Biol Chem*, 281, 11347–11356, 2006.

126. Nagata, D. et al., Molecular mechanism of the inhibitory effect of aldosterone on endothelial NO synthase activity, *Hypertension*, 48, 165–171, 2006.

127. Fransen, M. et al., Analysis of human Pex19p's domain structure by pentapeptide scanning mutagenesis, *J Mol Biol*, 346, 1275–1286, 2005.

128. Couchoux, H. et al., Loss of caveolin-3 induced by the dystrophy-associated P104L mutation impairs L-type calcium channel function in mouse skeletal muscle cells, *J Physiol*, 580, 745–754, 2007.

129. Mukai, H.Y. et al., Transgene insertion in proximity to the c-myb gene disrupts erythroid-megakaryocytic lineage bifurcation, *Mol Cell Biol*, 26, 7953–7965, 2006.

130. McCaig, C. et al., The role of matrix metalloproteinase-7 in redefining the gastric microenvironment in response to Helicobacter pylori, *Gastroenterology*, 130, 1754–1763, 2006.

131. Chudotvorova, I. et al., Early expression of KCC2 in rat hippocampal cultures augments expression of functional GABA synapses, *J Physiol*, 566, 671–679, 2005.

132. Lardi-Studler, B. et al., Vertebrate-specific sequences in the gephyrin E-domain regulate cytosolic aggregation and postsynaptic clustering, *J Cell Sci*, 120, 1371–1382, 2007.

133. Uchida, Y. et al., Semaphorin3A signalling is mediated via sequential Cdk5 and GSK3beta phosphorylation of CRMP2: Implication of common phosphorylating mechanism underlying axon guidance and Alzheimer's disease, *Genes Cells*, 10, 165–179, 2005.

134. de Lartigue, G. et al., Cocaine- and amphetamine-regulated transcript: Stimulation of expression in rat vagal afferent neurons by cholecystokinin and suppression by ghrelin, *J Neurosci*, 27, 2876–2882, 2007.

135. Sacha, J.B. et al., Gag-specific CD8+ T lymphocytes recognize infected cells before AIDS-virus integration and viral protein expression, *J Immunol*, 178, 2746–2754, 2007.

136. Xenariou, S. et al., Using magnetic forces to enhance nonviral gene transfer to airway epithelium in vivo, *Gene Ther*, 13, 1545–1552, 2006.

137. Hellwig, N. et al., Ultrasound-enhanced microbubble-magnetofection: A new approach for targeted delivery of nucleotides in vivo, *J Vasc Res*, 42, 86–87, 2005.

138. Galuppo, L.D. et al., Gene expression in synovial membrane cells after intraarticular delivery of plasmid-linked superparamagnetic iron oxide particles—a preliminary study in sheep, *J Nanosci Nanotechnol*, 6, 2841–2852, 2006.

139. Hellstern, D. et al., Systemic distribution and elimination of plain and with Cy3.5 functionalized poly(vinyl alcohol) coated superparamagnetic maghemite nanoparticles after intraarticular injection in sheep in vivo, *J Nanosci Nanotechnol*, 6, 3261–3268, 2006.

140. Schulze, K. et al., Uptake and biocompatibility of functionalized poly(vinylalcohol) coated superparamagnetic maghemite nanoparticles by synoviocytes in vitro, *J Nanosci Nanotechnol*, 6, 2829–2840, 2006.

141. Pagliara, P. et al., Differentiation of monocytic U937 cells under static magnetic field exposure, *Eur J Histochem*, 49, 75–86, 2005.

142. Rosen, A.D., Mechanism of action of moderate-intensity static magnetic fields on biological systems, *Cell Biochem Biophys*, 39, 163–173, 2003.

143. Flipo, D. et al., Increased apoptosis, changes in intracellular Ca^{2+}, and functional alterations in lymphocytes and macrophages after in vitro exposure to static magnetic field, *J Toxicol Environ Health A*, 54, 63–76, 1998.

144. Sabo, J. et al., Effects of static magnetic field on human leukemic cell line HL-60, *Bioelectrochem*, 56, 227–231, 2002.

145. Goodman, R. and Blank, M., Insights into electromagnetic interaction mechanisms, *J Cell Physiol*, 192, 16–22, 2002.

146. Hiraoka, M. et al., Induction of c-fos gene expression by exposure to a static magnetic field in HeLaS3 cells, *Cancer Res*, 52, 6522–6524, 1992.

147. Hirose, H. et al., Static magnetic field with a strong magnetic field gradient (41.7 T/m) induces c-Jun expression in HL-60 cells, *In Vitro Cell Dev Biol Anim*, 39, 348–352, 2003.

148. Potenza, L. et al., Effects of a static magnetic field on cell growth and gene expression in *Escherichia coli*, *Mutat Res*, 561, 53–62, 2004.

149. Babincova, M., Babinec, P., and Bergemann, C., High-gradient magnetic capture of ferrofluids: Implications for drug targeting and tumor embolization, *Z Naturforsch [C]*, 56, 909–911, 2001.

150. Babincova, M. et al., Site-specific in vivo targeting of magnetoliposomes using externally applied magnetic field, *Z Naturforsch [C]*, 55, 278–281, 2000.

151. Babincova, M. et al., AC-magnetic field controlled drug release from magnetoliposomes: Design of a method for site-specific chemotherapy, *Bioelectrochemistry*, 55, 17–19, 2002.

152. Klibanov, A.L., Ultrasound contrast agents: Development of the field and current status, *Top Curr Chem*, 222, 73–106, 2002.

153. Unger, E.C. et al., Therapeutic applications of lipid-coated microbubbles, *Adv Drug Deliv Rev*, 56, 1291–314, 2004.

154. Liu, Y., Miyoshi, H., and Nakamura, M., Encapsulated ultrasound microbubbles: Therapeutic application in drug/gene delivery, *J Control Release*, 114, 89–99, 2006.

155. van Wamel, A. et al., Vibrating microbubbles poking individual cells: Drug transfer into cells via sonoporation, *J Control Release*, 112, 149–155, 2006.

156. Chen, S. et al., Optimization of ultrasound parameters for cardiac gene delivery of adenoviral or plasmid deoxyribonucleic acid by ultrasound-targeted microbubble destruction, *J Am Coll Cardiol*, 42, 301–308, 2003.

157. Skyba, D.M. et al., Direct in vivo visualization of intravascular destruction of microbubbles by ultrasound and its local effects on tissue, *Circulation*, 98, 290–293, 1998.

158. Price, R.J. et al., Delivery of colloidal particles and red blood cells to tissue through microvessel ruptures created by targeted microbubble destruction with ultrasound, *Circulation*, 98, 1264–1267, 1998.

159. Hynynen, K. et al., Noninvasive MR imaging-guided focal opening of the blood-brain barrier in rabbits, *Radiology*, 220, 640–646, 2001.

160. Bekeredjian, R. et al., Ultrasound-targeted microbubble destruction can repeatedly direct highly specific plasmid expression to the heart, *Circulation*, 108, 1022–1026, 2003.

161. Danialou, G. et al., Ultrasound increases plasmid-mediated gene transfer to dystrophic muscles without collateral damage, *Mol Ther*, 6, 687–693, 2002.

162. Endoh, M. et al., Fetal gene transfer by intrauterine injection with microbubble-enhanced ultrasound, *Mol Ther*, 5, 501–508, 2002.

163. Frenkel, P.A. et al., DNA-loaded albumin microbubbles enhance ultrasound-mediated transfection in vitro, *Ultrasound Med Biol*, 28, 817–822, 2002.

164. Lawrie, A. et al., Microbubble-enhanced ultrasound for vascular gene delivery, *Gene Ther*, 7, 2023–2027, 2000.

165. Miller, D.L., Pislaru, S.V., and Greenleaf, J.E., Sonoporation: Mechanical DNA delivery by ultrasonic cavitation, *Somat Cell Mol Genet*, 27, 115–134, 2002.

166. Shohet, R.V. et al., Echocardiographic destruction of albumin microbubbles directs gene delivery to the myocardium, *Circulation*, 101, 2554–2556, 2000.

167. Teupe, C. et al., Vascular gene transfer of phosphomimetic endothelial nitric oxide synthase (S1177D) using ultrasound-enhanced destruction of plasmid-loaded microbubbles improves vasoreactivity, *Circulation*, 105, 1104–1109, 2002.

168. Tsutsui, J.M., Xie, F., and Porter, R.T., The use of microbubbles to target drug delivery, *Cardiovasc Ultrasound*, 2, 23, 2004.

169. Vannan, M. et al., Ultrasound-mediated transfection of canine myocardium by intravenous administration of cationic microbubble-linked plasmid DNA, *J Am Soc Echocardiogr*, 15, 214–218, 2002.

170. Lentacker, I. et al., Ultrasound-responsive polymer-coated microbubbles that bind and protect DNA, *Langmuir*, 22, 7273–7278, 2006.

171. Soetanto, K. and Watarai, H., Development of magnetic microbubbles for drug delivery system (DDS), *Jpn J Appl Phys*, 39, 3230–3232, 2000.

172. Hellwig, N. et al., Ultrasound-enhanced microbubble-magnetofection: A new approach for efficient targeted delivery of nucleotides in vivo, *J Vasc Res*, 43, 549–549, 2006.

173. Plank, C. et al., Localized nucleic acid delivery using magnetic nanoparticles, *Eur Cell Mater*, 10, 8, 2005.

174. Vlaskou, D. et al., Magnetic microbubbles: New carriers for localized gene and drug delivery, *Mol Ther*, 13, S290, 2006.

18 Biolistic and Other Needle-Free Delivery Systems

Mark A.F. Kendall

CONTENTS

18.1 INTRODUCTION

Immunotherapeutics (e.g., vaccines, allergens) are most commonly administered using a needle and syringe, a method first invented in 1853. The needle and syringe is effective, but unpopular, and creates a risk of iatrogenic disease from needle-stick injury or needle reuse as a consequence of the billions of administrations each year. Further, the needle and syringe does not deliver the vaccine ingredients optimally to the antigen-presenting cells (APCs), which alone can respond to the combination of antigen and adjuvant (innate immune stimulus) that makes a successful vaccine.

The provision of safe and efficient routes of delivery of immunotherapeutics to the immunologically sensitive dendritic cells in the skin (and mucosa) has the potential to enhance strategies in the treatment of major disease. Examples of these include DNA vaccines and the immunotherapy of allergies. The application of physical methods to achieving this goal presents unique engineering challenges in the physical transport of immunotherapeutic biomolecules (e.g., polynucleotides) to these cells.

In this chapter, the physiology, immunology, and material properties of the skin are examined in the context of the physical cell-targeting requirements of the viable epidermis. Selected cell-targeting technologies engineered to meet these needs are briefly presented. The operating principles of these approaches are described, together with a discussion of their effectiveness for the noninvasive targeting of viable epidermis cells and the DNA vaccination against major diseases.

The focus then moves to one of these methods, called biolistics, that ballistically delivers millions of microparticles coated with biomolecules to outer skin layers. The engineering of these devices is presented, beginning with earlier prototypes before examining a more advanced system configured for clinical use. Then, follows a theoretical and experimental analysis of the ballistic microparticle impact process, including the examination of induced cell death. Finally, the results of applying this technology to key human clinical trials are presented.

18.2 IMPORTANCE OF TARGETING SKIN AND MUCOSAL CELLS

Why are outer skin cells important targets in the treatment of disease? The answer is found from a consideration of skin structure, shown schematically in Figures 18.1 and 18.3. Human skin can be subdivided into a number of layers: the outer stratum corneum (SC, 10–20 μm in depth), the viable epidermis (50–100 μm), and the dermis (1–2 mm) [1,2]. The SC is the effective physical barrier of dead cells in a "bricks

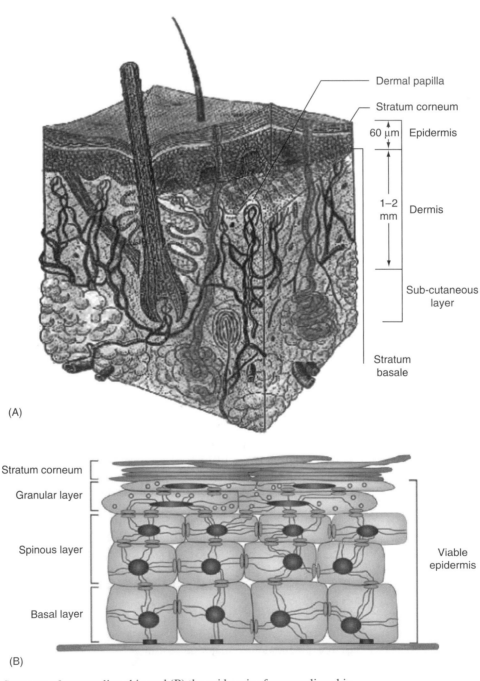

(A)

(B)

FIGURE 18.1   (A) Structure of mammalian skin and (B) the epidermis of mammalian skin

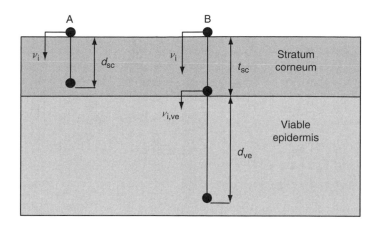

(C)

FIGURE 18.1 (continued) (C) the corresponding bilayer approximation of the epidermis used for the theoretical penetration model. Penetration case A denotes particle delivery into the SC (d_{sc}), whereas, in case B, the SC is fully breached (t_{sc}), and the final particle location is within the viable epidermis (d_{ve}). The impact velocity is v_i, while the input velocity for the viable epidermis is $v_{i,ve}$. (Adapted from Kendall, M.A.F., Rishworth, S., Carter, F.V., and Mitchell, T.J., *J. Invest. Dermatol.*, 122, 739, 2004. With permission.)

and mortar" structure [3,4]. The underlying viable epidermis is composed of cells, such as immunologically sensitive Langerhans cells, keratinocytes, stem cells, and melanocytes [2]. Unlike the dermis below, the viable epidermis lacks blood vessels and sensory nerve endings—important characteristics of a site for pain-free delivery with minimal damage.

In the viable epidermis, the skin has evolved a highly competent immunological function, with an abundance of Langerhans cells (500–1000 cells mm^{-2}) [5–7], often serving as the first line of defense against many pathogens [8]. In particular, Langerhans cells (illustrated in Figure 18.3) are extremely effective APCs, responsible for the uptake and processing of foreign materials in order to generate an effective immune response. Such cells are reported to be up to 1000-fold more effective than keratinocytes, fibroblasts, and myoblasts at eliciting a variety of immune responses [9–12]. Effective *in situ* (*in vivo*) targeting of Langerhans cells and other epidermal cells with polynucleotides or antigens will open up novel applications in disease control [9], including vaccination against major viruses/diseases, such as human immunodeficiency virus (HIV) and cancer.

18.3 ENGINEERING OF PHYSICAL APPROACHES FOR THE TARGETING OF SKIN AND MUCOSAL CELLS

Within the viable epidermis, the location of Langerhans cells—as a delivery target for immunotherapeutics—is tightly defined by:

- Vertical position at a consistent suprabasal location [13]
- Spatial distribution in the horizontal plane evenly distributed throughout the skin [14]
- Constitution of 2% of the total epidermal cell population [15] (in human skin)

How can these and other epidermal skin cells be targeted? Despite its recognized potential, the viable epidermis has only recently been viewed as a feasible cellular targeting site with the emergence of new biological and physical technologies. The challenge is the effective penetration of the SC and precise targeting of the cells of interest.

18.3.1 MECHANICAL PROPERTIES OF THE SC BARRIER

The SC is a semipermeable barrier that—owing to its variable mechanical properties—is challenging to breach, in a minimally invasive manner, to target the viable epidermal cells below. Mechanically, the SC is classified as a bioviscoelastic solid and shows highly variable properties. Obvious differences include the huge variation in thickness and composition with the skin site and the age of an individual [16]. However, there are more subtle and equally important variations in SC properties to consider when configuring targeting methods.

As one example, the SC mechanical breaking stress is strongly influenced by the ambient humidity/moisture content [17–21]—the relative humidity range from 0% to 100% results in a decrease in excised human SC breaking stress from 22.5 to 3.2 MPa [22]. Similarly, an increase in ambient temperature also results in an SC breaking stress decrease by an order of magnitude [23].

More recently, with indentation studies using small probes (diameters of 2 and 5 μm) fitted to a NANO indenter [24], we have found even more complexity and variation in key SC—and underlying viable epidermis—mechanical properties. Specifically:

- Storage modulus and mechanical breaking stress both dramatically decrease through the SC (Figure 18.2A and B).

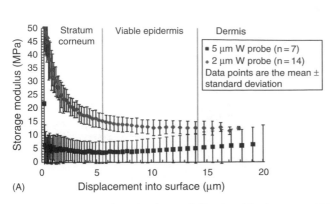

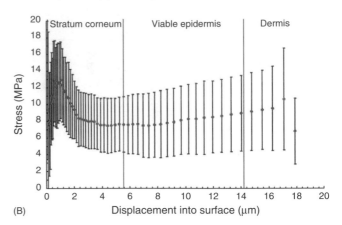

FIGURE 18.2 (See A in color insert following blank page 206.) Mechanical properties (mean ± standard deviation) as a function of displacement obtained with microprobes indented into murine ears. (A) Storage modulus with 5 and 2 μm microprobes. (B) Stress with a 2 μm microprobe. (Adapted from Kendall, M.A.F., Chong, Y., and Cock, A., *Biomaterials*, 28, 4968, 2007. With permission.)

- At a given depth within the SC and VE, decreasing the probe size significantly increases the storage modulus (Figure 18.2A).

These and other sources of variability in the SC mechanical properties present challenges in configuring approaches to breach, in a minimally invasive manner, the SC and effectively deliver biomolecules (e.g., polynucleotides, antigens or allergens) to the underlying cells.

18.3.2 BIOLOGICAL APPROACHES

Although the focus of this chapter is on physical approaches to target epidermal cells, it is also important to highlight biological approaches. A powerful biological approach to the transport

of biomolecules to epidermal (and other) cells, *in vivo*, exploits the evolved function of viruses in the transport to cells. In gene delivery, researchers have made use of genetically engineered viruses in the DNA vaccination and gene therapy of major diseases with encouraging results. However, viral gene delivery is hindered by safety concerns, a limited DNA-carrying capacity, production and packaging problems, and a high cost [25,26].

18.3.3 PHYSICAL CELL-TARGETING APPROACHES

Alternatively, many physical technologies are being developed. Potentially, they can overcome some limitations of biological approaches using needle-free mechanisms to breach the SC barrier to facilitate drug and vaccine administration directly to epidermal cells. Figure 18.3 illustrates schematically

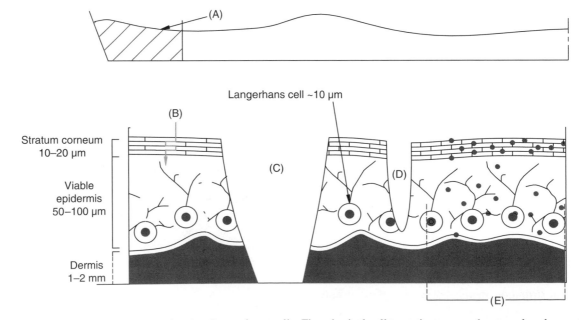

FIGURE 18.3 Cross-section of the skin showing Langerhans cells. Five physical cell-targeting approaches are also shown. (A) A half-section of a small gauge needle and syringe; (B) route of diffusion from patches; (C) penetration from a liquid jet injector; (D) a hole from a microinjector; and (E) distribution of microparticles following biolistic injection. (From Kendall, M.A.F., *Vaccine*, 24, 4651, 2006. With permission.)

key physical targeting approaches relative to the scale of typical skin and the Langerhans cell layer of interest.

18.3.3.1 Needle and Syringe

For the illustration of the most common physical delivery method, a small gauge needle and syringe is shown in half-section in Figure 18.3A. Although this approach easily breaches the SC, precise targeting of the Langerhans cell-rich viable epidermis cannot be practically achieved. Hence, the needle and syringe is used for intradermal or intramuscular injection. This inefficient, indirect targeting of dendritic cells with DNA has resulted in modest immune responses [27]. Other disadvantages of the needle and syringe include risks due to needle-stick injuries [28] and needle phobia [1].

18.3.3.2 Diffusion/Permeation Delivery

Perhaps the least invasive method of breaching the SC is by permeation through it, driven by diffusion from patches applied to the skin (Figure 18.3B) [29]. However, currently, the general view is that this mode of delivery is best suited to smaller biomolecules (<500 Da [29])—considerably smaller than oligonucleotides and antigens. This view is being challenged, with a recent study showing that very large recombinant antigens of ~1 MDa can be delivered to elicit systemic responses by diffusion from patches [30]. The transport of larger biomolecules through the SC can be further enhanced by simple approaches, including tape stripping with an adhesive tape, brushing with sandpaper [31,32], or the application of depilatory agents [26,33,34]. Amongst the more advanced technologies are electroporation [35,36], ablation by laser or heat, radiofrequency high-voltage currents [37], iontophoresis [38–40], sonophoresis, and microporation [8]. Many of these approaches remain untested for complex entities such as vaccines and immunotherapies. Permeation through the SC can also be enhanced by the coating of plasmid DNA on nanoparticles (~100 nm) for DNA vaccination [41].

18.3.3.3 Liquid Jet Injectors

Interest in using high-speed liquid jet injectors arose in the mid-twentieth century because of its needle-free approach [42]. This technique has seen a recent resurgence, with liquid delivered around the Langerhans cells in gene transfer and DNA vaccination experiments [42], and the delivery of drugs [43]. As shown in Figure 18.3C, current liquid jet injectors typically disrupt the skin in the epidermal and dermal layer. To target exclusively the viable epidermal cells, such as Langerhans cells, the challenge of more controlled delivery needs to be addressed. With the dermal disruption induced by administration, liquid jet injectors are also reported to cause pain to patients.

18.3.3.4 Microneedle Arrays/Patches

Researchers have overcome some of the disadvantages described by fabricating arrays of micrometre-scale projections to breach the SC and to deliver naked DNA to several cells in live animals [44]. Similar microprojection devices are

used to increase the permeability of drugs [45] and "conventional" protein antigen vaccines [45,46]. Figure 18.3D shows that, unlike current liquid jet injectors, these microneedles can accurately target the viable epidermis. Furthermore, they are as simple to use as patches, while overcoming the SC diffusion barrier to many molecules. Moreover, compared with both the needle and syringe and liquid jet injectors, these microneedle methods are pain-free because of epidermal targeting. By drawing upon a range of manufacturing techniques, McAllister et al. [46] have shown that these microneedle arrays can be made from a range of materials, including silicon, metal, and biodegradable polymers. This advantage makes microneedle patches a promising practical and cost-effective method of delivering oligonucleotides to epidermal cells for DNA vaccination [47].

18.3.4 Biolistics Microparticle Delivery

Currently, the most established physical method of DNA vaccination is biolistic microparticle delivery, otherwise known as gene guns (Figure 18.3E). Biolistic delivery is the focus of the remainder of this chapter.

18.4 BIOLISTICS MICROPARTICLE DELIVERY

18.4.1 Biolistics Operating Principle

In this needle-free technique, pharmaceutical or immuno-modulatory agents, formulated as particles, are accelerated in a supersonic gas jet to sufficient momentum to penetrate the skin (or mucosal) layer and to achieve a pharmacological effect.

Sanford and Klein [48] pioneered this innovation with systems designed to deliver DNA coated metal particles (of diameter of the order of 1 μm) into plant cells for genetic modification, using pistons accelerated along the barrels of adapted guns. The concept was extended to the treatment of humans with particles accelerated by entrainment in a supersonic gas flow [49]. Prototype devices embodying this concept have been shown to be effective, painless, and applicable to pharmaceutical therapies ranging from protein delivery [50] to conventional vaccines [51] and DNA vaccines [9,52,53].

Different embodiments of the concept (e.g., in Figures 18.4 and 18.6) all have a similar procedure of operation. Consider the prototype shown schematically in Figure 18.4A as one example.

Prior to operation, the gas canister is filled with helium or nitrogen to 2–6 MPa and the vaccine cassette, comprising two 20 μm diaphragms, is loaded with a powdered pharmaceutical payload of 0.5–2 mg. The pharmaceutical material is placed on the lower diaphragm surface. Operation commences when the valve in the gas canister is opened to release gas into the rupture chamber, where the pressure builds up until the two diaphragms retaining the vaccine particles sequentially burst. The rupture of the downstream diaphragm initiates a shock which propagates down the converging–diverging nozzle. The ensuing expansion of stored gas results in a short-duration flow in which the drug particles are entrained and accelerated

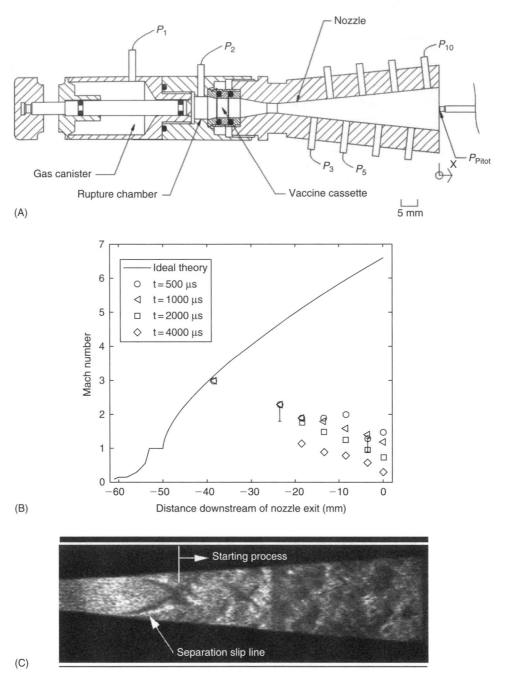

FIGURE 18.4 (A) Simplified prototype vaccine device instrumented for Pitot and static pressure measurements. The static pressure transducers are labeled P_1–P_{10}. (B) Experimental and ideal axial Mach number within the conical nozzle of investigation. The profiles are provided after the starting process. (C) Sample Schlieren image within the nozzle. (From Kendall, M.A.F., Quinlan, N.J., Thorpe, S.J., Ainsworth, R.W., and Bellhouse, B.J., *Exp. Fluids*, 37, 128, 2004. With permission.)

through the device. After leaving the device, particles impact on the skin and penetrate to the epidermis to deliver a pharmacological effect.

18.4.2 ENGINEERING OF HAND-HELD BIOLISTIC DEVICES FOR CLINICAL USE

Biolistic delivery of immunotherapeutics is an application of transonic flow technology that is otherwise applied to aerospace applications. In this section we introduce prototype devices and discuss the key engineering challenges in applying this aerospace technology to clinical biomedical applications. Key parameters used to guide the engineering of biolistic devices are

1. Nominally uniform, controlled, and quantified microparticle velocity and spatial distribution impacting the tissue target. Further, the impact momentum

is to be within the range needed for delivery to particular locations (e.g., the Langerhans cells for DNA vaccines).

2. Sufficient "footprint" on the tissue to deliver sufficient payload and target the appropriate number of cells.
3. Noise levels within the user guidelines, for both the operator and patient.
4. Device is to be hand-held.
5. For long-term stability, the pharmaceutical is to be stored within a sealed environment.
6. Device is to be produced from biocompatible materials.
7. Devices, manufactured in large numbers, are to be cost competitive with other relevant technologies.

18.4.2.1 Earlier Generation Systems

Early attempts to address these parameters were with a prototype device family generated from empirical studies. A schematic of one of these devices, using a convergent–divergent nozzle design, is shown in Figure 18.4A [54]. Working with these devices, the challenge was to establish the gas-particle dynamics behavior of the systems. A significant research programme was directed at this goal.

A suite of methods were used to characterize the gas and particle dynamics of these systems. Quinlan et al. [55] performed static pressure measurements to interrogate the gas flow, together with time-integrated Doppler global velocimetry (DGV) measurements of drug particle velocity. These measurements were very useful, but gave an incomplete description of the predominantly unsteady flow in the device.

In subsequent, broader studies, the transient gas and particle flow within the device were interrogated with Pitot–static pressure measurement (as instrumented in Figure 18.4A), together with Schlieren imaging, and time resolved DGV [55] and computational fluid dynamics (CFD) modeling [56]. The findings of this study are summarized with measured axial Mach number profiles through the nozzle (Figure 18.4B) and a single Schlieren image (Figure 18.4C).

The axial profiles of Mach number at various times after termination of the starting process (based on total-static and Pitot–static pressure measurements) are compared with the theoretical Mach number profile for steady isentropic quasi-one-dimensional supersonic flow (with the assumption of a choked throat) in Figure 18.4B. Pitot and static pressure measurements (P_2 and P_3, respectively, in Figure 18.4A) suggest that $500\,\mu s$ after diaphragm rupture, the flow $38.5\,mm$ upstream of the nozzle exit is supersonic and close to the isentropic ideal. Further downstream, however, the overexpanded nozzle flow is processed through an oblique shock system which induces flow separation. Consequently, the experimentally determined Mach number (determined from Pitot and static pressure) gradually falls from between 2 and 2.5 ($23.5\,mm$ upstream of the exit plane) to 1.5 at the exit plane. The Mach number in the downstream region of the nozzle decays with time as the shock system moves upstream.

Sequences of Schlieren images such as the sample shown in Figure 18.4C ($t = 132\,\mu s$) reveal the structure of the evolving flowfield with greater detail and clarity [54].

The oblique shocks visible have evolved to form at least three oblique shock cells that have interacted with the boundary layer and separated the nozzle flow.

DGV images show particles were entrained in the nozzle starting process and the separated nozzle flow—regimes with large variations in gas density and velocity—giving rise to large variations in particle velocity (200–$800\,ms^{-1}$) and spatial distributions [54]. Clearly, parameter (1) from above is not satisfied with this geometry.

Furthermore, the gas flow throughout much of the nozzle (Figure 18.4C) is highly sensitive to variations in the nozzle boundary condition imposed by inserting a tissue target and a silencer—because the boundary condition information can be communicated upstream. This means that this silenced device applied to the tissue target would have considerably lower and more variable impact velocities. In some cases, it is questionable whether these subsonic nozzle flow silenced devices would deliver particles with a sufficient momentum to reach the target tissue layer.

18.4.2.2 Improved Devices for Clinical Use

To overcome the large variations in particle impact conditions in earlier described devices—and meet the other important criteria of a practical clinical system (outlined above)—a next generation biolistic device, called the contoured shock tube (CST), was conceived and developed [57–62]. The devices operate with the principle of delivering a payload of microparticles to the skin with a narrow range of velocities, by entraining the drug payload in a quasi one dimensional, steady supersonic flowfield.

In experiments with simple prototype CST devices, it was shown that the desired gas flow was achieved repeatedly [57]. Importantly, further work with particle payloads measured a variation in free-jet particle velocity of ±4% [57]. In this research, measurements were made with particle image velocimetry (PIV). A sample PIV result is shown in Figure 18.5. Similar PIV images at a range of times after diaphragm rupture were processed to extract the mean centerline axial particle velocity profiles. Importantly, these PIV measurements show particle payloads do achieve near uniform exit-plane velocities at the device exit over the time interval studied. This CST device prototype was a benchtop prototype, not addressing the key criteria for a practical, hand-held clinical immunotherapeutic system.

An embodiment of the CST configured to meet these clinical needs is shown in Figure 18.6, with the key components labeled. The device was fabricated from biocompatible materials and the device wall thickness was kept relatively constant to meet autoclave sterilization requirements.

To reduce the overall system length, the bottle reservoir (which operates by an actuation pin) is located within the driver annulus. A challenge of this coaxial arrangement was to maintain integrity of transonic gas flow within the driver

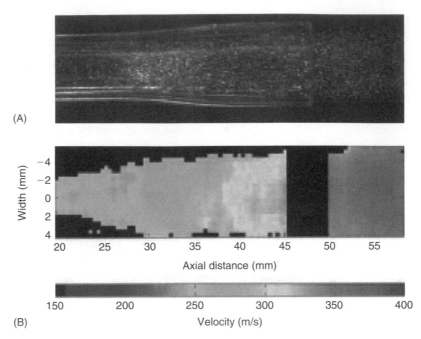

FIGURE 18.5 (See color insert following blank page 206.)   (A) Raw image and (B) derived PIV velocity map of the instantaneous particle flowfield of a CST prototype, taken 225 μs after diaphragm rupture. The payload was 2.2 mg of 39 μm diameter polystyrene spheres. (Adapted from Kendall, M.A.F., *Shock Waves J.*, 12, 22, 2002. With permission.)

initiated after diaphragm rupture. This challenge was met by carefully contouring the driver and obstacle of the mounting arrangement [60]. Possible fragments from opening of the aluminium gas bottle are contained by a sealed filter at the bottle head.

The powdered pharmaceutical is enclosed and sealed by a cassette created by the inclusion of additional diaphragms upstream of the particle payload. In this case, the cassette houses two jets designed to mix the particles into a cloud, hence reducing the dependence on the initial particle location [57,59]. Therefore, a nominally uniform spatial distribution of particles is released within the quasi-steady flow through the shock tube and nozzle. Repeated *in vitro* and *in vivo* experiments show that polycarbonate diaphragm fragments do not damage the target.

Elements of the silencing system are also shown in Figure 18.6. The primary shock initiated by diaphragm rupture, reflected from the target, is identified as the main source of sound to be attenuated. This shock is collapsed into compression waves by a series of compressions–expansions induced by an array of orifices and saw-tooth baffles, resulting in appropriate sound levels for the operator and patient.

The device lift-off force is also to be well within user constraints. A peak lift-off force of 13 N is achieved by the careful selection of endbell contact diameter, silencer volume, flowrates through the reservoir, and silencer geometry. This peak was for only a very short time within a gas flow lasting only ~200 μs (with a helium driver gas). The point of contact between the device and skin target was selected to maintain a target seal and to minimize the lift-off force, while not adversely

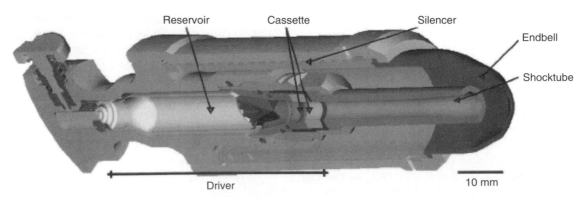

FIGURE 18.6   CST prototype configured for clinical biolistic delivery. (Adapted from Kendall, M.A.F., *Shock Waves J.*, 12, 22, 2002. With permission.)

affecting the impact velocities of the particles The effect of silencing was also minimized by maintaining a supersonic gas flow transporting particles through the nozzle—so changes in the nozzle boundary condition were not fed upstream.

The range of impact conditions for the CST platform was achieved by the selection of appropriate helium/nitrogen mixtures within the gas bottle driver/driven area ratios.

18.4.3 BALLISTICS MICROPARTICLE DELIVERY TO SKIN

We now examine delivery of microparticles from these quantified and highly controlled biolistics devices, impacting the skin. Shown in Figure 18.1, this skin is a highly variable, bio-viscoelastic material.

The described biolistic devices have been applied to a range of tissue targets for immunotherapeutic applications, including the skin of rodents [65], pigs [67], dogs [68], and humans [66]. Typically, two classes of particles are delivered to the tissue. In the powder delivery of conventional vaccines and allergens for allergy immunotherapy, particles of 10–20 μm in radius are delivered to the epidermis of the skin to achieve a therapeutic effect [65]. DNA vaccination, however, is an application in which smaller (radius 0.5–2 μm) gold particles coated with a DNA construct are targeted at the nuclei of key immunologically sensitive cells within the epidermis [53].

18.4.3.1 Theoretical Model for Ballistic Impact into Skin

In these particle impact studies, the mechanisms of particle impact were explored with a theoretical model based on a representation first proposed by Dehn [63]. The model attributes the particle resistive force (D) to plastic deformation and target inertia

$$D = \frac{1}{2} \rho_t A v^2 + 3 A \sigma_y \qquad (18.1)$$

where

ρ_t and σ_y are the density and yield stress of the target
A is the particle cross-sectional area
v is the particle velocity

The yield stress (sometimes known as the breaking stress) is the stress at which the tissue begins to exhibit plastic behavior. Equation 18.1 may be integrated to obtain the penetration depth as a function of particle impact and target parameters. The key parameters of the skin used in the model are summarized in Table 18.1. Note that these parameters have all been obtained at low, quasi-static strain-rates, and not the high ballistic strain-rates.

The theoretical model of particle penetration into the epidermis using Equation 18.1 in a two layer model is shown in Figure 18.1C. Equation 18.1 shows that the yield stress and density of the SC and viable epidermis are important in the ballistic delivery of particles to the epidermis.

In the case of particle delivery only to the SC (labeled "A" in Figure 18.1C), the particle depth into the SC (d_{sc}) is obtained by the integration of Equation 18.1.

$$d_{sc} = \frac{4\rho_p r_p}{3\rho_{sc}} \left\{ \ln\left(\frac{1}{2} \rho_{sc} v_i^2 + 3\sigma_{sc} \right) - \ln(3\sigma_{sc}) \right\} \qquad (18.2)$$

where the subscripts sc and p denote the SC and particle properties, respectively. Also, v_i and σ_{sc}, respectively, are the particle impact velocity and SC yield stress.

If the particle impact momentum is sufficient to breach the SC (labeled "B" in Figure 18.1C), Equation 18.2 is rearranged to obtain the velocity of the particle at the SC–viable epidermis boundary ($v_{i,ve}$), i.e.,

$$v_{i,ve} = \left\{ \left(v_i^2 + \frac{6\sigma_{sc}}{\rho_{sc}} \right) e^{\frac{-3\rho_{sc} t_{sc}}{4\rho_p r_p}} - \frac{6\sigma_{sc}}{\rho_{sc}} \right\}^{\frac{1}{2}} \qquad (18.3)$$

where t_{sc} is the thickness of the SC.

TABLE 18.1

Parameters and Assigned Values Used in the Theoretical Calculations of the Particle Penetration Depth as a Function of the Relative Humidity

Skin Region	Parameter	Value	Source
SC	σ_{sc} (MPa)	22.5–3.2 (0%–100% RH)	Wildnauer et al. (1971)
	ρ_{sc} (kg m^{-3})	1500	Duck (1990)
	t_{sc} (μm)	10–15.6 (0%–93% RH)	Blank et al. (1984) (measurement)
Viable epidermis	σ_{ve} (MPa)	2.2	Kishino and Yanagida (1988) (actin tensile)
		10	Mitchell et al. (2003) (epithelium)
	ρ_{ve} (kg m^{-3})	1150	Duck (1990)

Source: From Kendall, M.A.F., Rishworth, S., Carter, F.V., and Mitchell, T.J., *J. Invest. Dermatol.*, 122, 739, 2004.

The subsequent particle penetration in the viable epidermis (d_{ve}) is then calculated using Equation 18.2, using instead the material properties of the viable epidermis and $v_{i,ve}$. The total particle penetration depth (d_t) is thus

$$d_t = t_{sc} + d_{ve} \qquad (18.4)$$

An alternative fully numerical discrete element model approach has also been applied [64], but will not be discussed here.

18.4.3.2 Locations of Microparticles into Skin

As one example, particle delivery to excised human skin is shown for both classes of particles (Figure 18.7) [66]. In Figure 18.7A, a glass particle of 20 μm radius delivered to the skin at a nominal entry velocity of 260 m/s is shown. Note the variation in both the SC and epidermal thicknesses. Histological sampling of the three skin sites from the backs of cadavers. Measured SC and epidermal thickness compared very well with previous reports from the literature. Over 1800 readings of the deepest particle edge and size of the particles were made on similar histological sections with polystyrene, stainless steel, and glass particles, selected for different density and size ranges.

In Figure 18.7B, a histological section is shown after the impact of gold particles with a measured mean radius of 1 ± 0.2 μm on the skin with a mean calculated impact velocity of 580 ± 50 m/s. A sample particle depth measurement is labeled as d_i. Over 1200 readings of the deepest edge and size of the gold particles were made on similar histological sections. All the raw data collected from the histology sections (such as in

Figure 18.7) are plotted as a function of the particle impact parameter, ρvr where ρ is the density, v is the velocity, and r is the radius, in Figure 18.8. The variability of penetration as shown in Figure 18.8 is typical of results obtained with other tissues.

Some insights into the sources of scatter in the penetration data of Figure 18.8 can be gained when the data are grouped and processed. Consider, for instance, the gold particles shown in Figure 18.7 grouped by particle radius as shown in Figure 18.9. The error bars correspond to one standard deviation in collapsed particle penetration depth and ρvr. Note the trend indicating that for a given value of ρvr, an increase in radius (and hence a decrease in impact velocity) corresponds to a decrease in penetration depth. These data, together with other (unpublished) work, show the different particle sizes and the cell matrix results in different penetration depths. For instance, the gold particles are smaller than the average cell size, and, during deceleration through the skin tissue, are more likely to penetrate through individual cell membranes. For the larger particles, however, the tissue would primarily fail between the cell boundaries. Indeed, these ballistic penetration data are qualitatively consistent with findings from microprobe indentation studies [24], albeit at considerably higher strain rates.

Corresponding calculated penetration profiles using the theoretical model are also shown in Figure 18.8, and illustrate a similar trend with good agreement. Importantly, in this case the yield stress was held constant at 40 MPa, to achieve the closest fit with the data. This is considerably higher than the quasi-static yield stresses reported in the literature (summarized in Table 18.1 and Figure 18.2). This discrepancy is

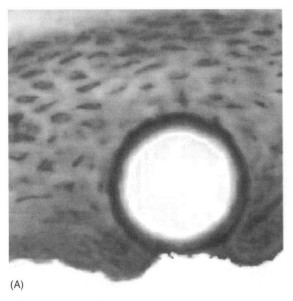

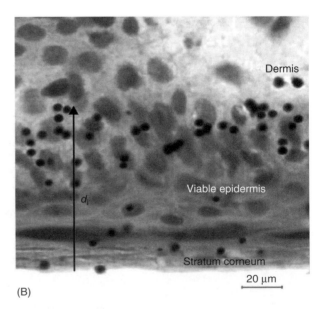

(A)   (B)

FIGURE 18.7   Photomicrographs of particles delivered to human skin. (A) A 20 μm radius glass sphere delivered at 260 m/s and (B) gold particles (1.0 ± 0.2 μm radius) delivered at 580 ± 50 m/s are shown. (Adapted from Kendall, M.A.F., Mitchell, T.J., and Wrighton-Smith, P., *J. Biomech.*, 37, 1733, 2004. With permission.)

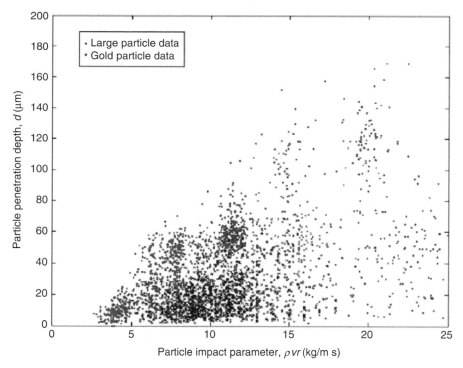

FIGURE 18.8 (See color insert following blank page 206.) Raw gold and larger particle penetration into excised human skin as a function of the particle radius, density, and impact velocity. (Adapted from Kendall, M.A.F., Mitchell, T.J., and Wrighton-Smith, P., *J. Biomech.*, 37, 1733, 2004. With permission.)

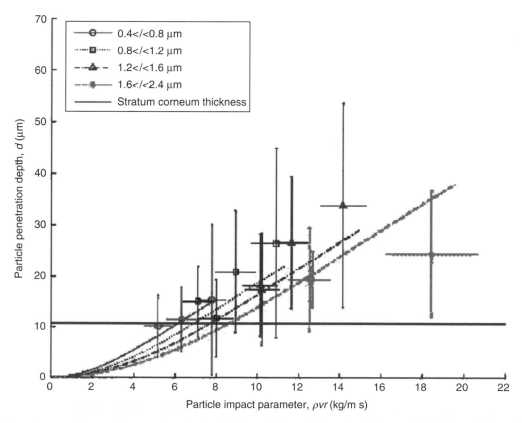

FIGURE 18.9 Impact parameters and penetration depth of gold particles within excised human skin. (Adapted from Kendall, M.A.F., Mitchell, T.J., and Wrighton-Smith, P., *J. Biomech.*, 37, 1733, 2004. With permission.)

attributed to a huge strain-rate effect: the ballistic impact of the microparticle has a peak strain-rate of ~10^6 s^{-1}. In a subsequent, more refined study [22], these strain-rate effects are further elucidated.

In addition to the described scale and strain-rate effects, another source of variability stems from the high sensitivity in SC mechanical properties to hydration and temperature, deriving from variation in ambient conditions (detailed in [22]). Increasing the RH from 15% to 95% (temperature at 25°C) led to a particle penetration increase by a factor of 1.8. Temperature increases from 20°C to 40°C (RH at 15%) enhanced particle penetration twofold. In both cases, these increases were sufficient to move the target layer from the SC to the viable epidermis. In immunotherapeutic applications, this is the difference between the ineffectual delivery of particles to the SC and the targeted delivery of specific cells in the viable epidermis.

These collective data show the momentum range obtained from the described biolistic devices primarily translate into delivery within targeted viable epidermis and SC. With the precise delivery conditions achieved from these devices, we have obtained new insights into the important biological variability in microparticle impact. This variability, together with more obvious differences in tissue thicknesses (with the tissue site of target, age, and gender), must be considered when

selecting device conditions for clinical biolistic immunotherapeutic delivery.

18.4.3.3 Skin Cell Death from Ballistic Impact

Of great importance in biolistics applications is the biological responses induced by biolistic impact. When delivered to the tissue surface, the microparticles undergo a tremendous deceleration—peaking at ~10^{10} g—and coming to rest within ~100–200 ns. Such deceleration induces shock and stress waves within the tissue, and it is important to determine under which conditions skin cells are killed. This was investigated in mice, where following the delivery of gold microparticles, the cell death was assayed with mixtures of ethidium bromide and acridine orange and images noninvasively with multiphoton microscopy (MPM) [69]. The data are summarized in Figures 18.10 and 18.11. It was found that each direct impact of a gold microparticle resulted in cell death. Further, even in cases where microparticles passed within ~10 μm of the cell surface—but not touching the cell—cell death resulted. A sufficiently high number density in the tissue can result in complete cell death within the viable epidermis. Clearly, this is important when considering the biological responses induced by microparticle impact.

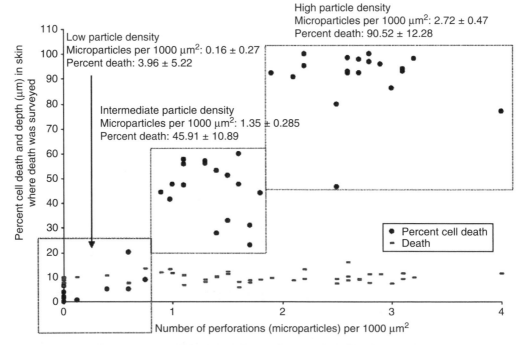

FIGURE 18.10 Viability of epidermal cells targeted by biolistic microparticle delivery. Data for this plot were generated by first enumerating the number of perforations per 1000 μm^2 in the SC caused by microparticle penetration (Figure 18.11 slide A). The perforations were equated to microparticles. Second, percent cell death was calculated in the viable epidermis below the SC using the acridine orange/ethidium bromide assay for discriminating live and dead cells (see Figure 18.11 slide B–D). (From Raju, P.A., Truong, N.K., and Kendall, M.A.F., *Vaccine*, 24, 4644, 2006. With permission.)

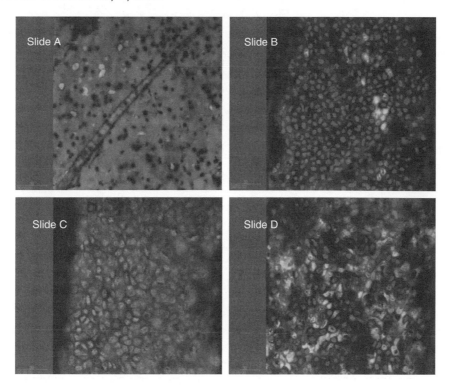

FIGURE 18.11 (See color insert following blank page 206.) Murine tissue stained for live–dead cell discrimination with a cocktail of acridine orange/ethidium bromide and imaged with near-infrared two-photon excitation. Cells emitting red fluorescence are dead whereas cells emitting green fluorescence are alive. In slide A, the dark spots are perforations in the SC caused by microparticle bombardment. Slide B is the corresponding viable epidermis 11.3 μm below the SC. Slides A and B represent an image set of "high particle density" shown in Figure 18.10. Slide C is the viable epidermis of the control (9.7 μm below the SC) where particle delivery to the tissue was not made. Slice D is representative of the viable epidermis of tissue with intermediate particle densities (10.5 μm below the SC). Images were collected with LaserSharp software (Carl Zeiss, Hertfordshire, U.K.). (From Raju, P.A., Truong, N.K., and Kendall, M.A.F., *Vaccine*, 24, 4644, 2006. With permission.)

18.4.4 CLINICAL RESULTS AND COMMERCIAL APPLICATION

18.4.4.1 Commercial Application

Biolistics is a platform technology for delivering a broad range of drugs and immunotherapeutics. Currently, the technology is progressing commercially in two streams:

- Delivery of lidocaine local anesthetic to the skin (the larger class of particles shown in Figure 18.8), approved by the FDA for market application (Zingo, Anesiva)
- Delivery of DNA vaccines on gold-microparticles (PowderMed, Pfizer), undergoing phase III clinical trials

18.4.4.2 Clinical Results

Although strong results are achieved in other immunotherapeutics such as allergy immunotherapy of the animal model [65] and lidocaine for anesthesia, the key clinical progress with DNA vaccines is discussed in this section.

The DNA plasmid that forms the active component DNA vaccines is precipitated onto microscopic gold particles

(typically 2 μg DNA on 1 mg gold). Microscopic elemental gold particles (mean particle diameter ~2 μm) are used as the plasmid DNA carrier, because it is inert and has the appropriate density needed to deliver the vaccine directly into the target epidermal immunologically sensitive cells, including Langerhans cells.

Following delivery into the APC, the DNA elutes off the gold particle and is transcribed into RNA. The RNA in turn is translated into the relevant antigen, which is then processed and presented on the cell surface as if it were an intracellular viral protein. An efficient cellular and humoral immune response is thus induced.

A series of clinical trials have been conducted to assess the immunogenicity and safety of a prophylactic hepatitis B virus DNA vaccine [52,70,71]. These studies have demonstrated that biolistic DNA vaccination can elicit antigen-specific humoral and T-cell responses. In the study by [52], DNA vaccination with 1–4 μg of hepatitis B surface antigen elicited measurable cytotoxic T-cell responses and Th cell responses in all 12 healthy adults who had not previously been immunized with a hepatitis B vaccine [52]. Furthermore, all 12 previously nonvaccinated subjects also seroconverted with levels of hepatitis B-specific antibody ranging from 10 to over

TABLE 18.2

Serum Antibody Responses, Seroconversion, and Seroprotection Rate

Group	Day	GMT (Range)	Seroconversion[a] (%)	Seroprotection[b] (%)	Mean GMT Increase (Fold)
1	0	16 (15–40)	—	17 (2/12)	—
	14	23 (5–160)	8 (1/12)	42 (5/12)	1.4
	21	28 (10–240)	17 (2/12)	33 (4/12)	1.7
	56	44 (10–320)	33 (4/12)	58 (7/12)	**2.8**
2	0	17 (5–40)	—	33 (4/12)	—
	14	29 (10–60)	17 (2/12)	50 (6/12)	1.7
	21	36 (20–80)	8 (1/12)	58 (7/12)	2.1
	56	65 (20–320)	**67 (8/12)**	**92 (11/12)**	**3.9**
3	0	12 (5–40)	—	8 (1/12)	—
	14	21 (5–80)	17 (2/12)	25 (3/12)	1.8
	21	40 (10–160)	33 (4/12)	67 (8/12)	3.4
	56	97 (40–640)	**64 (7/11)**	**100 (11/11)**	**8.1**

Source: From Drape, R.J., Macklin, M.D., Barr, L.J., Jones, S., Haynes, J.R., and Dean, H.J., *Vaccine,* 24, 4475, 2006. With permission.

Note: Values meeting CPMP criteria are bold.

[a] Seroconversion is defined as either a negative pre-vaccination titer (≤ 10) to a post-vaccination titer ≥ 40, or a significant increase in antibody titer, i.e., at least a fourfold increase between pre- and post-vaccination titers where the pre-vaccination titer is ≥ 10.

[b] Seroprotection rate is defined as the proportion of subjects achieving a titer ≥ 40.

5000 mIU/mL. This is of particular significance as intramuscular delivery of DNA—using the needle and syringe—with up to 1000-fold more DNA has generated only low or no antibody responses [72,73]. The same biolistic hepatitis B DNA vaccine was also shown to increase serum antibody titers in 7 of 11 subjects who had previously failed to seroconvert after three or more doses of conventional vaccination with licensed recombinant protein vaccine [70]. Finally this plasmid DNA construct has been used to successfully bridge between the earlier bulky experimental device and the simple, hand held disposable device that will be used for product commercialization [71].

A phase I study [74] has been carried out to investigate the safety and immunogenicity of biolistic administration of an influenza prophylactic plasmid, which encodes a single HA antigen of influenza A/Panama/2007/99 (H3N2). A total of 36 healthy subjects with low preexisting serological responses to this strain received a vaccination of either 1, 2, or 4 μg of DNA at a single administration session. The antibody response was then assessed according to the CHMP criteria for the approval of annual flu vaccines in the European Union. Table 18.2 summarizes these humoral responses, determined as a hemagglutination inhibition titre elicited on days 0 (pre-dose), 14, 21, and 56. Time points, where responses met the levels required by the CHMP guidelines for licensing of annual influenza vaccine, are shown in bold.

The 4 μg dose group met the CHMP criteria at day 21, demonstrating the ability of biolistic DNA vaccination to stimulate serological responses equivalent to those seen in protein based approaches. Furthermore, the responses in all groups continued to increase up to day 56 (the last day monitored) indicating that responses to biolistic vaccination may show a more sustained increase than is typically seen with protein vaccines. By day 56 100% of those subjects vaccinated with the 4 μg dose were seroprotected.

Overall vaccination was well tolerated and local reactogenicity results were typical of those seen in other biolistics studies.

18.5 CONCLUSION

Many immunotherapeutics (e.g., vaccines) can be radically improved by targeted delivery to particular immunologically sensitive cells within the outer skin layers. The push is on to develop a range of technologies to meet this need, either using physical or biological targeting approaches. One of these methods, called biolistics, ballistically delivers biomolecule-coated gold microparticles to the outer layers of the skin. The method of particle acceleration relies heavily on approaches usually applied to the aerospace industry. Consequently, many unique challenges had to be overcome in engineering biolistic devices for clinical use. Research with the resultant devices has yielded unique insights into the skin at microscale dynamic loading—both from mechanical and biological perspectives. Important progress is also being made in clinical trials using biolistic devices to deliver DNA vaccines in the following fields: hepatitis B, influenza, genital herpes, human papilloma virus, HIV/AIDS, Hantaan virus, melanoma, and a variety of other cancers.

REFERENCES

1. Givens, B., Oberle, S., and Lander, J. Taking the jab out of needles. *Can Nursing Journal*, 89, 37–40, 1993.
2. Fuchs, E. and Raghavan, S. Getting under the skin of epidermal morphogenesis. *Nat Rev Genet*, 3, 199–209, 2002.
3. Menton, D.N. and Eisen, A.Z. Structure and organization of mammalian stratum corneum. *J Ultrastruct Res*, 35(3), 247–264, 1971.
4. Nemes, Z. and Steinert, P.M. Bricks and mortar of the epidermal barrier. *Exp Mol Med*, 31, 5–19, 1999.
5. Stenn, K.S., Goldenhersh, M.A., and Trepeta, R.W. *Structure and Functions of the Skin. The Skin*, Vol. 9, Churchill Livingstone, 1–14, 1992.
6. Chen, H., Yuan, J., Wang, Y., and Silvers, W.K. Distribution of ATPase-positive Langerhans cells in normal adult human skin. *Br J Dermatol*, 113(6), 707–711, 1985.
7. Berman, B., Chen, V.L., France, D.S., Dotz, W.I., and Petroni, G. Anatomical mapping of epidermal Langerhans cell densities in adults. *Br J Dermatol*, 109(5), 553–558, 1983.
8. Babiuk, S., Baca-Estrada, M., Babiuk, L.A., Ewen, C., and Foldvari, M. Cutaneous vaccination: The skin as an immunologically active tissue and the challenge of antigen delivery. *J Control Release*, 66, 199–214, 2000.
9. Chen, D., Maa, Y., and Haynes, J.R. Needle-free epidermal powder immunization. *Expert Rev Vaccines*, 1(3), 89–100, 2002.
10. Banchereau, J. and Steinman, R.M. Dendritic cells and the control of immunity. *Nature*, 19(392), 245–252, 1998.
11. McKinney, E.C. and Streilein, J.W. On the extraordinary capacity of allogeneic epidermal Langerhans cells to prime cytotoxic T cells in vivo. *J Immunol*, 143, 1560–1564, 1989.
12. Timares, L., Takashima, A., and Johnston, S.A. Quantitative analysis of the immunopotency of genetically transfected dendritic cells. *Proc Natl Acad Sci U S A*, 95, 13147–13152, 1998.
13. Hoath, S.B. and Leahy, D.G. Formation and function of the stratum corneum. In: Marks R, Levenge J, Voegli R (Eds.), *The Essential Stratum Corneum*. London: Martin Dunitz, 2002.
14. Numahara, T., Tanemura, M., Nakagawa, T., and Takaiwa, T. Spatial data analysis by epidermal Langerhans cells reveals an elegant system. *J Dermatol Sci*, 25, 219–228, 2001.
15. Bauer, J., Bahmer, F.A., Worl, J., Neuhuber, W., Schuler, G., and Fartasch, M. A strikingly constant ratio exists between Langerhans cells and other epidermal cells in human skin. A stereologic study using the optical dissector method and the confocal laser-scanning microscope. *J Invest Dermatol*, 116, 313–318, 2001.
16. Hopewell, J.W. The skin: Its structure and response to ionizing radiation. *Int J Radiat Biol*, 57(4), 751–773, 1990.
17. Wildnauer, R.H., Bothwell, J.W., and Douglas, A.B. Stratum corneum properties I. Influence of relative humidity on normal and extracted stratum corneum. *J Invest Dermatol*, 56, 72–78, 1971.
18. Christensen, M.S., Hargens, C.W., Nacht, S., and Gans, E.H. Viscoelastic properties of intact human skin: Instrumentation, hydration effects and the contribution of the stratum corneum. *J Invest Dermatol*, 69, 282–286, 1977.
19. Rawlings, A., Harding, C., Watkinson, A., Banks, J., Ackerman, O., and Sabin, R. The effect of glycerol and humidity on desmosome degradation in the stratum corneum. *Arch Dermatol Res*, 287, 457–464, 1995.
20. Dobrev, H. In vivo non-invasive study of the mechanical properties of the human skin after single application of topical corticosteroids. *Folia Med (Plovdiv)*, 38, 7–11, 1996.
21. Nicolopoulos, C.S., Giannoudis, P.V., Glaros, K.D., and Barbanel, J.C. In vitro study of the failure of skin surface after influence of hydration and reconditioning. *Arch Dermatol Res*, 290, 638–640, 1998.
22. Kendall, M.A.F., Rishworth, S., Carter, F.V., and Mitchell, T.J. The effects of relative humidity and ambient temperature on the ballistic delivery of micro-particles into excised porcine skin. *J Invest Dermatol*, 122(3), 739–746, 2004.
23. Papir, Y.S., Hsu, K.-H., and Wildnauer, R.H. The mechanical properties of stratum corneum, I. The effect of water and ambient temperature on the tensile properties of newborn rat stratum corneum. *Biochim Biophys Acta*, 399, 170–180, 1975.
24. Kendall, M.A.F., Chong, Y., and Cock, A. The mechanical properties of the skin epidermis in relation to targeted gene and drug delivery. *Biomaterials*, 28, 4968–4977, 2007.
25. Lu, B., Federoff, H.J., Wang, Y., Goldsmith, L.A., and Scott, G. Topical application of viral vectors for epidermal gene transfer. *J Invest Dermatol*, 108, 803–808, 1997.
26. Tang, D.-C., Shi, Z., and Curiael, D.T. Vaccination onto bare skin. *Nature*, 388, 729–730, 1997.
27. Mumper, R.J. and Ledebur, H.C. Dendritic cell delivery of plasmid DNA: Application for controlled nucleic acid-based vaccines. *Mol Biotech*, 19, 79–95, 2001.
28. World Health Organisation. In: Safety of injections, Facts & Figures Fact Sheet N° 232, 1999.
29. Glenn, G.M., Kenney, R.T., Ellingsworth, L.R., Frech, S.A., Hammond, S.A., and Zoeteweweij, J.P. Transcutaneous immunization and immunostimulant strategies: Capitalizing on the immunocompetence of the skin. *Expert Rev Vaccines*, 2(2), 253–267, 2003.
30. Guerena-Burgueno, F., Hall, E.R., and Taylor, D.N. Safety and immunogenicity of a prototype enterotoxigenic *Escherichia coli* vaccine administered transcutaneously. *Infect Immun*, 70(4), 1874–1880, 2002.
31. Liu, L.J., Watabe, S., Yang, J., Hamajima, K., Ishii, N., Hagiwara, E., Onari, K., Xin, K.Q., and Okuda, K. Topical application of HIV DNA vaccine with cytokine-expression plasmids induces strong antigen-specific immune responses. *Vaccine*, 20(1–2), 42–48, 2001.
32. Watabe, S., Xin, K.Q., Ihata, A., Liu, L.J., Honsho, A., Aoki, I., Hamajima, K., Wahren, B., and Okuda, K. Protection against influenza virus challenge by topical application of influenza DNA vaccine. *Vaccine*, 19(31), 4434–4444, 2001.
33. Shi, Z., Zeng, M., Yang, G., Siegel, F., Cain, L.J., Van Kampen, K.R., Elmets, C.A., and Tang, D.C. Protection against tetanus by needle-free inoculation of adenovirus-vectored nasal and epicutaneous vaccines. *J Virol*, 75(23), 11474–11482, 2001.
34. Shi, Z., Curiel, D.T., and Tang, D.C. DNA-based non-invasive vaccination onto the skin. *Vaccine*, 17(17), 2136–2141, 1999.
35. Widera, G., Austin, M., Rabussay, D., Goldbeck, C., Barnett, S.W., Chen, M., Leung, L., Otten, G.R., Thudium, K., Selby, M.J., and Ulmer, J.B. Increased DNA vaccine delivery and immunogenicity by electroporation in vivo. *J Immunol*, 164, 4635–4640, 2000.
36. Zucchelli, S., Capone, S., Fattori, E., Folgori, A., Di Marco, A., Casimiro, D., Simon, A.J., Laufer, R., La Monica, N., Cortese, R., and Nicosia, A. Enhancing B- and T-cell immune response to a hepatitis C virus E2 DNA vaccine by intramuscular electrical gene transfer. *J Virol*, 74, 11598–11607, 2000.
37. Sintov, A.C., Krymberk, I., Daniel, D., Hannan, T., Sohn, Z., and Levin, G. Radiofrequency-driven skin microchanneling as a new way for electrically assisted transdermal delivery of hydrophilic drugs. *J Control Release*, 89(2), 311–320, 2003.

38. Alexander, M.Y. and Akhurst, R.J. Liposome-mediated gene transfer and expression via the skin. *Hum Molec Genet*, 4, 2279–2285, 1995.

39. Li, L. and Hoffman, R.M. The feasibility of targeted selective gene therapy of the hair follicle. *Nat Med*, 1, 705–706, 1995.

40. Domashenko, A., Gupta, S., and Cotsarelis, G. Efficient delivery of transgenes to human hair follicle progenitor cells using topical lipoplex. *Nat Biotechnol*, 18, 420–423, 2000.

41. Cui, Z. and Mumper, R.J. Dendritic cell-targeted genetic vaccines engineered from novel microemulsion precursors. *Mol Ther*, 3, S352, 2001.

42. Furth, P.A., Shamay, A., and Henninghausen, L. Gene transfer into mammalian cells by jet injection. *Hybridoma*, 14, 149–152, 1995.

43. Bremseth, D.L. and Pass, F. Delivery of insulin by jet injection: Recent observations. *Diabetes Technol Ther*, 3, 225–232, 2001.

44. Mikszta, J.A., Alarcon, J.B., Brittingham, J.M., Sutter, D.E., Pettis, R.J., and Harvery, N.G. Improved genetic immunization via micromechanical disruption of skin-barrier function and targeted epidermal delivery. *Nature*, 8(4), 415–419, 2002.

45. Matriano, J.A., Cormier, M., Johnson, J., Young, W.A., Buttery, M., Nyam, K., and Daddona, P.E. Macroflux micro-projection array patch technology: A new and efficient approach for intracutaneous immunization. *Pharm Res*, 19(1), 63–70, 2002.

46. McAllister, D.V., Wang, P.M., Davis, S.P., Park, J.H., Canatella, P.J., Allen, M.G., and Prausnitz, M.R. Microfabricated needles for transdermal delivery of macromolecules and nanoparticles: Fabrication methods and transport studies. *Proc Natl Acad Sci U S A*, 25(100), 13755–13760, 2003.

47. Kendall, M.A.F. Device for Delivery of Bioactive Materials and Other Stimuli. U.S. Patent Application, US 11/496,053, filed August 2006.

48. Sanford, J.C. and Klein, M.C. Delivery of substances into cells and tissues using a particle bombardment process. *Particulate Sci Tech*, 5, 27–37, 1987.

49. Bellhouse, B.J., Sarphie, D.F., and Greenford, J.C. Needleless syringe using supersonic gas flow for particle delivery. Int patent Wo94/24263, 1994.

50. Burkoth, T.L., Bellhouse, B.J., Hewson, G., Longridge, D.J., Muddle, A.G., and Sarphie, D.F. Transdermal and transmucosal powdered drug delivery. *Crit Rev Ther Drug Carrier Syst*, 16(4), 331–384, 1999.

51. Chen, D.X., Endres, R.L., Erickson, C.A., Weis, K.F., McGregor, M.W., Kawaoka, Y., and Payne, L.G. Epidermal immunization by a needle-free powder delivery technology: Immunogenicity of influenza vaccine and protection in mice. *Nature Medicine*, 6, 1187–1190, 2000.

52. Roy, M.J., Wu, M.S., Barr, L.J., Fuller, J.T., Tussey, L.G., Speller, S., Culp, J., Burkholder, J.K., Swain, W.F., Dixon, R.M., Widera, G., Vessey, R., King, A., Ogg, G., Gallimore, A., Haynes, J.R., and Heydenburg Fuller, D. Induction of antigen-specific CD8$^+$ T cells, T helper cells, and protective levels of antibody in humans by particle mediated administration of a hepatitis B virus DNA vaccine. *Vaccine*, 19, 764–778, 2000.

53. Lesinski, G.B., Smithson, S.L., Srivastava, N., Chen, D.X., Widera, G., and Westerink, J.A. DNA vaccine encoding a peptide mimic of *Streptococcus pneumoniae* serotype 4 capsular polysaccharide induces specific anti-carbohydrate antibodies in Balb/c mice. *Vaccine*, 19, 1717–1726, 2001.

54. Kendall, M.A.F., Quinlan, N.J., Thorpe, S.J., Ainsworth, R.W., and Bellhouse, B.J. Measurements of the gas and particle flow within a converging–diverging nozzle for high speed powdered vaccine and drug delivery. *Exp Fluids*, 37, 128–136, 2004.

55. Quinlan, N.J., Kendall, M.A.F., Bellhouse, B.J., and Ainsworth, R.W. Investigations of gas and particle dynamics in first generation needle-free drug delivery devices. *Int J Shock Waves*, 10(6), 395–404, 2001.

56. Liu, Y. and Kendall, M.A.F. Numerical study of a transient gas and particle flow in a high-speed needle-free ballistic particulate vaccine delivery system. *J Mech Med Biol*, 4(4), 1–20, 2004.

57. Kendall, M.A.F. The delivery of particulate vaccines and drugs to human skin with a practical, hand-held shock tube-based system. *Shock Waves J*, 12(1), 22–30, 2002.

58. Liu, Y. and Kendall, M.A.F. Numerical simulation of heat transfer from a transonic jet impinging on skin for needle-free powdered drug and vaccine delivery. *J Mech Engineer Sci, Proc Instit Mech Engineers*, 218(C), 1373–1383, 2004.

59. Truong, N.K., Liu, Y., and Kendall, M.A.F. Gas-particle dynamics characterization of a pre-clinical contoured shock tube for vaccine and drug delivery. *Shock Waves*, 15, 149–164, 2006.

60. Marrion, M., Kendall, M.A.F., and Liu, Y. The gas-dynamic effects of a hemisphere-cylinder obstacle in a shock-tube driver. *Exp Fluids*, 38, 319–327, 2005.

61. Hardy, M.P. and Kendall, M.A.F. Mucosal deformation from an impinging transonic gas jet and the ballistic impact of microparticles. *Phys Med Biol*, 50, 4567–4580, 2005.

62. Liu, Y., Truong, N.K., Kendall, M.A.F., and Bellhouse, B.J. Characteristics of a micro-biolistic system for murine immunological studies. *Biomed Microdevices*, 9, 465–474, 2007.

63. Dehn, J. A unified theory of penetration. *Int J Impact Eng,* 5, 239–248, 1976.

64. Mitchell, T.M. The ballistics of micro-particles into the mucosa and skin. DPhil Thesis, Engineering Science, University of Oxford, 2003.

65. Kendall, M.A.F., Mitchell, T.J., Costigan, G., Armitage, M., Lenzo, J.C., Thomas, J.A., Von Garnier, C., Zosky, G.R., Turner, D.J., Stumbles, P.A., Sly, P.D., Holt, P.G., and Thomas, W.R. Down regulation of IgE allergic responses in the lung by epidermal biolistic micro-particle delivery. *Allergy Clin Immunol J*, 117(2), 275–282, 2006.

66. Kendall, M.A.F., Mitchell, T.J., and Wrighton-Smith, P. Intradermal ballistic delivery of micro-particles into excised human skin for drug and vaccine applications. *J Biomechanics*, 37(11), 1733–1741, 2004.

67. Kendall, M.A.F., Rishworth, S., Carter, F.V., and Mitchell, T.J. The effects of relative humidity and ambient temperature on the ballistic delivery of micro-particles into excised porcine skin. *J Invest Dermat*, 122(3), 739–746, 2004.

68. Mitchell, T.J., Kendall, M.A.F., and Bellhouse, B.J. A ballistic study of micro-particle penetration to the oral mucosa. *Int J Impact Eng*, 28, 581–599, 2003.

69. Raju, P.A., Truong, N.K., and Kendall, M.A.F. Assessment of epidermal cell viability by near infra-red two-photon microscopy following ballistic delivery of gold micro-particles. *Vaccine*, 24(21), 4644–4647, 2006.

70. Rottinghaus, S.T., Poland, G.A., Jacobson, R.M., Barr, L.J., and Roy, M.J. Hepatitis B DNA vaccine induces protective antibody responses in human non-responders to conventional vaccination. *Vaccine*, 21(31), 4604–4608, 2003.

71. Roberts, L.K., Barr, L.J., Fuller, D.H., McMahon, C.W., Leese, P.T., and Jones, S. Clinical safety and efficacy of a powdered Hepatitis B nucleic acid vaccine delivered to the epidermis by a commercial prototype device. *Vaccine*, 23(40), 4867–4878, 2005.

72. MacGregor, R.R., Boyer, J.D., Ugen, K.E., Lacy, K.E., Gluckman, S.J., Bagarazzi, M.L., Chattergoon, M.A., Baine, Y., Higgins, T.J., Ciccarelli, R.B., Coney, L.R., Ginsberg, R.S., and Weiner, D.B. First human trial of a DNA-based vaccine for treatment of human immunodeficiency virus type 1 infection: Safety and host response. *J Infect Dis*, 178(1), 92–100, 1998.

73. MacGregor, R.R., Ginsberg, R., Ugen, K.E., Baine, Y., Kang, C.U., Tu, M., Higgins, T., Weiner, D.B., and Boyer, J.D. T-cell responses induced in normal volunteers immunized with a DNA-based vaccine containing HIV-1 env and rev. *AIDS*, 16(16), 2137–2143, 2002.

74. Drape, R.J., Macklin, M.D., Barr, L.J., Jones, S., Haynes, J.R., and Dean, H.J. Epidermal DNA vaccine for influenza is immunogenic in humans. *Vaccine*, 24(21), 4475–4481, 2006.

75. Kendall, M.A.F. Engineering of needle-free physical methods to target epidermal cells for DNA vaccination. *Vaccine*, 24(21), 4651–4656, 2006.

19 Optimization of Nonviral Gene Therapeutics

Nancy Smyth Templeton

CONTENTS

19.1 INTRODUCTION

Many investigators are focused on the production of effective nonviral gene therapeutics and on creating improved delivery systems that mix viral and nonviral vectors. Use of improved liposome formulations for delivery *in vivo* is valuable for gene therapy and would avoid several problems associated with viral delivery. Delivery of nucleic acids using liposomes is promising as a safe and nonimmunogenic approach to gene therapy. Furthermore, gene therapeutics composed of artificial reagents can be standardized and regulated as drugs rather than as biologics. Cationic lipids have been used for efficient delivery of nucleic acids to cells in tissue culture for several years [1,2]. Much effort has also been directed toward developing cationic liposomes for efficient delivery of nucleic acids in animals and in humans [3–12]. Most frequently, the formulations that are best to use for transfection of a broad range of cell types in culture are not optimal for achieving efficacy in small and large animal disease models.

Much effort has been devoted to the development of nonviral delivery vehicles due to the numerous disadvantages of viral vectors that have been used for gene therapy. Following viral delivery, *in vivo* immune responses are generated to express viral proteins that subsequently kill the target cells required to produce the therapeutic gene product; an innate humoral immune response can be produced to certain viral vectors due to previous exposure to the naturally occurring virus; random integration of some viral vectors into the host chromosome could occur and cause activation of proto-oncogenes resulting in tumor formation; clearance of viral vectors delivered systemically by complement activation can occur, viral vectors can be inactivated upon re-administration by the humoral immune response, and potential for recombination of the viral vector with DNA sequences in the host chromosome that generates a replication-competent infectious virus also exists. Specific delivery of viral vectors to target cells can be difficult because two distinct steps in engineering viral envelopes or capsids must be achieved. First, the virus envelope or capsid must be changed to inactivate the natural tropism of the virus to enter specific cell types. Then sequences must be introduced that allow the new viral vector to bind and internalize through a different cell surface receptor. Other disadvantages of viral vectors include the inability to administer certain viral vectors more than once, the high costs for producing large amounts of high-titer viral stocks for use in the clinic, and the limited size of the nucleic acid that can be packaged and used for viral gene therapy. Attempts are being made to overcome the immune responses produced by viral vectors after administration in immune competent animals and in humans, such as the use of gutted adenoviral vectors or encapsulation of viral vectors in liposomes [13]. However, complete elimination of all immune responses to viral vectors may be impossible.

Use of liposomes for gene therapy provides several advantages. A major advantage is the lack of immunogenicity after *in vivo* administration including systemic injections. Therefore, the nucleic acid–liposome complexes can be re-administered without harm to the patient and without compromising the efficacy of the nonviral gene therapeutic. Improved formulations of nucleic acid–liposome complexes can also evade complement inactivation after *in vivo* administration. Nucleic acids of unlimited size can be delivered ranging from single nucleotides to large mammalian artificial chromosomes. Furthermore, different types of nucleic acids can be delivered including plasmid DNA, RNA, oligonucleotides, DNA–RNA chimeras, synthetic ribozymes, antisense molecules, RNAi, viral nucleic acids, and others. Certain cationic formulations can also encapsulate and deliver viruses [13], proteins or partial proteins with a low isoelectric point (pI), and mixtures of nucleic acids and proteins of any pI. Creation of nonviral vectors for targeted delivery to specific cell types, organs, or tissues is relatively simple. Targeted delivery involves elimination of nonspecific charge interactions with nontarget cells and addition of ligands for binding and internalization through target cell surface receptors. Other advantages of nonviral vectors include the low cost and relative ease in producing nucleic acid–liposome complexes in large scale for use in the clinic. In addition, greater safety for patients is provided using nonviral delivery vehicles due to few or no viral sequences present in the nucleic acids used for delivery thereby precluding generation of an infectious virus. The disadvantage of nonviral delivery systems had been the low levels of delivery and gene expression produced by first-generation complexes. However, recent advances have dramatically improved transfection efficiencies and efficacy of liposomal vectors [14–18]. Reviews of other *in vivo* delivery systems and improvements using cationic liposomes have been published recently [19,20].

Cationic liposome–nucleic acid complexes can be administered via numerous delivery routes *in vivo*. Routes of delivery include direct injection (e.g., intratumoral), intravenous, intraperitoneal, intra-arterial, intrasplenic, mucosal (nasal, vaginal, rectal), intramuscular, subcutaneous, transdermal, intradermal, subretinal, intratracheal, intracranial, and others. Much interest has focused on noninvasive intravenous administration because many investigators believe that this route of delivery is the "holy grail" for the treatment or cure of cancer, cardiovascular, and other inherited or acquired diseases. Particularly for the treatment of metastatic cancer, therapeutics must reach not only the primary tumor but also the distant metastases.

Optimization of cationic liposomal complexes for *in vivo* applications and therapeutics is complex, involving many distinct components. These components include nucleic acid purification, plasmid design, formulation of the delivery vehicle, administration route and schedule, dosing, detection of gene expression, and others. Often I make the analogy of liposome optimization to a functional car. Of course the engine of the car, analogous to the liposome delivery vehicle, is extremely important. However, if the car does not have wheels, adequate tires, etc., the motorist will not be able to drive the vehicle to its destination. This chapter focuses on optimization of these distinct components for use in a variety of *in vivo* applications. Optimizing all components of the delivery system will allow broad use of liposomal complexes to treat or cure human diseases or disorders.

19.2 OPTIMIZATION OF CATIONIC LIPOSOME FORMULATIONS FOR USE *IN VIVO*

Much research has been directed toward the synthesis of new cationic lipids. Some new formulations led to the discovery of more efficient transfection agents for cells in culture. However, their efficiency measured *in vitro* did not correlate with their ability to deliver DNA after administration in animals. Functional properties defined *in vitro* do not assess the stability of the complexes in plasma or their pharmacokinetics and biodistribution, all of which are essential for optimal activity *in vivo*. Colloidal properties of the complexes in addition to the physicochemical properties of their component lipids also determine these parameters. In particular, in addition to efficient transfection of target cells, nucleic acid–liposome complexes must be able to traverse tight barriers *in vivo* and penetrate throughout the target tissue to produce efficacy for the treatment of disease. These are not issues for achieving efficient transfection of cells in culture with the exception of polarized tissue culture cells. Therefore, we are not surprised that optimized liposomal delivery vehicles for use *in vivo* may be different than those used for efficient delivery to cells in culture.

In summary, *in vivo* nucleic acid–liposome complexes that produce efficacy in animal models of disease have extended half-life in the circulation, are stable in serum, have broad biodistribution, efficiently encapsulate various sizes of nucleic acids, are targetable to specific organs and cell types, penetrate across tight barriers in several organs, penetrate evenly throughout the target tissue, are optimized for nucleic acid/lipid ratio and colloidal suspension *in vivo*, can be size fractionated to produce a totally homogenous population of complexes prior to injection, and can be repeatedly administered. Recently, we demonstrated efficacy of a robust liposomal delivery system in small and large animal models for lung [15], breast [17], head and neck, and pancreatic cancers [16], and for hepatitis B and C. On the basis of efficacy in these animal studies, this liposomal delivery system has been used successfully in a phase I clinical trial to treat end-stage nonsmall cell lung carcinoma patients who have failed to respond to chemotherapy (Lu et al., *AACR Meeting*, Los Angeles 2007). These patients have prolonged life spans and have demonstrated objective responses including tumor regression. This liposomal delivery system will also be used in upcoming clinical trials to treat other types of cancer including pancreatic, breast, and head and neck cancers. Our studies have demonstrated broad efficacy in the use of liposomes to treat disease and have dispelled several myths that exist concerning the use of liposomal systems.

19.3 LIPOSOME MORPHOLOGY AND EFFECTS ON GENE DELIVERY AND EXPRESSION

Efficient *in vivo* nucleic acid–liposome complexes have unique features including their morphology, mechanisms for crossing the cell membrane and entry into the nucleus, ability to be targeted for delivery to specific cell surface receptors, and ability to penetrate across tight barriers and throughout target tissues. Liposomes have different morphologies based upon their composition and the formulation method. Furthermore, the morphology of complexes can contribute to their ability to deliver nucleic acids *in vivo*. Formulations frequently used for the delivery of nucleic acids are lamellar structures which include small unilamellar vesicles (SUVs), multilamellar vesicles (MLVs), or bilamellar invaginated vesicles (BIVs) recently developed in our laboratory (Figure 19.1). Several investigators have developed liposomal delivery systems using hexagonal structures, however, they have demonstrated efficiency primarily for the transfection of some cell types in culture and not for *in vivo* delivery. SUVs condense nucleic acids on the surface and form "spaghetti and meatballs" structures [21]. DNA–liposome complexes made using SUVs produce little or no gene expression upon systemic delivery although these complexes transfect numerous cell types efficiently *in vitro* [1,2]. Furthermore, SUV liposome–DNA complexes cannot be targeted efficiently. SUV liposome–DNA complexes also have a short half life within the circulation, generally about 5–10 min. Polyethylene glycol (PEG) has been added to liposome formulations to extend their half-life [22–24], however, PEGylation created other problems that have not been resolved. PEG seems to hinder delivery of cationic liposomes into cells due to its sterically hindering ionic interactions, and it interferes with optimal condensation of nucleic acids onto the cationic delivery vehicle. Furthermore, extremely long half-life in the circulation, e.g., several days, has caused problems for patients because the bulk of the PEGylated liposomal formulation doxil that encapsulates the cytotoxic agent, doxorubicin, accumulates in the skin, hands, and feet. For example, patients contract mucositis and hand and foot syndrome [25,26] that cause extreme discomfort to the patient. Attempts to add ligands to doxil for delivery to specific cell surface receptors have not resulted in much cell-specific delivery, and the majority of the injected targeted formulation still accumulates in the skin, hands, and feet. Addition of PEG into formulations developed in our laboratory also caused steric hindrance in the bilamellar invaginated structures that did not encapsulate DNA efficiently, and gene expression was substantially diminished.

Some investigators have loaded nucleic acids within SUVs using a variety of methods; however, the bulk of the DNA does not load or stay within the liposomes. Furthermore, most

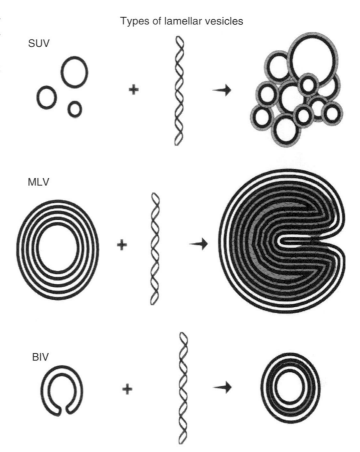

Types of lamellar vesicles

FIGURE 19.1 Diagrams drawn from cryo-electron micrographs of cross-sections through vitrified films of various types of liposomes and DNA–liposome complexes. SUVs are small unilamellar vesicles that condense nucleic acids on the surface and produce spaghetti and meatballs structures. MLVs are multilamellar vesicles that appear as Swiss Rolls after mixing with DNA. BIVs are bilamellar invaginated vesicles produced using a formulation developed in our laboratory. Nucleic acids are efficiently encapsulated between two bilamellar invaginated structures (BIVs). (From Templeton, N.S. et al., *Nat. Biotechnol.*, 15, 647, 1997.)

of the processes used for loading nucleic acids within liposomes are extremely time consuming and not cost effective. Therefore, SUVs are not the ideal liposomes for creating nonviral vehicles for targeted delivery.

Complexes made using MLVs appear as "Swiss rolls" when viewing cross-sections by cryo-electron microscopy [27]. These complexes can become too large for systemic administration or deliver nucleic acids inefficiently into cells due to inability to "unravel" at the cell surface. Addition of ligands onto MLV liposome–DNA complexes further aggravates these problems. Therefore, MLVs are not useful for the development of targeted delivery of nucleic acids.

Using a formulation developed in our laboratory, nucleic acids are efficiently encapsulated between two BIVs [14]. We created these unique structures using 1,2-bis(oleoyloxy)-3-(trimethylamino)propane (DOTAP) and cholesterol (Chol) and a novel formulation procedure. This procedure is different because it includes a brief, low frequency

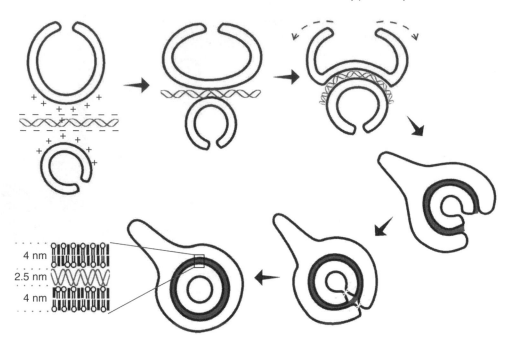

FIGURE 19.2 Proposed model showing cross-sections of extruded DOTAP:Chol:liposomes (BIVs) interacting with nucleic acids. Nucleic acids adsorb onto a BIV via electrostatic interactions. Attraction of a second BIV to this complex results in further charge neutralization. Expanding electrostatic interactions with nucleic acids cause inversion of the larger BIV and total encapsulation of the nucleic acids. Inversion can occur in these liposomes because of their excess surface area, which allows them to accommodate the stress created by the nucleic acid–lipid interactions. Nucleic acid binding reduces the surface area of the outer leaflet of the bilayer and induces the negative curvature due to lipid ordering and reduction of charge repulsion between cationic lipid headgroups. Condensation of the internalized nucleic acid–lipid sandwich expands the space between the bilayers and may induce membrane fusion to generate the apparently closed structures. The enlarged area shows the arrangement of nucleic acids condensed between two 4 nm bilayers of extruded DOTAP:Chol.

sonication followed by manual extrusion through filters of decreasing pore size. The 0.1 and 0.2 μm filters used are made of aluminum oxide and not polycarbonate that is typically used by other protocols. Aluminum oxide membranes contain more pores per surface area, evenly spaced and sized pores, and pores with straight channels. During the manual extrusion process the liposomes are passed through each of four different sized filters only once. This process produces 88% invaginated liposomes. Use of high frequency sonication or mechanical extrusion produces only SUVs.

The BIVs produced condense unusually large amounts of nucleic acids of any size (Figure 19.2) or viruses (Figure 19.3). Furthermore, addition of other DNA condensing agents including polymers is not necessary. For example, condensation of plasmid DNA onto polymers first before encapsulation in the BIVs did not increase condensation or subsequent gene expression after transfection *in vitro* or *in vivo*. Encapsulation of nucleic acids by these BIVs alone is spontaneous and immediate, and therefore, cost effective requiring only one step of simple mixing. The extruded DOTAP:Chol–nucleic acid complexes are also large enough so that they are not cleared rapidly by Kupffer cells in the liver and yet extravasate across tight barriers, including the endothelial cell barrier of the lungs in a normal mouse, and diffuse through target organs efficiently [15]. Our recent work demonstrating efficacy for

treatment of nonsmall cell lung cancer [15] showed that only BIV DOTAP:Chol-p53–DNA:liposome complexes produced efficacy, and SUV DOTAP:Chol-p53–DNA:liposome complexes produced no efficacy. Therefore, the choice of lipids

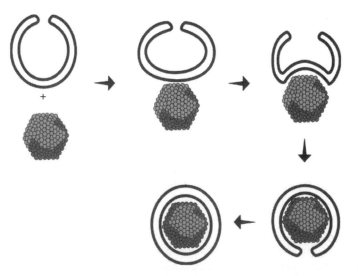

FIGURE 19.3 Proposed model showing cross-sections of an extruded DOTAP:Chol:liposome (BIV) interacting with adenovirus. Adenovirus interacts with a BIV causing negative curvature and wrapping around the virus particle.

alone is not sufficient for optimal DNA delivery, and the morphology of the complexes is essential.

19.4 OPTIMAL LIPIDS AND LIPOSOME MORPHOLOGY: EFFECTS ON GENE DELIVERY AND EXPRESSION

Choosing the best cationic lipids and neutral lipids are also essential for producing the optimal *in vivo* formulation. For example, using our novel manual extrusion procedure does not produce BIVs using the cationic lipid dimethyldioctadecy lammonium bromide (DDAB). Furthermore, DOTAP is biodegradable, whereas DDAB is not biodegradable. Use of biodegradable lipids is preferred for use in humans. Furthermore, only DOTAP and not DDAB containing liposomes produced highly efficient gene expression *in vivo* [14]. DDAB did not produce BIVs and was unable to encapsulate nucleic acids. Apparently, DDAB and DOTAP containing SUVs produce similar efficiency of gene delivery *in vivo*; however, these SUVs are not as efficient as BIV DOTAP:Chol [14]. In addition, use of L-α dioleoyl phosphatidylethanolamine (DOPE) as a neutral lipid creates liposomes that cannot wrap or encapsulate nucleic acids. Several investigators have reported efficient transfection of cells in culture using DOPE in liposomal formulations. However, our data showed that formulations consisting of DOPE were not efficient for producing gene expression *in vivo* [14].

Investigators must also consider the source and lot of certain lipids purchased from companies. For example, different lots of cholesterol from the same vendor can vary dramatically and will affect the formulation of liposomes. Recently, we are using synthetic cholesterol (Sigma, St. Louis, MO). Synthetic cholesterol instead of natural cholesterol that is purified from the wool of sheep is preferred by the Food and Drug Administration for use in producing therapeutics for injection into humans.

Our BIV formulations are also stable for a few years as liquid suspensions. Freeze-dried formulations can also be made that are stable indefinitely even at room temperature. Stability of liposomes and liposomal complexes is also essential particularly for the commercial development of human therapeutics.

19.5 LIPOSOME ENCAPSULATION, FLEXIBILITY, AND OPTIMAL COLLOIDAL SUSPENSIONS

A common belief is that artificial vehicles must be 100 nm or smaller to be effective for systemic delivery. However, this belief is most likely true only for large, inflexible delivery vehicles. Blood cells are several microns (up to 7000 nm) in size, and yet have no difficulty circulating in the blood including through the smallest capillaries. However, sickle cell blood cells, that are rigid, do have problems in the circulation. Therefore, we believe that flexibility is a more important issue than small size. In fact, BIV DNA–liposome complexes in the size range of 200–450 nm produced the highest levels of gene expression in all tissues after intravenous injection [14]. Delivery vehicles, including nonviral vectors and viruses, that are

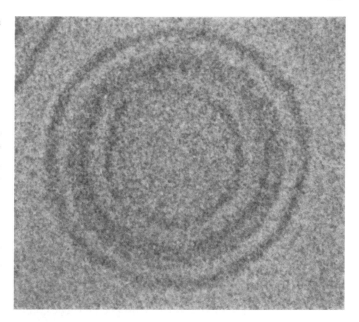

FIGURE 19.4 Cryo-electron micrograph of BIV DOTAP:Chol–DNA:liposome complexes. The plasmid DNA is encapsulated between two BIVs.

not PEGylated and are smaller than 200 nm are cleared quickly by the Kupffer cells in the liver. Therefore, increased size of liposomal complexes could extend their circulation time particularly when combined with injection of high colloidal suspensions. BIVs are able to encapsulate nucleic acids and viruses apparently due to the presence of cholesterol in the bilayer (Figure 19.4). Whereas, formulations including DOPE instead of cholesterol could not assemble nucleic acids by a "wrapping type" of mechanism (Figure 19.5) and produced little gene

FIGURE 19.5 Cryo-electron micrograph of extruded DOTAP: DOPE liposomes complexed to plasmid DNA. Although these liposomes were prepared by the same protocol that produces BIV DOTAP:Chol, these vesicles cannot wrap and encapsulate nucleic acids. The DNA condenses on the surfaces of the liposomes shown.

expression in the lungs and no expression in other tissues after intravenous injections. Because the extruded DOTAP:Chol BIV complexes are flexible and not rigid, are stable in high concentrations of serum, and have extended half-life, they do not have difficulty circulating efficiently in the bloodstream.

We believe that colloidal properties of nucleic acid–liposome complexes also determine the levels of gene expression produced after *in vivo* delivery [14,28] These properties include the DNA/lipid ratio that determines the overall charge density of the complexes and the colloidal suspension that is monitored by its turbidity. Complex size and shape, lipid composition and formulation, and encapsulation efficiency of nucleic acids by the liposomes also contribute to the colloidal properties of the complexes. The colloidal properties affect serum stability, protection from nuclease degradation, blood circulation time, and biodistribution of the complexes.

Our *in vivo* transfection data showed that an adequate amount of colloids in suspension was required to produce efficient gene expression in all tissues examined [14]. The colloidal suspension is assessed by measurement of adsorbance at 400 nm using a spectrophotometer optimized to measure turbidity. Our data showed that transfection efficiency in all tissues corresponded to OD400 of the complexes measured prior to intravenous injection.

19.6   OVERALL CHARGE OF COMPLEXES AND ENTRY INTO THE CELL

In addition, our delivery system is efficient because the complexes deliver DNA into cells by fusion with the cell membrane and avoid the endocytic pathway Figure 19.6. Cells are negatively charged on the surface, and specific cell types vary in their density of negative charge. These differences in charge density can influence the ability of cells to be transfected. Cationic complexes have nonspecific ionic charge interactions with cell surfaces. Efficient transfection of cells by cationic complexes is, in part, contributed by adequate charge interactions. In addition, recent publications report

that certain viruses have a partial positive charge around key subunits of viral proteins on the virus surface responsible for binding to and internalization through target cell surface receptors. Therefore, this partial positive charge is required for virus entry into the cell. Thus, maintenance of adequate positive charge on the surface of targeted liposome complexes is essential for optimal delivery into the cell. Different formulations of liposomes interact with cell surfaces via a variety of mechanisms. Two major pathways for interaction are by endocytosis or by direct fusion with the cell membrane [27,29–34]. Preliminary data suggest that nucleic acids delivered *in vitro* and *in vivo* using complexes developed in our laboratory enter the cell by direct fusion (Figure 19.6). Apparently, the bulk of the nucleic acids do not enter endosomes, and therefore, far more nucleic acid enters the nucleus. Cell transfection by direct fusion produced increased levels of gene expression, and numbers of cells transfected versus cells transfected through the endocytic pathway by several orders of magnitude.

We believe that maintenance of adequate positive charge on the surface of complexes is essential to drive cell entry by direct fusion. Therefore, we create targeted delivery of our complexes *in vivo* without the use of PEG. These ligand-coated complexes also reexpose the overall positive charge of the complexes as they approach the target cells. Through ionic interactions or covalent attachments, we have added monoclonal antibodies, Fab fragments, proteins, partial proteins, peptides, peptide mimetics, small molecules, and drugs to the surface of our complexes after mixing. These ligands efficiently bind to the target cell surface receptor, and maintain entry into the cell by direct fusion. Using novel methods for addition of ligands to the complexes for targeted delivery results in further increased gene expression in the target cells after transfection. Therefore, we design targeted liposomal delivery systems that retain predominant entry into cells by direct fusion versus the endocytic pathway. Figure 19.7 shows our optimized strategy to achieve targeted delivery, deshielding, fusion with the cell membrane, entry of nucleic acids into

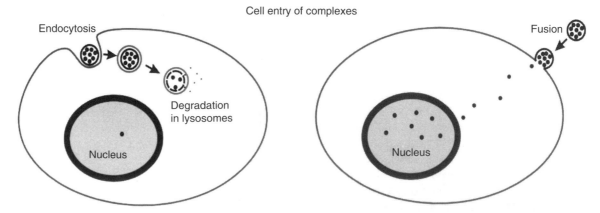

Cell entry of complexes

FIGURE 19.6   Mechanisms for cell entry of nucleic acid–liposome complexes. Two major pathways for interaction are by endocytosis or by direct fusion with the cell membrane. Complexes that enter the cell by direct fusion allow delivery of more nucleic acids to the nucleus because the bulk of the nucleic acids do not enter endosomes.

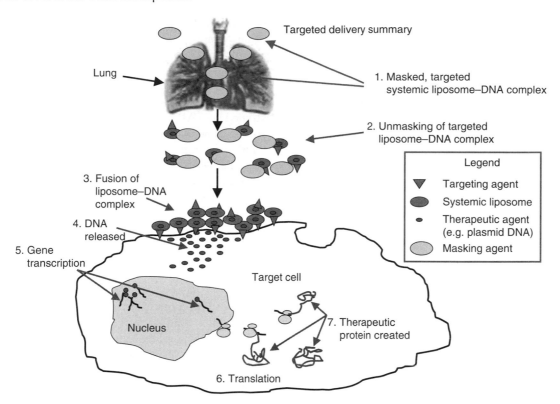

Targeted delivery summary

Lung

1. Masked, targeted systemic liposome–DNA complex

2. Unmasking of targeted liposome–DNA complex

Legend

▼ Targeting agent

⬬ Systemic liposome

• Therapeutic agent (e.g. plasmid DNA)

⬭ Masking agent

3. Fusion of liposome–DNA complex

4. DNA released

5. Gene transcription

Nucleus

Target cell

7. Therapeutic protein created

6. Translation

FIGURE 19.7 (See color insert following blank page 206.)   Optimized strategy for delivery and gene expression in the target cell. Optimization of many steps is required to achieve targeted delivery, shielding from nonspecific uptake in nontarget organs and tissues, deshielding, fusion with the cell membrane, entry of nucleic acids into the cell and to the nucleus, and production of gene expression of a cDNA cloned in a plasmid.

the cell and to the nucleus, and production of gene expression of a cDNA cloned in a plasmid.

19.7   SERUM STABILITY OF OPTIMIZED NUCLEIC ACID–LIPOSOME COMPLEXES FOR USE *IN VIVO*

Serum stability of cationic complexes is complicated and cannot be assessed by simply performing studies at a random concentration of serum. Figure 19.8 shows results from serum stability studies of DNA–liposome complexes that have been optimized in our laboratory for systemic delivery. Serum stability of these complexes was studied at 37°C out to 24 h at concentrations of serum ranging from 0% to 100%. Two different serum stability assays were performed. The first assay measured the OD400 of BIV DOTAP:Chol–DNA:liposome complexes added into tubes containing a different concentration of serum in each tube, ranging from 0% to 100%. The tubes were incubated at 37°C and small aliquots from each tube were removed at various time points out to 24 h. The OD400 of each aliquot was measured on a spectrophotometer calibrated to accurately measure turbidity. Previous work in our laboratory demonstrated that the OD400 predicted both the stability of the complexes and the transfection efficiency results obtained for multiple organs after intravenous injections [14,28]. Percent stability for this assay is defined as the transfection efficiency that is obtained at a particular OD400 of the

complexes used for intravenous injections. Therefore, this assay is rigorous because slight declines in OD400 of these complexes result in obtaining no transfection *in vivo*. Declines in the OD400 also measure precipitation of the complexes.

A second assay was performed to support the results obtained from the OD400 measurements described above.

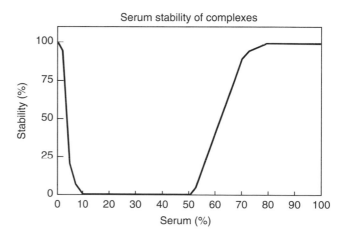

Serum stability of complexes

FIGURE 19.8   Serum stability profile for DNA–liposome complexes optimized for systemic delivery. Serum stability of these complexes was studied at 37°C out to 24 h at concentrations of serum ranging from 0% to 100%. Serum stability at the highest concentrations of serum, about 70%–100%, that are physiological concentrations of serum found in the bloodstream is required.

A different concentration of serum, ranging from 0% to 100%, was placed into each well of a 96-well microtiter dish. BIV DOTAP:Chol–DNA:liposome complexes were added to the serum in the wells, and the plate was incubated at 37°C. The plate was removed at various time points out to 24 h and complexes in the wells were observed under the microscope. Precipitation of complexes in the wells was assessed. 100% stability was set at no precipitation observed. Results from this assay were compared to those obtained in the first assay. 100% stability of complexes was set at no decline of OD400 in assay #1 and no observed precipitation in assay #2 at each percent serum concentration, and the results were plotted (Figure 19.8).

The results showed serum stability at the highest concentrations of serum, about 70%–100%, that are physiological concentrations of serum found in the bloodstream. In addition, these complexes were also stable in no or low concentrations of serum, whereas, the complexes were unstable at 10%–50% serum, perhaps due to salt bridging. Therefore, *in vitro* optimization of serum stability for formulations of cationic complexes must be performed over a broad range of serum concentration to be useful for applications *in vivo*.

19.8 OPTIMIZED HALF-LIFE IN THE CIRCULATION

As stated above, the extruded BIV DOTAP:Chol–nucleic acid complexes are large enough so that they are not cleared rapidly by Kupffer cells in the liver and yet extravasate across tight barriers and diffuse through the target organ efficiently. Further addition of ligands to the surface of extruded BIV DOTAP:Chol–nucleic acid complexes does not significantly increase the mean particle size. Extravasation and penetration through the target organ and gene expression produced after transfection are not diminished. These modified formulations are positively charged and deliver nucleic acids efficiently into cells *in vitro* and *in vivo*. Because extruded BIV DOTAP: Chol–nucleic acid complexes with or without ligands have a 5 h half-life in the circulation, these complexes do not accumulate in the skin, hands, or feet. Extended half-life in the circulation is provided primarily by the formulation, preparation method, injection of optimal colloidal suspensions, and optimal nucleic acid/lipid ratio used for mixing complexes, serum stability, and size (200 to 450 nm). Therefore, these BIVs are ideal for use in the development of effective, targeted nonviral delivery systems that clearly require encapsulation of nucleic acids.

19.9 BROAD BIODISTRIBUTION OF OPTIMIZED LIPOSOME FORMULATIONS

Our "generic" BIV nucleic acid–liposome formulation transfects many organs and tissues efficiently after intravenous injection [14] and has demonstrated efficacy in animal models for lung cancer [15], pancreatic cancer [17], breast cancer [16],

hepatitis B and C, and cardiovascular diseases [35]. Therefore, optimization of the morphology of the complexes, the lipids used, flexibility of the liposomes and complexes, colloidal suspension, overall charge, serum stability, and half-life in circulation allows for efficient delivery and gene expression in many organs and tissues other than the lung. Apparently, these extruded DOTAP:Chol BIV–nucleic acid:liposome complexes can overcome the tendency to be adsorbed only by the endothelial cells lining the circulation surrounding the lungs described by other investigators [36]. However, as discussed above and below, we can further direct delivery to specific target tissues or cells by our targeted delivery strategies in combination with reversible masking used to bypass nonspecific transfection.

19.10 OPTIMIZATION OF TARGETED DELIVERY

Much effort has been made to specifically deliver nucleic acid–liposome complexes to target organs, tissues, or cells. Ligands that bind to cell surface receptors are usually attached to PEG and then attached to the cationic or anionic delivery vehicle. Due to shielding the positive charge of cationic complexes by PEG, delivery to the specific cell surface receptor can be accomplished by only a small fraction of complexes injected systemically. Furthermore, delivery of PEGylated complexes into the cell occurs predominantly through the endocytic pathway, and subsequent degradation of the bulk of the nucleic acid occurs in the lysosomes. Thus, gene expression is generally lower in the target cell than using the nonspecific delivery of highly efficient cationic complexes.

As discussed above, the vast majority of the injected PEGylated complexes bypasses the target cell. Apparently, the PEGylated complexes cannot utilize critical charge interactions for optimal transfection into cells by direct fusion. Inability to expose positive charge on the surface of optimized complexes results in the transfection of fewer cells. PEGylation was first used to increase the half-life of complexes in the circulation and to avoid uptake in the lung. However, this technology also destroys the ability to efficiently transfect cells. We were able to increase the half-life in circulation of BIVs to 5 h without the use of PEG. Because the extended half-life of BIVs is not too long, this delivery system does not result in the accumulation of complexes in nontarget tissues that circulate for 1–3 days. Some investigators have now reported targeted delivery that produces increased gene expression in the target cell over their nontargeted complexes. However, these nontargeted and targeted delivery systems are inefficient [37] compared to efficient delivery systems such as the BIVs.

In using the extruded BIV DOTAP:Chol–nucleic acid: liposome complexes, we produced an optimal half-life in the circulation without the use of PEG [14]. Extended half-life was produced primarily by the formulation, preparation method, injection of optimal colloidal suspensions, serum stability, and optimal nucleic acid/lipid ratio used for mixing complexes, and size (200 to 450 nm). Furthermore, we avoid uptake in the lungs using the negative charge of the ligands and shielding/deshielding compounds that can be added to the complexes used for targeting just prior to injection or administration *in vivo*. Our strategy to bypass nonspecific transfection is called reversible masking (U.S. Patent No. 7,037,520 B2). By addition of ligands using the novel approaches that we developed, adequate overall positive charge on the surface of complexes is preserved. In summary, we achieve optimal circulation time of the complexes, reach and deliver to the target organ, avoid uptake in nontarget tissues, and efficiently interact with the cell surface to produce optimal transfection.

19.11 EFFICIENT DISSEMINATION THROUGHOUT TARGET TISSUES AND MIGRATION ACROSS TIGHT BARRIERS

A primary goal for efficient *in vivo* delivery is to achieve extravasation into and penetration throughout the target organ/tissue ideally by noninvasive systemic administration. Without these events therapeutic efficacy is highly compromised for any treatment including gene and drug therapies. Achieving this goal is difficult due to the many tight barriers that exist in animals and people. Furthermore, many of these barriers become tighter in the transition from neonates to becoming adults. Penetration throughout an entire tumor is further hindered due to the increased interstitial pressure within most tumors [38–40]. We believe that nonviral systems can play a pivotal role in achieving target organ extravasation and penetration needed to treat or cure certain diseases. Our preliminary studies have shown that extruded BIV DOTAP:Chol–nucleic acid:liposome complexes can extravasate across tight barriers and penetrate evenly throughout entire target organs, and viral vectors cannot cross identical barriers. These barriers include the endothelial cell barrier in a normal mouse, the posterior blood retinal barrier in adult mouse eyes, complete and even diffusion throughout large tumors [15], and penetration through several tight layers of smooth muscle cells in the arteries of pigs [35]. Diffusion throughout large tumors was measured by expression of β-galactosidase or the pro-apoptotic gene p53 in about half of the p53-null tumor cells after a single injection of BIV DOTAP:Chol–DNA:liposome complexes into the center of a tumor. Transfected cells were evenly spread throughout the tumors. Tumors injected with complexes encapsulating plasmid DNA encoding p53 showed apoptosis in almost all of the tumor cells by TUNEL staining. Tumor cells expressing p53 mediate a bystander effect on neighboring cells perhaps due to upregulation by Fas ligand that causes non-transfected tumor cells to undergo apoptosis. Currently, we are investigating the mechanisms used by extruded DOTAP:Chol–nucleic acid:liposome complexes to cross barriers and penetrate throughout target organs. By knowing more about these mechanisms, we hope to develop more robust nonviral gene therapeutics.

19.12 OPTIMIZATION OF PLASMIDS FOR *IN VIVO* GENE EXPRESSION

Delivery of DNA and subsequent gene expression may be poorly correlated [18,41]. Investigators may focus solely on the delivery formulation as the source of poor gene expression. In many cases, however, the delivery of DNA into the nucleus of a particular cell type may be efficient, although little or no gene expression is achieved. The causes of poor gene expression can be numerous. The following issues should be considered independent of the delivery formulation including suboptimal promoter–enhancers in the plasmid, poor preparation of plasmid DNA, and insensitive detection of gene expression.

Plasmid expression cassettes typically have not been optimized for animal studies. For example, many plasmids lack a full-length CMV promoter–enhancer. Over 100 variations of the CMV promoter–enhancer exist, and some variations produce greatly reduced or no gene expression in certain cell types [18]. Even commercially available plasmids contain suboptimal CMV promoters–enhancers, although these plasmids are advertised for use in animals. Furthermore, upon checking the company data for these plasmids, one would discover that these plasmids have never been tested in animals and have been tested in only one or two cultured cell lines. Conversely, plasmids that have been optimized for overall efficiency in animals may not be best for transfection of certain cell types *in vitro* or *in vivo*. For example, many investigators have shown that optimal CMV promoters–enhancers produce gene expression at levels several orders of magnitude less in certain cell types. In addition, one cannot assume that a CMV promoter that expresses well within the context of a viral vector, such as adenovirus, will function as well in a plasmid-based transfection system for the same cell context. Virus proteins produced by the viral vector are required for producing high levels of mRNA by the CMV promoter in specific cell nuclei.

Ideally, investigators design custom promoter–enhancer chimeras that produce the highest levels of gene expression in their target cells of interest. Recently, we designed a systematic approach for customizing plasmids used for breast cancer gene therapy using expression profiling [18]. Gene therapy clinical trials for cancer frequently produce inconsistent results. We believe that some of this variability could result from differences in transcriptional regulation that limit expression of therapeutic genes in specific cancers. Our systemic liposomal delivery of a nonviral plasmid DNA showed efficacy in animal models for several cancers. However, we observed large differences in the levels of gene expression from a CMV promoter–enhancer between lung and breast cancers. To optimize gene expression in breast cancer cells *in vitro* and *in vivo*, we created a new promoter–enhancer chimera to regulate gene expression. Serial analyses of gene expression (SAGE) data from a panel of breast carcinomas and normal breast cells predicted promoters that are highly active in breast cancers, for example the glyceraldehyde 3-phosphate dehydrogenase (GAPDH) promoter. Furthermore, GAPDH is upregulated by hypoxia which is common in tumors. We added the GAPDH promoter, including the hypoxia enhancer sequences, to our *in vivo* gene expression plasmid. The novel CMV-GAPDH promote–enhancer showed up to 70-fold increased gene expression in breast tumors compared to the optimized CMV promoter–enhancer alone. No significant increase in gene expression was observed in other tissues. These data demonstrate tissue-specific effects on gene expression after nonviral delivery and suggest that gene delivery systems may require plasmid modifications for the treatment of different tumor types. Furthermore, expression profiling can facilitate the design of optimal expression plasmids for use in specific cancers.

Several reviews have stated that nonviral systems are intrinsically inefficient compared to viral systems. However, as discussed above, one must separate issues of the delivery vehicle versus the plasmid that is delivered. Case in point, we have shown that our extruded liposomes optimized for systemic delivery could out-compete delivery using a lentivirus. For example, we have compared SIVmac239, a highly noninfectious virus, with nonviral delivery of SIVmac239 DNA complexed to BIVs in adult rhesus macaques after injection into the saphenous vein of the leg. Our data showed that the monkeys injected with SIV DNA encapsulated in DOTAP:Chol BIVs were infected 4 days postinjection, and high levels of infection were produced in these monkeys at 14 days postinjection. Furthermore, higher levels of SIV RNA in the blood were produced using our BIV liposomes for delivery versus using the SIV virus. CD4 counts were measured before and after injections. CD4 levels dropped in all monkeys to the lowest levels ever detected in the macaques in any experiment by 28 days postinjection, the first time point at which these counts were measured postinjection. All monkeys had clinical SIV infections and lost significant weight by day 28. These results were surprising because SIVmac239 is not highly infectious, and monkeys become sick with SIV infection only after several months or years postinjection with SIVmac239 virus. Therefore, we were able to induce SIV infection faster using our nonviral delivery of SIV plasmid DNA. In this case, we delivered a replication competent plasmid so that gene expression increased over time post-transfection. Our delivery system was highly efficient and exceeded that of the lentivirus. The critical feature in this nonviral experiment was the plasmid DNA that was delivered.

Plasmids can be engineered to provide for specific or long-term gene expression, replication, or integration. Persistence elements, such as the inverted terminal repeats from adenovirus or adeno-associated virus, have been added to plasmids to prolong gene expression *in vitro* and *in vivo*. Apparently, these elements bind to the nuclear matrix thereby retaining the plasmid in cell nuclei. For regulated gene expression, many different inducible promoters are used that promote expression only in the presence of a positive regulator or in the absence of a negative regulator. Tissue specific promoters have been used for the production of gene expression exclusively in the target cells. As discussed in the previous paragraph, replication competent plasmids or plasmids containing sequences for autonomous replication can be included that provide prolonged gene expression. Other plasmid-based strategies produce site-specific integration or homologous recombination within the host cell genome (Reviewed in Ref. [42]). Integration of a cDNA into a specific "silent site" in the genome could provide long-term gene expression without disruption of normal cellular functions. Homologous recombination could correct genetic mutations upon integration of wild-type sequences that replace mutations in the genome. Plasmids that contain fewer bacterial sequences and that produce high yield upon growth in *Escherichia coli* are also desirable.

19.13   OPTIMIZATION OF PLASMID DNA PREPARATIONS

The transfection quality of plasmid DNA is dependent on the preparation protocol and training of the person preparing the DNA. For example, we performed a blinded study asking three people to make DNA preparations of the same plasmid from the same box of a Qiagen Endo-Free plasmid preparation kit. One person then mixed all of the DNA–liposome complexes on the same morning using a single vial of liposomes. One person performed all tail vein injections, harvesting of tissues, preparation of extracts from tissues, and reporter gene assays on the tissue extracts. *In vivo* gene expression differed 30-fold among these three plasmid DNA preparations.

One source for this variability is that optimized methods to detect and remove contaminants from plasmid DNA preparations have not been available. We have identified large amounts of contaminants that exist in laboratory and clinical grade preparations of plasmid DNA. These contaminants co-purify with DNA by anion exchange chromatography and by cesium chloride density gradient centrifugation. Endotoxin removal does not remove these contaminants. HPLC cannot detect these contaminants. Therefore, we developed three proprietary methods for the detection of these contaminants in plasmid DNA preparations. We can now make clinical grade (GMP) DNA that does not contain these contaminants. To provide the greatest efficacy and levels of safety, these contaminants must be assessed and removed from plasmid DNA preparations. These contaminants belong to a class of molecules known to inhibit both DNA and RNA polymerase activities. Therefore, gene expression post-transfection can be increased by orders of magnitude if these contaminants are removed from DNA preparations. The presence of these contaminants in DNA also precludes high dose delivery of DNA–liposome complexes intravenously. Our group and other investigators have shown that intravenous injections of high doses of improved liposomes alone cause no adverse effects in small and large animals.

Some investigators have removed the majority of CpG sequences from their plasmids and report reduced toxicity after intravenous injections of cationic liposomes complexed to these plasmids [43]. However, only low doses containing up to 16.5 μg of DNA per injection into each mouse were shown to reduce toxicity. To achieve efficacy for cancer metastases, particularly in mice bearing aggressive tumors, most investigators are interested in injecting higher doses in the range of 50–150 μg of DNA per mouse. Therefore, removal of CpG sequences from plasmid based gene therapy vectors will not be useful for these applications because no difference in toxicity was shown after intravenous injections of these higher doses of plasmids, with or without reduced CpG sequences, complexed to liposomes [43]. Therefore, we believe that removal of the other contaminants in current DNA preparations, discussed above, is the major block to the safe intravenous injection of high doses of DNA–liposome complexes.

19.14 DETECTION OF GENE EXPRESSION

Thought should also be given to choosing the most sensitive detection method for every application of nonviral delivery rather than using the method that seems most simple. For example, detection of β-galactosidase expression is far more sensitive than that for the green fluorescent protein (GFP). Specifically, 500 molecules of β-galactosidase (β-gal) per cell are required for detection using X-gal staining. Whereas, about one million molecules of GFP per cell are required for direct detection. Furthermore, detection of GFP may be impossible if the fluorescence background of the target cell or tissue is too high. Detection of chloramphenicol acetyltransferase (CAT) is extremely sensitive with little or no background detected in untransfected cells. Often, assays for CAT expression can provide more useful information than using β-gal or GFP as reporter genes.

Few molecules of luciferase in a cell can be detected by luminescence assays of cell or tissue extracts post-transfection. The sensitivity of these assays is highly dependent on the type of instrument used to measure luminescence. However, luciferase results may not predict the therapeutic potential of a nonviral delivery system. For example, if several hundred or thousand molecules per cell of a therapeutic gene are required to produce efficacy for a certain disease, then production of only few molecules will not be adequate. If only few molecules of luciferase are produced in the target cell using a specific nonviral delivery system, then the investigator may be misled in using this system for therapeutic applications.

Furthermore, noninvasive detection of luciferase expression *in vivo* is not as sensitive as luminescence assays of cell or tissue extracts post-transfection. Recently, some colleagues of mine tried cooled charge coupled device (CCD) camera imaging on live mice after intravenous injection of other cationic liposomes complexed to plasmid DNA encoding luciferase, and they were not able to detect any transfection. However, these liposomal delivery systems had been used to detect luciferase by luminescence assays of organ extracts. My colleagues detected luciferase expression by CCD imaging after intravenous injections of BIV DOTAP:Chol–luciferase DNA: liposome complexes [44]. Because the luciferase protein is short-lived, maximal expression was detected at 5 h post-transfection. Whereas, detection of HSV-TK gene expression

using microPET imaging in the same mice was highest at 24 h posttransfection. In contrast to luciferase, the CAT protein accumulates over time, and therefore, the investigator is not restricted to a narrow time frame for assaying gene expression. Furthermore, detection of CAT seems to be more sensitive than CCD imaging of luciferase following intravenous injections of DNA–liposome complexes. However, the animals must be sacrificed in order to perform CAT assays on tissue or organ extracts. In summary, further work is still needed to develop *in vivo* detection systems that have high sensitivity and low background.

19.15 OPTIMIZATION OF DOSE AND FREQUENCY OF ADMINISTRATION

To establish the maximal efficacy for the treatment of certain diseases or for the creation of robust vaccines, injections, or administrations of the nonviral gene therapeutic, etc., via different routes may be required. For particular treatments, one should not assume that one delivery route is superior to others without performing the appropriate animal experiments. In addition, people with the appropriate expertise should perform the injections and administrations. In our experience, only a minority of people who claim expertise in performing tail vein injections can actually perform optimal injections.

The optimal dose should be determined for each therapeutic gene or other nucleic acid that is administered. The investigator should not assume that the highest tolerable dose is optimal for producing maximal efficacy. The optimal administration schedule should also be determined for each therapeutic gene or other nucleic acid. To progress faster, some investigators have simply used the same administration schedule that they used for chemotherapeutics. The investigator should perform *in vivo* experiments to determine when gene expression or efficacy drops significantly. Most likely, re-administration of the nonviral gene therapeutic is not necessary until this drop occurs. Loss of the therapeutic gene product will vary with the half-life of the protein produced. Therefore, if a therapeutic protein has a longer half-life, then the gene therapy could be administered less frequently.

19.16 SUMMARY

Overcoming some hurdles remain in the broad application of nonviral delivery, however, we are confident that we will successfully accomplish the remaining challenges soon. Furthermore, we predict that eventually the majority of gene therapies will utilize artificial reagents that can be standardized and regulated as drugs rather than biologics. We will continue to incorporate the molecular mechanisms of viral delivery that produce efficient delivery to cells into artificial systems. Therefore, the artificial systems, including liposomal delivery vehicles, will be further engineered to mimic the most beneficial parts of the viral delivery systems while circumventing their limitations. We will also maintain the numerous benefits of the liposomal delivery systems discussed in this chapter.

REFERENCES

1. Felgner, P.L. et al., Lipofection: A highly efficient lipid-mediated DNA transfection procedure, *Proc Natl Acad Sci U S A*, 84, 7413, 1987.
2. Felgner, J.H. et al., Enhanced gene delivery and mechanism studies with a novel series of cationic lipid formulations, *J Biol Chem*, 269, 2550, 1994.
3. Zhu, N. et al., Systemic gene expression after intravenous DNA delivery in adult mice, *Science*, 261, 209, 1993.
4. Philip, R. et al., In Vivo gene delivery: Efficient transfection of T lymphocytes in adult mice, *J Biol Chem*, 268, 16087, 1993.
5. Solodin, I. et al., A novel series of amphiphilic imidazolinium compounds for in vitro and in vivo gene delivery, *Biochemistry*, 34, 13537, 1995.
6. Liu, Y. et al., Cationic liposome-mediated intravenous gene delivery, *J Biol Chem*, 270, 24 864, 1995.
7. Liu, F. et al., Factors controlling the efficiency of cationic lipid-mediated transfection in vivo via intravenous administration, *Gene Ther*, 4, 517, 1997.
8. Liu, Y. et al., Factors influencing the efficiency of cationic-liposome mediated intravenous gene delivery, *Nat Biotechnol*, 15, 167, 1997.
9. Thierry, A.R. et al., Systemic gene therapy: Biodistribution and long-term expression of a transgene in mice, *Proc Natl Acad Sci U S A*, 92, 9742, 1995.
10. Tsukamoto, M. et al., Gene transfer and expression in progeny after intravenous DNA injection into pregnant mice, *Nat Genet*, 9, 243, 1995.
11. Aksentijevich, I. et al., In vitro and in vivo liposome-mediated gene transfer leads to human MDR1 expression in mouse bone marrow progenitor cells, *Hum Gene Ther*, 7, 1111, 1996.
12. Xu, Y. and Szoka, F.C., Mechanism of DNA release from cationic liposome/DNA complexes used in cell transfection, *Biochemistry*, 35, 5616, 1996.
13. Yotnda, P. et al., Bilamellar cationic liposomes protect adenovectors from preexisting humoral immune responses, *Mol Ther*, 5, 233, 2002.
14. Templeton, N.S. et al., Improved DNA: Liposome complexes for increased systemic delivery and gene expression, *Nat Biotechnol*, 15, 647, 1997.
15. Ramesh, R. et al., Successful treatment of primary and disseminated human lung cancers by systemic delivery of tumor suppressor genes using an improved liposome vector, *Mol Ther*, 3, 337, 2001.
16. Tirone, T.A. et al., Insulinoma induced hypoglycemic death in mice is prevented with beta cell specific gene therapy, *Ann Sur*, 233, 603, 2001.
17. Shi, H.Y. et al., Inhibition of breast tumor progression by systemic delivery of the maspin gene in a syngeneic tumor model, *Mol Ther*, 5, 755, 2002.
18. Lu, H. et al., Enhanced gene expression in breast cancer cells in vitro and tumors in vivo, *Mol Ther*, 6, 783, 2002.
19. Li, S. et al., Targeted delivery via lipidic vectors, in *Vector Targeting for Therapeutic Gene Delivery*, D.T. Curiel and J.T. Douglas, Editors, Wiley-Liss Inc.: Hoboken, New Jersey, 2002, p. 17.
20. Pirollo, K.F. et al., Immunoliposomes: A targeted delivery tool for cancer treatment, in *Vector Targeting for Therapeutic Gene Delivery*, D.T. Curiel and J.T. Douglas, Editors, Wiley-Liss Inc.: Hoboken, New Jersey, 2002, p. 33.
21. Sternberg, B., Morphology of cationic liposome/DNA complexes in relation to their chemical composition, *J Liposome Res*, 6, 515, 1996.
22. Senior, J. et al., Influence of surface hydrophilicity of liposomes on their interaction with plasma protein and clearance from the circulation: Studies with poly(ethylene glycol)-coated vesicles, *Biochim Biophys Acta*, 1062, 77, 1991.
23. Papahadjopoulos, D. et al., Sterically stabilized liposomes: Improvements in pharmacokinetics and antitumor therapeutic efficacy, *Proc Natl Acad Sci U S A*, 88, 11460, 1991.
24. Gabizon, A. et al., Prolonged circulation time and enhanced accumulation in malignant exudates of doxorubicin encapsulated in polyethylene-glycol coated liposomes, *Cancer Res*, 54, 987, 1994.
25. Gordon, K.B. et al., Hand-foot syndrome associated with liposome-encapsulated doxorubicin therapy, *Cancer*, 75, 2169, 1995.
26. Uziely, B. et al., Liposomal doxorubicin: antitumor activity and unique toxicities during two complementary phase I studies, *J Clin Oncol*, 13, 1777, 1995.
27. Gustafsson, J. et al., Complexes between cationic liposomes and DNA visualized by cryo-TEM, *Biochim Biophys Acta*, 1235, 305, 1995.
28. Templeton, N.S. and Lasic, D.D., New directions in liposome gene delivery, *Mol Biotechnol*, 11, 175, 1999.
29. Behr, J.-P. et al., Efficient gene transfer into mammalian primary endocrine cells with lipopolyamine-coated DNA, *Proc Natl Acad Sci U S A*, 86, 6982, 1989.
30. Felgner, P.L. and Ringold, G.M., Cationic liposome-mediated transfection, *Nature*, 337, 387, 1989.
31. Pinnaduwage, P. and Huang, L., The role of protein-linked oligosaccharide in the bilayer stabilization activity of glycophorin A for dioleoylphosphatidylethanolamine liposomes, *Biochim Biophys Acta*, 986, 106, 1989.
32. Leventis, R. and Silvius, J.R., Interactions of mammalian cells with lipid dispersions containing novel metabolizable cationic amphiphiles, *Biochim Biophys Acta*, 1023, 124, 1990.
33. Rose, J.K. et al., A new cationic liposome reagent mediating nearly quantitative transfection of animal cells, *Biotechniques*, 10, 520, 1991.
34. Loeffler, J.P. and Behr, J.-P., Gene transfer into primary and established mammalian cell lines with lipopolyamine-coated DNA, *Methods Enzymol*, 217, 599, 1993.
35. Templeton, N.S. et al., Non-viral vectors for the treatment of disease, *Keystone Symposia on Molecular and Cellular Biology of Gene Therapy*, Salt Lake City, Utah, 1999.
36. Mislick, K.A. and Baldeschwieler, J.D., Evidence for the role of proteoglycans in cation-mediated gene transfer, *Proc Natl Acad Sci U S A*, 93, 12349, 1996.
37. Hood, J.D. et al., Tumor regression by targeted gene delivery to the neovasculature, *Science*, 296, 2404, 2002.
38. Jain, R.K., Barriers to drug delivery in solid tumors, *Sci Am*, 271, 58, 1994.
39. Jain, R.K., Haemodynamic and transport barriers to the treatment of solid tumours, *Int J Radiat Biol*, 60, 85, 1991.
40. Jain, R.K., Transport of molecules, particles, and cells in solid tumors, *Annu Rev Biomed Eng*, 1, 241, 1999.
41. Handumrongkul, C. et al., Distinct sets of cellular genes control the expression of transfected, nuclear-localized genes, *Mol Ther*, 5, 186, 2002.
42. Gene targeting protocols, in *Methods in Molecular Biology*, E.B. Kmiec, Editor, Vol. 133, Humana Press: Totowa, New Jersey, 2000.
43. Yew, N.S. et al., CpG depleted plasmid DNA vectors with enhanced safety and long-term gene expression in vivo, *Mol Ther*, 5, 731, 2002.
44. Iyer, M. et al., Noninvasive imaging of cationic lipid-mediated delivery of optical and PET reporter genes in living mice, *Mol Ther*, 6, 555, 2002.

20 Photochemical Internalization: A Technology for Efficient and Site-Specific Gene Delivery

Anette Bonsted, Anders Høgset, and Kristian Berg

CONTENTS

20.1 INTRODUCTION

Photochemical internalization (PCI) is a technology that enables light-directed delivery of a variety of therapeutic molecules into the cell cytosol. The technique can be used *in vitro* and *in vivo* for site-specific delivery of proteins, DNA carried by nonviral and viral vectors, peptide nucleic acids (PNA), and small interfering RNA (siRNA). PCI is based on light activation of amphiphilic photosensitizers that localize in the membranes of the endocytic vesicles. When photosensitizer-containing cells or tissues are exposed to light at wavelengths absorbed by the photosensitizers, photochemical reactions are induced that rupture the membranes of the endocytic vesicles. Hence, the constituents of these vesicles (e.g., drugs or nucleic acids) can be released into the cytosol where they may act on their target or further translocate to the nucleus. As the effect of the photochemical treatment is dependent on light exposure, the enhancement of drug- or gene delivery is achieved only at illuminated regions. This chapter presents the background of PCI and its role for obtaining efficient delivery of nucleic acids.

20.2 PRINCIPLES OF PHOTOCHEMICAL INTERNALIZATION

The PCI technology derives from the field of photodynamic therapy (PDT). In PDT, the phototoxic effects induced by a photosensitizer, light, and oxygen are used therapeutically [1]. It is a minimally invasive treatment with great potential in malignant disease and premalignant conditions. PDT involves the application of a photosensitizer in combination with the targeted delivery of light, at a wavelength specific for the activation of the photosensitizer. Usually, wavelengths in the 600–800 nm range are used for optimal therapeutic effects. Upon light-dependent activation, the photosensitizers undergo photophysical reactions that lead to generation of reactive oxygen species (ROS) under aerobic conditions, most predominantly $^1O_2$. In PDT, the photochemical generation of ROS results in apoptosis or necrosis, harnessed therapeutically for the purpose of eliminating malignant tissue. However, as $^1O_2$ has a short lifetime ($<0.04\,\mu s$) and a short range of action (10–20 nm) in cells [2], only structures very close to the photosensitizer will be affected after light exposure. The

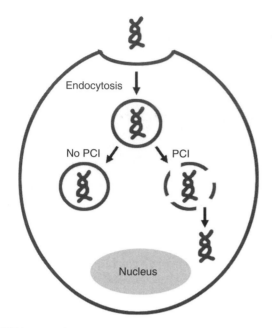

FIGURE 20.1 PCI-mediated delivery of nucleic acids. The nucleic acids are endocytosed and localize in endocytic vesicles. If not translocated, the nucleic acids are degraded inside the endocytic vesicles. The PCI treatment (i.e., light exposure of photosensitizer-treated cells) permeabilizes the endocytic membranes, leading to the release of the nucleic acids into the cytosol.

affected structures depend thus on the localization of the photosensitizer in the cell. PCI is based on the observation that light exposure of cells preloaded with amphiphilic photosensitizers that localize to the membranes of the endocytic vesicles leads to permeabilization of the vesicles [3]. Hence, the PCI treatment can induce a release of endocytosed compounds into the cell cytosol (Figure 20.1). This finding is of great importance because the translocation of genes from the endocytic vesicles into the cytosol is an important step and a possible barrier in the gene delivery process. Since the first publication in 1999, PCI has been shown to significantly enhance the delivery of a variety of macromolecules to cells and tissues.

To date, aluminium phthalocyanine with two sulfonate groups on adjacent phenyl rings (AlPcS$_{2a}$) and meso-tetraphenylporphine with two sulfonate groups on adjacent phenyl rings (TPPS$_{2a}$) are the most efficient photosensitizers for enhancing gene delivery (4; Figure 20.2). Photosensitizers that localize to other cellular structures than the endocytic vesicles are not efficient for inducing the PCI effect [4–6]. The light source used to induce the photochemical reactions may be any source emitting light of wavelengths absorbed by the photosensitizer. Typically, a red light source (peak wavelength at 670 nm) is used for the excitation of AlPcS$_{2a}$ and a blue light source (peak wavelength at 420 nm) for TPPS$_{2a}$.

20.2.1 PHOTOCHEMICAL INTERNALIZATION METHOD

The PCI procedure must be performed in subdued light to avoid unwanted uncontrollable activation of the photosensitizer and to protect the cells from undesirable photochemical damage. As an example, in *in vitro* assays, the cells are first incubated with cell culture medium containing the photosensitizer (either AlPcS$_{2a}$ or TPPS$_{2a}$) for 18 h at 37°C in a CO$_2$ incubator. Then, the cells are washed three times in photosensitizer-free medium, and chased for 4 h at 37°C prior to exposure to red or blue light at light doses inducing around 50%–70% cell survival. The DNA, PNA, or siRNA is added at different time points prior to light exposure, so that they localize in the endocytic vesicles at the time of light exposure (see Figure 20.3 for suggested time intervals).

An alternative PCI technique for obtaining increased transfection is to expose photosensitizer-treated cells to light before the gene vector is added [7]. The mechanism behind this technique is not fully clarified. However, it has been proposed that fusion events between the photochemically disrupted endocytic vesicles and intact vesicles containing the vector may lead to cytosolic release of the vector and increased transfection levels [7]. Illumination before gene delivery may be advantageous in some clinical situations. For example, if PCI is employed as an adjuvant treatment of residual disease after surgical removal of tumors, it would be an advantage to be able to apply both the light treatment and the therapeutic molecule at the end of surgical procedures and immediately following each other.

For *in vivo* treatment procedures, AlPcS$_{2a}$ is the preferred photosensitizer due to its higher extinction coefficient compared

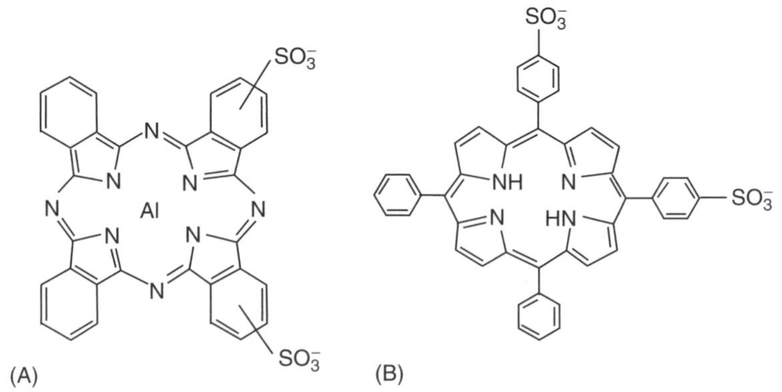

(A)   (B)

FIGURE 20.2 The photosensitizers AlPcS$_{2a}$ (A) and TPPS$_{2a}$ (B).

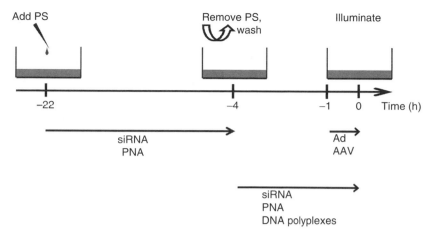

FIGURE 20.3 Experimental scheme for PCI treatment *in vitro*. The cells are incubated with the photosensitizer (PS) for 18 h. Then, the cells are washed and incubated in photosensitizer-free medium for 4 h prior to light exposure to disrupt the membranes of the endocytic vesicles. To perform PCI of siRNA or PNA, the cells can be incubated with the siRNA or PNA molecules at the same time as the photosensitizer and for a total of 18 h. Alternatively, siRNA or PNA can be added to the cells 4 h prior to light exposure. To perform PCI of DNA polyplexes, the DNA polyplexes are added to the cells 4 h prior to light exposure. Finally, for PCI of adenovirus and adeno-associated virus, the cells are chased for 3 h in photosensitizer-free medium prior to the adding of the virus to the cells.

to TPPS$_{2a}$ in the wavelength region where light penetration through tissue is high (above 600 nm). In order to perform PCI treatment on mouse xenograft models, the photosensitizer AlPcS$_{2a}$ is first injected intraperitoneally, intravenously, or intratumorally at 10 mg/kg [8,9]. Following the injection of the photosensitizer, the animals must be kept in the dark for at least 4 days in order to avoid undesirable photochemical damage to tissues [9]. Forty-eight hours after the injection of the photosensitizer, an intratumoral injection of the gene vector is performed under subdued light. After 6 h, the animals are covered in aluminium foil except above the tumor area, and the tumors are exposed to red light, typically from a diode laser. The photochemical doses must be optimized for the laser and the tumor model. Typically, light doses around 20–40 J/cm^2 are delivered to tumors around 100 mm^3 from a light source with an irradiance of less than 150 mW/cm^2.

20.3 EFFECTS OF PCI ON THE DELIVERY OF GENES AND OLIGONUCLEOTIDES *IN VITRO*

20.3.1 PCI WITH NONVIRAL GENE VECTORS

The physicochemical properties of plasmid DNA impede cell membrane translocation. Therefore, a variety of nonviral delivery systems have been proposed. Nonviral-based carriers, like cationic lipoplexes and polyplexes, are usually taken up by cells through charge-based binding and endocytosis. Once inside the endosomes, these complexes need to efficiently escape into the cytosol prior to acidification, in order to avoid lysosomal degradation of the DNA. Some vector materials are known to be able to destabilize the endosomal membrane, by acting as a proton sponge or a fusogen (e.g., polyethylenimine

[PEI], lipids) [10]. Nevertheless, the effect of PCI on gene delivery has been demonstrated both for gene vectors with low efficiency for endosomal release, such as polylysine (PLL), and for gene vectors able to induce endosomal release on their own, such as cationic lipids and PEI [4,11].

The first documentation of the PCI effect was obtained using a plasmid carrying the enhanced green fluorescent protein (EGFP) gene and complexed with PLL [5]. Further, it has been shown that the PCI-induced enhancement of transgene expression depends on the photochemical dose and the cell line, indicating the possible involvement of cellular processes or pathways that depending on the cell type might be differently affected by the photochemical treatment [12]. Moreover, the effect of PCI on transfection of DNA/cationic lipid complexes was more variable than found for PLL, and it was highly dependent on the composition of the lipid vector [11,13]. PCI also improved the efficiency of various PEI-based vectors, although PEI is one of the most potent nonviral vectors known [14,15]. However, the relative effect of PCI was most pronounced at doses of DNA/PEI that were suboptimal on their own. For example, PCI efficiently enhanced transfection of DNA/PEI polyplexes at N/P ratios ≤6. PEI displays only moderate endosomolytic activity in such formulations [16]. This is probably because the concentration of PEI inside the endocytic vesicles is too low to perform its endosomolytic function at these ratios, and therefore, the transfection may be enhanced by PCI. For *in vivo* gene therapy approaches, the observation that PCI is particularly suited to enhance gene delivery when the amount of PEI is low might be important. Due to the *in vivo* extracellular barriers, the amount of vector reaching the target cells will often be very low and probably often considerably below the threshold where PEI is able to induce efficient endosomal escape.

20.3.2 PCI with Viral Gene Vectors

In a variety of cancer cell lines, the transgene expression from several adenovirus vectors was substantially increased in PCI-treated cells compared to photochemically untreated cells. The photochemical effect on transduction was most pronounced at lower virus doses, where in some cases the same transduction efficiency could be achieved with a 20 times lower virus dose than with conventional infection. Moreover, the photochemical effect was dependent on the photochemical dose delivered to the cells and on the cell line tested [17].

Adenoviral transduction is believed to be an efficient process [18,19]. Nevertheless, the photochemical enhancement of transduction was not due to a change in the uptake mechanism of the adenoviral particles into the cells, as the viral infection occurred through clathrin-dependent, receptor-mediated endocytosis in both photochemically treated and untreated cells [20]. Results using real-time polymerase chain reaction (PCR) and fluorescence in situ hybridization (FISH) showed that the level of viral DNA in the nucleus of photochemically treated samples was increased compared to photochemically untreated samples. In addition, real-time PCR was also applied to measure the photochemically induced increase in the level of transgene mRNA. The increase in transgene mRNA level after photochemical treatment was higher than the fold increase observed at both the DNA and the protein level. This was not due to the CMV promoter, as adenovirus carrying the SV40-promoter instead of the CMV promoter responded in a similar manner to PCI. Finally, the cellular localization of the viral particles at the time of illumination was important to achieve increased transgene expression [6].

Cells lacking expression of the coxsackie- and adenovirus receptor (CAR) are resistant to adenoviral infection. However, as the adenovirus particles are negatively charged, they can be coated with positively charged polymers (e.g., polylysine) to transduce CAR-negative cells. Transduction is thus enhanced due to improved binding to the negatively charged cell surfaces. Photochemical treatment of cells infected with adenovirus/polylysine complexes enabled up to 38-fold increase in the percentage of β-galactosidase positive cells compared to photochemically untreated cells infected with uncomplexed virus [21].

The effect of PCI on transduction with adeno-associated virus (AAV) has been evaluated and compared to that of adenovirus or plasmid/polylysine complexes in two glioblastoma cell lines, U-87 Mg and GaMg [22]. PCI had a similar effect on transfection with plasmid/polylysine complexes in these cell lines. Therefore, it was of interest to use these cell lines to compare the effect of PCI on transduction with different viral vectors. Adenoviral vectors have often been preferred for cancer gene therapy. However, recent studies have pointed out that AAV vectors seem highly suitable for gene delivery to malignant tumors [23,24]. PCI increased the transgene expression from AAV in GaMg, but not in U-87 Mg cells. The lack of effect in U-87 Mg was independent of viral dose, photochemical dose, and treatment sequence (e.g., photochemical treatment prior to or after infection). In contrast, PCI enhanced transgene expression from a recombinant adenovirus vector in both cell lines, although the effect was more pronounced in U-87 Mg than in GaMg. These results support findings from others, describing that different steps of intracellular trafficking may be limiting for AAV-based vectors depending on the cell line tested [25–27]. Of note, the PCI effect of AAV5 was the highest in the cell line that demonstrated the lowest efficiency of AAV5 transduction without PCI, perhaps due to a less efficient AAV5 escape from endosomes in the GaMg cell line compared to the U-87 Mg cell line. However, further studies are needed to identify different barriers in intracellular trafficking between cell lines and viral vectors.

20.3.3 Double Targeting

PCI can be combined with biologically targeted gene vectors in order to obtain both specific binding of the gene vector to target cells and light-enhanced release from the endocytic vesicles. Such dual targeting may reduce the risk of side effects due to inadvertent expression of transgenes in nontarget cells. PCI has been combined with several epidermal growth factor receptor (EGFR)-targeted compounds, such as DNA polyplexes [15], adenovirus [28], and the protein toxin saporin [29]. EGFR is a frequently used target due to its overexpression in many human tumors, and antibodies towards EGFR and Her2/neu are approved for clinical use.

The magnitude of enhancement of transgene expression due to photochemical treatment depended on the formulation of the targeted compound, the cell line, and the photochemical dose. In a study by Kloeckner and colleagues, nonviral DNA polyplexes containing polyethylenimine (PEI) conjugated with the EGF protein as a cell-binding ligand for EGFR-mediated endocytosis and polyethylene glycol (PEG) for masking the polyplex surface charge were used in combination with PCI treatment [15]. PCI of such targeted PEG-PEI/DNA polyplexes enabled high and EGFR-specific gene transfer activity in cells. In another study by Bonsted and colleagues, the importance of PEG in such targeted complexes was confirmed [28]. Recombinant adenoviral particles were coated with pDMAEMA-based polymers coupled with ligands towards EGFR. An additional incorporation of polyethylene glycol (PEG)-conjugated pDMAEMA into the viral complexes provided charge shielding of the viral complexes and increased the specificity of EGFR-mediated transduction. Furthermore, transgene expression was efficiently enhanced by photochemical treatment of cells infected with the PEGylated and EGFR-targeted complexes [28]. These findings have implications for the possible use of PCI of targeted Ad *in vivo*, as conjugation of PEG to viral vectors extends the circulation time and reduces the toxicity and immunogenicity of the vectors.

A double targeting effect may also be obtained by the use of promoters that are activated by the photochemical treatment, such as heat-shock proteins (HSPs). Consequently, it has been investigated whether PCI also could activate HSPp, a promoter from the HSP family gene, and thereby stimulate transcription of a HSPp-controlled gene, the HSP70. It was shown that the photochemical treatment could enhance expression of cellular HSP70, which correlated with a photochemically enhanced expression (approximately twofold, at PCI-optimal doses) of the HSPp-controlled gene integrated in the genome. Furthermore, PCI also enhanced expression of the HSPp-controlled episomal transgene delivered as a plasmid. However, in plasmid-based transfection, the PCI-mediated enhancement with HSPp did not exceed the enhancement achieved with the constitutively active CMV promoter [30].

20.3.4 PCI OF THERAPEUTIC GENES

PCI has been evaluated for the delivery of therapeutic genes to cells. Plasmids carrying genes encoding p53 or HSV-tk (herpes simplex virus thymidine kinase) and adenovirus carrying a gene encoding TRAIL (tumor necrosis factor related apoptosis inducing ligand) have been used for this approach [14,31,32]. Enhanced transgene expression and biological activity of all three transgenes was demonstrated after photochemical treatment of cells. Thus, genes delivered by PCI are not only expressed, but are also able to perform a therapeutic function resulting in therapeutic effects in photochemically treated cells.

20.3.5 PCI OF OLIGONUCLEOTIDES

PNAs are oligonucleotide mimics that can specifically hybridize to DNA or mRNA, thus inhibiting the transcription and translation of their complementary target sequence. As PNAs are taken up into cells by endocytosis and tend to accumulate in endosomes, PCI is a candidate for enhancing their delivery into the cytosol. Folini and colleagues were the first to demonstrate a PCI-induced release of PNA into the cytoplasm of cells [33]. In their approach, they used naked PNA targeting the catalytic component of human telomerase reverse transcriptase (hTERT-PNA). After photochemical treatment, cells treated with the hTERT-PNA showed a marked inhibition of telomerase activity and a reduced cell survival, which was not observed after treatment with hTERT-PNA alone. Moreover, in a direct comparison, the PCI technology proved to be more efficient to internalize the hTERT-PNA than an alternative strategy based on the HIV-Tat internalization protein. In another study, Shiraishi and Nielsen showed that PCI could enhance the antisense effects (cytosolic/nuclear) of different peptide nucleic acid–peptide conjugates (Tat, Arg7, KLA) up to two orders of magnitude [34]. In addition, efficient gene silencing of the *S100A4* gene by PNA delivered by PCI has also been demonstrated [35]. These results emphasize the importance of endosomal release for cellular activity of this type of drug delivery.

SiRNA molecules, which are functional mediators of RNA interference, also need to enter the cytoplasm of cells in order to be able to interrupt translation of a specific protein by inducing posttranscriptional gene silencing. PCI has been employed to facilitate the escape of siRNA molecules complexed with lipofectamine from endocytic vesicles [36]. Combining antiepidermal growth factor receptor (EGFR) siRNA treatment

with PCI induced a 10-fold increased efficiency in knockdown of EGFR compared to siRNA treatment alone. Thus, lower doses of siRNA can be used when PCI is employed to augment its delivery. Lowering the doses of siRNA would prevent saturation of the RNAi machinery and reduce off-target effects. In addition, local light exposure of target tissue would enhance siRNA delivery in the desired cells, only, which further increases the specificity of the treatment.

20.4 EFFECTS OF PCI ON DELIVERY OF GENES *IN VIVO*

Although PCI is a useful tool for *in vitro* drug and gene delivery, the most interesting use of PCI is as a method for site-specific *in vivo* treatment. PCI-studies in Balb/c mice on subcutaneously growing tumors (colon carcinomas and sarcomas) have been performed with the protein toxin gelonin, the cytotoxic agent bleomycin and a plasmid carrying the p53 gene [8,9,37,38]. The *in vivo* studies performed on PCI of gelonin show that the inhibition on tumor growth and the depth of necrosis was much stronger using gelonin in combination with PCI as compared to using gelonin alone or the photochemical treatment alone. PCI of gelonin led to complete tumor regression in 50%–80% of the animals.

PCI with genes is a more complex process than with proteins, as the gene has to be transcribed and translated before it can exhibit a desirable function. However, Ndoye and colleagues [9] performed iterative PCI treatment with a nonviral vector encoding the p53 protein in a head and neck xenograft with mutated p53. Successful tumor cure was achieved in 83% of the mice after seven rounds of PCI treatment. Compared to data reported in the literature, the PCI *in vivo* data are highly promising because in gene therapy several groups report upon tumor regression, while tumor eradication is rarely reported even in the cases where viral vectors have been applied.

20.5 ADVANTAGES AND DRAWBACKS OF PCI

PCI is a versatile method that can be applied to enhance the delivery of a variety of molecules with endocytic uptake mechanism in a site-specific manner, both *in vitro* and *in vivo*. The photochemical treatment substantially enhances the expression of transgenes and the biological effect of PNAs and siRNAs in cells. For example, in some cell lines, 20 times less adenoviral particles can be applied in combination with PCI to obtain similar levels of transduction as with adenovirus infection in the absence of photochemical treatment [17]. Because the concentration of the gene vector drops rapidly from the application point in tissues, this may be an important improvement of gene delivery protocols.

The main drawbacks of the technology are, for certain applications, the limited light penetration through tissue and the cytotoxicity induced by the photochemical treatment itself. The light penetration is tissue dependent and decays approximately (e^{-1}) for every 2–3 mm of depth, and the maximum depth of necrosis observed for PDT is 1 cm [39]. PCI is expected to improve treatment of additional sublayers of tumors than those killed by PDT because of the additional effect of the transgene or drug delivered into the cells [8,9,37,38]. Nevertheless, in applications where tissue removal is the main goal, the limited light penetration in tissue can be counteracted by performing the treatment several times [9]. The cytotoxicity induced by the photochemical treatment may be a drawback for some gene therapy applications, but it has been successfully exploited in PDT of cancer as well as for the treatment of age-related macular degeneration [1]. It is possible to reduce the toxicity by reducing the photochemical dose, but this would also decrease the effect of the photochemical treatment on gene delivery. Approaches are being made to reduce the cellular toxicity of the treatment. For example, a system for DNA delivery in which the DNA and a novel dendritic photosensitizer (DP) are assembled into one structure has been developed [40]. This is a ternary complex composed of a core containing DNA packaged with cationic peptides and enveloped in the anionic DP. The ternary complex showed significantly enhanced transgene expression *in vitro* and *in vivo* with minimal unfavorable phototoxicity.

20.6 CONCLUSION

The future of gene therapy may rely on approaches that combine various concepts for enhancing the efficiency and specificity of gene transfer. The photochemical internalization technology described is a versatile method that can be applied for enhancing delivery and biological effect of various agents and gene vectors. Moreover, photochemical treatments are already in clinical use for cancer therapy, and they are very specific and have few side effects. On the basis of preclinical studies, we conclude that PCI has a clear potential for improving both the efficiency and the specificity of gene therapy protocols, making it possible to achieve efficient site-specific gene delivery in tissue.

REFERENCES

1. Brown, S.B., Brown, E.A., and Walker, I., The present and future role of photodynamic therapy in cancer treatment. *Lancet Oncol.*, 5, 497, 2004.
2. Moan, J. and Berg, K., The photodegradation of porphyrins in cells can be used to estimate the lifetime of singlet oxygen. *Photochem. Photobiol.*, 53, 549, 1991.
3. Berg, K. and Moan, J., Lysosomes as photochemical targets, *Int. J. Cancer*, 59, 814, 1994.
4. Prasmickaite, L., Høgset, A., and Berg, K., Evaluation of different photosensitizers for use in photochemical gene transfection. *Photochem. Photobiol.*, 73, 388, 2001.
5. Berg, K. et al., Photochemical internalization: A novel technology for delivery of macromolecules into cytosol. *Cancer Res.*, 59, 1180, 1999.
6. Engesaeter, B.O. et al., Photochemical treatment with endosomally localized photosensitizers enhances the number of adenoviruses in the nucleus. *J. Gene Med.*, 8, 707, 2006.
7. Prasmickaite, L. et al., Photochemical disruption of endocytic vesicles before delivery of drugs: A new strategy for cancer therapy. *British J. Cancer*, 86, 652, 2002.

8. Selbo, P.K. et al., *In vivo* documentation of photochemical internalization, a novel approach to site specific cancer therapy. *Int. J. Cancer*, 92, 761, 2001.
9. Ndoye, A. et al., Eradication of p53-mutated head and neck squamous cell carcinoma xenografts using nonviral p53 gene therapy and photochemical internalization. *Mol. Ther.*, 13, 1156, 2006.
10. Medina-Kauwe, L.K., Xie, J., and Hamm-Alvarez, S., Intracellular trafficking of nonviral vectors. *Gene Ther.*, 12, 1734, 2005.
11. Prasmickaite, L. et al., Role of endosomes in gene transfection mediated by photochemical internalisation (PCI). *J. Gene Med.*, 2, 477, 2000.
12. Høgset, A. et al., Photochemical transfection: A new technology for light-induced, site-directed gene delivery. *Hum. Gene Ther.*, 11, 869, 2000.
13. Hellum, M. et al., Photochemically enhanced gene delivery with cationic lipid formulations. *Photochem. Photobiol. Sci.*, 2, 407, 2003.
14. Prasmickaite, L. et al, Photochemically enhanced gene transfection increases the cytotoxicity of the herpes simplex virus thymidine kinase gene combined with ganciclovir. *Cancer Gene Ther.*, 11, 514, 2004.
15. Kloeckner, J. et al., Photochemically enhanced gene delivery of EGF receptor-targeted DNA polyplexes. *J. Drug Target.*, 12, 205, 2004.
16. Ogris, M. et al., The size of DNA/transferrin-PEI complexes is an important factor for gene expression in cultured cells. *Gene Ther.*, 5, 1425, 1998.
17. Høgset, A. et al., Light-induced adenovirus gene transfer, an efficient and specific gene delivery technology for cancer gene therapy. *Cancer Gene Ther.*, 9, 365, 2002.
18. Greber, U.F. et al., Stepwise dismantling of adenovirus 2 during entry into cells. *Cell*, 75, 477, 1993.
19. Leopold, P.L. et al., Fluorescent virions: Dynamic tracking of the pathway of adenoviral gene transfer vectors in living cells. *Hum. Gene Ther.*, 9, 367, 1998.
20. Engesaeter, B.Ø. et al., PCI-enhanced adenoviral transduction employs the known uptake mechanism of adenoviral particles. *Cancer Gene Ther.*, 12, 439, 2005.
21. Bonsted, A. et al., Transgene expression is increased by photochemically mediated transduction of polycation-complexed adenoviruses. *Gene Ther.*, 11, 152, 2004.
22. Bonsted, A. et al., Photochemical enhancement of gene delivery to glioblastoma cells is dependent on the vector applied. *Anticancer Res.*, 25, 291, 2005.
23. Enger, P.O. et al., Adeno-associated viral vectors penetrate human solid tumor tissue *in vivo* more effectively than adenoviral vectors. *Hum. Gene Ther.*, 13, 1115, 2002.
24. Ponnazhagan, S. and Hoover, F., Delivery of DNA to tumor cells *in vivo* using adeno-associated virus. *Methods Mol. Biol.*, 246, 237, 2004.
25. Douar, A.M. et al., Intracellular trafficking of adeno-associated virus vectors: Routing to the late endosomal compartment and proteasome degradation. *J. Virol.*, 75, 1824, 2001.
26. Hansen, J., Qing, K., and Srivastava, A., Adeno-associated virus type 2-mediated gene transfer: Altered endocytic processing enhances transduction efficiency in murine fibroblasts. *J. Virol.*, 75, 4080, 2001.
27. Pajusola, K. et al., Cell-type-specific characteristics modulate the transduction efficiency of adeno-associated virus type 2 and restrain infection of endothelial cells. *J. Virol.*, 76, 11530, 2002.
28. Bonsted, A. et al., Photochemically enhanced transduction of polymer-complexed adenovirus targeted to the epidermal growth factor receptor. *J. Gene Med.*, 8, 286, 2006.

29. Weyergang, A., Selbo, P.K., and Berg, K., Photochemically stimulated drug delivery increases the cytotoxicity and specificity of EGF-saporin. *J. Control. Release*, 111, 165, 2006.

30. Prasmickaite, L. et al., Photochemical internalization of transgenes controlled by the heat-shock protein 70 promoter. *Photochem. Photobiol.*, 82, 809, 2006.

31. Ndoye, A. et al., Enhanced gene transfer and cell death following p53 gene transfer using photochemical internalisation of glucosylated PEI-DNA complexes. *J. Gene Med.*, 6, 884, 2004.

32. Engesaeter, B.O. et al., Photochemically mediated delivery of AdhCMV-TRAIL augments the TRAIL-induced apoptosis in colorectal cancer cell lines. *Cancer Biol. Ther.*, 5, 1511, 2006.

33. Folini, M. et al., Photochemical internalization of a peptide nucleic acid targeting the catalytic subunit of human telomerase. *Cancer Res.*, 63, 3490, 2003.

34. Shiraishi, T. and Nielsen, P.E., Photochemically enhanced cellular delivery of cell penetrating peptide-PNA conjugates. *FEBS Lett.*, 580, 1451, 2006.

35. Bøe, S. and Hovig, E., Photochemically induced gene silencing using PNA-peptide conjugates. *Oligonucleotides*, 16, 145, 2006.

36. Oliveira, S. et al., Photochemical internalization enhances silencing of epidermal growth factor receptor through improved endosomal escape of siRNA. *Biochim. Biophys. Acta*, 1768, 1211, 2007.

37. Dietze, A. et al., Enhanced photodynamic destruction of a transplantable fibrosarcoma using photochemical internalisation of gelonin. *Br. J. Cancer*, 92, 2004, 2005.

38. Berg, K. et al., Site-specific drug delivery by photochemical internalization enhances the antitumor effect of bleomycin. *Clin. Cancer Res.*, 11, 8476, 2005.

39. Berg, K. and Moan, J., Photodynamic tumor therapy, 2nd and 3rd generation photosensitizers, in Moser, J.G. (Ed.), *Optimization of Wavelengths in Photodynamic Therapy*, Harwood Academic Publishers, London, 1998.

40. Nishiyama, N. et al., Light-induced gene transfer from packaged DNA enveloped in a dendrimeric photosensitizer. *Nat. Mater.*, 4, 934, 2005.

21 Interactions in Macromolecular Complexes Used as Nonviral Vectors for Gene Delivery

Rudolf Podgornik, Daniel Harries, Jason DeRouchey, Hemult H. Strey, and V. Adrian Parsegian

CONTENTS

21.1 INTRODUCTION

Designed by nature for information, valued by molecular biologists for manipulation, DNA is also a favorite molecule of physical chemists and physicists [1]. Its mechanical properties [2], interactions with other molecules [3], and modes of packing [4] present tractable but challenging problems whose answers have *in vivo* and *in vitro* consequences. In the context of DNA transfection and gene therapy [5], what has been learned through molecular mechanics, interaction and packing might teach us how to package DNA for more effective gene transfer. Among these modes of *in vitro* packaging are association with proteins, treatment with natural or synthetic cationic "condensing agents," and combination with synthetic positively charged lipids [6].

In vivo, DNA is tightly held, not at all like the dilute solution form often studied *in vitro* (see Figure 21.1). This tight assembly necessarily incurs huge energetic costs of confinement, which create a tension under which DNA is expected to ravel or to unravel its message. Through direct measurement of forces between DNA molecules [7] and direct observation of its modes of packing [8], we might see not only how to use concomitant energies to design better DNA transfer systems but also to reason better about the sequences of events by which DNA is read in cells.

What binds these structures? To first approximation, for large, flexible biological macromolecules, the relevant interactions resemble those found among colloidal particles [9] where the size of the molecule (such as DNA molecules, lipid membranes, actin bundles) distinguishes it from simpler, smaller species (such as small solutes or salt ions). On the colloidal scale of tens of nanometers (1 nm = 10^{-9} m), only the interactions between macromolecules are evaluated explicitly, while

the small molecular species only "dress" the large molecules and drive the interactions between them.

The electrical charge patterns of multivalent ions, such as Mn^{2+}, Co^{3+}, or $spermine^{4+}$, with cation binding to negative DNA create attractive electrostatic and solvation forces that move DNA double helices to finite separations despite the steric knock of thermal Brownian motion of the DNA [10]. Solvation patterns about the cation-dressed structures create solvation forces: DNA–DNA repulsion because of water clinging to the surface and attraction from the release of the solvent [11]. Positively charged histones spool DNA into carefully distributed skeins, which are arrayed for systematic unraveling and reading [12]. Viral capsids encase DNA, stuffed against its own DNA–DNA electrostatic and solvation repulsion, to keep it under pressure for release upon infection [13]. In artificial preparations, the glue of positively charged and neutral lipids can lump negative DNA into ordered structures that can move through lipids and through water solutions [14].

Changes in the suspending medium can modulate intermolecular forces. One example is the change in van der Waals charge-fluctuation forces (see Section 21.2.4) between lipid bilayers when small sugars modifying the dielectric dispersion properties of water are added to the solution [15]. More dramatic, the addition of salt to water can substantially reduce electrostatic interactions between charged molecules such as DNA or other charged macromolecules bathed by an aqueous solution [16]. These changes can modify the behavior of macromolecules quantitatively or induce qualitatively new features into their repertoire, the most notable among these being the precipitation of DNA by addition of organic polycations to the solution [10].

Similar observations can be made about the small molecules essential to practically every aspect of interaction between macromolecules. In case of water, a network is built: its dielectric

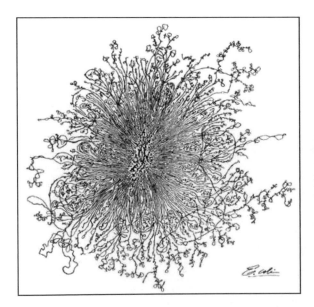
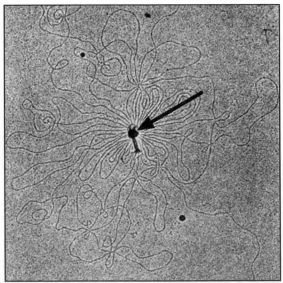

FIGURE 21.1 *In vivo* DNA is highly compacted. The figure shows *Esherichia coli* DNA and *T2* bacteriophage DNA after osmotic shock in distilled water has allowed them to expand from their much more compact *in vivo* configurations. (*E. coli* picture courtesy of Ruth Kavenoff, Bluegenes Inc., Los Angeles (1994); T2 picture from Kleinschmidt et al. *Biophys. Biochim. Acta*, 61, 252, 1962. With permission.)

contstant or dielectric permittivity facilitates electrostatic interactions, its pH enables its interaction in charging equilibria, and its fundamental molecular geometry creates the hydrogen bond network topology around simple solutes. This is, of course, the network of water molecule [17]. In what follows, we will limit ourselves to only three basic properties of macromolecules—charge, polarity (solubility), and conformational flexibility—that appear to govern the plethora of forces encountered in biological milieus. It is no surprise that highly ordered biological structures, such as the quasicrystalline spooling of DNA in viral heads or the multilamellar stacking of lipid membranes in visual receptor cells (Figure 21.2), can, in fact, be explained by the properties of a very small number of fundamental forces acting between macromolecules. Detailed experimental as well as theoretical investigations have identified hydration, electrostatic, van der Waals or dispersion, conformational fluctuation, and polyelectrolyte bridging forces as the most fundamental interactions governing the fate of biological macromolecules.

Our intent here is to sketch the measurements of these operative forces and to dwell upon concepts that rationalize them. It is from these concepts, with their insight into what controls organizing forces, that we expect people to learn to manipulate and to package DNA in more rewarding ways.

21.2 MOLECULAR FORCES

21.2.1 ORIGIN AND MEASUREMENT OF MOLECULAR FORCES

We divide these forces into two broad categories, both of which can be either attractive or repulsive. First, there are interactions that are connected with fields emanating from sources within or on the macromolecules themselves [16]; for example, electrostatic fields pointing from the fixed charge distributions on macromolecules into the surrounding space and fields of connectivity of hydrogen bond networks extending from the macromolecular surfaces into the bulk solution that are seen in hydration interactions. Second, there are the forces due to fluctuations that originate either in thermal Brownian motion or quantum jitter [15]. Consequent interactions include the van der Waals or dispersion forces that originate from thermal as well as quantum mechanical fluctuations of electromagnetic (EM) fields in the space between and within the interacting molecules and conformation–fluctuation forces from thermal gyrations of the macromolecule when thermal agitation pushes against the elastic energy resistance of the molecule and confinement imposed by neighboring macromolecules [16].

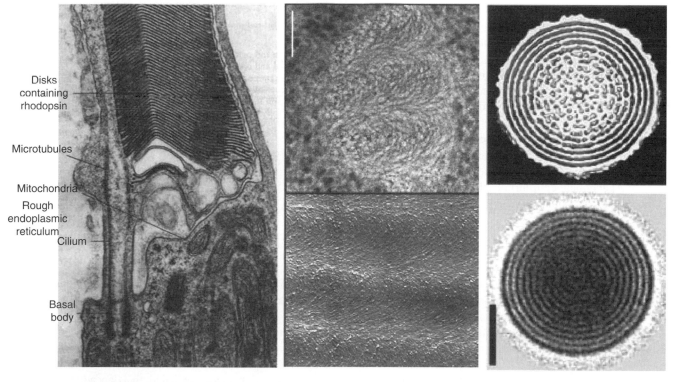

Disks containing rhodopsin

Microtubules

Mitochondria

Rough endoplasmic reticulum

Cilium

Basal body

FIGURE 21.2 Highly ordered assemblies, ubiquitous among biological structures, can be explained through the properties of a very small number of fundamental forces acting between macromolecules. On the left-hand side, electron micrograph of a part of a human eye rod cell showing multilamellar bilayer aggregate. (From Kessel, R.G. and Kardon, R.H., *Tissues and Organs*, W.H. Freeman and Co., San Francisco, CA, 1979.) In the middle, electron micrograph of an *in vivo* cholesteric phase of a wild type *E. coli* DNA. (Adapted from Frankiel Krispin, D. et al., *EMBO J.*, 2001, 20, 1184.) For comparison, we show the same type of structure for DNA *in vitro* below. (Adapted from Leforestier, A. and Livolant, F., *Biophys. J.*, 1993, 65, 56.) On the right-hand side, cryo-micrographs and computer-processed images of T7 phage heads showing ordered DNA spooling within the viral heads. (From Cerritelli, M.E., Cheng, N., Rosenberg, A.H., McPherson, C.E., Booy, F.P., and Steven, A.C., *Cell*, 91, 271, 1997. With permission.)

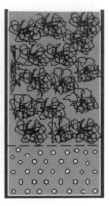

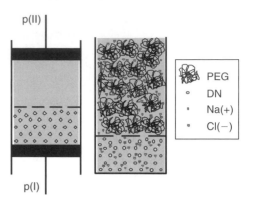

FIGURE 21.3 Osmotic pressure in macromolecular arrays. Dissolved polymers such as PEG exert an osmotic pressure on the part of the solution from which they are excluded (shown schematically by the weight). Instead of exerting osmotic pressure directly on the macromolecular subphase such as DNA or lipid arrays (small circles), one can equilibrate it with a solution of PEG at a set concentration and the PEG itself will exert osmotic stress on the macromolecular subphase. Osmotic weighing of polymers one against the other (the one with the known, set osmotic pressure against the unknown one) is the essence of the osmotic stress technique of measuring interactions in macromolecular solutions.

There are many ways to detect interactions between macromolecules. In this chapter, we consider only macromolecules interacting in ordered arrays that are particularly relevant for investigations of the packing and energetics of DNA–lipid complexes.

A fundamental concept in macromolecular arrays is that of osmotic pressure (Figure 21.3). It is equal to the pressure needed to hold a macromolecular array together against the forces acting between its constituent macromolecules. It can be applied either mechanically across a semipermeable membrane or via the osmotic stress of a high molecular weight (e.g., polyethylene glycol [PEG], polyvinyl pyrrolidone [PVP], dextran) polymer solution. At chemical equilibrium, the osmotic pressure of one solution (macromolecular array) balances that of the other (the bathing polymer solution). The chemical equilibrium can be maintained either via a semipermeable membrane or simply because the bathing polymer solution phase separates from the macromolecular array, as is many times the case with PEGs, PVP, and dextran. This osmotic balancing of different molecular solutions is the basis of the "osmotic stress method" to measure the equation of state of macromolecular arrays [18].

The equation of state of a macromolecular solution is defined as the dependence of its osmotic pressure on the density of the array (see Figure 21.4). By equilibrating the macromolecular array versus a solution of high molecular weight polymer with a known osmotic pressure, one can set the osmotic pressure in the macromolecular array itself [18]. If in addition, one measures the concurrent density of the macromolecular array, either via x-ray scattering or direct densitometry, one gets the dependence of the osmotic pressure of the array on its density, i.e., its equation of state. This is the essence of the osmotic stress method.

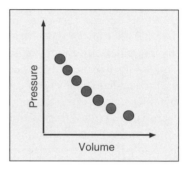

FIGURE 21.4 (See color insert following blank page 206.) Osmotic stress method. (From Parsegian, V.A., Rand, R.P., Fuller, N.L., and Rau, D.C., *Methods Enzymol.*, 1986, 127, 400.) DNA liquid crystals are equilibrated against solutions of a neutral polymer (such as PEG or PVP, depicted as disordered coils). These solutions are of known osmotic pressure, pH, temperature, and ionic composition. (From Rau, D.C., Lee, B.K., and Parsegian, V.A., *Proc. Natl. Acad. Sci. USA*, 1984, 81, 2621.) Equilibration of DNA under the osmotic stress of external polymer solution is effectively the same as exerting mechanical pressure on the DNA subphase with a piston that passes water and small solutes but not DNA. After equilibration under this known stress, DNA separation is measured either by x-ray scattering, if the DNA subphase is sufficiently ordered, or by densitometry. (From Strey, H.H., Parsegian, V.A., and Podgornik, R., *Phys. Rev. Lett.,* 1997, 78, 895.) DNA density and osmotic stress thus determined immediately provide an equation of state (osmotic pressure as a function of the density of the DNA subphase) to be codified in analytic form over an entire phase diagram.

21.2.2 HYDRATION FORCES

The hydration force is connected with a very simple observation that it takes increasing amounts of work to remove water from between electrically neutral lipids in multilamellar arrays or from between ordered arrays of polymers at large polymer concentrations [18]. Direct measurements of this work show that it increases exponentially with the diminishing separation between colloid surfaces with a decay length that depends as much on the bulk properties of the solvent as on the detailed characteristics of the interacting surfaces. There is nevertheless some profound universality in the interactions between macromolecular surfaces at close distances (see Figure 21.5)—whether they are charged, zwitterionic, or uncharged—that strongly suggest that water is essential to maintaining the stability of biological matter at high densities.

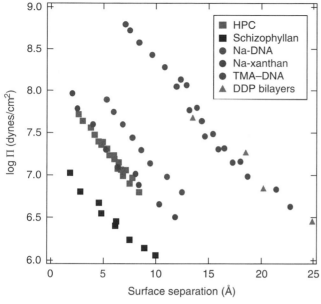

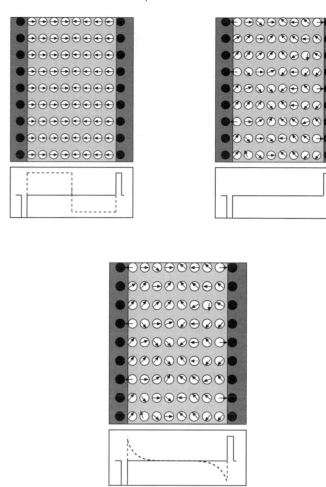

FIGURE 21.5 (See color insert following blank page 206.)   Interactions between biological macromolecules show striking universality at close surface-to-surface separations (or equivalently at very large densities). Hydroxypropyl cellulose, schizophyllan, different DNA salts, xanthan, and DDP bilayers at small intermolecular separations (given in terms of the separation between effective molecular surfaces of the interacting molecules) all show a strong repulsive interactions decaying with about the same characteristic decay length. The log-linear plot is thus more or less a straight line. (Composite data, courtesy D.C. Rau.)

FIGURE 21.6   The hydration force. Marčelja, S. and Radić, N., *Chem. Phys. Lett.*, 1976, 42, 129, introduced an order parameter P that would capture the local condition, or local ordering, of solvent molecules between the surfaces. We represent it as an arrow (that has magnitude and direction) on each water molecule that is trapped between the two apposing surfaces and is being acted upon by the surface fields, depicted schematically with a bold line below each of the three drawings. Minimizing the energy corresponding to a spatial profile of P leads to a configuration where P points (for example) away from both surfaces and there is thus mismatch at the midplane (the dotted line below the leftmost drawing). The entropy on the other hand would favor completely disordered configurations with no net value of P (the dotted line below the rightmost drawing). The free energy strikes a compromise between the two extrema, leading to a smooth profile of P, varying continuously as one goes from one surface to the other (the dotted line below the bottom drawing). As the two surfaces approach the nonmonotonic profile of the order parameter P, it leads to repulsive forces between them.

Hydration forces can be understood in different terms with no consensus yet on mechanism [11]. Marãelja and Radiç [19] first proposed the idea that colloid surfaces perturb the vicinal water and that the exponential decay of the hydration force is due to the weakening of the perturbation of the solvent as a function of the distance between the interacting surfaces (Figure 21.6). They introduced an order parameter $P(z)$ as a function of the transverse coordinate z, between the surfaces located at $z = D/2$ and $z = -D/2$, that would capture the local condition or local ordering of solvent molecules between the surfaces. The detailed physical nature of this order parameter is left unspecified, but since the theory builds on general principles of symmetry and perturbation expansions, molecular details are not needed. All one needs to know about P is that within the bulk water $P = 0$ and close to a macromolecular surface P remains nonzero. As a mnemonic, one can envision P as an arrow associated with each water molecule. In bulk water, the arrows point in all directions with equal probability. Close to a bounding macromolecular surface, they point preferentially toward or away from the surface (Figure 21.6) depending on the surface-orienting fields.

If we envisage solvent molecules between two perturbing surfaces, we can decompose the total free energy F of their configuration into its energy W and entropy S parts via the well-known thermodynamic definition $F = W - TS$, where T is the absolute temperature. Energetically, it would be most favorable for the surface-induced order to persist away from the surfaces, but that would create conflict between the apposing surfaces (see Figure 21.3). Entropy fights any type of ordering and wants to eliminate all orderly configurations between the two surfaces, creating a homogeneous state of molecular disorder characterized by $P = 0$. Energy and entropy compromise to create a nonuniform profile of the order parameter between the surfaces; surface-induced order propagates but progressively decreases away from the surfaces.

From the free energy, we can derive the repulsive hydration osmotic pressure p acting between the surfaces because, by definition, it is proportional to the derivative of the free energy with respect to the separation D. Osmotic pressure between two apposed lipid surfaces has been measured extensively for different lipids [20] and has been found to have the form $p = p_0 \exp(-D/\lambda)$. This is consistent with the previous theoretically derived form of the hydration free energy if one assumes that $p_0 \sim P^2$ ($z = D/2$). The corresponding interaction free energy per unit surface area of the interacting surfaces, $F(D)$, would thus behave as $F(D) \sim p_0 \lambda \exp(-D/\lambda)$. From these experiments, one can deduce the magnitude of the prefactor p_0, which determines the absolute magnitude of the hydration repulsion for a great variety of lipids and lipid mixtures within an interval 10^{12} to 10^{10} dynes/cm^2 or equivalently in pressures of hundreds of atmospheres.

As already noted, in this simple theoretical approach, the hydration decay length depends only on the bulk properties of the solvent and not on the properties of the surface. In order to generalize this simplification, Kornyshev and Leikin [21] formulated a variant of the hydration force theory also to take into account explicitly the nature of surface ordering. They derived a modified hydration decay length that clearly shows how surface order couples with the bare hydration decay length. Without going too deeply into this theory, we note that if the interacting surfaces have two-dimensional ordering patterns characterized by a wave vector $Q = 2\lambda$, where λ is the characteristic scale of the spatial variations of these patterns, then the effective hydration force decay length would be $\lambda_{KL} = 1/2\lambda$. Inserting numbers for the case of DNA, where the "surface" structure has a characteristic scale of 1–2 Å, we realize that the hydration decay length in this case would be almost entirely determined by the surface structure and not the bulk solvent properties. Given the experimentally determined variety of forces between phospholipids [20], it is indeed quite possible that even in the simplest cases, the measured decay lengths are not those of the water solvent itself but instead also include the surface properties in the characteristic scale of the surface ordering.

The other important facet of this theory is that it predicts that in certain circumstances, the hydration forces can become attractive [11]. This is particularly important in the case of interacting DNA molecules where this hydration attraction connected with condensing agents can hold DNAs into an ordered array even though the van der Waals forces themselves would be unable to accomplish [22]. This attraction is always an outcome of nonhomogeneous surface ordering and arises in situations where apposing surfaces have complementary checkerboard like order [11]. Unfortunately, in this situation, many mechanisms can contribute to attractions; it is difficult to argue for one strongest contribution.

21.2.3 Electrostatic Forces

Electrostatic forces between charged colloid bodies are among the key components of the force equilibria in (bio)colloid systems [23]. At larger separations, they are the only forces that can counteract van der Waals attractions and thus stabilize colloid assembly. The crucial role of the electrostatic interactions in (bio)colloid systems is well documented and explored following the seminal realization of Bernal and Fankuchen [24] that electrostatic interaction is the stabilizing force in tobacco mosaic virus (TMV) arrays.

Though the salient features of electrostatic interactions of fixed charges in a sea of mobile countercharges and salt ions are intuitively straightforward to understand, they are difficult to evaluate. These difficulties are clearly displayed by the early ambiguities in the sign of electrostatic interactions between two equally charged bodies that was first claimed to be attractive (Levine), then repulsive (Verwey-Overbeek), and finally realized that it is usually repulsive except if the counterions or the salt ions are of higher valency [25].

In this section, we introduce the electrostatic interaction on an intuitive footing (see Figure 21.7). Assume we have two equally charged bodies with counterions in between. Clearly the minimum of electrostatic energy W_E [28], which for the electrostatic field configuration at the spatial position $\mathbf{r}$, $\mathbf{E}(\mathbf{r})$, is proportional to the integral of $\mathbf{E}^2(\mathbf{r})$ over the whole space where one has nonzero electrostatic field, would correspond to the adsorption of counterions to the charges, leading to their complete neutralization. The equilibrium electrostatic field would be thus entirely concentrated right next to the surface. However, at a finite temperature, it is not the electrostatic energy but rather the free energy [26], $F = W_E - TS$, containing also the entropy S of the counterion distribution, that should be minimized. The entropy of the mobile particles with the local density $\rho_i(r)$ (we assume there is more than one species of mobile particles, for example, counterions and salt ions, tracked through the index i) is taken as an ideal gas entropy [26], which is proportional to the volume integral of $\sum_i[\rho_i(r) \ln(\rho_i(r)/\rho_0) - \rho_i(r) - \rho_{i0}]$, where ρ_{i0} is the density of the mobile charges in a reservoir that is in chemical equilibrium with the confined system under investigation. Entropy by itself would clearly lead to a uniform distribution of counterions between the charged bodies, $\rho_i(r) = \rho_{i0}$, while together with the electrostatic energy, it obviously leads to a nonmonotonic profile of the mobile charge distribution between the surfaces, minimizing the total free energy of the mobile ions.

The above discussion, though being far from rigorous, contains all the important theoretical underpinnings known under the title of "Poisson–Boltzmann theory" [27]. In order to arrive at the central equation corresponding to the core of this theory, one simply has to formally minimize the free energy $F = W_E - TS$, just as in the case of structural interactions, together with the basic electrostatic equation [28] (the Poisson equation) that connects the sources of the electrostatic field with the charge densities of different ionic species. The standard procedure now is to minimize the free energy, take into account the Poisson equation, and what follows is the well-known Poisson–Boltzmann (PB) equation, the solution of which gives the nonuniform profile of the mobile charges between the surfaces with fixed charges. This equation can be solved explicitly for some particularly simple geometries [27]. For two charged planar surfaces, the solution gives a screened

Energy minimization

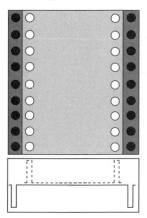

Entropy minimization

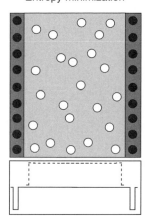

Free energy minimization

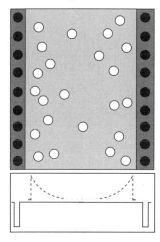

FIGURE 21.7 Pictorial exposition of the main ideas behind the PB theory of electrostatic interactions between (bio)colloidal surfaces. Electrostatic energy by itself would favor adsorption of counterions (white circles) to the oppositely charged surfaces (black circles). The equilibrium profile of the counterions in this case is presented by the dotted line below the leftmost drawing. Entropy, on the contrary, favors a completely disordered configuration, i.e., a uniform distribution of counterions between the surfaces, presented by the dotted line below the rightmost drawing. The free energy works a compromise between the two principles leading to a nonmonotonic profile of the counterion density (From Verwey, E.J.W. and Overbeek, J.T.G., *Theory of the Stability of Lyophobic Colloids*, Elsevier, New York, 1948), varying smoothly in the intersurface region. As the two surfaces are brought close, the overlapping counterion distributions originating at the fixed charge at the surfaces (the bold line below each of the drawings) create repulsive forces between them.

electrostatic potential that decays exponentially away from the walls. It is thus smallest in the middle of the region between the surfaces and largest at the surfaces. The spatial variation of the electrostatic interaction is just as in the case of structural interactions described with a characteristic decay length, termed the Debye length in this case, which for uni–uni valent salts assumes the value of $\lambda_D = 3\,\text{Å}/\sqrt{I}$, where I is the ionic strength of the salt in moles per liter. A 0.1 M solution of uni–uni valent salt, such as NaCl, would thus have the characteristic

decay length of about 9.5 Å. Beyond this separation, the charged bodies do not feel each other any more. By adding or removing salt from the bathing solution, we are thus able to regulate the range of electrostatic interactions.

The exponential decay of the electrostatic field away from the charged surfaces with a characteristic length independent (to the lowest order) of the surface charge is one of the most important results of the PB theory.

Obviously as the surfaces come closer together, their decaying electrostatic potentials begin to interpenetrate [25]. The consequence of this interpenetration is a repulsive force between the surfaces that again decays exponentially with the intersurface separation and a characteristic length again equal to the Debye length. For two planar surfaces at a separation D, bearing sufficiently small charges, characterized by the surface charge density σ, so that the ensuing electrostatic potential is never larger than $k_B T/e$, where k_B is the Boltzmann's constant and e is the elementary electron charge, one can derive the expression $F(D) \sim \sigma^2 \lambda \exp(-D/\lambda_D)$ [27] for the interaction free energy per unit surface area $F(D)$. The typical magnitude of the electrostatic interaction in different systems of course depends on the magnitude of the surface charge. It would not be unusual in lipids to have surface charge densities in the range of one elementary charge per 50 to 100 Å² surface area [29]. For this range of surface charge densities, the constant prefactor in the expression for the osmotic pressure would be of the order 0.4–1.2×10^7 N/m.

The same type of analysis would apply also to two charged cylindrical bodies, e.g., two molecules of DNA, interacting across an electrolyte solution. What one evaluates in this case is the interaction free energy per unit length of the cylinders [30], $g(R)$, where R is the separation between the cylinders, which can be obtained in the approximate form $g(R) \sim \mu^2 \exp(-R/\lambda_D)$. It is actually possible to get also an explicit form [30] of the interaction energy between two cylinders even if they are skewed by an angle θ between them. In this case, the relevant quantity is the interaction free energy itself (if θ is nonzero, then the interaction energy does not scale with the length of the molecules) that can be obtained in a closed form as $F(R,\theta) \sim \mu^2 \lambda_D R^{1/2} \exp(-R/\lambda_D)/\sin\theta$.

The predictions for the forces between charged colloid bodies have been reasonably well borne out for electrolyte solutions of uni–uni valent salts [31]. In that case, there is near quantitative agreement between theory and experiment. However, for higher valency salts, the PB theory does not only give the wrong numerical values for the strength of the electrostatic interactions, but also misses their sign. In higher valency salts, the correlations among mobile charges between charged colloid bodies due to thermal fluctuations in their mean concentration lead effectively to attractive interactions [32] that are in many respects similar to van der Waals forces that we analyze next.

21.2.4 VAN DER WAALS FORCES

van der Waals charge fluctuation forces are special in the sense that they are a consequence of thermodynamic as well as quantum mechanical fluctuations of the EM fields [15]. They exist even if the average charge, dipole moment, or higher multipole moments on the colloid bodies are zero. This is in stark contrast to electrostatic forces that require a net charge or a net polarization to drive the interaction. This also signifies that van der Waals forces are much more general and ubiquitous than any other force between colloid bodies [9].

There are many different approaches to van der Waals forces [15,33,34]. For small molecules interacting at a relatively large distance, one can distinguish different contributions to the van der Waals force, stemming from thermally averaged dipole–dipole potentials (the Keesom interaction), dipole-induced dipole interactions (the Debye interaction), and induced dipole–induced dipole interactions (the London interaction) [35]. They are all attractive and their respective interaction energy decays as the sixth power of the separation between the interacting molecules. The magnitude of the interaction energy depends on the EM absorption (dispersion) spectrum of interacting bodies, hence also the term dispersion forces.

For large colloidal bodies composed of many molecules, the calculation of the total van der Waals interactions is not trivial [15,34], even if we know the interactions between individual molecules composing the bodies. Hamaker assumed that one can simply add the interactions between composing molecules in a pairwise manner. It turned out that this was a very crude and simplistic approach to van der Waals forces in colloidal systems, as it does not take into account the highly nonlinear nature of the van der Waals interactions in condensed media. Molecules in a condensed body interact among themselves, thus changing their properties, hence their dispersion spectrum, which in turn modifies the van der Waals forces between them.

Lifshitz, following work of Casimir [9,15,34], realized how to circumvent this difficulty and formulated the theory of van der Waals forces in a way that already includes all these nonlinearities. The main assumption of this theory is that the presence of dielectric discontinuities, as in colloid surfaces, modifies the spectrum of EM field modes between these surfaces (see Figure 21.8). As the separation between colloid bodies varies, so do the eigenmode frequencies of the EM field between and within the colloid bodies. It is possible to deduce the change in the free energy of the EM modes due to the changes in the separation between colloid bodies coupled to their dispersion spectral characteristics [36].

Based on the work of Lifshitz, it is now clear that the van der Waals interaction energy is just the change of the free energy of field harmonic oscillators at a particular eigenmode frequency ω as a function of the separation between the interacting bodies D and temperature T, $\omega = \omega(D,T)$. With this equivalence in mind, it is quite straightforward to calculate the van der Waals interaction free energy between two planar surfaces at a separation D and temperature T; the dielectric permittivity between the two surfaces, ε, and within the surfaces, ε', must both be known as a function of the frequency ω of the EM field [36]. This is a consequence of the fact that in general, the dielectric media comprising the surfaces as well as the

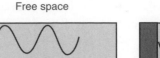
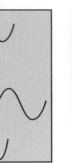
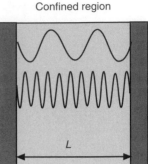

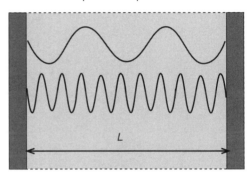

FIGURE 21.8 Pictorial introduction to the theory of Lifshitz–van der Waals forces between colloid bodies. Empty space is alive with EM field modes that are excited by thermal as well as quantum mechanical fluctuations. Their frequency is unconstrained and follows the black body radiation law. Between dielectric bodies only those EM modes survive that can fit into a confined geometry. As the width of the space between the bodies varies, so do the allowed EM mode frequencies. Every mode can be treated as a separate harmonic oscillator, each contributing to the free energy of the system. Since this free energy depends on the frequency of the modes that in turn depend on the separation between the bodies, the total free energy of the EM modes depends on the separation between the bodies. This is an intuitive description of the Lifshitz–van der Waals force. (From Mahanty, J. and Ninham, B.W., *Dispersion Forces*, Academic Press, London, 1976.)

space between them are dispersive, which means that their dielectric permittivities depend on the frequency of the EM field, i.e., $\varepsilon = \varepsilon(\omega)$. With this in mind, one can derive the interaction free energy per unit surface area of the interacting surfaces in the form $F(D) = A(D)12\pi D^2$, where the so called Hamaker coefficient A depends on the difference between the dielectric permittivities of the interacting materials at different imaginary frequencies $\iota\xi$. It can, in general, be split into two terms: the first term in the Hamaker coefficient is due to thermodynamic fluctuations, such as Brownian rotations of the dipoles of the molecules composing the media or the averaged dipole-induced dipole forces and depends on the static ($\omega = 0$) dielectric response of the interacting media, while the second term is purely quantum mechanical in nature [15]. The imaginary argument of the dielectric constants is not that odd since $\varepsilon(\iota\xi)$ is an even function of ξ, which makes $\varepsilon(\iota\xi)$ also a purely real quantity [36].

In order to evaluate the magnitude of the van der Waals forces, one has to know the dielectric dispersion $\varepsilon(\omega)$, or more appropriately $\varepsilon(\iota\xi)$, of all the media involved. This is no simple task and can be accomplished only for very few materials [34]. Experiments seem to be a much more straightforward way to proceed. The values for the Hamaker coefficients of different materials interacting across water are between 0.3 and 2.0 × 10^{-20} J. Specifically for lipids, the Hamaker constants are quite close to theoretical expectations except for the phosphatidyl-ethanolamines that show a much larger attractive interactions probably due to headgroup alignment [31]. Evidence from direct measurements of attractive contact energies as well as direct force measurements suggests that van der Waals forces are more than adequate to provide attraction between bilayers for them to form multilamellar systems [37].

For cylinders, the same type of argument applies except that due to the geometry, the calculations are a bit more tedious [38]. Here the relevant quantity is not the free energy per unit area but the interaction free energy per unit length of the two cylinders of radius a, $g(R)$, considered to be parallel at a separation R. The calculation [39] leads to the following form: $g(R) \sim Aa^4/R^5$ where the constant A again depends on the differences between dielectric permittivities, $\varepsilon_{\parallel}$ and $\varepsilon_{\perp}$, respectively, the parallel and the perpendicular components of the dielectric permittivity of the dielectric material of the cylinders, and ε_m, the dielectric permittivity of the bathing medium.

If, however, the two interacting cylinders are skewed at an angle θ, then the interaction free energy $G(R,\theta)$, this time not per length, is obtained [39] in the form $G(R) \sim (A + B\cos^2\theta)(a^4/R^4\sin\theta)$. The constants A and B describe the dielectric mismatch between the cylinder and the bathing medium at different imaginary frequencies. The same correspondence between the thermodynamic and quantum mechanical parts of the interactions as for two parallel cylinders applies also to this case. Clearly the van der Waals force between two cylinders has a profound angular dependence that, in general, creates torques between the two interacting molecules.

Taking the numerical values of the dielectric permittivities for two interacting DNA molecules, one can calculate that the van der Waals forces are quite small, typically one to two orders of magnitude smaller than the electrostatic repulsions between them, and in general cannot hold the DNAs together in an ordered array. Other forces, leading to condensation phenomena in DNA [10], clearly have to be added to the total force balance in order to get a stable array. There is as yet still no consensus on the exact nature of these additional attractions. It seems that they are due to the fluctuations of counterions atmosphere close to the molecules.

21.2.5 DLVO MODEL

The popular Derjaguin–Landau–Verwey–Overbeek (DLVO) [9,25] model assumes that electrostatic double layer and van der Waals interactions govern colloid stability. Applied with a piety not anticipated by its founders, this model actually does surprisingly work rather well in many cases. Direct osmotic stress measurements of forces between lipid bilayers show that at separations less than ~10 Å, there are qualitative deviations from the DLVO thinking [40]. For micron-sized objects and for macromolecules at greater separations, electrostatic double-layer forces and sometimes van der Waals forces tell us what we need to know about interactions governing movement and packing.

21.2.6 GEOMETRIC EFFECTS

Forces between macromolecular surfaces are most easily analyzed in plane-parallel geometry. Because most of the interacting colloid surfaces are not planar, one must either evaluate molecular interactions for each particular geometry or devise a way to connect the forces between planar surfaces with forces between surfaces of a more general shape. The Derjaguin approximation [9] assumes that interactions between curved bodies can be decomposed into interactions between small plane-parallel sections of the curved bodies (see Figure 21.9). The total interaction between curved bodies would be thus equal to a sum where each term corresponds to a partial interaction between quasiplane-parallel sections of the two bodies.

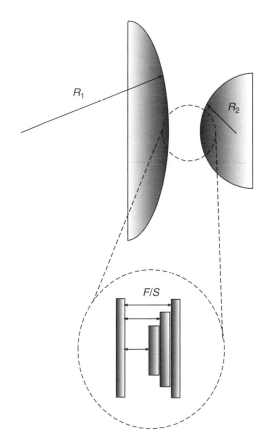

FIGURE 21.9 The Derjaguin approximation. Formulating forces between oppositely curved bodies (e.g., cylinders, spheres, etc.) is very difficult. But it is often possible to use an approximate procedure. Two curved bodies (two spheres of unequal radii in this case) are approximated by a succession of planar sections, interactions between which can be calculated relatively easily. The total interaction between curved bodies is obtained through a summation over these planar sections.

This idea can be given a completely rigorous form and leads to a connection between the interaction free energy per unit area of two interacting planar surfaces, $F(D)$, and the force acting between two spheres at minimal separation D, $f(D)$, one with the mean radius of curvature R_1 and the other one with R_2. The formal equivalence can be written as follows: $f(D) = 2\pi(R_1 R_2/(R_1 + R_2))F(D)$. A similar equation can also be obtained for two cylinders in the form, $f(D) = 2\pi(R_1 R_2)^{1/2}F(D)$.

These approximate relations clearly make the problem of calculating interactions between bodies of general shape tractable. The only caveat here is that the radii of curvature should be much larger than the proximal separation between the two interacting bodies, effectively limiting the Derjaguin approximation to sufficiently small separations.

Using the Derjaguin formula or evaluating the interaction energy explicitly for those geometries for which it is not an insurmountable task, one can now obtain a whole range of DLVO expressions for different interaction geometries (see Figure 21.10). The salient features of all these expressions are that the total interaction free energy always has a primary minimum that can only be eliminated by strong short-range

hydration forces and a secondary minimum due to the compensation of screened electrostatic repulsion and van der Waals–Lifshitz attraction. The position of the secondary minimum depends as much on the parameters of the forces (Hamaker constant, fixed charges, and ionic strength) as well as on the interaction geometry. One can state generally that the range of interaction between the bodies of different shapes is inversely proportional to their radii of curvature.

Thus the longest-range forces are observed between planar bodies, and the shortest between small (point-like) bodies. What we have not indicated on Figure 21.7 is that the interaction energy between two cylindrical bodies, skewed at a general angle θ and not just for parallel or crossed configurations, can be obtained in an explicit form. It follows simply from these results that the configuration of two interacting rods with minimal interaction energy is the one corresponding to $\theta = \pi/2$, i.e., corresponding to crossed rods.

21.2.7 FLUCTUATION FORCES

The term "fluctuation forces" is a bit misleading in this context because clearly van der Waals forces already are fluctuation

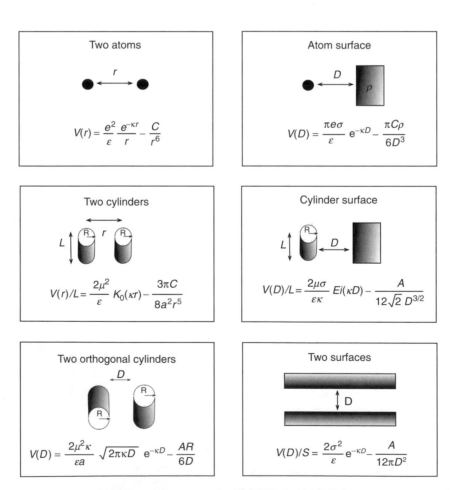

FIGURE 21.10 A representative set of DLVO interaction expressions for different geometries most commonly encountered in biological milieus: Two small particles, a particle and a wall, two parallel cylinders, a cylinder close to a wall, two skewed cylinders, and two walls. The DLVO interaction free energy is always composed of a repulsive electrostatic part (calculated from a linearized PB theory) and an attractive van der Waals part. Charge: e, charge per unit length of a cylinder: μ, charge per unit surface area of a wall: σ, C are geometry-dependent constants, ε the dielectric constant, κ the inverse Debye length, and ρ the density of the wall material. The functions $K_0(x)$ (the Bessel function K_0) and $Ei(x)$ (the exponential integral function) both depend essentially exponentially on their respective argument.

forces. What we have in mind is thus a generalization of the van der Waals forces to situations where the fluctuating quantities are not EM fields but other quantities subject to thermal fluctuations. No general observation as to the sign of these interactions can be made; they can be either repulsive or attractive and are as a rule of thumb comparable in magnitude to the van der Waals forces.

The most important and ubiquitous force in this category is the undulation or Helfrich force [41]. It has a very simple origin and operates among any type of deformable bodies as long as their curvature moduli are small enough (comparable to thermal energies). It was shown to be important for multilamellar lipid arrays [42] as well as in hexagonal polyelectrolyte arrays [43] (see Figure 21.11).

FIGURE 21.11 (See color insert following blank page 206.)   Thermally excited conformational fluctuations in a multilamellar membrane array (small molecules are waters and long chain molecules are phospholipids) or in a tightly packed polyelectrolyte chain array (the figure represents a hexagonally packed DNA array) leading to collisions between membranes or polyelectrolyte chains. These collisions contribute an additional repulsive contribution to the total osmotic pressure in the array, a repulsion that depends on the average spacing between the fluctuating objects. (Graphics courtesy of B.B. Brooks and M. Hodoscek, NIH)

The mechanism is simple. The shape of deformable bodies fluctuates because of thermal agitation (Brownian motion) [26]. If the bodies are close to each other, the conformational fluctuations of one will be constrained by the fluctuations of its neighbors. Thermal motion makes the bodies bump into each other, which creates spikes of repulsive force between them. The average of this force is smooth and decays continuously with the mean separation between the bodies.

One can estimate this steric interaction for multilamellar lipid systems and for condensed arrays of cylindrical polymers (Figure 21.11). The only quantity entering this calculation is the elastic energy of a single bilayer that can be written as the square of the average curvature of the surface, summed over the whole area of the surface, multiplied by the elastic modulus of the membrane, K_C. K_C is usually between 10 and $50 k_B T$ [44] for different lipid membranes. If the instantaneous deviation of the membrane from its overall planar shape in the plane is now introduced as u, the presence of neighboring membranes introduces a constraint on the fluctuations of u that basically demands that the average of the square of u must be proportional to D^2, where D is the average separation between the membranes in a multilamellar stack. Thus we should have $u^2 \sim D^2$. The free energy associated with this constraint can now be derived in the form $F(D) \sim (k_B T)^2 (K_C D^2)$, and is seen to decay in inverse proportion to the separation between bilayers squared [41].

It has thus obviously the same dependence on D as the van der Waals force. This is, however, not a general feature of undulation interactions as the next example clearly shows. Also we only indicated the general proportionality of the interaction energy. The calculation of the prefactors can be a difficult [45] especially because the elastic bodies usually do not interact with idealized hard repulsions but rather through soft potentials that have both attractive as well as repulsive regimes.

The same line of thought can now be applied to flexible polymers in a condensed array [43]. This system is a one-dimensional analog of the multilamellar membrane system. For polymers, the elastic energy can be written similarly to the membrane case as the square of the local curvature of the polymer, multiplied by the elastic modulus of the polymer, integrated over its whole length. The elastic modulus K_C is usually expressed through a persistence length $L_p = K_C/(k_B T)$. The value of the persistence length tells us how long a polymer can be before the thermal motion forces it to fluctuate wildly. For DNA, this length is about 50 nm. It spans, however, the whole range of values between about 10 nm for hyaluronic acid and all the way to 3 mm for microtubules. Using now the same constraint for the average fluctuations of the polymer away from the straight axis, one derives the relationship for the free energy change due to this constraint, $F(D) \sim (k_B T)(L_p^{1/3} D^{2/3})$ [43].

Clearly the D dependence for this geometry is very much different from the one for van der Waals force, which would be D^{-5}. There is thus no general connection between the van der Waals force and the undulation fluctuation force. Here again one has to indicate that if the interaction potential between fluctuating bodies is described by a soft potential, with no discernible hard core, the fluctuation interaction can have a profoundly different dependence on the mean separation [43].

Apart from the undulation fluctuation force, there are other fluctuation forces. The most important among them appears to be the monopolar charge fluctuation force [46], recently investigated in the context of DNA condensation. It arises from transient charge fluctuations along the DNA molecule due to constant statistical redistributions of the counterion atmosphere.

The theory of charge fluctuation forces is quite intricate and mathematically demanding [47]. Let us just quote a rather interesting result: if two point charges interact via a "bare" potential $V_0(R)$, where R is the separation between them, then the effect of the thermal fluctuations in the number of counterions surrounding these charges would lead to an effective interaction of the form $V(R) \sim k_B T(V_0(R))^2$. The fluctuation interaction in this case would thus be attractive and proportional to the square of the bare interaction.

This simple result already shows one of the salient features of the interaction potential for monopolar charge fluctuation forces, i.e., it is screened with half the Debye screening length (because of $V^2(R)$). If there is no screening, however, the monopolar charge fluctuation force becomes the strongest and longest ranged among all the fluctuation forces. It is, however, much less general than the related van der Waals force and at present, it is still not clear what the detailed conditions should be for its appearance, the main difficulty being the question whether charge fluctuations in the counterion atmosphere are constrained or not.

21.2.8 ATTRACTIVE ELECTROSTATIC FORCES: STRONG COUPLING AND POLYELECTROLYTE BRIDGING

As we pointed out in the section on the electrostatic forces, the predictions of the PB theory for interactions between charged macroions in electrolyte solutions of univalent salts conform well with osmotic stress experiments on ordered DNA arrays to the extent of a near quantitative agreement between theory and experiment [43]. The effective surface charge on DNA obtained from osmotic stress experiments is close, but nevertheless consistently somewhat smaller from the theoretical predictions based on the Manning counterion condensation theory. Theoretical arguments based on the more detailed analysis of the counterion condensation in the presence of salt [48] generally agree that the effective charge on DNA in an ionic solution should be smaller than the one based on the estimate of Manning condensation theory.

When the same analysis of the osmotic stress experiments is furthermore applied to salts containing at least one higher valency counterion, such as Mn^{2+}, $Co(NH_3)_6^{3+}$, or various polyamines, the theoretical predictions based on the PB theory tend to lose agreement with experiment. Not only does the PB theory give the wrong numerical values for the strength of the electrostatic interactions, but also and more importantly misses their sign since experiments point to the existence of electrostatic attractions [22]. This attraction is deduced from the shape of the osmotic pressure as a function of density of DNA, i.e., there are regions of DNA density where the corresponding osmotic pressure in a DNA array remains constant [49] (see left panel of Figure 21.12 for $Co(NH_3)_6^{3+}$ concentrations of 12 and 6 mM).

This is quite similar to the pressure versus volume isotherms in the case of a liquid–gas transition [50] (see right panel of Figure 21.12). In that case due to attractive van der Waals interactions between gas molecules, the gaseous phase condenses into a liquid phase at a certain condensation pressure that depends on the temperature. As we reduce the volume in the region of liquid–gas coexistence, more of the gas condenses

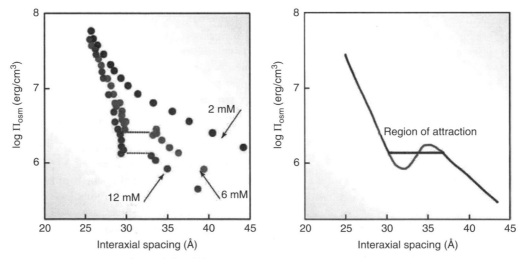

FIGURE 21.12 Left-hand side: Osmotic pressure as a function of DNA concentration in a DNA array with monovalent salt (0.25 M NaCl) with added trivalent counterion CoHex ($Co(NH_3)_6^{3+}$) at concentration from 0 to 20 mM. Above a sufficiently large value of CoHex concentration (17 mM), DNA spontaneously precipitates. For smaller values of CoHex concentration, e.g., 6 mM, the osmotic pressure dependence on the interaxial spacing shows a horizontal transition line between two regimes of repulsive forces. Right-hand side: A schematic explanation of the 6 mM CoHex concentration line. The dependence of the osmotic pressure on the interaxial spacing is in fact nonmonotonic due to the presence of attractive interactions in the region depicted in red. Because of the condition of stability, just as in the case of the van der Waals isotherm, the regions of attraction can only be traversed via a horizontal transition line.

into liquid, while the applied force required to keep the gas in place is unchanged. As a result, the part of the isotherm between the start and the end of the condensation process is flat (see right panel of Figure 21.12). In the DNA case, the role of inverse temperature is played (roughly) by the concentration of the polyvalent counterion. For sufficiently large concentration, for example, of $Co(NH_3)_6^{3+}$, the DNA array spontaneously precipitates or condenses into an ordered high density phase. One thus concludes that the polyvalent counterion should confer some kind of attractive interactions between nominally equally charged DNA molecules. What is so special about multivalent counterions, such as Mn^{2+}, $Co(NH_3)_6^{3+}$, or various polyamines, that leads to a complete breakdown of the simple PB framework?

In order to understand this anomaly, we should first relinquish a seemingly obvious explanation—that we are dealing with the effects of DNA–DNA van der Waals interactions as is the case in condensation of gases (see Section 21.2.4). These forces are much too small to account for the strong attractions seen with polyvalent counterions [34]. So how are we to rationalize the polyvalent counterion-mediated DNA attractions?

When dealing with small univalent counterions in the PB framework, we actually assume that they can be described collectively, via their number or charge density, without acknowledging that we have individual charges (see, for example, the chapter by Andelman in Ref. [27]). This approach is usually referred to as the mean-field approach in statistical thermodynamics and is used in a plethora of contexts. It works only if the concentration of the counterions or salt ions, in general, is high enough. PB theory is, in fact, a mean-field theory. When we go to polyvalent counterions of valency Z, and for the sake of the argument, assume that $Z \gg 1$, we have fewer counterions to satisfy the overall electroneutrality of the system. In an extreme case, we would be dealing with just a few of them as represented schematically in Figure 21.13. In this case, the mean-field description would break down miserably. Why? Because there is no proper "mean-field" to speak of. We have to deal with each of the counterions individually. This demands a completely different approach that has to be set apart and formulated in a completely different language than the popular PB (mean-field) theory. This alternative approach has been formulated by various people and goes under the strong-coupling approach [51] or the strong-correlation approach [52].

Let us try to describe this alternative approach without invoking the heavy analytical machinery on which it relies. For instance, assume that we have only one giant polyvalent counterion between two fixed charges of opposite sign (see right panel of Figure 21.13). The charge of this giant counterion would have to be large indeed in order to neutralize the two surfaces with fixed charge of opposite sign, but this is the assumption. The overall force between the fixed charges is composed of direct repulsion between the fixed charges, since they are assumed to be equally charged plus the attraction between the left fixed charge and the counterion (remember, they bear charges of opposite sign), assumed

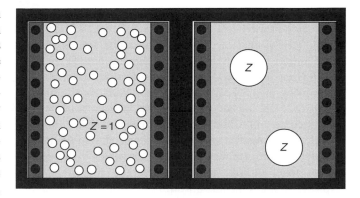

FIGURE 21.13 Monovalent counterions (small white circles) assumed to be positive between two oppositely charged surfaces (small black circles) versus polyvalent positive counterions (large white circles) in the same geometry. In the idealized case when the valency (Z) of the counterion is assumed to be very large ($Z \gg 1$), we can treat just a single counterion (of the two depicted in the right panel) in the space between the two surfaces bearing fixed charges. Direct electrostatic interactions between negatively charged surfaces are repulsive, but the interactions between the (single) counterion and the surfaces are attractive since the counterion bears a charge opposite in sign to those on the surfaces. The sum of the two is also net attractive. This is the physical origin of the correlation attraction in polyvalent counterion systems such as DNA with trivalent $Co(NH_3)_6^{3+}$ counterion.

to be in the middle of the space between the fixed charges and the attraction between the right fixed charge and the counterion (again remember, they bear charges of opposite sign). Summing together all these contribution, we get an overall attraction. Thus in this case, like-charged surfaces do not repel as goes the common wisdom, in fact they attract. This intuitive argument can be made exact within the strong-coupling approximation that supersedes the PB description in the case of polyvalent counterions. The strong coupling approach can be formulated also in an alternative form of the s.c. correlated Coulomb fluid theory [52], but is always reduced to the assumption that polyvalent counterions interact with the fixed charges individually and not collectively as in the PB framework. The attraction usually outweighs repulsion only at small separation between the surfaces bearing fixed charges, and as the separation increases the idealization invoked above becomes increasingly less realistic, we move smoothly from the strong-coupling attraction to the standard PB repulsion.

An alternative interpretation of the same effect [22] would be that the polyvalent counterion adsorbs onto the charged surface and thus changes its hydration pattern by interacting much more strongly with the water molecules in its vicinity than with the fixed charges on the surface. This modulation of the hydration pattern on both of the apposed surfaces could also induce structural attractions, qualitatively similar to the strong coupling electrostatic interactions. Both effects are short-ranged and are thus difficult to disentangle. Strong counterion specificity in the magnitude of attractive interactions shows that at least a part of it has an origin in counterion properties other than its valency.

Until now, when invoking polyvalent counterions what we had in mind were relatively small charged particles without any inherent structure. We now move forward. By increasing the valency of the counterions, thus adding more and more charge to them, ions do not only grow in size but in fact usually become more and more polymer-like. Their inherent chemical structure becomes increasingly chain-like. In fact, we refer to them as polyelectrolytes [53], which in this context means long, usually positively charged polymer chains. Typical examples include polyamines, such as spermine (Sp), and polypeptides such as poly-L-lysine (PL) and poly-L-arginine (PA). Mixing these long flexible polycations with DNA in an ionic solution leads again to ordered DNA arrays that allow the application of the osmotic stress technique, resulting in equation of state, i.e., osmotic pressure as a function of the DNA density, just as in the case of simple salts [54]. Strong attractions between DNA molecules have been measured also in solutions containing such polycations. In this case, however, it would be difficult to invoke the previous argument based on the strong-coupling picture, since the charged polymer chains have very extended configurations that do not allow us to use gedanken experiments based on point charge models as we did before, where the total interaction was composed of direct repulsion between the fixed charges plus the attraction between the fixed charges and the counterion. So what would be an appropriate conceptual picture to explain attractions between surfaces with fixed negative charges mediated by these long polycations?

We again give a simple description of what is going on without going into complicated mathematical details [55]. Imagine the situation of two cylindrical macroions with fixed negative charge together with a polycation chain and possibly simple salt and counterions (as depicted in Figure 21.14). Because the polyelectrolyte chain is oppositely charged from the macroions, it would like to neutralize them due to electrostatic attraction. In the case of sufficiently long polycations, the flexible chain can wrap around both macroions with fixed charges, creating a polyelectrolyte bridge between them (see the schematic representation on the left panel of Figure 21.14). This bridge is trying to pull the two fixed charges together by an entropic elastic force due to a tendency to maximize the possible number of polymer conformations, thus creating an attractive force between them. However, in this case, there is no strong-coupling or correlation effect. The chain draws the two macroions together simply because of its connectivity [56]. The total force in this case is composed of the direct electrostatic repulsion between the macroions with partially neutralized charges and the elastic term corresponding to the part of the polyelectrolyte chain between the macroions. Note the difference, however. In the strong coupling viewpoint, the attraction is still electrostatic in origin, stemming from the sharing of the simple polyvalent counterion between the two fixed charges. In the case of polyelectrolyte bridging on the other hand, the attraction is only distantly electrostatic in origin, stemming more directly from the elasticity, connectivity, and conformational flexibility of the polycation chain. Similar to the situation with simple, not polymeric, counterions, the polyelectrolyte bridging attraction is also usually short-ranged except in arrays of macromolecules where it can be also long(er) ranged [57].

Both mechanisms, strong coupling attraction as well as polyelectrolyte bridging, have a profound effect on the balance of forces in DNA arrays as well as for conformations of a single DNA in a very dilute solution. In the former case, they have been observed directly in the DNA–polycation complexes analyzed by the osmotic stress technique that we will describe in more detail later [54], whereas in the latter case, they are responsible for the phenomenon of DNA condensation that we will not deal with specifically in this chapter.

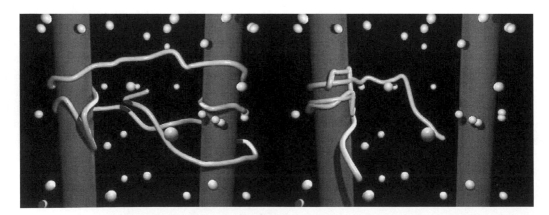

FIGURE 21.14 A schematic presentation of two cylindrical macroions with fixed (negative) charges (cylinders) with a polycation chain (wires) and explicit salt and counterions (small spheres) in between. The polycation chain tends to neutralize the fixed charges and thus wraps around both cylindrical macroions in a bridging configuration. Left-hand side: For small enough separation between the counterions, the polycation can bridge the space between them resulting in attractive bridging interactions. Right-hand side: For larger separations, the polycation cannot bridge the space between macroions and the bridging attraction is not present. This polycation-mediated bridging attraction is usually of short range but much stronger than the van der Waals interaction.

21.2.9 LESSONS

Molecular forces apparently convey a variety that is surprising considering the fact that they are all to some extent or another just a variant of electrostatic interactions. Quantum and thermal fluctuations apparently modify the underlying electrostatics, leading to qualitatively novel and unexpected features. Electrostatic interactions mediated by polyvalent counterions show a quite surprising feature of being actually attractive even between nominally equally charged macroions such as two DNA molecules. These attractions can be due to either complicated correlation or hydration effects, or polyelectrolyte bridging attractions if the counterions have long flexible charged chains. The menagerie of forces obtained in this way is what one has to deal with and understand when trying to make them work for us.

21.3 DNA MESOPHASES

21.3.1 POLYELECTROLYTE PROPERTIES OF DNA

We can define several levels of DNA organization similar to Ref. [1]. Its primary structure is the sequence of base pairs. Its secondary structure is the famous double helix that can exist in several conformations. In solution, the B-helical structure dominates [58]. The bases are perpendicular to the axis of the molecule and are 0.34 nm apart, and 10 of them make one turn of the helix. These parameters can vary for DNA in solution where up to 10.5 base pairs can make a whole turn of the double helix [59]. In the A structure, the bases are tilted with respect to the direction of the helix and this arrangement yields an internal hole, wider diameter, and closer packing (see Figure 21.15). Other conformations, such as the left-handed Z form, are rare. In solution, DNA's tertiary structure includes the many bent and twisted conformations in three dimensions.

DNA lengths can reach macroscopic dimensions. For instance, the human genome is coded in approximately 3 billion base pairs with a collective linear stretch on the order of a meter. Obviously, this molecule must undergo extensive compaction in order to fit in the cell nucleus. In natural environments, DNA is packaged by basic proteins, which form chromatin structures to keep DNA organized. In the test tube, DNA can be packaged into very tight and dense structures as well, primarily by various "condensing" agents. Their addition typically induces a random coil to globule transition. At large concentrations, DNA molecules, like lipids, form ordered liquid crystalline phases [10] that have been studied extensively at different solution conditions [8].

In vitro, at concentrations above a critical value [60], polyelectrolyte DNA self-organizes in highly ordered mesophases [8]. In this respect, it is a lyotropic liquid crystal. But contrary to the case of lipid mesophases, where the shape of constituent molecules plays a determining role, the organization of DNA in condensed phases is primarily a consequence of its relatively large stiffness [8]. The orientational ordering of DNA at high concentrations is promoted mostly by the interplay between entropically favored disorder or misalignment and the consequent price in terms of the high interaction

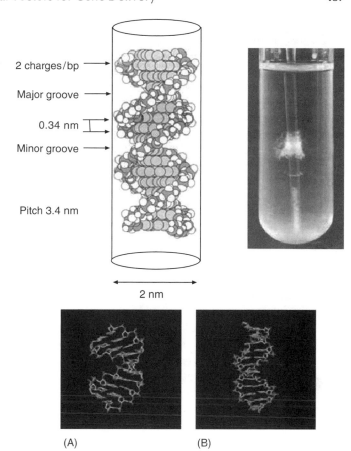

Polymer	Solvent	Persistence length (nm)
Microtubule	Water	3,000,000
F-actin	Water	15,000
Schizophyllan	Water	200
Xanthan	0.1 M NaCl	120
DNA	0.2 M NaCl	50
Hyaluronic acid	0.2 M NaCl	7

FIGURE 21.15 Structural parameters of a DNA molecule. The two relevant configurations of the DNA backbone. (A) DNA, common at small hydrations or high DNA densities. (B) DNA, common in solution at large hydrations and lower DNA densities. The test tube holds ethanol-precipitated DNA in solution. Its milky color is due to the light scattering by thermal conformational fluctuations in the hexatic phase (see main text). Box: Typical persistence lengths for different (bio)polymer chains in nanometers. DNA persistence length was first inferred from light scattering experiments in 1953 by Peterlin. (From Peterlin, A., *Nature*, 171, 259, 1953.)

energy. The mechanism of orientational ordering is thus the same as in standard short nematogens [61], with the main difference arising from the extended length of the polymeric chains. The discussion that follows will concentrate mostly on very long, on the order of 1000 persistence lengths, microns long, DNA molecules.

21.3.2 FLEXIBILITY OF DNA MOLECULES IN SOLUTION

In isotropic solutions, DNA can be in one of the several forms. For linear DNA, individual molecules are effectively straight over the span of a persistence length that can be defined also as the exponential decay length for the loss of angular correlation between two positions along the molecule, while for longer lengths they form a worm-like random coil. The persistence length of DNA is about 50 nm [1]. The persistence length has been determined by measuring the diffusion coefficient of different-length DNA molecules using dynamic light scattering and by enzymatic cyclization reactions [62]. It depends only weakly on the base-pair sequence and ionic strength.

DNA can also be circular as in the case of a plasmid. The closed form of a plasmid introduces an additional topological constraint on the conformation that is given by the linking number Lk [2]. The linking number gives the number of helical turns along a circular DNA molecule. Because plasmid DNA is closed, Lk has to be an integer number. By convention, Lk of a closed right-handed DNA helix is positive. The most frequent DNA conformation for plasmids in cells is negatively supercoiled. This means that for such plasmids, Lk is less than it would be for a torsionally relaxed DNA circle; negatively supercoiled DNA is underwound. This is a general phenomenon with important biological consequences. It seems that free energy of negative supercoiling catalyzes processes that depend on DNA untwisting such as DNA replication and transcription, which rely on DNA [63]. While the sequence of bases in exons determines the nature of proteins synthesized, it is possible that such structural features dictate the temporal and spatial evolution of DNA-encoded information.

21.3.3 LIQUID CRYSTALS

The fact that DNA is intrinsically stiff makes it form liquid crystals at high concentration [8]. Known for about 100 years, the simplest liquid crystals are formed by rodlike molecules. Solutions of rods exhibit a transition from an isotropic phase with no preferential orientation to a nematic phase, a fluid in which the axes of all molecules point on average in one direction (see Figure 21.11). The unit vector in which the molecules point is called the nematic director **n**. Nematic order is orientational order [61], in contrast to positional order that distinguishes between fluid and crystalline phases. Polymers with intrinsic stiffness can also form liquid crystals. This is because a long polymer with persistence length L_p acts much like a solution of individual rods that are all one persistence length long, thus the term "polymer nematics" [64].

If the molecules that comprise the liquid crystal are chiral, have a natural twist such as double helical DNA, then their orientational order tends to twist. This twist originates from the interaction between two molecules that are both of the same handedness. This chiral interaction is illustrated in Figure 21.16 for two helical or screw-like molecules. For steric reasons, two helices pack best when tilted with respect to each other. Instead of a nematic phase, chiral molecules form a cholesteric phase [61]. The cholesteric phase is a twisted nematic phase in which the nematic director twists continuously around the so-called cholesteric axis as shown in Figure 21.16. Using the same arguments as for plain polymers, chiral polymers form polymer cholesterics.

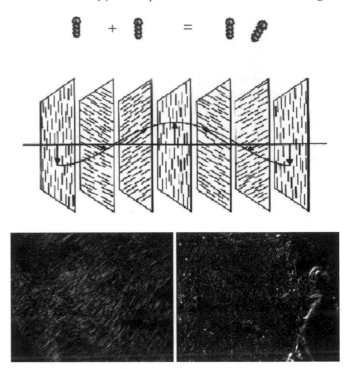

FIGURE 21.16 "Chiral" interaction for two helical or screw-like molecules. For steric reasons, two helices just as two screws (depicted on the figure) pack best when slightly tilted with respect to each other. Since DNA because of its double-stranded, helical nature is a type of molecular screw, it too exhibits chiral interactions. Instead of a nematic phase depicted in Figure 21.11, characterized by the average constant direction of molecules, chiral molecules form a cholesteric phase. (From De Gennes, P.G. and Prost, J., *The Physics of Liquid Crystals*, 2nd ed., Oxford University Press, Oxford, 1993.) The cholesteric phase is a twisted nematic phase in which the nematic director twists continuously around a "cholesteric axis" depicted on the middle drawing. Under crossed polarizers (bottom), the DNA cholesteric phase creates a characteristic striated texture. For long DNA molecules, the striations appear disordered.

Both cholesteric and hexagonal liquid crystalline DNA phases were identified in the 1960s. This discovery was especially exciting because both phases were also found in biological systems. The hexagonal liquid crystalline phase can be seen in bacterial phages and the cholesteric phase can be seen in cell nuclei of dinoflagellates [8].

21.3.4 MEASUREMENTS OF FORCES BETWEEN DNA MOLECULES

Liquid crystalline order lets us measure intermolecular forces directly. With the osmotic stress method, DNA liquid crystals are equilibrated against neutral polymer (such as PEG or PVP) solutions of known osmotic pressure, pH, temperature, and ionic composition [65]. Equilibration of DNA under osmotic stress of external polymer solution is effectively the same as exerting mechanical pressure on the DNA subphase with a piston (see Figure 21.4). In this respect, the osmotic stress technique is formally very much similar to the Boyle experiment where one compresses a gas with mechanical pistons and measures the ensuing pressure. After equilibration under this known stress, DNA separation is measured either by x-ray scattering, if the DNA subphase is sufficiently ordered, or by straightforward densitometry [66]. Known DNA density and osmotic stress immediately provide an equation of state (osmotic pressure as a function of the density of the DNA subphase) to be codified in analytic form for the entire phase diagram. Then, with the local packing symmetry derived from x-ray scattering [7,65], and sometimes to correct for DNA motion [42], it is possible to extract the bare interaxial forces between molecules which can be compared with theoretical predictions as developed in Chapter 2. *In vivo* observation of DNA liquid crystals [67] shows that the amount of stress needed for compaction and liquid crystalline ordering is the same as for DNA *in vitro*.

21.3.5 Interactions Between DNA Molecules

Direct force measurements performed on DNA in univalent salt solutions reveal two types of purely repulsive interactions between DNA double helices [4]:

1. At interaxial separations less than ~3 nm (surface separation ~1 nm), an exponentially varying "hydration" repulsion is thought to originate from partially ordered water near the DNA surface.
2. At surface separations greater than 1 nm, measured interactions reveal electrostatic double layer repulsion presumably from negative phosphates along the DNA backbone.

Measurements give no evidence for a significant DNA–DNA attraction expected on theoretical grounds [68]. Though charge fluctuation forces must certainly occur, they appear to be negligible at least for liquid crystal formation in monovalent-ion solutions. At these larger separations, the double layer repulsion often couples with configurational fluctuations to create exponentially decaying forces whose decay length is significantly larger than the expected Debye screening length [42].

Bare short-range molecular interactions between DNA molecules appear to be insensitive to the amount of added salt. This has been taken as evidence that they are not electrostatic in origin, as attested also by similar interactions between completely uncharged polymers such as schizophyllan (Figure 21.5). The term "hydration force" associates these forces with perturbations of the water structure around the DNA surface [65]. Alternatively, short-range repulsion has been viewed as a consequence of the electrostatic force specific to high DNA density and counterion concentration [69].

21.3.6 High-Density DNA Mesophases

Ordering of DNA can be induced by two alternative mechanisms. First of all, attractive interactions between different DNA segments can be enhanced by adding multivalent counterions thought to promote either counterion-correlation forces [70] or electrostatic [71] and hydration attraction [22]. In these cases, DNA aggregates spontaneously. Alternatively, one can add neutral crowding polymers to the bathing solution that phase separate from DNA and exert osmotic stress on the DNA subphase [72]. In this case, the intersegment repulsions in DNA are simply counteracted by the large externally applied osmotic pressure. DNA is forced in this case to condense under externally imposed constraints. This latter case is formally (but only formally) analogous to a Boyle gas pressure experiment but with osmotic pressure playing the role of ordinary pressure. The main difference being that ordinary pressure is set mechanically while osmotic pressure has to be set through the chemical potential of water, which is in turn controlled by the amount of neutral crowding polymers (such as PEG, PVP, or dextran) in the bathing solution [66].

At very high DNA densities, where the osmotic pressure exceeds 160 atm, DNA can exist only in a (poly)crystalline state [73]. Nearest neighbors in such an array are all oriented in parallel and show correlated (nucleotide) base stacking between neighboring duplexes (see Figures 21.11 and 21.17). This means that there is a long-range correlation in the positions of the backbone phosphates between different DNA molecules in the crystal. The local symmetry of the lattice is monoclinic. Because of the high osmotic pressure, DNA is actually forced to be in an A conformation characterized by a somewhat larger outer diameter as well as a somewhat smaller pitch than in the canonical B conformation (see Figure 21.15), which persists at smaller densities. If the osmotic pressure of such a crystal is increased above 400 atm, the helix begins to crack and the sample loses structural homogeneity [73].

Lowering the osmotic pressure does not have a pronounced effect on the DNA crystal until it is down to ~160 atm. Then the crystal as a whole simultaneously expands while individual DNA molecules undergo an A to B conformational transition (see Figure 21.17) [73]. This phase transformation is thus first order, and besides being a conformational transition for single DNA, is connected also with the melting of the base stacking as well as positional order of the helices in the lattice. The ensuing low-density mesophase, where DNA is in the B conformation, is therefore characterized by short-range base stacking order, short-range 2D positional order, and long-range bond orientational order (see Figure 21.18) [74]. This order is connected with the spatial direction of the nearest neighbors [75]. It is for this reason that the phase has been termed a "line hexatic" phase. Hexatics usually occur only in 2D systems. They have crystalline bond orientational order but liquid-like positional order. There might be a hexatic to hexagonal columnar transition somewhere along the hexatic line though a direct experimental proof is lacking.

The difference between the two phases is that the hexagonal columnar phase has also a crystalline positional order and is thus a real 2D crystal (see Figure 21.18) [76]. It is the long-range bond orientational order that gives the line hexatic

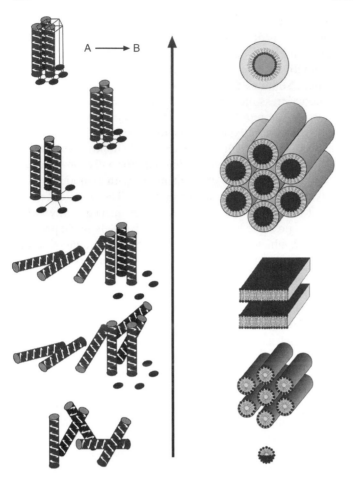

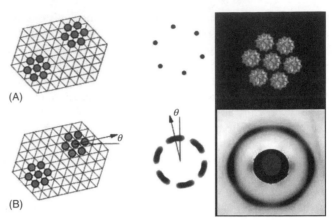

FIGURE 21.18 Bond orientational or hexatic order. With a real crystal, if one translates part of the crystal by a lattice vector, the new position of the atoms completely coincides with those already there. (Adapted from Chaikin, P.M. and Lubensky, T.C., *Principles of Condensed Matter Physics*, Cambridge University Press, Cambridge, MA, 1995.) In a hexatic phase, the directions to the nearest neighbors (bond orientations) coincide (after rotation by 60°), but the positions of the atoms do not coincide after displacement in one of the six directions. Consequently, a real crystal gives a series of very sharp Bragg peaks in x-ray scattering (upper half of box) whereas a hexatic gives hexagonally positioned broad spots. The pattern of x-ray scattering by high-density DNA samples gives a fingerprint of a hexatic phase. The densitogram of the scattering intensity (right half of figure) shows six pronounced peaks that can be Fourier decomposed with a marked sixth-order Fourier coefficient, another sign that that the scattering is due to long-range bond orientational order. (From Podgornik, R., Strey, H.H., Gawrisch, K., Rau, D.C., Rupprecht, A., and Parsegian, V.A., *Proc. Natl. Acad. Sci. USA*, 1996, 93, 4261.)

FIGURE 21.17 Schematic phase diagrams for DNA (left) and lipids (right). In both cases, the arrow indicates increasing density in both cases. DNA starts (bottom) as a completely disordered solution. It progresses through a sequence of "blue" phases characterized by cholesteric pitch in two perpendicular directions (From Leforestier, A. and Livolant, F., *Mol. Cryst. Liquid Cryst.*, 1994, 17, 651.) and then to a cholesteric phase with pitch in only one direction. At still larger densities, this second cholesteric phase is succeeded by a hexatic phase, characterized by short-range liquid-like positional order and long-range crystal-like bond orientational (or hexatic order, indicated by lines). At highest densities, there is a crystalline phase, characterized by long-range positional order of the molecules and long-range base stacking order in the direction of the long axes of the molecules. Between the hexatic and the crystalline forms, there might exist a hexagonal columnar liquid-crystalline phase that is similar to a crystal, but with base stacking order only on short scales. The lipid phase diagram (From Small, D.M., *The Physical Chemistry of Lipids: From Alkanes to Phospholipids*, Plenum Press, New York, 1986.) is a composite of results obtained for different lipids. It starts from a micellar solution and progresses through a phase of lipid tubes to a multilamellar phase of lipid bilayers. This is followed by an inverted hexagonal columnar phase of water cylinders and possibly goes to an inverted micellar phase. Most lipids show only a subset of these possibilities. Boundaries between the phases shown here might contain exotic cubic phases not included in this picture.

phase some crystalline character [77]. The DNA duplexes are still packed in parallel, while the local symmetry perpendicular to the long axes of the molecules is changed to hexagonal. The directions of the nearest neighbors persist through mac-

roscopic dimensions (on the order of mm) while their positions tend to become disordered already after several (typically 5 to 10) lattice spacings. This mesophase has a characteristic x-ray scattering fingerprint (see Figure 21.18). If the x-ray beam is directed parallel to the long axis of the molecules, it shows a hexagonally symmetric diffraction pattern of broad liquid-like peaks [78].

Typical lattice spacings in the line hexatic phase are between 25 and 35 Å (i.e., between 600 and 300 mg/mL of DNA) [74]. The free energy in this mesophase is mostly a consequence of the large hydration forces stemming from removal of water from the phosphates of the DNA backbone. Typically independent of the ionic strength of the bathing solution, these hydration forces [65] depend exponentially on the interhelical separation and decay with a decay length of about 3 Å [11] at these large densities. This value of the hydration decay length seems to indicate that it is determined solely by the bulk properties of the solvent, i.e., water.

It is interesting to note that the behavior of short-fragment DNA in this range of concentrations is different from the long DNA [76]. The short-fragment DNA, typically the nucleosomal DNA fragment of 146 bp, forms a two-dimensional hexagonal phase at interaxial spacing of ~30 Å, which progressively orders into a three-dimensional hexagonal phase on decrease of the interaxial spacing to ~23 Å [76].

At still larger concentrations, the short-fragment DNA forms a three-dimensional orthorhombic crystal, with a deformed hexagonal unit cell perpendicular to the c-axis. Concurrently to this symmetry transformation, the helical pitch of the condensed phase decreases continuously from 34.6 to 30.2 Å [76]. The reasons for this fundamental difference between the behaviors of long as opposed to short-fragment DNA are still not well understood.

When the osmotic pressure is lowered to about 10 atm (corresponding to interaxial spacing of about 35 Å or DNA density of about 300 mg/mL), the characteristic hexagonal x-ray diffraction fingerprint of the line hexatic mesophase disappears continuously. This disappearance suggests the presence of a continuous, second-order transition into a low-density cholesteric [74]. It is characterized by short-range (or effectively no) base stacking order, short-range positional order, short-range bond orientational order, but long-range

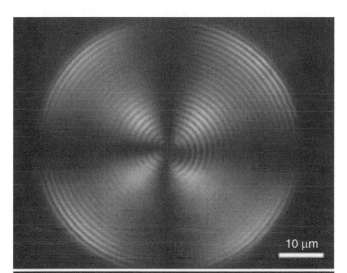

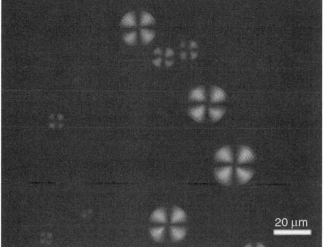

FIGURE 21.19 Texture of small drops of DNA cholesteric phase (spherulites) in a PEG solution under crossed polarizers. These patterns reveal the intricacies of DNA orientational packing when its local orientation is set by a compromise between interaction forces and the macroscopic geometry of a spherulite. The change from a bright to a dark stripe indicates that the orientation of the DNA molecule has changed by 90°.

cholesteric order, manifested in a continuing rotation of the long axis of the molecules in a preferred direction. In this sense, the cholesteric DNA mesophase would retain the symmetry of a one-dimensional crystal. X-ray diffraction pattern of the DNA in the cholesteric phase is isotropic and has the form of a ring. Crossed polarizers, however, reveal the existence of long-range cholesteric order just as in the case of short chiral molecules. The texture of small drops of DNA cholesteric phase (spherulites) under crossed polarizers (see Figure 21.19) reveals the intricacies of orientational packing of DNA where its local orientation is set by a compromise between interaction forces and macroscopic geometry of a spherulite. It is thus only at these low densities that the chiral character of the DNA finally makes an impact on the symmetry of the mesophase. It is not yet fully understood why the chiral order is effectively screened from the high-density DNA mesophases.

At still smaller DNA densities, the predominance of the chiral interactions in the behavior of the system remains. Recent work on the behavior of low-density DNA mesophases indicates [79] that the cholesteric part of the phase diagram might end with a sequence of blue phases that would emerge as a consequence of the loosened packing constraints coupled to the chiral character of the DNA molecule. At DNA density of about 10 mg/mL, the cholesteric phase line would end with DNA reentering the isotropic liquid solution where it remains at all subsequent densities except perhaps at very small ionic strengths [80].

21.3.7 DNA Equation of State

The free energy of the DNA cholesteric mesophase appears to be dominated by the large elastic shape fluctuations of its constituent DNA molecules [81] that leave their imprint in the very broad x-ray diffraction peak [66]. Instead of showing the expected exponential decay characteristic of screened electrostatic interactions [82], where the decay length is equal to the Debye length, it shows a fluctuation-enhanced repulsion similar to the Helfrich force existing in the flexible smectic multilamellar arrays [41]. Fluctuations not only boost the magnitude of the existing screened electrostatic repulsion but also extend its range through a modified decay length equal to four times the Debye length. The factor-of-four enhancement in the range of the repulsive force is a consequence of the coupling between the bare electrostatic repulsions of exponential type and the thermally driven elastic shape fluctuations described through elastic curvature energy that is proportional to the square of the second derivative of the local helix position [42]. In the last instance, it is a consequence of the fact that DNAs in the array interact via an extended, soft-screened electrostatic potential and not through hard bumps as assumed in the simple derivation in Section 21.2.7.

The similarity of the free energy behavior of the smectic arrays with repulsive interactions of Helfrich type and the DNA arrays in the cholesteric phase, which can as well be understood in the framework of the Helfrich-type enhanced repulsion, satisfies a consistency test for our understanding of flexible supermolecular arrays.

21.4 LIPID MESOPHASES

21.4.1 AGGREGATION OF LIPIDS IN AQUEOUS SOLUTIONS

Single-molecule solutions of biological lipids exist only over a negligible range of concentrations; virtually all interesting lipid properties are those of aggregate mesophases such as bilayers and micelles. Lipid molecules cluster into ordered structures to maximize hydrophilic and minimize hydrophobic interactions [83,84]. These interactions include negative free energy contribution from the solvation of polar heads and van der Waals interactions of hydrocarbon chains, competing with positive contributions such as steric, hydration, and electrostatic repulsions between polar heads. The "hydrophobic effect," which causes segregation of polar and nonpolar groups, is said to be driven by the increase of the entropy of the surrounding medium.

Intrinsic to the identity of surfactant lipids is the tension between water-soluble polar groups and lipid-soluble hydrocarbon chains. There is no surprise then that the amount of water available to an amphiphile is a parameter pertinent to its modes of packing and to its ability to incorporate foreign bodies.

These interactions therefore force lipid molecules to self-assemble into different ordered microscopic structures, such as bilayers, micelles (spherical, ellipsoidal, rodlike, or disklike), which can, especially at higher concentrations, pack into macroscopically ordered phases such as lamellar, hexagonal, inverted hexagonal, and cubic. The morphology of these macroscopic phases changes with the balance between attractive van der Waals and ion correlation forces versus electrostatic, steric, hydration, and undulation repulsion [85].

21.4.2 LIPID BILAYER

The workhorse of all lipid aggregates is the bilayer (see Figure 21.20) [84]. This sandwich of two monolayers, with nonpolar hydrocarbon chains tucked in toward each other and polar groups facing water solution, is only about 20 to 30 Å thick. Yet it has the physical resilience and the electrical resistance to form the "plasma" membrane that divides "in" from "out" in all biological cells. Its mechanical properties have been measured in terms of bending and stretching moduli. These strengths together with measured interactions between bilayers in multilamellar stacks have taught us to think quantitatively about the ways in which bilayers are formed and maintain their remarkable stability.

With some lipids, such as double-chain phospholipids, when there is the need to encompass voluminous hydrocarbon components compared with the size of polar groups, the small surface-to-volume ratio of spheres, ellipsoids, or even cylinders cannot suffice even at extreme dilution. Bilayers in this case are the aggregate form of choice. These may occur as single "unilamellar" vesicles, as onion-like multilayer vesicles, or multilamellar phases of indefinite extent. *In vivo*, bilayer-forming phospholipids create the flexible but tightly sealed plasma–membrane matrix that defines the inside from the outside of a cell. *In vitro*, multilayers are often chosen as a

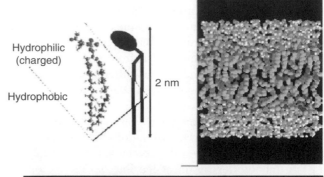

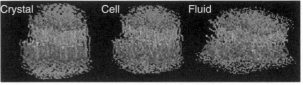

Lipid	Bending regidity $(10^{-19} J)$	Area compression (mN/m)
DMPC	1.15	145
SOPC	0.9	200
EggPC	0.4–2	
Plasma membrane	0.2–2	700
Red blood cell	0.13–0.3	450

FIGURE 21.20 (See color insert following blank page 206.) The lipid bilayer. A lipid molecule has a hydrophilic and a hydrophobic part (here shown is the phosphatidylserine molecule that has a charged headgroup). At high enough densities, lipid molecules assemble into a lipid bilayer. Together with membrane proteins as its most important component, the lipid bilayer is the underlying structural component of biological membranes. The degree of order of the lipids in a bilayer depends drastically on temperature and goes through a sequence of phases (see main text): Crystalline, gel, and fluid, as depicted in the middle drawing. The box at the bottom gives sample values of bilayer bending rigidity and area compressibility for some biologically relevant lipids and one well-studied cell membrane. (Graphics courtesy of B.B Brooks, and M. Hodoscek, NIH).

matrix of choice for the incorporation of polymers. Specifically, there are tight associations between positively charged lipids that merge with negatively charged DNA in a variety of forms (see Section 21.5.1).

The organization of lipid molecules in the bilayer itself can vary [84]. At low enough temperatures or dry enough conditions, the lipid tails are frozen in an all-*trans* conformation that minimizes the energy of molecular bonds in the alkyl tails of the lipids. Also the positions of the lipid heads along the surface of the bilayer are frozen in 2D positional order, making the overall conformation of the lipids in the bilayer crystal (L_C). The chains can either be oriented perpendicular to the bilayer surface (L_β and $L_{\beta'}$) or be tilted (crystalline phase L_C or ripple phase P_β). Such a crystalline bilayer cannot exist by itself but assembles with others to make a real 3D crystal.

Upon heating, various rearrangements in the 2D crystalline bilayers occur, first the positional order of the headgroups melts leading to a loss of 2D order ($L_{\beta'}$) and tilt (L_β) and then at the gel–liquid crystal phase transition, the untilted or rippled (P_β phase) bilayer changes into a bilayer membrane with disordered polar heads in two dimensions and conformationally frozen hydrocarbon chains, allowing them to spin around the long axes of the molecules, the so-called L_α phase. At still higher temperatures, the thermal disorder finally destroys also the ordered configuration of the alkyl chains, leading to a fluidlike bilayer phase. The fluid bilayer phase creates the fundamental matrix that according to the fluid mosaic model [83] contains different ingredients of biological membranes, e.g., membrane proteins, channels, etc.

Not only bilayers in multilamellar arrays but also liposome bilayers can also undergo such phase transitions; electron microscopy has revealed fluid phase, rippled and crystalline phase in which spherical liposomes transform into polyhedra due to very high values of bending elasticity of crystallized bilayers [86].

The fluid phase of the lipid bilayer is highly flexible. This flexibility makes it prone to pronounced thermal fluctuations resulting in large excursions away from a planar shape. This flexibility of the bilayer is essential for understanding the range of equilibrium shapes that can arise in closed bilayer (vesicles) systems [87]. Also, just as in the case of flexible DNA, it eventually leads to configurational entropic interactions between bilayers that have been crammed together [41]. Bilayers and linear polyelectrolytes thus share a substantial amount of fundamentally similar physics that allows us to analyze their behavior in the same framework.

21.4.3 Lipid Polymorphism

Low-temperature phases [88] are normally lamellar with frozen hydrocarbon chains tilted (crystalline phase L_C or ripple phase P_β) or nontilted (L_β and $L_{\beta'}$ form 3D, 2D, or 1D crystalline or gel phases) with respect to the plane of the lipid bilayers. Terminology from thermotropic liquid crystals phenomenology [61] can be used efficiently in this context: these phases are smectic, and SmA describes 2D fluid with no tilt while a variety of SmC phases with various indices encompass tilted phases with various degrees of 2D order. Upon melting, liquid crystalline phases with 1D (lamellar L_α), 2D (hexagonal II), or 3D (cubic) positional order can form.

The most frequently formed phases are micellar, lamellar, and hexagonal (Figure 21.17). Normal hexagonal phase consists of long cylindrical micelles ordered in a hexagonal array, while in the inverse hexagonal II (H_{II}) phase, water channels of inverse micelles are packed hexagonally with lipid tails filling the interstices. In excess water, such arrays are coated by a lipid monolayer. The morphology of these phases can be maintained upon their (mechanical) dispersal into colloidal dispersions. Despite the requirement that energy has to be used to generate dispersed mesophases, relatively stable colloidal dispersions of particles with lamellar, hexagonal, or cubic symmetry can be formed.

Many phospholipids found in lamellar cell membranes after extraction, purification, and resuspension prefer an inverted hexagonal geometry (Figure 21.21) [88]. Under

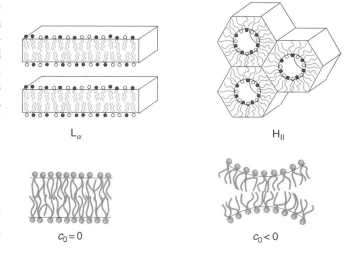

FIGURE 21.21 Different lipids are strained to different degrees when forced into lamellar packing. Relaxation of this strain contributes to the conditions for lamellar-to-inverted hexagonal phase transitions that depend on temperature, hydration, and salt concentration (for charged lipids).

excess-water conditions, different lipids assume different most-favored spontaneous radii for the water cylinder of this inverted phase [89]. An immediate implication is that different lipids are strained to different degrees when forced into lamellar packing. There are lamellar-inverted hexagonal phase transitions that occur with varied temperature, hydration, and salt concentration (for charged lipids) that form in order to alleviate this strain (see Figure 21.21).

In the presence of an immiscible organic phase, emulsion droplets can assemble [90]. In regions of phase diagram, which are rich in water, oil-in-water emulsions and microemulsions ($c > 0$) can be formed, while in oil-rich regions, these spherical particles have negative curvature and are therefore water-in-oil emulsions. The intermediate phase between the two is a bicontinuous emulsion that has zero average curvature and an anomalously low value of the surface tension (usually brought about the use of different cosurfactants) between the two immiscible components. Only microemulsions can form spontaneously (analogously to micelle formation) while for the formation of a homogeneous emulsion, some energy has to be dissipated into the system.

The detailed structure of these phases as well as the size and shape of colloidal particles are probably dominated by

- Average molecular geometry of lipid molecules
- Their aqueous solubility and effective charge
- Weaker interactions such as intra- and intermolecular hydrogen bonds
- Stereoisomerism as well as interactions within the medium

All depend on the temperature, lipid concentration, and electrostatic and van der Waals interactions with the solvent and solutes. With charged lipids, counterions, especially anions, may also be important. Ionotropic transitions have been observed with negatively charged phospholipids in the presence of metal ions leading to aggregation and fusion [91]. In cationic amphiphiles, it was shown that simple exchange of counterions can induce micelle–vesicle transition. Lipid polymorphism is very rich and even single-component lipid systems can form a variety of other phases, including ribbon-like phases, coexisting regions and various stacks of micelles of different shapes.

21.4.4 Forces in Multilamellar Bilayer Arrays

Except for differences in dimensionality, forces between bilayers are remarkably similar to those between DNA. At very great separations between lamellae, the sheet-like structures flex and "crumple" because of (thermal) Brownian motion [41]. Just as an isolated flexible linear polymer can escape from its one linear dimension into the three dimensions of the volume in which it is bathed, so can two-dimensional flexible sheets. In the most dilute solution, biological phospholipids typically form huge floppy closed vesicles; these vesicles enjoy flexibility while satisfying the need to keep all greasy nonpolar chains comfortably covered by polar groups rather than exposed at open edges. For this reason, in very dilute solution, the interactions between phospholipid bilayers are usually space wars of collision and volume occupation. This steric competition is always seen for neutral lipids; it is not always true for charged lipids [85].

Especially in the absence of any added salt, planar surfaces emit far-ranging electrostatic fields [27] that couple to thermally excited elastic excursions to create very long-range repulsion [44,92]. As with DNA, this repulsion is a mixture of direct electrostatic forces and soft collisions mediated by electrostatic forces rather than by actual bilayer contact. In some cases, electrostatic repulsion is strong enough to snuff out bilayer bending when bilayers form ordered arrays with periodicities as high as hundreds of angstroms [93].

Almost always bilayers align into well-formed stacks when their concentration approaches ~50 to 60 wt.% and their separation is brought down to a few tens of angstroms. In this region, charged layers are quite orderly with little lamellar undulation. In fact, bilayers of many neutral phospholipids often spontaneously fall out of dilute suspension to form arrays with bilayer separations between 20 and 30 Å. These spontaneous spacings are thought to reflect a balance between van der Waals attraction and undulation-enhanced hydration repulsion [85]. One way to test for the presence of van der Waals forces has been to add solutes such as ethylene glycol, glucose, or sucrose to the bathing solutions. It is possible then to correlate the changes in spacing with changes in van der Waals forces due to the changes in dielectric susceptibility as described above [94]. More convincing, there have been direct measurements of the work to pull apart bilayers that sit at spontaneously assumed spacings. This work of separation is of the magnitude expected for van der Waals attraction [95].

Similar to DNA, multilayers of charged or neutral lipids, subjected to strong osmotic stress, reveal exponential variation in osmotic pressure versus bilayer separation [85]. Typically at separations between dry "contact" and 20 Å, exponential decay lengths are 2–3 Å in distilled water or in salt solution, whether phospholipids are charged or neutral. Lipid bilayer repulsion in this range is thought to be due to the work of polar group dehydration sometimes enhanced by lamellar collisions from thermal agitation [96]. Normalized per area of interacting surface, the strength of hydration force acting in lamellar lipid arrays and DNA arrays is directly comparable.

Given excess water, neutral lipids usually find the above-mentioned separation of 20 to 30 Å at which this hydration repulsion is balanced by van der Waals attraction. Charged lipids, unless placed in solutions of high salt concentration, swell to take up indefinitely high amounts of water. Stiff charged bilayers repel with exponentially varying electrostatic double layer interactions, but most charged bilayers undulate at separations where direct electrostatic repulsion has weakened. In that case, similar to what has been described for DNA, electrostatic repulsion is enhanced by thermal undulations [97].

21.4.5 Equation of State of Lipid Mesophases

Lipid polymorphism shows much less universality than DNA. This is of course expected since lipid molecules come in many different varieties [84] with strong idiosyncrasies in terms of the detailed nature of their phase diagrams. One thus cannot achieve the same degree of generality and universality in the description of lipid phase diagram and consequent equations of state as was the case for DNA.

Nevertheless, recent extremely careful and detailed work on PCs by Nagle and his group [98] points strongly to the conclusion that at least in the lamellar part of the phase diagram of neutral lipids, the main features of the DNA and lipid membrane assembly physics indeed is the same [96]. This statement, however, demands qualification. The physics is the same provided one first disregards the dimensionality of the aggregates—one dimensional in the case of DNA and two dimensional in the case of lipid membranes—and takes into account the fact that while van der Waals forces in DNA arrays are negligible, they are essential in lipid membrane force equilibria. One of the reasons for this state of affairs is the large difference, unlike in the case of DNA, between the static dielectric constant of hydrophobic bilayer interior, composed of alkyl lipid tails, and the aqueous solution bathing the aggregate.

We have already pointed out that in the case of DNA arrays, quantitative agreement between theory, based on hydration and electrostatic forces augmented by thermal undulation forces, and experiment has been obtained and extensively tested [7,42]. The work on neutral lipids [96] claims that the same level of quantitative accuracy can also be achieved in lipid membrane assemblies if one takes into account hydration and van der Waals forces again augmented by thermal undulations. Of course, the nature of the fluctuations in the two systems is different and is set by the dimensionality of the fluctuating aggregates: one-versus two-dimensional.

The case of lipids adds an additional twist to the quantitative link between theory and experiments. DNA in the line hexatic as well as cholesteric phases (where reliable data for the equation of state exist) is essentially fluid as far as positional order is concerned and thus has unbounded positional fluctuations. Lipid membranes in the smectic multilamellar phase on the other hand are quite different in this respect. They are not really fluid as far as positional order is concerned but show something called quasilong range (QLR) order, meaning that they are in certain respects somewhere between a crystal and a fluid [61,78]. The QLR positional order makes itself recognizable through the shape of the x-ray diffraction peaks in the form of persistent (Caille) tails [78].

In a crystal, one would ideally expect infinitely sharp peaks with Gaussian broadening only because of finite accuracy of the experimental setup. Lipid multilamellar phases, however, show peaks with very broad, non-Gaussian, and extended tails that are one of the consequences of QLR positional order. The thickness of these peaks for different orders of x-ray reflections varies in a characteristic way with the order of the reflection [78]. It is this property that allows us to measure not only the average spacing between the molecules but also the amount of fluctuation around this average spacing. Luckily the theory predicts this property also and without any free parameters (all of them being already determined from the equation of state), the comparison between predicted and measured magnitude in positional fluctuations of membranes in a multilamellar assembly is more than satisfactory [96].

In summing up, the level of understanding of the equation of state reached for DNA and neutral lipid membrane arrays is pleasing.

21.5 DNA–LIPID INTERACTIONS

Mixed in solution with cationic lipids (CLs), DNA spontaneously forms CL–DNA aggregates of submicron size. These DNA–lipid aggregates, sometimes called "lipoplexes," [99] are routinely used for cell transfection *in vitro*. More important, they are used primarily as potential gene delivery vehicles for *in vivo* gene therapy (for recent reviews, see Refs. [100–105] and references therein). Under appropriate conditions, these aggregates reveal complex underlying thermodynamic phase behavior. There is a practical paradox here. We use stable equilibrium structures to reveal the forces that cause aggregation and assembly; we use this knowledge of forces to create the unstable preparations likely to be most efficient in transfection.

Lipoplexes for transfection were first proposed by Felgner and coworkers [106,107]. The guiding idea was to overcome the electrostatic repulsion between cell membranes (containing negatively charged lipids) and negative DNA by complexing DNA with positively charged CL. Preliminary experimental data showed that at least some lipoplexes deliver

DNA through direct fusion with the cell membrane [108,109]. More often, however, lipoplex internalization probably proceeds through endocytosis after initial interaction with the cell's membrane.

Prior to the attempts to utilize lipoplexes for transfection, studies of DNA aggregated with multivalent cations and coated with negatively charged liposomes were also explored as possible vectors. It was hoped that CL–DNA complexes would no longer require an additional complexing agent, and that also, the transfection efficiency would be higher. The complex's lipid coating could protect the tightly packed DNA cargo during its passage to the target cells. In recent years, this strategy has been slightly modified to complex DNA and anionic (or even neutral) membranes in ordered lamellar phases that should have lower cytotoxicity than the alternative CLs. An unresolved problem of this approach is the inefficient association between the ALs and DNA molecules, which is attributed to their like-charge electrostatic repulsion [110–114].

While not confronted with the immunological response, risked by the alternative viral vector strategy, the use of lipoplexes in gene therapy is still hampered by toxicity of the CL and low *in vivo* transfection efficiency despite the *in vitro* efficiency of some CL formulations. This discrepancy can be attributed to the multistage and multibarrier process the complexes must endure before transfection is achieved. These steps typically include passage in the serum, interaction with target and other cells, internalization, complex disintegration in the cytoplasm, transport of DNA into the nucleus, and ultimately expression.

In the search for increasingly more potent gene delivery vectors, the intimate relationship between the lipoplex's phase structure (or morphology) and its transfection efficiency probably serves as the greatest motivation for their study. How is transfection affected by lipoplex morphology? How may this structure be controlled? Experiment and theory of the past decade shed some light on such fundamental questions. They may give perspective for future strategies to design CL-based nonviral vectors.

To this end, we present here our current understanding of the structure and phase behavior of CL–DNA complexes. We review the relation of structure to transfection efficiency, and more specifically, to the way the complex formation overcomes one barrier to DNA release into the cytoplasm.

21.5.1 STRUCTURE OF CL–DNA COMPLEXES

In general, the structures of CL–DNA composite phases can be viewed as morphological hybrids of familiar pure-lipid and pure-DNA phases. A first example is the lamellar-like structure initially proposed by Lasic et al. [115,116]. The first comprehensive and unambiguous evidence for this structure came from a series of studies by Rädler et al. [117–121]. From high-resolution synchrotron x-ray diffraction and optical microscopy, they reported the existence of novel lamellar CL–DNA phase morphologies. In particular, one complex structure was shown to consist of lamellar multilayer. In this case,

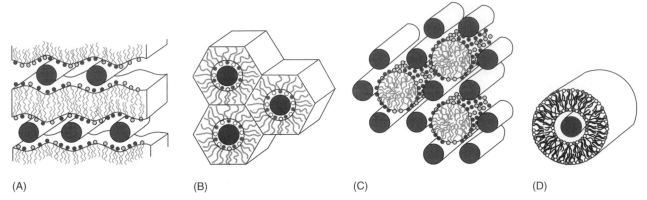

FIGURE 21.22 Schematic illustration of some possible structures of DNA–mixed lipid (cationic/nonionic) complexes. (A) The sandwich-like (L_α^c) lamellar complex composed of parallel DNA molecules intercalated between lipid bilayers. (B) The honeycomb-like (H_{II}^c) hexagonal complex, composed of a hexagonally packed bundle of monolayer-coated DNA strands. (C) Two interpenetrating hexagonal lattices, one of DNA, the other of micelles. (D) Spaghetti-like complex, composed of bilayer-coated DNA. (From May, S., Harries, D., and Ben-Shaul, A., *Biophys. J.*, 79, 1747, 2000. With permission.)

smectic-like stacks of mixed bilayers, each composed of a mixture of CL (e.g., dioleoyltrimethylammonium propane (DOTAP)) and neutral "helper" lipid (e.g., dioleoylphosphatidylcholine (DOPC)), with monolayers of DNA strands intercalated within the intervening water gaps (see Figure 21.22A) like a multilipid bilayer L_α phase [122]. Helper lipids (HLs) are often added for their fusogenic properties. Dioleoylphosphatidylethanol amine (DOPE), for example, is conjectured to promote transfection. In addition, because pure (synthetically derived) CLs often tend to form micelles in solution, HLs facilitate the formation of membranes.

In this L_α^c complex geometry, the DNA strands within each gallery are parallel to each other, exhibiting a well-defined repeat distance d. While d depends on the CL/DNA and CL/HL concentration ratios, the spacing between two apposed lipid monolayers is nearly constant at ~26 Å corresponding to the diameter of double stranded B-DNA, ca. 20 Å, surrounded by a thin hydration shell. This L_α^c lamellar (sandwich) complex is stabilized by the electrostatic attraction between the negatively charged DNA and the CL bilayer. Because of strong electrostatic repulsion between the charged bilayers (particularly at low salt conditions), the lamellar lipid phase is unstable without DNA.

Quite different equilibrium ordered phase morphologies were found to occur from other choices of neutral HL. In the case of DOPE, or lecithin, for example, inverted hexagonal ("honeycomb" or H_{II}^c) organization of the lipid, with stretches of double-stranded DNA laying in the aqueous solution regions, were found to form (see Figure 21.22B) [106,118,123]. The H_{II}^c structure may be regarded as the inverse-hexagonal (H_{II}) lipid phase, with DNA strands wrapped within its water tubes. Here too, the diameter of the water tubes is only slightly larger than the diameter of the DNA "rods." The presence of DNA is crucial for stabilizing the hexagonal structure. Without it, strong electrostatic repulsion will generally drive the lipids to organize themselves into planar bilayers. In fact, the most abundant aggregate structure of pure CL and HL mixtures,

from which hexagonal complexes are subsequently formed, is single-bilayer liposomes.

Other CL–DNA phases have also been observed. One of the earliest studies probing the structure of lipoplexes showed some evidence for a hexagonal arrangement of rodlike micelles intercalated between hexagonally packed DNA (Figure 21.22C) [124,125]. More recently, such structures have been unambiguously characterized in complexes composed of lipids that possess large, typically polyvalently charged headgroups [126]. There is also evidence that some lipid mixtures promote the formation of cubic phases that are also able to transfect [127].

The number of possibilities is even larger if one also considers metastable intermediates. The "spaghetti" structure (see Figure 21.22D), observed using freeze-fracture electron microscopy, has been predicted by theory to probably be one such metastable morphology [128,129]. In this structure, each (possibly supercoiled) DNA strand is coated by a cylindrical bilayer of the CL/HL lipid mixture [130,131]. Early proposed models of the CL–DNA complexes suggested a "beads on a string" type complex, in which the DNA is wrapped around or in between lipid vesicles (and even spherical micelles). While this may not turn out to be an equilibrium structure, such aggregates are sometimes found and may also serve as unstable intermediates [132–134]. Other structures, such as the bilamellar invaginated liposomes (BIV) made of DOTAP-Chol, have been proposed and demonstrated to be efficient vectors [108,135]. These structures resemble to some degree the L_α^c phase. However, formed from extruded liposomes, the BIVs are most probably metastable.

What factors determine which of these phases (or possibly several coexisting structures) actually form in solution? To what degree can we control and predict them? Control can first be achieved through the choice of type of CL and HL, and the ratio between the two used in forming liposomes. This in turn will determine such basic properties as the lipid bilayer's bending rigidity, spontaneous curvature, and surface charge density of the water–lipid aggregate interface. An additional experimentally controllable parameter is the ratio between the lipid and DNA content in solution. We show that both these parameters have significant effects on the phases that are formed.

21.5.2 COUNTERION RELEASE

From the start, it was realized that the expected condensation of DNA with oppositely charged lipids could be used to package and send DNA to transfect targeted cells. The expectation that the DNA and lipids would aggregate was intuitively based on the notion that oppositely charged bodies attract. Early experiments confirmed the aggregation of DNA and lipids. However, the mechanism by which CL and DNA were found to associate—previously termed in the context of macromolecular association "counterion release" [136]—is more intricate than the "opposites attract" mechanism that may be naively expected.

Prior to association, DNA and lipids are bathed in the aqueous solutions containing their respective counterions, so that the solutions are overall electrostatically neutral. The counterions are attracted to the oppositely charged macromolecules, thus gaining electrostatic energy. Here, in addition to DNA, we shall also refer to the preformed CL liposomes as a "macromolecules" since they typically retain their integrity in solution, even upon association with other charged macromolecules. The counterions are therefore confined to the vicinity of the oppositely charged macromolecules at the compromise of greater translational entropy in solution.

Upon association, the two oppositely charged macromolecules condense to form CL–DNA complexes (see Figure 21.23). Many (possibly all) previously confined counterions

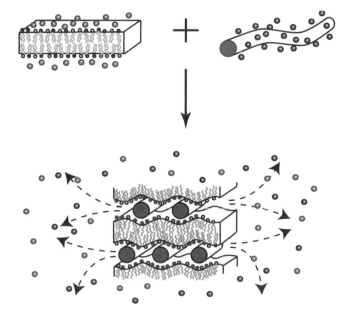

FIGURE 21.23 Schematic illustration of the condensation of DNA and lipid bilayers (liposomes) into CL–DNA complexes. In the process, the previously confined counterions are released into the bathing solution, thereby gaining translational entropy.

can now be expelled into the bulk solution from the lipoplex interior, thus gaining translational entropy. While the translational entropy of the paired macromolecules is reduced by (typically) only a few $k_B T$s, due to loss of conformational and translational entropy, many released counterions can now favorably contribute to a gain in entropy, each by a comparable amount. For this reason, it is sometimes stated that the DNA–lipid condensation is "entropically driven." The electrostatic energy can also contribute somewhat to stabilizing the lipoplexes. However, it has been well argued, both experimentally and theoretically, that the cardinal contribution to the association free energy of CL–DNA complexes is the entropy gain associated with counterion release [137–139].

Further support was given by counting released ions using conductivity measurements of the supernatant. It was possible to determine that a maximal number of counterions were released when the number of "fixed" charges on the DNA and lipid was exactly equal.

Calorimetric measurements confirm this finding and furthermore reveal that the association could in fact be endothermic so that it is only favorable for entropic reasons [140,141]. The special point at which the number of positive and negative fixed charges is equal has been termed the "isoelectric point." At this point, the (charging) free energy of the complex is minimal: the fixed charges of opposite signs fully compensate each other, thus allowing essentially *all* the counterions to be released into solution. Note, that by "counterions," we do not refer here to added salt ions. Ions of added salt will span the entire solution including the lipoplex interior. Thus, the salt content changes the thermodynamic phase behavior and the value of the adsorption free energy, mainly because a high ambient salt concentration lowers the entropic gain associated with releasing a counterion.

Theoretical predictions and estimates from calorimetry show that for a salt solution of concentration $n^0 = 4$ mM, and a 1:1 CL/HL mole ratio, the gain in free energy upon adsorption at the isoelectric point is a bemusingly large ~$7.5 k_B T$ per fixed charge pair (DNA and CL) [138–141]. This value translates to over $2000 k_B T$ when considering the energy per persistence length of DNA (about 50 nm), carrying approximately 300 charges.

21.5.3 LAMELLAR DNA–LIPID COMPLEXES AND OVERCHARGING

Many degrees of freedom with competing contributions are expected to ultimately determine the free energy minimum for equilibrium DNA/membrane structures. Typically these include (but are not limited to) electrostatic energy, elastic bending, solvation, van der Waals, ion mixing, and lipid mixing. Therefore, considering the lipoplex phase behavior, we begin, for simplicity, by discussing systems where only L_α^c complexes are found. This can be expected when the lipid membranes are rather rigid, such as in the case of mixtures of DOTAP/DOPC [100,118] or DMPC (dimyristoyl phosphatidylcholine)/DC-Chol [142]. The main structural parameter for the L_α^c phase is the DNA–DNA distance, reflecting the DNA packing density within the complex. A series of x-ray measurements by Rädler et al. revealed how the DNA–DNA spacings d vary with the ratio ρ of the number of lipid charges to the total number of charges on DNA. The measurements were repeated for each of several different lipid compositions defined by the ratio of charged to overall number of lipids, ϕ. It was found that for a lipid mixture of a given composition ϕ, the spacings are constant throughout the low ρ range where the complex coexists with excess DNA. In the high ρ range, where the complex coexists with excess lipid, the spacings are also nearly constant. In between these limits, there exists a "single-phase" region, where all the DNA and lipids participate in forming lipoplexes. This region is generally found to include the isoelectric point where, by definition, $\rho = 1$ (see Figure 21.24).

Several theoretical studies have been proposed to account for this phase behavior [137,143,144]. It was found that it is possible to account for most of the experimental observations within the scope of the nonlinear PB equation [144]. In this theoretical model, elastic deformations of the DNA and lipid bilayers were neglected, treating them as rigid macromolecules. On the other hand, the lipid's lateral (in plane) mobility in the membrane layer was explicitly taken into account. This turns out to be an important degree of freedom in mixed fluid bilayers, enabling the system to greatly enhance the free energy gain upon complexation, with respect to the case where no lipid mobility is allowed. This adds to the stability of the L_α^c complex. Generally, it was found that lipid mobility favors optimal (local) charge matching of the apposed DNA and lipid membrane. This is the state in which a maximal number of mobile counterions are expelled from the interaction zone, implying a maximal gain in free energy upon complex formation [145]. However, the tendency for charge matching (hence migration of lipid to and from the region of proximity) is opposed by an unfavorable local lipid demixing entropy loss. This entropic penalty will somewhat suppress the membrane's tendency to polarize in the vicinity of the DNA molecule. The extent to which the membrane will polarize is determined by the intricate balance between the

electrostatic and lipid mixing entropy contributions to the free energy of the complex. The contribution of lipid demixing to the stabilization of the complex is most pronounced when the membrane's average composition is far from that of the DNA, namely for low ϕ. Here, the system can gain most out of the polarization so as to come close to local charge matching.

The tendency of charged lipids to segregate in the vicinity of adsorbed rigid macromolecules has gained some experimental support from nuclear magnetic resonance (NMR) studies [146] although many systems may display a more complex behavior. Molecular dynamic simulations of L_α^c complexes, for a lipid mixture of DMTAP and DMPC, showed evidence for a favorable pairing of DMPC and DMTAP lipid molecules through the (partial) negative charge on DOPC and an interaction of the (remaining) positive charge of the zwiterionic DOPC with the DNA. In contrast to the model discussed above, this implies a nonideal lipid demixing: these lipid molecules preferentially move in pairs [147]. This may be anticipated since it is well known that lipids do not generally mix ideally even in free (unassociated) membranes [148]. Furthermore, there is evidence that to some extent neutral lipids also interact directly with DNA [149].

Figure 21.25 shows the experimental results and theoretical calculations for the dependence of d on ρ for several values of ϕ. For a specific value of ϕ (say $\phi = 0.5$), the three-phase

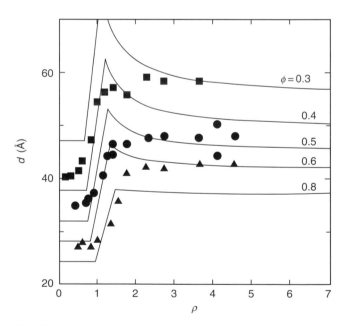

FIGURE 21.25 DNA–DNA spacing as a function of ρ in a series of theoretical and experimental results. The theoretical results correspond to (top to bottom) $\phi = 0.3$, 0.4, 0.5, 0.6, 0.8; all results are presented for a screening length of 50 Å (corresponding to ca. 4 nm of bathing salt solution). The experimental results correspond to $\phi = 0.3$ (squares), 0.5 (circles), 0.7 (triangles), and were performed with no added salt. (Theoretical results adapted from Harries, D., May, S., Gelbart, W.M., and Ben-Shaul, A., *Biophys. J.,* 1998, 75, 159 and Harries, D., PhD dissertation, The Hebrew University, Jerusalem, Israel, 2001; experimental results adapted from Rädler, J.O., Koltover, I., Salditt, T. et al., *Science,* 275, 810, 1997.)

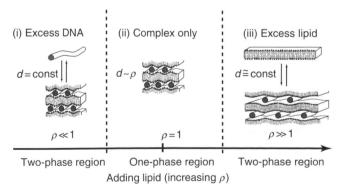

FIGURE 21.24 Schematic illustration of the phase evolution of the L_α^c complexes, for a constant lipid composition (cationic to nonionic lipid ratio). As lipid is added (ρ increases), the system evolves from a two-phase (complex and excess DNA) region through a one-phase (complex only) region, and finally to a two-phase (complex and excess lipid) region. The isoelectric point is generally contained within the one-phase region.

regimes can clearly be seen. As ρ increases, d changes from $\approx 35\,\text{Å}$ (in the excess DNA regime, $\rho \ll 1$) to $\approx 47\,\text{Å}$ (in the excess lipid regime, $\rho \gg 1$). Both theory and experiment show that for a wide range of lipid composition, ϕ, there exists a one-phase, complex-only region at ρ values somewhat larger and smaller than the isoelectric point. This implies that complexes may become either negatively or positively "overcharged" so that the total number of fixed positive and negative charges is not equal. Hence, the complex accommodates either an excess number of lipids or else an excess amount of DNA. The complex's free energy is thus not at its minimum, which occurs at isoelectricity ($\rho = 1$). The interplay between possible phases to minimize the total system's free energy dictates that the complex moves away from its minimal free energy. The alternative would be to expel the excess lipid ($\rho > 1$) or excess DNA ($\rho < 1$) into solution. The charge densities on these "free" unneutralized macromolecules would be very large, rendering this scenario highly unfavorable. Using a simple model based on this overcharging phenomenon, it was possible to account for the considerable extent of this one-phase region [144]. Within this model, only the uncompensated charges on apposed (DNA–DNA or bilayer–bilayer) surfaces of a L_α^c unit cell ("box") were considered in estimating the complex's free energy. Figure 21.25 also shows that as the membrane becomes enriched in CL (ϕ increases) the DNA–DNA distance is systematically reduced, reflecting the fact that smaller amounts of lipid membrane are needed to achieve isoelectricity.

Salt has a significant effect on the phase behavior. In general, added salt causes a significant decrease in d, presumably due to a screening of the repulsive DNA–DNA interaction. This effect is most pronounced when divalent salts are added in increasing amounts. A sharp decrease in the d value is observed for a certain salt molar concentration, resulting in very highly condensed DNA in each gallery [89,130]. Another interesting observation is that the identity of the CL's counterion used changes the (endothermic) association enthalpy considerably, particularly in the excess DNA region [140]. This probably reflects the nonelectrostatic interaction energies of different ions with membranes, which may influence the thermotropic behavior of the lipid membranes [150,151].

21.5.4 DNA Adsorption on Lipid Membranes

Further insight into the in-plane DNA ordering in L_α^c complexes has been gained through the atomic force microscopy (AFM) study by Fang and Yang [152,153] of DNA adsorption on supported lipid bilayers. In these experiments, DNA was first adsorbed on dipalmitoyldimethylammoniumylpropane (DPDAP) or distearoyl-DAP (DSDAP) CL bilayers, assumed to be in the gel phase. After equilibration and saturation of the surface, the DNA bulk solution was removed, and the surface was put in contact with solution of various concentrations of NaCl. After further equilibration, the salt solution was removed and the surface imaged by AFM. Plasmid and linear DNA similarly treated showed similar results.

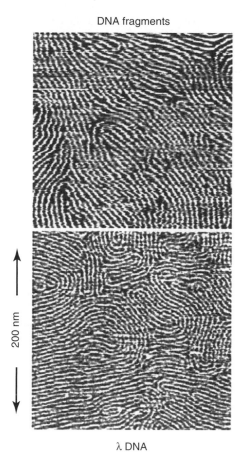

DNA fragments

200 nm

λ DNA

FIGURE 21.26 AFM images of DNA from different sources (see figure for details) condensed on DPDAP bilayers at room temperature in 20 mM NaCl. Striking fingerprint-like order is apparent, with a domain size of the order of the persistence length (ca. 50 nm). (Courtesy of J. Yang, University Vermont.)

Striking, fingerprint-like images of DNA adsorbed on the surface were revealed (Figure 21.26). The typical domain size for the aligned, smectic-like order is usually several hundred angstroms, reflecting the DNA's intrinsic persistence length. These structures are expected to be like those found in L_α^c complexes: the domain size, inferred from x-ray scattering is quite similar [119,120]. Furthermore, it was found that the surfaces are often overcharged when DNA is adsorbed, i.e., the number of DNA fixed charges exceeds the number of lipid charges. This can be anticipated on the basis of theoretical studies of a similar problem: adsorption of charged globular proteins (yet another macroion) on oppositely charged membranes [154]. In both cases the driving force for adsorption is similar to that driving lipoplex formation, namely counterion release. In L_α^c complex formation, much of the DNA can interact with the two sandwiching bilayers. In contrast, topology dictates that adsorbates on a single lipid bilayer will always possess a part proximal and a part distal to the interaction zone. If both parts are charged, as is the case with DNA, complete counterion release cannot be achieved since the distal part does not interact significantly with the underlying bilayer. Therefore, although charges on the lipid membrane are fully cancelled by charges on adsorbed DNA macroions, still the portion of DNA away from the contact zone imparts a net surface charge, i.e., overcharging of the DNA-covered membrane.

Yet another interesting feature is the dependence of the DNA–DNA distance on salt concentration. As the NaCl concentration was varied between 20 and 1000 mM, this distance grew from around 45 Å to almost 60 Å. At first this may seem baffling: adding salt should be expected to decrease the DNA–DNA electrostatic repulsion, and hence lower the distance between neighboring interacting strands. This is indeed the general trend that has been observed in L_α^c complexes [117,144]. However, because the DNA was primarily allowed to saturate the surface, and only subsequently treated with the salt solution (which was later also washed away), adsorption here was not at equilibrium. In fact, when faced with a neat salt solution, the adsorbed DNA can only detach and will not generally readsorb onto the surface. It is therefore hard to give full theoretical reasoning for the trend.

Theoretical explanations have previously been offered to account for this salt-dependent behavior based on a balance between membrane-mediated effective attraction (that may be the result of the DNA perturbation of the lipid bilayer) and electrostatic repulsion between DNA strands [155]. The predicted DNA–DNA spacing as a function of screening length is nonmonotonic: increasing first for low screening lengths and decreasing for high values. An alternative to this approach is related to the free energy gain upon adsorption and how it changes with the addition of salt. In the presence of added salt, the adsorption free energy can be expected to be lower since the gain in entropy upon release of counterions becomes very small when releasing an ion from an adsorbed layer into a bathing solution with a comparable concentration. Assuming that unbinding would occur when the free energy gain per persistence length is $\approx k_B T$, we can estimate from a simple model that the thickness of the confined layer is $l_{eff} \approx 5$ Å, rather close to the screening length in solution (3–4 Å) [138–141]. Thus, the lower binding free energy may cause some of the DNA strands to dissociate from the lipid surface once the system is exposed to salt. Allowing DNA to rearrange on the surface would then lead to an increase in the average DNA–DNA distance.

When multivalent salt is used, a crowding of DNA molecules is first observed as salt is added (in accordance with the observations in the L^c complexes), and then starts to grow for higher concentrations [100,156]. This may be a manifestation of the two competing forces as salt is added: lessened repulsion between strands versus weakened adsorption energy.

21.5.5 From Lamellar to Hexagonal Complexes

So far, we have discussed the L_α^c lipoplexes formed from lipid membranes that are rigid (bending rigidity much greater than $k_B T$) and tend to a planar geometry. Other lipoplex structures may ensue when the lipids possess a spontaneous curvature which is nonplanar, or when the membranes are soft enough to be deformed under the influence of the apposed macroion.

The lipid membrane thus responds to the presence of DNA by deforming elastically and by locally changing its composition ϕ.

Membrane elasticity may be varied substantially either by changing the lipid CL/HL composition, changing the lipid species, or by adding other agents, such as alcohols, to the membrane [157,158]. In contrast, double-stranded DNA generally remains rather stiff, with a typical persistence length of ≈ 500 Å. Hence, the lipoplex geometries are restricted to structures in which DNA remains linear on these large length scales. Usually, it is the interplay between the elastic (spontaneous curvature and bending rigidity) and electrostatic (charge density) properties of the membrane that will determine the optimum lipoplex geometry at equilibrium.

Often, the membrane elasticity and electrostatic contribution to the free energy display opposing tendencies. For example, the hexagonal H_{II}^c complex is electrostatically favored due to the cylindrical wrapping of the DNA by the lipid monolayer. This allows better contact between the two macromolecular charged surfaces. However, the highly curved lipid geometry may incur substantial elastic (curvature deformation) energy. The price to pay will be lower when the lipid (monolayer's) spontaneous curvature matches closely the DNA intrinsic (negative) curvature or when it has low bending rigidity. Under such conditions, the H_{II}^c complex may become more stable than the L_α^c phase. Usually, a neutral HL is used for adjusting the spontaneous curvature to the required negative curvature, since pure CLs typically tend to form uncurved or positively curved aggregates. Use of more HL in the mixed membranes may, on one hand, lower the elastic penalty, while, on the other hand, lower the monolayer's charge density, compromising the electrostatic energy gain upon association.

These qualitative notions were elegantly demonstrated by experiments in which the elastic properties of the lipid monolayers were controlled by changing the nature of the lipid mixture. The spontaneous curvature of the lipid bilayer was modified by changing the identity of HL. It was found that when using a mixture of DOTAP/DOPE, H_{II}^c was the preferred structure, while DOTAP/DOPC mixtures promoted the formation of the L_α^c phase. This is consistent with the fact that pure DOPE forms the inverted hexagonal phase, H_{II}, due to its high negative spontaneous curvature [159–161], while DOPC self-assembles into planar bilayer. In addition, by adding hexanol to the DOTAP/DOPC–DNA lipid mixture, the bending rigidity could be diminished by about one order of magnitude [157,158]. This induced a clear first-order $L_\alpha^c \rightarrow H_{II}^c$ phase transition [118].

Additional complexity can be expected when accounting for the coexistence of more than one phase in solution. A theoretical study of the phase equilibrium took into account the bare lipid phases L_α and H_{II}, the naked DNA and the complex L_α^c and H_{II}^c phases [162]. The phase diagram of the system was evaluated by minimization of the total free energy, which included electrostatic, elastic, and lipid demixing contributions. Several systems of different compositions were considered. Figure 21.27 shows the predicted phase coexistence corresponding to the simplest case already discussed of rigid

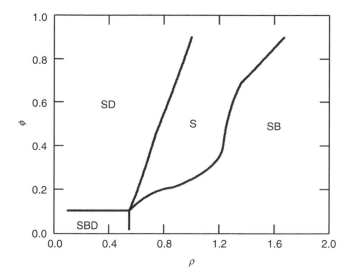

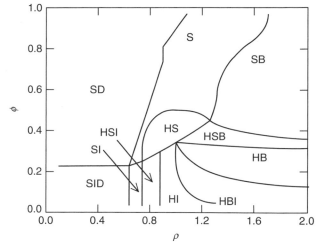

FIGURE 21.27 Phase diagram of a lipid–DNA mixture for lipids, which self-assemble into rigid planar membranes. The phase diagram was calculated for membranes characterized by a bending rigidity in the range of $4 < \kappa < \infty \, k_B T$ and a spontaneous curvature of $c = 0\,\text{Å}^{-1}$ for both helper and CLs. The symbols S, B, and D denote, respectively, the L_α^c, L_α, and uncomplexed (naked) DNA phases. (From May, S., Harries, D., and Ben-Shaul, A., *Biophys. J.* 2000, 79, 1747. Reprinted by permission from Biophysical Society.)

FIGURE 21.28 Phase diagram of a lipid–DNA mixture involving "curvature-loving" helper lipid: the spontaneous curvature of the helper lipid is $c = -1/25\,\text{Å}^{-1}$. For the CL the spontaneous curvature taken is $c = 0\,\text{Å}^{-1}$. The bending rigidity for both lipids is $\kappa = 10 \, k_B T$. The symbols S, H, B, I, and D denote, respectively, the L_α^c, H_{II}^c, L_α, H_{II}, and uncomplexed (naked) DNA phases. The broken line marks the single H_{II}^c phase. (From May, S., Harries, D., and Ben-Shaul, A., *Biophys. J.*, 79, 1747, 2000. Reprinted by permission from Biophysical Society.)

planar membranes. Results are presented for lipid membranes with a bending rigidity of $\kappa = 10 k_B T$ per monolayer and spontaneous curvature $c = 0\,\text{Å}^{-1}$ (typical for many bilayer forming lipids [122]) for which only lamellar complexes are expected to form. As the overall lipid composition is enriched in CL (higher ϕ) the one phase persists over a wider range of ρ. This indicates that for higher CL content, the complex may be expected to be more stable toward addition of either DNA or lipid (hence moving away from the isoelectric point).

The Gibbs phase rule allows for up to three phases to coexist concomitantly for this three-component (DNA, HL, and CL) system. Figure 21.28 shows the theoretical prediction for the phase diagram for a system in which the HL has a strong negative spontaneous curvature ($\kappa = 10 k_B T$ and $c = 1/25\,\text{Å}^{-1}$) [162]. For high ϕ values, the phase behavior resembles that of the previously discussed system. However, for lower values of ϕ, a multitude of regions of (up to three) different phases coexisting together can be found. In some regions, lamellar and hexagonal complexes appear coexisting side by side. A similarly complex diagram results when the membranes are soft (bending rigidity of $\approx k_B T$) as might be expected for membranes with added alcohols [162].

A more subtle demonstration of the underlying balance of forces can be found within the realm of the L_α^c complex. Thus far, the theoretical models considered for the lipid membranes in this lamellar phase assumed them to be perfectly planar slabs. However, this need not be so. When membranes are sufficiently soft (yet not soft enough to favor the H_{II}^c phase) or if one of the CL/HL has a propensity to form curved surfaces, the membrane may corrugate so as to

optimize its contact with DNA (see Figure 21.22A). If the membrane is further softened, finally, a transition may occur to the H_{II}^c phase. In this respect, the membrane corrugation in the L_α^c complex may be regarded as a further stabilization of the lamellar complex, and a delay to the onset of the $L_\alpha^c \to H_{II}^c$ transition.

A possible consequence of membrane corrugation in the L_α^c phase is an induced locking between neighboring galleries. This follows the formation of "troughs" in a gallery, induced by the interaction of the membrane with DNA in adjacent galleries. This imposes "adsorption sites" for the DNA in the two neighboring galleries, which propagates the order on. The formation of these troughs, as well as a very weak electrostatic interaction between galleries, may thus correlate between the positions of DNA in different galleries [147,163,164]. Limited experimental evidence supports this notion. In cryo-transmission electron microscopy (cryo-TEM) studies of the L_α^c phase, spatial correlations were found between DNA strands in different galleries [165]. In another series of x-ray studies, the corrugation and charge density modulation in an L_α^c-like complex, in which the membranes are in the gel phase, were measured [166]. Further support for the possible formation of corrugations is gained from computer simulations of lipid–DNA complexes [147].

In order to assess the extent of membrane corrugation, a balance of forces between many degrees of freedom should be taken into account. The free energy minimum now depends on the local membrane composition—dictating membrane properties such as local charge density, spontaneous curvature, and bending elasticity—and the extent of local deformation around the DNA. Theoretical predictions show that for a wide range of conditions, both stiff and soft membranes can show corrugations that are stable with respect to thermal undulations of the membranes [164]. The spacings between galleries and between DNA molecules are also predicted to change somewhat with respect to the case where no corrugations are allowed [163]. For the conditions in which the troughs are shallow, or absent altogether, one may anticipate the formation of phases where DNA in different galleries are positionally uncorrelated, while orientational order is preserved. These structures were predicted theoretically and termed "sliding phases" [119,120,165,167–169].

21.5.6 Lipoplex Structure and Transfection Efficiency

In recent years, a large number of CL–DNA formulations have been proposed as vectors. However, the fate of the CL–DNA complex once administered, its interaction with the cell membrane, and entry into the cell and subsequently into the cell nucleus, is likely complex and largely unresolved. The poorly understood [170–172] process of DNA release once in the cell interior must be important. For example, it has been shown from action in the nucleus that DNA expression is diminished when it is tightly complexed with lipids [175]. Hints to the mechanism of the intracellular release of lipoplexes come from experimental evidence *in vitro* showing that other added polyelectrolytes may compete with DNA and subsequently replace it in the complex [173]. This kind of replacement, by natural polyelectrolytes, may be one way in which DNA is released in cells [174]. Another possible mechanism is the fusion of complex lipids with lipid membranes in the cell [100,120].

Only a limited number of experiments have probed the relationship between the structure of CL–DNA complexes and the transfection efficiency. One emergent theme attributes an important role to complex frustration and destabilization in promoting transfection.

Experimental studies show that the two ordered complex structures, L_α^c and H_{II}^c, behave differently inside living cells. Furthermore, a correlation was found between the structure of the lipoplexes formed and the transfection efficiency. The structure formed depends in turn on the specific choice and relative amount of HL, CL, and DNA. The H_{II}^c complex was found (in the studied cases) to be a more potent vector than L_α^c [176,177]. Further information is gained from fluorescence studies of cell cultures with both complex types internalized in fibroblast L-cells. These indicate that the L_α^c complex is more stable inside the cells, while the H_{II}^c more readily disintegrates—its lipids fusing with the cell's own (endosomal or plasma) membranes—resulting in DNA release. This is in accord with the theoretical findings that the L_α^c complex structure is rather flexible toward changes in the system's compositional parameters due to its ability to tune both the membrane composition and the DNA–DNA spacing, while this tuning is more limited in the H_{II}^c phase.

The picture is further substantiated by a series of studies by Barenholtz and coworkers [101,171,172,178]. In general, it was shown that maximal transfection efficiency could be achieved in complexes that were formed in the excess lipid regime (with ρ in the range of 2–5). This correlated well with the point of maximal size heterogeneity of the complexes. These instabilities were shown to occur concomitantly with an increase in the amount of membrane defects that were in turn mainly attributed to the appearance of several coexisting structures in solution (e.g., H_{II}^c and L_α^c in DOTAP/DOPE lipoplexes, or micellar and lamellar phases in DOSPA/DOPE-based lipoplexes). This is in accordance with the theoretical prediction that the regions of most phase diversity and the largest number of coexisting phases occurs at high ρ (and low ϕ) values (see Figure 21.28 [101,132,159]).

Other evidence seems to agree with these notions. For example, some successful formulations, such as BIV, are also probably metastable [108,115,128]. This may suggest that it is in fact their instability, which helps them in releasing their DNA cargo once they are inside the cell. Attempts have also been made to destabilize lipoplexes more specifically once they are already internalized in the cells (rather than en route in the serum). Reduction-sensitive CLs were designed, and the subsequent lipoplexes that are formed were shown to undergo large structural changes when exposed to the cytoplasmic reductive systems. The lipoplexes are thus destabilized and the previously packaged DNA is released into the cytosol [103,179–181]. A decrease in the toxicity of the CL and increased transfection efficiency are thus achieved [182].

Recent experiments have shed additional light on the relation between complex structure and transfection efficiency. These studies concentrate on lamellar L_α^c complexes that are the ones most often encountered. It has been shown that the membrane charge density of the CL vector, rather than, say, the large valency of the CL is an important parameter that governs transfection efficiency. Specifically, the path for complex uptake is distinctly different for high versus low lipid charge density. While at both high and low charge densities, complexes are found to enter through endocytosis, at low lipid charge density, DNA was trapped in complexes, and those in turn were trapped inside endosomes, while at high density, endosomal entrapment does not seem to be a significant limiting factor. At very high lipid charge density, the transfection efficiency is again reduced. These findings have led to a new model of the intracellular release of DNA from lamellar complexes, through activated fusion with endosomal membranes. It has been suggested that complexes escape the endosome by fusion with the endosomal membrane. This fusion is favored by (attractive) electrostatic interaction energy that is higher for highly charged membranes [183–186].

In contrast, the transfection efficiency of the H_{II}^c complexes does not seem to depend on lipid charge density, perhaps because their structure leads to a distinctly different mechanism of cell entry. It seems that these complexes undergo rapid fusion with cellular membranes. The curvature-loving properties of the hexagonal complexes favor rapid fusion and escape of DNA from the endosome, and this process no longer limits the transfection rate.

Destabilizing lipoplexes is not the only barrier to transfection. For example, entry of DNA into the nucleus through the nuclear pore complex is inefficient for large pieces of DNA. It has been shown that the cell's own nuclear import machinery may be utilized to increase transfection efficiency dramatically by attaching a peptide containing a nuclear localization signal (NLS) to the DNA [187,188]. Furthermore, the size of the complexes also seems to play a crucial role in determining transfection efficiencies [101,102,108,115]. Here, the repulsive interaction between like-surface charge of the complex due to over/under charging (excess lipid or DNA) can aid in stabilizing the complexes, once they are formed, from fusing further.

Another strategy to controlling the interaction between aggregates and the stability of the aggregate *in vivo* is to modify the composition of the outer wrapping sheath of the lipoplex. The caveat is that the lipoplexes are not stabilized to such a degree that they can no longer disintegrate once inside the cells. For example, short-chain lipids possessing a PEG headgroup (or a derivative thereof) have been used to increase the stability of the lipoplexes in the blood stream due to repulsive steric interactions, while not interfering with the endosomal unwrapping once the lipoplexes are internalized in cells. These "PEGylated" lipids can reduce the transfection efficiency, but also increase the stability of complexes en route to their targets [189–191]. The PEG chains that are attached to the lipids are present inside as well as at the surface of CL–DNA complexes. This gives rise to polymer-mediated forces, such as depletion attraction between DNA strands that arises because polymer disfavors confinement of its degrees of freedom when it is between DNA rods.

More generally, we can expect that understanding how to control and manipulate the formation of specific phases on the one hand, while better understanding the multistage transfection mechanism, and the parameters (conditions) affecting it on the other, should aid in the design of more potent lipid-based gene delivery vectors in the future. These, together with control over the coating and targeting of the complexes, may render these vectors as useful vehicles in gene therapy.

21.6 DNA–POLYCATION INTERACTIONS

Stiff polyelectrolytes like DNA are readily capable of spontaneously forming complexes with various oppositely charged macromolecules to form submicron-sized complexes [53]. It is primarily thought that the formation of these self-assemblies is driven by Coulombic electrostatic and polyelectrolyte bridging interactions as well as the entropic gains derived upon the release of bound water and counterions. These forces compensate for the entropic loss resulting from a close packing of the stiff polyelectrolyte chains. Due to their potential for protecting and delivering genetic information, condensation of DNA with polycations is of particular interest. The polycation–DNA complex, often called "polyplex" [99], forms spontaneously upon mixing positively charged polymers with negatively charged nucleic acids. The advantage of polymer formulations lies in their ability to be generated economically in large quantities while offering versatility of synthetic chemistry and allowing easy tailoring of various chemical structures, molecular weights, and topologies with low immunogenicity [192].

Similar to lipoplexes, the primary goal of polycation research is the condensation of DNA into particles of virus-like dimensions that can migrate through the blood stream and into target tissue, overcome the electrostatic repulsions of the cell membrane, yet protect the nucleic acid from degradation and undesired interactions [193]. A wide variety of natural and synthetic polycations have been investigated, most of them based on amine chemistry. Factors influencing DNA compaction include the number of charges per chain, the type of charge (e.g., primary, secondary), charge spacing along the chain, chain architecture (linear, branched, dendritic), and chain hydrophobicity. External conditions such as solution ionic strength, concentration, positive-to-negative charge ratio of polycations to DNA, and mixing conditions also are observed to influence how polyplexes form. Observed differences in transfection efficiency to date are not easily correlated to polycation chemistry or structure. Large transfection differences are observed between various polycations as well as with the same polycation but of varied molecular weight or polymer chain architecture. Aggregation is generally diffusion limited upon mixing. Depending on the technical process of mixing one can form fibers; large micron-sized aggregates; or small rod, toroidal, or spherical aggregates [194–197]. Critical to increasing binding to the negatively charged cell membrane, polyplexes are usually overcharged by the polycation resulting in a net-positive colloidal aggregate.

In an ideal polyplex-mediated gene delivery, the polycations must not only compact DNA but also be able to overcome a wide variety of multibarrier processes to achieve successful transfection. Similar to a virus capsid, vector delivery would occur in a highly cell-specific manner, facilitate cellular uptake as well as endosomal release, and be able to maneuver through the cytoplasmic environment, disassociate, and then localize the desired nucleic acid vector into the cell nucleus ready for transcription. In addition, this ideal polycation would also be nontoxic, nonimmunogenic, and biodegradable. Obviously, this is a large list; in reality, no one polycation is likely to satisfy all these conditions. Consequently, a variety of additional functional elements have also been included in polyplex formulation to try to improve delivery, efficiency, targeting, and lower cytotoxicity. To date, no predictions of the correlation between polyplex structure and transfection efficiency can be made without extensive experiments. Clearly a balance of the molecular forces within and between the polyplexes is needed to create particles stable enough to carry the DNA to the nucleus yet sufficiently unstable to release the DNA when needed. In this chapter, we introduce our current understanding of the internal structure, phase behavior, and compressibility of polyplexes. We end with a brief discussion of recent advances in the understanding of polyplex gene delivery.

21.6.1 STRUCTURE OF POLYPLEXES

Early work investigating the internal packing and ordering of DNA–polycation complexes used small-angle x-ray scattering (SAXS) to look at pulled DNA fibers kept at

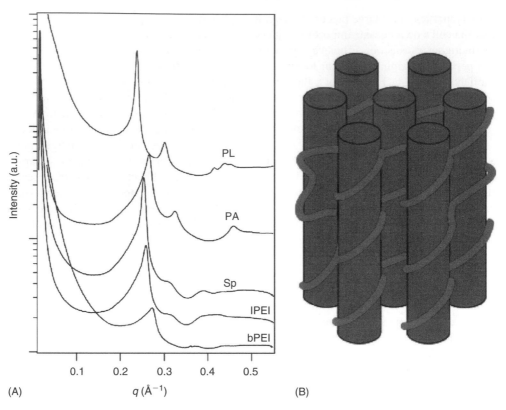

FIGURE 21.29 (A) Synchrotron small-angle x-ray intensity profiles for five different polycation–DNA complexes shifted in intensity for clarity. All samples were made with an excess of cationic charge to DNA phosphate and equilibrated in TE buffer without additional salt. (B) Schematic drawing of a close-packed hexagonal ordering of DNA rods held together by electrostatically bound polycations chains.

constant humidity levels [198–200]. More recent work has used SAXS to look at the internal packing and ordering of DNA–polycation complexes in buffered solutions [22,54]. Figure 21.29 shows a series of high-resolution synchrotron SAXS intensity profiles, shifted in intensity for clarity, for calf thymus DNA packaged with PL, PA, Sp, and linear and branched polyethylene imine (lPEI and bPEI, respectively) in TE buffer. Here, all samples are made with an excess of polycations and then washed to remove unreacted polycations.

All systems show a similar hexagonal close-packing of rods. A simple schematic view of this hexagonal ordering is shown in Figure 21.29. For simplicity, we represent DNA as a stiff rod to emphasize that on small length scales, such as those observable via SAXS, DNA behaves as a rigid polymer with a large persistence length ($L_p \sim 50\,\text{nm}$) in stark contrast to the highly flexible polycation chains ($L_p \sim 1\,\text{nm}$). Here, the flexible polycations act as a simple, electrostatically driven linker molecule, which can both wrap around a single DNA as well as bridge between DNA strands. The DNA packing dominates the scattering profiles here as DNA is significantly higher in electron density compared to the polycations. A dependence of the $d_{\text{DNA–DNA}}$ spacing, as well as small effects on the DNA pitch, is observed to depend on the chemical nature of the complexed polycation. Interhelical distances in buffer or low salt are typically of the order of 26–30 Å, equivalent to ~6–10 Å space between DNA rod surfaces. Due to the

kinetic nature of polyplex formation, sample preparation is also observed to affect the polyplex internal structure. Typical conditions used to form gene therapy vectors (dilute solutions mixed under low salt conditions) were observed to give similar hexagonal packaging of the DNA as shown in Figure 21.29, but with significantly poorer long-range order in the DNA array. These particles were found to be kinetically trapped in nonequilibrium structures. Once formed, they equilibrate extremely slowly with significant, internal spacing rearrangements (~5%) observed on the timescale of several months as the chains try to rearrange themselves to reach their thermodynamic equilibrium spacing. Qualitatively, samples with natural amino acids, lysine and arginine, show a higher degree of long-range order compared to the short natural polyamine (Sp) and the synthetic PEI samples. This loss of long-range order may be due to the shortness of the chain for Sp and charge mismatching in the synthetic PEI polycations. Better ordering is observed for linear compared to branched PEI. This suggests that the chain architecture as well as the chemical nature of the polycations plays some role in determining the internal packing of the polyplexes. Only small effects in interaxial spacings (1–3%) are seen upon changing polycation molecular weight to significantly higher molecular weights, increased nitrogen to phosphate (N/P) charge ratio (above charge neutrality), or changing pH after polyplex formation. An exception is that d spacings were observed to change some 10% with increasing N/P ratio for the PEI samples. However, as PEI is not a fully charged chain at neutral pH, the calculated N/P charge ratio does not reflect the true N/P ratio such that N/P significantly higher than one is necessary to achieve charge neutrality in the complex, consistent with known PEI behavior. Estimates for PEI place neutrality at N/P ~ 2.5 [201].

21.6.2 POLYPLEX PHASE BEHAVIOR

The structure and phase behavior of lipoplexes have been investigated in some detail showing a wealth of possibilities tuned by varying intermolecular forces through changing lipid chemistry or chain shape. In contrast, polyplexes have not shown much variability and have not been explored in as much detail. To understand the phase behavior, polyplexes of calf thymus DNA condensed with Sp, polylysine, polyarginine, and linear and branched PEI were investigated as a function of external monovalent salt concentration [54]. The addition of monovalent salt can weaken the molecular interactions of the polycations with DNA through screening of the electrostatic and bridging interactions, displacement of bound multivalents, and/or chloride binding to multivalent ions (see Figure 21.30).

In this work, universal phase behavior was observed for all polyplexes with increasing salt concentration. Initially all samples are observed to form a "tight bundle" phase of hexagonally close-packed DNA rods as depicted in Figure 21.29. The Bragg reflection in the SAXS curves not only gives information about the interaxial spacing between the rods in the array but also the peak width indicates the long-range in-plane ordering of the array through a correlation length ξ. ξ is seen to depend on the polycations and sample preparation; at low salt concentrations, ξ is observed to be on the order of 15–30 DNA repeats for the various polyplexes studied. At low salt concentrations, this hexagonal packing shows simple linear swelling behavior with the observed scattering while maintaining the in-plane correlations. At a critical salt concentration, c_s^*, dependent on polycation, the onset of a coexistence regime is observed (shaded regions in Figure 21.30). The coexistence regime occurs over a relatively narrow range of salt concentrations and is characterized by an overlapping of the initial sharp Bragg reflection with a new broad peak at lower q spacings in the SAXS measurements. In this phase, a significant fraction of the polycations has been displaced from the DNA, and a salt-induced melting transition is observed. Here, the polycations are more loosely associated with the DNA. This results in a phase with a wider distribution of interaxial spacings and a poor in-plane packing of the DNA and identified as a "loose bundle" phase. A high local concentration of the DNA is maintained through the bridging interactions of the polycations.

Due to the intrinsic stiffness or persistence length of DNA, it maintains some order similar in nature to the liquid crystalline phases observed in pure DNA phases at high concentration [8]. For an intuition on the strength of these interactions, c_s^* ranged from ~300 mM for Sp to as much as 1.6 M NaCl for polyarginine. With still more added salt, this broad peak of the loose bundle phase is observed simultaneously to shift to smaller q, or larger d spacings, and to broaden significantly, corresponding to decreasing in-plane correlations of the DNA arrays. Interestingly, if all the polyplexes are normalized with respect to c_s^*, the swelling behavior in both regimes and ξ above c_s^* is observed to collapse to a single curve (see Figure 21.30). At c_s^*, ξ is observed to drop sharply to ~6–7 DNA repeats, independent of polycations; ξ continues to decrease to ~3 DNA repeats at the highest observed salt concentrations. At large polymer/DNA concentrations, a network phase is observed to form and to grow at the expense of the loose bundle phase. At still higher salt, or lower polymer–DNA concentrations, this network dissociates completely.

A simple model [54] was proposed using a free energy function balancing only the electrostatic attraction and entropic repulsion between the polymer chains for a hexagonal bundle and a network phase. In the bundle phase, condensation is driven by attractive electrostatic interactions. The entropic gains from releasing the bound counterions and water from the DNA compensate for the entropic loss from a close packing of the polymer chains parallel. The network phase, in contrast, is stabilized by highly localized bridge points between polymers and dominates at high salt concentration or low polymer densities. Building on established theory, the potentials for the electrostatic interactions and entropic repulsions can be used to estimate the free energy expressions for both the bundle and network phases [202,203]. This simple balance of electrostatic attraction and entropic repulsion quantitatively and qualitatively describes the transition from a hexagonal loose bundle to a network phase at high salt concentrations on the order of 1 M NaCl, where the network phase grows at the expense of the bundle phase. To induce a phase separation between loose and tight bundles, additional nonelectrostatic attractive forces have to be invoked. Furthermore, this model does not include short-ranged specific interactions, which must arise from chemically distinct polycations as observed experimentally.

21.6.3 EQUATION OF STATE FOR POLYPLEXES

Insight into the intermolecular forces within condensed DNA arrays can be obtained through osmotic stress experiments. Stressing solutions of PEG exert a known osmotic pressure that is balanced by the intermolecular repulsion between helices. Measurement of the interhelical spacing at each osmotic pressure furnishes the DNA equation of state. Very few data have been collected on polyplexes, but we expect many similarities with the more extensively studied short condensing agents. Various condensing agents, ranging from divalent Mn salt to longer polycations such as spermidine and protamine, show many common features [22,48]. Equilibrium spacings, at zero osmotic pressure, depend on the polycations inducing assembly and do not change significantly at low pressures. The equilibrium spacing is typically 8–12 Å between DNA surfaces, indicating a balance of attraction and repulsion between helices. These surfaces can be brought

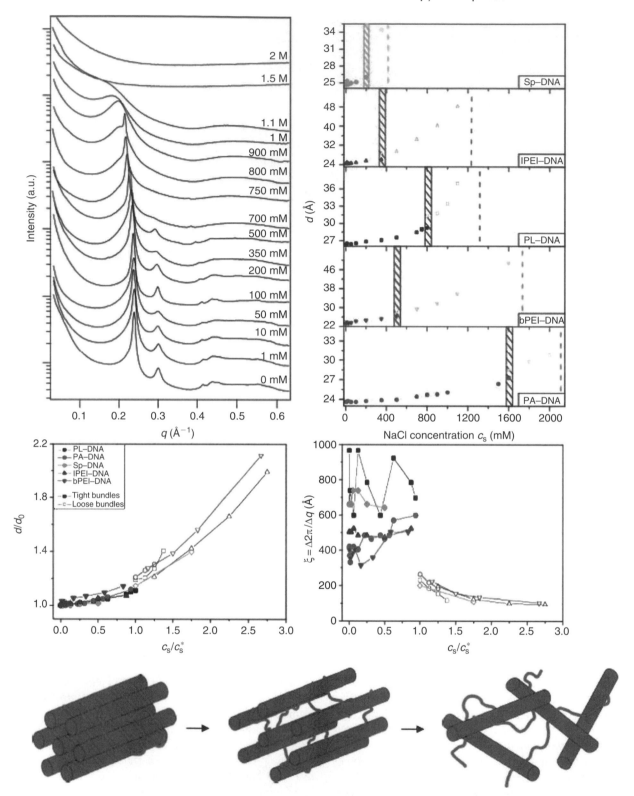

FIGURE 21.30 (See color insert following blank page 206.) Phase behavior of polyplexes with increasing monovalent salt concentration. Universal phase behavior is observed for all systems with an initial swelling of a tightly packed hexagonal array of DNA rods. As a critical salt concentration, c_s^*, dependent on polycation, the onset of a coexistence regime is observed between tight and loose bundles. In-plane correlation lengths, ξ, representative of long-range order within the arrays show a sharp decrease upon crossing c_s^*. When scaled with respect to c_s^*, both swelling ratio (d/d_o) and ξ collapse onto universal behavior. With increasing salt, the loose bundles lose both positional and orientational order with an increase in a network structure at the expense of the loose bundle regime. At sufficiently high salt concentration, or dilute polymer concentration, all Coulombic interactions are screened and the samples disassociate completely into the dilute phase.

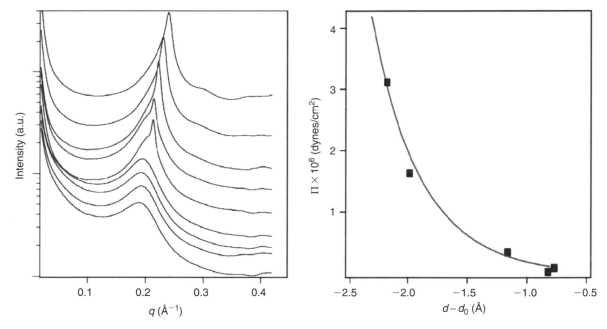

FIGURE 21.31 SAXS intensity curves of the osmotic stress induced reordering of PLL–DNA at 900 mM NaCl. With increasing pressure, the broad diffuse peak indicative of liquid crystalline–like "loose bundle" ordering is forced to reorder showing clear coexistence and then finally scattering from a purely tight-bundled phase is observed. Compression at low osmotic pressure in this loose bundle phase was predicted to scale as $1/d$ using a simple calculated energy balance of long-range electrostatic attraction against entropic repulsion. This $1/d$ scaling is shown in the inset and is found to agree well with the experimental results. (From DeRouchey, J., Netz, R.R., and Rädler, J.O., *Eur. Phys. J. E.*, 16, 17, 2005.)

closer together with osmotic pressure and show net hydration repulsion modulated by the presence of the polycations. The force needed to push helices closer than the equilibrium spacing depends on the polyvalent ion identity. At high pressures, however, all samples show exponentially increasing forces with decay lengths of 1.5–2 Å independent of the counterion species. The decay lengths of these repulsive interactions are approximately half that for an overall net hydration repulsion observed in pure DNA arrays and are insensitive to external salt concentrations. While decay length is only weakly cation dependent, there are significant differences in the magnitudes of the forces. The exact nature of these forces is not completely understood. The data do set limits that show inconsistency with several proposed ionic fluctuating models such as direct ionic bridging, ionic fluctuation, and van der Waals attraction balanced against hydration/electrostatic repulsion.

Available data for polyplexes show similar osmotic stress force curves. Equilibrium spacings for polylysine or polyarginine polyplexes are similar to spacings of DNA condensed with small cations and curves appear to converge to similar exponential limits. Osmotic stress was used to investigate a complex phase diagram for PL–DNA involving two coexisting phases [54]. Using high salt concentrations, where at $\Pi = 0$, the samples are found to be in a pure loose bundle phase, osmotic stress was applied using a high molecular weight PEG. Initially, simple compression of the loose bundles is observed with the average interaxial spacing between DNA inside the array getting smaller. With increasing osmotic

pressure, the system rearranges and the characteristic scattering for the coexistence regime is observed. Increasing the osmotic pressure further still results in the sample passing through the coexistence regime until scattering due solely to a tight bundle regime is observed. SAXS curves showing PLL–DNA at 900 mM NaCl, corresponding to loose bundles at $\Pi_{osm} = 0$, with increasing osmotic pressure are shown in Figure 21.31. Using osmotic pressure to stress from a loose bundle to a tight bundle phase showed that these two phases are in equilibrium.

21.6.4 Polyplex Transfection

Typical transfection formulations are mixed in low salt conditions with an excess of positive to negative charge ratio of polymer to DNA to limit aggregation, resulting in uniform nanometer-sized particles (hydrodynamic radii of the order of 20–50 nm). Unfortunately, free polycations, typically amines, are toxic because they destabilize cellular membranes. Optimal charge ratios for gene therapy are a balance between achieving small, stable aggregates while minimizing the toxic effects. Incorporation of uncharged hydrophilic polymers, such as PEG, has proven to be an effective method to shield polyplexes, lower particle surface charge, and reduce unspecific interactions with salt and blood components [204]. However, shielding typically results in reduced transfection efficiency. This reduction can, in part, be overcome by incorporation of targeting ligands or bioreversible shielding into the particles [205,206].

Cellular uptake and delivery is not well understood but is known to depend strongly on particle size. Polyplex nanoparticles are small enough to be brought into the cell through clathrin- and caveolae-dependent endocytosis [207,208]. However, clathrin- and caveolae-independent endocytosis has also been observed including macropinocytosis and phagocytosis [209,210]. Once inside the cell, single-particle tracking techniques suggest that polyplexes are actively transported to the perinuclear region by microtubule transport [211]. More recently, new single-particle tracking results suggest that polyplexes of poly(ethylene imine) (PEI/DNA) have many diffusive behaviors. Interestingly, in early stages, these polyplexes are seen to bind to the cell membrane and freely diffuse on the cell surface while inducing a progressive accumulation of syndecans, resulting in an actin cytoskeleton–mediated endocytosis. Once endocytosed, the active transport of polyplexes inside vesicles by molecular motors along microtubules filaments was observed [212].

One of the primary causes of poor gene delivery with synthetic formulations is believed to be inefficient endosomal release. PEI, one of the most frequently studied systems due to its low cost and excellent transfection efficiencies *in vitro* and significant transfections *in vivo*, shows significant improvement in endosomal release over other common polycations [213,214]. Endosomal escape is believed to proceed through the "proton sponge" effect where charges along the chain with a pK_a slightly below physiological pH gets activated upon acidification in the endosome, resulting in an influx of both protons and charge neutralizing Cl^- ions, inducing an osmotic stress that swells and destabilizes the endosome, releasing its contents into the cytoplasm [215,216]. These studies strongly suggest that the high transfection activity of PEI vectors is due to their unique ability to avoid acidic lysosomes. The subsequent steps, complex disassociation, and final nuclear transport of the plasmid are still not mechanistically well understood. While improved active transport across the nuclear membrane has been reported through the incorporation of NLS [217], clear evidence has also shown that passive DNA entry into the nucleus during cell division when the nuclear membrane is temporarily disrupted is important. Highest efficiencies are reported in dividing populations of cells [218]. While new techniques and studies are beginning to shed light on the complicated multiple barriers, both thermodynamic and kinetic, involved in successful gene delivery, clearly more work must be done to understand, balance, and use the molecular forces involved for improved delivery vectors.

21.7 RETROSPECT AND PROSPECT

Structural elucidation of the DNA–CL and DNA–polycation complexes and realization of the extent to which they share the structural features of pure-DNA or pure-lipid polymorphism have advanced notably in the past few years. Some old questions have been answered and new questions raised. It is these new questions that challenge our knowledge of the intricacies of interactions between macromolecules.

The DNA–lipid and DNA–polycation complexes found so far are only a sample of the much wider set of structures that will be seen on a full DNA–complex phase diagram. We argue that this larger set of possibilities be approached by firmly established methods to measure the energies of these structures at the same time that they are determined and located on a phase diagram. Built on principles of direct molecular interactions, recognizing the consequences of thermal agitation, this line of observation and analysis can lead to an understanding of the energetic "whys" and preparative "hows" of complex structures.

Forces so delineated are already knowledgeably applied in new preparations. Precisely how the structure of DNA–lipid and DNA–polycation aggregates will affect their efficacy in transfection remains to be seen. So far, the ideas we have are too general and have been learned from studying analytically tractable but technically inadequate preparations. General principles do not lead to specific results. Molecules are too interesting to allow easy success in clinical design. Still there is little doubt of a practical link between the energy and structure of these complexes and their viability in a technological application.

Even the present general understanding of forces, even the cartoon ideas of the directions in which forces act in macromolecular complexes, can tutor the bench scientist on how to improve preparations. There is enough known for a healthy iteration between experimental attempt and theoretical reason. Experimental successes and failures become the data for molecular force analyses. Various DNA–lipid and DNA–polycation assemblies reflect the various actions of competing forces. Molecular theorists can define and delineate these forces as they act to create each form; they can provide a logic to design variations in preparation. Basic scientists and clinicians are already in a position to help each other to improve their ways.

ACKNOWLEDGMENTS

We would like to thank Milan Hodoscek and Matjaz Licer for their help with molecular graphics in preparing some of the figures, and John Nagle and Stephanie Tristram-Nagle for their comments on an earlier version of this chapter. We would like to thank Don Rau for instructive discussions. R.P. and D.H. thank the support of the Israeli and Slovenian ministries of science and technology through a joint Slovenian–Israeli research grant. This research was supported by the Intramural Research Program of the NICHD, NIH.

REFERENCES

1. Bloomfield VA, Crothers DM, and Tinoco I. *Nucleic Acids: Structures, Properties and Functions*. Mill Valley: University Science Books, 1998.
2. Vologotskii AV. *Topology and Physics of Circular DNA (Physical Approaches to DNA)*. Boca Raton, FL: CRC Press, 1992.
3. Strey HH, Podgornik R, Rau DC, and Parsegian VA. Colloidal DNA. *Curr Opin Coll Interf Sci* 1998; 3: 534–539.

4. Podgornik R, Strey HH, and Parsegian VA. DNA–DNA interactions. *Curr Opin Struc Biol* 1998; 8: 309–313.

5. Kabanov VA, Flegner P, and Seymour LW. *Self-Assembling Complexes for Gene Delivery*. New York: Wiley, 1998.

6. Lasic DD. *Liposomes in Gene Delivery*. Boca Raton, FL: CRC Press, 1997.

7. Podgornik R, Rau DC, and Parsegian VA. Parametrization of direct and soft steric-undulatory forces between DNA double helical polyelectrolytes in solutions of several different anions and cations. *Biophys J* 1994; 66: 962–971.

8. Livolant F and Leforestier A. DNA mesophases. A structural analysis in polarizing and electron microscopy. *Mol Cryst Liq Cryst* 1992; 215: 47–56.

9. Derjaguin BV, Churaev NV, and Muller VM. *Surface Forces*. New York: Plenum, 1987.

10. Bloomfield VA. DNA condensation. *Curr Opin Struc Biol* 1996; 6: 334–341.

11. Leikin S, Parsegian VA, Rau DC, and Rand RP. Hydration forces. *Annu Rev Phys Chem* 1993; 44: 369–395.

12. Darnell J, Lodish H, and Baltimore D. *Molecular Cell Biology*, 2nd ed. New York: Scientific American Books, 1990.

13. Cerritelli ME, Cheng N, Rosenberg AH, McPherson CE, Booy FP, and Steven AC. Encapsidated conformation of bacteriophage T7 DNA. *Cell* 1997; 91: 271–280.

14. Safinya CR, Koltover I, and Rädler J. DNA at membrane surfaces: An experimental overview. *Curr Opin Colloid Interf Sci* 1998; 1: 69–77.

15. Mahanty J and Ninham BW. *Dispersion Forces*. London: Academic Press, 1976.

16. Safran SA. *Statistical Thermodynamics of Surfaces, Interfaces and Membranes*. New York: Addison Wesley, 1994.

17. Eisenberg D and Kauzmann W. *The Structure and Properties of Water*. Oxford: Clarendon Press, 1969.

18. Parsegian VA, Rand RP, Fuller NL, and Rau DC. Osmotic stress for the direct measurement of intermolecular forces. *Methods Enzymol* 1986; 127: 400–416.

19. Marãclja S and Radiç N. Repulsion of interfaces due to boundary water. *Chem Phys Lett* 1976; 42: 129–130.

20. Rand RP and Parsegian VA. Hydration forces between phospholipid bilayers. *Biochim Biophys Acta* 1989; 988: 351–376.

21. Kornyshev AA and Leikin S. Fluctuation theory of hydration forces: the dramatic effects of inhomogeneous boundary conditions. *Phys Rev A* 1998; 40: 6431–6437.

22. Rau DC and Parsegian VA. Direct measurement of the intermolecular forces between counterion-condensed DNA double helices. Evidence for long range attractive hydration forces. *Biophys J* 1992; 61: 246–259.

23. Parsegian VA. Long-range physical forces in the biological milieu. *Annu Rev Biophys Bioeng* 1973; 2: 221–255.

24. Bernal JD and Fankuchen I. X-ray and crystallographic studies of plant virus preparations. *J General Physiol* 1942; 25: 111–165.

25. Verwey EJW and Overbeek JTG. *Theory of the Stability of Lyophobic Colloids*. New York: Elsevier, 1948.

26. Hill TL. *Statistical Mechanics: Principles and Selected Applications*. New York: Dover, 1956.

27. Andelman D. Electrostatic properties of membranes: The Poisson–Boltzmann theory. In: Lipowsky R and Sackmann E, eds. *Structure and Dynamics of Membranes*. Vol. 1B. Amsterdam: Elsevier, 1995, pp. 603–642.

28. Landau LD and Lifshitz EM. *The Classical Theory of Fields*, 4th ed. Butterworth-Heinemann, 1986.

29. McLaughlin S. Electrostatic potential at membrane solution interfaces. *Curr Top Membrane Transp* 1985; 4: 71–144.

30. Brenner SL and McQuarrie DA. Force balances in systems of cylindrical polyelectrolytes. *Biophys J* 1973; 13: 301–331.

31. Parsegian VA, Rand RP, and Fuller NL. Direct osmotic stress measurements of hydration and electrostatic double-layer forces between bilayers of double-chained ammonium acetate surfactants. *J Phys Chem* 1991; 95: 4777–4782.

32. Kjellander R. Ion–ion correlations and effective charges in electrolyte and macroion systems. *Ber Bunsenges Phys Chem* 1996; 100: 894–904.

33. Hunter RJ. *Foundations of Colloid Science*. New York: Oxford University Press, 1987.

34. Parsegian VA, *Van der Waals Forces: A Handbook for Biologists, Chemists, Engineers, and Physicists*. Cambridge, MA: Cambridge University Press, 2005.

35. Parsegian VA. Long range van der Waals forces. In: van Olphen H. and Mysels KL, eds. *Physical Chemistry: Enriching Topics from Colloid and Interface Science*. 1975: 27–72.

36. Landau LD and Lifshitz EM. *Statistical Physics Part 2*. Butterworth-Heinemann, 1986.

37. Parsegian VA and Rand RP. Interaction in membrane assemblies. In: Lipowsky R and Sackmann E, eds. *Structure and Dynamics of Membranes*, Vol. 1B. Amsterdam: Elsevier, 1995, pp. 643–690.

38. Parsegian VA. Non-retarded van der Waals between anisotropic long thin rods at all angles. *J Chem Phys* 1972; 56: 4393–4397.

39. Brenner SL and Parsegian VA. A physical method for deriving the electrostatic interaction between rod-like polyions at all mutual angles. *Biophys J* 1974; 14: 327–334.

40. Parsegian VA, Fuller N, and Rand RP. Measured work of deformation and repulsion of lecithin bilayers. *Proc Natl Acad Sci USA* 1979; 76: 2750–2754.

41. Lipowsky R. Generic interactions of flexible membranes. In: Lipowsky R and Sackmann E, eds. *Structure and Dynamics of Membranes*, Vol. 1B. Amsterdam: Elsevier, 1995, pp. 521–596.

42. Helfrich W. Steric interactions of fluid membranes in multilayer systems. *Z Naturforsch* 1978; 33a: 305–315.

43. Strey HH, Parsegian VA, and Podgornik R. Equation of state for polymer liquid crystals: Theory and experiment. *Phys Rev E* 1998; 59: 999–1008.

44. Seifert U and Lipowsky R. Morphology of vesicles. In: Lipowsky R and Sackmann E, eds. *Structure and Dynamics of Membranes*, Vol. 1A. Amsterdam: Elsevier, 1995, p. 403.

45. Podgornik R and Parsegian VA. Thermal–mechanical fluctuations of fluid membranes in confined geometries: The case of soft confinement. *Langmuir* 1992; 8: 557–562.

46. Podgornik R and Parsegian VA. Charge-fluctuation forces between rodlike polyelectrolytes: Pairwise summability reexamined. *Phys Rev Lett* 1998; 80: 1560–1563.

47. Ha BJ and Liu AJ. Counterion-mediated attraction between two like charged rods. *Phys Rev Lett* 1997; 79: 1289–1292.

48. Trizac E and Tellez G. Onsager–Manning–Oosawa condensation phenomenon and the effect of salt. *Phys Rev Lett* 2006; 96: 38302.

49. Rau DC and Parsegian VA. Direct measurement of temperature-dependent solvation forces between DNA double helices. *Biophys J* 1992; 61: 260–271.

50. Hill TL. *An Introduction to Statistical Thermodynamics*. New York: Dover Publications, 1986.

51. Naji A, Jungblut S, Moreira AG, et al. Electrostatic interactions in strongly coupled soft matter. *Physica A* 2005; 352: 131–170.

52. Grosberg AY, Nguyen TT, and Shklovskii BI. Colloquium: The physics of charge inversion in chemical and biological systems. *Rev Mod Phys* 2002; 74: 329–345.

53. Oosawa F. *Polyelectrolytes*. New York: Marcel Dekker, 1971.

54. DeRouchey, J, Netz RR, and Rädler JO. Structural investigations of DNA–polycation complexes. *Eur Phys J E* 2005; 16: 17–28.

55. Podgornik R. Polyelectrolyte-mediated bridging interactions. *J Polym Sci, Polym Phys* 2004; 42: 3539–3556.

56. Podgornik R. Two-body polyelectrolyte mediated bridging interactions. *J Chem Phys* 2003; 118: 11286–11296.

57. Podgornik R and Saslow WM. Long-range many-body polyelectrolyte bridging interactions. *J Chem Phys* 2005; 122: 204902.

58. Saenger W. *Principles of Nucleic Acid Structure*. New York: Springer Verlag, 1984.

59. Rhodes D and Klug A. Helical periodicity of DNA determined by enzyme digestion. *Nature* 1980; 286: 573–578.

60. Rill RL. Liquid crystalline phases in concentrated DNA solutions. In: Pifa-Mrzljak G. ed. *Supramolecular Structure and Function*. New York: Springer, 1988, pp. 166–167.

61. De Gennes PG and Prost J. *The Physics of Liquid Crystals*, 2nd ed. Oxford: Oxford University Press, 1993.

62. Hagerman PJ. Flexibility of DNA. *Annu Rev Biophys Biophys Chem* 1988; 17: 265–286.

63. Pruss GJ and Drlica K. DNA supercoiling and prokaryotic transcription. *Cell* 1989; 56: 521–523.

64. Grosberg AY and Khokhlov AR. *Statistical Physics of Macromolecules* (AIP Series in Polymers and Complex Materials). Washington, DC: American Institute of Physics, 1994.

65. Rau DC, Lee BK, and Parsegian VA. Measurement of the repulsive force between polyelectrolyte molecules in ionic solution: Hydration forces between parallel DNA double helices, *Proc Natl Acad Sci USA* 1984; 81: 2621–2625.

66. Strey HH, Parsegian VA, and Podgornik R. Equation of state for DNA liquid crystals: Fluctuation enhanced electrostatic double layer repulsion. *Phys Rev Lett* 1997; 78: 895–898.

67. Reich Z, Wachtel EJ, and Minsky A. In vivo quantitative characterization of intermolecular interaction. *J Biol Chem* 1995; 270: 7045–7046.

68. Barrat JL and Joanny JF. Theory of polyelectrolyte solutions. In: Prigogine I and Rice SA, eds. *Advances in Chemical Physics*. New York: Wiley, 1995, pp. 1–66.

69. Lyubartsev AP and Nordenskiold L. Monte Carlo simulation study of ion distribution and osmotic pressure in hexagonally oriented DNA. *J Phys Chem* 1995; 99: 10373–10382.

70. Oosawa F. *Polyelectrolytes*. New York: Marcel Dekker, 1971.

71. Rouzina I and Bloomfield VA. Macro-ion attraction due to electrostatic correlation between screening counterions. *J Phys Chem* 1996; 100: 9977–9989.

72. Podgornik R, Strey HH, Rau DC, and Parsegian VA. Watching molecules crowd: DNA double helices under osmotic stress. *Biophys Chem* 1995; 26: 111–121.

73. Lindsay SM, Lee SA, Powell JW, Weidlich T, Demarco C, Lewen GD, Tao NJ, and Rupprecht A. The origin of the A to B transition in DNA fibers and films. *Biopolymers* 1988; 17: 1015–1043.

74. Podgornik R, Strey HH, Gawrisch K, Rau DC, Rupprecht A, and Parsegian VA. Bond orientational order, molecular motion, and free energy of high-density DNA mesophases. *Proc Natl Acad Sci USA* 1996; 93: 4261–4266.

75. Strandberg D. *Bond-Orientational Order in Condensed Matter Systems*. New York: Springer, 1992.

76. Durand D, Doucet J, and Livolant F. A study of the structure of highly concentrated phases of DNA by X-ray diffraction. *J Phys II France* 1992; 2: 1769–1783.

77. Kamien RD. Liquids with chiral bond order. *J Phys II France* 1996; 6: 461–475.

78. Chaikin PM and Lubensky TC. *Principles of Condensed Matter Physics*. Cambridge, MA: Cambridge University Press. 1995.

79. Leforestier A and Livolant F. DNA liquid-crystalline blue phases—electron-microscopy evidence and biological implications. *Mol Cryst Liquid Cryst* 1994; 17: 651–658.

80. Wang L and Bloomfield VA. Small-angle X-ray scattering of semidilute rodlike DNA solutions: Polyelectrolyte behavior. *Macromolecules* 1991; 24: 5791–5795.

81. Podgornik R, Rau DC, and Parsegian VA. The action of interhelical forces on the organization of DNA double helices: Fluctuation enhanced decay of electrostatic double layer and hydration forces. *Macromolecules* 1989; 22: 1780–1786.

82. Frank-Kamenetskii MD, Anshelevich VV, and Lukashin AV. Polyelectrolyte model of DNA. *Sov Phys Usp* 1987; 4: 317–330.

83. Tanford C. *The Hydrophobic Effect. Formation of Micelles and Biological Membranes*. New York: Wiley, 1980.

84. Cevc G and Marsh D. *Phospholipid Bilayers: Physical Principles and Models* (Cell Biology: A Series of Monographs, Vol 5). New York: Wiley-Interscience, 1987.

85. Parsegian VA and Evans EA. Long and short range intermolecular and intercolloidal forces. *Curr Opin Coll Interf Sci* 1996; 1: 53–60.

86. Duzgunes N, Willshut L, Hong K, Fraley R, Perry C, Friends DS, James TL, and Papahadjopoulos D. Physicochemical characterization of large unilamellar phospholipid vesicles prepared by reverse-phase evaporation. *Biophys Biochim Acta* 1983; 732: 289–299.

87. Seifert U. Configurations of fluid membranes and vesicles. *Adv Phys* 1997; 46: 13–137.

88. Small DM. *The Physical Chemistry of Lipids: From Alkanes to Phospholipids*. New York: Plenum Press, 1986.

89. Gruner SM, Parsegian VA, and Rand RP. Directly measured deformation energy of phospholipid H2 hexagonal phases. *Faraday Disc* 1986; 81: 213–221.

90. Daoud M and Williams CE. *Soft Matter Physics*. New York: Springer, 1999.

91. Lasic DD. *Liposomes: From Physics to Applications*. Amsterdam: Elsevier, 1993.

92. Parsegian VA and Podgornik R. Surface-tension suppression of lamellar swelling on solid substrates. *Colloids Surf A Physicochem Eng Asp* 1997; 129–130, 345–364.

93. Roux D and Safinya CR. A synchrotron X-ray study of competing undulation and electrostatic interlayer interactions in fluid multimembrane lyotropic phases. *J Phys—Paris* 1988; 49: 307–318.

94. Leneveu DM, Rand RP, Gingell D, and Parsegian VA. Apparent modification of forces between lecithin bilayers. *Science* 1976; 191: 399–400.

95. Parsegian VA. Reconciliation of van der Waals force measurements between phosphatidylcholine bilayers in water and between bilayer coated mica surfaces. *Langmuir* 1993; 9: 3625–3628.

96. Gouliaev N and Nagle JF. Simulations of interacting membranes in soft confinement regime. *Phys Rev Lett* 1998; 81: 2610–2613.

97. Rand RP and Parsegian, VA. Hydration forces between phospholipid bilayers. *Biochim Biophys Acta* 1989; 988: 351–376.

98. Petrache HI, Gouliaev N, Tristram-Nagle S, Zhang R, Suter RM, and Nagle JF. Interbilayer interactions from high-resolution X-ray scattering. *Phys Rev E* 1998; 57: 7014–7024.

99. Felgner PL, Barenholz Y, Behr JP, et al. Nomenclature for synthetic gene delivery systems. *Hum Gene Ther* 1997; 8: 511–512.

100. Safinya CR. Structure of lipid–DNA complexes: Supermolecular assembly and gene delivery. *Curr Opin Struc Biol* 2001; 11: 440–448.

101. Barenholz Y. Liposome application: Problems and prospects. *Curr Opin Coll Inter Sci* 2001; 6: 66–77.

102. Rädler JO. Structure and phase behavior of cationic-lipid DNA complexes. In: Holm C, Kékicheff P, and Podgornik R, eds. *Electrostatic Effects in Soft Matter and Biophysics.* Dordrecht: Kluwer, 2001, pp. 441–458.

103. Ilies MA and Balaban AT. Recent developments in cationic lipid-mediated gene delivery and gene therapy. *Expert Opin Ther Patents* 2001; 11: 1729–1751.

104. Ilies MA, William AS, and Balaban AT. Cationic lipids in gene delivery: Principles, vector design and therapeutical applications. *Curr Pharm Des* 2002; 8: 125–133.

105. Audouy S and Hoekstra D. Cationic-mediated transfection in vitro and in vivo. *Mol Membr Biol* 2001; 18: 129–143.

106. Felgner PL, Gadeck T, Holen M, et al. Lipofectin: A highly efficient lipid-mediated DNA transfection procedure. *Proc Acad Sci USA* 1987; 84: 7413–7417.

107. Felgner PL and Ringold GM. Cationic liposome-mediated transfection, *Nature* 1989; 337: 387–388.

108. Tempelton NS. Developments in liposomal gene delivery systems. *Expert Opin Biol Ther* 2001; 1: 1–4.

109. Fraley R, Subramani S, Berg P, and Papahadjopoulos D. Introduction of liposome-encapsulated SV40 DNA into cells. *J Biol Chem* 1980; 255: 10431–10435.

110. Fillion P, Desjardins A, Sayasith K, and Lagace J. Encapsulation of DNA in negatively charged liposomes and inhibition of bacterial gene expression with fluid liposome-encapsulated antisense oligonucleotides. *Biochim Biophys Acta Biomembr* 2001; 1515: 44–54.

111. Lakkaraju A, Dubinsky JM, Low WC, and Rahman YE. Neurons are protected from excitotoxic death by p53 antisense oligonucleotides delivered in anionic liposomes. *J Biol Chem* 2001; 276: 32000–32007.

112. Patil SD and Rhodes DG. Conformation of oligodeoxynucleotides associated with anionic liposomes. *Nucleic Acids Res* 2000; 28: 4125–4129.

113. Liang H, Harries D, and Wong GCL. Polymorphism of DNA-anionic liposome complexes reveals hierarchy of ion-mediated interactions. *Proc Natl Acad Sci USA* 2005; 102: 11173–11178.

114. Pisani M, Bruni P, Caracciolo G, et al. Structure and phase behavior of self-assembled DPPC–DNA–metal cation complexes *J Phys Chem B* 2006; 110: 13203–13211.

115. Lasic DD and Templeton NS. Liposomes in gene therapy. *Adv Drug Deliv Rev* 1996; 20: 221–266.

116. Lasic DD, Strey H, Stuart MCA, Podgornik R, and Frederik PM. The structure of DNA–liposome complexes. *J Am Chem Soc* 1997; 119: 832–833.

117. Rädler JO, Koltover I, Salditt T, et al. Structure of DNA–cationic liposome complexes: DNA intercalation in multilamellar membranes in distinct interhelical packing regimes. *Science* 1997; 275: 810–814.

118. Koltover I, Salditt T, Rädler JO, et al. An inverted hexagonal phase of cationic liposome–DNA complexes related to DNA release and delivery. *Science* 1998; 281: 78–81.

119. Salditt T, Koltover I, Rädler O, et al. Self-assembled DNA–cationic–lipid complexes: Two-dimensional smectic ordering, correlations, and interactions. *Phys Rev E* 1998; 58: 889–904.

120. Salditt T, Koltover I, Rädler J, and Safinya CR. Two dimensional smectic ordering of linear DNA chains in self-assembled DNA–cationic liposome mixtures. *Phys Rev Lett* 1997; 79: 2582–2585.

121. Rädler JO, Koltover I, Jamieson A, Salditt T, and Safinya CR. Structure and interfacial aspects of self-assembled cationic lipid–DNA gene carrier complexes. *Langmuir* 1998; 14: 4272–4283.

122. Sackmann E. Physical basis of self-organization and function of membranes: Physics of vesicles. In: Lipowsky R and Sackmann E, eds. *Structure and Dynamics of Membranes.* Amsterdam: Elsevier, 1995, pp. 213–304.

123. Tarahovsky YS, Khusainova RS, Gorelov AV, et al. DNA initiates polymorphic structural transitions lecithin. *FEBS Lett* 1996; 390: 133–136.

124. Ghirlando R, Wachtel EJ, Arad T, and Minsky A. DNA packaging induced by micellar aggregates: A novel in vitro DNA condensation system. *Biochemistry* 1992; 31: 7110–7119.

125. Gershon H, Ghirlando R, Guttman SB, and Minsky A. Mode of formation and structural features of DNA–cationic liposome complexes used for transfection. *Biochemistry* 1993; 32: 7143–7151.

126. Ewert KK, Evans HM, Zidovska A, et al. A columnar phase of dendritic lipid-based cationic liposome–DNA complexes for gene delivery: Hexagonally ordered cylindrical micelles embedded in a DNA honeycomb lattice. *J Am Chem Soc* 2006; 128: 3998–4006.

127. Koynova R, Rosenzweig HS, Wang L, et al. Novel fluorescent cationic phospholipid, O-4-napthylimido-1-butyl-DOPC, exhibits unusual foam morphology, forms hexagonal and cubic phases in mixtures, and transfects DNA. *Chem Phys Lipids* 2004; 129: 183–194.

128. May S and Ben-Shaul S. DNA–lipid complexes: Stability of honeycomb-like and spaghetti-like structures. *Biophys J* 1997; 73: 2427–2440.

129. Dan N. The structure of DNA complexes with cationic liposomes—cylindrical or lamellar? *Biophys Biochim Acta* 1998; 1369: 34–38.

130. Sternberg B, Sorgi FL, and Huang L. New structures in complex-formation between DNA and cationic liposomes visualized by freeze-fracture electron-microscopy. *FEBS Lett* 1994; 356: 361–366.

131. Sternberg B. New structures in complex formation between DNA and cationic liposomes visualized by freeze-fracture electron microscopy. *FEBS Lett* 1994; 356: 361–366.

132. Simberg D, Danino D, Talmon Y, Minsky A, Ferrari ME, Wheeler CJ, and Barenholz Y. Phase behavior, DNA ordering, and size instability of cationic lipoplexes. *J Biol Chem* 2001; 276: 47453–47459.

133. Pitard B, Aguerre O, Airiau M, Lachages AM, et al. Virus-sized self-assembling lamellar complexes between plasmid DNA and cationic micelles promote gene transfer. *Proc Natl Acad Sci USA* 1997; 94: 14412–14417.

134. Boukhnikashvili T, Aguerre-Chariol O, Airiau M, Lesieur S, Ollivon M, and Vacus J. Structure of in-serum transfecting DNA–cationic lipid complexes. *FEBS Lett* 1997; 409: 188–194.

135. Templeton NS, Lasic DD, Frederik PM, Strey HH, Roberts DD, and Pavlakis GN. Improved DNA: Liposome complexes for increased systemic delivery and gene expression. *Nature Biotech* 1997; 15: 647–652.

136. Record TM Jr, Anderson CF, and Lohman TM. Thermodynamic analysis of ion effects on the binding and conformational equilibria of proteins and nucleic acids: The roles of ion association or release, screening, and ion effects on water activity. *Quart Rev Biophys* 1978; 11: 103–178.

137. Bruinsma R. Electrostatics of DNA–cationic lipid complexes: Isoelectric instability. *Eur Phys J B* 1998; 4: 75–88.

138. Wagner K, Harries D, May S, et al. Counterion release upon cationic lipid–DNA complexation. *Langmuir* 2000; 16: 303–306.

139. Bordi F, Cametti C, Sennato S, et al. Counterion release in overcharging of polyion–liposome complexes. *Phys Rev E* 2006; 74: Art. No. 030402.

140. Barreleiro PCA, Olofsson G, and Alexandridis P. Interaction of DNA with cationic vesicles: A calorimetric study. *J Phys Chem B* 2000; 104: 7795–7802.

141. Kennedy MT, Pozharski EV, Rakhmanova VA, and MacDonald RC. Factors governing the assembly of cationic phospholipid–DNA complexes. *Biophys J* 2000; 78: 1620–1633.

142. Huebner S, Battersby BJ, Grimm R, and Cevc G. Lipid mediated complex formation: Reorganization and rupture of lipid vesicles in the presence of DNA as observed by cryoelectron microscopy. *Biophys J* 199; 76: 3158–3166.

143. Dan N. Multilamellar structures of DNA complexes with cationic liposomes. *Biophys J* 1997; 73: 1842–1846.

144. Harries D, May S, Gelbart WM, and Ben-Shaul A. Structure, stability and thermodynamics of lamellar DNA–lipid complexes. *Biophys J* 1998; 75: 159–173.

145. Parsegian VA and Gingell D. On the electrostatic interaction across a salt solution between two bodies bearing unequal charges. *Biophys J* 1972; 12: 1192–1204.

146. Mitrakos P and Macdonald PM. DNA-induced lateral segregation of cationic amphiphiles in lipid bilayer membranes as detected via ^{2}H NMR. *Biochemistry* 1996; 35: 16714–16722.

147. Bandyopadhyay S, Tarek M, and Klein ML. Molecular dynamics study of a lipid–DNA complex. *J Phys Chem B* 1999; 103: 1007–1008.

148. Garidel P, Johann C, and Blume A. Thermodynamics of lipid organization and domain formation in phospholipid bilayers. *J Liposome Res* 2000; 10: 131–158.

149. Koltover I, Wagner K, and Safinya CR. DNA condensation in two dimensions. *Proc Natl Aca Sci USA* 2000; 97: 14046–14051.

150. Dubois M, Zemb Th, Fuller N, Rand RP, and Parsegian VA. Equation of state of a charged bilayer system: Measure of the entropy of the lamellar–lamellar transition in DDABr. *J Chem Phys* 1998; 18: 7855–7869.

151. Bostrom M, Williams DRM, and Ninham B. Specific ion effects: Why DLVO theory fails. *Phys Rev Lett* 2001; 87: Art No. 168103.

152. Fang Y and Yang J. Two-dimensional condensation of DNA molecules on cationic lipid membranes. *J Phys Chem B* 1997; 101: 441–449.

153. Mou J, Czajkowsky DM, Zhang Y, and Shao Z. High-resolution atomic force microscopy of DNA: The pitch of the double helix. *FEBS Lett* 1995; 371: 279–282.

154. May S, Harries D, and Ben-Shaul A. Lipid demixing and protein–protein interactions in the adsorption of charged proteins on mixed membranes. *Biophys J* 2000; 79: 1747–1760.

155. Dan N. Formation of ordered domains in membrane-bound DNA. *Biophys J* 1996; 71: 1267–127.

156. Fang Y and Yang J. Effect of cationic strength and species on 2-D condensation of DNA. *J Phys Chem B* 1997; 101: 3453–3456.

157. Safinya CR, Sirota EB, Roux D, and Smith GS. Universality in interacting membranes: The effect of cosurfactants on the interfacial rigidity. *Phys Rev Lett* 1989; 62: 1134–1137.

158. Szleifer I, Kramer D, Ben-Shaul A, Roux D, and Gelbart MW. Curvature elasticity of pure and mixed surfactant films. *Phys Rev Lett* 1998; 60: 1966–1969.

159. Gawrisch K, Parsegian VA, Hajduk DA, Tate MW, Gruner SM, Fuller NL, and Dan N. Energetics of a hexagonal–lamellar–hexagonal-phase transition sequence in dioleoylphosphatidylethanolamine membranes. *Biochemistry* 1992; 31: 2856–2864.

160. Kozlov MM, Leikin S, and Rand RP. Bending, hydration, and intersticial energies quantitatively account for the hexagonal-lamellar-hexagonal reentrant phase transition in dioleoylphosphatidylethanolamine. *Biophys J* 1994; 67: 1603–1611.

161. Chen Z and Rand RP. Comparative study of the effects of several *n*-alkanes on phospholipid hexagonal phases. *Biophys J* 1998; 74: 944–952.

162. May S, Harries D, and Ben-Shaul A. The phase behavior of cationic lipid–DNA complexes. *Biophys J* 2000; 78: 1681–1697.

163. Schiessel H and Aranda-Espinoza H. Electrostatically induced undulations of lamellar DNA–lipid complexes. *Eur Phys J E* 2001; 5: 499–506.

164. Harries D. Electrostatic interaction between macromolecules and mixed lipid membranes. PhD dissertation. The Hebrew University, Jerusalem, Israel, 2001. Available from www.fh.huji.ac.il/~daniel/.

165. Battersby BJ, Grimm R, Huebner S, and Cevc G. Evidence for three-dimensional interlayer correlations in cationic lipid–DNA complexes as observed by cryo-electron microscopy. *Biochim Biophys Acta* 1998; 1372: 379–383.

166. Artzner F, Zantl R, Rapp G, and Rädler JO. Observation of a rectangular columnar phase in condensed lamellar cationic lipid–DNA complexes. *Phys Rev Lett* 1998; 81: 5015–5018.

167. O'Hern CS and Lubensky TC. Sliding columnar phase of DNA lipid complexes. *Phys Rev Lett* 1998; 80: 4345–4348.

168. Golubovic L and Golubovic M. Fluctuations of quasi-two-dimensional smectics intercalated between membranes in multilamellar phases of DNA cationic lipid complexes. *Phys Rev Lett* 1998; 80: 4341–4344.

169. Podgornik R and Žekž B. Coupling between smectic and twist modes in polymer intercalated smectics. *Phys Rev Lett* 1998; 80: 305–308.

170. Hui SW, Langner M, Zhao Y-L, Patrick R, Hurley E, and Chan K. The role of helper lipids in cationic liposome–mediated gene transfer. *Biophys J* 1996; 71: 590–59.

171. Zuidam NJ, Lerner DH, Margulies S, and Barenholz Y. Lamellarity of cationic liposomes and mode of preparation of lipoplexes affect transfection efficiency. *Biophys Biochim Acta* 1999; 1419: 207–220.

172. Meidan VM, Cohen JS, Amariglio N, Hirsch-Lerner D, and Barenholz Y. Interaction of oligonucleotides with cationic lipids: The relationship between electrostatics, hydration and state of aggregation. *Biochem Biophys Acta* 2000; 1464: 251–261.

173. Artzner F, Zantl R, and Rädler JO. Lipid–DNA and lipid–polyelectrolyte mesophases: Structure and exchange kinetics. *Cell Molec Biol* 2000; 46: 967–978.

174. Wiethoff CM, Smith JG, Koe GS, and Middaugh CR. The potential role of proteoglycans in cationic lipid-mediated gene delivery—studies of the interaction of cationic lipid–DNA complexes with model glycosaminoglycans. *J Biol Chem* 2001; 276: 32806–32813.

175. Zabner J, Fasbender AJ, Moninger T, Poellinger KA, and Welsh MJ. Cellular and molecular barriers to gene transfer by a cationic lipid. *J Biol Chem* 1995; 270: 18997–19007.

176. Lin AJ, Slack NL, Ahmad A, Koltover I George CX, Samuel CE, and Safinya CR. Structure and structure–function studies of lipid/plasmid DNA complexes. *J Drug Targeting* 2000; 8: 13–27.

177. Zuhorn IS, Bakowsky U, Polushkin E, Visser WH, Stuart MC, Engberts JB, and Hoekstra D. Nonbilayer phase of lipoplex–membrane mixture determines endosomal escape of genetic cargo and transfection efficiency. *Mol Ther* 2005; 11:801–10.

178. Hirsch-Lerner D and Barenholz Y. Probing DNA–cationic lipid interactions with the fluorophore trimethylammonium diphenyl-hexatrien (TMADPH). *Biochim Biophys Acta* 1998; 1370: 17–30.

179. Tang F and Hughes JA. Use of dithiodilycolic acid as a tether for cationic lipids decreases the cytotoxicity and increases transgene expression of plasmid DNA in vitro. *Bioconjug Chem* 1999; 10: 791–796.

180. Byk G, Wetzer B, Frederic M, et al. Reduction sensitive lipopolyamines as a novel nonviral gene delivery system for modulated release of DNA with improved transgene expression. *J Med Chem* 2000; 43: 4377–4387.

181. Balasubramaniam RP, Bennett MJ, Aberle AM, Malone JG, Nantz MH, and Malone RW. Structural and functional analysis of cationic transfection lipids: The hydrophobic domain. *Gene Ther* 1996; 3: 163–172.

182. Weltzer B, Byk G, Frederic M, et al. Reducible cationic lipids for gene transfer. *Biochem J* 2001; 356: 747–756.

183. Lin AJ, Slack NL, Ahmad A, George CX, Samuel CE, and Safinya CR. Three-dimensional imaging of lipid gene-carriers: Membrane charge density controls universal transfection behavior in lamellar cationic liposome–DNA complexes. *Biophys J* 2003; 84: 3307–3316.

184. Ewert KK, Ahmad A, Evans HM, Ahmad A, Slack NL, Lin AJ, Martin-Herranz A, and Safinya CR. Cationic lipid–DNA complexes for non-viral gene therapy: Relating supramolecular structures to cellular pathways. *Expert Opin Biol Therapy* 2005; 5: 33–53.

185. Lin AJ, Slack NL, Ahmad A, et al. Three-dimensional imaging of lipid gene-carriers: Membrane charge density controls universal transfection behavior in lamellar cationic liposome–DNA complexes. *Biophys J* 2003; 84: 3307–3316.

186. Ewert K, Slack NL, Ahmad A, Evans HM, Lin AJ, Samuel CE, and Safinya CR. Cationic lipid–DNA complexes for gene therapy: Understanding the relationship between complex structure and gene delivery pathways at the molecular level. *Curr Med Chem* 2004; 11: 133–149.

187. Zanta MA, Belguise-Valladier P, and Behr JP. Gene delivery: A single nuclear localization signal peptide is sufficient to carry DNA to the cell nucleus. *Proc Natl Acad Sci USA* 1999; 96: 91–96.

188. Cartier R and Reszka R. Utilization of synthetic peptides containing nuclear localization signal for nonviral gene transfer systems. *Gene Ther* 2002; 9: 157–167.

189. Wheeler JJ, Palmer L, Ossanlou M, et al. Stabilized plasmid–lipid particles: Construction and characterization. *Gene Ther* 1999; 6: 271–281.

190. Martin-Herranz A, Ahmad A, Evans HM, Ewert K, Schulze U, and Safinya CR. Surface functionalized cationic lipid–DNA complexes for gene delivery: PEGylatedlamellar complexes exhibit distinct DNA–DNA interaction regimes. *Biophys J* 2004; 86: 1160–1168.

191. Schulze, U, Schmidt, H-W, and Safinya, CR. Synthesis of novel cationic poly(ethylene glycol) containing lipids. *Bioconjugate Chem* 1999; 10: 548–552.

192. Duncan R. The dawning era of polymer therapeutics. *Nat Rev Drug Discovery* 2003; 2: 347–360.

193. Wagner E and Kloeckner J. Gene delivery using polymer therapeutics. *Adv Polymer Sci* 2006; 192: 135–173.

194. Maurstad G, Danielsen S, and Stokke BT. Analysis of compacted semiflexible polyanions visualized by atomic force microscopy: Influence of chain stiffness on the morphologies of polyelectrolyte complexes. *J Phys Chem B* 2003; 107: 8172–8180.

195. Dunlap DD, Maggi A, Soria MR, et al. Nanoscopic structure of DNA condensed for gene delivery. *Nucleic Acids Res* 1997; 25: 3095–3101.

196. Golan R, Pietrasanta LI, Hsieh W, et al. DNA toroids: Stages in condensation. *Biochemistry* 1999; 38: 14069–14076.

197. Liu D, Wang C, Lin Z, et al. Visualization of the intermediates in a uniform DNA condensation system by tapping mode atomic force microscopy. *Surface Interface Anal* 2001; 32: 15–19.

198. Suwalsky M. Comparative X-ray study of a nucleoprotamine and DNA complexes with polylysine and polyarginine. *Biopolymers* 1972; 11: 2223–2231.

199. Suwalsky M. An X-ray study of interaction of DNA with spermine. *J Mol Biol* 1969; 42: 363–373.

200. Azorin F. Interaction of DNA with lysine-rich polypeptides and proteins—the influence of polypeptide composition and secondary structure. *J Mol Biol* 1985; 185: 371–387.

201. Boeckle S, von Gersdorff K, van der Piepen S, et al. Purification of polyethylenimine polyplexes highlights the role of free polycations in gene transfer. *J Gene Med* 2004; 6: 1102–1111.

202. Netz RR. Variational charge renormalization in charged systems. *Eur Phys J E* 2003; 11: 301–311.

203. Odijk T. Undulation-enhanced electrostatic forces in hexagonal polyelectrolyte gels. *Biophys Chem* 1993; 46: 69–75.

204. Ogris M, Brunner S, Schuller S, Kircheis R, and Wagner E. PEGylated DNA/transferrin–PEI complexes: Reduced interaction with blood components, extended circulation in blood and potential for systemic gene delivery. *Gene Ther* 1999; 6: 595–605.

205. Walker GF, Fella C, Pelisek J, Fahrmeir J, Boeckle S, Ogris M, and Wagner E. Toward synthetic viruses: Endosomal pH-triggered deshielding of targeted polyplexes greatly enhances gene transfer in vitro and in vivo. *Mol Ther* 2005; 11: 418–425.

206. Schatzlein AG. Targeting of synthetic gene delivery systems. *J Biomed Biotechnol* 2003; 2: 149–158.

207. von Gersdorff K, Sanders NN, Vandenbroucke R, et al. The internalization route resulting in successful gene expression depends on polyethylenimine both cell line and polyplex type. *Molec Ther* 2006; 14: 745–753.

208. Rejman J, Oberle V, Zuhorn IS, et al. Size-dependent internalization of particles via the pathways of clathrin- and caveolae-mediated endocytosis. *Biochem J* 2004; 377: 159–169.

209. Kopatz I, Remy JS, and Behr JP. A model for non-viral gene delivery: Through syndecan adhesion molecules and powered by actin. *J Gene Med* 2004; 6: 769–776.

210. Goncalves C, Mennesson E, Fuchs R, et al. Macropinocytosis of polyplexes and recycling of plasmid via the clathrin-dependent pathway impair the transfection efficiency of human hepatocarcinoma cells. *Molec Ther* 2004; 10: 373–385.

211. Suh J, Wirtz D, and Hanes J. Efficient active transport of gene nanocarriers to the cell nucleus. *Proc Natl Acad Sci USA* 2003; 100: 3878–3882.

212. Bausinger R, von Gersdorff K, Braeckmans K, et al. The transport of nanosized gene carriers unraveled by live-cell imaging. *Ang Chem Int Ed* 2006; 45: 1568–1572.

213. Furgeson DY and Kim SW. Recent advances in poly(ethylene imine) gene carrier design. *ACS Symp Ser* 2006; 923: 182–197.

214. Neu M, Fischer D, and Kissel T. Recent advances in rational gene transfer vector design based on poly(ethylene imine) and its derivatives. *J Gene Med* 2005; 7: 992–1009.

215. Boussif O, Lezoualch F, Zanta MA, et al. A versatile vector for gene and oligonucleotide transfer into cells in culture and in vivo—polyethyleneimine. *Proc Natl Acad Sci USA* 1995; 92: 7297–7301.

216. Akinc A, Thomas M, Klibanov AM, et al. Exploring polyethylenimine-mediated DNA transfection and the proton sponge hypothesis. *J Gene Med* 2005; 7: 657–663.

217. Wilson GL, Dean BS, Wang G, et al. Nuclear import of plasmid DNA in digitonin-permeabilized cells requires both cytoplasmic factors and specific DNA sequences. *J Biol Chem* 1999; 274: 22025–22032.

218. Dean DA, Strong DD, and Zimmer WE. Nuclear entry of nonviral vectors. *Gene Ther* 2005; 12: 881–890.

219. Peterlin A. Light scattering by very stiff chain molecules, *Nature* 1953; 171: 259–260.

Part II

Other Therapeutic Strategies

22 *In Vivo* Applications of Morpholino Oligonucleotides

Jeff S. Isenberg, William A. Frazier, and David D. Roberts

CONTENTS

22.1 INTRODUCTION

The first demonstration that gene expression could be suppressed in a specific manner by targeting a messenger RNA (mRNA) using complementary hybridizing DNA sequences was done in 1978 [1,2]. However, several technical obstacles needed to be overcome to develop therapeutic applications based on this discovery. Unmodified oligonucleotides are sensitive to enzymatic degradation *in vivo*, and their negative charge limits their entry into target cells without use of a vehicle to transport them across the plasma membrane. Several chemical modifications of the phosphorodiester–deoxyribosyl backbone have been made to increase the stability of antisense oligonucleotides. One of these involves replacement of the deoxyribose moiety with a substituted morpholine, which maintains the base-specific hybridization of the parent oligonucleotide (Figure 22.1) [3,4]. Phosphorodiamidate morpholino oligonucleotides (morpholinos) are neutral antisense oligomers of this class that have become widely used as a research tool in developmental biology.

Morpholinos are currently the standard reagent to temporarily but efficiently suppress expression of target genes in developing zebrafish embryos [5–7]. Similar applications are emerging for developmental analysis of gene function in other animals for which transgenic technologies are not currently available. In mice and other animals where transgenesis is available, morpholinos are also useful to complement transgenic approaches to study the function of embryonic lethal genes at specific stages of oocyte and embryo differentiation.

Morpholinos are designed to hybridize with complementary nucleotide sequences in the mRNA of a target gene, but unlike RNA interference approaches, morpholinos do not generally induce RNA degradation [8]. Rather, they can be designed to prevent gene expression by two mechanisms (Figure 22.2). Translation-blocking morpholinos are designed to hybridize with the 5′ region of mature mRNA to prevent its efficient translation [9]. Splice-blocking morpholinos are designed to hybridize with an exon–intron junction in a specific nuclear pre-mRNA to prevent the splicing required to generate a mature functional mRNA [10]. The latter approach can be used to inhibit expression if the junction is chosen judiciously so that alternative splicing of the nuclear transcript cannot bypass the targeted splice junction to generate a functional mRNA. Conversely, splice-blocking morpholinos have also found use to force generation of alternatively spliced mRNAs that can rescue function of certain mutated genes.

The stability of these synthetic oligonucleotides and the duration of their activity to suppress gene expression have prompted attempts to apply morpholinos to therapeutically

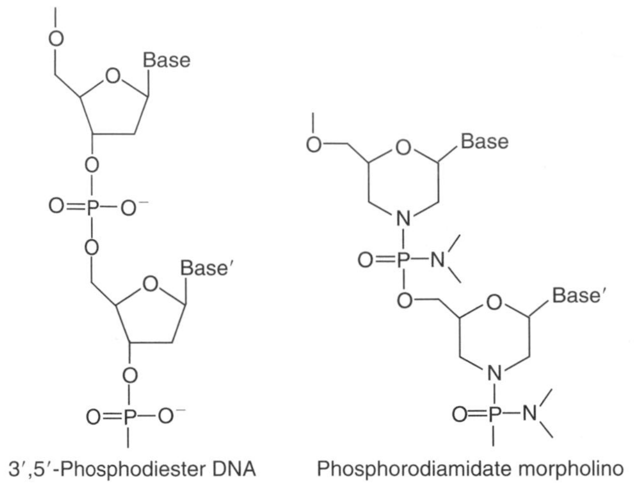

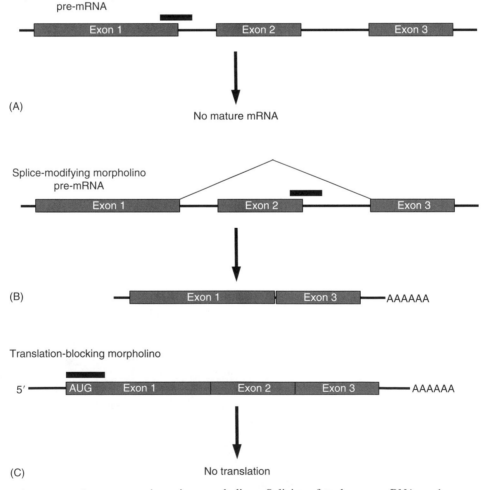

FIGURE 22.1 General structure of morpholinos. In morpholinos, the 3′,5′-phosphodiester linkage of DNA is replaced by the enzymatically stable phosphorodiamidate and the deoxyribose is replaced by a substituted morpholino.

suppress gene expression in adult animals [11,12]. We will review the results to date from these efforts and discuss the advantages of morpholinos for *in vivo* control of gene expression as well as the barriers that must be overcome to bring this approach to broad clinical use.

One of the central theoretical advantages of a morpholino-based therapeutic is its ability to specifically target the gene responsible for a specific pathology or disease state. The morpholino, because it is an electrostatically neutral molecule that engages the target RNA via Watson–Crick base pairing, may have very few if any off-target effects [13]. Some off-target activities can occur and could involve interactions of the morpholino oligonucleotide and various extracellular, membrane, and intracellular structures whose function could be perturbed by such binding. The current generation of morpholino oligonucleotides replaces the five-membered sugar ring backbone of natural DNA with an uncharged phosphorodiamidate linkage. The advantages of this derivative include better biologic stability, water solubility, excellent hybridization with RNA, and complete resistance to enzymatic

FIGURE 22.2 Strategies to control gene expression using morpholinos. Splicing of nuclear pre-mRNA can be prevented using a morpholino complementary to a splice donor sequence (A), which prevents generation of the mature processed mRNA. In some cases, splicing to generate an alternative mature mRNA can be forced by using a morpholino complementary to an internal exon (B). Translation-blocking morpholinos are typically designed to hybridize near the translation initiation sequence in a mature mRNA (C).

degradation. These agents also provide some advantages over other products used in gene silencing such as S-DNA and small-interfering RNA (siRNA), which were found to have some characteristic off-target effects, multiple protein interactions, and could stimulate both coagulation and immune system effects due their net negative charge, some being lethal in animal models [13–15].

The minimum effective length of the present generation morpholinos for target inhibition has been found to be 14 to 15 bases. This is true of both splice- and translation-blocking morpholinos. As such, once within the cell, the morpholino has free access to both the cytosol and nucleus, increasing the target pool of mRNAs upon which it can exert its therapeutic effects [13].

22.2 DELIVERY TECHNIQUES

22.2.1 *In Vitro* Delivery

The current phosphorodiamidate mopholinos are very effective at gene silencing, preventing protein production in at least 80%–95% of reported targets. However, such success is clearly dependent upon adequate delivery of the agent to the intracellular space. Most experience to date and research into drug delivery with morpholinos has employed cell culture systems [16,17]. In two-dimensional mammalian cell cultures, efficient delivery has been obtained using physical means to temporarily disrupt integrity of the plasma membrane. The two most effective physical methods are cell scraping and cell trituration [17].

Two nonphysical methods of cell delivery of morpholinos into cultured cells are available from Gene Tools, LLC. One is an ethoxylated polyethylenimine solution [18]. For this purpose, morpholinos are first complexed with a complementary conventional oligonucleotide, which creates an anionic heteroduplex. Then, this heteroduplex is complexed with positively charged polyethyleneimine to form a delivery complex. As has been reported for polyethyleneimine complexes with plasmids [19], these complexes deliver well to certain types of cultured cells including PC3, HT-29, U20S, HL-60, and U937 cells and primary endothelial, vascular smooth muscle, and B cells, but not to some other cell types [20–31].

The second commercial delivery reagent consists of a cell-penetrating peptide (CPP) (Endoporter), which does not directly associate with the morpholino but effectively delivers morpholinos or other macromolecules in a nonspecific manner into cells by inducing endocytosis and subsequent release of the endocytic contents into the cytoplasm [32]. Unlike the polyethyleneimine delivery vehicle, the CPP can be used in the presence of serum [16,33]. As described in the manufacturer's patent, the active component is a 19- to 37-residue two-face amphiphilic peptide, with one face composed predominantly (80%–100%) of aliphatic lipophilic amino acids, and the other face composed predominantly (80%–100%) of basic amino acids, with at least 70% of the amino acids of the weak-base face being histidines [34]. Following endocytosis, acidification of the endocytic vesicle by the cell induces a conformation change that causes the peptide to disrupt the membrane of the vesicle and release the morpholino into the cytoplasm. The commercial delivery vehicle contains 1 mM of this peptide in dimethylsulfoxide and results in efficient delivery of morpholinos into many types of cultured mammalian cells with suppression of target gene expression [35–47].

22.2.2 *In Vivo* Delivery

22.2.2.1 Nontargeted *in Vivo* Delivery Vehicles

In vivo delivery of morpholinos remains an area of evolving technology and therapeutic development. Local/regional infiltrative injection, central administration via intravenous injection of morpholinos, and topical application have been reported [48–51]. Topical delivery to the skin has been found to be safe and demonstrated tissue localization of the oligonucleotide [48,49]. The vehicle for delivery was a mixture of water, propylene glycol, and linoleic acid. Analysis of the underlying skin demonstrated the morpholino both at the site of application, but morpholino was also detected distantly in the liver. Target protein suppression was found after topical delivery.

Morpholinos have also been delivered locally to disease sites via direct injection with therapeutic effect [52–55]. This approach was especially useful in suppressing solid tumor growth in tumor xenograft models in nude mice [55,56]. Tumors were injected multiple times over several weeks. Intracranial injection of a morpholino was found to alter neuroendocrine responses in a rat model of stress-driven hyperprolactinemia [53]. In this application, a single intracerebroventricular injection was employed. Direct injection of a morpholino intramuscularly, again as a single injection, was found to suppress abnormal muscle cell protein production at the site of injection in a murine model of muscular dystrophy [57]. In these reports, the specific morpholino was delivered in phosphate-buffered saline (PBS) or normal saline and through direct deposition of the morpholino in the region of the site of pathology, enhancing target delivery of the morpholino and most likely increasing concentration of the morpholino at the disease site. Nonetheless, local delivery of morpholinos was associated with demonstration of the morpholino in distant tissue sites. Presumably the hydrostatic pressure of injection and the physical trauma of the needle transecting the tissues facilitate the local uptake of the morpholino at the cellular level, but the mechanism of uptake at distant sites remains a mystery.

Nontargeted administration of morpholinos has also been accomplished by systemic injection either through intravenous injection [58,59] or intraperitoneal (i.p.) injection [60–62]. Studies using i.p. injection dosage schedules have varied from a single dose [60] to inhibit Ebola viral infection to 30 doses over 5 weeks in a murine model of colonic neoplasia [61] to multiple-dose schedules in combination or concurrently with chemotherapy in a Lewis lung cancer model [63]. In these *in vivo* models, the morpholino was delivered in PBS or saline vehicle. Tail vein injection of a morpholino targeting the hepatitis C virus

(HCV) internal ribosome entry site, once more suspended in PBS, was given to mice with a xenotransplant of human HCV, and viral translation was monitored *in vivo* by a luciferase gene reporter assay [59]. Alter et al. also evaluated systemic delivery via tail vein injection once per week over three consecutive weeks and demonstrated systemic alterations in target protein [58]. Though other delivery modalities have been reported in experimental models including via gastrointestinal instillation and through the lungs by inhalation with other types of oligonucleotides, application of these approaches to morpholinos has not been reported.

Significant empirical experience has been gained in using morpholinos *in vitro* to suppress protein production. However, systematic *in vivo* pharmacokinetic data for morpholinos in mammals, regardless of the method of delivery, remain quite limited [13,64]. Tissue distribution may be determined by the means of delivery of the morpholino, with serum concentration and systemic distribution faster via intravenous injection as compared to other delivery methods [65]. Regardless of the *in vivo* delivery method, morpholinos are distributed to essentially all tissues in a manner that is not sequence-dependent [4,33]. However, the ability to find morpholinos in tissues following local or systemic administration does not equate with intracellular delivery of the drug [65]. Various cell types including hepatocytes, macrophages, vascular cells, and splenocytes have been found to take up morpholinos following *in vivo* administration, though intracellular distribution remains nonuniform following uptake similar to *in vitro* results.

22.2.2.2 Peptide-Mediated and Targeted *in Vivo* Delivery Vehicles

To enhance specificity and cell uptake of morpholinos, several novel delivery methods have been described [11,52,57,65–67]. In a model of intracranial glioblastoma, morpholinos were injected in a complex with Polycefin, reportedly to target a specific tumor cell surface protein [66]. Given the number of variables in the experimental design of this report, clear conclusions could not be made. Others have recently reported encouraging results to enhance *in vivo* responses by complexing the morpholino with a CPP [67,68]. Targeting an essential bacterial gene, morpholino-CPP complexes given via i.p. injection significantly decreased bacteremia in mice [67,68]. However, the general effectiveness of CPP for delivery of morpholinos is not clear. Recent reports from the same investigators showed that the peptide, particularly if the peptide had aminohexanoic residues, was subject to rapid degradation. Also, the entire complex of CPP–PMO localized to endosomes and lysosomes, and it remains unclear the manner by which such complexes escape these intracellular compartments to gain access to the nucleus [16].

In a novel approach to achieve a higher local concentration of morpholino with less systemic distribution, morpholinos were complexed to perfluorocarbon microbubbles and given intravenously at the time of coronary artery stent placement

[69]. This morpholino complex was found to localize to the vessel site of injury and suppress the target protein with minimal systemic presence.

22.3 *IN VIVO* THERAPEUTIC APPLICATIONS

Morpholinos have demonstrated therapeutic effects in a number of animal models of disease processes. Local and diffuse systemic diseases have been treated by either local or central routes of administration, although no general guidelines or standards are yet in place. This section reviews the results reported to date for several disease models in animals.

22.3.1 MUSCULAR DYSTROPHY

A morpholino targeting the dystrophin exon 23 donor splice site was given via injection in saline into the skeletal muscle. This treatment was associated with an increase in normal protein in an *mdx* mouse model of Duchene muscular dystrophy [52]. In this report, the oligonucleotide was annealed to a sense strand leash, creating a heteroduplex to facilitate cell penetration. The agents were then dissolved in normal saline and delivered via direct intramuscular injection in single doses or via i.p. injection as a single dose or multiple doses. Regardless of the agent or delivery method, target protein alteration was demonstrated.

Alter et al. [58] injected 2 μg of a morpholino targeting the sequence M23D + 07–18, specific for removal of exon 23 of the mouse dystrophin gene, into the tibialis anterior skeletal muscle of dystrophic mice. They found an increase in dystrophin-positive muscle fibers. They also administered 2 mg of morpholino via tail vein injection and found mRNA changes in muscle after 2 weeks. Given the temporary effects of morpholinos on cell culture protein levels, it is not clear how changes were sustained after 2 weeks, although the slower proliferation of cells *in vivo* may extend the duration of suppression.

22.3.2 ANTIVIRAL ACTIVITY

Burrer et al. [11] reported the minimization of hepatitis infection in mice treated with a morpholino targeting the 5′ terminus of coronavirus mRNA or murine hepatitis virus. Several formulations of these agents were employed including morpholino alone and morpholino covalently complexed by a thioether or an amide linkage to an arginine-rich peptide sequence. Both peptide-bound and unbound phosphorodiamidate morpholinos reduced viral titers in animals and reduced viral tissue damage. However, in a group inoculated with a high viral dose, several animals demonstrated toxicity that was absent in control uninfected animals who received the same treatment.

Peptide-conjugated morpholinos targeting the 5′ terminus of the viral genome or an essential 3′ RNA element required for genome cyclization displayed a broad spectrum of antiflavivirus activity, suppressing West Nile virus, Japanese encephalitis virus, and St. Louis encephalitis virus by 3–5 log units in cell culture [70]. Both of the conjugated morpholinos at 100 and 200 μg/day partially protected mice from West Nile virus disease with minimal to no toxicity. However, a higher treatment dose (300 μg/day) caused toxicity. In contrast to some other *in vivo* applications, unconjugated morpholinos at 3 mg/day showed neither efficacy nor toxicity, suggesting the importance of the peptide-conjugate for the observed antiviral activity.

Similar activity in mice was observed using a CPP-conjugated morpholino targeting a sequence in the 3′ portion of the Coxsackievirus B3 internal ribosomal entry site [65]. Coxsackievirus B3 is a primary cause of viral myocarditis. A/J mice received intravenous administration of the CPP-morpholino once prior to and once after Coxsackievirus B3 infection and showed an approximately 100-fold decreased viral titer in the myocardium at 7 days postinfection and a significantly less cardiac tissue damage, compared to the controls.

The same CPP conjugation strategy proved effective for protecting mice from Ebola virus infection [60]. A 22-mer CPP-morpholino complementary to the translation start site region of Ebola virus VP35 positive-sense RNA inhibited viral amplification in cell culture and provided complete protection to mice when administered before or after an otherwise lethal infection of Ebola virus. In this prophylaxis model, three nonconjugated morpholinos targeting the same VP35 sequence also protected the mice from Ebola virus challenge.

22.3.3 Cancer

Recently, attempts to combine morpholino delivery with increased uptake and specificity following intravenous administration have been applied to several cancer models. Ljubimova et al. [71] described a new polymer-derived drug delivery system. The scaffold incorporates antibodies specific for the target tumor tissue, the morpholino, and a fluorescent marker to track localization. In two murine xenograft models of cancer, the agent localized to the tumors in greater concentrations than other tissues. However, no therapeutic results were reported. Liu et al. [64,72] also reported relative specificity of drug delivery to tumors. However, they used the morpholino as a targeting agent (not as a primary therapeutic) to concentrate radioactivity in tumor tissue. A morpholino conjugated to the anti-CEA IgG antibody MN14 was injected via tail vein into mice bearing a colon cancer xenograft. This was then followed by a radiolabeled complimentary morpholino. Tumor growth rates were markedly inhibited compared to controls. The MN14–morpholino complex localized to the tumor cells and allowed for site specific delivery of the radiotherapy via the complimentary morpholino. Here the tissue specificity appears to depend on the localization of the bound antitumor antibody. This study supports the specificity of morpholino-antibody targeting but does not add new insight into the ability of morpholinos to enter tissues or cells *in vivo* to control gene expression.

22.3.4 Cardiovascular Disease

Morpholino oligonucleotides have also been employed in cardiovascular disease models in higher mammals including pigs [69,73] and rabbits [74] with some reported therapeutic success. In a porcine model of the coronary artery disease, several studies have employed a c-myc antisense morpholino in an effort to reduce intimal hyperplasia secondary to vascular smooth muscle cells (VSMC) proliferation [12,69,75]. The agents were locally delivered through catheters and/or implanted stents. Both delivery techniques resulted in decreased coronary lesion size in response to injury.

22.3.5 Ischemic Injury

Recently, we have developed a morpholino targeting a conserved sequence in human and murine CD47 to treat ischemic injuries. CD47, also known as integrin-associated protein, is a cell surface receptor for the secreted protein thrombospondin-1 [76]. Thrombospondin-1 is a potent inhibitor of nitric oxide/cGMP signaling in vascular cells [77,78]. The nitric oxide signaling pathway acutely controls blood flow into tissues by promoting vasodilation, and in the longer term promotes angiogenic remodeling of the vasculature (Figure 22.3). Transgenic mice lacking either thrombospondin-1 or CD47 show enhanced angiogenic responses as well as superior acute maintenance of blood flow and tissue oxygenation in ischemic tissues and improved long-term survival of ischemic injuries in cutaneous skin flap and hindlimb models [35,79]. These studies established that thrombospondin-1/CD47 signaling plays a critical role in acutely limiting blood flow and the long-term inhibition of angiogenic remodeling in ischemic tissues. Therefore, we sought therapeutic approaches to selectively block this pathway to treat ischemic injuries.

One effective approach to treat fixed tissue ischemia in several murine models, including random dorsal myocutaneous flaps and hind limb vascular occlusion, was to temporarily suppress CD47 expression using morpholinos [35]. The morpholino used here was first shown to be effective to suppress CD47 expression in cultured VSMC. Immunohistochemistry of tissue sections from treated mice also demonstrated a substantial reduction in CD47 protein expression *in vivo* following morpholino treatment. Local morpholino injection into the flaps and wound bed markedly improved survival of ischemic tissues in McFarland dorsal skin flaps (Figure 22.4A). Remarkably, the morpholino was equally effective at improving survival in ischemic hindlimb models (Figure 22.4B). The loss of target expression and clinical responses to these treatments following local injection indicate that delivery of morpholinos to the dermis in the flap model and to skeletal muscle in the hind limb model is relatively efficient.

We recently applied the same CD47 morpholino treatment to skin grafts. In a murine model of full thickness skin graft healing, direct injection of a CD47-targeting morpholino into both the graft and underlying wound bed resulted in a dramatic increase

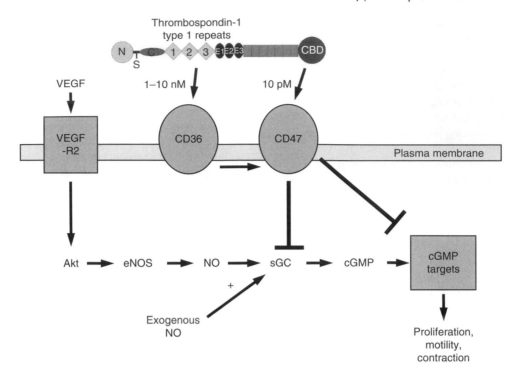

FIGURE 22.3 Thrombospondin-1 antagonism of nitric oxide signaling to regulate tissue perfusion and angiogenesis. Thrombospondin-1 inhibits nitric oxide signaling at the level of soluble guanylate cyclase (sGC) and downstream of cGMP by binding to its cell surface receptors CD36 and CD47. Of these, CD47 is the essential thrombospondin-1 receptor, and its genetic deletion, blocking with antibodies, or temporary suppression using a morpholino relieves the NO/cGMP signaling cascade from inhibition by thrombospondin-1. In endothelial cells, this pathway regulates proangiogenic signaling downstream of vascular endothelial growth factor (VEGF), which induces NO synthesis by increasing phosphorylation of endothelial nitric oxide synthase (eNOS) by the kinase Akt. In VSMC, exogenous NO produced by adjacent endothelial cells activates sGC leading to cGMP-mediated dephosphorylation of myosin light chain, which causes relaxation of the smooth muscle cells and increased tissue perfusion.

in skin graft survival [80]. Saline and control sequence treated skin grafts demonstrated approximately 100% tissue necrosis. In contrast, grafts treated with the CD47 morpholino showed near complete survival (Figure 22.5).

Interestingly, such a therapeutic response in the mouse does not require use of the proprietary delivery peptide that enhances cellular uptake via phagocytosis. We compared responses for mice injected with the CD47 morpholino in saline versus the morpholino-delivery peptide mixture in saline. We saw equivalent responses in terms of tissue survival of full thickness skin grafts and random myocutaneous flaps (Figure 22.6 and [80]). This implies that morpholinos can efficiently gain entry into the relevant target cells in a surgical wound independent of any cell delivery agent. We propose that surgical wounds may mimic the conditions of physical stress that are known to stimulate morpholino uptake in cell cultures, either due to the local physical stress on the tissue or in response to inflammatory cytokines that are released during the physiological wound response and may stimulate phagocytic uptake of morpholinos by the target cells. Thus, cells in a surgical field may be intrinsically more efficient for uptake of morpholinos. However, further research is needed to quantify the relative efficiencies of morpholino uptake into tissues in response to surgical stresses and other stressors

such as inflammation and to define the signaling pathways that regulate this uptake.

Murine models of tissue injury are economical and readily reproducible based on the genetic background homogeneity of laboratory mouse strains. However, mice differ substantially in their soft tissue vascular anatomy compared to other higher mammals including humans. The pig offers a model of soft tissue vascular anatomy that more closely parallels that found in people. Using the Yucatan white hairless pig, we are currently evaluating the ability of CD47 targeting morpholinos to enhance ischemic tissue survival in a random tissue flap model. As in murine studies of comparable ischemic tissue injury, local administration of the CD47 morpholino results in enhanced tissue survival [81].

In both the murine and porcine models of fixed ischemic tissue injury, the CD47 morpholino has been delivered via local subcutaneous and intradermal injection. Under these conditions, we have found that tissue survival was increased by the target morpholino but not by a control morpholino. These data support the possibility that the same approach can be effective to treat ischemic injuries in humans [82]. The ease of delivery via local injection at the site of soft tissue injury and vascular compromise will allow for ready therapeutic applications during surgical procedures.

Random myocutaneous flaps

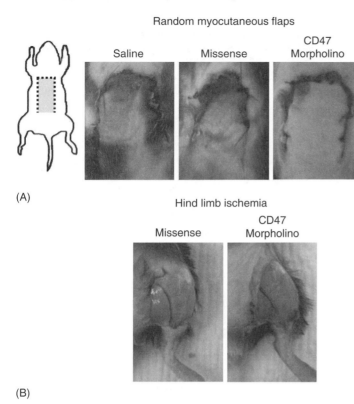

(A)

Hind limb ischemia

(B)

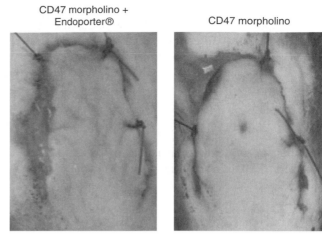

FIGURE 22.6 (See color insert following blank page 206.) CPP does not enhance the efficacy of a CD47 morpholino for ischemic injury. Wild-type C57BL/6 age- and sex-matched mice underwent 1 × 2 cm random myocutaneous flaps. Flaps were treated at the time of surgery with a targeted CD47 morpholino plus the manufacture's recommended cell uptake agent (Endoporter®) in saline or a CD47 morpholino in saline alone and tissue survival determined.

FIGURE 22.4 (See color insert following blank page 206.) Improved survival of fixed ischemic injury following CD47 morpholino treatment of dorsal skin flaps and ischemic hind limbs in C57BL/6 mice. Wild-type C57BL/6 age- and sex-matched mice underwent 1 × 2 cm random myocutaneous flaps. Flaps were treated at the time of surgery with saline, a mismatched morpholino control or a target CD47 morpholino, and tissue survival determined (A). Wild-type C57BL/6 age- and sex-matched mice underwent ligation of the femoral artery at the inguinal ligament. Limbs were treated at the time of surgery by direct intramuscular injection of vehicle (saline), a mismatched control morpholino sequence, or a CD47 morpholino.

22.4 CLINICALLY ORIENTED APPLICATIONS

Several studies in humans involving morpholino therapy are currently in phase I or II clinical trials [54,83,84]. An oligonucleotide morpholino targeting c-myc (AVI-4126), which was effective in decreasing tumor burden in a murine xenograft model of bladder cancer, has been administered to human volunteers in a phase I trial as a single intravenous bolus with no reported adverse events [83]. Plasma concentrations were reported to approach baseline within several hours of administration. This same agent is now in phase II trials [84]. Other agents are in development targeting coronary vascular stenosis, polycystic kidney disease, and drug metabolism via cytochrome P450 enzymes.

22.5 SUMMARY

Morpholino oligonucleotides represent an evolving and promising means of manipulating gene expression that could overcome several persistent limitations of alternative oligonucleotide-based technologies. Extensive use as a research tool in developmental biology has proven their efficacy for temporarily suppressing gene expression with minimal off-target effects. Their stability and lack of charge are clearly advantages to therapeutic applications. Morpholinos have demonstrated therapeutic activities in a number of animal species and disease models. However, many areas deserve continued research. In particular, efforts should be made to more fully describe the molecular mechanism by which the administered drug reaches the target mRNA in order to optimize drug delivery. Further development is also needed to optimize delivery protocols in both injured and uninjured tissues and to develop

Full thickness skin grafts

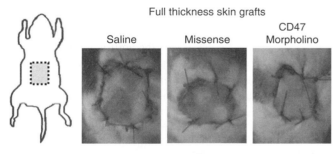

FIGURE 22.5 (See color insert following blank page 206.) Improved survival of full thickness skin grafts following CD47 morpholino treatment in C57BL/6 mice. Wild-type C57BL/6 age- and sex-matched mice underwent 1 × 1 cm full thickness skin grafts. At the time of surgery, grafts and wound beds were treated by direct injection with saline (vehicle), a missense control morpholino, or a CD47 morpholino (10 μM), recognizing the murine sequence of the protein. Tissue survival was assessed on postoperative day 7.

effective targeting strategies to permit selective suppression of gene expression in tumors and other diseased tissues that are not accessible to direct injection. Further studies are also needed to clarify the pharmacokinetics and pharmacodynamics of morpholinos, both for systemic delivery and in target tissues following local delivery to optimize their activities.

ACKNOWLEDGMENTS

The preparation of this manuscript was supported by the Intramural Research Program of the National Institutes of Health, National Cancer Institute, Center for Cancer Research (D.D.R.), and NIH grants HL54390 and GM57573 (W.A.F.).

REFERENCES

1. Zamecnik, P.C. and Stephenson, M.L., Inhibition of Rous sarcoma virus replication and cell transformation by a specific oligodeoxynucleotide, *Proc Natl Acad Sci USA*, 75, 280–284, 1978.
2. Stephenson, M.L. and Zamecnik, P.C., Inhibition of Rous sarcoma viral RNA translation by a specific oligodeoxyribonucleotide, *Proc Natl Acad Sci USA*, 75, 285–288, 1978.
3. Karkare, S. and Bhatnagar, D., Promising nucleic acid analogs and mimics: Characteristic features and applications of PNA, LNA, and morpholino, *Appl Microbiol Biotechnol*, 71, 575–586, 2006.
4. Amantana, A. and Iversen, P.L., Pharmacokinetics and biodistribution of phosphorodiamidate morpholino antisense oligomers, *Curr Opin Pharmacol*, 5, 550–555, 2005.
5. Pickart, M.A. et al., Functional genomics tools for the analysis of zebrafish pigment, *Pigment Cell Res*, 17, 461–470, 2004.
6. Iversen, P.L. and Newbry, S., Manipulation of zebrafish embryogenesis by phosphorodiamidate morpholino oligomers indicates minimal non-specific teratogenesis, *Curr Opin Mol Ther*, 7, 104–108, 2005.
7. Berman, J.N., Kanki, J.P., and Look, A.T., Zebrafish as a model for myelopoiesis during embryogenesis, *Exp Hematol*, 33, 997–1006, 2005.
8. Iversen, P.L., Phosphorodiamidate morpholino oligomers: Favorable properties for sequence-specific gene inactivation, *Curr Opin Mol Ther*, 3, 235–238, 2001.
9. Arora, V., Devi, G.R., and Iversen, P.L., Neutrally charged phosphorodiamidate morpholino antisense oligomers: Uptake, efficacy and pharmacokinetics, *Curr Pharm Biotechnol*, 5, 431–439, 2004.
10. Morcos, P.A., Achieving targeted and quantifiable alteration of mRNA splicing with morpholino oligos, *Biochem Biophys Res Commun*, 358, 521–527, 2007.
11. Burrer, R. et al., Antiviral effects of antisense morpholino oligomers in murine coronavirus infection models, *J Virol*, 81, 5637–5648, 2007.
12. Kipshidze, N.N. et al., Intramural coronary delivery of advanced antisense oligonucleotides reduces neointimal formation in the porcine stent restenosis model, *J Am Coll Cardiol*, 39, 1686–1691, 2002.
13. Summerton, J.E., Morpholino, siRNA, and S-DNA compared: Impact of structure and mechanism of action on off-target effects and sequence specificity, *Curr Top Med Chem*, 7, 651–660, 2007.
14. Qin, G., Taylor, M., Ning, Y.Y., Iversen, P., and Kobzik, L., In vivo evaluation of a morpholino antisense oligomer directed against tumor necrosis factor-alpha, *Antisense Nucleic Acid Drug Dev*, 10, 11–16, 2000.
15. Ghosh, C., Stein, D., Weller, D., and Iversen, P., Evaluation of antisense mechanisms of action, *Methods Enzymol*, 313, 135–143, 2000.
16. Youngblood, D.S., Hatlevig, S.A., Hassinger, J.N., Iversen, P.L., and Moulton, H.M., Stability of cell-penetrating peptide-morpholino oligomer conjugates in human serum and in cells, *Bioconjug Chem*, 18, 50–60, 2007.
17. Ghosh, C. and Iversen, P.L., Intracellular delivery strategies for antisense phosphorodiamidate morpholino oligomers, *Antisense Nucleic Acid Drug Dev*, 10, 263–274, 2000.
18. Siddall, L.S., Barcroft, L.C., and Watson, A.J., Targeting gene expression in the preimplantation mouse embryo using morpholino antisense oligonucleotides, *Mol Reprod Dev*, 63, 413–421, 2002.
19. Lungwitz, U., Breunig, M., Blunk, T., and Gopferich, A., Polyethylenimine-based non-viral gene delivery systems, *Eur J Pharm Biopharm*, 60, 247–266, 2005.
20. Zhou, L., Isenberg, J.S., Cao, Z., and Roberts, D.D., Type I collagen is a molecular target for inhibition of angiogenesis by endogenous thrombospondin-1, *Oncogene*, 25, 536–545, 2006.
21. Arora, V., Cate, M.L., Ghosh, C., and Iversen, P.L., Phosphorodiamidate morpholino antisense oligomers inhibit expression of human cytochrome P450 3A4 and alter selected drug metabolism, *Drug Metab Dispos*, 30, 757–762, 2002.
22. Braun, S. et al., Nrf2 transcription factor, a novel target of keratinocyte growth factor action which regulates gene expression and inflammation in the healing skin wound, *Mol Cell Biol*, 22, 5492–5505, 2002.
23. Munshi, C.B., Graeff, R., and Lee, H.C., Evidence for a causal role of CD38 expression in granulocytic differentiation of human HL-60 cells, *J Biol Chem*, 277, 49453–49458, 2002.
24. Yan, Y., Shirakabe, K., and Werb, Z., The metalloprotease Kuzbanian (ADAM10) mediates the transactivation of EGF receptor by G protein-coupled receptors, *J Cell Biol*, 158, 221–226, 2002.
25. Nisoli, E. et al., Mitochondrial biogenesis in mammals: The role of endogenous nitric oxide, *Science*, 299, 896–899, 2003.
26. Garl, P.J. et al., Perlecan-induced suppression of smooth muscle cell proliferation is mediated through increased activity of the tumor suppressor PTEN, *Circ Res*, 94, 175–183, 2004.
27. Mukhopadhyay, D., Houchen, C.W., Kennedy, S., Dieckgraefe, B.K., and Anant, S., Coupled mRNA stabilization and translational silencing of cyclooxygenase-2 by a novel RNA binding protein, CUGBP2, *Mol Cell*, 11, 113–126, 2003.
28. Li, F. et al., Plasmodium ookinete-secreted proteins secreted through a common micronemal pathway are targets of blocking malaria transmission, *J Biol Chem*, 279, 26635–26644, 2004.
29. Young, P.J. et al., The Ewing's sarcoma protein interacts with the Tudor domain of the survival motor neuron protein, *Brain Res Mol Brain Res*, 119, 37–49, 2003.
30. Acosta-Rodriguez, E.V. et al., Galectin-3 mediates IL-4-induced survival and differentiation of B cells: Functional cross-talk and implications during *Trypanosoma cruzi* infection, *J Immunol*, 172, 493–502, 2004.
31. Pak, J.H., Manevich, Y., Kim, H.S., Feinstein, S.I., and Fisher, A.B., An antisense oligonucleotide to 1-cys peroxiredoxin causes lipid peroxidation and apoptosis in lung epithelial cells, *J Biol Chem*, 277, 49927–49934, 2002.

32. Deshayes, S., Morris, M.C., Divita, G., and Heitz, F., Cell-penetrating peptides: Tools for intracellular delivery of therapeutics, *Cell Mol Life Sci*, 62, 1839–1849, 2005.

33. Amantana, A. et al., Pharmacokinetics, biodistribution, stability and toxicity of a cell-penetrating peptide-morpholino oligomer conjugate, *Bioconjug Chem*, 18, 1325–1331, 2007.

34. Summerton, J., Peptide composition and method for delivering substances into the cytosol of cells, U.S. Patent 7,084,248, Issued August 1, 2006.

35. Isenberg, J.S. et al., Thrombospondin-1 limits ischemic tissue survival by inhibiting nitric oxide-mediated vascular smooth muscle relaxation, *Blood*, 109, 1945–1952, 2007.

36. Sands, W.A., Woolson, H.D., Milne, G.R., Rutherford, C., and Palmer, T.M., Exchange protein activated by cyclic AMP (Epac)-mediated induction of suppressor of cytokine signaling 3 (SOCS-3) in vascular endothelial cells, *Mol Cell Biol*, 26, 6333–6346, 2006.

37. Mandayam, S., Huang, R., Tarnawski, A.S., and Chiou, S.K., Roles of survivin isoforms in the chemopreventive actions of NSAIDS on colon cancer cells, *Apoptosis*, 12, 1109–1116, 2007.

38. Kikuno, N. et al., Knockdown of astrocyte-elevated gene-1 inhibits prostate cancer progression through upregulation of FOXO3a activity, *Oncogene*, 26, 7647–7655, 2007.

39. Ammons, M.C., Siemsen, D.W., Nelson-Overton, L.K., Quinn, M.T., and Gauss, K.A., Binding of pleomorphic adenoma gene-like 2 to the tumor necrosis factor (TNF)-alpha-responsive region of the NCF2 promoter regulates p67(phox) expression and NADPH oxidase activity, *J Biol Chem*, 282, 17941–17952, 2007.

40. Yeh, T.C. et al., Genistein induces apoptosis in human hepatocellular carcinomas via interaction of endoplasmic reticulum stress and mitochondrial insult, *Biochem Pharmacol*, 73, 782–792, 2007.

41. Du, L., Pollard, J.M., and Gatti, R.A., Correction of prototypic ATM splicing mutations and aberrant ATM function with antisense morpholino oligonucleotides, *Proc Natl Acad Sci USA*, 104, 6007–6012, 2007.

42. Sample, R. et al., Inhibition of iridovirus protein synthesis and virus replication by antisense morpholino oligonucleotides targeted to the major capsid protein, the 18 kDa immediate-early protein, and a viral homolog of RNA polymerase II, *Virology*, 358, 311–320, 2007.

43. Madhavan, M. et al., The role of Pax-6 in lens regeneration, *Proc Natl Acad Sci USA*, 103, 14848–14853, 2006.

44. Chih, B., Gollan, L., and Scheiffele, P., Alternative splicing controls selective trans-synaptic interactions of the neuroligin–neurexin complex, *Neuron*, 51, 171–178, 2006.

45. Boyden, E.D. and Dietrich, W.F., Nalp1b controls mouse macrophage susceptibility to anthrax lethal toxin, *Nat Genet*, 38, 240–244, 2006.

46. Tyson-Capper, A.J. and Europe-Finner, G.N., Novel targeting of cyclooxygenase-2 (COX-2) pre-mRNA using antisense morpholino oligonucleotides directed to the 3′ acceptor and 5′ donor splice sites of exon 4: Suppression of COX-2 activity in human amnion-derived WISH and myometrial cells, *Mol Pharmacol*, 69, 796–804, 2006.

47. Masaki, M. et al., Smad1 protects cardiomyocytes from ischemia-reperfusion injury, *Circulation*, 111, 2752–2759, 2005.

48. Pannier, A.K., Arora, V., Iversen, P.L., and Brand, R.M., Transdermal delivery of phosphorodiamidate morpholino oligomers across hairless mouse skin, *Int J Pharm*, 275, 217–226, 2004.

49. Arora, V., Hannah, T.L., Iversen, P.L., and Brand, R.M., Transdermal use of phosphorodiamidate morpholino oligomer AVI-4472 inhibits cytochrome P450 3A2 activity in male rats, *Pharm Res*, 19, 1465–1470, 2002.

50. Sodhi, D. et al., Morpholino oligonucleotide-triggered beta-catenin knockdown compromises normal liver regeneration, *J Hepatol*, 43, 132–141, 2005.

51. Sazani, P. et al., Systemically delivered antisense oligomers upregulate gene expression in mouse tissues, *Nat Biotechnol*, 20, 1228–1233, 2002.

52. Fletcher, S. et al., Dystrophin expression in the mdx mouse after localised and systemic administration of a morpholino antisense oligonucleotide, *J Gene Med*, 8, 207–216, 2006.

53. Fujikawa, T. et al., Prolactin receptor knockdown in the rat paraventricular nucleus by a morpholino-antisense oligonucleotide causes hypocalcemia and stress gastric erosion, *Endocrinology*, 146, 3471–3480, 2005.

54. Devi, G.R. et al., In vivo bioavailability and pharmacokinetics of a c-MYC antisense phosphorodiamidate morpholino oligomer, AVI-4126, in solid tumors, *Clin Cancer Res*, 11, 3930–3938, 2005.

55. Takei, Y., Kadomatsu, K., Yuasa, K., Sato, W., and Muramatsu, T., Morpholino antisense oligomer targeting human midkine: Its application for cancer therapy, *Int J Cancer*, 114, 490–497, 2005.

56. London, C.A. et al., A novel antisense inhibitor of MMP-9 attenuates angiogenesis, human prostate cancer cell invasion and tumorigenicity, *Cancer Gene Ther*, 10, 823–832, 2003.

57. Gebski, B.L., Mann, C.J., Fletcher, S., and Wilton, S.D., Morpholino antisense oligonucleotide induced dystrophin exon 23 skipping in mdx mouse muscle, *Hum Mol Genet*, 12, 1801–1811, 2003.

58. Alter, J. et al., Systemic delivery of morpholino oligonucleotide restores dystrophin expression bodywide and improves dystrophic pathology, *Nat Med*, 12, 175–177, 2006.

59. McCaffrey, A.P., Meuse, L., Karimi, M., Contag, C.H., and Kay, M.A., A potent and specific morpholino antisense inhibitor of hepatitis C translation in mice, *Hepatology*, 38, 503–508, 2003.

60. Enterlein, S. et al., VP35 knockdown inhibits Ebola virus amplification and protects against lethal infection in mice, *Antimicrob Agents Chemother*, 50, 984–993, 2006.

61. Roy, H.K. et al., Down-regulation of SNAIL suppresses MIN mouse tumorigenesis: Modulation of apoptosis, proliferation, and fractal dimension, *Mol Cancer Ther*, 3, 1159–1165, 2004.

62. Ko, Y.J. et al., Androgen receptor down-regulation in prostate cancer with phosphorodiamidate morpholino antisense oligomers, *J Urol*, 172, 1140–1144, 2004.

63. Knapp, D.C., Mata, J.E., Reddy, M.T., Devi, G.R., and Iversen, P.L., Resistance to chemotherapeutic drugs overcome by c-myc inhibition in a Lewis lung carcinoma murine model, *Anticancer Drugs*, 14, 39–47, 2003.

64. Liu, G. et al., Successful radiotherapy of tumor in pretargeted mice by 188Re-radiolabeled phosphorodiamidate morpholino oligomer, a synthetic DNA analogue, *Clin Cancer Res*, 12, 4958–4964, 2006.

65. Yuan, J. et al., Inhibition of coxsackievirus B3 in cell cultures and in mice by peptide-conjugated morpholino oligomers targeting the internal ribosome entry site, *J Virol*, 80, 11510–11519, 2006.

66. Fujita, M. et al., Inhibition of laminin-8 in vivo using a novel poly(malic acid)-based carrier reduces glioma angiogenesis, *Angiogenesis*, 9, 183–191, 2006.

67. Tilley, L.D., Mellbye, B.L., Puckett, S.E., Iversen, P.L., and Geller, B.L., Antisense peptide-phosphorodiamidate morpholino oligomer conjugate: Dose–response in mice infected with *Escherichia coli*, *J Antimicrob Chemother*, 59, 66–73, 2007.

68. Geller, B.L., Deere, J., Tilley, L., and Iversen, P.L., Antisense phosphorodiamidate morpholino oligomer inhibits viability of *Escherichia coli* in pure culture and in mouse peritonitis, *J Antimicrob Chemother*, 55, 983–988, 2005.

69. Kipshidze, N.N. et al., Systemic targeted delivery of antisense with perflourobutane gas microbubble carrier reduced neo-intimal formation in the porcine coronary restenosis model, *Cardiovasc Radiat Med*, 4, 152–159, 2003.

70. Deas, T.S. et al., In vitro resistance selection and in vivo efficacy of morpholino oligomers against West Nile virus, *Antimicrob Agents Chemother*, 51, 2470–2482, 2007.

71. Ljubimova, J.Y. et al., Nanoconjugate based on polymalic acid for tumor targeting, *Chem Biol Interact*, 171, 195–203, 2007.

72. Liu, G. et al., Radiolabeling of MAG3-morpholino oligomers with 188Re at high labeling efficiency and specific radioactivity for tumor pretargeting, *Appl Radiat Isot*, 64, 971–978, 2006.

73. McClorey, G., Moulton, H.M., Iversen, P.L., Fletcher, S., and Wilton, S.D., Antisense oligonucleotide-induced exon skipping restores dystrophin expression in vitro in a canine model of DMD, *Gene Ther*, 13, 1373–1381, 2006.

74. Kipshidze, N. et al., Local delivery of c-myc neutrally charged antisense oligonucleotides with transport catheter inhibits myointimal hyperplasia and positively affects vascular remodeling in the rabbit balloon injury model, *Catheter Cardiovasc Interv*, 54, 247–256, 2001.

75. Iversen, P.L., Kipshidze, N., Moses, J.W., and Leon, M.B., Local application of antisense for prevention of restenosis, *Methods Mol Med*, 106, 37–50, 2005.

76. Brown, E.J. and Frazier, W.A., Integrin-associated protein (CD47) and its ligands, *Trends Cell Biol*, 11, 130–135, 2001.

77. Isenberg, J.S. et al., Thrombospondin-1 inhibits endothelial cell responses to nitric oxide in a cGMP-dependent manner, *Proc Natl Acad Sci USA*, 102, 13141–13146, 2005.

78. Isenberg, J.S., Wink, D.A., and Roberts, D.D., Thrombospondin-1 antagonizes nitric oxide-stimulated vascular smooth muscle cell responses, *Cardiovasc Res*, 71, 785–793, 2006.

79. Isenberg, J.S. et al., Increasing survival of ischemic tissue by targeting CD47, *Circ Res*, 100, 712–720, 2007.

80. Isenberg, J.S. et al., Blockade of thrombospondin-1-CD47 interactions prevents necrosis of full thickness skin grafts, *Ann Surgery*, 247, 180–190, 2008.

81. Isenberg, J.S. et al., Gene silencing of CD47 and antibody ligation of thrombospondin-1 enhance ischemic tissue survival in a porcine model, *Ann Surgery*, in press.

82. Kaczorowski, D.J. and Billiar, T.R., Targeting CD47: NO limit on therapeutic potential, *Circ Res*, 100, 602–603, 2007.

83. Iversen, P.L., Arora, V., Acker, A.J., Mason, D.H., and Devi, G.R., Efficacy of antisense morpholino oligomer targeted to c-myc in prostate cancer xenograft murine model and a phase I safety study in humans, *Clin Cancer Res*, 9, 2510–2519, 2003.

84. Stephens, A.C., Technology evaluation: AVI-4126, AVI BioPharma, *Curr Opin Mol Ther*, 6, 551–558, 2004.

23 Antisense Oligonucleotide-Based Therapeutics

C. Frank Bennett, Eric Swayze, Richard Geary, Scott Henry, Lloyd Tillman, and Greg Hardee

CONTENTS

23.1 INTRODUCTION

Antisense oligonucleotides are short synthetic oligonucleotides, usually between 15 and 25 bases in length designed to hybridize to RNA through Watson–Crick base pairing (Figure 23.1). Upon binding to the target RNA, the oligonucleotide prevents expression of the encoded protein product in a sequence-specific manner. As the rules for Watson–Crick base pairing are well characterized [1], antisense oligonucleotides represent, in principle, one of the few examples of rational drug design. In practice, exploitation of antisense oligonucleotides for therapies has presented a unique set of challenges, some anticipated and others unanticipated. Nevertheless, antisense oligonucleotides are showing promise as therapeutic agents broadly applicable for the treatment of human diseases. Currently there is one approved antisense product on the market and approximately 30 agents in clinical trials, several of which are in advanced stages of development (Table 23.1). In this review, we will summarize the properties of antisense oligonucleotides in terms of their application as therapeutic agents. Because of the wealth of information available on first-generation antisense oligonucleotides, i.e., phosphorothioate oligodeoxynucleotides, they will serve as a benchmark for comparison with some of the newer modified

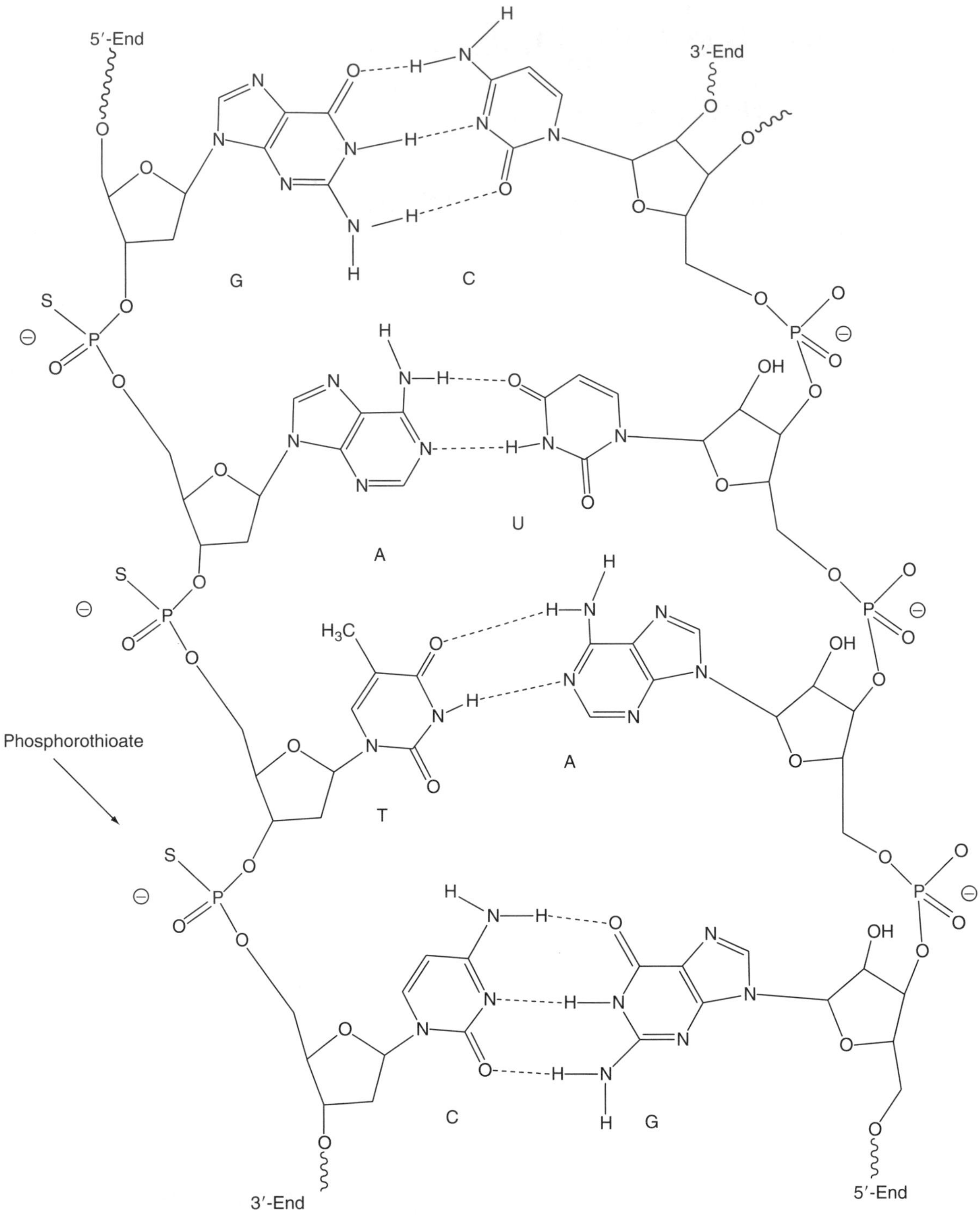

FIGURE 23.1 Phosphorothioate antisense oligodeoxynucleotide targeting an RNA receptor. Watson–Crick base pairing rules are indicated: nucleobase adenosine hydrogen bonds to nucleobase uracil and nucleobase cytosine hydrogen bonds to nucleobase guanine.

oligonucleotides. However, as discussed in this chapter, second-generation antisense chemistries exhibit a number of improved properties compared to first-generation oligonucleotides and should be viewed as the current standard.

23.2 ANTISENSE MECHANISM OF ACTION

Antisense oligonucleotides are small synthetic oligonucleotides designed to bind to mRNA through Watson–Crick

hybridization. Although most antisense oligonucleotides currently target protein coding mRNAs or pre-mRNAs, they can also be used to inhibit the function of noncoding RNAs [2–6]. Upon binding to the RNA, the oligonucleotide may inhibit function of the RNA through either inducing cleavage of the RNA by endogenous RNases or by occupancy of critical regulatory sites on the RNA, disrupting its function (Figure 23.2). Although intuitively one would anticipate that an antisense oligonucleotide that worked through a catalytic mechanism

TABLE 23.1
Antisense Oligonucleotides Approved or Currently in Clinical Development

Oligonucleotide	Molecular Target	Disease Indication	Chemistry	Route of Administration	Status	Sponsor
Vitravene (fomivirsen, ISIS 2922)	Human cytomegalovirus IE-2 gene	CMV retinitis	Phosphorothioate oligodeoxynucleotide	Intravitreal	Marketed	Novartis Ophthalmic/ISIS Pharmaceuticals
Oblimersen (Genasense, G3139)	BCL-2	Cancer	Phosphorothioate oligodeoxynucleotide	Intravenous	Phase III	Genta
Alicaforsen (ISIS 2302)	ICAM-1	Ulcerative colitis	Phosphorothioate oligodeoxynucleotide	Enema	Phase II	Atlantic Healthcare/ISIS Pharmaceuticals
GTI-2040	Ribonucleotide reductase R1 subunit	Cancer	Phosphorothioate oligodeoxynucleotide	Intravenous	Phase II	Lorus Therapeutics
GTI-2501	Ribonucleotide reductase R2 subunit	Cancer	Phosphorothioate oligodeoxynucleotide	Intravenous	Phase II	Lorus Therapeutics
AP12009	TGF-B2	Malignant glioma	Phosphorothioate oligodeoxynucleotide	Intravenous	Phase II	Antisense Pharma
TPI ASM 8	IL-3, IL-5, GM-CSF, rantes, eotaxin, MCP-3, MCP-4 receptors (2 drugs)	Asthma	Phosphorothioate oligodeoxynucleotide	Aerosol	Phase II	Topigen
OGX-011 (ISIS 112989)	Clusterin	Cancer	Phosphorothioate 2'-O-methoxyethyl/oligodeoxynucleotide chimera	Intravenous/subcutaneous	Phase II	Oncogenix/ISIS Pharmaceuticals
ISIS 113715	PTP-1B	Diabetes	Phosphorothioate 2'-O-methoxyethyl/ oligodeoxynucleotide chimera	Subcutaneous	Phase II	ISIS Pharmaceuticals
ATL1102 (ISIS 107248)	CD49D (alpha subunit of VLA4)	Multiple sclerosis	Phosphorothioate 2'-O-methoxyethyl/oligodeoxynucleotide chimera	Subcutaneous	Phase II	Antisense Therapeutics Ltd/Isis Pharmaceuticals
ISIS 301012	Apo B	Hyperlipidemia	Phosphorothioate 2'-O-methoxyethyl/oligodeoxynucleotide chimera	Subcutaneous	Phase II	Isis Pharmaceuticals
LY2181308	Survivin	Cancer	Phosphorothioate 2'-O-methoxyethyl/oligodeoxynucleotide chimera	Intravenous	Phase II	Eli Lilly/ Isis Pharmaceuticals
AEG35156	XIAP	Cancer	Phosphorothioate 2'-O-methyl/oligodeoxynucleotide chimera	Intravenous	Phase II	Aegera
Resten-NG	c-myc	Restenosis	Morpholino	Catheter delivery-intra-arterial	Phase II	AVI BioPharma
Bevasiranib	VEGF	Macular degeneration	Unmodified siRNA	Intravitreal	Phase II	Acuity Pharmaceuticals
AGN211745	VEGF receptor	Macular degeneration	Modified siRNA	Intravitreal	Phase II	Allergan

(continued)

TABLE 23.1 (continued)
Antisense Oligonucleotides Approved or Currently in Clinical Development

Oligonucleotide	Molecular Target	Disease Indication	Chemistry	Route of Administration	Status	Sponsor
GRN163L	Telomerase	Cancer	$N3' \rightarrow P5'$ thio-phosphoramidate	Intravenous	Phase II	Geron
RTP801i-14	RTP801	Macular degeneration	siRNA	Intravitreal	Phase II	Pfizer/ Quark Pharmaceuticals
Oncomyc-NG	c-myc	Cancer	Morpholino	Unknown	Phase I/II	AVI BioPharma
LY2275796	eIF-4E	Cancer	Phosphorothioate 2′-O-methoxyethyl/oligodeoxy-nucleotide chimera	Intravenous	Phase I	Eli Lilly/ Isis Pharmaceuticals
AVI-5126	c-myc	CABG	Morpholino	Intravenous	Phase I	AVI BioPharma
AVI-4557	Cytochrome P450 (CYP3A4)	Inhibit drug metabolism	Morpholino	Intravenous	Phase I	AVI BioPharma
SPC-2996	Bcl-2	Cancer	LNA/oligodeoxynucleotide chimera	Intravenous	Phase I	Santaris Pharma
SPC-2968	Hif-1α	Cancer	LNA/oligodeoxynucleotide chimera	Intravenous	Phase I	Enzon/Santaris Pharma
iCo-07	C-raf kinase	Macular degeneration	Phosphorothioate 2′-O-methoxyethyl/oligodeoxy-nucleotide chimera	Intravitreal	Phase I	iCo/ Isis Pharmaceuticals
ISIS 325568	Glucagon receptor	Type 2 diabetes	Phosphorothioate 2′-O-methoxyethyl/oligodeoxy-nucleotide chimera	Subcutaneous	Phase I	Ortho McNeil/Isis Pharmaceuticals
ISIS 377131	Glucocorticoid receptor	Type 2 diabetes	Phosphorothioate 2′-O-methoxyethyl/oligodeoxy-nucleotide chimera	Subcutaneous	Phase I	Ortho McNeil/Isis Pharmaceuticals
ISIS 369645	IL-4 receptor alpha	Asthma	Phosphorothioate 2′-O-methoxyethyl/oligodeoxy-nucleotide chimera	Aerosol	Phase I	Altair/Isis Pharmaceuticals
ALN-RSV-01	Respiratory synctial virus	RSV	Unmodified siRNA	Aerosol	Phase II	Alnylam Pharmaceuticals
AKIi-5	p53	Acute renal injury	siRNA	Intravenous	Phase I	Quark Pharmaceuticals

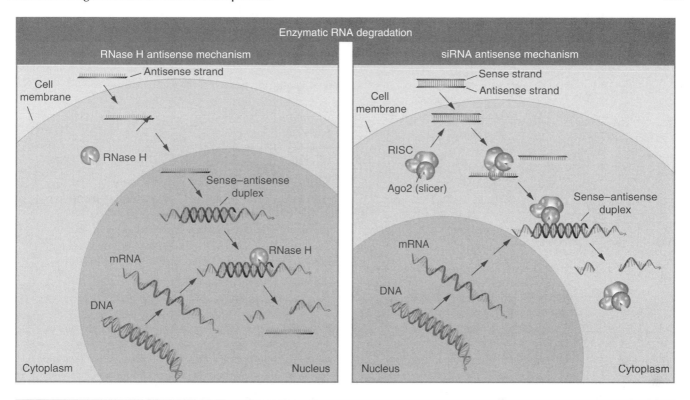

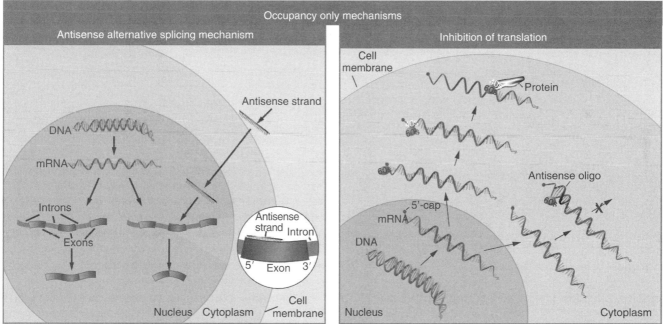

FIGURE 23.2 (See color insert following blank page 206.) Antisense mechanisms of action. Cartoon depicting three different mechanisms by which an antisense oligonucleotide can inhibit expression of a targeted gene product by hybridization to the mRNA or pre-mRNA, which codes for the gene product.

such as RNase H or RNA interference (RNAi) would result in more potent oligonucleotides, this has not necessarily been the case (Figure 23.3).

23.2.1 RNA DEGRADATION

Several studies have documented that phosphorothioate oligodeoxynucleotides promote cleavage of the targeted RNA by a mechanism consistent with RNase H cleavage [7–12]. RNase H is a ubiquitously expressed enzyme, which cleaves the RNA strand of an RNA–DNA heteroduplex [11,13–15]. If the antisense oligonucleotide contains at least five consecutive DNA nucleotides, it is capable of directing RNase H to specifically cleave the target RNA upon binding.

Another RNase-dependent antisense mechanism that has recently received much attention is RNAi [16–21]. Introduction

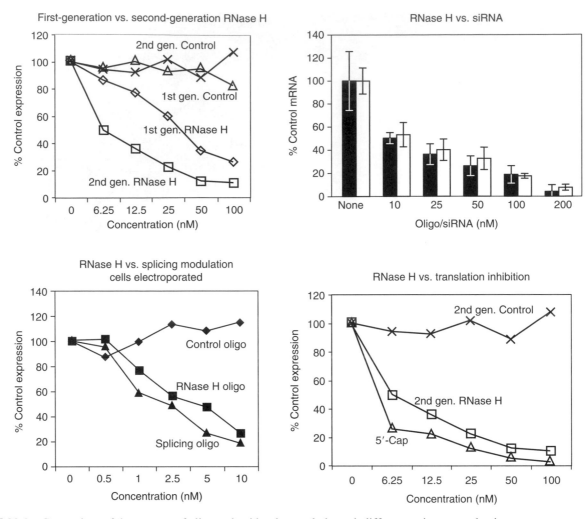

FIGURE 23.3   Comparison of the potency of oligonucleotides that work through different antisense mechanisms.

of long double-stranded RNA (dsRNA) into eukaryotic cells leads to the sequence-specific degradation of homologous gene transcripts. The long dsRNA molecules are metabolized to small 21–23 nucleotide-interfering RNAs (siRNAs) by the action of an endogenous ribonuclease [22–24]. The siRNA molecules bind to a protein complex, termed RNA-induced silencing complex (RISC), which contains a helicase activity that unwinds the two strands of RNA molecules, allowing the antisense strand to bind to the targeted RNA molecule [20,25]. The RISC contains an endonuclease, Ago 2, which hydrolyzes the target RNA by an RNase H-like mechanism at the site where the antisense strand is bound [26,27]. Therefore, RNAi is an antisense mechanism, as ultimately a single-strand RNA molecule binds to the target RNA molecule by Watson–Crick base pairing rules and recruits a ribonuclease that degrades the target RNA. More in depth discussion on RNAi is provided elsewhere in this book.

There are other RNases present in cells that may be exploited in manner similar to RNase H or the Ago 2 associated with RNAi. As an example, another RNase enzyme that has been

exploited for antisense applications is RNase L [28]. RNase L is ribonuclease activated by $2'$–$5'$-linked oligoadenylates generated in response to interferon activation. Selective linkage of $2'$–$5'$oligoadenylate to an antisense oligonucleotide has been reported to promote selective cleavage of the targeted mRNA [28–30].

It should be noted that not all oligonucleotides designed to hybridize to a target RNA effectively inhibit target gene expression [7,31–33]. This is thought to be due to inaccessibility of some regions of the RNA to the oligonucleotide due to secondary or tertiary structure or to protein interactions with the RNA. At this time, there are no good predictive algorithms for predicting antisense oligonucleotide binding sites on a target RNA. Thus some serendipity is still involved in the process of identifying and optimizing potent and effective antisense inhibitors.

23.2.2 Occupancy Only

Early on, it was thought that occupancy of the RNA (the receptor for the antisense oligonucleotide) by the oligonucleotide would be sufficient to block translation of the RNA, i.e., translation arrest [34]. Subsequent studies have documented that oligonucleotides are not efficient at blocking translation of mRNA if they bind 3′ to the AUG translation initiation codon. Furthermore, we have found that only certain sites in the 5′-untranslated region of an mRNA are effective target sites for an antisense oligonucleotide. In particular, the 5′-terminus of a transcript appears to be a good target site for oligonucleotides for some molecular targets in that occupancy of this region prevents assembly of the ribosome on the RNA [35]. It should be noted that occupancy of the receptor (RNA) and steric blocking factor of binding by high-affinity oligonucleotides can be an efficient mechanism for blocking gene expression. For the example cited above, the steric blocking oligonucleotide was approximately 10-fold more potent than an oligonucleotide that supports RNase H activity. These results suggest that catalytic turnover of the target RNA is not the rate-limiting step for antisense oligonucleotides.

Another process through which noncatalytic oligonucleotides can alter gene expression is by regulating RNA processing. Most mammalian RNAs undergo multiple post- or cotranscriptional processing steps including addition of a 5′-cap structure, splicing, and polyadenylation. Because single-stranded antisense oligonucleotides localize to the cell nucleus [36–39], they have the potential of regulating these processes. Several studies have been published documenting that antisense oligonucleotides can be used to regulate RNA splicing in both cell-based assays and in rodent tissues [40–50]. Oligonucleotides can be used to modulate alternative splicing by promoting use of cryptic splice sites as was exemplified for β-thalassemia and SMN-2 [40,41,48,51], by enhancing use of an alternative splice site [45]. Oligonucleotide binding to the pre-mRNA can also be exploited to mask polyadenylation signals on the pre-mRNA, forcing the cell to utilize alternative poly A sites [52]. Finally, oligonucleotides can regulate RNA function by sterically preventing factors from binding or changing the structure of the RNA such that it is no longer recognized by the factor. Thus there are multiple mechanisms by which oligonucleotides can be utilized to inhibit or modulate expression of a target gene product. No single mechanism is vastly superior to other mechanisms, thus one should select the antisense mechanism that best addresses the specific biological need.

23.3 ANTISENSE OLIGONUCLEOTIDE CHEMISTRY

The most advanced oligonucleotide chemistry used for antisense drugs is phosphorothioate oligodeoxynucleotides. These differ from natural DNA in that one of the nonbridging oxygen atoms in phosphodiester linkage is substituted with sulfur (Figure 23.1). Phosphorothioate oligodeoxynucleotides are commercially available, easily synthesized, support RNase H activity, exhibit acceptable pharmacokinetics for systemic and local delivery, and have not exhibited major toxicities, which would prevent their use in humans. There have been significant resources employed to identify chemical modifications that further improve upon the properties of phosphorothioate oligodeoxynucleotides. The primary objectives of the effort

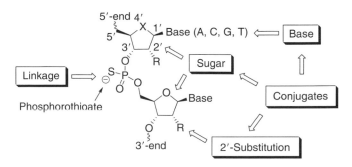

FIGURE 23.4 Positions, which have been chemically modified, for antisense oligonucleotides.

are similar to medicinal chemistry efforts for other types of pharmacological agents, i.e., to increase potency, improve pharmacokinetics, and to decrease toxicity.

A dimer of an oligonucleotide depicting subunits that may be modified to enhance oligonucleotide drug properties is depicted in Figure 23.4. In naturally occurring nucleic acids, these subunits are composed of heterocycles, carbohydrate sugars, and phosphodiester-based linkages between the sugars. The combination of the carbohydrate sugar (ribose in RNA, 2′-deoxyribose in DNA) and the linkage provides the backbone of the oligonucleotide polymer. Many modifications have been made on the individual base, sugar, and linkage subunits, and the sugar–phosphate backbone has been completely replaced with an appropriate substitute. Additionally, many diverse moieties have been conjugated to various positions in the subunits, mainly in an attempt to alter the biophysical properties of the polymer. Finally, prodrug modifications may be employed to enhance drug properties. Most of the positions available in a nucleoside dimer (approximately 25 positions for each dimer that do not directly interfere with Watson–Crick base pair hydrogen bonding) have been modified and studied for their effects on the properties of the resulting oligonucleotides.

The nucleobases or heterocycles of nucleic acids provide the recognition points for the Watson–Crick base pairing rules and any oligonucleotide modification must maintain these specific hydrogen bonding interactions. Thus, the scope of heterocyclic modifications is somewhat limited. The relevant heterocyclic modifications can be grouped into two structural classes: those that enhance base stacking and those that provide additional hydrogen bonding. The primary objective of heterocyclic modifications is to enhance hybridization, resulting in increased affinity (Figure 23.4). Modifications that enhance base stacking by expanding the π-electron cloud are represented by conjugated, lipophilic modifications in the 5-position of pyrimidines, such as propynes, hexynes, azoles, and simply a methyl group [53–56] and the 7-position of 7-deaza-purines position including iodo, propynyl, and cyano groups [57–59]. Investigators have continued to build out of the 5-position of cytosine by going from the propynes to five-membered heterocycles to tricyclic fused systems emanating from the 4,5-positions of cytosine clamps (Figure 23.5) [60–63]. A second type of heterocycle modification is represented by

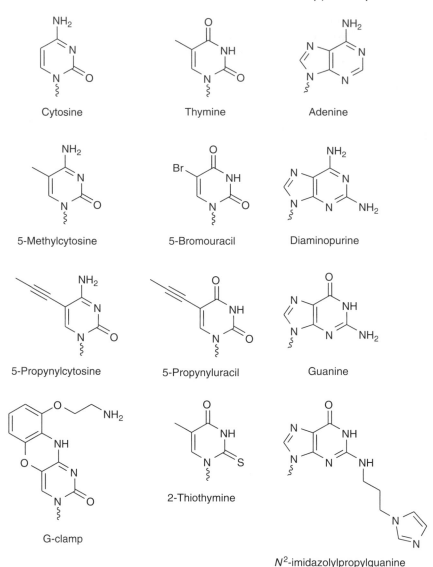

FIGURE 23.5   Examples of different heterocycle modifications, which support antisense activity.

the 2-amino-adenine (Figure 23.4), where the additional amino group provides another hydrogen bond in the A-T base pair, analogous to the three hydrogen bonds in a G-C base pair. Heterocycle modifications providing a combination of effects are represented by 2-amino-7-deaza-7-modified A [59] and the tricyclic cytosine analog having hydrogen bonding capabilities in the major groove of heteroduplexes [62] (Figure 23.4). Furthermore, N2-modified 2-amino adenine-modified oligonucleotides have exhibited interesting binding properties [64,65]. All of these modifications are positioned to lie in the major or minor groove of the heteroduplex, do not affect sugar conformation of the heteroduplex, provide little nuclease resistances, but will generally support an RNase H cleavage mechanism.

Modifications in the ribofuranosyl moiety have provided the most value in the quest to enhance oligonucleotide drug properties (Figure 23.5). In particular, certain 2′-*O*- modifications

have greatly increased binding affinity, nuclease resistance, altered pharmacokinetics, and are potentially less toxic [66]. Preorganization of the sugar into a 3′-*endo* pucker conformation is responsible for the increased binding affinity [67–69]. The 2′-*O*-methoxyethoxy (MOE) and 2′-*O*-methyl modifications (Figure 23.5) are the most advanced of the 2′-modified series having entered clinical trials and showing promising results in the clinic [70,71]. The cationic 2′-*O*-aminopropyl [72] and 2′-*O*-(dimethylaminooxyethyl) [73,74] have shown favorable binding affinity, with dramatically improved nuclease resistance. In an attempt to extend on the increased nuclease resistance of these cationic modifications to the high affinity seen with MOE, a dimethylaminoethoxyethyl version (DMAEOE) was prepared. This modification displays hybridization properties equal to or superior to that those of 2′-*O*-methoxyethyl, and nuclease resistance equal to that of the 2′-*O*-aminopropyl modification.

The sugar modification showing the largest known improvement in binding affinity is a bicyclic system having the 4'-carbon tethered to the 2'-hydroxyl group. As this modification "locks" the conformation of the ribose sugar into an RNA-like (3'-endo) conformation, it is referred to as locked nucleic acid (LNA) [75,76]. LNA shows dramatically improved hybridization affinity with regard to a reference DNA:RNA duplex and has increased nuclease resistance compared to RNA or DNA. LNA oligonucleotides have shown utility in several antisense oligonucleotide designs and have increased potency compared to first-generation oligonucleotides in several mouse models [76–79]. However, several oligonucleotides tested in mice exhibited profound hepatoxicity, resulting in a decrease in the therapeutic index [79]. Two different oligonucleotide-containing LNA nucleosides are in phase I clinical trials for the treatment of cancer (Table 23.1). While extremely promising from early biophysical and *in vitro* data, it remains to be seen whether these properties will translate into oligonucleotides with acceptable therapeutic index for broad therapeutic applications.

It is now well known that uniformly 2'-O-modified-oligonucleotides do not support an RNase H mechanism [80].

The heteroduplex formed has been shown to present a structural conformation that is recognized by the enzyme but cleavage is not supported [15,81,82]. Thus, uniformly modified, "RNA-like" oligonucleotides (3'-endo sugar conformation) will be unable to effect cleavage of the target mRNA and must therefore exert their effects via other means. This has led to the development of a chimeric strategy [8,80,83–85], which focuses on the design of high-affinity, nuclease-resistant antisense oligonucleotides that contain a "gap" of contiguous phosphorothioate-modified oligodeoxynucleotides (Figure 23.6). On hybridization to target RNA, a heteroduplex is presented, which supports an RNase H-mediated cleavage of the RNA strand via interaction with the 2'-deoxy gap region. The stretch of the modified oligonucleotide–RNA heteroduplex, which is recognized by RNase H may be placed anywhere within the modified oligonucleotide. The modifications in the flanking regions of the gap should not only provide nuclease resistance to exo- and endonucleases, but also not compromise binding affinity and base pair specificity. There are several types of structures that have been successfully developed (Figure 23.6), with the most advance being "gapmers," having a 7 to 10 base oligodeoxynucleotide gap flanked by 2 regions of

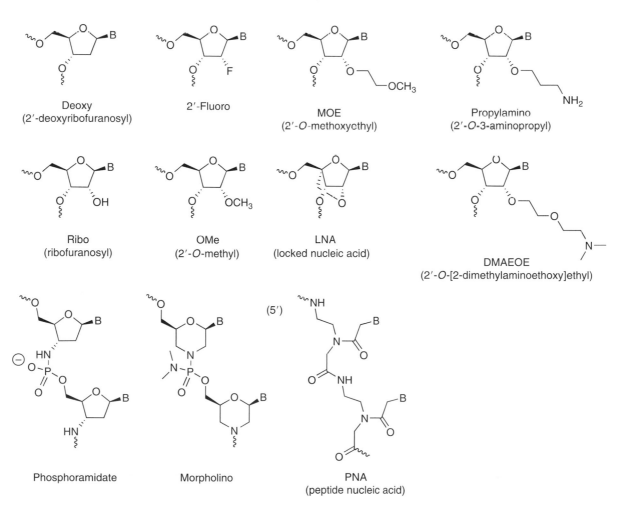

FIGURE 23.6 Examples of different sugar and backbone modifications that support antisense activity.

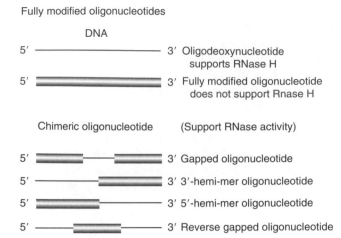

Fully modified oligonucleotides

FIGURE 23.7 Examples of different oligonucleotide structures.

TABLE 23.2
Attributes of Various Modified Oligonucleotides

Attribute	Examples
Increased affinity for RNA	2′-O-methyl, 2′-fluoro, MOE, DMAEOE, LNA, 5-MeC, 5-propynyl, phenoxazine G-clamp, PNA, phosphoramidate, others
Increased nuclease resistance	MOE, DMAEOE, LNA, PNA, phosphoramidate, morpholino, others
Alter tissue distribution	MOE, PNA, cholesterol conjugate, phosphoramidate, morpholino, others
Decrease toxicity	2′-O-methyl, MOE, 5-MeC, morpholino, others

2′-modified nucleosides (Figure 23.7). These oligomers, in particular 2′-MOE modified, show reduced toxicity, increased potency, and superior pharmacokinetics relative to the parent-unmodified 20-mer phosphorothioate oligodeoxynucleotide [85–89].

Several possible mechanisms exist for uniformly modified, non–RNase H activating oligonucleotides to show efficacy, such as prevention of assembly of the ribosome through binding in the 5′–UTR, "translation arrest" or ribosome stalling by blocking the reading of the mRNA ribosome, and modulation of splicing events by binding to splice junctions. While all of these strategies have been pursued, no uniformly modified oligonucleotides have advanced beyond RNase H oligonucleotides. However, much recent progress has been made with non–RNase H active oligonucleotides and there remains much potential for these modifications. LNA and MOE have been used in a uniform context in addition to the gapmer strategy (Figure 23.7), and early studies show promise. Another interesting uniform modification is the phosphoramidiate modification, which substitutes an amino group for the 3′ oxygen atom of the deoxyribose sugar of DNA. This results in a preference for the RNA–like (3′ endo) sugar conformation, and results in increased affinity as is seen with the 2′-O-alkyl modifications [90,91].

One of the most intriguing backbone oligonucleotide modifications is peptide nucleic acid (PNA). PNA is unique in that the sugar–phosphate backbone is completely replaced with a peptide-based backbone (Figure 23.5) [92]. This results in a polymer with a neutral backbone that has high affinity for complimentary nucleic acids. PNA has been extensively investigated as an antisense agent, but these efforts have generally been frustrated by the poor cellular penetration and *in vivo* pharmacokinetic properties of PNA [93]. A four-lysine peptide conjugated to a PNA was found to provide robust *in vivo* activity when targeted to a splice junction [47]. Extension of this work has lead to additional PNA motifs, which display activity in cell culture in the absence of a transfection agent and promising pharmacokinetics in animals [94,95]. Although these data are encouraging, initial pharmacological evaluation in animals has not yielded reproducible or robust effects. Thus more investment is required before the promise of PNA as an antisense therapeutic agent can be realized.

The most advanced uniform modification is the "morpholino" modification (Figure 23.6), which is currently in phase II clinical trials (Table 23.1). The morpholino modification simultaneously replaces the ribofuranosyl sugar with a morpholine ring and the negatively charged phosphate ester with a neutral phosphorodiamidate linkage [96,97]. Morpholinos are generally used around the translation initiation start codon and are believed to function via translation arrest. A morpholino oligonucleotide has shown *in vivo* activity [98], as well as oral bioavailability in rats [99], which would be a major advance if studies proved general and translated to larger mammalian species.

In addition to heterocycle, backbone, and sugar modifications discussed above, various pendant groups have been attached to oligonucleotides such as cholesterol, folic acid, fatty acids, etc. to alter pharmacokinetic properties [100,101]. The reader is referred to several reviews, which discuss the chemistry of oligonucleotides in more detail [102–106]. It should be noted that there is no single modification that covers all the desired properties for a modified oligonucleotide. Modifications have been identified, which increase hybridization affinity of the oligonucleotide for its target RNA, increase nuclease resistance, decrease toxicity, and alter the pharmacokinetics (Table 23.2). Furthermore, the ideal oligonucleotide will differ for different applications. Therefore, it is important to be able to mix and match the various modifications so as to obtain the optimal oligonucleotide for the task at hand.

23.4 PHARMACOKINETICS OF OLIGONUCLEOTIDES

23.4.1 CELLULAR PHARMACOKINETICS

Cellular uptake of phosphorothioate oligonucleotides has been documented to occur in most mammalian cells [107–114]. Cellular uptake of oligonucleotide is time- and temperature-dependent. It is also influenced by cell type, cell culture conditions and media, and the length/sequence of the oligonucleotide itself [107]. No obvious correlation between the lineage of cell, whether the cells are transformed or virally infected, and uptake has been identified. Cellular uptake appears to be an active process, i.e., oligonucleotide will accumulate in greater concentration intracellular than in the medium and is energy-dependent. Despite the fact that mammalian cells in culture will readily accumulate oligonucleotides, it has been necessary to further facilitate cytosolic delivery for many, but not all cells with transfection agents such as cationic lipids, dendrimers, fusogenic peptides, electroporation, etc. [38,115–119,46]. In the absence of these facilitators, it has been difficult to demonstrate true antisense effects in cultured cells although there are some exceptions. However, *in vivo* this is not the case. It has become apparent that *in vitro* cell uptake studies do not predict *in vivo* cell uptake and pharmacokinetics of oligonucleotides [107,158,120–124]. Our understanding of cellular and subcellular uptake has evolved as superior analytical tools have been developed. These advances include development of immunohistochemical techniques utilizing oligonucleotide-specific antibodies [125] or in situ hybridization with labeled oligonucleotide probes.

Our understanding of cellular and subcellular distribution and pharmacokinetics of oligonucleotides in whole animals is emerging. In our laboratories, we have utilized more specific tools for qualification and even quantification of intact oligonucleotide [121,125–127]. Phosphorothioate oligonucleotides rapidly distribute to whole tissue with distribution half-lives ranging from 30 to 60 min *in vivo*. Approximately half of the oligonucleotide associated with the liver (as an example) is intracellular in both parenchymal and nonparenchymal cells by 4 h after intravenous administration [121,128]. The other half of the organ-associated oligonucleotide appears to be associated with extracellular matrix, interstitium, or loosely bound to the cell membrane. Consistent with this observation, others have shown that phosphorothioates have been localized to connective tissue and can bind to various proteins within these matrices such as laminin and fibronectin [125,129,130]. Some of this matrix-associated oligonucleotide will diffuse to cells over time or be lost to efflux from the organ [125]. It is likely that both of these processes are functioning up to 24 h after administration of oligonucleotide. By 24 h after injection of phosphorothioate oligonucleotide, very little is seen to be associated with extracellular matrix [125]. Thus, whole-organ pharmacokinetic evaluation after the initial 24 h distribution phase parallels cellular clearance kinetics.

Although the *in vitro* studies fail to predict well which cell types will take up oligonucleotide *in vivo*, the general trend of variability from cell type to cell type continues to be observed *in vivo* [125]. Based upon these results, one would not expect to uniformly inhibit expression of a targeted gene product within a tissue or whole organism, resulting in differential sensitivity of different tissues and cells within tissue to the antisense effect. Subcellular distribution has been shown to be broad and the extent of cytosolic and nuclear distribution differs between cells [121]. In general, the total number of oligonucleotide molecules is greatest in the cytosol. However,

because of the much smaller volume of the nucleus, the nucleus may often contain a higher concentration of oligonucleotide than cytosol.

Nuclease metabolism has been shown to account for the clearance of phosphorothioate oligonucleotide from organs of distribution. Within the cells, the pattern of metabolites appears to be quite similar between cell types and the subcellular compartments (membrane associated, cytosolic, and nuclear). Increasing doses from 5 to 50 mg/kg only moderately decreased metabolism inside cells consistent with whole-organ data [121].

Several studies have suggested that active uptake processes including receptor-mediated endocytosis and pinocytosis are involved in uptake of oligonucleotides *in vivo*. At very low doses (less than 1 mg/kg), competition of binding for scavenger receptors *in vivo* altered the whole-organ distribution of oligonucleotides in liver but not kidney [131–133]. However, distribution studies conducted in scavenger receptor knockout mice did not show significantly altered intracellular and whole-organ distribution of phosphorothioate oligonucleotides [134].

Distribution in the kidney has been more thoroughly studied, and drug has been shown to be present in Bowman's capsule, the proximal convoluted tubule, the brush border membrane, and within renal tubular epithelial cells [125,135]. These data suggested that the oligonucleotides are filtered by the glomerulus and then reabsorbed by the proximal convoluted tubule epithelial cells. Moreover, the authors identified a specific protein in the brush border that may mediate uptake. In subsequent studies, the authors have purified the 45 kDa protein, reconstituted it in phospholipid vesicles, and demonstrated that it served as a channel allowing nucleic acid to pass through phospholipid bilayers [136]. In separate studies, other investigators have shown that although some oligonucleotide is taken up from the tubular lumen brush border, the distribution to the tubule epithelial cells is predominantly from the capillary serosal side [137]. The uptake from capillary circulation may not be receptor-mediated. In summary, it is likely that there are multiple processes involved in the uptake of oligonucleotides into cells *in vivo*. Additional research will be required to further elucidate these mechanisms.

23.4.2 Whole-Animal Oligonucleotide Pharmacokinetics

23.4.2.1 Phosphorothioate Oligodeoxynucleotides

The plasma pharmacokinetics of phosphorothioate oligodeoxynucleotides are characterized by rapid and dose-dependent clearance (30–60 min half-life) driven primarily by distribution to tissue and secondarily by metabolism. Only a minor fraction of a parenteral dose of phosphorothioate oligonucleotide is excreted in urine or feces by 24 h after administration [138–141]. Dose-dependent distribution kinetics from plasma is predominantly a function of saturable tissue distribution [140,142]. Metabolism has been shown to be unchanged in plasma over a large dose range (1–50 mg/kg) and after repeated administration up to 1 month, suggesting that metabolism is neither inhibited or induced by repeat administration [143].

The plasma pharmacokinetics are quite similar between animals and humans, and they scale from one species to the next on the basis of body weight, not surface area [138, 143–146]. For example, it is possible to show that, when dosed on the basis of body weight, the concentrations of oligonucleotides in plasma administered by a 2 h constant intravenous infusion are similar between humans and monkeys. Thus it has been possible to predict plasma concentrations in humans from nonclinical pharmacokinetic data.

Phosphorothioate oligonucleotides bind to circulating plasma proteins such as albumin and α-2-macroglobulin [147,148]. The apparent affinity for human serum albumin is low (10–30 µM). Therefore, plasma protein-binding provides a repository for these drugs preventing rapid renal excretion. Because serum protein binding is saturable at high concentrations, intact oligonucleotide may be found in urine in increasing amounts as dose rate and/or amount is increased [143, 149,150].

Phosphorothioate oligonucleotides are rapidly and extensively absorbed after intradermal, subcutaneous, intramuscular, or intraperitoneal administration [120,142,151,152]. Nonparenteral absorption has been characterized for pulmonary and oral routes of administration. Estimates of bioavailability range from 3% to 20% following intratracheal dosing, and <1% by the oral route [153,154]. Although permeability in the intestine is low, stability of these compounds in the intestine (prior to absorption) may be a rate-limiting factor to oral absorption [155,156]. As discussed below, some chemical modifications to the oligonucleotide enhance oral absorption. The metabolic half-life of a 20-mer phosphorothioate oligonucleotide in the rat intestine (*in vivo*) is less than 1 h (data shown in formulation section of this chapter).

Phosphorothioates are broadly distributed to all peripheral tissues. Highest concentrations of oligonucleotides are found in the liver, kidney, spleen, lymph nodes, and bone marrow with no measurable distribution to the brain [120,142,143,149,155]. Many other tissues take up smaller amounts of oligonucleotide, resulting in lower tissue concentrations. Phosphorothioate oligonucleotides are primarily cleared from tissues by nuclease metabolism. Rate of clearance differs between tissues with the spleen, lymph nodes, and liver generally clearing more rapidly than kidney, for example. In general, the clearance rates result in half-lives of elimination ranging from 2 to 5 days in rodents and primates [140,146].

In summary, pharmacokinetic studies of phosphorothioate oligonucleotides demonstrate that they are well absorbed from parenteral sites, distribute broadly to all peripheral tissues, do not cross the blood–brain barrier, and are eliminated primarily by slow metabolism. In short, once-a-day or every-other-day systemic dosing should be feasible. In general, the pharmacokinetic properties of this class of compounds appear to be largely driven by chemistry rather than sequence. Additional studies are required to determine whether there are subtle sequence-specific effects on the pharmacokinetic profile of this class of drugs.

23.4.2.2 Second-Generation Oligonucleotides

The plasma pharmacokinetics of 2′-*O*-methyl-, 2′-*O*-propyl-, or 2′-*O*-methoxyethyl-modified oligonucleotides do not differ significantly from their phosphorothioate oligodeoxynucleotide congeners [87,88,147,157]. Since metabolism plays only a minor role in the plasma distribution kinetics, 2′-ribose modifications do very little to alter the distribution and excretion kinetics. Early studies in our laboratory indicate that the binding affinity to serum albumin may be somewhat lessened by 2′-ribose sugar modifications, but the overall capacity of the plasma proteins to bind these oligonucleotides is not significantly changed (Table 23.3). Therefore, urinary excretion remains a minor route of elimination and these compounds are broadly distributed to peripheral tissues.

Several of the 2′-ribose sugar modification produces enough of an increase in nuclease resistance that it is possible to produce relatively stable oligonucleotides with phosphodiester linkages (Table 23.2). Thus this modification allows for elimination or reduction in the number of sulfurs contained in the internucleotide bridge, but these compounds are less stable than their 2′-modified phosphorothioate congeners [158]. In addition, as sulfur is removed, plasma protein binding is greatly decreased and rapid removal from plasma by filtration in the kidney increases significantly. This pharmacokinetic characteristic may limit the use of phosphodiester second-generation modified oligonucleotides intended for treatment of systemic disease [87]. Alternatively, this pharmacokinetic profile may be ideal for locally administered oligonucleotides since it limits the accumulation of systemically absorbed drug.

Absorption for parenterally administered modified oligonucleotides is consistently rapid and nearly complete. Some of the second-generation modified oligonucleotides have exhibited improved intestinal permeability [155] as well as significantly improved stability in the intestine [156]. This combination of improved biochemical characteristics has led

---

TABLE 23.3

Serum Albumin Affinity, Whole Plasma Fraction Bound to Proteins (F_b), and Fraction of Dose Excreted in Urine ($f_{excreted}$, 0–24 h) Following Intravenous Administration At 3 mg/kg: Comparison of First- and Second-Generation Chemistries

Compound No.	Chemistry	K_d (µM)	F_b (%)	$f_{excreted}$
ISIS 2302	PS ODN	17.7	99.2	0.003
ISIS 11159	PS 2′-MOE	29.3	95.5	0.032
ISIS 16952	PO 2′-MOE	>500	79.6	0.45

Note: PS ODN, phosphorothioate oligodeoxynucleotide; PS 2′-MOE, 2′-*O*-methoxyethyl ribose modified phosphorothioate (all nucleotides were modified); PO 2′-MOE, 2′-*O*-methoxyethyl ribose modified phosphodiester (all nucleotides were modified).

TABLE 23.4

Summary of Observed Organ Clearance Half-Lives (in Days) Comparing First- and Second-Generation Chemistries

Organ	2'-Modified Phosphoro-thioate Oligonucleotide[a]	Phosphorothioate Oligodeoxynucleotide
Kidney cortex	20–30	5.0
Kidney medulla	10–20	3.1
Liver	14–30	2.8
Spleen	8–15	3.3
Lymph nodes	16–24	0.9
Bone marrow	12–16	1.3

[a] Observed half-life range for more than five different chimeric 2'-MOE/2'-deoxy modified phosphorothioate oligonucleotide; differences apparently due to sequence. (From Krutzfeldt, J., Rajewsky, N., Braich, R., Rajeev, K. G., Tuschl, T., Manoharan, M., and Stoffel, M. *Nature*, 685, 438, 2005.)

to the observation of improved oral bioavailability [155] for this class of oligonucleotide compounds.

The distribution pattern of the 2'-ribose-modified phosphorothioate oligonucleotides is similar to first-generation phosphorothioates and similarly not altered by changes in sequence. Kidney, liver, spleen, bone marrow, and lymph nodes are the major sites of distribution. The most exciting difference in pharmacokinetics is not surprisingly manifested in prolonged terminal elimination half-lives from tissues of distribution. The elimination half-lives appear to be increased nearly 5–10-fold suggesting that once-weekly systemic dosing may be feasible (Table 23.4). Subsequent nonclinical and clinical testing have confirmed the effectiveness of once weekly dosing frequency for 2'-MOE-modified antisense oligonucleotides [70,127,141,159–164].

In summary, pharmacokinetic studies of 2'-modified ribose phosphorothioate oligonucleotides demonstrate that they are well absorbed from parenteral sites, may have improved oral absorption attributes, and distribute broadly to all peripheral tissues. Although stability has been greatly enhanced, nuclease metabolism remains the primary mechanism for ultimate elimination of these modified oligonucleotides [139,163]. In short, once-a-week systemic dosing should be feasible and oral administration may be possible in the near future. Additional studies are required to determine whether there are substantial sequence-specific effects on the pharmacokinetic profile of this class of drugs.

23.5 TOXICOLOGY OF OLIGONUCLEOTIDES

Phosphorothioate oligodeoxynucleotides have been examined extensively in a full range of acute, chronic, and reproductive studies in rodents, lagomorphs, and primates. At high doses, there is a distinctive pattern of toxicity that is common to all phosphorothioate oligodeoxynucleotides [165–168]. The remarkable similarity in toxicity with different phosphorothioate

oligodeoxynucleotides suggests that, for this class of antisense compounds, toxicity is independent of sequence and is the result of nonantisense-mediated mechanisms. The most probable mechanism of the observed toxicities is the binding of phosphorothioate oligodeoxynucleotides to proteins. These nonantisense-mediated pathways are thought to be responsible for most, if not all, of the toxicities associated with the administration of these compounds to laboratory animals. This conclusion is strengthened by studies in which little or no differences in toxicity is observed between pharmacologically active and inactive sequences. Different patterns of toxicity exist between rodents and primates. Understanding the mechanisms behind these differences is crucial to understanding which species best predicts the potential human effects. A comparison of the toxicological profiles of phosphorothioate oligodeoxynucleotides with that of the second-generation phosphorothioate oligodeoxynucleotides suggests that some of the chemical class-related toxicities of phosphorothioate oligodeoxynucleotides can be ameliorated by chemical modification.

A number of phosphorothioate oligodeoxynucleotides have been examined in one or more of the following battery of genotoxicity assays: Ames test, *in vitro* chromosomal aberrations, *in vitro* mammalian mutation (hypoxanthine-guanine phosphoribosyl transferase [HGPRT] locus and mouse lymphoma), *in vitro* DNA synthesis tests, and *in vivo* mouse micronucleus. In all of these assays, the results were negative and there was no evidence of mutagenicity or clastogenicity of these compounds [169].

23.5.1 ACUTE TOXICITIES

23.5.1.1 Phosphorothioate Oligodeoxynucleotides

In rodents, the acute toxicity of phosphorothioate oligodeoxynucleotides has been characterized as part of an effort to determine the maximum tolerated dose for *in vivo* genotoxicity assays. The doses of three phosphorothioate oligodeoxynucleotides required to produce 50% lethality (LD_{50}) were estimated to be approximately 750 mg/kg [169].

In primates, the acute dose-limiting toxicities are a transient inhibition of the clotting cascade and the activation of the complement cascade [165,170,171]. Both of these toxicities are thought to be related to the polyanionic nature of the molecules and the binding of these compounds to specific protein factors in plasma.

Prolongation of clotting times following administration of different phosphorothioate oligodeoxynucleotides is characterized by a concentration-dependent prolongation of activated partial thromboplastin times (aPTT) [168,172–174]. The prolongation of aPTT is highly transient and directly proportional to plasma concentrations of phosphorothioate oligodeoxynucleotide and therefore parallels the plasma drug concentration curves with various dose regimens. As drug is cleared from plasma, the inhibition diminishes such that there is complete reversal within hours of dosing. With repeated administration, there is no evidence of residual inhibition. Prolongation of aPTT has been observed in all species examine to date, including human, monkey, and rat. The mechanism of prolongation of aPTT by phosphorothioate oligodeoxynucleotides is thought to be a result of the interaction of the oligonucleotides with proteins. It is well known that polyanions are inhibitors of clotting, and phosphorothioate oligodeoxynucleotides may act through similar mechanisms. If these oligonucleotides inhibit the clotting cascade as a result of their polyanionic properties, then binding and inhibition of thrombin would be a likely mechanism of action. However, the greater sensitivity of the intrinsic pathway to inhibition by phosphorothioate oligodeoxynucleotides suggests that there are other clotting factors specific to this pathway that may be inhibited as well. Recent data suggest that there is a specific allosteric inhibition of the tenase complex as well as binding to thrombin [171,175].

Activation of the complement cascade by phosphorothioate oligodeoxynucleotides has the potential to produce the most profound acute toxicological effects. In primates, treatment with high doses over short infusion times (i.e., generating high C_{max} values) resulted in marked hematological effects and marked hemodynamic changes that are thought to be secondary to complement activation. Hematological changes are characterized by transient reduction in neutrophil counts, presumably due to margination, followed by neutrophilia with abundant immature, nonsegmented neutrophils [166,170]. In a small fraction of monkeys, complement activation was accompanied by marked reductions in heart rate, blood pressure, and subsequently cardiac output. In some animals, these hemodynamic changes were lethal [165,170,176].

There is an association between cardiovascular collapse and complement activation. That is, all monkeys demonstrating some degree of cardiovascular collapse or hemodynamic changes had markedly elevated levels of complement split products. However, the converse is not true, in that only a fraction of the animals with activated complement had cardiovascular functional changes [169]. Thus, this observation suggests that there may be sensitive subpopulations or predisposing factors within individual animals that make them susceptible to the physiologic sequelae of complement activation. Because of these observed hemodynamic changes, primate studies to monitor for these effects have become part of the normal evaluation of these compounds [177,178]. Although complement activation at high doses is consistent and predictable between animals, there is currently little appreciation for the variability in the severity of the associated hemodynamic changes. While the split product Bb can be used to monitor complement activation, it is C5a (complement split product) that is the most biologically active split product. Preliminary data obtained relating response to complement split product levels indicate that C5a levels are elevated more significantly in some of the more affected animals [169].

The goal of toxicity studies is to characterize the toxicity of compounds and to establish a framework upon which clinical safety studies can be designed. In this regard, it is useful to examine the relationship between plasma concentrations of oligonucleotides and the activation of complement. When Bb concentrations were plotted against the concurrent plasma concentrations of oligodeoxynucleotides in primates, it was

apparent that complement was only activated at concentrations of phosphorothioate oligodeoxynucleotides that exceed a threshold value of 40–50 μg/mL [170]. Bb levels remained unchanged from control values at plasma concentrations below the threshold. Remarkably, this threshold concentration is similar for three 20-mer phosphorothioate oligodeoxynucleotides and for an 8-mer phosphorothioate oligodeoxynucleotide that forms a tetrad complex [179,180]. Recent data demonstrate that human serum may be less sensitive to activation than monkey serum, suggesting a species difference in sensitivity. Regardless of small differences, it is clear that clinical dose regimens should be designed to avoid plasma oligodeoxynucleotide concentrations that exceed 40–50 μg/mL. To this end, the similarities in plasma pharmacokinetics between monkeys and humans have allowed the design of dose regimens that achieve desired plasma concentration profiles.

The most direct approach for staying below the plasma thresholds for complement activation is to reduce the dose rate by substituting prolonged infusions for bolus injections. In clinical trials with phosphorothioate oligodeoxynucleotides, the drugs are administered either as 2 h infusions or as constant 24 h infusions. At a rate of infusion of 2 mg/kg over 2 h, the C_{max} was 8–15 μg/mL, still well below the threshold for complement activation [144]. Phosphorothioate oligodeoxynucleotides have been administered by intravenous infusion to more than 3000 patients and volunteers without any significant indication of activation of the alternative complement cascade.

23.5.1.2 Modified Oligonucleotides

Replacement of phosphorothioate linkages with natural occurring phosphodiester linkages in oligodeoxynucleotides may reduce the potential to activate complement. However, as noted above, reduction of phosphorothioate linkages dramatically reduces distribution of oligonucleotide to tissues, diminishing the pharmacological effects. As a result, for antisense drugs that work inside of cells, replacement of phosphorothioate linkages with phosphodiester linkages may not be a viable approach for systemically administered drugs. Second-generation oligonucleotides containing 2′-O-methoxyethyl modifications appear to have a lower potency for complement activation in primates and anticoagulant effects [168,169,181]. The plasma concentration thresholds for complement activation are markedly increased for 2′-O-methoxyethyl-modified oligonucleotides, which may be due to the decrease in protein interactions this modification confers onto an oligonucleotide. Although the safety profile of phosphorothioate oligodeoxynucleotides has proven satisfactory, the acute safety profile of the next generation of oligonucleotides appears to be improved by the 2′-O-methoxyethyl modification.

23.5.2 Toxicological Effects Associated with Chronic Exposure

23.5.2.1 Phosphorothioate Oligodeoxynucleotides

One of the characteristic toxicities observed with repeated exposure of rodents to phosphorothioate oligodeoxynucleotides is a profile of effects that can be described as immune stimulation. The profile is characterized by splenomegaly, lymphoid hyperplasia, and diffuse multiorgan mixed mononuclear cell infiltrates [168]. The severity of these changes is dose-dependent and most notable at doses equal to or exceeding 10 mg/kg. The mixed mononuclear cell infiltrates consisted of monocytes, lymphocytes, and fibroblasts and were particularly notable in liver, kidney, heart, lung, thymus, pancreas, and periadrenal tissues [167,182–184].

Although immune stimulation in rodents is thought to be a class effect of phosphorothioate oligodeoxynucleotides and not dependent on hybridization, sequence is an important factor in determining immunostimulatory potential [185–188]. Immunostimulatory motifs have been described in the literature and involve palindromic sequences and CpG (cytosine–guanosine) motifs [188].

Among the most remarkable features of oligodeoxynucleotide-induced immune stimulation are the species differences. Rodents are highly susceptible to this generalized immune stimulation, whereas primates appear to be relatively insensitive to the effect at equivalent doses. Even 6 months of treatment of cynomolgus monkeys with 10 mg/kg of a 20-mer oligodeoxynucleotide, ISIS 2302, given every other day produced only a relatively mild increase in B cell numbers in spleen and lymph nodes of the primates with no change in organ weights. The mixed mononuclear cellular infiltrates in liver and other organs that are so characteristic of the response in rodents are absent even after long-term exposure in monkeys [168]. It is evident that there are both species and sequence differences involved in immune stimulation and specific sequences should, if possible, be excluded from oligodeoxynucleotides.

Morphologic changes in the bone marrow of mice were observed after 2 weeks of treatment (3 doses/week) with 100–150 mg/kg phosphorothioate oligodeoxynucleotide. There was a reduction in number of megakaryocytes that were accompanied by a reduction of approximately 50% in circulating platelet counts [183]. Reductions in platelets have been observed in rats treated with 21.7 mg/kg ISIS 2105 given every other day [167], but were not observed in primates administered 10 mg/kg. Similarly, a reduction in platelets was observed in mice, but not in monkeys treated for 4 weeks with ISIS 2302 at doses of 100 and 50 mg/kg every other day, respectively. Similar observations were made for ISIS 5132 with reductions in platelets at 20 and 100 mg/kg in mice and no observed effect in monkeys up to 10 mg/kg [182]. These data suggest that the mouse may be more sensitive to these subchronic effects on platelets than nonhuman primates. However, in acute studies in primates, transient reductions in platelets are occasionally observed. These transient reductions in platelets occur acutely during 2 h infusions at doses of 10 mg/kg, reverse after completion of the infusion, and have not been associated with any measurable change in platelet number 24 to 48 h after subchronic or chronic treatment regimens [169]. Thrombocytopenia has been reported in AIDS patients treated with GEM 91, a 27-mer phosphorothioate oligodeoxynucleotide [189].

Tissue distribution studies have shown that the liver and kidney are major sites of deposition of phosphorothioate oligodeoxynucleotide. In toxicity studies with phosphorothioate oligodeoxynucleotides, a variety of hepatic changes have been observed. The immune-mediated cellular infiltrates in rodent livers were discussed above. With high-dose administration of oligodeoxynucleotides in all species examined, there was a hypertrophic change in Kupffer cells accompanied by inclusions of basophilic material that was observed with hematoxylin and eosin staining. These basophilic granules have been identified as inclusions of oligodeoxynucleotide [125]. Furthermore, it was demonstrated that the presence of these inclusions was related to dose.

Hepatocellular changes were not a prominent feature of toxicity in primates. In cynomolgus monkeys, 50 mg ISIS 2302 per kg administered every other day for 4 weeks by intravenous injection produced no morphologic indication of liver toxicity, although there was a slight (1.5-fold) increase in aspartate aminotransferase (AST) in this group [190]. Following subcutaneous doses of ISIS 3521 and ISIS 5132 of up to 80 mg/kg every other day for four doses, there was Kupffer cell hypertrophy and periportal cell vacuolation, but no indication of necrosis and only a very slight increase in alanine aminotransferase (ALT) [169]. After 4 weeks of alternate-day dosing with 10 mg/kg via 2 h intravenous infusion of either ISIS 3521 or ISIS 5132, there were no alterations in AST or ALT, suggesting that at clinically relevant doses of these compounds, there was no evidence for hepatic pathology or transaminemia. In clinical trials with ISIS 2302, ISIS 3521, and ISIS 5132 at doses of 2 mg/kg administered by 2 h infusion on alternate days for 3–4 weeks, there was no indication of hepatic dysfunction nor was there any evidence of transaminemia.

Like Kupffer cells in the liver, renal proximal tubule epithelial cells take up oligodeoxynucleotide, as demonstrated by autoradiographic studies and immunohistochemistry as discussed previously [125,129,191,192] and by the use of special histologic stains [166]. The appearance of basophilic inclusions is dose-dependent in proximal tubule cells. Significant renal toxicity can be induced by extremely high doses. Doses of 80 mg/kg in rats and monkeys have induced both histologic and serum chemistry changes in the kidney [193]. At clinically relevant doses, however, there was no indication of renal dysfunction. In 4-week or 6-month toxicity studies with phosphorothioate oligodeoxynucleotides, we observed a much more subtle type of morphologic change in the kidney. At a dose of 10 mg/kg on alternate days, there was a decrease in the height of the brush border and enlarged nuclei in some proximal tubule cells. These changes have been characterized as minimal to mild tubular atrophic and regenerative changes. At a dose of 3 mg/kg and below, these changes were only infrequently observed if at all.

An important aspect of dose-dependent effects is characterization of exposure concentrations and their relationship to morphological changes. To assess exposure, concentrations of oligodeoxynucleotides have been measured in the renal cortex obtained in subchronic and chronic toxicity studies. Renal concentrations increase with increasing doses. The concentration of total oligodeoxynucleotide in the renal cortex associated with minimal to mild (although not clinically relevant) renal tubular atrophy or regenerative changes is approximately 1000 μg/g of tissue. The cortex concentrations of total oligodeoxynucleotides that are associated with moderate degenerative changes after subcutaneous doses of 40–80 mg/kg are greater than 2000 μg/g. At a clinically relevant dose of 3 mg/kg every other day, the steady-state concentration of total oligodeoxynucleotide in the kidney is in the range of 400–500 μg/g, thus demonstrating a significant margin of safety between the clinical doses and the doses associated with even the most minimal morphologic renal changes. Application of clearance and steady-state pharmacokinetic models suggests that continued administration of oligodeoxynucleotide at this dose should never achieve the renal concentrations associated with dysfunction [143]. These models have been confirmed in 6-month chronic toxicology studies where tissue concentrations measured at the end of 6 months of every other day dosing were no different than levels observed after 4 weeks of dosing at a similar or equivalent dose.

23.5.2.2 Chemical Modification of Oligodeoxynucleotides

Chemical modifications of oligodeoxynucleotides have been shown to reduce the potency of immune stimulation. The simplest modification with remarkable activity for reducing the immunostimulatory effects of oligodeoxynucleotides is the replacement of cytosine with 5-methyl cytosine. Methylation of a cytosine not in a CpG motif did not reduce the immunostimulatory potential [194]. In our experience with mice, when sequences with 5-methyl cytosine are compared with the same sequence without methylation, the methylated sequence has a lower potency for inducing immune stimulation, as determined by spleen weights and immune cell activation [195,196]. The addition of 2′-O-methyl and 2′-O-methoxyethyl substituents also reduced immunostimulatory potential [86,197]. In particular, we have found that 2′-O-methoxyethyl modification dramatically reduces signaling through Toll-like 9 receptors.

23.6 OLIGONUCLEOTIDE FORMULATIONS

23.6.1 Physical–Chemical Properties

Due to the presence of a mixture of diastereisomers, phosphorothioate oligodeoxynucleotides are amorphous solids possessing the expected physical properties of hygroscopicity, low-bulk density, electrostatic charge pick up, and poorly defined melting point prior to decomposition. Their good chemical stability allows storage in the form of a lyophilized powder, spray-dried powder, or a concentrated, sterile solution; more than 3 years of storage is possible at refrigerated temperatures.

Due to their polyanionic nature, phosphorothioate oligodeoxynucleotides are readily soluble in neutral and basic conditions. Drug-product concentrations are limited (in select applications) only by an increase in solution viscosity. The counterion composition, ionic strength, and pH also influence the apparent solubility. Phosphorothioate oligodeoxynucleotides have an apparent pK_a in the vicinity of 2 and will come out of solution in acidic environments, i.e., the stomach. This precipitation is readily reversible with increasing pH or by acid-mediated hydrolysis.

Instability of phosphorothioate oligodeoxynucleotides has been primarily attributed to two degradation mechanisms: oxidation and acid-catalyzed hydrolysis. Oxidation of the (P=S) bond in the backbone has been observed at elevated temperatures and under intense UV light, leading to partial phosphodiesters (still pharmacologically active) and are readily monitored by anion-exchange high-performance liquid chromatography. Under acidic conditions, hydrolysis reactions followed by chain-shortening depurination reactions have been documented by length-sensitive electrophoretic techniques.

23.6.2 Parenteral Injections

Given the excellent solution stability and solubility possessed by phosphorothioate oligodeoxynucleotides, it has been relatively straightforward to formulate the first-generation drug products in support of early clinical trials. Simple, buffered solutions have been successfully used in clinical studies by i.v., intradermal, and subcutaneous injections. And recently, the intravitreal route was approved for the first antisense drug application.

23.6.3 Oral Delivery

Of the numerous barriers proposed by Nicklin et al. [152] to the oral delivery of oligonucleotides, our experience has confirmed that two factors stand out as critical: instability in the gastrointestinal (GI) tract and low permeability across the intestinal mucosa. Given the formidable nature of these two barriers, it is not surprising that oral delivery of oligonucleotides has been considered impossible, or at best difficult, as is the case with proteins, which has necessitated the latter's nonenteral administration in order to achieve systemic concentrations considered therapeutic. Nevertheless, progress has been made to address and understand each of these barriers with respect to oligonucleotides. (P=S)-oligonucleotides have a distinct advantage over proteins in that the former does not rely on secondary structure for activity. This provides freedom from concern over secondary structure destabilization and allows for (P=S)–oligonucleotide structural modifications to address both presystemic and systemic metabolism.

Natural DNA and RNA are rapidly digested by the ubiquitous nucleases found within the gut. As a consequence, oligonucleotides need to be stabilized in order to achieve a reasonable GI residence time to allow for absorption to occur. Surprisingly, phosphorothioate oligodeoxynucleotides were found to be rapidly degraded by nucleases found in the GI tract, therefore additional protection from nuclease degradation is required to achieve significant oral bioavailability. Oligonucleotides that are uniformly modified or modified on the 3′-end (gapmers or 3′-hemimers; Figure 23.6) with nuclease-resistant modification have the potential to exhibit increased oral bioavailability. This was demonstrated for both backbone modifications (methylphosphonates) and sugar modified (2′-O-methyl) oligonucleotides [155,198]. We have found that 2′-O-methoxyethyl-modified oligonucleotides also exhibit increased oral absorption compared to phosphorothioate oligodeoxynucleotides [156,199,200].

The physicochemical properties of phosphorothioate oligodeoxynucleotides present a significant barrier to their GI absorption into the systemic circulation or the lymphatics. These factors include their large size and molecular weight (i.e., up to 6.5 kDa for 20-mers), hydrophilic nature (log $D_{o/w}$ approximating −3.5), and multiple ionization pk_a values (e.g., unpublished titration data), using a Sirius GlpKa instrument on a 20-mer sequence, noted over 17 pk_a values for phosphorothioate oligodeoxynucleotide and over 32 pk_a values for the 2′-O-methoxyethyl hemi-mer form (Figure 23.7). The use of formulations can improve upon GI permeability. Oligonucleotide drug formulations designed to improve oral bioavailability need to consider the mechanism of oligonucleotide absorption—either paracellular via the epithelial tight junctions or, transcellular, by direct passage through the lipid membrane bilayer. By using paracellular and transcellular models appropriate for water-soluble, hydrophilic macromolecules, it was determined that oligonucleotides predominantly traverse GI epithelium via the paracellular route. In this regard, formulation design considerations involve the selection of those penetration enhancers (PEs), which facilitate paracellular transport and meet other formulation criteria including suitable biopharmaceutics, safety considerations, manufacturability, physical and chemical stability, and practicality of the product configuration (i.e., regarding production costs, dosing regimen, and patient compliance, etc.). Work is in progress optimizing oligonucleotide chemistry with various permeation enhancers [156,199–202]. Preliminary data are encouraging and support continued investment of resources on this endeavor.

23.6.4 LIPOSOME FORMULATIONS

Early in the development of antisense oligonucleotides, it was speculated that cellular delivery of large polyanionic molecules across a plasma membrane was going to be challenging and therefore considerable effort was invested in delivery technologies such as liposomes [203–206]. Fortunately, single-stranded phosphorothioate oligonucleotides do, in fact, distribute to a variety of tissues and cell types within the tissues when administered systemically or locally [120,125,146, 149,154,164,207]. More importantly, it was demonstrated that the oligonucleotides produced a sequence-specific reduction in targeted RNA supporting that some of the oligonucleotides in the cells within the tissue was pharmacologically active [122,207–214]. Thus, the need for liposome formulations to enable antisense technology did not transpire. In contrast to single-stranded antisense oligonucleotides, double-stranded oligonucleotides that work through the RNAi pathway do

appear to require formulations to exhibit activity following systemic administration, at least as currently designed. There are marked physical–chemical differences between single-stranded DNA-like oligonucleotides and dsRNA-like oligonucleotides, which may account for these findings. More in depth coverage of formulations for delivery of siRNAs are covered elsewhere in this volume.

Although liposomes or other types of formulations are not obligatory to enable single-stranded antisense oligonucleotides, they may offer value to further improve the profile of antisense drugs. As an example, preliminary data suggest that not all of the oligonucleotide within tissues is available for binding to the target RNA. Therefore, using formulations as a means to deliver more antisense oligonucleotide to cellular compartments that contain the mRNA could increase the potency of the antisense drug. Secondly, attaching targeting ligands to liposome or other nanoparticulate carriers could change the tissue distribution and cellular distribution within tissues, resulting in more antisense drug accumulating in target tissue and less in nontarget tissues, increasing therapeutic index. In the design of such delivery systems, it is important to consider the value added by the formulation more than offsets the extra expense and complexity of formulating and characterization of the drug as well as demonstrate that the toxicological profile inherent in the formulation is an improvement over the oligonucleotide alone. Finally, increased cost for preparing the liposome formulation and different routes of administration have to be considered.

23.7 CLINICAL EXPERIENCE WITH ANTISENSE OLIGONUCLEOTIDES

Approximately 30 different antisense oligonucleotides are currently in clinical trials or approved for use in humans (Table 23.1). Similar to any other class of drugs, it can be expected that there will be failures in the clinic due to a variety of reasons such as selection of the wrong molecular target resulting in lack of efficacy, incorrect dosing, marketing consideration, toxicity, etc. It is hoped that because of the generic pharmacokinetics and chemical class-specific toxicity that the failure rates for antisense oligonucleotides will be lower than other classes of agents. However, this remains to be seen. Following is a brief summary of clinical status of several antisense oligonucleotides. An exhaustive review of clinical data on antisense oligonucleotides currently in development is beyond the scope of this review.

23.7.1 USE OF ANTISENSE OLIGONUCLEOTIDES IN ANTIVIRAL THERAPY

The most advanced antisense product is Vitravene (fomivirsen, ISIS 2922), which is marketed in the United States and the rest of the world for the treatment of patients with cytomegalovirus (CMV) retinitis. Fomivirsen was identified from a screen of a series of phosphorothioate oligodeoxynucleotides targeting human CMV (HCMV) DNA polymerase gene or to RNA transcripts of the major immediate-early regions 1 and 2 (IE1 and IE2) [215]. Fomivirsen is a 21-mer phosphorothioate oligodeoxynucleotide targeting the coding region of the IE2 gene. Fomivirsen inhibits viral protein expression, as measured by an enzyme-linked immunosorbent assay (ELISA) detecting an HCMV late protein product in fibroblasts with an EC_{50} value of $0.1\,\mu M$. Published studies in aggregate suggest that fomivirsen is a potent inhibitor of CMV replication, which is capable of inhibiting viral gene expression by an antisense mechanism of action, but also may inhibit viral replication by a nonantisense mechanism of action at higher concentrations [215–218]. Whether both mechanism of action are operational in the clinic remains to be elucidated.

Fomivirsen is approved for the local treatment of CMV retinitis in patients with acquired immunodeficiency syndrome, who are intolerant of or have a contraindication to other treatments of CMV retinitis [219,220]. The recommended dose is 330 μg every other week for two doses and then a maintenance dose administered every 4 weeks given as an intravitreal injection. The most frequently observed adverse event reported for fomivirsen is ocular inflammation (uveitis) including iritis and vitritis [221]. Ocular inflammation has been reported to occur in approximately 25% of the patients. Topical corticosteroids have been useful in treating the ocular inflammation. Open label, controlled clinical studies have been performed evaluating the safety and efficacy of fomivirsen in newly diagnosed CMV retinitis patients. Based upon assessment of fundus photographs, the median time to progression was approximately 80 days for patients treated with fomivirsen compared to 2 weeks for patients not receiving treatment [220]. Although the market for CMV retinitis is relatively small, this drug represents an important validation for the technology.

The other antiviral antisense oligonucleotide currently in clinical trials is ALN-RSV-01 for the treatment of respiratory synctial virus (RSV). RSV is a highly contagious virus that infects both upper and lower respiratory tract, predominantly in children. This antisense drug is a dsRNA that works through the RNAi mechanism [222]. Phase I trials in healthy volunteers demonstrated that ALN-RSV-01 was safe and well tolerated in over 100 volunteers [223]. ALN-RSV-01 is currently in phase II trials in which the safety, tolerability, and antiviral activity are determined in healthy volunteers experimentally challenged with RSV infection [223].

23.7.2 Use of Antisense Oligonucleotides for Cancer Therapy

Overexpression of bcl-2 is common in several cancers, in particular non-Hodgkin lymphoma, and may contribute to decreased sensitivity to chemotherapeutic agents [224,225]. An 18-mer phosphorothioate antisense oligodeoxynucleotide (oblimersen, also known as Genasense® or G3139) targeting the translation initiation codon of the bcl-2 gene was shown to inhibit the growth of lymphoma cells in severe combined immunodefecient (SCID) mice [226]. Follow-up studies demonstrated that oblimersen inhibited growth of lymphoma cells in severely immunocompromised SCID and NOD/SCID

mice, suggesting that the activity of the oligonucleotide was not secondary to an immunostimulatory effect [227]. Oblimersen has also demonstrated antitumor activity in preclinical models of various other cancers such as melanoma, prostate cancer, and gastric cancer [228–232]. Webb et al. conducted a phase I clinical trial of oblimersen at the Royal Marsden Hospital in London [233]. Genta 3139 was administered as a daily subcutaneous infusion for 14 days to patients with BCL-2-positive non-Hodgkin lymphoma. The dose of the drug given ranged from 4.6 to $73.6\,mg/m^2$. Other than local inflammation at the site of infusion, no treatment-related side effects were noted. In two patients, tomography scans revealed reductions in tumor size with one complete response. In two additional patients, the number of circulating lymphoma cells decreased during treatment. Reduced levels of bcl-2 protein expression in circulating lymphoma cells were detected in two out of five patients. These findings again demonstrate that phosphorothioate oligodeoxynucleotides can be safely administered to patients and also provide preliminary efficacy data with a bcl-2 antisense oligonucleotide. Several other phase I/II studies on oblimersen have been performed including studies in prostate cancer, breast cancer, colorectal cancer, AML, CML, multiple myeloma, and malignant melanoma [232,234]. Side effects associated with the use of oblimersen included thrombocytopenia, hypotension, fever, and hypoglycemia [232]. The sponsor completed positive phase III programs for oblimersen in advanced melanoma [235] and chronic lymphocytic leukemia [236]. Data from both trials were separately submitted to the U.S. Food and Drug Administration for marketing authorization and the melanoma data were submitted to the European Medicines Agency for approval. In all three cases, the regulatory agencies denied marketing approval for the drug. The company is continuing to conduct clinical studies with oblimersen to address deficiencies raised by the regulatory agencies.

Several additional first- and second-generation antisense oligonucleotides are currently in clinical trials for the treatment of a variety of malignancies (Table 23.1). One of the more advanced oligonucleotides in this group is OGX-011, a second-generation 2′-MOE modified gapped oligonucleotide targeting clusterin mRNA, a cytoprotective gene product [71,237]. In a phase I trial, prostate cancer patients scheduled to undergo prostectomy were pretreated with OGX-011 by 2 h intravenous infusions on days 1, 3, 5 and then weekly from days 8 to 29 at doses ranging from 40 to 640 mg combined with androgen ablative therapy [71]. Following prostectomy, tumor tissue was analyzed for drug, clusterin expression, and markers for apoptosis. The drug was well tolerated and demonstrated dose-dependent accumulation in tumor, dose-dependent decreases in clusterin expression, and increases in tumor cell apoptosis. Maximal effects were observed at doses between 480 and 640 mg per infusion. Although these data do not address clinical efficacy of OGX-011, they do demonstrate that second-generation antisense oligonucleotides can modulate gene expression in tumor tissue at a dose that is well tolerated. OGX-011 is currently in several phase II trials for prostate cancer, breast cancer, and small cell lung carcinoma. As noted in

Table 23.1, there are several additional antisense drugs being evaluated in clinical trials for treatment of malignancies. However, with limited published clinical studies on these drugs, it is difficult to determine how they are performing. It is anticipated that over the next couple of years, the trials will mature and additional information will become available.

23.7.3 USE OF ANTISENSE OLIGONUCLEOTIDES FOR THE TREATMENT OF INFLAMMATORY DISEASES

Intercellular adhesion molecule 1 (ICAM-1) is a member of the immunoglobulin gene family expressed at low levels on resting endothelial cells and can be markedly up regulated in response to inflammatory mediators such as tumor necrosis factor alpha (TNF-α), interleukin-1, and interferon-γ on a variety of cell types. ICAM-1 plays a role in the extravasation of leukocytes from the vasculature to inflamed tissue and activation of leukocytes in the inflamed tissue [238–241]. Alicaforsen (Isis 2302) was identified out of a screen of multiple first-generation phosphorothioate oligodeoxynucleotides targeting various regions of the human ICAM-1 [7,242]. Alicaforsen inhibits ICAM-1 expression in a variety of cell types by an RNase H–dependent mechanism of action [242–244]. Alicaforsen blocks both leukocyte adhesion to activated endothelial cells and costimulatory signals to T lymphocytes, both activities predicted based on previous studies with monoclonal antibodies to ICAM-1.

Numerous studies have been published demonstrating anti-inflammatory activity of the ICAM-1 antisense oligonucleotides in a variety of rodent models [184,245–254]. These data demonstrate that ICAM-1 antisense oligonucleotides administered systemically can attenuate an inflammatory response in animals and suggest that inhibition of ICAM-1 expression in humans could provide therapeutic benefit in patients suffering from a variety of inflammatory diseases.

Safety and pharmacokinetics of systemically administered alicaforsen was established in a phase I study performed at Guy's hospital in normal volunteers [144]. Alicaforsen was evaluated in a series of small phase IIa studies (20 to 40 patients in each trial) in patients with rheumatoid arthritis, psoriasis, Crohn's disease, ulcerative colitis, and renal transplant. With the exception of the psoriasis study, the trials were placebo-controlled, double-blinded in which the drug is administered as a 2 h intravenous infusion. In all trials, the drug was well tolerated. In the rheumatoid arthritis trial, alicaforsen failed to produce significant efficacy but showed positive trends [255]. The small sample size of the trial, 43 patients, did not allow definitive conclusions to be drawn. In the phase IIa Crohn's disease study, the drug was well tolerated and showed encouraging signs of efficacy [256]. Unfortunately, the initial results were not confirmed in several larger studies, therefore is no longer being studied in Crohn's disease [257–259].

Alicaforsen is currently being developed as an enema formulation for the treatment of ulcerative colitis. Several phase II studies have demonstrated efficacy of alicaforsen delivered as an enema [260–263]. Delivered as an enema, alicoforsen exhibits less than 1% bioavailability, suggesting that the beneficial effects of the drug are due to local effects on GI tissue [264]. Additional trials to support registration of alicaforsen are in progress.

ATL 1102 is a second-generation antisense oligonucleotide targeting the mRNA of another cell adhesion molecule, CD49d (the alpha subunit of VLA4). An antibody targeting VLA4, natilizumab, has demonstrated clinical benefit for the treatment of multiple sclerosis and Crohn's disease [265,266] and has been approved for the treatment of multiple sclerosis. VLA4 is an integrin expressed on lymphocytes, monocytes, and a subset of granulocytes and binds to vascular cell adhesion molecule 1 (VCAM-1) and fibronectin expressed on endothelium and other cell types. Similarly to ICAM-1, blocking VLA4 function or expression decreases extravasation of leukocytes into sites of inflammation and attenuates activation of leukocytes in inflamed tissue. A mouse-specific CD49d antisense oligonucleotide was shown to prevent and reverse clinical symptoms in a mouse model of multiple sclerosis [267]. ATL 1102 has completed phase I trials in normal volunteers and is currently in phase II trials for the treatment of multiple sclerosis (Table 23.1).

Two antisense drugs, both delivered by aerosol, are currently in clinical trials for the treatment of asthma. TPI ASM 8 is actually two different first-generation antisense oligonucleotides in a single formulation. One antisense oligonucleotide targets the common beta subunit of the IL-3, IL-5, and granulocyte-monocyte colony stimulatory factor (GM-CSF) receptor with the second antisense oligonucleotide targeting the CCR3 receptor. Early clinical trials with TPI ASM8 have generated encouraging data. This drug is currently in phase II trials. Another inhaled antisense oligonucleotide in clinical trials is ISIS 369645 (Table 23.1). ISIS 369645 is a second-generation antisense oligonucleotide targeting the IL-4 receptor alpha chain. Inhibiting the expression of the alpha chain blocks both IL-4 and IL-13 cytokine signaling, both of which have been implicated in contributing to the pathophysiology in asthma. Preclinical studies with a mouse specific IL-4 receptor alpha chain antisense oligonucleotide produced clinical benefit in both acute and chronic models of asthma when delivered by an aerosol [211]. ISIS 369645 is currently in phase I trials and if tolerated will be tested in asthmatic subjects.

23.8 CONCLUSION

As is to be expected with first-generation technology, undesirable properties have been identified for phosphorothioate oligodeoxynucleotides [168,169,268,269]. Despite these limitations, it is possible to use phosphorothioate oligodeoxynucleotides to selectively inhibit the expression of a targeted RNA in cell culture and in vivo. The pharmacokinetics of phosphorothioate oligodeoxynucleotides are similar across species and do not appear to exhibit major sequence-specific differences. When dosed at high levels, it is possible to identify toxicities in rodents and primates. However, at doses currently under evaluation in the clinic, phosphorothioate oligodeoxynucleotides have been well tolerated. In addition, there is evidence that phosphorothioate oligodeoxynucleotides provide clinical benefit to patients with viral infections, cancer, and inflammatory diseases. There are still several phosphorothioate oligodeoxynucleotides in late stage clinical trials, which will hopefully deliver more effective therapies for patients suffering from life-threatening and debilitating diseases.

Extensive medicinal chemistry efforts have been successfully focused on identifying improved antisense oligonucleotides, which address some of these issues. There are at least four areas in which chemistry can add value to first-generation drugs: increase potency, decrease toxicity, alter pharmacokinetics, and lower costs. As an example, numerous modified oligonucleotides have been identified, which have a higher affinity for target RNA than phosphorothioate oligodeoxynucleotides [92,97,101–103,270–272]. Oligonucleotide modifications, which exhibit increased resistance to serum and cellular nucleases, have been identified, enabling use of oligonucleotides that do not have phosphorothioate linkages. The tissue distribution of oligonucleotides may be altered with either chemical modifications or formulations [87,147,154,155, 157,198,272,273–275]. Preliminary data also suggest that oral delivery of antisense oligonucleotides may be feasible [155,199–201]. Finally a number of modified oligonucleotides have been described, which potentially exhibited less toxicities than first-generation phosphorothioate oligodeoxynucleotides [86,168,197]. The most advanced second-generation antisense chemistry is 2'-O-methoxyethyl, with over 700 subjects treated with oligonucleotides containing this modification [276]. To date, 2'-O-methoxyethyl-modified oligonucleotides have been well tolerated and have exhibited a marked increase in therapeutic index compared to first-generation phosphorothioate oligodeoxynucleotides.

In conclusion, first-generation phosphorothioate oligodeoxynucleotides have proven to be valuable pharmacological tools for the researcher and have produced new therapies for the patient. Identification of improved second- and third-generation oligonucleotides with novel formulation should be better therapies for patients. Although tremendous progress has been made for antisense technology during the past 14 years, there are far more questions that remain for the technology.

REFERENCES

1. Watson, J. and Crick, F. (1953) *Nature* **171**, 737.
2. Lanz, R. B., McKenna, N. J., Onate, S. A., Albrecht, U., Wong, J., Tsai, S. Y., Tsai, M. J., and O'Malley, B. W. (1999) *Cell* **97**(1), 17–27.
3. Esau, C., Kang, X., Peralta, E., Hanson, E., Marcusson, E. G., Ravichandran, L. V., Sun, Y., Koo, S., Perera, R. J., Jain, R., Dean, N. M., Freier, S. M., Bennett, C. F., Lollo, B., and Griffey, R. H. (2004) *J. Biol. Chem.* **279**, 52361–52365.
4. Esau, C., Davis, S., Murray, S. F., Yu, X. X., Pandey, S. K., Pear, M., Watts, L., Booten, S. L., Graham, M., McKay, R. A., Subramaniam, A., Propp, S., Lollo, B. A., Freier, S., Bennett, C. F., Bhanot, S., and Monia, B. P. (2006) *Cell Metab.* **3**, 87–98.
5. Krutzfeldt, J., Rajewsky, N., Braich, R., Rajeev, K. G., Tuschl, T., Manoharan, M., and Stoffel, M. (2005) *Nature* **438**, 685–689.
6. Prasanth, K. V., Prasanth, S. G., Xuan, Z., Hearn, S., Freier, S. M., Bennett, C. F., Zhang, M. Q., and Spector, D. L. (2005) *Cell* **123**(2), 249–263.
7. Chiang, M. Y., Chan, H., Zounes, M. A., Freier, S. M., Lima, W. F., and Bennett, C. F. (1991) *J. Biol. Chem.* **266**, 18162–18171.
8. Monia, B. P., Lesnik, E. A., Gonzalez, C., Lima, W. F., McGee, D., Guinosso, C. J., Kawasaki, A. M., Cook, P. D., and Freier, S. M. (1993) *J. Biol. Chem.* **268**(19), 14514–14522.
9. Giles, R. V., Spiller, D. G., and Tidd, D. M. (1995) *Antisense Res. Dev.* **5**, 23–31.
10. Condon, T. P. and Bennett, C. F. (1996) *J. Biol. Chem.* **271**(48), 30398–30403.
11. Lima, W. F., Wu, H., and Crooke, S. T. (2001) *Meth. Enzymol.* **341**, 430–440.
12. Wu, H., Lima, W. F., Zhang, H., Fan, A., Sun, H., and Crooke, S. T. (2004) *J. Biol. Chem.* **279**, 17181–17189.
13. Crouch, R. J. (1990) *New Biol.* **2**(9), 771–777.
14. Wu, H., Lima, W. F., and Crooke, S. T. (1998) *Antisense Nucleic Acid Drug Dev.* **8**, 53–61.
15. Wu, H., Lima, W. F., and Crooke, S. T. (1999) *J. Biol. Chem.* **274**(40), 28270–28278.
16. Fire, A., Xu, S., Montgomery, M. K., Kostas, S. A., Driver, S. E., and Mello, C. C. (1998) *Nature* **391**, 806–811.
17. Grishok, A. and Mello, C. C. (2002) *Adv. Genet.* **46**, 339–360.
18. Elbashir, S. M., Lendeckel, W., and Tuschl, T. (2001) *Genes Dev.* **15**, 188–200.
19. Harborth, J., Elbashir, S. M., Bechert, K., Tuschl, T., and Weber, K. (2001) *J. Cell Sci.* **114**, 4557–4565.
20. Zamore, P. D. (2002) *Science* **296**, 1265–1269.
21. McManus, M. T. and Sharp, P. A. (2002) *Nat. Rev. Genet.* **3**, 737–747.
22. Grishok, A., Tabara, H., and Mello, C. C. (2000) *Science* **287**, 2494–2497.
23. Bernstein, E., A. A., C., Hammond, S. M., and Hannon, G. J. (2001) *Nature* **409**, 295–296.
24. Knight, S. W. and Bass, B. L. (2001) *Science* **293**, 2269–2271.
25. Zamore, P. D., Tuschl, T., Sharp, P. A., and Bartel, D. P. (2000) *Cell* **101**, 25–33.
26. Meister, G., Landthaler, M., Patkaniowska, A., Dorsett, Y., Teng, G., and Tuschl, T. (2004) *Mol. Cell* **15**, 185–197.
27. Liu, J., Carmell, M. A., Rivas, F. V., Marsden, C. G., Thomson, J. M., Song, J.-J., Hammond, S. M., Joshua-Tor, L., and Hannon, G. J. (2004) *Science* **305**, 1437–1441.
28. Torrence, P. F., Maitra, R. K., Lesiak, K., Khamnei, S., Zhou, A., and Silverman, R. H. (1993) *Proc. Natl. Acad. Sci. U.S.A.* **90**, 1300–1304.
29. Maran, A., Maitra, R. K., Kumar, A., Dong, B., Xiao, W., Li, G., Williams, B. R. G., Torrence, P. F., and Silverman, R. H. (1994) *Science* **265**, 789–792.
30. Leaman, D. W., Longano, F. J., Okicki, J. R., Soike, K. F., Torrence, P. F., Silverman, R. H., and Cramer, H. (2002) *Virology* **292**, 70–77.
31. Bennett, C. F., Condon, T. P., Grimm, S., Chan, H., and Chiang, M. Y. (1994) *J. Immunol.* **152**(7), 3530–3540.
32. Dean, N. M., McKay, R., Condon, T. P., and Bennett, C. F. (1994) *J. Biol. Chem.* **269**, 16416–16424.
33. Monia, B. P., Johnston, J. F., Geiger, T., Muller, M., and Fabbro, D. (1996) *Nat. Med.* **2**(6), 668–675.
34. Helene, C. and Toulm, J.-J. (1990) *Biochim. Biophys. Acta Gene Struct. Express.* **1049**, 99–125.
35. Baker, B. F., Lot, S. S., Condon, T. P., Cheng-Flournoy, S., Lesnik, E. A., Sasmor, H. M., and Bennett, C. F. (1997) *J. Biol. Chem.* **272**, 11994–12000.

36. Chin, D. J., Green, G. A., Zon, G., Szoka, F. C., Jr., and Straubinger, R. M. (1990) *New Biol.* **2**, 1091–1100.

37. Leonetti, J. P., Mechti, N., Degols, G., Gagnor, C., and Lebleu, B. (1991) *Proc. Natl. Acad. Sci. U.S.A.* **88**, 2702–2706.

38. Bennett, C. F., Chiang, M. Y., Chan, H., Shoemaker, J. E. E., and Mirabelli, C. K. (1992) *Mol. Pharmacol.* **41**(6), 1023–1033.

39. Lorenz, P., Baker, B. F., Bennett, C. F., and Spector, D. L. (1998) *Mol. Biol. Cell* **9**, 1007–1023.

40. Dominski, Z. and Kole, R. (1994) *Mol. Cell. Biol.* **14**, 7445–7454.

41. Dominski, Z. and Kole, R. (1993) *Proc. Natl. Acad. Sci. U.S.A.* **90**, 8673–8677.

42. Sierakowska, H., Sambade, M. J., Agrawal, S., and Kole, R. (1996) *Proc. Natl. Acad. Sci. U.S.A.* **93**, 12840–12844.

43. Friedman, K. J., Kole, J., Cohn, J. A., Knowles, M. R., Silverman, L. M., and Kole, R. (1999) *J. Biol. Chem.* **274**, 36193–36199.

44. Sazani, P., Kang, S. H., Maier, M. A., Wei, C., Dillman, J., Summerton, J., Manoharan, M., and Kole, R. (2001) *Nucleic Acids Res.* **29**, 3965–3974.

45. Taylor, J., Zhang, Q., Wyatt, J., and Dean, N. (1999) *Nat. Biotech.* **17**, 1097–1100.

46. Karras, J. G., Maier, M. A., Lu, T., Watt, A., and Manoharan, M. (2001) *Biochem.* **40**, 7853–7859.

47. Sazani, P., Gemignani, F., Kang, S. H., Maier, M. A., Manoharan, M., Persmark, M., Bortner, D., and Kole, R. (2002) *Nat. Biotechnol.* **20**, 1228–1233.

48. Hua, Y., Vickers, T. A., Baker, B. F., Bennett, C. F., and Krainer, A. R. (2007) *PLOS Biology* **5**, e73.

49. Vickers, T. A., Zhang, H., Graham, M. J., Lemonidis, K. M., Zhao, C., and Dean, N. M. (2006) *J. Immunol.* **176**, 3652–3661.

50. Roberts, J., Palma, E., Sazani, P., Orum, H., and Kole, R. (2006) *Mol. Therap.* **14**, 471–475.

51. Suwanmanee, T., Sierakowaska, H., Fuchareon, S., and Kole, R. (2002) *Mol. Therap.* **6**, 718–726.

52. Vickers, T. A., Wyatt, J. R., Burckin, T., Bennett, C. F., and Freier, S. M. (2001) *Nucleic Acids Res.* **29**, 1293–1299.

53. Froehler, B. C., Jones, R. J., Cao, X., and Terhorst, T. J. (1993) *Tetrahedon Lett.* **34**(6), 1003–1006.

54. Gutierrez, A. J., Terhorst, T. J., Matteucci, M. D., and Frochler, B. C. (1994) *J. Am. Chem. Soc.* **116**(13), 5540–5544.

55. Guttierrez, A. J. and Froehler, B. C. (1996) *Tetrahedron Lett.* **37**(23), 3959–3962.

56. Lin, K. Y., Pudlo, J. S., Jones, R. J., Bischofberger, N., Matteucci, M. D., and Froehler, B. C. (1994) *Bioorg. Med. Chem. Lett.* **4**(8), 1061–1064.

57. Seela, F. and Thomas, H. (1995) *Helv. Chim. Acta* **78**(1), 94–108.

58. Seela, F., Zulauf, M., Rosemeyer, H., and Reuter, H. (1996) *J. Chem. Soc., Perkin Trans.* **2**(11), 2373–2376.

59. Balow, G., Mohan, V., Lesnik, E. A., Johnston, J. F., Monia, B. P., and Acevedo, O. L. (1998) *Nucleic Acids Res.* **26**, 3350–3357.

60. Flanagan, W. M., Wolf, J. J., Grant, D., Lin, K. Y., and Matteucci, M. D. (1999) *Nat. Biotechnol.* **17**, 48–52.

61. Lin, K.-Y., Jones, R. J., and Matteucci, M. (1995) *J. Am. Chem. Soc.* **117**, 3873–3874.

62. Lin, K. Y. and Matteucci, M. D. (1998) *J. Am. Chem. Soc.* **120**, 8531–8532.

63. Flanagan, W. M., Wolf, J. J., Olson, P., Grant, D., Lin, K. Y., Wagner, R. W., and Matteucci, M. D. (1999) *Proc. Natl. Acad. Sci. U.S.A.* **96**, 3513–3518.

64. Ramasamy, K. S., Zounes, M., Gonzalez, C., Freier, S. M., Lesnik, E. A., Cummins, L.L., Griffey, R. H., Monia, B. P., and Cook, P. D. (1994) *Tetrahedron Lett.* **35**(2), 215–218.

65. Manoharan, M., Ramasamy, K. S., Mohan, V., and Cook, P. D. (1996) *Tetrahedron Lett.* **37**(43), 7675–7678.

66. Cook, P. D. (1998) Antisense medicinal chemistry. In: Crooke, S. T. (ed.) *Antisense Research and Application*, Springer-Verlag, Berlin, Heidelberg.

67. Egli, M., Usman, N., and Rich, A. (1993) *Biochemistry* **32**, 3221–3237.

68. Tereshko, V., Portmann, S., Tay, E. C., Martin, P., Natt, F., Altmann, K.-H., and Egli, M. (1998) *Biochemistry* **37**, 10626–10634.

69. Teplova, M., Minasov, G., Tereshko, V., Inamati, G. B., Cook, P. D., Manoharan, M., and Egli, M. (1999) *Nat. Structural Biol.* **6**, 535–539.

70. Kastelein, J. J. P., Wedel, M. K., Baker, B. F., Su, J., Bradley, J. A., Yu, R. Z., Chuang, E., Graham, M. J., and Crooke, R. M. (2006) *Circulation* **114**, 1729–1735.

71. Chi, K. M., Eisenhauer, E., Fazli, L., Jones, E. C., Goldenberg, S. L., Powers, J., and Gleave, M. E. (2005) *J. Natl. Cancer Inst.* **97**, 1287–1296.

72. Griffey, R. H., Monia, B. P., Cummins, L. L., Freier, S., Greig, M. J., Guinosso, C. J., Lesnik, E., Manalili, S. M., Mohan, V., Owens, S., Ross, B. R., Sasmor, H., Wancewicz, E., Weiler, K., Wheeler, P. D., and Cook, P. D. (1996) *J. Med. Chem.* **39**, 5100–5109.

73. Prakash, T. P., Kawasaki, A. M., Vasquez, G., Fraser, A. S., Casper, M. D., Cook, P. D., and Manoharan, M. (1999) *Nucleos. Nucleot.* **18**, 1381–1382.

74. Prakash, T. P., Kawasaki, A. M., Fraser, A. S., Vasquez, G., and Manoharan, M. (2002) *J. Org. Chem.* **67**, 357–369.

75. Wengel, J. (1999) *Acc. Chem. Res.* **32**, 301–308.

76. Jepsen, J. S. and Wengel, J. (2004) *Curr. Opin. Drug Discov. Dev.* **7**, 188–194.

77. Fluiter, K., Frieden, M., Vreijling, J., Rosenbohm, C., de Wissel, M. B., Christensen, S. M., Koch, T., Orum, H., and Baas, F. (2005) *Chembiochem* **6**, 1–6.

78. Fluiter, K., ten Asbroek, A. L. M. A., de Wissel, M. B., Jakobs, M. E., Wissenbach, M., Olsson, H., Olsen, O., Oerum, H., and Baas, F. (2003) *Nucleic Acids Res.* **31**, 953–962.

79. Swayze, E. E., Siwkowski, A. M., Wancewicz, E. V., Migawa, M. T., Wyrzykiewicz, T. K., Hung, G., Monia, B. P., and Bennett, C. F. (2007) *Nucleic Acids Res.* **35**, 687–700.

80. Inoue, H., Hayase, Y., Iwai, S., and Ohtsuke, E. (1987) *FEBS Lett.* **215**, 327–330.

81. Lima, W. F., Mohan, V., and Crooke, S. T. (1997) *J. Biol. Chem.* **272**(29), 18191–18199.

82. Lima, W. F. and Crooke, S. T. (1997) *J. Biol. Chem.* **272**(44), 27513–27516.

83. Lamond, A. I. and Sproat, B. S. (1993) *FEBS Lett.* **325**, 123–127.

84. Yu, D., Iyer, R. P., Shaw, D. R., Lisziewicz, J., Li, Y., Jiang, Z., Roskey, A., and Agrawal, S. (1996) *Bioorg. Med. Chem.* **4**(10), 1685–1692.

85. McKay, R. A., Miraglia, L. J., Cummins, L. L., Owens, S. R., Sasmor, H., and Dean, N. M. (1999) *J. Biol. Chem.* **274**, 1715–1722.

86. Henry, S., Stecker, K., Brooks, D., Monteith, D., Conklin, B., and Bennett, C. F. (2000) *J. Pharm. Exp. Therap.* **292**, 468–479.

87. Geary, R. S., Watanabe, T. A., Truong, L., Freier, S., Lesnik, E. A., Sioufi, N. B., Sasmor, H., Manoharan, M., and Levin, A. A. (2001) *J. Pharmacol. Exp. Ther.* **296**, 890–897.

88. Geary, R. S., Khatsenko, O., Bunker, K., Crooke, R., Moore, M., Burckin, T., Truong, L., Sasmor, H., and Levin, A. A. (2001) *J. Pharmacol. Exp. Ther.* **296**, 898–904.

89. Yu, R. Z., Zhang, H., Geary, R. S., Graham, M., Masarjian, L., Lemonidis, K., Crooke, R., Dean, N. M., and Levin, A. A. (2001) *J. Pharmacol. Exp. Ther.* **296**, 388–395.

90. Gryaznov, S. M. (1999) *Biochim. Biophys. Acta,* **1489**, 131–147.

91. Faria, M., Spiller, D. G., Dubertret, C., Nelson, J. S., White, M. R. H., Scherman, D., Helene, C., and Giovannangeli, C. (2001) *Nat. Biotechnol.* **19**, 40–44.

92. Larsen, H. J., Bentin, T., and Nielsen, P. E. (1999) *Biochim. Biophys. Acta* **1489**, 159–166.

93. McMahon, B. M., Mays, D., Lipsky, J., Stewart, J. A., Fauq, A., and Richelson, E. (2002) *Antisense Nucleic Acid Drug Dev.* **12**, 65–73.

94. Maier, M. A., Esau, C. C., Siwkowski, A. M., Wancewicz, E. V., Albertshofer, K., Kingberger, G. A., Kabada, N. S., Watanabe, T., Manoharan, M., Bennett, C. F., Griffey, R. H., and Swayze, E. E. (2006) *J. Med. Chem.* **49**, 2534–2542.

95. Albertshofer, K., Siwkowski, A., Wancewicz, E. V., Esau, C. C., Watanabe, T., Nishihara, K. C., Kinberger, G. A., Malik, L., Eldrup, A. B., Manoharan, M., Geary, R. S., Monia, B. P., Swayze, E. E., Griffey, R. H., Bennett, C. F., and Maier, M. A. (2005) *J. Med. Chem.* **48**, 6741–6749.

96. Summerton, J. and Weller, D. (1997) *Antisense Nucleic Acid Drug Dev.* **7**, 187–195.

97. Summerton, J. (1999) *Biochim. Biophys. Acta,* **1489**, 141–158.

98. Arora, V., Knapp, D. C., Smith, B. L., Stadtfield, M. L., Stein, D. A., Reddy, M. T., Weller, D. D., and Iversen, P. L. (2000) *J. Pharm. Exp. Ther.* **292**, 921–930.

99. Arora, V., Knapp, D. C., Reddy, M. T., Weller, D. D., and Iversen, P. L. (2002) *J. Pharm. Sci.* **91**, 1009–1111.

100. Manoharan, M., Inamati, G. B., Lesnik, E. A., Sioufi, N. B., and Freier, S. M. (2002) *Chembiochem* **3**, 1257–1260.

101. Manoharan, M. (2002) *Antisense Nuleic Acid Drug Dev.* **12**, 103–128.

102. Cook, P. D. (2001) Medicinal chemistry of antisense oligonucleotides. In: Crooke, S. T. (ed). *Antisense Drug Technology: Principles, Strategies, and Applications*, Marcel Dekker, New York.

103. Matteucci, M. (1996) *Perspectives Drug Dis. Des.* **4**, 1–16.

104. Braasch, D. A. and Corey, D. R. (2002) *Biochemistry* **41**, 4503–4510.

105. Herdewijn, P. (2000) *Antisense Nucleic Acid Drug Dev.* **10**, 297–310.

106. Swayze, E. E. and Bhat, B. (2007) The medicinal chemistry of oligonucleotides. In: Crooke, S. T. (ed). *Antisense Drug Technology: Principles, Strategies and Applications*, 2nd ed., Taylor and Francis, Boca Raton, FL.

107. Crooke, R. M., Graham, M. J., Cooke, M. E., and Crooke, S. T. (1995) *J. Pharmacol. Exp. Ther.* **275**(1), 462–473.

108. Krieg, A. M. (1993) *Clin.Chem.* **39**, 710–712.

109. Stein, C. A., Tonkinson, J. L., Zhang, L.-M., Yakubov, L., Gervasoni, J., Taub, R., and Rotenberg, S. A. (1993) *Biochemistry* **32**, 4855–4861.

110. Tonkinson, J. L. and Stein, C. A. (1994) *Nucl. Acids Res.* **22**, 4268–4275.

111. Loke, S. L., Stein, C. A., Zhang, X. H., Mori, K., Nakanishi, M., Subasinghe, C., Cohen, J. S., and Neckers, L. M. (1989) *Proc. Natl. Acad. Sci. U.S.A.* **86**, 3474–3478.

112. Wu-Pong, S., Weiss, T. L., and Hunt, C. A. (1992) *Pharm. Res.* **9**, 1010–1017.

113. Beltinger, C., Saragovi, H. U., Smith, R. M., LeSauteur, L., Shah, N., DeDionisio, L., Christensen, L., Raible, A., Jarett, L., and Gewirtz, A. M. (1995) *J. Clin. Invest.* **95**, 1814–1823.

114. Iversen, P. L., Zhu, S., Meyer, A., and Zon, G. (1992) *Antisense Res. Dev.* **2**, 211–222.

115. Delong, R., Stephenson, K., Loftus, T., Fisher, M., Alahari, S., Nolting, A., and Juliano, R. L. (1997) *J. Pharm. Sci.* **86**, 762–764.

116. Bongartz, J.-P., Aubertin, A.-M., Milhaud, P. G., and Lebleu, B. (1994) *Nucleic Acids Res.* **22**, 4681–4688.

117. Leonetti, J. P., Degols, G., and Lebleu, B. (1990) *Bioconjug. Chem.* **1**, 149–153.

118. Zelphati, O. and Szoka, F. C., Jr. (1996) *Pharm. Res.* **13**(9), 1367–1372.

119. Benimetskaya, L., Guzzo-Pernell, N., Liu, S. T., Lai, J. C., Miller, P., and Stein, C. A. (2002) *Bioconjug. Chem.* **13**, 177–187.

120. Cossum, P. A., Sasmor, H., Dellinger, D., Truong, L., Cummins, L., Owens, S. R., Markham, P. M., Shea, J. P., and Crooke, S. (1993) *J. Pharmacol. Exp. Ther.* **267**, 1181–1190.

121. Graham, M. J., Crooke, S. T., Monteith, D. K., Cooper, S. R., Lemonidis, K. M., Stecker, K. K., Martin, M. J., and Crooke, R. M. (1998) *J. Pharm. Exp. Ther.* **286**(1), 447–458.

122. Dean, N. M. and McKay, R. (1994) *Proc. Natl. Acad. Sci. U.S.A.* **91**, 11762–11766.

123. Zhang, H., Cook, J., Nickel, J., Yu, R., Stecker, K., Myers, K., and Dean, N. M. (2000) *Nat. Biotech.* **18**, 862–867.

124. Butler, M., McKay, R. A., Popoff, I. J., Gaarde, W. A., Witchell, D., Murray, S. F., Dean, N. M., Bhanot, S., and Monia, B. P. (2002) *Diabetes* **51**, 1028–1034.

125. Butler, M., Stecker, K., and Bennett, C. F. (1997) *Lab. Invest.* **77**, 379–388.

126. Yu, R. Z., Baker, B., Chappell, A., Geary, R. S., Cheung, E., and Levin, A. A. (2002) *Anal. Biochem.* **304**, 19–25.

127. Sewell, L. K., Geary, R. S., Baker, B. F., Glover, J. M., Mant, T. G. K., Yu, R. Z., Tami, J. A., and Dorr, F. A. (2002) *J. Pharm. Exp. Therap.* **303**, 1334–1343.

128. Graham, M., Crooke, S. T., Lemonidis, K. M., Gaus, H. J., Templin, M. V., and Crooke, R. M. (2001) *Biochem. Pharmacol.* **62**, 297–306.

129. Plenat, F., Klein-Monhoven, N., Marie, B., Vignaud, J.-M., and Duprez, A. (1995) *Am. J. Pathol.* **147**, 124–135.

130. Benimetskaya, L., Tonkinson, J. L., Koziolkiewicz, M., Karwowski, B., Guga, P., Zeltser, R., Stec, W., and Stein, C. A. (1995) *Nucleic Acids Res.* **23**, 4239–4245.

131. Bijsterbosch, M. K., Manoharan, M., Rump, E. T., De Vrueh, R. L. A., van Veghel, R., Tivel, K. L., Biessen, E. A. L., Bennett, C. F., Cook, P. D., and van Berkel, T. J. C. (1997) *Nucleic Acids Res.* **25**(16), 3290–3296.

132. Biessen, E. A., Vietsch, H., Kuiper, J., Bijsterbosch, M. K., and Berkel, T. J. (1998) *Mol. Pharmacol.* **53**, 262–269.

133. Steward, A., Christian, R. A., Hamilton, K. O., and Nicklin, P. L. (1998) *Biochem. Pharmacol.* **56**, 509–516.

134. Butler, M., Crooke, R., Graham, M., Lougheed, M., Murray, S., Witchell, D. R., Steinbrecher, U., and Bennett, C. F. (2000) *J. Pharm. Exp. Ther.* **292**, 489–496.

135. Rappaport, J., Hanss, B., Kopp, J. B., Copeland, T. D., Bruggeman, L. A., Coffman, T. M., and Klotman, P. E. (1995) *Kidney Int.* **47**, 1462–1469.

136. Hanns, B., Leal-Pinto, E., Bruggeman, L. A., Copeland, T. D., and Klotman, P. E. (1998) *Proc. Natl. Acad. Sci. U.S.A.* **95**, 1921–1926.

137. Sawai, K., Takenori, M., Takakura, Y., and Hashida, M. (1995) *Antisense Res. Dev.* **5**, 279–287.

138. Geary, R. S., Leeds, J. M., Fitchett, J., Burckin, T., Truong, L., Spainhour, C., Creek, M., and Levin, A. A. (1997) *Drug Metab. Dispos.* **25**, 1272–1281.

139. Geary, R. S., Yu, R. Z., Watanabe, T., Henry, S. P., Hardee, G. E., Chappell, A., Matson, J. E., Sasmor, H., Cummins, L., and Levin, A. A. (2003) *Drug Metab. Dispos.* **31**, 1419–1428.

140. Levin, A. A., Geary, R. S., Leeds, J. M., Monteith, D. K., Yu, R., Templin, M. V., and Henry, S. P. (1998) The pharmacokinetics and toxicity of phosphorothioate oligonucleotides. In: Thomas, J.A. (ed). *Biotechnology and Safety Assessment*, Taylor & Francis, Boca Raton, FL.

141. Levin, A. A., Yu, R. Z., and Geary, R. S. (2007) Basic principles of the pharmacokinetics of antisense oligonucleotide drugs. In: Crooke, S. T. (ed). *Antisense Drug Technology: Principles, Strategies and Applications*, 2nd ed., Taylor and Francis, Boca Raton, FL.

142. Phillips, J. A., Craig, S. J., Bayley, D., Christian, R. A., Geary, R., and Nicklin, P. L. (1997) *Biochem. Pharmacol.* **54**, 657–668.

143. Geary, R. S., Leeds, J. M., Henry, S. P., Monteith, D. K., and Levin, A. A. (1997) *Anticancer Drug Design* **12**, 383–393.

144. Glover, J. M., Leeds, J. M., Mant, T. G. K., Amin, D., Kisner, D. L., Zuckerman, J. E., Geary, R. S., Levin, A. A., and Shanahan, W. R., Jr. (1997) *J. Pharmacol. Exp. Ther.* **282**, 1173–1180.

145. Yu, R. Z., Geary, R. S., Leeds, J. M., Watanabe, T., Moore, M., Fitchett, J., Matson, J., Burckin, T., Templin, M. V., and Levin, A. A. (2000) *J. Pharm. Sci.* **90**, 182–193.

146. Geary, R. S., Yu, R. Z., Leeds, J. M., Watanabe, T. A., Henry, S. P., Levin, A. A., and Templin, M. V. (2001) Pharmacokinetic properties in animals. In: Crooke, S. T. (ed). *Antisense Drug Technology: Principles, Strategies, and Applications*, Marcel Dekker, New York.

147. Crooke, S. T., Graham, M. J., Zuckerman, J. E., Brooks, D., Conklin, B. S., Cummins, L. L., Greig, M. J., Guinosso, C. J., Kornburst, D., Manoharan, M., Sasmor, H. M., Schleich, T., Tivel, K. L. and Griffey, R. H. (1996) *J. Pharmacol. Exp. Ther.* **277**, 923–937.

148. Watanabe, T. A., R. S., G., and Levin, A. A. (2006) *Oligonucleotides* **16**, 169–180.

149. Agrawal, S., Temsamani, J., and Tang, J. Y. (1991) *Proc. Natl. Acad. Sci. U.S.A.* **88**, 7595–7599.

150. Bishop, M. R., Iversen, P. L., Bayever, E., Sharp, J. G., Greiner, T. C., Copple, B. L., Ruddon, R., Zon, G., Spinolo, J., Arneson, M., Armitage, J. O., and Kessinger, A. (1996) *J. Clin. Oncol.* **14**, 1320–1326.

151. Cossum, P. A., Truong, L., Owens, S. R., Markham, P. M., Shea, J. P., and Crooke, S. T. (1994) *J. Pharmacol. Exp. Ther.* **269**(1), 89–94.

152. Nicklin, P. L., Craig, S. J., and Phillips, J. A. (1998) Pharmacokinetic properties of phosphorothioates in animals—absorption, distribution, metabolism and elimination. In: Crooke, S. T. (ed). *Antisense Research and Application*, Springer-Verlag, Berlin, Heidelberg.

153. Nicklin, P. L., Bayley, D., Giddings, J., Craig, S. J., Cummins, L. L., Hastewell, J. G., and Phillips, J. A. (1998) *Pharm. Res.* **15**, 583–591.

154. Templin, M. V., Levin, A. A., Graham, M. J., Aberg, P. M., Axelsson, B. I., Butler, M., Geary, R. S., and Bennett, C. F. (1999) *Antisense Nucleic Acid Drug Dev.* **10**, 359–368.

155. Agrawal, S., Zhang, X., Lu, Z., Zhao, H., Tamburin, J. M., Yan, J., Cai, H., Diasio, R. B., Habus, I., Jiang, Z., Iyer, R. P., Yu, D., and Zhang, R. (1995) *Biochem. Pharm.* **50**, 571–576.

156. Khatsenko, O., Morgan, R. A., Truong, L., York-Defalco, C., Sasmor, H., Conklin, B., and Geary, R. S. (2000) *Antisense Nucleic Acid Drug Dev.* **10**, 35–44.

157. Zhang, R., Lu, Z., Zhao, H., Zhang, X., Diasio, R. B., Habus, I., Jiang, Z., Iyer, R. P., Yu, D., and Agrawal, S. (1995) *Biochem. Pharmacol.* **50**, 545–556.

158. Agrawal, S., Jiang, Z., Zhao, Q., Shaw, D., Cai, Q., Roskey, A., Channavajjala, L., Saxinger, C., and Zhang, R. (1997) *Proc. Natl. Acad. Sci. U.S.A.* **94**, 2620–2625.

159. Graham, M. J., Lemonidis, K. M., Whipple, C. P., Subramaniam, S., Monia, B. P., Crooke, S. T., and Crooke, R. M. (2007) *J. Lipid Res.* **48**, 763–767.

160. Crooke, R. M., Graham, M. J., Lemonidis, K. M., Whipple, C. P., Koo, S., and Perera, R. J. (2005) *J. Lipid Res.* **46**, 872–884.

161. Sloop, K. W., Showalter, A. D., Cox, A. L., Cao, J. X.-C., Siesky, A. M., Zhang, H. Y., Bodenmiller, D. M., Jacobs, S. J., Irizarry, A. R., Murray, S. F., Booten, S. L., Finger, T., McKay, R. A., Monia, B. P., Bhanot, S., and Michael, M. D. (2007) *J. Biol. Chem.* **282**, 19113–19122.

162. Watts, L., Manchem, V., Leedom, T., Rivard, A., McKay, R. A., Bao, D., Neroladakis, T., Monia, B. P., Bodenmiller, D., Cao, J., Zhang, H., Cox, A., Jacobs, S., Michael, M., Sloop, K., and Bhanot, S. (2005) *Diabetes* **54**(6), 1846–1853.

163. Yu, R. Z., Kim, T. W., Hong, A., Watanabe, T. A., Gaus, H. J., and Geary, R. (2007) *Drug Metab. Dispos.* **35**, 460–468.

164. Geary, R. S., Yu, R. Z., Siwkowski, A., and Levin, A. A. (2007) Pharmacokinetic/pharmacodynamic properties of phosphorothioate 2′-O-(2-methoxyethyl)-modified antisense oligonucleotides in animals and man. In: Crooke, S. T. (ed). *Antisense Drug Technology: Principles, Strategies and Applications*, Taylor and Francis, Boca Raton, FL.

165. Galbraith, W. M., Hobson, W. C., Giclas, P. C., Schechter, P. J., and Agrawal, S. (1994) *Antisense Res. Dev.* **4**(3), 201–206.

166. Henry, S. P., Bolte, H., Auletta, C., and Kornbrust, D. J. (1997) *Toxicology* **120**(2), 145–155.

167. Henry, S. P., Grillone, L. R., Orr, J. L., Bruner, R. H., and Kornbrust, D. J. (1997) *Toxicology* **116**(1–3), 77–88.

168. Levin, A. A., Henry, S. P., Monteith, D., and Templin, M. V. (2001) Toxicity of antisense oligonucleotides. In: Crooke, S. T. (ed). *Antisense Drug Technology: Principles, Strategies, and Applications*, Marcel Dekker, New York.

169. Levin, A. A., Monteith, D. K., Leeds, J. M., Nicklin, P. L., Geary, R. S., Butler, M., Templin, M. V., and Henry, S. P. (1998) Toxicity of oligodeoxynucleotide therapeutic agents. In: Crooke, S. T. (ed). *Antisense Research and Application*, Springer-Verlag, Berlin, Heidelberg.

170. Henry, S. P., Giclas, P. C., Leeds, J., Pangburn, M., Auletta, C., Levin, A. A., and Kornbrust, D. J. (1997) *J. Pharmacol. Exp. Ther.* **281**, 810–816.

171. Sheehan, J. P. and Phan, T. M. (2001) *Biochemistry* **40**, 4980–4989.

172. Henry, S. P., Novotny, W., Leeds, J., Auletta, C., and Kornbrust, D. J. (1997) *Antisense Nucleic Acid Drug Dev.* **7**(5), 503–510.

173. Wallace, T. L., Bazemore, S. A., Kornbrust, D. J., and Cossum, P. A. (1996) *J. Pharm. Exp. Ther.* **278**, 1313–1317.

174. Nicklin, P. L., Ambler, J., Mitchelson, A., Bayley, D., Phillips, J. A., Craig, S. J., and Monia, B. P. (1997) *Nucleos. Nucleot.* **16**, 1145–1153.

175. Sheehan, J. P. and Lan, H.-C. (1998) *Blood* **92**, 1617–1625.

176. Cornish, K. G., Iversen, P., Smith, L., Arneson, M., and Bayever, E. (1993) *Pharmacol. Commun.* **3**, 239–247.

177. Black, L. E., Degeorge, J. J., Cavagnaro, J. A., Jordan, A., and Ahn, C. H. (1993) *Antisense Res. Dev.* **3**, 399–404.

178. Black, L. E., Degeorge, J. J., Cavagnaro, J. A., Jordan, A., and Ahn, C.-H. (1994) *Antisense Res. Dev.* **3**, 399–404.

179. Leeds, J. M., Henry, S. P., Truong, L., Zutshi, A., Levin, A. A., and Kornbrust, D. (1997) *Drug Metab. Dispos.* **25**, 921–926.

180. Leeds, J. M. and Geary, R. S. (1998) Pharmacokinetic properties of phosphorothioate oligonucleotides in humans. In: Crooke, S. T. (ed). *Antisense Research and Application*, Springer-Verlag, Berlin, Heidelberg.

181. Henry, S. P., Kim, T.-W., Kramer-Strickland, K., Zanardi, T. A., Fey, R. A., and Levin, A. A. (2007) Toxicological Properties of 2′-*O*-methoxyethyl chimeric antisense inhibitors in animals and man. In: Crooke, S. T. (ed). *Antisense Drug Technology: Principles, Strategies, and Applications*, 2nd ed., Taylor and Francis, Boca Raton, FL.

182. Monteith, D. K., Henry, S. P., Howard, R. B., Flournoy, S., Levin, A. A., Bennett, C. F., and Crooke, S. T. (1997) *Anticancer Drug Design* **12**(5), 421–432.

183. Sarmiento, U. M., Perez, J. R., Becker, J. M., and Narayanan, R. (1994) *Antisense Res. Dev.* **4**, 99–107.

184. Bennett, C. F., Kornbrust, D., Henry, S., Stecker, K., Howard, R., Cooper, S., Dutson, S., Hall, W., and Jacoby, H. I. (1997) *J. Pharmacol. Exp. Ther.* **280**(2), 988–1000.

185. Branda, R. F., Moore, A. L., Mathews, L., McCormack, J. J., and Zon, G. (1993) *Biochem. Pharmacol.* **45**, 2037–2043.

186. Krieg, A. M., Yi, A.-K., Matson, S., Waldschmidt, T. J., Bishop, G. A., Teasdale, R., Koretzky, G. A., and Klinman, D. M. (1995) *Nature* **374**, 546–549.

187. Kuramoto, E., Yano, O., Kimura, Y., Baba, M., Makino, T., Yamamoto, S., Yamamoto, T., Kataoka, T., and Tokunaga, T. (1992) *Jpn. J. Cancer Res.* **83**, 1128–1131.

188. Rankin, R., Pontarollo, R., Ioannou, X., Krieg, A. M., Hecker, R., Babiuk, L. A., and vanDrunen Littel-va den Hurk, S. (2001) *Antisense Nucleic Acid Drug Dev.* **11**, 333–340.

189. Plenat, F. (1996) *Mol. Med. Today* **2**, 250–257.

190. Henry, S. P., Monteith, D., Bennett, F., and Levin, A. A. (1997) *Anticancer Drug Design* **12**, 409–420.

191. Oberbauer, R., Schreiner, G. F., and Meyer, T. W. (1995) *Kidney Int.* **48**, 1226–1232.

192. Sands, H., Gorey-Feret, L. J., Cocuzza, A. J., Hobbs, F. W., Chidester, D., and Trainor, G. L. (1994) *Mol. Pharmacol.* **45**, 932–943.

193. Monteith, D. K., Horner, M. J., Gilett, N. A., Butler, M., Geary, R. S., Burckin, T., Ushiro-Watanabe, T., and Levin, A. A. (1999) *Toxicol. Pathol.* **27**, 307–317.

194. Krieg, A. M. and Stein, C. A. (1995) *Antisense Res. Dev.* **5**(4), 241.

195. Boggs, R. T., McGraw, K., Condon, T., Flournoy, S., Villiet, P., Bennett, C. F., and Monia, B. P. (1997) *Antisense Nucleic Acid Drug Dev.* **7**(5), 461–471.

196. Henry, S. P., Zuckerman, J. E., Rojko, J., Hall, W. C., Harman, R. J., Kitchen, D., and Crooke, S. T. (1997) *Anticancer Drug Design* **12**(1), 1–14.

197. Zhao, Q., Temsamani, J., Iadarola, P. L., Jiang, Z., and Agrawal, S. (1995) *Biochem. Pharmacol.* **51**, 173–182.

198. Zhang, R., Iyer, R. P., Yu, D., Tan, W., Zhang, X., Lu, Z., Zhao, H., and Agrawal, S. (1996) *J. Pharmacol. Exp. Ther.* **278**, 971–979.

199. Raoof, A. A., Chiu, P., Ramtoola, Z., Cumming, I. K., Teng, C. L., Weinbach, S. P., Hardee, G. E., Levin, A. A., and Geary, R. S. (2004) *J. Pharm. Sci.* **93**, 1431–1439.

200. Tillman, L. G., Geary, R. S., and Hardee, G. E. (2008) *J. Pharm. Sci.* 97, 225–236.

201. Raoff, A. A., Ramtoola, Z., McKenna, B., Yu, R. Z., Hardee, G. E., and Geary, R. S. (2002) *Eur. J. Pharm. Sci.* **17**, 131.

202. Gonzalez Ferreiro, M., Crooke, R. M., Tillman, L., Hardee, G., and Bodmeier, R. (2003) *Eur. J. Pharm. Biopharm.* **55**, 19–26.

203. Bartsch, M., Weeke-Kimp, A. H., Meijer, D. K., Scherphof, G. L., and Kamps, J. A. (2005) *J. Liposome Res.* **15**, 59–92.

204. Patil, S. D., Rhodes, D. G., and Burgess, D. J. (2005) *AAPS J.* **8**, E61–E77.

205. Fattal, E., Couvreur, P., and Dubernet, C. (2004) *Adv. Drug Deliv. Rev.* **56**, 931–946.

206. Akhtar, S., Hughes, M. D., Khan, A., Bibby, M., Hussain, M., Nawaz, Q., Double, J., and Sayyed, P. (2000) *Adv. Drug Deliv. Rev.* **44**, 3–21.

207. Smith, R. A., Miller, T. M., Yamanaka, K., Monia, B. P., Condon, T. P., Hung, G., Lobsiger, C. S., Ward, C. M., McAlonis-Downes, M., Wei, H., Wancewicz, E. V., Bennett, C. F., and Cleveland, D. W. (2006) *J. Clin. Invest.* **116**, 2290–2296.

208. Yazaki, T., Ahmad, S., Chahlavi, A., Zylber-Katz, E., Dean, N. M., Rabkin, S. D., Martuza, R. L., and Glazer, R. I. (1996) *Mol. Pharmacol.* **50**(2), 236–242.

209. Butler, M., McKay, R. A., Popoff, I., Gaarde, W., Witchell, D., Murray, S., Dean, N. M., Bhanot, S., and Monia, B. P. (2002) *Diabetes*, 1028–1034.

210. Zinker, B. A., Rondinone, C. M., Trevillyan, J. M., Gum, R. J., Clampit, J. E., Waring, J. F., Xie, N., Wilcox, D., Jacobson, P., Frost, L., Kroeger, P. E., Reilly, R. M., Koterski, S., Opgenorth, T. J., Ullrich, R. G., Crosby, S., Butler, M. M., Murray, S. F., McKay, R. A., Bhanot, S., Monia, B. P., and Jirousek, M. R. (2002) *Proc. Natl. Acad. Sci. U.S.A.* **99**, 11357–11362.

211. Karras, J. G., Crosby, J. R., Guha, M., Tung, D., Miller, D. A., Gaarde, W. A., Geary, R. S., Monia, B. P., and Gregory, S. A. (2007) *Am. J. Resp. Cell Mol. Biol.* **36**, 276–285.

212. Savage, D. B., Choi, C. S., Samuel, V. T., Liu, Z.-X., Zhang, D., Wang, A., Zhang, X. M., Cline, G. W., Yu, X. X., Geisler, J. G., Bhanot, S., Monia, B. P., and Shulman, G. I. (2006) *J. Clin. Invest.* **116**, 817–824.

213. Sloop, K., Cao, J.-C., Siesky, A., Zhang, H., Bodenmiller, D., Cox, A., Jacobs, S., Moyers, J., Owens, R., Showalter, A., Brenner, M., Raap, A., Gromada, J., Berridge, B., Monteith, D., Porksen, N., McKay, R. A., Monia, B. P., Bhanot, S., Watts, L., and Michael, M. (2004) *J. Clin. Invest.* **113**, 1571–1581.

214. Muse, E. D., Obici, S., Bhanot, S., Monia, B. P., McKay, R. A., Rajala, M. W., Scherer, P. E., and Rossetti, L. (2004) *J. Clin. Invest.* **114**, 232–239.

215. Azad, R. F., Driver, V. B., Tanaka, K., Crooke, R. M., and Anderson, K. P. (1993) *Antimicrobial Agents Chemother.* **37**(9), 1945–1954.

216. Azad, R. F., Brown-Driver, V., Buckheit, R. W. Jr., and Anderson, K. P. (1995) *Antiviral Res.* **28**, 101–111.

217. Anderson, K. P., Fox, M. C., Brown-Driver, V., Martin, M. J., and Azad, R. F. (1996) *Antimicrobial Agents Chemother.* **40**, 2004–2011.

218. Detrick, B., Nagineni, C. N., Grillone, L. R., Anderson, K. P., Henry, S. P., and Hooks, J. J. (2001) *Invest. Ophthmal. Vis. Sci.* **42**, 163–169.

219. Geary, R. S., Henry, S. P., and Grillone, L. R. (2002) *Clin. Pharmacokinet.* **41**, 255–260.

220. Vitravene Study Group (2002) *Am. J. Ophthalmol.* **133**, 475–483.

221. Vitravene Study Group (2002) *Am. J. Ophthalmol.* **133**, 484–498.

222. Bitko, V., Musiyenko, A., Shulyayeva, O., and Barik, S. (2005) *Nat. Med.* **11**, 50–55.

223. de Fougerolles, A. R. and Maraganore, J. M. (2007) Discovery and development of RNAi therapeutics. In: Crooke, S. T. (ed). *Antisense Drug Technology: Principles, strategy, and Applications*, Taylor and Francis, Boca Raton, FL.

224. Reed, J. C., Cuddy, M., Haldar, S., Croce, C., Nowell, P., Makover, D., and Bradley, K. (1990) *Proc. Natl. Acad. Sci. U.S.A.* **87**, 3660–3664.

225. Reed, J. C. (1998) *Oncogene* **17**, 3225–3236.

226. Cotter, F. E., Johnson, P., Hall, P., Pocock, C., Al Mahdi, N., Cowell, J. K., and Morgan, G. (1994) *Oncogene* **9**, 3049–3055.

227. Waters, J. S., Clarke, P. A., Cunningham, D., Millcr, B., Webb, A., Corbo, M., and Cotter, F. (2000) *Proc. Am. Soc. Clin. Oncol.* **19**, 14a.

228. Miayake, H., Tolcher, A. W., and Gleave, M. E. (2000) *J. Natl. Cancer Inst.* **92**, 34–41.

229. Jansen, B., Schlagbauer-Wadl, H., Brown, B. D., Bryan, R. N., van Elsas, A., Muller, M., Wolff, K., Eichler, H. G., and Pehamberger, H. (1998) *Nat. Med.* **4**, 232–234.

230. Wacheck, V., Heere-Ress, E., Halaschek-Weiner, J., Lucas, T., Meyer, H., Eichler, H. G., and Jansen, B. (2001) *J. Mol. Med.* **79**, 587–593.

231. Wacheck, V., Krepler, C., Strommer, S., Heere-Ress, E., Klem, R., Pehamberger, H., Eichler, H. G., and Jansen, B. (2002) *Antisense Nucleic Acid Drug Dev.* **12**, 359–367.

232. Klasa, R. J., Gillum, A. M., Klem, R. E., and Frankel, S. R. (2002) *Antisense Nucleic Acid Drug Dev.* **12**, 193–213.

233. Webb, A., Cunningham, D., Cotter, F., Clarke, P. A., di Stefano, F., Corbo, M., and Dziewanowska, Z. (1997) *Lancet* **349**, 1137–1141.

234. Chi, K. N., Gleave, M. E., Klasa, R., Murray, N., Bryce, C., Lopes de Menezes, D. E., D'Aloisio, S., and Tolcher, A. W. (2001) *Clin. Cancer Res.* **7**, 3920–3927.

235. Bedikian, A. Y., Millward, M., Pehamberger, H., Conry, R., Gore, M., Trefzer, U., Pavlick, A. C., Deconti, R., Hersh, E. M., Hersey, P., Kirkwood, J. M., and Haluska, F. G. (2006) *J. Clin. Oncol.* **24**, 4738–4745.

236. O'Brien, S., Moore, J. O., Boyd, T. E., Larratt, L. M., Skotnicki, A., Koziner, B., Chanan-Khan, A. A., Seymour, J. F., Bocick, R. G., Pavletic, S., and Rai, K. R. (2007) *J. Clin. Oncol.* **25**, 1114–1120.

237. Zellweger, T., Miyake, H., Cooper, S., Chi, K., Conklin, B. S., Monia, B. P., and Gleave, M. E. (2001) *J. Pharmacol. Exp. Ther.* **298**, 934–940.

238. Dustin, M. L. and Springer, T. A. (1991) *Annu. Rev. Immunol.* **9**, 27–66.

239. Springer, T. A. (1994) *Cell* **76**, 301–314.

240. Springer, T. A. (1990) *Nature* **346**, 425–434.

241. Butcher, E. C. (1991) *Cell* **67**, 1033–1036.

242. Bennett, C. F. (1994) *Clin.Chem.* **40**, 644–645.

243. Miele, M. E., Bennett, C. F., Miller, B. E., and Welch, D. R. (1994) *Exp.Cell Res.* **214**, 231–241.

244. Nestle, F. O., Mitra, R. S., Bennett, C. F., Chan, H., and Nickoloff, B. J. (1994) *J. Invest. Dermatol.* **103**, 569–575.

245. Stepkowski, S. M., Tu, Y., Condon, T. P., and Bennett, C. F. (1994) *J. Immunol.* **153**, 5336–5346.

246. Kumasaka, T., Quinlan, W. M., Doyle, N. A., Condon, T. P., Sligh, J., Takei, F., Beaudet, A. L., Bennett, C. F., and Doerschuk, C. M. (1996) *J. Clin. Invest.* **97**, 2362–2369.

247. Christofidou-Solomidou, M., Albelda, S. M., Bennett, C. F., and Murphy, G. F. (1997) *Am. J. Pathol.* **150**, 631–639.

248. Katz, S. M., Tian, L., Steopkowski, S. M., Phan, T., Bennett, C. F., and Kahan, B. D. (1997) *Transplant. Proc.* **29**, 748–749.

249. Musso, A., Condon, T. P., West, G. A., De La Motte, D., Strong, S. A., Levine, A. D., Bennett, C. F., and Fiocchi, C. (1999) *Gastroenterology* **117**, 546–556.

250. Klimuk, S. K., Semple, S. C., Nahirney, P. N., Mullen, M. C., Bennett, C. F., Scherrer, P., and Hope, M. J. (2000) *J. Pharm. Exp. Ther.* **292**, 480–488.

251. Stepkowski, S. M., Qu, X., Wang, M., Tian, L., Chen W., Wancewicz, E. V., Johnston, J., F., Bennett, C. F., and Monia, B. P. (2000) *Transplantation* **70**, 656–661.

252. Rijcken, E., Krieglstein, C. F., Anthoni, C., Laukoeter, M. G., Mcnnigcn, R., Spigcl, H. U., Scnninger, N., Bennett, C. F., and Schuermann, G. (2002) *Gut* **51**, 529–535.

253. Stepkowski, S. M., Wang, M.-E., Condon, T. P., Cheng-Flournoy, S., Stecker, K., Graham, M., Qu, X., Tian, L., Chen, W., Kahan, B. D., and Bennett, C. F. (1998) *Transplantation* **66**, 699–707.

254. Haller, H., Dragun, D., Miethke, A., Park, J. K., Weis, A., Lippoldt, A., Grob, V., and Luft, F. C. (1996) *Kidney Int.* **50**, 473–480.

255. Maksymowch, W. P., Blackburn, W. D., Tami, J. A., and Shanahan, W. R. (2002) *J. Rheumatol.* **29**, 447–453.

256. Yacyshyn, B. R., Bowen-Yacyshyn, M. B., Jewell, L., Tami, J. A., Bennett, C. F., Kisner, D. L., and Shanahan, W. R., Jr. (1998) *Gastroenterology* **114**, 1133–1142.

257. Yacyshyn, B. R., Chey, W. Y., Goff, J., Salzberg, B., Baerg, R., Buchman, A. L., Tami, J. A., Yu, R. Z., Gibiansky, E., and Shanahan, W. R. (2002) *Gut* **51**, 30–36.

258. Yacyshyn, B. R., Barish, C., Goff, J., Dalke, D., Gaspari, M., Yu, R. Z., Tami, J. A., Dorr, F. A., and Sewell, K. L. (2002) *Aliment Pharmacol. Ther.* **16**, 1761–1770.

259. Yacyshyn, B. R., Chey, W. Y., Wedel, M. K., Yu, R. Z., Paul, D., and Cheung, E. (2007) *Clin. Gastroenterol. Hepatol.* **5**, 215–220.

260. van Deventer, S. J. H., Tami, J. A., Wedel, M. K., and Group, E. C. S. (2004) *Gut* **53**, 1646–1651.

261. Miner, P., Wedel, M., Bane, B., and Bradley, J. (2004) *Aliment Pharmacol. Ther.* **19**, 281–286.

262. Miner, P. B., Jr., Wedel, M. K., Xia, S., Baker, B. F. (2006) *Aliment Pharmacol. Ther.* **23**, 1403 1413.

263. Van Deventer, S. J. H., Wedel, M. K., Baker, B. F., Xia, S., Chuang, E., Miner, P. B. Jr., (2006) *Aliment Pharmacol. Ther.* **23**, 1415–1425.

264. Miner, P. B., Jr., Wedel, M. K., Xia, S., Baker, B. F., Geary, R. S., and Matson, J. (2006) *Aliment Pharmacol. Ther.* **23**, 1427–1434.

265. Polman, C. H., O'Connor, P. W., Havrdova, E., Hutchinson, M., Kappos, L., Miller, D. H., Phillips, J. T., Lublin, F. D., Giovannoni, G., Wajgt, A., Toal, M., Lynn, F., Panzara, M., Sandrock, A. W., and Investigators, A. (2006) *New Engl. J. Med.* **354**, 899–910.

266. Targan, S. R., Feagan, B. G., Fedorak, R. N., Lashner, B. A., Panaccione, R., Present, D. H., Spehlmann, M. E., Rutgeerts, P. J., Tulassay, Z., Volfova, M., Wolf, D. C., Hernandez, C., Bornstein, J., Sandborn, W. J., and Group (2007) *Gastroenterology* **132**, 1672–1683.

267. Myers, K. J., Witchell, D. R., Graham, M. J., Koo, S., Butler, M., and Condon, T. P. (2003) *J. Neuroimm.* **160**, 12–24.

268. Stein, C. A. (1996) *TIBTECH* **14**, 147–149.

269. Bennett, C. F. (1998) *Biochem. Pharmacol.* **55**(1), 9–19.

270. Altmann, K.-H., Dean, N. M., Fabbro, D., Freier, S. M., Geiger, T., Haener, R., Huesken, D., Martin, P., and Monia, B. P. (1996) *Chimia* **50**(4), 168–176.

271. Rajwanshi, V. K., Kumar, R., Kofod-Hansen, M., and Wengel, J. (1999) *J. Chem. Soc., Perkin Trans.* **1**, 1407–1414.

272. Vester, B. and Wengel, J. (2004) *Biochemistry* **43**, 13233–13241.

273. Bennett, C. F., Zuckerman, J. E., Kornbrust, D., Henry, S., Sasmor, H., Leeds, J. M., and Crooke, S. T. (1996) *J. Cont. Rel.* **41**, 121–130.

274. Mehta, R. C., Stecker, K. K., Cooper, S. R., Templin, M. V., Tsai, Y. J., Condon, T. P., Bennett, C. F., and Hardee, G. E. (2000) *J. Invest. Derm.* **115**, 805–812.

275. Yu, R. Z., Geary, R. S., Leeds, J. M., Watanabe, T., Fitchett, J. R., Matson, J. E., Hardee, G. R., Templin, M. V., Huang, K., Newman, M. S., Quinn, Y., Uster, P., Zhu, G., Working, P. K., Nelson, J., and Levin, A. A. (1999) *Pharm. Res.* **16**, 1309–1315.

276. Kwoh, J. T. (2007) An overview of the clinical safety experience of first- and second-generation antisense oligonucleotides. In: Crooke, S. T. (ed). *Antisense Drug Technology: Principles, Strategies, and Applications*, Taylor and Francis, Boca Raton, FL.

24 Development of RNAi Therapeutics

Antonin R. de Fougerolles

CONTENTS

24.1 INTRODUCTION

Within less than a decade since its discovery, RNA interference (RNAi) as a novel mechanism to selectively silence messenger RNA (mRNA) expression has revolutionized the biological sciences in the postgenomic era. With RNAi, the target mRNA is enzymatically cleaved, leading to decreased levels of the corresponding protein. The specificity of this mRNA silencing is controlled very precisely at the nucleotide level. Given the identification and sequencing of the entire human genome, RNAi is a fundamental cellular mechanism that can also be harnessed to rapidly develop novel drugs against any disease target. The reduction in expression of pathological proteins through RNAi is applicable to all classes of molecular targets, including those that have been traditionally difficult to target with either small molecules or proteins including monoclonal antibodies. Numerous proof-of-concept studies in animal models of human disease have demonstrated the broad potential applicability of RNAi-based therapeutics. Further, RNAi therapeutics are now under clinical investigation for age-related macular degeneration (AMD) and respiratory syncytial virus (RSV) infection, with numerous other drug candidates poised to advance into clinic development in the years to come.

In this chapter, I will outline and discuss the various considerations that go into developing RNAi-based therapeutics starting from *in vitro* lead design and identification to *in vivo* preclinical drug delivery and testing, and lastly, to a review of clinical experiences to date with RNAi therapeutics.

24.2 RNAi OVERVIEW

The discovery of long double-stranded RNA (dsRNA)-mediated RNAi in the worm by Fire and Mello [1] and the subsequent publication by Tuschl and colleagues demonstrating that RNAi, mediated by small interfering RNA (siRNA), operates in mammalian cells [2] sparked an explosion of research to uncover new mechanisms of gene silencing, revolutionizing understanding of endogenous mechanisms of gene regulation and providing powerful new tools for biological research and drug discovery. While much is known about RNAi, this remains an area of intense research rich in new discoveries, with multiple classes of small noncoding RNA being discovered that play important roles in cell biology. At present, the majority of research into RNAi has focused on siRNAs and microRNAs (miRNAs), with the latter deriving from imperfectly paired noncoding hairpin RNA structures naturally transcribed by the genome [3,4]. Gene silencing can be induced by siRNA through

sequence-specific cleavage of perfectly complementary mRNA, while miRNAs mediate translational repression and transcript degradation for imperfectly complementary targets (Figure 24.1). In the endogenous pathway, RNAs containing stem-loops or short hairpin structures, encoded in intragenic regions or within introns, are processed in the nucleus and exported into the cytoplasm as precursor molecules, called pre-miRNA [4,5]. In the cytoplasm, the pre-miRNA is further shortened and processed by an RNAse III enzyme, called Dicer, to produce an imperfectly matched, double-stranded miRNA, while Dicer similarly processes long, perfectly matched dsRNA into siRNA (Figure 24.1). A multienzyme complex including Argonaute 2 (Ago2) and the RNA-induced silencing complex (RISC) binds either the miRNA duplex or the siRNA duplex and discards one

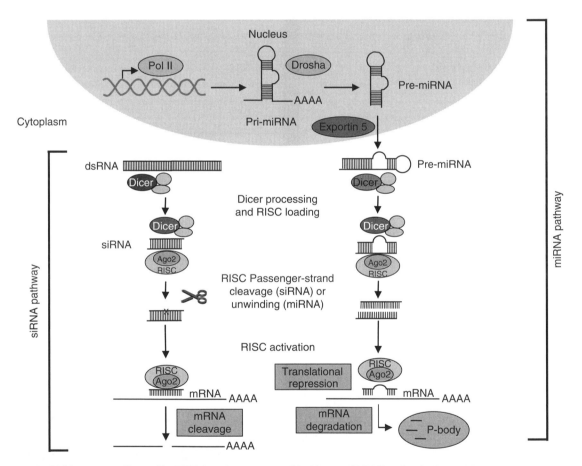

FIGURE 24.1 RNAi in mammalian cells. RNAi pathways are guided by small RNAs that include siRNA and miRNAs. The siRNA pathway begins with cleavage of long dsRNA by the Dicer enzyme complex into siRNA. These siRNA are incorporated into Ago2 and the RISC. When the RNA duplex loaded into RISC has perfect sequence complementarity, Ago2 cleaves the passenger (sense) strand so that active RISC is produced that contains the guide (antisense) strand. The siRNA guide strand recognizes target sites to direct mRNA cleavage, which is carried out by the catalytic domain of Ago2. RNAi therapeutics developed to harness the siRNA pathway typically involves delivery of synthetic siRNA into the cell cytoplasm. The miRNA pathway begins with endogenously encoded primary miRNA transcripts (pri-miRNAs) that are transcribed by RNA polymerase II (Pol II) and are processed by the Drosha enzyme complex to yield precursor miRNAs (pre-miRNAs). These precursors are then exported to the cytoplasm by exportin 5, and subsequently bind to the Dicer enzyme complex, which processes the pre-miRNA for loading into the Ago2/RISC complex. When the RNA duplex loaded into RISC has imperfect sequence complementarity, the passenger (sense) strand is unwound, leaving a mature miRNA bound to active RISC. The mature miRNA recognizes target sites (typically in the 3′ UTR) in the mRNA, leading to direct translational inhibition. Binding of miRNA to target mRNA may also lead to mRNA target degradation in P-bodies. (Adapted from de Fougerolles, A.R. et al., *Nat. Rev. Drug Discovery*, 6, 443, 2007.)

strand (it is thought the passenger or sense strand is cleaved in the case of siRNA and is released in the case of miRNA) to form an activated complex containing the guide or antisense strand [6]. The activated Ago2/RISC complex then seeks and binds an mRNA strand bearing a complementary sequence and inactivates its expression, either by blocking translation (if complementarity is less than perfect such as with miRNA) or by cleaving it between the nucleotides complementary to nucleotides 10 and 11 of the guide strand (relative to the 5′ end) (when complementarity is perfect or nearly perfect such as with siRNA) (Figure 24.1). In addition to translational repression, miRNAs may also mediate mRNA degradation which occurs in cytoplasmic compartments known as processing bodies (P-bodies) [7]. Gene silencing by mRNA cleavage is thought to be particularly potent, in part because mRNA cleavage leads to rapid nucleolytic degradation of the RNA fragments and frees the activated RISC complex to seek and destroy another target mRNA in a catalytic fashion [8]. While the mechanism by which an activated Ago2/RISC cleaves a target RNA is well understood, the molecular basis for how this complex finds its target RNA is less well understood. Recent studies have shown that the accessibility of the target site correlates directly with efficiency of cleavage, and that in the course of target recognition, the activated Ago2/RISC complex transiently contacts single-stranded RNA nonspecifically and promotes siRNA–target RNA annealing [9]. Furthermore, the 5′ part of the siRNA within RISC seems to create a thermodynamic threshold that then determines the stable association of RISC and the target RNA.

The goal of RNAi-based therapy is to activate selective mRNA cleavage for efficient gene silencing. It is possible to harness the endogenous pathway in one of two ways—either by using a viral vector to express short hairpin RNA (shRNA) that look like miRNA precursors or by introducing siRNAs that mimic the Dicer cleavage product into the cytoplasm. These siRNA can bypass the earlier steps in the RNAi silencing pathway and can be taken up directly into the Ago2/RISC complex. While delivery of synthetic siRNA and viral delivery of shRNA are being developed as potential RNAi-based therapeutic approaches, this chapter focuses solely on the development of synthetic siRNA as drugs. Synthetic siRNAs are readily defined drugs that harness the naturally occurring RNAi pathway in a manner that is consistent and predictable, thus making them particularly attractive as therapeutics. Moreover, since they enter the RNAi pathway later, siRNA is less likely to interfere with gene regulation by endogenous miRNAs [10,11]. As a consequence, siRNAs are the class of RNAi therapeutics that are most advanced in preclinical and clinical studies.

24.3 *IN VITRO* SELECTION OF LEAD CANDIDATES

This section highlights the various steps required to identify potent lead siRNA candidates starting from bioinformatics design through to *in vitro* characterization. The overall scheme is summarized in Figure 24.2. The three most important attributes to take into account when designing and selecting

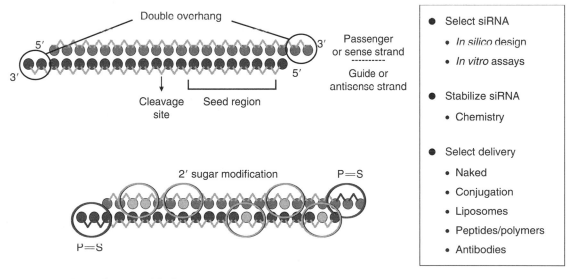

FIGURE 24.2 (See color insert following blank page 206.) Developing RNAi therapeutics. Steps involved in developing an RNAi therapeutic. This three-step process begins with *in silico* design and *in vitro* screening of target siRNA, is followed by incorporating stabilizing chemical modifications on lead siRNA as required, and ends with selection and *in vivo* evaluation of delivery technologies appropriate for the target cell type/organ and the disease setting. A schematic illustration of some of the important features of siRNA structure (two base pair overhangs, seed region, and mRNA cleavage site) is shown. An example of an optimized siRNA molecule, which incorporates chemical modifications to increase nuclease stability and to minimize off-targeting, is shown below. The phosphorothioate and 2′ base modifications shown are illustrative of exo-nuclease and endo-nuclease stabilizing chemistries, respectively; other chemistries also exist that confer similar properties on an siRNA duplex. (From Bumcrot, D. et al. *Nat. Chem. Biol.*, 2006, 2, 711; de Fougerolles, A.R. et al., *Nat. Rev. Drug Discovery*, 6, 443, 2007.)

siRNA are potency, specificity, and nuclease stability. With regard to specificity of siRNA, the two issues of "off-targeting" due to silencing of genes sharing partial homology to the siRNA and "immune stimulation" due to recognition of certain siRNAs by the innate immune system have been of special concern. With an increased understanding of the molecular and structural mechanism of RNAi, all issues around lead siRNA selection are better understood and also are now generally resolvable through the use of bioinformatics, chemical modifications, and empirical testing. Thus, it is now possible, in the span of several months, to very rapidly identify potent, specific and stable *in vitro* active lead siRNA candidates to any target of interest.

24.3.1 POTENCY

Work by Tuschl and colleagues [2] represented the first published study to demonstrate that RNAi could be mediated in mammalian cells through introduction of small fragments of dsRNA, termed siRNA, and that siRNAs had a specific architecture comprised of 21 nucleotides in a staggered 19 nucleotide duplex with a two-nucleotide 3′ overhang on each strand (Figures 24.1 and 24.2). siRNAs can be identified to silence any gene, often with *in vitro* activity at nanomolar or lower concentrations. Algorithms to predict effective siRNAs, developed based on common features of siRNAs identified by empiric testing, are available on the Web [12,13]. While algorithms can increase the chances of identifying active siRNA, they are imperfect and may sometimes miss the most potent siRNAs, which can only be identified by experimental testing. An alternative to using algorithms as a starting point to identify candidate siRNAs is to test all the tiled 21 nt sequences from the mRNA coding region experimentally and select a group of candidates that induce effective silencing at the lowest concentrations. One study suggests that somewhat longer siRNAs that need to be processed by Dicer before they can be incorporated into RISC can be used as "Dicer substrates" and maintain silencing activity, although these longer siRNAs are more complicated to synthesize and could have an increased propensity to activate untoward immune responses [14,15]. Loading of RISC with respect to the sense and antisense siRNA strands is not symmetrical [16–18]. The strand that is least tightly bound at its 5′-end is favored to bind into a deep pocket in the RISC to become the active strand. Strand selection could in fact be biased by making a single nucleotide substitution at the end of the duplex to alter the relative binding of the ends [16]. Thus, by designing an siRNA with a mismatch at the 5′-end of the intended active strand that favors its incorporation into RISC, it may be possible to increase the likelihood of identifying a potent duplex.

24.3.2 SPECIFICITY

RNAi-mediated silencing of gene expression has been shown to be exquisitely specific as evidenced by silencing fusion mRNA without affecting an unfused allele [19,20] and by

studies showing ability to silence point-mutated genes over wild-type sequence [21,22]. Nevertheless, along with on-target mRNA silencing, siRNA might have the potential to recognize nontarget mRNA, otherwise known as "off-target" silencing. Based on *in vitro* transcriptional profiling studies, siRNA duplexes have been reported to silence multiple genes in addition to the intended target gene under certain conditions. Not surprisingly, many of these observed off-target genes contain regions that are complementary to one of the two strands in the siRNA duplex [23–25]. More detailed bioinformatic analysis revealed that complementarity between the 5′ end of the guide strand and the mRNA was the key to off-targeting silencing, with the critical nucleotides being in positions 2 to 8 (from 5′ end of guide strand), which corresponds to the so-called "seed region" for miRNA recognition of endogenous target mRNAs [26,27] (Figure 24.2). Accordingly, one approach for minimizing off-target effects is careful design of siRNA to minimize exact seed pairings, although seed regions of 7 nucleotides are very short in length. Alternatively, chemical modifications of riboses in the guide strand can suppress most off-target effects, while maintaining target mRNA silencing [28,29]. In fact, incorporation of a single 2′-O-Me modification at nucleotide 2 was sufficient to suppress most off-target silencing of partially complementary mRNA transcripts by all siRNAs tested. Thus, in summary, bioinformatics design or position-specific, sequence-independent chemical modifications can be incorporated into siRNA that reduces off-target effects while maintaining target silencing.

A second mechanism whereby siRNA can also induce potential unwanted effects is by activating the innate immune response to dsRNAs, setting off defense systems usually used to combat viruses. One pathway involves recognition of dsRNAs by the PKR kinase (reviewed in Ref. [30]). This pathway is efficiently triggered only by dsRNAs > 30 nucleotides long. At higher concentrations, however, smaller siRNAs may be able to activate this pathway, resulting in global translational blockade and cell death. Perhaps of more concern is the potential to activate Toll-like receptors (TLRs), especially the dsRNA receptor TLR-7, in plasmacytoid dendritic cells to trigger the production of Type I interferons and proinflammatory cytokines and NF-κB activation [31]. In much the same way that certain CpG motifs in antisense oligonucleotides mediate TLR-9 immunostimulation, interferon induction by some siRNAs occurs principally via TLR-7, which is largely found within endosomes in immune antigen-presenting cells. Fortunately, TLR-7 binding is sequence-specific, favoring GU-rich sequences, and can be avoided by choosing sequences that are not recognized by the receptor. Additional work remains to identify immunostimulatory motifs and determine whether other receptors might also be involved. Several approaches exist to circumvent the immunostimulatory properties of certain siRNA duplexes. Candidate duplexes can be transfected into plasmacytoid dendritic cells to test whether they induce interferon expression [31,32], and those that do can be discarded. Moreover, the presence of 3′ overhang nucleotides and 2′-O-methyl modifications can be used to abrogate TLR-7 binding and thereby abolish immunostimulatory activity [31,33]. Since modifications of this sort also greatly reduce sequence-dependent off-target silencing, this type of modification may be particularly beneficial in enhancing siRNA target specificity. Potentially, TLR-7 signaling could also be evaded by using siRNA delivery strategies that avoid the cell types responsible for immune stimulation or that transport the siRNA directly to the cytoplasm, thereby avoiding TLR activation in the endosome.

24.3.3 STABILITY

Naked siRNAs are degraded in human plasma with a half-life on the order of minutes [34–36]. To convert siRNAs into optimized drugs, much effort has been expended to investigate chemical modifications that prolong half-life without jeopardizing biological activity. The starting point for these efforts has been chemistries already in use with antisense oligonucleotide and aptamer therapeutics. For instance, introduction of a phosphorothioate (P=S) backbone linkage at the 3'-end is used to protect against exonuclease degradation and 2'-sugar modification (2'-O-Me, 2'-F, others) is used for endonuclease resistance [34–38]. With respect to maintenance of RNAi silencing activity, exonuclease-stabilizing modifications are all very well tolerated. Introduction of internal sugar modifications to protect against endonucleases is also generally tolerated but can be more dependent on location of the modification within the duplex, with the sense strand being more amenable to modification than the antisense strand. Nevertheless, using simple, well-described modifications such as P=S, 2'-O-Me, and 2'-F it is possible in most instances to fully nuclease-stabilize an siRNA duplex and maintain mRNA silencing activity. Importantly, the degree of modifications required to fully stabilize the siRNA duplex can generally be limited in extent, thereby avoiding the toxicities associated with certain oligonucleotide chemistries.

Improved nuclease stability is especially important *in vivo* for siRNA duplexes, which are exposed to nuclease-rich environments (such as serum) and are formulated using excipients that do not themselves confer additional nuclease protection on the duplex. As might be expected in these situations, nuclease-stabilized siRNA shows improved pharmacokinetic properties *in vivo* [36,39]. In other situations, when delivering siRNA directly to more nuclease-amenable sites such as the lung or when delivering in conjunction with delivery agents such as liposomes, the degree of nuclease stabilization that is required can be reduced significantly. While the ability of an siRNA duplex to reach its target cell type intact is vitally important, whether nuclease protection confers a measurable benefit once an siRNA is inside the cell is now being more fully understood. *In vitro* comparisons of naked siRNA versus fully stabilized siRNA do not reveal significant differences in longevity of mRNA silencing whether rapidly dividing cells [34,36] or slowly dividing cells are used [36]. In fact, the most important factor affecting the *in vitro* longevity of siRNA-mediated silencing is dilution due to cell division and not

intracellular siRNA half-life [36,40]. Studies using rapidly dividing HeLa cells found target suppression lasting for around 1 week, while the same target is suppressed for up to 1 month in slowly dividing CCD-1074Sk cells [36]. Investigation into the role of chemical stabilization on *in vivo* silencing suggests that the stabilization advantages of nuclease-stabilized siRNA originate primarily from effects prior to and during internalization before the siRNA can interact with the intracellular machinery [36]. With the recent advent of fluorescence resonance energy transfer studies using siRNA [41], it should be possible in the near future to more completely understand the intracellular benefit of nuclease stabilization on longevity of RNAi-mediated silencing.

24.3.4 THERAPEUTIC CONSIDERATIONS

In designing siRNAs for therapeutic purposes, other considerations beyond an active target sequence exist. Where possible, it is desirable to identify target sequences that have identity across all the relevant species used in safety and efficacy studies, thus enabling development of a single drug candidate from research stage all the way through clinical trials. Other considerations in selecting a target sequence involve presence of single nucleotide polymorphisms and general ease of chemical synthesis.

Predicting the nucleotide sequence and chemical modifications required to yield an ideal RNAi therapeutic still remains a work in progress. While much progress has been made in understanding what attributes are required to identify an *in vitro* active and stable siRNA, much less is known about how well those attributes translate into identifying *in vivo* active siRNAs. For example, many of the issues around specificity are based on *in vitro* data and their *in vivo* relevance remains to be determined. For example, the range of off-target genes identified in tissue culture can differ dramatically depending upon the transfection method used to introduce siRNAs into cells [42]. Likewise, induction of innate immune responses by certain siRNAs has been shown to be cell-type-specific [15]. At present, in order to identify robust *in vitro* active lead candidate siRNAs suitable for subsequent *in vivo* study, the practical and prudent approach is to synthesize and screen a library of siRNA duplexes for potency, specificity, nuclease stability, and immunostimulatory activity.

24.4 *IN VIVO* DELIVERY

Effective delivery is the most challenging remaining consideration in the development of RNAi as a broad therapeutic platform. To date, animal studies using siRNA either have employed no additional formulation (i.e., "naked siRNA") or have delivered siRNA formulated as conjugates, as liposome/lipoplexes, or as complexes with peptides, polymers, or antibodies (Figure 24.3). The route of administration of siRNA has also ranged from local, direct delivery to systemic administration. Local delivery or "direct RNAi" has particular advantages for a developing technology in that, as with any

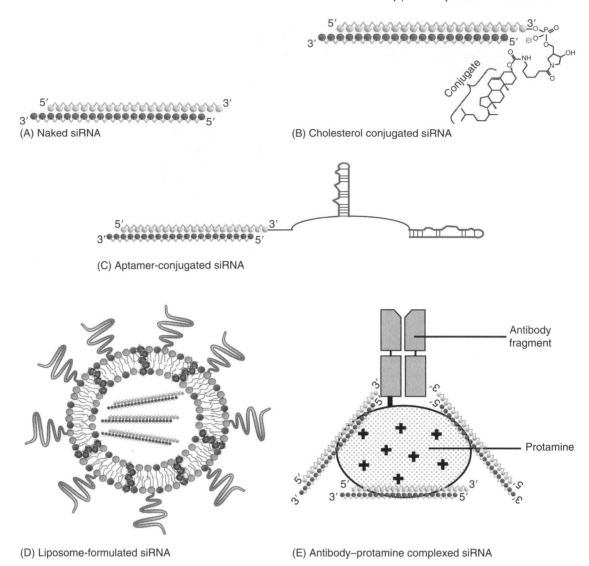

(A) Naked siRNA

(B) Cholesterol conjugated siRNA

(C) Aptamer-conjugated siRNA

(D) Liposome-formulated siRNA

(E) Antibody–protamine complexed siRNA

FIGURE 24.3 Delivery of siRNAs. Different strategies have been employed to deliver and achieve RNAi-mediated silencing *in vivo*. Direct injection of naked siRNA (unmodified or chemically modified) has proven efficacious in multiple contexts of ocular, respiratory, and CNS disease [46–66]. Direct conjugation of siRNA to a natural ligand such as cholesterol has demonstrated *in vivo* silencing in hepatocytes [68,69] and neuronal cells [70]. Liposome-mediated delivery of siRNA has been used to silence multiple targets following systemic administration [80–82,84]. Aptamer–siRNA conjugates have demonstrated specific delivery and silencing following local injection in a tumor xenograft model [75]. Lastly, antibody–protamine fusion proteins have been used to noncovalently bind siRNAs through charge interactions and deliver siRNA specifically to cells that express the surface receptor recognized by the antibody [115,116].

pharmacologic approach, doses of siRNA required for efficacy are substantially lower when siRNAs are injected into or administered at or near the target tissue. Direct delivery also allows for more focused delivery of siRNA, which might circumvent any theoretical undesired side effects resulting from systemic delivery. Systemic delivery of siRNA, especially with cholesterol conjugates and liposome formulations have also been widely explored with considerable success. This section will provide a review of the different delivery approaches utilized with siRNA. Many recent review publications offer comprehensive listings of successful *in vivo* efficacy studies [43–45].

24.4.1 NAKED siRNA

Many reports describing success with RNAi *in vivo* involve direct delivery of "naked" siRNA to tissues such as eye, lung, and central nervous system. It remains incompletely understood why certain cells, but not others, can directly take up siRNA into the cytoplasm where the RNAi machinery operates. As used here, the term "naked" siRNA refers to delivery of siRNA (unmodified or modified) in saline or other simple excipients such as 5% dextrose (D5W). The ease of formulation and administration using direct delivery of naked siRNA to tissues make this an attractive therapeutic approach.

Not surprisingly, the initial development of RNAi therapeutics has focused on disease targets and clinical indications (AMD and RSV infection) that allow for direct administration of siRNA to the diseased organ.

24.4.1.1 Ocular

Multiple examples of efficacious local delivery of siRNA in the eye exist, where proof-of-concept has been attained in animal models of ocular neovascularization and scarring using both saline- and lipid-based formulations [46–51]. Much evidence suggests that direct administration of naked siRNA is able to target cell types in the back of the eye and have profound disease-modifying effects. Using an optimized vascular endothelial growth factor (VEGF) targeting siRNA, we demonstrated robust specific inhibition of pathologic retinal neovascularization in a rat oxygen-induced model of retinopathy [51]. Following a single intravitreal injection of saline-formulated VEGF siRNA, over 75% inhibition of pathological neovascularization was achieved with no effect on the normal retinal vasculature. The inhibition seen was both dose-dependent and specific as a mismatched siRNA showed no inhibition. Separate earlier studies using lipid-formulated VEGF siRNA had shown a reduction of laser-induced choroidal neovascularization (CNV) in a mouse model of AMD [46]; this initial study was followed by nonhuman primate laser-induced CNV study where it was reported that intravitreal injection of a saline-formulated VEGF siRNA was well tolerated and efficacious [47]. Lastly, intravitreal injection of saline-formulated siRNA targeting VEGF receptor-1 was effective in reducing the area of ocular neovascularization by one-third to two-third in two mouse models [48]. These encouraging proof-of-concept studies in animal models have led to clinical trials of siRNA targeting the VEGF pathway in AMD.

24.4.1.2 Respiratory

A number of studies have demonstrated that intranasal or orotracheal local administration of siRNA can result in a significant target gene silencing and distinct phenotypes in the lung. Typically, siRNAs were administered in concentrations of ~100 μg per mouse and were directed against viral or endogenous disease-related targets. Most of the examples of successful direct delivery of siRNA to lung have involved delivery of naked siRNA either in saline or with excipients such as D5W or lung surfactants. In instances when delivery formulations such as Transit-TKO have been used, they are sometimes shown to be dispensable. However, the required dose or delivery to certain cell types in the lung for some indications will likely be improved by incorporating siRNA into a carrier. In one groundbreaking study, intranasal instillation of siRNAs (either unformulated or complexed with Transit-TKO) directed against viral targets reduced the viral load of RSV and parainfluenza virus (PIV), two relevant pathogens in pediatric and immune-compromised patients, in the lung by more than three orders of magnitude in a mouse model of viral infection at a dose as low as 70 μg/animal with

no observed adverse events [52]. In addition, other pathological features of RSV infection such as elevated leukotriene levels, pulmonary inflammation, as well as increased respiratory rate were reduced to baseline levels after siRNA administration. The RSV-specific siRNA-mediated reduction of viral load could be achieved in both prophylaxis and treatment paradigms. In a similar approach, siRNA formulated in D5W was administered intranasally in a nonhuman primate model of SARS corona virus (SCV) infection [53]. Macaques treated with siRNA prior to, simultaneously with, or with repeated doses after viral infection, showed a milder response to viral infection as judged by a reduced elevation of body temperature, a key indicator for the severity for SARS-like symptoms. In addition, siRNA treatment resulted in a significant reduction of interstitial infiltrates and pathological changes to the lung as well as in an inhibition of viral replication in the respiratory tract. Both studies demonstrate the potential of RNAi therapeutics to treat viral infection in the respiratory system.

A number of studies showed that siRNAs can achieve delivery in pulmonary tissue in certain disease settings. Lee and colleagues demonstrated that siRNA-mediated silencing of heme-oxygenase-1 (HO-1) after intranasal siRNA administration resulted in enhanced apoptosis in an ischemia-reperfusion (I-R) mouse model [54]. While I-R induced HO-1 in a number of organs, siRNA mediated HO-1 silencing following intranasal administration was restricted to the lung and resulted in a local elevation of FAS expression and caspase3 activity. In addition, biotinylated siRNA was detected histologically in lung parenchyma up to 16 h after instillation. To further understand the pathogenesis of acute lung injury (ALI), siRNA targeting the two chemokines, keratinocyte-derived chemokine (KC) and macrophage-inflammatory protein 2 (MIP-2), was administered by intratracheal instillation in a hemorrhage-induced and septic challenge model of ALI [55]. KC and MIP-2 lung mRNA levels were decreased by about 50%, resulting in a significant reduction of local IL-6 concentrations, and in the case of MIP-2 siRNA, treatment also resulted in reduced neutrophil influx, interstitial edema, and disruption of lung architecture. Silencing with Fas-siRNAs in the same animal model resulted in reduction of Fas to a level similar to control animals and ameliorated pulmonary apoptosis and inflammation [56]. Recently, in a hyperoxia-based mouse model of ALI, intranasal administration of an angiopoietin 2 (Ang2) siRNA in saline specifically and dramatically ameliorated hyperoxia-induced oxidant injury, cell death, inflammation, vascular permeability, and mortality [57]. Expression of Ang2 in this model is dramatically induced in lung epithelial cells and this was specifically inhibited by Ang2 siRNA but not control siRNA. Lastly, it was demonstrated that intranasal instillation of siRNA targeting discoidin domain receptor 1 (DDR-1) mRNA resulted in 60%–70% reduction of DDR1 protein in the lung of animals treated with bleomycin to induce pulmonary fibrosis [58]. DDR1 silencing was accompanied by an attenuation of infiltration of inflammatory cells and a reduction of cytokines. As a consequence, bleomycin-induced transforming growth factor (TGF)-beta up-regulation was suppressed and collagen deposition was significantly reduced. In sum, these studies exhibit the potential for direct instillation of naked siRNA to effectively silence endogenous lung genes and have disease-modifying effects.

24.4.1.3 Nervous System

The nervous system is a third area where direct instillation of siRNA in saline has proven successful in validating disease targets *in vivo*. Direct administration of saline-formulated siRNA by intracerebroventricular, intrathecal, or intraparenchymal infusion resulted in silencing of specific neuronal mRNA targets in multiple regions of the peripheral and central nervous system [59–62]. While the naked unformulated siRNA dose typically required for target silencing in these rodent studies is on the order of ~0.5 mg/day, use of polymer or lipid-based delivery systems seems to facilitate cellular uptake as effective *in vivo* doses are ~50 μg with these delivery systems [63–65].

24.4.1.4 Other

Other local administration paradigms are potentially amenable to injection with naked siRNA. Examples may include intramuscular, intradermal, and transtympanic injection of siRNA. In fact, a recent report demonstrated that intradermal injection of siRNA into a mouse footpad was able to specifically inhibit expression of a vector-based target mRNA [22,66].

24.4.2 Conjugation

Direct conjugation with molecular entities designed to help target or deliver drug into the appropriate target cell type is naturally appealing. Through conjugation, the therapeutic can be fixed as a one-component system thereby significantly reducing the complexity from a chemistry, manufacturing, and controls perspective. For siRNA, conjugation is especially attractive given the fact that only one of two strands in the duplex is active. Thus, conjugates can be attached to the sense strand without disrupting activity of the antisense strand. Typically, conjugates have been placed on either the 5′ or 3′ end of the sense strand, though they can in some instances also be tolerated on the antisense strand. To date, siRNA conjugates have been made using lipophilic molecules, proteins, peptides, aptamers, and even small-molecule antagonists.

24.4.2.1 Cholesterol

While the use of cholesterol conjugates to aid *in vivo* delivery to liver had been established using antisense oligonucleotides [67], the report in 2004 by Soutschek et al. [68] using a cholesterol–siRNA conjugate provided the first mechanistic *in vivo* proof-of-concept for RNAi. Effective silencing of the apolipoprotein B (ApoB) in mice was demonstrated following intravenous administration of cholesterol-conjugated siRNA duplexes. In these experiments, injections of cholesterol-conjugated ApoB siRNA at a relatively high dose of 50 mg/kg resulted in silencing of the ApoB mRNA by ~55%

in the liver and ~70% in the jejunum, the two principal sites for ApoB expression; control cholesterol-conjugated siRNA demonstrated no silencing activity. Critically, the authors went on to demonstrate that the reduction in ApoB mRNA was a result of RNAi, through 5′RACE detection of specific mRNA cleavage product. Paralleling the mRNA decrease in ApoB mRNA, ApoB protein levels in the plasma dropped by ~70%, and in addition, consistent with the biological function of ApoB, a 35%–40% decrease in serum cholesterol levels was also seen. Cholesterol conjugation imparted critical pharmacokinetic and cellular uptake properties to the duplex as evidenced by the fact that the unconjugated ApoB siRNA was rapidly cleared and unable to effect mRNA silencing. The mechanism by which cholesterol conjugates promote improved distribution and cellular uptake is an area of active has recently been shown to involve incorporation of the cholesterol–siRNA conjugate into circulating lipoprotein particles and then delivery through receptor-mediated processes into hepatocytes [69].

Whether other tissues or cell types can be efficiently targeted by cholesterol conjugates is also an area of intense interest. One recent publication has demonstrated that a single intrastriatal injection of cholesterol-conjugated siRNA targeting huntingtin in mice expressing a mutant huntingtin gene silences the target mRNA, attenuates neuronal pathology, and delays the abnormal behavioral phenotype observed in a rapid-onset, viral transgenic mouse model of Huntington's disease [70].

24.4.2.2 Other Natural Ligands

Beyond cholesterol, other natural ligands have been successfully directly conjugated to siRNA duplexes. Among some siRNA conjugates described are membrane-permeant peptides such as penetratin and transportan [71]. While these conjugates were successfully synthesized with RNAi silencing activity preserved, robust *in vitro* or *in vivo* evidence indicative of an impact on cellular delivery is still lacking. Nevertheless, work done using natural ligands as targeting agents for siRNA formulated in conjunction with delivery vehicles has been positive. Molecules such as transferrin [72], folate [73], and Arg-Gly-Asp (RGD) peptide [74] have been introduced into particle-based delivery vehicles to help provide specific targeting of siRNA to cells bearing the natural receptor.

24.4.2.3 Aptamers

The use of RNA aptamers as a conjugate for delivery and targeting of siRNA is a promising and elegant approach. In contrast to most of the other described delivery methods, this approach allows for specific targeting of particular cell types and yet, unlike a similar antibody-based approach, has an additional advantage that the therapeutic can be composed entirely of RNA (i.e., RNA aptamer linked to an siRNA). Recently, *in vitro* and *in vivo* proof of concept has been generated using aptamers to prostate-specific membrane antigen (PSMA), a cell-surface receptor overexpressed in prostate cancer cells and tumor vascular endothelium [75,76]. PSMA aptamers, when either directly linked to siRNA [75] or conjugated through a modular streptavidin bridge [76], were capable of promoting specific cellular uptake and RNAi-mediated silencing of target mRNA *in vitro*. Using siRNA directed against survival genes (plk1 and bcl-2) directly linked to PSMA aptamer, it was found these RNA chimeras were internalized by cells and resulted in RNAi-mediated target mRNA silencing and cell death [75].

One concern about aptamer–siRNA chimeras is the potential difficulties in synthesis of long RNA molecules. Short aptamers (25–35 bases) that bind to a wide variety of targets with high affinity have been described [77]. Thus, in theory, it is possible to design siRNA–aptamer chimeras that would have a long strand of 45–55 bases, a synthesis length that is within the range of technical and commercial feasibility.

24.4.2.4 Small Molecules

Beyond natural ligands and aptamers as targeting vehicles for siRNA, there also exists the possibility of using small-molecule drugs in this fashion. Conceptually, targeting siRNA to specific cell types through conjugation to small molecules (or as discussed below through complexation to antibodies) is appealing. Developing small-molecule conjugates capable of promoting entry of siRNA into cells requires selection of an appropriate small molecule that is capable of efficient cellular internalization and endosomal escape.

24.4.3 Liposomes and Lipoplexes

Drugs have traditionally been formulated in liposomes in order to provide increased pharmacokinetic properties or decreased toxicity profiles. With the advent of RNAi, research

in the use of liposomes to deliver drugs into cells has surged. Liposomes are vesicles that consist of an aqueous compartment enclosed in a phospholipid bilayer with drug typically entrapped in the central aqueous layer. When lipids simply complex with nucleic acids to form particles, they are known as lipoplexes and are typical of most commercial transfection agents, such as Lipofectamine 2000 and TKO. Liposomes are multicomponent lipid-based nanoparticle delivery systems typically comprising a lipid component (often containing a cationic or a fusogenic lipid), cholesterol, and a polyethylene glycol (PEG)–lipid. Each of these components have a critical function to play in the fusogenicity and pharmacokinetic properties of the liposome (Figure 24.4; for review see Refs. [78,79]).

Both liposomes and lipoplexes have been extensively utilized to deliver siRNA *in vitro* and *in vivo*. *In vitro* transfection of siRNA using lipid-based delivery agents is now a routine laboratory procedure. More recently, significant success has been demonstrated with both local and systemic administration of siRNA. One of the most important advances in the development of RNAi as a therapeutic came with the demonstration by Zimmermann et al. that systemically delivered siRNA in stable nucleic acid lipid particles (SNALPs [32]) was able to dramatically silence ApoB in mice and nonhuman primates [80]. Most importantly, a single intravenous dose of 2.5 mg/kg of SNALP-formulated siRNA in cynomolgus monkeys was able to reduce ApoB mRNA levels in the liver by >90%. As expected, accompanying ApoB mRNA silencing, levels of serum cholesterol and low-density lipoproteins were reduced by greater than 65 and 85%, respectively. Remarkably, the durability of silencing following a single intravenous dose of 2.5 mg/kg SNALP-formulated siRNA was shown to last for at least 11 days. Treatment with liposomally formulated siRNA was well tolerated in these

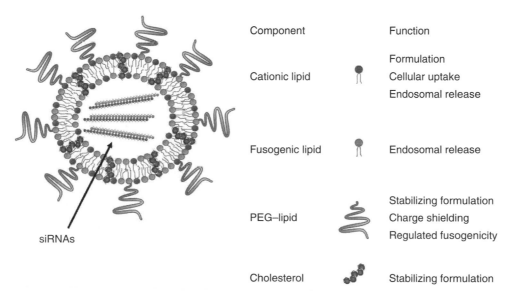

FIGURE 24.4 (See color insert following blank page 206.) Structure and function of a cationic liposome. Composition of a SNALP is shown. Other cationic liposome formulations also exist, which may or may not contain a fusogenic or PEG–lipid. Ratios of different components also can vary between formulations.

experiments, with transient increases in liver enzymes as the only reported evidence of toxicity, and these effects may be related to the targeting of ApoB itself. The general applicability of SNALP formulations for hepatic delivery of siRNA was also demonstrated in animal models of Hepatitis B virus and Ebola virus infection [81,82]. Other "non-SNALP" liposomes have also been demonstrated to be effective in delivering siRNA to different organs. Systemic delivery of a cationic liposome containing TNFα siRNA specifically inhibited TNFα production in knee joints and alleviated disease in a mouse model of rheumatoid arthritis [83], while a liposomally formulated caveolin-1 siRNA reduced target expression by 90% in mouse lung endothelia with concomitant expected physiological effects [84]. In summary, use of lipid-based formulations for systemic delivery of siRNA, especially to liver, represents one of the most promising near-term opportunities for development of RNAi therapeutics.

Local administration of liposomes and lipoplexes has also been successful in delivering siRNA to target cell types in the eye, nervous system, and tumors. Within the ocular context, subretinal injection of VEGF siRNA formulated in Transit-TKO was found to reduce CNV in a laser-induced mouse model [46], and subconjunctival injection of TGB-β receptor II siRNA in Transit-TKO to suppressed inflammation and fibrosis in a mouse subconjunctival scarring model [49]. With regards to the nervous system, intracranial delivery of lipid-complexed siRNA targeting the viral envelope genes protected mice against Japanese encephalitis virus– and West Nile Virus–induced encephalitis [65]. Similarly, intrathecal administration of cationic lipid formulated delta opioid receptor (DOR) siRNA facilitated delivery and specific target gene silencing in the spinal cord and dorsal root, resulting in blocked antinociception by a DOR-selective agonist [63]. Numerous publications have reported successful antitumoral effects following either direct or systemic tumoral injection of lipid-formulated siRNA (for reviews, see Refs. [38,43–45]). One recent example demonstrated that either intratumor injection or intraperitoneal injection of TransMessenger-formulated siRNA directed against the human papillomavirus E6 oncogene inhibited tumor growth, suppressed E6 expression, and induced tumor apoptosis in a subcutaneous cervical cancer xenograft model [85]. It is important to note that the necessity for lipid-based delivery systems (and this holds true for other delivery systems as well) in many of these direct RNAi applications must be assessed in a target cell- and disease-specific manner. For eye, lung, and nervous system, there are numerous examples of siRNA being successfully delivered in the absence of such agents and other examples where the converse is true.

Lastly, an area of increasing interest is direct application of lipoplexed siRNA to mucosal surfaces such as the vagina and intestine. Intravaginal delivery of lipid-complexed siRNA directed against herpes simplex 2 virus was found to protect mice when delivered before or after lethal herpes virus challenge [86]. Corroborating these results is evidence for specific lamin A/C and CCR5 silencing using the relevant targeting siRNA complexed with Lipofectamine 2000; similarly complexed irrelevant siRNA had no effect on silencing [87]. Direct delivery of a TNFα siRNA in a Lipofectamine formulation has also recently been demonstrated to not only reduce TNFα levels but also attenuate colonic inflammation after enema administration [87]. Intravaginal and intracolonic administration of siRNA using these lipid-based delivery systems was well tolerated in mice with no reported evidence of toxicity or activation of an interferon response [86,87]. Given the relatively easy physical access, the large number of potential unmet medical disorders, the general tolerability of the siRNA–lipid formulations, and importantly, the demonstrated robust *in vivo* delivery success, mucosal administration of lipid-formulated siRNA will be a fertile area for the development of future RNAi therapeutics.

24.4.4 PEPTIDES AND POLYMERS

Complexation of siRNA with positively charged peptides or polymers has been a field of increasing interest [88]. In general, cationic peptides and polymers are used to form complexes with the negatively charged phosphate backbone of the siRNA duplex. This noncovalent association is very stable and results in nanometer-sized particles. One concern with complexation approaches is the ability to prevent aggregation and thus control particle formation in a reproducible manner. This is most often done by either maintaining a net positive charge on the surface or incorporating molecules such as PEG that help stabilize the particle and prevent aggregation. PEG groups are also used with polymers to improve the pharmacokinetic profile, while agents such as folate, transferrin, antibodies, or sugars such as galactose and mannose can be incorporated for tissue targeting [89]. Beyond improving particle targeting *in vivo*, another essential aspect is to design particles that are capable of promoting efficient endosomal escape. Several pH-triggered polymers that facilitate endosomal escape have been identified [79,90]; data with one of these peptides have already been shown to improve RNAi-mediated silencing *in vitro* [91].

One of the most widely studied polymers for delivery of nucleic acids is polyethylenimine (PEI). PEI polymers are synthetic linear or branched structures with protonable amino groups and high cationic charge densities. Following complexation with siRNA, the cationic polyplexes are thought to interact with the cell surface through electrostatic interaction and are taken up by cells through endocytosis, where intracellularly, they act to buffer the low endosomal pH. Endosomal escape is hypothesized to occur due to a "proton sponge" effect, whereby PEI enhances the influx of protons and water, leading to endosomal destabilization and osmotic release of the polyplexes into the cytoplasm [92]. Multiple reports of use of siRNA–PEI complexes *in vivo* now exist. PEI complexed with influenza siRNA were reported to result in profound antiviral effects in infected mice [93], while similar targeting of the Ebola L gene resulted in protection against lethal Ebola infection in guinea pigs [82]. Profound antitumor activity was seen upon administration of PEI–siRNA complexes targeting pleiotrophin [94], VEGF [74], and HER2

[95]. Lastly, local injection of PEI–siRNA polyplexes targeting the NMDA receptor subunit NR2B significantly attenuated formalin-induced nociception in the rat [64]. One of the main concerns with the use of PEI as a therapeutic delivery vehicle is the extreme toxicity seen at higher doses. In an attempt to expand the safety margin, several groups are optimizing the physical structure of PEI to enable improved *in vivo* delivery of siRNA [96–98]. In addition to PEI, other synthetic polycations consisting of histidine and polylysine residues have also been evaluated for delivery of siRNA and appear to have improved *in vitro* efficacy as compared to standard PEI [99].

Four other polymer approaches that have yielded *in vivo* siRNA data are dynamic polyconjugates, chitosan nanoparticles, cyclodextrin-based nanoparticles, and atelocollagen nanoparticles. The first of these delivery approaches uses dynamic polyconjugates and has been shown *in vivo* in mice to effectively deliver siRNA and silence endogenous target mRNA (ApoB and PPAR-alpha) in hepatocytes [100]. This is a multicomponent polymer system whose key features include a membrane-active polymer to which siRNA is covalently coupled via a disulfide bond and where both PEG (for charge masking) and *N*-acetylgalactosamine (for hepatocyte targeting) groups are linked via pH-sensitive bonds [100]. Upon binding to the hepatocyte and entry into the endosome, the polymer complex disassembles in the low pH environment, with the polymer exposing its positive charge, leading to endosomal escape and cytoplasmic release of the siRNA from the polymer. While this is an elegant and exciting approach using both targeted delivery and endosomal escape mechanisms, further work remains to be done to understand whether it represents a viable drug delivery modality. Another interesting, though complex, polymer approach involves using transferrin-targeted cyclodextrin-containing polycation nanoparticles. These nanoparticles have demonstrated targeted silencing of the EWS–FLI1 gene product in transferrin receptor–expressing Ewing's sarcoma tumor cells [72] and siRNA formulated in these nanoparticles was well tolerated in nonhuman primates [101]. A third approach has used chitosan, which is a well-tolerated natural biodegradable polymer, to form cationic complexes with nucleic acids. Effective *in vivo* RNAi interference was achieved both in bronchiole epithelial cells of transgenic EGFP mice after intranasal administration of chitosan/siRNA formulations [102] and in subcutaneously implanted breast cancer cells of nude mice after intravenous administration of chitosan/RhoA siRNA complexes [103]. A fourth polymer approach, which has received substantial attention, uses a fragment of collagen known as atelocollagen. Systemic administration of atelocollagen–siRNA complexes had marked effects on subcutaneous tumor xenografts as well as on bone metastases [104,105]. Lastly, other related polymer approaches such as polyammidoamine (PAMAM) dendrimers and poly(DL-lactic-*co*-glycolic acid) (PLGA) nanoparticles are also being investigated as vehicles for siRNA delivery [106,107], though *in vivo* evidence for RNAi-mediated silencing remains to be demonstrated using these approaches.

Peptide-based approaches, including use of cell-penetrating peptides, to deliver oligonucleotides is an area of active research [90,108–110]. Several peptide-based gene delivery systems such as MPG [111], Penetratin [112], and cholesteryl oligo-D-arginine (Chol-R9) [113] have been shown to promote siRNA uptake *in vitro*. In the case of Chol-R9, local *in vivo* administration of complexed VEGF-targeting siRNA, but not complexed scrambled siRNA, led to tumor regression in a mouse model [113]. Another peptide-based system to demonstrate *in vivo* silencing was a synthetic chimeric peptide consisting of nonamer arginine residues (9R) added to the carboxy terminus of a 29–amino acid rabies virus glycoprotein (RVG) peptide, which specifically binds to the acetylcholine receptor expressed by neuronal cells [110]. Using this system, siRNA were complexed with the positively charged RVG-9R peptide and found to be capable of delivering siRNA to neuronal cells following intravenous administration, resulting in specific gene silencing [110]. Lastly, peptide-based approaches are often coupled with other delivery systems, such as liposomes to enable more targeted delivery of oligonucleotides [74,114].

24.4.5 ANTIBODIES

Antibody-based approaches for specific targeted delivery of siRNA hold much promise. Recently, elegant studies have demonstrated that a protamine–antibody fusion protein was able *in vitro* to selectively deliver siRNA to HIV-envelope expressing B16 melanoma cells or HIV-infected CD4 T cells [115]. Protamine is used for its nucleic acid–binding properties while the Fab fragment was used to mediate receptor-specific binding to only those cells expressing the gp160 HIV envelope protein. Importantly, using gp160-B16 cells as a tumor model, it was shown that siRNA–antibody–protamine complexes when delivered either intratumorally or intravenously are able to specifically deliver siRNA *in vivo* and retard tumor growth. Recently fusion proteins that target all human white blood cells or just activated leukocytes using conformation-sensitive, single-chain antibodies to the integrin LFA-1 have demonstrated the exquisite selectivity of this targeting strategy both *in vitro* and *in vivo* [116]. These studies demonstrate the potential for antibodies to direct siRNA selectively into cells *in vivo*.

24.5 CLINICAL TRIALS

RNAi has rapidly advanced from research discovery to clinical trials. Four different RNAi therapeutics are under clinical investigation, with several more poised to enter trials soon. Initial trials have focused on well-validated therapeutic targets, such as the VEGF pathway for the wet form of AMD and the RSV genome, for the treatment of RSV infection. The status of the clinical-stage programs is summarized below.

24.5.1 OCULAR

Ophthalmic indications have historically been attractive for oligonucleotide-based therapeutics. The only approved oligonucleotide-based drugs are used in the eye—Vitravene® to treat CMV retinitis [117] and Macugen® for AMD [118]. Direct injection into the vitreal cavity efficiently targets oligonucleotide drugs to the retina. Another advantage of the ocular compartment is that it is relatively free of nucleases, compared to serum. While this permits unmodified siRNA to be used in the vitreal cavity, exonuclease-stabilizing chemistries can dramatically improve the persistence of intact siRNA following intravitreal injection [51]. The ease of intraocular drug delivery, combined with the clinical validation of VEGF as an attractive therapeutic target for proliferative vascular retinopathies, led to the rapid development of several RNAi therapeutics.

Intravitreous injection of Bevasiranib or Cand5, an unmodified siRNA targeting all VEGF-A spliced isoforms, has completed a phase II trial in patients with serious progressive wet AMD and has been reported to provide dose-related benefits against several endpoints including near vision and lesion size (Opko Pharmaceuticals press release). Cand5 is currently being tested in phase III clinical trials for wet AMD.

Sirna-027, a chemically modified siRNA targeting the VEGF receptor (VEGF-R1), completed phase I trials in patients with wet AMD and was reported to be well tolerated. It was also reported to stabilize or improve visual acuity in a subset of patients (Sirna Pharmaceuticals press release). Development of this molecule has transitioned to Merck Pharmaceuticals following their acquisition of Sirna Therapeutics.

RTP801i-14, a chemically modified siRNA targeting the hypoxia-inducible gene RTP801, is currently in phase I/II trials for the treatment of wet AMD. Intravitreal injection of RTP801i-14 resulted in inhibition of CNV and vessel leakage in mouse and primate models (Quark press release). In preclinical studies, RTP 801i-14 was shown to work in cooperation or synergy with VEGF-based drugs.

24.5.2 RESPIRATORY

There are many advantages to delivering siRNA therapeutics directly to the lungs for targeting lung epithelial cells. Lung epithelial cells are capable of naked siRNA uptake and RNAi-mediated silencing [52,54,57]. Moreover, the local lung environment is relatively nuclease-free compared to serum. Delivery to the lung by inhalation is noninvasive and directly targets the tissue epithelium, which improves drug concentrations at the target tissue without first-pass metabolism, reduces drug dosing, and decreases the likelihood of systemic side effects. While in many instances, naked siRNA in saline appears to be effective, other approaches to optimize lung delivery may need to be developed for different indications.

The first pulmonary studies are directed at treating RSV, a serious neonatal respiratory infection. Two phase I intranasal trials with ALN-RSV01, an siRNA targeting the viral nucleocapsid (N) gene, have been completed in over 100 healthy adult volunteers. ALN-RSV01 was found to be safe and well tolerated (Alnylam Pharmaceuticals press release).

Clinical development of ALN-RSV01 is progressing with experimental infection studies ongoing.

24.6 SUMMARY

RNAi has advanced from research discovery to clinical trials in a very short span of time, starting with the now Nobel Prize winning discovery in 1998 by Fire and Mello that long dsRNA could mediate RNAi in *Caernorhabditis elegans* [1], to evidence in 2001 by Tuschl and colleagues that short synthetic siRNA could induce RNAi in mammalian cells [2], to the present day when four different RNAi therapeutics are in human clinical trials and numerous others poised to enter. Since discovery of its utility in mammalian cells six years ago, RNAi has revolutionized biomedical research and has spawned a veritable explosion of scientific knowledge covering all aspects of *in vitro* and *in vivo* RNAi biology. The principal challenge that remains in achieving the broadest application of RNAi therapeutics is the hurdle of delivery. Tremendous progress has been made on this front through the use of different approaches, such as conjugation, complexation, and lipid-based approaches, though it is clear that delivery is yet to be solved for all cell types and tissues. Delivery solutions as they arrive will likely need to be tailored for target cell type and the disease indication. In identifying and optimizing delivery solutions, the use of reporter systems (either endogenous cell-type-specific genes or transgenes) can greatly expedite the process. Likewise in optimizing formulations, it is vital that the key parameters of efficacy and safety be assessed in parallel in order to rapidly arrive at the optimal formulations, which have the greatest therapeutic window. As the challenge of siRNA delivery is met, it will be possible to rapidly advance RNAi therapeutics against potentially any disease target into clinical studies. The ongoing clinical trials with siRNA for the treatment of AMD and RSV infection will be the first indicator of the potential for RNAi therapeutics to be a revolutionary new class of drug molecules.

REFERENCES

1. Fire, A. et al. Potent and specific genetic interference by double-stranded RNA in Caenorhabditis elegans. *Nature* 391, 806–811 (1998).
2. Elbashir, S.M. et al. Duplexes of 21-nucleotide RNAs mediate RNA interference in cultured mammalian cells. *Nature* 411, 494–498 (2001).
3. Meister, G. and Tuschl, T. Mechanisms of gene silencing by double-stranded RNA. *Nature* 431, 343–349 (2004).
4. Kim, D.H. and Rossi, J.J. Strategies for silencing human disease using RNA interference. *Nat Rev Genet* 8, 173–184 (2007).
5. He, L. and Hannon, G.J. MicroRNAs: Small RNAs with a big role in gene regulation. *Nat Rev Genet* 5, 522–531 (2004).
6. Matranga, C. et al. Passenger-strand cleavage facilitates assembly of siRNA into Ago2-containing RNAi enzyme complexes. *Cell* 123, 607–620 (2005).
7. Liu, J. et al. MicroRNA-dependent localization of targeted mRNAs to mammalian P-bodies. *Nat Cell Biol* 7, 719–723 (2005).

8. Hutvagner, G. and Zamore, P.D. A microRNA in a multiple-turnover RNAi enzyme complex. *Science* 297, 2056–2060 (2002).

9. Ameres, S.L. et al. Molecular basis for target RNA recognition and cleavage by human RISC. *Cell* 130, 101–112 (2007).

10. Grimm, D. et al. Fatality in mice due to oversaturation of cellular microRNA/short hairpin RNA pathways. *Nature* 441, 537–541 (2006).

11. John, M. et al. Effective RNAi-mediated gene silencing without interruption of the endogenous microRNA pathway. *Nature* 449, 745–747 (2007).

12. Pei, Y. and Tuschl, T. On the art of identifying effective and specific siRNAs. *Nature Methods* 3, 670–676 (2006).

13. Reynolds, A. et al. Rational siRNA design for RNA interference. *Nat Biotech* 22, 326–330 (2004).

14. Kim, D.H. et al. Synthetic dsRNA Dicer substrates enhance RNAi potency and efficacy. *Nat Biotechnol* 23, 222–226 (2005).

15. Reynolds, A. et al. Induction of the interferon response by siRNA is cell type- and duplex length-dependent. *RNA* 6, 988–993 (2006).

16. Schwarz, D.S. et al. Asymmetry in the assembly of the RNAi enzyme complex. *Cell* 115, 199–208 (2003).

17. Khvorova, A., Reynolds, A., and Jayasena, S.D. Functional siRNAs and miRNAs exhibit strand bias. *Cell* 115, 209–216 (2003).

18. Aza-Blanc, P. et al. Identification of modulators of TRAIL-induced apoptosis via RNAi-based phenotypic screening. *Mol Cell* 12, 627–637 (2003).

19. Heidenreich O. et al. AML1/MTG8 oncogene suppression by small interfering RNAs supports myeloid differentiation of t(8;21)-positive leukemic cells. *Blood* 101, 3157–3163 (2003).

20. Wohlbold L. et al. Inhibition of *bcr-abl* gene expression by small interfering RNA sensitizes for imatinib mesylate (STI571). *Blood* 102, 2236–2239 (2003).

21. Schwarz, D.S. et al. Designing siRNA that distinguish between genes that differ by a single nucleotide. *PLoS Genetics* 2, 1307–1318 (2006).

22. Hickerson R.P. et al. Single-nucleotide-specific siRNA targeting in a dominant-negative skin model. *J Invest Dermatol* 128, 594–605 (2008).

23. Jackson, A.L. et al. Expression profiling reveals off-target gene regulation by RNAi. *Nat Biotech* 21, 635–637 (2003).

24. Lin, X. et al. siRNA-mediated off-target gene silencing triggered by a 7 nt complementation. *Nucleic Acids Res* 33, 4527–4535 (2005).

25. Qiu, S., Adema, C.M., and Lane, T. A computational study of off-target effects of RNA interference. *Nucleic Acids Res* 33, 1834–1847 (2005).

26. Jackson, A.L. et al. Widespread siRNA off-target transcript silencing mediated by seed region sequence complementarity. *RNA* 12, 1179–1187 (2006).

27. Birmingham, A. et al. 3′ UTR seed matches, but not overall identity, are associated with RNAi off-targets. *Nat Methods* 3, 199–204 (2006).

28. Jackson, A.L. et al. Position-specific chemical modification of siRNAs reduces off-target transcript silencing. *RNA* 12, 1197–1205 (2006).

29. Fedorov, Y. et al. Off-target effects by siRNA can induce toxic phenotype. *RNA* 12, 1188–1196 (2006).

30. Schlee, M. et al. SiRNA and isRNA: Two edges of one sword. *Mol. Ther.* 14, 463–470 (2006).

31. Hornung, V. et al. Sequence-specific potent induction of IFN-alpha by short interfering RNA in plasmacytoid dendritic cells through TLR7. *Nat Med* 11, 263–270 (2005).

32. Judge, A.D. et al. Sequence-dependent stimulation of the mammalian innate immune response by synthetic siRNA. *Nat Biotech* 23, 457–462 (2005).

33. Judge, A.D. et al. Design of noninflammatory synthetic siRNA mediating potent gene silencing in vivo. *Mol Ther* 13, 494–505 (2006).

34. Layzer, J.M. et al. In vivo activity of nuclease-resistant siRNAs. *RNA* 10, 766–771 (2004).

35. Choung, S. et al. Chemical modification of siRNAs to improve serum stability without loss of efficacy. *Biochem Biophys Res Commun* 342, 919–927 (2006).

36. Bartlett, D.W. and Davis, M.E. Effect of siRNA nuclease stability on the in vitro and in vivo kinetics of siRNA-mediated gene silencing. *Biotech Bioeng* 97, 909–921 (2007).

37. Allerson, C.R. et al. Fully 2′-modified oligonucleotide duplexes with improved in vitro potency and stability compared to unmodified small interfering RNA. *J Med Chem* 48, 901–904 (2005).

38. de Fougerolles, A.R. et al. RNA interference in vivo: Toward synthetic small inhibitory RNA-based therapeutics. *Methods Enzymol* 392, 278–296 (2005).

39. Morrissey, D.V. et al. Activity of stabilized short interfering RNA in a mouse model of hepatitis B virus replication. *Hepatology* 41, 1349–1356 (2005).

40. Bartlett, D.W. and Davis, M.E. Insights into the kinetics of siRNA-mediated gene silencing from live-cell and live-animal bioluminescent imaging. *Nucleic Acids Res* 34, 322–333 (2006).

41. Raemdonck, K. et al. In situ analysis of single-stranded and duplex siRNA integrity in living cells. *Biochemistry* 45, 10614–10623 (2006).

42. Fedorov, Y. et al. Different delivery methods-different expression profiles. *Nat Methods* 2, 241 (2005).

43. de Fougerolles, A.R. et al. Interfering with disease: A progress report on siRNA-based therapeutics. *Nat Rev Drug Discovery* 6, 443–453 (2007).

44. Aigner, A. Gene silencing through RNA interference (RNAi) in vivo: Strategies based on the direct application of siRNAs. *J Biotechnol* 124, 12–25 (2006).

45. Bumcrot, D. et al. RNAi therapeutics: A potential new class of pharmaceutical drugs. *Nat Chem Biol* 2, 711–719 (2006).

46. Reich, S.J. et al. Small interfering RNA (siRNA) targeting VEGF effectively inhibits ocular neovascularization in a mouse model. *Mol Vision* 9, 210–216 (2003).

47. Tolentino, M.J. et al. Intravitreal injection of vascular endothelial growth factor small interfering RNA inhibits growth and leakage in a nonhuman primate, laser-induced model of choroidal neovascularization. *Retina* 24, 132–138 and 660–661 (2004).

48. Shen, J. et al. Suppression of ocular neovascularization with siRNA targeting VEGF receptor 1. *Gene Ther* 13, 225–234 (2006).

49. Nakamura, H. et al. RNA interference targeting transforming growth factor-beta type II receptor suppresses ocular inflammation and fibrosis. *Mol Vision* 10, 703–711 (2004).

50. Campochiaro, P.A. et al. Potential applications for RNAi to probe pathogenesis and develop new treatments for ocular disorders. *Gene Ther* 13, 559–562 (2006).

51. de Fougerolles, A.R. and Maraganore, J.M. Discovery and development of RNAi therapeutics, in *Antisense Drug Technology*, Crooke, S.T. (Ed.), CRC Press, Boca Raton, FL, 2007, chap. 16.

52. Bitko, V. et al. Inhibition of respiratory viruses by nasally administered siRNA. *Nat Med* 11, 50–55 (2005).

53. Li, B.J. et al. Using siRNA in prophylactic and therapeutic regimens against SARS coronavirus in Rhesus macaque. *Nat Med* 11, 944–951 (2005).

54. Zhang, X. et al. Small interfering RNA targeting heme oxygenase-1 enhances ischemia-reperfusion-induced lung apoptosis. *J Biol Chem* 279, 10677–10684 (2004).

55. Lomas-Neira, J.L. et al. In vivo gene silencing (with siRNA) of pulmonary expression of MIP-2 versus KC results in divergent effects on hemorrhage-induced, neutrophil-mediated septic acute lung injury. *J Leukoc Biol* 77, 846–853 (2005).

56. Perl, M. et al. Silencing of Fas, but not caspase-8, in lung epithelial cells ameliorates pulmonary apoptosis, inflammation, and neutrophil influx after hemorrhagic shock and sepsis. *Am J Pathol* 167, 1545–1559 (2005).

57. Bhandari, V. et al. Hypoxia causes angiopoietin-2-mediated acute lung injury and necrotic cell death. *Nat Med* 12, 1286–1292 (2006).

58. Matsuyama, W. et al. Suppression of discoidin domain receptor 1 by RNA interference attenuates lung inflammation. *J Immunol* 176, 1928–1936 (2006).

59. Makimura, H. et al. Reducing hypothalamic AGRP by RNA interference increases metabolic rate and decreases body weight without influencing food intake. *BMC Neurosci* 3, 18 (2002).

60. Thakker, D.R. et al. Neurochemical and behavioral consequences of widespread gene knockdown in the adult mouse brain by using nonviral RNA interference. *Proc Natl Acad Sci USA* 101, 17270–17275 (2004).

61. Thakker, D.R. et al. siRNA-mediated knockdown of the serotonin transporter in the adult mouse brain. *Mol Psychiatry* 10, 782–789 (2005).

62. Dorn, G. et al. SiRNA relieves chronic neuropathic pain. *Nucleic Acids Res* 32, e49 (2004).

63. Luo, M.C. et al. An efficient intrathecal delivery of small interfering RNA to the spinal cord and peripheral neurons. *Mol Pain* 1, 29 (2005).

64. Tan, P.H. et al. Gene knockdown with intrathecal siRNA of NMDA receptor NR2B subunit reduces formalin-induced nociception in the rat. *Gene Ther* 12, 59–66 (2005).

65. Kumar, P. et al. A single siRNA suppresses fatal encephalitis induced by two different flaviviruses. *PLoS Med* 3, 505–514 (2006).

66. Wang, Q. et al. Delivery and inhibition of reporter genes by small interfering RNAs in a mouse skin model. *J Invest Dermatol* 127, 2577–2584 (2007).

67. Biessen, E.A. et al. Targeted delivery of oligodeoxynucleotides to parenchymal liver cells in vivo. *Biochem J* 340, 783–792 (1999).

68. Soutschek, J. et al. Therapeutic silencing of an endogenous gene by systemic administration of modified siRNAs. *Nature* 432, 173–178 (2004).

69. Wolfrum, C. et al. Mechanisms and optimization of in vivo delivery of lipophilic siRNAs. *Nat Biotechnol* 25, 1149–1157 (2007).

70. Difiglia, M. et al. Therapeutic silencing of mutant huntingtin with siRNA attenuates striatal and cortical neuropathology and behavioral deficits. *Proc Natl Acad Sci USA* 104, 17204–17209 (2007).

71. Muratovska, A. and Eccles, MR. Conjugate for efficient delivery of short interfering RNA (siRNA) into mammalian cells. *FEBS Lett* 558, 63–68 (2004).

72. Hu-Lieskovan, S. et al. Sequence-specific knockdown of EWS-FLI1 by targeted, nonviral delivery of small interfering RNA inhibits tumor growth in a murine model of metastatic Ewing's sarcoma. *Cancer Res* 65, 8984–8992 (2005).

73. Kim, S.H. et al. Target-specific gene silencing by siRNA plasmid DNA complexed with folate-modified poly(ethylenimine). *J Control Release* 104, 223–232 (2005).

74. Schiffelers, R.M. et al. Cancer siRNA therapy by tumor selective delivery with ligand-targeted sterically stabilized nanoparticle. *Nucleic Acids Res* 32, e149 (2004).

75. McNamara, J.O. et al. Cell type-specific delivery of siRNAs with aptamer-siRNA chimeras. *Nat Biotech* 24, 1005–1015 (2006).

76. Chu, T.C. et al. Aptamer mediated siRNA delivery. *Nucleic Acids Res* 34, e73 (2006).

77. Nimjee, S.M., Rusconi, C.P., and Sullenger B.A. Aptamers: An emerging class of therapeutics. *Annu Rev Med* 56, 555–583 (2005).

78. Torchilin, V.P. Recent approaches to intracellular delivery of drugs and DNA and organelle targeting. *Annu Rev Biomed Eng* 8, 343–375 (2006).

79. Li, W. and Szoka F.C. Lipid-based nanoparticles for nucleic acid delivery. *Pharm Res Adv Drug Deliv Rev* 24, 438–449 (2007).

80. Zimmermann, T.S. et al. RNAi-mediated gene silencing in non-human primates. *Nature* 441, 111–114 (2006).

81. Morrissey, D.V. et al. Potent and persistent in vivo anti-HBV activity of chemically modified siRNAs. *Nat Biotechnol* 23, 1002–1007 (2005).

82. Geisbert, T.W. et al. Postexposure protection of guinea pigs against lethal ebola virus challenge is conferred by RNA interference. *J Infect Dis* 193, 1650–1657 (2006).

83. Khoury, M. et al. Efficient new cationic liposome formulation for systemic delivery of small interfering RNA silencing tumor necrosis factor-alpha in experimental arthritis. *Arthritis Rheumatism* 54, 1867–1877 (2006).

84. Miyawaki-Shimizu, K. et al. SiRNA-induced caveolin-1 knockdown in mice increases lung vascular permeability via the junctional pathway. *Am J Physiol Lung Cell Mol Physiol* 290, L405–L413 (2006).

85. Niu, X.Y. et al. Inhibition of HPV 16 E6 oncogene expression by RNA interference in vitro and in vivo. *Int J Gynecol Cancer* 16, 743–751 (2006).

86. Palliser, D. et al. An siRNA-based microbicide protects mice from lethal herpes simplex virus 2 infection. *Nature* 439, 89–94 (2006).

87. Zhang, Y. et al. Engineering mucosal RNA interference in vivo. *Mol Ther* 4, 336–342 (2006).

88. Juliano, R.L. Peptide–oligonucleotide conjugates for the delivery of antisense and siRNA. *Curr Opin Mol Ther* 7, 132–136 (2005).

89. Merdan, T. et al. Prospects for cationic polymers in gene and oligonucleotide therapy against cancer. *Adv Drug Deliv Rev* 54, 715–758 (2002).

90. Li, W., Nicol, F., and Szoka F.C. GALA: A designed synthetic pH-responsive amphipathic peptide with applications in drug and gene delivery. *Adv Drug Deliv Rev* 56, 967–985 (2004).

91. Oliveira, S. et al. Fusogenic peptides enhance endosomal escape improving siRNA-induced silencing of oncogenes. *Int J Pharm* 331: 211–214 (2007).

92. Boussif, O. et al. A versatile vector for gene and oligonucleotide transfer into cells in culture and in vivo: Polyethylenimine. *Proc Natl Acad Sci USA* 92, 7297–7301 (1995).

93. Ge, Q. et al. Inhibition of influenza virus production in virus-infected mice by RNA interference. *Proc Natl Acad Sci USA* 101, 8676–8681 (2004).

94. Grzelinski, M. et al. RNA interference-mediated gene silencing of pleiotrophin through polyethylenimine-complexed small interfering RNAs in vivo exerts antitumoral effects in glioblastoma xenografts. *Hum Gene Ther* 17, 751–766 (2006).

95. Urban-Klein, B. et al. RNAi-mediated gene-targeting through systemic application of polyethylenimine (PEI)-complexed siRNA in vivo. *Gene Ther* 12, 461–466 (2005).

96. Thomas, M. et al. Full deacylation of polyethylenimine dramatically boosts its gene delivery efficiency and specificity to mouse lung. *Proc Natl Acad Sci USA* 102, 5679–5684 (2005).

97. Grayson, A.C. et al. Biophysical and structural characterization of polyethylenimine-mediated siRNA delivery in vitro. *Pharm Res* 23, 1868–1876 (2006).

98. Werth, S. et al. A low molecular weight fraction of polyethylenimine (PEI) displays increased transfection efficiency of DNA and siRNA in fresh or lyophilized complexes. *J Control Release* 112, 257–270 (2006).

99. Read, M.L. et al. A versatile reducible polycation-based system for efficient delivery of a broad range of nucleic acids. *Nucleic Acids Res* 33, e86 (2005).

100. Rozema, D.B. et al. Dynamic polyconjugates for targeted in vivo delivery of siRNA to hepatocytes. *Proc Natl Acad Sci USA* 104, 12982–12987 (2007).

101. Heidel, J.D. et al. Administration in non-human primates of escalating intravenous doses of targeted nanoparticles containing ribonucleotide reductase subunit M2 siRNA. *Proc Natl Acad Sci USA* 104, 5715–5721 (2007).

102. Howard, K.A. et al. RNA interference in vitro and in vivo using a novel chitosan/siRNA nanoparticle system. *Mol Ther* 14, 476–484 (2006).

103. Pille, J.Y. et al. Intravenous delivery of anti-RhoA small interfering RNA loaded in nanoparticles of chitosan in mice: Safety and efficacy in xenografted aggressive breast cancer. *Human Gene Ther* 17, 1019–1026 (2006).

104. Takei, Y. et al. A small interfering RNA targeting vascular endothelial growth factor as cancer therapeutics. *Cancer Res* 64, 3365–3370 (2004).

105. Takeshita, F. and Ochiya, T. Therapeutic potential of RNA interference against cancer. *Cancer Sci* 97, 689–696 (2006).

106. Khan, A. et al. Sustained polymeric delivery of gene silencing antisense ODNs, siRNA, DNAzymes and ribozymes: In vitro and in vivo studies. *J Drug Target* 12, 393–404 (2004).

107. Kang, H. et al. Tat-conjugated PAMAM dendrimers as delivery agents for antisense and siRNA oligonucleotides. *Pharm Res* 22, 2099–2106 (2005).

108. Zatsepin, T.S. et al. Conjugates of oligonucleotides and analogues with cell penetrating peptides as gene silencing agents. *Curr Pharm Des* 11, 3639–3654 (2005).

109. Meade, B.R. and Dowdy, S.F. Exogenous siRNA delivery using peptide transduction domains/cell penetrating peptides. *Adv Drug Deliv Rev* 59, 134–140 (2007).

110. Kumar, P. et al. Transvascular delivery of small interfering RNA to the central nervous system. *Nature* 448, 39–43 (2007).

111. Simeoni, F. et al. Insight into the mechanism of the peptide-based gene delivery system MPG: Implications for delivery of siRNA into mammalian cells. *Nucleic Acids Res* 31, 2717–2724 (2003).

112. Davidson, T.J. et al. Highly efficient small interfering RNA delivery to primary mammalian neurons induces microRNA-like effects before mRNA degradation. *J Neurosci* 24, 10040–10046 (2004).

113. Kim, W.J. et al. Cholesteryl oligoarginine delivering vascular endothelial growth factor siRNA effectively inhibits tumor growth in colon adenocarcinoma. *Mol Ther* 14, 343–350 (2006).

114. Longmuir, K.J. et al. Effective targeting of liposomes to liver and hepatocytes in vivo by incorporation of a plasmodium amino acid sequence. *Pharm Res* 23, 759–769 (2006).

115. Song, E. et al. Antibody mediated in vivo delivery of small interfering RNAs via cell-surface receptors. *Nat Biotechnol* 23, 709–717 (2005).

116. Peer, D. et al. Selective gene silencing in activated leukocytes by targeting siRNAs to the integrin lymphocyte function-associated antigen 1. *Proc Natl Acad Sci USA* 104, 4095–4100 (2007).

117. Jabs, D.A. et al. Fomivirsen for the treatment of cytomegalovirus retinitis. *Am J Ophthalmol* 133, 552–556 (2002).

118. Gragoudas, E.S. et al. Pegaptanib for neovascular age-related macular degeneration. *N Engl J Med* 351, 2805–2816 (2004).

25 Suicide Gene Therapy

Nikiforos Ballian, Bert W. O'Malley Jr., and F. Charles Brunicardi

CONTENTS

25.1   INTRODUCTION

Despite a trend toward decreasing incidence and mortality during the past decade, cancer is responsible for one in four deaths and approximately 1.5 million new cases annually in the United States [1]. Advances in surgical technique, as well as new forms of neoadjuvant chemotherapy and radiotherapy, are increasing the number of patients who can be cured by surgery. On the other hand, cancer patients who present with advanced disease not amenable to curative resection are being treated with radiotherapy and chemotherapy with the goal of achieving local and systemic disease control. Responses to these forms of treatment are usually temporary and cause significant toxicity to healthy tissues. Efforts are being made to

direct the cytotoxic effects of chemotherapy and radiotherapy to malignant cells in order to avoid injury to nonneoplastic tissues. At the same time, improved understanding of tumor immunology and genetics is fueling the development of new treatment modalities that will achieve the same goals.

Although initially perceived as a potential treatment for inherited single-gene defects, gene therapy is attracting significant research interest as a cancer treatment that could achieve selective tumor cell toxicity. The therapeutic principle behind gene therapy is the correction of genetic abnormalities by the insertion and expression of normal genes in cells with inherited or acquired monogenic defects. In contrast to single-gene disorders such as cystic fibrosis, the genesis of an invasive neoplasm is characterized by multiple gene mutations.

By the time tumor volume is significant to cause clinical manifestations, it consists of multiple subclones of cells that have resulted from different combinations of mutations. Hence, attempts to treat tumors by restoring the function of mutated genes would be impractical.

As a result of these limitations, suicide gene therapy has evolved as a different approach to cancer gene therapy, which can be used alone or in combination with other treatment modalities. Since suicide gene therapy emerged in the 1980s, two forms of therapy have been suggested [2]. The first relies on inserting genes into tumor cells that encode cytotoxins, which directly lead to cell death once the gene is expressed. This form of treatment has not received significant attention and will not be further discussed. The other form of suicide gene therapy, also known as gene-directed enzyme prodrug therapy (GDEPT), virus-directed enzyme prodrug therapy (VDEPT), or gene–prodrug activation therapy (GPAT), consists of inserting a gene into malignant cells, which encodes an enzyme able to convert a nontoxic prodrug to a toxic compound, which will result in cell death [3]. In this chapter, the term "suicide gene therapy" will be used. Key steps for the success of suicide gene therapy are (1) efficient suicide gene delivery to as many tumor cells as possible and (2) expression of the suicide gene in all transfected cells and until enough toxic metabolite has been produced to cause cell death. At the same time, delivery and expression of the suicide gene to normal cells should be avoided or minimized to prevent toxicity to healthy host tissues. Based on these principles, a number of "suicide" gene–prodrug combinations have been developed.

This chapter will discuss the physiologic principles of suicide gene therapy, the suicide gene–prodrug combinations currently in use, the different vectors used to deliver suicide genes, the mechanisms responsible for the bystander effect, and the current status of clinical trials of suicide gene therapy.

25.2 PRODRUGS, ENZYMES, AND SAFETY PRINCIPLES

A number of features are responsible for the different efficacy and safety of prodrugs and enzymes used in suicide gene therapy. Ideal prodrugs are stable compounds with minimal or no toxicity to host tissues that diffuse through the interstitial space and cell membrane to reach tumor cells in high concentrations. In addition, their activation should include as few steps and enzymes as possible to reduce the chance of enzyme mutation leading to reduced prodrug activation and resistance to treatment [3]. Prodrugs can be divided into two groups depending on their activation kinetics. Directly linked prodrugs are those that require chemical transformation to directly release an active, toxic compound. The second prodrug class, self-immolating prodrugs, produces an unstable intermediate during the activation process, which is converted to the final cytotoxic metabolite in a number of subsequent steps. Such prodrugs are characterized by a site of activation that differs from the site where the cytotoxic metabolite is released [2].

In contrast to prodrugs, enzymes have a large number of parameters that determine their efficacy and safety. At present, enzymes used in suicide gene therapy can be divided into two categories. The first consists of enzymes foreign to mammalian cells, with or without mammalian counterparts, such as viral thymidine kinase and bacterial cytosine deaminase (CD) that are part of metabolic pathways occurring exclusively in bacteria, viruses, or fungi. Structural differences must exist between these enzymes and their mammalian counterparts to ensure different prodrug affinity and hence selective prodrug activation in malignant cells. This is analogous to the selective effects of antimicrobials on bacteria, viruses, and fungi due to their ability to interfere with metabolic pathways specific to these organisms. The fact that such enzymes are foreign proteins makes them immunogenic. The second class of enzymes are normally present in mammalian cells but are not expressed in neoplastic cells. These include deoxycytidine kinase (dCK), carboxypeptidase A (CPA), β-glucuronidase (β-Glu), and cytochrome P450 (CYP). Their main advantage is reduced immunogenicity. On the other hand, their presence in healthy host tissues increases their potential for prodrug activation in these cells, reducing their safety profile. To solve this problem, these enzymes require modification prior to use to ensure different substrate affinity in tumor cells [3].

Besides lack of immunogenicity, the efficacy of enzymes depends on several other characteristics. Advantageous enzymes produce significant metabolite concentrations even at low enzyme concentrations, i.e., when the suicide gene is inefficiently or transiently expressed. Furthermore, enzyme catalytic activity should not require cofactors [3]. In terms of classic enzyme kinetics, ideal enzymes should have a high maximal velocity of catalytic activity, V_{max}, and a low substrate concentration at which the reaction can proceed efficiently, as expressed by the Michaelis-Menten constant, K_m. Current kinetic data available for enzyme–prodrug combinations are insufficient to accurately compare the different systems [3]. Also, it is unknown whether a sudden release of toxic metabolites is beneficial in terms of tumor cell killing compared to a slow, steady release. Finally, the relative sensitivity of resting and dividing tumor cells to toxic metabolites is also not known [3].

The enzymes used in suicide gene therapy are also characterized by different safety profiles, which are directly related to mechanism of enzyme action. Most suicide gene–prodrug combinations result in the production of metabolites that inhibit nucleic acid synthesis. Hence, their effects are more pronounced in cells that are in the S phase of the cell cycle, when nucleic acid synthesis occurs. As a result, malignant cells that divide more rapidly than normal host cells will be predominantly killed upon exposure to the cytotoxic products of suicide gene action. It follows that suicide gene–prodrug combinations that lack selectivity for dividing cells are associated with increased risks of host cell toxicity. Although this mechanism protects host cells from toxicity because of their lower rate of proliferation compared to tumor cells, it has the disadvantage of targeting only malignant cells that are proliferating. Since only a fraction of tumor cells are proliferating at the time of prodrug administration, it follows that prolonged treatment would be required to kill all cells within a tumor or that combination therapy with other treatment modalities able to target quiescent tumor cells would be necessary.

25.3 SUICIDE GENE–PRODRUG COMBINATIONS

Although all suicide gene–prodrug combinations follow the principles of suicide gene therapy discussed above, they differ in mechanism of action, degree of tumor cell toxicity, safety, and presence of a bystander effect (Table 25.1). For comparison of the efficiency of different prodrug–enzyme systems, two parameters are in use: the potential of activation of a given system and its degree of activation. The activation potential indicates the maximal killing efficiency of a given system toward a tumor cell line. It is defined as the ratio of the prodrug concentration required to kill 50% of cells in culture (known as IC_{50}) to the IC_{50} of the active drug in a nontransfected tumor cell line. The degree of activation is the ratio of the prodrug IC_{50} in the nontransfected cell line to the prodrug IC_{50} in the transfected cell line that expresses the activating enzyme. Both these parameters depend on the cell line's sensitivity to the drug, but the degree of activation also indicates the efficiency of the prodrug toward a specific cell line [2].

It should be noted that, for a number of reasons, efficient tumor cell killing *in vitro* does not always translate into equivalent *in vivo* results. In addition, cancers differ in their susceptibility to suicide gene therapy as a whole and to specific suicide gene–prodrug combinations. For example, HSV-TK, the most widely used form of suicide gene therapy, is less effective in the treatment of hemopoietic malignancies despite good results against most solid tumors [4]. This is most likely due to more rapid downregulation of suicide gene expression in hemopoietic malignancies, resulting in lower enzyme concentrations and inefficient prodrug conversion to its toxic metabolite [5].

25.3.1 HERPES SIMPLEX VIRUS-THYMIDINE KINASE

Herpes simplex virus-thymidine kinase (HSV-TK) has proven effective against a number of solid human tumors and is the most well-studied and widely applied suicide gene therapy strategy available. Since the first *in vivo* application of HSV-TK gene therapy in mice with gliomas, this form of suicide gene therapy has been tested in a number of human tumors including thoracic, head and neck, ovarian, and gastrointestinal tract malignancies [6–10]. HSV-TK-GCV is the "prototype" suicide gene therapy system and has been used to test new prodrugs, modified enzymes, multiple suicide gene therapy, combinations of suicide gene therapy with other cancer treatment modalities, and other techniques and ideas in the field of suicide gene therapy.

HSV-TK therapy relies on tumor cell transfection with the gene encoding the TK enzyme found in HSV. This enzyme monophosphorylates gancyclovir (GCV), an analog of acyclic guanosine that is in clinical use as an antiviral drug. GCV monophosphate is then further phosphorylated by mammalian cell kinases to GCV di- and triphosphate, which become incorporated into the replicating DNA chain in dividing tumor cells, inhibiting DNA polymerase, and terminating DNA production, resulting in apoptosis [11]. Although host cells express a mammalian thymidine kinase, which also monophosphorylates GCV, this occurs with 1000-fold lower efficiency than in tumor cells expressing the viral enzyme, leading to negligible host cell toxicity [12–16]. Importantly, phosphorylated GCV is a polar molecule that cannot cross the cell membrane and rapidly accumulates inside cells, enhancing its cytotoxicity to tumor cells in which it is produced [15,16]. However, it was

TABLE 25.1
Suicide Genes, Metabolites, Mechanisms of Action, and Bystander Effect

Gene	Final Metabolite(s)	Direct Cytotoxic Effect	Bystander Effect	Reference
HSV-TK	GCV TP	Disrupts DNA synthesis	Present	[12]
CD	5-FU-MP/TP	Disrupts DNA/RNA synthesis	Present	[25]
VZV-TK	Ara-ATP	Disrupts DNA synthesis	Low/absent	[29]
NTR	4-hydroxylamine metabolite of CB1954	Crosslinks DNA DNA breaks Disrupts DNA synthesis	Present	[32]
CYP2B1	Phosphoramide mustard Acrolein	DNA alkylation (crosslinks DNA) Crosslinks cellular proteins	Present	[39]
CPG2	Benzoic acid mustard	DNA alkylation (crosslinks DNA)	Present	[42]
XGRPT	6-Thioguanine MP	Disrupts DNA synthesis	Presumed	[45]
PNP	6-Methylpurine	Direct toxin/cellular necrosis	Present	[49]
HRP	IAA	Free radical formation	Present	[50]
CYP1A2	NABQI	Protein arylation/oxidation	Present	[65]
Mutated tyrosinase	Quinone compounds	Unknown	Unknown	[56]
UPRT	FUMP	Disrupts DNA/RNA synthesis	Minimal	[57]

demonstrated that labeled GCV can pass between cells via gap junctions, a property that depends on its small molecular weight [17]. As explained later, this is one mechanism responsible for the "bystander effect."

A relatively new strategy for targeting TK expression involves the use of tissue-specific promoters. These are gene elements that when activated by transcription factors stimulate gene expression. Inserting such a promoter upstream of a suicide gene to create a new "transgene" will stimulate gene expression in tissues expressing transcription factors that activate the promoter. TK has been the main suicide gene used to create transgenes with a number of tissue-specific promoters. An example of this technique is the rat insulin promoter (RIP)-TK transgene. This is a gene construct made by combining the RIP with the TK gene. The RIP is potently activated by pancreatic duodenal homeobox-1 (PDX-1), a transcription factor that activates the insulin gene and is expressed almost exclusively and in high levels by pancreatic β-cells. Hence, after systemic administration of this gene construct, TK gene expression will be potently stimulated in cells with abundant PDX-1. This strategy has been used against pancreatic adenocarcinoma in vitro and in vivo with encouraging results [18–20]. The same method has been used to treat metastatic prostate carcinoma using the osteocalcin (OC) gene promoter to create an OC-TK transgene. OC is expressed in human prostate cancer cells and surrounding stroma. However, due to its expression in normal osteoblasts, it is not entirely tumor-specific and cannot be given systemically. In a human clinical trial, OC-TK was injected into lymph node and bone prostate cancer metastases packaged in an adenoviral vector. This treatment was well tolerated and caused marked tumor cell apoptosis in seven of the eleven patients treated [21].

25.3.2 CYTOSINE DEAMINASE

CD is the second commonest suicide gene in clinical and preclinical application and shares a number of features with TK. It is also involved in nucleic acid metabolism and is absent from mammalian cells. CD is expressed in bacteria and fungi and normally catalyzes the deamination of cytosine to uracil. Its role in suicide gene therapy is due to its ability to convert the prodrug 5-fluorocytosine (5-FC) to 5-fluorouracil (5-FU) [22–24]. Due to the expression of CD in fungi and its absence from mammalian cells, 5-FC has been used as an antifungal. 5-FU is cytotoxic to mammalian cells without further phosphorylation, this being the basis of its use in cancer chemotherapy. However, in a step analogous to the phosphorylation of GCV monophosphate mentioned in the previous section, intracellular kinases present in both microbes and mammalian cells convert 5-FU to 5-fluorouridine 5'-triphosphate and 5-fluoro-2'-deoxyuridine 5'-monophosphate. These metabolic products are directly cytotoxic by disrupting RNA and DNA synthesis.

As with HSV-TK, CD-based suicide gene therapy results in a significant bystander effect. In this respect, CD is advantageous over HSV-TK due to the ability of 5-FU, the cytotoxic

product of this system, to cross cell membranes, enter neighboring nontransfected tumor cells and exert its lethal effects. In addition, 5-fluorouridine 5'-triphosphate and 5-fluoro-2'-deoxyuridine 5'-monophosphate have their own contribution to the bystander effect once they leak out of apoptotic tumor cells [25].

Like HSV-TK, CD has also been fused with tumor-specific promoters to create transgenes. Zhang and coworkers used the carcinoembryonic antigen (CEA) promoter to create a CEA-CD transgene, achieving specific in vitro expression of the CD gene by human colon cancer cell lines positive for CEA [26]. In a different study, the same group demonstrated tumor-specific expression of the CD suicide gene after adenoviral administration of CEA-CD to rats bearing hepatic tumors of human colon cancer cells [27]. CD has its own set of limitations that have to be addressed before it becomes more suitable for widespread clinical use. Although the effects of 5-FU are not S-phase-specific, it has increased toxicity on rapidly dividing cells, resulting in the implications on tumor cell toxicity described above. In addition, the intracellular concentrations of 5-FU required for cytotoxicity are relatively high, which translates into the need for efficient tumor cell transfection with CD, high levels of CD expression, and high doses of 5-FC. Due to CD expression by normal intestinal flora, high 5-FC doses result in bacterial 5-FU production that is released into the gut lumen and injures intestinal cells. Finally, the duration of 5-FU toxicity is approximately 10 min, requiring prolonged CD expression and 5-FC administration for 5-FU concentrations to reach toxic levels.

25.3.3 VARICELLA ZOSTER VIRUS-THYMIDINE KINASE

Varicella zoster (VZV) is another herpes virus that expresses its own TK, with a substrate specificity distinct from that of HSV-TK and mammalian enzymes. Once the gene is expressed after insertion into tumor cells, VZV-TK acts on the prodrug 9-(b-D-arabinofuranosyl)-6-methoxy-9H-purine, also known as araM [28]. AraM is initially phosphorylated by VZV-TK and then further metabolized by other enzymes that naturally occur in mammalian cells (AMP deaminase, AMP kinase, nucleoside diphosphate kinase, and adenylosuccinate synthetase lyase) into adenine arabinonucleoside triphosphate (araATP). AraATP is a highly toxic metabolite that kills tumor cells even when present in small concentrations. AraM cannot be metabolized by mammalian enzymes, ensuring exclusive araATP production in cells transfected with VZV-TK. This strategy is relatively new and shares the advantages and disadvantages of the HSV-TK-GCV. The safety and efficacy of VZV-TK-araM as well as the presence of a bystander effect are under investigation although in the in vitro bystander effect is minimal [29].

25.3.4 ESCHERICHIA COLI NITROREDUCTASE

E. coli nitroreductase (NTR) reduces dinitrophenylaziridine CB1954 to a 4-hydroxylamine metabolite that reacts with intracellular thioesters such as acetyl-CoA to produce an alkylating agent that is 10,000-fold more toxic than the original prodrug [30]. This product exerts its toxicity on dividing and nondividing cells through crosslinking of DNA strands, resulting in disruption of DNA synthesis and breaks in the DNA molecule [31]. A bystander effect has been shown for this system *in vitro* and *in vivo* [32]. The NTR-CB1954 system has a more rapid onset of cytolytic action than HSV-TK-GCV, which has been attributed to its lack of selectivity for dividing cells [31,33]. As mentioned previously, this is a particularly advantageous feature regarding tumor cell killing, but could also increase toxicity to normal host tissues. Hence, specific tumor cell targeting is especially important with this strategy. This limitation has precluded more widespread use of NTR-CB1954 until increased transfection efficiency and tumor-specific targeting are possible. As part of this effort, the NTR gene has been used to create transgenes with promoters of human telomerase reverse transcriptase (hTERT) and human telomerase (hTR), the two genes encoding for subunits of human telomerase. The latter enzyme is expressed in more than 85% of human cancers but is absent from other adult human cells with the exception of stem cells [34]. Bilsland and coworkers achieved suppression of tumor growth in animals bearing ovarian and cervical cancer xenografts after administration of hTR-NTR and hTERT-NTR [35]. In their study, tumor specificity of suicide gene expression was not assessed because the transgene was delivered locally [35,36].

25.3.5 Cytochrome P450 2B1

The antineoplastic activity of the nontoxic prodrug cyclophosphamide (CPA) is initiated by hepatic cytochrome P450 2B1 (CYP2B1) [37], which converts CPA to 4-hydroxy-CPA. The latter compound then spontaneously decomposes into acrolein, which covalently crosslinks cellular proteins, and phosphoramide mustard (PM), which alkylates DNA and causes DNA breaks during replication. Although both these compounds are theoretically toxic, the role of acrolein in killing tumor cells *in vivo* has not been demonstrated. It has been suggested that acrolein sensitizes tumor cells to PM [38]. On the other hand, PM is directly toxic to dividing and nondividing cells and shares the advantages and disadvantages of other suicide enzyme metabolites with this feature, as outlined above. In addition, 4-hydroxy-CPA is able to cross cell membranes and enter nontransduced tumor cells, resulting in a strong bystander effect [39].

The main concern with this form of suicide gene therapy is its safety profile. Despite low expression of CYP2B1 in tumor cells, hepatocytes express high levels of this enzyme, hence the potential for hepatotoxicity. In addition, lysis of hepatocytes releases intracellular toxic metabolites into the circulation, leading to systemic toxicity. Attempts to improve the safety of CYP2B1-CPA have focused on achieving high levels of CYP2B1 expression in tumor cells, which would decrease the CPA dose required for tumor cell killing and reduce hepatic and systemic injury. A recent phase I/II clinical trial evaluated the efficacy of CYP2B1 in the treatment of inoperable pancreatic tumors. Angiographic delivery of cells expressing CYP2B1 to the tumors, followed by systemic administration of ifosfamide, a CPA analog, led to tumor regression, tumor suppression, symptomatic improvement, and significantly prolonged survival [40].

25.3.6 Carboxypeptidase G2

The bacterial enzyme carboxypeptidase G2 (CPG2) acts by removing the glutamic acid moiety from 4-[(2-chloroethyl) (2-mesyloxyethylamino]bensoyl-L-glutamic acid (CMDA), releasing a toxic benzoic acid mustard derivative [41]. Unlike most other suicide enzyme metabolites, this product does not require further metabolic conversion for tumor cell toxicity. It owes its toxic effect to DNA alkylation and crosslinking, leading to apoptosis of both dividing and nondividing cells. The implications of this mechanism of action for tumor cytotoxicity and potential for systemic toxicity have already been discussed. A significant advantage of CPG2-CMDA is the one-step prodrug conversion process, which makes resistance to this form of therapy less likely than strategies that require multiple enzyme action. In the latter situation, a mutation in a single enzyme could prevent the formation of the final toxic product and lead to treatment resistance [42]. Finally, this system has one of the strongest bystander effects demonstrated *in vitro*, achieving significant tumor cell killing even with transfection of only 3.7% of tumor cells [43]. Clinically, this enzyme–prodrug combination is now almost exclusively used for antibody-directed enzyme prodrug therapy (ADEPT) with anti-CEA antibodies [44].

25.3.7 *E. coli* XGPRT

The *E. coli* gtp gene encodes the enzyme xanthine guanine phosphorybosyl transferase (XGPRT), which converts the xanthine analog, 6-thioxanthine (6-TX), into a weakly toxic purine analog that is subsequently phosphorylated into 6-thioxanthine monophosphate (6-XMP) by XGPRT. 6-XMP is then converted to 6-thioguanine monophosphate (6-GMP), a highly toxic compound. This form of suicide gene therapy has been studied in sarcomas and gliomas *in vitro* and *in vivo* [45,46]. Although only retroviral transfection has been attempted so far, adenoviruses could also prove suitable for this purpose.

25.3.8 *E. coli* PNP

The *E. coli* Deo gene encodes the enzyme purine nucleoside phosphorylase (PNP), which is normally absent from human cells and converts the prodrug 6-methyl purine deoxyriboside (MePdR) into the toxic compound 6-methylpurine (MeP). MeP can freely cross cell membranes and enter nontransduced tumor cells, resulting in a strong bystander effect. The efficacy of PNP–MePdR has been studied in human colon cancer, glioma, and melanoma cell lines *in vitro* and ovarian and glioma cell lines *in vivo* [47,48]. PNP–MePdR also had a potent effect against pancreatic tumors *in vitro* and *in vivo* [49].

25.3.9 HORSERADISH PEROXIDASE–INDOLEACETIC ACID

This is an enzyme–prodrug system described in 2000 [50]. Horseradish peroxidase (HRP) is a heme enzyme isolated from the roots of the horseradish plant, while indoleacetic acid (IAA) is also a plant product and is involved in the regulation of plant growth, cell division, and differentiation. IAA also occurs in mammals as a product of amino acid tryptophan metabolism by monoamine oxidase [51]. At physiological pH, HRP activates IAA to produce a toxic radical. It is unlikely that nonspecific activation of IAA would take place in mammalian tissues, since IAA is a poor substrate for mammalian peroxidases [52]. IAA was given to patients and volunteers in the 1950s as a potential treatment for diabetes, with no major toxicities reported [53]. A strong bystander effect is induced by HRP–IAA in air and hypoxia, requiring only 5% of transfected cells to kill 70% of cells, probably because the toxic metabolite of IAA is freely diffusible through cell membranes [54]. Hence, the HRP–IAA system has the potential for hypoxia-regulated gene therapy and has been tested against an urothelial cancer cell line *in vitro* [50].

25.3.10 CYP1A2–ACETAMINOPHEN

The CYP1A2 enzyme converts acetaminophen to the toxic metabolite *N*-acetyl-benzoquinoneimine. This system has been reported to have a potent bystander effect and has *in vitro* efficacy against ovarian and colon carcinoma [55].

25.3.11 MUTANT TYROSINASE–N-ACETYL-4S-CYSTEAMINYL PHENOL

This system has been reported to have a dose-dependent antiproliferative effect against breast adenocarcinoma, fibrosarcoma, and gliosarcoma [56].

25.3.12 E. COLI–URACIL

During DNA synthesis in mammalian cells, *E. coli* uracil phosphoribosyl transferase (UPRT) catalyzes the formation of uridine-5′-monophosphate from uracil. Its utility in suicide gene therapy comes from its ability to convert 5-FU to 5-fluorouridine-5′-monophosphate (FUMP). In mice-bearing tumors of human colon cancer cells transfected with the UPRT gene, 5-FU administration leads to significant tumor regression. A disadvantage of this system is its minimal bystander effect [57].

25.4 THE BYSTANDER EFFECT

At present, vectors used to introduce suicide genes in tumor cells *in vivo* can only transfect a minority of malignant cells comprising a tumor. Hence, to obtain significant antitumor effects, suicide gene therapy has to induce toxicity to cells not expressing the suicide gene. The bystander effect, defined as the toxicity of suicide gene therapy to tumor cells that have not been transfected with the suicide gene, is a property of most forms of this therapy. The mechanisms responsible for the bystander effect of suicide gene therapy are outlined below.

25.4.1 METABOLIC COOPERATION AND GAP JUNCTIONS

The observation that tumor cell killing was greater than could be accounted for by the number of transfected cells was first made by Moolten in the HSV-TK-GCV system [13]. Bi et al. later hypothesized that the mechanism of metabolic cooperation was responsible for this phenomenon [58]. Metabolic cooperation, a concept introduced in 1966, describes the ability of low molecular weight compounds to pass between adjacent cells via the gap junctional network [59]. This is a physiologic mechanism that promotes homeostasis by exchange of nutrients between cells. In the same way, toxic metabolites produced in a tumor cell transfected by the suicide gene enter and kill neighboring tumor cells (Figure 25.1). Interestingly, there is evidence that toxic metabolites from transduced tumor cells can also kill surrounding vascular endothelial cells by the same mechanism. This phenomenon could induce an indirect antitumor effect by causing tumor cell ischemia [60,61].

Gap junctions consist of proteins called connexins, coded for by a family of 12 genes [62]. These are small hexameric structures approximately 2 nm in diameter that span the membranes of adjacent cells and allow exchange of cytoplasmic molecules. Mesnil et al. demonstrated that HeLa cells, which do not express connexins and lack gap junctional communication, are not susceptible to a bystander effect with HSV-TK-GCV treatment [63]. However, when expression of the gap junctional protein connexin 43 was induced in the same cells, they became susceptible to bystander killing. These experiments provided the definitive proof for the role of metabolic cooperation and gap junctions in the bystander effect.

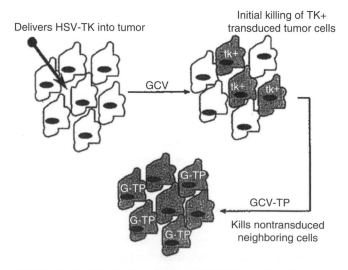

FIGURE 25.1 The bystander effect: GCV-TP produced in transfected cells has cytotoxic effects on neighboring nontransduced cells.

In addition, Andrade-Rozental et al. showed that the magnitude of the bystander effect depends on the type of connexins that make up the gap junctions of adjacent cells [64]. Hence, it seems the bystander effect depends on the interaction between toxic metabolites, gap junctions, and other tumor cell-specific parameters [61,64].

Interestingly, gap junctions are also responsible for a reciprocal phenomenon, named the "good samaritan effect." After production of toxic prodrug metabolite in transfected cells, gap junctions facilitate its diffusion into neighboring cells, hence lowering its concentration in the transfected cells and protecting them from toxicity. Despite this protection, persistence of these tumor cells can result in prolonged suicide gene transcription and toxic metabolite formation, which increases overall tumor cell toxicity [3,64,65].

25.4.2 Transfer of Toxic Metabolites via Apoptotic Vesicles or Direct Transmembrane Diffusion

As described in the previous section, toxic metabolites produced in cells where the suicide gene is expressed are cytotoxic to nontransduced tumor cells once they reach their cytoplasm. There is evidence that gap junctions are not the only method by which this can be achieved. As initially observed by Freeman et al., apoptotic tumor cells release apoptotic vesicles [66]. It has been demonstrated by using fluorescent tracking dye and fluorescence microscopy that these vesicles are phagocytosed by surrounding viable tumor cells. Hence, apoptotic vesicles contain toxic metabolites such as phosphorylated GCV or even the suicide gene itself and can release these molecules into neighboring tumor cells, thereby contributing to the bystander effect. However, other studies showed that phagocytosis of apoptotic vesicles occurs after exchange of toxic metabolites and does not contribute to the bystander effect [67]. Finally, the molecular characteristics of certain toxic metabolites such as 5-FU and MeP allow them to cross cell membranes and enter other tumor cells without the need of gap junctions or apoptotic vesicles, also enhancing the bystander effect.

25.4.3 Local Antitumor Immune Responses

Although the effects of toxic metabolites on nontransduced tumor cells are considered the most important mechanism responsible for the bystander effect, the immune system is also thought to contribute to this phenomenon. Initial evidence for this came from observations that, following *in vivo* suicide gene therapy, tumors became infiltrated by inflammatory cells [68,69]. Other studies showed reduced efficacy of HSV-TK-GCV therapy in T-cell-deficient mice compared to wild-type animals, revealing that T-cells could contribute to tumor cell killing [70,71]. Furthermore, treatment of intraperitoneal tumors with cells transfected with HSV-TK results in a cascade of cytokine release into the peritoneal fluid, providing more evidence for local activation of the immune system as a result of suicide gene therapy. CD/5-FU suicide gene therapy also causes *in vivo* cytokine release and increases

tumor-infiltrating CD4-, CD8-, and NK cells, leading to tumor regression in a rat model of liver metastases from colorectal carcinoma [72]. Research into this aspect of the bystander effect has attracted significant attention in recent years.

25.4.4 Systemic Antitumor Immune Response

In addition to local antitumor activity, the immune system appears to be systemically activated following suicide gene therapy, which also contributes to the bystander effect. From a physiological standpoint, suicide gene therapy can elicit an antitumor immune response. First, suicide genes encode microbial proteins that render transduced cells immunogenic [73]. Second, tumor cell lysis after suicide gene therapy could expose tumor-specific antigens to antigen-presenting cells (Figure 25.2). Initial evidence for the role of the immune system in the etiology of the bystander effect came from *in vivo* studies showing that, following HSV-TK-GCV therapy, tumor-bearing experimental animals were protected from developing new tumors when given additional doses of malignant cells [70,74]. Although this protective effect of HSV-TK GCV therapy was not permanent, T-cell-deficient mice were susceptible to engraftment of tumor cells when rechallenged, showing that T-cells had a role in this phenomenon.

An immune aspect of the bystander effect has also been demonstrated with CD suicide gene therapy. After *in vitro* expression of CD in adenocarcinoma and fibrosarcoma cell lines, administration of these cells to mice resulted in tumors susceptible to 5-FC therapy. In addition, mice became resistant to developing tumors when given wild-type cells of the same cell lines [73]. However, similar to observations made for HSV-TK therapy, this resistance was absent when the experimental animals were challenged with tumor cells of an antigenically different cell line, indicating an antigen-specific immune response had been induced.

There have been attempts to elucidate the role of different classes of immune cells in the bystander effect using antibodies to deplete specific populations of immune cells.

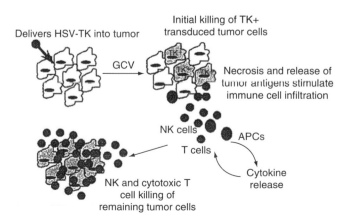

FIGURE 25.2 Local antitumor immune response following suicide gene therapy.

This showed that depletion of both granulocytes and CD8-positive lymphocytes, but not CD4-positive lymphocytes, decreased the level of tumor regression after CD suicide gene therapy. However, depletion of CD4-positive lymphocytes and CD8-positive lymphocytes did each independently reduce the protective effect of CD suicide gene therapy against developing tumors by readministration of tumor cells [69]. Despite these interesting observations, the exact mechanisms of immune-mediated bystander effect and how they differ between different systems remain unknown. Answering these questions might provide clues regarding the levels of antitumor immunity and its only temporary efficacy following suicide gene therapy.

Another benefit of the immune-mediated bystander effect is its potential utility in attacking metastases after treatment of the primary tumor. Initial studies in experimental animals did not show regression of synchronous tumors growing at a distant site from tumors of the same cell line treated with CD or HSV-TK [73]. However, Consalvo et al. showed significant regression of lung metastases from an orthotopic mammary tumor after CD suicide gene therapy [69]. In another animal model, hepatic metastases of colonic adenocarcinomas regressed after HSV-TK therapy [75]. Hence, it seems that in some cases, antitumor immunity induced by suicide gene therapy has the potency to treat metastatic lesions. Why then is this immunity ineffective against synchronous lesions? A hypothesis put forward to explain these findings is that synchronous subcutaneous lesions are immunologically protected due to decreased numbers of immune cells at that site compared to liver and lung, where cells of the reticuloendothelial system are more abundant. Alternatively, antigen presentation at these sites might be more efficient than that of subcutaneous tissue. The second possible etiology for failure of antitumor efficacy of the immune system against synchronous metastases is the difference in tumor burden between large synchronous lesions and the usually smaller metastases. Evidence for the role of tumor volume comes from the observation that antitumor immunity induced by suicide gene therapy is strong enough to inhibit formation of new tumors when animals are given additional doses of tumor cells, but too weak to cause regression of established synchronous tumors, which consist of a much larger volume of tumor cells. In recent years, systemic antitumor immune responses following suicide gene therapy have been shown for a number of different tumors and suicide gene strategies.

25.5 COMBINATION SUICIDE GENE THERAPY

As described earlier, there have been efforts to improve the results of "classic" one-system suicide gene therapy. Multiple approaches to this have been described, including engineering of enzymes and prodrugs, combining multiple prodrug–enzyme systems, using tissue-specific promoters to direct suicide gene expression and combining suicide gene therapy with other forms of cancer therapy.

Prodrugs can be modified in a number of ways to improve the efficacy of suicide gene therapy [3]. A primary aim is to design new prodrugs or alter existing compounds to improve kinetics of their respective enzyme–prodrug systems or confer other desirable features to the active drug, such as decreased toxicity to host tissues and enhanced ability to cross the blood–brain barrier for the treatment of central nervous system (CNS) tumors [3,76]. For example, despite its 10-fold greater potency compared to acyclovir, GCV has a reduced ability to cross the blood–brain barrier. Investigation into prodrugs with improved blood brain barrier permeability showed that administration of (E)-5-(2-bromovinyl)-2'-deoxyuridine (BVDU) in addition to GCV improved the efficacy of HSV-TK-GCV therapy [77,78].

Suicide genes and the enzymes they encode can also be modified. Suicide genes can be engineered for intra- or extracellular enzyme expression [43], extracellular enzymes becoming tethered to the outer surface of the cell membrane or secreted outside the cell. An advantage of extracellular suicide enzyme production is that prodrug does not need to enter malignant cells, leading to increased direct cytotoxicity and a potent bystander effect. On the other hand, a consequence of extracellular prodrug activation is the potential for systemic toxicity due to active metabolites entering the systemic circulation [43,79]. Further genetic engineering attempts to improve enzyme efficacy include the production of enzymes with improved affinity for their substrate [3]. In cases of gene–prodrug systems requiring multiple enzymatic reactions to produce the toxic metabolite, there have been efforts to transfect tumor cells with genes encoding more than on enzyme involved in the cascade. For example, in the HSV-TK-GCV system, tumor cell transfection with genes encoding kinases acting downstream of TK increased the yield of GCV triphosphate [80]. Interestingly, tumor cell transfection with two copies of the same gene has also been found to increase the production of active metabolite [81]. Finally, transfection of tumor cells with two different suicide genes followed by administration of their respective prodrugs improves *in vivo* and *in vitro* tumor cell killing compared to one enzyme–prodrug system [82,83].

It is established that tumor specific antigens present in the majority of cancers stimulate an antitumor immune response that usually lacks the potency to cause significant tumor regression. In addition, cytokines have a central role in effective antitumor immune responses induced by suicide gene therapy, as evidenced by alterations in cytokine levels following this treatment. Cytokines have been used to enhance this response in animal models and humans [84]. However, systemic administration of efficacious cytokine doses is associated with severe toxicity [85–87]. Cytokine gene therapy involves transfecting tumor cells with genes encoding for cytokines. The aim of this strategy is to achieve high and sustained cytokine concentrations within tumors to enhance tumor immunogenicity and at the same time avoid systemic cytokine toxicity. Importantly, this strategy has potential for synergism if combined with suicide gene therapy. Chen et al. treated colorectal cancer liver metastases with a combination of HSV-TK and interleukin (IL)-2 gene therapy, showing that intratumoral delivery of these genes lead to improved tumor regression compared to administration of either gene alone. Despite failing to prolong survival, this treatment did generate a temporary protective effect against a second challenge of wild-type tumor cells given at a distant site [88]. On the other hand, combined HSV-TK and IL-2 gene therapy had a synergistic effect on regression of head and neck squamous cell carcinomas in mice and also prolonged survival [89]. Also in this model, combined HSV-TK and IL-2 gene therapy protected animals from developing tumors when rechallenged with tumor cells [84]. Interestingly, tumors from mice transfected with IL-2 had more intratumoral lymphocytes than those from animals not receiving the gene, suggesting this might contribute to better results seen with combined therapy [89]. Other studies have investigated a combination of HSV-TK, IL-2, and granulocyte-monocyte colony stimulating factor (GM-CSF) gene therapy, which also enhanced tumor regression and improved survival and also induced protection against engraftment of wild-type tumor cells [90]. Investigators have suggested that the addition of GM-CSF acts by stimulating antigen-presenting cells and induces prolonged antitumor immunity mediated by CD4-positive tumor cells [91]. Unfortunately, not all tumors have proven susceptible to this triple scheme, most likely due to inherent differences in tumor type and site that affect response to GM-CSF [92].

Another approach has been to combine suicide gene therapy with the administration of cytotoxic drugs to obtain synergistic tumor cell toxicity. For example, ponicidin, a compound that activates HSV-TK has been found to increase tumor cell toxicity 3- to 87-fold when administered in addition to acyclovir or GCV [93]. A number of other drugs (butyrate, camptothecin, taxol, 7-hydroxystaurosporine) enhance the apoptotic effects of HSV-TK-GCV therapy without affecting enzyme function [94]. Also, hydroxyurea improves the toxicity of GCV-triphosphate by reducing the intracellular concentration of deoxyguanosine triphosphate, which competes with GCV triphosphate for incorporation into the DNA molecule [95]. Other compounds improve the tumor cell toxicity of suicide gene therapy by enhancing the bystander effect. Retinoids, apigenin, lovastatin, and cyclic AMP upregulate gap junctions, augmenting the bystander effect of HSV-TK-GCV therapy *in vitro* and *in vivo* [96–99].

The HSV-TK system has also proven useful in sensitizing tumor cells to the effects of radiotherapy. Tumor cells transfected with this enzyme and exposed to bromodeoxycytidine were 1.4- to 2.3-fold more sensitive to radiotherapy compared to tumor cells transfected with a control gene [100]. In this model, HSV-TK converts bromodeoxycytidine to bromodeoxyuridine, which is then incorporated into the DNA molecule.

25.6 METHODS OF SUICIDE GENE DELIVERY

For suicide gene therapy to become an effective form of cancer treatment in humans, limitations pertaining to the delivery of suicide genes to malignant cells remain significant. Efficient tumor cell targeting is more important with systemic gene delivery, where the aim is to transfect both cells of the primary tumor and metastatic cells in distant tissues. Targeting of tumor cells is less significant in trials that utilize intratumoral gene delivery [101]. The characteristics of the ideal gene delivery system include transfection of all malignant cells without the need for repeated treatments and lack of side effects. Gene delivery can be achieved with or without the use of viral vectors although combinations of viral and nonviral transfection methods have produced systems such as hemagglutinating virus of Japan liposomes and polycation-enhanced adenoviruses [102].

Nonviral gene delivery involves administration of the suicide gene itself and its control elements, such as promoters and enhancers, in the form of a circular double-stranded plasmid. Nonviral vectors include naked DNA, liposomes, peptides, and polymers [101]. Alternatively, suicide genes can be packaged into viral particles that utilize the natural process of inserting viral genes into the genome of mammalian cells. Viral methods achieve higher transfection efficiency and are being used in most human clinical trials of suicide gene therapy. However, they are far from being considered an ideal transfection method. The list of problems with viral vectors is long: poor transfection efficiency, random gene integration into the host cell genome, rapid clearance of systemically delivered vectors, reduced susceptibility of target cells to viral infection, generation of replication-competent particles, limited gene size that can be packaged in viral vectors, and inability to perform repeat treatments [103]. Importantly, safety concerns that were highlighted with the death of a patient following gene therapy with an adenoviral vector have increased interest in nonviral transfection methods [104,105]. In addition to the above issues, further problems need to be resolved before safe and efficient transfection vectors can be developed and used extensively in humans. First, development of new transfection methods using current animal models is unsatisfactory due to significant biological differences with human tissues. These include differences in susceptibility to infection by viral vectors and variations in vector doses that can be safely tolerated [106–108]. Finally, the optimal timing of vector administration with respect to prodrug administration is not well studied and depends on speed of transfection and suicide gene expression [101].

25.6.1 NONVIRAL TRANSFECTION

Nonviral transfection methods involve inserting a suicide gene and its regulatory elements in a double-stranded DNA molecule (plasmid) into target malignant cells. Gene regulatory elements, such as promoters and enhancers, also consist of double-stranded DNA and are inserted into the same plasmid. Other plasmid components can include engineered DNA sequences that control the processing and persistence of the suicide gene within the target cell. Plasmids dissolved in saline have been successfully expressed in muscle and thyroid after direct injection into these tissues and plasmid uptake by endocytosis [109,110]. However, this method is inefficient *in vivo*.

Currently, the optimal method of nonviral transfection is the use of liposomes or cationic lipids. These formulations have been designed to enhance plasmid uptake and can be modified for optimal DNA delivery according to the type of target tumor [111]. Additional advantages include lack of toxicity and immunogenicity, the ability to deliver genes of unlimited size, slow clearance of systemically delivered vector, the ability to perform repeat treatments, its ease of production, and low cost [103,104]. The poor transfection efficiency of first-generation liposomal vectors has been improved and liposomes can be administered intravenously, intramuscularly, subcutaneously, intradermally, transdermally, and via additional routes [103,112]. Clinically, liposomes have been used to deliver HSV-TK to patients with cerebral glioblastoma, confirming lack of toxicity and achieving significant reduction in tumor volume [113]. In addition, liposomal TK has been used to treat human pancreatic cancer xenografts in mice, also with lack of toxicity and antitumor efficacy [114,115]. Interestingly, liposomes have been used to conjugate suicide genes within adenoviral vectors. This method decreases the immunogenicity and toxicity of adenoviral vectors used alone and has been applied to the treatment of brain gliomas in mice [116].

There are several problems with DNA-mediated gene delivery at present. First, the maximal transfection it can achieve *in vivo* is 1%–3%, even when liposomes and cationic lipids are used. Although only a minority of tumor cells needs to express the suicide gene for clinically significant tumor regression, higher transfection efficiency rates do correlate with better antitumor effects. In general, transfection rates of less than 10% of cells within a tumor are not associated with tumor regression or delay in progression. The second major problem of nonviral transfection methods is that suicide gene expression is transient, regardless of the formulation used. This occurs because plasmids become quickly degraded or extruded from tumor cells, but fortunately has proved a less significant issue with suicide gene therapy than with other forms of cancer gene therapy as cells are generally killed before plasmids are eliminated. Both limitations of DNA-mediated gene delivery have been addressed by repeating the transfection process. Following nonviral transfection, permanent incorporation of suicide genes into the tumor cell genome rarely occurs *in vitro* and has not been reported to occur *in vivo*. Despite its limitations when prolonged gene expression is required in tumor cells, this phenomenon has the advantage of not causing a permanent alteration of the host cell genome. Hence, if healthy host cells are inadvertently transfected, the potential for toxicity is low.

25.6.2 Retrovirus-mediated Suicide Gene Delivery

Retroviral vectors can be divided to those derived from the murine leukemia virus and only replicate in dividing cells and to those derived from lentiviruses that do not have this limitation [117,118]. Historically, retroviruses were popular as vectors in early human trials of suicide gene therapy because they had a number of advantages. First, retroviral particles can be engineered to deliver the gene of interest without containing viral genes or viral gene products. Second, long-term gene expression is possible due to the ability of retroviruses to insert genes into host cell chromosomes. Third, retroviruses and the cell lines from which they are produced can be engineered for maximal host cell safety.

At the same time, limitations associated with retroviral vectors have prevented more extensive clinical use. As expected, insertion of viral genes into host cell chromosomes is associated with risk of carcinogenesis. Recently, Themis et al. reported a high incidence of tumorigenesis in mice transfected with a lentiviral vector, raising concerns about the safety of this vector family [119]. Furthermore, transfection efficiency of retroviral vectors remains poor for a number of reasons. First, because some of these vectors insert genes into actively dividing cells, only a fraction of tumor cells can be transfected. Second, it is difficult to achieve high titers of retroviral vectors. To overcome this problem, retroviral vectors themselves have to be delivered within their human or murine packaging cell line. Third, delivery of retroviruses within cells raises the issue of immunogenicity of murine cells, which can lead to murine cell killing within the host, thereby reducing transfection efficiency. Finally, the susceptibility of different tumors to retroviral infection varies.

25.6.3 Adenoviral Gene Transfer

Adenoviruses are a second family of viral particles that can be used to introduce suicide genes into tumor cells. In contrast to retroviruses, they do not integrate genes into host cell chromosomes, reducing the risk of mutagenesis. This is advantageous in terms of their safety profile but leads to shorter duration of suicide gene expression, lasting from one to several weeks [120,121]. As previously explained, duration of gene expression is not considered a significant limitation of the efficacy of suicide gene therapy [36]. Further advantages of adenoviral vectors are that they can be produced in 10,000-fold larger titers than retroviruses and their ability to transfect both dividing and nondividing tumor cells *in vivo* [122]. However, hemopoietic cells are not susceptible to adenoviral gene transfer.

First-generation adenoviral vectors were replication-incompetent but retained the ability to produce viral proteins after entering the host cell, which are immunogenic and can induce inflammation and cell lysis [102,123]. These disadvantages of adenoviral transfection were addressed by genetically engineering adenoviral particles to produce second-generation vectors that persist in the host cell, have reduced viral gene product expression, and tend to induce milder inflammatory reactions [124,125]. However, adenoviral transfection remains immunogenic, leading to antiviral antibody formation that limits the efficacy of repeat transfections.

25.6.4 Oncolytic Viruses

Until recently, viral vectors were all replication-incompetent to ensure safety to normal host tissues. However, the ability of some viruses to infect and lyse cells releasing numerous new viral particles is potentially useful in cancer therapy if directed specifically against malignant cells. Improved understanding of cancer genetics and viral replication have allowed the development of viruses able to selectively infect and kill malignant cells. Due to their ability to replicate and release more vectors that will infect new cells, oncolytic viruses offer improved gene delivery to tumor cells compared to replication-deficient viruses and are currently regarded as a superior transfection method [102,126]. Furthermore, they achieve tumor cell lysis independent of their ability to deliver suicide genes [102].

The Onyx-015 adenovirus, also known as dl1520, was the first oncolytic virus used in a human clinical trial. This vector is characterized by deletion of a gene coding for E1B-55 kDa, a p53-binding protein that permits viral replication in cells with intact copies of the p53 gene. Hence, mutant viral particles will not replicate in normal host cells, but will do so in the majority of human cancers where the p53 is mutated [127,128].

25.6.5 OTHER VIRAL VECTORS

Additional viruses might have potential as transfection vectors. Adeno-associated virus (AAV) is a nonpathogenic, naturally replication defective virus of the parvovirus family, with a single-stranded DNA genome [102]. Treatment of AAV can replace the majority of its genome by the therapeutic gene of choice [129], making such particles less immunogenic and able to transfect mammalian cells with prolonged expression of therapeutic genes. They have the added safety feature of integrating their genome at a specific locus on chromosome 19, leading to reduced potential for mutagenesis. Although engineered AAV have lost this property of predictable gene insertion, raising the possibility of carcinogenic potential, there is currently no evidence for AAV-mediated carcinogenesis [102,130,131]. A further factor that complicates the use of adeno-associated virus is the requirement for wild-type virus to be present during the vector production process. The recombinant vector then has to be purified from wild-type viral particles before amplification and *in vivo* use. This renders the vector production process inefficient and reduces the maximal viral titers that can be achieved. Furthermore, AAV does not specifically infect tumor cells and targeting techniques have to be used to obtain tumor specificity [102]. AAV has been used to transfect hepatoma cells *ex vivo* with TK and direct expression of this gene using the α-fetoprotein gene enhancer and albumin gene promoter [132]. Although growth of xenografts of such cells was suppressed following prodrug administration, the *in vivo* tumor specificity of the AAV vector was not evaluated [102].

HSV is another virus with potential use as a transfection vector and has the potential for indefinite intracellular persistence in a latent state. Replication-defective HSV particles able to express recombinant gene products in animals have been produced [133]. However, such particles retain the ability to produce viral gene products toxic to host cells, which has limited clinical use of HSV vectors.

Other viruses under investigation for use as transfection vectors include human papilloma virus, vaccinia virus, avipox virus, and baclovirus. Continued modification of existing viruses simultaneously with examination of new viral vectors will increase the number of viral vectors available for use in suicide gene therapy.

25.7 CURRENT SUICIDE GENE THERAPY TRIALS

Since the first human gene therapy trial started in 1989 and until January 2007, a total of 850 cancer gene therapy clinical trials had been initiated or completed worldwide. Excluding those that were withdrawn or never initiated, suicide gene therapy trials account for 105 of these. In turn, 74 of suicide gene therapy trials are being conducted in the United States.

With respect to vectors utilized, retroviruses and adenoviruses were the commonest, accounting for 51 and 42, respectively, of the 105 suicide gene therapy trials conducted. These trials have confirmed that both viral and nonviral transfection methods used in suicide gene therapy are safe and can achieve expression of suicide genes in tumor cells. However, despite mostly encouraging results, these have to be cautiously interpreted as phase I and phase I/II trials accounting for 64 and 32 of these trials, respectively [133].

REFERENCES

1. Jemal A, Siegel R, Ward E, Murray T, Xu J, Thun MJ. Cancer Statistics, 2007. *CA Cancer J Clin* 2007;57:43–66.
2. Springer CJ, Niculescu-Duvaz I. Prodrug-activating systems in suicide gene therapy. *J Clin Invest* 2000;105:1161–1167.
3. Niculescu-Duvaz I, Springer CJ. Introduction to the background, principles, and state of the art in suicide gene therapy. *Mol Biotechnol* 2005;30:71–88.
4. Moolten, FL, Wells JM. Curability of tumors bearing herpes thymidine kinase genes transferred by retroviral vectors. *J Natl Cancer Inst* 1990;82:297–300.
5. Abe A, Takeo T, Emi N, Tanimoto M, Ueda R, Yee JK, Friedman T, Saito H. Transduction of a drug sensitive toxic gene into human leukemia cell lines with a novel retroviral vector. *Proc Soc Exp Biol Med* 1993;203:354–359.
6. Culver KW, Ram Z, Wallbridge S, Ishii H, Oldfield EH, Blaese RM. In vivo gene transfer with retroviral vector-producer cells for treatment of experimental brain tumors. *Science* 1992;256:1550–1552.
7. Olfield EH, Ram Z, Culver KW, Blaese RM, DeVroom HL, Anderson WF. Gene therapy for the treatment of brain tumors using intra-tumoral transduction with the thymidine kinase gene and intravenous ganciclovir. *Hum Gene Ther* 1993;4:39–69.
8. Smyth WR, Hwang HC, Amin KM, et al. Use of recombinant adenovirus to transfer the herpes simplex virus thymidine kinase (HSVtk) gene to thoracic neoplasms: An effective in vitro sensitization system. *Cancer Res* 1994;54:2055–2059.
9. O'Malley BWJ, Chen S, Schwartz MR, Woo SLC. Adeno-virus-mediated gene therapy for human head and neck squamous cell cancer in a nude mouse mode. *Cancer Res* 55:1080–1085.
10. Freeman SM, McCune C, Robinson W, et al. Treatment of ovarian cancer using a gene-modified vaccine. *Hum Gene Ther* 1995;6:927–939.

11. Matthews T, Boehme R. Antiviral activity and mechanism of action of ganciclovir. *Rev Infect Dis* 1988;10:s490–s494.

12. St Clair MH, Lambe CU, Furman PA. Inhibition of ganciclovir of cell growth and DNA synthesis of cells biochemically transformed with herpesvirus genetic information. *Antimicrob Agents Chemother* 1987;31:844–849.

13. Moolten FL. Tumor chemosensitivity conferred by inserted herpes thymidine kinase genes: Paradigm for a prospective cancer control strategy. *Cancer Res* 1986;46:5276–5281.

14. Moolten FL. Drug sensitivity ("suicide") genes for selective cancer chemotherapy. *Cancer Gene Ther* 1994;4:279–287.

15. Elion GB, Furman PA, Pyfe JA, et al. Selectivity of action of an antiherpetic agent, 9-(2-hydroxyethoxmethly) guanine. *Proc Natl Acad Sci USA* 1997;74:5716–5720.

16. Elion GB. The chemotherapeutic exploitation of virus-specified enzymes. *Adv Enzyme Regul* 1980;18:53–60.

17. Simpson I, Rose B, Lowenstein WR. Size limit of moloxules permeating junctional membrane channels. *Science* 1977;195:294–296.

18. Liu SH, Davis A, Li Z, Ballian N, Davis E, Wang XP, Fisher W, Brunicardi FC. Effective ablation of pancreatic cancer cells in SCID mice using systemic adenoviral RIP-TK/GCV gene therapy. *J Surg Res* 2007;141:45–52.

19. Liu S, Wang XP, Brunicardi FC. Enhanced cytotoxicity of RIPTK gene therapy of pancreatic cancer via PDX-1 co-delivery. *J Surg Res* 2007;137:1–9.

20. Tirone TA, Wang XP, Templeton NS, Lee T, Nguyen L, Fisher W, Brunicardi FC. Cell-specific cytotoxicity of human pancreatic adenocarcinoma cells using rat insulin promoter thymidine kinase-directed gene therapy. *World J Surg* 2004;28:826–833.

21. Kubo H, Garder TA, Wada Y, et al. Phase I dose escalation clinical trial of adenovirus vector carrying osteocalcin promoter-driven herpes simplex virus thymidine kinase in localized and metastatic hormone-refractory prostate cancer. *Hum Gene Ther* 2003;14:227–241.

22. Austin EA, Huber BE. A first step in the development of gene therapy for colorectal carcinoma: Cloning sequencing and expression of *E. coli* cytosine deaminase. *Mol Pharmacol* 1992;43:380–387.

23. Mullen CA, Kilstrup M, Blaese M. Transfer of the bacterial gene for cytosine deaminase to mammalian cells confers lethal sensitivity to 5-flourocytosine: A negative selection system. *Proc Natl Acad Sci USA* 1992;89:33–37.

24. Huber BE, Austin EA, Good SS, Knick VC, Tibbels T, Richards CA. In vivo anti-tumor activity of 5-flourocytosine on human colorectal carcinoma cells genetically modified to express cytosine deaminase. *Cancer Res* 1995;53:4619–4626.

25. Huber BE, Austin EA, Richards CA, Davis ST, Good SS. Metabolism of 5-fluorocytosine to 5-fluorouracil in human colorectal tumor cells transduced with the cytosine deaminase gene: Significant antitumor effects when only a small percentage of tumor cells express cytosine deaminase. *Proc Natl Acad Sci USA* 1994;91:8302–8306.

26. Nyati MK, Sreekuman A, Li S, Zhang M, et al. High and selective expression of yeast cytosine deaminase under a carcinoembryonic antigen promoter-enhancer. *Cancer Res* 2001;62:237–2342.

27. Zhang M, Li S, Nyati MK, DeRemer S, et al. Regional delivery and selective expression of a high-activity yeast cytosine deaminase in an intrahepatic colon cancer model. *Cancer Res* 2003;63:658–663.

28. Pinedo HM, Peters GFJ. Flourouracil: Biochemistry and pharmacology. *J Clin Oncol* 1988;6:1653–1644.

29. Grignet-Debrus C, Cool V, Baudon N, Velu T, Caleberg-Bacq CM. The role of celluslar- and prodrug-associated factors in the bystander effect induced by the Varicella zoster and Herpes simplex viral thymidine kinases in suicide gene therapy. *Cancer Gene Ther* 2007;7:1456–1468.

30. Knox RJ, Friedlos P, Boland MP. The bioactivation of CB 1954 and its use as a prodrug in antibody-directed enzyme prodrug therapy (ADEPT). *Cancer Metastasis Rev* 1993;12:195–212.

31. Clark AJ, et al. Selective cell ablation in transgenic mice expressing *E. coli* nitroreductase. *Gene Ther* 1997;4:101–110.

32. Westphal Em, Ge J, Catchpole JR, Ford M, Kenney SC. The nitroreductase/CB1954 combination in Epstein–Barr virus-positive B-cell lines: Induction of bystander killing in vitro and in vivo. *Cancer Gene Ther* 2000;7:97–106.

33. Bridgewater JA, et al. Expression of the bacterial nitroreductase enzyme in mammalian cells renders them selectively sensitive to killing by the prodrug CB1954. *Eur J Cancer* 1995;31A:2362–2370.

34. Groot-Wassink T, Aboagye EO, Wang Y, Lemoine NR, Keith WN, Vassauz G. Noninvasive imaging of the transcriptional activities of human telomerase promoter fragments in mice. *Cancer Res* 2004;64:4906–4911.

35. Bilsland AE, Anderson CJ, Fletcher-Monaghan AJ, et al. Selective ablation of human cancer cells by telomerase-specific adenoviral suicide gene therapy vectors expressing bacterial nitroreductase. *Oncogene* 2003;22:370–380.

36. Schepelmann S, Springer CJ. Viral vectors for gene-directed enzyme prodrug therapy. *Curr Gene Ther* 2006;6: 647–670.

37. Clarke L, Waxman DJ. Oxidation metabolism of cyclophosphamide: Identification of the hepatic monooxygenase catalysts of drug activation. *Cancer Res* 1989;49:2344–2350.

38. Chen L, Waxman DL, Chen D, Kufe DW. Sensitization of human cancer cells to cyclophosphamide and ifosfamide by transfer of a liver cytochrome P450 gene. *Cancer Res* 1996;56:1331–1340.

39. Wei MX, Tamiya T, Rhee RJ, Breakefield XO, Chiocca EA. Diffusible cytotoxic metabolites contribute to the in vitro bystander effect associated with the cyclophosphamide/cytochrome P450 2B1 cancer gene therapy paradigm. *Clin Cancer Res* 1995;1:1171–1177.

40. Salmons B, Löhr M, Günzberg WH. Treatment of inoperable pancreatic carcinoma using a cell-based local chemotherapy: Results of a phase I/II clinical trial. *J Gastroenterol* 2003;15:78–84.

41. Springer CJ, Antoniw P, Bagshawe KD, et al. Novel prodrugs which are activated to cytoxic alkylating agents by cartoxypeptidase G2. *J Chem* 1990;33:677–681.

42. Niculescu-Duvaz I, Springer CJ. Gene-directed enzyme prodrug therapy (GDEPT): Choice of prodrugs. In: K.D. Bagshaw (ed.) *Gene-Directed Enzyme Prodrug Therapy.* Elsevier Science, Amsterdam, 1995.

43. Marais R, Spooner RA, Light Y, Martin J, Springer CJ. Gene-directed enzyme prodrug therapy with a mustard prodrug/carboxypeptidase G2 combination. *Cancer Res* 1996;56(20):4735–42.

44. Pedley RB, Sharma SK, Boxer GM, Boden R, et al. Enhancement of antibody-directed enzyme prodrug therapy in colorectal xenografts by an antivascular agent. *Cancer Res* 1999;59:3998–4003.

45. Tamiya T, Ono Y, Wei MX, Mroz PJ, Moolter FL, Chiocca EA. *Escherichia coli* gpt gene sensitizes rat glioma cells to killing by 6-thioxanthine or 6-thioguanine. *Cancer Gene Ther* 1996;3:155–162.

46. Mroz PJ, Moolten FL. Retrovirally transfected *Escherichia coli* gpt genes combine selectability with chemosensitivity capable of mediating tumor eradication. *Hum Gen Ther* 1993;4:4735–4742.

47. Sorscher EJ, Peng S, Bebok Z, et al. Tumor cell bystander killing in colonic carcinoma utilizing the *Escherichia coli* DeoD gene to generate toxic purinea. *Gene Ther* 1994;1:233–238.

48. Parker WB, King SA, Allan PW, et al. In vivo gene therapy of cancer with *E. coli* purine nucleoside phosphorylase. *Hum Gene Ther* 1997;8:1637–1644.

49. Dearvengt S, Wack S, Uhring M, Aprahamian M, Hajri A. Suicide gene/prodrug therapy for pancreatic adenocarcinoma by *E. coli* purine nucleoside phosphorylase and 6-methylpurine 2′-deoxyriboside. *Pancreas* 2004;28:E54–E64.

50. Greco O, Folkes LK, Wardman P, Tozer GM, Dachs GU. Development of a novel enzyme/prodrug combination for gene therapy of cancer: Horseradish peroxidase/indole-3-acetic acid. *Cancer Gene Ther* 2000;7:1414–1420.

51. Dachs GU, Tupper J, Tozer GM. From bench to bedside for gene-directed enzyme prodrug therapy of cancer. *Anticancer Drugs* 2005;16:349–359.

52. Folkes LK, Candeias LP, Wardman P. Toward targeted "oxidation therapy" of cancer: Peroxidase-catalysed cytotoxicity of indole-3-acetic acids. *Int J Radiat Oncol Biol Phys* 1998;42:917–920.

53. Diengott D, Mirsky IA. Hypoglycemic action of indole-3-acetic acid by mouth in patients with diabetes mellitus. *Proc Soc Exp Biol Med* 1956;93:109–110.

54. Greco O, Rossiter S, Kanthou C, et al. Horseradish peroxidase-mediated gene therapy: Choice of prodrugs in oxic and anoxic tumor conditions. *Mol Cancer Ther* 2001;1:151–160.

55. Thatcher NJ, Edwards RJ, Lemoine NR, et al. The potential of acetaminophen as a prodrug in gene-directed enzyme prodrug therapy. *Cancer Gene Ther* 2000;7:521–525.

56. Simonova M, Wall A, Weissleder R, Bogdanov A. Tyrosinase mutants are capable of prodrug activation in transfected non-melanotic cells. *Cancer Res* 2000;60:6656–6662.

57. Kawamur K, Tasaki K, Hamada H, et al. Expression of *Escherichia coli* uracil phosphoribosyltransferase gene in murine colon carcinoma cells augments the antitumoral effect of 5-flourouracil and induces protective immunity. *Cancer Gene Ther* 2000;7:637–643.

58. Bi WL, Parysek LM, Warnick R, Stambrook PJ. In vitro evidence that metabolic cooperation is responsible for the bystander effect observed with HSV tk retroviral gene therapy. *Hum Gene Ther* 1993;4:725–731.

59. Subak-Sharpe JH, Burk RR, Pitts JD. Metabolic cooperation by cell to cell transfer between genetically different mammalian cells. *Heredity* 1966;21:342–343.

60. Arafat WO, Casado E, Wang M, Alvarez RD, et al. Genetically modified CD34+ cells exert a cytotoxic bystander effect on human endothelial and cancer cells. *Clin Cancer Res* 2000;6:4442–4448.

61. van Dillen IJ, Mulder NH, Vaalburg W, de Vries EF, Hospers GA. Influence of the bystander effect on HSV-tk/GCV gene therapy. A review. *Curr Gene Ther* 2002;2:307–322.

62. Beyer EC. Gap junctions. *Int Rev Cytol* 1993;137C:1–37.

63. Mesnil M, Krutovskikh V, Piccoli C, et al. Negative growth control of HeLa cells by connexin genes: Connexin species specificity. *Cancer Res* 1995;55:629–639.

64. Andrade-Rozental AF, Rozental R, Hopperstad MG, et al. Gap junctions: The "kiss of death" and the "kiss of life." *Brain Res Rev* 2000;32:308–315.

65. Wygoda MR, Wilson MR, Davis MA, et al. Protection of herpes simplex virus thymidine kinase-transduced cells from ganciclovir-mediated cytotoxicity by bystander cells: The Good Samaritan effect. *Cancer Res* 1997;57:1699–1703.

66. Freeman SM, Abboud CN, Whartenby KA, et al. The "bystander effect": Tumor regression when a fraction of the tumor mass is genetically modified. *Cancer Res* 1999;53:5274–5283.

67. Hamel W, Magnelli L, Chiarugi VP, Israel MA. Herpes simplex virus thymidine kinase/ganciclovir-mediated apoptotic death of bystander cells. *Cancer Res* 1996;56:2697–2702.

68. Caruso M, Panis Y, Gagandeep S, et al. Regression of established microscopic liver metastases after in situ transduction of a suicide gene. *Proc Natl Acad Sci USA* 1993;90:7024–7028.

69. Consalvo M, Mullen CA, Modesti A, et al. 5-Flourocytosine induced eradication of murine adenocarcinomas engineered to express the cytosine deaminase suicide gene requires host immune competence and leaves an efficient memory. *J Immunol* 1995;154:5302–5312.

70. Vile RG, Nelson JA, Castelden S, et al. Systemic gene therapy of murine melanoma using tissue specific expression of HSVtk gene involves an immune component. *Cancer Res* 1994;54:6223–6224.

71. Gagandeep S, Brew R, Green B, et al. Prodrug-activated gene therapy: Involvement of an immunological component in the bystander effect. *Cancer Gene Ther* 1996;3:83–88.

72. Bertin S, Neves S, Gavelli A, et al. Cellular and molecular events associated with the antitumor response induced by the cytosine deaminase/5-flourocytosine suicide gene therapy system in a rat liver metastasis model. *Cancer Gene Ther* 2007; June 22 e-pub ahead of print.

73. Mullen CA, Coale MM, Lowe R, Blaese RM. Tumors expressing the cytosine deaminase suicide gene can be eliminated in vivo with 5-flourocytosine and induce protective immunity to wild type tumor. *Cancer Res* 1994;54:1503–1506.

74. Barba D, Hardin J, Sadelain M, Gage FH. Development of antitumor immunity following thymidine kinase mediated killing of experimental brain tumors. *Proc Natl Acad Sci USA* 1994;91:4348–4352.

75. Misawa T, Chiang M, Scotzco L, et al. Induction of systemic responses against hepatic metastases by HSV1-TK ganciclovir treatment in a rat model. *Cancer Gene Ther* 1995;2:332–337.

76. Hasegawa Y, Nishiyama Y, Imaizumi K, Ono N, Kinoshita T, Hatano S, Saito H, Shimokata K. Avoidance of bone marrow suppression using A-5021 as a nucleoside analog for retrovirus-mediated herpes simplex virus type I thymidine kinase gene therapy. *Cancer Gene Ther* 2000;7:557–62.

77. Hamel W, Zirkel D, Mehdorn HM, Westphal M, Israel MA. (*E*)-5-(2-bromovinyl)-2′-deoxyuridine potentiates ganciclovir-mediated cytotoxicity on herpes simplex virus-thymidine kinase-expressing cells. *Cancer Gene Ther* 2001;8:388–396.

78. Grignet-Debrus C, Cool V, Baudson N, Degrève B, Balzarini J, De Leval L, Debrus S, Velu T, Calberg-Bacq CM. Comparative in vitro and in vivo cytotoxic activity of (*E*)-5-(2-bromovinyl)-2′-deoxyuridine (BVDU) and its arabinosyl derivative, (*E*)-5-(2-bromovinyl)-1-beta-D-arabinofuranosyluracil (BVaraU), against tumor cells expressing either the Varicella zoster or the herpes simplex virus thymidine kinase. *Cancer Gene Ther* 2000;7:215–23.

79. Marais R, Spooner RA, Stribbling SM, Light Y, Martin J, Springer CJ. A cell surface tethered enzyme improves efficiency in gene-directed enzyme prodrug therapy. *Nat Biotechnol* 1997;15:1373–1377.

80. Blanche F., Cameron B., Coudet M., Crouzet J., Enzyme combinations for destroying proliferative cells. U.S. Patent W09735024, 1997. Rhone Poulenc Roerer:1–61.

81. Kim YG, Bi W, Feliciano ES, Drake RR, Stambrook PJ. Ganciclovir-mediated cell killing and bystander effect is enhanced in cells with two copies of the herpes simplex virus thymidine kinase gene. *Cancer Gene Ther* 2000;7:240–246.

82. Kammertoens T, Gelbmann W, Karle P, et al. Combined chemotherapy of murine mammary tumors by local activation of the prodrugs ifosfamide and 5-flourocytosine. *Cancer Gene Ther* 2000;7:629–636.

83. Rogulski KR, Wing MS, Paielli DL, et al. Double suicide gene therapy augments the antitumor activity of a replication-competent lytic adenovirus through enhanced cytotoxicity and radiosensitization. *Hum Gene Ther* 2000;11:67–76.

84. O'Malley BW, Sewell DA, Li D, et al. The role of interleukin-2 in combination adenovirus gene therapy for head and neck cancer. *Mol Endocrinol* 1997;11:667–673.

85. Rosenberg SA, Lotze M, Maui LM. A progress report on the treatment of 157 patients with advanced cancer using lymphokine-activated killer cells and interleukin-2 or high-dose interleukin-2 alone. *N Engl J Med* 1987;316:889–897.

86. West WH, Touer KW, Yanelli JR. Constant infusion of recombinant interleukin-2 in adoptive immunotherapy of advanced cancer. *N Engl J Med* 1987;316:898–905.

87. West WH, Touer KW, Yunelli Jr, et al. Constant infusion interleukin-2 in adoptive cellular therapy of cancer. *Proc ASCI* 1987;6:929–933.

88. Chen S, Li Chen XH, Wang Y, et al. Combination gene therapy for liver metastasis of colon carcinoma in vivo. *Proc Natl Acad Sci USA* 1995;92:2577–2581.

89. O'Malley BW, Cope KA, Chen S, et al. Combination gene therapy for oral cancer in a murine model. *Cancer Res* 1996;56:1737–1741.

90. Chen SH, Kosui K, Xu B, et al. Combination suicide and cytokine gene therapy for hepatic metastases of colon carcinoma: Sustained antitumor immunity prolongs animal survival. *Cancer Res* 1996;56:3758–3762.

91. Dranoff G, Jaffe EM, Lazenby A, et al. Vaccination with irradiated tumor cells engineered to secrete murine granulocyte-macrophage colony-stimulating factor stimulates potent, specific, and long lasting anti-tumor immunity. *Proc Natl Acad Sci USA* 1993;90:3539–3543.

92. Duy K, Li D, Duan L, O'Malley BW. Granulocyte-macrophage colony-stimulating factor is a combination gene therapy strategy for head and neck cancer. *Laryngoscope* 2001;111:801–806.

93. Hayashi K, Hayashi T, Sun HD, Takeda Y. Potentiation of ganciclovir toxicity in the herpes simplex virus thymidine kinase/ganciclovir administration system by ponicidin. *Cancer Gene Ther* 2000;7:45–52.

94. McMasters RA, Wilbert TN, Jones KE, Pitlyk K, Saylors RL, Moyer MP, Chambers TC, Drake RR. Two-drug combinations that increase apoptosis and modulate bak and bcl-X(L) expression in human colon tumor cell lines transduced with herpes simplex virus thymidine kinase. *Cancer Gene Ther* 2000;7(4):563–573.

95. Rubsam LZ, Davidson BL, Shewach DS. Superior cytotoxicity with ganciclovir compared with acyclovir and 1-beta-D-arabinofuranosylthymine in herpes simplex virus-thymidine kinase-expressing cells: A novel paradigm for cell killing. *Cancer Res* 1998;58:3873–3882.

96. Park JY, Elshami AA, Amin K, et al. Retinoids augment the bystander effect in vitro and in vivo in herpes simplex virus thymidine kinase/ganciclovir-mediated gene therapy. *Gene Ther* 1997;4:909–917.

97. Touraine RL, Vahanian N, Ramsey WJ, Blaeses RM. Enhancement of the herpes simplex virus thymidine kinase/ganciclovir bystander effect and its antitumor efficacy in vivo by pharmacologic manipulation of gap junctions. *Hum Gene Ther* 1998;9:2385–2391.

98. Carystinos GD, Katabi MM, Laird DW, Galipeau J, Chan H, Alaoui-Jamali MA, Batist G. Cyclic-AMP induction of gap junctional intercellular communication increases bystander effect in suicide gene therapy. *Clin Cancer Res* 1999;5:61–68.

99. Mesnil M, Yamasaki H. Bystander effect in herpes simplex virus-thymidine kinase/ganciclovir cancer gene therapy: Role of gap-junctional intercellular communication. *Cancer Res* 2000;60:3989–3999.

100. Brust D, Feden J, Farnsworth J, et al. Radiosensitization of rat glioma with bromodeoxycytidine and adenovirus expressing herpes simplex virus-thymidine kinase delivered by slow, rate-controlled positive pressure infusion. *Cancer Gene Ther* 2000;7:778–788.

101. Springer CJ. Introduction to vectors for suicide gene therapy. *Methods Mol Med* 2004;90:29–45.

102. Schepelmann S, Springer CJ. Viral vectors for gene-directed enzyme prodrug therapy. *Curr Gene Ther* 2006;6:647–670.

103. Tempelton NS. Cationic liposome-mediated gene delivery in vivo. *Biosci Rep* 2002;22:283–295.

104. Reszka RC, Jacobs A, Voges J. Liposome-mediated suicide gene therapy in humans. *Methods Enzymol* 2005;391:200–208.

105. Hollon T. Gene therapy—A loss of innocence. *Nat Med* 2000;6:1–2.

106. Raper SE, Chirmule N, Lee FS, et al. Fatal systemic inflammatory response syndrome in a ornithine transcarbamylase deficient patient following adenoviral gene transfer. *Mol Genet Metab* 2003;80:148–158.

107. Ryan PC, Jakubczak JF, Stewart DA, et al. Antitumor efficacy and tumor-selective replication with a single intravenous injection of OAS403, an oncolytic adenovirus dependent on two prevalent alterations in human cancer. *Cancer Gene Ther* 2004;11:555–569.

108. Russell SJ. RNA viruses as vibrotherapy agents. *Cancer Gene Ther* 2002;9:961–966.

109. Wolff JA, Malone RW, Williams P, Chong W, Acsadi G, Jani A, Felgner PL. Direct gene transfer into mouse muscle in vivo. *Science.* 1990;247(4949 Pt 1):1465–1468.

110. Sikes ML, O'Malley BWJ, Finegold MJ, et al. In vivo gene transfer into rabbit thyroid follicular cells by direct DNA injection. *Hum Gene Ther* 1994;5:837–884.

111. Felgner PL, Gadek TR, Holm M, et al. Lipfection: A highly efficient, lipid-mediated DNA-transfection procedure. *Proc Natl Acad Sci USA* 1984;84:7413–7417.

112. Ramesh R, Saeki T, Templeton NS, et al. Successful treatment of primary and disseminated human lung cancers by systemic delivery of tumor suppressor genes using an improved liposome vector. *Mol Ther* 2001;3:337–350.

113. Voges J, Reszka R, Gossmann A, et al. Imaging-guided convection-enhanced delivery and gene therapy of glioblastoma. *Ann Neurol* 2003;54:479–487.

114. Wang XP, Yazawa K, Templeton NS, Yang J, Liu S, Li Z, Li M, Yao Q, Chen C, Brunicardi FC. Intravenous delivery of liposome-mediated nonviral DNA is less toxic than intraperitoneal delivery in mice. *World J Surg* 2005;29(3):339–343.

115. Hajri A, Wack S, Lehn P, et al. Combined suicide gene therapy for pancreatic peritoneal carcinomatosis using BGTC liposomes. *Cancer Gene Ther* 2004;11:16–27.

116. Mizuno M, Ryuke Y, Yoshida J. Cationic liposomes conjugation to recombinant adenoviral vectors containing herpes simplex virus thymidine kinase gene followed by ganciclovir

treatment reduces viral antigenicity and maintains antitumor activity in mouse experimental glioma models. *Cancer Gene Ther* 2002;9:825–829.

117. Naldini L, Blömer U, Gallay P, et al. In vivo gene delivery and stable transduction of nondividing cells by a lentiviral vector. *Science* 1996;272:263–267.

118. Zufferey R, Nagy D, Mandel RJ, et al. Multiply attenuated lentiviral vector achieves efficient gene delivery in vivo. *Nat Biotechnol* 1997;15:871–875.

119. Themis M, Waddington SN, Schmidt M, et al. Oncogenesis following delivery of a nonprimate lentiviral gene therapy vector to fetal and neonatal mice. *Mol Ther* 2005;12:763–771.

120. Roth JA, Cristiano RJ. Gene therapy for cancer: What have we done and where are we going? *J Natl Cancer Inst* 1997;89:21–39.

121. Crystal RG. Transfer of genes to humans: Early lessons and obstacles to success. *Science* 1995;270:404–410.

122. O'Malley BW, Ledley FD. Somatic gene therapy: Methods for the present and future. *Arch Otolaryngol Head Neck Surg* 1993;119:1100–1107.

123. Engelhardt FJ, Ye X, Doranz B, Wilson JM. Ablation of E3A in recombinant adenoviruses improves transgene persistence and decreases inflammatory response in mouse liver. *Proc Natl Acad Sci USA* 1994;91:6196–6200.

124. Wang Q, Finer MH. Second generation adenovirus vectors. *Nature* 1996;2:714–716.

125. Ichikawa T, Chiocca EA. Comparative analyses of transgene delivery and expression in tumors inoculated with a replication-conditional or -defective viral vector. *Cancer Res* 2001;15:5336–5339.

126. McCormick F. Interactions between adenovirus proteins and the p53 pathway: The development of ONYX-015. *Semin Cancer Biol* 2000;10:453–459.

127. Kirn D, Niculescu-Duvaz I, Hallden G, Springer CJ. The emerging fields of suicide gene therapy and virotherapy. *Trends Mol Med* 2002;8(4 Suppl):S68–S73.

128. Samulski RJ, Chang LS, Shenk T. Helper-free stocks of recombinant adeno-associated viruses: Normal integration does not require viral gene expression. *J Virol* 1989;63:3822–3828.

129. Check E. Harmful potential of viral vectors fuels doubts over gene therapy. *Nature* 2003;423:573–574.

130. Hulbert CL, Alexander IE, Wolgamot GM, Miller AD. Adenoassociated virus vectors transduce primary cells much less efficiently than immortalized cells. *J Virol* 1995;69:1479.

131. Su H, Lu R, Chang JC, Kan YW. Tissue-specific expression of herpes simplex virus thymidine kinase gene delivered by adeno-associated virus inhibits the growth of human hepatocellular carcinoma in athymic mice. *Proc Natl Acad Sci USA* 1997;94:13891–13896.

132. Geller AI, Keyomarsi K, Bryan J, Pardee AI. An efficient deletion mutant packaging system for defective herpes simplex virus vectors: Potential application to human gene therapy and neuronal physiology. *Proc Natl Acad Sci USA* 1990;87:8950–8954.

133. J Gene Med Clinical Trial Site, http://www.wiley.co.uk/gene therapy/clinical/ (accessed July 31, 2007).

26 Selectable Markers for Gene Therapy

Jean-Pierre Gillet, Chava Kimchi-Sarfaty, Shiri Shinar, Thomas Licht, Caroline Lee, Peter Hafkemeyer, Christine A. Hrycyna, Ira Pastan, and Michael M. Gottesman

CONTENTS

26.1 INTRODUCTION

26.1.1 USE AND CHOICE OF SELECTABLE MARKERS

One of the major problems with current approaches to gene therapy is the instability of expression of genes transferred into recipient cells. Homologous recombination or use of artificial chromosomes can stabilize sequences with wild-type regulatory regions. Several studies have recently demonstrated the potential of such approaches to gene therapy using human cell lines and mice models [1–5]. However, clinical trials have not been undertaken yet. In most high-efficiency DNA transfer in current use in intact organisms, selectable markers must be used to maintain transferred sequences; in the absence of selection, the transferred DNAs or their expression is rapidly lost.

There are several different selectable markers that might be used for *in vivo* selection, including genes whose expression has been associated with resistance of cancers to anticancer drugs. Examples include: (a) methotrexate (MTX) resistance due to mutant dihydrofolate reductase (*DHFR*) [6]; (b) alkylating agent resistance due to expression of methylguanine methyltransferase

(*MGMT*) [7]; and (c) the expression of the multidrug transporting proteins P-glycoprotein (P-gp, the product of the *ABCB1/MDR1* gene) [8], *ABCC1/MRP1* (multidrug resistance associated protein 1) [9], and *ABCG2/BCRP* (breast cancer resistant protein) [10] (see Table 26.1). In this chapter, we will detail our experience with the *ABCB1/MDR1* gene.

The resistance of many cancers to anticancer drugs is due, in many cases, to the overexpression of several different adenosine triphosphate (ATP)-dependent transporters (ABC transporters), including the human multidrug resistance gene *ABCB1/MDR1* [23], ABCC1/*MRP1*, the multidrug resistance-associated protein [24,25], other MRP family members [26,27], and *ABCG2/MXR* [28] (for reviews, see Szakacs et al. [29] and Gillet et al. [30]). *ABCB1/MDR1* encodes P-gp, which is a 12-transmembrane (TM) domain glycoprotein composed of two homologous halves, each containing six TM domains and one ATP-binding site (see Figure 26.1) [31–33]. P-gp recognizes a large number of structurally unrelated hydrophobic and amphipathic molecules, including many chemotherapeutic agents, and removes them from the cell via an ATP-dependent transport process [31–33].

555

TABLE 26.1

Use of Drug Resistance Genes to Confer Resistance on Bone Marrow

Gene	Selection	Reference
Multidrug resistance gene 1 (*MDR1/ABCB1*)	Multiple cytotoxic natural product drugs	Gottesman et al. (1995) [8]
Multidrug ssociated protein 1 (MRP1/ABCC1)	Multiple cytotoxic natural product drugs	Omori et al. (1999) [11]
Breast cancer resistance protein (BCRP/ABCG2)	Multiple cytotoxic natural product drugs	Ujhelly et al. (2003) [10]
Dihydrofolate reductase (DHFR)	MTX and trimethotrexate	Flasshove et al. (1998) [6]; Warlick et al. (2002) [12]
Reduced folate carrier (RFC)	MTX and trimethotrexate	Rothem et al. (2005) [13]
Cytidine deaminase	Cytosine arabinoside	Momparler et al. (1996) [15]; Eliopoulos et al. (2002) [14]
Blasticidin S deaminase	Blasticidin S	Freitas et al. (2002) [16]
Glutathione transferase Yc	Melphalan, mechlorethamine, chlorambucil	Letourneau et al. (1996) [17]
Aldehyde dehydrogenase	Cyclophosphamide	Magni et al. (1996) [18]; Moreb et al. (1996) [19]
O^6-methylguanine methyltransferase (O^6-MGMT)	Nitrosourea (BCNU)	Allay et al. (1995) [20]
Gamma-glutamylcysteine synthetase	L-buthionine-*S,R*-sulfoximine (BSO)	Lorico et al. (2005) [21]; Rappa et al. (2007) [22]

ABCB1 has many obvious advantages for use as a selectable marker in gene therapy. It is a cell surface protein that can be easily detected by flow cytometry or immunohistochemistry. Cells expressing P-gp on their surfaces can be enriched using cell sorting or magnetic bead panning technologies. The very broad range of cytotoxic substrates recognized by P-gp makes it a pharmacologically flexible system, allowing the investigator to choose among many different selection regimens with differential toxicity for different tissues and different pharmacokinetic properties. Furthermore,

as will be discussed in detail in this chapter, *ABCB1* can be mutationally modified to increase resistance to specific substrates and alter inhibitor sensitivity.

26.1.2 LESSONS FROM TRANSGENIC AND KNOCKOUT MICE

Certain evidence supports the concept of using *ABCB1* as a selectable marker in human gene therapy (see Figure 26.2). Transgenic mice expressing the *ABCB1* gene in their bone

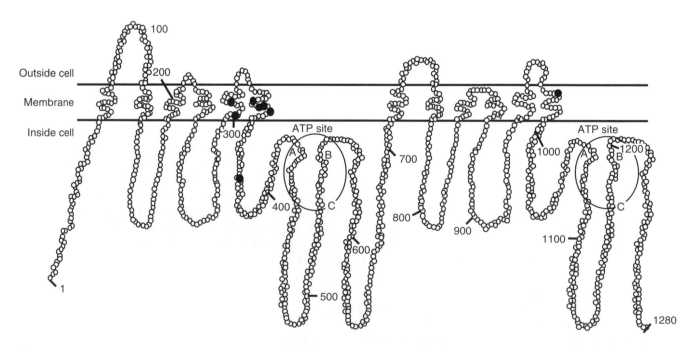

FIGURE 26.1 Hypothetical 2-D model of human P-gp, based on hydropathy analysis of the amino acid sequence and its functional domains. In this diagram, each circle represents an amino acid residue, with black solid circles showing the positions of mutations that alter the substrate specificity of P-gp (for clarity, only the mutations discussed in the manuscript are shown). The ATP sites are circled, with Walker A, B, and C regions indicated as A, B, and C, respectively.

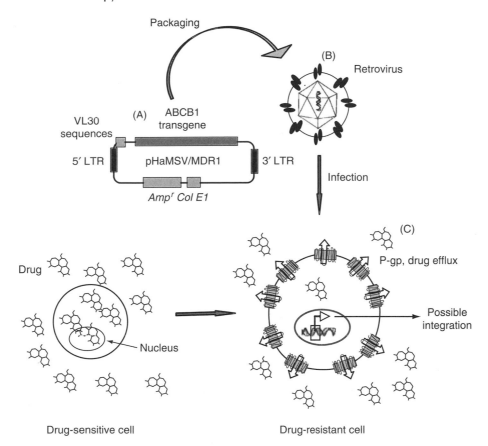

FIGURE 26.2 ABCB1-transduced cell using a retrovirus. (A) The pHaMSV/*MDR1* retroviral vector is composed of: (i) sequences from bacteria, including an origin of DNA replication (*colEI*) and an antibiotic resistance gene (*amp*^{r}); (ii) VL30 sequences containing *cis*-acting packaging elements; and (iii) an *ABCB1* transgene cassette. (B) The retrovirus consists of an envelope, a matrix, and a capsid, which contains the RNA genome. (C) In the host cell nucleus, the DNA either integrates into a specific site on chromosome 19, persists episomally, or integrates randomly into other chromosomes. In this example, *ABCB1* encoding the multidrug transporter P-gp confers the ability to the cell to be resistant to a wide panel of drugs.

marrow are resistant to the cytotoxic effects of many different anticancer drugs [34–38]. *ABCB1* transgenic bone marrow can be transplanted into drug-sensitive mice and the transplanted marrow is resistant to cytotoxic drugs [39,40]. Mice transplanted with bone marrow transduced with the human *ABCB1* cDNA and exposed to taxol show enrichment of the *ABCB1*-transduced cells [41–43], and this transduced marrow can be serially transplanted and remains drug-resistant [41]. Recently, this ability to select transduced bone marrow with taxol has been demonstrated in canine and mice bone marrow transplantation models [44,45].

The mouse *mdr1a* and *mdr1b* genes have been insertionally inactivated in mice [46–49]. These animals, although otherwise normal, are hypersensitive to cytotoxic substrates of P-gp. This hypersensitivity is due in part to the abrogation of the *mdr1a*-based blood brain/nerve barriers [50–53] and to enhanced absorption and decreased excretion of *mdr1* substrates [54]. These studies demonstrate the critical role that P-gp plays in drug distribution and pharmacokinetics and argue that specific targeting of P-gp to tissues that do not ordinarily express it (as in gene therapy) will protect such tissues from cytotoxic *mdr1* substrates.

26.1.3 Nanosystems for Gene–Drug Delivery: Emergence of a New Technology

A critical factor in gene therapy is the vector used for delivery of the transgene. Viruses have been extensively studied for use as vectors. Although they are able to mediate gene transfer with high efficiency, the size of the transgene that viruses can carry [55], the method used for the virus production and shedding of viral proteins are all serious concerns and challenges [56–59]. Nonviral methods using chemical or physical approaches have also been thoroughly investigated. However, most of these methods are less efficient or not always reproducible compared to virus-based delivery vectors and have limitations for *in vivo* gene delivery; for a review, see Gao et al. [60].

An expanding, multidisciplinary field has developed in the last few years based on nanotechnologies that apply engineering at the molecular level in a range from 1 to 100 nm. In medicine, nanoparticles have broad applications including molecular diagnostics, molecular imaging, targeted therapy, and gene therapy. These applications and particularly drug delivery devices are currently in clinical trials (see National Nanotechnology initiatives: http://www.nano.gov).

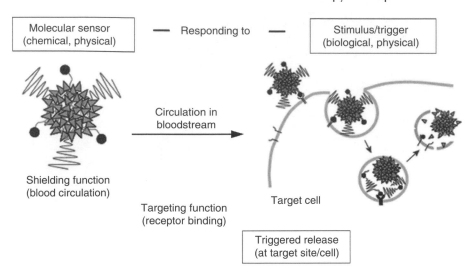

FIGURE 26.3 Nanosystems for programmed drug delivery. Preprogramming of nanoparticle properties by molecular sensors that respond to biological or physical stimuli during the drug delivery process. Timely operation of multiple tasks (e.g., circulation in blood, target binding, cell entry, and intracellular transport and drug release) is required. (From Wagner, E., *Expert Opin. Biol. Ther.*, 7, 588, 2007. With permission.)

Programmed nanostructured systems, a novel and elegant approach to gene therapy, have the ability to deliver both drug and gene in a controlled fashion. These systems are able to respond in a timely way to biological (i.e., pH variations of biological compartments, difference in redox potentials through the cell membrane, etc.) or physical stimuli (i.e., heat, light, ultrasound, electric or magnetic fields) (see Figure 26.3) (for a review, see Wagner [61]). Progammed nanoparticles can be easily used to deliver a transgene such as *ABCB1* to protect bone marrow, reducing technical and safety issues associated with the use of viruses.

26.2 SELECTABLE MARKERS FOR GENE THERAPY OF MALIGNANT HEMATOPOIETIC SYSTEM AND NONMALIGNANT DISEASES

Protection of chemosensitive cells from toxic compounds may be particularly helpful in the case of the hematopoietic system because most cells in blood and bone marrow are highly susceptible to antineoplastic compounds. Hematopoietic progenitor cells do not express glutathione-S-transferases [62] and only very low levels of endogenous *ABCB1* gene are expressed in myeloid and erythroid progenitor cells [63,64]. These low expression levels are not capable of providing protection from the cytotoxicity of anticancer drugs. Conversely, the high susceptibility of normal hematopoietic cells to cytotoxic agents allows selection strategies exploiting drug resistance genes if sufficient levels of resistance can be conferred.

Gene therapy, although thought to bear the potential of curing genetically determined diseases, has long been limited by low gene transfer efficiency to hematopoietic stem cells. In diseases where overexpression of the therapeutic gene does not confer a survival advantage, bicistronic vectors encoding

the therapeutic gene and a selectable marker gene such as drug resistance genes can provide a selective *in vivo* survival advantage.

26.2.1 *ABCB1* Gene as a Selectable Marker

As briefly mentioned in Section 26.1, several studies on human *ABCB1* transgenic mice established that constitutive overexpression of this gene protects animals from antineoplastic agents and thus could be useful as a selectable marker in human gene therapy. Indeed, it has been demonstrated that drugs can be administered safely at dose levels several-fold higher than those possible with mice of the respective background strains [34,36,37]. To demonstrate the specificity of this protection, verapamil, an inhibitor of P-gp, was coadministered, resulting in reversal of drug resistance [38]. Other studies have shown that retroviral transduction with a full-length *ABCB1* cDNA promoted by long-terminal repeats (LTR) of the Harvey sarcoma virus protects normal, clonogenic hematopoietic precursors, or erythroleukemia cells from anticancer drugs [65,66]. Furthermore, transduced cells were found to be resistant to multiple drugs including taxol, colchicine, and daunomycin. Murine hematopoietic stem cells originating from fetal liver [67], peripheral blood following mobilization with the use of growth factors [68], or from bone marrow [69] were also efficiently transduced with retroviral *ABCB1* vectors. Finally, it was shown that transplantation of *ABCB1*-transduced murine bone marrow cells into mast-cell-deficient (W/W^v) mutant mice [43] or lethally irradiated normal syngeneic mice [42] resulted in significant gene expression in the bone marrow of recipient animals. In both of these studies, investigators detected elevated levels of *ABCB1* expression after treatment of recipient mice with taxol, favoring the idea of a selective *in vivo* advantage of hematopoietic cells overexpressing the *ABCB1* transgene. This observation was in marked contrast to previous studies with selectable markers such as genes conferring resistance to neomycin, puromycin, or hygromycin. Because of their pharmacological or pharmacokinetic characteristics, such compounds cannot be used for *in vivo* selection. Further support for the potential usefulness of drug resistance genes for *in vivo* selection was provided by experiments in which *ABCB1*-transduced bone marrow was first transplanted into recipient mice [41]. After taxol treatment of recipient mice, their bone marrow was then retransplanted into a second generation of recipient mice. In several cycles of retransplantation and taxol treatment of recipient animals, increasingly higher levels of drug resistance were generated *in vivo*. Mice of the fifth and sixth generation survived doses of taxol (ranging from 15 to 20 mg/kg) that were lethal for mice that had not undergone bone marrow transplantation. Recently, Maier et al. reported that *ABCB1* overexpression may protect normal tissues from radiation- or chemotherapy-induced damage during tumor treatment. Indeed, several genes implicated in detoxification and exocytosis showed an increase from 1.4- to 4-fold in *ABCB1*-transduced cells compared to control counterparts. In addition, proapoptosis genes were downregulated, while the antiapoptotic gene *AKT3* was upregulated [70]. They subsequently corroborated the effect of *ABCB1* overexpression on apoptosis signaling in functional assays by showing reduced rates of apoptosis in response to irradiation in TK6 cells transduced with *ABCB1*. Therefore, the resistant phenotype of *ABCB1*-mediated P-gp-overexpressing cells seems to be associated with differential expression of genes coding for metabolic and apoptosis-related proteins.

Bunting et al. [71] reported that transduction of murine bone marrow cells with pHaMDR1 retroviral vector enables *ex vivo* stem cell expansion, which might help account for the ability of transduced cells to survive multiple cycles of transplantation. However, the biological safety of expansion of transduced stem cells is currently under scrutiny. When *ABCB1*-transduced progenitor cells are expanded with growth factors for extended periods (up to 12 days), uncontrolled proliferation occurs, as has been observed in one study [71]. The authors concluded that the finding was related to the *ABCB1* transgene because the development of a myeloproliferative syndrome was not observed either with a control DHFR vector or mock-transduced cells. Subsequent data confirmed and extended these earlier observations and provided further evidence that enforced *ABCB1* transporter function can alter the replicative behavior of stem cells [72]. Previous studies have not observed this adverse effect after *ABCB1* transfer to murine hematopoietic cells [42,43,73]. More importantly, comparable studies in nonhuman primates, using identical vectors but less efficient gene transfer conditions, did not reveal perturbations of myelopoiesis despite systemic challenge with growth factors [74]. However, the vector used by Bunting and colleagues in their stem cell expansion experiments was modified to reduce cryptic mRNA splicing within the *ABCB1* coding sequence [75,76] and likely expressed greater amounts of P-gp than the vectors used in the other studies. In a recent study, Modlich et al. [77] hypothesized that retroviral vector insertions near a proto-oncogene or another locus involved in regulating cell growth and proliferation may contribute to leukemogenesis, especially when collaborating with signal alterations evoked by the retrovirally encoded transgene product [78,79]. They found that when there are four or more copies of the transgene, *ABCB1*-associated leukemias are more likely to develop. *ABCB1* overexpression alone did not lead to this development. In addition, one leukemia developed after transduction with multiple copies of a vector expressing a fluorescent protein, strongly suggesting that combinatorial vector insertions can be sufficient to induce carcinogenesis [77]. The exact contribution of *ABCB1* to leukemia incidence under conditions of dose escalation and the possible synergy of *ABCB1* overexpression with insertional gene alterations have not been established.

Predictive preclinical models are thus required, both for risk assessment and prevention. The work of Modlich et al. mentioned above has revealed that studies involving dose-escalation of vectors are an important step in this direction, potentially allowing the establishment of a therapeutic index for retroviral transgene delivery. Du et al. reported that insertional mutagenesis can identify genes that promote the immortalization of hematopoietic cells, which normally have only limited self-renewal [80]. Taking advantage of this study, Modlich et al. developed new mutagenesis assays based on primary murine bone marrow cells that were transduced with a known multiplicity of infection (MOI) [81]. Self-inactivating (SIN) retroviral vectors do not contain the terminal repetition of the enhancer/promoter (at the opposite of their LTR counterparts), theoretically attenuating the interaction with neighboring cellular genes. Their study revealed that SIN retroviral vectors carrying a strong internal enhancer/promoter may transform primary hematopoietic cells by insertional mutagenesis though with significantly lower frequency than their LTR-driven counterparts usually used for stable gene transduction into hematopoietic cells [81].

Based on experiments in tissue culture and animal models, early clinical trials on transfer of the *ABCB1* gene to hematopoietic progenitor cells have been conducted [82–84]. Bone marrow or peripheral blood progenitor cells from patients suffering from advanced neoplastic diseases were retrovirally transduced and reinfused after high-dose chemotherapy [85–87]. These studies revealed that transduction efficiencies using *ABCB1* vectors as detected in bone marrow or peripheral blood of patients tended to be low and varied from one patient to another. Notably, in two studies enrichment of *ABCB1*-transduced cells was observed following treatment with etoposide or paclitaxel [88,89]. These studies confirm that the human multidrug resistance gene can serve as a drug-selectable marker gene *in vivo* in the hematopoietic system. However, gene transfer procedures and selection strategies need to be improved to efficiently protect human hematopoietic cells from the cytotoxicity of drug treatment.

These concerns have prompted new investigations with vector systems other than retroviruses. For instance, SV40 pseudovirions allow for highly efficient *ABCB1* gene transfer to hematopoietic cells [90,91]. Alternatively, Epstein–Barr virus-based (EBV) vectors can be designed that contain the

ABCB1 cDNA episomally in target cells [92]. With such vector systems, lifelong expression cannot be achieved because they fail to integrate into the genome. Conversely, for protection of hematopoietic cells during a series of chemotherapy cycles, sustained expression may not be required. Loss of *ABCB1* expression after the period of chemotherapy might in fact increase the safety of *ABCB1* gene therapy (see Section 26.3).

26.2.2 OTHER DRUG RESISTANCE GENES AS SELECTABLE MARKERS

Different patterns of chemoresistance can be attributed to various drug resistance genes. For instance, the *ABCC1* gene is genetically and functionally related to *ABCB1*. Retroviral transfer of *ABCC1* resulted in resistance to doxorubicin, etoposide, and vincristine [9]. However, since binding and transport of inhibitors to *ABCB1* may be different from *ABCC1*, transfer of this gene may be useful if naturally occurring resistance due to *ABCB1* overexpression in cancer cells has to be overcome to allow for effective chemotherapy of an *ABCB1*-expressing cancer.

As has been seen with *ABCB1*, chemoprotection of hematopoietic progenitor cells and a selective advantage *in vitro* were demonstrated following transduction by mutated DHFR cDNAs that confer resistance to MTX and trimetrexate (TMTX) [6,93–99]. Williams et al. [100], Cline et al. [101], and Vinh and McIvor [102] demonstrated the protection of recipient animals from lethal doses of MTX. These studies cannot be carried out in humans because the high doses necessary could be toxic.

Mice transplanted with DHFR-transduced hematopoietic cells display resistance against high levels of TMTX, which can be transferred to a second generation of transplant recipients [12]. Interestingly, mice can be transplanted with low amounts of DHFR-transduced cells following mild total body irradiation at a reduced dose of 1 Gy (Gray) and acquire MTX resistance by treatment with this drug for 60 days following transplantation [103]. Retransplantation experiments performed with DHFR [104] gave results comparable to those obtained with *ABCB1* [41]; both genes facilitate increased levels of resistance after several cycles of transplantation and drug treatment of recipient animals.

In vivo selection of retrovirally transduced hematopoietic cells has convincingly been demonstrated with the DHFR as a selectable marker [105]. In this study, mice were transplanted with DHFR-transduced bone marrow cells. Drug treatment resulted in significantly increased expression in granulocytes, erythrocytes, platelets, and T- and B-lymphocytes. Secondary recipients revealed that selection had occurred at the stem cell level. Recently, a complete tumor regression has been shown in mice after transplantation with mutant DHFR (L22Y)-transduced hematopoietic stem cells (HSC) followed by combined chemotherapy (MTX) and immunotherapy (anti-CD137) treatment [106].

Resistance to alkylating agents is multicausative and several genes may be useful as selectable markers. Retroviral transfer of a rat glutathione-S-transferase Yc cDNA to hematopoietic cells conveyed moderate resistance to melphalan, mechlorethamine, and chlorambucil [17]. Resistance to cyclophosphamide or 4-hydroperoxycyclophosphamide, respectively, could be conferred on hematopoietic cells by transfer of aldehyde dehydrogenase with the use of retroviral vectors [18,19]. Leukemic or primary hematopoietic cells were rendered resistant to 1,3-bis-(2-chloroethyl)-1-nitrosourea (BCNU) by retroviral transfer of a human *MGMT* cDNA, a gene encoding O^6-alkylguanine-DNA-alkyltransferase (AGT) [7,20,107]. Transplantation of transduced bone marrow cells rescued recipient animals from the toxicity of nitrosoureas [108]. In particular, nitrosourea-induced severe immunodeficiency can be overcome by transduction of immature progenitor cells [108,109]. However, Allay et al. demonstrated that the survival advantage to BCNU treatment in *MGMT*-transduced cells is modest and thus limits its therapeutic potential [20]. Indeed, in this report, *MGMT*-transduced cells were only twofold more resistant than the mock-transfected cells to BCNU treatment. In the meantime, it has been shown that G156A and P140A *MGMT* mutations are able to render the protein 240- and 40-fold more resistant to O^6-benzylguanine (BG), an inhibitor of *MGMT*, than the wild-type protein in cell-free extract [110], while P140K MGMT mutated protein has been shown to be over 1000-fold more resistant to BG [111]. These results have been confirmed using retroviral transduction of this mutant into human stem cells, which confers an increased resistance to combined treatment with BG and BCNU [112]. Furthermore, an *in vivo* study performed on mice demonstrated that transplantation of G156A *MGMT*-transduced bone marrow cells resulted in animal survival from lethal doses of BG and BCNU and in prolonged enrichment of the transduced cells [113]. Another mutation (P140K *MGMT*) [114] showed similar effects after transfer of the mutant gene into hematopoietic stem cells [115,116].

Upon overexpression in target cells, drug-resistant genes may also protect them from environmental toxins such as carcinogens in addition to amelioration of anticancer chemotherapy [117]. For instance, transfer of *MGMT* increases repair of DNA damage in sensitive cells. *In vitro* and *in vivo* studies confirmed this aspect of the function of drug resistance genes [118,119]. Liu et al. [120] showed that rapid repair of O^6-methylguanine–DNA adducts in transgenic mice protected them from *N*-methyl-nitrosourea-induced thymic lymphomas. This protection from carcinogens can be targeted to other organs such as the liver or the skin by suitable promoter systems [121,122].

To widen the range of anticancer drugs to be inactivated by gene therapy, vectors have been constructed for coexpression of two different drug resistance genes. For instance, a vector containing *ABCB1* (*or ABCC1*) and *MGMT* rendered human erythroleukemia cells resistant to the ABCB1 substrates, colchicine and doxorubicin, as well as to alkylating agents, *N*-methyl-*N*-nitrosourea and temozolomide [123–127]. Other examples include the use of *ABCB1* with DHFR, which confers MTX resistance [128], *ABCB1* plus *MGMT*, which confers resistance to certain alkylating agents, and *ABCC1* plus gamma-glutamylcysteine synthetase, which confers resistance to alkylating agents as well [129]. More recently, studies demonstrated the effect of mutant *DHFR*/mutant *thymidylate synthase* [130] and mutant *DHFR*/*cytidine deaminase* [131] fusion genes in conferring resistance to pemetrexed and both MTX and cytarabine, respectively.

A different approach to utilize vectors that allow for coexpression of two genes is to include a second gene that may enhance the efficacy of a selectable marker gene. This has been shown by construction of a vector that contained an *ABCC1*–cDNA and a cDNA encoding γ-glutamyl-cysteine synthetase (GCS), the rate-limiting enzyme of glutathione biosynthesis [129]. The transduced cells were resistant to substrates of ABCC1 and also to alkylating agents including melphalan, chlorambucil, and cisplatin [129]. A follow-up study showed that ABCC1 overexpression results in an intracellular decrease of glutathione (GSH) levels and in an increased GSH synthetic rate, increased steady-state GSH intracellular levels and normalization of the turnover time of GSH [132]. Furthermore, the report highlighted that ABCC1-overexpressing cells were hypersensitive to ʟ-buthionine-*S,R*-sulfoximine (BSO) [132], an inhibitor of γ-GCS, enhancing the cytotoxicity activity of alkylating agents, platinum compounds, and irradiation [133]. However, a resistance to BSO treatment rather than hypersensitivity has been shown in ABCC1/catalytic subunit of γ-GCS transduced cells [132]. These results, along with previous reports discussing the ability of BSO to reverse drug resistance, especially doxorubicin resistance, mediated by ABCC1 overexpressing experimental tumors [134,135], suggest that treatment with BSO could exert an anticancer effect and also modulated the anticancer activity of ABCC1 substrates and alkylating agents in ABCC1-overexpressing tumors.

26.2.3 DRUG RESISTANCE GENES AS SELECTABLE MARKERS FOR GENE THERAPY OF NONMALIGNANT DISEASES

Selectable bicistronic vectors provide great flexibility in coordinating expression of a selectable marker, such as *ABCB1*, and a therapeutic gene. The low translation efficiency of the internal ribosome entry site (IRES) results in asymmetric expression of genes positioned before and after the IRES. This asymmetric expression pattern makes it possible to alter the relative expression level of a therapeutic gene and P-gp to achieve maximum therapeutic effects while applying minimal selective pressure using a cytotoxic drug. By choosing different configurations, i.e., placing *ABCB1* before or after the IRES, we can select cells expressing a therapeutic gene at either a low level (*ABCB1* before the IRES) or a high level (*ABCB1* after IRES).

Lysosomal storage deficiencies such as the Gaucher and Fabry disorders are well-characterized single gene disorders and are not subject to complex regulation mechanisms. In addition, an enzymatic activity around 20% of the normal level is sufficient for clinical efficacy. These characteristics make lysosomal storage disorders good candidates for gene therapy.

Gaucher disease is characterized by accumulation of a glucosylceramide in glucocerebrosidase (GC)-deficient hematopoietic cells. These patients suffer from skeletal lesions, severe hepatosplenomegaly, anemia, and disorders of the central nervous system. While it is possible to efficiently transduce a GC cDNA to hematopoietic progenitor cells [136,137], expression levels tend to decrease after several weeks or months *in vivo* because of silencing or the limited lifespan of the transduced cells' progeny. To increase expression of GC *in vivo*, Aran et al. [138] constructed a transcriptional fusion between *ABCB1* and the GC gene. Increased expression of this gene was achieved by selection with cytotoxic substrates of P-gp. Appropriate selection strategies allowed complete restoration of the underlying genetic defect in cells from Gaucher patients [139]. Transduction of such bicistronic vectors into hematopoietic stem cells might allow treatment of patients by chemotherapeutic elimination of nontransduced cells that continue to synthesize or store glucosylceramide. Moreover, following chemotherapy, the numbers of genetically corrected hematopoietic progenitor cells should increase in the bone marrow to maintain physiological numbers of mature granulocytes, monocytes, and lymphocytes in peripheral blood. Havenga et al. demonstrated an *in vivo* selection for cells expressing GC with a vector containing the selectable marker gene, DHFR [140].

To demonstrate coexpression of a dominant selectable marker with a therapeutic gene using a bicistronic vector, the authors have coexpressed P-gp with GC [138,141], α-galactosidase A [142], adenosine deaminase (ADA) [143], a subunit of the NADPH oxidase complex [45,144], the shared gamma chain of the interleukin receptors [145], and a hammerhead ribozyme targeted to the U5 region of HIV-1 LTR [146]. In those experiments, *ABCB1* served as a selectable marker linked to the target gene by an IRES from encephalomyocarditis virus (EMCV) and constructed in a retroviral vector containing Harvey sarcoma virus LTR [147]. Two configurations, in which *ABCB1* is placed either before or after the IRES, have been examined in some cases. As demonstrated in those experiments, P-gp and the target gene are coexpressed in the cells selected using cytotoxic P-gp substrates, such as colchicine or vincristine; the expressed target proteins are functional as detected using *in vitro* or *ex vivo* analysis. In one case, using subcellular fractionation, we have demonstrated that P-gp and GC are translocated separately to the cell plasma membrane and lysosomes, indicating correct intracellular protein trafficking [138]. The demonstration that a noncoding RNA, such as a hammerhead ribozyme, can function even though tethered to an mRNA encoding a functional *ABCB1* provides an additional powerful way to use bicistronic vectors [146].

Using an IRES to generate a bicistronic mRNA ensures coexpression of two different proteins. However, IRES-dependent mRNA translation (or cap-independent translation) is less efficient than cap-dependent translation so that the two proteins are not expressed in equal amounts. It has been shown that in a monocistronic vector, insertion of an IRES upstream from an open reading frame of either *ABCB1* or *DHFR* reduces the translation efficiency by 2- to 10-fold [148,149]. Using a bicistronic vector, expression of *neo* in the position downstream from the IRES is 25% to 50% of that observed when *neo* is in the upstream position [150]. The asymmetric expression pattern of the bicistronic vector results in a significant difference in *ABCB1* transducing titer between a configuration with *ABCB1* placed before the IRES and a configuration in which *ABCB1* is placed after the IRES. We have found that the apparent titer of a bicistronic vector containing *ADA-IRES-ABCB1* was only 7% of the titer of a bicistronic vector containing *ABCB1-IRES-ADA* [143]. Similar reductions in *ABCB1* transducing titer and in expression of the nonselected downstream gene were seen with *ABCB1*-β-galactosidase bicistronic vectors [142]. The apparent *ABCB1* transducing titer of the retrovirus is based on the drug resistance conferred by expression of P-gp as the result of retroviral infection; thus the viral titer is proportional to the P-gp expression level. Insufficient expression of P-gp is unable to protect the cells from cytotoxic drug selection. To achieve P-gp expression at the same level, the lower efficiency of translation would have to be compensated for by a higher level of transcription, which can occur only in a minority of the cells in the transduced population. This may account for the apparent lower *ABCB1*-transducing titer of bicistronic vectors with a configuration of P-gp placed after the IRES. On the other hand, when cells express P-gp at the same level (i.e., the cells survived vincristine or colchicine selection at the same concentration), ADA expressed from *ADA-IRES-ABCB1* is 15-fold higher than the ADA expressed from *ABCB1-IRES-ADA*. This difference is probably due to a combination of the lower translation efficiency of ADA located downstream from the IRES and the high transcription level of *ADA-IRES-ABCB1* as the result of vincristine selection. A similar asymmetric expression of P-gp and human β-galactosidase A is also observed in NIH3T3 cells, where the difference is about eightfold.

IRES-dependent translation is a complex process in which mRNA-containing IRES interacts with various cellular proteins, including IRES-transacting factors (reviewed by Hellen and Sarnow [151]). The efficiency of IRES-dependent translation can be affected by the cell type [152], IRES origin [153,154], and the size and structure of a particular mRNA molecule. We have found that the titer of retrovirus containing pHa-*ABCB1* was higher than pHa-*ABCB1-IRES-ADA*, even though P-gp translation was cap-dependent in both cases. P-gp expressed from pHa-*ABCB1* was also at a higher level in a vincristine-resistant cell population than the P-gp expressed from pHa-*ABCB1-IRES-ADA*. A possible explanation for the relatively low retroviral titers observed is RNA instability or alternative splicing, since no DNA rearrangement was detected by Southern blot analysis of the transduced cells using an *ABCB1* probe.

High expression of the target gene can be selected using cytotoxic drugs, cytotoxic drugs combined with chemosensitizers, or the vector configured to place the target gene before the IRES. However, those approaches also reduce the overall number of cells that can survive the selection. Nevertheless, using a minimum concentration of drug, the selectable bicistronic vector provides options for selecting a large population of cells with low expression of the target gene, or a small population of cells with high expression of the target gene. Both options may be useful for gene therapy. For instance, ADA levels in normal individuals occur over a very broad range. Heterozygous carriers can be immunologically normal even with as little as 10% of the normal amount of ADA (reviewed by Blaese [155]). Expression of ADA at a low level in a large number of cells may prove sufficient to treat severe combined immunodeficiency (SCID). On the other hand, high ADA-expressing lymphoid cells, even through present as a small percentage of total cells, are also able to correct the SCID syndrome due to a beneficial bystander effect [156].

A different strategy to exploit the *ABCB1* gene as a drug selectable marker for correction of ADA deficiency was described by Germann et al. [157]. In this study, both genes were fused to a single cDNA encoding a bifunctional chimeric protein. This approach, however, cannot be used if the two proteins are physiologically located in different cellular compartments.

In gene therapy applications, the choice of the approach depends on the therapeutic strategy for a specific disease. Experiments on animal models are essential to prove the concepts that underlie gene therapy using selectable markers such as *ABCB1*. Detection of the function of transferred genes may be difficult if normal animals are utilized because of the activity of the respective endogenous gene product. To circumvent this difficulty, "knock-out" animals whose gene has been inactivated by targeted disruption can serve as useful models. For instance, mice whose β-galactosidase gene has been disrupted may be helpful to characterize a bicistronic vector in which *ABCB1* is combined with the respective human gene for correction of Fabry disease [158]. Another alternative is to use marking genes that are not physiologically expressed at high levels in normal tissues. To characterize bicistronic vectors containing *ABCB1*, this gene has been coexpressed with a green fluorescent protein or β-galactosidase [159]. These model systems should help to improve protocols for efficient drug selection and to identify strategies for selection at limited systemic toxicity. For instance, addition of P-gp inhibitors at low concentration to cytotoxic drugs may increase the stringency of drug selection, thereby allowing the use of anticancer drugs at low concentrations for selection [141]. Indeed, in the presence of a P-gp reversing agent, most P-gp-expressing cells are killed by the cytotoxic drug unless they express a large amount of P-gp to overcome the inhibitory effects. Using a combination of cytotoxic drug and chemosensitizer allows selection of cells expressing the therapeutic gene at a high level without need for a high concentration of cytotoxic drug. This strategy is especially desirable for an *in vivo* selection in which avoiding systemic toxicity is essential.

Another system in which selectable markers may be useful is in the skin. It is possible to grow keratinocytes in culture and introduce the *ABCB1* gene via retroviral vectors. Such keratinocytes are resistant to MDR drugs *in vitro*, and when transplanted on keratinocyte "rafts" to recipient animals, they remain resistant to colchicine, which can be applied as an ointment. If colchicine is withdrawn, transplanted keratinocytes are gradually replaced by nontransduced host skin; in the presence of selection, the transplanted keratinocyte graft is maintained. It should be possible in such a system to introduce other nonselectable genes via bicistronic vectors to serve as a source of protein to treat a genetic defect in the skin or elsewhere in the host [160,161].

26.3 AAV, SV40, AND EBV: EPISOMAL VECTORS EXPRESSING SELECTABLE MARKERS

Efficient delivery of a therapeutic gene to the appropriate target cells and its subsequent maintenance and expression are important steps for successful gene therapy. Genes introduced into cells are rapidly lost unless there is a mechanism to retain these genes within the nucleus and to ensure that the genes are also replicated and partitioned into daughter cells during cell division. Long-term expression of the transgene within cells can be achieved either via the integration of the transferred DNA into the host genome or maintenance of the introduced DNA as an autonomously replicating extrachromosomal element or episome. In either case, inclusion of a drug-selectable marker, such as the *ABCB1* gene, in the construct would ensure that rapidly dividing cells containing the transgene are given a selective growth advantage.

Delivery modalities can be viral or nonviral. Retroviral gene transfer, one of the most exploited systems for gene transfer into actively dividing cells, has been discussed earlier in this chapter while liposomal gene delivery will be discussed later in the chapter. In this section, nonretroviral and/or episomal vectors expressing selectable markers will be described.

26.3.1 AAV

In addition to retroviruses, adeno-associated virus (AAV) can also facilitate integration of the transgene into the host genome. Unlike retroviruses, AAV was found to integrate preferentially into a specific site on chromosome 19 [162]. AAV is a naturally defective, nonpathogenic, single-strand human DNA parvovirus. For productive infection and viral replication, coinfection with helper viruses, e.g., adenovirus, herpes virus, or vaccinia virus are required. In the absence of a helper virus, AAV establishes latency in the host by integrating itself into the host genome. AAV has a broad host range and is also able to infect both dividing and nondividing cells [163]. Hence recombinant AAV (rAAV) vectors have been exploited as alternative vehicles for gene therapy.

AAV-based vectors [164] are simple to construct, requiring only that the viral inverted terminal repeat (ITR) (which are 145 nucleotides each) is upstream from the gene of interest. Other important viral genes such as rep (involved in replication and integration) and cap (encoding structural genes) can then be supplied in *trans*. One disadvantage with such rAAV vectors is that site-specific integration of the gene of interest into the host genome is not observed [165]. This is probably because the rep gene, which is important for

mediating site-specific integration in the absence of helper viruses, is not included in the construct with the gene of interest. Nonetheless, rAAV has been successfully applied to the delivery of various genes into a variety of tissues and persistence of transgene expression in these nondividing tissues was reported [166–171]. Baudard et al. [165] demonstrated that in rapidly dividing cells, continuous selective pressure is necessary to sustain gene expression in cells. *ABCB1* was used as the selectable marker in this study. Being among the smallest DNA animal viruses (~20 nm in diameter), another disadvantage of the AAV system is its limited packaging capacity since it can accommodate only approximately 4.7 kb of the gene of interest. As such, a small and efficient promoter would be required to drive the expression of large genes. One such promoter is the AAV p5 promoter, which, together with the ITR, forms a 263-base pair cassette capable of mediating efficient expression in a CF bronchial epithelial cell line [166,167]. Baudard et al. further demonstrated that the reduction of the p5 promoter-ITR cassette to 234 bp was also able to promote efficient gene expression [165].

As mentioned above, although AAV has many features that render it a useful gene transfer vehicle, its small packaging capacity poses a challenge for AAV-mediated gene therapy in many inherited diseases such as cystic fibrosis (CF), hemophilia A, or Duchenne muscular dystrophy. In these diseases, the therapeutic expression cassette exceeds the viral packaging limit. Transsplicing AAV (tsAAV) vectors is a newly developed technology to double AAV packaging capacity. In this approach, the therapeutic gene is split into two parts and each part is carried by an AAV virion. After coinfection, the full-length gene is reconstituted through head-to-tail vector genome recombination. Therapeutic protein is expressed after the viral junction is removed from the mature mRNA by the cellular splicing machinery (reviewed by Duan et al. [172]). Lai et al. generated a set of *trans*-splicing vectors and achieved widespread expression of a truncated but functional dystrophin gene (exon 17–48) in skeletal muscle of mdx mice, a model for Duchenne muscular dystrophy. The dystrophic phenotype was ameliorated in both adult and aged mice [173]. The same group has recently demonstrated for the first time whole-body transduction using tsAAV vectors [174].

26.3.2 SV40

Vectors that facilitate extrachromosomal replication have some advantages; among these is high gene expression. This could be a result of vector amplification, promotion of nuclear localization and retention, as well as transcriptional activation by viral genes involved in episomal replication. However, selective pressure using selectable markers such as the *ABCB1* gene is necessary to maintain these episomes in actively dividing cells. In some cases, i.e., SV40 and retroviruses, chromosomal replication does occur followed by integration into the genome. This integration can produce mutations. At least two participants involved in a recent gene therapy trial for SCID developed leukemia after retroviral insertion [175]. Thus, another potential advantage of using episomally replicating vectors is the ability to extinguish expression at will by withdrawing selective pressure to replicating cells. One such example is the expression of the human *ABCB1* cDNA, which can be repressed by tetracycline and induced in its absence [176]. Episomally replicating vectors can be easily created by the inclusion of replicons into the vector design. These replicons can be derived from DNA viruses such as Simian virus 40 (SV40) [177], EBV [178], BK virus [179–181], vaccinia virus, canarypox virus, and Newcastle disease RNA virus [175]. The replicons are usually comprised of a viral origin of replication as well as a viral gene product that is important for maintaining extrachromosomal replication.

SV40 is a DNA polyomavirus that is a nonenveloped virus. Wild-type SV40 (wt SV40) virus is a double-stranded circular 5.2 kb DNA simian virus. The SV40 capsid is composed of 72 pentamers of the major capsid protein VP1 that are connected through their carboxy termini. The minor capsid proteins VP2 and VP3, which share 234 amino acids at their carboxy termini, connect the minichromosome core to the axial cavities of VP1. It has been suggested that correct interpentamer bonding is facilitated by host chaperones [182]. This interaction of the minor capsid proteins VP2 and VP3 with the major capsid protein VP1 is integral for viral infection. Amino acid residues of the three capsid proteins of SV40 have been studied and the residues required for protein interaction and infection of cells have been defined [183].

SV40 infection begins with the virus binding to its primary receptor, the major histocompatibility complex class I (MHC class I) without internalization of the receptors. The virus also binds to GM1 gangliosides, which act as plasma membrane receptors and travel with it to the endoplasmic reticulum (ER) [184]. After binding, SV40 enters cells through pinocytosis [185]. Transport of the virus to the ER, where the virus is disassembled, occurs in coated carrier vesicles (named COPI and COPII, [186]). Beta-COP, a component of COPI vesicles, is required for the retrograde retrieval pathway from the Golgi apparatus to the ER [187]. This transport to the ER is mediated by caveolae and facilitated by ER chaperones. Following disassembly, nuclear pore complexes assist in delivery of the viral DNA to the nucleus for replication. Nuclear membrane disassembly is not required for this delivery [187]. Infection of nonmonkey cells with wt SV40 can result in the integration of viral DNA into the host chromosome, permitting transmission of expression to daughter cells [188]. While human cells are semipermissive to wt SV40 replication, rodent cells are nonpermissive to viral replication; therefore, no progeny virions can be produced in these cells [189]. Nonlytic infection with wt SV40 enables the virus genome to persist as a self-replicating minichromosome due to the presence of the T antigen (Tag) [190].

Alternatively, when SV40 infects its host, it undergoes a lytic replication cycle, through which Tag is produced and is necessary for viral DNA replication and thus for the production of the capsid proteins. The binding of the large T antigen (LT) to the tumor suppressor proteins p53 and pRb may induce cellular immortalization *in vivo* [191].

The two major SV40 delivery systems are vectors that use SV40 sequences or the wt virus as a helper prepared *in vivo* (the SV40 delivery systems that are prepared using cell lines are defined here "*in vivo*" and those without helper virus and packaging cell lines are defined "*in vitro*"), and vectors that are packaged *in vitro*, with no SV40 sequences and in which the wild-type virus is not present. In *in vivo* preparation, replacing the late or early region with a reporter gene can generate SV40 recombinant viral particles (rSV40 [185,188, 192–198]). These are then propagated using either wild-type, or a temperature-sensitive mutant of SV40 as helper via a viral producer cell line, COS7. The packaging cell line COS7 stably expresses an origin-defective SV40 mutant and is capable of supporting the lytic cycle of SV40. Present SV40 vectors prepared *in vivo* have most of the viral coding sequences removed, retaining only the packaging sequences, the polyadenylation signal, and the early promoter of the virus, thus increasing the capacity for DNA to 5.3 kb. "Gutless" vectors, devoid of the SV40 sequence [190], can also be packaged in COS7 cells, for those do not require a helper virus or cotransfection with other plasmids. In this approach, the maximal DNA packaged in the rSV40 vector is 5.7 kb. Arad et al. [199] reported the use of COT18, a packaging cell line, with minimal sequence identity to the wild-type vector that unlike the Tag replacement vectors does not result in reacquisition of the Tag gene. An immune response and recombination with wt SV40 sequences are thus prevented.

The *in vitro* method of preparing helper-free SV40 vectors utilizes the SV40 viral late proteins, VP1, VP2, VP3, and agno or VP1 only [200–203]. The inclusion of VP1 in the vectors is essential and, due to its calcium-binding residues, VP1 participates in the assembly of the capsid and maintains viral integrity. Li et al. [204] demonstrated the involvement of calcium ions in viral entry to the cell and nucleus, controlling virion formation and structural alterations. In this method of *in vitro* packaging (IVP), nuclear extracts of VP1-baculovirus-transduced *Spodoptera frugiperda* (Sf9) insect cells are incubated with supercoiled plasmid DNA lacking SV40 sequences (i.e., the origin of replication, 182) or with siRNA [205] in the presence of 8 mM $MgCl_2$, 1 mM $CaCl_2$, and 5 mM ATP [90]. Such *in vitro* assembly allows larger DNA plasmids (up to 17.6 kb) to be packaged very efficiently with no need for SV40 sequences. In this delivery system, it is possible to package large genes together with the *ABCB1* gene (4.7 kDa) as a selective marker. Recently, new conditions for IVP have been defined and tested also for *ABCB1*: a scaled-up procedure for IVP of plasmid DNA, a procedure for concentrating virion volume without loss of activity, and a method that yields a great number of transduced cells [206].

SV40 has numerous advantages as a gene-delivery vehicle [91,194,207]:

1. *Tropism*. The virus is able to infect a wide variety of mammalian cells, including human cells, and to express its genes in these cells. Pseudovirions can transfer the gene of interest to a variety of cells (including hematopoietic cells) with high efficiency.

When using the SV40 *in vivo* delivery system, injections of the vectors into a specific organ or an organ's blood supply leads to the transduction of other organs as well [190]. However, when using the pseudomonas exotoxin (PE)-SV40 *in vitro* delivery system to reduce the size of adenocarcinomas in mice, there was no evidence of toxin expression in other tissues [208].

2. *Nucleic acid delivery.* The vector system is able to deliver untranslated RNA products [205].

3. *Expression persistence.* The gene expression may be transient or stable in cell lines depending on the specific SV40 system that is used. *in vivo* preparation results in long term expression: Strayer et al. [188] found evidence for integration of the recombinant gene or parts of it, a few days after transduction in random sites, which might explain the long-term expression. It has been demonstrated that when the large Tag gene is replaced with a reporter gene, replication-deficient recombinant SV40 viruses can be produced and can mediate gene transfer *in vivo*. Reporter gene expression was detectable for about 3 months without selection. Through IVP one can achieve short-term expression; selection with *ABCB1* can prolong expression. While expression of *ABCB1* and green fluorescence protein (GFP) are transient, lasting up to 21 days, it can be extended to 3 months under colchicine selection for *ABCB1*. When selection is withdrawn, expression is lost [90].

4. *High titers.* Multiple infections, in the preparation of the *in vivo* system, result in higher titers of the virus—up to 10^{10} infectious units/ml. However, the *in vitro* system produces a maximum 10^6 infectious units/ml. It should be noted that a comparison of rSV40 vector titers is problematic because there is no consensus on the titration method. While some researchers use a modified in situ polymerase chain reaction (PCR) [196], others use a real-time quantitative PCR method of titration [209] and yet others determine titers by quantifying luciferase activity in vectors carrying this transgene [199].

5. *Expression levels.* Using the SV40 *in vivo* system, episomal replication in the SV40 virus requires the SV40 replication origin as well as the large Tag, which activates the replication origin. Such episomal replication can generate more than 105 copies per cell of recombinant plasmids [210]. Introduction of the *in vitro*-packaged genes at 24 h intervals also results in greater expression than a single transduction [206].

6. *Inflammation and immunogenicity.* The replacement of the large Tag with firefly luciferase, controlled by the SV40 early promoter, results in expression of luciferase without causing inflammation [211]. Unlike other viral vectors, rSV40 was not reported to elicit antibody formation. Since rSV40 vectors lack Tag, which induces cell-mediated immunity, and often lack capsid genes encoding for capsid proteins that produce antibody-mediated immunity, probably only slight immune response is generated. It should also be noted that the entry pathway for SV40 is different from that of most viruses. It does not fuse with the cell membrane because it does not contain an envelope, nor is it phagocytosed. The viral capsid proteins, the potential targets for the immune system, are only processed at a later stage [185]. Transport of the virus to the ER in coated carrier vesicles and caveosomes bypasses endosomes and the Golgi apparatus [212]. Damm et al. [213] describe an alternative, faster entry pathway through which the virus travels to the ER without clathrin or caveolin and thus bypasses endocytic organelles. *in vitro* blocking of the ER continues to show GFP expression, thus suggesting that other pathways of viral entry to the nucleus exist [214]. In an attempt to identify the limiting step in the *in vitro* entry pathway of SV40 pseudovirions into the nucleus, Kimchi-Sarfaty et al. [214] showed that most of the DNA is not incorporated into the nucleus, but is trapped in the ER. Some of it moves back to the cytosol and is possibly degraded [215].

7. *Capacity. In vitro*-packaged SV40 vectors that do not contain an SV40 sequence permit packaging of plasmid DNA up to 17 kb in size. In order to determine the cloning capacity of IVP SV40 vectors, expression of ABC transporter genes carried by packaged plasmids ranging from 4.2 to 17.7 kb was measured. In addition to delivery of the *GFP* gene as a reporter, very efficient delivery of the *ABCB1*, *ABCG2*, and *ABCC1* genes was achieved along with multidrug resistance on virtually all cell types (human, murine, and monkey cell lines, [216]). The expression of both *ABCB1* and *GFP* genes is dose-dependent. The alteration in the level of expression suggests that MHC class I receptors play an important role in determining the efficiency of transduction. However, in IVP system high levels of MHC class I receptors are not necessary, nor are they sufficient for viral expression [216].

8. *Safety. In vitro*-packaged SV40 vectors that contain only capsid proteins and no viral sequences are as safe as nonviral vectors [216].

Application of SV40 in gene delivery is widely reported in the literature. Rund et al. [91] demonstrated very efficient delivery [217] of the drug-selectable marker, *ABCB1*, into various murine and human cell types including primary human bone marrow cells. *ABCB1* constructs that carried a promoter with an intron demonstrated higher expression than those without the intron. In low-expressing MHC class I cell lines, the CMV promoter expressed more P-gp compared with the SV40 promoter. *In vitro*-packaged *GFP* vectors that carried the CMV promoter consistently confirmed higher expression than those that carried the SV40 promoter [216].

Gene expression through IVP is usually limited in its duration. The short-term expression of the SV40/*ABCB1 in vitro* vectors may be an advantage for use in chemoprotection. Long-term expression beyond the chemotherapy period is undesirable, and may put patients at risk for treatment-induced myelodysplasia or secondary leukemia. The SV40/*ABCB1* vectors that are prepared *in vitro* may provide not only a safe vehicle for gene delivery but will also potentially avoid the problem of persistent bone marrow drug resistance in cancer patients.

Resting cells as well as dividing cells can express genes delivered through the SV40 delivery system [190]. Human mesenchymal stem cells (MSCs) in various stages of differentiation can be efficiently transduced by SV40 IVP vectors. When these cells are grown in medium containing factors that promote differentiation, high levels of GFP expression are achieved. While mouse-enriched bone marrow cells transduced with a retrovirus express MDR on the cell surface, a significantly greater amount of MDR is expressed by SV40-transduced cells 35 days posttransduction. This implies that *ABCB1* transduction of patient bone marrow through the SV40 delivery system may provide a strategy for chemoprotection during chemotherapy (unpublished data).

The involvement of SV40 in cancer therapy prompted the idea of delivery of lethal genes through the SV40 delivery system. SV40 pseudovirion delivery of Pseudomonas exotoxin A (PE38) was found to be effective in the treatment of human adenocarcinomas growing in mice either by direct injection or systemically. Using a combined treatment of SV40-PE38 with doxorubicin reduces the side effects of chemotherapy [208].

SV40 vectors efficiently deliver HIV-1-inhibitory RNAs using pol II or III promoters [218]. Vectors that encode a variable fragment antibody that recognizes HIV-1 integrase inhibit HIV-1 infection in SCID mice [218]. Neuronal apoptosis induced by HIV-1 gp120 can be reduced *in vivo* through the delivery of superoxide dismutase (SOD1) and glutathione peroxidase (GPx1) by rSV40 injected into rat brains [218]. While antiretroviral therapy penetrates the CNS poorly, rSV40 delivery to microglial cells is effective and can be used for the delivery of anti-HIV therapy, thus protecting from HIV-induced brain injury [219].

In conclusion, rSV40-based vectors have great potential in gene therapy. rSV40 delivery systems are highly efficient vehicles for the delivery of chemoprotective agents, such as members of the ABC transporter genes, into a wide variety of cells. The use of the *ABCB1* selective marker in these vectors along with large genes (total plasmid size of 5.7 kb in *in vivo* "gutless" SV40 delivery and 17.7 kb in *in vitro*-packaged delivery, lacking the SV40 sequence) enables well-controlled, long-term as well as short-term gene expression.

26.3.3 EBV AND OTHER EPISOMAL VECTORS

Episomal vectors based on EBV are also being developed for gene therapy purposes. EBV is a human B-lymphotropic herpesvirus that resides asymptomatically in more than 90% of the adult human population by establishing latency and maintaining its genome episomally [220]. The life cycle of EBV comprises two phases, a lytic and a latent phase. During the lytic phase, EBV DNA replicates via a rolling circle intermediate to achieve a 1000-fold increase in copy number. The origin of replication, Ori Lyt, and the transacting element ZEBRA are required for the lytic replication. Rolling circle replication results in the formation of linear head-to-tail concatamers. The presence of the EBV terminal repeat (TR) sequence causes cleavage of the concatemerized DNA to molecules of about 150–200 kbp, which are then packaged into virions. Upon infection into a permissive cell, the viral DNA circularizes by ligation of TR. Latency is established in the cells by episomal replication of the circular DNA.

Episomal replication in EBV is maintained by two elements interacting to ensure that the viral genome is retained within the nucleus, efficiently replicated and partitioned into daughter cells. Although the copy numbers of episomal viral DNA vary from 1 to 800, only between 4 and 10 episomal copies per cell are usually observed using vectors containing EBV OriP and EBNA-1 [221]. Unlike other episomal vector systems, very low rates of spontaneous mutation have been observed with EBV-based episomal vectors [222]. The *cis*-acting element responsible for episomal replication is a 1.8 kb OriP while the transacting element is EBNA-1. OriP comprises two distinct sequence motifs, the dyad symmetry (DS) motif from which replication is initiated and the family of repeats (FR) that serve as a replication fork barrier. Interaction of EBNA-1 with DS initiates bidirectional replication, while binding of EBNA-1 to FR enhances transcription from the episome and terminates DNA replication. EBNA is reported not to be oncogenic nor immunogenic. It evades the host immune system via the presence of the repeat motif, Gly-Ala, which was found to interfere with antigen processing and MHC class I-restricted presentation [223]. These EBV episomal vectors replicate once per cell cycle [224] and are capable of stably maintaining human genomic inserts of sizes between 60 and 330 kb for at least 60 generations [225].

Vos and colleagues [226] developed a helper-dependent infectious recombinant EBV to evaluate the feasibility of using such a vector system to correct hereditary syndromes in B-lymphocytes already harboring the EBV virus latently. The EBV-containing target B-lymphocytes will supply EBNA-1 in *trans* for the episomal maintenance of the transgene. Hence only minimal *cis*-EBV elements for episomal replication (OriP), viral amplification (Ori Lyt), and packaging (TR) are included in their construct. The hygromycin resistance gene was included as a selectable marker in their vector. Infectious virions are generated by the producer cell line HH514. They demonstrated successful transfer of such infectious virions carrying the therapeutic gene, Fanconi anemia group C (FA-C) cDNA, into HSC536, an FA-C patient cell line. Upon selection with hygromycin, long-term (at least 6 months) correction of the Fanconi phenotype *in vitro* was observed, as determined by cellular resistance to the crosslinking agent, diepoxybutane. They also observed that in the absence of selective pressure, their episomal vector is retained in rapidly dividing cells at a rate of 98% per cell division translating to a half-life of 30 days in cells doubling every 20 hours.

The authors have explored the use of EBV episomal vectors containing only the OriP and EBNA-1 and carrying the selectable marker *ABCB1* as potential gene therapy vectors. Using the liposome formulation, DOGS/DOPE (1:1) [227], we successfully delivered the vector to various cultured cells as well as human stem cells. *ABCB1* was found to be expressed at a higher level in the episomal vector compared to its non-episomal counterpart and more drug colonies were obtained upon selection. Episomal plasmids could be recovered in drug-selected cells for many weeks [92].

Other episomally replicating vectors can be derived from BPV viruses [228] or BK virus [180]. Unfortunately, BPV vectors cannot be reliably maintained as episomes, as they exhibit a high spontaneous mutation rate (~1%), frequently undergoing integration, deletion, recombination, and rearrangements [229]. Furthermore, BPV has a limited host range and BPV vectors cannot be efficiently maintained in human cells. Not much is known about BK-virus-derived episomal vectors. Nonetheless, successful stable maintenance of episomal gene expression was reported in human transitional carcinoma cells using BK-based vectors but not EBV-based vectors, probably due to the differential tropism of BK and EBV viruses for human uroepithelial cells [179].

Various chimeric viruses have been developed to improve the efficiency of gene transfer as well as the maintenance of gene expression within target cells. These chimeric virus systems attempt to combine the favorable attributes of each vector system and overcome the limitations associated with each system. The episomal replication ability of EBV was exploited to produce both rapid and long-term high-titer recombinant retroviruses (up to 10^7 TU/ml) for efficient gene transfer into human hematopoietic progenitor cells [230,231]. A novel adenoviral/retroviral chimeric vector was also reported in which an adenoviral delivery system was utilized to efficiently deliver both the retroviral vector and its packaging components, thereby inducing the target cells to function as transient retroviral producers capable of infecting neighboring cells. This system capitalizes on the superior efficiency of adenoviruses to deliver genes *in vivo* and the integrative ability of retroviruses to achieve stable gene expression [232]. An EBV/HSV-1 amplicon vector system was also described that combines the efficiency of HSV-1 virus to transfer DNA into various mammalian cells, including the postmitotic neuronal cells and the ability of EBV to maintain genes episomally. This vector system contains the HSV-1 origin of DNA replication (oriS) and a packaging signal, which allow replication and packaging of the amplicon into HSV-1 virions in the presence of HSV-1 helper functions as well as EBV oriP and EBNA-1 [233]. Another report describes the use of a similar HSV-1 amplicon system for efficient gene transfer, but AAV was included in their vector to achieve stable expression. This HSV/AAV hybrid vector contains OriS and packaging sequences from HSV-1, a transgene cassette that is flanked by AAV ITRs as well as an AAV rep gene residing outside the transgene cassette to mediate amplification and genomic integration of ITR-flanked sequences [234]. An HVJ-liposome vector system reported by Dzau et al. [235] was utilized to improve the efficiency of liposome-mediated transfer of an EBV-episomally maintained transgene [236,237]. This system exploits the fusigenic properties of the hemagglutinating virus of Japan (HVJ or Sendai virus) since envelope proteins of inactivated HVJ were found to mediate liposome–cell membrane fusion and facilitate cellular uptake of packaged plasmid DNA, bypassing endocytosis and lysosomal degradation.

One of the limitations with using viral episomal systems is the limited host range of such vectors. Although EBV episomal vectors replicate well in various human and primate cells, they are unable to replicate in rodent cells, limiting their utility in gene therapy since they cannot be tested in rodent models. Nonetheless, it was found that large fragments of human genomic DNA (between 10 and 15 kb) can mediate autonomous replication if there is also a mechanism to retain them in the nucleus [238]. Such vectors based on a human origin of replication were also found to be capable of replicating in rodent cells [239], probably due to the common host factors that drive their replication. A hybrid class of vectors was thus developed, which employs a human origin of replication to mediate vector replication as well as the EBV FR and EBNA-1 gene product to provide nuclear retention functions (see Calos [240]). EBNA-1 binding to the FR of the vector DNA causes the adherence of this complex to the chromosomal scaffold in a noncovalent fashion, thus retaining the vector DNA in the nucleus [241]. These vectors were reported to replicate somewhat in synchrony with chromosomal DNA once per cell cycle. Maintenance of these vectors within cells is related to the frequency of cell division [240]. Such vectors have been reported to persist in cells for at least 2 months under no selective pressure [238,242].

Ultimately, the development of a true mammalian artificial chromosome (MAC) without dependence on viral elements will be the key to obtaining stable episomal replication without dependence on selective pressure. Functional elements in mammalian cells important for maintaining DNA episomally as a minichromosome include a replication origin to promote autonomous replication, telomeres to protect ends of linear DNA and replicate DNA termini, and a centromere to facilitate correct segregation of the construct during mitotic division. Various mammalian chromosomal DNA replication initiation sites have been identified (reviewed in DePamphilis [243]) and found to comprise a 0.5–11 kb primary origin of bidirectional replication flanked by an initiation zone of about 6–55 kb. These sequences show characteristics of DNA unwinding, a densely methylated island, attachment sites to the nuclear matrix, and some palindromic sequences.

Vectors utilizing human genomic sequences that promote extrachromosomal vector replication have already been successfully applied as mentioned above. Telomeres that are required for the stability and integrity of the eukaryotic chromosome have been well characterized. In mammalian cells, the telomeric tracts comprise 2–50 kb of tandem TTAGGG repeats. Human centromeres, necessary for proper chromosome segregation at mitosis and meiosis, have been localized cytogenetically as primary constrictions of the chromosomes.

They are thought to consist of up to several megabases of highly repetitive DNA belonging to the alpha satellite DNA family and are attached to microtubules [244]. Until recently, the functional isolation of the centromere has been a great hurdle in the progress toward the construction of an MAC. The group of Willard et al. developed the first generation of human artificial microchromosomes (HAC) by creating synthetic alpha satellite arrays ~1 Mb in size [245]. They found that such an HAC, which is about 6–10 Mb in size, is mitotically and cytogenetically stable for up to 6 months in culture in the absence of selective pressure. Nonetheless, the technical challenge of assembling an MAC is still formidable, as cloning and manipulating such large constructs are not trivial using conventional bacterial cloning systems, and transfer to mammalian cells is difficult.

26.4 NANOPARTICLE-BASED DRUG AND GENE DELIVERY: A NEW PROMISE FOR GENE THERAPY

Occasionally, a new technology/field comes along that represents an emerging and novel scientific trend. Although nanotechnology has become very popular in the last few years, the concept is not new. Richard P. Feynman pioneered the field in the late 1950s when he introduced the concept of manipulating individual atoms: "there is plenty of room at the bottom." Developing nanoparticles depends on understanding the limitations and advantages of nanoscale studies. Despite Feynman's early vision, working at nanometer scale has only recently been conceivable.

Nanotechnology is a multidisciplinary field ranging from materials science to personal care application. Among these disciplines, nanomedicine has the potential to revolutionize diagnostics and therapeutics, two main applications that currently are in clinical trials (see National Nanotechnology initiatives: http://www.nano.gov). As mentioned in Section 26.1, a critical factor in gene therapy is the delivery vector. Therapeutics-based nanomedicine raises the opportunity to circumvent the issues related to present systems (including low transducing efficiency, toxicity, etc.) or at least to improve the gene and/or drug delivery process (reviewed by Emerich and Thanos [246] and Nie et al. [247]).

One type of nanoparticle is the liposome, which consists of a lipid shell surrounding a core containing a therapeutic gene or drug. Liposome-mediated gene transfer is a safe and noninvasive method of DNA delivery into cells. The use of lipids, specifically cationic lipids, as tools for the delivery of nucleic acids into cells has certain advantages over other gene delivery methods such as viral transfection vectors. These advantages include safety, simple preparation and handling, robust production of vectors, and the ability to inject large lipid:DNA complexes [248]. When liposomes are used for gene delivery, cationic lipids interact with the DNA spontaneously by noncovalent binding to form lipoplexes. These can infect a wide variety of cells in vitro with varied efficiency. Some have been reported to infect more than 70% of the cells [249]. Lipoplexes do not limit the size of the DNA transferred, unlike viral vectors, and were not shown to induce an immunological response, thus allowing multiple dose regimens [250].

The major shortcoming of liposomal methods for gene delivery that makes them inferior to viral vectors is their low and variable efficiency and nonstable expression in vivo. A possible explanation for this is that organism complexity and systemic barriers alter the cationic lipid transfection efficiency observed in vitro. Systemic barriers can be overcome by peptoids (oligomers with N-substituted glycines [251]) and lipoplex formulation methodologies. The incorporation of nonlipid substances, such as proteins or peptides [250], into the lipoplex can also assist in transport through cell barriers. Pharmacokinetic properties of lipoplexes can be altered by the use of PEGylated liposomes (attachment of polyethylene glycol). While expression is enhanced and renal clearance reduced, the efficacy and safety of these systems are compromised. Repeated in vivo administrations cause a long-term antibody response against PEG [252]. This response may be minimized by modifying the alkyl chain of the PEG–lipid conjugate. Through the change in pharmacokinetic properties, more constant and sustained plasma concentration can be achieved. In turn, this results in increased clinical effectiveness and decreased side effects [253]. Lipoplexes are characterized by their size, surface charge, DNA and lipid organization, DNA accessibility, and colloidal stability. It has been shown that certain lipoplex characteristics can influence transfection efficiency [249]. Of these, lipoplex size is worth mentioning [254], as well as charge ratio of the cation to anion and cell type.

Like liposomes, calcium phosphate nanoparticles are an alternative to viral gene delivery systems. Experimentation with this system has yielded high vector transfection efficiency, with no adverse effects from the carrier particles [255].

The use of the human ABCB1 gene as a selectable marker may allow for the selection and enrichment of the recipient cells and may be useful in the future for the long-term maintenance of the cationic liposome:DNA complex. Previous studies have shown that a liposomal delivery system can mediate successful ABCB1 transfection of mouse bone marrow cells and in vivo expression of functional P-gp in hematopoietic cells [256]. The introduction via liposomes into hematopoietic cells of an ABCB1 gene driven by Harvey murine sarcoma virus LTR sequences (Ha-MSV-LTR) can be achieved in two ways: either "directly" by intravenous administration into mice or "indirectly" by adoptive transplantation of previously in vitro-transfected bone marrow cells. In these studies, using a cationic liposome complex consisting of dioctadecylamidoglycyl spermidine (DOGS) and dioleoylphosphophosphatidyl ethanolamine (DOPE, a popular colipid), ABCB1 transfection was detected in up to 30% of unselected and 66% of vincristine-preselected murine bone marrow cells. This was demonstrated by drug resistance in an in vitro, colony-forming unit assay. Transfection into human bone marrow cells is likely to be much less efficient. However, obtaining drug-selectable mouse bone marrow progenitor cells through liposome-mediated gene transfer may eventually prevent bone marrow toxicity in cancer patients undergoing chemotherapy. Liposome-mediated gene transfer can also be used for *in vivo* delivery of adeno-associated-vectors (AAV)-*ABCB1*-based vectors. Baudard et al. [165] report drug-selected coexpression of both P-gp and GC with an AAV vector containing the *ABCB1*-IRES-GC fusion delivered to NIH 3T3 cells by lipofection. Moreover, 7 weeks after a single intravenous injection of this bicistronic vector complexed with cationic liposomes, GC and *ABCB1* sequences were detected by PCR in all recipient mice organs tested.

For nonintegrating DNA vectors such as EBV-based systems (see Section 26.4) and the AAV system [165], liposome-based gene delivery usually results in transient transgene expression. This can be explained by the episomal nature of the transfected plasmid and the loss of the plasmid when the cells proliferate [257,258]. Use of a selectable marker such as *ABCB1* may make it possible to maintain nonintegrated episomal forms in proliferating cells (see Section 26.4). Combining liposomes with AAV- or EBV-based vectors and *ABCB1* as a selectable marker may make it possible to expand the population of expressing cells by *ABCB1*-drug selection.

Polymer nanospheres are also being investigated (reviewed by Uchegbu [259]). Perez et al. constructed poly(lactic acid)/poly(ethylene glycol) (PEG) composite nanospheres (diameter <300 nm) and showed modulation of release with the addition of either poly(vinyl alcohol) or poly(vinylpyrrolidone) alone or in combination [260]. Putnam et al. demonstrated the complexation of polyhistidine–PEG nanoparticles to DNA by forming micelles of DNA/polyhistidine surrounded by a hydrated shell of PEG and showed that they could achieve similar efficiency of transfection to polylysine DNA conjugates, but lower than polyethyleneimine [261]. Dendritic and hyperbranched polymers are other nanoparticles developed for gene delivery (reviewed by Paleos et al. [262]). These polymers form electrostatic interactions between the negatively charged phosphate group on the DNA backbone and the positively charged amino groups on the polymer. Polyamidoamines (PANAM) are the most often used and characterized dendrimers for gene delivery and several groups have demonstrated their efficacy. One of them, Maksimenko et al., reported successful transfection of several cell lines by pCMV β-gal plasmid-dendrimer complexes in the presence of anionic oligomers including oligonucleotides or dextran sulfate [263]. The size and the charge of the additive affected the transfection efficiency, with 35–50 phosphate group oligonucleotides providing the highest efficiency.

Although further investigation is required to study the potential toxicity of nanoparticles, this novel technology opens interesting avenues for gene therapy. Nanoparticles could soon replace viral systems for gene delivery and circumvent some of the present difficulties involved in gene therapy.

26.5 ENGINEERING VECTORS TO IMPROVE EFFICIENCY OF DRUG SELECTION

26.5.1 NOVEL SEQUENCES AND DELIVERY VECTORS

Almost all coding eukaryotic mRNA molecules are monocistronic. Initiation of translation of eukaryotic mRNA is mediated by a cap-binding protein that recognizes a methylated guanosine cap at the 5′ terminus of mRNA. However, some viral mRNA molecules transcribed in eukaryotic cells are polycistronic. They can use a cap-independent mechanism to initiate translation in the middle of mRNA molecules. For picornavirus, this cap-independent internal initiation of translation is mediated through a unique IRES within the mRNA molecule [264,265]. Identification of IRES sequences has led to the development of bicistronic vectors that allow coexpression of two different polypeptides from a single mRNA molecule in eukaryotic cells [148,150]. Using a bicistronic vector containing an IRES to coexpress a target gene and a selectable marker has several advantages. First, since two polypeptides are translated from the same mRNA molecule, the bicistronic vector guarantees coexpression of a selectable marker and a second protein. Secondly, bicistronic mRNA allows two polypeptides to be translated separately. Thus, this system does not compromise the correct intracellular trafficking of proteins directed to different subcellular compartments. In addition, using a bicistronic vector, expression of a target gene is proportionate to the expression of a selectable marker. Hence, expression of a target protein can be achieved quantitatively by applying selections of different stringencies.

In addition to IRESes derived from viruses, several IRES elements have been identified in human genes. Those IRESes play important roles in cell cycle-dependent or stress–response translation regulation (reviewed in Sachs [266]). In contrast to viral IRESes, human IRESes are shorter and are complementary to 18s rRNA (reviewed in Mauro and Edelman [267]). It has been found that a 9-nt sequence from the 5′-UTR of the mRNA encoding the Gtx homeodomain protein can function as an IRES. Ten linked copies of the 9-nt sequence are 3- to 63-fold more active than the classical EMCV IRES in all 11 cell lines tested [268]. Similarly, an IRES isolated from the human EIF 4G gene also exhibits 100-fold more IRES activity than EMCV IRES in four different cell lines [269]. In addition to higher efficiency and smaller size, translation from a human IRES can be regulated by cellular events [269], which may be advantageous for certain cancer gene therapies.

Another approach to the use of *ABCB1*-based bicistronic vectors is to develop "suicide" vectors for cancer gene therapy. Using *ABCB1* to protect bone marrow cells from cytotoxic drugs represents a promising approach to improve cancer chemotherapy. However, contaminating cancer cells may be inadvertently transduced with *ABCB1*, or transduced bone marrow cells may accidentally develop new tumors. In those cases, overexpression of P-gp could cause multidrug resistance in inadvertently transduced tumor cells that contaminate bone marrow, or in any transduced cells that later become malignant. A bicistronic "suicide" vector developed by the authors links P-gp expression with herpes simplex virus thymidine kinase (TK) expression [270,271]. Thus the cells containing this vector can be eliminated through ganciclovir treatment.

Using the multidrug resistance gene, Metz et al. [272] showed that retroviral vectors derived from Harvey viruses can be substantially shortened without reduction of gene transfer efficiency, thereby increasing the maximum size of the packaged gene of interest. By systematic analysis of the U3-region of various 5′-LTR, Baum et al. [273] optimized *ABCB1* transfer to hematopoietic cells. Notably, transfer to immature hematopoietic progenitor cells, which are generally difficult to transduce, was improved [274]. Subsequent improvements of posttranscriptional processing led to a vector that reliably ensured *ABCB1* expression and drug efflux in human hematopoietic cells following an *in vivo* passage in immunodeficient mice [217]. Other vector systems used for chemoresistance gene transfer to hematopoietic cells include AAV vectors [165], liposomes [256], and new nanosystems (see Sections 26.3 and 26.4).

New vector constructs increase the efficiency of gene transfer to hematopoietic cells but do not necessarily ensure gene expression for sustained periods. A major obstacle to long-term gene expression is the limited lifespan of some transduced cell clones. Since only hematopoietic stem cells have the capability of self-renewal, the lifespan of progeny generated by more differentiated progenitor cells is limited. Berger et al. [275] have shown that expansion of cells with cytokines, particularly with interleukin-3 (IL-3), reduces the frequency of long-term culture-initiating cells (LTC-IC), which correlated with reduction of Rhodamine-123 efflux from immature progenitor cells. In accord with these findings, Schiedlmeier et al. [276] reported that IL-3-stimulated hematopoietic cells engrafted more poorly than cells grown in the presence of other growth factor combinations. Both studies resulted in efficient retroviral *ABCB1* transfer to primitive human progenitor/stem cells.

Alternatively, a dominant-positive selectable marker gene can be coexpressed with a negative selectable marker such as thymidine kinase from Herpes simplex virus (HSV-TK) [270,271]. The latter approach allows selective elimination of transduced cells. Such an approach may increase the safety of gene transfer if cancer cells contaminating hematopoietic cell preparations are inadvertently rendered drug-resistant, or if transduced cells become malignant [71,277]. Selective killing of *ABCB1*-HSV-TK transduced cells *in vivo* has been demonstrated [271]. TK may not only facilitate selective killing of cancer cells but instead increase the efficacy of certain selectable marker genes. A bicistronic vector in which TK was combined with DHFR displayed enhanced resistance as compared to a construct that contained a neomycin phosphotransferase instead of TK [278].

To increase the safety of gene therapy of cancer, drug resistance genes may be combined with cDNAs that specifically eliminate cancer cells. This has been demonstrated for chronic myeloid leukemia (CML), which is characterized by a specific molecular marker, the BCR/ABL gene fusion. A vector has been constructed that combined a MTX-resistant DHFR with an antiBCR/ABL antisense sequence [279]. Transfer of this vector to CML cells led to the restoration of normal cellular function of BCR/ABL cDNA + cells due to reduced levels of transcripts while conferring drug resistance.

26.5.2 ABCB1 Mutations and SNP Studies Lead to Improved Drug Treatment Efficacy

One of the hallmark characteristics of the multidrug transporter is its extremely broad substrate specificity. Over the past several years, the identification of specific domains and amino acid residues involved in substrate recognition has contributed to our present understanding of the mechanism of action of P-gp. The major sites of interaction have been shown to reside in TM domains 5 and 6 in the N-terminal half of the protein and in TMs 11 and 12 in the C-terminal half and the loops that conjoin them [280–284]. For the purposes of chemoprotection, the design of a P-gp that has increased resistance to chemotherapeutic agents compared to endogenous P-gp and/or decreased sensitivity to inhibitors would be most useful because increased doses of the agent could be administered without harming the bone marrow cells expressing the exogenous P-gp molecule, and naturally occurring resistance of cancer cells could be reversed while still allowing protection of bone marrow by transfected P-gp. This increased resistance and decreased sensitivity to inhibitors can be achieved by generating ABCB1-mutant transporters.

Mutations in TM domains of P-gps from both rodents and humans have demonstrated significant alterations in substrate specificity [8,285,286]. An F338A mutation in hamster P-gp enhances resistance to vincristine, colchicine, and daunorubicin but has little impact on resistance to actinomycin D [287,288]. An F339P mutation in the same molecule only increases actinomycin D resistance. However, the double F338A/F339P mutant demonstrates an increased level of resistance to actinomycin D and vincristine but a lowered level of resistance to colchicine and daunorubicin [287,288]. Of these mutants, the F338A may prove most useful because it confers increased resistance to a wider range of chemotherapeutic agents. In human P-gp, however, a homologous mutation at F335 confers greater resistance to colchicine and doxorubicin but causes a severe reduction in resistance to vinblastine and actinomycin D [289,290]. Additionally, cells expressing a Val > Ala mutation at position 338 also exhibit preferential resistance to colchicine and doxorubicin but are severely impaired for vinblastine [290]. Resistance to actinomycin D, however, is unaffected. Mutation of proline 223 to alanine in the fourth TM segment of human P-gp decreased the affinity for colchicine, whereas the affinity for vinblastine remained unchanged [289]. Alanine scanning of TM 11 in mouse P-gp encoded by *mdr*1a revealed that two mutants, M944A and F940A, show an increase in resistance to doxorubicin and colchicine while maintaining wild-type levels of resistance to vinblastine and actinomycin D [291]. For certain treatment protocols, it is conceivable that increased resistance to certain agents would be desirable, and the reduction in levels of resistance to other compounds would not be problematic, especially if a well-defined chemotherapy regimen was being employed.

Although the majority of residues that increase resistance to various chemotherapeutic agents reside in the TM domains, a number of residues in the putative cytoplasmic loops also have been implicated in defining drug resistance profiles for cytotoxic drugs. The best characterized of these mutations is the G185 V mutant that confers an increased resistance to colchicine and etoposide but decreased resistance to actinomycin D, vinblastine, doxorubicin, vincristine, and taxol [292–295]. Interestingly, and perhaps relevant clinically, when this mutation is made in conjunction with an Asn > Ser mutation at residue 183, increased resistance to actinomycin D, vinblastine, and doxorubicin is achieved without loss of increase in colchicine resistance [293]. Mutations of Gly-141, 187, 288, 812, or 830 to Val in human P-gp increase the relative resistance of NIH3T3 cells to colchicine and doxorubicin but do not alter resistance to vinblastine [296]. Only the mutations at positions 187, 288, and 830 confer decreased resistance to actinomycin D to cells in culture.

Due to its broad substrate specificity, P-gp not only interacts with chemotherapeutic compounds but also with reversing agents and inhibitors. In combination chemotherapies, reversing agents increase the efficacy of cytotoxic agents in ABCB1-expressing cancers. Two of the most potent reversing agents currently in use or in clinical trials are cyclosporin A and its nonimmunosuppressive analog PSC833. Recently, a number of mutants have been described that affect sensitivity to these agents. Cells expressing a human P-gp containing a deletion at Phe335 or Phe334 are substantially resistant to cyclosporin A and PSC-833 [297]. A similar phenotype has been observed for a transporter containing five mutations in the region including TM 5 and TM 6, namely Ile299Met, Thr319Ser, Leu322Ile, Gly324Lys, and Ser351Asn [298] (see Figure 26.1). Additionally, in hamster P-gp, the substitution of an alanine at position 339 with proline results in a transporter that confers lowered sensitivity to cyclosporin A [288]. From these studies, it appears that TM 6 plays an important role in the recognition of cyclosporin A and its analogs. The decreased sensitivity to these reversing agents observed in cells expressing the TM 6 mutations could help protect bone marrow stem cells transduced with the mutant ABCB1 gene from the toxic effects of chemotherapy given with reversing agents to sensitize ABCB1-expressing tumors.

The cis- and trans-isomers of flupentixol, a dopamine receptor antagonist, have also been shown to inhibit drug transport and reverse drug resistance mediated by P-gp [299,300]. The substitution of a single phenylalanine residue with alanine (F983A) in TM 12 affects inhibition of P-gp-mediated drug transport by both isomers of flupentixol [301–303]. Both isomers were found to be less effective at reversing P-gp-mediated drug transport of daunorubicin and bisantrene. However, the inhibitory effects of other reversing agents such as cyclosporin A were not affected. The reduced sensitivity of the F983A mutant to this compound coupled to the apparent lack of clinical toxicity of (trans)-flupentixol [299], suggests that this mutant may be useful in combining ABCB1 gene therapy with chemotherapy including trans-flupentixol as a chemosensitizer. This approach, in theory, should allow for effective treatment at lower doses of chemotherapeutic agents while maintaining bone marrow protection.

The assembly of gene delivery vectors that incorporate the ABCB1 gene as a selectable marker should take into consideration common polymorphisms along the gene that are known to alter P-gp expression or function. This knowledge is of great importance when one attempts to create a P-gp molecule that enables high resistance or sensitivity to MDR1 substrates. Different single-nucleotide polymorphisms (SNPs) are linked to drug efficacy and disease susceptibility.

Five nonsynonymous SNPs in positions N21D, F103L, S400N, A893S, and A998T showed no differences in P-gp expression and function in vitro when compared to wild type [304]. In contrast, three SNPs in positions 12/C1236T, 21/G2677T, and 26/C3435T, frequently appearing in various populations, might have clinical implications. Although no differences in mRNA or protein levels are observed in transient infection–transfection studies, there is evidence that this haplotype is associated with a conformational change that results in altered substrate and inhibitor recognition. The haplotype protein has been shown to be relatively resistant to the inhibitory effects of cyclosporin A and verapamil. This is of particular interest, since two of these three polymorphisms are synonymous and do not change the amino acid sequence. The mechanism of this resistance appears to be altered protein conformation resulting from a change in the kinetics of translation of the polymorphic P-gp [305]. In lymphoblastoid cell lines derived from HIV positive Japanese patients, higher concentrations of nelfanivir (protease inhibitor, PI) were found in individuals with the T/T genotype in location 3435, when compared to the C/C genotype [306]. This suggests higher P-gp expression for the C wild-type allele. Furthermore, this SNP is associated with a predisposition to refractory Crohn's disease and ulcerative colitis and to a reduction in the risk of developing Parkinson's disease [307]. Another common SNP influencing P-gp expression is in exon 21, 2677G > T. This SNP, much like the former, has been correlated with a greater risk for resistance to chemotherapy [308]. It has also been linked to tacrolimus-induced neurotoxicity in Japanese liver transplant patients [309]. Moreover, chemotherapy efficiency in acute myeloid leukemia patients has been associated with this SNP, i.e., homozygous patients (2677GG/TT) exhibit a worse prognosis than heterozygous patients [310]. Similarly, clinical implications have been reported for the SNP at location 12/1236C > T. This conformation has been linked to a greater export of protease inhibitors in the treatment of HIV and to CD4 cell count recovery during PI treatment [306]. These studies suggest that one should take into consideration possible SNP-genotype-phenotype correlations when constructing vectors carrying the ABCB1 gene as a selectable marker.

26.6 CONCLUSIONS AND FUTURE PROSPECTS

We have argued in this review that drug-selectable marker genes may be helpful for gene therapy in two ways: first, to protect bone marrow progenitor cells (and other sensitive cells) from the cytotoxicity of anticancer drugs, thereby allowing safe chemotherapeutic treatment at reduced risk of severe side effects, and second, to enrich the expression of otherwise non selectable genes in drug-sensitive cells to overcome low or unstable gene expression *in vivo*. Given the current instability of expression of genes from existing vectors, especially episomal vectors, such selectable markers may be an essential component of gene therapy protocols. The variety of selective markers with different drug resistance profiles that are now available make it possible for gene therapy to be used to overcome myelotoxicity in cancer chemotherapy.

We are still in the early stages of vector development, and until transduction efficiencies into human tissues such as bone marrow are improved and shown to be safe, long-term human gene therapy will not be feasible. The combination of more efficient gene transfer, targeted vector systems and effective, relatively nontoxic selection systems to maintain gene expression may make long-term correction of human genetic defects feasible and safe.

Finally, a new era for drug/gene delivery has arrived with the development of nanotechnology. Nanoparticles could soon replace viral systems for gene delivery and circumvent some of the present difficulties involved in gene therapy. Although further investigation is required to study the potential toxicity of nanoparticles, this novel technology has the potential to revolutionize medicine and especially, gene therapy. Moreover, the analysis of SNPs or engineered mutants, coupled with our knowledge of the current set of anticancer and reversing agents, offers an opportunity to begin designing novel and promising vectors.

REFERENCES

1. Gomez-Sebastian, S., et al., Infectious delivery and expression of a 135 kb human FRDA genomic DNA locus complements Friedreich's ataxia deficiency in human cells, *Mol. Ther.*, 15, 248–254, 2007.
2. Lim, F., et al., Functional recovery in a Friedreich's ataxia mouse model by frataxin gene transfer using an HSV-1 amplicon vector, *Mol. Ther.*, 15, 1072–1078, 2007.
3. Hibbitt, O. C., et al., Delivery and long-term expression of a 135 kb LDLR genomic DNA locus in vivo by hydrodynamic tail vein injection, *J. Gene Med.*, 9, 488–497, 2007.
4. Kanzaki, S., et al., Transgene correction maintains normal cochlear structure and function in 6-month-old Myo15a mutant mice, *Hearing Res.*, 214, 37–44, 2006.
5. Suda, T., et al., Heat-regulated production and secretion of insulin from a human artificial chromosome vector, *Biochem. Biophys. Res. Commun.*, 340, 1053–1061, 2006.
6. Flasshove, M., et al., Retroviral transduction of human CD34 + umbilical cord blood progenitor cells with a mutated dihydrofolate reductase cDNA, *Hum. Gene Ther.*, 9, 63–71, 1998.
7. Wang, G., et al., Retrovirus-mediated transfer of the human O6-methylguanine-DNA methyltransferase gene into a murine hematopoietic stem cell line and resistance to the toxic effects of certain alkylating agents, *Biochem. Pharmacol.*, 51, 1221–1228, 1996.
8. Gottesman, M. M., et al., Genetic analysis of the multidrug transporter, *Ann. Rev. Genet.*, 29, 607–649, 1995.
9. D'Hondt, V., Caruso, M., and Bank, A., Retrovirus-mediated gene transfer of the multidrug resistance-associated protein (MRP) cDNA protects cells from chemotherapeutic agents, *Hum. Gene Ther.*, 8, 1745–1751, 1997.
10. Ujhelly, O., et al., Application of a human multidrug transporter (ABCG2) variant as selectable marker in gene transfer to progenitor cells, *Hum. Gene Ther.*, 14, 403–412, 2003.
11. Omori, F., et al., Retroviral-mediated transfer and expression of the multidrug resistance protein 1 gene (MRP1) protect human hematopoietic cells from antineoplastic drugs, *J. Hematother. Stem Cell Res.*, 8, 503–514, 1999.
12. Warlick, C. A., et al., In vivo selection of antifolate-resistant transgenic hematopoietic stem cells in a murine bone marrow transplant model, *J. Pharmacol. Exp. Ther.*, 300, 50–56, 2002.
13. Rothem, L., et al., The reduced folate carrier gene is a novel selectable marker for recombinant protein overexpression, *Mol. Pharmacol.*, 68, 616–624, 2005.
14. Eliopoulos, N., et al., Human cytidine deaminase as an ex vivo drug selectable marker in gene-modified primary bone marrow stromal cells, *Gene Ther.*, 9, 452–462, 2002.
15. Momparler, R. L., et al., Resistance to cytosine arabinoside by retrovirally mediated gene transfer of human cytidine deaminase into murine fibroblast and hematopoietic cells, *Cancer Gene Ther.*, 3, 331–338, 1996.
16. Freitas, A. C., et al., Modified blasticidin S resistance gene (bsrm) as a selectable marker for construction of retroviral vectors, *J. Biotechnol.*, 95, 57–62, 2002.
17. Letourneau, S., Greenbaum, M., and Cournoyer, D., Retrovirus-mediated gene transfer of rat glutathione S-transferase Yc confers in vitro resistance to alkylating agents in human leukemia cells and in clonogenic mouse hematopoietic progenitor cells, *Hum. Gene Ther.*, 7, 831–840, 1996.
18. Magni, M., et al., Induction of cyclophosphamide-resistance by aldehyde-dehydrogenase gene transfer, *Blood*, 87, 1097–1103, 1996.
19. Moreb, J., et al., Overexpression of the human aldehyde dehydrogenase class I results in increased resistance to 4-hydroperoxycyclophosphamide, *Cancer Gene Ther.*, 3, 24–30, 1996.
20. Allay, J. A., et al., Retroviral transduction and expression of the human alkyltransferase cDNA provides nitrosourea resistance to hematopoietic cells, *Blood*, 85, 3342–3351, 1995.
21. Lorico, A., et al., Gamma-glutamylcysteine synthetase and L-buthionine-(S,R)-sulfoximine: A new selection strategy for gene-transduced neural and hematopoietic stem/progenitor cells, *Hum. Gene Ther.*, 16, 711–724, 2005.
22. Rappa, G., et al., Gamma-glutamylcysteine synthetase-based selection strategy for gene therapy of chronic granulomatous disease and graft-vs.-host disease, *Eur. J. Haematol.*, 78, 440–448, 2007.
23. Gottesman, M. M. and Ling, V., The molecular basis of multidrug resistance in cancer: The early years of P-glycoprotein research, *FEBS Lett.*, 580, 998–1009, 2006.
24. Deeley, R. G., Westlake, C., and Cole, S. P., Transmembrane transport of endo- and xenobiotics by mammalian ATP-binding cassette multidrug resistance proteins, *Physiol. Rev.*, 86, 849–899, 2006.
25. Bakos, E. and Homolya, L., Portrait of multifaceted transporter, the multidrug resistance-associated protein 1 (MRP1/ABCC1), *Pflugers Arch.*, 453, 621–641, 2007.
26. Kruh, G. D., et al., ABCC10, ABCC11, and ABCC12, *Pflugers Arch.*, 453, 675–684, 2007.
27. Borst, P., de Wolf, C., and van de Wetering, K., Multidrug resistance-associated proteins 3, 4, and 5, *Pflugers Arch.*, 453, 661–673, 2007.
28. Robey, R. W., et al., ABCG2: Determining its relevance in clinical drug resistance, *Cancer Metastasis Rev.*, 26, 39–57, 2007.

29. Szakacs, G., et al., Targeting multidrug resistance in cancer, *Nat. Rev. Drug Discov.,* 5, 219–234, 2006.
30. Gillet, J.-P., Efferth, T., and Remacle, J., Chemotherapy-induced resistance by ATP-binding cassette transporter genes, *Biochim. Biophys. Acta,* 1775, 237–262, 2007.
31. Ambudkar, S. V., Kim, I. W., and Sauna, Z. E., The power of the pump: Mechanisms of action of P-glycoprotein (ABCB1), *Eur. J. Pharm. Sci.,* 27, 392–400, 2006.
32. Ambudkar, S. V., et al., P-glycoprotein: From genomics to mechanism, *Oncogene,* 22, 7468–7485, 2003.
33. Sauna, Z. E. and Ambudkar, S. V., About a switch: How P-glycoprotein (ABCB1) harnesses the energy of ATP binding and hydrolysis to do mechanical work, *Mol. Cancer Ther.,* 6, 13–23, 2007.
34. Guo, C. and Jin, X., Chemoprotection effect of multidrug resistance 1 (MDR1) gene transfer to hematopoietic progenitor cells and engrafted in mice with cancer allows intensified chemotherapy, *Cancer Invest.,* 24, 659–668, 2006.
35. Carpinteiro, A., et al., Genetic protection of repopulating hematopoietic cells with an improved MDR1-retrovirus allows administration of intensified chemotherapy following stem cell transplantation in mice, *Int. J. Cancer,* 98, 785–792, 2002.
36. Galski, H., et al., Expression of a human multidrug resistance cDNA (MDR1) in the bone marrow of transgenic mice: Resistance to daunomycin-induced leukopenia, *Mol. Cell. Biol.,* 9, 4357–4363, 1989.
37. Mickisch, G. H., et al., Chemotherapy and chemosensitization of transgenic mice which express the human multidrug resistance gene in bone marrow: Efficacy, potency, and toxicity, *Cancer Res.,* 51, 5417–5424, 1991.
38. Mickisch, G. H., et al., Transgenic mice that express the human multidrug-resistance gene in bone marrow enable a rapid identification of agents that reverse drug resistance, *Proc. Natl Acad. Sci. USA,* 88, 547–551, 1991.
39. Mickisch, G. H., et al., Transplantation of bone marrow cells from transgenic mice expressing the human MDR1 gene results in long-term protection against the myelosuppressive effect of chemotherapy in mice, *Blood,* 79, 1087–1093, 1992.
40. May, C., Gunther, R., and McIvor, R. S., Protection of mice from lethal doses of methotrexate by transplantation with transgenic marrow expressing drug-resistant dihydrofolate reductase activity, *Blood,* 86, 2439–2448, 1995.
41. Hanania, E. G. and Deisseroth, A. B., Serial transplantation shows that early hematopoietic precursor cells are transduced by MDR-1 retroviral vector in a mouse gene therapy model, *Cancer Gene Ther.,* 1, 21–25, 1994.
42. Podda, S., et al., Transfer and expression of the human multiple drug resistance gene into live mice, *Proc. Natl Acad. Sci. USA,* 89, 9676–9680, 1992.
43. Sorrentino, B. P., et al., Selection of drug-resistant bone marrow cells in vivo after retroviral transfer of human MDR1, *Science,* 257, 99–103, 1992.
44. Licht, T., et al., Drug selection with paclitaxel restores expression of linked IL-2 receptor gamma -chain and multidrug resistance (MDR1) transgenes in canine bone marrow, *Proc. Natl Acad. Sci. USA,* 99, 3123–3128, 2002.
45. Sugimoto, Y., et al., Drug-selected co-expression of P-glycoprotein and gp91 in vivo from an MDR1-bicistronic retrovirus vector Ha-MDR-IRES-gp91, *J. Gene Med.,* 5, 366–376, 2003.
46. Schinkel, A. H., et al., Normal viability and altered pharmacokinetics in mice lacking mdr1-type (drug-transporting) P-glycoproteins, *Proc. Natl. Acad. Sci. USA,* 94, 4028–4033, 1997.

47. Schinkel, A. H., et al., Disruption of the mouse mdr1a P-glycoprotein gene leads to a deficiency in the blood–brain barrier and to increased sensitivity to drugs, *Cell,* 77, 491–502, 1994.
48. Schinkel, A. H., et al., P-glycoprotein in the blood–brain barrier of mice influences the brain penetration and pharmacological activity of many drugs, *J. Clin. Invest.,* 97, 2517–2524, 1996.
49. Smit, J. W., et al., Hepatobiliary and intestinal clearance of amphiphilic cationic drugs in mice in which both mdr1a and mdr1b genes have been disrupted, *Br. J. Pharmacol.,* 124, 416–424, 1998.
50. de Lange, E. C., et al., BBB transport and P-glycoprotein functionality using MDR1A (–/–) and wild-type mice. Total brain versus microdialysis concentration profiles of rhodamine-123, *Pharm. Res.,* 15, 1657–1665, 1998.
51. Saito, T., et al., Homozygous disruption of the mdr1a P-glycoprotein gene affects blood–nerve barrier function in mice administered with neurotoxic drugs, *Acta Otolaryngol.,* 121, 735–742, 2001.
52. Saito, T., et al., Inhibitory effect of cyclosporin A on p-glycoprotein function in peripheral nerves of mice treated with doxorubicin and vinblastine, *Acta Otolaryngol.,* 124, 313–317, 2004.
53. Grauer, M. T. and Uhr, M., P-glycoprotein reduces the ability of amitriptyline metabolites to cross the blood brain barrier in mice after a 10-day administration of amitriptyline, *J. Psychopharmacol.,* 18, 66–74, 2004.
54. van Asperen, J., et al., Comparative pharmacokinetics of vinblastine after a 96-hour continuous infusion in wild-type mice and mice lacking mdr1a P-glycoprotein, *J. Pharmacol. Exp. Ther.,* 289, 329–333, 1999.
55. Romano, G., et al., Latest developments in gene transfer technology: Achievements, perspectives, and controversies over therapeutic applications, *Stem Cells,* 18, 19–39, 2000.
56. Hacein-Bey-Abina, S., et al., A serious adverse event after successful gene therapy for X-linked severe combined immunodeficiency, *N. Engl. J. Med.,* 348, 255–256, 2003.
57. Hacein-Bey-Abina, S., et al., LMO2-associated clonal T cell proliferation in two patients after gene therapy for SCID-X1, *Science,* 302, 415–419, 2003.
58. Marshall, E., Gene therapy death prompts review of adenovirus vector, *Science,* 286, 2244–2245, 1999.
59. Raper, S. E., et al., Fatal systemic inflammatory response syndrome in an ornithine transcarbamylase deficient patient following adenoviral gene transfer, *Mol. Genet. Metab.,* 80, 148–158, 2003.
60. Gao, X., Kim, K. S., and Liu, D., Nonviral gene delivery: What we know and what is next, *AAPS J.,* 9, E92–E104, 2007.
61. Wagner, E., Programmed drug delivery: Nanosystems for tumor targeting, *Expert Opin. Biol. Ther.,* 7, 587–593, 2007.
62. Czerwinski, M., Kiem, H. P., and Slattery, J. T., Human CD34 + cells do not express glutathione S-transferases alpha, *Gene Ther.,* 4, 268–270, 1997.
63. Drach, D., et al., Subpopulations of normal peripheral blood and bone marrow cells express a functional multidrug resistant phenotype, *Blood,* 80, 2729–2734, 1992.
64. Klimecki, W. T., et al., P-glycoprotein expression and function in circulating blood cells from normal volunteers, *Blood,* 83, 2451–2458, 1994.
65. DelaFlor-Weiss, E., et al., Transfer and expression of the human multidrug resistance gene in mouse erythroleukemia cells, *Blood,* 80, 3106–3111, 1992.

66. McLachlin, J. R., et al., Expression of a human complementary DNA for the multidrug resistance gene in murine hematopoietic precursor cells with the use of retroviral gene transfer, *J. Natl. Cancer Inst.,* 82, 1260–1263, 1990.

67. Richardson, C., et al., Mouse fetal liver cells lack functional amphotropic retroviral receptors, *Blood,* 84, 433–439, 1994.

68. Bodine, D. M., et al., Efficient retrovirus transduction of mouse pluripotent hematopoietic stem cells mobilized into the peripheral blood by treatment with granulocyte colony-stimulating factor and stem cell factor, *Blood,* 84, 1482–1491, 1994.

69. Licht, T., et al., Efficient expression of functional human MDR1 gene in murine bone marrow after retroviral transduction of purified hematopoietic stem cells, *Blood,* 86, 111–121, 1995.

70. Maier, P., et al., Overexpression of MDR1 using a retroviral vector differentially regulates genes involved in detoxification and apoptosis and confers radioprotection, *Radiation Res.,* 166, 463–473, 2006.

71. Bunting, K. D., et al., Transduction of murine bone marrow cells with an MDR1 vector enables ex vivo stem cell expansion, but these expanded grafts cause a myeloproliferative syndrome in transplanted mice, *Blood,* 92, 2269–2279, 1998.

72. Bunting, K. D., et al., Enforced P-glycoprotein pump function in murine bone marrow cells results in expansion of side population stem cells in vitro and repopulating cells in vivo, *Blood,* 96, 902–909, 2000.

73. Hanania, E. G., et al., Resistance to taxol chemotherapy produced in mouse marrow cells by safety-modified retroviruses containing a human MDR-1 transcription unit, *Gene Ther.,* 2, 279–284, 1995.

74. Sellers, S. E., et al., The effect of multidrug-resistance 1 gene versus neo transduction on *ex vivo* and in vivo expansion of rhesus macaque hematopoietic repopulating cells, *Blood,* 97, 1888–1891, 2001.

75. Galipeau, J., et al., A bicistronic retroviral vector for protecting hematopoietic cells against antifolates and P-glycoprotein effluxed drugs, *Hum. Gene Ther.,* 8, 1773–1783, 1997.

76. Sorrentino, B. P., et al., Expression of retroviral vectors containing the human multidrug resistance 1 cDNA in hematopoietic cells of transplanted mice, *Blood,* 86, 491–501, 1995.

77. Modlich, U., et al., Leukemias following retroviral transfer of multidrug resistance 1 (MDR1) are driven by combinatorial insertional mutagenesis, *Blood,* 105, 4235–4246, 2005.

78. Baum, C., et al., Side effects of retroviral gene transfer into hematopoietic stem cells, *Blood,* 101, 2099–2114, 2003.

79. Li, Z., et al., Murine leukemia induced by retroviral gene marking, *Science,* 296, 497, 2002.

80. Du, Y., Jenkins, N. A., and Copeland, N. G., Insertional mutagenesis identifies genes that promote the immortalization of primary bone marrow progenitor cells, *Blood,* 106, 3932–3939, 2005.

81. Modlich, U., et al., Cell-culture assays reveal the importance of retroviral vector design for insertional genotoxicity, *Blood,* 108, 2545–2553, 2006.

82. Deisseroth, A. B., Kavanagh, J., and Champlin, R., Use of safety-modified retroviruses to introduce chemotherapy resistance sequences into normal hematopoietic cells for chemoprotection during the therapy of ovarian cancer: A pilot trial, *Hum. Gene Ther.,* 5, 1507–1522, 1994.

83. Hesdorffer, C., et al., Human MDR gene transfer in patients with advanced cancer, *Hum. Gene Ther.,* 5, 1151–1160, 1994.

84. O'Shaughnessy, J. A., et al., Retroviral mediated transfer of the human multidrug resistance gene (MDR-1) into hematopoietic stem cells during autologous transplantation after intensive chemotherapy for metastatic breast cancer, *Hum. Gene Ther.,* 5, 891–911, 1994.

85. Cowan, K. H., et al., Paclitaxel chemotherapy after autologous stem-cell transplantation and engraftment of hematopoietic cells transduced with a retrovirus containing the multidrug resistance complementary DNA (MDR1) in metastatic breast cancer patients, *Clin. Cancer Res.,* 5, 1619–1628, 1999.

86. Hanania, E. G., et al., Results of MDR-1 vector modification trial indicate that granulocyte/macrophage colony-forming unit cells do not contribute to posttransplant hematopoietic recovery following intensive systemic therapy, *Proc. Natl. Acad. Sci. USA,* 93, 15346–15351, 1996.

87. Hesdorffer, C., et al., Phase I trial of retroviral-mediated transfer of the human MDR1 gene as marrow chemoprotection in patients undergoing high-dose chemotherapy and autologous stem-cell transplantation, *J. Clin. Oncol.,* 16, 165–172, 1998.

88. Abonour, R., et al., Efficient retrovirus-mediated transfer of the multidrug resistance 1 gene into autologous human long-term repopulating hematopoietic stem cells, *Nat. Med.,* 6, 652–658, 2000.

89. Moscow, J. A., et al., Engraftment of MDR1 and NeoR gene-transduced hematopoietic cells after breast cancer chemotherapy, *Blood,* 94, 52–61, 1999.

90. Kimchi-Sarfaty, C., et al., In vitro-packaged SV40 pseudovirions as highly efficient vectors for gene transfer, *Hum. Gene Ther.,* 13, 299–310, 2002.

91. Rund, D., et al., Efficient transduction of human hematopoietic cells with the human multidrug resistance gene 1 via SV40 pseudovirions, *Hum. Gene Ther.,* 9, 649–657, 1998.

92. Lee, C. G., et al., An episomally maintained MDR1 gene for gene therapy, *Hum. Gene Ther.,* 12, 945–953, 2001.

93. Hock, R. A. and Miller, A. D., Retrovirus-mediated transfer and expression of drug resistance genes in human haematopoietic progenitor cells, *Nature,* 320, 275–277, 1986.

94. Li, M. X., et al., Development of a retroviral construct containing a human mutated dihydrofolate reductase cDNA for hematopoietic stem cell transduction, *Blood,* 83, 3403–3408, 1994.

95. Miller, A. D., Law, M. F., and Verma, I. M., Generation of helper-free amphotropic retroviruses that transduce a dominant-acting, methotrexate-resistant dihydrofolate reductase gene, *Mol. Cell. Biol.,* 5, 431–437, 1985.

96. Zhao, S. C., et al., Long-term protection of recipient mice from lethal doses of methotrexate by marrow infected with a double-copy vector retrovirus containing a mutant dihydrofolate reductase, *Cancer Gene Ther.,* 1, 27–33, 1994.

97. Flasshove, M., et al., Increased resistance to methotrexate in human hematopoietic cells after gene transfer of the Ser31 DHFR mutant, *Leukemia,* 9 Suppl 1, S34–S37, 1995.

98. Isola, L. M. and Gordon, J. W., Systemic resistance to methotrexate in transgenic mice carrying a mutant dihydrofolate reductase gene, *Proc. Natl. Acad. Sci. USA,* 83, 9621–9625, 1986.

99. James, R. I., et al., Transgenic mice expressing the tyr22 variant of murine DHFR: Protection of transgenic marrow transplant recipients from lethal doses of methotrexate, *Exp. Hematol.,* 25, 1286–1295, 1997.

100. Williams, D. A., et al., Protection of bone marrow transplant recipients from lethal doses of methotrexate by the generation of methotrexate-resistant bone marrow, *J. Exp. Med.,* 166, 210–218, 1987.

101. Cline, M. J., et al., Gene transfer in intact animals, *Nature,* 284, 422–425, 1980.

102. Vinh, D. B. and McIvor, R. S., Selective expression of metho-trexate-resistant dihydrofolate reductase (DHFR) activity in mice transduced with DHFR retrovirus and administered methotrexate, *J. Pharmacol. Exp. Ther.*, 267, 989–996, 1993.

103. James, R. I., et al., Mild preconditioning and low-level engraft-ment confer methotrexate resistance in mice transplanted with marrow expressing drug-resistant dihydrofolate reductase activity, *Blood*, 96, 1334–1341, 2000.

104. Corey, C. A., et al., Serial transplantation of methotrexate-resistant bone marrow: Protection of murine recipients from drug toxicity by progeny of transduced stem cells, *Blood*, 75, 337–343, 1990.

105. Allay, J. A., et al., In vivo selection of retrovirally transduced hematopoietic stem cells, *Nat. Med.*, 4, 1136–1143, 1998.

106. McMillin, D. W., et al., Complete regression of large solid tumors using engineered drug-resistant hematopoietic cells and anti-CD137 immunotherapy, *Hum. Gene Ther.*, 17, 798–806, 2006.

107. Moritz, T., et al., Retrovirus-mediated expression of a DNA repair protein in bone marrow protects hematopoietic cells from nitrosourea-induced toxicity in vitro and in vivo, *Cancer Res.*, 55, 2608–2614, 1995.

108. Maze, R., et al., Increasing DNA repair methyltransferase levels via bone marrow stem cell transduction rescues mice from the toxic effects of 1,3-bis(2-chlorocthyl)-1-nitrosourca, a chemotherapeutic alkylating agent, *Proc. Natl. Acad. Sci. USA*, 93, 206–210, 1996.

109. Maze, R., et al., Reversal of 1,3-bis(2-chloroethyl)-1-nitro-sourea-induced severe immunodeficiency by transduction of murine long-lived hemopoietic progenitor cells using O6-methylguanine DNA methyltransferase complementary DNA, *J. Immunol.*, 158, 1006–1013, 1997.

110. Crone, T. M., et al., Mutations in human O6-alkylguanine-DNA alkyltransferase imparting resistance to O6-benzylguanine, *Cancer Res.*, 54, 6221–6227, 1994.

111. Xu-Welliver, M., Kanugula, S., and Pegg, A. E., Isolation of human O6-alkylguanine-DNA alkyltransferase mutants highly resistant to inactivation by O6-benzylguanine, *Cancer Res.*, 58, 1936–1945, 1998.

112. Reese, J. S., et al., Retroviral transduction of a mutant meth-ylguanine DNA methyltransferase gene into human CD34 cells confers resistance to O6-benzylguanine plus 1,3-bis (2-chloroethyl)-1-nitrosourea, *Proc. Natl. Acad. Sci. USA*, 93, 14088–14093, 1996.

113. Davis, B. M., et al., Selection for G156A O6-methylguanine DNA methyltransferase gene-transduced hematopoietic pro-genitors and protection from lethality in mice treated with O6-benzylguanine and 1,3-bis(2-chloroethyl)-1-nitrosourea, *Cancer Res.*, 57, 5093–5099, 1997.

114. Davis, B. M., et al., Characterization of the P140K, PVP (138–140)MLK, and G156A O6-methylguanine-DNA meth-yltransferase mutants: Implications for drug resistance gene therapy, *Hum. Gene Ther.*, 10, 2769–2778, 1999.

115. Jansen, M., et al., Hematoprotection and enrichment of trans-duced cells in vivo after gene transfer of MGMT(P140K) into hematopoietic stem cells, *Cancer Gene Ther.*, 9, 737–746, 2002.

116. Neff, T., et al., Methylguanine methyltransferase-mediated in vivo selection and chemoprotection of allogeneic stem cells in a large-animal model, *J. Clin. Invest.*, 112, 1581–1588, 2003.

117. Gottesman, M. M., Multidrug resistance during chemical carcinogenesis: A mechanism revealed? *J. Natl. Cancer Inst.*, 80, 1352–1353, 1988.

118. Ishizaki, K., et al., Transfer of the *E. coli* O6-methylguanine methyltransferase gene into repair-deficient human cells and restoration of cellular resistance to *N*-methyl-*N'*-nitro-*N*-nitrosoguanidine, *Mutat. Res.*, 166, 135–141, 1986.

119. Zaidi, N. H., et al., Transgenic expression of human MGMT protects against azoxymethane-induced aberrant crypt foci and G to A mutations in the K-ras oncogene of mouse colon, *Carcinogenesis*, 16, 451–456, 1995.

120. Liu, L., et al., Rapid repair of O6-methylguanine–DNA adducts protects transgenic mice from *N*-methylnitrosourea-induced thymic lymphomas, *Cancer Res.*, 54, 4648–4652, 1994.

121. Becker, K., Gregel, C. M., and Kaina, B., The DNA repair protein O6-methylguanine–DNA methyltransferase protects against skin tumor formation induced by antineoplastic chlo-roethylnitrosourea, *Cancer Res.*, 57, 3335–3338, 1997.

122. Nakatsuru, Y., et al., O6-methylguanine–DNA methyltrans-ferase protects against nitrosamine-induced hepatocarcino-genesis, *Proc. Natl. Acad. Sci. USA*, 90, 6468–6472, 1993.

123. Jelinek, J., et al., A novel dual function retrovirus expressing multidrug resistance 1 and O6-alkylguanine-DNA-alkyltransferase for engineering resistance of haemopoietic progenitor cells to mul-tiple chemotherapeutic agents, *Gene Ther.*, 6, 1489–1493, 1999.

124. Southgate, T. D., et al., Dual agent chemoprotection by retroviral co-expression of either MDR1 or MRP1 with the P140K mutant of O6-methylguanine-DNA-methyl transferase, *J. Gene Med.*, 8, 972–979, 2006.

125. Suzuki, M., et al., Retroviral coexpression of two different types of drug resistance genes to protect normal cells from combination chemotherapy, *Clin. Cancer Res.*, 3, 947–954, 1997.

126. Suzuki, M., Sugimoto, Y., and Tsuruo, T., Efficient protection of cells from the genotoxicity of nitrosoureas by the retrovirus-mediated transfer of human O6-methylguanine-DNA methyl-transferase using bicistronic vectors with human multidrug resistance gene 1, *Mutat. Res.*, 401, 133–141, 1998.

127. Wang, J., et al., Improvement of combination chemotherapy tolerance by introduction of polycistronic retroviral vector drug resistance genes MGMT and MDR1 into human umbili-cal cord blood CD34 + cells, *Leuk. Res.*, 26, 281–288, 2002.

128. Zhang, S., et al., pHaMDR-DHFR bicistronic expression system for mutational analysis of P-glycoprotein, *Methods Enzymol.*, 292, 474–480, 1998.

129. Rappa, G., et al., Novel bicistronic retroviral vector expressing gamma-glutamylcysteine synthetase and the multidrug resis-tance protein 1 (MRP1) protects cells from MRP1-effluxed drugs and alkylating agents, *Hum. Gene Ther.*, 12, 1785–1796, 2001.

130. Capiaux, G. M., et al., Protection of hematopoietic stem cells from pemetrexed toxicity by retroviral gene transfer with a mutant dihydrofolate reductase-mutant thymidylate synthase fusion gene, *Cancer Gene Ther.*, 11, 767–773, 2004.

131. Budak-Alpdogan, T., et al., Methotrexate and cytarabine inhibit progression of human lymphoma in NOD/SCID mice carrying a mutant dihydrofolate reductase and cytidine deam-inase fusion gene, *Mol. Ther.*, 10, 574–584, 2004.

132. Rappa, G., et al., Retroviral transfer of MRP1 and gamma-glutamyl cysteine synthetase modulates cell sensitivity to L-buthionine-*S*,*R*-sulphoximine (BSO): New rationale for the use of BSO in cancer therapy, *Eur. J. Cancer*, 39, 120–128, 2003.

133. Bailey, H. H., L-*S*,*R*-buthionine sulfoximine: Historical devel-opment and clinical issues, *Chem. Biol. Interact.*, 111–112, 239–254, 1998.

134. Vanhoefer, U., et al., D,L-buthionine-(S,R)-sulfoximine potentiates in vivo the therapeutic efficacy of doxorubicin against multidrug resistance protein-expressing tumors, *Clin. Cancer Res.*, 2, 1961–1968, 1996.

135. Versantvoort, C. H., et al., Resistance-associated factors in human small-cell lung-carcinoma GLC4 sub-lines with increasing adriamycin resistance, *Int. J. Cancer*, 61, 375–380, 1995.

136. Havenga, M., et al., Development of safe and efficient retroviral vectors for Gaucher disease, *Gene Ther.*, 4, 1393–1400, 1997.

137. Xu, L. C., et al., Growth factors and stromal support generate very efficient retroviral transduction of peripheral blood CD34 + cells from Gaucher patients, *Blood*, 86, 141–146, 1995.

138. Aran, J. M., Gottesman, M. M., and Pastan, I., Drug-selected coexpression of human glucocerebrosidase and P-glycoprotein using a bicistronic vector, *Proc. Natl. Acad. Sci. USA*, 91, 3176–3180, 1994.

139. Aran, J. M., et al., Construction and characterization of a selectable multidrug resistance-glucocerebrosidase fusion gene, *Cytokines Mol. Ther.*, 2, 47–57, 1996.

140. Havenga, M., et al., In vivo methotrexate selection of murine hemopoietic cells transduced with a retroviral vector for Gaucher disease, *Gene Ther.*, 6, 1661–1669, 1999.

141. Aran, J. M., et al., Complete restoration of glucocerebrosidase deficiency in Gaucher fibroblasts using a bicistronic MDR retrovirus and a new selection strategy, *Hum. Gene Ther.*, 7, 2165–2175, 1996.

142. Sugimoto, Y., et al., Retroviral coexpression of a multidrug resistance gene (MDR1) and human alpha-galactosidase A for gene therapy of Fabry disease, *Hum. Gene Ther.*, 6, 905–915, 1995.

143. Zhou, Y., et al., Co-expression of human adenosine deaminase and multidrug resistance using a bicistronic retroviral vector, *Hum. Gene Ther.*, 9, 287–293, 1998.

144. Sokolic, R. A., et al., A bicistronic retrovirus vector containing a picornavirus internal ribosome entry site allows for correction of X-linked CGD by selection for MDR1 expression, *Blood*, 87, 42–50, 1996.

145. Kleiman, S. E., et al., Characterization of an MDR1 retroviral bicistronic vector for correction of X-linked severe combined immunodeficiency, *Gene Ther.*, 5, 671–676, 1998.

146. Lee, C. G., et al., Efficient long-term coexpression of a hammerhead ribozyme targeted to the U5 region of HIV-1 LTR by linkage to the multidrug-resistance gene, *Antisense Nucleic Acid Drug Dev.*, 7, 511–522, 1997.

147. Pastan, I., et al., A retrovirus carrying an MDR1 cDNA confers multidrug resistance and polarized expression of P-glycoprotein in MDCK cells, *Proc. Natl. Acad. Sci. USA*, 85, 4486–4490, 1988.

148. Kaufman, R. J., et al., Improved vectors for stable expression of foreign genes in mammalian cells by use of the untranslated leader sequence from EMC virus, *Nucleic Acids Res.*, 19, 4485–4490, 1991.

149. Sugimoto, Y., et al., Efficient expression of drug-selectable genes in retroviral vectors under control of an internal ribosome entry site, *Biotechnology*, 12, 694–698, 1994.

150. Adam, M. A., et al., Internal initiation of translation in retroviral vectors carrying picornavirus 5′ nontranslated regions, *J. Virol.*, 65, 4985–4990, 1991.

151. Hellen, C. U. and Sarnow, P., Internal ribosome entry sites in eukaryotic mRNA molecules, *Genes Dev.*, 15, 1593–1612, 2001.

152. Toyoda, H., et al., Host factors required for internal initiation of translation on poliovirus RNA, *Arch. Virol.*, 138, 1–15, 1994.

153. Borman, A. M., et al., Picornavirus internal ribosome entry segments: Comparison of translation efficiency and the requirements for optimal internal initiation of translation in vitro, *Nucleic Acids Res.*, 23, 3656–3663, 1995.

154. Brown, E. A., Zajac, A. J., and Lemon, S. M., In vitro characterization of an internal ribosomal entry site (IRES) present within the 5′ nontranslated region of hepatitis A virus RNA: Comparison with the IRES of encephalomyocarditis virus, *J. Virol.*, 68, 1066–1074, 1994.

155. Blaese, R. M., Development of gene therapy for immunodeficiency: Adenosine deaminase deficiency, *Pediatr. Res.*, 33, S49–S53; discussion S53–S45, 1993.

156. Kantoff, P. W., et al., Correction of adenosine deaminase deficiency in cultured human T and B cells by retrovirus-mediated gene transfer, *Proc. Natl. Acad. Sci. USA*, 83, 6563–6567, 1986.

157. Germann, U. A., et al., Retroviral transfer of a chimeric multidrug resistance-adenosine deaminase gene, *Faseb J.*, 4, 1501–1507, 1990.

158. Ohshima, T., et al., Alpha-galactosidase A deficient mice: A model of Fabry disease, *Proc. Natl. Acad. Sci. USA*, 94, 2540–2544, 1997.

159. Aran, J. M., Gottesman, M. M., and Pastan, I., Construction and characterization of bicistronic retroviral vectors encoding the multidrug transporter and beta-galactosidase or green fluorescent protein, *Cancer Gene Ther.*, 5, 195–206, 1998.

160. Pfutzner, W., et al., Topical colchicine selection of keratinocytes transduced with the multidrug resistance gene (MDR1) can sustain and enhance transgene expression in vivo, *Proc. Natl. Acad. Sci. USA*, 99, 13096–13101, 2002.

161. Pfutzner, W. and Vogel, J. C., Topical colchicine selection of keratinocytes transduced with the multidrug resistance gene (MDR1) can sustain and enhance transgene expression in vivo, *Cells Tissues Organs*, 177, 151–159, 2004.

162. Kotin, R. M., et al., Site-specific integration by adeno-associated virus, *Proc. Natl. Acad. Sci. USA*, 87, 2211–2215, 1990.

163. Flotte, T. R., Afione, S. A., and Zeitlin, P. L., Adeno-associated virus vector gene expression occurs in nondividing cells in the absence of vector DNA integration, *Am. J. Respir. Cell. Mol. Biol.*, 11, 517–521, 1994.

164. Flotte, T. R. and Carter, B. J., Adeno-associated virus vectors for gene therapy, *Gene Ther.*, 2, 357–362, 1995.

165. Baudard, M., et al., Expression of the human multidrug resistance and glucocerebrosidase cDNAs from adeno-associated vectors: Efficient promoter activity of AAV sequences and in vivo delivery via liposomes, *Hum. Gene Ther.*, 7, 1309–1322, 1996.

166. Flotte, T. R., et al., Stable in vivo expression of the cystic fibrosis transmembrane conductance regulator with an adeno-associated virus vector, *Proc. Natl. Acad. Sci. USA*, 90, 10613–10617, 1993.

167. Flotte, T. R., et al., Expression of the cystic fibrosis transmembrane conductance regulator from a novel adeno-associated virus promoter, *J. Biol. Chem.*, 268, 3781–3790, 1993.

168. Kaplitt, M. G., et al., Long-term gene expression and phenotypic correction using adeno-associated virus vectors in the mammalian brain, *Nat. Genet.*, 8, 148–154, 1994.

169. Kessler, P. D., et al., Gene delivery to skeletal muscle results in sustained expression and systemic delivery of a therapeutic protein, *Proc. Natl. Acad. Sci. USA*, 93, 14082–14087, 1996.

170. Snyder, R. O., et al., Persistent and therapeutic concentrations of human factor IX in mice after hepatic gene transfer of recombinant AAV vectors, *Nat. Genet.*, 16, 270–276, 1997.

171. Xiao, X., et al., Gene transfer by adeno-associated virus vectors into the central nervous system, *Exp. Neurol.*, 144, 113–124, 1997.

172. Duan, D., Yan, Z., and Engelhardt, J., Expanding the capacity of AAV vectors, in *Parvoviruses*, Bloom M. E, et al. (Eds.), Hodder Arnold, New York, pp. 525–532, 2006.

173. Lai, Y., et al., Efficient in vivo gene expression by *trans*-splicing adeno-associated viral vectors, *Nat. Biotechnol.*, 23, 1435–1439, 2005.

174. Ghosh, A., et al., Efficient whole-body transduction with *trans*-splicing adeno-associated viral vectors, *Mol. Ther.*, 15, 1220, 2007.

175. Gottesman, M. M., Cancer gene therapy: An awkward adolescence, *Cancer Gene Ther.*, 10, 501–508, 2003.

176. Aleman, C., et al., P-glycoprotein, expressed in multidrug resistant cells, is not responsible for alterations in membrane fluidity or membrane potential, *Cancer Res.*, 63, 3084–3091, 2003.

177. Tsui, L. C., et al., Persistence of freely replicating SV40 recombinant molecules carrying a selectable marker in permissive simian cells, *Cell*, 30, 499–508, 1982.

178. Yates, J. L., Warren, N., and Sugden, B., Stable replication of plasmids derived from Epstein–Barr virus in various mammalian cells, *Nature*, 313, 812–815, 1985.

179. Cooper, M. J. and Miron, S., Efficient episomal expression vector for human transitional carcinoma cells, *Hum. Gene Ther.*, 4, 557–566, 1993.

180. Milanesi, G., et al., BK virus-plasmid expression vector that persists episomally in human cells and shuttles into *Escherichia coli*, *Mol. Cell. Biol.*, 4, 1551–1560, 1984.

181. Sabbioni, S., et al., A BK virus episomal vector for constitutive high expression of exogenous cDNAs in human cells, *Arch. Virol.*, 140, 335–339, 1995.

182. Khalili, K., *Human Polyomaviruses*, Wiley-Liss Inc., New York, 2001.

183. Nakanishi, A., et al., Identification of amino acid residues within simian virus 40 capsid proteins Vp1, Vp2, and Vp3 that are required for their interaction and for viral infection, *J. Virol.*, 80, 8891–8898, 2006.

184. Tsai, B., et al., Gangliosides are receptors for murine polyoma virus and SV40, *EMBO J.*, 22, 4346–4355, 2003.

185. Strayer, D. S., Gene therapy using SV40-derived vectors: What does the future hold? *J. Cell Physiol.*, 181, 375–384, 1999.

186. Richards, A. A., et al., Inhibitors of COP-mediated transport and cholera toxin action inhibit simian virus 40 infection, *Mol. Biol. Cell*, 13, 1750–1764, 2002.

187. Norkin, L. C., et al., Caveolar endocytosis of simian virus 40 is followed by brefeldin A-sensitive transport to the endoplasmic reticulum, where the virus disassembles, *J. Virol.*, 76, 5156–5166, 2002.

188. Strayer, D. S., Zern, M. A., and Chowdhury, J. R., What can SV40-derived vectors do for gene therapy? *Curr. Opin. Mol. Ther.*, 4, 313–323, 2002.

189. Chia, W. and Rigby, P. W., Fate of viral DNA in nonpermissive cells infected with simian virus 40, *Proc. Natl. Acad. Sci. USA*, 78, 6638–6642, 1981.

190. Strayer, D. S., et al., What they are, how they work and why they do what they do? The story of SV40-derived gene therapy vectors and what they have to offer, *Curr. Gene Ther.*, 5, 151–165, 2005.

191. Kao, C., et al., Role of SV40 T antigen binding to pRB and p53 in multistep transformation in vitro of human uroepithelial cells, *Carcinogenesis*, 14, 2297–2302, 1993.

192. Dalyot-Herman, N., et al., The simian virus 40 packaging signal ses is composed of redundant DNA elements which are partly interchangeable, *J. Mol. Biol.*, 259, 69–80, 1996.

193. Dalyot-Herman, N., Rund, D., and Oppenheim, A., Expression of beta-globin in primary erythroid progenitors of beta-thalassemia patients using an SV40-based gene delivery system, *J. Hematother. Stem Cell Res.*, 8, 593–599, 1999.

194. Oppenheim, A., et al., Efficient introduction of plasmid DNA into human hemopoietic cells by encapsidation in simian virus 40 pseudovirions, *Proc. Natl. Acad. Sci. USA*, 83, 6925–6929, 1986.

195. Oppenheim, A., Peleg, A., and Rachmilewitz, E. A., Efficient introduction and transient expression of exogenous genes in human hemopoietic cells, *Ann. N Y Acad. Sci.*, 511, 418–427, 1987.

196. Strayer, D. S., Gene delivery to human hematopoietic progenitor cells to address inherited defects in the erythroid cellular lineage, *J. Hematother. Stem Cell Res.*, 8, 573–574, 1999.

197. Strayer, D. S., Effective gene transfer using viral vectors based on SV40, *Methods Mol. Biol.*, 133, 61–74, 2000.

198. Strayer, D. S., et al., Use of SV40-based vectors to transduce foreign genes to normal human peripheral blood mononuclear cells, *Gene Ther.*, 4, 219–225, 1997.

199. Arad, U., et al., A new packaging cell line for SV40 vectors that eliminates the generation of T-antigen-positive, replication-competent recombinants, *Virology*, 304, 155–159, 2002.

200. Sandalon, Z., et al., In vitro assembly of SV40 virions and pseudovirions: Vector development for gene therapy, *Hum. Gene Ther.*, 8, 843–849, 1997.

201. Sandalon, Z. and Oppenheim, A., Self-assembly and protein–protein interactions between the SV40 capsid proteins produced in insect cells, *Virology*, 237, 414–421, 1997.

202. Sandalon, Z. and Oppenheim, A., Production of SV40 proteins in insect cells and in vitro packaging of virions and pseudovirions, *Methods Mol. Biol.*, 165, 119–128, 2001.

203. Wrobel, B., et al., Production and purification of SV40 major capsid protein (VP1) in *Escherichia coli* strains deficient for the GroELS chaperone machine, *J. Biotechnol.*, 84, 285–289, 2000.

204. Li, P. P., et al., Role of simian virus 40 Vp1 cysteines in virion infectivity, *J. Virol.*, 74, 11388–11393, 2000.

205. Kimchi-Sarfaty, C., et al., Efficient delivery of RNA interference effectors via in vitro-packaged SV40 pseudovirions, *Hum. Gene Ther.*, 16, 1110–1115, 2005.

206. Kimchi-Sarfaty, C., et al., Transduction of multiple cell types using improved conditions for gene delivery and expression of SV40 pseudovirions packaged in vitro, *BioTechniques*, 37, 270–275, 2004.

207. Strayer, D. S., et al., Efficient gene transfer to hematopoietic progenitor cells using SV40 derived vectors, *Gene Ther.*, 7, 886–895, 2000.

208. Kimchi-Sarfaty, C., et al., SV40 Pseudovirion gene delivery of a toxin to treat human adenocarcinomas in mice, *Cancer Gene Ther.*, 13, 648–657, 2006.

209. Vera, M., et al., Factors influencing the production of recombinant SV40 vectors, *Mol. Ther.*, 10, 780–791, 2004.

210. Mellon, P., et al., Identification of DNA sequences required for transcription of the human alpha 1-globin gene in a new SV40 host-vector system, *Cell*, 27, 279–288, 1981.

211. Strayer, D. S., SV40 as an effective gene transfer vector in vivo, *J. Biol. Chem.*, 271, 24741–24746, 1996.

212. Norkin, L. C. and Kuksin, D., The caveolae-mediated sv40 entry pathway bypasses the golgi complex en route to the endoplasmic reticulum, *Virology J.,* 2, 38, 2005.

213. Damm, E. M., et al., Clathrin- and caveolin-1-independent endocytosis: Entry of simian virus 40 into cells devoid of caveolae, *J. Cell Biol.,* 168, 477–488, 2005.

214. Kimchi-Sarfaty, C., Garfield, S., Alexander, N. S., Ali, S., Cruz, C., Chinnasamy, D., and Gottesman, M. M., The pathway of uptake of SV40 pseudovirions packaged in vitro: From MHC class I receptors to the nucleus, *Gene Ther. Mol. Biol.,* 8, 439–450, 2004.

215. Mannova, P. and Forstova, J., Mouse polyomavirus utilizes recycling endosomes for a traffic pathway independent of COPI vesicle transport, *J. Virol.,* 77, 1672–1681, 2003.

216. Kimchi-Sarfaty, C., et al., High cloning capacity of in vitro packaged SV40 vectors with no SV40 virus sequences, *Hum. Gene Ther.,* 14, 167–177, 2003.

217. Knipper, R., et al., Improved post-transcriptional processing of an MDR1 retrovirus elevates expression of multidrug resistance in primary human hematopoietic cells, *Gene Ther.,* 8, 239–246, 2001.

218. Agrawal, L., et al., Antioxidant enzyme gene delivery to protect from HIV-1 gp120-induced neuronal apoptosis, *Gene Ther.,* 13, 1645–1656, 2006.

219. Cordelier, P. and Strayer, D. S., Using gene delivery to protect HIV-susceptible CNS cells: Inhibiting HIV replication in microglia, *Virus Res.,* 118, 87–97, 2006.

220. Kieff, E. and Liebowitz, D., Epstein–Barr virus and its replication, in Fields, B.N., Knipe, D. M., (Eds.), *Virology,* 2nd Ed., Raven Press, New York, pp. 1889–1920, 1990.

221. Margolskee, R. F., Epstein–Barr virus based expression vectors, *Curr. Top. Microbiol. Immunol.,* 158, 67–95, 1992.

222. Sarasin, A., Shuttle vectors for studying mutagenesis in mammalian cells, *J. Photochem. Photobiol., B,* 3, 143–155, 1989.

223. Levitskaya, J., et al., Inhibition of antigen processing by the internal repeat region of the Epstein–Barr virus nuclear antigen-1, *Nature,* 375, 685–688, 1995.

224. Haase, S. B. and Calos, M. P., Replication control of autonomously replicating human sequences, *Nucleic Acids Res.,* 19, 5053–5058, 1991.

225. Sun, T. Q., Fernstermacher, D. A., and Vos, J. M., Human artificial episomal chromosomes for cloning large DNA fragments in human cells, *Nat. Genet.,* 8, 33–41, 1994.

226. Banerjee, S., Livanos, E., and Vos, J. M., Therapeutic gene delivery in human B-lymphoblastoid cells by engineered nontransforming infectious Epstein–Barr virus, *Nat. Med.,* 1, 1303–1308, 1995.

227. Lee, C., et al., Delivery systems for the MDR1 gene, in Boulikas, T., (Ed.), *Gene Therapy and Molecular Biology,* Gene Therapy Press, Athens, Greece, pp. 241–251, 1998.

228. DiMaio, D., Treisman, R., and Maniatis, T., Bovine papilltomavirus vector that propagates as a plasmid in both mouse and bacterial cells, *Proc. Natl. Acad. Sci. USA,* 79, 4030–4034, 1982.

229. Mecsas, J. and Sugden, B., Replication of plasmids derived from bovine papilloma virus type 1 and Epstein–Barr virus in cells in culture, *Annu. Rev. Cell Biol.,* 3, 87–108, 1987.

230. Grignani, F., et al., High-efficiency gene transfer and selection of human hematopoietic progenitor cells with a hybrid EBV/retroviral vector expressing the green fluorescence protein, *Cancer Res.,* 58, 14–19, 1998.

231. Kinsella, T. M. and Nolan, G. P., Episomal vectors rapidly and stably produce high-titer recombinant retrovirus, *Hum. Gene Ther.,* 7, 1405–1413, 1996.

232. Feng, M., et al., Stable in vivo gene transduction via a novel adenoviral/retroviral chimeric vector, *Nat. Biotechnol.,* 15, 866–870, 1997.

233. Wang, S., et al., A novel herpesvirus amplicon system for in vivo gene delivery, *Gene Ther.,* 4, 1132–1141, 1997.

234. Fraefel, C., et al., Gene transfer into hepatocytes mediated by helper virus-free HSV/AAV hybrid vectors, *Mol. Med.,* 3, 813–825, 1997.

235. Dzau, V. J., et al., Fusigenic viral liposome for gene therapy in cardiovascular diseases, *Proc. Natl. Acad. Sci. USA,* 93, 11421–11425, 1996.

236. Hirai, H., et al., Use of EBV-based vector/HVJ–liposome complex vector for targeted gene therapy of EBV-associated neoplasms, *Biochem. Biophys. Res. Commun.,* 241, 112–118, 1997.

237. Satoh, E., et al., Efficient gene transduction by Epstein–Barr-virus-based vectors coupled with cationic liposome and HVJ-liposome, *Biochem. Biophys. Res. Commun.,* 238, 795–799, 1997.

238. Krysan, P. J., Haase, S. B., and Calos, M. P., Isolation of human sequences that replicate autonomously in human cells, *Mol. Cell. Biol.,* 9, 1026–1033, 1989.

239. Krysan, P. J. and Calos, M. P., Epstein–Barr virus-based vectors that replicate in rodent cells, *Gene,* 136, 137–143, 1993.

240. Calos, M. P., The potential of extrachromosomal replicating vectors for gene therapy, *Trends Genet.,* 12, 463–466, 1996.

241. Jankelevich, S., et al., A nuclear matrix attachment region organizes the Epstein–Barr viral plasmid in Raji cells into a single DNA domain, *EMBO J.,* 11, 1165–1176, 1992.

242. Wohlgemuth, J. G., et al., Long-term gene expression from autonomously replicating vectors in mammalian cells, *Gene Ther.,* 3, 503–512, 1996.

243. DePamphilis, M., (Ed.), Eukaryotic replication origins, in *DNA Replication in Eukaryotic Cells,* Cold Spring Harbor Laboratory Press, Cold Spring Harbor, NY, pp. 983–1004, 1996.

244. Bloom, K., The centromere frontier: Kinetochore components, microtubule-based motility, and the CEN-value paradox, *Cell,* 73, 621–624, 1993.

245. Harrington, J. J., et al., Formation of de novo centromeres and construction of first-generation human artificial microchromosomes, *Nat. Genet.,* 15, 345–355, 1997.

246. Emerich, D. F. and Thanos, C. G., The pinpoint promise of nanoparticle-based drug delivery and molecular diagnosis, *Biomol. Eng.,* 23, 171–184, 2006.

247. Nie, S., et al., Nanotechnology applications in cancer, *Annu. Rev. Biomed. Eng.,* 9, 257–288, 2007.

248. Karmali, P. P. and Chaudhuri, A., Cationic liposomes as nonviral carriers of gene medicines: Resolved issues, open questions, and future promises, *Med. Res. Rev.,* 27, 696–722, 2006.

249. Rao, N. M. and Gopal, V., Cell biological and biophysical aspects of lipid-mediated gene delivery, *Bioscience Rep.,* 26, 301–324, 2006.

250. Zuhorn, I. S., Engberts, J. B., and Hoekstra, D., Gene delivery by cationic lipid vectors: Overcoming cellular barriers, *Eur. Biophys. J.,* 36, 349–362, 2007.

251. Burkoth, T. S., et al., Incorporation of unprotected heterocyclic side chains into peptoid oligomers via solid-phase submonomer synthesis, *J. Am. Chem. Soc.,* 125, 8841–8845, 2003.

252. Judge, A., et al., Hypersensitivity and loss of disease site targeting caused by antibody responses to PEGylated liposomes, *Mol. Ther.,* 13, 328–337, 2006.

253. Harris, J. M., Martin, N. E., and Modi, M., Pegylation: A novel process for modifying pharmacokinetics, *Clin. Pharmacokinet.,* 40, 539–551, 2001.

254. Goncalves, E., Debs, R. J., and Heath, T. D., The effect of liposome size on the final lipid/DNA ratio of cationic lipoplexes, *Biophys. J.*, 86, 1554–1563, 2004.

255. Bisht, S., et al., pDNA loaded calcium phosphate nanoparticles: Highly efficient non-viral vector for gene delivery, *Int. J. Pharm.*, 288, 157–168, 2005.

256. Aksentijevich, I., et al., In vitro and in vivo liposome-mediated gene transfer leads to human MDR1 expression in mouse bone marrow progenitor cells, *Hum. Gene Ther.*, 7, 1111–1122, 1996.

257. Philip, R., et al., Efficient and sustained gene expression in primary T lymphocytes and primary and cultured tumor cells mediated by adeno-associated virus plasmid DNA complexed to cationic liposomes, *Mol. Cell. Biol.*, 14, 2411–2418, 1994.

258. Philip, R., et al., In vivo gene delivery. Efficient transfection of T lymphocytes in adult mice, *J. Biol. Chem.*, 268, 16087–16090, 1993.

259. Uchegbu, I. F., Pharmaceutical nanotechnology: Polymeric vesicles for drug and gene delivery, *Expert Opin. Drug Del.*, 3, 629–640, 2006.

260. Perez, C., et al., Poly(lactic acid)–poly(ethylene glycol) nanoparticles as new carriers for the delivery of plasmid DNA, *J. Control. Release*, 75, 211–224, 2001.

261. Putnam, D., et al., Polyhistidine-PEG:DNA nanocomposites for gene delivery, *Biomaterials*, 24, 4425–4433, 2003.

262. Paleos, C. M., Tsiourvas, D., and Sideratou, Z., Molecular engineering of dendritic polymers and their application as drug and gene delivery systems, *Mol. Pharmaceut.*, 4, 169–188, 2007.

263. Maksimenko, A. V., et al., Optimisation of dendrimer-mediated gene transfer by anionic oligomers, *J. Gene Med.*, 5, 61–71, 2003.

264. Pelletier, J. and Sonenberg, N., Internal initiation of translation of eukaryotic mRNA directed by a sequence derived from poliovirus RNA, *Nature*, 334, 320–325, 1988.

265. Trono, D., et al., Translation in mammalian cells of a gene linked to the poliovirus 5′ noncoding region, *Science*, 241, 445–448, 1988.

266. Sachs, A. B., Cell cycle-dependent translation initiation: IRES elements prevail, *Cell*, 101, 243–245, 2000.

267. Mauro, V. P. and Edelman, G. M., The ribosome filter hypothesis, *Proc. Natl. Acad. Sci. USA*, 99, 12031–12036, 2002.

268. Chappell, S. A., Edelman, G. M., and Mauro, V. P., A 9-nt segment of a cellular mRNA can function as an internal ribosome entry site (IRES) and when present in linked multiple copies greatly enhances IRES activity, *Proc. Natl. Acad. Sci. USA*, 97, 1536–1541, 2000.

269. Wong, E. T., Ngoi, S. M., and Lee, C. G., Improved co-expression of multiple genes in vectors containing internal ribosome entry sites (IRESes) from human genes, *Gene Ther.*, 9, 337–344, 2002.

270. Sugimoto, Y., et al., Coexpression of a multidrug-resistance gene (MDR1) and herpes simplex virus thymidine kinase gene as part of a bicistronic messenger RNA in a retrovirus vector allows selective killing of MDR1-transduced cells, *Clin. Cancer Res.*, 1, 447–457, 1995.

271. Sugimoto, Y., et al., Coexpression of a multidrug resistance gene (MDR1) and herpes simplex virus thymidine kinase gene in a bicistronic retroviral vector Ha-MDR-IRES-TK allows selective killing of MDR1-transduced human tumors transplanted in nude mice, *Cancer Gene Ther.*, 4, 51–58, 1997.

272. Metz, M. Z., Best, D. M., and Kane, S. E., Harvey murine sarcoma virus/MDR1 retroviral vectors: Efficient virus production and foreign gene transduction using MDR1 as a selectable marker, *Virology*, 208, 634–643, 1995.

273. Baum, C., et al., Novel retroviral vectors for efficient expression of the multidrug resistance (mdr-1) gene in early hematopoietic cells, *J. Virol.*, 69, 7541–7547, 1995.

274. Eckert, H. G., et al., High-dose multidrug resistance in primary human hematopoietic progenitor cells transduced with optimized retroviral vectors, *Blood*, 88, 3407–3415, 1996.

275. Berger, F., et al., Efficient retrovirus-mediated transduction of primitive human peripheral blood progenitor cells in stroma-free suspension culture, *Gene Ther.*, 8, 687–696, 2001.

276. Schiedlmeier, B., et al., Quantitative assessment of retroviral transfer of the human multidrug resistance 1 gene to human mobilized peripheral blood progenitor cells engrafted in nonobese diabetic/severe combined immunodeficient mice, *Blood*, 95, 1237–1248, 2000.

277. Bunting, K. D., et al., Effects of retroviral-mediated MDR1 expression on hematopoietic stem cell self-renewal and differentiation in culture, *Ann. N.Y. Acad. Sci.*, 872, 125–140, discussion 140–121, 1999.

278. Mineishi, S., et al., Co-expression of the herpes simplex virus thymidine kinase gene potentiates methotrexate resistance conferred by transfer of a mutated dihydrofolate reductase gene, *Gene Ther.*, 4, 570–576, 1997.

279. Zhao, R. C., et al., Gene therapy for chronic myelogenous leukemia (CML): A retroviral vector that renders hematopoietic progenitors methotrexate-resistant and CML progenitors functionally normal and nontumorigenic in vivo, *Blood*, 90, 4687–4698, 1997.

280. Bruggemann, E. P., et al., Characterization of the azidopine and vinblastine binding site of P-glycoprotein, *J. Biol. Chem.*, 267, 21020–21026, 1992.

281. Greenberger, L. M., Major photoaffinity drug labeling sites for iodoaryl azidoprazosin in P-glycoprotein are within, or immediately C-terminal to, transmembrane domains 6 and 12, *J. Biol. Chem.*, 268, 11417–11425, 1993.

282. Greenberger, L. M., et al., Domain mapping of the photoaffinity drug-binding sites in P-glycoprotein encoded by mouse mdr1b, *J. Biol. Chem.*, 266, 20744–20751, 1991.

283. Morris, D. I., et al., Localization of the forskolin labeling sites to both halves of P-glycoprotein: Similarity of the sites labeled by forskolin and prazosin, *Mol. Pharmacol.*, 46, 329–337, 1994.

284. Zhang, X., Collins, K. I., and Greenberger, L. M., Functional evidence that transmembrane 12 and the loop between transmembrane 11 and 12 form part of the drug-binding domain in P-glycoprotein encoded by MDR1, *J. Biol. Chem.*, 270, 5441–5448, 1995.

285. Gottesman, M. M., Pastan, I., and Ambudkar, S. V., P-glycoprotein and multidrug resistance, *Curr. Opin. Genet. Dev.*, 6, 610–617, 1996.

286. Ambudkar, S. V., et al., Biochemical, cellular, and pharmacological aspects of the multidrug transporter, *Ann. Rev. Pharmacol. Toxicol.*, 39, 361–398, 1999.

287. Devine, S. E., Ling, V., and Melera, P. W., Amino acid substitutions in the sixth transmembrane domain of P-glycoprotein alter multidrug resistance, *Proc. Natl. Acad. Sci. USA*, 89, 4564–4568, 1992.

288. Ma, J. F., Grant, G., and Melera, P. W., Mutations in the sixth transmembrane domain of P-glycoprotein that alter the pattern of cross-resistance also alter sensitivity to cyclosporin A reversal, *Mol. Pharmacol.*, 51, 922–930, 1997.

289. Loo, T. W. and Clarke, D. M., Functional consequences of phenylalanine mutations in the predicted transmembrane domain of P-glycoprotein, *J. Biol. Chem.*, 268, 19965–19972, 1993.

290. Loo, T. W. and Clarke, D. M., Mutations to amino acids located in predicted transmembrane segment 6 (TM6) modulate the

activity and substrate specificity of human P-glycoprotein, *Biochemistry,* 33, 14049–14057, 1994.

291. Hanna, M., et al., Mutagenesis of transmembrane domain 11 of P-glycoprotein by alanine scanning, *Biochemistry,* 35, 3625–3635, 1996.

292. Choi, K. H., et al., An altered pattern of cross-resistance in multidrug-resistant human cells results from spontaneous mutations in the mdr1 (P-glycoprotein) gene, *Cell,* 53, 519–529, 1988.

293. Currier, S. J., et al., Identification of residues in the first cytoplasmic loop of P-glycoprotein involved in the function of chimeric human MDR1–MDR2 transporters, *J. Biol. Chem.,* 267, 25153–25159, 1992.

294. Kioka, N., et al., P-glycoprotein gene (MDR1) cDNA from human adrenal: Normal P-glycoprotein carries Gly185 with an altered pattern of multidrug resistance, *Biochem. Biophys. Res. Commun.,* 162, 224–231, 1989.

295. Safa, A. R., et al., Molecular basis of preferential resistance to colchicine in multidrug-resistant human cells conferred by Gly-185–Val-185 substitution in P-glycoprotein, *Proc. Natl. Acad. Sci. USA,* 87, 7225–7229, 1990.

296. Loo, T. W. and Clarke, D. M., Functional consequences of glycine mutations in the predicted cytoplasmic loops of P-glycoprotein, *J. Biol. Chem.,* 269, 7243–7248, 1994.

297. Chen, G., et al., Multidrug-resistant human sarcoma cells with a mutant P-glycoprotein, altered phenotype, and resistance to cyclosporins, *J. Biol. Chem.,* 272, 5974–5982, 1997.

298. Shoshani, T., et al., Analysis of random recombination between human MDR1 and mouse mdr1a cDNA in a pHaMDR-dihydrofolate reductase bicistronic expression system, *Mol. Pharmacol.,* 54, 623–630, 1998.

299. Ford, J. M., Prozialeck, W. C., and Hait, W. N., Structural features determining activity of phenothiazines and related drugs for inhibition of cell growth and reversal of multidrug resistance, *Mol. Pharmacol.,* 35, 105–115, 1989.

300. Yang, J. M., et al., Characteristics of P388/VMDRC.04, a simple, sensitive model for studying P-glycoprotein antagonists, *Cancer Res.,* 54, 730–737, 1994.

301. Dey, S., et al., A single amino acid residue contributes to distinct mechanisms of inhibition of the human multidrug transporter by stereoisomers of the dopamine receptor antagonist flupentixol, *Biochemistry,* 38, 6630–6639, 1999.

302. Hafkemeyer, P., et al., Contribution to substrate specificity and transport of nonconserved residues in transmembrane domain 12 of human P-glycoprotein, *Biochemistry,* 37, 16400–16409, 1998.

303. Hafkemeyer, P., et al., Chemoprotection of hematopoietic cells by a mutant P-glycoprotein resistant to a potent chemosensitizer of multidrug-resistant cancers, *Hum. Gene Ther.,* 11, 555–565, 2000.

304. Kimchi-Sarfaty, C., Gribar, J. J., and Gottesman, M. M., Functional characterization of coding polymorphisms in the human MDR1 gene using a vaccinia virus expression system, *Mol. Pharmacol.,* 62, 1–6, 2002.

305. Kimchi-Sarfaty, C., et al., A "silent" polymorphism in the MDR1 gene changes substrate specificity, *Science,* 315, 525–528, 2007.

306. Zhu, D., et al., Influence of single-nucleotide polymorphisms in the multidrug resistance-1 gene on the cellular export of nelfinavir and its clinical implication for highly active antiretroviral therapy, *Antiviral Ther.,* 9, 929–935, 2004.

307. Tan, E. K., et al., Effect of MDR1 haplotype on risk of Parkinson disease, *Arch. Neurol.,* 62, 460–464, 2005.

308. Goreva, O. B., et al., Possible prediction of the efficiency of chemotherapy in patients with lymphoproliferative diseases based on MDR1 gene G2677T and C3435T polymorphisms, *Bull. Exp. Biomed.,* 136, 183–185, 2003.

309. Yamauchi, A., Oishi, R., and Kataoka, Y., Tacrolimus-induced neurotoxicity and nephrotoxicity is ameliorated by administration in the dark phase in rats, *Cell. Mol. Neurobiol.,* 24, 695–704, 2004.

310. van den Heuvel-Eibrink, M. M., et al., MDR1 gene-related clonal selection and P-glycoprotein function and expression in relapsed or refractory acute myeloid leukemia, *Blood,* 97, 3605–3611, 2001.

27 Embryonic Stem Cells and Their Application in Regenerative Medicine and Tissue Engineering

Nicholas D. Evans and Julia M. Polak

CONTENTS

27.1 INTRODUCTION

Humans have been fascinated by the possibility of biological regeneration since ancient times. The ancient Greeks may have known something about it, as they named the liver, one of the few organs that has the capacity to regenerate, *hēpar* after the Greek word for "repairable." And this is evident, of course, in the famous myth of Prometheus, who was forever condemned to have his liver pecked out by an eagle, only for it to regenerate by the next day [1].

We can see evidence all around us to suggest that other organisms can undergo substantial regeneration; for example, if the limbs of amphibians such as salamanders and newts are amputated, they will simply grow new ones. We too can regenerate—most parts of our bodies can heal and regenerate to a certain degree, but not as spectacularly as we might hope. For this reason, natural philosophers, physicians, and more recently, scientists have tried to understand how the body heals, how they can encourage it to heal, and how we may devise new ways of making substitute body parts.

Transplantation is one of the more familiar ways to achieving this and there are sporadic reports of transplantation throughout history, some undoubtedly mythical, but others which are bona fide. For example, an Italian surgeon, Tagliacozzi, was performing autografts in the sixteenth century [2]. But it was not until just over 100 years ago that organ transplantation was pioneered by the French Nobel Prize winning surgeon, Alexis Carrel, who took lessons from a seamstress in order to perfect his work. Problems with rejection, however, meant that it was not until 1954 that the first successful organ transplant was performed [3]. In the interim, transplantation has become a routine and successful procedure for the replacement of many defective tissue and organs. Its success now has almost become a burden: there simply are not enough replacement body parts to go round.

In addition to transplantation, surgeons have investigated synthetic materials as potential replacement body parts. For example, a group in Brazil recently showed that the ancient Incas used gold plates to repair skull defects [4]. Many medical implants have been developed in more modern times, such as artificial heart valves and stents, cardiac pacemakers, and artificial joints. Such developments have brought a substantial improvement in the quality of life for many, but this success is tempered by the often limited lifespan of the implants.

So, to improve on these developments, a new area of research has emerged in recent years—tissue engineering—which seeks to "apply the principles of engineering and life sciences toward the development of biological substitutes for the repair or regeneration of tissue or organ functions" [5]. In other words, tissue engineers are faced with the challenge of making tissues and organs from scratch, using cells, scaffolds, and matrices as building materials. This approach brings modern scientists and engineers from many disciplines together. For example, material scientists must design novel, biocompatible scaffolds, on which biologists are faced with growing cells, and eventually tissues, that can function in the intended manner. Increasingly, biologists are investigating stem cells. These cells have the potential to "self-renew" or to differentiate into more than one cell type, like those cells that give rise new limbs in salamanders. The recent isolation of embryonic stem cells (ESCs) from early human embryos [6] has suggested tantalizingly that we may be able to make any cell type for therapeutic purposes since these cells ultimately develop naturally into all of the cells and tissues found in the adult.

Here we will review the need for cell therapy and tissue engineering, how cells, and in particular ESCs, can be used in cell therapy, and we will discuss some of the materials and scaffolds that may be combined with these cells to create replacement body parts.

27.2 WHY IS REGENERATIVE MEDICINE NEEDED?

Tissue engineering and cell therapy aims to provide therapies for any disease in which a tissue or organ is lost and so its remit covers a wide range of degenerative diseases. The need for tissue engineering strategies is becoming especially marked in the developed world, where an ageing population means that tissues and organs become worn out and need replacing to ensure quality of life. Similarly lifestyles, which increase the risk of other diseases such as heart disease and diabetes, are also contributing to burdens on health services, which increase the need for novel therapies.

27.2.1 NEURODEGENERATIVE DISEASES

Age is one of the foremost risk factors for neurodegenerative conditions. For example, more than 100,000 people in the United Kingdom suffer from Parkinson's disease, and this costs the National Health Service €1 billion annually [7]. This disease involves the loss of the cells of the substantia nigra of the brain, which secrete dopamine. Currently, this disease is treated with drugs that increase dopamine in the brain, but these drugs are not ideal, having often unpleasant side effects and benefits that decline with use [8]. The transplantation of cells from aborted fetuses has been used, but there are ethical questions concerning this procedure; effects are often variable and tissue is limited. Similarly, Huntington's disease, Alzheimer's disease, and stroke (the third highest killer in industrialized countries) [9] all result in significant neuronal death in the brain, which could be partially rectified by cell transplantation. And, of course, spinal cord injuries are particularly debilitating and are usually irreversible, causing lifelong paralysis. Axons do have the capacity to regenerate, but are prevented from doing so by scarring at the injury site [10]. Cell transplantation may help to improve the prognosis of people with these injuries.

Another degenerative neural disease that is increasing as our societies age is age-related macular degeneration (ARMD). It is estimated that 30 million people worldwide suffer from this condition and it is the leading cause of blindness in people over 60 in the United States [11]. This disease is caused by the death of the photoreceptors in the retina due to a decrease in blood supply, but there is interest in using cell transplantation to help correct these diseases, although research is at an early stage.

27.2.2 CARDIOVASCULAR DISEASE

Cardiovascular disease causes 40% of all deaths in the United States each year [12]. Many treatments, including pharmacological, lifestyle, and surgical, are available for coronary heart disease, but as yet there is no treatment other than transplantation that can cause repair of damaged heart muscle. The survival of patients aged 45 years and over following myocardial infarction was only around 60% in the United Kingdom when measured in 2005 [13], although the number is declining slightly. Alternatives to transplant are being considered, including xenotransplantation and artificial mechanical hearts [14], but new treatments that can repair or replace damaged tissue in the heart are desperately needed.

27.2.3 DIABETES

A report in 1997 estimated that there were over 120 million diabetic people worldwide and predicted that this would increase to 220 million before the year 2010 [15]. The increased incidence of diabetes is especially marked in the developing world, with obesity and decreased physical activity being high risk factors for the development of type 2 diabetes. Aside from the poorer health and quality of life that diabetic people suffer, the economic cost for society is huge. In most countries, more than 3% of the annual health budget is spent on diabetes, of which around 75% is accounted for by the associated long-term complications [16]. Studies in the United Kingdom forecast significant increases in these costs due to the aging population [17]. As the incidence of diabetes increases, so do the associated diseases that it causes. For example, end-stage renal disease therapy cost $25 billion in the United States in 2002. Cell therapies that replace pancreatic β cells in type 1 diabetes and augment their effect in type 2 diabetes would lighten the economical and social burden considerably.

27.2.4 MUSCULOSKELETAL DISEASE

Although bone has a remarkable capacity to heal itself, disease or injury often results in severe damage that must be corrected surgically. Autologous grafting is considered the "gold standard" treatment, but it is not ideal: these treatments can cause pain and tissue damage at the donor site [18]. Allogenic grafts are also used, but are from cadavers. This brings the risk of poor quality or necrotic tissue, which may carry infectious agents [19]. Artificial joint replacements have had great success in the treatment of arthritis and joint disease and injury over the past few decades, and currently nearly 500,000 knee and hip replacements take place in the United States each year [20]. But these implants have a limited lifespan, which is a particular problem in the treatment of young patients. New treatments that replace damaged cartilage and bone are urgently required.

27.2.5 OTHER DEGENERATIVE DISEASES

Lung and liver diseases are two other prominent diseases that may benefit from stem cells and tissue engineering. It is estimated that one in seven people in the United Kingdom are affected by lung disease, and for many chronic conditions, transplantation is the only eventual cure. Similarly, although the liver has a substantial capacity for self-repair, in many cases, such as in cirrhosis and hepatitis, liver tissue must be replaced to ensure survival.

27.3 CELL SOURCES

Cell therapies and tissue engineering, which aim to replace cells that are lost in the diseases described above, obviously require a source of transplantable cells. In the ideal situation, the relevant cells would be isolated from the patient, grown in culture to a volume sufficient to correct the disease, and implanted back into the patient. But in practice, these cells may be reduced or absent, difficult to isolate, or difficult to grow or maintain in culture. In addition, most somatic cells in an organism are terminally differentiated, which means that they cannot divide further. Others, such as fibroblasts, may become senescent, which makes the growth of sufficient number of cells with the correct phenotype difficult to achieve. For these reasons, and because many tissues like skin, liver, and bone have the capacity to heal and regenerate, researchers have speculated that stem cells or progenitor cells exist in the body that can contribute to new tissue formation after injury.

27.3.1 HEMATOPOIETIC STEM CELLS

Hematologists can probably lay claim to being the pioneers of stem cell research. During the 1950s and 1960s, cells were discovered in the bone marrow, which had the potential to generate a new immune system in mice in which the old one had been destroyed. These cellular entities were designated hematopoetic stem cells (HSCs) [21], and shortly later, Mathé et al. showed that transplants of these cells could replace a cancerous blood system in humans and hence cure leukemia [22]. The precise identity of these cells has been somewhat elusive, and there has been a great deal of debate on how they are actually identified. For instance, it is not possible to culture a pure population of HSCs. But there is now some consensus on the pattern of surface markers that these powerful stem cells express.

27.3.2 MESENCHYMAL STEM CELLS

As with HSCs, there does not appear to be a consensus in the literature on what precisely constitutes a mesenchymal stem cell (MSC). In most instances, the term "marrow stromal cell" (which conveniently has the same initials) is a more accurate term. They were first noticed by Friedenstein et al., who saw that rare cells (c0.001%) in bone marrow aspirates adhered to the culture substratum and formed colonies [23]. These cells were found to be highly proliferative and could be maintained in tissue culture for 20 to 40 passages before senescence. Like HSCs, these cells are multipotent and can differentiate readily into cells of the connective tissue such as osteoblasts, adipocytes, and chondrocytes [24] under the correct culture conditions. More recently, other groups [25,26] have shown that they can differentiate into many other lineages, not limited to mesoderm tissue. MSCs are intrinsically primary cell cultures, and are not clonal, raising the question that the observed differentiation might be because the initial cell population is mixed, containing cells at various stages of commitment. But a journal article published in 1999 [27] demonstrated, by isolating individual cells from MSC cultures, expanding them, and differentiating these new cell culture "strains" into three lineages, that at least some of these cells do have true multipotency. MSCs are hotly investigated cells in muscoskeletal tissue engineering since they can be isolated as autologous cells for tissue transplantation. But it is important to note that they do have limited proliferative potential (which decreases with age), and it might not be possible to expand enough cells for regenerating large tissue defects, as is often the case in this field of medicine.

27.3.3 TISSUE-SPECIFIC STEM CELLS

Stem cells are thought to exist in many other niches throughout the body apart from the bone marrow, but are perhaps best studied in the brain and the skin. Reynolds and Wiess [28] and Richards et al. [29] demonstrated in the 1990s that single cells from the brains of adult rodents could be grown in suspension culture as clusters of cells, which were named "neurospheres." Neural stem cells (NSCs), as they became to be known, could be isolated from these neurospheres and selectively differentiated into the three main cells of the brain—neurons, glial cells (astrocytes), and oligodendrocytes (see Ref. [30] for an excellent review). More recently, it has been shown that NSCs can also differentiate into other cells types, for instance blood [31], and when they are combined with early embryos, they can contribute to the formation of an organism, subsequently being found in tissue such as kidney, heart, and liver [32]. NSCs may have an important role to play in the treatment of many neural diseases. Transplantation of fetal cells, which are in scarce supply, is known to improve the condition of people with Parkinson's disease [33,34], and recently Redmond et al. have shown that undifferentiated NSCs can be effective in a primate model of Parkinson's disease [35].

Skin also has a population of stem or progenitor cells and the molecular processes that govern the renewal and differentiation of these cells are well studied (see Ref. [36] for a review). Stem cells are also thought to occur in niches in other tissues in the body such as liver [37] and kidney [38], which act to maintain the population of functional, terminally differentiated cells in these tissues. These cells are rather different from HSCs and MSCs, however, in that they exist in small numbers *in vivo*, but cannot be isolated or expanded in large numbers, so their potential in regenerative medicine and tissue engineering is not yet clear.

27.3.4 EMBRYONIC STEM CELLS

ESCs have the potential to grow indefinitely and can differentiate into all cell types of the adult. They have had and continue to have widespread public appeal, probably because of the controversial way in which they are generated, but more importantly because of the significant potential that they have in medicine. They are derived from the preimplantation blastocyst, a ball of around 50 cells, which develops several days after conception in humans. The blastocyst is a spherical fluid-filled structure, composed of a layer of outer cells called the trophectoderm that goes on to form the placenta, and a ball of cells attached to the inner wall of this structure called the inner cell mass that goes on to form the embryo proper. It is these latter cells from which ESCs are derived. These cells have the advantage over other cell types of being truly pluripotent, i.e., they have the potential to develop into all three "germ layers" of tissues found in the adult organism. It is this property, which suggests that we may be able to derive any cell type from them, that interests biologists. The germ layers comprise the ectoderm, which goes on to form the neural tissue and the skin; the mesoderm, which gives rise to the

bones, blood, muscles, and other connective tissues; and the endoderm, which gives rise to the cells that line the gut as well as the internal organs, including liver, lungs, and pancreas (Figure 27.1).

ESC research has its origins in the rather obscure study of tumors called teratomas. These morbidly fascinating tumors occur predominantly in the germline tissues and may be comprised of a wide range of misplaced tissues such as teeth and hair. In the most bizarre cases, even more advanced tissue may be present, such as in fetaform teratoma, where highly developed structures such as rudimentary limbs sometimes develop [39]. Several groups found that cells from teratomas, which were named embryonal carcinoma (EC) cells, could be isolated and grown indefinitely in culture [40]. When these cells were then reintroduced into adult mice, further teratomas were generated and, most intriguingly, when introduced into mouse blastocysts, these cells could survive, differentiate, and eventually contribute to some of the tissues found in the adult mouse. Because it seemed possible to grow these cells in their undifferentiated state *in vitro*, the scientists working on them reasoned that it may be possible to isolate and expand cells from the early mouse embryo. And in 1981, Evans and Kaufman at Cambridge [41], and Martin [42], who had previously worked with Evans at Cambridge, reported the derivation and extended culture of pluripotent cell lines from mouse blastocysts. Most tellingly, three years later, Bradley et al. [43] demonstrated that when these cells were injected into mouse blastocysts, they could contribute to all of the tissue types present in the adult, including the germline. This last point was particularly important since this new technology meant that it became theoretically possible to make a mouse, albeit in the second generation, which was wholly derived from *in vitro* cultured cells. And this was the advance that lead to transgenic mice: growing ESCs could be genetically manipulated *in vitro*, introduced into a mouse blastocyst, grown to adulthood, and mated to create heterozygous mice and eventually homozygous mice [44]. This approach has proved invaluable in many areas of research, particularly in identifying the functions of genes through "knockout" mice over the last 15–20 years.

There has also been a great deal of interest in the molecular mechanisms that keep a cell from differentiating. Initially, it was necessary to culture mouse ESCs either in a medium conditioned by EC cells or, like EC cells, on a cell layer derived from embryonic fibroblasts. The discovery of a molecule called leukemia inhibitory factor (LIF) by two groups in 1988 [45,46] paved the way for feeder-independent culture of ESCs, and this also led to further identification of the genes involved in maintenance of cells in the undifferentiated state, such as *Oct3/4* and *Nanog*. More recently, it has been found that ESCs can be grown in the absence of serum if bone morphogenic protein (BMP) is included in the growth medium [47]. Identification of the genes and molecular pathways involved in pluripotency led in 2006 to the genetic reprogramming of adult cells so that they are virtually indistinguishable from ESCs. In these studies, Yamanaka's group showed that the retroviral introduction of only four genes, *Oct3/4*, *Sox2*,

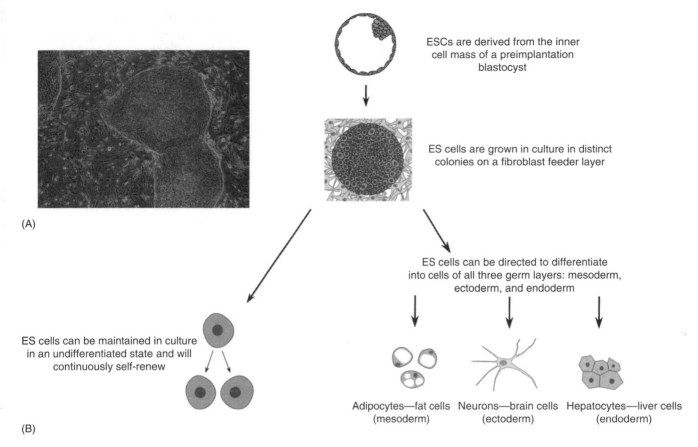

(A)

(B)

ESCs are derived from the inner cell mass of a preimplantation blastocyst

ES cells are grown in culture in distinct colonies on a fibroblast feeder layer

ES cells can be directed to differentiate into cells of all three germ layers: mesoderm, ectoderm, and endoderm

ES cells can be maintained in culture in an undifferentiated state and will continuously self-renew

Adipocytes—fat cells (mesoderm)   Neurons—brain cells (ectoderm)   Hepatocytes—liver cells (endoderm)

FIGURE 27.1   (A) Phase contrast microscope image of undifferentiated human ESCs (H1 cell line) grown on a layer of embryonic fibroblast feeder cells. (B) Schematic illustration of derivation, differentiation, and maintenance of ESCs. ESCs are able to self-renew or to differentiate into cells of all three mammalian germ layers.

c-Myc, and *Klf4* (albeit at multiple copy number), caused embryonic and adult fibroblasts to revert to an "induced pluripotent" state [48]. Initially, these cells could not form chimaeras in mice, but more recently this technique has been refined so that, remarkably, it now appears possible to make not only chimaeras, but also mice that are comprised wholly of cells that were, previously, ordinary fibroblasts growing in culture [49,50].

Probably due to the lack of available embryos, and also because of the intrinsic reticence of the cells themselves, it was not until 1998 that the first human ESCs were derived [6]. Like mouse ESCs, human ESCs can be grown in an undifferentiated state *in vitro* and, when implanted in an immunodeficient mouse, form teratomas comprised of all three germ layers. But there remain important differences between hESCs and mESCs. For instance, there are differences in the proteoglycan cell surface markers that they express and hESCs cannot be maintained in an undifferentiated state through the addition of LIF (interestingly LIF appears to be an evolutionary adaptation to allow mouse embryos to survive in the uterus when implantation is delayed by the presence of a still-suckling litter [51]). Instead hESCs can be maintained in an undifferentiated state with the addition of two other signaling molecules, activin and fibroblast growth factor (FGF) [52]. From a technical perspective, one of the most infuriating

aspects of hESC culture for scientist, or a business, working on them is that they propagate poorly as single cells and must be manually dissected rather than by trypsin addition. The reason for these differences has remained something of an enigma, though Brons et al. have recently shed some light on the matter [53]. They found that cells isolated from postimplantation mouse blastocysts (named "epiblast stem cells" [EpiSCs]) behaved similarly to hESCs, indicating that the mESCs that have been studied for the past 25 years may represent a developmentally earlier state than hESCs, and possibly that human development is more advanced at implantation. The study of how cells control differentiation is not only academically interesting, but also has great bearing on the real therapeutic applications that ESCs might have, the ability to differentiate them into functional, mature tissue for transplantation.

27.4   DIFFERENTIATION OF ESCs *IN VITRO*

As with derivation, the control of differentiation of ESCs owes much to work done in the 1970s on EC cells. Like with EC cells, differentiation can be induced by the culture of ESCs as cellular aggregates called "embryoid bodies" (EBs) in suspension on nonadherent substrata [54]. In this way, intercellular communication appears to spontaneously induce differentiation in a way analogous to that occurring in the embryo in its natural environment. Initially in mESCs, an outer layer of endoderm develops in the EB, followed by an interior rim of ectoderm [55]. Later, mesodermal cells also begin to differentiate and when EBs are plated onto tissue culture plastic, cells with the properties of many different cells, including cardiomyocytes, neurons, and hepatocytes, begin to appear. But unlike in embryos, the tissues that form in EBs are disorganized and no polarity is established. The main aim of differentiating ESCs *in vitro* is to reduce this complexity, or at least to be able to control it predictably so that cells of the required type can be grown and isolated for their intended use.

In vivo, the process of development from a single cell to a mature organism is, of course, an enormously complicated process and is much better understood in simple model organisms than in mammals. In the nematode worm, *Caenorhabditis elegans*, Sulston and Horvitz [56] used light microscopy to painstakingly map the fate of every cell that arose through cell division. In this way, they were able to generate a "family tree" of all the 959 specialized cells that arose, and to begin to analyze the precise spatial and chemical cues that were required to generate a particular type of cell. The process is certainly much more complex in the mammal. Delineating the precise developmental path of every single cell may well prove impossible. Biologists have, however, used a variety of techniques to preferentially generate cell types of interest. These include culturing cells in a medium designed for the target cell type; growing the cells on a specific extracellular matrix (ECM); coculturing ESCs with another cell type; and introducing transcription factor or selection genes into cells to either promote differentiation or to enable cell selection.

27.4.1 ECTODERM TISSUES

Arguably, differentiation of ESCs into cell types of the ectodermal lineage, skin, and nerve tissue has had the most success. Ectodermal differentiation appears, in fact, to be the default differentiation pathway of ESCs [57]. In 1995, several groups [58,59] found that neuronal cell types, including glial cells and neurons, could be generated from EBs. This differentiation was induced by retinoic acid (RA), previously known to be an important signaling molecule in the central nervous system (CNS) [60]. Serum removal [61,62], coculture with stromal cells [63], and addition of growth factors such as basic FGF and epidermal growth factor [64], as well as incubation

with neuronal survival factors [65], were also found to increase the number of neuronal cells derived from ESCs. Furthermore, the complexity and ill-defined nature of EB-based differentiation have been circumvented by the development of more controlled monolayer differentiation protocols [66]. A genetic approach was also used, where an antibiotic selection marker was introduced under the control of the neuroepithelial-specific *Sox2* promoter. In this way, all cells that did not express this gene could be eliminated in culture, resulting in reasonably pure populations of neuronal cell types.

In these studies, all three of the main neural cell types, neurons, astrocytes, and oligodendrocytes, were found in the expanded populations. More recent efforts have sought to more precisely control which of these cell types are isolated. Kim et al. [67] engineered an ESC line that over expressed *Nurr1* and showed that midbrain dopaminergic neurons, the type of cells that are lost in Parkinson's disease, could be selectively generated. And Wichterle et al. [68] used an ESC line with enhanced green fluorescence protein (EGFP) under the control of the *HB9* promoter to isolate cells with the characteristics of motor neurons.

Similar approaches have been used by Smith's group to find genes that are implicated in neuronal differentiation from ES cells. For instance, by producing a knock-in ESC line with GFP under the control of the *Sox1* promoter, and using it to screen many possible genes for neuroinduction, these authors demonstrated that inhibition of the Wnt signaling pathway could promote neural differentiation [69].

In parallel with these developments on mESCs, many groups have also shown that human ESCs have the potential to differentiate into neuronal cells and that hESC derived neural progenitors can be selectively differentiated into the three main neural cell types [70]. As with mESCs, coculture, combinations of growth factors, and ECM proteins can all increase NSC differentiation, and that, like with NSCs, human ESC-derived neural progenitors could be maintained in culture as neurospheres [71].

Other ectodermal cell types have also been generated in culture. Bagutti et al. [72] demonstrated that kerantinocytes could be generated from mESCs and later, Green et al. [73] have shown cytokeratin-expressing cells can be generated from human ESCs. In another study, Coraux et al. [74] have shown that, by culturing ESCs on a matrix derived from a skin cell line, a tissue with the complexity and organization of embryonic skin can be generated (Figure 27.2).

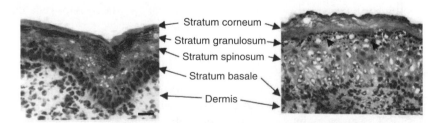

FIGURE 27.2 (See color insert following blank page 206.) Hematoxylin/eosin staining of sections of 17.5-day-old mouse embryo skin (left) and of tissue formed from ESCs grown on a matrix secreted by a skin cell line (right). (From Coraux, C., Hilmi, C., Rouleau, M., Spadafora, A., Hinnrasky, J., Ortonne, J.P., et al. *Curr. Biol.*, 13, 849, 2003. With permission.)

27.4.2 MESODERM TISSUES

Similar principles to induce differentiation have been used for cells of the mesoderm lineage, which include connective tissues such as cartilage and bone, as well as muscle and blood. For example, by initially aggregating ESCs as EBs, plating them, and then growing them in a medium containing β-glycerophosphate, ascorbate, and dexamethasone (usually used for the osteogenic differentiation of cell lines), Buttery et al. [75] and Bielby et al. [76] have demonstrated that ESCs form calcified deposits on the cell culture substratum called bone nodules, and preferentially express bone-specific genes (Figure 27.3). Others have investigated further the means by which this happens, finding that vitamin D [77] and compactin, [78] as well as BMPs [79,80] and coculture with a liver carcinoma cell line [81] can regulate bone formation. In similar ways, others have demonstrated formation of chondrocytes, adipocytes, and muscle.

There is a glut of papers in the literature on cardiac differentiation, probably because cell therapies for cardiac disease are considered to be urgently needed. Cardiomyocytes commonly develop in EBs in the presence of serum and can be easily identified by the fact that they beat. They have also been characterized and express many of the proteins found in adult cardiomyocytes, such as sarcomeric proteins [82], and they are similar electrophysiologially [82,83]. These cells have been isolated, like in neural cells, by modifying them to express either antibiotic resistance genes or fluorescent proteins under the transcription control of cardiac-specific promoters. For example, Klug et al. [84] engineered an ESC line to express an α-MHC-dependent neomycin gene, while Muller et al. [85] engineered GFP under the control of the myosin light chain 2v promoter. Such fluorescent cells could then be isolated by fluorescence-activated cell sorting (FACS). The factors that induce cardiomyogenesis have also been widely studied. BMP2 and FGF2 [86] and ascorbate [87] appear to promote cardiomyocyte differentiation. Mummery's group has shown that coculture with a visceral endoderm cell line, END-2, can promote cardiomyogenic differentiation [88,89]. As with mESCs, human ESCs also differentiate into cardiomyocytes under similar conditions, although these cells appear to have properties of fetal cells rather than adult cells [90].

Differentiation to cells of the blood system has also been well studied, mostly because ESCs offer an ideal model system to investigate development of the blood system, but also because it was suspected ESCs might offer an unlimited source of HSCs for transplants. Keller has written an extensive review on this subject [91]. As early as 1985, Doetschman et al. noticed that red hemoglobin-containing cells were evident in EBs [55]. Later, Keller and colleagues showed that under the correct culture conditions, the majority of ESCs could be induced to express Flk-1, a marker of vascular and hematopoetic cells [92]. Furthermore, specific cell types of the blood are evident in hematopoesis, such as erythrocytes, macrophages, and neutrophils. Despite this, however, the isolation of hematopoietic stem cells from ESCs (which have the critical property of ability to reconstitute the immune system of an irradiated animal) has yet to be achieved reproducibly (see Ref. [91] for an exhaustive review).

The early events that control mesoderm development are only beginning to be understood, but by investigating the effects of various growth factors in the absence of serum, researchers are starting to precisely define the series of steps that occur. It is especially important to determine precisely when and what changes in gene expression occur when differentiating ESCs become committed to each of the three germ layers. This process occurs at "gastrulation," a series of organizing events in which the specification and position of cells becomes determined. At this point in development, the cells that are committed to the mesoderm and the endoderm begin to express a characteristic series of genes. In an attempt to mark cells committed to the mesoderm lineage, Fehling et al. made an ESC line expressing GFP under the control of the *Brachyury* gene, thought to be one of the earliest markers of mesoderm [93]. They were then able to demonstrate that the hemangioblast, the putative precursor of both the blood cells and endothelial cells, could be derived from Brachyury [+] cells in the primitive streak. Surprisingly, these Brachyury [+] cells were also found to have the potential to form endoderm, with a greater capacity when activin was included in the differentiation media and when serum was omitted [94]. In fact, serum, which is an undefined mixture of growth factors and other chemicals, was found to inhibit endoderm formation, a point that illustrated the problems with using ill-defined factors in a supposedly reductionist approach of understanding a biological process. Subsequently, the Brachyury–GFP cell line has been further modified to include the CD4 surface receptor protein under the control of the *Foxa2* gene, which is thought to be an early endoderm marker [95]. This has allowed Gadue et al. to establish signaling pathways and positional differences

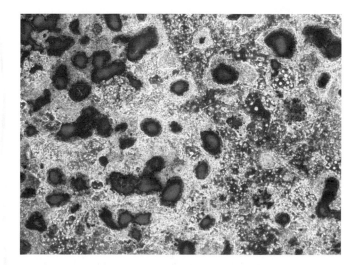

FIGURE 27.3 (See color insert following blank page 206.) Phase contrast and fluorescence composite image of murine ESCs grown for 21 days in the presence of osteogenic supplements (β-glycerophosphate, ascorbate, and dexamethasone). Bone nodules are indicated by areas that stain positively for alizarin red S, which fluoresces when bound to calcified ECM.

involved in the very early development of mesoderm and endoderm by isolating these cells by FACS.

27.4.3 Endoderm Tissues

The differentiation of ESCs to the endodermal lineage has mainly focused on generating hepatocytes and pancreatic β cells for therapies treating liver disease and type 1 diabetes, respectively. Again, these tissues were known to form in teratomas when ESCs were introduced into severe combined immunodeficient (SCID) mice, and therefore researchers used similar protocols to those used for mesoderm and ectoderm. An early protocol for generating pancreatic beta cells involved a complex five-step protocol in which ESCs were grown for four days as EBs, transferred to adherent culture in the absence of serum, then following the expression of nestin in these cells, FGF was added. Its removal several days later resulted in the onset on pancreatic differentiation [96]. But since this paper was published, doubt has been cast on the conclusions that were drawn; another group demonstrated that the insulin-positive cells in this experiment may, in fact, have been absorbed by the cells in culture [97]. Drug resistance genes inserted under the control of the insulin promoter have been used to select cells that have subsequently been shown to have some of the properties of pancreatic β cells, such as glucose-dependent insulin release [98]. Other genes have also been investigated: the transcription factors *Pax4* [99] or *Pdx1* [100] are necessary for the development of β cells. Overexpression of these genes has been investigated as a mechanism to create β cells from ESCs with some moderate success [101,102]. When Micallef et al. [103] attempted to select *Pdx1*+cells from cultures of ESCs using GFP as a marker, they found that these cells did indeed appear to be of the endoderm lineage, but did not express any other markers of pancreatic maturation. More recently, Jiang et al. [104] have employed a serum-free differentiation protocol to generate cultures of cells of which 15% express C-peptide, a β-cell-specific marker, which must be expressed endogenously and cannot be taken up from the growth medium.

Both liver and lung cells have also been shown to develop from ESCs. Hamazaki et al. [105] showed early and late hepatic markers in differentiating ESCs and subsequently the differentiation of liver cells to bile duct epithelial and oval cells has been demonstrated. Recently Soto-Gutierrez et al. [106] demonstrated that these ESC-derived liver cells could metabolize chemicals such as ammonia at a rate two-thirds that of primary liver cells. Alveolar cells of the lung have also been derived from ESCs by selecting cells that express GFP under the control of the *Spc* gene [107,108].

27.5 ESC THERAPIES

ESCs have several advantages that make them attractive as potential cellular therapies. As previously noted, they can divide indefinitely, so are potentially unlimited as a tissue source, and have been shown to differentiate into many different tissues. But like any transplanted cell source, ESCs would first need to be tissue-matched to a potential donor to avoid the problems of rejection. Because ESC cells are derived from embryos, this raised the possibility that cloned ESC could be derived by somatic cell nuclear transfer (SCNT). This involves introducing DNA from an adult cell into an ovum from which the host DNA has been removed and growing the embryo to blastocyst stage. ESCs can then be derived from the inner cell mass and, because the DNA is from an adult cell, these stem cell lines might be used to treat the person from whom the DNA was originally isolated. Hwang et al. reported the derivation of such lines in 2004 and 2005 [109,110] to worldwide media attention, but this work was later found to be fraudulent. This approach remains viable, however, but the likely costs and time involved in such a procedure are likely to be too prohibitive for SCNT to be used as routine treatments. Others have suggested that ESCs might be genetically engineered to reduce their immunogenicity [111] or ESC-derived white blood cells might be derived and implanted prior to a transplant to modulate any immune response [112]. More realistically, however, ESCs, like blood cells, might be banked and could then be selected based on their human leukocyte antigen (HLA) type. An interesting recent study by Taylor et al. [113] has shown that a bank of only 150 ESC lines, which is a realistic possibility in the light of the recent establishment of the U.K. stem cell bank, would provide an acceptable HLA match for around 80% of the population (Figure 27.4).

It is, however, something of an understatement to say that several safety issues and technical difficulties with ESCs need to be overcome before they can be used in a clinical setting. For instance, many researchers consider that the derivation and growth of ESCs in the presence of animal products, such as serum, feeder cells, and other reagents, is an obstacle to therapeutic use. For instance, one ESC line culture in serum was found to express a mouse-derived sialic acid [114]. This has been used as an argument to promote the further derivation of ESCs in a precisely defined, xeno-free environment [115]. Another problem with ESCs is their potential to form teratomas when implanted into animals. And, of course, any cell type derived from ESCs must be functional and able to do its job *in vivo*. For these reasons, there is a wealth of information in the literature on the implantation of ESC-derived cells into animal models of disease.

27.5.1 Cardiac Disease

Many studies have investigated the effect of bone marrow transplantation on heart repair in humans. For example, Bittira et al. [116] have shown that MSCs can contribute to heart repair by homing to the site of injury. Because of safety issues, to date there have been no studies investigating the effect of ESC transplantation for heart disease. But there have been several animal studies. In 1996, Klug et al. [84] were able to select ESC-derived cardiomyocytes by fusion of the cardiac-specific α-MHC gene with a drug resistance gene. These cells were then implanted into mice, were able to integrate into the host tissue, and were able to survive for at least 8 weeks. Since this study, other groups have investigated ES transplantation

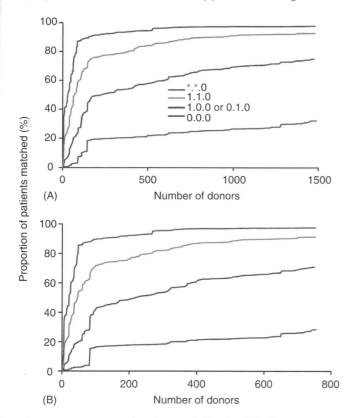

(A) Number of donors

(B) Number of donors

Proportion of patients matched (%)

FIGURE 27.4 (See color insert following blank page 206.) Graphs of estimates of how many ESC donors are required to provide different degrees of HLA matching in potential recipients. (0.0.0) indicates a zero HLA-A, HLA-B, and HLA-DR mismatch; (1.0.0 or 0.1.0) indicates a zero HLA-DR mismatch with no more than a single HLA-A or HLA-B mismatch; (1.1.0) indicates a zero HLA-DR mismatch with no more than a single HLA-A and a single HLA-B mismatch; and (**. 0) indicates a zero HLA-DR mismatch. (A) was obtained using data from 1500 consecutive cadaveric organ donors and (B) 760 consecutive blood-group-O cadaveric organ donors. (From Zhan, X., Dravid, G., Ye, Z., Hammond, H., Shamblott, M., Gearhart, J., et al., *Lancet*, 364, 163, 2004. With permission.)

and have shown improvements in cardiac function. While most studies have investigated the effect of implanting cells differentiated *in vitro*, others have looked at the implantation of undifferentiated cells [117]. The expected outcome of this treatment would be the formation of a teratoma, but remarkably in these studies, the host environment appears to direct the cells to a cardiac phenotype. These findings are contested by others who have shown teratoma formation under similar conditions [118,119]. There has been significantly less work published on the implantation of human ESCs, although recently Laflamme et al. reported that a combination of cardiomyocytes enriched from ESCs and a cocktail of "survival" factors could restore heart function and remuscularize myocardial infarcts in mice [120].

27.5.2 Type 1 Diabetes

There have been several examples of the implantation ESC-derived β-cell-like cells into animal models of diabetes, such as mice treated with streptazotocin. In a report in 2000,

Soria et al. reported the correction of hyperglycemia in mice implanted with such cells [98]. This is a commonly used method of testing the functionality of β cells produced by many groups working on differentiation strategies. There has been some controversy in this area of research; some researchers suggest that the insulin-producing cells obtained in such experiments may be more closely related to nerve cells than to endocrine cells [121]. The most recent study at the time of writing claimed that euglycemia was restored in 30% of streptozotocin-treated mice implanted with ESC-derived β cells [104]. Until the characteristics of the "pancreatic" tissue produced from ESCs are better defined, the application of ESCs for clinical correction of type 1 diabetes remains a distant prospect.

27.5.3 Musculoskeletal Disease

There are few reports of the use of ESC-derived cells to correct disease models of musculoskeletal disease, and it is not yet clear whether ESCs could give rise to functional bone or cartilage tissue. It is important to bear in mind that, more so than in many other stem cell–based therapies, it is the functional properties of the tissue as a whole rather than the individual cells, which is of primary importance. This means that stem cell research into the generation of bone and cartilage is linked intrinsically to tissue engineering. Most stem cell–based research in this field has been concentrated on MSCs, which can differentiate into both osteoblasts and chondrocytes, and various studies have shown that a combination of MSCs and a suitable scaffold can rectify bone injuries in animals (for example, Ref. [122]). But MSCs do have their disadvantages, including limited replication potential especially in the elderly, and so the importance of ESCs cannot be understated, even if their application is still some time away yet. A demonstration that ESCs could be used to correct an animal model of bone or cartilage disease would be a significant step forward in this field.

27.5.4 Neurodegenerative Disease

Early studies by Brustle et al. [123,124] demonstrated that ESC-derived neural cells could form all three brain cell types in rat brains and that oligodendrocytes could form myelin sheaths around host neurons in an animal model of multiple sclerosis. In another study, Kim et al. [67] found that ESCs overexpressing the transcription factor *Nurr1* could cause a significant improvement in the condition of a rat model of Parkinson's disease. Several other studies have had similar findings, although there are still risks of tumor formation that should be taken into account before such transplants can be considered in people. More recently, Bühnemann et al. [125] have shown that ESC-derived neural cells can repopulate previously ischemic areas of rat brain (a model of stroke), with cell survival for 12 weeks and differentiation into cells expressing many neural markers, while in a similar study Yanagisawa et al. [126] have shown an improvement in the behavior of recipient rats. There have also been studies investigating the implantation of ESCs for motor neurone disease and spinal cord injury. Lee et al. [127] showed that ESC-derived neurons could engraft successfully in chick embryos, and most importantly, project axons into peripheral host tissues a significant distance away from the area of engraftment. In a rat model of spinal injury, Deshpande et al. [128] showed that ESCs, engrafted at the site of damage, could send out axonal projections to muscle, partially restoring movement in these animals.

27.5.5 OTHER DISEASES

ESC-derived hepatocytes have also been shown to be able to contribute to host tissue following implantation into a host animal. Wang et al. [129] isolated fluorescent cells from cultures of ESCs expressing GFP under the control of the albumin gene promoter and showed engraftment in the injured liver, while Moriya et al. [130] showed an improvement in liver fibers in ESC-treated animals. In another study, an implanted bioartificial liver containing ESC-derived hepatocytes was shown to improve whole-animal liver function following a 90% hepatectomy [106]. Recently, Lane et al. [131] have injected pneumatocyte-enriched ESCs systemically into mouse models of lung disease and shown that these cells selectively engraft in the lungs when compared to other organs. This suggests that ESCs might be used as a cell source for such diseases as cystic fibrosis.

27.6 ESCs FOR SCREENING

ESCs are not only important for their application in cell therapy and tissue engineering, but also as a source of cells for testing for the toxicity of compounds produced in the pharmaceutical industry. Currently new drugs must be screened either using animal models that are expensive, which the governments in Europe are trying to limit for ethical and political reasons, or using human primary cell lines. Of particular interest are cardiac and liver cells lines; two of the main causes of failure in preclinical drug trials are from cardio- and hepatotoxicity. But these cells lines are often limited because of a lack of healthy donors and may have variations from batch to batch, which limit the accuracy of such tests. ESCs may be useful in such circumstances as they are unlimited in supply and culture protocols and testing procedures might be standardized (see Ref. [132] for a review). Several companies are focusing on deriving and differentiating human ESCs under defined conditions for toxicity testing; for example, Cellartis in Sweden is planning to sell cardiomyoctyes for testing while Geron in the United States has produced hepatocytes. Another advantage that ESCs have is that they might intrinsically carry, or be made to carry, specific genetic disease. For example, Pickering et al. have derived an ESC line that carries a genetic lesion associated with cystic fibrosis [133]. Although technically challenging currently, there is no reason to suggest that in time, it will become easier to genetically engineer human ESCs, making it possible to study the effect of drugs on

many genetic lesions, which cause disease in people. Hwang's studies in the mid-2000s initially seemed to show it was possible to create ESCs that were clones of adult donors. This work was, of course, subsequently shown to be faked, but in theory, this technique remains possible. Experimentation on SCNT is limited by the supply of human eggs, so to circumvent this problem researchers in the United Kingdom asked for permission to create hybrid ESC lines, i.e., using animal eggs rather than human eggs. After some initial controversy and political posturing [134], this license was recently granted [135]. If successful, this may mean that ESC lines will be produced that carry a whole range of human diseases and that it will be possible to model the effect of drugs on these diseases.

27.7 SCAFFOLDS

In many cases, ESC therapies might be delivered by cell transplantation alone. For example, ESC-derived β cells or dopaminergic neurons could potentially be directly injected into a recipient for the treatment of type 1 diabetes or Parkinson's disease, respectively. Cell therapy may also benefit a host of other diseases where injected cells might engraft and improve function in damaged tissue such as heart, liver, and lung disease. Often the area of degeneration may require the transplant of a significant volume of tissue, including the diseases mentioned above, but particularly in musculoskeletal disease. In these cases, it may be necessary to generate a significant volume of ESC-derived tissue prior to implantation. Tissue engineering emerged as a discipline in the 1980s [5] in response to this need and is focused on developing implants that combine materials with a source of cells. Skin tissue engineering is an example of where this approach may provide a simple, reliable therapy, and it is today available commercially as a treatment for burns and ulcers (Dermagraft, Apligraft). In these examples, human fibroblasts, derived from neonate foreskin, are expanded *in vitro* and seeded onto a collagenous matrix, a simple cell "scaffold," which is then implanted in the recipient. Luckily, keratinocytes and fibroblasts lack MHC antigens and are therefore immunoprivileged, so rejection is minimized. Similarly, Atala et al. [136] recently generated functional, artificial bladders. In this case, biopsies of the patients' bladders were taken, cells were expanded on bladder-shaped collagen scaffolds, and the resulting constructs were implanted successfully back into the patients as autologous transplants.

Skin and bladder tissue engineering have the advantage of requiring the growth of relatively thin sheets of cells, which can be easily grown in conventional *in vitro* culture systems. But many applications may require replacement tissue that is much thicker and more complex, and so tissue engineers have begun to devise many ways of achieving this.

27.7.1 CONVENTIONAL SCAFFOLDS

Scaffolds must provide a suitable three-dimensional environment in which tissue can grow and develop. Historically, biomedical implants were designed to be bioinert, but more recently for tissue engineering purposes, scaffolds are being designed that actively contribute to the development of a tissue, the so-called "bioactive" scaffolds. In some cases, scaffolds may even carry no cellular component; the structure and composition of the scaffolds are designed to be sufficient to encourage cellular in-growth and the formation of new tissue *in vivo* [137].

In cases where cells are added to scaffolds and grown before implantation, it is critical that cells throughout the depth of the scaffold have access to oxygen and nutrients. For this reason, scaffolds have generally been designed to incorporate a porous or permeable structure. Also the scaffolds are usually chosen so that they do not provoke a morbid immune response and provide a physical environment approaching that of the tissue to be replaced. Many materials can be chosen and manipulated in this way. For example, polyesters such as polyglycolide can be foamed up and hardened, or can be molded around grains of salt that can subsequently be dissolved out in aqueous solution, leaving a connected porous network [138]. Many consider these polymers to have the advantage of biodegradability; these materials were originally used in dissolving sutures and often dissolve to yield products such as lactic acid that can be metabolized in the body. Perhaps because of these qualities, there is a vast literature on polymer scaffolds. But these polymers are often not very hydrophilic and do not encourage cell attachment, spreading, and infiltration [139,140], but they can be modified by protein coating, for example, with fibronectin [141,142] or combination with other materials, such as collagen [143]. Such scaffolds have been considered for many applications, including bone, cartilage, liver, and lung.

Because bone has a high mineral content, inorganic materials such as calcium phosphates and hydroxyapatite-based materials have been considered for the tissue engineering of bone. Such materials can be ground up to form small particles, which can then be sintered, i.e., heated and partially melted so that each particle melts slightly into the next. In this way, porous networks are formed [144]. These scaffolds can then be seeded with cells and potentially implanted as scaffolds. Such scaffolds usually are brittle but have high mechanical strength when combined with polymers or natural organic material [145] and so could be beneficial in the tissue engineering of load-bearing bones. Bioactive glasses, composed of various combinations of CaO, P_2O_5, and SiO_2, have also been considered for bone tissue engineering. Like calcium phosphate–based materials, they can be sintered [144], or alternatively can be foamed with a surfactant to form sol gels [146]. It is also well-established that bioactive glass enhances osteogenic activity in several types of cells and can accelerate bone healing.

Artificial and natural hydrogels show great promise in soft tissue engineering. The advantage of this approach is that cells can be mixed in a solution, which can subsequently be gelled and molded without killing the cells. For example, cells can be mixed with polyethylene glycol (PEG) and a crosslinker agent and UV light can be used to solidify the gel [147]. Similarly alginate, a polymer of guluronic and manuronic

acids, which can be isolated from seaweed, can be hardened by the addition of divalent, nontoxic ions such as Ca^{2+} [148]. Like PEG, this material has been considered as an encapsulant in cell therapy for diabetes and has also been used as a space filler to allow in situ formation of tissue. Stevens et al. injected alginate into the periosteum of bone and observed de novo bone formation [137]. Collagen also has been used in this fashion; gelation of collagen can be achieved by increasing pH. And there are a host of studies investigating collagen gels in many different types of tissue engineering, but particularly in bone, cartilage, and tendon.

One of the main disadvantages with porous scaffolds is that it will prove very difficult to make thick tissues as in an *in vitro* setting, there is no blood supply to ensure the delivery of nutrients to the center of a construct and the removal of waste. As previously mentioned, this problem has largely been tackled by developing porous scaffolds. But pore size is generally a trade-off between maintaining a three-dimensional tissue structure and ensuring nutrient diffusion. That is, if pore size is too great, it is arguable whether cells experience a three-dimensional environment; they merely grow on the curved surfaces of pores in the same way that cells grow on tissue culture plastic. If pore size is too small, on the other hand, nutrients cannot diffuse into the center of the construct. To increase nutrient diffusion, bioreactors or spinner flasks have been used in which either the medium surrounding a construct is forced to flow or the medium is forcibly pumped through the scaffold. Another problem with such scaffolds is that because of the way they are constructed as bulk objects, their precise structure cannot be controlled and so they are poor organizers of tissue growth. For these reasons, many groups are investigating novel scaffolds that can be patterned at a microscale resolution to recreate vascular systems and tissue structure, and at the molecular level to direct cell differentiation and function.

27.7.2 MICRO AND NANO SCAFFOLDS

Photolithographic techniques, introduced from the microelectronics industry, have recently been applied to tissue engineering to allow a form of scaffold patterning at a micrometer scale. Here, raised relief patterns can be produced on flat photosensitive silicon wafers called photoresists by selectively exposing areas to UV light using a mask. Patterned masks of any sort can be made easily on computer drawing programs. A curable polymer, such as polydimethylsiloxane (PDMS), can then be poured over the raised relief pattern, allowed to harden, and then removed. This technique has been employed to great success in the field of microfluidics, where the PDMS is patterned with channels, adhered to glass slides before allowing liquid to flow through the channels (see Ref. [149] for a review). But recently, several groups have realized that other curable material that are more biocompatible, such as collagen, alginate, or PEG, can be first combined with cells, poured over the silicon (or PDMS) molds, and then cured, creating porous, patterned scaffolds (Figure 27.5) [150,151]. In this way, some groups have simulated vascular-like structure, where a central, wide inlet "artery" branches into many

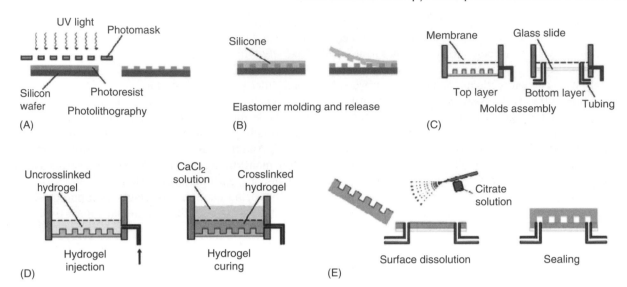

FIGURE 27.5 (See color insert following blank page 206.)   Process by which patterned hydrogel scaffolds can be constructed. Silicon photoresists are first patterned by exposure to UV light (A). A curable gel, such as silicone can then be molded over the photoresist (B) and used as a mold to create structure in a biocompatible hydrogel such as alginate that may contain cells (C and D). Channels in the hydrogel can then be perfused with a liquid such as growth medium (E). (From Cabodi, M., Choi, N.W., Gleghorn, J.P., Lee, C.S.D., Bonassar, L.J., and Stroock, A.D., *J. Am. Chem. Soc.*, 127, 13788, 2005. With permission.)

"capillaries" before recombining into another wide "vein" (Figure 27.6) [152,153]. Such scaffolds are advantageous because they can supply cells deeply embedded within a scaffold with nutrients in an organized manner, although it is unclear whether these channels are structurally durable and whether or not there would be an even delivery of nutrients to all parts of the scaffold.

Curable polymers can also be patterned without the need for a mold. In these cases, a layer of cell-containing PEG is exposed to UV light in a pattern determined by a mask, the uncured areas of polymer are removed, and then another layer of cell–polymer mixture can be added above it. In this way, different shapes, possibly containing different cell types, can be patterned to simulate tissue architecture [154].

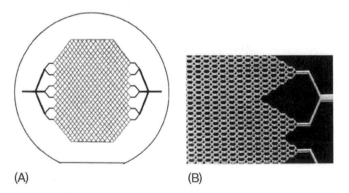

FIGURE 27.6 Patterns can be produced easily on photoresists to simulate vascular architecture. (A) Illustrates a macroscopic view of such a construct and (B) shows a magnified image of the branching structure. (From Kaihara, S., Borenstein, J., Koka, R., Lalan, S., Ochoa, E.R., Ravens, M., et al., *Tissue Eng.*, 6, 105, 2000. With permission.)

Aside from compartmentalizing cells in this manner, others are investigating scaffolds that are textured or patterned at a nanometer scale. In their natural environment, cells are usually surrounded on all sides by other cells and proteins of the ECM; they live in a truly three-dimensional environment. But in scaffolds with large pores, cells may spread out with growth medium on one side and the material of the scaffold on the other. To try and create a more physiological environment, several novel types of scaffold are under investigation. In one approach, materials, which can be an artificial polymer or a more natural material such as collagen, are extruded rapidly through a fine nozzle under the influence of a high electric field to generate very fine threads of material from around 5 nm to 5 μm in cross-section diameter [155]. Meshes of this material, which resembles cotton wool, can then be seeded with cells and maintained in culture. Cell-binding ligands, such as the triamino acid motif Arginine–Glycine Aspartate (RGD), can be attached to such materials to encourage cell binding and spreading the three dimensions. Another ingenious scaffold technology under investigation includes so-called amphiphilic self-assembling peptides [156]. These molecules function using much the same principle as soap. Each molecule has a highly hydrophobic tail and a hydrophilic amino acid head. In aqueous solution, the hydrophobic tails attract one another and remain hidden; the hydrophilic heads protrude out into the aqueous solution (Figure 27.7). Because of the structure of the center of the molecules, rather than forming spherical micelles, these peptides form long cylindrical fibers, which when viewed macroscopically, resemble a sticky gel. A particular advantage of these peptides is that they can be added to

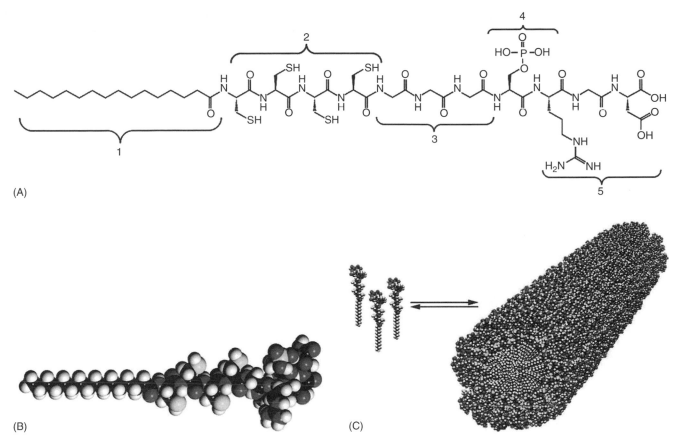

FIGURE 27.7 (See color insert following blank page 206.) Amphiphilic molecules can be designed with a hydrophilic peptide "head" (5) and a hydrophobic tail (1) (A). These peptides self-assemble to form thread-like nanofibres (B and C). (From Hartgerink, J.D., Beniash, E., and Stupp, S.I., *Science*, 294, 1684, 2001. With permission.)

a solution that contains cells so that the fiber network forms around the cells rather than vice versa, and that specific cell-recognition peptides can be introduced as the fiber head so that cell function and differentiation can be chemically controlled. The ability of a scaffold to directly control the differentiation of cells and the development of tissue is a particularly important factor when considering stem cells in tissue engineering.

27.8 SCAFFOLDS AND STEM CELLS

In the case of stem cells, it is particularly important that the scaffold should play some role in the differentiation of the cells that grow on and around it and that tissue should form in an organized manner. Perhaps because of the reason that stem cell biologists have concentrated on finding ways to make a specific type of cell from ESCs, there have been surprisingly few studies on the differentiation of ESCs on scaffolds and how these cells differ in three dimensions. Recently, however, there have been a few interesting examples. Levenberg et al. [157,158] have shown that the tissues and cells that form when ESCs differentiate in an ECM-based scaffold, made of Matrigel and polyester, express much higher levels of proteins associated with differentiation than those

differentiated on two-dimensional surfaces. Another group has shown that the chondrogenic differentiation of ESCs cultured in a PEG hydrogel is dependent upon whether or not the cells are exposed to RGD-binding ligands [159]. Gerecht et al. have also investigated the growth of ESCs in three dimensions using either hyaluronic acid or a photocurable hydrogel made of poly(glycerolco-sebacate)-acrylate [160,161].

In future, it will probably be beneficial to investigate the effect of more bioactive scaffolds on ESC differentiation in three dimensions. For example, several groups have begun to investigate bioactive scaffolds on other types of stem cell. Kim et al. [162] created a polyester-based scaffold that could slowly secrete ascorbate and β-glycerophosphate. This could enhance the differentiation of MSCs to osteoblasts. Another group has used a similar principle to make a scaffold that slowly exudes BMP, a growth factor that promotes osetogenic differentiation in MSCs [163]. Interestingly, two groups have recently begun to investigate amphiphilic peptide scaffolds to stem cell differentiation, Silva et al. introduced a laminin-specific peptide at the hydrophilic end of the amphiphiles and demonstrated an increased neural differentiation of cells culture in three dimensions in this scaffold compared with if it was used as a substrate [164]. Such scaffolds may have exciting applications in the study of ESC differentiation in the future.

27.9 CONCLUSIONS AND FUTURE DIRECTIONS

ESCs present us with a possible cure to many diseases in which tissue is lost. Studies on the fundamental processes that govern cellular differentiation, as well as the development of new techniques, which will allow us to produce functional differentiated cells, are vital if we are to produce therapies from these versatile cells. Similarly, if we wish to engineer new tissues for implantation, an understanding of how cells respond to a three-dimensional environment will be necessary. New biomaterials and cell culture techniques are vital to develop this understanding, and in the past two decades, material scientists and cell biologists have begin to collaborate. We are only beginning to see the benefits of this new field of research, and in the future, we may be encouraged that the construction of three-dimensional, intricate replicas of human tissue in the laboratory may become a reality.

REFERENCES

1. Schneider MD. Regenerative medicine: Prometheus unbound. *Nature* 2004, 432: 451–453.
2. Santoni-Rugiu P and Mazzola R. Leonardo Fioravanti (1517–1588): A barber-surgeon who influenced the development of reconstructive surgery. *Plast Reconstr Surg* 1997, 99: 570–575.
3. McAlister VC. Clinical kidney transplantation: A 50th anniversary review of the first reported series. *Am J Surg* 2005, 190: 485–488.
4. Marino R, Jr. and Gonzales-Portillo M. Preconquest Peruvian neurosurgeons: A study of Inca and pre-Columbian trephination and the art of medicine in ancient Peru. *Neurosurgery* 2000, 47: 940–950.
5. Langer R and Vacanti JP. Tissue engineering. *Science* 1993, 260: 920–926.
6. Thomson JA, Itskovitz-Eldor J, Shapiro SS, Waknitz MA, Swiergiel JJ, Marshall VS et al. Embryonic stem cell lines derived from human blastocysts. *Science* 1998, 282: 1145–1147.
7. Findley L, Aujla M, Bain PG, Baker M, Beech C, Bowman C et al. Direct economic impact of Parkinson's disease: A research survey in the United Kingdom. *Mov Disord* 2003, 18: 1139–1145.
8. Lindvall O, Kokaia Z, and Martinez-Serrano A. Stem cell therapy for human neurodegenerative disorders—how to make it work. *Nat Med* 2004, 10 Suppl: S42–S50.
9. Bacigaluppi M, Pluchino S, Martino G, Kilic E, and Hermann DM. Neural stem/precursor cells for the treatment of ischemic stroke. *J Neurol Sci* 2007, 265: 73–77.
10. Yiu G and He Z. Glial inhibition of CNS axon regeneration. *Nat Rev Neurosci* 2006, 7: 617–627.
11. Lund RD, Wang S, Klimanskaya I, Holmes T, Ramos-Kelsey R, Lu B et al. Human embryonic stem cell-derived cells rescue visual function in dystrophic RCS rats. *Cloning Stem Cells* 2006, 8: 189–199.
12. Rosamond W, Flegal K, Friday G, Furie K, Go A, Greenlund K et al. Heart disease and stroke statistics—2007 update: A report from the American Heart Association Statistics Committee and Stroke Statistics Subcommittee. *Circulation* 2007, 115: e69–e171.
13. Department of Health. Measurement of healthcare output and productivity; Survival rates for patients admitted to hospital with myocardial infarction. http://www.dh.gov.uk/prod_consum_dh/idcplg?IdcService=GET_FILE&dID=16183&Rendition=Web 5886[5]. Department of Health, UK (accessed September 1, 2007).
14. Keon WJ. Heart transplantation in perspective. *J Card Surg* 1999, 14: 147–151.
15. Amos AF, McCarty DJ, and Zimmet P. The rising global burden of diabetes and its complications: Estimates and projections to the year 2010. *Diabet Med* 1997, 14 Suppl 5: S1–S85.
16. Jonsson B, Henriksson UFJ, and Lindgren P. The social cost of diabetes mellitus and the cost-effectiveness of interventions. In *Textbook of Diabetes*, Pickup JC and Williams G (eds). Oxford: Blackwell Science, 2003.
17. Currie CJ, Kraus D, Morgan CL, Gill L, Stott NC, and Peters JR. NHS acute sector expenditure for diabetes: The present, future, and excess in-patient cost of care. *Diabet Med* 1997, 14: 686–692.
18. Arrington ED, Smith WJ, Chambers HG, Bucknell AL, and Davino NA. Complications of iliac crest bone graft harvesting. *Clin Orthop Relat Res* 1996, 329: 300–309.
19. Kainer MA, Linden JV, Whaley DN, Holmes HT, Jarvis WR, Jernigan DB et al. Clostridium infections associated with musculoskeletal-tissue allografts. *N Engl J Med* 2004, 350: 2564–2571.
20. Vats A, Bielby RC, Tolley NS, Nerem R, and Polak JM. Stem cells. *Lancet* 2005, 366: 592–602.
21. Ford CE, Hamerton JL, Barnes DW, and Loutit JF. Cytological identification of radiation-chimaeras. *Nature* 1956, 177: 452–454.
22. Mathé G, Amiel JL, Schwarzenberg L, Cattan A, and Schneider M. Haematopoietic chimera in man after allogenic (homologous) bone-marrow transplantation. (Control of the secondary syndrome. Specific tolerance due to the chimerism). *Br Med J* 1963, 5373: 1633–1635.
23. Friedenstein AJ, Chailakhjan RK, and Lalykina KS. The development of fibroblast colonies in monolayer cultures of guinea-pig bone marrow and spleen cells. *Cell Tissue Kinet* 1970, 3: 393–403.
24. Haynesworth SE, Goshima J, Goldberg VM, and Caplan AI. Characterization of cells with osteogenic potential from human marrow. *Bone* 1992, 13: 81–88.
25. Hofstetter CP, Schwarz EJ, Hess D, Widenfalk J, El MA, Prockop DJ et al. Marrow stromal cells form guiding strands in the injured spinal cord and promote recovery. *Proc Natl Acad Sci USA* 2002, 99: 2199–2204.
26. Liechty KW, MacKenzie TC, Shaaban AF, Radu A, Moseley AM, Deans R et al. Human mesenchymal stem cells engraft and demonstrate site-specific differentiation after in utero transplantation in sheep. *Nat Med* 2000, 6: 1282–1286.
27. Pittenger MF, Mackay AM, Beck SC, Jaiswal RK, Douglas R, Mosca JD et al. Multilineage potential of adult human mesenchymal stem cells. *Science* 1999, 284: 143–147.
28. Reynolds BA and Weiss S. Generation of neurons and astrocytes from isolated cells of the adult mammalian central nervous system. *Science* 1992, 255: 1707–1710.
29. Richards LJ, Kilpatrick TJ, and Bartlett PF. De novo generation of neuronal cells from the adult mouse brain. *Proc Natl Acad Sci USA* 1992, 89: 8591–8595.
30. Kornblum HI. Introduction to neural stem cells. *Stroke* 2007, 38: 810–816.
31. Bjornson CR, Rietze RL, Reynolds BA, Magli MC, and Vescovi AL. Turning brain into blood: A hematopoietic fate adopted by adult neural stem cells in vivo. *Science* 1999, 283: 534–537.

32. Clarke DL, Johansson CB, Wilbertz J, Veress B, Nilsson E, Karlstrom H et al. Generalized potential of adult neural stem cells. *Science* 2000, 288: 1660–1663.

33. Leker RR and McKay RD. Using endogenous neural stem cells to enhance recovery from ischemic brain injury. *Curr Neurovasc Res* 2004, 1: 421–427.

34. Lindvall O, Sawle G, Widner H, Rothwell JC, Bjorklund A, Brooks D et al. Evidence for long-term survival and function of dopaminergic grafts in progressive Parkinson's disease. *Ann Neurol* 1994, 35: 172–180.

35. Redmond DE, Jr, Bjugstad KB, Teng YD, Ourednik V, Ourednik J, Wakeman DR et al. Behavioral improvement in a primate Parkinson's model is associated with multiple homeostatic effects of human neural stem cells. *Proc Natl Acad Sci USA* 2007, 104: 12175–12180.

36. Fuchs E. Scratching the surface of skin development. *Nature* 2007, 445: 834–842.

37. Piscaglia AC, Shupe TD, Oh SH, Gasbarrini A, and Petersen BE. Granulocyte-colony stimulating factor promotes liver repair and induces oval cell migration and proliferation in rats. *Gastroenterology* 2007, 133: 619–631.

38. Oliver JA. Adult renal stem cells and renal repair. *Curr Opin Nephrol Hypertens* 2004, 13: 17–22.

39. Weiss JR, Burgess JR, and Kaplan KJ. Fetiform teratoma (homunculus). *Arch Pathol Lab Med* 2006, 130: 1552–1556.

40. Kahan BW and Ephrussi B. Developmental potentialities of clonal in vitro cultures of mouse testicular teratoma. *J Natl Cancer Inst* 1970, 44: 1015–1036.

41. Evans MJ and Kaufman MH. Establishment in culture of pluripotential cells from mouse embryos. *Nature* 1981, 292: 154–156.

42. Martin GR. Isolation of a pluripotent cell line from early mouse embryos cultured in medium conditioned by teratocarcinoma stem cells. *Proc Natl Acad Sci USA* 1981, 78: 7634–7638.

43. Bradley A, Evans M, Kaufman MH, and Robertson E. Formation of germ-line chimaeras from embryo-derived teratocarcinoma cell lines. *Nature* 1984, 309: 255–256.

44. Mansour SL, Thomas KR, and Capecchi MR. Disruption of the proto-oncogene int-2 in mouse embryo-derived stem cells: A general strategy for targeting mutations to non-selectable genes. *Nature* 1988, 336: 348–352.

45. Smith AG, Heath JK, Donaldson DD, Wong GG, Moreau J, Stahl M et al. Inhibition of pluripotential embryonic stem cell differentiation by purified polypeptides. *Nature* 1988, 336: 688–690.

46. Williams RL, Hilton DJ, Pease S, Willson TA, Stewart CL, Gearing DP et al. Myeloid leukaemia inhibitory factor maintains the developmental potential of embryonic stem cells. *Nature* 1988, 336: 684–687.

47. Ying QL, Nichols J, Chambers I, and Smith A. BMP induction of Id proteins suppresses differentiation and sustains embryonic stem cell self-renewal in collaboration with STAT3. *Cell* 2003, 115: 281–292.

48. Takahashi K and Yamanaka S. Induction of pluripotent stem cells from mouse embryonic and adult fibroblast cultures by defined factors. *Cell* 2006, 126: 663–676.

49. Okita K, Ichisaka T, and Yamanaka S. Generation of germline-competent induced pluripotent stem cells. *Nature* 2007, 448: 313–317.

50. Wernig M, Meissner A, Foreman R, Brambrink T, Ku M, Hochedlinger K et al. In vitro reprogramming of fibroblasts into a pluripotent ES-cell-like state. *Nature* 2007, 448: 318–324.

51. Chambers I and Smith A. Self-renewal of teratocarcinoma and embryonic stem cells. *Oncogene* 2004, 23: 7150–7160.

52. Vallier L, Alexander M, and Pedersen RA. Activin/Nodal and FGF pathways cooperate to maintain pluripotency of human embryonic stem cells. *J Cell Sci* 2005, 118: 4495–4509.

53. Brons IG, Smithers LE, Trotter MW, Rugg-Gunn P, Sun B, Chuva de Sousa Lopes SM et al. Derivation of pluripotent epiblast stem cells from mammalian embryos. *Nature* 2007, 448: 191–195.

54. Martin GR and Evans MJ. Differentiation of clonal lines of teratocarcinoma cells: Formation of embryoid bodies in vitro. *Proc Natl Acad Sci USA* 1975, 72: 1441–1445.

55. Doetschman TC, Eistetter H, Katz M, Schmidt W, and Kemler R. The in vitro development of blastocyst-derived embryonic stem cell lines: Formation of visceral yolk sac, blood islands and myocardium. *J Embryol Exp Morphol* 1985, 87: 27–45.

56. Sulston JE and Horvitz HR. Post-embryonic cell lineages of the nematode, *Caenorhabditis elegans*. *Dev Biol* 1977, 56: 110–156.

57. Vallier L, Reynolds D, and Pedersen RA. Nodal inhibits differentiation of human embryonic stem cells along the neuroectodermal default pathway. *Dev Biol* 2004, 275: 403–421.

58. Bain G, Kitchens D, Yao M, Huettner JE, and Gottlieb DI. Embryonic stem cells express neuronal properties in vitro. *Dev Biol* 1995, 168: 342–357.

59. Fraichard A, Chassande O, Bilbaut G, Dehay C, Savatier P, and Samarut J. In vitro differentiation of embryonic stem cells into glial cells and functional neurons. *J Cell Sci* 1995, 108(Pt 10): 3181–3188.

60. Maden M, Keen G, and Jones GE. Retinoic acid as a chemotactic molecule in neuronal development. *Int J Dev Neurosci* 1998, 16: 317–322.

61. Okabe S, Forsberg-Nilsson K, Spiro AC, Segal M, and McKay RD. Development of neuronal precursor cells and functional postmitotic neurons from embryonic stem cells in vitro. *Mech Dev* 1996, 59: 89–102.

62. Tropepe V, Hitoshi S, Sirard C, Mak TW, Rossant J, and van der KD. Direct neural fate specification from embryonic stem cells: A primitive mammalian neural stem cell stage acquired through a default mechanism. *Neuron* 2001, 30: 65–78.

63. Kawasaki H, Mizuseki K, Nishikawa S, Kaneko S, Kuwana Y, Nakanishi S et al. Induction of midbrain dopaminergic neurons from ES cells by stromal cell-derived inducing activity. *Neuron* 2000, 28: 31–40.

64. Lee SH, Lumelsky N, Studer L, Auerbach JM, and McKay RD. Efficient generation of midbrain and hindbrain neurons from mouse embryonic stem cells. *Nat Biotechnol* 2000, 18: 675–679.

65. Rolletschek A, Chang H, Guan K, Czyz J, Meyer M, and Wobus AM. Differentiation of embryonic stem cell-derived dopaminergic neurons is enhanced by survival-promoting factors. *Mech Dev* 2001, 105: 93–104.

66. Ying QL and Smith AG. Defined conditions for neural commitment and differentiation. *Methods Enzymol* 2003, 365: 327–341.

67. Kim JH, Auerbach JM, Rodriguez Gomez JA, Velasco I, Gavin D, Lumelsky N et al. Dopamine neurons derived from embryonic stem cells function in an animal model of Parkinson's disease. *Nature* 2002, 418: 50–56.

68. Wichterle H, Lieberam I, Porter JA, and Jessell TM. Directed differentiation of embryonic stem cells into motor neurons. *Cell* 2002, 110: 385–397.

69. Aubert J, Dunstan H, Chambers I, and Smith A. Functional gene screening in embryonic stem cells implicates Wnt antagonism in neural differentiation. *Nat Biotechnol* 2002, 20: 1240–1245.

70. Carpenter MK, Inokuma MS, Denham J, Mujtaba T, Chiu CP, and Rao MS. Enrichment of neurons and neural precursors from human embryonic stem cells. *Exp Neurol* 2001, 172: 383–397.

71. Reubinoff BE, Itsykson P, Turetsky T, Pera MF, Reinhartz E, Itzik A et al. Neural progenitors from human embryonic stem cells. *Nat Biotechnol* 2001, 19: 1134–1140.

72. Bagutti C, Wobus AM, Fassler R, and Watt FM. Differentiation of embryonal stem cells into keratinocytes: Comparison of wild-type and beta 1 integrin-deficient cells. *Dev Biol* 1996, 179: 184–196.

73. Green H, Easley K, and Iuchi S. Marker succession during the development of keratinocytes from cultured human embryonic stem cells. *Proc Natl Acad Sci USA* 2003, 100: 15625–15630.

74. Coraux C, Hilmi C, Rouleau M, Spadafora A, Hinnrasky J, Ortonne JP et al. Reconstituted skin from murine embryonic stem cells. *Curr Biol* 2003, 13: 849–853.

75. Buttery LD, Bourne S, Xynos JD, Wood H, Hughes FJ, Hughes SP et al. Differentiation of osteoblasts and in vitro bone formation from murine embryonic stem cells. *Tissue Eng* 2001, 7: 89–99.

76. Bielby RC, Boccaccini AR, Polak JM, and Buttery LD. In vitro differentiation and in vivo mineralization of osteogenic cells derived from human embryonic stem cells. *Tissue Eng* 2004, 10: 1518–1525.

77. zur Nieden NI, Kempka G, and Ahr HJ. In vitro differentiation of embryonic stem cells into mineralized osteoblasts. *Differentiation* 2003, 71: 18–27.

78. Phillips BW, Belmonte N, Vernochet C, Ailhaud G, and Dani C. Compactin enhances osteogenesis in murine embryonic stem cells. *Biochem Biophys Res Commun* 2001, 284: 478–484.

79. zur Nieden NI, Kempka G, Rancourt DE, and Ahr HJ. Induction of chondro-, osteo- and adipogenesis in embryonic stem cells by bone morphogenetic protein-2: Effect of cofactors on differentiating lineages. *BMC Dev Biol* 2005, 5: 1.

80. Kawaguchi J, Mee PJ, and Smith AG. Osteogenic and chondrogenic differentiation of embryonic stem cells in response to specific growth factors. *Bone* 2005, 36: 758–769.

81. Hwang YS, Randle WL, Bielby RC, Polak JM, and Mantalaris A. Enhanced derivation of osteogenic cells from murine embryonic stem cells after treatment with HepG2-conditioned medium and modulation of the embryoid body formation period: Application to skeletal tissue engineering. *Tissue Eng* 2006, 12: 1381–1392.

82. Boheler KR, Czyz J, Tweedie D, Yang HT, Anisimov SV, and Wobus AM. Differentiation of pluripotent embryonic stem cells into cardiomyocytes. *Circ Res* 2002, 91: 189–201.

83. Hescheler J, Fleischmann BK, Lentini S, Maltsev VA, Rohwedel J, Wobus AM et al. Embryonic stem cells: A model to study structural and functional properties in cardiomyogenesis. *Cardiovasc Res* 1997, 36: 149–162.

84. Klug MG, Soonpaa MH, Koh GY, and Field LJ. Genetically selected cardiomyocytes from differentiating embryonic stem cells form stable intracardiac grafts. *J Clin Invest* 1996, 98: 216–224.

85. Muller M, Fleischmann BK, Selbert S, Ji GJ, Endl E, Middeler G et al. Selection of ventricular-like cardiomyocytes from ES cells in vitro. *FASEB J* 2000, 14: 2540–2548.

86. Kawai T, Takahashi T, Esaki M, Ushikoshi H, Nagano S, Fujiwara H et al. Efficient cardiomyogenic differentiation of embryonic stem cell by fibroblast growth factor 2 and bone morphogenetic protein 2. *Circ J* 2004, 68: 691–702.

87. Takahashi T, Lord B, Schulze PC, Fryer RM, Sarang SS, Gullans SR et al. Ascorbic acid enhances differentiation of embryonic stem cells into cardiac myocytes. *Circulation* 2003, 107: 1912–1916.

88. Mummery C, Ward D, van den Brink CE, Bird SD, Doevendans PA, Opthof T et al. Cardiomyocyte differentiation of mouse and human embryonic stem cells. *J Anat* 2002, 200: 233–242.

89. Mummery C, Ward-van OD, Doevendans P, Spijker R, van den BS, Hassink R et al. Differentiation of human embryonic stem cells to cardiomyocytes: Role of coculture with visceral endoderm-like cells. *Circulation* 2003, 107: 2733–2740.

90. Kehat I, Kenyagin-Karsenti D, Snir M, Segev H, Amit M, Gepstein A et al. Human embryonic stem cells can differentiate into myocytes with structural and functional properties of cardiomyocytes. *J Clin Invest* 2001, 108: 407–414.

91. Keller G. Embryonic stem cell differentiation: Emergence of a new era in biology and medicine. *Genes Dev* 2005, 19: 1129–1155.

92. Kabrun N, Buhring HJ, Choi K, Ullrich A, Risau W, and Keller G. Flk-1 expression defines a population of early embryonic hematopoietic precursors. *Development* 1997, 124: 2039–2048.

93. Fehling HJ, Lacaud G, Kubo A, Kennedy M, Robertson S, Keller G et al. Tracking mesoderm induction and its specification to the hemangioblast during embryonic stem cell differentiation. *Development* 2003, 130: 4217–4227.

94. Kubo A, Shinozaki K, Shannon JM, Kouskoff V, Kennedy M, Woo S et al. Development of definitive endoderm from embryonic stem cells in culture. *Development* 2004, 131: 1651–1662.

95. Gadue P, Huber TL, Paddison PJ, and Keller GM. Wnt and TGF-beta signaling are required for the induction of an in vitro model of primitive streak formation using embryonic stem cells. *Proc Natl Acad Sci USA* 2006, 103: 16806–16811.

96. Lumelsky N, Blondel O, Laeng P, Velasco I, Ravin R, and McKay R. Differentiation of embryonic stem cells to insulin-secreting structures similar to pancreatic islets. *Science* 2001, 292: 1389–1394.

97. Rajagopal J, Anderson WJ, Kume S, Martinez OI, and Melton DA. Insulin staining of ES cell progeny from insulin uptake. *Science* 2003, 299: 363.

98. Soria B, Roche E, Berna G, Leon-Quinto T, Reig JA, and Martin F. Insulin-secreting cells derived from embryonic stem cells normalize glycemia in streptozotocin-induced diabetic mice. *Diabetes* 2000, 49: 157–162.

99. Sosa-Pineda B, Chowdhury K, Torres M, Oliver G, and Gruss P. The Pax4 gene is essential for differentiation of insulin-producing beta cells in the mammalian pancreas. *Nature* 1997, 386: 399–402.

100. Jonsson J, Carlsson L, Edlund T, and Edlund H. Insulin-promoter-factor 1 is required for pancreas development in mice. *Nature* 1994, 371: 606–609.

101. Soria B. In-vitro differentiation of pancreatic beta-cells. *Differentiation* 2001, 68: 205–219.

102. Blyszczuk P, Czyz J, Kania G, Wagner M, Roll U, St-Onge L et al. Expression of Pax4 in embryonic stem cells promotes differentiation of nestin-positive progenitor and insulin-producing cells. *Proc Natl Acad Sci USA* 2003, 100: 998–1003.

103. Micallef SJ, Janes ME, Knezevic K, Davis RP, Elefanty AG, and Stanley EG. Retinoic acid induces Pdx1-positive endoderm in differentiating mouse embryonic stem cells. *Diabetes* 2005, 54: 301–305.

104. Jiang W, Shi Y, Zhao D, Chen S, Yong J, Zhang J et al. In vitro derivation of functional insulin-producing cells from human embryonic stem cells. *Cell Res* 2007, 17: 333–344.

105. Hamazaki T, Iiboshi Y, Oka M, Papst PJ, Meacham AM, Zon LI et al. Hepatic maturation in differentiating embryonic stem cells in vitro. *FEBS Lett* 2001, 497: 15–19.

106. Soto-Gutierrez A, Kobayashi N, Rivas-Carrillo JD, Navarro-Alvarez N, Zhao D, Okitsu T et al. Reversal of mouse hepatic failure using an implanted liver-assist device containing ES cell-derived hepatocytes. *Nat Biotechnol* 2006, 24: 1412–1419.

107. Rippon HJ, Polak JM, Qin M, and Bishop AE. Derivation of distal lung epithelial progenitors from murine embryonic stem cells using a novel three-step differentiation protocol. *Stem Cells* 2006, 24: 1389–1398.

108. Wang D, Haviland DL, Burns AR, Zsigmond E, and Wetsel RA. A pure population of lung alveolar epithelial type II cells derived from human embryonic stem cells. *Proc Natl Acad Sci USA* 2007, 104: 4449–4454.

109. Hwang WS, Ryu YJ, Park JH, Park ES, Lee EG, Koo JM et al. Evidence of a pluripotent human embryonic stem cell line derived from a cloned blastocyst. *Science* 2004, 303: 1669–1674.

110. Hwang WS, Roh SI, Lee BC, Kang SK, Kwon DK, Kim S et al. Patient-specific embryonic stem cells derived from human SCNT blastocysts. *Science* 2005, 308: 1777–1783.

111. Bradley JA, Bolton EM, and Pedersen RA. Stem cell medicine encounters the immune system. *Nat Rev Immunol* 2002, 2: 859–871.

112. Zhan X, Dravid G, Ye Z, Hammond H, Shamblott M, Gearhart J et al. Functional antigen-presenting leucocytes derived from human embryonic stem cells in vitro. *Lancet* 2004, 364: 163–171.

113. Taylor CJ, Bolton EM, Pocock S, Sharples LD, Pedersen RA, and Bradley JA. Banking on human embryonic stem cells: Estimating the number of donor cell lines needed for HLA matching. *Lancet* 2005, 366: 2019–2025.

114. Martin MJ, Muotri A, Gage F, and Varki A. Human embryonic stem cells express an immunogenic nonhuman sialic acid. *Nat Med* 2005, 11: 228–232.

115. Rajala K, Hakala H, Panula S, Aivio S, Pihlajamaki H, Suuronen R et al. Testing of nine different xeno-free culture media for human embryonic stem cell cultures. *Hum Reprod* 2007, 22: 1231–1238.

116. Bittira B, Shum-Tim D, Al-Khaldi A, and Chiu RC. Mobilization and homing of bone marrow stromal cells in myocardial infarction. *Eur J Cardiothorac Surg* 2003, 24: 393–398.

117. Hodgson DM, Behfar A, Zingman LV, Kane GC, Perez-Terzic C, Alekseev AE et al. Stable benefit of embryonic stem cell therapy in myocardial infarction. *Am J Physiol Heart Circ Physiol* 2004, 287: H471–H479.

118. Laflamme MA and Murry CE. Regenerating the heart. *Nat Biotechnol* 2005, 23: 845–856.

119. Nussbaum J, Minami E, Laflamme MA, Virag JA, Ware CB, Masino A et al. Transplantation of undifferentiated murine embryonic stem cells in the heart: Teratoma formation and immune response. *FASEB J* 2007, 21: 1345–1357.

120. Laflamme MA, Chen KY, Naumova AV, Muskheli V, Fugate JA, Dupras SK et al. Cardiomyocytes derived from human embryonic stem cells in pro-survival factors enhance function of infarcted rat hearts. *Nat Biotechnol* 2007, 25: 1015–1024.

121. Hansson M, Tonning A, Frandsen U, Petri A, Rajagopal J, Englund MC et al. Artifactual insulin release from differentiated embryonic stem cells. *Diabetes* 2004, 53: 2603–2609.

122. Meinel L, Betz O, Fajardo R, Hofmann S, Nazarian A, Cory E et al. Silk based biomaterials to heal critical sized femur defects. *Bone* 2006, 39: 922–931.

123. Brustle O, Spiro AC, Karram K, Choudhary K, Okabe S, and McKay RD. In vitro-generated neural precursors participate in mammalian brain development. *Proc Natl Acad Sci USA* 1997, 94: 14809–14814.

124. Brustle O, Jones KN, Learish RD, Karram K, Choudhary K, Wiestler OD et al. Embryonic stem cell-derived glial precursors: A source of myelinating transplants. *Science* 1999, 285: 754–756.

125. Buhnemann C, Scholz A, Bernreuther C, Malik CY, Braun H, Schachner M et al. Neuronal differentiation of transplanted embryonic stem cell-derived precursors in stroke lesions of adult rats. *Brain* 2006, 129: 3238–3248.

126. Yanagisawa D, Qi M, Kim DH, Kitamura Y, Inden M, Tsuchiya D et al. Improvement of focal ischemia-induced rat dopaminergic dysfunction by striatal transplantation of mouse embryonic stem cells. *Neurosci Lett* 2006, 407: 74–79.

127. Lee H, Shamy GA, Elkabetz Y, Schofield CM, Harrsion NL, Panagiotakos G et al. Directed differentiation and transplantation of human embryonic stem cell-derived motoneurons. *Stem Cells* 2007, 25: 1931–1939.

128. Deshpande DM, Kim YS, Martinez T, Carmen J, Dike S, Shats I et al. Recovery from paralysis in adult rats using embryonic stem cells. *Ann Neurol* 2006, 60: 32–44.

129. Wang X, Ge S, McNamara G, Hao QL, Crooks GM, and Nolta JA. Albumin-expressing hepatocyte-like cells develop in the livers of immune-deficient mice that received transplants of highly purified human hematopoietic stem cells. *Blood* 2003, 101: 4201–4208.

130. Moriya K, Yoshikawa M, Saito K, Ouji Y, Nishiofuku M, Hayashi N et al. Embryonic stem cells develop into hepatocytes after intrasplenic transplantation in CCl4-treated mice. *World J Gastroenterol* 2007, 13: 866–873.

131. Lane S, Rippon HJ, and Bishop AE. Personal communication, 2007.

132. Thomson H. Bioprocessing of embryonic stem cells for drug discovery. *Trends Biotechnol* 2007, 25: 224–230.

133. Pickering SJ, Minger SL, Patel M, Taylor H, Black C, Burns CJ et al. Generation of a human embryonic stem cell line encoding the cystic fibrosis mutation deltaF508, using preimplantation genetic diagnosis. *Reprod Biomed Online* 2005, 10: 390–397.

134. An unwieldy hybrid. *Nature* 2007, 447: 353–354.

135. Animal–human hybrid–embryo research. *Lancet* 2007, 370: 909.

136. Atala A, Bauer SB, Soker S, Yoo JJ, and Retik AB. Tissue-engineered autologous bladders for patients needing cystoplasty. *Lancet* 2006, 367: 1241–1246.

137. Stevens MM, Marini RP, Schaefer D, Aronson J, Langer R, and Shastri VP. In vivo engineering of organs: The bone bioreactor. *Proc Natl Acad Sci USA* 2005, 102: 11450–11455.

138. Murphy WL, Dennis RG, Kileny JL, and Mooney DJ. Salt fusion: An approach to improve pore interconnectivity within tissue engineering scaffolds. *Tissue Eng* 2002, 8: 43–52.

139. Wang S, Cui W, and Bei J. Bulk and surface modifications of polylactide. *Anal Bioanal Chem* 2005, 381: 547–556.

140. Pompe T, Keller K, Mothes G, Nitschke M, Teese M, Zimmermann R et al. Surface modification of poly(hydroxybutyrate) films to control cell–matrix adhesion. *Biomaterials* 2007, 28: 28–37.

141. Nagai M, Hayakawa T, Makimura M, and Yoshinari M. Fibronectin immobilization using water-soluble carbodiimide on poly-L-lactic acid for enhancing initial fibroblast attachment. *J Biomater Appl* 2006, 21: 33–47.

142. Yang XB, Roach HI, Clarke NM, Howdle SM, Quirk R, Shakesheff KM et al. Human osteoprogenitor growth and differentiation on synthetic biodegradable structures after surface modification. *Bone* 2001, 29: 523–531.

143. Liu B, Cai SX, Ma KW, Xu ZL, Dai XZ, Yang L et al. Fabrication of a PLGA–collagen peripheral nerve scaffold and investigation of its sustained release property in vitro. *J Mater Sci Mater Med* 2007, 19: 1127–1132.

144. Jones JR, Ehrenfried LM, and Hench LL. Optimising bioactive glass scaffolds for bone tissue engineering. *Biomaterials* 2006, 27: 964–973.

145. Weir MD and Xu HH. High-strength, in situ-setting calcium phosphate composite with protein release. *J Biomed Mater Res A* 2007, 85: 388–396.

146. Sepulveda P, Jones JR, and Hench LL. Bioactive sol–gel foams for tissue repair. *J Biomed Mater Res* 2002, 59: 340–348.

147. Mann BK, Gobin AS, Tsai AT, Schmedlen RH, and West JL. Smooth muscle cell growth in photopolymerized hydrogels with cell adhesive and proteolytically degradable domains: Synthetic ECM analogs for tissue engineering. *Biomaterials* 2001, 22: 3045–3051.

148. Simpson NE, Stabler CL, Simpson CP, Sambanis A, and Constantinidis I. The role of the $CaCl_2$–guluronic acid interaction on alginate encapsulated betaTC3 cells. *Biomaterials* 2004, 25: 2603–2610.

149. Falconnet D, Csucs G, Grandin HM, and Textor M. Surface engineering approaches to micropattern surfaces for cell-based assays. *Biomaterials* 2006, 27: 3044–3063.

150. Cabodi M, Choi NW, Gleghorn JP, Lee CSD, Bonassar LJ, and Stroock AD. A microfluidic biomaterial. *J Am Chem Soc* 2005, 127: 13788–13789.

151. Tan W and Desai TA. Microfluidic patterning of cells in extracellular matrix biopolymers: Effects of channel size, cell type, and matrix composition on pattern integrity. *Tissue Eng* 2003, 9: 255–267.

152. Shin M, Matsuda K, Ishii O, Terai H, Kaazempur-Mofrad M, Borenstein J et al. Endothelialized networks with a vascular geometry in microfabricated poly(dimethyl siloxane). *Biomed Microdevices* 2004, 6: 269–278.

153. Kaihara S, Borenstein J, Koka R, Lalan S, Ochoa ER, Ravens M et al. Silicon micromachining to tissue engineer branched vascular channels for liver fabrication. *Tissue Eng* 2000, 6: 105–117.

154. Liu VA and Bhatia SN. Three-dimensional photopatterning of hydrogels containing living cells. *Biomed Microdevices* 2002, 4: 257–266.

155. Pham QP, Sharma U, and Mikos AG. Electrospinning of polymeric nanofibers for tissue engineering applications: A review. *Tissue Eng* 2006, 12: 1197–1211.

156. Hartgerink JD, Beniash E, and Stupp SI. Self-assembly and mineralization of peptide–amphiphile nanofibers. *Science* 2001, 294: 1684–1688.

157. Levenberg S, Huang NF, Lavik E, Rogers AB, Itskovitz-Eldor J, and Langer R. Differentiation of human embryonic stem cells on three-dimensional polymer scaffolds. *Proc Natl Acad Sci USA* 2003, 100: 12741–12746.

158. Levenberg S, Burdick JA, Kraehenbuehl T, and Langer R. Neurotrophin-induced differentiation of human embryonic stem cells on three-dimensional polymeric scaffolds. *Tissue Eng* 2005, 11: 506–512.

159. Elisseeff J, Ferran A, Hwang S, Varghese S, and Zhang Z. The role of biomaterials in stem cell differentiation: Applications in the musculoskeletal system. *Stem Cells Dev* 2006, 15: 295–303.

160. Gerecht S, Townsend SA, Pressler H, Zhu H, Nijst CL, Bruggeman JP et al. A porous photocurable elastomer for cell encapsulation and culture. *Biomaterials* 2007, 28: 4826–4835.

161. Gerecht S, Burdick JA, Ferreira LS, Townsend SA, Langer R, and Vunjak-Novakovic G. Hyaluronic acid hydrogel for controlled self-renewal and differentiation of human embryonic stem cells. *Proc Natl Acad Sci USA* 2007, 104: 11298–11303.

162. Kim H, Suh H, Jo SA, Kim HW, Lee JM, Kim EH et al. In vivo bone formation by human marrow stromal cells in biodegradable scaffolds that release dexamethasone and ascorbate-2-phosphate. *Biochem Biophys Res Commun* 2005, 332: 1053–1060.

163. Yang XB, Whitaker MJ, Sebald W, Clarke N, Howdle SM, Shakesheff KM et al. Human osteoprogenitor bone formation using encapsulated bone morphogenetic protein 2 in porous polymer scaffolds. *Tissue Eng* 2004, 10: 1037–1045.

164. Silva GA, Czeisler C, Niece KL, Beniash E, Harrington DA, Kessler JA et al. Selective differentiation of neural progenitor cells by high-epitope density nanofibers. *Science* 2004, 303: 1352–1355.

28 Polymeric Nano- and Microparticles for the Delivery of Antisense Oligonucleotides and siRNA

Elias Fattal and Giuseppe De Rosa

CONTENTS

28.1 INTRODUCTION

Nano and microtechnologies support a number of biomedical applications including drug targeting. Among the several micron and submicron particles developed for drug targeting, biodegradable polymeric nanoparticles have been widely used to carry different types of drugs including oligonucleotides (ODNs) and small-interfering RNA (siRNA). These short gene fragments are able to control gene expression once they get inside the cell and therefore potentially active to treat oncogene-related cancers, viral infections, or inflammatory diseases. Because of their low intracellular penetration and poor stability in biological fluids, the use of a carrier able to overcome these issues is crucial to exploit nucleic acids in therapeutics. Polymeric nanoparticles offer good potentialities as carriers because of their chemical stability and their ability to achieve tissue and cellular targeting. Besides nanotechnologies, microtechnologies were also applied to the delivery of nucleic acids. Since they produce long-term drug release, they can be used for the local delivery of nucleic acids in body regions that can be damaged if frequent administrations are given (e.g., intraocular delivery). This chapter will describe the different strategies for designing polymeric particulate systems for optimal administration of ODNs and siRNA.

28.2 BASIC CONCEPT AND MECHANISM OF ODNs AND siRNA

ODNs can be designed to modulate gene expression by different strategies. Antisense ODNs (AS-ODNs) are certainly the more largely investigated ODNs. AS-ODNs are synthetic single-stranded DNA fragments, from 13 to about 25 nucleotides, able to bind to specific sequence (sense) of intracellular messenger RNA (mRNA) strands, forming a short double helix. The resulting short double helix hinders mRNA translation with consequent inhibition of protein synthesis expressed by the targeted gene [1]. The ability of an AS-ODN to form a hybrid depends on its binding affinity and sequence specificity. Binding affinity is a function of the number of hydrogen bonds formed between the AS-ODN and the targeted sequence. The affinity can be measured by determining the melting temperature (T_m) at which 50% of the double-stranded material is dissociated into single strands. T_m depends on the concentration of the AS-ODN, the nature of the base pairs, and the ionic strength of the solvent in which hybridization occurs [2]. Among several recognized mechanisms, one commonly described is the so-called translational arrest (Figure 28.1). In this mechanism, the AS-ODN binds to the single-strand mRNA by Watson–Crick base pairing forming a double helix hybrid, sterically blocking the translation of this transcript into a protein [3]. In addition, the hybridized mRNA prevents the binding of factors that initiate or modulate the translation and may block the movement of ribosomes along the mRNA [4]. Another mechanism that was widely described involves RNase-H-mediated cleavage of the target mRNA (Figure 28.1). RNase-H is a ribonuclease that recognizes RNA–DNA duplexes and selectively cleaves the RNA strand. This mechanism is catalytic: once an RNA molecule is cleaved, the AS-ODN dissociates from the duplex and becomes available to bind a second target mRNA molecule [5]. However, this recognition by RNase-H is restricted to a few compounds since it does not occur with most of the chemically modified ODNs.

Direct DNA targeting is also possible by using ODNs (antigen ODNs) (Figure 28.1). These single-stranded ODNs can form triple helices with DNA. The third strand binds in the major groove of the double helix specifically recognizing homopolypurine:homopolypyrimidine sequences present in a number of eukaryotic genes. The third strand of polypyrimidine

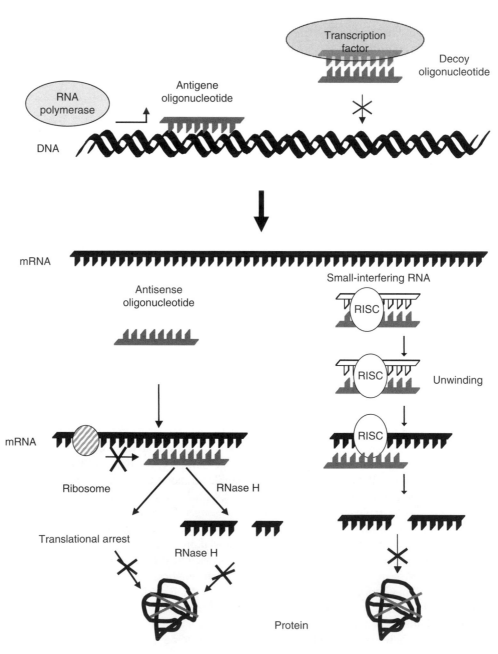

FIGURE 28.1 Mechanisms of action of ODNs and siRNA.

binds parallel to the polypurine strand of the duplex DNA by Hoogsteen hydrogen bonds, whereas the third strand of polypurine, which is typically G-rich and antiparallel to the polypurine tract, binds to the purine strand of the duplex DNA via reverse-Hoogsteen hydrogen bonds. Formation of this triplex construct lead to the blockage of the transcription of the target gene [6,7].

An alternative way to modulate gene expression by ODNs is the transcription factor decoy strategy (Figure 28.1). This experimental approach is based on the competition for transcription factors between endogenous *cis*-elements present within regulatory regions of target genes and exogenously added DNA sequences (double-stranded ODNs) mimicking the specific *cis*-elements. The binding of decoy ODN to transcription factors results in an inhibition of the expression of specific proteins [8].

SiRNA was discovered by showing that the introduction of long double-stranded RNA (dsRNA) into a variety of hosts could induce post-transcriptional silencing of all homologous host genes or transgenes by a mechanism that is summarized in Figure 28.1 [9–12]. Within the intracellular compartment, the long dsRNA molecules are metabolized to small 21–23 nucleotide-interfering RNAs by the action of an endogenous ribonuclease: dsRNA-specific RNAse III enzyme Dicer [11,13,14]. The siRNA molecules then assemble to a multiprotein complex, termed RNA-induced silencing complex (RISC), which contains a helicase activity that unwinds the two strands of RNA molecules, allowing the antisense strand to bind to the targeted RNA molecule [12] and an endonuclease activity, which hydrolyzes the target mRNA homologous at the site where the antisense strand is bound. RNA interference has an antisense mechanism of action, as ultimately a single-strand RNA molecule binds to the target RNA molecule by Watson–Crick base pairing rules and recruits a ribonuclease that degrades the target RNA [15]. Although there are similarities between AS-ODNs and siRNA, very few molecules of siRNA are needed to inhibit gene expression compared to AS-ODNs [16]. As suggested by Corey [17], the reason might be that, unlike siRNA, there is no evolved mechanism for promoting antisense strand recognition and it is likely that AS-ODNs must find their target unassisted [17].

28.3 CHEMICAL MODIFICATIONS OF ODNs AND siRNA

ODNs and siRNAs are natural phosphodiester compounds (Figure 28.2). However, critical drawbacks such as poor stability versus nuclease activity *in vitro* and *in vivo*, low intracellular penetration, and low bioavailability have limited their use in therapeutics [18,19]. As a result, clinical applications of both types of nucleic acids required chemical modifications with the aim of retaining their capacity to knock down protein expression while increasing stability and cellular penetration.

Replacement of the nonbridging oxygen of the phosphodiester backbone by sulfur resulted in the synthesis of phosphorothioate ODN or siRNA (Figure 28.2) with enhanced stability to enzymatic degradation. For ODNs, although the duplex formed with the target RNA has a lower melting temperature (T_m) (i.e., lower affinity) than the phosphodiester compound, it remains a substrate for RNase H. With siRNA, such

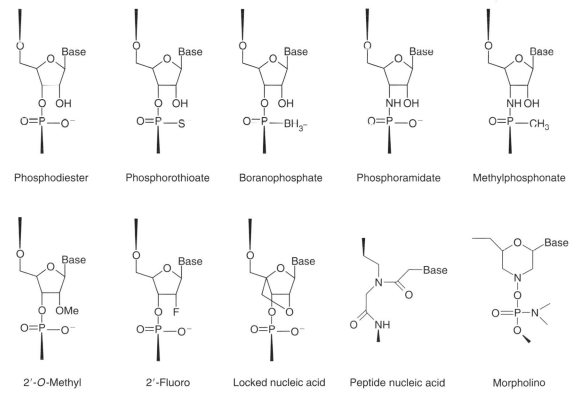

FIGURE 28.2 Chemical structure of natural and modified monomer in ODNs and siRNA.

a modification did not significantly affect silencing efficiency [20]. However, the main disadvantage of the phosphorothioate modification is their binding to certain proteins, inducing undesirable effects [21,22]. Moreover, some authors have shown that these linkages do not significantly enhance siRNA stability versus nucleases and reduce the melting temperatures of the duplexes as compared with unmodified RNA [23]. In another study, a phosphorothioate modification was shown to clearly reduce siRNA activity [24]. Thus, the importance of phosphorothioate modification might have been overestimated.

Other chemical approaches tested on ODNs have been the replacement of the nonbridging oxygen by a methyl group, which results in a methylphosphonate ODN (Figure 28.2) or the replacement of the same oxygen by a borane in ODNs [25] and siRNA [26] (Figure 28.2). Methylphosphonate ODNs have a greater hydrophobicity due to the loss of the negative charge. However, in this case, RNase H activation is also reduced [27]. Boranophosphate ODNs are highly resistant to nucleases, hybridize with both DNA and RNA, and activate RNase H cleavage of complementary RNA [25,28]. Moreover, for these ODNs, no demonstrable toxicity has been shown. Similarly, boranophosphate siRNAs were shown to be more effective than natural or phosphorothioate siRNAs mostly because they are at least 10 times more nuclease-resistant than unmodified siRNAs [26,29]. Other modifications made on the phosphate group have led to the synthesis of phosphoramidates (Figure 28.2). In these molecules, the 3'-oxygen is substituted by a 3'-amino group. This modification creates a highly nuclease-resistant molecule with an ability to form very stable duplexes with RNA by Watson–Crick base pairing. In contrast to phosphodiesters, the phosphoramidates also form stable triplexes with double-stranded DNA under near physiological conditions. Although the ability of phosphoramidates to activate RNase-H is weak, they effectively block translation because of the high stability of the DNA/RNA hybrids formed [30,31].

ODN or siRNA sugar moiety can be modified at the 2' position of the ribose by linkage of O-methyl (2'-OMe), O-methoxyethyl (2'-OMeEt), fluoro (2'-F), and locked nucleic acid (LNA) (Figure 28.2). 2'-O-methyl and 2'-O-methoxyethyl ODNs have raised a large interest. The RNA/RNA duplex formed with these ODNs is highly stable as indicated by the elevated T_m. However, none of the 2'-O-alkyl ODNs derivative can induce RNase-H cleavage of target RNA. For this reason, their antisense effect can only be due to a physical blockage of ribosomal machinery, with consequent reduction of their potency [32,33]. The siRNA motif consisting of 2'-O-Me and 2'-F has enhanced plasma stability and increased in vivo potency as well [34]. This modification has also been shown to increase the nuclease resistance of siRNA duplexes [24]. Nevertheless, this high potency is being discussed by several authors. Indeed, it was recently shown that siRNA duplexes containing full 2'-O-Me-modified sense strands display comparable activity to the unmodified analog of similar sequence [35]. Another interesting study has shown that even though the 2'-F modified siRNAs greatly improve

resistance to nuclease degradation in plasma, this increase in stability did not translate into enhanced or prolonged inhibitory activity of target gene reduction in mice following tail vein injection [36].

LNA, also referred to as inaccessible RNA, is a family of conformationally locked nucleotide analogs that display unprecedented hybridization affinity toward complementary DNA and RNA [37,38]. Commonly used LNA contains a methylene bridge connecting the 2' oxygen with the 4' carbon of the ribose ring (Figure 28.2). This bridge locks the ribose ring in the 3'-endo conformation characteristic of RNA [39,40]. LNA ODNs possess a high stability in blood serum and cell extracts and are able to activate RNase H [37]. LNA ODNs have been shown to inhibit tumor growth in a murine xenograft model [41]. However, they also displayed profound hepatotoxicity as measured by serum transaminases, as well as by organ and body weights [42]. LNA were also compatible with siRNA intracellular machinery, preserving molecular integrity while offering several improvements that are relevant to the development of siRNA technology [23,43].

Chemical modifications have also led to the synthesis of an important class, which is the peptide nucleic acids (PNAs) [44] (Figure 28.2). In this case, the phosphodiester linkage is completely replaced by a polyamide (peptide) backbone composed of (2-aminoethyl) glycine units. These ODNs have high hybridization properties and a good biological stability versus nucleases. Nevertheless, they present some drawbacks such as their poor aqueous solubility and low cellular uptake. In addition, they do not induce target RNA cleavage by RNase-H. PNAs seem to be nontoxic as they are uncharged molecules with a low affinity for extracellular proteins. Another family of nonionic DNA analogs is morpholino ODNs (Figure 28.2) that consist of nucleic acids in which the ribose is replaced by a morpholino moiety and phosphoroamidate intersubunit linkages are used as a substitute for phosphodiester bonds. These ODNs are interesting because their affinity for their target is similar to that of natural ODNs and they display a low toxicity [45,46].

28.4 CELLULAR UPTAKE OF ODNs AND siRNA

In order to downregulate gene expression, ODNs and siRNAs must penetrate into the targeted cells and reach the cytoplasm. The exact mechanisms involved in ODN penetration are so far unclear. Uptake occurs through active transport, which in turn depends on temperature [47], structure, concentration of ODN [48], and of course cell type. At present, it is believed that adsorptive endocytosis and fluid phase pinocytosis are the major mechanisms of ODN internalization, with the relative proportions of internalized material depending on ODN concentration. At relatively low ODN concentrations, it is likely that internalization occurs via interaction with a membrane-bound receptor [47]. However, only a limited amount of ODN that enters into cells can escape from endosomes. Also in the case of siRNAs, the mechanism of internalization into cells is unclear. It is known that they are readily taken up by invertebrate cells such as *Caenorhabditis elegans*. However, they are not taken up by most mammalian cells in a way that preserves their activity. Actually, siRNAs do not readily cross the cellular membrane because of their negative charge and size [49] and therefore the simple addition of naked, unmodified siRNAs to the culture media that overlies mammalian cells does not result in effective knockdown of the target gene [19].

28.5 PHARMACOKINETIC AND BIODISTRIBUTION OF ODNs AND siRNA

The first pharmacokinetics studies were conducted with a 25-mer natural phosphodiester ODN that was injected intravenously in monkeys. Analysis of blood sample by polyacrylamide gel electrophoresis or capillary gel electrophoresis showed that after 15 min, all the ODNs was degraded [50], with a blood half-life estimated to be about 5 min [50]. Similar results were reported in rats for a 20-mer phosphodiester ODN [51]. The majority of the chemically modified ODNs can improve pharmacokinetics parameters. Indeed in rodent models (rats and mice), phosphorothioate ODNs showed distribution half-lives ranging from 30 to 60 min [52–55]. For 2'-*O*-methyl derivatives, the pharmacokinetics properties were close to those of phosphorothioates [55,56] while the PNAs did not display any increase in the distribution half-life [57]. The plasma clearance depended in some cases on the animal model. In mice, at equivalent doses of phosphorothioates ODNs, plasma clearance was at least twofold more rapid than in monkeys [58]. Plasma clearance was largely attributed to ODN distribution into peripheral tissues. ODNs were mainly distributed to the liver, kidney, spleen, lymph nodes, and bone marrow with no measurable distribution to the brain [50,52,53,59]. The biodistribution pattern depended on the nature of the ODN backbone. The phosphodiester ODNs accumulated highly in the liver, whereas high kidney uptake dominated the phosphorothioates, 2'-*O*-methyl phosphodiester [60], and PNA patterns [57]. ODNs do not only distribute to tissues but also accumulate within cells in tissues [61–63]. At early time points after injection, phosphorothioate ODNs appear to be associated with extracellular matrix and within cells. However, by 24 h, almost all of the ODNs are found within cells in tissues [61]. The mechanisms by which ODNs accumulate within the cells following parenteral administration is currently unknown [64]. Regarding siRNAs, their half-life *in vivo* is as short as for ODNs (seconds to minutes) [65]. This is predominantly due to their rapid elimination by kidney filtration because of their small size (~7 kDa) [65]. Like ODNs, they can also be degraded by endogenous serum RNases with a serum half-life of 60 min. The biodistribution of radiolabeled siRNA in mice showed an accumulation primarily in the liver and kidneys [66]. They were also detected in the heart, spleen, and lung. Both intravenous and intraperitoneal delivery resulted in similar behavior of siRNA *in vivo*. SiRNAs levels were relatively stable from 1 to 4 h following injection and were still detectable for up to 72 h [66]. These patterns closely resemble the characteristics of ODNs.

28.6 NANO AND MICROTECHNOLOGIES FOR THE DELIVERY OF NUCLEIC ACIDS

Despite the continuous synthesis of new entities of ODNs or more recently siRNA, a few issues remain to be solved to optimize their delivery. It is very important to meet all the critical parameters corresponding to an optimal activity: improved stability, increased blood half-life, tissue and cellular targeting, improved cellular penetration, and release of the nucleic acids in the right intracellular compartment. A number of research groups have focused on the design of nanotechnologies that could meet all the requirements mentioned above for an optimal delivery of nucleic acids. Two main approaches were utilized. First, due to their negative charges, ODNs can be complexed through electrostatic interactions with cationic polymers or nanoparticles. Second, the nucleic acids are delivered by nanoparticles that are able to trap them in their inner structure. Moreover, microparticles containing nucleic acids have been investigated extensively mainly for their long-term release effect.

28.6.1 PARTICULATE SYSTEMS BY COMPLEX FORMATION WITH ODNs AND siRNA

Several systems have been developed to enhance the intracellular delivery of ODNs through complex formation. Nucleic acids are negatively charged at physiological pH and can electrostatically interact with cationic charges to form complexes simply by mixing the components at optimal conditions. The positive charges can either come from soluble cationic polymers or cationic nanoparticles. As the complex possesses generally a global positive surface charge, they get easily attached to negatively charged cell surface with subsequent endocytosis. Complexes have shown to improve the cellular uptake of nucleic acids with variable success *in vitro* and *in vivo* [67–69].

28.6.1.1 Poly(ethyleneimine)

Polyethyleneimine (PEI) is a polymer, available in different molecular weights and in linear or branched form, characterized by a high density of positive charges. PEI is largely known to be capable of effective nucleic acid transfer to mammalian cells [67,70] (Figure 28.3). The high transfection efficiency of PEI has been attributed to the buffering effect or the "proton sponge effect" of the polymer caused by the presence of amino groups in the molecule. The strong buffering effect and the consequent endosome rupture should help in rapid and massive diffusion of ODN into the cytoplasm [71]. Cytotoxicity and transfection efficiency of PEI are directly proportional to its molecular weight [72]. Many studies have aimed to reduce the toxicity by synthesis of PEI with graft copolymers such as linear poly(ethylene glycol) (PEG) [73],

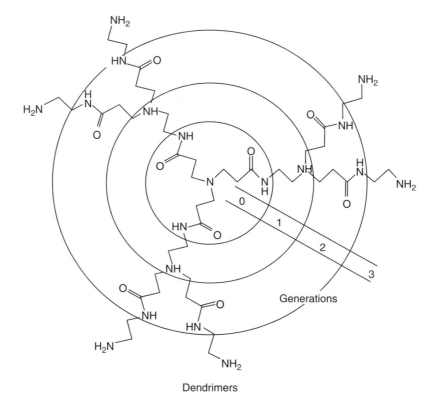

Polyethyleneimine

Dendrimers

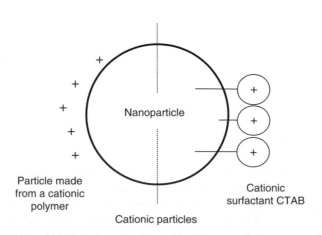

FIGURE 28.3 Main cationic polymers and nanospheres used for complexation of nucleic acids.

incorporation of low–molecular weight PEI, and PEI glycosylation [74]. *In vitro* efficacy of PEI to deliver ODNs and siRNAs was clearly demonstrated on different cell populations including nondividing cells [67,75,76]. The medium in which the complex is formulated is of utmost importance to obtain nanosized complexes and subsequent optimal cellular delivery of ODNs [76]. Other authors have tried to reduce complex aggregation by micelle formation using a PEI–PEG conjugate [77]. It was shown that the molecular weight of PEI and the substitution degree of PEG are critical parameters for stability of the complexes with ODNs [78]. More stable complexes are obtained with low–molecular weight PEI containing 10 mol% PEG [78]. A higher degree of substitution leads to a quick decomplexation of the ODNs/PEI complexes [79].

Successful siRNA delivery was observed within a very narrow window of conditions and only with the 25 KDa branched PEI. The ability of PEI to improve the *in vitro* and *in vivo* efficacy of siRNA has been studied by a few authors. While the zeta potential and size of PEI/siRNA complexes correlated to transfection efficacy in some cases, complex instability may have also reduced nucleic acid efficacy [80]. It was shown that complexation of chemically unmodified siRNAs with PEI leads to the formation of complexes that condense and completely cover siRNAs as determined by atomic force microscopy [81]. More importantly, upon complexation with PEI, siRNAs were efficiently protected against degradation both *in vitro* in the presence of RNase and *in vivo* in the presence of serum nucleases [82]. In tissue culture, PEI/siRNA complexes are internalized by tumor cells within a few hours, leading to the intracellular release of siRNA molecules, which display full bioactivity. Grzelinski et al. [81] tested with success the delivery of an siRNA against the growth factor pleiotrophin. The complex was able to generate antitumoral effects in an orthotropic mouse glioblastoma model with U87 cells growing intracranially. In a subcutaneous mouse tumor model, the intraperitoneal administration of complexed, but not of naked siRNAs, led to the delivery of the intact siRNAs into the tumors [82]. The intraperitoneal injection of PEI complexes targeting the HER-2 receptor also resulted in a marked reduction of tumor growth through siRNA-mediated HER-2 downregulation [82]. Ligand-targeted PEI has been used to improve the efficiency of siRNA delivery. Self-assembling nanoparticles with siRNA were constructed with PEI that was PEGylated with an Arg-Gly-Asp (RGD) peptide ligand attached at the distal end of the PEG. The complex was used to target tumor neovasculature expressing integrins and deliver siRNA inhibiting vascular endothelial growth factor receptor-2 expression blocking tumor angiogenesis. Cell delivery and activity of PEGylated PEI was found to depend on the presence of peptide ligand and could be competed by free peptide. Intravenous administration into tumor-bearing mice gave selective tumor uptake, siRNA sequence-specific inhibition of protein expression within the tumor, and inhibition of both tumor angiogenesis and growth rate [83].

As well as for ODNs, the stability of siRNA/PEI complexes is of critical importance for efficient siRNA delivery to the cytoplasm. The effect of PEGylation on stability was tested by Mao et al. [84]. The stability and size of PEI/siRNA complexes were clearly influenced by PEI–PEG structure, and high degrees of substitution resulted in large (300–400 nm), diffuse complexes, which showed condensation behavior only at high nitrogen-to-phosphorus ratios (N/P) [84]. The stability of siRNA polyplexes against heparin displacement and RNase digestion could be improved by PEGylation. In knockdown experiments using NIH/3T3 fibroblasts stably expressing beta-galactosidase, it was shown that PEG chain length had a significant influence on biological activity of siRNA. PEGylated complexes with siRNA yielded knockdown efficiencies of around 70% higher than with PEI alone. These results were explained by the fact that the high siRNA condensation of PEI might prevent siRNA release in the cytosol whereas the condensation was lower in the case of PEGylated PEI [84].

28.6.1.2 Dendrimers

Starburst polyamidoamine (PAMAM) dendrimers are synthetic branched polymers characterized by a spherical shape and a high charge density surface. Dendrimers are produced in an iterative sequence of reaction steps, in which each additional iteration leads to a higher generation dendrimer (theoretically from G1 to Gn) (Figure 28.3). Transfections of AS-ODNs using dendrimers into clones generated from D5 mouse melanoma and Rat2 embryonal fibroblast cell lines expressing luciferase cDNA resulted in a specific and dose-dependent inhibition of luciferase expression [85]. This inhibition caused approximately 25%–50% reduction of baseline luciferase activity [85]. These results were confirmed by others using G5 dendrimers [86]. Helin et al. [87] studied the intracellular distribution of a fluorescein isothiocyanate-labeled AS-ODN delivered by PAMAM and found that intracellular ODN distribution was dependent on the phase of the cell cycle, with a nuclear localization predominantly in the G2/M phase.

Dendrimers can condense siRNA into nanoscale particles. They protect siRNA from enzymatic degradation and achieve substantial release of siRNA over an extended period of time for efficient gene silencing [88]. The ability of siRNA–dendrimer complexes to deliver siRNA to cells was tested using a model of endogenous gene knockdown. For these assays, A549Luc cells stably expressing the GL3 luciferase gene were used. Efficient gene silencing was observed with the specific GL3Luc siRNA–G7 complex while neither the naked siRNA nor the nonspecific GL2Luc siRNA–G7 complex showed any gene silencing effect [88]. The siRNA–dendrimer complex showed generation-dependent gene silencing behavior: the higher the dendrimer generation, the more efficient the gene silencing was found to be. The best gene silencing results were obtained with G7 at an N/P ratio of 10 to 20 and siRNA concentration of 100 nM [88]. For targeting purposes, PAMAM G5 dendrimer was conjugated to Tat peptide, a cell-penetrating peptide that should increase cellular delivery. Delivery of both AS-ODNs and siRNAs designed to inhibit MDR1 gene expression in NIH/3T3 cell line was achieved [89]. MDR1 gene expression was partially inhibited by the antisense complex and weakly inhibited by the siRNA complex when both were tested at nontoxic levels of the dendrimer. Conjugation with Tat peptide did not improve the delivery efficiency of the dendrimer [89].

28.6.1.3 Complexation by Cationic Nanospheres

Nanospheres (nanoparticles consisting of spherical polymer matrix) can be used to support complexation of nucleic acids. In this case, the use of cationic nanospheres, made of cationic polymers (Figure 28.3) or nanospheres that are rendered positive through the adsorption of a cationic surfactant (Figure 28.3), is required. In this latest case, nanospheres are composed of synthetic biodegradable polymers such as poly(alkylcyanoacrylate) (PACA) [90] or poly(lactic acid) (PLA) [91,92]. Since ODNs have no affinity for the polymeric matrix, association with nanospheres has been achieved by ion pairing with a cationic surfactant, cetyltrimethylammonium bromide (CTAB) adsorbed onto the nanosphere surface [90]. AS-ODNs bound to PACA nanosphere in this way were protected from nucleases *in vitro* [93] and their intracellular uptake was increased [68]. In addition, nanospheres were able to concentrate intact AS-ODNs in the liver and spleen [94]. Using such a formulation, AS-ODNs were able to specifically inhibit mutated Ha-*ras*-mediated cell proliferation and tumorigenicity in nude mice after intratumoral administration [95].

28.6.2 Particulate Systems Encapsulating ODNs and siRNA

The association of nucleic acids to the surface of nanoparticles involves a number of issues. In most cases, the use of an ion pair agent is required to confer a positive charge to particle surface. However, this kind of association was found to be rapidly reversible in the presence of protein-rich biological medium [94]. Moreover, when associated through electrostatic forces, nucleic acids adopt conformations that make them unlikely to hybridize with their biological target [96]. Finally, the use of surfactants, such as CTAB, can contribute to the cytotoxicity of the formulation, as demonstrated by Delie et al. [97]. Thus, for optimal activity, ODNs should not be adsorbed, but entrapped within nano or microparticles, such as nano/microcapsules and nano/microspheres. Nano and microcapsules are vesicular systems in which the drug is confined to a cavity (an oily or aqueous core) surrounded by a unique polymeric membrane (Figure 28.4) whereas for nano and microspheres, the drug is dispersed throughout the polymeric matrix (Figure 28.4). Nanospheres or nanocapsules have a diameter less than 1 μm whereas the diameter of microspheres and microcapsules is larger than 1 μm. Different methods have been developed to prepare micro and nanoparticles for the encapsulation of hydrophilic macromolecules such as nucleic acid. These techniques can be classified into two main categories according to whether the formation of nanoparticles requires a polymerization reaction or it is achieved directly from a macromolecule or a preformed polymer.

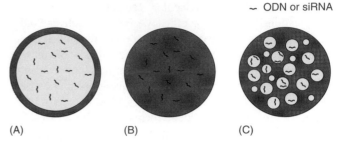

FIGURE 28.4 Polymeric particles with a capsular (A) or matrix (B,C) structure. In nanocapsules or microcapsules, the nucleic acid is solubilized in the liquid core surrounded by a polymeric membrane (A). In nanospheres or microspheres, the nucleic acid is dispersed in the polymeric matrix (B) or localized into cavities present within the polymeric matrix (C).

28.6.2.1 Particles Obtained by Polymerization of a Monomer

al Khouri-Fallouh et al. [98] have developed a method to produce PACA nanocapsules made of an internal oily core. They actually proposed to solubilize the monomer in an alcohol phase containing oil, this phase being then dispersed into an aqueous phase containing surfactants. The monomer, insoluble in water, polymerizes at the interface of the phases to form the wall of the nanocapsules. This process, which is simple to apply, was designed to encapsulate lipophilic drugs [99]. To encapsulate hydrophilic drugs, Lambert et al. [100] have designed a method to produce polyisobutylcyanoacrylate (PIBCA) nanocapsules containing an aqueous core for the encapsulation of nucleic acids, such as AS-ODN or siRNA. In this case, an aqueous solution containing the nucleic acids is emulsified within an oily phase containing a surfactant, forming a water-in-oil microemulsion. Following the slow addition of the monomer, polymerization occurs at the water-in-oil interface, thus producing nanocapsule shell (Figure 28.5) [100–102]. PIBCA nanocapsules containing an aqueous core were successfully used for the encapsulation of water-soluble molecules such as peptide and nucleic acids [100,103,104].

AS-ODNs encapsulation into the aqueous core of nanocapsules improved their stability against enzymatic degradation. Their half-life in serum was increased from 3 to 6 h when comparing naked AS-ODNs and AS-ODNs adsorbed onto PIBCA nanospheres, respectively [100]. *In vitro* AS-ODNs release from nanocapsules is generally high in the first 30 min, followed by a linear and almost complete release after 3 h [105]. This effect was explained by the presence of two nanocapsule population with different membrane fragilities. The initial release could be due to a rapid disruption of the polymeric wall occurring in the case of one of the population (probably the larger nanocapsules). The following linear release was instead attributed to the progressive degradation of the more resistant nanocapsule wall by serum esterase. It can therefore be assumed that AS-ODN degradation in serum occurs only after release from nanocapsules [105].

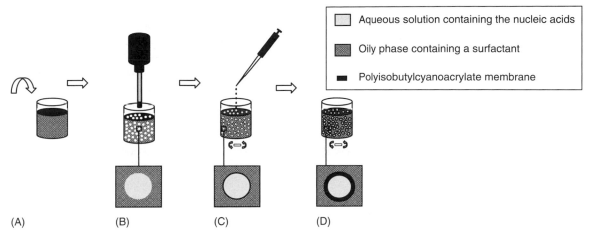

FIGURE 28.5 Preparation of poly(isobutylcyanoacrylate) nanocapsules containing an aqueous core obtained by interfacial polymerization. The aqueous solution containing the nucleic acids is emulsified within an oily phase containing a surfactant (generally sorbitan monooleate) (A) to form a water-in-oil microemulsion (B); polymerization occurs for slow addition of the monomer at the water-in-oil interface (C); after about 4 h under stirring, the aqueous droplets are surrounded by a poly(isobutylcyanoacrylate) membrane (nanocapsules) (D).

Encapsulation into PIBCA nanocapsules importantly increased cell uptake of both AS-ODN and siRNA. Toub et al. [106] investigated the subcellular distribution of AS-ODNs when incubated under naked or encapsulated form [106]. Incubation of naked AS-ODNs with cultured vascular smooth muscle cells resulted in membrane-bound species higher than the internalized fraction, while the ODN encapsulation led to an inverted profile with an internalized fraction 10 times higher as compared to naked ODN. Toub et al. [106] confirmed that the intracellular fraction of naked AS-ODNs was mainly found in vesicular fraction. AS-ODNs delivered by PIBCA nanocapsules drastically changed the subcellular distribution: less than 40% of ODN remained entrapped into vesicles, whereas about 50% was found in the nuclear fraction. This suggests that PIBCA nanocapsules were able to deliver a substantial amount of its cargo to the cytosolic compartment with further diffusion of the AS-ODNs to the nucleus. Similar AS-ODNs distribution into cells was observed using either fluorescence or radioactivity measurements, confirming the integrity of the internalized compounds [106]. Finally, AS-ODNs accumulation into the nucleus was found to increase as a function of the time, as shown in confocal microscopy experiments performed at different time intervals, ranging from 2 to 15 h [107].

The mechanism by which encapsulated AS-ODNs can escape from endosomes is not yet fully understood. However, detergents such as sorbitan monooleate used in their formulation may play a role in endosomal membrane destabilization [108]. Similar results were obtained for siRNA encapsulated into PIBCA with an aqueous core [104]. Confocal microscopy studies showed that when a fluorescently labelled naked siRNA was added to the culture medium, only a low fluorescence localized on the extracellular matrix was found. On the contrary, when siRNA was encapsulated in nanocapsules, a punctuate fluorescence was observed into the cells, suggesting an endosomal localization of the siRNA [104].

In vivo, AS-ODNs-containing nanocapsules were mainly tested on a murine model of Ewing sarcoma-related tumors targeting EWS Fli-1, a fusion gene resulting from a translocation that is found in 90% of both Ewing sarcoma and primitive neuroectodermal tumor [105]. All treatments were given by the intratumoral route. At a cumulative ODN dose of 14.4 nmol, only AS-ODNs encapsulated within nanocapsules led to a significant inhibition of tumor growth, while only a slight effect was observed with naked AS-ODNs at the same dose. Tanaka et al. [109] observed a significant reduction of tumor growth with injection of 500 nmol naked AS-ODNs. Thus, the use of nanocapsules allowed to obtain a 35-fold reduction of the AS-ODN administered dose. In the same experimental model, the efficiency of AS-ODN-containing nanocapsules was compared with that of the same AS-ODNs adsorbed onto poly(iso hexylcyanoacrylate) (PIHCA) nanosphere through interactions with preadsorbed CTAB [110]. In this case, a completely phosphorothioate AS-ODNs and a chimeric phosphorothioate/phosphodiester derivative were tested. Both nanoparticulate systems led to an efficient inhibition of the tumor volume from about 66% to 82%. The antisense effect of the delivered AS-ODN was demonstrated by downregulation of EWS-Fli-1 mRNA. The authors speculate that in the case of the nanoparticles, CTAB can permeate the endosomal membrane, thus inducing a more efficient cytoplasm delivery of the AS-ODN. Using again the same *in vivo* experimental model, Toub et al. showed that PIBCA nanocapsules were efficient to deliver siRNAs into tumor [104]. In particular, a dramatic inhibition of tumor growth was observed, especially when a higher cumulative dose was employed. In particular, a reduction of the tumor volume from about 43% to about 80% with a cumulative dose of 0.8 and 1.1 mg/kg, respectively, were observed allowing with a lower daily and cumulative dose to obtain a better efficacy than for encapsulated AS-ODNs. For both doses, naked siRNA had no inhibitory effect on tumor growth.

28.6.2.2 Particles Obtained by Preformed Polymer

To avoid limitations due to the use of monomers such as the toxicity of eventual residual monomers or the possibility of drug inactivation through the formation of covalent link between the drug and the polymer, particulate delivery systems based on preformed polymers have been proposed. Among the preformed polymers, PLA and its copolymers with glycolic acid (PLGA) have been extensively used in the last three decades. This family of polymers is present in the market at different molecular weights, lactide-to-glycolide ratios, and hydrophilicity (i.e., capped and uncapped endgroups), offering the possibility to choose among very different biodegradation rates. Phosphodiester and phosphothioate AS-ODNs were firstly loaded onto films made of PLA, obtaining a sustained release and an efficient protection against serum nucleases as well as a maintained hybridization capability [111]. However, the use of such polymeric devices needs a surgical implantation. The possibility to formulate these polymers in the form of microspheres offers the advantage to directly inject the delivery system into the administration site. Also when encapsulated into microsphere, AS-ODNs are efficiently protected by degradation upon incubation in serum [112,113].

Encapsulation of ODNs into PLA or PLGA particles was mainly achieved by the water-in-oil-in-water (or multiple) emulsion/solvent evaporation or extraction technique. In these techniques, an aqueous solution of the ODN is emulsified in an organic polymer solution to form a water-in-oil dispersion. This primary emulsion is then dispersed into an aqueous solution containing a hydrocolloid, generally poly(vinyl alcohol), to obtain a multiple emulsion. Solvent evaporation allows polymer hardening with a consequent formation of matrix-like particles in which the ODN is mainly localized in internal cavities of the first emulsion (Figure 28.6). In the case of solvent extraction, a cosolvent is added to the external aqueous phase to accelerate polymer hardening. ODN length and chemistry (phosphodiester or phosphorothioate), ODN loading as well as microspheres size strongly affect encapsulation

efficiency and release properties [112]. Higher drug loading is generally associated with a larger size [114] but is independent of ODN length [112]. *In vitro* release profile of AS-ODN from particles was found to be affected by microsphere size, drug loading, and ODN length and polymer molecular weight. Small microparticles (1–2 μm) released most of the entrapped AS-ODNs within the first 4 days, compared with 40 days for larger particles (10–20 μm) [114]. In the case of larger particles, a triphasic release was observed: a burst release in the first 48 h attributed to the rapid release of the AS-ODN present on or close to the microsphere surface; a sustained release phase (about 25 days) in which an AS-ODN diffusion through pores present in the polymer matrix should occur; a final more rapid release phase (up to a complete release at about 60 days) due to the polymer degradation that should accelerate ODN release rate [112]. In the case of smaller particles, the increased surface area-to-volume ratio results in a greater concentration of ODN at or near to the surface with the consequent more rapid release [112]. An increase of ODN loading also results in a higher burst effect [115,116] and a more rapid release rate [112]. Release profile is only modestly dependent on nucleic acid length, with a release rate higher as the AS-ODN molecular weight is lowered [117]. Finally, a high burst effect as well as the rapid *in vitro* release rate has been related to microsphere porosity. Pores on microsphere surface originate when an osmotically active agent, such as AS-ODN, is added to the internal aqueous phase. The use of an osmotic agent, such as NaCl, into the external aqueous phase allows to reduce microsphere surface porosity, thus reducing the initial ODN burst effect as well as the *in vitro* release rate [113,118,119]. *In vitro* decoy ODN release from PLGA microparticles is dramatically modified at pH different than the physiological value [116]. In particular, it was found that at pH values of 5 and 10, the overall released decoy ODN dramatically reduced. The pH effect was reduced by the incorporation of PEG into the microparticles [116].

AS-ODNs were also coencapsulated with the anticancer drug 5-fluorouracil with a similar loading efficiency. However,

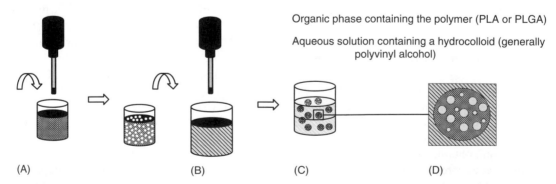

Organic phase containing the polymer (PLA or PLGA)

Aqueous solution containing a hydrocolloid (generally polyvinyl alcohol)

(A)   (B)   (C)   (D)

FIGURE 28.6 Preparation of PLA/PLGA particles containing nucleic acids by water-in-oil-in-water emulsion/solvent evaporation or extraction technique. An aqueous solution containing the nucleic acid is emulsified in an organic polymer solution (A) to form a water-in-oil dispersion; this primary emulsion is then dispersed in an aqueous solution containing a hydrocolloid (generally polyvinyl alcohol) (B) to obtain a multiple emulsion (C). Solvent evaporation or extraction allows polymer hardening with a consequent formation of matrix-like particles in which the nucleic acid is mainly localized in the internal cavities (D).

the release profiles were slower when the two drugs were coencapsulated, suggesting an interaction between ODN and 5-fluorouracil [120]. Besides single-stranded AS-ODNs, also double-stranded ODN have been successfully encapsulated within PLGA microspheres [121]. In particular, microspheres encapsulating a decoy ODN against the transcription factor κB (NF-κB) showed a negligible burst effect and a very gradual release profile, with the decoy ODN completely released after about 40 days [121]. A double-stranded siRNA was also encapsulated into PLGA nanoparticles [122] and microparticles [117]. For the same formulation, the *in vitro* release of siRNA was found to be much slower and with a very limited burst effect, compared with that obtained with a single-stranded AS-ODN. The difference between the release profiles was ascribed to the different chemical nature of the two nucleic acids, while only a modest effect was attributed to the different molecular weights [117].

Increase of ODN encapsulation efficiency into PLGA or PLA nano and microparticles, as well as optimization of their *in vitro* release properties, has been achieved by using cationic additives able to complex the nucleic acid within the matrix. The *in vitro* release rate of AS-ODN from PLGA microspheres was slowed down by complexation with zinc acetate [123]. A limited burst effect together with a linear release was observed (70% of the loaded ODN released over 9 days). Our group used the same rationale to design a "Trojan" delivery system that combines the advantages of microspheres (long term release) and of PEI/AS-ODNs complexes that is an enhanced ODN intracellular penetration [113,119,124,125]. Several different types of AS-ODNs were entrapped with the microparticles and depending on the ODNs characteristics and on the experimental conditions in which ODN and PEI are mixed, liquid [124], or solid complexes can be encapsulated [119,125]. ODN complexation with PEI improved AS-ODN encapsulation efficiency [119,124,125], probably by an electrostatic interaction between the cationic ODN/PEI complex and the anionic PLGA [124]. Release rate was shown to be dependent on the porosity of the microparticles, the higher the porosity, the faster was the release and the higher was the burst effect. Using an analytical solution of Fick's second law of diffusion, it was shown that the early phase of phosphorothioate AS-ODN and phosphorothioate AS-ODN-PEI complex release was primarily controlled by pure diffusion irrespective of the type of microspheres [119].

When incubated with cells, the *in vitro* fate of the encapsulated drug can vary according to particle size. In particular, small PLGA particles (including nanoparticles with a mean diameter lower than 1 μm and microparticles with a mean diameter between 1 and 5 μm) are generally developed to be taken up by cells. It was therefore shown that AS-ODNs entrapped within 1–2 μm-sized microparticles displayed 10-fold improved cell association in murine macrophages, compared with naked AS-ODNs [114]. Cell uptake was enhanced when macrophages were activated with interferon-γ or lipopolysaccharide (LPS) treatment, but decreased significantly in the presence of metabolic and phagocytosis inhibitors. PLGA microparticles loaded with AS-ODNs against the rat

tenascin mRNA was tested on smooth muscle cell proliferation and migration [126]. The study demonstrated a dose-dependent growth inhibition with the AS-ODN-loaded microparticles. Microparticles were found to be able to deliver a fluorescently labeled phosphorothioate AS-ODNs into "nonacidic compartments" of DU145 cells, presumably cytoplasm and nucleus, rather than in the endosome, where naked AS-ODNs are preferentially found [127]. Microparticles with a mean diameter of about 3.5 μm and encapsulating ODN/PEI complexes were efficiently internalized into HeLa cells [128]. In this case, the global positive charge of particles was thought to contribute to the cell/particle interaction. Moreover, formulation with the lower ODN release rate exhibited the better ODNs accumulation into the nucleus [128]. The use of nanoparticles (diameter lower than 1 μm) allowed to improve cell internalization and cytoplasm localization of a platelet-derived growth factor (PDGF) β-receptor AS-ODN [129]. A dose–response antiproliferative effect was found with AS-ODN-encapsulated nanoparticle on rat smooth muscle cells at the dose of 0.5 and 1 μM, while the effect was the same at higher concentration (2 μM). The antiproliferative effect was significantly lower with naked ODN at the same concentration [129]. A higher uptake of nanoparticles containing AS-ODNs against vascular endothelial growth factor (VEGF) was observed by Aukunuru et al. [130] in human retinal pigment epithelial cells. In this case, AS-ODN uptake increased about 4.3-fold by using nanoparticles, compared with naked AS-ODN. These data were confirmed by the efficient inhibition of VEGF mRNA expression together with a significant reduction of VEGF secretion by the cells. Efficient uptake of PLGA microparticles encapsulating adjuvant ODN containing unmethylated CpG motifs was also showed on dendritic cells *in vitro* by confocal microscopy, fluorescence-activated cell sorting (FACS), and scanning electron microscopy (SEM) [131]. In this study, dendritic cells incubated with PLGA nanoparticles containing tetanus toxoid induced significantly higher T cell proliferation when coencapsulated with CpG ODN rather than associated to the adjuvant at the same dose in solution [131].

Only a few studies have been focused on the *in vitro* fate of AS-ODN when delivered with large particles (diameter higher than 5 μm) that are unable to be taken up by cells. Of course, the main aim in this case is to obtain a long-term delivery of the ODN, thus reducing the need of frequent administrations. However, the release rate can play an important role in cellular uptake. In recent studies, we have demonstrated that, at a fixed concentration, ODN cell uptake is enhanced by slow release, compared with ODN administered in naked form [113,121,124]. Confocal microscopy studies showed a significant higher uptake of a single-stranded ODN when slowly released form microspheres of about 30 μm, compared with naked ODN [124]. In another study, a double-stranded decoy ODN against NF-κB was encapsulated in PLGA microspheres with a mean diameter of about 25 μm. Also in this case, an increased ODN uptake into cells was observed by confocal microscopy [121]. The increased decoy ODN uptake fitted with an increased inhibition of NF-κB activation (80-fold higher than with naked

Decoy ODN), as suggested by decreased NO production, iNOS, and tumor necrosis factor alpha (TNF-α) expression as well as reduced NF-κB/DNA binding [121]. The contemporary release of AS-ODN and PEI resulted in a further improvement of ODN cell uptake with accumulation into the nucleus [113,124]. All these results could be explained by the fact that ODN entry into cells prevalently by a pinocytotic process [132] that is saturated at high concentrations of naked ODN exposing a high dose fraction to degradation by exonucleases. When encapsulated into PLGA microspheres, ODN is efficiently protected from the enzymatic activity; then, when a little amount of ODN is released from microspheres, pinocytotic saturation is not expected to occur and a higher percentage of ODN can enter into cells. For siRNA, only a recent report has used PLGA nanoparticles containing a green fluorescent protein (GFP) gene silencing siRNA with GFP-expressing 293T cells [122]. Significant silencing effect was observed after 1 day incubation and was still persistent 1 week later.

Different studies have investigated the *in vivo* potentialities of PLA/PLGA nano- and microparticles for the delivery of AS-ODN compared with AS-ODN administered in a naked form. The therapeutic effect of PLGA microspheres encapsulating an AS-ODN targeted to the proto-oncogene *c-myc* was demonstrated in two different animal tumor models, human xenograft leukemia and melanoma [123]. In particular, the antitumor effect of microspheres encapsulating ODN/zinc complexes, administered by subcutaneous route, was compared with the efficacy of either subcutaneously or intravenously administered naked ODN at the same doses. The microsphere-based formulation was more effective as shown by reduced tumor growth, decreased number of metastases, reduced *c-myc* expression, and increased survival in melanoma model. A decreased metastatic potential and an increased survival were also observed in the leukemia model [123]. Different AS-ODN distribution in neostriatum of rat brain was observed by comparing a fluorescently labeled AS-ODN administered in naked or in microencapsulated form [133]. In particular, in the case of naked AS-ODN, a punctuate fluorescence was found in the cell cytosol at 24h postinjection, while a weak signal was observed at 48h. In contrast, when administered in microencapsulated form, fluorescently labeled AS-ODNs were still well visible in the neuronal tissue [133]. Biodistribution of tritium-labeled ODN, administered in naked form or encapsulated into microspheres, was investigated after subcutaneous administration in Balb-c mice [117]. In the case of naked AS-ODN, the majority of the radioactivity associated with the nucleic acid was eliminated within 24h. The use of AS-ODNs entrapped within microspheres resulted in high levels of radioactivity at the injection site and a lower level through the body after 24h from the inoculation. After 7 days, the level of radioactivity remains high at the administration site and considerably lower in the liver, kidney, and spleen. Microautoradiography analysis of liver and kidney showed similar biodistribution for naked and microencapsulated AS-ODN [117]. In a rat carotid *in vivo* model, PLGA nanoparticles containing an AS-ODN against PDGF receptor

showed an antirestenotic effect after 14 days from the treatment [129]. In this case, however, the same effect was observed with naked AS-ODN administered at the same dose. It is worthy to note that in the case of nanoparticles only a little amount of the loaded AS-ODN should be released during 14 days; consequently, the naked AS-ODN was effective compared with the encapsulated AS-ODN, but at higher doses [129]. The higher effect of AS-ODN, when administered with PLA/PLGA particles, was also investigated with PLGA nanoparticles containing CpG ODN (used as immunostimulant) [131]. Strong antigen-specific T cell proliferation was observed in Balb/c mice receiving immunogen tetanus toxoid associated to the adjuvant CpG ODN. A comparable stimulation index was observed with naked CpG ODN at a dose of 5 µg or with CpG ODN encapsulated into PLGA nanoparticles at a dose of 0.05 µg [131]. The *in vivo* potential of PLGA microspheres as delivery system for a decoy ODN to NF-κB has been recently tested in a rat model of chronic inflammation. Subcutaneous injection of PLGA microspheres releasing decoy ODN inhibited λ-carrageenin caused leukocyte infiltration and formation of granulation tissue up to 15 days whereas naked AS-ODN, at the same administered dose, showed a similar effect only from 1 day up to 5 days (G. De Rosa, personal communication). The so-called "Trojan delivery system" consisting of PLGA microparticles releasing AS-ODN/PEI complexes was tested *in vivo* [119]. In this study, an AS-ODN against the TGF-β2 was tested for wound healing after filtration surgery in an animal model of glaucoma. The cryosection of rabbit conjunctiva showed that after 1 day from the subconjunctival injection of microspheres containing AS-ODN/PEI complexes, the AS-ODN was located in the conjunctival stroma. Six days after the injection, AS-ODN was located in conjunctival cells, with accumulation into the nucleus. No conjunctival hyperemia was observed for 6 weeks after the injection. Rabbit treatment with PLGA microspheres releasing anti-TGF β2 AS-ODN prolong bleb survival for 28 days (50% of the treated eyes) following trabeculectomy compared to control. This effect was significantly higher in the case of PLGA microspheres releasing AS-ODN/PEI complexes with a bleb survival of 42 days on 100% of the treated eyes [119].

28.7 CONCLUSION

It is clear from this review that particle engineering has been extensively applied to the delivery of AS-ODNs and siRNAs. Major successes were obtained *in vitro* and *in vivo* for the delivery of these compounds. However, one issue remains to be solved. How can nanotechnology address specifically its content to the diseased cell? There is therefore a need to develop more targeted systems. A successful tissue-targetable system should have three properties: nucleic acid release into the cell cytosol, protection from nonspecific interactions, and tissue targeting that provides cell uptake. No single material has all of these required properties. Thus, the alternative approach could be the design of nanocapsule material that assembles these multiple properties.

REFERENCES

1. Loke, S. L., Stein, C. A., Zhang, X. H., Mori, K., Nakanishi, M., Subasinghe, C., Cohen, J. S., and Neckers, L. M., Characterization of oligonucleotide transport into living cells, *Proc Natl Acad Sci USA* 86(10), 3474–3478, 1989.
2. Breslauer, K. J., Frank, R., Blocker, H., and Marky, L. A., Predicting DNA duplex stability from the base sequence, *Proc Natl Acad Sci USA* 83(11), 3746–3750, 1986.
3. Baker, B. F., Lot, S. S., Condon, T. P., Cheng-Flournoy, S., Lesnik, E. A., Sasmor, H. M., and Bennett, C. F., 2′-O-(2-Methoxy)ethyl-modified anti-intercellular adhesion molecule 1 (ICAM-1) oligonucleotides selectively increase the ICAM-1 mRNA level and inhibit formation of the ICAM-1 translation initiation complex in human umbilical vein endothelial cells, *J Biol Chem* 272(18), 11994–2000, 1997.
4. Dias, N., Dheur, S., Nielsen, P. E., Gryaznov, S., van Aerschot, A., Herdewijn, P., Helene, C., and Saison-Behmoaras, T. E., Antisense PNA tridecamers targeted to the coding region of Ha-ras mRNA arrest polypeptide chain elongation, *J Mol Biol* 294(2), 403–416, 1999.
5. Walder, R. Y. and Walder, J. A., Role of RNase H in hybrid-arrested translation by antisense oligonucleotides, *Proc Natl Acad Sci USA* 85(14), 5011–5015, 1988.
6. Moser, H. E. and Dervan, P. B., Sequence-specific cleavage of double helical DNA by triple helix formation, *Science* 238(4827), 645–650, 1987.
7. Guntaka, R. V., Varma, B. R., and Weber, K. T., Triplex-forming oligonucleotides as modulators of gene expression, *Int J Biochem Cell Biol* 35(1), 22–31, 2003.
8. Cho-Chung, Y. S., Park, Y. G., and Lee, Y. N., Oligonucleotides as transcription factor decoys, *Curr Opin Mol Ther* 1(3), 386–392, 1999.
9. Fire, A., Xu, S., Montgomery, M. K., Kostas, S. A., Driver, S. E., and Mello, C. C., Potent and specific genetic interference by double-stranded RNA in *Caenorhabditis elegans*, *Nature* 391(6669), 806–811, 1998.
10. Kennerdell, J. R., Yamaguchi, S., and Carthew, R. W., RNAi is activated during Drosophila oocyte maturation in a manner dependent on aubergine and spindle-E, *Genes Dev* 16(15), 1884–1889, 2002.
11. Elbashir, S. M., Harborth, J., Lendeckel, W., Yalcin, A., Weber, K., and Tuschl, T., Duplexes of 21-nucleotide RNAs mediate RNA interference in cultured mammalian cells, *Nature* 411(6836), 494–498, 2001.
12. Vaucheret, H. and Fagard, M., Transcriptional gene silencing in plants: Targets, inducers and regulators, *Trends Genet* 17(1), 29–35, 2001.
13. Timmons, L. and Fire, A., Specific interference by ingested dsRNA, *Nature* 395(6705), 854, 1998.
14. Ketting, R. F., Fischer, S. E., Bernstein, E., Sijen, T., Hannon, G. J., and Plasterk, R. H., Dicer functions in RNA interference and in synthesis of small RNA involved in developmental timing in *C. elegans*, *Genes Dev* 15(20), 2654–2659, 2001.
15. Zamore, P. D., Tuschl, T., Sharp, P. A., and Bartel, D. P., RNAi: Double-stranded RNA directs the ATP-dependent cleavage of mRNA at 21 to 23 nucleotide intervals, *Cell* 101(1), 25–33, 2000.
16. Bertrand, J. R., Pottier, M., Vekris, A., Opolon, P., Maksimenko, A., and Malvy, C., Comparison of antisense oligonucleotides and siRNAs in cell culture and in vivo, *Biochem Biophys Res Commun* 296(4), 1000–1004, 2002.
17. Corey, D. R., RNA learns from antisense, *Nat Chem Biol* 3(1), 8–11, 2007.
18. Opalinska, J. B. and Gewirtz, A. M., Nucleic-acid therapeutics: Basic principles and recent applications, *Nat Rev Drug Discov* 1(7), 503–514, 2002.
19. Dykxhoorn, D. M. and Lieberman, J., Knocking down disease with siRNAs, *Cell* 126(2), 231–235, 2006.
20. Harborth, J., Elbashir, S. M., Vandenburgh, K., Manninga, H., Scaringe, S. A., Weber, K., and Tuschl, T., Sequence, chemical, and structural variation of small interfering RNAs and short hairpin RNAs and the effect on mammalian gene silencing, *Antisense Nucleic Acid Drug Dev* 13(2), 83–105, 2003.
21. Stein, C. A., Phosphorothioate antisense oligodeoxynucleotides: Questions of specificity, *Biotechnology* 14(5), 47–149, 1996.
22. Amarzguioui, M., Holen, T., Babaie, E., and Prydz, H., Tolerance for mutations and chemical modifications in a siRNA, *Nucleic Acids Res* 31(2), 589–595, 2003.
23. Braasch, D. A., Jensen, S., Liu, Y., Kaur, K., Arar, K., White, M. A., and Corey, D. R., RNA interference in mammalian cells by chemically modified RNA, *Biochemistry* 42(26), 7967–7975, 2003.
24. Chiu, Y. L. and Rana, T. M., siRNA function in RNAi: A chemical modification analysis, *RNA* 9(9), 1034–1048, 2003.
25. Summers, J. S. and Shaw, B. R., Boranophosphates as mimics of natural phosphodiesters in DNA, *Curr Med Chem* 8(10), 1147–1155, 2001.
26. Hall, A. H., Wan, J., Spesock, A., Sergueeva, Z., Shaw, B. R., and Alexander, K. A., High potency silencing by single-stranded boranophosphate siRNA, *Nucleic Acids Res* 34(9), 2773–2781, 2006.
27. Agrawal, S., Importance of nucleotide sequence and chemical modifications of antisense oligonucleotides, *Biochim Biophys Acta* 1489(1), 53–68, 1999.
28. Wang, X., Dobrikov, M., Sergueev, D., and Shaw, B. R., RNase H activation by stereoregular boranophosphate oligonucleotide, *Nucleosides Nucleotides Nucleic Acids* 22(5–8), 1151–1153, 2003.
29. Hall, A. H., Wan, J., Shaughnessy, E. E., Ramsay Shaw, B., and Alexander, K. A., RNA interference using boranophosphate siRNAs: Structure–activity relationships, *Nucleic Acids Res* 32(20), 5991–6000, 2004.
30. Gryaznov, S. M., Lloyd, D. H., Chen, J. K., Schultz, R. G., DeDionisio, L. A., Ratmeyer, L., and Wilson, W. D., Oligonucleotide N3′→P5′ phosphoramidates, *Proc Natl Acad Sci USA* 92(13), 5798–5802, 1995.
31. Gryaznov, S., Skorski, T., Cucco, C., Nieborowska-Skorska, M., Chiu, C. Y., Lloyd, D., Chen, J. K., Koziolkiewicz, M., and Calabretta, B., Oligonucleotide N3′→P5′ phosphoramidates as antisense agents, *Nucleic Acids Res* 24(8), 1508–1514, 1996.
32. Kurreck, J., Antisense technologies. Improvement through novel chemical modifications, *Eur J Biochem* 270(8), 1628–1644, 2003.
33. Urban, E. and Noe, C. R., Structural modifications of antisense oligonucleotides, *Farmaco* 58(3), 243–258, 2003.
34. Allerson, C. R., Sioufi, N., Jarres, R., Prakash, T. P., Naik, N., Berdeja, A., Wanders, L., Griffey, R. H., Swayze, E. E., and Bhat, B., Fully 2′-modified oligonucleotide duplexes with improved in vitro potency and stability compared to unmodified small interfering RNA, *J Med Chem* 48(4), 901–904, 2005.
35. Kraynack, B. A. and Baker, B. F., Small interfering RNAs containing full 2′-O-methylribonucleotide-modified sense strands display Argonaute2/eIF2C2-dependent activity, *RNA* 12(1), 163–176, 2006.

36. Layzer, J. M., McCaffrey, A. P., Tanner, A. K., Huang, Z., Kay, M. A., and Sullenger, B. A., In vivo activity of nuclease-resistant siRNAs, *RNA* 10(5), 766–771, 2004.

37. Wahlestedt, C., Salmi, P., Good, L., Kela, J., Johnsson, T., Hokfelt, T., Broberger, C., Porreca, F., Lai, J., Ren, K., Ossipov, M., Koshkin, A., Jakobsen, N., Skouv, J., Oerum, H., Jacobsen, M. H., and Wengel, J., Potent and nontoxic antisense oligonucleotides containing locked nucleic acids, *Proc Natl Acad Sci USA* 97(10), 5633–5638, 2000.

38. Vester, B. and Wengel, J., LNA (locked nucleic acid): High-affinity targeting of complementary RNA and DNA, *Biochemistry* 43(42), 13233–13241, 2004.

39. Bondensgaard, K., Petersen, M., Singh, S. K., Rajwanshi, V. K., Kumar, R., Wengel, J., and Jacobsen, J. P., Structural studies of LNA:RNA duplexes by NMR: Conformations and implications for RNase H activity, *Chemistry* 6(15), 2687–2695, 2000.

40. Braasch, D. A. and Corey, D. R., Locked nucleic acid (LNA): Fine-tuning the recognition of DNA and RNA, *Chem Biol* 8(1), 1–7, 2001.

41. Jepsen, J. S., Pfundheller, H. M., and Lykkesfeldt, A. E., Downregulation of p21(WAF1/CIP1) and estrogen receptor alpha in MCF-7 cells by antisense oligonucleotides containing locked nucleic acid (LNA), *Oligonucleotides* 14(2), 147–156, 2004.

42. Swayze, E. E., Siwkowski, A. M., Wancewicz, E. V., Migawa, M. T., Wyrzykiewicz, T. K., Hung, G., Monia, B. P., and Bennett, C. F., Antisense oligonucleotides containing locked nucleic acid improve potency but cause significant hepatotoxicity in animals, *Nucleic Acids Res* 35(2), 687–700, 2007.

43. Elmen, J., Thonberg, H., Ljungberg, K., Frieden, M., Westergaard, M., Xu, Y., Wahren, B., Liang, Z., Orum, H., Koch, T., and Wahlestedt, C., Locked nucleic acid (LNA) mediated improvements in siRNA stability and functionality, *Nucleic Acids Res* 33(1), 439–447, 2005.

44. Hanvey, J. C., Peffer, N. J., Bisi, J. E., Thomson, S. A., Cadilla, R., Josey, J. A., Ricca, D. J., Hassman, C. F., Bonham, M. A., Au, K. G., et al., Antisense and antigene properties of peptide nucleic acids, *Science* 258(5087), 1481–1485, 1992.

45. Summerton, J. and Weller, D., Morpholino antisense oligomers: Design, preparation, and properties, *Antisense Nucleic Acid Drug Dev* 7(3), 187–195, 1997.

46. Summerton, J., Morpholino antisense oligomers: The case for an RNase H-independent structural type, *Biochim Biophys Acta* 1489(1), 141–158, 1999.

47. Yakubov, L. A., Deeva, E. A., Zarytova, V. F., Ivanova, E. M., Ryte, A. S., Yurchenko, L. V., and Vlassov, V. V., Mechanism of oligonucleotide uptake by cells: Involvement of specific receptors? *Proc Natl Acad Sci USA* 86(17), 6454–6458, 1989.

48. Vlassov, V. V., Balakireva, L. A., and Yakubov, L. A., Transport of oligonucleotides across natural and model membranes, *Biochim Biophys Acta* 1197(2), 95–108, 1994.

49. Aagaard, L. and Rossi, J. J., RNAi therapeutics: Principles, prospects and challenges, *Adv Drug Deliv Rev* 59(2–3), 75–86, 2007.

50. Agrawal, S., Temsamani, J., Galbraith, W., and Tang, J., Pharmacokinetics of antisense oligonucleotides, *Clin Pharmacokinet* 28(1), 7–16, 1995.

51. Sands, H., Gorey-Feret, L. J., Cocuzza, A. J., Hobbs, F. W., Chidester, D., and Trainor, G. L., Biodistribution and metabolism of internally 3H-labeled oligonucleotides. I. Comparison of a phosphodiester and a phosphorothioate, *Mol Pharmacol* 45(5), 932–943, 1994.

52. Agrawal, S., Temsamani, J., and Tang, J. Y., Pharmacokinetics, biodistribution, and stability of oligodeoxynucleotide phosphorothioates in mice, *Proc Natl Acad Sci USA* 88(17), 7595–7599, 1991.

53. Cossum, P. A., Sasmor, H., Dellinger, D., Truong, L., Cummins, L., Owens, S. R., Markham, P. M., Shea, J. P., and Crooke, S., Disposition of the ^{14}C-labeled phosphorothioate oligonucleotide ISIS 2105 after intravenous administration to rats, *J Pharmacol Exp Ther* 267(3), 1181–1190, 1993.

54. Zhang, R., Diasio, R. B., Lu, Z., Liu, T., Jiang, Z., Galbraith, W. M., and Agrawal, S., Pharmacokinetics and tissue distribution in rats of an oligodeoxynucleotide phosphorothioate (GEM 91) developed as a therapeutic agent for human immunodeficiency virus type-1, *Biochem Pharmacol* 49(7), 929–939, 1995.

55. Geary, R. S., Watanabe, T. A., Truong, L., Freier, S., Lesnik, E. A., Sioufi, N. B., Sasmor, H., Manoharan, M., and Levin, A. A., Pharmacokinetic properties of 2′-O-(2-methoxyethyl)-modified oligonucleotide analogs in rats, *J Pharmacol Exp Ther* 296(3), 890–897, 2001.

56. Sewell, K. L., Geary, R. S., Baker, B. F., Glover, J. M., Mant, T. G., Yu, R. Z., Tami, J. A., and Dorr, F. A., Phase I trial of ISIS 104838, a 2′-methoxyethyl modified antisense oligonucleotide targeting tumor necrosis factor-alpha, *J Pharmacol Exp Ther* 303(3), 1334–1343, 2002.

57. McMahon, B. M., Mays, D., Lipsky, J., Stewart, J. A., Fauq, A., and Richelson, E., Pharmacokinetics and tissue distribution of a peptide nucleic acid after intravenous administration, *Antisense Nucleic Acid Drug Dev* 12(2), 65–70, 2002.

58. Yu, R. Z., Geary, R. S., Leeds, J. M., Watanabe, T., Moore, M., Fitchett, J., Matson, J., Burckin, T., Templin, M. V., and Levin, A. A., Comparison of pharmacokinetics and tissue disposition of an antisense phosphorothioate oligonucleotide targeting human Ha-ras mRNA in mouse and monkey, *J Pharm Sci* 90(2), 182–193, 2001.

59. Geary, R. S., Leeds, J. M., Henry, S. P., Monteith, D. K., and Levin, A. A., Antisense oligonucleotide inhibitors for the treatment of cancer: 1. Pharmacokinetic properties of phosphorothioate oligodeoxynucleotides, *Anticancer Drug Des* 12(5), 383–393, 1997.

60. Lendvai, G., Velikyan, I., Bergstrom, M., Estrada, S., Laryea, D., Valila, M., Salomaki, S., Langstrom, B., and Roivainen, A., Biodistribution of 68Ga-labelled phosphodiester, phosphorothioate, and 2′-O-methyl phosphodiester oligonucleotides in normal rats, *Eur J Pharm Sci* 26(1), 26–38, 2005.

61. Butler, M., Stecker, K., and Bennett, C. F., Cellular distribution of phosphorothioate oligodeoxynucleotides in normal rodent tissues, *Lab Invest* 77(4), 379–388, 1997.

62. Graham, M. J., Crooke, S. T., Monteith, D. K., Cooper, S. R., Lemonidis, K. M., Stecker, K. K., Martin, M. J., and Crooke, R. M., In vivo distribution and metabolism of a phosphorothioate oligonucleotide within rat liver after intravenous administration, *J Pharmacol Exp Ther* 286(1), 447–458, 1998.

63. Yu, R. Z., Zhang, H., Geary, R. S., Graham, M., Masarjian, L., Lemonidis, K., Crooke, R., Dean, N. M., and Levin, A. A., Pharmacokinetics and pharmacodynamics of an antisense phosphorothioate oligonucleotide targeting Fas mRNA in mice, *J Pharmacol Exp Ther* 296(2), 388–395, 2001.

64. Butler, M., Crooke, R. M., Graham, M. J., Lemonidis, K. M., Lougheed, M., Murray, S. F., Witchell, D., Steinbrecher, U., and Bennett, C. F., Phosphorothioate oligodeoxynucleotides distribute similarly in class A scavenger receptor knockout and wild-type mice, *J Pharmacol Exp Ther* 292(2), 489–496, 2000.

65. Soutschek, J., Akinc, A., Bramlage, B., Charisse, K., Constien, R., Donoghue, M., Elbashir, S., Geick, A., Hadwiger, P., Harborth, J., John, M., Kesavan, V., Lavine, G., Pandey, R. K., Racie, T., Rajeev, K. G., Rohl, I., Toudjarska, I., Wang, G., Wuschko, S., Bumcrot, D., Koteliansky, V., Limmer, S., Manoharan, M., and Vornlocher, H. P., Therapeutic silencing of an endogenous gene by systemic administration of modified siRNAs, *Nature* 432(7014), 173–178, 2004.

66. Braasch, D. A., Paroo, Z., Constantinescu, A., Ren, G., Oz, O. K., Mason, R. P., and Corey, D. R., Biodistribution of phosphodiester and phosphorothioate siRNA, *Bioorg Med Chem Lett* 14(5), 1139–1143, 2004.

67. Boussif, O., Lezoualc'h, F., Zanta, M. A., Mergny, M. D., Scherman, D., Demeneix, B., and Behr, J. P., A versatile vector for gene and oligonucleotide transfer into cells in culture and in vivo: Polyethylenimine, *Proc Natl Acad Sci USA* 92(16), 7297–7301, 1995.

68. Chavany, C., Saison-Behmoaras, T., Le Doan, T., Puisieux, F., Couvreur, P., and Helene, C., Adsorption of oligonucleotides onto polyisohexylcyanoacrylate nanoparticles protects them against nucleases and increases their cellular uptake, *Pharm Res* 11(9), 1370–1378, 1994.

69. Demeneix, B. and Behr, J. P., Polyethylenimine (PEI), *Adv Genet* 53PA, 215–230, 2005.

70. de Semir, D., Petriz, J., Avinyo, A., Larriba, S., Nunes, V., Casals, T., Estivill, X., and Aran, J. M., Non-viral vector-mediated uptake, distribution, and stability of chimeraplasts in human airway epithelial cells, *J Gene Med* 4(3), 308–322, 2002.

71. Boussif, O., Zanta, M. A., and Behr, J. P., Optimized galenics improve in vitro gene transfer with cationic molecules up to 1000-fold, *Gene Ther* 3(12), 1074–1080, 1996.

72. Godbey, W. T. and Mikos, A. G., Recent progress in gene delivery using non-viral transfer complexes, *J Control Release* 72(1–3), 115–125, 2001.

73. Petersen, H., Fechner, P., Fischer, D., and Kissel, T., Synthesis, characterization, and biocompatibility of polyethylenimine-*graft*-poly(ethylene glycol) block copolymers, *Macromolecules* 35, 6867–6874, 2002.

74. Leclercq, F., Dubertret, C., Pitard, B., Scherman, D., and Herscovici, J., Synthesis of glycosylated polyethylenimine with reduced toxicity and high transfecting efficiency, *Bioorg Med Chem Lett* 10(11), 1233–1235, 2000.

75. Dheur, S., Dias, N., van Aerschot, A., Herdewijn, P., Bettinger, T., Remy, J. S., Helene, C., and Saison-Behmoaras, E. T., Polyethylenimine but not cationic lipid improves antisense activity of 3′-capped phosphodiester oligonucleotides, *Antisense Nucleic Acid Drug Dev* 9(6), 515–525, 1999.

76. Gomes dos Santos, A. L., Bochot, A., Tsapis, N., Artzner, F., Bejjani, R. A., Thillaye-Goldenberg, B., de Kozak, Y., Fattal, E., and Behar-Cohen, F., Oligonucleotide–polyethylenimine complexes targeting retinal cells: Structural analysis and application to anti-TGFbeta-2 therapy, *Pharm Res* 23(4), 770–781, 2006.

77. Vinogradov, S. V., Bronich, T. K., and Kabanov, A. V., Self-assembly of polyamine–poly(ethylene glycol) copolymers with phosphorothioate oligonucleotides, *Bioconjug Chem* 9(6), 805–812, 1998.

78. Glodde, M., Sirsi, S. R., and Lutz, G. J., Physiochemical properties of low and high molecular weight poly(ethylene glycol)-grafted poly(ethylene imine) copolymers and their complexes with oligonucleotides, *Biomacromolecules* 7(1), 347–356, 2006.

79. Fischer, D., Osburg, B., Petersen, H., Kissel, T., and Bickel, U., Effect of poly(ethylene imine) molecular weight and PEGylation on organ distribution and pharmacokinetics of polyplexes with oligodeoxynucleotides in mice, *Drug Metab Dispos* 32(9), 983–992, 2004.

80. Grayson, A. C., Doody, A. M., and Putnam, D., Biophysical and structural characterization of polyethylenimine-mediated siRNA delivery in vitro, *Pharm Res* 23(8), 1868–1876, 2006.

81. Grzelinski, M., Urban-Klein, B., Martens, T., Lamszus, K., Bakowsky, U., Hobel, S., Czubayko, F., and Aigner, A., RNA interference-mediated gene silencing of pleiotrophin through polyethylenimine-complexed small interfering RNAs in vivo exerts antitumoral effects in glioblastoma xenografts, *Hum Gene Ther* 17(7), 751–766, 2006.

82. Urban-Klein, B., Werth, S., Abuharbeid, S., Czubayko, F., and Aigner, A., RNAi-mediated gene-targeting through systemic application of polyethylenimine (PEI)-complexed siRNA in vivo, *Gene Ther* 12(5), 461–466, 2005.

83. Schiffelers, R. M., Ansari, A., Xu, J., Zhou, Q., Tang, Q., Storm, G., Molema, G., Lu, P. Y., Scaria, P. V., and Woodle, M. C., Cancer siRNA therapy by tumor selective delivery with ligand-targeted sterically stabilized nanoparticle, *Nucleic Acids Res* 32(19), e149, 2004.

84. Mao, S., Neu, M., Germershaus, O., Merkel, O., Sitterberg, J., Bakowsky, U., and Kissel, T., Influence of polyethylene glycol chain length on the physicochemical and biological properties of poly(ethylene imine)-*graft*-poly(ethylene glycol) block copolymer/SiRNA polyplexes, *Bioconjug Chem* 17(5), 1209–1218, 2006.

85. Bielinska, A., Kukowska-Latallo, J. F., Johnson, J., Tomalia, D. A., and Baker, J. R., Jr., Regulation of in vitro gene expression using antisense oligonucleotides or antisense expression plasmids transfected using starburst PAMAM dendrimers, *Nucleic Acids Res* 24(11), 2176–2182, 1996.

86. Yoo, H., Sazani, P., and Juliano, R. L., PAMAM dendrimers as delivery agents for antisense oligonucleotides, *Pharm Res* 16(12), 1799–1804, 1999.

87. Helin, V., Gottikh, M., Mishal, Z., Subra, F., Malvy, C., and Lavignon, M., Cell cycle-dependent distribution and specific inhibitory effect of vectorized antisense oligonucleotides in cell culture, *Biochem Pharmacol* 58(1), 95–107, 1999.

88. Zhou, J., Wu, J., Hafdi, N., Behr, J. P., Erbacher, P., and Peng, L., PAMAM dendrimers for efficient siRNA delivery and potent gene silencing, *Chem Commun (Camb)* (22), 2362–2364, 2006.

89. Kang, H., DeLong, R., Fisher, M. H., and Juliano, R. L., Tat-conjugated PAMAM dendrimers as delivery agents for antisense and siRNA oligonucleotides, *Pharm Res* 22(12), 2099–2106, 2005.

90. Fattal, E., Vauthier, C., Aynie, I., Nakada, Y., Lambert, G., Malvy, C., and Couvreur, P., Biodegradable polyalkylcyanoacrylate nanoparticles for the delivery of oligonucleotides, *J Control Release* 53(1–3), 137–143, 1998.

91. Singh, M., Ugozzoli, M., Briones, M., Kazzaz, J., Soenawan, E., and O'Hagan, D. T., The effect of CTAB concentration in cationic PLG microparticles on DNA adsorption and in vivo performance, *Pharm Res* 20(2), 247–251, 2003.

92. Oster, C. G., Kim, N., Grode, L., Barbu-Tudoran, L., Schaper, A. K., Kaufmann, S. H., and Kissel, T., Cationic microparticles consisting of poly(lactide-*co*-glycolide) and polyethylenimine as carriers systems for parental DNA vaccination, *J Control Release* 104(2), 359–377, 2005.

93. Chavany, C., Le Doan, T., Couvreur, P., Puisieux, F., and Helene, C., Polyalkylcyanoacrylate nanoparticles as polymeric carriers for antisense oligonucleotides, *Pharm Res* 9(4), 441–449, 1992.

94. Nakada, Y., Fattal, E., Foulquier, M., and Couvreur, P., Pharmacokinetics and biodistribution of oligonucleotide adsorbed onto poly(isobutylcyanoacrylate) nanoparticles after intravenous administration in mice, *Pharm Res* 13(1), 38–43, 1996.

95. Schwab, G., Chavany, C., Duroux, I., Goubin, G., Lebeau, J., Helene, C., and Saison-Behmoaras, T., Antisense oligonucleotides adsorbed to polyalkylcyanoacrylate nanoparticles specifically inhibit mutated Ha-*ras*-mediated cell proliferation and tumorigenicity in nude mice, *Proc Natl Acad Sci USA* 91(22), 10460–10464, 1994.

96. Ganachaud, F., Elaissari, A., Pichot, C., Laayoun, A., and Cros, P., Adsorption of single-stranded DNA fragments onto cationic aminated latex particles, *Langmuir* 13(4), 701–707, 1997.

97. Delie, F., Berton, M., Allemann, E., and Gurny, R., Comparison of two methods of encapsulation of an oligonucleotide into poly (D,L-lactic acid) particles, *Int J Pharm* 214(1–2), 25–30, 2001.

98. al Khouri-Fallouh, N., Fessi, H., Roblot-Treupel, L., Devissaguet, J. P., and Puisieux, F., [An original procedure for preparing nanocapsules of polyalkylcyanoacrylates for interfacial polymerization], *Pharm Acta Helv* 61(10–11), 274–281, 1986.

99. Aboubakar, M., Puisieux, F., Couvreur, P., Deyme, M., and Vauthier, C., Study of the mechanism of insulin encapsulation in poly(isobutylcyanoacrylate) nanocapsules obtained by interfacial polymerization, *J Biomed Mater Res* 47(4), 568–576, 1999.

100. Lambert, G., Fattal, E., Pinto-Alphandary, H., Gulik, A., and Couvreur, P., Polyisobutylcyanoacrylate nanocapsules containing an aqueous core as a novel colloidal carrier for the delivery of oligonucleotides, *Pharm Res* 17(6), 707–714, 2000.

101. Lambert, G., Fattal, E., Brehier, A., Feger, J., and Couvreur, P., Effect of polyisobutylcyanoacrylate nanoparticles and lipofectin loaded with oligonucleotides on cell viability and PKC alpha neosynthesis in HepG2 cells, *Biochimie* 80(12), 969–976, 1998.

102. Lambert, G., Fattal, E., Pinto-Alphandary, H., Gulik, A., and Couvreur, P., Polyisobutylcyanoacrylate nanocapsules containing an aqueous core for the delivery of oligonucleotides, *Int J Pharm* 214(1–2), 13–16, 2001.

103. Watnasirichaikul, S., Davies, N. M., Rades, T., and Tucker, I. G., Preparation of biodegradable insulin nanocapsules from biocompatible microemulsions, *Pharm Res* 17(6), 684–689, 2000.

104. Toub, N., Bertrand, J. R., Tamaddon, A., Elhamess, H., Hillaireau, H., Maksimenko, A., Maccario, J., Malvy, C., Fattal, E., and Couvreur, P., Efficacy of siRNA nanocapsules targeted against the EWS-Fli1 oncogene in Ewing sarcoma, *Pharm Res* 23(5), 892–900, 2006.

105. Lambert, G., Bertrand, J. R., Fattal, E., Subra, F., Pinto-Alphandary, H., Malvy, C., Auclair, C., and Couvreur, P., EWS fli-1 antisense nanocapsules inhibits ewing sarcoma-related tumor in mice, *Biochem Biophys Res Commun* 279(2), 401–406, 2000.

106. Toub, N., Angiari, C., Eboue, D., Fattal, E., Tenu, J. P., Le Doan, T., and Couvreur, P., Cellular fate of oligonucleotides when delivered by nanocapsules of poly(isobutylcyanoacrylate), *J Control Release* 106(1–2), 209–213, 2005.

107. Toub, N., Bertrand, J. R., Malvy, C., Fattal, E., and Couvreur, P., Antisense oligonucleotide nanocapsules efficiently inhibit EWS-Fli1 expression in a Ewing's sarcoma model, *Oligonucleotides* 16(2), 158–168, 2006.

108. Torchilin, V. P., Zhou, F., and Huang, L., pHsensitive liposomes, *J Liposome Res* 3, 201–255, 1993.

109. Tanaka, K., Iwakuma, T., Harimaya, K., Sato, H., and Iwamoto, Y., EWS-Fli1 antisense oligodeoxynucleotide inhibits proliferation of human Ewing's sarcoma and primitive neuroectodermal tumor cells, *J Clin Invest* 99(2), 239–247, 1997.

110. Maksimenko, A., Malvy, C., Lambert, G., Bertrand, J. R., Fattal, E., Maccario, J., and Couvreur, P., Oligonucleotides targeted against a junction oncogene are made efficient by nanotechnologies, *Pharm Res* 20(10), 1565–1567, 2003.

111. Lewis, K. J., Irwin, W. J., and Akhtar, S., Biodegradable poly(L-lactic acid) matrices for the sustained delivery of antisense oligonucleotides, *J Control Release* 37, 173–183, 1995.

112. Lewis, K. J., Irwin, W. J., and Akhtar, S., Development of a sustained-release biodegradable polymer delivery system for site-specific delivery of oligonucleotides: Characterization of P(LA-GA) copolymer microspheres in vitro, *J Drug Target* 5(4), 291–302, 1998.

113. De Rosa, G., Quaglia, F., Bochot, A., Ungaro, F., and Fattal, E., Long-term release and improved intracellular penetration of oligonucleotide–polyethylenimine complexes entrapped in biodegradable microspheres, *Biomacromolecules* 4(3), 529–536, 2003.

114. Akhtar, S. and Lewis, K. J., Antisense oligonucleotide delivery to cultured macrophages is improved by incorporation into sustained-release biodegradable polymer microspheres, *Int J Pharm* 151, 57–67, 1997.

115. Kilic, A. C., Capan, Y., Vural, I., Gursoy, R. N., Dalkara, T., Cuine, A., and Hincal, A. A., Preparation and characterization of PLGA nanospheres for the targeted delivery of NR2B-specific antisense oligonucleotides to the NMDA receptors in the brain, *J Microencapsul* 22(6), 633–641, 2005.

116. Zhu, X., Lu, L., Currier, B. L., Windebank, A. J., and Yaszemski, M. J., Controlled release of NFkappaB decoy oligonucleotides from biodegradable polymer microparticles, *Biomaterials* 23(13), 2683–2692, 2002.

117. Khan, A., Benboubetra, M., Sayyed, P. Z., Ng, K. W., Fox, S., Beck, G., Benter, I. F., and Akhtar, S., Sustained polymeric delivery of gene silencing antisense ODNs, siRNA, DNAzymes and ribozymes: In vitro and in vivo studies, *J Drug Target* 12(6), 393–404, 2004.

118. Freytag, T., Dashevsky, A., Tillman, L., Hardee, G. E., and Bodmeier, R., Improvement of the encapsulation efficiency of oligonucleotide-containing biodegradable microspheres, *J Control Release* 69(1), 197–207, 2000.

119. Gomes dos Santos, A. L., Bochot, A., Doyle, A., Tsapis, N., Siepmann, J., Siepmann, F., Schmaler, J., Besnard, M., Behar-Cohen, F., and Fattal, E., Sustained release of nano-sized complexes of polyethylenimine and anti-TGF-beta 2 oligonucleotide improves the outcome of glaucoma surgery, *J Control Release* 112(3), 369–381, 2006.

120. Hussain, M., Beale, G., Hughes, M., and Akhtar, S., Co-delivery of an antisense oligonucleotide and 5-fluorouracil using sustained release poly (lactide-*co*-glycolide) microsphere formulations for potential combination therapy in cancer, *Int J Pharm* 234(1–2), 129–138, 2002.

121. De Rosa, G., Maiuri, M. C., Ungaro, F., De Stefano, D., Quaglia, F., La Rotonda, M. I., and Carnuccio, R., Enhanced intracellular uptake and inhibition of NF-kappaB activation by decoy oligonucleotide released from PLGA microspheres, *J Gene Med* 7(6), 771–781, 2005.

122. Yuan, X., Li, L., Rathinavelu, A., Hao, J., Narasimhan, M., He, M., Heitlage, V., Tam, L., Viqar, S., and Salehi, M., SiRNA drug delivery by biodegradable polymeric nanoparticles, *J Nanosci Nanotechnol* 6(9–10), 2821–2828, 2006.

123. Putney, S. D., Brown, J., Cucco, C., Lee, R., Skorski, T., Leonetti, C., Geiser, T., Calabretta, B., Zupi, G., and Zon, G., Enhanced anti-tumor effects with microencapsulated c-myc antisense oligonucleotide, *Antisense Nucleic Acid Drug Dev* 9(5), 451–458, 1999.

124. De Rosa, G., Quaglia, F., La Rotonda, M. I., Appel, M., Alphandary, H., and Fattal, E., Poly(lactide-*co*-glycolide) microspheres for the controlled release of oligonuclcotide/polyethylenimine complexes, *J Pharm Sci* 91(3), 790–799, 2002.

125. De Rosa, G., Bochot, A., Quaglia, F., Besnard, M., and Fattal, E., A new delivery system for antisense therapy: PLGA microspheres encapsulating oligonucleotide/polyethyleneimine solid complexes, *Int J Pharm* 254(1), 89–93, 2003.

126. Cleek, R. L., Rege, A. A., Denner, L. A., Eskin, S. G., and Mikos, A. G., Inhibition of smooth muscle cell growth in vitro by an antisense oligodeoxynucleotide released from poly(DL-lactic-*co*-glycolic acid) microparticles, *J Biomed Mater Res* 35(4), 525–530, 1997.

127. Berton, M., Benimetskaya, L., Allemann, E., Stein, C. A., and Gurny, R., Uptake of oligonucleotide-loaded nanoparticles in prostatic cancer cells and their intracellular localization, *Eur J Pharm Biopharm* 47(2), 119–123, 1999.

128. Ungaro, F., De Rosa, G., Quaglia, F., Fattal, E., and La Rotonda, M. I., Controlled release of oligonucleotide/polyethylenimine complexes from PLGA-based microspheres: Potential of spray-drying technique, *J Drug Delivery Sci Technol* 15(2), 113–192, 2005.

129. Cohen-Sacks, H., Najajreh, Y., Tchaikovski, V., Gao, G., Elazer, V., Dahan, R., Gati, I., Kanaan, M., Waltenberger, J., and Golomb, G., Novel PDGFbetaR antisense encapsulated in polymeric nanospheres for the treatment of restenosis, *Gene Ther* 9(23), 1607–1616, 2002.

130. Aukunuru, J. V., Ayalasomayajula, S. P., and Kompella, U. B., Nanoparticle formulation enhances the delivery and activity of a vascular endothelial growth factor antisense oligonucleotide in human retinal pigment epithelial cells, *J Pharm Pharmacol* 55(9), 1199–1206, 2003.

131. Diwan, M., Elamanchili, P., Lane, H., Gainer, A., and Samuel, J., Biodegradable nanoparticle mediated antigen delivery to human cord blood derived dendritic cells for induction of primary T cell responses, *J Drug Target* 11(8–10), 495–507, 2003.

132. Lebedeva, I., Benimetskaya, L., Stein, C. A., and Vilenchik, M., Cellular delivery of antisense oligonucleotides, *Eur J Pharm Biopharm* 50(1), 101–119, 2000.

133. Khan, A., Sommer, W., Fuxe, K., and Akhtar, S., Site-specific administration of antisense oligonucleotides using biodegradable polymer microspheres provides sustained delivery and improved subcellular biodistribution in the neostriatum of the rat brain, *J Drug Target* 8(5), 319–334, 2000.

Part III

Gene Expression and Detection

29 Deliberate Regulation of Therapeutic Transgenes

Nuria Vilaboa and Richard Voellmy

CONTENTS

29.1 INTRODUCTION

The expression of transgenes encoding powerful biological mediators or effectors should be limited to the optimal therapeutic window. Multiple gene switches that can be utilized to regulate the temporal expression of transgenes were developed. They typically consist of an artificial (chimeric) transcription factor that is activated or inactivated by a small-molecule ligand and a promoter that is activated or silenced by the transcription factor and is functionally linked to the transgene to be regulated. All systems appear to be, in principle, capable of functioning as on/off gene switches. However, they differ in respect of the level of maximal transgene expression they support, pharmacodynamic and kinetic properties of the activating or inactivating small-molecule ligands required, degree of humanization achieved or potentially achievable, and the nature of the transcription factor, i.e., whether it is monomeric, or homo- or hetero-oligomeric. Other differences relate to the level of characterization and consequently, useful information available to the research community.

Perhaps even more important than temporal regulation is spatial restriction of transgene expression. Although efforts have been undertaken to produce vectors that have specificity for certain target tissues, standard viral and nonviral vectors have limited ability or no ability to deliver transgenes exclusively to tissues, organs, or regions in need of therapy. Several approaches have been pursued to restrict expression of therapeutic genes or replication of viruses such as oncolytic viruses to an intended site of action.

Historically, the foremost strategy has been employment of tissue-specific or tissue-restricted promoters for controlling transgene expression [1,2]. Although simple in principle, this approach to transcriptional targeting is associated with several difficulties relating to less than perfect tissue specificity, fixed level of activity of a given promoter, and a need for using different promoters for different targets. Another strategy takes advantage of disease-associated biological or biochemical differences. Hypoxia-inducible promoters have been used to limit transgene expression to ischemic tissues including ischemic portions of solid tumors [3,4]. Promoters that are selectively active in proliferating cells have been exploited for confining transgene activity to tumor tissues [5,6]. A glucose-sensitive promoter has been employed for controlling insulin expression [7]. E1A- and E1B-deleted oncolytic viruses were investigated for their ability to selectively replicate in certain tumor tissues. Perhaps, the best known example of such a

619

virus is ONYX-015, an E1B-deleted oncolytic adenovirus that has undergone advanced clinical studies [8]. While the latter approaches are expected to be more generally applicable than the tissue-specific promoters discussed before, they may suffer from some of the same disadvantages, i.e., uncertainties relating to specificity of gene regulation as well as to uniformity of gene expression throughout diseased tissues or organs.

Another strategy that is expected to be essentially free of the limitations discussed before relies on the use of promoters, which are activated in essentially all tissues by directed physical force. A major portion of this chapter will be devoted to a discussion of this strategy. Promoters of this type include heat shock protein (hsp) gene promoters and radiation-induced promoters. Transgenes controlled by such promoters can be activated and their expression restricted to desired target regions by directed radiation or ultrasound, respectively. It is noted that the kinetics of activation and subsequent deactivation of the latter inducible promoters are defined by endogenous signaling mechanisms and therefore cannot be readily adapted to yield a desirable temporal profile of expression of a therapeutic transgene or virus replication. More complex systems were built and tested that combine a radiation- or ultrasound-activated promoter and a gene switch comprising a small-molecule-dependent transactivator. Such systems allow for deliberate control of both location and duration of transgene expression.

29.2 SMALL-MOLECULE-DEPENDENT GENE SWITCHES TO ACHIEVE CONTROL OVER DURATION OF TRANSGENE EXPRESSION

29.2.1 GENE SWITCHES COMPRISING TETRACYCLINE REPRESSOR–DERIVED TRANSACTIVATORS

Tetracycline-responsive gene switches were shown to effectively control transgene expression in cultured cells as well as in yeast, *Dictyostelium*, *Drosophila*, *Candida*, amphibia, rodents, and nonhuman primates. The *Escherichia coli* Tn10 tetracycline resistance operator consists of tetracycline repressor protein (TetR) and a specific DNA-binding site, the tetracycline operator (TetO) [9]. TetO is also referred to as tetracycline-response element (TRE) [10]. TetR forms dimers that bind to TRE. Tetracycline and derivatives including doxycycline interact with TetR and induce a conformational change that causes dissociation from TRE. A first tetracycline-dependent eukaryotic gene switch employed a fusion protein consisting of TetR and an activation domain from herpes simplex virus (HSV) protein VP16. This fusion protein was termed tTA. *In vitro*, tTA enhanced the activity of an artificial promoter consisting of a minimal cytomegalovirus (CMV) promoter and seven TRE sequences four orders of magnitude better in the absence than in the presence of tetracycline. This gene switch is referred to as the Tet-Off system (Figure 29.1A). The Tet-Off system was also shown to function in transgenic mice containing a tTA-encoding gene and a luciferase or β-galactosidase reporter gene [11]. To decrease the likelihood

that tTA interacted with cellular factors, the VP16 activation domain was reduced subsequently to three minimal repeats of 12 amino acids each [12].

Silencing of a transgene controlled by the Tet-Off system requires continuous administration of tetracycline or tetracycline derivative. This feature of the system may preclude its clinical application, especially when vectors are employed that result in an extended residence time of a therapeutic gene. A mutant of TetR was identified whose TRE-binding ability was induced rather than inhibited by tetracycline. This mutant was used to design the so-called Tet-On gene switch that incorporated an antibiotic-activated transactivator, referred to as rtTA (Figure 29.1B) [13]. As would have been expected, activation of the Tet-On switch was considerably more rapid than that of the Tet-Off switch. However, rtTA was less sensitive to doxycycline than tTA. This raised a concern that full transgene activation may not be achievable in tissues in which elevated concentrations of antibiotic cannot be attained. In addition, the Tet-On system exhibited instability that, presumably, was due to a residual affinity of rtTA for TRE. Several avenues were pursued to produce an improved version of the Tet-On system. A new transactivator, rtTA2^S-M2, was engineered through random mutagenesis and codon optimization. rtTA2^S-M2 had an improved sensitivity to inducer as well as a reduced residual activity in the absence of inducer [14]. More recently, viral evolution approaches were used for obtaining different variants of rtTA that also were more sensitive to inducer and, in addition, were more active than the original rtTA [15,16].

Residual rtTA activity could also be suppressed by means of a tetracycline-controlled transcriptional silencer (tTS) (Figure 29.1C) [17]. tTS is capable of binding TRE-containing promoters, preventing their activation by rtTA in the absence of doxycycline. When doxycycline is added, tTS is inactivated, rtTA is fully activated, and transcription from TRE-containing promoters is fully enabled. Zhu et al. [18] prepared tTS^{Kid}, a fusion between TetR and the KRAB domain of human kidney protein Kid-1. Gene switches that included both rtTA and tTS^{Kid} were able to stringently control expression of genes for toxic products. Transgenic mice that expressed tTS^{Kid} and rtTA under the control of a lung-specific promoter and interleukin-13 (IL-13) under the control of an artificial promoter containing TRE sequences were established. tTS^{Kid} suppressed residual IL-13 expression in the absence of doxycycline. In the presence of inducer, rtTA effectively stimulated IL-13 expression. To optimize delivery, the genes for rtTA, tTS^{Kid}, and IL-13 were introduced in a single adenoviral or episomally replicating vector. These systems performed well *in vitro* and *in vivo* [19–23]. However, not all experimentation with tTS-containing systems was successful. Lamartina et al. [24] reported that addition of tTS to a Tet-On switch resulted in impaired transgene activation. Barde et al. [25] found that target gene activation in CHO cells transduced with a lentiviral vector expressing rtTA2^S-M2 and tTS^{Kid} was progressively lost with time and repeated induction.

Yet another strategy for minimizing residual rtTA activity in the absence of inducer was to regulate rtTA expression by

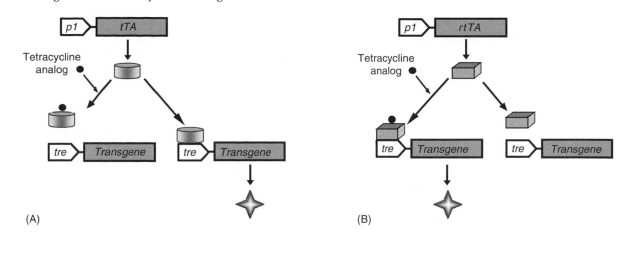

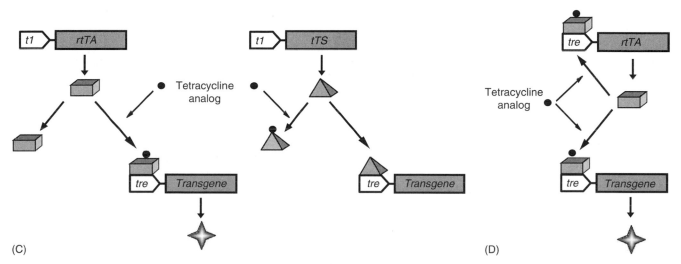

FIGURE 29.1 Tetracycline-responsive gene switches. (A) Tet-Off switch. (B) Tet-On switch. (C) Tet-On switch incorporating a tetracycline-controlled transcriptional silencer. (D) Tet-On switch comprising an autoactivated gene for rtTA. tTA and rtTA refer to tetracycline-inactivated and tetracycline-activated transactivators, respectively. tTS is a tetracycline-inactivated transcriptional silencer. tre, TRE sequence-containing (tetracycline-responsive) promoter; t1, tissue-restricted promoter.

means of an autoactivated promoter (Figure 29.1D) [26]. Such a system was tested in a panel of human primary and established cell lines transduced with lentiviral vectors and was found to exhibit low background activity in the absence of inducer and strong induction in the presence of inducer. Induced gene expression levels surpassed those produced by a CMV promoter [27].

To achieve prostate cell-specific transgene expression, a DiSTRES vector was recently designed by Woraratanadharm et al. [28]. Several gene cassettes were subcloned into the genome of a replication-incompetent adenovirus vector. To develop a prostate-specific and tTA-inducible promoter, a TRE was introduced upstream of a probasin promoter. The resulting *TRE-ARR2PB* promoter was used to control the expression of a Lac repressor (*lacI*) gene. A second component consisted of a probasin promoter linked to a *tTA* gene that was cloned immediately opposite of the *TRE-ARR2PB-lacI*

cassette such that the TRE had bidirectional activity. A third component contained a green fluorescence protein (*gfp*) gene under the control of a tetracycline-responsive promoter. In addition, a gene cassette containing a newly developed Lac repressor-responsive promoter was cloned into the left-end of the genome and controlled expression of a *tTS* gene. In a prostate cell, both *tTA* and *lacI* genes would be active, because they are under the control of prostate-specific TRE-ARR2PB promoters. In the absence of tetracycline or a derivative, tTA would then transactivate the *gfp* gene. The tTA protein would also bind to the TREs within its own *TRE-ARR2PB* promoter, thereby establishing a feedforward loop of tTA amplification. Lac repressor would bind to the Lac repressor-responsive promoter of the *tTS* gene, thereby suppressing *tTS* gene transcription. Neither tTA nor Lac repressor would be expressed in nonprostate cells. In the absence of Lac repressor, the *tTS* gene is actively transcribed, resulting in transcriptional silencing of all tetracycline-regulated genes, including the gene of interest (here the *gfp* gene). Note that tetracycline or its derivatives are not required for cell-specific performance of this system. Green fluorescent protein (GFP) expression from DiSTRES adenovector was more than 30-fold higher than from Ad/CMV-*GFP* in LNCaP cells.

Weber et al. [29] constructed a synthetic mammalian gene network that processes different input signals to control tightly the expression of a specific target gene. Three components of the switch were controlled by constitutive promoters: the first component of the circuitry was a VP16 transactivation domain fused to an AVITAG biotinylation signal; the second component was *E. coli* biotin ligase BirA, and the third component was streptavidin fused to tetracycline repressor (TetR-SA). A fourth component consisted of a TRE-containing promoter that controlled a human placental–secreted alkaline phosphatase target gene. In the presence of biotin and in the absence of tetracycline, VP16 was biotinylated by BirA and interacted with TetR-SA that was bound to the TRE-containing promoter, resulting in target gene expression. Upon removal of biotin, biotinylated VP16 was eliminated by degradation and target gene expression ceased. In the absence of biotin, VP16 was not biotinylated and the target gene promoter remained silent. Addition of tetracycline induced dissociation of TetR-SA from the TRE-containing promoter, disabling the target gene expression.

In summary, the newer tetracycline-controlled gene switches are capable of tightly regulating transgene expression *in vitro* and *in vivo*. Fully activated transgene expression compares well with that achievable by other small-molecule-dependent gene switches [30,31]. However, some concerns remain. TetR is of bacterial origin. Hence, there is a possibility of immune responses to TetR-derived regulators. In fact, rtTA transactivators were shown to be immunogenic in monkeys, an effect that appeared to be related to short-lived expression of transgenes under their control [32,33]. Furthermore, therapy with a transgene under the control of a tetracycline-responsive gene switch may result in the development of resistance to tetracycline or other antibiotically active derivatives. Finally, there appear to exist cellular factors such as interferon α–stimulated gene factors and GATA factors that have an affinity for standard tetracycline-responsive promoters. Evidence was obtained that binding of these factors affects the activity of tetracycline-responsive systems [34,35].

29.2.2 Other Gene Switches Employing Bacterial-Derived Transactivators

Gene switches that responded to streptogramins, such as pristinamycin, or macrolide antibiotics, such as erythromycin, were described [36–41]. Like the tetracycline-responsive systems, these gene switches employed transactivator proteins of bacterial origin. It is noted that pristinamycin and erythromycin are approved for human use. Cronin et al. [42] reengineered the *lacI* gene to avoid silencing and to ensure widespread expression in mouse tissues. The repressor was found capable of controlling effectively a tyrosinase transgene

in a transgenic mouse model. The transgene was active in the presence but not in the absence of inducer isopropyl-β-D-thiogalactopyranoside (administered at a nontoxic dose). Transgene expression resulted in readily visible increases in fur and eye pigmentation. Ryan and Scrable [43] reported another application of this system. The regulatory mechanisms of the *p-cym* operon of *Pseudomonas putida* were exploited for the development of several mammalian gene switches incorporating the CymR repressor that is responsive to small-molecule inducer cumate [44]. A concern common to these systems is that they involve expression of bacterial protein sequences. Therefore, the possibility of adverse immune reactions exists that may render these systems unsuitable for controlling long-term expression of transgenes.

29.2.3 Gene Switches Employing Steroid Receptor–Derived Transactivators

29.2.3.1 Ecdysone Receptor-Based Gene Switches

Insect ecdysone receptor (EcR) forms a heterodimeric transactivator with ultraspiracle protein (USP). USP is the insect ortholog of vertebrate retinoid X receptor (RXR) [45]. Binding of ecdysteroids (insect steroids) to EcR is increased in the presence of USP. Binding stabilizes EcR/USP complex and enhances the affinity of the complex for ecdysteroid response elements (EcRE) present in promoters of ecdysteroid response genes [46]. A mammalian homolog of EcR does not appear to exist. Ecdysteroids are not detected in vertebrates. Like mammalian steroids, ecdysteroids readily distribute to all important tissues and are quickly metabolized and cleared.

No et al. [47] described a gene switch consisting of chimeric receptor VpEcR and a target gene promoter. VpEcR is a fusion protein comprising *Drosophila melanogaster* EcR (*Dm*EcR) and an HSV VP16 transactivation region. In human 293 cells, VpEcR effectively forms a complex with endogenous RXR in the presence of an ecdysteroid such as ponasterone A. VpEcR/RXR complex binds to EcRE-containing promoters and activates downstream genes. Highly induced reporter gene activity was observed in 293 cells expressing VpEcR in the presence of muristerone A. An early improvement consisted of replacing the EcRE sequences in the target gene promoter with synthetic E/GRE sequences and of mutagenizing VpEcR to enable it to bind the latter synthetic sequences. E/GRE is a hybrid between a glucocorticoid response element and a response element for a type II receptor (e.g., EcR, RXR). This improvement was aimed at preventing possible interference by endogenous receptors such as farnesoid X receptors. The improved gene switch consisting of VgEcR (mutated VpEcR) and a target gene under the control of an E/GRE-containing promoter exhibited lower basal activity and higher inducibility than a tetracycline-regulated switch in 293 cells [47].

Despite the fact that RXR is expressed in many mammalian cell types, human 293 cells were found to be uniquely capable of supporting efficient formation of EcR/RXR complex. In other cells, extra RXR needed to be supplied from

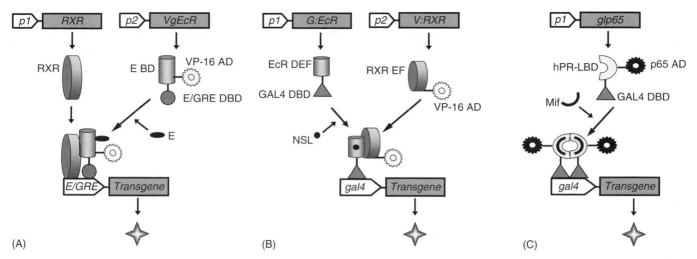

FIGURE 29.2 Gene switches employing steroid receptor–derived transactivators. (A) System based on coexpression of retinoid X receptor (RXR) and an ecdysone receptor (EcR)-derived transactivator (VgEcR) consisting of an E/GRE DNA-binding domain (E/GRE DBD), ecdysteroid-binding domain (E BD) and VP16 transcriptional activation domain (VP-16 AD). (B) EcR/RXR-based gene switch in two-hybrid format. G:EcR comprises a yeast GAL4 DNA-binding domain (GAL4 DBD) and EcR D, E and F domains (EcR DEF), and V:RXR includes a VP16 transcriptional activation domain and RXR E and F domains (RXR EF). (C) Progesterone receptor–based gene switch. The *glp65* gene encodes a mifepristone-activated transactivator comprising a yeast GAL4 DNA-binding domain, a human progesterone receptor ligand-binding domain (hPR-LBD) and a human P65 transcriptional activation domain (p65 AD). E, Ecdysteroid; NSL, nonsteroidal ligand; Mif, mifepristone.

exogenous genes (Figure 29.2A) [45,46,48]. To achieve coordinate expression of VgEcR and RXR, Wyborski et al. [49] used a CMV promoter-driven bicistronic expression cassette. Galimi et al. [50] transduced mouse hematopoietic progenitor cells with two lentiviral vectors, the first expressing the VgEcR–RXR cassette and the second harboring a reporter gene encoding GFP. Four months after transplantation into mice, exposure to an ecdysteroid increased the number of peripheral blood cells expressing the *gfp* reporter gene. This result showed that the gene switch could be employed in a gene therapy setting.

Further engineering resulted in versions of EcR-based gene switches that did not require cointroduction of an RXR gene. A modified *Bombyx mori* EcR transactivated effectively in the presence of ecdysone agonist tebufenozide [51]. Hoppe et al. [52] constructed *DB*EcR, a chimeric *Drosophila/Bombyx* ecdysone receptor that combined activation and DNA-binding domains of VgEcR with hinge region and ligand-binding domain of *B. mori* EcR. *DB*EcR was a potent transactivator in the presence of ecdysteroid ponasterone A in mammalian cells or of nonsteroidal ecdysone agonist GS-E ([*N*-(3-methoxy-2-ethylbenzoyl)-*N*′-(3,5-dimethylbenzoyl)-*N*″-*tert*-butylhydrazine]) both *in vitro* and *in vivo*. *DB*EcR-based gene switches have the advantages of being encoded by a relatively short nucleic acid, as only a single transactivator subunit needs to be supplied, and of not causing large increases in RXR concentration, which may have adverse biological effects.

More recently, EcR-based gene switches were also developed in a two-hybrid format (Figure 29.2B). Palli et al. [53] prepared a G:*Cf*E (DEF) fusion receptor comprising the D, E, and F domains of *Choristoneura fumiferana* EcR (*Cf*EcR)

and a yeast GAL4 DNA-binding domain. Two-hybrid partner was V:*Mm*R (EF) comprising E and F domains of mouse RXR and a VP16 activation domain. The heterodimeric, chimeric receptor transactivated a reporter gene controlled by a *gal4* promoter almost 9000-fold better in the presence than in the absence of ligand. Nonsteroidal ligands were more effective than ecdysteroids. This system, now known as RheoSwitch, performed well in stable cell lines and a bladder cancer xenograft model [54]. Use of the ligand binding domain of a new mutant of *Cf*EcR (GEvy) enhanced the sensitivity of the system such that maximal activation occurred at nanomolar concentrations of nonsteroidal ligands [55]. Substitution of mouse RXR sequences with corresponding sequences from *Locusta migratoria* RXR (*Lm*RXR) also increased ligand sensitivity. Two-hybrid transactivator G:*Cf*E (DEF)–V:*Lm*R (EF) stimulated a reporter gene in 3T3 cells at lower GS-E concentrations than G:*Cf*E (DEF)–V:*Hs*R (EF) [56]. A further improved RXR chimera was prepared that contained helices 1–8 from *Hs*RXR and 9–12 from *Lm*RXR and tested in combination with G:*Cf*E (DEF). Nanomolar concentrations of GS-E sufficed to induce reporter activity in 3T3 cells. This two-hybrid gene switch also performed well in mouse muscle subsequent to *in vivo* electroporation. The development of multiplexed gene switches is also notable, which switches exploited differences in affinities of different EcR forms for steroidal and nonsteroidal ligands [57].

In summary, EcR-based gene switches are now available that exhibit high ligand sensitivity. Different systems preferentially respond to either steroidal or nonsteroidal ligands. The systems are characterized generally by low basal and high induced activities. One possible drawback of employment of these systems in mammalian organisms relates to the fact that they necessarily include nonmammalian protein sequences. The possibility of immune responses that limit their usefulness cannot be excluded. Although EcR-based systems should not influence or be influenced by endogenous steroid regulation, a recent report by Oehme et al. [58] raises some concern. The authors observed that FasL- and TRAIL-induced apoptosis in colon carcinoma cells was inhibited by ecdysteroid muristerone A.

29.2.3.2 Progesterone Receptor-Based Gene Switches

Mifepristone is a synthetic steroid that acts as a competitive antagonist of progesterone receptor. The drug substance that was formerly known as RU486 is approved for human use. Mifepristone may be administered orally or, in animal studies, intraperitoneally (i.p.). To avoid repetitive administration, implanted mifepristone timed release pellets may be utilized [59].

A mutant human progesterone receptor lacking 42 carboxy-terminal amino acids failed to bind progesterone but interacted with mifepristone and enhanced transcription of reporter genes [60]. A chimeric transactivator was constructed by replacement of the original DNA-binding domain with a GAL4 DNA-binding domain and addition of a VP16 transcriptional activation domain at the amino terminus [61]. The chimeric receptor transactivated reporter genes controlled by a promoter containing 17-mer GAL4 upstream activating sequences in the presence of very low concentrations (0.1–1 nM) of mifepristone both *in vitro* and *ex vivo*. A more effective transactivator was obtained, in which the progesterone receptor ligand-binding domain was extended and the VP16 activation domain was added at the carboxy terminus [62]. To partially humanize the transactivator, the VP16 activation domain was replaced with an activation domain from the P65 subunit of human NF-κB [63]. As already discussed in the context of Tet-On systems, placing transactivator expression under feedforward regulation by means of a promoter containing GAL4 response elements reduced basal activity [64]. Finally, shortening the GAL4 DNA-binding domain resulted in transactivator GS4 that has a reduced ability to homodimerize and, consequently, to transactivate in the absence of a ligand [65]. See Figure 29.2C for a depiction of a progesterone receptor-based gene switch.

Wang et al. [66] produced transgenic mice containing a gene for a mifepristone-activated, chimeric transactivator under the control of a "liver-specific" promoter and a human growth hormone (hGH) reporter gene driven by a *gal4* promoter. A single dose of mifepristone induced hGH expression, resulting in transiently elevated serum levels. Minimal induction was observed in the absence of mifepristone. Expression of hGH was reversible and could be re-induced by administration of a second dose of ligand. Similarly, transgenic mice were described that were capable of expressing inhibin under mifepristone control [59]. Mifepristone-dependent expression *in vivo* was also observed subsequent to the delivery of

transactivator and target gene by means of adenovirus vectors [63,67,68]. Mifepristone-dependent transgene expression in the brain resulted from delivery of gene switch and target gene by means of an HSV vector [69].

Generally, mifepristone-activated gene switches are characterized by low basal activity and high degree of inducibility. However, Xu et al. [31] reported that induced levels of target gene expression tend to be lower than those produced by tetracycline-responsive or dimerizer-activated gene switches. Possible concerns relate to the fact that the currently available mifepristone-dependent gene switches contain yeast GAL4 protein sequences. Hence, the mifepristone-activated transactivators may be immunogenic, limiting the usefulness of the gene switches in human genetic therapy. Furthermore, although mifepristone-dependent gene switches can be triggered by subclinical concentrations of ligand, whether these low concentrations are, in fact, without effect on the ovarian cycle appears not to be known.

29.2.3.3 Estrogen Receptor-Based, Fully Humanized Gene Switch

Roscilli et al. [70] engineered a gene switch based on chimeric transactivator HEA-3. This transactivator is comprised of the DNA-binding domain of human hepatocyte nuclear factor-1α (HNF1α), the G521R mutant of the ligand-binding domain of human estrogen receptor and activation domains from the P65 subunit of human NF-κB. Transcription factor HNF1α is expressed in the liver but not in the muscle. The G521R domain binds antiestrogen 4-hydroxytamixofen but not estradiol. The expression of a reporter gene in cells expressing HEA-3 was three orders of magnitude greater in the presence than in the absence of 4-hydroxytamoxifen. Induction of about 100-fold was measured in mouse muscle into which transactivator and reporter genes had been electroinjected. Expression in muscle of a HEA-3-controlled erythropoietin (*epo*) gene could be maintained for about 300 days in mice dosed daily with tamoxifen, a validated drug substance that is converted to 4-hydroxytamoxifen *in vivo*.

The tamoxifen-activated gene switch was capable of effectively controlling target gene expression *in vitro* and *in vivo*. The system is fully humanized. Hence, limiting immune reactions in human therapy are not expected. Potential concerns include that activation of the system requires elevated doses of tamoxifen. Tamoxifen has been known to induce thromboembolic phenomena, a low increased risk of endometrial cancer and hot flashes (discussed in Ref. [71]). Furthermore, the DNA-binding domain of HEA-3 originates from a human transcription factor. It cannot be excluded that HEA-3 may modulate endogenous gene expression, particularly when delivered by means of standard, untargeted vectors.

29.2.4 DIMERIZER-ACTIVATED GENE SWITCHES

These gene switches take advantage of known biological properties of drug substances such as rapamycin and FK506. To begin with rapamycin, this macrolide is approved for human use. It causes inhibition of T cell proliferation and immunosuppression. Of interest in the context of gene regulation is that the molecule distributes systemically, including in the brain. Rapamycin is known to interact with abundant immunophilin FKBP12. FKBP12–rapamycin complex associates with mTOR/FRAP and inhibits its enzymatic activity. FKBP rapamycin-associated protein (FRAP) is a phosphoinositide-3-kinase homolog that is involved in the control of cell growth and division (reviewed in Ref. [71]). The concept underlying the rapamycin-based gene switches is that essential components of a transcription factor, i.e., DNA-binding site and activation domain(s), are distributed between two fusion proteins comprising FKBP12 and sequences required for docking of rapamycin–FKBP12 complex, respectively. In the presence of rapamycin, the two fusion proteins are joined. This dimerized factor, but not the individual fusion proteins, is capable of functioning as a transcription factor.

A first rapamycin-based gene switch was described by Rivera et al. (Figure 29.3) [72]. The first fusion protein contained DNA-binding fusion ZFHD1 and three tandem copies of FKBP12. ZFHD1 includes human zinc finger and homeodomain DNA-binding domains that bind tightly to composite DNA response elements with the consensus sequence TAATTANGGGNG [73]. The second fusion protein included FKBP12-rapamycin-associated protein (FRB), a 100-residue domain of FRAP that is sufficient for binding rapamycin–FKBP12 complex, and an activation domain from P65. The gene to be regulated was under the control of a minimal *il-2* promoter supplemented with 12 ZFHD1 binding sites. This system was tested in stably transfected cells and mice [72,74]. Strict rapamycin-dependent regulation as well as high levels of induced expression were reported. Maximal induced target gene activity was found to be limited by the nature of the transactivation domain used. The current version of the system that is considerably more active than the original version employs a combination of activation domains from P65 and human heat shock factor 1 [75].

Stringent rapamycin-dependent gene expression was observed following delivery of transactivator and target genes by means of onco-retroviral and lentiviral vectors [75,76], adenoviral vectors [77–79], HSV amplicon vectors [80], and adeno-associated viral (AAV) vectors [77,81–83]. Long-term expression following AAV-mediated delivery was investigated extensively. Johnston et al. [84] introduced an *epo* gene under the control of a rapamycin-dependent gene switch into muscle of β-thalassemic mice. Substantial increases in hematocrit values were measured upon periodic activation of

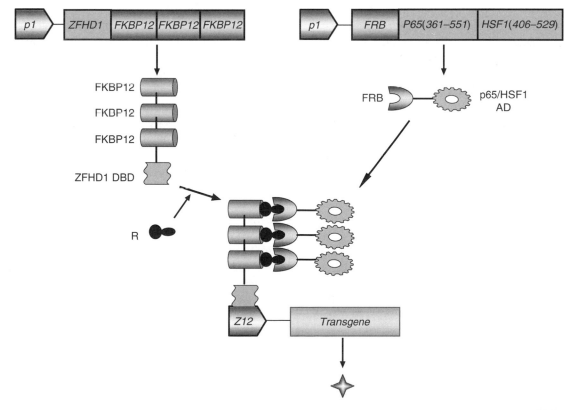

FIGURE 29.3 Rapamycin-activated gene switch. A first fusion protein encodes a ZFHD1 DNA-binding domain (ZFHD1 DBD) and three copies of FKBP12 arranged in tandem. The second fusion protein comprises the 100-residue FRB domain and a composite transcriptional activation domain made from activation domains from human P65 protein and human heat shock transcription factor HSF1 (P65/HSF1 AD). Note that numbers refer to amino acid positions in P65 and HSF1, respectively. *p1*: promoter that drives expression of fusion proteins; *Z12*: promoter containing 12 recognition sequences for ZFHD1; R: rapamycin.

the rapamycin-controlled *epo* gene. Importantly, the severe toxicity (polycythemia) known to be caused by uncontrolled overexpression of expression of erythropoietin (EPO) from constitutively expressed transgenes was not observed. Notably, the *epo* gene could still be activated reversibly half a year after delivery. Rivera et al. [85] introduced AAV vectors containing a rapamycin-dependent gene switch and an *epo* target gene into leg muscle of Rhesus monkeys. Remarkably, EPO expression could be induced reversibly over a period of more than five years. Induced levels of EPO were sufficient to produce significantly increased hematocrit values. Lebherz et al. [86] published results of a similar study, in which long-term inducible expression in the eyes of Rhesus monkeys was obtained.

Several studies compared optimized rapamycin-activated systems with Tet-On or Tet-Off and steroid receptor-based systems [30,31]. The consensus emerged that transgene expression was controlled more stringently by rapamycin-activated gene switches than any of the other systems. This conclusion is not surprising if one considers that rapamycin-activated gene switches are based on two proteins containing necessary components of a transcription factor, each of which has an affinity for rapamycin but not the other protein. Perhaps, the most remarkable demonstration of the degree of regulation achievable was provided by Pollock et al. [75]. These authors were able to establish a cell line, in which a rapamycin-activated gene switch controlled expression of the highly toxic diphtheria toxin A chain protein. Rapamycin-activated and Tet-On/Tet-Off systems were reported to support higher levels of target gene expression than steroid receptor-based systems.

The main concern with rapamycin-based systems relates to the immunosuppressive properties of rapamycin. However, several derivatives of rapamycin, so-called rapalogs, were identified that have greatly reduced immunosuppressive activity but still can trigger rapamycin-activated gene switches, provided that the FRB domain present in one of the two transactivator halves contains a compensatory Thr-to-Leu substitution at position 2098 [83,87]. Two such rapalogs, AP22565 and AP21967, were shown to be competent activators *in vitro* and in mice [79,83]. Rivera et al. [85] tested activation of an *epo* gene controlled by a rapamycin-activated gene switch by other rapalogs in muscle of Rhesus monkeys. AP22594 (28-epi-rapamycin), a rapalog that is about 100 times less immunosuppressive than rapamycin, was a similarly potent activator of EPO expression as the prototype drug.

Clemons et al. [88] designed an FK506-activated gene switch. This gene switch relies on interactions of FK506 and FKBP12, and of FK506–FKBP12 complex and calcium-dependent protein phosphatase calcineurin that results in immunosuppression. FK506 is a well-characterized molecule that has been used clinically. The authors succeeded in constructing a fusion protein comprising sequences from both calcineurin subunits that was capable of binding FK506–FKBP12 complex. They also were able to prepare nonimmunosuppressive derivatives of FK506 that interacted with mutant forms of the FKBP12 fusion protein. These developments enabled the construction of a novel gene switch that is activated by nonimmunosuppressive derivatives of FK506.

29.3 SPATIAL RESTRICTION OF TRANSGENE EXPRESSION: PROMOTERS RESPONDING TO EXOGENOUS STIMULI

29.3.1 RADIATION-INDUCED PROMOTERS

Irradiation of cells is known to result in changes in gene expression. A group of genes, so-called "immediate early genes," react rapidly to ionizing radiation (IR). Protein synthesis is not required for this induction. Immediate early genes include genes for transcription factors such as c-FOS, c-JUN, AP-1, NF-κB, and EGR-1 (reviewed in Ref. [89]). Early growth response-1 (EGR-1) binds to consensus sequence GCGGGGGCG and regulates genes for tumor necrosis factor α (TNF-α), IL-1, platelet-derived growth factor α, and basic fibroblast growth factor. Activation of the *egr-1* promoter by IR is transient: it occurs within 15 min of irradiation and the promoter remains active for about 3 h. The primary mechanism of activation involves P38 kinase and JUNK/SAPK pathways that mediate phosphorylation of ternary complex factor ELK-1 and binding of serum response factor/ELK-1 complex to sequence elements comprising the core sequence $CC(A/T)_6GG$ in the *egr-1* promoter. Alternatively, ELK-1 phosphorylation may be achieved by a mechanism that involves activated protein kinase C.

The use of focused IR for spatially restricting transgene expression was advocated by Weichselbaum et al. [90]. Their studies largely focused on tumor therapy by TNF-α and on controlling expression of TNF-α by means of an *egr-1* promoter and directed IR [91]. TNF-α was an attractive choice for an effector both because of its demonstrated antitumor activity and its known systemic toxicity. Regarding systemic toxicity, studies by Marr et al. [92,93] had shown significant toxicity of TNF-α expressed from a constitutively active gene delivered using an adenoviral vector. Concerning antitumor activity, TNF-α was known for its cytotoxicity, antiangiogenic effects, and enhancement of antitumor immunity. Considering the current limitations to gene delivery, the fact that TNF-α is a secreted molecule could be expected to contribute significantly to therapy efficacy.

Hallahan et al. [94] experimented with a nude mouse model onto which a radio-resistant human SQ-20B tumor was grafted. A pure IR therapy was compared with a combination IR/TNF-α therapy. Combination therapy involved intratumoral administration of an adenovector containing an *egr-1* promoter-controlled *tnf-α* gene (Ad.Egr-TNF). Animals were subjected to fractionated IR (50 Gy). Regression of tumors was observed in the course of a 60-day period. Regression was more pronounced for the combination therapy than the pure IR therapy. Immunochemical analysis of tumor samples revealed that irradiation produced an eightfold increase of TNF-α expression. It is noteworthy that serum concentration of TNF-α remained far below the known toxic level. As far as the mechanism of regression is concerned, Hallahan et al. observed that the combination therapy produced significant tumor cell apoptosis and necrosis in the tumor bed. Work by others indicated that the therapy resulted in selective damage to the tumor vasculature [95,96]. An antiangiogenic effect of TNF-α was also described [97].

Two phase I clinical trials investigated combination therapy with IR and TNF-α. These trials were completed and were reviewed in Mezhir et al. [98]. Generally, the trials showed improved tumor responses compared to historical controls. TNF-α was provided by means of a recombinant adenovirus referred to as TNFerade. TNFerade consisted of a second generation adenovirus vector containing E1, E3, and E4 deletions that harbored a *tnf-α* gene functionally linked to an *egr-1* promoter. In one of these phase I studies, patients having different types of solid tumors were treated by intratumoral injections of TNFerade over a period of 6 weeks (4×10^7 to 4×10^{11} particle units) and IR (30–70 Gy) [99]. Twenty-one of the 30 evaluable patients showed objective tumor responses. The study also included five patients with synchronous lesions. Both lesions were subjected to IR, but only one received TNFerade. In four of the five patients, the combination treatment showed an improved response compared to IR alone. In the other study, patients with soft tissue sarcomas of the extremities were injected intratumorally with TNFerade over a 5-week period (4×10^9 to 4×10^{11} particle units) and received IR (36–50.4 Gy) [100]. Tumor responses, including two complete responses, were recorded in 11 of the 13 evaluable patients. No dose-limiting toxicity of TNFerade was observed. Several phase II trials are under way.

Other combinations of radiation-induced promoters and transgenes were also tested. Kawashita et al. [101] examined effects of a suicide therapy in a hepatocellular carcinoma model. The therapy involved delivery of an HSV thymidine kinase *(tk)* gene under the control of an *egr-1* promoter. Tsurushima et al. [102] recently tested a plasmid harboring the radiation sensitive *egr-1* promoter linked to a gene for tumor necrosis factor–related apoptosis-inducing ligand (TRAIL) for the treatment of malignant gliomas. *In vivo* electroporation of U251 gliomas growing subcutaneously in nude mice followed by IR led to significant decreases in tumor volumes. Worthington et al. [103] prepared a construct comprising an inducible nitric oxide synthase *(inos)* gene controlled by a *p21/waf1* promoter and utilized this construct to transfect rat tail arterial sections. After a 4 Gy IR dose, a fivefold increase in iNOS level could be measured, whose increase correlated with a dramatically reduced contractile response to phenylephrin. The same group also advocated use of a *p21* promoter for controlling *inos* gene therapy of cancer. The combination of iNOS and IR showed efficacy in models of both p53 wild-type and p53 mutant tumors [104]. The *p21* promoter, transduced by rAAV vectors, was found to be remarkably responsive to very low doses of IR [105]. This high sensitivity of the *p21* promoter to IR may be exploited for the design of safer, radiation-based cancer gene therapies.

The *egr-1* promoter is only weakly inducible by IR [106]. However, significant improvements were shown to be possible. Manipulation of the CArG units of the *egr-1* promoter was found to enhance the response to IR. Improved IR responsiveness was obtained after increasing the number of CArG elements or after specific alteration of the core A/T sequences [107]. Combination of optimal number and sequence of CArG elements resulted in the best IR induction levels attained to date, both *in vitro* and *in vivo*. Scott et al. [108] designed a two-component gene switch. The first component was a gene for CRE recombinase under the control of a synthetic *egr-1*-like promoter. The second component consisted of a gene for an enhanced GFP (EGFP) linked to a CMV promoter. Promoter and gene were separated by a stop cassette flanked by LOX sites. IR induced expression of CRE recombinase from the first construct. CRE recombinase removed the stop cassette from the second construct, enabling EGFP expression. EGFP expression was found to be induced about 40-fold by IR. It is noted that target gene expression is irreversibly activated by IR in this system, limiting its range of possible applications. A gene switch of this type was recently employed by Greco et al. [109] in a model of IR-induced HSV *tk* suicide tumor therapy.

As reviewed herein, radiation-inducible promoters were used successfully to regulate different transgenes, and therapeutic effects, without limiting toxicity, were obtained. As therapies become more effective, the low inducibility, a significant basal activity of many radiation-induced promoters, is expected to become an obstacle to their therapeutic use. However, it has already been demonstrated that systems with substantially increased inducibility can be engineered. Hence, improved IR-induced systems may be available when the need arises. What remains are potential dangers inherent in IR. IR-inducible gene switches may be utilized primarily in situations in which IR therapy has some demonstrated effectiveness and addition of an IR-induced transgene can be expected to improve upon the therapeutic outcome.

29.3.2 Heat-Induced Promoters

Heat can be administered to a subject in a focused fashion. Heat delivery may be achieved simply by contact with a heated surface or liquid. Microwave devices may be utilized for heating selective body regions. Devices that should be suitable for such applications have been constructed and are being marketed for hyperthermic treatment of tumors. Infrared radiation allows for heating of areas close to the body surface, while ultrasound can heat deep-seated areas. Combining focused ultrasound and magnetic resonance imaging (MRI)-based, real-time temperature mapping resulted in an approach that was shown to be capable of achieving controlled heating of narrowly defined regions within the body [110,111].

Heat shock or stress protein *(hsp)* genes are a small group of conserved genes encoding proteins (HSPs) of different subunit molecular weights. Different HSPs are named HSP90, HSP70, HSP60, etc., the numbers referring to approximate subunit molecular weights in kDa. While HSPs are normally present and play important physiological roles, they accumulate transiently to elevated levels in cells exposed to certain physical and chemical stresses. Elevated levels of stress proteins are known to tolerize cells to the latter stresses. HSPs are typically expressed from multiple genes, of which at least one is induced when cells are exposed to a mild to moderate physical or chemical insult. Heat appears to be the most powerful inducer of these genes. One of the most highly inducible *hsp* genes appears to be the human *hsp70B* gene. First described in 1985, the promoter of this gene (hereinafter the *hsp70B* promoter) has a very low basal activity, which increases several thousand fold upon heat induction [112–114].

The only recognizable feature common to heat-inducible *hsp* promoters is the presence of one or more so-called heat shock elements (HSEs). These HSEs typically comprise three or more modules of the sequence NGAAN that are joined in alternating orientations. HSEs are binding sites for heat shock factors (HSFs). The *hsp70B* promoter contains three HSEs located within about 300 bp from the transcription start site. All three elements appear to contribute to the heat inducibility of the promoter [113]. Note that "minimal *hsp70B* promoter" refers to a truncated *hsp70B* promoter that includes the latter three HSEs but little additional upstream sequence.

Induced expression of *hsp* genes is mediated by HSFs that interact with HSEs present in their promoters. Although mammalian cells express at least three different types of HSFs, the factor known as HSF1 appears to be the most important. This conclusion is based on the observations that (1) heat-induced HSE DNA-binding activity can be traced back to HSF1 (reviewed in Ref. [115]), (2) heat or chemical stresses do not induce *hsp* genes in HSF1-deficient mice [116], and (3) a constitutively active form of HSF1 is capable of inducing enhanced expression of inducible HSPs [117]. HSF1 is ubiquitously expressed and is normally present in an inactive form. Transient activation of HSF1 occurs when a cell is exposed to heat or another insult. This activation is a complex process that involves homotrimerization of the factor (reviewed in Ref. [118]).

As was discussed above, heat can be delivered in a focused fashion. Consequently, it should be possible to use *hsp* promoters for spatially controlling transgene expression, a concept that has been discussed since the 1990s. *hsp* promoters have several additional attributes that should increase their usefulness. Perhaps, the most important among these additional features is that the promoters can be activated in most cell types. Hence, it should be possible to activate a therapeutic gene throughout an organ comprised of different tissues or a complex tumor, eliminating the concern associated with the use of tissue-restricted promoters that certain cells or tissues within an organ or complex tumor may escape therapy. From the point of view of convenience, only one transgene construct would need to be prepared. This construct could be expected to be functional in all possible target tissues. Furthermore, the level of induced activity of an *hsp* promoter is correlated with the intensity of the activating heat dose. Therefore, it should be possible to express in a target tissue a therapeutic gene product at the level that produces optimal therapeutic results. Moreover, activated *hsp* promoters support higher levels of target gene expression than many tissue-restricted promoters. Finally, heat activation of *hsp* promoters is reversible. The promoters are silenced within a period of at most a few hours from the activating heat treatment. Hence, they provide an element of therapeutic safety. However, as is discussed in more detail below, this feature of *hsp* promoters also limits

their usefulness as they may not be suitable for directing therapies requiring transgene activity for more than the latter short time window.

Several studies demonstrated that the *hsp70B* promoter can be utilized for local control of transgene expression. In one such study, an *egfp* reporter gene linked to a minimal human *hsp70B* promoter was introduced into rat C6 glioblastoma cells and a stable cell line was selected [119]. This line was introduced subcutaneously in nude mice. After sizeable tumors had developed, the tumor cells were heated by means of a heated needle that was inserted in the tumors. Heating was either at 50°C for 3 min or at 43°C for 30 min. Strong induction of EGFP expression occurred in heated tumors but not in unheated tumors. The shorter heat treatment at the higher temperature resulted in a better definition of gene activity within the tumors. Time courses of transgene activation and silencing were defined *in vitro*: Peak EGFP level was attained one day after thermal treatment. Two days later, EGFP concentration had returned to background level.

Huang et al. [120] prepared *hsp70B-egfp*, *hsp70B-tnf-α*, and *hsp70B-il-12* gene cassettes and introduced them in replication-deficient adenovirus vectors. Initial experiments in which 4T1 mouse mammary tumor cells were infected with the different viruses revealed that all three transgenes were highly induced by heat treatment (41–43°C for 20 min). EGFP level, detected by fluorescence-activated cell sorter (FACS), increased 500–1000-fold, and levels of TNF-α and IL-12 6.8×10^5 and 13,600-fold, respectively. EGFP level was maximal 18–24 h after heat treatment. At 72 h, EGFP was no longer detectable. Induction *in vivo* was tested following injection of recombinant adenoviruses into B16F10 melanomas growing subcutaneously in the hindleg of C57BL/6 mice and heat treatment by immersion for 40 min in a 42°C waterbath. Induction factors of 835 and 33 were determined for TNF-α and IL-12, respectively. In the absence of heat treatment, expression of the two cytokines was not detected.

Adenovectors containing an *hsp70B*-luciferase (Ad-*hsp70B-luc*) or an *hsp70B*-Fas ligand gene cassette (Ad-*hsp70B-fasl*) were prepared by Smith et al. [121]. The constructs were first tested *in vitro* in smooth muscle cells. The *hsp70B* promoter-driven luciferase gene was essentially silent in the absence of a heat treatment. Heat treatment resulted in a rapid rise of luciferase activity. Consistent with observations made in the previous study and elsewhere, luciferase activity declined to an undetectable level within about 48 h after heat treatment. To test the *hsp70B-luciferase* cassette *in vivo*, Ad-*hsp70B-luc* was injected intradermally in immunodeficient mice or rats. Localized heating was achieved by means of a portable therapeutic ultrasound device. Luciferase activity was assayed 12 h after heating. No luciferase activity above background was detected in unheated animals, whereas substantial activity was measured in heat-treated animals. Induction factors were 212-fold in mice and 540-fold in rats. Ad-*hsp70B-fasl* was introduced through the rat tail vein. The abdominal region was heated by directed ultrasound. Elevated levels of liver enzyme alanine aminotransferase were detected in locally heat-treated animals, where as only normal levels were found in not-heated animals. This result was taken to suggest that FasL expression also was stringently controlled *in vivo* by the *hsp70B* promoter.

Guilhon et al. [122] explored the benefits of using online magnetic resonance (MR) thermometry for enhancing the spatial accuracy of tissue heating by focused ultrasound. The authors were able to specifically heat subcutaneous C6 glioblastoma-derived tumors containing an *hsp70B*-driven *egfp* gene, inducing strong expression of the reporter protein. It is noted that to obtain optimal definition of the heated area, heating was intense but of short duration.

The feasibility of using focused ultrasound-induced hyperthermia for precisely inducing expression of a target gene under the control of an *hsp70B* promoter in liver, a deep-seated parenchymal organ, was demonstrated by Plathow et al. [123]. An adenovector containing an *hsp70B-gfp* gene cassette was injected into the tail vein of Copenhagen rats. Twenty four hours later, acoustic power at 1.92 W was directed to the rat livers. As indicated by noninvasive magnetic resonance imaging (MRI)-based temperature mapping, a relatively uniform focal temperature reaching 42.5°C within liver parenchyma was achieved. One day later, animals were euthanized and liver sections were examined. Fluorescence microscopy indicated that the *hsp70B* promoter-controlled gene was specifically expressed in the heated areas. In another study conducted by Sicox et al. [124], an adenovirus harboring an *hsp70B*-luciferase gene cassette was directly administered into both lobes of the prostate of three beagles. Two days later, the left lobe of the prostate was heated by MRI-guided ultrasound. High levels of luciferase expression could be measured that were restricted to the areas that had been heated by ultrasound.

In another study of directed ultrasound heating, an *hsp70B*-luciferase gene cassette was introduced in gastrocnemius muscles of mice by electroinjection [125]. The muscles were heated by focused ultrasound. Heating was feedback-controlled by means of a thermocouple inserted in the muscles. Heat-induced luciferase expression was maximal after 24 h. Two days later, activity had declined to background level. The optimal in situ heat dose was surprisingly low (39°C for 30 min). Perhaps, the muscles had been mildly injured by the electroinjection procedure and had not recovered by the time of heat treatment, which was administered only one day later.

The *hsp70B* promoter was used to control therapeutic transgenes in several model studies of cancer therapy. Huang et al. [120] injected adenoviruses containing *hsp70B*-controlled *il-12* or *egfp* genes into B16F10 tumors growing subcutaneously in hindlegs of C57BL/6 mice. Two virus doses were administered 1 week apart. After each virus dose, hindlegs were heated by immersion for 40 min in a 42.5°C waterbath. Tumor growth was delayed significantly in the experimental group (*il-12* gene- and heat-treated) but not in control groups. In a related study, adenovirus vector containing an *hsp70B-il-12* gene cassette was injected into B16F10 melanomas or 4T1 mammary adenocarcinomas growing in C57BL/6 mice [126]. Heating was done as before, i.e., by immersion in a water bath for 40 min at 42.5°C. Twenty four hours later,

substantial levels of IL-12 were detected in the tumors, while systemic toxicity was negligible. IL-12 levels in liver were significantly lower than those detected in animals that had been injected intratumorally with an adenovirus carrying an *il-12* gene controlled by a constitutively active promoter. In a study by Siddiqui et al. [127], intratumoral injection of an adenoviral vector containing an *hsp* promoter-linked murine *il-12* gene followed by heat treatment produced antiangiogenic effects in 4T1 tumors growing in Balb/C mice. A replication-deficient adenoviral vector was recently designed for delivering the feline *il-12* gene under the control of the human *hsp70B* promoter [128]. This vector was tested in a phase I clinical trial in cats with spontaneously arising soft tissue sarcoma. Intratumoral injection of adenovirus followed by heating for 1 h at 41°C resulted in high levels of *il-12* mRNA in the tumors and limited systemic toxicity [129].

Brade et al. [130] constructed a modified *hsp70B* promoter, *HSE.70b*, by addition of three synthetic HSE sequences to a minimal *hsp70B* promoter. The *HSE.70b* promoter could be induced by lower heat doses than the original *hsp70B* promoter. The *HSE.70b* promoter was utilized in a subsequent study for controlling the expression of a gene encoding a modified Gibbon ape leukemia virus envelope protein (GALV FMG) [131]. GALV FMG induces cell–cell fusion in permissive cells [132,133]. An *HSE.70b-galv fmg* gene cassette was introduced into HT-1080 tumor cells and stable line HG5 was isolated. In a model tumor therapy experiment, HG5 and control line CI1 were injected in hind limb footpads of severe combined immunodeficiency (SCID) mice. After sizeable tumors had formed, hind limbs were heat-treated by immersion in a waterbath for 30 min at 44°C. Animals were monitored and sacrificed at tumor volumes of 150 mm³. Median time to sacrifice of mice containing an HG5 tumor was 52 days for heat-treated animals versus 4.3 days for not-heated animals. Only a minor effect of heat treatment was observed in mice with control tumors.

Braiden et al. [134] delivered in several passages an HSV *tk* gene under the control of the *hsp70B* promoter or a control gene to FM3A tumors (mouse mammary carcinoma) growing subcutaneously in nude mice using HVJ (Sendai virus) liposomes. Animals were heated every other day by immersion in a waterbath for 30 min at 43°C. Gancyclovir (GCV) was administered daily during a 2-week period. Animals whose tumor had been administered the *hsp70B*-HSV *tk* gene cassette and were heat-treated showed tumor regression. Animals whose tumor was not heat-treated or had received a control gene instead of the *hsp70B*-HSV *tk* gene cassette continued to grow. In a further experiment, intraabdominal FM3A tumors were treated with *hsp70B*-HSV *tk* liposomes, GCV, and heat exposures. This treatment resulted in a survival rate of 80% at the end of the observation period. Control mice did not survive. A recent study extended these observations to a model of gastric cancer [135].

The studies reviewed above demonstrated convincingly that the *hsp70B* promoter was capable of stringently controlling expression of transgenes *in vivo*. Regional administration of heat resulted in gene expression that was confined to tumors, organs, or limbs. Focused ultrasound guided by MRI was capable of restricting heating, and consequently, transgene expression to narrowly defined body regions (of about 1 cm in diameter) [110,122]. Several studies investigated time courses of appearance and subsequent disappearance of gene products expressed from *hsp70B* promoter-controlled genes. Typically, gene products accumulated to maximal levels about 1 day after activating heat treatment. About 2 days later, concentrations had fallen to undetectable levels. Although single-activating heat treatments produced therapeutic effects in some of the tumor therapy experiments summarized above, the expectation is that the therapeutic window represented by the time of activity of an *hsp* promoter subsequent to an activating heat treatment is generally too short for effective therapy. Hence, effective clinical protocols may be envisioned to resemble that used in the Braiden study [134], in which elevated levels of transgene product were maintained through multiple heat treatments administered at regular intervals. Obviously, such protocols would be burdensome for patient, physician, and hospital. A concern particular to the use of *hsp* promoters in the context of cancer therapies has been that activating heat treatments would not only enhance expression of a therapeutic protein but also that of endogenous HSPs. Elevated levels of HSPs, which are known to have cytoprotective and antiapoptotic properties, could neutralize the effects of the therapeutic gene product. However, in the experiments reported to date, heat treatments administered for the activation of *hsp70B* promoter-controlled transgenes did not appear to interfere with therapy efficacy. Surprisingly, *in vitro* experiments reported by the Braiden study [134] even furnished evidence for a substantial synergistic effect of heat on tumor cell killing by TK/GCV. Heat treatment of mouse mammary carcinoma FM3A cells constitutively expressing an HSV *tk* gene reduced the IC_{50} of GCV by about 15-fold. This effect was ascribed to increased levels of Fas and FasL in the heated cells. Other studies demonstrated that cell killing induced by powerful mediators of apoptosis was potentiated by heat treatment or introduction of a constitutively active form of HSF1 [136,137].

29.4 *hsp70* PROMOTER-BASED GENE SWITCHES FOR DELIBERATE SPATIAL AND TEMPORAL REGULATION OF TRANSGENES

As discussed before, *hsp* promoters remain active for at most a few hours after heat activation. At the level of protein product of a transgene controlled by an *hsp* promoter, synthesis of the protein will continue after the controlling *hsp* promoter has returned to an inactive state. However, continued expression may not last much longer than one day, if that long. This narrow time window of active synthesis of therapeutic gene product will not be compatible with therapies requiring the continued presence of an effective level of a therapeutic protein over a longer period of time. In addition, the short period of activity of *hsp* promoters caps peak levels of gene product that can be attained. Repetitive heat treatments administered

at an appropriate interval could compensate at least partially for these shortcomings. However, as also alluded to before, a clinical protocol involving multiple, carefully timed heat treatments may not be accepted readily by patients and care providers.

A more important problem is involuntary activation of *hsp* promoter-controlled therapeutic genes that could occur subsequent to therapy in target as well as nontarget tissues. When occurring in the wake of cytotoxic, cytostatic, or angiogenic therapies, such involuntary activation of the therapeutic gene could compromise severely patient safety. The problem arises from the fact that current technology is not capable of targeted delivery of transgenes. Transgenes inevitably are disseminated systemically even if initially administered directly to a target tissue such as a tumor [138]. A rise in body temperature caused by disease, strenuous exercise, or certain pharmacological interventions, or by ischemic events [139–146] could cause inadvertent activation of an *hsp* promoter-controlled transgene, resulting in systemic expression of transgene product. Ingestion of heavy metals and other toxicants could also prompt activation (see, e.g., Ref. [147]). While it should be possible to avoid such inadvertent systemic activation during therapy in a hospital setting, this would be difficult or impossible to achieve subsequent to hospital discharge or in the context of outpatient care, i.e., when patients are not or no longer monitored carefully (or do not wish to be restricted in their customary activities).

We hypothesized that both problems associated with uses of *hsp* promoters, i.e., insufficiently long periods of transgene expression and the possibility of inadvertent, systemic expression of transgenes encoding highly potent mediators/effectors could be avoided by a regulatory circuit that properly combined an *hsp70B* promoter and a small-molecule-dependent transactivator [148]. Figure 29.4 outlines components and operation of such a gene switch. The first component is a gene for a small-molecule-dependent transactivator that is under the control of a dual-responsive promoter or promoter cassette. This promoter cassette or promoter incorporates an *hsp70B* promoter and a transactivator-responsive promoter or combines aspects thereof. The second component is a transactivator-responsive promoter that is functionally linked to a target gene. In a cell-containing gene switch and target gene, neither transactivator nor target gene is expected to be expressed in the absence of small-molecule ligand and without an activating heat treatment. Heat treatment will trigger activation of endogenous HSF1, which transactivates the transactivator gene. Transactivator will be synthesized. However, in the absence of small-molecule ligand, transactivator will remain inactive and the target gene will not be expressed. In the presence of small-molecule ligand alone, the minute amounts of transactivator present in the unheated cell will be insufficient to cause autoactivation of the transactivator gene. When the cell is both exposed to small-molecule ligand and subjected to a transient heat treatment, activated HSF1 will transactivate the transactivator gene. Newly made transactivator will be bound by small-molecule ligand, become transcriptionally competent, and transactivate the

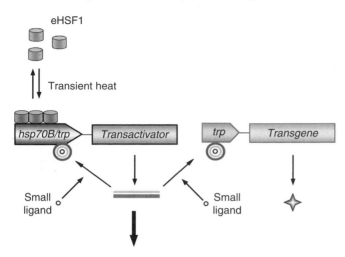

FIGURE 29.4 Heat-activated and small-molecule ligand-dependent gene switch comprising a transactivator gene controlled by a promoter comprising components of an *hsp70B* promoter and a transactivator-responsive promoter (*trp*) or a promoter cassette including an *hsp70B* promoter and a *trp* promoter, and a *trp* promoter to which a transgene of interest is linked. The solid arrow indicates transactivator turnover. eHSF1, endogenous heat shock transcription factor 1.

target gene as well as autoactivate its own gene. Even though HSF1 will revert to its inactive state shortly after heat treatment and the *hsp70B* promoter or promoter-component will become silent, transactivator level is expected to be maintained by the latter feedforward mechanism. Target gene expression should continue for as long as small-molecule ligand is present. Removal of ligand should return the gene switch to its initial inactive state.

The elements of an actual regulatory circuit are depicted in Figure 29.5. Its first component included a gene for GLP65,

a partially humanized, mifepristone-activated, chimeric transactivator containing a GAL4 DNA-binding domain [63] (see also Section 29.2.3.2). This gene was controlled by a promoter cassette that combined an *hsp70B* promoter and a *gal4* promoter. The second component was a *gal4* promoter that was linked to a target gene, e.g., a luciferase gene.

This gene switch was tested extensively *in vitro* and *in vivo* [148]. The firefly luciferase gene was utilized as the target gene for most of these experiments. Because of the low stability of mRNA and protein product, luciferase activity served as a reasonable indicator of transcriptional rate. Stability and performance was examined in a human cell line stably containing gene switch and luciferase target gene. The key findings are represented by the data shown in Figure 29.6. The system had very low basal activity in the absence of an activating heat treatment and mifepristone. Addition of mifepristone as well as continued exposure to mifepristone for 6 days did not

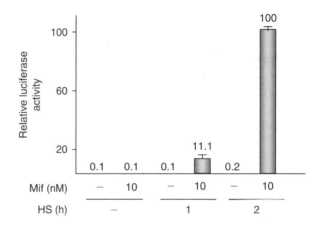

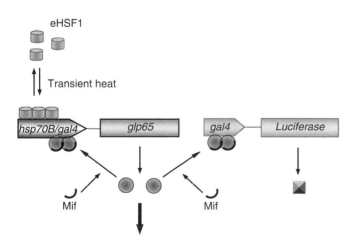

FIGURE 29.5 Heat-activated and mifepristone-dependent gene switch comprising transactivator gene *glp65* that is under the control of cassette promoter *hsp70B/gal4*, which responds to activated endogenous heat shock factor 1 (eHSF1) as well as GLP65, and GLP65-responsive promoter *gal4* that controls the linked luciferase transgene. The solid arrow indicates transactivator turnover. Mif, mifepristone.

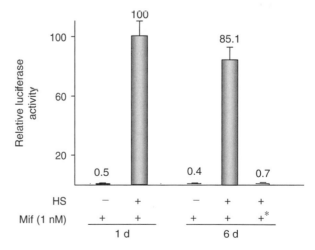

FIGURE 29.6 Performance of the mifepristone-activated gene switch shown in Figure 29.5 in a cell line stably transfected with gene switch and luciferase reporter/target gene. Top panel: Reporter gene activity one day after activating heat treatment (HS) at 43°C. Bottom panel: Reporter gene activity 1 day (1 d) and 6 days (6 d) after heat activation (43°C/2 h) in the presence of mifepristone, and reversiblity of activation. *Mifepristone was washed away 1 day after heat activation. Mif: mifepristone. See Vilaboa, N. et al., *Mol. Ther.*, 12, 290, 2005 for a more complete data set.

cause a significant increase in luciferase activity. Heat treatment in the absence of mifepristone was without effect. However, in the presence of mifepristone, a 2 h heat treatment at 43°C resulted in a luciferase level that was about 1000 times higher than basal activity. Luciferase expression was proportional to the activating heat dose (cf. activity levels induced by 1 or 2 h heat treatments). Finally, high-level, induced expression of luciferase continued throughout the entire observation period of 6 days. Removal of mifepristone 1 day after activating heat treatment led to rapid decline in luciferase level; background level was attained 3 days later. A similar gene switch and luciferase target gene were electroinjected into hindleg muscles of mice [148]. Basal luciferase activity in the muscles was extremely low. A transient heat treatment administered to the hindleg in the presence of systemic mifepristone resulted in a 1000-fold increase in luciferase activity. As in the *in vitro* experiments, the system proved essentially insensitive to heat alone or mifepristone alone. Furthermore, induced activity persisted at a high level during the 6-day observation period during which mifepristone was maintained by daily i.p. injections. Discontinuation of mifepristone administration resulted in deactivation of the gene switch.

These studies demonstrated that a gene switch that combines an *hsp70B* promoter and a small-molecule-dependent transactivator could overcome the shortcomings of *hsp* promoter-based transgene regulation. A single-activating heat treatment produced sustained high-level expression of a transgene. Moreover, transgene expression was dependent on the continuous presence of transactivator ligand. Hence, the danger of inadvertent systemic transgene expression subsequent to therapy can be avoided by using such a system.

29.5 CONCLUDING REMARKS

Until vehicles become available that deliver transgenes with exquisite selectivity to any tissue or organ afflicted by a disease susceptible to gene therapy, achievement of effective transcriptional targeting remains an important goal.

As illustrated in the preceding section, versatile *hsp*-based gene switches can be assembled that are expected to be capable of activation in essentially all cell types. The gene switches allow for deliberate control over spatial as well as temporal aspects of transgene expression. Transgene activity can be restricted to the tissue or organ of interest by means of a focused, transient heat treatment that initiates expression of a small-molecule-dependent transactivator. Transgene expression commensurate with a particular therapeutic window can be achieved by an appropriate protocol of small-molecule administration. In addition, transgene activity may be adjusted to a desired level by modulation of the activating heat dose. Methods for accurate heating of narrowly defined regions within the body were developed as evidenced by the studies reviewed in Section 29.3. Hence, clinical exploration of *hsp*-based gene switches should now be feasible.

The example *hsp*-based gene switch discussed in Section 29.4 incorporated a mifepristone-activated gene switch. In principle, any small-molecule-controlled gene switch may be utilized, including those described under Section 29.2. The choice of small-molecule-controlled gene switch will impact the properties of the resulting *hsp*-based gene switch. As discussed in Section 29.2, small-molecule-dependent gene switches can differ in respect of both maximal levels and basal, uninduced levels of transgene expression. Additional differences originate from the nature of the small-molecule ligands that activate or repress particular transactivators. The ligands may or may not distribute to all tissues, and they will have different *in vivo* half-lives and toxicities. Furthermore, they may or may not affect endogenous gene expression, and consequently cause unintended physiological reactions. Finally, transactivators may differ in the degree of humanization and, therefore, may pose a greater or lesser danger of untoward immune reactions. Given these differences, the optimal choice of small-molecule-controlled gene switch to be incorporated in an *hsp*-based gene switch can be expected to be influenced by multiple factors, including the potency of the therapeutic gene product in question, the therapeutic indication targeted, and the anticipated therapy protocol. Certainly, future improvements that may be made to different ligand-dependent gene switches will also affect this choice.

Radiation-inducible promoters are encumbered with similar limiting features as *hsp* promoters. They are inactivated within hours of irradiation. As with *hsp* promoters, there exists a danger of inadvertent activation. Radiation-inducible promoters can be expected to be activated by many drugs, including several widely used chemotherapy drugs [98]. Because of the similarity of problems, the principles used for the design of *hsp*-based gene switches should also be applicable to the preparation of IR-inducible promoter-based gene switches. Such gene switches would not only increase the safety of IR-induced therapy but may also allow for the use of therapy protocols that cannot presently be envisaged, e.g., sequential radiation-induced genetic therapy and classical chemotherapy of cancer.

ACKNOWLEDGMENTS

N.V. was supported by grant FIS 01/3027 from the Fondo de Investigaciones Sanitarias and by the Fundación Mutua Madrileña.

REFERENCES

1. Beck, C. et al., Tissue-specific targeting for cardiovascular gene transfer. Potential vectors and future challenges, *Curr. Gene Ther.*, 4, 457, 2004.
2. Melo, L.G. et al., Endothelium-targeted gene and cell-based therapies for cardiovascular disease, *Artherioscler. Thromb. Vasc. Biol.*, 24, 1761, 2004.
3. Su, H. et al., Adeno-associated viral vector delivers cardiac-specific and hypoxia-inducible VEGF expression in ischemic mouse hearts, *Proc. Natl. Acad. Sci. USA*, 101, 16280, 2004.
4. Marignol, L. et al., Achieving hypoxia-inducible gene expression in tumors, *Cancer Biol. Ther.*, 4, 359, 2005.

5. Takeda, T. et al., Tumor-specific gene therapy for undifferentiated thyroid carcinoma utilizing the telomerase reverse transcriptase promoter, *J. Clin. Endocrinol. Metab.*, 88, 3531, 2003.

6. Lo, H.W., Day, C.P., and Hung, M.C., Cancer-specific gene therapy, *Adv. Genet.*, 54, 235, 2005.

7. Burkhardt, B.R. et al., Glucose transporter-2 (GLUT2) promoter mediated transgenic insulin production reduces hyperglycemia in diabetic mice, *FEBS Lett.*, 579, 5759, 2005.

8. McCormick, F., Cancer-specific viruses and the development of ONYX-015, *Cancer Biol. Ther.*, 2, S157, 2003.

9. Hillen, W. and Wissman, A., Protein–nucleic acid interaction: Tet repressor- tet operator interaction, In: W. Saeger and U. Heinemann (Eds.), *Topics in Molecular and Structural Biology*, Macmillan, London, 1989, Vol. 10, p. 143.

10. Gossen, M. and Bujard, H., Tight control of gene expression in mammalian cells by tetracycline-responsive promoters, *Proc. Natl. Acad. Sci. USA*, 89, 5547, 1992.

11. Furth, P.A. et al., Temporal control of gene expression in transgenic mice by a tetracycline-responsive promoter, *Proc. Natl. Acad. Sci. USA*, 91, 9302, 1994.

12. Baron, U., Gossen, M., and Bujard, H., Tetracycline-controlled transcription in eukaryotes: Novel transactivators with graded transactivation potential, *Nucleic Acids Res.*, 25, 2723, 1997.

13. Gossen, M. et al., Transcriptional activation by tetracyclines in mammalian cells, *Science*, 268, 1766, 1995.

14. Urlinger, S. et al., Exploring the sequence space for tetracycline-dependent transcriptional activators: Novel mutations yield expanded range and sensitivity, *Proc. Natl. Acad. Sci. USA*, 97, 7963, 2000.

15. Das, A.T. et al., Viral evolution as a tool to improve the tetracycline-regulated gene expression system, *J. Biol. Chem.*, 279, 18776, 2004.

16. Zhou, X. et al., Optimization of the Tet-On system for regulated gene expression through viral evolution, *Gene Ther.*, 13, 1382, 2006.

17. Freundlieb, S., Schirra-Muller, C., and Bujard, H., A tetracycline-controlled activation/repression system with increased potential for gene transfer into mammalian cells, *J. Gene Med.*, 1, 4, 1999.

18. Zhu, Z. et al., Use of the tetracycline-controlled transcriptional silencer (tTS) to eliminate transgene leak in inducible overexpression transgenic mice, *J. Biol. Chem.*, 276, 25222, 2001.

19. Salucci, V. et al., Tight control of gene expression by a helper-dependent adenovirus vector carrying the rtTA2(s)-M2 tetracycline transactivator and repressor system, *Gene Ther.*, 9, 1415, 2002.

20. Mizuguchi, H. et al., Tight positive regulation of transgene expression by a single adenovirus vector containing the rtTA and tTS expression cassettes in separate genome regions, *Hum. Gene Ther.*, 14, 1265, 2003.

21. Bornkamm, G.W. et al., Stringent doxycycline-dependent control of gene activities using an episomal one-vector system, *Nucleic Acids Res.*, 33, e137, 2005.

22. Rubinchik, S. et al., New complex Ad vectors incorporating both rtTA and tTS deliver tightly regulated transgene expression both in vitro and in vivo, *Gene Ther.*, 12, 504, 2005.

23. Xiong, W. et al., Regulatable gutless adenovirus vectors sustain inducible transgene expression in the brain in the presence of an immune response against adenoviruses, *J. Virol.*, 80, 27, 2006.

24. Lamartina, S. et al., Construction of an rtTA2(s)-m2/tts(kid)-based transcription regulatory switch that displays no basal activity, good inducibility, and high responsiveness to doxycycline in mice and non-human primates, *Mol. Ther.*, 7, 271, 2003.

25. Barde, I. et al., Efficient control of gene expression in the hematopoietic system using a single tet-on inducible lentiviral vector, *Mol. Ther.*, 13, 382, 2006.

26. Chtarto, A. et al., Tetracycline-inducible transgene expression mediated by a single AAV vector, *Gene Ther.*, 10, 84, 2003.

27. Markusic, D. et al., Comparison of single regulated lentiviral vectors with rtTA expression driven by an autoregulatory loop or a constitutive promoter, *Nucleic Acids Res.*, 33, e63, 2005.

28. Woraratanadharm, J. et al., Novel system uses probasin-based promoter, transcriptional silencers and amplification loop to induce high-level prostate expression, *BMC Biotechnol.*, 12, 7, 2007.

29. Weber, W. et al., A synthetic time-delay circuit in mammalian cells and mice, *Proc. Natl. Acad. Sci. USA*, 104, 2643, 2007.

30. Go, W.Y. and Ho, S.N., Optimization and direct comparison of the dimerizer and reverse tet transcriptional control systems, *J. Gene Med.*, 4, 258, 2002.

31. Xu, Z.L. et al., Regulated gene expression from adenovirus vectors: A systematic comparison of various inducible systems, *Gene*, 309, 145, 2003.

32. Favre, D. et al., Lack of an immune response against the tetracycline-dependent transactivator correlates with long-term doxycycline regulated transgene expression in nonhuman primates after intramuscular injection of recombinant adeno-associated virus, *J. Virol.*, 76, 11605, 2002.

33. Latta-Mahieu, M. et al., Gene transfer of a chimeric transactivator is immunogenic and results in short-lived transgene expression, *Hum. Gene Ther.*, 13, 1611, 2002.

34. Rang, A. and Will, H., The tetracycline-responsive promoter contains functional interferon-inducible response elements, *Nucleic Acids Res.*, 28, 1120, 2000.

35. Gould, D.J. and Chernajovsky, Y., Endogenous GATA factors bind the core sequence of the tetO and influence gene regulation with the tetracycline system, *Mol. Ther.*, 10, 127, 2004.

36. Fussenegger, M. et al., Streptogramin-based gene regulation systems for mammalian cells, *Nat. Biotechnol.*, 18, 1203, 2000.

37. Roberts, M., Resistance to tetracycline, macrolide-lincosamide-streptogramin, trimethoprim and sulfonamide drug classes, *Mol. Biotechnol.*, 20, 261, 2002.

38. Weber, W. et al., Macrolide-based transgene control in mammalian cells and mice, *Nat. Biotechnol.*, 20, 901, 2002.

39. Weber, W. et al., Novel promoter/transactivator configurations for macroloide- and streptogramin-responsive transgene expression in mammalian cells, *J. Gene Med.*, 4, 676, 2002.

40. Fux, C. et al., Novel macrolide-adjustable bidirectional expression modules for coordinated expression of two different transgenes in mice, *J. Gene Med.*, 5, 1067, 2003.

41. Mitta, B., Weber, C.C., and Fussenegger, M., In vivo transduction of HIV-1-derived lentiviral particles engineered for macrolide-adjustable transgene expression, *J. Gene Med.*, 7, 1400, 2005.

42. Cronin, C.A., Gluba, W., and Scrable, H., The lac operator-repressor system is functional in the mouse, *Genes Dev.*, 15, 1506, 2001.

43. Ryan, A. and Scrable, H., Visualization of the dynamics of gene expression in the living mouse, *Mol. Imaging*, 3, 33, 2004.

44. Mullick, A. et al., The cumate gene-switch: A system for regulated expression in mammalian cells, *BMC Biotechnol.*, 6, 43, 2006.

45. Thomas, H.E., Stunnenberg, H.G., and Stewart, A.F., Heterodimerization of the Drosophila ecdysone receptor with retinoid X receptor and ultraspiracle, *Nature*, 362, 471, 1993.

46. Yao, T.P. et al., Functional ecdysone receptor is the product of EcR and Ultraspiracle genes, *Nature*, 366, 476, 1993.

47. No, D., Yao, T.P., and Evans, R.M., Ecdysone-inducible gene expression in mammalian cells and transgenic mice, *Proc. Natl. Acad. Sci. USA*, 93, 3346, 1996.

48. Yao, T.P. et al., Drosophila ultraspiracle modulates ecdysone receptor function via heterodimer formation, *Cell*, 71, 63, 1992.

49. Wyborski, D.L., Bauer, J.C., and Vaillancourt, P., Bicistronic expression of ecdysone-inducible receptors in mammalian cells. *Biotechniques*, 31, 618, 2001.

50. Galimi, F. et al., Development of ecdysone-regulated lentiviral vectors, *Mol. Ther.*, 11, 142, 2005.

51. Suhr, S.T. et al., High level transactivation by a modified Bombyx ecdysone receptor in mammalian cells without exogenous retinoid X receptor, *Proc. Natl. Acad. Sci. USA*, 95, 7999, 1998.

52. Hoppe, U.C., Marban, E., and Johns, D.C., Adenovirus-mediated inducible gene expression in vivo by a hybrid ecdysone receptor, *Mol. Ther.*, 1, 159, 2000.

53. Palli, S.R. et al., Improved ecdysone receptor-based inducible gene regulation system, *Eur. J. Biochem.*, 270, 1308, 2003.

54. Karns, L.R. et al., Manipulation of gene expression by an ecdysone-inducible gene switch in tumor xenografts, *BMC Biotechnol.*, 1, 11, 2001.

55. Karzenowski, D., Potter, D.W., and Padidam, M., Inducible control of transgene expression with ecdysone receptor: Gene switches with high sensitivity, robust expression, and reduced size, *Biotechniques*, 39, 191, 2005.

56. Palli, S.R., Kapitskaya, M.Z., and Potter, D.W., The influence of heterodimer partner ultraspiracle/retinoid X receptor on the function of ecdysone receptor, *FEBS J.*, 272, 5979, 2005.

57. Kumar, M.B. et al., Highly flexible ligand binding pocket of ecdysone receptor: A single amino acid change leads to discrimination between two groups of nonsteroidal ecdysone agonists, *J. Biol. Chem.*, 279, 27211, 2004.

58. Oehme, I., Bosser, S., and Zornig, M., Agonists of an ecdysone-inducible mammalian expression system inhibit Fas Ligand- and TRAIL-induced apoptosis in the human colon carcinoma cell line RKO, *Cell Death Differ.*, 13, 189, 2006.

59. Pierson, T.M. et al., Regulable expression of inhibin A in wild-type and inhibin alpha null mice, *Mol. Endocrinol.*, 14, 1075, 2000.

60. Vegeto, E. et al., The mechanism of RU486 antagonism is dependent on the conformation of the carboxy-terminal tail of the human progesterone receptor, *Cell*, 69, 703, 1992.

61. Wang, Y. et al., A regulatory system for use in gene transfer, *Proc. Natl. Acad. Sci. USA*, 91, 8180, 1994.

62. Wang, Y. et al., Positive and negative regulation of gene expression in eukaryotic cells with an inducible transcriptional regulator, *Gene Ther.*, 4, 432, 1997.

63. Burcin, M.M. et al., Adenovirus-mediated regulable target gene expression in vivo, *Proc. Natl. Acad. Sci. USA*, 96, 355, 1999.

64. Abruzzese, R.V. et al., Ligand-dependent regulation of vascular endothelial growth factor and erythropoietin expression by a plasmid-based, autoinducible GeneSwitch system, *Mol. Ther.*, 2, 276, 2000.

65. Nordstrom, J.L., Antiprogestin-controllable transgene regulation in vivo, *Curr. Opin. Biotechnol.*, 13, 453, 2002.

66. Wang, Y. et al., Ligand-inducible and liver-specific target gene expression in transgenic mice, *Nat. Biotechnol.*, 15, 239, 1997.

67. Wang, L. et al., Prolonged and inducible transgene expression in the liver using gutless adenovirus: A potential therapy for liver cancer, *Gastroenterology*, 126, 278, 2004.

68. Schillinger, K.J. et al., Regulatable atrial natriuretic peptide gene therapy for hypertension, *Proc. Natl. Acad. Sci. USA*, 102, 13789, 2005.

69. Oligino, T. et al., Drug inducible transgene expression in brain using a herpes simplex virus vector, *Gene Ther.*, 5, 491, 1998.

70. Roscilli, G. et al., Long-term and tight control of gene expression in mouse skeletal muscle by a new hybrid human transcription factor, *Mol. Ther.*, 6, 653, 2002.

71. Abraham, R.T., Mammalian target of rapamycin: Immunosuppressive drugs uncover a novel pathway of cytokine recepetor signaling, *Curr. Opin. Immunol.*, 10, 330, 1998.

72. Rivera, V.M. et al., A humanized system for pharmacologic control of gene expression, *Nat. Med.*, 2, 1028, 1996.

73. Pomerantz, J.L., Sharp, P.A., and Pabo, C.O., Structure-based design of transcription factors, *Science*, 267, 93, 1995.

74. Magari, S.R. et al., Pharmacologic control of a humanized gene therapy system implanted into nude mice, *J. Clin. Invest.*, 100, 2865, 1997.

75. Pollock, R. et al., Delivery of a stringent, dimerizer-regulated gene expression system in a single retroviral vector, *Proc. Natl. Acad. Sci. USA*, 97, 13221, 2000.

76. Luce, M.J. and Reiser, J., Development of regulated lentiviral vectors, *Mol. Ther.*, 5, S32, 2002.

77. Rivera, V.M. et al., Long-term regulated expression of growth hormone in mice after intramuscular gene transfer, *Proc. Natl. Acad. Sci. USA*, 96, 8657, 1999.

78. Auricchio, A. et al., Constitutive and regulated expression of processed insulin following *in vivo* hepatic gene transfer, *Gene Ther.*, 9, 963, 2002.

79. Chong, H. et al., A system for small-molecule control of conditionally replication-competent adenoviral vectors, *Mol. Ther.*, 5, 195, 2002.

80. Wang, S., Petravicz, J., and Breakefield, X.O., Single HSV-amplicon vector mediates drug-induced gene expression via dimerizer system, *Mol. Ther.*, 7, 790, 2003.

81. Ye, X. et al., Regulated delivery of therapeutic proteins after in vivo somatic cell gene transfer, *Science*, 283, 88, 1999.

82. Auricchio, A. et al., Pharmacological regulation of protein expression from adeno-associated viral vectors in the eye, *Mol. Ther.*, 6, 238, 2002.

83. Pollock, R. and Clackson, T., Dimerizer-regulated gene expression, *Curr. Opin. Biotechnol.*, 13, 459, 2002.

84. Johnston, J. et al., Regulated expression of erythropoietin from an AAV vector safely improves the anemia of β-thalassemia in a mouse model, *Mol. Ther.*, 7, 493, 2003.

85. Rivera, V.M. et al., Long-term pharmacologically regulated expression of erythropoietin in primates following AAV-mediated gene transfer, *Blood*, 105, 1424, 2005.

86. Lebherz, C. et al., Long-term inducible gene expression in the eye via adeno-associated virus gene transfer in nonhuman primates, *Hum. Gene Ther.*, 16, 178, 2005.

87. Toniatti, C. et al., Gene therapy progress and prospects: Transcription regulatory systems, *Gene Ther.*, 11, 649, 2004.

88. Clemons, P.A. et al., Synthesis of calcineurin-resistant derivatives of FK506 and selection of compensatory receptors, *Chem. Biol.*, 9, 49, 2002.

89. Chastel, C., Jiricny, J., and Jaussi, R., Activation of stress-responsive promoters by ionizing radiation for deployment in targeted gene therapy, *DNA Repair*, 3, 201, 2004.

90. Weichselbaum, R.R. et al., Gene therapy targeted by ionizing radiation, *Int. J. Radiat. Oncol. Biol. Phys.*, 24, 565, 1992.

91. Weichselbaum, R.R. et al., Radiation-induced tumor necrosis factor-α expression: Clinical application of physical and transcriptional targeting of gene therapy, *Lancet Oncol.*, 3, 665, 2002.

92. Marr, R.A. et al., Tumour therapy in mice using adenovirus vectors expressing human TNFa, *Int. J. Oncol.*, 12, 509, 1998.

93. Marr, R.A. et al., A p75 tumor necrosis factor receptor-specific mutant of murine tumor necrosis factor alpha expressed from an adenovirus vector induces an antitumor response with reduced toxicity, *Cancer Gene Ther.*, 6, 465, 1999.

94. Hallahan, D.E. et al., Spatial and temporal control of gene therapy using ionizing radiation, *Nat. Med.*, 1, 786, 1995.

95. Seung, L.P. et al., Genetic radiotherapy overcomes tumor resistance to cytotoxic agents, *Cancer Res.*, 55, 5561, 1995.

96. Mauceri, H.J. et al., Tumor necrosis factor alpha (TNF alpha) gene therapy targeted by ionizing radiation selectively damages tumor vasculature, *Cancer Res.*, 56, 4311, 1996.

97. Mauceri, H.J. et al., Combined effects of angiostatin and ionizing radiation in antitumour therapy, *Nature*, 394, 287, 1998.

98. Mezhir, J.J. et al., Ionizing radiation: A genetic switch for cancer gene therapy, *Cancer Gene Ther.*, 13, 1, 2006.

99. Senzer, N. et al., TNFerade biologic, an adenovector with a radiation-inducible promoter, carrying the human tumor necrosis factor alpha gene: A phase I study in patients with solid tumors, *J. Clin. Oncol.*, 22, 592, 2004.

100. Mundt, A.J. et al., A phase I trial of TNFerade biologic in patients with soft tissue sarcoma in the extremities, *Clin. Cancer Res.*, 10, 5747, 2004.

101. Kawashita, Y. et al., Regression of hepatocellular carcinoma in vitro and in vivo by radiosensitizing suicide gene therapy under the inducible and spatial control of radiation, *Hum. Gene Ther.*, 10, 1509, 1999.

102. Tsurushima H. et al., Radioresponsive tumor necrosis factor-related apoptosis-inducing ligand (TRAIL) gene therapy for malignant brain tumors, *Cancer Gene Ther.*, 14, 706, 2007.

103. Worthington, J. et al., Modification of vascular tone using inos under the control of a radiation-inducible promoter, *Gene Ther.*, 7, 1126, 2000.

104. McCarthy H.O. et al., p21(WAF1)-mediated transcriptional targeting of inducible nitric oxide synthase gene therapy sensitizes tumours to fractionated radiotherapy, *Gene Ther.*, 14, 246, 2007.

105. Nenoi, M. et al., Low-dose radiation response of the p21WAF1/CIP1 gene promoter transduced by adeno-associated virus vector, *Exp. Mol. Med.*, 38, 553, 2006.

106. Manome, Y. et al., Transgene expression in malignant glioma using a replication-defective adenoviral vector containing the *EGR-1* promoter: Activation by ionizing radiation or uptake of radioactive iododeoxyuridine, *Hum. Gene Ther.*, 9, 1409, 1998.

107. Scott, S.D., Joiner, M.C., and Marples, B., Optimizing radiation-responsive gene promoters for radiogenetic cancer therapy, *Gene Ther.*, 9, 1396, 2002.

108. Scott, S.D. et al., A radiation-controlled molecular switch for use in gene therapy of cancer, *Gene Ther.*, 7, 1121, 2000.

109. Greco, O. et al., Hypoxia- and radiation-activated Cre/loxP "molecular switch" vectors for gene therapy of cancer, *Gene Ther.*, 13, 206, 2006.

110. Rome, C., Couillaud, F., and Moonen, C.T., Spatial and temporal control of expression of therapeutic genes using heat shock protein promoters, *Methods*, 35, 188, 2005.

111. Moonen, C.T., Spatio-temporal control of gene expression and cancer treatment using magnetic resonance imaging-guided focused ultrasound, *Clin. Cancer Res.*, 13, 3482, 2007.

112. Voellmy, R. et al., Isolation and functional analysis of a human 70,000 dalton heat shock protein gene segment, *Proc. Natl. Acad. Sci. USA*, 82, 4949, 1985.

113. Schiller, P. et al., *Cis*-acting elements involved in the regulated expresion of a human hsp70 gene, *J. Mol. Biol.*, 203, 97, 1988.

114. Dreano, M. et al., High-level, heat-regulated synthesis of proteins in eukaryotic cells, *Gene*, 49, 1, 1986.

115. Voellmy, R., On mechanisms that control heat shock transcription factor activity in metazoan cells, *Cell Stress Chaperones*, 9, 122, 2004.

116. Christians, E.S. and Benjamin, I.J., Heat shock response: Lessons from mouse knockouts, *Handb. Exp. Pharmacol.*, 172, 139, 2006.

117. Xia, W. et al., Modulation of tolerance by mutant heat shock transcription factors, *Cell Stress Chaperones*, 4, 8, 1999.

118. Voellmy, R., Feedback regulation of the heat shock response, *Handb. Exp. Pharmacol.*, 172, 43, 2006.

119. Vekris, A. et al., Control of transgene expression using local hyperthermia in combination with a heat-sensitive promoter, *J. Gene Med.*, 2, 89, 2000.

120. Huang, Q. et al., Heat-induced gene expression as a novel targeted cancer gene therapy strategy, *Cancer Res.*, 60, 3435, 2000.

121. Smith, R.C. et al., Spatial and temporal control of transgene expression through ultrasound-mediated induction of the heat shock protein 70B promoter in vivo, *Hum. Gene Ther.*, 13, 697, 2002.

122. Guilhon, E. et al., Spatial and temporal control of transgene expression in vivo using a heat-sensitive promoter and MRI-guided focused ultrasound, *J. Gene Med.*, 5, 333, 2003.

123. Plathow, C. et al., Focal gene induction in the liver of rats by a heat-inducible promoter using focused ultrasound hyperthermia: Preliminary results, *Invest. Radiol.*, 40, 729, 2005.

124. Silcox, C.E. et al., MRI-guided ultrasonic heating allows spatial control of exogenous luciferase in canine prostate, *Ultrasound Med. Biol.*, 31, 965, 2005.

125. Xu, L. et al., Regulation of transgene expression in muscles by ultrasound-mediated hyperthermia, *Gene Ther.*, 11, 894, 2004.

126. Lohr, F. et al., Systemic vector leakage and transgene expression by intratumorally injected recombinant adenovirus vectors, *Clin. Cancer Res.*, 7, 3625, 2001.

127. Siddiqui, F. et al., Anti-angiogenic effects of interleukin-12 delivered by a novel hyperthermia induced gene construct, *Int. J. Hyperthermia*, 22, 587, 2006.

128. Siddiqui, F. et al., Characterization of a recombinant adenovirus vector encoding heat-inducible feline interleukin-12 for use in hyperthermia-induced gene-therapy, *Int. J. Hyperthermia*, 22, 117, 2006.

129. Siddiqui, F. et al., A phase I trial of hyperthermia-induced interleukin-12 gene therapy in spontaneously arising feline soft tissue sarcomas, *Mol. Cancer Ther.*, 6, 380, 2007.

130. Brade, A.M. et al., Heat-directed gene targeting of adenoviral vectors to tumor cells, *Cancer Gene Ther.*, 7, 1566, 2000.

131. Brade, A.M. et al., Heat-directed tumor cell fusion, *Hum. Gene Ther.*, 14, 447, 2003.

132. Bateman, A. et al., Fusogenic membrane glycoproteins as a novel class of genes for the local and immune-mediated control of tumor growth, *Cancer Res.*, 60, 1492, 2000.

133. Diaz, R.M. et al., A lentiviral vector expressing a fusogenic glycoprotein for cancer gene therapy, *Gene Ther.*, 7, 1656, 2000.

134. Braiden, V. et al., Eradication of breast cancer xenografts by hyperthermic suicide gene therapy under the control of the heat shock protein promoter, *Hum. Gene Ther.*, 11, 2453, 2000.

135. Isomoto, H. et al., Heat-directed suicide gene therapy mediated by heat shock promoter for gastric cancer, *Oncol. Rep.*, 15, 629, 2006.

136. Xia, W., Voellmy, R., and Spector, N., Sensitization of tumor cells to fas kiling through overexpression of heat-shock transcription factor 1, *J. Cell Physiol.*, 183, 425, 2000.

137. Xia, W. et al., Concurrent exposure to heat shock and H7 synergizes to trigger breast cancer cell apoptosis while sparing normal cells, *Breast Cancer Res. Treat.*, 77, 233, 2003.

138. Bramson, J.L. et al., Pre-existing immunity to adenovirus does not prevent tumor regression following intratumoral administration of a vector expressing IL-12 but inhibits virus dissemination, *Gene Ther.*, 4, 1069, 1997.

139. Mehta, H.B., Popovich, B.K., and Dillmann, W.H., Ischemia induces changes in the level of mRNAs coding for stress protein 71 and creatine kinase M, *Circ. Res.*, 63, 512, 1988.

140. Benjamin, I.J., Kroger, B., and Williams, R.S., Activation of the heat shock transcription factor by hypoxia in mammalian cells, *Proc. Natl. Acad. Sci. USA*, 87, 6263, 1990.

141. Locke, M. et al., Activation of heat-shock factor in rat heart after heat shock and exercise, *Am. J. Physiol.*, 268, C1387, 1995.

142. Venkataseshan, V.S. and Marquet, E., Heat shock protein 72/73 in normal and diseased kidneys, *Nephron*, 73, 442, 1996.

143. Salminen, W.F., Jr, Voellmy, R., and Roberts, S.M., Differential heat shock protein induction by acetaminophen and a nonhepatotoxic regioisomer, 3'-hydroxyacetanilide, in mouse liver, *J. Pharmacol. Exp. Ther.*, 282, 1533, 1997.

144. Moseley, P.L., Heat shock proteins and the inflammatory response, *Ann. N.Y. Acad. Sci.*, 856, 206, 1998.

145. Bajramovic, J.J. et al., Differential expression of stress proteins in human adult astrocytes in response to cytokines, *J. Neuroimmunol.*, 106, 14, 2000.

146. Shastry, S., Toft, D.O., and Joyner, M.J., hsp70 and hsp90 expression in leucocytes after exercise in moderately trained humans, *Acta Physiol. Scand.*, 175, 138, 2002.

147. Zou, J. et al., Correlation between glutathione oxidation and trimerization of heat shock factor 1, an early step in the stress induction of the Hsp response, *Cell Stress Chaperones*, 3, 130, 1998.

148. Vilaboa, N. et al., Novel gene switches for targeted and timed expression of proteins of interest, *Mol. Ther.*, 12, 290, 2005.

30 Monitoring Gene and Cell Therapies in Living Subjects with Molecular Imaging Technologies

Sunetra Ray, Sandip Biswal, and Sanjiv Sam Gambhir

CONTENTS

30.1   INTRODUCTION

Human clinical trials for gene- and cell-based therapeutic strategies can be significantly aided by the ability to determine the locations, magnitude, and time variation of transgene expression. Gene therapy vectors have to be held accountable for their actions and recent developments in noninvasively imaging gene expression in living subjects are certain to help in this regard. Following delivery of a therapeutic vector into a

patient, a gene therapist wants to know the answers to some fundamental questions: Did the vector get delivered to the desired target organ, tumor, or cell population or have inappropriate tissues been transfected? Did gene transfer or transcription take place? How much gene product is being generated from the transgene? How much control do I have in the expression of this vector? What happens to expression levels as the patient responds (or fails) to therapy? Current clinical methods have difficulty in answering these questions: They rely upon serum markers, tissue sampling followed by histochemical analysis or autoradiography, anatomical-based imaging, and physical examination, most of which are inefficient, invasive, or inadequate in this era of targeted gene therapy and molecular medicine. Answering these questions rapidly, easily, and effectively during a patient's clinical course is clearly desirable not only to tailor an individual patient's needs and improve clinical outcomes but also to promote gene therapy for routine clinical use.

Recognizing this need, molecular imaging researchers have advanced the principles of reporter gene technology for use in intact, viable subjects. To the molecular biologists, reporter gene technology is well-trodden territory and they have been familiar with the process of coupling a reporter gene to a therapeutic gene as a means of measuring the efficacy of gene delivery and tracking gene expression. Commonly used reporter genes for *in vitro* or *ex vivo* analyses include: (a) chloramphenicol acetyltransferase, which transfers (radioactive) acetyl groups to chloramphenicol and detection is rendered through thin layer chromatography and autoradiography [1–3]; (b) β-galactosidase (GAL), which hydrolyzes colorless galactosides to yield colored precipitate, is detected via tissue extracts or histochemical techniques [4–6]; (c) β-lactamase, which catalyzes hydrolysis of cephalosporin, is monitored by a change in fluorescence emission of a substrate [7,8]; (d) green fluorescent protein (GFP), which fluoresces upon radiation, is examined with epifluorescence microscopy or utilized in cell sorting techniques [9–11]; (e) Firefly luciferase (FL), which oxidizes D-luciferin to generate bioluminescence, can be visualized using luminometry or microscopes fitted to specialized cameras [12,13]; and (f) Renilla luciferase, which catalyzes oxidation of coelenterazine, leading to bioluminescence [14]. While some of these technologies cannot make the jump into studies of living subjects (e.g., chloramphenicol acetyltransferase, β-galactosidase) because of technical limitations, the optically based technologies, such as those based on GFP and the luciferases, have been applied to intact, small living subjects using sensitive optical detection systems [15–18].

Furthermore, there has been an exciting surge in the development of reporter gene technologies that are compatible with current clinical imaging modalities such as the radionuclide-based techniques (positron emission tomography [PET] and single-photon emission computed tomography [SPECT]), magnetic resonance imaging (MRI), and magnetic resonance spectroscopy (MRS). To understand these clinical technologies does not require a large leap of faith. In fact, they are conceptually similar to the optical systems described above. All *in vivo* reporter techniques, including the aforementioned optical techniques, depend on the detection of electromagnetic homing signals that arise either from the probe itself or the specific interaction of a reporter protein and its corresponding reporter probe. Just as GFP emits photons in the visible range of electromagnetic spectrum, which can be seen with the naked eye or cameras, radionuclide-based reporter techniques emit photons in the gamma ray-wave range that are detected by specialized scintillation crystals or more recently solid state detectors. Table 30.1 lists the various reporter systems available for *in vivo* imaging. In this chapter,

TABLE 30.1
Summary of Reporter Gene Systems Used in Living Subjects

Method	Imaging	Reporter Gene	Action	Imaging Agent/Substrate	Ref.
In vitro (selected few)	Chromatography and autoradiography	Chloramphenicol acetyltransferase	Acetylation	Radioactive acetyl group	[2]
	Light microscopy	β-Galactosidase	Hydrolysis	Metabolized galactosides	[6]
		β-glucuronidase	Hydrolysis	Metabolized glucuronides	[217]
	Epifluorescence microscopy	β-Lactamase	Hydrolysis	Metabolized cephalosporin	[8]
		GFP	Light excitation	None (irradiation)	[9]
		Red fluorescent protein	Light excitation	None (irradiation)	[87]
		Enhanced GFP	Light excitation	None (irradiation)	[80]
	MRI	Tyrosinase	Hydroxylation	Melanin production	[162]
Living subjects	Optical: Fluorescence	GFP	Light excitation	None (irradiation)	[18]
		Red fluorescent protein (dsRed)	Light excitation	None (irradiation)	[218]
		Monomeric red fluorescent protein (mRFP1 and mPLUM)	Light excitation	None (irradiation)	[88]
	Optical: Fluorescence activation	Cathepsin D	Enzymatic cleavage	Quenched NIRF fluorochromes	[19]
		Matrix metalloproteinase-2 (MMP-2)	Enzymatic cleavage	Quenched NIRF fluorochromes	[96]
	Optical: Bioluminescence	Luciferase (firefly)	Luciferase-D-Luciferin action in presence of ATP, Mg^{2+}, and O_2	D-Luciferin	[78]

TABLE 30.1 (continued)

Summary of Reporter Gene Systems Used in Living Subjects

Method	Imaging	Reporter Gene	Action	Imaging Agent/Substrate	Ref.
		Thermostable firefly luciferase	Luciferase-D-Luciferin action in presence of ATP, Mg^{2+}, and O$_2$	D-Luciferin	[100]
		Luciferase (Renilla)	Luciferase–substrate action in presence of O$_2$ only	Coelenterazine	[15]
		Red-shifted Renilla	Luciferase–substrate action in presence of O$_2$ only	Coelenterazine	[102]
		Luciferase (Gaussia)	Luciferase–substrate action in presence of O$_2$ only	Coelenterazine	[21]
	Radionuclide Imaging: PET	Herpes simplex virus type 1 thymidine kinase (gene) (HSV1-tk)	Phosphorylation	[^{124}I]FIAU [^{18}F]GCV [^{18}F]PCV [^{18}F]FHPG [^{18}F]FHBG [^{18}F]FEAU	[219]
		Mutant HSV1-tk (HSV1-sr39-tk)	Phosphorylation	[^{18}F]PCV [^{18}F]FHBG	[122]
		Truncated human mitochondrial thymidine kinase 2 (ΔhTk2)	Phosphorylation	[^{18}F]FEAU [^{124}I]FIAU	[123]
		Varicella-zoster virus thymidine kinase gene (VZV-tk)	Phosphorylation	Radiolabeled BCNA	[124]
		Cytosine deaminase (CD)	Enzymatic conversion	[^{18}F]-5-fluorocytosine (5-FC)	[138]
		L-Amino acid decarboxylase (AADC)	Enzymatic conversion	6-[^{18}F]fluoro-L-*m*-tyrosine	[220]
		Dopamine-2-receptor (D2R)	Receptor–ligand	[^{18}F]FESP	[108]
		Mutant D2R	Receptor–ligand	[^{18}F]FESP	[107]
		Sodium/iodide symporter (NIS)	Ion pump	^{124}I	[134]
	Radionuclide Imaging: Gamma camera/ SPECT	Somatostatin (SS) receptor	Affinity binding	[^{99m}Tc]SS analogue P2045 [^{111}In]DTPA-d-Phe-octreotide [^{188}Re]SS analogue	[150]
		HSV1-tk	Phosphorylation	[^{131}I]FIAU [^{14}C]FIAU [^{125}I]FIAU	[150] [221]
		NIS	Ion pump	^{123}I, ^{131}I, ^{99m}Tc pertechnetate	[131]
		Gastrin-releasing peptide receptor	Receptor–ligand	[^{125}I]mIP-Des-Met^{14}-bombesin (7–13)NH$_2$ [^{125}I]Bombesin [^{99m}Tc]Bombesin analog	[222]
	MRI	β-Galactosidase	Hydrolysis	EgadMe	[168]
		Engineered transferrin receptor (ETR)	Receptor–ligand internalization	Tf-MION Tf-CLIO	[154] [157]
		Tyrosinase	Enzymatic conversion	DOPA	[162,163]
		Ferritin	Iron storage	Magnetoferrin molecules	[165,166]
	MRS	Creatine kinase (CK)	Chemical shift spectroscopy	Phosphocreatine	[169]
		Arginine kinase (AK)	Chemical shift spectroscopy	Phosphoarginine	[170]
		CD	Spectroscopy	Production of 5-fluorouracil (5-FU)	[138]

we will discuss the basic principles of the various tools now available, display several examples of the images generated from such tools, and at the end, provide a strategy to help scientists or gene therapists interested in using *in vivo* reporter techniques to choose optimal strategies.

30.2   IMAGING INSTRUMENTATION FOR LIVING SUBJECTS

30.2.1   BASIC PRINCIPLES OF OPTICAL IMAGING (FLUORESCENCE AND BIOLUMINESCENCE)

Although photons emitted in the visible light range of the electromagnetic spectrum face considerable obstacles traveling through layers of tissues, notable advances in light sensor technology have permitted the use of optical reporter genes in intact organisms. There are fundamentally two different types of optically based imaging systems: (a) fluorescence imaging, which uses emitters such as GFP, wavelength-shifted GFP mutants, red fluorescent protein (RFP), fluorescent small molecules, "smart" near-infrared fluorescent (NIRF) probes and (b) bioluminescence imaging, which utilizes systems such as firefly luciferase-D-Luciferin or Renilla luciferase/Gaussia luciferase-coelenterazine [15,19–21]. Each of these systems will be discussed in further detail in Section 30.3.1. Emission of light from fluorescent markers requires external light excitation while bioluminescence systems generate light de novo after an injectable substrate is introduced and the appropriate conditions are met (Figure 30.1). In both cases, light emitted from either system can be detected with a thermoelectrically cooled charge-coupled device (CCD) camera since they emit light in the visible light range (400–750 nm) to near-infrared (NIR) range (~800 nm). Cooled to −120°C to −150°C, these cameras can detect weakly luminescent sources within a light-tight chamber. Being exquisitely sensitive to light, these desktop camera systems allow for quantitative analysis of the data. The method of imaging bioluminescence sources in living subjects with a CCD camera is relatively straightforward: the animal is anesthetized, subsequently injected intravenously or intraperitoneally with the substrate, and placed in the light-tight chamber for a few seconds to minutes. A standard light photographic image of the animal is obtained, followed by a bioluminescence image captured by the cooled CCD camera positioned above the subject within the confines of the dark chamber. A computer subsequently superimposes the two images on one another and relative location of luciferase activity is inferred from the composite image. An adjacent color scale quantitates relative or absolute number of photons detected. This scale does not reflect the color (wavelength) of the emitted photons, but only the number of such photons, measured as photons/cm²/s/steradian. Here steradian is a measurement of solid angle. Differences between fluorescence and bioluminescence systems are discussed in a later section of this chapter.

Comparison of optical-based imaging systems with the other imaging modalities, such as the radionuclide-based or

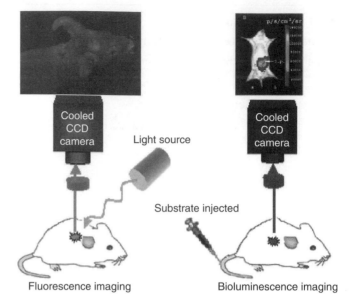

FIGURE 30.1 (See color insert following blank page 206.) Basic principles of optical CCD imaging (fluorescence/bioluminescence). There are fundamentally two different types of optically based imaging systems: fluorescence imaging, which uses emitters such as GFP, wavelength-shifted GFP mutants, RFP, smart probes and near-infrared fluoreseent (NIRF) probes, and bioluminescence imaging, which utilize systems such as firefly luciferase/D-Luciferin or Renilla luciferase/coelenterazine. Emission of light from fluorescent markers requires external light excitation while bioluminescent systems generate light de novo when the appropriate substrates/cofactors are made available. In both cases, light emitted from either system can be detected with a thermoelectrically cooled CCD camera since they emit light in the visible light range (400–700 nm) to NIR range (~800 nm). Cooled to −120°C to −150°C, these cameras can detect weakly luminescent sources within a light-tight chamber. Being exquisitely sensitive to light, these desktop camera systems allow for quantitative analysis of the data. Image shown above the "fluorescence imaging" schematic is a representative image obtained from a glioma model, which expresseses RFP (image used with permission from Anticancer, Inc.). The method of imaging bioluminescent sources living subjects with a CCD camera is relatively straightforward: The animal is anesthetized, subsequently injected with the substrate, and immediately placed in the light-tight chamber. A light photographic image of the animal is obtained which is followed by a bioluminescence image captured by the cooled CCD camera positioned above the subject within the confines of the dark chamber. The two images are subsequently superimposed on one another by a computer, and relative location of luciferase activity is inferred from the composite image. An adjacent color scale confers relative concentration of luciferase activity. Sample image above the "Bioluminescence imaging" schematic is a typical image obtained with this technology. In this specific example, image was obtained after intravenous injection of coelenterazine into a mouse containing intraperitoneal Renilla luciferase-expressing tumor cells. Significant bioluminescence is detected from the region of the xenograft. (From Bhaumik, S. and Gambhir, S.S., *Proc. Natl. Acad. Sci. USA*, 99, 377, 2002. With permission.)

MRI-based systems, reveals important differences. Advantages of optical-based reporter systems are that they are at least an order of magnitude more sensitive than the radionuclide-based techniques but at limited depths [22]. Furthermore, the direct and indirect costs are generally less than radionuclide-based techniques or MRI. However, there is significantly less spatial information obtained from optical imaging and signal obtained from light-emitting reporter systems is limited by the tissue depth from which it arises. Skin autofluorescence is also a major obstacle to imaging at visible light wavelengths. The Maestro *in-Vivo* Imaging System from Cambridge Research and Instrumentation helps to resolve some of these issues by allowing simultaneous imaging at multiple spectra with a tunable filter that can operate at wavelengths from 500 to 950 nm. This multispectral acquisition, unmixing, and analysis approach dramatically improves sensitivity by detecting and minimizing autofluorescence noise. The system also provides multicolor flexibility and quantitative accuracy for both visible and NIR labels [23].

In the recent past, significant progress has been made to localize fluorescent signals tomographically to obtain distribution of fluorochromes in deep tissues [24]. Currently there are three approaches for achieving optical tomography: continuous wave, time-domain (TD), and frequency-domain. In continuous wave fluorescence tomography, the subject is trans-illuminated by a constant intensity light source and transmitted light is collected at different points on the surface surrounding the subject [25]. The fluorescence molecular tomography (FMT) platform from VisEn Medical uses this approach to quantify fluorescent signals from living animals at different depths and at multiple NIR wavelengths. FMT captures approximately 50,000–100,000 source–detector pair measurements rapidly within a scan time on the order of 2–4 min [26]. These large data sets are then processed by a sophisticated tomographic reconstruction algorithm taking into account the optical properties of the tissue being imaged and of the probe being used [23].

The second approach to accurately determine the localization and concentration of fluorophores is based on the principles of TD imaging. Most biologically relevant fluorophores exhibit characteristic decay times ranging from picoseconds to nanoseconds. In TD optical imaging, photons are classified based on the time at which they emerge from the tissue. The eXplore OptixTM system from ART, Inc. exploits this concept to facilitate detection, localization, and tracking of fluorescent probes noninvasively and in real-time. Typically, the subject is illuminated with a short, subnanosecond duration laser pulse and a fast detector measures the arrival distribution of photons as a function of time at different locations. A light pulse propagating through the tissue is broadened and attenuated due to scattering and absorption [27]. A fluorophore embedded in tissue is excited by this pulse of light and emits a fluorescence pulse with lifetime decay typical of the fluorophore. The generated time-of-flight distribution (generally referred to as temporal-point-spread-function) is a statistical distribution of all possible photon paths between the point of illumination and the point where the light exits the tissue. This distribution is then used to determine the optical characteristics of the specimen and to discriminate scattering from absorption [28]. Theoretically, fluorescence lifetime is capable of distinguishing between exogenous and endogenous fluorescence, resulting in an increase in the signal-to-noise ratio, hence of the sensitivity. Apart from measuring the fluorescence intensity, tissue transport properties, or local fluorophore microenvironment, the system is also capable of tomographic reconstruction [29]. Some other advantages include the ability to distinguish and quantify two different fluorophores simultaneously based on their overlapping decay kinetics and to distinguish when certain classes of fluorophores are in a given chemical environment based on variations in fluorescence lifetime.

The various optical imaging strategies have been recently compared. Results indicate that the eXplore Optix system can be more sensitive with significantly higher detection depth and spatial resolution compared to the IVIS200 continuous wave imaging system [27]. However, unlike the IVIS200 system, major drawbacks of the TD imaging system are increased acquisition time and the inability to image multiple animals simultaneously.

Another strategy, which utilizes frequency domain photon migration imaging, is also under active investigation [30,31] but no commercial systems are yet available. In this approach, the incident light is modulated at a frequency of 10 mHz–1 GHz (range for physiological tissues being 30–200 mHz). The modulated excitation light then propagates within the tissue and generates a redshifted and intensity-modulated emission wave on reaching a fluorescent target. The emitted light then propagates to the tissue surface and is collected and passed through optical filters to reject the excitation portion of the reemitted light signal. These frequency domain systems may prove to have some advantages including enhanced depth penetration, lower noise, and higher sensitivity compared to the continuous wave method [31].

More effective *in vivo* optical imaging has become possible now because of the development of far-red and NIR probes (700–900 nm) that can overcome the unique obstacles posed by living tissues. The principal absorbers of light in the biological systems, like hemoglobin, water and lipid, are known to have their lowest absorbance at the NIR region. An additional advantage of imaging at NIR region involves low-tissue autofluorescence, which helps to further improve the target/background ratio.

30.2.2 Basic Principles of Positron Emission Tomography

With recent breakthroughs in molecular/cell biology and target discovery, it is now possible to design specific markers to image events noninvasively in small animals and humans with PET. Natural biological molecules such as glucose, peptides, proteins, and a variety of other structures can either be labeled with a radioisotope or slightly modified to accommodate a radioisotope. In the jargon of molecular imagers, these radiolabeled molecules are referred to as molecular probes, reporter probes, markers, or tracers. The term contrast agent should be avoided as it usually refers to a nonspecific imaging agent in distinction to molecular probes. In PET imaging, molecules are labeled with isotopes that emit positrons from their nucleus. This is in contrast to gamma camera or SPECT imaging (discussed in Section 30.2.3), where molecules are "tagged" with radioisotopes that emit gamma rays. When a molecular probe (tracer) is injected into a living subject and then placed into a PET scanner, the image we acquire is a snapshot of the physiologic distribution and concentration of that tracer.

Let us take the example of 2-deoxy-2-[¹⁸F] fluoro-D-deoxyglucose (FDG) [32–34], a glucose analog labeled with the positron-emitting isotope, ¹⁸F. In the synthesis of this tracer, PET radiochemists have devised a method to replace one of the hydrogens with the ¹⁸F radioisotope. Following intravenous injection, this particular probe distributes readily from the intravascular compartments to the extravascular and intracellular compartments. FDG becomes sequestered in cellular populations and tissues, which have a predilection for glucose,

i.e., cells, which possess a larger number of glucose transporters or hexokinase II activity. When FDG encounters hexokinase II, it becomes phosphorylated and subsequently trapped within the cell because it cannot diffuse out due to a negative charge. In contrast, FDG clears from cells or tissues that lack the ability to transport or phosphorylate FDG. The positron-emitting moiety of FDG, in this example, ¹⁸F, decays by emitting a positron from its nucleus. This positron eventually collides with a nearby electron, resulting in an annihilation event where two 511,000 eV photons in the form of gamma rays are emitted ~180° apart. The two emitted photons travel extracorporeally and are detected nearly simultaneously as they interact with the PET camera, a ring array of detectors (composed of scintillation crystals and photomultiplier tubes) surrounding the subject (Figure 30.2). Detection of a single annihilation event results in the activation of detectors opposing one another, which is recorded as a coincident event, thus defining a set of coincident lines [35]. The recording of multiple detector pair combinations yields a large number of these coincident lines. Sophisticated mathematical analyses of the coincident lines, which include filtered back projection and attenuation correction (correction for gamma rays that do not

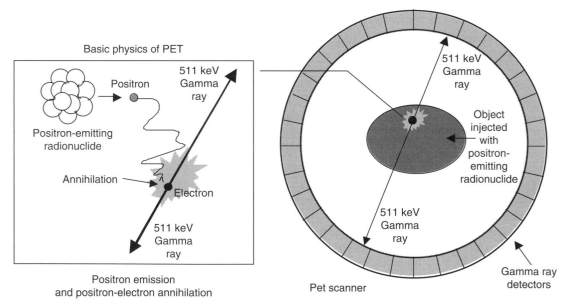

Basic physics of PET

Positron

511 keV
Gamma
ray

Positron-emitting
radionuclide

Annihilation

Electron

511 keV
Gamma
ray

Positron emission
and positron-electron annihilation

511 keV
Gamma
ray

Object
injected
with
positron-
emitting
radionuclide

511 keV
Gamma
ray

Gamma ray
detectors

Pet scanner

FIGURE 30.2 Basic principles of PET imaging. Biologically active molecules such as glucose, peptides, and proteins can be radiolabeled with positron emitting radioisotopes. This radiolabeled molecule is referred to as a probe or tracer. The positron emitting isotope decays by emitting a positron from its nucleus. This positron eventually collides with a nearby electron resulting in an annihilation event where two 511,000 eV photons in the form of gamma rays are emitted ~180° apart. The two emitted photons travel extracorporeally and are detected nearly simultaneously as they interact with a ring of detectors (composed of scintillation crystals and photomultiplier tubes) surrounding the subject. Detection of a single annihilation event results in the "activation" of detectors opposing one another, which is recorded as a "coincident event." The recording of multiple detector pair combinations yields a large number of these coincident lines. Sophisticated mathematical analyses of the coincident lines, which include filtered back projection and attenuation correction, yields the location of cell populations or tissues that contain the molecule labeled with the positron emitter. Tomographic images of relative probe concentration can be reconstructed in the conventional sagittal, coronal, and transverse imaging planes or, actually, in any arbitrary plane. The resultant image depicts the distribution and concentration of the radiolabeled tracer. Sensitivity of PET is in the range of 10^{-11} to 10^{-12} m/L and is independent of the location depth of the tracer of interest. It is also important to note that all positron-emitting radioisotopes produce two gamma rays of the same energy, so if two molecular probes, each with a different positron-emitting isotope, are injected simultaneously, there is no way for the PET camera to distinguish between the two molecular probes. Therefore, to perform studies which look at two or more distinct molecular events (e.g., suicide gene therapy and imaging apoptosis, cardiac gene therapy and perfusion ¹³N ammonia imaging, etc.), one has to inject molecular probes separately, which allows decay of the isotope.

penetrate through due to absorption), yield the location of cell populations or tissues that have accumulated FDG. Tomographic images of relative probe concentration can be reconstructed in the sagittal, coronal, and transverse imaging planes. Quantitative information obtained from the images is, in turn, related to the underlying biochemical process.

Radiolabeling molecules is not just limited to ^{18}F. A collection of positron-emitting isotopes are available for use, which include the more commonly utilized isotopes, ^{15}O, ^{13}N, ^{11}C, and ^{18}F, and the less commonly used ^{14}O, ^{64}Cu, ^{62}Cu, ^{124}I, ^{76}Br, ^{82}Rb, and ^{68}Ga. Most of these isotopes are created in a cyclotron, a device used to accelerate charged particles to create the relatively short-lived positron-emitting isotopes (for example, ^{18}F, the half-life of which is 110 min) [36]. Automated synthesizers can then couple the isotope to a molecule of interest to produce the molecular probe (tracer). Given the relatively short half-life of positron emitters, the process of producing isotopically labeled molecules has to be performed with great efficiency and in relatively close proximity to the hospital, clinic, or animal research facility. In this regard, a modest number of PET radiopharmacies are available worldwide, producing PET tracers on a daily basis.

Clinical PET scanners have been around for several decades and, in the last decade, PET cameras for small animals have been developed for the purpose of developing molecular imaging assays in small rodents prior to their application in humans. The spatial resolution of most clinical PET scanners is $\sim(6-8)^3$ mm^3 with the more recent scanners achieving $\sim(2)^3$ mm^3 capabilities. By comparison, most small animal PET scanners (microPET) have a resolution of $\sim(2)^3$ mm^3 with newer generation scanners attaining $\sim(1)^3$ mm^3 [37–41]. When compared to other modalities, the sensitivity of PET is relatively high, on the order of 10^{-11} to 10^{-12} m/L (MRI's intrinsic sensitivity is $\sim10^{-4}$ to 10^{-5} M) [42]. Furthermore, the location or depth of the tracer of interest does not affect sensitivity. In contrast, the imaging of many optical imaging probes is significantly affected by tissue depth. Under appropriate conditions, the smallest cluster of cells that can be visualized by a clinical PET scanner is 10^6 to 10^9 in number. Thus, radiotracer imaging techniques afford the detailed locations, magnitude, and persistence of probes or tracers for *in vivo* use in animals and humans.

30.2.3 BASIC PRINCIPLES OF THE GAMMA CAMERA AND SINGLE-PHOTON EMISSION COMPUTED TOMOGRAPHY IMAGING

Imaging with a gamma camera is similar to PET, but the radiolabel emits gamma rays instead of positrons. A variety of radiolabels, each emitting at characteristic photon energies, can be attached to molecules including ^{111}In (171–245 keV), ^{125}I (27–35 keV), ^{131}I (364 keV), and ^{99m}Tc (140 keV). Once introduced into the body, detection of these radiolabeled probes is performed with a gamma camera, a scintillation detector consisting of collimator, a sodium iodide crystal, and a set of photomultiplier tubes (Figure 30.3).

Upon decay, these radionuclides emit a gamma ray at their characteristic energies in different directions. Some of the gamma rays will scatter or lose energy and others may never interact with the camera. Since the gamma camera is situated only on one side of the subject, only rays directed toward the camera will be captured. Furthermore, only those gamma

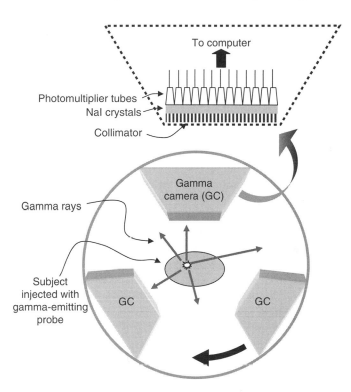

FIGURE 30.3 Basic principles of gamma camera/SPECT imaging. Imaging with a gamma camera is similar to PET, but the radiolabel emits gamma rays instead of positrons. A variety of radioisotopes, each emitting at characteristic photon energies, can be attached to a variety of molecules. Examples of isotopes include ^{111}In (171–245 keV), ^{125}I (27–35 keV), ^{131}I (364 keV), and ^{99m}Tc (140 keV). Once introduced into the body, detection of these radiolabled probes is performed with a gamma camera, a scintillation detector consisting of collimator, a sodium iodide crystal, and a set of photomutliplier tubes. Upon decay, these radionuclides emit a gamma ray at their characteristic energies in different directions. Some of the gamma rays will scatter or lose energy and others may never interact with the camera. Since the gamma camera is situated only on one side of the subject, only rays directed toward the camera will potentially be "captured." Furthermore, only those gamma rays, which arise parallel to collimator, will be detected since scattered gamma rays will be absorbed by the collimator. Those rays, which successfully reach the crystal and stopped by it, will be converted into photons of light. In turn, the photomultiplier tubes convert the light into an electrical signal that is proportional to the incidental gamma ray. Gamma rays, which arrive at detector lower than the expected characteristic energy, are thought to be the result of scattering and summarily rejected from the analysis. Since gamma cameras acquire data in a single plane, the resultant images are a two-dimensional representation of a three-dimensional subject (referred to as "planar imaging"). SPECT acquires volumetric data by rotating a gamma camera around the subject and/or using multidetector systems (shown above). (From Rosenthal, M.S., Cullom, J., Hawkins, W., Moore, S.C., Tsui, B.M., and Yester, M., *J. Nucl. Med.*, 36, 1489, 1995.)

rays that arise parallel to the collimator will be detected since the collimator will absorb scattered gamma rays. Those rays that successfully reach the crystal will be converted into photons of light. In turn, the photomultiplier tubes convert the light into an electrical signal that is proportional to the incidental gamma ray. Gamma rays, which arrive at the detector lower than the expected characteristic energy, are thought to be the result of scattering and summarily rejected from the analysis. Since gamma cameras acquire data in a single plane, the resultant images are a two-dimensional representation of a three-dimensional subject (referred to as planar imaging).

While more affordable and accessible than PET, the limitations of gamma camera imaging are obvious: (1) diminished sensitivity, since many decay events are either rejected or never captured due to collimation, (2) decreased signal-to-noise since overlapping foci of activity are not delineated, and (3) lower spatial resolution. Alleviating some of these problems is SPECT, which acquires volumetric data by rotating the gamma camera around the subject or using multidetector systems [43] (Figure 30.3). As with microPET, small-animal SPECT devices (microSPECT) have been created to study the use of gamma-emitting radiolabeled reporter probes in animal models of cancer and gene therapy. These instruments have resolutions on the order of $1 \, mm^3$. Thus, advantages of the SPECT systems are that they generally allow for better spatial resolution. Another advantage of SPECT is that two radioisotopes of different energies can be imaged simultaneously, allowing for the concurrent study of two distinctly radiolabeled molecules (e.g., one radiolabeled with ^{99m}Tc and another radiolabeled with ^{125}I). With PET imaging, such simultaneous imaging is not possible since all positron-emitting events lead to 511 keV gamma rays. A disadvantage for SPECT, however, is that it is an order of magnitude lower in sensitivity than PET. However, small-animal SPECT imaging systems continue to be developed and can achieve very high spatial resolution [44,45].

30.2.4 BASIC PRINCIPLES OF MAGNETIC RESONANCE IMAGING

With regard to the electromagnetic spectrum, MRI works with longer wavelengths (lower energies) than those used by the radionuclide or optical techniques. A comprehensive review of MRI is beyond the scope of this chapter; there are several excellent reviews and books available [46–49]. A very brief, simplistic description follows here: Spinning charged nuclei generate a magnetic field and MRI depends on these charged atomic nuclei, which contain an odd number of protons. Those nuclei that contain an odd number of protons always have an unpaired proton, which gives the atom a net magnetic field or "magnetic dipole moment" (MDM). In contrast, atoms, which have an even number of protons, have a net magnetic field of zero. Thus, the nucleus of hydrogen (^1H), which possesses a single unpaired proton as its nucleus and is present in abundance (in the form of H_2O and $-CH_2-$) in cells and tissues, is primarily used for MRI.

The majority of available MR scanners are fitted with a superconducting magnet in the shape of a long hollow tube. The magnet creates an external magnetic field (B_0) and within the hollow center of the tube-shaped magnet, the magnetic field is nearly homogeneous and parallel to the long axis of the tube. When a subject is placed within the hollow confines of the magnet, the MDMs of the hydrogen atoms align themselves with the main magnetic field B_0, much like the way iron filings behave when placed in the vicinity of a magnet.

Once equilibrium has been established in the external magnetic field, B_0, a magnetic pulse (an electromagnetic wave), otherwise known as radiofrequency (RF) pulse is introduced perpendicular to B_0. It is called an RF pulse because the frequency (and energy) of the pulse is in the RF range of the electromagnetic spectrum (3–100 MHz). This causes the hydrogen nuclei to transiently orient their MDM parallel to the new magnetic field (perpendicular to B_0). After the RF pulse, they realign (relax) their MDMs to the main magnetic field B_0, and, in the process give off energy in the form of RF waves that can be detected by receiver coil, which typically surrounds the subject. To reiterate, both RF pulse and magnetic field are used to perturbthe underlying subject, with an RF wave being generated in the process to produce an image. Also, the time it takes for the hydrogen nuclei to relax to equilibrium (or a fraction thereof) can be measured. The rate at which a hydrogen nucleus relaxes is dependent upon the nature of its parent molecule such as a freely mobile water proton versus the rigidly attached proton of hydrocarbon backbone of a fatty acid. Water protons, which randomly tumble in aqueous solution, take longer to regain equilibrium with the main magnetic field, B_0, than those protons associated with much larger, more fixed molecules. The measurements of relaxation rates can be converted into a value, which translates into image pixel value, with each pixel representing a small, representative unit volume of the subject (voxel). On a certain MRI protocol called a T1-weighted sequence, a voxel composed mostly of fatty (hydrocarbons) protons will have a high (bright) signal since the rate of relaxation is rapid. Compare this to the voxel, which contains a large number of water protons: this voxel will have a low (dark) signal on T1-weighted MRI since the rate of relaxation is much longer. Each MR image is of 256×256 or 512×512 pixels, each a representative slice through the subject (Figure 30.4).

Certain exogenous or endogenous atoms or molecules, like gadolinium (paramagnetic) and iron/hemosiderin (superparamagnetic), respectively, can influence the local magnetic field by their powerful magnetic properties, significantly alter the rate of relaxation of the protons, and therefore generate contrast in the image. Paramagnetic substances have unpaired electrons. They become magnetized in the presence of an external magnetic field and contribute to an increase in the effective magnetic field. Examples include the rare-earth element, gadolinium (Gd) (seven unpaired electrons), deoxyhemoglobin (four unpaired electrons), and methemoglobin (five unpaired electrons). Hemosiderin, an end-stage by-product of hemorrhage, has more than 10,000 unpaired electrons and, thus, belongs to superparamagnetic

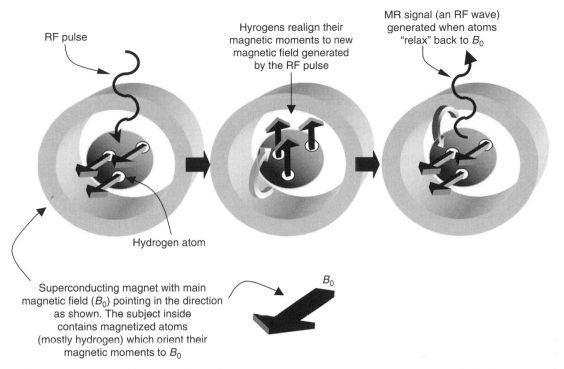

FIGURE 30.4 Basic principles of MRI. In MRI, subjects are placed in a strong, external magnetic field, B_0, produced by a hollow, cylindrical magnet. The B_0 field is nearly uniform and points parallel to the long axis of the magnet. Imaging with MRI is dependent upon atomic nuclei with an odd number of protons, such as hydrogen (^{1}II). Such atoms have their own net magnetic field (known as MDM) and their moments align accordingly when placed in this external magnetic field. Once equilibrium has been achieved between the subject and the magnet, energy can be added to the system in the form of an RF pulse. In most cases, this pulse, which generates its own magnetic field, can change the alignment of the hydrogen atoms such that their moments are now perpendicular to B_0. Once the RF pulse is "turned off," the hydrogen atoms realign or "relax" to B_0 and give up energy in the form of RF waves during the relaxation period. Receivers located in the magnet capture this RF wave. One of the calculations made from the captured information is the rate at which the hydrogen atoms relax to equilibrium. Image construction and image contrast are possible with MRI because hydrogen atoms associated with macromolecules like fat and proteins have a significantly different relaxation rate than the hydrogen atoms of bulk water. The measurements of relaxation rates can be converted into a value, which translates into image pixel value, with each pixel representing a small, representative, unit volume of the subject (voxel). On a certain MR imaging protocol called a "T1-weighted" sequence, a voxel composed mostly of fatty (hydrocarbons) protons will have a high (bright) signal since the rate of relaxation is rapid. In contrast, voxels, which contain a large number of water protons, will have a low (dark) signal on T1-weighted MR imaging since the rate of relaxation is much longer.

group of substances. The magnetic susceptibility of superparamagnetic substances is 100 to 1000 times stronger than paramagnetic substances. Both super- and paramagnetic substances help localize reporter gene expression during MRI (see also Figure 30.10). As we will see in Section 30.3.3, the properties of paramagnetic and superparamagnetic compounds are exploited for MR-based reporter systems.

MRI techniques offer phenomenal spatial resolution (voxel resolutions of ~10 μm³ *in vitro* and ~50 μm³ in small animals) but are several orders of magnitude less sensitive than optical- and radionuclide-based techniques. Sensitivity of MRI is on the order of 10^{-3} M while PET imaging is 10^{-12} M, and thus, substantially more MRI probe has to be injected into the living subject in order to provide sufficient contrast [50]. However, in MRI, subjects are not exposed to radiation, which poses as one of the biggest advantage of this modality.

Hyperpolarized gases have also been used extensively in MRI as MR contrast agents or for enhancing temporal reso-

lution and to enhance sensitivity mostly in the lungs [51]. A low proton density and a lack of tissue homogeneity degrade the magnetic resonance signal and therefore interfere with image acquisition in the lungs with conventional MRI [52]. Inert noble gases like helium (^{3}He) upon inhalation can diffuse rapidly to fill airspaces of the lungs and permit visualization and measurement of the ventilated airways and alveolar spaces [53,54]. However, the density of gases under normal conditions is too low to produce a detectable MR signal. This drawback is overcome by artificially increasing the amount of polarization per unit volume, a process called hyperpolarization. This procedure causes the normal random spin distribution of atoms to be altered and a population of atoms with spins in parallel configuration dominates causing a log-order increase in signal and a favorable signal-to-noise ratio. Hyperpolarization can provide nuclear polarizations that are five orders of magnitude higher than those achieved at thermal equilibrium [55]. The most common methods for hyperpolarization of noble gases are based on laser optical pumping, and include alkali–metal spin-exchange or meta-stability exchange. Due to its relatively long stability and its inert gas properties, ^{3}He became the most commonly applied gas for MRI, with applications including high-resolution functional lung imaging to detect diseases related to smoking or asthma, cystic fibrosis, etc. [53,56,57]. Recently hyperpolarized ^{13}C has shown a lot of promise in angiography and other medical imaging approaches [58,59]. Use of bis-1, 1-(hydroxymethyl)-1-^{13}C-cyclopropane-D8, a water-soluble compound, demonstrated long relaxation times and excellent resolution in angiographic studies [60]. Other applications of hyperpolarized ^{13}C include angiographic studies with ^{13}C-labeled urea, or more recently, real-time imaging of metabolism with ^{13}C-labeled pyruvate as a metabolic marker [61].

30.2.5 COMPUTED TOMOGRAPHY

Computed tomography (CT) deals in the x-ray range of the electromagnetic spectrum. In CT, the subject is placed in the center of a ring of detectors. A rotating focused x-ray source emits radiation, which penetrates the subject and reaches a set of detectors on the other side of the patient. The amount of x-ray reaching the detector depends upon the amount absorbed by the patient and is inversely proportional to the density of tissues encountered as it passes through the patient. The amount of radiation reaching a detector is given a value, and, through a complex set of back calculations, tomographic images in the transverse plane can be constructed.

While this modality is not currently utilized to monitor gene expression, efforts are under way. Rather, CT's main role in human gene therapy will grow significantly as it currently serves as an anatomical adjunct to PET and SPECT imaging. Both CT and PET (or SPECT) data sets can be coregistered, and because of its superior spatial resolution, CT gives PET information more specificity. Clinical CT scanners have recently achieved spatial resolutions <1 mm. Small animal CT scanners (microCT) have attained resolutions of 50 μm [62]. Already showing its prowess in oncology, PET/CT clinical machines as well as small animal machines should help to show gene expression coupled to anatomy. More details about PET/CT can be found elsewhere [63]. The increasing use of microCT and hybrid systems that combine PET and SPECT with x-ray CT in small-animal imaging has prompted the development of novel imaging probes. Lately, several nanoscale carriers like nanoparticles, liposomes, water-soluble polymers, micelles, and dendrimers have been developed for targeted delivery of cancer diagnostic and therapeutic agents and can also behave as CT-contrast agents. Rabin et al. described a polymer-coated bismuth sulfide nanoparticle preparation as an injectable CT imaging agent. According to their initial results, this preparation shows excellent stability at high concentrations, fivefold higher x-ray absorption compared to iodine-based agents, >2 h long circulation times, and much greater safety than iodinated imaging agents [64]. Such agents can have a huge impact on applications involving

angiogenesis imaging, lymph-node imaging, and developing targeted imaging probes.

30.2.6 ULTRASOUND

Ultrasound (US) imaging uses US waves (frequency > 20 kHz) to produce either static or dynamic images of body organs and to measure velocity of blood flow. US probes are essentially acoustic transducers emitting wave pulses that are reflected as echoes bouncing off the interfaces between materials of different acoustic impedances. The echo signal provides information on the interface depth and the nature of the materials at the interface. Although the echo signals received by the probes are converted into single dimensional signals, it is possible to develop 2D or 3D images by combining the signals obtained from different angles around an organ under investigation [65].

Ultrasonography is being extensively used in small-animal research and is proving to be a valuable tool in gene therapy. In gene therapy, US microbubbles can be employed to successfully deliver genes in living subjects. There is evidence that when US is applied to microbubbles, they are disrupted and this can cause small perforations in the target cells (sonoporation), which allows the DNA to enter. US has been shown to enhance gene transfer into cells both *in vitro* and *in vivo* [66–70]. Enhanced gene transfer was observed when pure plasmid DNA was attached to microbubble shells [71]. In 2003, researchers showed that an innovative combination of US and microbubbles helped in delivering DNA in mice skeletal muscles and the efficiency of gene therapy was improved by about 10 times [72]. In a recent study, researchers were able to transfect prostate tumors in mice efficiently by a combination of therapeutic US and contrast agent, Optison, following intratumoral administration [73].

US-enhanced gene therapy is a rapidly evolving field. The procedure is very safe since the exposure levels required to destroy microbubbles lie in the diagnostic range. Moreover, targeting ligands can be attached to the surface of the microbubbles (i.e., targeted microbubbles), which would assist in directing the therapeutic transgenes to specific tissues [74–76]. In a recent study by Willmann et al., vascular endothelial growth factor receptor 2 (VEGFR2)-targeted microbubble allowed noninvasive visualization of VEGFR2 receptors in two different murine tumor models by contrast-enhanced US imaging [77]. Such tools can be extremely useful in delivering therapeutic genes, peptides, or drugs to the neovasculature bed of a growing tumor.

30.3 REPORTER GENES AND REPORTER PROBES

30.3.1 OPTICAL REPORTER SYSTEMS

There are essentially four different types of optical reporter genes currently in use for studies in living animals. These genes either encode (a) a protein, which contains a chromophore (a short, internal peptide segment that contributes to the protein's fluorescent capabilities following obligatory posttranslational modifications), which fluoresces when externally irradiated (fluorescence imaging), (b) an enzyme, which can convert an exogenously added optically quiescent substrate into a fluorescent complex, or an enzyme, which changes the conformation of a substrate such that it fluoresces a different color (fluorescence imaging), (c) luciferases that generate light when presented with the appropriate substrate (bioluminescence imaging), or (d) fusion proteins, which couple transgene products with fluorescence or bioluminescence optical reporters with a peptide linker (discussed in greater detail in Section 30.3.1). When compared to other imaging modalities, fluorescence and bioluminescence imaging techniques hold tremendous potential for study of small living animals because of its affordability, relative ease of use, high assay sensitivity, and low requirement for specialized support personnel. In contrast to the radionuclide-based techniques described earlier, all optical reporter gene/probe systems are forms of indirect imaging. The measured light emissions generated from these reporter systems may or may not correlate to the amount of therapeutic gene product present. Briefly reviewed below, extensive work in the past several years has been dedicated to mutating existing or cloning new fluorescent or bioluminescence genes, which are more compatible for use in living subjects.

Mammalian tissues pose a number of obstacles for the propagation of light and, thus, are a challenge for the evaluation of fluorescence- or bioluminescence-based reporter genes in living subjects. Fluorescent-based techniques, in particular, face additional challenges since both the light used to excite the reporter probe and the light emitted from the probe are subject to absorption, scatter, and other optical tribulations. More specifically, the excitation light is not only limited by its ability to penetrate nontransparent tissue, but also contributes to background autofluorescence (from hemoglobin and cytochromes) especially when excitation wavelengths are in the blue and green portions of the spectrum, which is generally the case when GFP (and its close relatives such as blue fluorescent proteins) is used as reporter genes [22]. As a result, fluorescence imaging has relatively poorer signal-to-noise when compared to bioluminescent techniques.

Light emitted from either fluorescence- or bioluminescence-based techniques are subject to tissue absorption and scattering. In mammalian tissues, the blameworthy structures responsible for light absorption are predominantly molecules of hemoglobin, which absorb wavelength emissions of 400–600 nm. To a lesser extent, melanin and other pigmented macromolecules also contribute to light absorption, and, thus, experiments using white-furred or hairless subjects are preferred when using optical reporter systems. Light absorption by water molecules is also another important factor, but not until wavelengths approach >900 nm range. Light scatter is another important confounding factor in the detection of low-level photon emissions; the interfaces at the surface membranes of cells and organelles are largely the cause for this occurrence [22].

Yet, despite these impediments, recent advances in the fields of optics and sensor technology make it possible to detect relatively low emissions events with great sensitivity and generate remarkable images [78,79]. Furthermore, optical reporters are being developed to operate with longer wavelengths; that is, away from the blue–green part of the spectrum and towards the red (also called redshifted, between 600 and 900 nm) so as to maximize transmission, minimize absorption, and background autofluorescence. Additionally, the fluorescence efficiency of each of the fluorescent proteins is being optimized through site-directed mutagenesis. The intrinsic physical and chemical properties of the mutants are altered. Characteristics such as protein stability, extinction coefficients, fluorescence quantum yield, tendency to dimerize or form multimers, requirement for oxygen, efficiency of fluorochrome formation, and susceptibility to photoisomerization and photobleaching can be modulated through mutagenesis, and, thus, the amount and rate of light photons emitted from such structures can be optimized [80].

For example, wild-type *Aequorea* GFP, a 238 amino acid polypeptide (27–30 kD) specifically isolated from the Pacific jellyfish (*Aequorea victoria*), is a highly fluorescent molecule with excitation peaks at 395 (largest peak) and 475 nm [81]. GFP emission occurs at 509 nm, which is in lower green portion of the visible spectrum and is therefore highly susceptible to tissue attenuation. As a result, imaging of GFP expression in living subjects often requires surgical measures like creating skin or skull windows for optimal visualization of GFP in deeper structures such as the pancreas or brain [82]. Such tissue-window–based intravital microscopy methods allow high-resolution imaging of implanted labeled cells, extracellular matrix (ECM) components, and local blood vessels in the living animal over days and weeks [83,84]. Additionally, GFP fluorescence is not immediate and only detectable at about 7 h after injection of recombinant adenovirus carrying GFP (vAd-CMV-GFP). The rate-limiting step in GFP "maturation" appears to be a necessary oxidation step in chromophore formation [85].

Extensive work has been dedicated to creating GFP mutants since wild-type GFP possess a few compromising factors for its use in living subjects. The list of available GFP mutants, each having their own characteristics, is quite lengthy. An excellent review is available [80]. One of the more thoroughly studied mutant S65T, also known as enhanced GFP (EGFP), has some impressive advantages. Ser65, one of the amino acids of the chromophore, is replaced by Thr in this mutant. The wild-type 395 nm excitation peak is suppressed and the 475 nm peak is enhanced five- to sixfold in amplitude (sixfold increase in brightness), the peak is shifted to 489–490 nm [85]. It is fourfold faster during the rate-limiting oxidation step, not subject to photoisomerization and exhibits very slow photobleaching [86]. Examples of the use of EGFP in living subjects are shown in Figure 30.5.

GFP mutants come in all kinds of colors: blue, cyan, yellow, and green. However, the maximum emission peak attained by the GFPs is 529 nm. Cloned red fluorescent protein (dRFP), a 28 kDa protein responsible for red coloration

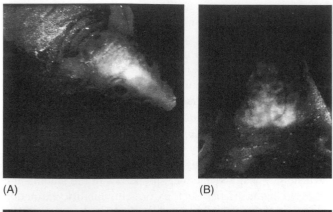

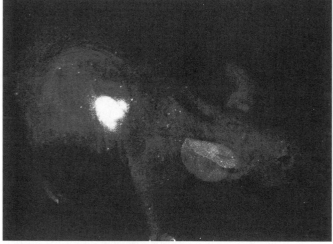

FIGURE 30.5 (See color insert following blank page 206.) Optical (fluorescence) imaging of transgene expression. The adenoviral (vAd) vector AdCMV5GFPAE1/AE3 [vAd-GFP] (Quantum, Montreal, Canada) constitutively expresses an EGFP, which is driven by a cytomegalo virus (CMV) promoter. The vector was delivered to the brain after an upper midline scalp incision and creation of a parietal skull window. Twenty microliters containing 8×10^{10} plaque-forming units (pfu)/ml vAd-GFP per mouse was injected into the brain. Twenty four hours later, fluorescence imaging of the entire animal (lower magnification) was carried out in a light box illuminated by blue light fiber optics, which provided the external excitation wavelength and imaged by using the cooled color CCD camera. Emitted fluorescence was collected through a long-pass filter GG475 on a three-chip thermoelectrically cooled, color CCD camera. Images of 1,024 × 724 pixels are captured directly on a personal computer or continuously through video output on a high-resolution video recorder. Images are subsequently processed for contrast and brightness and analyzed with imaging software. Higher magnification images (not shown here) can be accomplished by using a fluorescence stereomicroscope equipped with a 50 W mercury lamp. In this scenario, selective excitation of GFP is produced through a D425/60 bandpass filter and 470 DCXR dichroic mirror. Emitted fluorescence are captured and processed as described above. Images (A) and (B) demonstrate GFP transgene expression following adenoviral delivery to the brain. Image (C) demonstrates Ad-CMV-GFP delivery to the livery via portal vein cannulation. (Courtesy of Anticancer, Inc. and from Yang, M., Baranov, E., Moossa, A.R., Penman, S., and Hoffman, R.M., *Proc. Natl. Acad. Sci. USA,* 97, 12278, 2000. With permission.)

seen in the coral *Discosoma*, has broken the 529 nm barrier and excitation and emission maxima are at 558 and 583 nm, respectively [87]. The relatively high extinction coefficient and fluorescence quantum yield indicate that the brightness of the mature, well-folded protein is comparable to any other fluorescent protein. Furthermore, a commercially available mutant, DsRed, is resistant to photobleaching and has been further redshifted to 602 nm, which reduces the tissue absorption of the emitted light photon. Unfortunately, there are some significant limitations to the use of dRFP. It is an obligate tetramer and it is quite slow in its maturation; it takes days for it to mature from green to red. Mutations attempted to alleviate this problem resulted in a monomeric form designated as mRFP1 that matures more than 10 times faster than the native DsRed protein [88]. In addition, the excitation and emission peaks of mRFP1, 584 and 607 nm, are approximately 25 nm redshifted from DsRed, which should confer greater tissue penetration and spectral separation from autofluorescence and other fluorescent proteins. A recently discovered far redshifted variant of mRFP1, mPLUM (emission 649 nm), will be very useful in optical imaging experiments in living animals [89].

Apart from fluorescent reporter genes, many fluorescent reporter probes like nanoparticles or "smart probes" are also being investigated actively. Fluorescent organic and inorganic nanoparticles like quantum dots (QDs) that interface with biological systems have recently attracted widespread interest in the fields of biology and medicine. They have the potential as intravascular or cellular probes for both diagnostic (imaging) and therapeutic purposes (gene delivery) in gene therapy. Target-specific gene delivery and early diagnosis in cancer treatment using these probes can play a critical role in medicine [90]. QDs are essentially semiconductor nanocrystals with a core/shell structure of 2–8 nm in diameter with size-dependent fluorescence emission. The unique optical properties of QDs that make it suitable for *in vivo* optical imaging include high absorbency, high quantum yield, narrow emission bands, large Stokes shifts, and high resistance to photobleaching [91]. QDs can also allow multiplex detection of multiple targets in a single experiment. There are some excellent reviews that summarize the synthesis, bioconjugation chemistry, optical features, and applications of QDs for *in vivo* imaging [92–94].

Two recently developed smart fluorescent probes have been developed, which are significantly redshifted to the extent that they work in the NIR portion of the electromagnetic spectrum (excitation wavelength 673 nm; emissions wavelength of 689 nm), making them ideal fluorescent agents for use in intact organisms. Although they are being exploited to measure endogenous protein levels, they are worth mentioning here since they have the potential to be activated by exogenously delivered genes and be used to monitor gene therapy [19]. This new breed of fluorescent probe is dependent upon the close proximity of multiple near-infrared fluorochromes, Cy5.5, bound to a synthetic graft copolymer consisting of a cleavable backbone (partially methoxy poly(ethylene glycol) modified poly-L-lysine) [95]. When placed in close proximity, a pair of these fluorochromes will quench each other and, therefore, not be detectable. Upon enzymatic cleavage of the backbone with a protease that has lysine–lysine specificity, the fluorochromes are spatially dissociated and will begin to fluoresce.

Because tumor progression and angiogenesis necessarily produce certain proteases, these clever or smart biocompatible autoquenched near-infrared fluorescent probes can be used to detect tumors, which are known to upregulate certain proteases. Cathepsins B and H are tumor proteases, which have lysine–lysine specificity and have been shown to activate this fluorescent probe in tumor xenografts. Other known tumor-enhanced proteases, Cathepsin D and MMP-2, which are dependent on other specific peptide sequences for their action, can also activate this probe if the NIR fluorochromes are attached to the backbone via Cathepsin D-sensitive or MMP-2-sensitive sequences; these enzyme-specific probes have been demonstrated in living subjects [19,96].

Bioluminescent optical systems are increasingly being used in studying living subjects because of their inherently low background (no excitation/irradiation needed that would otherwise cause background autofluorescence). The most commonly used bioluminescence reporter gene is the Firefly luciferase gene (*Fluc*), which encodes for a 550 amino acid, 61 kDa monomeric protein (FL) derived from *Photinus pyralis*, the North American firefly [79]. Photon emission is achieved through oxidation of its native substrate, D-luciferin (D-(-)-2-(6′-hydroxy-2′-benzothiazolyl)-thiazone-4-carboxylic acid) into oxyluciferin in a reaction that requires ATP, Mg^{2+}, and O_2. The reaction produces a broad spectral emission that peaks at 560 nm. A number of modifications to the gene since its discovery have facilitated its use in mammalian tissues, which include amino acid substitutions that redshift the emissions peak above 600 nm, optimized mammalian codon language, removal of a peroxisome targeting site for increased expression levels, and cytoplasmic localization [22,97]. Once produced, luciferase does not require post-translational processing for enzymatic activity and it can immediately function as a genetic reporter. However, the enzyme is unstable and loses activity rapidly even at room temperature (*in vitro* half-life of 2 to 3 min at 37°C and *in vivo* half-life of 1 to 4 h) [98]. This property can potentially compromise sensitivity and precision in analytical experiments involving firefly luciferase as a reporter gene. To overcome this problem, a luciferase mutant (*tfl*) containing four point mutations (bearing the mutations E354K, I232A, T214A, and F295L) was generated that showed remarkably greater thermostability (*in vitro* half-life of 15 min) relative to the wild type [99,100]. Examples of an adenoviral-mediated firefly luciferase gene delivery are given in Figures 30.6 and 30.7. Figure 30.6 is an example of cardiac gene therapy and Figure 30.7 is an example of how tissue-specific transgene expression can be achieved using tissue-specific promoters in living subjects.

A second bioluminescence system, which utilizes the Renilla luciferase gene (*Rluc*) that encodes a 36 kDa monomeric protein (RL), has recently been tested in small rodents [15]. Its peak emission displays a blue–green bioluminescence at 480 nm when Renilla luciferase interacts with its substrate, coelenterazine. In comparison to FL, RL does not need cofactors or ATP to oxidize its substrate, and therefore will be less taxing on the cell in which it is expressed. It also has much more rapid kinetics in terms of light production so that it can potentially be used simultaneously with FL through the injection of both substrates and multiple time-point imaging [15].

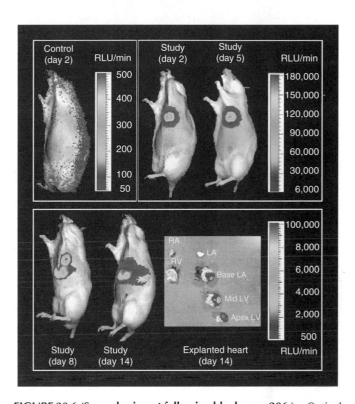

FIGURE 30.6 (See color insert following blank page 206.) Optical (bioluminescence) imaging of cardiac reporter gene expression. Replication-defective adenovirus carrying firefly luciferase (*Fluc*) driven by a constitutive cytomegalovirus (CMV) promoter (Ad-CMV-*Fluc*, 1×10^9 pfu) was injected directly into the myocardium (anterolateral wall) of a rat. Images obtained 2, 5, 8, and 14 days later from a cooled CCD camera demonstrate significant cardiac emissions from FL activity ($P < 0.05$ vs. control). By day 8, luciferase activity is seen in the liver, which is probably from spillover of adenoviral vector into the systemic circulation and subsequent hepatic transfection of the virus via coxsackie-adenovirus receptors on hepatocytes. On day 14, the heart of the same rat was explanted after whole-body imaging was performed and sliced into three sections (bottom right). Firefly luciferase activity is localized at anterolateral wall of left ventricle along the site of virus injection. Control rats, which received an intracardiac injection of Ad-CMV-HSV1-sr39tk (1×10^9 pfu), demonstrate no significant firefly luciferase activity 2 days after the injection (upper left). Please note the bioluminescent scales are different for control rat, study rat days 2 to 5, and study rat days 8 to 14 to account for the wide range of cardiac firefly luciferase activity observed. Scales are a quantitative indicator of light photons detected (relative light units [RLU]/minute [min]). RA indicates right atrium; RV, right ventricle; and LV, left ventricle. (Reproduced from Wu, J.C., Inubushi, M., Sundaresan, G., Schelbert, H.R., and Gambhir, S.S., *Circulation*, 105, 1631, 2002. With permission.)

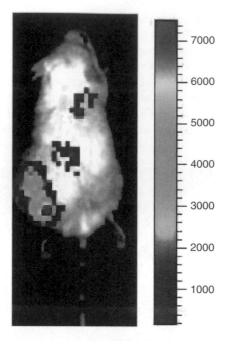

FIGURE 30.7 (See color insert following blank page 206.) Optical (bioluminescence) imaging of targeted transgene expression using a tissue-specific promoter. One way to target gene therapy is through the use of tissue-specific promoters. However, most tissue-specific promoters yield low levels of transcription. In this example, certain key regulatory elements of the promoter and enhancer of prostate-specific antigen (PSA) have been multimerized to yield a construct, PSE-BC, which is 20-fold more active than the native PSA promoter/enhancer. Following incorporation into an adenovirus vector (AdPSE-BC-luc) and subsequent intratumoral injection into a human prostate cancer xenograft model (LAPC series), firefly luciferase expression can be seen in the main tumor xenograft (left flank) as well as other extratumoral sites, such as the back and chest, in this male SCID mice 21 days after vector delivery. Detailed histologic analysis of the xenograft and extratumoral sites demonstrates that firefly luciferase expression is restricted to the prostate tumor and prostate metastases, respectively. The metastases, in this case, are located in the spine and lung. By comparison, CCD imaging and histologic analysis of xenografts injected with AdCMV-luc show markedly diminished expression of firefly luciferase in the xenograft and increased nonspecific expression in the liver at 21 days postinjection (figure not shown). Results from this study indicate that tissue-specific transgene expression is possible and that CCD imaging can be used to track firefly luciferase-marked tumor cells. Scale indicates the number of photons detected (RLU/min). (Reproduced from Adams, J.Y., Johnson, M., Sato, M. et al., *Nat. Med.*, 8, 891, 2002. With permission.)

Recently, RLuc was mutated at eight sites to develop an optimized protein (RLUC8) which, compared with the parental enzyme, is 200-fold more resistant to inactivation in murine serum and exhibits a fourfold improvement in light output. Based on mutation analysis, a double mutant showed half the resistance to inactivation in serum of the native enzyme while yielding a fivefold improvement in light output, and hence more suitable as a reporter gene [101]. RLUC8 has been mutated further to obtain Renilla luciferase variants with

bathochromic (red) shifts of up to 66 nm (547 nm peak) that also had greater stability and higher light emission than native enzyme [102]. Such redshifted forms will be immensely helpful for imaging gene expression from deep tissues.

A third example of a bioluminescent reporter gene is the *Gaussia* luciferase enzyme from the marine copepod *Gaussia princeps*. *Gaussia* luciferase (*GLuc*; 185 aa, 19.9 kDa) is the smallest luciferase known and is naturally secreted. This luciferase emits light at a peak of 480 nm with a broad emission spectrum extending to 600 nm when it reacts with its substrate, coelenterazine. The humanized form of the protein when expressed in mammalian cells from an HSV-amplicon vector showed about 100-fold higher bioluminescent signal intensity compared to other well-studied luciferases like *Rluc* or *Fluc* under most conditions. Like other coelenterazine luciferases, *Gluc* does not require ATP for activity, but unlike other luciferases, it is a secretory protein and therefore can report from the cells themselves as well as their immediate environment [21]. The secretory nature of the protein however poses a problem while trying to correlate signal intensity with the number of transduced cells. Another major disadvantage involves rapid loss of signal through diffusion away from transduced cells. Regardless, because of its small size, *Gluc* will likely have many potential uses as reporter genes in gene therapy using vectors in which payload size is limited.

Recently, activatable luciferase molecules have been developed for imaging apoptosis. In principle, the luciferase activity is silenced by flanking the protein moiety at both termini with in-frame caspase-3 substrate sequences and the estrogen receptor regulatory domain [103]. The attenuated activity of the fusion protein can be unleashed on induction of caspase-3 in cells undergoing apoptosis and thus can be imaged with bioluminescence imaging. A different strategy was employed to build a new caspase-3 sensor fusion protein (MTF) using three different reporter genes to allow multimodality imaging of apoptosis. This fusion protein contains a fluorescent reporter-mRFP1, a bioluminescence reporter-FL, and a PET reporter-HSV1-sr39truncated thymidine kinase (protein) (TK)] fused by the caspase-3-recognizable polypeptide linker [Asp-Glu-Val-Asp(DEVD)]. Upon cleavage by activated caspase-3 in apoptotic cells, the fusion protein with attenuated activity for each component showed significant increases in the activities of all the reporter genes both in mammalian cells and in mouse tumor models (Ray et al., unpublished data).

Comparing the technical and practical aspects between fluorescent and bioluminescence systems reveals substantial differences, which are briefly mentioned here. Fluorescence imaging benefits greatly from relatively high photon yield from its proteins and dyes and can be imaged with conventional photographic equipment rather than the more expensive cooled CCD cameras. By comparison, the photon yield of bioluminescence systems is significantly lower, and imaging of living subjects always requires a cooled CCD camera. Fluorescence imaging can potentially be used to image simultaneous signals of different colors, whereas multiplexing different and simultaneous signals in bioluminescence imaging is difficult, owing to the differential enzyme kinetics and time to peak photon flux between varying luciferases.

On the other hand, bioluminescence imaging has certain advantages over the fluorescence systems. For example, autofluorescence, a source of background signal in the image, is significantly less in bioluminescence for reasons discussed earlier. Also, bioluminescence reporters can be utilized as soon as they are synthesized. In contrast, the fluorescent proteins usually have a requisite period, some substantially longer than others, of posttranslational processing and maturation prior to their function as a fluorescent reporter [80]. In some cases, where temporal resolution is an important issue in the monitoring of gene therapy, the use of fluorescence techniques may be limiting in this regard. Bioluminescence proteins also take some finite time to mature, but anecdotal evidence suggests that it is not as long as the fluorescent proteins; future studies are sure to address this issue.

Imaging with bioluminescence techniques permits rapid, repetitive, or prolonged imaging periods so long as the required substrates and cofactors are available. The physical phenomena of photobleaching, a property seen in many fluorescent proteins where the fluorochrome is permanently extinguished after light excitation or ultraviolet light exposure, makes imaging difficult if prolonged excitation periods or rapid repetitive imaging is needed; repeat imaging can only take place once fresh fluorescent protein has been generated. Among the different fluorescent probes, QDs are much more resistant to photobleaching. These inorganic nanoparticles possess a well-encapsulated semiconductor core, which protects the QDs from the photochemical reactions that cause photobleaching. Therefore, QDs can be employed in continuous tracking studies over a long period of time [104]. However, photobleaching is not a significant problem in bioluminescence imaging. Having stated the above, no formal comparison has yet been made between fluorescence and bioluminescence using the same animal model, fusion reporter/individual reporters, and the same CCD camera. Future studies addressing these issues should help better define the more ideal reporter.

30.3.2 RADIONUCLIDE-IMAGING REPORTER SYSTEMS

Currently, PET or gamma/SPECT reporter genes encode a receptor, enzyme or ion pump, which bind a radiolabeled ligand, interact with a radiolabeled substrate, or facilitate intracellular translocation of ionic radioisotopes, respectively. The reporter gene product is designed to sequester the probe intracellularly and/or on the surface of cells expressing these genes. Ideally, those cells lacking the transgene will be unable to trap the reporter probe. The amount of probe or tracer used is typically in the nanogram range and does not lead to any pharmacological effect [105]. In contrast, bioluminescence strategies, NIRF and MRI require mass amounts of probe (usually micrograms to milligrams) and, as a result, may produce a pharmacologic effect and have a greater potential for toxicity.

Radionuclide imaging has significant advantages in that it permits quantitative and repetitive imaging in the same subject over time [106–109]. Perhaps the main drawback of radionuclide-based techniques is that unsequestered probe usually circulates in the enterohepatic system or is excreted in the kidneys. As a result, probe can collect in the gut, kidneys, urinary bladder, or gall bladder, making it difficult to specifically evaluate these organs. Discussed below are two fundamentally different approaches to radionuclide-based reporter imaging: direct and indirect imaging.

30.3.2.1 Direct Radionuclide Imaging: HSV1-TK, Mutant HSV1-TK, Sodium Iodide Symporter, and Cytosine Deaminase

One of the major advantages of using the radionuclide techniques is that it allows the delivered therapeutic gene to be imaged directly (i.e., direct imaging). In direct imaging, there is no need to couple the expression of a therapeutic gene with an additional reporter gene since reporter probes already exist for the therapeutic gene. The HSV1-tk gene when used under certain conditions is an exemplary model of direct imaging.

Mammalian TKs phosphorylate thymidine for its normal incorporation into DNA during replication. The HSV1-tk, gene on the other hand, is normally employed as a suicide gene for the therapy of cancer. The gene product, with its broad substrate specificity, is able to phosphorylate acycloguanosine, guanosine, and thymidine derivatives and subsequently trap these substrates intracellularly. Therefore, when delivered with a prodrug such as acyclovir, gancyclovir, and penciclovir, the HSV1-tk acts as a suicide gene. The enzyme converts the prodrug to phosphorylated nucleoside analogs, which when trapped inside the cell at high concentrations leads to cell death due to premature chain termination. Transfected cell populations particularly affected by this type of therapy are those which have a high mitotic index. When radiolabeled acycloguanosines (e.g., 9-[(3-[^{18}F] fluoro-1-hydroxy-2-propoxy)methyl]guanine ([^{18}F]FHPG); 9-(4-[^{18}F] fluoro-3-hydroxymethylbutyl)guanine ([^{18}F]FHBG)), guanosines (e.g., fluorogangciclovir ([^{18}F]FPGV; fluoropenciclovir ([^{18}F]FPCV)), and thymidine derivatives (2'-[^{124}I] fluoro-2'deoxy-1-β-D-arabinofuranosyl-5-iodouracil ([^{124}I]FIAU)), 2'-[^{18}F]fluoro-5-ethyl-1-beta-D-arabinofuranosyluracil ([^{18}F]-FEAU) are used in nonpharmacologic (trace) doses, they can serve as PET reporter probes. Similarly, [^{131}I]FIAU and [$^{123/125}$I]FIAU can be used as gamma camera or SPECT reporter probes. The pyrimidine analog FEAU is a strong antiherpetic agent with much lower toxicity against host cells. Compared to other pyrimidine analogs, FEAU exhibited a much higher affinity for viral TKs than for host cell enzymes, has primarily renal clearance, leading to a low background signal in the abdomen. This led to the growing interest in radiolabeled FEAU as a PET tracer for the HSV1-tk reporter gene [110–112]. Details of the synthesis and kinetics of these and other similar agents have been reviewed [113–119].

Endogenous TK, HSV1-tk and a mutated form of HSV1-tk, HSV1-sr39tk, each demonstrate different substrate specificity, which can be exploited for therapeutic and imaging purposes. For example, endogenous TK demonstrates narrow substrate specificity and cannot efficiently phosphorylate radiolabeled prodrugs. Although endogenous TK can phosphorylate the prodrugs/probes to a minor degree, nontransfected cells cannot accumulate significant amounts of the radiolabeled prodrugs. Instead, these radiolabeled prodrugs specifically localize to cells, tissues, or tumors that express the HSV1-tk gene.

Comparison of the HSV1-tk probes demonstrate that [^{18}F]FEAU displays favorable pharmacokinetics (greater sensitivity, better contrast, less background noise) and, thus, the best imaging potential. Cell uptake studies showed that [^{3}H]FEAU has a higher specificity compared to [^{13}C]FIAU. This is consistent with another study with C6 cells where [^{18}F]FEAU showed the highest specificity ratio of 84.5 compared to 10.3 for [^{18}F]FIAU and 40.8 for [^{18}F]FHBG [110]. Studies with the fluorine-18-labeled tracer in mice with HT29 xenografts showed rapid clearance from the blood within 2 h and a fivefold higher uptake in HSV1-tk expressing tumor compared to the wild-type tumor. Majority of the activity was retained in the tumor while all other organs demonstrated low accumulation at 2 h except the bladder owing to urinary excretion. Uptake in wild-type tumors in the left thigh was as low as the background activity at 2 h [112].

Among the other tracers, [^{124}I]FIAU may also be a good choice in some cases because the half-life of ^{124}I is 4.2 days, allowing for longer systemic clearance. On the other hand, this longer half-life may also prove problematic in cases when frequent repeat imaging is needed since a minimum 10 to 12 days between imaging sessions is needed to allow clearance of prior probe administration. Further testing has shown that [^{18}F]FHPG and [^{18}F]FHBG are inferior agents for HSV1-TK. This variability seen amongst HSV1-TK reporter probes is in part due to differences in biologic half-life, stability, substrate competition, degree of nonspecific binding, specific retention, rates of cellular transport, method of transfer (viral-mediated vs. stable transfection), and routes of clearance [120].

Mutated versions of the suicide gene, HSV1-tk, have been created to enhance its killing potential by increasing its ability to phosphorylate prodrugs [121]. From a library of site-directed mutants, it has been determined that the product of mutant HSV1-sr39tk suicide gene is more adept at phosphorylating ganciclovir and less efficient in phosphorylating endogenous thymidine when compared to wild-type HSV1-TK. In light of this, mutant kinase has been exploited as a reporter gene since radiolabeled reporter probes already exist and are identical to those used for the wild-type HSV1-TK. As expected, HSV1-sr39TK efficiently phosphorylates [^{18}F]FGCV, [^{18}F]FPCV, [^{18}F]FHPG, and [^{18}F]FHBG, with [^{18}F]FHBG as the most effective substrate for the mutant TK. The ability to trap the reporter probe is improved by at least a factor of 2 to 3 in mutant TK-expressing tumor xenografts when compared with wild-type TK. Significantly improved uptake of the reporter probe is seen in the liver following systemic delivery of a recombinant adenoviral vector carrying the mutated transgene, resulting in enhanced sensitivity for imaging this transgene *in vivo* when compared to wild-type HSV1-tk [122]. An example of the use of HSV1-sr39tk as a potential reporter gene for cardiac gene therapy is provided (Figure 30.8). As an aside, an example of an optical reporter for use in cardiac gene therapy (using a nearly identical adenoviral vector and the firefly luciferase reporter gene) is provided for side-by-side comparison to better demonstrate the spatial advantages of microPET imaging (Figure 30.8).

Ponomarev et al. recently described a novel, human derived, nonimmunogenic reporter gene, the human mitochondrial thymidine kinase 2 (hTK2) [123]. A truncated version of the protein, ΔhTK2, was capable of phosphorylating and trapping PET probes FIAU and FEAU both *in vitro* and in animals with high specificity. Because of its human origin, such a reporter gene has the potential of being applied in preclinical and clinical trials to follow molecular events noninvasively.

Other viral TKs have also been explored as reporter genes. The varicella-zoster virus TK gene (VZV-tk) was explored as a potential PET imaging gene recently. The protein product of VZV-tk gene is capable of phosphorylating bicyclic deoxynucleoside analogs (BCNAs) and trapping them in cells. Thus BCNAs labeled with a positron-emitting radioisotope in combination with the VZV-tk gene can result in a new PET reporter gene/probe system [124].

Another therapeutic transgene, which lends itself to direct imaging in living subjects, is the sodium/iodide symporter (NIS) gene (Figure 30.9). NIS is an intrinsic membrane protein, which is responsible for translocating and concentrating iodide within thyroid follicular cells. In this normal, physiologic situation, iodide is eventually used to make the thyroid hormones [125]. Recent cloning of the symporter has permitted the investigation of its role as a suicide gene [126]. Because the symporter can concentrate high intracellular levels of iodide (including radioiodide, ^{131}I), targeted cells expressing the symporter can be killed in this form of targeted radiotherapy by the accumulation of radioiodide. It is estimated that a dose of up to 50,000 cGy of ionizing radiation can be achieved in targeted cancer cells [127–130]. The lethal effects of the NIS have been cleverly demonstrated *in vitro* in a variety of cell lines including melanoma, colon cancer, and ovarian carcinoma, and in a murine model of a transfected melanoma xenograft following intraperitoneal injection of ^{131}I [131]. Also, tissue-specific expression of NIS transgene is possible by fusing a prostate-specific antigen (PSA) promoter fragment with the NIS gene, the result being tissue-specific expression of NIS transgene and subsequent accelerated death of an androgen-dependent prostatic carcinoma xenograft after injection with a therapeutic dose of radioiodine [132]. In another elegant study, researchers used a novel oncolytic vesicular stomatitis virus expressing the NIS gene to treat radiosensitive multiple myeloma with ^{131}I. In this study, NIS also acted as a reporter gene and helped in monitoring the intratumoral spread of the infection noninvasively by serial gamma camera imaging of ^{123}I-iodide biodistribution [133].

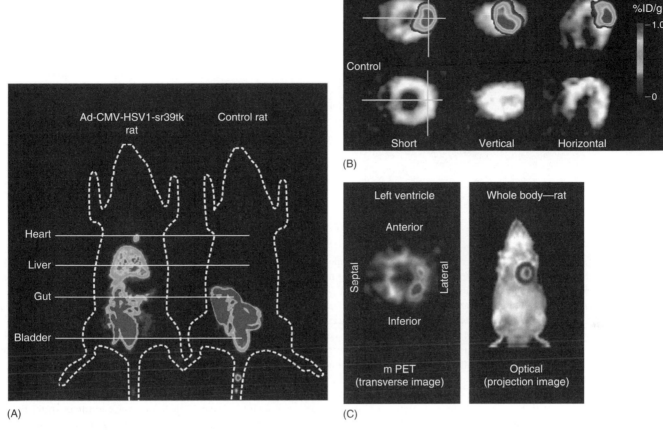

FIGURE 30.8 (See color insert following blank page 206.) Micropet and optical (bioluminescence) imaging of cardiac reporter gene delivery. (A) Imaging cardiac gene expression using adenoviral-mediated mutant thymidine kinase (HSV1-sr39tk) as PET reporter gene and [18F]FHBG as PET reporter probe. Trapping of tracer occurs only in cells expressing the reporter gene. At day 4, whole-body microPET image of a rat shows focal cardiac [18F]FHBG activity at the site of intramyocardial Ad-CMV-HSV1-sr39tk injection. Liver [18F]FHBG activity is also seen because of systemic adenoviral leakage with transduction of hepatocytes. Control rat injected with Ad-CMV-*Fluc* shows no [18F]FHBG activity in either the cardiac of hepatic regions. Radiolabeled probe is always "visible" with radionuclide imaging regardless of whether it has localized to its target or not. As a result, radionuclide-based images will exhibit a certain degree of nonspecific tracer localization since "unbound" reporter probe is metabolized through either the enterohepatic or urinary system or both. In this example, nonspecific reporter probe activity gut and bladder activities are seen for both study and control rats because of route of [18F]FHBG clearance. (B) Tomographic views of cardiac microPET images. The [13N]NH3 (gray scale) images of perfusion are superimposed on [18F]FHBG images (color scale), demonstrating HSV1-sr39tk reporter gene expression. [18F]FHBG activity is seen in the anterolateral wall for experimental rat compared with background signal in control rat. Perpendicular lines represent the axis for vertical and horizontal cuts. Color scale is expressed as % ID/g. (C) Comparison of typical images obtained with PET (left) and optical imaging (right). The optical method is more sensitive (at limited depths), easier to perform, and demonstrates minimal background noise. With PET, we can see that the transgene was delivered to the anterolateral aspect of the left ventricle. Such spatial resolution is not afforded by *in vivo* optical imaging at this time. (Reproduced from Wu, J.C., Inubushi, M., Sundaresan, G., Schelbert, H.R., and Gambhir, S.S., *Circulation*, 105, 1631, 2002; Wu, J.C., Inubushi, M., Sundaresan, G., Schelbert, H.R., and Gambhir, S.S., *Circulation*, 106, 180, 2002. With permission.)

Imaging of the NIS transgene is relatively straightforward since the therapeutic agent, ^{131}I, can also be used as an imaging agent with a gamma camera or SPECT. Similarly, ^{123}I or ^{125}I, both of which are commercially available, can also be used as reporter probes. A significant advantage of this reporter system is that the reporter probe is relatively simple to produce and commercially available. Specialized radiochemistry, such as that required for HSV1-tk reporter probes, is not needed here. Another significant advantage of this system is that this reporter system can readily be imaged with PET using [124]I as the positron-emitting reporter probe, which is already available [134–136]; the very same living subject carrying the NIS transgene can be imaged with either SPECT or PET depending upon the choice of reporter probe. In a recent study, the combined [124]I -PET/CT technology was employed to follow NIS gene expression in living animals with great accuracy [137]. The main confounding issue with using this reporter system, on the other hand, is radioiodine will not only localize to target cells, but it will also accumulate in normal tissues such as the thyroid, salivary glands, breast, and stomach, all of which express physiologic levels of endogenous NIS. Also, the use of radioiodide requires specialized environmental safety precautions as iodine is easily aerosolized and can pose a hazard to researchers or health care workers.

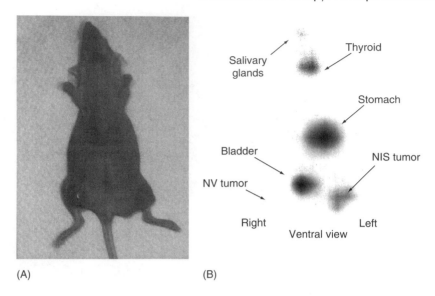

(A)   (B)

FIGURE 30.9 Gamma camera imaging of the sodium iodide symporter transgene. Recently cloned rat and human sodium iodide symporter genes (rNIS and hNIS, respectively) are increasingly showing their potential as a novel suicide gene therapy for a variety of cancer models. Once transduced to a cancer cell line, tumor or target organ via recombinant viral transfection or liposomal-mediated techniques, the expressed membrane proteins facilitate the active intracellular transport of iodide (I^-) into targeted cells. Exogenously administered radiotracers ^{123}I, ^{125}I, or ^{99m}Tc-pertechnetate results in intracellular accumulation of these tracers in cells that are transduced with this symporter. The distribution of these tracers can be imaged with a gamma camera or SPECT, and therefore be used as a means of localizing cells that have been transduced with NIS. Similarly, targeted brachytherapy can be performed by the administration of ^{131}I. (A) A retroviral vector carrying the rNIS gene was used to transduce A375 human melanoma cell line. Transduced (NIS) and nontransduced (NV) tumor xenografts were subcutaneously implanted into the left and right flank of the photographed mouse, respectively. (B) By 30 days, the tumor had reached approximately 10 mm in diameter. An intraperitoneal dose of ^{131}I was administered and a gamma camera image was obtained after a 1 h incubation period. rNIS-transduced xenograft (left flank) demonstrates radioiodide uptake while nontransduced tumor (right flank) does not. The thyroid, stomach and, to a lesser extent, the salivary glands endogenously express sodium iodide symporters and, thus, normal, physiologic radioiodide uptake is seen in these organs. (Reproduced from Mandell, R.B., Mandell, L.Z., and Link, C.J., Jr., *Cancer Res.*, 59, 661, 1999. With permission.)

The CD transgene, another suicide gene, encodes an enzyme, which converts 5-FC to the toxic 5-FU. Efforts to image this transgene, which have been performed with radiolabeled fluorocytosine as a probe, have largely been hampered by suboptimal pharmacokinetics: poor tracer uptake and poor retention of the toxic metabolite [138]. Extended periods of up to 48 h are required to see differential accumulation of tracer between transfected and control cells. While targeted delivery of CD remains a viable method of suicide gene therapy [139], alternative substrates for CD, which are rapidly transported, deaminated, and trapped intracellularly, will have to be developed in order to use this transgene as a reporter gene.

30.3.2.2 Indirect Radionuclide Imaging: D2R, Mutant D2R, and Somatostatin-2 Receptor

In cases where a reporter probe does not already exist for a delivered transgene, a number of strategies exist for indirect transgene imaging. Indirect imaging involves the simultaneous coexpression of the therapeutic gene and reporter gene with both driven by the same or identical promoter. Because reporter gene expression can directly correlate with transgene expression, such approaches have the potential to give us

valuable information on the quantity and localization of transgene expression [140]. It should be noted, however, that it is not given that they directly correlate; one can hope that they do but each approach/application has to be tested. Further details regarding such indirect approaches are provided in a later section.

In radionuclide imaging, one way to perform indirect imaging is to couple the therapeutic gene to a reporter gene, which encodes a receptor that can bind, and, therefore, trap radiolabeled ligands. Following transfection, targeted cells may express the therapeutic gene product and receptor in proportional amounts. Subsequently, the ectopically expressed receptor will either localize to the cell membrane or remain intracellular. Following exposure to radiolabeled ligand probe, cells expressing the receptor will specifically bind the probe resulting in a complex, which can be detected by PET/SPECT/ gamma camera imaging. The intensity of activity on the PET image is directly proportional to the number of these receptor–ligand complexes, and, therefore, correlated to the amount of therapeutic gene expressed. Ideally, those cells not producing the receptor will be devoid of tracer signal.

An example of a receptor-based reporter gene is the dopamine type 2 receptor reporter gene (D2R) [108]. D2R is normally an endogenous, cell-surface receptor predominantly expressed in the striatum. When activated, it causes a G-protein-coupled reduction of cyclic adenosine monophosphate (cAMP) via its inhibition of adenylate cyclase. When D2R is used as a reporter gene, a radiolabeled D2R antagonist, spiperone [3-(2′ [18F]fluoroethyl)spiperone ([18F]FESP)], serves as the receptor's ligand and accumulates intracellularly and on the cell surface of D2R-expressing tissue. Radiolabeled spiperone, originally used to monitor levels of endogenous levels of striatal D2 receptors in vivo, binds to D2R with high affinity and is able to cross the blood–brain barrier (BBB). To overcome potential deleterious effects of ectopic D2R activation by circulating endogenous ligands, mutant D2R reporter genes, D2R80A and D2R194A, have been created, which are disengaged from downstream transduction events while maintaining a high binding affinity for the ligand probe [107].

A few advantages of the D2R-based reporter system are worth commenting. [18F]FESP's ability to cross the BBB and cell membranes favors its use in the central nervous system (CNS) relative to other reporter systems such as the HSV1-tk (mutant or not) reporter system since [18F]FHBG is not as efficient in crossing this important barrier. Furthermore, [18F]FESP has a relatively easier time of localizing to target, which is a cell-surface and intracellular receptor. In contrast, [18F]FHBG has to cross the cell membrane in order to interact with the target enzyme and is, therefore, subject to transport kinetics. Also, D2R, an endogenous protein, is not immunogenic compared to HSV1-TK and therefore probably more appropriate for repeated imaging during longitudinal studies. Interestingly, despite these relative advantages of the D2R/[18F]FESP system, equivalent sensitivities are reported between the D2R/[18F]FESP and mutant HSV1-tk/[18F]FHBG PET reporter gene imaging systems in the liver (~20% injected dose (ID)/g in the liver when used with adenoviral delivery systems carrying constitutive CMV-based expression of the reporter gene). One must also note that endogenous D2R expression in the striatum will produce background noise and therefore [18F]FESP imaging will only be useful outside the striatum.

Another receptor-based reporter system takes advantage of the SS membrane receptors (SSTR), which also belong to the family of G-protein-coupled receptors. Under normal conditions, the interaction between SSTR and its ligand, SS is known to have a variety of biologic effects including a role in vasoconstriction, immunomodulation, and an inhibitory effect on endocrine and exocrine secretory functions. More recently, SSTR-activated signal transduction pathways have been implicated in the induction of apoptosis and inhibition of cell growth [141]. SSTRs have also been found in normal and hyperplastic human endothelium where they are believed to exert negative effects on angiogenesis. Five different SS receptor subtypes have been described thus far (SSTR1 through SSTR5) and their respective genes have been cloned [142–144].

Historically, the development of this reporter gene centered on probes which were already in existence and were being used clinically to identify diseased states characterized by upregulated SSTR receptors. Radiolabeled SS analogs, for example, have been clinically used to identify a number of primary human cancers and their metastases where elevated levels of SS receptors are seen [145]. Neuroendocrine tumors (including carcinoid, islet cell tumors, small-cell lung cancers, pheocromocytomas, gastrinoma, paragangliomas, and medullary thyroid cancers), pituitary gland tumors as well as sarcomas, meningiomas, low-grade astroctomas, lymphomas, some breast cancers, and metastatic prostate cancers are known to express high levels of SSTR, particularly SSTR2. In fact, the combination of high SSTR receptor density seen in some tumors and the antiproliferative, antiangiogenic, and antisecretory effects of SS analogs form the premise for SS analog therapy for cancer patients [146,147].

The favorable binding kinetics of SS-SSTR are the basis for use of the sstr2 gene as reporter gene. [111In]-DTPA-D-Phe1-octreotide (Octreoscan) and [99mTc]-depreotide (Neotect), radiolabeled SS analogs have been in routine clinical use for the past several years for the imaging of SSTR-positive tumors using gamma cameras or SPECT, i.e., SSTR scintigraphy [148,149]. These agents have also been useful in the imaging of non-neoplastic conditions, which are associated with SSTR upregulation such as a variety of autoimmune and granulomatous diseases.

The potential of using sstr2 as a reporter gene has been realized. Figure 30.10 demonstrates adenoviral delivery via intratumoral injection of human sstr2 gene into a tumor xenograft of a mouse. The transfected tumor is subsequently detected with a gamma camera following an intravenous injection of [99mTc]P2045, an SS analog [150]. Recently, a positron-emitting SS analog [94mTc] Demotate-1 could be used to determine adenovirus-mediated SSTR2 gene transfer by microPET imaging, a modality that can improve the sensitivity of the SSTR2 reporter gene system [151]. For sstr2 to serve as a true reporter gene, however, it will have to be uncoupled from signal transduction since it may have undesirable effects in targeted and surrounding cells in its current state.

On a side note, sstr2 is being tried as a double-edged suicide transgene for cancer therapy. Not only does SSTR primarily mediate antimetabolic effects as described earlier, but it can also be utilized for its ability to internalize and retain SS analogs. Efforts are being made to deliver cytotoxic agents to targeted cells by using the receptor as a courier of toxic SS analogs, such as [90Y]-DOTA-D-Phe1-Tyr3-octreotide ([90Y]-SMT 487). 90Y exerts its lethal effects by local irradiation via emitted β-particle. When SSTR–null tumor xenografts were injected with recombinant adenovirus encoding the SSTR2 receptor, they become susceptible to systemically administered [90Y]-SMT 487 [152]. This version of targeted radiotherapy is able to significantly reduce quadrupling times of the xenograft. When the sstr2 transgene is used in this manner it can be directly imaged using Octreoscan.

Indirect imaging can also be accomplished by coupling the therapeutic gene to a reporter gene that encodes an enzyme, which converts freely dispersible radiolabeled substrate probes into sequestered products [153]. For example,

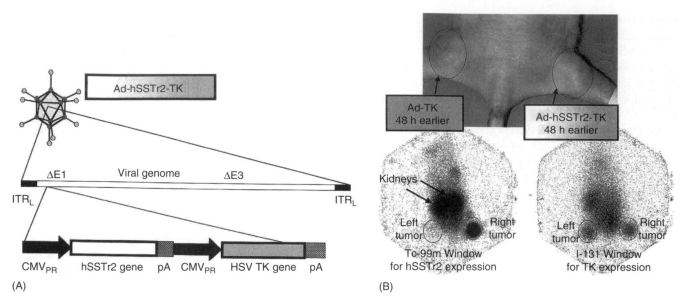

FIGURE 30.10 Gamma camera imaging of a dual promoter construct. (A) To measure target (therapeutic) gene expression, several strategies are employed to "link" the therapeutic gene with a reporter gene on a single vector. In one strategy, indirect measurements of a target gene expression can be made by a downstream reporter gene with both genes driven by separate and identical promoters (i.e., the "dual-promoter" construct). Replication-incompetent adenovirus encoding human type 2 somatostatin receptor (hsstr2) and the herpes simplex TK enzyme (Ad-CMV-hsstr2-CMV-tk) is an example of this construct. Both transgenes are driven by the CMV promoter element. (B) Human nonsmall cell lung cancer tumor xenografts were subcutaneously implanted in the right and left flank of a mouse. The left tumor was injected with Ad-CMV-tk and the right tumor was injected with Ad-CMV-hsstr2-CMV-tk. Forty-eight hours later, the mouse was simultaneously injected (IV) with both ^{99m}Tc-P2045, which is an SS receptor peptide ligand, to detect expression of hsstr2 and radioiodinated ^{131}I-2'-deoxy-2'-fluoro-β-D-arabinofuranosyl-5-iodouracil (FIAU) to detect TK expression. Imaging was performed with an Anger gamma camera 5 h after injection of the radiotracers. The gamma camera can be "tuned" to select for gamma rays that fall within a defined range. It is this property of the gamma camera which can discriminate between the activity of different radioisotopes such as ^{99m}Tc (140 keV) and ^{131}I (364 keV). Thus, two images can be obtained from the same animal by changing the window settings in the gamma camera: one for ^{99m}Tc (for hsstr2 expression, lower left image) and the other for ^{131}I (for TK expression, lower right image). The right flank xenograft, which was injected with the dual promoter construct (Ad-CMV-hsstr2-CMV-tk), demonstrates uptake of both radiotracers, while the left flank xenograft (injected with Ad-CMV-tk) demonstrates [^{131}I]FIAU uptake only. These findings support the feasibility of the dual promoter approach for tracking transgene delivery. (Reproduced from Zinn, K.R., Chaudhuri, T.R., Krasnykh, V.N. et al., *Radiology*, 223, 417, 2002. With permission.)

previously described HSV1-tk or HSV1-sr39tk can be used as reporter gene by using [^{124}I]FIAU or [^{18}F]FHBG at tracer levels (subpharmacologic, nontherapeutic dose). In this manner, the HSV1-tk or mutant counterpart can be utilized strictly as a reporter gene.

30.3.3 MRI REPORTER GENES

Imaging transgene expression using MRI depends on reporter genes that encode receptors or enzymes, which specifically interact with probes which are attached or chemically modified to accommodate paramagnetic or supraparamagnetic substances. Once localized to their targets, these probes alter the local magnetic field, which changes the relaxivity of nearby protons, and, concomitantly, effects a change in the RF signal detected by the receiver. A number of MRI reporter systems have been proven in animal studies: Engineered transferrin receptor (ETR)-dependent reporter systems, the iron-binding metalloproteins (e.g., the tyrosinase enzyme and the ferritin protein), and the *lacZ*-EgadMe system.

Iron, a superparamagnetic ion, can cause significant changes of the local magnetic field, which can be detected by MRI if supraphysiologic concentrations can be achieved. Transferrin (Tf), an iron-binding protein, and its cell-surface receptor (Tf-R), ubiquitously present on most cell types, mediate normal cellular iron metabolism and regulation. Normal intracellular stores of iron are dependent upon internalization kinetics of this receptor–ligand complex. Removal of the mRNA destabilization motifs in the 3' untranslated region and the iron-regulatory region of the Tf-R gene resulted in an ETR that can be constitutively overexpressed and liberated from feedback regulatory control [154,155]. As expected, ETR-transfected cells accumulate approximately 500% more probe (holo-Tf) than control cells [156]. To further augment the difference between transfected and control cells, the reporter probe itself has been modified to possess even greater magnetic susceptibility characteristics. It involves the synthesis of a 3 nm monocrystalline iron oxide nanoparticle (MION) that is surrounded by a layer of low molecular-weight dextran to which holotransferrin is covalently bound (Tf-MION). On the average, each MION particle contains approximately 2000 superparamagnetic Fe atoms (compared to the 2 Fe atoms present in a single molecule of the paramagnetic chelate, holotransferrin). As expected, T2-weighted gradient echo MR imaging (1.5 T; imaging time 3–7 min per sequence; voxel resolution $300 \times 300 \times 700 \mu m$) reveal significantly lower signal intensity in ETR$^+$ tumor xenografts than control following intravenous administration of Tf-MION [154]. Recently, a second superparamagnetic reporter probe, a dextran cross-linked iron-oxide (CLIO) superparamagnetic particle conjugated to transferrin (Tf-CLIO) has also been effective in identifying ETR$^+$ tissues with MRI [157]. Such superparamagnetic iron-oxide (SPIO) nanoparticles have shown much promise as cell-labeling agents giving a negative contrast in MR images wherever the labeled cells are present. The small size of the particles assists in easy transport across cell membranes, and low toxicity of the particles allows for labeling with huge iron loads (25 pg/cell) [158,159]. Stem cells (e.g., embryonic or neural stem cells) can be easily labeled by incubating them *in vitro* with SPIO particles and visualized *in vivo* using MRI [160,161]. However, a fundamental drawback of these particles is that they cannot be distinguished from a void in the MR image. Moreover these negative contrast agents also suffer from partial volume effects. Recent developments in the field allow cells labeled with SPIO to be imaged with positive contrast [159]. Spectrally selective RF pulses are employed to excite and refocus the off-resonance water surrounding the labeled cells, thereby allowing only the tissues and fluid around the cells to be detected in the MR image. Phantom studies showed a significant linear correlation between the number of cells and the signal observed ($r = 0.87, p < 0.005$). This will help in quick assessment of the location of injected cells and also for quantifying the volume of labeled cells in cell-based therapeutic approaches.

Another example of an MR reporter gene is the tyrosinase enzyme. Tyrosinase catalyzes the hydroxylation of tyrosine-yielding dioxyphenylalanine (DOPA) and its subsequent oxidation to DOPAquinone. DOPAquinone is subsequently converted to melanin, which in turn shows a high affinity for iron. Thus over-expression of tyrosinase enzyme results in cellular accumulation of iron, which accounts for the iron-induced T1 hyperintensities in MR images. Such significant increase in signal intensities is well documented in cells [162,163]. However, toxicity issues can limit the application of this system. It has been found that melanin and melanin precursors produce highly reactive oxygen species, and thus exhibit significant toxic effects [164].

Ferritin, another metalloprotein, serves as the body's iron depot and its main function is to store iron inside its protein shell in a biochemically safe crystal configuration. Native ferritin is, however, a weak contrast agent. In an effort to improve its relaxivity, the native iron oxyhydroxide core was removed and the protein shell was reconstituted with a superparamagnetic core. These so-called magnetoferritin molecules have proven to be effective T_2 contrast agents both *in vitro* and *in vivo* at 1.5 T [165,166]. In an effort to use ferritin as a reporter gene, an adenovirus vector encoding the protein was injected into the mouse brain. After injection, the transfected cells showed a marked hypointensity on T_2-weighted MR images [167].

A different MR imaging method shown to be compatible with living subjects relies on an enzymatic amplification strategy to monitor gene expression. As mentioned before, gadolinium (Gd) is a rare-earth element with the largest number of unpaired electrons. With seven unpaired electrons, Gd is a strong paramagnetic substance, collectively affecting the spins of water protons immediately surrounding it. The increased signal (for T1-weighted sequences) seen in Gd-enhanced MR imaging is afforded by the increased relaxation rate of intimately associated water protons surrounding a Gd atom. Gd's ability to affect the relaxation rate of protons varies inversely with the distance between the paramagnetic ion and water protons.

Based on these principles, a reporter probe, (1-(2-(β-galac topyranosyloxy)propyl)-4,7,10-tris(carboxymethyl)-1,4,7,10-t etraazacyclododecane)gadolinium(III) (EgadMe), has been cleverly formed by encasing Gd in a water-proof package, an artificial barrier or cage, which is designed to keep water molecules away so as not to be affected by the Gd's magnetic properties (Figure 30.11A) [168]. In this configuration, water has no access to the paramagnetic ion and therefore this probe is silent on MR imaging. As it turns out, part of the physical barrier is composed of a sugar- a galactopyranose cap, which has been attached to the cage by a β-galactosidase-cleavable linker. If *lacZ* is used as a reporter gene, subsequent enzyme cleavage releases the cap and allows water access to the Gd ion, thus activating this novel MR contrast agent. β-galactosidase activity, following introduction of linearized plasmid cDNA encoding *lacZ* into a specific subset of cells in a *Xenopus laevis* embryo, has been imaged after an intracellular injection of EgadMe (Figure 30.11B and C).

In its current state, EgadMe has difficulty crossing cell membranes and as a result has to be directly injected intracellularly to maximize detection *in vivo*. Furthermore, relatively slow kinetics of cleavage for this agent is perhaps suboptimal for imaging gene expression [168]. Regardless, these relatively recent developments will be refined and indicate great potential for MRI of transgene expression.

30.3.4 MAGNETIC RESONANCE SPECTROSCOPY REPORTER GENES

Certain metabolites produced physiologically from endogenous enzymes or uniquely from exogenous enzymes have unique chemical signatures that can be detected using MRS also called nuclear magnetic resonance (NMR) spectroscopy. These enzymes can be overexpressed in target tissues and can be used as MRS reporter genes to successfully track gene expression in transgenic models, transfected tumor xenografts, or viral-mediated gene transfer experiments. At this time, MRS does not produce true, spatial pictures, and,

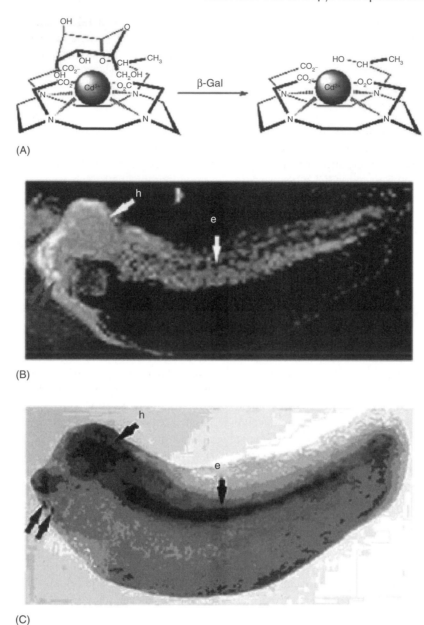

FIGURE 30.11 MRI of β-galactosidase-activated MRI reporter probe (EgadMe). One can obtain image contrast in MRI by using paramagnetic substances, which change the local magnetic field and, thereby, increase the relaxation rate of nearby water protons. Gd is an example of a paramagnetic substance and relatively high local concentration of this agent translates into enhanced brightness as seen on T1-weighted images. (A) A reporter probe, EgadMe, has been formed by "encasing" Gd in a molecular casing, an artificial barrier, which is designed to keep water molecules away so as not to be affected by the Gd's magnetic effects. In this configuration, water has no access to the paramagnetic ion and therefore this probe is "silent" on MR imaging. Part of the physical barrier is composed of a sugar, a galactopyranose cap, which has been attached to the cage by a β-galactosidase-cleavable linker. If *lacZ* is used as a reporter gene, subsequent enzyme cleavage releases the cap and allows water access to the Gd ion, thus, "activating" this novel MR contrast agent. (B) EgadMe permits MRI detection of *lacZ* gene expression. Linearized plasmid cDNA encoding *lacZ* is injected into one of the cells of the two-cell stage *Xenopus laevis* embryo. EgadMe is injected into both cells of the two-cell stage. Subsequent enzyme expression is on one side of the animal since the two cells represent the future right and left sides of the animal. MR imaging of the embryos has been obtained at approximately the 100,000-cell stage using a 11.7 T magnet. As expected, β-galactosidase activity is seen in one half of the animal as depicted as areas of high signal intensity within the endoderm (e) and head (h). (C) Light microscopic images of same embryo fixed and stained with X-gal. Areas of X-gal staining follow regions of high signal intensity on MR image. (Reproduced from Louie, A.Y., Huber, M.M., Ahrens, E.T. et al., *Nat. Biotechnol.*, 18, 321, 2000. With permission.)

instead, shows spectral tracings of the various metabolites it is able to identify. It is a sensitive, quantitative, and relatively fast technique when compared to MRI [42].

One particular MRS-sensitive metabolic reaction produces ATP, a process catalyzed by CK:

$$H^+ + PCr + ADP \Leftrightarrow ATP + Cr$$

where PCr is phosphocreatine and Cr is creatine. More specifically, ^{31}P-MRS identifies amounts of PCr, ATP, ADP, and free phosphorus in the reaction. This molecule can be detected readily in the heart, muscle, and brain since they are produced in great quantities in these organs. The liver, on the other hand, has very low levels and thus can serve as a background for situations where CK is overexpressed. A transgenic mouse model that overexpresses this enzyme in the liver has shown that it can generate MRS-detectable levels of PCr [169]. This technique is however invasive in nature as a tissue window has to be created to minimize background noise from the overlying muscle and other surrounding structures rich in CK. Future developments in ^{31}P 3D spectroscopy may eventually prove helpful.

A related study employs an invertebrate analog of CK as its reporter gene. *Drosophila melanogaster* AK has been cloned and, when introduced into mammalian muscle, produces phosphoarginine (PArg) in the following reaction [170]:

$$H^+ + PArg + ADP \Leftrightarrow ATP + Arg$$

PArg provides a unique phosphorus signal, which is amplified when the transgene for AK is delivered to mammalian tissues (Figure 30.12). One important consideration in the use of *Drosophila* AK is it can act as an ATP buffer in mammalian tissues and thus the consequences of this need to be explored prior to its widespread use as an *in vivo* gene reporter gene.

Another MRS-friendly system is the CD reporter system [171]. Using ^{19}F MRS, the conversion of the relatively benign 5-FC to the cytotoxic agent (5-FU) driven by this enzyme can be detected. Tumor xenografts transfected to express yeast CD have been shown to produce 5-FU with MRS.

30.4 STRATEGIES FOR TRACKING VEHICLES AND TRANSDUCED CELLS

The list is quite long for the array of vehicles being developed, both viral and nonviral, for the delivery of gene vectors. The details of such carriers are, of course, provided elsewhere throughout this book. There are two fundamentally different ways to track the biodistribution of delivered agents. One method is to track the sites of successful gene transfer/expression (various methods described in this chapter) and the other method relies on directly labeling vehicles, DNA, or cells with radionuclides, fluorescent dyes, or MR-compatible contrast agents. The last method may give better spatiotemporal information with respect to distribution of the vehicles or DNA material used for therapy, but does not give any information with regard to the success of gene transfer. The latter method has been reviewed recently and includes the direct labeling of herpes virus with ^{111}In, adenoviral knob with ^{99m}Tc, liposomes with ^{111}In, double-stranded DNA by peptide-based chelates ([^{99m}Tc]PBC), genetically modified mesothelioma cells with technitium ([^{99m}Tc]PA1-STK), myoblasts with technetium, and DNA delivery systems with MRI-detectable DNA-binding chelates [172–179].

For those involved in the development of novel viral or nonviral delivery vehicles, targeting and efficiency of gene transfer are primary concerns, and, thus, examples of the former method are given here. The success of exogenous gene expression is dependent on its ability to at least survive the following series of stringent events: the DNA–vehicle complex has to bind specific cell-surface receptors, undergo receptor-mediated endocytosis, survive endosomal lysis, be released from endosomal captivity, endure the cytoplasmic environment, be destined for targeted entry of the nucleus, and ultimately released from carrier molecules to facilitate gene expression [180]. Whether it is pseudotyped lentiviruses, modified PEI complexes, liposomes, etc., one major role reporter genes are expected to provide is the monitoring of localization, biodistribution, and gene transfer efficiency of these delivery vehicles in living subjects. Noninvasive localization of retroviral [181], adenoviral [182], herpes viral vector [183] mediated HSV1-tk gene transfer has been performed. Figure 30.13 is an example of replication-conditional, oncolytic herpes simplex virus–mediated gene delivery. Constitutively expressed optical or PET reporter genes have been helpful in studying the distribution of nonviral vehicles such

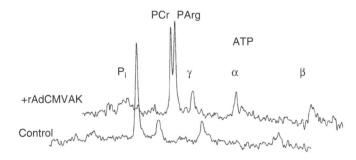

FIGURE 30.12 MRS of transgene expression. ^{31}P-MRS has the ability to detect phosphorus NMR signals. A recombinant adenovirus (rAdCMVAK) can be constructed to deliver AK, an enzyme unique to invertebrates, into muscle. Once introduced into mammalian muscle, the enzyme catalyzes the production of PArg, which has a unique spectral signature (both in magnitude and location on the spectrum) and which can be detected by MRS. The figure represents *in vivo* basal ^{31}P spectra from the hind limbs of a 6-month-old mouse. ^{31}P-MRS spectra from the rAdCMVAK-injected limb (upper spectrum) reveal a ^{31}P resonance at the chemical shift for PArg that is not present in the contralateral control limb (lower spectrum). (Reproduced from Walter, G., Barton, E.R., and Sweeney, H.L., *Proc. Natl. Acad. Sci. USA*, 97, 5151, 2000. With permission.)

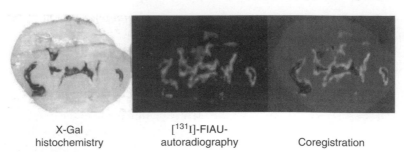

| X-Gal histochemistry | [131I]-FIAU-autoradiography | Coregistration |

FIGURE 30.13 (See color insert following blank page 206.) Tracking HSV infection with [131I]FIAU using autoradiography. Tracking wild-type HSV-1 infection with radionuclide-based techniques can be accomplished using the virus' native TK gene and a reporter probe such as radiolabeled [131I]FIAU. To help corroborate imaging findings with histochemistry findings, a replication conditional, oncolytic recombinant HSV-1 virus vector, hrR3, containing a *lacZ* insertional mutation within the *RR* gene locus, has been prepared. Following injection of the vector into rat gliosarcoma xenografts, tumors were processed for tissue-sectioning, autoradiography, and β-galactosidase-stained histology. Image coregistration of tumor histology, HSV-1-tk related radioactivity (assessed by [131I]FIAU autoradiography), and *lacZ* gene expression (assessed by β-galactosidase staining) demonstrated a characteristic pattern of gene expression around the injection sites. A narrow band of *lacZ* gene expression immediately adjacent to necrotic tumor areas is observed, and this zone is surrounded by a rim of HSV-1-tk-related radioactivity, primarily in viable-appearing tumor tissue. PET images (not shown) of injected tumors in the intact animal have also been performed using [124I]FIAU as a reporter probe; the areas of PET-labeled probe uptake correlate well with the β-galactosidase-stained photomicrographs. (Reproduced from Jacobs, A., Tjuvajev, J.G., Dubrovin, M. et al., *Cancer Res.,* 61, 2983, 2001. With permission.)

as PEI polyplexes, which have been covalently modified with Tf to facilitate targeting (Figure 30.14) and cationic lipid–DNA complexes (Figure 30.15).

30.5 NONINVASIVE IMAGING OF CELL TRAFFICKING

Stem cell–based therapeutic approaches hold great promise in the treatment of degenerative diseases, cardiovascular diseases, and certain genetically inherited deficiencies. For example, transplantation of embryonic stem cells into an ailing heart is gradually emerging as a promising therapeutic option for coronary heart disease. Similarly, neural stem cell–based approaches have been successfully employed in several CNS pathologies like Parkinson's disease, traumatic lesions of the spinal cord, ischemic and hemorrhagic brain lesions, and multiple sclerosis [184]. On a different note, elimination of tumor cells by T-cell transfusion, a process termed as adoptive immunotherapy, is being rigorously pursued as a promising treatment option for cancer. All these cell-based therapeutic approaches can hugely benefit from ways to repeatedly monitor the locations of engrafted cells to determine their viability, performance, and expansion noninvasively [185]. This can be achieved by either transfecting these cells stably with a suitable reporter gene or by labeling the cells by a suitable contrast agent prior to systemic administration. For long-term tracking, the reporter gene approach is more desirable. Reporter genes integrated in to the cell genome would be extremely useful in assessing survival status of the implanted cells because the reporter will be expressed as long as the cells are alive and will be

passed to daughter cells upon cell division. However, long-term expression of the reporter gene can be subjected to silencing by epigenetic mechanisms such as promoter methylation or histone deacetylation. Krishnan et al. showed that luciferase expression in a rat embryonic cardiomyoblast cell line H9C2 stably transfected with firefly luciferase reporter gene decreased with every passage, such that at passage 8 it was only 0.01% of that of passage 1 [186]. When cells were treated with an antimethylating agent, 5-Aza, they showed a higher luciferase signal in living animals compared to the untreated counterparts even 8 days postimplantation. Such loss of reporter gene expression poses a difficult challenge for molecular imaging of reporter gene-labeled cells.

30.5.1 NONINVASIVE IMAGING OF STEM CELL BASED THERAPY

Since stem cells are emerging as promising tools to replace damaged organs and to rejuvenate diseased organs in many human disorders, ways to track these cells once they enter the body have become a very important field of study. Different imaging modalities and contrast agents have been employed to monitor cell transplantation, transduction, and migration, and for evaluation of cell trafficking, correction mechanisms, and efficacy of new therapeutic strategies in different animal models of human diseases [184]. Such imaging approaches may not only improve our understanding of therapeutic mechanisms in preclinical studies but may also facilitate rapid translation of cell-based therapies into clinical practice.

Stem cells can be labeled directly by contrast agents such as 2-[F-18]-fluoro-2-deoxy-D-glucose ([F-18]FDG) for PET, [In-111]oxine for SPECT, and SPIO particles for MRI. This

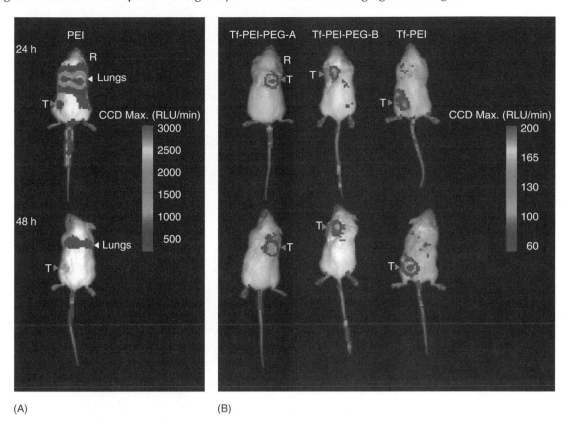

(A)   (B)

FIGURE 30.14 (See color insert following blank page 206.)   Tracking Tf targeted polyethylenimine (PEI)-mediated gene delivery using optical bioluminescence imaging. Delivery of the bioluminescence reporter gene, firefly luciferase (*Fluc*), by CMV-*Fluc* DNA/PEI polyplexes and subsequent *Fluc* expression can be imaged in living mice using a cooled CCD camera. Additionally, the biodistribution of modified PEI polycation complexes, altered with molecules such as Tf or polyethylene glycol (PEG), can be studied in this manner. Tf targeting has been shown to improve the transfection efficiency in certain tumor cell lines and PEG modification has been shown to improve circulation times of DNA/PEI complexes and prevent their nonspecific uptake by the reticuloendothelial system. All CCD images are of living mice carrying N2A xenograft 24 or 48 h after intravenous injection of various DNA/PEI polyplexes. Site of tumor is indicated (T). (A) PEI (positive control) treated animals show relatively high *Fluc* expression (using 1× D-Luciferin) in the lungs as compared with the tumor. The activity on the left-hind limb is from the N2A cell tumor (T). Nonspecific tail activity occurs at the DNA/PEI polyplex injection site. All *Fluc* expression decreases at 48 h. (B) Tf-PEI-PEG-A, Tf-PEI-PEG-B, and Tf-PEI treated mice show *Fluc* expression (using 2× D-luciferin) in the tumor (T) and tail regions, but no detectable signal in the lungs. For each formulation, expression in the tumor varied over 24 to 48 h. All images are quantitated as indicated by the two scales (RLU/min). (Reproduced from Hildebrandt, I.J., Iyer, M., Wagner, E., and Gambhir, S.S., *Gene Ther.*, in press. With permission.)

method does not involve extensive manipulation of the cells and, therefore, is often preferred for clinical applications. However, there are a couple of limitations: labels are diluted upon cell division, making these cells invisible and labels may efflux from cells or may be degraded over time. Furthermore, even cells that are nonviable can lead to signal till the dead cells are removed by the macrophages.

As described earlier, reporter genes can be used to label stem cells. Imaging of genetically labeled cells was performed using rat cardiomyoblasts, which were infected *ex vivo* with adenovirus carrying the HSV1-sr39tk and luciferase reporter genes (detectable number of cells was 5×10^5 cells by optical imaging and 3×10^6 cells by PET) [187]. Cell-specific *in vivo* optical and micro-PET imaging was feasible for up to 2 weeks after direct injection of cells into the myocardium of nude rats. Similarly mouse embryonic stem cells labeled with a trifusion reporter gene (*fluc-mrfp-ttk*)

helped in monitoring the kinetics of ES cell survival, proliferation, and migration following cardiac delivery (Figure 30.16) [188].

The ideal imaging modality should be the one that can provide integrated information related to the entire process of cell engraftment, survival, and functional outcome with high sensitivity and better resolution. Both radionuclide technology and MRI have the advantage of clinical applicability, with MRI having some advantages in terms of spatial resolution. While studies using conventional MRI scanners could detect up to 105 iron-oxide labeled hematopoietic bone marrow–derived and mesenchymal stem cells, use of high-field magnets (11.7 T) enhanced the sensitivity many fold and showed the ability to track single cells containing a single iron particle [189,190]. Compared to MRI, direct labeling of cells with radionuclides can provide high sensitivity due to low background. After intraventicular injection,

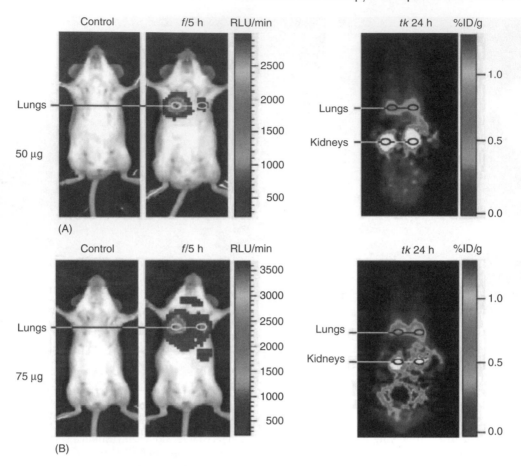

FIGURE 30.15 (See color insert following blank page 206.) Tracking cationic lipid-mediated reporter gene delivery using optical (bioluminescence) and PET imaging. Cationic lipids associate with negatively charged DNA to form complexes that bind to cell surfaces by way of electrostatic interaction, thereby allowing a nonviral means of gene transfer. Distribution of systemic administration of DNA–lipid complexes in mice is demonstrated by delivering prepared DNA–lipid complexes that carry optical and PET reporter genes. CMV-*fl* plasmid DNA (CMV promoter driving expression of firefly luciferase [*fl*] gene) was mixed with cationic lipid, 1,2-dioleoyl-3-trimethylammonium-propoane (DOTAP) and cholesterol, to form *fl* DNA–lipid complexes. Similar procedure was used to prepare HSV1-sr39tk DNA–lipid complex (tk DNA–lipid complex). (A, B) show images following administration of 50 and 75 μg of each *fl* and tk DNA–lipid complexes via tail vein injection into CD-1 mice, respectively. Bioluminescent images (left images) were obtained 5 h after injection of the vector and 5 min after intraperitoneal injection of D-Luciferin. MicroPET images (right images) were obtained 24 h after vector delivery and 1 h after [^{18}F]FHBG injection. Control mice (left) optical images obtained prior to administration of D-luciferin. Optical and PET images demonstrate that lungs are primary organs for transgene expression. Increased dose of DNA–lipid complex results in greater pulmonary transgene expression. Activity seen in the kidneys in the microPET images is the result of excreted, unsequestered reporter probe, [^{18}F]FHBG. (Reproduced from Iyer, M., Berenji, M., Templeton, N.S., and Gambhir, S.S., *Mol. Ther.*, 6, 555, 2002. With permission.)

a small number of indium (In)-111–labeled endothelial and hematopoietic progenitor cells were found to accumulate in infarcted rat myocardium by SPECT [191]. In conclusion, molecular and cellular imaging retains a fundamental role not only in preclinical studies, but also in clinical practice, since accurate delivery or homing of cells to target tissues is crucial for the clinical success of such cell-based therapeutic applications [184].

30.5.2 NONINVASIVE IMAGING OF T-CELL TRAFFICKING

Immunotherapy with tumor-targeted lymphocytes is an active area of basic and clinical research. For better therapeutic outcome, it will be highly beneficial if the kinetics, specificity, and longevity of tumor-targeted lymphocytes

could be repetitively monitored in cancer patients receiving cellular immunotherapy. Dubey et al. infected T-cells from mice carrying Moloney murine sarcoma virus/Moloney murine leukemia virus (M-MSV/M-MuLV) tumors with a retrovirus encoding for both GFP and HSV1-sr39tk [192]. The modified lymphocyte population once injected into animals bearing M-MSV/M-MuLV tumors could be followed repetitively by PET and showed specific accumulation at the target M-MSV/M-MuLV tumors (Figure 30.17). In a similar experiment, Epstein–Barr virus (EBV)-specific T cells programmed to encode HSV1-TK enzyme selectively accumulated radiolabeled FIAU after adoptive transfer in animals and the HSV1-TK + T cells could be tracked noninvasively in SCID mice bearing human tumor by PET [193]. These T cells selectively accumulated in EBV + target tumors but

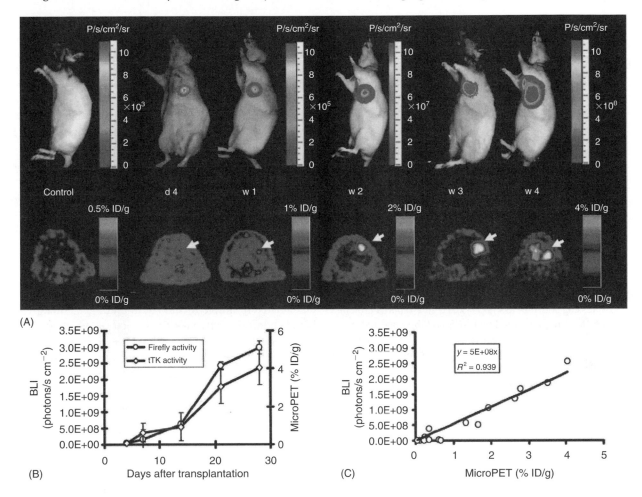

FIGURE 30.16 (See color insert following blank page 206.) Molecular imaging of transplanted ES cells with bioluminescence and PET imaging. (A) Embryonic stem cells stably transduced with a triple fusion reporter gene mrfp-fluc-ttk were transplanted in the heart of athymic rats. Animals were subsequently imaged for 4 weeks to assess cell survival. A representative study animal is shown in the figure. Significant bioluminescence (top) and PET (bottom) signals were observed at day 4, week 1, week 2, week 3, and week 4 following transplantation. In contrast, control animals had background activities only. (B) Quantification of imaging signals showed a drastic increase of fluc and ttk activities from week 2 to week 4. Extracardiac signals were observed during subsequent weeks. (C) Quantification of cell signals showed a robust *in vivo* correlation between bioluminescence and PET imaging ($r^2 = 0.92$). BLI indicates bioluminescence. (Reproduced from Cao, F., Lin, S., Xie, X. et al., *Circulation*, 113, 1005, 2006. With permission.)

not in control tumors. Moreover, the radiolabeled transduced T cells retained their capacity to eliminate targeted tumors selectively.

Tracking of cells with MR provides detailed anatomical information and is therefore more informative for the evaluation of the localization of therapeutic cells after injection. However, it has been relatively difficult to label cells with contrast agents at high enough concentrations to monitor cell trafficking by MR [194]. In a recent report, ovalbumin-specific splenocytes (OT-1) were labeled with anionic γ-Fe$_2$O$_3$ SPIO nanoparticles and adoptively transferred into mice with ovalbumin-expressing tumors [195]. OT-1 cells could be tracked *in vivo* with high resolution using a 7 T MRI and showed significant negative enhancement of the spleen at 24 h, and of the tumor at 48 and 72 h after labeled cell injection.

Bioluminescence imaging has been employed to study T-cell trafficking. Luciferase-expressing human papillomavirus type 16 (HPV-16) E7-specific CD8(+) T cells were followed noninvasively over time. The injected E7-specific T cells not only showed preferential migration to the E7-expressing target tumor but also proliferated at the tumor site [196]. In a different approach, optical imaging was employed to study the kinetics of distribution and homing of luciferase-expressing cytokine-induced killer (CIK) cells, following intravenous administration into tumor-bearing mice [197]. The study showed that CIK cells exhibit early localization to the lungs, followed by the liver and spleen and subsequently the tumor.

The different imaging approaches can be safely adopted to noninvasively monitor tumor targeting of antigen-specific cell-based cancer therapies. These studies will be helpful in determining various parameters like routes of administration, cellular, and systemic factors affecting T-cell survival, residence time, and activation to maximize the therapeutic outcome.

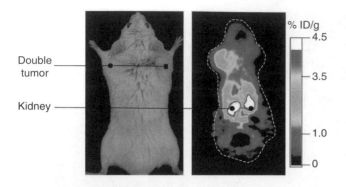

FIGURE 30.17 (See color insert following blank page 206.) Noninvasive imaging of T-cell trafficking following adoptive transfer. Tumors were implanted subcutaneously on the shoulders of the mouse in the figure: A Moloney murine sarcoma virus–Moloney murine leukemia virus (M-MSV/MMuLV) tumor on its left shoulder and a control P815 tumor on its right shoulder. T-cells from an animal carrying an M-MSV/M-MuLV tumor were transfected with a retrovirus expressing both the HSV1sr39Tk and GFP reporter proteins. Transfected cells were first sorted by FACS and then injected intraperitoneally into the mouse bearing the M-MSV/M-MuLV and P815 tumor, and the animals were imaged by micoPET following [18F]FHBG injection. Homing of the cells to the M-MSV/M-MuLV tumor, but not the P815 tumor, was apparent within day 13 of injection. (Reproduced from Dubey, P., Su, H., Adonai, N. et al., *Proc. Natl. Acad. Sci. USA,* 100, 1232, 2003. With permission.)

30.5.3 NONINVASIVE IMAGING OF T-CELL ACTIVATION

In addition to elucidating the trafficking pathways of immune cells *in vivo*, means to visualize their differentiation, and antigen-dependent activation status will be immensely beneficial for a better understanding of how the immune system interacts with cancer. In an effort to image T-cell activation, Ponomarev et al. created a human T-cell line expressing an HSV1-TK–GFP fusion protein under the control of the T-cell receptor (TCR)-dependent nuclear factor of activated T cells (NFAT)-responsive promoter [198]. Mice bearing subcutaneously implanted cells were treated with anti-CD28 and anti-CD3 antibody to initiate TCR mediated T-cell activation. This in turn triggered expression from the NFAT promoter, which was subsequently measured by PET analysis of [124I]FIAU accumulation by the HSV1-tk gene.

With the advent of new probes and new ways to label T-cells, repetitive and noninvasive monitoring of both tumor-specific lymphocytes homing and antigen-dependent activation of these cells at their targets will be highly feasible.

30.6 MONITORING GENE THERAPY LEVELS

Precise localization and quantitative assessment of the magnitude and temporal variation of transgene expression is a necessary component of any gene therapy trial. Direct imaging with a transgene-specific imaging probe is ideal but not feasible or practical in most cases. To develop a specific probe for each individual transgene is not always technically possible;

furthermore, it is necessarily labor- and cost-intensive. Using indirect imaging methods by linking a portable reporter gene to a therapeutic gene allows for more flexibility as a variety of transgenes can be individually monitored by cloning in a reporter gene into appropriate sites of the vector. Coexpression of the therapeutic gene product and reporter gene product in a coordinated and regulated manner enables a correlative and quantitative relationship between the two genes. Thus, levels of therapeutic gene expression can be inferred by the amount measured from reporter genes, provided that the expression of both genes remains coupled. Several such approaches, ranging from the more straightforward, like the dual vector approach, to the more sophisticated, such as the bidirectional transcriptional approach, are currently being developed and briefly discussed below [153].

30.6.1 COVECTOR ADMINISTRATION

One relatively simple method to monitor gene therapy *in vivo* is to coadminister two different vectors, which are identical except for the transgene they are carrying: one vector would encode the therapeutic gene and the other would encode the reporter gene and both genes would be driven by the same promoter. This approach has been validated using the two PET reporter genes, HSV-sr39tk and D2R, each cloned into distinct adenoviral vectors and both driven by the same CMV promoter [140]. While individual cell differences in expression levels may be seen, macroscopic measurements made at the tissue culture or organ level (adenoviral-mediated hepatic transfer) correlate quite well ($r^2 \geq 0.93$). The technique may prove useful in specific experimental situations.

30.6.2 SINGLE VECTOR APPROACHES

The use of an internal ribosomal entry site (IRES) signifies the bicistronic approach to coupling genes [199–201]. In a bicistronic expression cassette, an IRES sequence is interpositioned between the therapeutic and reporter genes, usually the first and second cistron, respectively. Both genes are under the control of the same promoter and transcription of this construct results in a single mRNA molecule. Initiation of translation of the first cistron is by way of the usual cap-dependent manner, but translation of the second cistron is facilitated by the IRES sequence in a cap-independent mechanism, which allows translation by a second ribosome. This approach has been verified in a few studies. For example, an IRES derived from an encephalomyocarditis virus has been used to construct a bicistronic vector from which both D2R and HSV1-sr39tk reporter genes are coexpressed from a common CMV promoter (pCMV-D2R-IRES-HSV1-sr39tk) [106]. The levels of D2R and HSV1-sr39tk activity demonstrate a high degree of correlation ($r^2 = 0.97$) using [18F]FESP and [18F]FHBG as imaging probes, respectively. Another vector, which encodes Renilla luciferase (*Rluc*) in a bicistronic configuration, pCMV-*Rluc*-IRES-sr39tk, or pCMV-sr39tk-IRES-*Rluc*, also shows excellent correlation

[106]. Similar relationships have also been seen with the use of an HSV1-tk gene that has been IRES-linked to the *lacZ* gene [202]; imaging with iodinated FIAU (SPECT reporter probe for HSV1-tk) correlates well with β-galactosidase activity seen by light microscopy. These studies corroborate the use of radionuclide and optical reporter genes as means of quantitatively determining relative levels of target gene expression.

One interesting finding in this approach, however, is that expression levels from the gene upstream to IRES sequence is consistently more robust than the levels seen from the gene downstream to the IRES [106]. This may have to with cell-specific differential translation from the IRES sequence, but nonetheless emphasizes the need for a highly sensitive reporter system as levels of the reporter gene product will be significantly diminished compared to the upstream gene. Further understanding and exploitation of regulatory modules recently found within the IRES may help circumvent this problem in

the future [203]. Alternatively, two different genes expressed from two distinct, but identical, promoters within a single vector can be done to avoid the attenuation problem and tissue variation issues experienced with the IRES-based approach. This is a variation of the bicistronic approach and is otherwise known as the dual-promoter approach [150]. Strong correlation between two reporter genes, hsstr2 and HSV1-tk, each driven by an independent but identical CMV promoter, has been exhibited (Figure 30.10).

In some situations, gene therapists will want to externally control the levels of transgene expression and, additionally, will need to verify the extent of control with imaging techniques. One particularly novel indirect imaging method makes this entire scenario possible by the use of a single inducible bidirectional tetracycline-responsive element (TRE) and two flanking minimal CMV promoters (Figure 30.18A). The fusion transactivator protein rtetR-VP16, which is constitutively expressed and can potentially

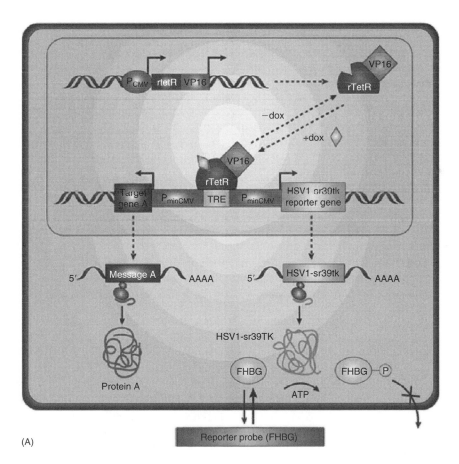

(A)

FIGURE 30.18 (See color insert following blank page 206.) Indirect PET imaging using a bi-directional transcriptional approach. (A) Target (therapeutic) gene expression can be measured indirectly by imaging reporter gene expression if expression of the two genes is linked. Both genes can be simultaneously expressed from two minimal CMV promoters that are regulated by a single bidirectional TRE. The rTetR–VP16 fusion protein is produced constitutively from a CMV promoter. When the rTetR–VP16 fusion protein binds doxycycline, this complex binds to the TRE regulatory sequence and substantially enhances expression from the two minimal CMV promoters. The target gene A in one coding region and a reporter gene (for example, a reporter kinase such as HSV1-sr39tk) in the alternative coding region are transcribed simultaneously into two mRNA molecules. Translation of the two mRNA molecules yields two distinct proteins in amounts that are directly correlated with each other.

(*continued*)

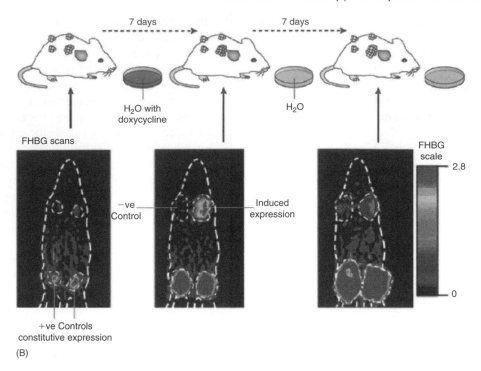

FIGURE 30.18 (continued) (B) Quantitative imaging of the locations and magnitude of PET reporter gene expression by trapping of a PET tracer inside the cell (for example, by phosphorylation of [^{18}F]FHBG by the HSV1-sr39TK reporter protein) provides an indirect measure of target gene expression. Sequential microPET imaging studies of a nude mouse carrying four tumors. Four tumor cell lines, two positive controls (constitutive reporter gene expression), one negative control, and one inducible line (reporter gene expression) induced by doxycycline, were injected subcutaneously into four separate sites in a single mouse. When tumors reached a size of at least 5 mm, the mouse was imaged with 9-(4-[^{18}F] fluoro-3-hydroxymethylbutyl)guanine ([^{18}F]FHBG). Doxycycline was then added to the water supply for 7 days. The mouse was then scanned again with [^{18}F]FHBG. Doxycycline was removed from the water supply for the next 7 days, and the mouse was again scanned with [^{18}F]FHBG. The locations of the four tumors and the mouse outline are shown by the dotted regions of interest. All images are 1–2 mm coronal sections through the four tumors. The % ID/g (% injected dose per gram tissue) scale for [^{18}F]FHBG is shown on the right. The negative control tumors show no gene expression and the positive control tumors show increased expression over the time course. The tumor on the top right, with inducible gene expression, initially does not accumulate [^{18}F]FHBG, then at 7 days after addition of doxycycline, induction of reporter gene expression traps [^{18}F]FHBG. Seven days after withdrawal of doxycycline, there is decreased induction and minimal trapping of [^{18}F]FHBG. The [^{18}F]FHBG image signal correlates well with target gene expression (not shown). (Reproduced from Gambhir, S.S., *Nat. Rev. Cancer*, 2, 683, 2002. With permission.)

be incorporated into the same bidirectional vector, binds to TRE only in the presence of tetracycline or one of its analogs. By varying levels of an exogenously added inducer such as doxycycline, a gene therapist can control transcription and magnitude of expression that can be verified by the accompanying reporter gene. Proof of principle has been shown in rat xenograft models using the reporter genes D2R and HSV1-sr39tk (Figure 30.18B) [204].

The fusion gene–protein approach is yet another powerful means of indirect monitoring of gene therapy. Constructs in this protocol contain two or more genes linked together within the same reading frame so that a single protein is translated. The resultant hybrid or fusion protein will have therapeutic and reporter properties and the expression of the fused gene can be closely monitored since the expression of the therapeutic component is stoichiometrically coupled to the reporter component of the protein. HSV1-tk-*gfp*, HSV1-tk-*Fluc*, HSV1-sr39tk-*Rluc, CD*-HSV1-tk are a few successful examples [22,205–208]. Various triple fusion proteins encoded by

fusion genes composed of a bioluminescent, a fluorescent, and a therapeutic/reporter gene are also reported for imaging gene expression from a single cell to living animals. For example, HSV1-sr39tk-*mrfp1-Rluc*, Δ45HSV1-tk-*gfp-Fluc*, and the most recent m*TFLuc-mrfp1*-HSV1-tk (m*TFLuc* stands for mutated Thermostabile *Fluc*) [98,209,210]. The fusion gene is engineered such that the proteins are linked via a short peptide spacer. However, while appealing in concept, it is quite challenging in practice to produce generalizable proteins since fusion proteins are often inactive or less active than their individual counterparts; furthermore, fusion proteins also may not localize to appropriate compartments since appropriate signaling mechanisms are either masked or unavailable [153]. Using this technique for every therapeutic gene developed would prove to be a daunting task since each newly generated fusion protein has its own peculiarities. Future technological improvements in the physical linkage of two proteins may help minimize discrepant behavior and allow more flexibility for this technique.

30.7 CLINICAL APPLICATIONS

While human gene therapy trials have been in effect for several years, the imaging of transgene expression in humans has only started. Transgenes, which can serve as both therapeutic gene and reporter gene (i.e., direct imaging), will most likely have an easier transition toward clinical application when compared to indirect imaging methods. More specifically, the kinetics, biodistribution, stability, dosimetry (needed for radionuclide approaches), and safety of the transgene and reporter probe have to be determined in direct imaging reporter protocols. In comparison, an additional verification and characterization of the reporter gene product has to be performed in indirect imaging systems.

Utilizing a direct imaging method, as is the case with the HSV1-tk suicide/reporter gene and one of its reporter probes, [18F]FHBG, where the requisite pharmacokinetics, biosafety, and other related testing have been established in human volunteers [117]. The study indicated that while [18F]FHBG demonstrates acceptable characteristics and should prove to be an acceptable probe for HSV1-tk imaging, [18F]FHBG does not readily cross the BBB. Methods used to disrupt the BBB during [18F]FHBG administration may aid in this regard if needed. Furthermore, imaging HSV1-tk reporter gene expression near the gall bladder, kidneys, and bladder with [18F]FHBG may prove to be difficult since nonsequestered [18F]FHBG passes through these organs, contributing to the background signal seen in these structures. In a recent clinical trial in patients with hepatocarcinoma undergoing adenoviral-HSV1-tk gene therapy, transgene expression could be successfully followed by [18F]FHBG and PET (Figure 30.19) [211–213]. Interestingly, transgene expression in the tumor was largely dependent on the injected dose of the adenovirus and was detectable in all patients who received ≥10^12 viral particles. However, 9 days postinjection, no expression could be observed. Most importantly, no specific expression of the transgene could be detected in distant organs or in the surrounding cirrhotic tissue in any of the cases studied.

In another study, which also employed a suicide/reporter gene system, a small group of patients suffering from recurrent glioblastomas were treated with suicide gene therapy (HSV1-tk) [214]. The construct was delivered with a cationic liposomal vector and among the five patients treated with this therapy, one showed accumulation of the reporter probe [124I]FIAU in a portion of the tumor. Following a 2 week course of ganciclovir, FDG PET and methionine (MET) PET scans at the end of treatment indicated that necrosis occurred in the region where [124I]FIAU uptake was previously shown, suggesting successful suicide gene transfer to a portion of the tumor. Those patients who could not accumulate detectable levels of [124I]FIAU in their infused tumors were found to have a low mitotic index in their tumors.

With this small yet novel study, we are able to learn significant amounts about human gene therapy using *in vivo* imaging methods. For example, the investigators were able to identify and localize successful gene transfer in one patient

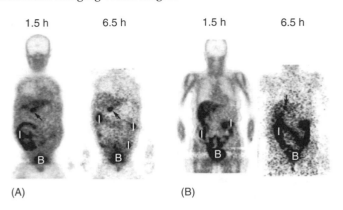

1.5 h 6.5 h 1.5 h 6.5 h

(A) (B)

FIGURE 30.19 [18F]FHBG-PET whole-body imaging in patients with Ad-CMV-HSV1-tk adenovirus shows that the expression of the transgene is circumscribed to the injection site and also that higher vector titers do not necessarily produce higher transgene expression. Whole-body images show that specific accumulation of the tracer (and hence HSV1-tk expression) is restricted to the vector injection site in the liver tumoral nodule (arrows). At later times, the tracer accumulates in the bladder (B) and the intestines (I) due to physiological elimination. In addition, this figure shows that higher vector titers do not necessarily produce higher transgene expression in the tumor as evidenced by [18F]FHBG accumulation. Coronal [18F]FHBG-PET images obtained from patients treated with 10^12 vp (panel a) and with 2 × 10^12 vp (panel b) starting 1.5 h and 6.5 h after injection of the tracer. For the patient treated with the lower adenoviral vector dose, specific [18F]FHBG accumulation in the treated lesion can readily be ascertained in the early images while it can only be seen 6.5 h p.i. for the patient treated with the higher dose. (From Penuelas, I., Haberkorn, U., Yaghoubi, S., and Gambhir, S.S., *Eur. J. Nucl. Med. Mol. Imaging*, 32(Suppl. 2), S384, 2005. With permission.)

68 h after tumor infusion. Also, in the patient where [124I]FIAU uptake was seen, only a fraction of the tumor was sequestering [124I]FIAU. This suggests that the rest of the tumor was either not transduced or that the transduction was too weak to be detected. Additional follow-up scans of the patient would help to determine the outcome of the heterogeneously treated tumor and help determine importance of reporter probe uptake. In the patients who did not demonstrate [124I]FIAU uptake, the failure to see [124I]FIAU reporter probe accumulation may be due to sheer lack of transducible cells or failure of gene transfer. Future studies addressing these and other issues are sure to be extremely helpful in understanding the effectiveness of gene therapy.

30.8 DECISION ALGORITHM FOR REPORTER GENE TECHNOLOGY

The road to clinical gene therapy trials is an arduous one but can be helped with the variety of reporter gene technologies described in this chapter. As with any drug testing, there is a prerequisite period of preclinical testing, much of which takes place in animal models. As expected, there will be a strong inclination toward optical and radionuclide approaches since these approaches are the most accessible and more rigorously tested among the imaging options. Furthermore, it is likely that a single investigator or a single group of investigators working on a specific gene therapy will most likely have to utilize a combination of modalities (e.g., optical- and radionuclide-based technologies) to efficiently test gene therapy vectors and to facilitate their use in higher organisms. As testing progresses in higher species, efforts need to be made toward the radionuclide approaches and substantial investment of time and money may be needed to develop PET reporter probes. Eventually, human trials will employ either PET, gamma/SPECT and, in the future, MR-based methods. The decision and timing to use certain reporter systems over others is a relatively complex one, and we will attempt to simplify the decision making process by providing the set of guidelines we use to make such decisions.

The first decision depends upon whether you are attempting to study the biodistribution/pharmacokinetics of a gene delivery vehicle or whether you are interested in monitoring gene expression. From an earlier discussion (Section 30.4), the distribution of a vehicle can be directly imaged by directly labeling the vehicle with radioisotope-, fluorescent- or MR-compatible markers. Distribution of injected vehicle can also be inferred from a reporter gene coupled to a constitutive promoter. However, this method is less sensitive since the rate of gene transfer is always less than 100%, and, in fact, is a minority in most cases.

For those interested in monitoring transgene expression, a search for reporter probes that may already exist for the transgene is mandatory. For example, for those who use the suicide gene, HSV1-tk, a number of reporter probes already exist and direct imaging protocols can be performed using radionuclide techniques. In the example of HSV1-tk, the decision to use [18F]FEAU over other probes such as [18F]FHBG is based on preliminary favorable evidence toward [18F]FEAU described earlier. However, the selection of the appropriate reporter probe is not trivial, and, as further tests are being performed, it is likely that the other reporter probes may prove to be just as or more efficient than FEAU depending on specific cell type, local pharmacokinetics, mode of delivery of the transgene (viral vs. nonviral), etc. As we learn more about each reporter system, selection of specific transgene–reporter probe combinations will be disease-specific; it is likely that each reporter combination will have certain strengths and weaknesses depending on the disease model and target tissues. The investigators will have to adjust accordingly. Future studies are bound to address these and other related issues.

If no reporter probe for the transgene exists, then indirect imaging methods are needed. This requires the coupling of the therapeutic genes with an optical, radionuclide-based or MR-based reporter genes. As described earlier, a number of indirect methods are available including the dual promoter approach, bidirectional approach, bicistronic approach, etc. The selection of one of these promoter configurations will largely depend upon the nature of the promoter of the therapeutic gene. If a robust, constitutive promoter drives the therapeutic gene, then the dual promoter, the bicistronic (IRES-mediated) or bidirectional approaches may suffice. However, if a weak or tissue-specific promoter drives the therapeutic gene, then amplification strategies for the reporter gene (and possibly the therapeutic gene) will need to be employed. Amplification strategies for tissue-specific promoters using the VP16 transactivation domain fused to the yeast GAL4 DNA-binding domain has been verified for use with reporter genes [215]. A recently developed Gal4-VP16-activated bidirectional promoter is capable of linking the amplified expression of a therapeutic gene from a weak tissue-specific promoter to that of a reporter gene, thereby facilitating noninvasive imaging of transgene expression in gene therapy applications [216].

A variety of other factors also dictate the selection of reporter gene systems. For example, if a research group is intent on bringing a specific gene therapy to human application, it requires them to use multimodality reporter gene systems, i.e., use fused reporter genes such as HSV1-tk-*Fluc* (keeping in mind that HSV1-tk is not being used as a therapeutic gene in this case when the reporter probe is administered in nonpharmacologic amounts). By coupling a therapeutic gene to a fused reporter gene, it will be possible to move quickly between the preclinical verification phase (optical imaging of FL) and the clinical phase (radionuclide-based imaging of HSV1-tk). Alternatively, if a research group strictly deals with small animals, then the use of optical reporter genes only may suffice. Along this line of reasoning, the use of large animals precludes the use of optical methods, and as a result, radionuclide-based or MR-based technologies become more important.

Imaging disease processes in the CNS also limits the selection of reporter genes. Some of the radionuclide-based technologies, such as HSV1-tk and its mutants, are limited since their reporter probes do not readily cross the BBB. On the other hand, [18F]FESP, the reporter probe for D2R, easily crosses the BBB and can be used to a certain extent in the CNS. Use of the D2R-[18F]FESP system in the CNS is limited in the striatum where endogenous dopamine receptors are present. Optical methods, both fluorescence and bioluminescence, can be utilized if specific spatial localization is not needed. The substrates of the bioluminescence methods, D-luciferin and coelenterazine, can easily cross the BBB thereby facilitating the use of these reporter genes in the CNS.

In the same token, imaging in the lung, connective tissue or cortical bone, may limit the use of superparamagnetic-labeled agents in MR imaging since it will be difficult to differentiate between signal from the contrast agent and the signal of these anatomic structures since they are identical in certain MR sequences.

Other important factors in the selection of reporter genes include the need for good spatial resolution (preference given to radionuclide and MRI-based imaging), repetitive imaging (preference given to radionuclide approaches), and image quantitation (preference given to radionuclide, optical, and MRS-based methods). Cost, institutional infrastructure, requirement for support personnel, and physical space required can also be significant factors, which favorably inclined toward the optical and gamma camera methods and less toward the PET and MR-based methods. Thus, a large number of factors have to be considered prior to the selection and implementation of reporter genes in living subjects. With careful planning, an optimal imaging strategy can be followed and can be enormously helpful in the study of gene therapy and disease entities.

30.9 SUMMARY AND FUTURE DIRECTIONS

A wide array of tools are becoming available for the evaluation of gene therapy in living subjects. These tools will be extremely helpful toward the advancement of human gene therapy by providing a means of continuous monitoring of locations, magnitude, and temporal variation of gene delivery and expression. Viral and nonviral transfer of genetic material is bound to improve and these developments will be supported by direct and indirect reporter gene technologies. The ability to monitor gene expression through a variety of bicistronic vectors and control gene expression, either through inducible promoters or through tissue-specific amplification techniques, is likely to play an increasing role in gene therapy as we begin to move away from constitutive expression in some cases.

In the next decade, PET/CT and SPECT/CT will probably be the major workhorses for human gene therapy trials. PET-based technologies particularly have an advantage given their greater sensitivity as well as the ability to use biological molecules that nearly mimic their parent molecule after being radiolabeled. Future technical improvements are certain to aid optical and MR-based protocols for human application. Continued alliances between gene therapists, molecular biologists, engineers, chemists, physicists, and pharmacologists will help build the next generation of molecular imaging technologies to expand its capabilities in gene and cell therapy.

REFERENCES

1. Westphal H, Overbeek PA, Khillan JS et al. 1985. Promoter sequences of murine alpha A crystallin, murine alpha 2(I) collagen or of avian sarcoma virus genes linked to the bacterial chloramphenicol acetyl transferase gene direct tissue-specific patterns of chloramphenicol acetyl transferase expression in transgenic mice. *Cold Spring Harb Symp Quant Biol* 50: 411–416.
2. Zhou D, Zhou C, and Chen S. 1997. Gene regulation studies of aromatase expression in breast cancer and adipose stromal cells. *J Steroid Biochem Mol Biol* 61: 273–280.
3. Leite JP, Niel C, and D'Halluin JC. 1986. Expression of the chloramphenicol acetyl transferase gene in human cells under the control of early adenovirus subgroup C promoters: Effect of E1A gene products from other subgroups on gene expression. *Gene* 41: 207–215.

4. Naciff JM, Behbehani MM, Misawa H, and Dedman JR. 1999. Identification and transgenic analysis of a murine promoter that targets cholinergic neuron expression. *J Neurochem* 72: 17–28.
5. Forss-Petter S, Danielson PE, Catsicas S et al. 1990. Transgenic mice expressing beta-galactosidase in mature neurons under neuron-specific enolase promoter control. *Neuron* 5: 187–197.
6. Lee JH, Federoff HJ, and Schoeniger LO. 1999. G207, modified herpes simplex virus type 1, kills human pancreatic cancer cells in vitro. *J Gastrointest Surg* 3: 127–131; discussion 132–133.
7. Sauvonnet N and Pugsley AP. 1996. Identification of two regions of *Klebsiella oxytoca* pullulanase that together are capable of promoting beta-lactamase secretion by the general secretory pathway. *Mol Microbiol* 22: 1–7.
8. Zlokarnik G, Negulescu PA, Knapp TE et al. 1998. Quantitation of transcription and clonal selection of single living cells with beta-lactamase as reporter. *Science* 279: 84–88.
9. Chalfie M, Tu Y, Euskirchen G, Ward WW, and Prasher DC. 1994. Green fluorescent protein as a marker for gene expression. *Science* 263: 802–805.
10. Naylor LH. 1999. Reporter gene technology: The future looks bright. *Biochem Pharmacol* 58: 749–757.
11. Leffel SM, Mabon SA, and Stewart CN, Jr. 1997. Applications of green fluorescent protein in plants. *Biotechniques* 23: 912–918.
12. Karp M, Akerman K, Lindqvist C, Kuusisto A, Saviranta P, and Oker-Blom C. 1992. A sensitive model system for in vivo monitoring of baculovirus gene expression in single infected insect cells. *Biotechnology* 10: 565–569.
13. Gnant MF, Noll LA, Irvine KR et al. 1999. Tumor-specific gene delivery using recombinant vaccinia virus in a rabbit model of liver metastases. *J Natl Cancer Inst* 91: 1744–1750.
14. Lorenz WW, Cormier MJ, O'Kane DJ, Hua D, Escher AA, and Szalay AA. 1996. Expression of the *Renilla reniformis* luciferase gene in mammalian cells. *J Biolumin Chemilumin* 11: 31–37.
15. Bhaumik S and Gambhir SS. 2002. Optical imaging of Renilla luciferase reporter gene expression in living mice. *Proc Natl Acad Sci USA* 99: 377–382.
16. Contag CH, Contag PR, Mullins JI, Spilman SD, Stevenson DK, and Benaron DA. 1995. Photonic detection of bacterial pathogens in living hosts. *Mol Microbiol* 18: 593–603.
17. Contag CH, Jenkins D, Contag PR, and Negrin RS. 2000. Use of reporter genes for optical measurements of neoplastic disease in vivo. *Neoplasia* 2: 41–52.
18. Yang M, Baranov E, Moossa AR, Penman S, and Hoffman RM. 2000. Visualizing gene expression by whole-body fluorescence imaging. *Proc Natl Acad Sci USA* 97: 12278–12282.
19. Tung CH, Bredow S, Mahmood U, and Weissleder R. 1999. Preparation of a cathepsin D sensitive near-infrared fluorescence probe for imaging. *Bioconjug Chem* 10: 892–896.
20. Contag CH and Ross BD. 2002. It's not just about anatomy: In vivo bioluminescence imaging as an eyepiece into biology. *J Magn Reson Imaging* 16: 378–387.
21. Tannous BA, Kim DE, Fernandez JL, Weissleder R, and Breakefield XO. 2005. Codon-optimized Gaussia luciferase cDNA for mammalian gene expression in culture and in vivo. *Mol Ther* 11: 435–443.
22. Contag CH and Bachmann MH. 2002. Advances in in vivo bioluminescence imaging of gene expression. *Annu Rev Biomed Eng* 4: 235–260.
23. Eisenstein M. 2006. Helping cells to tell a colorful tale. *Nature Methods* 3: 647–655.

24. Ntziachristos V, Tung CH, Bremer C, and Weissleder R. 2002. Fluorescence molecular tomography resolves protease activity in vivo. *Nat Med* 8: 757–760.

25. Graves EE, Ripoll J, Weissleder R, and Ntziachristos V. 2003. A submillimeter resolution fluorescence molecular imaging system for small animal imaging. *Med Phys* 30: 901–911.

26. http://www.visenmedical.com/technologies/fmt.html.

27. Keren S, Gheysens O, Levin CS, and Gambhir SS. 2007. A comparison between a time domain and continuous wave small animal optical imaging system. *IEEE Trans. Med. Imaging* 27: 58–63.

28. Long W and Vernon M. 2004. http://www.art.ca/en/products/INOPaper040129.pdf.

29. Elson D, Requejo-Isidro J, Munro I et al. 2004. Time-domain fluorescence lifetime imaging applied to biological tissue. *Photochem Photobiol Sci* 3: 795–801.

30. Hawrysz DJ and Sevick-Muraca EM. 2000. Developments toward diagnostic breast cancer imaging using near-infrared optical measurements and fluorescent contrast agents. *Neoplasia* 2: 388–417.

31. Houston JP, Thompson AB, Gurfinkel M, and Sevick-Muraca EM. 2003. Sensitivity and depth penetration of continuous wave versus frequency-domain photon migration near-infrared fluorescence contrast-enhanced imaging. *Photochem Photobiol* 77: 420–430.

32. Avril N and Propper D. 2007. Functional PET imaging in cancer drug development. *Future Oncol* 3: 215–228.

33. Weber WA, Schwaiger M, and Avril N. 2000. Quantitative assessment of tumor metabolism using FDG-PET imaging. *Nucl Med Biol* 27: 683–687.

34. Israel O and Kuten A. 2007. Early detection of cancer recurrence: 18F-FDG PET/CT can make a difference in diagnosis and patient care. *J Nucl Med* 48 Suppl 1: 28S–35S.

35. Phelps ME, Hoffman EJ, Mullani NA, and Ter-Pogossian MM. 1975. Application of annihilation coincidence detection to transaxial reconstruction tomography. *J Nucl Med* 16: 210–224.

36. Strijckmans K. 2001. The isochronous cyclotron: Principles and recent developments. *Comput Med Imaging Graph* 25: 69–78.

37. Chatziioannou AF. 2002. Molecular imaging of small animals with dedicated PET tomographs. *Eur J Nucl Med Mol Imaging* 29: 98–114.

38. Chatziioannou A, Tai YC, Doshi N, and Cherry SR. 2001. Detector development for microPET II: A 1 microl resolution PET scanner for small animal imaging. *Phys Med Biol* 46: 2899–2910.

39. McVeigh ER. 2006. Emerging imaging techniques. *Circ Res* 98: 879–886.

40. Sossi V and Ruth TJ. 2005. Micropet imaging: In vivo biochemistry in small animals. *J Neural Transm* 112: 319–330.

41. Weber S and Bauer A. 2004. Small animal PET: Aspects of performance assessment. *Eur J Nucl Med Mol Imaging* 31: 1545–1555.

42. Louie AY, Duimstra JA, and Meade TJ. 2002. Mapping gene expression by MRI. In: Toga AW, Mazziotta JC (eds.) *Brain Mapping: The Methods*. Elsevier Science, Amsterdam, pp. 819–828.

43. Rosenthal MS, Cullom J, Hawkins W, Moore SC, Tsui BM, and Yester M. 1995. Quantitative SPECT imaging: A review and recommendations by the Focus Committee of the Society of Nuclear Medicine Computer and Instrumentation Council. *J Nucl Med* 36: 1489–1513.

44. Beekman F and van der Have F. 2007. The pinhole: Gateway to ultra-high-resolution three-dimensional radionuclide imaging. *Eur J Nucl Med Mol Imaging* 34: 151–161.

45. Funk T, Despres P, Barber WC, Shah KS, and Hasegawa BH. 2006. A multipinhole small animal SPECT system with submillimeter spatial resolution. *Med Phys* 33: 1259–1268.

46. Hashemi RH and Bradley WG. 1997. *MRI: The Basics*. Williams & Wilkins, Baltimore, MD.

47. Allport JR and Weissleder R. 2001. In vivo imaging of gene and cell therapies. *Exp Hematol* 29: 1237–1246.

48. Storey P. 2006. Introduction to magnetic resonance imaging and spectroscopy. *Methods Mol Med* 124: 3–57.

49. Gibby WA. 2005. Basic principles of magnetic resonance imaging. *Neurosurg Clin N Am* 16: 1–64.

50. Massoud TF and Gambhir SS. 2003. Molecular imaging in living subjects: Seeing fundamental biological processes in a new light. *Genes Dev* 17: 545–580.

51. Oros AM and Shah NJ. 2004. Hyperpolarized xenon in NMR and MRI. *Phys Med Biol* 49: R105–R153.

52. Hopkins SR, Levin DL, Emami K et al. 2007. Advances in magnetic resonance imaging of lung physiology. *J Appl Physiol* 102: 1244–1254.

53. Fain SB, Korosec FR, Holmes JH, O'Halloran R, Sorkness RL, and Grist TM. 2007. Functional lung imaging using hyperpolarized gas MRI. *J Magn Reson Imaging* 25: 910–923.

54. Kauczor HU. 2003. Hyperpolarized helium-3 gas magnetic resonance imaging of the lung. *Top Magn Reson Imaging* 14: 223–230.

55. Salerno M, Altes TA, Mugler JP, 3rd, Nakatsu M, Hatabu H, and de Lange EE. 2001. Hyperpolarized noble gas MR imaging of the lung: Potential clinical applications. *Eur J Radiol* 40: 33–44.

56. Hoffman EA and van Beek E. 2006. Hyperpolarized media MR imaging—expanding the boundaries? *Acad Radiol* 13: 929–931.

57. van Beek EJ, Hill C, Woodhouse N et al. 2007. Assessment of lung disease in children with cystic fibrosis using hyperpolarized 3-Helium MRI: Comparison with Shwachman score, Chrispin-Norman score and spirometry. *Eur Radiol* 17: 1018–1024.

58. Mansson S, Johansson E, Magnusson P et al. 2006.^{13}C imaging—a new diagnostic platform. *Eur Radiol* 16: 57–67.

59. Olsson LE, Chai CM, Axelsson O, Karlsson M, Golman K, and Petersson JS. 2006. MR coronary angiography in pigs with intraarterial injections of a hyperpolarized ^{13}C substance. *Magn Reson Med* 55: 731–737.

60. Johansson E, Mansson S, Wirestam R et al. 2004. Cerebral perfusion assessment by bolus tracking using hyperpolarized ^{13}C. *Magn Reson Med* 51: 464–472.

61. Golman K and Petersson JS. 2006. Metabolic imaging and other applications of hyperpolarized ^{13}C1. *Acad Radiol* 13: 932–942.

62. Paulus MJ, Gleason SS, Kennel SJ, Hunsicker PR, and Johnson DK. 2000. High resolution X-ray computed tomography: An emerging tool for small animal cancer research. *Neoplasia* 2: 62–70.

63. Gambhir SS. 2002. Molecular imaging of cancer with positron emission tomography. *Nat Rev Cancer* 2: 683–693.

64. Rabin O, Manuel Perez J, Grimm J, Wojtkiewicz G, and Weissleder R. 2006. An X-ray computed tomography imaging agent based on long-circulating bismuth sulphide nanoparticles. *Nat Mater* 5: 118–122.

65. Lecchi M, Ottobrini L, Martelli C, Del Sole A, and Lucignani G. 2007. Instrumentation and probes for molecular and cellular imaging. *Q J Nucl Med Mol Imaging* 51: 111–126.

66. Iwanaga K, Tominaga K, Yamamoto K et al. 2007. Local delivery system of cytotoxic agents to tumors by focused sonoporation. *Cancer Gene Ther* 14: 354–363.

67. Bednarski MD, Lee JW, Callstrom MR, and Li KC. 1997. In vivo target-specific delivery of macromolecular agents with MR-guided focused ultrasound. *Radiology* 204: 263–268.

68. Bao S, Thrall BD, Gies RA, and Miller DL. 1998. In vivo transfection of melanoma cells by lithotripter shock waves. *Cancer Res* 58: 219–221.

69. Newman CM and Bettinger T. 2007. Gene therapy progress and prospects: Ultrasound for gene transfer. *Gene Ther* 14: 465–475.

70. Xenariou S, Griesenbach U, Liang HD et al. 2007. Use of ultrasound to enhance nonviral lung gene transfer in vivo. *Gene Ther* 14: 768–774.

71. Lawrie A, Brisken AF, Francis SE, Cumberland DC, Crossman DC, and Newman CM. 2000. Microbubble-enhanced ultrasound for vascular gene delivery. *Gene Ther* 7: 2023–2027.

72. Lu QL, Liang HD, Partridge T, and Blomley MJ. 2003. Microbubble ultrasound improves the efficiency of gene transduction in skeletal muscle in vivo with reduced tissue damage. *Gene Ther* 10: 396–405.

73. Duvshani-Eshet M and Machluf M. 2007. Efficient transfection of tumors facilitated by long-term therapeutic ultrasound in combination with contrast agent: From in vitro to in vivo setting. *Cancer Gene Ther* 14: 306–315.

74. Price RJ and Kaul S. 2002. Contrast ultrasound targeted drug and gene delivery: An update on a new therapeutic modality. *J Cardiovasc Pharmacol Ther* 7: 171–180.

75. Dayton PA and Ferrara KW. 2002. Targeted imaging using ultrasound. *J Magn Reson Imaging* 16: 362–377.

76. Lien YH and Lai LW. 2004. Gene therapy for renal disorders. *Expert Opin Biol Ther* 4: 919–926.

77. Willmann J, Paulmurugan R, Chen K, Gheysens O, Rodriguez-Porcel M, Lutz AM, Chen IY, Chen X, and Gambhir SS. 2007. Ultrasonic imaging of tumor angiogenesis with contrast microbubbles targeted to vascular endothelial growth factor receptor type 2 in mice. *Radiology* 246: 508–518.

78. Contag CH, Spilman SD, Contag PR et al. 1997. Visualizing gene expression in living mammals using a bioluminescent reporter. *Photochem Photobiol* 66: 523–531.

79. Contag PR, Olomu IN, Stevenson DK, and Contag CH. 1998. Bioluminescent indicators in living mammals. *Nat Med* 4: 245–247.

80. Tsien RY. 1998. The green fluorescent protein. *Annu Rev Biochem* 67: 509–544.

81. Welsh S and Kay SA. 1997. Reporter gene expression for monitoring gene transfer. *Curr Opin Biotechnol* 8: 617–622.

82. Pfeifer A, Kessler T, Yang M et al. 2001. Transduction of liver cells by lentiviral vectors: Analysis in living animals by fluorescence imaging. *Mol Ther* 3: 319–322.

83. Raikwar SP, Temm CJ, Raikwar NS, Kao C, Molitoris BA, and Gardner TA. 2005. Adenoviral vectors expressing human endostatin–angiostatin and soluble Tie2: Enhanced suppression of tumor growth and antiangiogenic effects in a prostate tumor model. *Mol Ther* 12: 1091–1100.

84. Oh P, Borgstrom P, Witkiewicz H et al. 2007. Live dynamic imaging of caveolae pumping targeted antibody rapidly and specifically across endothelium in the lung. *Nat Biotechnol* 25: 327–337.

85. Cubitt AB, Heim R, Adams SR, Boyd AE, Gross LA, and Tsien RY. 1995. Understanding, improving and using green fluorescent proteins. *Trends Biochem Sci* 20: 448–455.

86. Prasher DC. 1995. Using GFP to see the light. *Trends Genet* 11: 320–323.

87. Baird GS, Zacharias DA, and Tsien RY. 2000. Biochemistry, mutagenesis, and oligomerization of DsRed, a red fluorescent protein from coral. *Proc Natl Acad Sci USA* 97: 11984–11989.

88. Campbell RE, Tour O, Palmer AE et al. 2002. A monomeric red fluorescent protein. *Proc Natl Acad Sci USA* 99: 7877–7882.

89. Wang L, Jackson WC, Steinbach PA, and Tsien RY. 2004. Evolution of new nonantibody proteins via iterative somatic hypermutation. *Proc Natl Acad Sci USA* 101: 16745–16749.

90. Liu Y, Miyoshi H, and Nakamura M. 2007. Nanomedicine for drug delivery and imaging: A promising avenue for cancer therapy and diagnosis using targeted functional nanoparticles. *Int J Cancer* 120: 2527–2537.

91. Rao J, Dragulescu-Andrasi A, and Yao H. 2007. Fluorescence imaging in vivo: Recent advances. *Curr Opin Biotechnol* 18: 17–25.

92. Ballou B, Ernst LA, and Waggoner AS. 2005. Fluorescence imaging of tumors in vivo. *Curr Med Chem* 12: 795–805.

93. Michalet X, Pinaud FF, Bentolila LA et al. 2005. Quantum dots for live cells, in vivo imaging, and diagnostics. *Science* 307: 538–544.

94. Medintz IL, Uyeda HT, Goldman ER, and Mattoussi H. 2005. Quantum dot bioconjugates for imaging, labelling and sensing. *Nat Mater* 4: 435–446.

95. Weissleder R, Tung CH, Mahmood U, and Bogdanov A, Jr. 1999. In vivo imaging of tumors with protease-activated near-infrared fluorescent probes. *Nat Biotechnol* 17: 375–378.

96. Bremer C, Bredow S, Mahmood U, Weissleder R, and Tung CH. 2001. Optical imaging of matrix metalloproteinase-2 activity in tumors: Feasibility study in a mouse model. *Radiology* 221: 523–529.

97. Sherf B and Wood K. 1994. Firefly luciferase engineered for improved genetic reporting. *Promega Notes* 49: 14–21.

98. Ray P, Tsien R, and Gambhir SS. 2007. Construction and validation of improved triple fusion reporter gene vectors for molecular imaging of living subjects. *Cancer Res* 67: 3085–3093.

99. White PJ, Squirrell DJ, Arnaud P, Lowe CR, and Murray JA. 1996. Improved thermostability of the North American firefly luciferase: Saturation mutagenesis at position 354. *Biochem J* 319 (Pt 2): 343–350.

100. Tisi L, White PJ, Squirrell DJ, Murphy MJ, Lowe CR, and Murray JAH. 2002. Development of a thermostable firefly luciferase. *Anal Chim Acta* 457: 115–123.

101. Loening AM, Fenn TD, Wu AM, and Gambhir SS. 2006. Consensus guided mutagenesis of Renilla luciferase yields enhanced stability and light output. *Protein Eng Des Sel* 19: 391–400.

102. Loening AM, Wu AM, and Gambhir SS. 2007. Red-shifted Renilla reniformis luciferase variants for imaging in living subjects. *Nat Methods* 4: 641–643.

103. Laxman B, Hall DE, Bhojani MS et al. 2002. Noninvasive real-time imaging of apoptosis. *Proc Natl Acad Sci USA* 99: 16551–16555.

104. Morgan NY, English S, Chen W et al. 2005. Real time in vivo non-invasive optical imaging using near-infrared fluorescent quantum dots. *Acad Radiol* 12: 313–323.

105. Gambhir SS, Herschman HR, Cherry SR et al. 2000. Imaging transgene expression with radionuclide imaging technologies. *Neoplasia* 2: 118–138.

106. Yu Y, Annala AJ, Barrio JR et al. 2000. Quantification of target gene expression by imaging reporter gene expression in living animals. *Nat Med* 6: 933–937.

107. Liang Q, Satyamurthy N, Barrio JR et al. 2001. Noninvasive, quantitative imaging in living animals of a mutant dopamine D2 receptor reporter gene in which ligand binding is uncoupled from signal transduction. *Gene Ther* 8: 1490–1498.

108. MacLaren DC, Gambhir SS, Satyamurthy N et al. 1999. Repetitive, non-invasive imaging of the dopamine D2 receptor as a reporter gene in living animals. *Gene Ther* 6: 785–791.

109. Herschman HR, MacLaren DC, Iyer M et al. 2000. Seeing is believing: Non-invasive, quantitative and repetitive imaging of reporter gene expression in living animals, using positron emission tomography. *J Neurosci Res* 59: 699–705.

110. Buursma AR, Rutgers V, Hospers GA, Mulder NH, Vaalburg W, and de Vries EF. 2006. ^{18}F-FEAU as a radiotracer for herpes simplex virus thymidine kinase gene expression: In vitro comparison with other PET tracers. *Nucl Med Commun* 27: 25–30.

111. Soghomonyan S, Hajitou A, Rangel R et al. 2007. Molecular PET imaging of HSV1-tk reporter gene expression using [^{18}F]FEAU. *Nat Protoc* 2: 416–423.

112. Alauddin MM, Shahinian A, Park R, Tohme M, Fissekis JD, and Conti PS. 2007. In vivo evaluation of 2′-deoxy-2′-[(18)F]fluoro-5-iodo-1-beta-D-arabinofuranosyluracil ([(18)F]FIAU) and 2′-deoxy-2′-[(18)F]fluoro-5-ethyl-1-beta-D-arabinofuranosyluracil ([(18)F]FEAU) as markers for suicide gene expression. *Eur J Nucl Med Mol Imaging* 34: 822–829.

113. Namavari M, Barrio JR, Toyokuni T et al. 2000. Synthesis of 8-[(18)F]fluoroguanine derivatives: In vivo probes for imaging gene expression with positron emission tomography. *Nucl Med Biol* 27: 157–162.

114. Alauddin MM and Conti PS. 1998. Synthesis and preliminary evaluation of 9-(4-[18F]-fluoro-3-hydroxymethylbutyl)guanine ([18F]FHBG): A new potential imaging agent for viral infection and gene therapy using PET. *Nucl Med Biol* 25: 175–180.

115. Alauddin MM, Conti PS, Mazza SM, Hamzeh FM, and Lever JR. 1996. 9-[(3-[18F]-fluoro-1-hydroxy-2-propoxy)methyl]guanine ([18F]-FHPG): A potential imaging agent of viral infection and gene therapy using PET. *Nucl Med Biol* 23: 787–792.

116. Iyer M, Barrio JR, Namavari M et al. 2001. 8-[18F]Fluoropenciclovir: An improved reporter probe for imaging HSV1-tk reporter gene expression in vivo using PET. *J Nucl Med* 42: 96–105.

117. Yaghoubi S, Barrio JR, Dahlbom M et al. 2001. Human pharmacokinetic and dosimetry studies of [(18)F]FHBG: A reporter probe for imaging herpes simplex virus type-1 thymidine kinase reporter gene expression. *J Nucl Med* 42: 1225–1234.

118. Tjuvajev JG, Finn R, Watanabe K et al. 1996. Noninvasive imaging of herpes virus thymidine kinase gene transfer and expression: A potential method for monitoring clinical gene therapy. *Cancer Res* 56: 4087–4095.

119. Alauddin MM, Shahinian A, Park R, Tohme M, Fissekis JD, and Conti PS. 2007. In vivo evaluation of 2′-deoxy-2′-[(18)F]fluoro-5-iodo-1-beta-D-arabinofuranosyluracil ([(18)F]FIAU) and 2′-deoxy-2′-[(18)F]fluoro-5-ethyl-1-beta-D-arabinofuranosyluracil ([(18)F]FEAU) as markers for suicide gene expression. *Eur J Nucl Med Mol Imaging* 34: 822–829.

120. MacLaren DC, Toyokuni T, Cherry SR et al. 2000. PET imaging of transgene expression. *Biol Psychiatry* 48: 337–348.

121. Black ME, Newcomb TG, Wilson HM, and Loeb LA. 1996. Creation of drug-specific herpes simplex virus type 1 thymidine kinase mutants for gene therapy. *Proc Natl Acad Sci USA* 93: 3525–3529.

122. Gambhir SS, Bauer E, Black ME et al. 2000. A mutant herpes simplex virus type 1 thymidine kinase reporter gene shows improved sensitivity for imaging reporter gene expression with positron emission tomography. *Proc Natl Acad Sci USA* 97: 2785–2790.

123. Ponomarev V, Doubrovin M, Shavrin A et al. 2007. A human-derived reporter gene for noninvasive imaging in humans:

124. Chitneni SK, Deroose CM, Balzarini J et al. 2007. Synthesis and preliminary evaluation of ^{18}F- or ^{11}C-labeled bicyclic nucleoside analogues as potential probes for imaging varicella-zoster virus thymidine kinase gene expression using positron emission tomography. *J Med Chem* 50: 1041–1049.

125. Riedel C, Dohan O, De la Vieja A, Ginter CS, and Carrasco N. 2001. Journey of the iodide transporter NIS: From its molecular identification to its clinical role in cancer. *Trends Biochem Sci* 26: 490–496.

126. Dwyer RM, Schatz SM, Bergert ER et al. 2005. A preclinical large animal model of adenovirus-mediated expression of the sodium-iodide symporter for radioiodide imaging and therapy of locally recurrent prostate cancer. *Mol Ther* 12: 835–841.

127. Spitzweg C, Dietz AB, O'Connor MK et al. 2001. *In vivo* sodium iodide symporter gene therapy of prostate cancer. *Gene Ther* 8: 1524–1531.

128. Dwyer RM, Bergert ER, O'Connor MK, Gendler SJ, and Morris JC. 2005. *In vivo* radioiodide imaging and treatment of breast cancer xenografts after MUC1-driven expression of the sodium iodide symporter. *Clin Cancer Res* 11: 1483–1489.

129. Dwyer RM, Bergert ER, O'Connor MK, Gendler SJ, and Morris JC. 2006. Adenovirus-mediated and targeted expression of the sodium-iodide symporter permits *in vivo* radioiodide imaging and therapy of pancreatic tumors. *Hum Gene Ther* 17: 661–668.

130. Dwyer RM, Bergert ER, O'Connor MK, Gendler SJ, and Morris JC. 2006. Sodium iodide symporter-mediated radioiodide imaging and therapy of ovarian tumor xenografts in mice. *Gene Ther* 13: 60–66.

131. Mandell RB, Mandell LZ, and Link CJ, Jr. 1999. Radioisotope concentrator gene therapy using the sodium/iodide symporter gene. *Cancer Res* 59: 661–668.

132. Spitzweg C, O'Connor MK, Bergert ER, Tindall DJ, Young CY, and Morris JC. 2000. Treatment of prostate cancer by radioiodine therapy after tissue-specific expression of the sodium iodide symporter. *Cancer Res* 60: 6526–6530.

133. Goel A, Carlson SK, Classic KL et al. 2007. Radioiodide imaging and radiovirotherapy of multiple myeloma using VSV(Delta51)-NIS, an attenuated vesicular stomatitis virus encoding the sodium iodide symporter gene. *Blood* 110: 2342–2350.

134. Groot-Wassink T, Aboagye EO, Glaser M, Lemoine NR, and Vassaux G. 2002. Adenovirus biodistribution and noninvasive imaging of gene expression in vivo by positron emission tomography using human sodium/iodide symporter as reporter gene. *Hum Gene Ther* 13: 1723–1735.

135. Niu G, Krager KJ, Graham MM, Hichwa RD, and Domann FE. 2005. Noninvasive radiological imaging of pulmonary gene transfer and expression using the human sodium iodide symporter. *Eur J Nucl Med Mol Imaging* 32: 534–540.

136. Groot-Wassink T, Aboagye EO, Wang Y, Lemoine NR, Reader AJ, and Vassaux G. 2004. Quantitative imaging of Na/I symporter transgene expression using positron emission tomography in the living animal. *Mol Ther* 9: 436–442.

137. Dingli D, Kemp BJ, O'Connor MK, Morris JC, Russell SJ, and Lowe VJ. 2006. Combined I-124 positron emission tomography/computed tomography imaging of NIS gene expression in animal models of stably transfected and intravenously transfected tumor. *Mol Imaging Biol* 8: 16–23.

138. Haberkorn U, Oberdorfer F, Gebert J et al. 1996. Monitoring gene therapy with cytosine deaminase: In vitro studies using tritiated-5-fluorocytosine. *J Nucl Med* 37: 87–94.

Mitochondrial thymidine kinase type 2. *J Nucl Med* 48: 819–826.

139. Yazawa K, Fisher WE, and Brunicardi FC. 2002. Current progress in suicide gene therapy for cancer. *World J Surg* 26: 783–789.

140. Yaghoubi SS, Wu L, Liang Q et al. 2001. Direct correlation between positron emission tomographic images of two reporter genes delivered by two distinct adenoviral vectors. *Gene Ther* 8: 1072–1080.

141. Lamberts SW, de Herder WW, and Hofland LJ. 2002 Somatostatin analogs in the diagnosis and treatment of cancer. *Trends Endocrinol Metab* 13: 451–457.

142. Patel YC. 1999 Somatostatin and its receptor family. *Front Neuroendocrinol* 20: 157–198.

143. Lahlou H, Guillermet J, Hortala M et al. 2004. Molecular signaling of somatostatin receptors. *Ann N Y Acad Sci* 1014: 121–131.

144. Oberg K. 2004. Future aspects of somatostatin-receptor-mediated therapy. *Neuroendocrinology* 80 Suppl 1: 57–61.

145. Ell PaG SS. 2004. *Nuclear Medicine in Clinical Diagnosis and Treatment.* Churchill Livingstone, New York, p. 1924.

146. Shojamanesh H, Gibril F, Louie A et al. 2002. Prospective study of the antitumor efficacy of long-term octreotide treatment in patients with progressive metastatic gastrinoma. *Cancer* 94: 331–343.

147. Arnold R, Simon B, and Wied M. 2000. Treatment of neuroendocrine GEP tumours with somatostatin analogues: A review. *Digestion* 62 Suppl 1: 84–91.

148. Vallabhajosula S, Moyer BR, Lister-James J et al. 1996. Preclinical evaluation of technetium-99 m-labeled somatostatin receptor-binding peptides. *J Nucl Med* 37: 1016–1022.

149. Breeman WA, de Jong M, Kwekkeboom DJ et al. 2001. Somatostatin receptor-mediated imaging and therapy: Basic science, current knowledge, limitations and future perspectives. *Eur J Nucl Med* 28: 1421–1429.

150. Zinn KR, Chaudhuri TR, Krasnykh VN et al. 2002. Gamma camera dual imaging with a somatostatin receptor and thymidine kinase after gene transfer with a bicistronic adenovirus in mice. *Radiology* 223: 417–425.

151. Rogers BE, Parry JJ, Andrews R, Cordopatis P, Nock BA, and Maina T. 2005. MicroPET imaging of gene transfer with a somatostatin receptor-based reporter gene and (94m) Tc-Demotate 1. *J Nucl Med* 46: 1889–1897.

152. Rogers BE, Zinn KR, Lin CY, Chaudhuri TR, and Buchsbaum DJ. 2002. Targeted radiotherapy with [(90)Y]-SMT 487 in mice bearing human nonsmall cell lung tumor xenografts induced to express human somatostatin receptor subtype 2 with an adenoviral vector. *Cancer* 94: 1298–1305.

153. Gobalakrishnan S and Gambhir SS. 2002. Radionuclide imaging of reporter gene expression. In: Toga AW and Mazziotta JC (eds.). *Brain Mapping: The Methods.* Elsevier Science, Amsterdam, pp. 799–818.

154. Weissleder R, Moore A, Mahmood U et al. 2000. In vivo magnetic resonance imaging of transgene expression. *Nat Med* 6: 351–355.

155. Moore A, Basilion JP, Chiocca EA, and Weissleder R. 1998. Measuring transferrin receptor gene expression by NMR imaging. *Biochim Biophys Acta* 1402: 239–249.

156. Moore A, Josephson L, Bhorade RM, Basilion JP, and Weissleder R. 2001. Human transferrin receptor gene as a marker gene for MR imaging. *Radiology* 221: 244–250.

157. Ichikawa T, Hogemann D, Saeki Y et al. 2002. MRI of transgene expression: Correlation to therapeutic gene expression. *Neoplasia* 4: 523–530.

158. Bowen CV, Zhang X, Saab G, Gareau PJ, and Rutt BK. 2002. Application of the static dephasing regime theory to superparamagnetic iron-oxide loaded cells. *Magn Reson Med* 48: 52–61.

159. Cunningham CH, Arai T, Yang PC, McConnell MV, Pauly JM, and Conolly SM. 2005. Positive contrast magnetic resonance imaging of cells labeled with magnetic nanoparticles. *Magn Reson Med* 53: 999–1005.

160. Bulte JW, Zhang S, van Gelderen P et al. 1999. Neurotransplantation of magnetically labeled oligodendrocyte progenitors: Magnetic resonance tracking of cell migration and myelination. *Proc Natl Acad Sci USA* 96: 15256–15261.

161. Watson DJ, Walton RM, Magnitsky SG, Bulte JW, Poptani H, and Wolfe JH. 2006. Structure-specific patterns of neural stem cell engraftment after transplantation in the adult mouse brain. *Hum Gene Ther* 17: 693–704.

162. Weissleder R, Simonova M, Bogdanova A, Bredow S, Enochs WS, and Bogdanov A, Jr. 1997. MR imaging and scintigraphy of gene expression through melanin induction. *Radiology* 204: 425–429.

163. Alfke H, Stoppler H, Nocken F et al. 2003. In vitro MR imaging of regulated gene expression. *Radiology* 228: 488–492.

164. Gilad AA, Winnard PT, Jr., van Zijl PC, and Bulte JW. 2007. Developing MR reporter genes: Promises and pitfalls. *NMR Biomed* 20: 275–290.

165. Bulte JW, Douglas T, Mann S et al. 1994. Magnetoferritin: Characterization of a novel superparamagnetic MR contrast agent. *J Magn Reson Imaging* 4: 497–505.

166. Bulte JW, Douglas T, Mann S, Vymazal J, Laughlin PG, and Frank JA. 1995. Initial assessment of magnetoferritin biokinetics and proton relaxation enhancement in rats. *Acad Radiol* 2: 871–878.

167. Genove G, DeMarco U, Xu H, Goins WF, and Ahrens ET. 2005. A new transgene reporter for in vivo magnetic resonance imaging. *Nat Med* 11: 450–454.

168. Louie AY, Huber MM, Ahrens ET et al. 2000. In vivo visualization of gene expression using magnetic resonance imaging. *Nat Biotechnol* 18: 321–325.

169. Koretsky AP, Brosnan MJ, Chen LH, Chen JD, and Van Dyke T. 1990. NMR detection of creatine kinase expressed in liver of transgenic mice: Determination of free ADP levels. *Proc Natl Acad Sci USA* 87: 3112–3116.

170. Walter G, Barton ER, and Sweeney HL. 2000. Noninvasive measurement of gene expression in skeletal muscle. *Proc Natl Acad Sci USA* 97: 5151–5155.

171. Stegman LD, Rehemtulla A, Beattie B et al. 1999. Noninvasive quantitation of cytosine deaminase transgene expression in human tumor xenografts with in vivo magnetic resonance spectroscopy. *Proc Natl Acad Sci USA* 96: 9821–9826.

172. de Marco G, Bogdanov A, Marecos E, Moore A, Simonova M, and Weissleder R. 1998. MR imaging of gene delivery to the central nervous system with an artificial vector. *Radiology* 208: 65–71.

173. Min JJ and Gambhir SS. 2004. Gene therapy progress and prospects: noninvasive imaging of gene therapy in living subjects. *Gene Ther* 11: 115–125.

174. Schellingerhout D, Rainov NG, Breakefield XO, and Weissleder R. 2000. Quantitation of HSV mass distribution in a rodent brain tumor model. *Gene Ther* 7: 1648–1655.

175. Zinn KR, Douglas JT, Smyth CA et al. 1998. Imaging and tissue biodistribution of 99mTc-labeled adenovirus knob (serotype 5). *Gene Ther* 5: 798–808.

176. Bogdanov A, Jr., Tung CH, Bredow S, and Weissleder R. 2001. DNA binding chelates for nonviral gene delivery imaging. *Gene Ther* 8: 515–522.

177. Harrington KJ, Mohammadtaghi S, Uster PS et al. 2001. Effective targeting of solid tumors in patients with locally advanced cancers by radiolabeled pegylated liposomes. *Clin Cancer Res* 7: 243–254.

178. Harrison LH, Jr., Schwarzenberger PO, Byrne PS, Marrogi AJ, Kolls JK, and McCarthy KE. 2000. Gene-modified PA1-STK cells home to tumor sites in patients with malignant pleural mesothelioma. *Ann Thorac Surg* 70: 407–411.

179. Colombo FR, Torrente Y, Casati R et al. 2001. Biodistribution studies of 99mTc-labeled myoblasts in a murine model of muscular dystrophy. *Nucl Med Biol* 28: 935–940.

180. Ameri K and Wagner E. 2000. Receptor-mediated gene transfer. In: Templeton NS and Lasic DD (eds.). *Gene Therapy: Therapeutic Mechanisms and Strategies*. Marcel Dekker, New York, pp. 141–164.

181. Tjuvajev JG, Stockhammer G, Desai R et al. 1995. Imaging the expression of transfected genes in vivo. *Cancer Res* 55: 6126–6132.

182. Gambhir SS, Barrio JR, Phelps ME et al. 1999. Imaging adenoviral-directed reporter gene expression in living animals with positron emission tomography. *Proc Natl Acad Sci USA* 96: 2333–2338.

183. Jacobs A, Tjuvajev JG, Dubrovin M et al. 2001. Positron emission tomography-based imaging of transgene expression mediated by replication-conditional, oncolytic herpes simplex virus type 1 mutant vectors in vivo. *Cancer Res* 61: 2983–2995.

184. Politi LS. 2007. MR-based imaging of neural stem cells. *Neuroradiology* 49: 523–534.

185. Herschman HR. 2004. PET reporter genes for noninvasive imaging of gene therapy, cell tracking and transgenic analysis. *Crit Rev Oncol Hematol* 51: 191–204.

186. Krishnan M, Park JM, Cao F et al. 2006. Effects of epigenetic modulation on reporter gene expression: Implications for stem cell imaging. *Faseb J* 20: 106–108.

187. Wu JC, Chen IY, Sundaresan G et al. 2003. Molecular imaging of cardiac cell transplantation in living animals using optical bioluminescence and positron emission tomography. *Circulation* 108: 1302–1305.

188. Cao F, Lin S, Xie X et al. 2006. In vivo visualization of embryonic stem cell survival, proliferation, and migration after cardiac delivery. *Circulation* 113: 1005–1014.

189. Kraitchman DL, Tatsumi M, Gilson WD et al. 2005. Dynamic imaging of allogeneic mesenchymal stem cells trafficking to myocardial infarction. *Circulation* 112: 1451–1461.

190. Shapiro EM, Skrtic S, Sharer K, Hill JM, Dunbar CE, and Koretsky AP. 2004. MRI detection of single particles for cellular imaging. *Proc Natl Acad Sci USA* 101: 10901–10906.

191. Chin BB, Nakamoto Y, Bulte JW, Pittenger MF, Wahl R, and Kraitchman DL. 2003. [111]In oxine labelled mesenchymal stem cell SPECT after intravenous administration in myocardial infarction. *Nucl Med Commun* 24: 1149–1154.

192. Dubey P, Su H, Adonai N et al. 2003. Quantitative imaging of the T cell antitumor response by positron-emission tomography. *Proc Natl Acad Sci USA* 100: 1232–1237.

193. Koehne G, Doubrovin M, Doubrovina E et al. 2003. Serial in vivo imaging of the targeted migration of human HSV-TK-transduced antigen-specific lymphocytes. *Nat Biotechnol* 21: 405–413.

194. Herschman HR. 2004. Noninvasive imaging of reporter gene expression in living subjects. *Adv Cancer Res* 92: 29–80.

195. Smirnov P, Lavergne E, Gazeau F et al. 2006. In vivo cellular imaging of lymphocyte trafficking by MRI: A tumor model approach to cell-based anticancer therapy. *Magn Reson Med* 56: 498–508.

196. Kim D, Hung CF, and Wu TC. 2007. Monitoring the trafficking of adoptively transferred antigen-specific CD8-positive T cells in vivo, using noninvasive luminescence imaging. *Hum Gene Ther* 18: 575–588.

197. Edinger M, Cao YA, Verneris MR, Bachmann MH, Contag CH, and Negrin RS. 2003. Revealing lymphoma growth and the efficacy of immune cell therapies using in vivo bioluminescence imaging. *Blood* 101: 640–648.

198. Ponomarev V, Doubrovin M, Lyddane C et al. 2001. Imaging TCR-dependent NFAT-mediated T-cell activation with positron emission tomography in vivo. *Neoplasia* 3: 480–488.

199. Jang SK, Davies MV, Kaufman RJ, and Wimmer E. 1989. Initiation of protein synthesis by internal entry of ribosomes into the 5′ nontranslated region of encephalomyocarditis virus RNA in vivo. *J Virol* 63: 1651–1660.

200. Jang SK, Krausslich HG, Nicklin MJ, Duke GM, Palmenberg AC, and Wimmer E. 1988. A segment of the 5′ nontranslated region of encephalomyocarditis virus RNA directs internal entry of ribosomes during in vitro translation. *J Virol* 62: 2636–2643.

201. Pelletier J and Sonenberg N. 1989. Internal binding of eucaryotic ribosomes on poliovirus RNA: Translation in HeLa cell extracts. *J Virol* 63: 441–444.

202. Tjuvajev JG, Joshi A, Callegari J et al. 1999. A general approach to the non-invasive imaging of transgenes using cis-linked herpes simplex virus thymidine kinase. *Neoplasia* 1: 315–320.

203. Chappell SA, Edelman GM, and Mauro VP. 2000. A 9-nt segment of a cellular mRNA can function as an internal ribosome entry site (IRES) and when present in linked multiple copies greatly enhances IRES activity. *Proc Natl Acad Sci USA* 97: 1536–1541.

204. Sun X, Annala AJ, Yaghoubi SS et al. 2001. Quantitative imaging of gene induction in living animals. *Gene Ther* 8: 1572–1579.

205. Loimas S, Wahlfors J, and Janne J. 1998. Herpes simplex virus thymidine kinase-green fluorescent protein fusion gene: New tool for gene transfer studies and gene therapy. *Biotechniques* 24: 614–618.

206. Strathdee CA, McLeod MR, and Underhill TM. 2000. Dominant positive and negative selection using luciferase, green fluorescent protein and beta-galactosidase reporter gene fusions. *Biotechniques* 28: 210–214.

207. Ray P, Wu A, and Gambhir SS. 2003. Optical bioluminescence and positron emission tomography (PET) imaging of a novel fusion reporter gene in tumor xenografts of living mice. *Cancer Res* 63: 1160–1165.

208. Hackman T, Doubrovin M, Balatoni J et al. 2002. Imaging expression of cytosine deaminase-herpes virus thymidine kinase fusion gene (CD/TK) expression with [124I]FIAU and PET. *Mol Imaging* 1: 36–42.

209. Ray P, De A, Min JJ, Tsien RY, and Gambhir SS. 2004. Imaging tri-fusion multimodality reporter gene expression in living subjects. *Cancer Res* 64: 1323–1330.

210. Ponomarev V, Doubrovin M, Serganova I et al. 2004. A novel triple-modality reporter gene for whole-body fluorescent, bioluminescent, and nuclear noninvasive imaging. *Eur J Nucl Med Mol Imaging* 31: 740–751.

211. Penuelas I, Haberkorn U, Yaghoubi S, and Gambhir SS. 2005. Gene therapy imaging in patients for oncological applications. *Eur J Nucl Med Mol Imaging* 32(Suppl 2): S384–S403.

212. Yaghoubi SS, Barrio JR, Namavari M et al. 2005. Imaging progress of herpes simplex virus type 1 thymidine kinase suicide gene therapy in living subjects with positron emission tomography. *Cancer Gene Ther* 12: 329–339.

213. Penuelas I, Mazzolini G, Boan JF et al. 2005. Positron emission tomography imaging of adenoviral-mediated transgene expression in liver cancer patients. *Gastroenterology* 128: 1787–1795.

214. Jacobs A, Voges J, Reszka R et al. 2001. Positron-emission tomography of vector-mediated gene expression in gene therapy for gliomas. *Lancet* 358: 727–729.

215. Iyer M, Wu L, Carey M, Wang Y, Smallwood A, and Gambhir SS. 2001. Two-step transcriptional amplification as a method for imaging reporter gene expression using weak promoters. *Proc Natl Acad Sci USA* 98: 14595–14600.

216. Ray S, Paulmurugan R, Hildebrandt I et al. 2004. Novel bidirectional vector strategy for amplification of therapeutic and reporter gene expression. *Hum Gene Ther* 15: 681–690.

217. Hirt H. 1991. A novel method for in situ screening of yeast colonies with the beta-glucuronidase reporter gene. *Curr Genet* 20: 437–439.

218. Handler AM and Harrell RA, 2nd. 2001. Polyubiquitin-regulated DsRed marker for transgenic insects. *Biotechniques* 31: 820, 824–828.

219. Gambhir SS, Barrio JR, Herschman HR, and Phelps ME. 1999. Assays for noninvasive imaging of reporter gene expression. *Nucl Med Biol* 26: 481–490.

220. Bankiewicz KS, Eberling JL, Kohutnicka M et al. 2000. Convection-enhanced delivery of AAV vector in parkinsonian monkeys; in vivo detection of gene expression and restoration of dopaminergic function using pro-drug approach. *Exp Neurol* 164: 2–14.

221. Bengel FM, Anton M, Avril N et al. 2000. Uptake of radiolabeled 2′-fluoro-2′-deoxy-5-iodo-1-beta-D-arabinofuranosyluracil in cardiac cells after adenoviral transfer of the herpesvirus thymidine kinase gene: The cellular basis for cardiac gene imaging. *Circulation* 102: 948–950.

222. Rogers BE, Curiel DT, Mayo MS, Laffoon KK, Bright SJ, and Buchsbaum DJ. 1997. Tumor localization of a radiolabeled bombesin analogue in mice bearing human ovarian tumors induced to express the gastrin-releasing peptide receptor by an adenoviral vector. *Cancer* 80: 2419–2424.

223. Wu JC, Inubushi M, Sundaresan G, Schelbert HR, and Gambhir SS. 2002. Optical imaging of cardiac reporter gene expression in living rats. *Circulation* 105: 1631–1634.

224. Adams JY, Johnson M, Sato M et al. 2002. Visualization of advanced human prostate cancer lesions in living mice by a targeted gene transfer vector and optical imaging. *Nat Med* 8: 891–897.

225. Wu JC, Inubushi M, Sundaresan G, Schelbert HR, and Gambhir SS. 2002. Positron emission tomography imaging of cardiac reporter gene expression in living rats. *Circulation* 106: 180–183.

226. Hildebrandt IJ, Iyer M, Wagner E, and Gambhir SS. Optical imaging of transferrin targeted PEI/DNA complexes in living subjects. *Gene Ther* 10: 758–764.

227. Iyer M, Berenji M, Templeton NS, and Gambhir SS. 2002. Noninvasive imaging of cationic lipid-mediated delivery of optical and PET reporter genes in living mice. *Mol Ther* 6: 555–562.

31 Nonviral Genome Modification Strategies for Gene Therapy: Transposon, Integrase, and Homologous Recombination Systems

Lauren E. Woodard and Michele P. Calos

CONTENTS

31.1 INTRODUCTION

31.1.1 GENE THERAPY

Gene therapy involves a new class of pharmaceuticals that uses a nucleic acid drug to treat a medical condition. By introducing a corrected copy of a defective gene or disrupting the production of a harmful protein, gene therapy treats a disease at its source. The ideal gene therapy varies depending on the indication, but in every case, it should be nontoxic, nonimmunogenic, safe, and free of detrimental side effects. For DNA vaccines or some cancer gene therapy approaches, short-term gene expression may be adequate to control dosing and length of expression of the gene product. By contrast, for recessive monogenic diseases, the therapy should give persistent and sufficient levels of expression of the missing gene product for the life of the patient. Dominant negative diseases could also be cured in theory by disrupting production of the disease-causing protein. Cancer is a somatic genetic disease, and gene therapy treatments for cancer may include either long-term production of a secreted protein drug in normal cells or short-term intracellular manufacture of proteins that cause cell death in cancer cells. We argue that many genetic diseases have the potential to be addressed clinically with nonviral, integrating systems.

31.1.2 LIMITATIONS OF VIRAL VECTORS

Viral systems were the first to be developed for gene therapy and since that time have been tested extensively in clinical trials. Viruses are naturally adapted for cellular entry and transgene expression. However, the human body has evolved under selection to implement complicated and extensive antiviral defenses that make genetically engineered virus-derived systems imperfect tools for human gene therapy. Viral vectors are costly to prepare and may be contaminated during preparation. Additionally, there is some risk of inadvertently creating a self-replicating engineered virus. Systemic inflammatory response syndrome and insertional mutagenesis have been responsible for deaths in two clinical trials [1,2]. The number of new gene therapy clinical trials increased until 1999, when the first death occurred, from which point the numbers have been dropping steadily [1,3].

31.1.3 ADVANTAGES OF NONVIRAL GENE THERAPY SYSTEMS

While some investigators attempt to remedy the flaws that have been uncovered in using viral systems for gene therapy by modifying the viral vectors, another viable strategy is to use nonviral gene therapy systems. As drugs, these systems are predicted to be more clinically acceptable than viral vectors have been to date. Nonviral vectors are expected to have a lower cost of preparation, a greater ease of handling, and fewer side effects from toxicity and immunogenicity than viruses. There is little to no risk of creation of a self-replicating agent, either before administration or after the therapy is given to the host. Nonviral vectors are often less complicated than viral vectors to design, many having no size limit, promoter flexibility, and requiring short, if any, DNA sequences for integration. While some viruses have specific tissue preferences and constitutive promoters, nonviral DNA may be delivered to particular organs and can be engineered to express transgenes with tissue-specific promoters.

31.1.4 INTEGRATING VS. NONINTEGRATING NONVIRAL GENE THERAPY

All nonviral gene therapy clinical trials to date have used plasmid DNA without integrating systems. In part due to silencing or loss of plasmids from degradation or during cell division, these trials have been able to demonstrate only limited clinical benefit. Silencing may occur due to viral promoters, bacterial sequences, or CpG methylation [4]. Optimized plasmids, including minicircles, have been demonstrated that allow for expression of transgene products episomally in some tissues without silencing [5,6]. This is encouraging for settings in which plasmid loss due to cell division is not a major issue, such as delivery to the brain or muscle, or when expression is only desired short term, such as for DNA vaccines or the production of proteins to combat cancer. Nonintegrating systems carry a low risk of insertional mutagenesis, because random integration occurs at a very low rate, and this rate would be similar for any nonviral system [7]. Another advantage of nonintegrating plasmid systems is that there is no size limit for cargo genes [8]. However, repeated administration to nonquiescent tissues may be prohibitively expensive or difficult over the lifetime of the patient, making integrating technologies that permanently modify the genome preferable. Nonintegrating plasmids may be an improvement over recombinant protein therapy for treatment of recessive genetic disorders. However, approaches that merely potentiate a temporary benefit are expected to be replaced by technologies that do not require repeated administration, such as integrating vectors.

31.1.5 THREE PROMISING NONVIRAL INTEGRATING GENE THERAPY SYSTEMS

There are currently three leading methods under development for nonviral integrating gene therapy. The first to be developed were transposon systems, followed by sequence-specific phage integrases, and then zinc finger nucleases (ZFNs) that catalyze homologous recombination. These three methods were developed independently, but share some common features.

All require cellular production of a foreign DNA-modifying protein and the presence of a donor plasmid containing therapeutic gene sequences. The application of each of these systems for gene therapy is entirely dependent on getting the nucleic acids into cells, and all three will benefit greatly from advances in DNA transfection. Cells may be modified *ex vivo* and then reintroduced to the patient, either after selection of individual colonies so that integration sites can be identified and transgene product assessed, or as pools of cells. Electroporation, nucleofection, and lipophilic reagents are all possible transfection methods. The success of *in vivo* transfection is dependent on the tissue type, but general methods include hydrodynamic delivery, electroporation, and cationic lipids. One advantage of *in vivo* approaches may be a lower cost to the patient since *in vivo* protocols could be performed in the same way on many patients. *Ex vivo* approaches carry with them the need for costly amplification and manipulation of patient cells in cell culture under strict, highly regulated conditions. However, some of the *in vivo* methods that have been proposed are invasive, possibly carrying a greater risk to the patient than *ex vivo* methods. The feasibility of translation into the clinic of most of the transfection methods currently used in research animals *in vivo* is yet to be determined.

There are three major risks associated with nonviral integrating systems for gene therapy. The first is oncogenic transformation, resulting in a cancer. Genomic integration near a proto-oncogene can lend the potency of the transgene promoter to the endogenous coding regions near the insertion, causing dysregulation of protein expression [9]. Toward the same end, insertion into a tumor suppressor gene can disrupt the activity of genes that prevent cancer [10]. Double strand breaks (DSBs) are also well-known causative agents in tumor formation [11]. The second risk, immunogenicity, is expected to be minimal for any nonviral vector because they lack an immunogenic viral capsid. Lastly, toxicity must be kept within limits that are tolerable, ideally less than the therapy that is currently available, if any. The acceptable level of toxicity may depend on the disease indication. For example, the chemotherapies used to treat cancer are notoriously toxic. Safety profiles are important, because they often determine the success or failure of an effective drug. These three risks vary depending on the system and will be discussed further individually.

31.2 TRANSPOSON SYSTEMS

31.2.1 BACKGROUND

Transposons are naturally occurring mobile genetic elements, also known as "jumping genes." Originally discovered by Barbara McClintock, they are responsible for the multiple corn kernel colors found in patchwork corn ears by disrupting pigment-producing enzymes [12]. Natural transposable elements act as mutagens. By excising, sometimes replicating, and inserting themselves elsewhere in the host genome, transposons are capable of accelerating evolution and expanding genetic possibilities. Transposons are one of the most important forces in shaping the genomes of higher organisms; genomes are littered with the inactive remains of mobile genetic elements called transposon fossils [13]. These transposon fossils can create duplications during meiosis due to repeated sequences, causing errors during homologous recombination. In some cases, transposition-induced mutations may have negative effects. In humans, transposons are to blame for a small fraction of genetic diseases, which are caused by transposons inserting themselves in or near important genes.

There are two major classes of transposons: retrotransposons and DNA transposons. Retrotransposons replicate via a RNA intermediate, which allows the coding sequence to copy and insert itself into the genome multiple times [14]. DNA of retrotransposon origin is abundant in the human genome. In contrast, transposon systems suitable for gene therapy are DNA transposons, which naturally mobilize without replication, by excision and reinsertion of the DNA into the host genome [15]. The largest family of DNA transposons is the *Tc1/mariner* family [16]. Systems from this family that have been applied to gene therapy applications include *Sleeping Beauty* (*SB*), *Frog Prince*, and *Himar1* [17–19]. Transposons from other families that have been shown to function in mammalian cells include *hAT*-like *Tol2* and *piggyBac* [20,21].

The *SB* system will be the central focus of this discussion because it is the leading and most widely used transposon system for gene therapy. However, it should be noted that recent work in cell culture has shown that *piggyBac* may outperform *SB* in integration frequency and that *Tol2* does not show overexpression inhibition, an effect of transposase to transposon ratios that limits translation between applications [22].

31.2.2 *SB* MOBILE GENETIC ELEMENT

Similar inactive transposable element sequences from fish were aligned and compared to create a consensus sequence [23]. The consensus yielded a functioning synthetic transposon, and since it was awakened after a 10 million year long evolutionary "nap," the element was aptly named *Sleeping Beauty* [23,24]. *SB* is a two-component system consisting of transposon and transposase. The transposon is a DNA molecule, containing two terminal inverted repeat (IR)/direct repeat (DR) elements separated by a stretch of DNA that in nature would code for the transposase, but in engineered systems may contain any gene of choice, commonly a selectable marker and/or therapeutic gene (Figure 31.1). The transposase protein binds to the DR regions to mediate insertion of the transposon into thymidine–adenine dinucleotides (TA) in another DNA sequence [23]. For genomic engineering, the transposase may be coded for on the same plasmid as the transposon (outside of the two inverted terminal repeat [ITR] regions), on a separate plasmid, or by an mRNA [25,26].

Transposition occurs in a three step process: binding, cutting, and pasting. First, four molecules of transposase bind to the transposon at the DR elements to form a protein–DNA complex [27,28]. Second, three cleavages occur, resulting in TA dinucleotide overhangs in the host DNA and GTC overhangs at both ends of the transposon. After excision, host cofactors assist in nonhomologous end joining to ligate the transposon into the host DNA [29–31]. Other host cofactors then complete the reaction by filling in single-stranded DNA gaps to repair the DNA.

31.2.2.1 Improvements to the Transposon

The originally isolated *SB* transposase was named "SB10," and the transposon was called "pT." This first system was capable of transposition in 0.3% of immortalized human cells, a frequency 20–40 times the background rate of random integration [23]. Improvements of the IR/DR sequences have resulted in pT2, pT3, and pMSZ transposon vectors, with rates of transposition about two to four times that seen with pT when SB10 is used as the transposase [28,32–34]. A recent improvement to the transposase is the HSB17 enzyme, which represents a 30-fold improvement of transposition frequency in human tissue culture cells when used in conjunction with pMSZ over the original SB10/pT system [35]. However, after

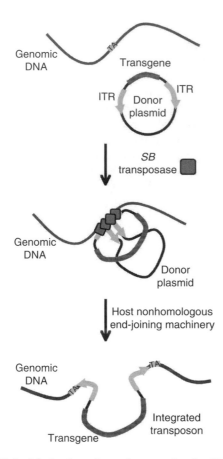

FIGURE 31.1 Mechanism of gene therapy using the *SB* transposon. A transposon donor plasmid, along with a source of *SB* transposase protein, is introduced into the cell by any transfection method. The donor plasmid contains two IR/DR ITR regions oriented in opposite directions from the transgene insert. The remainder of the donor plasmid may contain selectable markers or code for the transposase. Two molecules of transposase bind to each ITR element and the genomic DNA. *SB* transposase cleaves the DNA in three places in a staggered manner, twice at the ends of the ITRs and once across the TA dinucleotide in the genomic DNA. Nonhomologous end-joining factors provided by the host repair the DNA, and single-stranded gaps are filled in. The TA dinucleotide is repeated as a consequence of staggered cleavage. The promoter/enhancer and transgene are permanently inserted, along with the flanking ITR sequences, into the genomic DNA.

hydrodynamic tail vein injection for liver-specific transgene expression, the HSB17 enzyme was slightly less than twofold above SB10 when both were used with pMSZ [35]. This is less than the sevenfold increase in expression observed in liver when the HSB2 enzyme was used with either the pT or pT3 transposon [34].

31.2.2.2 Factors Affecting Transposition

Knowledge of how the transposase functions in mammalian cells may also allow investigators to increase transposition by better understanding the role of the host environment. HMGB1, a DNA-binding protein and pleiotropic cytokine, was found to interact with *SB* transposase *in vivo* and may

play a role in DNA interaction during transposition [36]. The Ku protein, a member of the nonhomologous end-joining pathway of double-strand break (DSB) repair, was also shown to interact with *SB* transposase [37]. In addition, the transposase binds to the transcription factor Miz1, causing an extension of the G1 phase of the cell cycle [38]. This knowledge may be useful in predetermining the fitness of a particular tissue for transposition by evaluating the levels of these host proteins. Although it has been suggested that it may be beneficial to transiently introduce extra doses of one or more of these host proteins to increase transposition rates, such a strategy increases the complexity of the system and may be difficult or unsafe in practice.

There are several limiting factors to *SB* transposition beyond host cofactor requirements. Both the size of the transgene within the transposon, as well as the overall size of the transposon-delivering plasmid, have been shown to affect efficiency [39,40]. There is a decrease in transposition as the size of the transgene increases, and this inverse relationship becomes more pronounced when prokaryotic DNA is used as the transposon cargo instead of eukaryotic DNA [33,39,40]. Similarly, it has also been shown that transposition increases when the transposon ends are brought closer to each other, either by decreasing the size of the transposon or the non-transposon portion of the plasmid vector [40]. The methylation state of the transposon also plays a role in transposition. Methylation of the transposon at CpG sites was found to increase overall transposition rates by 11-fold in mouse ES cells [41]. Yusa et al. [41] hypothesized that the DNA condensation brought about by histone binding to the methylated transposon sequence may bring the ends of the transposon together physically. Thus it seems that some of the limiting factors for *SB* transposition may be at least partially explained by steric properties, since it appears that transposition increases whenever the transposon ends are closer in physical proximity to one another.

A related issue is the phenomenon of overexpression inhibition, which was first identified in *mariner* transposons [42]. Because dimers of *SB* binding both ends of the transposon must come together to form a tetrameric complex for excision to occur, it is thought that an abundance of *SB* transposase may result in the formation of tetrameric complexes containing only one transposon end [25]. Thus, an increase in transposase protein levels beyond the optimal amount results in a decrease in transposition efficiency [33]. Since different cell types and transfection methods result in varying steady-state levels of transposase protein, the transposase-to-transposon ratios must be optimized for each application of the *SB* system [17]. This optimization limits comparisons between studies and necessitates a substantial input of time and resources whenever new methods using the *SB* system are tested.

31.2.2.3 Integration Site Preferences

The integration preference of the *SB* transposon into the host genome is essentially random within host genomes; all that is absolutely required is a TA dinucleotide sequence [43].

The few statistically significant preferences that have been identified are for AT-rich DNA, microsatellite repeats, and a small bias toward genes and regulatory sequences [44]. *SB* transposase has been demonstrated to have an affinity for heterochromatic regions and the efficiency of transposition is increased 100-fold when DNA is made heterochromatic [45]. Within those insertions into genes, transcriptional activity did not play a significant role [44]. Slightly longer consensus sequences (AYATATRT or (AT)^4) have been identified from rescued genomic integration sites [43,44,46]. Also, highly favored sites have been shown to occur during integration into plasmids, although this consensus did not match the genomic consensus sequences [47]. A common theme from each of these integration site studies is that *SB* transposase favors TA dinucleotides that are surrounded by highly deformable sequence. While the random nature of *SB* transposition is preferable to the retroviral preference for promoter regions, random integration is also used as a tool for insertional mutagenesis [48–50]. This feature indicates that gene therapy with the *SB* system could theoretically result in oncogenic transformation as well. This risk should not be taken lightly. Treatment with a nonspecifically integrating retrovirus resulted in insertion events near the LMO2 proto-oncogene, resulting in four cases of leukemia (one of which was lethal) after a clinical trial involving 11 patients [2,51,52].

31.2.2.4 Strategies to Make *SB* More Specific

To reduce the risk of insertional mutagenesis, steps have been taken to improve the *SB* system by attempting to make it more site-specific in order for gene therapy to proceed toward the clinic, especially in tissues that are at high risk for transformation. The idea that a chimeric fusion protein could be produced, which would direct transposition to a more specific site in the genome, is appealing. While it has been noted that directing transposition to a specific site should also be coupled with a mechanism to decrease insertion into random sites, experiments to date have only focused on the former [53]. Wilson et al. first tested the feasibility of this approach in cell culture by expressing a fusion protein containing a zinc finger (ZF) protein fused to the C-terminus of SB12 transposase [54]. Transposition activity was fully preserved and overexpression inhibition was abolished. However, it was not demonstrated that the ZF portion actually redirected transposition activity toward its DNA-binding site [54]. In a second study, a hyperactive version of the transposase, HSB5, was fused to a DNA-binding domain at the N-terminus and was shown to be 15-fold more specific than wild-type HSB5 in an extrachromosomal assay in HeLa cells [53]. A third group also made a fusion protein in which a ZF protein was tethered to the N-terminus of the transposase [55]. While transposition activity was maintained, targeting was not successful. Two other strategies were also tested in which a separate tethering protein was expressed that contained two fused domains. One domain bound to a specific chromosomal region, while the second domain bound either a DNA sequence in the transposon

or the transposase protein directly. *SB* transposase was also introduced without modification. Using the targeting protein that bound to the transposase, about 10% of integration events were observed at the expected genomic DNA sequence in cultured human cells, a 10^7-fold enrichment compared to transposase alone [55]. This approach may be preferable to the creation of transposase fusion proteins because these chimeric enzymes often lose activity [53–55]. Also, it may be possible to further refine the tethering protein system to reduce random integration. However, an increase in specificity of *SB* is yet to be demonstrated *in vivo*.

31.2.2.5 Bioengineering with *SB*

The *SB* system has proven useful for a number of genomic engineering techniques in addition to gene therapy. For example, *SB* has been shown to be active when expressed in the cells of many vertebrates [40]. It has been used effectively as a tool for the creation of transgenic organisms such as fish (zebrafish and medaka), frogs, and rats [56–59]. The random integration feature of *SB* makes it useful for gene discovery to identify the expression profiles of genes in zebrafish [60]. Another successful application of the *SB* system is cancer gene discovery by insertional mutagenesis in mouse and rat [49,50,61]. Transposon systems are an improvement over the traditional retroviral method of cancer gene discovery, because transposons may be designed to mutagenize any tissue of choice by expressing the transposase under a tissue-specific promoter [62,63].

31.2.2.6 Strategies for Transposon Gene Therapy

To achieve gene therapy, all nonviral integrating systems must be coupled with delivery methods to introduce the component parts into the cell. Because the transposon DNA must be introduced, it is often easiest to introduce the transposase gene as DNA as well, either outside the transposon region on the same plasmid (*cis*) or on a different plasmid (*trans*) [64]. It is theoretically possible to use purified transposase protein for gene therapy, although it has not yet been demonstrated. It has been shown that UTR-stabilized *SB* transposase mRNA is capable of mediating transposition in mouse liver [26]. Regardless of the mode of delivery, each new method used must be optimized for transposon-to-transposase ratios since an excess of transposase is detrimental due to overexpression inhibition.

Nonviral transfection methods frequently use either lipophilic reagents or physical methods of delivery. The first use of *SB* for gene therapy in mice used hydrodynamic tail vein injection, a procedure for *in vivo* transfection of hepatocytes in which a large volume of a dilute solution of naked nucleic acids in saline is introduced into the tail vein of the mouse [17,65,66]. Both the human alpha-1 antitrypsin (hAAT) gene and the human Factor IX (hFIX) genes were used as transgenes in normal mice, and *SB* groups had significantly higher levels of transgene product from several months after administration until the termination of the experiment [17]. Also, in a mouse model of hemophilia B, after treatment with the *SB* system, bleeding time was decreased to about twice that of normal mice, indicating a substantial amelioration of the disease in the mouse model [17]. Since then, hydrodynamic tail vein injection has been used with *SB* successfully for therapeutic correction of tyrosemia type I, diabetes, hemophilia A, and mucopolysaccharidosis type I [67–70].

DNA–polyethylenimine (PEI) complexes containing transposon DNA carrying the luciferase gene were found to transfect mouse lung cells, and luciferase expression was prolonged when the *SB* transposase was available to mediate genomic integration [71]. The most common type of cell to express luciferase in this study was a type II pneumocyte [71]. It was also possible to target expression to endothelial cells using DNA–PEI complexes together with an endothelial-specific promoter, and it was shown that secreted alkaline phosphatase levels in serum were an order of magnitude higher when appropriate *SB* transposase levels were provided with the transposon [32]. This system was applied to a mouse model of hemophilia A and delivery of an *SB* transposon carrying the B-domain deleted Factor VIII gene corrected the bleeding disorder when expressed in lung endothelial cells [72]. DNA–PEI complexes have also been used for lung delivery of the indoleamine 2,3-dioxygenase in rat, with the aim of creating a gene therapy for obliterative bronchiolitis after lung transplantation [73]. In a similar study, the endothelial nitric oxide synthase gene was delivered in the same way to reduce the effect of subsequent pulmonary hypertension in rats [74]. The other major use for DNA–PEI complexes has been in cancer gene therapy. *SB* has been used in a mouse model of glioblastoma to prolong transgene expression, resulting in tumor regression and increased survival times when therapeutic genes were delivered [75–77].

SB was first used for *ex vivo* gene therapy for functional correction of junctional epidermolysis bullosa by introducing the *LAMB3* gene into human primary keratinocytes derived from junctional epidermolysis bullosa patients and grafted onto immune-deficient mice [78]. Peripheral blood leukocytes have also been manipulated using nucleofection and the *SB* system to stably express a suicide gene for cancer gene therapy [64]. Both immortalized K-562 hematopoetic cells and CD34+ hematopoetic stem cells were nucelofected with the *SB* system carrying the reporter gene GFP [79]. While both cell types showed high levels of transposition, the hematopoetic stem cells engrafted at extremely low numbers after nucleofection [79]. A lipophilic reagent has also been used to transfect K-562 cells with an erythroid-specific *SB* system to mediate stable beta-globin expression with the eventual goal of treating sickle cell disease [80]. Blood outgrowth endothelial cells were transfected by a commercially available lipophilic reagent with an *SB* system to secrete hAAT, hFIX, and hFVIII [81].

The transposon system can also be used as a delivery vehicle for the cellular manufacture of short-interfering RNAs or short hairpin RNAs for RNA interference to drastically decrease the cellular levels of a protein by preventing the translation of its mRNA. Both *SB* and *Frog Prince* have been used to deliver RNA interference constructs in tissue culture to successfully knock down genes such as lamin A and huntingtin [82,83]. The toxicity of short hairpin RNAs has

been shown to be variable when expressed strongly in liver cells, resulting in the deaths of some mice, so *in vivo* application will require optimization of the construct not only for effectiveness, but also safety [84].

31.3 PHAGE INTEGRASE SYSTEMS

31.3.1 Background

Bacteriophages are ubiquitous viral parasites of bacteria that have been studied for decades. A staggering diversity and abundance of phages exist in the world. It is estimated that 10^{27} bacteriophages are found in the ocean alone and that these tiny particles sequester about 20 times as much carbon as all living whales combined [85]. Temperate phages are able to enter either the lytic or lysogenic cycle based on the environment of the host upon infection. If the lysogenic cycle is favored, the phage DNA integrates into the bacterial genome to emerge at a later date when conditions change. Natural selection favoring those phages that are capable of infecting a diversity of bacteria is expected to occur, so it is not surprising that some phages do not require host-specific cofactors for integration [86]. Integration is generally performed by a site-specific recombinase. There are two major classes of site-specific recombinases: tyrosine and serine, which refers to the catalytic residue used to cut the DNA. Tyrosine recombinases include lambda integrase, as well as resolvases such as Cre and FLP. Both of these resolvases are more efficient at excision reactions since recombination takes place between, and generates, the same recognition sites. Serine integrases are better suited to gene therapy due to the unidirectional mechanism of integration between unlike recognition sites, together with their lack of cofactor requirements [86]. Serine integrases include phiC31, phiBT1, R4, and TP901-1 [87–90]. Several serine integrases have been tested for their ability to integrate efficiently into native sites in mammalian genomes, and of these, phiC31 integrase was found to have the most favorable properties for gene therapy [88,91]. Therefore, the following discussion will focus on phiC31 as the hallmark of the phage integrase systems currently under development.

31.3.2 PhiC31 Integrase

PhiC31 integrase was originally isolated from the actinophage phiC31 of *Streptomyces*, a soil bacterium that produces a diversity of antibiotic metabolites and has a complex life cycle [92–94]. The semipalindromic DNA sequences that the serine integrase attaches to are termed *attB* for bacterial attachment site and *attP* for phage attachment site. The most likely mechanism of recombination, as shown by supershift studies of phiC31 and the structure of family member gamma–delta resolvase, is that two molecules of phiC31 attach to each *att* site and come together to form a tetrameric complex that positions both strands of DNA next to each other (Figure 31.2) [95,96]. The formation of the tetramer allows the DNA to be cleaved at the core. Simultaneously, a flat, hydrophobic surface forms perpendicular to the DNA strands between the halves of each original dimer [96]. Rotation at the hydrophobic surface allows for recombination and the

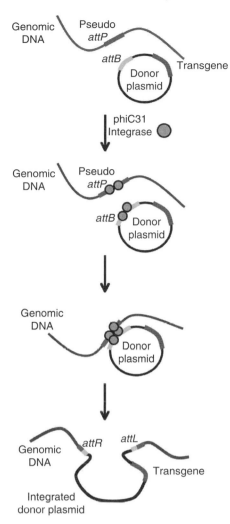

FIGURE 31.2 Mode of action of phiC31 integrase-mediated gene therapy. A donor plasmid containing a short *attB* sequence and the desired transgene is introduced into cells, together with a source of phiC31 integrase. A phiC31 integrase molecule binds to each half-site of both the *attB* site on the donor plasmid and endogenous pseudo *attP* sites naturally present in the genome for a total of four integrase molecules. The two dimers come together to form a tetrameric complex including the pseudo *attP* site in the genomic DNA and the *attB* site on the donor plasmid. This complex cleaves both the pseudo *attP* and the *attB* sites in a staggered manner at the core sequence, simultaneously swapping the DNA strands. The new hybrid sites, *attR* and *attL*, are formed, and the entire donor plasmid is now integrated into the genomic DNA.

formation of new sites *attR* and *attL*, named for their right or left proximity to the phage sequence after integration into the bacterial genome. Presumably because phiC31 integrase cannot form a stable synapse between the hybrid *attR* and *attL* sites, the reaction is unidirectional [94,95].

31.3.2.1 Use of PhiC31 Integrase in Mammalian Cells

The inherent properties of phiC31 integrase seemed to indicate that it might fulfill the needs of gene therapists for a nonviral, site-specific enzyme for genomic integration. It was first shown to be capable of recombination in mammalian cells between

extrachromosomal *attB* and *attP* sites [87]. The length of the minimal sites for recombination was determined in *Escherichia coli* to be 34 bp for *attB* and 39 bp for *attP* [87]. For human gene therapy, an integrating enzyme must be able to find suitable preexisting genomic sites. For site-specific recombinases, such sites are generally termed pseudo sites, or sites that share common sequence features with the actual recognition site, but may be degenerate. PhiC31 integrase was demonstrated to mediate recombination into a chromosomally placed *attP* site in a human cell line [97]. The next step was to determine what, if any, pseudo sites existed. Integration was tested by using an *attB*-containing plasmid to find pseudo *attP* sites and an *attP*-containing plasmid to search for pseudo *attB* sites. Surprisingly, phiC31 integrase has a definite preference for integration into pseudo *attP* sites rather than pseudo *attB* sites [97]. This bias may be related to the lower GC content of the *attP* site; possibly either because methylated CpGs inhibit the reaction or simply due to the general rarity of GC-rich regions in mammalian genomes [8]. Therefore, for use in transgenic organisms and gene therapy, the system usually consists of a coding region for the phiC31 integrase introduced as a plasmid or mRNA, as well as a separate donor plasmid containing the *attB* site and transgene of choice.

31.3.2.2 Integration Site Analysis of PhiC31 Integrase in the Human Genome

The ideal gene augmentation strategy would involve integration at one safe site in the genome. The integration specificity of phiC31 integrase has been evaluated extensively by plasmid rescue in three human cell lines [98]. It was found that integration is indeed sequence-specific, enabling the identification of a ~30 bp consensus pseudo *attP* sequence. Nineteen pseudo *attP* sites were found more than once, representing just over half of all integration events. The most common pseudo site, present on chromosome 19, was found in 7.5% of genomic integrations. With regard to safety, while none of the rescued events showed any reasonable chance of causing insertional mutagenesis after comparison to a list of known cancer genes, about 10% of rescues involved aberrant events, either translocations or large deletions of the chromosomes [98]. The translocation events that were rescued in this study were incapable of producing fusion proteins, because the two chromosome arms were separated by the donor plasmid, which is several kilobases long. In general, phiC31 integrase appears to be safer than other integrating systems, although more experiments are needed to address the cancer risk of a phiC31 integrase-mediated gene therapy. However, there is already a growing body of evidence that phiC31 integrase does not mediate chromosomal translocations at a detrimental rate *in vivo*. No tumors have been found in normal mice, rats, or rabbits given phiC31 integrase-based gene therapies. Also, phiC31 integrase has been expressed constitutively in mice, and these strains exhibit normal fertility and development [99,100]. Additionally, other experiments are under way to mutate the enzyme to better recognize one of the chromosomal locations that was a preferred integration site. We predict that phiC31 integrase will evolve over the coming years into a more ideal enzyme with higher efficiency in mammalian cells and more specificity in the human genome.

31.3.2.3 Genetic Engineering with PhiC31 Integrase

In addition to its utility for gene therapy, the unique properties of phiC31 integrase have made it popular for applications in transgenic organisms. In mice, the integrase has been transiently expressed for transgene integraton into embryonic genomes [101]. Additionally, two separate strains of mice have been created in which phiC31 integrase is constitutively expressed for genomic manipulation by crossing [99,100]. In *Drosophila*, phiC31 integrase has been used for both the creation of transgenics and cassette exchange [102,103]. Other genomes in which chromosomal manipulations with phiC31 integrase have been successful include fission yeast, tobacco plastid, mosquito, silkworm, *Xenopus laevis*, Chinese hamster ovary cells, mouse, rat, rabbit synovial cells, and bovine fibroblasts [19,104–111]. It has been reported that phiC31 integrase was incapable of finding pseudo *attP* sites in chicken fibroblasts; however, it is unknown exactly why integration was unsuccessful [112].

31.3.2.4 Gene Therapy Mediated by PhiC31 Integrase

In the gene therapy arena, phiC31 integrase has been shown to be effective in mediating long-term gene expression in many preclinical treatment models. Hydrodynamic tail vein injection has been used as a method to transfect hepatocytes with the phiC31 integrase system, resulting in permanent, high levels of transgene product [113–115]. Fumarylacetoacetate hydrolase, hFIX, and hAAT have been successfully used as transgenes. Protein levels were consistently greater in groups that received integrase in comparison to groups that received no active integrase. In the 2002 study, some mice in both hFIX groups were given partial hepatectomies to encourage liver regeneration [113]. A decrease in hFIX levels was only seen in the nonintegrase group, indicating that maintenance after cell division was specific to those hepatocytes receiving integrase as well as the donor plasmid. In the same study, the mouse mpsL1 pseudo *attP* site was identified, and phiC31 integrase-mediated integration at this site was shown to have occurred in every liver sample tested [113]. Using this *in vivo* hepatocyte transfection technique, phiC31 integrase was demonstrated to be capable of imparting significantly higher levels of hFIX than the *SB* system when comparable backbone sequences were used for the hFIX-containing plasmid [115]. Notably, after liver regeneration was induced by a two-thirds partial hepatectomy, phiC31 integrase-treated hepatocytes maintained hFIX expression better than *SB*-treated hepatocytes, indicating that phiC31 integrase may also be the better choice for long-term therapeutic gene product expression [115].

Additionally, phiC31 integrase has proven effective for genomic integration of transgenes after *in vivo* electroporation [116–118]. Rats were injected subretinally with a solution containing a luciferase donor plasmid and a plasmid coding for phiC31 integrase, then subjected to *in situ* electroporation.

This treatment resulted in long-term luciferase expression in the retinal pigment epithelium for the duration of the study (4.5 months) [116]. In other studies, long-term transgene expression in the muscles of juvenile mice was seen after *in vivo* electroporation of donor and integrase plasmids for treatment of muscular dystrophy and peripheral ischemia [117,118].

In a rabbit model of arthritis, joints were intra-articularly injected with naked DNA with and without phiC31 integrase. When phiC31 integrase was present, transgene levels were significantly higher; however, transfection was relatively inefficient [119].

Ex vivo approaches using phiC31 integrase are also promising. The first use of phiC31 integrase for *ex vivo* gene therapy was for recessive dystrophic epidermolysis bullosa (RDEB) [120]. RDEB patients have defects in both copies of their collagen VII genes. Human keratinocytes obtained from an RDEB patient registry were stably modified by phiC31 integrase to express collagen VII. After regenerating skin from cells engrafted onto recipient *scid* mice, the skin was biopsied and shown to be pathologically indistinguishable from skin generated from normal keratinocytes at time points up to 14 weeks post-transplant [120]. To further explore potential treatments for RDEB, transfection of fibroblasts with plasmids expressing collagen VII and phiC31 integrase, followed by intradermal injection into grafts of RDEB skin on immune-deficient mice, was also successful [121]. While wild-type cells were incapable of correcting the RDEB phenotype, phiC31-mediated correction of fibroblasts resulted in levels of collagen VII that exceeded wild-type, allowing the regeneration of normal skin from these fibroblasts [121]. Robust expression of transgenes placed in the genome is a commonly observed feature of the phiC31 integrase system, possibly due to genomic placement at chromatin regions that are accessible, not only to phiC31 integrase but also to proteins involved in transcription [98,121].

Patients with junctional epidermolysis bullosa are lacking laminin V expression. PhiC31 integrase was used to mediate integration of the laminin V gene into patient keratinocytes, resulting in functional correction of the disease when grafts were generated on immune-deficient mice [122]. Another *ex vivo* approach used electroporation in a human T-cell line derived from an X-linked severe combined immune deficiency patient, which was functionally corrected by genomic integration of the common cytokine receptor gamma chain gene with phiC31 integrase [123]. Muscle precursor cells were nucleofected with the dystrophin gene and phiC31 integrase, resulting in long-term dystrophin expression after engraftment [124,125].

31.4 HOMOLOGOUS RECOMBINATION FOR GENE CORRECTION

31.4.1 BACKGROUND

Homologous recombination, a process of exchanging similar DNA sequences between sister chromatids, naturally occurs in organisms ranging from bacteria to humans. The process has been selected for over time for two purposes: (1) to prevent catastrophic somatic mutations, which could result in cell death or cancer, by repairing DNA that has lost its molecular integrity through a DSB and (2) to increase the diversity of individuals in a population by recombining paternal and maternal chromosomes during meiosis to create new hybrid chromosomes that are passed on through gametes to the next generation. Homologous recombination has also been taken advantage of as a method for genomic manipulation of organisms such as bacteria, yeast, plants, and mice. Recently, the natural host machinery has also been harnessed for gene therapy applications, allowing for gene correction rather than simply gene addition.

There are two major strategies that have been devised for gene therapy via homologous recombination, specially designed oligonucleotides and ZFNs. Because spontaneous homologous recombination occurs at a very low background rate (in about one in a million somatic cells after DNA introduction), DNA manipulations of either the genomic DNA strand or the introduced template molecule are necessary to stimulate the homologous recombination machinery to act on the targeted area of genomic DNA. Chimeraplasts are short RNA–DNA nucleic acid hybrid molecules that are introduced into cells, triggering homologous recombination by virtue of their abnormal structure [126]. Unfortunately, gene therapy studies using chimeraplasty have been notable for their inconsistent results [127]. While the laboratories of Kmiec and Steer have published accounts of chimeraplasty working well some of the time, many accounts of failure have also been documented [127,128]. This form of gene therapy through homologous recombination, and the similar single- or triple-stranded DNA oligonucleotide correction vehicles, may be better understood in the future. Although they are appealing in principle, because the genomic DNA is left intact until correction occurs, an alternative route to homologous recombination induction in human cells using ZFNs is currently under development and will be discussed in Section 31.4.2. It should be noted that ZFNs constitute a gene therapy technology that is less developed than either the *SB* or phiC31 integrase systems.

31.4.2 ZINC FINGER NUCLEASES

31.4.2.1 ZF Motifs

A ZF motif is a 30 amino acid folded protein domain that chelates a zinc ion. It is capable of binding DNA by virtue of an alpha helix that inserts itself into the major groove of the DNA [129]. Certain residues along this alpha helix have been found to determine the exact DNA sequence that the ZF binds to [129–131]. These ZFs may be linked together like beads on a string to extend the length of the recognized DNA sequence [132–134]. Other domains can also be fused to the ZF motifs, including the nonspecific *FokI* cleavage domain, to create a ZFN [135–138].

So far the ability to design ZFNs is limited by the availability of ZF motifs that recognize triplet sequences [139].

Because multiple ZF modules are linked together, the long recognition site desired to increase specificity of the ZFN must be balanced against the availability of ZF motifs that recognize the sequence. Since for genetic correction, it is desirable to make a DSB as close to the mutation site as possible, it would be helpful to ZF designers to have a ZF motif for every possible triplet DNA sequence [139].

31.4.2.2 Restriction Endonuclease *FokI*

Most restriction enzymes cleave DNA at a simple, palindromic recognition site, because cleavage and site recognition are coupled events within the same protein domain. *FokI* is a different type of enzyme, in that it cleaves the DNA 9/13 nucleotides downstream of the nonpalindromic recognition site. This unique cleavage pattern occurs because *FokI* consists of two separate domains, a N-terminal sequence-recognition domain and a separate C-terminal nuclease domain [140,141]. Normally, the N-terminal domain must recognize the sequence before the C-terminal domain can cut the DNA due to protein–protein interactions between the two domains [142]. However, it has been shown that single amino acid mutations can disrupt those interactions, decoupling the two processes [143,144]. Additionally, the endonuclease domain requires dimerization for cleavage of the DNA to occur [145]. This dimerization requirement also applies to fusion proteins containing this domain, such as ZFNs [136,146]. Proof of concept for chimeric proteins containing the *FokI* endonuclease domain first occurred after fusion to the *Drosophila Ubx* homeodomain, which showed cleavage at the *Ubx* recognition sequence [147]. Since then, other DNA-binding motifs have been fused to the cleavage domain of *FokI*, including transcription factors [148].

31.4.2.3 Designing ZFNs

The mechanism of the ZFN method of homologous recombination for gene substitution is relatively straightforward in principle, but far from trivial to achieve. First, the investigator designs right-side and left-side ZFNs containing at least three ZF motifs that will each recognize one half of the sequence flanking the cut site. Practically, this involves a deep understanding of the ZF motifs available for use and an application of this knowledge to the immediate sequence surrounding the disease-causing mutation. Then, optimization of the alpha helices must be performed to maintain protein structure integrity [149]. Putative designed ZFNs are made and tested for activity. Cells are transfected with both a donor plasmid carrying the correct DNA sequence and another plasmid encoding the ZFN protein. Within the cell, the ZFN cleaves the DNA at the target site, creating a DSB (Figure 31.3). This DSB recruits the homologous recombination machinery to the site. The donor plasmid is available at high numbers in the cell and is matched with the damaged chromosomal sequence and used as a template. The treated cells are then analyzed for gene correction. Correction efficiencies of 18% have been found without selection, including 7% of cells that were homozygous for the desired modification [149].

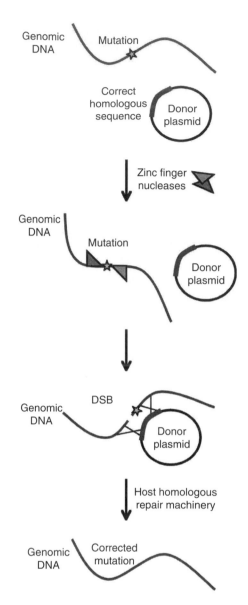

FIGURE 31.3 Gene correction by induction of homologous recombination via ZFNs. A known mutation in the genomic DNA is identified and ZFNs are designed for each half of the sequence flanking the mutation. A source of both ZFN proteins is introduced into the cell, as well as a plasmid containing wild-type DNA that is homologous to the sequence surrounding the DNA mutation. The ZFNs bind to their specific DNA sequences, determined by the ZF motifs selected and then dimerize at the *FokI* nuclease domain. The DNA is then cleaved near the harmful mutation by the ZFNs. Host homologous repair machinery initiates strand invasion and the cleaved DNA is repaired using the correct sequence found on abundant donor plasmids as a template. The genomic DNA is now permanently modified to carry the correct DNA sequence instead of a harmful genetic mutation.

31.4.2.4   Gene Therapy with ZFNs

So far, only two gene therapy applications of ZFNs have been demonstrated, both in cultured cells and chosen to highlight the potential of the technology. First, a therapy for X-linked severe combined immune deficiency was devised

targeting a common mutation in exon 5 of the *IL2R*γ gene in both K-562 immortalized hematopoetic cells and primary human CD4 +T-cells [149]. In human T-cells, observed levels of gene correction were 5.3% corresponding to 20% of the cells that were successfully transfected by nucleofection [149]. Other ZFNs have also been made against the *IL2R*γ gene [150]. Additionally, ZFNs exist that target the human CD8α gene, although a therapeutic application of these is unclear [150]. Sickle cell anemia is a disease caused by a defect in the β-globin gene, which could potentially be cured by gene correction with a ZFN. Toward this end, a ZFN was designed and demonstrated to target the β-globin gene in human cells [150].

ZFN-mediated gene therapy is still in its infancy. Much work remains to be done before homologous recombination will be ready for the clinic. First, it is difficult to design ZFNs and one must be an experienced ZF specialist to understand how to make a working ZFN [151]. Second, much of the promising technology, including ZF libraries, ZFNs, and patents, is owned by one company, Sangamo, which, in order to protect its intellectual property, does not freely distribute its techniques and products. This makes it difficult for academic scientists to gain access to these materials. Third, ZFNs must be shown to mediate long-term, safe gene therapy when used in human cells *ex vivo* that are reintroduced to irradiated *scid* mice. Fourth, ZFNs must be optimized to mediate gene correction at a usable frequency, while limiting off-target DSBs. For greater safety, it would be helpful to make ZF motifs more specific [152].

31.5   COMPARISON OF INTEGRATING SYSTEMS

All three of the developed systems for nonviral integration have their advantages and disadvantages. It is possible that all three will eventually be put to clinical use for different indications. Nonintegrating plasmid vectors may be adequate for gene therapy in quiescent tissues such as brain, for DNA vaccines, and for short-term therapies, on the order of days to months, in proliferating tissues. Transposon systems may be best suited for gene therapy in tissues such as muscle, where oncogenesis is rare. Theoretically, they may also be useful for dominant negative genetic diseases for which a DNA-binding domain is already known, so that a tethering fusion protein could be constructed. *Ex vivo* screening could possibly allow the isolation of stem cells with insertions that disrupt and replace the defective gene. Phage integrases, with their limited number of insertion sites *in vivo*, may represent the best choice for *in vivo* transfection protocols to proliferating cells for treatment of monogenic diseases, as well as long-term cancer therapies. Treatment for recessive diseases is an obvious application, but treatment for dominant negative diseases may also be possible by designing a vector that would carry an RNA interference component to reduce levels of the defective gene product, along with a substantially different, engineered DNA sequence encoding the correct endogenous gene. This approach would couple knockdown of the defective gene to expression of the correct gene. The strategy would only work if the RNA interference could be designed to be nontoxic and a small amount of defective gene product would not be an issue, because RNA interference usually reduces steady-state protein levels by only about 90%. Additionally, phage integrases are preferred for many *ex vivo* approaches, because the identification of the insertion site can be accomplished by a simple PCR for the most common insertion sites. While it may be possible to engineer a phage integrase to preferentially integrate at one selected site by using mutagenesis screens, it will most likely not be possible to design a different phage integrase for every dominant negative gene due to constraints on sequence recognition [153]. ZFNs are well suited to clonal *ex vivo* approaches for correction of dominant negative diseases, because they edit and correct genetic mutations. The modular nature of this approach may allow adaptation to a number of genetic diseases. However, due to the potentially oncogenic nature of DSBs, this approach may not have a favorable risk/benefit profile for *in vivo* transfection of cells that commonly become transformed into cancers. A second disadvantage of this approach is that for every new disease application, a new ZFN must be designed, made, and tested. However, the possibility of correcting a genetic disease by actually replacing the defective DNA with correct, native DNA, leaving no leftover traces of the therapeutic machinery, is an approach of as yet unparalleled elegance in the field of gene therapy [151]. Table 31.1 compares and contrasts the

features of nonreplicating episomal, transposon, integrase and ZFN systems for gene therapy.

Knowledge of the limitations as well as abilities of each system will be key to thoughtfully evaluating which approaches should advance to the clinic. Hopefully in the coming decades, gene therapy will find a routine place in human medicine. However, it is crucial to success that the cure is not worse than the disease. Systems should be rigorously evaluated in sensitive models to assess the contribution of the gene therapy to cancer formation. Gene therapy in humans would require administration of the therapy to a greater number of cells, since humans are much larger than the rodent models commonly used in gene therapy experiments [152]. Other important differences exist between rodent models and humans, such as tumor spectrum, genomic composition, and immune response. There is no way to be absolutely sure that a gene therapy system is safe prior to clinical trials in human patients, but the predicted risk to the patient should not exceed the expected benefit.

In conclusion, while the promise of gene therapy is yet to be realized, new strategies are under development to address the problems that have been encountered to date in the clinic. Nonviral genome modifying systems may soon emerge as revolutionary new human medicines. Many of the strategies proposed go beyond simple administration of a pharmaceutical and couple surgery or stem cells with a permanent gene transfer system to mediate potential disease cures.

TABLE 31.1
Comparison of Nonviral Systems for Gene Therapy

	Episomal	*SB* Transposon	PhiC31 Integrase	Zinc Finger Nuclease
Is there prolonged gene expression (up to a few months)?	In some tissues	Yes	Yes	Yes
Is expression theoretically permanent for the life of the cell and its progenitors?	No	Yes	Yes	Yes
What levels of transgene product have been observed?	High, tapering over time	Low to high	Usually high	Endogenous levels
Is an immune response expected?	No	No	No	No
Is the cost to produce probably lower than that of viral vectors?	Yes	Yes	Yes	Yes
Has the system been used in a model system for dominant negative disorders?	No	No	No	Yes
Is the dose critical to success?	No	Yes, due to overexpression inhibition	No	Possibly, due to off-target DSB toxicity
Has targeting to known genomic sites been demonstrated?	N/A	Yes, in modified systems in cell culture	Yes, in cell culture and *in vivo*	Yes, in cell culture
Are there free DSBs?	No	No	No	Yes
Have chromosomal recombination and/or deletion events been detected?	No	No	Yes, in cell culture	Not reported, but likely, due to DSBs
What is the expected risk of oncogenesis?	Lowest	Higher, because randomly integrating	Lower, because integration is sequence-specific	Higher, because of off-target DSBs
How toxic would the therapy be?	Low	Low	Low	Higher if administered *in vivo*; lower if administered *ex vivo*

REFERENCES

1. Raper, S.E., et al., Fatal systemic inflammatory response syndrome in a ornithine transcarbamylase deficient patient following adenoviral gene transfer, *Mol Genet Metab*, 80, 148, 2003.

2. Hacein-Bey-Abina, S., et al., LMO2-associated clonal T cell proliferation in two patients after gene therapy for SCID-X1, *Science*, 302, 415, 2003.

3. Edelstein, M.L., et al., Gene therapy clinical trials worldwide 1989–2004—an overview, *J Gene Med*, 6, 597, 2004.

4. Miao, C.H., A novel gene expression system: Non-viral gene transfer for hemophilia as model systems, *Adv Genet*, 54, 143, 2005.

5. Darquet, A.M., et al., Minicircle: An improved DNA molecule for in vitro and in vivo gene transfer, *Gene Ther*, 6, 209, 1999.

6. Rodriguez, E.G., Nonviral DNA vectors for immunization and therapy: Design and methods for their obtention, *J Mol Med*, 82, 500, 2004.

7. Murnane, J.P., Yezzi, M.J., and Young, B.R., Recombination events during integration of transfected DNA into normal human cells, *Nucleic Acids Res*, 18, 2733, 1990.

8. Calos, M.P., The phiC31 integrase system for gene therapy, *Curr Gene Ther*, 6, 633, 2006.

9. Pall, M.L., Gene-amplification model of carcinogenesis, *Proc Natl Acad Sci USA*, 78, 2465, 1981.

10. Sager, R., Genetic suppression of tumor formation: A new frontier in cancer research, *Cancer Res*, 46, 1573, 1986.

11. Khanna, K.K. and Jackson, S.P., DNA double-strand breaks: Signaling, repair and the cancer connection, *Nat Genet*, 27, 247, 2001.

12. McClintock, B., The origin and behavior of mutable loci in maize, *Proc Natl Acad Sci* 36, 344, 1950.

13. Waterston, R.H., et al., Initial sequencing and comparative analysis of the mouse genome, *Nature*, 420, 520, 2002.

14. Baltimore, D., Retroviruses and retrotransposons: The role of reverse transcription in shaping the eukaryotic genome, *Cell*, 40, 481, 1985.

15. Essner, J.J., McIvor, R.S., and Hackett, P.B., Awakening gene therapy with Sleeping Beauty transposons, *Curr Opin Pharmacol*, 5, 513, 2005.

16. Miskey, C., et al., DNA transposons in vertebrate functional genomics, *Cell Mol Life Sci*, 62, 629, 2005.

17. Yant, S.R., et al., Somatic integration and long-term transgene expression in normal and haemophilic mice using a DNA transposon system, *Nat Genet*, 25, 35, 2000.

18. Miskey, C., et al., The Frog Prince: A reconstructed transposon from *Rana pipiens* with high transpositional activity in vertebrate cells, *Nucleic Acids Res*, 31, 6873, 2003.

19. Keravala, A., et al., Hyperactive Himar1 transposase mediates transposition in cell culture and enhances gene expression in vivo, *Hum Gene Ther*, 17, 1006, 2006.

20. Koga, A., et al., The medaka fish Tol2 transposable element can undergo excision in human and mouse cells, *J Hum Genet*, 48, 231, 2003.

21. Ding, S., et al., Efficient transposition of the piggyBac (PB) transposon in mammalian cells and mice, *Cell*, 122, 473, 2005.

22. Wu, S.C., et al., PiggyBac is a flexible and highly active transposon as compared to sleeping beauty, Tol2, and Mos1 in mammalian cells, *Proc Natl Acad Sci U S A*, 103, 15008, 2006.

23. Ivics, Z., et al., Molecular reconstruction of Sleeping Beauty, a Tc1-like transposon from fish, and its transposition in human cells, *Cell*, 91, 501, 1997.

24. Ivics, Z., et al., Identification of functional domains and evolution of Tc1-like transposable elements, *Proc Natl Acad Sci USA*, 93, 5008, 1996.

25. Hackett, P.B., et al., Sleeping beauty transposon-mediated gene therapy for prolonged expression, *Adv Genet*, 54, 189, 2005.

26. Wilber, A., et al., Messenger RNA as a source of transposase for sleeping beauty transposon-mediated correction of hereditary tyrosinemia Type I, *Mol Ther*, 15, 1280, 2007.

27. Izsvak, Z., et al., Involvement of a bifunctional, paired-like DNA-binding domain and a transpositional enhancer in Sleeping Beauty transposition, *J Biol Chem*, 277, 34581, 2002.

28. Cui, Z., et al., Structure–function analysis of the inverted terminal repeats of the sleeping beauty transposon, *J Mol Biol*, 318, 1221, 2002.

29. Plasterk, R.H., Izsvak, Z., and Ivics, Z., Resident aliens: The Tc1/mariner superfamily of transposable elements, *Trends Genet*, 15, 326, 1999.

30. Fischer, S.E., Wienholds, E., and Plasterk, R.H., Regulated transposition of a fish transposon in the mouse germ line, *Proc Natl Acad Sci USA*, 98, 6759, 2001.

31. Yant, S.R. and Kay, M.A., Nonhomologous-end-joining factors regulate DNA repair fidelity during sleeping beauty element transposition in mammalian cells, *Mol Cell Biol*, 23, 8505, 2003.

32. Liu, L., et al., Endothelial targeting of the sleeping beauty transposon within lung, *Mol Ther*, 10, 97, 2004.

33. Geurts, A.M., et al., Gene transfer into genomes of human cells by the sleeping beauty transposon system, *Mol Ther*, 8, 108, 2003.

34. Yant, S.R., et al., Mutational analysis of the N-terminal DNA-binding domain of sleeping beauty transposase: Critical residues for DNA binding and hyperactivity in mammalian cells, *Mol Cell Biol*, 24, 9239, 2004.

35. Baus, J., et al., Hyperactive transposase mutants of the sleeping beauty transposon, *Mol Ther*, 12, 1148, 2005.

36. Zayed, H., et al., The DNA-bending protein HMGB1 is a cellular cofactor of sleeping beauty transposition, *Nucleic Acids Res*, 31, 2313, 2003.

37. Izsvak, Z. and Ivics, Z., Sleeping beauty transposition: Biology and applications for molecular therapy, *Mol Ther*, 9, 147, 2004.

38. Walisko, O., et al., Sleeping beauty transposase modulates cell-cycle progression through interaction with Miz-1, *Proc Natl Acad Sci USA*, 103, 4062, 2006.

39. Karsi, A., et al., Effects of insert size on transposition efficiency of the sleeping beauty transposon in mouse cells, *Mar Biotechnol (NY)*, 3, 241, 2001.

40. Izsvak, Z., Ivics, Z., and Plasterk, R.H., Sleeping beauty, a wide host-range transposon vector for genetic transformation in vertebrates, *J Mol Biol*, 302, 93, 2000.

41. Yusa, K., Takeda, J., and Horie, K., Enhancement of sleeping beauty transposition by CpG methylation: Possible role of heterochromatin formation, *Mol Cell Biol*, 24, 4004, 2004.

42. Hartl, D.L., et al., What restricts the activity of mariner-like transposable elements, *Trends Genet*, 13, 197, 1997.

43. Vigdal, T.J., et al., Common physical properties of DNA affecting target site selection of sleeping beauty and other Tc1/mariner transposable elements, *J Mol Biol*, 323, 441, 2002.

44. Yant, S.R., et al., High-resolution genome-wide mapping of transposon integration in mammals, *Mol Cell Biol*, 25, 2085, 2005.

45. Ikeda, R., et al., Sleeping beauty transposase has an affinity for heterochromatin conformation, *Mol Cell Biol*, 27, 1665, 2007.

46. Carlson, C.M., et al., Transposon mutagenesis of the mouse germline, *Genetics*, 165, 243, 2003.

47. Liu, G., et al., Target-site preferences of sleeping beauty transposons, *J Mol Biol*, 346, 161, 2005.

48. Schroder, A.R., et al., HIV-1 integration in the human genome favors active genes and local hotspots, *Cell*, 110, 521, 2002.

49. Collier, L.S., et al., Cancer gene discovery in solid tumours using transposon-based somatic mutagenesis in the mouse, *Nature*, 436, 272, 2005.

50. Dupuy, A.J., et al., Mammalian mutagenesis using a highly mobile somatic Sleeping Beauty transposon system, *Nature*, 436, 221, 2005.

51. McCormack, M.P., et al., The LMO2 T-cell oncogene is activated via chromosomal translocations or retroviral insertion during gene therapy but has no mandatory role in normal T-cell development, *Mol Cell Biol*, 23, 9003, 2003.

52. Baum, C., What are the consequences of the fourth case?, *Mol Ther*, 15, 1401, 2007.

53. Yant, S.R., et al., Site-directed transposon integration in human cells, *Nucleic Acids Res*, 35, e50, 2007.

54. Wilson, M.H., Kaminski, J.M., and George, A.L., Jr., Functional zinc finger/sleeping beauty transposase chimeras exhibit attenuated overproduction inhibition, *FEBS Lett*, 579, 6205, 2005.

55. Ivics, Z., et al., Targeted sleeping beauty transposition in human cells, *Mol Ther*, 15, 1137, 2007.

56. Davidson, A.E., et al., Efficient gene delivery and gene expression in zebrafish using the sleeping beauty transposon, *Dev Biol*, 263, 191, 2003.

57. Grabher, C., et al., Transposon-mediated enhancer trapping in medaka, *Gene*, 322, 57, 2003.

58. Sinzelle, L., et al., Generation of trangenic xenopus laevis using the Sleeping Beauty transposon system, *Transgenic Res*, 15, 751, 2006.

59. Lu, B., et al., Generation of rat mutants using a coat color-tagged sleeping beauty transposon system, *Mamm Genome*, 18, 338, 2007.

60. Balciunas, D., et al., Enhancer trapping in zebrafish using the sleeping beauty transposon, *BMC Genomics*, 5, 62, 2004.

61. Kitada, K., et al., Transposon-tagged mutagenesis in the rat, *Nat Methods*, 4, 131, 2007.

62. Starr, T.K. and Largaespada, D.A., Cancer gene discovery using the sleeping beauty transposon, *Cell Cycle*, 4, 1744, 2005.

63. Dupuy, A.J., Jenkins, N.A., and Copeland, N.G., Sleeping beauty: A novel cancer gene discovery tool, *Hum Mol Genet*, 15 Spec No 1, R75, 2006.

64. Huang, X., et al., Stable gene transfer and expression in human primary T cells by the sleeping beauty transposon system, *Blood*, 107, 483, 2006.

65. Liu, F., Song, Y., and Liu, D., Hydrodynamics-based transfection in animals by systemic administration of plasmid DNA, *Gene Ther*, 6, 1258, 1999.

66. Zhang, G., Budker, V., and Wolff, J.A., High levels of foreign gene expression in hepatocytes after tail vein injections of naked plasmid DNA, *Hum Gene Ther*, 10, 1735, 1999.

67. Montini, E., et al., In vivo correction of murine tyrosinemia type I by DNA-mediated transposition, *Mol Ther*, 6, 759, 2002.

68. He, C.X., et al., Insulin expression in livers of diabetic mice mediated by hydrodynamics-based administration, *World J Gastroenterol*, 10, 567, 2004.

69. Ohlfest, J.R., et al., Phenotypic correction and long-term expression of factor VIII in hemophilic mice by immunotolerization and nonviral gene transfer using the sleeping beauty transposon system, *Blood*, 105, 2691, 2005.

70. Aronovich, E.L., et al., Prolonged expression of a lysosomal enzyme in mouse liver after sleeping beauty transposon-mediated gene delivery: Implications for non-viral gene therapy of mucopolysaccharidoses, *J Gene Med*, 9, 403, 2007.

71. Belur, L.R., et al., Gene insertion and long-term expression in lung mediated by the sleeping beauty transposon system, *Mol Ther*, 8, 501, 2003.

72. Liu, L., Mah, C., and Fletcher, B.S., Sustained FVIII expression and phenotypic correction of hemophilia A in neonatal mice using an endothelial-targeted sleeping beauty transposon, *Mol Ther*, 13, 1006, 2006.

73. Liu, H., et al., Sleeping beauty-based gene therapy with indoleamine 2,3-dioxygenase inhibits lung allograft fibrosis, *Faseb J*, 20, 2384, 2006.

74. Liu, L., et al., Sleeping beauty-mediated eNOS gene therapy attenuates monocrotaline-induced pulmonary hypertension in rats, *Faseb J*, 20, 2594, 2006.

75. Ohlfest, J.R., et al., Integration and long-term expression in xenografted human glioblastoma cells using a plasmid-based transposon system, *Mol Ther*, 10, 260, 2004.

76. Ohlfest, J.R., et al., Combinatorial antiangiogenic gene therapy by nonviral gene transfer using the sleeping beauty transposon causes tumor regression and improves survival in mice bearing intracranial human glioblastoma, *Mol Ther*, 12, 778, 2005.

77. Wu, A., et al., Transposon-based interferon gamma gene transfer overcomes limitations of episomal plasmid for immunogene therapy of glioblastoma, *Cancer Gene Ther*, 14, 550, 2007.

78. Ortiz-Urda, S., et al., Sustainable correction of junctional epidermolysis bullosa via transposon-mediated nonviral gene transfer, *Gene Ther*, 10, 1099, 2003.

79. Hollis, R.P., et al., Stable gene transfer to human CD34(+) hematopoietic cells using the sleeping beauty transposon, *Exp Hematol*, 34, 1333, 2006.

80. Zhu, J., et al., Erythroid-specific expression of beta-globin by the sleeping beauty transposon for sickle cell disease, *Biochemistry*, 46, 6844, 2007.

81. Kren, B.T., et al., Blood outgrowth endothelial cells as a vehicle for transgene expression of hepatocyte-secreted proteins via sleeping beauty, *Endothelium*, 14, 97, 2007.

82. Heggestad, A.D., Notterpek, L., and Fletcher, B.S., Transposon-based RNAi delivery system for generating knockdown cell lines, *Biochem Biophys Res Commun*, 316, 643, 2004.

83. Chen, Z.J., et al., Sleeping Beauty-mediated down-regulation of huntingtin expression by RNA interference, *Biochem Biophys Res Commun*, 329, 646, 2005.

84. Grimm, D., et al., Fatality in mice due to oversaturation of cellular microRNA/short hairpin RNA pathways, *Nature*, 441, 537, 2006.

85. Wilhelm, S.W. and Shuttle, C.A., Viruses and nutrient cycles in the sea, *Bioscience*, 49, 781, 1999.

86. Smith, M.C. and Thorpe, H.M., Diversity in the serine recombinases, *Mol Microbiol*, 44, 299, 2002.

87. Groth, A.C., et al., A phage integrase directs efficient site-specific integration in human cells, *Proc Natl Acad Sci USA*, 97, 5995, 2000.

88. Olivares, E.C., Hollis, R.P., and Calos, M.P., Phage R4 integrase mediates site-specific integration in human cells, *Gene*, 278, 167, 2001.

89. Chen, L. and Woo, S.L., Complete and persistent phenotypic correction of phenylketonuria in mice by site-specific genome integration of murine phenylalanine hydroxylase cDNA, *Proc Natl Acad Sci USA*, 102, 15581, 2005.

90. Stoll, S.M., Ginsburg, D.S., and Calos, M.P., Phage TP901–1 site-specific integrase functions in human cells, *J Bacteriol*, 184, 3657, 2002.

91. Keravala, A., et al., A diversity of serine phage integrases mediate site-specific recombination in mammalian cells, *Mol Genet Genomics*, 276, 135, 2006.

92. Hopwood, D.A., The Leeuwenhoek lecture, 1987. Towards an understanding of gene switching in streptomyces, the basis of sporulation and antibiotic production, *Proc R Soc Lond B Biol Sci*, 235, 121, 1988.

93. Kuhstoss, S. and Rao, R.N., Analysis of the integration function of the streptomycete bacteriophage phiC31, *J Mol Biol*, 222, 897, 1991.

94. Rausch, H. and Lehmann, M., Structural analysis of the actinophage phiC31 attachment site, *Nucleic Acids Res*, 19, 5187, 1991.

95. Thorpe, H.M., Wilson, S.E., and Smith, M.C., Control of directionality in the site-specific recombination system of the streptomyces phage phiC31, *Mol Microbiol*, 38, 232, 2000.

96. Li, W., et al., Structure of a synaptic gammadelta resolvase tetramer covalently linked to two cleaved DNAs, *Science*, 309, 1210, 2005.

97. Thyagarajan, B., et al., Site-specific genomic integration in mammalian cells mediated by phage phiC31 integrase, *Mol Cell Biol*, 21, 3926, 2001.

98. Chalberg, T.W., et al., Integration specificity of phage phiC31 integrase in the human genome, *J Mol Biol*, 357, 28, 2006.

99. Belteki, G., et al., Site-specific cassette exchange and germline transmission with mouse ES cells expressing phiC31 integrase, *Nat Biotechnol*, 21, 321, 2003.

100. Raymond, C.S. and Soriano, P., High-efficiency FLP and PhiC31 site-specific recombination in mammalian cells, *PLoS ONE*, 2, e162, 2007.

101. Hollis, R.P., et al., Phage integrases for the construction and manipulation of transgenic mammals, *Reprod Biol Endocrinol*, 1, 79, 2003.

102. Groth, A.C., et al., Construction of transgenic drosophila by using the site-specific integrase from phage phiC31, *Genetics*, 166, 1775, 2004.

103. Bateman, J.R., Lee, A.M., and Wu, C.T., Site-specific transformation of drosophila via phiC31 integrase-mediated cassette exchange, *Genetics*, 173, 769, 2006.

104. Thomason, L.C., Calendar, R., and Ow, D.W., Gene insertion and replacement in schizosaccharomyces pombe mediated by the streptomyces bacteriophage phiC31 site-specific recombination system, *Mol Genet Genomics*, 265, 1031, 2001.

105. Lutz, K.A., et al., A novel approach to plastid transformation utilizes the phiC31 phage integrase, *Plant J*, 37, 906, 2004.

106. Kittiwongwattana, C., et al., Plastid marker gene excision by the phiC31 phage site-specific recombinase, *Plant Mol Biol*, 64, 137, 2007.

107. Nimmo, D.D., et al., High efficiency site-specific genetic engineering of the mosquito genome, *Insect Mol Biol*, 15, 129, 2006.

108. Nakayama, G., et al., Site-specific gene integration in cultured silkworm cells mediated by phiC31 integrase, *Mol Genet Genomics*, 275, 1, 2006.

109. Allen, B.G. and Weeks, D.L., Transgenic xenopus laevis embryos can be generated using phiC31 integrase, *Nat Methods*, 2, 975, 2005.

110. Thyagarajan, B. and Calos, M.P., Site-specific integration for high-level protein production in mammalian cells, *Methods Mol Biol*, 308, 99, 2005.

111. Ma, Q.W., et al., Identification of pseudo *attP* sites for phage phiC31 integrase in bovine genome, *Biochem Biophys Res Commun*, 345, 984, 2006.

112. Malla, S., et al., Rearranging the centromere of the human Y chromosome with phiC31 integrase, *Nucleic Acids Res*, 33, 6101, 2005.

113. Olivares, E.C., et al., Site-specific genomic integration produces therapeutic factor IX levels in mice, *Nat Biotechnol*, 20, 1124, 2002.

114. Held, P.K., et al., In vivo correction of murine hereditary tyrosinemia type I by phiC31 integrase-mediated gene delivery, *Mol Ther*, 11, 399, 2005.

115. Ehrhardt, A., et al., A direct comparison of two nonviral gene therapy vectors for somatic integration: In vivo evaluation of the bacteriophage integrase phiC31 and the Sleeping Beauty transposase, *Mol Ther*, 11, 695, 2005.

116. Chalberg, T.W., et al., phiC31 integrase confers genomic integration and long-term transgene expression in rat retina, *Invest Ophthalmol Vis Sci*, 46, 2140, 2005.

117. Bertoni, C., et al., Enhancement of plasmid-mediated gene therapy for muscular dystrophy by directed plasmid integration, *Proc Natl Acad Sci USA*, 103, 419, 2006.

118. Portlock, J.L., et al., Long-term increase in mVEGF164 in mouse hindlimb muscle mediated by phage phiC31 integrase after nonviral DNA delivery, *Hum Gene Ther*, 17, 871, 2006.

119. Keravala, A., et al., PhiC31 integrase mediates integration in cultured synovial cells and enhances gene expression in rabbit joints, *J Gene Med*, 8, 1008, 2006.

120. Ortiz-Urda, S., et al., Stable nonviral genetic correction of inherited human skin disease, *Nat Med*, 8, 1166, 2002.

121. Ortiz-Urda, S., et al., Injection of genetically engineered fibroblasts corrects regenerated human epidermolysis bullosa skin tissue, *J Clin Invest*, 111, 251, 2003.

122. Ortiz-Urda, S., et al., PhiC31 integrase-mediated nonviral genetic correction of junctional epidermolysis bullosa, *Hum Gene Ther*, 14, 923, 2003.

123. Ishikawa, Y., et al., Phage phiC31 integrase-mediated genomic integration of the common cytokine receptor gamma chain in human T-cell lines, *J Gene Med*, 8, 646, 2006.

124. Quenneville, S.P., et al., Nucleofection of muscle-derived stem cells and myoblasts with phiC31 integrase: Stable expression of a full-length-dystrophin fusion gene by human myoblasts, *Mol Ther*, 10, 679, 2004.

125. Quenneville, S.P., et al., Dystrophin expression in host muscle following transplantation of muscle precursor cells modified with the phiC31 integrase, *Gene Ther*, 14, 514, 2007.

126. Richardson, P.D., et al., Gene repair and transposon-mediated gene therapy, *Stem Cells*, 20, 105, 2002.

127. De Semir, D. and Aran, J.M., Targeted gene repair: The ups and downs of a promising gene therapy approach, *Curr Gene Ther*, 6, 481, 2006.

128. Taubes, G., Gene therapy. The strange case of chimeraplasty, *Science*, 298, 2116, 2002.

129. Pavletich, N.P. and Pabo, C.O., Zinc finger-DNA recognition: Crystal structure of a Zif268–DNA complex at 2.1 A, *Science*, 252, 809, 1991.

130. Shi, Y. and Berg, J.M., A direct comparison of the properties of natural and designed zinc-finger proteins, *Chem Biol*, 2, 83, 1995.

131. Elrod-Erickson, M. and Pabo, C.O., Binding studies with mutants of Zif268. Contribution of individual side chains to binding affinity and specificity in the Zif268 zinc finger–DNA complex, *J Biol Chem*, 274, 19281, 1999.

132. Liu, Q., et al., Design of polydactyl zinc-finger proteins for unique addressing within complex genomes, *Proc Natl Acad Sci USA*, 94, 5525, 1997.

133. Beerli, R.R., et al., Toward controlling gene expression at will: Specific regulation of the erbB-2/HER-2 promoter by using polydactyl zinc finger proteins constructed from modular building blocks, *Proc Natl Acad Sci USA*, 95, 14628, 1998.

134. Kim, J.S. and Pabo, C.O., Getting a handhold on DNA: Design of poly-zinc finger proteins with femtomolar dissociation constants, *Proc Natl Acad Sci USA*, 95, 2812, 1998.

135. Kim, Y.G., Cha, J., and Chandrasegaran, S., Hybrid restriction enzymes: Zinc finger fusions to Fok I cleavage domain, *Proc Natl Acad Sci USA*, 93, 1156, 1996.

136. Smith, J., et al., Requirements for double-strand cleavage by chimeric restriction enzymes with zinc finger DNA-recognition domains, *Nucleic Acids Res*, 28, 3361, 2000.

137. Mani, M., et al., Design, engineering, and characterization of zinc finger nucleases, *Biochem Biophys Res Commun*, 335, 447, 2005.

138. Alwin, S., et al., Custom zinc-finger nucleases for use in human cells, *Mol Ther*, 12, 610, 2005.

139. Durai, S., et al., Zinc finger nucleases: Custom-designed molecular scissors for genome engineering of plant and mammalian cells, *Nucleic Acids Res*, 33, 5978, 2005.

140. Li, L., Wu, L.P., and Chandrasegaran, S., Functional domains in Fok I restriction endonuclease, *Proc Natl Acad Sci USA*, 89, 4275, 1992.

141. Li, L. and Chandrasegaran, S., Alteration of the cleavage distance of Fok I restriction endonuclease by insertion mutagenesis, *Proc Natl Acad Sci USA*, 90, 2764, 1993.

142. Waugh, D.S. and Sauer, R.T., Single amino acid substitutions uncouple the DNA binding and strand scission activities of Fok I endonuclease, *Proc Natl Acad Sci USA*, 90, 9596, 1993.

143. Wah, D.A., et al., Structure of the multimodular endonuclease FokI bound to DNA, *Nature*, 388, 97, 1997.

144. Wah, D.A., et al., Structure of FokI has implications for DNA cleavage, *Proc Natl Acad Sci USA*, 95, 10564, 1998.

145. Bitinaite, J., et al., FokI dimerization is required for DNA cleavage, *Proc Natl Acad Sci USA*, 95, 10570, 1998.

146. Mani, M., et al., Binding of two zinc finger nuclease monomers to two specific sites is required for effective double-strand DNA cleavage, *Biochem Biophys Res Commun*, 334, 1191, 2005.

147. Kim, Y.G. and Chandrasegaran, S., Chimeric restriction endonuclease, *Proc Natl Acad Sci USA*, 91, 883, 1994.

148. Ruminy, P., et al., Long-range identification of hepatocyte nuclear factor-3 (FoxA) high and low-affinity binding sites with a chimeric nuclease, *J Mol Biol*, 310, 523, 2001.

149. Urnov, F.D., et al., Highly efficient endogenous human gene correction using designed zinc-finger nucleases, *Nature*, 435, 646, 2005.

150. Porteus, M.H., Mammalian gene targeting with designed zinc finger nucleases, *Mol Ther*, 13, 438, 2006.

151. Kaiser, J., Gene therapy. Putting the fingers on gene repair, *Science*, 310, 1894, 2005.

152. Porteus, M.H., Connelly, J.P., and Pruett, S.M., A look to future directions in gene therapy research for monogenic diseases, *PLoS Genet*, 2, e133, 2006.

153. Sclimenti, C.R., Thyagarajan, B., and Calos, M.P., Directed evolution of a recombinase for improved genomic integration at a native human sequence, *Nucleic Acids Res*, 29, 5044, 2001.

Part IV

Disease Targets and Therapeutic Strategies

32 Hematopoietic Progenitor Cells

Kuan-Yin Karen Lin, Olga Sirin, and Margaret A. Goodell

CONTENTS

32.1 INTRODUCTION

Hematopoietic stem cells (HSCs) maintain hematopoiesis or the generation of blood cells throughout an organism's lifetime. This is accomplished by the capacity of HSCs to self-renew as well as differentiate into the various cell types of the blood. In view of this ability, the HSCs are a promising target for gene therapy, as genetically altering the HSC results in modifications of the myeloid and lymphoid lineages. Furthermore, the availability of surface markers and dyes allows for isolation of pure HSCs with potential to be manipulated for gene therapy. Despite the advantages of HSCs, there are drawbacks. Efficient gene transfer to HSCs has proven to be difficult. Limitations in our knowledge of regulators of the self-renewal and differentiation process has inhibited our ability to fully manipulate the HSCs, such as expansion of HSC *ex vivo*, which would have tremendous advantages for the clinical use of HSCs and bone marrow transplantations.

In this chapter, we will review basic biology of murine and human HSCs, the latest advances made in the understanding of molecular regulators of HSC function, and the recent clinical attempts in HSC gene therapy.

The presence of HSCs was first demonstrated with bone marrow transplantation experiments in which lethally irradiated mice were injected with whole bone marrow (WBM) and rescue was achieved [1,2]. Since then, it has become apparent that there are different populations of progenitors with varying degrees of "stemness." Based on the extent of the self-renewing ability, they are referred to as "long-term HSC" (LT-HSC), "short-term HSC" (ST-HSC), or multipotent progenitor (MMP) [3–5] (Figure 32.1). These progenitors comprise a hierarchy according to their ability to self-renew and to give rise to multilineage blood progenies. The LT-HSC has, in theory, an infinite self-renewing and repopulating ability, while the ST-HSC and MMP has respectively less. Furthermore, while the LT-HSC is mainly quiescent, the ST-HSC is actively cycling [5,6].

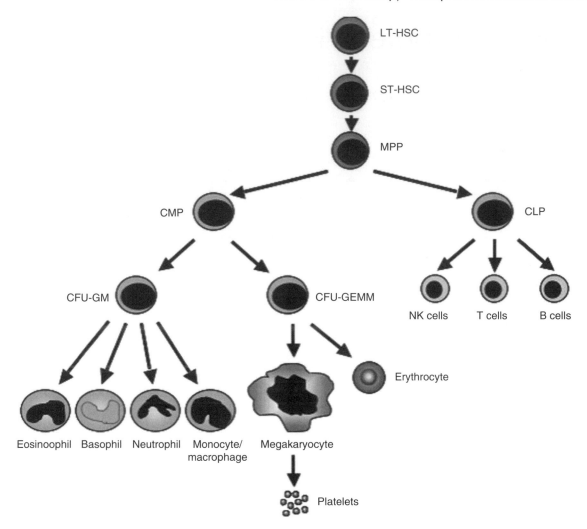

FIGURE 32.1 Hematopoietic differentiation: The hierarchy of the HSC, hematopoietic progenitors, and blood progenies. The hematopoiesis tree starts from LT-HSCs, which are the most primitive stem cells that are able to indefinitely self-renew as well as differentiate. The self-renewal capacity of progenitors degenerates when cells are differentiating toward the end of the hematopoiesis tree. ST-HSCs possess a limited capacity and remain multipotent, while the multipotent progenitors only have the capacity to differentiate into all the blood lineages.

32.2   ISOLATION OF HSCs

32.2.1   SOURCES

The existence of HSCs was first suggested by Ford et al. when they rescued lethally irradiated mice with WBM cells [1,7]. Several approaches such as sedimentation rates and fluorescence activated cell sorting (FACS) were then exploited to identify the regenerating precursors in the adult bone marrow. In addition to WBM, peripheral blood is another source for HSCs in the adult. Brecher and Cronkite first discovered that lethally irradiated mice can be rescued by nonirradiated partners through parabiosis, that is, surgically joining of the circulatory systems [8]. It was further established by Goodman and Hodgson that lethally irradiated mice can be rescued by transplanting more than 5 million peripheral blood cells [9]. However, the requirement of high peripheral blood cell numbers and the low engraftment ability suggest HSCs in the peripheral blood are of low frequency as well as less primitive.

In humans, HSCs are present in ample amounts in WBM while the HSCs in peripheral blood are infrequent and less potent [10,11]. However, because the procedure to acquire peripheral blood is less complicated, and the infusion of granulocyte-macrophage colony stimulating factor (GM-CSF), a cytokine, can greatly increase the frequency of HSCs in human peripheral blood [12]. GM-CSF mobilized peripheral blood is a frequently used source for HSCs for bone marrow transplantations.

Human cord blood (CB) is highly enriched with the HSCs relative to normal peripheral blood. Because CB-HSCs are found to be primitive and resistant to extensive handling such as cryopreservation, as performed in cord blood banks and for cell culture, they have shown great promise for cell-based therapy. In 1989, Gluckman et al. reported the first case of successful hematopoietic reconstitution of a pediatric patient with cord blood cells to correct Fanconi's anemia [13]. Since then, thousands of successful transplantations in children have been performed to cure both malignant and benign blood diseases. Through these studies, and the increasing interest in national CB banking, CB transplantation has increased. CB is primarily used as a source of HSC for pediatric patients due to the limited total volume of CB that can be obtained: one unit of CB does not provide a sufficient number of HSC to transplant a typical adult-sized patient. However, numerous strategies to overcome this limitation are being examined, such as limited expansion of the CB-HSC or combining CB units from different sources into the same patient [14].

During early embryonic development, two waves of hematopoiesis occur [15,16]. The first wave of hematopoiesis, which gives rise to primitive (embryonic) erythroblasts (EryP), occurs in yolk sac around day 7 of gestation in mice [16]. The second wave of hematopoiesis occurs around day 9–10 of gestation from the aorta-gonad-mesonephros (AGM), a mesoderm-derived region, which gives rise to definitive (adult) hematopoietic progenitors [17,18]. It has been found that the embryonic precursors from yolk sac are not able to reconstitute lethally irradiated recipients, while the definitive precursors from AGMs are capable of engrafting. The definitive HSCs are later found in the fetal liver (at day 11 of gestation), where they are thought to have migrated from the AGM. Ultimately, the HSC migrate to the bone marrow where they reside in the adult. Additionally, Gekas et al. has found the placenta as a site of definitive hematopoiesis, which emerges at day 12.5–13.5 of gestation after the AGM and during the fetal liver hematopoiesis [19].

Mouse and human embryonic stem (ES) cells also generate hematopoietic cells during *in vitro* differentiation. The differentiation pattern is thought to recapitulate the hematopoiesis in early embryonic development [20]. The idea that ES cells can be indefinitely expanded and then made to differentiate into desired numbers of potent HSCs has drawn scientists to the attempt of deriving HSC from ES cells. Doetschman et al. were the first to generate blood cells *in vitro* from ES cell suspension cultures (with no feeder layers or the supply of leukemia inhibitory factor) in 1985 [21]. However, ES-derived blood cells are found to resemble embryonic hematopoietic progenitors that are destined to produce embryonic erythrocytes and lack the ability to produce lymphocytes [20]. Furthermore, these differentiated hematopoietic cells were reported to possess limited ability to engraft in adult animals [20,22], which presents a major drawback for the idea of using ES-derived HSCs as a clinical source. However, a few studies have been able to reconstitute certain hematopoietic components in immune-compromised mice by infusion with HSC-enriched ES-derived hematopoietic cells [23,24]. These studies indicate that although adult hematopoietic precursors occur transiently and are of low frequency, one may utilize strategies to improve the generation of adult hematopoietic precursors from ES cells. Therefore, several approaches, such as supplementing with hematopoietic cytokines and gene targeting, have been undertaken to improve the derivation of adult HSCs from ES cells. Enforced expression of transcription factors such as BCR/Abl [25], Hoxb4 [26], or Cdx4 [27] enhances the generation of definitive (adult) hematopoiesis to allow engraftment in lethally irradiated mice. Moreover, a study showed that by combining strategies, culturing cells with hematopoietic cytokines, and purifying hematopoietic progenitors with flow cytometry, ES-derived HSCs are able to engraft in major histocompatibility complex (MHC)-mismatched recipient mice [28]. Thus, with further development long-term, we may see ES-derived HSCs moved into the clinic.

32.2.2 Cell Surface Markers and Dyes

32.2.2.1 Cell Surface Phenotype of Murine and Human HSC

Since bone marrow is a heterogeneous mixture of cells, identification of the cells within marrow that are responsible for engraftment of irradiated hosts has been a focus of interest for some time. Physical properties, such as sedimentation rates, of bone marrow (BM) cell subsets were initially exploited to enrich for HSCs, followed by antigenic or dye staining properties that could be identified by FACS.

Class I major histocompatibility antigens (MHC I) were one of the first molecules exploited for fractionating WBM [29]. Depletion of activity assays, which utilized antibodies capable of fixing complement and directing the lysis of target cells, were employed to show that the murine HSC did not express MHC I [29]. In contrast, human primitive progenitors do express MHC I [30], establishing the first of many phenotypic differences between murine and human HSCs. Ultimately, it was shown that murine progenitors and HSCs do not express MHC I, while human progenitors express high levels of MHC I, and human HSCs express low levels [31–34].

Cell "panning" has also been employed to fractionate WBM, where an antibody is fixed to the surface of a tissue culture plate and BM cells, which adhere to the antibodies, are collected and assayed for the enrichment or depletion of primitive hematopoietic potential. Cell panning helped identify c-Kit [35] expression by HSC as well as their ability to bind wheat germ agglutinin [36].

The development of flow cytometry allowed multiple antigens or fluorescent dye efflux characteristics to be utilized together for improved purification. A major advance was to use a cocktail of antibodies against antigens specific for the differentiated lineages to deplete the BM of the 90% differentiated cells [37]. Such "lineage"-negative BM is depleted of T cells, B cells, granulocytes, macrophages, and erythrocyte progenitors. This lineage cocktail is commonly used in purification schemes of murine HSC. In human progenitor purifications, antibodies against CD33 are used, as it is expressed by progenitors but not the HSC [38]. Similarly, CD38 is present on human progenitors but absent from HSC [39] so it can be used for negative selection of differentiated cells.

Another major breakthrough in HSC purification entailed the use of the CD34 antigen for positive selection of human HSC [33,40]. Primate studies ushered in the use of CD34 as a tool for isolating HSC for bone marrow transplantation [41], and CD34 became the most clinically exploited molecule expressed by HSC.

In the studies of HSC biology using the mouse as a model, multiple purification schemes are exploited. The first strategy that utilizes the Thy-1 antigen was reported by Spangrude et al. to enrich for murine HSCs [37]. Thereafter, different combinations of cell surface makers were discovered by various laboratories in the attempt to further enrich for HSCs from bone marrow. Although the strategies among different laboratories vary, there is a core element that utilizes the expression of Sca-1 (stem cell antigen), c-Kit, and the absent expression of lineage markers, resulting in an HSC-enriched cell population usually called the c-Kit + Sca-1 + Lineage- (KSL) cells. The KSL cell population comprises ~0.1% of murine WBM cells and is a heterogeneous population that contains lineage-primed progenitors as well as stem cells [42]. Therefore, several additional markers such as Thy1.1 [3], CD34 [43], and Flk-2 [4] were individually utilized to exclude differentiated progenitors by various laboratories to advance the enrichment scheme, which as a result lead to a variety of cell surface marker combinations. The resulting cell populations from these schemes are thought to be highly overlapping with each other if not identical. More recently, Kiel et al. used a new set of cell surface markers, CD150, CD244, and CD48, which belong to the SLAM family, to purify HSCs [44]. However, it is still to be determined how similar and homogeneous SLAM-purified cells are compared with other well-characterized populations.

Although multiple surface markers are expressed on murine HSCs, the expression of these markers does not always translate into human studies. For example, although Sca-1 is routinely used to purify murine HSC, a human homolog for Sca-1 has never been identified. Likewise, high expression of CD34 is critical for purification of human HSCs, but CD34 is expressed at very low levels on murine HSCs [45]. C-kit expression, on the other hand, has also been used to purify both human and murine HSC [42,46–49].

In summary, murine and human HSC can be selected on the basis of the expression of a number of antigens in combination: Sca-1, lineage[neg] c-Kit[pos] with either Thy1.1 (low), CD34[low], or Flk-2 in the mouse, and CD34+, CD38–, lineage[neg], and c-kit[pos] in the human.

32.2.2.2 Vital Dye Enrichments

Vital dyes have also been used alone or in combination with cell surface markers. Most WBM cells stain brightly with the DNA-binding dye Hoechst 33342, but progenitors stain dimly [50–52]. Hoechst-based enrichment for stem cells was refined with the observation that with the use of two emission wavelengths, multiple populations could be defined. A small population can be observed on the side of the profile (the so-called side population, or SP, Figure 32.2) [53], which contains all of the hematopoietic reconstituting potential of mouse bone marrow. An SP has been observed in multiple species [45] and appears to be due to a multidrug resistance (MDR)-like mediated efflux of the Hoechst dye [53]. The efflux of Hoechst dye by HSC is thought to be due to the transporter ABCG2 [54], but additional drug pumps are likely also involved. While the

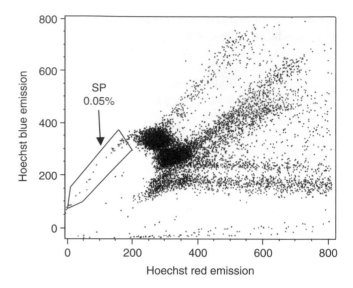

FIGURE 32.2 Representative Hoechst 33342 staining profile of murine bone marrow. When mouse bone marrow is stained with the DNA-binding dye Hoechst 33342, multiple populations are observed. All the hematopoietic reconstituting cells are found in the small population on the side, the "side population" or SP (boxed).

natural substrate of the dye efflux is currently unknown, we hypothesize that it may be a molecule that otherwise would cause the cells to differentiate. FACS purification using the SP profile alone affords 1000- to 3000-fold enrichment. When used in combination with additional cell surface marker expression such as c-Kit, lineage markers, and Sca-1, another 15% or so of non-stem cells can be eliminated, resulting in a remarkably pure and homogenous HSC population "SPKLS" (pronounced as "sparkle") in which one out of every three cells has functional stem cell activity, which approaches a theoretical maximum [2,45,53,55,56].

Rhodamine 123 (Rh123) is another vital dye used for HSC enrichment. Murine and human rhodamine[Low] WBM cells are enriched for HSCs [57,58]. When Hoechst[Low] WBM cells were fractionated into rhodamine[Medium] and rhodamine[Low] populations [52], Rhodamine[Low] cells were shown to be more primitive [52]. Rh123 binds mitochondria and therefore Rh123 fluorescence is thought to directly reflect cellular metabolic activity [59]. However, since Rh123 is also a substrate for MDR transporters, low Rh123 may simply reflect dye efflux, as is the case with Hoechst 33342.

32.3 MOLECULAR REGULATION OF HSC FUNCTIONS

Under normal conditions, HSCs remain in a quiescent state. However, when given appropriate signals such as stress, they have the ability to self-renew, differentiate to generate mature blood cells, or undergo apoptosis (Figure 32.3). What molecularly dictates this fate determining process is poorly understood; however, to date several key players have been identified (Figure 32.4).

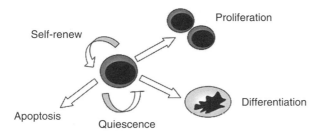

FIGURE 32.3 HSC cell fate decisions during its lifespan. Multiple mechanisms in regulating cell cycle and cell survival are found to be involved in HSC cell fate decisions.

32.3.1 INTRACELLULAR FACTORS MEDIATING SELF-RENEWAL

The cyclin-dependent kinase inhibitors (CKIs) such as p21cip/waf1, p27 kip1, and p18 INK4A have been shown to play a role in the regulation of stem cell quiescence. p21 KO mice have an increased number of HSCs with elevated numbers progressing though cell cycle [60], suggesting that p21 negatively regulates HSC cell cycle entry. P27 has been shown to play a similar role; however, rather than increasing the HSC pool and the number of proliferating HSCs, it seems to affect the number of mature progenitors without changing the number of HSCs [61]. Furthermore, p18 has been shown to regulate the frequency of self-renewing HSCs [62] suggested by an improved long-term engraftment of p18 KO bone marrow cells in transplantation assays. This group of cell cycle checkpoint regulators demonstrates the complexity of the HSC self-renewal regulation. Though the function of these three CKIs is to regulate cell cycle, they have distinct roles in regulating HSCs kinetics.

Several transcription factors have been identified to regulate HSC self-renewal. Among these are the homeobox genes. The Hox genes, well studied for their role in embryogenesis, have recently been implicated to play an important role in hematopoiesis. Several independent studies have shown that HoxB4 promotes HSC expansion both *in vitro* and *in vivo* [63,64]. Overexpression of HoxB4 was associated with an engraftment advantage compared to that of wild-type [63]. Similarly, Hoxb3 has been shown to play a role in regulating stem cell regeneration [65]. Pbx-1, a cofactor for Hox transcription factors, has been shown to be essential for the maintenance rather than the initiation of hematopoiesis [66].

Another group of genes implicated in HSC self-renewal are antiapoptotic genes such as Bcl2 and Mcl-1. Mice overexpressing Bcl2 were found to have increased numbers of HSCs [67]. Loss of Mcl-1 expression in mice resulted in loss of progenitors and HSCs [68]. Together these findings suggest that apoptosis plays a significant role in controlling HSC homeostasis.

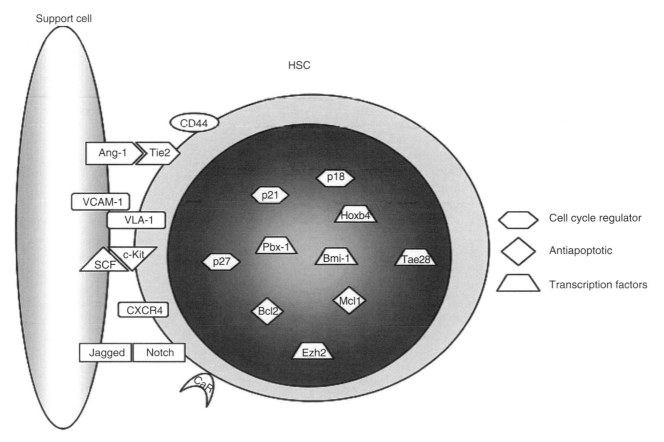

FIGURE 32.4 Key players that regulate HSC self-renewal. Examples are the intrinsic factors that regulate HSC cell cycle, survival, and epigenetic programming, and the extrinsic factors needed for the expression of ligands on the support cells (osteoblasts and stromal cells) to stimulate the downstream pathways.

32.3.2 EXTRACELLULAR FACTORS/SIGNALING PATHWAYS MEDIATING SELF-RENEWAL

Signaling pathways such as the Notch, Wnt, and Hedgehog are classically known for their involvement in embryonic development and have recently been shown to also play critical roles in HSC self-renewal. The evolutionarily well-conserved Notch receptors and their ligands have been shown to play different roles in hematopoiesis both in the expansion of HSCs as well as the differentiation process [69–74]. Though there have been conflicting results, several studies have implicated members of the Wnt pathway such as β-catenin in maintaining hematopoiesis homeostasis [75–81]. Overexpressing β-catenin has been shown to improve long-term engraftment. Furthermore, when the Wnt signaling pathway was blocked, a reduction in HSC growth was observed both *in vitro* and *in vivo* [75]. As with the Notch pathway, activating the Hedgehog pathway has recently been demonstrated to control HSC pool by inhibiting differentiation [82–85]. Moreover it seems that the Hedgehog signaling pathway regulates the pool of pluripotent hematopoietic repopulating cells by utilizing the BMP signaling pathway.

Recently, researchers have uncovered another level of regulation of HSC homeostasis. Various laboratories have demonstrated the importance of epigenetic modifications in HSC fate determination [86–91]. The Polycomb group (PcG) chromatin-modifying proteins have been implicated in the epigenetic regulation of HSC self-renewal, differentiation, and quiescence. Bmi-1, a member of the PcG family, has been shown to increase HSC self-renewal when it is overexpressed, while loss of its expression was associated with an overall decrease in HSC self-renewal [88]. Other PcG proteins such as Rae28 have shown to be necessary in maintaining hematopoiesis by regulating HSC activity [89], and Ezh2, involved in histone methylation and deacetylation, has been shown to prevent HSC exhaustion and fully conserve long-term repopulating potential of HSCs [90,91]. An alternative approach to demonstrate the importance of epigenetic modification in HSC homeostasis was taken by Milhem et al. [87]. They chemically blocked DNA methylation and histone deacetylation by using 5azaD and TSA and observed changes in HSC and progenitors cell fate.

As our understanding of HSC development and its basic regulators deepens, we increase the number of potential targets for gene therapy. Once we know the players involved, we may be able to manipulate the pathways to expand HSC numbers and achieve therapeutic levels of desired cell types.

32.3.3 HSC NICHE

The interaction of an HSC with its microenvironment "niche" is critical for HSC homeostasis. The role of the HSC niche is proposed to regulate a balance between differentiation and self-renewal. Schofield in 1978 was the first to propose the idea of a niche [92]. Nowadays the "HSC niche" most commonly refers to the BM encapsulated by bone or more specifically osteoblasts and stromal cells. Direct evidence of the important roles

the niche exerts in nurturing HSC comes from one study that found by increasing the number of osteoblastic cells in the trabecular region of murine hind legs, the absolute number of HSCs increased accordingly [93]. The microenvironment may provide direct cell–cell contact to HSCs through surface molecules as well as provide growth factors such as stem cell factor (SCF), the ligand to c-Kit on HSCs [94], Wnt molecules, ligands of Wingless pathway [95], and Jagged1, a ligand of Notch [72], which will be discussed in further detail below. In addition, a more recent study found that osteoblasts also secrete a small glycoprotein, osteopontin, to regulate HSC cell numbers [96].

As discussed in the self-renewal portion, Notch signaling is involved in HSC expansion. HSCs express Notch and the osteoblasts express their ligand Jagged. The osteoblasts are therefore directly involved in regulating HSC differentiation and self-renewal. C-Kit is another receptor expressed on HSCs. It is relatively well characterized and has been shown to be essential for HSC survival. The absence of c-Kit expression in HSC [97] or its ligand SCF that is expressed by the support cells in the niche lead to hematopoietic failures in both murine and human, again demonstrating the importance of the communication between the HSC and its niche [94].

Tie2 is also a receptor expressed on the surface of HSCs. The ligand Ang-1, expressed on osteoblasts, has been shown to allow the adhesion of HSCs to osteoblasts. Furthermore, this receptor pair has been shown to maintain HSC quiescence [98]. Another receptor, which was more recently discovered to be involved in confining HSCs to the niche, is CaR. It has been found that loss of CaR is associated with a loss of HSC attachment to the osteoblasts in the niche even though they were able to home to bone marrow properly [99].

Besides providing direct contact with the HSCs, osteoblasts produce hematopoietic growth factors that help regulate HSC differentiation [100–102]. Some of these factors are IL-1, IL-6 [103–105], GM-CSF, M-CSF [106,107], and G-CSF [108]. Though it has been clearly demonstrated that these factors are needed for HSC survival and expansion *in vivo*, when added *in vitro*, these factors are not enough to maintain HSCs in culture for long periods of time.

In the context of BM transplantations, transplanted HSC must successfully migrate to their niche in order to repopulate the ablated bone marrow. How the HSCs find their way to the niche, so called "home" to the niche, is not well understood. Nevertheless, a number of players have been identified. Adhesion molecules and their respective receptors, integrins, and selectins as well as chemokines have been shown to be involved in HSC homing to the BM [109,110]. Specifically, the involvement SDF-1, a potent HSC chemoattractant, and its receptor CXCR4 (present on the HSCs), was found to be necessary not only for HSC mobilization and homing but also HSC survival [111–116]. SDF-1 from endothelial cells and osteoblasts attracts HSCs to migrate into the niche through a process that is suggested to be mediated by E and P selectins [110,115,117,118].

Another adhesion molecule shown to be involved in the homing of HSCs is VLA-4 and its interaction with vascular cell adhesion molecule-1 (VCAM-1) (expressed on endothelial and stromal cells). It has been demonstrated that by blocking this interaction with antibodies, HSCs remain mobilized in peripheral blood and are not able to home to the bone marrow niche [109,119–121]. CD44 and RHAMM, two receptors for HA (hyaluronan) expressed on the HSCs, have also been implicated in HSC mobilization. Mice lacking CD44 have decreased mobilization. Furthermore, inhibiting CD44 signaling has been shown to block homing of HSCs to the BM [122–125].

32.3.4 HSC AND AGING

Under homeostasis, the pool of HSCs is maintained to support an individual throughout a lifetime. HSCs therefore have to be present in ample amounts in both young and old mice, which is also for humans with a much longer lifespan than mice. In fact, the HSC compartment of older mice is expanded relative to that of younger mice [126–128]. Even though HSCs derived from both old and young mice have the ability to reconstitute and rescue a lethally irradiated recipient, there is a decline of HSC function on a per-HSC basis [128]. This suggests that the "stemness" of HSCs decreases with age or possibly with each self-renewal division [129–132]. Another indication of this phenomenon is seen among different strains of mice. For instance, C57Bl/6 mice live longer than DBA mice, and if one compares the HSCs function of the two strains, C57Bl/6-derived HSCs outcompete HSCs derived from DBA mice. Furthemore, the DBA HSCs have been shown to divide faster than that of C57Bl/6 and might therefore be "aging" faster [133,134]. Again, this suggests that increased division leads to faster cell exhaustion.

32.4 GENETIC MANIPULATION OF HSC

As discussed earlier, HSCs have the ability to give rise to all the different blood cell types. Any genetic manipulation made to the HSC is therefore passed on to its progeny, making HSCs ideal targets for gene therapy. Thus far, HSCs have been targeted using different vectors, some of which are listed below and reviewed in more detail elsewhere in this book. It is important to note that there is no one good system for targeting, and depending on the type of "disease," the vector system needs to be chosen accordingly.

32.4.1 RETROVIRAL AND LENTIVIRAL VECTORS

Retrovirus and lentivirus have been extensively used in preclinical and clinical gene therapy trials. These viruses are particularly advantageous because they integrate into the host genome, allowing for stable long-term expression. In the absence of integration, the viral DNA would be lost as the stem and progenitor cells proliferate and differentiate. However, with the advantage of stable chromosomal integration comes the risk of potential insertional mutagenesis, which has been one of the major setbacks in using these vector systems in human clinical trials [135–138]. Another disadvantage to retroviral and lentiviral gene transfer is the inability, or compromised ability, to infect nondividing cells. When

aiming to target HSCs, this proves to be a major problem since most HSCs are quiescent. However, this problem can be overcome by stimulation of the HSCs to proliferate. When using these vectors, there is also a limit to the insert size as the retro- and lentiviruses can only package approximately 10 kb of genetic material.

A second major advantage of using one of these two vector systems is the ability "pseudotype" the viruses. This means that the envelope of these viruses can be replaced with the envelope of another, allowing for a broader host range. This technique has also been widely utilized to make viruses that specifically or preferentially infect a certain cell type, a desired feature for gene therapy.

32.4.2 ADENO-ASSOCIATED VIRUS

Adeno-associated virus (AAV) is a defective virus requiring coinfection with a helper virus (i.e., Ad). AAV can persist as an episome or in the absence of helper virus, it can stably integrate into a specific site on chromosome 19 of the host genome where it remains dormant until rescued by a helper virus. A major advantage of AAV is that it can infect both diving and nondividing cells, making it a useful tool for HSC targeting. One major disadvantage; however, is its minimal capacity. AAV can only accommodate ~4.7 kb, limiting the number of possible therapeutic genes. Furthermore, the difficulty of achieving high titers has kept AAV from being more widely used in clinical trials.

Despite its limitations, AAV possesses qualities desirable for gene therapy. AAV does not appear to be immunogenic, allowing for multiple administrations; furthermore it has a broad host range. AAV has been shown to stably transduce HSCs; however, the extent of its efficacy has been controversial where independent studies have shown transduction efficiency varying from 4% to 80% [139–142]. Nonetheless, it has proven to be a potential tool for targeting HSCs.

32.4.3 ADENOVIRUS

Long-term expression of the gene of interest is not always desired, in which case Adenovirus (Ad) has a great advantage. This vector has a large packaging capacity and therefore allows for large inserts unlike AAV. Because Ad does not integrate, it allows for transient transductions and expression. One major drawback is that Ad viral proteins have been shown to be immunogenic. There does appear to be a dose-dependent toxicity so repeated administration needs to be avoided.

Similar to AAV, it can infect both dividing and nondividing cells, which is again advantageous for targeting HSCs. The first demonstration of Ad's ability to transduce hematopoietic cells was as shown early as 1985 [142]. Another major advantage of using Ad is that it allows for high expression levels of the therapeutic gene. Furthermore, with Ad, one can achieve high titers and thereby sufficient virus for large-scale transductions as with clinical trials.

32.4.4 Nonviral Alternatives

During the recent years, nonviral alternatives for gene delivery have become more attractive. Some general advantages that the nonviral delivery methods mentioned below have in common are that they allow for a wide range of cells to be targeted, as they are not limited by receptors, nor do they have a major limitation of DNA insert size. Furthermore, because there is no major toxicity issues as with the viral particles, repeated administrations can be performed.

Injection of naked DNA is conceptually the simplest method for gene delivery. The major disadvantage of this technique; however, is that it only allows for targeting of one cell at a time. Though this may be desirable for certain applications, it is not ideal for targeting the large number of cells commonly needed to achieve for therapeutic targeting. Furthermore, though this method has proven to be efficient, it is slow and rather laborious.

Lipofection is also a well-established technique used as an alternative to viral gene delivery [143]. It involves the delivery of RNA and/or DNA to cells by the use of positively charged lipids. This method takes advantage of the negatively charged nucleic acids. The electrostatic interaction between the positively charged macromolecules and DNA/RNA creates complexes that are more easily taken up by the cells through fusion with the cell membrane. Lipofection provides an efficient alternative to viral delivery. It has been shown to work for transduction of a wide range of cell types, however, only provides a transient transduction.

Another nonviral method for gene delivery is electroporation, which involves the exposure of cells to a high-voltage electrical current that creates pores in the cell membrane, allowing for entry of nucleic acids, proteins, and even certain drugs. Electroporation has been used not only for gene delivery *in vitro* but also *in vivo*. The main target cell for *in vivo* studies has been skin cells, however, as the technology develops, tissues that reside deeper in the body may be targeted. As with the lipofection and injection of naked DNA, the expression of inserted DNA is transient. However, combining these techniques with homologous recombination [144] could allow for long-term correction. Unlike the naked DNA injection, electroporation allows for targeting a large number of cells at once and is significantly faster and more effortless. Electroporation had been used in a gene therapy trial for *ex vivo* gene transfer of IL-2 to natural killer cells for treatment of certain cancers [145]. As the technology develops, its use will most likely expand. Recently, a modification of electroporation has been made, referred to as "nucleofection." Nucleofection allows for DNA/RNA to enter directly into the nucleus and therefore increases the efficacy of gene expression. This method has already proven to be a promising tool for gene therapy, as it is tailored for certain cell types, allowing for transduction of various primary cells and cell lines that have in the past shown to be difficult to transfect [146].

32.5 THERAPEUTIC APPLICATIONS

The ability of HSCs to consistently supply blood progeny throughout one's lifespan has been utilized in human clinical treatments. Hematopoietic stem cell transplantation (HSCT) is one of the most potent therapeutic strategies to correct hematopoietic diseases with blood cell malfunction such as immunodeficiency, autoimmune diseases, anemia, and leukemia. Many of these diseases are corrected by grafting HSCs enriched from healthy donors into patients. However, the most successful allogenic blood engraftment has matched MHC molecules between donor and recipients, and very often it is hard to find a perfect match for each patient. On the other hand, HSC are among the few tissue cells that can be easily isolated, gene modified by virus transduction, and reinfused into patients. It has therefore been successful to combine gene targeting in HSCs and autologous transplantation to correct hematopoietic diseases with genetic defects. In addition, with the development of umbilical cord blood banks, it has been proposed to utilize self-umbilical cord blood that is enriched for HSCs as a source for stem cell gene targeting. Discussed below are different strategies currently used to treat a range of hematopoietic disorders.

32.5.1 Hematopoietic Disorders and HSC Transplantation

The first successful bone marrow transplantation was reported by E. Donnall Thomas in 1975 [147]. After this report, bone marrow transplantation was gradually accepted to treat blood-borne malignancy. More recently, it was discovered that peripheral blood could be a rich source for HSCs when treating donors with G-CSF to mobilize the HSC from bone marrow. With the combination of FACS or magnetic beads in conjunction with antibodies to HSC surface markers such as CD34, HSCs can be further purified. To date, HSCT has been utilized in combination with chemotherapy regimens to treat malignant blood diseases such as myeloid and lymphoid leukemia [148–152]. Recently, HSCTs have also been performed to treat autoimmune diseases [153].

32.5.2 Genetic Defects in Hematopoietic Disease and Potential Gene Therapy

Allogeneic transplantations are often accompanied by side effects such as graft-versus-host disease in which the engrafted donor immune cells attack host cells and therefore present a certain risk. With some diseases such as Wiskott–Aldrich syndrome (WAS), a genetic defect resulting in immunodeficiency, autologous transplantations have been demonstrated to be the best option to avoid high mortality [154,155]. Fortunately, HSCs are relatively easy to isolate, genetically manipulated, and reinfused. Gene targeting in HSCs have been proposed to correct some genetic hematopoietic defects. There are a number of mouse models for

TABLE 32.1
Animal and *in Vitro* Models of Gene Therapy of HSC Disorders

Deficiency	Vector	Outcome	References
JAK-3 mutations	Retrovirus	Complete correction *in vivo* in mouse	[161,162]
WAS	Lentivirus	Complete restoration of WASP expression in blood cells *in vivo* in mouse. Restore human T cell function *in vitro*	[164,165,167,188]
CGD	Lentvirus	Complete correction *in vitro* with human cells; partial restoration of phox expression and neutrophil generation in NOD/SCID mouse with human cell	[174]
Fanconi anemia	Retrovirus; AAV (*in vitro*)	Complete correction *in vivo* in mouse; complete correction *in vitro* with human cells	[177,178]
Sickle cell anemia	Lentivirus	Complete correction *in vivo* in mouse	[179]
β-Thalassemia	Lentivirus	Complete correction *in vivo* in mouse	[180]

human immunodeficiencies that have served as templates for gene therapy trials in human (Table 32.1) and these are discussed below.

32.5.2.1 Animal and *in Vitro* Models for Gene Therapy

JAK-3 is a protein tyrosine kinase that is involved in cytokine signaling [156]. It associates with the common gamma chain, which is used by the cytokines IL-2, 4, 7, 9, and 15 [156]. Lack of JAK-3 function results in low T and NK cell numbers, leading to opportunistic infections and often death [157]. It is estimated that about 7% of human severe combined immunodeficiency (SCID) cases are due to defective JAK-3 expression. JAK-3 null mice have a similar phenotype to humans, although they have low levels of B cells as well [158–160]. One group has used retrovirus expressing JAK-3 to transduce murine HSC, which are then transplanted into JAK-3 null mice. These studies have shown that restoration of JAK-3 by this method increases the numbers of B and T cells. Most importantly, these mice were able to generate normal immune responses after immunization [161] as well as after challenge with influenza A virus [162]. Since human JAK-3-deficient patients are often prone to virus infections, these studies show that the use of gene therapy of HSC to restore JAK-3 expression may be applicable in humans.

Another immunodeficiency in which gene therapy may be useful is WAS. WAS is an X-linked disorder that is characterized by thrombocytopenia, eczema, and abnormalities in cell-mediated and antibody immune responses [163]. The protein involved in this disorder is the WAS protein (WASP). Studies have shown that retroviral-mediated restoration of WASP in B cell lines, T cell lines, and primary T cells from WAS patients can partially restore normal function in these cells [164,165]. More recently, the WASP-null mice have been used to set up a therapeutic protocol, which combines an improved vector design and a HSC transduction setting. In this series of studies, WASP is expressed under control of its autologous promoter and purified precursor cells were transduced and subsequently

generate blood cells expressing WASP protein. These strategies have successfully restored T cell proliferation, immune synapse formation, and cytokine secretion, which are important to correct the T cell defect in the WAS syndrome [166,167]. However, how the animals respond to pathogens is yet to be determined.

Chronic granulomatous disease (CGD) results from a mutation in one of four proteins that make up nicotinamide adenine dinucleotide phosphate (NADPH) oxidase. NADPH oxidase generates large amounts of superoxide in the respiratory burst. CGD patients have diminished NADPH activity; as a result, they are susceptible to bacterial and fungal infections [168]. Several investigators have used mouse models for CGD to test restoration of NADPH oxidase activity by retroviral transfer of the defective protein subunit [169–172]. In these studies, WBM (or bone marrow partially enriched for HSC) was transduced with a retroviral vector carrying a functional NADPH oxidase subunit gene. The cells were then transplanted into irradiated recipients, which lack that gene. In all cases, NADPH oxidase activity was only partially restored, as protein levels of the expressed gene was detected at low amounts. However, this partial restoration of NADPH activity was sufficient to achieve marked improvement in defense against pathogens that would decimate mice lacking the gene entirely. One study [173] has shown that retroviral transduction of a phox gene into human CD34+ cells from CGD patients results in restoration of about half the phox activity in these cells. Protein levels of the transduced gene were similar to those in wild-type cells. More recently, two studies have further showed successful restoration of phox genes in CD34 human progenitors cells *ex vivo* with lentiviral vectors, and these corrected cells can further differentiate into neutrophil cells *in vivo* in NOD/SCID mice [174,175].

Fanconi anemia (FA) is a disorder in which cells are hypersensitive to DNA cross-linking agents such as mitomycin C [176]. The disease, which is due to mutations in the FANC-C gene, often results in bone marrow failure and predisposition to malignancies [176]. A retrovirus carrying a functional FANC-C gene was used to transduce bone marrow from

FANC-C deficient mice [177]. The cells were then transplanted into FANC-C deficient mice and subjected to treatment with mitomycin C. Mice receiving the corrected gene exhibited normal blood counts, while mice not receiving the gene exhibited decreased blood counts, leading to death. Therefore, phenotypic correction was established by retroviral transfer of the FANC-C gene. In addition, proof-of-principle studies using human CD34 cells from FA patients have demonstrated that AAV [178] or retroviral [178] delivery of the human FANC-C gene can restore resistance to mitomycin C *in vitro*. Similarly, sickle cell anemia and β-thalassemia have been corrected in mouse models by lentiviral transduction of HSC [179,180].

In summary, gene therapy of HSC in mouse models of human diseases has provided proof-of-principle evidence that gene therapy can be used to correct some genetic disorders, although the translation from mouse to human studies can be quite difficult. The use of mouse models for human disease does provide a template to design and test gene therapy protocols for future human trials.

32.5.2.2 Clinical Trials of Gene Therapy for Human Hematopoietic Disorders

There are many human hematopoietic disorders that could potentially be treated using gene therapy (Table 32.2). In most of these diseases, transplantation of bone marrow from a healthy donor could potentially cure the patient. However, a suitable donor is not always available and the transplantation procedure itself can be toxic. Therefore, the use of gene therapy as an alternative treatment would be a significant advance. The first disorder in which a gene therapy protocol was used in clinical trials was ADA-SCID, which occurs due to deficiency in adenosine deaminase (ADA) expression, resulting in SCID. Patients can be successfully treated by receiving exogenous ADA (PEG-ADA), but must continue this regimen for life. Early gene therapy trials used a retrovirus expressing an ADA minigene to transduce umbilical cord blood or bone marrow cells or peripheral blood lymphocytes [165,181,182]. Although some transduced cells persisted and were functional, patients were still dependent on PEG-ADA treatment; removal of PEG-ADA resulted in a relapse into immunodeficiency.

Correction of ADA-SCID by gene therapy has now been improved with better HSC transduction and transplantation protocols [183]. Two patients were reinfused with their transduced bone marrow after a nonmyeloablative-conditioning regimen. Normalization of the patient's immune system in the absence of PEG-ADA was achieved and had been sustained for about 1 year at the time of the report [183].

The first successful human gene therapy trials were performed on patients with X-linked SCID (SCID-X1). SCID-X1 is caused by mutations in the common gamma chain, which is a signaling chain used by several cytokines. Lack of this chain results in drastically reduced T- and NK-cell development, leading to SCID. CD34 cells from SCID-X1 patients were transduced with a retrovirus carrying the common gamma chain and infused, without myeloablation, into the patients. Nearly 1 year after this, immune function was normal and patients were living normal lives at home [184,185]. However, two of the 10 patients treated in this manner developed a leukemia-like disease due to integration of the retrovirus into the LMO2 locus, activating this powerful known proto-oncogene [186], raising questions as to the safety of integrating viruses for gene therapy.

Retroviral transduction of HSC has also been conducted for disorders such as CGD and FA as discussed earlier [187]. Although these studies have shown long-term expression of the defective genes, the expression level and transduction rates have been too low to show any lasting clinical benefit.

While the successful trials with SCID-X1 and ADA-SCID patients reflect the promise of gene therapy, failures of other trials represents the obstacles needed to be overcome. One of these obstacles is the frequent problem of low expression levels of the transgene following transduction. However, as demonstrated by the mouse CGD experiments described earlier, low transgene expression may be sufficient to restore function to the affected cells. In other situations, even low levels of transgene expression may be sufficient for preferential selection of the transduced cells, bypassing the need for high transduction efficiencies and high gene expression levels. Another question is how many HSCs must be transduced in order to achieve lasting therapeutic effect. It is likely that these questions must be empirically answered for each disorder being treated.

The retroviral-linked leukemia in also outlines another problem in gene therapy that needs to be addressed [186]. Studies regarding the safety of different integrating retroviruses, and the study of stable expression via nonintegrating episomes may provide more information on the plausibility of safe gene therapy treatments. Although gene therapy of

TABLE 32.2
Clinical Trials Using Gene Therapy of HSC

Disorder	Vector Used	Result	References
ADA-SCID	Retrovirus	Partial correction	[181,182,189,190]
SCID-X1	Retrovirus	Complete correction	[184,185]
CGD	Retrovirus	Very few cells corrected	[191]
FA	Retrovirus	Partial transient correction	[187]

HSC has shown great promise, these questions must be answered before this type of treatment can be used on a regular basis.

ACKNOWLEDGMENTS

We thank Shannon McKinney-Freeman and Steven Bradfute for their contributions to the original version of this chapter.

REFERENCES

1. Ford CE, Hamerton JL, Barnes DWH, Loutit JF. Cytological identification of radiation chimaeras. *Nature* 1956; 177:452–454.
2. Camargo FD, Chambers SM, Drew E, McNagny KM, Goodell MA. Hematopoietic stem cells do not engraft with absolute efficiencies. *Blood* 2006;107:501–507.
3. Morrison SJ, Weissman IL. The long-term repopulating subset of hematopoietic stem cells is deterministic and isolatable by phenotype. *Immunity* 1994;1:661–673.
4. Christensen JL, Weissman IL. Flk-2 is a marker in hematopoietic stem cell differentiation: A simple method to isolate long-term stem cells. *Proc Natl Acad Sci USA* 2001;98:14541–14546.
5. Passegue E, Wagers AJ, Giuriato S, Anderson WC, Weissman IL. Global analysis of proliferation and cell cycle gene expression in the regulation of hematopoietic stem and progenitor cell fates. *J Exp Med* 2005;202:1599–1611.
6. Cheshier SH, Morrison SJ, Liao X, Weissman IL. In vivo proliferation and cell cycle kinetics of long-term self-renewing hematopoietic stem cells. *Proc Natl Acad Sci USA* 1999; 96:3120–3125.
7. Ford CE, Hamerton JL, Barnes DW, Loutit JF. Cytological identification of radiation-chimaeras. *Nature* 1956;177: 452–454.
8. Brecher G, Cronkite EP. Post-radiation parabiosis and survival in rats. *Proc Soc Exp Biol Med* 1951;77:292–294.
9. Goodman JW, Hodgson GS. Evidence for stem cells in the peripheral blood of mice. *Blood* 1962;19:702–714.
10. Barr RD, Whang-Peng J, Perry S. Hemopoietic stem cells in human peripheral blood. *Science* 1975;190:284–285.
11. McCredie KB, Hersh EM, Freireich EJ. Cells capable of colony formation in the peripheral blood of man. *Science* 1971;171:293–294.
12. Socinski MA, Cannistra SA, Elias A, Antman KH, Schnipper L, Griffin JD. Granulocyte-macrophage colony stimulating factor expands the circulating haemopoietic progenitor cell compartment in man. *Lancet* 1988;1:1194–1198.
13. Gluckman E, Broxmeyer HA, Auerbach AD, Friedman HS, Douglas GW, Devergie A, Esperou H, Thierry D, Socie G, Lehn P, et al. Hematopoietic reconstitution in a patient with fanconi's anemia by means of umbilical-cord blood from an hla-identical sibling. *N Engl J Med* 1989;321:1174–1178.
14. Hofmeister CC, Zhang J, Knight KL, Le P, Stiff PJ. *Ex vivo* expansion of umbilical cord blood stem cells for transplantation: Growing knowledge from the hematopoietic niche. *Bone Marrow Transplant* 2007;39:11–23.
15. Medvinsky AL, Dzierzak EA. Development of the definitive hematopoietic hierarchy in the mouse. *Dev Comp Immunol* 1998;22:289–301.
16. Lensch MW, Daley GQ. Origins of mammalian hematopoiesis: In vivo paradigms and in vitro models. *Curr Top Dev Biol* 2004;60:127–196.
17. Medvinsky A, Dzierzak E. Definitive hematopoiesis is autonomously initiated by the agm region. *Cell* 1996;86:897–906.
18. Muller AM, Medvinsky A, Strouboulis J, Grosveld F, Dzierzak E. Development of hematopoietic stem cell activity in the mouse embryo. *Immunity* 1994;1:291–301.
19. Gekas C, Dieterlen-Lievre F, Orkin SH, Mikkola HK. The placenta is a niche for hematopoietic stem cells. *Dev Cell* 2005;8:365–375.
20. Kyba M, Daley GQ. Hematopoiesis from embryonic stem cells: Lessons from and for ontogeny. *Exp Hematol* 2003;31:994–1006.
21. Doetschman TC, Eistetter H, Katz M, Schmidt W, Kemler R. The in vitro development of blastocyst-derived embryonic stem cell lines: Formation of visceral yolk sac, blood islands and myocardium. *J Embryol Exp Morphol* 1985;87:27–45.
22. Hole N, Graham GJ, Menzel U, Ansell JD. A limited temporal window for the derivation of multilineage repopulating hematopoietic progenitors during embryonal stem cell differentiation in vitro. *Blood* 1996;88:1266–1276.
23. Muller AM, Dzierzak EA. Es cells have only a limited lymphopoietic potential after adoptive transfer into mouse recipients. *Development* 1993;118:1343–1351.
24. Potocnik AJ, Nerz G, Kohler H, Eichmann K. Reconstitution of b cell subsets in rag deficient mice by transplantation of in vitro differentiated embryonic stem cells. *Immunol Lett* 1997;57:131–137.
25. Perlingeiro RC, Kyba M, Daley GQ. Clonal analysis of differentiating embryonic stem cells reveals a hematopoietic progenitor with primitive erythroid and adult lymphoid-myeloid potential. *Development* 2001;128:4597–4604.
26. Kyba M, Perlingeiro RC, Daley GQ. Hoxb4 confers definitive lymphoid-myeloid engraftment potential on embryonic stem cell and yolk sac hematopoietic progenitors. *Cell* 2002;109:29–37.
27. Wang Y, Yates F, Naveiras O, Ernst P, Daley GQ. Embryonic stem cell-derived hematopoietic stem cells. *Proc Natl Acad Sci USA* 2005;102:19081–19086.
28. Burt RK, Verda L, Kim DA, Oyama Y, Luo K, Link C. Embryonic stem cells as an alternate marrow donor source: Engraftment without graft-versus-host disease. *J Exp Med* 2004;199:895–904.
29. Basch RS, Janossy G, Greaves MF. Murine pluripotential stem cells lack ia antigen. *Nature* 1977;270:520–522.
30. Fitchen JH, Le Fevre C, Ferrone S, Cline MJ. Expression of ia-like and hla-a,b antigens on human multipotential hematopoietic progenitor cells. *Blood* 1982;59:188–190.
31. Falkenburg JH, Fibbe WE, Goselink HM, Van Rood JJ, Jansen J. Human hematopoietic progenitor cells in long-term cultures express hla-dr antigens and lack hla-dq antigens. *J Exp Med* 1985;162:1359–1369.
32. Griffin JD, Sabbath KD, Herrmann F, Larcom P, Nichols K, Kornacki M, Levine H, Cannistra SA. Differential expression of hla-dr antigens in subsets of human cfu-gm. *Blood* 1985;66:788–795.
33. Sutherland HJ, Eaves CJ, Eaves AC, Dragowska W, Lansdorp PM. Characterization and partial purification of human marrow cells capable of initiating long-term hematopoiesis in vitro. *Blood* 1989;74:1563–1570.
34. Huang S, Terstappen LW. Lymphoid and myeloid differentiation of single human cd34+, hla-dr+, cd38– hematopoietic stem cells. *Blood* 1994;83:1515–1526.
35. Papayannopoulou T, Brice M, Broudy VC, Zsebo KM. Isolation of c-kit receptor-expressing cells from bone marrow,

peripheral blood, and fetal liver: Functional properties and composite antigenic profile. *Blood* 1991;78:1403–1412.

36. Visser JW, Bauman JG, Mulder AH, Eliason JF, de Leeuw AM. Isolation of murine pluripotent hemopoietic stem cells. *J Exp Med* 1984;159:1576–1590.

37. Spangrude GJ, Heimfeld S, Weissman IL. Purification and characterization of mouse hematopoietic stem cells. *Science* 1988;241:58–62.

38. Andrews RG, Takahashi M, Segal GM, Powell JS, Bernstein ID, Singer JW. The l4f3 antigen is expressed by unipotent and multipotent colony-forming cells but not by their precursors. *Blood* 1986;68:1030–1035.

39. Terstappen LW, Huang S, Safford M, Lansdorp PM, Loken MR. Sequential generations of hematopoietic colonies derived from single nonlineage-committed cd34+, cd38– progenitor cells. *Blood* 1991;77:1218–1227.

40. Civin CI, Strauss LC, Brovall C, Fackler MJ, Schwartz JF, Shaper JH. Antigenic analysis of hematopoiesis. III. A hematopoietic progenitor cell surface antigen defined by a monoclonal antibody raised against kg-1a cells. *J Immunol* 1984;133:157–165.

41. Berenson RJ, Andrews RG, Bensinger WI, Kalamasz D, Knitter G, Buckner CD, Bernstein ID. Antigen cd34+ marrow cells engraft lethally irradiated baboons. *J Clin Invest* 1988;81:951–955.

42. Okada S, Nakauchi H, Nagayoshi K, Nishikawa S, Miura Y, Suda T. In vivo and in vitro stem cell function of c-kit- and sca-1-positive murine hematopoietic cells. *Blood* 1992; 80:3044–3050.

43. Osawa M, Hanada K, Hamada H, Nakauchi H. Long-term lymphohematopoietic reconstitution by a single cd34-low/negative hematopoietic stem cell. *Science* 1996;273:242–245.

44. Kiel MJ, Yilmaz OH, Iwashita T, Yilmaz OH, Terhorst C, Morrison SJ. Slam family receptors distinguish hematopoietic stem and progenitor cells and reveal endothelial niches for stem cells. *Cell* 2005;121:1109–1121.

45. Goodell MA, Rosenzweig M, Kim H, Marks DF, DeMaria M, Paradis G, Grupp SA, Sieff CA, Mulligan RC, Johnson RP. Dye efflux studies suggest that hematopoietic stem cells expressing low or undetectable levels of cd34 antigen exist in multiple species. *Nat Med* 1997;3:1337–1345.

46. Ogawa M, Matsuzaki Y, Nishikawa S, Hayashi S, Kunisada T, Sudo T, Kina T, Nakauchi H. Expression and function of c-kit in hemopoietic progenitor cells. *J Exp Med* 1991;174: 63–71.

47. Ikuta K, Weissman IL. Evidence that hematopoietic stem cells express mouse c-kit but do not depend on steel factor for their generation. *Proc Natl Acad Sci USA* 1992;89:1502–1506.

48. Briddell RA, Broudy VC, Bruno E, Brandt JE, Srour EF, Hoffman R. Further phenotypic characterization and isolation of human hematopoietic progenitor cells using a monoclonal antibody to the c-kit receptor. *Blood* 1992;79:3159–3167.

49. Moore T, Huang S, Terstappen LW, Bennett M, Kumar V. Expression of cd43 on murine and human pluripotent hematopoietic stem cells. *J Immunol* 1994;153:4978–4987.

50. Pallavicini MG, Summers LJ, Dean PN, Gray JW. Enrichment of murine hemopoietic clonogenic cells by multivariate analyses and sorting. *Exp Hematol* 1985;13:1173–1181.

51. Neben S, Redfearn WJ, Parra M, Brecher G, Pallavicini MG. Short- and long-term repopulation of lethally irradiated mice by bone marrow stem cells enriched on the basis of light scatter and Hoechst 33342 fluorescence. *Exp Hematol* 1991;19:958–967.

52. Wolf NS, Kone A, Priestley GV, Bartelmez SH. In vivo and in vitro characterization of long-term repopulating primitive hematopoietic cells isolated by sequential Hoechst 33342-rhodamine 123 facs selection. *Exp Hematol* 1993;21:614–622.

53. Goodell MA, Brose K, Paradis G, Conner AS, Mulligan RC. Isolation and functional properties of murine hematopoietic stem cells that are replicating in vivo. *J Exp Med* 1996;183:1797–1806.

54. Zhou S, Schuetz JD, Bunting KD, Colapietro AM, Sampath J, Morris JJ, Lagutina I, Grosveld GC, Osawa M, Nakauchi H, Sorrentino BP. The abc transporter bcrp1/abcg2 is expressed in a wide variety of stem cells and is a molecular determinant of the side-population phenotype. *Nat Med* 2001; 7:1028–1034.

55. Goodell MA. Stem cell identification and sorting using the Hoechst 33342 side population (sp); in Robinson JP, Darzynkiewicz Z, Dean PN, Hibbs AR, Orfao A, Rabinovitch PS, Wheeless LL (eds). *Current Protocols in Cytometry*, vol 2, Wiley, New York, 2002, pp. 9.18.11–19.18.11.

56. Lin KK, Goodell MA. Purification of hematopoietic stem cells using the side population. *Methods Enzymol* 2006; 420:255–264.

57. Chaudhary PM, Roninson IB. Expression and activity of p-glycoprotein, a multidrug efflux pump, in human hematopoietic stem cells. *Cell* 1991;66:85–94.

58. Li CL, Johnson GR. Rhodamine 123 reveals heterogeneity within murine lin-, sca-1 + hemopoietic stem cells. *J Exp Med* 1992;175:1443–1447.

59. Johnson LV, Walsh ML, Chen LB. Localization of mitochondria in living cells with rhodamine 123. *Proc Natl Acad Sci USA* 1980;77:990–994.

60. Cheng T, Rodrigues N, Shen H, Yang Y, Dombkowski D, Sykes M, Scadden DT. Hematopoietic stem cell quiescence maintained by p21cip1/waf1. *Science* 2000;287:1804–1808.

61. Cheng T, Rodrigues N, Dombkowski D, Stier S, Scadden DT. Stem cell repopulation efficiency but not pool size is governed by p27(kip1). *Nat Med* 2000;6:1235–1240.

62. Yuan Y, Shen H, Franklin DS, Scadden DT, Cheng T. In vivo self-renewing divisions of haematopoietic stem cells are increased in the absence of the early g1-phase inhibitor, p18ink4c. *Nat Cell Biol* 2004;6:436–442.

63. Sauvageau G, Thorsteinsdottir U, Eaves CJ, Lawrence HJ, Largman C, Lansdorp PM, Humphries RK. Overexpression of hoxb4 in hematopoietic cells causes the selective expansion of more primitive populations in vitro and in vivo. *Genes Dev* 1995;9:1753–1765.

64. Thorsteinsdottir U, Sauvageau G, Humphries RK. Enhanced in vivo regenerative potential of hoxb4-transduced hematopoietic stem cells with regulation of their pool size. *Blood* 1999;94:2605–2612.

65. Bjornsson JM, Larsson N, Brun AC, Magnusson M, Andersson E, Lundstrom P, Larsson J, Repetowska E, Ehinger M, Humphries RK, Karlsson S. Reduced proliferative capacity of hematopoietic stem cells deficient in hoxb3 and hoxb4. *Mol Cell Biol* 2003;23:3872–3883.

66. DiMartino JF, Selleri L, Traver D, Firpo MT, Rhee J, Warnke R, O'Gorman S, Weissman IL, Cleary ML. The hox cofactor and proto-oncogene pbx1 is required for maintenance of definitive hematopoiesis in the fetal liver. *Blood* 2001;98:618–626.

67. Domen J, Cheshier SH, Weissman IL. The role of apoptosis in the regulation of hematopoietic stem cells: Overexpression of bcl-2 increases both their number and repopulation potential. *J Exp Med* 2000;191:253–264.

68. Opferman JT, Iwasaki H, Ong CC, Suh H, Mizuno S, Akashi K, Korsmeyer SJ. Obligate role of anti-apoptotic mcl-1 in the survival of hematopoietic stem cells. *Science* 2005;307:1101–1104.

69. Dallas MH, Varnum-Finney B, Martin PJ, Bernstein ID. Enhanced t-cell reconstitution by hematopoietic progenitors expanded *ex vivo* using the notch ligand delta1. *Blood* 2007;109:3579–3587.

70. Karanu FN, Murdoch B, Gallacher L, Wu DM, Koremoto M, Sakano S, Bhatia M. The notch ligand jagged-1 represents a novel growth factor of human hematopoietic stem cells. *J Exp Med* 2000;192:1365–1372.

71. Karanu FN, Murdoch B, Miyabayashi T, Ohno M, Koremoto M, Gallacher L, Wu D, Itoh A, Sakano S, Bhatia M. Human homologues of delta-1 and delta-4 function as mitogenic regulators of primitive human hematopoietic cells. *Blood* 2001;97:1960–1967.

72. Stier S, Cheng T, Dombkowski D, Carlesso N, Scadden DT. Notch1 activation increases hematopoietic stem cell self-renewal in vivo and favors lymphoid over myeloid lineage outcome. *Blood* 2002;99:2369–2378.

73. Carlesso N, Aster JC, Sklar J, Scadden DT. Notch1-induced delay of human hematopoietic progenitor cell differentiation is associated with altered cell cycle kinetics. *Blood* 1999;93:838–848.

74. Pui JC, Allman D, Xu L, DeRocco S, Karnell FG, Bakkour S, Lee JY, Kadesch T, Hardy RR, Aster JC, Pear WS. Notch1 expression in early lymphopoiesis influences b versus t lineage determination. *Immunity* 1999;11:299–308.

75. Reya T, Duncan AW, Ailles L, Domen J, Scherer DC, Willert K, Hintz L, Nusse R, Weissman IL. A role for wnt signalling in self-renewal of haematopoietic stem cells. *Nature* 2003;423:409–414.

76. Wang H, Gilner JB, Bautch VL, Wang DZ, Wainwright BJ, Kirby SL, Patterson C. Wnt2 coordinates the commitment of mesoderm to hematopoietic, endothelial, and cardiac lineages in embryoid bodies. *J Biol Chem* 2007;282:782–791.

77. Kirstetter P, Anderson K, Porse BT, Jacobsen SE, Nerlov C. Activation of the canonical wnt pathway leads to loss of hematopoietic stem cell repopulation and multilineage differentiation block. *Nat Immunol* 2006;7:1048–1056.

78. Scheller M, Huelsken J, Rosenbauer F, Taketo MM, Birchmeier W, Tenen DG, Leutz A. Hematopoietic stem cell and multilineage defects generated by constitutive beta-catenin activation. *Nat Immunol* 2006;7:1037–1047.

79. Cobas M, Wilson A, Ernst B, Mancini SJ, MacDonald HR, Kemler R, Radtke F. Beta-catenin is dispensable for hematopoiesis and lymphopoiesis. *J Exp Med* 2004;199:221–229.

80. Murdoch B, Chadwick K, Martin M, Shojaei F, Shah KV, Gallacher L, Moon RT, Bhatia M. Wnt-5a augments repopulating capacity and primitive hematopoietic development of human blood stem cells in vivo. *Proc Natl Acad Sci USA* 2003;100:3422–3427.

81. Austin TW, Solar GP, Ziegler FC, Liem L, Matthews W. A role for the wnt gene family in hematopoiesis: Expansion of multilineage progenitor cells. *Blood* 1997;89:3624–3635.

82. Trowbridge JJ, Scott MP, Bhatia M. Hedgehog modulates cell cycle regulators in stem cells to control hematopoietic regeneration. *Proc Natl Acad Sci USA* 2006;103:14134–14139.

83. Gering M, Patient R. Hedgehog signaling is required for adult blood stem cell formation in zebrafish embryos. *Dev Cell* 2005;8:389–400.

84. Dyer MA, Farrington SM, Mohn D, Munday JR, Baron MH. Indian hedgehog activates hematopoiesis and vasculogenesis and can respecify prospective neurectodermal cell fate in the mouse embryo. *Development* 2001;128:1717–1730.

85. Bhardwaj G, Murdoch B, Wu D, Baker DP, Williams KP, Chadwick K, Ling LE, Karanu FN, Bhatia M. Sonic hedgehog induces the proliferation of primitive human hematopoietic cells via bmp regulation. *Nat Immunol* 2001;2:172–180.

86. Lessard J, Sauvageau G. Polycomb group genes as epigenetic regulators of normal and leukemic hemopoiesis. *Exp Hematol* 2003;31:567–585.

87. Milhem M, Mahmud N, Lavelle D, Araki H, DeSimone J, Saunthararajah Y, Hoffman R. Modification of hematopoietic stem cell fate by 5aza 2′deoxycytidine and trichostatin a. *Blood* 2004;103:4102–4110.

88. Iwama A, Oguro H, Negishi M, Kato Y, Morita Y, Tsukui H, Ema H, Kamijo T, Katoh-Fukui Y, Koseki H, van Lohuizen M, Nakauchi H. Enhanced self-renewal of hematopoietic stem cells mediated by the polycomb gene product BMI-1. *Immunity* 2004;21:843–851.

89. Ohta H, Sawada A, Kim JY, Tokimasa S, Nishiguchi S, Humphries RK, Hara J, Takihara Y. Polycomb group gene rae28 is required for sustaining activity of hematopoietic stem cells. *J Exp Med* 2002;195:759–770.

90. Kamminga LM, Bystrykh LV, de Boer A, Houwer S, Douma J, Weersing E, Dontje B, de Haan G. The polycomb group gene ezh2 prevents hematopoietic stem cell exhaustion. *Blood* 2006;107:2170–2179.

91. Vire E, Brenner C, Deplus R, Blanchon L, Fraga M, Didelot C, Morey L, Van Eynde A, Bernard D, Vanderwinden JM, Bollen M, Esteller M, Di Croce L, de Launoit Y, Fuks F. The polycomb group protein ezh2 directly controls DNA methylation. *Nature* 2006;439:871–874.

92. Schofield R. The relationship between the spleen colony-forming cell and the haemopoietic stem cell. *Blood Cells* 1978;4:7–25.

93. Calvi LM, Adams GB, Weibrecht KW, Weber JM, Olson DP, Knight MC, Martin RP, Schipani E, Divieti P, Bringhurst FR, Milner LA, Kronenberg HM, Scadden DT. Osteoblastic cells regulate the haematopoietic stem cell niche. *Nature* 2003;425:841–846.

94. Heissig B, Hattori K, Dias S, Friedrich M, Ferris B, Hackett NR, Crystal RG, Besmer P, Lyden D, Moore MA, Werb Z, Rafii S. Recruitment of stem and progenitor cells from the bone marrow niche requires mmp-9 mediated release of kit-ligand. *Cell* 2002;109:625–637.

95. Moore KA. Recent advances in defining the hematopoietic stem cell niche. *Curr Opin Hematol* 2004;11:107–111.

96. Stier S, Ko Y, Forkert R, Lutz C, Neuhaus T, Grunewald E, Cheng T, Dombkowski D, Calvi LM, Rittling SR, Scadden DT. Osteopontin is a hematopoietic stem cell niche component that negatively regulates stem cell pool size. *J Exp Med* 2005;201:1781–1791.

97. Bernstein SE, Russell ES. Implantation of normal bloodforming tissue in genetically anemic mice, without X-irradiation of host. Proceedings of the Society for Experimental Biology and Medicine Society for Experimental Biology and Medicine, New York, 1959;101:769–773.

98. Arai F, Hirao A, Ohmura M, Sato H, Matsuoka S, Takubo K, Ito K, Koh GY, Suda T. Tie2/angiopoietin-1 signaling regulates hematopoietic stem cell quiescence in the bone marrow niche. *Cell* 2004;118:149–161.

99. Adams GB, Chabner KT, Alley IR, Olson DP, Szczepiorkowski ZM, Poznansky MC, Kos CH, Pollak MR, Brown EM, Scadden DT. Stem cell engraftment at the endosteal niche is specified by the calcium-sensing receptor. *Nature* 2006;439:599–603.

100. Taichman R, Reilly M, Verma R, Ehrenman K, Emerson S. Hepatocyte growth factor is secreted by osteoblasts and cooperatively permits the survival of haematopoietic progenitors. *Br J Haematol* 2001;112:438–448.

101. Taichman RS, Reilly MJ, Emerson SG. Human osteoblasts support human hematopoietic progenitor cells in vitro bone marrow cultures. *Blood* 1996;87:518–524.

102. Jung Y, Wang J, Havens A, Sun Y, Jin T, Taichman RS. Cell-to-cell contact is critical for the survival of hematopoietic progenitor cells on osteoblasts. *Cytokine* 2005;32:155–162.

103. Hanazawa S, Amano S, Nakada K, Ohmori Y, Miyoshi T, Hirose K, Kitano S. Biological characterization of interleukin-1-like cytokine produced by cultured bone cells from newborn mouse calvaria. *Calcif Tissue Int* 1987;41:31–37.

104. Feyen JH, Elford P, Di Padova FE, Trechsel U. Interleukin-6 is produced by bone and modulated by parathyroid hormone. *J Bone Miner Res* 1989;4:633–638.

105. Ishimi Y, Miyaura C, Jin CH, Akatsu T, Abe E, Nakamura Y, Yamaguchi A, Yoshiki S, Matsuda T, Hirano T, et al. Il-6 is produced by osteoblasts and induces bone resorption. *J Immunol* 1990;145:3297–3303.

106. Horowitz MC, Einhorn TA, Philbrick W, Jilka RL. Functional and molecular changes in colony stimulating factor secretion by osteoblasts. *Connect Tissue Res* 1989;20:159–168.

107. Elford PR, Felix R, Cecchini M, Trechsel U, Fleisch H. Murine osteoblastlike cells and the osteogenic cell mc3t3-e1 release a macrophage colony-stimulating activity in culture. *Calcif Tissue Int* 1987;41:151–156.

108. Felix R, Elford PR, Stoerckle C, Cecchini M, Wetterwald A, Trechsel U, Fleisch H, Stadler BM. Production of hemopoietic growth factors by bone tissue and bone cells in culture. *J Bone Miner Res* 1988;3:27–36.

109. Papayannopoulou T, Nakamoto B. Peripheralization of hemopoietic progenitors in primates treated with anti-vla4 integrin. *Proc Natl Acad Sci USA* 1993;90:9374–9378.

110. Frenette PS, Subbarao S, Mazo IB, von Andrian UH, Wagner DD. Endothelial selectins and vascular cell adhesion molecule-1 promote hematopoietic progenitor homing to bone marrow. *Proc Natl Acad Sci USA* 1998;95:14423–14428.

111. Dar A, Goichberg P, Shinder V, Kalinkovich A, Kollet O, Netzer N, Margalit R, Zsak M, Nagler A, Hardan I, Resnick I, Rot A, Lapidot T. Chemokine receptor cxcr4-dependent internalization and resecretion of functional chemokine sdf-1 by bone marrow endothelial and stromal cells. *Nat Immunol* 2005;6:1038–1046.

112. Lapidot T, Dar A, Kollet O. How do stem cells find their way home? *Blood* 2005;106:1901–1910.

113. Kawabata K, Ujikawa M, Egawa T, Kawamoto H, Tachibana K, Iizasa H, Katsura Y, Kishimoto T, Nagasawa T. A cell-autonomousrequirement for cxcr4 in long-term lymphoid and myeloid reconstitution. *Proc Natl Acad Sci USA* 1999;96:5663–5667.

114. Ma Q, Jones D, Springer TA. The chemokine receptor cxcr4 is required for the retention of b lineage and granulocytic precursors within the bone marrow microenvironment. *Immunity* 1999;10:463–471.

115. Kortesidis A, Zannettino A, Isenmann S, Shi S, Lapidot T, Gronthos S. Stromal-derived factor-1 promotes the growth, survival, and development of human bone marrow stromal stem cells. *Blood* 2005;105:3793–3801.

116. Peled A, Grabovsky V, Habler L, Sandbank J, Arenzana-Seisdedos F, Petit I, Ben-Hur H, Lapidot T, Alon R. The chemokine sdf-1 stimulates integrin-mediated arrest of cd34(+) cells on vascular endothelium under shear flow. *J Clin Invest* 1999;104:1199–1211.

117. Huang Y, Kucia M, Rezzoug F, Ratajczak J, Tanner MK, Ratajczak MZ, Schanie CL, Xu H, Fugier-Vivier I, Ildstad ST. Flt3-ligand-mobilized peripheral blood, but not flt3-ligand-expanded bone marrow, facilitating cells promote establishment of chimerism and tolerance. *Stem Cells* 2006;24:936–948.

118. Naiyer AJ, Jo DY, Ahn J, Mohle R, Peichev M, Lam G, Silverstein RL, Moore MA, Rafii S. Stromal derived factor-1-induced chemokinesis of cord blood cd34(+) cells (long-term culture-initiating cells) through endothelial cells is mediated by e-selectin. *Blood* 1999;94:4011–4019.

119. Papayannopoulou T, Craddock C, Nakamoto B, Priestley GV, Wolf NS. The vla4/vcam-1 adhesion pathway defines contrasting mechanisms of lodgement of transplanted murine hemopoietic progenitors between bone marrow and spleen. *Proc Natl Acad Sci USA* 1995;92:9647–9651.

120. Papayannopoulou T, Priestley GV, Nakamoto B, Zafiropoulos V, Scott LM, Harlan JM. Synergistic mobilization of hemopoietic progenitor cells using concurrent beta1 and beta2 integrin blockade or beta2-deficient mice. *Blood* 2001;97:1282–1288.

121. Scott LM, Priestley GV, Papayannopoulou T. Deletion of alpha4 integrins from adult hematopoietic cells reveals roles in homeostasis, regeneration, and homing. *Mol Cell Biol* 2003;23:9349–9360.

122. Schmits R, Filmus J, Gerwin N, Senaldi G, Kiefer F, Kundig T, Wakeham A, Shahinian A, Catzavelos C, Rak J, Furlonger C, Zakarian A, Simard JJ, Ohashi PS, Paige CJ, Gutierrez-Ramos JC, Mak TW. Cd44 regulates hematopoietic progenitor distribution, granuloma formation, and tumorigenicity. *Blood* 1997;90:2217–2233.

123. Pilarski LM, Pruski E, Wizniak J, Paine D, Seeberger K, Mant MJ, Brown CB, Belch AR. Potential role for hyaluronan and the hyaluronan receptor rhamm in mobilization and trafficking of hematopoietic progenitor cells. *Blood* 1999;93:2918–2927.

124. Avigdor A, Goichberg P, Shivtiel S, Dar A, Peled A, Samira S, Kollet O, Hershkoviz R, Alon R, Hardan I, Ben-Hur H, Naor D, Nagler A, Lapidot T. Cd44 and hyaluronic acid cooperate with sdf-1 in the trafficking of human cd34+ stem/progenitor cells to bone marrow. *Blood* 2004;103:2981–2989.

125. Vermeulen M, Le Pesteur F, Gagnerault MC, Mary JY, Sainteny F, Lepault F. Role of adhesion molecules in the homing and mobilization of murine hematopoietic stem and progenitor cells. *Blood* 1998;92:894–900.

126. Sudo K, Ema H, Morita Y, Nakauchi H. Age-associated characteristics of murine hematopoietic stem cells. *J Exp Med* 2000;192:1273–1280.

127. Morrison SJ, Wandycz AM, Akashi K, Globerson A, Weissman IL. The aging of hematopoietic stem cells. *Nat Med* 1996;2:1011–1016.

128. Chambers SM, Shaw CA, Gatza C, Fisk CJ, Donehower LA, Goodell MA. Aging hematopoietic stem cells decline in function and exhibit epigenetic dysregulation. *PLoS Biology* 2007;5:e201.

129. Cudkowicz G, Upton AC, Shearer GM, Hughes WL. Lymphocyte content and proliferative capacity of serially transplanted mouse bone marrow. *Nature* 1964;201:165–167.

130. Harrison DE, Astle CM, Delaittre JA. Loss of proliferative capacity in immunohemopoietic stem cells caused by serial transplantation rather than aging. *J Exp Med* 1978;147:1526–1531.

131. Siminovitch L, Till JE, McCulloch EA. Decline in colony-forming ability of marrow cells subjected to serial transplantation into irradiated mice. *J Cell Physiol* 1964;64:23–31.

132. Harrison DE, Stone M, Astle CM. Effects of transplantation on the primitive immunohematopoietic stem cell. *J Exp Med* 1990;172:431–437.

133. de Haan G, Nijhof W, van Zant G. Mouse strain-dependent changes in frequency and proliferation of hematopoietic stem cells during aging: Correlation between lifespan and cycling activity. *Blood* 1997;89:1543–1550.

134. Chen J, Astle CM, Harrison DE. Genetic regulation of primitive hematopoietic stem cell senescence. *Exp Hematol* 2000;28:442–450.

135. Hacein-Bey-Abina S, Von Kalle C, Schmidt M, McCormack MP, Wulffraat N, Leboulch P, Lim A, Osborne CS, Pawliuk R, Morillon E, Sorensen R, Forster A, Fraser P, Cohen JI, de Saint Basile G, Alexander I, Wintergerst U, Frebourg T, Aurias A, Stoppa-Lyonnet D, Romana S, Radford-Weiss I, Gross F, Valensi F, Delabesse E, Macintyre E, Sigaux F, Soulier J, Leiva LE, Wissler M, Prinz C, Rabbitts TH, Le Deist F, Fischer A, Cavazzana-Calvo M. Lmo2-associated clonal T cell proliferation in two patients after gene therapy for scid-x1. *Science* 2003;302:415–419.

136. Hacein-Bey-Abina S, von Kalle C, Schmidt M, Le Deist F, Wulffraat N, McIntyre E, Radford I, Villeval JL, Fraser CC, Cavazzana-Calvo M, Fischer A. A serious adverse event after successful gene therapy for X-linked severe combined immunodeficiency. *N Engl J Med* 2003;348:255–256.

137. Schroder AR, Shinn P, Chen H, Berry C, Ecker JR, Bushman F. HIV-1 integration in the human genome favors active genes and local hotspots. *Cell* 2002;110:521–529.

138. Baum C, Dullmann J, Li Z, Fehse B, Meyer J, Williams DA, von Kalle C. Side effects of retroviral gene transfer into hematopoietic stem cells. *Blood* 2003;101:2099–2114.

139. Goodman S, Xiao X, Donahue RE, Moulton A, Miller J, Walsh C, Young NS, Samulski RJ, Nienhuis AW. Recombinant adeno-associated virus-mediated gene transfer into hematopoietic progenitor cells. *Blood* 1994;84:1492–1500.

140. Chatterjee S, Li W, Wong CA, Fisher-Adams G, Lu D, Guha M, Macer JA, Forman SJ, Wong KK, Jr. Transduction of primitive human marrow and cord blood-derived hematopoietic progenitor cells with adeno-associated virus vectors. *Blood* 1999;93:1882–1894.

141. Miller JL, Donahue RE, Sellers SE, Samulski RJ, Young NS, Nienhuis AW. Recombinant adeno-associated virus (raav)-mediated expression of a human gamma-globin gene in human progenitor-derived erythroid cells. *Proc Natl Acad Sci USA* 1994;91:10183–10187.

142. Karlsson S, Humphries RK, Gluzman Y, Nienhuis AW. Transfer of genes into hematopoietic cells using recombinant DNA viruses. *Proc Natl Acad Sci USA* 1985;82:158–162.

143. Felgner PL, Gadek TR, Holm M, Roman R, Chan HW, Wenz M, Northrop JP, Ringold GM, Danielsen M. Lipofection: A highly efficient, lipid-mediated DNA-transfection procedure. *Proc Natl Acad Sci USA* 1987;84:7413–7417.

144. Hatada S, Nikkuni K, Bentley SA, Kirby S, Smithies O. Gene correction in hematopoietic progenitor cells by homologous recombination. *Proc Natl Acad Sci USA* 2000;97:13807–13811.

145. Schmidt-Wolf IG, Finke S, Trojaneck B, Denkena A, Lefterova P, Schwella N, Heuft HG, Prange G, Korte M, Takeya M, Dorbic T, Neubauer A, Wittig B, Huhn D. Phase I clinical study applying autologous immunological effector cells transfected with the interleukin-2 gene in patients with metastatic renal cancer, colorectal cancer and lymphoma. *Br J Cancer* 1999;81:1009–1016.

146. Gresch O, Engel FB, Nesic D, Tran TT, England HM, Hickman ES, Korner I, Gan L, Chen S, Castro-Obregon S, Hammermann R, Wolf J, Muller-Hartmann H, Nix M, Siebenkotten G, Kraus G, Lun K. New nonviral method for gene transfer into primary cells. *Methods* 2004;33:151–163.

147. Thomas E, Storb R, Clift RA, Fefer A, Johnson FL, Neiman PE, Lerner KG, Glucksberg H, Buckner CD. Bone-marrow transplantation (first of two parts). *N Engl J Med* 1975;292:832–843.

148. Armitage JO. Bone marrow transplantation. *N Engl J Med* 1994;330:827–838.

149. Ljungman P, Urbano-Ispizua A, Cavazzana-Calvo M, Demirer T, Dini G, Einsele H, Gratwohl A, Madrigal A, Niederwieser D, Passweg J, Rocha V, Saccardi R, Schouten H, Schmitz N, Socie G, Sureda A, Apperley J. Allogeneic and autologous transplantation for haematological diseases, solid tumours and immune disorders: Definitions and current practice in Europe. *Bone Marrow Transplant* 2006;37:439–449.

150. Collaboration CAT. Autologous stem cell transplantation in chronic myeloid leukaemia: A meta-analysis of six randomized trials. *Cancer Treat Rev* 2007;33:39–47.

151. Kharfan-Dabaja MA, Anasetti C, Santos ES. Hematopoietic cell transplantation for chronic lymphocytic leukemia: An evolving concept. *Biol Blood Marrow Transplant* 2007;13:373–385.

152. Tallman MS, Nabhan C, Feusner JH, Rowe JM. Acute promyelocytic leukemia: Evolving therapeutic strategies. *Blood* 2002;99:759–767.

153. van Laar JM, Tyndall A. Adult stem cells in the treatment of autoimmune diseases. *Rheumatology (Oxford)* 2006; 45:1187–1193.

154. Filipovich AH, Stone JV, Tomany SC, Ireland M, Kollman C, Pelz CJ, Casper JT, Cowan MJ, Edwards JR, Fasth A, Gale RP, Junker A, Kamani NR, Loechelt BJ, Pietryga DW, Ringden O, Vowels M, Hegland J, Williams AV, Klein JP, Sobocinski KA, Rowlings PA, Horowitz MM. Impact of donor type on outcome of bone marrow transplantation for Wiskott–Aldrich syndrome: Collaborative study of the international bone marrow transplant registry and the national marrow donor program. *Blood* 2001;97:1598–1603.

155. Antoine C, Muller S, Cant A, Cavazzana-Calvo M, Veys P, Vossen J, Fasth A, Heilmann C, Wulffraat N, Seger R, Blanche S, Friedrich W, Abinun M, Davies G, Bredius R, Schulz A, Landais P, Fischer A. Long-term survival and transplantation of haemopoietic stem cells for immunodeficiencies: Report of the European experience 1968–99. *Lancet* 2003;361:553–560.

156. Imada K, Leonard WJ. The JAK-STAT pathway. *Mol Immunol* 2000;37:1–11.

157. Buckley RH, Schiff RI, Schiff SE, Markert ML, Williams LW, Harville TO, Roberts JL, Puck JM. Human severe combined immunodeficiency: Genetic, phenotypic, and functional diversity in one hundred eight infants. *J Pediatr* 1997;130:378–387.

158. Nosaka T, van Deursen JM, Tripp RA, Thierfelder WE, Witthuhn BA, McMickle AP, Doherty PC, Grosveld GC, Ihle JN. Defective lymphoid development in mice lacking JAK3. *Science* 1995;270:800–802.

159. Park SY, Saijo K, Takahashi T, Osawa M, Arase H, Hirayama N, Miyake K, Nakauchi H, Shirasawa T, Saito T. Developmental defects of lymphoid cells in JAK3 kinase-deficient mice. *Immunity* 1995;3:771–782.

160. Thomis DC, Gurniak CB, Tivol E, Sharpe AH, Berg LJ. Defects in B lymphocyte maturation and T lymphocyte activation in mice lacking jak3. *Science* 1995;270:794–797.

161. Bunting KD, Sangster MY, Ihle JN, Sorrentino BP. Restoration of lymphocyte function in Janus kinase 3-deficient mice by retroviral-mediated gene transfer. *Nat Med* 1998;4:58–64.

162. Bunting KD, Flynn KJ, Riberdy JM, Doherty PC, Sorrentino BP. Virus-specific immunity after gene therapy in a murine model

of severe combined immunodeficiency. *Proc Natl Acad Sci USA* 1999;96:232–237.

163. Thrasher AJ, Kinnon C. The Wiskott–Aldrich syndrome. *Clin Exp Immunol* 2000;120:2–9.

164. Wada T, Jagadeesh GJ, Nelson DL, Candotti F. Retrovirus-mediated wasp gene transfer corrects Wiskott–Aldrich syndrome T-cell dysfunction. *Hum Gene Ther* 2002;13:1039–1046.

165. Huang MM, Tsuboi S, Wong A, Yu XJ, Oh-Eda M, Derry JM, Francke U, Fukuda M, Weinberg KI, Kohn DB. Expression of human Wiskott–Aldrich syndrome protein in patients' cells leads to partial correction of a phenotypic abnormality of cell surface glycoproteins. *Gene Ther* 2000;7:314–320.

166. Dupre L, Marangoni F, Scaramuzza S, Trifari S, Hernandez RJ, Aiuti A, Naldini L, Roncarolo MG. Efficacy of gene therapy for Wiskott–Aldrich syndrome using a was promoter/CDNA-containing lentiviral vector and nonlethal irradiation. *Hum Gene Ther* 2006;17:303–313.

167. Charrier S, Stockholm D, Seye K, Opolon P, Taveau M, Gross DA, Bucher-Laurent S, Delenda C, Vainchenker W, Danos O, Galy A. A lentiviral vector encoding the human Wiskott–Aldrich syndrome protein corrects immune and cytoskeletal defects in wasp knockout mice. *Gene Ther* 2005;12:597–606.

168. Johnston SL, Unsworth DJ, Dwight JF, Kennedy CT. Wiskott–Aldrich syndrome, vasculitis and critical aortic dilatation. *Acta Paediatr* 2001;90:1346–1348.

169. Bjorgvinsdottir H, Ding C, Pech N, Gifford MA, Li LL, Dinauer MC. Retroviral-mediated gene transfer of gp91phox into bone marrow cells rescues defect in host defense against aspergillus fumigatus in murine x-linked chronic granulomatous disease. *Blood* 1997;89:41–48.

170. Mardiney M, 3rd, Jackson SH, Spratt SK, Li F, Holland SM, Malech HL. Enhanced host defense after gene transfer in the murine p47phox-deficient model of chronic granulomatous disease. *Blood* 1997;89:2268–2275.

171. Dinauer MC, Gifford MA, Pech N, Li LL, Emshwiller P. Variable correction of host defense following gene transfer and bone marrow transplantation in murine X-linked chronic granulomatous disease. *Blood* 2001;97:3738–3745.

172. Dinauer MC, Li LL, Bjorgvinsdottir H, Ding C, Pech N. Long-term correction of phagocyte NADPH oxidase activity by retroviral-mediated gene transfer in murine X-linked chronic granulomatous disease. *Blood* 1999;94:914–922.

173. Becker S, Wasser S, Hauses M, Hossle JP, Ott MG, Dinauer MC, Ganser A, Hoelzer D, Seger R, Grez M. Correction of respiratory burst activity in X-linked chronic granulomatous cells to therapeutically relevant levels after gene transfer into bone marrow cd34+ cells. *Hum Gene Ther* 1998;9:1561–1570.

174. Roesler J, Brenner S, Bukovsky AA, Whiting-Theobald N, Dull T, Kelly M, Civin CI, Malech HL. Third-generation, self-inactivating gp91(phox) lentivector corrects the oxidase defect in nod/scid mouse-repopulating peripheral blood-mobilized cd34+ cells from patients with X-linked chronic granulomatous disease. *Blood* 2002;100:4381–4390.

175. Brenner S, Whiting-Theobald NL, Linton GF, Holmes KL, Anderson-Cohen M, Kelly PF, Vanin EF, Pilon AM, Bodine DM, Horwitz ME, Malech HL. Concentrated rd114-pseudotyped mfgs-gp91phox vector achieves high levels of functional correction of the chronic granulomatous disease oxidase defect in nod/scid/beta -microglobulin-/- repopulating mobilized human peripheral blood cd34+ cells. *Blood* 2003;102:2789–2797.

176. Grompe M, D'Andrea A. Fanconi anemia and DNA repair. *Hum Mol Genet* 2001;10:2253–2259.

177. Gush KA, Fu KL, Grompe M, Walsh CE. Phenotypic correction of fanconi anemia group c knockout mice. *Blood* 2000;95:700–704.

178. Walsh CE, Nienhuis AW, Samulski RJ, Brown MG, Miller JL, Young NS, Liu JM. Phenotypic correction of fanconi anemia in human hematopoietic cells with a recombinant adeno-associated virus vector. *J Clin Invest* 1994;94: 1440–1448.

179. Pawliuk R, Westerman KA, Fabry ME, Payen E, Tighe R, Bouhassira EE, Acharya SA, Ellis J, London IM, Eaves CJ, Humphries RK, Beuzard Y, Nagel RL, Leboulch P. Correction of sickle cell disease in transgenic mouse models by gene therapy. *Science* 2001;294:2368–2371.

180. May C, Rivella S, Callegari J, Heller G, Gaensler KM, Luzzatto L, Sadelain M. Therapeutic haemoglobin synthesis in beta-thalassaemic mice expressing lentivirus-encoded human beta-globin. *Nature* 2000;406:82–86.

181. Blaese RM, Culver KW, Miller AD, Carter CS, Fleisher T, Clerici M, Shearer G, Chang L, Chiang Y, Tolstoshev P, et al. T lymphocyte-directed gene therapy for ada-scid: Initial trial results after 4 years. *Science* 1995;270:475–480.

182. Bordignon C, Notarangelo LD, Nobili N, Ferrari G, Casorati G, Panina P, Mazzolari E, Maggioni D, Rossi C, Servida P, et al. Gene therapy in peripheral blood lymphocytes and bone marrow for ada-immunodeficient patients. *Science* 1995;270:470–475.

183. Aiuti A, Slavin S, Aker M, Ficara F, Deola S, Mortellaro A, Morecki S, Andolfi G, Tabucchi A, Carlucci F, Marinello E, Cattaneo F, Vai S, Servida P, Miniero R, Roncarolo MG, Bordignon C. Correction of ada-SCID by stem cell gene therapy combined with nonmyeloablative conditioning. *Science* 2002;296:2410–2413.

184. Cavazzana-Calvo M, Hacein-Bey S, de Saint Basile G, Gross F, Yvon E, Nusbaum P, Selz F, Hue C, Certain S, Casanova JL, Bousso P, Deist FL, Fischer A. Gene therapy of human severe combined immunodeficiency (SCID)-x1 disease. *Science* 2000;288:669–672.

185. Hacein-Bey-Abina S, Le Deist F, Carlier F, Bouneaud C, Hue C, De Villartay JP, Thrasher AJ, Wulffraat N, Sorensen R, Dupuis-Girod S, Fischer A, Davies EG, Kuis W, Leiva L, Cavazzana-Calvo M. Sustained correction of X-linked severe combined immunodeficiency by *ex vivo* gene therapy. *N Engl J Med* 2002;346:1185–1193.

186. Marshall E. Clinical research. Gene therapy a suspect in leukemia-like disease. *Science* 2002;298:34–35.

187. Liu JM, Kim S, Read EJ, Futaki M, Dokal I, Carter CS, Leitman SF, Pensiero M, Young NS, Walsh CE. Engraftment of hematopoietic progenitor cells transduced with the fanconi anemia group c gene (fancc). *Hum Gene Ther* 1999;10:2337–2346.

188. Charrier S, Dupre L, Scaramuzza S, Jeanson-Leh L, Blundell MP, Danos O, Cattaneo F, Aiuti A, Eckenberg R, Thrasher AJ, Roncarolo MG, Galy A. Lentiviral vectors targeting wasp expression to hematopoietic cells, efficiently transduce and correct cells from was patients. *Gene Ther* 2007;14:415–428.

189. Kohn DB, Hershfield MS, Carbonaro D, Shigeoka A, Brooks J, Smogorzewska EM, Barsky LW, Chan R, Burotto F, Annett G, Nolta JA, Crooks G, Kapoor N, Elder M, Wara D, Bowen T, Madsen E, Snyder FF, Bastian J, Muul L, Blaese RM, Weinberg K, Parkman R. T lymphocytes with a normal ada gene accumulate

after transplantation of transduced autologous umbilical cord blood cd34+ cells in ada-deficient SCID neonates. *Nat Med* 1998;4:775–780.

190. Kohn DB, Weinberg KI, Nolta JA, Heiss LN, Lenarsky C, Crooks GM, Hanley ME, Annett G, Brooks JS, el-Khoureiy A, et al. Engraftment of gene-modified umbilical cord blood cells in neonates with adenosine deaminase deficiency. *Nat Med* 1995;1:1017–1023.

191 Malech HL, Maples PB, Whiting-Theobald N, Linton GF, Sekhsaria S, Vowells SJ, Li F, Miller JA, DeCarlo E, Holland SM, Leitman SF, Carter CS, Butz RE, Read EJ, Fleisher TA, Schneiderman RD, Van Epps DE, Spratt SK, Maack CA, Rokovich JA, Cohen LK, Gallin JI. Prolonged production of nadph oxidase-corrected granulocytes after gene therapy of chronic granulomatous disease. *Proc Natl Acad Sci USA* 1997;94:12133–12138.

33 Gene Therapy for Hematopoietic Disorders

Elizabeth M. Kang, Mitchell E. Horwitz, and Harry L. Malech

CONTENTS

33.1 INTRODUCTION

Clinical gene transfer to hematopoietic cells began on May 22, 1989 when autologous T-lymphocytes from a cancer patient were gene marked and transplanted back into the patient [1]. Since then, over 500 patients have been enrolled in hematopoietic cell gene marking or gene therapy protocols, and studies such as this one have played an important role in the evolution of the field of gene therapy (Table 33.1). This chapter will review the practical issues involved in the conduct of a clinical gene therapy trial, as well as the published trials to date involving actual transfusion of genetically modified cells.

There are two general strategies to clinical gene transfer protocols for hematopoietic cells. The first strategy is to target the pluripotent hematopoietic stem cell (HSC) with the goal of permanent gene transfer to all three hematopoietic cell lineages. Clinical HSC trials have involved transfer of both marker genes as well as therapeutic genes. The second strategy involves gene transfer into a single, terminally differentiated

TABLE 33.1

Summary of the Published Clinical Gene Transfer Trials Involving Hematopoietic Cells

Gene	Vector Type	Cellular Target	Reference
Neomycin resistance[a]	Retroviral	HSC	[2–7]
P47 (chronic granulomatous disease [CGD])	Retroviral	HSC	[8]
GP91 (CGD)	Retroviral	HSC	[9,10]
Adenosine deaminase (ADA) (severe combined immunodeficiency [SCID])	Retroviral	HSC	[11–16]
Multidrug resistance 1 [MDR1]	Retroviral	HSC	[17–20]
Glucocerebrosidase (Gaucher's disease)	Retroviral	HSC	[21]
CD18 (leukocyte adhesion deficiency)	Retroviral	HSC	[22]
Rev responsive element decoy	Retroviral	HSC	[23]
Common gamma chain (X-linked severe combined immunodeficiency)	Retroviral	HSC	[24–27]
Transdominant rev protein (HIV)	Retroviral	HSC	[28]
Neomycin resistance[a]	Retroviral	Lymphocyte	[1,29–31]
ADA (SCID)	Retroviral	Lymphocyte	[12,14,32]
Herpes simplex virus-thymidine kinase[b]	Retroviral	Lymphocyte	[33–35]
Transdominant rev protein (HIV)	Retroviral	Lymphocyte	[36,37]
Anti-Tat 1 ribozyme (HIV)	Retroviral	Lymphocyte	[38]
Anti-MART-1 TCR (Melanoma)	Retroviral	Lymphocyte	[39]
HIV envelope antisense protein (HIV)	Lentiviral	Lymphocyte	[40]

[a] Marker gene studies.

[b] Suicide gene studies.

hematopoietic cell lineage. With this strategy, persistence of gene corrected cells is dependent on the lifespan of the target cell. At present, this approach has been applied clinically to allogeneic and autologous lymphocytes but other cell types such as dendritic cells are being explored. There are three broad categories of genes that have been used in clinical trials targeting hematopoietic cells: (1) marker genes, (2) suicide genes, and (3) therapeutic genes. Marker genes provide a mechanism for detection of the transduced cells by introducing a gene product that is easily detectable by flow cytometry, enzymatic activity, or by conferring a drug resistance phenotype. Antibiotic resistance genes such as the bacterial neomycin phosphotransferase gene are most commonly used as markers. Suicide genes allow for *in vivo* elimination of transduced cells. The herpes thymidine kinase, for example, confers sensitivity of the transduced cell to the drug ganciclovir. There are a number of different types of therapeutic transgenes that have been utilized in clinical trials. Some are designed to interfere with the life cycle of the human immunodeficiency virus, some are designed to protect normal HSCs from chemotherapy, some are designed to augment tumor killing, and others are designed to correct inherited monogenic disorders.

33.2 MURINE-BASED RETROVIRAL VECTORS

All clinical gene therapy trials to date targeting hematopoietic cells have utilized murine-based retroviral vectors from the oncoviridae subfamily. As discussed in Chapter 1, the simplicity of the murine retroviral genome has facilitated the

development of a class of replication-incompetent vectors that can be produced by specially engineered producer cell lines. However, murine retroviral vectors can efficiently transfer genes only to actively replicating target cells [41]. Relatively quiescent target cells such as HSCs, which reside predominantly in the G0/G1 phase of the cell cycle, are inefficiently transduced by murine retroviruses. Most retroviral vectors that have been used clinically for hematopoietic cell gene transfer are based on the oncovirus called Moloney murine leukemia virus (MoMLV) [42], though other vectors used clinically have been derived from either the Harvey murine sarcoma virus (HaMSV), murine stem cell virus (MSCV), or the spleen focus forming virus (SFFV). HaMSV, MSCV, and SFFV are similar to MoMLV in overall structure and function, as are the modifications incorporated to make them replication-incompetent and safe to use in humans. The HaMSV and MSCV long terminal repeats (LTRs) may function better in HSCs and may be less subject to silencing, but this has not been proven in humans or nonhuman primates. The SFFV has been used in some trials due to its purportedly increased expression in myeloid lines. Modifications to the MoMLV T1 LTR have been made by some investigators to achieve a similar goal of preventing silencing related to methylation by altering methylation-sensitive sites [43]. The general description below of MoMLV also applies to the other related oncoviridae. The MoMLV is composed of two copies of RNA ranging in size from 2 to 9 kb. In order to render these vectors replication-incompetent, the *gag* (core proteins), *pol* (reverse transcriptase and integrase), and *env* (envelope protein determining the host cell range or tropism of the retrovirus) genes are deleted. What remains are the 5′ and 3′ LTR sequences on each end of the construct along with the packaging (Ψ) site. The 5′ LTR functions as the promoter and enhancer region and the 3′ LTR contains the poly-A signal. The Ψ region serves as a binding site for the *gag* polyprotein, which packages the RNA into a viral core. The transgene is cloned into a site downstream of this region. In some vectors, splice donor and acceptor sites are retained or deliberately engineered into the Ψ region to generate, from a portion of the full-length mRNA, a subgenomic mRNA that more efficiently translates the downstream inserted cDNA open reading frame. Replication-incompetent viral particles are produced in specifically engineered "packaging cells." Packaging cell lines such as PA317 [44], (NIH3T3-derived murine fibroblasts), AM12 [45], Ψ crip [46], or 293SPA (293 derived human embryonal kidney cells) [47] constitutively express the *gag*, *pol*, and *env* proteins and therefore secrete empty virions into the culture media. When plasmid DNA of the retroviral vector is transfected into the packaging cells, clones of producer cells can be selected that secrete into the culture media replication-incompetent but infectious virions containing the transgene. The culture media, known as "viral supernatant" can then be collected and used *ex vivo* to infect the desired target cells. An important issue regarding retrovirus vector transduction of hematopoietic cells is that stem cells and lymphocytes express relatively low levels of the surface receptor required for binding of the most widely used amphotropic envelope packaging element [48]. Some studies have suggested that packaging lines using the envelope proteins of the Gibbon ape leukemia virus (GALV) or the feline endogenous virus (RD114) may be advantageous for targeting both stem cells and T lymphocytes because these cells have higher levels of receptor for this envelope [49,50]. There is considerable current interest in bringing lentivirus vectors to the clinic due to a number of desirable characteristics, including their ability to transduce nondividing cells (although even lentivectors may poorly transduce highly quiescent cells in G_0), as well as evidence that their insertion site pattern relative to gamma retrovirus vectors may be less prone to causing leukemogenesis [51–55]. Although a number of groups are planning and a few are conducting clinical studies using a lentivirus based vector, the current requirement for a four plasmid transfection-mediated transient production of replication incompetent vector has made large-scale production of lentiviral vectors cumbersome. Despite numerous attempts, there are yet no available stable high titer producer cell lines for lentiviral production and current safety requirements for most "third-generation" systems designed to limit possible combinatorial events, require four plasmids for transfection. To date, only one group has published a clinical study using a lentiviral vector and that notably involves only patients already infected with HIV [40] (as discussed in Chapter 9). There are many groups working toward improved methods of production and it is likely that production issues will be resolved over the next few years.

33.3 PRACTICAL CONSIDERATIONS OF HEMATOPOIETIC CELL GENE THERAPY

The following section outlines the three phases of a retro-viral-based HSC and lymphocyte gene transfer protocol: (1) hematopoietic cell procurement, (2) *ex vivo* gene transfer, and (3) infusion of the corrected cells (Figure 33.1).

33.3.1 GENE TRANSFER TO HSCS

33.3.1.1 HSC Procurement

Collection of large numbers of autologous HSCs is the first step in an HSC gene therapy protocol. The three sites from which HSC can be harvested are: (1) bone marrow, (2) cytokine or chemotherapy/cytokine-mobilized peripheral blood, and (3) umbilical cord blood. Although the precise phenotype of a true HSC is unknown, large numbers of HSCs are contained in a population of cells expressing the CD34 antigen [56]. CD34 cells make up 0.5%–5% of nucleated cells in the bone marrow and only a fraction of these are HSCs. The actual function of CD34 is unknown and investigators have been exploring other markers such as AC133 or characteristics such as the ability to extrude specific dyes to try to select for more primitive cells. Studies using these markers include a description of one patient receiving autologous CD133-positive cells postchemotherapy; however to date, all clinical gene therapy protocols have used CD34 to characterize their progenitor population [57–61]. HSCs can be safely aspirated from bone marrow of the posterior superior iliac crest in a minor operative procedure. The major side effects of this procedure include mild discomfort at the aspiration site and an occasional hematoma. For smaller individuals, symptomatic anemia may require a blood transfusion, which, if anticipated in advance, can be an autologous unit. The

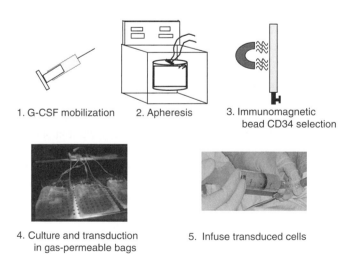

1. G-CSF mobilization   2. Apheresis   3. Immunomagnetic bead CD34 selection

4. Culture and transduction in gas-permeable bags   5. Infuse transduced cells

FIGURE 33.1 Clinical scale HSC gene therapy scheme.

major disadvantage of large volume bone marrow aspiration is that it must be done under general anesthesia. Because of these issues, repeated large-scale marrow harvests are not desirable.

In most individuals, administration of granulocyte-colony stimulating factor (G-CSF) at 10–16 μg/kg per day subcutaneously for 5 or 6 days results in a transient 20- to 100-fold increase in the absolute number of CD34 hematopoietic progenitor and stem cells in the circulation (from a baseline of about 1.4 CD34 positive cells per microliter to an average of 78 CD34 positive cells per microliter; this phenomenon termed stem cell mobilization). Other growth factors have been investigated for their potential to induce mobilization including Flt-3-ligand (Flt3-L) [62–64], granulocyte-macrophage colony stimulating factor (GM-CSF) [65–68], and stem cell factor (SCF) [69,70], though none are as safe and effective as G-CSF. Food and Drug Administration approved formulations only exist for human recombinant G-CSF or GM-CSF, and in the standard clinical setting, only G-CSF or (in poor mobilizers) the combination of G-CSF with GM-CSF is used for mobilization of stem cells. For gene therapy, it may be desirable to obtain larger numbers of CD34-positive stem cells from an individual patient than can be mobilized, collected, and purified from one cycle of mobilization. As well, patients with CGD and possibly other chronic illnesses may mobilize less efficiently than healthy volunteers [71]. However, the mobilization process may be safely repeated at 4- to 8-week intervals. Thus, very large numbers of CD34 cells (exceeding in total $10–15 \times 10^6$ cells/kg patient body weight) may be obtained by cryopreserving and then pooling cells collected from several repeated cycles of mobilization and apheresis at 6-week intervals. Unlike the procurement of bone marrow, no operative procedure is needed and the entire process can be done without a hospital admission.

Dunbar et al. [72] have demonstrated improved gene marking in nonhuman primates when peripheral blood HSC mobilized with G-CSF and SCF are compared to bone marrow HSC. The same group later showed superior *in vivo* gene marking of nonhuman primate HSC mobilized with G-CSF and SCF compared to HSC mobilized with G-CSF and Flt-3 ligand or G-CSF alone [73]. High-resolution cell cycle analysis of cytokine-mobilized peripheral blood HSC has revealed that these cells are more likely to have entered cell cycle and express higher levels of the amphotropic receptor mRNA [74–76]. While certainly easier to collect, human gene marking studies have yet to demonstrate unequivocally that cytokine-mobilized peripheral blood HSCs are better targets for retroviral transduction than bone marrow HSCs. However, the logistical advantages to obtaining very large numbers of peripheral blood HSCs dictate these as the preferred source for gene therapy and other transplant purposes except when the donor has a poor CD34 mobilization response to cytokine administration.

Recently there has been significant interest in the stem cell mobilizing activity of an experimental new small molecule in advanced stage of clinical testing for HSC mobilization, known as AMD3100, this bicyclam molecule initially developed for its ability to act as a competitive inhibitor of CXCR4 for use in HIV treatment (CXCR4 is one of the several known coreceptors for HIV entry). CXCR4 is a G-protein-coupled receptor found on primitive HSCs. It is the cognate receptor for SDF-1 (chemokine CXCL12) expressed on and secreted by marrow stromal cells. SDF-1 binding to CXCR4 mediates homing and retention of HSCs in the bone marrow. G-CSF administration appears to facilitate a transient reduction in CXCR4 on HSCs and increased degradation of SDF-1 that requires about 5 days to reach peak effect on mobilization of HSC from marrow to peripheral blood. However, AMD3100 acts by directly binding to CXCR4 without activating the receptor, effectively blocking interaction with SDF-1, achieving maximum effect on mobilization of HSCs from the marrow to peripheral blood within 6 hours. There is animal data proving the reconstitutive potential of AMD3100-mobilized cells as well as clinical data showing synergy in mobilization using G-CSF and AMD3100 in combination. AMD3100 may be particularly useful for those patients who do not respond adequately to GCSF alone. Studies are still in progress to determine its exact dosing and best indication for use at this time [77–81].

Umbilical cord blood contains a higher concentration of primitive hematopoietic progenitors than bone marrow [82]. Recent data suggest that the HSCs from umbilical cord blood may also express higher levels of the amphotropic retrovirus receptor [83], which may result in more efficient transduction with amphotropic retroviral vectors. On average 20×10^6 CD34 cells can be collected from the placenta at the time of delivery, which is approximately 10-fold less than what can be collected from mobilized peripheral blood of an adult. Because of the low efficiency of HSC transduction with current techniques, clinical application for gene-corrected umbilical cord blood stem cells may only be practical in neonates.

33.3.1.2 *Ex Vivo* Gene Transfer

Optimization of *ex vivo* retroviral transduction conditions for HSCs has proven to be a formidable task. It appears that 48 to 72 h of culture in growth factors is required for quiescent lineage negative CD34 cells to enter the cell cycle and thus become receptive targets for retroviral transduction [84]. However, studies have shown that prolonged *ex vivo* culture of HSCs results in loss of long-term repopulating ability as a consequence of lineage commitment and of an acquired defect in the ability of cycling cells to engraft [85,86]. This loss of repopulating potential with *ex vivo* culture may be gradual and to some degree reversible. Takatoku et al. manipulated the *ex vivo* culture conditions and growth factor combinations such that HSCs from nonhuman primates were first induced into cell cycle by using a combination of active cytokines, thereby facilitating retroviral transduction. Then, before the cells were transplanted back into the animal, they were returned to a quiescent state by incubating the cells in media containing only SCF. This method of "resting" stem cells prior to infusion resulted in improved engraftment of gene-marked cells [87]. Measurable gene transfer into HSCs in the clinical setting has been reported with *ex vivo* transduction periods ranging from 6 to 72 h [3,15]. While the report using a 6 h regimen appeared to succeed in achieving measurable gene transfer without use of growth factors [3], most investigators have found that growth factors and an *ex vivo* culture period of 48–96 h are required for optimum transduction. Growth factors are also essential to prevent apoptosis of HSCs during prolonged *ex vivo* culture [21,88–90]. Many studies have focused on determination of growth factor combinations used without stromal layers that can achieve highest transduction while maintaining reconstitutive potential of the transduced HSC. Flt3-L [91] and thrombopoietin (TPO) [91,92] have emerged as important agents to add to the *ex vivo* culture in relatively high concentrations (100–300 ng/mL) to achieve these dual goals. These growth factors work optimally in synergy with other growth factors. The presence of SCF in the culture at more modest concentrations of 50–100 ng/mL appears to be important as well to maintaining viability and achieving cycling of the most primitive HSCs. While interleukin 3 (IL-3) and IL-6 have also been widely used *ex vivo* in the transduction culture in clinical trials to enhance cycling, there is controversy regarding the necessity, utility, and helpful effects of these cytokine growth factors. In fact, use of concentrations of IL-3 higher than 30–50 ng/mL may be detrimental to preservation of cells with long-term engraftment potential. Low doses of IL-3 in the culture in the range of 1–20 ng/mL may be sufficient to achieve synergy with other factors in enhancing cycling of primitive HSCs without also inducing undesired loss of engraftment potential. Given current concerns about vector insertional mutagenesis and the likelihood that there could be correlation between gene activation at the time of transduction and vector gene insertion targeting, it is probably best to use the lowest concentration of all growth factors in the transduction culture that achieves the desired level of transduction of primitive HSCs. Other factors essential for maintenance and development of lymphoid progenitors from stem cells, such as IL-7, have not been used clinically, but may in the future prove to be useful for the transduction of lymphocytes [93,94].

A number of techniques have been devised to encourage the interaction of hematopoietic progenitors with viral particles during the *ex vivo* transduction period. Most investigators have opted to transduce a cell population enriched for CD34 expression. CD34 cell enrichment enables an improved stem cell/viral particle ratio, thereby optimizing the multiplicity of infection while using less of the valuable clinical grade retroviral supernatant. A variety of stem cell selection devices that use monoclonal antibodies specific for the CD34 antigen have been employed in experimental clinical protocols (reviewed in Refs. [95,96]). These devices are able to select large numbers of CD34 cells from a bone marrow or mobilized peripheral blood apheresis graft at 50%–80% efficiency yielding a product, which consists of 60%–90% CD34 cells. For reasons that are not well defined, centrifugation of target cells during incubation in a retrovirus vector supernatant increases transduction

efficiency, a technique that has been termed "spinocculation" [97]. The g-forces employed to achieve the effect are as low as 1200 g for 20 min, making it unlikely that the effect is due to sedimentation of individual virus particles. Cocultivation of the target cells on a confluent layer of retrovirus producer cells has also been shown to enhance transduction. However, regulatory issues related to the safety of cocultivation of HSC with producer lines make this approach impractical for clinical application. One of the more exciting techniques to be described is the finding that a specific proteolytic fragment of fibronectin facilitates stem cell–retroviral particle interaction when this peptide is used to coat the surface of the culture vessel [98]. Fibronectin, a prominent component of the extracellular matrix of bone marrow stromal cells, has numerous hematopoietic cell-binding domains. Moritz et al. have demonstrated binding of both viral particles and hematopoietic target cells to a proteolytic fragment of fibronectin that contains the CS1 binding site [99]. The CS1 binding site of fibronectin interacts with the VLA4 adhesion molecule found on HSCs [100]. Thus, when HSCs are incubated with retroviral particles in the presence of this specific fibronectin fragment, transduction efficiency is improved. The availability of a clinical grade recombinant human C-terminal fibronectin fragment (CH-296; RetroNectin™) has facilitated its use in many clinical gene therapy trials targeting HSCs. Prior to the use of this fibronectin fragment, autologous stroma was used to help maintain the cells during the transduction period in many animal studies [88,90,101], however this is technically and logistically much more difficult than use of this fibronectin fragment to coat the culture vessel. Rhesus studies have shown equivalence if not improvement in using fibronectin fragment CH-296 compared to stroma and now, even with retrovirus vectors of modest titer, acceptable transduction of CD34 cells can be achieved [102]. Importantly, with retrovirus vectors at titers 5×10^6 infective particles per mL, the use of CH-296 fibronectin fragment-coated culture vessels can achieve transduction of 50%–70% of CD34 cells routinely at clinical scale. It is also possible that use of fibronectin fragment coating may help to preserve the long-term engraftment potential of cultured HSC [103] and this has become the standard in most human clinical HSC transduction methods.

33.3.1.3 Transplantation of Transduced HSC

Following *ex vivo* transduction, the extensively washed HSCs are resuspended in physiologic saline and infused into a peripheral vein of the patient. Within 24 h, the majority of the HSCs have homed to the bone marrow. Experience with stem cell transplantation for treatment of hematological malignancies has shown that the bone marrow can be completely reconstituted by transplanted HSCs (autologous or allogeneic) following myeloablative doses of chemotherapy or radiation. Because loss of long-term repopulating ability may occur during *ex vivo* transduction of HSC, it is unethical to rescue hematopoiesis in a myeloablated patient with *ex vivo* manipulated HSCs only. However, preclinical and clinical studies suggest that some degree of cytoreductive therapy administered prior to infusion is required to establish clinically relevant levels of gene marking. Using clinically applicable tools, long-term marking at levels of 5%–10% can be achieved in nonhuman primates following the administration of high-dose total body irradiation [78,104,105]. Aiuti and colleagues have been the first to employ nonmyeloablative bone marrow conditioning in a human gene therapy study for children with ADA-deficient SCID (discussed in more detail in Section 33.4.1.3) [73].

33.3.2 Gene Transfer to Lymphocytes

33.3.2.1 Lymphocyte Procurement

With few exceptions, large numbers of lymphocytes circulate in the peripheral blood and are therefore easily harvested from gene therapy candidates using apheresis. Lymphocytes may also be harvested from special sites such as tumors. These cells are of particular interest since they may possess unique antitumor properties.

33.3.2.2 Ex Vivo Gene Transfer and Infusion

Compared to HSC, fewer hurdles exist in the quest to optimize retroviral gene transfer of lymphocytes. Since these cells are terminally differentiated, loss of phenotype during *ex vivo* manipulation is not a concern. T-lymphocytes, which are the most common target for lymphocyte-based gene therapy, are expanded in culture with agents such as IL-2 or monoclonal antibodies to CD3. While being cultured, many of the cells are stimulated into active phase of the cell cycle and become susceptible to permanent retroviral integration. It has been observed that with long-term culture of T-lymphocytes, enrichment of CD8 T-lymphocytes relative to CD4 T-lymphocytes develops. This issue may be addressed by altering the ratio of cells added to the initial culture mixture [32]. Techniques that have been shown to improve lymphocyte transduction efficiency include: (1) the use of retroviral vectors pseudotyped with the GALV envelope protein, (2) upregulation of amphotropic or GALV retroviral receptor expression by growth in phosphate depleted media, and (3) transduction in a culture vessel coated with the CS1 fibronectin fragment [106,107]. Incorporation of these techniques together in the same protocol can yield transduction efficiency of 50%–60%. Given that lymphocytes are the natural target of HIV and other viruses, lentiviral-based vectors pseudotyped with vesicular stomatitis virus G (VSV-G) are a particularly attractive option for T cell mediated gene therapy. Infusion of the gene-corrected lymphocytes takes place as would any routine infusion of cell products. Although it has previously been assumed that cytoreductive conditioning of the recipient is not necessary to achieve persistence of the transplanted lymphocytes, where the therapeutic gene may provide a survival advantage, there clearly are clinical settings in which modest cytoreductive chemotherapy conditioning aimed at marrow stem cells (or in some cases at T lymphocytes) may significantly enhance the clinical outcome from gene therapy. This will be discussed in more detail in later sections of this chapter.

33.3.3 Clinical Scale-Up

The transition from a laboratory-based gene transfer assay to one that is ready for inclusion in a clinical protocol can be quite challenging. The most obvious differences relate to the number of hematopoietic progenitors that must be transduced at one time. Large volumes of retroviral supernatant must be produced in a facility licensed to provide clinical grade material (GMP). It is common to find a decrement in the viral titer of the clinical material compared to titers that can be achieved at small scale in the laboratory. Besides the requisite sterility and endotoxin testing, the product must always be tested for the presence of replication-competent retrovirus. Regulatory and proprietary issues regarding use of reagents or devices often hinder the ability to replicate in the clinical setting what is done with ease in the laboratory. Performing retroviral transduction in a clinically approved facility may require modifications of a laboratory-optimized assay [108]. Use of closed-system, gas-permeable, culture containers compatible with the standard sterile transfer techniques used by blood banks is one method that has been adapted for this purpose [8].

33.4 HUMAN GENE TRANSFER TRIALS

33.4.1 HSC Gene Transfer Studies

33.4.1.1 Marker Gene Studies

The insertion of marker genes into HSCs has been useful in the evolution of gene transfer technology and has led to a better understanding of autologous stem cell transplantation. The first and still pivotal gene marking study was reported by Brenner et al. in 1993 [3]. This study enrolled 20 patients under the age of 20 who were candidates for autologous bone marrow transplantation for acute myelogenous leukemia or neuroblastoma. Bone marrow was harvested as the patient recovered from a cycle of cytotoxic chemotherapy. Two-thirds of the harvest was immediately frozen and the remaining one-third was transduced with a retroviral vector containing the neomycin resistance gene. *In vitro* transduction efficiency, measured as a percentage of hematopoietic colonies resistant to G418 (neomycin analog), ranged from 2%–14%. The transduced and unmanipulated bone marrow cells were infused after the administration of high-dose chemotherapy. Using polymerase chain reaction (PCR) analysis, the marker gene was detected in approximately 5% of bone marrow mononuclear cells 1 month following transplantation. G418 resistance was observed in 5%–20% of bone marrow colony forming units (CFU) in 5 of 5 evaluable patients at 1 year, and in 2 patients at 18 months.

Three important principles emerged from this study. First, the study proved that long-term repopulating cells could be successfully gene marked *ex vivo*. Second, the study demonstrated that autologous marrow infusion following high-dose chemotherapy participates in the marrow recovery. Finally, the authors also reported that tumor cells obtained from patients who relapsed after the gene marked, autologous BMT contained the neomycin resistance gene, suggesting tumor cell contamination of the autologous stem cell graft [4]. Deisseroth et al. performed a similar gene marked, autologous, BMT study in patients with chronic myelogenous leukemia where gene marking of normal bone marrow CFU was demonstrated 6 months postinfusion. The investigators also demonstrated gene-marked tumor cells at relapse, again suggesting tumor contamination of the autograft [5]. Two other studies were unsuccessful at demonstrating tumor contamination of autografts, which may be a consequence of much lower rates of *ex vivo* gene transfer [2,109]. Dunbar et al. employed HSC transduction conditions optimized by others with animal models in an attempt to improve the low *ex vivo* transduction efficiency reported by Brenner et al. [3,4]. This gene-marking study also set out to compare the engraftment capabilities of bone marrow and peripheral blood stem cells (PBSC) and their suitability as targets for retroviral transduction. Bone marrow and chemotherapy/cytokine-mobilized peripheral blood mononuclear cells were procured from breast cancer and multiple myeloma patients who were candidates for autologous stem cell transplantation. Two-thirds of each product was cryopreserved without manipulation. CD34 + cells were purified from the remaining one-third of the mobilized peripheral blood, and bone marrow product and retroviral transduction of each product was performed using one of the two molecularly distinct retroviral vectors containing the neomycin resistance gene. Since the bone marrow and peripheral blood CD34 cells were not transduced with the same vector, the contribution of each to engraftment following autologous stem cell transplantation could be tracked using PCR. *Ex vivo* transduction conditions consisted of a 72 h culture of the target cells with retroviral supernatant supplemented with the hematopoietic cytokines IL-3, SCF, and IL-6 (IL-6 was omitted from the culture of CD34 cells from multiple myeloma patients). All transduced and unmanipulated cells were pooled and infused following high-dose chemotherapy administration. The *ex vivo* transduction efficiency as measured by a clonogenic assay was 18%–24%. Following transplantation, gene-marked bone marrow and peripheral blood cells were detected in 10 of 10 evaluable patients at a frequency of 0.1%–1% using a semiquantitative PCR technique. At 600 days postinfusion, only one patient had detectable levels of gene-marked cells. The authors did not identify any significant difference between bone marrow or PBSC as targets for retroviral gene transduction [6].

33.4.1.2 X-Linked SCID

The majority of patients with SCID have an X-linked form caused by mutations in the common gamma chain of the receptors for interleukin 2, 4, 7, 9, 15, and 21. Without a functional gamma chain, there is complete absence of T-cell and NK cell development and failure to complete B-cell maturation, resulting in a profound immunodeficiency in cell-mediated and humeral immunity that is usually fatal in the first year of life. Standard therapy for infants with an HLA-identical sibling is a T-cell-depleted bone marrow transplant with no (or very low level) cytoreductive conditioning that often results in restoration of the T-cell compartment and in many cases also some prolonged restoration of B-cell and NK cell functions. For infants with no HLA-matched sibling, it has become common practice to transplant such children with a T-cell-depleted HLA-haploidentical graft from a parent (usually the mother), usually without cytoreductive conditioning. Surprisingly, in 70%–90% of children, there is sufficient engraftment of donor T-cells to provide adequate prolonged T-cell function, but there is seldom restoration of B-cell or NK cell function. Almost all of these patients require intravenous gamma globulin and in some cases, there is slow decrease in the functionality of donor T-cell function, leading to recurrent infections, gastrointestinal dysfunction, poor growth, and alloimmune phenomena. As noted, B-cell function is often deficient in patients who are transplanted without marrow conditioning. Such patients are dependent on lifelong immunoglobulin supplementation to remain healthy. The residual humoral immune defect, along with risks of graft versus host disease (GVHD) and conditioning-related toxicity for patients treated with a traditional stem cell transplant led investigators from the Long-Necker hospital in Paris to conduct the first stem cell gene transfer trial for X-linked SCID. To date, CD34-selected bone marrow cells from a total of 11 affected boys were transduced with an MoMLV (MFG) packaged in the crip producer line and infused without bone marrow conditioning. In 10 of 11 patients, normal functioning gene-corrected T-cells and NK cells were detected in peripheral circulation approximately 3 months after the transplant. Gene-corrected cells were never detected in one of the patients and he was subsequently successfully treated with allogeneic stem cell transplantation [112]. As a result of the selective advantage for gene-corrected T-cells, total T-cell numbers reached normal levels within 6 months. Serum immunoglobulin levels normalized in 9 of 11 patients and were low but detectable in the fourth patient. All patients with successful engraftment of gene-corrected stem cell showed resolution of all stigmata of profound immunodeficiency, making this the first gene therapy trial to provide clear-cut clinical benefit.

Unfortunately, the unequivocal clinical benefit that accrued to these patients and the appropriate enthusiasm and excitement surrounding the success of this trial has been tempered by the development of T-cell leukemia in four patients in the Paris cohort. Analysis from the tumor cells of three of the four patients revealed that malignant transformation most likely occurred at least in part as a consequence of insertional mutagenesis of LMO-2, a gene linked to T-cell leukemia as well as other genes [113,114]. Gaspar et al. have published results on clinically beneficial gene therapy in four infants with X-linked SCID and have additional positive results in additional infants using a GALV (as opposed to amphotrophic) pseudotyped MFG gamma retrovirus vector [27]. They also used bone marrow as the stem cell source and all patients were less than 3 years of age with the oldest patient of 33 months having the longest time to T-cell recovery. The authors have also noted that the kinetics of their reconstitution in general were slower than that seen in the study published by Cavazzano-Calvo et al. Of note, despite some suggestion from animal studies that the common gamma chain of IL-2 itself has oncogenic properties, there have been no reports of leukemogenesis in this or any other X-SCID gene therapy trials [115,116] even with follow-up times similar to the trial from Long-Necker. Some differences between the Paris and London cohorts in the *ex vivo* transduction conditions were some very slight differences in the vector backbone sequence, the use of serum-free (London) versus fetal calf serum (Paris) in the medium, GALV envelope (London) versus amphotropic envelope (Paris), and the use of lower concentrations of growth factors particularly IL-3 in the London versus Paris cohort. It is completely speculative whether any of these factors might have made a difference in the observed frequency of leukemic events.

Interestingly, gene therapy for X-SCID has not been as successful in older patients. Two trials reported to date (U.S. and European cohorts) have shown only benefit in one of the five patients treated (one of the three in the U.S. cohort) [75,76]. Both groups used PBSC as the stem cell source and a retroviral vector encoding the common gamma chain pseudotyped with the GALV enveloped for patients with X-linked SCID. In the trial in the United States by Chinen et al., only patients having failed a haploidentical transplant were eligible and of the three patients treated (aged 10–14 years old), only one patient (the youngest) had any significant prolonged clinical improvement with partial restoration of immunity postgene therapy. One of the patients had, in fact, rejected four attempts at transplantation from both his father and mother, but due to an aberrant mutation was able to produce a sufficient level of protein to allow survival, but with continued problems of recurrent infections and autoimmunity. Although there were detectable numbers of marked cells in all treated patients, including in the myeloid lineages, there was no proliferative outgrowth of corrected cells except in the youngest patient treated [26]. In the European trial of two patients with X-SCID performed by Thrasher et al., results were also limited with poor T cell development in the two patients, ages 20 and 15, who were treated. Of these patients, one again had an aberrant mutation able to produce some IL-2 receptor common gamma chain on T cells, but insufficient for normal immunity. The other had graft failure post a haploidentical transplant. [25] It is possible that the lack of thymic function in these older patients is a barrier to success as well as a replete, albeit dysfunctional T cell compartment, from either a poorly functional, but persistent graft or the patients' own T cells, which prevent expansion of the genetically modified autologous T lymphocytes.

33.4.1.3 Adenosine Deaminase Deficiency SCID

The inherited deficiency of ADA is responsible for approximately one-quarter of the cases of SCID. In the absence of ADA, lymphocytes accumulate high levels of 2′-deoxyadenosine, which is converted to the toxic compound deoxyadenosine triphosphate. The result is a patient with profound T and B lymphocyte dysfunction that is susceptible to infections with opportunistic pathogens. ADA deficiency is an attractive target for gene therapy because corrected lymphocytes have a survival advantage over noncorrected, ADA-negative lymphocytes. Multiple gene therapy trials for ADA deficiency SCID (ADA-SCID) targeting HSC as well as lymphocytes have been performed [12–16,117]. Discussion in this section will focus on HSCs as the target for gene therapy.

Van Beusechem et al. laid the groundwork for HSC-based, ADA gene transfer by demonstrating in nonhuman primates that prolonged low-level expression of human ADA from peripheral blood cells was possible after bone marrow cells were transduced with retrovirus containing the ADA cDNA [118]. In 1995, Bordignon et al. in Milan, Italy, published results of a clinical trial in which both peripheral blood lymphocytes and HSCs were transduced with two molecularly distinct retroviral vectors carrying the ADA and neomycin resistance genes and then reinfused into the patient without cytoreductive preconditioning [12]. One patient received nine injections of gene-corrected cells over a two-year period. The other patient received five injections over 10 months. Prolonged detection of both transgenes was observed in both bone marrow and peripheral blood myeloid and lymphoid cells for the entire 35-month period of follow-up. Until the 35-month time point, the vector used for HSC correction was detected only in circulating T cells and bone marrow cells but not in the lymphocytes. Gene-marked lymphocytes contained only the vector used to transduce lymphocytes. At 35 months, however, lymphocytes derived from the gene-corrected HSCs were first detected and subsequently became more prevalent than those that carried the lymphocyte-specific vector. For ethical reasons, all patients treated in this study were treated concomitantly with polyethylene glycol (PEG)-ADA, thereby blunting the inherent selective advantage that exists for the gene-corrected cells. In one patient however, immune function became progressively dysfunctional despite PEG-ADA therapy and sustained long-term expression of the ADA gene in genetically modified lymphocytes. This provided the investigators with the rational for discontinuation of PEG-ADA therapy, with hope of restoring a selective advantage to the corrected cells and improving immune function [98]. Prior to discontinuation of the PEG-ADA, the patient received additional gene-corrected autologous lymphocytes. The results of this intervention were positive. There was an increase in total peripheral blood T-cell numbers, an increase in intracellular ADA levels, and normalization of *in vitro* T-cell responsiveness to anti-CD3 and alloantigens. Finally, the patient showed evidence of improved humoral immune function as measured by generation of anti-X174 antibodies following vaccination. However, the investigators also measured a concomitant increase in the toxic metabolites of adenosine in red blood cells, suggesting that despite the encouraging clinical effects of PEG-ADA withdrawal, repair of the underlying metabolic defect was incomplete.

Kohn et al. transduced CD34 cells isolated from umbilical cord blood of three ADA-deficient neonates with an MoMLV-based retroviral vector containing human ADA cDNA as well as the cDNA encoding the bacterial neomycin resistance gene [15]. Cells were returned to the unconditioned recipient on the fourth day of life. *Ex vivo* transduction efficiency as measured by percent G418-resistant CFU ranged from 12% to 19%. Clonogenic myeloid precursors from the bone marrow were assayed at 1 year and found to be G418-resistant at a frequency of 4%–6%. Concurrent with the infusion of gene-corrected HSCs, the patients began ADA replacement therapy with PEG-ADA. As a result, the selective advantage of ADA-positive clones was partially blunted. By 4 years of follow-up, the levels of gene-corrected cells in the peripheral blood mononuclear cell fraction had increased 50- to 100-fold in all the three patients [14]. The frequency of gene-corrected T-lymphocytes increased to 1%–10%. This rate of increase was far greater than that observed in the granulocyte series (0.01%–0.03%). In an attempt to assess clinical efficacy of gene transfer, PEG-ADA administration was discontinued in the patient with the highest level of gene-corrected T-lymphocytes. This resulted in reduction of S-adenosylhomocysteine hydrolase activity and loss of antigen specific T-lymphocyte reactivity, suggesting inadequate ADA production by the transduced cells. After 2 months, PEG-ADA replacement was restarted when oral monilia, sinusitis, and an upper respiratory tract infection developed. Hence, despite the inability to demonstrate clinical benefit, the persistence of the transgene for over 4 years in both the lymphoid and the myeloid cell lines suggests that gene transfer into true long-term, repopulating HSCs is possible from cells collected from umbilical cord blood.

It became clear from gene therapy studies of ADA-SCID that whether the target of transduction was T lymphocytes, marrow stem cells, or cord blood stem cells, there appeared to be insufficient expansion and activity of autologous gene-corrected T lymphocytes to achieve significant and lasting clinical benefit. Using improved transduction methodology and for the first time in the setting of gene therapy for a non-malignant disorder, the use of low-intensity bone marrow conditioning (Busulfan 4 mg/kg), the Italian group of Aiuti and coworkers in collaboration with investigators from Israel initially transplanted 2 ADA-SCID patients with gene-corrected HSC [11]. These patients were either unable to obtain or did not tolerate PEG-ADA treatment and for this reason were not treated with PEG-ADA in addition to receiving low-dose busulfan conditioning. Using this approach, the patients demonstrated an impressive increase in the number of peripheral blood T-cells within 2 months following gene therapy, obviating the need for PEG-ADA therapy and thus preserving the selective advantage of the corrected cells. Concurrent with increased T-cell numbers was normalization of T cell proliferative responses, increased levels of IgM and IgA, and in one patient, a robust humoral response to vaccination with tetanus toxoid. The most notable finding from this trial is that one patient had sustained, high-level (5%–15%) gene-marking in multilineage bone marrow and

peripheral blood cells as measured by real-time PCR. Based on this data, other groups have begun to use nonmyeloablative conditioning prior to gene therapy as well as holding the PEG-ADA. The London group from Great Ormond Street Hospital has published their results of one patient using melphalan (140 mg/m^2) conditioning and also stopping the PEG-ADA 1 month prior to the gene therapy. This patient appears to be doing well with evidence of T cell normalization and biochemical detoxification [16]. A total of 12 patients to date have been treated for ADA-SCID by various groups using nonmyeloablative conditioning regimens with clinical success in 11 of the 12 (as presented at platform session at the American Society for Gene Therapy National Meeting in June 2007).

33.4.1.4 Chronic Granulomatous Disease

CGD is an inherited immune deficiency caused by genetic mutations in any of the four subunits of the phagocyte NADPH oxidase (p47phox, p67phox, gp91phox, p22phox), resulting in the inability of phagocytes to produce microbicidal superoxide and hydrogen peroxide. CGD patients are prone to recurrent bacterial and fungal infections as well as granuloma formation [119]. Stem cell transplantation can cure CGD, indicating that stem cells are the appropriate target for gene therapy [120–124]. Preclinical work demonstrated that normal oxidase-positive neutrophils differentiate from CGD stem cells transduced with retroviral vectors encoding the corrective normal oxidase subunit cDNA [125–127].

Based on this work, investigators at the National Institutes of Health have undertaken a number of clinical gene transfer studies to treat CGD. However, unlike the two diseases discussed previously, there is no proliferative advantage conferred upon the corrected cells. The first study targeted CGD patients with deficiency of the p47phox subunit [8]. The p47phox cDNA was cloned into the MoMLV-based MFGS retrovirus vector and this vector was packaged by the murine crip amphotropic packaging cell line. Mobilized PBSC were apheresed from patients after treatment with G-CSF followed by CD34 enrichment using an immuno-paramagnetic bead-selection device. Retroviral transduction of the PBSC was performed in serum-free media for 6h in the presence of hematopoietic cytokines for 3 consecutive days. The transduced PBSC were then transfused into the patient without marrow conditioning. *Ex vivo* transduction efficiency was determined by several methods, including quantitative assessment of oxidase-positive (Nitroblue tetrazolium dye test) myeloid colonies as well as determination of vector copy number using Southern blot hybridization. The percent of oxidase-positive colonies ranged from 9% to 29%. This was closely correlated with the vector copy number of 0.05 to 0.18 (equivalent to 5%–18% of cells, assuming a copy number of one per transduced cell) found in transduced and cultured PBSC. Following intravenous infusion of these transduced autologous CD34 cells into patients, it was possible to demonstrate the appearance of oxidase-normal neutrophils in the peripheral blood as assessed using a highly sensitive flow cytometry assay based on the increased fluorescence of dihydrorhodamine-123 oxidized by hydrogen peroxide [128]. Oxidase-normal neutrophils first appeared 2 weeks after transplantation and peak correction occurred after 3–6 weeks. The frequency of corrected neutrophils ranged from 0.004% to 0.05% though it is important to note that the oxidase activing in these rare gene-corrected neutrophils was equivalent to that seen in oxidase-normal neutrophils from healthy individuals.

The NIH investigators then performed a follow-up study that targeted CGD patients with the X-linked form of CGD caused by mutations in the transmembrane gp91phox subunit. All five patients enrolled in this trial had the gp91phox null form of X-linked CGD and had previously received allogeneic normal granulocytes as treatment for severe infection. A high-titer MFGS-gp91phox retroviral vector was prepared using the human-derived 293 (8) amphotropic envelope packaging cell line. The mobilized selected CD34-positive PBSC were exposed to retroviral supernatant in gas-permeable flexible plastic bag containers, which were coated with a clinical-grade preparation of the CH-296 fibronectin fragment (RetroNectin). As in the first trial, cells were then infused into the patient without marrow conditioning. Two or three cycles of gene therapy were administered at approximately 50-day intervals. *Ex vivo* transduction efficiency was assessed by flow cytometry using an anti-gp91 monoclonal antibody as well as by nitroblue tetrazolium dye testing for oxidase-positive colonies. Seventy-five percent to 85% of CD34 + CD38-, Lin-, HLA DR-cells were positive for the gp91 as detected by flow cytometry after 7 days of *in vitro* culture. In the case of myeloid CFUs, 50% were found to be oxidase-positive. Gene-corrected neutrophils were detected in three of the five patients beginning 2–3 weeks following the infusion of cells. One patient peaked at almost 0.15% oxidase-positive neutrophils. Similar peaks of corrected neutrophils were seen after each cycle, but the size of the peak correlated with the dose of gene-corrected cells that were infused. Gene-corrected cells gradually disappeared from the peripheral circulation approximately 3–6 months after the last cycle. In one patient, corrected neutrophils were recovered from the pus of a liver abscess that predated the time of enrollment in the study (gene therapy was administered partly in the hope for benefit in treating this infection). The inability to detect corrected neutrophils in two of the five patients may be a result of immune-mediated elimination of gene-corrected cells in this protein-null, inherited disorder as the therapeutic gene product might have been perceived by the patient immune system as a neoantigen.

In 2004, a group in Germany began a study using an SFFV based vector for transduction of G-CSF mobilized PBSC for patients with CGD and the results from two patients were published in 2006 [10]. Similar to the studies for ADA-SCID, the patients also received busulfan conditioning prior to the stem cell infusion this time at a dose of 8 mg/kg given as divided doses over 2 days. As noted, the vector used was an SFFV vector containing the GP91 transgene. The bulk transduction efficiency in the *ex vivo* culture was 39.5%–45.2%

and the number of cells infused ranged from 3.6 to 5.1×10^6 CD34s/kg. Of note is that there was resolution of underlying infections existing in both patients at the time of gene therapy. Early after gene therapy, 15% to 20% of circulating neutrophils were gene-marked and oxidase positive, but gene marking increased to greater than 40% and 60% in each patient, respectively, after 6 months. However, the majority of gene marking in both patients was due to contribution from only a few clones and was limited to the myeloid lineage. Analysis of the different insertion sites showed a predilection for the transgene to insert into myeloid-promoting genes, specifically SETBP1, PRDM16, and MDS/EVI1. Despite the high levels of gene marking seen, the expression of the therapeutic transgene protein was limited, resulting in only 14% and 30%, respectively, of normal levels oxidase production per oxidase positive neutrophil. As recently communicated by these investigators at several international meetings, by 2 years after gene therapy, the high level of gene marking that persisted had become dominated by one or a few clones producing neutrophils that appear to have lost or very significantly diminished oxidase activity. More recently these investigators noted that both of their reported patients may have had the slow emergence of a clone of myeloid cells manifesting monosomy 7. One patient has succumbed to a severe infection, while the other patient appears to be stable. A third patient has also been treated and has not appeared to have this same clonal outgrowth noted in the first two published patients, but the marking is also significantly lower at less than 1% to date.

The investigators at NIH have also undertaken a new study using a nonmyeloablative conditioning regimen consisting of busulfan 10 mg/kg in divided doses over 2 days prior to gene therapy. Enrolling only patients who have an underlying infection unresponsive to conventional therapy and who also do not have an HLA-matched healthy sibling who could serve as a stem cell donor for transplant, the aim of the trial is to use gene therapy as a treatment for the infection. Other than the addition of nonmyeloablative busulfan conditioning, this new clinical trial remains very similar to the previously described NIH study, including using the exactly same amphotropic MFGS-gp91phox vector for the transduction. The RetroNectin-coated gas permeable bags with serum-free medium is used for transduction in the presence of SCF, Flt-3 ligand, TPO, and low-dose IL-3 as growth factors. This resulted in 73% bulk transduction in the *ex vivo* culture after 4 days of transduction of the first patient's PBSC. To date, the first patient treated has had early appearance of almost 24% oxidase-corrected neutrophils in peripheral blood. This slowly decreased over the first 6 months, stabilizing at about 1% of circulating neutrophils that were oxidase normal between 6 and 10 months post gene therapy. Thus, using the same vector and similar methods of transduction, the use of busulfan conditioning resulted in a 10- to 100-fold improvement from the prior trials. Most important is that there was clinical cure of a massive liver abscess that had previously been unresponsive to conventional therapy. In addition, throughout the follow-up period, the oxidase activity per gene-corrected cell remained equivalent to the fully normal level of oxidase activity seen in neutrophils from a healthy donor. Neither oligoclonality of gene marking, nor outgrowth of gene marked clones, nor predilection for vector insertion in particular genes has been observed to date in follow-up.

33.4.1.5 Multidrug Resistance Gene Transfer

Some of the other chapters in this volume will discuss in great detail the use of certain mammalian genes that confer resistance that allowed for *in vivo* selective enhancement of gene marking. So this topic will not be discussed in detail in this chapter. However, the MDR1 gene as a selective marker is one of the first mammalian genes tried for this purpose and serves as a model for this concept and is not discussed in great detail elsewhere in this volume. Therefore, it will be reviewed here. The MDR1 gene encodes a drug efflux pump called P-glycoprotein (reviewed in Refs. [129–131]). This pump confers resistance to a variety of anticancer agents including doxorubicin, mitoxantrone, vincristine, etoposide, and taxol. MDR1 can be used as both a therapeutic gene and as a selectable marker for other therapeutic genes [132].

In most cases, myelosuppression is the dose-limiting toxicity of chemotherapeutic agents used to treat cancer. Therefore, if the transfer of MDR1 into HSC is able to attenuate myelosuppression, higher and potentially more effective doses of chemotherapy could be delivered safely to the patient. Proof of this principle was provided in transgenic mice engineered to constitutively express MDR1 [133]. These mice could tolerate several-fold higher doses of taxol and daunomycin than their wild-type counterparts [134,135]. A similar effect has been demonstrated using retroviral gene transfer of MDR1 into HSC [136–138].

Results from the early studies were severely compromised by low levels of *ex vivo* transduction of HSCs, a low transplanted dose of transduced HSCs, and by the advanced stage of the cancer patients who were treated [19,20]. Subsequent studies were notable for the fact that they provided evidence that in a few select patients, engrafted MDR1-transduced HSCs could reduce the intensity of the neutrophil nadir following high-dose chemotherapy and that repeated cycles of chemotherapy provided them with a selective advantage [17,18,139]. The most notable MDR1 clinical trial came from Abonour et al. [17]. The transgene vector consisted of an HMSV backbone packaged in the AM12 amphotropic packaging cell line. Autologous peripheral blood CD34 cells were transduced on dishes coated with the fibronectin derivative CH-296 in media containing fetal calf serum and cytokines (SCF and IL6 or G-CSF/mega-karyocyte growth and development factor [MGDF]/SCF) and infused into the patient following conditioning with VP16 and carboplatin. Transgenes containing peripheral blood cells were detected by quantitative real-time PCR in seven of 12 patients, with one patient as high as 5.6% 2 weeks following the transplant. No definitive evidence of hematopoietic chemoprotection was demonstrated when the gene therapy patients were treated with further chemotherapy. However, there was an increase in the percentage of transgene-containing cells in six patients following

additional chemotherapy, suggesting a selective advantage for these cells. Importantly, no unexpected adverse events occurred in the study participants. Safety issues regarding the use of MDR1 as a selectable marker have been brought into question by a recent study that demonstrates development of a myeloproliferative syndrome in mice transplanted with MDR1-transduced cells. However, here the transplanted cells had been subjected to prolonged *ex vivo* culture [140]. Further studies in large animal models are needed to confirm this possible toxicity. In addition to MDR1, other promising selectable mammalian genes that can be used for *in vivo* selection include mutant versions of dihydrofolate reductase and the 6-*O*-benzylguanine resistant P140K mutant of the DNA alkylation repair enzyme methylguanine methyltransferase (MGMT). As noted, the latter selective marker has attracted much recent interest and features in two ongoing clinical trials as will be discussed in Chapter 26.

33.4.1.6 Fanconi Anemia

Fanconi anemia (FA) is an autosomal recessive disorder manifested by aplastic anemia, physical malformations, and cancer susceptibility [141]. Eight separate genotypic groups of FA have been described (FA-A though FA-H) [142]. Cells carrying the FA mutation are hypersensitive to DNA-damaging agents such as mitomycin C [143]. Since stem cell transplantation successfully treats the hematological manifestations of FA [144,145], it is a logical candidate for stem cell gene therapy. In addition, there is evidence to suggest that normal stem cells may have a selective growth advantage over HSCs with the mutated FA gene [146]. After demonstrating *in vitro* that HSCs from patients with the FA group C (FAC), gene mutation could be functionally corrected by retroviral gene transfer of the normal FAC gene, Liu et al. initiated a clinical trial [147]. Three children and one adult with the FAC mutation were treated with sequential cycles of autologous, G-CSF-mobilized CD34 selected progenitor cells that were transduced with an MoMLV-based retroviral vector containing the cDNA for FAC and neomycin resistance. The cells were transduced in the presence of hematopoietic cytokines IL-3, IL-6, and SCF for 72 h and returned to the patients without cytotoxic bone marrow conditioning. The cell dose for patients 1 and 2 was extremely low due to poor CD34 cell mobilization in response to G-CSF. FAC vector sequence was detected by PCR in the peripheral blood and bone marrow of all three patients at levels that ranged from 0.01% to 3%. Multilineage, peripheral blood marking was detected in one patient for a 16-month period during which the patient received four cycles of gene transfer. The other patients had only transient bone marrow and peripheral blood positivity despite repeated cycles of gene therapy. Bone marrow sampling of each patient following the infusion of gene-corrected cells revealed an increase in the number of colonies resistant to the DNA-damaging effects of mitomycin C. Radiation therapy delivered to patient 4 approximately 50 days following gene therapy for treatment of squamous cell carcinoma of the vulva resulted in detection, for the first time, of gene-marked cells in the peripheral blood. These gene-corrected cells were undetectable prior to radiation therapy, suggesting that they were afforded a selective advantage by the radiation treatments [148]. A large animal study using the same vector design as that used for the clinical study was also performed. The animals underwent low-dose radiation prior to infusion of the genetically modified cells, but were otherwise hematologically normal. In this setting, the modified cell did not have a selective advantage but the marking levels, while low, were detectable for greater than 1 year, suggesting that again as in the ADA-SCID trial, with low-dose conditioning and the selective advantage of corrected cells in the setting of abnormal cells, gene therapy may be more efficacious [149].

33.4.1.7   Gaucher's Disease

Gaucher's disease is an autosomal recessive disorder that results in a deficiency of the lysosomal enzyme glucocerebrosidase [150]. This leads to the accumulation of glucosylcerebroside in macrophages throughout the reticular endothelial system. Although the clinical course is quite variable, most patients develop hepatosplenomegaly and painful, lytic, bone lesions. Conventional treatment includes stem cell transplantation or glucocerebrosidase supplementation. Dunbar et al. published the first clinical gene therapy trial for Gaucher's disease [21]. Three patients received CD34-selected bone marrow or mobilized PB transduced on autologous stroma/cytokines with a retroviral vector containing the human glucocerebrosidase cDNA. Transduction efficiencies were low, ranging from 1% to 10% using a semiquantitative PCR assay. Transgene was detected in peripheral blood mononuclear cells (0.02%) for 3 months in the patient whose cells were transduced with the highest efficiency.

33.4.1.8   Wiskott–Aldrich Syndrome

Wiskott-Aldrich syndrome is a primary immunodeficiency characterized by recurrent sinobacterial infections, autoimmunity typically manifesting as eczema and thrombocytopenia with dysfunctional platelets. The spectrum of the disease is variable, but the majority of patients have thrombocytopenia and varying levels of immunocompromise. The WASP protein is a cytoskeletal protein found on all hematopoietic cells and necessary for actin polymerization. As it can be cured with allogeneic transplantation, WA is also a good candidate for gene therapy. Two patients have been reportedly treated using transduced HSCs; however as with many other gene therapies, the levels of marking have been low as the cells were infused without any prior bone marrow conditioning. A number of other trials are being planned internationally, and most are now designed to include some form of conditioning.

33.4.1.9   Human Immunodeficiency Virus

HSC and mature T-lymphocytes are targets for anti-HIV gene therapy. Strategies include transfer of gene-encoding

ribozyme, which target the viral RNA genome, RNA decoys, and mutant transactivator genes that interfere with viral gene expression [38,151–153]. These approaches, along with a review of the clinical trials, are discussed in Chapter 41.

33.4.1.10   Summary

Techniques for gene transfer to HSC have advanced to the point where definitive clinical benefit has been achieved by *ex vivo* HSC gene therapy in ADA-SCID, X-CID, and X-linked CGD. Sustained engraftment of transduced HSC whose progeny do not have a selective advantage is best demonstrated in the gene transfer studies where cytotoxic bone marrow conditioning was administered prior to the transplant [3,4,6,11]. Besides reducing the number of resident naive stem cells, marrow conditioning results in a bone marrow microenvironment that is more conducive for engraftment of transduced HSC. Improvement in retroviral transduction conditions that utilize cytokines such as SCF, FLT-3 ligand, TPO, and limited concentration of IL-3 have resulted in impressive rates of HSC gene transfer in myeloablated nonhuman primates [86,104] and in the human clinical trials of gene therapy for ADA-SCID, X-SCID, and X-CGD discussed above. Alternative vectors such as those derived from lentivirus hold considerable promise for improvements in future HSC-gene-transfer strategies, assuming that their safety can be maximized and production issues can be resolved. Given the newly described cases of insertional mutagenesis that has complicated one of the X-SCID trials and the insertion-mediated outgrowth of myeloid cells seen in one of the X-CGD trials, the risk versus benefit of gene therapy will need to be reevaluated for each candidate stem cell disorder and for different vector systems and transduction conditions.

33.4.2   Lymphocyte Gene Transfer Studies

33.4.2.1   Gene Marking Studies

The first clinical gene therapy trial involved gene transfer into lymphocytes and was undertaken in the late 1980s as a means of characterizing tumor-infiltrating lymphocytes (TIL) [1]. TIL were isolated from tumors of patients with metastatic melanoma. These cells were expanded and then transduced with a retroviral vector containing the neomycin resistance gene. Following reinfusion, the marker gene was used to track the migration of the marked lymphocytes. TIL were consistently found in the peripheral blood as well as tumor deposits for up to 2 months. A similar approach was employed to study the persistence of Epstein–Barr virus (EBV)-specific cytotoxic T-lymphocytes (CTL) generated *ex vivo* and infused as treatment of post bone marrow transplant EBV-related lymphoproliferation [30,31]. EBV-specific CTLs marked with a neomycin-resistance vector were detectable in the peripheral circulation for 10 weeks postinfusion.

33.4.2.2 ADA Deficiency

Blaese et al. were the first to use a therapeutic gene in a retroviral gene transfer trial for ADA-deficient SCID patients [32]. Two patients with an incomplete response to PEG-ADA therapy were infused with T-lymphocytes transduced with an MoMLV-based retroviral vector containing the cDNA coding for ADA. The transduction efficiency of lymphocytes prior to infusion ranged from 0.1% to 10%. Patients were treated with multiple cycles of transduced T-cells over a period of 1–2 years. Gene-corrected cells were detected in circulation for 2 years following the final cycle of gene therapy, demonstrating a T-lymphocyte lifespan much longer than what was predicted. There was clear-cut evidence of improved cellular and humoral immune response following gene therapy in one of the two patients treated. This discrepancy in patient response was attributed to superior *ex vivo* transduction efficiency obtained in the responding patient. No toxicity was attributed to the conduct of the protocol. Using the identical experimental design and retroviral vector, Onodera et al. accomplished similar results in one patient treated in Japan [117]. These studies, in conjunction with the previously described Bordignon study [12], demonstrate the potential of lymphocyte-based gene therapy to provide prolonged clinical benefit.

33.4.2.3 Suicide Gene Transfer

The use of herpes simplex virus thymidine kinase (HSV-TK) gene and other "suicide" genes for gene therapy is discussed for the phosphorylation of nucleoside analogs such as ganciclovir. Once phosphorylated, ganciclovir becomes toxic to the cell as it is incorporated into DNA. There are three published hematopoietic cell gene therapy trials utilizing the HSV-TK. Lymphocytes were the target cells in these studies [33–35]. In an attempt to make HIV therapy with autologous cytotoxic T-cells safer, Riddell et al. transduced HIV-specific cytotoxic T-cells with the HSV-TK gene [33,34]. The transduced cells could then be eliminated with ganciclovir treatment if toxicity were to arise from their presence. The vector used to transduce the lymphocytes also carried the hygromycin phosphotransferase gene. Unexpectedly, the transduced lymphocytes were rejected by a brisk host CTL response against the transduced cells and thus efficacy could not be assessed.

In allogeneic bone marrow transplantation, lymphocytes taken from the bone marrow donor are often infused into the recipient to treat tumor relapse via the graft versus tumor effect, or to aid in immune reconstitution. Significant toxicity may arise if transplanted lymphocytes mount an immunological attack against the recipient (graft vs. host disease). As in the Riddell study, Bonini et al. studied HSV-TK-transduced donor lymphocyte infusions as a method of protecting against GVHD [33]. The vector used for transduction also carried the marker gene that coded for a truncated form of the human low-affinity receptor for nerve growth factor as well as the neomycin resistance gene. Gene-marked lymphocytes were

detected in seven of eight patients available for analysis with a range of 0.01% to 13.4% of the total circulating lymphocytes. Three of the patients developed GVHD and were treated with ganciclovir. The percent of genetically modified lymphocytes decreased dramatically with complete resolution of GVHD in two of the three patients. Partial resolution of GVHD occurred in the other. While there was no evidence of an immune-mediated elimination of the transduced cells, these bone marrow transplant patients were likely immunocompromised even more profoundly than the HIV patients studied by Riddell et al. Using a similar strategy, Tiberghien et al. infused donor lymphocytes transduced with an MoMuLV-derived retroviral vector carrying the genes for HSV-TK and neomycin resistance into recipients of a T-cell-depleted allogeneic stem cell transplantation [35]. Of the 12 patients treated, four developed clinically significant GVHD, prompting ganciclovir administration. Ganciclovir administration resulted in resolution of GVHD in three of the four patients and, as in the earlier study, correlated with a drop in the number of gene-marked lymphocytes.

33.4.2.4 Cancer Therapy

A number of investigators have been engineering lymphocytes and other immune system targets to more effectively target cancers. A common technique is to transduce dendritic cells to express tumor antigens and thereby expand *ex vivo* reactive T cells for infusion into the patient. (This is a technique that is also being used to generate virus specific T cells, as discussed further in Chapters 36 and 47.) Other groups have modified the T cells themselves. Morgan et al. treated 17 patients with treatment refractory melanoma using an HLA A*0201 TCR receptor specific for the MART-1 antigen. Peripheral blood lymphocytes were transduced with genes encoding the alpha and beta chains of the anti-Mart-1 TCR using a retroviral vector. The initial three patients had less than 2% positive cells 50 days postinfusion. As a result, the next 10 patients, including one patient from the first cohort, were treated with cells cultured for a shorter period of time. Here >9% positive cells were detectable 1 and 4 weeks posttreatment. For the third cohort of four patients, the cells underwent a second expansion to increase the cell dose. All patients received lympho-depleting chemotherapy with fludarabine prior to the infusion of cells. Two patients, one from the second cohort and the other from the third, had a clinical response with complete regression of tumor [39].

33.5 CONCLUSION

To date, all published clinical trials involving transduction of hematopoietic cells have utilized murine-based retroviral vectors. When lymphocytes are the target cell, modest clinical benefit was observed in both the ADA and HSV-TK trials. This is attributable to the relatively high level of transduction efficiency using the latest techniques. Though clinical applicability of HSC gene transfer is much broader than for lymphocytes, long-term repopulating HSCs are less receptive to retroviral-based gene transfer. While alternate vectors such as lentivirus and adeno-associated viral vectors are promising, a number of studies have proven that the retroviral vector can successfully target the long term repopulating cell, albeit with an extremely low efficiency. To improve the ability to transduce HSC, further knowledge of stem cell biology such as their cell cycle characteristics, their trigger for self-replication versus lineage commitment, and the optimal *ex vivo* growth conditions will aid in the ability to transduce these cells. The production of high-titer, replication-defective, retroviral supernatant is another technique that must be perfected. It is unclear to what extent host immune rejection of cells expressing the transgene will hinder progress of gene therapy for protein-null disorders or disorders where a heterologous gene product is produced.

And finally, it is becoming clear that some degree of cytoreductive bone marrow conditioning will be necessary to achieve clinically significant levels of HSC gene transfer.

REFERENCES

1. Rosenberg, S.A., et al., Gene transfer into humans—immunotherapy of patients with advanced melanoma using tumor-infiltrating lymphocytes modified by retroviral gene transduction. *N Engl J Med*, 1990, **323**(9): 570–578.
2. Bachier, C.R., et al., Hematopoietic retroviral gene marking in patients with follicular non-Hodgkin's lymphoma. *Leuk Lymphoma*, 1999, **32**(3–4): 279–288.
3. Brenner, M.K., et al., Gene marking to determine whether autologous marrow infusion restores long-term haemopoiesis in cancer patients. *Lancet*, 1993, **342**(8880): 1134–1137.
4. Brenner, M.K., et al., Gene-marking to trace origin of relapse after autologous bone-marrow transplantation. *Lancet*, 1993, **341**(8837): 85–86.
5. Deisseroth, A.B., et al., Genetic marking shows that Ph + cells present in autologous transplants of chronic myelogenous leukemia (CML) contribute to relapse after autologous bone marrow in CML. *Blood*, 1994, **83**(10): 3068–3076.
6. Dunbar, C.E., et al., Retrovirally marked CD34-enriched peripheral blood and bone marrow cells contribute to long-term engraftment after autologous transplantation. *Blood*, 1995, **85**(11): 3048–3057.
7. Emmons, R.V., et al., Retroviral gene transduction of adult peripheral blood or marrow- derived CD34 + cells for six hours without growth factors or on autologous stroma does not improve marking efficiency assessed in vivo. *Blood*, 1997, **89**(11): 4040–4046.
8. Malech, H.L., et al., Prolonged production of NADPH oxidase-corrected granulocytes after gene therapy of chronic granulomatous disease. *Proc Natl Acad Sci USA*, 1997, **94**(22): 12133–12138.
9. Malech, H.L., et al., Extended production of oxidase normal neutrophils in X-linked chronic granulomatous disease (CGD) following gene therapy with gp91phos transduced CD34 + cells. *Blood*, 1998, **92**(10 Suppl. 1): 690a.
10. Ott, M.G., et al., Correction of X-linked chronic granulomatous disease by gene therapy, augmented by insertional activation of MDS1-EVI1, PRDM16 or SETBP1. *Nat Med*, 2006, **12**(4): 401–409.
11. Aiuti, A., et al., Correction of ADA-SCID by stem cell gene therapy combined with nonmyeloablative conditioning. *Science*, 2002, **296**(5577): 2410–2413.
12. Bordignon, C., et al., Gene therapy in peripheral blood lymphocytes and bone marrow for ADA- immunodeficient patients. *Science*, 1995, **270**(5235): 470–475.
13. Hoogerbrugge, P.M., et al., Bone marrow gene transfer in three patients with adenosine deaminase deficiency. *Gene Ther*, 1996, **3**(2): 179–183.
14. Kohn, D.B., et al., T lymphocytes with a normal ADA gene accumulate after transplantation of transduced autologous umbilical cord blood CD34 + cells in ADA- deficient SCID neonates. *Nat Med*, 1998, **4**(7): 775–780.
15. Kohn, D.B., et al., Engraftment of gene-modified umbilical cord blood cells in neonates with adenosine deaminase deficiency. *Nat Med*, 1995, **1**(10): 1017–1023.
16. Gaspar, H.B., et al., Successful reconstitution of immunity in ADA-SCID by stem cell gene therapy following cessation of PEG-ADA and use of mild preconditioning. *Mol Ther*, 2006, **14**(4): 505–513.
17. Abonour, R., et al., Efficient retrovirus-mediated transfer of the multidrug resistance 1 gene into autologous human long-term repopulating hematopoietic stem cells [see comments]. *Nat Med*, 2000, **6**(6): 652–658.
18. Cowan, K.H., et al., Paclitaxel chemotherapy after autologous stem-cell transplantation and engraftment of hematopoietic cells transduced with a retrovirus containing the multidrug resistance complementary DNA (MDR1) in metastatic breast cancer patients. *Clin Cancer Res*, 1999, **5**(7): 1619–1628.
19. Hanania, E.G., et al., Results of MDR-1 vector modification trial indicate that granulocyte/macrophage colony-forming unit cells do not contribute to posttransplant hematopoietic recovery following intensive systemic therapy. *Proc Natl Acad Sci USA*, 1996, **93**(26): 15346–15351.
20. Hesdorffer, C., et al., Human MDR gene transfer in patients with advanced cancer. *Hum Gene Ther*, 1994, **5**(9): 1151–1160.
21. Dunbar, C.E., et al., Retroviral transfer of the glucocerebrosidase gene into CD34 + cells from patients with Gaucher disease: in vivo detection of transduced cells without myeloablation. *Hum Gene Ther*, 1998, **9**(17): 2629–2640.
22. Bauer, T.R., Jr. and D.D. Hickstein, Gene therapy for leukocyte adhesion deficiency. *Curr Opin Mol Ther*, 2000, **2**(4): 383–388.
23. Kohn, D.B., et al., A clinical trial of retroviral-mediated transfer of a rev-responsive element decoy gene into CD34(+) cells from the bone marrow of human immunodeficiency virus-1-infected children. *Blood*, 1999, **94**(1): 368–371.
24. Cavazzana-Calvo, M., et al., Gene therapy of human severe combined immunodeficiency (SCID)-X1 disease [see comments]. *Science*, 2000, **288**(5466): 669–672.
25. Thrasher, A.J., et al., Failure of SCID-X1 gene therapy in older patients. *Blood*, 2005, **105**(11): 4255–4257.
26. Chinen, J., et al., Gene therapy improves immune function in pre-adolescents with X-linked severe combined immunodeficiency. *Blood*, 2007, 111(1): 67–73.
27. Gaspar, H.B., et al., Gene therapy of X-linked severe combined immunodeficiency by use of a pseudotyped gammaretroviral vector. *The Lancet*, 2004, **364**(9452): 2181–2187.
28. Kang, E.M., et al., Nonmyeloablative conditioning followed by transplantation of genetically modified HLA-matched peripheral blood progenitor cells for hematologic malignancies in patients with acquired immunodeficiency syndrome. *Blood*, 2002, **99**(2): 698–701.

29. Merrouche, Y., et al., Clinical application of retroviral gene transfer in oncology: results of a French study with tumor-infiltrating lymphocytes transduced with the gene of resistance to neomycin. *J Clin Oncol*, 1995, **13**(2): 410–418.

30. Rooney, C.M., et al., Use of gene-modified virus-specific T lymphocytes to control Epstein–Barr-virus-related lymphoproliferation. *Lancet*, 1995, **345**(8941): 9–13.

31. Rooney, C.M., et al., Infusion of cytotoxic T cells for the prevention and treatment of Epstein–Barr virus-induced lymphoma in allogeneic transplant recipients. *Blood*, 1998, **92**(5): 1549–1555.

32. Blaese, R.M., et al., T lymphocyte-directed gene therapy for ADA- SCID: initial trial results after 4 years. *Science*, 1995, **270**(5235): 475–480.

33. Bonini, C., et al., HSV-TK gene transfer into donor lymphocytes for control of allogeneic graft-versus-leukemia. *Science*, 1997, **276**(5319): 1719–1724.

34. Riddell, S.R., et al., T-cell mediated rejection of gene-modified HIV-specific cytotoxic T lymphocytes in HIV-infected patients [see comments]. *Nat Med*, 1996, **2**(2): 216–223.

35. Tiberghien, P., et al., Administration of herpes simplex-thymidine kinase-expressing donor T cells with a T-cell-depleted allogeneic marrow graft. *Blood*, 2001, **97**(1): 63–72.

36. Ranga, U., et al., Enhanced T cell engraftment after retroviral delivery of an antiviral gene in HIV-infected individuals. *Proc Natl Acad Sci USA*, 1998, **95**(3): 1201–1216.

37. Woffendin, C., et al., Expression of a protective gene-prolongs survival of T cells in human immunodeficiency virus-infected patients. *Proc Natl Acad Sci USA*, 1996, **93**(7): 2889–2894.

38. Macpherson, J.I., et al., Long-term survival and concomitant gene expression of ribozyme-transduced CD4 + T-lymphocytes in HIV-infected patients. *J Gene Med*, 2005, **7**(5): 552–564.

39. Morgan, R.A., et al., Cancer regression in patients after transfer of genetically engineered lymphocytes. *Science*, 2006, **314**(5796): 126–129.

40. Levine, B.L., et al., Gene transfer in humans using a conditionally replicating lentiviral vector. *Proc Natl Acad Sci USA*, 2006, **103**(46): 17372–17377.

41. Roe, T., et al., Integration of murine leukemia virus DNA depends on mitosis. *Embo J*, 1993, **12**(5): 2099–2108.

42. Karlsson, S., Treatment of genetic defects in hematopoietic cell function by gene transfer. *Blood*, 1991, **78**(10): 2481–2492.

43. Challita, P.M., et al., Multiple modifications in cis elements of the long terminal repeat of retroviral vectors lead to increased expression and decreased DNA methylation in embryonic carcinoma cells. *J Virol*, 1995, **69**(2): 748–755.

44. Miller, A.D. and C. Buttimore, Redesign of retrovirus packaging cell lines to avoid recombination leading to helper virus production. *Mol Cell Biol*, 1986, **6**(8): 2895–2902.

45. Markowitz, D., S. Goff, and A. Bank, A safe packaging line for gene transfer: separating viral genes on two different plasmids. *J Virol*, 1988, **62**(4): 1120–1124.

46. Danos, O. and R.C. Mulligan, Safe and efficient generation of recombinant retroviruses with amphotropic and ecotropic host ranges. *Proc Natl Acad Sci USA*, 1988, **85**(17): 6460–6464.

47. Davis, J.L., et al., Retroviral particles produced from a stable human-derived packaging cell line transduce target cells with very high efficiencies. *Hum Gene Ther*, 1997, **8**(12): 1459–1467.

48. Orlic, D., et al., The level of mRNA encoding the amphotropic retrovirus receptor in mouse and human hematopoietic stem cells is low and correlates with the efficiency of retrovirus transduction. *Proc Natl Acad Sci USA*, 1996, **93**(20): 11097–11102.

49. Kelly, P.F., et al., Highly efficient gene transfer into cord blood nonobese diabetic/severe combined immunodeficiency repopulating cells by oncoretroviral vector particles pseudotyped with the feline endogenous retrovirus (RD114) envelope protein. *Blood*, 2000, **96**(4): 1206–1214.

50. Kiem, H.P., et al., Gene transfer into marrow repopulating cells: comparison between amphotropic and gibbon ape leukemia virus pseudotyped retroviral vectors in a competitive repopulation assay in baboons. *Blood*, 1997, **90**(11): 4638–4645.

51. Leurs, C., et al., Comparison of three retroviral vector systems for transduction of nonobese diabetic/severe combined immunodeficiency mice repopulating human CD34 + cord blood cells. *Hum Gene Ther*, 2003, **14**(6): 509–519.

52. Hematti, P., et al., Distinct genomic integration of MLV and SIV vectors in primate hematopoietic stem and progenitor cells. *PLoS Biology*, 2004, **2**(12): e423.

53. De Palma, M., et al., Promoter trapping reveals significant differences in integration site selection between MLV and HIV vectors in primary hematopoietic cells. *Blood*, 2005, **105**(6): 2307–2315.

54. Wu, X., et al., Transcription start regions in the human genome are favored targets for MLV integration. *Science*, 2003, **300**(5626): 1749–1751.

55. Schroder, A.R., et al., HIV-1 integration in the human genome favors active genes and local hotspots. *Cell*, 2002, **110**(4): 521–529.

56. Andrews, R.G., J.W. Singer, and I.D. Bernstein, Precursors of colony-forming cells in humans can be distinguished from colony-forming cells by expression of the CD33 and CD34 antigens and light scatter properties. *J Exp Med*, 1989, **169**(5): 1721–1731.

57. Yin, A.H., et al., AC133, a novel marker for human hematopoietic stem and progenitor cells. *Blood*, 1997, **90**(12): 5002–5012.

58. Gallacher, L., et al., Isolation and characterization of human CD34-Lin- and CD34 + Lin- hematopoietic stem cells using cell surface markers AC133 and CD7. *Blood*, 2000, **95**(9): 2813–2820.

59. Barfield, R.C., et al., Autologous transplantation of CD133 selected hematopoietic progenitor cells for treatment of relapsed acute lymphoblastic leukemia. *Pediatric Blood Cancer*, 2007, **48**(3): 349–353.

60. Goodell, M.A., et al., Isolation and functional properties of murine hematopoietic stem cells that are replicating in vivo. *J Exp Med*, 1996, **183**(4): 1797–1806.

61. Zhang, J.L., et al., Long-term transgene expression and survival of transgene-expressing grafts following lentivirus transduction of bone marrow side population cells. *Transplantation*, 2005, **79**(8): 882–888.

62. Gabbianelli, M., et al., Multi-level effects of flt3 ligand on human hematopoiesis: expansion of putative stem cells and proliferation of granulomonocytic progenitors/monocytic precursors. *Blood*, 1995, **86**(5): 1661–1670.

63. Rusten, L.S., et al., The FLT3 ligand is a direct and potent stimulator of the growth of primitive and committed human CD34 + bone marrow progenitor cells in vitro. *Blood*, 1996, **87**(4): 1317–1325.

64. Sudo, Y., et al., Synergistic effect of FLT-3 ligand on the granulocyte colony-stimulating factor-induced mobilization of hematopoietic stem cells and progenitor cells into blood in mice. *Blood*, 1997, **89**(9): 3186–3191.

65. Brugger, W., et al., Reconstitution of hematopoiesis after high-dose chemotherapy by autologous progenitor cells generated ex vivo. *N Engl J Med*, 1995, **333**(5): 283–287.

66. Haas, R., et al., Successful autologous transplantation of blood stem cells mobilized with recombinant human granulocyte-macrophage colony-stimulating factor. *Exp Hematol*, 1990, **18**(2): 94–98.

67. Lane, T.A., et al., Harvesting and enrichment of hematopoietic progenitor cells mobilized into the peripheral blood of normal donors by granulocyte-macrophage colony-stimulating factor (GM-CSF) or G-CSF: potential role in allogeneic marrow transplantation. *Blood*, 1995, **85**(1): 275–282.

68. Socinski, M.A., et al., Granulocyte-macrophage colony stimulating factor expands the circulating haemopoietic progenitor cell compartment in man. *Lancet*, 1988, **1**(8596): 1194–1198.

69. Andrews, R.G., et al., In vivo synergy between recombinant human stem cell factor and recombinant human granulocyte colony-stimulating factor in baboons enhanced circulation of progenitor cells. *Blood*, 1994, **84**(3): 800–810.

70. Moskowitz, C.H., et al., Recombinant methionyl human stem cell factor and filgrastim for peripheral blood progenitor cell mobilization and transplantation in non-Hodgkin's lymphoma patients—results of a phase I/II trial. *Blood*, 1997, **89**(9): 3136–3147.

71. Sekhsaria, S., et al., Granulocyte colony-stimulating factor recruitment of CD34 + progenitors to peripheral blood: impaired mobilization in chronic granulomatous disease and adenosine deaminase–deficient severe combined immunodeficiency disease patients. *Blood*, 1996, **88**(3): 1104–1112.

72. Dunbar, C.E., et al., Improved retroviral gene transfer into murine and Rhesus peripheral blood or bone marrow repopulating cells primed in vivo with stem cell factor and granulocyte colony-stimulating factor. *Proc Natl Acad Sci USA*, 1996, **93**(21): 11871–11876.

73. Hematti, P., et al., Retroviral transduction efficiency of G-CSF + SCF-mobilized peripheral blood CD34 + cells is superior to G-CSF or G-CSF + Flt3-L-mobilized cells in nonhuman primates. *Blood*, 2003, **101**(6): 2199–2205.

74. Bregni, M., et al., Mobilized peripheral blood CD34 + cells express more amphotropic retrovirus receptor than bone marrow CD34 + cells. *Haematologica*, 1998, **83**(3): 204–208.

75. Horwitz, M.E., et al., Granulocyte colony-stimulating factor mobilized peripheral blood stem cells enter into G1 of the cell cycle and express higher levels of amphotropic retrovirus receptor mRNA. *Exp Hematol*, 1999, **27**(7): 1160–1167.

76. Lemoli, R.M., et al., Biological characterization of CD34 + cells mobilized into peripheral blood. *Bone Marrow Transplant*, 1998, **22 Suppl 5**: S47–S50.

77. Liles, W.C., et al., Mobilization of hematopoietic progenitor cells in healthy volunteers by AMD3100, a CXCR4 antagonist. *Blood*, 2003, **102**(8): 2728–2730.

78. Hess, D.A., et al., Human progenitor cells rapidly mobilized by AMD3100 repopulate NOD/SCID mice with increased frequency in comparison to cells from the same donor mobilized by granulocyte colony stimulating factor. *Biol Blood Marrow Transplant*, 2007, **13**(4): 398–411.

79. Larochelle, A., et al., AMD3100 mobilizes hematopoietic stem cells with long-term repopulating capacity in nonhuman primates. *Blood*, 2006, **107**(9): 3772–3778.

80. Broxmeyer, H.E., et al., Rapid mobilization of murine and human hematopoietic stem and progenitor cells with AMD3100, a CXCR4 antagonist. *J Exp Med*, 2005, **201**(8): 1307–1318.

81. Flomenberg, N., et al., The use of AMD3100 plus G-CSF for autologous hematopoietic progenitor cell mobilization is superior to G-CSF alone. *Blood*, 2005, **106**(5): 1867–1874.

82. Broxmeyer, H.E., et al., Growth characteristics of marrow hematopoietic progenitor/precursor cells from patients on a phase I clinical trial with purified recombinant human granulocyte-macrophage colony-stimulating factor. *Exp Hematol*, 1988, **16**(7): 594–602.

83. Orlic, D., et al., Identification of human and mouse hematopoietic stem cell populations expressing high levels of mRNA encoding retrovirus receptors. *Blood*, 1998, **91**(9): 3247–3254.

84. Jordan, C.T., G. Yamasaki, and D. Minamoto, High-resolution cell cycle analysis of defined phenotypic subsets within primitive human hematopoietic cell populations. *Exp Hematol*, 1996, **24**(11): 1347–1355.

85. Gothot, A., et al., Cell cycle-related changes in repopulating capacity of human mobilized peripheral blood CD34(+) cells in non-obese diabetic/severe combined immune-deficient mice. *Blood*, 1998, **92**(8): 2641–2649.

86. Tisdale, J.F., et al., Ex vivo expansion of genetically marked rhesus peripheral blood progenitor cells results in diminished long-term repopulating ability. *Blood*, 1998, **92**(4): 1131–1141.

87. Takatoku, M., et al., Avoidance of stimulation improves engraftment of cultured and retrovirally transduced hematopoietic cells in primates. *J Clin Invest*, 2001, **108**(3): 447–455.

88. Moore, K.A., et al., Stromal support enhances cell-free retroviral vector transduction of human bone marrow long-term culture-initiating cells. *Blood*, 1992, **79**(6): 1393–1399.

89. Nolta, J.A., E.M. Smogorzewska, and D.B. Kohn, Analysis of optimal conditions for retroviral-mediated transduction of primitive human hematopoietic cells. *Blood*, 1995, **86**(1): 101–110.

90. Wells, S., et al., The presence of an autologous marrow stromal cell layer increases glucocerebrosidase gene transduction of long-term culture initiating cells (LTCICs) from the bone marrow of a patient with Gaucher disease. *Gene Ther*, 1995, **2**(8): 512–520.

91. Luens, K.M., et al., Thrombopoietin, kit ligand, and flk2/flt3 ligand together induce increased numbers of primitive hematopoietic progenitors from human CD34 + Thy-1 + Lin- cells with preserved ability to engraft SCID-hu bone. *Blood*, 1998, **91**(4): 1206–1215.

92. Borge, O.J., et al., Ability of early acting cytokines to directly promote survival and suppress apoptosis of human primitive CD34 + CD38- bone marrow cells with multilineage potential at the single-cell level: key role of thrombopoietin. *Blood*, 1997, **90**(6): 2282–2292.

93. Peschon, J.J., et al., Early lymphocyte expansion is severely impaired in interleukin 7 receptor-deficient mice. *J Exp Med*, 1994, **180**(5): 1955–1960.

94. Plum, J., et al., Interleukin-7 is a critical growth factor in early human T-cell development. *Blood*, 1996, **88**(11): 4239–4245.

95. Cagnoni, P.J. and E.J. Shpall, Mobilization and selection of CD34-positive hematopoietic progenitors. *Blood Rev*, 1996, **10**(1): 1–7.

96. Shpall, E.J., et al., Peripheral blood stem cell harvesting and CD34-positive cell selection. *Cancer Treat Res*, 1997, **77**: 143–157.

97. Kotani, H., et al., Improved methods of retroviral vector transduction and production for gene therapy. *Hum Gene Ther*, 1994, **5**(1): 19–28.

98. Hanenberg, H., et al., Colocalization of retrovirus and target cells on specific fibronectin fragments increases genetic transduction of mammalian cells. *Nat Med*, 1996, **2**(8): 876–882.

99. Moritz, T., et al., Fibronectin improves transduction of reconstituting hematopoietic stem cells by retroviral vectors: evidence of direct viral binding to chymotryptic carboxy-terminal fragments. *Blood*, 1996, **88**(3): 855–862.

100. Hurley, R.W., J.B. McCarthy, and C.M. Verfaillie, Direct adhesion to bone marrow stroma via fibronectin receptors

inhibits hematopoietic progenitor proliferation. *J Clin Invest*, 1995, **96**(1): 511–519.

101. Nolta, J.A., M.B. Hanley, and D.B. Kohn, Sustained human hematopoiesis in immunodeficient mice by cotransplantation of marrow stroma expressing human interleukin-3: analysis of gene transduction of long-lived progenitors. *Blood*, 1994, **83**(10): 3041–3051.

102. Wu, T., et al., Prolonged high-level detection of retrovirally marked hematopoietic cells in nonhuman primates after transduction of CD34 + progenitors using clinically feasible methods. *Molec Ther*, 2000, **1**(3): 285–293.

103. Dao, M.A., et al., Adhesion to fibronectin maintains regenerative capacity during ex vivo culture and transduction of human hematopoietic stem and progenitor cells. *Blood*, 1998, **92**(12): 4612–4621.

104. Kiem, H.P., et al., Improved gene transfer into baboon marrow repopulating cells using recombinant human fibronectin fragment CH-296 in combination with interleukin-6, stem cell factor, FLT-3 ligand, and megakaryocyte growth and development factor. *Blood*, 1998, **92**(6): 1878–1886.

105. Rosenzweig, M., et al., Efficient and durable gene marking of hematopoietic progenitor cells in nonhuman primates after nonablative conditioning. *Blood*, 1999, **94**(7): 2271–2286.

106. Bunnell, B.A., et al., High-efficiency retroviral-mediated gene transfer into human and nonhuman primate peripheral blood lymphocytes. *Proc Natl Acad Sci USA*, 1995, **92**(17): 7739–7743.

107. Lam, J.S., et al., Improved gene transfer into human lymphocytes using retroviruses with the Gibbon ape leukemia virus envelope. *Hum Gene Ther*, 1996, **7**(12): 1415–1422.

108. Bosse, R., et al., Good manufacturing practice production of human stem cells for somatic cell and gene therapy. *Stem Cells*, 1997, **15 Suppl 1**: 275–280.

109. Cornetta, K., et al., Retroviral gene transfer in autologous bone marrow transplantation for adult acute leukemia. *Hum Gene Ther*, 1996, **7**(11): 1323–1329.

110. Hughes, P.F., et al., Retroviral gene transfer to primitive normal and leukemic hematopoietic cells using clinically applicable procedures. *J Clin Invest*, 1992, **89**(6): 1817–1824.

111. Nolta, J.A. and D.B. Kohn, Comparison of the effects of growth factors on retroviral vector-mediated gene transfer and the proliferative status of human hematopoietic progenitor cells. *Hum Gene Ther*, 1990, **1**(3): 257–268.

112. Hacein-Bey-Abina, S., et al., Sustained correction of X-linked severe combined immunodeficiency by ex vivo gene therapy. *N Engl J Med*, 2002. **346**(16): 1185–1193.

113. Hacein-Bey-Abina, S., et al., A serious adverse event after successful gene therapy for X-linked severe combined immunodeficiency. *N Engl J Med*, 2003, **348**(3): 255–256.

114. Hacein-Bey-Abina, S., et al., LMO2-associated clonal T cell proliferation in two patients after gene therapy for SCID-X1. *Science*, 2003, **302**(5644): 415–419.

115. Dave, U.P., N.A. Jenkins, and N.G. Copeland, Gene therapy insertional mutagenesis insights. *Science*, 2004, **303**(5656): 333.

116. Woods, N.-B., et al., Gene therapy: therapeutic gene causing lymphoma. *Nature*, 2006, **440**(7088): 1123–1123.

117. Onodera, M., et al., Successful peripheral T-lymphocyte-directed gene transfer for a patient with severe combined immune deficiency caused by adenosine deaminase deficiency. *Blood*, 1998, **91**(1): 30–36.

118. van Beusechem, V.W., et al., Long-term expression of human adenosine deaminase in rhesus monkeys transplanted with retrovirus-infected bone-marrow cells. *Proc Natl Acad Sci USA*, 1992, **89**(16): 7640–7644.

119. Klempner, M. and H.L. Malech, Phagocytes:normal and abnormal neutrophil host defenses, in *Infectious Diseases*, Gorbach, S.L. and Blacklow, N. editors. 1997, W.B. Saunders: Philadelphia, PA. pp. 41–46.

120. Calvino, M.C., et al., Bone marrow transplantation in chronic granulomatous disease. *Eur J Pediatr*, 1996, **155**(10): 877–879.

121. Ho, C.M., et al., Successful bone marrow transplantation in a child with X-linked chronic granulomatous disease. *Bone Marrow Transpl*, 1996, **18**(1): 213–215.

122. Horwitz, M.E., et al., Treatment of chronic granulomatous disease with nonmyeloablative conditioning and a T-cell-depleted hematopoietic allograft. *N Engl J Med*, 2001, **344**(12): 881–888.

123. Ozsahin, H., et al., Successful treatment of invasive aspergillosis in chronic granulomatous disease by bone marrow transplantation, granulocyte colony-stimulating factor-mobilized granulocytes, and liposomal amphotericin-B. *Blood*, 1998, **92**(8): 2719–2724.

124. Seger, R.A., et al., Treatment of chronic granulomatous disease with myeloablative conditioning and an unmodified hemopoietic allograft: a survey of the European experience, 1985–2000. *Blood*, 2002, **100**(13): 4344–4350.

125. Li, F., et al., CD34 + peripheral blood progenitors as a target for genetic correction of the two flavocytochrome b558 defective forms of chronic granulomatous disease. *Blood*, 1994, **84**(1): 53–58.

126. Sekhsaria, S., et al., Peripheral blood progenitors as a target for genetic correction of p47phox-deficient chronic granulomatous disease. *Proc Natl Acad Sci USA*, 1993, **90**(16): 7446–7450.

127. Weil, W.M., et al., Genetic correction of p67phox deficient chronic granulomatous disease using peripheral blood progenitor cells as a target for retrovirus mediated gene transfer. *Blood*, 1997, **89**(5): 1754–1761.

128. Vowells, S.J., et al., Flow cytometric analysis of the granulocyte respiratory burst: a comparison study of fluorescent probes. *J Immunol Methods*, 1995, **178**(1): 89–97.

129. Endicott, J.A. and V. Ling, The biochemistry of P-glycoprotein-mediated multidrug resistance. *Annu Rev Biochem*, 1989, **58**: 137–171.

130. Gottesman, M.M. and I. Pastan, Biochemistry of multidrug resistance mediated by the multidrug transporter. *Annu Rev Biochem*, 1993, **62**: 385–427.

131. Pastan, I. and M. Gottesman, Multiple-drug resistance in human cancer. *N Engl J Med*, 1987, **316**(22): 1388–1393.

132. Licht, T., et al., The multidrug-resistance gene in gene therapy of cancer and hematopoietic disorders. *Ann Hematol*, 1996, **72**(4): 184–193.

133. Galski, H., et al., Expression of a human multidrug resistance cDNA (MDR1) in the bone marrow of transgenic mice: resistance to daunomycin-induced leukopenia. *Mol Cell Biol*, 1989, **9**(10): 4357–4363.

134. Mickisch, G.H., et al., Chemotherapy and chemosensitization of transgenic mice which express the human multidrug resistance gene in bone marrow: efficacy, potency, and toxicity. *Cancer Res*, 1991, **51**(19): 5417–5424.

135. Mickisch, G.H., et al., Transgenic mice that express the human multidrug-resistance gene in bone marrow enable a rapid identification of agents that reverse drug resistance. *Proc Natl Acad Sci USA*, 1991, **88**(2): 547–551.

136. Hanania, E.G. and A.B. Deisseroth, Serial transplantation shows that early hematopoietic precursor cells are transduced by MDR-1 retroviral vector in a mouse gene therapy model. *Cancer Gene Ther*, 1994, **1**(1): 21–25.

137. Podda, S., et al., Transfer and expression of the human multiple drug resistance gene into live mice. *Proc Natl Acad Sci USA*, 1992, **89**(20): 9676–9680.

138. Sorrentino, B.P., et al., Selection of drug-resistant bone marrow cells in vivo after retroviral transfer of human MDR1. *Science*, 1992, **257**(5066): 99–103.

139. Moscow, J.A., et al., Engraftment of MDR1 and NeoR gene-transduced hematopoietic cells after breast cancer chemotherapy. *Blood*, 1999, **94**(1): 52–61.

140. Bunting, K.D., et al., Transduction of murine bone marrow cells with an MDR1 vector enables ex vivo stem cell expansion, but these expanded grafts cause a myeloproliferative syndrome in transplanted mice. *Blood*, 1998, **92**(7): 2269–2279.

141. Fanconi, G., Familial constitutional panmyelocytopathy, Fanconi's anemia (F.A.). I. Clinical aspects. *Semin Hematol*, 1967, **4**(3): 233–240.

142. Joenje, H., et al., Evidence for at least eight Fanconi anemia genes. *Am J Hum Genet*, 1997, **61**(4): 940–944.

143. Liu, J.M., et al., Fanconi anemia and novel strategies for therapy. *Blood*, 1994, **84**(12): 3995–4007.

144. Gluckman, E., et al., Bone marrow transplantation for Fanconi anemia. *Blood*, 1995, **86**(7): 2856–2862.

145. Gluckman, E., et al., Hematopoietic reconstitution in a patient with Fanconi's anemia by means of umbilical-cord blood from an HLA-identical sibling. *N Engl J Med*, 1989, **321**(17): 1174–1178.

146. Lo Ten Foe, J.R., et al., Somatic mosaicism in Fanconi anemia: molecular basis and clinical significance. *Eur J Hum Genet*, 1997, **5**(3): 137–148.

147. Liu, J.M., et al., Retroviral mediated gene transfer of the Fanconi anemia complementation group C gene to hematopoietic progenitors of group C patients. *Hum Gene Ther*, 1997, **8**(14): 1715–1730.

148. Liu, J.M., et al., Engraftment of hematopoietic progenitor cells transduced with the Fanconi anemia group C gene (FANCC). *Hum Gene Ther*, 1999, **10**(14): 2337–2346.

149. Kang, E.M., et al., Persistent low-level engraftment of rhesus peripheral blood progenitor cells transduced with the fanconi anemia c gene after conditioning with low-dose irradiation. *Mol Ther*, 2001, **3**(6): 911–919.

150. Beutler, E., Gaucher's disease. *N Engl J Med*, 1991, **325**(19): 1354–1360.

151. Leavitt, M.C., et al., Transfer of an anti-HIV-1 ribozyme gene into primary human lymphocytes. *Hum Gene Ther*, 1994, **5**(9): 1115–1120.

152. Malim, M.H., et al., Stable expression of transdominant Rev protein in human T cells inhibits human immunodeficiency virus replication. *J Exp Med*, 1992, **176**(4): 1197–1201.

153. Sullenger, B.A., et al., Overexpression of TAR sequences renders cells resistant to human immunodeficiency virus replication. *Cell*, 1990, **63**(3): 601–608.

34 Use of Genetically Modified Stem Cells in Experimental Gene Therapies

Thomas P. Zwaka

CONTENTS

34.1 INTRODUCTION

Gene therapy is a novel therapeutic branch of modern medicine. Its emergence is a direct consequence of the revolution heralded by the introduction of recombinant DNA methodology in the 1970s. Gene therapy is still highly experimental, but has the potential to become an important treatment regimen. In principle, it allows the transfer of genetic information into patient tissues and organs. Consequently, diseased genes can be eliminated or their normal functions rescued. Furthermore, the procedure allows the addition of new functions to cells, such as the production of immune system mediator proteins that help to combat cancer and other diseases.

Originally, monogenic inherited diseases (inherited single gene defects), such as cystic fibrosis, were considered primary targets for gene therapy. For instance, in pioneering studies, the correction of adenosine deaminase deficiency, a lymphocyte-associated severe combined immunodeficiency (SCID), was attempted (Mullen et al., 1996). Although no modulation of immune function was observed, data from this study, together with other early clinical trials, demonstrated the potential feasibility of gene transfer approaches as effective therapeutic strategies. The first successful clinical trials using gene therapy to treat a monogenic disorder involved a different type of SCID, caused by mutation of an X chromosome-linked lymphocyte growth factor receptor (Hacein-Bey-Abina et al., 2002). While the positive therapeutic outcome was celebrated as a breakthrough for gene therapy, a serious drawback subsequently became evident. By February 2005, 3 children out of 17 who had been successfully treated for X-linked SCID developed leukemia because the vector inserted near an oncogene (a cancer-causing gene) inadvertently caused it to be inappropriately expressed in the genetically engineered lymphocyte target cell (Hacein-Bey-Abina et al., 2003a,b). On a more positive note, a small number of patients with adenosine deaminase-deficient SCID have been successfully treated by gene therapy without any adverse side effects (Aiuti et al., 2002).

A minority of more recent gene therapy clinical trials, however, are concerned with monogenic disorders. Out of the 1000 or so recorded clinical trials (January 2005), less than 10% target these diseases (Figure 34.1). The majority of current clinical trials focus on polygenic diseases, particularly cancer (66% of all trials).

Gene therapy relies on similar principles as traditional pharmacological therapy; specifically, regional specificity for the targeted tissue, specificity of the introduced gene function in relation to disease, and stability and controllability of expression of the introduced gene. To integrate all these aspects into a successful therapy is an exceedingly complex process that requires expertise from many disciplines, including molecular and cell biology, genetics, and virology, in addition to bioprocess manufacturing capability and clinical laboratory infrastructure.

34.2 TWO PATHS TO GENE THERAPY

Gene therapy can be performed either by direct transfer of genes into the patient or by using living cells as vehicles to transport the genes of interest. Both modes have certain advantages and disadvantages.

On the one hand, direct gene transfer is particularly attractive because of its relative simplicity. In this scenario, genes are directly delivered into a patient's tissues or bloodstream by

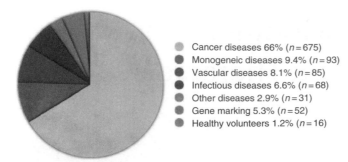

Cancer diseases 66% (*n* = 675)
Monogeneic diseases 9.4% (*n* = 93)
Vascular diseases 8.1% (*n* = 85)
Infectious diseases 6.6% (*n* = 68)
Other diseases 2.9% (*n* = 31)
Gene marking 5.3% (*n* = 52)
Healthy volunteers 1.2% (*n* = 16)

The Journal of Gene Medicine, © 2005 John Wiley & Sons Ltd   www.wiley.co.uk/genmed/clinical

FIGURE 34.1 (See color insert following blank page 206.) Percentage of diseases targeted by clinical gene therapy trials. (http://www.wiley.co.uk/genmed/clinical/)

packaging into liposomes (spherical vessels made up of the molecules that form the membranes of cells) or other biological microparticles. Alternatively, the genes are packaged into genetically engineered viruses, such as retroviruses or adenoviruses. Because of biosafety concerns, the viruses are typically altered so that they are not toxic or infectious (i.e., they are replication incompetent). These basic tools of gene therapists have been extensively optimized over the past 10 years. Their biggest strength, simplicity, however, is simultaneously their biggest weakness. In many cases, direct gene transfer does not allow very sophisticated control over the therapeutic gene. This is because the transferred gene either randomly integrates into the patient's chromosomes or persists unintegrated for a relatively short period of time in the targeted tissue. Additionally, the targeted organ or tissue is not always easily accessible for direct application of the therapeutic gene.

On the other hand, therapeutic genes can be delivered using living cells. This procedure is relatively complex in comparison to direct gene transfer and can be divided into three major steps. In the first step, cells from the patient or other sources are isolated and propagated in the laboratory. Secondly, the therapeutic gene is introduced into these cells, applying methods similar to those used in direct gene transfer. Finally, the genetically modified cells are returned to the patient. The use of cells as gene transfer vehicles has certain advantages. In the laboratory dish (*in vitro*), cells can be manipulated much more precisely than in the body (*in vivo*). Some of the cell types that continue to divide under laboratory conditions may be expanded significantly before reintroduction into the patient. Moreover, some cell types are able to localize to particular regions of the human body, such as hematopoietic (blood-forming) stem cells, which return to the bone marrow. This "homing" phenomenon may be useful for applying the therapeutic gene with regional specificity. A major disadvantage is the additional biological complexity brought into systems by living cells. Isolation of a specific cell type requires not only extensive knowledge of biological markers, but also insight into the requirements for that cell type to

stay alive *in vitro* and continue to divide. Unfortunately, specific biological markers are not known for many cell types, and the majority of normal human cells cannot be maintained for long periods of time *in vitro* without acquiring deleterious mutations.

34.3 STEM CELLS AS VEHICLES FOR GENE THERAPY

Stem cells can be classified as embryonic or adult, depending on their tissue of origin. The role of adult stem cells is to sustain an established repertoire of mature cell types in essentially steady-state numbers over the lifetime of the organism. Although adult tissues with a high turnover rate, such as blood, skin, and intestinal epithelium, are maintained by tissue-specific stem cells, the stem cells themselves rarely divide. However, in certain situations such as during tissue repair after injury or following transplantation, stem cell divisions can become more frequent.

The prototypic example of adult stem cells, the hematopoietic stem cell, has already been demonstrated to be of utility in gene therapy (Aiuti et al., 2002; Hacein-Bey-Abina et al., 2002). Although they are relatively rare in the human body, they can be readily isolated from bone marrow or after mobilization into peripheral blood. Specific surface markers allow the identification and enrichment of hematopoietic stem cells from a mixed population of bone marrow or peripheral blood cells. After *in vitro* manipulation, these cells can be transplanted back into patients by injection into the bloodstream, where they travel automatically to their place in the bone marrow in which they are functionally active. Hematopoietic stem cells that are explanted, *in vitro* manipulated, and re-transplanted into the same patient (autologous transplantation) or a different patient (allogeneic transplantation) retain the ability to contribute to all mature blood cell types of the recipient for an extended period of time (when patients' cells are temporarily grown "outside the body" before being returned to them, the *in vitro* process is typically referred to as an *ex vivo* approach).

Another adult bone marrow–derived stem cell type with potential use as a vehicle for gene transfer is the mesenchymal stem cell that has the ability to form cartilage, bone, adipose (fat) tissue, and marrow stroma (the bone marrow microenvironment) (Gregory et al., 2005). Recently, a related stem cell type, the multipotent adult progenitor cell, has been isolated from bone marrow that can differentiate into multiple lineages, including neurons, hepatocytes (liver cells), endothelial cells (such as the cells that form the lining of blood vessels), and other cell types (Jiang et al., 2002). Other adult stem cells have been identified, such as those in the central nervous system and heart, but these are less well characterized and not as easily accessible (Stocum, 2005).

The traditional way to introduce a therapeutic gene into hematopoietic stem cells from bone marrow or peripheral blood involves the use of a vector derived from a certain class of virus called the retrovirus. One type of retroviral vector was initially employed to show proof-of-principle that a foreign gene (in that instance the gene was not therapeutic, but was used as a molecular tag to genetically mark the cells) introduced into bone marrow cells may be stably maintained for several months (Brenner et al., 1993). These particular retroviral vectors are only capable of transferring the therapeutic gene into actively dividing cells. Since most adult stem cells divide at a relatively slow rate, efficiency was rather low. Vectors derived from other types of retroviruses (the so-called lentiviruses) and adenoviruses have the potential to overcome this limitation, since they target nondividing cells as well. The major drawback of these methods is that the therapeutic gene frequently integrates more or less randomly into the chromosomes of the target cell. In principle, this is dangerous, because the gene therapy vector can potentially modify the activity of neighboring genes (positively or negatively) in close proximity to the insertion site or even inactivate host genes by integrating into them, a phenomenon referred to as "insertional mutagenesis." In extreme cases, like in the X-linked SCID gene therapy trials, these mutations contribute to the malignant transformation of the targeted cells, ultimately resulting in cancer.

Another major limitation of using adult stem cells is that it is still relatively difficult to maintain their stem cell state during *ex vivo* manipulations. Under current suboptimal conditions, adult stem cells tend to become more specialized giving rise to mature cell types through a process termed differentiation, concomitantly losing their stem cell properties. Recent advances in supportive culture conditions for mouse hematopoietic stem cells may ultimately facilitate more effective use of human hematopoietic stem cells in gene therapy applications (Reya et al., 2003; Willert et al., 2003).

34.4 EMBRYONIC STEM CELL: "THE ULTIMATE STEM CELL"

Embryonic stem cells are capable of unlimited self-renewal while maintaining the potential to differentiate into derivatives of all three germ layers. Even after months and years of growth in the laboratory, they retain the ability to form any cell type in the body. These properties reflect their origin from cells of the early embryo at a stage during which the cellular machinery is geared toward the rapid expansion and diversification of cell types.

Murine (mouse) embryonic stem cells were isolated over 20 years ago (Evans and Kaufman, 1981; Martin, 1981) and paved the way for the isolation of nonhuman primate and finally human embryonic stem cells (Thomson et al., 1998). Much of the anticipated potential surrounding human embryonic stem cells is an extrapolation from pioneering experiments in the mouse system. Experiments performed with human embryonic stem cells in the last couple of years indicate that these cells have the potential to make an important impact on medical science, at least in certain fields. In particular, this includes (a) differentiation of human embryonic stem cells into various cell types, such as neurons, cardiac,

vascular, hematopoietic, pancreatic, hepatic, and placental cells, (b) the derivation of new cell lines under alternative conditions, and (c) the establishment of protocols that allow the genetic modification of these cells.

34.5 POTENTIAL OF HUMAN EMBRYONIC STEM CELLS FOR GENE THERAPY

Due to their rapid growth, remarkable stability, ability to mature *in vitro* into multiple cell types of the body, and especially because after derivation, they are easily accessible for controlled and specific genetic manipulation, human embryonic stem cells are attractive potential tools for gene therapy. There are two possible scenarios whereby human embryonic stem cells may benefit the field of gene therapy.

One way is that human embryonic stem cells could be genetically manipulated to introduce the therapeutic gene. This gene may either be active or awaiting later activation once the modified embryonic stem cell has differentiated into the desired cell type. Recently published reports establish the feasibility of such an approach (Rideout et al., 2002). Skin cells from an immunodeficient mouse were used to generate cellular therapy that partially restored immune function in the mouse. In these experiments, embryonic stem cells were generated from an immunodeficient mouse by nuclear transfer technology. The nucleus of an egg cell was replaced with that from a skin cell of an adult mouse with the genetic immunodeficiency. The egg was developed to the blastula stage at which embryonic stem cells were derived. The genetic defect was corrected by a genetic modification strategy designated "gene targeting." These "cured" embryonic stem cells were differentiated into hematopoietic "stem" cells and transplanted into immunodeficient mice. Interestingly, the immune function in these animals was partially restored. In principle, this approach may be employed for treating human patients with immunodeficiency or other diseases correctable by cell transplantation. However, significant advances would need to be made: The levels of immune system reconstitution observed were quite modest (<1% normal) while the methodology employed to achieve hematopoietic engraftment, which involved using a more severely immunodeficient mouse as recipient (which also had the mouse equivalent of the human X-linked SCID mutation) and genetically engineering the hematopoietic engrafting cells with a potential oncogene prior to transplantation, would not be clinically feasible.

Embryonic stem cells may additionally be indirectly beneficial for cellular gene therapy. Since these cells can be differentiated *in vitro* into many cell types, including presumably tissue-specific stem cells, they may provide a constant *in vitro* source of cellular material. Such "adult" stem cells derived from embryonic stem cells may thus be utilized to optimize protocols for propagation and genetic manipulation techniques (Barberi et al., 2005). To acquire optimal cellular material from clinical samples in larger quantities for experimental and optimization purposes is usually rather difficult since access to these samples is limited.

34.6 GENETIC MANIPULATION OF STEM CELLS

The therapeutic gene needs to be introduced into the cell type used for therapy. Genes can be introduced into cells by transfection or transduction.

Transfection utilizes chemical or physical methods to introduce new genes into cells. Usually, small molecules, such as liposomes, as well as other cationic-lipid based particles are employed to facilitate the entry of DNA encoding the gene of interest into the cells. Brief electric shocks are additionally used to facilitate DNA entry into living cells. All these techniques have been applied to various stem cells, including human embryonic stem cells. However, the destiny of the introduced DNA is relatively poorly controlled using these procedures. In most cells, the DNA disappears after days or weeks, and in rare cases, integrates randomly into host chromosomal DNA. *In vitro* drug selection strategies allow the isolation and expansion of cells that are stably transfected, as long as they significantly express the newly introduced gene.

Transduction utilizes viral vectors for DNA transfer. Viruses, by nature, introduce DNA or RNA into cells very efficiently. Engineered viruses can be used to introduce almost any genetic information into cells. However, there are usually limitations in the size of the introduced gene. Additionally, some viruses (particularly retroviruses) only infect dividing cells effectively, whereas others (lentiviruses) do not require actively dividing cells. In most cases, the genetic information carried by the viral vector is stably integrated into the host cell genome (the total complement of chromosomes in the cell). An important parameter that has to be carefully monitored is this random integration into the host genome, since on the one hand, this process can induce mutations leading to malignant transformation or serious gene dysfunction, while on the other hand, several copies of the therapeutic gene integrated into the genome help to bypass positional effects and gene silencing. Positional effects are caused by certain areas within the genome and directly influence the activity of the introduced gene. Gene silencing refers to the phenomenon whereby over time, most artificially introduced active genes are turned off by the host cell, a mechanism that is not currently well understood. In these cases, integration of several copies may help to achieve stable gene expression, since a subset of the introduced genes may integrate into favorable sites. In the past, gene silencing and positional effects were a particular problem in mouse hematopoietic stem cells. These problems led to the optimization of retroviral and lentiviral vector systems, for example by addition of genetic control elements referred to as chromatin domain insulators and scaffold/matrix attachment regions into the vectors, resulting in more robust expression in differentiating cell systems, including human embryonic stem cells (Ma et al., 2003).

In some gene transfer systems, the foreign transgene does not integrate at a high rate and remains separate from the host genomic DNA, a status denoted "episomal." Specific proteins stabilizing these episomal DNA molecules have been identified as well as viruses (adenovirus) that persist stably for some time in an episomal condition. Recently, episomal systems have been applied to embryonic stem cells (Aubert et al., 2002).

An elegant way to circumvent positional effects and gene silencing is to introduce the gene of interest specifically into a defined region of the genome by the gene targeting technique referred to previously (Kyba et al., 2002). The gene targeting technique takes advantage of a cellular DNA repair process known as homologous recombination (Smithies, 2001). Homologous recombination provides a precise mechanism for defined modifications of genomes in living cells and has been used extensively with mouse embryonic stem cells to investigate gene function and create mouse models of human diseases. Recombinant DNA is altered *in vitro* and the therapeutic gene is introduced into a copy of the genomic DNA that is targeted during this process. Next, recombinant DNA is introduced by transfection into the cell, where it recombines with the homologous part of the cell genome. This in turn results in the replacement of normal genomic DNA with recombinant DNA containing genetic modifications. Homologous recombination is a very rare event in cells, and thus a powerful selection strategy is necessary to identify the cells in which it occurs. Usually, the introduced construct has an additional gene coding for antibiotic resistance (referred to as a selectable marker), allowing cells that have incorporated the recombinant DNA to be positively selected in culture. However, antibiotic resistance only reveals that the cells have taken up recombinant DNA and incorporated it somewhere in the genome. To select for cells in which homologous recombination has occurred, the end of the recombination construct often includes the thymidine kinase gene from the herpes simplex virus. Cells that randomly incorporate recombinant DNA usually retain the entire DNA construct, including the herpes virus thymidine kinase gene. In cells displaying homologous recombination between the recombinant construct and cellular DNA, an exchange of homologous DNA sequences is involved and the nonhomologous thymidine kinase gene at the end of the construct is eliminated. Since cells expressing the thymidine kinase gene are killed by the antiviral drug (negative selection), ganciclovir, those undergoing homologous recombination are unique in that they are resistant to both the antibiotic and ganciclovir, allowing effective selection with these drugs (Figure 34.2).

Gene targeting by homologous recombination has recently been applied to human embryonic stem cells (Zwaka and Thomson, 2003). This is important for studying gene functions *in vitro* for lineage selection and marking. For therapeutic applications in transplantation medicine, the controlled modification of specific genes should be useful for purifying specific embryonic stem cell-derived, differentiated cell types from a mixed population, altering the antigenicity of embryonic stem cell derivatives and adding defined markers that allow the identification of transplanted cells. Additionally, since the therapeutic gene can now be introduced into defined regions of the human genome, controlled expression of the therapeutic gene should be possible. This also significantly reduces the risk of insertional mutagenesis.

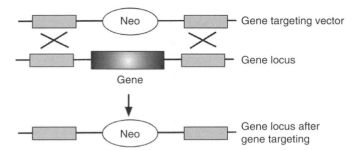

FIGURE 34.2 Homologous recombination: endogenous locus is replaced by targeting vector.

34.7 FUTURE CHALLENGES FOR STEM CELL-BASED GENE THERAPY

Despite very promising scientific results with genetically modified stem cells, some major problems remain to be overcome. The more specific and extensive the genetic modification, the longer the stem cells have to remain *in vitro*. Although human embryonic stem cells in the culture dish remain remarkably stable, the cells could accumulate genetic and epigenetic changes that might be harmful to the patient (epigenetic changes regulate gene activity without altering the genetic blueprint of the cell). Indeed sporadic chromosomal abnormalities in human embryonic stem cell culture have been reported and these may occur more frequently when the cells are passaged as bulk populations, which reinforces the necessity to further optimize culture conditions, explore new human embryonic stem cell lines, and monitor the existing cell lines (Cowan et al., 2004; Draper et al., 2004; Mitalipova et al., 2005). Additionally undifferentiated embryonic stem cells have the potential to form a type of cancer called a teratocarcinoma. Safety precautions are therefore necessary, and currently, protocols are being developed to allow the complete depletion of any remaining undifferentiated embryonic stem cells (Gerecht-Nir and Itskovitz-Eldor, 2004). This may be achieved by rigorous purification of embryonic stem cell derivatives or introducing suicide genes that can be externally controlled.

Another issue is the patient's immune system response. Transgenic genes as well as vectors introducing these genes (such as those derived from viruses) potentially trigger immune system responses. If stem cells (for instance, human embryonic stem cells) are not autologous, they eventually cause immuno-rejection of the transplanted cell type. Strategies to circumvent these problems, such as the expression of immune system-modulating genes by stem cells, creation of chimeric, immunotolerable bone marrow or suppression of human leukocyte antigen (HLA) genes have been suggested (Gerecht-Nir and Itskovitz-Eldor, 2004).

The addition of human embryonic stem cells to the experimental gene therapy arsenal offers great promise in overcoming many of the existing problems of cellular-based gene therapy that have been encountered in clinic trials. Further research is essential to determine the full potential of both adult and embryonic stem cells in this exciting new field.

REFERENCES

Aiuti A., Slavin S., Aker M., Ficara F., Deola S., Mortellaro A., Morecki S., Andolfi G., Tabucchi A., Carlucci F. et al. 2002. Correction of ADA-SCID by stem cell gene therapy combined with nonmyeloablative conditioning. *Science* 296, 2410–2413.

Aubert, J., Dunstan, H., Chambers, I., and Smith, A. 2002. Functional gene screening in embryonic stem cells implicates Wnt antagonism in neural differentiation. *Nat Biotechnol* 20, 1240–1245.

Barberi T., Willis, L. M., Socci, N. D., and Studer, L. 2005. Derivation of multipotent mesenchymal precursors from human embryonic stem cells. *PLoS Med* 2, e161.

Brenner, M. K., Rill, D. R., Holladay, M. S., Heslop, H. E., Moen, R. C., Buschle, M., Krance, R. A., Santana, V. M., Anderson, W. F., and Ihle, J. N. 1993. Gene marking to determine whether autologous marrow infusion restores long-term haemopoiesis in cancer patients. *Lancet* 342, 1134–1137.

Cowan, C. A., Klimanskaya, I., McMahon, J., Atienza, J., Witmyer, J., Zucker, J. P., Wang, S., Morton, C. C., McMahon, A. P., Powers, D. et al. 2004. Derivation of embryonic stem-cell lines from human blastocysts. *N Engl J Med* 350, 1353–1356.

Draper, J. S., Smith, K., Gokhale, P., Moore, H. D., Maltby, E., Johnson, J., Meisner, L., Zwaka, T. P., Thomson, J. A., and Andrews, P. W. 2004. Recurrent gain of chromosomes 17q and 12 in cultured human embryonic stem cells. *Nat Biotechnol* 22, 53–54.

Evans, M. J. and Kaufman, M. H. 1981. Establishment in culture of pluripotential cells from mouse embryos. *Nature* 292, 154–156.

Gerecht-Nir, S. and Itskovitz-Eldor, J. 2004. Cell therapy using human embryonic stem cells. *Transpl Immunol* 12, 203–209.

Gregory, C. A., Prockop, D. J., and Spees, J. L. 2005. Non-hematopoietic bone marrow stem cells: Molecular control of expansion and differentiation. *Exp Cell Res* 306, 330–335.

Hacein-Bey-Abina, S., Le Deist, F., Carlier, F., Bouneaud, C., Hue, C., De Villartay, J. P., Thrasher, A. J., Wulffraat, N., Sorensen, R., Dupuis-Girod, S. et al. 2002. Sustained correction of X-linked severe combined immunodeficiency by *ex vivo* gene therapy. *N Engl J Med* 346, 1185–1193.

Hacein-Bey-Abina, S., von Kalle, C., Schmidt, M., Le Deist, F., Wulffraat, N., McIntyre, E., Radford, I., Villeval, J. L., Fraser, C. C., Cavazzana-Calvo, M. et al. 2003a. A serious adverse event after successful gene therapy for X-linked severe combined immunodeficiency. *N Engl J Med* 348, 255–256.

Hacein-Bey-Abina S., Von Kalle C., Schmidt M., McCormack M. P., Wulffraat N., Leboulch P., Lim A., Osborne C.S., Pawliuk R., Morillon E. et al. 2003b. LMO2-associated clonal T cell proliferation in two patients after gene therapy for SCID-X1. *Science* 302, 415–419.

Jiang, Y., Jahagirdar, B. N., Reinhardt, R. L., Schwartz, R. E., Keene, C. D., Ortiz-Gonzalez, X. R., Reyes, M., Lenvik, T., Lund, T., Blackstad, M. et al. 2002. Pluripotency of mesenchymal stem cells derived from adult marrow. *Nature* 418, 41–49.

Kyba, M., Perlingeiro, R. C., and Daley, G. Q. 2002. HoxB4 confers definitive lymphoid–myeloid engraftment potential on embryonic stem cell and yolk sac hematopoietic progenitors. *Cell* 109, 29–37.

Ma, Y., Ramezani, A., Lewis, R., Hawley, R. G., and Thomson, J. A. 2003. High-level sustained transgene expression in human embryonic stem cells using lentiviral vectors. *Stem Cells* 21, 111–117.

Martin, G. R. 1981. Isolation of a pluripotent cell line from early mouse embryos cultured in medium conditioned by teratocarcinoma stem cells. *Proc Natl Acad Sci USA* 78, 7634–7638.

Mitalipova, M. M., Rao, R. R., Hoyer, D. M., Johnson, J. A., Meisner, L. F., Jones, K.L., Dalton, S., and Stice, S. L. 2005. Preserving the genetic integrity of human embryonic stem cells. *Nat Biotechnol* 23, 19–20.

Mullen, C. A., Snitzer, K., Culver, K. W., Morgan, R. A., Anderson, W. F., and Blaese, R. M. 1996. Molecular analysis of T lymphocyte-directed gene therapy for adenosine deaminase deficiency: Long-term expression in vivo of genes introduced with a retroviral vector. *Hum Gene Ther* 7, 1123–1129.

Reya, T., Duncan, A. W., Ailles, L., Domen, J., Scherer, D. C., Willert, K., Hintz, L., Nusse, R., and Weissman, I. L. 2003. A role for Wnt signalling in self-renewal of haematopoietic stem cells. *Nature* 423, 409–414.

Rideout, W. M., 3rd, Hochedlinger, K., Kyba, M., Daley, G. Q., and Jaenisch, R. 2002. Correction of a genetic defect by nuclear transplantation and combined cell and gene therapy. *Cell* 109, 17–27.

Smithies, O. 2001. Forty years with homologous recombination. *Nat Med* 7, 1083–1086.

Stocum, D.L. 2005. Stem cells in CNS and cardiac regeneration. *Adv Biochem Eng Biotechnol* 93, 135–159.

Thomson, J. A., Itskovitz-Eldor, J., Shapiro, S. S., Waknitz, M. A., Swiergiel, J. J., Marshall, V. S., and Jones, J. M. 1998. Embryonic stem cell lines derived from human blastocysts. *Science* 282, 1145–1147.

Willert, K., Brown, J. D., Danenberg, E., Duncan, A. W., Weissman, I. L., Reya, T., Yates, J. R., 3rd, and Nusse, R. 2003. Wnt proteins are lipid-modified and can act as stem cell growth factors. *Nature* 423, 448–452.

Zwaka, T. P. and Thomson, J. A. 2003. Homologous recombination in human embryonic stem cells. *Nat Biotechnol* 21, 319–321.

35 Cardiovascular Gene and Cell Therapy

Kazuhiro Oka

CONTENTS

35.1   INTRODUCTION

The concept of gene therapy was introduced with great promise and high expectations. Manipulating gene expression to treat patients was novel and created the prospect to revolutionize clinical practice in genetic disorders as well as common ailments. For the last decade, the field of gene therapy has been transformed from treatments for rare monogenetic disorders to more complex polygenetic disorders. Despite advances in primary and secondary interventions, cardiovascular disease remains a major cause of morbidity and mortality in industrial countries, so successful gene therapy could have a profound impact.

The field of gene therapy has been rapidly evolving with a better understanding of pathophysiology and advances in gene delivery system. However, we should be aware of limitations of gene and cell therapy strategies, including technical issues and the development of more conventional treatments for the intended diseases. For example, vascular gene therapy to prevent postangioplasty restenosis, once an attractive gene therapy target, is no longer an issue because of development of drug-eluting vascular stents [1]. The possibility of cell therapy to

repair damaged heart muscle also has generated tremendous excitement. However, its clinical efficacy has not been as evident in large-scale randomized trials, perhaps due to our imperfect understanding of the biological process [2]. Careful evaluation of existing knowledge and diligence at integrating advances in the field may allow these therapies to become reality.

The topic of cardiovascular gene and cell therapy is broad. This chapter will review general strategies of gene and cell manipulation in the cardiovascular system and give an overview of potential gene therapy targets. It will also review examples of experimental gene therapy in animal and human trials.

35.2   GENETIC MANIPULATION OF CARDIOVASCULAR TISSUE

35.2.1   Gene and Cell Therapy Strategies

35.2.1.1   Manipulation of Gene Expression

Gene therapy is a method that manipulates gene expression levels for the purpose of treating disease. This is achieved through the introduction of therapeutic genetic materials (DNA or RNA) into cells in a process referred to as transduction or transfection. Transduction is the transfer of genetic information to a cell using a viral vector. Transfection is the uptake of genetic information by eukaryotic cells, often mediated by liposome–DNA or –RNA complexes. Gene transfer allows replacement of a missing gene product, augmentation of endogenous gene function, or interference with gene expression.

Gene replacement or augmentation involves the transfer of a gene that is missing from a cell, defective or underexpressed. The expressed protein may be active within the target cell or secreted to act in a paracrine, endocrine, or autocrine manner. Some proteins require expression levels comparable to normal, while other gene products can function at lower levels to yield sufficient therapeutic results. In certain conditions, excessive expression of a therapeutic gene may unintentionally cause harmful effects.

In some cases, monogenic disease is caused by the expression of dominant negative mutant or excessive expression of a gene; in these cases, the gene requires silencing. There are four major strategies for gene silencing (Figure 35.1) [3]: (a) antisense oligodeoxynucleotides (ODNs), (b) catalytically active RNA ribozymes, (c) double-stranded small interfering RNAs (siRNA) that induce RNA degradation through RNA interference (RNAi), and (d) transcriptional decoys.

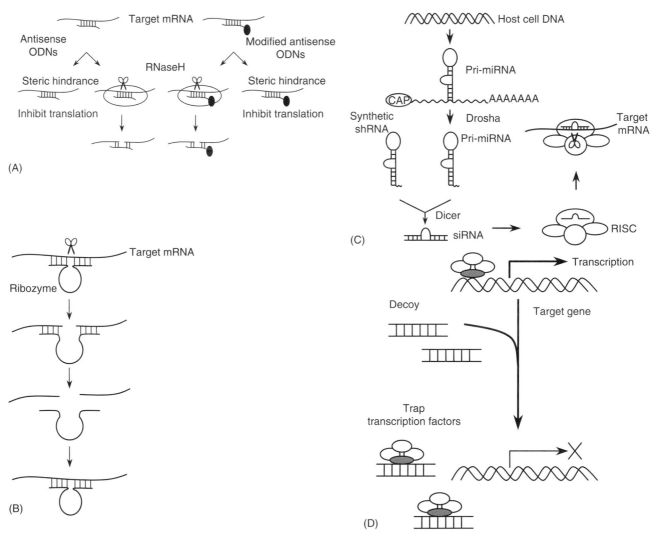

FIGURE 35.1 Strategies for gene silencing. There are four major categories for gene silencing. (A) Antisense ODNs are short single-stranded deoxynucleotides that have a base sequence complementary to a segment of target RNAs, which inhibit translation or recruit RNase H to cleave the target mRNA. Most ODNs are modified to enhance their stability. (B) Ribozymes are RNA with catalytic activity to cleave a target mRNA. Ribozymes bind the complementary target RNA through base-pairing. After cleaving the substrate, they release the cleaved products and are recycled like enzymes. (C) RNA interference (RNAi) is a naturally occurring gene regulatory mechanism. RNAi is mediated by short double-stranded RNA (dsRNA) termed small interfering RNA (siRNA) duplexes. These siRNA duplexes are generated by cleavage of long dsRNA or short hairpin RNA (shRNA) by Dicer. One strand of the siRNA duplex, referred to as the guide strand, is incorporated into a nuclease complex known as the RNA-induced silencing complex (RISC) while the second passenger strand is released and degraded. Upon incorporation into RISC, the siRNA guides the RISC to the complementary target mRNA for cleavage. MicroRNAs (miRNAs) are endogenous single-stranded noncoding RNAs. The primary transcripts of miRNAs, pri-miRNA, are transcribed from introns of protein-coding genes. The RNase Drosha processes the pri-miRNA to hairpin-shaped pre-miRNA. Pre-miRNA is exported to the cytoplasm where it is cleaved by Dicer. (D) Another strategy to interfere with gene transcription is a decoy for transcription factor. Decoys are double-stranded ODNs designed to mimic the chromosomal binding sites of transcription factors. They bind up the available transcription factors, which prevents the binding of transcription factor complexes to target genes and inhibit transcription.

1. *Antisense ODNs.* Antisense ODNs are short [15–20 nucleotides (nt) in length] single-stranded DNAs complementary to a target mRNA. Upon cellular uptake, the ODNs hybridize to the corresponding segment of mRNA to form a DNA–RNA duplex. Translation into protein is inhibited either by direct steric hindrance of ribosome translation machinery, interference with the splicing of pre-mRNA, or by recruiting RNase H to cleave the target mRNA. Major problems with antisense ODNs are their rapid degradation by ubiquitous DNase, the high dose required for a biological response, and the difficult delivery into target cells. Unmodified ODNs are degraded rapidly *in vivo*. Therapeutic ODNs are routinely modified to enhance their stability, mostly by chemical modification within their sugar phosphate backbone (i.e., phosphotioate, peptide nucleic acid, morpholino phosphramidate and locked nucleic acid) [4].

2. *Ribozymes.* Certain RNA molecules catalyze RNA cleavage in the absence of protein cofactors. Ribozymes contain a catalytic region that cleaves other RNA molecules in a sequence-specific manner and an adjacent sequence conferring specificity. Hammerhead hairpin ribozymes are composed of small RNA of 40–160 nt in length complementary to target RNA through base-pairing. After cleaving the substrate, they release the cleaved products and are recycled, which allow cleavage of multiple substrates. Ribozymes can be administered to the patient either by vector mediated gene transfer or by direct injection of synthetic ribozymes. The limitations of this approach are variability in efficacy depending on the target cell [5].

3. *RNAi.* RNAi is an evolutionarily conserved gene regulatory mechanism. RNAi is mediated by 21–24 nt dsRNA having two nt 3′ overhangs, termed short-interfering RNA (siRNA) duplexes. These siRNA duplexes can be generated by cleavage of long dsRNA or shRNA by the dsRNA-specific RNase III enzyme, Dicer. One strand of the siRNA duplex, referred to as the guide strand, is incorporated into a nuclease complex known as the RNA-induced silencing complex (RISC), while the second passenger strand is released and degraded. Upon incorporation into RISC, the siRNA guides the RISC to the complementary target sites within mRNA, which is cleaved by the RISC component, Argonaute2, leading to their degradation.

   miRNAs are endogenous ~22 nt single-stranded noncoding RNAs, which appear to be functionally identical to siRNAs in mammalian cells. They are transcribed from introns of protein-coding genes. Since dsRNA occurs mostly in viruses, the RNAi pathway could have been evolved as a host defense against virus infection. Over the last years, RNAi has become the most widely used approach for gene knockdown due to its potency.

The requirements for successful gene knockdown using siRNA are well known. These include methods of delivery, enhanced stability, minimization of off-target effects, and identification of effective sites in the target RNA. The Web sites for the design of siRNAs are available for free access to the public at http://jura.wi.mit.edu/bioc/siRNAext/home.php (Whitehead Institute) and http://www.ambion.com/techlib/misc/siRNA_finder.html (Ambion). However, the algorithms are not perfect and several target-specific siRNA sequences should be tested in cells. Synthetic siRNA or expressing plasmid DNAs can be transfected into cells using liposome. However, a more effective approach is to continuously express siRNA in the relevant cells using vectors derived from adenovirus (Ad), retrovirus (RV), or lentivirus (LV). The RNAi consortium (TRC) is a collaborative effort based at the Broad Institute of MIT and Harvard; it has created lentiviral shRNA libraries targeting 15,000 human and 15,000 mouse annotated genes. The shRNA for a specific gene can be searched by gene symbol, GenBank accession number, or clone ID, and be purchased through Open Biosystems at http://www.openbiosystems.com/ or Sigma-Aldrich at http://www.sigmaaldrich.com/.

4. *Decoy.* An additional approach to gene regulation is the blockade of transcription factors. Transcription factors regulate gene expression by binding to chromosomal DNA at regulatory sequences or by forming transcription regulatory complexes; these interactions affect the activity of the RNA Polymerase II complex. Double-stranded ODNs are designed to mimic the chromosomal binding sites of these transcription factors and act as decoys that titrate the available transcription factors, preventing them from binding to target genes.

35.2.1.2 Cell-Based Therapy

Cell-based therapy refers to treatments using cells instead of DNA or RNA. The most well-known procedure is bone marrow transplantation (BMT). The interest for cell-based therapy for cardiovascular disease has increased tremendously in recent years since the demonstration of homing and incorporation of BM-derived endothelial progenitor cells (EPCs) into ischemic tissues [6] and regeneration of infarcted myocardium by transplanted BM cells [7,8]. The most widely used cell sources in clinical trials have been skeletal myoblasts, BM mononuclear cells, peripheral blood cells, and mesenchymal stem cells [9]. Other potential cells for this type of treatment include embryonic stem cells, stem cells in cord blood, and cardiac stem cells. These cells can be genetically modified prior to transplantation. The use of stem cells to generate replacement cells for damaged myocardium, vessels, and valves holds great promise for treatments of acute myocardial infarction (MI), restenosis, chronic heart failure, and other cardiovascular repairs.

35.2.2 GENE DELIVERY APPROACH

There are two approaches for how to deliver therapeutic genes (Figure 35.2). The *ex vivo* approach refers to manipulation of

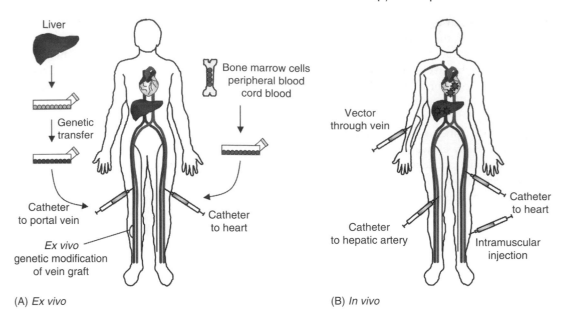

FIGURE 35.2 Gene and cell delivery approaches. (A) *Ex vivo* approach. In *ex vivo* approach, cells or tissues are isolated from the body and genetically modified *in vitro* before transplantation. Cells or tissues can be autologous (cells originated from the host) or be from donors. (B) *In vivo* approach. In this approach, gene transfer vectors are directly injected into the body.

cells or tissues outside of the body, where they can be genetically modified before transplantation. The advantage is that the gene delivery vector does not circulate *in vivo* and transduction efficiency can be monitored before transplantation. It is also possible to select cells with a desired phenotype. Most cell-based therapies, including stem cell therapy, use this approach.

The *in vivo* approach is the most direct strategy for gene transfer. The vector containing therapeutic gene is directly administered to the organ of interest or into blood circulation. When specific physical or biological targeting methods are available, this is a preferred approach. However, engineering targeted vectors is a challenge [10]. Physical targeting includes catheter-based gene delivery, direct intramuscular injection, and injection of biodegradable microspheres or nanoparticles coated with gene delivery vectors. Diagnostic ultrasound in conjunction with microbubble contrast agent complexed with gene delivery vector may improve the efficiency of targeted gene transfer [11].

35.2.3 Vectors for Gene Transfer

The underlying concept of gene therapy is that transfer of genetic information as DNA will modify a disease, yielding therapeutic results. The "holy grail" of gene therapy is a vector that delivers genetic materials efficiently and safely to the target cells. An ideal vector must overcome the following hurdles: (1) The therapeutic gene must reach the target cell within the desired organs; (2) it must enter the cell, travel through the plasma membrane, traffic through cytoplasm, and reach the nucleus; (3) it must recruit the endogenous transcriptional machinery to unleash its encrypted information; and (4) it cannot have detrimental consequences on cell function other than intended effects.

Gene transfer vectors can be divided into two categories: nonviral and viral. Nonviral methods require direct transfection of naked DNA into cells or tissues. Unfortunately, it is relatively inefficient. To achieve gene transfer, the gene is usually complexed with liposomes, coupled with nanoparticles, or incorporated in a recombinant viral vector.

35.2.3.1 Physical and Chemical Delivery of Nonviral DNA

The use of naked plasmid DNA has obvious advantages including the relative ease of DNA preparation and storage. It can be handled as a chemical instead of as a biological substance. Furthermore, it is less immunogenic and allows repeated administration. Gene expression is low with local injection of plasmid DNA into the muscle, liver, skin, or lung airway [12]. Nevertheless, nonviral DNA vector may have greater utility in the treatment of acute disease and DNA-based immunization. The major limitation for transfection of the liver by naked DNA is poor transduction efficiency by systemic administration. To overcome this problem, a hydrodynamic delivery method has been developed. The procedure involves the rapid injection of plasmid DNA in a large volume (approximately 8%–12% of the body weight) within 5–8 s. This results in a transient increase in intravascular pressure, leading to more than a 1000-fold increased transgene expression, mainly in the liver [13]. Increased hepatic transgene expression with this method has been reported with lentiviral [14] and adenoviral vectors [15]. This method has recently been adopted for hepatic injection of plasmid DNA or Ad into large animal models using a balloon catheter [16,17]. Other physical methods for plasmid DNA delivery include electroporation and ballistic delivery [18].

35.2.3.2 Nonviral Vectors

The expression of therapeutic genes delivered by nonviral vectors is mostly transient and cannot be maintained over an extended time period. To prolong transgene expression, recently developed nonviral vectors have incorporated an integration system. Insertional mutagenesis is the primary concern in viral vectors. However, the integration of a transgene into a predetermined safe site in the host chromosome could avoid this problem, at least in theory.

1. *Phage integrase.* Phage integrases are enzymes that mediate unidirectional site-specific recombination between two DNA recognition sequences: the phage attachment site, *attP*, and the bacterial attachment site, *attB* [19]. Some integrases such as ΦC31 or R4 integrase do not require host cofactors, while others act with the aid of phage cofactors or bacterial host factors. ΦC31 integrase is functional in mammalian cells and it can integrate plasmid DNA containing an *attB* site into pseudo-*attP* sites in the mammalian genome. The ΦC31 integrase-mediated gene transfer has been used in small animal experiments [20–22]. The integrase system has no apparent upper size limit and repeated injections can be made without inducing a host immune response to the vector. Multiple coinjections of a plasmid expressing phiBT1 integrase and a phenylalanine hydroxylase have been reported to correct the phenotype, a mouse model of phenylketonuria [23].

   Plasmid DNA is degraded rapidly if it is not integrated. This might be ideal for a plasmid coding for integrase since it should be eliminated from cells when recombination is complete. Although the *attP* recognition sequence is 39 base pair (bp) in length, 101 integration sites have been identified in human cell lines [24]. Despite a more restricted pattern of integration sites compared with Sleeping Beauty transposon [25], chromosomal abnormalities including chromosome translocation have been reported with the ΦC31 integrase-mediated integration [26,27]. This may be due to DNA breakage catalyzed by the integrase.

2. *Adeno-associated virus Rep protein-mediated integration.* Adeno-associated viruses (AAVs) are the only known mammalian virus capable of site-specific integration in a human chromosome. During latency in humans, AAV2 preferentially integrates at a site on chromosome 19q13.3-qtr, the so-called "AAVS1" site [28]. This site-specific integration is mediated by AAV Rep protein by targeting a sequence composed of a *rep*-binding element and terminal resolution site identical to the viral-inverted terminal repeats (ITRs) [29]. Similar sites have been identified in an African Green Monkey kidney cell line, CV-1 [30], and on mouse chromosome 7 [31]. Despite its proximity to the muscle-specific myosin-binding subunit 85 gene, integration into the AAVS1 site has not been associated

with any diseases. The absence of Rep function allows the transgene flanked by AAV ITRs to randomly integrate at chromosome breaks. This integration has no apparent strict size limitation [32]. The 138 bp p5 promoter called p5 integration efficiency element enhances integration efficiency and is sufficient to mediate AAVS1 integration without ITRs [33]. It should be noted that Rep protein arrests the cell cycle [34].

3. *Transposon-mediated integration of transgene.* Another nonviral integrating system is based on DNA transposable elements. The Sleeping Beauty transposon has been used for gene delivery in animal models [22,35]. Integration mediated by transposon is efficient, but occurs randomly. In order to achieve site-specific integration, DNA-binding domains such as custom-designed zinc finger proteins (ZFPs) have been fused to a transposase. Although this modification increases site-specific targeting, it may interfere with the activity of transposase [36,37]. Nevertheless, this approach has potential for safe transgene integration.

35.2.3.3 Viral Vectors

Viruses are highly evolved biological machines whose sole purpose is to gain access to the host cell and replicate using the host's cellular machinery. Therefore, viruses have evolved highly efficient gene transfer mechanisms, fully tested in nature. There are two major groups of viruses according to the essence of their genetic materials: DNA virus and RNA virus. The sequences of RNA viruses are reverse transcribed into the host chromosome before the viral gene is expressed. Although attempts to identify new viral species for vector development are in progress, AAV, Ad, RV, LV, and foamy virus (FV) are the best characterized vectors for cardiovascular gene therapy. Key features of each viral system are summarized in Table 35.1. Each viral system has its advantage and limitations. It is important to choose the best fit system for the application.

35.2.3.3.1 DNA Viral Vectors

AAV. AAVs are single-stranded DNA viruses, which were first described as a contaminant of a tissue culture-grown simian Ad [38,39]. AAV is a member of the *Parvoviridae* family and has a capsid with icosahedral symmetry approximately 20 nm in diameter [40]. Taking advantage of the natural diversity and evolution of AAV in primate species, more than 100 unique AAV capsid sequences have been identified [41]. Several AAV receptors and coreceptors have been identified, which include heparan sulfate proteoglycan [42], $\alpha_V\beta_5$ integrins [43], fibroblast growth factor receptor-1 [44], platelet-derived growth factor (PDGF) receptor [45], and sialic acid [46]. The 37/67 kDa laminin receptor has recently been identified as a receptor for serotypes 2, 3, 8, and 9 [47]. The presence of these receptors or coreceptors in tissues determines cell type- or tissue-specific transduction. Species-dependent tropism of AAV serotypes

TABLE 35.1

Main Features of Viral Vectors

Vector	Genetic Material	Particle Size (nm)	Cloning Capacity (kb)	Transduction	Integration[a]	Immune Problems	Special Feature
AAV	ssDNA	20–25	−4.7 (~10)[b]	Dividing cells Nondividing cells	Mainly episomal	Low	Nonpathogenic
FGAd	dsDNA	70–100	8–10	Dividing cells Nondividing cells	Episomal	Extensive	
HDAd	dsDNA	70–100	~37	Dividing cells Nondividing cells	Essentially Episomal[c]	moderate	
RV	RNA	100	7–10	Dividing cells	Integrated	Low	
LV	RNA	100	9–10	Dividing cells Nondividing cells	Integrated	Low	
FV	RNA	100–140	−9.2	Dividing cells Nondividing cells	Integrated	Low	Nonpathogenic

AAV, adeno-associated virus; FGAd, first generation adenovirus; HDAd, helper-dependent adenovirus.

[a] The major advantage of integrating vectors is a persistent gene transfer. However, this feature can be disadvantageous as it may induce oncogenesis in some applications.

[b] This capacity can be achieved by an intermolecular joining method using two AAV vectors.

[c] Minimal but significantly higher integration frequency of HDAd vector than FGAd has been reported.

has also been reported [48]. This is an important issue for AAV-mediated gene delivery. To date, 11 AAV serotypes have been characterized for gene therapy. AAV vectors express the transgene mostly in an episomal form (>90%), but also in an integrated form (<10%) [49]. One disadvantage of AAV vectors is their limited cloning capacity (4.7 kb), though this size limitation can be partly overcome by an intermolecular joining of two AAV vectors [50–52]. AAV vectors package a single-stranded genome and require host cell synthesis of the complementary strand for transduction. This causes a relatively long latent period between vector delivery and initiation of transgene expression. An AAV vector containing self-complementary genome can overcome this problem; however, it reduces the vector-cloning capacity for transgene to half [53].

In contrast to the cellular pathogenic response induced by an Ad vector infection, AAV vectors induce a nonpathogenic response affecting antiproliferative genes [54]. There is no known human disease associated with an AAV infection, making this vector a popular gene therapy vehicle. However, the potential for oncogenesis may exist. In the absence of *rep* gene, AAV genome has been shown to integrate randomly into the mouse genome with propensity toward gene regulatory sequences, albeit at low frequency [55]. Although studies involving large numbers of mice have failed to confirm an association between AAV vector treatment and tumor formation [56,57], there have been reports of a potential role of AAV2 in spontaneous abortion [58] and hepatocellular carcinoma in mice [59]. Donsante et al. have determined AAV integration sites in mouse hepatocellular carcinoma and mapped two of the insertions to chromosome 12 close to miRNA transcript. Furthermore, microarray analysis revealed upregulation of genes adjacent and telomeric to the AAV vector proviruses. These results have raised safety concerns over the human use because a syntenic locus on human chromosome 14 has been linked to several cancers [60].

Preexisting immunity against AAV has not been considered a serious problem until recently. The majority of the population is infected with AAV during childhood. Although AAV has a tremendously safety record in animal studies in diverse species and in humans, preexisting immunity may challenge its efficacy for *in vivo* transduction of tissues in humans. Transient self-limited liver toxicity has been reported in hemophilia B clinical trial [61]. The subsequent study of cytotoxic T cell (CTL) responses in one patient revealed that the expansion and contraction of the capsid-specific CD8+ T cell population paralleled the rise and fall of serum transaminase in the patient, while no such expansion occurred in mice. T-cell responses to AAV apparently targeted the conserved peptide sequence of the AAV capsid [62]. Preexisting CD8+ T cells to AAV capsid neither targeted AAV-transduced hepatocytes in mice [63,64] nor prevented AAV8-mediated Factor IX expression in hemophilia B dogs [65]. These studies highlighted critical differences in immunological responses against AAV among animal species.

Adenoviral vector. Adenoviruses (Ads) are nonenveloped, dsDNA viruses of ~70–100 nm in diameter. They have an icosahedral protein shell surrounding a protein core that contains the linear viral genome of ~35 kb. Ad infects cells independent of the cell cycle. The primary attachment sites of the Ad fiber protein subgroup C to cell surface receptors are "coxackievirus and adenovirus receptor" (CAR) [66]. The internalization of Ad vector is mediated by interaction of the penton base protein with integrin $\alpha_V\beta_3$ or $\alpha_V\beta_5$ as the secondary internalization receptors [67]. Heparan sulfate proteoglycans have been also reported as cellular attachment sites for this subgroup of Ads [68]. In contrast, subgroup B, exemplified by serotype 11 and 35, uses CD46 as the entry site [69,70]. Of the over 50 different Ad serotypes, subgroup B serotypes 2 and 5 are the most characterized.

First-generation Ad vectors (FGAd) contain all of the essential Ad genome except E1A, which encodes the key regulators of early gene expression. FGAd is replication-defective. FGAd vectors induce strong host innate and adaptive immune responses, which is characterized by the generation of Ad-specific MHC Class I restricted CD8+ CTL [71]. Moreover leaky Ad gene expression induces adaptive immunity, resulting in destruction of Ad-transduced hepatocytes [72,73]. In immunocompetent hosts, this response causes transient transgene expression, which lasts for less than a month. In order to improve the safety profile, second- and third-generation Ad vectors have been developed. There is only marginal improvement in the *in vivo* toxicity or duration of transgene expression in these vectors. Nevertheless, these early generation Ad vectors have been useful in proof-of-principle experiments and remain a useful vector for cancer gene therapy and vaccination.

Helper-dependent adenoviral vector (HDAd). HDAds are the latest generation Ad vectors, which have a greatly improved safety profile and allow for sustained transgene expression *in vivo* [74]. HDAds have a large cloning capacity, accommodating inserts of up to 37 kb. The HDAd lacks all viral protein-coding genes and encompasses only the ITR for replication and the packaging signal. Helper virus, an FGAd, provides the necessary viral proteins *in trans* for packaging in cultured cells. Thus, chronic hepatotoxicity seen in early generation Ads has been eliminated in purified HDAds. However, minimal acute toxicity still exists. The innate immune response to HDAds underlies this residual toxicity [75,76], which is caused by the direct interaction of viral particles with innate immune effector cells upon systemic vector administration, commonly used for hepatocyte transduction. Transient depletion of Kupffer cells is effective in reducing toxicity as well as to increase hepatic transduction by HDAds [77].

Apart from innate immune responses, HDAds have been considered to evade most of the adaptive immunity and thus sustain long-term transgene expression [76]. In animal models, HDAd-mediated hepatic gene delivery exhibits stable, long-term transgene expression associated with reduced hepatotoxicity [17,78–82]. However, both FGAd- and HDAd-transduced human dendritic cells (DCs) efficiently and stimulated the proliferation of autologous T cells, suggesting that preexisting immunity is directed primarily against the viral capsid [83]. A significant proportion of humans acquire respiratory tract infections caused by various Ads. Over 85% of human peripheral blood mononuclear cells (PBMCs) exhibited T cell response against human Ad5 [84]. Therefore, cell-mediated immunity could remain a challenge even with HDAd gene therapy.

35.2.3.3.2 RNA Viral Vectors

The Molony murine leukemia RV was the first gene transfer vector developed for gene therapy. Other RNA viral vectors include LV and FV. These RNA viral vectors integrate the transgene into the host chromosome, resulting in a sustained transgene expression.

Oncoretroviral vector (RV). RVs are single-stranded RNA viruses. The current vector is pseudotyped with the vesicular stomatitis virus glycoprotein (VSV-G). This confers a broad host range and stabilization of the vector particles, allowing the vector stock to be concentrated to high titers by ultracentrifugation. The packaging capacity of this vector is about 8 kb. Transgene expression persists due to integration into the host chromosome. However, the vector genome is integrated randomly, rendering insertional mutagenesis and oncogenesis as rare but definite possibilities. Another drawback is that RVs work only in cells that are actively dividing because they can access the nucleus only if the nuclear membrane has been broken down. RVs are mainly used in *ex vivo* gene therapy and are effective in transducing hematopoietic stem cells [49].

Lentiviral vector (LV). LVs were derived from human immunodeficiency virus (a member of the RV family) and are mostly pseudotyped with VSV-G. LV can penetrate an intact nuclear membrane and transduce nondividing cells. Thus, LV can be used in both *ex vivo* and *in vivo* applications. The deletion of most of the parental genome in the vector and in packaging constructs minimizes the possibility of generating a replication competent virus. As with RVs, most LVs now in use are self-inactivating, lacking the regulatory elements in the downstream long-terminal repeat (LTR), thus eliminating transcription of the packaging signal. Few immunological problems have been reported for both RV and LV. RV favors transgene integration near transcription start sites and shows little preference for active transcription units, while LV preferentially integrates into active genes [85,86].

Foamy viral vector (FV). FVs are the most recently developed gene transfer vectors. FVs comprise a genus of retroviruses but have distinct properties from oncoretroviruses and lentiviruses [87]. They have the largest genome size (up to 9.2 kb) of all retroviruses and are considered apathogenic in their natural animal hosts and after zoonotic transmission into humans [88]. The FV genome contains canonical retroviral *gag, pol,* and *env* genes, but replication is dependent on a transcriptional activator, Tas, encoded by the *bel*1 gene [89]. FV has a broad tropism and infects both dividing and nondividing cells. Although they show a preference for integration near transcription sites and for CpG islands (methylated DNA), they do not integrate preferentially within genes [90]. They are effective in transducing hematopoietic stem cells of mice and humans and human embryonic stem cells [91–93].

35.3 GENE THERAPY FOR HYPERLIPIDEMIA

Over the last half century, the mortality rate of coronary heart disease (CHD) has continued to decline [94]. However, an aging population and an increasing prevalence of overweight and obese individuals (see recent trend at http://www.cdc.gov/nccdphp/dnpa/obesity/trend/index.htm) could reverse the gain in treatment of CHD. Inherited and acquired forms of hyperlipidemia are an independent risk factor for the development of atherosclerosis. Most of diet-related hyperlipidemia and heterozygous familial hypercholesterolemia (FH) can be treated with lipid-lowering drugs such as statins (3-hydroxy-3-methylglutaryl-coenzyme A [HMG CoA] reductase inhibitors), bile acid binding resins, ezetimibe to inhibit cholesterol absorption in intestine, or combinations of the above.

FH is an autosomal-dominant disorder caused by mutations of the low-density lipoprotein (LDL) receptor (LDLR) gene. Close to 2000 sequence variants have been reported (www.ucl.ac.uk/fh). FH is one of most common inherited disorders that affect 1 in 500 individuals in the heterozygous form. Homozygous FH patients exhibit severe phenotype in which the plasma cholesterol levels can range from 700 to 1200 mg/dl. Many homozygous FH patients develop angina in childhood due to the aortic stenosis and coronary atheroma. MI has been reported as early as age 2 years, and most patients do not survive beyond the third decades of life if not treated [95]. A subset of heterozygous FH and the most severe forms of homozygous FH do not respond to the conventional therapy. Several other therapeutic options are liver transplantation, LDL apheresis, biliary diversion, and portacaval anastomosis [96]. The lack of conventional interventions has attracted interest for gene replacement therapy.

35.3.1 Ex Vivo Approach

LDLR facilitates the uptake of LDL in the liver and disposal of cholesterol into the bile to maintain normal cholesterol level (Figure 35.3). Unregulated hepatic overexpression of LDLR has not been associated with significant adverse events in animal studies, although hepatic steatosis is a primary concern. In the Watanabe heritable hyperlipidemic (WHHL) rabbit, a model of FH, Chowdhury et al. developed an ex vivo LDLR replacement gene therapy method [97]. Autologous hepatocytes were harvested by a partial hepatectomy and transduced in culture with RV carrying normal LDLR. The modified hepatocytes were then transplanted via the portal vein into the liver of the animal from which the cells were originally derived. This treatment resulted in a 30%–50% reduction of plasma cholesterol, which persisted for 6.5 months. These encouraging results led to the first pilot clinical gene therapy trial for FH. A total of five patients were treated. Although the procedure was safe and feasible, it was concluded that the variable metabolic responses due to the low gene transfer precluded broad application [98,99]. This study uncovered important practical implications for human application. It required isolation of autologous hepatocytes through an invasive partial liver dissection, cell proliferation, transduction with a vector that could potentially induce genotoxicity, and handling of large number of hepatocytes.

35.3.2 In Vivo Approach

Researchers have demonstrated efficient gene transfer to the liver by intravenous injection of Ad vector. Several groups have reported complete phenotypic correction of elevated LDL cholesterol levels in mouse and rabbit models of FH after a single intravenous injection of FGAd vector carrying

LDLR or very-low-density lipoprotein receptor (VLDLR) [100–103]. VLDLR is a multiligand apoE receptor, which is normally synthesized in LDLR-deficient mice. The effect of gene transfer with Ad was transient and associated with hepatotoxicity. However, HDAd-mediated LDLR and VLDLR gene transfer into LDLR-deficient mice conferred long-term phenotypic correction and protection against diet-induced atherosclerosis lasting up to 2 years [79,104]. LDLR was more effective, though it produced anti-LDLR antibodies at high dose. Therefore, in the context of LDLR protein deficiency, VLDLR may be preferable. Lebherz et al. treated LDLR-deficient mice with an intraportal vein injection of AAV2, AAV7, and AAV8 carrying human LDLR gene. AAV7 and AAV8 treatment normalized serum lipid for at least 21 weeks and significantly protected LDLR-deficient mice against atherosclerosis [105].

In vivo administration of RV has also been tested for the long-term reversal of hypercholesterolemia in a WHHL rabbit [106]. A 20% reduction of plasma cholesterol was observed 52 weeks after treatment. However, the requirement of both a 10% liver resection to induce cell proliferation and repeated vector injection combined with plasmid/liposome-mediated thymidine kinase gene transfer to induce cell division makes this impractical for clinical application. LV is a more efficient in vivo integrating vector. Kankkonen et al. treated WHHL rabbits with an intraportal vein injection of a third-generation LV carrying LDLR. The treatment reduced serum cholesterol by 34% at 2 years after the treatment without major safety issues [107]. No clinical trial for FH using this in vivo approach has been performed.

Apolipoprotein (apo) E-deficient mice develop spontaneous atherosclerosis. In this mouse model, a single injection of HDAd carrying apoE gene and its liver enhancer element has been shown to reverse hypercholesterolemia for their lifetime [78]. AAV7 and AAV8 expressing apoE also have been used to prevent atherosclerosis in apoE-deficient mice [108].

There are other therapeutic genes that can be exploited to treat dyslipidemia. Apolipoprotein A-I (apoAI) is a major protein component of high-density lipoproteins (HDL). HDL cholesterol is inversely correlated with the risk of CHD. Raising HDL cholesterol by elevated apoAI has attracted an attention for prevention of atherosclerosis. Long-term hepatic expression of apoAI inhibited a progression of atherosclerosis and remodeled unstable plaques to a stable phenotype in LDLR-deficient mice [109]. Similarly, AAV-mediated apoAI expression was stable for at least 1 year [110].

Another potential target is apoB100, which is a protein component of LDL. Teng et al. used an FGAd carrying apoB mRNA editing enzyme catalytic subunit-1 (apobec-1) to induce apoB100 mRNA editing and reduced plasma cholesterol in LDLR-deficient mice [111]. This is possible because resulting apoB48 containing lipoproteins are more efficiently taken up by the liver LDLR via apoE ligand. Enjoji et al. designed a hammerhead ribozyme to cleave human apoB100 mRNA and introduced it by FGAd-mediated transfer into

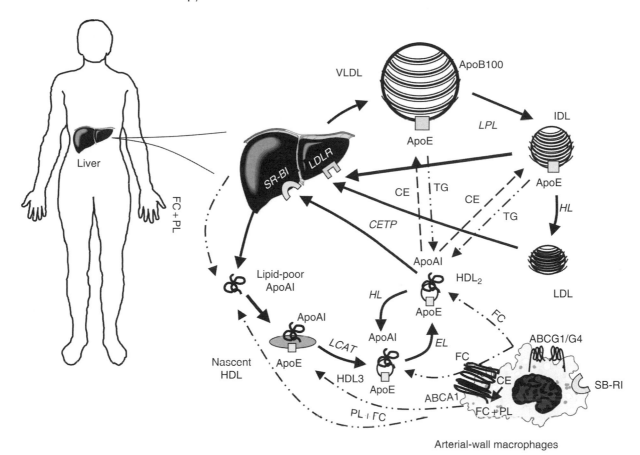

FIGURE 35.3 Overview of metabolism of liver-secreted lipoproteins and reverse cholesterol transfer via macrophages in humans. apoB100 containing triglyceride (TG)-rich VLDL is secreted from the liver. TGs on VLDL are hydrolyzed to free fatty acids and glycerol by the action of lipoprotein lipase (LPL). This process remodels VLDL to a smaller, denser, intermediate-density lipoprotein (IDL). IDL is taken up by the LDLR or other apoE receptors through an apoE ligand and is further remodeled by hepatic lipase (HL). LDL particles are taken up by LDLR through an apoB100 ligand. Lipid-free or lipid-poor apoA-I is secreted from the liver and serves as an acceptor for ATP binding cassette transporter-1 (ABCA1)-mediated lipid efflux from hepatocytes as well as macrophages. ABCA1-mediated efflux of free cholesterol (FC) and phospholipids (PL) to apoAI forms nascent or pre-β-high-density-lipoprotein (β-HDL) which is further modified by lecithin-cholesterol acyltransferase (LCAT). In humans, the resulting large, less dense HDL₂ and smaller, more dense HDL₃ can serve as acceptors for ABC subfamily G member 1 (ABCG1)/ABCG4-mediated cholesterol efflux. This pathway is favored by the presence of apoE in HDL. HL also hydrolyzes PL in HDL₂ and, therefore, is involved in the conversion of HDL₂ to HDL₃. Endothelial lipase (EL) is considered to act on HDL-PL. HDL cholesterol is taken up by the liver and is secreted into the bile for disposal through the scavenger receptor class B type I (SR-BI). Cholesteryl ester transfer protein (CETP) transfers CE from HDL to TG rich lipoproteins in exchange for TG. LDLR-related protein (LRP) in liver is considered the receptor for diet-derived chylomicron remnants, which is not shown. There are substantial differences in lipoprotein metabolism between humans and mice. In mice, liver-secreted VLDL contains both apoB100 and apoB48 due to the presence of apoB's mRNA editing activity. TGs on apoB48-VLDL are hydrolyzed by LPL and are remodeled to smaller, denser apoB48-VLDL remnants. These particles are taken up through an interaction with apoE receptors. Mouse plasma does not have CETP activity. As a result, major cholesterol carrying lipoproteins in humans are LDL, while they are HDL in mice.

apoB100 transgenic apobec-1-deficient mice. This resulted in reduced apoB100 and plasma cholesterol [112]. Similarly, siRNA has also been used to reduce plasma cholesterol by decreasing apoB mRNAs [113,114]. Recently, antisense ODNs against apoB have been used in a small phase II double-blind randomized, placebo-controlled, dose-escalation clinical trial for mild dyslipidemia [115].

Cell-based approaches also have shown some success in modifying dyslipidemia, mostly through transplantation of genetically modified BM cells. Macrophages synthesize and secrete apoE. Transplantation of BM cells transduced with RV- or LV-expressing apoE has been reported to protect apoE-deficient mice against atherosclerosis [116,117]. Also, apoAI expression in macrophages has been reported to attenuate development of atherosclerosis in apoE-deficient mice [118]. This is likely through the apoAI-mediated promotion of cholesterol efflux from macrophages.

Successful gene therapy for FH requires high-level long-term stable expression of therapeutic genes and evasion of the host immune system. Progress in gene therapy for FH has been the standard to inspire treatments for other systemic acquired or genetic diseases.

35.4 GENE THERAPY FOR VASCULAR DISEASE

35.4.1 Restenosis

Stenosis is the narrowing of arteries leading to decreased perfusion of tissues. Restenosis is a recurrent reduction in luminal size following intraarterial procedures including angioplasty and atherectomy because of injury to the arteries. Grüntzig pioneered the nonsurgical technique for the treatment of symptomatic coronary artery disease known as percutaneous transluminal coronary angioplasty (PTCA) [119]. This has revolutionized the treatment of acute coronary syndrome by cardiologists. However, on a cellular level, normally quiescent smooth muscle cells (SMCs) within the treated vessel can be activated to migrate and proliferate as a part of the normal wound-healing process. This has been documented following percutaneous coronary interventions (PCIs) including coronary stenting and balloon angioplasty. A post-PTCA restenosis occurs in 30%–40% of treated coronary lesions in 6–7 months. Gene therapy has been explored as a potential single long-lasting treatment. However, vascular gene therapy to prevent postangioplasty restenosis has experienced a significant challenge in the past few years. The combination of coronary stenting and aggressive antithrombotic and antiplatelet therapy and development of drug-eluting stents have significantly reduced the incidence of restenosis. Future studies using gene therapy approach for the treatment of restenosis should be combined with drug-eluting stents to further improve outcomes. Such strategies have been reported in rabbit vascular injury models [120,121].

Targets of gene therapy. The physiological response to vascular injury is divided into four stages: (i) the early phase; (ii) the thrombogenic phase (formation of thrombi due to local hemorrhage and thrombosis); (iii) neointima hyperplasia (the proliferation and migration of medial and intimal vascular SMCs); and (iv) remodeling (an adaptive growth of a fibrocellular lesion composed of vascular SMCs and extracellular matrix with a more synthetic state). Growth factors and cytokines are believed to play a key role in the stimulation of vascular SMCs during neointima hyperplasia. It was considered that a key target for gene therapy should be to decrease growth factors, thus inhibiting the pathologic proliferation of SMCs and preventing restenosis. However, this approach was not fruitful due to redundancy in functions among growth factors.

Effective treatments to prevent restenosis are based on the knowledge of regulation of cell cycle in cancer cells [2]. Medial vascular SMCs are normally quiescent and exist in G0 phase. In response to vascular injury or stimulation by growth factors, they progress to G1. This transition is regulated by cyclins and cyclin-dependent kinases (CDKs). The progression of the cell cycle depends on the phosphorylation and dephosphorylation of CDKs. The CDK inhibitor, p27, plays a predominant role in the regulation of the transition from G1 to S phase, and is an essential checkpoint in the progression through the cell cycle. Mice deficient in p27 display enhanced growth and multiorgan hyperplasia [122]. Rapamycin is a pharmacological approach to this problem since rapamycin-coated stents inhibit intimal hyperplasia by increasing p27 [123]. Using a gene therapy approach, Ad-mediated overexpression of p21 inhibits neointima formation in a rat model of balloon angioplasty [124].

Another key regulator of cell cycle progression is the product of the retinoblastoma gene (pRB), a tumor suppressor. Inactivation of RB is lethal in mice, causing multiple defects in hematopoiesis and neurogenesis [125] while overexpression of the RB gene in transgenic mice causes dwarfism [126]. pRB is a nuclear phosphoprotein that arrests cell cycle at the G1 phase by repressing genes required for transition to the S phase. This is accomplished through formation of a complex with the transcription factor E2F. This complex formation prevents E2F's activation. As CDKs activity increases, pRB becomes phosphorylated and dissociates from E2F, activating the S phase genes. Transcription decoys bearing the E2F consensus-binding sequence inhibit E2F effects on its target genes [127]. In the ITALICS trial (Randomized Investigation by the Thoraxcenter of Antisense DNA using local delivery and IVUS after Coronary Stenting), proto-oncogene c-myc was targeted by ODNs. Phosphorothioate-modified 15-mer antisense ODNs were delivered into the coronary arteries after stent implantation. Unfortunately, the treatment did not reduce neointimal volume obstruction or restenosis [128].

35.4.2 Vein Graft Failure

In contrast to gene therapy for postangioplasty stenosis, vein graft stenosis may be a more amenable target for gene therapy because of direct access to the vein during surgery. The vein graft can be manipulated *ex vivo* prior to implantation [129]. Matrix metalloproteinases (MMPs) play a role in extracellular matrix degradation, allowing SMC migration during neointima formation. MMPs are inhibited by endogenous inhibitors called TIMPs (the "tissue inhibitors for MMP"). Studies have shown the expression of TIMP 1–3 in human saphenous vein (SV) cultures inhibits migration of SMCs. Ultrasound exposure of vein grafts in the presence of microbubbles complexed with plasmid DNA expressing TIMP-3 inhibited neointimal formation in a pig model [130]. *Ex vivo* approaches have proven successful in animal models of neointimal hyperplasia: Ad-mediated expression of the carboxyl terminus of the β-adrenergic (β-AR) receptor kinase in a aortocoronary SV graft in a canine model [131] and Ad expression of antisense transforming growth factor β1 (TGF β1), a chemotactic and mitogenic factor, to inhibit SMC migration, endothelial cell proliferation, and extracellular matrix synthesis in rats [132].

Nitric oxide (NO) plays a key role in maintaining vascular homeostasis by regulation of platelet adhesion and aggregation, vascular SMC migration, and proliferation, as well as inhibition of endothelial activation and inflammation [133]. West et al. have shown that exposure to Ad, expressing nitric oxide synthase (NOS), during jugular-carotid bypass surgery in rabbits improved endothelial cell function, blocked SMC proliferation, and reduced vascular superoxide production [134]. Endothelial NOS (eNOS) was more effective in preventing intimal hyperplasia, suggesting the potential of this approach for prevention of restenosis [135]. In another study, the expression of the tumor suppressor gene p53 caused apoptosis of SMCs, thereby blocking their migration and inhibiting neointimal formation in human SV coronary artery grafts [136]. Similarly, Ad-mediated p53 gene transfer into the luminal surface of porcine SV grafts prior to grafting resulted in increased lumen size and inhibition of neointimal formation [137]. Although most studies have used Ads, the treatment for vein graft stenosis may require long-term therapeutic gene expression and AAV may have more utility in this approach [138].

Despite early reports suggesting that an E2F decoy can block cellular proliferation in vein grafts [139], this treatment was ineffective in preventing graft failure in two large multicenter randomized double-blind placebo-controlled studies of 2400 [140] and 1404 patients [141]. Perhaps, siRNA technology could be more effective than decoys to silence harmful factors in vein grafts. Midkine is a heparin-binding growth factor and deficient mice exhibit a reduction in neointima formation in a restenosis model. ShRNA to midkine was mixed with atelocollagen and was shown to reduce intimal hyperplasia up to 90% in jugular vein-to-carotid artery interposition vein grafts in rabbits 4 weeks after treatment [142].

35.4.3 THERAPEUTIC ANGIOGENESIS FOR ISCHEMIA

A study by Folkman et al. suggested that angiogenic growth factors are required for induction of new blood vessel formation and tumor growth [143]. The growth of blood vessels is essential for organ growth and repair. Development of the vascular system occurs via vasculogenesis (angioblast mobilization), angiogenesis (sprouting), and arteriogenesis (maturation) (Figure 35.4) [144].

Angiogenesis. Angiogenesis is the sprouting of new capillaries from postcapillary venules. In adults, it is stimulated mainly by tissue hypoxia via activation of hypoxia-induced factor (HIF)-1α expression. Among many genes induced by HIF-1α, angiongensis-specific genes include the vascular endothelial growth factor (VEGF) family, angiopoietins (Ang), and inducible nitric oxide synthase (iNOS). The VEGF family is composed of five closely related genes: VEGF-A, -B, -C, -D, and PIGF. There are several isoforms of VEGF-A that differ by their amino acid length and their ability to bind

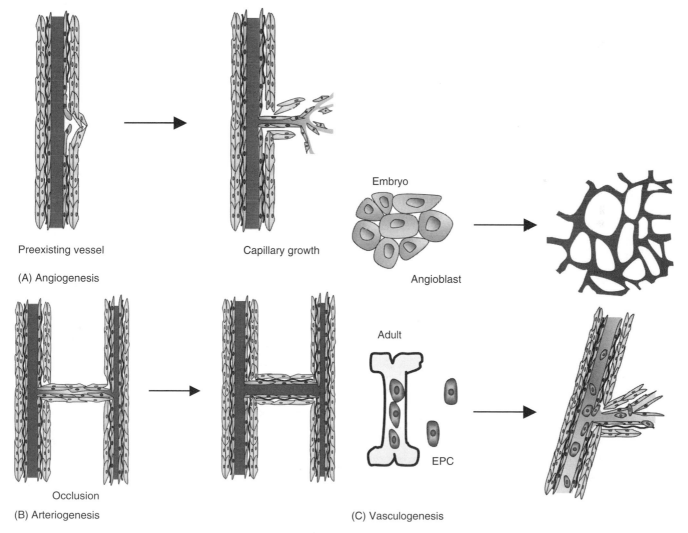

FIGURE 35.4 Vascular growth. (A) Angiogenesis is the process by which preexisting vascular structures sprout to new capillary network, operated by the in situ proliferation and migration of preexisting endothelial cells. (B) Arteriogenesis is maturation or de novo growth of collateral conduits. It occurs frequently outside the area of ischemia in response to local change in shear stress. (C) Vasculogenesis. In embryo, angioblasts (EPCs) assemble in a primitive network. In adults, bone marrow derived EPCs are activated in response to vascular trauma, home to the site of neovascularlization and are incorporated into blood and lymphatic vessels.

to heparan sulfate. The long isoforms, $VEGF_{204}$ and $VEGF_{186}$, bind to heparan sulfate tightly and do not diffuse through extracelluar matrix. In contrast, a short isoform, $VEGF_{121}$, lacks binding to heparan sulfate and shows wide diffusibility. $VEGF_{165}$ preserves both properties and thus possesses increased ability to stimulate VEGF receptors [145]. This isoform is the most promising for promotion of vascular growth [146]. Placenta growth factor (PIGF) is a unique member of the VEGF family. Its arteriogenic effects and ability to release EPCs from the BM have been reported [147,148]. Angs is a family of four genes involved in the regulation of vessel stability and remodeling. Ang-1 counteracts the VEGF-induced angiogenesis that causes the vessels to be more permeable and tightens the permeability barrier by enhancing the interaction between endothelial cells, pericytes, and the surrounding matrix [149].

Arteriogenesis. Arteriogenesis occurs frequently outside the area of ischemia in response to local accumulation of blood-derived mononuclear cells at the sites of arterial stenosis, where shear-stress is increased [150]. Angiographic studies in rodent hind limb ischemic models have demonstrated remodeling of preexisting vessels [151]. However, it remains a hotly debated issue whether arteriogenesis occurs de novo or represents enlargement of preexisting vascular structures. Both angiogenesis and arteriogenesis increase blood flow, which has attracted attention as a potential treatment for CHD and peripheral arterial disease (PAD). Fibroblast growth factors (FGFs) are the best-studied family of arteriogenic growth factors and are potent inducers of cell growth and migration through their interaction with four tyrosine kinase receptors and syndecan-4. Those with the most pronounced angiogenic activity include FGF-1, -2, -4, and -5 [152]. However, FGF activity is regulated mainly at the level of FGF receptor expression in target tissues [153]. Other contributors to arteriogenic response include PDGF and hepatocyte growth factors [146].

Vasculogenesis. Circulating BM-derived EPCs are activated in response to stimuli such as cytokines or tissue ischemia. They home to sites of neovascularization and are incorporated into the foci [154]. This phenomenon of vasculogenesis was originally described for embryonic neovascularization but also has been found in adults [6,155]. BM-derived EPCs functionally contribute to vasculogenesis during wound healing [156], limb ischemia [6,157], postmyocardial infarction [158], endothelialization, and microvessel formation from vascular grafts [159,160], and physiological and pathological neovascularization [157].

Therapeutic myocardial angiogenesis is an attractive treatment option for patients with end-stage coronary disease who have failed angioplasty and surgical methods of revascularization. Many potential genes have been identified in proof-of-principle experiments using small or large animals. The most popular model for therapeutic angiogenesis has been the hind limb ischemia in rodents, rabbits, or pigs [161–164]. Direct injection of naked DNA or viral vectors expressing VEGF has been reported to be effective in inducing collateralization and improving perfusion [165–167]. Most studies of gene therapy for ischemic disease have used *in vivo* delivery of single genes, usually growth factors, which initiate a cascade of angiogenesis. Kondoh et al. have tested an *ex vivo* approach with dual gene expression. In a rabbit model of hind limb ischemia, autologous fibroblasts were infected with Ad expressing VEGF or basic FGF, alone or combination, and injected via the left internal iliac artery. In the combination group, collateral conductance showed synergistic effects while additive effects were found *in vivo* blood flow and SMC-positive vessel density [168].

Other approaches include the use of multiple active genes or transcription factors that impact on the angiogenesis pathways. The injection of skeletal myoblasts infected with Ad bicistronic vector expressing $VEGF_{165}$ and Ang-1 stimulated functional neovascularization of hind limb [169]. Whitlock et al. used Ad expression of a VEGF cDNA/genomic hybrid that allows the expression of three major VEGF isoforms 121, 165, and 189 for greater restoration of blood flow [170]. Another approach has been to upregulate endogenous VEGF expression by engineering ZFP transcription factors. Ad-expressing VEGF-activating ZFP was delivered to the adducter muscle of the ischemic hind limb in a mouse model. The treatment improved blood flow and limb salvage, demonstrating the feasibility of designing ZFPs to therapeutically regulate gene expression [171].

Despite encouraging results in early small open-labeled trials, randomized controlled trials employing angiogenic gene therapy have not shown a clinical benefit [1,172]. In one phase II study, patients with lower limb ischemia were recruited for treatment with Ad-$VEGF_{165}$ or plasmid-$VEGF_{165}$. The primary endpoint was increased vascularity measured by angiography at 3 months; this was significantly improved but the restenosis rate was not [173]. In a small double-blinded placebo-controlled study, patients with critical limb ischemia received intramuscular injection of plasmid-carrying $VEGF_{165}$. Although significant improvement in pressure indices (ankle-to-brachial index, toe-to-brachial index) and clinical improvement (skin, pain, and quality of life score) were detected, the amputation rate was not reduced [174]. In the regional angiogenesis with vascular endothelial growth factor (RAVE) trial, 105 patients with intermittent claudication were treated with intramuscular injection of low- or high-dose Ad-$VEGF_{121}$ or placebo. Unfortunately, the treatment did not improve exercise performance or quality of life so the results did not support local delivery of Ad-$VEGF_{121}$ as a treatment for patients with intermittent claudication. In fact, it was noted that the Ad-$VEGF_{121}$ administration was associated with increased peripheral edema [175]. However, in another phase II randomized controlled trial of Ad-$VEGF_{121}$ (REVASC), there was improvement in exercise-induced ischemia at 26 weeks in patients who received intramyocardial delivery of Ad-$VEGF_{121}$ [176]. Therapeutic angiogenesis was also tested for patients with a previous MI or angina. Ad-mediated transfer of $VEGF_{165}$ or FGF-4 has been reported to be effective in increasing myocardial perfusion [177,178] while the treatment with plasmid-carrying VEGF$_{165}$ was ineffective [179].

35.4.4 HYPERTENSION

High blood pressure (BP) is associated with an increased risk of mortality and morbidity from stroke, CHD, congestive heart failure (CHF), and end-stage renal disease. Worldwide prevalence is estimated to be approximately 1 billion individuals [180]. More than 75 antihypertensive agents in nine classes are available. Much progress has been made in recent years in the prevention, detection, and treatment of high BP. More emphasis is now placed on lifestyle/behavior modification (obesity, high dietary intake of fat and sodium, physical activity, smoking, excessive alcohol intake, low dietary potassium intake) to control BP as well as to improve the efficacy of pharmacologic treatment of high BP. Also, new classes of antihypertensive agents along with new compounds in established drug classes are widening the repertoire that physicians have to combat hypertension. Some examples include aldosterone receptor blockers, direct renin inhibitors, endothelin receptor antagonists, and dual endopeptidase inhibitors [181]. Although these pharmacologic solutions make gene therapy out to be an unlikely treatment option, part of the challenge in controlling BP is patient noncompliance. Patients with hypertension tend to have low compliance rate because this disorder is asymptomatic. The conventional antihypertensive therapies do not cure the problem and some patients find it difficult to continue with lifestyle modifications. Thus, long-acting gene therapy could be an unexpected treatment option.

RNAi- and antisense ODN-based strategies may be useful for the control of hypertension. Targeting of the renin–angiotensin system appears to be a viable strategy [182]. A single intracardiac administration of RV-containing angiotensin II type I receptor antisense gene resulted in prolonged BP lowering and prevention of cardiac hypertrophy in hypertensive rodent models. However, this treatment did not attenuate endothelial dysfunction [183,184]. Tyrosine hydroxylase (TH) is a rate-limiting enzyme for catecholamine synthesis. Inhibition of TH may have similar effects to that of β-AR receptor blockers. Tail vein injection of antisense ODNs against TH caused hypotensive effects in SHR with a concomitant reduction of TH activity, epinephrine, and norepinephrine levels in the adrenal medulla [185].

Increased levels of reactive oxygen species have been shown in blood vessels and in tissues in experimental models of hypertension [186]. Increased superoxide may contribute to hypertension by inactivation of vascular NO that induces vascular relaxation [187]. Thus, gene transfer of extracellular superoxide dismutase (ecSOD) may reduce arterial pressure. Intravenous injection of Ad-expressing ecSOD but not ecSOD lacking the heparin-binding domain reduced BP and cumulative sodium balance in SHR [188].

Other vasodilatory approaches may be viable as well. Atrial natriuretic peptide (ANP) is a 27 amino acid peptide with potent vasodilatory, natriuretic, and diuretic effects [189].

Prolonged infusion of synthetic ANP to humans with essential hypertension causes a reduction in BP and enhanced natriuresis [190]. Similar results have been reported by injection of naked ANP plasmid into SHR or Ad expressing ANP into Dahl salt-sensitive rats [191,192]. Interestingly, the treatment also reduced stroke-induced mortality in Dahl salt-sensitive rats signifying that the important cardiovascular outcomes could be affected [192]. To achieve long-term ANP expression, Schillinger et al. have used HDAd-mediated mifepristone-inducible ANP gene transfer into the BHP/2 mouse model, and demonstrated decreased BP, increased urinary cGMP output, and decreases in heart weight [193]. Other therapeutic genes that might decrease BP by inducing vasodilation are kallikrein, adrenomedullin, and eNOS [194,195].

Primary pulmonary hypertension is a rare but life-threatening disease characterized by progressive pulmonary hypertension, leading to right ventricular failure and premature death [196]. Prostacyclin is a potent vasodilator that inhibits platelet adhesion and cell growth. Impaired prostacyclin production has been linked to the development of pulmonary hypertension [197]. Intratracheal transfer of cDNA coding for human prostacyclin synthase into the lungs of monocrotaline (MCT)-induced pulmonary hypertensive rats has been reported to attenuate pulmonary arterial pressure and pulmonary resistance and improve survival of rats [198].

Factors that inhibit SMC proliferation and induce apoptosis also are potential therapeutic genes. Survivin is an inhibitor of apoptosis found primarily in cancer cells. It has been reported that balloon-mediated arterial injury in rabbits results in expression of survivin in vascular cells where it is involved in pathological vessel-wall remodeling [199]. Survivin is also expressed in pulmonary arteries of patients with pulmonary hypertension and rats with MCT-induced pulmonary hypertension. Inhibition of survivin by inhaled Ad expressing the dominant negative mutant reversed MCT-induced pulmonary arterial hypertension and prolonged survival by 25% in rats [200].

35.4.5 THROMBOSIS

Deep vein thrombosis is a common condition, affecting 1% to 2% of the population. More than 200,000 new cases occur in the United States annually. Of these, 30% of patients die within 30 days of the event and one-fifth will suffer sudden death due to pulmonary embolism [201]. Drug treatment strategies to prevent thrombosis include antiplatelet and systemic anticoagulation. Vascular cyclooxygenase-1 (COX-1) is the rate-limiting enzyme in prostacyclin synthesis. Injection of Ad-expressing COX-1 into angioplasty-injured porcine carotid arteries was able to inhibit thrombus formation [202]. Other genes tested for their utility are tissue plasminogen activator [203], thrombomodulin [204], tissue factor pathway inhibitor [205], C-type natriuretic peptide [206], and ectonucleoside triphosphate diphosphohydrolase [207]. Direct injection of plasmid-expressing VEGF into thrombus also was shown to improve thrombus recanalization [208].

35.5   GENE THERAPY FOR HEART FAILURE

Heart failure remains one of major causes of morbidity and mortality in industrial societies, worsening as the population ages. The anatomical structure of the heart and advances in percutaneous methods to access the heart make it possible to target this organ for local gene or cell therapy. The failed heart may contain both nondiseased and dysfunctional cardiomyocytes and the goal of gene therapy is to improve the contractile function of cardiomyocytes. Cell therapy attempts to replace lost cardiomyocytes.

The main targets of gene therapy for heart failure are enhancement of contractility via intracellular calcium and β-AR pathways, inhibition of cardiac hypertrophy, remodeling and fibrosis, and prevention of arrhythmias. Enhancement of contractility of cardiac myocytes can be achieved through manipulation of intracellular calcium levels. The sarcoplasmic reticulum (SR) plays a central role in the regulation of intracellular calcium in cardiac muscles during excitation–contraction coupling. The SR ATPase pump is responsible for most of the Ca^{2+} removal during relaxation, and relaxation abnormalities related to deficient SR Ca^{2+} uptake have been documented in heart failure [209]. The catheter-based direct injection of Ad-expressing SR Ca^{2+}-ATPase into left ventricle has been reported to improve left-ventricular function in a rat model of heart failure [210]. Other genes tested to sequester calcium include parvalbumin [211], a mutant of phospholamban, which is a key regulator of cardiac SR Ca^{2+} cycling [212], and Ca^{2+}-binding protein S100A1 [213].

The β-AR system mediates the inotrophic and chronotrophic effects of epinephrine and norepinephrine and therefore plays a crucial role in the regulation of cardiac function. Intracoronary delivery of Ad expressing β2-AR has been reported to increase contractility in rabbits [214]. Because of our rapidly progressing understanding of cardiomyocyte signal transduction, new potential targets have been emerged. The protein kinase C (PKC) family is involved in hypertrophy, dilated cardiomyopathy, ischemic injury, and mitogen stimulation [215]. Inhibition of PKCα by expression of a dominant-negative mutant or overexpression of p38 kinase have been reported to enhance cardiac contractility, restore pump function, and increase cell survival in a rat model of postinfarction cardiomyopathy [216,217]. However, a more direct approach to rescue failing cardiac myocytes may be to inhibit apoptosis through a blockade of caspase activation [218]. Ischemia and oxidative stress are the major mechanisms of tissue injury. The heme oxygenase system is a regulator of endothelial cell (EC) integrity and oxidative stress. Direct delivery of the cytoprotective heme oxygenase gene into the rat myocardium using AAV has been reported to protect the myocardium from ischemia/reperfusion injury in rat [219]. Myocardial ischemia is the most common cause of CHF and enhancement of myocardial perfusion could improve myocardial dysfunction. Ad-VEGF$_{121}$ treatment has been reported to enhance myocardial function in a swine model of pacing-induced heart failure [220].

An interesting development in treatment of heart failure is the use of biological pacemakers. Although electronic pacemakers have been used to manage symptomatic bradycardia, implantable devices are associated with multiple complications. In recent years, gene and cell therapy approaches have been explored to replace or complement electronic devices with biological pacemakers to treat sick sinus syndrome or atrioventricular conduction block [221]. There are three strategies: gene transfer to existing cardiomyocytes; cellular transplantation; and delivery of genetically modified cells to the heart. The normal cardiac rhythm originates in the sinoatrial node in the right atrium. Proper pacemaker function is encoded by the hyperpolarized-activated cyclic nucleotide-gated channel (HCN) gene family, whose gene products generate the current. Ad-mediated gene transfer of synthetic HCN into the left ventricular myocardium has been reported to induce pacemaker activity with spontaneous action potential oscillations in myocardium [222–224]. Long-term studies with AAV or integrating viral vectors are clearly needed to evaluate its effectiveness. Much remains to be learned at the preclinical level and fine tuning of this approach is necessary before clinical trials are designed.

35.6   CELL THERAPY FOR MYOCARDIAL INFARCTION

Prolonged interruption of myocardial blood flow initiates events that lead to the cumulative loss of cardiac myocytes. MI, ischemic cardiomyopathy, and CHF remain the predominant causes of morbidity and mortality throughout the world. Currently, no medication or procedure has shown efficacy in replacing damaged myocardium. Therefore, reports suggesting that the transfer of BM cells could regenerate infarcted myocardium and improve cardiac function by neovascularization have generated tremendous excitement [7,8]. All studies of cell transfer into damaged cardiac tissues in small animal models have reported beneficial effects. However, how those transplanted cells repair the damage has not been fully determined [225]. The initial suggestion for trans-differentiation of donor BM cells into myocardium [7] has been supported by evidence of phenotypic conversion of progenitor cells [226–228], while others failed to demonstrate this phenomenon [229–231].

In contrast, cell fusion is a well-recognized phenomenon in several tissues [232–234]. Another potential mechanism for repair of damaged cardiac muscle is cell integration, which refers to the ability of certain cell types to integrate into the cardiac syncytium without transdifferentiation into cardiomyocytes [232,235]. Whether there is true functional integration or a paracrine effect by transplanted cells requires further study. Transplanted cells may secrete various cytokines, chemokines, and growth factors, which could indirectly promote survival of cardiomyocytes, mobilization of endogenous progenitor cells, or neovascularlization [236–240]. Although the prospect of regeneration of cardiac tissues is exciting, so far animal studies have challenged the idea that BM cells can generate cardiomyocytes efficiently [229,231] and clinical studies have shown that only a small fraction of infused BM cells are retained in the heart [241].

Strauer et al. used catheter-based intracoronary administration of autologous BM cells in patients with acute MI [242]. The results of this study and other small studies using similar methods were encouraging [243–245]. In the REPAIR-AMI (Reinfusion of Enriched Progenitor Cells and Infarct Remodeling in Acute Myocardial Infarction) trial, Schächinger et al. reported a significant improvement of left ventricular ejection fraction (LVEF) 4 months after cell infusion by angiography. The benefit was greatest in patients with the worst LVEF at baseline [245]. However, three randomized trials in LVEF have been disappointing [246–248]. The bone marrow transfer to enhanced ST-elevation infarct regeneration (BOOST) trial reported that the improvement in LVEF seen at 6 months after cell infusion was no longer significant at 18 months [247]. Lunde et al. also did not find a significant improvement in LVEF 6 months after infarction [248]. In the TOPCARE-CHD trial (transplantation of progenitor cells and recovery of LV function in patients with chronic ischemic heart disease), Assmus et al. evaluated the transplantation of progenitor cells derived from BM cells or circulating progenitor cells (CPC) in patients with chronic ventricular dysfunction and reported a significant improvement in LVEF 3 months after treatment. Patients received BM cells had significantly greater effects than those who received CPC, but the benefit was modest [249].

35.7 CONCLUSION

Research in gene and cell therapy for cardiovascular disease has exploded in the last several years. Our knowledge of vascular biology and pathogenesis of cardiovascular diseases has identified many therapeutic targets. Despite encouraging proof-of-concept experiments and small-scale clinical trials, the early promise of efficacy has not been confirmed in large-scale double-blinded randomized clinical trials. Some of this failure is due to limitations seen in all gene therapy approaches. For gene therapy, the challenges remain to include vector systems, route of administration, pharmacokinetics, and immunity against both the vector and therapeutic gene. In cell therapy, sufficient understanding of biological processes leading to the disease may be lacking. Some key issues that must be addressed are cell types to be used, optimum dosing and timing, routes of administration, and suitable patient population. Our ultimate success or failure will rest on the clinical efficacy and the effects of treatment on mortality and morbidity.

ACKNOWLEDGMENTS

The author is grateful to Drs. S.L. Samson and K.I. Oka for their comments and to Dr. T. Terashima for his assistance in manuscript preparation. This work was supported in part by HL73144 and HL59314.

REFERENCES

1. Rissanen, T.T. and Yla-Herttuala, S. 2007. Current status of cardiovascular gene therapy. *Mol Ther 15*, 1233–1247.
2. Fuster, V. and Sanz, J. 2007. Gene therapy and stem cell therapy for cardiovascular diseases today: A model for translational research. *Nat Clin Pract Cardiovasc Med 4 Suppl 1*, S1–S8.
3. Scherer, L.J. and Rossi, J.J. 2003. Approaches for the sequence-specific knockdown of mRNA. *Nat Biotechnol 21*, 1457–1465.
4. Chen, X., Dudgeon, N., Shen, L., and Wang, J.H. 2005. Chemical modification of gene silencing oligonucleotides for drug discovery and development. *Drug Discov Today 10*, 587–593.
5. Akashi, H., Matsumoto, S., and Taira, K. 2005. Gene discovery by ribozyme and siRNA libraries. *Nat Rev Mol Cell Biol 6*, 413–422.
6. Asahara, T., Murohara, T., Sullivan, A., Silver, M., van der Zee, R., Li, T., Witzenbichler, B., Schatteman, G., and Isner, J.M. 1997. Isolation of putative progenitor endothelial cells for angiogenesis. *Science 275*, 964–967.
7. Orlic, D., Kajstura, J., Chimenti, S., Jakoniuk, I., Anderson, S.M., Li, B., Pickel, J., McKay, R., Nadal-Ginard, B., Bodine, D.M., Leri, A., and Anversa, P. 2001. Bone marrow cells regenerate infarcted myocardium. *Nature 410*, 701–705.
8. Kocher, A.A., Schuster, M.D., Szabolcs, M.J., Takuma, S., Burkhoff, D., Wang, J., Homma, S., Edwards, N.M., and Itescu, S. 2001. Neovascularization of ischemic myocardium by human bone-marrow-derived angioblasts prevents cardiomyocyte apoptosis, reduces remodeling and improves cardiac function. *Nat Med 7*, 430–436.
9. Sanchez, P.L., Sanchez-Guijo, F.M., Villa, A., del Canizo, C., Arnold, R., San Roman, J.A., and Fernandez-Aviles, F. 2007. Launching a clinical program of stem cell therapy for cardiovascular repair. *Nat Clin Pract Cardiovasc Med 4 Suppl 1*, S123–S129.
10. Waehler, R., Russell, S.J., and Curiel, D.T. 2007. Engineering targeted viral vectors for gene therapy. *Nat Rev Genet 8*, 573–587.
11. Yla-Herttuala, S. and Alitalo, K. 2003. Gene transfer as a tool to induce therapeutic vascular growth. *Nat Med 9*, 694–701.
12. Gao, X., Kim, K.S., and Liu, D. 2007. Nonviral gene delivery: What we know and what is next. *Aaps J 9*, E92–E104.
13. Liu, F., Song, Y., and Liu, D. 1999. Hydrodynamics-based transfection in animals by systemic administration of plasmid DNA. *Gene Ther 6*, 1258–1266.
14. Condiotti, R., Curran, M.A., Nolan, G.P., Giladi, H., Ketzinel-Gilad, M., Gross, E., and Galun, E. 2004. Prolonged liver-specific transgene expression by a non-primate lentiviral vector. *Biochem Biophys Res Commun 320*, 998–1006.
15. Brunetti-Pierri, N., Palmer, D.J., Mane, V., Finegold, M., Beaudet, A.L., and Ng, P. 2005. Increased hepatic transduction with reduced systemic dissemination and proinflammatory cytokines following hydrodynamic injection of helper-dependent adenoviral vectors. *Mol Ther 12*, 99–106.
16. Alino, S.F., Herrero, M.J., Noguera, I., Dasi, F., and Sanchez, M. 2007. Pig liver gene therapy by noninvasive interventionist catheterism. *Gene Ther 14*, 334–343.
17. Brunetti-Pierri, N., Stapleton, G.E., Palmer, D.J., Zuo, Y., Mane, V.P., Finegold, M.J., Beaudet, A.L., Leland, M.M., Mullins, C.E., and Ng, P. 2007. Pseudo-hydrodynamic delivery of helper-dependent adenoviral vectors into nonhuman primates for liver-directed gene therapy. *Mol Ther 15*, 732–740.

18. Glover, D.J., Lipps, H.J., and Jans, D.A. 2005. Towards safe, non-viral therapeutic gene expression in humans. *Nat Rev Genet 6*, 299–310.

19. Groth, A.C. and Calos, M.P. 2004. Phage integrases: Biology and applications. *J Mol Biol 335*, 667–678.

20. Ortiz-Urda, S., Thyagarajan, B., Keene, D.R., Lin, Q., Fang, M., Calos, M.P., and Khavari, P.A. 2002. Stable non-viral genetic correction of inherited human skin disease. *Nat Med 8*, 1166–1170.

21. Held, P.K., Olivares, E.C., Aguilar, C.P., Finegold, M., Calos, M.P., and Grompe, M. 2005. In vivo correction of murine hereditary tyrosinemia type I by phiC31 integrase-mediated gene delivery. *Mol Ther 11*, 399–408.

22. Ehrhardt, A., Xu, H., Huang, Z., Engler, J.A., and Kay, M.A. 2005. A direct comparison of two nonviral gene therapy vectors for somatic integration: In vivo evaluation of the bacteriophage integrase phiC31 and the Sleeping Beauty transposase. *Mol Ther 11*, 695–706.

23. Chen, L. and Woo, S.L. 2005. Complete and persistent phenotypic correction of phenylketonuria in mice by site-specific genome integration of murine phenylalanine hydroxylase cDNA. *Proc Natl Acad Sci USA 102*, 15581–15586.

24. Chalberg, T.W., Portlock, J.L., Olivares, E.C., Thyagarajan, B., Kirby, P.J., Hillman, R.T., Hoelters, J., and Calos, M.P. 2006. Integration specificity of phage phiC31 integrase in the human genome. *J Mol Biol 357*, 28–48.

25. Yant, S.R., Wu, X., Huang, Y., Garrison, B., Burgess, S.M., and Kay, M.A. 2005. High-resolution genome-wide mapping of transposon integration in mammals. *Mol Cell Biol 25*, 2085–2094.

26. Liu, J., Jeppesen, I., Nielsen, K., and Jensen, T.G. 2006. Phi c31 integrase induces chromosomal aberrations in primary human fibroblasts. *Gene Ther 13*, 1188–1190.

27. Ehrhardt, A., Engler, J.A., Xu, H., Cherry, A.M., and Kay, M.A. 2006. Molecular analysis of chromosomal rearrangements in mammalian cells after phiC31-mediated integration. *Hum Gene Ther 17*, 1077–1094.

28. Kotin, R.M., Siniscalco, M., Samulski, R.J., Zhu, X.D., Hunter, L., Laughlin, C.A., McLaughlin, S., Muzyczka, N., Rocchi, M., and Berns, K.I. 1990. Site-specific integration by adeno-associated virus. *Proc Natl Acad Sci USA 87*, 2211–2215.

29. Weitzman, M.D., Kyostio, S.R., Kotin, R.M., and Owens, R.A. 1994. Adeno-associated virus (AAV) Rep proteins mediate complex formation between AAV DNA and its integration site in human DNA. *Proc Natl Acad Sci USA 91*, 5808–5812.

30. Amiss, T.J., McCarty, D.M., Skulimowski, A., and Samulski, R.J. 2003. Identification and characterization of an adeno-associated virus integration site in CV-1 cells from the African green monkey. *J Virol 77*, 1904–1915.

31. Dutheil, N., Yoon-Robarts, M., Ward, P., Henckaerts, E., Skrabanek, L., Berns, K.I., Campagne, F., and Linden, R.M. 2004. Characterization of the mouse adeno-associated virus AAVS1 ortholog. *J Virol 78*, 8917–8921.

32. McCarty, D.M., Young, S.M., Jr., and Samulski, R.J. 2004. Integration of adeno-associated virus (AAV) and recombinant AAV vectors. *Annu Rev Genet 38*, 819–845.

33. Philpott, N.J., Gomos, J., and Falck-Pedersen, E. 2004. Transgene expression after rep-mediated site-specific integration into chromosome 19. *Hum Gene Ther 15*, 47–61.

34. Berthet, C., Raj, K., Saudan, P., and Beard, P. 2005. How adeno-associated virus Rep78 protein arrests cells completely in S phase. *Proc Natl Acad Sci USA 102*, 13634–13639.

35. Yant, S.R., Meuse, L., Chiu, W., Ivics, Z., Izsvak, Z., and Kay, M.A. 2000. Somatic integration and long-term transgene expression in normal and haemophilic mice using a DNA transposon system. *Nat Genet 25*, 35–41.

36. Yant, S.R., Huang, Y., Akache, B., and Kay, M.A. 2007. Site-directed transposon integration in human cells. *Nucleic Acids Res 35*, e50.

37. Ivics, Z., Katzer, A., Stuwe, E.E., Fiedler, D., Knespel, S., and Izsvak, Z. 2007. Targeted Sleeping Beauty transposition in human cells. *Mol Ther 15*, 1137–1144.

38. Atchison, R.W., Casto, B.C., and Hammon, W.M. 1965. Adenovirus-associated defective virus particles. *Science 149*, 754–756.

39. Hoggan, M.D., Blacklow, N.R., and Rowe, W.P. 1966. Studies of small DNA viruses found in various adenovirus preparations: Physical, biological, and immunological characteristics. *Proc Natl Acad Sci USA 55*, 1467–1474.

40. Carter, B.J. 2004. Adeno-associated virus and the development of adeno-associated virus vectors: A historical perspective. *Mol Ther 10*, 981–989.

41. Gao, G., Alvira, M.R., Somanathan, S., Lu, Y., Vandenberghe, L.H., Rux, J.J., Calcedo, R., Sanmiguel, J., Abbas, Z., and Wilson, J.M. 2003. Adeno-associated viruses undergo substantial evolution in primates during natural infections. *Proc Natl Acad Sci USA 100*, 6081–6086.

42. Summerford, C. and Samulski, R.J. 1998. Membrane-associated heparan sulfate proteoglycan is a receptor for adeno-associated virus type 2 virions. *J Virol 72*, 1438–1445.

43. Summerford, C., Bartlett, J.S., and Samulski, R.J. 1999. AlphaVbeta5 integrin: A co-receptor for adeno-associated virus type 2 infection. *Nat Med 5*, 78–82.

44. Qing, K., Mah, C., Hansen, J., Zhou, S., Dwarki, V., and Srivastava, A. 1999. Human fibroblast growth factor receptor 1 is a co-receptor for infection by adeno-associated virus 2. *Nat Med 5*, 71–77.

45. Di Pasquale, G., Davidson, B.L., Stein, C.S., Martins, I., Scudiero, D., Monks, A., and Chiorini, J.A. 2003. Identification of PDGFR as a receptor for AAV-5 transduction. *Nat Med 9*, 1306–1312.

46. Walters, R.W., Yi, S.M., Keshavjee, S., Brown, K.E., Welsh, M.J., Chiorini, J.A., and Zabner, J. 2001. Binding of adeno-associated virus type 5 to 2,3-linked sialic acid is required for gene transfer. *J Biol Chem 276*, 20610–20616.

47. Akache, B., Grimm, D., Pandey, K., Yant, S.R., Xu, H., and Kay, M.A. 2006. The 37/67-kilodalton laminin receptor is a receptor for adeno-associated virus serotypes 8, 2, 3, and 9. *J Virol 80*, 9831–9836.

48. Jiang, H., Couto, L.B., Patarroyo-White, S., Liu, T., Nagy, D., Vargas, J.A., Zhou, S., Scallan, C.D., Sommer, J., Vijay, S., Mingozzi, F., High, K.A., and Pierce, G.F. 2006. Effects of transient immunosuppression on adenoassociated, virus-mediated, liver-directed gene transfer in rhesus macaques and implications for human gene therapy. *Blood 108*, 3321–3328.

49. Thomas, C.E., Ehrhardt, A., and Kay, M.A. 2003. Progress and problems with the use of viral vectors for gene therapy. *Nat Rev Genet 4*, 346–358.

50. Duan, D., Yue, Y., Yan, Z., and Engelhardt, J.F. 2000. A new dual-vector approach to enhance recombinant adeno-associated virus-mediated gene expression through intermolecular cis activation. *Nat Med 6*, 595–598.

51. Nakai, H., Storm, T.A., and Kay, M.A. 2000. Increasing the size of rAAV-mediated expression cassettes in vivo by intermolecular joining of two complementary vectors. *Nat Biotechnol 18*, 527–532.

52. Sun, L., Li, J., and Xiao, X. 2000. Overcoming adeno-associated virus vector size limitation through viral DNA heterodimerization. *Nat Med 6*, 599–602.

53. McCarty, D.M., Monahan, P.E., and Samulski, R.J. 2001. Self-complementary recombinant adeno-associated virus (scAAV) vectors promote efficient transduction independently of DNA synthesis. *Gene Ther 8*, 1248–1254.

54. Stilwell, J.L. and Samulski, R.J. 2004. Role of viral vectors and virion shells in cellular gene expression. *Mol Ther 9*, 337–346.

55. Nakai, H., Montini, E., Fuess, S., Storm, T.A., Grompe, M., and Kay, M.A. 2003. AAV serotype 2 vectors preferentially integrate into active genes in mice. *Nat Genet 34*, 297–302.

56. Bell, P., Wang, L., Lebherz, C., Flieder, D.B., Bove, M.S., Wu, D., Gao, G.P., Wilson, J.M., and Wivel, N.A. 2005. No evidence for tumorigenesis of AAV vectors in a large-scale study in mice. *Mol Ther 12*, 299–306.

57. Bell, P., Moscioni, A.D., McCarter, R.J., Wu, D., Gao, G., Hoang, A., Sanmiguel, J.C., Sun, X., Wivel, N.A., Raper, S.E., Furth, E.E., Batshaw, M.L., and Wilson, J.M. 2006. Analysis of tumors arising in male B6C3F1 mice with and without AAV vector delivery to liver. *Mol Ther 14*, 34–44.

58. Tobiasch, E., Rabreau, M., Geletneky, K., Larue-Charlus, S., Severin, F., Becker, N., and Schlehofer, J.R. 1994. Detection of adeno-associated virus DNA in human genital tissue and in material from spontaneous abortion. *J Med Virol 44*, 215–222.

59. Donsante, A., Vogler, C., Muzyczka, N., Crawford, J.M., Barker, J., Flotte, T., Campbell-Thompson, M., Daly, T., and Sands, M.S. 2001. Observed incidence of tumorigenesis in long-term rodent studies of rAAV vectors. *Gene Ther 8*, 1343–1346.

60. Donsante, A., Miller, D.G., Li, Y., Vogler, C., Brunt, E.M., Russell, D.W., and Sands, M.S. 2007. AAV vector integration sites in mouse hepatocellular carcinoma. *Science 317*, 477.

61. Manno, C.S., Pierce, G.F., Arruda, V.R., Glader, B., Ragni, M., Rasko, J.J., Ozelo, M.C., Hoots, K., Blatt, P., Konkle, B., Dake, M., Kaye, R., Razavi, M., Zajko, A., Zehnder, J., Rustagi, P.K., Nakai, H., Chew, A., Leonard, D., Wright, J.F., Lessard, R.R., Sommer, J.M., Tigges, M., Sabatino, D., Luk, A., Jiang, H., Mingozzi, F., Couto, L., Ertl, H.C., High, K.A., and Kay, M.A. 2006. Successful transduction of liver in hemophilia by AAV-Factor IX and limitations imposed by the host immune response. *Nat Med 12*, 342–347.

62. Mingozzi, F., Maus, M.V., Hui, D.J., Sabatino, D.E., Murphy, S.L., Rasko, J.E., Ragni, M.V., Manno, C.S., Sommer, J., Jiang, H., Pierce, G.F., Ertl, H.C., and High, K.A. 2007. CD8(+) T-cell responses to adeno-associated virus capsid in humans. *Nat Med 13*, 419–422.

63. Li, H., Murphy, S.L., Giles-Davis, W., Edmonson, S., Xiang, Z., Li, Y., Lasaro, M.O., High, K.A., and Ertl, H.C. 2007. Pre-existing AAV capsid-specific CD8+ T cells are unable to eliminate AAV-transduced hepatocytes. *Mol Ther 15*, 792–800.

64. Wang, L., Figueredo, J., Calcedo, R., Lin, J., and Wilson, J.M. 2007. Cross-presentation of adeno-associated virus serotype 2 capsids activates cytotoxic T cells but does not render hepatocytes effective cytolytic targets. *Hum Gene Ther 18*, 185–194.

65. Wang, L., Calcedo, R., Nichols, T.C., Bellinger, D.A., Dillow, A., Verma, I.M., and Wilson, J.M. 2005. Sustained correction of disease in naive and AAV2-pretreated hemophilia B dogs: AAV2/8-mediated, liver-directed gene therapy. *Blood 105*, 3079–3086.

66. Bergelson, J.M., Cunningham, J.A., Droguett, G., Kurt-Jones, E.A., Krithivas, A., Hong, J.S., Horwitz, M.S., Crowell, R.L., and Finberg, R.W. 1997. Isolation of a common receptor for Coxsackie B viruses and adenoviruses 2 and 5. *Science 275*, 1320–1323.

67. Nemerow, G.R. 2000. Cell receptors involved in adenovirus entry. *Virology 274*, 1–4.

68. Dechecchi, M.C., Melotti, P., Bonizzato, A., Santacatterina, M., Chilosi, M., and Cabrini, G. 2001. Heparan sulfate glycosaminoglycans are receptors sufficient to mediate the initial binding of adenovirus types 2 and 5. *J Virol 75*, 8772–8780.

69. Gaggar, A., Shayakhmetov, D.M., and Lieber, A. 2003. CD46 is a cellular receptor for group B adenoviruses. *Nat Med 9*, 1408–1412.

70. Segerman, A., Atkinson, J.P., Marttila, M., Dennerquist, V., Wadell, G., and Arnberg, N. 2003. Adenovirus type 11 uses CD46 as a cellular receptor. *J Virol 77*, 9183–9191.

71. Liu, Q. and Muruve, D.A. 2003. Molecular basis of the inflammatory response to adenovirus vectors. *Gene Ther 10*, 935–940.

72. Yang, Y., Nunes, F.A., Berencsi, K., Furth, E.E., Gonczol, E., and Wilson, J.M. 1994. Cellular immunity to viral antigens limits E1-deleted adenoviruses for gene therapy. *Proc Natl Acad Sci USA 91*, 4407–4411.

73. Yang, Y., Jooss, K.U., Su, Q., Ertl, H.C., and Wilson, J.M. 1996. Immune responses to viral antigens versus transgene product in the elimination of recombinant adenovirus-infected hepatocytes in vivo. *Gene Ther 3*, 137–144.

74. Kochanek, S. 1999. High-capacity adenoviral vectors for gene transfer and somatic gene therapy. *Hum Gene Ther 10*, 2451–2459.

75. Brunetti-Pierri, N., Palmer, D.J., Beaudet, A.L., Carey, K.D., Finegold, M., and Ng, P. 2004. Acute toxicity after high-dose systemic injection of helper-dependent adenoviral vectors into nonhuman primates. *Hum Gene Ther 15*, 35–46.

76. Muruve, D.A., Cotter, M.J., Zaiss, A.K., White, L.R., Liu, Q., Chan, T., Clark, S.A., Ross, P.J., Meulenbroek, R.A., Maelandsmo, G.M., and Parks, R.J. 2004. Helper-dependent adenovirus vectors elicit intact innate but attenuated adaptive host immune responses in vivo. *J Virol 78*, 5966–5972.

77. Schiedner, G., Hertel, S., Johnston, M., Dries, V., van Rooijen, N., and Kochanek, S. 2003. Selective depletion or blockade of Kupffer cells leads to enhanced and prolonged hepatic transgene expression using high-capacity adenoviral vectors. *Mol Ther 7*, 35–43.

78. Kim, I.H., Jozkowicz, A., Piedra, P.A., Oka, K., and Chan, L. 2001. Lifetime correction of genetic deficiency in mice with a single injection of helper-dependent adenoviral vector. *Proc Natl Acad Sci USA 98*, 13282–13287.

79. Nomura, S., Merched, A., Nour, E., Dieker, C., Oka, K., and Chan, L. 2004. Low-density lipoprotein receptor gene therapy using helper-dependent adenovirus produces long-term protection against atherosclerosis in a mouse model of familial hypercholesterolemia. *Gene Ther 11*, 1540–1548.

80. Toietta, G., Mane, V.P., Norona, W.S., Finegold, M.J., Ng, P., McDonagh, A.F., Beaudet, A.L., and Lee, B. 2005. Lifelong elimination of hyperbilirubinemia in the Gunn rat with a single injection of helper-dependent adenoviral vector. *Proc Natl Acad Sci USA 102*, 3930–3935.

81. Brunetti-Pierri, N., Ng, T., Iannitti, D.A., Palmer, D.J., Beaudet, A.L., Finegold, M.J., Carey, K.D., Cioffi, W.G., and Ng, P. 2006. Improved hepatic transduction, reduced systemic vector dissemination, and long-term transgene expression by delivering helper-dependent adenoviral vectors into the surgically isolated liver of nonhuman primates. *Hum Gene Ther 17*, 391–404.

82. Oka, K., Belalcazar, L.M., Dieker, C., Nour, E.A., Nuno-Gonzalez, P., Paul, A., Cormier, S., Shin, J.K., Finegold, M., and Chan, L. 2007. Sustained phenotypic correction in a mouse model of hypoalphalipoproteinemia with a helper-dependent adenovirus vector. *Gene Ther 14*, 191–202.

83. Roth, M.D., Cheng, Q., Harui, A., Basak, S.K., Mitani, K., Low, T.A., and Kiertscher, S.M. 2002. Helper-dependent adenoviral vectors efficiently express transgenes in human dendritic cells but still stimulate antiviral immune responses. *J Immunol 169*, 4651–4656.

84. Perreau, M. and Kremer, E.J. 2005. Frequency, proliferation, and activation of human memory T cells induced by a non-human adenovirus. *J Virol 79*, 14595–14605.

85. Schroder, A.R., Shinn, P., Chen, H., Berry, C., Ecker, J.R., and Bushman, F. 2002. HIV-1 integration in the human genome favors active genes and local hotspots. *Cell 110*, 521–529.

86. Wu, X., Li, Y., Crise, B., and Burgess, S.M. 2003. Transcription start regions in the human genome are favored targets for MLV integration. *Science 300*, 1749–1751.

87. Lecellier, C.H. and Saib, A. 2000. Foamy viruses: Between retroviruses and pararetroviruses. *Virology 271*, 1–8.

88. Schweizer, M., Turek, R., Hahn, H., Schliephake, A., Netzer, K.O., Eder, G., Reinhardt, M., Rethwilm, A., and Neumann-Haefelin, D. 1995. Markers of foamy virus infections in monkeys, apes, and accidentally infected humans: Appropriate testing fails to confirm suspected foamy virus prevalence in humans. *AIDS Res Hum Retroviruses 11*, 161–170.

89. Verma, I.M. and Weitzman, M.D. 2005. Gene therapy: Twenty-first century medicine. *Annu Rev Biochem 74*, 711–738.

90. Trobridge, G.D., Miller, D.G., Jacobs, M.A., Allen, J.M., Kiem, H.P., Kaul, R., and Russell, D.W. 2006. Foamy virus vector integration sites in normal human cells. *Proc Natl Acad Sci USA 103*, 1498–1503.

91. Vassilopoulos, G., Trobridge, G., Josephson, N.C., and Russell, D.W. 2001. Gene transfer into murine hematopoietic stem cells with helper-free foamy virus vectors. *Blood 98*, 604–609.

92. Leurs, C., Jansen, M., Pollok, K.E., Heinkelein, M., Schmidt, M., Wissler, M., Lindemann, D., Von Kalle, C., Rethwilm, A., Williams, D.A., and Hanenberg, H. 2003. Comparison of three retroviral vector systems for transduction of nonobese diabetic/severe combined immunodeficiency mice repopulating human CD34+ cord blood cells. *Hum Gene Ther 14*, 509–519.

93. Gharwan, H., Hirata, R.K., Wang, P., Richard, R.E., Wang, L., Olson, E., Allen, J., Ware, C.B., and Russell, D.W. 2007. Transduction of human embryonic stem cells by foamy virus vectors. *Mol Ther 15*, 1827–1833.

94. Rosamond, W., Flegal, K., Friday, G., Furie, K., Go, A., Greenlund, K., Haase, N., Ho, M., Howard, V., Kissela, B., Kittner, S., Lloyd-Jones, D., McDermott, M., Meigs, J., Moy, C., Nichol, G., O'Donnell, C.J., Roger, V., Rumsfeld, J., Sorlie, P., Steinberger, J., Thom, T., Wasserthiel-Smoller, S., and Hong, Y. 2007. Heart disease and stroke statistics—2007 update: A report from the American Heart Association Statistics Committee and Stroke Statistics Subcommittee. *Circulation 115*, e69–e171.

95. Durrington, P. 2003. Dyslipidaemia. *Lancet 362*, 717–731.

96. Kermani, T. and Frishman, W.H. 2005. Nonpharmacologic approaches for the treatment of hyperlipidemia. *Cardiol Rev 13*, 247–255.

97. Chowdhury, J.R., Grossman, M., Gupta, S., Chowdhury, N.R., Baker, J.R., Jr., and Wilson, J.M. 1991. Long-term improvement of hypercholesterolemia after *ex vivo* gene therapy in LDLR-deficient rabbits. *Science 254*, 1802–1805.

98. Grossman, M., Raper, S.E., Kozarsky, K., Stein, E.A., Engelhardt, J.F., Muller, D., Lupien, P.J., and Wilson, J.M. 1994. Successful *ex vivo* gene therapy directed to liver in a patient with familial hypercholesterolaemia. *Nat Genet 6*, 335–341.

99. Grossman, M., Rader, D.J., Muller, D.W., Kolansky, D.M., Kozarsky, K., Clark, B.J., 3rd, Stein, E.A., Lupien, P.J., Brewer, H.B., Jr., Raper, S.E. et al. 1995. A pilot study of *ex vivo* gene therapy for homozygous familial hypercholesterolaemia. *Nat Med 1*, 1148–1154.

100. Ishibashi, S., Brown, M.S., Goldstein, J.L., Gerard, R.D., Hammer, R.E., and Herz, J. 1993. Hypercholesterolemia in low density lipoprotein receptor knockout mice and its reversal by adenovirus-mediated gene delivery. *J Clin Invest 92*, 883–893.

101. Kozarsky, K.F., McKinley, D.R., Austin, L.L., Raper, S.E., Stratford-Perricaudet, L.D., and Wilson, J.M. 1994. In vivo correction of low density lipoprotein receptor deficiency in the Watanabe heritable hyperlipidemic rabbit with recombinant adenoviruses. *J Biol Chem 269*, 13695–13702.

102. Kozarsky, K.F., Jooss, K., Donahee, M., Strauss, J.F., 3rd, and Wilson, J.M. 1996. Effective treatment of familial hypercholesterolaemia in the mouse model using adenovirus-mediated transfer of the VLDL receptor gene. *Nat Genet 13*, 54–62.

103. Kobayashi, K., Oka, K., Forte, T., Ishida, B., Teng, B., Ishimura-Oka, K., Nakamuta, M., and Chan, L. 1996. Reversal of hypercholesterolemia in low density lipoprotein receptor knockout mice by adenovirus-mediated gene transfer of the very low density lipoprotein receptor. *J Biol Chem 271*, 6852–6860.

104. Oka, K., Pastore, L., Kim, I.H., Merched, A., Nomura, S., Lee, H.J., Merched-Sauvage, M., Arden-Riley, C., Lee, B., Finegold, M., Beaudet, A., and Chan, L. 2001. Long-term stable correction of low-density lipoprotein receptor-deficient mice with a helper-dependent adenoviral vector expressing the very low-density lipoprotein receptor. *Circulation 103*, 1274–1281.

105. Lebherz, C., Gao, G., Louboutin, J.P., Millar, J., Rader, D., and Wilson, J.M. 2004. Gene therapy with novel adeno-associated virus vectors substantially diminishes atherosclerosis in a murine model of familial hypercholesterolemia. *J Gene Med 6*, 663–672.

106. Pakkanen, T.M., Laitinen, M., Hippelainen, M., Kallionpaa, H., Lehtolainen, P., Leppanen, P., Luoma, J.S., Tarvainen, R., Alhava, E., and Yla-Herttuala, S. 1999. Enhanced plasma cholesterol lowering effect of retrovirus-mediated LDL receptor gene transfer to WHHL rabbit liver after improved surgical technique and stimulation of hepatocyte proliferation by combined partial liver resection and thymidine kinase–ganciclovir treatment. *Gene Ther 6*, 34–41.

107. Kankkonen, H.M., Vahakangas, E., Marr, R.A., Pakkanen, T., Laurema, A., Leppanen, P., Jalkanen, J., Verma, I.M., and Yla-Herttuala, S. 2004. Long-term lowering of plasma cholesterol levels in LDL-receptor-deficient WHHL rabbits by gene therapy. *Mol Ther 9*, 548–556.

108. Kitajima, K., Marchadier, D.H., Miller, G.C., Gao, G.P., Wilson, J.M., and Rader, D.J. 2006. Complete prevention of atherosclerosis in apoE-deficient mice by hepatic human apoE gene transfer with adeno-associated virus serotypes 7 and 8. *Arterioscler Thromb Vasc Biol 26*, 1852–1857.

109. Belalcazar, L.M., Merched, A., Carr, B., Oka, K., Chen, K.H., Pastore, L., Beaudet, A., and Chan, L. 2003. Long-term stable expression of human apolipoprotein A-I mediated by helper-dependent adenovirus gene transfer inhibits atherosclerosis progression and remodels atherosclerotic plaques in a mouse model of familial hypercholesterolemia. *Circulation 107*, 2726–2732.

110. Kitajima, K., Marchadier, D.H., Burstein, H., and Rader, D.J. 2006. Persistent liver expression of murine apoA-1 using vectors based on adeno-associated viral vectors serotypes 5 and 1. *Atherosclerosis 186*, 65–73.

111. Teng, B., Ishida, B., Forte, T.M., Blumenthal, S., Song, L.Z., Gotto, A.M., Jr., and Chan, L. 1997. Effective lowering of plasma, LDL, and esterified cholesterol in LDL receptor-knockout mice by adenovirus-mediated gene delivery of ApoB mRNA editing enzyme (Apobec1). *Arterioscler Thromb Vasc Biol 17*, 889–897.

112. Enjoji, M., Wang, F., Nakamuta, M., Chan, L., and Teng, B.B. 2000. Hammerhead ribozyme as a therapeutic agent for hyperlipidemia: Production of truncated apolipoprotein B and hypolipidemic effects in a dyslipidemia murine model. *Hum Gene Ther 11*, 2415–2430.

113. Soutschek, J., Akinc, A., Bramlage, B., Charisse, K., Constien, R., Donoghue, M., Elbashir, S., Geick, A., Hadwiger, P., Harborth, J., John, M., Kesavan, V., Lavine, G., Pandey, R.K., Racie, T., Rajeev, K.G., Rohl, I., Toudjarska, I., Wang, G., Wuschko, S., Bumcrot, D., Koteliansky, V., Limmer, S., Manoharan, M., and Vornlocher, H.P. 2004. Therapeutic silencing of an endogenous gene by systemic administration of modified siRNAs. *Nature 432*, 173–178.

114. Rozema, D.B., Lewis, D.L., Wakefield, D.H., Wong, S.C., Klein, J.J., Roesch, P.L., Bertin, S.L., Reppen, T.W., Chu, Q., Blokhin, A.V., Hagstrom, J.E., and Wolff, J.A. 2007. Dynamic polyconjugates for targeted in vivo delivery of siRNA to hepatocytes. *Proc Natl Acad Sci USA 104*, 12982–12987.

115. Kastelein, J.J., Wedel, M.K., Baker, B.F., Su, J., Bradley, J.D., Yu, R.Z., Chuang, E., Graham, M.J., and Crooke, R.M. 2006. Potent reduction of apolipoprotein B and low-density lipoprotein cholesterol by short-term administration of an antisense inhibitor of apolipoprotein B. *Circulation 114*, 1729–1735.

116. Yoshida, H., Hasty, A.H., Major, A.S., Ishiguro, H., Su, Y.R., Gleaves, L.A., Babaev, V.R., Linton, M.F., and Fazio, S. 2001. Isoform-specific effects of apolipoprotein E on atherogenesis: Gene transduction studies in mice. *Circulation 104*, 2820–2825.

117. He, W., Qiang, M., Ma, W., Valente, A.J., Quinones, M.P., Wang, W., Reddick, R.L., Xiao, Q., Ahuja, S.S., Clark, R.A., Freeman, G.L., and Li, S. 2006. Development of a synthetic promoter for macrophage gene therapy. *Hum Gene Ther 17*, 949–959.

118. Su, Y.R., Ishiguro, H., Major, A.S., Dove, D.E., Zhang, W., Hasty, A.H., Babaev, V.R., Linton, M.F., and Fazio, S. 2003. Macrophage apolipoprotein A-I expression protects against atherosclerosis in ApoE-deficient mice and up-regulates ABC transporters. *Mol Ther 8*, 576–583.

119. Gruntzig, A. 1978. Transluminal dilatation of coronary-artery stenosis. *Lancet 1*, 263.

120. Walter, D.H., Cejna, M., Diaz-Sandoval, L., Willis, S., Kirkwood, L., Stratford, P.W., Tietz, A.B., Kirchmair, R., Silver, M., Curry, C., Wecker, A., Yoon, Y.S., Heidenreich, R., Hanley, A., Kearney, M., Tio, F.O., Kuenzler, P., Isner, J.M., and Losordo, D.W. 2004. Local gene transfer of phVEGF-2 plasmid by gene-eluting stents: An alternative strategy for inhibition of restenosis. *Circulation 110*, 36–45.

121. Sharif, F., Hynes, S.O., McMahon, J., Cooney, R., Conroy, S., Dockery, P., Duffy, G., Daly, K., Crowley, J., Bartlett, J.S., and O'Brien, T. 2006. Gene-eluting stents: Comparison of adenoviral and adeno-associated viral gene delivery to the blood vessel wall in vivo. *Hum Gene Ther 17*, 741–750.

122. Nakayama, K., Ishida, N., Shirane, M., Inomata, A., Inoue, T., Shishido, N., Horii, I., and Loh, D.Y. 1996. Mice lacking p27(Kip1) display increased body size, multiple organ hyperplasia, retinal dysplasia, and pituitary tumors. *Cell 85*, 707–720.

123. Marx, S.O. and Marks, A.R. 2001. Bench to bedside: The development of rapamycin and its application to stent restenosis. *Circulation 104*, 852–855.

124. Chang, M.W., Barr, E., Lu, M.M., Barton, K., and Leiden, J.M. 1995. Adenovirus-mediated over-expression of the cyclin/cyclin-dependent kinase inhibitor, p21 inhibits vascular smooth muscle cell proliferation and neointima formation in the rat carotid artery model of balloon angioplasty. *J Clin Invest 96*, 2260–2268.

125. Lee, E.Y., Chang, C.Y., Hu, N., Wang, Y.C., Lai, C.C., Herrup, K., Lee, W.H., and Bradley, A. 1992. Mice deficient for Rb are nonviable and show defects in neurogenesis and haematopoiesis. *Nature 359*, 288–294.

126. Bignon, Y.J., Chen, Y., Chang, C.Y., Riley, D.J., Windle, J.J., Mellon, P.L., and Lee, W.H. 1993. Expression of a retinoblastoma transgene results in dwarf mice. *Genes Dev 7*, 1654–1662.

127. Morishita, R., Gibbons, G.H., Horiuchi, M., Ellison, K.E., Nakama, M., Zhang, L., Kaneda, Y., Ogihara, T., and Dzau, V.J. 1995. A gene therapy strategy using a transcription factor decoy of the E2F binding site inhibits smooth muscle proliferation in vivo. *Proc Natl Acad Sci USA 92*, 5855–5859.

128. Kutryk, M.J., Foley, D.P., van den Brand, M., Hamburger, J.N., van der Giessen, W.J., deFeyter, P.J., Bruining, N., Sabate, M., and Serruys, P.W. 2002. Local intracoronary administration of antisense oligonucleotide against c-myc for the prevention of in-stent restenosis: Results of the randomized investigation by the thoraxcenter of antisense DNA using local delivery and IVUS after coronary stenting (ITALICS) trial. *J Am Coll Cardiol 39*, 281–287.

129. Turunen, P., Puhakka, H.L., Heikura, T., Romppanen, E., Inkala, M., Leppanen, O., and Yla-Herttuala, S. 2006. Extracellular superoxide dismutase with vaccinia virus anti-inflammatory protein 35K or tissue inhibitor of metalloproteinase-1: Combination gene therapy in the treatment of vein graft stenosis in rabbits. *Hum Gene Ther 17*, 405–414.

130. Akowuah, E.F., Gray, C., Lawrie, A., Sheridan, P.J., Su, C.H., Bettinger, T., Brisken, A.F., Gunn, J., Crossman, D.C., Francis, S.E., Baker, A.H., and Newman, C.M. 2005. Ultrasound-mediated delivery of TIMP-3 plasmid DNA into saphenous vein leads to increased lumen size in a porcine interposition graft model. *Gene Ther 12*, 1154–1157.

131. Petrofski, J.A., Hata, J.A., Gehrig, T.R., Hanish, S.I., Williams, M.L., Thompson, R.B., Parsa, C.J., Koch, W.J., and Milano, C.A. 2004. Gene delivery to aortocoronary saphenous vein grafts in a large animal model of intimal hyperplasia. *J Thorac Cardiovasc Surg 127*, 27–33.

132. Wolff, R.A., Ryomoto, M., Stark, V.E., Malinowski, R., Tomas, J.J., Stinauer, M.A., Hullett, D.A., and Hoch, J.R. 2005. Antisense to transforming growth factor-beta1 messenger RNA reduces vein graft intimal hyperplasia and monocyte chemotactic protein 1. *J Vasc Surg 41*, 498–508.

133. Ahanchi, S.S., Tsihlis, N.D., and Kibbe, M.R. 2007. The role of nitric oxide in the pathophysiology of intimal hyperplasia. *J Vasc Surg 45 Suppl A*, A64–A73.

134. West, N.E., Qian, H., Guzik, T.J., Black, E., Cai, S., George, S.E., and Channon, K.M. 2001. Nitric oxide synthase (nNOS) gene transfer modifies venous bypass graft remodeling: Effects on vascular smooth muscle cell differentiation and superoxide production. *Circulation 104*, 1526–1532.

135. Cooney, R., Hynes, S.O., Sharif, F., Howard, L., and O'Brien, T. 2007. Effect of gene delivery of NOS isoforms on intimal hyperplasia and endothelial regeneration after balloon injury. *Gene Ther 14*, 396–404.

136. George, S.J., Angelini, G.D., Capogrossi, M.C., and Baker, A.H. 2001. Wild-type p53 gene transfer inhibits neointima formation in human saphenous vein by modulation of smooth muscle cell migration and induction of apoptosis. *Gene Ther 8*, 668–676.

137. Wan, S., George, S.J., Nicklin, S.A., Yim, A.P., and Baker, A.H. 2004. Overexpression of p53 increases lumen size and blocks neointima formation in porcine interposition vein grafts. *Mol Ther 9*, 689–698.

138. Eslami, M.H., Gangadharan, S.P., Sui, X., Rhynhart, K.K., Snyder, R.O., and Conte, M.S. 2000. Gene delivery to in situ veins: Differential effects of adenovirus and adeno-associated viral vectors. *J Vasc Surg 31*, 1149–1159.

139. Mann, M.J., Whittemore, A.D., Donaldson, M.C., Belkin, M., Conte, M.S., Polak, J.F., Orav, E.J., Ehsan, A., Dell'Acqua, G., and Dzau, V.J. 1999. *Ex-vivo* gene therapy of human vascular bypass grafts with E2F decoy: The PREVENT single-centre, randomised, controlled trial. *Lancet 354*, 1493–1498.

140. Alexander, J.H., Ferguson, T.B., Jr., Joseph, D.M., Mack, M.J., Wolf, R.K., Gibson, C.M., Gennevois, D., Lorenz, T.J., Harrington, R.A., Peterson, E.D., Lee, K.L., Califf, R.M., and Kouchoukos, N.T. 2005. The project of *ex-vivo* vein graft engineering via transfection IV (PREVENT IV) trial: Study rationale, design, and baseline patient characteristics. *Am Heart J 150*, 643–649.

141. Conte, M.S., Bandyk, D.F., Clowes, A.W., Moneta, G.L., Seely, L., Lorenz, T.J., Namini, H., Hamdan, A.D., Roddy, S.P., Belkin, M., Berceli, S.A., DeMasi, R.J., Samson, R.H., and Berman, S.S. 2006. Results of PREVENT III: A multi-center, randomized trial of edifoligide for the prevention of vein graft failure in lower extremity bypass surgery. *J Vasc Surg 43*, 742–751; discussion 751.

142. Banno, H., Takei, Y., Muramatsu, T., Komori, K., and Kadomatsu, K. 2006. Controlled release of small interfering RNA targeting midkine attenuates intimal hyperplasia in vein grafts. *J Vasc Surg 44*, 633–641.

143. Folkman, J., Merler, E., Abernathy, C., and Williams, G. 1971. Isolation of a tumor factor responsible for angiogenesis. *J Exp Med 133*, 275–288.

144. Carmeliet, P. 2000. Mechanisms of angiogenesis and arteriogenesis. *Nat Med 6*, 389–395.

145. Ferrara, N. 2004. Vascular endothelial growth factor: Basic science and clinical progress. *Endocr Rev 25*, 581–611.

146. Yla-Herttuala, S., Rissanen, T.T., Vajanto, I., and Hartikainen, J. 2007. Vascular endothelial growth factors: Biology and current status of clinical applications in cardiovascular medicine. *J Am Coll Cardiol 49*, 1015–1026.

147. Pipp, F., Heil, M., Issbrucker, K., Ziegelhoeffer, T., Martin, S., van den Heuvel, J., Weich, H., Fernandez, B., Golomb, G., Carmeliet, P., Schaper, W., and Clauss, M. 2003. VEGFR-1-selective VEGF homologue PlGF is arteriogenic: Evidence for a monocyte-mediated mechanism. *Circ Res 92*, 378–385.

148. Babiak, A., Schumm, A.M., Wangler, C., Loukas, M., Wu, J., Dombrowski, S., Matuschek, C., Kotzerke, J., Dehio, C., and Waltenberger, J. 2004. Coordinated activation of VEGFR-1 and VEGFR-2 is a potent arteriogenic stimulus leading to enhancement of regional perfusion. *Cardiovasc Res 61*, 789–795.

149. Thurston, G., Rudge, J.S., Ioffe, E., Zhou, H., Ross, L., Croll, S.D., Glazer, N., Holash, J., McDonald, D.M., and Yancopoulos, G.D. 2000. Angiopoietin-1 protects the adult vasculature against plasma leakage. *Nat Med 6*, 460–463.

150. Resnick, N., Einav, S., Chen-Konak, L., Zilberman, M., Yahav, H., and Shay-Salit, A. 2003. Hemodynamic forces as a stimulus for arteriogenesis. *Endothelium 10*, 197–206.

151. Helisch, A. and Schaper, W. 2003. Arteriogenesis: The development and growth of collateral arteries. *Microcirculation 10*, 83–97.

152. Horowitz, A., Tkachenko, E., and Simons, M. 2002. Fibroblast growth factor-specific modulation of cellular response by syndecan-4. *J Cell Biol 157*, 715–725.

153. Cao, R., Brakenhielm, E., Pawliuk, R., Wariaro, D., Post, M.J., Wahlberg, E., Leboulch, P., and Cao, Y. 2003. Angiogenic synergism, vascular stability and improvement of hind-limb ischemia by a combination of PDGF-BB and FGF-2. *Nat Med 9*, 604–613.

154. Eguchi, M., Masuda, H., and Asahara, T. 2007. Endothelial progenitor cells for postnatal vasculogenesis. *Clin Exp Nephrol 11*, 18–25.

155. Shi, Q., Rafii, S., Wu, M.H., Wijelath, E.S., Yu, C., Ishida, A., Fujita, Y., Kothari, S., Mohle, R., Sauvage, L.R., Moore, M.A., Storb, R.F., and Hammond, W.P. 1998. Evidence for circulating bone marrow-derived endothelial cells. *Blood 92*, 362–367.

156. Gill, M., Dias, S., Hattori, K., Rivera, M.L., Hicklin, D., Witte, L., Girardi, L., Yurt, R., Himel, H., and Rafii, S. 2001. Vascular trauma induces rapid but transient mobilization of VEGFR2(+)AC133(+) endothelial precursor cells. *Circ Res 88*, 167–174.

157. Asahara, T., Masuda, H., Takahashi, T., Kalka, C., Pastore, C., Silver, M., Kearne, M., Magner, M., and Isner, J.M. 1999. Bone marrow origin of endothelial progenitor cells responsible for postnatal vasculogenesis in physiological and pathological neovascularization. *Circ Res 85*, 221–228.

158. Shintani, S., Murohara, T., Ikeda, H., Ueno, T., Honma, T., Katoh, A., Sasaki, K., Shimada, T., Oike, Y., and Imaizumi, T. 2001. Mobilization of endothelial progenitor cells in patients with acute myocardial infarction. *Circulation 103*, 2776–2779.

159. Bhattacharya, V., McSweeney, P.A., Shi, Q., Bruno, B., Ishida, A., Nash, R., Storb, R.F., Sauvage, L.R., Hammond, W.P., and Wu, M.H. 2000. Enhanced endothelialization and microvessel formation in polyester grafts seeded with CD34(+) bone marrow cells. *Blood 95*, 581–585.

160. Kaushal, S., Amiel, G.E., Guleserian, K.J., Shapira, O.M., Perry, T., Sutherland, F.W., Rabkin, E., Moran, A.M., Schoen, F.J., Atala, A., Soker, S., Bischoff, J., and Mayer, J.E., Jr. 2001. Functional small-diameter neovessels created using endothelial progenitor cells expanded *ex vivo*. *Nat Med 7*, 1035–1040.

161. Takeshita, S., Zheng, L.P., Brogi, E., Kearney, M., Pu, L.Q., Bunting, S., Ferrara, N., Symes, J.F., and Isner, J.M. 1994. Therapeutic angiogenesis. A single intraarterial bolus of vascular endothelial growth factor augments revascularization in a rabbit ischemic hind limb model. *J Clin Invest 93*, 662–670.

162. Taniyama, Y., Morishita, R., Hiraoka, K., Aoki, M., Nakagami, H., Yamasaki, K., Matsumoto, K., Nakamura, T., Kaneda, Y., and Ogihara, T. 2001. Therapeutic angiogenesis induced by human hepatocyte growth factor gene in rat diabetic hind limb ischemia model: Molecular mechanisms of delayed angiogenesis in diabetes. *Circulation 104*, 2344–2350.

163. Pipp, F., Boehm, S., Cai, W.J., Adili, F., Ziegler, B., Karanovic, G., Ritter, R., Balzer, J., Scheler, C., Schaper, W., and Schmitz-Rixen, T. 2004. Elevated fluid shear stress enhances postocclusive collateral artery growth and gene expression in the pig hind limb. *Arterioscler Thromb Vasc Biol 24*, 1664–1668.

164. Xie, D., Li, Y., Reed, E.A., Odronic, S.I., Kontos, C.D., and Annex, B.H. 2006. An engineered vascular endothelial growth factor-activating transcription factor induces therapeutic angiogenesis in ApoE knockout mice with hindlimb ischemia. *J Vasc Surg 44*, 166–175.

165. Tsurumi, Y., Takeshita, S., Chen, D., Kearney, M., Rossow, S.T., Passeri, J., Horowitz, J.R., Symes, J.F., and Isner, J.M. 1996. Direct intramuscular gene transfer of naked DNA encoding vascular endothelial growth factor augments collateral development and tissue perfusion. *Circulation 94*, 3281–3290.

166. Rivard, A., Silver, M., Chen, D., Kearney, M., Magner, M., Annex, B., Peters, K., and Isner, J.M. 1999. Rescue of diabetes-related impairment of angiogenesis by intramuscular gene therapy with adeno-VEGF. *Am J Pathol 154*, 355–363.

167. Vajanto, I., Rissanen, T.T., Rutanen, J., Hiltunen, M.O., Tuomisto, T.T., Arve, K., Narvanen, O., Manninen, H., Rasanen, H., Hippelainen, M., Alhava, E., and Yla-Herttuala, S. 2002. Evaluation of angiogenesis and side effects in ischemic rabbit hindlimbs after intramuscular injection of adenoviral vectors encoding VEGF and LacZ. *J Gene Med 4*, 371–380.

168. Kondoh, K., Koyama, H., Miyata, T., Takato, T., Hamada, H., and Shigematsu, H. 2004. Conduction performance of collateral vessels induced by vascular endothelial growth factor or basic fibroblast growth factor. *Cardiovasc Res 61*, 132–142.

169. Niagara, M.I., Haider, H., Ye, L., Koh, V.S., Lim, Y.T., Poh, K.K., Ge, R., and Sim, E.K. 2004. Autologous skeletal myoblasts transduced with a new adenoviral bicistronic vector for treatment of hind limb ischemia. *J Vasc Surg 40*, 774–785.

170. Whitlock, P.R., Hackett, N.R., Leopold, P.L., Rosengart, T.K., and Crystal, R.G. 2004. Adenovirus-mediated transfer of a minigene expressing multiple isoforms of VEGF is more effective at inducing angiogenesis than comparable vectors expressing individual VEGF cDNAs. *Mol Ther 9*, 67–75.

171. Yu, J., Lei, L., Liang, Y., Hinh, L., Hickey, R.P., Huang, Y., Liu, D., Yeh, J.L., Rebar, E., Case, C., Spratt, K., Sessa, W.C., and Giordano, F.J. 2006. An engineered VEGF-activating zinc finger protein transcription factor improves blood flow and limb salvage in advanced-age mice. *Faseb J 20*, 479–481.

172. Gaffney, M.M., Hynes, S.O., Barry, F., and O'Brien, T. 2007. Cardiovascular gene therapy: Current status and therapeutic potential. *Br J Pharmacol 152*, 175–188.

173. Makinen, K., Manninen, H., Hedman, M., Matsi, P., Mussalo, H., Alhava, E., and Yla-Herttuala, S. 2002. Increased vascularity detected by digital subtraction angiography after VEGF gene transfer to human lower limb artery: a randomized, placebo-controlled, double-blinded phase II study. *Mol Ther 6*, 127–133.

174. Kusumanto, Y.H., van Weel, V., Mulder, N.H., Smit, A.J., van den Dungen, J.J., Hooymans, J.M., Sluiter, W.J., Tio, R.A., Quax, P.H., Gans, R.O., Dullaart, R.P., and Hospers, G.A. 2006. Treatment with intramuscular vascular endothelial growth factor gene compared with placebo for patients with diabetes mellitus and critical limb ischemia: A double-blind randomized trial. *Hum Gene Ther 17*, 683–691.

175. Rajagopalan, S., Mohler, E.R., 3rd, Lederman, R.J., Mendelsohn, F.O., Saucedo, J.F., Goldman, C.K., Blebea, J., Macko, J., Kessler, P.D., Rasmussen, H.S., and Annex, B.H. 2003. Regional angiogenesis with vascular endothelial growth factor in peripheral arterial disease: A phase II randomized, double-blind, controlled study of adenoviral delivery of vascular endothelial growth factor 121 in patients with disabling intermittent claudication. *Circulation 108*, 1933–1938.

176. Stewart, D.J., Hilton, J.D., Arnold, J.M., Gregoire, J., Rivard, A., Archer, S.L., Charbonneau, F., Cohen, E., Curtis, M., Buller, C.E., Mendelsohn, F.O., Dib, N., Page, P., Ducas, J., Plante, S., Sullivan, J., Macko, J., Rasmussen, C., Kessler, P.D., and Rasmussen, H.S. 2006. Angiogenic gene therapy in patients with nonrevascularizable ischemic heart disease: A phase 2 randomized, controlled trial of AdVEGF(121) (AdVEGF121) versus maximum medical treatment. *Gene Ther 13*, 1503–1511.

177. Hedman, M., Hartikainen, J., Syvanne, M., Stjernvall, J., Hedman, A., Kivela, A., Vanninen, E., Mussalo, H., Kauppila, E., Simula, S., Narvanen, O., Rantala, A., Peuhkurinen, K., Nieminen, M.S., Laakso, M., and Yla-Herttuala, S. 2003. Safety and feasibility of catheter-based local intracoronary vascular endothelial growth factor gene transfer in the prevention of postangioplasty and in-stent restenosis and in the treatment of chronic myocardial ischemia: Phase II results of the Kuopio Angiogenesis Trial (KAT). *Circulation 107*, 2677–2683.

178. Grines, C.L., Watkins, M.W., Mahmarian, J.J., Iskandrian, A.E., Rade, J.J., Marrott, P., Pratt, C., and Kleiman, N. 2003. A randomized, double-blind, placebo-controlled trial of Ad5FGF-4 gene therapy and its effect on myocardial perfusion in patients with stable angina. *J Am Coll Cardiol 42*, 1339–1347.

179. Kastrup, J., Jorgensen, E., Ruck, A., Tagil, K., Glogar, D., Ruzyllo, W., Botker, H.E., Dudek, D., Drvota, V., Hesse, B., Thuesen, L., Blomberg, P., Gyongyosi, M., and Sylven, C. 2005. Direct intramyocardial plasmid vascular endothelial growth factor-A165 gene therapy in patients with stable severe angina pectoris a randomized double-blind placebo-controlled study: The Euroinject One trial. *J Am Coll Cardiol 45*, 982–988.

180. Chobanian, A.V., Bakris, G.L., Black, H.R., Cushman, W.C., Green, L.A., Izzo, J.L., Jr., Jones, D.W., Materson, B.J., Oparil, S., Wright, J.T., Jr., and Roccella, E.J. 2003. The seventh report of the joint national committee on prevention, eetection, evaluation, and treatment of high blood pressure: the JNC 7 report. *Jama 289*, 2560–2572.

181. Israili, Z.H., Hernandez-Hernandez, R., and Valasco, M. 2007. The future of antihypertensive treatment. *Am J Ther 14*, 121–134.

182. Phillips, M.I. and Kimura, B. 2005. Gene therapy for hypertension: Antisense inhibition of the renin-angiotensin system. *Methods Mol Med 108*, 363–379.

183. Pachori, A.S., Numan, M.T., Ferrario, C.M., Diz, D.M., Raizada, M.K., and Katovich, M.J. 2002. Blood pressure-independent attenuation of cardiac hypertrophy by AT(1)R-AS gene therapy. *Hypertension 39*, 969–975.

184. Reaves, P.Y., Beck, C.R., Wang, H.W., Raizada, M.K., and Katovich, M.J. 2003. Endothelial-independent prevention of high blood pressure in L-NAME-treated rats by angiotensin II type I receptor antisense gene therapy. *Exp Physiol 88*, 467–473.

185. Kumai, T., Tateishi, T., Tanaka, M., Watanabe, M., Shimizu, H., and Kobayashi, S. 2001. Tyrosine hydroxylase antisense gene therapy causes hypotensive effects in the spontaneously hypertensive rats. *J Hypertens 19*, 1769–1773.

186. McIntyre, M., Bohr, D.F., and Dominiczak, A.F. 1999. Endothelial function in hypertension: The role of superoxide anion. *Hypertension 34*, 539–545.

187. Benkusky, N.A., Lewis, S.J., and Kooy, N.W. 1998. Attenuation of vascular relaxation after development of tachyphylaxis to peroxynitrite in vivo. *Am J Physiol 275*, H501–H508.

188. Chu, Y., Iida, S., Lund, D.D., Weiss, R.M., DiBona, G.F., Watanabe, Y., Faraci, F.M., and Heistad, D.D. 2003. Gene transfer of extracellular superoxide dismutase reduces arterial pressure in spontaneously hypertensive rats: Role of heparin-binding domain. *Circ Res 92*, 461–468.

189. D'Souza, S.P., Davis, M., and Baxter, G.F. 2004. Autocrine and paracrine actions of natriuretic peptides in the heart. *Pharmacol Ther 101*, 113–129.

190. Cusson, J.R., Thibault, G., Cantin, M., and Larochelle, P. 1990. Prolonged low dose infusion of atrial natriuretic factor in essential hypertension. *Clin Exp Hypertens A 12*, 111–135.

191. Lin, K.F., Chao, J., and Chao, L. 1995. Human atrial natriuretic peptide gene delivery reduces blood pressure in hypertensive rats. *Hypertension 26*, 847–853.

192. Lin, K.F., Chao, J., and Chao, L. 1999. Atrial natriuretic peptide gene delivery reduces stroke-induced mortality rate in dahl salt-sensitive rats. *Hypertension 33*, 219–224.

193. Schillinger, K.J., Tsai, S.Y., Taffet, G.E., Reddy, A.K., Marian, A.J., Entman, M.L., Oka, K., Chan, L., and O'Malley, B.W. 2005. Regulatable atrial natriuretic peptide gene therapy for hypertension. *Proc Natl Acad Sci USA 102*, 13789–13794.

194. Miller, W.H., Brosnan, M.J., Graham, D., Nicol, C.G., Morecroft, I., Channon, K.M., Danilov, S.M., Reynolds, P.N., Baker, A.H., and Dominiczak, A.F. 2005. Targeting endothelial cells with adenovirus expressing nitric oxide synthase prevents elevation of blood pressure in stroke-prone spontaneously hypertensive rats. *Mol Ther 12*, 321–327.

195. Wang, T., Li, H., Zhao, C., Chen, C., Li, J., Chao, J., Chao, L., Xiao, X., and Wang, D.W. 2004. Recombinant adeno-associated virus-mediated kallikrein gene therapy reduces hypertension and attenuates its cardiovascular injuries. *Gene Ther 11*, 1342–1350.

196. Rich, S., Dantzker, D.R., Ayres, S.M., Bergofsky, E.H., Brundage, B.H., Detre, K.M., Fishman, A.P., Goldring, R.M., Groves, B.M., Koerner, S.K., and et al. 1987. Primary pulmonary hypertension. A national prospective study. *Ann Intern Med 107*, 216–223.

197. Christman, B.W., McPherson, C.D., Newman, J.H., King, G.A., Bernard, G.R., Groves, B.M., and Loyd, J.E. 1992. An imbalance between the excretion of thromboxane and prostacyclin metabolites in pulmonary hypertension. *N Engl J Med 327*, 70–75.

198. Nagaya, N., Yokoyama, C., Kyotani, S., Shimonishi, M., Morishita, R., Uematsu, M., Nishikimi, T., Nakanishi, N., Ogihara, T., Yamagishi, M., Miyatake, K., Kaneda, Y., and Tanabe, T. 2000. Gene transfer of human prostacyclin synthase ameliorates monocrotaline-induced pulmonary hypertension in rats. *Circulation 102*, 2005–2010.

199. Blanc-Brude, O.P., Yu, J., Simosa, H., Conte, M.S., Sessa, W.C., and Altieri, D.C. 2002. Inhibitor of apoptosis protein survivin regulates vascular injury. *Nat Med 8*, 987–994.

200. McMurtry, M.S., Archer, S.L., Altieri, D.C., Bonnet, S., Haromy, A., Harry, G., Puttagunta, L., and Michelakis, E.D. 2005. Gene therapy targeting survivin selectively induces pulmonary vascular apoptosis and reverses pulmonary arterial hypertension. *J Clin Invest 115*, 1479–1491.

201. Heit, J.A. 2002. Venous thromboembolism epidemiology: Implications for prevention and management. *Semin Thromb Hemost 28 Suppl 2*, 3–13.

202. Zoldhelyi, P., McNatt, J., Xu, X.M., Loose-Mitchell, D., Meidell, R.S., Clubb, F.J., Jr., Buja, L.M., Willerson, J.T., and Wu, K.K. 1996. Prevention of arterial thrombosis by adenovirus-mediated transfer of cyclooxygenase gene. *Circulation 93*, 10–17.

203. Waugh, J.M., Kattash, M., Li, J., Yuksel, E., Kuo, M.D., Lussier, M., Weinfeld, A.B., Saxena, R., Rabinovsky, E.D., Thung, S., Woo, S.L., and Shenaq, S.M. 1999. Gene therapy to promote thromboresistance: Local overexpression of tissue plasminogen activator to prevent arterial thrombosis in an in vivo rabbit model. *Proc Natl Acad Sci USA 96*, 1065–1070.

204. Waugh, J.M., Yuksel, E., Li, J., Kuo, M.D., Kattash, M., Saxena, R., Geske, R., Thung, S.N., Shenaq, S.M., and Woo, S.L. 1999. Local overexpression of thrombomodulin for in vivo prevention of arterial thrombosis in a rabbit model. *Circ Res 84*, 84–92.

205. Zoldhelyi, P., McNatt, J., Shelat, H.S., Yamamoto, Y., Chen, Z.Q., and Willerson, J.T. 2000. Thromboresistance of balloon-injured porcine carotid arteries after local gene transfer of human tissue factor pathway inhibitor. *Circulation 101*, 289–295.

206. Ohno, N., Itoh, H., Ikeda, T., Ueyama, K., Yamahara, K., Doi, K., Yamashita, J., Inoue, M., Masatsugu, K., Sawada, N., Fukunaga, Y., Sakaguchi, S., Sone, M., Yurugi, T., Kook, H., Komeda, M., and Nakao, K. 2002. Accelerated reendothelialization with suppressed thrombogenic property and neointimal hyperplasia of rabbit jugular vein grafts by adenovirus-mediated gene transfer of C-type natriuretic peptide. *Circulation 105*, 1623–1626.

207. Furukoji, E., Matsumoto, M., Yamashita, A., Yagi, H., Sakurai, Y., Marutsuka, K., Hatakeyama, K., Morishita, K., Fujimura, Y., Tamura, S., and Asada, Y. 2005. Adenovirus-mediated transfer of human placental ectonucleoside triphosphate diphosphohydrolase to vascular smooth muscle cells suppresses platelet aggregation in vitro and arterial thrombus formation in vivo. *Circulation 111*, 808–815.

208. Waltham, M., Burnand, K., Fenske, C., Modarai, B., Humphries, J., and Smith, A. 2005. Vascular endothelial growth factor naked DNA gene transfer enhances thrombus recanalization and resolution. *J Vasc Surg 42*, 1183–1189.

209. Gwathmey, J.K., Slawsky, M.T., Hajjar, R.J., Briggs, G.M., and Morgan, J.P. 1990. Role of intracellular calcium handling in force-interval relationships of human ventricular myocardium. *J Clin Invest 85*, 1599–1613.

210. Miyamoto, M.I., del Monte, F., Schmidt, U., DiSalvo, T.S., Kang, Z.B., Matsui, T., Guerrero, J.L., Gwathmey, J.K., Rosenzweig, A., and Hajjar, R.J. 2000. Adenoviral gene transfer of SERCA2a improves left-ventricular function in aortic-banded rats in transition to heart failure. *Proc Natl Acad Sci USA 97*, 793–798.

211. Szatkowski, M.L., Westfall, M.V., Gomez, C.A., Wahr, P.A., Michele, D.E., DelloRusso, C., Turner, II, Hong, K.E., Albayya, F.P., and Metzger, J.M. 2001. In vivo acceleration of heart relaxation performance by parvalbumin gene delivery. *J Clin Invest 107*, 191–198.

212. Hoshijima, M., Ikeda, Y., Iwanaga, Y., Minamisawa, S., Date, M.O., Gu, Y., Iwatate, M., Li, M., Wang, L., Wilson, J.M., Wang, Y., Ross, J., Jr., and Chien, K.R. 2002. Chronic suppression of heart-failure progression by a pseudophosphorylated mutant of phospholamban via in vivo cardiac rAAV gene delivery. *Nat Med 8*, 864–871.

213. Most, P., Pleger, S.T., Volkers, M., Heidt, B., Boerries, M., Weichenhan, D., Loffler, E., Janssen, P.M., Eckhart, A.D., Martini, J., Williams, M.L., Katus, H.A., Remppis, A., and Koch, W.J. 2004. Cardiac adenoviral S100A1 gene delivery rescues failing myocardium. *J Clin Invest 114*, 1550–1563.

214. Maurice, J.P., Hata, J.A., Shah, A.S., White, D.C., McDonald, P.H., Dolber, P.C., Wilson, K.H., Lefkowitz, R.J., Glower, D.D., and Koch, W.J. 1999. Enhancement of cardiac function after adenoviral-mediated in vivo intracoronary beta2-adrenergic receptor gene delivery. *J Clin Invest 104*, 21–29.

215. Molkentin, J.D. and Dorn, I.G., 2nd. 2001. Cytoplasmic signaling pathways that regulate cardiac hypertrophy. *Annu Rev Physiol 63*, 391–426.

216. Hambleton, M., Hahn, H., Pleger, S.T., Kuhn, M.C., Klevitsky, R., Carr, A.N., Kimball, T.F., Hewett, T.E., Dorn, G.W., 2nd, Koch, W.J., and Molkentin, J.D. 2006. Pharmacological- and gene therapy-based inhibition of protein kinase Calpha/beta enhances cardiac contractility and attenuates heart failure. *Circulation 114*, 574–582.

217. Tenhunen, O., Soini, Y., Ilves, M., Rysa, J., Tuukkanen, J., Serpi, R., Pennanen, H., Ruskoaho, H., and Leskinen, H. 2006. p38 Kinase rescues failing myocardium after myocardial infarction: Evidence for angiogenic and anti-apoptotic mechanisms. *Faseb J 20*, 1907–1909.

218. Laugwitz, K.L., Moretti, A., Weig, H.J., Gillitzer, A., Pinkernell, K., Ott, T., Pragst, I., Stadele, C., Seyfarth, M., Schomig, A., and Ungerer, M. 2001. Blocking caspase-activated apoptosis improves contractility in failing myocardium. *Hum Gene Ther 12*, 2051–2063.

219. Melo, L.G., Agrawal, R., Zhang, L., Rezvani, M., Mangi, A.A., Ehsan, A., Griese, D.P., Dell'Acqua, G., Mann, M.J., Oyama, J., Yet, S.F., Layne, M.D., Perrella, M.A., and Dzau, V.J. 2002. Gene therapy strategy for long-term myocardial protection using adeno-associated virus-mediated delivery of heme oxygenase gene. *Circulation 105*, 602–607.

220. Leotta, E., Patejunas, G., Murphy, G., Szokol, J., McGregor, L., Carbray, J., Hamawy, A., Winchester, D., Hackett, N., Crystal, R., and Rosengart, T. 2002. Gene therapy with adenovirus-mediated myocardial transfer of vascular endothelial growth factor 121 improves cardiac performance in a pacing model of congestive heart failure. *J Thorac Cardiovasc Surg 123*, 1101–1113.

221. Rosen, M.R., Brink, P.R., Cohen, I.S., and Robinson, R.B. 2004. Genes, stem cells and biological pacemakers. *Cardiovasc Res 64*, 12–23.

222. Bucchi, A., Plotnikov, A.N., Shlapakova, I., Danilo, P., Jr., Kryukova, Y., Qu, J., Lu, Z., Liu, H., Pan, Z., Potapova, I., KenKnight, B., Girouard, S., Cohen, I.S., Brink, P.R., Robinson, R.B., and Rosen, M.R. 2006. Wild-type and mutant HCN channels in a tandem biological-electronic cardiac pacemaker. *Circulation 114*, 992–999.

223. Tse, H.F., Xue, T., Lau, C.P., Siu, C.W., Wang, K., Zhang, Q.Y., Tomaselli, G.F., Akar, F.G., and Li, R.A. 2006. Bioartificial sinus node constructed via in vivo gene transfer of an engineered pacemaker HCN channel reduces the dependence on electronic pacemaker in a sick-sinus syndrome model. *Circulation 114*, 1000–1011.

224. Kashiwakura, Y., Cho, H.C., Barth, A.S., Azene, E., and Marban, E. 2006. Gene transfer of a synthetic pacemaker channel into the heart: A novel strategy for biological pacing. *Circulation 114*, 1682–1686.

225. De Muinck, E.D., Thompson, C., and Simons, M. 2006. Progress and prospects: Cell based regenerative therapy for cardiovascular disease. *Gene Ther 13*, 659–671.

226. Badorff, C., Brandes, R.P., Popp, R., Rupp, S., Urbich, C., Aicher, A., Fleming, I., Busse, R., Zeiher, A.M., and Dimmeler, S. 2003. Transdifferentiation of blood-derived human adult endothelial progenitor cells into functionally active cardiomyocytes. *Circulation 107*, 1024–1032.

227. Deb, A., Wang, S., Skelding, K.A., Miller, D., Simper, D., and Caplice, N.M. 2003. Bone marrow-derived cardiomyocytes are present in adult human heart: A study of gender-mismatched bone marrow transplantation patients. *Circulation 107*, 1247–1249.

228. Kajstura, J., Rota, M., Whang, B., Cascapera, S., Hosoda, T., Bearzi, C., Nurzynska, D., Kasahara, H., Zias, E., Bonafe, M., Nadal-Ginard, B., Torella, D., Nascimbene, A., Quaini, F., Urbanek, K., Leri, A., and Anversa, P. 2005. Bone marrow cells differentiate in cardiac cell lineages after infarction independently of cell fusion. *Circ Res 96*, 127–137.

229. Balsam, L.B., Wagers, A.J., Christensen, J.L., Kofidis, T., Weissman, I.L., and Robbins, R.C. 2004. Haematopoietic stem cells adopt mature haematopoietic fates in ischaemic myocardium. *Nature 428*, 668–673.

230. Wagers, A.J., Sherwood, R.I., Christensen, J.L., and Weissman, I.L. 2002. Little evidence for developmental plasticity of adult hematopoietic stem cells. *Science 297*, 2256–2259.

231. Murry, C.E., Soonpaa, M.H., Reinecke, H., Nakajima, H., Nakajima, H.O., Rubart, M., Pasumarthi, K.B., Virag, J.I., Bartelmez, S.H., Poppa, V., Bradford, G., Dowell, J.D., Williams, D.A., and Field, L.J. 2004. Haematopoietic stem cells do not transdifferentiate into cardiac myocytes in myocardial infarcts. *Nature 428*, 664–668.

232. Reinecke, H., Minami, E., Poppa, V., and Murry, C.E. 2004. Evidence for fusion between cardiac and skeletal muscle cells. *Circ Res 94*, e56–e60.

233. Nygren, J.M., Jovinge, S., Breitbach, M., Sawen, P., Roll, W., Hescheler, J., Taneera, J., Fleischmann, B.K., and Jacobsen, S.E. 2004. Bone marrow-derived hematopoietic cells generate cardiomyocytes at a low frequency through cell fusion, but not transdifferentiation. *Nat Med 10*, 494–501.

234. Terada, N., Hamazaki, T., Oka, M., Hoki, M., Mastalerz, D.M., Nakano, Y., Meyer, E.M., Morel, L., Petersen, B.E., and Scott, E.W. 2002. Bone marrow cells adopt the phenotype of other cells by spontaneous cell fusion. *Nature 416*, 542–545.

235. Rubart, M., Soonpaa, M.H., Nakajima, H., and Field, L.J. 2004. Spontaneous and evoked intracellular calcium transients in donor-derived myocytes following intracardiac myoblast transplantation. *J Clin Invest 114*, 775–783.

236. Rehman, J., Li, J., Orschell, C.M., and March, K.L. 2003. Peripheral blood "endothelial progenitor cells" are derived from monocyte/macrophages and secrete angiogenic growth factors. *Circulation 107*, 1164–1169.

237. Kinnaird, T., Stabile, E., Burnett, M.S., and Epstein, S.E. 2004. Bone-marrow-derived cells for enhancing collateral development: Mechanisms, animal data, and initial clinical experiences. *Circ Res 95*, 354–363.

238. Yoon, C.H., Hur, J., Park, K.W., Kim, J.H., Lee, C.S., Oh, I.Y., Kim, T.Y., Cho, H.J., Kang, H.J., Chae, I.H., Yang, H.K., Oh, B.H., Park, Y.B., and Kim, H.S. 2005. Synergistic neovascularization by mixed transplantation of early endothelial progenitor cells and late outgrowth endothelial cells: The role of angiogenic cytokines and matrix metalloproteinases. *Circulation 112*, 1618–1627.

239. Kupatt, C., Horstkotte, J., Vlastos, G.A., Pfosser, A., Lebherz, C., Semisch, M., Thalgott, M., Buttner, K., Browarzyk, C., Mages, J., Hoffmann, R., Deten, A., Lamparter, M., Muller, F., Beck, H., Buning, H., Boekstegers, P., and Hatzopoulos, A.K. 2005. Embryonic endothelial progenitor cells expressing a broad range of proangiogenic and remodeling factors enhance vascularization and tissue recovery in acute and chronic ischemia. *Faseb J 19*, 1576–1578.

240. Gnecchi, M., He, H., Liang, O.D., Melo, L.G., Morello, F., Mu, H., Noiseux, N., Zhang, L., Pratt, R.E., Ingwall, J.S., and Dzau, V.J. 2005. Paracrine action accounts for marked protection of ischemic heart by Akt-modified mesenchymal stem cells. *Nat Med 11*, 367–368.

241. Hofmann, M., Wollert, K.C., Meyer, G.P., Menke, A., Arseniev, L., Hertenstein, B., Ganser, A., Knapp, W.H., and Drexler, H. 2005. Monitoring of bone marrow cell homing into the infarcted human myocardium. *Circulation 111*, 2198–2202.

242. Strauer, B.E., Brehm, M., Zeus, T., Kostering, M., Hernandez, A., Sorg, R.V., Kogler, G., and Wernet, P. 2002. Repair of infarcted myocardium by autologous intracoronary mononuclear bone marrow cell transplantation in humans. *Circulation 106*, 1913–1918.

243. Assmus, B., Schachinger, V., Teupe, C., Britten, M., Lehmann, R., Dobert, N., Grunwald, F., Aicher, A., Urbich, C., Martin, H., Hoelzer, D., Dimmeler, S., and Zeiher, A.M. 2002. Transplantation of Progenitor Cells and Regeneration Enhancement in Acute Myocardial Infarction (TOPCARE-AMI). *Circulation 106*, 3009–3017.

244. Wollert, K.C., Meyer, G.P., Lotz, J., Ringes-Lichtenberg, S., Lippolt, P., Breidenbach, C., Fichtner, S., Korte, T., Hornig, B., Messinger, D., Arseniev, L., Hertenstein, B., Ganser, A., and Drexler, H. 2004. Intracoronary autologous bone-marrow cell transfer after myocardial infarction: the BOOST randomised controlled clinical trial. *Lancet 364*, 141–148.

245. Schächinger, V., Erbs, S., Elsasser, A., Haberbosch, W., Hambrecht, R., Holschermann, H., Yu, J., Corti, R., Mathey, D.G., Hamm, C.W., Suselbeck, T., Assmus, B., Tonn, T., Dimmeler, S., and Zeiher, A.M. 2006. Intracoronary bone marrow-derived progenitor cells in acute myocardial infarction. *N EnglJ Med 355*, 1210–1221.

246. Janssens, S., Dubois, C., Bogaert, J., Theunissen, K., Deroose, C., Desmet, W., Kalantzi, M., Herbots, L., Sinnaeve, P., Dens, J., Maertens, J., Rademakers, F., Dymarkowski, S., Gheysens, O., Van Cleemput, J., Bormans, G., Nuyts, J., Belmans, A., Mortelmans, L., Boogaerts, M., and Van de Werf, F. 2006. Autologous bone marrow-derived stem-cell transfer in patients with ST-segment elevation myocardial infarction: Double-blind, randomised controlled trial. *Lancet 367*, 113–121.

247. Meyer, G.P., Wollert, K.C., Lotz, J., Steffens, J., Lippolt, P., Fichtner, S., Hecker, H., Schaefer, A., Arseniev, L., Hertenstein, B., Ganser, A., and Drexler, H. 2006. Intracoronary bone marrow cell transfer after myocardial infarction: Eighteen months' follow-up data from the randomized, controlled BOOST (BOne marrOw transfer to enhance ST-elevation infarct regeneration) trial. *Circulation 113*, 1287–1294.

248. Lunde, K., Solheim, S., Aakhus, S., Arnesen, H., Abdelnoor, M., Egeland, T., Endresen, K., Ilebekk, A., Mangschau, A., Fjeld, J.G., Smith, H.J., Taraldsrud, E., Grogaard, H.K., Bjornerheim, R., Brekke, M., Muller, C., Hopp, E., Ragnarsson, A., Brinchmann, J.E., and Forfang, K. 2006. Intracoronary injection of mononuclear bone marrow cells in acute myocardial infarction. *N Engl J Med 355*, 1199–1209.

249. Assmus, B., Honold, J., Schachinger, V., Britten, M.B., Fischer-Rasokat, U., Lehmann, R., Teupe, C., Pistorius, K., Martin, H., Abolmaali, N.D., Tonn, T., Dimmeler, S., and Zeiher, A.M. 2006. Transcoronary transplantation of progenitor cells after myocardial infarction. *N Engl J Med 355*, 1222–1232.

36 Gene Therapy for Cancer

Karsten Brand

CONTENTS

36.1 INTRODUCTION

Gene therapy for cancer has generated great interest for nearly two decades and intensive experimental and clinical investigations are under way. In general, it comprises different technologies to deliver a cDNA of choice to cancer cells or to normal tissue for a variety of diagnostic and therapeutic applications. Based on the complex nature of cancer, these technologies are very heterogeneous, as demonstrated by a variety of concepts such as immunomodulation, the "suicide strategy" (i.e., transfer of the cDNA of a prodrug converting enzyme); gene replacement strategies like transfer of a tumor

suppressor or an antioncogene; viral oncolysis; antiangiogenic and antiproteolytic gene therapy; or the delivery of drug resistance genes into hematopoietic precursor cells. Many experimental approaches are now in clinical evaluation and more than 1280 trials have been published or initiated or are pending. More than two-thirds of these trials are directed against cancer. In the first part of this contribution, I have tried to summarize the current state of the art of the gene therapeutic vectors now available and review gene therapy principles strictly from the point of view of cancer therapy. In the second part, I have described experimental and clinical approaches for several types of cancer. The peculiarities of the respective cancer types allow or require customized gene therapeutic approaches, particularly in a clinical setting.

This contribution cannot give a complete picture of all current gene therapy approaches and I apologize for any omissions. Cross-references to literature reviews and to the corresponding chapters in this book are given. Helpful tools for getting an overview of gene therapy trials are found at http://www.wiley.co.uk/genetherapy/clinical/ at the Genetic Medicine Clinical Trials Database and at http://www4.od.nih.gov/oba/RAC/GeMCRIS/GeMCRIS.htm, a Web site provided by the NIH, Office of Biotechnology.

36.2 THERAPEUTIC PRINCIPLES

36.2.1 IMMUNOMODULATION

It has long been recognized that tumors exhibit to a certain extent of immunogenicity (Ottgen 1991). As demonstrated by several studies, the human immune system seems to respond to this immunogenicity by recognizing specific tumor antigens and, consequently, by mounting humoral and cellular responses. However, during cancer development, this response is only of limited intensity and duration (Pardoll 1994; Sahin et al. 1995). As the mechanisms that cancer cells use to escape detection by the immune system have been elucidated, many strategies have been developed to reconstitute an effective antitumor immune response. The tremendous increase in the knowledge of the immunobiology of cancer has made immunological approaches, e.g., immunomodulation, the most dominant strategy in cancer gene therapy during recent years (Blankenstein et al. 1996). In general, immunomodulation studies can be categorized according to (a) the target cells (tumor cells, host cells, T-cells, or antigen-presenting cells such as dendritic cells [DCs] or other cells); (b) the mode of gene delivery (which vector, *in vitro*, *ex vivo*, *in vivo*); or (c) the transferred transgenes (cytokines, costimulatory molecules, tumor-associated antigens). Four examples from the multiplicity of these approaches will be described in more detail.

One of the most attractive target cell types for genetic modification is T-lymphocytes. The application of cytokine-transduced tumor-infiltrating T-lymphocytes (TILs) (Rosenberg et al. 1990; Friedmann 1991; Merrouche et al. 1995; Riddell and Greenberg 1995) was among the earliest clinical protocols for gene therapy. More recently, T-lymphocytes have been the target for *ex vivo* genetic modification by cytokine gene transfer and for redirection by tumor antigen–specific T-cell receptors or chimeric receptor genes (extracellular domain: antigen binding, intracellular domain: cell signaling). They have also been isolated from genetically modified tumors or their draining lymph nodes (Greenberg 1991; Yee et al. 1996; Rosenberg 1997). Other approaches enhance T-cell reactivity with antibodies that are targeted directly at the respective receptors on T-cells (Leach et al. 1996; Melero et al. 1997). These approaches are complemented by different methods to enhance antigen recognition on the surface of tumor cells.

Another immunologic approach to generate a local inflammatory response is the use of short-range communications between immune and nonimmune cells. Cytokine-transfected tumor cells or fibroblasts are transferred, which are then supposed to directly activate specific as well as nonspecific immune cells. In this context, the continuous local release of cytokines has been shown to increase the therapeutic index. Additionally or alternatively, costimulatory molecules can be transferred. In tumor cells where major histocompatibility complex (MHC) class I or class II molecules or costimulatory molecules such as HLA-B7 are downregulated, the transfer of the corresponding wild-type cDNA can reactivate antigen recognition on the surface of the transduced tumor cells.

During the last decade, many tumor-derived antigens have been defined. These became a very interesting target for gene transfer approaches (van der Bruggen et al. 1991). In addition to applications of the corresponding peptides, some of these antigens have been transferred by viral vectors or as naked DNA either directly to the tumors or via DCs, evoking an immune response against the tumor. Detection of new tumor antigens with new powerful methods, such as serological analysis of tumor antigens by recombinant expression (SEREX) cloning (Sahin et al. 1995, 1997) or methods for detection of differential gene expression such as microarray technologies, hold great promise for the future.

Finally, antibody-based immunotherapy should be mentioned in this context (Riethmuller et al. 1994), particularly since recombinant antibodies have finally shown marked clinical benefit. The underlying concept of this strategy is to raise monoclonal antibodies against soluble tumor antigens or against antigens that are expressed on the surface of the malignant cells or on the tumor stroma. A cancer gene therapeutic use of antibody genes is probably only a matter of time.

36.2.2 PRODRUG CONVERTING ENZYMES (SUICIDE STRATEGY)

The suicide strategy in oncology combines classical cytotoxic chemotherapy with gene transfer technology. The underlying concept is to limit the action of a known cytotoxic drug to the local area of the tumor lesion. To this end, the cDNA of a prodrug converting enzyme is delivered into the tumor by a vector system of choice. This is followed by regional or systemic application of the corresponding nontoxic prodrug. Once the prodrug reaches the tumor and is taken up by the tumor cells, which express the prodrug converting enzyme, it is converted into the cytotoxic drug. In conventional chemotherapy, toxic and myeloablative side effects can be dose-limiting. In contrast, when using the suicide gene concept, the cytotoxic effects of the converted drug are mainly restricted to the area of tumor infiltration and the time of action is limited to the presence of the cancer cells expressing the prodrug-converting enzyme. In addition, the efficacy of the suicide strategy is enhanced by the "bystander effect." This molecular mechanism allows the killing of even uninfected tumor cells in the neighborhood of infected cells due to intercellular communication mediated by gap junctions (for example). However, this mechanism is still not fully understood. There are many different prodrug-converting enzymes under experimental or clinical investigation. For a more detailed overview of the application of the suicide strategy in cancer gene therapy, please see *Gene and Cell Therapy* (Chapter 25).

The prodrug-converting enzyme most often used for clinical purposes is the herpes simplex virus thymidine kinase gene (HSV tk). The enzyme tk phosphorylates the prodrug ganciclovir (GCV) to GCV-monophosphate, which is then further phosphorylated to toxic GCV-triphosphate. Inhibition of the DNA polymerase by GCV-triphosphate finally leads to cell death. As demonstrated in various studies, tumor eradication has been achieved even if only 10% of the tumor mass is transduced (Moolten et al. 1990). This effect is probably due to a very potent bystander effect. In preclinical animal experiments, a vector carrying the HSV tk gene is usually applied intratumorally. In several reports using immunocompromised animals, a reduction of tumor volume of more than 50% of the controls was achieved in various experimental settings and even complete remissions were frequently observed. In addition to these experimental studies in immunocompromised animals, reports in immunocompetent animals suggest that the immune system may play a supportive role in the efficacy of this approach (Barba et al. 1994). Promising results of tumor growth inhibition are not without side effects induced by the HSV tk system. Using an *in vivo* model of adenovirus-mediated gene transfer to the liver of mice and rats, liver toxicity was observed due to unwanted transduction of normal tissue (Brand et al. 1997). This can be eliminated either by targeted vectors or tissue-specific gene expression (Brand et al. 1997, 1998).

As in other gene therapy strategies, the capability of the suicide strategy to kill cancer cells in experimental and clinical studies is limited by overall low efficacy of gene transfer *in vivo*. This is probably the main reason for the overall limited clinical efficacy of the suicide strategy in several phase I, II, and III clinical trials for several types of cancer. Even with the most efficient vector systems, direct intratumoral vector application only leads to a partial transduction of tumors with a nonhomogenous intratumoral vector distribution. Consequently, low antitumor efficacy is seen, particularly in humans who present much larger tumor burdens than rodents in a typical laboratory situation. Vector distribution could, however, be improved by changing the anatomical route of vector application (intravesical, intraperitioneal, intrathecal, intraventricular, repeated, and bulk flow); through the use of conditionally replication competent vectors (see *Gene and Cell Therapy* (Chapter 8)); by enhancing the bystander effect through the transfer of connexins, which increase the number of gap junctions; and by receptor targeting (see *Gene and Cell Therapy* (Chapter 4)).

36.2.3 Tumor Suppressor Genes and Antioncogenes

During the last decade, an increasing number of genes has been identified, which become dysregulated during carcinogenesis. During a complex and multifactorial process leading ultimately to the macroscopic presence of cancer, genes become dysregulated by different molecular mechanisms including gene deletion, mutation, or promotor silencing. At the end of these processes of genetic alteration, which result in activation or inactivation of multiple genes, the cancer cell proliferates in an uncontrolled fashion. It is not able to go into apoptosis and achieves invasive potential. Genes promoting these processes, like oncogenes, become activated. Genes suppressing the processes, like tumor-suppressor genes, become inactivated.

Current approaches in this area include the inactivation of overexpressed oncogenes by antisense molecules or dominant negative mutants or, alternatively, the reconstitution of cells with tumor suppressor genes, which have been lost or have mutated. The rationale is not so much the reversion of tumor cells back into normal cells. This task would be difficult to solve because more then one mutation is usually acquired by the tumor cell during the transformation process. The aim is rather to define the weak point in the cell's regulatory balance and consequently to identify the gene or combination of genes, which would have the highest impact on the exertion of cell cycle arrest or better apoptosis. To this end, it is of crucial importance to acquire a sufficient understanding of the cell's balance with respect to signal transduction, cell cycle regulation, and finally susceptibility to apoptosis. The targeting of the genes known to be dysfunctional in the respective tumor is usually the most efficient procedure. In some cases, a combination of several genes could increase specificity and efficacy. The main obstacle to this therapeutic approach is the need for a particularly high gene transfer efficacy. Although a mild bystander effect has been reported in the context of p53 gene transfer (Bouvet et al. 1998), a high-transfer efficacy is usually necessary for eradication of tumors—at least in those cases where the immune system as an adjuvant is not dramatically activated. Such high-transfer efficacies are the exception with current vectors. Therefore, a breakthrough for this highly tumor-specific approach particularly depends on future vector development. Among several tumor suppressor genes, which have shown *in vivo* efficacy, p53 has made it into the clinic and is discussed in more detail (e.g., in Section 36.4.8). Besides proapoptotic bax (Bargou et al. 1996) and bcl-x$_s$ (Ealovega et al. 1996), caspases (Shinoura, Saito et al. 2000), PTEN and BRCA1, those genes involved in the regulation of the G1 phase of the cell cycle have been evaluated particularly closely in animal experiments. Tumor growth could be inhibited by transfer of wild-type and truncated pRb (Riley et al. 1996; Xu et al. 1996), which in its active form binds E2F-1 and prevents entry into S-phase, as well as by the cdk/cyclin inhibitors p16 (Jin et al. 1995; Sandig et al. 1997) and p21 (Yang et al. 1995), which mainly keep pRb in its hypophosphorylated, active state. It remains to be shown whether cell cycle arrest is sufficient for therapy of established tumors or whether a strong induction of apoptosis as sometimes seen by transfer of p16 and p27 (Jin et al. 1995) is required. Undoubtedly, restoration of tumor cells with p53 or p21 can increase the sensitivity of cells to radiation (Gallardo et al. 1996; Spitz et al. 1996; Chang 1997) and chemotherapy (Fujiwara et al. 1994; Li et al. 1997; Skladanowski and Larsen 1997; Blagosklonny and El-Deiry

1998). Also, combinations of tumor suppressor genes could prove to have overadditive tumoricidal effects as has been shown for the combination of p53 and p16 (Sandig et al. 1997).

Gene therapy–directed against oncogenes or apoptosis suppressors like bcl-2 or bcl-x$_l$ has targeted several types of cancer and several clinical trials are under way. These are discussed in more detail in Section 36.4 and in Chapter 42. The antioncogene approach is in principle supported by the recent success of small-molecular inhibitors of oncogenes: inhibitors of the epidermal growth factor (EGF) receptor, the Rat sarcoma (RAS) pathway, and the ABL gene (Gleevec). Gene therapy has the potential to compete with small molecules because small molecules still efficiently target only enzymes and receptors whereas transdominant antioncogenes, antisense oligonucleotides, and (recently) the very promising RNA-interfering nucleotides (Borkhardt 2002; Brummelkamp et al. 2002) can inhibit virtually any oncogene of relevance.

36.2.4 TUMOR LYSIS BY RECOMBINANT VIRUSES

Since its early beginnings, cancer therapy by viral oncolysis has been one of the most challenging strategies to treat human malignancies (Kovesdi et al. 1997). The idea that viruses may be used as selective anticancer agents dates back almost a century (Dock 1904). In 1957, four years after the discovery of replicating adenoviruses, they were used in cancer therapy. Of patients with advanced cervical carcinoma receiving intratumoral or intra-arterial injections of wild-type human adenoviruses, 65% had a marked to moderate local tumor response and only three patients on steroids out of 30 patients had a viral syndrome of short duration (Smith 1956). The underlying concept of this strategy is to inject the virus directly into the tumor, leading to the transduction of a certain number of cells in which viral replication takes place. Consequently, the infected cells are finally disrupted and viral progeny are released, allowing the spread of the infection and an increase in transfer efficacy. Interest and research in this area of molecular biology has exploded, and the available virus production techniques and purification techniques have been markedly improved. Thus, much larger amounts of adenovirus can now be applied. However, one of the major concerns has been how to limit the viral replication to the site of the tumor. One strategy to confer tumor-specific replication has been developed which utilizes the dependence of the replicating adenovirus on the genetic status of p53 in the infected host cell. Like many DNA viruses, adenoviruses have also developed specific gene products that seem to counteract apoptosis-inducing molecules like p53. Among the adenoviral genes, the E1B 55 kD gene blocks p53, the best known inducer of apoptosis and indirect inducer of cell cycle arrest. In an intriguing approach (Bischoff et al. 1996), the E1B gene was deleted from the adenoviral genome, allowing viral replication only in p53-negative tumor cells but creating apoptosis upon viral infection in p53-positive normal cells (Bischoff et al. 1996; Heise et al. 1997). Since more than 50% of common solid tumors lack functional p53, this approach was thought to be widely

applicable. In the mean time, both the lack of replication in all p53-positive cells and the potent replication in all p53-negative cells (Hall et al. 1998; McCormick 2001) have been questioned. It seems that defects further down the p53 pathway, as well as functions of E1B55K independent from inactivation of p53, need to be taken into account. They may perhaps moderately limit the applicability of the approach. Independent of the outcome of this debate, this concept has already stimulated virologists and cancer biologists to develop further generations of conditionally replication-competent adenoviruses. Adenoviral E1A inactivates pRB, the well-known tumor suppressor and inducer of cell cycle arrest, and facilitates adenoviral replication by allowing cellular replication. In an approach similar to the one described above, E1A was mutated so as to no longer bind and inactivate pRb; this should lead to viral replication only in tumor cells, which typically have defects in the pRb pathway. This concept has been successfully applied *in vitro* and *in vivo* (Doronin et al. 2000; Fueyo et al. 2000; Heise et al. 2000).

Another way to achieve cancer-specific replication is through the use of tissue-specific or tumor-specific gene expression. The adenoviral E1A region, which controls viral replication through several mechanisms, was placed under the control of a tissue- or cancer-specific promoter instead of the E1A promoter. This approach allowed specific replication in the respective cancer cells and tumors when prostate-specific antigen (PSA) or probasin promoters (Rodriguez et al. 1997; Yu et al. 2001); TCF-responsive elements, which are preferentially activated in colon cancer; or the MUC1 promoter (which is active in breast carcinoma, Kurihara et al. 2000) were used to drive the E1A gene (Brunori et al. 2001).

Besides adenoviruses, other oncolytic viruses have been tested in animal models. Like adenoviruses, HSV, a neurotrophic DNA virus, directly lyses cells during viral shedding. First-generation HSV contains a 360 bp deletion in the tk gene; this seems to prevent replication in quiescent cells but allows replication in rapidly proliferating cells (Martuza et al. 1991; Jia et al. 1994). Second-generation viruses with additional mutations have been generated to increase tumor specific replication and reduce neurovirulence (Kirn 1999). Newcastle disease virus (NDV) is a chicken paramyxovirus associated with minimal disease in humans. Cytotoxicity for numerous human tumor cell lines and resistance of several human fibroblast lines have been reported (Reichard et al. 1992). The mechanism that causes tumor selectivity is not fully resolved but a failure to mount a protective interferon response may be causally involved. Tumor cell killing may involve virus replication and direct cell lysis or induction of tumor necrosis factor (TNF) secretion as well as increased sensitivity of tumor cells to TNF-mediated killing (Cassel and Garrett 1965; Lorence et al. 1988). Recently, a clinical trial of intravenous injection of NDV into 67 patients with solid tumors was published. Objective responses occurred at higher doses (Pecora et al. 2002).

Autonomous parvoviruses are DNA viruses with small genomes, which depend on helper viruses or specific cellular functions for replication. Their cytopathic effect seems to be dependent on DNA replication as well as on the expression of nonstructural gene products and enhanced sensitivity to their effects (Kirn 1999).

In the future, the tremendous increase in knowledge in the field of tumor and virus biology can be used to create appropriate vectors. Several factors that potentially influence clinical efficacy have already been defined. Among these are (a) characteristics of the viruses, such as the size, the time between infection and lysis, as well as the number of virions produced or the induction of a humoral or cellular immune response and (b) characteristics of the tumor such as distribution of the vasculature or physical barriers to virus spread such as fibrosis. An example for how to surmount such a natural barrier is the demonstration that inactivation of particular adenoviral serotypes by preformed antibodies can be circumvented if other serotypes are used (Kass-Eisler et al. 1996).

36.2.5 ANTIANGIOGENIC AND ANTIPROTEOLYTIC GENE THERAPY

These types of gene therapies differ markedly from the majority of the other nonimmunologic gene cancer therapeutic approaches because the products of the delivered genes act extracellularly. This fact obviates the need to transduce the majority of tumor cells and even allows one to target normal tissue, which maybe easier to transduce than the tumor. In addition, these approaches are usually appropriate for long-term treatments. Therefore, the conventional demands of cancer gene therapy—high efficacy of gene transfer, acceptance of immunogenicity, and sufficiency of short duration of gene expression—are replaced by requirements more typical for the treatment of monogenetic disease. These include low immunogenicity of vector and transgene, long-duration gene expression, and achievement of at least threshold levels of gene product in the target tissue or in the blood.

Inhibition of angiogenesis has become a very promising target for cancer therapy. Advantages of targeting the growth of new blood vessels as opposed to other types of therapy include the avoidance of tumor cell resistance mechanisms, the broad and generally applicable mechanism, and easy access to targets within the vasculature. More than 20 endogenous inhibitors of angiogenesis and nearly as many stimulators, which can serve as targets for small inhibitory molecules, have been characterized (Scappaticci 2002). Interestingly, several standard chemotherapeutic drugs have been found to exert their antitumor effect in part through inhibition of angiogenesis.

There are over 80 antiangiogenic agents (proteins as well as small molecules) in clinical trials. In general, antitumor efficacy has been observed, although overall success rates have not lived up to expectations. One important reason is perhaps the choice of disease stage. Although antiangiogenic therapy will most likely be of highest efficacy at the early stages of tumor growth or metastasis, patients with excessive tumor burdens have been mostly treated so far.

Antiangiogenic gene therapy has not yet been tested in patients, but several preclinical experiments have shown promise (Scappaticci 2002). Among the first genes examined were the genes for the soluble form of endothelial cell receptor proteins, Tie2 and FLT-1. These proteins interact with the angiogenetic factors angiopoietin-1 and vascular endothelial growth factor (VEGF), respectively. These genes were transferred in vivo or ex vivo and resulted in significant inhibition of tumor growth (Folkman 1998; Goldman et al. 1998; Lin et al. 1998). Other strategies include the transfer of antiangiogenic proteolytic fragments like angiostatin or endostatin, immunomodulatory genes with antiangiogenic properties like interferons and IL-12, or other molecules better known for other functions in the body like tissue inhibitor of metalloproteinases (TIMPs), p53, or p16.

There are several advantages of molecular therapy over direct application of the effector substance. Firstly, the use of targeted vectors allows increased intratumoral concentrations of antiangiogenic factors without the risk of potential side effects on wound healing, endometrial maturation, or embryo growth (Klauber et al. 1997). Secondly, the cost of gene delivery may be lower than prolonged protein therapy. Thirdly, experimental data suggest that effective antiangiogenic therapy requires the continuous presence of the inhibitor in the blood, which may perhaps be more efficiently achieved by gene therapy than by bolus protein therapy.

Antiangiogenic therapy itself is only in its infancy and the future development of this treatment modality will obviously have a large impact on the corresponding gene therapy approach. Because of its mainly tumoristatic nature, it seems likely that this form of therapy will have its main impact before or after surgery with a view to facilitating surgery and preventing recurrence of distant metastases, as well as in combination with conventional chemotherapy, radiotherapy, or immunotherapy.

Excessive degradation and remodeling of the extracellular matrix (ECM) is one of the hallmarks of cancer progression at nearly every step from the first breakdown of the basal membrane of a primary tumor up to the extended growth of established metastases. There are four known classes of proteases implicated in ECM turn over: matrix metalloproteinases (MMPs), serine proteinases, ADAMs, and the BMP1 family. The naturally occurring inhibitors of the MMPs, the four known TIMPs, and the inhibitors of serine proteases the PAIs have been preclinically evaluated for antitumor efficacy and are the most prominent examples of antiproteolytic gene therapy. Intratumoral injection of first-generation TIMP-2 adenovirus into subcutaneously growing tumors of mammary, colorectal, and bronchial origin reduced tumor size by 60%–80% as compared to the controls, and the incidence of lung metastases was reduced by 90% (Li et al. 2001). A similar efficacy was achieved by adenovirus-mediated transfer of TIMP-3 into subcutaneously growing melanomas (Ahonen et al. 2002). Systemic delivery of TIMP-2 by adenovirus application into the tail vein of immunocompetent mice bearing orthotopically implanted breast tumors resulted in 50% growth inhibition (Hajitou et al. 2001), and the intraperitoneal injection of TIMP-2-carrying liposomes for the treatment of spontaneously arising breast tumors allowed for a 40% size reduction and a dramatic reduction of the incidence of lung metastases (Sacco et al. 2001). An intraperitoneal approach was taken to treat peritoneally disseminated pancreatic cancer and adenoviral delivery of either TIMP-1 or TIMP-2 resulted in a 75%–85% reduction of tumor loads and a significant prolongation of survival (Rigg and Lemoine 2001).

A prevention of colorectal liver metastasis of up to 95% was achieved if Ad-TIMP-2 was injected into the tail vein of mice, leading to a preferential transduction of the liver with the antitumor gene (Brand et al. 2000). Administration of TIMP-2 vector to mice bearing established liver metastases resulted in a 77% growth inhibition. When this same concept of gene transfer into the host tissue was applied for the treatment of lymphatic liver metastases, adenoviral transfer of TIMP-1 reduced metastatic deposits by 93% (Elezkurtaj 2004). The peculiarity of this approach is the transfer of anticancer defense genes into the normal noncancerous tissue, which was termed the "impregnation" approach.

In contrast to these findings of a marked antitumor efficacy of TIMP-1, -2, and -3 with three different types of animal models and modes of vector administration, systemic transfer of TIMP-4 by intramuscular electro-adjuvated injection of naked plasmid for the prevention of orthotopically growing breast tumors resulted in a threefold increase in size and up to sevenfold increase in the incidence of primary tumors (Jiang et al. 2001). None of the cited papers has reported any TIMP-related toxicity. Mild and probably vector-related toxicity to the liver was occasionally seen when first-generation adenovirus was delivered systemically.

Transfection of tumor cells with PAI-1, an inhibitor of serine proteases, reduced metastasis. Treatment with adenovirus-mediated PAI-1 gene transfer reduced the growth of uveal primary tumors as well as the incidence of metastases (Andreasen et al. 2000). Similarly, gene transfer of PAI-2 had unequivocal antitumoral effects (Andreasen et al. 2000). In contrast and unexpectedly, transgenic mice overexpressing PAI-1 developed pulmonary metastases at the same rate as wild-type mice in a melanoma model (Ma et al. 1997). Even more paradoxically, PAI-1 knock-out mice displayed a dramatically reduced local invasion and tumor vascularization of transplanted malignant keratinocytes and administration of PAI-1 by adenoviral gene transfer restored invasion and associated angiogenesis (Bajou et al. 1998). These data indicate that antiproteolytic therapy with PAIs, and to a lesser extent with TIMPs, can have dramatic antitumor efficacy but that paradoxical tumor-promoting effects can occur. Similar to the MMPs and TIMPs, the PAs and PAIs have been shown to display effects independent from the degradation of the ECM (Chapman 1997), which may be partly responsible for the unexpected findings. Therefore, increased knowledge of either protease/inhibitor system may help us to better understand the underlying mechanisms, resolve some unexpected findings, and to draw conclusions with therapeutic relevance.

36.2.6   Drug Resistance Genes

Instead of killing tumor cells, this strategy is aimed at preventing toxic side effects of modern chemotherapy, for example, by making normal cells resistant to chemotherapeutic toxicity. This approach tries to solve the problem that the cytotoxicity of chemotherapy cannot usually discriminate between proliferating normal and cancer cells. Proliferating normal cells like hematopoietic precursor cells are especially affected, as evidenced by the fact that bone marrow suppression is still dose-limiting in most high-dose courses of chemotherapy. To overcome this limitation, different experimental approaches have been developed. One strategy to make hematopoietic precursor cells resistant to chemotherapy is the direct gene transfer of drug resistance genes like transporters that extrude drugs across the plasma membrane. These include the multiple drug resistance gene 1 (MDR1), the multiple drug resistance proteins (MRPs), or enzymes that render drugs resistant to alkylating agents such as gluthathion-S-transferase and O^6-alkylguanine DNA alkyltransferase or resistant to antimetabolites such as dihydrofolate reductase (DGFR) (Licht and Peschel 2002). The MDR1 gene encodes for the P-glycoprotein, a cell membrane multidrug transporter that effluxes a broad spectrum of hydrophobic and amphipathic compounds, including several chemotherapeutic drugs currently used in clinical studies (Gottesman et al. 1995). *Ex vivo* transfer and expression of the MDR1 gene in hematopoietic progenitor cells has been shown to increase the resistance of hematopoietic precursor cells to chemotherapy in rodent models (Mickisch et al. 1992; Podda et al. 1992; Sorrentino et al. 1992; Licht et al. 1995; Sorrentino et al. 1995). In the first several clinical trials, low level expression of the transferred gene was reported only occasionally. More recently, however, prompt hematopoietic recovery and long term *in vivo* vector expression was seen (Moscow et al. 1999; Abonour et al. 2000). We may see further clinical improvements if efficient transduction is achieved (especially of very immature cells) and drug selection strategies are further optimized. The encouraging report of *in vivo* correction of X-linked severe combined immunodeficiency (SCID) (Cavazzana-Calvo et al. 2000; Hacein-Bey-Abina 2002) with at least 17 out of 20 patients now benefiting from this life-saving gene therapy also indicates that transduction of human hematopoietic stem cells is now feasible (Cavazzana-Calvo and Fischer 2007). Clinical benefit may result, in particular, in situations where corrected cells have an *in vivo* selective advantage.

36.2.7   Marker Genes

Six percent of all clinical gene therapy protocols use marker genes for several purposes (Rosenberg et al. 2000). Marker trials have confirmed the assumption that contamination of reinfused bone marrow with tumor cells causes tumor recurrence after autologous bone marrow transplantation (Rill et al. 1994). In other studies, hematopoietic stem cells or T-cells were marked and traced over a longer duration, providing important insights into their long-term survival.

36.3   VECTORS FOR CANCER GENE THERAPY

36.3.1   Virus-Based Gene Transfer Vectors

In this section, the various gene transfer vectors (e.g., adenovirus- and retrovirus-based vectors) will be discussed with respect to their specific relevance for efficacy and toxicity in cancer gene therapy. For the general biology, development and design of the vectors, the reader is referred to the corresponding chapters in this textbook, which focus on particular vectors.

One of the most widely used vectors for *in vivo* application in cancer gene therapy is the replication-deficient recombinant adenovirus. First-generation adenoviral vectors (Ad vector) can be generated to the highest titers (up to 10^{12}/plaque forming [i.e., infectious] units/mL) among viral vectors (Wilson 1996). Moreover, they easily infect cells of epithelial origin including cancer cells, most likely because these cells express the appropriate receptors, the coxsackie adenovirus receptor (Bergelson et al. 1997) and integrins (Wickham et al. 1993) for binding and internalization, respectively. Since Ad vectors infect dividing and nondividing cells, dormant tumor cells, which can make up a considerable fraction of the tumor mass, can also be killed. These properties have made first-generation adenoviruses one of the vectors of choice for clinical studies targeting cancer. Of the 1280 clinical gene therapy protocols initiated to date, 326 (24.9%) use Ad vectors, the majority of those for cancer treatment. Commonly, serotype C adenoviruses are used but the discovery of CD46 as the cellular receptor for serotyoe B adenoviruses has made these vectors promising alternatives (Gaggar et al. 2003). Ad vector-mediated gene transfer can be accompanied by toxic side effects on normal tissue mainly due to residual adenoviral gene expression in occasionally transduced normal cells (Nevins 1981; Spergel and Chen-Kiang 1991; Yang et al. 1994; Lieber et al. 1997; Amalfitano et al. 1998; Brand et al. 1999). This toxicity can be reduced using Ad vectors of the second generation where adenoviral genes such as the E4 region or the E2a gene are deleted (Wang et al. 1995; Bramson et al. 1996; Gao et al. 1996; Wang and Finer 1996; Wilson 1996; Yeh et al. 1996) or helper-dependent (i.e., minimal or gutless) adenoviral vectors (HD-Ads), which are completely devoid of all viral genes (Haecker et al. 1996; Parks et al. 1996; Parks and Graham 1997). The side effects of first-generation Ad-vectors such as toxicity and immunity can be turned into a desired adjuvant situation for cancer therapy as demonstrated by the group of Rosenberg (Chen et al. 1996). However, in cancer gene therapy where vector toxicity to the tumor cells at the first glance may be desirable, this toxicity may be of disadvantage in some situations. For instance, leakage of intratumorally injected vectors carrying a prodrug-converting enzyme (suicide gene strategy) into the host organ has been shown to produce severe side effects (Brand et al. 1997). Regional and systemic application of vectors with the aim of targeting metastases will inevitably transduce normal tissue. Other anticancer strategies are even explicitly based on transduction of normal tissue for its use as a bioreactor as in the antiproteolytic impregnation concept, antiangiogenic, and some immunological approaches. Therefore, cancer gene therapy (like gene therapy of other diseases) will depend enormously on safer vectors. In addition to increased safety (Morsy et al. 1998), HD-Ads have some other advantages over first-generation vectors including an extremely high packaging capacity, which allows for a better adaptation to the tumor

cell's particularities, as well as the potential for long-term gene expression (e.g., Schiedner et al. 1998), apparently higher gene expression (Maione et al. 2000), and better retained tumor-specific gene expression (Shi et al. 2002). The two main disadvantages of HD-Ads have been contamination with helper virus and difficulties in obtaining high titer preparations. Helper contamination has been significantly improved and titers of up to 2×10^{11} bfu (blue forming units) are possible with modern techniques. HD-Ads have been tested in several preclinical models including cancer models (Rodicker et al. 2001) but no clinical trials have been conducted so far.

Another very popular vehicle for gene transfer, accounting for more than one-third of all trials and half of all patients treated, is retrovirus-based vectors. Most of these are derived from murine leukemia virus (MLV). Retroviral vectors were the first vectors for which all the viral genes were provided by vector producer cells (VPCs). The retroviruses themselves contain only the LTRs, the packaging signal, and the transgene. In the field of cancer therapy, these vectors have dominated trials utilizing *ex vivo* transfer of markers, drug resistance, and immunomodulatory genes. For *in vivo* gene transfer, the VPCs themselves were often directly applied into the tumors due to the initially low achievable titers of 10^6/mL in the retroviral supernatant with the rationale that these VPCs will deliver retroviruses until they are eliminated by the immune system (Culver et al. 1992). As for adenoviral vectors, many animal studies were carried out, including *in vivo* gene transfer of retroVPCs or retroviral vectors into established tumors, and complete remissions of microscopic tumors have been reported (Culver et al. 1992). Current packaging cell lines give titers of greater than 10^7 infectious particles per mL (Cosset et al. 1995). Moreover, the use of different envelope proteins, such as the G protein from vesicular-stomatitis virus, has improved titers following concentration to greater than 10^9/mL (Cosset et al. 1995), although these titers are still far below what is achievable with adenoviruses. Especially from a safety standpoint, their property of only infecting dividing cells may prevent toxicity to nondividing normal tissue. On the other hand, these vectors leave out dormant cancer cells, which may decrease their antitumor efficacy.

In contrast to MLV-based vectors, lentiviral retroviruses based on pseudotyped HIV-1 are able to infect nondividing cells (Blomer et al. 1996; Naldini et al. 1996a,b; Robbins et al. 1998), although there is some debate about whether hematopoetic stem cells should preferentially not be in G0 (Sutton et al. 1999) and whether hepatocytes are much more efficiently transduced *in vivo* if proliferating (Park et al. 2000; Follenzi et al. 2002; VandenDriessche et al. 2002). Lentiviral vectors may therefore be of interest for cancer gene therapeutic approaches, which rely on long-term transduction of noncancerous tissue for cancer therapy. Mainly due to (perhaps overestimated) safety concerns with respect to the generation of wild-type virus, only nine clinical trials have been initiated yet. However, the use for example, of apathogenic animal lentiviruses (Stitz et al. 2001; Kuate et al. 2002) may soon circumvent this concern.

Based on their tropism for neuronal tissues, HSVs are in principle highly suitable for neurological cancer therapy. In general, they can infect dividing and nondividing cells. Similar to adenoviruses, cytotoxic genes must be deleted from the viral genome and genes must be provided from helper viruses or cell lines in order to prevent unwanted toxicity if normal tissue is transduced. Successful attempts in this direction have been made (Fink et al. 1995; Marconi et al. 1996; Wu et al. 1996; Zhu et al. 1996; Fink and Glorioso 1997a,b). HSV has already been used in 43 clinical trials and the intratumoral injection mode at least did not cause severe toxicity (see Section 36.4.4 on Glioblastoma).

A fourth type of vector, which has become very interesting for gene therapy in general during the last years, is adeno-associated virus (AAV)–based vectors. These helper virus (e.g., adenovirus)-dependent vectors induce only a minor cellular immune response and have the potential to integrate into the host genome. Therefore, they are very promising candidate vectors when long-term expression is required (for instance, for the protection of hematopoietic precursor cells during high-dose chemotherapy). With respect to clinical trials, AAV are fairly advanced. Forty-eight trials have already been initiated, most of them, however, for noncancerous disease. For the general biology of AAV-based gene transfer vectors, please refer to *Gene and Cell Therapy* (chapter 6).

The poxviruses fowlpox virus and vaccinia virus infect a wide variety of cells types. In addition, they are safe and have a high packaging capacity (Scheiflinger et al. 1992). Due to their high intrinsic immunogenicity, they have mainly been used as an adjuvant and transfer vehicle for immunogene therapy of cancer. An impressive number of 125 clinical studies using pox virus have been initiated.

In addition, several other viruses have been used for the transfer of genes into cancer cells. They usually have some interesting features but also bring certain disadvantages, which have so far prevented widespread use. EBV viruses have a natural tropism for B-cells, but the potential development of wild-type transforming viruses by homologous recombination raises a safety concern. Baculoviruses are insect viruses and cannot replicate in human cells but they have been shown to efficiently infect a variety of human cell lines (Hofmann et al. 1995; Sandig et al. 1996). However, the human complement system rapidly inactivates these vectors such that extensive modification of the cell surface is required (Sandig et al. 1996; Hofmann and Strauss 1998; Huser et al. 2001) to improve gene transfer *in vivo*. Alpha viruses like sindbis virus (Xiong et al. 1989) and Semliki-forest-virus (Berglund et al. 1993) can efficiently multiply their RNA genome in target cells, allowing for very high transgene expression. Further improvements in their packaging systems could make these vectors very interesting agents for the future, especially for gene delivery to the brain, antitumor therapy by induction of apoptosis, or intratumoral replication and immunogene transfer (Lundstrom 2001).

Autonomous parvoviral vectors (Maxwell et al. 1996) such as H1, MVM, and LIII can only replicate during the S-phase of the host cell and have been used as replication-competent viruses for cancer gene therapy *in vitro* or in preclinical settings. Other viruses which have been evaluated for gene therapy include NDV, reovirus, poliovirus, and vesicular stomatitis virus (Wildner 2001).

Chimeric vectors, which are constructed by the use of two or more viruses, may have the advantages and disadvantages of both vectors. To date, several types of hybrid vectors have been published including first-generation adenovirus/AAV (Herz and Gerard 1993; Fisher et al. 1996; Liu et al. 1999), HD-Ad/AAV (Goncalves et al. 2002; Shayakhmetov et al. 2002) HD-Ad/transposon (Yant et al. 2002), HD-Ad/retrotransposon (Soifer et al. 2001) and first-generation adenovirus/retrovirus (Feng et al. 1997; Caplen et al. 1999; Roberts et al. 2002). These chimeras were evaluated mainly to combine the high efficacy of gene transfer with the long duration of gene expression; to facilitate AAV production; and to overcome the limited cloning capacity of AAV. HSV/AAV (Johnston et al. 1997; Heister et al. 2002) chimeras or baculovirus/AAV chimeras (Palombo et al. 1998) were constructed to combine cell specificity with stable gene expression.

With respect to long-term safety of viral vectors, occasional reports of possible oncogenicity have been described in animals for AAV (Donsante et al. 2001). Due to the development of therapy-associated leukemias, most likely due to insertional oncogenesis, safety concerns are higher for MLV-derived retroviruses (Li et al. 2002). Three cases of leukemia occurred in a trial for SCID that was very successful to that point (Hacein-Bey-Abina et al. 2003). A careful case-by-case risk–benefit analysis is clearly required (Buchholz and Cichutek 2006). Oncogenicity has not been a major concern for adenoviruses, which, however, can integrate at a low rate as well (Hillgenberg et al. 2001; Mitani and Kubo 2002).

Gene silencing as a reason for limited duration of gene expression despite vector integration has been described extensively for retrovirus and also for AAV (Chen et al. 1997) but inclusion of matrix attachment regions (MAR) or scaffold attachment regions (SAR) can improve the problem (Agarwal et al. 1998).

In summary, it seems that with respect to viral vectors it is possible to exploit nature's evolutionary achievements to our benefit and even to deviate markedly from nature's path without deleterious consequences. It is sometimes quite mysterious that even dramatic modifications of viral genomes still give viable and effective vectors. Since it seems that there is a workaround for nearly every problem, as long as time and money are provided, one can be confident that the vector problem, still the main issue in gene therapy, can be solved as long as the target diseases justify the effort and alternative ways of treatment are not provided.

36.3.2 NONVIRUS-BASED GENE TRANSFER VECTORS

In general, nonviral vector systems have some advantages over viral systems. They are less toxic, less immunogenic, and easier to prepare. On the other hand, they have a much lower efficacy of gene transfer, particularly *in vivo*, and a more limited duration of gene expression than several viral vectors. Roughly 30% of clinical trials have utilized nonviral vectors. Two-thirds of these used naked DNA plasmids and one-third cationic liposomes, which are probably the most important nonviral vectors to date. Attempts to increase efficacy of gene transfer with cationic lipids and polymers will have to target three major barriers: stability during and after the manufacturing process, extracellular barriers, and intracellular barriers. An important step in improving extracorporal stability of positively charged liposomes and polymers (which have a tendency to aggregate and precipitate) was the inclusion of hydrophilic polymers like polyethylene glycol (PEG). This provides steric stabilization. PEGylation seems also to reduce interactions of cationic particles with blood components such as albumine, which present major extracellular barriers to nonviral gene transfer. Modifications to decrease unwanted preferential uptake of liposomes by the reticulo endothelial system include the addition of sialic acid to make "stealth" liposomes (Lasic et al. 1991). The most challenging and perhaps most rewarding task will be the elucidation of intracellular barriers to nonviral gene delivery. Cationic DNA complexes bind to the negatively charged cell membrane and are taken up by endocytosis. As for viral vectors, targeting by addition of ligands for cellular receptors has been used to circumvent extracellular barriers. For example, transferrin is a common ligand used to target tumor cells (Kircheis et al. 2001; Mao et al. 2001). Endocytosed DNA is largely retained in the endosomal compartment and it is important that endosomal release is completed before lysosomal degradation initiates. Fusogenic lipids like Dioleylphosphatidylethanolamine (DOPE), fusogenic peptides from viral vectors (Wagner 1999), or substances with high buffering capacity (which in addition can swell like polyethylenimine [PEI]) have been used to release DNA from endosomes. After vesicle escape, the nucleic acid must traffic to the nucleus. DNA is not very stable in the cytoplasm, which may be due to the presence of cytoplasmic nucleases. It is therefore particularly important to define the optimal time point to release DNA from protection by the endosome or from coverage from vector after endosomal release. Recently, nuclease inhibitors have shown to improve naked DNA-mediated gene transfer (Walther et al. 2005). Once the vector arrives at the nucleus, another hurdle has to be overcome since nuclear entry of DNAs larger than 250 bp is hardly possible. A promising attempt to circumvent this problem is the inclusion of nuclear localization signals (NLS, Kaneda et al. 1989; Dean 1997).

Naked DNA is most interesting for immunological approaches in which efficacy of gene transfer is not the most crucial criterion. Interestingly, skeletal muscle cells are fairly susceptible to transfection with directly injected plasmid DNA. However, in mice, systemic application of naked DNA has led to astonishing high rates of gene transfer in mice if administered systemically in extremely high volumes and with generation of high pressure in target organs (Budker et al. 1996, 1998). Transient membrane defects are probably the mechanism how naked DNA enters the cells by this hydrodynamic method (Zhang et al. 2004).

In addition to liposomes and naked DNA, cationic polymers (e.g., linear PEI, LPEI) have gained attention during recent years although no clinical trials yet exist. Other modalities include electroporation (limited spatial distribution *in vivo*), ultrasound-facilitated gene transfer, and particle bombardment (Yang et al. 1990; Burkholder et al. 1993; Cheng et al. 1993; Gao et al. 2007). In this latter technique, 1–3 μm small gold or tungsten particles are covered with plasmid DNA accelerated in an electrical field and fired onto the target tissue. Due to a depth of penetration of up to 50 cell layers, this approach has been applied for vaccination into uninvolved skin of melanoma patients (Cassaday et al. 2007).

36.3.3 TRANSGENE

A gene therapeutic expression cassette consists of the regulatory elements, the transgene, and the poly A signal. The transgene is usually a wild-type cDNA but other structures are possible. Among these are mutated transgenes, such as dominant negative oncogenes to inhibit oncogenes, or mutated cDNAs, which lead to other defects favorable for cancer gene therapy. The availability of high-capacity vectors like HD-Ad or HSV has allowed the inclusion of full-length genes including introns and other features, which may eventually be as interesting for cancer gene therapy as they already are for the treatment of monogenetic disease. Finally, recent developments of small double-stranded DNA oligonucleotides, which apparently can interfere with RNA, may dramatically influence cancer gene therapy. These oligonucleotides have already been included in viral vectors and antitumor efficacy has been reported after *ex vivo* gene transfer (Brummelkamp et al. 2002).

36.3.4 CANCER-SPECIFIC GENE EXPRESSION (TRANSCRIPTIONAL TARGETING)

If one wants to exploit differences between tumor and normal tissue for tumor therapy, tumor-specific gene expression is a particularly rewarding area. Transcriptional targeting restricts gene expression to the target tissue even if undesired tissues have been transduced as well, thereby providing a significant improvement in safety. Tumor-specific promoters are only active in those target tissues naturally expressing their respective tumor-specific endogenous genes. If one engineers an expression cassette in which any therapeutic gene is placed behind that tumor-specific promoter, it will, due to its promiscuous nature, express this transgene in a tumor-specific fashion as well. A huge variety of tumor-specific promoters have been examined during the last years for their use in gene therapy. This is basically because every tumor type has its own more or less tumor-specific genes. Tumor-specific gene expression can obviously only be as specific as the degree of differential gene expression of the respective endogenous gene. Therefore, it has certain limitations. In other words, residual gene expression in undesired tissues is frequently observed. Among the promoters used are tissue-specific promoters, which will also be active in the corresponding normal tissues, and proliferation-specific promoters, which will be active in proliferating nonmalignant tissue. In addition, the expression levels are often only gradually higher in the tumor than in the normal tissue. One must carefully examine how these gradual differences translate into the therapeutic situation.

Tumor-specific promoters are particularly interesting in the situation of metastatic cancer where the tissue-specific promoter allows transgene expression in the metastatic tumor cell but not in the surrounding normal tissue, even if it is transduced by the vector. For instance, the carcinoembryonic antigen (CEA), which is physiologically expressed predominantly in colon tissue but not in liver tissue, allows transgene expression predominantly in colon cancer cells metastatic to the liver. Toxicity to the liver seen with the ubiquitous CMV promoter that expresses a suicide gene (Brand et al. 1997; van der Eb et al. 1998) can be prevented if the CEA promoter is used instead (Brand et al. 1998). Other examples of such a promoter strategy include the alpha-feto-protein (AFP) promoter for hepatocellular carcinoma (HCC); the promoters of erbB2, an oncogene often found in breast tumors; the PSA; or the promoter of the tyrosinase gene, which has specificity for melanoma.

However, the maintenance of tissue specificity can be a problem after inclusion into a vector. For adenoviral vectors, loss as well as maintenance of tissue specificity has been reported (Chen et al. 1995; Connelly et al. 1996; Imler et al. 1996; Sandig et al. 1996; Siders et al. 1996; Tanaka et al. 1997; Brand et al. 1998). In this context, the orientation of the expression cassette can play an important role. For instance, the transcriptional activity of the E1A enhancer, which is not deleted in E1-deficient first-generation adenoviruses can induce loss of tissue specificity. This type of promoter interference can even stem from regions located downstream from the expression cassette (Steinwaerder and Lieber 2000). This can even occur if the expression cassette is inserted in the reverse orientation to the main adenoviral reading frame (own observations). To circumvent this problem, an insulator element derived from the chicken gamma-globin locus (HS-4) was employed to shield an inducible promoter from viral enhancers. Induction ratios could be improved up to 40-fold as compared to vectors with an uninsulated promoter (Steinwaerder and Lieber 2000). In a similar approach, transcriptional terminators have been implemented to restrict read-through of the E1A enhancer (Buvoli et al. 2002). Finally, insertion of a specific promoter at a site distant from the E1A enhancer (e.g., close to the right ITR) has retained higher specificity and lower background (Rubinchik et al. 2001).

In contrast to first-generation adenovirus, HD-Ads seem to allow full maintenance of promoter specificity. This has recently been demonstrated for the PSA and the tyrosinase promoters (Shi et al. 2002).

Also in contrast to first-generation adenovirus, insertion of expression cassettes in reverse orientation into retroviral vectors normally maintains tissue specificity. Orientation in frame, however, puts the transgene under the influence of the strong retroviral LTR with subsequent loss of specificity (Wu et al. 1996). Alternatively, self-inactivating (SIN) vectors, which do not possess any promoter on the 3′ or 5′ end of their genome after integration into the host cell genome (Yu et al. 1986), can be used. The LTR can also be replaced by the tissue-specific promoter. The problems of tissue-specific expression seen in adenovirus or retrovirus vectors do not seem to occur in AAV vectors (Su et al. 1996; Zhou et al. 1996). This is probably because the flanking ITRs possess no regulative activity. Recently, maintenance of tissue-specific expression was reported for the PSA promoter in a lentiviral vector (Wu et al. 2001).

A major problem with tumor-specific promoters, whose specificity may be satisfying, can be insufficient promoter strength. An ingenious strategy to compensate for the weakness of tumor-specific promoters has been made possible through the introduction of the CRE lox system into the concept of transcriptional targeting. On one of the constructs, the tumor-specific promoter regulates the expression of the CRE recombinase. This enzyme then recognizes loxP sites (on a second construct), which flank a spacer region separating a strong viral promoter and a therapeutic transgene. Cleavage of the loxP sites excises the spacer, thereby inducing transcription of the transgene. Expression of the transgene with this system is supposed to be higher than with simple tumor-specific expression because the stability of the CRE enzyme will amplify the tumor-specific effect. The utility of this system has been proven using the AFP promoter in vitro (Sakai et al. 2001) and the CEA promoter in vivo (Kijima et al. 1999). Other strategies to compensate for promoter weakness have used transcriptional transactivators under tissue-specific control, which then have activated a minimal promoter for transgene expression (Koch et al. 2001; Qiao et al. 2002; Zhang et al. 2002). So far, all these systems have used two types of adenoviruses for either promoter/transgene construct. However, incorporation of both expression cassettes on one vector should be possible.

As in other fields of gene therapy, the packaging size of the vector to be used for gene transfer is one of the most important limitations for the construction of cancer-specific expression cassettes. So far, vectorologists have mainly worked with the intron-free cDNA coding for the transgene. Recently, the availability of vectors with substantially increased packaging capacity, such as HD adenoviruses, has allowed the inclusion of whole minigenes into a viral vector (Mitani et al. 1995; Fisher et al. 1996; Kochanek et al. 1996; Parks et al. 1996). This provides the necessary space for longer and potentially even more specific regulatory elements.

In contrast to constitutive promoters, tumor specificity in patients who are resistant to chemotherapy could be mediated by therapy-inducible promoters. In this respect, the examination of gene regulation mechanisms in cancer cells like the multidrug resistance gene (MDR1), the X-irradiation-induced tissue-type plasminogen activator (t-PA), the early growth response gene (Egr-1), the human heat shock protein HSP 70, or the glucose-regulated protein (GRP78) led to the discovery of a class of promoter sequences that are involved in such stress responses. These promoters carry responsive elements that are inducible either by radiotherapy, cytostatic drugs, or hyperthermia (all conventional treatment modalities). It has already been shown that the expression of therapeutic genes could be enhanced and the efficacy increased if placed under the control of therapy-inducible promoters (Joki et al. 1995). The combination of therapeutic genes under the control of therapy-inducible promoters used with conventional cancer treatment methods could enhance overall treatment efficacy and also retain specificity.

36.3.5 Targeted Vectors (Transductional Targeting)

Currently, the predominantly used viral vectors for gene transfer can infect a broad variety of target cells and tissues. Although this is of interest for cell type–independent gene expression, it is a disadvantage when tissue-specific gene expression is required as in cancer gene therapy. Therefore, the surface of the vector needs to be modified to infect the cancer cell in a more specific manner (retargeting).

Several approaches have been developed to redirect the tropism of adenoviral vectors in favor of cancer cells. The efficacy of binding to the surface of a given target cell by an adenovirus depends on the presence of specific receptors like the coxsackie and adenovirus receptor (CAR) and to a lesser extent the alpha2 domain of MHC class I on the cell surface, which interact with the adenoviral fiber knob protein (Wickham et al. 1993; Bergelson et al. 1997; Hong et al. 1997; Tomko et al. 1997). After binding, the adenovirus is internalized through clathrin-coated pits. This process is mediated through an interaction between RGD motifs on adenoviral penton base protein loops and integrins $\alpha_v\beta_3$ and $\alpha_v\beta_5$. Two major strategies have been used to redirect adenoviral infection: conjugate-based strategies and genetic targeting strategies. Conjugate-based strategies include bispecific recombinant proteins, peptides, or chemical conjugates to redirect adenoviral infection. Examples include bispecific conjugates consisting of the Fab fragment of a neutralizing antiknob monoclonal antibody covalently linked to folate (Douglas et al. 1996; Wickham et al. 1996; Goldman et al. 1997); polymer-mediated ligand coupling to adenoviral capsid, or avidin bridging between biotinylated adenovirus; and biotinylated potential ligands for cellular receptors. Increases of 100-fold and more in infectability have been described in several *in vitro* studies as compared to wild-type adenovirus. The best results reported so far for *in vivo* systemic vascular delivery of two-component targeted vectors were obtained using adenovirus with bispecific antibodies. These blocked the CAR-binding domain and redirected the vector either to FGF receptors on tumor cells, which led to substantially increased survival using the HSV tk/GCV approach (Gu et al. 1999) or to angiotensin-converting enzyme on pulmonary capillary epithelium, which enhanced gene transfer to the lung at least 20-fold. The same group has recently initiated the first clinical study using retargeted adenoviruses for the treatment of cancer with a suicide gene approach (Barnett et al. 2002).

Conjugate-based approaches have two main advantages: the wide range of cellular receptors, which can be targeted and the fact that there is no need to make structural changes to the adenovirus itself. The disadvantages of the system include the larger size and irregular shape of the particles, as well as regulatory issues such as the need to obtain approval of all components of a two- or three-component system.

These problems can be circumvented by genetic targeting approaches, which genetically modify fiber, penton base, and hexon capsid proteins. By far the most extensively studied alterations have been done on fiber knob, including fiber knob pseudotyping, genetic incorporation of targeting ligands into the fiber protein, and genetic fiber replacement strategies. Fortunately, two sites in the fiber protein have been identified where ligands can be inserted safely without disrupting fiber knob trimerization. These are the C-terminal end of the fiber tail, which projects from the surface of the virion, and the so-called HI loop within the fiber knob. Incorporation of targeting ligands led to increased infectabilitiy of target cells *in vitro* of up to three orders of magnitude (Amalfitano and Parks 2002; Nicklin and Baker 2002). Even large reporter enzyme genes for imaging can be inserted into the C-terminus of protein IX (Li et al. 2005, 2006). The advantage of entire replacement of fiber over just incorporating ligands is the possibility of achieving real retargeting because only then can initial affinity of adenovirus to CAR be abrogated. To date, attempts have been successfully made to abrogate infection of CAR positive cells, but infectability of retargeted adenoviruses in general was lowered 20-fold (van Beusechem et al. 2000; Magnusson et al. 2001). A pseudotyped adenoviral vector has recently been utilized for targeting replication competent adenovirus to liver metastases and an increase of the transduction rate from 0% to 8% was observed (Shayakhmetov et al. 2002).

A combination of conjugate-based strategies and genetic targeting has recently been described. Efficient and selective gene transfer into human brain tumor spheroids has been achieved using antibody-targeted vectors; native tropism to both CAR and integrin were abolished (van Beusechem et al. 2002).

A variety of retargeting attempts have recently been undertaken for AAV. The primary attachment receptor for AAV is heparan sulfate proteoglycan (Summerford and Samulski 1998), and coreceptors are fibroblast growth factor receptor-1 (Qing et al. 1999) and $\alpha_v\beta_5$ integrins (Summerford et al. 1999). These receptors are widely expressed in the human body. Retargeting is therefore feasible especially as AAV is an integrating vector. In addition, it seems that despite efficient internalization, nuclear trafficking of AAV may be inefficient. Retargeting to new receptors could allow for new and more efficient ways of trafficking. As for adenoviruses, genetic and nongenetic approaches have been used to retarget AAV. The former strategy has been more popular by far, however. *In vivo* gene transfer has been performed with AAV-2 pseudotyped with AAV-5 capsid or AAV-8 capsid; this produced higher levels of transduction than wild-type AAV-2 in injected skeletal mouse muscle or liver, respectively (Gao et al. 2002).

A lot of effort has recently been put into genetic incorporation of targeting ligands into the AAV capsid. In contrast to adenovirus, the structure of the AAV capsid has not yet been elucidated. This will eventually enable pinpointing of the capsid epitopes, which mediate cell surface receptor binding and enable specific site-directed mutagenesis for ablation of wild-type AAV-2 tropism. Despite this drawback, by modeling the AAV capsid on the known structure of canine parvovirus, several studies have already demonstrated that genetic incorporation of peptide epitopes into the AAV capsid can modify tropism.

The first genetically targeted HD-Ad has been published (Biermann et al. 2001). The wild types of commonly used retroviruses such as ecotropic MLV strains do not recognize a receptor on human cells. They have therefore been pseudotyped with natural viral fusion proteins such as the amphotrophic MLV env protein or the vesicular stomatitis virus (VSV)-G glycoproteins (Yee et al. 1994). These pseudotyped retroviruses now infect a variety of human cells. This is not always desirable, as in the case of adenovirus and AAV. Therefore, retargeting of these vectors to specific cell types has been attempted. In addition, studies have shown that several cell types including hematopoietic stem cells are still insufficiently infected by pseudotyped vectors. Some specificity is achieved just by using other viruses for pseudotyping (such as gibbon ape leukemia virus [GaLV] [von Kalle et al. 1994], human foamy virus [HFV] [Lindemann et al. 1997], simian immunodeficiency virus [Indraccolo et al. 1998], and HIV-1 [Mammano et al. 1997]).

More was expected from direct targeting approaches such as inclusion of a wide variety of ligands into the MLV glycoproteins. Efficient retargeting was achieved in several cases. Unfortunately, upon binding most of the chimeras were unable to trigger the highly complicated fusion events of viral envelope and target cell membrane necessary for efficient viral transduction. The most straightforward approach to circumvent this problem was the creation of tethering or escorting ligands. These were again cloned into the MLV env glycoproteins, but at a site which does not abrogate wild-type receptor interaction and disturb fusion. This type of retargeting has already proven successful for preclinical cancer therapy. Systemic application of an ECM-targeted retrovirus allowed enhanced transduction of tumor tissue and significantly prolonged survival of mice as compared to a nontargeted vector (Gordon et al. 2001). During these studies, an observation was made that was paradoxical at first glance: some receptor ligands conferred excellent binding to target cells but due to lysosomal sequestration, no gene transfer was achieved (Cosset et al. 1995; Fielding et al. 1998; Russell and Cosset 1999). This phenomenon, which has been called inverse targeting, can cause the selective transduction of EGF-receptor negative tissue if the vector has been engineered to contain EGF and if the EGF-receptor negative tissue is principally infectable by the wild-type vector (as has been demonstrated by inverse targeting of lentiviral vectors to the spleen) (Peng et al. 2001). This type of inverse targeting can be reversed into true targeting by inclusion of a protease cleavage site inserted between the EGF domain and the amphotrophic receptor. In this case, EGF serves as a tethering molecule but extracellular proteases

present on the target cells cleave it off and efficient infection occurs. This approach has proven to discriminate between (matrix metallo) protease rich and poor in EGF receptor-positive tumors in nude mice (Peng et al. 1999).

Another innovative way to achieve retargeting without losing fusion uses proteins of nonretroviral enveloped viruses instead of the natural receptors. These proteins are able to trigger membrane fusion. MLV vectors coated with the influenza virus hemagglutinin (HA) glycoprotein, including a cell surface-specific ligand, could be selectively targeted to and incorporated into cells expressing the expected target cell surface molecules (Hatziioannou et al. 1999). Finally, similarly to adenovirus and AAV, bifunctional bridge proteins have been examined for targeting with success *in vitro* (Boerger et al. 1999).

The benefit of combining transcriptional targeting and tropism-modified targeting was recently demonstrated using the endothelial cell-specific FLT-1 promoter and an anti-ACE antibody. This combination achieved an impressive 300,000-fold improvement in adenoviral transduction of the lung (Reynolds et al. 2001).

36.4 CLINICAL TARGETS FOR CANCER GENE THERAPY

36.4.1 Bladder Cancer

Superficial transitional cell cancer (TCC) of the bladder is the fifth most common solid malignancy in the United States (Greenlee et al. 2000). Seventy to eighty percent of patients with bladder cancer present with these low-grade, noninvasive tumors confined to mucosa. Although most superficial cancers can be managed with periodic transurethral resection, this is not an ideal situation because recurrence is the rule (Dalbagni and Herr 2000). Even with close surveillance and follow-up, at least 50% will eventually require a cystectomy. Intravesical chemotherapy or installation of Bacille Calmette–Guérin can prolong the disease, but still 30% of patients will die of recurrent metastatic bladder cancer within 15 years (Dalbagni and Herr 2000).

Intravesical gene therapy is a promising new approach for the treatment of refractory superficial bladder cancer. Several approaches have proven preclinical efficacy, including transfer of immunomodulatory genes such as IL-2 (Horiguchi et al. 2000), IFN-γ (Hashimura et al. 1997), and IFN-α2b (Adam et al. 2007); installation of tumor cells transfected with GM-CSF, IL-2, or HLA B7-1; HSV tkHSV tk-based suicide approaches (Jin et al. 1995; Sandig et al. 1997); oncolytic herpes virus, which was interestingly efficient against metastases if injected systemically (Gallardo et al. 1996) and transfer of p16 (Fujiwara et al. 1994), pRb (Spitz et al. 1996), or p53 (Melero et al. 1997).

A clinical trial of installation of empty vaccinia virus did not lead to any systemic toxicity but recruitment of lymphocytes and induction of a brisk local inflammatory response were seen (Gomella et al. 2001). Intravesical installation (but not intratumoral injection) of Adeno-p53 led to transgene expression and induction of the p53 target gene p21 in patients with bladder cancer awaiting total bladder resection. No dose-limiting toxicity was observed at doses of up to 7.5×10^{13} particles (Kuball et al. 2002). The superficial nature of TCC and the unique opportunity of intravesical vector installation make this disease particularly suited for cancer gene therapy at the current state of development.

36.4.2 BREAST CANCER

One of the most important target malignancies for cancer gene therapy is breast cancer. Breast cancer will affect one in every nine women in the United States and a similar incidence is seen in Europe (Harris et al. 993). Conventional treatments such as surgery, radiotherapy, and adjuvant systemic therapy allow disease-free survival for many years. However, the loco regional recurrence rate and the rate of disseminated disease are high, and even 10%–40% of patients without obvious axillary lymph node involvement at the time of surgery relapse (Miller et al. 1993). In these cases, curability by conventional methods is very unlikely.

Based on the unsatisfying outcome of classical strategies to improve cancer treatment, several gene transfer approaches are under experimental and clinical investigation. These include immunological approaches or the transfer of tumor suppressor genes or prodrug-activating enzymes.

Several immunological strategies are based on the use of tumor-specific antigens to improve recognition of breast cancer cells by the effector cells of the immune system. In this respect, several tumor-specific antigens are under experimental investigation to set up a specific vaccination strategy for patients with breast cancer. Candidate antigens are mucin 1 (MUC-1), MAGE-1, CEA, and members of the erbB gene family of cell surface receptors (Ruppert et al. 1997). A recombinant vaccinia virus expressing MUC1 and IL-2 has been used in a phase I/II study in nine patients with advanced breast cancer (Scholl et al. 2000); one patient had a concomitant decrease in CEA serum levels and remained clinically stable for 10 weeks.

Among the first immunological approaches for breast cancer patients was a phase I study of immunotherapy of cutaneous metastasis using allogenic (A2, HLA-B13) and xenogenic (HLA- H-2K (k)) MHC–DNA liposome complexes (Hui et al. 1997). Several partial and complete responses of the injected tumorous nodules were observed.

Intratumoral adenovirus-mediated gene transfer of the 4-1BB costimulatory molecule-ligand (ADV/4-1BBL) to liver metastases in a syngeneic animal model of breast cancer induced a dramatic regression of pre-established tumor (Martinet et al. 2002). Excellent tumor regressions were also observed with a combination of IL-2- and IL-12-expressing adenoviruses in a mouse model of mammary adenocarcinoma; 63% of animals underwent complete regression of both treated and untreated tumors (Addison et al. 1998). Direct *in vivo* IFN-β gene delivery into established tumors generated high local concentrations of IFN-β, inhibited tumor growth, and in many cases caused complete tumor regression. Because the mice

were immune-deficient, it is likely that the antitumor effect was exerted primarily through direct interferon-mediated inhibition of tumor cell proliferation and survival (Qin et al. 1998).

Gene transfer strategies for breast cancer patients can also be used to improve classical strategies in breast cancer therapy like chemotherapy. In this field, purging techniques in high-dose chemotherapy are under intensive experimental investigation. Before initiation of high-dose chemotherapy for women with breast cancer, autologous stem cell transplants are collected from the patient and are given back after the therapeutic protocol is completed. Because these preparations can still be contaminated by tumor cells, different strategies aim at purging these stem cell transplants from contamination. However, magnetic purging techniques are expensive and not very efficient while pharmacological purging is very effective but toxic not only to the tumor (Hildebrandt et al. 1997). Therefore, to limit the toxicity of pharmacological purging of contaminating cancer cells, either hematopoietic precursor cells of the autologous stem cell transplants need to be made resistant to chemotherapy, or contaminating tumor cells need to be made more sensitive to chemotherapy. Thus, one genetic approach tries to infect only hematopoietic cells and make them resistant to chemotherapy through transfer of a multidrug resistant gene (Frey et al. 1998). An example of the other strategy of making contaminating tumor cells more sensitive to chemotherapy is adenovirus-mediated gene transfer of a prodrug-activating enzyme like HSV-tk or CD (Chen et al. 1995; Yee et al. 1996; Garcia-Sanchez et al. 1998). This approach exploits the fact that breast cancer cells, which contaminate autologous steam cell transplants, express more adenovirus-internalizing integrins than hematopoietic precursor cells (Seth et al. 1996; Wroblewski et al. 1996; Li 1997). Preclinical studies have already demonstrated that the concept of adenovirus-mediated gene transfer into contaminating breast cancer cells of autologous stem cells is a real alternative strategy for direct clinical application (Wolff et al. 1998; Lillo et al. 2002). Purging of hematopoietic stem cells from breast cancer cells has also been successfully performed with conditionally replication-competent HSV (Wu et al. 2001) and a complete lack of toxicity has been shown for a p53 adenovirus (Hirai et al. 2001). Bone marrow cells themselves are the target of retroviral transfer of MDR genes. Recent clinical trials showed a selective enrichment of these cells by subsequent chemotherapy and the maintenance of transduced cells for more than 1 year has been reported (Takahashi et al. 2006).

Another strategy, which has been adapted for gene therapy of breast cancer, is the use of antioncogene and tumor suppressor genes. *In vivo* administration of BRCA1 led to tumor growth retardation comparable to that obtained with p53 or p21 (Randrianarison et al. 2001).

Hydrodynamic-based gene delivery of a secreted form of the TNF-related apoptosis-inducing ligand, TRAIL, led to the regression of a human breast tumor established in SCID mice (Wu et al. 2001).

Adenovirus-mediated transfer of the cDNA of the pro-apoptotic bcl-x$_s$ into breast cancer cells led to a significant reduction of tumor growth after the transduced cells were transplanted into immunodeficient mice (Clarke et al. 1995; Ealovega et al. 1996). As described in the literature on colorectal cancer, breast cancer also develops by a succession of genetic alterations (Mars 1990; Cho and Vogelstein 1992; Parsons et al. 1993; Cox et al. 1994; Devilee and Cornelisse 1994). Although the degree of genetic heterogeneity is particularly high in breast cancer, these multiple genetic changes may interfere with just a few critical cell cycle regulatory pathways and therefore represent suitable targets for corrective gene therapy (Lukas et al. 1995). Successful *in vitro* approaches include oligonucleoutide-mediated transfer of antisense myc (McManaway et al. 1990; Watson et al. 1991), ErbB-2 (Bertram et al. 1994; Casalini et al. 1997), cyclin D1 (Zhou et al. 1995), and TGF alpha (Kenney et al. 1993); the transfer of genes for intracellular antibodies, which prevent growth factor receptors to reach the cell surface, as reported for ErbB-2 (Wright et al. 1997); and other transgenes such as PKA, EGFR, TGF-beta, IGFIR, P12, MDM2, BRCA, Bcl-2, ER, VEGF, MDR, ferritin, transferrin receptor, IRE, C-fos, HSP27, C-myc, C-raf, and metallothionin genes (Head et al. 2002). Some of these have already been tested *in vivo*. For example, excellent inhibition of breast tumor growth *in vivo* was shown by retroviral transfer of antisense C-FOS (Arteaga and Holt 1996).

The best existing *in vivo* data so far has been seen with adenoviral transfer of the tumor suppressor p53, which argues for the fruitfulness of combining a highly efficient gene transfer vehicle with a nearly universal apoptosis inducer (Nielsen et al. 1997). In addition, good efficacy of tumor growth reduction was achieved using liposomes for the transfer of a p53 cDNA (Xu et al. 1997). The importance of the p53 status for the efficacy of chemotherapeutic drugs is a matter of intense debate. Whereas in the majority of studies, the loss of p53 has been associated with decreased sensitivity to chemotherapy, the opposite was reported for the chemotherapeutic drug taxol (Wahl et al. 1996). An interesting approach to counteract the MDM2-protein, which is often elevated in wtp53 positive tumors, was conducted by generating a recombinant adenovirus expressing a p53 variant (i.e., deleted for the amino acid sequence necessary for MDM2 binding). Apoptotic activity of rAd-p53-expressing the mutated p53 with that of a recombinant adenovirus expressing wild-type p53 in cell lines that differed in endogenous p53 status caused higher levels of apoptosis in p53 wild-type tumor lines compared with wild-type p53 treatment, showed apoptotic activity similar to that seen with wild-type p53 treatment in p53-altered tumor lines, and showed greater antitumor activity in an established p53 wild-type (hepatocellular) tumor compared with treatment with wild-type p53 (Atencio et al. 2001). With a similarly mutated adenovirus, tumor regressions were also observed in a sarcoma model (Bougeret et al. 2000). In combination with chemotherapy, coapplication of docorubicin did show an overadditive effect (Lebedeva et al. 2001) whereas cyclophosphamide did not (Nielsen 2000).

Stable preclinical tumor regressions were also achieved by intratumoral injection of an adenovirus harboring the adenoviral E1A gene under the control of the tumor-specific MUC1 promoter with and without an additional TNF-α transgene expression cassette (Kurihara et al. 2000). In a clinical trial, an E1A gene complexed with DCC-E1A cationic liposome was injected once a week into the thoracic or peritoneal cavity of 18 patients with advanced cancer of the breast ($n = 6$) or ovary ($n = 12$) and led to HER-2/neu downregulation, increased apoptosis, and reduced proliferation (Hortobagyi et al. 2001).

Injection of a plasmid harboring the CD gene under the control of the tumor-specific erbB-2 promoter in 12 patients with breast cancer showed the safety of the approach; expression of the transgene was seen in 90% of cases although no antitumor efficacy was reported (Pandha et al. 1999). Retroviral transfer of the suicide gene P450 led to partial responses and stable disease in patients with cutaneous metastases upon application of cyclophosphamid (Braybrooke et al. 2005).

Injection of adenovirus-harboring dominant negative mutants of estrogen receptors into pre-established T47D tumors in nude mice induced tumor regressions (Lee et al. 2001).

Finally, *in vivo* studies of reovirus breast cancer therapy revealed that viral administration could cause tumor regression in an MDA-MB-435S mammary fat pad model in SCID mice. Reovirus could also effect regression of tumors remote from the injection site in an MDA-MB-468 bilateral tumor model, raising the possibility of systemic therapy of breast cancer by the oncolytic agent (Norman et al. 2002).

In summary, although metastatic breast cancer poses high demands on current cancer gene therapists, nearly all therapeutic principles have been evaluated in animals and some are already in clinical evaluation.

36.4.3 COLORECTAL CANCER

Colorectal cancer is the third most frequently occurring cancer in the United States (Parker et al. 1997). Surgical removal of the primary tumor is the established first-line therapy. This strategy can be a curative approach if all cancer cells are eliminated. However, already at a very early stage of tumor development, colorectal cancer metastasizes to the liver. In 60% of all cases, the only manifestation of distant metastasis (Kemeny and Seiter 1993) occurs in this organ. Consequently, liver metastasis of colorectal cancer is one of the very few indications where treatment of metastasis can lead to a significant improvement of the prognosis of the disease.

Single liver metastasis of colorectal cancer represents a promising target for intratumoral or regional gene therapy. Studies in experimental animals demonstrated significant tumor reduction by gene transfer of the tumor suppressor gene p53 (Harris et al. 1996). However, neither a clinical trial applying p53 adenovirus into the hepatic artery (Venook 1998; Chung-Faye et al. 2000) nor intratumoral injections (Habib et al. 1995) have revealed any clinical response. The suicide approach by gene transfer of prodrug-activating enzymes like cytosine deaminase, HSV-tk, or nitroreductase followed by systemic application of the corresponding prodrugs 5-fluorocytosine, GCV, or CB1954 has led to good results in animal experiments (Kemeny et al. 1993; Hirschowitz et al. 1995; Djeha et al. 2001).

A single dose of adenovirus transferring the HSV-tk gene, followed by a 10-day intraperitoneal GCV treatment, led to a reduced tumor growth of more than 90% and a significant reduction of the tumor volume (O'Malley et al. 1997; Schrewe et al. 1990). This experimental approach was extended by the combination of adenovirus vectors carrying the transgene of different cytokines with the adenovirus encoding for HSV-tk. The results so far have suggested that the HSV-tk/GCV effect can be increased by simultaneous cytokine gene expression (O'Malley et al. 1997). However, in an orthotopic model of colon carcinoma metastatic to the liver, the HSV-tk strategy was compromised by severe hepatic toxicity and the death of several animals (Brand et al. 1997). Similar toxicity was seen if the vector was applied intraportally for the treatment of HCC in mice or rats (van der Eb et al. 1998). In contrast, no toxicity was observed with retroviral vectors (Caruso et al. 1996) (own findings). The liver toxicity observed using adenovirus vectors could be abrogated if the CMV promoter was replaced by the colon-specific CEA promoter in these vectors (Brand et al. 1998). The generation of tumor cell-specific transgene expression in colorectal liver metastases by adenovirus vectors can also be accomplished by modification of the adenoviral fiber protein by inclusion of the CEA-receptor (Curiel et al. 1992). Recently, the results of an Adeno-HSV tk/GCV phase I trial with intratumoral injection of an RSV.HSV tk-adenovirus were reported. Ultrasound-guided percutaneous injection of up to 10^{13} adenoviral vector particles only caused mild hepatic toxicity upon GCV treatment; this may indicate that extensive leakage into the liver tissue was prevented. No clinical benefit was observed (Sung et al. 2001).

A clinical phase I trial with conditionally oncolytic replication-competent E1B55K deleted adenovirus was performed (Habib et al. 2001). It was possible to safely deliver virus by direct intratumoral, intraarterial, and intravenous injection up to a dose of 3×10^{11} infectious particles. In a phase II trial for patients with colorectal liver metastases, virus was administered intraarterially together with the chemotherapeutic 5-FU. After 3 months, stable disease was demonstrated in six of seven patients (Habib et al. 2001). More recently, the combination of RCA with chemotherapy again showed overadditive effects when 25% of patients with colorectal liver metastases displayed marginal to partial response (Reid et al. 2002). The sole application of Onyx-15 RCA into the hepatic artery of patients with colorectal liver metastases, which had failed 5-FU therapy, lead to increased survival in a phase II study (Reid et al. 2005). Among other potential oncolytic viruses HSV showed promising effects in animal models (Bennett et al. 2002) and is now used in a clinical trial for patients with gastrointestinal cancer metastatic to the liver.

Antiangiogenic and antiproteolytic strategies have been implemented but are still in the preclinical phase. A prevention of colorectal liver metastasis of up to 95% was achieved when Ad-TIMP-2 or Ad-TIMP-1 were injected into the tail vein of mice; this leads to a preferential transduction and quasi-impregnation of the liver with the antitumor gene (Brand et al. 2000; Elezkurtaj et al. 2004)). Good preclinical results have been obtained with the proangiogenic endostatin gene as well, which was transferred in models for primary and metastatic liver cancer either alone or in combination with chemotherapy (for review, see Hernandez-Alcoceba et al. 2006).

Multiple liver metastases are also targets for immunological gene therapy strategies. In this respect, the CEA is one of the most promising candidates under investigation. In two clinical studies of recombinant vaccinia virus–CEA immunization, T-cell responses were demonstrated (Conry et al. 1995; Tsang et al. 1995). In a recently published clinical trial of a dual expression plasmid encoding CEA and hepatitis B surface antigen (HBsAg) in 17 patients with metastatic colorectal carcinoma, four patients developed lymphoproliferative responses to CEA after vaccination although no objective clinical responses to the DNA vaccine were observed (Conry et al. 2002). A combination of CEA and costimulatory molecules has revealed a CEA-specific immune response and some positive clinical results (for review, see Mohr et al. 2007) such that subsequent trials are active at Duke and Georgtown Universities. In addition, significant tumor growth inhibition of xenografts from colon tumor cells in established animal models has also been accomplished using different approaches. These include vaccinia virus-mediated transfer of B7-1 and IL-12 (Rao et al. 1996); adenovirus-mediated transfer of IL-12 (Caruso et al. 1996); liposomal transfer of MHC class I molecules (Plautz et al. 1993); or the transfer of fibroblasts, which had been transduced in vitro with IL-2 (Fakhrai et al. 1995) (this last result has led to the initiation of a clinical phase I study) (Sobol et al. 1995). The results of these experimental studies indicate that the dormant or suppressed immunogenicity of colon tumor cells can be evoked by several immunomodulatory mechanisms. Moreover, the good antitumor efficacy achieved even with vector systems, which traditionally suffer from low gene transfer efficacy, suggests that a certain level of gene transfer may be sufficient to induce an immune response.

In summary, colorectal liver metastases are unique. Their cure could mean a cure for the patient. Therefore, they have served as the model disease for gene therapy of metastatic disease. Initial, encouraging clinical results have been obtained.

36.4.4 GLIOBLASTOMA

Glioblastoma is the most common primary brain tumor. Despite advances in diagnosis and treatment, the median survival time is still only 1 year from the time of diagnosis (Mahaley et al. 1989). This tumor rarely metastasizes to distant organs, suggesting that improvements in local treatment could be of great benefit. Therefore, glioblastoma has been one of the model diseases for gene therapy with suicide genes. Reviews on this topic are provided by Rainov et al. (2001) and Lam and Breakefield (2001). The first preclinical studies for cancer gene therapy were performed using retrovirus-mediated transfer of the HSV-tk gene into established intracranial glioblastomas in Fisher rats (Culver et al. 1992). Based on these early promising results, other vector systems have been tested for the HSV-tk/GCV approach including adenoviruses,

TABLE 36.1

Selected Recent Reviews and Books on Some of the Covered Topics

Topic	References
Cancer gene therapy in general	McCormick (2001), Lattime (2002), Cross and Burmester (2006)
Oncolytic viruses	Wildner (2001)
Angiogenesis	Scappaticci (2002)
Antiproteolysis	Brand (2002)
Non viral vectors general	Nishikawa and Huang (2001), Davis (2002), Liu and Huang (2002), Gao et al. (2007)
Cationic polymers	Merdan et al. (2002)
Targeting	
Adenovirus	Amalfitano and Parks (2002), Barnett et al. (2002), Nicklin and Baker (2002); Nicklin et al. (2005), Campos and Barry (2007)
Retrovirus	Lavillette et al. (2001)
AAV	Nicklin and Baker (2002)
Lentivirus	Nicklin and Baker (2002)
General	Russ and Wagner (2007)
RNAi	Zhang et al. (2006)
Cancer types	
Bladder	Ardelt and Bohle (2002)
Bone	Witlox et al. (2007)
Breast	Plunkett and Miles (2002), Takahashi et al. (2006)
Brain	Lam and Breakefield (2001), Rainov and Kramm (2001), Lesniak (2006)
Colon	Chung-Faye et al. (2000), Havlik et al. (2002), Mohr et al. (2007)
Head and neck	Nemunaitis and O'Brien (2002), St John et al. (2006)
Hematological	Brenner (2001), Russell and Dunbar (2001), Buttgereit and Schmidt-Wolf (2002), Schmidt-Wolf and Schmidt-Wolf (2002), Larsen and Rasko (2005)
Liver	Hernandez-Alcoceba et al. (2006)
Lung	Roth et al. (2001), Albelda et al. (2002)
Melanoma	Sotomayor et al. (2002)
Ovary	Wolf and Jenkins (2002), Kimball et al. (2006)
Pancreas	Kasuya et al. (2001), Gilliam and Watson (2002), Bhattacharyya and Lemoine (2006)
Prostate	Mabjeesh et al. (2002)

liposomes, AAV, or HSV. All these vectors have been successfully used in animal studies where it was frequently possible to observe complete remissions with long-term survival (Table 36.1). Consequently, several clinical phase I, II, and III trials were initiated making the gene therapy of glioblastoma by the HSV-tk/GSV approach the most advanced system for cancer gene therapy. The suicide gene approach has been combined with surgery. As much malignant tissue as possible is removed, leaving the local infiltrating parts for multiple vector injections. This strategy has to leave out the large isles of healthy tissue within the tumor network, which is a characteristic of this tumor. This is the reason why complete resections can be rarely performed. As demonstrated by results from a multicenter phase II trial in Germany, there is no clearcut clinical outcome so far. In a 1 year follow-up report of 10 patients with recurrent glioblastoma multiforme, where retroviral vector packaging cells were administered into the tumor followed by application of GCV, 4 of the 10 patients died because of tumor progression. Of the other six patients, one presented a complete remission at 12 months, and five had progressive disease but with a significant increase in quality

of life (Weber et al. 1997). Other reports demonstrated responses in the CT scan; a clear enhancement was visible in the areas were the retroviral VPCs carrying the transgene had been injected (Ram et al. 1997). However, clinical responses were rare and not marked. Unfortunately, the first phase III trial with HSV tk retroVPCs did not have any treatment benefit (Rainov et al. 2001). Some responders were observed in trials using adenoviral vectors but therapy was accompanied by severe neurological symptoms (Alavi et al. 1998; Trask et al. 2000). As discussed in the context of colorectal liver metastases, toxicity can in principle be abrogated by tissue-specific expression of the transgene, e.g., by the nestin (Kurihara et al. 2000) or myelin basic protein (Shinoura et al. 2000) promoter or targeted vectors (Rainov et al. 2001). Several phase I trials with conditionally replication-competent HSVs have been performed to date. Toxicities related to the vector were not reported and occasionally exciting antitumor efficacy was seen (Markert et al. 2000; Rampling et al. 2000; Papanastassiou et al. 2002; Harrow et al. 2004). Conditionally replication-competent adenovirus (e.g., Onyx-015) has been evaluated in clinical trials as well (Chiocca et al. 2004) and most recent clinical data indicates that the injection of the virus is well tolerated and occasional regressions can occur.

Besides the strategy of using prodrug-converting enzymes, gene transfer of the tumor suppressor gene p53 has also been commonly tested in experimental models with success either alone (Badie et al. 1995; Cirielli et al. 1999) or in combination with radiotherapy and chemotherapy (Trepel et al. 1998; Biroccio et al. 1999; Broaddus et al. 1999). Among the cell cycle inhibitory proteins, p16 is very often inactivated in glioblastoma. Adenovirus-mediated gene transfer of p16 generated significant tumor growth reduction in p16-negative glioblastomas (Fueyo et al. 1996). For the transcription factor E2F-1, it could be demonstrated that adenovirus-mediated overexpression resulted in a tumor growth reduction in p53 wild-type expressing glioblastoma cells (Fueyo et al. 1998). Other apoptosis-inducing genes (caspases [Shinoura et al. 2000; Jacobson et al. 2000], TNF-α [Walczak et al. 1999; Niranjan et al. 2000]) or antioncogenes (ras, Shu et al. 1999 meningioma) have been successfully evaluated in animal models. In addition to the use of molecules regulating cell cycle and apoptosis, several immunomodulatory genes have been tested to treat experimental glioblastoma tumors either alone and in combination (Sampson et al. 1997; Glick et al. 1999; DiMeco et al. 2000; Herrlinger et al. 2000). Growth inhibition was seen in most cases, which indicates that although the brain is an immunoprivileged site, this barrier could effectively be surmounted at least in some tumors. A clinical trial using interferon-β (Eck et al. 2001) has been initiated. In several preclinical trials the combination of gene/viral therapy with chemotherapy has shown overadditive effects (for review, see Lesniak (2006)).

In summary, although the HSV tk suicide gene trials probably represented one of the earliest sobering results in gene therapy, the disease is certainly a very good candidate for gene therapy. This is especially so if targeted vectors will distinguish between normal brain tissue and the deeply infiltrating and surgically inaccessible tumor branches of glioblastoma multiforme.

36.4.5 Head and Neck Cancer

Each year, in the United States, approximately 40,000 individuals are diagnosed with carcinoma of the head and neck (HNSCC) and upper aerodigestive tract (Society 1993). More than two-thirds of the individuals with HNSCC present with stage II or IV of the disease (Dimery and Hong 1993) and 50–60% of these patients ultimately develop local recurrence despite optimal local therapy. These patients may therefore obtain significant benefit from local or regional gene therapeutic approaches.

A high percentage of these tumors are negative for p53 and several clinical trials with percutanous or endobronchial p53 gene transfer are under way or already closed. Several clinical studies have used adenoviral vectors, while one trial used retroviruses. In summary, no major toxicity was observed. Indications of antitumor efficacy included increased survival times compared to historical controls in four of 15 patients with recurrent disease; dose-related improvements in time-to-disease

progression in a large phase II trial; and partial response or at least stable disease in two-thirds of patients in a trial with 52 patients who had previously failed conventional treatment. Consequently, one p53 adenoviral vector (INGN 201) is now in phase III testing. Most strikingly, the first gene therapeutic cancer medicine to obtain regulatory approval (2003, China) is such an p53 adenovirus (Guo and Xin 2006). Since approval, thousands of patients have been treated with this virus and reports on large-scale efficacy are eagerly awaited.

Despite these findings of clinical efficacy, improvements of the strategy are necessary, mainly because of the requirement of highly efficient gene transfer. Based on the bulky mass of head and neck cancer, it is unlikely that the large tumor burdens, which remain even after radical surgery, can be sufficiently transduced even by the highly efficacious replication-deficient adenoviral vectors. Therefore, and because of the good accessibility, head and neck cancer has become the model disease for therapy with selectively replication-competent adenoviruses. The underlying concept of the use of replication-competent recombinant adenoviruses is based on the capability of adenovirus to induce cell lysis by its progeny inside of an infected target cell (Horwitz 1996; Shenk 1996). To limit the cytotoxicity only to tumor cells and not to normal cells, researchers exploited the fact that about 50% of tumors carry a mutation in the tumor suppressor gene p53. A recombinant replication-competent adenovirus was constructed, which should specifically replicate only in p53 negative tumor cells (Bischoff et al. 1996). For further explanation of this concept, please see the Section 36.2.4. Although the concept underlying the mode of action of the vector is not yet completely understood, several clinical trials have been performed. The published data, mainly with the vector ONYX-015, are among the most encouraging in clinical cancer gene therapy. With respect to toxicity, results suggest that the intratumoral injection of ONYX-15 limits proliferation of the vector to malignant tissue; it is well tolerated except for transient low-grade fever and injection site pain. Three clinical trials have been formally published so far (Ganly et al. 2000; Khuri et al. 2000; Nemunaitis et al. 2001): A phase I trial, a phase II trial for intratumoral injection of ONYX-15 alone, and one trial for the combination of virus and cisplatin. Five of 36 patients in the phase II trial showed partial or even complete (n = 3) responses (Nemunaitis et al. 2001). In the combinatorial approach, 63% as compared to 30%–40% in the historical controls showed partial or complete responses (Khuri et al. 2000). Replication-competent NDVs have led to a complete response in one patient of squamous cell carcinoma of the tonsils (Pecora et al. 2002).

Besides these dominant strategies for HNSCC, immune modulatory approaches have already been tested clinically. Partial responses have been observed in four out of nine patients with advanced HNSCC upon intratumoral injection of HLA-B7 plasmid (Allovectin-7) (Wickham et al. 1993). Among several approaches showing promising preclinical efficacy, including transfer of suicide genes, Bcl-2, superoxid dismutase (SOD), and EGF-receptor, transfer of adenovirus E1A in a liposomal formulation has been tested clinically.

It showed only minimal toxicity and modest clinical efficacy with one out of 24 patients with recurrent, unresectable HNSCC having a complete response; two patients having a minor response; and seven patients having stable disease (Villaret et al. 2002).

With respect to immunogene therapy, liposomal-mediated gene transfer of HLA-B7 costimulatory molecules has led to stable disease and some partial responses in phase I and phase II trials for HNSCC patients (Gleich et al. 2001).

Head and neck cancer is probably the tumor type with the best clinical results in cancer gene therapy obtained so far. This may be due to the good accessibility for intratumoral approaches; the low tendency for metastazation, with the opportunity for long time follow-ups and (perhaps) a favorable immunologic and intracellular molecular situation. These first successes have encouraged many researchers to believe that less favorable starting points in other cancer types can be overcome, leading to clinical efficacy.

36.4.6 HEMATOLOGIC AND LYMPHATIC MALIGNANCIES

These malignancies are the domain of chemotherapy. Due to the development of advanced protocols, the initial rate of remission and the rate of long-term survivors have dramatically increased. In all stages, the primary therapeutic intention is curative. The usual treatment regime consists of several cycles of intensive or high-dose chemotherapy up to a full eradication of the patient's bone marrow followed by autologous or allogeneic bone marrow transplantation. Due to the good susceptibility of leukemic and lymphatic cells for chemotherapy, conventional treatment usually leaves only minimal residual disease. This residual tumor load can, however, lead to relapses in up to 80% of patients with leukemia (Braun et al. 1997) and has usually a bad prognosis. Gene therapy could try to prevent this situation.

Several immunologic approaches have been tried. Preclinical in vivo gene transfer of TNF-alpha was used for T-cell lymphoma (Gillio et al. 1996) and myeloma (Cao et al. 1998), B7 for several lymphomas (Chen et al. 1994; Martin-Fontecha et al. 1996) and GM-CSF for T-cell leukemia (Hsieh et al. 1997). Transfer of IL-2 into lymphomas has been clinically examined (Hersh et al. 1996).

Human primary hematologic malignant cells are unfortunately highly resistant to transduction by most available vectors. Only recently, modifications of vectors and ex vivo transduction protocols have allowed increased gene transfer at least into cultured allogeneic vaccine lines with adenovirus, HSV, liposomes, and retroviral vectors. Another problem is the considerable phenotypic heterogeneity of the cancer, which can cause an incomplete representation of potential antigens of the whole malignant cell population in the population selected for gene transfer and vaccination.

Despite these problems, a considerable number of cancer gene therapy trials using ex vivo cytokine-transduced vaccines have been conducted. Among several cytokines, IL-4 and GM-CSF protected best against tumor challenge, and GM-CSF-transduced cells could also inhibit further progression of preestablished lymphatic tumors (Levitsky et al. 1996) in a mouse model. Most clinical trials used IL-2, but in others IL-4, 7, 12, IFN-γ, and GM-CSF vaccines were applied.

Vaccinations with the DNA of tumor antigens have also been carried out. Gene therapy provides an advantage here over the use of the respective protein when this is difficult to obtain in the required amount or in the correctly glycosylated form. Encouraging results were obtained in a clinical phase I study of the vaccination of patients with chronic lymphatic leukemia (CLL) with CLL cells adenovirally transduced to overexpress the CD 40 ligand (CD 154); a phase II trial has been initiated (Kipps et al. 2000).

Ex vivo gene transfer into effector cells has also been examined clinically. In a phase I clinical trial, autologous cytokine-induced killer (CIK) cells *ex vivo* transduced with IL-2 were infused in patients with metastatic disease. One out of 10 patients had a complete response of his lymphoma (Schmidt-Wolf et al. 1999).

The generation of chimeric T-cell receptors (TCRs) has been another strategy to generate antitumor immunity for B cell lymphoma. T cells are directed against target cells by grafting an antibody V region of desired specificity onto the TCR and the constant regions. The specificity of this approach *in vitro* has been shown (Gross et al. 1995).

The transfer of donor leukocytes (Mackinnon et al. 1995) or EBV-specific cytotoxic T lymphocytes (Rooney et al. 1995) has been a successful approach for the treatment of B cell lymphoma. A potential problem of this strategy, however, could be graft versus host disease (GvHD). An ingenious way to circumvent this problem is the transduction of the donor T cells with a suicide gene like HSV-tk to be able to kill the donor T cells by GCV if signs of GvHD appear (Verzeletti et al. 1998; Bonini et al. 1997). In a clinical trial, GvHD resolved in two of three patients upon application of GCV. A review of the pros and cons of this strategy can be found in Sadelain and Riviere (2002). Gene marking was used to track the *in vivo* persistence and fate of EBV-specific CTLs in 14 patients with relapsed Hodgkin lymphoma. Infused cells expanded by approximately 1000-fold *in vivo* and five patiens were in complete remission at up to 40 months (Larsen and Rasko 2005).

One sustained response in a clinical trial was observed upon the application of TILs engineered to secrete TNF (Brenner 1998). For therapy of multiple myeloma, direct intratumoral gene transfer of the TNF gene was performed preclinically and a substantial gene transfer in nearly 50% of tumors was seen (Wright et al. 1998). Anecdotally, one woman received an injection of an IL-2 adenovirus into an occipital myeloma lesion and gene transfer could be demonstrated although no tumor regression was observed (Stewart et al. 1998). In fact, myeloma cells seem to be among the easiest transducable hematologic cells at least for adenoviruses (Wattel et al. 1996) and AAV (Wendtner et al. 1997).

Lymphomas and leukemias have also been targeted with oligonucleotides against myc (McManaway et al. 1990; Skorski et al. 1995, 1996), bcl-2 (Madrigal et al. 1997), myb (Ratajczak et al. 1992), bcr-abl (Skorski et al. 1995, 1996) and bcr-abl ribozymes (Lange et al. 1993). Transfer efficacy of antisense oligonucleotides *in vitro* and *in vivo* is highly controversial, but a clinical trial using a bcl-2 antisense oligonucleotide (oblimersen, Genasense; Genta Inc. Berkeley Heights, NJ) demonstrates modest clinical activity in patients with refractory/relapsed CLL (O'Brien et al. 2005).

Therapy-associated side effects like infections and hemorrhages account for 70% of the deaths of adult patients with acute leukemia (Mertelsmann 1998). The transfer of MDR genes into hematopoietic stem cells for chemoprotection holds promise and clinical studies of the transduction of the MDR gene in patients with relapsed and resistant lymphomas are already under way. This approach, however, runs the risk that tumor cells are transduced, making them resistant to chemotherapy. Only the future will show whether this problem can be neglected or circumvented (e.g., by targeted vectors).

In the area of oncolytic viruses, attenuated measles virus, which has a natural tropism for lymphoid tissue, has been examined preclinically. The incorporation of an anti-CD20 fusion peptide into the viral capsid retarded the growth of CD20-positive tumors in mice (Bucheit et al. 2003).

In summary, it comes as no surprise that hematologic malignancies as a systemic disease are the domain of immunologic approaches, which have demonstrated clinical success in several cases.

36.4.7 HEPATOCELLULAR CARCINOMA

Although HCC is of moderate epidemiologic relevance in the Western world, it is the most common cancer in large areas in Asia. HCC often remains localized to the liver, but only a minority of patients is amenable for local therapy such as surgery, liver transplantation, or cryoablation. Standard chemotherapy is largely ineffective. Therefore, HCC is a suitable target for intratumoral application of therapeutic genes. As demonstrated, adenovirus vectors can easily infect and express different types of genes in tumor cells of HCC (Sandig et al. 1997; Brand et al. 1999). In addition, the alpha-fetoprotein (AFP)-promoter is extremely HCC-specific (Ido et al. 1995) for achieving tumor-specific gene expression. Using adenovirus-mediated gene transfer, different experimental approaches were studied demonstrating significant tumor volume reductions by IL-12 (Bui et al. 1997), HSV-tk/GCV (Qian et al. 1997; Sa Cunha et al. 2002), and oncolytic adenovirus (Takahashi et al. 2002); combinatorial expression of p53 and p16 (Sandig et al. 1997) or p21 and GM-CSF using an EGF-receptor targeted adenovirus (Liu et al. 2002).

A pilot clinical study to assess the therapeutic potential of percutaneous intratumoral injection of wild-type p53 (wt-p53) in patients with primary hepatocellular carcinoma was initiated in 1996. Nine patients with primary hepatocellular carcinoma and six patients with colorectal liver metastases received percutaneous injections of a wild-type p53 DNA–liposome complex. In contrast to nonresponders with colorectal metastases, four of nine patients with HCC showed a reduction of tumor volume and a significant decrease in serum AFP levels (Habib 1995; Habib et al.

1996). A phase I trial was recently published, which compared intratumoral injection of ethanol with that of conditionally oncolytic replication-competent E1B55K-deleted adenovirus. In the gene therapy group, one patient showed a partial response and four patients had progressive disease. In the ethanol-treated group, two patients had stable disease and three patients showed disease progression (Habib et al. 2002). When hepatobiliary tumors were treated in a different trial, 50% of patients displayed a transient reduction of tumor markers (Makower et al. 2003). Besides RCAs other oncolytic viruses have been investigated. Excellent results with vesicular stomatitis virus (VSV)-derived vectors have been obtained in HCC-bearing rats and these vectors are currently investigated in the clinic (Shinozaki et al. 2004).

An antiangiogenic gene therapy approach has been pursued with a recently identified antiangiogenic factor, the pigment epithelium-derived factor (PEDF) the transfer of which resulted in antitumor efficacy in a murine HCC model (Matsumoto et al. 2004).

With respect to immunologic approaches transfer of the IL-12 cytokines to HCC gave good preclinical results (Barajas et al. 2001; Waehler et al. 2005). The results of a subsequent clinical phase I trial demonstrate the safety and feasibility of intratumoral injection of a first-generation adenoviral vector expressing IL-12 in primary and metastatic liver cancer patients (Sangro et al. 2004).

36.4.8 LUNG AND PLEURAL CANCER

In the Western world, lung cancer has become the most frequent tumor for males. The main reason for this is smoking; more than 90% of these patients are or were smokers. The prognosis of this cancer is bad. Due to regional or systemic metastases, tumors can be resected in only 25% of the patients. Only one-fourth of these patients will survive 5 years; only 6% of the patients are curable. Because of the tendency for early metastasis, gene therapy approaches aiming at a systemic response such as immunogene therapy are needed. A potent evocation of an immune response has only recently been reported in animal experiments (Esandi et al. 1998). Mesothelioma is a rare type of cancer originating from the pleura.

Because lung cancer metastasizes early, this tumor is a very difficult target for gene therapy. However, the high epidemiologic relevance of lung cancer creates an urgent need to develop alternatives to the standard therapeutic approaches. This is especially true for nonsmall cell lung cancer (NSCLC). Because NSCLC is in most of the cases highly resistant to any kind of chemotherapy and well-documented studies about genetic defects in NSCLC exist, this tumor was from the beginning an attractive target for the use of tumor suppressor genes, e.g., p53. Impressive results were obtained in preclinical studies: Significant tumor growth inhibition in subcutaneous and orthotopic animal models (Fujiwara et al. 1994; Zhang et al. 1994). Therefore, one of the first cancer gene therapy trials was initiated to treat NSCLC by gene transfer of p53. A retroviral vector containing the wild-type p53 gene under control of a beta-actin promoter was used. Nine patients for whom conventional treatments had failed received direct injections into the tumor. Despite a low efficacy of gene transfer, and lacking evidence for an involvement of T-cell mediated immunity, partial tumor regression was noted in three patients and tumor growth was stabilized in three other patients (Roth et al. 1996). To improve the *in vivo* gene transfer efficacy, adenoviral vectors were used for p53 gene transfer in three additional studies (Schuler et al. 1998; Swisher et al. 1999; Kubba et al. 2000), and in three more studies p53 adenovirus was combined with chemotherapy (Nemunaitis et al. 2000; Weill et al. 2000; Schuler et al. 2001). In several patients in these trials, transient local control was observed. This was clearly attributed to the gene therapy treatment. Several more tumor suppressor genes have now been evaluated on preclinical models including FUS I (tumor suppressor, now in phase I), p16, and pRB (both targeting the cell cycle G1-restiction point), fragile histidine triad (FHIT) and proapoptotic TRAIL (for review, see Toloza et al. 2006).

Lung cancer has also been a model disease for antioncogene therapy. The K-ras oncogene is frequently overexpressed in lung cancer. Intratracheal transfer of retroviruses carrying an antisense K-ras construct markedly reduced tumor size and the number of lung tumors in nude mice (Georges et al. 1993). Another candidate for antisense therapy is the ErbB2 transmembrane protein kinase receptor whose aberrant expression has been shown to contribute to malignant transformation and progression. The promoter of the ErbB2 gene could be used for expression of toxic transgenes. Promoters that are activated by ionizing radiation (Weichselbaum et al. 1994) have also been used for this purpose. Researchers hope that sufficient specificity for tumor deposits will be achieved in the future, enabling even metastases to be efficiently transduced without toxicity for the surrounding normal tissue. Under these circumstances, the combination of gene therapy with conventional therapy could be a particularly attractive strategy as discussed above for p53 and cisplatin.

Using the powerful siRNA technology, preclinical antitumor efficacy was obtained by knock-down of the cell cycle regulator skp-2 by intratumoral injection of adenoviral vectors harboring the respective skp-2 siRNA for the treatment of subcutaneously established SCLC tumors (Sumimoto et al. 2005). Adenoviral vectors were also used for gene transfer of dominant negative insulin-like growth factor I receptor, which demonstrated antitumor effects in a preclinical model of lung cancer (Lee et al. 2003).

To date, there has been no published trial using suicide-gene therapy in patients with lung cancer. However, there are two reports of the treatment of patients with pleural melanoma (Molnar-Kimber et al. 1998; Sterman et al. 2000). In both trials, HSV tk-adenovirus was injected intratumorally and only minimal side effects were observed. A marked difference between the trials was that in one study, a high dose of glucocorticosteroids was applied to counteract vector-related immune responses (Lin et al. 1998). In this study, partial tumor regressions were observed in some patients; two of them even remained tumor free 3 years after treatment.

Several phase I and II immunogene therapeutic clinical trials for the treatment of lung cancer have been conducted using the adenovirally GM-CSF-transduced allogenic NSCLC cell line GVAX (Simons et al. 1997). An interim analysis of 30 patients from a multicenter phase I and II trial demonstrated a major response rate in 18% of the patients, 10 of whom showed complete tumor remissions (Albelda et al. 2002). Complete clinical responses with this vector were observed in 3 out of 33 patients with advanced stage NSCLC (Nemunaitis et al. 2004). Less favorable outcomes were observed in a subsequent trial in which autologous tumor cells were admixed with allogeneic GMCSF-secreting tumor cells (Nemunaitis et al. 2006).

Partial regression of a lung mass was observed in one of 10 patients treated with IL-2-transfected TILs (Tan et al. 1996). Another immunogene therapeutic approach with promising preclinical results is the application of adenovirally IL-7-transduced DCs for the treatment of lung cancer (Miller et al. 2000). A clinical trial of direct intratumoral injections of vaccinia virus harboring IL-2 for the treatment of pleural mesothelioma led to diverse immunological responses although no significant tumor regressions were seen (Robinson et al. 1998).

Based on the observation that there is an imbalance of proangiogenic ELC+ and antiangiogenic ELC-CXC-chemokines in NSCLC samples, the ELC-CXC-chemokine MIG was transferred to NSCLC tumors, resulting in an inhibition of tumor growth and reduced vessel density in preclinical models (Addison et al. 2000).

In summary, some clinical efficacy with cancer gene therapy has been demonstrated. However, this is clearly neither satisfactory nor sufficient for this devastating disease.

36.4.9 MELANOMA

Malignant melanoma is a tumor of average incidence worldwide but with extremely high incidence in certain areas such as Australia. Worldwide, an annual increase in incidence of 6–7% was reported (Monson 1989). Melanoma is a tumor with a very high resistance to treatment. Treatment schedules including radiation, chemotherapy, and combinations of both have no significant impact on the overall survival of the patients (Ahmann et al. 1989; Ho and Sober 1990). Although the primary tumor can usually easily be excised, distant metastases cause the death of nearly all patients. Early and fatal metastasation is one reason why this disease is an important target for approaches like immunotherapeutic therapies from which a systemic antitumor efficacy can be expected. Fortunately, melanoma displays a naturally high immunogenicity, which facilitates the recognition of the tumor by the immune system.

Basically, all principles of immunomodulation described at the beginning of this chapter have been applied to melanoma. Trials in melanoma alone account for 54% of all open immunogene therapy trials in the United States. At the time of writing, the number of clinical immunogene therapy studies for melanoma worldwide exceeds 50. The earliest trials were based on the transfer of TNF-transduced TILs (Rosenberg et al. 1990). Phase I trials are now closed, but due to minor success, no further trials entirely focusing on TILs have been initiated. However, in a later trial, TILs obtained from a patient's tumor nodules were injected with the HLA-B7 molecule, expanded, and reinfused, and direct immunological effects were demonstrated (DeBruyne et al. 1996). The use of chimeric T-cell receptors with cytokine or melanoma antigen domains is an interesting approach in the area of adoptive gene therapy with genetically modified immune cells. In this approach, CD8+ T-cells are redirected to the tumor and create a favorable cytokine-rich environment (Yee et al. 1997).

Recently, much has been expected from adoptive transfer of DCs, which are probably the most potent antigen-presenting cells in the body. They are also easy to transfect with several vectors. DCs have been transduced *ex vivo* with tumor antigens and costimulatory molecules. MART-1 antigen-transfected DCs in particular generated MART-1 specific immunity and arrested the growth of established tumors (Ribas et al. 1997; Butterfield et al. 1999). In a clinical trial transfer of T-cells engineered to express a MART 1 targeted T cell receptor led to objective responses in two out of 15 patients (Morgan et al. 2006). Recently, it has been clearly shown that CD8+ T cells receive CD4 help directly through the receptor CD40 and that this interaction is fundamental for CD8+ T cell memory generation (Bourgeois et al. 2002). Intratumoral injections of DCs, which had been transduced *in vitro* with CD40-ligand, led to regression of injected as well as distant B16 melanoma nodules (Kikuchi et al. 2000).

The majority of preclinical and clinical data for melanoma immunogene therapy has been generated in vaccination trials. Vaccination is performed traditionally with autologous tumor cells, but allogeneic vaccines consisting of autologous fibroblast or standardized gene-transduced cell lines are used as well (reducing cost and time). Many human trials using either cytokines (e.g., IL-2, -4, -6, -7, -12, IFN-α, and IFN-γ) or costimulatory molecules have been completed or are still ongoing. The best results in clinical vaccination trials have probably been obtained with granulocyte-macrophage colony-stimulating factor (GM-CSF); this had already shown the greatest degree of systemic immunity among 10 different cytokines in preclinical models (Dranoff et al. 1993). In a melanoma vaccination trial with GM-CSF-transduced autologous melanoma cells, extensive tumor destruction and pronounced immunologic responses were observed in 11 of 16 patients (Soiffer et al. 1998). GM-CSF was also transferred intratumorally using vaccinia virus (Mastrangelo et al. 1999) or transfected into tumor cells to prime tumor-derived lymph node cells in a trial of adoptive gene transfer (Chang et al. 2000). Excellent responses, including complete remissions, have been obtained in both types of clinical studies.

More than 50 patients in five different studies have received intratumoral injections of HLA-B7 DNA using liposomes (Hersh 1996; Vogelzang et al. 1996). DNA or protein was detected in the great majority of the patients. Toxicity not attributable to the mechanical irritations was the exception, and local or even general responses were seen in one-third of the cases.

Published results of a trial involving multiple intratumoral injections into melanoma metastases of a retroviral vector carrying the IFN-gene showed either stable disease or partial or complete response of the injected lesions (Fujii et al. 2000). In another trial, adenovirus was used to package the IL-24 (MDA-7) cytokine, which can induce cancer cell apoptosis. Following on two phase I trials, which reported systemic immune activation and local apoptosis (Tong et al. 2005) and even complete or partial response in 2/28 patients, a phase II trial is now under way.

Besides the immunological approaches illustrated above, which clearly dominate gene therapy of melanoma, preclinical suicide gene (Table 36.1), and tumor suppressor gene (p53) (Table 36.1) therapies have also been performed. Combinatorial approaches (e.g., cotransfer of suicide genes and cytokines) have resulted in additive tumor growth inhibition (Bonnekoh et al. 1995; Cao et al. 1998). A clinical phase I/II trial (Klatzmann et al. 1998) for suicide therapy of metastatic melanoma with retroviral HSV tk vectors resulted in significant tumor necrosis in three of eight patients despite a low transfection efficacy of less than 1%. Intratumoral injections of a p53 adenoviral vector for the treatment of patients with breast and melanoma metastases were associated with minimal toxicity and the detection of p53 mRNA in biopsies of injected tumors (Dummer et al. 2000).

In the preclinical stage, treatment of melanoma cells with oncogene Ha-ras (Jansen et al. 1995) or c-myc (Putney et al. 1999) specific antisense oligonucleotides led to good antitumor efficacy in SCID mice. The application of dominant negative Stat3 cancer-related gene led to tumor regressions due to massive apoptosis (Putney et al. 1999). In one clinical trial for the treatment of malignant melanoma with dacarbacin and antisense-bcl-2, one complete remission has been observed (Lebedeva 2001).

In summary, the preclinical and in particular the clinical results obtained with immunogene therapy of melanoma are among the most encouraging findings in the field of cancer gene therapy.

36.4.10 OSTEOSARCOMA

Osteosarcoma primarily afflicts young people within the first decades of life and accounts for 5% of all childhood malignancies (Hudson et al. 1990). The overall 2-year metastasis-free survival rate approaches 66% (Link et al. 1986; Goorin et al. 1987). Metastases, mainly to the lung, are the predominant cause for mortality. Therefore, immunological approaches such as the *in vivo* transfer of the B7 gene (Hayakawa et al. 1997) may be particularly fruitful. However, the osteocalcin promoter constitutes an available tissue-specific promoter, which potentially allows systemic or regional treatment with cytotoxic genes. It has already been proven to be efficient *in vivo* in an adenoviral context (Ko et al. 1996). The osteocalcin promoter expressing the adenoviral E1A gene to allow for tumor-specific replication has been proposed for intravenous application in a clinical trial for the treatment of metastasized osteosarcoma (Benjamin et al. 2001). Other successful preclinical approaches include aerosolic adenovirus-IL-12 (Worth et al. 2000), PEI-IL-12 (Jia et al. 2002), or PEI–p53 complexes for the treatment of lung metastases; adenovirus-IL-3-β gene transfer for isolated limb perfusion (de Wilt et al. 2001); and intratumoral injection of retrovirus VPCs-HSV tk (Charissoux et al. 1999).

36.4.11 Ovarian Cancer

Ovarian cancer is the leading cause of death from gynecological malignancies in women (Ko et al. 1996). Due to improvements in surgery, radiation, and chemotherapy techniques, the 5-year survival rate has improved over the last 20 years. However, over two-thirds of the patients have advanced stage disease at presentation; despite transient responses, the long-term survival of these patients rarely exceeds 15%–30%. Although even patients with advanced stages often have their disease confined to their abdomen for extended periods of time, intraperitoneal chemotherapies have only moderate success. Basically, this is because they do not provide the reduced toxicity profiles initially hoped for. Therefore, ovarian cancer has become one of the model diseases for gene therapeutic approaches with intracavital vector applications.

Several clinical studies have been initiated with HSV-tk as transgene in which vectors or modified cells were applied intraperitoneally. Transfer of adeno-HSV tk at doses up to 1×10^{11} pfus led to stable disease in 5 of 14 patients and transient fever in four patients. Transient hematologic grade 3 and 4 toxicities were observed if an adeno-HSV tk was combined with topotecan chemotherapy, but since toxicity was not related to vector dose, these side effects are more likely related to the chemotherapy (Hasenburg et al. 2001). In a vaccine/suicide trial with patients receiving irradiated PAI-1 ovarian cancer cells, *ex vivo* transfected with HSV tkHSV tk, a high incidence of fever and abdominal pain was seen in several patients along with resolution of ascites or decreased CA-125 levels (Buttgereit and Schmidt-Wolf 2002). Results of a trial using retroviral VPCs (Link et al. 1996) have not yet been published.

Phase I trials have also been initiated for adenoviral delivery of an anti-erbB-2 single chain antibody gene (Deshane et al. 1995; Alvarez and Curiel 1997). This intrabody approach could become a potent alternative to antisense strategies. In a clinical trial, adeno-anti-erbB-2 was injected intraperitoneally. Despite the relatively low dose of up to 1×10^{11} pfus, 5 of 13 evaluable patients had stable disease (Hasenburg et al. 2001).

The results of two trials using p53-adenovirus have been published. In one trial, 2 of 11 patients had a partial response, four patients had stable disease, and no major toxicity was reported despite multiple injections of up to 3×10^{12} viral particles (Alvarez et al. 2000). In the second phase I/II trial enrolling 36 patients, adeno-p53 was administered either alone or combined with chemotherapy at single or multiple doses. A comparison with historical controls led the authors to suggest one of two explanations: either the combination of chemotherapy and gene therapy had a synergistic effect or gene therapy changes the nature of the cancer, contributing to a chronicity of the disease that favorably impacts survival (Buller et al. 2002a,b).

A trial with replication-competent adenovirus showed significant toxicity but no efficacy (Vasey et al. 2002).

Finally, intraperitoneal injection of a retrovirus harboring a BRCA-1 splice-variant showed partial responses in 3 of 12 patients in a phase I study. However, the lack of any response led to an early termination of a subsequent phase II study (Tait et al. 1997, 1999). The authors discuss differences in tumor burden and immune system status of the patients as possible reasons for the differing response to BRCA1 gene therapy. Liposomal transfer of adenoviral E1A in a reduction of the tumor marker CA-125 and one stable disease but grade 3 and 4 toxicity in 9/12 patients (Madhusudan et al. 2004).

In summary, intracavital gene therapy of ovarian cancer showed clinical antitumor effects with immunologic and nonimmunologic approaches.

36.4.12 Pancreatic Cancer

This disease has a low incidence but a very bad prognosis, mainly because the primary tumors are not resectable at the time of discovery. It accounts for only 2% of all newly diagnosed cancers in the United States but for 5% of all cancer deaths (Kasuya et al. 2001). Moreover, the tumors are highly chemoresistant. Partial remissions of up to 15% of the cases have been seen with the new drug gemcitabine (Mertelsmann 1998), which remains the standard chemotherapy at present.

More than 80% of pancreatic tumors contain a mutation of the ras oncogene in position 12 (Abrams et al. 1996). Since this mutation (as well as others) is recognized by the immune system, ras-peptide based immunotherapy has been successful in mice. Clinical studies have been performed and an association between prolonged survival and an immune response against the vaccine was seen (Gjertsen et al. 1996; Abrams et al. 1997; Gjertsen et al. 2001). Recently, antigen-presenting cells, genetically modified to express tumor antigen, were shown to induce antigen-specific cytotoxic T cell responses *in vitro* and *in vivo* (Kubuschok et al. 2002). H-ras antisense oligos have also been used in a phase II clinical trial and clinical stabilization was seen in two out of 16 patients (Perez et al. 2001). The combination of this oligo with gemcitabine in a phase II study including 48 patients with metastatic pancreatic cancer was well tolerated and lead to one complete remission and one partial remission (for review, see Bhattacharyya and Lemoine, 2006) More recently, the combination of genetic expression of K-ras and IFN-α has shown overadditive preclinical efficacy (Hatanaka et al. 2004). Clinical efficacy has been observed by the combination of gemcitabine with the oncolytic adenovirus Onyx 015 (Hecht et al. 2003) whereas application of Onyx 015 alone had no antitumor efficacy in a number of clinical trials (Kirn 2000). In preclinical studies, a different oncolytic adenovirus alone inhibited tumor growth, which was attributed to an inhibition of VEGF production and subsequent inhibition of angiogenesis (Saito et al. 2006).

Effective preclinical gene therapy for pancreatic cancer was also seen by combining oncolytic conditional replication-competent adenoviruses and IL-2 or IL-12 adenoviruses (Motoi et al. 2000). Replication-deficient IL-2 adenoviruses were used in a phase I/II trial for digestive cancer and a single intraoperative intratumoral injection did not cause any adverse effects (Gilly et al. 1999). Mild liver toxicity and mild antitumor effects were observed if patients with advanced GI-cancer including seven patients with pancreatic cancer were given intratumoral injections of IL-12 adenovirus (replication defective) (Sangro et al. 2004).

In a phase I study, allogenic pancreas cells engineered to secrete GM-CSF were intradermally applied and three patients who had received higher doses of cells seemed to have increased survival times (Jaffee et al. 2001). A combination of a tumor antigen, several immunostimulatory genes harbored by two sequentially applied immunogenic viruses (vaccinia and fowlpox) (PANVAC-VF, Petrulio and Kaufman 2006) is probably in the most advanced stage of clinical examination since recently a phase III trial has been completed. This indicates that at least in immunogene therapy, a combination of synergistic principles may be highly beneficial.

The cell cycle inhibitor p16 (G1/S transition) is frequently mutated in pancreatic cancer. Although adenoviral expression of p16 in pancreatic cell lines induced significant suppression of proliferation (Kobayashi et al. 1999), apparently no preclinical or clinical studies have been published. Larger phase I and II trials have been initiated to treat pancreatic carcinoma patients with a retroviral vector transferring a gene designed to interfere with the cell cycle driver cyclin G1 (Rexin-G). In a preceding phase I trial, three out of three patients experienced tumor growth arrest with two patients experiencing stable disease (Gordon et al. 2004). This drug is the first injectable gene therapeutic agent to achieve orphan drug status from the U.S. Food and Drug Administration (Gordon and Hall 2005).

The tumor suppressor smad4, which is inactivated in more than 50% of pancreatic cancers, has shown tumor growth inhibition in immunodeficient mice after adenoviral gene transfer (Duda et al. 2003).

Because the local disease is often life-limiting in pancreatic cancer, the disease is attractive for cytotoxic gene therapy approaches (Aoki et al. 1995; Block et al. 1997; Evoy et al. 1997). A combination of HSV oncolysis and HSV tk/GCV cytotoxicity increased the percentage of mice with peritoneal pancreatic tumors obtaining significantly prolonged lifespan to 70% as compared to 40% with oncolysis alone (Kasuya et al. 1999). A very interesting trial used *ex vivo* suicide gene-transduced autologous tumor cells for localized activation of

chemotherapy. Genetically modified allogeneic cells, which expressed a cytochrome P450 enzyme, were encapsulated in cellulose sulfate and delivered by supraselective angiography to the tumor vasculature. These cells locally activated systemically administered ifosfamide. The tumors of four of 14 patients regressed after treatment and those of the other 10 individuals who completed the study remained stable. Median survival was doubled in the treatment group by comparison with historic controls and 1-year survival rate was three times better (Lohr et al. 2001).

siRNA technology has recently been utilized for knockdown of the antiapoptotic bcl-2 oncogene, which is overexpressed in most pancreatic cancers. Intraperitoneal administration of bcl-2 siRNA delayed the growth of pancreatic cancer xenografts (Ocker et al. 2005). Also directly targeting tumor cell apoptosis adenoviral gene transfer of TRAIL (tumor necrosis factor-related apoptosis-inducing ligand) either alone (Katz et al. 2003) or in combination with gemcitabine (Jacob et al. 2005) significantly suppressed pancreatic tumor growth. The transgene is driven by the tumor-specific human telomerase transcripatase (hTERT) promoter in this vector.

Another approach aiming at tumor-specific transgene expression is the use of radiation-inducible promoters. The TNFerade vector is an adenovirus, which delivers the TNF-α cytokine under the control of such a radiation-inducible promoter. Based on response rates of 85% inkling, two complete responses in patients with soft tissue sarcoma (Mundt et al. 2004), TNFerade is now in late phase II clinical trials. Larger studies are conducted for other cancers including pancreatic cancers as well (Senzer et al. 2004; McLoughlin et al. 2005).

36.4.13 Prostate Carcinoma

Prostate cancer is the most frequently diagnosed cancer in men in the United States and the second leading cause of death from malignancy (Carter and Coffey 1990; Boring et al. 1994). Locally restricted tumors can be treated by surgical resection or radiotherapy. Androgen-ablative therapy often induces dramatic responses, but virtually all patients progress to an androgen refractory state with a median survival of 12–18 months. The prostate is a unique accessory organ and expresses several hundred unique gene products as potential targets for gene therapy especially for tumor-specific gene expression. Prostate cancer can therefore serve as a model disease for tumor-specific gene therapy. The efficacy and specificity of the PSA promoter has been proven in animal experiments (Gotoh et al. 1998). This promoter has also been used to confer prostate-specific replication of conditionally replication-competent adenoviruses (Rodriguez et al. 1997). A clinical trial using this vector for injection into locally recurrent prostate carcinoma provided relative safety with no grade 4 toxicity and a more than 50% reduction in PSA levels in all patients treated with the highest dose (DeWeese et al. 2001).

Prostate cancer has proven to be highly responsive to immunomodulatory gene therapy. The most advanced drug is probably GVAX™, a GM-CSF secreting vaccine created with lethally radiated prostate cancer cell lines. Currently, a phase III trial is ongoing in which GVAX™ is compared to the standard chemotherapy for metastatic hormone-refractive prostate cancer (docetaxel plus prednisolon). Initiation of such a trial was possible because a number of phase I/II trials had shown survival times greater than in historical controls (Simons and Sacks 2006), declines of PSA levels, and stabilization of bone scans (Simons et al. 2006).

A second approach that had extensive clinical testing with encouraging clinical results (for review, see Stanizzi and Hall 2007) is the use of immunogenic viruses such as vaccinia virus, fowlpox virus, or adenovirus expressing candidate antigens such as PSA either alone or in combination with costimulatory molecules or cytokines.

Other immunogene therapy trials include direct intratumoral injection of liposomal IL-2 plasmid led to decreases in serum PSA levels in 80% of patients (Pantuck et al. 2000); Vaccinia virus MUC-1-IL-2 also decreased PSA levels (Pantuck et al. 2000). A large-scale phase II study based on a fowlpox/Muc-1 virus is under way (Liu et al. 2004).

As an example of gene replacement therapy, a trial with direct injection of retrovirus-BRCA-1 should be mentioned: while it did not reveal any antitumor efficacy, it did at least demonstrate the safety of the approach (Steiner and Gingrich 2000).

Clear objective responses were observed when an adenovirus-HSV tk vector was injected into the tumorous prostate. At the highest treatment dose, PSA levels declined by 50%. However, reversible grade 4 thrombocytopenia and grade 3 hepatotoxicity occurred (Herman et al. 1999). Importantly, an adenoviral vector harboring the HSV tk gene under the control of the prostate cancer cell-specific osteoclacin promoter has made it into a clinical trial, which demonstrated the absence of severe toxicity and necrosis, apoptosis, and the indiction of T-cells in 64% of treated patients (Kubo et al. 2003). Such a vector utilizing tissue-specific gene expression holds the promise to widen the therapeutic window.

When combined with radiation therapy, Ad-HSV tk approaches proved to be superior to historical controls (Teh et al. 2001), which apparently is due to HSV tk/GCV acting as a radiation sensitizer. If combination therapy included a replication-competent adenoviral vector harboring the HSV tk gene, PSA half-life was significantly shorter than reported for patients treated with radiotherapy alone (Freytag et al. 2003). The promising toxicity profile was further improved by using second generation adenoviruses (Freytag et al. 2007).

Successful preclinical experiments have been reported for the transfer of the anti-IGF-1 receptor (Burfeind et al. 1996), p21 (Eastham et al. 1995), p53 (Eastham et al. 1995; Ko et al. 1996), p16, and antisense myc (Steiner et al. 1998). Three clinical trials have already been initiated with antisense myc.

Immunogene therapy of prostate cancer is in the forefront of candidates for approval for clinical application in the Western world. The future use of prostate-specific regulatory elements for expression of suicide genes or conditional replication-competent viral vectors holds promise in the treatment of disseminated disease, the major cause of death in prostate cancer patients. Treatment will eventually be possible if systemic application of the vectors harboring cytotoxic genes can be done without major toxicity.

36.5 CONCLUSIONS

We have now seen nearly two decades of clinical gene therapy. A wide variety of vectors and transgenes have been successfully tested in animals. Several types of vectors and many different therapeutic principles have been evaluated clinically, producing more than 850 studies of cancer gene therapy.

The general outcome of these clinical studies indicates that gene therapeutic vectors are usually tolerated with mild to medium acute toxicity. The toxicity has been reversible in all cases. No cases of long-term toxicity have been reported. In the majority of clinical trials, at least some efficacy was seen. In some trials, a clear benefit for the patient was observed in comparison with either historical controls or (in some exceptional cases) with internal controls. Apparently, in all cases where the best results in clinical trials have been achieved, the cancer target was naturally most well suited to current cancer gene therapeutics. Examples of this are the good accessibility of superficial bladder cancer for intravesically installed viruses, the easy access to manifestations of head and neck cancer for the intratumoral transfer of replication competent vectors, and the good infectability of HCC by adenoviral vectors or of glioblastoma by HSV (which is a neurotrophic virus by nature). Particularly encouraging results have been obtained in many immunogene therapeutic trials, which do not depend as heavily on efficacy of gene transfer as nonimmunologic approaches. The sometimes astonishing modulatibility of vectors gives rise to the hope that cancers that are less well-suited for cancer gene therapy will also have particular vectors designed against them.

Although most clinical studies were designed as phase I or phase I/II trials with the aim of dose finding and toxicity evaluation, the lack of a real breakthrough in clinical efficacy has been disappointing at least for those who had envisioned a very rapid development of gene therapy into a routine treatment modality. The discrepancy between excellent preclinical data and the clinical outcome has been particularly difficult to understand, especially for the public. There are several reasons for this discrepancy. First of all, animal models are artificial models, which have inherent simplifications as compared to the clinical situation. A well-known argument is that animal tumors are usually developed from one fairly homogenous cell line, whereas human tumors are often very heterogeneous though they have also originated from a single clone. Consequently, a strategy, which targets one specific molecular event in an animal tumor, may only be effective against a fraction of tumor cells in a patient. But there are less sophisticated reasons such as the mere difference in size of tumors between a rodent and a human. This may also dramatically affect clinical outcome. Another explanation is that dose-limiting toxicity has not been reached in most of the trials performed so far. In addition, mice are probably more robust and will probably tolerate higher doses. It is evident that some kinds of treatments well tolerated in mice would hardly ever be considered for humans.

Other setbacks for cancer gene therapy have included the first severe adverse events in gene therapy: the death of a young man in a trial for the metabolic disorder OTC and the occurrence of three cases of leukemia in what was a very successful trial for SCID up to that point.

All these considerations lead to the conclusion that the therapeutic window of most cancer gene therapeutic strategies is apparently still to narrow to allow prominent antitumor efficacy. However, nearly all kinds of medical therapy have been confronted with this problem during their development into routine clinical applications. This therapeutic window can be broadened by either creating less toxic medicines, which allows one to increase the dose, or by creating more efficient medicines, which allows one to obtain efficacy without the necessity of increasing the dose. Both avenues have been followed in cancer gene therapy. An example of a less toxic vector is HD-Ad as compared to first-generation Ad. An example of a more efficient therapeutic principle is the combination of conventional suicide gene strategies with replication-competent viral vectors to increase the efficacy of gene transfer. A third strategy, which allows a reduction in the dose without losing efficacy, is vector targeting, either in the form of transductional or transcriptional targeting. This extremely promising approach will eventually reveal the real potential of gene therapy: the design of intelligent drugs that are capable of overcoming intelligent cancer.

But will not the design of small molecular drugs be intelligent enough to solve the cancer problem? Do we need gene therapy if small molecules can do the job? Especially if one considers the recent genomics-related dramatic increase in knowledge about potential drug targets? Maybe, in some cases. Admittedly, the use of a vehicle to transfer a gene into the human body is less direct and usually more complicated than conventional methods of cancer therapy. However, the interposition of several steps between the application of an active substance and the generation of an observable effect allows an unusual high degree of freedom of regulation. In cancer gene therapy, one can comply with the biology and especially the weak spots of the cancer cells to a degree never seen before. Fortunately, nature seems to be predictable to a certain degree, which makes such complex therapeutic strategies possible. Since we are still far away from such a tight adaptation of our vectors and transgenes to the specificities of the cancer cells; it comes as no surprise that to date clinical efficacy of cancer gene therapy has not been convincing. However, the potential fruitfulness of a new method cannot be judged by its initial success but rather by the general limitations of the whole concept. Since in the area of cancer gene therapy such limitations have not yet been seen, a broad breakthrough in therapeutic efficacy will most likely be a matter of time.

REFERENCES

Abonour, R., D. A. Williams, et al. 2000. Efficient retrovirus-mediated transfer of the multidrug resistance 1 gene into autologous human long-term repopulating hematopoietic stem cells. *Nat Med* 6(6): 652–658.

Abrams, S. I., P. H. Hand, et al. 1996. Mutant ras epitopes as targets for cancer vaccines. *Semin Oncol* 23(1): 118–134.

Abrams, S. I., S. N. Khleif, et al. 1997. Generation of stable CD4+ and CD8+ T cell lines from patients immunized with ras oncogene-derived peptides reflecting codon 12 mutations. *Cell Immunol* 182(2): 137–151.

Adam, L., P. C. Black, et al. 2007. Adenoviral mediated interferon-alpha 2b gene therapy suppresses the pro-angiogenic effect of vascular endothelial growth factor in superficial bladder cancer. *J Urol* 177(5): 1900–1906.

Addison, C. L., D. A. Arenberg, et al. 2000. The CXC chemokine, monokine induced by interferon-gamma, inhibits non-small cell lung carcinoma tumor growth and metastasis. *Hum Gene Ther* 11(2): 247–261.

Addison, C. L., J. L. Bramson, et al. 1998. Intratumoral coinjection of adenoviral vectors expressing IL-2 and IL-12 results in enhanced frequency of regression of injected and untreated distal tumors. *Gene Ther* 5(10): 1400–1409.

Agarwal, M., T. W. Austin, et al. 1998. Scaffold attachment region-mediated enhancement of retroviral vector expression in primary T cells. *J Virol* 72(5): 3720–3728.

Ahmann, D. L., E. T. Creagan, et al. 1989. Complete responses and long-term survivals after systemic chemotherapy for patients with advanced malignant melanoma. *Cancer* 63(2): 224–227.

Ahonen, M., R. Ala-Aho, et al. 2002. Antitumor activity and bystander effect of adenovirally delivered tissue inhibitor of metalloproteinases-3. *Mol Ther* 5(6): 705–715.

Alavi, J. B., Judy, K., Alavi, A., Hackney, D., Phillips, P., Smith, J., Recio, A., Wilson, J., and Eck, S. 1998. Phase I trial of gene therapy in primary brain tumors. In: American Society of Gene Therapy 1st Annual Meeting, Seattle, WA.

Albelda, S. M., R. Wiewrodt, et al. 2002. Gene therapy for lung neoplasms. *Clin Chest Med* 23(1): 265–277.

Alvarez, R. D., M. N. Barnes, et al. 2000. A cancer gene therapy approach utilizing an anti-erbB-2 single-chain antibody-encoding adenovirus (AD21): A phase I trial. *Clin Cancer Res* 6(8): 3081–3087.

Alvarez, R. D. and D. T. Curiel 1997. A phase I study of recombinant adenovirus vector-mediated intraperitoneal delivery of herpes simplex virus thymidine kinase (HSV-TK) gene and intravenous ganciclovir for previously treated ovarian and extraovarian cancer patients. *Hum Gene Ther* 8(5): 597–613.

Amalfitano, A., M. A. Hauser, et al. 1998. Production and characterization of improved adenovirus vectors with the E1, E2b, and E3 genes deleted. *J Virol* 72(2): 926–933.

Amalfitano, A. and R. J. Parks 2002. Separating fact from fiction: Assessing the potential of modified adenovirus vectors for use in human gene therapy. *Curr Gene Ther* 2(2): 111–133.

Andreasen, P. A., R. Egelund, et al. 2000. The plasminogen activation system in tumor growth, invasion, and metastasis. *Cell Mol Life Sci* 57(1): 25–40.

Aoki, K., T. Yoshida, et al. 1995. Liposome-mediated in vivo gene transfer of antisense K-ras construct inhibits pancreatic tumor dissemination in the murine peritoneal cavity. *Cancer Res* 55(17): 3810–3816.

Ardelt, P. and A. Bohle 2002. Molecular aspects of bladder cancer IV: Gene therapy of bladder cancer. *Eur Urol* 41(4): 372–380; discussion 380–381.

Arteaga, C. L. and J. T. Holt 1996. Tissue-targeted antisense c-fos retroviral vector inhibits established breast cancer xenografts in nude mice. *Cancer Res* 56(5): 1098–1103.

Atencio, I. A., J. B. Avanzini, et al. 2001. Enhanced apoptotic activity of a p53 variant in tumors resistant to wild-type p53 treatment. *Mol Ther* 4(1): 5–12.

Badie, B., K. E. Drazan, et al. 1995. Adenovirus-mediated p53 gene delivery inhibits 9L glioma growth in rats. *Neurol Res* 17(3): 209–216.

Bajou, K., A. Noel, et al. 1998. Absence of host plasminogen activator inhibitor 1 prevents cancer invasion and vascularization. *Nat Med* 4(8): 923–928.

Barajas, M., G. Mazzolini, et al. 2001. Gene therapy of orthotopic hepatocellular carcinoma in rats using adenovirus coding for interleukin 12. *Hepatology* 33(1): 52–61.

Barba, D., J. Hardin, et al. 1994. Development of anti-tumor immunity following thymidine kinase-mediated killing of experimental brain tumors. *Proc Natl Acad Sci USA* 91(10): 4348–4352.

Bargou, R. C., C. Wagener, et al. 1996. Overexpression of the death-promoting gene bax-alpha which is downregulated in breast cancer restores sensitivity to different apoptotic stimuli and reduces tumor growth in SCID mice. *J Clin Invest* 97(11): 2651–2659.

Barnett, B. G., C. J. Crews, et al. 2002. Targeted adenoviral vectors. *Biochim Biophys Acta* 1575(1–3): 1–14.

Benjamin, R., L. Helman, et al. 2001. A phase I/II dose escalation and activity study of intravenous injections of OCaP1 for subjects with refractory osteosarcoma metastatic to lung. *Hum Gene Ther* 12(12): 1591–1593.

Bennett, J. J., K. A. Delman, et al. 2002. Comparison of safety, delivery, and efficacy of two oncolytic herpes viruses (G207 and NV1020) for peritoneal cancer. *Cancer Gene Ther* 9(11): 935–945.

Bergelson, J. M., J. A. Cunningham, et al. 1997. Isolation of a common receptor for Coxsackie B viruses and adenoviruses 2 and 5. *Science* 275(5304): 1320–1323.

Berglund, P., M. Sjoberg, et al. 1993. Semliki Forest virus expression system: Production of conditionally infectious recombinant particles. *Biotechnology (N Y)* 11(8): 916–920.

Bertram, J., M. Killian, et al. 1994. Reduction of erbB2 gene product in mamma carcinoma cell lines by erbB2 mRNA-specific and tyrosine kinase consensus phosphorothioate antisense oligonucleotides. *Biochem Biophys Res Commun* 200(1): 661–667.

Bhattacharyya, M. and N. R. Lemoine 2006. Gene therapy developments for pancreatic cancer. *Best Pract Res Clin Gastroenterol* 20(2): 285–298.

Biermann, V., C. Volpers, et al. 2001. Targeting of high-capacity adenoviral vectors. *Hum Gene Ther* 12(14): 1757–1769.

Biroccio, A., D. D. Bufalo, et al. 1999. Increase of BCNU sensitivity by wt-p53 gene therapy in glioblastoma lines depends on the administration schedule. *Gene Ther* 6(6): 1064–1072.

Bischoff, J. R., D. H. Kirn, et al. 1996. An adenovirus mutant that replicates selectively in p53-deficient human tumor cells. *Science* 274(5286): 373–376.

Blagosklonny, M. V. and W. S. El-Deiry 1998. Acute overexpression of wt p53 facilitates anticancer drug-induced death of cancer and normal cells. *Int J Cancer* 75(6): 933–940.

Blankenstein, T., S. Cayeux, et al. 1996. Genetic approaches to cancer immunotherapy. *Rev Physiol Biochem Pharmacol* 129: 1–49.

Block, A., S. H. Chen, et al. 1997. Adenoviral-mediated herpes simplex virus thymidine kinase gene transfer: Regression of hepatic metastasis of pancreatic tumors. *Pancreas* 15(1): 25–34.

Blomer, U., L. Naldini, et al. 1996. Applications of gene therapy to the CNS. *Hum Mol Genet* 5 Spec No: 1397–1404.

Boerger, A. L., S. Snitkovsky, et al. 1999. Retroviral vectors preloaded with a viral receptor-ligand bridge protein are targeted to specific cell types. *Proc Natl Acad Sci USA* 96(17): 9867–9872.

Bonini, C., G. Ferrari, et al. 1997. HSV-TK gene transfer into donor lymphocytes for control of allogeneic graft-versus-leukemia. *Science* 276(5319): 1719–1724.

Bonnekoh, B., D. A. Greenhalgh, et al. 1995. Inhibition of melanoma growth by adenoviral-mediated HSV thymidine kinase gene transfer in vivo. *J Invest Dermatol* 104(3): 313–317.

Boring, C. C., T. S. Squires, et al. 1994. Cancer statistics, 1994. *CA Cancer J Clin* 44(1): 7–26.

Borkhardt, A. 2002. Blocking oncogenes in malignant cells by RNA interference—New hope for a highly specific cancer treatment? *Cancer Cell* 2(3): 167.

Bougeret, C., A. Virone-Oddos, et al. 2000. Cancer gene therapy mediated by CTS1, a p53 derivative: Advantage over wild-type p53 in growth inhibition of human tumors overexpressing MDM2. *Cancer Gene Ther* 7(5): 789–798.

Bourgeois, C., B. Rocha, et al. 2002. A role for CD40 expression on CD8+ T cells in the generation of CD8+ T cell memory. *Science* 297(5589): 2060–2063.

Bouvet, M., L. M. Ellis, et al. 1998. Adenovirus-mediated wild-type p53 gene transfer down-regulates vascular endothelial growth factor expression and inhibits angiogenesis in human colon cancer. *Cancer Res* 58(11): 2288–2292.

Bramson, J., M. Hitt, et al. 1996. Construction of a double recombinant adenovirus vector expressing a heterodimeric cytokine: In vitro and in vivo production of biologically active interleukin-12. *Hum Gene Ther* 7(3): 333–342.

Brand, K. 2002. Cancer gene therapy with tissue inhibitors of metalloproteinases (TIMPs). *Curr Gene Ther* 2(2): 255–271.

Brand, K., W. Arnold, et al. 1997. Liver-associated toxicity of the HSV-tk/GCV approach and adenoviral vectors. *Cancer Gene Ther* 4(1): 9–16.

Brand, K., A. H. Baker, et al. 2000. Treatment of colorectal liver metastases by adenoviral transfer of tissue inhibitor of metalloproteinases-2 into the liver tissue. *Cancer Res* 60(20): 5723–5730.

Brand, K., R. Klocke, et al. 1999. Induction of apoptosis and G2/M arrest by infection with replication-deficient adenovirus at high multiplicity of infection. *Gene Ther* 6(6): 1054–1063.

Brand, K., P. Loser, et al. 1998. Tumor cell-specific transgene expression prevents liver toxicity of the adeno-HSV tk/GCV approach. *Gene Ther* 5(10): 1363–1371.

Braun, S. E., K. Chen, et al. 1997. Gene therapy strategies for leukemia. *Mol Med Today* 3(1): 39–46.

Braybrooke, J. P., A. Slade, et al. 2005. Phase I study of MetXia-P450 gene therapy and oral cyclophosphamide for patients with advanced breast cancer or melanoma. *Clin Cancer Res* 11(4): 1512–1520.

Brenner, M. 1998. Applications of gene transfer. *Recent Results Cancer Res* 144: 60–69.

Brenner, M. K. 2001. Gene transfer and the treatment of haematological malignancy. *J Intern Med* 249(4): 345–358.

Broaddus, W. C., Y. Liu, et al. 1999. Enhanced radiosensitivity of malignant glioma cells after adenoviral p53 transduction. *J Neurosurg* 91(6): 997–1004.

Brummelkamp, T., R. Bernards, et al. 2002. Stable suppression of tumorigenicity by virus-mediated RNA interference. *Cancer Cell* 2(3): 243.

Brunori, M., M. Malerba, et al. 2001. Replicating adenoviruses that target tumors with constitutive activation of the wnt signaling pathway. *J Virol* 75(6): 2857–2865.

Bucheit, A. D., S. Kumar, et al. 2003. An oncolytic measles virus engineered to enter cells through the CD20 antigen. *Mol Ther* 7(1): 62–72.

Buchholz, C. J. and K. Cichutek 2006. Is it going to be SIN?: A European Society of Gene Therapy commentary. Phasing-out the clinical use of non self-inactivating murine leukemia virus vectors: Initiative on hold. *J Gene Med* 8(10): 1274–1276.

Budker, V., G. Zhang, et al. 1996. Naked DNA delivered intraportally expresses efficiently in hepatocytes. *Gene Ther* 3(7): 593–598.

Budker, V., G. Zhang, et al. 1998. The efficient expression of intravascularly delivered DNA in rat muscle. *Gene Ther* 5(2): 272–276.

Bui, L. A., L. H. Butterfield, et al. 1997. In vivo therapy of hepatocellular carcinoma with a tumor-specific adenoviral vector expressing interleukin-2. *Hum Gene Ther* 8(18): 2173–2182.

Buller, R. E., I. B. Runnebaum, et al. 2002a. A phase I/II trial of rAd/p53 (SCH 58500) gene replacement in recurrent ovarian cancer. *Cancer Gene Ther* 9(7): 553–566.

Buller, R. E., M. S. Shahin, et al. 2002b. Long term follow-up of patients with recurrent ovarian cancer after Ad p53 gene replacement with SCH 58500. *Cancer Gene Ther* 9(7): 567–572.

Burfeind, P., C. L. Chernicky, et al. 1996. Antisense RNA to the type I insulin-like growth factor receptor suppresses tumor growth and prevents invasion by rat prostate cancer cells in vivo. *Proc Natl Acad Sci USA* 93(14): 7263–7268.

Burkholder, J. K., J. Decker, and N. S. Yang, 1993. Rapid transgene expression in lymphocyte and macrophage primary cultures after particle bombardment-mediated gene transfer. *J Immunol Methods* 165: 149–156.

Butterfield, L. H., A. Ribas, and J. S. Economou 1999. DNA and dendritic cell-based genetic immunization against cancer. In: E. C. Lattime and S. L. Gerson (eds.), *Gene Therapy of Cancer*, Academic Press, New York, pp. 285–298.

Buttgereit, P. and I. G. Schmidt-Wolf 2002. Gene therapy of lymphoma. *J Hematother Stem Cell Res* 11(3): 457–467.

Duvoli, M., S. J. Langer, et al. 2002. Potential limitations of transcription terminators used as transgene insulators in adenoviral vectors. *Gene Ther* 9(3): 227–231.

Campos, S. K. and M. A. Barry 2007. Current advances and future challenges in adenoviral vector biology and targeting. *Curr Gene Ther* 7(3): 189–204.

Cao, X., D. W. Ju, et al. 1998. Adenovirus-mediated GM-CSF gene and cytosine deaminase gene transfer followed by 5-fluorocytosine administration elicit more potent antitumor response in tumor-bearing mice. *Gene Ther* 5(8): 1130–1136.

Caplen, N. J., J. N. Higginbotham, et al. 1999. Adeno-retroviral chimeric viruses as in vivo transducing agents. *Gene Ther* 6(3): 454–459.

Carter, H. B. and D. S. Coffey 1990. The prostate: An increasing medical problem. *Prostate* 16(1): 39–48.

Caruso, M., K. Pham-Nguyen, et al. 1996. Adenovirus-mediated interleukin-12 gene therapy for metastatic colon carcinoma. *Proc Natl Acad Sci USA* 93(21): 11302–11306.

Casalini, P., S. Menard, et al. 1997. Inhibition of tumorigenicity in lung adenocarcinoma cells by c-erbB-2 antisense expression. *Int J Cancer* 72(4): 631–636.

Cassaday, R. D., P. M. Sondel, et al. 2007. A phase I study of immunization using particle-mediated epidermal delivery of genes for gp100 and GM-CSF into uninvolved skin of melanoma patients. *Clin Cancer Res* 13(2 Pt 1): 540–549.

Cassel, W. A. and R. E. Garrett 1965. Newcastle Disease virus as an antineoplastic agent. *Cancer* 18: 863–868.

Cavazzana-Calvo, M. and A. Fischer 2007. Gene therapy for severe combined immunodeficiency: Are we there yet? *J Clin Invest* 117(6): 1456–1465.

Cavazzana-Calvo, M., S. Hacein-Bey, et al. 2000. Gene therapy of human severe combined immunodeficiency (SCID)-X1 disease. *Science* 288(5466): 669–672.

Chang, A. E., Q. Li, et al. 2000. Immunogenetic therapy of human melanoma utilizing autologous tumor cells transduced to secrete granulocyte-macrophage colony-stimulating factor. *Hum Gene Ther* 11(6): 839–850.

Chang, E. H., Y. J. Jang, et al. 1997. Restoration of the G1 checkpoint and the apoptotic pathway mediated by wild-type p53 sensitizes squamous cell carcinoma of the head and neck to radiotherapy. *Arch Otolaryngol Head Neck Surg* 123(5): 507–512.

Chapman, H. A. 1997. Plasminogen activators, integrins, and the coordinated regulation of cell adhesion and migration. *Curr Opin Cell Biol* 9(5): 714–724.

Charissoux, J. L., L. Grossin, et al. 1999. Treatment of experimental osteosarcoma tumors in rat by herpes simplex thymidine kinase gene transfer and ganciclovir. *Anticancer Res* 19(1A): 77–80.

Chen, L., D. Chen, et al. 1995. Breast cancer selective gene expression and therapy mediated by recombinant adenoviruses containing the DF3/MUC1 promoter. *J Clin Invest* 96(6): 2775–2782.

Chen, L., P. McGowan, et al. 1994. Tumor immunogenicity determines the effect of B7 costimulation on T cell-mediated tumor immunity. *J Exp Med* 179(2): 523–532.

Chen, P. W., M. Wang, et al. 1996. Therapeutic antitumor response after immunization with a recombinant adenovirus encoding a model tumor-associated antigen. *J Immunol* 156(1): 224–231.

Chen, W. Y., E. C. Bailey, et al. 1997. Reactivation of silenced, virally transduced genes by inhibitors of histone deacetylase. *Proc Natl Acad Sci USA* 94(11): 5798–5803.

Cheng, L., Ziegelhoffer, P. R., and Yang, N. S. 1993. In vivo promoter activity and transgene expression in mammalian somatic tissues evaluated by using particle bombardment. *Proc Natl Acad Sci USA* 90: 4455–4459.

Chiocca, E. A., K. M. Abbed, et al. 2004. A phase I open-label, dose-escalation, multi-institutional trial of injection with an E1B-attenuated adenovirus, ONYX-015, into the peritumoral region of recurrent malignant gliomas, in the adjuvant setting. *Mol Ther* 10(5): 958–966.

Cho, K. R. and B. Vogelstein 1992. Suppressor gene alterations in the colorectal adenoma-carcinoma sequence. *J Cell Biochem Suppl* 166: 137–141.

Chung-Faye, G. A., D. J. Kerr, et al. 2000. Gene therapy strategies for colon cancer. *Mol Med Today* 6(2): 82–87.

Cirielli, C., K. Inyaku, et al. 1999. Adenovirus-mediated wild-type p53 expression induces apoptosis and suppresses tumorigenesis of experimental intracranial human malignant glioma. *J Neurooncol* 43(2): 99–108.

Clarke, M. F., I. J. Apel, et al. 1995. A recombinant bcl-x$_s$ adenovirus selectively induces apoptosis in cancer cells but not in normal bone marrow cells. *Proc Natl Acad Sci USA* 92(24): 11024–11028.

Connelly, S., J. M. Gardner, et al. 1996. High-level tissue-specific expression of functional human factor VIII in mice. *Hum Gene Ther* 7(2): 183–195.

Conry, R., et al. 1995. Breaking tolerance to CEA with rV virus vaccine in man (abstr.). *Proc Am Assoc Cancer Res* 36, 492.

Conry, R. M., D. T. Curiel, et al. 2002. Safety and immunogenicity of a DNA vaccine encoding carcinoembryonic antigen and hepatitis B surface antigen in colorectal carcinoma patients. *Clin Cancer Res* 8(9): 2782–2787.

Cosset, F. L., F. J. Morling, et al. 1995. Retroviral retargeting by envelopes expressing an N-terminal binding domain. *J Virol* 69(10): 6314–6322.

Cosset, F. L., Y. Takeuchi, et al. 1995. High-titer packaging cells producing recombinant retroviruses resistant to human serum. *J Virol* 69(12): 7430–7436.

Cox, L. A., G. Chen, and E. Y. Lee, 1994. Tumor suppressor genes and their roles in breast cancer. *Breast Cancer Res Treat* 32: 19–38.

Cross, D. and J. K. Burmester 2006. Gene therapy for cancer treatment: Past, present and future. *Clin Med Res* 4(3): 218–227.

Culver, K. W., Z. Ram, et al. 1992. In vivo gene transfer with retroviral vector-producer cells for treatment of experimental brain tumors. *Science* 256(5063): 1550–1552.

Curiel, D. T., E. Wagner, et al. 1992. High-efficiency gene transfer mediated by adenovirus coupled to DNA–polylysine complexes. *Hum Gene Ther* 3(2): 147–154.

Dalbagni, G. and H. W. Herr 2000. Current use and questions concerning intravesical bladder cancer group for superficial bladder cancer. *Urol Clin North Am* 27(1): 137–146.

Davis, M. E. 2002. Non-viral gene delivery systems. *Curr Opin Biotechnol* 13(2): 128–131.

de Wilt, J. H., A. Bout, et al. 2001. Adenovirus-mediated interleukin 3 beta gene transfer by isolated limb perfusion inhibits growth of limb sarcoma in rats. *Hum Gene Ther* 12(5): 489–502.

Dean, D. A. 1997. Import of plasmid DNA into the nucleus is sequence specific. *Exp Cell Res* 230(2): 293–302.

DeBruyne, L. A., A. E. Chang, et al. 1996. Direct transfer of a foreign MHC gene into human melanoma alters T cell receptor V beta usage by tumor-infiltrating lymphocytes. *Cancer Immunol Immunother* 43(1): 49–58.

Deshane, J., G. P. Siegal, et al. 1995. Targeted tumor killing via an intracellular antibody against erbB-2. *J Clin Invest* 96(6): 2980–2989.

Devilee, P. and C. J. Cornelisse 1994. Somatic genetic changes in human breast cancer. *Biochim Biophys Acta* 1198(2–3): 113–130.

DeWeese, T. L., H. van der Poel, et al. 2001. A phase I trial of CV706, a replication-competent, PSA selective oncolytic adenovirus, for the treatment of locally recurrent prostate cancer following radiation therapy. *Cancer Res* 61(20): 7464–7472.

DiMeco, F., L. D. Rhines, et al. 2000. Paracrine delivery of IL-12 against intracranial 9L gliosarcoma in rats. *J Neurosurg* 92(3): 419–427.

Dimery, I. W. and W. K. Hong 1993. Overview of combined modality therapies for head and neck cancer. *J Natl Cancer Inst* 85(2): 95–111.

Djeha, A. H., T. A. Thomson, et al. 2001. Combined adenovirus-mediated nitroreductase gene delivery and CB1954 treatment: A well-tolerated therapy for established solid tumors. *Mol Ther* 3(2): 233–240.

Dock, G. 1904. *Am J Med Sci* 127: 563.

Donsante, A., C. Vogler, et al. 2001. Observed incidence of tumorigenesis in long-term rodent studies of rAAV vectors. *Gene Ther* 8(17): 1343–1346.

Doronin, K., K. Toth, et al. 2000. Tumor-specific, replication-competent adenovirus vectors overexpressing the adenovirus death protein. *J Virol* 74(13): 6147–6155.

Douglas, J. T., B. E. Rogers, et al. 1996. Targeted gene delivery by tropism-modified adenoviral vectors. *Nat Biotechnol* 14(11): 1574–1578.

Dranoff, G., E. Jaffee, et al. 1993. Vaccination with irradiated tumor cells engineered to secrete murine granulocyte-macrophage colony-stimulating factor stimulates potent, specific, and long-lasting anti-tumor immunity. *Proc Natl Acad Sci USA* 90(8): 3539–3543.

Duda, D. G., M. Sunamura, et al. 2003. Restoration of SMAD4 by gene therapy reverses the invasive phenotype in pancreatic adenocarcinoma cells. *Oncogene* 22(44): 6857–6864.

Dummer, R., J. Bergh, et al. 2000. Biological activity and safety of adenoviral vector-expressed wild-type p53 after intratumoral injection in melanoma and breast cancer patients with p53-overexpressing tumors. *Cancer Gene Ther* 7(7): 1069–1076.

Ealovega, M. W., P. K. McGinnis, et al. 1996. bcl-x$_s$ gene therapy induces apoptosis of human mammary tumors in nude mice. *Cancer Res* 56(9): 1965–1969.

Eastham, J. A., S. J. Hall, et al. 1995. In vivo gene therapy with p53 or p21 adenovirus for prostate cancer. *Cancer Res* 55(22): 5151–5155.

Eck, S. L., J. B. Alavi, et al. 2001. Treatment of recurrent or progressive malignant glioma with a recombinant adenovirus expressing human interferon-beta (H5.010CMVhIFN-beta): A phase I trial. *Hum Gene Ther* 12(1): 97–113.

Elezkurtaj, S., C. Kopitz, et al. 2004. Adenovirus-mediated overexpression of tissue inhibitor of metalloproteinases-1 in the liver: Efficient protection against T-cell lymphoma and colon carcinoma metastasis. *J Gene Med* 6(11): 1228–1237.

Esandi, M. C., G. D. van Someren, et al. 1998. IL-1/IL-3 gene therapy of non-small cell lung cancer (NSCLC) in rats using "cracked" adenoproducer cells. *Gene Ther* 5(6): 778–788.

Evoy, D., E. A. Hirschowitz, et al. 1997. In vivo adenoviral-mediated gene transfer in the treatment of pancreatic cancer. *J Surg Res* 69(1): 226–231.

Fakhrai, H., D. L. Shawler, et al. 1995. Cytokine gene therapy with interleukin-2-transduced fibroblasts: Effects of IL-2 dose on anti-tumor immunity. *Hum Gene Ther* 6(5): 591–601.

Feng, M., W. H. Jackson, Jr., et al. 1997. Stable in vivo gene transduction via a novel adenoviral/retroviral chimeric vector. *Nat Biotechnol* 15(9): 866–870.

Fielding, A. K., M. Maurice, et al. 1998. Inverse targeting of retroviral vectors: Selective gene transfer in a mixed population of hematopoietic and nonhematopoietic cells. *Blood* 91(5): 1802–1809.

Fink, D. J. and J. C. Glorioso 1997a. Engineering herpes simplex virus vectors for gene transfer to neurons. *Nat Med* 3(3): 357–359.

Fink, D. J. and J. C. Glorioso 1997b. Herpes simplex virus-based vectors: Problems and some solutions. *Adv Neurol* 72: 149–156.

Fink, D. J., R. Ramakrishnan, et al. 1995. Advances in the development of herpes simplex virus-based gene transfer vectors for the nervous system. *Clin Neurosci* 3(5): 284–291.

Fisher, K. J., H. Choi, et al. 1996. Recombinant adenovirus deleted of all viral genes for gene therapy of cystic fibrosis. *Virology* 217(1): 11–22.

Folkman, J. 1998. Antiangiogenic gene therapy. *Proc Natl Acad Sci USA* 95(16): 9064–9066.

Follenzi, A., G. Sabatino, et al. 2002. Efficient gene delivery and targeted expression to hepatocytes in vivo by improved lentiviral vectors. *Hum Gene Ther* 13(2): 243–260.

Frey, B. M., N. R. Hackett, et al. 1998. High-efficiency gene transfer into ex vivo expanded human hematopoietic progenitors and precursor cells by adenovirus vectors. *Blood* 91(8): 2781–2792.

Freytag, S. O., B. Movsas, et al. 2007. Phase I trial of replication-competent adenovirus-mediated suicide gene therapy combined with IMRT for prostate cancer. *Mol Ther* 15(5): 1016–1023.

Freytag, S. O., H. Stricker, et al. 2003. Phase I study of replication-competent adenovirus-mediated double-suicide gene therapy in combination with conventional-dose three-dimensional conformal radiation therapy for the treatment of newly diagnosed, intermediate- to high-risk prostate cancer. *Cancer Res* 63(21): 7497–7506.

Friedmann, T. 1991. Genetically modified tumor-infiltrating lymphocytes for cancer therapy. *Cancer Cells* 3(7): 271–274.

Fueyo, J., C. Gomez-Manzano, et al. 1996. Adenovirus-mediated p16/CDKN2 gene transfer induces growth arrest and modifies the transformed phenotype of glioma cells. *Oncogene* 12(1): 103–110.

Fueyo, J., C. Gomez-Manzano, et al. 1998. Overexpression of E2F-1 in glioma triggers apoptosis and suppresses tumor growth in vitro and in vivo. *Nat Med* 4(6): 685–690.

Fueyo, J., C. Gomez-Manzano, et al. 2000. A mutant oncolytic adenovirus targeting the Rb pathway produces anti-glioma effect in vivo. *Oncogene* 19(1): 2–12.

Fujii, S., S. Huang, et al. 2000. Induction of melanoma-associated antigen systemic immunity upon intratumoral delivery of interferon-gamma retroviral vector in melanoma patients. *Cancer Gene Ther* 7(9): 1220–1230.

Fujiwara, T., D. W. Cai, et al. 1994. Therapeutic effect of a retroviral wild-type p53 expression vector in an orthotopic lung cancer model. *J Natl Cancer Inst* 86(19): 1458–1462.

Fujiwara, T., E. A. Grimm, et al. 1994. Induction of chemosensitivity in human lung cancer cells in vivo by adenovirus-mediated transfer of the wild-type p53 gene. *Cancer Res* 54(9): 2287–2291.

Gaggar, A., D. M. Shayakhmetov, et al. 2003. CD46 is a cellular receptor for group B adenoviruses. *Nat Med* 9(11): 1408–1412.

Gallardo, D., K. E. Drazan, et al. 1996. Adenovirus-based transfer of wild-type p53 gene increases ovarian tumor radiosensitivity. *Cancer Res* 56(21): 4891–4893.

Ganly, I., D. Kirn, et al. 2000. A phase I study of Onyx-015, an E1B attenuated adenovirus, administered intratumorally to patients with recurrent head and neck cancer. *Clin Cancer Res* 6(3): 798–806.

Gao, G. P., M. R. Alvira, et al. 2002. Novel adeno-associated viruses from rhesus monkeys as vectors for human gene therapy. *Proc Natl Acad Sci USA* 99(18): 11854–11859.

Gao, G. P., Y. Yang, et al. 1996. Biology of adenovirus vectors with E1 and E4 deletions for liver-directed gene therapy. *J Virol* 70(12): 8934–8943.

Gao, X., K. S. Kim, et al. 2007. Nonviral gene delivery: What we know and what is next. *Aaps J* 9(1): E92–E104.

Garcia-Sanchez, F., G. Pizzorno, et al. 1998. Cytosine deaminase adenoviral vector and 5-fluorocytosine selectively reduce breast cancer cells 1 million-fold when they contaminate hematopoietic cells: A potential purging method for autologous transplantation. *Blood* 92(2): 672–682.

Georges, R. N., T. Mukhopadhyay, et al. 1993. Prevention of orthotopic human lung cancer growth by intratracheal instillation of a retroviral antisense K-ras construct. *Cancer Res* 53(8): 1743–1746.

Gilliam, A. D. and S. A. Watson 2002. Emerging biological therapies for pancreatic carcinoma. *Eur J Surg Oncol* 28(4): 370–378.

Gillio, T. A., Cignetti, A., Rovera, G., and Foa, R. 1996. Retroviral vector-mediated transfer of the tumor necrosis factor alpha gene into human cancer cells restores an apoptotic cell death program and induces a bystander-killing effect. *Blood* 87: 2486–2495.

Gilly, F. N., A. Beaujard, et al. 1999. Gene therapy with Adv-IL-2 in unresectable digestive cancer: Phase I–II study, intermediate report. *Hepatogastroenterology* 46(Suppl 1): 1268–1273.

Gjertsen, M. K., A. Bakka, et al. 1996. Ex vivo ras peptide vaccination in patients with advanced pancreatic cancer: Results of a phase I/II study. *Int J Cancer* 65(4): 450–453.

Gjertsen, M. K., T. Buanes, et al. 2001. Intradermal ras peptide vaccination with granulocyte-macrophage colony-stimulating factor as adjuvant: Clinical and immunological responses in patients with pancreatic adenocarcinoma. *Int J Cancer* 92(3): 441–450.

Gleich, L. L., J. L. Gluckman, et al. 2001. Clinical experience with HLA-B7 plasmid DNA/lipid complex in advanced squamous cell carcinoma of the head and neck. *Arch Otolaryngol Head Neck Surg* 127(7): 775–779.

Glick, R. P., T. Lichtor, et al. 1999. Prolongation of survival of mice with glioma treated with semiallogeneic fibroblasts secreting interleukin-2. *Neurosurgery* 45(4): 867–874.

Goldman, C. K., R. L. Kendall, et al. 1998. Paracrine expression of a native soluble vascular endothelial growth factor receptor inhibits tumor growth, metastasis, and mortality rate. *Proc Natl Acad Sci U S A* 95(15): 8795–8800.

Goldman, C. K., B. E. Rogers, et al. 1997. Targeted gene delivery to Kaposi's sarcoma cells via the fibroblast growth factor receptor. *Cancer Res* 57(8): 1447–1451.

Gomella, L. G., M. J. Mastrangelo, et al. 2001. Phase I study of intravesical vaccinia virus as a vector for gene therapy of bladder cancer. *J Urol* 166(4): 1291–1295.

Goncalves, M. A., I. Van Der Velde, et al. 2002. Efficient generation and amplification of high capacity adeno-associated virus/ adenovirus hybrid vectors. *J Virol* 76(21): 10734–10744.

Goorin, A. M., A. Perez-Atayde, et al. 1987. Weekly high-dose methotrexate and doxorubicin for osteosarcoma: The Dana-Farber Cancer Institute/the Children's Hospital—study III. *J Clin Oncol* 5(8): 1178–1184.

Gordon, E. M., Z. H. Chen, et al. 2001. Systemic administration of a matrix-targeted retroviral vector is efficacious for cancer gene therapy in mice. *Hum Gene Ther* 12(2): 193–204.

Gordon, E. M., G. H. Cornelio, et al. 2004. First clinical experience using a "pathotropic" injectable retroviral vector (Rexin-G) as intervention for stage IV pancreatic cancer. *Int J Oncol* 24(1): 177–185.

Gordon, E. M. and F. L. Hall 2005. Nanotechnology blooms, at last [Review]. *Oncol Rep* 13(6): 1003–1007.

Gotoh, A., S. C. Ko, et al. 1998. Development of prostate-specific antigen promoter-based gene therapy for androgen-independent human prostate cancer. *J Urol* 160(1): 220–229.

Gottesman, M. M., C. A. Hrycyna, et al. 1995. Genetic analysis of the multidrug transporter. *Annu Rev Genet* 29: 607–649.

Greenberg, P. D. 1991. Adoptive T cell therapy of tumors: Mechanisms operative in the recognition and elimination of tumor cells. *Adv Immunol* 49: 281–355.

Greenlee, R. T., T. Murray, et al. 2000. Cancer statistics, 2000. *CA Cancer J Clin* 50(1): 7–33.

Gross, G., S. Levy, et al. 1995. Chimaeric T-cell receptors specific to a B-lymphoma idiotype: A model for tumour immunotherapy. *Biochem Soc Trans* 23(4): 1079–1082.

Gu, D. L., A. M. Gonzalez, et al. 1999. Fibroblast growth factor 2 retargeted adenovirus has redirected cellular tropism: Evidence for reduced toxicity and enhanced antitumor activity in mice. *Cancer Res* 59(11): 2608–2614.

Guo, J. and H. Xin 2006. Chinese gene therapy. Splicing out the West? *Science* 314(5803): 1232–1235.

Habib, N. A., S. F. Ding, et al. 1996. Preliminary report: The short-term effects of direct p53 DNA injection in primary hepatocellular carcinomas. *Cancer Detect Prev* 20(2): 103–107.

Habib N. A., D. S. El-Masry, et al. 1995. Contrasting effects of direct p53 injection in primary and secondary liver tumours. *Tumour Targeting* 1: 295–298.

Habib, N., H. Salama, et al. 2002. Clinical trial of E1B-deleted adenovirus (dl1520) gene therapy for hepatocellular carcinoma. *Cancer Gene Ther* 9(3): 254–259.

Habib, N. A., C. E. Sarraf, et al. 2001. E1B-deleted adenovirus (dl1520) gene therapy for patients with primary and secondary liver tumors. *Hum Gene Ther* 12(3): 219–226.

Hacein-Bey-Abina, S., F. Le Deist, et al. 2002. Sustained correction of X-linked severe combined immunodeficiency by ex vivo gene therapy. *N Engl J Med* 346(16): 1185–1193.

Hacein-Bey-Abina, S., C. Von Kalle, et al. 2003. LMO2-associated clonal T cell proliferation in two patients after gene therapy for SCID-X1. *Science* 302(5644): 415–419.

Haecker, S. E., H. H. Stedman, et al. 1996. In vivo expression of full-length human dystrophin from adenoviral vectors deleted of all viral genes. *Hum Gene Ther* 7(15): 1907–1914.

Hajitou, A., N. E. Sounni, et al. 2001. Down-regulation of vascular endothelial growth factor by tissue inhibitor of metalloproteinase-2: Effect on in vivo mammary tumor growth and angiogenesis. *Cancer Res* 61(8): 3450–3457.

Hall, A. R., B. R. Dix, et al. 1998. p53-dependent cell death/apoptosis is required for a productive adenovirus infection. *Nat Med* 4(9): 1068–1072.

Harris, J. R., M. Morrow, and E. A. Bonadonng, (1993). Cancer of the breast. In: V. DeVita, S. Hellman, and S. A. Rosenberg (eds.), *Cancer: Principles and Practice of Oncology*, Lippincott Williams and Wilkins, Philadelphia, 4th edn., pp. 1264–1332.

Harris, M. P., S. Sutjipto, et al. 1996. Adenovirus-mediated p53 gene transfer inhibits growth of human tumor cells expressing mutant p53 protein. *Cancer Gene Ther* 3(2): 121–130.

Harrow, S., V. Papanastassiou, et al. 2004. HSV1716 injection into the brain adjacent to tumour following surgical resection of high-grade glioma: Safety data and long-term survival. *Gene Ther* 11(22): 1648–1658.

Hasenburg, A., X. W. Tong, et al. 2001. Adenovirus-mediated thymidine kinase gene therapy in combination with topotecan for patients with recurrent ovarian cancer: 2.5-year follow-up. *Gynecol Oncol* 83(3): 549–554.

Hashimura, T., T. Ueda, et al. 1997. Gene therapy by in vivo interferon-gamma gene transfer to murine bladder tumor. *Hinyokika Kiyo* 43(11): 809–813.

Hatanaka, K., K. Suzuki, et al. 2004. Interferon-alpha and antisense K-ras RNA combination gene therapy against pancreatic cancer. *J Gene Med* 6(10): 1139–1148.

Hatziioannou, T., E. Delahaye, et al. 1999. Retroviral display of functional binding domains fused to the amino terminus of influenza hemagglutinin. *Hum Gene Ther* 10(9): 1533–1544.

Havlik, R., L. R. Jiao, et al. 2002. Gene therapy for liver metastases. *Semin Oncol* 29(2): 202–208.

Hayakawa, M., S. Kawaguchi, et al. 1997. B7–1-transfected tumor vaccine counteracts chemotherapy-induced immunosuppression and prolongs the survival of rats bearing highly metastatic osteosarcoma cells. *Int J Cancer* 71(6): 1091–1102.

Head, J. F., R. L. Elliott, et al. 2002. Gene targets of antisense therapies in breast cancer. *Expert Opin Ther Targets* 6(3): 375–385.

Hecht, J. R., R. Bedford, et al. 2003. A phase I/II trial of intratumoral endoscopic ultrasound injection of ONYX-015 with intravenous gemcitabine in unresectable pancreatic carcinoma. *Clin Cancer Res* 9(2): 555–561.

Heise, C., T. Hermiston, et al. 2000. An adenovirus E1A mutant that demonstrates potent and selective systemic anti-tumoral efficacy. *Nat Med* 6(10): 1134–1139.

Heise, C., A. Sampson-Johannes, et al. 1997. ONYX-015, an E1B gene-attenuated adenovirus, causes tumor-specific cytolysis and antitumoral efficacy that can be augmented by standard chemotherapeutic agents. *Nat Med* 3(6): 639–645.

Heister, T., I. Heid, et al. 2002. Herpes simplex virus type 1/adeno-associated virus hybrid vectors mediate site-specific integration at the adeno-associated virus preintegration site, AAVS1, on human chromosome 19. *J Virol* 76(14): 7163–7173.

Herman, J. R., H. L. Adler, et al. 1999. In situ gene therapy for adenocarcinoma of the prostate: A phase I clinical trial. *Hum Gene Ther* 10(7): 1239–1249.

Hernandez-Alcoceba, R., B. Sangro, et al. 2006. Gene therapy of liver cancer. *World J Gastroenterol* 12(38): 6085–6097.

Herrlinger, U., A. Jacobs, et al. 2000. Helper virus-free herpes simplex virus type 1 amplicon vectors for granulocyte-macrophage colony-stimulating factor-enhanced vaccination therapy for experimental glioma. *Hum Gene Ther* 11(10): 1429–1438.

Hersh, E. M., G. Nabel, H. Silver, et al. 1996. Intratumoral injection of plasmid DNA in cationic lipid vectors for cancer gene therapy. *Cancer Gene Ther* 3: 11.

Herz, J. and R. D. Gerard 1993. Adenovirus-mediated transfer of low density lipoprotein receptor gene acutely accelerates cholesterol clearance in normal mice. *Proc Natl Acad Sci USA* 90: 2812–2816.

Hildebrandt, M., M. Y. Mapara, et al. 1997. Reverse transcriptase-polymerase chain reaction (RT-PCR)-controlled immunomagnetic purging of breast cancer cells using the magnetic cell separation (MACS) system: A sensitive method for monitoring purging efficiency. *Exp Hematol* 25(1): 57–65.

Hillgenberg, M., H. Tonnies, et al. 2001. Chromosomal integration pattern of a helper-dependent minimal adenovirus vector with a selectable marker inserted into a 27.4-kilobase genomic stuffer. *J Virol* 75(20): 9896–9908.

Hirai, M., D. LaFace, et al. 2001. Ex vivo purging by adenoviral p53 gene therapy does not affect NOD-SCID repopulating activity of human CD34+ cells. *Cancer Gene Ther* 8(12): 936–947.

Hirschowitz, E. A., A. Ohwada, et al. 1995. In vivo adenovirus-mediated gene transfer of the *Escherichia coli* cytosine deaminase gene to human colon carcinoma-derived tumors induces chemosensitivity to 5-fluorocytosine. *Hum Gene Ther* 6(8): 1055–1063.

Ho, V. C. and A. J. Sober 1990. Therapy for cutaneous melanoma: An update. *J Am Acad Dermatol* 22(2 Pt 1): 159–176.

Hofmann, C., V. Sandig, et al. 1995. Efficient gene transfer into human hepatocytes by baculovirus vectors. *Proc Natl Acad Sci USA* 92(22): 10099–10103.

Hofmann, C. and M. Strauss 1998. Baculovirus-mediated gene transfer in the presence of human serum or blood facilitated by inhibition of the complement system. *Gene Ther* 5(4): 531–536.

Hong, S. S., L. Karayan, et al. 1997. Adenovirus type 5 fiber knob binds to MHC class I alpha2 domain at the surface of human epithelial and B lymphoblastoid cells. *Embo J* 16(9): 2294–2306.

Horiguchi, Y., W. A. Larchian, et al. 2000. Intravesical liposome-mediated interleukin-2 gene therapy in orthotopic murine bladder cancer model. *Gene Ther* 7(10): 844–851.

Hortobagyi, G. N., N. T. Ueno, et al. 2001. Cationic liposome-mediated E1A gene transfer to human breast and ovarian cancer cells and its biologic effects: A phase I clinical trial. *J Clin Oncol* 19(14): 3422–3433.

Horwitz, M. S. 1996. Adenoviruses. In: B. N. Fields, D. M. Knipe, P. M. Howley, et al. (eds.), *Fields Virology*, 3rd edn., Philadelphia, PA: Lippincott-Raven, pp. 2149–2171.

Hsieh, C. L., V. F. Pang, et al. 1997. Regression of established mouse leukemia by GM-CSF-transduced tumor vaccine: Implications for cytotoxic T lymphocyte responses and tumor burdens. *Hum Gene Ther* 8(16): 1843–1854.

Hudson, M., M. R. Jaffe, et al. 1990. Pediatric osteosarcoma: Therapeutic strategies, results, and prognostic factors derived from a 10-year experience. *J Clin Oncol* 8(12): 1988–1997.

Hui, K. M., P. T. Ang, et al. 1997. Phase I study of immunotherapy of cutaneous metastases of human carcinoma using allogeneic and xenogeneic MHC DNA-liposome complexes. *Gene Ther* 4(8): 783–790.

Huser, A., M. Rudolph, et al. 2001. Incorporation of decay-accelerating factor into the baculovirus envelope generates complement-resistant gene transfer vectors. *Nat Biotechnol* 19(5): 451–455.

Ido, A., K. Nakata, et al. 1995. Gene therapy for hepatoma cells using a retrovirus vector carrying herpes simplex virus thymidine kinase gene under the control of human alpha-fetoprotein gene promoter. *Cancer Res* 55(14): 3105–3109.

Imler, J. L., F. Dupuit, et al. 1996. Targeting cell-specific gene expression with an adenovirus vector containing the lacZ gene under the control of the CFTR promoter. *Gene Ther* 3(1): 49–58.

Indraccolo, S., S. Minuzzo, et al. 1998. Pseudotyping of Moloney leukemia virus-based retroviral vectors with simian immunodeficiency virus envelope leads to targeted infection of human CD4+ lymphoid cells. *Gene Ther* 5(2): 209–217.

Jacob, D., J. J. Davis, et al. 2005. Suppression of pancreatic tumor growth in the liver by systemic administration of the TRAIL gene driven by the hTERT promoter. *Cancer Gene Ther* 12(2): 109–115.

Jacobson, M. D., M. Brown, J. Bertin, L. Chiang, B. Horsburgh, J. Bakowska, X. O. Breakefield, and P. DiStefano. 2000. Herpes simplex virus amplicons as tools to analyze pro-apoptotic gene function in primary neurons. *Soc Neurosci Abstract* 227(17).

Jaffee, E. M., R. H. Hruban, et al. 2001. Novel allogeneic granulocyte-macrophage colony-stimulating factor-secreting tumor vaccine for pancreatic cancer: A phase I trial of safety and immune activation. *J Clin Oncol* 19(1): 145–156.

Jansen, B., H. Wadl, et al. 1995. Phosphorothioate oligonucleotides reduce melanoma growth in a SCID-hu mouse model by a nonantisense mechanism. *Antisense Res Dev* 5(4): 271–277.

Jia, S. F., L. L. Worth, et al. 2002. Eradication of osteosarcoma lung metastases following intranasal interleukin-12 gene therapy using a nonviral polyethylenimine vector. *Cancer Gene Ther* 9(3): 260–266.

Jia, W. W., M. McDermott, et al. 1994. Selective destruction of gliomas in immunocompetent rats by thymidine kinase-defective herpes simplex virus type 1. *J Natl Cancer Inst* 86(16): 1209–1215.

Jiang, Y., M. Wang, et al. 2001. Stimulation of mammary tumorigenesis by systemic tissue inhibitor of matrix metalloproteinase 4 gene delivery. *Cancer Res* 61(6): 2365–2370.

Jin, X., D. Nguyen, et al. 1995. Cell cycle arrest and inhibition of tumor cell proliferation by the p16INK4 gene mediated by an adenovirus vector. *Cancer Res* 55(15): 3250–3253.

Johnston, K. M., D. Jacoby, et al. 1997. HSV/AAV hybrid amplicon vectors extend transgene expression in human glioma cells. *Hum Gene Ther* 8(3): 359–370.

Joki, T., M. Nakamura, et al. 1995. Activation of the radiosensitive EGR-1 promoter induces expression of the herpes simplex virus thymidine kinase gene and sensitivity of human glioma cells to ganciclovir. *Hum Gene Ther* 6(12): 1507–1513.

Kaneda, Y., K. Iwai, et al. 1989. Increased expression of DNA cointroduced with nuclear protein in adult rat liver. *Science* 243(4889): 375–378.

Kass-Eisler, A., L. Leinwand, et al. 1996. Circumventing the immune response to adenovirus-mediated gene therapy. *Gene Ther* 3(2): 154–162.

Kasuya, H., Y. Nishiyama, et al. 1999. Intraperitoneal delivery of hrR3 and ganciclovir prolongs survival in mice with disseminated pancreatic cancer. *J Surg Oncol* 72(3): 136–141.

Kasuya, H., S. Nomoto, et al. 2001. Gene therapy for pancreatic cancer. *Hepatogastroenterology* 48(40): 957–961.

Katz, M. H., D. E. Spivack, et al. 2003. Gene therapy of pancreatic cancer with green fluorescent protein and tumor necrosis factor-related apoptosis-inducing ligand fusion gene expression driven by a human telomerase reverse transcriptase promoter. *Ann Surg Oncol* 10(7): 762–772.

Kemeny, N., J. A. Conti, et al. 1993. A pilot study of hepatic artery floxuridine combined with systemic 5-fluorouracil and leucovorin. A potential adjuvant program after resection of colorectal hepatic metastases. *Cancer* 71(6): 1964–1971.

Kemeny, N. and Seiter, K. 1993. Treatment option for patients with metastatic colorectal cancer. In: J. Niederhuber (ed.), *Current Therapy in Oncology*, St. Louis, MO: Mosby Year Book, pp. 447–457.

Kenney, N. J., T. Saeki, et al. 1993. Expression of transforming growth factor alpha antisense mRNA inhibits the estrogen-induced production of TGF alpha and estrogen induced proliferation of estrogen-responsive human breast cancer cells. *J Cell Physiol* 156(3): 497–514.

Khuri, F. R., J. Nemunaitis, et al. 2000. A controlled trial of intratumoral ONYX-015, a selectively-replicating adenovirus, in combination with cisplatin and 5-fluorouracil in patients with recurrent head and neck cancer. *Nat Med* 6(8): 879–885.

Kijima, T., T. Osaki, et al. 1999. Application of the Cre recombinase/loxP system further enhances antitumor effects in cell type-specific gene therapy against carcinoembryonic antigen-producing cancer. *Cancer Res* 59(19): 4906–4911.

Kikuchi, T., M. A. Moore, et al. 2000. Dendritic cells modified to express CD40 ligand elicit therapeutic immunity against pre-existing murine tumors. *Blood* 96(1): 91–99.

Kimball, K. J., T. M. Numnum, et al. 2006. Gene therapy for ovarian cancer. *Curr Oncol Rep* 8(6): 441–447.

Kipps, T. J., P. Chu, et al. 2000. Immunogenetic therapy for B-cell malignancies. *Semin Oncol* 27(6 Suppl 12): 104–109.

Kircheis, R., T. Blessing, et al. 2001. Tumor targeting with surface-shielded ligand–polycation DNA complexes. *J Control Release* 72(1–3): 165–170.

Kirn, D. 1999. Selectively replicating viruses as therapeutic agents against cancer. In: E. C. Lattime and S. L. Gerson (eds.), *Gene Therapy of Cancer*. New York: Academic Press, pp. 235–247.

Kirn, D. 2000. Replication-selective oncolytic adenoviruses: Virotherapy aimed at genetic targets in cancer. *Oncogene* 19(56): 6660–6669.

Klatzmann, D., C. A. Valery, et al. 1998. A phase I/II study of herpes simplex virus type 1 thymidine kinase suicide gene therapy for recurrent glioblastoma. Study Group on Gene Therapy for Glioblastoma. *Hum Gene Ther* 9(17): 2595–2604.

Klauber, N., R. M. Rohan, et al. 1997. Critical components of the female reproductive pathway are suppressed by the angiogenesis inhibitor AGM-1470. *Nat Med* 3(4): 443–446.

Ko, S. C., J. Cheon, et al. 1996. Osteocalcin promoter-based toxic gene therapy for the treatment of osteosarcoma in experimental models. *Cancer Res* 56(20): 4614–4619.

Ko, S. C., A. Gotoh, et al. 1996. Molecular therapy with recombinant p53 adenovirus in an androgen-independent, metastatic human prostate cancer model. *Hum Gene Ther* 7(14): 1683–1691.

Kobayashi, S., H. Shirasawa, et al. 1999. P16INK4a expression adenovirus vector to suppress pancreas cancer cell proliferation. *Clin Cancer Res* 5(12): 4182–4185.

Koch, P. E., Z. S. Guo, et al. 2001. Augmenting transgene expression from carcinoembryonic antigen (CEA) promoter via a GAL4 gene regulatory system. *Mol Ther* 3(3): 278–283.

Kochanek, S., P. R. Clemens, et al. 1996. A new adenoviral vector: Replacement of all viral coding sequences with 28 kb of DNA independently expressing both full-length dystrophin and beta-galactosidase. *Proc Natl Acad Sci USA* 93(12): 5731–5736.

Kovesdi, I., D. E. Brough, et al. 1997. Adenoviral vectors for gene transfer. *Curr Opin Biotechnol* 8(5): 583–589.

Kuate, S., R. Wagner, et al. 2002. Development and characterization of a minimal inducible packaging cell line for simian immunodeficiency virus-based lentiviral vectors. *J Gene Med* 4(4): 347–355.

Kuball, J., S. F. Wen, et al. 2002. Successful adenovirus-mediated wild-type p53 gene transfer in patients with bladder cancer by intravesical vector instillation. *J Clin Oncol* 20(4): 957–965.

Kubba S, A. S., Schiller J, et al. 2000. Phase 1 trial of adenovirus p53 in brochioalveolar lung carcinoma (BAC) administered by bronchoalveolar lavage. Abstract 1904. *Proceedings of the American Society of clinical Oncology* 19.

Kubo, H., T. A. Gardner, et al. 2003. Phase I dose escalation clinical trial of adenovirus vector carrying osteocalcin promoter-driven herpes simplex virus thymidine kinase in localized and metastatic hormone-refractory prostate cancer. *Hum Gene Ther* 14(3): 227–241.

Kubuschok, B., R. Schmits, et al. 2002. Use of spontaneous Epstein–Barr virus-lymphoblastoid cell lines genetically modified to express tumor antigen as cancer vaccines: Mutated p21 ras oncogene in pancreatic carcinoma as a model. *Hum Gene Ther* 13(7): 815–827.

Kurihara, H., A. Zama, et al. 2000. Glioma/glioblastoma-specific adenoviral gene expression using the nestin gene regulator. *Gene Ther* 7(8): 686–693.

Kurihara, T., D. E. Brough, et al. 2000. Selectivity of a replication-competent adenovirus for human breast carcinoma cells expressing the MUC1 antigen. *J Clin Invest* 106(6): 763–771.

Lam, P. Y. and X. O. Breakefield 2001. Potential of gene therapy for brain tumors. *Hum Mol Genet* 10(7): 777–787.

Lange, W., E. M. Cantin, et al. 1993. In vitro and in vivo effects of synthetic ribozymes targeted against BCR/ABL mRNA. *Leukemia* 7(11): 1786–1794.

Larsen, S. R. and J. E. Rasko 2005. Lymphoproliferative disorders: Prospects for gene therapy. *Pathology* 37(6): 523–533.

Lasic, D. D., F. J. Martin, et al. 1991. Sterically stabilized liposomes: A hypothesis on the molecular origin of the extended circulation times. *Biochim Biophys Acta* 1070(1): 187–192.

Lavillette, D., S. J. Russell, et al. 2001. Retargeting gene delivery using surface-engineered retroviral vector particles. *Curr Opin Biotechnol* 12(5): 461–466.

Le, L. P., H. N. Le, et al. 2006. Dynamic monitoring of oncolytic adenovirus in vivo by genetic capsid labeling. *J Natl Cancer Inst* 98(3): 203–214.

Leach, D. R., M. F. Krummel, et al. 1996. Enhancement of antitumor immunity by CTLA-4 blockade. *Science* 271(5256): 1734–1736.

Lebedeva IV S.C., Antisense downregulation of the apoptosis-related and Bcl-xl proteins: A new approach to cancer therapy. In: Lattime, E.C and Gerson, S.L., eds., *Gene Therapy of Cancer* 2nd Ed., Academic Press, 2001, pp. 315–330.

Lebedeva, S., S. Bagdasarova, et al. 2001. Tumor suppression and therapy sensitization of localized and metastatic breast cancer by adenovirus p53. *Hum Gene Ther* 12(7): 763–772.

Lee, C. T., K. H. Park, et al. 2003. Recombinant adenoviruses expressing dominant negative insulin-like growth factor-I receptor demonstrate antitumor effects on lung cancer. *Cancer Gene Ther* 10(1): 57–63.

Lee, E. J., M. Jakacka, et al. 2001. Adenovirus-directed expression of dominant negative estrogen receptor induces apoptosis in breast cancer cells and regression of tumors in nude mice. *Mol Med* 7(11): 773–782.

Lesniak, M. S. 2006. Gene therapy for malignant glioma. *Expert Rev Neurother* 6(4): 479–488.

Levitsky, H. I., J. Montgomery, et al. 1996. Immunization with granulocyte-macrophage colony-stimulating factor-transduced, but not B7-1-transduced, lymphoma cells primes idiotype-specific T cells and generates potent systemic antitumor immunity. *J Immunol* 156(10): 3858–3865.

Li, H., F. Lindenmeyer, et al. 2001. AdTIMP-2 inhibits tumor growth, angiogenesis, and metastasis, and prolongs survival in mice. *Hum Gene Ther* 12(5): 515–526.

Li, J., L. Le, et al. 2005. Genetic incorporation of HSV-1 thymidine kinase into the adenovirus protein IX for functional display on the virion. *Virology* 338(2): 247–258.

Li, W. W., J. Fan, et al. 1997. Overexpression of p21waf1 leads to increased inhibition of E2F-1 phosphorylation and sensitivity to anticancer drugs in retinoblastoma-negative human sarcoma cells. *Cancer Res* 57(11): 2193–2199.

Li, Z., J. Dullmann, et al. 2002. Murine leukemia induced by retroviral gene marking. *Science* 296(5567): 497.

Li, Z., Shanmugam, N., Katayose, D., Huber, B., Srivastava, S., Cowan, K., and Seth, P. 1997. Enzyme/prodrug gene therapy approach for breast cancer using a recombinant adenovirus expressing *Escherichia coli* cytosine deaminase. *Cancer Gene Ther* 4: 113–117.

Licht, T., I. Aksentijevich, et al. 1995. Efficient expression of functional human MDR1 gene in murine bone marrow after retroviral transduction of purified hematopoietic stem cells. *Blood* 86(1): 111–121.

Licht, T. and C. Peschel 2002. Restoration of transgene expression in hematopoietic cells with drug-selectable marker genes. *Curr Gene Ther* 2(2): 227–234.

Lieber, A., C. Y. He, et al. 1997. The role of Kupffer cell activation and viral gene expression in early liver toxicity after infusion of recombinant adenovirus vectors. *J Virol* 71(11): 8798–8807.

Lillo, R., M. Ramirez, et al. 2002. Efficient and nontoxic adenoviral purging method for autologous transplantation in breast cancer patients. *Cancer Res* 62(17): 5013–5018.

Lin, P., J. A. Buxton, et al. 1998. Antiangiogenic gene therapy targeting the endothelium-specific receptor tyrosine kinase Tie2. *Proc Natl Acad Sci USA* 95(15): 8829–8834.

Lindemann, D., M. Bock, et al. 1997. Efficient pseudotyping of murine leukemia virus particles with chimeric human foamy virus envelope proteins. *J Virol* 71(6): 4815–4820.

Link, C. J., Jr., D. Moorman, et al. 1996. A phase I trial of in vivo gene therapy with the herpes simplex thymidine kinase/ganciclovir system for the treatment of refractory or recurrent ovarian cancer. *Hum Gene Ther* 7(9): 1161–1179.

Link, M. P., A. M. Goorin, et al. 1986. The effect of adjuvant chemotherapy on relapse-free survival in patients with osteosarcoma of the extremity. *N Engl J Med* 314(25): 1600–1606.

Liu, F. and L. Huang 2002. Development of non-viral vectors for systemic gene delivery. *J Control Release* 78(1–3): 259–266.

Liu, M., B. Acres, et al. 2004. Gene-based vaccines and immunotherapeutics. *Proc Natl Acad Sci USA* 101(Suppl 2): 14567–14571.

Liu, X., P. Tian, et al. 2002. Enhanced antitumor effect of EGF R-targeted p21WAF-1 and GM-CSF gene transfer in the established murine hepatoma by peritumoral injection. *Cancer Gene Ther* 9(1): 100–108.

Liu, X. L., K. R. Clark, et al. 1999. Production of recombinant adeno-associated virus vectors using a packaging cell line and a hybrid recombinant adenovirus. *Gene Ther* 6(2): 293–299.

Lohr, M., A. Hoffmeyer, et al. 2001. Microencapsulated cell-mediated treatment of inoperable pancreatic carcinoma. *Lancet* 357(9268): 1591–1592.

Lorence, R. M., P. A. Rood, et al. 1988. Newcastle disease virus as an antineoplastic agent: Induction of tumor necrosis factor-alpha and augmentation of its cytotoxicity. *J Natl Cancer Inst* 80(16): 1305–1312.

Lukas, J., L. Aagaard, et al. 1995. Oncogenic aberrations of p16INK4/CDKN2 and cyclin D1 cooperate to deregulate G1 control. *Cancer Res* 55(21): 4818–4823.

Lundstrom, K. 2001. Alphavirus vectors for gene therapy applications. *Curr Gene Ther* 1(1): 19–29.

Lv, W., C. Zhang, et al. 2006. RNAi technology: A revolutionary tool for the colorectal cancer therapeutics. *World J Gastroenterol* 12(29): 4636–4639.

Ma, D., R. D. Gerard, et al. 1997. Inhibition of metastasis of intraocular melanomas by adenovirus-mediated gene transfer of plasminogen activator inhibitor type 1 (PAI-1) in an athymic mouse model. *Blood* 90(7): 2738–2746.

Mabjeesh, N. J., H. Zhong, et al. 2002. Gene therapy of prostate cancer: Current and future directions. *Endocr Relat Cancer* 9(2): 115–139.

Mackinnon, S., E. B. Papadopoulos, et al. 1995. Adoptive immunotherapy using donor leukocytes following bone marrow transplantation for chronic myeloid leukemia: Is T cell dose important in determining biological response? *Bone Marrow Transplant* 15(4): 591–594.

Madhusudan, S., A. Tamir, et al. 2004. A multicenter phase I gene therapy clinical trial involving intraperitoneal administration of E1A-lipid complex in patients with recurrent epithelial ovarian cancer overexpressing HER-2/neu oncogene. *Clin Cancer Res* 10(9): 2986–2996.

Madrigal, M., M. F. Janicek, et al. 1997. In vitro antigene therapy targeting HPV-16 E6 and E7 in cervical carcinoma. *Gynecol Oncol* 64(1): 18–25.

Magnusson, M. K., S. S. Hong, et al. 2001. Genetic retargeting of adenovirus: Novel strategy employing "deknobbing" of the fiber. *J Virol* 75(16): 7280–7289.

Mahaley, M. S., Jr., C. Mettlin, et al. 1989. National survey of patterns of care for brain-tumor patients. *J Neurosurg* 71(6): 826–836.

Maione, D., M. Wiznerowicz, et al. 2000. Prolonged expression and effective readministration of erythropoietin delivered with a fully deleted adenoviral vector. *Hum Gene Ther* 11(6): 859–868.

Makower, D., A. Rozenblit, et al. 2003. Phase II clinical trial of intralesional administration of the oncolytic adenovirus ONYX-015 in patients with hepatobiliary tumors with correlative p53 studies. *Clin Cancer Res* 9(2): 693–702.

Mammano, F., F. Salvatori, et al. 1997. Truncation of the human immunodeficiency virus type 1 envelope glycoprotein allows efficient pseudotyping of Moloney murine leukemia virus particles and gene transfer into CD4+ cells. *J Virol* 71(4): 3341–3345.

Mao, H. Q., K. Roy, et al. 2001. Chitosan-DNA nanoparticles as gene carriers: Synthesis, characterization and transfection efficiency. *J Control Release* 70(3): 399–421.

Marconi, P., D. Krisky, et al. 1996. Replication-defective herpes simplex virus vectors for gene transfer in vivo. *Proc Natl Acad Sci USA* 93(21): 11319 11320.

Markert, J. M., M. D. Medlock, et al. 2000. Conditionally replicating herpes simplex virus mutant, G207 for the treatment of malignant glioma: Results of a phase I trial. *Gene Ther* 7(10): 867–874.

Mars, W. M and G. F. Saunders 1990. Chromosomal abnormalities in human breast cancer. *Cancer Metastasis Rev* 9: 35–43.

Martin-Fontecha, A., F. Cavallo, et al. 1996. Heterogeneous effects of B7-1 and B7-2 in the induction of both protective and therapeutic anti-tumor immunity against different mouse tumors. *Eur J Immunol* 26(8): 1851–1859.

Martinet, O., C. M. Divino, et al. 2002. T cell activation with systemic agonist antibody versus local 4-1BB ligand gene delivery combined with interleukin-12 eradicate liver metastases of breast cancer. *Gene Ther* 9(12): 786–792.

Martuza, R. L., A. Malick, et al. 1991. Experimental therapy of human glioma by means of a genetically engineered virus mutant. *Science* 252(5007): 854–856.

Mastrangelo, M. J., H. C. Maguire, Jr., et al. 1999. Intratumoral recombinant GM-CSF-encoding virus as gene therapy in patients with cutaneous melanoma. *Cancer Gene Ther* 6(5): 409–422.

Matsumoto, K., H. Ishikawa, et al. 2004. Antiangiogenic property of pigment epithelium-derived factor in hepatocellular carcinoma. *Hepatology* 40(1): 252–259.

Maxwell, I. H., Spitzer, A. L., Long, C. J., and Maxwell, F. 1996. Autonomous parvovirus transduction of a gene under control of tissue-specific or inducible promoters. *Gene Ther* 3: 28–36.

McCormick, F. 2001. Cancer gene therapy: Fringe or cutting edge? *Nat Rev Cancer* 1(2): 130–141.

McLoughlin, J. M., T. M. McCarty, et al. 2005. TNFerade, an adenovector carrying the transgene for human tumor necrosis factor alpha, for patients with advanced solid tumors: Surgical experience and long-term follow-up. *Ann Surg Oncol* 12(10): 825–830.

McManaway, M. E., L. M. Neckers, et al. 1990. Tumour-specific inhibition of lymphoma growth by an antisense oligodeoxynucleotide. *Lancet* 335(8693): 808–811.

Melero, I., W. W. Shuford, et al. 1997. Monoclonal antibodies against the 4-1BB T-cell activation molecule eradicate established tumors. *Nat Med* 3(6): 682–685.

Merdan, T., J. Kopecek, et al. 2002. Prospects for cationic polymers in gene and oligonucleotide therapy against cancer. *Adv Drug Deliv Rev* 54(5): 715–758.

Mertelsmann, R. 1998. Blutbildendes and Lymphatisches System. In: Weihrauch , T. R., ed., Internistische Therapie 98/99. urban and Schwarzenberg München, pp. 689–737.

Merrouche, Y., S. Negrier, et al. 1995. Clinical application of retroviral gene transfer in oncology: Results of a French study with tumor-infiltrating lymphocytes transduced with the gene of resistance to neomycin. *J Clin Oncol* 13(2): 410–418.

Mickisch, G. H., I. Aksentijevich, et al. 1992. Transplantation of bone marrow cells from transgenic mice expressing the human MDR1 gene results in long-term protection against the myelosuppressive effect of chemotherapy in mice. *Blood* 79(4): 1087–1093.

Miller, B. A., E. J. Feuer, et al. 1993. Recent incidence trends for breast cancer in women and the relevance of early detection: An update. *CA Cancer J Clin* 43(1): 27–41.

Miller, P. W., S. Sharma, et al. 2000. Intratumoral administration of adenoviral interleukin 7 gene-modified dendritic cells augments specific antitumor immunity and achieves tumor eradication. *Hum Gene Ther* 11(1): 53–65.

Mitani, K., F. L. Graham, et al. 1995. Rescue, propagation, and partial purification of a helper virus-dependent adenovirus vector. *Proc Natl Acad Sci USA* 92(9): 3854–3858.

Mitani, K. and S. Kubo 2002. Adenovirus as an integrating vector. *Curr Gene Ther* 2(2): 135–144.

Mohr, A., M. Lyons, et al. 2007. Gene therapy strategies for colorectal cancer. *Dtsch Med Wochenschr* 132(11): 567–570.

Molnar-Kimber, K. L., D. H. Sterman, et al. 1998. Impact of preexisting and induced humoral and cellular immune responses in an adenovirus-based gene therapy phase I clinical trial for localized mesothelioma. *Hum Gene Ther* 9(14): 2121–2133.

Monson, J. R. 1989. Malignant melanoma: A plague of our times. *Br J Surg* 76(10): 997–998.

Moolten, F. L., J. M. Wells, et al. 1990. Lymphoma regression induced by ganciclovir in mice bearing a herpes thymidine kinase transgene. *Hum Gene Ther* 1(2): 125–134.

Morgan, R. A., M. E. Dudley, et al. 2006. Cancer regression in patients after transfer of genetically engineered lymphocytes. *Science* 314(5796): 126–129.

Morsy, M. A., M. Gu, et al. 1998. An adenoviral vector deleted for all viral coding sequences results in enhanced safety and extended expression of a leptin transgene. *Proc Natl Acad Sci USA* 95(14): 7866–7871.

Moscow, J. A., H. Huang, et al. 1999. Engraftment of MDR1 and NeoR gene-transduced hematopoietic cells after breast cancer chemotherapy. *Blood* 94(1): 52–61.

Motoi, F., M. Sunamura, et al. 2000. Effective gene therapy for pancreatic cancer by cytokines mediated by restricted replication-competent adenovirus. *Hum Gene Ther* 11(2): 223–235.

Mundt, A. J., S. Vijayakumar, et al. 2004. A phase I trial of TNFerade biologic in patients with soft tissue sarcoma in the extremities. *Clin Cancer Res* 10(17): 5747–5753.

Naldini, L., U. Blomer, et al. 1996a. Efficient transfer, integration, and sustained long-term expression of the transgene in adult rat brains injected with a lentiviral vector. *Proc Natl Acad Sci USA* 93(21): 11382–11388.

Naldini, L., U. Blomer, et al. 1996b. In vivo gene delivery and stable transduction of nondividing cells by a lentiviral vector. *Science* 272(5259): 263–267.

Nemunaitis, J., T. Jahan, et al. 2006. Phase 1/2 trial of autologous tumor mixed with an allogeneic GVAX vaccine in advanced-stage non-small-cell lung cancer. *Cancer Gene Ther* 13(6): 555–562.

Nemunaitis, J., F. Khuri, et al. 2001. Phase II trial of intratumoral administration of ONYX-015, a replication-selective adenovirus, in patients with refractory head and neck cancer. *J Clin Oncol* 19(2): 289–298.

Nemunaitis, J. and J. O'Brien 2002. Head and neck cancer: Gene therapy approaches. Part II: Genes delivered. *Expert Opin Biol Ther* 2(3): 311–324.

Nemunaitis, J., D. Sterman, et al. 2004. Granulocyte-macrophage colony-stimulating factor gene-modified autologous tumor vaccines in non-small-cell lung cancer. *J Natl Cancer Inst* 96(4): 326–331.

Nemunaitis, J., S. G. Swisher, et al. 2000. Adenovirus-mediated p53 gene transfer in sequence with cisplatin to tumors of patients with non-small-cell lung cancer. *J Clin Oncol* 18(3): 609–622.

Nevins, J. R. 1981. Mechanism of activation of early viral transcription by the adenovirus E1A gene product. *Cell* 26(2 Pt 2): 213–220.

Nicklin, S. A. and A. H. Baker 2002. Tropism-modified adenoviral and adeno-associated viral vectors for gene therapy. *Curr Gene Ther* 2(3): 273–293.

Nicklin, S. A., E. Wu, et al. 2005. The influence of adenovirus fiber structure and function on vector development for gene therapy. *Mol Ther* 12(3): 384–393.

Nielsen, L. L. 2000. Combination therapy with SCH58500 (p53 adenovirus) and cyclophosphamide in preclinical cancer models. *Oncol Rep* 7(6): 1191–1196.

Nielsen, L. L., J. Dell, et al. 1997. Efficacy of p53 adenovirus-mediated gene therapy against human breast cancer xenografts. *Cancer Gene Ther* 4(2): 129–138.

Niranjan, A., S. Moriuchi, et al. 2000. Effective treatment of experimental glioblastoma by HSV vector-mediated TNF alpha and HSV-tk gene transfer in combination with radiosurgery and ganciclovir administration. *Mol Ther* 2(2): 114–120.

Nishikawa, M. and L. Huang 2001. Nonviral vectors in the new millennium: Delivery barriers in gene transfer. *Hum Gene Ther* 12(8): 861–870.

Norman, K. L., M. C. Coffey, et al. 2002. Reovirus oncolysis of human breast cancer. *Hum Gene Ther* 13(5): 641–652.

O'Brien, S. M., C. C. Cunningham, et al. 2005. Phase I to II multicenter study of oblimersen sodium, a Bcl-2 antisense oligonucleotide, in patients with advanced chronic lymphocytic leukemia. *J Clin Oncol* 23(30): 7697–7702.

O'Malley, B. W., Jr., D. A. Sewell, et al. 1997. The role of interleukin-2 in combination adenovirus gene therapy for head and neck cancer. *Mol Endocrinol* 11(6): 667–673.

Ocker, M., D. Neureiter, et al. 2005. Variants of bcl-2 specific siRNA for silencing antiapoptotic bcl-2 in pancreatic cancer. *Gut* 54(9): 1298–1308.

Ottgen, H. F. and Old, L. J. 1991. The history of cancer immunotherapy. In: De Vita, V. T., Hellmann, S., and Rosenberg, S. A. (eds.), *Biological Therapy of Cancer: Principles and Practice*, JB Lippincott, Philadelphia, pp. 87–111.

Palombo, F., A. Monciotti, et al. 1998. Site-specific integration in mammalian cells mediated by a new hybrid baculovirus-adeno-associated virus vector. *J Virol* 72(6): 5025–5034.

Pandha, H. S., L. A. Martin, et al. 1999. Genetic prodrug activation therapy for breast cancer: A phase I clinical trial of erbB-2-directed suicide gene expression. *J Clin Oncol* 17(7): 2180–2189.

Pantuck, A. J., A. Zisman, et al. 2000. Gene therapy for prostate cancer at the University of California, Los Angeles: Preliminary results and future directions. *World J Urol* 18(2): 143–147.

Papanastassiou, V., R. Rampling, et al. 2002. The potential for efficacy of the modified (ICP 34.5 (-)) herpes simplex virus HSV1716 following intratumoural injection into human malignant glioma: A proof of principle study. *Gene Ther* 9(6): 398–406.

Pardoll, D. M. 1994. Tumour antigens. A new look for the 1990s. *Nature* 369(6479): 357.

Park, F., K. Ohashi, et al. 2000. Efficient lentiviral transduction of liver requires cell cycling in vivo. *Nat Genet* 24(1): 49–52.

Parker, S. L., T. Tong, et al. 1997. Cancer statistics, 1997. *CA Cancer J Clin* 47(1): 5–27.

Parks, R. J., L. Chen, et al. 1996. A helper-dependent adenovirus vector system: Removal of helper virus by Cre-mediated excision of the viral packaging signal. *Proc Natl Acad Sci USA* 93(24): 13565–13570.

Parks, R. J. and F. L. Graham 1997. A helper-dependent system for adenovirus vector production helps define a lower limit for efficient DNA packaging. *J Virol* 71(4): 3293–3298.

Parsons, R., G. M. Li, et al. 1993. Hypermutability and mismatch repair deficiency in RER+ tumor cells. *Cell* 75(6): 1227–1236.

Pecora, A. L., N. Rizvi, et al. 2002. Phase I trial of intravenous administration of PV701, an oncolytic virus, in patients with advanced solid cancers. *J Clin Oncol* 20(9): 2251–2266.

Peng, K. W., L. Pham, et al. 2001. Organ distribution of gene expression after intravenous infusion of targeted and untargeted lentiviral vectors. *Gene Ther* 8(19): 1456–1463.

Peng, K. W., R. Vile, et al. 1999. Selective transduction of protease-rich tumors by matrix-metalloproteinase-targeted retroviral vectors. *Gene Ther* 6(9): 1552–1557.

Perez, R. P., J. W. III. Smith, S. R. Alberts, et al. 2001. Phase II trials of ISIS 2503, antisense inhibitor of H-ras, in patients (pts) with advanced pancreatic carcinoma (CA). *J Clin Oncol* 20: 628.

Petrulio, C. A. and H. L. Kaufman 2006. Development of the PANVAC-VF vaccine for pancreatic cancer. *Expert Rev Vaccines* 5(1): 9–19.

Plautz, G. E., Z. Y. Yang, et al. 1993. Immunotherapy of malignancy by in vivo gene transfer into tumors. *Proc Natl Acad Sci USA* 90(10): 4645–4649.

Plunkett, T. A. and D. W. Miles 2002. 13. New biological therapies for breast cancer. *Int J Clin Pract* 56(4): 261–266.

Podda, S., M. Ward, et al. 1992. Transfer and expression of the human multiple drug resistance gene into live mice. *Proc Natl Acad Sci USA* 89(20): 9676–9680.

Putney, S. D., J. Brown, et al. 1999. Enhanced anti-tumor effects with microencapsulated c-myc antisense oligonucleotide. *Antisense Nucleic Acid Drug Dev* 9(5): 451–458.

Qian, C., M. Idoate, et al. 1997. Gene transfer and therapy with adenoviral vector in rats with diethylnitrosamine-induced hepatocellular carcinoma. *Hum Gene Ther* 8(3): 349–358.

Qiao, J., M. Doubrovin, et al. 2002. Tumor-specific transcriptional targeting of suicide gene therapy. *Gene Ther* 9(3): 168–175.

Qin, X. Q., N. Tao, et al. 1998. Interferon-beta gene therapy inhibits tumor formation and causes regression of established tumors in immune-deficient mice. *Proc Natl Acad Sci USA* 95(24): 14411–14416.

Qing, K., C. Mah, et al. 1999. Human fibroblast growth factor receptor 1 is a co-receptor for infection by adeno-associated virus 2. *Nat Med* 5(1): 71–77.

Rainov, N. G., C. Fels, et al. 2001. Temozolomide enhances herpes simplex virus thymidine kinase/ganciclovir therapy of malignant glioma. *Cancer Gene Ther* 8(9): 662–668.

Rainov, N. G. and C. M. Kramm 2001. Vector delivery methods and targeting strategies for gene therapy of brain tumors. *Curr Gene Ther* 1(4): 367–383.

Ram, Z., K. W. Culver, et al. 1997. Therapy of malignant brain tumors by intratumoral implantation of retroviral vector-producing cells. *Nat Med* 3(12): 1354–1361.

Rampling, R., G. Cruickshank, et al. 2000. Toxicity evaluation of replication-competent herpes simplex virus (ICP 34.5 null mutant 1716) in patients with recurrent malignant glioma. *Gene Ther* 7(10): 859–866.

Randrianarison, V., D. Marot, et al. 2001. BRCA1 carries tumor suppressor activity distinct from that of p53 and p21. *Cancer Gene Ther* 8(10): 759–770.

Rao, J. B., R. S. Chamberlain, et al. 1996. IL-12 is an effective adjuvant to recombinant vaccinia virus-based tumor vaccines: Enhancement by simultaneous B7-1 expression. *J Immunol* 156(9): 3357–3365.

Ratajczak, M. Z., J. A. Kant, et al. 1992. In vivo treatment of human leukemia in a scid mouse model with c-myb antisense oligodeoxynucleotides. *Proc Natl Acad Sci USA* 89(24): 11823–11827.

Reichard, K. W., R. M. Lorence, et al. 1992. Newcastle disease virus selectively kills human tumor cells. *J Surg Res* 52(5): 448–453.

Reid, T., E. Galanis, et al. 2002. Hepatic arterial infusion of a replication-selective oncolytic adenovirus (dl1520): Phase II viral, immunologic, and clinical endpoints. *Cancer Res* 62(21): 6070–6079.

Reid, T. R., S. Freeman, et al. 2005. Effects of Onyx-015 among metastatic colorectal cancer patients that have failed prior treatment with 5-FU/leucovorin. *Cancer Gene Ther* 12(8): 673–681.

Reynolds, P. N., S. A. Nicklin, et al. 2001. Combined transductional and transcriptional targeting improves the specificity of transgene expression in vivo. *Nat Biotechnol* 19(9): 838–842.

Ribas, A., L. H. Butterfield, et al. 1997. Genetic immunization for the melanoma antigen MART-1/Melan-A using recombinant adenovirus-transduced murine dendritic cells. *Cancer Res* 57(14): 2865–2869.

Riddell, S. R. and P. D. Greenberg 1995. Principles for adoptive T cell therapy of human viral diseases. *Annu Rev Immunol* 13: 545–586.

Riethmuller, G., E. Schneider-Gadicke, et al. 1994. Randomised trial of monoclonal antibody for adjuvant therapy of resected Dukes' C colorectal carcinoma. German Cancer Aid 17-1A Study Group. *Lancet* 343(8907): 1177–1183.

Rigg, A. S. and N. R. Lemoine 2001. Adenoviral delivery of TIMP1 or TIMP2 can modify the invasive behavior of pancreatic cancer and can have a significant antitumor effect in vivo. *Cancer Gene Ther* 8(11): 869–878.

Riley, D. J., A. Y. Nikitin, et al. 1996. Adenovirus-mediated retinoblastoma gene therapy suppresses spontaneous pituitary melanotroph tumors in Rb+/− mice. *Nat Med* 2(12): 1316–1321.

Rill, D. R., V. M. Santana, et al. 1994. Direct demonstration that autologous bone marrow transplantation for solid tumors can return a multiplicity of tumorigenic cells. *Blood* 84(2): 380–383.

Robbins, P. D., H. Tahara, et al. 1998. Viral vectors for gene therapy. *Trends Biotechnol* 16(1): 35–40.

Roberts, M. L., D. J. Wells, et al. 2002. Stable micro-dystrophin gene transfer using an integrating adeno-retroviral hybrid vector ameliorates the dystrophic pathology in mdx mouse muscle. *Hum Mol Genet* 11(15): 1719–1730.

Robinson, B. W., S. A. Mukherjee, et al. 1998. Cytokine gene therapy or infusion as treatment for solid human cancer. *J Immunother* 21(3): 211–217.

Rodicker, F., T. Stiewe, et al. 2001. Therapeutic efficacy of E2F1 in pancreatic cancer correlates with TP73 induction. *Cancer Res* 61(19): 7052–7055.

Rodriguez, R., E. R. Schuur, et al. 1997. Prostate attenuated replication competent adenovirus (ARCA) CN706: A selective cytotoxic for prostate-specific antigen-positive prostate cancer cells. *Cancer Res* 57(13): 2559–2563.

Rooney, C. M., C. A. Smith, et al. 1995. Use of gene-modified virus-specific T lymphocytes to control Epstein–Barr-virus-related lymphoproliferation. *Lancet* 345(8941): 9–13.

Rosenberg, S. A. 1997. Cancer vaccines based on the identification of genes encoding cancer regression antigens. *Immunol Today* 18(4): 175–182.

Rosenberg, S. A., P. Aebersold, et al. 1990. Gene transfer into humans—Immunotherapy of patients with advanced melanoma, using tumor-infiltrating lymphocytes modified by retroviral gene transduction. *N Engl J Med* 323(9): 570–578.

Rosenberg, S. A., R. M. Blaese, et al. 2000. Human gene marker/ therapy clinical protocols. *Hum Gene Ther* 11(6): 919–979.

Roth, J. A., S. F. Grammer, et al. 2001. Gene therapy approaches for the management of non-small cell lung cancer. *Semin Oncol* 28(4 Suppl 14): 50–56.

Roth, J. A., D. Nguyen, et al. 1996. Retrovirus-mediated wild-type p53 gene transfer to tumors of patients with lung cancer. *Nat Med* 2(9): 985–991.

Rubinchik, S., S. Lowe, et al. 2001. Creation of a new transgene cloning site near the right ITR of Ad5 results in reduced enhancer interference with tissue-specific and regulatable promoters. *Gene Ther* 8(3): 247–253.

Ruppert, J. M., M. Wright, et al. 1997. Gene therapy strategies for carcinoma of the breast. *Breast Cancer Res Treat* 44(2): 93–114.

Russ, V. and E. Wagner 2007. Cell and tissue targeting of nucleic acids for cancer gene therapy. *Pharm Res* 24(6): 1047–1057.

Russell, S. J. and F. L. Cosset 1999. Modifying the host range properties of retroviral vectors. *J Gene Med* 1(5): 300–311.

Russell, S. J. and C. E. Dunbar 2001. Gene therapy approaches for multiple myeloma. *Semin Hematol* 38(3): 268–275.

Sa Cunha, A., E. Bonte, et al. 2002. Inhibition of rat hepatocellular carcinoma tumor growth after multiple infusions of recombinant Ad.AFPtk followed by ganciclovir treatment. *J Hepatol* 37(2): 222–230.

Sacco, M. G., E. M. Cato, et al. 2001. Systemic gene therapy with anti-angiogenic factors inhibits spontaneous breast tumor growth and metastasis in MMTVneu transgenic mice. *Gene Ther* 8(1): 67–70.

Sadelain, M. and I. Riviere 2002. Sturm und Drang over Suicidal Lymphocytes. *Mol Ther* 5(6): 655–657.

Sahin, U., O. Tureci, et al. 1997. Serological identification of human tumor antigens. *Curr Opin Immunol* 9(5): 709–716.

Sahin, U., O. Tureci, et al. 1995. Human neoplasms elicit multiple specific immune responses in the autologous host. *Proc Natl Acad Sci U S A* 92(25): 11810–11813.

Saito, Y., M. Sunamura, et al. 2006. Oncolytic replication-competent adenovirus suppresses tumor angiogenesis through preserved E1A region. *Cancer Gene Ther* 13(3): 242–252.

Sakai, Y., S. Kaneko, et al. 2001. Gene therapy for hepatocellular carcinoma using two recombinant adenovirus vectors with alpha-fetoprotein promoter and Cre/lox P system. *J Virol Methods* 92(1): 5–17.

Sampson, J. H., D. M. Ashley, et al. 1997. Characterization of a spontaneous murine astrocytoma and abrogation of its tumorigenicity by cytokine secretion. *Neurosurgery* 41(6): 1365–1372; discussion 1372–1373.

Sandig, V., K. Brand, et al. 1997. Adenovirally transferred p16INK4/ CDKN2 and p53 genes cooperate to induce apoptotic tumor cell death. *Nat Med* 3(3): 313–319.

Sandig, V., C. Hofmann, et al. 1996. Gene transfer into hepatocytes and human liver tissue by baculovirus vectors. *Hum Gene Ther* 7(16): 1937–1945.

Sandig, V., P. Loser, et al. 1996. HBV-derived promoters direct liver-specific expression of an adenovirally transduced LDL receptor gene. *Gene Ther* 3(11): 1002–1009.

Sangro, B., G. Mazzolini, et al. 2004. Phase I trial of intratumoral injection of an adenovirus encoding interleukin-12 for advanced digestive tumors. *J Clin Oncol* 22(8): 1389–1397.

Scappaticci, F. A. 2002. Mechanisms and future directions for angiogenesis-based cancer therapies. *J Clin Oncol* 20(18): 3906–3927.

Scheiflinger, F., F. Dorner, et al. 1992. Construction of chimeric vaccinia viruses by molecular cloning and packaging. *Proc Natl Acad Sci USA* 89(21): 9977–9981.

Schiedner, G., N. Morral, et al. 1998. Genomic DNA transfer with a high-capacity adenovirus vector results in improved in vivo gene expression and decreased toxicity. *Nat Genet* 18(2): 180–183.

Schmidt-Wolf, G. D. and I. G. Schmidt-Wolf 2002. Immunomodulatory gene therapy for haematological malignancies. *Br J Haematol* 117(1): 23–32.

Schmidt-Wolf, I. G., S. Finke, et al. 1999. Phase I clinical study applying autologous immunological effector cells transfected with the interleukin-2 gene in patients with metastatic renal cancer, colorectal cancer and lymphoma. *Br J Cancer* 81(6): 1009–1016.

Scholl, S. M., J. M. Balloul, et al. 2000. Recombinant vaccinia virus encoding human MUC1 and IL2 as immunotherapy in patients with breast cancer. *J Immunother* 23(5): 570–580.

Schrewe, H., J. Thompson, et al. 1990. Cloning of the complete gene for carcinoembryonic antigen: Analysis of its promoter indicates a region conveying cell type-specific expression. *Mol Cell Biol* 10(6): 2738–2748.

Schuler, M., R. Herrmann, et al. 2001. Adenovirus-mediated wild-type p53 gene transfer in patients receiving chemotherapy for advanced non-small-cell lung cancer: Results of a multicenter phase II study. *J Clin Oncol* 19(6): 1750–1758.

Schuler, M., C. Rochlitz, et al. 1998. A phase I study of adenovirus-mediated wild-type p53 gene transfer in patients with advanced non-small cell lung cancer. *Hum Gene Ther* 9(14): 2075–2082.

Senzer, N., S. Mani, et al. 2004. TNFerade biologic, an adenovector with a radiation-inducible promoter, carrying the human tumor necrosis factor alpha gene: A phase I study in patients with solid tumors. *J Clin Oncol* 22(4): 592–601.

Seth, P., U. Brinkmann, et al. 1996. Adenovirus-mediated gene transfer to human breast tumor cells: an approach for cancer gene therapy and bone marrow purging. *Cancer Res* 56(6): 1346–1351.

Shayakhmetov, D. M., C. A. Carlson, et al. 2002. A high-capacity, capsid-modified hybrid adenovirus/adeno-associated virus vector for stable transduction of human hematopoietic cells. *J Virol* 76(3): 1135–1143.

Shayakhmetov, D. M., Z. Y. Li, et al. 2002. Targeting of adenovirus vectors to tumor cells does not enable efficient transduction of breast cancer metastases. *Cancer Res* 62(4): 1063–1068.

Shenk, T. 1996. Adenoviridae: The viruses and their replication. In: B. N. Fields, D. M. Knipe, P. M. Howley, et al. (eds.), *Fields Virology*, 3rd edn., Philadelphia, PA: Lippincott-Raven, pp. 2111–2148.

Shi, C. X., M. Hitt, et al. 2002. Superior tissue-specific expression from tyrosinase and prostate-specific antigen promoters/ enhancers in helper-dependent compared with first-generation adenoviral vectors. *Hum Gene Ther* 13(2): 211–224.

Shinoura, N., K. Saito, et al. 2000. Adenovirus-mediated transfer of bax with caspase-8 controlled by myelin basic protein promoter exerts an enhanced cytotoxic effect in gliomas. *Cancer Gene Ther* 7(5): 739–748.

Shinozaki, K., O. Ebert, et al. 2004. Oncolysis of multifocal hepatocellular carcinoma in the rat liver by hepatic artery infusion of vesicular stomatitis virus. *Mol Ther* 9(3): 368–376.

Shu, J., J. H. Lee, et al. 1999. Adenovirus-mediated gene transfer of dominant negative Ha-Ras inhibits proliferation of primary meningioma cells. *Neurosurgery* 44(3): 579–587; discussion 587–588.

Siders, W. M., P. J. Halloran, et al. 1996. Transcriptional targeting of recombinant adenoviruses to human and murine melanoma cells. *Cancer Res* 56(24): 5638–5646.

Simons, J. W., M. A. Carducci, et al. 2006. Phase I/II trial of an allogeneic cellular immunotherapy in hormone-naive prostate cancer. *Clin Cancer Res* 12(11 Pt 1): 3394–3401.

Simons, J. W., E. M. Jaffee, et al. 1997. Bioactivity of autologous irradiated renal cell carcinoma vaccines generated by ex vivo granulocyte-macrophage colony-stimulating factor gene transfer. *Cancer Res* 57(8): 1537–1546.

Simons, J. W. and N. Sacks 2006. Granulocyte-macrophage colony-stimulating factor-transduced allogeneic cancer cellular immunotherapy: The GVAX vaccine for prostate cancer. *Urol Oncol* 24(5): 419–424.

Skladanowski, A. and A. K. Larsen 1997. Expression of wild-type p53 increases etoposide cytotoxicity in M1 myeloid leukemia cells by facilitated G2 to M transition: Implications for gene therapy. *Cancer Res* 57(5): 818–823.

Skorski, T., M. Nieborowska-Skorska, et al. 1995. Leukemia treatment in severe combined immunodeficiency mice by antisense oligodeoxynucleotides targeting cooperating oncogenes. *J Exp Med* 182(6): 1645–1653.

Skorski, T., M. Nieborowska-Skorska, et al. 1996. Antisense oligodeoxynucleotide combination therapy of primary chronic myelogenous leukemia blast crisis in SCID mice. *Blood* 88(3): 1005–1012.

Smith, R., Huebner, R. J., Rowe, W. P., Schatten, W. E., and Thomas, L. P., (1956). Studies on the use of viruses in the treatment of carcinoma of the cervix. *Cancer* 9: 1211–1218.

Sobol, R. E., I. Royston, et al. 1995. Injection of colon carcinoma patients with autologous irradiated tumor cells and fibroblasts genetically modified to secrete interleukin-2 (IL-2). A phase I study. *Hum Gene Ther* 6(2): 195–204.

Society, A. C. 1993. *American Cancer Society Facts and Figures.* Washington, DC: American Cancer Society: Publ. No. 93-400.

Soifer, H., C. Higo, et al. 2001. Stable integration of transgenes delivered by a retrotransposon-adenovirus hybrid vector. *Hum Gene Ther* 12(11): 1417–1428.

Soiffer, R., T. Lynch, et al. 1998. Vaccination with irradiated autologous melanoma cells engineered to secrete human granulocyte-macrophage colony-stimulating factor generates potent antitumor immunity in patients with metastatic melanoma. *Proc Natl Acad Sci USA* 95(22): 13141–13146.

Sorrentino, B. P., S. J. Brandt, et al. 1992. Selection of drug-resistant bone marrow cells in vivo after retroviral transfer of human MDR1. *Science* 257(5066): 99–103.

Sorrentino, B. P., K. T. McDonagh, et al. 1995. Expression of retroviral vectors containing the human multidrug resistance 1 cDNA in hematopoietic cells of transplanted mice. *Blood* 86(2): 491–501.

Sotomayor, M. G., H. Hu, et al. 2002. Advances in gene therapy for malignant melanoma. *Cancer Control* 9(1): 39–48.

Spergel, J. M. and S. Chen-Kiang 1991. Interleukin 6 enhances a cellular activity that functionally substitutes for E1A protein in transactivation. *Proc Natl Acad Sci USA* 88(15): 6472–6476.

Spitz, F. R., D. Nguyen, et al. 1996. Adenoviral-mediated wild-type p53 gene expression sensitizes colorectal cancer cells to ionizing radiation. *Clin Cancer Res* 2(10): 1665–1671.

St. John, M. A., E. Abemayor, et al. 2006. Recent new approaches to the treatment of head and neck cancer. *Anticancer Drugs* 17(4): 365–375.

Stanizzi, M. A. and S. J. Hall 2007. Clinical experience with gene therapy for the treatment of prostate cancer. *Rev Urol* 9(Suppl 1): S20–S8.

Steiner, M. S., C. T. Anthony, et al. 1998. Antisense c-myc retroviral vector suppresses established human prostate cancer. *Hum Gene Ther* 9(5): 747–755.

Steiner, M. S. and J. R. Gingrich 2000. Gene therapy for prostate cancer: Where are we now? *J Urol* 164(4): 1121–1136.

Steinwaerder, D. S. and A. Lieber 2000. Insulation from viral transcriptional regulatory elements improves inducible transgene expression from adenovirus vectors in vitro and in vivo. *Gene Ther* 7(7): 556–567.

Sterman, D. H., K. Molnar-Kimber, et al. 2000. A pilot study of systemic corticosteroid administration in conjunction with intrapleural adenoviral vector administration in patients with malignant pleural mesothelioma. *Cancer Gene Ther* 7(12): 1511–1518.

Stewart, A. K., A. D. Schimmer, et al. 1998. In vivo adenoviral-mediated gene transfer of interleukin-2 in cutaneous plasmacytoma. *Blood* 91(3): 1095–1097.

Stitz, J., M. D. Muhlebach, et al. 2001. A novel lentivirus vector derived from apathogenic simian immunodeficiency virus. *Virology* 291(2): 191–197.

Su, H., J. C. Chang, et al. 1996. Selective killing of AFP-positive hepatocellular carcinoma cells by adeno-associated virus transfer of the herpes simplex virus thymidine kinase gene. *Hum Gene Ther* 7(4): 463–470.

Sumimoto, H., S. Yamagata, et al. 2005. Gene therapy for human small-cell lung carcinoma by inactivation of Skp-2 with virally mediated RNA interference. *Gene Ther* 12(1): 95–100.

Summerford, C., J. S. Bartlett, et al. 1999. AlphaVbeta5 integrin: A co-receptor for adeno-associated virus type 2 infection. *Nat Med* 5(1): 78–82.

Summerford, C. and R. J. Samulski 1998. Membrane-associated heparan sulfate proteoglycan is a receptor for adeno-associated virus type 2 virions. *J Virol* 72(2): 1438–1445.

Sung, M. W., H. C. Yeh, et al. 2001. Intratumoral adenovirus-mediated suicide gene transfer for hepatic metastases from colorectal adenocarcinoma: Results of a phase I clinical trial. *Mol Ther* 4(3): 182–191.

Sutton, R. E., M. J. Reitsma, et al. 1999. Transduction of human progenitor hematopoietic stem cells by human immunodeficiency virus type 1-based vectors is cell cycle dependent. *J Virol* 73(5): 3649–3660.

Swisher, S. G., J. A. Roth, et al. 1999. Adenovirus-mediated p53 gene transfer in advanced non-small-cell lung cancer. *J Natl Cancer Inst* 91(9): 763–771.

Tait, D. L., P. S. Obermiller, et al. 1997. A phase I trial of retroviral BRCA1sv gene therapy in ovarian cancer. *Clin Cancer Res* 3(11): 1959–1968.

Tait, D. L., P. S. Obermiller, et al. 1999. Ovarian cancer BRCA1 gene therapy: Phase I and II trial differences in immune response and vector stability. *Clin Cancer Res* 5(7): 1708–1714.

Takahashi, M., T. Sato, et al. 2002. E1B-55K-deleted adenovirus expressing E1A-13S by AFP-enhancer/promoter is capable of highly specific replication in AFP-producing hepatocellular carcinoma and eradication of established tumor. *Mol Ther* 5(5 Pt 1): 627–634.

Takahashi, S., Y. Ito, et al. 2006. Gene therapy for breast cancer—Review of clinical gene therapy trials for breast cancer and MDR1 gene therapy trial in Cancer Institute Hospital. *Breast Cancer* 13(1): 8–15.

Tan, Y., M. Xu, et al. 1996. IL-2 gene therapy of advanced lung cancer patients. *Anticancer Res* 16(4A): 1993–1998.

Tanaka, T., F. Kanai, et al. 1997. Adenovirus-mediated gene therapy of gastric carcinoma using cancer-specific gene expression in vivo. *Biochem Biophys Res Commun* 231(3): 775–779.

Teh, B. S., E. Aguilar-Cordova, et al. 2001. Phase I/II trial evaluating combined radiotherapy and in situ gene therapy with or without hormonal therapy in the treatment of prostate cancer—A preliminary report. *Int J Radiat Oncol Biol Phys* 51(3): 605–613.

Toloza, E. M., M. A. Morse, et al. 2006. Gene therapy for lung cancer. *J Cell Biochem* 99(1): 1–22.

Tomko, R. P., R. Xu, et al. 1997. HCAR and MCAR: The human and mouse cellular receptors for subgroup C adenoviruses and group B coxsackieviruses. *Proc Natl Acad Sci USA* 94(7): 3352–3356.

Tong, A. W., J. Nemunaitis, et al. 2005. Intratumoral injection of INGN 241, a nonreplicating adenovector expressing the melanoma-differentiation associated gene-7 (mda-7/IL24): Biologic outcome in advanced cancer patients. *Mol Ther* 11(1): 160–172.

Trask, T. W., R. P. Trask, et al. 2000. Phase I study of adenoviral delivery of the HSV-tk gene and ganciclovir administration in patients with current malignant brain tumors. *Mol Ther* 1(2), 195–203.

Trepel, M., P. Groscurth, et al. 1998. Chemosensitivity of human malignant glioma: Modulation by p53 gene transfer. *J Neurooncol* 39(1): 19–32.

Tsang, K. Y., S. Zaremba, et al. 1995. Generation of human cytotoxic T cells specific for human carcinoembryonic antigen epitopes from patients immunized with recombinant vaccinia-CEA vaccine. *J Natl Cancer Inst* 87(13): 982–990.

van Beusechem, V. W., J. Grill, et al. 2002. Efficient and selective gene transfer into primary human brain tumors by using single-chain antibody-targeted adenoviral vectors with native tropism abolished. *J Virol* 76(6): 2753–2762.

van Beusechem, V. W., A. L. van Rijswijk, et al. 2000. Recombinant adenovirus vectors with knobless fibers for targeted gene transfer. *Gene Ther* 7(22): 1940–1946.

van der Bruggen, P., C. Traversari, et al. 1991. A gene encoding an antigen recognized by cytolytic T lymphocytes on a human melanoma. *Science* 254(5038): 1643–1647.

van der Eb, M. M., S. J. Cramer, et al. 1998. Severe hepatic dysfunction after adenovirus-mediated transfer of the herpes simplex virus thymidine kinase gene and ganciclovir administration. *Gene Ther* 5(4): 451–458.

VandenDriessche, T., L. Thorrez, et al. 2002. Lentiviral vectors containing the human immunodeficiency virus type-1 central polypurine tract can efficiently transduce nondividing hepatocytes and antigen-presenting cells in vivo. *Blood* 100(3): 813–822.

Vasey, P. A., L. N. Shulman, et al. 2002. Phase I trial of intraperitoneal injection of the E1B-55-kd-gene-deleted adenovirus ONYX-015 (dl1520) given on days 1 through 5 every 3 weeks in patients with recurrent/refractory epithelial ovarian cancer. *J Clin Oncol* 20(6): 1562–1569.

Venook, A., 1998. Gene therapy of colorectal liver metastases using a recombinant adenovirus encoding wt p53 (Sch 58500) via hepatic artery infusion: A phase I study (abstr.). *American society of clinical Oncology (ASCO)* Proceedings 17: 431a.

Verzeletti, S., C. Bonini, et al. 1998. Herpes simplex virus thymidine kinase gene transfer for controlled graft-versus-host disease and graft-versus-leukemia: Clinical follow-up and improved new vectors. *Hum Gene Ther* 9(15): 2243–2251.

Villaret, D., B. Glisson, et al. 2002. A multicenter phase II study of tgDCC-E1A for the intratumoral treatment of patients with recurrent head and neck squamous cell carcinoma. *Head Neck* 24(7): 661–669.

Vogelzang, N. J., Sudakoff, G., Hersh, E. M., et al. 1996. Clinical experience in phase I and phase II testing of direct intratumoral administration with Allovectin-7: A gene-based immunotherapeutic agent. *Proc Am Soc Clin Oncol* 15: 235.

von Kalle, C., H. P. Kiem, et al. 1994. Increased gene transfer into human hematopoietic progenitor cells by extended in vitro exposure to a pseudotyped retroviral vector. *Blood* 84(9): 2890–2897.

Waehler, R., H. Ittrich, et al. 2005. Low-dose adenoviral immunotherapy of rat hepatocellular carcinoma using single-chain interleukin-12. *Hum Gene Ther* 16(3): 307–317.

Wagner, E. 1999. Application of membrane-active peptides for nonviral gene delivery. *Adv Drug Deliv Rev* 38(3): 279–289.

Wahl, A. F., K. L. Donaldson, et al. 1996. Loss of normal p53 function confers sensitization to Taxol by increasing G2/M arrest and apoptosis. *Nat Med* 2(1): 72–79.

Walczak, H., R. E. Miller, et al. 1999. Tumoricidal activity of tumor necrosis factor-related apoptosis-inducing ligand in vivo. *Nat Med* 5(2): 157–163.

Walther, W., U. Stein, et al. 2005. Use of the nuclease inhibitor aurintricarboxylic acid (ATA) for improved non-viral intratumoral in vivo gene transfer by jet-injection. *J Gene Med* 7(4): 477–485.

Wang, Q. and M. H. Finer 1996. Second-generation adenovirus vectors. *Nat Med* 2(6): 714–716.

Wang, Q., X. C. Jia, et al. 1995. A packaging cell line for propagation of recombinant adenovirus vectors containing two lethal gene-region deletions. *Gene Ther* 2(10): 775–783.

Watson, P. H., R. T. Pon, et al. 1991. Inhibition of c-myc expression by phosphorothioate antisense oligonucleotide identifies a critical role for c-myc in the growth of human breast cancer. *Cancer Res* 51(15): 3996–4000.

Wattel, E., M. Vanrumbeke, et al. 1996. Differential efficacy of adenoviral mediated gene transfer into cells from hematological cell lines and fresh hematological malignancies. *Leukemia* 10(1): 171–174.

Weber, F., Bojar, H., Priesack, H. B., Floeth, F., Lenartz, D., Kiwit, J., and Bock, W. 1997. Gene therapy of glioblastoma—One year clinical experience with ten patients. *J Mol Med* 75: B40.

Weichselbaum, R. R., D. E. Hallahan, et al. 1994. Gene therapy targeted by radiation preferentially radiosensitizes tumor cells. *Cancer Res* 54(16): 4266–4269.

Weill, D., M. Mack, et al. 2000. Adenoviral-mediated p53 gene transfer to non-small cell lung cancer through endobronchial injection. *Chest* 118(4): 966–970.

Wendtner, C. M., A. Nolte, et al. 1997. Gene transfer of the costimulatory molecules B7-1 and B7-2 into human multiple myeloma cells by recombinant adeno-associated virus enhances the cytolytic T cell response. *Gene Ther* 4(7): 726–735.

Wickham, T. J., P. Mathias, et al. 1993. Integrins alpha v beta 3 and alpha v beta 5 promote adenovirus internalization but not virus attachment. *Cell* 73(2): 309–319.

Wickham, T. J., D. M. Segal, et al. 1996. Targeted adenovirus gene transfer to endothelial and smooth muscle cells by using bispecific antibodies. *J Virol* 70(10): 6831–6838.

Wildner, O. 2001. Oncolytic viruses as therapeutic agents. *Ann Med* 33(5): 291–304.

Wilson, J. M. 1996. Adenoviruses as gene-delivery vehicles. *N Engl J Med* 334(18): 1185–1187.

Witlox, M. A., M. L. Lamfers, et al. 2007. Evolving gene therapy approaches for osteosarcoma using viral vectors: Review. *Bone* 40(4): 797–812.

Wolf, J. K. and A. D. Jenkins 2002. Gene therapy for ovarian cancer (review). *Int J Oncol* 21(3): 461–468.

Wolff, G., I. J. Korner, et al. 1998. Ex vivo breast cancer cell purging by adenovirus-mediated cytosine deaminase gene transfer and short-term incubation with 5-fluorocytosine completely prevents tumor growth after transplantation. *Hum Gene Ther* 9(15): 2277–2284.

Worth, L. L., S. F. Jia, et al. 2000. Intranasal therapy with an adenoviral vector containing the murine interleukin-12 gene eradicates osteosarcoma lung metastases. *Clin Cancer Res* 6(9): 3713–3718.

Wright, M., J. Grim, et al. 1997. An intracellular anti-erbB-2 single-chain antibody is specifically cytotoxic to human breast carcinoma cells overexpressing erbB-2. *Gene Ther* 4(4): 317–322.

Wright, P., C. Zheng, et al. 1998. Intratumoral vaccination of adenoviruses expressing fusion protein RM4/tumor necrosis factor (TNF)-alpha induces significant tumor regression. *Cancer Gene Ther* 5(6): 371–379.

Wroblewski, J. M., L. T. Lay, et al. 1996. Selective elimination (purging) of contaminating malignant cells from hematopoietic stem cell autografts using recombinant adenovirus. *Cancer Gene Ther* 3(4): 257–264.

Wu, A., A. Mazumder, et al. 2001. Biological purging of breast cancer cells using an attenuated replication-competent herpes simplex virus in human hematopoietic stem cell transplantation. *Cancer Res* 61(7): 3009–3015.

Wu, L., J. Matherly, et al. 2001. Chimeric PSA enhancers exhibit augmented activity in prostate cancer gene therapy vectors. *Gene Ther* 8(18): 1416–1426.

Wu, N., S. C. Watkins, et al. 1996. Prolonged gene expression and cell survival after infection by a herpes simplex virus mutant defective in the immediate-early genes encoding ICP4, ICP27, and ICP22. *J Virol* 70(9): 6358–6369.

Wu, X., Y. He, et al. 2001. Regression of human mammary adenocarcinoma by systemic administration of a recombinant gene encoding the hFlex-TRAIL fusion protein. *Mol Ther* 3(3): 368–374.

Wu, X., J. Holschen, et al. 1996. Retroviral vector sequences may interact with some internal promoters and influence expression. *Hum Gene Ther* 7(2): 159–171.

Xiong, C., R. Levis, et al. 1989. Sindbis virus: An efficient, broad host range vector for gene expression in animal cells. *Science* 243(4895): 1188–1191.

Xu, H. J., Y. Zhou, et al. 1996. Enhanced tumor suppressor gene therapy via replication-deficient adenovirus vectors expressing an N-terminal truncated retinoblastoma protein. *Cancer Res* 56(10): 2245–2249.

Xu, M., D. Kumar, et al. 1997. Parenteral gene therapy with p53 inhibits human breast tumors in vivo through a bystander mechanism without evidence of toxicity. *Hum Gene Ther* 8(2): 177–185.

Yang, N. S., Burkholder, J., Roberts, B., Martinell, B., and McCabe, D. 1990. In vivo and in vitro gene transfer to mammalian somatic cells by particle bombardment. *Proc Natl Acad Sci USA* 87: 9568–9572.

Yang, Y., F. A. Nunes, et al. 1994. Inactivation of E2a in recombinant adenoviruses improves the prospect for gene therapy in cystic fibrosis. *Nat Genet* 7(3): 362–369.

Yang, Z. Y., N. D. Perkins, et al. 1995. The p21 cyclin-dependent kinase inhibitor suppresses tumorigenicity in vivo. *Nat Med* 1(10): 1052–1056.

Yant, S. R., A. Ehrhardt, et al. 2002. Transposition from a gutless adeno-transposon vector stabilizes transgene expression in vivo. *Nat Biotechnol* 20(10): 999–1005.

Yee, C., M. J. Gilbert, et al. 1996. Isolation of tyrosinase-specific CD8+ and CD4+ T cell clones from the peripheral blood of melanoma patients following in vitro stimulation with recombinant vaccinia virus. *J Immunol* 157(9): 4079–4086.

Yee, C., S. R. Riddell, et al. 1997. Prospects for adoptive T cell therapy. *Curr Opin Immunol* 9(5): 702–708.

Yee, D., S. E. McGuire, et al. 1996. Adenovirus-mediated gene transfer of herpes simplex virus thymidine kinase in an ascites model of human breast cancer. *Hum Gene Ther* 7(10): 1251–1257.

Yee, J. K., A. Miyanohara, et al. 1994. A general method for the generation of high-titer, pantropic retroviral vectors: Highly efficient infection of primary hepatocytes. *Proc Natl Acad Sci USA* 91(20): 9564–9568.

Yeh, P., J. F. Dedieu, et al. 1996. Efficient dual transcomplementation of adenovirus E1 and E4 regions from a 293-derived cell line expressing a minimal E4 functional unit. *J Virol* 70(1): 559–565.

Yu, D. C., Y. Chen, et al. 2001. Antitumor synergy of CV787, a prostate cancer-specific adenovirus, and paclitaxel and docetaxel. *Cancer Res* 61(2): 517–525.

Yu, S. F., T. von Ruden, et al. 1986. Self-inactivating retroviral vectors designed for transfer of whole genes into mammalian cells. *Proc Natl Acad Sci USA* 83(10): 3194–3198.

Zhang, G., X. Gao, et al. 2004. Hydroporation as the mechanism of hydrodynamic delivery. *Gene Ther* 11(8): 675–682.

Zhang, L., H. Akbulut, et al. 2002. Adenoviral vectors with E1A regulated by tumor-specific promoters are selectively cytolytic for breast cancer and melanoma. *Mol Ther* 6(3): 386.

Zhang, W. W., X. Fang, et al. 1994. High-efficiency gene transfer and high-level expression of wild-type p53 in human lung cancer cells mediated by recombinant adenovirus. *Cancer Gene Ther* 1(1): 5–13.

Zhou, P., W. Jiang, et al. 1995. Antisense to cyclin D1 inhibits growth and reverses the transformed phenotype of human esophageal cancer cells. *Oncogene* 11(3): 571–580.

Zhou, S. Z., Q. Li, et al. 1996. Adeno-associated virus 2-mediated transduction and erythroid cell-specific expression of a human beta-globin gene. *Gene Ther* 3(3): 223–229.

Zhu, Z., N. A. DeLuca, et al. 1996. Overexpression of the herpes simplex virus type 1 immediate-early regulatory protein, ICP27, is responsible for the aberrant localization of ICP0 and mutant forms of ICP4 in ICP4 mutant virus-infected cells. *J Virol* 70(8): 5346–5356.

37 Molecular Therapy for Type 1 and Type 2 Diabetes

Susan Leanne Samson, Vijay Yechoor, and Lawrence Chan

CONTENTS

The prevalence of diabetes worldwide has been on the rise over the last century. According to the International Diabetes Federation (IDF), the disease affects 246 million people worldwide in 2006. The IDF projects that, if nothing is done, the total number will skyrocket to 380 million by 2025 [1]. In the United States, the rise in incidence has been especially steep in the last decade [2]. The increase in incidence is also accompanied by a relative decline in the risk of death among diabetic patients, at least in men [3], which further increases the overall prevalence of the disease. It is projected that if the rate of increase continues, almost 50 million people in the United States will carry the diagnosis of diabetes by the year 2050 [4].

Most diabetic patients do not die of the acute complications such as diabetic ketoacidosis, which are becoming increasingly uncommon as a cause of death, as they are mostly preventable by modern therapy. Patients with diabetes live longer lives and are increasingly being afflicted by the chronic complications [5]. These include the microvascular complications of retinopathy, nephropathy, and neuropathy, and the macrovascular complications resulting from the consequences of accelerated atherosclerosis, including myocardial infarction and stroke.

Diabetic complications are thus the major cause of morbidity and mortality among type 1 and type 2 diabetics. The pathogenesis of the chronic complications is complex. There is good evidence that inadequately controlled hyperglycemia is an all important causative factor. The best way to prevent or delay the onset of complications is the prevention or reversal of hyperglycemia.

Recent developments in therapy have greatly expanded our armamentarium in tackling the disease. As a consequence, blood glucose control of most diabetic patients today is much better than that accomplished only a decade ago. However, despite the introduction of different kinds of diabetes medications, it is still impossible to consistently reverse hyperglycemia in diabetes without occasional or frequent hypoglycemic episodes [6]. The problem is that it is very difficult, if not impossible, to attain physiological blood glucose control using oral agents or parenteral medications.

A biomechanical artificial pancreas is one way of restoring euglycemia. In theory, a well-designed closed loop system can accomplish this without significant risk of hypoglycemia. Unfortunately, a portable, user-friendly, and foolproof system is not yet available despite intense research in industry over the last few decades [7]. In the meantime, academic investigators and industry have initiated research programs to develop cell and gene therapy for diabetes based on increasing understanding of islet development and immune dysregulation in type 1 diabetes and the interacting signaling networks and metabolic interactions of different organs underlying the insulin resistance and β cell failure in type 2 diabetes (Figure 37.1).

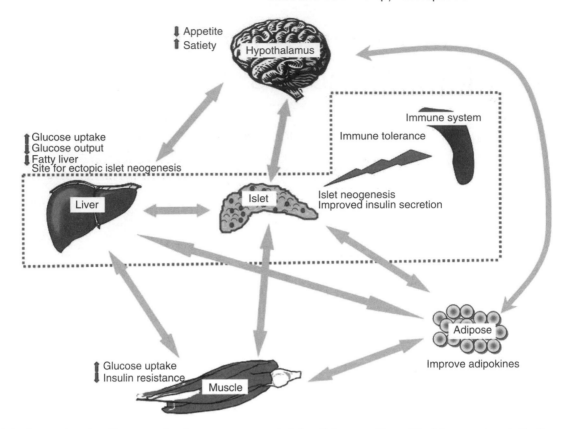

FIGURE 37.1 Interacting signaling networks of organ systems underlying diabetes mellitus. All of these are targets for therapy for type 2 diabetes. Type 1 diabetes targets, represented by the dashed box, are primarily limited to the pancreas, liver, and the immune system. The desired therapeutic effects in each organ are listed.

37.1 GENE THERAPY FOR TYPE 1 DIABETES MELLITUS

This chapter deals exclusively with *in vivo* gene therapy, as it is the method of choice at the present time because the regimen can be delivered to a recipient like a pharmaceutical reagent. A major challenge is the development of safe efficient vectors for *in vivo* gene delivery, which is the subject of other chapters in this volume.

There are two major problems to be addressed in any curative treatment of type 1a diabetes, the autoimmune form of diabetes that is the cause of severe β cell destruction in the vast majority of type 1 patients leading to absolute insulin deficiency and hyperglycemia. The approaches for *in vivo* gene therapy to correct the underlying metabolic derangements in diabetes, all with the ultimate objective of restoration of euglycemia [8], can be broadly divided into three strategies that involve the delivery of genes to express (1) glucose-regulatable insulin, (2) proteins that facilitate glucose utilization or inhibit hepatic glucose production, and (3) developmental/transcription factors that induce the production of new β cells. The last approach must be coupled with a strategy to address the autoimmune β cell destruction to protect the newly formed islets from persistent immune attack. In fact, there is recent evidence that even in long-standing type 1a diabetes, the pancreas continues to make valiant though futile

attempts at regeneration [9]. Thus, gene therapy to correct, or circumvent, the autoimmunity, could by itself enable β cell regeneration to take place, leading to improvements in islet function. Modulation of autoimmunity has been tried both as a primary and secondary prevention strategy in mouse models; at this time, it could have potential practical application only as a secondary prevention approach in humans as discussed later in this chapter.

37.1.1 GENE THERAPY TO ADDRESS THE HYPERGLYCEMIA AND METABOLIC DERANGEMENTS

Some of the strategies discussed in this section can be applicable in type 2 diabetes and those discussed in the section on type 2 diabetes may also find application in type 1 diabetes therapy. Many of the strategies reviewed below have been tested in type 1 insulin-deficient animal models.

37.1.1.1 Transfer of Glucose-Responsive Insulin Transgene

The vast majority of publications on diabetes gene therapy involve the delivery into liver cells of versions of the insulin gene, which have been modified (i) either to make the proinsulin expressed susceptible to processing into mature insulin or to obviate the need for processing and (ii) to render gene expression responsive to changes in blood glucose concentration [10,11]. As liver cells do not produce the islet prohormone convertases Pc1/3 and Pc2, many investigators have introduced new proteolytic cleavage sites into the proinsulin molecules that are recognized by furin, a protease present in many tissues including liver cells [12–15]. Alternatively, the insulin gene can be modified to encode single-chain insulin [16], which has 20%–40% of the activity of normal mature insulin [17]. The most challenging part of insulin gene therapy is to confer glucose responsiveness to insulin transgene expression that mimics that of normal β cells, wherein, the onset of hyperglycemia is met with an almost instantaneous burst of insulin (in less than a minute) from primed preformed insulin-containing vesicles, the first phase response followed by a more sustained delayed response resulting from transcriptional and translational regulation, the second phase response. The most popular strategy to accomplish this objective has been the use of different glucose-responsive promoters, e.g., promoters from phosphoenolpyruvate carboxykinase (PEPCK, [18]) gene, elements from the L-pyruvate kinase gene [19], the glucose-6-phosphatase [20], and other genes (reviewed in Refs. [10,21]), which allow the insulin transgene transcription to be upregulated by hyperglycemia within 1–2 h, and stimulated insulin protein production and secretion within 3–4 h. The increased insulin production and secretion also take a long time to be turned off when blood glucose is normal or low. Because of the lag in the secretory response, glycemic control by transcriptionally regulated insulin transgenes is erratic and hypoglycemia is a major complication. Another approach is to control secretion at the level of endoplasmic reticulum by drug-induced protein disaggregation [22]. However, any manipulation requiring pharmacological agents defeats the purpose of gene therapy, as there is considerable flexibility with the wide array of pharmacological therapy using the different forms of insulin formulations that are currently available.

37.1.1.2 Transfer of Genes Other Than Insulin to Lower Blood Glucose

There are two types of noninsulin transgenes, which have been used to lower blood glucose: (i) transgenes that inhibit glucose production in the liver and (ii) those that enhance glucose utilization by the liver or skeletal muscle.

In the first group, glucokinase (Gck) gene transfer in rodents has been used by many different groups [23–27]. Although Gck has been categorized as a transgene that lowers glucose production in the liver [28], hepatic glucose production was not measured in any of these studies and it is likely that a major downstream effect of Gck is increased glucose utilization [24]. High-dose Gck gene transfer causes hyperlipidemia and fatty liver [24,27], and Gck gene transfer is best used as adjuvant treatment (a debatable role for gene therapy) to complement insulin therapy [27]. Gck regulatory protein gene transfer produces a very similar effect as Gck gene transfer [29]. A mutant form of 6-phosphofructo-2-kinase/fructose-2,6-bisphosphatase has been used to activate

phosphofructokinase-1 and to simultaneously inhibit fructose-1,6-bisphosphatase to downregulate gluconeogenesis. Its overexpression has been shown to downregulate glucose-6-phosphatase and upregulate Gck levels, stimulating glucose disposal and inhibiting hepatic glucose production in a mouse model of type 2 diabetes [30]. Another way to downregulate hepatic glucose production is to divert the glucose to glycogen by overexpression of protein targeting to glycogen (PTG) [31,32]. PTG is a member of the family of glycogen-targeting subunits of protein phosphatase-1 that regulate glycogen metabolism. Adenovirus-mediated transfer of PTG stimulates glycogen synthesis in the liver and lowers blood glucose in rats, and represents a potential therapeutic approach to diabetes gene therapy [31]. As alluded to earlier, hepatic overexpression of Gck appears to enhance glucose utilization and thus is one of the potential therapeutic genes that simulate glucose disposal. Interestingly, skeletal muscle overexpression of Gck also stimulates glucose disposal and protects against hyperglycemia in streptozotocin (STZ)-induced diabetic mice [33].

In summary, there are different strategies to modulate blood glucose by targeting genes (other than insulin) that affect glucose production and utilization. However, most of these represent adjuvant therapy. Gene therapy may not be the best strategy for upregulating enzymes that modulate glucose metabolism; a better strategy to target enzyme activity is the screening and development of small molecular weight compounds (e.g., see Ref. [34]), a time-honored method that has worked well in pharmaceutical companies.

In addition to the enzyme overexpression approach summarized above, different gene therapy strategies to downregulate gene specific gene expression have also been tested. Most of the targets are applicable to both type 1 and type 2 diabetes and will be discussed later in the section on type 2 diabetes.

37.1.1.3 Transfer of Developmental/Transcription Factors to Induce the Production of β Cells in the Liver

An increasing understanding of the ontogeny of pancreatic and islet organogenesis has enabled investigators to initiate studies on gene therapy–based induced islet neogenesis as a curative therapy for diabetes. In comparison with insulin transgene delivery, induction of new islets/β cell formation has many potential advantages (Table 37.1). Strategies to induce new β cells, either in the pancreas or more commonly in the liver, have been tested in different laboratories in the last few years with some encouraging results.

To understand the rationale underlying gene therapy-induced islet/β cell neogenesis as a form of therapy, it is necessary to briefly review the development of the pancreas and the pancreatic islets. The pancreas develops during embryogenesis by cell-type specification and subsequent differentiation orchestrated, in the gut endoderm, by various transcription factors including, among others, Hlxb9, Sox9, Pdx1 (also known as Ipf1), Hex, Isl1, Ptf1a, Hes1, and Oc1 [35,36]. A distinct multipotent progenitor cell population in the developing

TABLE 37.1

Limitations of Insulin Gene Therapy as Compared to Native or Induced Islets/β Cells

Islet/β Cells	Insulin Gene Therapy
Regulated by the insulin promoter	Usually by a glucose regulated promoter
Post-transcriptional regulation	None
Post-translational regulation	None
Have a first phase and second phase response due to primed preformed insulin granules	Only one phase that lags in response
Other islet cell contribution	None

pancreas that is Pdx1⁺ Ptf1a⁺ cMyc^High^Cpa1⁺ and negative for differentiated lineage markers ultimately produces the exocrine, endocrine, and duct cells [37]. Once the ventral and dorsal pancreatic buds develop, the endocrine cell specification is initiated in them by the bHLH transcription factor neurogenin3 (Ngn3) within the domain of Pdx1-positive cells [38–40], which is accompanied by a suppression of notch signaling in these cells. This results in the determination of the islets of Langerhans that ultimately produce important hormones that control glucose homeostasis, including insulin from β cells. Ngn3 turns on a downstream cascade of transcription factors including Neurod1, Pax6, Isl1, Pax4, Nkx2.2, Nkx6.1, and MafA in β cells and Arx and Brn4 to specify glucagons-secreting α cells. Investigators interested in induced islet neogenesis as a therapeutic approach have postulated that in competent lineages, induced expression of some of these transcription factors may cause the target cells to differentiate into islet or β cell types.

Attempts to use the pancreas as the target organ to induce islet neogenesis have been limited primarily by the technical difficulty of getting the gene into the target cell type. Recent work demonstrated that adenoviral vectors may be used to transfer gene payloads to the pancreas either retrograde through the pancreatic duct, or by intravenous injection via systemic circulation, or via the duodenal vein [41–43]. In contrast to the pancreas, the liver is a much more accessible target for various vectors, e.g., adenovirus "naturally" targets the liver efficiently [44], and, alternatively, hepatic gene delivery can be readily accomplished for almost any vector by catheter-based delivery via the hepatic artery or portal vein.

Pdx-1 is the master regulator of pancreatic organogenesis, the lack of which leads to agenesis of the pancreas [45,46]. Pdx-1 expression thus defines the pancreatic (both exocrine and endocrine) lineage. Investigators have attempted to take advantage of this in using Pdx-1 to induce islet/β cells. There was initial success at lowering blood glucose levels up to 10 days in insulin-deficient STZ-induced diabetic mice that were treated intravenously with first-generation adenovirus expressing Pdx-1 (FGAd-PDX-1). The glucose-lowering effect was ascribed to the induction of rare insulin positive cells in the liver though the blood insulin level remained uncorrected [47]. Subsequently, it was noted by others that there was coexpression of trypsin, a pancreatic exocrine enzyme, in the

insulin-producing cells in the liver. With expression of Pdx-1 in the liver at a level that would influence glucose homeostasis, the concomitant production of exocrine enzymes was found to lead to severe morbidity and substantial mortality resulting from a fulminant hepatitis caused by ectopic pancreatic enzyme production in the absence of a normal pancreatic ductal system to divert the locally produced digestive enzymes into the intestine [48]. In a different model, conditional transgenic expression of Pdx-1 in albumin-producing cells in the liver was also found to lead to considerable morbidity and mortality associated with hyperbilirubinemia and severe hepatic dysmorphogenesis, caused by the concomitant production of exocrine enzymes (elastase and chymotrypsinogen 1B) in cells in the liver that also expressed endocrine markers (islet hormones including insulin) [49]. This was not an unexpected outcome as Pdx-1 is a master developmental factor that specifies both endocrine and exocrine lineages in the gut endoderm from which both liver and pancreas develop. Hence, application of Pdx-1 as the sole transgene would not be the best strategy for inducing β cells in the liver *in vivo*.

To avoid the problems with Pdx-1, others have used endocrine lineage-specific factors that are downstream of Pdx-1, such as Ngn3, Neurod1, and MafA. These have been used singly, in combination, and along with Pdx-1. Hepatic delivery of hybrid adeno-associated virus (AAV) vectors carrying Ngn3, Neurod1, and MafA has been shown to lower blood glucose levels in diabetic mice [50], with expression of insulin and other islet markers in the liver. Hepatic delivery of a combination of Pdx-1 with Neurod1 and MafA produced similar results in another laboratory [51]. The latter group also tested a modified version of Pdx-1, one fused with the strong transactivation domain of herpes simplex virus protein VP-16, and noted that, when delivered with Ngn3 or Neurod1, the modified Pdx-1 effectively reversed hyperglycemia in insulin-deficient mice [52]. All studies reviewed above involving Pdx-1 and downstream factors consisted of short-term observations that were made within 2 weeks of gene delivery. One recent study used hyprdrodynamic delivery to coinjected naked DNA constructs of Ngn3 with an adenovirus carrying an unrelated gene (which by itself had no effect) in STZ-diabetic mice led to correction of hyperglycemia in these animals up to 2 months after treatment [53]. The most promising data to date involved the hepatic delivery of Neurod1 by helper-dependent adenovirus (HDAd), by itself or in combination with an islet growth factor (beta cellulin), in STZ-diabetic mice. The treatment induced the appearance of clusters of insulin-positive cells in the liver, a complete reversal of hyperglycemia, and restoration of plasma insulin and glucose tolerance for over 4 months of observation [48]. The liver cells isolated from the treated mice exhibited pancreatic β cell markers and functions, including glucose-stimulated insulin secretion. They also responded appropriately to glibenclamide, a sulfonylurea class of drug used in diabetic patients, similar to the response seen with pancreatic β cells. However, the newly formed insulin positive cells coexpressed other islet hormones, in contrast to adult islet cells that each normally expresses only a single hormone. The occurrence of neoislet cells that express multiple islet hormones suggests that they are immature islet cells, as insulin-glucagon double positive cells are seen commonly during embryonic islet development.

The function of many transcription factors involved in β cell development has been determined by the use of knockout mouse models, and many were found to play key roles in normal islet organogenesis and β cell function; their effects when overexpressed ectopically often have been short of expectation. This probably stems from the fact that these transcription factors act in lineage-competent cells during embryonic development. It is unclear if they are competent in inducing the complete *trans*-differentiation of already differentiated cells like hepatocytes into a fully differentiated endocrine pancreas lineage. *Trans*-differentiation has been defined as the lineage switching from one differentiated state to another with an intervening undifferentiated intermediate stage. Although *trans*-differentiation was claimed in several of the reports cited above, none of the aforementioned studies have rigorously examined whether lineage switching from the treated hepatocytes to an islet lineage has indeed happened.

Mechanistic experiments are clearly needed for future studies on induced islet neogenesis to determine if true *trans*-differentiation has occurred. Unlike induced differentiation from a progenitor cell population, long-term stable induced *trans*-differentiation from another differentiated lineage is a tall order for a transcription factor and for gene therapy. Thus, if we want to generate new islets to treat diabetes, the identification of receptive progenitor cells that are competent to differentiate into an islet cell lineage is an important component of any gene therapy regimen utilizing lineage-defining transcription factors.

37.1.2 GENE THERAPY TO MODULATE IMMUNE DERANGEMENT

Immune modulation is an important therapeutic strategy in type 1a diabetes. It finds application in the disease at multiple levels: (1) As alluded to earlier, there is evidence for persistent foci of islet neogenesis in the pancreas of most type 1 diabetics, even after many decades of disease. It is conceivable that long-term success with correction of autoimmunity in a type 1a diabetic might eventually lead to restoration of β cell function. Even a small improvement in insulin reserve may convert a brittle diabetic to a much more stable one, as exemplified by the improvements shown by some type 1a patients who have been treated by islet transplantation who exhibit minor improvement in insulin requirement but marked improvement in the quality of life because of reversal of their "brittleness" [54]. (2) It can be used to circumvent the toxic immunosuppression cocktails given to type 1 diabetics after islet transplantation. (3) It may be an essential adjuvant therapy for induced islet neogenesis. Regimens that successfully induce islet neogenesis in STZ-mice have been found to work very poorly in autoimmune models due to the inflammatory response to the neo-islets (Yechoor et al. unpublished results). One approach is to use standard immunosuppressive therapy to prepare the diabetic animal (or eventually, patient) for the islet induction, but an attractive strategy is to modulate the immune system, either locally or systemically. There have been some primary prevention strategies that have been attempted in mouse models of autoimmune diabetes that may hold promise as secondary prevention strategies for immune modulation in concert with inducing new islets/β cells. These have predominantly focused on genes to modulate the known β cell death-inducing cytokine pathways such as IFN-γ, TNF-α, IL-1β, and methods to induce tolerance [55–63].

Immune modulation by gene therapy is discussed in detail elsewhere in this book; we will discuss below specific aspects that are applicable to type 1 autoimmune diabetes. Immune modulation by inducing tolerance to known antigens or by expressing known autoimmune-resistant MHC Class II genotype related proteins has had some success in both preventing diabetes in nonobese diabetic (NOD) mice and also in reversing some of the insulitis associated with established diabetes. In these studies, GAD 500–585 (glutamic acid decarboxylase, an autoantigen) transferred via an AAV vector intramuscularly [64] or transferred as a fusion protein with an immunoglobulin chain using a retroviral vector *ex vivo*, to stimulated splenocytes and then transferred into prediabetic NOD mice [65], decreased insulitis and the incidence of diabetes significantly in treated mice by eliminating autoreactive T cells and enhancing the protective suppressor T regulatory (Treg) cells. Similarly, proinsulin fused with the cholera B toxin and expressed by a recombinant vaccinia virus in prediabetic NOD mice decreased the incidence of diabetes significantly [66]. Unfortunately, none of these approaches have demonstrated long-term success nor have they been shown to be effective as a secondary prevention measure in established diabetes, making their application to human autoimmune diabetes untenable at this time. Another strategy is to express the autoimmunity-resistant genotype-associated MHC Class II I-Aβ molecule on autologous bone marrow stem cells *ex vivo* by retroviral transduction; although effective for primary prevention in NOD diabetes, it has not been tested for treating clinically overt autoimmune diabetic animals [67].

Other successful strategies for primary prevention include approaches that target the autoreactive T cells using systemic IL-10, IL-4, and TGF-β to change the immune response from a destructive Th1 response to a protective Th2 response. Systemic overexpression of IL-10 administered as a single intramuscular injection using rAAV vector induces Treg (CD4 + CD25+) cells *in vivo* and significantly reduces insulitis in treated animals while completely preventing diabetes in NOD mice [56]. Interestingly, gene therapy of diabetic NOD mice with FGAd-TGF-β not only induces native pancreatic islets to regenerate but also protects syngenic islets after transplantation from being destroyed by autoimmunity [68]. An increase in Tregs was once again seen with this approach. An *ex vivo* approach by infecting dendritic cells with adenovirus carrying IL-4 [69] or converting antigen-specific diabetogenic T cells into Tregs by retroviral transduction of Treg-defining transcription factor FoxP3 [70] *ex vivo* and subsequently infusing them into prediabetic NOD mice was effective in primary prevention of diabetes. Similar studies abound with similar results; we have selected the examples above as representative of the approaches and preliminary outcomes [71].

In summary, many approaches have been used to induce tolerance with the common pathway being the induction of Tregs and a switch from a destructive Th1 response to a protective Th2 response. Studies testing secondary prevention of autoimmunity in animal models are few and wanting in rigor in establishing efficacy and durability of response. In human diabetes, there are no available markers that identify at risk patients definitively before the onset of overt diabetes, making it impractical to implement immune modulation as a primary prevention strategy. However, if an *in vivo* gene therapy induced islet neogenesis proves efficacious, it is paramount that adjuvant strategies be developed to induce tolerance to these neo-islets. Hence, gene therapy for type 1a diabetes would involve a combination of new islet induction to correct the β cell deficit and immune modulation of autoimmune-mediated suppression of Treg cells to allow the neo-islets to survive. The ideal scenario is to administer genes to accomplish both objectives in the same gene therapy regimen such that one can reverse diabetes in individuals with overt type 1a diabetes in a single treatment.

37.2   GENE THERAPY FOR TYPE 2 DIABETES MELLITUS

As discussed above, a major goal of gene therapy for type 1 diabetes mellitus is to replenish insulin production and secretion, and the most successful approach may be through the development of "neo-islets" using expression of pancreatic developmental transcription factors. The importance of immune destruction in the pathogenesis of type 1 diabetes also requires that new avenues are explored to circumvent or dampen autoimmunity to protect the newly formed β cells.

Gene therapy for type 2 diabetes mellitus (Dm2) has its own unique challenges because it is the culmination of two pathologic processes: insulin resistance and β cell failure [72]. Through most stages of this disease, there is hyperinsulinemia, rather than the absolute insulin deficiency seen in type 1 diabetes. Hyperglycemia occurs because the insulin levels are inadequate due to peripheral insulin resistance.

Even though high levels of insulin can be secreted under these circumstances, diabetes occurs because of relative β cell failure, since the β cells are unable to secrete enough insulin to meet physiologic requirements and maintain glucose homeostasis. Over the long term, the overtaxed β cells gradually will lose their glucose-sensing and insulin-secretory abilities, and absolute insulin deficiency will become an issue. Boosting insulin production through islet neo-genesis or cell replacement therapies, as described for type 1 diabetes, potentially could assist in reducing the relative insulin deficiency of β cell failure in Dm2. However, this does not correct the pathogenic contributions of insulin resistance, which is the stress that drove the endogenous β cells to "exhaustion" in the first place. Further, intrinsic defects in the native β cells are now believed to exist in many of the Dm2 patients, which contributed to their failure; replacement β cells or neo-islets may suffer a similar fate under the challenge of insulin resistance. Therefore, a comprehensive gene therapy approach to Dm2 requires a two-pronged tactic: namely, that both problems of insulin resistance and β cell failure be addressed (Figure 37.2).

37.2.1   GENE THERAPY TO COMBAT INSULIN RESISTANCE

Identification of Dm2-associated genes has the potential to provide new targets for pharmacologic and gene therapies to prevent or reverse Dm2. However, Dm2 is considered by many to be a polygenic disease, so that replacement of a single "deficient" gene product may not have significant effects overall. Also, there are large contributions by environmental and epigenetic factors, which contribute to the development of diabetes [73]. Perhaps the most crucial of these factors is obesity. The dramatic increase in the prevalence of Dm2 parallels that of obesity in developing as well as developed nations, emphasizing its contribution to the development of Dm2 mostly through increased insulin resistance. However, not all obese individuals develop diabetes, reminding us that both genetic and environmental factors combine to cause Dm2.

37.2.1.1   Targets for Peripheral and Central Control of Obesity

The mechanisms by which obesity leads to insulin resistance are multifold and are the focus of many research laboratories. In simple terms, with excess adipose tissue, there is an increase in circulating free fatty acids, which are directly responsible for inducing insulin resistance, likely through their intracellular accumulation in key metabolic tissues, liver and muscle, leading to impaired insulin signaling [74–76]. Adipose tissue is not an inert storage tissue, but is actually an endocrine organ, secreting both adipokines (such as adiponectin, leptin, and resistin), as well as cytokines (such as TNFα and IL-6), which are dysregulated and also have pathologic effects on insulin signaling [76,77]. Therefore, reducing adipose tissue mass can abrogate a key factor in the pathogenesis of insulin resistance leading to Dm2.

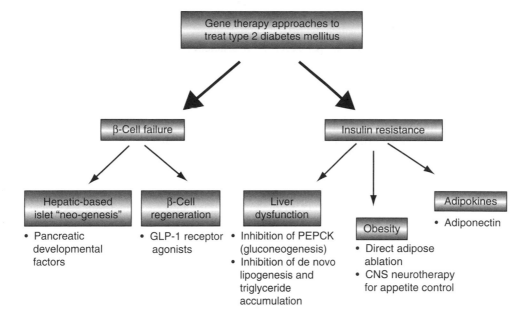

FIGURE 37.2 Variety of gene therapy approaches have been applied to treat type 2 diabetes, obesity, and β cell failure in rodent models. (Abbreviations: CNS, central nervous system; GLP-1, glucagon-like peptide-1; PEPCK, phosphoenolpyruvate carboxykinase.)

37.2.1.1.1 Peripheral Adipose Tissue Ablation

To tackle the problem of obesity, the gene therapy approaches can be direct, attacking the adipocytes themselves, or indirect by inducing a decrease in energy intake and increase in energy expenditure to result in a negative energy balance and weight loss. The direct approach has been demonstrated using molecular therapy with targeted ablation of adipose tissue in an obese mouse model. Using phage-display techniques, Kolonin et al. [78] identified a novel peptide motif, which targets adipose blood vessels specifically via specific surface proteins or "zip-codes" on the vasculature. By linking the fat "homing-peptide" to an apoptotic signal (called KLAKLAK), which causes cell death, peptide treatment of $Lep^{Ob/Ob}$ mice resulted in a sustained reduction in adipose tissue mass, decreased lipid accumulation in muscle and liver, and increased energy expenditure. Importantly, there were no phenotypic manifestations of lipodystrophy, namely insulin resistance and dyslipidemia, and fat ablation actually improved glucose homeostasis. In this respect, if this form of treatment proves safe and efficacious in humans, it may be superior to surgical liposuction. A well-controlled study involving obese women with or without diabetes showed that surgical liposuction does not improve glucose or lipid homeostasis, or ameliorate diabetes, despite the removal of over 20 kg of fat [79]. A major difference between adipose reduction by liposuction and targeted apoptosis of adipose-specific vasculature is that liposuction selectively removes subcutaneous fat whereas targeted vasculature apoptosis simultaneously ablates visceral and subcutaneous fat, because adipose depots in different parts of the body appear to share similar "zip-codes" in their vascular supply. Visceral fat differs markedly from subcutaneous fat in its metabolism and secretory profile [76,80] and visceral fat

expansion is associated with much more deleterious metabolic consequences than is subcutaneous fat expansion [80,81]. Targeted adipose tissue vasculature ablation as a form of molecular therapy for obesity and Dm2 is being tested in nonhuman primates.

37.2.1.1.2 Central Appetite Regulation

Another approach to induce weight loss is by decreasing food intake or increasing energy expenditure, or both. A large portion of the integration of peripheral nutrient and energy signals takes place in the central nervous system (CNS), mainly in the hypothalamus and the brainstem [82]. The hypothalamus is housed adjacent to the third cerebral ventricle and contains important neural nuclei involved in appetite regulation [82,83]. The neuropeptide Y (NPY) neurons in the arcuate nucleus (ARC) express NPY, agouti-related peptide (AgRP), and GABA, which stimulate appetite (orexigenic). Other neurons are anorexic and express the product of the cocaine- and amphetamine-regulated transcript (CART) and the pro-opiomelanocortin (POMC) peptide, which is processed to produce α- and β-melanocyte stimulating hormone (MSH) [84]. MSH acts to decrease appetite through melanocortin receptors (MCR3 and 4) in the paraventricular nucleus (PVN), dorsomedial (DH), ventromedial (VMH), and lateral (LH) regions of the hypothalamus, as well as the nucleus of the solitary tract in the brainstem [82,85]. These two opposing sets of orexigenic and anorexigenic neurons are regulated by signals from the periphery.

Leptin is secreted from adipose tissue and acts centrally at receptors on the hypothalamic neurons to increase satiety by inhibition of NPY neurons and activation of POMC neurons [84,86]. Activation of the leptin receptor on hypothalamic neurons activates several intracellular signal transduction pathways including JAK/STAT3, phosphoinositide-3-kinase (PI3K), and mitogen-activated protein kinase (MAPK) [82,87]. STAT3 phosphorylation increases transcription of the POMC gene but reduces that of NPY. An opposing peripheral signal is ghrelin, which is released from specialized oxyntic cells in the stomach to stimulate appetite through activation of NPY neurons and inhibition of POMC neurons. These peripheral signals are able to reach the hypothalamus through a permissive blood brain barrier (fenestrated capillaries) or specific transport mechanisms [88–90].

The centralized nature of appetite control and energy expenditure provides a target for gene therapy. However, one challenge is that systemic approaches, such as liver-based expression of anorectic peptides, may not be potent enough to have effects in the hypothalamus. A physiologic example of this problem is that of leptin. Leptin levels increase proportionately with adipose tissue mass, which would be expected to inhibit central appetite centers, causing decreased energy intake in the face of obesity. In genetically obese mice lacking leptin ($Lep^{ob/ob}$), adenovirus-mediated peripheral leptin therapy is partially effective in causing decreased food intake and weight loss [91,92]. This mouse model is a monogenic deficiency of the Ob (leptin) gene, and direct replacement of leptin is expected to be successful. However, the promise of leptin therapy for the treatment of generalized obesity has not been realized due to central leptin resistance [86,93,94]. In high-fat diet (HFD) rodent models, there is decreased STAT3 phosphorylation with CNS leptin injections compared to nonobese controls, suggesting that there is a postreceptor defect leading to leptin resistance [95]. Moreover, the higher peripheral leptin levels that accompany obesity are unable to compensate because leptin transport mechanisms appear to be saturable, providing a second mechanism for leptin resistance [88,89].

There may also be decreased local leptin expression from hypothalamic neurons, further contributing to central leptin "insufficiency" in obesity [86].

If peripheral therapies are unlikely to succeed to control appetite, one solution is to directly target appetite centers in the CNS using "neurotherapy." Vectors derived from AAV are the most common gene therapy tool used for CNS transduction. AAV serotypes 2, 3, and 5 have tropism for neuronal tissue and have been shown to transduce brain cells when directly injected into CNS regions of interest [96–98]. These vectors are replication-deficient and virtually nonimmunogenic [97]. The minimal spread of the AAV vectors throughout the brain may be a disadvantage when more global CNS gene therapy is desired, such as for Alzheimer's disease or lysosomal storage disorders, but it is sufficient for transduction of discrete areas in the hypothalamus for appetite regulation. Compared to AAV, FGAds may have less utility due to their immunogenicity and the wide distribution of their receptor, the coxsackie and adenovirus receptor (CAR) [99]. To solve this problem, more cell-type restricted transgene expression is accomplished using neuronal cell-specific promoters [99]. HDAds also have been studied, which are less toxic and much better tolerated. Canine adenovirus-2 (CAV-2) is able to transduce neural tissues and transgene expression is detectable in striatal neurons, the substantia nigra, and the basal nuclei of Maynert 1 year after striaital injections [100] suggesting that further development of these vectors may have important applications for long-term gene therapy in the CNS.

Directed injection of gene therapy vectors into the CNS of rodents is accomplished under anesthesia using specialized stereotaxic injection equipment and specific coordinates (Figure 37.3; [97]). Both intracerebroventricular and direct hypothalamic injections have been successful for studies of

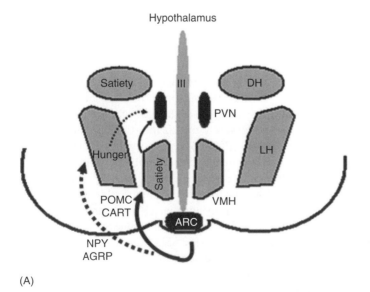

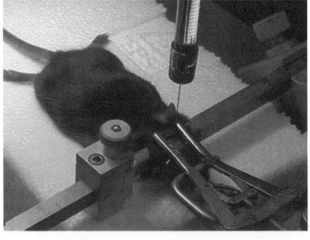

(A)   (B)

FIGURE 37.3 (A) Schematic diagram of the hypothalamus including the ARC, the VMH, LH, DH in relation to the third ventricle (III) and the PVN. (B) Anesthetized mouse is undergoing intracerebroventricular injection using stereotaxic equipment. (Courtesy of Dr. K. Oka.)

appetite regulation using AAV in rodent models [101–103]. Overexpression of leptin in the hypothalamus decreases appetite and weight gain in most rodent models of obesity, unlike with systemic leptin gene therapy [86,104,105]. The expressed leptin also is able to override an orexigenic stimulus by exogenously administered ghrelin [105]. AAV-mediated leptin expression in the hypothalamus was shown to decrease body weight in regular chow and HFD fed rats [106,107]. Only some of the weight loss was explained by reduced food intake because pair-fed control rats had higher weights at the end of the experiment [106]. One explanation was the finding that expression levels of uncoupling protein-1 (UCP-1) were upregulated in the brown adipose tissue (BAT) of central leptin-treated rats, causing increased thermogenesis to promote energy dissipation [106,107]. AAV–leptin treatment of HFD-fed rats also displayed improvements in triglycerides, free fatty acids, and reduced insulin levels compared to pair-fed mice, suggesting that central leptin expression improves peripheral metabolic parameters beyond the effects of weight loss [107]. However, in this study, HFD feeding was started at the time of AAV treatment and was more of a prevention than a treatment study [107]. Other groups who have studied central AAV–leptin gene therapy administered after chronic feeding with HFD have not observed such marked effects [108]. With long-term HFD, only control chow fed and HFD-resistant rats showed appropriate hypothalamic STAT3 phosphorylation and increased BAT UCP-1 expression in response to leptin, while HFD susceptible rats failed to show these responses to central leptin gene therapy [108].

There may be other approaches to circumvent leptin resistance. Leukemia inhibitory factor (LIF) is a member of the IL-6 cytokine family and has anorectic effects [82]. The LIF receptor shares homology with the CNS leptin receptor (Ob-R) and is present on neurons in the rat hypothalamus including the PVN and ARC nucleus [109]. The LIF receptor is part of the class I cytokine receptor family and causes phosphorylation of STAT-3 similar to leptin [82]. A single intracerebroventricular injection of AAV-LIF has been shown to decrease food intake and body weight in rats over several weeks, suggesting that LIF and its receptor may constitute another possible target to combat obesity [110].

Central gene therapy using downstream mediators of leptin, such as POMC, also has been successful for the regulation of appetite in rodent models of obesity. Zucker fatty (fa/fa) rats are deficient in leptin, like $Lep^{Ob/Ob}$ mice. AAV-mediated POMC overexpression in the ARC nucleus of fa/fa rats results in decreased food intake and visceral adiposity with a concomitant improvement in glucose homeostasis and insulin resistance [101]. During aging, increased weight gain and leptin resistance are observed in rodent models on regular diet, but central AAV–leptin gene therapy is able to abrogate the effects of aging in F344/BN rats [111]. One caveat is that the central POMC gene therapy does not have similar effects in young, lean rodents. In fact, POMC gene therapy paradoxically increased diet-induced obesity when lean rats were challenged with an HFD [112]. Although the underlying mechanisms are not fully understood, the findings were attributed to MCR

desensitization or overriding orexigenic signals in lean animals. AAV-POMC expression in the brainstem nucleus solitarius also results in similar, but more prolonged, anorectic effects in aging obese rats compared to hypothalamic expression, which may be due to a lack of counter-regulatory orexigenic neurons in this region [85].

Our knowledge of appetite regulation is increasing exponentially through experiments in animal models of hyperphagia and obesity. In addition, the repertoire of available anorexigenic peptides is expanding. There are at minimum over 28 neuropeptides known to inhibit appetite (Table 37.2) [87]. Many of these peptides are produced in the periphery by the gastrointestinal tract including cholecystokinin (CCK), glucagon-like peptide 1 (GLP-1), oxyntomodulin, amylin, and peptide tyrosin-tyrosine (PYY_{3-36}) [87,90]. Others are produced centrally in the hypothalamus: brain-derived neurotrophic factor (BDNF), melanin-concentrating hormone (MCH), prolactin-releasing peptide (PrRP), and galanin-like peptide (GALP) [113]. As their mechanisms of action become better defined, these peptides may also prove efficacious for central gene therapy approaches to appetite and obesity. The major challenge in applying this approach to combat obesity is to devise methods that ensure sustainability over time, and to develop treatment methods that circumvent repeated CNS deliveries.

37.2.1.2 Other Targets for Glycemic/Metabolic Control

37.2.1.2.1 Adiponectin Gene Therapy
The majority of adipokines and cytokines secreted by adipose tissue have negative effects on insulin sensitivity and glucose homeostasis. With increased adiposity, elevated secretion of these molecules is pathogenic, leading to the metabolic dysregulation seen with obesity. One exception is the adipokine, adiponectin (Acrp30), which circulates in high levels in the serum (µg/ml quantities), with a concentration that is inversely correlated with adipose tissue mass and insulin resistance [77,114]. Adiponectin is one of few endogenous molecules that can be exploited to increase insulin sensitivity. The impact of adiponectin on glucose homeostasis and metabolism is significant. Enhanced secretion and circulation of adiponectin in a transgenic mouse model is able to rescue the insulin-resistant phenotype caused by leptin deficiency in $Lep^{Ob/Ob}$ mice even though the treatment causes increased weight gain and fat mass [115]. Several mechanisms contribute to the positive effects of adiponectin on glucose homeostasis and insulin sensitivity. These include increased lipid oxidation in skeletal muscle, decreased free fatty acid flux and uptake by the liver, and decreased hepatic glucose production [77,114,116]. Systemic overexpression of adiponectin using gene therapy technology has had promising results in animal models of obesity and insulin resistance. For liver-based expression and secretion, intraportal injection of AAV-adiponectin into rats decreased food intake and weight, with an improvement in glucose tolerance [117]. Similar results were obtained in a diet-induced obesity mouse model with minicircle gene

TABLE 37.2

Peptides Known to Modify Appetite

Major Secretory Organ/Tissue	Anorexigenic	Orexigenic
Adipose tissue	• Leptin • TNFα • IL-1β	
Gastrointestinal[a] (stomach intestine, liver, pancreatic islet)	• GLP-1 • CCK • Peptide YY$_{3-36}$ (PYY) • Insulin-like growth factors-I and II • Oxyntomodulin • Neuromedin B/Bombesin • Insulin • Somatostatin • Amylin • Enterostatin	• Ghrelin • VGF • Motilin
CNS (mostly hypothalamic)	• α-MSH • Cocaine- and amphetamine-related peptide • Neurotensin • Corticotropin-releasing hormone • Urocortin • Thyrotropin releasing hormone • Ciliary neurotrophic factor • Neuropeptide K • Calcitonin-gene related peptide • PrRP • Neuromedin B and Neuromedin U • Neuropeptide B and Neuropeptide W • Somatostatin • Anorectin	• NPY • Agouti-related peptide (AgRP) • MCH • Hypocretins/orexins A and B • Growth hormone releasing hormone • Dynorphin • ß-Endorphin • 26RFa • Galanin • Galanin-like peptide (GALP)

[a] Many gastrointestinal peptides can also be found in the CNS.

delivery of adiponectin, which demonstrated improved fasting glucose and insulin sensitivity as measured by homeostatic model assessment [118].

37.2.1.2.2 Hepatic Enzyme Gene Therapy

Liver is a metabolically active tissue, which responds to nutrient signals mediated by insulin and its counter-regulatory hormones, such as glucagon. Elevated glucose, such as after a meal, induces insulin secretion, which will suppress hepatic gluconeogenesis and glycogenolysis and promote glucose storage as glycogen. The effects of insulin are direct, at the level of the hepatocyte, as well as indirect, through inhibition of glucagon secretion from the pancreatic islet. Both mechanisms are impaired in insulin resistance and type 2 diabetes, contributing to hyperglycemia.

One approach to decreasing hepatic glucose output in both type 1 and type 2 diabetes is to directly inhibit the expression of enzymes required for hepatic gluconeogenesis. However, inhibiting gene expression by gene or molecular therapy is more challenging than simple overexpression. The synthesis of a protein can be inhibited by repression of

translation or degradation of its message using interfering RNA (RNAi) [119]. Once a specific RNAi is confirmed to be effective, its sequence can be expressed *in vivo* using plasmid-based or viral vector approaches. PEPCK is the rate-limiting enzyme for the initial step in gluconeogenesis. Gomez-Valades et al. [120] used hydrodynamic systemic injection of a plasmid to express a short-hairpin RNA (shRNA), which significantly decreased hepatic PEPCK expression and activity, was found to normalize glucose tolerance in STZ-diabetic mice.

Glucokinase is the glucose-sensor for β cells and the first step in metabolism of glucose by hepatocytes (see discussion earlier in chapter under therapy for type 1 diabetes). Glucokinase (GcK) activity in the liver is controlled by binding to its regulatory protein (GKRP), which sequesters GcK in the nucleus [121]. GKRP has been described as an inhibitor of GcK activity, which releases GcK to translocate to the cytoplasm only in the presence of high glucose. However, GKRP appears to protect and stabilize GcK so that adenovirus overexpression of GKRP in the liver of HFD mice improved glucose metabolism, with decreased fasting glucose and insulin as discussed earlier in this chapter (under type 1 diabetes therapy) [29].

Additional effects were decreases in circulating free fatty acids and triglycerides.

Other possible hepatic targets have been revealed by the use of antisense oligonucleotides to repress the expression of enzymes involved in hepatic lipid metabolism. Obesity leads to hepatic steatosis because of increased free fatty acid flux from adipose tissue and uptake by the liver; there is also increased de novo lipogenesis and reesterification of fatty acids to be stored as triglycerides [122]. Hepatic steatosis then completes the circle by interfering with insulin signaling. If the accumulation of hepatic lipid can be prevented, hepatic insulin sensitivity can be restored. For example, diacylglycerol acyltransferase (DGAT2) is the final enzyme in the production and storage of liver triglycerides [123,124]. Inhibition of DGAT2 expression resulted in decreased liver triglyceride accumulation along with improved hepatic and peripheral insulin sensitivity measured as increased glucose uptake and insulin signaling [124]. Similarly, inhibition of other enzymes, such as acetyl CoA carboxylase [125], and stearoyl-CoA desaturase-1 [126] has led to parallel decreases in hepatic fat accumulation and insulin resistance.

37.2.2 Gene Therapy for β Cell Failure

Although obesity and ensuing insulin resistance contribute to the development of type 2 diabetes, hyperglycemia would not develop if the β cells could compensate for the increased insulin requirements. The familial nature of many cases of Dm2 implies that there is an important inherited component. Although obesity is a risk factor, not all obese individuals develop diabetes, suggesting that those who do have diabetes, an intrinsic and likely heritable defect that allows β cell failure [72,127]. The identification and characterization of genes involved in β cell failure will be an important step toward preventing and curing Dm2.

Over the last decade, geneticists have attempted to identify single genes associated with Dm2, with the greatest success from genome-wide association and candidate gene studies [128–130]. These methods have identified variants in genes encoding the potassium inwardly rectifying channel (*KCNJ11*), involved in the function of the β cell sulfonylurea receptor, and peroxisome proliferator-activated receptor-γ, the adipocyte differentiation transcription factor. Other recently identified genes include variants of transcription factor-7-like 2 (TCF7L2) involved in WNT signaling [130,131], Calpain-10, which has a role in islet cell apoptosis [132], and others with less well-characterized functions [128].

Of all of the genes identified by genome-wide analysis, the association of the TCF7L2 gene to Dm2 is the strongest. TCF7L2 was identified through classical linkage analysis and genotyping of microsatellite markers among a cohort of type 2 diabetics in Iceland; this was further confirmed with cohorts from Denmark and the United States [133]. Homozygosity and heterozygosity for the "at risk" T allele increased the risk of type 2 diabetes by 2.41 and 1.45, respectively. The importance of TCF7L2 was further substantiated by genotyping

patients from the Diabetes Prevention Program [134], and the Scandanavian MPP and Botnia studies [135] with an increased risk for progression from impaired glucose tolerance to Dm2. Importantly, these individuals did not have increased insulin resistance, but instead had decreased insulin secretion, suggesting that TCF7L2 gene product influences β cell function [134]. Characterization of the β cell defects revealed a decreased "incretin" response [135]. Incretins are hormones released from specialized L cells in the intestine in response to an oral glucose load or a meal. The incretin hormones are GLP-1 and glucose-dependent insulinotropic peptide (GIP). Upon release, these peptides augment glucose-regulated insulin secretion from the β cell so that an oral nutritional load causes greater insulin secretion than parenteral glucose administration. GLP-1 receptor agonists have attracted much attention with their recent debut for clinical use in human diabetes. In long-standing Dm2, the first phase of insulin release can be dampened, but GLP-1 agonists improve the early β cell response, decreasing hyperglycemia. Significantly, TCF7L2 is involved in the transcription regulation of the pro-glucagon gene, which encodes GLP-1 and GIP. TCF7L2 also is implicated in WNT signaling, which has a role in β cell proliferation [135,136]. Although the consequences of TCF7L2 polymorphisms need further characterization, it seems reasonable that differences in incretin levels or action could regulate β cells and their ability to compensate for increased insulin resistance. In this way, incretins could be an important avenue to improve β cell function in Dm2.

It is convenient that GLP-1 is a peptide hormone, which makes it amenable to expression from gene therapy vector systems. One disadvantage is the short half-life of the peptide, which is degraded in minutes by peptidases, especially dipeptidyl peptidase IV (DPPIV). For this reason, attempts have been made to stabilize the GLP-1 peptide and to extend its half-life by rendering it resistant to DPPIV action. These approaches have included engineering amino acid substitutions in the GLP-1 sequence [137] or use of the similarly active but DPPIV-resistant exenatide (exendin 4) sequence from gila monster (*Heloderma suspectum*). Others have used peptide dimerization or noncovalent or covalent binding to albumin for extended systemic circulation [138].

In addition to their immediate effects on insulin secretion, GLP-1 receptor agonists are documented to increase functional β cell mass, at least in rodent model systems, through a decrease in apoptosis and an increase in β cell proliferation [139–141]. GLP-1 also contributes to regeneration of β cells after partial pancreatectomy [140]. For type Dm2, this property could be exploited to prevent β cell loss and to rebuild β cell mass after failure due to prolonged hyperglycemia. In both toxin-induced (STZ) insulin-deficient mouse model and an obesity model (*db/db*), adenoviral-mediated systemic expression of GLP-1 resulted in remission of diabetes, with restoration of insulin levels and increased insulin-staining cells in the pancreas [137,142]. Thus, prolonged expression of GLP-1 receptor agonists, such as from gene therapy vectors, may have potential to restore functioning β cells.

TABLE 37.3

Potential Contributions of GLP-1 to Improve Glucose Homeostasis and Insulin Sensitivity in Type 2 Diabetes

Tissue	GLP-1 Action
Islet	• Increased β cell mass
	• Augmented glucose-regulated insulin secretion (incretin effect)
Liver	• Increased insulin signaling
	• Decreased hepatic steatosis
	• Decreased gluconeogenesis
Muscle	• Increased glucose uptake
	• Increased insulin signaling
CNS	• Satiety and decreased food intake
Adipose tissue	• Weight loss

These peptides also have numerous other properties that make them attractive for Dm2 gene therapy (Table 37.3). They decrease the inappropriately high glucagon levels seen in type 2 diabetes, reducing the contributions of hepatic gluconeogenesis to hyperglycemia [143]. In addition to their desirable hormonal effects on insulin and glucagon secretion, GLP-1 receptor agonists appear to improve insulin sensitivity. Some of this effect may be promotion of weight loss since GLP-1 receptor agonists are documented to decrease food intake and body weight in both humans and rodents. Two different groups [137,144] have reported decreased food intake and decreased weight gain in hyperphagic rodent models of *fa/fa* Zucker rats and *Lep*^*ob/ob*^ mice, respectively, when they treated them with an FGAd expressing a GLP-1 receptor agonist. The decreased food intake is attributed to a slowing of gastric emptying and increased satiety [145]. Although the precise mechanisms remain to be elucidated, some of the properties may be mediated through effects on CNS feeding centers [146] such as through leptin [147], POMC [148], or CART [149].

In addition to effects on obesity, GLP-1 and its analogs also directly increase insulin sensitivity, which has been documented in several rodent systems. The insulin-sensitizing effects seen with GLP-1 receptor agonist infusion also is seen with the steady-state expression attained by gene therapy [144,150,151]. Adenoviral expression of GLP-1 in *Lep*^*ob/ob*^ mice improves insulin tolerance testing, an *in vivo* test of insulin sensitivity. Surrogate markers of insulin sensitivity also are improved, with increased insulin-stimulated phosphorylation of IRS-1 and PKC activity in muscle and liver and increased Akt phosphorylation in muscle [144]. Although one argument could be that improved insulin sensitivity is a direct result of decreased weight gain and fat mass, the improved insulin sensitivity weight in *fa/fa* rats was reported to be independent of changes in body weight [151].

From the above discussion, it is clear that GLP-1 receptor agonists have great potential for Dm2 gene therapy because they address the two key pathologic processes that lead to Dm2: β cell failure and insulin resistance. They accomplish this through multiple mechanisms which include: (1) promoting increased β cell mass, (2) increasing insulin sensitivity through indirect (decreased weight) and direct mechanisms, and (3) decreasing hepatic glucose output.

37.3 CONCLUDING REMARKS

Type 1a diabetes is the clinical consequence of autoimmune destruction of β cells, whereas type 2 diabetes is a manifestation of a severely compromised β cell function in the presence of insulin resistance, often caused by obesity. Herein, we have summarized the recent advances in the molecular therapy of type 1a and type 2 diabetes. To date, all therapies are still at the preclinical phase. A major hurdle to moving these regimens toward clinical trial is the availability of safe and efficient vectors that are suitable for *in vivo* gene delivery in humans, an issue that is discussed in detail in other chapters. Despite this limitation, we have witnessed important developments in diabetes gene therapy in the last two decades, including many exciting proof-of-concept experiments. Diabetes is a serious growing health problem worldwide that justifies a continued search for novel treatment options, including gene therapy.

ACKNOWLEDGMENTS

Research performed in the authors' laboratories that was discussed in the chapter was supported by U.S. National Institutes of Health Grants R01-HL51586, DK068037, R21-DK075002, K08-DK68391, and by the Betty Rutherford Chair in Diabetes Research at Baylor College of Medicine and St. Luke's Episcopal Hospital in Houston, the Iacocca Family Foundation, and the T.T. and W.F. Chao Global Foundation.

REFERENCES

1. Diabetes Atlas. 3rd ed. International Diabetes Federation; 2006.
2. Fox CS, Pencina MJ, Meigs JB, Vasan RS, Levitzky YS, and D'Agostino RB, Sr. Trends in the incidence of type 2 diabetes mellitus from the 1970s to the 1990s: The Framingham Heart Study. *Circulation* 2006; 113(25):2914–2918.

3. Gregg EW, Gu Q, Cheng YJ, Narayan KM, and Cowie CC. Mortality trends in men and women with diabetes, 1971 to 2000. *Ann Intern Med* 2007; 147(3):149–155.

4. Narayan KM, Boyle JP, Geiss LS, Saaddine JB, and Thompson TJ. Impact of recent increase in incidence on future diabetes burden: U.S., 2005–2050. *Diabetes Care* 2006; 29(9):2114–2116.

5. *Diabetes in America*, 2nd ed. National Institutes of Health, Bethesda, MD; 1995.

6. Kilpatrick ES, Rigby AS, Goode K, and Atkin SL. Relating mean blood glucose and glucose variability to the risk of multiple episodes of hypoglycaemia in type 1 diabetes. *Diabetologia* 2007; 50(12):2553–2561.

7. Hovorka R. Continuous glucose monitoring and closed-loop systems. *Diabet Med* 2006; 23(1):1–12.

8. Chan L, Fujimiya M, and Kojima H. In vivo gene therapy for diabetes mellitus. *Trends Mol Med* 2003; 9(10):430–435.

9. Butler AE, Galasso R, Meier JJ, Basu R, Rizza RA, and Butler PC. Modestly increased beta cell apoptosis but no increased beta cell replication in recent-onset type 1 diabetic patients who died of diabetic ketoacidosis. *Diabetologia* 2007; 50(11):2323–2331.

10. Dong H and Woo SLC. Hepatic insulin production for type 1 diabetes. *Trends Endocrinol Metab* 2001; 12:441–446.

11. Yoon JW and Jun HS. Recent advances in insulin gene therapy for type 1 diabetes. *Trends Mol Med* 2002; 8(2):62–68.

12. Short DK, Okada S, Yamauchi K, and Pessin JE. Adenovirus-mediated transfer of a modified human proinsulin gene reverses hyperglycemia in diabetic mice. *Am J Physiol Endocrinol Metab* 1998; 275:E748–E756.

13. Muzzin P, Eisensmith RC, Copeland KC, and Woo SLC. Hepatic insulin gene expression as treatment for type 1 diabetes mellitus in rats. *Mol Endocrinol* 1997; 11:833–837.

14. Auricchio A, Gao G-P, Yu QC et al. Constitutive and regulated expression of processed insulin following in vivo hepatic gene transfer. *Gene Ther* 2002; 9:963–971.

15. Ren B, O'Brien BA, Swan MA et al. Long-term correction of diabetes in rats after lentiviral hepatic insulin gene therapy. *Diabetologia* 2007; 50(9):1910–1920.

16. Hui H and Perfetti R. Pancreas duodenum homeobox-1 regulates pancreas development during embryogenesis and islet cell function in adulthood. *Eur J Endo* 2002; 146:129–141.

17. Lee HC, Kim S-J, Shin H-C, and Yoon J-W. Remission in models of type 1 diabetes by gene therapy using a single-chain insulin analogue. *Nature* 2000; 408:483–488.

18. Lu D, Tamemoto H, Shibata H, Saito I, and Takeuchi T. Regulatable production of insulin from primary-cultured hepatocytes: Insulin production is up-regulated by glucagon and cAMP and down-regulated by insulin. *Gene Ther* 1998; 5:888–895.

19. Thule PM, Liu J, and Phillips LS. Glucose regulated production of human insulin in rat hepatocytes. *Gene Ther* 2000; 7:205–214.

20. Chen R, Meseck ML, and Woo SLC. Auto-regulated hepatic insulin gene expression in type 1 diabetic rats. *Mol Ther* 2001; 3:584–590.

21. Vaulont S, Vasseur-Cognet M, and Kahn A. Glucose regulation of gene transcription. *J Biol Chem* 2000; 275:31555–31558.

22. Rivera VM, Wang X, and Wardwell S et al. Regulation of protein secretion through controlled aggregation in the endoplasmic reticulum. *Science* 2000; 287:826–830.

23. Ferre T, Pujol A, Riu E, Bosch F, and Valera A. Correction of diabetic alterations by glucokinase. *Proc Natl Acad Sci USA* 1996; 93:7225–7230.

24. O'Doherty RM, Lehman DL, Telemaque-Potts S, and Newgard CB. Metabolic impact of glucokinase overexpression in liver. *Diabetes* 1999; 48:2022–2027.

25. Shiota M, Postic C, Fujimoto Y et al. Glucokinase gene locus transgenic mice are resistant to the development of obesity-induced type 2 diabetes. *Diabetes* 2001; 50:622–629.

26. Desai UJ, Slosberg ED, Boettcher BR et al. Phenotypic correction of diabetic mice by adenovirus-mediated glucokinase expression. *Diabetes* 2001; 50:2287–2295.

27. Morral N, McEvoy R, Dong H et al. Adenovirus-mediated expression of glucokinase in the liver as an adjuvant treatment for type 1 diabetes. *Hum Gene Ther* 2002; 13:1561–1570.

28. Morral N. Novel targets and therapeutic strategies for type 2 diabetes. *Trends Endocrinol Metab* 2003; 14:169–175.

29. Slosberg ED, Desai UJ, Fanelli B et al. Treatment of type 2 diabetes by adenoviral-mediated overexpression of the glucokinase regulatory protein. *Diabetes* 2001; 50:1813–1820.

30. Wu C, Okar DA, Newgard CB, and Lange AJ. Increasing fructose 2,6-bisphosphate overcomes hepatic insulin resistance of type 2 diabetes. *Am J Physiol Endocrinol Metab* 2002; 252:E38–E45.

31. O'Doherty RM, Jensen PB, Anderson P et al. Activation of direct and indirect pathways of glycogen synthesis by hepatic overexpression of protein targeting to glycogen. *J Clin Invest* 2000; 105:479–488.

32. Newgard CB, Brady MJ, O'Doherty RM, and Saltiel AR. Organizing glucose disposal. Emerging roles of the glycogen targeting subunits of protein phosphatase-1. *Diabetes* 2000; 49:1967–1977.

33. Otaegui PJ, Ferre T, Pujol A, Riu E, Jimenez R, and Bosch F. Expression of glucokinase in skeletal muscle: A new approach to counteract diabetic hyperglycemia. *Human Gene Ther* 2000; 11:1543–1552.

34. Grimsby J, Sarabu R, Corbett WL et al. Allosteric activators of glucokinase: Potential role in diabetes therapy. *Science* 2003; 301:370–373.

35. Collombat P, Hecksher-Sorensen J, Serup P, and Mansouri A. Specifying pancreatic endocrine cell fates. *Mech Dev* 2006; 123(7):501–512.

36. Jorgensen MC, Hnfelt-Ronne J, Hald J, Madsen OD, Serup P, and Hecksher-Sorensen J. An illustrated review of early pancreas development in the mouse. *Endocr Rev* 2007; 28(6):685–705.

37. Zhou Q, Law AC, Rajagopal J, Anderson WJ, Gray PA, and Melton DA. A multipotent progenitor domain guides pancreatic organogenesis. *Dev Cell* 2007; 13(1):103–114.

38. Gu G, Dubauskaite J, and Melton DA. Direct evidence for the pancreatic lineage: NGN3 + cells are islet progenitors and are distinct from duct progenitors. *Development* 2002; 129(10):2447–2457.

39. Smith SB, Watada H, and German MS. Neurogenin3 activates the islet differentiation program while repressing its own expression. *Mol Endocrinol* 2004; 18(1):142–149.

40. Apelqvist A, Li H, Sommer L et al. Notch signalling controls pancreatic cell differentiation. *Nature* 1999; 400(6747):877–881.

41. Taniguchi H, Yamato E, Tashiro F, Ikegami H, Ogihara T, and Miyazaki J. Beta cell neogenesis induced by adenovirus-mediated gene delivery of transcription factor pdx-1 into mouse pancreas. *Gene Ther* 2003; 10(1):15–23.

42. Ayuso E, Chillon M, Agudo J et al. In vivo gene transfer to pancreatic beta cells by systemic delivery of adenoviral vectors. *Hum Gene Ther* 2004; 15(8):805–812.

43. Ayuso E, Chillon M, Garcia F et al. In vivo gene transfer to healthy and diabetic canine pancreas. *Mol Ther* 2006; 13(4):747–755.

44. Campos SK and Barry MA. Current advances and future challenges in Adenoviral vector biology and targeting. *Curr Gene Ther* 2007; 7(3):189–204.

45. Jonsson J, Carlsson L, Edlund T, and Edlund H. Insulin-promoter-factor 1 is required for pancreas development in mice. *Nature* 1994; 371(6498):606–609.

46. Ahlgren U, Jonsson J, and Edlund H. The morphogenesis of the pancreatic mesenchyme is uncoupled from that of the pancreatic epithelium in IPF1/PDX1-deficient mice. *Development* 1996; 122(5):1409–1416.

47. Ferber S, Halkin A, Cohen H et al. Pancreatic and duodenal homeobox gene 1 induces expression of insulin genes in liver and ameliorates streptozotocin-induced hyperglycemia. *Nat Med* 2000; 6(5):568–572.

48. Kojima H, Fujimiya M, Matsumura K et al. NeuroD-betacellulin gene therapy induces islet neogenesis in the liver and reverses diabetes in mice. *Nat Med* 2003; 9(5):596–603.

49. Miyatsuka T, Kaneto H, Kajimoto Y et al. Ectopically expressed PDX-1 in liver initiates endocrine and exocrine pancreas differentiation but causes dysmorphogenesis. *Biochem Biophys Res Commun* 2003; 310(3):1017–1025.

50. Song YD, Lee EJ, Yashar P, Pfaff LE, Kim SY, and Jameson JL. Islet cell differentiation in liver by combinatorial expression of transcription factors neurogenin-3, BETA2, and RIPE3b1. *Biochem Biophys Res Commun* 2007; 354(2):334–339.

51. Kaneto H, Matsuoka TA, Nakatani Y et al. A crucial role of MafA as a novel therapeutic target for diabetes. *J Biol Chem* 2005; 280(15):15047–15052.

52. Kaneto H, Nakatani Y, Miyatsuka T et al. PDX-1/VP16 fusion protein, together with NeuroD or Ngn3, markedly induces insulin gene transcription and ameliorates glucose tolerance. *Diabetes* 2005; 54(4):1009–1022.

53. Wang AY, Ehrhardt A, Xu H, and Kay MA. Adenovirus transduction is required for the correction of diabetes using Pdx-1 or Neurogenin-3 in the liver. *Mol Ther* 2007; 15(2):255–263.

54. Shapiro AM, Ricordi C, Hering BJ et al. International trial of the Edmonton protocol for islet transplantation. *N Engl J Med* 2006; 355(13):1318–1330.

55. Goudy K, Song S, Wasserfall C et al. Adeno-associated virus vector-mediated IL-10 gene delivery prevents type 1 diabetes in NOD mice. *Proc Natl Acad Sci USA* 2001; 98(24):13913–13918.

56. Goudy KS, Burkhardt BR, Wasserfall C et al. Systemic overexpression of IL-10 induces CD4 + CD25 + cell populations in vivo and ameliorates type 1 diabetes in nonobese diabetic mice in a dose-dependent fashion. *J Immunol* 2003; 171(5):2270–2278.

57. Jun HS, Chung YH, Han J et al. Prevention of autoimmune diabetes by immunogene therapy using recombinant vaccinia virus expressing glutamic acid decarboxylase. *Diabetologia* 2002; 45(5):668–676.

58. Cameron MJ, Arreaza GA, Waldhauser L, Gauldie J, and Delovitch TL. Immunotherapy of spontaneous type 1 diabetes in nonobese diabetic mice by systemic interleukin-4 treatment employing adenovirus vector-mediated gene transfer. *Gene Ther* 2000; 7(21):1840–1846.

59. Flodstrom-Tullberg M, Yadav D, Hagerkvist R et al. Target cell expression of suppressor of cytokine signaling-1 prevents diabetes in the NOD mouse. *Diabetes* 2003; 52(11):2696–2700.

60. Herold KC, Hagopian W, Auger JA et al. Anti-CD3 monoclonal antibody in new-onset type 1 diabetes mellitus. *N Engl J Med* 2002; 346(22):1692–1698.

61. Ryu S, Kodama S, Ryu K, Schoenfeld DA, and Faustman DL. Reversal of established autoimmune diabetes by restoration of endogenous beta cell function. *J Clin Invest* 2001; 108(1):63–72.

62. Kodama S, Kuhtreiber W, Fujimura S, Dale EA, and Faustman DL. Islet regeneration during the reversal of autoimmune diabetes in NOD mice. *Science* 2003; 302(5648):1223–1227.

63. Bottino R, Lemarchand P, Trucco M, and Giannoukakis N. Gene- and cell-based therapeutics for type I diabetes mellitus. *Gene Ther* 2003; 10(10):875–889.

64. Han G, Li Y, Wang J et al. Active tolerance induction and prevention of autoimmune diabetes by immunogene therapy using recombinant adenoassociated virus expressing glutamic acid decarboxylase 65 peptide GAD(500–585). *J Immunol* 2005; 174(8):4516–4524.

65. Song L, Wang J, Wang R et al. Retroviral delivery of GAD-IgG fusion construct induces tolerance and modulates diabetes: A role for CD4 + regulatory T cells and TGF-beta? *Gene Ther* 2004; 11(20):1487–1496.

66. Denes B, Yu J, Fodor N, Takatsy Z, Fodor I, and Langridge WH. Suppression of hyperglycemia in NOD mice after inoculation with recombinant vaccinia viruses. *Mol Biotechnol* 2006; 34(3):317–327.

67. Tian C, Bagley J, Cretin N, Seth N, Wucherpfennig KW, and Iacomini J. Prevention of type 1 diabetes by gene therapy. *J Clin Invest* 2004; 114(7):969–978.

68. Luo X, Yang H, Kim IS et al. Systemic transforming growth factor-beta1 gene therapy induces Foxp3 + regulatory cells, restores self-tolerance, and facilitates regeneration of beta cell function in overtly diabetic nonobese diabetic mice. *Transplantation* 2005; 79(9):1091–1096.

69. Feili-Hariri M, Falkner DH, Gambotto A et al. Dendritic cells transduced to express interleukin-4 prevent diabetes in nonobese diabetic mice with advanced insulitis. *Hum Gene Ther* 2003; 14(1):13–23.

70. Peng J, Dicker B, Du W et al. Converting antigen-specific diabetogenic CD4 and CD8 T cells to TGF-beta producing nonpathogenic regulatory cells following FoxP3 transduction. *J Autoimmun* 2007; 28(4):188–200.

71. Goudy KS and Tisch R. Immunotherapy for the prevention and treatment of type 1 diabetes. *Int Rev Immunol* 2005; 24(5–6):307–326.

72. Wajchenberg BL. β cell failure in diabetes and preservation by clinical treatment. *Endocrine Rev* 2007; 28(2):187–218.

73. Ozanne SE and Constancia M. Mechanisms of disease: The developmental origins of disease and the role of the epigenotype. *Nat Clin Pract Endocrinol Metab* 2007; 3:539–546.

74. Shulman GI. Cellular mechanisms of insulin resistance. *J Clin Invest* 2000; 106:171–176.

75. Kahn SE, Hull RL, and Utzschneider KM. Mechanisms linking obesity to insulin resistance and type 2 diabetes. *Nature* 2006; 444(7121):840–846.

76. Qatanani M and Lazar MA. Mechanisms of obesity-associated insulin resistance: Many choices on the menu. *Genes Dev* 2007; 21(12):1443–1455.

77. Kershaw EE and Flier JS. Adipose tissue as an endocrine organ. *J Clin Endocrinol Metab* 2004; 89:2548–2556.

78. Kolonin MG, Saha PK, Chan L, Pasqualini R, and Arap W. Reversal of obesity by targeted ablation of adipose tissue. *Nat Med* 2004; 10:625–632.

79. Klein S, Fontana L, Young VL et al. Absence of an effect of liposuction on insulin action and risk factors for coronary heart disease. *N Engl J Med* 2004; 350(25):2549–2557.

80. Fontana L, Eagon JC, Trujillo ME, Scherer PE, and Klein S. Visceral fat adipokine secretion is associated with systemic inflammation in obese humans. *Diabetes* 2007; 56(4):1010–1013.

81. Bergman RN. Orchestration of glucose homeostasis: From a small acorn to the California oak. *Diabetes* 2007; 56(6):1489–1501.

82. Gao Q and Horvath TL. Neurobiology of feeding and energy expenditure. *Annu Rev Neurosci* 2007; 30:367–398.

83. Pritchard LE and White A. Neuropeptide processing and its impact on melanocortin pathways. *Endocrinology* 2007; 148(9):4201–4207.

84. Kalra SP and Kalra PS. Gene-transfer technology: A preventive neurotherapy to curb obesity, ameliorate metabolic syndrome and extend life expectancy. *Trends Pharmacol Sci* 2005; 26(10):488–495.

85. Li G, Zhang Y, Rodrigues E et al. Melanocortin activation of nucleus of the solitary tract avoids anorectic tachyphylaxis and induces prolonged weight loss. *Am J Endocrinol Metab* 2007; 293:E252–E258.

86. Boghossian S, Lecklin A, Torto R, Kalra PS, and Kalra SP. Suppression of fat deposition for the life time with gene therapy. *Peptides* 2005; 26(8):1512–1519.

87. Sahu A. Minireview: A hypothalamic role in energy balance with special emphasis on leptin. *Endocrinology* 2004; 145(6):2613–2620.

88. Caro JF, Kolaczynski JW, Nyce MR et al. Decreased cerebrospinal-fluid/serum leptin ratio in obesity: A possible mechanism for leptin resistance. *Lancet* 1996; 348(9021):159–161.

89. Schwartz MW, Peskind E, Raskind M, Boyko EJ, and Porte D. Cerebrospinal fluid leptin levels: Relationship to plasma levels and to adiposity in humans. *Nat Med* 1996; 2:589–593.

90. Wynne K and Bloom SR. The role of oxyntomodulin and peptide tyrosine-tyrosine (PYY) in appetite control. *Nat Clin Pract Endocrinol Metab* 2006; 2:612–620.

91. Muzzin P, Eisensmith R, Copeland KC, and Woo SL. Correction of obesity and diabetes in genetically obese mice by leptin gene therapy. *Proc Natl Acad Sci USA* 1996; 93(25):14804–14808.

92. Lundberg C, Jungles SJ, and Mulligan RC. Direct delivery of leptin to the hypothalamus using recombinant adeno-associated virus vectors results in increased therapeutic efficacy. *Nat Biotech* 2001; 19(2):169–172.

93. Widdowson PS, Upton R, Buckingham R, Arch J, and Williams G. Inhibition of food response to intracerebroventricular injection of leptin is attenuated in rats with diet-induced obesity. *Diabetes* 1997; 46:1782–1785.

94. Levin BE and Dunn-Meynell AA. Reduced central leptin sensitivity in rats with diet-induced obesity. *Am J Physiol Regul Integr Comp Physiol* 2002; 283(4):R941–R948.

95. El Haschimi K, Pierroz DD, Hileman SM, Bjorbak C, and Flier JS. Two defects contribute to hypothalamic leptin resistance in mice with diet-induced obesity. *J Clin Invest* 2000; 105(12):1827–1832.

96. Okada T, Nomoto T, Shimazaki K et al. Adeno-associated virus vectors for gene transfer to the brain. *Methods* 2002; 28(2):237–247.

97. Ruitenberg MJ, Eggers R, Boer GJ, and Verhaagen J. Adeno-associated viral vectors as agents for gene delivery: Application in disorders and trauma of the central nervous system. *Methods* 2002; 28(2):182–194.

98. Shevtsova Z, Malik JMI, Michel U, Bahr M, and Kugler S. Promoters and serotypes: Targeting of adeno-associated virus vectors for gene transfer in the rat central nervous system in vitro and in vivo. *Exp Physiol* 2005; 90(1):53–59.

99. Hermening S, Kugler S, Bahr M, and Isenmann S. Improved high-capacity adenoviral vectors for high-level neuron-restricted gene transfer to the CNS. *J Virol Methods* 2006; 136(1–2):30–37.

100. Soudais C, Skander N, and Kremer EJ. Long-term in vivo transduction of neurons throughout the rat CNS using novel helper-dependent CAV-2 vectors. *FASEB J* 2004; 18:391–393.

101. Li G, Mobbs CV, and Scarpace PJ. Central pro-opiomelanocortin gene delivery results in hypophagia, reduced visceral adiposity, and improved insulin sensitivity in genetically obese zucker rats. *Diabetes* 2003; 52(8):1951–1957.

102. Kalra SP and Kalra PS. NPY and cohorts in regulating appetite, obesity and metabolic syndrome: Beneficial effects of gene therapy. *Neuropeptides* 2004; 38(4):201–211.

103. Zhang Y and Scarpace PJ. Circumventing central leptin resistance: Lessons from central leptin and POMC gene delivery. *Peptides* 2006; 27(2):350–364.

104. Kalra SP. Circumventing leptin resistance for weight control. *Proc Natl Acad Sci USA* 2001; 98(8):4279–4281.

105. Ueno N, Dube MG, Inui A, Kalra PS, and Kalra SP. Leptin modulates orexigenic effects of ghrelin and attenuates adiponectin and insulin levels and selectively the dark-phase feeding as revealed by central leptin gene therapy. *Endocrinology* 2004; 145(9):4176–4184.

106. Dhillon H, Kalra SP, and Kalra PS. Dose-dependent effects of central leptin gene therapy on genes that regulate body weight and appetite in the hypothalamus. *Mol Ther* 2001; 4:139–145.

107. Dube MG, Beretta E, Dhillon H, Ueno N, Kalra PS, and Kalra SP. Central leptin gene therapy blocks high-fat diet-induced weight gain, hyperleptinemia, and hyperinsulinemia: Increase in serum ghrelin levels. *Diabetes* 2002; 51(6):1729–1736.

108. Wilsey J, Zolotukhin S, Prima V, and Scarpace PJ. Central leptin gene therapy fails to overcome leptin resistance associated with diet-induced obesity. *Am J Physiol Regul Integr Comp Physiol* 2003; 285(5):R1011–R1020.

109. Yamakuni H, Minami M, and Satoh M. Localization of mRNA for leukemia inhibitory factor receptor in the adult rat brain. *J Neuroimmunol* 1996; 70(1):45–53.

110. Beretta E, Dhillon H, Kalra PS, and Kalra SP. Central LIF gene therapy suppresses food intake, body weight, serum leptin and insulin for extended periods. *Peptides* 2002; 23(5):975–984.

111. Li G, Zhang Y, Wilsey JT, and Scarpace PJ. Hypothalamic pro-opiomelanocortin gene delivery ameliorates obesity and glucose intolerance in aged rats. *Diabetologia* 2005; 48:2376–2385.

112. Li G, Zhang Y, Cheng KY, and Scarpace PJ. Lean rats with hypothalamic pro-opiomelanocortin overexpression exhibit greater diet-induced obesity and impaired central melanocortin responsiveness. *Diabetologia* 2007; 50:1490–1499.

113. Arora S and Anubhuti V. Role of neuropeptides in appetite regulation and obesity—A review. *Neuropeptides* 2006; 40(6):375–401.

114. Whitehead JP, Rechards AA, Hickman IJ, Macdonald GA, and Prins JB. Adiponectin—a key adipokine in the metabolic syndrome. *Diabetes Obes Metab* 2006; 8:264–280.

115. Kim JY, van de WE, Laplante M et al. Obesity-associated improvements in metabolic profile through expansion of adipose tissue. *J Clin Invest* 2007; 117(9):2621–2637.

116. Berg AH, Combs TP, Du X, Brownlee M, and Scherer PE. The adipocyte-secreted protein Acrp30 enhances insulin action. *Nat Med* 2001; 7:947–953.

117. Shklyaev S, Aslanidi G, Tennant M et al. Sustained peripheral expression of transgene adiponectin offsets the development of diet-induced obesity in rats. *Proc Natl Acad Sci USA* 2003; 100(24):14217–14222.

118. Park JH, Lee M, and Kim SW. Non-viral adiponectin gene therapy into obese type 2 diabetic mice ameliorates insulin resistance. *J Control Release* 2006; 114:118–125.

119. Kim DH and Rossi JJ. Strategies for silencing human disease using RNA interference. *Nat Rev Genet* 2007; 8(3):173–184.

120. Gomez-Valades AC, Vidal-Alabro A, Molas M et al. Overcoming diabetes-induced hyperglycemia through inhibition of hepatic phosphoenolpyruvate carboxykinase(GTP) with RNAi. *Mol Ther* 2006; 13:401–410.

121. Brown KS, Kalinowski SS, Megill JR, Durham SK, and Mookhtiar KA. Glucokinase regulatory protein may interact with glucokinase in the hepatocyte nucleus. *Diabetes* 1997; 46(2):179–186.

122. Perlemuter G, Bigorgne A, Cassard-Doulcier AM, and Naveau S. Nonalcoholic fatty liver disease: From pathogenesis to patient care. *Nat Clin Pract Endocrinol Metab* 2007; 3:458–469.

123. Yu XX, Murray SF, Pandey SK et al. Antisense oligonucleotide reduction of DGAT2 expression improves hepatic steatosis and hyperlipidemia in obese mice. *Hepatology* 2005; 42(2):362–371.

124. Choi CS, Savage DB, Kulkarni A et al. Suppression of diacylglycerol acyltransferase-2 (DGAT2), but not DGAT1, with antisense oligonucleotides reverses diet-induced hepatic steatosis and insulin resistance. *J Biol Chem* 2007; 282(31):22678–22688.

125. Savage DB, Choi CS, Samuel VT et al. Reversal of diet-induced hepatic steatosis and hepatic insulin resistance by antisense oligonucleotide inhibitors of acetyl-CoA carboxylases 1 and 2. *J Clin Invest* 2006; 116(3):817–824.

126. Gutierrez-Juarez R, Pocai A, Mulas C et al. Critical role of stearoyl-CoA desaturase-1 (SCD1) in the onset of diet-induced hepatic insulin resistance. *J Clin Invest* 2006; 116(6):1686–1695.

127. Prentki M and Nolan CJ. Islet beta cell failure in type 2 diabetes. *J Clin Invest* 2006; 116:1802–1812.

128. Frayling TM. Genome-wide association studies provide new insights into type 2 diabetes aetiology. *Nat Rev Genet* 2007; 8(9):657–662.

129. Watanabe RM, Black MH, Xiang AH, Allayee H, Lawrence JM, and Buchanan TA. Genetics of gestational diabetes mellitus and type 2 diabetes. *Diabetes Care* 2007; 30(Suppl 2):S134–S140.

130. Hattersley AT. Prime suspect: The TCF7L2 gene and type 2 diabetes risk. *J Clin Invest* 2007; 117:2077–2079.

131. Nauck MA and Meier JJ. The enteroinsular axis may mediate the diabetogenic effects of TCF7L2 polymorphisms. *Diabetologia* 2007; 50:2413–2416.

132. Horikawa Y. Calpain-10 (NIDDM1) as a susceptibility gene for common type 2 diabetes. *Endocr J* 2006; 53:567–576.

133. Grant SFA, Thorleifsson G, Reynisdottir I et al. Variant of transcription factor 7-like 2 (TCF7L2) gene confers risk of type 2 diabetes. *Nat Genet* 2006; 38(3):320–323.

134. Florez JC, Jablonski KA, Bayley N et al. TCF7L2 polymorphisms and progression to diabetes in the Diabetes Prevention Program. *N Engl J Med* 2006; 355(3):241–250.

135. Lyssenko V, Lupi R, Marchetti P et al. Mechanisms by which common variants in the TCF7L2 gene increase risk of type 2 diabetes. *J Clin Invest* 2007; 117(8):2155–2163.

136. Rulifson IC, Karnik SK, Heiser PW et al. Wnt signaling regulates pancreatic beta cell proliferation. *Proc Natl Acad Sci USA* 2007; 104(15):6247–6252.

137. Parsons GB, Souza DW, Wu H et al. Ectopic expression of glucagon-like peptide 1 for gene therapy of type II diabetes. *Gene Ther* 2007; 14(1):38–48.

138. Kumar M, Hunag Y, Glinka Y, Prud'Homme GJ, and Wang Q. Gene therapy of diabetes using a novel GLP-1//IgG1-Fc fusion construct normalizes glucose levels in db//db mice. *Gene Ther* 2006; 14(2):162–172.

139. Stoffers DA, Kieffer TJ, Hussain MA et al. Insulinotropic glucagon-like peptide 1 agonists stimulate expression of homeodomain protein IDX-1 and increase islet size in mouse pancreas. *Diabetes* 2000; 49(5):741–748.

140. De Leon DD, Deng S, Madani R, Ahima RS, Drucker DJ, and Stoffers DA. Role of endogenous glucagon-like peptide-1 in islet regeneration after partial pancreatectomy. *Diabetes* 2003; 52:365–371.

141. Brubaker PL and Drucker DJ. Minireview: Glucagon-like peptides regulate cell proliferation and apoptosis in the pancreas, gut, and central nervous system. *Endocrinology* 2004; 145(6):2653–2659.

142. Liu MJ, Shin S, Li N et al. Prolonged remission of diabetes by regeneration of β cells in diabetic mice treated with recombinant adenoviral vector expressing glucagon-like peptide-1. *Mol Ther* 2007; 15(1):86–93.

143. Burcelin R, Dolci W, and Thorens B. Long-lasting antidiabetic effect of a dipeptidyl peptidase IV-resistant analog of glucagon-like peptide-1. *Metabolism* 1999; 48(2):252–258.

144. Lee YS, Shin S, Shigihara T et al. Glucagon-like peptide-1 gene therapy in obese diabetic mice results in long-term cure of diabetes by improving insulin sensitivity and reducing hepatic gluconeogenesis. *Diabetes* 2007; 56(6):1671–1679.

145. Naslund E, Gutniak M, Skogar S, Rossner S, and Hellstrom PM. Glucagon-like peptide 1 increases the period of postprandial satiety and slows gastric emptying in obese men. *Am J Clin Nutr* 1998; 68(3):525–530.

146. Schick RR, Zimmermann JP, vorm Walde T, and Schusdziarra V. Peptides that regulate food intake: Glucagon-like peptide 1-(7-36) amide acts at lateral and medial hypothalamic sites to suppress feeding in rats. *Am J Physiol Regul Integr Comp Physiol* 2003; 284(6):R1427–R1435.

147. Williams DL, Baskin DG, and Schwartz MW. Leptin regulation of the anorexic response to glucagon-like peptide-1 receptor stimulation. *Diabetes* 2006; 55(12):3387–3393.

148. Ma X, Zubcevic L, Bruning JC, Ashcroft FM, and Burdakov D. Electrical inhibition of identified anorexigenic POMC neurons by orexin/hypocretin. *J Neurosci* 2007; 27(7):1529–1533.

149. Aja S, Ewing C, Lin J, Hyun J, and Moran TH. Blockade of central GLP-1 receptors prevents CART-induced hypophagia and brain c-Fos expression. *Peptides* 2006; 27(1): 157–164.

150. Zander M, Madsbad S, Madsen JL, and Holst JJ. Effect of 6-week course of glucagon-like peptide 1 on glycaemic control, insulin sensitivity, and β cell function in type 2 diabetes: A parallel-group study. *Lancet* 2002; 359(9309):824–830.

151. Gedulin BR, Nikoulina SE, Smith PA et al. Exenatide (exendin-4) improves insulin sensitivity and beta cell mass in insulin-resistance obese fa/fa Zucker rats independent of glycemia and body weight. *Endocrinol* 2005; 146:2069–2076.

38 Progress toward Gene Therapy for Cystic Fibrosis

Ruth S. Everett and Larry G. Johnson

CONTENTS

38.1 INTRODUCTION

Cystic fibrosis (CF) is a common monogenic disorder affecting approximately one in 2500 Caucasian births in the United States [1]. The disease affects epithelia throughout the body including the lungs, pancreas, and the digestive system, but lung disease is the leading cause of morbidity and mortality [2]. Located on chromosome 7 [3–6], the CF gene encodes the CF transmembrane conductance regulator (CFTR) protein [7]. CFTR is a cyclic AMP-regulated chloride channel that is expressed in epithelial cells of the airways, sweat ducts, pancreatic ducts, intestine, and the vas deferens [8]. Over 1400 mutations within the CF gene have been identified that are associated with varying severities of disease [10]. The most common mutation, a deletion of phenylalanine (F) at position 508 of the protein product (ΔF508), is found in 70% of CF chromosomes worldwide [9,10]. This mutation leads to a misfolded CFTR that is degraded in the endoplasmic reticulum such that little or no CFTR traffics to the apical membrane of affected epithelia. The lack of CFTR at the apical membrane prevents secretion of salt and water into the luminal airway surface liquid periciliary layer. Because wild-type CFTR inhibits epithelial sodium channels (ENaC) in the apical membrane, absence of CFTR leads to uninhibited ENaC activity and enhanced sodium absorption from the lumen. The inability to secrete chloride combined with enhanced sodium absorption leads to a decrease in the height and volume of the periciliary liquid layer on the luminal surface airway epithelia, resulting in decreased mucociliary and cough clearance of secretions, bacterial infection, inflammation, airways obstruction, lung destruction, and premature death.

Since CF heterozygotes exhibit a normal phenotype, introduction of a single normal copy of the gene into the defective CF epithelial cells should result in the restoration of the CFTR-mediated Cl⁻ transport function and a normal airway epithelial phenotype. *In vitro* studies suggest that as little as 6%–10% of normal CFTR function may be required to reverse the chloride channel defect [11,12], although enhanced sodium transport in CF airways may require correction of all the cells within the epithelium. These and other *in vitro* studies have established the feasibility of gene transfer for CF following introduction of a normal CFTR into CF airway epithelial cells using viral and nonviral vectors [13–17].

Identifying the cellular targets for CF gene transfer in humans has been controversial. The site where CF lung disease begins remains controversial with both the superficial columnar epithelial cells lining the lumen of the small airways and the serous cells of submucosal glands having been identified as potential sites. Clinical data support the theory that the disease begins in the small airways [16,17], whereas the submucosal glands were previously reported to be the predominant site of CFTR expression [18]. A subsequent study of CFTR expression in the lung with immunohistochemical and biochemical techniques demonstrated localization of wild-type CFTR protein to the apical membrane of ciliated cells within the superficial epithelium and gland ducts, with little CFTR expression in the gland acinus and alveolar epithelial cells [19]. Current approaches have almost exclusively focused on luminal delivery of gene transfer vectors.

Despite its initial role as the prototypical disease for direct delivery of vectors for human gene transfer, progress toward development of gene therapy for CF has been disappointing due to a variety of host- and vector-specific barriers. In this chapter, we review the progress and pitfalls of gene transfer systems used in CF gene delivery, the barriers to gene transfer for CF, and the progress made in the development and application of these gene delivery systems to enhancing efficiency and overcoming these barriers.

38.2 CYSTIC FIBROSIS CLINICAL TRIALS UPDATE

More than 20 clinical gene transfer safety and efficacy trials have been performed in CF subjects with a variety of vectors [22]. Two clinical trials have been published in the past 4 years. One trial was based on the use of compacted DNA nanoparticles to deliver the CFTR transgene whereas the other trial used an adeno-associated virus-2 (AAV-2) vector to deliver the transgene to lower airways.

Konstan et al. performed a double-blinded, dose escalation, phase I clinical gene transfer safety and efficacy trial of intranasal administration of compacted DNA nanoparticles or placebo (saline) in 12 CF subjects. The vector in this trial consisted of single molecules of plasmid DNA carrying the CFTR gene compacted into DNA nanoparticles with 10 kDa polyethylene glycol (PEG) substituted poly-L-lysine peptides [20]. All subjects received either compacted DNA or saline in one nostril and saline (vehicle) in the other nostril (2 mL of each). However, because of spillage of vector into the contralateral nostril during vector administration, the saline nostril could not be used as a placebo control. The doses of the nanoparticles administered were 0.8, 2.67, or 8.0 mg. Primary end points of the trial were safety and tolerability with efficacy evaluation as a secondary endpoint.

Safety evaluation included serum and nasal washings for measurement of cytokine levels, complement, and C reactive protein. One subject demonstrated a transient elevation in serum IL-6, but no significant local or systemic toxicity was otherwise detected. No significant changes in complement or C reactive protein levels were measured either. Significant levels of vector DNA were observed in the nasal epithelial cells at 2 weeks following gene transfer. Efficacy as assessed by nasal potential difference measurements demonstrated partial to complete restoration of CFTR chloride channel function in 8 out of 12 subjects. The correction persisted for 6 days after gene transfer and for up to 28 days in one subject. Raised sodium transport as defined by a change in basal nasal potential difference (PD) with vector administration was not detected.

A randomized, double-blinded, placebo-controlled phase II clinical trial of repeated aerosol delivery of an AAV2-CFTR

vector to the lower respiratory tract of CF subjects was performed by Moss et al. [21]. The primary end points were safety and tolerability of repetitive administration of aerosolized AAV2-CFTR vectors. The vector or placebo was delivered to the lungs of 37 CF patients with mild lung disease by inhalation of three aerosolized doses. Twenty subjects received at least one dose of the vector and 17 subjects received placebo. Thirty days after vector administration, a small improvement in FEV1 was measured in the subjects who received the tgAAVCFTR as compared to placebo controls. No significant toxicity was detected in this study and a significant decrease in IL-8 levels in subjects randomized to the tgAAV2-CFTR was detected on day 14 as compared to an increase in IL-8 levels in subjects who received the placebo. No differences in IL-8 and IL-10 were measured on days 45 and 75. Vector DNA was detected in bronchial cells from bronchial brushings of all six subjects who received vector, but gene expression (vector-specific CFTR mRNA) was not detected. Although the results from this trial of multiple doses of aerosolized AAV-2 CFTR demonstrated both safety and tolerability, evidence for efficacy was limited.

These latter clinical studies for CF gene transfer were consistent with the prior studies that demonstrated evidence for gene transfer based on the surrogate markers that were measured, but highly variable and limited evidence for efficacy.

38.3 VECTORS FOR CF GENE TRANSFER

Both viral and nonviral vector systems have been administered to subjects in CF clinical trials. An ideal vector system for airway gene transfer should have an adequate capacity for the CFTR cDNA and a suitable promoter, efficiently transduce airway cells from the lumen, be nonimmunogenic and noninflammatory, and produce either long-term transgene expression or have the ability to be readministered without loss of gene transfer efficiency. No vector to date has been identified that meets all of these criteria.

38.3.1 Viral Vectors

Adenovirus (Ad) and AAV vectors have each been used to deliver the CFTR transgene in clinical human gene transfer trials. Lentivirus vectors have undergone extensive preclinical development, but have not yet been entered into CF clinical trials.

1. *Ad vectors.* Replication-deficient adenoviral vectors have been extensively developed as tools for gene transfer into mammalian cells [22]. Ads are double-stranded DNA viruses that have a 36 kbp genome consisting of a series of early genes that are responsible for virus replication, antigen presentation, and surveillance, and a series of late genes that encode viral structural proteins [23–26]. Several generations of Ad vectors have been developed based on Ad serotypes 2 (Ad2) and 5 (Ad5). First generation vectors have had the early region one (E1) genes deleted to make the vector replication-defective [27,28], whereas deletion of the E3 region creates room to insert therapeutic cDNAs with a suitable promoter.

E1-deleted second generation vectors have been developed in which the E2a region has been mutated to form a temperature-sensitive mutant virus that replicates in 293 cells at 32°C, but not at 39°C, bringing an additional safety feature to this vector during production [29–31]. E1-deleted second generation vectors with deletions in most (except for open reading frame six, ORF6) or all of the E4 region to limit late viral gene expression [32–34] have also been developed. All viral genes have been deleted from third generation Ad vectors, retaining only a small packaging signal and the inverted terminal repeats [35–41]. These vectors can accept insert cDNAs or even genomic DNA in excess of 30 kb [25], but may require a stuffer sequence for adequate packaging of vector constructs containing smaller cDNAs [39,41]. Because these completely deleted vectors require coinfection of producer cells with an Ad helper virus for production of viral structural proteins, they are commonly referred to as gutless or helper-dependent (HD) Ad vectors. Although the use of Ad helper virus for production may contaminate vector production stocks, the advent of Cre-lox recombination with newer virus production techniques has reduced helper contamination to as low as 0.01% [42–44].

Ad binds to a high-affinity coxsackie and adenoviral 2/5 receptor (CAR) [45] and is internalized through an $\alpha_v\beta_{3/5}$ integrin-mediated vesicular (endocytic) process [46]. As a result of endosomolytic properties mediated by the Ad penton base, Ad avoids lysosomal degradation and efficiently translocates to the nucleus where it exists as an episome (extrachromosomal DNA) mediating expression of therapeutic genes (cDNAs). Because Ad vectors do not integrate at high frequencies, transient expression occurs so repetitive administration will be required for CF gene therapy.

2. *AAV vectors.* AAV vectors were derived from the naturally defective and nonpathogenic wild-type human parvoviruses based on serotypes 2 and 3, AAV-2 and AAV-3 [47–51]. AAV requires the presence of a helper virus, e.g., Ad or herpes simplex virus, to replicate or to cause a lytic infection. In the absence of a helper virus, AAV integrates into the host cell genome and becomes latent. Upon a subsequent wild-type Ad or wild-type herpesvirus infection, the AAV genome can be rescued (excised) from the chromosome to generate a lytic infection [47–51]. AAV-2 is a single-stranded DNA virus with a genome size of ~4.7 kb that consists of the following: (1) inverted terminal repeats at the 5′ and 3′ ends of the molecule [47–50], which play a role in replication and are important for integration into the host cell genome and (2) the viral genes *rep* and *cap*, which mediate viral replication and nucleocapsid formation. Deletion of the *rep* and *cap* genes permits 4.5 kb for DNA insertion into the AAV-2 vector, This small insert size limits the cloning capacity for large coding sequences such as the CFTR cDNA (4.5 kb) and the regulatory elements required to drive CFTR transcription [52].

The receptor for AAV-2 is a membrane-associated heparan sulfate proteoglycan with fibroblast growth factor 1 and $\alpha_v B_5$

integrins serving as coreceptors for AAV-2 entry [53–55]. Similar to CAR, heparan sulfate proteoglycans and the coreceptors are localized predominantly to the basolateral surface of airway epithelial cells [56]. The ability of AAV vectors to transduce nondividing cells makes them a good candidate for CFTR gene transfer. Although AAV-2 vectors exhibit poor entry across the apical membrane of well-differentiated (WD) human airway epithelial (HAE) cells following apical application, a variety of serotypes may offer hope for CFTR gene transfer.

3. *Lentiviral vectors.* Retroviruses are members of a large group of viruses called the Retroviridae [57,58]. Several different classification schemes have been delineated for retroviruses including those based on biology as well as those based on morphology. Despite these classification schemes, most members of the retroviral family can generally be characterized by the either simple or complex composition of their individual genomes. Simple retroviruses are RNA viruses whose genomes consist of two viral long terminal repeats (LTRs), a packaging signal, and a series of structural genes, gag, pol and env. These structural genes encode the capsid protein, reverse transcriptase, protease, an integrase, and the envelope glycoprotein. Deletion of the structural genes gag, pol, and env makes the virus replication defective and enables the insertion of therapeutic cDNAs plus promoter elements into the retroviral genome, forming a replication-defective retroviral vector [57]. The genomes of lentiviruses are more complex, encoding a variety of regulatory accessory proteins and pathogenesis factors that are not present in the genomes of simple retroviruses [58–61]. Furthermore, genes encoding proteins that utilize the cellular nuclear import machinery e.g., MA, IN, and Vpr of HIV-1, to target the preintegration complex to the nucleus are also encoded within this complex genome. Major deletions in these accessory and pathogenesis factor genes have enabled the development of lentiviral vectors for gene transfer [58–61]. Exogenous (internal) promoters have also been included within the sequences of the inserted gene cassette since transcription from the viral LTR may constitute a safety hazard. Deletions in the LTR to prevent transcription (self-inactivating vectors) have been introduced as a safety feature of retroviral and lentiviral vectors [62,63].

The envelope glycoproteins of wild-type retroviruses and lentiviruses bind to cell surface receptors to facilitate entry of the virus into the cytoplasm where the viral RNA is reverse transcribed to form a cDNA, the provirus. This provirus is translocated to the nucleus where it integrates into the host cell chromosomes and through the normal process of DNA transcription, encodes new viral proteins and new viral RNA, which are assembled at the cell surface into new viral particles. Replication-defective lentiviral vectors infect cells by similar mechanisms, but unlike wild-type viruses, the integrated provirus from these vectors encodes the therapeutic gene and viral particles are not produced. As a safety measure, the endogenous envelope proteins of lentiviral vectors have been deleted and replaced by envelope proteins from other viruses, a process known as pseudotyping. The vesicular stomatitis virus glycoprotein (VSV-G) has been among the most commonly used envelope [64]. Initial studies demonstrated that pseudotyped lentivirus vectors can transfect differentiated airway epithelial cells in some systems, but much work is required *in vivo* to move them toward clinical trials. Lentiviral vectors, derived from the human immunodeficiency viruses (HIV), feline immunodeficiency virus (FIV), and equine infectious anemia virus (EIAV) have each been developed as gene transfer vectors for CF.

38.3.2 NONVIRAL VECTORS

Since viral vectors can elicit immune and inflammatory responses and the induction of neutralizing antibodies potentially preventing vector readministration [65], some investigators, particularly the United Kingdom Gene Therapy Consortium, have focused their efforts on developing nonviral vectors as gene transfer tools. Nonviral vectors explored for CF gene transfer to date include naked DNA, cationic liposomes, and polymers.

1. *Naked DNA*. Naked plasmid DNA has been used as a gene transfer vehicle in airway epithelial cells both *in vitro* and *in vivo*, but has generally been ineffective at delivering genes to CF airways. Although delivery of naked DNA to airways is a simple and safe technique, gene transfer efficiency in prior studies was inadequate for further development of this vector system for lung gene delivery [66].

2. *Cationic liposomes*. Cationic liposomes have been extensively evaluated in clinical trials. Cationic liposomes are composed of cationic lipids mixed in varying molar ratios with cholesterol and dioleoylphosphatidylethanolamine (DOPE), a neutral phospholipid [67,68]. Commonly used cationic lipids for gene transfer include N[1-(2,3-dioleoxy)propyl]-N,N,N-trimethylammonium (DOTMA), 1,2-dimyristyloxypropyl-3-dimethylhydroxyethylammonium bromide (DMRIE), or 3β [N,N',N'-dimethylamino ethane-carbamoyl] cholesterol (DC-Chol), N[1-(2,3-dioleoxy)propyl]-N,N,N-trimethylammonium methyl sulfate (DOTAP), p-ethyl dimyristoyl phosphatidyl choline (EDMPC) cholesterol, and N^4-sperminine cholesteryl carbamate (GL-67). Cationic liposomes bind to negatively charged plasmid DNA to form DNA–liposome complexes, which may, under conditions of excess molar DNA, have a net negative charge. DNA–liposome complexes (also known as lipoplexes) enter cells primarily by endocytosis [69,70], although the mechanism and specificity of binding to the cell surface has not been clearly delineated. Cationic liposomes also promote the release of plasmid DNA from the endosome into the cytoplasm [70]. Like Ads, cationic liposomes do not integrate into the host cell genome, so that expression may be lost with cell division. Thus, repetitive administration of lipoplexes will be required for CF gene therapy. Lipoplexes have been shown to induce host cytotoxic responses associated with bacterial unmethylated cytosine-pyrimidine-guanine (CpG) sequences present within the plasmid DNA [71].

3. *Polymers*. Complexes of cationic polymers with plasmid DNA (polyplexes) that often condense the DNA and prevent it

from degradation have also been developed. These complexes enter epithelial cells on binding to specific cell surface receptors followed by receptor-mediated endocytosis. Poly-L-lysine, polyethyleneimine (PEI), and polyamidoamine have been commonly considered as tools for gene transfer research [72], but only poly-L-lysine has actually been developed sufficiently to form a gene transfer product for human gene transfer. The cationic polymer PEI when complexed to plasmid DNA has been reported to transfect up to 5% of pulmonary cells after intravenous administration [73], but has not undergone significant development for clinical trials in CF. Polyplexes based on poly-L-lysine conjugated to a Fab fragment of IgG against human secretory component have been shown to deliver reporter genes to the airway epithelia *in vitro* and *in vivo* following intravenous administration [74,75], but may suffer from immune surveillance due to the protein component. A PEG-substituted poly-L-lysine has been used to generate compact DNA into nanoparticles for gene transfer to airways *in vivo* and a phase II trial has been completed with this vector (see Section 38.2).

38.4 BARRIERS TO GENE TRANSFER FOR CF

Multiple barriers to airway gene transfer exist in the lung. Nonspecific host barriers include airway mucus, cell surface glycoconjugates, and the inflammatory milieu, while vector-specific barriers include factors affecting cell binding and entry, nuclear translocation, and factors limiting transgene expression postnuclear entry. These barriers combine to make *in vivo* gene transfer to human airways inefficient.

38.4.1 HOST BARRIERS TO GENE TRANSFER

1. *Luminal contents*. Luminal components of the chronic pulmonary infection and inflammation in the CF airway have been shown to inhibit gene transduction by a variety of vectors. Fresh sputum obtained from CF patients has been shown to inhibit liposomal (DC-Chol/DOPE) and adenoviral gene transfer *in vitro* [76]. Sterilization of the sputum with ultraviolet light demonstrated that active infectious organisms alone were not required. Rather the simulation of this inhibitory effect on gene transfer by pretreatment of cells with genomic DNA prior to transduction combined with reversibility of CF sputum-induced effects on gene transfer by pretreatment with recombinant DNAse (rDNAse) suggested that excessive DNA could also inhibit liposomal and adenoviral gene transfer [76].

Separation of sputum collected from CF patients with acute exacerbations receiving antibiotics and rDNAse into aqueous (sol) and gel components by ultracentrifugation demonstrated that the aqueous components or sol of the airway lumen could inhibit Ad gene transfer to airways [77]. This effect arose from the presence of Ad-specific antibodies in the sol of CF sputum, an effect that could be overcome by heat inactivation of CF sol. BAL fluid from CF patients has been shown to inhibit AAV gene transfer *in vitro* [78]. This effect was reversible when CF BAL fluid was incubated with alpha 1-antitrypsin (α_1AT). Subsequent experiments with neutrophil elastase (NE) and purified human neutrophil peptide (HNP) demonstrated that HNP mediated the inhibitory effect on AAV gene transfer.

The inflammatory milieu induced by Pseudomonas is a formidable barrier to transduction [79]. Murine studies in which mice inoculated with *Pseudomonas aeruginosa*-laden agarose beads develop chronic bronchopulmonary infection exhibit a greater than twofold reduction in gene transfer efficiency following nasal instillation of an Ad-*lacZ* vector as compared to mice that had sterile beads or Ad vector alone instilled and a 10-fold reduction on Ad gene transfer efficiency following delivery of Ad-*lacZ* vector to nasal airways of Pseudomonas (PAO1 strain)-infected mice as compared to noninfected nasal airways is consistent with this concept [79,80].

Airway and alveolar macrophages may play a major role in the rapid loss of vector genomes following luminal airway delivery of gene transfer vectors. Worgall et al. demonstrated a 70% loss of Ad-*lacZ* genomes in both immunocompetent and immunodeficient mice within 24 h of transtracheal administration of an Ad-*lacZ* vector [81]. Elimination of macrophages from murine lungs by pretreatment with liposomes containing dichloromethylene biphosphonate to eliminate macrophages prior to Ad vector administration overcame this effect. Alveolar macrophages have also been shown to inhibit retrovirus-mediated gene transfer to airway epithelia *in vitro* [82,83]. Transduction of HAE cells by an amphotropic enveloped retroviral vector was inhibited ~40% by alveolar macrophages and by more than 60% by lipopolysaccharide (LPS)-activated alveolar macrophages. Dexamethasone (1 μM) partially reversed this inhibition of retroviral transduction. Consistent with the Ad studies, a rapid loss of vector retroviral DNA was detected consistent with rapid degradation. Thus, luminal gene transfer approaches targeting the superficial airway epithelium will have to overcome the barriers induced by the inflammatory response in lumen to gain access to the airway epithelium.

2. *Cell surface components.* The glycocalyx may also have inhibitory effects on airway gene transfer. This structure forms a matrix of complex carbohydrate moieties, glycolipids, and glycoproteins on the apical surfaces of airways. At least five mucin glycoproteins have been localized to airway surfaces: MUC1, MUC4, and MUC16, which are tethered mucins and the secreted mucins MUC5ac and Muc5b [84–87]. Overexpression of the mucin MUC1 *in vitro* has been linked to inefficient Ad gene transfer to Madin Darby canine kidney (MDCK) and bronchial epithelial cells that could be reversed by removal of sialic acid residues from the apical surface by neuraminidase [87–89]. The role of the glycocalyx as a barrier to gene transfer has been further explored by overexpression of glycosylphosphatidyl inositol (gpi)-linked CAR, which localizes to the apical membrane of polarized cells [90,91]. Ad5-mediated transduction of polarized MDCK cells overexpressing gpi-linked CAR was inefficient following apical application, but was enhanced by removal of sialic residues with neuraminidase [90]. A similar finding was reported in MUC1-deficient mice *in vivo* [92]. In contrast, apical application of an Ad2 vector in a prior study efficiently transduced polarized HAE cells overexpressing gpi-linked CAR, an effect not enhanced by neuraminidase pretreatment [91]. This incongruity may result from different levels of expression of glycocalyx components in the cell cultures, rather than differences in vector serotype.

38.4.2 Vector-Specific Barriers to Luminal Airway Gene Transfer

Barriers specific to each vector considered or tested in human clinical gene transfer trials for CF to date have been identified. These vector-specific barriers to airway gene transfer are discussed below.

1. *Ad vectors.* Preclinical studies and clinical trials have established that luminal Ad gene transfer to airways is inefficient [93–95]. This inefficiency of Ad gene transfer results from localization of the coxsackie B and adenovirus receptor (CAR) and $\alpha_v\beta_{3/5}$ integrins, which mediate uptake of Ad vectors [46], to the basolateral rather than the apical membrane, of columnar cells limiting vector binding and entry and hence efficient gene transfer [90,94,96,97]. Significant concern still remains for the adaptive immune response to Ad, even for HDAd vector, which may have a longer duration of expression, but still require vector readministration for CF gene therapy [98].

2. *AAV vectors.* Several barriers to AAV-mediated gene transfer remain. The small insert size, which is at the upper size limit for insertion of full-length wild-type CFTR (coding region of ~4.5 kb) driven by an exogenous promoter, is one potential significant barrier for CFTR gene transfer. The generation of small promoters to drive transcription and the development of shorter versions of CFTR (minigenes) may overcome this limitation, but may not be ideal.

Lack of functional apical membrane receptors due to the basolateral localization of the membrane-associated heparan sulfate proteoglycans receptor and its coreceptors fibroblast growth factor-1 (FGF-1) and $\alpha_v\beta_5$ integrins has been a limitation for AAV-2-mediated gene transfer to airways [53–56,99]. However, binding studies of radiolabeled AAV-2 support the existence of nonfunctional apical membrane receptors for AAV-2 [56,99]. Recent developments in AAV serotypes may soon overcome the limitations of AAV-2.

A postentry barrier to transduction is the persistence of AAV-2 genomes as single-stranded episomes (ssDNA), which are inefficiently converted to double-stranded DNA (dsDNA), a requirement for transgene expression [100,101]. This limitation leads to a delay in the onset of transgene expression, which may be overcome by waiting long enough for maximal transgene expression to occur (~4 weeks) or by the use of DNA damaging agents, topoisomerase inhibitors, and Ad early gene products [102–104]. Thus, barriers to AAV-mediated transduction of CFTR into CF airways include small insert size, decreased binding, and uptake of vector due to absent or decreased functional receptor expression on the apical membrane, and inefficient single-strand to double-strand conversion of AAV genomes.

3. *Lentiviral vectors.* The development of HIV, EIAV, and FIV vectors [58,59,61,63,105,106], which can transduce non-dividing airway cells, has overcome potential limitations caused by the low rate of cell proliferation in human airway epithelia [107]. Advances in retroviral production techniques and pseudotyping of vectors to create stable envelopes that permit concentration of vector stocks have also overcome the limitations of titer [108]. However, titers of retroviruses remain ~1–2 logs lower than that of Ad and AAV vectors.

Apical membrane barriers to efficient transduction of WD airway cells are also limiting. *In vitro* and *in vivo* studies of polarized airway epithelial cells have localized low levels of functional amphotropic receptors (RAM-1 or Pit-2) to the basolateral surface of polarized WD airway cells. *In vivo* studies have confirmed low levels of expression of RAM-1 or Pit-2 in murine lung [109]. Similar findings have been reported for lentiviral vectors pseudotyped with the VSV envelope glycoprotein (G). VSV-G pseudotyped HIV, EIAV, and FIV vectors have been shown to transduce polarized WD HAE cells *in vitro* and *in vivo* from the basolateral surface [105,106,110–112].

4. *Nonviral vectors.* Barriers to nonviral gene transfer have been identified at several steps. Nuclear entry was identified as the rate-limiting factor for efficient liposome-mediated gene transfer into cell lines resistant to transfection [113,114], whereas gene transfer to WD airway epithelial cells has been limited by both inefficient nuclear entry and failure of DNA–liposome complexes to enter the cell [115]. Loss of phagocytic entry mechanisms, decreased cell surface binding, and decreased uptake in differentiated airway epithelial cells (central cells) as compared to poorly differentiated cells (edge cells) were the reasons for inefficient transduction. Subsequent studies confirmed these observations by demonstrating decreased amounts of cell-associated lipoplexes in differentiated airway epithelia as compared to poorly differentiated epithelia [113,114]. Liposome-mediated gene transfer into proliferating cells was also enhanced relative to quiescent cells, raising the possibility of enhanced nuclear transport of DNA during mitosis due to breakdown of the nuclear envelope. Thus, barriers to efficient transduction of WD airway cells by lipoplexes include decreased binding and uptake of lipoplexes and poor nuclear translocation of DNA.

Molecular conjugates are typically designed to target specific receptors on cell surfaces to improve binding and uptake. However, poor nuclear translocation with trafficking of significant portions of vector to lysosomal compartments for degradation remains a concern for molecular conjugates.

38.5 STRATEGIES FOR IMPROVING GENE TRANSFER EFFICIENCY

The apical membrane of WD airway cells remains the major barrier to efficient transduction by all the current gene transfer vectors such that strategies still focus primarily on this site. Current strategies have not fully addressed the problem with luminal contents, i.e., neutrophils, bacteria, excess DNA, and inflammation, focusing instead on improving apical membrane binding and entry. The hope is that pretreatment with α_1AT, rDNase, or antibiotic therapy will limit the effects of luminal contents on gene transfer.

Two general strategies have been proposed for overcoming barriers to airway epithelial gene transfer: (1) modification of the host airway to enhance gene transfer and (2) modification of the vectors to target receptors expressed on the apical membrane of airways *in vivo*.

38.5.1 HOST MODIFICATION

1. *Modulation of paracellular permeability.* Treatment of the airway epithelium with agents that transiently increase permeability through the tight junctions to facilitate access of the vectors to the basolateral membrane is one strategy for enhancing gene transfer efficiency to airway epithelia. Pretreatment of nasal airways with polidocanol enhanced gene transfer mediated by an Ad-*lacZ* vector and facilitated partial correction of the Cl⁻ transport defect in the nasal epithelium of CF mice following a single dose of vector [80]. The single dose of Ad-CFTR vector used following pretreatment with the surface agent polidocanol generated the same degree of CFTR correction previously reported by Grubb et al. in which four doses of an Ad vector were required to generate a 40%–50% correction of Cl⁻ transport [93].

Luminal or apical application of the calcium chelator ethylene glycol bis-(β-aminoethyl ether)-*N,N,N′*, *N′*-tetraacetic acid (EGTA) in hypotonic solutions followed by luminal application of an amphotropic retroviral vector has enabled investigators to correct the Cl⁻ transport defect in WD CF HAE cells stimulated to proliferate with keratinocyte growth factor (KGF) and has also enhanced luminal FIV-based lentiviral gene transfer to rabbit airways *in vivo* [106,112]. EGTA/hypotonic solution has also been reported to enhance AAV and Ad gene transfer to WD HAE cells *in vitro* and Ad gene transfer to airway epithelia *in vivo* following luminal application [56,116,117]. However, much higher concentrations of EGTA (0.1–0.4 M) may be required for optimal enhancement of Ad gene transfer to nasal and lower airway epithelia of mice *in vivo* [117].

Medium chain fatty acids have also been shown to enhance airway gene transfer [118–120]. The sodium salt of capric acid (sodium caprate) has been shown to more effectively enhance Ad gene transfer than EGTA *in vitro* and to enhance Ad gene transfer to murine airways *in vivo* [118–121].

Water-induced hypoosmotic shock alone may transiently permeabilize tight junctions temporarily to enhance vector uptake [122]. PEI–pDNA polyplexes nebulized in hypoosmotic distilled water were shown to enhance transgene expression by 57-fold more than polyplexes in 5% glucose, and by 185-fold than those in Hepes-buffered saline. However, no formal studies of tight junction function were performed in the study and the size of the PEI polyplexes in water was significantly smaller than those measured in glucose or saline.

Lysophosphatidyl-choline (LPC), a natural component found in lung surfactant, has also been used to increase vector access to basolateral membrane receptors. A prior study demonstrated that treatment of the airways with LPC resulted in enhanced long-term gene transfer *in vivo* with a VSV-G pseudotyped lentivirus and partial correction of CFTR function in the nasal epithelia of CF deficient mice [123]. Formulating a HD Ad*lacZ* vector in LPC also enhanced transduction efficiency in rabbit airways with detection of transgene expression in the small and large airway epithelial cells, from trachea to terminal bronchioles [122]. All cell types of the surface epithelium were transduced and extensive transduction of the epithelium was obtained with virus formulated in isotonic 0.1% LPC.

The apical application of the occludin peptide OP_{90-103} in WD HAE cells has also shown to enhance Ad and AAV viral vector transduction by transient disruption of the epithelial tight junctions resulting in alterations in paracellular permeability in a concentration-dependent manner [124]. AdLacZ and AAV2GFP vectors each showed significantly higher levels of transduction in WD HAE cells pretreated apically with an OP_{90-103} [124]. Relatively specific and transient redistribution of tight junction proteins were detected with minimal toxicity *in vitro* in this study.

These studies established the feasibility of transient permeabilization of the paracellular path to enhance airway gene transfer mediated by vectors with receptors that are localized on the basolateral membrane. However, major concerns have been raised in the CF community regarding the safety of modulating paracellular permeability to enhance airway gene transfer efficiency. Chelating agents, medium chain fatty acids, polidocanol, and LPC have each been associated with some toxicity [80,120,122,123]. Efforts to develop agents to that target proteins in the apical junctional complex, e.g., occludin or one of the claudins, with high specificity offer promise, but much work needs to be done to perfect the technology.

2. *Enhancing endosomal processing.* The significantly lower transduction efficiency of AAV transduction from the apical surface of airway epithelial cells as opposed to basolateral application appears to be due to differences in endosomal processing and nuclear trafficking of apically or basolaterally internalized virions [56,99,125,126]. Since AAV capsid proteins are ubiquitinated after endocytosis, gene transfer using these vectors was shown to be significantly enhanced by proteasome or ubiquitin ligase inhibitors [127,128]. Tripeptidyl aldehyde proteasome inhibitors increased luminal AAV transduction efficiency by >200-fold, to a level almost equivalent to that observed with basolateral application of the vector. *In vivo* application of these proteasome inhibitors in murine lungs increased AAV-mediated gene transfer from undetectable levels to a mean of 10.4% ± 1.6% of the epithelial cells in large bronchioles. These results support potential use of proteasome inhibitors for AAV-mediated gene transfer to circumvent the barriers of endosomal processing and nuclear trafficking.

3. *Increasing dwell time.* Formulation of gene transfer vectors such as Ad and AAV with viscoelastic gels such as carboxymethyl-cellulose (CMC) and methylcellulose (MC) to increase the residence time of the vectors on the epithelial cells may result in enhanced gene transfer efficiency. Mucociliary clearance of gene transfer vectors from the surface of airway epithelial cells by the coordinated beating of the cilia may reduce contact time of the gene transfer vectors with the epithelial surface, which could prevent uptake of the vector by the epithelial cells. Previously, recombinant Ad formulated in a thixotropic solution containing 1% Avicel RC-591, a mixture of microcrystalline cellulose and CMC, enhanced gene transfer efficiency in airway epithelial cells by inhibiting mucociliary clearance [129,130]. The viscoelastic gels CMC, MC, and poloxamer 407 (P407) were tested for their ability to enhance gene transfer of adenoviral, AAV5, and lentiviral vectors in the airways of mice. MC and CMC enhanced the efficiency of gene transfer to the airways mediated by Ad, AAV5, and GP64-pseudotyped FIV gene transfer. The safety of this approach has not been carefully evaluated, but may raise concerns since decreased clearance can lead to impaired lung function in CF patients.

38.5.2 VECTOR MODIFICATION

Modifications of specific gene transfer vectors are an alternative approach for enhancing gene transfer efficiency. Targeting the vector to receptors that are endogenously expressed on the apical membrane of WD HAE cells is the goal of this approach.

1. *Ad vectors.* While proof-of-concept studies have shown the feasibility of retargeting Ad vectors to alternative receptors on airway apical surfaces, little meaningful progress has been made [90,131,132]. Rather the advances in gene transfer efficiency with HDAd vectors in rodent, rabbit, and primate lungs can be attributed to the use of agents that modulate paracellular permeability [65,98,133].

Studies of chimeric Ad vectors that target apical membrane of airway cells have been proposed. Zabner et al. screened 12 adenoviral serotypes from Ad subgroups A–F for their ability to bind and infect the apical surface of polarized WD HAE cells in culture [134]. Wild-type Ad17 bound to WD HAE cells more efficiently than wild-type Ad2. Subsequently chimeric Ad2 vectors with Ad17 fiber transduced WD HAE cells ~100-fold more efficiently than the original Ad2 vector. These data suggested that it may be possible to generate chimeric Ad vectors that more efficiently transduced HAE cells following luminal application. However, development of chimeric vectors would not be able to eliminate potential problems with neutralizing antibodies.

To overcome the immunological barriers of neutralizing antibodies that prevent Ad vector readministration, nonhuman Ad vectors developed from primate Ad serotypes, to which humans do not possess neutralizing antibodies, have been evaluated [135–137]. A rotation of Ad vector serotypes (serotype switching) could prolong the time between readministration of the same serotype in an attempt to avoid preexisting immunity to common Ad serotypes. However as with human Ad vectors, primate vectors still need to gain entry to the receptors located on the basolateral surface of the epithelial cells. In the first study involving readministration of HD Ad vectors to the lung, a second administration of these vectors to murine lungs was effective, with minimal loss of transgene expression [133]. In contrast, reduced transgene expression was detected in mice that were administered a first generation Ad vector followed by repeat delivery of a HD Ad vector. While these data suggest that HD Ad vectors can be readminstered at least once, multiple administrations were not performed.

2. *AAV vectors.* Some progress has been made in addressing barriers to AAV2-mediated gene transfer to CF airway epithelia. The restriction on insert size that limits efficient packaging of wild-type CFTR with a suitable promoter remains a significant concern. Although trans-splicing strategies have been considered in which different portions of the CFTR carried on different vectors can splice together to form a complete CFTR molecule, the complexity of achieving efficient transduction of two different vectors into the same cell to permit trans-splicing has slowed progress in this area [138]. However, spliceosome-mediated RNA trans-splicing (SmaRT) with AAV vectors has been explored as a means for functionally correcting endogenous ΔF508 CFTR transcripts both *in vitro* in human CF polarized airway epithelia and in human CF bronchial xenografts [139]. Previously, Ad vectors encoding a pre-trans-splicing molecule (PTM) targeted to intron 9 of CFTR-induced partial correction of endogenous ΔF508 messenger RNA and protein and human CF bronchial xenografts infected with Ad. CFTR-PTM also demonstrated partial correction of CFTR-mediated chloride channel activity suggesting the feasibility of SmaRT-mediated repair of mutant endogenous CFTR. The low level of correction was probably due to the inefficient infection of these vectors from the apical surface of the airway epithelium. Recently the SmaRT technique has been applied to AAV vectors, resulting in correction of CFTR function following luminal application in human CF airway epithelia [139]. Adeno-associated viral vectors carrying a PTM targeted to bind intron 9 of CFTR pre-mRNA and trans-splice the normal sequence for human CFTR into the endogenous pre-mRNA were generated. Human CF airway epithelia were infected from the luminal surface with rAAV2 or rAAV5 CFTR-PTM vectors and cells were treated with proteasome-modulating agents to enhance the efficiency of transduction. Partial CFTR restoration was observed with both the rAAV2 and rAAV5 CFTR-PTM vectors and wild-type CFTR transcripts in CFTR-PTM-corrected epithelia were detected by RT-PCR analysis. These data suggested that the SmaRT technology with its ability to repair endogenous CFTR mRNA could overcome the insert size limitation of AAV due to its smaller CFTR-PTM cassette.

Concerns also remain about which promoters might be best for CFTR gene transfer mediated by AAV vectors.

A pseudotyped AAV5-CFTR vector containing a CFTR minigene and a chicken β-actin promoter with a CMV enhancer has been shown to significantly enhance transgene expression in primates [140]. Aerosolization of this pseudotyped rAAV5-CFTR or rAAV5-GFP genes to nonhuman primate lungs resulted in CFTR and GFP expression in the airway epithelium, suggesting that the CFTR minigene construct could potentially be a promising candidate for CF gene transfer promoters. Another study, which evaluated the effects of promoter sequences on transduction rates and gene expression levels in the lung, showed that the Rous sarcoma virus (RSV) promoter performed significantly better than a human cytomegalovirus (CMV) promoter in the airway epithelium [141]. A hybrid promoter consisting of a CMV enhancer, beta-actin promoter and splice donor, and a beta-globin splice acceptor (CAG promoter) showed even higher expression than either of the strong viral promoters alone, with a 38-fold increase in protein expression over the RSV promoter. Vectors containing either the RSV or CAG promoter also showed good expression in the nasal and tracheal epithelium, suggesting that efficient transduction in airways can be achieved by using the appropriate AAV serotypes and promoter sequences.

Retargeting AAV2 vectors to enhance uptake via the apical surface of airway epithelia can be accomplished by introducing ligands for receptors on the apical membrane into the viral capsid or by using capsids of alternative serotypes to target the apical membrane. Current research efforts to improve apical membrane targeting of AAV2 vectors have focused on identifying wild-type AAV vectors of other serotypes. Once identified, the alternative serotypes can be developed into new gene transfer vectors or alternatively, their capsid proteins can serve as pseudotypes for AAV2 vectors. Over 40 different AAV serotypes have been considered for gene transfer [142] and more than nine AAV serotypes have been tested as gene transfer vectors [142,143]. The most efficient for airway epithelia have been AAV5 [144,145], AAV6 [146,147], and AAV9 [148]. A prior study demonstrated that rAAV-5 vectors were able to transduce murine lungs more effectively than AAV-2 vectors. Intranasal delivery of pseudotyped AAV-2 genomes with AAV-5 capsids (AAV2/5) resulted in a 250-fold-higher level of transgene expression in murine lungs than rAAV-2 capsid-mediated infection [144,145]. Previous reports show that the AAV6 pseudotyped vectors can mediate high transduction efficiency in large and small airways, with up to 80% transduction in some airways [146]. While both AAV2 and AAV6 serotypes transduce alveolar epithelia with similar efficiencies, the transduction efficiency of bronchial and distal airway epithelial cells with AAV6 was significantly higher than AAV2.

In more recent studies, a human AAV2/9 vector transduced murine nasal and lung airway epithelia *in vivo* with persistence of expression for 9 months. The AAV2/9 vectors could be readministered as early as 30 days following initial exposure with a minimal effect on transgene expression despite the presence of high levels of neutralizing antibodies in the serum [148].

Nevertheless, most investigators accept that readministration of AAV vectors will be required for lifelong therapy of CF. One strategy to circumvent the problem of repeat administration hinges on modification of the surface of AAV vectors with PEG to mask them from the immune system. PEGylation of the viral capsids with tresyl chloride (TMPEG) resulted in enhanced efficacy of AAV2 vectors in murine lungs by five-fold following repeat administration [149], suggesting that TMPEG provided protection from AAV neutralizing antibodies in vivo.

Serotype switching is another method that has been considered for overcoming immune inhibition of AAV gene transfer. The use of nonhuman primate serotypes has been explored as a method to overcome neutralizing antibodies that limit the efficiency of gene transfer. A recombinant primate AAV vector containing a capsid derived from the rhesus macaques and the AAV2 internal terminal repeats (AAVrh.10) has been shown to boost alpha-1 antitrypsin levels when administered intrapleurally to mice preimmunized to AAV2/5 vectors, thus circumventing immunity to AAV [150]. The efficiency of this vector in airways has not been reported in detail to date, but it may offer yet another serotype that could be used to circumvent humoral immunity to other AAV serotypes.

The limitations of single-strand to double-strand conversion previously mentioned can be overcome by the application of self-complementing AAV vectors (sc-AAV) to CF gene transfer. The sc-AAV vector includes both the sense strand as well as the reverse-complement sequence of the promoter-transgene-polyadenylation insert. An AAV2-hairpin loop located between the sense and reverse complement enables the vector DNA to immediately fold back upon itself into a double-strand form as soon as it is uncoated. This configuration reduces the time for second strand conversion of vector genomes and the onset of transgene expression is accelerated [126,151,152].

3. *Lentiviral vectors.* Transduction of airway cells following luminal delivery remains rate-limiting for lentiviral vectors. Lentiviral vectors pseudotyped with the VSV-G can transduce nondividing cells, but these pseudotyped vectors transduce airway epithelia via receptors that are located predominantly on the basolateral surface of the airway epithelium. As a result, their transduction efficiency is low. However, altering the envelope glycoprotein by pseudotyping lentiviral vectors with the Ebola Zaire envelope glycoprotein (EboZ) has been shown to significantly increase uptake of the vector when applied luminally, since the receptors required for binding and entry of the EboZ virus are located on the apical surface of the airway epithelial cells [126,151–154]. Recently, production of lentiviral vectors pseudotyped with the avian plague virus hemaglutinin was significantly improved by coexpression of influenza virus M2 in vector-producing cells. These vectors efficiently transduced murine tracheal epithelia cells *in vitro* and murine tracheas *in vivo* following luminal application [155].

An FIV, pseudotyped with a baculovirus GP64 envelope glycoprotein resulted in persistent expression for up to 50 weeks in murine nasal epithelia. This pseudotyped lentivirus recognizes a receptor that is almost 10 times more abundant on the apical surface than the basolateral surface of the airway epithelium [153]. However, efficient transduction occurred only when the virus was mixed with the viscoelastic gel MC, which increases the residence time of the vector [130]. Under these conditions, FIV vectors pseudotyped with the baculovirus GP64 envelope were more effective than FIV vectors pseudotyped with a modified Ebola virus envelope [156].

4. *Paramyxovirus systems.* A novel viral vector system under development is the paramyxovirus vector [157,158]. Sendai virus (SeV), also known as murine parainfluenza virus type 1, is a negative sense, single-stranded RNA virus of the Paramyxoviridae family that mediates cell attachment and entry via the envelope glycoproteins hemaglutinin-neuraminidase (HN) and fusion protein (F). Using reverse genetics, a nontransmissable SeV vector system has been developed from replication-competent SeV vectors [158]. Second-generation SeV, which are transmission-incompetent and potentially safer than first generation SeV, have been shown to reverse the defective chloride transport in CF knockout mice [159]. These recombinant SeV vectors recognize a sialic acid-containing receptor on the surface of airway epithelial cells. SeV vectors containing the reporter genes such as lacZ and luciferase were shown to transduce airway epithelium *in vivo* with high efficiency [158]. Recently, it was shown that the 4.5 kb CFTR cDNA can be inserted within the SeV vector and that the use of this SeV-CFTR vector for gene transfer to airway epithelial cells both *in vitro* and *in vivo* resulted in the expression of a functional CFTR protein [159]. However, the use of this vector elicited airway inflammation. Furthermore, transgene expression using these vectors is transient, and hence repeat administration, which has so far proven to be inefficient, will be required.

Respiratory syncytial virus and parainfluenza virus-3 (PIV) are other members of the paramyxoviridae family of viruses that have been explored as potential gene transfer vectors for CF [160–162]. *In vitro* studies using a replication-competent respiratory syncytial virus-GFP vector showed efficient transduction of human airway epithelia following apical application. Similar results have been reported with PIV-3 vectors [162]. These vectors specifically target ciliated airway epithelial cells from the luminal surface, and *in vitro* studies have shown restoration of CFTR function following transduction with a PIV-CFTR vector [162].

5. *Nonviral vectors.* Recent evaluations of nonviral gene transfer agents including naked DNA, PEI–pDNA polyplexes, PEG-substituted poly-L-lysine pDNA polyplexes, and a cationic lipid 67 (GL67) showed that delivery of a human CFTR plasmid complexed with GL67 into ovine lungs resulted in the production of equal amounts of human CFTR compared to levels of endogenous ovine CFTR with slightly higher efficacy than the other gene transfer agents [163]. These data have led to a clinical trial program by the United Kingdom CF Gene Transfer Consortium.

Since lipoplexes have been shown to induce toxicity with low efficacy in nondividing cells, their safety profile has been improved by removing unmethylated CpG sequences within the plasmid that are known to elicit a toxic response [164]. Following the elimination of 270 out of 526 CpG sequences in cationic lipid:plasmid DNA complexes, the CpG-reduced vector was found to be significantly less immunostimulatory compared to the unmodified vector as evidenced by a 40% to 75% reduction in the levels of IL-12, IFN-gamma, and IL-6 in the serum 24h after intravenous delivery in mice. Similar reductions in cytokine levels were also observed in the bronchoalveolar lavage fluids after intranasal administration in mice, while the levels of reporter gene expression were not affected by these modifications. Inhibitors of the CpG signaling pathways, such as chloroquine and quinacrine, also significantly reduced the induction of IL-12 from mouse spleen cells *in vitro* and inhibited cytokine production in the lung by approximately 50% without affecting gene expression. These results demonstrate that the elimination of CpG sequences in a plasmid DNA vector or inhibitors of CpG immunostimulation can reduce toxicity associated with cationic lipid and plasmid DNA complexes. A CpG-reduced vector has been selected for the United Kingdom CF Gene Transfer Consortium Trial.

Poor apical membrane binding, entry, and nuclear translocation must be overcome to improve the efficiency of nonviral gene transfer in human airways. In general, polyplexes have been more successful at delivering genes to airways than lipoplexes. Branched chain PEI–DNA complexes can deliver genes to airways *in vivo* following luminal delivery [165], but are not biodegradable and induce inflammatory responses [165–167]. Other investigators have focused on PEG-substituted poly-L-lysine plasmid DNA polyplexes that form compacted DNA nanoparticles [149,150]. Although not targeted, these nanoparticles have a small size (~25 nm) that promotes entry across the apical membrane of airway epithelia and enhances nuclear translocation and transcription [168,169]. PEGylation also appears to protect them from inflammatory responses due to unmethylated CpG motifs in plasmid DNA. Correction (often complete) of the CFTR chloride transport defect in the nasal epithelia in CF mice has been reported and partial correction was reported in a clinical gene safety and efficacy trial [20,170].

38.6 SUMMARY

Although progress has been made in CF gene transfer techniques both with viral and nonviral vectors, several challenges still remain and several concerns need to be addressed. An ideal gene transfer vector system that is both effective and safe still needs to be developed, and the problems associated with vector immunogenicity and readministration remain unresolved. A better understanding of the interaction of viral capsid proteins with the innate immune system is required to address the problem of viral vector immunogenicity. Nonviral gene delivery systems also have important issues that need to be addressed. Nonviral vectors that can achieve long-term

expression and specifically target lung epithelia need to be exploited. Research directed toward these issues and toward developing a new generation of vectors with better performance profiles is currently under way and provides hope for success in gene transfer for CF lung disease.

ACKNOWLEDGMENT

The authors wish to acknowledge support from the Arkansas Biosciences Institute in the preparation of this manuscript.

REFERENCES

1. Bobadilla JL, Macek M, Jr., Fine JP, and Farrell PM. Cystic fibrosis: A worldwide analysis of CFTR mutations—Correlation with incidence data and application to screening. *Hum Mutat* 2002;19(6):575–606.
2. Brennan AL and Geddes DM. Cystic fibrosis. *Curr Opin Infect Dis* 2002;15(2):175–182.
3. Knowlton RG, Cohen-Haguenauer O, Van Cong N, et al. A polymorphic DNA marker linked to cystic fibrosis is located on chromosome 7. *Nature* 1985;318(6044):380–382.
4. Wainwright BJ, Scambler PJ, Schmidtke J, et al. Localization of cystic fibrosis locus to human chromosome 7cen-q22. *Nature* 1985;318(6044):384–385.
5. Rommens JM, Iannuzzi MC, Kerem B, et al. Identification of the cystic fibrosis gene: Chromosome walking and jumping. *Science* 1989;245(4922):1059–1065.
6. White R, Woodward S, Leppert M, et al. A closely linked genetic marker for cystic fibrosis. *Nature* 1985;318(6044):382–384.
7. Riordan JR, Rommens JM, Kerem B, et al. Identification of the cystic fibrosis gene: Cloning and characterization of complementary DNA. *Science* 1989;245(4922):1066–1073.
8. Bear CE, Li CH, Kartner N, et al. Purification and functional reconstitution of the cystic fibrosis transmembrane conductance regulator (CFTR). *Cell* 1992;68(4):809–818.
9. Kerem B, Rommens JM, Buchanan JA, et al. Identification of the cystic fibrosis gene: Genetic analysis. *Science* 1989;245(4922):1073–1080.
10. Rowe SM, Miller S, and Sorscher EJ. Cystic fibrosis. *N Engl J Med* 2005;352(19):1992–2001.
11. Johnson LG, Boyles SE, Wilson J, and Boucher RC. Normalization of raised sodium absorption and raised calcium-mediated chloride secretion by adenovirus-mediated expression of cystic fibrosis transmembrane conductance regulator in primary human cystic fibrosis airway epithelial cells. *J Clin Invest* 1995;95(3):1377–1382.
12. Johnson LG, Olsen JC, Sarkadi B, Moore KL, Swanstrom R, and Boucher RC. Efficiency of gene transfer for restoration of normal airway epithelial function in cystic fibrosis. *Nat Genet* 1992;2(1):21–25.
13. Drumm ML, Pope HA, Cliff WH, et al. Correction of the cystic fibrosis defect in vitro by retrovirus-mediated gene transfer. *Cell* 1990;62(6):1227–1233.
14. Rich DP, Anderson MP, Gregory RJ, et al. Expression of cystic fibrosis transmembrane conductance regulator corrects defective chloride channel regulation in cystic fibrosis airway epithelial cells. *Nature* 1990;347(6291):358–363.
15. Olsen JC, Johnson LG, Stutts MJ, et al. Correction of the apical membrane chloride permeability defect in polarized cystic fibrosis airway epithelia following retroviral-mediated gene transfer. *Hum Gene Ther* 1992;3(3):253–266.

16. Rosenfeld MA, Yoshimura K, Trapnell BC, et al. In vivo transfer of the human cystic fibrosis transmembrane conductance regulator gene to the airway epithelium. *Cell* 1992; 68(1):143–155.

17. Egan M, Flotte T, Afione S, et al. Defective regulation of outwardly rectifying Cl- channels by protein kinase A corrected by insertion of CFTR. *Nature* 1992;358(6387):581–584.

18. Engelhardt JF, Yankaskas JR, Ernst SA, et al. Submucosal glands are the predominant site of CFTR expression in the human bronchus. *Nat Genet* 1992;2(3):240–248.

19. Kreda SM, Mall M, Mengos A, et al. Characterization of wild-type and deltaF508 cystic fibrosis transmembrane regulator in human respiratory epithelia. *Mol Biol Cell* 2005;16(5):2154–2167.

20. Konstan MW, Davis PB, Wagener JS, et al. Compacted DNA nanoparticles administered to the nasal mucosa of cystic fibrosis subjects are safe and demonstrate partial to complete cystic fibrosis transmembrane regulator reconstitution. *Hum Gene Ther* 2004;15(12):1255–1269.

21. Moss RB, Rodman D, Spencer LT, et al. Repeated adeno-associated virus serotype 2 aerosol-mediated cystic fibrosis transmembrane regulator gene transfer to the lungs of patients with cystic fibrosis: A multicenter, double-blind, placebo-controlled trial. *Chest* 2004;125(2):509–521.

22. Graham FL and Prevec L. Methods for construction of adenovirus vectors. *Mol Biotechnol* 1995;3(3):207–220.

23. Berkner KL. Development of adenovirus vectors for the expression of heterologous genes. *Biotechniques* 1988;6(7): 616–629.

24. Ginsberg HS, Lundholm-Beauchamp U, Horswood RL, et al. Role of early region 3 (E3) in pathogenesis of adenovirus disease. *Proc Natl Acad Sci USA* 1989;86(10):3823–3827.

25. Kovesdi I, Brough DE, Bruder JT, and Wickham TJ. Adenoviral vectors for gene transfer. *Curr Opin Biotechnol* 1997;8(5):583–589.

26. Rich DP, Couture LA, Cardoza LM, et al. Development and analysis of recombinant adenoviruses for gene therapy of cystic fibrosis. *Hum Gene Ther* 1993;4(4):461–476.

27. Engelhardt JF, Simon RH, Yang Y, et al. Adenovirus-mediated transfer of the CFTR gene to lung of nonhuman primates: Biological efficacy study. *Hum Gene Ther* 1993;4(6):759–769.

28. Engelhardt JF, Yang Y, Stratford-Perricaudet LD, et al. Direct gene transfer of human CFTR into human bronchial epithelia of xenografts with E1-deleted adenoviruses. *Nat Genet* 1993;4(1):27–34.

29. Engelhardt JF, Litzky L, and Wilson JM. Prolonged transgene expression in cotton rat lung with recombinant adenoviruses defective in E2a. *Hum Gene Ther* 1994; 5(10):1217–1229.

30. Goldman MJ, Litzky LA, Engelhardt JF, and Wilson JM. Transfer of the CFTR gene to the lung of nonhuman primates with E1-deleted, E2a-defective recombinant adenoviruses: A preclinical toxicology study. *Hum Gene Ther* 1995;6(7):839–851.

31. Yang Y, Nunes FA, Berencsi K, Gonczol E, Engelhardt JF, and Wilson JM. Inactivation of E2a in recombinant adenoviruses improves the prospect for gene therapy in cystic fibrosis. *Nat Genet* 1994;7(3):362–369.

32. Wang Q, Jia XC, and Finer MH. A packaging cell line for propagation of recombinant adenovirus vectors containing two lethal gene-region deletions. *Gene Ther* 1995;2(10): 775–783.

33. Kaplan JM, Armentano D, Sparer TE, et al. Characterization of factors involved in modulating persistence of transgene

34. Armentano D, Sookdeo CC, Hehir KM, et al. Characterization of an adenovirus gene transfer vector containing an E4 deletion. *Hum Gene Ther* 1995;6(10):1343–1353.

35. Parks RJ, Chen L, Anton M, Sankar U, Rudnicki MA, and Graham FL. A helper-dependent adenovirus vector system: Removal of helper virus by Cre-mediated excision of the viral packaging signal. *Proc Natl Acad Sci USA* 1996;93(24):13565–13570.

36. Fisher KJ, Choi H, Burda J, Chen SJ, and Wilson JM. Recombinant adenovirus deleted of all viral genes for gene therapy of cystic fibrosis. *Virology* 1996;217(1):11–22.

37. Hardy S, Kitamura M, Harris-Stansil T, Dai Y, and Phipps ML. Construction of adenovirus vectors through Cre-lox recombination. *J Virol* 1997;71(3):1842–1849.

38. Hartigan-O'Connor D, Amalfitano A, and Chamberlain JS. Improved production of gutted adenovirus in cells expressing adenovirus preterminal protein and DNA polymerase. *J Virol* 1999;73(9):7835–7841.

39. Kochanek S, Clemens PR, Mitani K, Chen HH, Chan S, and Caskey CT. A new adenoviral vector: Replacement of all viral coding sequences with 28 kb of DNA independently expressing both full-length dystrophin and beta-galactosidase. *Proc Natl Acad Sci USA* 1996;93(12):5731–5736.

40. Lieber A, He CY, Kirillova I, and Kay MA. Recombinant adenoviruses with large deletions generated by Cre-mediated excision exhibit different biological properties compared with first-generation vectors in vitro and in vivo. *J Virol* 1996;70(12):8944–8960.

41. Morsy MA, Gu M, Motzel S, et al. An adenoviral vector deleted for all viral coding sequences results in enhanced safety and extended expression of a leptin transgene. *Proc Natl Acad Sci USA* 1998;95(14):7866–7871.

42. Ng P, Parks RJ, and Graham FL. Preparation of helper-dependent adenoviral vectors. *Methods Mol Med* 2002;69: 371–388.

43. Ng P, Evelegh C, Cummings D, and Graham FL. Cre levels limit packaging signal excision efficiency in the Cre/loxP helper-dependent adenoviral vector system. *J Virol* 2002;76(9):4181–4189.

44. Ng P, Parks RJ, Cummings DT, Evelegh CM, Sankar U, and Graham FL. A high-efficiency Cre/loxP-based system for construction of adenoviral vectors. *Hum Gene Ther* 1999;10(16):2667–2672.

45. Bergelson JM, Cunningham JA, Droguett G, et al. Isolation of a common receptor for Coxsackie B viruses and adenoviruses 2 and 5. *Science* 1997;275(5304):1320–1323.

46. Wickham TJ, Mathias P, Cheresh DA, and Nemerow GR. Integrins alpha v beta 3 and alpha v beta 5 promote adenovirus internalization but not virus attachment. *Cell* 1993;73(2): 309–319.

47. Flotte TR, Solow R, Owens RA, Afione S, Zeitlin PL, and Carter BJ. Gene expression from adeno-associated virus vectors in airway epithelial cells. *Am J Respir Cell Mol Biol* 1992;7(3):349–356.

48. Flotte TR, Afione SA, Conrad C, et al. Stable in vivo expression of the cystic fibrosis transmembrane conductance regulator with an adeno-associated virus vector. *Proc Natl Acad Sci USA* 1993;90(22):10613–10617.

49. Flotte TR, Afione SA, Solow R, et al. Expression of the cystic fibrosis transmembrane conductance regulator from a novel adeno-associated virus promoter. *J Biol Chem* 1993;268(5):3781–3790.

expression from recombinant adenovirus in the mouse lung. *Hum Gene Ther* 1997;8(1):45–56.

50. Flotte TR. Prospects for virus-based gene therapy for cystic fibrosis. *J Bioenerg Biomembr* 1993;25(1):37–42.

51. Flotte TR and Carter BJ. Adeno-associated virus vectors for gene therapy. *Gene Ther* 1995;2(6):357–362.

52. Zhang L, Wang D, Fischer H, et al. Efficient expression of CFTR function with adeno-associated virus vectors that carry shortened CFTR genes. *Proc Natl Acad Sci USA* 1998;95(17):10158–10163.

53. Qing K, Mah C, Hansen J, Zhou S, Dwarki V, and Srivastava A. Human fibroblast growth factor receptor 1 is a co-receptor for infection by adeno-associated virus 2. *Nat Med* 1999;5(1):71–77.

54. Summerford C, Bartlett JS, and Samulski RJ. AlphaVbeta5 integrin: A co-receptor for adeno-associated virus type 2 infection. *Nat Med* 1999;5(1):78–82.

55. Summerford C and Samulski RJ. Membrane-associated heparan sulfate proteoglycan is a receptor for adeno-associated virus type 2 virions. *J Virol* 1998;72(2):1438–1445.

56. Duan D, Yue Y, Yan Z, McCray PB, Jr., and Engelhardt JF. Polarity influences the efficiency of recombinant adenoassociated virus infection in differentiated airway epithelia. *Hum Gene Ther* 1998;9(18):2761–2776.

57. Miller AD and Rosman GJ. Improved retroviral vectors for gene transfer and expression. *Biotechniques* 1989;7(9):980–982, 984–986, 989–990.

58. Naldini L. Lentiviruses as gene transfer agents for delivery to non-dividing cells. *Curr Opin Biotechnol* 1998;9(5):457–463.

59. Naldini L, Blomer U, Gage FH, Trono D, and Verma IM. Efficient transfer, integration, and sustained long-term expression of the transgene in adult rat brains injected with a lentiviral vector. *Proc Natl Acad Sci USA* 1996;93(21):11382–11388.

60. Naldini L, Blomer U, Gallay P, et al. In vivo gene delivery and stable transduction of nondividing cells by a lentiviral vector. *Science* 1996;272(5259):263–267.

61. Zufferey R, Nagy D, Mandel RJ, Naldini L, and Trono D. Multiply attenuated lentiviral vector achieves efficient gene delivery in vivo. *Nat Biotechnol* 1997;15(9):871–875.

62. Delviks KA, Hu WS, and Pathak VK. Psi-vectors: Murine leukemia virus-based self-inactivating and self-activating retroviral vectors. *J Virol* 1997;71(8):6218–6224.

63. Zufferey R, Dull T, Mandel RJ, et al. Self-inactivating lentivirus vector for safe and efficient in vivo gene delivery. *J Virol* 1998;72(12):9873–9880.

64. Goldman MJ, Lee PS, Yang JS, and Wilson JM. Lentiviral vectors for gene therapy of cystic fibrosis. *Hum Gene Ther* 1997;8(18):2261–2268.

65. Kushwah R, Cao H, and Hu J. Potential of helper-dependent adenoviral vectors in modulating airway innate immunity. *Cell Mol Immunol* 2007;4(2):81–89.

66. Zabner J, Cheng SH, Meeker D, et al. Comparison of DNA–lipid complexes and DNA alone for gene transfer to cystic fibrosis airway epithelia in vivo. *J Clin Invest* 1997;100(6):1529–1537.

67. Ledley FD. Nonviral gene therapy: The promise of genes as pharmaceutical products. *Hum Gene Ther* 1995;6(9):1129–1144.

68. Scherman D, Bessodes M, Cameron B, et al. Application of lipids and plasmid design for gene delivery to mammalian cells. *Curr Opin Biotechnol* 1998;9(5):480–485.

69. Legendre JY and Szoka FC, Jr. Delivery of plasmid DNA into mammalian cell lines using pH-sensitive liposomes: Comparison with cationic liposomes. *Pharm Res* 1992;9(10):1235–1242.

70. Zhou X and Huang L. DNA transfection mediated by cationic liposomes containing lipopolylysine: Characterization and mechanism of action. *Biochim Biophys Acta* 1994;1189(2):195–203.

71. Yew NS, Wang KX, Przybylska M, et al. Contribution of plasmid DNA to inflammation in the lung after administration of cationic lipid:pDNA complexes. *Hum Gene Ther* 1999;10(2):223–234.

72. Boussif O, Lezoualc'h F, Zanta MA, et al. A versatile vector for gene and oligonucleotide transfer into cells in culture and in vivo: Polyethylenimine. *Proc Natl Acad Sci USA* 1995;92(16):7297–72301.

73. Zou SM, Erbacher P, Remy JS, and Behr JP. Systemic linear polyethylenimine (L-PEI)-mediated gene delivery in the mouse. *J Gene Med* 2000;2(2):128–134.

74. Ferkol T, Kaetzel CS, and Davis PB. Gene transfer into respiratory epithelial cells by targeting the polymeric immunoglobulin receptor. *J Clin Invest* 1993;92(5):2394–2400.

75. Ferkol T, Perales JC, Eckman E, Kaetzel CS, Hanson RW, and Davis PB. Gene transfer into the airway epithelium of animals by targeting the polymeric immunoglobulin receptor. *J Clin Invest* 1995;95(2):493–502.

76. Stern M, Caplen NJ, Browning JE, et al. The effect of mucolytic agents on gene transfer across a CF sputum barrier in vitro. *Gene Ther* 1998;5(1):91–98.

77. Perricone MA, Rees DD, Sacks CR, Smith KA, Kaplan JM, and St George JA. Inhibitory effect of cystic fibrosis sputum on adenovirus-mediated gene transfer in cultured epithelial cells. *Hum Gene Ther* 2000;11(14):1997–2008.

78. Virella-Lowell I, Poirier A, Chesnut KA, Brantly M, and Flotte TR. Inhibition of recombinant adeno-associated virus (rAAV) transduction by bronchial secretions from cystic fibrosis patients. *Gene Ther* 2000;7(20):1783–1789.

79. van Heeckeren A, Ferkol T, and Tosi M. Effects of bronchopulmonary inflammation induced by pseudomonas aeruginosa on adenovirus-mediated gene transfer to airway epithelial cells in mice. *Gene Ther* 1998;5(3):345–351.

80. Parsons DW, Grubb BR, Johnson LG, and Boucher RC. Enhanced in vivo airway gene transfer via transient modification of host barrier properties with a surface-active agent. *Hum Gene Ther* 1998;9(18):2661–2672.

81. Worgall S, Leopold PL, Wolff G, Ferris B, Van Roijen N, and Crystal RG. Role of alveolar macrophages in rapid elimination of adenovirus vectors administered to the epithelial surface of the respiratory tract. *Hum Gene Ther* 1997;8(14):1675–1684.

82. McCray PB, Jr., Wang G, Kline JN, et al. Alveolar macrophages inhibit retrovirus-mediated gene transfer to airway epithelia. *Hum Gene Ther* 1997;8(9):1087–1093.

83. Bosch A, McCray PB, Jr., Chang SM, et al. Proliferation induced by keratinocyte growth factor enhances in vivo retroviral-mediated gene transfer to mouse hepatocytes. *J Clin Invest* 1996;98(12):2683–2687.

84. Bernacki SH, Nelson AL, Abdullah L, et al. Mucin gene expression during differentiation of human airway epithelia in vitro. Muc4 and muc5b are strongly induced. *Am J Respir Cell Mol Biol* 1999;20(4):595–604.

85. Livraghi A and Randell SH. Cystic fibrosis and other respiratory diseases of impaired mucus clearance. *Toxicol Pathol* 2007;35(1):116–129.

86. Meerzaman D, Shapiro PS, and Kim KC. Involvement of the MAP kinase ERK2 in MUC1 mucin signaling. *Am J Physiol Lung Cell Mol Physiol* 2001;281(1):L86–L91.

87. Arcasoy SM, Latoche JD, Gondor M, Pitt BR, and Pilewski JM. Polycations increase the efficiency of adenovirus-mediated gene transfer to epithelial and endothelial cells in vitro. *Gene Ther* 1997;4(1):32–38.

88. Arcasoy SM, Latoche J, Gondor M, et al. The effects of sialo-glycoconjugates on adenovirus-mediated gene transfer to epithelial cells in vitro and in human airway xenografts. *Chest* 1997;111(6 Suppl):142S–143S.

89. Arcasoy SM, Latoche J, Gondor M, et al. MUC1 and other sialoglycoconjugates inhibit adenovirus-mediated gene transfer to epithelial cells. *Am J Respir Cell Mol Biol* 1997;17(4): 422–435.

90. Pickles RJ, Fahrner JA, Petrella JM, Boucher RC, and Bergelson JM. Retargeting the coxsackievirus and adenovirus receptor to the apical surface of polarized epithelial cells reveals the glyco-calyx as a barrier to adenovirus-mediated gene transfer. *J Virol* 2000;74(13):6050–6057.

91. Walters RW, van't Hof W, Yi SM, et al. Apical localization of the coxsackie-adenovirus receptor by glycosyl-phosphatidylinositol modification is sufficient for adenovirus-mediated gene transfer through the apical surface of human airway epithelia. *J Virol* 2001;75(16):7703–7711.

92. Stonebraker JR, Wagner D, Lefensty RW, et al. Glyco-calyx restricts adenoviral vector access to apical receptors expressed on respiratory epithelium in vitro and in vivo: Role for tethered mucins as barriers to lumenal infection. *J Virol* 2004;78(24):13755–13768.

93. Grubb BR, Pickles RJ, Ye H, et al. Inefficient gene transfer by adenovirus vector to cystic fibrosis airway epithelia of mice and humans. *Nature* 1994;371(6500):802–806.

94. Pickles RJ, Barker PM, Ye H, and Boucher RC. Efficient adenovirus-mediated gene transfer to basal but not colum-nar cells of cartilaginous airway epithelia. *Hum Gene Ther* 1996;7(8):921–931.

95. Dupuit F, Bout A, Hinnrasky J, et al. Expression and localiza-tion of CFTR in the rhesus monkey surface airway epithe-lium. *Gene Ther* 1995;2(2):156–163.

96. Zabner J, Freimuth P, Puga A, Fabrega A, and Welsh MJ. Lack of high affinity fiber receptor activity explains the resistance of ciliated airway epithelia to adenovirus infection. *J Clin Invest* 1997;100(5):1144–1149.

97. Pickles RJ, McCarty D, Matsui H, Hart PJ, Randell SH, and Boucher RC. Limited entry of adenovirus vectors into well-differentiated airway epithelium is responsible for inefficient gene transfer. *J Virol* 1998;72(7):6014–6023.

98. Koehler DR, Frndova H, Leung K, et al. Aerosol delivery of an enhanced helper-dependent adenovirus formulation to rabbit lung using an intratracheal catheter. *J Gene Med* 2005;7(11):1409–1420.

99. Duan D, Yue Y, Yan Z, Yang J, and Engelhardt JF. Endosomal processing limits gene transfer to polarized airway epithelia by adeno-associated virus. *J Clin Invest* 2000;105(11):1573–1587.

100. Halbert CL, Alexander IE, Wolgamot GM, and Miller AD. Adeno-associated virus vectors transduce primary cells much less efficiently than immortalized cells. *J Virol* 1995;69(3): 1473–1479.

101. Ferrari FK, Samulski T, Shenk T, and Samulski RJ. Second-strand synthesis is a rate-limiting step for efficient transduc-tion by recombinant adeno-associated virus vectors. *J Virol* 1996;70(5):3227–3234.

102. Alexander IE, Russell DW, and Miller AD. DNA-damaging agents greatly increase the transduction of nondividing cells by adeno-associated virus vectors. *J Virol* 1994;68(12):8282–8287.

103. Russell DW, Alexander IE, and Miller AD. DNA synthe-sis and topoisomerase inhibitors increase transduction by adeno-associated virus vectors. *Proc Natl Acad Sci USA* 1995;92(12):5719–5723.

104. Rabinowitz JE and Samulski J. Adeno-associated virus expression systems for gene transfer. *Curr Opin Biotechnol* 1998;9(5):470–475.

105. Olsen JC. Gene transfer vectors derived from equine infec-tious anemia virus. *Gene Ther* 1998;5(11):1481–1487.

106. Wang G, Slepushkin V, Zabner J, et al. Feline immunodefi-ciency virus vectors persistently transduce nondividing air-way epithelia and correct the cystic fibrosis defect. *J Clin Invest* 1999;104(11):R55–R62.

107. Leigh MW, Kylander JE, Yankaskas JR, and Boucher RC. Cell proliferation in bronchial epithelium and submucosal glands of cystic fibrosis patients. *Am J Respir Cell Mol Biol* 1995;12(6):605–612.

108. Burns JC, Friedmann T, Driever W, Burrascano M, and Yee JK. Vesicular stomatitis virus G glycoprotein pseudotyped retroviral vectors: Concentration to very high titer and efficient gene transfer into mammalian and nonmammalian cells. *Proc Natl Acad Sci USA* 1993;90(17):8033–8037.

109. Zsengeller ZK, Halbert C, Miller AD, Wert SE, Whitsett JA, and Bachurski CJ. Keratinocyte growth factor stimulates transduction of the respiratory epithelium by retroviral vec-tors. *Hum Gene Ther* 1999;10(3):341–353.

110. Johnson LG, Mewshaw JP, Ni H, Friedmann T, Boucher RC, and Olsen JC. Effect of host modification and age on airway epithelial gene transfer mediated by a murine leukemia virus-derived vector. *J Virol* 1998;72(11):8861–8872.

111. Johnson LG, Olsen JC, Naldini L, and Boucher RC. Pseudo-typed human lentiviral vector-mediated gene transfer to air-way epithelia in vivo. *Gene Ther* 2000;7(7):568–574.

112. Wang G, Davidson BL, Melchert P, et al. Influence of cell polarity on retrovirus-mediated gene transfer to differentiated human airway epithelia. *J Virol* 1998;72(12):9818–9826.

113. Zabner J, Fasbender AJ, Moninger T, Poellinger KA, and Welsh MJ. Cellular and molecular barriers to gene transfer by a cationic lipid. *J Biol Chem* 1995;270(32):18997–19007.

114. Fasbender A, Zabner J, Zeiher BG, and Welsh MJ. A low rate of cell proliferation and reduced DNA uptake limit cationic lipid-mediated gene transfer to primary cultures of ciliated human airway epithelia. *Gene Ther* 1997;4(11):1173–1180.

115. Matsui H, Johnson LG, Randell SH, and Boucher RC. Loss of binding and entry of liposome–DNA complexes decreases transfection efficiency in differentiated airway epithelial cells. *J Biol Chem* 1997;272(2):1117–1126.

116. Wang G, Zabner J, Deering C, et al. Increasing epithelial junc-tion permeability enhances gene transfer to airway epithelia in vivo. *Am J Respir Cell Mol Biol* 2000;22(2):129–138.

117. Chu Q, St George JA, Lukason M, Cheng SH, Scheule RK, and Eastman SJ. EGTA enhancement of adenovirus-mediated gene transfer to mouse tracheal epithelium in vivo. *Hum Gene Ther* 2001;12(5):455–467.

118. Coyne CB, Kelly MM, Boucher RC, and Johnson LG. Enhanced epithelial gene transfer by modulation of tight junctions with sodium caprate. *Am J Respir Cell Mol Biol* 2000;23(5):602–609.

119. Coyne CB, Ribeiro CM, Boucher RC, and Johnson LG. Acute mechanism of medium chain fatty acid-induced enhancement of airway epithelial permeability. *J Pharmacol Exp Ther* 2003;305(2):440–450.

120. Johnson LG, Vanhook MK, Coyne CB, Haykal-Coates N, and Gavett SH. Safety and efficiency of modulating paracellular permeability to enhance airway epithelial gene transfer in vivo. *Hum Gene Ther* 2003;14(8):729–747.

121. Gregory LG, Harbottle RP, Lawrence L, Knapton HJ, Themis M, and Coutelle C. Enhancement of adenovirus-mediated gene transfer to the airways by DEAE dextran and sodium caprate in vivo. *Mol Ther* 2003;7(1):19–26.

122. Rudolph C, Schillinger U, Ortiz A, et al. Aerosolized nanogram quantities of plasmid DNA mediate highly efficient gene delivery to mouse airway epithelium. *Mol Ther* 2005;12(3):493–501.

123. Limberis M, Anson DS, Fuller M, and Parsons DW. Recovery of airway cystic fibrosis transmembrane conductance regulator function in mice with cystic fibrosis after single-dose lentivirus-mediated gene transfer. *Hum Gene Ther* 2002;13(16):1961–1970.

124. Everett RS, Vanhook MK, Barozzi N, Toth I, and Johnson LG. Specific modulation of airway epithelial tight junctions by apical application of an occludin peptide. *Mol Pharmacol* 2006;69(2):492–500.

125. Duan D, Sharma P, Yang J, et al. Circular intermediates of recombinant adeno-associated virus have defined structural characteristics responsible for long-term episomal persistence in muscle tissue. *J Virol* 1998;72(11):8568–8577.

126. Ding W, Yan Z, Zak R, Saavedra M, Rodman DM, and Engelhardt JF. Second-strand genome conversion of adeno-associated virus type 2 (AAV-2) and AAV-5 is not rate limiting following apical infection of polarized human airway epithelia. *J Virol* 2003;77(13):7361–7366.

127. Yan Z, Zak R, Luxton GW, Ritchie TC, Bantel-Schaal U, and Engelhardt JF. Ubiquitination of both adeno-associated virus type 2 and 5 capsid proteins affects the transduction efficiency of recombinant vectors. *J Virol* 2002;76(5):2043–2053.

128. Yan Z, Zak R, Zhang Y, et al. Distinct classes of proteasome-modulating agents cooperatively augment recombinant adeno-associated virus type 2 and type 5-mediated transduction from the apical surfaces of human airway epithelia. *J Virol* 2004;78(6):2863–2874.

129. Seiler MP, Luner P, Moninger TO, Karp PH, Keshavjee S, and Zabner J. Thixotropic solutions enhance viral-mediated gene transfer to airway epithelia. *Am J Respir Cell Mol Biol* 2002;27(2):133–140.

130. Sinn PL, Shah AJ, Donovan MD, and McCray PB, Jr. Viscoelastic gel formulations enhance airway epithelial gene transfer with viral vectors. *Am J Respir Cell Mol Biol* 2005;32(5):404–410.

131. Drapkin PT, O'Riordan CR, Yi SM, et al. Targeting the urokinase plasminogen activator receptor enhances gene transfer to human airway epithelia. *J Clin Invest* 2000;105(5):589–596.

132. Kreda SM, Pickles RJ, Lazarowski ER, and Boucher RC. G-protein-coupled receptors as targets for gene transfer vectors using natural small-molecule ligands. *Nat Biotechnol* 2000;18(6):635–640.

133. Koehler DR, Martin B, Corey M, et al. Readministration of helper-dependent adenovirus to mouse lung. *Gene Ther* 2006;13(9):773–780.

134. Zabner J, Chillon M, Grunst T, et al. A chimeric type 2 adenovirus vector with a type 17 fiber enhances gene transfer to human airway epithelia. *J Virol* 1999;73(10):8689–8695.

135. Parks R, Evelegh C, and Graham F. Use of helper-dependent adenoviral vectors of alternative serotypes permits repeat vector administration. *Gene Ther* 1999;6(9):1565–1573.

136. Morral N, O'Neal W, Rice K, et al. Administration of helper-dependent adenoviral vectors and sequential delivery of different vector serotype for long-term liver-directed gene transfer in baboons. *Proc Natl Acad Sci USA* 1999;96(22):12816–12821.

137. Hashimoto M, Boyer JL, Hackett NR, Wilson JM, and Crystal RG. Induction of protective immunity to anthrax lethal toxin with a nonhuman primate adenovirus-based vaccine in the presence of preexisting anti-human adenovirus immunity. *Infect Immun* 2005;73(10):6885–6891.

138. Yan Z, Zhang Y, Duan D, and Engelhardt JF. Trans-splicing vectors expand the utility of adeno-associated virus for gene therapy. *Proc Natl Acad Sci USA* 2000;97(12):6716–6721.

139. Liu X, Luo M, Zhang LN, et al. Spliceosome-mediated RNA trans-splicing with recombinant adeno-associated virus partially restores cystic fibrosis transmembrane conductance regulator function to polarized human cystic fibrosis airway epithelial cells. *Hum Gene Ther* 2005;16(9):1116–1123.

140. Fischer AC, Smith CI, Cebotaru L, et al. Expression of a truncated cystic fibrosis transmembrane conductance regulator with an AAV5-pseudotyped vector in primates. *Mol Ther* 2007;15(4):756–763.

141. Halbert CL, Lam SL, and Miller AD. High-efficiency promoter-dependent transduction by adeno-associated virus type 6 vectors in mouse lung. *Hum Gene Ther* 2007;18(4):344–354.

142. Gao G, Vandenberghe LH, Alvira MR, et al. Clades of Adeno-associated viruses are widely disseminated in human tissues. *J Virol* 2004;78(12):6381–6388.

143. Wu Z, Asokan A, and Samulski RJ. Adeno-associated virus serotypes: Vector toolkit for human gene therapy. *Mol Ther* 2006;14(3):316–327.

144. Zabner J, Seiler M, Walters R, et al. Adeno-associated virus type 5 (AAV5) but not AAV2 binds to the apical surfaces of airway epithelia and facilitates gene transfer. *J Virol* 2000;74(8):3852–3858.

145. Auricchio A, O'Connor E, Weiner D, et al. Noninvasive gene transfer to the lung for systemic delivery of therapeutic proteins. *J Clin Invest* 2002;110(4):499–504.

146. Halbert CL, Allen JM, and Miller AD. Adeno-associated virus type 6 (AAV6) vectors mediate efficient transduction of airway epithelial cells in mouse lungs compared to that of AAV2 vectors. *J Virol* 2001;75(14):6615–6624.

147. Halbert CL and Miller AD. AAV-mediated gene transfer to mouse lungs. *Methods Mol Biol* 2004;246:201–212.

148. Limberis MP and Wilson JM. Adeno-associated virus serotype 9 vectors transduce murine alveolar and nasal epithelia and can be readministered. *Proc Natl Acad Sci USA* 2006;103(35):12993–12998.

149. Le HT, Yu QC, Wilson JM, and Croyle MA. Utility of PEGylated recombinant adeno-associated viruses for gene transfer. *J Control Release* 2005;108(1):161–177.

150. De BP, Heguy A, Hackett NR, et al. High levels of persistent expression of alpha1-antitrypsin mediated by the nonhuman primate serotype rh.10 adeno-associated virus despite preexisting immunity to common human adeno-associated viruses. *Mol Ther* 2006;13(1):67–76.

151. McCarty DM, Monahan PE, and Samulski RJ. Self-complementary recombinant adeno-associated virus (scAAV) vectors promote efficient transduction independently of DNA synthesis. *Gene Ther* 2001;8(16):1248–1254.

152. Wang Z, Ma HI, Li J, Sun L, Zhang J, and Xiao X. Rapid and highly efficient transduction by double-stranded adeno-associated virus vectors in vitro and in vivo. *Gene Ther* 2003;10(26):2105–2111.

153. Medina MF, Kobinger GP, Rux J, et al. Lentiviral vectors pseudotyped with minimal filovirus envelopes increased gene transfer in murine lung. *Mol Ther* 2003;8(5):777–789.

154. Kobinger GP, Weiner DJ, Yu QC, and Wilson JM. Filovirus-pseudotyped lentiviral vector can efficiently and stably transduce airway epithelia in vivo. *Nat Biotechnol* 2001;19(3):225–230.

155. McKay T, Patel M, Pickles RJ, Johnson LG, and Olsen JC. Influenza M2 envelope protein augments avian influenza hemagglutinin pseudotyping of lentiviral vectors. *Gene Ther* 2006;13(8):715–724.

156. Sinn PL, Burnight ER, Hickey MA, Blissard GW, and McCray PB, Jr. Persistent gene expression in mouse nasal epithelia following feline immunodeficiency virus-based vector gene transfer. *J Virol* 2005;79(20):12818–12827.

157. Yonemitsu Y, Kitson C, Ferrari S, et al. Efficient gene transfer to airway epithelium using recombinant Sendai virus. *Nat Biotechnol* 2000;18(9):970–973.

158. Li HO, Zhu YF, Asakawa M, et al. A cytoplasmic RNA vector derived from nontransmissible Sendai virus with efficient gene transfer and expression. *J Virol* 2000;74(14):6564–6569.

159. Ferrari S, Griesenbach U, Iida A, et al. Sendai virus-mediated CFTR gene transfer to the airway epithelium. *Gene Ther* 2007;14(19):1371–1379.

160. Hall CB. Respiratory syncytial virus and parainfluenza virus. *N Engl J Med* 2001;344(25):1917–1928.

161. Zhang L, Peeples ME, Boucher RC, Collins PL, and Pickles RJ. Respiratory syncytial virus infection of human airway epithelial cells is polarized, specific to ciliated cells, and without obvious cytopathology. *J Virol* 2002;76(11):5654–5666.

162. Zhang L, Bukreyev A, Thompson CI, et al. Infection of ciliated cells by human parainfluenza virus type 3 in an in vitro model of human airway epithelium. *J Virol* 2005;79(2):1113–1124.

163. McLachlan G, Baker A, Tennant P, et al. Optimizing aerosol gene delivery and expression in the ovine lung. *Mol Ther* 2007;15(2):348–354.

164. Yew NS and Cheng SH. Reducing the immunostimulatory activity of CpG-containing plasmid DNA vectors for non-viral gene therapy. *Expert Opin Drug Deliv* 2004;1(1):115–125.

165. Gautam A, Densmore CL, Golunski E, Xu B, and Waldrep JC. Transgene expression in mouse airway epithelium by aerosol gene therapy with PEI–DNA complexes. *Mol Ther* 2001;3(4):551–556.

166. Uduehi AN, Stammberger U, Kubisa B, Gugger M, Buehler TA, and Schmid RA. Effects of linear polyethylenimine and polyethylenimine/DNA on lung function after airway instillation to rat lungs. *Mol Ther* 2001;4(1):52–57.

167. Gautam A, Densmore CL, and Waldrep JC. Pulmonary cytokine responses associated with PEI–DNA aerosol gene therapy. *Gene Ther* 2001;8(3):254–257.

168. Liu G, Li D, Pasumarthy MK, et al. Nanoparticles of compacted DNA transfect postmitotic cells. *J Biol Chem* 2003;278(35):32578–32586.

169. Ziady AG, Gedeon CR, Miller T, et al. Transfection of airway epithelium by stable PEGylated poly-L-lysine DNA nanoparticles in vivo. *Mol Ther* 2003;8(6):936–947.

170. Ziady AG, Kelley TJ, Milliken E, Ferkol T, and Davis PB. Functional evidence of CFTR gene transfer in nasal epithelium of cystic fibrosis mice in vivo following luminal application of DNA complexes targeted to the serpin–enzyme complex receptor. *Mol Ther* 2002;5(4):413–419.

39 Gene Therapy for Neurological Diseases

William J. Bowers, Suresh de Silva, and Howard J. Federoff

CONTENTS

39.1 INTRODUCTION

Highly effective methodologies for therapeutic gene delivery to cells of the central nervous system (CNS) have been developed from the exponential gains achieved in the area of molecular biology over recent decades. This capability has created promise for therapeutic intervention in neurodegenerative disorders. Presently, the most widely employed treatments for neurodegenerative diseases are based upon symptom alleviation through the use of pharmacological compounds, but these approaches offer little curative, or even neuroprotective properties. An ideal therapy for a neurodegenerative disorder would stabilize and repair afflicted cells and prevent other cells from succumbing to the pathophysiological process.

TABLE 39.1

Viral Vector Platforms Employed in CNS Gene Transfer

Viral Vector	Payload Capacity	Tropism	Duration Expression	Other Properties
Adenovirus	7–8 kb (Gen[a] 1/2); 30 kb (Gutless)	Glial/neuronal	Months (Gen1/2); year (Gutless)	Episomal; immunogenic (Gen1/2); less immunogenic (Gutless)
AAV	4.5 kb	Neuronal	Years	Episomal and integrative; minimal immune response
HSV recombinant	~10 kb	Neuronal > glial	Months	Episomal; some cytotoxicity
HSV amplicon	~130 kb	Neuronal > glial	Months to a year	Episomal; all viral genes deleted; little immune response
HSV/AAV hybrid amplicon	>20 kb	Neuronal	Months to a year	Episomal and integrative; all viral genes deleted
Lentivirus	8–10 kb	Neuronal	Years	Integrative

[a] Gen1/2, Generation 1 and 2.

Using gene transfer-based methodologies, it may be possible to create such therapies. The characteristics of presently available virus-based vectors are summarized in Table 39.1. Each vector platform exhibits a series of advantages and disadvantages that need to be assessed. However, prior to their implementation in the treatment of neurological disorders, a detailed understanding of the disease-specific mechanisms is crucial. These issues will be addressed in greater detail below.

39.1.1 THERAPEUTIC TRANSCRIPTIONAL UNIT

A thorough understanding of the molecular mechanisms underlying the neurodegenerative disease process is requisite for selection of a proposed therapeutic gene. In general, the therapeutic transcriptional unit should have one or more of the following properties: (1) neuroprotective, (2) neuroregenerative, (3) neurotrophic, or (4) antiapoptotic.

Equally important is choosing promoters and regulatory elements to drive expression of the desired genes. The choice of these components will determine the specificity, duration, and regulation of transgene expression. Once the transcriptional unit has been designed, a suitable vector for delivering the DNA must be selected. The vector must be capable of delivering and expressing the desired genes in the appropriate cells to be targeted.

39.1.2 VECTOR SELECTION

Subsequent to selection of the gene therapeutic effector gene, whether it possesses a neuroaugmentative or neuroprotective mechanism of action, the appropriate gene delivery vector can be chosen. The following points must be considered for informed selection of an appropriate gene therapeutic vehicle for the treatment of neurodegenerative disorders: vector capacity, vector specificity, vector genome maintenance, vector-mediated transgene expression duration and levels, and vector safety profile. The length of the therapeutic transcription unit is sometimes used as a first criterion to limit vector choice. This category also includes a given vector's ability to harbor multiple transcription units, thereby potentially affording recapitulation of a complex biochemical pathway (i.e., dopamine [DA] biosynthesis in the case of Parkinson's disease [PD]). Potential applications for several presently available vectors are restricted by insertion size limitations and are sometimes excluded if multiple gene delivery is required for therapy. As demonstrated with adeno-associated vector (AAV), the coinjection of multiple vectors, where each expresses a different therapeutic gene, represents one approach to circumvent this issue [1].

Cell type specificity is also an important issue when developing a gene-based therapeutic intervention for neurodegenerative disorders. It would be most beneficial if the vector of choice could transduce and express in cell types that comprise the disease-affected neural pathway. Vector tropism can be regulated through modulation of cellular receptor interactions by one or more of the following approaches: alteration of virus docking proteins, utilization of alternate serotypes, pseudotyping, and introduction of cellular receptors into the viral envelope. Once a vector is optimally targeted to the brain region of interest, therapeutic transgene expression can be restricted to selected cell populations via the utilization of cell type-specific promoters or transcriptional elements [2–6]. Strict spatial control of transgene expression is important to ensure that the correct cells manufacture the gene product. This control, in turn, reduces the risk that ectopic transgene expression will occur and lead to untoward effects on adjacent neurological pathways.

Because many neurodegenerative disorders are protracted in duration, gene-based modalities will be required to impart therapeutic benefit for several decades of an individual's lifetime. To this end, a vector genome should be stably maintained within the transduced cell of the neural pathway for extended periods of time. Vector genome maintenance is, therefore, an important factor in selection of an appropriate gene therapy vehicle for such disorders. Vector genomes can exist episomally or as integrated forms within nuclei of transduced cells. The postmitotic property of CNS neuronal populations does not exclude the utilization of episomal vectors because genomes can be maintained without the fear of progressive loss due to mitosis. Integrating vectors circumvent this issue but their use augments safety concerns including potential to transactivate nearby proto-oncogenes and to disrupt essential host genes via insertional mutagenesis. This fear has apparently become a reality as evidenced in the case of the children in Europe treated for X-linked severe combined immune deficiency disease (XSCID) with a murine retroviral vector expressing the T cell growth factor, GammaC. Two of ten enrolled infants developed similar leukemia-like illnesses and the trial was halted to determine the genesis of the adverse event [7].

Another similar issue regarding vector selection relates to the desired levels and duration of gene product expression for treatment of neurodegenerative disorders. Depending upon the vector and transcriptional elements chosen, pharmacologic or physiologic levels of transgene expression can be achieved for short or long-term periods of duration. As with other selection criteria, the decision of which level or duration of expression is preferred rests heavily on the underlying molecular mechanisms to be targeted and at which time during the disease course the intervention is to be implemented. Early interventions may require maintenance of long-term physiologic levels of transgene expression (i.e., neuroprotective strategies) since the neuronal system is likely to be intact at this time. A vector–promoter combination that safely and stably maintains gene expression at nearly physiologic levels in the CNS would serve as a potential candidate for such early treatment approaches. Treatment modalities that are implemented after presentation of clinical neurodegenerative disease symptoms may require long-term pharmacologic levels of transgene product to restore neurologic function to a brain region decimated by disease.

Safety is of utmost concern regarding the application of novel gene therapeutic strategies to the treatment of neurodegenerative disorders. Many presently available vectors trigger immunogenic or inflammatory responses when introduced into the CNS. These responses are known to arise from the humoral or cell-mediated arms of the immune system, and the magnitude differs depending upon which vector type is employed. For example, repeat administration of early generation viral vectors has been shown to lead to lower transgene expression and serious inflammation, likely the result of a primed immune system [8]. Therefore, a vector that is stably maintained and that can express its encoded transgene for extended periods of time would likely prove to be a more favorable choice as a gene therapeutic vehicle for neurodegenerative disorders. Another aspect that is often overlooked regarding gene therapy safety is the role of transgene products in the elaboration of immune responses and toxicity. Transgene products that are of foreign origin, ectopically expressed, or pharmacologically expressed possess the potential to induce cytotoxicity or immune responses. Research addressing these issues is imperative to elucidate the role of transgene products in the elaboration of these potentially harmful responses, and how such responses can be successfully circumvented. Utilization of regulation-competent transcriptional or post-transcriptional elements in delivery vectors to provide "fine-tuning" of therapeutic transgene expression levels is a way to minimize harmful clinical outcomes.

39.1.3 VECTOR DELIVERY

Once a suitable vector has been chosen for treating a particular neurological disease, a safe means for delivery must be established to provide optimal therapeutic benefit. With recent refinements of stereotactic surgical procedures, highly precise and reproducible delivery to specific regions of the brain is now performed (reviewed in Ref. [9]). By placing the patient's head within a rigid frame and using three-dimensional cartesian reference points, delivery of vectors or cells can be made to a defined space within the coordinate system. Gene therapeutic vectors can be introduced into the brain via direct or indirect means. Direct gene transfer involves either local or global delivery of a selected vector to the brain. Direct local delivery using stereotactic methods is highly suitable for treatment of certain neurological diseases due to the fairly circumscribed region that may be afflicted. Figure 39.1 illustrates the finely tuned and restricted delivery of a herpes simplex virus

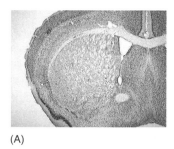 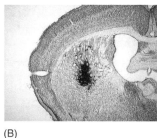 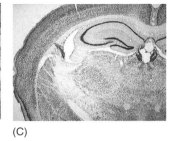

(A)   (B)   (C)

FIGURE 39.1 (See color insert following blank page 206.) Stereotactic delivery of an HSV amplicon vector expressing β-galactosidase into the mouse striatum. Mice were injected with 1×10^5 transduction units of HSVlac using a microprocessor-controlled pump. Animals were sacrificed and perfused 4 days post-transduction and X-gal histochemistry was performed on 40 μm sections. Sections representative of the injection site (Panel A), a site anterior of the injection (Panel B), and a site posterior of the injection (Panel C). All sections were counterstained with thionin and acquired at a magnification of 2.5×. The photomicrographs indicate that focal delivery of a viral vector can be achieved within the brain. (From Brooks, A.I., et al., *J. Neurosci. Methods*, 80(2), 137, 1998. © 1998 Elsevier Science B.V. With permission.)

(HSV) amplicon vector expressing the β-galactosidase gene within the striatum of a mouse [10].

Other investigative groups have utilized more global approaches to CNS gene delivery that have broader applications for many neurological diseases where the volume of affected brain region is beyond the feasibility of conventional stereotactic methods. Convection-enhanced delivery (CED) has been developed to distribute homogeneous tissue concentrations of vectors over a large region of rodent or primate brain [11–14]. This method has been utilized extensively in the development of gene therapeutic strategies for neurodegenerative disorders, but further research is required to determine if global distribution of vectors produces deleterious effects on neural pathway functioning. For example, global expression of vector-derived neurotrophic factors may induce uncontrolled neuritic sprouting and subsequently, altered neuronal activity and physiology.

Indirect gene transfer, or *ex vivo* therapy, utilizes transplantation of genetically altered (vector-transduced) or unaltered cells capable of restoring functionality to a diseased region. Although *ex vivo* therapy includes the use of neural stem cells for repopulation of the denervated dopaminergic system and restoration of pathway function, this type of therapy will not be discussed in this chapter. Vectors used to genetically modify transplanted cells typically express either secreted trophic factors for neuroprotection/neuroaugmentation of surrounding host tissues or cell-intrinsic survival factors for protection of grafted cells from the stress of transplantation. Vectors delivered via *ex vivo* therapy must be able to integrate into the host cell genome as transplanted cell populations are expanded prior to grafting. Utilization of gene transfer technologies may allow for the expansion of the graft cell population or

increased graft survival via expression of a growth or survival factor thereby minimizing the amount of starting fetal tissue [15,16].

39.2 NEURODEGENERATIVE DISEASES AMENABLE TO GENE THERAPY

The following discussion will illustrate the most common forms of neurodegenerative diseases (summarized in Table 39.2) and potential gene therapy approaches that have been employed based on current understandings of disease mechanisms.

39.2.1 PARKINSON'S DISEASE

PD is the second most common chronic neurodegenerative disease in humans. The incidence of PD was estimated in 1995 to be between 1:100 and 1:500 individuals [17,18]. This incidence translates to approximately 1%–2% of the population over the age of 65 [19–21]. The disorder, initially described in six patients by Dr. James Parkinson, is typified clinically by symptoms including bradykinesia, resting tremor, rigidity, and gait abnormalities followed by postural instability, dementia, and autonomic dysfunction. Pathologically, PD patients experience specific degeneration of dopaminergic neurons in the substantia nigra pars compacta (SNpc) as well as dopaminergic ventral tegmental area neurons and noradrenergic neurons of the locus coeruleus. Furthermore, neuronal loss has been reported in other brain areas such as the cerebral cortex, anterior thalamus, hypothalamus, amygdala, and basal forebrain. In addition to neuronal loss, accumulation of proteinaceous cytoplasmic inclusions called Lewy bodies is a neuropathologic hallmark of PD. The exact role of Lewy

TABLE 39.2
Neurological Disorders Amenable to Gene Therapy

Disease	Age Onset	Disease Mechanism	Basis
Parkinson's	50–70s	Loss pigmented DA neurons in midbrain	*Genetic* (rare): α-Synuclein, C-terminal ubiquitin hydrolase; parkin
			Environmental: Pesticides, fungicides, neurotoxicants
Alzheimer's	50–80s	Aβ accumulation; tau hyperphosphorylation; synapse loss	*Genetic* (infrequent): APP, PS (susceptibility): ApoE4
			Environmental: Viral infection/Diet
Lysosomal storage	Infancy	Enzyme deficiency leading to lysosomal protein accumulation and eventual peripheral and CNS degeneration	*Genetic*: Lysosomal enzyme/transport gene deletion/mutation
Huntington's	20–50s	Loss of striatal medium spiny neurons	*Genetic*: Polyglutamine expansion in Huntingtin (Htt) gene locus
Stroke	Varies	Hypoxic insult leads to necrotic and apoptotic cell death	Potential genetic and environmental susceptibility interaction
Epilepsy	Childhood	Aberrant hypersynchronization of local or global neuronal networks	Unknown; potential early lifetime subclinical event
Motor neuron disorders	40–60s	Loss of spinal motor neurons	*Genetic*: missense mutations (ALS); gene deletions/aberrant splicing (SMA)

bodies in PD is unclear, but other neurodegenerative disorders also exhibit intracellular and extracellular protein aggregates. Understanding the molecular mechanisms underlying protein aggregates in neurodegenerative diseases may assist to illuminate a common target for gene therapy.

Although more than 180 years have passed since Parkinson's first description, the disease etiology is still largely unknown. Since the evolution of successful gene therapeutic strategies will rely heavily on a detailed understanding of the molecular and cellular processes governing the clinical presentations of PD, exploring the pathophysiology of PD is crucial. The following section will summarize possible mechanisms of PD and how methods in gene delivery and expression might be applied to interdict the pathogenic pathways of this neurodegenerative disorder.

39.2.1.1 Mechanisms of Disease

PD exists as both a sporadic and familial disorder. Although the exact etiology of PD is unknown, the common pathway of both sporadic and familial PD is a loss of DA neurons. Importantly, this decline in DA neuron number below a critical threshold produces early symptomatic PD (reviewed in Refs. [22–24]). Given that environmental factors such as pesticides, herbicides, and industrial chemicals have been identified as potential risk factors for PD and genetic mutations have also been identified, it is likely that either alone or in combination, these triggers will produce a clinical syndrome similar to PD (reviewed in Ref. [24]). Figure 39.2 outlines a common pathway for PD and suggests several targets for gene therapy without necessarily interdicting the

initiating mechanism [367]. In this "common pathway" model for PD, multiple triggering mechanisms such as genetic, toxicant, and environmental trigger plus genetic vulnerability converge on a shared common pathway to cell death. The first step along this pathway may encompass presynaptic injury and dysfunction followed by cellular compensation and metabolic stress. This preclinical injury and damage would result in DA deficiency and cell death. Gene therapy treatment opportunities would include (1) targeting specific triggering mechanisms, (2) targeting shared early pathways prior to presynaptic dopaminergic dysfunction, (3) targeting shared later pathways when dysfunction occurs, and (4) restoring DA biosynthesis in the denervated striatum. As stated above, PD is likely the result of a combination of environmental, toxin, and genetic triggering factors. Because of this, it is difficult to review the impact of each individually but we will attempt to briefly outline the potential impact of these factors on the etiology of PD and summarize their convergence on the pathophysiology of this disease.

39.2.1.2 Environmental Factors

Several findings support an etiologic role for exogenous factors in PD. The earliest observation was that a synthetic by-product of meperidine production, 1,2,3,6-methylphenyl-tetrahydropyridine (MPTP), produced a syndrome similar to PD. MPTP treatment of mice and monkeys has become a common method to achieve dopaminergic neuronal loss and an animal "model" of PD. The toxic compound MPTP is converted in glia to the pyridinium ion (MPP+) by monoaminoxidase type B (MAO-B) and subsequently taken up by DA

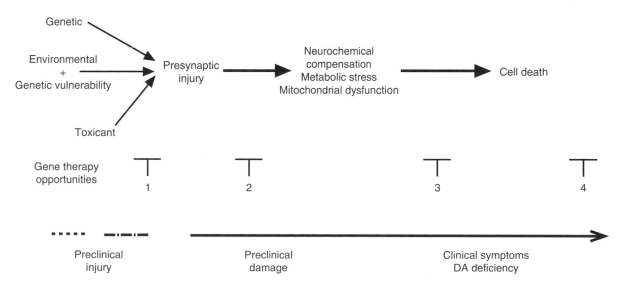

1. Targeting specific triggering mechanisms
2. Targeting shared early pathway step prior to presynaptic DA dysfunction
3. Targeting shared later pathway step when dopaminergic neuron dysfunction occurs
4. Restoring DA biosysnthesis in denervated striatum

FIGURE 39.2 Schematic depicting the common pathway model for PD and points along this pathway where gene-based therapy may be applied. (From Bowers, W.J., et al., *Clin. Neurosci. Res.*, 1, 483, 2001. © 2001 Elsevier Science B.V. With permission.)

neurons via the DA transporter. MPP+ is then actively transported to the mitochondria where it inhibits complex I, interfering with mitochondrial respiration and resulting in increased production of the superoxide anion. [25].

Recent reports lend biological plausibility to pesticide exposure as both an alternative model of nigrostriatal dopaminergic degeneration and a risk factor for PD [26–28]. Greenamyre's group reported that chronic, systemic treatment of rats with the lipophilic pesticide, rotenone, resulted in highly selective nigrostriatal dopaminergic degeneration [27,29]. This degeneration is associated with PD-like behavioral changes including hypokinesia and rigidity. Neuropathologically, these animals accumulate fibrillar cytoplasmic inclusions that contain both ubiquitin and α-synuclein. This treatment represents an important alternate PD model since these inclusions resemble Lewy bodies, a hallmark of PD. Of equal importance is the suggestion from this model that mitochondrial dysfunction in dopaminergic neurons, in particular complex I inhibition, plays an important role in the pathophysiology of Lewy body formation and some cases of idiopathic PD. Other groups have implicated a combination of herbicide and fungicide exposure as a potent risk factor for PD [26,28]. Mice treated with a combination of paraquat and maneb demonstrated a sustained decrease in motor activity and reduced tyrosine hydroxylase immununoreactivity in the dorsal striatum. These effects were greater in combination than either paraquat or maneb alone, suggesting that multiple compound exposure alone or in concert with genetic vulnerability may trigger a cascade of events leading to a PD-like syndrome.

39.2.1.3 Genetic Factors

Several genetic factors have recently been identified to play a central role in the etiology of familial PD (summarized in Table 39.3). In 1996, genetic linkage strategies were applied to a large Italian family with early onset PD [17,30]. The susceptibility gene was identified within the *PARK1* locus on the

long arm of chromosome 4q21-q23, and positional cloning identified a missense mutation at position 53 (A53T) in the *SNCA* gene, which encodes for the α-synuclein protein [31]. Another mutation in α-synuclein, A30P, was identified in a German family [32]. Recently, a third missense mutation, E46K, was identified in a Spanish family with an autosomal dominant form of PD [33]. As of yet, the physiological role of α-synuclein has been elusive, but the A53T and A30P mutant forms of α-synuclein have been shown to aggregate *in vitro*, with the propensity to form oligomeric conformers capable of accelerating the progression of the disease [34]. The E46K mutation has also proven to accelerate α-synuclein filament formation, and unlike the A53T and A30P mutant forms, it harbors an increased ability to bind phospholipids *in vitro* [35]. Furthermore, overproduction of the wild-type α-synuclein protein is sufficient to cause PD as evidenced by a patient of the Contursi kindred who harbored a triplication of the α-synuclein locus [36]. Although mutant α-synuclein is responsible for a small number of PD cases, it has vaulted to the forefront of PD research since wild-type α-synuclein has been identified as a key component of Lewy bodies. In fact, α-synuclein staining is now widely utilized as a neuropathological criterion for PD [37–42].

Two other genetic mutations related to proteasome function have been identified in familial PD. One mutation in the *parkin* gene (*PARK2*) was discovered in an early onset juvenile autosomal recessive form of parkinsonism (AR-JP) that presents with mild symptoms, slow progression, and the absence of Lewy bodies [43–46]. Homozygous deletions and point mutations in *parkin* also account for the majority of autosomal recessive inherited PD [47–53]. A second mutation in the ubiquitin carboxy-terminal-hydrolase-L1 (UCH-L1) gene has been identified in the *PARK5* locus in a German family with PD [54]. Both Parkin and UCH-L1 are involved in the regulation of the ubiquitin–proteasome pathway that suggests dysregulation of protein processing in the pathophysiology of PD [21,55,56].

More recently, two other genes believed to be involved in cellular oxidative stress response have been implicated by linkage analysis to rare familial forms of PD. In 2003, Bonifati et al. discovered mutations in the DJ-1 gene (*PARK7*) to be associated with an autosomal recessive early-onset form of human parkinsonism [57]. Although the exact function of DJ-1 is unclear, it is proposed to act as a redox-regulated molecular chaperone important in combating oxidative stress. Two homozygous mutations in PINK1 (PTEN-induced kinase 1), a mitochondrial resident protein kinase, have also been linked with a hereditary early-onset form of PD mapped to the *PARK6* locus in a large Sicilian family [58]. Cell culture studies have speculated PINK1 to function during cellular oxidative stress to protect against mitochondrial dysfunction and apoptosis of neurons. In 2004, the mutant gene responsible for *PARK8*-linked autosomal dominant form of PD was cloned and identified as leucine-rich repeat kinase 2 (LRRK2), which encodes for the dardarin protein throughout the adult brain and proposed to function as a tyrosine kinase [59,60]. In some kindreds of Middle Eastern origins, a prevalent mutation in

TABLE 39.3
Identified Genes Associated with Familial PD

PARK	Locus	Gene	Proposed Function	References
PARK 1	4q21–22	α-Synuclein	Presynaptic	[31]
PARK 2	6q25–27	Parkin	E3 ubiquitin ligase	[46]
PARK 5	4p14	UCH-L1	Ubiquitin recycling enzyme	[54]
PARK 6	1p25–36	Pink1	Mitochondrial	[57]
PARK 7	1p36	DJ1	Oxidative stress response	[58]
PARK 8	12p11.2q13.1	LRRK2	Kinase	[60]

Source: Adapted from Fleming, S.M., et al., *NeuroRx*, 2(3), 495, 2005. With permission.

LRRK2 (G2019S) has been described as a common cause of disease. This multifunctional protein appears to suffer mutations, which deregulated its catalytic activity and cause a dominantly inherited form of PD [61–65]. Very recent data, obtained from whole genome scans, have identified several genes that confer increase risk for the development of PD [66–68]. Whether the mechanisms underlying their conveyance of increased vulnerability will be therapeutically relevant remains to be established.

As evidenced above, rigorous genetic linkage studies are identifying genes implicated in rare familial forms of PD, which will undoubtedly facilitate unraveling the etiology of PD and lead the way for timely diagnosis and novel gene-based therapeutics in the future.

39.2.1.4 Potential Gene-Based Therapies for PD

As outlined in Figure 39.2 and discussed above, toxicant and environmental triggers combined with genetic vulnerability are likely to converge on a common pathway toward cell death in PD [69]. These common downstream events that culminate in neuronal cell death are all possible targets for gene therapy in the treatment of PD. Teaming the appropriate gene with the most powerful vector system will enable the best experimental therapeutic approaches for patients with PD.

39.2.1.4.1 Ad Vectors and PD

Adenovirus-based therapies for PD have been tested extensively in rodent models. A first-generation Ad vector encoding the TH gene, when introduced into the striatum of 6-OHDA-lesioned rats, led to a reduction in amphetamine-induced rotational behavior [70]. Vector-directed TH gene expression was observed for 1 to 2 weeks following gene transfer but was accompanied by a vigorous inflammatory response, gliosis, and local tissue damage. Ad vectors expressing the TGFβ family member, glial cell line-derived neurotrophic factor (GDNF) have also been tested in 6-OHDA-lesioned rats where protection of the dopaminergic phenotype from chemical induced damage is observed for up to 6 weeks postlesion [71]. The protection afforded in a prophylactic setting was of equivalent magnitude irrespective of the site of vector instillation: striatum or the SN [72,73]. Furthermore, using a glial fibrillary acidic protein (GFAP) promoter to restrict expression of GDNF to astrocytes, Do Thi et al. demonstrated neuroprotective effects of DA neurons up to 3 months post-transduction in a 6-OHDA rat model [74]. Additionally, the modified E1, E3/E4-defective, Ad-GFAP-GDNF vector employed in this study demonstrated a reduction in vector-induced cytotoxicity compared to "first-generation" Ad vectors, an important observation with potential relevance for the translational extension from preclinical studies. Connor et al. demonstrated that Ad-mediated GDNF delivery was only protective in aged rats when the virus was delivered to the striatum, whereas the dopaminergic system of young 6-OHDA-lesioned rats could be protected by delivery to either the SN or striatum [75]. This group hypothesized, based upon a thesis put forth by Zigmond et al., that compensatory changes occur in the CNS of the aged rat that likely increases its

sensitivity to Parkinsonian lesioning [76]. In these preclinical studies, Ad-GDNF was delivered shortly before or after the drug-induced lesion in the rodent model, which inadequately reflects a realistic clinical setting where the nigrostriatal pathway has substantially deteriorated. Therefore, Zheng et al. administered an Ad-GDNF vector 4 weeks postlesion into the striatum of a 6-OHDA rat model and observed an increase in DA levels and TH-positive neurons 10 weeks postinjection, thus demonstrating the potential of GDNF-based gene therapy even at a later time point of the disease in the rodent model [77].

As mentioned previously, oxidative stress may play a significant role in PD etiology in terms of exerting downstream effects on DA neurons, leading to their degeneration in the nigrostriatal pathway. Cu/Zn superoxide dismutase (SOD1) is a key enzyme in a cells' response to oxidative stress, serving to scavenge and mitigate the harmful effects of reactive oxygen species (ROS). Barkats et al. more recently demonstrated a neuroprotective role of SOD1 on DA neurons within the nigrostriatal pathway by overexpressing SOD1 in the striatum of an MPP^+-lesioned rat model utilizing an Ad vector [78].

39.2.1.4.2 AAV Vectors and PD

Recombinant AAV vectors exhibit great promise in the arena of PD gene therapy. For example, Kaplitt et al. injected AAV vectors expressing either β-galactosidase or human tyrosine hydroxylase (hTH) into the brains of 6-OHDA-lesioned rats and demonstrated long-term transgene expression (3 months) and functional recovery [79]. Mandel et al. demonstrated longer-term transgene expression (at least 1 year) in the rat striatum [80]. In this series of experiments, AAV vectors were constructed to express either hTH or human GTP-cyclohydrolase I (hGTPCHI) in the 6-OHDA lesioned rat striatum. Elevated levels of L-DOPA were observed in animals receiving both vectors, but disappointingly, no reduction in apomorphine-induced rotational behavior was apparent in these animals. CED-mediated delivery of AAV expressing the aromatic amino acid decarboxylase (AADC) gene to the striata of MPTP-lesioned monkeys resulted in stable L-DOPA-regulated DA production persisting for more than 2 years post-therapy in these Parkinsonian animals [11,81]. An illustration of an experiment from this study is shown in Figure 39.3. This study had profound clinical relevance for PD gene therapy, since it demonstrated for the first time the sustained expression of a transgene delivered via AAV vectors (up to 6 years) in a nonhuman primate. A phase I safety study using AAV serotype 2 vectors to deliver AADC to the striatum of advanced PD patients has been initiated and continues to accrue patients (NIH OBA Human Gene Transfer Clinical Trials Database entry number 593-(2003/07)). Another study using AAV vectors to deliver hTH and AADC genes to the brains of MPTP-treated monkeys showed that AAV vectors could direct long-term transgene expression devoid of significant toxicity [82]. Moreover, Shen et al. have utilized a triple AAV vector administration approach to preclinically assess therapeutic efficacy in a PD model [1]. These vectors direct the expression of TH, AADC, and GTPCHI and were shown

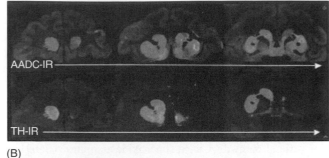

(A)

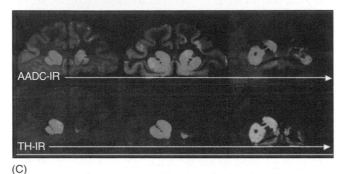

(B)

(C)

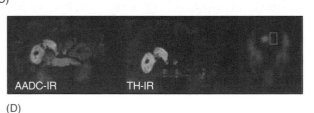

(D)

FIGURE 39.3 (See color insert following blank page 206.) Immunohistochemistry of AADC in hemilesioned monkeys before and after AAV-hAADC treatment. (A) Three [18F]FMT PET coronal images of an AAV-hAADC-treated monkey striatum at 18 months show the strongest AADC transgene expression in targeted area (midstriatum) and correlate with AADC immunostaining. (B–D) In (B) and (C), color-coded immunostaining against TH and AADC in progressive anterior to posterior coronal slices in two representative AAV-hAADC monkeys is shown at 3 years after AAV delivery. (D) shows TH and AADC staining in a representative AAV-LacZ monkey at 2 years after AAV delivery. The pattern of TH and AADC immunoreactivity, normally colocalized in intact dopaminergic pathways, is very different in these MPTP-lesioned animals. Note the dramatic reduction of anti-TH staining in the right stratum of all animals compared to the intact left side, confirming profound loss of dopaminergic fibers and terminals on the lesioned side (B–D). Anti-AADC staining shows endogenous AADC in the nonlesioned, left striatum and almost complete restoration of the enzyme within the right striatum of AAV-hAADC animals, demonstrating widespread distribution of AADC transgene expression (B and C). No such AADC staining or [18F]FMT PET signal is seen on right side of control animals (D). The box in the PET image in (D) indicates the region of infusion of AAV-LacZ. (From Bankiewicz, K.S., et al., *Mol. Ther.*, 14(4), 564, 2006. © 2006 The American Society of Gene Therapy. With permission.)

in combination in rats to enhance levels of tetrahydrobiopterin (BH4) and DA production, as well as to improve apomorphine-induced rotational behavior for up to 1 year. Other studies have also demonstrated extended expression in the CNS using AAV vectors [83–85]. The complexity of this approach, however, mitigates against its rapid translation to clinical trials.

In 2002, Luo and colleagues [86] described the use of AAV vectors to express two isoforms of glutamic acid decarboxylase (GAD) gene, GAD_{65} and GAD_{67}, in the setting of a rat toxicant model of PD. The conceptual approach was intended to phenoconvert excitatory neurons (glutamate neurotransmitter-releasing) to inhibitory neurons (GABA neurotransmitter-releasing). The rationale behind this novel approach relates to the altered neurotransmission that occurs within the basal ganglia during PD. In this pathophysiological state, dopaminergic neuron projections from the SNpc to the striatum degenerate. The observed reduction in DA release in the striatum results in reduced GABA neurotransmission, also called "disinhibition," in the subthalamic nucleus (STN). This state, coupled with direct disinhibition within the striatum itself, results in increased glutamate neurotransmission and thus excessive excitation of the substantia nigra pars reticulata (SNr) and the globus pallidum internal segment (GPi). Delivery of AAV expressing $GAD_{65/67}$ to the STN led to increased GABA release in the SNr, shifted SNr single-unit recording responses from primarily excitatory to inhibitory, and when delivered prior to 6-OHDA lesioning protected rats from the neurotoxin. Based on these encouraging preclinical results, the first gene-therapy phase I clinical trial was launched in 2003 at New York Weill Cornell Medical Center (NIH OBA Human Gene Transfer Clinical Trials Database entry number 469-(2001/04)). The phase I clinical trial entailed delivery of an AAV-GAD vector to the subthalamus of 12 advanced PD patients refractory to standard pharmacological treatment. It will be interesting to determine what the long-term effects of neuronal phenoconversion may have on the basal ganglia circuitry of these patients.

More recently, Ceregene has completed a phase I, open-label safety study of intrastriatal delivery of an AAV serotype 2 vector expressing the neurotrophic factor neurturin (NTN) to subjects with idiopathic PD (NIH OBA Human Gene Transfer Clinical trials Database entry number 689-(2005/01)). These patients did not manifest any SAEs and although the trial design was of an open label type marked reductions in the motor subscale of the unified PD rating scale (UPDRS) were reported. (Cite the RAC presentation). A phase II double-blinded sham surgical controlled trial with Cere120 has recently begun to enroll patients.

As previously mentioned, Parkin and DJ-1 are genes linked to rare inherited forms of PD and function to safeguard cells from the adverse effects of oxidative stress. Paterna et al. used recombinant AAV vectors to deliver and overexpress Parkin and DJ-1 in the substantia nigra of an MPTP mouse model of sporadic PD [87]. When compared to control mice that received injections of AAV-enhanced green fluorescent protein virus, both AAV-Parkin and AAV-DJ-1 virus injected mice showed inhibition of DA neuron death, 1 week following MPTP insult. A caveat in this study was that unilateral expression of DJ-1 in nigral dopaminergic neurons caused reduced DA receptor signaling in normal mice evidenced by apomorphine-induced ipsilateral turning toward the AAV-DJ-1 injected side. Additionally, neither Parkin nor DJ-1 was able to protect against MPTP-induced loss of catecholamines (DA, DOPAC, HVA) in the striatum of these MPTP-lesioned mice. Therefore, thorough investigation into the effects of overexpressing Parkin and DJ-1 on presynaptic neurotransmission as well as DA biosynthesis is required before considering them as candidates for PD-gene therapy.

39.2.1.4.3 Lentivirus Vectors and PD

Lentiviral vector-based therapies for PD have been tested extensively in rodent and nonhuman primate models. Injection of a self-inactivating lentivirus vector expressing GDNF has been shown to be protective in the 6-OHDA rat and MPTP nonhuman primate models of PD and in nonlesioned aged rhesus monkeys [88–90]. An example of one of these studies is shown in Figure 39.4. Long-term striatal overexpression of this potent factor by a lentiviral vector in nonlesioned rats has been shown to markedly downregulate TH expression [91]. Furthermore, a dose-dependent downregulation of TH-positive staining in the rat substantia nigra was observed 6 weeks after injecting a GDNF overexpressing lentiviral vector (rLV-GDNF) [92]. The mechanism responsible for GDNF-mediated repression of TH expression remains speculative, but proposals made by Georgievska and colleagues suggest that TH repression either at the protein level or enzyme activity level may offset the hyperstimulatory effects mediated by high levels of GDNF, which ultimately sustains long term benefit to DA neuron function. These observations warrant caution in implementation of gene transfer approaches clinically that involve long-term, uncontrolled GDNF expression. Addressing this concern, Georgievska et al. have developed a lentiviral vector system, which temporally regulates the expression of GDNF in a tetracycline-dependent manner [93]. GDNF expression was induced in the rat striatum by administering doxycycline (a tetracycline analog) in the drinking water, which transactivates the rtTAS2-M2 regulator element and permits GDNF production (Tet-on system). Consequently, TH

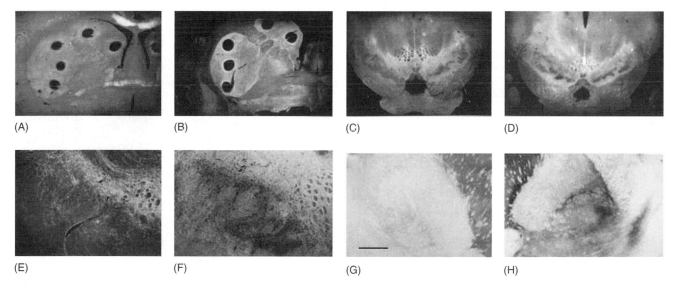

(A)   (B)   (C)   (D)

(E)   (F)   (G)   (H)

FIGURE 39.4 (See color insert following blank page 206.) Delivery of a lentiviral vector expressing the GDNF gene in Parkinsonian nonhuman primates. Rhesus monkeys received the dopaminergic toxin MPTP unilaterally (right side) to establish a parkinsonian state. One week later, the animals received ipsilateral infusions of lentiviral vector expressing GDNF (lenti-GDNF) or one expressing the reporter protein β-galactosidase (lenti-βGal). Three months following treatment the animals were sacrificed and immunohistochemistry was performed. Panels A and B depict low-power dark-field photomicrographs through the right striatum of TH-immunostained sections of MPTP-treated monkeys treated with lenti-βGal (Panel A) or lenti-GDNF (Panel B). There appeared to be a comprehensive diminution of TH immunoreactivity in the caudate and putamen of lenti-βGal-treated animals, while a nearly normal level of TH immunoreactivity was seen in the animals receiving lenti-GDNF. Low power (Panels C and D) and medium power (Panels E and F) photomicrographs are shown of a TH-immunostained section through the substantia nigra of animals treated with lenti-βGal (Panels C and E) and lenti-GDNF (Panels D and F). Note the loss of TH-immunoreactive neurons in the lenti-βGal-treated animals on the side of the MPTP infusion. TH-immunoreactive sprouting fibers, as well as an above normal number of TH-positive nigral perikarya, are observed in lenti-GDNF-treated animals on the side of the MPTP injection. Panels G and H depict bright-field low-power photomicrographs of a TH-immunostained section from a lenti-GDNF-treated monkey. Note the normal TH-immunoreactive fiber density through the globus pallidus on the intact side that was not treated with lenti-GDNF (Panel G). In contrast, an enhanced network of TH-immunoreactive fibers is seen on the side treated with both MPTP and lenti-GDNF. Scale bar in (G) represents the following magnifications: Panels A, B, C, and D at 3500 μm; Panels E, F, G, and H at 1150 μm. (From Kordower, J.H., et al., *Science*, 290(5492), 767, 2000. © 2000 AAAS. With permission.)

protein expression was reduced during doxycycline-induced GDNF expression, but gradual recovery of TH protein levels was observed 2–8 weeks after discontinuation of doxycycline. Hence, the downregulation of TH associated with the above therapeutic approach cannot be averted by temporally regulating vector-directed GDNF expression.

In a recent study, Vercammen et al. assessed the neuroprotective capacity of the E3 ubiquitin ligase, Parkin, the mutant gene linked to autosomal recessive juvenile PD. Wild-type Parkin was overexpressed via a lentiviral vector to the substantia nigra of a 6-OHDA rat model 2 weeks prior to lesioning [94]. Results from this study showed significant inhibition of DA neuron loss at 1 and 3 weeks postlesion accompanied by behavioral improvements lasting 20 weeks as evidenced by amphetamine-induced rotation test and cylinder test.

Lentiviral vectors harboring different transgenes have demonstrated preclinical efficacy in models of PD. The nigrostriatal loss of DA biosynthetic capacity has been approached with lentiviral vectors by the transduction of multiple enzymatic components to restore DA within the striatum. Due to the moderate size capacity of lentiviral vectors, multicistronic versions of the vector platform have been developed. Azzouz et al. recently created a lentiviral vector that coexpresses aromatic L-AADC, tyrosine hydroxylase, and GTP cyclohydrolase I [95]. Delivery of this vector to the striata of 6-OHDA lesioned rats led to stable DA biosynthesis and functional improvement for up to 5 months post-treatment.

39.2.1.4.4 HSV Vectors and PD

Dependent upon the transgene size capacity of a given gene transfer vector, the use of cellular promoters to direct transgene expression appears to be another promising strategy for development of a gene-based therapy for PD, as these promoters tend to yield longer term, cell-specific expression. This has been demonstrated using HSV amplicons equipped with either the preproenkephalin or 9 kb tyrosine hydroxylase (TH) promoter *in vivo*. Kaplitt et al. showed extended expression duration and striatal cell specificity using a version of the preproenkephalin promoter inserted into an HSV amplicon [96]. Using 9 kb rat TH promoter to drive expression of the β-galactosidase (*lacZ*) reporter gene in the mouse striatum, Jin et al. observed expression of *lacZ* in TH-positive dopaminergic neurons in the substantia nigra due to retrograde transport of amplicon virions [97]. In some of the initial studies utilizing amplicons as the chosen gene therapy vector, During et al. treated 6-OHDA lesioned rats with an HSV amplicon expressing hTH [98]. Behavioral and biochemical recovery was maintained for 1 year following vector introduction. Since that initial observation, Sun et al. exploited the large transgene capacity of HSV-1 amplicon vectors as well as the helper virus-free packaging technology to coexpress four therapeutic genes: tyrosine hydroxylase (TH), aromatic AADC, GTP cyclohydrolase (GTP CHI), and vesicular monoamine transporter 2 (VMAT2), which are involved in DA biosynthesis and transport in striatal

neurons [99]. Long-term, neuron-specific expression of the four genes was achieved by a modified neurofilament gene promoter in the striatum of 6-OHDA neurotoxicant rat model of PD. The rats that received the four-gene-vector demonstrated expression for 14 months accompanied by correction of apomorphine-induced rotational behavior up to 6 months. Biochemical analysis revealed extracellular levels of DA and dihydroxyphenylacetic acid (DOPAC) comparable to normal rats proximal to the injection sites. This study further confirms the versatility of the HSV-1 vector in delivering multitherapeutic transcription units for the treatment of PD. Furthermore, Sun et al. compared the neuroprotective effects of GDNF and brain-derived neurotrophic factor (BDNF) on dopamineric neurons of the nigrostriatal pathway using the HSV-1 vector. Their study concluded that GDNF was more potent in behavioral and biochemical recovery of the 6-OHDA lesioned mice when compared to either BDNF alone or when GDNF and BDNF were coexpressed.

39.2.1.4.5 Nonviral Vectors and PD

The use of nonviral means of DNA delivery for treatment of PD is appealing due to the inherent lack of immunogenic or cytotoxic viral gene products in the system. Nonviral delivery of genes includes the following means of transfer: "naked" DNA, DNA encapsulated within cationic lipids or polycationic polymers, or DNA attached to positively charged metal particles and introduced via particle bombardment [3,100–102]. Until recently, successful implementation of this gene transfer modality has largely been impeded by the low transfection efficiencies observed in neurons. *In vivo* nonviral means of DNA transfer were initially demonstrated to occur in muscle cells with surprising efficiency [103]. DNA transfer using polycationic lipid formulations to glia and neurons has been also demonstrated, albeit at low efficiencies [10,100]. Martinez-Fong et al. recently described their use of a neurotensin-SPDP-poly-L-lysine conjugate that was competent to bind and transfer plasmid DNA to neurotensin receptor-expressing cell lines (N1E-115 and HT-29, [104]). Because the high-affinity neurotensin receptor is expressed by a subset of neurons of the nigrostriatal and mesolimbic dopaminergic systems, this nonviral gene delivery modality may have the potential to be extended from preclinical studies. Additionally, the 25 kD cationic polymer polyethylenimine was shown by Abdallah et al. to mediate transfer of a luciferase-expressing plasmid to neurons and glia of adult mice [105].

Zhang and colleagues developed another nonviral gene delivery modality that has been recently tested in the 6-OHDA rat model of PD [106]. A TH-expressing plasmid, encapsulated in PEGylated immunoliposomes that were targeted to the brain with a rat transferrin receptor-specific monoclonal antibody, was administered intravenously. The plasmid was shown to effectively cross the blood–brain barrier (BBB) and lead to transient normalization of TH levels within the striatum of 6-OHDA lesioned rats. If issues regarding transgene expression silencing can be resolved and strict targeting to the striatum can be achieved using brain-specific gene promoters, this approach may become promising.

39.2.2 ALZHEIMER'S DISEASE

39.2.2.1 Introduction

As the most common cause of senile dementia, Alzheimer's disease affects millions of people worldwide. Clinically, patients experience progressive cognitive impairment leading to dementia and ultimately death. The neuropathological hallmarks of AD are senile neuritic plaques (NPs) and neurofibrillary tangles (NFTs). NPs are extracellular aggregations of protein, including the fibrillar peptide, β-amyloid. NFTs are neuronal inclusions of filamentous structures containing hyperphosphorylated forms of the microtubule associated protein tau. While these tissue and cellular abnormalities are well described the role of each in producing neuronal demise is still actively debated.

AD is divided into two types based on age of symptom onset, early (before 60 years) and late (after 60 years). Early-onset Alzheimer's disease (EOAD) is primarily an inheritable form of the disease. Genetic linkage studies of several families exhibiting EOAD identified a locus on the long arm of chromosome 21 near the amyloid precursor protein (APP) gene [107]. Further studies revealed that mutations in the human APP gene increased the production of β-amyloid (a major constituent of senile plaques) [108]. The APP linkage represents <1% of early onset cases suggesting other genetic loci exist. Two loci, one on chromosome 14q and the other on chromosome 1 containing the presenilin (PS) 1 and 2 genes, have also been identified in linkage studies [109,110]. Many mutations in PS-1 have been identified in EOAD cases, which account for >50% of all EOAD cases. The function of the PSs is currently emerging and appears to involve signal transduction [111–113].

A genetic locus on chromosome 19q is implicated in late-onset Alzheimer's disease (LOAD) [114,115]. The apolipoprotein E gene is located in this region, but no mutations have been found in AD, however one of three isoforms of ApoE, E4, is a significant risk factor for LOAD [108]. In 2004, Dodart et al. utilized a lentiviral vector to express the human ApoE4 isoform in the hippocampus of an AD mouse model (PDAPP), which overexpresses a human APP mutation (V717F) under the transcriptional regulation of the platelet-derived growth factor (PDGF) promoter, and also lacks the mouse ApoE gene (ApoE$^{-/-}$) [116]. Their results revealed increased amyloid burden and Aβ_{1-42} levels 5 weeks postinjection in the hippocampi of these mice (PDAPP- ApoE$^{-/-}$) compared to control mice that received the lenti-GFP control vector. Other potential LOAD-linked genes have been identified [117–123] and further study is needed to substantiate the importance of their linkage. The vast majority of LOAD cases are sporadic with no identified genetic component. Studies using microarray gene expression profile analysis are beginning to identify altered gene expression in LOAD compared to normally aged brains (P. Coleman, personal communication, [124]). Such analyses will lead to further studies characterizing the role of candidate genes in AD possibly allowing for a genetic screen to identify the best therapeutic strategy.

39.2.2.2 Putative Mechanisms Underlying Alzheimer's Disease

Based on the neuropathological findings of AD, NPs, and NFTs, extensive research efforts have focused on their role in the pathogenesis of AD. Many hypotheses are under examination: the acetylcholine hypothesis [125,126], the amyloid cascade hypothesis [127–129], the neuroinflammatory hypothesis [130], energy-metabolism hypothesis [131], and the oxidative stress hypothesis [132]. A comprehensive review of the research behind each of these hypotheses is beyond the scope of this chapter; instead, they will be discussed generally as potential target areas for gene therapy.

39.2.2.3 Potential Gene-Based Therapies for Alzheimer's Disease

One of the affected brain regions is the basal forebrain cholinergic complex as cholinergic deficiency correlates with both the magnitude of pathological severity and degree of dementia [125,133–136]. One therapeutic approach is to augment cholinergic function by increasing activity of the biosynthetic enzyme for acetylcholine, choline acetyltransferase (ChAT) to the affected area. Nerve growth factor (NGF) upregulates expression of ChAT [137] and promotes survival and maintenance of the septo-hippocampal pathway that is a major pathway for memory and learning [138]. Studies reporting that NGF increases β-amyloid production *in vitro* contrast with *in vivo* studies suggesting that NGF delivery does not increase plaque formation in primates [139]. Grafting of fibroblasts genetically modified to produce NGF via a Moloney murine leukemia virus vector has been shown to promote restoration and survival of the septohippocampal pathway [140–144]. Figure 39.5 depicts the bioactivity of transduced grafts *in vivo*. These initial *ex vivo* studies together with extensive preclinical studies in primates set the stage for the first, phase 1 clinical trial of *ex vivo* NGF gene therapy. Results from this trial revealed an improvement in the rate of cognitive decline in 6 patients, and no apparent detrimental effects of NGF expression from the grafted fibroblasts 22 months post-therapy [145]. Two patients had severe adverse events (SAEs), which were attributed to the neurosurgical procedure. As previously mentioned, *ex vivo* gene therapy has been shown to be transiently efficacious, and therefore focus has now been shifted to *in vivo* based strategies for protracted expression of NGF. In 2006, Ceregene launched a phase 1 clinical trial using an adeno-associated viral vector expressing NGF (AAV-βNGF) (NIH OBA Human Gene Transfer Clinical trials Database entry number 623-(2004/01)). NGF clearly has therapeutic potential but further clinical trials are warranted to evaluate its efficacy in subverting disease processes or ameliorating symptoms of AD.

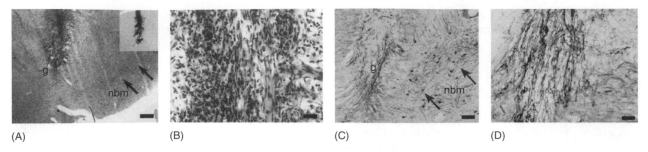

(A)   (B)   (C)   (D)

FIGURE 39.5 (See color insert following blank page 206.)   Trophic response to NGF in the human brain. (A,B) Nissl stain of autologous, NGF-secreting cell implant in brain of individual with Alzheimer's disease 5 weeks after treatment. Graft (g) adjacent to nucleus basalis of Meynert (nbm; arrows). Inset, robust mRNA encoding NGF by in situ hybridization within graft. Scale bar in (A) 247 μm; in (B) 24 μm. Note proximity of graft to nbm seen in similar perspective in c at higher magnification. (C,D) Immunocytochemistry for cholinergic neurons (p75) shows graft implant on left (g) and adjacent neurons of nbm (arrows). Higher magnification (d) shows dense penetration of cholinergic axons into graft. Scale bar in (C) 82 μm; in (D) 11 μm. (From Tuszynski, M.H., et al., *Nat. Med.*, 11(5), 551, 2005. © 2005 Nature Publishing Group. With permission.)

Amyloid-containing plaques may be a cause of AD or a by-product of the disease process. Identified mutations in both the APP and the PS genes correlate with increased production of β-amyloid and potentially plaques. PSs are transmembrane proteins localized predominantly in endosomes and Golgi apparati and are thought to promote the aggregation of β-amyloid by increasing the activity of γ-secretase, the enzyme responsible for liberating the Aβ 1–42 fragment from APP subsequent to β-secretase (BACE1) cleavage [113,146]. Gene products that may modify the activity of β- or γ-secretase or proteins capable of disrupting β-amyloid aggregation may potentially slow progression of the disease. Proof-of-concept for such a strategy was provided in 2005 by Singer et al., where they utilized a short interfering RNA (siRNA)-expressing lentiviral vector to target and knockdown the expression of β-secretase (BACE1) in an APP transgenic mouse model of AD [147], since embryonic disruption of γ-secretase activity via a traditional knock-out strategy has been shown to be lethal [148]. Results from this study showed a reduction in APP processing, evidenced by reduced $A\beta_{1-42}$ peptide levels 4 weeks postinjection in the hippocampi of APP transgenic mice that received the lenti-siBACE1–6 vector. Furthermore, siBACE1–6 expressing transgenic mice demonstrated less neurodegeneration in the hippocampus as evidenced by immunohistochemical analyses. This pathological correlate was substantiated functionally by improved performance in spatial learning and memory compared to untreated APP transgenic mice using the Morris water maze behavioral test. In 2006, Hong et al. constructed and tested a replication defective HSV vector capable of downregulating the expression of APP using an RNA interference (RNAi) strategy [149]. Delivery of this vector to the hippocampus of a novel AD mouse model (LV-APP^{Sw}) resulted in a significant reduction in $A\beta_{40}$ peptide levels 4 weeks postinjection. Furthermore, this group examined the efficacy of neprilysin in Aβ peptide clearance utilizing the same LV-APP^{Sw} mouse model. Neprilysin is an extracellular metalloendopeptidase, which Iwata et al. identified as an enzyme capable of degrading extracellular Aβ and thus preventing its accumulation in the brain [150]. Codelivery of an HSV vector expressing neprilysin together with the LV-APP^{Sw} construct into the hippocampus of C57BL/6J mice resulted in a significant reduction in $A\beta_{40}$ expression as evidenced by immunohistochemistry 4 weeks postinjection [149]. These findings corroborate those previously described by Marr et al. in which they demonstrated a significant reduction in Aβ deposition in the hippocampus of PDAPP transgenic mice one month after being stereotactically injected with a lentiviral vector expressing neprilysin [151].

Other areas of therapeutic interest for AD are inflammation and oxidative stress. Many mediators of inflammation have been detected in postmortem AD brain (reviewed in Refs. [130,152]). Whether inflammation is promoted by the production of plaques and tangles or inflammation initiates production of plaques and tangles, which in turn propagates an inflammatory response is not clear. It is thought that the inflammatory response of glia contributes to neuronal demise [130]. Although there is no unequivocal evidence directly linking inflammation and AD etiology, Janelsins et al. propose that inflammation may play an early role in the progression of AD based on their observations in a novel triple-transgenic mouse model (3xTg-AD), which recapitulates the disease progression and pathological hallmarks of AD [153]. Their findings reveal a spatial and temporal increase in the mRNA transcript levels of inflammatory mediators, TNF-α and MCP-1, coupled with augmented numbers of microglia and macrophages in the entorhinal cortex of 6 month old 3xTg-AD mice compared to 2 month old 3xTg-AD mice. Epidemiological studies with antioxidants and nonsteroidal anti-inflammatory drugs suggest that decreasing free radicals and inhibiting inflammatory processes may confer protection against AD or slow the rate of cognitive decline seen in AD (discussed in Ref. [132]). The delivery of gene products capable of scavenging free radicals and blocking inflammatory processes may also prove to be an effective therapeutic approach.

Another promising therapeutic approach in combating AD is Aβ-based immunotherapy [154–156]. Numerous investigative teams have reported diminution in AD-like pathology (i.e., Aβ deposition) and behavioral improvements in different animal models of the disease as a result of Aβ peptide immunization or via the administration of Aβ-specific antibodies. The biological mechanisms by which these therapies act are still incompletely understood. However, two nonmutually exclusive hypotheses have emerged: (1) Aβ-specific antibodies whether raised intrinsically or passively administered act at the BBB to alter the equilibrium of soluble $Aβ_{1-42}$ concentrations to favor its clearance from the brain. This is posited to prevent additional Aβ peptide deposition and perhaps promote aggregate dissolution. (2) Aβ-specific antibodies cross a compromised BBB and bind to Aβ within the brain parenchyma where antibody–Aβ complexes are postulated to be bound by complement, and recognized by microglia, which dissolve existing amyloid deposits. Both mechanisms of vaccine action can lead to possibly harmful side effects, with the most feared being systemic autoimmune disease and CNS inflammation as illuminated by recent findings from an Elan Pharmaceuticals peptide vaccine phase II clinical trial conducted in Europe (commentary in Refs. [157,158]). More extensive 4 year clinical follow-up from this Elan vaccine trial has been recently reported. No additional patients developed meningoencephalitis and, interestingly, there was blunted cognitive deterioration in the vaccine group. These data compel the development of safer vaccination approaches.

To circumvent such deleterious side effects caused by synthetic Aβ-peptide based vaccinations/immunizations, researchers have utilized and tested recombinant adeno-associated virus, adenoviral- and HSV-1 derived amplicon vectors to deliver Aβ-peptide encoding open-reading frames in the presence and absence of molecular adjuvants to various transgenic mouse models of AD [159–161]. These attempts have revealed a reduction in Aβ deposition accompanied by a Th2 type anti-inflammatory response against Aβ in transgenic mice receiving the Aβ-peptide expressing viral vector. Vector-based gene transfer technology due to its inherent versatility may allow for regulated antigen presentation and even code-livery of immunomodulatory gene products that could lead to safer and more efficacious vaccines for AD.

Recently, the use of anti-Aβ single chain variable fragment (scFv) antibodies to reduce or disrupt amyloid plaque formation in AD mouse models has been pursued as a passive immunotherapeutic modality for AD. The rationale for using only the variable fragment (Fv) of the anti-Aβ antibody stems from recent clinical trial results, where adverse immune responses were elicited and believed to be mediated by T cell or by the Fc portion of the antibody [162]. scFv antibodies constitute a single polypeptide chain bearing a heavy chain variable domain attached via a flexible linker to a light chain variable domain. In an early study, Fukuchi et al. utilized a recombinant AAV to deliver an anti-Aβ single chain antibody, termed scFv59, to the cortex and hippocampus of Tg2576 mice (hAPPswe-expressing) and demonstrated via immuno-histochemistry a reduction in Aβ deposition at the injection

site compared to PBS-injected Tg2576 control mice 6 months postinjection [163]. Moreover, immunohistochemical analysis did not reveal any cytotoxic effects upon scFv expression in the treated Tg2576 mice. Recently, Levites et al. conducted a similar study using three different scFv antibodies derived from anti-Aβ monoclonal antibodies raised against Aβ1–16, Aβ40, and Aβ42 peptides [164]. Delivery of these anti-Aβ scFv-expressing AAV vectors (AAV-scFv) into the ventricle of newborn Aβ-overexpressing CRND8 transgenic mice resulted in a decrease in amyloid plaque deposition accompanied by attenuated Aβ40 and Aβ42 levels 3 months post-therapy (see Figure 39.6). These preclinical anti-Aβ scFv studies offer an attractive therapeutic modality for the development of "safer" AD-targeted immunotherapies that are potentially incapable of eliciting deleterious immune responses.

39.2.3 Lysosomal Storage Diseases

39.2.3.1 Introduction

Lysosomal storage diseases (LSDs) are a diverse group of more than 40 disorders that originate mainly from a deficiency of a lysosomal enzyme, which results in lysosomal distention followed by accumulation of partially degraded substrates and cellular dysfunction. The lysosome, which is the cellular reservoir for many pH-dependent acid hydrolases, is critical for the degradation and clearance of unwanted cellular or extracellular components: the end-products of endocytic, phagocytic, and autophagic processes in the cell. Many of the LSDs exhibit moderate to severe deleterious effects on somatic tissues as well as the CNS. These degenerative disorders differ from many of the neurodegenerative diseases in that the pathogenic mechanism responsible for most LSDs is known and results from loss of an enzymatic function (i.e., lysosomal enzyme deficiency, defects in cofactors/activators, or transport proteins). There are examples of LSDs, however, for which the missing lysosomal function or protein activity has not been identified (e.g., a subset of the ceroid lipofuscinosis diseases; reviewed in Ref. [165]). Depending upon the type of LSD, enzyme replacement therapy (ERT) by direct infusion of the missing enzyme into peripheral tissues has been a clinically useful strategy [166]. Such an approach has been viable due to a phenomenon referred to as "cross-correction," where the majority of the exogenously supplied enzyme is sorted in the Golgi apparatus and targeted to the lysosome via the mannose 6-phosphate receptor-mediated pathway, while a small fraction is secreted into the extracellular space enabling its uptake by neighboring cells and leading to disease correction at distal sites (reviewed in Ref. [167]). This is no more evidence than in the case of Gaucher disease, which results from a deficiency of glucocerebrosidase. Loss of glucocerebrosidase, also referred to as β-glucosidase, leads to accumulation of glucocerebroside, a by-product of sphingolipid degradation, in macrophages resulting in spleen and liver enlargement, bone malformation, and pulmonary dysfunction [166]. Infusion of recombinant glucocerebrosidase leads to significant correction of peripheral tissue disease

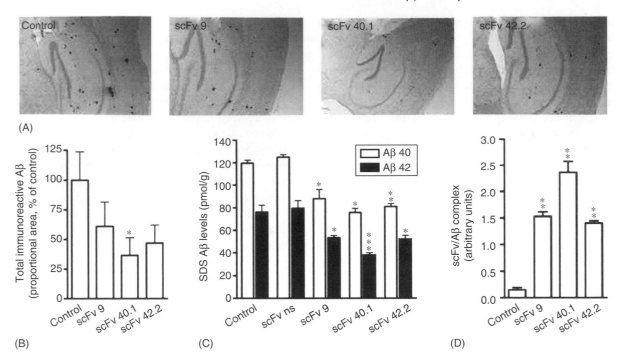

FIGURE 39.6 Anti-Aβ scFvs attenuate Aβ deposition in 3-month-old CRND8 mice. Newborn CRND8 mice were injected intracerebro-ventricularly with AAV1 expressing scFv9, scFv40.1, and scFv42.2. Control mice received AAV1-scFv ns or PBS. Three months later, mice were killed after treatment. One hemibrain was used for immunohistochemistry and the other was used for biochemical analysis. (A) Representative immunostained sections for amyloid plaques from brains of scFv-treated CRND8 mice. Magnification, 40×. (B) Quantitative image analysis of amyloid plaque burden in the neocortex of scFv-treated CRND8 mice. $*p < 0.05$ versus control. (C) Aβ levels in SDS-soluble extracts. (D) An Aβ-scFv complex in plasma was detected by ELISA with a capture antibody specific to the free end of Aβ (for scFv, mAb40.1; for scFv40.1 and scFv42.2, mAb9) and anti-myc-HRP as detection. $n = 7$; $*p < 0.05$ versus nonspecific scFv; $**p < 0.01$ versus nonspecific scFv; $***p < 0.005$ versus nonspecific scFv. Error bars indicate SEM. (From Levites, Y., et al., *J. Neurosci.*, 26(46), 11923, 2006. © 2006 Society for Neuroscience. With permission.)

[168]. A similar approach has proven successful in the treatment of Fabry disease, another sphingolipid disorder that primarily affects kidney, heart, and skin (reviewed in Ref. [169]). In clinical trials, intravenous delivery of α-galactosidase A was found to be safe and led to marked reductions in plasma glycosphingolipid levels and microvascular endothelial deposits in major organs [170,171]. Other LSDs being considered candidates for ERT are mucopolysaccharidosis I (MPS I), MPS II, MPS VI, and Pompe disease.

Other LSDs exhibit both systemic and CNS involvement. The CNS component of the mucopolysaccharidosis (MPS) storage disorders and other LSDs has proven to be a more difficult task to correct using standard ERT. This difficulty lies in the inability of peripherally infused lysosomal enzymes to traverse the BBB. Another caveat to CNS-directed ERT is the possibility of triggering a deleterious immune response against the exogenously delivered protein in the brain by repeated administration, as observed in preclinical studies conducted by Kakkis and coworkers [172,173]. Development of methodologies to effectively deliver and distribute the deficient enzyme throughout the brain with minimal to no immune response involvement would represent a major advance in therapies for LSDs. To

that end, the implementation of gene-based technology in this endeavor has been vigorously pursued and has demonstrated initial promise.

39.2.3.2 Potential Gene-Based Therapies for Correcting CNS Dysfunction Caused by LSDs

Numerous mouse models for the various LSDs have been developed and have proven extremely useful in assessing novel therapies. *In vivo* therapeutic strategies have been developed employing adeno-associated virus, adenovirus, and lentivirus-based vectors to deliver the wild-type lysosomal enzyme-encoding gene to the CNS of its respective LSD mouse model to assess its efficacy in attenuating or reversing disease symptoms. Gus^mps mice, which exhibit the β-glucuronidase enzyme deficiency associated with Sly disease (MPS VII), undergo progressive lysosomal accumulation of nondegraded glycosaminoglycans in multiple organs including the brain. Initial studies by the Sands laboratory demonstrated that AAV vector-mediated delivery of β-glucuronidase into the anterior cortex and hippocampus of newborn MPS VII mice led to a reduction in glycosaminoglycan deposition and concomitant improvement in cognitive function as measured in the Morris Water Maze paradigm [174]. Although transgene expression levels were generally greater than normal β-glucuronidase levels, particularly at the injection site, no overt toxicity was observed. Recently, the same group utilized an AAV type 4 vector to specifically transduce and deliver the β-glucuronidase gene to the ependymal cells lining the ventricle of adult MPS VII mice with pre-established disease [175]. The treated mice demonstrated a significant improvement in biochemical and behavioral symptoms 4 weeks postinjection compared to untreated MPS VII mice. Their findings also revealed the ependyma to be a viable point source for continuous enzyme secretion with potential for cross-corrective distribution of the enzyme to distal areas of the brain, hence precluding the need for repeated vector administration. Several other groups

have also demonstrated the utility of AAV vector-mediated β-glucuronidase delivery in gus^mps mice [176–178].

Brooks and colleagues have utilized recombinant viral vectors based on feline immunodeficiency virus (FIV) to deliver the β-glucuronidase gene to the brains of MPS VII mice to determine if restoration of this enzyme diminished pre-established lysosomal accumulations and corrected associated CNS deficits [179]. FIV vector-mediated bilateral delivery of β-glucuronidase via the striatum led to bilateral correction of protein deposits and a reversal of spatial learning and memory impairments. The effect of FIV vector-mediated expression of β-glucuronidase on lysosomal storage is illustrated in Figure 39.7. Perhaps shedding light on potential mechanisms of action, gene expression profiling indicated significant increases in genes associated with neuronal plasticity mediation. An interesting extension of these studies was performed by Elliger et al. [180]. The coding sequences for the Igκ secretion and HIV-1 TAT uptake signals were engineered into the β-glucuronidase open reading frame in order to effect therapeutic benefit in more distal organs. This modified transgene was delivered intrathecally via an AAV recombinant vector to newborn gus^mps mice. Treated mice were found to be more active, exhibited less stunted growth, and did not show evidence for abnormal storage deposits in the brain or liver or in tissues not harboring AAV vector genomes. Recently, a phase I study for lentiviral vector treatment of Sly's disease was reviewed by the Recombinant DNA Advisory Committee (RAC).

The Twitcher mouse, which harbors a genetic disruption in the galactocerebrosidase (GALC) locus, serves as an informative murine model for human globoid cell leukodystrophy (Krabbe disease). Loss of GALC activity in the CNS leads to the accumulation of galactolipids, such as psychosine, which

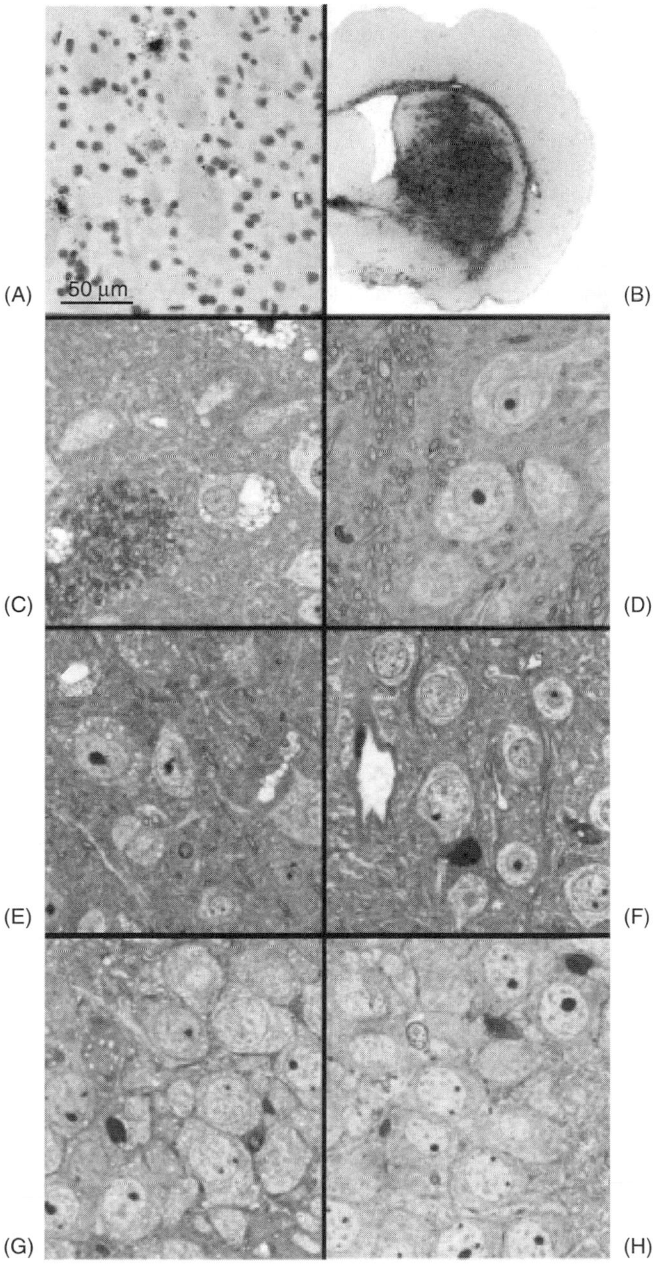

(A) 50 μm (B)
(C) (D)
(E) (F)
(G) (H)

FIGURE 39.7 β-Glucuronidase expression following FIV-mediated gene transfer into the striata of MPS VIII mice and its effect on lysosomal storage. Eight-week-old MPS VII (gus^mps) mice were injected unilaterally into the striatum with 1×10^6 transduction units of an FIV vector expressing β-glucuronidase (FIVβgluc). Six weeks later, the animals were sacrificed and analyzed for transgene expression, β-glucuronidase enzyme activity, and lysosomal storage profiles. Transgene-positive cells were detected near the injection site as revealed by in situ RNA analyses (Panel A). β-Glucuronidase activity in the brain of an MPS VII mouse injected with FIVβgluc was found to encompass a wide volume of the brain as determined by histological staining (red staining; Panel B). Representative examples of lysosomal storage in the striatum (Panel C), cortex (Panel E), and hippocampus (Panel G) in nontreated 8–12-week-old MPS VII mice are shown. Noticeable lysosomal storage is evident at this age. Analysis of age-matched, FIVβglu-treated MPS VII mice shows significant correction of the storage deficit in the contralateral striatum (Panel D), cortex (Panel F), and hippocampus (Panel H). In data not shown, treated animals exhibited improved learning and memory behavior, indicating that lentivirus-based delivery of β-glucuronidase to the CNS of an animal with pre-established LSD can reverse the neurological deficits caused by the disease. (From Brooks, A.I., et al., *Proc. Natl. Acad. Sci. USA*, 99(9), 6216, 2002. © 2002 National Academy of Sciences. With permission.)

results in apoptotic death of myelin-producing cells and severe pathological aberrations and premature death in humans [181]. Intraventricular infusion of a recombinant adenovirus vector expressing GALC to newborn Twitcher mice led to marked reduction in lysosomal storage pathology [182]. However, unlike in the case of gene delivery to MPS VII mice, treatment of Twitcher mice with pre-established disease had no significant effect on disease pathology. These results indicate that timing of interventions for Krabbe disease is crucial or adenovirus vectors, due to their inherent immunogenicity and reduced duration of expression, are not as useful for treating this LSD. Moreover, various recombinant AAV serotype vectors have also been used to deliver mouse GALC to the CNS of neonatal Twitcher mice resulting in attenuated disease progression and increased life span compared to untreated mice [183,184]. It is interesting to note that there exist autosomal recessive forms of Krabbe disease that have been identified in dogs and rhesus monkeys [185]. As other gene transfer vector approaches are applied to this disease, these higher order mammalian models will become invaluable in the stepwise progression toward clinical application to the human form of Krabbe disease.

Until recently, the lack of appropriate mouse models for neuronal ceroid lipofuscinoses (NCLs) has hindered the development of therapeutic modalities for this premature death-causing family of LSDs. In 2004, Sleat et al. generated and characterized a CLN2 knockout mouse (CLN2$^{-/-}$), which recapitulates the neurological deficits observed in classical late infantile NCL (cLINCL) [186]. In humans, CLN2 encodes the lysosomal enzyme tripeptidyl peptidase-1 (TPP-1), and deficiency of the enzyme results in accumulation of autofluorescent storage material predominantly in the CNS, which is the pathological hallmark of all NCL family members. cLINCL in humans is characterized by ataxia, loss of motor function, brain atrophy, and axonal degeneration, resulting in premature death between the ages of 7 and 15 years [187]. The monogenic nature of cLINCL prompted Passini et al. to deliver the functional CLN2-encoding gene via AAV2 and AAV5 viral vectors into the CNS of a CLN2$^{-/-}$ mouse model in order to ameliorate or reverse the pathological hallmarks associated with the disease. Their findings revealed that wild-type human CLN2 expression in the motor cortex, thalamus, and cerebellum of these CLN2$^{-/-}$ knockout mice resulted in elevated TPP-1 activity in these regions accompanied by a significant reduction of autofluorescent storage material at the injection site with limited spread in adjacent regions [188]. In addition, the Sands laboratory has also demonstrated a similar reduction of autofluorescent storage material in palmitoyl protein thioesterase-1 null mice (PPT-1$^{-/-}$), a murine model of infantile NCL (INCL), following treatment using an AAV2 vector expressing wild-type human PPT-1 [189,190]. Although, these studies observed limited cross-correction of the disease in their respective murine models, they provide ample proof-of-principle for the further development of gene therapy strategies using recombinant AAV vectors for combating NCLs.

39.2.4 Huntington's Disease

39.2.4.1 Introduction

Huntington's disease (HD) is a fully penetrant genetic neurodegenerative disorder that is inherited as an autosomal dominant mutation of the huntingtin (Htt) gene [191]. HD is one of a family of trinucleotide repeat disorders, which includes spinal and bulbar muscular atrophy, spinocerebellar ataxia (types 1, 2, 3, 5, and 7), and dentatorubropallidolusian atrophy. Affected individuals begin to exhibit symptoms in the third to fifth decade of life with some cases of juvenile (under 20 years old) and late onset (over 65 years old) [192,193]. Clinical manifestations of the disease include progressive chorea, emotional disturbances, and dementia. Neuropathological features of HD include an extensive loss of neurons and astrogliosis in the striatum, primarily the caudate nucleus. Histologically, intracellular inclusions consisting of ubiquitinated polyglutamine aggregates have been found in neurons of the striatum and other less affected areas such as neocortex.

The Htt gene encoding the Htt protein was first localized to chromosome 4p16.3 in 1983 [194]. The normal function of Htt is speculative at present, however, cloning of the normal and diseased gene revealed the mutant form of the gene contains an increased number of glutamine-encoding CAG repeats in the first exon [194]. Normal individuals have 11 to 34 repeats whereas affected individuals have 40 or more repeats [195,196]. Such an expansion in polyglutamine residues results in an atypical conformation of the mutant protein and predisposes it to form intracellular aggregates. An initial study by Ashizawa et al. [196], as well as a recent study by Rosenblatt et al. [197], provided evidence that CAG repeat number correlated with age of onset and clinical symptom severity. However, the mechanism by which the expression of this mutant protein leads to cell death remains under investigation.

39.2.4.2 Possible Mechanisms of Disease

Understanding the normal function of Htt may prove pivotal in understanding how the mutant form leads to the selective loss of γ-aminobutyric acid-containing (GABAergic) medium spiny neurons in the striatum. Recently, several lines of evidence implicate wild-type Htt in vesicular trafficking of extracellular bound proteins, and this transport process is disrupted when the mutant form of the protein is expressed in HD models [198,199]. While the physiological role of wild-type Htt is still being elucidated, studies have suggested possible mechanisms by which the mutant HD gene product causes cell death and provided a basis for developing therapeutic strategies.

One proposed mechanism involves a pathway of cellular protein degradation. The ubiquitin/proteasome pathway is the major protein degradation pathway of the cell. The proteasome, a cylindrical peptidase-containing complex that cleaves ubiquitinated proteins into their amino acid constituents, is thought to be involved in the degradation of Htt [200,201]. In HD, ubiquitinated polyglutamine aggregates have been identified suggesting the incompletely degraded mutant form of Htt forms aggregates, leading to an apparent apoptotic cell death [202,203]. Although downstream events leading to neuronal demise remain elusive, studies have shown that mutant Htt in its soluble or aggregated form is capable of recruiting/sequestering specific transcription factors. These include factors involved in chromatin remodeling (e.g., CREB-binding protein), and thereby are believed to instigate transcriptional dysregulation within the cell (reviewed in Refs. [204,205]).

Htt has been shown to interact with Htt associated protein 1 (HAP-1) [206], Htt interacting protein 1 and 2 (HIP-1, -2) [207,208], glyceraldehydes-3-phosphate dehydrogenase (GAPDH) [209], and calmodulin [210]. HAP-1 binding to Htt is enhanced by increased glutamine repeat length [206]. This factor has been shown to be involved in vesicular trafficking of various membrane receptors, further reinforcing Htt's role in intracellular protein transport [211–213]. Unlike HAP-1, HIP-1 binding to Htt is inversely related to the polyglutamine residue number. The loss of this interaction in a HD neuron may affect the cytoskeletal architecture given HIP-1's similarity to cytoskeletal proteins [208]. HIP-2 is an ubiquitin-conjugating enzyme that may be involved in the proteasomal degradation or aggregation of mutant Htt [214]. The glycolytic enzyme, GAPDH, binds to cleaved Htt fragments in vitro [209]. The increased binding of GAPDH to cleaved Htt fragments may alter cellular energy production leading to membrane depolarization, increased intracellular calcium, and cell death. Furthermore, Bae et al. have demonstrated that the nuclear translocation of mutant Htt is reliant on its interaction with GAPDH and Siah1, an ubiquitin-E3-ligase, using a mouse neuroblastoma cell line [215]. Htt can from a complex with calmodulin in a Ca^{2+}-dependent manner, but the mutant form binds independent of calcium [210]. Subsequent activation of downstream targets of calmodulin, such as those activated in excitotoxic cell death, may lead to cell death.

Adding to the putative mechanisms of striatal neuron death in HD is the ability of mutant Htt to aberrantly influence mitochondrial function and energy metabolism. This hypothesis stems from what is known about mitochondrial complex II function and protein levels found altered in the striatum of postmortem HD brains [216,217]. Recently, evidence for the above hypothesis was provided by in vitro cell culture studies using rat primary striatal neurons. Downregulation of mitochondrial complex II expression or function induced selective striatal neuron death upon lentiviral vector-mediated expression of a mutant form of the Htt protein (Htt171–82Q), which encompasses the first 171 amino acids of the protein including an expanded polyglutamine tract [217,218]. Furthermore, Cui et al. reported that PGC-1α mRNA levels are decreased in the caudate of postmortem brains of presyptomatic HD patients, and further demonstrated that mutant Htt can inhibit the expression of peroxisome proliferator-activated receptor gamma coactivator-1α (PGC-1α) in striatal neurons of HD mice

[219]. PGC-1α is a transcriptional coactivator protein critical for proper mitochondrial processes and energy metabolism. Thus, transcriptional repression of PGC-1α has been linked to mitochondrial dysfunction and the demise of striatal neurons.

Further understanding of the interactions of Htt with the aforementioned proteins as well as its ability to transcriptionally dysregulate endogenous genes in its mutant form will facilitate the development of targeted gene therapy. Based on our current understanding, therapies designed to block apoptosis, relieve oxidative stress, and prevent the expression and accumulation of degraded mutant Htt may help relieve or prevent symptoms in affected patients.

39.2.4.3 Potential Gene-Based Therapies for HD

Yamamoto et al. [220] demonstrated that continuous expression of mutant protein is necessary to maintain intracellular inclusions and symptoms. Thus, taking into account the monogenic nature of HD, therapies designed to decrease mutant gene expression via RNAi strategies are being extensively pursued with promising preclinical results in mouse models of HD. In this strategy, a viral vector expressing an siRNA or short hairpin RNA (shRNA) to mutant Htt is delivered to the striatum of the HD mouse. These small RNA species specifically bind to the mutant Htt mRNA and instigate degradation of the transcript via the cells RNAi machinery. Several groups have used adeno-associated viral vectors to deliver anti-Htt siRNA into the striatum of various mouse models of HD, which express different forms of the mutated human Htt protein. These studies revealed that partial repression of mutant htt protein levels in the striatum of these mice via RNAi is sufficient to ameliorate neuropathological and behavioral aberrations linked to HD [221–224]. Prior to implementing such a strategy in HD patients, it will be crucial to design and test siRNAs capable of exclusively targeting the mutant Htt mRNA, and also preclude off-target suppression of the wild-type Htt mRNA and other bystander transcripts. Alternatively, if a technology were available to reduce the glutamine repeats or replace the mutant allele with a normal allele, a permanent reversal of the disease might be accomplished. The use of chimeroplasts, target-specific RNA/DNA oligonucleotides, could possibly correct the genetic deficit, but this may not be technically feasible for such a large mutation [225–227].

A second approach takes advantage of the well-characterized neurotrophic factors, endogenous soluble proteins that regulate survival, growth, morphological plasticity or synthesis of proteins for differentiated functions of neurons [227]. The delivery of the neurotrophic factors GDNF, BDNF, or CNTF have shown neuroprotective effects when given prior to a quinolinic acid challenge in rodent and primate models of HD [228–232]. The neuroprotective effect of lentiviral vector-mediated delivery of CNTF in the quinolinic acid model for HD is illustrated in Figure 39.8 [368]. Further analysis is needed to evaluate the full potential of neurotrophic factor treatment for HD.

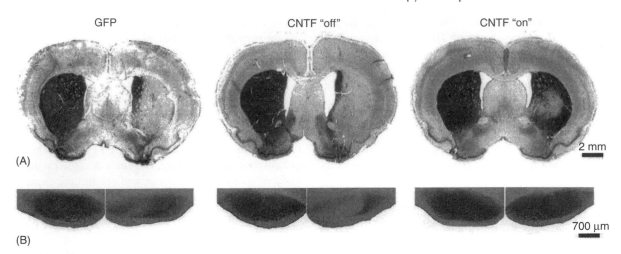

FIGURE 39.8 Use of a tetracycline-regulated lentiviral vector expressing ciliary neurotrophic factor (CNTF) in the quinolinic acid rat model of HD has a dose-dependent neuroprotective effect. Rats received intrastriatal injections of a tetracycline-regulated lentiviral vector expressing either CNTF (TRE-CNTF) or the reporter protein GFP (TRE-GFP). The vector-injected rats were treated with saline ("off") or doxycycline ("on") to regulate transgene expression. Quinolinic acid was infused into the striatum of these mice and they were subsequently sacrificed and brains analyzed by immunocytochemistry for the striatal marker DARPP-32. Representative photomicrographs showing DARPP-32-immunostained striatal (Panel A) and nigral sections (Panel B). The quinolinic acid-induced lesion is clearly indentified by the loss of DARPP-32 staining in the GFP- and CNTF-off groups, whereas a significant protective effect is observed when CNTF expression is switched on. (From Regulier, E., et al., *Hum. Gene Ther.*, 13(16), 1981, 2002. © 2002 Mary Ann Liebert, Inc. With permission.)

A third approach would be to target mechanisms of apoptotic cell death and oxidative stress. Similar to neurotrophic factors, this approach may delay the progression of the disease, but does not permanently correct the genetic deficit. The antiapoptotic gene, *bcl-2*, has been shown to inhibit neuronal apoptosis both *in vitro* and *in vivo* [233–243]. Delivery of the *bcl-2* gene to affected neurons may prevent or delay cell death that occurs in HD. If oxidative free radicals contribute to cell death, the expression of gene products capable of directly or indirectly lowering or removing free radicals (e.g., SOD, catalase, glutathione reductase [GR], glutathione peroxidase [GPX], and glutathione [GSH]) might prove useful in combating neuronal loss. Recently, the previously described coactivator PGC-1α has been shown to elicit neuroprotective effects on ROS-mediated cytotoxicity by regulating the expression of genes involved in oxidative stress responses (reviewed in Ref. [244]). Lentiviral delivery of PGC-1α to the striatum of R6/2 transgenic HD mice leads to the protection of striatal neurons 3.5 weeks postinjection [219]. Keeping in mind that PGC-1α protein is expressed in postmortem HD brains and also in the striatum of an established mouse model of HD, further investigation into the therapeutic potential of PGC-1α is warranted.

Another strategy devised to prevent mutant Htt-induced cytotoxicity and potentially retarding the disease progression could be to inhibit its aggregation and thus preclude cytotoxicity in striatal neurons (reviewed in Ref. [245]). In 2004, Colby et al. demonstrated that a single-domain intracellular antibody specific for Htt was able to inhibit aggregation in a mammalian cell model of HD [246]. The application of intracellular antibodies (intrabodies) for such an endeavor

seems coherent given that scFv antibodies are currently are currently being tested in Alzheimer's disease models, where disruption of protein aggregation has led to reduced amyloid deposition in the brain [164].

39.2.5 STROKE

39.2.5.1 Introduction

Hypoxic injuries, such as stroke, are the genesis of substantial morbidity and mortality in neonates and older individuals. Oxygen deprivation can lead to profound effects on motor function and cognition [247], leading to severe disability and diminished quality of life. Stroke is most commonly the result of embolic obstruction that results in either reduced or complete loss of blood flow to downstream fields. Two types of cell death appear to occur following a stroke. At the ischemic core, necrosis is readily apparent and has been linked to elevated extracellular glutamate and intracellular calcium levels. Necrotic cell death involves nonspecific DNA degradation, nuclear pyknosis, diminished membrane integrity, and mitochondrial swelling. Neuronal degeneration within this area following blood vessel obstruction occurs very rapidly (minutes) following the ischemic insult, and due to this time limitation, does not represent a viable target for gene-based therapeutics. In brain areas more distal to the ischemic core, termed the prenumbra, neurons undergo a more delayed form of cell death (hours to days) that exhibits dependence upon de novo gene expression [248–250]. This delayed neuronal death, or apoptosis, is capable of producing damage equivalent to the acute necrotic lesion observed soon after severe ischemia.

Cells undergoing apoptosis exhibit characteristic fingerprints, including chromosome condensation, internucleosomal DNA fragmentation, and membrane blebbing. A thorough understanding of the gene expression profiles and signaling events that occur in the prenumbra immediately following an ischemic event is likely to yield several molecular/cellular targets that are amenable to pharmacological and gene-based therapeutics.

39.2.5.2 Possible Mechanisms of Disease

Ischemia promotes adaptive and pathologic responses in the neuronal compartment. As illustrated in Figure 39.9 extrinsic and intrinsic perturbations in the ischemic brain appear to activate neuronal ischemic sensors, which in turn promotes adaptive and pathologic gene expression [369]. One of the

ischemic sensors believed to trigger de novo gene expression during oxygen deprivation is the transcription regulator, hypoxia-inducible factor-1α (HIF-1α). During hypoxia, the otherwise rapidly turned-over HIF-1α is stabilized and leads to the transcriptional activation of numerous genes including vascular endothelial growth factor (VEGF), erythropoietin (EPO), and glucose transporter 1 (GLUT-1) (reviewed in Ref. [251]). In 2005, the Barlow laboratory conditionally knocked-out HIF-1α in the mouse brain and subjected these mice to acute hypoxic insult [252]. Their findings suggest a more complex ischemic sensory network is present during hypoxia, including the activation of HIF-1α and its downstream targets. Surprisingly, the lack of HIF-1α in the brains of conditional HIF-1α$^{-/-}$ mice appears neuroprotective when these mice are subjected to acute hypoxic insult compared to control animals. Previous studies have proposed other ischemic sensors,

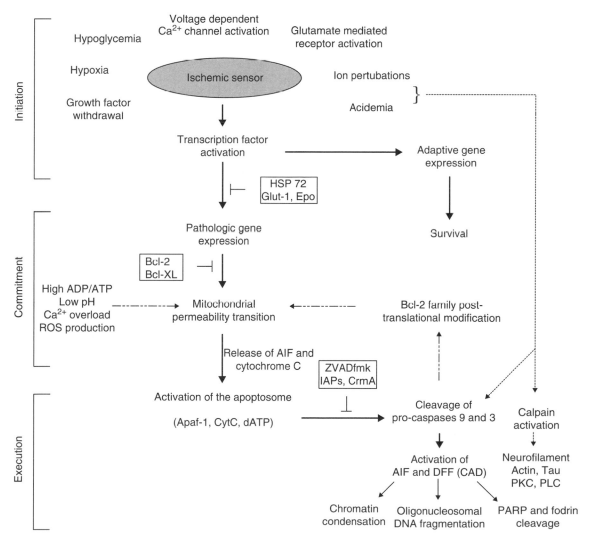

FIGURE 39.9 Ischemia promotes adaptive and pathologic responses in the neuronal compartment. Extrinsic and intrinsic perturbations in the ischemic brain activate a neuronal ischemic sensor, which in turn promotes adaptive and pathologic gene expression (solid lines). In addition, select stimuli can activate neuronal death independent of de novo gene expression (dashed lines). Specific gene products, which have demonstrated the ability to block the initiation, commitment, or execution phases of the programmed death pathway, are also included (boxed items). (From Halterman, M.W., et al., *J. Neurosci.*, 19(16), 6818, 1999. © 1999 Elsevier Science B.V. With permission.)

including HIF-2α [253,254], play integral roles during oxygen deprivation, further highlighting the complex nature of events that occur during the pathological and adaptive response to stroke.

It is the complex interplay amongst the numerous molecular signaling cascades and time sensitivity of the disorder that makes development of gene-based approaches for stroke both an exciting and daunting task. The ischemic sensors activate a series of transcription factors that, depending upon their respective stoichiometry and subcellular localization, act to initiate either an adaptive or pathologic signaling cascade. Insights into candidate gene products that confer protection against ischemic exposure can be gleaned from prior preconditioning studies [255]. The gene products involved in adaptive processes include those that inactivate ROS, and those that participate in DNA repair and cytoplasmic calcium regulation [256,257]. Hypoxic insult, as previously described in context with HIF-1α, stimulates the secretion of factors like EPO and VEGF, which can act via a paracrine manner to effect neuroprotection [258,259]. Other gene products, including heat shock protein 72 (HSP 72) and GLUT-1, have been identified that block the initiation of pathologic gene expression [260–263]. Such factors may provide sufficient protection during the early stages of cellular responses to ischemia in order to subvert potential downstream apoptotic signaling.

39.2.5.3   Potential Gene-Based Therapies for Stroke

Many early experimental gene therapy approaches targeted either intermediate or the late stages of the apoptotic process. The commitment to apoptosis, which is believed to involve mitochondrial permeability transition, can be averted by overexpression of a subset of Bcl-2 protein family members and a neuronal apoptosis inhibitory protein (NAIP). HSV-based vectors overexpressing Bcl-2 reduces the incidence of apoptosis and improves neuronal survival both *in vitro* and *in vivo* [239,264]. In 2003, the Steinberg laboratory provided direct evidence for the potential use of BCL-*w*, a novel Bcl-2 antiapoptotic family member, in protecting neurons from hypoxia-induced injury [265]. In this study, a recombinant AAV vector capable of overexpressing BCL-*w* was delivered to the cortex and striatum of adult rats 3 weeks prior to focal cerebral ischemia induced by MCA occlusion. The BCL-*w* overexpressing rats demonstrated reduced infarct volumes 24 h and 1-week following focal ischemia, as well as improved neurological function compared to control rats. Additionally, Kilic et al. demonstrated similar neuronal protection following MCAO in mice pretreated with an adenoviral vector expressing Bcl-X_L, which also belongs to the antiapoptotic Bcl family [266]. Thus, both studies exemplify the therapeutic benefit afforded by prosurvival Bcl-2 family members in the treatment of stroke. Furthermore, adenoviral vector-mediated expression of NAIP resulted in reduced levels of activated caspase-3 and diminished neuronal degeneration following transient forebrain ischemia [267,268]. Other strategies involve inhibition of cellular caspases that act during later stages of

apoptosis through the use of cell-permeable peptides to block the morphologic features of apoptosis [248]. Additionally, vector-mediated expression of anti-inflammatory molecules (i.e., interleukin-1 receptor antagonist) has been employed to minimize poststroke inflammatory responses within the brain [269–272]. However, since ischemia induces global disruptions of cellular processes upstream of mitochondrial commitment, caspase activation, and resultant inflammation, it is unclear whether these strategies can effectively interrupt the initiation of apoptosis or restore function to neurons endangered by this cellular process [273,274]. It is also important to note that many of the stroke-related gene transfer approaches performed to date have been employed prior to the onset of experimental ischemia. For gene transfer to be clinically applicable for ischemic disorders, treatments must be assessed following the ischemic event (realistically at times 1–2 h postinjury, [275])

Gene-based neuroprotective approaches that employ neurotrophic and angiogenic factors have also been extensively examined. Tsai et al. utilized AAV vectors to express GDNF in the setting of experimental stroke as an approach to minimize neuronal damage caused by transient ischemia [276]. Rats receiving AAV-GDNF immediately following bilateral common carotid artery ligation and middle cerebral artery occlusion (MCAO) exhibited significantly reduced infarct volumes as compared to animals receiving a β-galactosidase-expressing AAV control vector (see Figure 39.10) [370]. Zhang et al. performed studies using adenovirus-expressed GDNF to examine the therapeutic window following transient MCAO in rats [275]. The protective effect afforded by GDNF overexpression (i.e., reduced infarct size and inhibition of caspase-3 expression) was evident only if Ad-GDNF was administered at the time of reperfusion, but the effects of the gene transfer were minimal at 1 h postreperfusion. Additionally, Harvey et al. exploited the neurotropism and rapid expression profile of the HSV amplicon to overexpress GDNF and assess its efficacy in a focal ischemic stroke model [277]. Their findings revealed that pretreatment with HSV-GDNF, 4 days prior to MCA occlusion, afforded neuroprotection and behavioral improvements compared to rats receiving the neurotrophic vector 3 days post-MCAO. However, as previously highlighted, the viral vector of choice for therapeutic gene delivery plays a pivotal role in the efficacy of potential gene-based modalities for CNS disorders. Shirakura et al. showed that administration of negative-strand RNA virus-derived Sendai vectors carrying GDNF (SeV/GDNF), as well as other neurotrophic factor genes NGF and BDNF, at 4 or 6 h postischemic injury were capable of impeding hippocampal neuron death in gerbils [278] (see Figure 39.11). Taken together, these results underscore the importance of early postischemia interventions in minimizing neurodegeneration.

Restoration of energy stores in hypoxic neurons via overexpression of the rat GLUT-1 has been an approach that has shown promise in experimental models of ischemia and brain injury. HSV vector-mediated delivery of GLUT-1 in three models of injury (transient focal cerebral ischemia,

PBS   rAAV-lacZ   rAAV-GDNF

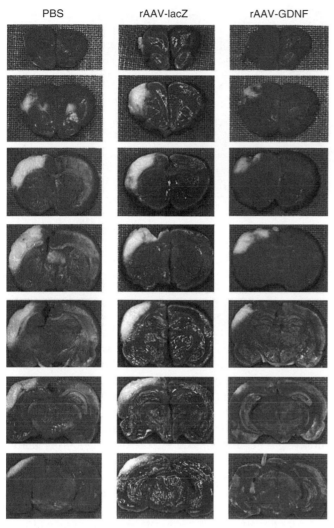

FIGURE 39.10 (See color insert following blank page 206.) Injection of rAAV-GDNF markedly reduces cortical infarction induced by middle cerebral arterial ligation in rats. The right middle cerbral artery and bilateral common carotid arterial were occluded for 90 min. Animals received a PBS, rAAV-lacZ (10^{10} viral particles), or rAAV-GDNF (10^{10} viral particles) unilateral infusion during arterial occlusion, were sacrificed 72 h later, and their brains were coronally sectioned (2 mm thickness) for TTC staining. White areas represent infarcted zones in the cerebral cortex. Rats receiving rAAV-GDNF exhibited a marked reduction in infarct size as compared to animals receiving the rAAV-lacZ or PBS control. (From Tsai, T.H., et al., *ExNeurol.*, 166(2), 266, 2000. © 2000 Elsevier Science B.V. With permission.)

kainic acid, and 3-acetylpyridine) led to localized increased uptake of glucose, reduced neuron loss, blunted decline in ATP concentrations and metabolism, and decreased glutamate release and cytosolic calcium levels [262,263,279]. Approaches that more directly address the detrimental cytosolic calcium excess observed following ischemia include overexpression of calcium-binding proteins. Viral vector-mediated delivery of the calcium-binding protein calbindin D28K exhibits calcium-buffering activity and is protective in conditions of hypoglycemia and experimental stroke [280–282].

As with many neurodegenerative disorders, the augmented generation of ROS during cerebral ischemia contributes significantly to the ultimate demise of neurons within the oxygen-deprived area of the brain. Hence, transient overexpression of enzymatic antioxidant genes, SOD1 and glutathione peroxidase (GPX), via HSV-1 amplicon vectors pre- and post-transient ischemia have been shown to protect striatal neurons by inhibiting pathways leading to apoptotic cell death [283,284]. Although these studies appear promising, infarct volumes were unaffected in either study, thereby bringing into question the therapeutic mechanisms of action underlying antioxidant gene therapy.

During times of cellular stress, resident proteins can become misfolded, a process which can result in diminution of protein function and activity and potentially lead to intracellular aggregation (reviewed in Ref. [285]). Stress response factors, such as HSPs, have been shown to act as molecular chaperones to assist in protein folding and may represent another approach to support a cell under ischemic attack. Several reports of vector-mediated overexpression of HSPs (e.g., HSP72) in ischemic animal models demonstrate a neuroprotective role for these factors [286,287]. HSV amplicon-mediated delivery of HSP72 to rats 30 min after MCAO resulted in higher numbers of surviving neuron numbers as compared to animals receiving a β-galactosidase-expressing amplicon [286].

In aggregate, gene transfer applications for stroke have shown preclinical promise. For such approaches to gain merit as viable treatment options in humans, major issues need to be addressed. One consideration relates to means of vector delivery. Focal administration of viral vectors following an ischemic event will likely have minimal benefit due to the large areas of the brain that are typically affected. Convection-enhanced delivery may provide the means in which to widely distribute a given therapeutic vector, but remains largely untested in ischemic paradigms [11]. A second issue, which is not entirely unrelated to the first, concerns the window of therapeutic opportunity. As described above, a multitude of signaling events occur immediately after ischemia that determine whether the compromised neuron follows an adaptive or a pathologic set of molecular instructions. This time window is extremely limited (1–2 h poststroke), which makes the implementation of stereotactic means of gene therapeutic vector delivery nearly improbable. However, efforts have been made to prolong this opportunistic period by mild postischemic hypothermia in an experimental model of stroke [288]. Another approach being investigated to exploit this timeframe for therapy is the development of a hypoxia-inducible vector system capable of instantaneously "switching-on" neuroprotective genes during an ischemic event, thus, minimizing the extent of the hypoxic injury [289,290]. A comprehensive understanding of the molecular signals and their temporal expression profiles will likely identify targets at the earliest of times within this restricted therapeutic window. Until these gaps in disease process knowledge are filled and technical hurdles overcome, clinical gene therapy interventions for stroke will remain an impractical potential alternative to pharmacologic compound-based therapeutics.

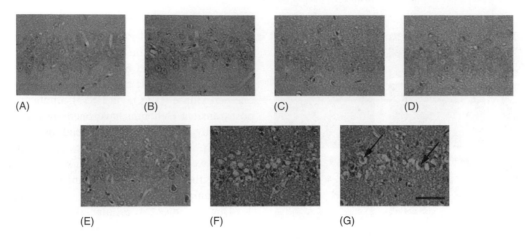

FIGURE 39.11 (See color insert following blank page 206.) Representative photographs of pyramidal neurons in the hippocampal CA1 regions after treatment (4 and 6 h after ischemia) with SeV vector following ischemic injury. Sham-operated gerbils (A), SeV/GDNF (B,E), SeV/NGF (C,F), or SeV/GFP (D,G) were administered intraventricularly 4 h (B,C,D) or 6 h (E,F,G) after ischemic insult. The sections were stained with hematoxylin and eosin. Scale bar = 50 μm. (From Shirakura, M., et al., *Gene Ther.*, 11(9), 784, 2004. © 2004 Nature Publishing Group. With permission.)

39.2.6 Epilepsy

39.2.6.1 Introduction

The epilepsies constitute a group of neurologic disorders characterized clinically by various seizure syndromes and at the cellular level by hyperexcitability. The inherent cell biologic problem is that of pathologic shifting of the normally well-regulated balance between inhibition and excitation towards the latter. The etiology of acquired epilepsy is unknown although an emerging viewpoint is that early lifetime events, perhaps subclinical in nature, contribute to or trigger the development of the neural substrate underlying the hyperexcitable state (reviewed in Ref. [291]). Since the processes responsible for inducing epileptogenesis are not fully elucidated, present treatment strategies are therefore directed at controlling symptoms (seizures). The seizures are the clinical manifestations of aberrant hypersynchronization of neural networks that can remain local, spread to other locations, or involve all cortical regions at once. Neocortical or hippocampal circuits are uniformly activated during an epileptic seizure. Synaptically connected regions may be recruited in a manner to augment or attenuate the discharge. When epileptic discharges travel unimpeded through neural networks, the network functions are subverted, temporarily rendering them incapable of executing normal tasks. With increasing duration of disease, particularly in epilepsies involving the temporal lobe, the neural substrate is scarred, presumably from chronic exposure to excitatory neurotransmitters (e.g., mesial temporal sclerosis).

For those epilepsies that are unresponsive to pharmacologic therapies, the only present means of relief for some individuals is neurosurgical resection. Due to the highly complex nature of the mammalian CNS and the unknown etiology of epilepsy, the execution of complicated neurosurgery to treat this debilitating disorder is approached with extreme caution. Such medically refractory and surgically difficult cases may prove to be suitable targets for gene therapy. Gene transfer would provide a means to locally alter synaptic transmission so as to synchronize excitatory and inhibitory signals, and in doing so, disrupt the hyperexcitable state. Of course, a number of considerations must be addressed, including the cell types to be targeted, localization of the transgene product (e.g., cell intrinsic vs. secreted), transgene expression control (e.g., constitutive vs. regulated), and effects on neighboring normal neuronal circuitry. These concerns were highlighted during an attempt to treat focal seizures by downregulating levels of the *N*-methyl-D-aspartic acid receptor subunit by AAV-mediated antisense therapy, where different promoter elements transcriptionally driving an identical antisense oligonucleotide segment surprisingly elicited opposite effects on seizure sensitivity *in vivo* [292]. However, novel approaches for treating epilepsy using gene-based means have begun to exhibit promise as illustrated below.

39.2.6.2 Potential Gene-Based Therapies for Epilepsy

The development of animal models for epilepsy has greatly enhanced the means to assess the possible etiology, and in turn, new therapies for this disorder. Extant in the literature is the observation that accumulation of *N*-acetyl-L-aspartate (NAA) in the brain leads to induction of seizure-like syndromes. NAA is found predominantly in neurons of the CNS and has served as a useful marker of neuronal injury or death in several neurodegenerative diseases. Intraventricular infusions of NAA produce absence-like seizures with signature hippocampal electroencephalograms (EEG; spike-wave-like complexes) in rats [293]. A deletion mutation in the aspartoacylase (ASPA) gene, which encodes for the enzyme responsible for metabolizing NAA into aspartate and acetate in glial cells, has been discovered in tremor rats, a model of human petit mal epilepsy [294]. The tremor rat model represents a useful *in vivo* system in which to test novel antiepilepsy gene transfer approaches [295]. Seki et al. utilized a replication-defective recombinant adenoviral vector to deliver the rat ASPA gene to the tremor rat [296]. Injection of the ASPA-expressing vector into the brains of 7-week-old tremor rats significantly reduced the appearance of absence-like seizures as compared to rats receiving a β-galactosidase expressing control adenovirus vector. Subsequently, they delivered the ASPA-expressing recombinant adenoviral vector into an 11-week-old spontaneously epileptic rat (SER: tm/tm, zi/zi), in order to assess its efficacy in preventing tonic convulsions observed in this animal model [297]. The SER rat is a double mutant derived from the tremor rat and exhibits tonic convulsions along with absence-like seizures at 8–9 weeks of age [298,299], thus making it a compelling animal model for epilepsy. Administration of the adenoviral ASPA vector reduced the number of tonic convulsions 7 days after treatment. However, the therapeutic benefit afforded by ASPA expression via the adenoviral vector was transient in nature and failed to prolong survival times compared to control animals. This study brings into question the efficacy of first-generation adenoviral vectors in the treatment of epilepsy and warrants further investigation into the usage of other viral vectors with enhanced expression duration profiles for the delivery of ASPA in future preclinical studies.

Another animal model that can be utilized to assess novel gene-based therapies for epilepsy is the protein-L-isoaspartyl methyltransferase (PIMT) deficient mouse [300,301]. PIMT is a ubiquitously expressed enzyme that repairs proteins that have undergone isomerization of aspartate residues or deamidation of asparagine residues. Failure to repair these spontaneously occurring protein alterations leads to mislocalization and diminished protein function [302]. In fact, PIMT-deficient mice, which are devoid of both splice variant forms of PIMT, accumulate isoaspartate (IsoAsp) within their brains. One obvious phenotype of this deficiency is the progressive development of fatal epileptic seizures. Intraventricular delivery of both splice variants (PIMT-I and -II) via *exo-utero* means to E14.5 PIMT-deficient mice led to improved growth, diminished IsoAsp accumulation, enhanced survival times, and marked reduction in occurrence of epileptic seizures [303]. Sustained therapeutic gene expression and concomitant clinical benefit was evident at times up to 7 weeks postinjection.

In recent years, specific neuroactive peptides present in the CNS have proven to be anticonvulsant and antiepileptogenic during seizures and have been deemed promising candidates for the treatment of epilepsy as corroborated by the following preclinical accounts. In 2003, Haberman et al. constructed a rAAV vector capable of expressing the neuropeptide, galanin (GAL), which was modified for *in vivo* secretion by fusing the fibronectin (FIB) secretion signal sequence to its N-terminus [304]. GAL is posited to exert its anticonvulsive effects by inhibiting glutamate release in the hippocampus, thus reducing the hyperexcitabilty associated with epilepsy [305]. Delivery of the AAV-FIB-GAL vector to the inferior collicular cortex in a rat model of acute, focal seizure resulted in a significant attenuation of seizures over a 4-week period compared to animals receiving a control AAV vector expressing a secretory form of the green fluorescent protein (AAV-FIB-GFP) [304]. Furthermore, infusion of the AAV-FIB-GAL vector into the hippocampus of a rat model of temporal lobe epilepsy (TLE) resulted in neuroprotection of the otherwise compromised hilar neurons (see Figure 39.12), but was futile in suppressing limbic seizure activity induced by kainic acid administration [304]. Subsequently, delivery of the AAV-FIB-GAL vector to the piriform cortex of the TLE rat model 7 days prior to kainic acid administration resulted in a significant

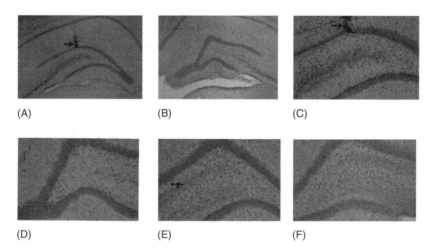

(A) (B) (C)

(D) (E) (F)

FIGURE 39.12 Effects of unilateral AAV-FIB-GAL or AAV-FIB-GFP infusion on hippocampal hilar neurons 2 weeks before peripheral administration of kainic acid. (A,C) Two weeks after kainic acid administration, hilar cells appear normal in the area of AAV-FIB-GAL infusion (arrows indicate injection tract; cresyl violet stain). (B,D) Contralateral hippocampus shows significant cell loss 2 weeks after the kainic acid administration (uninjected, damaged side exhibits a deformity of the dentate gyrus; $P < 0.05$ by paired t-test). (E,F) Infusion of AAV-GAL into the hippocampus (E; arrow indicates injector tip) was not protective against kainic acid treatment 2 weeks later. The injected side (E) exhibits the same amount of damage as the contralateral uninjected hippocampus (F). This damage is in marked contrast to (C). (From Haberman, R.P., et al., *Nat. Med.*, 9(8), 1076, 2003. © 2003 Nature Publishing Group. With permission.)

decrease in limbic seizure activity in rats previously exposed to seizure activity, thus, making this treatment modality more clinically applicable [306].

Neuropeptide Y (NPY), is a 36 amino acid polypeptide capable of exerting anticonvulsive properties during seizures (reviewed in Refs. [307,308]). In 2004, Richichi et al. overexpressed human NPY via a rAAV vector in the hippocampus of rats prior to intrahippocampal kainic acid administration and observed a striking reduction in EEG seizures with a delay in their onset as well as disruption of kindling epileptogenesis compared to control animals receiving an empty vector [309]. These promising preliminary results warrant further preclinical investigation in more established animal models of epilepsy.

In addition to these novel gene therapeutic modalities for treating epilepsy, Kanter-Shlifke et al. recently demonstrated the seizure-suppressant activity of the glial-cell derived neurotrophic factor (GDNF) using two models of TLE [310]. In their study, intrahippocampal delivery of a GDNF-overexpressing rAAV vector did not impact epileptogenesis, but significantly attenuated generalized seizures and increased seizure induction thresholds in kindled and status epilepticus (SE) rats. Their results also suggest that GDNF overexpression induces changes in the levels of neuropeptide Y, which could potentially underlie the antiepileptic effects elicited with GDNF treatment.

In aggregate, the promising results garnered from these epilepsy gene transfer studies, along with the discovery of potential antiepileptogenic gene targets from recent microarray data obtained from a rat model of TLE [311], evoke enthusiasm for further development of gene-based therapies for those epileptic syndromes refractory to standard symptomatic treatments.

39.2.7 Motor Neuron Disease

39.2.7.1 Introduction

The motor neuron diseases are a group of disorders characterized by predominant degeneration of the motor neurons. The most commonly recognized are familial spastic paraplegia (FSP, upper motor neuron disease), spinal muscular atrophy (SMA), spinobulbar muscular atrophy (SBMA, or Kennedy's disease) (lower motor neuron diseases), and amyotrophic lateral sclerosis (ALS; combination of upper and lower motor neuron disease). These disease syndromes can range in severity from debilitating (spastic paraplegia; Kennedy's disease) to lethal (some spinal muscular atrophies; ALS).

The clinical features of these conditions are well known to neurologists, but the recognition of the molecular basis, and the possibility of molecular confirmation of diagnosis, has clarified the extent and relationship of some of these conditions. A genetic basis has now been identified for all these diseases. Different genetic mechanisms, including missense mutations (ALS), gene deletions or aberrant gene splicing (SMA), and trinucleotide repeat expansions (SBMA), are responsible for the pathological findings and clinical presentations

of these motor neuron disorders. The molecular findings are of variable value as diagnostic tests. More importantly, it is hoped that understanding the underlying molecular basis of these conditions will ultimately usher forward specifically targeted treatments.

39.2.7.2 Models of Motor Neuron Disease

A number of mouse models of lower motor neuron degeneration or neuronopathy are currently available for experimental therapeutics. These include a number of naturally occurring mutant mice: wobbler (wr), progressive motor neuronopathy (pmn), neuromuscular degeneration (nmd), wasted (wst), motor neuron degeneration (mnd and mnd2), motor neuron disease (mneu), and muscle deficient (mdf). In all of these mice, the trait is autosomal recessive reflecting a presumed loss of function of the disease gene. To date, disease genes underlying these spontaneous mutations have not been identified [312]. Within the past decade, transgenic mouse models have been created whose features closely model human motor neuron degenerative disease. These include mice with engineered abnormalities of one of the neurofilament subunits, and mice overexpressing mutant Cu/Zn SOD1 that has been shown to underlie a subset of familial ALS cases [313–317]. In the case of most of the transgenic mouse models, lower motor neuron pathology or degeneration is most likely the result of a gain-of-function related to the neurofilament or mutant SOD1. The comparable gene knockout strategies result in more subtle pathology in these mice [318,319]. The underlying mechanism of mutant SOD1-mediated neurodegeneration in ALS models is still under investigation. However, recent cell-based models of ALS propose that mutant SOD1-expressing astrocytes, one of the principal glial cell populations in the CNS, may have a detrimental effect on motor neurons in a noncell autonomous manner [320,321]. These in vitro studies suggest that soluble neurotoxic factors released from mouse astrocytes harboring the mutant SOD1 gene are capable of selectively reducing the life span of wild-type and mutant SOD1-expressing motor neurons derived from various transgenic mutant SOD-1 mice. The task of identifying the toxic soluble factors responsible for motor neuron death in this culture model is currently in progress. Nevertheless, activation of proapoptotic pathways, such as the Bax-dependent cascade, has been implicated in downstream events, leading to motor neuron death in vitro and in vivo [321,322]. The identification of such cellular contributors plus the pathways leading to ultimate motor neuron degeneration will facilitate the development of more efficacious gene therapeutic modalities for ALS and other motor neuron diseases.

39.2.7.3 Potential Gene-Based Therapies for Motor Neuron Disease

Although the genetic basis for certain familial motor neuron diseases has been elucidated, the complex network of downstream events that eventually lead to motor neuron demise has not been fully understood. Thus, developing gene-based therapies for motor neuron disease remains a challenge, but recent preclinical studies have begun to show promise and offer promising new therapeutic targets for treating ALS.

Putative therapeutic approaches for motor neuron disease have been categorized based on hypothetical pathogenetic mechanisms or cell death pathways. Evidence that excess glutamate excitotoxicity can lead to motor neuron death mediated via non-NMDA receptors suggests that pharmacologic agents targeting the non-NMDA receptors might provide a promising strategy. Motor neuron death has been linked to excitotoxicity involving non-NMDA glutamate receptors, the glial glutamate transporter EAAT2 (also called GLT-1) [323–326], and related calcium-dependent pathways [327,328] in ALS and in *in vitro* models. Compared to other neurons, motor neurons possess a lower Ca^{2+}-buffering capacity due to reduced levels of calcium-binding proteins such as calbindin-D28K and parvalbumin [329] as well as a higher density of Ca^{2+}-permeable alpha-amino-3-hydroxy-5-methyl-4-isoxazolepro-prionic acid (AMPA) receptors [330], which renders them more susceptible to excitotoxic death. These same non-NMDA receptor-directed excitotoxic phenomena have now been implicated in mutant SOD1-mediated motor neuron death [331,332]. Related evidence supports a secondary role for voltage-gated calcium channels and downstream intracellular calcium dysregulation in motor neuron death [332]. In an *in vitro* model, expression of the cytoplasmic calcium-binding protein, calbindin-D28K, was able to rescue mouse spinal motor neurons from death downstream of mutant SOD1 expression, exogenous glutamate, or paraquat [332]. Additionally, overexpression of parvalbumin in spinal motor neurons in mutant SOD1 transgenic mice resulted in delayed onset of motor neuron degeneration and prolonged survival times compared to control mice [333]. Some beneficial effects were observed with administration of two antiglutamatergic agents, riluzole and gabapentin, to the transgenic mice [334]. Riluzole is currently the only medication for which effect on human disease has been demonstrated [335].

Accumulated neuronal damage resulting from oxidative stress over the course of decades may be a common pathogenetic mechanism shared by neurodegenerative disorders including ALS and PD. The fact that approximately 20% of all familial ALS cases are associated with gain-of-function mutations in the gene encoding the vital cellular antioxidant enzyme, SOD1, bears compelling evidence to the role of oxidative stress in motor neuron degeneration (reviewed in Ref. [336]). Oxidative damage has been documented at the molecular level in motor neuron disease [337,338]. Agents, which counteract oxidative stress, have had mixed success in preventing motor neuron degeneration in *in vitro* models [332], transgenic mice, and ALS patients themselves. Some beneficial effects in transgenic mice (either delayed onset, prolonged survival, or improved function) were observed with dietary supplementation with vitamin E [334], a modified catalase, given subcutaneously [339], and treatment with trientine and ascorbate [340]. *N*-acetyl-L-cysteine reduced lower motor neuron degeneration in wr [341,342]. Nitric oxide synthase inhibitors failed to alter disease course in the SOD1 mutant transgenic mice [343]. Recently a manganese porphyrin,

AEOL 10150, with antioxidant and free radical scavenging properties was tested in an early phase clinical trial after preclinical studies conducted in a mutant SOD1 transgenic mouse revealed therapeutic benefit of the compound when administered at symptom onset [344]. Adding to this list is the cellular energy substrate, pyruvate, which Park et al. have shown to retard disease progression in an SOD1 transgenic mouse model possibly due to its inherent antioxidant properties [345]. Collectively, these data suggest enzymes/compounds capable of reducing oxidative free radicals (as described in aforementioned neurodegenerative disorders) may be useful in treating motor neuron diseases.

Apoptosis plays a key role in development and morphogenesis. Interplay between antiapoptotic genes such as *bcl-2* and proapoptotic genes such as *bax* and *bad* may determine the activation of the enzymatic cascade of caspases leading to condensation and cleavage of nuclear DNA and the hallmark death of the cell with blebbing of the membrane and cytosolic contents. Increasing evidence gleaned from model systems suggest that motor neurons undergo apoptosis when subjected to pathological conditions related to motor neuron disease [346,347]. Thus, antiapoptotic strategies have been used in *in vitro* and transgenic mouse models to prevent motor neuron degeneration. Delayed onset of disease was observed in double transgenic (mutant SOD1 + *bcl-2* overexpression) [348] or in transgenic mice injected intraspinally with a recombinant AAV vector encoding the antiapoptotic gene *bcl-2* [349]. Furthermore, overexpression of another antiapoptotic family member, Bcl-$_X$L, was shown to protect cultured rat E15 motor neurons from glutamate excitotoxicity when transduced via a rAAV vector [350]. Some beneficial effects were also seen with the caspase inhibitor, zVAD-fmk [351]. Ultimately, protection against apoptosis may not be sufficient in preventing motor neuron death [352]. The contribution of apoptosis to the etiology of motor neuron death is still being debated [353].

The use of neurotrophic factors has been the most aggressively pursued approach for therapy in the past decade. Members of the NGF family, NGF itself, NT3, NT4/5-BDNF, CNTF, LIF, and GDNF, as well as other trophic factors like IGF-1 (myotrophin) and VEGF, have all been tested for their efficacy in supporting survival or regeneration of motor neurons in *in vitro* and mouse models. These factors may provide trophic support to the neuronal cell body, promote neurite elongation/sprouting, or reinnervation of the neuromuscular junction. Figure 39.13 depicts the neuroprotective effects of a VEGF-expressing lentiviral vector in a mouse model for ALS [354]. Ad vectors have been employed to introduce CNTF, BDNF, and NT-3 in rodent models of motor neuron degeneration [343,346,347]. Based on data from the earliest of these studies, CNTF, BDNF, GDNF, and myotrophin have all been employed at various stages of clinical trial evaluation in ALS patients [355,356]. In one approach, microencapsulated cells, genetically engineered to secrete CNTF, were implanted in lumbar intrathecal space [357]. Results from this study and others clinical trials have been discouraging [358,359]. Since then, alternate gene therapeutic strategies have been developed using AAV, lentiviral, and

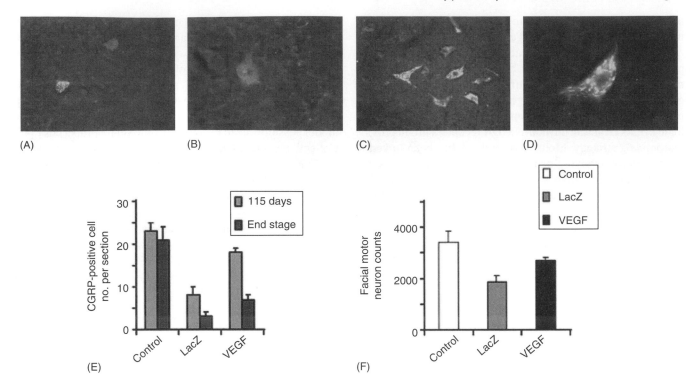

FIGURE 39.13 (See color insert following blank page 206.)   VEGF gene therapy protects spinal and brainstem motor neurons in SOD1^{G93A} transgenic mice. (A)–(D) Immunohistochemistry showing CGRP-positive neurons in lumbar spinal cord of EIAV-LacZ (A,B) and EIAV-VEGF injected animals (C,D). (E) Cell counts of surviving lumbar spinal cord motor neurons in control (wild-type), EIAV-LacZ, and EIAV-VEGF–treated SOD1^{G93A} mice at 115 days of age (blue) and at the end stage of disease (red). (F) Quantification of facial nucleus motor neurons in animals injected with EIAV-LacZ, EIAV-VEGF, and control animals at the end stage of disease. (From Azzouz, M., et al., *ExNeurol.*, 141(2), 225, 1996. © 2004 Nature Publishing Group. With permission.)

EBV vectors, which have exhibited therapeutic benefit in mouse models of ALS. One such strategy employed intramuscular grafting of myoblasts that were genetically modified to secrete the neurotrophic factor, GDNF, using a hybrid EBV/retroviral vector. Familial ALS transgenic mice treated by this method demonstrated reduced motor neuron loss and reduced disease progression [360]. Another approach exploited the retrograde transport capability of lentiviral and AAV vectors with the intention of delivering IGF-1- and VEGF-encoding genes to motor neurons of the spinal cord following intramuscular infusion of the viral vector. Both IGF-1 and VEGF delivered via AAV and lentiviral vectors, respectively, were able to delay the onset of symptoms and prolong the life expectancy in a mutant SOD1 mouse model of ALS [354,361].

A strategy to combat motor neuron degeneration in familial cases of ALS caused by the aberrant SOD1 enzyme would be to eliminate its cellular presence by silencing the dominantly inherited, mutant SOD1 encoding gene. In 2005, Ralph et al. employed a lentiviral-based RNAi strategy to specifically downregulate the expression of mutant human SOD1 (G93A) gene in a mouse model of ALS [362]. Transgenic mice that received the hSOD1-targeted shRNA vector demonstrated a significant reduction in hSOD1 expression, which translated into a significant delay in the onset of ALS symptoms and prolonged survival compared to mice receiving an empty

vector. Furthermore, efforts have also been made to design siRNAs that can specifically silence the mutated SOD1 allele, whilst permitting expression from the wild-type copy in a transgenic mouse model of ALS [363]. However, such a therapeutic approach would be limited to familial ALS patients in whom the specific mutation within the SOD1 allele is known.

Taken together, these studies suggest that the most effective therapies for motor neuron degeneration will include agents drawn from several of the classes described above. Further basic research addressing issues of vector, mode of delivery, and upregulation and control of expression of the therapeutic agent (such as neurotrophic factor) is crucial to the design of improved gene therapy strategies for motor neuron diseases [364–366].

39.3   FUTURE OF GENE THERAPY AND NEUROLOGICAL DISEASES

The ability to deliver and express genes *in vivo* has revolutionized the conduct of biological research and created potential new therapeutic approaches. Gene therapy offers potential advantages over traditional pharmacological therapy, a permanent correction of genetic anomaly thereby precluding the need for repetitive dosing. The past decade has seen significant strides in the development of novel gene therapeutic modalities to combat a wide spectrum of human genetic disorders. However, the challenges that present themselves are also numerous due to the complex etiologies of most neurologic diseases and limitations in current gene delivery technology. Neurodegenerative disorders affect millions of people, account for billions of dollars in health care costs annually, and as a group of diseases are without available curative therapy. As technology improves, this will lead to a better understanding of the underlying molecular mechanisms that constitute the pathogenesis of neurodegenerative disease and will facilitate the generation of superior disease recapitulating animal models. These models will thus expedite the discovery of additional efficacious therapies, which can easily be translated into a clinical setting for treating patients. With this, the full potential of gene therapy will be realized.

REFERENCES

1. Shen, Y., et al., Triple transduction with adeno-associated virus vectors expressing tyrosine hydroxylase, aromatic-L-amino-acid decarboxylase, and GTP cyclohydrolase I for gene therapy of Parkinson's disease. *Hum Gene Ther*, 2000, 11(11): 1509–19.

2. Zatloukal, K., et al., Transferrinfection: A highly efficient way to express gene constructs in eukaryotic cells. *Ann N Y Acad Sci*, 1992, 660: 136–53.

3. Wagner, E., et al., Coupling of adenovirus to transferrin-polylysine/DNA complexes greatly enhances receptor-mediated gene delivery and expression of transfected genes. *Proc Natl Acad Sci USA*, 1992, 89(13): 6099–103.

4. Zabner, J., et al., A chimeric type 2 adenovirus vector with a type 17 fiber enhances gene transfer to human airway epithelia. *J Virol*, 1999, 73(10): 8689–95.

5. Toyoda, K., et al., Cationic polymer and lipids enhance adenovirus-mediated gene transfer to rabbit carotid artery. *Stroke*, 1998, 29(10): 2181–8.

6. Beer, S.J., et al., Poly (lactic-glycolic) acid copolymer encapsulation of recombinant adenovirus reduces immunogenicity in vivo. *Gene Ther*, 1998, 5(6): 740–6.

7. Hacein-Bey-Abina, S., et al., A serious adverse event after successful gene therapy for X-linked severe combined immunodeficiency. *N Engl J Med*, 2003, 348(3): 255–6.

8. Byrnes, A.P., et al., Immunological instability of persistent adenovirus vectors in the brain: Peripheral exposure to vector leads to renewed inflammation, reduced gene expression, and demyelination. *J Neurosci*, 1996, 16: 3045–55.

9. Kelly, P.J., Stereotactic procedures for molecular neurosurgery. *Exp Neurol*, 1997, 144(1): 157–9.

10. Brooks, A.I., et al., Reproducible and efficient murine CNS gene delivery using a microprocessor-controlled injector. *J Neurosci Methods*, 1998, 80(2): 137–47.

11. Bankiewicz, K.S., et al., Convection-enhanced delivery of AAV vector in parkinsonian monkeys: In vivo detection of gene expression and restoration of dopaminergic function using pro-drug approach. *Exp Neurol*, 2000, 164(1): 2–14.

12. Bankiewicz, K.S., et al., Application of gene therapy for Parkinson's disease: Nonhuman primate experience. *Adv Pharmacol*, 1998, 42: 801–6.

13. Cunningham, J., et al., Distribution of AAV-TK following intracranial convection-enhanced delivery into rats. *Cell Transplant*, 2000, 9(5): 585–94.

14. Finberg, J.P., et al., Influence of selective inhibition of monoamine oxidase A or B on striatal metabolism of L-DOPA in hemiparkinsonian rats. *J Neurochem*, 1995, 65(3): 1213–20.

15. Barkats, M., et al., Adenovirus in the brain: Recent advances of gene therapy for neurodegenerative diseases. *Prog Neurobiol*, 1998, 55(4): 333–41.

16. Barkats, M., et al., Intrastriatal grafts of embryonic mesencephalic rat neurons genetically modified using an adenovirus encoding human Cu/Zn superoxide dismutase. *Neuroscience*, 1997, 78(3): 703–13.

17. Polymeropoulos, M.H., Genetics of Parkinson's disease. *Ann N Y Acad Sci*, 2000, 920: 28–32.

18. Schapira, A.H., Oxidative stress in Parkinson's disease. *Neuropathol Appl Neurobiol*, 1995, 21(1): 3–9.

19. Lang, A.E. and A.M. Lozano, Parkinson's disease. Second of two parts. *N Engl J Med*, 1998, 339(16): 1130–43.

20. Lang, A.E. and A.M. Lozano, Parkinson's disease. First of two parts. *N Engl J Med*, 1998, 339(15): 1044–53.

21. Zhang, Y., et al., Oxidative stress and genetics in the pathogenesis of Parkinson's disease. *Neurobiol Dis*, 2000, 7(4): 240–50.

22. Duvoisin, R.C., Overview of Parkinson's disease. *Ann N Y Acad Sci*, 1992, 648: 187–93.

23. Fahn, S., Future strategies for the treatment of Parkinson's disease. *Adv Neurol*, 1993, 60: 636–40.

24. Olanow, C.W. and W.G. Tatton, Etiology and pathogenesis of Parkinson's disease. *Annu Rev Neurosci*, 1999, 22: 123–44.

25. Singer, T.P., et al., Biochemical events in the development of parkinsonism induced by 1-methyl-4-phenyl-1,2,3,6-tetrahydropyridine. *J Neurochem*, 1987, 49(1): 1–8.

26. Brooks, A.I., et al., Paraquat elicited neurobehavioral syndrome caused by dopaminergic neuron loss. *Brain Res*, 1999, 823(1–2): 1–10.

27. Betarbet, R., et al., Chronic systemic pesticide exposure reproduces features of Parkinson's disease. *Nat Neurosci*, 2000, 3(12): 1301–6.

28. Thiruchelvam, M., et al., Potentiated and preferential effects of combined paraquat and maneb on nigrostriatal dopamine systems: Environmental risk factors for Parkinson's disease? *Brain Res*, 2000, 873(2): 225–34.

29. Greenamyre, J.T., et al., Response: Parkinson's disease, pesticides and mitochondrial dysfunction. *Trends Neurosci*, 2001, 24(5): 247.

30. Polymeropoulos, M.H., et al., Mapping of a gene for Parkinson's disease to chromosome 4q21–q23. *Science*, 1996, 274(5290): 1197–9.

31. Polymeropoulos, M.H., et al., Mutation in the alpha-synuclein gene identified in families with Parkinson's disease. *Science*, 1997, 276(5321): 2045–7.

32. Kruger, R., et al., Ala30Pro mutation in the gene encoding alpha-synuclein in Parkinson's disease. *Nat Genet*, 1998, 18(2): 106–8.

33. Zarranz, J.J., et al., The new mutation, E46K, of alpha-synuclein causes Parkinson and Lewy body dementia. *Ann Neurol*, 2004, 55(2): 164–73.

34. Conway, K.A., et al., Acceleration of oligomerization, not fibrillization, is a shared property of both alpha-synuclein mutations linked to early-onset Parkinson's disease: Implications for pathogenesis and therapy. *Proc Natl Acad Sci USA*, 2000, 97(2): 571–6.

35. Choi, W., et al., Mutation E46K increases phospholipid binding and assembly into filaments of human alpha-synuclein. *FEBS Lett*, 2004, 576(3): 363–8.

36. Singleton, A.B., et al., Alpha-synuclein locus triplication causes Parkinson's disease. *Science*, 2003, 302(5646): 841.

37. Mezey, E., et al., Alpha synuclein is present in Lewy bodies in sporadic Parkinson's disease. *Mol Psychiatry*, 1998, 3(6): 493–9.

38. Spillantini, M.G., et al., Alpha-synuclein in Lewy bodies. *Nature*, 1997, 388(6645): 839–40.

39. Spillantini, M.G., et al., Alpha-synuclein in filamentous inclusions of Lewy bodies from Parkinson's disease and dementia with lewy bodies. *Proc Natl Acad Sci USA*, 1998, 95(11): 6469–73.

40. Trojanowski, J.Q. and V.M. Lee, Aggregation of neurofilament and alpha-synuclein proteins in Lewy bodies: Implications for the pathogenesis of Parkinson disease and Lewy body dementia. *Arch Neurol*, 1998, 55(2): 151–2.

41. Wakabayashi, K., et al., Accumulation of alpha-synuclein/NACP is a cytopathological feature common to Lewy body disease and multiple system atrophy. *Acta Neuropathol (Berl)*, 1998, 96(5): 445–52.

42. Wakabayashi, K., et al., Alpha-synuclein immunoreactivity in glial cytoplasmic inclusions in multiple system atrophy. *Neurosci Lett*, 1998, 249(2–3): 180–2.

43. Ishikawa, A. and S. Tsuji, Clinical analysis of 17 patients in 12 Japanese families with autosomal-recessive type juvenile parkinsonism. *Neurology*, 1996, 47(1): 160–6.

44. Kitada, T., et al., Progress in the clinical and molecular genetics of familial parkinsonism. *Neurogenetics*, 2000, 2(4): 207–18.

45. Mizuno, Y., et al., Genetic and environmental factors in the pathogenesis of Parkinson's disease. *Adv Neurol*, 1999, 80: 171–9.

46. Kitada, T., et al., Mutations in the parkin gene cause autosomal recessive juvenile parkinsonism. *Nature*, 1998, 392(6676): 605–8.

47. Abbas, N., et al., A wide variety of mutations in the parkin gene are responsible for autosomal recessive parkinsonism in Europe. French Parkinson's Disease Genetics Study Group and the European Consortium on Genetic Susceptibility in Parkinson's Disease. *Hum Mol Genet*, 1999, 8(4): 567–74.

48. Hilker, R., et al., Positron emission tomographic analysis of the nigrostriatal dopaminergic system in familial parkinsonism associated with mutations in the parkin gene. *Ann Neurol*, 2001, 49(3): 367–76.

49. Lucking, C.B., et al., Association between early-onset Parkinson's disease and mutations in the parkin gene. French Parkinson's Disease Genetics Study Group. *N Engl J Med*, 2000, 342(21): 1560–7.

50. Riess, O. and R. Kruger, Parkinson's disease—a multifactorial neurodegenerative disorder. *J Neural Transm Suppl*, 1999, 56: 113–25.

51. Valente, E.M., et al., Localization of a novel locus for autosomal recessive early-onset parkinsonism, PARK6, on human chromosome 1p35-p36. *Am J Hum Genet*, 2001, 68(4): 895–900.

52. van de Warrenburg, B.P., et al., Clinical and pathologic abnormalities in a family with parkinsonism and parkin gene mutations. *Neurology*, 2001, 56(4): 555–7.

53. Zarate-Lagunes, M., et al., Parkin immunoreactivity in the brain of human and non-human primates: An immunohistochemical analysis in normal conditions and in Parkinsonian syndromes. *J Comp Neurol*, 2001, 432(2): 184–96.

54. Wintermeyer, P., et al., Mutation analysis and association studies of the UCHL1 gene in German Parkinson's disease patients. *Neuroreport*, 2000, 11(10): 2079–82.

55. Leroy, E., et al., Deletions in the Parkin gene and genetic heterogeneity in a Greek family with early onset Parkinson's disease. *Hum Genet*, 1998, 103(4): 424–7.

56. Tanaka, Y., et al., Inducible expression of mutant alpha-synuclein decreases proteasome activity and increases sensitivity to mitochondria-dependent apoptosis. *Hum Mol Genet*, 2001, 10(9): 919–926.

57. Bonifati, V., et al., Mutations in the DJ-1 gene associated with autosomal recessive early-onset parkinsonism. *Science*, 2003, 299(5604): 256–9.

58. Valente, E.M., et al., Hereditary early-onset Parkinson's disease caused by mutations in PINK1. *Science*, 2004, 304(5674): 1158–60.

59. Paisan-Ruiz, C., et al., Cloning of the gene containing mutations that cause PARK8-linked Parkinson's disease. *Neuron*, 2004, 44(4): 595–600.

60. Di Fonzo, A., et al., A frequent LRRK2 gene mutation associated with autosomal dominant Parkinson's disease. *Lancet*, 2005, 365(9457): 412–5.

61. Nichols, W.C., et al., Genetic screening for a single common LRRK2 mutation in familial Parkinson's disease. *Lancet*, 2005, 365(9457): 410–2.

62. Paisan-Ruiz, C., et al., Testing association between LRRK2 and Parkinson's disease and investigating linkage disequilibrium. *J Med Genet*, 2006, 43(2): e9.

63. Paisan-Ruiz, C., et al., LRRK2 gene in Parkinson disease: Mutation analysis and case control association study. *Neurology*, 2005, 65(5): 696–700.

64. West, A.B., et al., Parkinson's disease-associated mutations in LRRK2 link enhanced GTP-binding and kinase activities to neuronal toxicity. *Hum Mol Genet*, 2007, 16(2): 223–32.

65. Xiromerisiou, G., et al., Screening for SNCA and LRRK2 mutations in Greek sporadic and autosomal dominant Parkinson's disease: Identification of two novel LRRK2 variants. *Eur J Neurol*, 2007, 14(1): 7–11.

66. Hauser, M.A., et al., Expression profiling of substantia nigra in Parkinson disease, progressive supranuclear palsy, and frontotemporal dementia with parkinsonism. *Arch Neurol*, 2005, 62(6): 917–21.

67. Li, Y.J., et al., Investigation of the PARK10 Gene in Parkinson Disease. *Ann Hum Genet*, 2007, 71: 639–47.

68. Noureddine, M.A., et al., Genomic convergence to identify candidate genes for Parkinson disease: SAGE analysis of the substantia nigra. *Mov Disord*, 2005, 20(10): 1299–309.

69. Maguire-Zeiss, K.A. and H.J. Federoff, Convergent pathobiologic model of Parkinson's disease. *Ann N Y Acad Sci*, 2003, 991: 152–66.

70. Horellou, P., et al., Direct intracerebral gene transfer of an adenoviral vector expressing tyrosine hydroxylase in a rat model of Parkinson's disease. *Neuroreport*, 1994, 6(1): 49–53.

71. Choi-Lundberg, D.L., et al., Dopaminergic neurons protected from degeneration by GDNF gene therapy. *Science*, 1997, 275(5301): 838–41.

72. Choi-Lundberg, D.L., et al., Behavioral and cellular protection of rat dopaminergic neurons by an adenoviral vector encoding glial cell line-derived neurotrophic factor. *Exp Neurol*, 1998, 154(2): 261–75.

73. Bilang-Bleuel, A., et al., Intrastriatal injection of an adenoviral vector expressing glial-cell line-derived neurotrophic factor prevents dopaminergic neuron degeneration and behavioral impairment in a rat model of Parkinson disease. *Proc Natl Acad Sci USA*, 1997, 94(16): 8818–23.

74. Do Thi, N.A., et al., Delivery of GDNF by an E1, E3/E4 deleted adenoviral vector and driven by a GFAP promoter prevents dopaminergic neuron degeneration in a rat model of Parkinson's disease. *Gene Ther*, 2004, 11(9): 746–56.

75. Connor, B., et al., Differential effects of glial cell line-derived neurotrophic factor (GDNF) in the striatum and substantia nigra of the aged Parkinsonian rat. *Gene Ther*, 1999, 6(12): 1936–51.

76. Zigmond, M., et al., Neurochemical responses to 6-hydroxy-dopamine and L-DOPA therapy: Implications for PD, in *Neurotoxins and Neurodegenerative Disease*, W. Langston and A. Young, Editors. 1992, New York Academy of Sciences: New York. pp. 71–86.

77. Zheng, J.S., et al., Delayed gene therapy of glial cell line-derived neurotrophic factor is efficacious in a rat model of Parkinson's disease. *Brain Res Mol Brain Res*, 2005, 134(1): 155–61.

78. Barkats, M., et al., 1-Methyl-4-phenylpyridinium neurotoxicity is attenuated by adenoviral gene transfer of human Cu/Zn superoxide dismutase. *J Neurosci Res*, 2006, 83(2): 233–42.

79. Kaplitt, M.G., et al., Long-term gene expression and phenotypic correction using adeno-associated virus vectors in the mammalian brain. *Nat Genet*, 1994, 8(2): 148–54.

80. Mandel, R.J., et al., Characterization of intrastriatal recombinant adeno-associated virus-mediated gene transfer of human tyrosine hydroxylase and human GTP-cyclohydrolase I in a rat model of Parkinson's disease. *J Neurosci*, 1998, 18(11): 4271–84.

81. Bankiewicz, K.S., et al., Long-term clinical improvement in MPTP-lesioned primates after gene therapy with AAV-hAADC. *Mol Ther*, 2006, 14(4): 564–70.

82. During, M.J., et al., In vivo expression of therapeutic human genes for dopamine production in the caudates of MPTP-treated monkeys using an AAV vector. *Gene Ther*, 1998, 5(6): 820–7.

83. Klein, R.L., et al., Long-term actions of vector-derived nerve growth factor or brain-derived neurotrophic factor on choline acetyltransferase and Trk receptor levels in the adult rat basal forebrain. *Neuroscience*, 1999, 90(3): 815–21.

84. Klein, R.L., et al., Prevention of 6-hydroxydopamine-induced rotational behavior by BDNF somatic gene transfer. *Brain Res*, 1999, 847(2): 314–20.

85. Klein, R.L., et al., Neuron-specific transduction in the rat septohippocampal or nigrostriatal pathway by recombinant adeno-associated virus vectors. *Exp Neurol*, 1998, 150(2): 183–94.

86. Luo, J., et al., Subthalamic GAD gene therapy in a Parkinson's disease rat model. *Science*, 2002, 298(5592): 425–9.

87. Paterna, J.C., et al., DJ-1 and Parkin modulate dopamine-dependent behavior and inhibit MPTP-induced nigral dopamine neuron loss in mice. *Mol Ther*, 2007, 15(4): 698–704.

88. Bensadoun, J.C., et al., Lentiviral vectors as a gene delivery system in the mouse midbrain: Cellular and behavioral improvements in a 6-OHDA model of Parkinson's disease using GDNF. *Exp Neurol*, 2000, 164(1): 15–24.

89. Kordower, J.H., et al., Neurodegeneration prevented by lentiviral vector delivery of GDNF in primate models of Parkinson's disease. *Science*, 2000, 290(5492): 767–73.

90. Deglon, N., et al., Self-inactivating lentiviral vectors with enhanced transgene expression as potential gene transfer system in Parkinson's disease. *Hum Gene Ther*, 2000, 11(1): 179–90.

91. Rosenblad, C., et al., Long-term striatal overexpression of GDNF selectively downregulates tyrosine hydroxylase in the intact nigrostriatal dopamine system. *Eur J Neurosci*, 2003, 17(2): 260–70.

92. Georgievska, B., et al., Overexpression of glial cell line-derived neurotrophic factor using a lentiviral vector induces time- and dose-dependent downregulation of tyrosine hydroxylase in the intact nigrostriatal dopamine system. *J Neurosci*, 2004, 24(29): 6437–45.

93. Georgievska, B., et al., Regulated delivery of glial cell line-derived neurotrophic factor into rat striatum, using a tetracycline-dependent lentiviral vector. *Hum Gene Ther*, 2004, 15(10): 934–44.

94. Vercammen, L., et al., Parkin protects against neurotoxicity in the 6-hydroxydopamine rat model for Parkinson's disease. *Mol Ther*, 2006, 14(5): 716–23.

95. Azzouz, M., et al., Multicistronic lentiviral vector-mediated striatal gene transfer of aromatic L-amino acid decarboxylase, tyrosine hydroxylase, and GTP cyclohydrolase I induces sustained transgene expression, dopamine production, and functional improvement in a rat model of Parkinson's disease. *J Neurosci*, 2002, 22(23): 10302–12.

96. Kaplitt, M.G., et al., Preproenkephalin promoter yields region-specific and long-term expression in adult brain after direct in vivo gene transfer via a defective herpes simplex viral vector. *Proc Natl Acad Sci USA*, 1994, 91(19): 8979–83.

97. Jin, B.K., et al., Prolonged in vivo gene expression driven by a tyrosine hydroxylase promoter in a defective herpes simplex virus amplicon vector. *Hum Gene Ther*, 1996, 7: 2015–24.

98. During, M.J., et al., Long-term behavioral recovery in parkinsonian rats by an HSV vector expressing tyrosine hydroxylase. *Science*, 1994, 266(5189): 1399–403.

99. Sun, M., et al., Coexpression of tyrosine hydroxylase, GTP cyclohydrolase I, aromatic amino acid decarboxylase, and vesicular monoamine transporter 2 from a helper virus-free herpes simplex virus type 1 vector supports high-level, long-term biochemical and behavioral correction of a rat model of Parkinson's disease. *Hum Gene Ther*, 2004, 15(12): 1177–96.

100. Schwartz, B., et al., Gene transfer by naked DNA into adult mouse brain. *Gene Ther*, 1996, 3(5): 405–11.

101. Yang, K., et al., Gene therapy for central nervous system injury: The use of cationic liposomes: An invited review. *J Neurotrauma*, 1997, 14(5): 281–97.

102. Jiao, S., et al., Particle bombardment-mediated gene transfer and expression in rat brain tissues. *Biotechnology (N Y)*, 1993, 11(4): 497–502.

103. Wolff, J.A., et al., Direct gene transfer into mouse muscle in vivo. *Science*, 1990, 247(4949 Pt 1): 1465–8.

104. Martinez-Fong, D., et al., Neurotensin-SPDP-poly-L-lysine conjugate: A nonviral vector for targeted gene delivery to neural cells. *Brain Res Mol Brain Res*, 1999, 69(2): 249–62.

105. Abdallah, B., et al., A powerful nonviral vector for in vivo gene transfer into the adult mammalian brain: Polyethylenimine. *Hum Gene Ther*, 1996, 7(16): 1947–54.

106. Zhang, Y., et al., Intravenous nonviral gene therapy causes normalization of striatal tyrosine hydroxylase and reversal of motor impairment in experimental parkinsonism. *Hum Gene Ther*, 2003, 14(1): 1–12.

107. St George-Hyslop, P.H., et al., The genetic defect causing familial Alzheimer's disease maps on chromosome 21. *Science*, 1987, 235(4791): 885–90.

108. Van Broeckhoven, C., Alzheimer's disease: Identification of genes and genetic risk factors. *Prog Brain Res*, 1998, 117: 315–25.

109. Sherrington, R., et al., Cloning of a gene bearing missense mutations in early-onset familial Alzheimer's disease. *Nature*, 1995, 375(6534): 754–60.

110. Van Broeckhoven, C., Presenilins and Alzheimer disease. *Nat Genet*, 1995, 11(3): 230–2.

111. Price, D.L. and S.S. Sisodia, Mutant genes in familial Alzheimer's disease and transgenic models. *Annu Rev Neurosci*, 1998, 21: 479–505.

112. Haass, C., et al., Proteolytic processing of Alzheimer's disease associated proteins. *J Neural Transm Suppl*, 1998, 53: 159–67.

113. Thinakaran, G., The role of presenilins in Alzheimer's disease. *J Clin Invest*, 1999, 104(10): 1321–7.

114. Corder, E.H., et al., Protective effect of apolipoprotein E type 2 allele for late onset Alzheimer disease. *Nat Genet*, 1994, 7(2): 180–4.

115. Strittmatter, W.J. and A.D. Roses, Apolipoprotein E and Alzheimer disease. *Proc Natl Acad Sci USA*, 1995, 92(11): 4725–7.

116. Dodart, J.C., et al., Gene delivery of human apolipoprotein E alters brain Abeta burden in a mouse model of Alzheimer's disease. *Proc Natl Acad Sci USA*, 2005, 102(4): 1211–6.

117. Kamboh, M.I., et al., APOE*4-associated Alzheimer's disease risk is modified by alpha 1-antichymotrypsin polymorphism [published erratum appears in *Nat Genet* 199511(1): 104]. *Nat Genet*, 1995, 10(4): 486–8.

118. Okuizumi, K., et al., Genetic association of the very low density lipoprotein (VLDL) receptor gene with sporadic Alzheimer's disease. *Nat Genet*, 1995, 11(2): 207–9.

119. Xia, Y., et al., Genetic studies in Alzheimer's disease with an NACP/alpha-synuclein polymorphism. *Ann Neurol*, 1996, 40(2): 207–15.

120. Saitoh, T., et al., The CYP2D6B mutant allele is overrepresented in the Lewy body variant of Alzheimer's disease. *Ann Neurol*, 1995, 37(1): 110–2.

121. Tycko, B., et al., Polymorphisms in the human apolipoprotein-J/clusterin gene: Ethnic variation and distribution in Alzheimer's disease [published erratum appears in Hum Genet 1998 Apr;102(4):496]. *Hum Genet*, 1996, 98(4): 430–6.

122. Li, T., et al., Allelic functional variation of serotonin transporter expression is a susceptibility factor for late onset Alzheimer's disease. *Neuroreport*, 1997, 8(3): 683–6.

123. Curran, M., et al., HLA-DR antigens associated with major genetic risk for late-onset Alzheimer's disease. *Neuroreport*, 1997, 8(6): 1467–9.

124. Blalock, E.M., et al., Harnessing the power of gene microarrays for the study of brain aging and Alzheimer's disease: Statistical reliability and functional correlation. *Ageing Res Rev*, 2005, 4(4): 481–512.

125. Bartus, R.T., et al., The cholinergic hypothesis of geriatric memory dysfunction. *Science*, 1982, 217(4558): 408–14.

126. Whitehouse, P.J., et al., Alzheimer's disease and senile dementia: Loss of neurons in the basal forebrain. *Science*, 1982, 215(4537): 1237–9.

127. Hardy, J. and D. Allsop, Amyloid deposition as the central event in the aetiology of Alzheimer's disease. *Trends Pharmacol Sci*, 1991, 12(10): 383–8.

128. Kang, J., et al., The precursor of Alzheimer's disease amyloid A4 protein resembles a cell-surface receptor. *Nature*, 1987, 325(6106): 733–6.

129. Selkoe, D.J., Physiological production of the beta-amyloid protein and the mechanism of Alzheimer's disease. *Trends Neurosci*, 1993, 16(10): 403–9.

130. Akiyama, H., et al., Inflammation and Alzheimer's disease. *Neurobiol Aging*, 2000, 21(3): 383–421.

131. Hoyer, S., Age as risk factor for sporadic dementia of the Alzheimer type? *Ann N Y Acad Sci*, 1994, 719: 248–56.

132. Behl, C., Alzheimer's disease and oxidative stress: Implications for novel therapeutic approaches. *Prog Neurobiol*, 1999, 57(3): 301–23.

133. Coyle, J.T., et al., Alzheimer's disease: A disorder of cortical cholinergic innervation. *Science*, 1983, 219(4589): 1184–90.

134. Masliah, E., et al., Cortical and subcortical patterns of synaptophysinlike immunoreactivity in Alzheimer's disease. *Am J Pathol*, 1991, 138(1): 235–46.

135. Perry, E.K., et al., Changes in brain cholinesterases in senile dementia of Alzheimer type. *Neuropathol Appl Neurobiol*, 1978, 4(4): 273–7.

136. Perry, E.K., et al., Correlation of cholinergic abnormalities with senile plaques and mental test scores in senile dementia. *Br Med J*, 1978, 2(6150): 1457–9.

137. Hefti, F., et al., Chronic intraventricular injections of nerve growth factor elevate hippocampal choline acetyltransferase activity in adult rats with partial septo-hippocampal lesions. *Brain Res*, 1984, 293(2): 305–11.

138. Dutar, P., et al., The septohippocampal pathway: Structure and function of a central cholinergic system. *Physiol Rev*, 1995, 75(2): 393–427.

139. Blesch, A., et al., Neurotrophin gene therapy in CNS models of trauma and degeneration. *Prog Brain Res*, 1998, 117: 473–84.

140. Blesch, A. and M. Tuszynski, *Ex vivo* gene therapy for Alzheimer's disease and spinal cord injury. *Clin Neurosci*, 1995, 3(5): 268–74.

141. Tuszynski, M.H. and F.H. Gage, Bridging grafts and transient nerve growth factor infusions promote long-term central nervous system neuronal rescue and partial functional recovery. *Proc Natl Acad Sci USA*, 1995, 92(10): 4621–5.

142. Tuszynski, M.H. and F.H. Gage, Potential use of neurotrophic agents in the treatment of neurodegenerative disorders. *Acta Neurobiol Exp (Warsz)*, 1990, 50(4–5): 311–22.

143. Tuszynski, M.H., et al., Somatic gene transfer to the adult primate central nervous system: In vitro and in vivo characterization of cells genetically modified to secrete nerve growth factor. *Neurobiol Dis*, 1994, 1(1–2): 67–78.

144. Rosenberg, M.B., et al., Grafting genetically modified cells to the damaged brain: Restorative effects of NGF expression. *Science*, 1988, 242(4885): 1575–8.

145. Tuszynski, M.H., et al., A phase 1 clinical trial of nerve growth factor gene therapy for Alzheimer disease. *Nat Med*, 2005, 11(5): 551–5.

146. Wolfe, M.S., et al., Two transmembrane aspartates in presenilin-1 required for presenilin endoproteolysis and gamma-secretase activity. *Nature*, 1999, 398(6727): 513–7.

147. Singer, O., et al., Targeting BACE1 with siRNAs ameliorates Alzheimer disease neuropathology in a transgenic model. *Nat Neurosci*, 2005, 8(10): 1343–9.

148. Shen, J., et al., Skeletal and CNS defects in Presenilin-1-deficient mice. *Cell*, 1997, 89(4): 629–39.

149. Hong, C.S., et al., Herpes simplex virus RNAi and neprilysin gene transfer vectors reduce accumulation of Alzheimer's disease-related amyloid-beta peptide in vivo. *Gene Ther*, 2006, 13(14): 1068–79.

150. Iwata, N., et al., Metabolic regulation of brain Abeta by neprilysin. *Science*, 2001, 292(5521): 1550–2.

151. Marr, R.A., et al., Neprilysin gene transfer reduces human amyloid pathology in transgenic mice. *J Neurosci*, 2003, 23(6): 1992–6.

152. Wyss-Coray, T., Inflammation in Alzheimer disease: Driving force, bystander or beneficial response? *Nat Med*, 2006, 12(9): 1005–15.

153. Janelsins, M.C., et al., Early correlation of microglial activation with enhanced tumor necrosis factor-alpha and monocyte chemoattractant protein-1 expression specifically within the entorhinal cortex of triple transgenic Alzheimer's disease mice. *J Neuroinflammation*, 2005, 2: 23.

154. Morgan, D., et al., A beta peptide vaccination prevents memory loss in an animal model of Alzheimer's disease. *Nature*, 2000, 408(6815): 982–5.

155. Miravalle, L., et al., Substitutions at codon 22 of Alzheimer's abeta peptide induce diverse conformational changes and apoptotic effects in human cerebral endothelial cells. *J Biol Chem*, 2000, 275(35): 27110–6.

156. Janus, C., et al., A beta peptide immunization reduces behavioural impairment and plaques in a model of Alzheimer's disease. *Nature*, 2000, 408(6815): 979–82.

157. Bowers, W.J. and H.J. Federoff, Amyloid immunotherapy-engendered CNS inflammation. *Neurobiol Aging*, 2002, 23(5): 675–6; discussion 683–4.

158. Imbimbo, B.P., Toxicity of beta-amyloid vaccination in patients with Alzheimer's disease. *Ann Neurol*, 2002, 51(6): 794.

159. Zhang, J., et al., A novel recombinant adeno-associated virus vaccine reduces behavioral impairment and beta-amyloid plaques in a mouse model of Alzheimer's disease. *Neurobiol Dis*, 2003, 14(3): 365–79.

160. Kim, H.D., et al., Induction of anti-inflammatory immune response by an adenovirus vector encoding 11 tandem repeats of Abeta1–6: Toward safer and effective vaccines against Alzheimer's disease. *Biochem Biophys Res Commun*, 2005, 336(1): 84–92.

161. Bowers, W.J., et al., HSV amplicon-mediated Abeta vaccination in Tg2576 mice: Differential antigen-specific immune responses. *Neurobiol Aging*, 2005, 26(4): 393–407.

162. Orgogozo, J.M., et al., Subacute meningoencephalitis in a subset of patients with AD after Abeta42 immunization. *Neurology*, 2003, 61(1): 46–54.

163. Fukuchi, K., et al., Anti-Abeta single-chain antibody delivery via adeno-associated virus for treatment of Alzheimer's disease. *Neurobiol Dis*, 2006, 23(3): 502–11.

164. Levites, Y., et al., Intracranial adeno-associated virus-mediated delivery of anti-pan amyloid beta, amyloid beta40, and amyloid beta42 single-chain variable fragments attenuates plaque pathology in amyloid precursor protein mice. *J Neurosci*, 2006, 26(46): 11923–8.

165. Wraith, J.E., Lysosomal disorders. *Semin Neonatol*, 2002, 7(1): 75–83.

166. Barton, N.W., et al., Replacement therapy for inherited enzyme deficiency—macrophage-targeted glucocerebrosidase for Gaucher's disease. *N Engl J Med*, 1991, 324(21): 1464–70.

167. Sands, M.S. and B.L. Davidson, Gene therapy for lysosomal storage diseases. *Mol Ther*, 2006, 13(5): 839–49.

168. Barranger, J.A. and E. O'Rourke, Lessons learned from the development of enzyme therapy for Gaucher disease. *J Inherit Metab Dis*, 2001, 24 Suppl 2: 89–96; discussion 87–8.

169. Fabry, H., Angiokeratoma corporis diffusum—Fabry disease: Historical review from the original description to the introduction of enzyme replacement therapy. *Acta Paediatr Suppl*, 2002, 91(439): 3–5.

170. Eng, C.M., et al., Safety and efficacy of recombinant human alpha-galactosidase A—Replacement therapy in Fabry's disease. *N Engl J Med*, 2001, 345(1): 9–16.

171. Schiffmann, R., et al., Enzyme replacement therapy in Fabry disease: A randomized controlled trial. *Jama*, 2001, 285(21): 2743–9.

172. Kakkis, E., et al., Successful induction of immune tolerance to enzyme replacement therapy in canine mucopolysaccharidosis I. *Proc Natl Acad Sci USA*, 2004, 101(3): 829–34.

173. Shull, R.M., et al., Enzyme replacement in a canine model of Hurler syndrome. *Proc Natl Acad Sci USA*, 1994, 91(26): 12937–41.

174. Frisella, W.A., et al., Intracranial injection of recombinant adeno-associated virus improves cognitive function in a murine model of mucopolysaccharidosis type VII. *Mol Ther*, 2001, 3(3): 351–8.

175. Liu, G., et al., Functional correction of CNS phenotypes in a lysosomal storage disease model using adeno-associated virus type 4 vectors. *J Neurosci*, 2005, 25(41): 9321–7.

176. Skorupa, A.F., et al., Sustained production of beta-glucuronidase from localized sites after AAV vector gene transfer results in widespread distribution of enzyme and reversal of lysosomal storage lesions in a large volume of brain in mucopolysaccharidosis VII mice. *Exp Neurol*, 1999, 160(1): 17–27.

177. Sferra, T.J., et al., Recombinant adeno-associated virus-mediated correction of lysosomal storage within the central nervous system of the adult mucopolysaccharidosis type VII mouse. *Hum Gene Ther*, 2000, 11(4): 507–19.

178. Bosch, A., et al., Long-term and significant correction of brain lesions in adult mucopolysaccharidosis type VII mice using recombinant AAV vectors. *Mol Ther*, 2000, 1(1): 63–70.

179. Brooks, A.I., et al., Functional correction of established central nervous system deficits in an animal model of lysosomal storage disease with feline immunodeficiency virus-based vectors. *Proc Natl Acad Sci USA*, 2002, 99(9): 6216–21.

180. Elliger, S.S., et al., Enhanced secretion and uptake of beta-glucuronidase improves adeno-associated viral-mediated gene therapy of mucopolysaccharidosis type VII mice. *Mol Ther*, 2002, 5(5 Pt 1): 617–26.

181. Haq, E., et al., Molecular mechanism of psychosine-induced cell death in human oligodendrocyte cell line. *J Neurochem*, 2003, 86(6): 1428–40.

182. Shen, J.S., et al., Intraventricular administration of recombinant adenovirus to neonatal twitcher mouse leads to clinicopathological improvements. *Gene Ther*, 2001, 8(14): 1081–7.

183. Lin, D., et al., AAV2/5 vector expressing galactocerebrosidase ameliorates CNS disease in the murine model of globoid-cell leukodystrophy more efficiently than AAV2. *Mol Ther*, 2005, 12(3): 422–30.

184. Rafi, M.A., et al., AAV-mediated expression of galactocerebrosidase in brain results in attenuated symptoms and extended life span in murine models of globoid cell leukodystrophy. *Mol Ther*, 2005, 11(5): 734–44.

185. Baskin, G.B., et al., Genetic galactocerebrosidase deficiency (globoid cell leukodystrophy, Krabbe disease) in rhesus monkeys (*Macaca mulatta*). *Lab Anim Sci*, 1998, 48(5): 476–82.

186. Sleat, D.E., et al., A mouse model of classical late-infantile neuronal ceroid lipofuscinosis based on targeted disruption of the CLN2 gene results in a loss of tripeptidyl-peptidase I activity and progressive neurodegeneration. *J Neurosci*, 2004, 24(41): 9117–26.

187. Hofmann, S.L. and L. Peltonen, The neuronal ceroid lipofuscinosis, in *The Metabolic and Molecular Basis of Inherited Disease*, C.R. Scriver, A.L. Beaudet, W.S. Sly, and D. Valle, Editors. 2001, McGraw-Hill: New York. pp. 3877–94.

188. Passini, M.A., et al., Intracranial delivery of CLN2 reduces brain pathology in a mouse model of classical late infantile neuronal ceroid lipofuscinosis. *J Neurosci*, 2006, 26(5): 1334–42.

189. Griffey, M.A., et al., CNS-directed AAV2-mediated gene therapy ameliorates functional deficits in a murine model of infantile neuronal ceroid lipofuscinosis. *Mol Ther*, 2006, 13(3): 538–47.

190. Griffey, M., et al., Adeno-associated virus 2-mediated gene therapy decreases autofluorescent storage material and increases brain mass in a murine model of infantile neuronal ceroid lipofuscinosis. *Neurobiol Dis*, 2004, 16(2): 360–9.

191. The Huntington's Disease Collaborative Research Group, A novel gene containing a trinucleotide repeat that is expanded and unstable on Huntington's disease chromosomes. *Cell*, 1993, 72(6): 971–83.

192. Conneally, P.M., Huntington disease: Genetics and epidemiology. *Am J Hum Genet*, 1984, 36(3): 506–26.

193. Farrer, L.A. and P.M. Conneally, A genetic model for age at onset in Huntington disease. *Am J Hum Genet*, 1985, 37(2): 350–7.

194. Gusella, J.F., et al., A polymorphic DNA marker genetically linked to Huntington's disease. *Nature*, 1983, 306(5940): 234–8.

195. Gusella, J.F., et al., The genetic defect causing Huntington's disease: Repeated in other contexts? *Mol Med*, 1997, 3(4): 238–46.

196. Ashizawa, T., et al., CAG repeat size and clinical presentation in Huntington's disease. *Neurology*, 1994, 44(6): 1137–43.

197. Rosenblatt, A., et al., The association of CAG repeat length with clinical progression in Huntington disease. *Neurology*, 2006, 66(7): 1016–20.

198. Strehlow, A.N., et al., Wild-type huntingtin participates in protein trafficking between the Golgi and the extracellular space. *Hum Mol Genet*, 2007, 16(4): 391–409.

199. Li, S. and X.J. Li, Multiple pathways contribute to the pathogenesis of Huntington disease. *Mol Neurodegener*, 2006, 1: 19.

200. Wyttenbach, A., et al., Effects of heat shock, heat shock protein 40 (HDJ-2), and proteasome inhibition on protein aggregation in cellular models of Huntington's disease. *Proc Natl Acad Sci USA*, 2000, 97(6): 2898–903.

201. Krobitsch, S. and S. Lindquist, Aggregation of huntingtin in yeast varies with the length of the polyglutamine expansion and the expression of chaperone proteins. *Proc Natl Acad Sci USA*, 2000, 97(4): 1589–94.

202. Alves-Rodrigues, A., et al., Ubiquitin, cellular inclusions and their role in neurodegeneration. *Trends Neurosci*, 1998, 21(12): 516–20.

203. Saudou, F., et al., Huntingtin acts in the nucleus to induce apoptosis but death does not correlate with the formation of intranuclear inclusions. *Cell*, 1998, 95(1): 55–66.

204. Li, S.H. and X.J. Li, Huntingtin–protein interactions and the pathogenesis of Huntington's disease. *Trends Genet*, 2004, 20(3): 146–54.

205. Sugars, K.L. and D.C. Rubinsztein, Transcriptional abnormalities in Huntington disease. *Trends Genet*, 2003, 19(5): 233–8.

206. Li, X.J., et al., A huntingtin-associated protein enriched in brain with implications for pathology. *Nature*, 1995, 378(6555): 398–402.

207. Wanker, E.E., et al., HIP-I: A huntingtin interacting protein isolated by the yeast two-hybrid system. *Hum Mol Genet*, 1997, 6(3): 487–95.

208. Kalchman, M.A., et al., HIP1, a human homologue of S. cerevisiae Sla2p, interacts with membrane-associated huntingtin in the brain. *Nat Genet*, 1997, 16(1): 44–53.

209. Burke, J.R., et al., Huntingtin and DRPLA proteins selectively interact with the enzyme GAPDH. *Nat Med*, 1996, 2(3): 347–50.

210. Bao, J., et al., Expansion of polyglutamine repeat in huntingtin leads to abnormal protein interactions involving calmodulin. *Proc Natl Acad Sci USA*, 1996, 93(10): 5037–42.

211. Block-Galarza, J., et al., Fast transport and retrograde movement of huntingtin and HAP 1 in axons. *Neuroreport*, 1997, 8(9–10): 2247–51.

212. McGuire, J.R., et al., Interaction of Huntingtin-associated protein-1 with kinesin light chain: Implications in intracellular trafficking in neurons. *J Biol Chem*, 2006, 281(6): 3552–9.

213. Kittler, J.T., et al., Huntingtin-associated protein 1 regulates inhibitory synaptic transmission by modulating gamma-aminobutyric acid type A receptor membrane trafficking. *Proc Natl Acad Sci USA*, 2004, 101(34): 12736–41.

214. Kalchman, M.A., et al., Huntingtin is ubiquitinated and interacts with a specific ubiquitin-conjugating enzyme. *J Biol Chem*, 1996, 271(32): 19385–94.

215. Bae, B.I., et al., Mutant huntingtin: Nuclear translocation and cytotoxicity mediated by GAPDH. *Proc Natl Acad Sci USA*, 2006, 103(9): 3405–9.

216. Gu, M., et al., Mitochondrial defect in Huntington's disease caudate nucleus. *Ann Neurol*, 1996, 39(3): 385–9.

217. Benchoua, A., et al., Involvement of mitochondrial complex II defects in neuronal death produced by N-terminus fragment of mutated huntingtin. *Mol Biol Cell*, 2006, 17(4): 1652–63.

218. Zala, D., et al., Progressive and selective striatal degeneration in primary neuronal cultures using lentiviral vector coding for a mutant huntingtin fragment. *Neurobiol Dis*, 2005, 20(3): 785–98.

219. Cui, L., et al., Transcriptional repression of PGC-1alpha by mutant huntingtin leads to mitochondrial dysfunction and neurodegeneration. *Cell*, 2006, 127(1): 59–69.

220. Yamamoto, A., et al., Reversal of neuropathology and motor dysfunction in a conditional model of Huntington's disease. *Cell*, 2000, 101(1): 57–66.

221. Wang, Y.L., et al., Clinico-pathological rescue of a model mouse of Huntington's disease by siRNA. *Neurosci Res*, 2005, 53(3): 241–9.

222. Harper, S.Q., et al., RNA interference improves motor and neuropathological abnormalities in a Huntington's disease mouse model. *Proc Natl Acad Sci USA*, 2005, 102(16): 5820–5.

223. Rodriguez-Lebron, E., et al., Intrastriatal rAAV-mediated delivery of anti-huntingtin shRNAs induces partial reversal of disease progression in R6/1 Huntington's disease transgenic mice. *Mol Ther*, 2005, 12(4): 618–33.

224. Machida, Y., et al., rAAV-mediated shRNA ameliorated neuropathology in Huntington disease model mouse. *Biochem Biophys Res Commun*, 2006, 343(1): 190–7.

225. Yoon, K., et al., Targeted gene correction of episomal DNA in mammalian cells mediated by a chimeric RNA.DNA oligonucleotide. *Proc Natl Acad Sci USA*, 1996, 93(5): 2071–6.

226. Alexeev, V. and K. Yoon, Stable and inheritable changes in genotype and phenotype of albino melanocytes induced by an RNA–DNA oligonucleotide. *Nat Biotechnol*, 1998, 16(13): 1343–6.

227. Hefti, F., et al., Neurotrophic factors: What are they and what are they doing? in *Neurotrophic Factors*, S. Loughlin and J. Fallon, Editors. 1993, Academic Press: New York. pp. 25–50.

228. Bemelmans, A.P., et al., Brain-derived neurotrophic factor-mediated protection of striatal neurons in an excitotoxic rat model of Huntington's disease, as demonstrated by adenoviral gene transfer. *Hum Gene Ther*, 1999, 10(18): 2987–97.

229. Emerich, D.F., et al., Protective effect of encapsulated cells producing neurotrophic factor CNTF in a monkey model of Huntington's disease. *Nature*, 1997, 386(6623): 395–9.

230. Perez-Navarro, E., et al., Intrastriatal grafting of a GDNF-producing cell line protects striatonigral neurons from quinolinic acid excitotoxicity in vivo. *Eur J Neurosci*, 1999, 11(1): 241–9.

231. Kells, A.P., et al., AAV-mediated gene delivery of BDNF or GDNF is neuroprotective in a model of Huntington disease. *Mol Ther*, 2004, 9(5): 682–8.

232. McBride, J.L., et al., Viral delivery of glial cell line-derived neurotrophic factor improves behavior and protects striatal neurons in a mouse model of Huntington's disease. *Proc Natl Acad Sci USA*, 2006, 103(24): 9345–50.

233. Anton, R., et al., Neural transplantation of cells expressing the anti-apoptotic gene bcl-2. *Cell Transplant*, 1995, 4(1): 49–54.

234. Batistatou, A., et al., Bcl-2 affects survival but not neuronal differentiation of PC12 cells. *J Neurosci*, 1993, 13(10): 4422–8.

235. Boise, L.H., et al., bcl-x, a bcl-2-related gene that functions as a dominant regulator of apoptotic cell death. *Cell*, 1993, 74(4): 597–608.

236. Greenlund, L.J., et al., Role of BCL-2 in the survival and function of developing and mature sympathetic neurons. *Neuron*, 1995, 15(3): 649–61.

237. Hockenbery, D., et al., Bcl-2 is an inner mitochondrial membrane protein that blocks programmed cell death. *Nature*, 1990, 348(6299): 334–6.

238. Kane, D.J., et al., Bcl-2 inhibition of neural death: Decreased generation of reactive oxygen species. *Science*, 1993, 262 (5137): 1274–7.

239. Linnik, M.D., et al., Expression of bcl-2 from a defective herpes simplex virus-1 vector limits neuronal death in focal cerebral ischemia. *Stroke*, 1995, 26(9): 1670–4; discussion 1675.

240. Mah, S.P., et al., The protooncogene bcl-2 inhibits apoptosis in PC12 cells. *J Neurochem*, 1993, 60(3): 1183–6.

241. Martinou, J.C., et al., Overexpression of BCL-2 in transgenic mice protects neurons from naturally occurring cell death and experimental ischemia. *Neuron*, 1994, 13(4): 1017–30.

242. Zhong, L.T., et al., bcl-2 inhibits death of central neural cells induced by multiple agents. *Proc Natl Acad Sci USA*, 1993, 90(10): 4533–7.

243. Zhong, L.T., et al., BCL-2 blocks glutamate toxicity in neural cell lines. *Brain Res Mol Brain Res*, 1993, 19(4): 353–5.

244. McGill, J.K. and M.F. Beal, PGC-1alpha, a new therapeutic target in Huntington's disease? *Cell*, 2006, 127(3): 465–8.

245. Miller, T.W. and A. Messer, Intrabody applications in neurological disorders: Progress and future prospects. *Mol Ther*, 2005, 12(3): 394–401.

246. Colby, D.W., et al., Potent inhibition of huntingtin aggregation and cytotoxicity by a disulfide bond-free single-domain intracellular antibody. *Proc Natl Acad Sci USA*, 2004, 101(51): 17616–21.

247. Vannucci, R.C., Mechanisms of perinatal hypoxic-ischemic brain damage. *Semin Perinatol*, 1993, 17(5): 330–7.

248. Gottron, F.J., et al., Caspase inhibition selectively reduces the apoptotic component of oxygen-glucose deprivation-induced cortical neuronal cell death. *Mol Cell Neurosci*, 1997, 9(3): 159–69.

249. Korsmeyer, S.J., Regulators of cell death. *Trends Genet*, 1995, 11(3): 101–5.

250. Reed, J.C., Double identity for proteins of the Bcl-2 family. *Nature*, 1997, 387(6635): 773–6.

251. Wenger, R.H., Cellular adaptation to hypoxia: O_2-sensing protein hydroxylases, hypoxia-inducible transcription factors, and O_2-regulated gene expression. *Faseb J*, 2002, 16(10): 1151–62.

252. Helton, R., et al., Brain-specific knock-out of hypoxia-inducible factor-1alpha reduces rather than increases hypoxic-ischemic damage. *J Neurosci*, 2005, 25(16): 4099–107.

253. Wiesener, M.S., et al., Widespread hypoxia-inducible expression of HIF-2alpha in distinct cell populations of different organs. *Faseb J*, 2003, 17(2): 271–3.

254. Brusselmans, K., et al., Heterozygous deficiency of hypoxia-inducible factor-2alpha protects mice against pulmonary hypertension and right ventricular dysfunction during prolonged hypoxia. *J Clin Invest*, 2003, 111(10): 1519–27.

255. Gage, A.T. and P.K. Stanton, Hypoxia triggers neuroprotective alterations in hippocampal gene expression via a heme-containing sensor. *Brain Res*, 1996, 719(1–2): 172–8.

256. Kirino, T., et al., Ischemic tolerance. *Adv Neurol*, 1996, 71: 505–11.

257. Sakaki, T., et al., Brief exposure to hypoxia induces bFGF mRNA and protein and protects rat cortical neurons from prolonged hypoxic stress. *Neurosci Res*, 1995, 23(3): 289–96.

258. Goldberg, M.A., et al., Regulation of the erythropoietin gene: Evidence that the oxygen sensor is a heme protein. *Science*, 1988, 242(4884): 1412–5.

259. Maxwell, P.H., et al., Inducible operation of the erythropoietin 3′ enhancer in multiple cell lines: Evidence for a widespread oxygen-sensing mechanism. *Proc Natl Acad Sci USA*, 1993, 90(6): 2423–7.

260. Ho, D.Y., et al., Defective herpes simplex virus vectors expressing the rat brain glucose transporter protect cultured neurons from necrotic insults. *J Neurochem*, 1995, 65(2): 842–50.

261. Fink, S.L., et al., An adenoviral vector expressing the glucose transporter protects cultured striatal neurons from 3-nitropropionic acid. *Brain Res*, 2000, 859(1): 21–5.

262. Lawrence, M.S., et al., Overexpression of the glucose transporter gene with a herpes simplex viral vector protects striatal neurons against stroke. *J Cereb Blood Flow Metab*, 1996, 16(2): 181–5.

263. Dash, R., et al., A herpes simplex virus vector overexpressing the glucose transporter gene protects the rat dentate gyrus from an antimetabolite toxin. *Exp Neurol*, 1996, 137(1): 43–8.

264. Lawrence, M.S., et al., Overexpression of Bcl-2 with herpes simplex virus vectors protects CNS neurons against neurological insults in vitro and in vivo. *J Neurosci*, 1996, 16(2): 486–96.

265. Sun, Y., et al., Adeno-associated virus-mediated delivery of BCL-w gene improves outcome after transient focal cerebral ischemia. *Gene Ther*, 2003, 10(2): 115–22.

266. Kilic, E., et al., Adenovirus-mediated Bcl-X(L) expression using a neuron-specific synapsin-1 promoter protects against disseminated neuronal injury and brain infarction following focal cerebral ischemia in mice. *Neurobiol Dis*, 2002, 11(2): 275–84.

267. Xu, D.G., et al., Elevation of neuronal expression of NAIP reduces ischemic damage in the rat hippocampus. *Nat Med*, 1997, 3(9): 997–1004.

268. Xu, D., et al., Attenuation of ischemia-induced cellular and behavioral deficits by X chromosome-linked inhibitor of apoptosis protein overexpression in the rat hippocampus. *J Neurosci*, 1999, 19(12): 5026–33.

269. Pang, L., et al., Reduction of inflammatory response in the mouse brain with adenoviral-mediated transforming growth factor-ss1 expression. *Stroke*, 2001, 32(2): 544–52.

270. Masada, T., et al., Attenuation of intracerebral hemorrhage and thrombin-induced brain edema by overexpression of interleukin-1 receptor antagonist. *J Neurosurg*, 2001, 95(4): 680–6.

271. Yang, G.Y., et al., Attenuation of ischemic inflammatory response in mouse brain using an adenoviral vector to induce overexpression of interleukin-1 receptor antagonist. *J Cereb Blood Flow Metab*, 1998, 18(8): 840–7.

272. Tsai, T.H., et al., Gene treatment of cerebral stroke by rAAV vector delivering IL-1ra in a rat model. *Neuroreport*, 2003, 14(6): 803–7.

273. Bortner, C.D., et al., A primary role for K+ and Na+ efflux in the activation of apoptosis. *J Biol Chem*, 1997, 272(51): 32436–42.

274. Bortner, C.D. and J.A. Cidlowski, A necessary role for cell shrinkage in apoptosis. *Biochem Pharmacol*, 1998, 56(12): 1549–59.

275. Zhang, W.R., et al., Therapeutic time window of adenovirus-mediated GDNF gene transfer after transient middle cerebral artery occlusion in rat. *Brain Res*, 2002, 947(1): 140–5.

276. Tsai, T.H., et al., Gene therapy for treatment of cerebral ischemia using defective recombinant adeno-associated virus vectors. *Methods*, 2002, 28(2): 253–8.

277. Harvey, B.K., et al., HSV amplicon delivery of glial cell line-derived neurotrophic factor is neuroprotective against ischemic injury. *Exp Neurol*, 2003, 183(1): 47–55.

278. Shirakura, M., et al., Postischemic administration of Sendai virus vector carrying neurotrophic factor genes prevents delayed neuronal death in gerbils. *Gene Ther*, 2004, 11(9): 784–90.

279. Lawrence, M.S., et al., Herpes simplex virus vectors overexpressing the glucose transporter gene protect against seizure-induced neuron loss. *Proc Natl Acad Sci USA*, 1995, 92(16): 7247–51.

280. Yenari, M.A., et al., Calbindin d28k overexpression protects striatal neurons from transient focal cerebral ischemia. *Stroke*, 2001, 32(4): 1028–35.

281. Meier, T.J., et al., Increased expression of calbindin D28k via herpes simplex virus amplicon vector decreases calcium ion mobilization and enhances neuronal survival after hypoglycemic challenge. *J Neurochem*, 1997, 69(3): 1039–47.

282. Meier, T.J., et al., Gene transfer of calbindin D28k cDNA via herpes simplex virus amplicon vector decreases cytoplasmic calcium ion response and enhances neuronal survival following glutamatergic challenge but not following cyanide. *J Neurochem*, 1998, 71(3): 1013–23.

283. Davis, A.S., et al., Gene therapy using SOD1 protects striatal neurons from experimental stroke. *Neurosci Lett*, 2007, 411(1): 32–6.

284. Hoehn, B., et al., Glutathione peroxidase overexpression inhibits cytochrome C release and proapoptotic mediators to protect neurons from experimental stroke. *Stroke*, 2003, 34(10): 2489–94.

285. Beck, F.X., et al., Molecular chaperones in the kidney: Distribution, putative roles, and regulation. *Am J Physiol Renal Physiol*, 2000, 279(2): F203–15.

286. Hoehn, B., et al., Overexpression of HSP72 after induction of experimental stroke protects neurons from ischemic damage. *J Cereb Blood Flow Metab*, 2001, 21(11): 1303–9.

287. Yenari, M.A., et al., Gene therapy with HSP72 is neuroprotective in rat models of stroke and epilepsy. *Ann Neurol*, 1998, 44(4): 584–91.

288. Zhao, H., et al., Mild postischemic hypothermia prolongs the time window for gene therapy by inhibiting cytochrome C release. *Stroke*, 2004, 35(2): 572–7.

289. Tang, Y.L., et al., A hypoxia-inducible vigilant vector system for activating therapeutic genes in ischemia. *Gene Ther*, 2005, 12(15): 1163–70.

290. Shen, F., et al., Adeno-associated viral-vector-mediated hypoxia-inducible vascular endothelial growth factor gene expression attenuates ischemic brain injury after focal cerebral ischemia in mice. *Stroke*, 2006, 37(10): 2601–6.

291. McNamara, J.O., Emerging insights into the genesis of epilepsy. *Nature*, 1999, 399(6738 Suppl): A15–22.

292. Haberman, R., et al., Therapeutic liabilities of in vivo viral vector tropism: Adeno-associated virus vectors, NMDAR1 antisense, and focal seizure sensitivity. *Mol Ther*, 2002, 6(4): 495–500.

293. Akimitsu, T., et al., Epileptic seizures induced by N-acetyl-L-aspartate in rats: In vivo and in vitro studies. *Brain Res*, 2000, 861(1): 143–50.

294. Kitada, K., et al., Accumulation of *N*-acetyl-L-aspartate in the brain of the tremor rat, a mutant exhibiting absence-like seizure and spongiform degeneration in the central nervous system. *J Neurochem*, 2000, 74(6): 2512–9.

295. Yamada, J., et al., Rats with congenital tremor and curled whiskers and hair. *Jikken Dobutsu*, 1985, 34(2): 183–8.

296. Seki, T., et al., Adenoviral gene transfer of aspartoacylase into the tremor rat, a genetic model of epilepsy, as a trial of gene therapy for inherited epileptic disorder. *Neurosci Lett*, 2002, 328(3): 249–52.

297. Seki, T., et al., Adenoviral gene transfer of aspartoacylase ameliorates tonic convulsions of spontaneously epileptic rats. *Neurochem Int*, 2004, 45(1): 171–8.

298. Sasa, M., et al., Effects of antiepileptic drugs on absence-like and tonic seizures in the spontaneously epileptic rat, a double mutant rat. *Epilepsia*, 1988, 29(5): 505–13.

299. Serikawa, T. and J. Yamada, Epileptic seizures in rats homozygous for two mutations, zitter and tremor. *J Hered*, 1986, 77(6): 441–4.

300. Yamamoto, A., et al., Deficiency in protein L-isoaspartyl methyltransferase results in a fatal progressive epilepsy. *J Neurosci*, 1998, 18(6): 2063–74.

301. Kim, E., et al., Deficiency of a protein-repair enzyme results in the accumulation of altered proteins, retardation of growth, and fatal seizures in mice. *Proc Natl Acad Sci USA*, 1997, 94(12): 6132–7.

302. Ota, I.M. and S. Clarke, Calcium affects the spontaneous degradation of aspartyl/asparaginyl residues in calmodulin. *Biochemistry*, 1989, 28(9): 4020–7.

303. Ogawara, M., et al., Adenoviral expression of protein-L-isoaspartyl methyltransferase (PIMT) partially attenuates the biochemical changes in PIMT-deficient mice. *J Neurosci Res*, 2002, 69(3): 353–61.

304. Haberman, R.P., et al., Attenuation of seizures and neuronal death by adeno-associated virus vector galanin expression and secretion. *Nat Med*, 2003, 9(8): 1076–80.

305. Mazarati, A.M., et al., Modulation of hippocampal excitability and seizures by galanin. *J Neurosci*, 2000, 20(16): 6276–81.

306. McCown, T.J., Adeno-associated virus-mediated expression and constitutive secretion of galanin suppresses limbic seizure activity in vivo. *Mol Ther*, 2006, 14(1): 63–8.

307. Vezzani, A., et al., Neuropeptide Y: Emerging evidence for a functional role in seizure modulation. *Trends Neurosci*, 1999, 22(1): 25–30.

308. Noe, F., et al., Gene therapy in epilepsy: The focus on NPY. *Peptides*, 2007, 28(2): 377–83.

309. Richichi, C., et al., Anticonvulsant and antiepileptogenic effects mediated by adeno-associated virus vector neuropeptide Y expression in the rat hippocampus. *J Neurosci*, 2004, 24(12): 3051–9.

310. Kanter-Schlifke, I., et al., Seizure suppression by GDNF Gene therapy in animal models of epilepsy. *Mol Ther*, 2007, 15: 1106–13.

311. Gorter, J.A., et al., Potential new antiepileptogenic targets indicated by microarray analysis in a rat model for temporal lobe epilepsy. *J Neurosci*, 2006, 26(43): 11083–110.

312. Fisher, E.M., Modelling motor neuron degenerative disease. *Neuropathol Appl Neurobiol*, 1998, 24(2): 90–6.

313. Rosen, D.R., et al., Mutations in Cu/Zn superoxide dismutase gene are associated with familial amyotrophic lateral sclerosis. *Nature*, 1993, 362(6415): 59–62.

314. Wong, P.C., et al., Familial amyotrophic lateral sclerosis and Alzheimer's disease. Transgenic models. *Adv Exp Med Biol*, 1998, 446: 145–59.

315. Morrison, B.M., et al., Superoxide dismutase and neurofilament transgenic models of amyotrophic lateral sclerosis. *J Exp Zool*, 1998, 282(1–2): 32–47.

316. Julien, J.P., et al., Transgenic mice in the study of ALS: The role of neurofilaments. *Brain Pathol*, 1998, 8(4): 759–69.

317. Borchelt, D.R., et al., Transgenic mouse models of Alzheimer's disease and amyotrophic lateral sclerosis. *Brain Pathol*, 1998, 8(4): 735–57.

318. Reaume, A.G., et al., Motor neurons in Cu/Zn superoxide dismutase-deficient mice develop normally but exhibit enhanced cell death after axonal injury. *Nat Genet*, 1996, 13(1): 43–7.

319. Eyer, J. and A. Peterson, Neurofilament-deficient axons and perikaryal aggregates in viable transgenic mice expressing a neurofilament-beta-galactosidase fusion protein. *Neuron*, 1994, 12(2): 389–405.

320. Nagai, M., et al., Astrocytes expressing ALS-linked mutated SOD1 release factors selectively toxic to motor neurons. *Nat Neurosci*, 2007, 10(5): 615–22.

321. Di Giorgio, F.P., et al., Non-cell autonomous effect of glia on motor neurons in an embryonic stem cell-based ALS model. *Nat Neurosci*, 2007, 10(5): 608–14.

322. Gould, T.W., et al., Complete dissociation of motor neuron death from motor dysfunction by Bax deletion in a mouse model of ALS. *J Neurosci*, 2006, 26(34): 8774–86.

323. Rothstein, J.D., et al., Selective loss of glial glutamate transporter GLT-1 in amyotrophic lateral sclerosis. *Ann Neurol*, 1995, 38(1): 73–84.

324. Rothstein, J.D., Excitotoxic mechanisms in the pathogenesis of amyotrophic lateral sclerosis. *Adv Neurol*, 1995, 68: 7–20; discussion 21–7.

325. Carriedo, S.G., et al., Motor neurons are selectively vulnerable to AMPA/kainate receptor-mediated injury in vitro. *J Neurosci*, 1996, 16(13): 4069–79.

326. Van Den Bosch, L., et al., The role of excitotoxicity in the pathogenesis of amyotrophic lateral sclerosis. *Biochim Biophys Acta*, 2006, 1762(11–12): 1068–82.

327. Carriedo, S.G., et al., Rapid Ca^{2+} entry through Ca^{2+}-permeable AMPA/Kainate channels triggers marked intracellular Ca^{2+} rises and consequent oxygen radical production. *J Neurosci*, 1998, 18(19): 7727–38.

328. Williams, T.L., et al., Calcium-permeable alpha-amino-3-hydroxy-5-methyl-4-isoxazole propionic acid receptors: A molecular determinant of selective vulnerability in amyotrophic lateral sclerosis. *Ann Neurol*, 1997, 42(2): 200–7.

329. Sasaki, S., et al., Parvalbumin and calbindin D-28k immunoreactivity in transgenic mice with a G93A mutant SOD1 gene. *Brain Res*, 2006, 1083(1): 196–203.

330. Van Den Bosch, L., et al., Ca^{2+}-permeable AMPA receptors and selective vulnerability of motor neurons. *J Neurol Sci*, 2000, 180(1–2): 29–34.

331. Siklos, L., et al., Intracellular calcium parallels motoneuron degeneration in SOD-1 mutant mice. *J Neuropathol Exp Neurol*, 1998, 57(6): 571–87.

332. Roy, J., et al., Glutamate potentiates the toxicity of mutant Cu/Zn-superoxide dismutase in motor neurons by postsynaptic calcium-dependent mechanisms. *J Neurosci*, 1998, 18(23): 9673–84.

333. Beers, D.R., et al., Parvalbumin overexpression alters immune-mediated increases in intracellular calcium, and delays disease onset in a transgenic model of familial amyotrophic lateral sclerosis. *J Neurochem*, 2001, 79(3): 499–509.

334. Gurney, M.E., et al., Benefit of vitamin E, riluzole, and gabapentin in a transgenic model of familial amyotrophic lateral sclerosis. *Ann Neurol*, 1996, 39(2): 147–57.

335. Miller, R.G., et al., Riluzole for amyotrophic lateral sclerosis (ALS)/motor neuron disease (MND). *Cochrane Database Syst Rev*, 2000(2): CD001447.

336. Barber, S.C., et al., Oxidative stress in ALS: A mechanism of neurodegeneration and a therapeutic target. *Biochim Biophys Acta*, 2006, 1762(11–12): 1051–67.

337. Ferrante, R.J., et al., Evidence of increased oxidative damage in both sporadic and familial amyotrophic lateral sclerosis. *J Neurochem*, 1997, 69(5): 2064–74.

338. Przedborski, S. and V. Jackson-Lewis, Experimental developments in movement disorders: Update on proposed free radical mechanisms. *Curr Opin Neurol*, 1998, 11(4): 335–9.

339. Reinholz, M.M., et al., Therapeutic benefits of putrescine-modified catalase in a transgenic mouse model of familial amyotrophic lateral sclerosis. *Exp Neurol*, 1999, 159(1): 204–16.

340. Nagano, S., et al., Benefit of a combined treatment with trientine and ascorbate in familial amyotrophic lateral sclerosis model mice. *Neurosci Lett*, 1999, 265(3): 159–62.

341. Henderson, J.T., et al., Reduction of lower motor neuron degeneration in wobbler mice by N-acetyl-L-cysteine. *J Neurosci*, 1996, 16(23): 7574–82.

342. Jaarsma, D., et al., The antioxidant N-acetylcysteine does not delay disease onset and death in a transgenic mouse model of amyotrophic lateral sclerosis. *Ann Neurol*, 1998, 44(2): 293.

343. Upton-Rice, M.N., et al., Administration of nitric oxide synthase inhibitors does not alter disease course of amyotrophic lateral sclerosis SOD1 mutant transgenic mice. *Ann Neurol*, 1999, 45(3): 413–4.

344. Crow, J.P., et al., Manganese porphyrin given at symptom onset markedly extends survival of ALS mice. *Ann Neurol*, 2005, 58(2): 258–65.

345. Park, J.H., et al., Pyruvate slows disease progression in a G93A SOD1 mutant transgenic mouse model. *Neurosci Lett*, 2007, 413(3): 265–9.

346. Durham, H.D., et al., Aggregation of mutant Cu/Zn superoxide dismutase proteins in a culture model of ALS. *J Neuropathol Exp Neurol*, 1997, 56(5): 523–30.

347. Martin, L.J., Neuronal death in amyotrophic lateral sclerosis is apoptosis: Possible contribution of a programmed cell death mechanism. *J Neuropathol Exp Neurol*, 1999, 58(5): 459–71.

348. Kostic, V., et al., Bcl-2: Prolonging life in a transgenic mouse model of familial amyotrophic lateral sclerosis. *Science*, 1997, 277(5325): 559–62.

349. Azzouz, M., et al., Increased motoneuron survival and improved neuromuscular function in transgenic ALS mice after intraspinal injection of an adeno-associated virus encoding Bcl-2. *Hum Mol Genet*, 2000, 9(5): 803–11.

350. Garrity-Moses, M.E., et al., Neuroprotective adeno-associated virus Bcl-xL gene transfer in models of motor neuron disease. *Muscle Nerve*, 2005, 32(6): 734–44.

351. Li, M., et al., Functional role of caspase-1 and caspase-3 in an ALS transgenic mouse model. *Science*, 2000, 288(5464): 335–9.

352. Houseweart, M.K. and D.W. Cleveland, Bcl-2 overexpression does not protect neurons from mutant neurofilament-mediated motor neuron degeneration. *J Neurosci*, 1999, 19(15): 6446–56.

353. Migheli, A., et al., Lack of apoptosis in mice with ALS. *Nat Med*, 1999, 5(9): 966–7.

354. Azzouz, M., et al., VEGF delivery with retrogradely transported lentivector prolongs survival in a mouse ALS model. *Nature*, 2004, 429(6990): 413–7.

355. Dittrich, F., et al., Pharmacokinetics of intrathecally applied BDNF and effects on spinal motoneurons. *Exp Neurol*, 1996, 141(2): 225–39.

356. Cudkowicz, M.E., et al., Intrathecal administration of recombinant human superoxide dismutase 1 in amyotrophic lateral sclerosis: A preliminary safety and pharmacokinetic study. *Neurology*, 1997, 49(1): 213–22.

357. Aebischer, P., et al., Intrathecal delivery of CNTF using encapsulated genetically modified xenogeneic cells in amyotrophic lateral sclerosis patients. *Nat Med*, 1996, 2(6): 696–9.

358. Borasio, G.D., et al., A placebo-controlled trial of insulin-like growth factor-I in amyotrophic lateral sclerosis. European ALS/IGF-I Study Group. *Neurology*, 1998, 51(2): 583–6.

359. A controlled trial of recombinant methionyl human BDNF in ALS: The BDNF Study Group (Phase III). *Neurology*, 1999, 52(7): 1427–33.

360. Mohajeri, M.H., et al., Intramuscular grafts of myoblasts genetically modified to secrete glial cell line-derived neurotrophic factor prevent motoneuron loss and disease progression in a mouse model of familial amyotrophic lateral sclerosis. *Hum Gene Ther*, 1999, 10(11): 1853–66.

361. Kaspar, B.K., et al., Retrograde viral delivery of IGF-1 prolongs survival in a mouse ALS model. *Science*, 2003, 301(5634): 839–42.

362. Ralph, G.S., et al., Silencing mutant SOD1 using RNAi protects against neurodegeneration and extends survival in an ALS model. *Nat Med*, 2005, 11(4): 429–33.

363. Xia, X., et al., Allele-specific RNAi selectively silences mutant SOD1 and achieves significant therapeutic benefit in vivo. *Neurobiol Dis*, 2006, 23(3): 578–86.

364. Bartlett, S.E., et al., Retrograde axonal transport of neurotrophins: Differences between neuronal populations and implications for motor neuron disease. *Immunol Cell Biol*, 1998, 76(5): 419–23.

365. Haase, G., et al., Therapeutic benefit of ciliary neurotrophic factor in progressive motor neuronopathy depends on the route of delivery. *Ann Neurol*, 1999, 45(3): 296–304.

366. Bledsoe, A.W., et al., Cytokine production in motor neurons by poliovirus replicon vector gene delivery. *Nat Biotechnol*, 2000, 18(9): 964–9.

367. Bowers, W.J., et al., Gene therapeutic approaches to the treatment of Parkinson's disease. *Clin Neurosci Res*, 2001, 1: 483–495.

368. Regulier, E., et al., Dose-dependent neuroprotective effect of ciliary neurotrophic factor delivered via tetracycline-regulated lentiviral vectors in the quinolinic acid rat model of Huntington's disease. *Hum Gene Ther*, 2002, 13(16): 1981–90.

369. Halterman, M.W., et al., Hypoxia-inducible factor-1α mediates hypoxia-induced delayed neuronal death that involves p53. *J Neurosci*, 1999, 19(16): 6818–6824.

370. Tsai, T.H., et al., Recombinant adeno-associated virus vector expressing glial cell line-derived neurotrophic factor reduces ischemia-induced damage. *Exp Neurol*, 2000, 166(2): 266–75.

40 Principles of Gene Therapy for Inborn Errors of Metabolism

Jon A. Wolff and Cary O. Harding

CONTENTS

40.1 INTRODUCTION

40.1.1 HISTORICAL PERSPECTIVE ON INBORN ERRORS OF METABOLISM

Inborn errors of metabolism (IEM) have played a central role in the formulation of modern genetics. The hallmark of IEM is the accumulation of a biochemical in a bodily tissue. With the development of chemical analytical techniques, it became possible to identify and measure these biochemicals and correlate them with specific diseases (Figure 40.1). Knowledge of metabolic pathways enabled enzymatic defects to be identified, which eventually led to discovery of the cognate proteins and genes.

The appreciation that inborn susceptibilities play important roles in diseases was first promulgated at the beginning of the twentieth century by Alfred Garrod. He formulated the concept of an inherited metabolic disease on the basis of his studies of patients with alkaptonuria, albinism, cystinuria, and pentosuria. Cognizant of the laws of Mendel, he postulated that the relevant biochemical accumulates due to a metabolic block that is inherited in a recessive process.

The one gene–one enzyme principle developed by Beadle and Tatum provided the next conceptual framework for understanding IEM. This principle provides that metabolic processes are the result of specific enzymatic steps, which are under the control of a single gene. A mutation in a gene leads to deficiency of the enzyme that catalyzes the specific step. The molecular basis for a defective enzyme was provided by Pauling and Ingram's experiments on sickle-cell anemia,

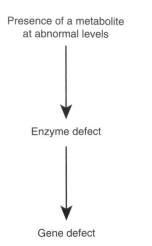

Presence of a metabolite
at abnormal levels

↓

Enzyme defect

↓

Gene defect

FIGURE 40.1 Flow of information in elucidating the genetic basis of metabolic disorders.

while the molecular basis for a defective gene was provided by Watson and Crick and subsequent elaboration of the central dogma. This dogma defines the flow of information as proceeding from DNA to RNA to protein. The objective of gene therapy is to modulate the flow of genetic information so as to attenuate the disease state.

40.1.2 Historical Perspective on Gene Therapy of IEM

IEM have also played a pivotal role in the formulation of gene therapy [1]. In fact, many of the first human clinical trials in gene therapy were for IEM. In the late 1960s, Rogers attempted to treat three siblings with arginase deficiency by injecting them with the Shope virus on the basis of the incorrect assumption that the virus contained an arginase gene. While being ahead of his time in anticipating the development of viral vectors, the injections had no effect on the subjects' arginine levels. In the more modern era of gene therapy, the first human trials for treating a disease involved children with severe combined immunodeficiency (SCID) caused by a deficiency in adenosine deaminase (ADA).

In addition to human gene therapy trials, IEM have played an important role in the development of gene therapy tools [1]. Cell lines deficient in hypoxanthine phosphoribosyl transferase (HPRT) and grown in hypoxanthine/aminopterin/thymidine (HAT) media enabled the selection for genetically modified cells that take up the HPRT gene in conjunction with other foreign genes. Similarly, cell lines deficient in thymidine kinase (TK) can be used for gene transfer selection. Many of the initial clinical trials in the gene therapy field revolved around IEM such as ADA deficiency, familial hypercholesterolemia (FH), ornithine transcarbomylase deficiency (OTC), and Canavans disease.

With the advent of positional cloning and the human genome project, disorders are being linked to defective genes without any understanding of how metabolism has been disrupted.

The current challenge will be to identify how the defective gene leads to a disturbance in development or homeostasis. Gene transfer and expression in animals and humans will provide critical tests for hypotheses of pathogenesis.

40.2 BASIS FOR GENE THERAPY FOR IEM

40.2.1 Types of IEM

One common type of IEM is caused by deficiency of an enzyme that catalyzes the conversion of one chemical to another (Figure 40.2A). Deficiency of a specific enzyme can cause disease through three separate mechanisms: (1) excessive accumulation of substrate to toxic levels, (2) deficiency of an essential product, or (3) metabolism of the substrate through alternative biochemical pathways leading to toxic secondary metabolites. Examples of such IEM include phenylketonuria (PKU) and methylmalonic aciduria.

IEM can also be caused by deficiency of protein that is involved in the transport of a metabolite (Figure 40.2B). Examples include the cystine transporter in cystinosis and the low-density lipoprotein (LDL) receptor in FH.

Other genes relevant to IEM are required for the proper formation of organelles (Figure 40.2C). Neonatal adrenoleukodystrophy and Zellweger syndrome are caused by defects in genes that are required for the proper formation of peroxisomes. Disorders affecting the biogenesis of organelles lead to the disruption of multiple enzymatic pathways and organelle functions.

40.2.2 Different Pathogenesis Models for IEM

The pathogenesis of IEM can be explained by several models (Figure 40.3). One major category includes IEM in which organ dysfunction occurs by a circulating toxic metabolite (Figure 40.3A). Another major category is organ dysfunction resulting from a cell-autonomous process (Figure 40.3B). Although these concepts are useful in formulating gene therapy approaches, it should be appreciated that they are only models and that our understanding of the pathogenesis for many IEM is incomplete. In fact, gene therapy trials may provide decisive information concerning the mechanism by which the metabolic defect leads to the diseased state.

40.2.3 Circulating Toxic Metabolite

In this class of disorder, a metabolite accumulates in one tissue as a result of an enzymatic deficiency (Figure 40.3A). This leads to increased metabolite levels in the blood and toxicity in other tissues. The prototype for this type of disorder is PKU in which deficiency of hepatic phenylalanine hydroxylase (PAH) leads to increased blood levels of phenylalanine and toxic effects to the developing brain. FH is another IEM that fits this model. Deficiency of the LDL receptor (LDLR) in the liver leads to increased levels of LDL and subsequent damage to the coronary arteries.

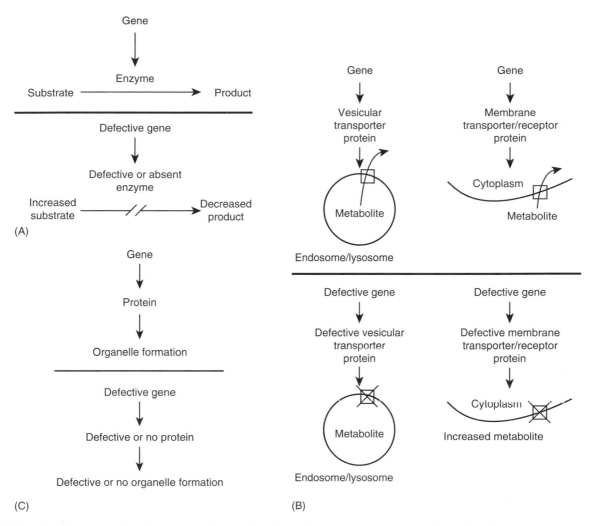

FIGURE 40.2 Types of normal functions that are disrupted in IEMs: (A) defects in enzymatic activity, (B) defects in the uptake or transport of metabolites, and (C) defects in the formation of organelles.

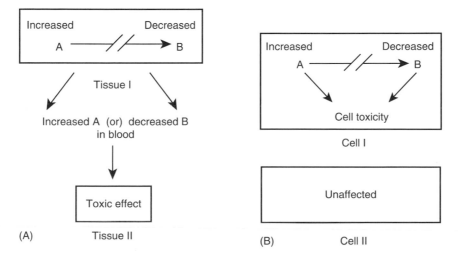

FIGURE 40.3 Two models by which IEM leads to cellular toxicity: (A) circulating toxic metabolite and (B) cell autonomous toxicity.

A corollary of this model is intraorgan toxicity from a metabolite that accumulates in an extracellular space within the affected tissue. It is particularly applicable to the central nervous system (CNS). Some IEM associated with neurological dysfunction may be caused by a toxic metabolite that accumulates within the brain and circulates in the cerebral spinal fluid (CSF). Gene therapy could then be predicated on providing gene expression in any cell within the brain as long as the expressed enzyme could lower levels of the toxic metabolite in the CNS. The CSF could provide the conduit for such exchange.

A metabolic defect in one tissue could also harm another tissue by decreasing the circulating level of a metabolite. For example, a defect in gluconeogenesis that occurs in the liver and muscle (e.g., glycogen storage disorder) can cause hypoglycemia and damage to the brain.

40.2.4 CELL AUTONOMOUS TOXICITY

In other IEM, the metabolic defect only leads to toxicity to the cell that has the metabolic deficiency (Figure 40.3B). Cellular toxicity results from either increased or decreased levels of a metabolite within the affected cell. For these disorders, gene therapy would be effective only if the normal gene is targeted to the dysfunctional cell.

40.2.5 METHOD OF GENE CORRECTION

A variety of parameters of expression are important determinants of the ability of gene therapy to treat specific disorders. Some generalizations can be made concerning the expression requirements for IEM (Table 40.1). Most IEM are recessive conditions and addition of a single gene copy is sufficient to correct the disease phenotype. In effect, gene addition converts the patient to a biochemical state analogous to that of a carrier. For those patients with single-point mutations, targeted gene correction using gene conversion or homologous recombination is a possible therapy, but gene correction is not necessary if a functional gene can be added. The obvious therapeutic gene to be added in IEM is the human gene that is defective in the disease state, but it is conceivable that a therapeutic effect could be achieved using another gene. For example,

a gene from another species could metabolize a toxic metabolite by a different mechanism.

40.2.6 REQUIREMENTS FOR EXPRESSION PERSISTENCE

For most IEM, gene expression does not have to be regulated and can be constant. Most genes involved in IEM are considered "housekeeping" genes. In contrast, in diabetes mellitus, insulin expression has to be regulated in response to blood glucose levels.

Given that IEM are chronic conditions, persistent expression is needed. It would be best if gene correction and therefore a "cure" could be done with one or few administrations. If expression cannot be persistent after one gene dose, then repetitive administrations are required. Repetitive administrations can be problematic for some vectors such as adenoviral and adenoassociated vectors that induce neutralizing antibodies. Loss of expression from vectors can be a result of removal of the foreign DNA, promoter suppression, or rejection of the foreign gene product.

Immune effects can arise even if the gene product is intracellular because all parts of proteins are presented to the immune system via the MHC I complex. The important issue is whether, in the disease state, the patient expresses any residual native protein and is immunologically tolerant to the normal gene product. One measure of this is whether tissues from the patient exhibit crossreactive material (CRM), protein that crossreacts with antibodies against the native protein. This is best determined by performing immunoblot (Western blot) analysis. Even if protein is not present, native protein could have been produced but be unstable. Expression of the foreign gene in such a patient may not induce an immune effect because the protein is not recognized as foreign. Further experience is necessary to determine whether the immune system will prevent stable expression of the normal gene in patients with IEM.

40.2.7 REQUIREMENTS FOR EXPRESSION LEVELS

The level of expression is a critical determinant for the success of a gene therapy. For most IEM, foreign gene expression only has to be greater than 5% of normal levels in order to

---

TABLE 40.1
Generalizations Concerning Expression Requirements for Gene Therapy of IEM

Expression Parameter	Requirement
Method of modification	Gene addition is sufficient
Therapeutic gene	Normal gene that is defective in patient
Duration	Persistent
Regulation	Not needed
Levels	>5% of normal levels
Target tissues	Liver, CNS, blood cells, muscle, heterologous expression possible in some disorders

attenuate the majority of the diseased state. This is based on clinical experience in which the percent of residual enzyme activity is correlated with the phenotype. In many IEM, people with more than 5% of normal enzymatic activity are free of symptoms. If enzymatic activity is between 1% and 5%, their clinical course is less severe than patients with 0% of enzymatic activity.

Although the total enzymatic activity is one measure, the percent of cells expressing the foreign gene may also be important. Overexpression in a few cells may not lead to a therapeutic effect if the expressed enzyme alone cannot completely produce the metabolic conversion. The protein deficient in the patient may be part of an enzymatic complex so that overexpression of one component would not necessarily lead to higher activity of the complete complex. Similarly, other enzymatic steps, cofactors, or transport of metabolites may limit the ability of the cell to perform the required metabolic conversion at a rate higher than the normal level. If so, the therapeutic gene has to be expressed in more than 5% of the target cells.

40.2.8　Target Tissues

Details of the pathogenesis for the IEM need to be understood, and the target tissue has to be tailored for each disorder. For IEM that fit the "circulating toxic metabolite" model, the therapeutic gene does not necessarily have to be targeted to the tissue that normally expresses the affected gene (Figure 40.4). Although correction of the deficient enzymatic activity in the affected organ would be most straightforward, expression within a heterologous tissue (different from that which normally expresses the enzyme) could clear the circulating toxic metabolite and attenuate the disease state. For this approach to be effective, the enzyme must be functional within the heterologous tissue. Restrictions on enzymatic function can include requirements for protein subunits, cofactors, substrate, and clearance of product. Given the ability for several gene transfer systems (e.g., plasmid DNA, adenoviral

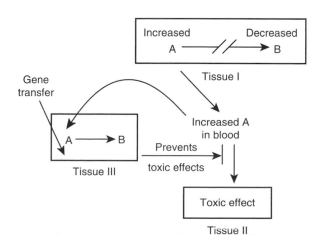

FIGURE 40.4 Heterologous tissue expression of a therapeutic gene to treat an IEM in which a circulating toxic metabolite causes the diseased state.

vectors, and AAV vectors) to express foreign genes stably in muscle, it will be a useful tissue for many heterologous gene therapy approaches. Blood cells derived from genetically modified stem cells are another candidate tissue for heterologous gene expression if the problems associated with stable foreign expression are solved.

For IEM that fit the "cell autonomous" model, expression within the affected cell is generally required. The exception is for the lysosomal storage disorders in which the enzyme can be transferred from one cell to another.

For IEM that affect the brain, "global" gene expression throughout the brain may be required. Alternatively, specific neurological symptoms could be treated by targeting specific regions of the brain. For example, in Lesch–Nyhan syndrome, choreoathetoid movements could be treated by targeting the basal ganglia.

40.2.9　Mitochondrial Disorders of Oxidative Phosphorylation

Several IEM are caused by defective oxidative phosphorylation within the respiratory chain complex of mitochondria. A unique feature of mitochondria is that 13 of the more than 80 respiratory chain subunits are encoded within the mitochondrial genome. The mitochondria also contain 22 transfer RNA (tRNA) and two ribosomal RNA that enable protein synthesis within the mitochondria. The remaining 70 or so respiratory chain subunits are encoded within the nuclear genome. These proteins are produced within the cytoplasm and contain an amino terminus that targets their entry into the mitochondria by interacting with a number of chaperone and transport proteins.

For disorders caused by mutations in the nuclear encoded respiratory chain genes, the gene therapy approaches described above are germane. However, disorders caused by mutations in the mitochondrial genome offer additional challenges for gene therapy. One approach would be to express the deficient subunit within the nucleus, regardless of its native mitochondrial origin. The subunit could be modified to contain an amino leader sequence to enable entry into the mitochondria.

The other approach of genetically modifying the mitochondrial genome is at an early conceptual stage. Toward this end, a peptide mitochondria-targeting sequence has been covalently attached to oligonucleotide to enable mitochondrial entry. The oligonucleotide could correct a point mutation by some type of gene conversion or recombination process. Point mutations occur in mitochondrial disorders such as Leber hereditary optic atrophy (LHON), myoclonic epilepsy and ragged-red fiber disease (MERRF), and mitochondrial encephalomyopathy, lactic acidosis, and strokelike episodes (MELAS). An alternative treatment approach would be the addition of functional tRNA genes to patients with mitochondrial disorders such as MERRF or MELAS that are caused by tRNA mutations. The treatment of mitochondrial DNA deletion diseases would require the delivery of larger DNA sequences (>5 kb), which would be more challenging. Deletions occur in disorders such as Kearns–Sayre syndrome. Another option would be to deliver normal mitochondria *in toto*.

Different mitochondria can proliferate in a tissue at different rates. This may explain why inborn and somatic (acquired) mitochondrial defects often present in later life. Any genetic modification of mitochondria must enable the corrected mitochondria to have a proliferation advantage over the abnormal mitochondria in order to achieve a permanent cure. A final challenge for mitochondrial disorders is that they often involve the nervous system, which is less accessible than other organs to therapeutic endeavors.

40.2.10 Newborn and Prenatal Screening

Gene therapy for IEM will have a significant impact on newborn screening programs and vice versa. Screening for IEM at birth enables gene therapy to be initiated prior to the onset of symptoms and any irreversible tissue damage and thereby increases the value of the gene therapy. Irreversible brain damage occurs in many IEM when a neonatal metabolic crisis is not prevented. For example, the extent of perinatal hyperammonemia in a urea cycle defect, OTC, has been directly correlated with intelligence in later life.

One criterion for the initiation of newborn screening for a particular disorder is whether an effective treatment exists. The development of effective gene therapy for a disorder could satisfy this criterion. Another criterion is the availability of a reliable, inexpensive laboratory method for disease detection.

Currently, most states in the United States and many other nations are screening for PKU and galactosemia. Screening for maple syrup urine disease or homocysteinemia is less common. Tandem mass spectroscopy procedures are being developed for analyzing blood spots in amino acids and organic acids conjugated to carnitine (acylcarnitines) in order to detect many of the disorders in amino acid and fat metabolism and organic acidurias. Such comprehensive newborn screening programs developed in conjunction with new gene therapies are having a major impact on the morbidity and mortality of IEM.

Many IEM can be reliably diagnosed in the prenatal period. As intrauterine gene therapy approaches are developed, IEM will be good candidates for such approaches. One potential advantage of prenatal approaches may be a decreased chance of an immune recognition of the therapeutic gene product.

40.3 GENE THERAPY OF SPECIFIC DISORDERS

A comprehensive review of the tremendous progress in the gene therapy for IEM is beyond the scope of this chapter. In fact, gene therapy studies have been conducted in almost every type of IEM. Instead, specific IEM were chosen either because they illustrate the above principles or for their important historical role.

40.3.1 Aminoacidopathies

The aminoacidopathies are a heterogeneous group of recessively inherited enzyme deficiencies that are associated with the accumulation of specific amino acids in blood and other tissues. The best known and most studied aminoacidopathy is PKU (Table 40.2) caused by deficiency of the liver enzyme PAH (Figure 40.5). PAH deficiency prevents the hydroxylation of phenylalanine to tyrosine and leads to excessive accumulation of phenylalanine in the body. If PKU is left untreated in an infant, poor brain and physical growth, seizures, and mental retardation will result from increased levels of the circulating toxic metabolite phenylalanine. Other examples of aminoacidopathies include tyrosinemia, maple syrup urine disease, homocystinuria, and ornithine transcarbamylase (OTC) deficiency (Table 40.2). Each of these enzyme deficiencies leads to the accumulation of a different specific substrate and causes a different symptom complex.

Contemporary therapy for these diseases is based on an understanding of the pathogenesis involved in each case, and the design of any gene therapy protocol must also be grounded upon a rational understanding of the specific disease pathophysiology. For example, high levels of phenylalanine in PKU are toxic to the developing brain and reducing blood phenylalanine levels are critical to successful treatment of PKU. However, some symptoms of PKU may be caused by deficiencies of specific neurotransmitters, such as dopamine, that are synthesized from tyrosine. In tyrosinemia 1 due to fumarylacetoacetate hydrolase (FAH) deficiency, elevated blood

TABLE 40.2
Summary of Select Aminoacidopathies

Disease	Deficient Enzyme	Elevated Blood Amino Acids	Animal Model
PKU	PAH	Phenylalanine	*Pah^{enu2}* mouse
Tyrosinemia type I	FAH	Tyrosine	FAH knockout mouse
OTC deficiency	OTC	Glutamine; elevated blood ammonia	Sparse fur *(spf)* mouse
Maple syrup urine disease (MSUD)	Branched-chain keto acid dehydrogenase	Leucine, valine, isoleucine	Hereford inbred calf MSUD
Homocystinuria	Cystathionine β-synthase	Homocystine	None

Typical metabolism

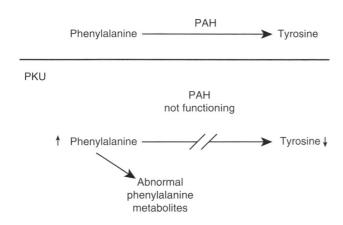

FIGURE 40.5 Pathogenesis of PKU.

tyrosine levels do not appear to be directly toxic, but accumulation of the toxic substrate fumarylacetoacetic acid and its conversion to an equally toxic secondary metabolite succinylacetone causes apoptosis in FAH-deficient hepatocytes. So, for PKU, any therapy that lowers blood phenylalanine and restores tyrosine levels will correct the disease phenotype. In tyrosinemia, clearance of tyrosine from blood is less important but FAH activity must be restored within individual hepatocytes to prevent the accumulation of toxic metabolites within the cell. These considerations must play a role in the design of any gene therapy protocol.

In many aminoacidopathies, the deficient enzyme is normally either exclusively or primarily expressed in liver; liver is the obvious target for gene transfer in these diseases [2]. However, for select disorders, circulating toxic metabolites may be effectively removed from the body by enzyme expressed in a tissue other than liver. The concept of expressing in an alternative tissue a protein that is normally restricted to a specific organ is known as heterologous gene therapy. As an example, PAH expression in skeletal muscle, if supplied with the necessary cofactors, might effectively clear phenylalanine from the circulation of a person with PKU. Gene targeting to the liver may not be essential for some aminoacidopathies. For other diseases, specific pathophysiological features limit the effectiveness of a heterologous gene therapy approach. In OTC deficiency, the substrate for OTC, carbamyl phosphate, is produced only locally in the liver and does not appear in the circulation. As we have already seen, hepatocellular damage in tyrosinemia type 1 can only be prevented if FAH activity is restored in hepatocytes. If possible, disease-specific pathophysiological features and the effectiveness of any gene therapy approach should be demonstrated in an animal model of the disease, if one is available, prior to application of the method in humans. Table 40.2 lists animal models available for the study of some aminoacidopathies.

Gene transfer experiments to treat PKU illustrate the difficulties and complexities of gene therapy for aminoacidopathies and other liver diseases [3]. The availability of a mouse

model, the *Pah^enu2* mouse, that accurately portrays human PAH deficiency has allowed significant advances in PKU gene therapy research. Experimental work with this animal model has demonstrated that PAH activity must be restored in at least 5% of hepatocytes before a substantial reduction in blood phenylalanine is achieved and early gene therapy attempts were limited by the inability to stably transduce sufficient numbers of hepatocytes [4]. The feasibility of hepatocyte-directed gene transfer was demonstrated using a recombinant Moloney murine leukemia virus vector to correct PAH deficiency in cultured hepatocytes from the animal model. However, portal venous infusion of cultured hepatocytes treated *ex vivo* with a β-galactosidase reporter gene-containing retroviral vector yielded only 1% β-galactosidase-expressing cells in the liver of a dog, a transduction frequency that would be insufficient to prevent hyperphenylalaninemia in PKU. As discussed below, a human clinical trial of this approach in subjects with FH yielded a similar transduction frequency and incomplete phenotypic correction [5]. This approach has, however, been employed successfully in the treatment of murine tyrosinemia type I [6]. In this mouse model, enzymatically corrected hepatocytes have a survival advantage over enzyme-deficient cells in the host; although they initially constitute only a very small fraction of the liver, retrovirus-treated enzyme-expressing hepatocytes gradually repopulate the entire liver. Recombinant retroviral-mediated gene therapy may be useful for disorders that require correction of only 1% to 2% of the liver to effect a phenotypic change or in situations where corrected cells have a competitive advantage over the native hepatocytes.

Higher transduction frequencies can be achieved using recombinant adenovirus vectors injected directly into the liver circulation. A recombinant adenovirus containing the PAH cDNA has been infused into the portal circulation of PKU mice [7]. In this experiment, hepatic PAH activity was reconstituted to 5% to 20% of control levels. Complete normalization of plasma phenylalanine levels occurred in animals with hepatic PAH activity equivalent to at least 10% of that in control animals. However, the effect had disappeared by 3 weeks following the treatment, and no hepatic PAH activity was detected following a second treatment with the adenoviral vector. The stability of expression from adenoviral vectors administered *in vivo* is limited by the immune response of the host against the vector or the reporter gene product. In experimental trials of adenovirus-mediated, liver-directed gene therapy employing a variety of different therapeutic genes to treat several different animal models, gene expression has been stable for 7 to 10 days and then has decreased to undetectable levels over the next few weeks. Preexisting immunity and vigorous inflammatory responses against adenovirus in humans further limit the applicability of this vector. In a clinical trial that administered a recombinant adenovirus containing the human OTC cDNA via the hepatic artery to humans with partial OTC deficiency, an overwhelming systemic inflammatory response and fatal hepatic failure occurred following infusion of the vector into a single subject [8].

Recombinant adeno-associated virus (rAAV) vectors have recently enjoyed widespread application in preclinical studies of gene therapy for IEM. AAV is a nonpathogenic single-stranded DNA parvovirus that is capable of stable integration without active cell division. Recombinant AAV vectors are capable of infecting multiple different cell types but pseudotyping with capsid proteins from different AAV serotypes can substantially alter tissue tropism. In comparison to adenovirus vectors, rAAV vector administration to animals has not triggered a vigorous inflammatory reaction. Unlike wild-type AAV, rAAV vector genomes remain predominantly episomal within the target cell; integration of the rAAV genome into the host genome occurs only rarely. Yet, in the absence of rapid target cell turnover and loss of episomal genomes, rAAV administration can lead to long-term sustained therapeutic gene expression. Sustained and complete correction of hyperphenylalaninemia in *Pah^enu2* mice following injection of PAH-expressing rAAV2/8 vector has been accomplished [9,10]. Concerns about tumorigenesis, hepatotoxicity, and immune complications require further study before an AAV-based clinical trial is initiated for PKU. Success with a naked plasmid DNA approach has also been achieved in the mouse model [11].

40.3.2 ORGANIC ACIDURIAS

Propionic aciduria and methylmalonic aciduria are two organic acidurias that have been well studied and are excellent candidates for gene therapy (Figure 40.6). In severe cases, patients present in the neonatal period with coma, metabolic ketoacidosis, and hyperammonemia. With vigorous medical support, they can survive this initial metabolic crisis, but they then must adhere to a strict diet restricted in protein intake. Despite dietary therapy, they continue to have metabolic crises that can be life-threatening. Given the inadequacy of dietary therapy, a gene therapy approach is needed.

Both disorders are caused by enzyme deficiencies in the metabolism of three-carbon species that are generated from the catabolism of amino acids and other metabolites (Figure 40.6). Methylmalonic aciduria is caused by a deficiency of methylmalonyl-CoA mutase activity that is a result of a defect either in the apoenzyme or in the active form of vitamin B_{12}. Patients with the latter defect often respond well to treatment with large amounts of vitamin B_{12} and are therefore in less need of gene therapy.

The prominent target tissue for both disorders is presumed to be the liver. Hepatorenal transplantation has been successfully employed in a patient with a severe form of methylmalonic aciduria. Nonetheless, the major pathology associated with these organic acidurias is due to circulating toxic metabolites. The associated enzymes are normally expressed in many tissues including leukocytes, muscle, and fibroblasts. Therefore, these heterologous tissues should be explored in gene therapy preclinical studies. Unfortunately, animal models for these disorders do not exist at the present time.

Five percent of normal enzymatic activity in either propionic aciduria or methylmalonic aciduria is associated with a benign clinical course, indicating that this level of expression in a gene therapy should be sufficient to realize a large clinical benefit. This level of expression may have to be distributed over approximately 5% of the cells because overexpression of the relevant genes may not lead to a proportional increase in metabolic flux through the three-carbon pathway. For example, in propionic aciduria, the propionyl-CoA carboxylase has two different subunits, α and β. In patients with a defect in the β-subunit, overexpression would require gene transfer with both subunits. However, in patients with a defect in the α-subunit, overexpression of the α subunit may be sufficient because the β-subunit is produced in a fivefold excess over that of the α-subunit.

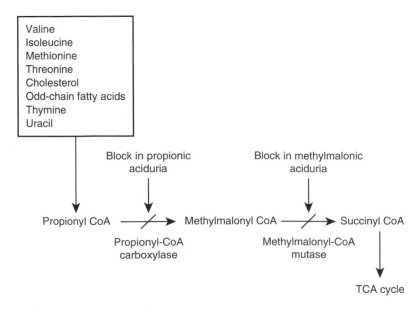

FIGURE 40.6 Enzymatic deficiencies in methylmalonic aciduria or propionic aciduria. TCA, tricarboxylic acid cycle.

40.3.3 LYSOSOMAL STORAGE DISEASES

The common feature of lysosomal storage diseases is the inappropriate accumulation of normal cellular components within lysosomes. This storage of material is visible in cells by light microscopy as very large lysosomes that displace a large part of the cytoplasm. This class of disorders is caused by deficiency of specific lysosomal enzymes that are required for the degradation and recycling of glycoproteins and other cellular components. Without a specific degradative enzyme, the substrate for the reaction accumulates and cannot be removed from the lysosome. The clinical phenotype of each disease is dependent on the tissue type most affected by storage and by the accumulation rate. Physical findings that are suggestive of lysosomal storage include enlargement of the liver and spleen, anemia, and thrombocytopenia due to replacement of normal bone marrow by stored material, destruction of bone, and for those enzymatic deficiencies that cause lysosomal storage in neurons, severe developmental regression, seizures, and other neurological symptoms. Not all of these problems are present in all lysosomal storage diseases; each different enzymatic deficiency presents with a specific phenotypic complex.

The challenges of gene therapy for lysosomal storage diseases are illustrated by the results of contemporary treatment with enzyme replacement therapy (ERT) or bone marrow transplantation. A major challenge to treating a lysosomal storage disease with gene therapy (in contrast to treatment of a liver enzymopathy) is the necessity of reversing lysosomal storage in multiple separate tissues. No currently available gene transfer technique is capable of delivering DNA to multiple target tissues efficiently. However, many lines of evidence demonstrate that lysosomal enzyme proteins can be produced in isolated tissues or even purified *ex vivo* and effectively delivered to most target tissues. For example, glucocerebrosidase, the enzyme deficient in Gaucher's disease can be produced *in vitro* using standard recombinant techniques, chemically modified to facilitate lysosomal targeting, and delivered to affected organs by simple intravenous infusion. This therapy if repeated periodically dramatically reduces liver and spleen size, corrects anemia and thrombocytopenia, and possibly prevents bone deterioration, all major debilitating features of Gaucher's disease. So, at least for gene therapy of Gaucher's disease, the enzyme would not need to be locally produced in all affected tissues. The enzyme could potentially be produced in a single target tissue, secreted into the circulation, and taken up by other diseased tissues. The major limitation of this approach is the difficulty of engineering a secreted form of the enzyme that would be efficiently taken up by other cells and incorporated into lysosomes.

Alternatively, the enzyme could be transferred from the site of production to diseased tissues via circulating blood cells. Seminal experiments demonstrated that functional lysosomal enzymes may be transferred directly from a normal cell to an enzyme-deficient cell in tissue culture. Bone marrow transplantation in the treatment of lysosomal storage diseases exploits this phenomenon. Replacement of enzyme-deficient host bone marrow with enzyme-sufficient donor bone marrow yields a population of circulating blood cells of the reticuloendothelial lineage that infiltrates tissues and transfers lysosomal enzyme to the native cells. Bone marrow transplantation has been employed successfully in Gaucher's disease and in select other storage diseases that do not exhibit brain involvement. Apparently either insufficient numbers of corrected cells penetrate the CNS or insufficient enzyme is transferred to neurons to successfully ameliorate the neurological phenotype of many lysosomal storage disorders. Presumably, difficulties with correcting enzyme deficiency in the brain will also be a major obstacle to successful gene therapy.

Gene therapy for lysosomal storage diseases has to date focused on gene transfer into bone marrow stem cells for the purpose of supplying enzyme via circulating reticuloendothelial cells [12]. Enzymatic correction of Gaucher's bone marrow cells in culture has been accomplished with recombinant retroviral vectors. Similar experiments using other lysosomal enzymes in both cultured bone marrow and fibroblasts have been successful. Persistent production of enzyme in circulating blood cells has been demonstrated in rodents. Phenotypic improvement following retroviral-mediated gene transfer into bone marrow has been shown in gus^{mps}/gus^{mps} mice, a β-glucuro-nidase-deficient mouse model of human mucopolysaccharidosis type VII. As expected, enzymatic correction of bone marrow resulted in amelioration of the somatic symptoms but did not arrest progressive neurological deterioration in this model. However, lysosomal storage in the brain did decrease in mice that had received intracerebral β-glucuronidase-expressing fibroblast implants. Clinical trials of retroviral-mediated bone marrow stem cell-directed gene therapy are under way in humans with Gaucher's disease and in patients with Hunter's syndrome (mucopolysaccharidosis type II) who have little CNS involvement.

40.3.4 LESCH–NYHAN SYNDROME

This X-linked syndrome is caused by a deficiency in HPRT, an enzyme required for salvaging purines (Figure 40.7). It is characterized clinically by increased blood and urine uric acid, mental retardation, choreoathetoid movements, and, most extraordinarily, self-mutilation. It is not understood how a deficiency in HPRT leads to these remarkable neurological sequelae. A genetic mouse model completely lacking HPRT activity does not exhibit any neurological dysfunction except when stressed with amphetamine administration or inhibition of adenine phosphoribosyl transferase (APRT) with 9-ethyladenine. The choreoathetoid movement disorder, however, is postulated to be due to dysfunction within the basal ganglion secondary to disturbed dopamine metabolism.

Although the hyperuric acidemia and its sequela can be controlled with allopurinol, the absence of treatment for the neurological symptoms has prompted the search for gene therapy approaches. Historically, Lesch–Nyhan syndrome has played an important role in the development of gene therapy. One of the first demonstrations of the ability of retroviral

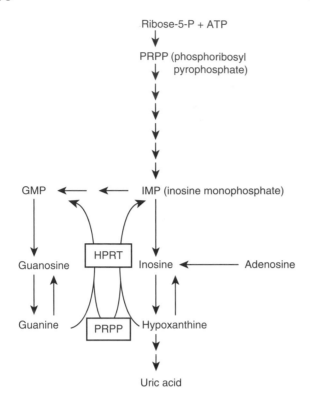

FIGURE 40.7   Enzymatic deficiency in Lesch–Nyhan syndrome.

vectors to correct a genetic mutation was done using the human HPRT gene. The first animal experiment in which a foreign gene was expressed in the brain was done by intracerebrally transplanting fibroblasts genetically modified to express the human HPRT [13]. Although HPRT is expressed in all cells, its high levels in the basal ganglia suggest that this area of the brain should be targeted for gene transfer. Prevention of the mental retardation may require more global expression within the brain.

The amount of normal gene expression required to effect relief can be extrapolated from clinical experience. Although it was previously believed that the severity of the syndrome was not correlated with residual enzymatic activity, it is now realized that its severity does correlate with the amount of HPRT activity in whole cells. Patients with 1.6% to 8% of normal activity had choreoathetosis but not mental retardation or self-mutilation.

In summary, this syndrome is an example of a genetic disorder in which therapy is lacking even when so much is known about its genetic and molecular basis. The development of effective gene transfer methods into the brain may not only provide a therapy but will be quite revealing about its pathogenesis.

40.3.5   Glycogen Storage Diseases

Glycogen storage diseases (GSD) are generally associated with hypoglycemia, hepatomegaly, skeletal and cardiac myopathy depending upon the specific enzymatic defect. Clinical problems arise from deficiency of product, glucose, in GSD type I and III or from lysosomal accumulation of the substrate, glycogen, in GSD type II. For those disorders such as GSD

type I (deficiency in glucose-6-phosphatase) and III (deficiency in glycogen debrancher enzyme) that are associated with hypoglycemia, the mainstay of therapy has been continuous gastric feedings or cornstarch therapy. But this has not prevented late stage liver tumors and renal disease, prompting the search for more definitive therapies such as gene therapy [14]. AAV vectors have been used to correct the metabolic disturbances (hypoglycemia and hypercholesterolemia) present in genetic mouse and dog models of GSD type I and prolonged survival.

GSD type II is caused by deficiency in the lysosomal acid alpha-glucosidase (GAA) enzyme. The infantile form of GSD II (Pompe disease) presents in infancy with generalized muscle weakness and hypertrophic cardiomyopathy, leading to cardiorespiratory failure and death. The juvenile and adult forms of Pompe do not exhibit cardiomyopathy but still have progressive skeletal muscle weakness and eventual respiratory failure. ERT is showing promise in both the infantile and late-onset forms but requires frequent infusions of large amounts of the enzyme, thus sparking interest in gene therapy approaches. Both adenoviral and AAV vectors have corrected the metabolic defect in a GAA-knock-out model for Pompe disease. Humoral antibodies against the GAA enzyme is problematic with both ERT and gene therapy approaches. In the mouse model, this problem has been avoided by the use of an AAV 2/8 vector expressing GAA from a liver-specific promoter.

40.3.6   Familial Hypercholesterolemia

Gene therapy has the potential to significantly improve the clinical status of patients with FH, which is caused by a defect in the LDLR (Figure 40.8). Deficiency in this receptor leads to reduced clearance of LDL by the liver and higher blood levels of LDL. In addition, affected individuals synthesize more cholesterol because the inhibitory effect of LDL on cholesterol synthesis is lost. This inhibition results from decreased HMG CoA reductase activity, the rate-limiting step in cholesterol synthesis.

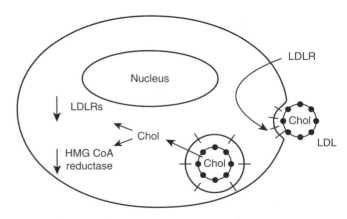

FIGURE 40.8   Pathogenesis of familial hypercholesterolemia. Chol, cholesterol; HMG, 3-hydroxymethylglutaryl; LDLR, low-density lipoprotein receptor.

Heterozygotes with LDLR deficiency occur at a frequency of 1:500 (as common as insulin-dependent diabetes mellitus), making it one of the most common genetic disorders in the United States, Europe, and Japan. Such patients have a two-fold elevation in plasma cholesterol levels (300–600 mg/dL) and may develop coronary artery disease by the fourth decade of life. Three percent to 6% of survivors of myocardial infarctions are heterozygotes for FH.

Homozygotes with LDLR deficiency occur much more infrequently (one in a million), but have much higher cholesterol levels (600–1000 mg/dL) and invariably die from coronary artery disease in their twenties. The severity of the sequelae is attenuated in the homozygote by a few percent residual LDLR activity. Deaths were much less frequent in those homozygotes who had at least 10% of normal LDLR activity. This indicates that clinical benefit could be achieved by a gene therapy in which only a small percentage of LDLR activity is restored. Furthermore, the severity of this disorder increases the benefit-to-risk ratio of clinical trials and thereby facilitates them. A gene therapy protocol can be first tested in the homozygotes (aided by Orphan Drug Status) and then extended to the more common heterozygotes.

Liver transplantation in children has proven that correction of the LDLR defect in the liver can normalize cholesterol levels. For this reason, gene therapy techniques for FH have been directed at the hepatocyte. Based on preclinical studies in mouse and rabbit LDLR-deficient models, *ex vivo* gene therapy in five homozygous FH patients using retrovirus-mediated LDLR gene transfer was performed. This technically challenging protocol yielded a highly variable metabolic response with some improvement in only one of the patients [5]. This study indicates that important modifications must be made to the *ex vivo* gene transfer method before gene therapy can be used as a general therapeutic procedure for such patients [15].

Given the borderline results of the human clinical trial, efforts were initiated with adenoviral vectors carrying the LDLR gene. *In vivo* adenovirus-mediated transfer of the LDLR was shown to be highly effective in reversing the hypercholesterolemia in LDLR knockout mice and WHHL rabbits [16]. The important limitation of adenoviral-mediated gene transfer remains the transient expression *in vivo* after infection of somatic cells with recombinant adenovirus. Nonetheless, these studies demonstrate the proof-of-principle for the gene therapy of FH by the transfer of the normal LDLR gene and highlight the inadequacies of current gene transfer methods.

Current therapy for hypercholesterolemia (not limited to homozygotic FH) includes the use of HMG-CoA reductase inhibitors, which work by secondarily inducing expression of the LDLR, thereby lowering plasma LDL levels. These agents not only lower serum cholesterol, but also lower all-cause mortality by at least 30% in men and women who have coronary disease and total cholesterol levels of 215 to 300 mg/dL. However, 2% of patients suffer liver toxicity and 0.2% develop muscle disease requiring cessation of drug administration. These drugs have to be taken once or twice every day for extended periods of time and compliance is often difficult. A gene therapeutic agent that is administered less than every month (even by intravenous injection) would offer substantial benefit to the patient.

At high efficiencies of liver gene transfer, LDLR gene transfer into the liver could be used to prevent coronary artery disease in the general population. Taking into account all types of hypercholesterolemias, the third National Health and Nutrition Examination Survey (NHANES III) concluded that lipid-lowering therapy was required for 29% of Americans over 20 years of age [17]. The Cholesterol and Recurrent Events study showed that patients with coronary artery disease but having "normal" LDL cholesterol levels benefited from treatment with a single statin therapy [18]. The positive correlation between LDL levels and coronary artery disease is a continuum. In addition, overexpression of the normal LDLR in the liver of transgenic mice (four to five times that of the endogenous receptor) prevented diet-induced hypercholesterolemia, suggesting that unregulated overexpression of the LDLR by liver gene therapy would be therapeutic in humans with hypercholesterolemia of various causes [19].

Many individuals develop coronary artery disease from other causes not amenable to statin therapy but that are potentially treatable by gene therapy. Liver gene therapy using the apoB mRNA editing enzyme (Apobec 1) or the very low density lipoproteins (VLDL) receptor genes could modify LDL cholesterol levels. Other lipoprotein factors besides LDL cholesterol levels influence the onset of coronary artery disease and are amenable to modulation by liver gene transfer. Additional expression of apoA-I in the liver by foreign gene transfer could raise high-density lipoprotein levels and prevent atherosclerosis, as has been demonstrated in mouse and rabbit models. Hypertension, a predisposing factor for coronary artery disease, could be treated by delivering the kalikrein gene to the liver.

Another approach is to block the expression of specific genes using antisense or siRNA molecules. Antisense against apoB has been reported to lower cholesterol levels in people. However, blockage of apoB may lead to steatosis. Other anti-hypercholesterolemic targets for gene knockdown include PCKS9 and acetylCoA-cholesterol transferase, which appear to lower blood cholesterol levels without causing steatosis. It is likely that the siRNA approaches will succeed antisense approaches given siRNA's greater potency and efficacy.

40.3.7 CRIGLER–NAJJAR HYPERBILIRUBINEMIA

Crigler–Najjar syndrome is caused by genetic mutations in the UGT1A gene that encodes UDP-glucuronosyltransferase 1. Deficiency in this enzyme causes unconjugated hyperbilirubinemia (jaundice) that can lead to bilirubin encephalopathy (kernicterus). Current therapy consists of phototherapy or liver transplantation [20].

Bilirubin is produced from the reticuloendothelial cells (macrophages) when they take up aged erythrocytes and degrade the heme moiety within hemoglobin. This unconjugated bilirubin is released into the circulation and taken up by hepatocytes where it is converted by UDP-glucuronosyltransferase into the conjugated form for secretion into the bile duct. Accordingly, adenoviral, AAV, lentiviral, and naked plasmid DNA approaches have been used to express the enzyme in hepatocytes and correct the hyperbilirubinemia present in the Gunn rat model for Crigler–Najjar syndrome [21–23]. A heterologous expression approach has been to express the enzyme in muscle [24]. The idea is that muscle can take up the unconjugated bilirubin, convert it to the conjugated form, and release the conjugated bilirubin into the circulation where it is either excreted in the urine or taken up by hepatocytes for secretion into the bile.

40.4 SUMMARY

The foundation for treating many IEM by gene therapy has been established. It is clear that the expression of the cognate gene can correct the metabolic disturbance and the disease state in most IEM. As in other types of disorders, clinical success has been thwarted by inefficiencies in the gene delivery and expression systems. As new vectors and expression systems are developed and improved over the next few years, it is anticipated that clinical efficacy will be demonstrated for an increasing number of IEM.

ACKNOWLEDGMENTS

This work was supported by National Institutes of Health grants DK065090 and DK059371 and the Propionic Acidemia Foundation.

REFERENCES

1. Wolff, JA and Lederberg, J 1994. An early history of gene transfer and therapy. *Hum Gene Ther* **5**: 469–480.
2. Eisensmith, RC and Woo, SL 1996. Somatic gene therapy for phenylketonuria and other hepatic deficiencies. *J Inherit Metab Dis* **19**: 412–423.
3. Ding, Z, Harding, CO, and Thony, B 2004. State-of-the-art 2003 on PKU gene therapy. *Mol Genet Metab* **81**: 3–8.
4. Hamman, K, Clark, H, Montini, E, Al-Dhalimy, M, Grompe, M, Finegold, M, et al. 2005. Low therapeutic threshold for hepatocyte replacement in murine phenylketonuria. *Mol Ther* **12**: 337–344.
5. Grossman, M, Rader, DJ, Muller, DW, Kolansky, DM, Kozarsky, K, Clark Jr, B, et al. 1995. A pilot study of *ex vivo* gene therapy for homozygous familial hypercholesterolaemia. *Nat Med* **1**: 1148–1154.
6. Overturf, K, Al-Dhalimy, M, Manning, K, Ou, CN, Finegold, M, and Grompe, M 1998. *Ex vivo* hepatic gene therapy of a mouse model of hereditary tyrosinemia type I. *Hum Gene Ther* **9**: 295–304.
7. Fang, B, Eisensmith, RC, Li, XHC, Finegold, MJ, Shedlovsky, A, Dove, W, et al. 1994. Gene therapy for phenylketonuria: Phenotypic correction in a genetically deficient mouse model by adenovirus-mediated hepatic gene therapy. *Gene Therap* **1**: 247–254.
8. Raper, SE, Chirmule, N, Lee, FS, Wivel, NA, Bagg, A, Gao, GP, et al. 2003. Fatal systemic inflammatory response syndrome in a ornithine transcarbamylase deficient patient following adenoviral gene transfer. *Mol Genet Metab* **80**: 148–158.
9. Ding, Z, Georgiev, P, and Thony, B 2006. Administration-route and gender-independent long-term therapeutic correction of phenylketonuria (PKU) in a mouse model by recombinant adeno-associated virus 8 pseudotyped vector-mediated gene transfer. *Gene Ther* **13**: 587–593.
10. Harding, CO, Gillingham, MB, Hamman, K, Clark, H, Goebel-Daghighi, E, Bird, A, et al. 2006. Complete correction of hyperphenylalaninemia following liver-directed, recombinant AAV2/8 vector-mediated gene therapy in murine phenylketonuria. *Gene Ther* **13**: 457–462.
11. Chen, L and Woo, SL 2005. Complete and persistent phenotypic correction of phenylketonuria in mice by site-specific genome integration of murine phenylalanine hydroxylase cDNA. *Proc Natl Acad Sci USA* **102**: 15581–15586.
12. Salvetti, A, Heard, JM, and Danos, O 1995. Gene therapy of lysosomal storage disorders. *Br Med Bull* **51**: 106–122.
13. Gage, FH, Wolff, JA, Rosenberg, MB, Xu, L, Yee, JK, Shults, C, et al. 1987. Grafting genetically modified cells to the brain: Possibilities for the future. *Neuroscience* **23**: 795–807.
14. Koeberl, DD, Kishnani, PS, and Chen, YT 2007. Glycogen storage disease types I and II: Treatment updates. *J Inherit Metab Dis* **30**: 159–164.
15. Brown, MS, Goldstein, JL, Havel, RJ, and Steinberg, D 1994. Gene therapy for cholesterol. *Nat Genet* **7**: 349–350.
16. Ishibashi, S, Brown, MS, Goldstein, JL, Gerard, RD, Hammer, RE, and Herz, J 1993. Hypercholesterolemia in low density lipoprotein receptor knockout mice and its reversal by adenovirus-mediated gene delivery. *J Clin Invest* **92**: 883–893.
17. Sempos, CT, Cleeman, JI, Carroll, MD, Johnson, CL, Bachorik, PS, Gordon, DJ, et al. 1993. Prevalence of high blood cholesterol among US adults. An update based on guidelines from the second report of the National Cholesterol Education Program Adult Treatment Panel. *Jama* **269**: 3009–3014.
18. Sacks, FM, Pfeffer, MA, Moye, LA, Rouleau, JL, Rutherford, JD, Cole, TG, et al. 1996. The effect of pravastatin on coronary events after myocardial infarction in patients with average cholesterol levels. Cholesterol and Recurrent Events Trial investigators. *N Engl J Med* **335**: 1001–1009.
19. Yokode, M, Hammer, RE, Ishibashi, S, Brown, MS, and Goldstein, JL 1990. Diet-induced hypercholesterolemia in mice: Prevention by overexpression of LDL receptors. *Science* **250**: 1273–1275.
20. Strauss, KA, Robinson, DL, Vreman, HJ, Puffenberger, EG, Hart, G, and Morton, DH 2006. Management of hyperbilirubinemia and prevention of kernicterus in 20 patients with Crigler–Najjar disease. *Eur J Pediatr* **165**: 306–319.
21. Jia, Z and Danko, I 2005. Single hepatic venous injection of liver-specific naked plasmid vector expressing human UGT1A1 leads to long-term correction of hyperbilirubinemia and prevention of chronic bilirubin toxicity in Gunn rats. *Hum Gene Ther* **16**: 985–995.
22. Seppen, J, Bakker, C, de Jong, B, Kunne, C, van den Oever, K, Vandenberghe, K, et al. 2006. Adeno-associated virus vector serotypes mediate sustained correction of bilirubin UDP glucuronosyltransferase deficiency in rats. *Mol Ther* **13**: 1085–1092.
23. van der Wegen, P, Louwen, R, Imam, AM, Buijs-Offerman, RM, Sinaasappel, M, Grosveld, F, et al. 2006. Successful treatment of UGT1A1 deficiency in a rat model of Crigler–Najjar disease by intravenous administration of a liver-specific lentiviral vector. *Mol Ther* **13**: 374–381.
24. Jia, Z and Danko, I 2005. Long-term correction of hyperbilirubinemia in the Gunn rat by repeated intravenous delivery of naked plasmid DNA into muscle. *Mol Ther* **12**: 860–866.

41 RNA Interference–Based Gene Therapy Strategies for the Treatment of HIV Infection

Lisa J. Scherer and John J. Rossi

CONTENTS

41.1 INTRODUCTION

Controlling HIV infection continues to be a major challenge both in underdeveloped and developed nations despite the fact that the currently employed drug cocktails have markedly changed the profile of progression to acquired immunodeficiency syndrome (AIDS) in HIV-infected individuals. Despite the successes, highly active antiretroviral therapy (HAART) treatment is not without significant problems and drawbacks [2–9]. Because of pharmacokinetic differences between individuals, there are multiple drug-related toxicities leading to nonadherence problems. There is a need for personalized dosing regimens and combinations and continued therapeutic monitoring of the drugs themselves. They continue to be drug failures for those on HAART as a consequence of viral resistance and other complications due to a lifelong regimen of chemotherapy [10,11]. Temporary interruption of HAART therapy is common; however, in marked contrast to earlier thinking, a recent study shows that compared to continuous antiretroviral therapy, episodic HAART leads to an increased risk of liver cirrhosis, myocardial infarction, renal failure, and stroke with no concomitant increase in quality of life [12]. The increase in side effects due to long-term conventional treatment is due in part to the improved lifespan brought on by the very success of antiretroviral therapies. In addition, treatment guidelines traditionally have not recommended initiating therapy in early stages of infection despite the risks associated with loss of immunological function, increased likelihood of transmission, and the development of a larger pool of viral subspecies that serve as a reservoir for potential resistance. However, there has also been a recent shift toward starting retroviral therapy earlier before CD4 counts drop below $200 \times 10^6/l$ [1,13]. As a possible means of circumventing some of the problems associated with HAART, a number of investigators are focusing their attention on gene therapy either as a stand alone approach or as an adjuvant to pharmacological drug regimens.

We will focus on the progress in developing RNA-based anti-HIV gene therapeutics for long-term applications, with particular attention to molecular targets and their mechanisms of action within the context of the special challenges posed by HIV. An important consideration is the issue of viral versus cellular targets. RNA antivirals can be designed with high specificity and HIV-1 products are the preferred target; however, many viral RNAs are highly abundant and viral escape is a major problem that can only be partially ameliorated by targeting highly conserved sequences. Cellular targets are far less prone to mutational escape and are often in lower abundance, but the side effects of downregulating cellular targets for the long-term are unknown. We evaluate RNA gene therapies in the light of the emerging consensus that combinatorial gene therapeutics has the greatest likelihood of success analogous to HAART therapy.

41.2 RNA INTERFERENCE AS A POWERFUL MECHANISM FOR TARGETED INHIBITION OF GENE EXPRESSION

RNAi is a regulatory mechanism of most eukaryotic cells that use small double-stranded RNA (dsRNA) molecules as triggers to direct homology-dependent control of gene activity [14]. Known as small interfering RNAs (siRNA) these ~21–22 bp long dsRNA molecules have characteristic two-nucleotide 3′ overhangs that allow them to be recognized by the enzymatic machinery of RNAi that eventually leads to homology-dependent degradation of the target mRNA. In mammalian cells, siRNAs are produced from cleavage of longer dsRNA precursors by the RNaseIII endonuclease Dicer [15]. Dicer is complexed with two RNA-binding proteins, the TAR-RNA-binding protein (TRBP) and protein kinase R-activating protein (PACT), which are involved in the hand off of siRNAs to the RNA-induced silencing complex (RISC) [16]. The core components of RISC are the Argonaute (Ago) family members. In humans, there are eight members of this family but only Ago-2 possesses an active catalytic domain for cleavage activity [17,18]. While siRNAs loaded into RISC are double-stranded, Ago-2 cleaves and releases the "passenger" strand, leading to an activated form of RISC with a single-stranded "guide" RNA molecule that directs the specificity of the target recognition by intermolecular base pairing [19]. Rules that govern selectivity of strand loading into RISC are based upon differential thermodynamic stabilities of the ends of the siRNAs [20,21]. The less thermodynamically stable end is favored for binding to the PIWI domain of Ago-2.

An important arm of RNAi involves the microRNAs (miRNAs). These are endogenous duplexes that regulate gene expression posttranscriptionally by complexing with RISC and binding to the 3′ untranslated regions (UTRs) of target sequences via short stretches of homology termed the "seed sequences" [22,23]. The primary mechanism of action of miRNAs is translational repression although this can be accompanied by message degradation [24].

The miRNA duplexes possess incomplete Watson–Crick base pairing and the antisense strand cannot be chosen by cleavage of the passenger strand as for siRNAs, therefore the antisense strand must be chosen by an alternative mechanism [25–27]. miRNAs are endogenous substrates for the RNAi machinery. They are initially expressed as long primary transcripts (pri-miRNAs), which are processed within the nucleus into 60–70 bp hairpins by the microprocessor complex, which consists of Drosha-DiGeorge syndrome critical region gene 8 protein (DGCR) [28,29]. The pre-miRNAs are further processed in the cytoplasm by Dicer and one of the two strands is loaded into RISC, presumably via interaction with PACT [16]. Importantly, it is possible to exploit this native gene silencing pathway for regulation of gene(s) of choice. If the siRNA effector is delivered to the cell it will "activate" RISC, resulting in potent and specific silencing of the targeted mRNA. Because of the potency and selectivity of RNAi, it has become the methodology of choice for silencing specific gene expression in mammalian cells.

Control of disease-associated genes makes RNAi an attractive choice for future therapeutics. Basically every human disease caused by activity from one or a few genes should be amenable for RNAi-based intervention. This list includes cancer, autoimmune diseases, dominant genetic disorders, and viral infections. RNAi can be triggered by two different pathways: (1) an RNA-based approach where synthetic effector siRNAs are delivered by various carriers to target cells as preformed 21 base duplexes; or (2) via DNA-based strategies in which the siRNA effectors are produced by intracellular processing of longer RNA hairpin transcripts (reviewed in Refs. [28,29]). The latter approach is primarily based on nuclear synthesis of short hairpin RNAs (shRNAs) that are transported to the cytoplasm via the miRNA export pathway and are processed into siRNAs by Dicer. While direct use of synthetic siRNA effectors is simple and usually results in potent gene silencing, the effect is transient. In a clinical setting, this would usually mean repeated treatments would have to be administered and these are large and costly drugs. In the case of HIV infection, this would be a lifelong treatment. DNA-based RNAi drugs on the other hand have the potential of being stably introduced when used in a gene therapy setting allowing in principle, a single treatment of viral vector delivered shRNA genes.

41.3 RNAi AND HIV THERAPEUTICS

HIV was the first infectious agent targeted by RNAi perhaps because the life cycle and pattern of gene expression of HIV is well understood. Synthetic siRNAs and expressed shRNAs have been used to target virtually all of the HIV-encoded RNAs in cell lines, including *tat, rev, gag, pol, nef, vif, env, vpr*, and the LTR [30–34]. Subsequent work showed a host of other viruses, including HBV, HCV, poliovirus, RSV, and others were targetable by RNAi (recently reviewed in Ref. [35]).

Despite the early successes of RNAi-mediated inhibition of HIV-encoded RNAs in cell lines, targeting the virus directly represents a substantial challenge for clinical applications because the high viral mutation rate will lead to mutants that can escape being targeted [36–39] although a clever recent strategy takes advantage of escape mutants in critical genes by targeting the mutants directly [40]. An alternative approach to avoid this problem is to target cellular transcripts that encode functions required for HIV-1 entry and replication. To this end, cellular cofactors such as NF kappa beta, the HIV receptor CD4, and the coreceptors CCR5 and CXCR4 have all been downregulated with the result of blocking viral replication or entry [32,33,41–43]. The macrophage-tropic CCR5 coreceptor holds particular promise as a target. This receptor is not essential for normal immune function and individuals homozygous for a 32 bp deletion in this gene are resistant to HIV infection whereas individuals who are heterozygous for this deletion have delayed progression to AIDS [44–46]. CXCR4 is essential for hematopoietic stem cell homing to marrow and subsequent T cell differentiation [47–49]; targeting this receptor is therefore not a good choice for anti-HIV therapy nor is targeting the essential CD4 receptor. A possible exception is in dendritic cells where the DC-SIGN receptor can be targeted by siRNAs to prevent infection [50]. Targeting only the CCR5 coreceptor may also present problems since HIV-1 switches to CXCR4 tropism during the course of AIDS, sometimes creating a more virulent infection [51]. Thus, there are drawbacks in solely targeting cellular HIV cofactors and viral targets will need to be included in any successful strategy using RNAi. It should be pointed out that it may be some-day be possible to use RNAi to prevent viral transmission by employing siRNAs as a microbicides [52].

Viral targets should be sequences that are highly conserved throughout the various clades to ensure efficacy against all viral strains and to minimize emergence of viral mutants resistant to RNAi. Multiplexing si/shRNAs by simultaneously targeting several sites in the virus is an option that should be fully explored and carefully examined for efficacy, inhibition of viral mutants, and potential toxicity. Since the shRNA pathway impinges on the endogenous miRNA pathway, there is ample opportunity for off-target effects and competition with miRNAs for loading into RISC [53,54]. An additional

potential concern is the putative inhibition of RNAi via HIV Tat and TAR. HIV-1 Tat has been demonstrated to bind and inhibit Dicer [55]. It should be pointed out that most investigators do not see inhibition of RNAi when targeting HIV though. TAR also binds TRBP, which is a Dicer cofactor and is a component of RISC [56]. Moreover, unlike other components of RISC, TRBP is made in limited amounts in the cell and hence binding to the TAR RNA could sequester TRBP from interacting with RISC and perhaps limit the effectiveness of an RNAi-based therapy. Binding of TRBP by TAR may also be a factor in the observed changes in miRNA profiles in HIV-infected cells [57].

An alternative approach to relying solely upon RNAi as an anti-HIV approach is mixing a single shRNA with other antiviral genes to provide a potent combinatorial approach (Figure 41.1). This has been successfully accomplished by coexpressing an anti *tat/rev* shRNA, a nucleolar localizing TAR decoy and an anti-CCR5 ribozyme in a single vector backbone [58]. A somewhat different combination used an shRNA with a dominant-negative Rev M10 protein in a coexpression system [59]. Perhaps other, more potent combinations

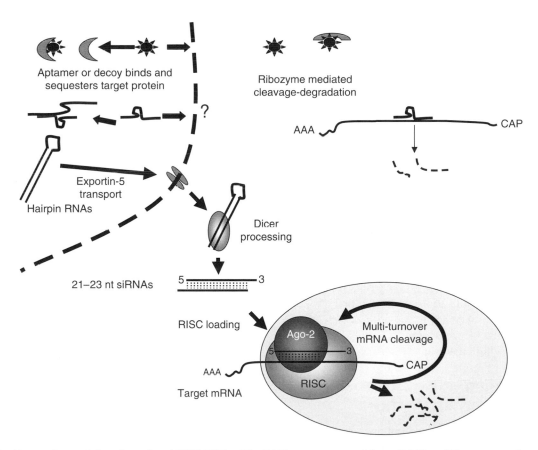

FIGURE 41.1 Expression and function of anti-HIV RNAs. The RNAs are expressed from Pol II or III promoters from viral vectors. Expressed hairpins are exported to the nucleus via the Exportin 5-Ran GTP carrier. In the cytoplasm, the RNAseIII enzyme Dicer processes the hairpins into siRNAs. One of the strands is chosen as the guide and enters the RISC where it serves to guide RISC for cleavage and functional destruction of targeted RNAs. The ribozymes are either exported to the cytoplasm via mRNA export pathways or are retained in the nucleus, as the RNAse P guide sequence. The ribozymes either cleave the target autonomously or use RNAse P to assist in the cleavage mechanism. The aptamers or decoys bind and sequester proteins that they are designed to interact with. This can take place either in the nucleus or cytoplasm.

of shRNAs with mixtures of non-shRNA antivirals will be developed in the near future for testing in preclinical settings.

41.3.1 T-Cells, Stem Cells, and Viral Vectors

Delivery of siRNAs or shRNAs to HIV-1 infected cells is also a challenging problem. The target cells are primarily T-lymphocytes, monocytes, dendritic cells, and macrophages. Since synthetic siRNAs will not persist for long periods in cells, they would have to be delivered repetitively for years to effectively treat the infection. Systemic delivery of siRNAs to T-lymphocytes is probably not feasible. Therefore, a potential method is to isolate T-cells from patients, followed by transduction, expansion of the transduced cells, and reinfusion. In an ongoing clinical trial, T-lymphocytes from HIV-infected individuals are transduced *ex vivo* with a lentiviral vector encoding an anti-HIV antisense RNA [60,61]. The transduced cells are subsequently expanded and reinfused into patients. This type of therapeutic approach could also be applicable to vectors harboring genes that encode siRNAs. A different approach is to transduce isolated hematopoietic progenitor or stem cells with vectors harboring the therapeutic genes. This approach has the advantage that following maturation and differentiation of these cells, all the hematopoietic lineages capable of being infected by the virus are transduced and protected. Hematopoietic stem cells are mobilized from the patients and transduced *ex vivo* prior to reinfusion. Two clinical trials where retroviral vectors expressing ribozymes were transduced into mobilized, autologous hematopoietic stem cells have demonstrated the feasibility of this approach [62,63]. Since RNAi is more potent than ribozyme or antisense approaches, movement of this technology to a human clinical trial for HIV treatment is expected to take place in the near future.

RNAi for the treatment of HIV-1 infection was first demonstrated in 2002. In just four years, there have been over 100 publications testing various siRNAs and shRNAs against different strains of the virus. Viral vector–mediated delivery to hematopoietic cells, including stem cells, is a feasible approach for shRNA gene delivery. Clearly, the barriers that initially confronted therapeutic applications of RNAi for HIV infection are rapidly being broken down, and one can expect to see this powerful cellular process applied clinically to HIV-1-infected patients within the year.

41.4 MULTITARGETING RNAi APPROACHES TO INHIBIT HIV REPLICATION

HIV-inhibitory RNAi therapeutics continues to be investigated because effective siRNAs mediate very potent target knockdown; both the limitations and strengths of using this approach are becoming clearer. Although RNAi can inhibit HIV-1 replication effectively in the initial stages of infection, a single base change in nearly any position within the target sequence may be sufficient to create siRNA/shRNA-resistant viruses after only a few days in culture, even in highly conserved

target sequences [36,38,39]. In addition, mutations outside the RNAi target sequence can allow viral escape to occur by evolving alternative RNA secondary structures [37]. Simultaneous targeting of both wild-type virus and RNAi-escape mutant variants by two or more shRNAs was ineffective in one study, presumably due to competition between shRNAs directed against the wild-type and major escape variant targets [66].

However, using multiple shRNAs to target separate conserved sites in HIV, akin to the HAART approach, is more promising, as it prevents crossresistance between different RNAi-effectors or between RNAi-effectors and conventional pharmaceuticals. Multiple RNAi-effectors would thus have the advantage of limiting escape and targeting a range of sequences as is found in different viral genotypes or quasispecies [64,65]. Viruses, which escape the antiviral effects of RNAi, can be reinhibited by targeting different sequences [66] and thus a multiple inhibitory approach should aim to target distinct genomic regions of HIV-1 or, alternatively, target host-derived factors, which contribute to viral replication.

A variety of approaches are being explored to express multiple siRNAs [67]. The target sequence of a typical siRNA is 21 nucleotides. Long hairpin RNAs (lhRNAS) greater than 50 bp in length can be expressed in cells and create multiple siRNAs via Dicer-mediated processing [68–70]. Expressed lhRNAs have been shown to be effective in cell culture against targets for HCV [71] and HIV [72–74] and *in vivo* for targets against HBV [69]. However, processing and knockdown efficacy of these substrates is asymmetric, being greatest at the base of the hairpin and tapering-off across the full length of the duplex [69]. A 50 bp U6 lhRNA against a conserved HIV-1 *int* region suppressed HIV replication in a variant resistant to a shorter shRNA in the same target [72]. A replication-competent *nef*-deleted HIV-1 variant with a 300 bp lhRNA targeted to *nef* showed significant inhibition of HIV-1 in *trans*, although intriguingly through a non-RNAi-mediated mechanism [73].

Another approach is the use of polycistronic shRNAs, where several hairpins are expressed as a single transcript. Unlike the lhRNAs, which typically target adjacent sequences, different short hairpins (usually 19 to 29-nt in length) in the primary transcript can be directed against widely separated targets. The individual units can be either simple hairpins or modeled on miRNAs, and potentially processed individually for better control of their individual activity.

Finally, multiple shRNAs may be expressed from individual promoters, typically the U6 and H1 polymerase III promoters. Recent studies show that it is possible to saturate the RNAi pathway by overexpressing shRNAS [53,54], resulting in cellular toxicity, particularly with the U6 promoter [53]. We recently described the use of tRNA^{Lys3}-shRNA chimeric cassettes that mediate graded shRNA knockdown, which may be valuable in multiplexing strategies [90], particularly if the principles can be extended to other tRNA isotypes. Use of lower multiplicities of infection when introducing multiplexed shRNAs in lentiviral backbones can alleviate toxicity; in one study, three separate shRNAs targeted to regions within *pol* and *gag* were capable of inhibiting HIV-1 with the addition of each shRNA-expressing cassette providing an additive inhibitory effect [75].

41.5 IMMUNOSTIMULATION BY RNAi

One general concern regarding the use of RNA-based gene therapy approaches remains the potential to induce unwanted immunostimulation. The intracellular presence of duplexed RNA often elicits components of the innate immune system and specifically the cytoplasmic receptors dsRNA-dependent protein kinase (PKR) and retinoic-acid-inducible gene-I (RIG-I), leading to a type 1 interferon (IFN) response (reviewed in Ref. [76]). Over 300 IFN-stimulated genes (ISGs) are activated by increased circulating levels of cytokines IFN-α and β [77]. In particular, the activation of 2′-5′-oligoadenylate synthetases (2–5-OAS) leads to apoptosis via the nonspecific degradation of cellular mRNAs by activated RNase L (reviewed in Refs. [76,78]). Exogenously introduced RNAs (ssRNAs and dsRNAs) are also capable of interacting with different endosomal Toll-like receptors (TLRs), leading to a signaling cascade, which elicits the IFN pathway. However, we have recently shown that endogenously (nuclear) expressed shRNAs evade detection by TLRs, RIG-1, and PKR when integrated in CD34 + progenitor hematopoietic stem cells [79]. The expression of lhRNAs (>30 bp) also appears to be well-tolerated and does not induce ISGs, whether transiently expressed in cells [68,70,74] or *in vivo* when injected in mice by hydrodynamic tail-vein injection [69]. These results suggest that expressed dsRNAs may mimic nuclear-derived natural RNAi precursors and augurs well for future gene therapy clinical trials.

41.6 LENTIVIRAL VECTOR TRIPLE HIV THERAPEUTIC MOVES TOWARD THE CLINIC

We have previously described a triple combination lentiviral construct comprised of a U6-driven TAR-RNA decoy appended upon a U16 snoRNA for nucleolar localization, a U6-promoted shRNA targeted to both *tat* and *rev* open reading frames, and a VA1 promoted, chimeric anti-CCR5 *trans*-cleaving hammerhead ribozyme [80]. The triple construct efficiently transduced human progenitor CD34 + cells and demonstrated improved suppression of HIV-1 over 42 days when compared to a single anti-*tat/rev* shRNA or double combinations of shRNA/ribozyme or decoy [80]. These three expression cassettes utilize a broad spectrum of the alternatives available in RNA therapeutics, demonstrating the potential power of this approach.

41.7 EXPANDING THE REPERTOIRE OF HIV TARGETS

Viral proteins and cellular partners involved in the early events in the emergence of HIV from latency to active replication have been a focus of drug development, since the problems associated with both emergence of viral resistance and perturbation of cellular metabolism as well as viral knockdown vastly increase after the onset of active HIV replication. The Tat–TAR interaction has received special attention due to its central role in the HIV transcriptional transactivation, but cellular cofactors such as NF-κB are also targets of drug development [81]. Not surprisingly, drugs against cellular cofactors have high toxicity, but are still being pursued as options in the event of failure of traditional chemotherapy. Other recently identified cofactors in activation of HIV transcription include the Werner's syndrome helicase [82] and p90 ribosomal S6 kinase 2 (RSK2) [83].

There has also been a great deal of interest regarding the TAR-RNA-binding protein (TRBP) and its effects on HIV replication. Initially identified as an activator of HIV transcription, its subsequent recognition as a partner of the RNAi enzyme Dicer [84] led to the suggestion that TAR subverts the RNAi pathway during HIV infection by sequestering TRBP [85]. However, HIV infection does not cause general downregulation of endogenous miRNAs; in fact, some are upregulated [86]. The effects of siRNA knockdown of TRBP indicate that the activation activities of TRBP predominate in HIV infection at several points during replication [87], consistent with observations that low endogenous expression of TRBP is directly linked to low HIV replication as in astrocytes [88,89].

41.8 PROSPECTS

The successful use of RNA mediators of anti-HIV activity in human hematopoietic cells has now been validated by many different investigators. With the development of genetically modified viral vectors that are capable of transducing hematopoietic cells with therapeutic RNA-encoding constructs, there will be more proof-of-concept studies in animal models within the next couple of years. Ongoing clinical trials using anti-HIV ribozymes and antisense RNAs have demonstrated the safety of hematopoietic-based gene therapies. The next step is to prove efficacy of these RNA-based inhibitors. Once efficacious use of antiviral RNAs in a gene therapy setting has been demonstrated, this will open the door for expanded clinical applications.

While a cure for AIDS does not appear to be immanent, the success of HAART therapy presents the very real possibility of maintaining HIV as a chronic condition. While many hurdles need to be overcome, the hope is that gene therapy will provide an added option when pharmacological drugs fail and perhaps an adjuvant to current therapies that promote longer and better quality of life.

REFERENCES

1. Deeks, S.G., Antiretroviral treatment of HIV infected adults, *BMJ*, 332, 1489, 2006.
2. Agrawal, L. et al., Anti-HIV therapy: Current and future directions, *Curr. Pharm. Des.*, 12, 2031, 2006.
3. Clerici, M., Immunomodulants for the treatment of HIV infection: The search goes on, *Expert Opin. Investig. Drugs*, 15, 197, 2006.

4. Hawkins, T., Appearance-related side effects of HIV-1 treatment, *AIDS Patient Care STDs,* 20, 6, 2006.

5. Justesen, U.S., Therapeutic drug monitoring and human immunodeficiency virus (HIV) antiretroviral therapy, *Basic Clin. Pharmacol. Toxicol.,* 98, 20, 2006.

6. Negredo, E., Bonjoch, A., and Clotet, B., Benefits and concerns of simplification strategies in HIV-infected patients, *J. Antimicrob. Chemother.,* 58, 235, 2006.

7. Piacenti, F.J., An update and review of antiretroviral therapy, *Pharmacotherapy,* 26, 1111, 2006.

8. Rodriguez-Novoa, S. et al., Overview of the pharmacogenetics of HIV therapy, *Pharmacogenomics J.,* 6, 234, 2006.

9. Slama, L. et al., [Adherence to antiretroviral therapy during HIV infection, a multidisciplinary approach], *Med. Mal. Infect.,* 36, 16, 2006.

10. Baba, M., Recent status of HIV-1 gene expression inhibitors, *Antiviral Res.,* 71, 301, 2006.

11. Yeni, P., Update on HAART in HIV, *J. Hepatol.,* 44, S100, 2006.

12. Podlekareva, D. et al., Factors associated with the development of opportunistic infections in HIV-1-infected adults with high CD4 + cell counts: A EuroSIDA study, *J. Infect. Dis.,* 194, 633, 2006.

13. Phillips, A.N. et al., When should antiretroviral therapy for HIV be started? *BMJ,* 334, 76, 2007.

14. Almeida, R. and Allshire, R.C., RNA silencing and genome regulation, *Trends Cell Biol.,* 15, 251, 2005.

15. Zhang, H. et al., Single processing center models for human Dicer and bacterial RNase III, *Cell,* 118, 57, 2004.

16. Lee, Y. et al., The role of PACT in the RNA silencing pathway, *EMBO J.,* 25, 522, 2006.

17. Meister, G. et al., Human Argonaute2 mediates RNA cleavage targeted by miRNAs and siRNAs, *Mol. Cell,* 15, 185, 2004.

18. Liu, J. et al., Argonaute2 is the catalytic engine of mammalian RNAi, *Science,* 305, 5689, 2004.

19. Tang, G., siRNA and miRNA: An insight into RISCs, *Trends Biochem. Sci.,* 30, 106, 2005.

20. Schwarz, D.S. et al., Asymmetry in the assembly of the RNAi enzyme complex, *Cell,* 115, 199, 2003.

21. Khvorova, A., Reynolds, A., and Jayasena, S.D., Functional siRNAs and miRNAs exhibit strand bias, *Cell,* 115, 209, 2003.

22. Bartel, D.P., MicroRNAs: Genomics, biogenesis, mechanism, and function, *Cell,* 116, 281, 2004.

23. Bartel, D.P. and Chen, C.Z., Micromanagers of gene expression: The potentially widespread influence of metazoan microRNAs, *Nat. Rev. Genet.,* 5, 396, 2004.

24. Bagga, S. et al., Regulation by let-7 and lin-4 miRNAs results in target mRNA degradation, *Cell,* 122, 553, 2005.

25. Leuschner, P.J. et al., Cleavage of the siRNA passenger strand during RISC assembly in human cells, *EMBO Rep.,* 7, 314, 2006.

26. Gregory, R.I. et al., Human RISC couples microRNA biogenesis and posttranscriptional gene silencing, *Cell,* 123, 631, 2005.

27. Matranga, C. et al., Passenger-strand cleavage facilitates assembly of siRNA into Ago2-containing RNAi enzyme complexes, *Cell,* 123, 607, 2005.

28. Hannon, G.J. and Rossi, J.J., Unlocking the potential of the human genome with RNA interference, *Nature,* 431, 371, 2004.

29. Scherer, L.J. and Rossi, J.J., Approaches for the sequence-specific knockdown of mRNA, *Nat. Biotechnol.,* 21, 1457, 2003.

30. Jacque, J.M., Triques, K., and Stevenson, M., Modulation of HIV-1 replication by RNA interference, *Nature,* 418, 435, 2002.

31. Lee, N.S. et al., Expression of small interfering RNAs targeted against HIV-1 rev transcripts in human cells, *Nat. Biotechnol.,* 20, 500, 2002.

32. Martinez, M.A., Clotet, B., and Este, J.A., RNA interference of HIV replication, *Trends Immunol.,* 23, 559, 2002.

33. Novina, C.D. et al., siRNA-directed inhibition of HIV-1 infection, *Nat. Med.,* 8, 681, 2002.

34. Coburn, G.A. and Cullen, B.R., Potent and specific inhibition of human immunodeficiency virus type 1 replication by RNA interference, *J. Virol.,* 76, 9225, 2002.

35. Leonard, J.N. and Schaffer, D.V., Antiviral RNAi therapy: Emerging approaches for hitting a moving target, *Gene Ther.,* 79, 3, 2005.

36. Boden, D. et al., Human immunodeficiency virus type 1 escape from RNA interference, *J. Virol.,* 77, 11531, 2003.

37. Westerhout, E.M. et al., HIV-1 can escape from RNA interference by evolving an alternative structure in its RNA genome, *Nucleic Acids Res.,* 33, 796, 2005.

38. Das, A.T. et al., Human immunodeficiency virus type 1 escapes from RNA interference-mediated inhibition, *J. Virol.,* 78, 2601, 2004.

39. Sabariegos, R. et al., Sequence homology required by human immunodeficiency virus type 1 to escape from short interfering RNAs, *J. Virol.,* 80, 571, 2006.

40. Brake, O. and Berkhout, B., A novel approach for inhibition of HIV-1 by RNA interference: Counteracting viral escape with a second generation of siRNAs, *J. RNAi Gene Silencing,* 1, 56, 2005.

41. Anderson, J. and Akkina, R., CXCR4 and CCR5 shRNA transgenic CD34 + cell derived macrophages are functionally normal and resist HIV-1 infection, *Retrovirology,* 2, 53, 2005.

42. Cordelier, P., Morse, B., and Strayer, D.S., Targeting CCR5 with siRNAs: Using recombinant SV40-derived vectors to protect macrophages and microglia from R5-tropic HIV, *Oligonucleotides,* 13, 281, 2003.

43. Surabhi, R.M. and Gaynor, R.B., RNA interference directed against viral and cellular targets inhibits human immunodeficiency Virus Type 1 replication, *J. Virol.,* 76, 12963, 2002.

44. Eugen-Olsen, J. et al., Heterozygosity for a deletion in the CKR-5 gene leads to prolonged AIDS-free survival and slower CD4 T-cell decline in a cohort of HIV-seropositive individuals, *Aids,* 11, 305, 1997.

45. Garred, P. et al., Dual effect of CCR5 delta 32 gene deletion in HIV-1-infected patients. Copenhagen AIDS Study Group, *Lancet,* 349, 1884, 1997.

46. Samson, M. et al., Resistance to HIV-1 infection in caucasian individuals bearing mutant alleles of the CCR-5 chemokine receptor gene, *Nature,* 382, 722, 1996.

47. Lapidot, T., Mechanism of human stem cell migration and repopulation of NOD/SCID and B2mnull NOD/SCID mice. The role of SDF-1/CXCR4 interactions, *Ann. N.Y. Acad. Sci.,* 938, 83, 2001.

48. Lapidot, T. and Kollet, O., The essential roles of the chemokine SDF-1 and its receptor CXCR4 in human stem cell homing and repopulation of transplanted immune-deficient NOD/SCID and NOD/SCID/B2m(null) mice, *Leukemia,* 16, 1992, 2002.

49. Kahn, J. et al., Overexpression of CXCR4 on human CD34 + progenitors increases their proliferation, migration, and NOD/SCID repopulation, *Blood,* 103, 2942, 2004.

50. Nair, M.P. et al., RNAi-directed inhibition of DC-SIGN by dendritic cells: Prospects for HIV-1 therapy, *Aaps J.,* 7, E572, 2005.

51. Arien, K.K. et al., Replicative fitness of CCR5-using and CXCR4-using human immunodeficiency virus type 1 biological clones, *Virology,* 347, 65, 2006.

52. Palliser, D. et al., An siRNA-based microbicide protects mice from lethal herpes simplex virus 2 infection, *Nature,* 439, 89, 2006.

53. An, D.S. et al., Optimization and functional effects of stable short hairpin RNA expression in primary human lymphocytes via lentiviral vectors, *Mol. Ther.*, 14, 494, 2006.

54. Grimm, D. et al., Fatality in mice due to oversaturation of cellular microRNA/short hairpin RNA pathways, *Nature*, 441, 537, 2006.

55. Bennasser, Y. et al., Evidence that HIV-1 encodes an siRNA and a suppressor of RNA silencing, *Immunity*, 22, 607, 2005.

56. Gatignol, A., Laine, S., and Clerzius, G., Dual role of TRBP in HIV replication and RNA interference: Viral diversion of a cellular pathway or evasion from antiviral immunity? *Retrovirology*, 2, 65, 2005.

57. Yeung, M.L. et al., Changes in microRNA expression profiles in HIV-1-transfected human cells, *Retrovirology*, 2, 81, 2005.

58. Li, M.J. et al., Long-term inhibition of HIV-1 infection in primary hematopoietic cells by lentiviral vector delivery of a triple combination of anti-HIV shRNA, anti-CCR5 ribozyme, and a nucleolar-localizing TAR decoy, *Mol. Ther.*, 12, 900, 2005.

59. Unwalla, H.J. et al., Novel Pol II fusion promoter directs human immunodeficiency virus type 1-inducible coexpression of a short hairpin RNA and protein, *J. Virol.*, 80, 1863, 2006.

60. Dropulic, B., Lentivirus in the clinic, *Mol. Ther.*, 4, 511, 2001.

61. Levine, B.L. et al., Gene transfer in humans using a conditionally replicating lentiviral vector, *Proc. Natl. Acad. Sci. USA*, 103, 17372, 2006.

62. Amado, R.G. et al., Anti-human immunodeficiency virus hematopoietic progenitor cell-delivered ribozyme in a phase I study: Myeloid and lymphoid reconstitution in human immunodeficiency virus type-1-infected patients, *Hum. Gene Ther.*, 15, 251, 2004.

63. Michienzi, A. et al., RNA-mediated inhibition of HIV in a gene therapy setting, *Ann. N. Y. Acad. Sci.*, 1002, 63, 2003.

64. Chang, L.J., Liu, X., and He, J., Lentiviral siRNAs targeting multiple highly conserved RNA sequences of human immunodeficiency virus type 1, *Gene Ther.*, 12, 1133, 2005.

65. Berkhout, B. and Haasnoot, J., The interplay between virus infection and the cellular RNA interference machinery, *FEBS Lett.*, 580, 2896, 2006.

66. Wilson, J.A. and Richardson, C.D., Hepatitis C virus replicons escape RNA interference induced by a short interfering RNA directed against the NS5b coding region, *J. Virol.*, 79, 7050, 2005.

67. Kim, D.H. and Rossi, J.J., Strategies for silencing human disease using RNA interference, *Nat. Rev. Genet.*, 8, 173, 2007.

68. Akashi, H. et al., Escape from the interferon response associated with RNA interference using vectors that encode long modified hairpin-RNA, *Mol. Biosyst.*, 1, 382, 2005.

69. Weinberg, M.S. et al., Specific inhibition of HBV replication in vitro and in vivo with expressed long hairpin RNA, *Mol. Ther.*, 15, 534, 2007.

70. Strat, A. et al., Specific and nontoxic silencing in mammalian cells with expressed long dsRNAs, *Nucleic Acids Res.*, 34, 3803, 2006.

71. Watanabe, T. et al., Intracellular-diced dsRNA has enhanced efficacy for silencing HCV RNA and overcomes variation in the viral genotype, *Gene Ther.*, 13, 883, 2006.

72. Nishitsuji, H. et al., Effective suppression of human immunodeficiency virus type 1 through a combination of short- or long-hairpin RNAs targeting essential sequences for retroviral integration, *J. Virol.*, 80, 7658, 2006.

73. Konstantinova, P. et al., Trans-inhibition of HIV-1 by a long hairpin RNA expressed within the viral genome, *Retrovirology*, 4, 15, 2007.

74. Konstantinova, P. et al., Inhibition of human immunodeficiency virus type 1 by RNA interference using long-hairpin RNA, *Gene Ther.*, 13, 1403, 2006.

75. ter Brake, O. et al., Silencing of HIV-1 with RNA interference: A multiple shRNA approach, *Mol. Ther.*, 14, 883, 2006.

76. Garcia-Sastre, A. and Biron, C.A., Type 1 interferons and the virus-host relationship: A lesson in detente, *Science*, 312, 879, 2006.

77. de Veer, M.J., Sledz, C.A., and Williams, B.R., Detection of foreign RNA: Implications for RNAi, *Immunol. Cell Biol.*, 83, 224, 2005.

78. Karpala, A.J., Doran, T.J., and Bean, A.G., Immune responses to dsRNA: Implications for gene silencing technologies, *Immunol. Cell Biol.*, 83, 211, 2005.

79. Robbins, M.A. et al., Stable expression of shRNAs in human CD34 + progenitor cells can avoid induction of interferon responses to siRNAs in vitro, *Nat. Biotechnol.*, 24, 566, 2006.

80. Li, M. and Rossi, J.J., Lentiviral vector delivery of siRNA and shRNA encoding genes into cultured and primary hematopoietic cells, *Methods Mol. Biol.*, 309, 261, 2005.

81. Stevens, M., De Clercq, E., and Balzarini, J., The regulation of HIV-1 transcription: Molecular targets for chemotherapeutic intervention, *Med. Res. Rev.*, 26, 595, 2006.

82. Sharma, A. et al., The Werner's syndrome helicase is a cofactor for HIV-1 LTR transactivation and retroviral replication, *J. Biol. Chem.*, 16, 12048, 2007.

83. Hetzer, C. et al., Recruitment and activation of RSK2 by HIV-1 Tat, *PLoS ONE*, 2, e151, 2007.

84. Haase, A.D. et al., TRBP, a regulator of cellular PKR and HIV-1 virus expression, interacts with Dicer and functions in RNA silencing, *EMBO Rep.*, 6, 961, 2005.

85. Bennasser, Y., Yeung, M.L., and Jeang, K.T., HIV-1 TAR RNA subverts RNA interference in transfected cells through sequestration of TAR RNA-binding protein, TRBP, *J. Biol. Chem.*, 281, 27674, 2006.

86. Triboulet, R. et al., Suppression of microRNA-silencing pathway by HIV-1 during virus replication, *Science*, 315, 1579, 2007.

87. Christensen, H.S. et al., siRNAs against the TAR RNA binding Protein, TRBP, a Dicer cofactor, inhibit HIV-1 long terminal repeat expression and viral production, *J. Virol.*, 81, 5121, 2007.

88. Bannwarth, S. et al., Cell-specific regulation of TRBP1 promoter by NF-Y transcription factor in lymphocytes and astrocytes, *J. Mol. Biol.*, 355, 898, 2006.

89. Ong, C.L. et al., Low TRBP levels support an innate human immunodeficiency virus type 1 resistance in astrocytes by enhancing the PKR antiviral response, *J. Virol.*, 79, 12763, 2005.

90. Scherer, L.J., Frank, R., and Rossi, J.J., Optimization and characterization of tRNA-shRNA expression constructs, *Nucleic Acids Res.*, 35, 2620, 2007.

42 Cutaneous Gene Transfer: Cell and Gene Therapy of Skin and Systemic Disorders

William Buitrago and Dennis R. Roop

CONTENTS

42.1   INTRODUCTION

Greater understanding about the skin's molecular biology and about the genetics of many skin and systemic disorders has led to significant progress in corrective gene therapy through cutaneous gene transfer. This progress has only been possible due to advances in the areas of vector design and efficiency of delivery, immune modulation, regulation of targeted gene replacement, correction and expression, and the development of appropriate animal models. The skin has a unique and complex structure and it presents remarkable advantages as a tissue for developing innovative gene therapy strategies. The skin is a readily accessible organ, which facilitates gene delivery, subsequent monitoring of transgene expression, and excision of small or large areas if required [1]. Epidermal keratinocytes and dermal fibroblasts can be readily expanded in culture. In addition, keratinocytes have high proliferative potential and inherent biological characteristics that allow them to synthesize mature proteins from a vast array of transgenes [2]. This rapid progress has made the skin a reasonable target to develop and test a variety of gene therapy approaches to both cutaneous and systemic diseases.

42.2 SKIN

42.2.1 STRUCTURE, FUNCTION, AND DEVELOPMENT

The skin is the most superficial and the largest organ in the body. Its functions are essential for the homeostatic balance of the organism. The skin provides protection against UV light, mechanical, thermal, and chemical insults, and also prevents excessive dehydration due to its relative water impermeability [3]. It also acts as a physical barrier against microorganisms and is involved in the coordination of multiple immune responses [4,5]. The skin is the major organ involved in sensory perception. For instance, it possesses tactile, pressure, pain, and temperature receptors [6,7]. In humans and in many other mammals, the skin is essential in thermoregulatory responses [5,8,9]. For example, heat conservation is aided by the presence of hair and adipose tissue in the hypodermis, and heat loss is increased by the evaporation of water through the skin's surface and by increasing the blood flux through the rich capillary plexus of the dermis [8]. The skin also plays a definite role in the body's metabolism as an important store of energy in the form of triglycerides and in the synthesis of vitamin D [5]. The skin presents regional variations with regard to thickness, coloration, and presence of adnexa. For instance, the skin is thicker, presents lower levels of coloration, and has no hair follicles in the palms and soles. However, the basic structure of the skin is maintained in all body areas. The external part of the skin, the epidermis, is formed by a stratified squamous epithelium, which is composed of keratinocytes in majority. Other cell types such as melanocytes, Langerhans cells, and Merkel cells are also found in this layer (Figure 42.1). The epidermis is divided into two major compartments: the basal (proliferative) and suprabasal (differentiation) compartments. The suprabasal compartment is further divided into different layers based upon microscopic characteristics: the spinous layer, the granular layer, and the cornified layer (Figure 42.1). The basal keratinocytes are attached to the basement membrane (BM) through specialized multiprotein junctional complexes called hemidesmosomes [10].

Early in development, the epidermis develops from the surface ectoderm, which is one cell layer thick and expresses keratin 18 (K18) up until its commitment to stratification [11]. The surface ectoderm shortly becomes a two-layer structure composed of an outer layer that differentiates into the periderm, which is a temporary structure, and an inner layer (basal layer), which gives rise to all the cells of the epidermis [11,12]. Once

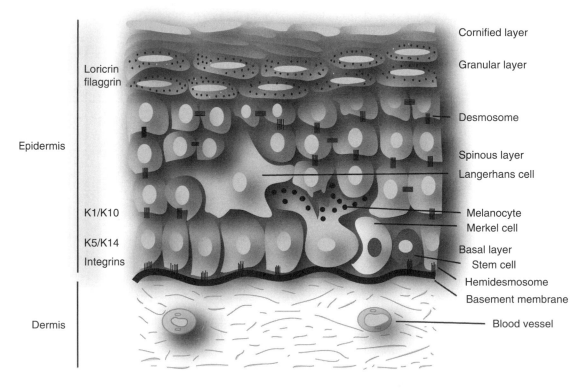

FIGURE 42.1 Schematic representation of the epidermis. The epidermis is a stratified epithelium consisting of basal, spinous, granular, and stratum corneum layers. The basal layer contains the stem cells and adheres to the BM via hemidesmosomes. K5, K14, and integrins are expressed by basal cells and their expression is downregulated as the cells detach from the BM and move upward to the suprabasal compartment where K1 and K10 become highly expressed. As the differentiating keratinocytes move toward the epidermal surface, the expression of other markers, such as filaggrin and loricrin, is detected. Other cell types such as melanocytes, Langerhans cells, and Merkel cells are also found in this layer.

the cells of the surface ectoderm commit to stratification, keratin 5 (K5) and keratin 14 (K14) are induced and K18 is downregulated starting at embryonic day (E) 9.5 [13]. The periderm is thought to provide protection to the developing embryo against amniotic fluid pending the formation of the epidermal barrier [14]. At E13.5, an intermediate layer starts developing between the basal layer and the periderm [15,16]. Further epidermal development takes place as the periderm is lost and the basal layer divides to produce an outer population of cells that will form the spinous layer, which in turn differentiates to form the granular and cornified layers [17]. The spinous cells are first seen at E15.5 and express keratin 1 (K1) and keratin 10 (K10) [17]. Between E16.5 and 17.5, loricrin and filaggrin begin to be expressed concomitant with the formation of the granular layer and stratum corneum [17]. The periderm is entirely lost and epidermal stratification is complete on E18.5 [18].

The expression of keratins, integrins, and involucrin is also regulated during keratinocyte differentiation in the mature epidermis. K5, K14, and integrins are expressed by basal cells and their expression is downregulated as the cells detach from the BM and move upward to the suprabasal compartment where K1 and K10 become highly expressed [13,19,20]. Subsequently, as keratinocytes migrate toward the epidermal surface, the expression of other markers, such as filaggrin and loricrin, is detected [17,19,21,22] (Figure 42.1). Most of the cellular organelles are degraded in the stratum granulosum. Terminally differentiated cells then undergo programmed cell death and these dead keratinocytes form the stratum corneum.

The dermis, a thick layer of fibroelastic dense connective tissue, which supports and nourishes the epidermis, is composed of numerous blood and lymphatic vessels, sensory elements, and different cell types such as fibroblasts, macrophages, and lymphocytes. The hypodermis (subcutaneous layer or subcutis) is localized under the dermis and contains adipose tissue and blood vessels. The adnexal structures such as sweat glands, sebaceous glands, and hair follicles are structures of ectodermal origin that form by invagination of the epidermal epithelium into the dermis, and sometimes the hypodermis.

42.2.2 Epidermal Stem Cells: Skin Renewal, Regeneration, and Repair

The epidermis, a tissue in constant turnover, is sustained by permanent mitotic divisions in the basal compartment where stem cells reside [23]. The cells in the basal layer undergo a series of maturational changes, and at some point, a given number move upward to the postmitotic suprabasal compartment where they undergo terminal differentiation. The basal layer is composed of a heterogeneous population of cells that can be classified according to their capacity for sustained growth into three subpopulations: (a) the holoclones or stem cells, which have the greatest reproductive capacity, (b) the paraclones or differentiated cells with a short replicative lifespan and limited growth capacity, and (c) the intermediate meroclones, which are the transitional stage between the holoclones and the paraclones [24].

Numerous studies provide strong evidence for the existence of stem cells in the upper outer root sheath (the hair follicle bulge region; bulge stem cells) in the basal layer of the interfollicular epidermis (interfollicular stem cells), and, potentially, a third stem cell population in the sebaceous gland (Figure 42.2). These stem cell populations are heterogeneous. Under normal homeostatic conditions, the bulge and interfollicular stem cells represent two distinct populations, but when the epidermis is damaged, it can be completely regenerated by bulge stem cells [25,26]. In addition, interfollicular stem cells

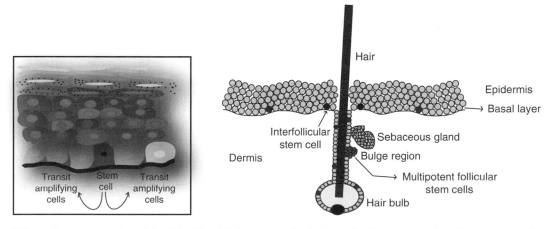

FIGURE 42.2 Schematic representation of the skin. The skin is composed of epidermis, dermis, and adnexal structures (sebaceous glands and hair follicles). The dermis supports and nourishes the epidermis and is composed of blood and lymphatic vessels, sensory elements, and different cell types such as fibroblasts, macrophages, and lymphocytes. The adnexal structures such as sweat glands, sebaceous glands, and hair follicles are structures of ectodermal origin that form by invagination of the epidermal epithelium into the dermis. The epidermal stem cells are found in the bulge region of the hair follicles, in the basal layer of the interfollicular epidermis, and a potentially (under debate) in the sebaceous glands. The interfollicular stem cells give rise to transient amplifying cells, which in turn generate the cells that undergo terminal differentiation and form the suprabasal layers (inset).

can be induced to form hair follicles and sebaceous glands when in contact with dermal papilla cells [27,28]. The presence of an intrinsic stem cell population in the sebaceous gland is still under much debate; it is possible that the sebaceous gland is maintained by bulge stem cells.

42.3 CUTANEOUS GENE TRANSFER

42.3.1 STRATEGIES

Multiple *ex vivo* and *in vivo* approaches are currently used with various degrees of success for gene delivery to the skin. In the *ex vivo* approach, the skin cells from the host are isolated and harvested after removal by biopsy. The cells are then grown *in vitro* where therapeutic gene transfer is performed. Finally, the altered cells are grafted back into the host (Figure 42.3A). This method offers some advantages because primary keratinocytes, including human, are receptive to gene modification, readily expanded in tissue culture under selective conditions, and easily grafted back into host [29–33]. However, the *ex vivo* approach is disadvantageous because of its labor intensity and potential scarring.

The *in vivo* approach consists of gene transfer through direct administration of genes by different modalities such as topical application, direct intradermal injection, particle-mediated gene transfer or liposome technology (Figure 42.3B). The genetic material maybe delivered in the form of naked DNA or packaged in vectors of viral or nonviral origins. This approach is favored over the *ex vivo* approach as the need for cell culture and surgery are bypassed making it technically, clinically, and economically advantageous [29]. In spite of these advantages, the direct *in vivo* approach is still limited by the low levels of transduction frequency for stem cells, leading to transient expression of the transferred gene.

One of the main strategies for genetic therapy is to achieve the overexpression of the correct protein after delivery and incorporation of the wild-type gene. However, most dominant disorders cannot be treated by simply overexpressing the correct

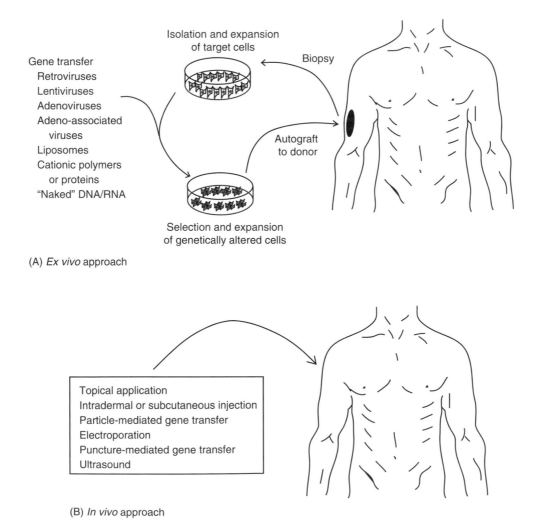

FIGURE 42.3 *Ex vivo* and *in vivo* approaches for cutaneous gene transfer. (A) In the *ex vivo* approach, the host skin cells are isolated and harvested after being removed by biopsy. The cells are grown *in vitro* where they can be genetically altered, selected, and expanded before being grafted back into the host. (B) In the *in vivo* approach, genetic transfer is accomplished by administration of genetic material to the cells in situ. This could be done by different modalities, such as naked DNA or in vectors of nonviral or viral origins. There is no need for cell culture and surgery; therefore, it is technically, clinically, and economically advantageous.

protein; thus, it is imperative to correct the mutant allele or to decrease or completely silence its expression. New gene therapy strategies have recently emerged in such cases. For example, RNA–DNA oligonucleotides have been used for targeted single-base gene correction through homologous recombination and mismatch DNA repair in skin cells [34–37]. Antisense DNA oligonucleotides (AS-ODNs), which are short-length synthetic DNA molecules that hybridize with specific mRNA sequences and silence gene expression, have been successfully used in cutaneous gene transfer studies [38–41]. Ribozymes, RNA molecules possessing a self-catalytic enzymatic function [42], have been used to cleave mutant target RNA and to repair mutant RNA through splicing [43–45]. Furthermore, ribozymes have been used to downregulate gene expression in the skin [46]. Small interfering RNAs (siRNAs) are short double-stranded RNA molecules capable of binding and inducing the degradation of specific mRNAs [47–49]. The first studies demonstrating the effective application of siRNA technology in the skin to suppress gene expression have been published [50,51]. However, the potential use of siRNAs in clinical applications for cutaneous gene therapy remains to be fully explored. In addition, the spliceosome, the site where introns are removed to produce mature RNA species through a process known as splicing, has been postulated as a target in efforts to achieve gene correction in the skin [52]. Splicing may take place by using alternative splicing sites when standard splice sites are mutated. In these cases, RNA–DNA oligonucleotides or ribozymes might be used to correct these mutations or AS-ODNs may be employed to induce suppression of alternative splice site usage. Recently, a related cellular mechanism known as trans-splicing whereby an intron of a pre-RNA interacts with an intron of another pre-RNA, enhancing the recombination of splice sites between two conventional pre-RNAs, has been exploited for gene correction [53–55]. This strategy is referred as spliceosome-mediated RNA transplicing (SMaRT) and it has been used in keratinocytes derived from generalized atrophic benign epidermolysis bullosa (GABEB) patients to correct the type XVII collagen (COL17A1) gene [56].

42.3.2 Gene Transfer Systems

The first step in effective gene therapy is to transfer the corrective genetic material to target tissue with high efficiency. The skin can be specifically targeted by either *in vivo* or *ex vivo* strategies. A variety of systems with successful application in skin gene transfer have been developed and they can be classified into two large categories: viral and nonviral.

42.3.2.1 Viral Gene Transfer Vectors

Currently, viruses provide the most efficient means for delivering genes to target cells. Replication-defective viral vectors are created by removal of viral genes. The gene of interest is inserted and the viral proteins required for cell entry and gene delivery are provided either by packaging cells lines or by other vectors. There are six main factors to consider in a viral system for efficient gene transfer: ability for high titer generation, cargo-carrying capacity, capacity to transduce dividing or nondividing cells, integration properties, vector antigenicity, the ease with which clinical-grade vectors can be prepared free of replication-competent virus, and the length of time of gene expression required [57]. Some disadvantages should also be considered when working with viral vectors. Since the expression of the delivered genes depends heavily on the integration site, it is important to understand that all the integrative viral vectors have the limitation of their lack of true site-specific integration. Furthermore, downregulation of the introduced genes by epigenetic mechanisms represents another significant obstacle. To overcome these obstacles, several strategies have been developed including the use of insulator elements and the bacterial tetracycline regulatory system [58–60]. Several classes of virus have been employed for gene delivery to skin cells and the choice of vector depends on the specific application (Table 42.1). Viral vectors are efficient for gene transfer in cell culture as part of the *ex vivo* approach and they have also yielded positive outcomes when administered topically or by direct injection during the *in vivo* approach [33,57,61–71].

TABLE 42.1
Viral Vectors Used in Cutaneous Gene Transfer

Vector	Advantages	Disadvantages
Retroviruses	Transduce almost all dividing mammalian cells; high transduction efficiency; long-term expression of transgene	Random integration into host genome; insertional mutagenesis known to cause target cell transformation; inability to transduce nondividing cells; only genes <6 Kb can be packaged
Lentiviruses	Transduce dividing and nondividing cells; high transduction efficiency; long-term expression of transgene	Random integration into host genome; virus is pathogenic; large-scale production is relatively costly and time-consuming; safety concerns in generating replication-competent viruses
Adenoviruses	Low risk of serious side effects; high cargo capacity; high titer easily obtained; no insertion into host genome; transduce dividing and nondividing cells	Highly antigenic; elicit cellular and humoral immune responses; most people already exposed to virus; transient expression of transgene; does not integrate into host genome; host range depends on serotype
Adenoassociated viruses	High safety profile; not associated with any known pathology in humans; transduce almost all mammalian cell types; transduce dividing and non-dividing cells	Low cargo capacity; transduction efficiency usually low; insert into host genome; induce antibody response; host range depends on serotype; difficult to produce in large quantities

42.3.2.1.1 Retroviruses

These are the most widely used viruses for gene delivery. They are capable of delivering genes to nearly any dividing mammalian cell type and integrate their genomes onto host cell chromosomes [72]. With the introduction of pantropic envelope proteins, "pseudotyped" retroviruses have been used to stably deliver genes to a much broader host range, including insect cells and nonmammalian eukaryotes [73]. Retroviral vectors can accommodate up to 6 Kb of foreign DNA and high-titer virus production is easily obtained [57]. Their transduction capacity is restricted to nonproliferating cells, therefore, their use in skin gene therapy has been limited to *ex vivo* approaches [29,57,65–67]. In addition, insertional mutagenesis has been reported to cause target cell transformation in humans [74–76]. Nonetheless, a recent landmark study has demonstrated the effective use of a retroviral vector for the gene therapy of junctional epidermolysis bullosa (JEB) in a patient [33]. The investigators isolated epidermal stem cells from a patient and transduced them with a retroviral vector carrying the LAMB3 cDNA encoding the laminin β3 subunit of laminin 5, a component of the BM. The corrected stem cells were expanded into epidermal grafts, which were transplanted into the patient, leading to functional correction of the phenotype *in vivo*. Of importance, the authors report no evidence of selection of specific integration events or clonal expansion.

42.3.2.1.2 Lentiviruses

One of the most important advances in retroviral vector applications is the use of lentiviruses. Lentiviruses have the unique capacity to infect nondividing as well as dividing cells [77]. This is of particular importance for gene therapy approaches in which target cells are nondividing or slow-dividing such as epidermal basal cells (including stem cells) or nervous tissue cells [67,70,71,78]. In fact, *in vivo* approaches have been successful in these tissues where stable, long-term production of proteins has been observed for genes delivered by lentiviral vectors [67,70,71,78,79]. Of significance, a lentiviral vector was employed to restore expression of the type VII collagen protein in keratinocytes and fibroblasts from dystrophic epidermolysis bullosa (DEB) patients using an *ex vivo* approach [80] and in a human DEB composite skin equivalent grafted onto immunodeficient mice by an *in vivo* approach [81]. These experiments provide proof-of-concept that gene correction in the skin is feasible using lentiviral vectors.

42.3.2.1.3 Adenoviruses

These vectors have been widely used for skin gene transfer. Most of these vectors have been rendered replication-deficient by deletion of their E1A and E1B essential genes [82,83]. The development of "gutless" adenoviral vectors, in which more viral genes are stripped, has increased their cargo capacity and minimized the toxicity of viral products to target cells [84]. Adenovirus vectors can carry up to 35 kb of foreign DNA and the viral particles can be produced at high titers. They can infect a wide variety of cell types and both replicating and nonreplicating cells. In addition, safety precedents already exist because adenovirus-based vaccines have been used in patients without any major side effect [85]. Adenoviral transgene expression has been detected in all cell types of both dermis and epidermis when injected subcutaneously [29,57,61,63,64]. However, adenoviruses are highly antigenic and because they do not integrate into the host genome, only transient expression of the delivered gene is achieved most of the time. Therefore, adenoviral vectors are not ideal for sustained expression in a regenerating tissue such as the epidermis, making their use to correct inherited skin diseases limited. Nonetheless, it has been reported that recombinant adenovirus vectors carrying the xeroderma pigmentosum (XP) A (XPA) and XPC genes achieved long-term expression and long-term restoration of biological activity (up to 2 months) in XPA and XPC immortalized and primary fibroblast cell lines [86]. Adenoviral vectors have been successfully used in applications such as DNA vaccination [87], anticancer therapy [88], and promotion of wound healing [89–91].

42.3.2.1.4 Adeno-Associated Viruses

Adeno-associated viruses (AAVs) are nonpathogenic viruses that were once considered poor vectors due to their limiting packaging size (about 4.0 kb). However, the requirement for a helper virus (adenovirus or herpesvirus) for productive infection makes them one of the safest viral vectors [92]. Currently, recombinant AAVs have a high safety profile because 96% of the AAV genome has been removed. AAVs can transduce a great variety of human cell types, both *in vitro* and *in vivo* [93]. AAVs can exist in both integrated and nonintegrated forms and are able to transduce replicating and nonreplicating cells [57,94]. AAVs do not induce a cytotoxic T cell response and long-term expression of the transgene is possible, but they are able to induce an antibody response and the transfection efficiencies are usually low [95]. In addition, they are difficult to produce in large quantities. Nonetheless, cutaneous gene transfer with long-term expression using these vectors has been reported [68,69].

42.3.2.2 Nonviral Gene Transfer Vectors

Although viral vectors are quite efficient for gene delivery, safety concerns have led to the development of other means for genetic transfer, the nonviral vectors. The use of nonviral vectors avoids some of the problems that occur with viral vectors, such as endogenous virus recombination, potential oncogenicity, and induction of an immune response [96]. However, low gene transfer rates and only transient expression are typically achieved in most of the cases. Nonviral reagents can be used to increase the stability and the efficient delivery and uptake of genetic material into skin cells. Overall, nonviral vector techniques for gene transfer can be classified into two groups: delivery by a chemical/biochemical vector and delivery of genetic material by physical methods [96,97].

42.3.2.2.1 Chemical or Biochemical Nonviral Vectors

The advantages of chemical or biochemical nonviral vectors include their lack of immunogenicity, absence of a theoretical

TABLE 42.2
Nonviral Vectors Used in Cutaneous Gene Transfer

Vector	Advantages	Disadvantages
Cationic polymers/proteins and cationic lipids	Lack of immunogenicity; no theoretical size limit for cargo capacity; can be used with naked DNA, RNA–DNA oligonucleotides, ribozymes or other genetic materials; low cost	Gene transfer less efficient than viral vectors; short-lived therapeutic gene expression; long-term selection usually needed
"Naked" DNA	Low immunogenicity; low biosafety risk	Low efficiency of gene transfer; short-term transgene expression
Φ31 bacteriophage integrase and engineered transposases	Precise integration of large DNA molecules into host genome; low biosafety risk; more efficient for stable gene transfer than plasmid-based systems; no theoretical size limit for cargo capacity	Further development needed for clinical applications

size limit for the therapeutic gene, use of one transfer agent for any desired gene, and the consequently ease of production, applicability, and low cost (Table 42.2). The two main nonviral vectors include cationic polymers or proteins (such as dendrimers, polylysine, DEAE-dextran, protamine, or polyethyleneimine) and liposomes [96–100]. These vectors are versatile and have been successfully applied in cutaneous gene transfer. Polymer and liposome formulations have been shown to be efficient vehicles for transdermal delivery of genes following topical application [101,102]. In the epidermis, ribozyme technology has been recently shown to be effective in downregulating gene expression using either an ethylene glycol-based vehicle or a liposome vector delivery system [46]. In a recent study, cationized gelatin microspheres were used to deliver a plasmid expressing an siRNA directed against vascular endothelial growth factor (VEGF) to squamous cell carcinoma (SCC) tumors in vivo [50]. The authors showed that this strategy is effective in decreasing vascularity and tumor growth. In another study, efficient lipid-mediated transfection of malignant melanoma (MM) cells in vitro of the microphthalmia-associated transcription factor (MITF)-specific siRNA was reported [51]. MITF is essential in the survival of MM cells and its downregulation by the MITF-specific siRNA induced apoptosis of these cells.

In addition, other means of gene delivery for cutaneous gene transfer have been developed in recent years. An engineered transposase, named Sleeping Beauty, originally isolated from fish binds to the inverted repeats of salmonoid transposons in a substrate-specific manner and mediates the precise cut-and-paste transposition in fish, as well as in mouse and human cells [103]. This is an active transposon system that can be used for the introduction of a therapeutic gene or construct. In fact, it has been used to drive the integration of the LAMB3 cDNA into holoclones from JEB patients [104]. These cells regenerated human skin with corrected levels of laminin 5 protein expression, hemidesmosome formation, and blistering. Additionally, the ΦC31 bacteriophage integrase, which is a site-specific recombinase that stably integrates large DNA molecules containing an attB phage attachment-site sequence into genomic "pseudo-attp sites" [105], has been recently used in studies targeting epidermal progenitor cells [106]. By implementing a Φ31 integrase-based gene transfer

system, Ortiz-Urda et al. were able to stably transfer the type VII collagen alpha-1 (COL7A1) cDNA into the genome of primary epidermal progenitor cells of recessive dystrophic epidermolysis bullosa (RDEB) patients and to achieve phenotypic correction in skin regenerated using these cells. Clearly, these two systems are advantageous because they lack the biosafety risks of viral vectors, they are more efficient than the stable transfer achieved with plasmid-based systems, and they allow for the stable integration of large DNA sequences into chromosomal DNA.

42.3.2.2.2 Physical Methods
In recent years, many physical techniques for in vivo cutaneous gene transfer have been developed (Table 42.3). Cutaneous gene transfer by topical application has been reported [46,101,102, 107,108]. This strategy has many advantages particularly for the treatment of large surface areas, but its efficiency is low and transgene expression is usually temporary. Direct intradermal injection of naked DNA has been shown to be an efficient mechanism for gene delivery to the skin. Using this technique, gene expression has been detected in both epidermal and dermal cells [2,67,109–112]. The advantages of this method are its simplicity and safety, but it is relatively inefficient and the coverage area tends to be small [2,96]. Therefore, additional physical methods in combination with other approaches have been tested in order to improve the efficiency of in vivo gene transfer. For example, topical application of antisense oligonucleotides and of DNA complexed with liposomes is effective in targeting the epidermal cells [113–115]. The use of epidermal enzymes, depilatory creams, or the stripping of the skin with tape to induce structural disruption of the stratum corneum can potentially increase the uptake of liposomal particles or oligonucleotides [29]. Electroporation, which is the use of electrical pulse fields to increase cell permeabilization [116], has been determined to increase gene uptake and transgene expression following injection of naked DNA in numerous tissues including the skin [71,117–122]. Puncture-mediated gene transfer, a method in which a device with a constantly high-frequency oscillating bundle of fine metal needles, leads to DNA transfer, and subsequent detection of reporter genes in skin cells [123]. Particle-mediated gene transfer or "gene gun" is also an effective method

TABLE 42.3
Physical Methods for Cutaneous Gene Transfer

Technique	Advantages	Disadvantages
Topical application	Practical; simple application of multiple treatments; advantageous for treatment of large surface areas	Low efficiency of gene transfer; transgene expression usually short-term
Direct intradermal injection	Simple; targets epidermal and dermal cells; safe	Low efficiency of gene transfer; small coverage area per application
Particle-mediated gene transfer (gene gun)	Targets both nonadherent cells and tissues; simple; practical; allows high expression of transferred genes	Short-term transgene expression; low transfection efficiency; deposits foreign material (gold microparticles) in tissue
Electroporation	Circumvents the use of viral vectors; high expression of transgene; long-term transgene expression may be achieved; area of gene transfer clearly delineated; protocols are easily optimized; recent technological developments have occurred	High risk of voltage-induced cellular injury; can induce tissue ischemia; may induce calcium-mediated protease activation leading to tissue damage; efficiency of transgene transfer needs to be optimized; limited effective range between electrodes
Puncture-mediated gene transfer (microseeding)	Does not deposit other foreign material on tissue; simple; practical; no labor-intensive preparation of genetic material required	Requires further optimization to be useful in clinical trials
Ultrasound	Noninvasive, can penetrate soft tissue; practical; large coverage area	Low transgene delivery efficiency; current value for cutaneous gene transfer is minimal
Hydrodynamic injection	Simple; reproducible; good for highly perfused tissues	Low transgene delivery efficiency; current value for cutaneous gene transfer is minimal

to deliver genetic material to skin cells, producing high transgene expression. In this method, small DNA-coated particles are accelerated into tissues and are capable of penetrating the cell by virtue of their small size; once inside the cell, the DNA dissociates and its expression ensues [96,124,125]. The major limitation of the gene gun technique is the degree of penetration into tissues, but it has been demonstrated that epidermal and dermal penetration is sufficient to induce high levels of transgene expression in the targeted areas [126–129]. In addition, other techniques such as ultrasound and hydrodynamic injection are being developed, but their current value for gene transfer to the skin is minimal [96,130,131].

42.4   CUTANEOUS GENE THERAPY

Many skin diseases are characterized by single gene mutations, either dominant or recessive and therefore can be corrected by destruction of the mutant gene product or by introduction of a wild-type gene product. In addition, taking advantage of the potent antigen-presenting dendritic cells in the epidermis (Langerhans cells) and dermis, the skin can be used as a route of immunization against tumor-associated or infection-associated antigens. Furthermore, the skin can be used as a delivery system to produce and secrete a variety of polypeptides, such as enzymes, growth factors, and cytokines, into the systemic circulation in order to aid in wound healing or to treat certain systemic diseases.

42.4.1   CUTANEOUS GENE THERAPY FOR INHERITED SKIN DISEASES

42.4.1.1 Characterization of Diseased Genes

The genetic basis of a number of inherited skin diseases has been elucidated in recent years. These include genetic lesions

leading to skin blistering, abnormal cutaneous cornification, and predisposition for cancer. In the epidermis, mutations in the genes expressed in the basal keratinocytes usually lead to skin fragility and blistering. Examples include K5 and K14 mutations in epidermolysis bullosa (EB) simplex (EBS) [132,133], mutation in genes encoding the BM-component laminin 5 (LAM5) in a subset of JEB [134,135], and mutations in the COL7A1 gene in DEB [136,137].

Mutations in genes expressed in the suprabasal keratinocytes lead to abnormal epidermal terminal differentiation and are often manifested as keratinization disorders known as ichthyoses, characterized by thickened and scaly skin [19]. Examples include transglutaminase 1 (TGM1) mutations in lamellar ichthyosis (LI) [138] and K1 and K10 mutations in epidermolytic hyperkeratosis (EHK) [139–141]. Genetic defects in genes expressed in the skin have also been linked to cancer predisposition, as in the cases of patched (PTC) mutations in basal cell nevus syndrome [142,143] and mutations in XP genes in XP [144].

42.4.1.2   Need to Correct Stem Cells and Other Considerations for Gene Therapy of Dominant and Recessive Inherited Skin Diseases

The structure of the epidermis is maintained by division of cells in the proliferative basal layer, which replaces cells in the outer layer that are sloughed into the environment. Only the stem cells persist for the lifetime of the epidermis [145]. Thus, in a renewing tissue, such as the epidermis, any permanent gene therapy approach must target the stem cell population.

Understanding of the molecular basis of disease pathogenesis provides the cornerstone in designing rational gene therapy strategies. Previous attempts to develop gene therapy for inherited skin disorders have focused mostly on recessive disorders. In recessive disorders, where mutations in both alleles are required to elicit the mutant phenotype, the introduction of one normal allele into epidermal stem cells would mimic the heterozygous "carrier state," and result in a corrected phenotype. Therefore, correction of recessive phenotypes, in principle, requires the correct expression of the wild-type gene product where it was previously absent or defective. Significant progress has been made in gene therapy for recessively inherited skin diseases in model systems. Furthermore, a recent study reports successful gene therapy on a patient with JEB [33]. Although these studies demonstrate the feasibility of genetically correcting these disorders, one concern is whether the corrected keratinocytes will be rejected when returned to patients. Since the missing protein was never synthesized by the patient, it may elicit an immune response. This potential problem is not a concern when considering dominantly inherited skin disorders because patients express normal protein from the wild-type allele. However, correction of dominant-negative phenotypes presents more of a challenge. In general, correction or suppression of the mutant gene or disruption of the dominant-negative gene product has to be achieved before the normal functions of the wild-type gene product can be restored.

42.4.1.3 Establishment of Disease Models

Besides facilitating the elucidation of the pathogenesis of a particular skin disease, an appropriate model has to be established to test possible gene therapy strategies before they are applied to patients. Therefore, it is desirable to utilize a preclinical animal model to determine the efficacy and safety of these approaches. Such a model can be either a human tissue/animal chimeric model or an entirely animal model.

For recessive diseases, the pathologic phenotypes develop only when both alleles are mutated or deleted from the genome. Although mutant cells retaining disease characteristics from patients can be used to test therapeutic approaches in vitro or grafted onto an animal [146], such material is often limited. The development of mouse embryonic stem (ES) cell techniques and the use of homologous recombination made it possible to obtain virtually unlimited material if a mouse model could be generated for a specific disease. The ES cell techniques and homologous recombination allow manipulation of the mouse genome in a finely controlled manner, where mutations can be made in a particular gene to observe the consequence of its alteration during development and differentiation. Mouse models for several recessive skin diseases have been made by making a null mutation in epidermal genes, such as TGM1 gene in LI [147] and type VII collagen (COL7A1) gene in RDEB [148]. These mouse models not only provided much information on the mechanisms of disease, but also became invaluable model systems to test gene therapy approaches for these diseases.

For dominant diseases caused by haploinsufficiency, as in the case of striate palmoplantar keratoderma [149,150], introduction of a second wild-type allele for desmoplakin (DSP) may eliminate the disease phenotype, as in the case for recessive diseases. For diseases caused by a dominant-negative mutation, traditional transgenic mouse models were very

informative in helping us understand the disease mechanisms. In this approach, a mutated transgene is introduced into the mouse genome and when it is expressed at a high enough level to compete with gene products from both wild-type alleles, a phenotype is observed. Mouse models were made in this manner for EBS [151] and EHK [152]. However, such mouse models cannot be used to test gene therapy approaches. First, because both wild-type alleles are present, the mutant gene product has to compete with wild-type product from both alleles. Second, the transgene is integrated into the genome randomly; therefore its expression level and consequently the severity of the phenotype are affected by the surrounding sequences. Thus the ratio of wild-type to mutant gene product, a crucial factor in judging the success of the therapy, remains variable in these models. Therefore, there is a need for more refined animal models with inducible and reversible transgene expression that can be controlled both spatially and temporally. In our laboratory, we took advantage of the mouse ES cell techniques and homologous recombination and generated mouse models for dominant skin diseases with these characteristics. "Hot spot" mutations in the 1A region of the rod domain in K14 and K10 have been linked to EBS and EHK, respectively [153,154]. Using a knock-in/replacement strategy, we replaced a wild-type K14 or K10 allele with a mutant allele containing a "hot spot" mutation [155,156]. Heterozygous mutant mice developed phenotypes similar to EBS and EHK, respectively, as expected for the dominant mutations. These mouse models mimic the diseases at both the genetic and phenotypic level and are ideal systems to test gene therapy approaches for EBS and EHK.

42.4.1.4 Recent Progress in Gene Therapy for Inherited Skin Disorders

42.4.1.4.1 EB Simplex

EB in general is a group of hereditary mechanobullous disorders with at least 11 distinct forms, seven of which are dominantly inherited. The main clinical feature is the skin fragility and blister formation even after mild mechanical trauma. The presentation of EB can vary from a relatively mild to a severe condition that could be fatal. EB is classified according to the level of blister formation in EB, JED, and DEB. In EBS, the blister formation occurs within the epidermis. In JEB, blister formation is seen at the lamina lucida within the BM zone and in DEB, the blistering is present below the lamina densa at the level of the papillary dermis.

The estimated incidence of EBS is 10 per 1 million births in the United States [157,158] with a considerable perinatal mortality due to electrolyte imbalance, marked protein loss, and sepsis. Blistering occurs within the epidermis and is due to lysis of basal keratinocytes, which show clumps of keratin-intermediate filament proteins upon ultrastructural examination. The majority of EBS cases are due to dominant keratin mutations. EBS was linked to the type I keratin gene cluster on chromosome 17 [132] and the type II keratin gene cluster on chromosome 12 [159]. Mutations have been identified in the basally expressed keratins K5 and K14, and are mostly found in the conserved parts of the rod domain of the protein

[153,154]. The more severe forms of EBS are caused by point mutations located at the beginning and end of the rod domain, whereas milder forms have been associated with mutations in less conserved regions either within or outside the rod domain [154]. EBS is classified according to its clinical presentation [160]. The most severe form of EBS, EB herpetiformis or Dowling-Meara (EBS-DM) presents at birth with generalized blistering [161]. Blisters occur characteristically in groups on the trunk and extremities, including palms and soles and usually heal without scarring. The development of hyperkeratoses often starts later in childhood. Interfamilial phenotypic variations are not uncommon, and have also been described among members of the same family. Approximately 70% of the reported mutations in EBS-DM occur at the same mutational "hot spot," codon 125, which encodes an arginine and is located at the beginning of the rod domain of K14.

On ultrastructural examination, EBS-DM cells have perinuclear aggregates of keratin filaments in the basal cells instead of keratin bundles throughout the cytoplasm in normal basal cells [158,162,163]. Because K14 is only expressed in the basal keratinocytes, the suprabasal cells appear normal and undergo normal terminal differentiation. EBS-Koebner (EBS-K) and EBS Weber-Cockayne (EBS-WC) forms have milder presentations. In EBS-K, the blisters are also generalized but less severe and in EBS-WC, the blisters are present mainly on the hands and feet [164].

Two mouse models were previously developed, including a transgenic model [151] and a K14 null model [165]. Although both models helped us understand the disease mechanisms, and the K14 null model also provided important insight into K15 functions, neither model mimics EBS-DM at the genetic level and therefore cannot be used to test gene therapy approaches for the disease. We recently developed a mouse model for EBS-DM, where a wild-type K14 allele was replaced with a mutant allele in the mouse germline [156]. The presence of a neomycin-resistance cassette in an intron-affected expression from the mutant K14 allele and the heterozygotes had subclinical phenotypes. But homozygous pups developed extensive blisters and died shortly after birth. When the neoselection cassette was removed by Cre-mediated excision, the resulting heterozygous pups developed large blisters, as was expected for this dominant mutation. This is the first mouse model that mimics EBS-DM at both the genetic and phenotypic levels. Unfortunately, the pups died because of severe blistering. To overcome such problems, we recently developed a transgenic mouse model that allows the focal deletion of a genomic sequence via Cre-mediated excision. A transgenic mouse line carries a Cre recombinase fused to a truncated progesterone receptor, driven by a K14 promoter (K14. CrePR1) [166]. These mice were crossed with mice heterozygous for the mutant K14 allele with the neocassette flanked by loxP sites (mtK14neo), and heterozygous mutant mice carrying the transgene were obtained. On topical treatment with an antiprogestin, the neocassette can be deleted in focal areas by activated Cre. We expected that in such areas, the mutant K14 expression would be comparable with that of the wild-type K14, causing blistering in the skin. We were indeed able to induce blisters using this system and the mice remained viable [156]. This transgenic mouse model is ideally suited to test gene therapy approaches for EBS-DM. In this mouse model, the mutant K14 allele can be focally activated in epidermal stem cells and following topical administration of an inducer, blisters develop in treated areas. However, after a few weeks, blisters heal and never reappear. Some skin disorders are characterized by a mosaic pattern, with alternating stripes of affected and unaffected skin that follow the lines of Blaschko, a term given to the pattern assumed by different inherited and acquired skin diseases on the human skin and mucosae. These lines likely represent the distribution of a clone of genetically altered cells. These nonrandom patterns are believed to be caused by postzygotic mutations that occur during embryogenesis. Interestingly, a mosaic form of EBS has never been reported. It has been suggested that basal stem cells carrying a postzygotic mutation in K5 or K14 would have a selective disadvantage and be rapidly displaced by wild-type basal stem cells, which can move laterally. Using laser capture microdissection (LCM), we have shown that the induced blisters healed by migration of surrounding nonphenotypic stem cells into the wound bed. We also demonstrated that EBS keratinocytes have a diminished capacity to repopulate graft sites compared to normal keratinocytes and that EBS stem cells have a selective disadvantage over normal stem cells in a graft environment (our unpublished data). Thus, our model predicts that if EBS stem cells could be corrected, they will have a selective growth advantage when introduced into areas prone to blistering. This observation provides an explanation for the lack of mosaic forms of EBS-DM. In addition, it has important implications for gene therapy since it suggests that defective EBS stem cells will be replaced by nondefective stem cells. Another unexpected observation from this mouse model was the discovery that mice expressing the mutant K14 allele at levels approximately 50% of wild-type K14 do not exhibit a skin phenotype. Previously, it had been assumed that gene therapy approaches for dominant disorders like EBS must aim to either correct the mutant allele or completely inhibit its expression. Our model predicts that the EBS phenotype may be eliminated by overexpression of the normal K14 allele or partial suppression of the mutant K14 allele, thus increasing the ratio of wild-type to mutant protein. Technically, it is much easier to design gene therapy strategies that would achieve a partial suppression of a mutant-dominant allele rather than a complete correction or suppression. Therefore, we designed gene therapy strategies aimed to correct the EBS-DM phenotype by altering the ratio of wild-type to mutant K14 by either increasing the expression of wild-type K14 or by decreasing the expression of the mutant allele. We developed lentiviral vectors carrying the wild-type K14 gene and others encoding siRNA constructs that specifically cleave the mutant K14 transcripts. We demonstrated that they effectively transduce keratinocytes with long-term transgene expression in tissue culture and *in vivo*. We have been able to correct the EBS phenotype by showing that corrected EBS keratinocytes form a normal epidermis when grafted onto nude mice (unpublished results).

42.4.1.4.2 Junctional Epidermolysis Bullosa

This disease is a subtype of EB characterized by blister formation at the level of the lamina lucida within the BM zone. There are two major forms of JEB, the Herlitz variant or EB letalis, and GABEB [167]. Both forms are transmitted in an autosomal recessive manner and both have onset at birth and are associated with marked skin fragility and generalized blister formation. While scarring and milia formation are usually absent in both forms of JEB, skin atrophy is a characteristic finding in GABEB patients who also tend to have marked dystrophic or absent nails, palmoplantar hyperkeratosis, and may have significant scarring alopecia of the scalp [167]. A characteristic feature of Herlitz disease is the development of large, non-healing areas of granulation tissue; common sites include the perioral and perinasal areas, trunk, and nape of the neck. In addition, extracutaneous involvement may occur in both forms of JEB. In Herlitz disease, findings may include oral blisters and erosions, dysplastic teeth, marked growth retardation, and severe anemia. In contrast, patients with GABEB have milder mucosal involvement (oral cavity, conjunctiva, and esophagus) and early loss of permanent teeth, but neither growth retardation nor anemia [167]. Unlike DEB, musculoskeletal abnormalities are absent in both forms of JEB. The dissociation of the dermal–epidermal junction occurs beneath the basal cell layer, but above the lamina densa in this disease. On examination by electron microscopy, JEB is caused by dissolution of the lamina lucida. Mutations in a number of genes encoding vital structural proteins, including laminin 5 components α3 (LAMA3), β3 (LAMB3), γ2 (LAMC2), [134,135], BP180 (type XVII collagen or BPAG2) [168], integrin β4 (ITGB4) [169], and plectin (PLEC1) [170] have been identified in JEB.

Several somatic gene therapy approaches have been tested to correct the JEB phenotype. In one study, keratinocytes from a patient with Herlitz JEB with a mutation in laminin 5 β3 were transduced with a LAMB3 transgene. The transduced keratinocytes synthesized β3 peptide that assembled with endogenous α3 and γ2 forming biologically active laminin 5, which was secreted, processed, and deposited into the extracellular matrix. Reexpression of laminin 5 induced cell spreading, nucleation of semidesmosomal-like structures, and enhanced adhesion to culture substrate [171]. Organotypic cultures with the transduced keratinocytes reconstituted epidermis closely adhering to the mesenchyme and presenting mature hemidesmosomes, bridging the cytoplasmic intermediate filaments of the basal cells to the anchoring filaments of the BM [171]. A mouse line with targeted disruption of LAMA3 was created a few years ago [172]. Although the mutation in homozygous pups caused neonatal lethality, cells isolated from these pups were used successfully to test therapeutic approaches in vitro. In a study carried out to restore BP180 function in cultured JEB patient keratinocytes and skin grafts through gene transduction [173], the transduced cells had normalization of their adhesion parameter. Of significance, a revertant mosaicism was reported in a patient with GABEB, representing a "natural gene therapy" [174].

The reversion of the affected genotype to carrier (heterozygote in a recessive disease) genotype in about 50% of the keratinocytes was sufficient for the normal functioning of the skin. Recent studies were successful in restoring and sustaining expression of LAMB3 in human JEB skin and led to the correction of the JEB phenotype in vitro and in vivo, respectively [175,176]. In addition, the engineered transposase Sleeping Beauty has been used to integrate the LAMB3 cDNA into holoclones from JEB patients [104]. The corrected cells regenerated human skin with normalized levels of laminin 5 protein expression, hemidesmosome formation, and blistering. In a recent landmark study [33], the researchers employed epidermal stem cells from the palms of the patient, which had a few unaffected areas, transduced them with a retroviral viral vector expressing LAMB3 cDNA, and used them to prepare genetically corrected cultured epidermal grafts. The grafts were then placed onto surgically prepared regions of the patient's legs. They report that after 8 days, there was complete epidermal regeneration on both legs and a normal adherent epidermis was maintained throughout the 1-year follow-up. The grafted skin did not have blisters, inflammation, or infections and there was absence of an immune response. A risk for reintroducing a therapeutic wild-type protein into patients with a recessive disease is that immunologic responses may be elicited. Extra caution has to be taken to decrease such possibility. However, in this case, the patient was a double heterozygote carrier of a null allele and a single-point mutation in the LAMB3 gene. This is a study of paramount importance because it demonstrates for the first time that gene therapy can cure a disease of solid tissue in patients.

42.4.1.4.3 Dystrophic Epidermolysis Bullosa

DEB is a group of blistering skin disorders with either autosomal-dominant or autosomal-recessive inheritance characterized by tissue separation, which occurs below the dermal–epidermal BM at the level of the anchoring fibrils due to mutations in the COL7A1 gene. A series of successful genetic corrective studies have been performed for DEB. In one study, the expression of type VII collagen was normalized in keratinocytes from recessive DEB patients [177]. Also, the expression of the type VII collagen protein in keratinocytes and fibroblasts from DEB patients using a lentiviral vector was restored in an ex vivo approach [80]. The corrected keratinocytes had long-term transgene expression and displayed normal morphology, normal behavior (i.e., motility, attachment and proliferative potential), and were capable of regenerate normal skin in vivo. The ΦC31 integrase-based gene transfer system was used by Ortiz-Urda et al. to stably transfer the COL7A1 cDNA into the genome of primary epidermal progenitor cells from four unrelated RDEB patients [106]. Collagen VII was correctly localized at the BM and phenotypic correction was achieved in skin regenerated using these cells. In another study by the same group, intradermal injection of genetically engineered RDEB fibroblasts overexpressing type VII collagen into reconstituted RDEB skin stably restored type VII collagen expression in vivo [178]. The type VII collagen protein was properly localized and the RDEB phenotype, including subepidermal blistering and anchoring fibril defects, was corrected. Also, intradermal injection of a lentiviral vector expressing human COL7A1 cDNA into athymic hairless mice and into a human DEB composite skin equivalent grafted onto immunodeficient mice was performed [81]. The vector transduced dermal cells, which expressed the protein. The collagen VII incorporated into the BM zone at the dermal–epidermal junction and remained stable for at least 3 months after a single intradermal injection. The lentivirus did not target keratinocytes; fibroblasts and endothelial cells were the target cells for the in vivo lentiviral vector transduction. Furthermore, the investigators showed that intradermal injection of the COL7A1 lentivector into the RDEB skin equivalents corrects the subepidermal blistering and restores the anchoring fibrils to a similar pattern seen in skin regenerated from normal controls.

42.4.1.4.4 Epidermolytic Hyperkeratosis

This disease is inherited in an autosomal-dominant mode, with an incidence of 1 in 200,000 to 300,000 newborns [158]. Up to 50% of the reported cases arise sporadically. Affected children present at birth with erythroderma, blistering, and peeling. Erythroderma and blistering diminish during the first year of life, and hyperkeratoses develop, predominantly over the flexural areas of the extremities. On histopathological examination, findings consist of hyperkeratosis and parakeratosis, lysis of the suprabasal keratinocytes, and perinuclear vacuolar degeneration. The basal keratinocytes appear normal but exhibit hyperproliferation [161]. The transit time for keratinocytes to move from the basal layer to the stratum corneum is remarkably shortened and takes only 4 days in EHK patients instead of the normal 4 weeks [179]. Mutations to K1 and K10 have been linked to EHK and a mutational "hot spot" at Arg 156 in K10 has been identified in most severe cases of EHK [141,180]. Three mouse models were previously generated, including two transgenic models [152,181] and K10 null model [182]. However, because they do not mimic EHK at the genetic level, none can be used as a model system to test gene therapy for this disease.

In our laboratory, we recently generated a mouse model that mimics EHK at both the genetic and phenotypic levels by replacing a wild-type K10 allele with one that has a point mutation at Arg 156 [155]. We have started strategies aimed at the correction of the EHK phenotype by altering the ratio of wild-type to mutant K10 by increasing expression of wild-type K10. To do this, we have collected epidermal stem cells from heterozygous mutant mice and transduced them with lentiviral vectors to overexpress the wild-type K10 allele. In addition, we are attempting to correct the EHK phenotype by suppressing the expression of the mutant K10 allele by transducing EHK cells with a viral vector expressing an siRNA construct specifically designed to cleave the mutant K10 transcripts. As opposed to EBS, in which a correction of the mutation in the basal keratins translates into a growth advantage, corrected and mutant stem cells in EHK will proliferate at similar rates because the mutant K10 is not expressed in stem cells, but only in the progeny of stem cells after they have differentiated and moved into the suprabasal layers.

There is no selection against epidermal stem cells with K10 mutations and these stem cells continue to give rise to defective differentiated progeny. It has recently been demonstrated that topical selection with colchicine can be used to select, in vivo, for human keratinocyte stem cells transduced with the multidrug resistance (MDR) gene [183]. Therefore, to efficiently amplify the corrected cell population, the MDR gene is included as a selection marker in our targeting strategy to ablate defective epidermal stem cells and allow their replacement with corrected stem cells, which then repopulate the epidermis.

42.4.1.4.5 Lamellar Ichthyosis

The autosomal-recessive ichthyoses are a clinically heterogeneous family of diseases characterized by abnormal cornification and comprise LI and congenital ichthyosiform erythroderma [184]. In LI, patients are born encased in a "collodian" membrane that is later shed and followed by development of large, thick scales and varying degrees of erythema. Palmar and plantar hyperkeratoses are often present.

Defects in the gene encoding keratinocyte TGM1 were identified in a number of patients with LI [138,185]. TGM1 is normally expressed in differentiated keratinocytes and catalyzes crosslinking of cornified envelope precursor molecules, such as involucrin, loricrin, and small proline-rich proteins [186,187]. With loss of TGM1 function in the formation of insoluble cornified envelope, the barrier function of the outer epidermis is disrupted [19,188]. Therefore restoration of the TGM1 enzymatic activity represents a possible strategy for correcting the LI disorder. A high-efficiency retroviral vector containing a wild-type TGM1 gene was used to transduce mutant keratinocytes from LI patients [30]. More than 98% of the primary cells expressed wild-type TGM1, as measured by the proportion of keratinocytes positive for the transferred gene compared with the total number of cells determined by propidium iodide counterstaining [30]. TGM1 enzymatic activity was restored to normal levels and was targeted to the membrane fraction. In addition, transduced keratinocytes also demonstrated restored involucrin crosslinking and normal cornification [30]. When these transduced keratinocytes were grafted to immunodeficient mice, the regenerated skin displayed restored TGM1 protein expression in vivo and was normalized at the levels of histology, clinical surface appearance, and barrier function [31]. However, TGM1 expression in the human skin graft was lost after 1 month because of silencing of the vector [32].

Besides transplantation of genetically modified cells, another way to deliver genes to the skin is through direct administration to the intact tissue. Choate and Khavari [189] regenerated skin from LI patients on nude mice to examine the corrective impact of direct injection of naked plasmid DNA. Regenerated LI skin received repeated in vivo injections with a TGM1 expression plasmid and restoration of TGM1 expression at the correct tissue location in the suprabasal epidermis was observed. However, unlike LI skin regenerated from keratinocytes first transduced in vitro with a retrovirus carrying TGM1 prior to grafting, directly injected LI skin displayed a nonuniform TGM1 expression pattern [189]. In addition, direct injection failed to correct the central histologic and functional abnormalities of LI. These results show that partial restoration of gene expression can be achieved through direct injection of naked DNA into human skin disease area, but underscore the need for new advances to achieve efficient and sustained plasmid-based gene delivery to the skin.

42.4.1.4.6 X-linked Ichthyosis

X-linked ichthyosis (XLI) is caused by a deficiency in steroid sulfatase (STS), which leads to the accumulation of cholesterol sulfate and results in abnormal scaling skin [190]. XLI is inherited in a pseudoautosomal mode, escaping X-inactivation in both humans and mice. Transduction using a retroviral vector expressing the STS gene led to restoration of protein expression as well as enzymatic activity [191]. In the same study, it was shown that transduced keratinocytes from XLI patients regenerated epidermis histologically indistinguishable from that formed by keratinocytes from healthy patients after grafting onto immunodeficient mice. In addition, the transduced XLI epidermis presented a return of barrier function parameters to normal.

42.4.1.4.7 Harlequin Ichthyosis

Harlequin ichthyosis (HI) is a genetic disorder with a devastating phenotype characterized by ectropion, eclabium, flattened ears, and thick scales in the whole body [184]. This disease is characterized by a high perinatal mortality rate associated with severe dehydration and infections due to the disruption of the skin barrier. Mutations in the adenosine-triphosphate-binding cassette A12 (ABCA12) gene were initially reported to be associated with an LI type 2 [192]. Just recently, two different groups found that mutations causing truncation or deletion of highly conserved regions of ABCA12 result in loss of the skin lipid barrier and cause HI [193,194]. ABCA12 is a lipid transporter present in keratinocytes. Its function is important in lamellar granule formation and lipid transport on the keratinocyte surface [193]. Recovery of lamellar granule-mediated lipid secretion has been demonstrated in cultured HI keratinocytes after corrective gene transfer of the ABCA12 gene [193].

42.4.1.4.8 Xeroderma Pigmentosum

XP is a group of autosomal-recessive disorders associated with defects in a number of DNA repair genes and are characterized by inadequacies in DNA repair after ultraviolet injury [144,195]. XP genes, which are involved in a specific mechanism of DNA repair called nucleotide excision repair (NER), fall into seven complementation groups (XPA to XPG) and all have been cloned [196]. XP affects both the basal and suprabasal compartments of the epidermis, including melanocytes. Thus, XP patients have increased susceptibility to epidermal neoplasms, such as basal cell carcinomas (BCC), SCC, and MM [197]. Treatment of this condition consists of avoidance to UV-light exposure, photoprotection, and surveillance and surgical resection of skin tumors when those arise. Since

every cell that is unable to repair the UV-induced DNA damage should be considered a potentially tumoral cell, any treatment to be curative, must target every cell in the epidermis including every stem cell and melanocytes. Although initial promising attempts to correct the genetic defects in this condition were performed [198–201], it became clear that XP represents a challenging condition for cutaneous targeted gene therapy because high-efficiency targeting of melanocytes remains to be developed. However, recent studies outline improved gene transfer efficiency of XP genes to different skin cells and show amelioration or correction of XP phenotypes. In one study, a functional XPC gene was introduced into primary keratinocytes of XPC patients [202]. Reexpression of the normal XPC protein corrected DNA repair capacity, UV-cell survival, and clonal transition capacity. Also, the XPC-corrected cells were able to participate in skin reconstitution in vitro. Marchetto et al. [203] used a recombinant adenovirus to deliver the human XPA cDNA to mouse embryo fibroblasts of Xpa-mice, which showed recovery of their defective DNA repair ability after transduction. The authors also demonstrated that subcutaneous injection of the XPA-carrying adenovirus was able to target basal keratinocytes and prevent tumor development after UV radiation in the treated mice. In another study, Marchetto et al. [204] tested the efficiency of lentiviral vectors carrying the XPA, XPC, and XPD genes to transduce XP fibroblasts. These vectors were efficient in transducing the XP fibroblasts in vitro. Furthermore, transduced cells showed correction of the XP cellular phenotypes such as normal UV survival and unscheduled DNA synthesis after UV radiation. Results from these recent studies suggest that significant progress in the area of XP gene therapy will be seen in the near future.

42.4.1.4.9 Other Diseases

In addition to the dominant-negative mutations in keratins that cause various blistering diseases and hyperkeratoses, several other diseases are inherited in a dominant mode through haploinsufficiency, such as palmoplantar keratoderma that results from DSP mutations [149,150]. Gene therapy strategies could involve introduction of a wild-type allele and host immunologic reactions will not be a concern in these cases.

42.4.2 Cutaneous Gene Therapy for Melanoma and SCC

The incidence of MM has been on the rise in the last years. SCC is a common malignancy of the epidermal keratinocytes. Due to the advantages that make the skin an amenable tissue for gene transfer, MM and SCC are frequently used as models for tumor-specific gene therapy. Different strategies have been applied for both conditions. In general, gene therapy efforts for MM have centered on genetically modifying host cells by transferring tumor suicide genes, genes that increase tumor rejection responses, or genes whose action interferes with signaling cascades involved in tumor development and growth [205]. The same principles apply to SCC gene therapy efforts. For instance, adenoviral-mediated transfer of the herpes simplex thymidine kinase "suicide" gene (tk) by direct intratumoral injection, followed by ganciclovir administration, was used to treat human MM established in nude mice [206]. In a different study using the B16 melanoma model, a synergistic effect was observed when combination therapy by adenovirus-mediated transfer of tk and interleukin 2 (IL2) or tk and granulocyte-macrophage colony stimulating factor (GM-CSF) was used [207]. Similar approaches have been employed in studies with SCC models [208,209]. Besides gene transfer mediated by adenoviral vectors, an *in vivo* liposomal-mediated approach with tk and IL-2, and a transgenic model constitutively expressing a costimulatory molecule have been performed for treatment of SCC [29]. In another study, the tk gene was cloned into an amplicon vector that was used effectively for its *in vivo* delivery on human MM xenografts [210]. Recently, a number of molecules have been targeted in gene therapy studies for MM and SCC. For example, VEGF is a key molecule in angiogenesis, which is an important process for tumor growth [211]. In one study, a vector expressing siRNA directed against VEGF was delivered by means of cationized gelatin microspheres to SCC tumors established by a xenograft model in mice [50]. The authors demonstrated that downregulation of VEGF induced a reduction in vascularity and tumor. Other studies demonstrated that administration of siRNA against the S phase kinase-interacting protein 2 (Skp2) gene, which encodes a protein overexpressed in MM and oral SCC, inhibited the cell growth of tumor cells *in vitro* and also suppressed tumor proliferation *in vivo* [212,213]. The Y-box binding (YB-1) protein was found to be upregulated in MM metastases [214]. Stable downregulation of YB-1 using siRNA in metastatic MM cells resulted in a reduced rate of proliferation, increased rate of apoptosis, and reduced the migration and invasion of MM cells. In addition, MM cells with a decreased YB-1 were more sensitive to chemotherapeutic agents [215]. MITF is essential in the survival of MM cells and its gene has been shown to be amplified in 20% of MM [216]. Nakai et al. [51] reported the downregulation of MITF on MM cells using an siRNA with a concomitant induction of apoptosis *in vitro*. Furthermore, the siRNA was delivered by electroporation into MM tumors established in mice by subcutaneous injection of B16 cells. The authors showed that the tumor cells transfected with the MITF-siRNA underwent apoptosis *in vivo* and that this strategy was effective in reducing tumor outgrowth.

42.4.3 CUTANEOUS GENE THERAPY FOR WOUND HEALING

The process of wound healing involves three stages: inflammatory reaction, formation of granulation tissue, and tissue remodeling [217]. All these events are known to be regulated by different cytokines and growth factors. Gene transfer techniques are intended to treat wound healing abnormalities or to enhance the wound healing rate. Genetic therapy for wound healing is eased because the epidermal barrier is defective, transgene expression for a short period is usually sufficient, and in most cases, the treated area is localized [218]. However,

the presence of inflammatory exudates, deposited proteins, and necrotic tissue may prevent the transfer of genetic material [218]. One strategy is to use genetically modified cells (such as keratinocytes or fibroblasts) and graft them to deliver a desired protein (e.g., growth factors) to the wound bead. For example, Eming et al. [219,220] introduced the genes encoding platelet-derived growth factor (PFGF) and insulin-like growth factor (IGF) isoforms into keratinocytes and grafted them on mice. The modified keratinocytes were shown to induce the formation of granulation tissue and to increase the proliferation of the modified cells, respectively. Similar approaches have been undertaken using dermal fibroblasts with successful results [221–225].

Multiple studies have investigated gene therapy for wound healing using *in vivo* approaches. Multiple groups have tested different modalities of gene transfer to skin wounds such as topical application, the gene gun approach, and the use of genetically modified cultured skin substitutes with promising results [126,226]. Tyrone et al. [227] successfully treated ischemic dermal ulcers in rabbits by topical application of PDGF-A or B DNA plasmids embedded within a collagen lattice. They showed a substantial increment in the formation of new granulation tissue, epithelialization, and wound closure after treatment. Subcutaneous injection of a liposomal IGF-1 cDNA construct was shown to effectively promote re-epithelialization of burn wounds by decreasing prolonged local inflammation through modulation of the expression of pro- and antiinflammatory cytokines [228]. Intradermal injection of an AVV vector expressing human VEGF-A to full-thickness excisional wounds in rats was found to induce new vessel formation and enhance wound healing rate [229]. Similar results were obtained by Romano Di et al. using topical application of an adenovirus vector to deliver VEGF on excisional wounds of streptozotocin-induced diabetic mice [230]. In other studies, liposomal keratinocyte growth factor (KFG) cDNA delivery to acute wounds in rats enhanced wound healing by increasing cell proliferation, re-epithelialization, and neovascularization, by reducing cell apoptosis, and by activating mesenchymal cells through the induction of IGF-1 expression [231,232]. Nitric oxide production at adequate levels is essential for adequate wound healing and is driven by the activity of nitric oxide synthase (NOS). Gene transfer by topical application of an adenovirus carrying the inducible NOS (iNOS) gene was shown to reverse the wound healing defects in iNOS-deficient mice [233]. Gene transfer studies of the human hepatocyte growth factor (HGF) have shown that it promotes re-epithelialization and neovascularization in the early phase of wound healing [234] and that it enhances wound healing and inhibits overscarring [235]. In addition to growth factor genes, gene transfer studies of proinflammatory cytokines [236], a protease inhibitor [237], and a transcription factor [238] have been reported to be effective in wound healing promotion. Furthermore, it has been proposed that signal transduction molecules such as Smads can be used to promote wound healing or to treat wound healing abnormalities [239].

42.4.4 Cutaneous Immunomodulation in Cancer Gene Therapy

Gene transfer is an effective approach to induce and enhance immune responses. Injection of naked DNA encoding antigenic epitopes can induce strong specific humoral immune responses [240]. The advantages of immunomodulation by genetic vaccination are evident; for example, it does not require isolation and purification of proteins and it circumvents the use of life or attenuated viral vaccines. The skin is rich in antigen-presenting cells that can initiate and control specific immune responses. In addition, cutaneous transfer of plasmid DNA or mRNA allows the delivery of not only genetic material encoding antigenic epitopes, but also the delivery of immunomodulators [241,242]. DNA vaccination has been used to induce immune responses as an strategy in antitumoral immunotherapy [243,244]. In a clinical trial conducted by Thurner et al., dendritic cells were isolated, propagated *ex vivo*, and pulsed with a Mage-3A1 tumor peptide. Subsequently, the cells were injected subcutaneously and intradermally to advanced MM patients; regression of some metastases was seen in 6 of 11 patients, and expansion of specific cytotoxic T cells was seen in 8 of 11 patients [245]. Sun et al. [243] demonstrated tumor regression and increased survival in mice-bearing subcutaneously implanted tumors after skin gene transfer of the antitumoral cytokines IL-2, IL-6, and interferon gamma (IFNG). Regression of established primary and metastatic murine tumors was documented after cutaneous gene gun delivery of a plasmid carrying the mouse IL-12 [246]. In the same study, the researchers were able to demonstrate that a tumor-specific immunological response had been induced in the treated mice. Recent clinical trials have shown that intralesional injection of an adenovirus carrying the IFNG cDNA is an effective therapy for cutaneous lymphomas [244,247]. In addition, a similar approach with IL-2 is also effective in patients with solid tumors including melanoma [244]. It is known that adenoviral vectors are capable of inducing an innate immune response and an interferon-mediated response as well. For example, Urosevic et al. [248] performed gene profiling of lymphomas before and after intralesional IFNG transfer using and adenoviral vector. They show that the IFN response is induced by the adenoviral vector itself and that it potentiates the response induced by the IFNG transgene. Furthermore, the authors suggest that this potentiation may have significant therapeutic value.

42.4.5 Cutaneous Gene Therapy for Systemic Diseases: Skin as a Bioreactor

Epidermal keratinocytes secrete a variety of proteins such as collagen VII, laminin, proteinases, proteinase inhibitors, growth factors, and cytokines [249]. Studies using cultured human keratinocytes, grafted onto athymic mice and rats, demonstrated that keratinocytes secrete proteins that reach the systemic circulation. One of the first studies of the secretory function of keratinocytes was carried out by Taichman

and colleagues [250]. They first monitored secretion of human apolipoprotein E (HuAPOE) in cultured human keratinocytes [250]. The protein was identified as APOE on the basis of molecular weight, isoform pattern, and immunoreactivity. When they grafted human keratinocytes onto athymic mice and rats, human APOE was detected in the systemic circulation of graft-bearing animals as long as the graft remained on the animals [251]. Within 24 h of graft removal, human APOE was not detected in the plasma, indicating that human APOE in the plasma resulted from continuous production of the protein by grafted human keratinocytes. These results showed that proteins as large as APOE (299 amino acids) can transverse the epidermal–dermal barrier and achieve systemic circulation.

Once it was established that keratinocyte-secreted proteins could reach the systemic circulation, genetically modified keratinocytes or fibroblasts were used to test whether they could deliver transgene products into the bloodstream. Subsequent experiments using both *in vivo* and *ex vivo* approaches have been successful in delivering different polypeptides, such as growth hormone (GH), erythropoietin (EPO), factors VIII and IX, leptin, phenylalanine hydroxylase (PAH), IL-10, GM-CSF, and others to the circulation. In our laboratory, we have further enhanced the usefulness of the skin as a bioreactor by developing a bigenic gene-switch system that allows focal induction of transgene expression via topical administration of an inducer [252]. Therefore, due of its capacity to deliver various polypeptides into the systemic circulation, its accessibility and abundant vascularization, added to the refinement of gene transfer systems (e.g., the gene-switch system), the skin is a very attractive tissue for test gene therapy studies for systemic conditions that respond to delivery of polypeptides into the circulation.

42.4.5.1 Progress in Cutaneous Gene Therapy of Specific Systemic Diseases

42.4.5.1.1 Hemophilia A

This is an X-linked inherited disease caused by deficiency of factor VIII and has an incidence of 1 in 5000 male live births [253]. This condition occurs in mild, moderate, and severe forms, reflecting the mutational heterogeneity seen in the factor VIII gene, and symptoms vary from excessive bleeding only after trauma or surgery to frequent episodes of spontaneous or excessive bleeding after minor trauma, particularly into joints and muscles [253,254]. In one study, factor VIII deficient-transgenic mice expressing human factor VIII under the control of the involucrin promoter were generated [255]. Plasma factor VIII activity and correction of the phenotype were seen in this mouse model. In the same study, skin explants from these transgenic mice were grafted into factor VIII double knockouts, which showed plasma factor VIII activity of 4% to 20% normal and had improved whole blood clotting [255]. Recently, a successful clinical trial with hemophilia A patients was performed [256]. Dermal fibroblasts from six patients were grown in culture, transfected with the factor VIII cDNA, and administered by laparoscopic injection in the omentum. Factor VIII plasma levels were significantly increased in four patients; a reduction in bleeding and the need for supplementation with growth factor VIII was seen.

42.4.5.1.2 Leptin Deficiency

Leptin, a 16 kDa protein hormone, is involved in the regulation of body weight in mammals [257,258]. It is secreted primarily by adipocytes and it has been shown to regulate food intake and neuroendrocrine function through its action in the hypothalamus (Friedman and Halaas, 1998). In accordance with the phenotype seen in the ob/ob mice [259], it has been determined that congenital leptin deficiency is associated with early-onset obesity in humans [260]. Leptin replacement therapy has provided encouraging results in clinical studies [261]; however, the need for repetitive dosing has prompted an alternative approach using gene therapy to correct this condition. In a recent study, a cutaneous gene therapy approach for leptin deficiency was successful in correcting the mouse ob/ob phenotype [262]. Here, immunodeficient ob/ob mice grafted with skin implants from mice overexpressing leptin reached body weight equivalent to that of wild-type animals. In addition, immunosupressed ob/ob mice that were transplanted with skin grafts made of human keratinocytes transduced with a leptin cDNA-carrying retroviral vector showed weight reduction concomitant with a decrease in blood glucose and food intake [262].

42.4.5.1.3 Anemia Due to EPO Deficiency

EPO is a kidney-produced glycoprotein that regulates red cell production. It binds to its receptor found in erythroid progenitor cells, activating a signaling pathway that leads to the increase of survival of these cells by inhibiting apoptosis, accelerating the release of reticulocytes from the bone marrow [263]. In 1987, recombinant human Epo (rHuEPO) was approved in the United States for the treatment of anemia of end-stage renal disease [263]. In addition, its therapeutic use has been extended to other conditions, such as anemia associated with Zidovudine treatment of patients with AIDS, anemia secondary to chemotherapy in the treatment of cancer, anemia of pregnancy, anemia of prematurity, myelodysplastic syndrome, and bone marrow transplantation [263]. Descamps et al. [264] transduced HeLa keratinocytes with an AAV vector containing the EPO gene and implanted them in nude mice. They observed a long-term increase in hematocrit, which correlated with the size of the induced tumor. This study provided proof-of-principle for the treatment of anemia using this approach. In another study, a lentiviral vector encoding HuEPO was delivered by single intracutaneous injection into human skin grafts on immunodeficient mice [70]. The investigators demonstrated that HuEPO was present in serum and its levels increased in a dose-dependent fashion. In addition, the hematocrit improved within 1 month after lentiviral injection and remained stable for almost 1 year. Siprashvili and Khavari [265] developed a lentiviral vector in which the expression of EPO can be controlled by topical administration of steroid

ligands. Also, in this system, the proviral insert can be excised by applying a topical inducer of a Cre recombinase. The investigators injected the lentivirus into human skin regenerated on immunodeficient mice and were able to control EPO and hematocrit levels over time. A clinical trial was performed in patients with chronic renal failure by *ex vivo* transduction of patient-derived dermal cores with an adenovirus containing the EPO gene [266]. EPO expression and reticulocyte counts were increased for 14 days. The researchers reported an increase in CD8 cytotoxic T cells, which could be the cause of the subsequent decrease in EPO expression. It is possible that this or a similar strategy can have increased efficiency if a different gene transfer technique is used.

42.4.5.1.4 GH Deficiency

Original experiments demonstrating the release of exogenous GH by transduced keratinocytes into the circulation were performed by Morgan et al. [267]. Using the recently developed bigenic gene-switch mouse model, we showed that after a single induction, high levels of human GH (huGH) were released from keratinocytes into the circulation [252]. The serum levels of huGH were dependent on the amount of inducer applied and repeated induction resulted in increased weight gain by transgenic versus control mice. Furthermore, physiological levels of huGH were detected in the serum of nude mice after topical induction of small transgenic skin grafts. These results clearly demonstrate the feasibility of using the gene-switch system to regulate the delivery of GH into the circulation for the treatment of GH deficiency. Bellini et al. [268] transduced primary human keratinocytes with the huGH gene and grafted them into immunodeficient dwarf mice (lit/scid). The graft-bearing mice had increased levels of the hormone that was associated with weight increase.

42.4.5.1.5 Contact Hypersensitivity

Contact hypersensitivity (CHS) responses are regulated by T cells, which release cytokines and attract other inflammatory cells after reacting with antigen. IL-10 has been known to be a key regulatory cytokine in both inflammatory and immune responses. Studies have demonstrated that IL-10 is involved in the regulation of the hypersensitivity response because recombinant IL-10 (rIL-10) prevented the elicitation of CHS in previously sensitized mice [269]. Meng et al. [270] injected a DNA plasmid containing human IL-10 into the dorsal skin of hairless rats. They detected local expression of mRNA and protein in a dose-dependent manner. They also showed the production and release of IL-10 into the circulation by transduced keratinocytes. Furthermore, they quantified a reduced response to challenge in distant areas from the injection site of previously sensitized animals.

42.4.5.1.6 Other Conditions

Additional studies testing the skin as a bioreactor approach and been performed. For example, two forms of APOE, both the endogenous HuAPOE and a recombinant form from a transfected vector, were detected in the serum of athymic mice-bearing grafts of modified human keratinocytes [250,271]. When human keratinocytes in culture were transduced with a retroviral vector carrying the factor IX gene, they secreted active form of the protein into the medium [272]. When they were grafted onto nude mice, small quantities of factor IX were detected in the bloodstream. In addition, other gene therapy studies for hemophilia B have been performed [273–275]. A recent study achieved the production of insulin from human keratinocytes and tissue-engineered skin substitutes as the initial steps for developing a potential strategy to treat diabetes [276]. Other examples include the genetic transfer of adenosine deaminase (ADA) gene to keratinocytes from ADA-deficient patients [277] and of ornithine-delta aminotransferase (OAT) in keratinocytes from patients with gyrate atrophy [278].

42.5 FUTURE PERSPECTIVES

The general idea in gene therapy is to introduce wild-type gene products or to destroy mutant gene products (including the mutant gene itself). However, the patient-specific nature of some diseases may impose a tremendous challenge that must be addressed on a case-by-case basis. During the last decade, we have seen the development of highly efficient gene transfer systems and the creation of better animal models for human disease. It has become clear that the skin has multiple advantages as a target organ for the development of gene therapy strategies. This is true not only for inherited skin diseases, but for systemic conditions as well. With the recent identification and characterization of the genetic basis and molecular biology of a significant number of diseases, we are today in an exceptional position to design successful gene therapy approaches for them. Clearly, to apply these advances in human clinical trials, more efficient gene transfer methods with increased and prolonged levels of expression, high safety profile, low cost, and ease of administration must be developed. The recent landmark study by Mavilio et al. [33] reports the effective gene therapy treatment of a patient suffering from JEB, demonstrating for the first time that gene therapy can cure a disease of solid tissue in patients. Recent advances in disease model systems, epidermal stem cell techniques, and gene delivery technology highlight the promise of successful gene therapy using cutaneous gene transfer approaches. Even though significant challenges lie ahead, the future in this field is bright.

REFERENCES

1. Greenhalgh, D.A., Rothnagel, J.A., and Roop, D.R., Epidermis: An attractive target tissue for gene therapy, *J. Invest. Dermatol.*, 103, 63S, 1994.
2. Sawamura, D., Akiyama, M., and Shimizu, H., Direct injection of naked DNA and cytokine transgene expression: Implications for keratinocyte gene therapy, *Clin. Exp. Dermatol.*, 27, 480, 2002.
3. Lucas, R.M. and Ponsonby, A.L., Ultraviolet radiation and health: Friend and foe, *Med. J. Aust.*, 177, 594, 2002.
4. Williams, I.R. and Kupper, T.S., Immunity at the surface: Homeostatic mechanisms of the skin immune system, *Life Sci.*, 58, 1485, 1996.
5. Spellberg, B., The cutaneous citadel: A holistic view of skin and immunity, *Life Sci.*, 67, 477, 2000.
6. Gardner, E.P., Martin, J.H., and Jessell, T.M., The bodily senses, in *Principles of Neural Science*. Kandell, E.R., Schwartz, J.H., Jessell, T.M. 4th ed, New York. McGraw-Hill, Health Professions Division, pp. 430–450, 2000.
7. Griffin, J.W., McArthur, J.C., and Polydefkis, M., Assessment of cutaneous innervation by skin biopsies, *Curr. Opin. Neurol.*, 14, 655, 2001.
8. Junqueira, L.C., Carneiro, J., and Kelley, R.O., Skin, in *Basic Histology*. 9th ed., Stamford, Conn. Appleton & Lange, pp. 374–359, 1998.
9. Austin, K.G., et al., Thermoregulation in burn patients during exercise, *J. Burn Care Rehabil.*, 24, 9, 2003.
10. Borradori, L. and Sonnenberg, A., Structure and function of hemidesmosomes: More than simple adhesion complexes, *J. Invest. Dermatol.*, 112, 411, 1999.
11. Moll, R., et al., The catalog of human cytokeratins: Patterns of expression in normal epithelia, tumors and cultured cells, *Cell*, 31, 11, 1982.
12. M'Boneko, V. and Merker, H.J., Development and morphology of the periderm of mouse embryos (days 9–12 of gestation), *Acta Anat. (Basel)*, 133, 325, 1988.
13. Byrne, C., Tainsky, M., and Fuchs, E., Programming gene expression in developing epidermis, *Development*, 120, 2369, 1994.
14. Hardman, M.J., et al., Barrier formation in the human fetus is patterned, *J. Invest. Dermatol.*, 113, 1106, 1999.
15. Weiss, L.W. and Zelickson, A.S., Embryology of the epidermis: Ultrastructural aspects. 1. Formation and early development in the mouse with mammalian comparisons, *Acta Derm. Venereol.*, 55, 161, 1975.
16. Weiss, L.W. and Zelickson, A.S., Embryology of the epidermis: Ultrastructural aspects. II. Period of differentiation in the mouse with mammalian comparisons, *Acta Derm. Venereol.*, 55, 321, 1975.
17. Bickenbach, J.R., et al., Loricrin expression is coordinated with other epidermal proteins and the appearance of lipid lamellar granules in development, *J. Invest. Dermatol.*, 104, 405, 1995.
18. Koster, M.I. and Roop, D.R., The role of p63 in development and differentiation of the epidermis, *J. Dermatol. Sci.*, 34, 3, 2004.
19. Roop, D., Defects in the barrier, *Science*, 267, 474, 1995.
20. Fuchs, E. and Segre, J.A., Stem cells: A new lease on life, *Cell*, 100, 143, 2000.
21. Steven, A.C., et al., Biosynthetic pathways of filaggrin and loricrin—two major proteins expressed by terminally differentiated epidermal keratinocytes, *J. Struct. Biol.*, 104, 150, 1990.
22. Hohl, D., et al., Characterization of human loricrin. Structure and function of a new class of epidermal cell envelope proteins, *J. Biol. Chem.*, 266, 6626, 1991.
23. Watt, F.M., Epidermal stem cells: Markers, patterning and the control of stem cell fate, *Philos. Trans. R. Soc. Lond. B Biol. Sci.*, 353, 831, 1998.
24. Barrandon, Y. and Green, H., Three clonal types of keratinocyte with different capacities for multiplication, *Proc. Natl. Acad. Sci. USA*, 84, 2302, 1987.
25. Ito, M., et al., Stem cells in the hair follicle bulge contribute to wound repair but not to homeostasis of the epidermis, *Nat. Med.*, 11, 1351, 2005.
26. Cotsarelis, G., Epithelial stem cells: A folliculocentric view, *J. Invest. Dermatol.*, 126, 1459, 2006.

27. Niemann, C. and Watt, F.M., Designer skin: Lineage commitment in postnatal epidermis, *Trends Cell Biol.*, 12, 185, 2002.

28. Owens, D.M. and Watt, F.M., Contribution of stem cells and differentiated cells to epidermal tumours, *Nat. Rev. Cancer*, 3, 444, 2003.

29. Trainer, A.H. and Alexander, M.Y., Gene delivery to the epidermis, *Hum. Mol. Genet.*, 6, 1761, 1997.

30. Choate, K.A., et al., Transglutaminase 1 delivery to lamellar ichthyosis keratinocytes, *Hum. Gene Ther.*, 7, 2247, 1996.

31. Choate, K.A., et al., Corrective gene transfer in the human skin disorder lamellar ichthyosis, *Nat. Med.*, 2, 1263, 1996.

32. Choate, K.A. and Khavari, P.A., Sustainability of keratinocyte gene transfer and cell survival in vivo, *Hum. Gene Ther.*, 8, 895, 1997.

33. Mavilio, F., et al., Correction of junctional epidermolysis bullosa by transplantation of genetically modified epidermal stem cells, *Nat. Med.*, 12, 1397, 2006.

34. Alexeev, V., et al., Localized in vivo genotypic and phenotypic correction of the albino mutation in skin by RNA–DNA oligonucleotide, *Nat. Biotechnol.*, 18, 43, 2000.

35. Alexeev, V. and Yoon, K., Gene correction by RNA–DNA oligonucleotides, *Pigment Cell Res.*, 13, 72, 2000.

36. Santana, E., et al., Different frequency of gene targeting events by the RNA–DNA oligonucleotide among epithelial cells, *J. Invest. Dermatol.*, 111, 1172, 1998.

37. Igoucheva, O. and Yoon, K., Targeted single-base correction by RNA–DNA oligonucleotides, *Hum. Gene Ther.*, 11, 2307, 2000.

38. Kim, H.M., Choi, D.H., and Lee, Y.M., Inhibition of wound-induced expression of transforming growth factor-beta 1 mRNA by its antisense oligonucleotides, *Pharmacol. Res.*, 37, 289, 1998.

39. Choi, B.M., et al., Control of scarring in adult wounds using antisense transforming growth factor-beta 1 oligodeoxynucleotides, *Immunol. Cell Biol.*, 74, 144, 1996.

40. Citro, G., et al., c-myc antisense oligodeoxynucleotides enhance the efficacy of cisplatin in melanoma chemotherapy in vitro and in nude mice, *Cancer Res.*, 58, 283, 1998.

41. Leonetti, C., et al., Increase of cisplatin sensitivity by c-myc antisense oligodeoxynucleotides in a human metastatic melanoma inherently resistant to cisplatin, *Clin. Cancer Res.*, 5, 2588, 1999.

42. Cech, T.R., The chemistry of self-splicing RNA and RNA enzymes, *Science*, 236, 1532, 1987.

43. Welch, P.J., Barber, J.R., and Wong-Staal, F., Expression of ribozymes in gene transfer systems to modulate target RNA levels, *Curr. Opin. Biotechnol.*, 9, 486, 1998.

44. Watanabe, T. and Sullenger, B.A., RNA repair: A novel approach to gene therapy, *Adv. Drug Deliv. Rev.*, 44, 109, 2000.

45. Sullenger, B.A. and Gilboa, E., Emerging clinical applications of RNA, *Nature*, 418, 252, 2002.

46. Cserhalmi-Friedman, P.B., Panteleyev, A.A., and Christiano, A.M., Recapitulation of the hairless mouse phenotype using catalytic oligonucleotides: Implications for permanent hair removal, *Exp. Dermatol.*, 13, 155, 2004.

47. Shi, Y., Mammalian RNAi for the masses, *Trends Genet.*, 19, 9, 2003.

48. Hannon, G.J., RNA interference, *Nature*, 418, 244, 2002.

49. Paddison, P.J., Caudy, A.A., and Hannon, G.J., Stable suppression of gene expression by RNAi in mammalian cells, *Proc. Natl. Acad. Sci. USA*, 99, 1443, 2002.

50. Matsumoto, G., et al., Cationized gelatin delivery of a plasmid DNA expressing small interference RNA for VEGF inhibits murine squamous cell carcinoma, *Cancer Sci.*, 97, 313, 2006.

51. Nakai, N., et al., Therapeutic RNA interference of malignant melanoma by electrotransfer of small interfering RNA targeting Mitf, *Gene Ther.*, 14, 357, 2007.

52. Pulkkinen, L., et al., Compound heterozygosity for novel splice site mutations in the BPAG2/COL17A1 gene underlies generalized atrophic benign epidermolysis bullosa, *J. Invest. Dermatol.*, 113, 1114, 1999.

53. Puttaraju, M., et al., Spliceosome-mediated RNA trans-splicing as a tool for gene therapy, *Nat. Biotechnol.*, 17, 246, 1999.

54. Mansfield, S.G., et al., Repair of CFTR mRNA by spliceosome-mediated RNA trans-splicing, *Gene Ther.*, 7, 1885, 2000.

55. Puttaraju, M., et al., Messenger RNA repair and restoration of protein function by spliceosome-mediated RNA trans-splicing, *Mol. Ther.*, 4, 105, 2001.

56. Dallinger, G., et al., Development of spliceosome-mediated RNA trans-splicing (SMaRT) for the correction of inherited skin diseases, *Exp. Dermatol.*, 12, 37, 2003.

57. Ghazizadeh, S. and Taichman, L.B., Virus-mediated gene transfer for cutaneous gene therapy, *Hum. Gene Ther.*, 11, 2247, 2000.

58. Gossen, M. and Bujard, H., Tight control of gene expression in mammalian cells by tetracycline-responsive promoters, *Proc. Natl. Acad. Sci. USA*, 89, 5547, 1992.

59. Hofmann, A., Nolan, G.P., and Blau, H.M., Rapid retroviral delivery of tetracycline-inducible genes in a single autoregulatory cassette, *Proc. Natl. Acad. Sci. USA*, 93, 5185, 1996.

60. Paulus, W., et al., Self-contained, tetracycline-regulated retroviral vector system for gene delivery to mammalian cells, *J. Virol.*, 70, 62, 1996.

61. Lu, B., et al., Topical application of viral vectors for epidermal gene transfer, *J. Invest. Dermatol.*, 108, 803, 1997.

62. Klatzmann, D., et al., A phase I/II dose-escalation study of herpes simplex virus type 1 thymidine kinase "suicide" gene therapy for metastatic melanoma. Study Group on Gene Therapy of Metastatic Melanoma, *Hum. Gene Ther.*, 9, 2585, 1998.

63. Tang, D.C., Shi, Z., and Curiel, D.T., Vaccination onto bare skin, *Nature*, 388, 729, 1997.

64. Bramson, J., et al., Enabling topical immunization via microporation: A novel method for pain-free and needle-free delivery of adenovirus-based vaccines, *Gene Ther.*, 10, 251, 2003.

65. Deng, H., Lin, Q., and Khavari, P.A., Sustainable cutaneous gene delivery, *Nat. Biotechnol.*, 15, 1388, 1997.

66. Kolodka, T.M., Garlick, J.A., and Taichman, L.B., Evidence for keratinocyte stem cells in vitro: Long term engraftment and persistence of transgene expression from retrovirus-transduced keratinocytes, *Proc. Natl. Acad. Sci. USA*, 95, 4356, 1998.

67. Ghazizadeh, S., Harrington, R., and Taichman, L.B., In vivo transduction of mouse epidermis with recombinant retroviral vectors: Implications for cutaneous gene therapy, *Gene Ther.*, 6, 1267, 1999.

68. Donahue, B.A., et al., Selective uptake and sustained expression of AAV vectors following subcutaneous delivery, *J. Gene Med.*, 1, 31, 1999.

69. Hengge, U.R. and Mirmohammadsadegh, A., Adeno-associated virus expresses transgenes in hair follicles and epidermis, *Mol. Ther.*, 2, 188, 2000.

70. Baek, S.C., et al., Sustainable systemic delivery via a single injection of lentivirus into human skin tissue, *Hum. Gene Ther.*, 12, 1551, 2001.

71. Kuhn, U., et al., In vivo assessment of gene delivery to keratinocytes by lentiviral vectors, *J. Virol.*, 76, 1496, 2002.

72. Coffin, J.M., Hughes, S.H., and Varmus, H.H., *Retroviruses.* Cold Spring Harbor Laboratory Press, New York, 1997.

73. Yee, J.K., Friedmann, T., and Burns, J.C., Generation of high-titer pseudotyped retroviral vectors with very broad host range, *Methods Cell Biol.*, 43 Pt A, 99, 1994.

74. Hacein-Bey-Abina, S., et al., A serious adverse event after successful gene therapy for X-linked severe combined immunodeficiency, *N. Engl. J. Med.*, 348, 255, 2003.

75. Hacein-Bey-Abina, S., et al., LMO2-associated clonal T cell proliferation in two patients after gene therapy for SCID-X1, *Science*, 302, 415, 2003.

76. Marshall, E., Gene therapy. Second child in French trial is found to have leukemia, *Science*, 299, 320, 2003.

77. Naldini, L., Lentiviruses as gene transfer agents for delivery to non-dividing cells, *Curr. Opin. Biotechnol.*, 9, 457, 1998.

78. Naldini, L., et al., In vivo gene delivery and stable transduction of nondividing cells by a lentiviral vector, *Science*, 272, 263, 1996.

79. Nolan, G.P. and Shatzman, A.R., Expression vectors and delivery systems, *Curr. Opin. Biotechnol.*, 9, 447, 1998.

80. Chen, M., et al., Restoration of type VII collagen expression and function in dystrophic epidermolysis bullosa, *Nat. Genet.*, 32, 670, 2002.

81. Woodley, D.T., et al., Intradermal injection of lentiviral vectors corrects regenerated human dystrophic epidermolysis bullosa skin tissue in vivo, *Mol. Ther.*, 10, 318, 2004.

82. Shenk, T., Group C adenoviruses for gene therapy, in *Viral Vectors: Gene Therapy and Neuroscience Applications.* Kaplitt, M.G. and Loewy, A.D. San Diego. Academic Press Inc. pp. 43–51, 1995.

83. Danthinne, X. and Imperiale, M.J., Production of first generation adenovirus vectors: A review, *Gene Ther.*, 7, 1707, 2000.

84. Hardy, S., et al., Construction of adenovirus vectors through Cre-lox recombination, *J. Virol.*, 71, 1842, 1997.

85. Kovesdi, I., et al., Adenoviral vectors for gene transfer, *Curr. Opin. Biotechnol.*, 8, 583, 1997.

86. Muotri, A.R., et al., Complementation of the DNA repair deficiency in human xeroderma pigmentosum group a and C cells by recombinant adenovirus-mediated gene transfer, *Hum. Gene Ther.*, 13, 1833, 2002.

87. Falo, L.D., Jr., Targeting the skin for genetic immunization, *Proc. Assoc. Am. Physicians*, 111, 211, 1999.

88. Schmutzler, C. and Koehrle, J., Innovative strategies for the treatment of thyroid cancer, *Eur. J. Endocrinol.*, 143, 15, 2000.

89. Jaakkola, P., et al., Transcriptional targeting of adenoviral gene delivery into migrating wound keratinocytes using FiRE, a growth factor-inducible regulatory element, *Gene Ther.*, 7, 1640, 2000.

90. Campbell, C., et al., Green fluorescent protein-adenoviral construct as a model for transient gene therapy for human cultured keratinocytes in an athymic mouse model, *J. Trauma*, 54, 72, 2003.

91. Chandler, L.A., et al., Matrix-enabled gene transfer for cutaneous wound repair, *Wound. Repair Regen.*, 8, 473, 2000.

92. Samulski, R.J., Chang, L.S., and Shenk, T., Helper-free stocks of recombinant adeno-associated viruses: Normal integration does not require viral gene expression, *J. Virol.*, 63, 3822, 1989.

93. Grimm, D. and Kleinschmidt, J.A., Progress in adeno-associated virus type 2 vector production: Promises and prospects for clinical use, *Hum. Gene Ther.*, 10, 2445, 1999.

94. Rabinowitz, J.E. and Samulski, J., Adeno-associated virus expression systems for gene transfer, *Curr. Opin. Biotechnol.*, 9, 470, 1998.

95. Mountz, J.D., Adeno-associated viruses: Monkey see, monkey do, *Gene Ther.*, 10, 194, 2002.

96. Niidome, T. and Huang, L., Gene therapy progress and prospects: Nonviral vectors, *Gene Ther.*, 9, 1647, 2002.

97. Scherman, D., et al., Application of lipids and plasmid design for gene delivery to mammalian cells, *Curr. Opin. Biotechnol.*, 9, 480, 1998.

98. Nishikawa, M. and Hashida, M., Nonviral approaches satisfying various requirements for effective in vivo gene therapy, *Biol. Pharm. Bull.*, 25, 275, 2002.

99. Scherer, F., et al., Nonviral vector loaded collagen sponges for sustained gene delivery in vitro and in vivo, *J. Gene Med.*, 4, 634, 2002.

100. Pelisek, J., et al., Optimization of nonviral transfection: Variables influencing liposome-mediated gene transfer in proliferating vs. quiescent cells in culture and in vivo using a porcine restenosis model, *J. Mol. Med.*, 80, 724, 2002.

101. Raghavachari, N. and Fahl, W.E., Targeted gene delivery to skin cells in vivo: A comparative study of liposomes and polymers as delivery vehicles, *J. Pharm. Sci.*, 91, 615, 2002.

102. Domashenko, A., Gupta, S., and Cotsarelis, G., Efficient delivery of transgenes to human hair follicle progenitor cells using topical lipoplex, *Nat. Biotechnol.*, 18, 420, 2000.

103. Ivics, Z., et al., Molecular reconstruction of Sleeping Beauty, a Tc1-like transposon from fish, and its transposition in human cells, *Cell*, 91, 501, 1997.

104. Ortiz-Urda, S., et al., Sustainable correction of junctional epidermolysis bullosa via transposon-mediated nonviral gene transfer, *Gene Ther.*, 10, 1099, 2003.

105. Thyagarajan, B., et al., Site-specific genomic integration in mammalian cells mediated by phage phiC31 integrase, *Mol. Cell Biol.*, 21, 3926, 2001.

106. Ortiz-Urda, S., et al., Stable nonviral genetic correction of inherited human skin disease, *Nat. Med.*, 8, 1166, 2002.

107. Li, L. and Hoffman, R.M., The feasibility of targeted selective gene therapy of the hair follicle, *Nat. Med.*, 1, 705, 1995.

108. Fan, H., et al., Immunization via hair follicles by topical application of naked DNA to normal skin, *Nat. Biotechnol.*, 17, 870, 1999.

109. Meuli, M., et al., Efficient gene expression in skin wound sites following local plasmid injection, *J. Invest. Dermatol.*, 116, 131, 2001.

110. Rao, K.V., He, Y.X., and Ramaswamy, K., Suppression of cutaneous inflammation by intradermal gene delivery, *Gene Ther.*, 9, 38, 2002.

111. Hengge, U.R., Walker, P.S., and Vogel, J.C., Expression of naked DNA in human, pig, and mouse skin, *J. Clin. Invest.*, 97, 2911, 1996.

112. Hengge, U.R., et al., Cytokine gene expression in epidermis with biological effects following injection of naked DNA, *Nat. Genet.*, 10, 161, 1995.

113. White, P.J., et al., Live confocal microscopy of oligonucleotide uptake by keratinocytes in human skin grafts on nude mice, *J. Invest. Dermatol.*, 112, 887, 1999.

114. Boulikas, T., Liposome DNA delivery and uptake by cells, *Oncol. Rep.*, 3, 989, 1996.

115. Nakamura, H., et al., Prevention and regression of atopic dermatitis by ointment containing NF-kB decoy oligodeoxynucleotides in NC/Nga atopic mouse model, *Gene Ther.*, 9, 1221, 2002.

116. Vanbever, R. and Preat, V.V., In vivo efficacy and safety of skin electroporation, *Adv. Drug Deliv. Rev.*, 35, 77, 1999.

117. Somiari, S., et al., Theory and in vivo application of electroporative gene delivery, *Mol. Ther.*, 2, 178, 2000.

118. Glasspool-Malone, J., et al., Efficient nonviral cutaneous transfection, *Mol. Ther.*, 2, 140, 2000.

119. Drabick, J.J., et al., Cutaneous transfection and immune responses to intradermal nucleic acid vaccination are significantly enhanced by in vivo electropermeabilization, *Mol. Ther.*, 3, 249, 2001.

120. Maruyama, H., et al., Skin-targeted gene transfer using in vivo electroporation, *Gene Ther.*, 8, 1808, 2001.

121. Zhang, L., et al., Enhanced delivery of naked DNA to the skin by non-invasive in vivo electroporation, *Biochim. Biophys. Acta*, 1572, 1, 2002.

122. Neumann, E., et al., Gene transfer into mouse lyoma cells by electroporation in high electric fields, *EMBO J.*, 1, 841, 1982.

123. Ciernik, I.F., Krayenbuhl, B.H., and Carbone, D.P., Puncture-mediated gene transfer to the skin, *Hum. Gene Ther.*, 7, 893, 1996.

124. Lin, M.T., et al., The gene gun: Current applications in cutaneous gene therapy, *Int. J. Dermatol.*, 39, 161, 2000.

125. Lin, M.T., et al., Differential expression of tissue-specific promoters by gene gun, *Br. J. Dermatol.*, 144, 34, 2001.

126. Nanney, L.B., et al., Boosting epidermal growth factor receptor expression by gene gun transfection stimulates epidermal growth in vivo, *Wound. Repair Regen.*, 8, 117, 2000.

127. Dileo, J., et al., Gene transfer to subdermal tissues via a new gene gun design, *Hum. Gene Ther.*, 14, 79, 2003.

128. Eming, S.A., et al., Particle-mediated gene transfer of PDGF isoforms promotes wound repair, *J. Invest. Dermatol.*, 112, 297, 1999.

129. Davidson, J.M., Krieg, T., and Eming, S.A., Particle-mediated gene therapy of wounds, *Wound. Repair Regen.*, 8, 452, 2000.

130. Yang, L., et al., Microbubble-enhanced ultrasound for gene transfer into living skin equivalents, *J. Dermatol. Sci.*, 40, 105, 2005.

131. Kim, K.S. and Park, Y.S., Antitumor effects of angiostatin K1–3 and endostatin genes coadministered by the hydrodynamics-based transfection method, *Oncol. Res.*, 15, 343, 2005.

132. Bonifas, J.M., Rothman, A.L., and Epstein, E.H., Jr., Epidermolysis bullosa simplex: Evidence in two families for keratin gene abnormalities, *Science*, 254, 1202, 1991.

133. Coulombe, P.A., et al., Point mutations in human keratin 14 genes of epidermolysis bullosa simplex patients: Genetic and functional analyses, *Cell*, 66, 1301, 1991.

134. Aberdam, D., et al., Herlitz's junctional epidermolysis bullosa is linked to mutations in the gene (LAMC2) for the gamma 2 subunit of nicein/kalinin (LAMININ-5), *Nat. Genet.*, 6, 299, 1994.

135. Pulkkinen, L., et al., Mutations in the gamma 2 chain gene (LAMC2) of kalinin/laminin 5 in the junctional forms of epidermolysis bullosa, *Nat. Genet.*, 6, 293, 1994.

136. Christiano, A.M., et al., A missense mutation in type VII collagen in two affected siblings with recessive dystrophic epidermolysis bullosa, *Nat. Genet.*, 4, 62, 1993.

137. Hilal, L., et al., A homozygous insertion-deletion in the type VII collagen gene (COL7A1) in Hallopeau-Siemens dystrophic epidermolysis bullosa, *Nat. Genet.*, 5, 287, 1993.

138. Huber, M., et al., Mutations of keratinocyte transglutaminase in lamellar ichthyosis, *Science*, 267, 525, 1995.

139. Cheng, J., et al., The genetic basis of epidermolytic hyperkeratosis: A disorder of differentiation-specific epidermal keratin genes, *Cell*, 70, 811, 1992.

140. Rothnagel, J.A., et al., Mutations in the rod domains of keratins 1 and 10 in epidermolytic hyperkeratosis, *Science*, 257, 1128, 1992.

141. Rothnagel, J.A., et al., A mutational hot spot in keratin 10 (KRT 10) in patients with epidermolytic hyperkeratosis, *Hum. Mol. Genet.*, 2, 2147, 1993.

142. Hahn, H., et al., Mutations of the human homolog of *Drosophila* patched in the nevoid basal cell carcinoma syndrome, *Cell*, 85, 841, 1996.

143. Johnson, R.L., et al., Human homolog of patched, a candidate gene for the basal cell nevus syndrome, *Science*, 272, 1668, 1996.

144. Kraemer, K.H., Xeroderma pigmentosum knockouts, *Lancet*, 347, 278, 1996.

145. Cairns, J., Mutation selection and the natural history of cancer, *Nature*, 255, 197, 1975.

146. Kim, Y.H., et al., Recessive dystrophic epidermolysis bullosa phenotype is preserved in xenografts using SCID mice: Development of an experimental in vivo model, *J. Invest. Dermatol.*, 98, 191, 1992.

147. Matsuki, M., et al., Defective stratum corneum and early neonatal death in mice lacking the gene for transglutaminase 1 (keratinocyte transglutaminase), *Proc. Natl. Acad. Sci. USA*, 95, 1044, 1998.

148. Heinonen, S., et al., Targeted inactivation of the type VII collagen gene (Col7a1) in mice results in severe blistering phenotype: A model for recessive dystrophic epidermolysis bullosa, *J. Cell Sci.*, 112 (Pt 21), 3641, 1999.

149. Whittock, N.V., et al., Striate palmoplantar keratoderma resulting from desmoplakin haploinsufficiency, *J. Invest. Dermatol.*, 113, 940, 1999.

150. Armstrong, D.K., et al., Haploinsufficiency of desmoplakin causes a striate subtype of palmoplantar keratoderma, *Hum. Mol. Genet.*, 8, 143, 1999.

151. Vassar, R., et al., Mutant keratin expression in transgenic mice causes marked abnormalities resembling a human genetic skin disease, *Cell*, 64, 365, 1991.

152. Fuchs, E., Esteves, R.A., and Coulombe, P.A., Transgenic mice expressing a mutant keratin 10 gene reveal the likely genetic basis for epidermolytic hyperkeratosis, *Proc. Natl. Acad. Sci. USA*, 89, 6906, 1992.

153. Corden, L.D. and McLean, W.H., Human keratin diseases: Hereditary fragility of specific epithelial tissues, *Exp. Dermatol.*, 5, 297, 1996.

154. Irvine, A.D. and McLean, W.H., Human keratin diseases: The increasing spectrum of disease and subtlety of the phenotype-genotype correlation, *Br. J. Dermatol.*, 140, 815, 1999.

155. Arin, M.J., et al., Focal activation of a mutant allele defines the role of stem cells in mosaic skin disorders, *J. Cell Biol.*, 152, 645, 2001.

156. Cao, T., et al., An inducible mouse model for epidermolysis bullosa simplex: Implications for gene therapy, *J. Cell Biol.*, 152, 651, 2001.

157. Marinkovich, M.P., et al., Hereditary Epidermolysis Bullosa in *Fitzpatrick's Dermatology in General Medicine*, Freedberg, I.M., Eisen, A.Z., Wolff, K., Austen, K.F., Goldsmith, L.A., Katz, S.I., Fitzpatrick, T.B., eds. 5th ed, New York. McGraw-Hill, Health Professions Division, pp. 690–702, 1999.

158. Marinkovich, M.P., et al., Inherited epidermolysis bullosa, in *Dermatology in General Medicine*. Freedberg, I.M., Eisen, A.Z., Wolff, K., Austen, K.F., Goldsmith, L.A., Katz, S.I., et al., eds. 6th ed., New York. McGraw-Hill, Health Professions Division, pp. 65, 2003.

159. Ryynanen, M., Knowlton, R.G., and Uitto, J., Mapping of epidermolysis bullosa simplex mutation to chromosome 12, *Am. J. Hum. Genet.*, 49, 978, 1991.

160. Fine, J.D., et al., Revised clinical and laboratory criteria for subtypes of inherited epidermolysis bullosa. A consensus report by the Subcommittee on Diagnosis and Classification of the National Epidermolysis Bullosa Registry, *J. Am. Acad. Dermatol.*, 24, 119, 1991.

161. Anton-Lamprecht, I. and Schnyder, U.W., Epidermolysis bullosa herpetiformis Dowling-Meara. Report of a case and pathomorphogenesis, *Dermatologica*, 164, 221, 1982.

162. Niemi, K.M., et al., Epidermolysis bullosa simplex. A new histologic subgroup, *Arch. Dermatol.*, 119, 138, 1983.

163. McGrath, J.A., et al., Epidermolysis bullosa simplex (Dowling-Meara). A clinicopathological review, *Br. J. Dermatol.*, 126, 421, 1992.

164. Haneke, E. and Anton-Lamprecht, I., Ultrastructure of blister formation in epidermolysis bullosa hereditaria: V. Epidermolysis bullosa simplex localisata type Weber-Cockayne, *J. Invest. Dermatol.*, 78, 219, 1982.

165. Lloyd, C., et al., The basal keratin network of stratified squamous epithelia: Defining K15 function in the absence of K14, *J. Cell Biol.*, 129, 1329, 1995.

166. Berton, T.R., et al., Characterization of an inducible, epidermal-specific knockout system: Differential expression of lacZ in different Cre reporter mouse strains, *Genesis*, 26, 160, 2000.

167. Kero, M. and Niemi, K.M., Epidermolysis bullosa, *Int. J. Dermatol.*, 25, 75, 1986.

168. McGrath, J.A., et al., Mutations in the 180-kD bullous pemphigoid antigen (BPAG2), a hemidesmosomal transmembrane collagen (COL17A1), in generalized atrophic benign epidermolysis bullosa, *Nat. Genet.*, 11, 83, 1995.

169. Vidal, F., et al., Integrin beta 4 mutations associated with junctional epidermolysis bullosa with pyloric atresia, *Nat. Genet.*, 10, 229, 1995.

170. Smith, F.J., et al., Plectin deficiency results in muscular dystrophy with epidermolysis bullosa, *Nat. Genet.*, 13, 450, 1996.

171. Vailly, J., et al., Corrective gene transfer of keratinocytes from patients with junctional epidermolysis bullosa restores assembly of hemidesmosomes in reconstructed epithelia, *Gene Ther.*, 5, 1322, 1998.

172. Ryan, M.C., et al., Targeted disruption of the LAMA3 gene in mice reveals abnormalities in survival and late stage differentiation of epithelial cells, *J. Cell Biol.*, 145, 1309, 1999.

173. Seitz, C.S., et al., BP180 gene delivery in junctional epidermolysis bullosa, *Gene Ther.*, 6, 42, 1999.

174. Jonkman, M.F., et al., Revertant mosaicism in epidermolysis bullosa caused by mitotic gene conversion, *Cell*, 88, 543, 1997.

175. Dellambra, E., et al., Corrective transduction of human epidermal stem cells in laminin-5-dependent junctional epidermolysis bullosa, *Hum. Gene Ther.*, 9, 1359, 1998.

176. Robbins, P.B., et al., In vivo restoration of laminin 5 beta 3 expression and function in junctional epidermolysis bullosa, *Proc. Natl. Acad. Sci. USA*, 98, 5193, 2001.

177. Compton, S.H., et al., Stable integration of large (>100 kb) PAC constructs in HaCaT keratinocytes using an integrin-targeting peptide delivery system, *Gene Ther.*, 7, 1600, 2000.

178. Ortiz-Urda, S., et al., Injection of genetically engineered fibroblasts corrects regenerated human epidermolysis bullosa skin tissue, *J. Clin. Invest.*, 111, 251, 2003.

179. Frost, P. and Van Scott, E.J., Ichthyosiform dermatoses. Classification based on anatomic and biometric observations, *Arch. Dermatol.*, 94, 113, 1966.

180. Syder, A.J., et al., Genetic mutations in the K1 and K10 genes of patients with epidermolytic hyperkeratosis. Correlation between location and disease severity, *J. Clin. Invest.*, 93, 1533, 1994.

181. Bickenbach, J.R., et al., A transgenic mouse model that recapitulates the clinical features of both neonatal and adult forms of the skin disease epidermolytic hyperkeratosis, *Differentiation*, 61, 129, 1996.

182. Porter, R.M., et al., Gene targeting at the mouse cytokeratin 10 locus: Severe skin fragility and changes of cytokeratin expression in the epidermis, *J. Cell Biol.*, 132, 925, 1996.

183. Pfutzner, W., et al., Selection of keratinocytes transduced with the multidrug resistance gene in an in vitro skin model presents a strategy for enhancing gene expression in vivo, *Hum. Gene Ther.*, 10, 2811, 1999.

184. Williams, M.L. and Elias, P.M., Genetically transmitted, generalized disorders of cornification. The ichthyoses, *Dermatol. Clin.*, 5, 155, 1987.

185. Russell, L.J., et al., Linkage of autosomal recessive lamellar ichthyosis to chromosome 14q, *Am. J. Hum. Genet.*, 55, 1146, 1994.

186. Rice, R.H., et al., Keratinocyte transglutaminase: Differentiation marker and member of an extended family, *Epithelial Cell Biol.*, 1, 128, 1992.

187. Kim, S.Y., et al., The structure of the transglutaminase 1 enzyme. Deletion cloning reveals domains that regulate its specific activity and substrate specificity, *J. Biol. Chem.*, 269, 27979, 1994.

188. Lavrijsen, A.P., et al., Barrier function parameters in various keratinization disorders: Transepidermal water loss and vascular response to hexyl nicotinate, *Br. J. Dermatol.*, 129, 547, 1993.

189. Choate, K.A. and Khavari, P.A., Direct cutaneous gene delivery in a human genetic skin disease, *Hum. Gene Ther.*, 8, 1659, 1997.

190. Yen, P.H., et al., Cloning and expression of steroid sulfatase cDNA and the frequent occurrence of deletions in STS deficiency: Implications for X–Y interchange, *Cell*, 49, 443, 1987.

191. Freiberg, R.A., et al., A model of corrective gene transfer in X-linked ichthyosis, *Hum. Mol. Genet.*, 6, 927, 1997.

192. Lefevre, C., et al., Mutations in the transporter ABCA12 are associated with lamellar ichthyosis type 2, *Hum. Mol. Genet.*, 12, 2369, 2003.

193. Akiyama, M., et al., Mutations in lipid transporter ABCA12 in harlequin ichthyosis and functional recovery by corrective gene transfer, *J. Clin. Invest.*, 115, 1777, 2005.

194. Kelsell, D.P., et al., Mutations in ABCA12 underlie the severe congenital skin disease harlequin ichthyosis, *Am. J. Hum. Genet.*, 76, 794, 2005.

195. Li, L., et al., Characterization of molecular defects in xeroderma pigmentosum group C, *Nat. Genet.*, 5, 413, 1993.

196. Magnaldo, T. and Sarasin, A., Genetic reversion of inherited skin disorders, *Mutat. Res.*, 509, 211, 2002.

197. Kraemer, K.H., Lee, M.M., and Scotto, J., Xeroderma pigmentosum. Cutaneous, ocular, and neurologic abnormalities in 830 published cases, *Arch. Dermatol.*, 123, 241, 1987.

198. Carreau, M., et al., Functional retroviral vector for gene therapy of xeroderma pigmentosum group D patients, *Hum. Gene Ther.*, 6, 1307, 1995.

199. Marionnet, C., et al., Recovery of normal DNA repair and mutagenesis in trichothiodystrophy cells after transduction of the XPD human gene, *Cancer Res.*, 56, 5450, 1996.

200. Myrand, S.P., Topping, R.S., and States, J.C., Stable transformation of xeroderma pigmentosum group A cells with an XPA minigene restores normal DNA repair and mutagenesis of UV-treated plasmids, *Carcinogenesis*, 17, 1909, 1996.

201. Zeng, L., et al., Retrovirus-mediated gene transfer corrects DNA repair defect of xeroderma pigmentosum cells of complementation groups A, B and C, *Gene Ther.*, 4, 1077, 1997.

202. Rnaudeau-Begard, C., et al., Genetic correction of DNA repair-deficient/cancer-prone xeroderma pigmentosum group C keratinocytes, *Hum. Gene Ther.*, 14, 983, 2003.

203. Marchetto, M.C., et al., Gene transduction in skin cells: Preventing cancer in xeroderma pigmentosum mice, *Proc. Natl. Acad. Sci. USA*, 101, 17759, 2004.

204. Marchetto, M.C., et al., Functional lentiviral vectors for xeroderma pigmentosum gene therapy, *J. Biotechnol.*, 126, 424, 2006.

205. Schadendorf, D., Gene-based therapy of malignant melanoma, *Semin. Oncol.*, 29, 503, 2002.

206. Bonnekoh, B., et al., Adenoviral-mediated herpes simplex virus-thymidine kinase gene transfer in vivo for treatment of experimental human melanoma, *J. Invest Dermatol.*, 106, 1163, 1996.

207. Bonnekoh, B., et al., Ex vivo and in vivo adenovirus-mediated gene therapy strategies induce a systemic anti-tumor immune defence in the B16 melanoma model, *J. Invest. Dermatol.*, 110, 867, 1998.

208. O'Malley, B.W., et al., Adenovirus-mediated gene therapy for human head and neck squamous cell cancer in a nude mouse model, *Cancer Res.*, 55, 1080, 1995.

209. O'Malley, B.W., et al., Combination gene therapy for oral cancer in a murine model, *Cancer Res.*, 56, 1737, 1996.

210. Wang, S., et al., Antitumor effects on human melanoma xenografts of an amplicon vector transducing the herpes thymidine kinase gene followed by ganciclovir, *Cancer Gene Ther.*, 9, 1, 2002.

211. Yancopoulos, G.D., et al., Vascular-specific growth factors and blood vessel formation, *Nature*, 407, 242, 2000.

212. Kudo, Y., et al., Small interfering RNA targeting of S phase kinase-interacting protein 2 inhibits cell growth of oral cancer cells by inhibiting p27 degradation, *Mol. Cancer Ther.*, 4, 471, 2005.

213. Katagiri, Y., Hozumi, Y., and Kondo, S., Knockdown of Skp2 by siRNA inhibits melanoma cell growth in vitro and in vivo, *J. Dermatol. Sci.*, 42, 215, 2006.

214. Hipfel, R., et al., Specifically regulated genes in malignant melanoma tissues identified by subtractive hybridization, *Br. J. Cancer*, 82, 1149, 2000.

215. Schittek, B., et al., The increased expression of Y box-binding protein 1 in melanoma stimulates proliferation and tumor invasion, antagonizes apoptosis and enhances chemoresistance, *Int. J. Cancer*, 120, 2110, 2007.

216. Garraway, L.A., et al., Integrative genomic analyses identify MITF as a lineage survival oncogene amplified in malignant melanoma, *Nature*, 436, 117, 2005.

217. Clark, R.A., Biology of dermal wound repair, *Dermatol. Clin.*, 11, 647, 1993.

218. Khavari, P.A., Rollman, O., and Vahlquist, A., Cutaneous gene transfer for skin and systemic diseases, *J. Intern. Med.*, 252, 1, 2002.

219. Eming, S.A., et al., Genetically modified human epidermis overexpressing PDGF-A directs the development of a cellular and vascular connective tissue stroma when transplanted to athymic mice–implications for the use of genetically modified keratinocytes to modulate dermal regeneration, *J. Invest. Dermatol.*, 105, 756, 1995.

220. Eming, S.A., et al., Targeted expression of insulin-like growth factor to human keratinocytes: Modification of the autocrine control of keratinocyte proliferation, *J. Invest. Dermatol.*, 107, 113, 1996.

221. Mesri, E.A., Federoff, H.J., and Brownlee, M., Expression of vascular endothelial growth factor from a defective herpes simplex virus type 1 amplicon vector induces angiogenesis in mice, *Circ. Res.*, 76, 161, 1995.

222. Breitbart, A.S., et al., Gene-enhanced tissue engineering: Applications for wound healing using cultured dermal fibroblasts transduced retrovirally with the PDGF-B gene, *Ann. Plast. Surg.*, 43, 632, 1999.

223. Breitbart, A.S., et al., Treatment of ischemic wounds using cultured dermal fibroblasts transduced retrovirally with PDGF-B and VEGF121 genes, *Ann. Plast. Surg.*, 46, 555, 2001.

224. Machens, H.G., et al., Genetically modified fibroblasts induce angiogenesis in the rat epigastric island flap, *Langenbecks Arch. Surg.*, 383, 345, 1998.

225. Breitbart, A.S., et al., Accelerated diabetic wound healing using cultured dermal fibroblasts retrovirally transduced with the platelet-derived growth factor B gene, *Ann. Plast. Surg.*, 51, 409, 2003.

226. Supp, D.M., et al., Genetic modification of cultured skin substitutes by transduction of human keratinocytes and fibroblasts with platelet-derived growth factor-A, *Wound Repair Regen.*, 8, 26, 2000.

227. Tyrone, J.W., et al., Collagen-embedded platelet-derived growth factor DNA plasmid promotes wound healing in a dermal ulcer model, *J. Surg. Res.*, 93, 230, 2000.

228. Spies, M., et al., Liposomal IGF-1 gene transfer modulates pro- and anti-inflammatory cytokine mRNA expression in the burn wound, *Gene Ther.*, 8, 1409, 2001.

229. Deodato, B., et al., Recombinant AAV vector encoding human VEGF165 enhances wound healing, *Gene Ther.*, 9, 777, 2002.

230. Romano Di, P.S., et al., Adenovirus-mediated VEGF(165) gene transfer enhances wound healing by promoting angiogenesis in CD1 diabetic mice, *Gene Ther.*, 9, 1271, 2002.

231. Jeschke, M.G., et al., Non-viral liposomal keratinocyte growth factor (KGF) cDNA gene transfer improves dermal and epidermal regeneration through stimulation of epithelial and mesenchymal factors, *Gene Ther.*, 9, 1065, 2002.

232. Jeschke, M.G., et al., Therapeutic success and efficacy of non-viral liposomal cDNA gene transfer to the skin in vivo is dose dependent, *Gene Ther.*, 8, 1777, 2001.

233. Yamasaki, K., et al., Reversal of impaired wound repair in iNOS-deficient mice by topical adenoviral-mediated iNOS gene transfer, *J. Clin. Invest.*, 101, 967, 1998.

234. Nakanishi, K., et al., Gene transfer of human hepatocyte growth factor into rat skin wounds mediated by liposomes coated with the sendai virus (hemagglutinating virus of Japan), *Am. J. Pathol.*, 161, 1761, 2002.

235. Ha, X., et al., Effect of human hepatocyte growth factor on promoting wound healing and preventing scar formation by adenovirus-mediated gene transfer, *Chin. Med. J. (Engl.)*, 116, 1029, 2003.

236. Erdag, G. and Morgan, J.R., Interleukin-1alpha and interleukin-6 enhance the antibacterial properties of cultured composite keratinocyte grafts, *Ann. Surg.*, 235, 113, 2002.

237. Terasaki, K., et al., Effects of recombinant human tissue inhibitor of metalloproteinases-2 (rh-TIMP-2) on migration of epidermal keratinocytes in vitro and wound healing in vivo, *J. Dermatol.*, 30, 165, 2003.

238. Elson, D.A., et al., Induction of hypervascularity without leakage or inflammation in transgenic mice overexpressing hypoxia-inducible factor-1alpha, *Genes Dev.*, 15, 2520, 2001.

239. Eming, S.A., Krieg, T., and Davidson, J.M., Gene therapy and wound healing, *Clin. Dermatol.*, 25, 79, 2007.

240. Yang, N.S. and Sun, W.H., Gene gun and other non-viral approaches for cancer gene therapy, *Nat. Med.*, 1, 481, 1995.

241. Larregina, A.T. and Falo, L.D., Jr., Generating and regulating immune responses through cutaneous gene delivery, *Hum. Gene Ther.*, 11, 2301, 2000.

242. Tang, D.C., DeVit, M., and Johnston, S.A., Genetic immunization is a simple method for eliciting an immune response, *Nature*, 356, 152, 1992.

243. Sun, W.H., et al., In vivo cytokine gene transfer by gene gun reduces tumor growth in mice, *Proc. Natl. Acad. Sci. USA*, 92, 2889, 1995.

244. Liu, M., et al., Gene-based vaccines and immunotherapeutics, *Proc. Natl. Acad. Sci. USA*, 101 Suppl 2, 14567, 2004.

245. Thurner, B., et al., Vaccination with mage-3A1 peptide-pulsed mature, monocyte-derived dendritic cells expands specific cytotoxic T cells and induces regression of some metastases in advanced stage IV melanoma, *J. Exp. Med.*, 190, 1669, 1999.

246. Rakhmilevich, A.L., et al., Gene gun-mediated skin transfection with interleukin 12 gene results in regression of established primary and metastatic murine tumors, *Proc. Natl. Acad. Sci. USA*, 93, 6291, 1996.

247. Dummer, R., et al., Adenovirus-mediated intralesional interferon-gamma gene transfer induces tumor regressions in cutaneous lymphomas, *Blood*, 104, 1631, 2004.

248. Urosevic, M., et al., Type I IFN innate immune response to adenovirus-mediated IFN-gamma gene transfer contributes to the regression of cutaneous lymphomas, *J. Clin. Invest.*, 117, 2834, 2007.

249. Cao, T., Wang, X.J., and Roop, D.R., Regulated cutaneous gene delivery: The skin as a bioreactor, *Hum. Gene Ther.*, 11, 2297, 2000.

250. Gordon, D.A., et al., Synthesis and secretion of apolipoprotein E by cultured human keratinocytes, *J. Invest. Dermatol.*, 92, 96, 1989.

251. Fenjves, E.S., et al., Systemic distribution of apolipoprotein E secreted by grafts of epidermal keratinocytes: Implications for epidermal function and gene therapy, *Proc. Natl. Acad. Sci. USA*, 86, 8803, 1989.

252. Cao, T., et al., The epidermis as a bioreactor. Topically regulated cutaneous delivery into the circulation, *Hum. Gene Ther.*, 13, 1075, 2002.

253. Mannucci, P.M. and Tuddenham, E.G., The hemophilias—From royal genes to gene therapy, *N. Engl. J. Med.*, 344, 1773, 2001.

254. Bowen, D.J., Haemophilia A and haemophilia B: Molecular insights, *Mol. Pathol.*, 55, 127, 2002.

255. Fakharzadeh, S.S., et al., Correction of the coagulation defect in hemophilia A mice through factor VIII expression in skin, *Blood*, 95, 2799, 2000.

256. Roth, D.A., et al., Nonviral transfer of the gene encoding coagulation factor VIII in patients with severe hemophilia A, *N. Engl. J. Med.*, 344, 1735, 2001.

257. Halaas, J.L. and Friedman, J.M., Leptin and its receptor, *J. Endocrinol.*, 155, 215, 1997.

258. Friedman, J.M. and Halaas, J.L., Leptin and the regulation of body weight in mammals, *Nature*, 395, 763, 1998.

259. Zhang, Y., et al., Positional cloning of the mouse obese gene and its human homologue, *Nature*, 372, 425, 1994.

260. Montague, C.T., et al., Congenital leptin deficiency is associated with severe early-onset obesity in humans, *Nature*, 387, 903, 1997.

261. Farooqi, I.S., et al., Effects of recombinant leptin therapy in a child with congenital leptin deficiency, *N. Engl. J. Med.*, 341, 879, 1999.

262. Larcher, F., et al., A cutaneous gene therapy approach to human leptin deficiencies: Correction of the murine ob/ob phenotype using leptin-targeted keratinocyte grafts, *FASEB J.*, 15, 1529, 2001.

263. Fisher, J.W., Erythropoietin: Physiology and pharmacology update, *Exp. Biol. Med. (Maywood.)*, 228, 1, 2003.

264. Descamps, V., et al., Keratinocytes as a target for gene therapy. Sustained production of erythropoietin in mice by human keratinocytes transduced with an adenoassociated virus vector, *Arch. Dermatol.*, 132, 1207, 1996.

265. Siprashvili, Z. and Khavari, P.A., Lentivectors for regulated and reversible cutaneous gene delivery, *Mol. Ther.*, 9, 93, 2004.

266. Lippin, Y., et al., Human erythropoietin gene therapy for patients with chronic renal failure, *Blood*, 106, 2280, 2005.

267. Morgan, J.R., et al., Expression of an exogenous growth hormone gene by transplantable human epidermal cells, *Science*, 237, 1476, 1987.

268. Bellini, M.H., Peroni, C.N., and Bartolini, P., Increases in weight of growth hormone-deficient and immunodeficient (lit/scid) dwarf mice after grafting of hGH-secreting, primary human keratinocytes, *FASEB J.*, 17, 2322, 2003.

269. Ferguson, T.A., Dube, P., and Griffith, T.S., Regulation of contact hypersensitivity by interleukin 10, *J. Exp. Med.*, 179, 1597, 1994.

270. Meng, X., et al., Keratinocyte gene therapy for systemic diseases. Circulating interleukin 10 released from gene-transferred keratinocytes inhibits contact hypersensitivity at distant areas of the skin, *J. Clin. Invest.*, 101, 1462, 1998.

271. Fenjves, E.S., et al., Systemic delivery of secreted protein by grafts of epidermal keratinocytes: Prospects for keratinocyte gene therapy, *Hum. Gene Ther.*, 5, 1241, 1994.

272. Gerrard, A.J., et al., Towards gene therapy for haemophilia B using primary human keratinocytes, *Nat. Genet.*, 3, 180, 1993.

273. Alexander, M.Y., et al., Circulating human factor IX produced in keratin-promoter transgenic mice: A feasibility study for gene therapy of haemophilia B, *Hum. Mol. Genet.*, 4, 993, 1995.

274. Page, S.M. and Brownlee, G.G., An ex vivo keratinocyte model for gene therapy of hemophilia B, *J. Invest. Dermatol.*, 109, 139, 1997.

275. White, S.J., et al., Long-term expression of human clotting factor IX from retrovirally transduced primary human keratinocytes in vivo, *Hum. Gene Ther.*, 9, 1187, 1998.

276. Lei, P., et al., Efficient production of bioactive insulin from human epidermal keratinocytes and tissue-engineered skin substitutes: Implications for treatment of diabetes, *Tissue Eng.*, 13, 2119, 2007.

277. Fenjves, E.S., et al., Keratinocyte gene therapy for adenosine deaminase deficiency: A model approach for inherited metabolic disorders, *Hum. Gene Ther.*, 8, 911, 1997.

278. Jensen, T.G., et al., Retrovirus-mediated gene transfer of ornithine-delta-aminotransferase into keratinocytes from gyrate atrophy patients, *Hum. Gene Ther.*, 8, 2125, 1997.

43 Gene Therapy for Childhood Onset Blindness

Matthew C. Canver and Jean Bennett

CONTENTS

43.1 GENE THERAPY FOR EARLY ONSET BLINDNESS

43.1.1 BACKGROUND: GENE THERAPY FOR EARLY ONSET BLINDNESS

In recent years, mutations in a large number of different genes have been implicated in the pathophysiology of retinal degenerative diseases [1]. Among these are mutations, which lead to particularly severe early onset blinding diseases. With the recent development of methods for *in vivo* retinal gene delivery (reviewed in Refs. [2–4], it has become possible to evaluate efficacy of gene therapy in animal models for these blinding diseases. We selected a canine model to study the efficacy of gene therapy for a particularly debilitating disease, Leber congenital amaurosis (LCA). LCA is a disease that is usually diagnosed in infancy as it severely handicaps its subjects, leaving them with severely impaired visual function and abnormal ocular movements (nystagmus). This is a disease that affects all ethnic groups and results in blindness in approximately 10,000 individuals in the United States (www.blindness.org). There are at least nine different genes, which, when mutated, can lead to LCA [1,5,6]. One of these, *RPE65*, is named for the evolutionarily conserved retinal pigment epithelium (RPE)-specific 65 kDa protein which it encodes [7,8] and accounts for approximately 7%–16% of these cases [8–10]. Mutations in *RPE65* can also lead to other severe and early onset blinding diseases including retinitis pigmentosa (RP) and cone-rod dystrophy [1,8,10–13].

43.1.2 CANINE MODEL FOR LCA: A UNIQUE RESOURCE

An advantage of selecting the *RPE65* gene defect for study is the availability of a large animal model for the LCA disease: the *RPE65* mutant dog. This animal, first described by Narfstrom et al. [14], suffers from an autosomal recessive inheritance of a retinal disease with severe visual deficits but a relatively slow degenerative component [15]. The canine *RPE65* gene defect is a homozygous 4 bp deletion [15,16]. The deletion results in a frame shift, leading to a premature stop codon. This eliminates more than two-thirds of the wild-type polypeptide. Histopathology in homozygotes shows prominent RPE inclusions and slightly abnormal rod photoreceptor morphology present within the first year of life [17]. There is also a lack of immunohistochemically detectable *RPE65* protein [15]. As the dogs age, a slowly progressive photoreceptor degeneration becomes apparent and is readily detectable in older (~5-year-old) dogs [17]. The *RPE65* protein was identified as the retinoid isomerase where it assists in vitamin A metabolism in the retina [18].

Without *RPE65* protein, there is a deficiency of 11-*cis* retinal in the retina. This compound is necessary for the generation of rhodopsin in rod photoreceptors. Without this, the cells cannot respond to light and initiate the series of biochemical/electrophysiologic events leading to vision. This accounts for the abnormal qualitative and quantitative measures of visual function that are apparent early in life in *RPE65*-deficient animals and humans.

43.1.3 Proof-of-Principle of Gene Therapy for Retinal Degeneration Has Been Established

Efficacy of gene therapy for inherited retinal degeneration may be efficiently demonstrated by studying animal models. Proof-of-principle for such treatments has been demonstrated using multiple strategies, viral vectors, and in a number of animal models of different species [2,3,19–51]. Successful strategies include delivery of wild-type cDNA in the case of loss-of-function disease [22,23,27–34,40,41,52,42–46], delivery of ribozymes, which specifically target the mutant mRNA in the case of gain-of-function disease [36,53,49,50], delivery of neurotrophic factors [24–26,37], and delivery of genes with antiapoptotic function [38,51]. Rescue has been achieved with a number of different vectors, including adenovirus, adeno-associated virus (AAV), lentivirus, and gutted adenovirus. Clinical end points in many of these studies have involved electrophysiological evidence of improvement in retinal function as assessed by electroretinograms (ERGs). In some instances, additional tests of visual function have been utilized. Pupillometry was used in conjunction with ERGs in a murine study [54]. The Morris water maze test has been implemented where mice were placed in a water tank with the goal of locating the escape route [39]. Another visual function test that has been employed is OptoMotry, which measures the optomotor response to a rotating sine-wave grating [40,55]. Finally, visual-evoked potential (VEP) has been utilized to test responses to external stimuli [41]. ERGs, pupillometry, and VEPs have also been used in large-animal studies [27,28,42,45,55]. Studies in dogs have also shown improvements in nystagmus [55,56] and in cortical vision [57]. All of these measures can be recorded in a noninvasive fashion. Interpretation of the results benefits from the use of the contralateral eye as a control. Results from retinal function testing have been supplemented with histological/immunohistochemical data obtained in terminal studies. In some of these paradigms, rescue has been detected for years after treatment [27,46]. One *RPE65* mutant dog that is being followed for life still enjoys vision more than 7 years after treatment (Figure 43.1; Bennett et al., unpublished data).

43.1.4 Treatment Strategy That Resulted in Rescue of Vision in the *RPE65* Mutant Dog

Restoration of vision in *RPE65* mutant dogs was first accomplished in July 2000 and was reported in 2001 [28]. Since

FIGURE 43.1 Seven years after subretinal delivery of AAV. *RPE65* in the right eye, Lancelot, an *RPE65* mutant dog, still enjoys vision. He is shown signing autographs at a research fundraiser. (October, 2006; K. Maguire, photographer.)

then, at least two other groups have confirmed these results [42,43,45,46]. In all of these studies, the experimental objective was to deliver wild-type copies of the disease-causing gene (*RPE65*) to cell populations primarily affected by the gene mutation (i.e., RPE). Delivery was achieved through subretinal injection of a recombinant AAV carrying the *RPE65* cDNA (Figure 43.2). This exposes cells bordering the subretinal space, RPE cells, photoreceptors, and Muller cells to the experimental reagent. Remarkably, the localized retinal detachment subsides within hours [58]. Additional controls assuring expression predominantly by RPE cells can be achieved through selection of the vector capsid or incorporation of RPE cell-specific promoters [27,42,55]. AAV2 targets RPE cells predominantly through this approach [27,55]. Production of *RPE65* protein by the transduced *RPE65* then restores the biochemical circuit by allowing production of 11-*cis* retinal.

43.2 LCA CLINICAL TRIAL: FROM BENCH TO BEDSIDE

43.2.1 Challenges in Planning a Gene Therapy Clinical Trial for LCA

Why does it take so long to embark on a human clinical trial?

In July 2000 (and then reported in May 2001), AAV-2-mediated delivery of a wild-type version of the defective *RPE65* gene was used to provide retinal function in a canine model of severe early onset retinal degeneration [28]. Subsequent studies have demonstrated safety and reproducibility of this

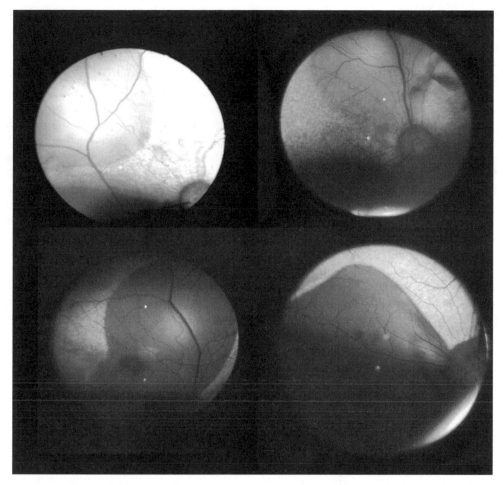

FIGURE 43.2 Appearance of the eyes of affected Briard dogs immediately after subretinal injection of 150 µl AAV2.hRPE65. In all of these eyes, the bleb (which prevents visualization of the speckled pattern of the underlying RPE) at least partially extends through the superior (tapetal) retina. The superior retina reflects light because of the tapetum. The inner retinal blood vessels are lifted up (with the rest of the neural retina) in the region of the bleb. The inferior (nontapetal) retina is the dark area below the optic nerve. (The entire optic nerve head is visible in the panel on the top right.) In panels going from top left to bottom right, the blebs are located as follows: superior, superior, mainly superior, and partially inferior (dumbbell-shaped), and equally superior and inferior. Photos were taken of eyes injected for studies described by Jacobson, S.G. et al., *Mol. Ther.*, 13, 1074, 2006. (Photographs were taken by Dr. A. Maguire.)

result [27,52,59,60]. Three clinical trials have been approved and one is already under way. Dr. Robin Ali of the Moorfields Eye Hospital and University College London began testing human LCA patients with a *RPE65* mutation in the spring of 2007. It is too early to assess the results of the patients that already had the procedure (http://www.newscientist.com/article/dn11765-first-trial-of-gene-therapy-to-restore-human-sight.html). Two other clinical trials are imminent for this disease. One will be under the direction of Dr. Samuel G. Jacobson (University of Pennsylvania) and will treat patients at University of Florida, Gainesville (under Dr. Barry J. Byrne) (http://clinicaltrials.gov/). The second trial will enroll pediatric subjects and will be carried out at the Children's Hospital of Philadelphia (CHOP) under the direction of Drs. Albert Maguire and Jean Bennett (with Dr. Katherine High, sponsor). Additional clinical trials, including one to be carried out in France, are likely to be initiated over the next few years (http://www.osnsupersite.com/view.asp?rID=21894). With the excellent safety and efficacy data, why did it take

nearly 7 years from the time of the initial experiments to get to the point of embarking on a human clinical trial?

There are many steps that must be taken before this treatment could be tested in humans with LCA. One of the challenges inherent in translational research is that the ideal construct (including species of origin of the cDNA, AAV capsid serotype, regulatory sequence, etc.) must be identified early on as all of the subsequent safety and efficacy testing using this vector is very costly and time-consuming. Safety testing must be extremely rigorous, especially since the ultimate goal is to treat a disease that is not lethal. Thus, we and others first evaluated a long list of variables for effects on safety and efficacy before selecting the clinical candidate vector. Special attention had to be paid to the possibility that the transgene could potentially be introduced to the brain via unwanted ganglion cell/optic nerve targeting [61–63]. The duration of the therapeutic effect needed to be characterized since LCA is a lifelong disease.

Vector selection was difficult because there was continued progress in developing viral vectors with slightly improved transduction qualities. At a certain point, a choice had to be made: Whether to select a particular vector as the candidate for formal toxicity studies (in anticipation of initiating a human clinical trial using this vector) or whether to do additional efficacy tests on newly described vectors. The toxicity studies are rigorous, time-consuming, expensive, and use a large number of animals. It is preferable not to have to repeat them using a second vector or even with slight modifications of the original vector (unless the additional benefits of that vector warrant the effort). Thus, two of the teams planning clinical trials for LCA (University of Florida, Gainesville and CHOP, Philadelphia) use the AAV vector and regulatory element (AAV2 and a chicken β actin promoter) for which the most data was available. The team conducting the clinical trial in the United Kingdom also uses AAV2, but instead of a constitutive promoter, uses an RPE-specific promoter. The team planning a trial in France has chosen AAV4 as its vector [42]. In the United States, clinical-grade vector must be generated using protocols and quality assurance assays that are in compliance with Food and Drug Administration (FDA) guidelines. Similar safety assurances must be met in Europe.

For the trials that were already approved, once the ideal vector was selected, toxicity testing was conducted to verify that the treatment did not cause any impairment to the eye or to the rest of the body. This had to be performed using the same (clinical grade) virus that was to be used in humans. Typically, toxicity studies entail both acute and chronic studies and often are carried out in animals of two different species. Most of the data is collected and evaluated by outside consultants, thereby assuring its integrity. Therefore, it can take a long time to complete the data analyses. Once the data are collected, they are submitted for both institutional (including Institutional Review Board and Institutional Biosafety Committee) and federal regulatory agency approvals. This process is quite lengthy.

43.2.2 How Can Potential Subjects for LCA-*RPE65* Clinical Trials Be Identified?

Once an individual with LCA is identified and their disease is clinically characterized, they must be shown to carry disease-causing *RPE65* mutations to be eligible to participate in a gene therapy trial for LCA due to *RPE65* mutations. This is a challenge as LCA is a rare disease. There are at most, an estimated 3000 individuals in the United States with LCA. Genotyping must be carried out by a Clinical Laboratory Improvement Amendments (CLIA) certified laboratory to ensure quality-controlled test results. Such testing is often expensive and tedious and is generally not covered by health insurance. A tremendous aid for this effort was provided by Project 3000 in the United States. This project, implemented by two major sports figures, aims to provide free testing for every man, woman, and child affected by LCA in the United States (http://carverlab.org/project3000/index.shtml). The testing will be carried out at the John and Marcia Carver Genetic Testing Laboratory, a CLIA-approved laboratory, under the direction of Dr. Edwin Stone.

43.2.3 What Is the Ideal Age for a Participant in a Human Gene Therapy Clinical Trial Targeting LCA-*RPE65*?

Recent studies have characterized and continue to characterize the progression pattern of the human disease (with respect to visual function measures and death of retinal cells) [64–69]. This will allow the selection of the optimal noninvasive outcome measures for evaluation of the effects of treatment in a human clinical trial. One question which arose was whether the first treatments, which are phase I safety and toxicity tests, should be performed on adult versus pediatric subjects. Two of the three different groups involved in the first set of human clinical trials (Directors Drs. R. Ali and S. Jacobson) argued that safety should be tested first in adult subjects. However, data from animals suggested improved efficacy in younger animals [56,64]. Also, photoreceptor cells in the retinas of animals and humans with *RPE65* mutations deteriorate over time, so that if one waits too long to treat the tissue, there will be few viable cells remaining. Finally, there is a vast amount of data demonstrating plasticity of the brain with respect to visual input in animals and humans that are young and a reduction in this plasticity with age [70]. Subjects with LCA-*RPE65* may have some light input at young ages, however, and so there may be more plasticity with respect to central nervous system (CNS) responses than is seen in other forms of vision deprivation [57]. After considering all of these issues, the team involved in the third clinical trial (Director Dr. A. Maguire) argued that treatment should be tested in pediatric subjects in order to optimize the risk–benefit ratio. Thus, the trial that is scheduled to be initiated at CHOP will enroll subjects aged 8 and older and will then include even younger subjects in follow-up studies (http://www.webconferences.com/nihoba/13_dec_2005.html).

43.2.4 What Are the Risks of a Gene Therapy Clinical Trial for LCA-*RPE65*?

There has been a vast amount of experience with AAV-mediated gene delivery to retinas of animal models of LCA-RPE. To date, the safety profile is excellent. Animals treated with AAV2.*RPE65* have remained healthy, active, and fertile. The most common clinical findings in safety/efficacy studies relate to the surgical delivery of the vector. This delivery leaves a small retinotomy scar, but this scar does not interfere with retinal/macular function. There are histologic changes after delivery to the fovea [60] and so in human studies, care must be taken to minimize toxicity to this region. The only significant inflammatory events reported to date occurred in animals that were injected subretinally with a research-grade virus preparation that was contaminated with foreign protein [27]. Despite the inflammation, animals treated subretinally with this lot of vector recovered vision. There have also been reports of low-level exposure of tear fluid, serum, optic nerve, and rare exposure to other organs after intraocular injection of AAV-*RPE65*. There have been no reports to date of toxic cell-mediated immune responses after this mode of delivery [52, Bennicelli, in press]. Because low doses will be administered in human clinical trials and because these doses will be delivered to a geographically enclosed compartment (the immune privileged eye), immunologic complications are not expected.

43.3 EXTRAPOLATION OF SUCCESS TO OTHER FORMS OF EARLY ONSET RETINAL DEGENERATION

43.3.1 EXTRAPOLATION OF THE SUCCESS IN THE CANINE LCA GENE THERAPY STUDIES TO OTHER FORMS OF EARLY ONSET RETINAL DEGENERATION

A major challenge is how to extrapolate the success in the *RPE65* mutant dog to successful treatment of other severe retinal degenerative diseases. One approach to meeting this challenge is to identify the conditions that are responsible for the success of the *RPE65* dog study. Besides the virus delivery characteristics, there may be elements of the *RPE65*-caused disease, which make it particularly amenable to treatment. Is the success due to the fact that the *RPE65* mutant retinal photoreceptors degenerate very slowly and thus there is a large therapeutic window during which the retina can be rescued? In some animal models (those with homozygous mutations in the beta subunit of rod cGMP phosphodiesterase, for example), the degeneration is so rapid that rod photoreceptors do not even have a chance to fully differentiate before they are lost [71]. In that disease, it is likely that the treatment will have to be administered very early in life while the target cells are still viable [31,32].

In animal models, which can have a rapidly progressive retinal degeneration, the AAV2 vector may not have a rapid enough onset to make any impact in the disease progression of rapidly progressive degenerative diseases. Thus, for demonstration of proof-of-concept, one might want to employ a viral vector that results in a rapid onset of transgene expression. AAV-2 is slow to reach peak levels of transgene expression *in vivo* [72]. In contrast, cargo packaged within AAV capsids of serotypes other than AAV2 deliver transgenes that are expressed within a few days of injection [47,73] (Bennett, personal communication). It should be noted, however, that the time-of-onset limitation may not be as important with respect to human treatment since the disease process occurs on a slower timescale than in animal models.

In particular diseases, there may be requirements of the disease model that will have to be met by developing alternative vectors/vector strategies. For example, one challenge is how to proceed with AAV if the cDNA or transgene cassette is too large to be packaged. The AAV cargo capacity is a maximum of 4.8 kb. In such a situation, one would have to consider other vectors that could carry the intact transgene cassette. Options include lentivirus, which can carry a cassette of ~8 kb, and gutted adenovirus, which can carry a cassette of >30 kb. If one wanted to use AAV even though the transgene cassette was too large, one could harness the AAV dual vector approach and split the transgene construct into two complementary vectors, which would be trans-spliced in the target cell [74]. Alternatively, particular novel or engineered AAV capsids may be able to package large DNA cargo [75].

Recently discovered LCA-causing mutations provide viable options for a gene therapy approach. Mutations in *LCA5*, which encodes for the protein lebercilin, have been shown to cause LCA [40]. Additionally, a mutation in *CEP290*, which causes premature protein truncation, has been demonstrated to cause LCA [76]. Finally, mutations in rod-derived cone viability factor (RdCVF), which is a protein expressed by photoreceptors, have been shown to disturb RdCVF's ability to maintain cone cell viability in LCA [77]. It may be possible to extrapolate the success of the *RPE65* mutation studies to any or all of these newly discovered LCA-causing mutations.

Finally, demonstration of safety of AAV vectors in therapy of human LCA will pave the way for treatment of other retinal diseases, including those targeting other cells types besides RPE and those involving gain-of-function disease such as autosomal dominant RP and macular degeneration. Once safety of the vector is demonstrated, these vectors can be used to test a host of other treatment targets/strategies. The latter include use of RNA interference as a treatment for gain-of-function disease and delivery of neurotrophic factors, which could enhance or prolong the function of the diseased target cells.

43.4 CONCLUSIONS

In summary, we and others have shown that efficacy of gene therapy for inherited retinal degeneration may be efficiently demonstrated studying animal models for the most clinically severe retinal degenerations, i.e., those that result in early onset blindness. AAV-2 was used to deliver corrective genes subretinally to a canine model of a severe retinal disease and efficacy of treatment was readily apparent. Success in this study led to identification of critical parameters for the success of gene therapy for retinal degeneration and also to human clinical trials designed to treat LCA. Safety and efficacy data derived from these clinical trials could pave the way for treatments of additional forms of LCA as well as other inherited and acquired forms of retinal degeneration.

ACKNOWLEDGMENTS

Since the article was submitted, initial results from two of the three clinical trials have been reported [78,79]. This study was supported by Children's Hospital of Philadelphia NIH grants EY10820, The Foundation Fighting Blindness; Research to Prevent Blindness; the Macular Vision Research Foundation, the Mackall Foundation Trust, and the F. M. Kirby Foundation.

REFERENCES

1. RetNet 2007.
2. Bennett, J. 2000. Gene therapy for retinitis pigmentosa. *Curr. Opin. Med. Ther.*, 2, 420.
3. Dejneka, N., et al. 2003. Gene therapy for Leber congenital amaurosis. *Adv. Exp. Med. Biol.*, 533, 415.
4. Allocca, M., et al. 2006. AAV-mediated gene transfer for retinal diseases. *Exp. Opin. Biol. Ther.*, 6, 1279.
5. Cremers, F. P., et al. 2002. Molecular genetics of Leber congenital amaurosis. *Hum. Mol. Genet.*, 11, 1169.
6. den Hollander, A., et al. 2007. Mutations in *LCA5*, encoding the novel ciliary protein lebercilin, cause Leber congenital amaurosis. *Nat. Genet.*, 39, 889.
7. Redmond, T. and Hamel, C. 2000. Genetic analysis of RPE65: From human disease to mouse model. *Meth. Enzymol.*, 317, 705.
8. Marlhens, F., et al. 1997. Mutations in RPE65 cause Leber's congenital amaurosis. *Nat. Genet.*, 17, 139.
9. Lotery, A., et al. 2000. Mutation analysis of 3 genes in patients with Leber congenital amaurosis. *Arch. Ophthalmol.*, 118, 538.
10. Morimura, H., et al. 1998. Mutations in the RPE65 gene in patients with autosomal recessive retinitis pigmentosa or Leber congenital amaurosis. *Proc. Natl. Acad. Sci. USA*, 95, 3088.
11. Thompson, D., et al. 2000. Genetics and phenotypes of RPE65 mutations in inherited retinal degeneration. *Invest. Ophthalmol. Vis. Sci.*, 41, 4293.
12. Gu, S. M., et al. 1997. Mutations in RPE65 cause autosomal recessive childhood-onset severe retinal dystrophy. *Nat. Genet.*, 17, 194.
13. Lorenz, B., et al. 2000. Early-onset severe rod-cone dystrophy in young children with RPE65 mutations. *Invest. Ophthalmol. Vis. Sci.*, 41, 2735.
14. Narfstrom, K., et al. 1989. The Briard dogs: A new animal model of congenital stationary night blindness. *Brit. J. Ophthalmol.*, 73, 750.
15. Aguirre, G., et al. 1998. Congenital stationary night blindness in the dog: Common mutation in the RPE65 gene indicates founder effect. *Mol. Vis.*, 4, 23.
16. Veske, A., et al. 1999. Retinal dystrophy of Swedish Briard/Briard-Beagle dogs is due to a 4-bp deletion in RPE65. *Genomics*, 57, 57.
17. Wrigstad, A. 1994. Hereditary dystrophy of the retina and the retinal pigment epithelium in a strain of Briard dogs: A clinical, morphological and electrophysiological study. Linkoping University Medical Dissertation, 1994.
18. Saari, J. 2000. Biochemistry of visual pigment regeneration. *Invest. Ophthalmol. Vis. Sci.*, 41, 337.
19. Min, S. H., et al. 2005. Prolonged recovery of retinal structure/function after gene therapy in an Rs1h-deficient mouse model of x-linked juvenile retinoschisis. *Mol. Ther.*, 12, 644.
20. Lebherz, C., et al. 2005. Long-term inducible gene expression in the eye via adeno-associated virus gene transfer in nonhuman primates. *Hum. Gene Ther.*, 16, 178.
21. Hong, D.-H., et al. 2004. Dominant, gain-of-function mutant produced by truncation of RPGR. *Invest. Ophthalmol. Vis. Sci*, 45, 36.
22. Vollrath, D., et al. 2001. Correction of the retinal dystrophy phenotype of the RCS rat by viral gene transfer of Mertk. *Proc. Natl. Acad. Sci. USA*, 98, 12584.
23. Sarra, G.-M., et al. 2001. Gene replacement therapy in the retinal degeneration slow (*rds*) mouse: The effect on retinal degeneration following partial transduction of the retina. *Hum. Mol. Genet.*, 10, 2353.
24. McGee Sanftner, L. H., et al. 2001. Glial cell line derived neurotrophic factor delays photoreceptor degeneration in a transgenic rat model of retinitis pigmentosa. *Mol. Ther.*, 4, 622.
25. Liang, F.-Q., et al. 2001. Long-term protection of retinal structure but not function using rAAV.CNTF in animal models of retinitis pigmentosa. *Mol. Ther.*, 4, 461.
26. Liang, F.-Q., et al. 2001. AAV-mediated delivery of ciliary neurotrophic factor prolongs photoreceptor survival in the rhodopsin knockout mouse. *Mol. Ther.*, 3, 241.
27. Acland, G. M., et al. 2005. Long-term restoration of rod and cone vision by single dose rAAV-mediated gene transfer to the retina in a canine model of childhood blindness. *Mol. Ther.*, 12, 1072.
28. Acland, G. M., et al. 2001. Gene therapy restores vision in a canine model of childhood blindness. *Nat. Gen.*, 28, 92.
29. Bemelmans, A., et al. 2006. Lentiviral gene transfer of RPE65 rescues survival and function of cones in a mouse model of Leber congenital amaurosis. *PLoS Med.*, 3, 347.
30. Dejneka, N., et al. 2004. Fetal virus-mediated delivery of the human RPE65 gene rescues vision in a murine model of congenital retinal blindness. *Mol. Ther.*, 9, 182.
31. Bennett, J., et al. 1996. Photoreceptor cell rescue in retinal degeneration (*rd*) mice by in vivo gene therapy. *Nat. Med.*, 2, 649.
32. Kumar-Singh, R. and Farber, D. 1998. Encapsidated adenovirus mini-chromosome-mediated delivery of genes to the retina: Application to the rescue of photoreceptor degeneration. *Hum. Mol. Genet.*, 7, 1893.
33. Ho, T., et al. 2002. Phenotypic rescue after adeno-associated virus-mediated delivery of 4-sulfatase to the retinal pigment epithelium of feline mucopolysaccharidosis VI. *J. Gene Med.*, 4, 613.
34. Takahashi, M., et al. 1999. Rescue from photoreceptor degeneration in the rd mouse by human immunodeficiency virus vector-mediated gene transfer. *J. Virol.* 73, 7812.
35. Gorbatyuk, M., et al. 2007. Preservation of photoreceptor morphology and function in P23H rats using an allele independent ribozyme. *Exp. Eye Res.*, 84, 44.
36. Lewin, A. S., et al. 1998. Ribozyme rescue of photoreceptor cells in a transgenic rat model of autosomal dominant retinitis pigmentosa. *Nat. Med.*, 4, 967.
37. Adamus, G., et al. 2003. Anti-apoptotic effects of CNTF gene transfer on photoreceptor degeneration in experimental antibody-induced retinopathy. *J. Autoimmunity*, 21, 121.
38. Bennett, J., et al. 1998. Adenovirus-mediated delivery of rhodopsin-promoted *bcl-2* results in a delay in photoreceptor cell death in the *rd/rd* mouse. *Gene Ther.*, 5, 1156.
39. Pang, J. J., et al. 2006. Gene therapy restores vision-dependent behavior as well as retinal structure and function in a mouse model of RPE65 Leber congenital amaurosis. *Mol. Ther.* 13, 565.
40. Alexander, J., et al. 2007. Restoration of cone vision in a mouse model of achromatopsia. *Nat. Med.*, 13, 685.
41. Nusinowitz, S., et al. 2006. Cortical visual function in the rd12 mouse model of Leber congenital amarousis (LCA) after gene replacement therapy to restore retinal function. *Vis. Res.*, 46, 3926.
42. Le Meur, G., et al. 2007. Restoration of vision in RPE65-deficient Briard dogs using an AAV serotype 4 vector that specifically targets the retinal pigmented epithelium. *Gene Ther.*, 14, 292.
43. Rolling, F., et al. 2006. Gene therapeutic prospects in early onset of severe retinal dystrophy: Restoration of vision in

RPE65 Briard dogs using an AAV serotype 4 vector that specifically targets the retinal pigmented epithelium. *Bulletin et memoires de l'Academie royale de medecine de Belgique*, 161, 497.

44. Narfstrom, K., et al. 2003. Functional and structural evaluation after AAV.RPE65 gene transfer in the canine model of Leber's congenital amaurosis. *Adv. Exp. Med. Biol.*, 533, 423.

45. Narfstrom, K., et al. 2003. Functional and structural recovery of the retina after gene therapy in the RPE65 null mutation dog. *Invest. Ophthalmol. Vis. Sci.*, 44, 1663.

46. Narfstrom, K., et al. 2003. In vivo gene therapy in young and adult RPE65-/- dogs produces long-term visual improvement. *J. Hered.*, 94, 31.

47. Auricchio, A., et al. 2001. Exchange of surface proteins impacts on viral vector cellular specificity and transduction characteristics: The retina as a model. *Hum. Mol. Genet.*, 10, 3075.

48. Reich, S., et al. 2003. Efficient trans-splicing in the retina expands the utility of adeno-associated virus as a vector for gene therapy. *Hum. Gene Ther.*, 14, 37.

49. O'Reilly, M., et al. 2007. RNA interference-mediated suppression and replacement of human rhodopsin in vivo. *Am. J. Hum. Genet.*, 81, 127.

50. Cashman, S., et al. 2005. Towards mutation-independent silencing of genes involved in retinal degeneration by RNA interference. *Gene Ther.*, 12, 1223.

51. Leonard, K. C., et al. 2007. XIAP protection of photoreceptors in animal models of retinitis pigmentosa. *PLoS ONE*, 2, e314.

52. Bennicelli, J., et al. 2008. Reversal of visual deficits in animal models of Leber congenital amaurosis within weeks after treatment with a serotype 2, optimized adeno-associated virus (AAV), *Mol. Ther.*, 16, 458.

53. LaVail, M. M., et al. 2000. Ribozyme rescue of photoreceptor cells in P23H transgenic rats: Long-term survival and late stage therapy. *Proc. Natl. Acad. Sci. USA*, 97, 11488.

54. Dejneka, N. S., et al. 2004. In utero gene therapy rescues vision in a murine model of congenital blindness. *Mol. Ther.*, 9, 182.

55. Bennicelli, J., et al. 2008. Reversal of visual deficits in animal models of Leber congenital amaurosis within weeks after treatment using optimized AAV2-mediated gene transfer. *Mol. Ther.*, 16, 458.

56. Jacobs, J., et al. 2006. Eye movement recordings as an effectiveness indicator of gene therapy in RPE65-deficient canines: Implications for the ocular motor system. *Invest. Ophthalmol. Vis. Sci.*, 47, 2865.

57. Aguirre, G. K., et al. 2007. Canine and human visual cortex intact and responsive despite early retinal blindness from RPE65 mutation. *PLoS Med.*, 4, e230.

58. Bennett, J., et al. 2000. Cross-species comparison of in vivo reporter gene expression after recombinant adeno-associated virus-mediated retinal transduction. *Meth. Enzymol.*, 316, 777.

59. Jacobson, S. G., et al. 2006. Safety of recombinant adeno-associated virus type 2-RPE65 vector delivered by ocular subretinal injection. *Mol. Ther.*, 13, 1074.

60. Jacobson, S. G., et al. 2006. Safety in nonhuman primates of ocular AAV2-RPE65, a candidate treatment for blindness in Leber congenital amaurosis. *Hum. Gene Ther.*, 17, 845.

61. Dudus, L., et al. 1999. Persistent transgene product in retina, optic nerve and brain after intraocular injection of rAAV. *Vis. Res.*, 39, 2545.

62. Provost, N., et al. 2005. Biodistribution of rAAV vectors following intraocular administration: Evidence for the presence and persistence of vector DNA in the optic nerve and in the brain. *Mol. Ther.*, 11, 275.

63. Guy, J., et al. 1999. Reporter expression persists 1 year after adeno-associated virus-mediated gene transfer to the optic nerve. *Arch. Ophthalmol.*, 117, 929.

64. Jacobson, S. G., et al. 2005. Identifying photoreceptors in blind eyes caused by RPE65 mutations: Prerequisite for human gene therapy success. *Proc. Natl. Acad. Sci. USA*, 102, 6177.

65. Lorenz, B., et al. 2004. Lack of fundus autofluorescence to 488 nanometers from childhood on in patients with early-onset severe retinal dystrophy associated with mutations in RPE65. *Ophthalmology*, 111, 1585.

66. Aleman, T. S., et al. 2004. Impairment of the transient pupillary light reflex in Rpe65(-/-) mice and humans with Leber congenital amaurosis. *Invest. Ophthalmol. Vis. Sci.*, 45, 1259.

67. Jacobson, S. G., et al. 2007. RDH12 and RPE65, visual cycle genes causing Leber congenital amaurosis, differ in disease expression. *Invest. Ophthalmol. Vis. Sci.*, 48, 332.

68. Paunescu, K., et al. 2005. Longitudinal and cross-sectional study of patients with early-onset severe retinal dystrophy associated with RPE65 mutations. *Graefe's Arch. Clin. Exp. Ophthalmol.* 243, 417.

69. Yzer, S., et al. 2003. A Tyr368His RPE65 founder mutation is associated with variable expression and progression of early onset retinal dystrophy in 10 families of a genetically isolated population. *J. Med. Genet.*, 40, 709.

70. Hubel, D. H., et al. 1977. Plasticity of ocular dominance columns in monkey striate cortex. *Phil. Trans. Royal Soc. Lond.*, 278, 377.

71. Buyukmihci, N., et al. 1980. Retinal degenerations in the dog II. Development of the retina in rod-cone dysplasia. *Exp. Eye Res.*, 30, 575.

72. Anand, V., et al. 2000. Additional transduction events after subretinal readministration of recombinant adeno-associated virus. *Hum. Gene Ther.*, 11, 449.

73. Lebherz, C., et al. (submitted for publication). Novel AAV serotypes for improved ocular gene transfer.

74. Reich, S., et al. 2002. Trans-splicing AAV vector system expands the packaging capacity of AAV gene therapy vectors for delivery of large transgenes to the retina. *Invest. Ophthalmol. Vis. Sci.*

75. Allocca, M., et al. 2008. Serotype-dependent packaging of large genes in adeno-associated viral vectors results in effective gene delivery in mice. *J. Clin. Invest.*, 118, 1955.

76. Menotti-Raymond, M., et al. 2007. Mutation in CEP290 discovered for cat model of human retinal degeneration. *J. Hered.*, 98, 211.

77. Hanein, S., et al. 2006. Disease-associated variants of the rod-derived cone viability factor (RdCVF) in Leber congenital amaurosis. Rod-derived cone viability variants in LCA. *Adv. Exp. Med. Biol.*, 572, 9.

78. Bainbridge, J.W., et al. 2008. Effect of gene therapy on visual function in Leber's Congenital Amaurosis, *N. Engl. J. Med.*, 358, 2231.

79. Maguire, A.M., et al. 2008. Safety and efficacy of gene transfer for Leber's Congenital Amaurosis, *N. Engl. J. Med.*, 358, 2240.

44 DNA Vaccines

Jian Yan and David B. Weiner

CONTENTS

44.1 IMPORTANCE OF VACCINES

44.1.1 HISTORICAL IMPORTANCE OF VACCINES

Vaccination is the deliberate introduction of foreign materials into humans or animals to elicit immune responses against the material that protect against disease [1]. The use of vaccines dates back to ancient times. In the seventh century, some Indian Buddhists drank snake venom to protect themselves from snake bites [1]. During the ninth century in China, a manuscript entitled *The Correct Treatment of Small Pox* was written by a Buddhist nun. The manuscript recommended a mixture of ground dried smallpox scabs and herb to be blown into the nostrils of children. In March 1718, upon learning of the Turkish practice of inoculating healthy children with a weakened strain of smallpox to confer immunity from the more virulent strains of the disease, Lady Mary Wortley Montague immediately had her son inoculated. Four years later, after returning home to England, she introduced this custom to the British nobility by having her daughter inoculated to demonstrate the practice. However, this procedure was risky as 1–2 of 100 inoculations resulted in death by smallpox. Worse, however, was that this practice could start an event of smallpox by infectious spread to uninoculated individuals. Even with such a long history, immunization was not widely used until Edward Jenner described deliberately injecting cowpox virus, vaccinia, into humans to protect them from the ravages of smallpox. He had observed that milkmaids were routinely infected with vaccinia from milking cows and these women were immune to smallpox, an immunologically cross-reactive viral infection. His procedure of vaccination was based on four principles of the immune response: (i) memory, those who received vaccinia were protected from smallpox for life; (ii) specificity, they were not protected from other infectious diseases such as polio; (iii) induceability, the protected state needed to be created by immunization; and (iv) self versus nonself, somehow the smallpox was targeted and not the person themselves. Fifty years later, Pasteur went on to build on these results by directly attenuating a pathogen and demonstrating its vaccine potential. Over the next 150 years, the development of vaccines against pathogenic microorganisms has become one of the most important advances in the history of medicine. Vaccines have not only provided protection from smallpox, but also from poliomyelitis, measles, mumps, rubella, yellow fever, pertussis (whooping cough), hepatitis A, hepatitis B, rotavirus, varicella, and now human papilloma virus (HPV) as well as others. These vaccines dramatically reduced morbidity from infectious agents and combined with antibiotics have directly protected more human lives than all other avenues of modern medicine combined. Yet, as increased standards for effectiveness and safety and the increased costs of developing and manufacturing vaccines have become more restrictive, the development of new vaccines has slowed. Furthermore, as new pathogens continue to emerge, it is important that novel methods for vaccine development be developed to meet the demanding requirements of the twenty-first century.

44.1.2 Vaccine Immunology

Traditional vaccines have relied on either live-replicating or nonliving preparations of microorganisms as their antigen preparation. The injected material functions as a vaccine by generating immunity against the inoculum and the resulting immune responses function to prevent disease by preventing or slowing pathogen infection so that the immune system can effectively destroy the pathogen. This type of induced immunity is referred to as protective immunity and results from the vaccine-activating specific B and T lymphocytes, which compose the lymphocyte subsets of the white blood cells of the immune system [2].

As the major components of humoral immune response, B cells are lymphocytes that develop in the fetal liver and subsequently mature in the bone marrow. Mature B cells carry surface immunoglobulins that act as their antigen-specific receptors. They move from the bone marrow through the circulation to secondary lymphoid tissues, the lymph nodes and spleen, where they respond to antigenic stimuli by dividing and differentiating into plasma cells under direction of cytokines produced by T cells. When they are activated, B lymphocytes become terminally differentiated to become plasma cells, which are entirely devoted to the production of secreted antibody. Antibodies are large water-soluble serum proteins, which are induced following contact with antigen. Antibodies bind to specific antigens (pieces of pathogens), which drive B cell expansion and antibody production. The induced antibodies can either directly neutralize or inactivate the pathogens or the antibodies can also direct other cells of the immune system such as macrophages and phagocytes to dispose of the antigen or infected cells. Furthermore, they can direct complement, highly toxic soluble immune mediators to bind to and destroy an invading pathogen.

T cells are lymphocytes that develop in the thymus. T cells acquire their antigen receptors in the thymus and differentiate into a number of subpopulations, which have separate functions and which can be recognized by their different cell surface markers. T lymphocytes develop as one of two subsets of white blood cells termed T helper cells or T cytotoxic or T killer cells. These cells are the basis of cell-mediated immune responses and function to eradicate pathogens in different ways. T helper cells, as their name implies, help and direct B cells to produce antibody, which targets and destroys extracellular pathogens, preventing their entry into host cells. T helper cells also cooperate with cytotoxic T lymphocytes (CTL) to expand them, aiding the destruction of virally infected cells. Activated T helper cells secrete small protein messengers termed cytokines, which activate and expand either or both humoral and cellular immune responses. New experimental data also indicates that CD4+ T-cell help is absolutely required for the generation of CD8+ T cells capable of efficient recall responses to antigen [3]. On the other hand, activated T cytotoxic cells seek out and destroy cells that have been infected with pathogens. These T killer cells bind and destroy allogeneic and virally infected cells, which display recognizable antigen-major histocompatibility complex (MHC) class I molecules. T killer cells induce these pathogen-infected cells to die either through the release of toxic proteins such as granzyme B or through initiating apoptosis or programmed cell death in the target cells. CD8+ cells can also mobilize the immune response at the point of antigen contact through the release of beta chemokines. In addition to defense against pathogens, CTL are particularly important in eradicating aberrant or pathologic host cells such as cancer cells. Humoral or cellular immunity can act independently or in concert to destroy the pathogenic organism within a vaccinated host. The T-cell response is initiated by unique antigen-presenting cells (APCs), the most efficient of these is the dendritic cell. Immature dendritic cells eat foreign invading organisms or their free antigens, and they process and present these to T cells in the context of the host MHC. The dendritic cell signals the T cell to expand and fight the invading pathogens through specific molecules that are presented along with the MHC pathogen peptide complex to the T cell termed costimulatory molecules. This stimulation is absolutely required for the induction of effective long-lived immunity. Therefore all vaccines must deliver antigen in some form to the APC, in particular the dendritic cell, for the induction of T-cell immunity and they must usually deliver spatially relevant antigens to B cells to induce a protective antibody-mediated vaccine response.

44.1.3 TRADITIONAL VACCINES

In the case of live vaccines, the infectious material has been weakened or attenuated so that it no longer induces disease in the general population. In the case of the nonlive preparations, the vaccine material has been manufactured to contain killed organisms or pieces of killed organisms that cannot grow when inoculated into a host. In some instances, specific components can be purified away from other portions of the microorganism or artificially manufactured in the laboratory to function as a subunit vaccine. Both categories of vaccines are presently utilized throughout the world to protect individuals against specific pathogens. Each vaccine has its own general characteristics for generating immunity and exhibits properties that can be beneficial or deleterious to an individual.

Live attenuated vaccines, such as the polio and smallpox vaccines, stimulate protective immunity as they replicate in the body of an immunized host. These vaccines emulate the natural infection of pathogens and generate a broad spectrum of immune responses. Because they are a weakened form of the pathogen, no disease occurs. This category of vaccine induces broad protective immunity with induction of both antibodies and activated T cells. More specifically, because CTLs are only induced if an infectious agent or vaccine actually is produced within host cells, live attenuated vaccines are the most effective inducer of CTL. Attenuated vaccines have an additional benefit in that they frequently provide lifelong immunity. In contrast, nonliving inactivated vaccines (including subunit preparations), such as the vaccine for hepatitis B virus, produce protective immunity that is limited to the generation of antibodies and helper T cells but they cannot induce killer T cells. Accordingly, the ability of these preparations to induce protection is limited to pathogens that can be destroyed by extracellular defenses. Unlike the live-attenuated vaccines, the protective immunity induced by inactivated vaccines is normally short-term and requires repeated booster injections to promote lifetime immunity. Based on these immunological characteristics, live attenuated vaccines immunologically represent the vaccines of choice. Still, they are not without problems. For instance, there are safety concerns related to the use of live attenuated vaccines. The potential exists for the attenuated vaccine to mutate back to the original disease-causing organism through a process is called reversion. Attenuated vaccines may also cause disease when inoculated into persons with weak or compromised immune systems, such as cancer patients receiving chemotherapy, acquired immunodeficiency disease (AIDS) patients, or in the elderly where the immune system deteriorates with age. Furthermore, live vaccines can infect individuals other than the inoculated individual and thus inadvertently expose a disease to a susceptible unknowing individual.

Even though inactivated vaccines are safer in theory than their live attenuated counterparts, certain problems also exist with some inactivated vaccines. For instance, the whole organism used as the inactivated vaccine can be contaminated with components from cell culture that are not removed during the manufacturing process. This contaminated material may be a catalyst for an autoimmune disease. Additional shortcomings with inactivated vaccines include contamination by components of the pathogen that are not important in the generation of protective immunity. These components may generate immune responses that are not relevant to protective immunity. Deleterious reactions, such as inflammation and allergic reactions, may also result from vaccination with the inactivated whole organism. These concerns regarding contaminants and the safety issues related to whole organism vaccines point to the use of purified subunit component vaccines.

In subunit vaccines, only the components of the microorganism involved in conferring protective immunity are included in the final formulation, while other portions of the microorganism are removed in an extensive purification process. This increases the cost of manufacturing the vaccine. Subunit component vaccines have an increased specificity that can target induction of the immune response in a very effective manner; again these vaccines elicit protective immunity by the generation of antibodies and limited T helper responses. However, if antibodies alone are insufficient to provide protective immunity against a particular pathogen, it becomes necessary to also involve the activated T lymphocyte component of the immune response and subunit preparations are not very efficient in this regard. As a need for vaccines against new pathogens emerges, safe vaccines that elicit both antibodies and activated T cells will have an advantage, particularly when the requirements for protective immunity in the host are not yet unknown. It would be a distinct advantage for vaccine development to have a technology that could induce the broad immunity normally associated with a live attenuated vaccine while exhibiting the safety and the focus of the subunit preparations. In addition, any simplification in manufacturing and increase in vaccine product stability are likely to positively impact on vaccine production for the developing world. These advantages are the promise of DNA vaccines.

44.2 DNA VACCINES

44.2.1 CONCEPTS OF DNA VACCINES

Pioneering work from a small number of laboratories, including that of the authors, has demonstrated that the injection of a DNA plasmid containing foreign genes encoding proteins of a pathogen or cancer antigens directly into a host can give rise to specific immunity. This injection results in the subsequent

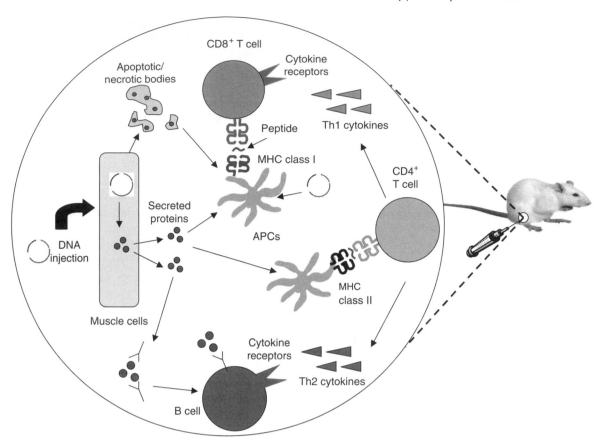

FIGURE 44.1	Induction of antigen-specific humoral and cellular immune responses following DNA immunization.

expression of the foreign genes in a host and the presentation of the specific encoded proteins to the immune system (Figure 44.1) [4]. DNA vaccine constructs can be produced as small circular vehicles or plasmids. These plasmids are constructed with a promoter site, which starts the transcription process, an antigenic DNA sequence, and a messenger RNA stop site containing the poly A tract necessary for conversion of the messenger RNA sequence into the antigen protein by the ribosomal protein manufacturing machinery (Figure 44.2). The concept of genetic immunization provides that both DNA and RNA that encode specific proteins can be used to generate specific immune responses. Since DNA and RNA are both nucleic acids, the term nucleic acid vaccine has also been used to describe this process.

44.2.2 HISTORY OF DNA VACCINES

The ability of genetic material to deliver genes for therapeutic purposes and its use in gene therapy has been appreciated for some time. Early experiments describing DNA inoculation into living cells were DNA transfer experiments performed by a number of investigators in the 1950s as well as 1960s [5–7]. These reports describe the ability of DNA preparations isolated from tumors or viral infections to induce tumors or virus infection following injection into animals. Importantly, many

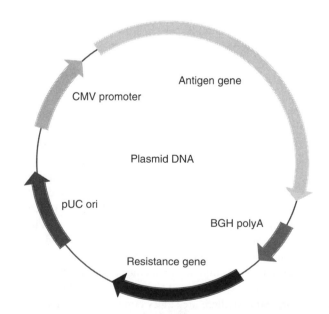

FIGURE 44.2	Diagram of a DNA vaccine construct consisting of a mammalian expression vector. The plasmid is constructed with a CMV promoter, an antigenic DNA sequence, and a messenger RNA stop site containing the poly A tract.

such inoculated animals developed antibody responses to the proteins encoded within the injected DNA sequences. Over the next 20 years, a number of scattered reports, all focusing on gene function or gene therapy techniques, provided evidence that injection of viral DNA or plasmids containing foreign gene resulted in antibody production was related to the DNA inoculations. In 1985, Dobensky reported that plasmids containing insulin DNA sequences could produce insulin following inoculation into a living animal for some period of time. Longer term expression of foreign genes was described in 1990 by Wolff et al. following plasmid inoculation *in vivo* [8]. These two separate observations demonstrated that DNA in the absence of viral vectors could deliver proteins that might have biological relevance. While both of these studies focused on the use of this technology for gene replacement strategies, studies were already under way in several laboratories using this same technology for vaccine applications.

In 1992, Johnston and colleagues at the Southwest Foundation reported that injection of DNA encoding human growth hormone into mice resulted in transient hormone production followed by the development of antibodies in the inoculated animals specific for the human growth hormone gene [9]. This important work utilized a "genetic gun" or gene gun to shoot gold particles covered with DNA through the skin layers of mice. While these investigators were actually studying the use of this technology for a gene therapy replacement strategy, they described this development of antibody responses due to this unusual immunization procedure as genetic immunization. Simultaneously with the publication by Johnston and colleagues, a critical vaccine meeting was being held at the Cold Spring Harbor Laboratory, Long Island, New York, in September, 1992. Talks by Margaret Liu (Merck), Harriet Robinson (UMass), and David Weiner (UPenn) described the use of DNA immunization to generate humoral and cellular immune responses against a human pathogen as well as protection from both tumors and viral challenges in animal systems.

Liu reported on the induction of immune responses to intramuscular (IM) injected plasmid encoding pathogen proteins. She observed that both antibody responses as well as CTL responses were induced to influenza viral gene products by this immunization technique. Furthermore vaccinated mice were able to resist lethal viral challenge. Robinson and her colleagues at the University of Massachusetts reported on the use of the gene gun to deliver influenza virus genes in DNA plasmids and reported that both antibody and T lymphocyte responses were produced in vaccinated mice and chickens. In challenge studies, these responses were protective. The use of the gene gun allowed investigators to deliver very low nanogram amount of DNA at the site of injection and still observe immune responses. Weiner and his colleagues at the University of Pennsylvania reported the direct injection of DNA encoding the genes for the human immunodeficiency virus (HIV). Again both antibodies and T lymphocyte responses specific for the viral gene products were observed in experimental animals. As HIV does not infect

mice, an *in vivo* mouse model was used where tumor cells, which are normally lethal to the mice, were constructed to express HIV or tumor proteins. Animals vaccinated with the specific DNA vaccine were demonstrated to be immune to implanted antigen-expressing tumor cells. While the audience was skeptical of the ability of nonliving genetic material to produce useful immune responses, the large amount of data presented by each of these groups representing several years of successful work in diverse systems could no longer be ignored by the scientific vaccine community. DNA vaccines were officially born.

Following these initial reports, DNA vaccination and the generation of antibody and T lymphocyte responses as well as protective responses in a variety of animal models have been reported in the scientific literature for many human pathogens such as hepatitis B virus (HBV), rabies virus, herpes simplex virus, hepatitis C virus, human T cell leukemia virus, HPV, and tuberculosis (TB) [10–12]. Now over 16 years later, there are thousands of articles that have been published utilizing DNA vaccine technology. Many clinical trials have been undertaken. Here we review some of the basic biology and application of this still emerging technology.

44.2.3 MECHANISM OF DNA VACCINES

The exact mechanism for immune induction by DNA immunization has been at the center of a major debate [13–15], but is likely to be similar to traditional antigen presentation models. In the body's immune system, APCs process and present antigenic peptides to lymphocytes in order to stimulate an antigen-specific immune response. Thus, antigen must be broken down and delivered to T lymphocytes by APCs [16]. Antigen presentation and recognition is a complex biological process that involves many interactions between APCs and T cells (Figure 44.3). There are four primary components that are critical in the professional APCs' ability to present the antigen to T cells and activate them for appropriate immune responses. These components are MHC–antigen complexes, costimulatory molecules (primarily CD80 and CD86), intracellular adhesion molecules, and soluble cytokines. Naive T cells circulate through the body across lymph nodes and secondary lymphoid organs such as the lymph nodes and spleen. Their migration is mediated by intercellular adhesion molecules and cytokines. As the T cells travel through immune organs, they bind to and dissociate from various APCs. This action is mediated through adhesion molecules. When a naive T cell binds to an APC expressing relevant MHC:peptide complex, the T cell expresses high levels of high-affinity IL-2 receptor. Only when this T cell receives a costimulatory signal through CD80/CD86–CD28 interaction does the T cell make soluble IL-2, which then drives the now-armed effector T cell to activate and proliferate initiating the immune response.

Antigen is expressed at significant levels in muscle following IM inoculation of plasmid DNA [8]. Using reporter gene injections in mice, various investigators have reported the detection of gene expression after IM injection of DNA

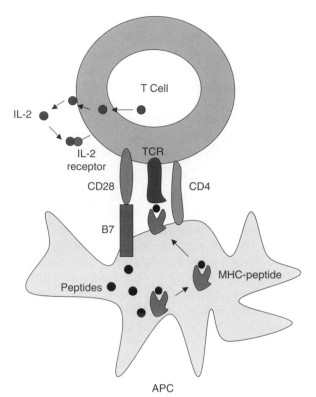

FIGURE 44.3 (See color insert following blank page 206.) Effective T cell activation by APC. The interaction between antigen-MHC and T cell receptor leads to the expression of IL-2 receptor. This T cell proliferates when the second signal is provided from APC's costimulatory proteins. CD28–CD80/CD86 ligation initiates the production of IL-2 production and leads to the proliferation of activated T cell.

expression cassettes [8]. Protein expression was detected in the quadriceps muscle of mice after injecting plasmid vectors encoding chloramphenicol acetyltransferase, luciferase, and β-galactosidase reporter genes into the muscle.

Muscle cells have several structural and functional features, which make them well suited for DNA uptake *in vivo* [17]. Muscle consists of multinucleated contractile muscle fibers with a cylindrical shape and tapered ends. These muscle fibers have mycogenic stem cells attached to them. When the muscle fibers are damaged or stressed, local stem cells become activated. The resulting myoblasts proliferate and eventually fuse to form new muscle fibers. It is believed that this continual activation and proliferation of the myoblasts allow a more opportunistic uptake of injected DNA. Because it has been shown that the uptake of the injected DNA and the subsequent production of protein occurs in muscle cells, they have been proposed as a potential site of antigen processing and presentation. However, the myocytes, which make up the muscle tissues, do not express CD80 or CD86 costimulatory molecules needed for efficient presentation. A recent study has identified an additional costimulatory molecule distinct from CD80 or CD86, which can be expressed in muscle cells [18]. However, if such molecules play a role in APC activity in the muscle requires more study. The question of ability of muscle cells to provide costimulatory signals

drives a current debate in the literature regarding the mechanism of antigen presentation following IM DNA immunization.

One potential mechanism is that the antigens produced in muscle are secreted from transfected muscle cells or released due to cell apoptosis [13,14]. Such exogenous antigen could then be taken up by professional APCs, locally which then migrate to, or shed antigen into the draining lymph nodes, where the antigen is processed via the MHC antigen presentation pathway on professional APC. These APCs are hypothesized to present the processed peptides to T cells. Recently, there have been reports that indicate that immune system has an inherent mechanism by which exogenous antigens access MHC class I molecules. One recent report identified dendritic cells as the potent mediator of such presentation antigen derived from phagocytosed apoptotic cells [19]. Immature dendritic cells engulf apoptotic cells and cross-present antigen from these sources to induce class I-restricted CTLs.

Another possible mechanism is the direct transfection of professional APCs by the injected DNA. Such a mechanism may be more probable in intradermal delivery of DNA because skin is rich in professional APCs, especially the dendritic cells. Condon et al. have reported that through DNA immunization into skin, they were able to show expression of proteins encoded by DNA plasmids [20]. On the other hand, such a mechanism is less efficient within the muscle tissue where there is a significantly lower presence of APCs. More recently, studies have reported that direct transfection of dendritic cells can occur following IM inoculation of DNA vaccine constructs, albeit at a lower level [21]. Studies support that macrophages and dendritic cells may be target cells for DNA transfection *in vivo* and that such a target might be important in driving immune responses [22]. A clear understanding of the role of APCs in DNA vaccination could have important implications for this technology.

44.2.4 POTENTIAL ADVANTAGES OF DNA VACCINES

As summarized in Table 44.1, DNA immunization affords several potential advantages over traditional vaccination strategies such as whole killed or live attenuated virus and recombinant protein-based vaccines without specific shortcomings and inherent risks associated with these vaccination methods. Like inactivated or subunit vaccines, DNA vaccines appear to be very well tolerated in humans. With several thousand persons having received DNA vaccines, the most common adverse effect has been injection-site irritation. Preclinical safety studies indicated that there was little evidence of plasmid integration which was an initial concern of scientists and regulators. The frequency of integration for DNA immunization by the IM, intradermal (ID), gene gun, or Biojet routes is at least three orders of magnitude below the spontaneous rate of gene-inactivating mutations [23–25]. In contrast to inactivated or subunit vaccines, DNA vaccine cassettes produce immunological responses that are more similar to live vaccine preparations. By directly introducing DNA into the host cell, the

TABLE 44.1
Potential Advantages of DNA Vaccines

Immunogenicity	Can induce both humoral and cellular immune responses
	Low effective dosages (micrograms) in animal models
Safety	Unable to revert into virulence unlike live vaccines
	Efficiency does not require the use of toxic treatments unlike some killed vaccines
Engineering	Plasmid vectors are simple to manipulate and can be tested rapidly
	Combination approaches are easily adapted
Manufacture	Conceptually low cost and reproducible large scale production and isolation
	Produced at high frequency in bacteria and easily isolated
Stability	More temperature-stable than conventional vaccines
	Long shelf life
Mobility	Ease of storage and transport
	Likely not to require a cold chain

host cell is essentially directed to produce the antigenic protein, mimicking viral replication or tumor cell antigen presentation in the host. Studies have shown that DNA can provide protection from infections from various pathogens, including viruses, bacteria, and parasites [26–29] in small animal models. DNA vaccines have been shown to be effective at eliciting antigen-specific CTLs, T helper cells, and antibodies similar to live infections. Unlike a live attenuated vaccine, the other major theoretical advantage of DNA vaccine is that they can be used for repeated administration as the efficacy of plasmid vectors are not influenced by preexisting neutralizing antibodies [10]. The ability to target multiple antigenic components may be a particularly important characteristic of DNA vaccines since multicomponent DNA vaccines can be engineered to include specific immunogens that can optimize and amplify desirable immunologic responses. Furthermore, DNA vaccines appear to be very stable and simple to produce. These features could allow for the rapid flow of DNA vaccines into use in developing countries. All these advantages of DNA vaccines make them appealing to both medical professionals and manufacturers in the health care industry.

44.3 STRATEGIES TO IMPROVE DNA VACCINES

Since the first DNA vaccines were described over 16 years ago, there has been great interest in developing a new generation of DNA vaccines that induce more potent immune responses. Recently, two veterinary DNA vaccines, a vaccine for West Nile Virus (WNV) in horses and a vaccine for infectious hematopoietic necrosis in salmon, have been licensed. Furthermore, license of a canine melanoma DNA vaccine is eminent. However, despite promising results in preclinical immunogenicity studies, DNA vaccines have been shown not to elicit robust antigen-specific immune responses in humans. Along with all of the advantages of DNA vaccines, an increase in potency would generate a vaccine that would appeal to the global health initiative. Thus, many strategies have been recently applied to improve the potency of the DNA vaccine platform.

44.3.1 IMPROVEMENT OF EXPRESSION AND IMMUNOGENICITY OF DNA VACCINES

Previous studies have demonstrated that the expression of viral or bacterial genes in mammalian cells can be hampered by the codon usage bias. For example, HIV viral RNA is AU-rich, in contrast to the high GC-rich content of the mammalian genome [30,31]. The tRNA pools required by HIV viral mRNA translation are different from those required by its host mammalian cells, resulting in poor expression of viral proteins in such cells. It is believed that the codon-optimized sequences for viral genes can be translated more efficiently using tRNAs in mammalian cells than the wild-type viral genes. For HIV plasmid vaccines, the issue of *rev* independence has appeared as another important problem. The structural genes of HIV are made as long unspliced transcripts that contain overlapping reading frames with small regulatory genes of HIV, *tat* and *rev*. These transcripts are retained in the nucleus and rapidly spliced to encode the small regulatory in the absence of the rev shuttle protein, thus preventing the transport of full-length message to the endoplasmic reticulum (ER) where structural gene translation can occur. Messages are inhibited from transport to the ER by both known and cryptic *rev*-dependent sequences. Previous studies suggest that optimization of codons to the usage of highly expressed human genes as well as deletion of residual inhibitory sequences significantly enhanced Gag expression, leading to potent augmentation of immune responses [32,33]. Therefore, codon optimization circumvents this problem. Wang et al. recently demonstrated that optimized DNA constructs were able to express much more envelope antigens than their wild-type counterparts, and one mechanism for higher expression was to have more newly synthesized, gp120-specific transcripts available for protein expression through increased stability of their mRNAs [34]. Scientists also found that optimized DNA vaccines could elicit more sustainable antibody responses than nonoptimized DNA vaccines [35,36].

RNA optimization suggested originally by the impact of HIV rev on HIV antigen expression appears to be another way to increase the expression/immunogenicity of DNA vaccines [37,38]. Research has shown that sequences with GC-rich sequence stretches are more likely to form secondary structures that can hamper protein translation. During the RNA optimization process, the negative cis-acting motifs such as internal TATA-boxes, ribosomal entry sites, repeat sequences, splice sites, and poly(A) signals are also removed in order to make protein translation more efficient.

Modification of native leader sequences can play an important role on protein translation efficiency. To date, several highly efficient leader sequences have been used to increase the expression levels of DNA vaccines, such as human tissue plasminogen activator leader sequence (tPA) [39,40] and immunoglobin leader sequence that have weak RNA secondary structure [41]. Previous studies indicated that the native signal sequence for HIV-1 envelope glycoproteins could inhibit gp120 folding inside the ER by delaying its own cleavage [42]. A significant increase in envelope protein production could be detected when the native leader sequence is replaced by a human tissue plasminogen activator leader sequence [39]. Data also suggest that the expression of IL-15 was enhanced dramatically when two native leader sequences were substituted by an IgE leader sequence [43]. Highly efficient leader sequences help to stabilize the mRNA and enhance the expression of DNA vaccines, thus exhibiting positive effects on immune responses.

Both cell-specific and ubiquitous promoters have been used to drive antigen-specific immune responses. Cell-specific promoters include the fascin promoter [44], the muscle creatine kinase promoter [45] and the keratinocyte K14 promoters [46]. While promising, the ubiquitous cytomegalovirus (CMV) promoter has been shown to be more effective at driving high levels of protein expression [47] and has been widely used to develop various DNA vaccines. The enhancement of antigen-specific immune responses was also observed in primates when the CMV promoter was used in a DNA vaccine cassette [48]. In addition, the CMV immediate early (IE) promoter has been shown to yield higher expression when it was tested in parallel with other promoters by both in vitro and in vivo experiments [40,49]. The intron A sequence, which is an enhancer, could cooperate with the CMV-IE promoter, yielding even higher expression levels [49]. A transcription factor-binding sequence that is thought to provide a promoter-specific regulatory element is located in the intron A sequence, and this sequence may contribute to the enhancing effect of intron A. However, limited results have been reported with this system. Studies on developing new efficient promoters, such as nonviral, synthetic, and chimeric promoters, capable of driving long-lasting transcription are also under way [50,51].

Besides the molecular strategies mentioned above, there are some other potential methods used to increase the expression of DNA vaccines. The inclusion of CpGs sequences in the backbone of plasmid DNA can enhance IL-12 and type I IFN secretion from murine bone marrow-derived DCs [52]. Toll-like receptor 9 (TLR9) recognizes unmethylated CpG motifs and CpGs activate TLR9-bearing APCs [53]. Studies

have also shown that constitutive transport elements (CTEs) increase the transport and stability of mRNA from the nucleus, thus enhancing the expression of DNA immunogens [54]. Taken together, all these approaches can contribute to the overall performance of the DNA vaccines through enhanced protein expression by increasing the efficiency of transcription and translation after DNA immunization. However, it is important to try several methods of optimization, as some can be inhibitory and others can be complementary.

Recent studies have indicated that improved immune responses came from an approach that combined and optimized several molecular strategies in a DNA vaccine design. Yan et al. [55] recently developed a novel engineered HIV-1 subtype B envelope DNA vaccine (pEY2E1-B) by incorporating several genetic modifications including codon optimization, RNA optimization, and the addition of a highly efficient immunoglobin leader sequence to increase the immunogenicity. A highly efficient IgE leader sequence was fused in frame upstream of the start codon to facilitate expression. The codon usage of this envelope gene was adapted to the codon bias of Homo sapiens genes. In addition, RNA optimization [37] was also performed: regions of very high (>80%) or very low (<30%) GC content and the cis-acting sequence motifs such as internal TATA boxes, chi-sites, and ribosomal entry sites were avoided. They found that this novel envelope immunogen elicited much stronger cellular immune responses than a codon-optimized only envelope immunogen. Moreover, they developed a novel engineered SIV env construct (pSM-optEnv) by using the same strategies used for pEY2E1-B design and found this novel SIV env construct could elicit about fourfold stronger cellular immune responses as compared to these induced by a partially optimized SIV env construct (pSIV-Env) (Figure 44.4A). Previous studies have shown that the cellular immune response to SIV env DNA antigen in monkeys was weak and inconsistent. Since the number of SIV env-specific IFN-γ producing CD8 effector cells was low, it was difficult to detect CD8 T-cell proliferation. In contrast, more than 20% env-specific CD8 T-cell proliferation was detected in primates immunized with pSM-Env (Figure 44.4B), supporting the improved immunogenicity of this combination approach.

44.3.2 INDUCTION OF CROSSREACTIVE IMMUNE RESPONSES

The genomic sequences of viruses such as HIV and HCV are very plastic. Highly error-prone genetic replication system, unique to each of these viruses, leads to the generation and accumulation of such genetic diversity. In HIV, this high mutation rate is driven by at least two mechanisms: the low fidelity of the viral reverse transcriptase (RT), resulting in at least one point mutation per replication cycle [56] and the dual effects of the antiretroviral cellular factor APOBEC3G gene and viral infectivity factor Vif accessory gene [57,58]. Genomes with every possible mutation and many double mutations are generated during every replication cycle, resulting in tremendous antigenic diversity. Research has shown that genetic differences among HIV-1 groups M, N, and O are

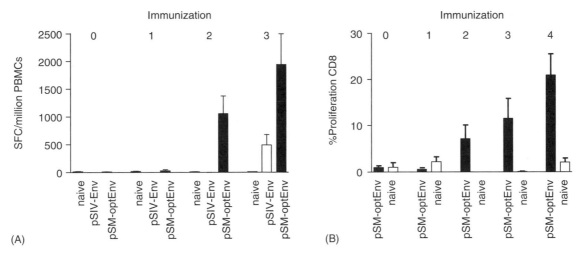

FIGURE 44.4 Improved potency of a genetically engineered optimized SIV envelope DNA vaccine. (A) SIV envelope-specific IFN-γ-producing cells following immunizations with the DNA vaccines. (B) SIV envelope-specific CD8 T-cell proliferation following immunizations with the pSM-optEnv DNA vaccine. Each group of macaques ($n = 5$) was immunized with 2 mg of DNA vaccine, and PBMCs were isolated and assessed for an SIV envelope-specific response by a standard IFN-γ ELISPOT assay and CFSE staining assay.

extensive, ranging from 30% to 50% in the *gag* and *env* genes, respectively [59–61]. The HIV-1 "major" (M) group viruses are primarily responsible for the AIDS pandemic and are further classified into nine genetically distinct subtypes (A, B, C, D, F, G, H, J, and K) [62]. The protein sequence differences between Env proteins from various subtypes differ up to 35% [63]. In addition, HIV-1 also frequently recombines among different subtypes to create circulating recombinant forms (CRFs) and novel recombinants [64]. Initial optimism that a vaccine designed from an HIV-1 primary isolate would eliminate disease caused by HIV-1 infection appears unfounded. Like other RNA viruses, HCV has a high mutation rate. The most divergent isolates sharing only 60% nucleotide sequence homology. The high level of genetic variability poses a major hurdle for developing vaccines against diseases caused by HIV, HCV, or other viruses. It has been argued that a candidate DNA vaccine derived from an individual isolate may not elicit sufficient crossreactivity to protect against diverse circulating viruses. In order to enhance the breadth of the elicited immune responses, scientists have applied an approach that includes using antigenic components from as many diverse HIV-1 isolates as possible in the vaccine, with the intension of inducing broader immune responses against divergent viral target antigens. Recently, in order to maximize the genetic similarity between the candidate vaccine and contemporary HIV-1 stains and to simplify immunogen production, new strategies, such as centralized sequence-based vaccines and polyvalent vaccines, are being developed.

44.3.2.1 Consensus-Based DNA Vaccines

A consensus sequence (CON) can be designed based on analysis of all circulating strains. In its simplest form, the CON is a sequence that has at each site the most frequent nucleotide or amino acid residue across an alignment of homologous sequences. The CON will be genetically closer to currently circulating strains than any given individual natural virus isolate. Previous phylogenetic analysis of more than 70 HIV-1 subtype C genomes suggested that the amino acid distances to the consensus sequence were significantly lower than distances between samples within all HIV-1 subtype C isolates [65]. The evaluation of the predictive power of the consensus sequence showed the additional benefit using the consensus-based vaccine. Only 3% to 8% amino acid differences were found when comparing the subtype C consensus sequence with the sets of subtype C sequences. The translated amino acid distances to the consensus sequence were significantly lower than distances between samples within all HIV-1 subtype C proteins. The high probability that a new sequence will be within the identified consensus sequence, and the consensus consistency and stability to the introduction of a new isolate strongly suggest that a consensus-based vaccine should be able to enhance the breadth of cellular immune responses against any single virus within a population.

Recently, Gao et al. developed an HIV-1 group M consensus DNA immunogen (CON6) and found that this artificial immunogen could fold into native conformation, preserving envelope antigenic epitopes. This construct could elicit weak neutralizing antibody responses [66]. Further study on this consensus immunogen indicated that vaccination with the pCON6 could induce a greater number of T-cell epitope responses than any single wild-type subtype A, B, and C *env* immunogen [63]. On the other hand, previous studies have shown that there were high rates of selection identified in different regions of subtype B and C envelope proteins [67]. This may be caused by immune pressures on different regions of the envelope protein in subtype B and C. Therefore, there may be advantages in using a subtype-specific CON envelope vaccines, as the immune responses to the vaccine and the circulating virus would share more antigenic domains. Yan et al. developed a novel HIV-1 subtype B consensus envelope DNA vaccine (pEY2E1-B) and peptide mapping results indicated that the EY2E1-B immunogen generated increased crossreactive cellular immune responses when compared to a primary envelope construct [55]. The use of consensus approaches is also being explored against other pathogens including influenza [68]. This is a promising area for further study of novel designed immunogens.

44.3.2.2 Ancestral Sequence-Based DNA Vaccines

Any consensus sequence may be biased by the limits of sampling and may associate polymorphisms in combinations not found in any natural virus, thus potentially resulting in improper protein conformations. Many scientists have proposed the use of ancestral sequence as a vaccine candidate to get around these issues [67,69–71]. The generation of an ancestral sequence involves reconstruction of the ancestral state of the antigen sequence [67]. An ancestral antigen that embodies the most recent common ancestor (MRCA) would have at least two advantages. First, it would be more genetically similar to all circulating strains than any other strains. Second, the protein generated from an ancestral sequence would more likely have native folding and function. Doria-Rose et al. constructed a full-length HIV-1 subtype B ancestral gene and demonstrated that the protein encoded by this gene could bind and fuse with cells expressing the HIV-1 coreceptor CCR5 [72]. When used as a DNA vaccine to immunize rabbits in a DNA prime-protein boost regimen, the artificial immunogen elicited neutralizing antibodies similar to those induced by a natural isolate, HIV-1 SF162. The cross-reactive cellular immune responses, however, were not determined due to the limitations in the rabbit model used. Further studies to explore whether these ancestor DNA immunogens could elicit broader cellular and humoral responses will be of great importance.

44.3.2.3 COT-Based DNA Vaccines

Gaschen et al. pointed out that an increasingly heterogeneous group of variants could lead to an ancestral sequence that is progressively more divergent from the variants [67]. Mullins and colleagues have recently developed a new hybrid method that produces a summary sequence that is tree-based while eliminating the influence of outlier sequences as seen in MRCA [73]. This new method identifies a point called the center of the tree (COT) on an unrooted phylogenic tree. Theoretically, the COT minimizes the average evolutionary distance to all sequences in the data set and also could potentially enhance the breadth of immune responses. Rolland et al. computationally derived COT sequences from circulating HIV-1 subtype B sequences for the *gag*, *tat*, and *nef* genes [73]. The results indicated that all these COT genes retained biological functions, COT gag generated virus-like particles, while COT tat transactivated gene expression from the HIV-1 LTR, and COT nef mediated downregulation of cell surface

MHC-1. In addition, these COT-based DNA vaccines were capable of eliciting antigen-specific cellular immune responses in mice. COT gag and nef could also induce cross-clade CTL responses when studied for reactivity against subtype A, B, or group M peptide pools. These data provide another promising avenue for globally relevant DNA vaccine design.

44.3.2.4 Polyvalent Vaccines

Recently, Korber and colleagues present the design of polyvalent vaccine antigen sets, focused on T-lymphocyte responses and optimized for either subtype B and C or for group M [74]. Mosaic proteins resemble natural proteins and a mosaic set maximizes the coverage of potential population. They found the coverage of viral diversity using polyvalent vaccines was significantly enhanced for both variable and conserved proteins. Four mosaic proteins matched 74% of 9-amino-acid potential epitopes in global gag sequences and 87% of potential epitopes matched at least eight of nine positions, while a single natural gag protein covered only 37% (9 of 9) and 67% (8 of 9). Based on these data, a polyvalent DNA vaccine may also be developed using mosaic DNA sequences encoding potential T-cell epitopes in global HIV-1 variants. Overall, the area of synthetic immunogen design is getting traction as the DNA platform is the most simple for initial testing of innovative antigen design strategies.

44.3.3 IMPROVEMENT IN DELIVERY OF DNA VACCINES

An ideal delivery system for *in vivo* gene transfer should be simple and safe, and provide an appropriate expression of transgenes at therapeutic levels. Traditionally, plasmid-based technology has been limited in scope as expression levels following naked DNA transfer have been low, only a fraction of that of viral-mediated gene transfer. Improvements in delivery are critical to improve the response rates of DNA vaccines for activation of both cellular and humoral responses in humans.

Among the approaches for plasmid delivery *in vivo*, electroporation (EP) is receiving significant attention. This delivery method has great potential for DNA vaccination applications since a more efficient level of plasmid DNA transfer and transgene expression has been reported [75]. This physical process exposes the target tissue to a brief electric field pulse that induces temporary and reversible pores in local cell membranes. During the period of membrane destabilization, molecules such as plasmids may be taken up, resulting in 100- to 1000-fold increases in protein expression [76–79]. However, EP may cause mild localized tissue damage, resulting in the release of cellular components, which might provide a "danger signal" and thus provide an adjuvant effect [80]. EP also increases the number of tissue-resident transfected mononuclear cells [81], which could be very important for the induction of effective immune responses.

EP has been studied in large-animal species such as dogs, pigs, cattle, and more recently nonhuman primates to deliver therapeutic genes that encode for a variety of hormones, cytokines, enzymes, or antigens [80]. Skeletal muscle cells have provided an ideal target for direct plasmid transfer for DNA vaccines and other applications [80]. The expression levels are increased by as much as three orders of magnitude over plasmid injection alone. For instance, a single injection of plasmid-encoding neuraminidase from influenza virus followed by EP in mice was able to provide long-term protection from influenza challenge [82]. However, similar results have been reported with other DNA delivery methods. Luckay et al. observed a 10- to 40-fold increase in HIV-1 antigen-specific responses in macaques immunized in combination with *in vivo* EP compared to those for macaques receiving a five-fold higher dose of vaccine without *in vivo* EP [83]. Importantly, *in vivo* EP was especially efficient at increasing the cellular immune responses against the less immunogenic antigens, such as nef, tat, and vif, resulting in a more balanced immune response overall. The data also suggested that *in vivo* EP could result in an approximate $2.5 - \log_{10}$ increase in antibody responses. Hirao et al. reported that the increased cellular and humoral immune responses were observed after IM delivery of plasmids by using a new constant-current device, including induction of polyfunctional responses. Data indicated that EP induced about four- to fivefold stronger cellular immune responses as compared to IM injection in mice (Figure 44.5). EP also induced between two- and fivefold stronger antibody responses as compared to IM injection and at DNA concentrations that were 20× less than the dose given by IM injection alone. Therefore, EP is an important method for further analysis as a DNA vaccine delivery technology [84,85].

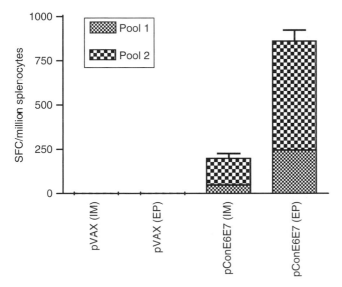

FIGURE 44.5 Comparison of cellular immune responses induced by pConE6E7 using two DNA delivery methods: IM injection versus EP. The splenocytes were isolated from individual immunized mice (five mice per group) and stimulated *in vitro* with overlapping HPV-16 E6/E7 peptides pools. HPV-16 E6/E7 specific IFN-γ production was measured by a standard IFN-γ ELISPOT assay. pVAX immunized mice were included as a negative control.

Although EP can markedly augment immune responses induce by DNA vaccines, the tolerability of EP has been an important concern for vaccine delivery into humans. In many studies, the applied voltage was 150 to 250 V/cm. In addition, the long pulse length was also used when EP was performed. To make this technology feasible for human vaccination studies, more benign methods such as lower voltage, shorter pulse length should be investigated. Furthermore, skin delivery may be appealing. However, the increased delivery of this method may require additional studies of biodistribution and plasmid persistence.

Gene gun delivery has been shown to be another promising delivery method. Unlike IM injection, gene gun DNA immunization directly transfects cells by depositing DNA-coated gold beads within the cell cytoplasma [9]. Robinson and colleagues found that DNA immunization by IM injection induced a predominantly Th1 response while gene gun DNA immunization elicited a predominantly Th2 response [86]. It has been demonstrated that gene gun immunization used 100- to 1000-fold less DNA than IM immunization to generate an equivalent Ab response [87]. These findings have very important implications for vaccine design.

44.3.4 Molecular Adjuvants

The primary goal of the first generation DNA vaccination studies was to demonstrate and evaluate the ability of DNA vaccines to elicit humoral and cellular responses *in vivo* in a safe and well-tolerated manner. As we explore the next generation of DNA vaccines, it would be desirable to refine current DNA vaccination expectations to demand that they elicit more clinically efficacious immune responses. In this regard, the next generation DNA vaccines may require better augmentation of the magnitude and direction (humoral or cellular) of the immune responses induced. Such modulation of immune responses may be accomplished by the use of genetic adjuvants encoded as part of the plasmid cocktail. Genetic or molecular adjuvants are different than the traditional adjuvants in that they are comprised of gene expression constructs encoding specific immunologically important molecules. These molecules include cytokines, heat shock proteins, chemokines, costimulatory molecules, and molecules that induce cell death among others. Molecular adjuvant constructs are coadministered along with immunogen constructs to modulate the immune responses induced by the vaccine cassettes themselves.

44.3.4.1 Cytokine Molecular Adjuvants

Many cytokines, including interleukins (IL), tumor necrosis factors (TNF), and interferons (IFN), have been evaluated as DNA vaccine adjuvants. The recent approval of therapeutic GHRH plasmid delivery for fetal wasting in pigs supports that this area is of importance. Here we describe several examples that illustrate findings and have potential implications for future vaccine and immune therapeutic exploitation.

Granulocyte-macrophage colony stimulating factor (GM-CSF) can recruit dendritic cells and promote their survival. It is one of the first cytokines to demonstrate clear potential to enhance both humor and cellular immune responses to a DNA vaccine in mice. Previous studies have shown that coimmunization of GM-CSF with HIV-1 envelope gp120 in mice was able to increase cellular immune responses and CD4 T-cell proliferation [88,89]. However, studies in primates and humans showed that there was no dramatic boosting of the immune responses when GM-CSF was used as a DNA vaccine adjuvant in these species. This illustrates the need for examination of products in several systems prior to drawing final conclusions regarding adjuvant immune potency.

IL-12 is mainly produced by activated APCs including macrophages, dendritic cells, and B cells. It induces Th1-type immune responses through inducing the maturation of Th1 cells and promotes natural killer (NK) cell activity [88,90]. It has been shown that coimmunization with IL-12 plasmid results increased T helper cell proliferation and a significant enhancement of CTL responses in both mice and primates [88,89]. IL-12 has also been successfully used in cancer immunotherapy since it has a critical role in inducing antitumor immune responses [91,92]. Kim et al. observed a significant enhancement of CTL response *in vivo* with the coadministration of murine IL-12 genes with three different HIV-1 DNA immunogens (*gag/pol, envelope, vif,* and *nef*). Importantly, recent data have demonstrated that stronger and earlier CTL and antibody responses were induced when IL-12 was used as a molecular adjuvant in rhesus monkeys. This promising adjuvant has been now moved forward to clinical examination. Preliminary results from these studies should be available late in 2008.

IL-2 is another important immune adjuvant. IL-2 can activate both CD4, CD8 T cells and NK cells [93,94] and has been shown to enhance both cellular and humoral immune responses to HBV surface antigen in mice [95]. Barouch et al. reported that monkeys vaccinated with DNA vaccines expressing SIVmac239 Gag and HIV-1 89.6P Env in combination with the purified fusion protein IL-2/Ig demonstrated augment secondary CTL responses, stable CD4 + T cell counts, very low viral loads, and no evidence of clinical disease by day 140 after a pathogenic SHIV-89.6P challenge [96]. Therefore, other adjuvants that target the same cytokine pathway may also be important in this context. This concept has been moved forward into clinical examination. In these studies, the IL-2/Ig is given 2 days following administration of the vaccine to maximize immune boosting.

The use of molecular adjuvant IL-15 has been considered as an important tool to increase the potency of DNA vaccines. Importantly, IL-15 has been reported to enhance memory CD8 + T cell proliferation as a DNA vaccine adjuvant [97]. It has been shown that there is a lack of memory phenotype CD8+ T cells in IL-15Rα- and IL-15-deficient mice [98]. An optimized IL-15 construct was generated and coimmunization with this construct increased HIV-1 gag-specific CD8+ T-cell proliferation, IFN-γ secretion, and longevity of CD8 T cells [99]. In an influenza DNA vaccine model, stronger memory T-cell responses that protected mice from a lethal mucosal influenza challenge were induced in mice immunized with pHA + pIL-15-Opt compared to these immunized

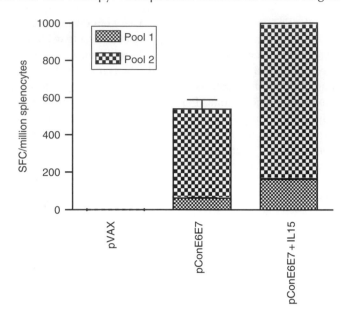

FIGURE 44.6 Comparison of cellular immune responses induced by pConE6E7 with or without using molecular adjuvant IL-15. The splenocytes were isolated from individual immunized mice (five mice per group) and stimulated *in vitro* with overlapping HPV-16 E6/E7 peptides pools. HPV-16 E6/E7 specific IFN-γ production was measured by a standard IFN-γ ELISPOT assay. pVAX immunized mice were included as a negative control.

with pHA [99]. pIL-15-Opt could restore CD8 secondary immune responses in the partial absence of CD4+ T cell help, suggesting that IL-15 may be particularly useful as a candidate adjuvant for HIV immunotherapy. Moreover, Boyer et al. reported that macaques immunized with SHIV DNA vaccine together with IL-15 have an improved ability to control viral challenge and exhibit undetectable viral loads even 1 year after challenge. In addition, IL-15 was also shown to enhance the *in vivo* antitumor activity of adoptively transferred CD8+ T cells [100]. Yan et al. found that the IL-15 plasmid as a molecular adjuvant can improve the antitumor immunity induced by the DNA immunogen in a tumor model as well (Figure 44.6). The E6/E7 transgenic mice immunized with a plasmid-encoded HPV16 E6 and E7 (pConE6E7) and IL-15 exhibited smaller tumors compared to these only immunized with pConE6E7. Nine out of 10 mice survived in pConE6E7 + IL15 group while only 4 of 10 mice survived in pConE6E7 group 60 days after tumor implantation, indicating that vaccination with pConE6E7 in combination with IL-15 increased the ability of pConE6E7 to overcome immune tolerance and could cause better tumor regression in the E6/E7 transgenic mice (Figure 44.7). Collectively the advances in the cytokine area are impressive. More study in this rapidly developing area is important.

44.3.4.2 Chemokine Molecular Adjuvants

Similar to cytokine gene codelivery, coimmunization with chemokine genes along with DNA immunogen constructs can modulate the direction and magnitude of induced immune

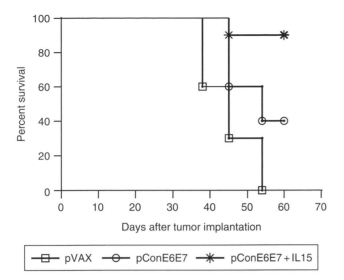

FIGURE 44.7 Vaccination in combination with IL-15 induced better tumor regression in C57BL/6 mice. On day 0, 5×10^4 TC-1 cells were implanted subcutaneously into C57BL/6 mice (10 mice/per group). Starting on day 3, each group of mice was immunized with pVAX, pConE6E7 and pConE6E7 + IL15 for three times at weekly intervals and tumors measurement was performed.

responses [101]. For example, coimmunization with IL-8 and macrophage inflammatory protein 1α (MIP-1α) genes increased the antibody response in a similar manner to IL-4 or GM-CSF coimmunization. IL-8 and RANTES (regulated on activation, normal T expressed and secreted; CCL5) as adjuvants could also enhance T helper proliferation. RANTES and monocyte chemoattractant protein-1 (MCP-1) coinjections resulted in a high level of CTL enhancement, almost as significant as IL-12, a potent CTL inducer for DNA vaccines. There is evidence that suggests some chemokine-adjuvanted vaccines require a direct fusion of the immunogen with the chemokine. For instance, immunization against HIV gp120 with MCP-3/CCL7 or macrophage-derived chemokine (MDC/CCL22) requires the direct fusion of the two for effective immune response induction. Accordingly, coimmunization revealed minimal immune enhancement [102]. Therefore, the use of these chemokine adjuvants could be particularly important as HIV vaccine modulators of β-chemokines. It has been reported that β-chemokines as vaccine adjuvants augmented β-chemokine production in a vaccine antigen-specific manner [103]. This aspect could be especially important for a development of a vaccine that modulates the earliest aspects of the inflammatory response.

44.3.4.3 Costimulatory Molecule Molecular Adjuvants

Professional APCs initiate T-cell activation through binding of antigenic peptide–MHC complexes to specific T-cell receptor molecules. In addition, the APCs provide critical costimulatory signals to T cells, which are required for the clonal expansion and differentiation of T cells. Among different costimulatory molecules, B7 molecules (CD80 and CD86) have been observed to provide potent immune signals [104,105]. They bind to their

receptors (CD28/CTLA-4) present on T cells. The CD80 and CD86 molecules are surface glycoproteins and members of immunoglobulin superfamily, which are expressed only on professional APCs [104–106]. The blocking of this additional costimulatory signal leads to T-cell anergy [107].

CD86 molecules play a prominent role in the antigen-specific induction of CD8+ CTLs when delivered as vaccine adjuvants [101]. Coadministration of CD86 cDNA along with DNA encoding HIV-1 antigens intramuscularly dramatically increased antigen-specific T-cell responses without a significant change to the level of the humoral response. This enhancement of CTL response was both MHC class I-restricted and CD8+ T-cell-dependent. These results have been extended by other investigators who also found that CD86, not CD80 coexpression resulted in the enhancement of T-cell-mediated immune responses [108,109]. However, Santra et al. reported that either CD86 or CD80 administered concurrently with a plasmid DNA vaccine can fully costimulate vaccine-elicited CTL responses [110].

It is possible that engineering of nonprofessional APCs such as muscle cells to express CD86 costimulatory molecules could empower them to prime CTL precursors. On the other hand, the enhancement effect of CD86 codelivery could also have been mediated through the direct transfection of a small number of professional APCs residing within the muscle tissue. Subsequently, these cells could have greater expression of costimulatory molecules and could in theory become more potent. This issue has been investigated using bone marrow chimeras, which demonstrated the ability to cross prime CTL immunity [107]. This method of engineering nonhematopoietic cells to be more efficient APCs could be especially important in cases where antigen alone fails to elicit a CTL response due to poor presentation by the host APCs. However, recent studies of B7.2 containing DNA vaccines in the nonhuman primate model have not been very encouraging. Engineering this pathway to generate useful levels of cellular immunity in nonhuman primates remains an important challenge.

In addition to the B7 family of costimulatory molecules, the immunology of CD40 and its ligand has also been implemented to enhance DNA vaccine potency. CD40 functions by interacting with CD40 ligand (CD40L) expressed on activated CD4+ or CD8+ T cells. The attachment of CD40L onto APCs, with or without T helper cells, have been shown to "condition" the APCs for antigen-specific CTL activation [111–113]. Additionally, their ligation has shown to enhance the expression of B7 costimulatory molecules on APCs, including dendritic cells [114–116]. Therefore, the engagement of CD40 with its ligand becomes advantageous for activation of CTLs during an immune priming response. Accordingly, coimmunization of plasmids coding for beta-galactosidase and CD40L has been reported to induce immune enhancement [117]. Importantly, the addition induced enhancement of CTL responses without suppressing the development of antibody responses was observed [118,119]. Interestingly, coinjection of CD40L was revealed to be more effective than CD40 [120], indicating that expression of CD40L on muscle cells may also induce the "licensing" of APCs for the activation of CTLs. However, the possibility of direct transfection into antigen specific infiltrating T-cells cannot be eliminated, although T-cells have not been yet established as an *in vivo* transfection target for DNA vaccines. Studies using multimeric forms of the CD40L molecule have been encouraging [121]. Such molecules maintain critical structure of the nature ligand. However, as with B7.2 initial studies in nonhuman primates have not been highly impressive.

Adhension molecules have also been studied in this context. These molecules are particularly important in the initiation of the host immune response. Antigen-specific T-cell responses can be enhanced by the coexpression of DNA immunogen and adhesion molecules intercellular adhesion molecule-1 (ICAM-1) and lymphocyte function-associated antigen 3 (LFA-3). Coexpression of ICAM-1 or LFA-3 molecules along with DNA immunogens resulted in a significant enhancement of Th cell proliferative responses. In addition, coimmunization with ICAM-1 (and more moderately with LFA-3) resulted in a dramatic enhancement of CD8-restricted CTL responses. Although vascular cell adhesion molecule-1 (VCAM-1) and ICAM-1 are similar in size, VCAM-1 coimmunization did not have any measurable effect on cell-mediated responses. Rather, these results imply that ICAM-1 and LFA-3 provide direct T-cell costimulation. These observations were further supported by the finding that coinjection with ICAM-1 dramatically enhanced the level of IFN-γ and β-chemokines MIP-1α, MIP-1β, and RANTES produced by stimulated T cells. ICAM-1/LFA-1T-cell costimulatory pathways appear independent of CD86/CD28 pathways, and they may synergistically expand T-cell responses *in vivo*. Furthermore, these studies indicate that CD8+ effector T cells at the site of inflammation can regulate the level of effector function through the expression of specific chemokines and adhesion molecules (Figure 44.8) [101,103]. Therefore, the end-stage effector T cells in the expansion phase of an antigen-specific immune response could direct their own destiny through coordinated expression and release of these molecules. This pathway can likely be exploited in the DNA vaccine setting.

44.3.4.4 Molecular Adjuvants That Induce Cell Death

In addition to powerful signaling, dendritic cells often function as scavenger cells by engulfing and processing apoptotic bodies. For instance, immature dendritic cells phagocytose apoptotic bodies by employing the receptors alphavbeta5 integrin and CD36 [122–124]. Subsequent engulfment of the

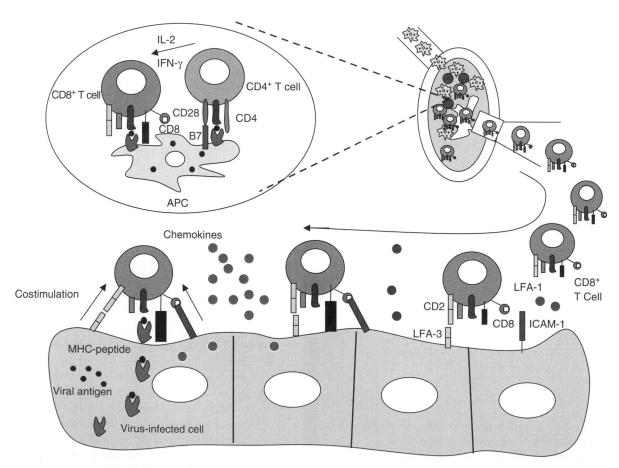

FIGURE 44.8 (See color insert following blank page 206.) Regulation of CD8+ T cell expansion by adhesion molecules and chemokines in the periphery. Specific adhesion molecules and chemokines provide modulatory signals to CD8+ T cells in effector stage. This network of cytokine, chemokine, costimulatory molecules, and adhesion molecules represents a coordinated regulation and maintenance of effector T cells in the periphery.

apoptotic body by both immature and mature dendritic cells induces viral and tumor immunogens to activate MHC class I restricted CD8+ CTLs [19,124]. Interestingly, *in vivo* depletion of CD11c+ and CD8+ cells in mice fail to crossprime antigens to prime naïve T-cells, specifying its essential role in inducing T-cell activation [125]. Both dendritic cells and macrophages have been shown to present apoptotic engulfed antigens but the latter fails to activate naïve T-cells, which becomes a vital step in the activation of adaptive immunity [19,126]. In addition, Rovere et al. [124] and Ronchetti et al. [127] have demonstrated that there is a quantitative dependency on apoptotic bodies by dendritic cells in inducing the secretion of proinflammatory cytokines TNF-α and IL-1β both *in vitro* and *in vivo*, respectively. Hence, an optimum strategy to develop potent vaccines would necessitate the activation of dendritic cells and the packaging of immunogens in these apoptotic bodies for uptake. Chattergoon et al. [128] employed a novel strategy whereby immunogen constructs were coimmunized with the death cell receptor Fas. In theory, this model induces the expression of both the immunogen and Fas within the same muscle cells and apoptosis would materialize through the interaction of Fas with its ligand, possibly via T-cell assistance. When these two constructs were coimmunized, there was a significant augmentation of immune responses as measured by enhanced CTLs and Th1 cytokines including IFN-γ and IL-12. Additionally, implementing Fas as the apoptosis receptor may also provide a compounding effect on dendritic cell maturation, as Fas engagement possesses multiple functional roles. For instance, Fas not only contributes with the induction of cell death, but also stimulates the maturation of dendritic cells when engaged with its ligand [129]. This is especially crucial, as direct *in vivo* transfection of dendritic cells have been proposed as a potential mechanism responsible for the induction of immune activation with respect to DNA vaccines [20–22]. More recent work by Sasaki et al. implemented mutant caspases to decrease apoptotic efficiency to aliquot ample time for immunogen expression, while still delivering apoptosis-mediated antigens to dendritic cells [130]. This raises an interesting question as these mutant caspases decreased apoptotic efficiency *in vitro* by nearly 10-fold compared to the native caspase proteases [130]. However, it is also currently understood that high quantities of apoptotic bodies are engulfed by scavenger macrophages and induce subsequent release of anti-inflammatory factors including IL-10, TGF-beta, PGE2, and PAF [131,132]. Accordingly, inflammation is observed when the clearance of apoptotic bodies become inefficient and results in their delayed removal from the surrounding environment [133]. This results in the development of postapoptotic necrosis, which may function to deliver additional signals for proinflammatory developments and dendritic cell maturation [133,134]. However, decrease in overall apoptotic quantity and potency with mutant caspases were able to provide enhanced immune response levels when compared to the more potent apoptotic signal [130]. Therefore, it is also likely that delaying the expression of the apoptotic signal or working upstream of caspases may validate the requirement

for the balance between immunogen expression versus apoptotic stimuli [130]. These studies reveal the necessity to delay the expression of apoptotic signals to maximize immunogen expression, while maintaining the potency of the death signal. It is also important to note that necrosis, not apoptosis, is traditionally the prime cell death mechanism by which inflammation and immune activation occurs [135]. Interestingly, recent work suggests several models may be involved to induce immune activation through cell death. For instance, one model suggests that necrosis in conjunction with apoptosis delivers the maturation signal to dendritic cells [134,136]. However, the role of postapoptotic necrosis as the stimulus is ruled out or insufficiently strong to deliver a maturation signal. Another model suggests that both primary necrotic and apoptotic cells are equivalent at inducing the maturation of dendritic cells and initiating immune activation [137]. Therefore, either channel of cell death may be competent to stimulate the immune system for activation. Lastly, it is suggested that cell death or injury releases adjuvanting properties from the cytoplasm and is able to effectively stimulate immune activation [138]. Therefore, components are actively secreted from dead cells that trigger the induction of immune responses. It is also evident that apoptosis and necrosis may function together to deliver immunogens to dendritic cells while inducing potent maturation signals [136]. However, further studies are crucial to validate the precise roles that these cell death components may play in the generation of inflammatory responses induced by DNA immunization.

44.3.5 RECOMBINANT PRIME/BOOST STRATEGIES IN PRECLINICAL AND CLINICAL STUDIES

The generation of potent CTL responses through the DNA immunization has become an attractive approach targeting many pathogens including HIV-1. The inefficient generation of antibodies possessing high crossneutralizing activity against different HIV strains has limited antibody based approaches for HIV [139,140]. On the other hand, an inverse correlation can be made between HIV viral load and CTL frequency, indicating its importance in both treatment and vaccine applications [141,142]. In fact, virus-specific CTLs attribute significantly to the control of acute phase infection and denote the mechanism by which viral loads are controlled [143–145]. Furthermore, studies on SIV infection models have also indicated that viremia of viral load could be effectively controlled by CD8+ T cells [145,146]. As a result, recent applications have concentrated on maximizing the cellular arm of the immune system by generating potent CTLs to target and eliminate virally infected host cells. Previously DNA vaccines showed efficacy in inbred primate models of animals challenged with nonpathogenic AIDS viruses [147,148] and against pathogenic SHIV-89.6P [96]. DNA vaccines were less effective in outbred animals and against stringent SIV challenge [149]. Heterologous prime boost regimens have been employed to improve DNA immune potency and to address the issue of preexisting vector immunity. Studies by differing groups have reported significant enhancement of CTLs by priming with DNA and boosting with attenuated viral vectors in SIV, HIV, heptitis B, and other model systems.

Studies have reported that the order of priming and boosting is important. For instance, vaccine regimens in reverse (virus priming and DNA boosting) or merely DNA alone exhibit minimal amplification in many studies and fail to provide protective immunity against subsequent challenges [150–152]. Historically, the order of application in attenuated prime/boost strategies has also exhibited similar results, as priming with a recombinant influenza virus and boosting with recombinant vaccinia resulted in enhanced CD8+ responses and protective immunity against malaria [153]. It was proposed that perhaps the influence of the viruses to preferentially migrate CTLs to sites favored by the boosting virus may influence the overall potency of the vaccine [154]. However, another study proposed that the immunogenicity might be correlated to the immunodominance of the boosting, as the antigen-specific memory responses from the priming may provide a greater focused isolation of the antigen of interest. Consequently, boosting with recombinant viruses concentrates the primed antigens from the recombinant and enhances its amplification [152–154]. This theory correlates with other reports indicating that homologous prime/boosting with the same virus with heterologous antigens fail to proliferate antigen specific CD8+ cells [155]. Meanwhile, heterologous boosting with the modified vaccinia virus Ankara (MVA) and fowlpox viruses, while maintaining the same immunogen, provided significant enhancement of CTLs [156]. It is also likely that the enhanced assembly of memory T cells that resulted from the primed immunization may augment the boost's proliferation by providing a larger memory T-cell pool.

In addition to its efficacy, attenuated virus boosting specifically with MVA provides a resilient history of safety, as it was employed to vaccinate 120,000 humans during the smallpox eradication campaign [152,157]. Furthermore, the inability of the virus to replicate in humans while maintaining efficient viral gene expression makes it a perfect component for recombinant boosting approaches [158,159]. Cytokine studies of these viruses indicate an elevated degree of Th1 versus Th2 responses, corresponding to the potent CTLs essential for an efficacious HIV-1 DNA vaccine [160,161]. Further reports on the immunology of the MVA recombinant virus both *in vitro* and *in vivo* demonstrate that potent CTLs are generated as indicated by enhanced CD8+ populations and secretion of proinflammatory cytokines interleukin 6 (IL-6) and tumor necrosis factor alpha (TNF-α) [162–165]. Consequently, the natural immune response of the virus correlates well with essential characteristics of vaccines by compounding the host's cellular responses.

To date, many different viruses have been studied to function as recombinant boosters for DNA vaccines including the modified vaccinia Ankara, fowlpox, adenovirus, AAV as well as alphaviruses-VSV [166–171]. While most studies have reported potent CTLs and enhanced Th1-biased cytokine expression, the studies focused on are recent studies implementing the DNA/MVA boost scheme, as well as the DNA/Adeno5 model. In the rhesus macaque model, these studies induced protective responses against subsequent pathogenic SHIV89.6P challenge by preventing clinical AIDS while displaying low to no viral loads. The overall comparisons indicate that the most promising model is the Ad5 through survival of higher T-cell counts and lower viral loads [171]. Egan et al. added an SIVgag plasmid DNA priming regimen to a series of rVSV vectored booster immunizations and found both SIVgag-specific cell-mediated immune responses and humoral responses were significantly increased [172]. The viral loads post-SHIV89.6P challenge were much lower compared to macaques receiving only the rVSV vectored immunizations. Moreover, the inclusion of the plasmid DNA priming regimen also increased the preservation of peripheral blood CD4+ cells and reduce the incidence of AIDS-like disease symptoms and death associated with SHIV89.6P infection. Mattapallil et al. recently demonstrated that systemic vaccination with a DNA-prime recombinant adenovirus boost regimen preserved CD4 T cells [173]. Letvin et al. [174] immunized monkeys with a DNA-prime recombinant adenovirus boost regimen and then challenged them with pathogenic SIV and their data indicated a prolonged survival. This survival was associated with preserved central memory CD4+ T lymphocytes. Shiver et al. reported that an effective DNA vaccine prime combined with an Ad5 booster resulted in a T-cell immune response that was similar to those observed with multiple high doses of an Ad5-based vaccine [175]. Taken together, these immune correlates of vaccine efficacy should be of great help to guide the development of AIDS vaccines in humans.

Despite the potential functional breadth of the antiviral T-cell response, the immunogenicity of HIV-specific T-cell responses and candidate T-cell-based vaccine constructs is largely determined by the production of IFN-γ as mentioned to date in ELISpot assays. However, other functional activities of CD8+ T cells may also play an important role in controlling HIV replication. The recent use of polychromatic flow cytometry to measure multiple phenotypic and functional parameters of T cells stimulated with vast libraries of peptides that represent the entire HIV proteome has revealed valuable immune correlates. Research has demonstrated that the functional profile of HIV-specific T cells in HIV-1 progressors is limited compared to that of nonprogressors, who consistently maintain polyfunctional T cells (as measured by the number of cytokines produced) [176,177]. Polyfunctional T-cells are a hallmark of existing vaccines such as the effective smallpox vaccine and are also believed to be important for an effective HIV vaccine. A most recent study indicated that the DNA prime/Ad5 boost regimen could elicit more polyfunctional HIV-specific CD4+ and CD8+ T cells than either candidate alone in macaques. Therefore, recombinant prime/boost immunizations, especially DNA prime/adeno boost immunization, may provide better immune responses against HIV-1 infection [178–180].

44.4 DNA VACCINES FOR INFECTIOUS DISEASES AND CANCER

44.4.1 DNA VACCINES AGAINST HIV-1

The WHO and UNAIDS estimated that approximately 35.5 million people living with human immunodeficiency virus-1 (HIV-1), 2.9 million individuals died as a result of AIDS-related diseases and 4.3 million people were newly infected with HIV-1 in the year 2006 [181]. HIV-1 is a retrovirus, which preferentially infects and kills CD4$^+$ T cells and macrophages, ultimately resulting in immune system failure and multipathogen infections. The development of antiretroviral therapies has resulted in a decrease in morbidity and mortality associated with AIDS in many developed countries. However, the costs and the stringent administration regimen requirements of these therapies make it clear that these therapies will only be effectively utilized in a limited part of the world population. Therefore, to address the worldwide problem of HIV-1 infection, there remains a need for a prophylactic vaccination strategy designed to control the epidemic through mass immunization campaigns.

There have been about 30 ongoing or completed human clinical trials to test the safety and immunogenicity of HIV-1 DNA vaccines in humans (Table 44.2). Several important conclusions can be made by inference. One is that the primate studies do not directly translate to similar levels of immunity in humans. Overall it happens that humans have less of an overall immune response induced to DNA vaccines than do macaques. However, clinical trials using DNA prime/boost strategies, such as DNA prime/protein boost by Lu et al. at Umass and by Barnett et al. from Novartis, and DNA prime MVA boost from the Karolynskia group and DNA prime/Ad5 boost by the VRC have all reported positive effects on the resulting immune response in these prime boost strategies. Similar positive effects of DNA prime/MVA boost in humans have been reported by Hill et al. at Oxford. Furthermore, efficacy trials are being planned by the VRC using the DNA prime/Ad5 boost approach. These are very exciting developments. We will learn a great deal from these studies on the clinical efficacy of such prime/boost strategies for HIV over the next several years.

44.4.2 DNA VACCINES AGAINST OTHER VIRAL DISEASES

There is a growing list of DNA vaccines targeted against other human diseases. One of the central attractions of the technologies is its flexibility in modulating immune responses. As a consequence, several vaccine cocktails have been created that specifically target highly immunogenic antigens, which results in potent antiviral immune responses.

Influenza is a good example of the potential utility of DNA vaccines. In fact, an initial study by Ulmer et al. targeted the nucleoprotein of influenza as its antigen and effectively generated CTLs to achieve protection against subsequent challenge with a heterologous strain of influenza A virus [182]

in mice. Protection was also achieved through immunization against the hemagglutinin antigen [183]. A recent phase I trial using a DNA vaccine encoded HA delivered by gene gun technology induced protective antibody responses after three immunizations [184]. While the ability of antibodies to prevent influenza infection is well established, the antibody responses may give no protection if the challenge virus is divergent from the virus strain used to generate vaccines. In contrast, cellular immune responses are less specific and far more crossreactive [68]. Subsequent studies detailing the crossreactivity induced by these DNA vaccines, in addition to the ability of these vaccines to show protection against broad lethal virus challenges, are currently in progress.

HBV affects 350 million people worldwide and there are no commercially available therapeutic vaccines for the treatment of chronic HBV infection. Studies have demonstrated that immunization of DNA vaccines encoding HBsAg elicit rapid, stronger and long-lasting cellular and humoral immune responses and the efficacy could be significantly enhanced by coexpression of IL-2 [95]. Moreover, a DNA vaccine encoding the HBV major and middle envelope proteins has been shown to induce strong antibody responses in chimpanzees [185]. In 2006, a DNA vaccine expressing HBsAg and HBcAg received U.S. IND approval together with approval from Singapore, Hong Kong, and Taiwanese and entered Phase I Clinical Trials. Assessments of immunogenicity and possible clinical response will be made soon.

Smallpox, a devastating disease caused by the variola virus, was eradicated in 1979 by smallpox vaccination. However, with renewed concerns regarding bioterrorism, the potential threat from an outbreak of smallpox through deliberate release has caused a re-examination of the vaccine status. Dryvax, a licensed smallpox live virus vaccine, carries a high risk of complications for people with suppressed immunity and other conditions. Since a DNA vaccine would not have the adverse side effects, recently many studies have been performed in the area of developing smallpox DNA vaccines. Hooper et al. have reported that vaccination with a DNA vaccine expressing four vaccinia virus genes (L1R, A27L, A33R, and B5R) could protect rhesus macaques against lethal challenge of monkeypox virus [186]. Therefore, DNA vaccines as a promising alternative approach may contribute greatly to vaccination strategies aimed at reducing the health hazards of the conventional smallpox vaccination.

The WNV is transmitted through the bite of an infected mosquito to humans, horses, and other animals by virus-carrying mosquitoes and it induces brain inflammation and death in endemic regions. Introduction of WNV into the United States in 1999 created major human and animal health concerns. There have been more than 16,000 reported cases of WNV in humans and more than 650 deaths since 1999. In addition, more than 21,000 WNV-infected cases in horses have been reported since 1999. As there is no specific therapy for the WNV infection at this moment, there is an increasing demand for the development of vaccine strategies to prevent disease from this virus. Encouraged by an initial study that

TABLE 44.2
Clinical Trials of Preventative HIV DNA Vaccines

Phase	Official Number	Antigen (Clade)	Vector	Organizer–Sponsor	Project Site	Start Date
I	96-I-0050	env (MN), rev (HXBc2)	DNA	NIAID–NIAID	United States	3/16/1996
I	AVEG 033	gag/pol (HXB2)	DNA	NIAID–NIAID	United States	1/21/1998
I	IAVI001	gag p17 and p24 (A); 25 CTL epitopes from gag, pol, nef, env (A)	DNA	MRC/Oxford–IAVI	University of Oxford, U.K.	8/1/2000
I	01-I-0079	gag, pol (B)	DNA	VRC–NIAID	United States	1/30/2001
I	IAVI005	gag p17 and p24 (A); 25 CTL epitopes from gag, pol, nef, env (A)	DNA prime, MVA boost	MRC/Oxford–IAVI	University of Oxford, U.K.	10/1/2001
I	FIT Bioteck	Nef (B)	DNA	FIT Biotech–FIT Bioteck	Finland	7/1/2002
I	VRC 004 (03-1-0022)	gag, pol, nef (B); env (A, B, C)	DNA	VRC–NIAID	United States	11/13/2002
I	IAVI009	gag p17 and p24 (A); 25 CTL epitopes from gag, pol, nef, env (A)	DNA prime, MVA boost	Uganda Virus Research Institute–IAVI	Uganda	1/13/2003
I	HVTN 045	gag, pol, env, tat, rev, and vpu (B)	DNA	Emory Vaccine Center, USA–HVTN	United States	1/23/2003
I	HVTN 048	21 CTL epitopes for Gag, Pol, Env, Nef, Rev, and Vpr, from all of the major subtypes and CRF	DNA	HVTN–HVTN Botswana	United States	4/1/2003
I	IAVI C001	gag, env (C); pol, nef, tat (C)	DNA	Aaron Diamond AIDS Research Center–IAVI	United States	12/1/2003
I	HVTN 052	gag, pol, nef (B); env (A, B, C)	DNA	HVTN–NIAID	United States	12/1/2003
I	HVTN 044	gag, pol, nef (B); env (A, B, C)	DNA + IL-2/Ig DNA adjuvant	HVTN–NIAID	United States	12/1/2003
I	UMMS vaccine	gag (C), env (A, B, C, E)	DNA prime, protein boost	HVDDT–UMMS	United States	4/1/2004
I	VRC 008 (05-I-0148)	gag, pol, nef (B), env (A, B, C); gag, pol (B), env (A, B, C)	DNA prime, adeno boost	NIAID–NIAID	United States	4/1/2004
I	VRC 007 (04-I-0254)	gag, pol, nef (B), env (A, B, C)	DNA	NIAID–NIAID	United States	8/1/2004
I	HVTN 049	gag and env DNA/PLG (B); Oligomeric gp 140 (B)	DNA/PLG prime, protein boost	NIAID–NIAID, Chiron	United States	1/1/2005
I	EnvDNA	env (A, B, C, D, E)	DNA	St. Jude, NIAID–St. Jude, NIAID	United States	2/1/2005
I	HIVIS 01	env (A, B, C), gag (A,B), RT (B), rev (B)	DNA	Karolinska Institute, Karolinska Institute and SIIDC, Vecura	Sweden	2/1/2005
I	EuroVac02	env, gag, pol, nef (C); env, gag, pol, nef (C)	DNA prime, NYVAC boost	European Vaccine Effort–European Vaccina Effect	Switzerland, U.K.	2/1/2005
I	HVTN 060	gag (B); gag (B) or env, gag, nef (B) multirepitope peptide	DNA+/ŠIL-12 prime, DNA or peptide boost	NIAID, Wyeth–Wyeth	United States	8/1/2005
I	HVTN 063	gag (B); gag (B) or env, gag, nef (B) multiepitope peptide	DNA+/ŠIL-15 prime, DNA +/ ŠIL-15, DNA +/ŠIL-12 or peptide boost	NIAID, Wyeth–Wyeth	United States, Brazil	9/1/2005
I	HVTN 064	gag, pol, vpr, nef (B); protein containing T-helper epitopes from env, gag, pol, vpu (B)	DNA, protein, or DNA+protein	NIAID, Pharmexa–Epimmune–Pharmexa-Epimmune	United States, Peru	1/1/2006
I/II	N/A	gag, RT, rev, tat, vpu, and env (B)	DNA prime, rFPV boost	National Centre in HIV Epidemiology and Clinical Research–Univ. New South Wales	Australia	6/1/2003
I/II	C060301	nef, rev, tat, gag, pol, env, CTL epitopes (B)	DNA	FIT Biotech-FIT Biotech, IAVI	Finland	2/1/2004
I/II	HIVIS 03	env (E), gag (A), pol (E)	DNA prime, MVA boost	Karolinska Institute–Karolinsa Institute, Vecura, USMHRP	Tanzania	1/1/2006

showed a DNA vaccine expressed the WN virus prM and E proteins could induce protective immunity [187], CDC and Wyeth Fort Dodge Animal Health Unit established a research and development agreement in 2001 to test a DNA vaccine and make it commercially available. On July 8, 2005, they were able to successfully develop the world's first licensed DNA vaccine that could protect horses from WNV with no adverse or major side effects. This is an exciting scientific breakthrough that has potential benefits far beyond preventing WNV infection in horses.

Almost at the same time, a DNA vaccine against infectious hematopoietic necrosis virus (IHNV) was licensed in Canada. IHNV, the causative agent of IHN, is endemic to the western coast of North America. The viral infection causes the deaths of large numbers of commercially raised salmon and trout and is an economic problem for commercial fisheries. The only method of reducing the risk IHNV poses to the Atlantic salmon farming industry is through the use of an efficacious vaccine. As of 1999, a DNA vaccine encoding the surface glycoprotein G of IHNV was being tested and the results indicated that immunization of this vaccine could cause protective immunity [188–191]. Thereafter, more than six million salmon were inoculated while this vaccine was evaluated by Canadian regulators. Tests showed the vaccine, called Apex-IHN, protected the fish without adverse effects. In July 2005, Canada licensed the DNA vaccine product for sale. Taken together, these two veterinarian DNA-based vaccines have become the first to make it from the laboratory into commercial use, buoying hopes for similar vaccines for human diseases, such as Ebola, HIV/AIDS, and SARS.

44.4.3 DNA Vaccines for Cancer

Although advances in science have led to countless theories and methods designed to combat human cancers, the battle is far from over. Surgical excision of tumors, drug therapies, and chemotherapy have been effective in certain cases but in other situations, particularly when the tumor has begun to metastasize, effective treatment is far more difficult and far less potent. Thus researchers are continually investigating novel and more effective treatment strategies for various forms of cancer. The concept of vaccinating against cancer has been around for many decades starting with nonspecific immunostimulatory approaches. The recent identification and characterization of genes encoding for tumor-associated antigens has made it possible to design antigen-specific cancer vaccines based on plasmid DNA and recombinant viral vectors [192].

Since the first human tumor antigen (MAGE-1) recognized by CD8+ T cells was identified, numerous tumor antigens have been confirmed as CD8+ T-cell targets. Studies also provided direct evidence that CD8+ T cells play a key role in tumor rejection in humans [193,194]. The capacity of DNA vaccines to induce CD8+ T-cell responses has been demonstrated in many preclinical and clinical studies in infectious diseases. However, it is important to note that immune responses to microbial antigens in healthy individuals are not comparable to immune responses to cancer antigens in cancer patients. Little is known as to how the host-antigenic tolerance to tumor antigens in patients with active cancers interferes with immune responses induced by DNA vaccines. Development of cancer DNA vaccines has to face several generic problems, such as how to break immune tolerance and how to avoid the negative impact of regulatory T cells on the induction of antitumor responses etc.

44.4.3.1 DNA Vaccines against Prostate Cancer

Prostate cancer is the most common form of cancer and the second most common cause of cancer-related death in American men. The appearance of prostate cancer is much more common in men over the age of 50 [195]. Three of the most widely used treatments are surgical excision of the prostate and seminal vesicles, external bean irradiation, and androgen deprivation. However, conventional therapies lose their efficacy once the tumor has metastasized, which is the case in more than half of initial diagnoses [196,197].

PSA is a serine protease and a human glandular kallikrein gene product of 240 amino acids, which is secreted by both normal and transformed epithelial cells of the prostate gland [198,199]. Because cancer cells secrete much higher levels of the antigen, PSA level is a particularly reliable and effective diagnostic indicator of the presence of prostate cancer [200]. PSA is also found in normal prostate epithelial tissue and its expression is highly specific [201].

The immune responses induced by a DNA vaccine encoding for human PSA has been investigated in a murine model [201]. The vaccine construct was constructed by cloning a gene for PSA into expression vectors under control of a CMV promoter. Following the injection of the PSA DNA construct (pCPSA), various assays were performed to measure both the humoral and cellular immune responses of the mice. PSA-specific immune responses induced in vivo by immunization were characterized by enzyme-linked immunosorbent assay (ELISA), T helper proliferation CTL, and flow cytometry assays. Strong and persistent antibody responses were observed against PSA for at least 180 days following immunization. In addition, a significant T helper cell proliferation was observed against PSA protein. Immunization with pCPSA also induced MHC Class I CD8+ T-cell-restricted cytotoxic T lymphocyte response against tumor cell targets expressing PSA. The induction of PSA-specific humoral and cellular immune responses following injection with pCPSA was also observed in rhesus macaques [202].

44.4.3.2 DNA Vaccines against Colon Cancer

Human carcinoembryonic antigen (CEA) is a 180 kDa glycoprotein expressed in elevated levels in 90% of gastrointestinal malignancies, including colon, rectal, stomach, and pancreatic tumors, 70% of lung cancers, and 50% of breast cancers [203,204]. CEA is also found in human fetal digestive organ tissue, hence the name carcinoembryonic antigen [205]. It has been discovered that CEA is expressed in normal adult colon epithelium as well, albeit at far lower levels [206,207].

Sequencing of CEA shows that it is associated with the human immunoglobulin gene superfamily and that it may be involved in the metastasizing of tumor cells [205].

The immune response to nucleic acid vaccination using a CEA DNA construct was characterized in a murine model. The CEA insert was cloned into a vector containing the CMV early promoter/enhancer and injected intramuscularly. CEA-specific humoral and cellular responses were detected in the immunized mice. These responses were comparable to the immune response generated by rV-CEA [206]. The CEA DNA vaccine was also characterized in a canine model, where sera obtained from dogs injected intramuscularly with the construct demonstrated an increase in antibody levels [208]. Cellular immune responses quantified using the lymphoblast transformation (LBT) assay also revealed proliferation of CEA-specific lymphocytes. Therefore a CEA DNA vaccine was able to induce both arms of the immune responses [208].

Although numerous cancer DNA vaccine candidates have been reported, only a few have been evaluated in clinical trials. Conry et al. [209] tested the safety and immunogenicity of a plasmid DNA vaccine encoding the CEA antigen and the HBV surface antigen (HbsAg) in a bicistronic vector. The results showed that this DNA vaccine induced HbsAg-specific antibodies but CEA-specific antibodies were not able to be detected. Four out of 17 patients developed lymphoproliferative responses against CEA after vaccination. No objective clinical responses were observed. Antigen-specific cellular immune responses were not observed in another clinical study testing a DNA vaccine expressing melanoma antigen (gp100) delivered either intradermally or intramuscularly [210]. Take together, cancer DNA vaccines are safe and could induce modest immunogenicity in cancer patients. More studies are needed to improve immunogenicity of cancer DNA vaccines.

44.4.3.3 DNA Vaccines against B-Cell Lymphoma

Several studies have shown that DNA immunization encoding the idiotypic determinants of a B cell lymphoma generates tumor-specific immunity. Although DNA vaccines containing single chain Fv sequences alone induced low levels of anti-idiotypic antibody responses, an enhanced antibody responses were able to be induced by fusing the gene for fragment C of tetanus toxin to the C terminus of human scFv [211]. Timmerman et al. performed a phase I/II clinical trial to study the safety and immunogenicity of DNA vaccines encoding chimeric idiotype in patients with follicular B-cell lymphoma [212]. The data showed that seven of 12 patients mounted both cellular and humoral immune responses. These results will help guide the design of such future DNA vaccine trials.

44.4.3.4 DNA Vaccines against Cervical Cancer

Cancers associated with the HPV pose a significant international health problem. Cervical cancer is the second most common cancer among women worldwide, but it is the most common cancer in developing countries [213]. Moreover, HPV-associated head and neck cancer affects as many as 115,000 people each year globally [214,215]. The development of these cancers is closely associated with infection by high-risk types of HPV [216]. Studies have shown that HPV-16 is the predominant subtype in both cancers, accounting for 46%–63% of cervical cancer cases and 90% of head and neck cancer cases [215,217].

There are two major families of vaccine that can be considered in the fight against HPV-related cancer. Prophylactic vaccines are aimed at inducing natural immunity against HPV infection in naïve individuals. Anti-L1 and L2-directed vaccines are efficient in this role. Preventive vaccines developed to date [218] are constructed of VLPs or even pentameric L1 capsomerases. When used as vaccines, they generate strong serum-neutralizing antibodies. HPV prophylactic vaccines have shown significant clinical efficacy. They can prevent about 70%–80% of all cervical cancers worldwide based on strains used in the vaccine [219]. However, prophylactic HPV vaccines would not be effective in controlling existing HPV infections or HPV-associated lesions. Therefore, an important need is to develop therapeutic HPV vaccines focused on women with active disease and aimed at eliminating or controlling existing infection or disease progression through the induction of a strong cell-mediated response. There has been a lot of work in this area.

Two HPV gene products, E6 and E7, are expressed in most and perhaps all HPV-related pre- and cancerous cells. These genes are responsible for the transformation of cells and are required for the maintenance of HPV associated malignancies, thus making them ideal immunotherapeutic targets [220,221]. Heat shock proteins have been found to induce tumor immunogenicity and their levels of expression is enhanced by highly immunogenic necrotic bodies [222–224]. In view of this, the HSP70 of mycobacterium TB was fused to the human HPV-16 E7 antigen to construct a chimeric DNA vaccine. The E7-HSP70 DNA vaccine induced significantly enhanced levels of cellular responses including a ratio of 435:14 (E7-HSP70 to E7) of E7 specific IFN-γ spot-forming CD8$^+$ T cells via ELISPOT assays [225]. In a similar fashion, another member of the HSP family that has been shown to augment the potency of DNA vaccines is calreticulin [112]. Accordingly, when calreticulin was fused to HPV-16 E7 antigen as a DNA vaccine, a potent, antitumor effect was provoked. The resulting response was attributed to both the enhanced immunogenicity against E7 and the generation of antiangiogenesis [226]. Although therapeutic vaccines that target either HPV-16 E6 or HPV-16 E7 have been previously described [227,228], few vaccines that target both oncoproteins have been developed and demonstrated to have effective antitumor activity. Research suggests that the combination of E6 and E7 DNA vaccines may have better antitumor potential than alone [229]. Yan et al. developed an engineered consensus-based HPV-16 DNA vaccine by using a multiphase design strategy [230]. When studied as a DNA vaccine, compared to a primary HPV-16 E7 DNA vaccine, this synthetic construct elicited strong cellular immune responses in both C57BL/6 and the E6/E7 transgenic mice. Prophylactic administration of this vaccine results in 100% protection against HPV E6 and E7-expressing tumors in both murine models. Therapeutic studies indicate that vaccination with this DNA vaccine induces tumor regression. Moreover, coadministration of IL-15 with this vaccine enhances its efficacy against established tumors.

Most animal models for HPV vaccine studies are based on analysis of fast-growing transplantable tumors; however, malignancies in cervical cancer patients in general develop much more slowly, which may lead to immune suppression or immune tolerance [231]. Therefore, whether HPV-specific vaccines are potent enough to overcome the tolerance becomes an important concern. In this regard, research has demonstrated that vaccination of HPV16 E7 protein-based vaccine reduced the growth of pre-established TC-1 tumors although there was no measurable CTL response detected [232]. Previous studies have also indicated that recombinant Semliki forest virus (SFV) expressing a fusion E6/E7 protein had the capacity to induce HPV16 E7-specific cytotoxic T cells in HPV-transgenic mice [231]. Furthermore, Souders et al. found that *Listeria*-based vaccines can partially break tolerance by expanding low avidity CD8+ T cells capable of causing tumor regression in a E6/E7 transgenic mouse model [233]. In contrast, several recent studies showed that immunization with HPV16 E6/E7 DNA vaccines was not able to induce HPV-specific CTL activity in E6/E7 transgenic mice and overcome E6- or E7-specific tolerance [231,234]. Yan et al. reported for the first time that a HPV E6/E7 DNA vaccine could have the ability to induce HPV-specific CTL activity and partially overcome immune tolerance in the E6/E7 transgenic mouse model system [230]. Building on these results will be important. Taken together, these data suggest further study of such DNA immunogens in the eventual context of immune therapy for select HPV-associated cancers are wanted.

44.4.3.5 DNA Vaccines against Melanoma

Canine malignant melanoma (CMM) is an aggressive form of cancer that typically appears in a dog's mouth, but also may appear in the nail bed, foot-pad, or other areas. CMM is the most common tumor in the dog, representing about 4% of all canine tumors [235,236]. To date, the most common treatments for CMM and advanced human melanoma (HM) have been radiation and surgery. Also, CMM and HM are chemo-resistant neoplasm [235,237]. Based on these similarities, CMM is a good clinical model for evaluating new treatments for advanced HM.

While controversial, some studies have indicated that immunization of mice with plasmids encoding cancer differentiation antigens is not effective when self-DNA is used, but effective tumor immunity can be induced by orthologous DNA from another species [238]. Wang et al. demonstrated that the xenogeneic human DNA encoding tyrosinase family proteins induced both cellular and humoral immune responses against syngeneic B16 melanoma cells in C57BL/6 mice [239]. Xenogenic DNA immunization protected tumor growth from syngeneic melanoma challenge and autoimmune hypopigmentation [240,241]. A phase I veterinary clinical trial conducted by Bergman and colleagues showed that xenogeneic human tyrosinase DNA vaccination of dogs was safe and potentially

efficacious. Only minimal local toxicity and a lack of systemic toxicity was detected and the clinical antitumor responses and remarkably prolonged median survival time was also observed after vaccination with this DNA vaccine. Importantly, the USDA has issued a conditional U.S. Veterinary Biological Product License for this therapeutic vaccine in May 2007. This conditional license is a response to an application and assurance of safety and purity, and a reasonable expectation of efficacy based on initial trials performed at MSKCC and AMC. This DNA vaccine will be used to improve the health and well-being of dogs, and this is the first time that the U.S. government has approved a therapeutic vaccine for the treatment of cancer in either animals or humans. Since humans and dogs develop this cancer in a similar manner, the use of this vaccine in dogs may result in improved cancer treatment in humans.

44.4.4 DNA Vaccines for Autoimmune Diseases

A breakdown in tolerance to self-antigens is often considered to result in autoimmune disease. However, the similar immunological reactions that control immune responses to foreign antigens are also involved during the progress of autoimmune disease. Th1 cells have been shown to be involved in many autoimmune diseases while Th2 cells may inhibit disease development [242,243]. Data also indicated that skin delivery of DNA by gene gun DNA immunization is associated with a rapid induction of Th2 responses. Therefore, the induction of Th2 responses by DNA vaccines may be one method whereby autoimmune responses can be inhibited. For example, many studies have reported the use of DNA vaccination as a means of preventing the autoimmune disease experimental allergic encephalomyelitis (EAE). Ramshaw et al. demonstrated that DNA immunization could induce MBP-specific antibody immune responses in mice [244]. Data also showed that the vaccination of a DNA vaccine encoding an immunodominant MBP peptide targeted to Fc of immunoglobulin G suppressed clinical and histopathological signs of EAE [245]. Taken together, DNA vaccines may find a place in the treatment of autoimmune diseases. However, this is an early area of investigation and more study is needed to define the themes that are most likely to bring successful interventions to this area.

44.5 CONCLUSIONS

DNA immunization holds great promise for providing safe and effective vaccines for many infectious diseases and cancers. The direct injection of foreign genes by genetic immunization has resulted in specific immune responses that exhibit characteristics of protective immunity against a number of infectious agents in a variety of animal models. For example, genetic vaccination cassettes targeting each of HIV-1's three major genes (*env*, *gag*, and *pol*), regulatory genes, and accessory genes have been developed and studied in small animals, primates, and humans. DNA vaccine constructs for cancers targeting tumor-specific antigens have also been studied in a variety of animal models. Developing successful vaccines for HIV-1 or cancers will likely involve targeting multiple antigenic components to direct and empower the immune system. Such a collection of immunization cassettes should be capable of stimulating broad immunity against both humoral and cellular epitopes, thus giving a vaccine the maximum ability to deal with viral immune escape or tumor growth. There is still considerable room for advancing the current technology. The combination of genetic modifications, including codon and RNA optimization, and addition of high efficient leader sequence and promotor have shown great impact on improvement of DNA vaccine immunogenicity. The "centralized" sequences provide a novel method to maximum the crossreactivity induced by DNA vaccines. In addition, the potential of the molecular adjuvant coadministration to dramatically enhance and regulate the antigen-specific humoral and cellular immune responses induced by DNA immunogens represents important new avenue for vaccine and immune therapeutic exploration. DNA vaccines can be combined with other vaccines including recombinant protein, poxvirus, adenovirus, as well as others to further enhance initial immune responses. In addition, improved transfection efficiency and delivery methods such as the gene gun and EP are also central areas in the continued development of the field. It is clear that there has been steady progress and recently there has been real excitement realized by the approvals in the veterinary area and the generation of immune responses with some potency in humans. It is likely that through a combination of technology enhancements and incremental advances, DNA vaccine technology will become a central strategy for prevention of infectious diseases and treatment of human cancers over the next 10 years. We are looking forward to seeing this important change in the field.

REFERENCES

1. Plotkin, S. A. and Mortimer, E. A. *Vaccines*. Philadelphia: W.B. Saunders, 1988.
2. Janeway, C. A. and Travers, P. *Immunobiology*. London: Current Biology Ltd/Garland, 1994.
3. Rocha, B. and Tanchot, C. Towards a cellular definition of CD8+ T-cell memory: The role of CD4+ T-cell help in CD8+ T-cell responses. *Curr. Opin. Immunol.* 2004;16:259–263.
4. Wang, B., Ugen, K. E., Srikantan, V., et al. Gene inoculation generates immune responses against human immunodeficiency virus type 1. *Proc. Natl. Acad. Sci. USA* 1993;90:4156–4160.
5. Stasney, J., Cantarow, A., and Paschkis, K. E. Production of neoplasms by injection of fractions of mammalian neoplasms. *Cancer Res.* 1955;11:775–782.
6. Paschkis, K. E., Cantarow, A., and Stasney, J. Induction of neoplasms by injection of tumor chromatin. *J. Natl. Cancer Inst.* 1955;15:1525–1532.
7. Ito, Y. A tumor-producing factor extracted by phenol from papillomatous tissue of cottontail rabbits. *Virology* 1960;12:596–601.
8. Wolff, J. A., Malone, R. W., Williams, P., et al. Direct gene transfer into mouse muscle in vivo. *Science* 1990;247.
9. Tang, D., DeVit, M., and Johnston, S. Genetic immunization is a simple method for eliciting an immune response. *Nature* 1992;356:152–154.
10. Chattergoon, M., Boyer, J., and Weiner, D. B. Genetic immunization: A new era in vaccines and immune therapies. *FASEB J.* 1997;11:753–763.
11. Kim, J. J. and Weiner, D. B. DNA/genetic vaccination for HIV. *Springer Sem. Immunopathol.* 1997;19:174–195.
12. Donnelly, J. J., Ulmer, J. B., Shiver, J. W., and Liu, M. A. DNA vaccines. *Annu. Rev. Immunol.* 1997;15:617–648.
13. Doe, B., Selby, M., Barnett, S., Baenziger, J., and Walker, C. M. Induction of cytotoxic T lymphocytes by intramuscular immunization with plasmid DNA is facilitated by bone-marrow-derived cells. *Proc. Natl. Acad. Sci. USA* 1996;93:8578–8583.
14. Corr, M., Lee, D. J., Carson, D. A., and Tighe, H. Gene vaccination with naked plasmid DNA: Mechanism of CTL priming. *J. Exp. Med.* 1996;184:1555–1560.
15. Pardoll, D. M. and Beckerleg, A. M. Exposing the immunology of naked DNA vaccines. *Immunity* 1995;3:165–169.
16. Brodsky, F. M. and Guagliardi, L. E. The cell biology of antigen processing and presentation. *Annu. Rev. Immunol.* 1991;9:707–744.
17. Goebels, N., Michaelis, D., Wekerle, M., and Hohlfeld, R. Human myoblasts as antigen-presenting cells. *J. Immunol.* 1992;149:661–667.
18. Behrens, L. M. K., Misgeld, T., Goebels, N., Wekerle, M., and Hohlfeld, R. Human muscle cells express a functional costimulatory molecule distinct from B7.1 (CD80) and B7.2 (CD86) in vivo and in inflammatory lesions. *J. Immunol.* 1998; 161:5943–5951.
19. Albert, M. L., Saulter, B., and Bhardwaj, N. Dendritic cells acquire antigen from apoptotic cells and induce class I-restricted CTLs. *Nature* 1998;392:86–89.
20. Condon, C., Watkins, S. C., Celluzzi, C. M., Thompson, K., and Falo, L. D. DNA-based immunization by in vivo transfection of dendritic cells. *Nat. Med.* 1996;10:1122–1128.
21. Casares, S., Inaba, K., Brumeanu, T. D., Steinman, R. M., and Bona, C. A. Antigen presentation by dendritic cells after immunization with DNA encoding a major histocompatibility complex class II-restricted viral epitope. *J. Exp. Med.* 1997; 186:1481–1486.
22. Chattergoon, M., Robinson, T. A., Boyer, J., and Weiner, D. B. Specific immune induction following DNA-based immunization through in vivo transfection and activation of macrophages. *J. Immunol.* 1998;160:5707–5718.
23. Martin, T., Parker, S. E., Hedstrom, R., et al. Plasmid DNA malaria vaccine: The potential for genomic integration after intramuscular injection. *Hum. Gene Ther.* 1999;10:759–768.
24. Nichols, W. W., Ledwith, B. J., Manam, S. V., and Triolo, P. J. Potential DNA vaccine integration into host cell genome. *Ann. N. Y. Acad. Sci.* 1995;772:30–39.
25. Wang, Z., Troilo, P. J., Wang, X., et al. Detection of integration of plasmid DNA into host genomic DNA following intramuscular injection and eletroporation. *Gene Ther.* 2004;11:711–721.
26. Walker, C. M., Scharton-Kersten, T., Rowton, E. D., et al. Genetic immunization with glycoprotein 63 cDNA results in a helper T cell type 1 immune response and protection in a murine model of leishmaniasis. *Hum. Gene Ther.* 1998;9:1899–1907.
27. Luke, C. J., Carner, K., Liang, X., and Barbour, A. G. An OspA-based DNA vaccine protects mice against infection with Borrelia burgdorferi. *J. Infect. Dis.* 1997;175:91–97.
28. Huygen, K., Content, J., Denis, O., et al. Immunogenicity adn protective efficacy of a tuberculosis DNA vaccine. *Nat. Med.* 1996;2.
29. Gonzalez, J. C., Morello, C. S., Cranmer, L. D., and Spector, D. H. DNA immunization confers protection against murine cytomegalovirus infection. *J. Virol.* 1996;70:7921–7928.
30. Kypr, J. and Mrazek, J. Unusual codon usage of HIV. *Nature* 1987;327:20–26.
31. Stephens, C. R. and Waelbroeck, H. Codon bias and mutability in HIV sequences. *J. Mol. Evol.* 1999;48:390–397.

32. Zur megede, J., Chen, M. C., Doe, B., et al. Increased expression and immunogenicity of sequence-modified human immunodeficiency virus type 1 gag gene. *J. Virol.* 2000;74:2628–2635.

33. Muthumani, K., Kudchodkar, S., Zhang, D., et al. Issues for improving multiplasmid DNA vaccines for HIV-1. *Vaccine* 2002;20:1999–2003.

34. Wang, S., Farfan-Arribas, D. J., Shen, S., et al. Relative contributions of codon usage, promoter efficiency and leader sequence to the antigen expression and immunogenicity of HIV-1 Env DNA vaccine. *Vaccine* 2006;24:4531–4540.

35. Uchijima, M., Yoshida, A., Nagata, T., and Koide, Y. Optimization of codon usage of plasmid DNA is required for the effective MHC class I-restricted T cell responses against intracellular bacterium. *J. Immunol.* 1998;161.

36. Deml, L., Bojak, A., Steck, S., et al. Multiple effects of codon usage optimization on expression and immunogenicity of DNA candidate vaccines encoding the human immunodeficiency virus type 1 Gag protein. *J. Virol.* 2001;75:10991–11001.

37. Schneider, R., Campbell, M., Nasioulas, G., Felber, B. K., and Pavlakis, G. N. Inactivation of the human immunodeficiency virus type 1 inhibitory elements allows Rev-independent expression of Gag and Gag/protease adn particle formation. *J. Virol.* 1997;71:4892–4903.

38. Muthumani, K. and Weiner, D. B. Novel engineered HIV-1 East African Clade A gp160 plamid constrct induces strong humoral and cell-mediated immune responses in vivo. *Virology* 2003;314:134–146.

39. Chapman, B. S., Thayer, R. M., Vincent, K. A., and Haigwood, N. L. Effect of intron A from human cytomegalovirus (Towne) immediate-early gene on heterologous expression in mammalian cells. *Nucleic Acids Res.* 1991;19:3979–3986.

40. Xu, Z. L., Mizuguchi, H., Ishii-Watabe, A., et al. Optimization of transcriptional regulatory elements for constructing plasmid vectors. *Gene* 2001;272:149–156.

41. Yang, J. S., Muthumani, K., Ayyavoo, V., Boyer, J., and Weiner, D. B. Induction of potent Th1-type immune responses from a novel DNA vaccine for West Nile virus New York isolate (WNV-NY1999). *J. Infect. Dis.* 2001;184:809–816.

42. Li, Y., Luo, L., Thomas, D. Y., and Kang, C. Y. The HIV-1 Env protein signal sequence retards its cleavage and down-regulates the glycoprotein folding. *Virology* 2000;272:417–428.

43. Laddy, D. J. and Weiner, D. B. From plasmids to protection: A review of DNA vaccines against infectious diseases. *Int. Rev. Immunol.* 2006;25:99–123.

44. Ross, R., Sudowe, S., Beisner, J., et al. Transcriptional targeting of dendritic cells for gene therapy using the promoter of the cytoskeletal protein fascin. *Gene Ther.* 2003;10:1035–1040.

45. Bojak, A., Hammer, D., Wolf, H., and Wagner, R. Muscle specific versus ubiquitous expression of Gag based HIV-1 DNA vaccines: A comparative analysis. *Vaccine* 2002;20:1975–1979.

46. Lin, M. T., Wang, F., Uitto, J., and Yoon, K. Differential expression of tissue-specific promoters by gene gun. *Br. J. Dermatol.* 2001;144:34–39.

47. Lee, A. H., Suh, Y. S., Sung, J. H., Yang, S. H., and Sung, Y. C. Comparison of various expression plasmids for the induction of immune responses by DNA immunization. *Mol. Cells* 1997; 7:495–501.

48. Galvin, T. A., Muller, J., and Khan, A. S. Effect of different promoters on immune responses elicited by HIV-1 gag/env multigene DNA vaccine in *Macaca mulatta* and *Macaca nemestrina*. *Vaccine* 2000;18:2566–2583.

49. Cheng, L., Ziegelhoffer, P. R., and Yang, N. S. In vivo promoter activity and transgene expression in mammalian somatic tissues evaluated by using particle bombardment. *Proc. Natl. Acad. Sci. USA* 1993;90:4455–4459.

50. Vanniasinkam, T., Reddy, S. T., and Ertl, H. C. DNA immunization using a non-viral promoter. *Virology* 2006;344:412–420.

51. Garg, S., Oran, A. E., Hon, H., and Jacob, J. The hybrid cytomegalovirus enhancer/chicken beta-actin promoter along with woodchuck hepatitis virus posttranscriptional regulatory element enhances the protective efficacy of DNA vaccines. *J. Immunol.* 2004;173:550–558.

52. Tudor, D., Dubuquoy, C., Gaboriau, V., et al. TLR9 pathway is involved in adjuvant effects of plasmid DNA-based vaccines. *Vaccine* 2005;23:1258–1264.

53. Hemmi, H., Takeuchi, O., Kawai, T., et al. A Toll-like receptor recognizes bacterial DNA. *Nature* 2000;408:740–745.

54. Wodrich, H., Schambach, A., and Krausslich, H. G. Multiple copies of the Mason-Pfizer monkey virus constitutive RNA transport element lead to enhanced HIV-1 Gag expression in a context-dependent manner. *Nucleic Acids Res.* 2000;28:901–910.

55. Yan, J., Yoon, H., Kumur, S., et al. Enhanced cellular immune responses elicited by an engineered HIV-1 subtype B consensus-based envelope DNA vaccine. *Mol. Ther.* 2007;15:411–421.

56. Mansky, L. M. and Temin, H. M. Lower in vivo mutation rate of human immunodeficiency virus type 1 than that predicted from the fidelity of purified reverse transcriptase. *J. Virol.* 1995; 69:5087–5094.

57. Marin, M., Rose, K. M., Kozak, S. L., and Kabat, D. HIV-1 Vif protein binds the editing enzyme APOBEC3G and induces its degradation. *Nat. Med.* 2003;9:1398–1403.

58. Sheehy, A. M., Gaddis, N. C., and Malim, M. H. The antiretroviral enzyme APOBEC3G is degraded by the proteasome in response to HIV-1 Vif. *Nat. Med.* 2003;9.

59. Gurtler, L. G., Hauser, P. H., Eberle, J., et al. A new subtype of human immunodeficiency virus type 1 (MVP-5180) from Cameroon. *J. Virol.* 1994;68:1581–1585.

60. Simon, F., Mauclere, P., Roques, P., et al. Identification of a new human immunodeficiency virus type 1 distinct from group M and group O. *Nat. Med.* 1998;4:1032–1037.

61. Vanden Haesevelde, M., Decourt, J. L., De Leys, R. J., et al. Genomic cloning and complete sequence analysis of a highly divergent African human immunodeficiency virus isolate. *J. Virol.* 1994,68:1586–1596.

62. Leitner, T., Foley, B., Hahn, B. H., et al. *HIV Sequence Compendium 2003.* Los Alamos National Laboratory, Los Alamos, New Mexico; U.S.A., 2004.

63. Weaver, E. A., Lu, Z., Camacho, Z. T., et al. Cross-subtype T-cell immune responses induced by a human immunodeficiency virus type 1 group M consensus Env immunogen. *J. Virol.* 2006;80:6745–6756.

64. Robertson, D. L., Anderson, J. P., Bradac, J. A., et al. HIV-1 nomenclature proposal. *Science* 2000;288:55–56.

65. Novitsky, V., Smith, U. R., Gilbert, P., et al. Human immunodeficiency virus type 1 subtype C molecular phylogeny: Consensus sequence for an AIDS vaccine design? *J. Virol.* 2002; 76:5435–5451.

66. Gao, F., Weaver, E. A., Lu, Z., et al. Antigenicity and immunogenicity of a synthetic human immunodeficiency virus type 1 group M consensus envelope glycoprotein. *J. Virol.* 2005; 79:1154–1163.

67. Gaschen, B., Taylor, J., Yusim, K., et al. Diversity considerations in HIV-1 vaccine selection. *Science* 2002;296: 2354–2360.

68. Laddy, D. J., Yan, J., Corbitt, N., et al. Immunogenicity of novel consensus-based DNA vaccines against avian influenza. *Vaccine* 2007;25:2984–2989.

69. Gao, F., Korber, B., Weaver, E. A., et al. Centralized immunogens as a vaccine strategy to overcome HIV-1 diversity. *Expert Rev. Vaccines* 2004;3:S161–S168.

70. Mullins, J. I., Nickle, D. C., Heath, L., Rodrigo, A. G., and Learn, G. H. Immungen sequence: The fourth tier of AIDS vaccine design. *Expert Rev. Vaccines* 2004;3:S151–S159.

71. Nickle, D. C., Jensen, M. A., Gottlieb, G. S., et al. Consensus and ancestral state HIV vaccines. *Science* 2003;299:1515–1518.

72. Doria-Rose, N. A., Learn, G. H., Rodrigo, A. G., et al. Human immunodeficiency virus type 1 subtype B ancestral envelope protein is functional and elicits neutralizing antibodies in rabbits similar to those elicited by a circulating subtype B envelope. *J. Virol.* 2005;79:11214–11224.

73. Rolland, M., Jensen, M. A., Nickle, D. C., et al. Reconstruction and function of ancestral central-of-tree (COT) HIV-1 proteins. *J. Virol.* 2007;81:8507–8514.

74. Fischer, W., Perkins, S., Theiler, J., et al. Polyvalent vaccines for optimal coverage of potential T-cell epitopes in global HIV-1 variants. *Nat. Med.* 2007;13:100–106.

75. Gehl, J. Electroporation: Theory and methods, perspectives for drug delivery, gene therapy and research. *Acta Physiol. Scand.* 2003;177:437–447.

76. Aihara, H. and Miyazaki, J. Gene transfer into muscle by electroporation in vivo. *Nat. Biotechnol.* 1998;16:867–870.

77. Mathiesen, I. Electropermeabilization of skeletal muscle enhances gene transfer in vivo. *Gene Ther.* 1999;6:508–514.

78. Rizzuto, G., Cappelletti, M., Maione, D., et al. Efficient and regulated erythropoietin production by naked DNA injection and muscle electroporation. *Proc. Natl. Acad. Sci. USA* 1999;96:6417–6422.

79. Widera, G., Austin, M., Rabussay, D., et al. Increased DNA vaccine delivery and immunogenicity by electroporation in vivo. *J. Immunol.* 2000;164:4635–4640.

80. Prud'homme, G. J., Glinka, Y., Khan, A. S., and Draghia-Akli, R. Electroporation-enhanced nonviral gene transfer for the prevention ro treatment of immunological, endocrine and neoplastic diseases. *Curr. Gene. Ther.* 2006;6:243–273.

81. Gronevik, E., Tollefsen, S., Sikkeland, L. I., et al. DNA transfection of mononuclear cells in muscle tissue. *J. Gene Med.* 2003;5:909–917.

82. Chen, J., Fang, F., Li, X., Chang, H., and Chen, Z. Protection against influenza virus infection in BalB/C mice immunized with a single dose of neuraminidase-expressing DNAs by electroporation. *Vaccine* 2005;23:4322–4328.

83. Luckay, A., Sidhu, M. K., Kjeken, R., et al. Effect of plasmid DNA vaccine design and in vivo electroporation on the resulting vaccine-specific immune responses in Rhesus macaques. *J. Virol.* 2007;81:5257–5269.

84. Medi, B. M. and Singh, J. Skin targeted DNA vaccine delivery using electroporation in rabbits. *Int. J. Pharmaceut.* 2005; 308:61–68.

85. Fredriksen, A. B., Sandlie, I., and Bogen, B. DNA vaccines increase immunogenicity of idiotypic tumor antigen by targeting novel fusion proteins to antigen presenting cells. *Mol. Ther.* 2006;13:776–785.

86. Feltquate, D. M., Heaney, S., Webster, R. G., and Robinson, H. L. Different T helper cell types and antibody isotypes generated by saline and gene gun DNA immunization. *J. Immunol.* 1997;158:2278–2284.

87. Pertmer, T. M., Eisenbraun, M. D., McCabe, D., et al. Gene gun-based nucleic acid immunization: Elicitation of humoral and cytotoxic T lymphocyte responses following epidermal delivery of nanogram quantities of DNA. *Vaccine* 1995;13:1427–1433.

88. Kim, J. J., Ayyavoo, V., Bagarazzi, M. L., et al. In vivo engineering of a cellular immune response by co-administration of IL-12 expression vector with a DNA immunogen. *J. Immunol.* 1997;158:816–826.

89. Kim, J. J., Ayyavoo, V., Bagarazzi, M. L., et al. Development of a multi-component candidate vaccine for HIV-1. *Vaccine* 1997;15:879–883.

90. Kim, J. J., Maguire, H. C., Nottingham, L. K., et al. Coadminstration of IL-12 or IL-10 expression cassettes drives immune responses toward a Th1 phenotypes. *J. Interferon Cytokine Res.* 1998;18:537–547.

91. Dias, S., Thomas, H., and Balkwill, F. Multiple molecular and cellular changes assocaited with tumor stasis and regression during IL-12 therapy of a murine breast cancer model. *Int. J. Cancer* 1998;75:151–157.

92. Rakhmilevich, A., Turner, J., and Ford, M. Gene gun-mediated skin transfection with interleukin-12 gene results in regression of established primary and metastatic murine tumors. *Proc. Natl. Acad. Sci. USA* 1996;93:6291–6296.

93. Smith, K. A. T-cell growth factor. *Immunol. Rev.* 1980;51:337–357.

94. Trinchieri, G., Matsumoto-Kobayashi, M., Clark, S. C., et al. Responses of resting human peripheral blood natural killer cells to interleukin 2. *J. Exp. Med.* 1984;160:1147–1169.

95. Chow, Y. H., Huang, W. L., Chi, W. K., Chu, Y. D., and Tao, M. H. Improvement of hepatitis B virus DNA vaccines by plasmids coexpressing hepatitis B surface antigen and interleukin-1. *J. Virol.* 1997;71:169–178.

96. Barouch, D. H., Santra, S., Schmitz, J. E., et al. Control of viremia and prevention of clinical AIDS in rhesus monkeys by cytokine-augmented DNA vaccination. *Science* 2000;290:486–492.

97. Zhang, X., Sun, S., Hwang, I., and Tougn, D. F. Potential selective stimulation of memory-phenotype CD8+ T cells in vivo by IL-15. *Immunity* 1998;8:591–599.

98. Schluns, K. S., Williams, K., Ma, A., Zheng, X. X., and Lefrancois, L. Cutting edge: Requirement for IL-15 in the generation of primary and memory antigen-specific CD8 T cells. *J. Immunol.* 2002;168:4827–4831.

99. Kutzler, M., Robinson, T. A., Chattergoon, M., et al. Coimmunization with an optimized IL-15 plasmid results in enhanced function and longevity of CD8 T cell that are partially independent of CD4 T cell help. *J. Immunol.* 2005;175:112–123.

100. Klebanoff, C. A., Finkelstein, S. E., and Surman, D. R. IL-15 enhances the in vivo antitumor activity of tumor-reactive CD8+ T cells. *Proc. Natl. Acad. Sci. USA* 2004;101:1969–1974.

101. Kim, J. J., Nottingham, L. K., Sin, J. I., et al. CD8 positive T cells controls antigen-specific immune responses through the expression of chemokines. *J. Clin. Invest.* 1998;102:1112–1124.

102. Biragyn, A., Belyakov, I. M., Chow, Y. H., et al. DNA vaccines encoding human immunodeficiency virus type 1 glycoprotein 120 fusions with proinflammatory chemoattractants induce systemic and mucosal immune responses. *Blood* 2002;100:1153–1159.

103. Kim, J. J., Tsai, A., Nottingham, L. K., et al. Intracellular adhesion molecule-1 (ICAM-1) modulates b-chemokines and provides costimulatory signals required for T cell activation and expansion in vivo. *J. Clin. Invest.* 1999;103:869–877.

104. Lanier, L. L., O'Fallon, S., Somoza, C., et al. CD80 (B7) and CD86 (B70) provide similar costimulatory signals for T cell proliferation, cytokine production, and generation of CTL. *J. Immunol.* 1995;154:97–105.

105. Linsley, P. S., Clark, E. A., and Ledbetter, J. A. The T cell antigen, CD28, mediates adhension with B cells by interacting with activation antigen, B7/BB-1. *Proc. Natl. Acad. Sci. USA* 1990;87:5031–5035.

106. June, C., Bluestone, J. A., Nadler, L. M., and Thompson, C. B. The B7 and CD28 receptor families. *Immunol. Today* 1994;15:321–333.

107. Agadjanyan, M. G., Kim, J. J., Trivedi, N., et al. CD86 (B7-2) can function to drive MHC-restricted antigen-specific cytotoxic T lymphocyte responses in vivo. *J. Immunol.* 1999;162:3417–3427.

108. Iwasaki, A., Stiernholm, B. J., Chan, A. K., Berstein, N. L., and Barber, B. H. Enhanced CTL responses mediated by plasmid DNA immunogens encoding costimulatory molecules and cytokines. *J. Immunol.* 1997;158:4591–4601.

109. Tsuji, T., Hamajima, K., Ishii, N., et al. Immunomodulatory effects of a plasmid expressing B7-2 on human immunodeficiency virus-1-specific cell-mediated immunity induced by a plasmid encoding the viral antigen. *Eur. J. Immunol.* 1997;27:782–787.

110. Santra, S., Barouch, D. H., Jackson, S. S., et al. Functional equivalency of B7-1 and B7-2 for costimulating plasmid DNA vaccine-elicited CTL responses. *J. Immunol.* 2000;165:6791–6795.

111. Ridge, J. P., Di Rose, F., and Matzinger, P. A conditioned dendritic cell can be a temporal bridge between a CD4+ T-helper and a T-killer cell. *Nature* 1998;393:474–478.

112. Bennett, S. R., Carbone, F. R., Karamalis, F., et al. Help for cytotoxic-T-cell responses is mediated by CD40 signaling. *Nature* 1998;393:478–480.

113. Schoenberger, S. P., Toes, R. E., van der Voort, E. I., Offringa, R., and Melief, C. J. T-cell help for cytotoxic T lymphocytes is mediated by CD40–CD40L interaction. *Nature* 1998;393:480–483.

114. Yang, Y. and Wilson, J. M. CD40 ligand-dependent T cell activation: Requirement of B7-CD28 signaling through CD40. *Science* 1996;273:1862–1864.

115. Cella, M., Scheidegger, D., Palmer-Lehmann, K., et al. Ligation of CD40 on dendritic cells triggers production of high levels of interleukin-12 and enhances T cell stimulatory capacity: T–T help via APC activation. *J. Exp. Med.* 1996;184:747–752.

116. Caux, C., Massacrier, C., Vanbervliet, B., et al. Activation of human dendritic cells through CD40 cross-linking. *J. Exp. Med.* 1994;180:1263–1272.

117. Mendoza, R. B., Cantwell, M. J., and Kipps, T. J. Immunostimulatory effects of a plasmid expressing CD40 ligand (CD154) on gene immunization. *J. Immunol.* 1997;159:5777–5781.

118. Ihata, A., Watabe, S., Sasaki, S., et al. Immunomodulatory effect of a plasmid expressing CD40 ligand on DNA vaccination against human immunodeficiency virus type-1. *J. Immunol.* 1999;98:432–462.

119. Gurunathan, S., Irvine, K. R., Wu, C. Y., et al. CD40 ligand/trimer DNA enhances both humoral and cellular immune responses and induces protective immunity to infectious and tumor challenge. *J. Immunol.* 1998;161:4563–4571.

120. Sin, J. I., Kim, J. J., Zhang, D., and Weiner, D. B. Modulation of cellular responses by plasmid CD40L:CD40L plasmid vectors enhance antigen-specific helper T cell type 1 CD4+ T cell-mediated protective immunity against herpes simplex virus type 2 in vivo. *Hum. Gene Ther.* 2001;12:1091–1102.

121. Stone, G. W., Barzee, S., Snarsky, V., et al. Multimeric soluble CD40 ligand as adjuvants for human immunodeficiency virus DNA vaccines. *J. Virol.* 2006;80:1762–1772.

122. Rubartelli, A., Poggi, A., and Zocchi, M. R. The selective engulfment of apoptotic bodies by dendritic cells is mediated by the alpha(V)beta3 integrin and requires intracellular and extracellular calcium. *Eur. J. Immunol.* 1997;27:1893–1900.

123. Albert, M. L., Pearce, S. F., Francisco, L. M., et al. Immature dendritic cells phagocytose apoptotic cells via alpha(V)beta5 and CD36, and cross-present antigens to cytotoxic T lymphocytes. *J. Exp. Med.* 1998;188:1359–1368.

124. Rovere, P., Vallinoto, C., Bondanza, A., et al. Bystander apoptosis triggers dendritic cell maturation and antigen-presenting function. *J. Immunol.* 1998;161:4467–4471.

125. Jung, S., Unutmaz, D., Wong, P., et al. In vivo dendritic of CD11c(+) dendritic cells abrogates priming of CD8(+) T cells by exogenous cell-associated antigens. *Immunity* 2002;17:211–220.

126. Bellone, M., Lezzi, G., Rovere, P., et al. Processing of engulfed apoptotic bodies yield T cell epitopes. *J. Immunol.* 1997;159:5391–5399.

127. Ronchetti, A., Rovere, P., Iezzi, G., et al. Immunogenicity of apoptotic cells in vivo: role of antigen load, antigen-presenting cells, and cytokine. *J. Immunol.* 1999;163:130–136.

128. Chattergoon, M., Kim, J. J., Yang, J. S., et al. Targeted antigen delivery to antigen-presenting cells including dendritic cells by engineered Fas-mediated apoptosis. *Nat. Biotechnol.* 2000;18:974–979.

129. Rescigno, M., Piguet, V., Valzasina, B., et al. Fas engagement induces the matuation of dendritic cells (DCs), the release of interleukin (IL)-1β, and the production of interferon γ in the absence of IL-12 during DC-T cell cogmate interaction: A new roel for fas ligand in inflammatory responses. *J. Exp. Med.* 2000;192:1661–1668.

130. Sasaki, S., Amara, R. R., Oran, A. E., Smith, J. M., and Robinson, H. L. Apoptosis-mediated enhancement of DNA-raised immune responses by mutant caspases. *Nat. Biotechnol.* 2001;19:543–547.

131. Voll, R. E., Herrmann, M., Roth, E. A., Stach, C., and Kalden, J. R. Immunosuppressive effects of apoptotic cells. *Nature* 1997;390:350–351.

132. Fadok, V. A., Bratton, D. L., Konowal, A., et al. Macrophages that have ingested apoptotic cells in vitro inhibit proinflammatory cytokine production through /paracrine mechanisms involving TGF-beta, PGE2, and PAF. *J. Clin. Invest.* 1998;101:890–898.

133. Rovere, P., Sabbadini, G. M., Vallinoto, C., et al. Delayed clearance of apoptotic lymphoma cells allows cross-presentation of intracellular antigens by mature dendritic cells. *J. Leukocyte Biol.* 1999;66:345–349.

134. Sauter, B., Albert, M. L., Francisco, L. M., et al. Consequences of cell death: Exposure to necrotic tumor cells, but not primary tissue cells or apoptotic cells, induces the maturation of immunostimulatory dendritic cells. *J. Exp. Med.* 2000;191:423–433.

135. Gallucci, S., Lolkema, M., and Matzinger, P. Natural adjuvants: Endogenous activators of dendritic cells. *Nat. Med.* 1999;5:1249–1255.

136. Pietra, G., Mortarini, R., Parmiani, G., and Anichini, A. Phases of apoptosis of melanoma cells, but not of normal melanocytes, differently affect maturation of myeloid dendritic cells. *Cancer Res.* 2001;61:8218–8226.

137. Kotera, Y., Shimizu, K., and Mule, J. J. Comparative analysis of necrotic and apoptotic tumor cells as a source of antigen(s) in dendritic cell-based immunization. *Cancer Res.* 2001;61:8105–8109.

138. Shi, Y., Zheng, W., and Rock, K. L. Cell injury releases endogenous adjuvants that stimulate cytotoxic T cell responses. *Proc. Natl. Acad. Sci. USA* 2000;97:14590–14595.

139. Robinson, H. L. DNA vaccine for immunodeficiency virus. *AIDS* 1997;11:S109–S115.

140. Letvin, N. L. Progress in the development of an HIV-1 vaccine. *Science* 1998;280:1875–1880.

141. Ogg, G. S., Xia, J., Bonhoeffer, S., et al. Quantitation of HIV-1-specific cytotoxic T lymphocytes and plasma load of viral RNA. *Science* 1998;279:2103–2106.

142. McMichael, A. J. T cell responses and viral escape. *Cell* 1998;93:673–676.

143. Wilson, J. D., Ogg, G. S., Allen, R. L., et al. Temporal association of cellular immune responses with the initial control of viremia in primary human immunodeficiency virus type 1 infection. *J. Virol.* 1994;68:4650–4655.

144. Borrow, P., Lewicki, H., Hahn, B. H., Shaw, G. M., and Oldstone, M. B. Virus-specific CD8+ T cytotoxic T-lymphocyte activity associated with control of virema in primary human immunodeficiency virus type 1infection. *J. Virol.* 1994;68:6103–6110.

145. Schmitz, J. E., Kuroda, M. J., Santra, S., et al. Control of viremia in simian immunodeficiency virus infection by CD8+ lymphocytes. *Science* 1999;283:857–860.

146. Gillimore, A., Cranage, M., Cook, N., et al. Early suppression of SIV replication by CD8+ nef-specific cytotoxic T cells in vaccinated macaques. *Nat. Med.* 1995;1:1167–1173.

147. Boyer, J., Ugen, K. E., Wang, B., et al. Protection of chimpanzees from high-dose heterologous HIV-1 challenge by DNA vaccination. *Nat. Med.* 1997;3:526–532.

148. Letvin, N. L., Montefiori, D. C., Yasutomi, Y., et al. Potent, protective anti-HIV immune responses generated by bimodal HIV envelope DNA plus protein vaccination. *Proc. Natl. Acad. Sci. USA* 1997;94:9378–9383.

149. Dale, C. J., Liu, X. S., De Rose, R., et al. Chimeric human papilloma virus-simian/human immunodeficiency virus virus-like particle vaccines: Immunogenicity and protective efficacy in macaques. *Virology* 2002;301:176–187.

150. Schneider, J., Gilbert, S. C., Blanchard, T. J., et al. Enhanced immunogenicity for CD8+ T cell induction and complete protective efficacy of malaria DNA vaccination by boosting with modified vaccinia virus Ankara. *Nat. Med.* 1998;4:397–402.

151. Sedegah, M., Jones, T. R., Kaur, M., et al. Boosting with recombinant vaccinia increases immunogenicity and protective efficacy of malaria DNA vaccine. *Proc. Natl. Acad. Sci. USA* 1998;95:7648–7653.

152. Schneider, J., Gilbert, S. C., Hannan, C. M., et al. Induction of CD8+ T cells using heterologous prime-boost immunization strategies. *Immunol. Rev.* 1999;170:29–38.

153. Li, S., Rodrigues, M., Rodriquez, D., et al. Priming with recombinant influenza virus followed by administration of recombinant vaccinia virus induces CD8+ T-cell mediacted protective immunity against malaria. *Proc. Natl. Acad. Sci. USA* 1993;90:5214–5218.

154. Gilbert, S. C., Schneider, J., Plebanski, M., et al. Ty virus-like particles, DNA vaccines and modified vaccinia virus Ankara: Comparisons and combinations. *Biol. Chem.* 1999;380:299–303.

155. Murata, K. and Garcia-Sastre, A. Characterization of in vivo primary and secondary CD8+ T cell responses induced by recombinant influenza and vaccinia viruses. *Cell Immunol.* 1996;173:96–104.

156. Irvine, K. R., Chamberlain, R. S., Shulman, E. P., et al. Enhancing efficacy of recombinant anti-cancer vaccines with prime/boost regimens that use two different vectors. *J. Natl. Cancer Inst.* 1997;89:1595–1601.

157. Mayr, A., Stickl, H., Muller, H. K., Danner, K., and Singer, H. The smallpox vaccination strain MVA: Marker, genetic structure, experience gained with parenteral vaccination and behavior in organisms with a debilitated defense mechanism. *Zentralbl. Bakteriol.* 1978;167:375–390.

158. Sutter, G. and Moss, B. Nonreplicating vaccinia vector efficiently expresses recombinant genes. *Proc. Natl. Acad. Sci. USA* 1992;89:10847–10851.

159. Drexler, I., Heller, K., Wahren, B., Erfle, V., and Sutter, G. Highly attenuated modified vaccinia virus Ankara replicates in baby hamster kidney cells, a potential host for virus propagation, but not in various human transformed and primary cells. *J. Gen. Virol.* 1998;79:347–352.

160. Ramshaw, I. A. and Ramsay, A. J. The priming-boost strategy: Exciting prospects for improved vaccination. *Immunol. Today* 2000;21:163–165.

161. Hanke, T., Blanchard, T. J., Schneider, J., et al. Enhancement of MHC class I-restricted peptide-specific T cell induction by a DNA prime/MVA boost vaccination regime. *Vaccine* 1998;16:439–445.

162. Seth, A., Ourmanov, I., Kuroda, M. J., et al. Recombinant modified vaccinia virus Ankara-simian immunodeficiency virus gag pol elicits cytotoxic T lymphocytes in rhesus monkeys detected by a major histocompatibility complex class I/peptide tetramer. *Proc. Natl. Acad. Sci. USA* 1998;95:10112–10116.

163. Belyakov, I. M., Wyatt, L. S., Ahlers, J. D., et al. Induction of a mucosal cytotoxic T-lymphocyte response by intrarectal immunization with a replication-deficient recombinant vaccinia virus expressing human immunodeficiency virus 89.6 envelope protein. *J. Virol.* 1998;72:8264–8272.

164. Ramirez, J. C., Gherardi, M. M., and Esteban, M. Biology of attenuated modified vaccinia virus Ankara recombinant vector in mice: virsu fate and activation of B- and T-cell immune responses in comparision with the Western Reserve strain and advantages as a vaccine. *J. Virol.* 2000;74:923–933.

165. Dorrell, L., O'Callaghan, C. A., Britton, W., et al. Recombinant modified vaccinia virus Ankara efficiently restimulates human cytotoxic T lymphocytes in vitro. *Vaccine* 2000;19:327–336.

166. Allen, T. M., Vogel, T. U., Fuller, D. H., et al. Induction of AIDS virus-specific CTL activity in fresh, unstimulated peripheral blood lymphocytes from rhesus macaques vaccinated with a DNA prime/modified vaccinia virus Ankara boost regimen. *J. Immunol.* 2000;164:4968–4978.

167. Amara, R. R., Villinger, F., Altman, J. D., et al. Control of a mucosal challenge and prevention of AIDS by a multiprotein DNA/MVA vaccine. *Science* 2001;292:69–74.

168. Hanke, T., Samuel, R. V., Blanchard, T. J., et al. Effective induction of simian immunodeficiency virus-specific cytotoxic T lymphocytes in macaques by using a multiepitope gene and DNA prime-modified vaccinia virus Ankara boost vaccination regimen. *J. Virol.* 1999;73:7524–7532.

169. Kent, S. J., Zhao, A., Best, S. J., et al. Enhanced T-cell immunogenicity and protective efficacy of a human immunodeficiency virus type 1 vaccine regimen consisting of consecutive priming with DNA and boosting with recombinant fowlpox virus. *J. Virol.* 1998;72:10180–10188.

170. Robinson, H. L., Montefiori, D. C., Johnston, R. P., et al. Neutralizing antibody-independent containment of immmunodeficiency virus challenges by DNA priming and recombinant pox virus booster immunizations. *Nat. Med.* 1999;5:526–534.

171. Shiver, J. W. and Fu, T. M. Replication-incompetent adenoviral vaccine vector elicits effective anti-immunodeficiency-virus immunity. *Nature* 2002;415:331–335.

172. Egan, M. A., Chong, S. Y., Megati, S., et al. Priming with plasmid DNAs expressing interleukin-12 and simian immunodeficiency Virsu Gag enhances the immunogenicity and efficacy of an experimental AIDS vaccine based on recombinant vesicular stomatitis Virus. *AIDS Res. Human Retroviruses* 2005; 21:629–643.

173. Mattapallil, J., Douek, D. C., Buckler-White, A., et al. Vaccination preserves CD4 memory T cells during acute simian immunodeficieicy virus challenge. *J. Exp. Med.* 2006;203:1533–1541.

174. Letvin, N. L., Mascola, J. R., Sun, Y., et al. Preserved CD4+ central memory T cells and survival in vaccinated SIV-challenged monkeys. *Science* 2006;312:1530–1533.

175. Casimiro, D., Tang, A., Chen, L., et al. Vaccine-induced immunity in baboons by using DNA and replication-incompetent adenovirus type 5 vectors expressing a human immunodeficiency virus type 1 gag gene. *J. Virol.* 2003;77:7663–7668.

176. Makedonas, G. and Betts, M. R. Polyfunctional analysis of human T cell responses: Importance in vaccine immunogencity and natural infection. *Springer Sem. Immunopathol.* 2006; 28:209–219.

177. Betts, M. R., Nason, M. C., West, S. M., et al. HIV nonprogressors preferentially maintain highly functional HIV-specific CD8+ T cells. *Blood* 2006;107:4781–4789.

178. Ulmer, J. B., Wahren, B., and Liu, M. A. DNA vaccines for HIV/AIDS. *Curr. Opin. HIV AIDS* 2006;1:309–313.

179. Wahren, B., Brave, A., Boberg, A., et al. Potent cellular and humoral immunity against HIV-1 elicited in mice by a DNA-prime/MVA-boost vaccine regimen intended for human use. *Retrovirology* 2006;3:83–88.

180. Wang, S., Arthos, J., Lawrence, J. M., et al. Enhanced immunogenicity of gp120 protein when combined with recombinant DNA priming to generate antibodies that neutralize the JR-FL primary isolate of human immunodeficiency virus type 1. *J. Virol.* 2005;79:7933–7937.

181. WHO AIDS epidemic update. 2006.

182. Ulmer, J. B., Donnelly, J. J., Parker, S. E., et al. Heterologous protection against influenza by injection of DNA encoding a viral protein. *Science* 1993;259:1745–1749.

183. Ross, T. M., Xu, Y., Bright, R. A., and Robinson, H. L. C3d enhancement of antibodies to hemagglutinin accelerates protection against influenza virus challenge. *Nat. Immunol.* 2000;1:127–131.

184. Drape, R. J., Macklin, M. D., Barr, L. J., et al. Epidermal DNA vaccine for influenza is immunogenic in humans. *Vaccine* 2006;24:4475–4481.

185. Davis, H. L., McCluskie, M. J., Gerin, J. L., and Purcell, R. H. DNA vaccine for hepatitis B: Evidence for immunogenicity in chimpanzees and comparison with other vaccines. *Proc. Natl. Acad. Sci. USA* 1996;93:7213–7218.

186. Hooper, J. W., Thompson, E., Wilhelmsen, C., et al. Smallpox DNA vaccine protects nonhuman primates against lethal monkeypox. *J. Virol.* 2004;78:4433–4443.

187. Davis, B., Chang, G. J., Croop, B., et al. West Nile virus recombinant DNA vaccine protects mouse and horse from virus challenge and expresses in vitro a noninfectious recombinant antigen that can be used in enzyme-linked immunosorbent assays. *J. Virol.* 2001;75:4040–4047.

188. Corbeil, S., La Patra, S. E., Anderson, E. D., et al. Evaluation of the protective immunogenicity of the N, P. M. NV and G proteins of infectious hematopoietic necrosis virus in rainbow trout (*Oncorhynchus mykiss*) using DNA vaccines. *Dis. Aquat. Organ.* 1999;39:29–36.

189. Traxler, G. S., Anderson, E., LaPatra, S. E., et al. Naked DNA vaccination of Atlantic salmon Salmo salar against IHNV. *Dis. Aquat. Organ.* 1999;38:183–190.

190. La Patra, S. E., Corbeil, S., Jones, G. R., Shewmaker, W. D., and Kurath, G. The dose-dependent effect on protection and humoral response to a DNA vaccine against IHN virus in subyearling rainbow trout. *J. Aquat. Anim. Health* 2000;2000:181–188.

191. La Patra, S. E., Corbeil, S., Jones, G. R., et al. Protection of rainbow trout against infectious hematopoietic necrosis virus four days after specific or semi-specific DNA vaccination. *Vaccine* 2001;19:4011–4019.

192. Sogn, J. A., Finerty, J. F., Heath, A. K., Shen, G. C., and Austin, F. C. Cancer vaccines: The perspective of the cancer immunology branch. *Ann. N. Y. Acad. Sci.* 1993;690:322–330.

193. Dudley, M. E., Wunderlich, J. R., and Rosenberg, E. S. Adoptive cell transfer therapy following non-myeloablative but lymphodepleting chemotherapy for the treatment of patients with refractory metastatic melanoma. *J. Clin. Oncol.* 2005;23:2346–2357.

194. Powell, D. J., Dudley, M. E., Hogan, K. A., Wunderlich, J. R., and Rosenberg, S. A. Adoptive transfer of vaccine-induced peripheral blood mononuclear cells to patients with metastatic melanoma following lymphodepletion. *J. Immunol.* 2006; 177:6527–6539.

195. Gilliland, F. D. and Keys, C. R. Male genital cancers. *Cancer* 1995;75:295–315.

196. Wei, C., Willis, R. A., Tilton, B. R., et al. Tissue-specific expression of the human prostate-specific antigen gene in transgenic mice: Implications for tolerance and immunotherapy. *Proc. Natl. Acad. Sci. USA* 1997;94:6369–6374.

197. Ko, S. C., Gotoh, A., Thalmann, G. N., et al. Molecular therapy with recombinant p53 adenovirus in an androgen-independent, metastatic human prostate cancer model. *Hum. Gene Ther.* 1996;7:1683–1691.

198. Wang, M. C., Kuriyama, M., Papsidera, L., D., et al. Prostate antigen of human cancer patients. *Methods Cancer Res.* 1982;19:179–197.

199. Watt, K. K., Lee, P. J., M'Timkulu, T., Chan, W. P., and Loor, R. M. Human prostate-specific antigen: Structural and functional similarity with serine proteases. *Proc. Natl. Acad. Sci. USA* 1986;83:3166–3170.

200. Labrie, F., DuPond, A., Suburu, R., et al. Serum prostate specific antigen as a pre-screening test for prostate cancer. *J. Urol.* 1992;151:1283–1290.

201. Kim, J. J., Trivedi, N., Wilson, D. M., et al. Molecular and immunological analysis of genetic prostate specific antigen (PSA) vaccine. *Oncogene* 1998;17:3125–3135.

202. Kim, J. J., Yang, J. S., Nottingham, L. K., et al. Induction of immune responses and safety profiles in rhesus macaques immunized with a DNA vaccine expressing human prostate specific antigen (PSA). *Oncogene* 2001;20:4497–4506.

203. Kelley, J. R. and Cole, D. J. Gene therapy strategies utilizing carcinoembryonic antigen as a tumor associated antigen for vaccination against solid malignancies. *Gene Ther. Mol. Biol.* 1998;2:14–30.

204. Zaremba, S., Barzaga, E., Zhu, M., et al. Identification of an enhancer agonist cytotoxic T lymphocyte peptide from human carcinoembryonic antigen. *Cancer Res.* 1997;57:4570–4577.

205. Foon, K. A., Chakraborty, M., John, W. J., et al. Immune response to the carcinoembryonic antigen in patients treated with an anti-idiotype antibody vaccine. *J. Clin. Invest.* 1997;96:334–342.

206. Conry, R. M., LoBuglio, A. F., Kantor, J., et al. Immune response to a carcinoembryonic antigen polynucleotide vaccine. *Cancer Res.* 1994;54:1164–1168.

207. Conry, R. M., LoBuglio, A. F., and Curiel, D. T. Polynucleotide-mediated immunization therapy of cancer. *Semin. Oncol.* 1996;23:135–147.

208. Smith, B. F., Baker, H. J., Curiel, D. T., Jiang, W., and Conry, R. M. Humoral and cellular immune responses of dogs immunized with a nucleic acid vaccne encoding human carcinoembryonic antigen. *Gene Ther.* 1998;5:865–868.

209. Conry, R. M., Curiel, D. T., Strong, T. V., et al. Safety and immunogenicity of a DNA vaccine encoding carcinoembryonic antigen and hepatitis B surface antigen in colorectal carcinoma patients. *Clin. Cancer Res.* 2002;8:2782–2787.

210. Rosenberg, E. S., Zhai, Y., Yang, J. C., et al. Immunizing patients with metastatic melanoma using recombinant adenoviruses encoding MART-1 or gp100 melanoma antigens. *J. Natl. Cancer Inst.* 1998;90:1894–1900.

211. Spellerberg, M. B., Zhu, D., Thompsett, A., et al. Promotion of anti-idiotypic antibody responses induced by single chain Fv genes by fusion to tetanus toxin fragment C. *J. Immunol.* 1997;159:1885–1892.

212. Timmerman, J. M., Singh, G., Hermanson, G., et al. Immunogenicity of a plasmid DNA vaccine encoding chimeric idiotype in patients with B-cell lymphoma. *Cancer Res.* 2002;62:5845–5852.

213. Franceschi, S., Rajkumar, T., Vaccarella, S., et al. Human papillomavirus and risk factors for cervical cancer in Chennai, India: A case-control study. *Int. J. Cancer* 2003;107:127–133.

214. Parkin, D. M., Bray, F., Ferlay, J., and Pisani, P. Global cancer statistics. *CA Cancer J. Clin.* 2005;55:74–108.

215. Gillison, M., Koch, W. M., and Capone, R. B. Evidence for a causal association between human papillomavirus and a subset of head and neck cancers. *J. Natl. Cancer Inst.* 2000;92:709–720.

216. Bosch, F. X., Manos, M. M., Munoz, N., et al. Prevalence of human papillomavirus in cervical cancer: A worldwide perspective. International biological study on cervical cancer (IBSCC) Study Group. *J. Natl. Cancer Inst.* 1995;87:796–802.

217. Munoz, N., Bosch, F. X., Castellsague, X., et al. Against which human papillomavirus types shall we vaccinate and screen? The international perspective. *Int. J. Cancer* 2004;111:278–285.

218. Roden, R. and Wu, T. C. Preventative and therapeutic vaccines for cervical cancer. *Expert Rev. Vaccines* 2003;2:495–516.

219. Roden, R. and Wu, T. C. How will HPV vaccines affect cervical cancer? *Nat. Rev. Cancer* 2006;6:753–763.

220. Munger, K. and Howley, P. M. Human papillomavirus immortalization and transformation functions. *Virus Res.* 2002;89:213–228.

221. Eiben, G. L., Velders, M. P., and Kast, W. M. The cell-mediated immune response to human papillomavirus-induced cervical cancer: Implications for immunotherapy. *Adv. Cancer Res.* 2002;86:113–148.

222. Tamura, Y., Peng, P., Liu, K., Daou, M., and Srivastava, P. K. Immunotherapy of tumors with autologous tumor-derived heat shock protein preparations. *Science* 1997;278:117–120.

223. Melcher, A., Todryk, S., Hardwick, N., et al. Tumor immunogenicity is determined by the mechanism of cell death via induction of heat shock protein expression. *Nat. Med.* 1998;4:581–587.

224. Basu, S., Binder, R. J., Suto, R., Anderson, K. M., and Srivastava, P. K. Necrotic but not apoptotic cell death releases heat shock proteins, which deliver a partial maturation signal to dendritic cells and activates the NF-kappa B pathway. *Int. Immunol.* 2000;12:1539–1546.

225. Chen, C. H., Wang, T. L., Hung, C. F., et al. Enhancement of DNA vaccine potency by linkage of antigen gene to an HSP70 gene. *Cancer Res.* 2000;60:1035–1042.

226. Cheng, W. F., Huang, C. F., Chai, C. Y., et al. Tumor-specific immunity and antiangiogenesis generated by a DNA vaccine encoding calreticulin linked to a tumor antigen. *J. Clin. Invest.* 2001;108:669–678.

227. Peng, S., Trimble, C., He, L., et al. Characterization of HLA-A2-restricted HPV-16 E7-specific CD8(+) T-cell immune responses induced by DNA vaccines in HLA-A2 transgenic mice. *Gene Ther.* 2006;13:67–77.

228. Lin, C. T., Tsai, Y. C., He, L., et al. A DNA vaccine encoding a codon-optimized human papillomavirus type 16 E6 gene enhances CTL response and anti-tumor activity. *J. Biomed. Sci.* 2006;13:481–488.

229. Peng, S., Tomson, T. T., Trimble, C., et al. A combination of DNA vaccines targeting human papillomavirus type 16 E6 and E7 generates potent antitumor effects. *Gene Ther.* 2006;13:257–265.

230. Yan, J., Reichenbach, D. K., Corbitt, N., et al. A novel HPV-16 DNA vaccine encoding a E6/E7 fusion consensus protein in combination with IL-15 enhances T cell immunity and modulates tumor growth in vivo. Submitted for publication.

231. Riezebos-Brilman, A., Regts, J., Freyschmidt, E. J., et al. Induction of human papilloma virus E6/E7-specific cytotoxic T-lymphocyte activity in immune-tolerant, E6/E7-transgenic mice. *Gene Ther.* 2005;12:1410–1414.

232. Gerard, C. M., Baudson, N., Kraemer, K., et al. Recombinant human papillomavirus type 16 E7 protein as a model antigen to study the vaccine potential in control and E7 transgenic mice. *Clin. Cancer Res.* 2001;7:838s–847s.

233. Souders, N. C., Sewell, D., Pan, Z. K., et al. Listeria-based vaccines can overcome tolerance by expanding low avidity CD8+ T cells capable of eradicating a solid tumor in a transgenic mouse model of cancer. *Cancer Immunity* 2007;7:1–12.

234. Michel, N., Osen, W., Gissmann, L., et al. Enhanced immunogenicity of HPV 16 E7 fusion proteins in DNA vaccination. *Virology* 2002;294:47–59.

235. MacEwen, E. G., Patnaik, A. K., Harvey, H. J., Hayes, A. A., and Matus, R. Canine oral melanoma: Comparison of surgery versus surgery plus *Corynebacterium parvum*. *Cancer Invest.* 1986;4:397–402.

236. Schwarz, P. D., Withrow, S. J., Curtis, C. R., Powers, B. E., and Straw, R. C. Mandibular resection as a treatment for oral cancer in 81 dogs. *J. Am. Anim. Hosp. Assoc.* 1991;27:601–610.

237. Bateman, K. E., Catton, P. A., Pennock, P. W., and Kruth, S. A. Radiation therapy for the treatment of canine oral melanoma. *J. Vet. Intern. Med.* 1994;8:267–272.

238. Weber, L. W., Bowne, W. B., Wolchok, J. D., et al. Tumor immunity and autoimmunity induced by immunization with homologous DNA. *J. Clin. Invest.* 1998;102:1258–1264.

239. Wang, S., Bartido, S., Yang, G., et al. A role for a melanosome transport signal in accessing the MHC class II presentation pathway and in eliciting CD4+ T cell responses. *J. Immunol.* 1999;163:2029–2042.

240. Bergman, P. J., McKnight, J. A., Novosad, A., et al. Long-term survival of dogs with advanced malignant melanoma after DNA vaccination with xenogeneic human tyrosinase: A phase I trial. *Clin. Cancer Res.* 2003;9:1284–1290.

241. Bergman, P. J., Camps-Palau, M. A., McKnight, J. A., et al. Development of a xenogeneic DNA vaccine program for canine maligmant melanoma at the Animal Medical Center. *Vaccine* 2006;24:4582–4585.

242. Liblau, R. S., Singer, S. M., and McDevitt, H. O. Th1 and Th2 CD4+ T cells in the pathogenesis of organ-specific autoimmune diseases. *Immunol. Today* 1995;16:34–38.

243. Katz, J. D., Benoist, C., and Mathis, D. T helper cell subsets in insulin-dependent diabetes. *Science* 1995;268:1185–1188.

244. Ramshaw, I. A., Fordham, S. A., Bernard, C. C., et al. DNA vaccines for the treatment of autoimmune disease. *Immunol. Cell Biology* 1997;75:409–413.

245. Lobell, A., Weissert, R., Stroch, M. K., et al. Vaccination with DNA encoding an immunodominant myelin basic protein peptide targeted to Fc of immunoglobulin G suppresses experimental antoimmune encephalomyelitis. *J. Exp. Med.* 1998;187:1543–1548.

45 Immunoisolation for CNS Cell Therapy

Dwaine F. Emerich and Christopher G. Thanos

CONTENTS

45.1 INTRODUCTION

Diseases of the central nervous system (CNS) are frequently characterized by the degeneration of specific cells and a consequent continuous deterioration of function. Advances in molecular biology, genetic engineering, proteomics, and genomics are identifying potentially efficacious proteins, peptides, and other compounds. Unfortunately, most of these compounds are not active following systemic administration, largely because the blood–brain barrier (BBB) restricts exchange between the vasculature and brain parenchyma [1,2]. A number of strategies have been described to circumvent the BBB including: (1) carrier-, or receptor-mediated transcytosis [3,4]; (2) osmotic opening [5,6]; (3) direct infusion with stereotactic guidance [7–9]; (4) osmotic pumps [10,11]; (5) sustained-release polymer systems [12–14]; (6) cell replacement/cell therapy [15–20]; and (7) direct gene therapy [21–25].

One form of cell-based therapy uses xenogeneic cells encased within a selectively permeable polymeric membrane. The polymer membrane containing the cells can be implanted into the brain allowing cell-based delivery of secreted products directly to the target site. This process, known as immunoisolation, is enabled because the encapsulated cells are protected from host rejection by the immunoisolatory, semipermeable membrane. The selective, semipermeable membrane barrier admits oxygen and required nutrients and releases bioactive cell secretions, but restricts passage of larger cytotoxic agents from the host immune defense system.

TABLE 45.1

Advantages and Disadvantages of Encapsulated Cells

	Advantages	Disadvantages
Unencapsulated cells or tissue	Permits anatomical integration between the host and transplanted tissue Good cell viability and neurochemical diffusion	Likely requires immunosuppression Limited tissue availability Difficult retrieval Societal and ethical issues
Microencapsulation	Permits use of allo- and xeno-grafts without immunosuppression Thin wall and spherical shape are optimal for cell viability and neurochemical diffusion	Mechanically and chemically fragile Multiple implant sites Limited retrievability
Macroencapsulation	Permits us of allo- and xeno-grafts without immunosuppression Reasonable mechanical stability Adequate cell viability and neurochemical diffusion Retrievable	Dimensions may limit neurochemical diffusion and cell viability Multiple implant sites May produce significant tissue damage/displacement during implantation

Immunoisolation therefore eliminates the need for chronic immunosuppression of the host and allows the implanted cells to be obtained from nonhuman sources, thus avoiding the constraints associated with cell sourcing, which have limited the clinical application of unencapsulated cell transplantation. This chapter discusses the preclinical and clinical evaluation of encapsulated cells across a range of CNS diseases highlighting the therapeutic potential of genetically modified, encapsulated cells for Alzheimer' disease (AD), Parkinson's disease (PD), and Huntington's disease (HD).

45.2   CELL IMMUNOISOLATION

There are generally two categories for cell encapsulation, micro- and macro-, each with some benefits and limitations shown in Table 45.1 [26–29]. This chapter focuses primarily on macroencapsulation (Figure 45.1). Macroencapsulation involves filling a hollow, usually cylindrical, selectively permeable membrane with cells, generally suspended in a matrix, and then sealing the ends to form a capsule. Polymers used for macroencapsulation are biodurable, with a thicker wall than

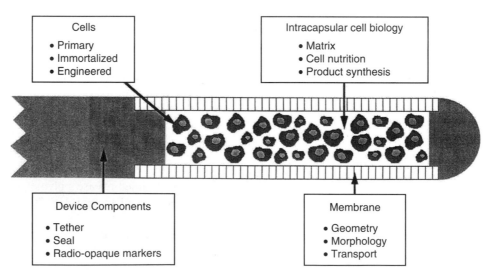

FIGURE 45.1   Diagram illustrating the different components of a macrocapsule contributing to successful implantation and cell viability. The manufacturing process involves several different aspects each with its own complexities. The initial choice of cell types includes primary, immortalized, or engineered. Intracapsular cell biology issues following encapsulation include a consideration of the need to use a compatible extracellular matrix (ECM) and other considerations specific to that cell type because they impact cell nutrition and product synthesis. A series of other device-related issues include membrane geometry, morphology, and transport of molecules into and out of the device. Finally, the device must be sealed and depending on the site of implantation could require a tether for subsequent retrieval or the inclusion radio-opaque markers for imaging purposes.

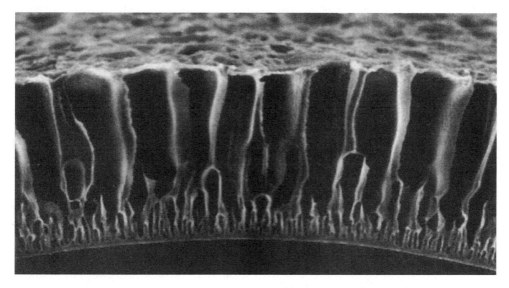

FIGURE 45.2   Scanning electron micrograph of a typical macrovoid-containing membrane used in cell encapsulation.

that found in microencapsulation. While thicker wall and larger implant diameters can enhance implant stability, these same features may also impair diffusion, compromise the viability of the tissue, and slow the release kinetics of desired factors. Macrocapsules can be retrieved from the recipient and replaced if necessary.

Macroencapsulation is generally achieved by filling preformed thermoplastic hollow fibers with a cell suspension. The hollow fiber is formed by pumping a polymer solution in a water-miscible solvent through an outer annular region of a nozzle while an aqueous solution is pumped through a central bore. The polymer precipitates upon contact with the water and forms a cylindrical hollow fiber with a permselective inner membrane. Further precipitation of the polymer occurs as the water moves through the polymer wall, forcing the organic solvent out and forming a trabecular wall structure. The hollow fiber is collected in an aqueous water bath, where complete precipitation of the polymer and dissolution of the organic solvent occur. The ends of the hollow fiber are then sealed to form a macrocapsule. This final step is not a trivial one, since reliably sealing the ends of capsules can be difficult, and provides the barrier paramount for successful immunoisolation. Figures 45.2 and 45.3 depict cross-sections of typical macrofibers.

A second method of macroencapsulation, called coextrusion, avoids the sealing problem by entrapping cells within the lumen of a hollow fiber during the fabrication process [30]. Pinching the fiber before complete precipitation of the polymer causes fusion of the walls, providing closure of the extremities while the cells are inside. The advantages of coextrusion over loading preformed capsules include distributing the cells more uniformly along the entire length of the fiber,

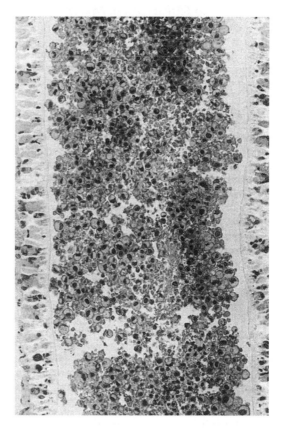

FIGURE 45.3   Example of encapsulated fibroblasts retrieved from monkey lateral ventricle (LV) illustrating many of the components listed above. Cells were modified to produce NGF and were encapsulated using Vitrogen as an ECM. Under these conditions, numerous H&E positive cells were visible and evenly distributed within the full length of the capsule one month following implantation.

reducing the shear stresses on the cells during the loading process and the potential for mass production.

While polymer capsules are relatively sturdy, it is critical to ensure that the encapsulating membrane is compliant enough to meet the dynamics of the surrounding tissue while remaining mechanically resilient to resist failure during device implantation/retrieval. To accomplish this, the macrocapsule can be designed to provide added strength and device integrity. One approach includes the addition of a crosslinked hydrogel, e.g., a 2% alginate solution, within the device. This modification enhances structural support during the implantation procedure [31]. However, with time, the hydrogel loses structural integrity and does not provide added strength, especially with regard to tensile strength, an important consideration for device retrieval. Mechanical supports can also be used for added strength. Such supports are better served from within the device to not impede diffusion between the encapsulating membrane and the surrounding tissue. Titanium wires and braided materials are examples of approaches that may be used to provide added tensile strength.

45.2.1 Cells and Extracellular Matrices Used in Encapsulation

Cells placed within encapsulation devices generally fall into one of three categories. The first category is primary postmitotic cells such as islets of Langerhans for diabetes, adrenal chromaffin cells for chronic pain, or hepatocytes for liver devices. Secondly, immortalized (or dividing) cells such as PC12 cells have been utilized to deliver dopamine for PD. The third category is cell lines genetically engineered to secrete a bioactive substance such as baby hamster kidney (BHK) cells to secrete factors such as nerve growth factor (NGF). Dividing tissue has advantages over postmitotic tissue; it can be expanded, banked, and easily tested for sterility and contaminants. However, dividing tissue is also constrained by the potential for overgrowth within the capsule environment, resulting in an accumulation of necrotic tissue that can diminish the membrane's permeability characteristics, reducing cell viability and neurochemical output.

In vivo, ECMs control cell function by regulating morphology, proliferation, differentiation, migration, and metastasis [32,33]. Within a capsule, ECMs were originally employed simply to prevent aggregation of cells (immobilization) and central necrosis, but are also beneficial to the viability and function of cells that require immobilization or scaffolding for anchorage. For example, immobilizing adrenal chromaffin cells in alginate prevents aggregation that, in turn, reduces central necrotic cores from forming [34]. The chromaffin cells appear to thrive in alginate whereas mitotically active fibroblasts do not. In this case, the use of alginate is essential to the optimal functioning of this device since some anchorage-dependent cells such as fibroblasts or endothelial cells are present with the adrenal chromaffin cells. In the absence of alginate or other immobilizing matrices, the fibroblasts expand and overgrow the capsule, resulting in a device deficient in bioactive factors produced from the chro-

maffin cells [35]. In contrast, fibroblastic BHK cells prefer collagen, while PC12 cells prefer distribution within precipitated chitosan, which provides a scaffolding structure on which the cells anchor [36].

In order to provide a substratum designed for more specific functions, the matrix material can be manipulated chemically or mechanically, which in turn may influence cell attachment, differentiation, and/or proliferation. For example, peptides, such as arginine–glycine–aspartic acid (RGD), have been immobilized on a variety of surfaces to promote cell adhesion [37]. Integrin receptors on the cell surface membrane interact with the RGD sequence, a known ligand for fibronectin receptors. Glass microbeads have been modified by attaching RGD or tyrosine–isoleucine–glycine–serine–arginine (YIGSR) motifs to provide sites for cell adhesion [37]. Spherical ferromagnetic beads have been coated with specific receptor ligands to mediate cell attachment [38]. With competitive binding assays and a mechanical stress testing apparatus, the endothelial cell's interaction with the ECM receptor, integrin beta-1, supported a force-dependent stiffening response, while nonadhesion receptors did not. A laminin-derived oligopeptide sequence, CDPG-YIGSR, has also been derivatized within an agarose hydrogel and permitted a dose-dependent increase in neurite outgrowth of neuronal cell bioassays [39]. Similarly, YIGSR and isoleucine-lysine-valine-alanine-valine (IKVAV), both of which are found in laminin, have been immobilized on surfaces to promote neuronal cell adhesion and neurite outgrowth [40]. Poly(ethylene oxide) (PEO)-star copolymers have been fabricated as a potential synthetic ECM [41]. The star copolymers provide many hydroxyl groups where various synthetic oligopeptides can be attached to desired specifications.

The survival and differentiation of encapsulated cells can be influenced by the matrix interactions. A variety of matrices for use in immunoisolatory devices, such as alginate, rat tail collagen extracts, gelatin shards, porous gelatin, or collagen microcarriers, carrageenan, chondroitin sulfate, fibrin, hyaluronic acid, the positively charged substrate chitosan, and an acrylamide-based thermoresponsive gel are available and were recently reviewed [42]. All in all, successful cell encapsulation involves the choice of the cells to be encapsulated, the type of intracapsular matrix used, and the ability to control membrane properties such as geometry, morphology, and transport. The interactions between the encapsulating membrane characteristics and the capsule core, or matrix within, should be rigorously characterized to determine the optimal configuration for each cell type.

45.2.2 Validation of the Concept of Immunoisolation

45.2.2.1 *In Vitro* Studies

The maintenance of immunoisolation can be easily confirmed *in vitro*. In one study, a polydisperse (10^3–10^6 g/mol) dextran solution was encapsulated into hollow fibers and the flux of the molecules across the semipermeable membrane into a surrounding reservoir was monitored over time. A dextran rejection curve was produced from the filtrate and reservoir concentrations measured using gel permeation chromatography [43]. With control devices that had been damaged, the large molecular weight dextran species rapidly escaped. In contrast, intact capsules retained the encapsulated dextran. Capsule integrity can also be confirmed using immunological assays such as measuring the protection of encapsulated cells against the cytotoxic killing of antibody (IgG)-mediated complement lysis [29]. With integral PC12 cell-loaded capsules, in the presence of antibody and complement, the membrane prevented antibody-mediated complement lysis (<10% cell death), while complete killing (100%) occurred with damaged capsules or unencapsulated cells.

45.2.2.2 *In Vivo* Studies

The importance of polymer capsule integrity for xenografted cell survival is illustrated in studies where unencapsulated PC12 cells, or cells encapsulated in intentionally damaged membranes, have been implanted in the brains of guinea pigs [30]. Intact PC12 cell-loaded capsules implanted into the guinea pig striatum showed no lymphocytic infiltration and a minimal astrocytic reaction assessed by glial fibrillary acidic protein (GFAP) staining. In contrast, cell survival was poor in damaged capsules, with marked inflammation and heavy lymphocytic invasion into the capsule. Parallel studies confirmed that unencapsulated PC12 cells do not survive following implantation into either the guinea pig or the monkey striatum while encapsulated PC12 cells were viable for 6 months in monkeys [44]. Similar results have been obtained in rats that received intraventricular implants of bovine adrenal chromaffin cells [45]. There was no evidence of elevated serum levels of rat antibovine adrenal chromaffin cell IgG or IgM with encapsulated xenogeneic adrenal chromaffin cells for nearly 1.5 years *in vivo*. In contrast, a robust host immune response was induced in animals after implantation of unencapsulated bovine adrenal chromaffin cells.

45.2.3 BIOCOMPATIBILITY

Transplant survival, with and without an encapsulating membrane, is mediated by many factors. The cellular/tissue reaction generated by a host in response to a foreign body (i.e., biocompatibility) impacts the success of the transplant. Factors affecting biocompatibility include the method of implantation, the implant site, and implant properties, such as composition of the polymer, potential residual processing agents, surface integrity and microgeometry, and the size and shape of the implant. The constituents of the implants should be assessed rigorously, both *in vitro* and *in vivo*, to determine the safety of the materials. The CNS tissue is not only privileged from an immunologic standpoint, but it also lacks the primary reactive cells, fibroblasts, and macrophages found in peripheral locations. The brain therefore offers a unique environment both in terms of the inflammatory response, as well as the cellular constituents that comprise the reactive cells. Immunospecific antibodies are available to delineate the roles of the brain reactive cells, the astroctyes and microglia, with respect to their reactivity. Nevertheless, few studies have systematically examined the reaction of host brain tissue to the presence of polymeric devices.

Early investigations utilized electron microscopy to characterize the brain tissue reaction to plastic-embedded metal electrodes and polymer implants [46–48]. Necrosis of the tissue surrounding the polymer capsules implanted into the striatum of rodents was minimal with small Nissl-positive cells and capillaries invading the open trabeculae in the capsules [48]. GFAP immunocytochemistry revealed local reactive astrocytes 1 to 2 weeks after implantation. The intensity of the GFAP reaction diminished rapidly such that by 4 weeks after implantation, the gliotic reaction surrounding the polymer implant was minimal. No significant changes in myelin basic protein reactive oligodendroglia were observed, and neuron specific enolase-reactive neurons were readily identifiable adjacent to the implant. Subsequent studies with an immunospecific antisera against rat microglia, OX-42, revealed a reaction in magnitude and time course similar to that seen for the astrocytes.

The lack of a significant host tissue reaction to the implant is crucial for the initial viability of the encapsulated cells as well as diffusion from the capsule. While capillary invasion into the capsule walls helps provide nutrients and oxygen in proximity to the encapsulated cells the process of angiogenesis for neovascularization typically evolves in a 4–7-day period [49]. The encapsulated cells must endure an initial period of nutrient and oxygen deprivation obtaining these essential factors only by diffusion. Moreover, since the only means of delivery of the desired cellular products from an encapsulated cell implant is by diffusion, any reaction around the capsule can diminish the diffusion of therapeutic products from the encapsulated cells. Studies demonstrating biocompatibility within the CNS therefore suggest that the bidirectional transport of low molecular weight solutes across the permselective membrane can be maintained *in vivo*. The inclusion of cells such as bovine adrenal chromaffin cells or PC12 cells does appear to significantly impact the host reaction to the polymer device.

It should be noted that in an effort to maintain or even further enhance the biocompatibility for cell line-containing implants, or reduce protein adsorption that may negatively impact the ability to maintain adequate long-term diffusive characteristics, several postsynthesis modifications have been attempted. Poly(acrylonitrile-*co*-vinyl chloride) (PAN/VC) hollow fiber membranes, which were surface modified by grafting PEO groups, exhibited improved biocompatibility in brain tissue over the unmodified PAN/VC controls [50]. Similar observations were made with PEO-modified poly(hydroxyethyl methacrylate-*co*-methyl methacrylate) membranes utilized extensively in cellular microencapsulation [26].

45.2.4 LONG-TERM PRODUCT SECRETION AND DELIVERY

Before patients suffering from chronic CNS diseases can be routinely implanted with encapsulated cells, long-term survival of the encapsulated cells and continued release of the therapeutic molecule must be demonstrated. Although effective immunoisolation should result in long-term survival of encapsulated cells, surprisingly few studies have examined implant viability for more than a few months. A few notable exceptions exist and provide compelling evidence about the potential for long-term survival and release of molecules from the cells. Following encapsulation, PC12 cells have been maintained *in vitro* and *in vivo* for at least 6 months, while maintaining a typical morphology and clustered arrangement along the lumen of the device [30,44,51]. The cells remain tyrosine hydroxylase (TH)-immunoreactive and mitotically active, with necrosis primarily in regions of high cell density. Electron microscopy confirmed the presence of numerous mitochondria, polysomes, and electron-dense secretory vesicles distributed within the cytoplasm. Spontaneous and evoked release of dopamine can be detected from capsules maintained both *in vitro* and following explantation from the CNS. Both rodent microdialysis and positron emission tomography (PET) studies in primates have confirmed that encapsulated PC12 cells continue to produce L-dopa/dopamine in situ [51,52].

Other cell types, including encapsulated bovine adrenal chromaffin cells, also survive for prolonged periods of time [45,53]. Intraventricular implants of encapsulated bovine chromaffin cell implants survived for nearly 1.5 years and continued to produce catecholamines and met-enkephalin [45]. Polymer-encapsulated, genetically modified cells also survived and continued to secrete NGF for 12–13.5 months in rats [54,55]. The cells remained viable and the NGF secreted from the encapsulated cells was 64% higher following removal from the rat LVs [55]. NGF transgene copy number was equivalent to preimplant levels, indicating NGF gene stability. No deleterious effects from long-term NGF were detectable on body weight, mortality rate, motor/ambulatory function, or cognitive function as assessed with the Morris water maze and delayed matching to position [55]. There was no evidence of NGF-induced hyperalgesia, although tests of somatosensory thresholds did reveal effects related to the NGF delivery. NGF from the encapsulated cells produced a marked hypertrophy of cholinergic neurons within the striatum and nucleus basalis as well as a robust sprouting of cholinergic fibers within the frontal cortex and lateral septum proximal to the implant site. Although no deleterious behavioral effects were observed, the profound anatomical changes and their relationship to functional alterations in normal and diseased brain warrant additional study.

45.2.4.1 Host-Specific Effects on Output of Encapsulated Cells

The available data suggest substantial variability in the *in vivo* performance of encapsulated cells even during the first few months. For example, there was a large range in dopamine and L-dopa output from explanted rodent-sized devices, from 0 to more than 50 pmol/device [56]. In fact, 15% of the devices from that study had no detectable output after only 4 months *in vivo*. Similar variability in device performance has been observed in the majority of *in vivo* studies. In a primate study that produced

therapeutic effects in two-thirds of the 1-methyl-4-phenyl-12,3,6 tetrahydropyrine (MPTP) monkeys implanted with PC12 cells, all five devices implanted in one monkey had virtually 0 output after explantation, while all five devices implanted in another monkey had relatively high output and all five devices in the third monkey had catecholamine output in the midrange [44]. These results suggest some of the variability in device performance may be attributable to individual differences between hosts, a result consistent with that reported for NGF output from encapsulated BHK cells implanted into the LV of rodents [57]. The exact mechanism for these individual differences remains undetermined but deserves serious attention.

45.2.5 DIFFUSION OF MOLECULES FROM POLYMER DEVICES

Tresco et al. [51] conducted a series of *in vivo* experiments to elucidate the relationship between diffusion of dopamine from encapsulated PC12 cells and behavioral recovery in dopamine-depleted rodents. Dopamine was detectable up to 200 ∝m from PC12 cell-loaded macrocapsules, as determined by microdialysis, in concentrations similar to those seen in unlesioned control striatum. In contrast, the levels of dopamine in the perfusate of animals that did not exhibit behavioral recovery were undetectable. Immunocytochemistry was used to estimate the diffusion of NGF from encapsulated cell implanted into the striatum of rats [58]. One month after implantation, the diffusion of NGF was approximately 1 mm. Ciliary neurotrophic factor (CNTF) has also been reported to be detectable in the cerebrospinal fluid (CSF) of amyotropic lateral sclerosis (ALS) patients receiving intrathecal implants of encapsulated CNTF-producing BHK cells [59]. In contrast, CSF levels of NGF were not detectable in nonhuman primates that received intraventricular grafts of NGF-producing cells [60]. At best, these data suggest that diffusion of molecules from encapsulated cells is limited. Moreover, the majority of degenerative CNS diseases will likely not be treatable by delivering drugs from the ventricular space given that diffusion of compounds in CNS tissue is generally severely limited when only governed by passive diffusion. Future work, particularly clinical studies, must seriously consider this issue when determining the optimal numbers and spacing of polymer devices.

45.3 CURRENT THERAPY USING ENCAPSULATED CELLS

Table 45.2 shows a summary of diseases currently being targeted by encapsulated cell therapy in different stages of investigation. While this list is quite extensive, we will concentrate the discussion on the CNS therapies with the highest level of characterization.

45.3.1 PARKINSON'S DISEASE

45.3.1.1 Effects of Catecholamine-Producing Cells

PD is an age-related neurodegenerative disorder characterized by hypokinesia, rigidity, and tremor secondary to the loss

TABLE 45.2

Diseases for Which Cell Encapsulation Is Being Applied

Disease/Model	Encapsulated Cell/Experimental Paradigm	Results
Hormonal and whole organ diseases		
Diabetes	Islets in rodents and dogs	Normoglycemia for 2 years
Hypoparathyroidism	Parathyroid tissue in rats	Normocalcemia for 30 weeks
Kidney failure	Orally delivered *Escherichia coli* bacteria to rats	Normalized urea metabolism
Growth hormone deficiency	Growth hormone producing cells in dogs	Hormone secretion for 1 year
Single gene diseases		
Hemophilia	Factor 9 cells in rats	Cell survival and secretion
Lysosomal storage disease	Beta-glucuronidase cells in mice	Behavioral normalization
Age-related/neurodegenerative diseases		
Age-related motor decline	Catecholamine and GDNF cells in rats	Improvement in motor function
Amyotropic lateral sclerosis	CNTF producing cells in mice	Protection of motor neurons
Alzheimer's disease	NGF cells in rat and primates	Protection of cholinergic neurons, improved memory
Huntington's disease	NGF and CNTF cells in rat and primates	Protection of neurons, improved behavior
Parkinson's disease	Catecholamine and GDNF cells in rat and primate brain	Improved behavior, protection of dopaminergic neurons
Retinitis pigmentosa	Human cells secreting hCNTF into the vitreous	Rod preservation
Spinal cord damage	BDNF cells in rats	Outgrowth of neurites
Oncology		
Colon cancer	iNOS cells (tet-regulated system) in mice	Enhanced survival
Glioblastoma	Endostatin cells in mice and rats	Reduced tumor growth, enhanced survival
HER-2/neu positive tumors	IL-2 fused with anti HER-2/neu antibody in mice	Modest survival benefit
Leukemia	Hybridoma producing antibodies to p15E in mice	Enhanced survival
Ovarian cancer	iNOS cells (tet-regulated system) in mice	Enhanced survival, cures
Other		
Acute and chronic pain	Chromaffin cells in rats	Reduced pain
Clinical trials		
Amyotropic lateral sclerosis	CNTF cells intrathecally	Sustained delivery, no toxicity
Chronic pain	Chromaffin cells in subarachnoid space	Prolonged cell survival, no pain reduction in phase II trials
Diabetes	Human islets intraperitoneally	Insulin independence for 9 months
Huntington's disease	CNTF cells into ventricles	Delivery for 6 months
Hypoparathyroidism	Parathyroid tissue	Successful in two patients
Pancreatic cancer	CYP2B1 cells in tumor vessel	Local tumor growth controlled, well tolerated
Retinitis pigmentosa	CNTF cells in eye	Ongoing

of dopaminergic neurons in the pars compacta of the substantia nigra. Replacing or increasing striatal dopamine levels with oral L-dopa significantly improves the motor deficits in the early stages of PD. Unfortunately, this pharmacological approach has limitations. Systemic administration results in drug distribution to extrastriatal dopamine receptors producing psychoses and vomiting. The therapeutic window of L-dopa's beneficial effects becomes progressively limited as the disease continues its degenerative course. Pulsatile delivery also seems to be associated with more adverse effects than continuous delivery [61]. Studies using pumps to deliver dopamine continuously and directly to the striatum in both rodent and primate models of PD have reported improved motor function with few adverse effects [62–64].

45.3.1.1.1 Rodent Studies

Given the promise of continuous, local L-dopa delivery to the striatum, several studies have detailed the effects of implanting encapsulated dopamine-producing cells into the striatum in rodents and primates (Table 45.3). Rats with PC12 cell-loaded capsules implanted in a 6-hydroxydopamine (6-OHDA) lesioned striatum exhibit fewer rotations after apomorphine administration than nonimplanted rats, which suggests that the devices are releasing catecholamines at levels sufficient to reduce the degree of synaptic supersensitivity that develops after dopamine-depleting lesions. Within 2 weeks following implantation, the number of apomorphine-induced rotations decreased 40%–50% and remained at that level for up to 6 months [51]. Reductions in rotational behavioral do not occur following

TABLE 45.3

**Behavioral Outcome from Encapsulated Catecholamine-Secreting
PC12 Cells in Models of PD**

Animal Model	Testing Paradigm	Outcome
Unilateral lesioned rats	Receptor-activated rotations	+
	T-maze alternation	+
	Somatosensory function	+
	Forelimb freezing/bracing	−
	Forelimb placing	−
	Forelimb akinesia	−
	Postural/locomotor forelimb use	−
Bilateral lesioned rats	Feeding/drinking	−
Aged rats	Activity levels	−
	Balance and coordination	+
MPTP-lesioned primates	Skilled hand use	+
	Clinical Rating Scale	+

+, Positive outcome; −, Negative outcome.

implants of empty polymer devices and are only evident in rats implanted in the denervated striatum and not in rats with devices implanted into the LVs [56,65]. The behavioral effects persist only as long as the devices remain in the striatum.

Measures of drug-induced rotations provide a convenient method for assessing potential efficacy, and significant information has been acquired using this initial preclinical screen. However, relying exclusively on changes in drug-induced rotations has limited clinical relevance and specificity. Accordingly, the effects of PC12 cells on a battery of non-drug-induced behaviors have been examined in 6-OHDA-lesioned rats. Neurological testing revealed behavioral deficits in the affected forelimb that were significantly attenuated by oral Sinemet® [56,66]. Since any transplantation procedure will be utilized as an adjunct to L-dopa administration, these data provided the opportunity to investigate the effects of PC12 cells on both relevant behavioral measures and on the therapeutic window of oral Sinemet. Rats with severe unilateral dopamine depletions received striatal implants of encapsulated PC12 cells and were evaluated on a series of behavioral tests over a range of doses of oral Sinemet. Delivery of L-dopa and dopamine from the encapsulated PC12 cells to the denervated striatum attenuated parkinsonian symptoms. The magnitude of the therapeutic effect produced by continuous, site-specific delivery of catecholamines was greater than the effect produced by acute, systemic oral Sinemet. The beneficial effects of oral Sinemet and striatal implants of PC12 cells were additive, but there were no adverse effects related to the implantation of the PC12 cells and these devices did not increase the adverse effects related to oral Sinemet [67]. Therefore, striatal implants of catecholamine-producing devices have direct therapeutic effects and perhaps more importantly and may widen the therapeutic window of oral Sinemet.

In addition to motor deficits produced experimentally by depleting nigrostriatal dopamine systems in young rodents, motor deficits are also observed with increased age. Age-related deficits in motor functions include deficits in balance, coordinated movement, and generalized locomotion. Enhancement of striatal dopamine function in aged animals by induction of dopamine receptor upregulation or administration of dopaminergic agonists can reduce age-related motor deficits. Since the motor deficits observed in aged rodents appear to be mediated partially by striatal dopamine systems, studies evaluated the potential efficacy of striatal implants of polymer-encapsulated PC12 cells on age-related motor dysfunction in rats [68]. In these studies, aged rats were significantly hypoactive relative to young animals. Moreover, compared to young rats the aged rats (1) remain suspended from a horizontal wire for less time, (2) are unable to descend a wooden pole covered with wire mesh in a coordinated manner, (3) fall more rapidly from a rotating rod, and (4) are unable to maintain their balance on a series of wooden beams with either a square or rounded top of varying widths. Following baseline testing, aged rats received bilateral striatal implants of empty capsules or PC12 cell-loaded capsules. Three weeks later, the aged rats that received PC12 cells showed a robust improvement in performance on the rotarod task and balance on the wooden beams.

45.3.1.1.2 Primate Studies

The studies conducted in the 6-OHDA rat model of PD generally support the clinical utility of encapsulated catecholamine-producing cells. Nonhuman primates are a more relevant model due to their size and complexity of the nervous system, which more closely approximates humans. In terms of tissue volume, diffusion through a rat brain is easier to accomplish than adequate diffusion through the much larger human brain. The nonhuman primate model allows the assessment of therapeutic potential at this level in a way that cannot be approximated in rodents. The potential efficacy of encapsulated PC12 cells has been evaluated in a unilateral MPTP-lesioned primate model. Cynomolgus monkeys trained to perform a task that involved picking food from small food wells that were unilaterally lesioned with an injection of MPTP in the right carotid artery. The resulting MPTP-induced lesion produced a significant and stable impairment in the ability of the animals to use the contralateral limb to retrieve food rewards from the wells. The times required to empty the wells were measured for 3 months postlesion, and the monkeys were then implanted with a U-shaped device that was immediately filled with a suspension of PC12 cells [69]. Following implantation of the cells, manual dexterity was improved and the time required for the monkeys to empty the tray using the impaired hand gradually decreased. Although the PC12 cells attenuated the parkinsonian deficit in this task, some tremor remained and the animals did not recover to prelesion levels. Prior to the implantation of cells, the monkey's left arm was essentially immobile. After implantation, the monkeys could consistently move both their arm and use their fingers. When the cells were flushed out of the device, performance declined to preimplant levels. Together, these data indicated that encapsulated cells survived, were functional, and promoted behavioral recovery even in primate models of PD.

Similar results were obtained in a study by Kordower et al. [44] where four cynomolgus primates were trained on a skilled reaching task similar to that described above and then rendered hemiparkinsonian with an intracarotid injection of MPTP. Three animals received implants of encapsulated PC12 cells into both the caudate and putamen and one animal received implants of empty capsules and served as a control. After a transient improvement, limb use in the control monkey dissipated and returned to post-MPTP levels of disability. Two of the three PC12 cell implanted monkeys recovered on the task to near normal levels for up to 6.5 months posttransplantation. Capsules retrieved from the monkeys who recovered limb function contained abundant viable PC12 cells that continued to release L-dopa and dopamine. In contrast, capsules retrieved from the monkey that did not recover contained few viable PC12 cells. Neuroanatomical and neurochemical evaluation of the implanted striatum failed to reveal any host-derived sprouting of catecholaminergic or indolaminergic fibers, which further suggested that the observed behavioral recovery was due to secretion of catecholamines from the encapsulated PC12 cells.

45.3.1.2 Effects of Neurotrophic Factor-Producing Cells

Several studies have investigated the ability of encapsulated neurotrophic factor-secreting cells to exert neurotrophic effects in rodent models of PD [70–72]. In an initial study [71], encapsulated cells releasing approximately 5 ng of glial derived neurotrophic factor (GDNF)/day were implanted immediately rostral to the substantia nigra. The medial forebrain

bundle was transected 1 week later and the ability of encapsulated GDNF-producing cells to minimize the behavioral effects of the lesion and prevent the degeneration of dopaminergic neurons was determined. GDNF treatment reduced the numbers of amphetamine-induced rotations in lesioned animals and attenuated the loss of neurons in the substantia nigra but had no effect on striatal dopamine levels. Using the same model system, neurturin-producing cells, a homologue of GDNF, were investigated for its neurotrophic activity [72]. Neurturin-treated animals had significantly more TH-positive neurons in the substantia nigra (51% compared to 16% in controls) but failed to show any behavioral improvement as measured by rotational behavior. Together, these data suggest that encapsulated cells may have a role in neurotrophic therapy for PD. However, additional studies in animal models are required to determine the relationship between the anatomical and behavioral consequences of cell-based delivery.

45.3.1.2.1 Encapsulated Neurotrophic Factor-Producing Cells to Enable Survival of Unencapsulated Cografts

The transplantation of encapsulated genetically modified cells also represents a potential means of delivering trophic factors to the brain to support the survival of cografted cells or tissue. In a series of studies, NGHF-secreting BHK cells were encapsulated and implanted into the left LV or the left striatum of hemiparkinsonian rats approximately 1.5 mm away from cografted unencapsulated adrenal medullary chromaffin cells [73]. Although the animals receiving adrenal medulla alone or adrenal medulla with intraventricular NGF-secreting cell grafting did not show recovery of apomorphine-induced rotational behavior, the animals receiving adrenal medulla with intrastriatal NGF-secreting cell implants showed a significant recovery of rotational behavior 2 and 4 weeks after transplantation. Histological analysis revealed that intraventricular NGF increased the number of surviving chromaffin cells five to six times above that seen in animals receiving adrenal medulla alone. Even more impressively, intrastriatal NGF-secreting cells increased the number of surviving chromaffin cells by more than 20 times than that in animals receiving adrenal medullary cells alone. The beneficial effects of NGF-producing cells were evident for as long as 12 months postgrafting (i.e., the longest time point examined) and were independent of the age of the chromaffin cell donor [54,74]. These results indicate the potential use of intrastriatal implantation of encapsulated NGF-secreting cells for augmenting the survival of cografted chromaffin cells as well as promoting the functional recovery of hemiparkinsonian rats.

45.3.2 Encapsulated NGF-Producing Cells in Animal Models of Alzheimer's Disease

AD affects approximately 5% of the population over the age of 65 and is the most prevalent form of adult-onset dementia. The most prominent feature of AD is a progressive deterioration of cognitive and mnemonic ability, which is at least partially related to the degeneration of basal forebrain cholinergic neurons. Treatments do not slow or prevent cholinergic neuron loss or the associated memory deficits. Several converging lines of evidence indicate that NGF has potent target-derived trophic and tropic effects upon cholinergic basal forebrain neurons [75,76].

45.3.2.1 Rodent Studies

Although no model faithfully recapitulates the complex etiology and time-dependent loss of cholinergic neurons seen in AD patients, model systems have been developed to determine if NGF prevents the death of damaged cholinergic neurons following trauma. Initial studies determined whether encapsulated NGF-secreting BHK cells (Figure 45.4) could prevent cholinergic neuron loss following aspiration of the fimbria/fornix [77]. Rats received lesions followed by intraventricular implants of either NGF-producing or control (nontransfected) cells. Control-implanted animals had an extensive loss (88%) of ChAT-positive cholinergic neurons ipsilateral to the lesion that was prevented by NGF cell implants (14% loss).

One of the cardinal behavioral symptoms of AD is a progressive loss of cognitive ability. Just as no animal model faithfully mimics the complex etiology and pathophysiology of AD, comparable behavioral abnormalities are difficult to reproduce in animal models. However, the aged rodent shows a progressive degeneration of basal forebrain cholinergic neurons together with marked cognitive impairments that are partly reversible by administering NGF. Lindner et al. [57] trained 3-, 18-, and 24-month-old rats on a spatial learning task in a Morris water maze (Figure 45.5). Cognitive function, as measured in this task, declined with age. Following training, animals received bilateral intraventricular implants of encapsulated NGF or control cells. The 18- and 24-month-old animals receiving NGF cells showed a significant improvement in cognitive function. No improvements or deleterious effects were observed in the young nonimpaired animals. There was no evidence that the NGF cells produced changes in mortality, body weights, somatosensory thresholds, potential hyperalgesia, or activity levels, suggesting that the levels of NGF produced were neither toxic nor harmful to the aged

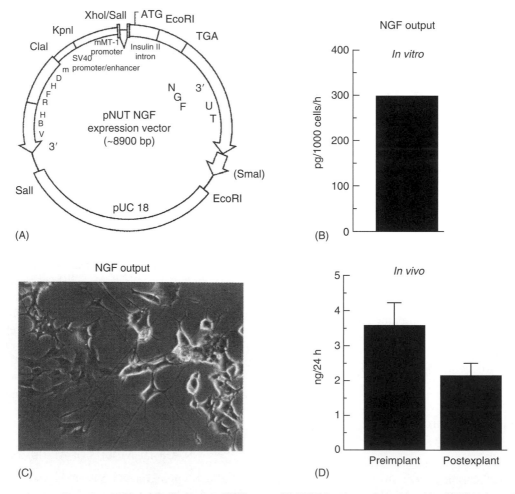

(A)

(B)

(C)

(D)

FIGURE 45.4 (A) Expression vector containing the human NGF gene. (B) NGF levels, as determined by ELISA, in unencapsulated (top) and encapsulated (bottom) BHK cells. The *in vivo* levels were determined from devices 3 months following explant from rodent striatum. (C) The biological activity of the NGF from encapsulated BHK cells is shown in phase-contrast photomicrographs of PC12 cells that exhibit extensive neurite processes. Original magnification = 25 μm.

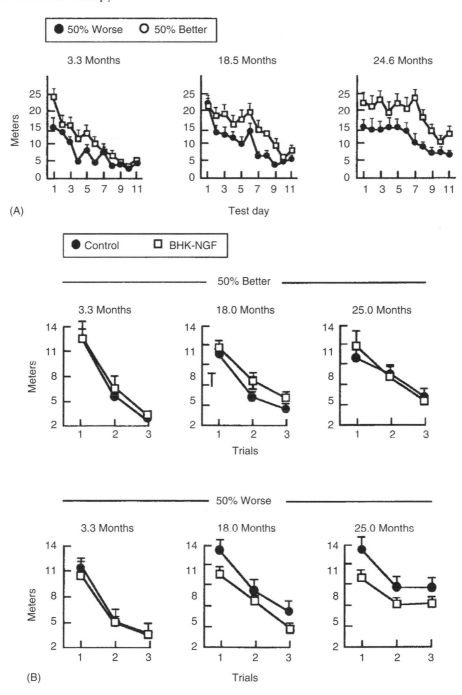

FIGURE 45.5 Cognitive function in young, middle-aged, and aged rats following implantation of encapsulated NGF-producing cells into the LVs. (A) Prior to implantation, animals were divided into the 50% worse and 50% better performers in a water maze task. (B) Following implantation, NGF was found to improve performance in the middle-aged and aged animals. Moreover, the improvements in cognitive performance were greatest in those animals that demonstrated the worst initial performance.

rats. Evidence of age-related atrophy of cholinergic neurons was observed in the striatum, medial septum, nucleus basalis, and vertical limb of the diagonal band. These anatomical changes were most severe in animals with the greatest cognitive impairments, suggesting a link between the two pathological processes. Anatomically, the NGF released from the encapsulated cells increased the size of the atrophied basal forebrain and striatal cholinergic neurons to the size of the neurons in the young healthy rats.

45.3.2.2 Primate Studies

Results similar to those obtained in rodents were obtained in nonhuman primates [60], an essential prerequisite to human clinical trials (Figure 45.6). Cynomolgus primates received transections of the fornix followed by placement of encapsulated NGF-secreting BHK or control cells into the LV. In the control animals, a significant reduction in the number of cholinergic neurons was observed in the medial septum and vertical limb of the diagonal band of Broca. Again, loss of

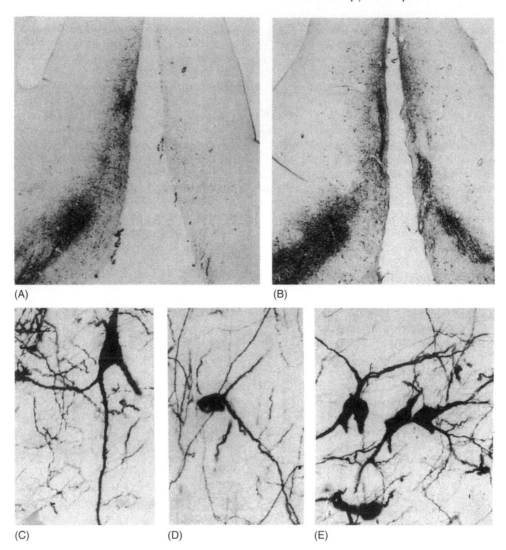

FIGURE 45.6 Photomicrographs through the septal diagonal band complex of cynomolgus monkeys. Monkeys received a fornix transection together with polymer-encapsulated implants of either BHK-control (A) or a BHK-NGF cells (B). Note the extensive loss of NGF receptor positive neurons ipsilateral to the lesion (right) in the control-implanted monkeys. In contrast, numerous NGF cholinergic neurons were observed in fornix transected monkeys receiving the BHK-NGF implant. High-power photomicrographs illustrating the morphology of cholinergic neurons within the medial septum of young monkeys receiving BHK-NGF (C–E) implants. Note the enlarged perikarya and extensive neuritic arbor displayed by monkeys receiving the BHK-NGF implants.

cholinergic neurons was prevented by implants of NGF-secreting cells. It also appeared that cholinergic neurons within the medial septum of NGF-treated animals were larger, more intensely labeled, and elaborated more extensive proximal dendrites than those displayed by BHK-control animals.

In addition to the effects on cell viability, NGF implants induced a robust sprouting of cholinergic fibers proximal to the implant site [60]. All monkeys receiving NGF implants displayed dense collections of NGF receptor-immunoreactive fibers throughout the dorsoventral extent of the lateral septum. This effect was unilateral as the contralateral side displayed only a few cholinergic fibers in a manner similar to that seen in control-implanted monkeys. The cholinergic nature of this sprouting was confirmed by an identical pattern of fibers that

were ChAT-immunoreactive and AChE-positive. These fibers ramified against the ependymal lining of the LV adjacent to the transplant site and were particularly prominent within the dorsolateral quadrant of the septum corresponding to the normal course of the fornix. The cell sparing and sprouting results have been replicated (Figure 45.7) in a group of aged nonhuman primates [78].

45.3.3 POLYMER-ENCAPSULATED CELLS TO DELIVER NEUROTROPHIC FACTORS IN ANIMAL MODELS OF HD

HD is an inherited, progressive neurological disorder characterized by a severe degeneration of basal ganglia neurons, particularly the intrinsic neurons of the striatum. Accompanying

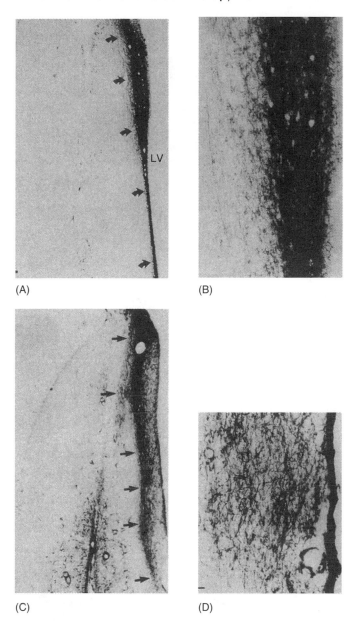

(A)

(B)

(C)

(D)

FIGURE 45.7 Sprouting of cholinergic fibers in young (A,B) and aged (C,D) NGF-treated monkeys. Low (A,C) and high (B,D) power photomicrographs of NGF receptor immunostained sections illustrating a dense plexus of cholinergic fibers on the side of NGF treatment (arrows) adjacent to the LV. Scale bar in (A) = 1000 μm for (A) and (C), scale bar in (D) = 50 μm for (B) and (D).

the pathological changes is a progressive dementia coupled with uncontrollable movements and abnormal postures. From the time of onset, an intractable course of mental deterioration and progressive motor abnormalities begins with death usually occurring within 15–17 years. Overall, the prevalence rate of HD in the United States is approximately 50 per 1,000,000. There is no treatment that effectively addresses the behavioral symptoms or slows the inexorable neural degeneration in HD.

Intrastriatal injections of excitotoxins such as quinolinic acid (QA) are a useful model of HD and can serve to evaluate

novel therapeutic strategies aimed at preventing, attenuating, or reversing neuroanatomical and behavioral changes associated with HD [79–82]. The use of trophic factors in a neural protection strategy may be particularly relevant for the treatment of HD. Unlike other neurodegenerative diseases, genetic screening can identify individuals at risk providing an opportunity to design treatment strategies to intervene prior to the onset of striatal degeneration.

45.3.3.1 Rodent Studies

Infusions of trophic factors such as NGF or implants of cells genetically modified to secrete NGF have proven effective in preventing the neuropathological sequelae, resulting from intrastriatal injections of excitotoxins including QA [83–86]. We [87,88] examined the ability of encapsulated trophic factor-secreting cells to affect central striatal neurons in a series of defined animal models of HD. In these experiments, rats received intraventricular implants of encapsulated NGF- or CNTF-producing cells (Figure 45.8). One week later, the same animals received unilateral injections of QA (225 nmol) or the saline vehicle into the ipsilateral striatum. An analysis of Nissl-stained sections demonstrated that the size of the lesion was significantly reduced in animals receiving NGF and CNTF cells relative to controls. Moreover, CNTF cells attenuated the extent of host neural damage produced by QA as assessed by a sparing of cholinergic, diaphorase-positive and GABAergic neurons (Figure 45.9). Neurochemical analyses confirmed the protection of multiple striatal cell populations using this strategy. Importantly, behavioral studies offer additional and compelling evidence of neuronal protection that can be produced in animal models of HD. Trophic factor-secreting cells have provided extensive behavioral protection as measured by tests that assess both gross and subtle movement abnormalities. Moreover, these same animals show improved performance on learning and memory tasks, indicating the anatomical protection afforded by trophic factors in this model is paralleled by a robust and relevant behavioral protection [89].

45.3.3.2 Primate Studies

The ability of cellularly delivered trophic factors to preserve neurons within the striatum in a rodent model of HD led to similar studies in nonhuman primates [90]. Polymer capsules containing CNTF-producing cells were grafted into the striatum of Rhesus monkeys. One week later, a QA injection was placed into the putamen and caudate proximal to the capsule implants. As seen in the rodent studies, the volume of striatal damage was decreased and both GABAergic and cholinergic neurons destined to degenerate were spared in CNTF grafted animals. Although all animals had significant lesions, there was a threefold and sevenfold increase in GABAergic neurons in the caudate and putamen, respectively, in CNTF-grafted animals relative to controls. Similarly, there was a 2.5-fold and fourfold increase in cholinergic neurons in the caudate and putamen, respectively, in CNTF-grafted animals (Figure 45.10).

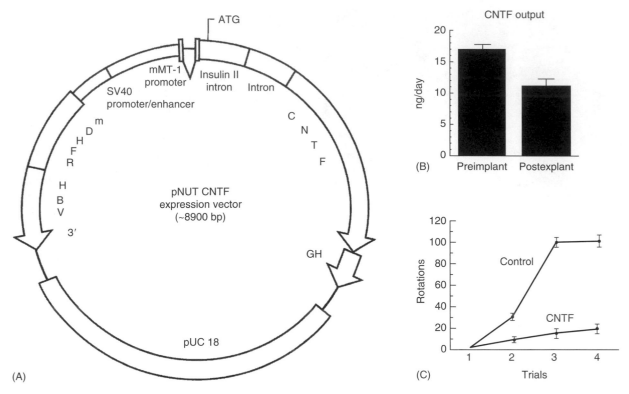

FIGURE 45.8 (A) Expression vector containing the human CNTF gene. (B) CNTF levels, as determined by ELISA, in encapsulated BHK cells immediately prior to implantation (left) and immediately following retrieval from rodent LV 70 days following implantation (right). (C) Implants of encapsulated CNTF producing cells reduce apomorphine rotations in rats after unilateral striatal injection of QA.

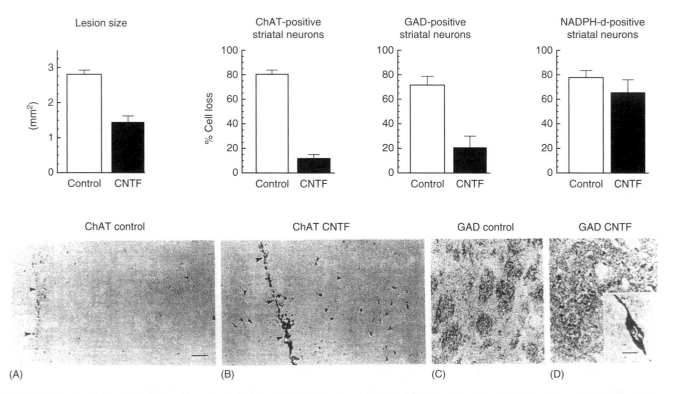

FIGURE 45.9 Lesion volume and neuronal cell counts in QA-lesioned rats. Control-implanted animals displayed a marked lesion volume (A) determined by Nissl staining and a significant loss of multiple types of striatal cell types including cholinergic (B), GABAergic (C), and diaphorase-positive neurons (D). The cholinergic and GABAergic neuronal losses were largely prevented in animals receiving CNTF implants, while the loss of diaphorase-positive neurons was not affected. In each case, data are presented as a percent loss of neurons on the lesioned/implanted side compared to intact contralateral side. Representative photomicrographs for cholinergic and GABAergic cells are shown for both control and CNTF-implanted animals. Note the appearance of numerous healthy appearing cholinergic and GABAergic neurons in the CNTF-treated animals. Scale bar in ChAT control = 500 μm for ChAT control/CNTF and scale bar in insert = 100 μm for GAD control/CNTF and 17 μm.

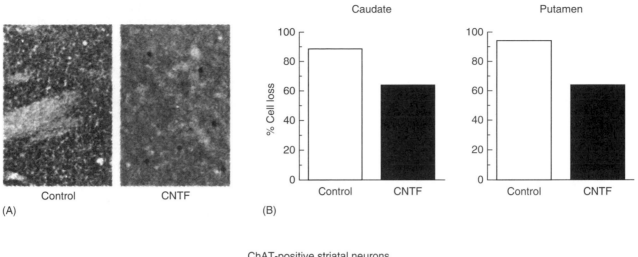

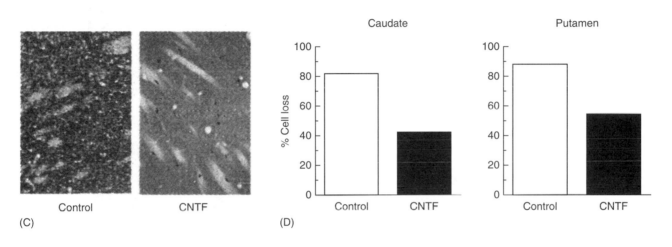

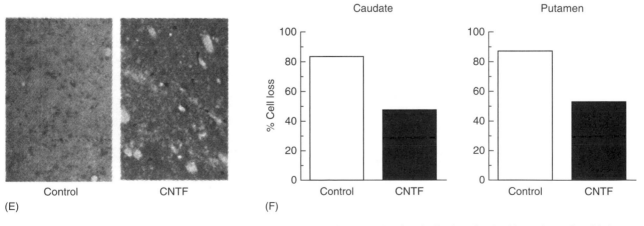

FIGURE 45.10 Neuronal cell counts in QA-lesioned monkeys. Control-implanted animals displayed a significant loss of multiple types of striatal neurons including GABAergic (A and B), cholinergic (C and D), and diaphorase-positive neurons (E and F). Although neuronal loss was still present in animals receiving CNTF implants, it was significantly attenuated in both the caudate and putamen. In each case, data are presented as a percent of neurons on the lesioned/implanted side compared to the intact contralateral side. Representative photomicrographs for all three cell types are shown for both control and CNTF-implanted animals (A = GABAergic, C = cholinergic, and E = diaphorase-positive neurons).

The ability to preserve GABAergic neurons in animals models of HD is an important, though not entirely sufficient, step to develop a useful therapeutic. If the perikarya are preserved without sustaining their innervation, then the experimental therapeutic strategy under investigation is not likely to yield significant value. The striatum is a central station in a series of loop circuits that receive inputs from all of the neocortex, projecting to a number of subcortical sites, and then returns information flow to the cerebral cortex. One critical part of this circuitry is the GABAergic projections to the globus pallidus and substantia nigra pars reticulata, the parts of the direct and indirect basal ganglia loop circuits. The integrity of this circuit can be evaluated immunocytochemically using an antibody that recognizes GABAergic terminals (DARPP-32) to determine if the preservation of GABAergic somata within the striatum also results in the preservation of the axons of these neurons to critical extrastriatal sites. Using quantitative morphological assessment of DARPP-32 optical density, monkeys receiving QA lesions had significant reductions in DARPP-32 immunoreactivity within the globus pallidus and substantia nigra. The lesion-induced decrease in GABAergic innervation for both of these regions was prevented in CNTF-grafted monkeys (Figure 45.11), demonstrating that this treatment strategy protected GABAergic neurons and sustained their normal projection systems [90].

The intrinsic striatal cytoarchitecture can be preserved in monkeys by CNTF grafts, and once exposed to these grafts, the cells apparently maintain their projections. But are the afferents to the striatum, specifically from the cerebral cortex, also influenced by these grafts? This may be particularly critical if some of the more devastating nonmotor symptoms seen in HD result from cortical changes secondary to striatal degeneration. Since layer V neurons from motor cortex send a dense projection to the postcommissural putamen, a region that was severely impacted by the QA lesion, the effects of QA lesions and CNTF implants on the number and size of cortical neurons in this region were examined. Although the QA lesion did not impact the number of neurons in this cortical area, layer V neurons were significantly reduced in cross-sectional area on the side ipsilateral to the lesion in control-grafted monkeys. This atrophy of cortical neurons was reversed by CNTF grafts [90].

A recent set of studies using CNTF producing cells in 3NP-treated monkeys have replicated and extended these results [91]. Following 10 weeks of 3NP treatment, monkeys displayed pronounced chorea and severe deficits in frontal lobe cognitive performance as assessed by the object retrieval detour test. Following implantation of CNTF-producing cells, a progressive and significant recovery of motor and cognitive recovery occurred. Histological analysis demonstrated that

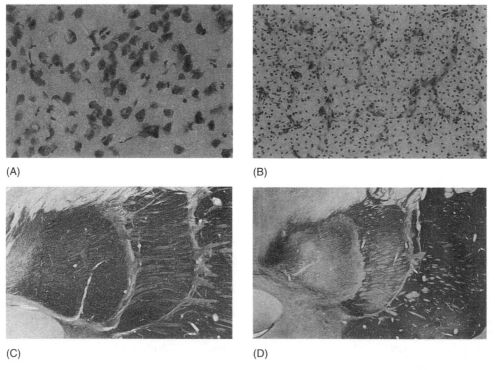

(A)   (B)

(C)   (D)

FIGURE 45.11 Photomicrographs of Nissl-stained sections through the striatum of monkeys that received QA injections into the striatum followed by implants of encapsulated CNTF producing (A) or control (B) BHK cells. A paucity of healthy neurons is observed in the striatum of control monkeys that is in stark contrast to the numerous healthy appearing neurons seen in the same region of the CNTF-treated monkeys. Together with the sparing of strital neurons is a preservation of the GABEergic projection from the striatum to the globus pallidus. DARPP-32 immunocytochemistry revealed an intense, normal appearing immunoreactivity within both the external and internal segments of the globus pallidus of CNTF-treated animals (C). In contrast, DARPP-32 immunoreactivity is reduced in control-implanted animals as a consequence of the lesion (D). Quantitative analysis confirmed the sparing of this projection in CNTF-treated monkeys.

CNTF was neuroprotective and spared NeuN and calbindin-positive cells in the caudate and putamen.

While the sparing of striatal neurons and maintenance of intrinsic circuitry are impressive, the effect is less than that seen in rodents. In primates, robust protection is limited to the area of the capsules and the total area of the lesion remains extensive, likely because the diffusion of CNTF from the capsule is not sufficient to protect more distant striatal regions undergoing degeneration. This concept is supported by a recent experiment examining the effects of intraventricular grafts of encapsulated CNTF grafts in the nonhuman primate model of HD [92]. In contrast to when the capsules were placed directly within brain parenchyma, intraventricular placements failed to protect any striatal cell types. The complete lack of neuroprotection provided by intraventricular implants in primates should be considered more carefully in clinical trials where encapsulated cells have to date been placed into the LVs of HD patients [93,94]. If human trials are to yield clinically relevant positive effects, the means of CNTF delivery utilized in these studies needs to be improved. Whether this entails grafting more capsules, enhancing the CNTF delivery from the cells by changing the vector system or cell type employed or changing the characteristics of the polymer membrane remains to be determined.

45.3.4 Encapsulated Cells for Treating Brain Tumors

Malignant tumors of the brain represent 2%–3% of all adult neoplasms occurring in up to 9 in 100,000 individuals. Gliomas account for approximately half of all primary brain tumors. Despite the continued refinement of neuroimaging techniques, progress in microsurgery and evolving chemotherapeutic drugs, the prognosis for glioma patients is poor with a median survival of around 1 year. Surgical debulking of the growing tumor mass is the primary treatment modality. While surgery provides short-term improvement for subsequent treatments it is ultimately ineffective because of the infiltrative nature of glioma resulting in tumor recurrence neighboring the resection margin.

If a means of delivering therapeutic molecules directly to the region of tumor recurrence could be developed, then the prognosis for glioma patients might be significantly improved. Cell encapsulation has recently been suggested as one means of achieving this goal [95–102]. While the majority of the data obtained so far utilizes microencapsulation techniques, the data sufficiently intriguing to be presented here. Of particular relevance to glioma, recent studies demonstrate that experimental gliomas could be successfully treated with endostatin-producing alginate encapsulated cells. In one study, BT4C glioma cells and endostatin-producing capsules were implanted into the brains of syngeneic rats [99]. The encapsulated cells survived well and secreted endostatin for up to 4 months *in vivo*. Survival was increased by 84% relative to control animals. Histological assessment of the transplant revealed that the endostatin-producing cells produced cellular apoptosis and the formation of large necrotic areas within the treated tumors. The treatment was otherwise reportedly well-tolerated with no serious side effects. These data were supported by a similar study in nude mice. Endostatin-producing BHK cells were encapsulated into alginate microspheres and injected 10 days prior to a subcutaneous injection of U87 glioma cells. At 21 days posttreatment, the growth of the gliobastoma xenografts was reduced with less neovascularization surrounding the tumor [96]. Similar results have been achieved using encapsulated kidney 293 cells modified to produce endostatin [103]. Intravital video microscopy revealed reductions in the density, diameter, and function of tumor-associated microvessels.

Another approach to treating glioma uses encapsulated cells to enable gene transfer *in vivo*. Martinet et al. [98] encapsulated mouse psi20VIK packaging cells into alginate microspheres. The capsules were then stereotaxically injected into established C6 glioblastomas and the animals were treated with gancyclovir. After 14 days, the tumors were harvested and it was determined that 3%–5% of the tumor cells had been transduced, resulting in a 45% reduction in tumor volume.

Together, these data represent a young but encouraging application of encapsulated cell therapy. The ability of cell-based therapies to provide continuous, local delivery of a wide range of therapeutic molecules will likely lead to further refinements in this area. Still, as recently pointed out [101], optimization of capsule biostability, tissue biocompatibility, cell choice and viability, dose-release, and biodistribution of therapeutic molecules represent unresolved areas of preclinical research.

45.3.5 Encapsulated Cells for Retinitis Pigmentosa

Retinitis pigmentosa (RP), an inherited degenerative disease of the retina, affecting roughly 1 out of 4000 individuals worldwide and is the sixth leading cause of blindness [104]. The disease progresses slowly from mild degeneration of the rod photoreceptors in the periphery to a more pronounced loss of vision involving the cones in the central retina. The end result is complete blindness that occurs on the order of months or years depending on the etiology of the disease.

RP is a disease with an extremely complex genetic profile, and over 100 rp-inducing mutations have been identified to date [105]. While there has been progress on this front, the influence of genetic medicine on the development of therapeutics for RP continues to be minimal. Current clinical strategies focus on optimization of the light path (lens replacements) and treatment of secondary effects of the disease. Work in the field of protein therapeutics, specifically neurotrophic factors, has led to the use of CNTF as a potential neuroprotective mediator of RP. For example, in the *rd* rodent model of retinal degeneration, it was shown that gene transfer of CNTF retarded the degeneration of photoreceptors [106,107]. Purified CNTF also slows photoreceptor degeneration in the *rd/rd* and *nr/nr* murine models of the disease [108]. Because of the relatively short half-life of the recombinant protein and the immunoprivileged environment of the eye, encapsulated cell technology is an extremely attractive mode of delivery for CNTF and other potential protein therapeutics to the eye.

More recently, retinal epithelial cells have been transfected with a plasmid containing the CNTF gene. CNTF is released *in vivo* at a rate of approximately 1.5 ng/day. Initial efficacy studies with the CNTF-producing cell line were carried out in s334ter-3 transgenic rats [109]. Naked cells were injected into the vitreous of rat eyes and after 20 days, eyes were enucleated and processed for histology. While untreated eyes or eyes injected with the nontransfected parental cell line showed severe progression of retinal degeneration (about one row of the outer nuclear layer [ONL] remaining), eyes treated with the CNTF-secreting cell line showed five to six times thicker ONL throughout the retina.

The canine *rcd1* model for retinal degeneration [110,111] was used to evaluate the device loaded with the CNTF-secreting cell line [109]. In this experiment, various amounts and combinations of parental cells and secreting cells were used to achieve five different doses between <100 pg/day to 15 ng/day. Again, as with the work in rodents, neuroprotection was shown to occur, and in this case, it was dose-dependent. Untreated eyes showed on average three layers of the ONL, while treated eyes showed between three and six layers. In all cases, the distribution of retinal preservation was homogenous throughout the retina.

45.4 INITIAL CLINICAL TRIALS

45.4.1 Amyotropic Lateral Sclerosis

Neuromuscular disorders such as ALS are marked by a progressive degeneration of spinal motor neurons. Different families of neurotrophic factors demonstrate therapeutic potential *in vitro* and in animal models of motor neuron disease [112–119]. The cytokine CNTF has neuroprotective effects for motor neurons in *wobbler* mice [116] and homozygote pmn (progressive motorneuropathy) mice [117,118]. The delivery of CNTF to motor neurons by peripheral administration proved difficult due to severe systemic side effects, short half-life of CNTF, and the inability of CNTF to cross the BBB [120–122].

Continuous intrathecal delivery of CNTF proximal to the nerve roots in the spinal cord is a practical alternative that could result in less side effects and better efficacy of CNTF in ALS patients. After safety, toxicology, and preclinical evaluation [123], a clinical trial to establish safety has been performed in ALS patients using polymer-encapsulated cells genetically modified to secrete CNTF [124]. A total of six ALS patients with early stage disease indicated by a forced vital capacity greater than 75% with no other major illness or treated with any investigational drugs for ALS were included. These patients were baseline tested for Tufts Quantitative Neuromuscular Evaluation, the Norris scale; blood levels of acute reactive proteins, and CNTF levels in the serum and CSF. BHK cells were encapsulated into 5 cm long by 0.6 mm diameter hollow membranes and implanted into the lumbar intrathecal space. The device included a silicone tether that was sutured to the lumbodorsal fascia and the skin was closed over the device. CNTF concentrations in the CSF

were not detectable prior to implantation, but were found in all six patients at 3–4 months postimplantation. All six explanted devices had viable cells and CNTF secretion of approximately 0.2–0.9 μg/day. No CNTF was detected in the serum. More recently, a phase I/II clinical trial was initiated in 12 ALS patients using the same approach. Again, CNTF was detectable for several weeks in the CSF of 9 out of 11 patients. Concurrent *in vitro* studies using CSF samples from these same patients revealed only a very weak antigenic immune response with bovine fetuin as the main antigenic component [125].

45.4.2 Chronic Pain

Numerous studies with rodent models of acute and chronic pain have suggested that adrenal chromaffin cells implanted into the intrathecal space and in the periaqueductal gray reliably produce significant analgesic effects [126,127]. While the majority of these studies have used unencapsulated cells, recent studies suggested that encapsulated cell implants also produce analgesia in rats. The analgesic effects of adrenal chromaffin cells in rodent model have provided the rationale to pursue clinical trials in patients with chronic pain. Small, open label trials demonstrated that the implantation procedure was minimally invasive and well tolerated [128,129]. Neurochemical and histological studies determined that the encapsulated cells survived and were biochemically functional for up to 1 year. Because reductions in morphine intake were noted following implantation (suggesting efficacy), larger scale, randomized studies were initiated in a collaborative study between CytoTherapeutics and Astra Pharmacueticals. The trials were recently halted because the efficacy achieved did not reach a level high enough to warrant further study [130].

Interestingly, several recent reports indicate encapsulated adrenal chromaffin cell implants may not produce efficacy as originally suggested [131–133]. Extensive studies in acute and chronic rodent pain models have failed to find any evidence of analgesia. This lack of effect occurred under conditions that were apparently designed to exactly reproduce previous testing procedures that did demonstrate efficacy. Among the variables examined were the location of implant (intrathecal vs. intraventricular), a wide range in cell preparation techniques, and an exhaustive battery of acute and chronic pain tests with and without nicotine stimulation. Importantly, the authors reported that systemic administration of morphine produced significant analgesia when tested in parallel in the same models. While subtle testing differences cannot be ruled out as contributing factors in the differences between these recent and previous studies, together with the only well controlled clinical trial conducted to date, it appears that, at the least, adrenal chromaffin cells do not produce analgesic effects as consistently as previously reported.

45.4.3 Huntington's Disease

Recently, clinical trials were completed to determine the safety and tolerability of CNTF-producing cells (0.15–0.50 μg

CNTF/day) implanted into the LV of HD patients [134,135]. HD symptoms were also analyzed using neuropsychological, motor, neurological, and neurophysiological tests and the striatal pathology monitored using MRI and PET scan imaging. Six subjects with stage 1 or 2 HD had one capsule implanted into the right LV; the capsule was retrieved and exchanged for a new one every 6 months over a total period of 2 years. No sign of CNTF-induced toxicity was observed; however, depression occurred in three subjects after removal of the last capsule, which may have correlated with the lack of any future therapeutic option. All retrieved capsules were intact but contained variable numbers of surviving cells, and CNTF release was low in 13 of 24 cases. Improvements in electrophysiological results were observed and were correlated with capsules releasing the largest amount of CNTF. This phase I study shows the safety, feasibility, and tolerability of this gene therapy procedure. Heterogeneous cell survival, however, stresses the need for improving the technique.

45.4.4 RETINITIS PIGMENTOSA

Phase I clinical studies have just been completed examining the safety of encapsulated CNTF-secreting cells surgically implanted into the vitreous of the eye. The trial restricted study participants to RP patients with 20/100 vision or less, a central visual field diameter of 40° or less, and a flicker ERG amplitude of 2 μV or less. Ten participants received CNTF implants in one eye. When the implants were removed after 6 months, they contained viable cells with minimal cell loss and gave CNTF output at levels previously shown to be therapeutic for retinal degeneration in rcd1 dogs. Although the trial was not powered to demonstrate clinical efficacy, of seven eyes for which visual acuity could be tracked by conventional reading charts, three eyes reached and maintained improved acuities of 10–15 letters, equivalent to two- to three-line improvement on standard Snellen acuity charts. This phase I trial indicated that CNTF is safe for the human retina even with severely compromised photoreceptors [136]. Neurotech is currently progressing to a multicenter phase II clinical trial.

45.5 CONCLUSIONS AND FUTURE DIRECTIONS

There is considerable promise of encapsulated cell therapy for treating a wide range of CNS disorders. Still, a number of research avenues exist, which are incompletely explored and deserve attention. In some preclinical studies, the extent of diffusion from the implants appears to limit the therapeutic effectiveness of the encapsulated cells [51,58]. Given the size of the human brain relative to the rodent and nonhuman primate brain, the potential problems related to limited tissue diffusion should be examined empirically.

Encapsulation allows the use of cells from human and animal sources with and without genetic modification. In theory, the capsule should isolate the cells from the surrounding tissue. Still, should a capsule rupture during implantation or retrieval, a deleterious host immunological response could

be induced. Though the host immune system should reject any released cells following capsule damage, the potential for tumor growth remains. Alterations in the ability of the host immune system to reject cells following damage to implants could also change upon long-term residence of the cells within the host. To date, no studies have systematically evaluated these risks, particularly with regard to the long-term effects of encapsulated cell implants. Again, large animal studies using intact and intentionally damaged devices would provide a useful starting point for evaluating these issues. These studies could use normal and immunosuppressed animals and evaluate potential tumorigenicity and changes in the host immune system over long periods of time.

Regulation of dosage is another area that deserves attention. In its most basic iteration, varying the numbers of cells within an implant, the size of the implant, or the use of multiple implants, may permit a range of doses to be delivered. While some long-term cell survival studies have been conducted, they tend to utilize only CSF-filled spaces and have not systematically examined cell survival and output of the desired molecule over long periods of time. Rather, studies have provided a "snapshot" of survival and output at a single time point. Large, long-term, well-controlled studies need to be conducted to examine the relationship between variables that include time, cell survival, gene expression (when modified cells are used), neurochemical output, the initial numbers of cell encapsulated, and the type of semipermeable membrane and ECM used for encapsulation. Obviously, such studies are time-consuming and expensive. But without them, the conditions optimal for successful cell encapsulation will remain speculative.

Another area that has attracted very little attention concerns the variability in the *in vivo* performance of encapsulated cells and the possible role that the host tissue environment plays in this variability. As discussed earlier, it appears that at least some of the variability in device performance is attributable to differences between hosts. While the mechanisms underlying these individual differences remain undetermined, several potential candidates exist including the variations in the general health of the animals, between animal differences in immune function and undetected microbreaches in the polymer membrane prior to or during implantation. The notion that the viability of grafted cells may depend in part on host-related variability in the CNS environment has only been suggested for encapsulated cells to date [56]. However, this emerging concept might also prove to be relevant for all CNS transplantation approaches that are cellular based. Indeed, the entire field of neural transplantation might benefit from this new perspective uncovered using encapsulated cells.

Finally, very few clinical studies have been conducted to date. While several small safety studies have been completed, only one large, controlled clinical study has been performed using encapsulation technology. This study evaluated the use of encapsulated adrenal chromaffin cells for the treatment of pain but failed to reveal analgesia sufficient enough to continue the trials. As we have already discussed in a previous section, the selection of pain as an initial indication for detailed study might have been an unfortunate choice given that recent preclinical data using encapsulated chromaffin cells are mixed at best. The only other clinical targets under investigation are ALS and HD and these are apparently modest efforts. Until, larger, controlled clinical trials are conducted, the potential of this technology will not be fully realized.

In conclusion, it appears that the implantation of encapsulated cells may provide an effective means of alleviating the symptoms of numerous human conditions/diseases. One particularly attractive avenue of research continues to be the application of trophic factors to minimize or halt the progression of neural degeneration or promote regeneration of damaged central nerves. However, caution must be applied when considering any novel therapy for treating brain disorders and the wide scale use of polymer neural implants should be considered only after rigorous scientific experimentation in animal models and their demonstrated efficacy and safety in human clinical trials.

REFERENCES

1. Cervos-Navarro, J., Kannuki, S., and Nakagawa, Y., Blood–brain barrier (BBB). Review from morphological aspect, *Histol. Histopathol.*, 3, 203, 1998.
2. Rubin, L.L. and Staddon, J.M., The cell biology of the blood–brain barrier, *Ann. Rev. Neurosci.*, 22, 11, 1999.
3. Friden, P.M. et al., Blood–brain barrier penetration and in vivo activity of an NGF conjugate, *Science*, 259, 373, 1993.
4. Friden, P.M., Receptor-mediated transport of therapeutics across the blood–brain barrier, *Neurosurgery*, 35, 294, 1994.
5. Jiao, S., Miller, P.J., and Lapchak, P.A., Enhanced delivery of [^{125}I] glial cell line-derived neurotrophic factor to the rat CNS following osmotic blood–brain barrier modification, *Neurosci. Lett.*, 220, 187, 1996.
6. Kroll, R.A. and Neuwelt, E.A., Outwitting the blood–brain barrier for therapeutic purposes: Osmotic opening and other means, *Neurosurgery*, 42, 1083, 1998.
7. Kordower, J.H. et al., Delivery of trophic factors to the primate brain, *Exp. Neurol.*, 124, 21, 1993.
8. Mufson, E. et al., Distribution and retrograde transport of trophic factors in the central nervous system: Functional implications for the treatment of neurodegenerative diseases, *Prog. Neurobiol.*, 57, 451, 1999.
9. Riddle, D.R., Katz, L.C., and Lo, D.C., Focal delivery of neurotrophins into the central nervous system using fluorescent latex microspheres, *Biotechniques*, 23, 928, 1997.
10. Olson, L. et al., Intraputaminal infusion of nerve growth factor to support adrenal medullary autografts in Parkinson's disease. One-year follow-up of first clinical trial, *Arch. Neurol.*, 48, 373, 1991.
11. Vahlsing, H.L. et al., An improved device for continuous intraventricular infusions prevents the introduction of pump-derived toxins and increases the effectiveness of NGF treatments, *Exp. Neurol.*, 105, 233, 1989.
12. Hoffman, D., Wahlberg, L., and Aebischer, P., NGF released from a polymer matrix prevents loss of ChAT expression in basal forebrain neurons following a fimbria-fornix lesion, *Exp. Neurol.*, 110, 39, 1990.
13. Langer, R. and Moses, M., Biocompatible controlled release polymers for delivery of polypeptides and growth factors, *J. Cell Biochem.*, 45, 340, 1991.
14. Winn, S.R., Wahlberg, L., Tresco, P.A., and Aebischer, P., An encapsulated dopamine-releasing polymer alleviates experimental Parkinsonism in rats, *Exp. Neurol.*, 105, 244, 1989.
15. Dunnett, S.B. and Bjorklund, A., *Functional Neural Transplantation*, Raven Press, New York, 1994.
16. Freed, W.J., Poltorak, M., and Becker, J.B., Intracerebral adrenal medulla grafts: A review, *Exp. Neurol.*, 110, 139, 1990.
17. Olanow, C.W. et al., A double-blind controlled trial of bilateral fetal nigral transplantation in Parkinson's disease, *Ann. Neurol.*, 5, 403, 2003.
18. Shoichet, M.S. and Winn, S.R., Cell delivery to the central nervous system, *Adv. Drug Deliv. Rev.*, 20, 81, 2000.
19. Winn, S.R. et al., Behavioral recovery following intrastriatal implantation of microencapsulated PC12 cells, *Exp. Neurol.*, 113, 322, 1991.
20. Yurek, D.M. and Sladek, J.R., Dopamine cell replacement: Parkinson's disease, *Ann. Rev. Neurosci.*, 13, 415, 1990.
21. Barkats, M. et al., Adenovirus in the brain: Recent advances of gene therapy for neuro degenerative diseases, *Prog. Neurobiol.*, 55, 333, 1998.
22. Bowers, W., Howard, D., and Federoff, H., Gene therapeutic strategies for neuroprotection: Implications for Parkinson's disease, *Exp. Neurol.*, 144, 58, 1997.
23. Kaplitt, M., Darakchiev, B., and During, M., Prospects for gene therapy in pediatric neurosurgery, *Ped. Neurosurg.*, 28, 3, 1998.
24. Kordower, J.H., In vivo gene delivery of glial cell line-derived neurotrophic factor for Parkinson's disease, *Ann. Neurol.*, 53, 120, 2003.
25. Zlokovic, B.V. and Apuzzo, M.L.J., Cellular and molecular neurosurgery: Part II: Vector systems and delivery methodologies for gene therapy of the central nervous system, *Neurosurgery*, 40, 805, 1997.
26. Crooks, C.A., Douglas, J.A., Broughton, R.L., and Sefton, M.V., Microencapsulation of mammalian cells in a HEMA–MMA copolymer: Effects on capsule mophology and permeability, *J. Biomed. Mater. Res.*, 24, 1241, 1990.
27. Gentile, F.T. et al., Polymer science for macroencapsulation of cells for central nervous system transplantation, *J. Reactive Polymers*, 25, 207, 1995.
28. Shoichet, M.S., Gentile, F.T., and Winn, S.R., The use of polymers in the treatment of neurological disorders, *Trends Poly. Sci.*, 3, 374, 1995.
29. Winn, S.R. and Tresco, P.A. Hydrogel applications for encapsulated cellular transplants. In Flanagan, T.F., Emerich, D.F., and Winn, S.R., eds., *Providing Pharmacological Access to the Brain*, Academic Press; Orlando, FL, 1994, p. 387.
30. Aebischer, P. et al., Long-term cross-species brain transplantation of a polymer encapsulated dopamine-secreting cell line, *Exp. Neurol.*, 111, 269, 1991.
31. Shoichet, M.S. and Rein, D.H., In vivo biostability of a polymeric hollow fiber membrane for cell encapsulation, *Biomaterials*, 17, 285, 1996.
32. Dunn, J.C.Y., Tompkins, R.G., and Yarmush, M.L., Long-term in vitro function of adult hepatocytes in a collagen sandwich configuration, *Biotechnol. Prog.*, 7, 237, 1991.
33. Emerich, D.F. et al., Polymer-encapsulated PC12 cells promote recovery of motor function in aged rats, *Exp. Neurol.*, 122, 37, 1993.
34. Aebischer, P., Tresco, P.A., Sagen, J., and Winn, S.R., Transplantation of microencapsulated bovine chromaffin cells reduces lesion-induced rotational asymmetry in rats, *Brain Res.*, 560, 43, 1991.
35. Tresco, P.A., Winn, S.R., and Aebischer, P., Polymer encapsulated neurotransmitter secreting cells: Potential treatment for Parkinson's disease, *ASAIO*, 38, 17, 1992.

36. Emerich, D.F. et al., Transplantation of polymer encapsulated PC12 cells: Use of chitosan as an immobilization matrix, *Cell Transpl.*, 2, 241, 1993.

37. Massia, S.P. and Hubbell, J.A., Covalent surface immobilization of Arg-Gly-Asp- and Tyr-Ile-Gly-Ser-Arg-containing peptides to obtain well-defined cell addhesive substrates, *Anal. Biochem.*, 187, 292, 1990.

38. Wang, N., Butler, J.P., and Ingber, D.E., Mechanotransduction across the cell surface and through the cytoskeleton, *Science*, 260, 1124, 1993.

39. Bellamkonda, R., Ranieri, J.P., and Aebischer, P., Laminin oligopeptide derivatized agarose gels allow three-dimensional neurite extension in vitro, *J. Neurosci. Res.*, 41, 501, 1995.

40. Tong, Y.W. and Shoichet, M.S., Peptide surface modification of poly(tetrafluoroethylene-*co*-hexafluoroethylene) enhances its interaction with central nervous system neurons, *J. Biomed. Mater. Res.*, 42, 85, 1998.

41. Cima, L.G., Lopina, S.T., Kaufamn, M., and Merrill, E.W., Polyethylene oxide hydrogels modified with cell attachment ligands, ASAIO National Meeting, San Francisco, CA, 1994.

42. Emerich, D.F. and Winn, S.R., Application of polymer-encapsulated cell therapy for CNS diseases. In, Dunnett, S.B., Boulton, A.A., and Baker, G.B., eds., *Neuromethods: Neural Transplantation Methods*, Humana Press, Totowa, NJ, 1999, p. 233.

43. Dionne, K.E. et al., Transport characterization of membranes for immunoisolation, *Biomaterials*, 17, 257, 1996.

44. Kordower, J.H., Liu, Y-T., Winn, S.R., and Emerich, D.F., Encapsulated PC12 cell transplants into hemiparkinsonian monkeys: A behavioral, neuroanatomical and neurochemical analysis, *Cell Transpl.*, 4, 155, 1995.

45. Lindner, M.D. et al., Intraventricular encapsulated bovine adrenal chromaffin cells: Viable for at least 500 days in vivo without detectable host immune sensitization or adverse effects on behavioral/cognitive function, *Restor. Neurol. Neurosci.*, 11, 21, 1997.

46. Rauch, H.C. et al., Histopathologic evaluation following chronic implantation of chromium and steel based metal alloys in the rabbit central nervous system, *J. Biomed. Mater. Res.*, 20, 1277, 1986.

47. Stensass, S.S. and Stensass, L.J., Histopathological evaluation of materials implanted into the cerebral cortex, *Acta Neuropathol.*, 4, 145, 1978.

48. Winn, S.R., Aebischer, P., and Galletti, P.M., Brain tissue reaction to permselective polymer capsules, *J. Biomed. Mater. Res.*, 23, 31, 1989.

49. Clark, R.A.F., *The Molecular and Cellular Biology of Wound Repair*, Plenum Press, New York, 1996.

50. Shoichet, M.S. et al., Poly(ethylene oxide)-grafted thermoplastic membranes for use as cellular hybrid bioartificial organs in the central nervous system, *Biotechnol. Bioeng.*, 43, 563, 1994.

51. Tresco, P.A. et al., Polymer-encapsulated PC12 cells: Long-term survival and associated reduction in lesioned-induced rotational behavior, *Cell Transpl.*, 1, 255, 1992.

52. Subramanian, T. et al., Polymer-encapsulated PC12 cells demonstrate high affinity uptake of dopamine in vitro and 18F-dopa uptake and metabolism after intracerebral implantation in nonhuman primate, *Cell Transpl.*, 6, 469, 1997.

53. Sagen, J., Wang, H., Tresco, P.A., and Aebischer, P., Transplants of immunologically isolated xenogeneic chromaffin cells provide a long-term source of pain-reducing neuroactive substances, *J. Neurosci.*, 13, 2415, 1993.

54. Date, I., Shingo, T., Ohmoto, T., and Emerich, D.F., Long-term enhanced chromaffin cell survival and behavioral recovery in hemiparkinsonian rats with co-grafted polymer-encapsulated human NGF-secreting cells, *Exp. Neurol.*, 147, 10, 1997.

55. Winn, S.R. et al., Polymer-encapsulated genetically-modified cells continue to secrete human nerve growth factor for over one year in rat ventricles: Behavioral and anatomical consequences, *Exp. Neurol.*, 140, 126, 1996.

56. Lindner, M.D. and Emerich, D.F., Therapeutic potential of a polymer-encapsulated L-DOPA and dopamine-producing cell line in rodent and primate models of Parkinsons disease, *Cell Transpl.*, 7, 65, 1998.

57. Lindner, M.D. et al., Effects of intraventricular encapsulated hNGF-secreting fibroblasts in aged rats, *Cell Transpl.*, 5, 205, 1996.

58. Kordower, J.H. et al., Intrastriatal implants of polymer-encapsulated cells genetically modified to secrete human NGF: Trophic effects upon cholinergic and noncholinergic neurons, *Neuroscience*, 72, 63, 1996.

59. Aebischer, P. et al., Intrathecal delivery of CNTF using encapsulated genetically modified xenogeneic cells in amyotrophic lateral sclerosis patients, *Nat. Med.*, 6, 696, 1996.

60. Emerich, D.F. et al., Implants of polymer-encapsulated human NGF-secreting cells in the nonhuman primate: Rescue and sprouting of degenerating cholinergic basal forebrain neurons, *J. Comp. Neurol.*, 349, 148, 1994.

61. Fahn, S., Fluctuations of disability in Parkinson's disease: Pathophysiological aspects. In: Marsden, C.D. and Fahn, S., eds., *Movement Disorders*, Butterworth Scientific, London, 1982, p. 123.

62. Becker, J. et al., Sustained behavioral recovery from unilateral nigrostriatal damage produced by the controlled release of dopamine from a silicone polymer pellet placed into the denervated striatum, *Brain Res.*, 508, 60, 1990.

63. DeYebens, J.G. et al., Intracerebroventricular infusion of dopamine and its agonists in rodents and primates: An experimental approach to the treatment of Parkinson's disease, *Trans. Am. Soc. Artif. Intern. Organs*, 34, 951, 1998.

64. Hargraves, R. and Freed, W.J., Chronic intrastriatal dopamine infusions in rats with unilateral lesions of the substantia nigra, *Life Sci.*, 40, 959, 1987.

65. Emerich, D.F., Winn, S.R., and Lindner, M.D., Continued presence of intrastriatal but not intraventricular polymer-encapsulated PC12 cells is required for alleviation of behavioral deficits in Parkinsonian rats, *Cell Transpl.*, 5, 589, 1997.

66. Lindner, M.D., Plone, M.A., Francis, J.M., and Emerich, D.F., Validation of a rodent model of Parkinson's disease: Evidence of a therapeutic window for oral Sinemet, *Brain Res. Bull.*, 39, 367, 1996.

67. Lindner, M.D. et al., Somatic delivery of catecholamines in the striatum attenuate parkinsonian symptoms and widen the therapeutic window or oral Sinemet in rats, *Exp. Neurol.*, 14, 130, 1997.

68. Emerich, D.F. et al., Polymer-encapsulated PC12 cells promote recovery of motor function in aged rats, *Exp. Neurol.*, 122, 37, 1993.

69. Aebischer, P., Goddard, M.B., Signore, P., and Timpson, R., Functional recovery in hemiparkinsonian primates transplanted with polymer encapsulated PC12 cells, *Exp. Neurol.*, 126, 1, 1994.

70. Hoane, M.R. et al., Mammalian-cell-produced Neurturin (NTN) is more potent that purified *Escherichia coli*-produced NTN, *Exp. Neurol.*, 162, 189, 2000.

71. Tseng, J.L., Baetge, E.E., Zurn, A.D., and Aebischer, P., GDNF reduces drug-induced rotational behavior after medial forebrain bundle transection by a mechanism not involving striatal dopamine, *J. Neurosci.*, 1, 325, 1997.

72. Tseng, J.L., Bruhn, S.L., Zurn, A.D., and Aebischer, P., Neurturin protects dopaminergic neurons following medial forebrain bundle axotomy, *NeuroReport*, 9, 1817, 1998.

73. Date, I. et al., Cografting with polymer-encapsulated human nerve growth factor-secreting cells and chromaffin cell survival and behavioral recovery in hemiparkinsonian rats, *J. Neurosurg.*, 84, 1006, 1996.

74. Date, I. et al., Chromaffin cell survival from both young and old donors is enhanced by co-grafts of polymer-encapsulated human NGF-secreting cells, *NeuroReport*, 7, 1813, 1996.

75. Hefti, F., Neurotrophic factor therapy for nervous system degenerative diseases, *J. Neurobiol.*, 25, 1418, 1994.

76. Hefti, F., Pharmacology of neurotrophic factors, *Annu. Rev. Pharmacol. Toxicol.*, 37, 239, 1997.

77. Winn, S.R. et al., Polymer-encapsulated cells genetically modified to secrete human nerve growth factor promote the survival of axotomized septal cholinergic neurons, *Proc. Natl. Acad. Sci. USA*, 91, 2324, 1994.

78. Kordower, J.H. et al., The aged monkey basal forebrain: Rescue and sprouting of axotomized basal forebrain neurons after grafts of encapsulated cells secreting human nerve growth factor, *Proc. Natl. Acad. Sci. USA*, 91, 10898, 1994.

79. Beal, M.F. et al., Replication of the neurochemical characteristics Huntington's disease by quinolinic acid, *Nature*, 321, 168, 1986.

80. Beal, M.F. et al., Somatostatin and neuropeptide Y concentrations in pathologically graded cases of Huntington's disease, *Ann. Neurol.*, 23, 562, 1988.

81. Beal, M.F. et al., Differential sparing of somatostatin-neuropeptide Y and cholinergic neurons following striatal excitotoxin lesions, *Synapse*, 3, 38, 1989.

82. Sanberg, P.R. et al., The quinolinic acid model of Huntington's disease: Locomotor abnormalities, *Exp. Neurol.*, 105, 45, 1989.

83. Frim, D.M. et al., Local response to intracerebral grafts of NGF-secreting fibroblasts: Induction of a peroxidative enzyme, *Soc. Neurosci. Abstr.*, 18, 1100, 1992.

84. Frim, D.M. et al., Striatal degeneration induced by mitochondrial blockade is prevented by biologically delivered NGF, *J. Neurosci. Res.*, 35, 452, 1993.

85. Frim, D.M. et al., Effects of biologically delivered NGF, BDNF, and bFGF on striatal excitotoxic lesions, *NeuroReport*, 4, 367, 1993.

86. Schumacher, J.M. et al., Intracerebral implantation of nerve growth factor-producing fibroblasts protects striatum against neurotoxic levels of excitatory amino acids, *Neuroscience*, 45, 561, 1991.

87. Emerich, D.F., Hammang, J.P., Baetge, E.E., and Winn, S.R., Implantation of polymer-encapsulated human nerve growth factor-secreting fibroblasts attenuates the behavioral and neuropathological consequences of quinolinic acid injections into rodent striatum, *Exp. Neurol.*, 130, 141, 1996.

88. Emerich, D.F. et al., Implants of encapsulated human CNTF-producing fibroblasts prevent behavioral deficits and striatal degeneration in a rodent model of Huntington's disease, *J. Neurosci.*, 1, 5168, 1996.

89. Emerich, D.F. et al., Cellular delivery of human CNTF prevents motor and cognitive dysfunction in a rodent model of Huntington's disease, *Cell Transpl.*, 6, 249, 1997.

90. Emerich, D.F. et al., Protective effects of encapsulated cells producing neurotrophic factor CNTF in a monkey model of Huntington's disease, *Nature*, 386, 395, 1997.

91. Mittoux, V. et al., Restoration of cognitive and motor function with ciliary neurotrophic factor in a primate model of Huntington's disease, *Hum. Gene Ther.*, 11, 1177, 2000.

92. Kordower, J.H., Isacson, O., and Emerich, D.F., Cellular delivery of trophic factors for the treatment of Huntington's disease: Is neuroprotection possible? *Exp. Neurol.*, 59, 4, 1999.

93. Bachoud-Levi, A.C. et al., Neuroprotective gene therapy for Huntington's disease using a polymer encapsulated BHK cell line engineered to secrete human CNTF, *Hum. Gene Ther.*, 11, 1723, 2000.

94. Emerich, D.F., Encapsulated CNTF-producing cells for Huntington's disease, *Cell Transpl.*, 8, 581, 2000.

95. Cirone, P., Bourgeois, J.M., Austin, R.C., and Chang, P.L., A novel approach to tumor suppression with microencapsulated recombinant cells, *Hum. Gene Ther.*, 13, 1157, 2000.

96. Joki, T., Machluf, M., and Atala, A., Continuous release of endostatin from microencapsulated engineered cells for tumor therapy, *Nat. Biotech.*, 19, 35, 2001.

97. Lang, M.S. et al., Immunotherapy with monoclonal antibodies directed against the immunosuppressive domain of p15E inhibits tumour growth, *Clin. Exp. Immunol.*, 102, 468, 1995.

98. Martnet, O., Schreyer, N., Reis, E.D., and Joseph, J.M., Encapsulation of packaging cell results in successful retroviral-mediated transfer of a suicide gene in vivo in an experimental model of glioblastoma, *Eur. J. Surg. Oncol.*, 29, 351, 2003.

99. Read, T.A. et al., Local endostatin treatment of gliomas administered by microencapsulated producer cells, *Nat. Biotech.*, 19, 29, 2001.

100. Thorsen, F. et al., Alginate-encapsulated producer cells: A potential new approach for the treatment of malignant brain tumors, *Cell Transpl.*, 9, 773, 2000.

101. Visted, T. and Lund-Johansen, M., Progress and challenges for cell encapsulation in brain tumor therapy, *Expert Opin. Biol. Ther.*, 3, 551, 2003.

102. Xu, W., Liu, L., and Charles, I.G., Microencapsulated iNOS-expressing cells cause tumor suppression in mice, *FASEB J.*, 16, 213, 2002.

103. Bjerkvig, R. et al., Cell therapy using encapsulated cells producing endostatin, *Acta Neurochir. Suppl.*, 88, 137, 2003.

104. Boughman, J.A., Conneally, P.M., and Nance, W.E., Population genetic studies of retinitis pigmentosa, *Am. J. Hum. Genet.*, 32, 223, 1980.

105. Farrar, G., Kenna, P., and Humphries, P., On the genetics of retinitis pigmentosa and on mutation-independent approaches to therapeutic intervention, *EMBO J.*, 21, 857, 2002.

106. Cayouette, M. and Gravel, C., Adenovirus-mediated gene transfer of ciliary neurotrophic factor can prevent photoreceptor degeneration in the retinal degeneration (rd) mouse, *Hum. Gene Ther.*, 8, 423, 1997.

107. Cayouette, M. et al., Intraocular gene transfer of ciliary neurotrophic factor prevents death and increases responsiveness of rod photoreceptors in the retinal degeneration slow mouse, *J. Neurosci.*, 18, 9282, 1998.

108. LaVail, M., Asumura, D., and Matthes, M., Protection of mouse photoreceptors by survival factors in retinal degenerations, *Invest. Ophthalmol. Vis. Sci.*, 39, 592, 1998.

109. Tao, W. et al., Encapsulated cell-based delivery of CNTF reduces photoreceptor degeneration in animal models of retinitis pigmentosa, *Invest. Ophthalmol. Vis. Sci.*, 43, 3292, 2002.

110. Aguirre, G., Farber, D., and Lolley, R., Retinal degenerations in the dog. III: Abnormal cyclic nucleotide metabolism in rod-cone dysplasia, *Exp. Eye Res.*, 35, 625, 1982.

111. Schmidt, S. and Aguirre, G., Reduction in taurine secondary to photoreceptor loss in Irish setters with rod-cone dysplasia, *Invest. Ophthalmol. Vis. Sci.*, 26, 679, 1985.

112. Henderson, C.E., GDNF: A potent survival factor of motoneurons present in peripheral nerve and muscle, *Science*, 266, 1062, 1994.

113. Hughes, R.A., Sendtner, M., and Thoenen, H., Members of several gene families influence survival of rat motoneurons in vitro and in vivo, *J. Neurosci. Res.*, 36, 663, 1993.

114. Kato, A.C. and Lindsay. R.M., Overlapping and additive effects of neurotrophins and CNTF on cultured human spinal cord neurons, *Exp. Neurol.*, 130, 196, 1994.

115. Lewis, M.E. et al., Insulin-like growth factor-I: Potential for treatment of motor neuronal disorders, *Exp. Neurol.*,124, 73, 1993.

116. Mitsumoto, H. et al., Arrest of motor neuron disease in wobbler mice co-treated with CNTF and BDNF, *Science*, 265, 1107, 1994.

117. Sagot, Y. et al., Polymer encapsulated cell lines genetically engineered to release ciliary neurotrophic factor can slow down progressive motor neuronopathy in the mouse, *Eur. J. Neurosci.*, 7, 1313, 1995.

118. Sendtner, M. et al., Brain-derived neurotrophic factor prevents the death of motor neurons in newborn rats after nerve section, *Nature*, 360, 757, 1992.

119. Sendtner, M. et al., Ciliary neurotrophic factor prevents degeneration of motor neurons in mouse mutant progressive motor neuronopathy, *Nature*, 358, 502, 1992.

120. Dittrich, F., Thoenen, H., and Sendtner, M.. Ciliary neurotrophic factor: Pharmacokinetics and acute-phase response in rat, *Ann. Neurol.*, 35, 151, 1994.

121. The ALS CNTF Treatment Study (ACTS) Phase I-II Study Group, The pharmacokinetics of subcutaneously administered recombinant human ciliary neurotrophic factor (rhCNTF) in patients with amytrophic lateral sclerosis: Relationship to parameters of the acute phase response, *Clin. Neuropharmacol.*, 18, 500, 1995.

122. The ALS CNTF Treatment Study (ACTS) Phase I-II Study Group, A phase I study of recombinant human ciliary neurotrophic factor (rHCNTF) in patients with amyotrophic lateral sclerosis, *Clin. Neuropharmacol.*, 18, 515, 1995.

123. Tan, S.A. et al., Rescue of motoneurons from axotomy-induced cell death by polymer encapsulated cells genetically engineered to release CNTF, *Cell Transpl.*, 5, 577, 1996.

124. Aebischer, P. et al., Intrathecal delivery of CNTF using encapsulated genetically modified xenogeneic cells in amyotrophic lateral sclerosis patients, *Nat. Med.*, 6, 696, 1996.

125. Zurn, A.D. et al., Evaluation of an intrathecal immune response in amyotrophic lateral sclerosis patients implanted with encapsulated genetically-engineered xenogeneic cells, *Cell Transpl.*, 9, 471, 2000.

126. Czech, K.A. and Sagen, J. Update on cellular transplantation into the rat CNS as a novel therapy for chronic pain, *Prog. Neurobiol.*, 46, 507, 1995.

127. Sagen, J., Cellular Transplantation for intractable pain, *Adv. Pharmacol.*, 42, 579, 1998.

128. Aebischer, P. et al., Transplantation in humans of encapsulated xenogeneic cells without immunosuppression—a preliminary report, *Transplantation*, 58, 1275, 1994.

129. Buschser, W. et al., Immunoisolated xenogenic chromaffin cell therapy for chronic pain. Initial clinical experience, *Anesthesiology*, 85, 1005, 1996.

130. CytoTherapeutics press release, Providence, RI, June 24, 1999.

131. Lindner, M.D. et al., Numerous adrenal chromaffin cell preparations fail to produce analgesic effects in the formalin test or in tests of acute pain even with nicotine stimulation, *Pain*, 8, 177, 1999.

132. Lindner, M.D. et al., The analgesic potential of intraventricular polymer-encapsulated adrenal chromaffin cells in a rodent model of chronic neuropathic pain, *Exp. Clin. Psychopharmacol.*, 8, 524, 2000.

133. Lindner, M.D., Francis, J.M., and Saydoff, J.A., Intrathecal polymer-encapsulated bovine adrenal chromaffin cells fail to produce analgesic effects in the hotplate and formalin test, *Exp. Neurol.*, 165, 370, 2000.

134. Bachoud-Levi, A.C., et al., Neuroprotective gene therapy for Huntington's disease using a polymer encapsulated BHK cell line engineered to secrete human CNTF, *Hum. Gene Ther.*, 11, 1723, 2000.

135. Bloche, J., et al. Neuroprotective gene therapy for Huntington's disease, using polymer-encapsulated cells engineered to secrete human ciliary neurotrophic factor: Results of a phase I study, *Hum. Gene Ther.*, 15, 968, 2004.

136. Sieving, P.A., et al., Ciliary neurotrophic factor (CNTF) for human retinal degeneration: Phase I trial of CNTF delivered by encapsulated cell intraocular implants, *Proc. Natl. Acad. Sci. USA*, 103, 3896, 2006.

Part V

Clinical Trials and Regulatory Issues

46 Cardiovascular Gene Therapy

Jörn Tongers, Jerome G. Roncalli, and Douglas W. Losordo

CONTENTS

46.1 BACKGROUND

Alteration of the vascular system leading to subsequent chronic structural and functional vessel damage is directly related to rising age and cardiovascular risk burden. Its consequences are primarily manifested in the heart (coronary artery disease [CAD]), lower extremities (peripheral vascular disease [PVD]), and the brain (cerebrovascular disease). Arteriosclerosis, atherothrombosis, endothelial dysfunction, inflammation, and thrombosis have been characterized as distinct key players in vascular pathology [1,2]. Research performed during the last few decades has provided insight into the molecular and cellular processes underlying vascular disease. In addition to strategies for prevention, the knowledge gained has enabled progress toward the gold standard of anti-arteriosclerotic, pleiotropic drug treatment to attenuate the progression or to promote the regression of vascular alterations.

Vascular diseases are the main causes of morbidity and mortality worldwide. In developed countries, projections indicate that the two main contributors to morbidity and mortality in 2020 will be ischemic heart disease and cerebrovascular disease [3]. Approximately 1.2 million individuals in the United States experience coronary events each year, 40% of whom will eventually develop chronic heart failure of ischemic origin [4]. In the peripheral vasculature, PVD predominates and leads to intractable ischemia, impaired mobility, ulceration, impaired wound healing, and amputation; 8 million Americans are currently affected by PVD-associated morbidity and mortality [5]. Occlusion of cerebral vessels can lead to stroke-induced neurological deficits and life-long disability; in 2004, the estimated prevalence of stroke in the United States was 5.7 million [4].

During the later stages of vascular disease, mechanical revascularization strategies can effectively alleviate ischemia. Surgical techniques designed to clear or bypass obstructed arteries have improved substantially over the years. Less invasive interventional therapies such as angioplasty and stent implantation may delay or potentially supplant operative revascularization. Both surgical and interventional techniques are well-established strategies for restoring tissue perfusion and alleviating the symptoms of vascular disease, particularly for CAD and PVD. However, the feasibility of conventional revascularization in severe vascular disease is limited.

Despite modern medical treatments and revascularization strategies, the consequences of vascular disease remain devastating. In addition, advancements in modern medicine have produced an aging, multimorbid society that suffers from diffuse, severe, and often occlusive vascular diseases. As a result, there is a growing subpopulation of individuals with symptomatic cardiovascular disease who have received numerous revascularization procedures and lack options for further treatment. In extreme cases, these individuals experience intractable chest pain, recurrent decompensation from unstable heart failure, disabling claudication, or nonhealing wounds. The fundamental cause for all of these afflictions is an inadequate blood supply during low levels of activity or even at rest.

Clinical experience indicates that myocardial tissue possesses only a limited capacity for regeneration, so the Holy Grail of cardiovascular medicine is the growth of new vasculature in critically ischemic tissue, thereby relieving symptoms, improving organ function, and rescuing at-risk tissues by re-establishing perfusion. The modern concept of cardiovascular gene therapy grew from pioneering work by Folkman et al. in the 1970s, who demonstrated that the development and maintenance of an adequate vascular supply were essential for the growth of normal as well as neoplastic tissue [6]. Subsequent insights into the underlying mechanisms included the identification and detailed characterization of angiogenic growth factors and led to modern treatment paradigms for promoting (via gene therapy) or inhibiting (via antiangiogenic tumor therapy) angiogenesis.

Gene therapy is an attractive treatment option for inducing new vascular growth in critically ischemic tissue. However, the development of gene therapy has been slowed by technical challenges; for example, a decade passed before sufficient amounts of growth factors could be produced for preclinical experiments [7]. After preclinical

studies of gene therapy in animal models of cardiovascular disease supported the concept of this treatment strategy, its feasibility and safety was established by early nonrandomized, uncontrolled, clinical studies [8,9]. Subsequent randomized, controlled clinical trials yielded variable results that may be explained by biological, technical, methodological, or disease-related factors. Gene therapy has also been used for other clinical conditions, such as heart failure, postinterventional vascular remodeling/restenosis, maintenance of vascular prosthesis and bypass grafts, and prevention of hyperlipidemia and hypertension [10].

The purpose of this chapter is to review the current status of cardiovascular gene therapy. Angiogenic therapy will be emphasized because it is the most clinically advanced application of gene therapy.

46.2 NEOVASCULARIZATION

In postnatal life, new blood vessels form by three distinct mechanisms: arteriogenesis, angiogenesis, and vasculogenesis. Neovascularization comprises the latter two mechanisms [11].

Collateral (macrovascular) vessel formation from conduit arteries is driven by biomechanical forces in response to significant vessel obstruction and occurs through the in situ enlargement and growth of muscular collateral vessels from pre-existing arteriolar anastomoses. A portion of the new vasculature may evolve through the proliferative remodeling of medium-sized arteries into larger collateral vessels [12]. It is unknown whether this remodeling is induced by growth factors or by the flow-mediated maturation of collateral conduits, which is referred to as arteriolization.

Angiogenesis is the growth of new blood vessels via the sprouting, bridging, intussusception, and/or enlargement of preexisting capillaries. Angiogenesis is driven by ischemia and is crucial for the preservation of tissue function on the microvascular level during critical ischemia. The underlying mechanism is characterized by the proliferation and migration of mature endothelial cells (ECs) from adjacent vessels in response to stimuli such as hypoxia, ischemia, mechanical stretching, and inflammation [11,13]. Angiogenesis plays an essential role in both physiological (e.g., wound healing) and pathological (e.g., neoplasm, proliferative diabetic retinopathy) processes.

Vasculogenesis was previously believed to occur only during embryonic vascular development as mesodermal precursor cells (angioblasts) differentiate into ECs, which subsequently assemble into a vascular labyrinth (the primary capillary plexus) [14]. Hematopoietic stem cells were first linked to postnatal angiogenesis in a seminal study by Asahara et al. [15], who showed that CD34+ progenitor cells isolated from mature tissue can differentiate into ECs and thereby demonstrated the importance of circulating stem and progenitor cells for postnatal angiogenesis [15]. The identification of bone marrow-derived endothelial progenitor cells (EPCs) in sites of ischemia-induced neovascularization further demonstrated that neovascularization in adults comprises both angiogenesis and vasculogenesis [16,17]. Recent reports from our laboratory and others indicate that the clinical application of stem cells is beneficial and associated with enhanced perfusion in ischemic heart and limb tissue [18,19].

46.3 THERAPEUTIC TARGETS

A variety of growth factors possess angiogenic potential. *In vivo*, these factors induce neovascularization by stimulating both angiogenesis and vasculogenesis. Therapeutic angiogenesis has been achieved by gene transfer of vascular endothelial growth factor (VEGF), fibroblast growth factor (FGF), platelet-derived growth factor (PDGF), hepatocyte growth factor (HGF), hypoxia-related transcription factor (e.g., hypoxia-inducible factor 1α [HIF1α]), and many other agents [20,21]. Investigations of VEGF and FGF predominate in both preclinical studies and clinical trials, which reflect the well-established link between VEGF receptor 2 or basic FGF and the differentiation of angioblasts during embryonic vasculogenesis [22–24]. In addition to these established candidates (Table 46.1), recent preclinical studies have identified stromal cell-derived factor 1 (SDF-1) [25–28] and sonic hedgehog (Shh) [25,29] as potential targets for future research, and several other angiogenic factors may soon be investigated clinically.

46.3.1 VASCULAR ENDOTHELIAL GROWTH FACTOR

In the complex process of new vessel formation, the role of VEGF is often crucial. It is involved in all types of vascular growth, is one of the key regulators in vascular embryogenesis [22,23], and is essential for neovascularization and vascular maintenance in both early and later adult life [30].

TABLE 46.1
Gene Therapeutic Targets

Therapeutic Goal	Underlying Mechanism	Potential Targets
Therapeutic angiogenesis	Angiogenesis Arteriogenesis Lymphangiogenesis	Adrenomedullin, Ang-1/-2, Del-1, EGR-1, eNOS, Ets-1, ezrin, FGF-1/-2/-4/-5, GM-CSF, HGF, HIF-1, Id1, IGF-1/-2, iNOS, kallikrein, MCP-1, PDGF-A/-B/-C/-D, PlGF, PDGF, netrin-1/-4, PR39, SDF-1, Shh, secretoneurin, thrombopoietin, thymidine phosphorylase, VEGF-A/-B/-C/-D/-E
Prevention of post-interventional vascular remodeling	Endothelial function Endothelial repair Re-endothelialization SMC proliferation matrix Production/degradation Apoptosis	β-adrenergic receptor kinase, β-interferon, activator protein-1, C-type natriuretic peptide, Bcl, catalase, cecropine A, CDC2 kinase, CDK-2, CGRP, COX, c-myc, CyA, E2F, cyclin B1/ G1, E2F, ecSOD, eNOS, Fas ligand, FGF-2, Forkhead, gax, Gβγ, HGF, HO-1, homeobox gene, gax, ICAM, iNOS, kallikrein, lipoprotein-associated phospholipase A2, MCP-1, midkine, NFkB, p16, p21, p27, p16 Chimera, p53, PAI-1, PCNA, PDGF-B, PDGFR-β, prostacyclin, PPAR, ras, RAD50, Ras, Rb, RB2/p130, Rho kinase, sdi-1, TGF-β 3, thymidine kinase, TIMP, TNF, TGF-β, soluble TGF-β type II receptor, VCAM, VEGF-A/-B/-C/-D/-E
Stabilization of vulnerable plaque and aneurysm	Cholesterol homeostasis Cholesterol uptake Inflammation	COX, Ets-1, LDL receptor in liver, MCP-1, PDGFs, soluble scavenger-receptor decoy, soluble VCAM, TGF-β, TIMP, TNFα
Treatment of hyperlipidemia	Cholesterol homeostasis	apoA-1, apoAI, apoB-100, apoB editing enzyme, apoE, LP(a), antioxidant enzymes, LCAT, LDL receptor, lipid transfer proteins, lipoprotein lipase, hepatic lipase, soluble scavenger-receptor decoy, soluble ICAM, soluble VCAM, VLDL receptor
Treatment of hypertension	Renin-angiotensin system Sympatoadrenergic system Vasoconstriction/-dilatation	β-adrenergic receptor, angiotensinogen, ACE, adrenomedullin, ATII R1, atrial natriuretic peptide, eNOS, kallikrein
Treatment of heart failure	Cardiomyocyte homeostasis Cellular contraction Hypertrophy Fibrosis Apoptosis	β-adrenergic receptor kinase, β2-adrenergic receptor, adenylyl cyclase VI, Ang-1, Bcl-2, cyclin A2, ecSOD, HGF, HO-1, kallikrein inhibitor-2, leukemia inhibitory factor, MCP-1, norepinephrine transporter uptake-1, p38 kinase, p35 phosphoinositide kinase domain, parvalbumin, phospholamban, PKCα, S100A1, SEK-1, SERCA1/2a, sorcin, TGF-β, TIMP-1, VEGFs, V2 receptor
Treatment of thrombotic state	Platelet degradation Fibrin formation Coagulation factors	C-type natriuretic peptide, COX, hirudin, tPA, TFPI, thrombomodulin, ectonucleoside triphosphate diphosphohydrolase, tissue factor
Modulation of arrhythmic activity	Spontaneous rhythmic	β2-adrenergic receptor, hyperpolarization-activated cyclic nucleotide-gated genes 1–4, hyperpolarization-activated nonselective channel, Kir2.1

Since the discovery of VEGFA (also called VEGF1) in 1989, VEGFC (VEGF2), VEGFB (VEGF3), VEGFD, VEGFE, and placental growth factor (PLGF) have also been identified as distinct members of the human VEGF superfamily. These proteins share homology but are encoded by distinct genetic loci located on different chromosomes. Whereas the diverse pro-angiogenic properties of VEGFA identify it as a prime candidate for therapeutic angiogenesis, the roles of the other VEGF superfamily members in vascular biology remain elusive. The involvement of VEGFA and VEGFC is critical during embryonic vascular development [23], but only VEGFA is abundant in fully vascularized tissue, which suggests that low VEGF levels are sufficient for vascular homeostasis in adult life [31]. Expression of VEGFA is upregulated postnatally during pathological vessel growth [32].

Alternative splicing of VEGFA pre-mRNA generates distinct isoforms identified by the number of amino acids: $VEGF_{121}$, $VEGF_{165}$, $VEGF_{189}$, and $VEGF_{206}$ [20]. These isoforms show angiogenic potential in animal models [33], and differences in their solubility and binding characteristics (e.g., to heparin and the extracellular matrix) generate variations in their abilities to bind target cells. The $VEGF_{165}$ isoform has the greatest therapeutic potential because of its high biological potency (100-fold higher than $VEGF_{121}$) and its ability to generate developmental gradients in tissue, which is required for the patterning of vessel growth [34]. Three distinct receptor subtypes (endothelial-specific, fms-like tyrosine kinases) are bound by VEGF in an overlapping pattern: VEGFR1 (Flt1), VEGFR2 (Flk1/KDR), and VEGFR3 (Flt4). Certain tumor cells and ECs also express cell-surface VEGF binding sites (neuropilins, NPR1, and NPR2) with distinct affinities [35].

ECs are the principal cellular target of VEGF. Expressions of both VEGF protein and VEGF receptor are upregulated in ECs in response to hypoxia [36], and VEGF functions as a survival factor for ECs both *in vitro* and *in vivo* [37,38]. The formation of ECs into tubes and subsequently into vessels is mediated by VEGFR1 [39], and VEGFR1 is also involved in the biological activity of stem and progenitor cells. The proliferation and migration of mature ECs [40,41] is mediated by VEGFR2, which thereby influences embryonic, neonatal, and pathological angiogenesis [38]. In contrast, the primary function of VEGFR3 appears to be in lymph angiogenesis [42]. Thus, the VEGF ligands and their receptors possess distinct functions in different tissue compartments.

The VEGF protein possesses several qualities that may facilitate effective gene therapy. First, the hydrophobic leader sequence comprises a secretory signal that induces VEGF secretion from intact cells, thereby triggering a cascade of paracrine effects [43]. Second, the high-affinity binding sites for VEGF are endothelial-specific, so its mitogenic effects are restricted to ECs (acidic and basic FGF induce mitogenesis of smooth muscle cells and fibroblasts in addition to ECs) [44,45]. Third, the protein sequence includes an autocrine loop that is shared by other angiogenic factors and modulates EC response. After activation by hypoxia, the autocrine loop amplifies the effect of exogenously administered VEGF, and factors secreted by hypoxic tissue also upregulate VEGFR expression.

This localized receptor expression explains why angiogenesis is generally limited to the site of ischemia. Finally, studies in both mice and humans indicate that the number and regenerative potential of circulating EPCs are augmented by VEGF gene transfer [46–48].

Although other members of the VEGF superfamily exhibit angiogenic potential, the evidence that identifies VEGFA as a key component of physiological and pathological angiogenesis is convincing. The bi-allelic requirement of VEGF during embryonic vasculogenesis, its hypoxia-inducible expression, and its robust induction of angiogenesis and of vascular permeability underscore the therapeutic potential of VEGF. For these reasons, VEGF may be considered the prototype angiogenic growth factor for use in gene therapy [10], and the great majority of cardiovascular gene therapy trials have been conducted with VEGF/VEGFA.

46.3.2 Fibroblast Growth Factors

The FGF family comprises 23 members including acidic FGF (FGF1), basic FGF (bFGF, FGF2), and FGFs 3–9. Acidic and basic FGF are the best-characterized family members. Although they are not crucial for embryogenesis [49], both have been successfully used for therapeutic angiogenesis [50]. Because they lack a signaling sequence, acidic and basic FGF are nonsecreted, intracellular peptides that enter the extracellular space passively after cell damage; however, FGF4 and FGF5 are efficiently secreted and consequently, may have therapeutic advantages over the intracellular FGFs. FGFs bind with high affinity to four tyrosine kinase receptors. Upon binding, the FGF/receptor complex is rapidly removed from the circulation and localized to cells or to the extracellular matrix.

The biological activity of FGF is not well understood. Although FGF proteins are potent mitogens, their activity is not limited to ECs, and they induce mitogenesis in other cell types, including vascular smooth muscle cells and fibroblasts. Like VEGF, secreted FGF stimulates the synthesis of proteases such as plasminogen activator and metalloproteinase. Proteases contribute to angiogenesis by digesting the extracellular matrix, thereby loosening intercellular structure and liberating sequestered angiogenic factors [51]. Basic FGF enhances the proliferative activity of ECs in response to hypoxia [52], and hypoxia also induces formation of an $HIF1\alpha$-dependent autocrine loop in FGF2, which subsequently drives angiogenesis [53]. However, since the most well-characterized forms of FGF lack a secretory signal sequence, FGF gene transfer requires either genetic modification of the common FGFs or the use of less-common FGF genes that contain a signal sequence [54–56].

46.3.3 Other Genetic Targets

Numerous pro-angiogenic proteins, growth factors, cytokines, and regulatory factors have been the focus of angiogenic research. The following discussion includes the more well-characterized therapeutic candidates.

46.3.3.1 Hypoxia-Related Transcription Factors

As an alternative to the transfer of genes that encode a single angiogenic growth factor, investigations have begun to assess the utility of upstream transcription factors that regulate multiple angiogenic genes. The metabolic stimuli associated with hypoxia and ischemia (e.g., hypoglycemia and low pH) induce expression of a variety of transcription factors that stimulate angiogenesis. Hypoxia-inducible factors (HIFs) and hypoxia-response elements (HREs) can function as master switches for an "angiogenesis program" that upregulates the expression of a number of growth factors and cytokines simultaneously, thereby mimicking a natural angiogenic response. On the cellular level, the detection of hypoxia is mediated by HIF1α [57]; HIF1α is degraded under normoxic conditions, but it accumulates during hypoxia and binds to HREs located in the promotor region of hypoxia-regulated genes. During postischemic vasculogenesis, the mobilization and trafficking of progenitor cells is critically dependent on HIF1α, and the effects are directly related to HIF1α regulation of SDF-1 expression [58]. However, HIF1α may also induce cell death [59], so its potential for vascular gene therapy must be carefully investigated, and HIF1α may be a significantly less potent angiogenic agent than VEGF [10].

46.3.3.2 Platelet-Derived Growth Factors

The role of PDGFs in the recruitment of pericytes to growing vessels [60] suggests that PDGFs could be added to growth-factor therapy to enhance angiogenesis. The combined application of PDGF and VEGF$_{165}$ via a polymeric system for growth factor delivery rapidly produced a mature and stable vascular network in mice [61], and a PDGF–FGF2 combination induced the formation of stable, long-term vessels in the corneas of rodents [62]. Recombinant PDGFC has been shown to mobilize EPCs, induce bone marrow cell differentiation, and boost neovascularization of ischemic tissue in mice [63].

46.3.3.3 Hepatocyte Growth Factor

HGF, the ligand of the receptor encoded by the c-met proto-oncogene, is an endothelium-specific growth factor with potent mitogenic activity in ECs. Its angiogenic capacity has been demonstrated in cell cultures, and gene transfer of human HGF improved perfusion in ischemic animal myocardium by activating the transcription factor ets, which is essential for angiogenesis [64].

46.3.3.4 Angiopoietins

Two members of this group, angiopoietin-1 (Ang1) and angiopoietin-2 (Ang2), modulate the maturation of blood vessels through agonist–antagonist interactions at the Tie2 receptor. By binding to the Tie2 receptor without activation, Ang2 antagonizes the activity of Ang1 and impairs angiogenesis [65]. The anti-inflammatory agonist Ang1 stabilizes mature vessels; although the mechanism is not known, it seems to prevent VEGF-induced vessel leakage [66]. Overall, there is evidence that a number of these pro-angiogenic factors influence angiogenesis, at least in part, by induction of VEGF [34].

46.3.4 Emerging Genetic Targets

46.3.4.1 Stromal Cell-Derived Factor 1

Stromal cell-derived factor 1 (CXCL12) is a 68-amino acid cytokine of the CXC superfamily. Since its identification, its activity as a growth factor has been primarily linked to lymphopoiesis and myelopoiesis [67]. Two isoforms, SDF-1α and SDF-1β, are encoded as splice variants of a single gene. The two are nearly identical, but SDF-1α is four amino-acids shorter [68]. It is expressed by both ECs and stem cells.

Although most investigations of the SDF-1/CXCR4 axis have focused on cell trafficking [69], it also mediates other biological functions, including vascular organogenesis [70]. Endothelial stem cells recruited by SDF-1 contribute to tumor vascularization [71], and SDF-1α is essential for post-ischemic recruitment and homing of stem and progenitor cells [28]; cell traffic to ischemic tissue follows an HIF1-induced SDF-1 gradient [58]. Results in our laboratory indicate that EPC integration after limb ischemia is augmented by local SDF-1α administration [26], and Grunewald et al. demonstrated the need for SDF-1α in close perivascular proximity during VEGF-induced postnatal neovascularization [72]. Because SDF-1α is essential for cell trafficking and maintenance during vasculogenesis, it is a promising candidate for gene therapy.

46.3.4.2 Sonic Hedgehog

Nearly three decades ago, the hedgehog (Hh) gene family was identified as a crucial regulator of cell-fate determination during embryogenesis [73], and it was subsequently shown to modify cell survival and proliferation [74]. Hh genes act as morphogens in various tissues during embryonic development, but most investigations have focused on their involvement in neurogenesis [75,76]. The secreted glycoprotein Shh is a vertebrate homologue of the Drosophila segment polarity gene Hh. Although its precise role remains unclear, the trophic or mitogenic activity of Shh seems to be required continuously during embryogenesis [73].

Our laboratory has demonstrated that recombinant Shh protein induces a robust angiogenic effect by upregulating multiple angiogenic factors, including VEGF, in interstitial mesenchymal cells [29]. This observation suggests that the embryonic pathway of vascular development could be reactivated postnatally to combat pathological conditions. Intramyocardial Shh gene transfer enhances the regeneration of ischemic myocardium by inducing expression of trophic factors such as SDF-1 and, consequently, by increasing the recruitment and incorporation of bone marrow-derived EPCs during neovascularization [25].

Because it appears to be capable of triggering a cascade of proangiogenic factors, Shh likely possesses considerable potential for vascular regeneration and is an appealing candidate for future research in cardiovascular gene therapy. In addition, Shh has recently been shown to accelerate wound healing in diabetic mice by enhancing EPC-mediated microvascular remodeling [77].

46.4 BASICS AND LIMITATIONS

In addition to the detailed information provided in the specific sections of this chapter, some general aspects of gene therapy with special relevance for cardiovascular applications must be addressed (Figure 46.1).

In general, the presence of each genetic component can be modified directly by delivering recombinant proteins or indirectly by transferring genes that are then overexpressed. To date, combined therapy that includes both recombinant protein delivery and gene transfer has also been tolerated by cardiovascular patients with multiple morbidities, but the preparation, delivery strategy, and dosage of each agent must be optimized with consideration for each individual patient's fragility.

Direct protein delivery enables the dose to be delivered more precisely, with well-defined dynamics, kinetics, and safety profiles. Some investigators also advocate direct protein delivery for practical reasons; it is particularly suitable for administration to the limb. However, recombinant protein is often administered systemically, which is subject to at least three limitations. First, adequate tissue uptake requires high plasma concentrations, with a consequently higher risk for adverse effects. Second, administered proteins do not produce a sustained effect because they are rapidly degraded by circulating proteinases [78,79]. Third, the production of recombinant human protein is complex and may be prohibitively expensive.

Gene transfer enables the attainment of high, sustained levels of gene expression accompanied by a lower probability of adverse reactions and lower costs. The effectiveness of gene transfer depends on the efficiency of transfection, which depends, in turn, on insertion of the transgene into the target tissue and the magnitude and endurance of its subsequent expression. The cellular penetration and intracellular trafficking of the transgene is facilitated by vectors.

There are two categories of vectors: viral and nonviral. Early attempts at cardiovascular gene therapy relied on nonviral, plasmid DNA because it was considered safe and easy to produce, and striated and cardiac muscle, the primary tissues of interest in cardiovascular medicine, had been shown to effectively take up and express naked plasmid DNA as well as transgenes encoded by viral vectors. However, later randomized, controlled trials did not support the use of plasmid DNA because of low gene transfer efficiency [80,81], and transfer efficiency was not improved by advancements in engineering (e.g., liposome polymers) or techniques designed

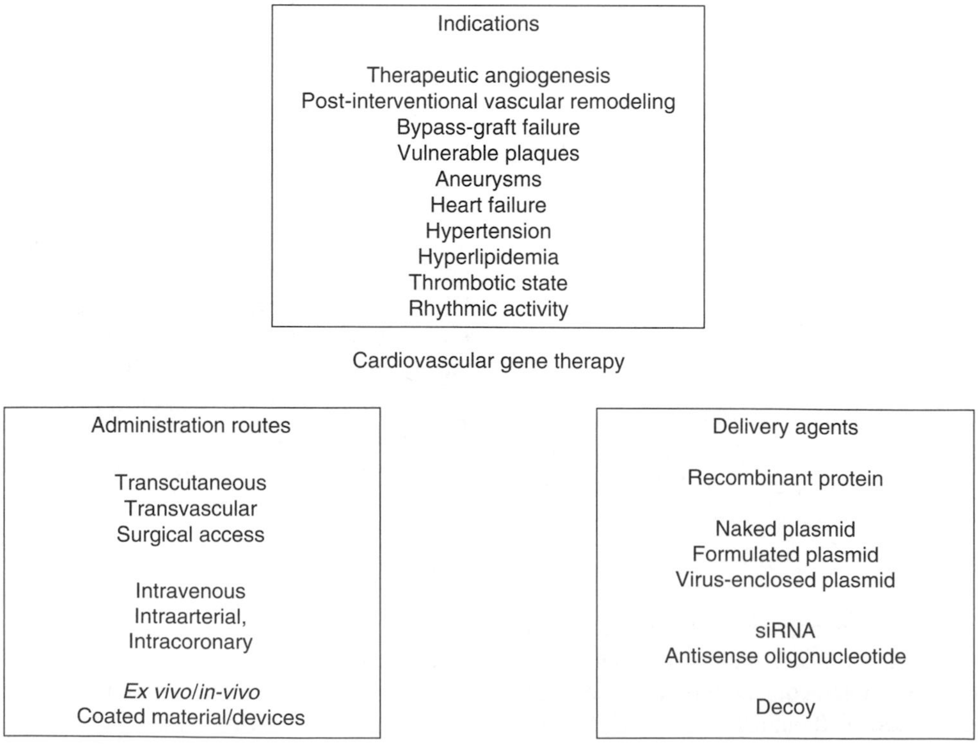

```
┌─────────────────────────────────────────────┐
│              Indications                      │
│                                               │
│         Therapeutic angiogenesis              │
│   Post-interventional vascular remodeling     │
│            Bypass-graft failure               │
│             Vulnerable plaques                │
│                Aneurysms                      │
│               Heart failure                   │
│               Hypertension                    │
│               Hyperlipidemia                  │
│              Thrombotic state                 │
│              Rhythmic activity                │
└─────────────────────────────────────────────┘
```

Cardiovascular gene therapy

```
┌─────────────────────────────┐        ┌─────────────────────────────┐
│    Administration routes    │        │       Delivery agents       │
│                             │        │                             │
│      Transcutaneous         │        │     Recombinant protein     │
│       Transvascular         │        │                             │
│      Surgical access        │        │        Naked plasmid        │
│                             │        │      Formulated plasmid     │
│        Intravenous          │        │    Virus-enclosed plasmid   │
│        Intraarterial,       │        │                             │
│        Intracoronary        │        │            siRNA            │
│                             │        │   Antisense oligonucleotide │
│      Ex vivo/in-vivo        │        │                             │
│    Coated material/devices  │        │            Decoy            │
└─────────────────────────────┘        └─────────────────────────────┘
```

FIGURE 46.1 Efficiency or failure of cardiovascular gene therapy is determined by the targeted disease, route of administration, delivery vector/agents, and gene of therapeutic interest (Table 46.1).

to improve cell penetration (e.g., electroporation and ultrasound) [82–84]. Unlike viral transfer, naked DNA often fails to produce a dose-dependent response [85,86], and although naked plasmid DNA is generally safe, it may cause infarction in peripheral muscle and myocardium [87,88].

Adenoviruses (Ads), adeno-associated viruses (AAVs), and retroviruses are the most frequently used vectors for cardiovascular gene therapy. Both achieve high transduction efficiency in vessel walls and in heart and skeletal muscle [82]. AD seems particularly effective for gene transfer to cardiomyocytes and skeletal myocytes [89], and the safety of Ads has been very good in cardiovascular clinical trials [80,90]. Natural tropisms that favor vascular smooth muscle cells, cardiomyocytes, and skeletal myocytes are among the more useful characteristics of AAVs, and the expression of AAV-transfected genes can endure for months.

Although retroviral vectors were among the first used for cardiovascular gene therapy, their popularity has declined because of safety issues. Lentiviral vectors have emerged more recently and appear to efficiently transduce blood vessels [91,92], but additional engineering is needed to optimize their efficiency. Small interfering RNA (siRNA) has begun to replace antisense oligonucleotides for inhibition studies and may soon provide a new option for nonviral gene modulation. Very high transfection efficiency may be achievable by administration of decoy receptors or antibodies that circumvent blocking factors in cardiovascular tissue.

In addition to receptor binding, angiogenic efficiency is influenced by two other variables. Local growth factor concentration is a key determinant of vascular growth [93] and growth factor spread is influenced by the binding properties of the matrix and the volume and number of injections [85]. Transfection efficiency is fivefold greater when genes are transferred into ischemic, rather than healthy, muscle [94,95]. Intramuscular gene transfer yields higher efficiency and angiogenesis with less ectopic gene expression than intravascular injection [96], but it is more feasible for peripheral muscle than for the heart.

The successful use of gene therapy in the clinical setting will require practical delivery systems and techniques that are specific to the most commonly targeted tissues: myocardium, skeletal muscle, and blood vessel tissue. Gene transfer can be performed by direct organ injection (local delivery), intravascular application, and *ex vivo* gene transfer. Local delivery is easily applied to peripheral tissue; intramyocardial and intramuscular injection is an efficient strategy for inducing robust gene expression. Local delivery to the heart initially required open-chest surgery or thoracoscopy, but the recent development of ultrasound-guided, catheter-based delivery systems (e.g., navigation and catheter mapping technology [NOGA]) made local transvascular delivery to the myocardium much more feasible [85]. The transvascular approach bypasses open operative procedures and enables selective delivery to the targeted organ; however, preclinical studies have suggested that intra-arterial transvascular delivery is insufficient unless the permeability of the endothelium is modified or gradient methods are applied [96,97].

Intravenous gene infusion is practical and inexpensive, can be administered to most patients, can be done repetitively, and does not require a unique set of skills, but the vector is widely distributed throughout the body, which may be a critical safety limitation, and the amount needed is unreasonably high. Genes can be transferred to vascular grafts with high efficiency via in situ incubation or *ex vivo* immersion in the gene transfer solution, and a gene-eluting stent has recently been introduced for intravascular administration of gene therapy [98].

46.5 CLINICAL AND PRECLINICAL EVIDENCE

Recent investigations have assessed the therapeutic potential of vascular gene therapy for preventing postinterventional vascular remodeling (restenosis, in-stent restenosis) and bypass-graft failure, for stabilizing vulnerable plaques and aneurysms, for modulating rhythmic activity, and for managing heart failure, hypertension, hyperlipidemia, and thrombotic states [10,99]. However, the bulk of investigative work has studied its use for therapeutic angiogenesis, which remains the most clinically advanced application of vascular gene therapy. For these reasons, most of the studies discussed below focused on the therapeutic potential of vascular gene therapy for angiogenesis.

46.5.1 THERAPEUTIC ANGIOGENESIS—CAD

Chronic heart failure is associated with the highest morbidity and mortality rates in the cardiovascular field and generates obvious socioeconomic consequences. As medical treatment of acute events has advanced, long-term survival after myocardial infarction has become a significant contributing factor to chronic heart failure. Therefore, strategies to prevent myocardial ischemia as well as to improve cardiac function after an ischemic event are urgently needed. In addition, patients with refractory angina sometimes require multiple procedures to achieve sufficient symptomatic relief, and because of the diffuse nature of CAD, more invasive techniques may not be suitable for a considerable number of patients (Tables 46.1 and 46.2).

46.5.1.1 VEGF

Recombinant $VEGF_{165}$ protein has been administered via a variety of routes to induce angiogenesis in ischemic myocardium. Phase 1 studies of both intracoronary and intravenous $VEGF_{165}$ protein injection in patients with symptomatic, inoperable CAD found encouraging improvements in anginal status and nuclear perfusion both at rest and when under stress [100–102].

The phase 2 VIVA trial tested the effect of recombinant VEGF protein in patients with CAD, angina, and viable myocardial tissue who were not candidates for revascularization. In this double-blind, placebo-controlled, multicenter, dose-escalation study, patients were randomized to one intracoronary injection or

TABLE 46.2

Randomized Controlled Phase II/III Clinical Trials in CAD—Therapeutic Angiogenesis

Trial	Entity	Agent	Route	Primary Endpoint	n	Follow-up	Outcome	Reference
VEGF								
Euroinject One	CAD angina CCS III–IV	Naked VEGF$_{165}$ plasmid	Percutaneous intramyocardial	Perfusion	74	3 months	Negative	[81]
Genasis	CAD angina CCS III–IV	Naked VEGF-2 (VEGF-C) Plasmid	Percutaneous intramyocardial	Exercise tolerance	295	3 months	Negative stopped before completion	(Unpublished)
KAT	CAD angina CCS II–IV	AdVEGF$_{165}$ liposome VEGF$_{165}$ plasmid	Intracoronary	Perfusion	103	6 months	Positive (AdVEGF$_{165}$) negative (VEGF$_{165}$ plasmid)	[80]
Northern	CAD angina CCS II–IV	AdVEGF$_{121}$	Percutaneous intramyocardial	Perfusion	120	12 weeks	(ongoing)	(Unpublished)
NOVA	CAD angina CCS II–IV	AdVEGF$_{121}$	Percutaneous intramyocardial	Exercise tolerance	129	26 weeks	(ongoing)	(Unpublished)
REVASC	CAD angina CCS II–IV	AdVEGF$_{121}$	Minimal-invasive intramyocardial	Ischemia in ECG exercise tolerance	67	26 weeks	Positive	[229]
FGF								
AGENT-2	CAD angina CCS II–IV	AdFGF-4	Intracoronary	Perfusion	52	8 weeks	Positive	[133]
AGENT-3	CAD angina CCS II–IV	AdFGF-4	Intracoronary	Exercise tolerance	416	12 weeks	Negative positive— subgroup analysis	[134]
AGENT-4	CAD angina CCS II–IV	AdFGF-4	Intracoronary	Exercise tolerance	116	12 weeks	Negative	[134]
AWARE	CAD angina CCS III–IV	AdFGF-4	Intracoronary	Exercise tolerance	300	6 months	(Ongoing)	(Unpublished)
Other targets								
Italics	CAD in-stent restenosis	c-myc antisense oligonucleotide	Local delivery	Neointimal vessel obstruction	85	6 months	Negative	[159]
PREVENT IV	CAD vein grafts	E2F transcription factor decoy	*Ex-vivo* delivery	Death vein graft stenosis (>75%)	2400	12–18 months	Negative	[178]

three dose-escalating intravenous infusions. Patients were followed for up to 1 year. Although the increase in exercise time did not differ significantly between the placebo group and either treated group after 60 days, there was a trend toward a lower angina class in VEGF-treated patients that could not be explained by differences in clinical status and diagnostic parameters. Angiographic and single photon emission computed tomography (SPECT) imaging were similar at follow-up [100,103]. Safety results indicated that hypotension was a significant, dose-limiting side effect among patients who received intravascular infusions. Hypotension had been previously observed in animal experiments with recombinant VEGF and may have been caused by VEGF-induced release of nitric oxide [104,105].

Because some delivery routes (e.g., intravenous, periadventitial) cannot produce sufficient plasma levels of the VEGF protein [106,107], the ineffectiveness of VEGF in some studies can likely be attributed to the method of administration. This assertion is supported by results from the phase 2 KAT trial. Here, 103 patients with CAD and angina (Canadian Cardiovascular Society [CCS] class II or III) were treated with a VEGF Ad vector, a VEGF liposome plasmid, or placebo via intracoronary injection while undergoing angioplasty and stenting. After 6 months, myocardial perfusion increased significantly in only the AdVEGF-treated group; the distribution of restenosis was similar in all three groups [80].

In some animal models of CAD, VEGF gene therapy has been successfully administered via naked plasmid DNA and viral vectors. Direct intramyocardial injection of AdVEGF$_{121}$ through a thoracoscopic access improved collateralization, tissue perfusion, and cardiac function after infarction in large animals [108,109], and intracoronary Ad-vector gene delivery yielded efficient gene transfer and VEGF expression in the myocardium [108]. However, local pericardial delivery of AdVEGF failed to improve collateral myocardial perfusion [79].

The first clinical study to transfer naked VEGF plasmid directly to targeted tissue was a phase 1, open-label, dose-escalating study initiated by our center that focused on safety and bioactivity. Preliminary results suggested that VEGF gene therapy administered by this method was safely and successfully transfected and yielded favorable clinical effects. Patients with refractory, stable angina when under exertion and with multivessel occlusive CAD and hibernating myocardium (determined via perfusion scanning) were selected. By using transesophageal echocardiographic control and a cardiac stabilizer, gene delivery was accomplished with no periprocedural complications. All patients experienced marked symptomatic improvement, including lower angina frequency and severity, reduced nitrate use, and improved exercise tolerance over a 1-year follow-up period. The clinical benefits were accompanied by enhanced perfusion in the ischemic area and recovery of wall motion abnormalities 60 days after gene transfer [110–113]. We observed similar results with VEGF2 plasmid DNA in an open-label, dose-escalating, multicenter, clinical trial in 30 patients with CAD and refractory class III–IV angina. All individuals had previously undergone bypass surgery, had sustained one to two episodes of acute ischemia, and were using more than two anti-anginal drugs. Gene transfer was performed safely and without complication. Twelve months later, anginal episode frequency and nitrate consumption were significantly lower, and mean exercise duration (determined by a standard treadmill protocol) increased by more than 2 minutes [114].

A number of other studies investigated direct myocardial VEGF gene transfer via an AdVEGF construct. The REVASC phase 2 trial was the largest investigation of AdVEGF$_{121}$ gene therapy in patients with ischemic CAD. Sixty-seven patients with CCS angina class II–IV were randomized to continue with optimal medical management or to receive intramyocardial AdVEGF$_{121}$ injection through a limited thoracotomy [115]. After 26 weeks, symptoms of ischemia occurred significantly later during treadmill testing in the AdVEGF$_{121}$-treatment group. Angina status and several domains of the Seattle Angina Questionnaire were also improved among patients who received VEGF$_{121}$ therapy; however, four patients experienced complications from the thoracotomy. Myocardial injection of AdVEGF$_{121}$ improved symptoms and exercise duration when administered to patients undergoing bypass surgery for advanced CAD and when administered without bypass surgery via minithoracotomy. Stress-induced nuclear perfusion images remained unchanged, and the vector was well-tolerated.

All of these early VEGF gene therapy studies in patients with myocardial ischemia were limited by the need for thoracotomic access. This procedure is associated with a small but relevant risk that can include substantial morbidity, and also precludes placebo-controlled investigations. More recent investigations have used a less invasive, catheter-based delivery system. The system integrates the NOGA mapping technique with a working catheter (Biosense-Webster, Warren, NJ). The distal catheter tip incorporates a 27-gauge needle used for injections into the myocardium. The mapping capabilities of the NOGA system are accompanied by three additional advantages when compared to an operative approach. First, the transgene can be delivered precisely to predetermined zones of ischemia, including sites that are less accessible via minithoracotomy; this, in turn, prevents accidental gene delivery to a myocardial scar. Second, catheter-based delivery eliminates the need for general anesthesia and operative dissection, so placebo-controlled, double-blind studies can be initiated. Third, the intervention can be performed as an outpatient procedure and repeated if necessary. Proof-of-concept, preclinical experiments confirmed the safety, reliability, reproducibility, and local accuracy of this approach [116,117].

We conducted a pilot study of percutaneous, catheter-based VEGF2 DNA gene transfer guided by the NOGA mapping system in six patients with symptomatic myocardial perfusion who were not candidates for revascularization. Twelve months after gene transfer, patients reported significant reductions in angina frequency and nitrate consumption, and improved myocardial perfusion was observed by both SPECT perfusion scanning and electromechanical mapping. After witnessing these encouraging results, we designed a double-blind, randomized, placebo-controlled, multicenter trial to evaluate catheter-based, VEGF2 gene transfer in patients with chronic myocardial ischemia not amenable to revascularization. Patients were randomized to receive six injections of VEGF2 plasmid (phVEGF2) or placebo. Gene transfer was well-tolerated and no relevant complications were reported. Twelve weeks later, the phVEGF2-treated group experienced significant improvement in angina severity class and other endpoints, such as exercise duration, functional improvement, and the Seattle Angina Questionnaire, exhibited strong trends favoring phVEGF2 over placebo treatment [118]. The study was closed by the U.S. Food and Drug Administration after 19 of the planned 27 patients had been enrolled, and larger, phase 2/3 trials were encouraged.

Only a few phase 2/3 trials have been initiated to date. In the Euroinject One phase 2 trial, 80 patients with CCS class III–IV CAD who lacked any other treatment options were randomly assigned to receive percutaneous, intramyocardial injection of naked VEGF$_{165}$ plasmid or placebo. Three months after plasmid delivery, myocardial perfusion (as assessed by SPECT) was similar in both groups, but regional wall motion (assessed both by NOGA and contrast ventriculography) was significantly better in VEGF-treated patients than in patients administered placebo [81]. The ongoing phase 2/3 Northern trial employed

AdVEGF$_{121}$ to therapeutically induce angiogenesis in no-option patients with CCS class III–IV angina. Change in myocardial perfusion was assessed 12 weeks after percutaneous, intramyocardial vector application. Results from 120 patients who have completed the trial are pending (ClinicalTrials.gov, NCT00143585). Patient recruitment for the phase 2/3 NOVA trial was recently halted because of substantial similarities with the Northern trial; the study design used percutaneous, intramyocardial application of AdVEGF$_{121}$ in patients with CAD and CCS class II–IV angina (ClinicalTrials.gov, NCT00215696). Finally, the GENASIS phase 2 study was designed to investigate the effect of naked VEGF2 plasmid applied by percutaneous, intramyocardial gene transfer on exercise tolerance in patients with CCS class II–IV CAD. The study was interrupted after 295 of the planned 404 patients had been enrolled because of complications stemming from the intramyocardial injection catheter (D. Losordo, unpublished data).

Taken together, clinical investigations of VEGF gene transfer for therapeutic angiogenesis are encouraging. Direct gene transfer into the myocardium is superior to intravascular applications and development of a catheter-based delivery system has revolutionized myocardial gene therapy.

46.5.1.2 FGF

Preclinical investigations in animal models of myocardial ischemia demonstrated that local FGF application improved myocardial perfusion and function and increased collateral blood flow [119–122]. Subsequent, small-scale phase 1 studies established the safety of recombinant FGF1 and FGF2 in patients with myocardial ischemia [123–127], and reported some evidence of clinical improvement and functional cardiac recovery in patients with CAD. Interpretation of these results is limited, however, because direct FGF injection to the myocardium was performed during coronary artery bypass graft surgery. Less invasive intravenous and intracoronary administration of recombinant FGF2 has been performed in phase 1 studies of patients with symptomatic CAD, but the results were mixed. Intravenous (but not intracoronary) injection alleviated angina, extended exercise time, and improved left ventricular function, but several patients experienced transient hypotension, conduction system disturbances, thrombocytopenia, or proteinuria [128–130].

The FIRST trial was an early phase 2 trial in 337 patients with inoperable CAD; patients received placebo or one of three doses of recombinant FGF2 by intracoronary injection. There were no significant differences between placebo and FGF2 treatment in any endpoint, including exercise duration and stress nuclear perfusion 90 days after treatment; however, there was a trend toward reduced anginal complaints in patients treated with FGF2, particularly among those who were older and more symptomatic [131].

Although the results achieved with recombinant FGF protein have been disappointing, FGF gene therapy remains promising. The Angiogenic Gene Therapy (AGENT) trial, conducted by Grines et al., was the first clinical investigation of FGF gene therapy for CAD. Patients with chronic stable

angina were randomized to receive placebo or escalating doses of AdFGF4 by intracoronary injection. The AdFGF4 infusion was well-tolerated, with few periprocedural adverse events (e.g., early transient fever, minor and self-limited liver enzyme elevation) and no significant difference in adverse event occurrence between AdFGF4- and placebo-treated individuals [132]. Four weeks after therapy, patients receiving AdFGF4 tended to exercise longer before manifestation of angina, and the difference between FGF4- and placebo-treated patients was significant in the subgroup of patients whose baseline exercise tolerance was less than 10 minutes. These results suggested that intracoronary infusion of AdFGF4 is safe and may exert favorable anti-ischemic effects in patients with severe myocardial ischemia.

In a subsequent double-blind, placebo-controlled study in 52 patients with stable angina and reversible ischemia (AGENT-2), AdFGF4 administered via intracoronary injection was associated with significantly smaller ischemic defect size (determined by SPECT) 8 weeks after treatment than was observed with placebo treatment [133]. However, the large-scale, multinational, double-blind AGENT-4 trial found no evidence of a robust effect on exercise tolerance among the 116 enrolled patients (ClinicalTrials.gov, NCT00185263). In the United States, patient enrollment in the AGENT-3 trial was terminated early because the primary endpoint was not achieved; however, a post hoc analysis of the AGENT-3 patients who were older (>55 years) and sicker found evidence of significant efficacy associated with FGF4 gene therapy, echoing the subgroup analyses performed in the FIRST trial. Recently, a pooled analysis of AGENT-3 and AGENT-4 patients revealed gender specificity in patient response to AdFGF4 gene therapy: men responded to placebo treatment but not to AdFGF4 therapy, and women experienced greater exercise time and reduced angina severity with FGF4 treatment but negligible response to placebo [134]. The therapeutic efficiency of FGF4 gene therapy in women is being investigated in the ongoing AWARE trial. This phase 3 multicenter study is designed to evaluate patient response to a one-time intracoronary FGF4 injection in 300 female patients with stable CAD (ClinicalTrials.gov, NCT00438867).

In summary, there is no evidence to support the use of intracoronary recombinant FGF protein for treatment of CAD, and the effectiveness of FGF gene therapy remains unknown. Further large-scale trials are needed to characterize the potential age- and sex-specific efficacy of FGF gene therapy.

46.5.2 Therapeutic Angiogenesis—PVD

In the consensus statement of the European Working Group on Critical Limb Ischemia, group members noted that no medical treatment has been shown to alter the course of critical limb ischemia. In a large proportion of patients with critical limb ischemia, the distribution and extent of arterial occlusion precludes proper revascularization. As a consequence, the quality of life for patients in later stages of the disease is comparable to that of patients with terminal cancer [135]. Ultimately, the chronic progression of critical limb ischemia

TABLE 46.3

Randomized Controlled Phase II/III Clinical Trials in Peripheral Arterial Disease—Therapeutic Angiogenesis

Trial	Entity	Agent	Route	Primary Endpoint	n	Follow-up	Outcome	Reference
VEGF								
Groningen	PAD critical limb ischemia	Naked VEGF$_{165}$ plasmid	Intramuscular	Amputation rate	54	100 days	Negative–positive— secondary endpoint	[148]
RAVE	PAD claudication	AdVEGF$_{121}$	Intramuscular	Walking time	105	12 weeks	Negative	[150]
VEGF peripheral vascular disease	PAD claudication	AdVEGF liposome VEGF$_{165}$ plasmid	Intraarterial	Vascularity	54	3 months	Positive	[90]
FGF								
PM 202	PAD critical limb ischemia	Naked FGF-1 plasmid	Intramuscular	po2	71		Negative	(Unpublished)
TALISMAN 201	PAD critical limb ischemia	Naked FGF-1 plasmid	Intramuscular	Ulcer healing	107	6 months	Negative–positive— secondary endpoint	(Unpublished)
Other targets								
DELTA-1	PAD claudication	Del-1 plasmid (poloxamer 188 formulated)	Intramuscular	Walking time	157	3 months	Negative	(Unpublished)
HGF-STAT	PAD critical limb ischemia	Naked HGF plasmid	Intramuscular	Wound healing amputation rate rest pain ABI	(48)		(Ongoing)	[158]
Prevent III	PAD critical limb ischemia vein graft	E2F transcription factor decoy	*Ex vivo* delivery	Time to graft failure: reintervention, amputation	1404	1 year	Negative–positive— secondary endpoint	[179]
WALK	PAD claudication	AdHIF-1α/VP16	Intramuscular		300	6 months	(Ongoing)	(Unpublished)

leads to limb loss, and of patients who undergo one amputation, 10% will undergo two or more [136–138]. Despite advances in conventional therapy, including interventional and operative techniques, there is an urgent need for treatment alternatives in patients with critical limb ischemia (Tables 46.1 and 46.3).

46.5.2.1 VEGF

The angiogenic efficiency of VEGF in critical limb ischemia was evident in early preclinical animal models. In rabbits, intra-arterial VEGF protein administration augmented collateral vessel and capillary development after severe, unilateral hind limb ischemia [139], and the angiogenic potency of VEGF protein was confirmed in a diverse set of other animal models [45,140]. However, clinical use was hampered by the same difficulties encountered with intravascular growth factor application. Cellular uptake of naked DNA after intravascular injection is poor, presumably because of prompt degradation by circulating nucleases, and the systemic distribution of intravascular gene transfer does not allow targeted treatment. Thoughtfully designed preclinical studies quickly established the feasibility and efficiency of site-specific, intramuscular VEGF gene transfer for treatment of critical limb ischemia. Meaningful biological outcomes were observed after direct injection of naked VEGF DNA into skeletal muscle, and blood pressure ratio and

Doppler-derived blood flow measurements correlated well with angiographic scores and capillary density [94,141].

The positive results of these and other preclinical studies encouraged the initiation of clinical studies. In a pilot clinical investigation, Jeffrey Isner, one of the pioneers of VEGF gene therapy in ischemic disease [142], found angiographic and histological evidence of angiogenesis after delivery of human VEGF$_{165}$ plasmid (phVEGF$_{165}$) in patients with ischemic peripheral arterial disease (PAD) [143]. In a small number of selected patients, transfer of naked phVEGF$_{165}$ (4000 µg) successfully induced VEGF expression and was associated with improvements in a number of objective clinical parameters as well as subjective clinical benefit [144]. In a subsequent trial, 55 patients with ischemic pain at rest and ischemic ulcers were locally treated with phVEGF$_{165}$ injection. Over a follow-up period lasting up to 36 months, 65.5% of treated patients achieved a favorable clinical outcome, defined as improved ankle-brachial index, collateralization, peripheral blood flow, and limb salvage. Rest pain and an age of less than 50 years were independent predictors of a favorable clinical outcome (determined by multiple logistic regression), while classic risk factors and the magnitude of the phVEGF$_{165}$ dose were not [145,146]. Adverse effects were limited to self-limiting lower-extremity edema, which developed in one third of patients; prolonged edema resolved with short-term diuretic therapy [147].

In the recently published, placebo-controlled, phase 1/2 Groningen trial, the amputation rate at day 100 among 54 diabetic patients (27 in each group) with PAD and critical ischemia was lower (though not significantly) in patients who received naked VEGF$_{165}$ plasmid by intramuscular injection rather than placebo, and predetermined secondary endpoints, including clinical condition and ankle-brachial index, were significantly improved [148]. The VEGF for PVD phase 2 trial comprehensively compared plasmid-liposome and AdVEGF$_{165}$ gene transfer in 54 patients with PAD and intermittent claudication who were selected for percutaneous transluminal angioplasty (PTA). Catheter-based, local, intra-arterial delivery of both preparations led to improved clinical parameters after three months, including vascular growth distal to the injection sites. Both treatments were safe and well tolerated [90].

In addition to patients with classic PAD, those with Buerger's disease who were unresponsive to conventional treatments for critical limb ischemia benefited from intramuscular phVEGF$_{165}$ therapy. Nocturnal rest pain, ankle-brachial index, and collateralization determined by serial angiography improved with treatment. These findings were consistent with healing limb ulcers [149].

Adenoviral VEGF$_{121}$ gene therapy has been investigated in disabled patients with PAD by Rajagopalan et al. In their initial phase 1 trial with five patients, intramuscular injection at sites where collateral vessel formation was desired improved lower extremity endothelial function and flow reserve [150]; however, the subsequent double-blind, randomized, controlled, phase 2, RAVE trial failed to reproduce these benefits. Twelve and 26 weeks after treatment, there was no significant difference in quality of life or any clinical parameter between patients administered AdVEGF$_{121}$ and those who received placebo. Thus, the authors did not advocate the use of AdVEGF$_{121}$ for AGENT in patients with PAD [151]. Treatment with AdVEGF$_{121}$ also resulted in transient edema, as we observed with the intramuscular injection of AdVEGF$_{165}$, but severe adverse effects were not reported.

Overall, local gene transfer of VEGF$_{165}$ appears to exhibit potential for treatment of patients with PAD accompanied by intermittent claudication and critical ischemia. Intravascular injection is not appropriate, however, because of striking limits in the methodology. The therapeutic application of VEGF$_{121}$ and VEGF2 has not been supported by clinical evidence, but only a few properly designed trials have been attempted thus far, so these constructs may still prove to be effective.

46.5.2.2 FGF

After preclinical studies demonstrated the improvement of ischemic hind limb perfusion by recombinant FGF2 application [152–154], an initial phase 1 study in patients with atherosclerotic PVD and claudication tested the feasibility, safety, and tolerability of intra-arterial FGF2 administration. In this double blind, placebo-controlled, dose-escalation trial, blood flow in the calf was greater in FGF2-treated patients than in

the control group, and the authors recommended the initiation of phase 2 trials [155]. The TRAFFIC trial evaluated the administration of recombinant FGF2 in patients with moderate-to-severe, intermittent claudication from PVD. A total of 190 patients were randomized to receive placebo, one dose (30 µg/kg), or two doses (30 µg/kg each, one month apart) of FGF2. After 90 days, the increase in peak walking time was higher in patients who received either FGF2 dosing regimen than in placebo-treated patients ($p = 0.034$ when all 190 patients were analyzed), but the second FGF2 dose provided no additional benefit. Serious adverse events were similarly distributed in all groups [156].

Gene therapy with FGF was initially performed in a small number of patients with ischemic rest pain or tissue necrosis due to end-stage peripheral arterial occlusive disease for whom mechanical revascularization was unsuitable. In this phase 1 study, naked plasmid DNA encoding for FGF1 was administered to the ischemic thigh or calf muscle by intramuscular injection in either escalating single or escalating repeated doses. Overall, the treatment was well tolerated and there were no FGF1-related adverse events. Plasmid biodistribution was limited with no increase in FGF1 serum level and none present in the urine, presumably because of metabolism by endogenous endonucleases. Administration of FGF1 gene therapy was associated with significant reductions in pain and ulcer size as well as increased oxygen pressure and ankle-brachial index at follow-up, but response was not dose-dependent [157]. Results from subsequent, unpublished, phase 2 trials that employed intramuscular injection of naked FGF plasmid were less encouraging. In the PM 202 trial in patients with critical limb ischemia, FGF therapy was not associated with significant improvement in the primary endpoint (ulcer healing) or in transcutaneous pO$_2$ [10], and there was no relevant difference in the primary endpoint (ulcer healing) of the TALISMAN 201 trial, although FGF therapy was associated with a significant reduction in amputation rate [10].

The findings described above suggest that recombinant FGF2 may be effective for therapeutic angiogenesis [156], but the scarce number of FGF gene therapy trials do not provide convincing support for administering the FGF gene to patients with PVD. Because conclusive evidence is lacking, additional large-scale studies are needed.

46.5.2.3 Other Factors

In addition to VEGF and FGF, a small number of other potential gene therapies are beginning to reach the clinical stage of investigation. Intramuscular injection of a Del-1 (developmentally regulated endothelial locus 1) plasmid formulated with poloxamer 188 in 157 patients with PAD and claudication did not substantially change the peak walking time after 3 months (DELTA-1 phase 2b trial, unpublished) [10]. Intramuscular application of AdHIF1α/VP16 is being investigated in the ongoing, phase 2, WALK trial; the primary study endpoint is peak walking time 6 months after treatment and the planned enrollment is 300 patients with severe intermittent claudication (ClinicalTrials.gov, NCT00117650). The effect of naked HGF plasmid injection into the skeletal musculature in patients with PAD and critical ischemia is being evaluated in the ongoing HGF-STAT trial [158]; wound healing, amputation rate, rest pain, and ankle-brachial index reflecting tissue perfusion are the designated study endpoints, and the planned enrollment is 100.

46.5.3 PREVENTION OF POSTINTERVENTIONAL REMODELING

One of the primary challenges remaining in cardiovascular medicine is postinterventional vascular remodeling, which can lead to acute thrombosis, restenosis, or in-stent stenosis. Even the most recent technical inventions often fail to overcome this shortcoming (Tables 46.1 and 46.4).

46.5.3.1 Coronary Artery System

Although drug-eluting stents dramatically reduce the occurrence of in-stent restenosis they also delay postinterventional endothelial recovery and increase the risk of fatality due to stent thrombosis. Brachytherapy has not provided a clinical breakthrough to date.

Acceleration of endothelial tissue formation (re-endothelialization) is known to attenuate the risk of restenosis after coronary stenting. Because VEGF is a potent stimulator of EC recovery, a VEGF-coated stent may reduce the risk of restenosis. In a preclinical model, rabbits were treated with uncoated or phVEGF2-coated stents. After 3 months, the coated stents were associated with significantly larger lumen cross-sectional areas and less cross-sectional narrowing; the underlying EC recovery was related to the greater EPC recruitment associated with VEGF gene-coated stenting [98]. A phase 1 clinical trial evaluating the safety of phVEGF2-coated stents has been initiated.

An alternative strategy for preventing restenosis that uses a specialized infiltration catheter to directly pre-infuse VEGF plasmid into the coronary plaque before stenting has been proposed. A technically similar approach was investigated in the Italics trial. Antisense oligonucleotides against c-myc were locally administered after stenting, but the neointimal vessel obstruction was similar in oligonucleotide- and saline-treated patients 6 months later [159].

46.5.3.2 Peripheral Arterial Vasculature

PTA is a successful therapy for alleviating arterial obstructions; however, the diffuse nature of arteriosclerosis and the increased risk of postinterventional restenosis make procedures performed in the peripheral vasculature more challenging and generally less effective than coronary procedures. This observation has not been satisfactorily explained, and limb vasculature in the adductor canal seems particularly prone to stenosis and restenosis. Whereas the success rate for standard PTA is up to 90%, restenosis may complicate the clinical course for as many as 60% of patients undergoing PTA of the superficial femoral artery [136–138]. Because efforts to limit the development of restenosis by nonmechanical means have not been effective, newer strategies attempt to speed restoration of endothelial integrity after the intervention. Results in our laboratory indicated that the administration of mitogens such as VEGF accelerates re-endothelialization in animals by promoting EC migration and/or proliferation, thereby reducing intimal thickening [160–163]. Other preclinical studies of re-endothelialization in different vascular beds after VEGF overexpression have yielded conflicting results [164–166].

In an unblinded, single-center, phase 1 pilot trial, we investigated whether VEGF gene therapy reduces restenosis after PTA-induced endothelial disruption by enhancing re-endothelialization. Gene transfer was intra-arterial, and transfer efficiency was established by a rise in plasma VEGF levels. Duration of exercise before claudication increased and Rutherford class improved with VEGF gene therapy, and there was a sustained improvement in ankle-brachial index, but the effect on the PAD severity score was unstable. During invasive follow-up with intravascular ultrasound, restenosis was less in patients administered VEGF gene therapy than in patients who did not receive the gene or in historical controls who had undergone brachytherapy. These results suggested that VEGF gene therapy could safely prevent restenosis by enhancing re-endothelialization [167].

Other target genes with vasoprotective potential, such as nitric oxide synthase (NOS) or prostacyclin synthase, have been successful in animal models [168,169], but these findings have yet to be confirmed in patients. Combining targeted gene therapy with drug-eluting stents may help overcome the clinical drawbacks of antiproliferative drug activity on the endothelium, and the costs of extended antithrombotic drug therapy may also be reduced. Gene-eluting stents have been engineered and tested in animal models [98], but stents that elute both drugs and genes are not yet available.

TABLE 46.4

Randomized Controlled Phase II/III Clinical Trials—Further Cardiovascular Indications

Trial	Entity	Agent	Route	Primary Endpoint	n	Follow-up	Outcome	Reference
ISIS	Familial hypercholesterolemia	apoB antisense oligonucleotide	Intravenous, followed by subcutaneous maintaining dose	LDL-cholesterol	36	12 weeks	Positive	[182]

46.5.4 PREVENTION OF GRAFT FAILURE

Because of the absence of established treatments and a desire for easier techniques, gene therapy could be a valuable option for prevention of graft alteration and subsequent occlusion (Tables 46.1 and 46.4). The concept is supported by promising results in animal models, and *ex vivo* gene transfer may be an attractive option before coronary, peripheral, and arteriovenous graft implantation [170–172]. To some extent, VEGF activity is vasculoprotective, and blockade of the VEGF cascade has been linked to cardiovascular events [173]. Animal studies suggest VEGF may exert a pro-arteriogenic effect [166,174], but adenoviral overexpression of VEGF in multiple clinical studies does not support this notion [34]. Because of its vasodilative, antithrombotic, and antiproliferative properties, NOS may be a suitable candidate gene for this application [175], and Ang1 could also serve as a target to reduce graft vasculopathy [176]. However, clinical attempts to reduce graft failure with any of these agents are rare.

After the in situ transfer of genes to grafts was demonstrated feasible and safe in a clinical pilot study [177], some large-scale trials were initiated to test the clinical relevance of this concept. In the PREVENT IV, phase 3 trial, 2400 patients who were undergoing coronary artery bypass surgery were implanted with venous grafts that had been treated with edifoligide (an E2F transcription factor decoy) or placebo. After 12 to 18 months, the failure rate of venous grafts (defined as $\geq 75\%$ vein graft stenosis) was similar in both groups; longer-term responses were not evaluated [178]. Similar venous grafts were used during bypass surgery for PAD in the phase 3 PREVENT III trial. In 1404 patients with critical limb ischemia who underwent infrainguinal revascularization, there was no significant difference in primary graft patency or limb salvage between patients who received treated or untreated grafts; however, there was significantly better secondary graft patency within 1 year [179]. These results indicate that graft pretreatment with antisense oligonucleotides may not be sufficiently effective to justify routine clinical use, although pretreatment with alternative agents (e.g., siRNA) may still prove beneficial.

46.5.5 THERAPY FOR HYPERLIPIDEMIA, HYPERTENSION, AND THROMBOSIS

Although the effectiveness of conventional therapy has improved substantially in recent decades, the cardiovascular burden of arteriosclerosis continues to rise. Among the many patients who have this disease, those with familial hypercholesterolemia or hypertension refractory to drug therapy may be particularly good candidates for targeted gene therapy (Tables 46.1 and 46.4).

In an early study, Grossman et al. injected autologous hepatocytes transfected to overexpress a functional LDL receptor (LDLr) into the portal vein of five patients with homozygous familial hypercholesterolemia. Although the therapy appeared safe and feasible, LDL levels were only modestly reduced, perhaps because the number of transfected

cells was small [180]. Subsequent animal studies demonstrated that excessive and unregulated AdLDLr overexpression created an imbalance between LDL uptake and metabolism, resulting in lipid accumulation in the liver [181]. In the first and only clinical study to use an antisense oligonucleotide inhibitor of apoB, 36 volunteers with mild dyslipidemia were treated by injection with a multiple-dosing regimen of inhibitor or placebo. Twelve weeks later, LDL-cholesterol and apolipoprotein B levels were significantly reduced by up to 35% and 50%, respectively. Injection-site erythema was common, but the therapy was otherwise safe [182].

Despite promising results in animal models [183–186], there have been no clinical investigations of gene therapy for treatment of hypertension and thrombosis. Hypertension could potentially be controlled by inhibiting the implicated genes or by overexpressing vasodilative genes; stable regulation of chronic hypertension might be accomplished with vectors that enable long-term gene expression. Experimental data suggest that thrombosis might be prevented by targeting antithrombotic genes such as tissue plasminogen activator and thrombomodulin.

46.5.6 THERAPY FOR HEART FAILURE AND MODULATION OF ELECTRICAL ACTIVITY

Long-term survival after myocardial infarction is now the most common cause of chronic heart failure, and despite significant medical advances, postischemic heart failure remains a primary cause of morbidity and mortality in the western world [187]. As available knowledge about the molecular mechanisms that lead to heart failure increases, gene therapy that targets these mechanisms could become a valuable therapeutic option. To improve contractile function, gene therapies will likely be designed to enhance tissue perfusion, increase Ca^{2+} transients, attenuate ventricular remodeling, and modulate electrical activity (Tables 46.1 and 46.4).

Gene therapy with VEGF has been shown to improve left-ventricular function in animal models of ischemic [34] and pacing-induced heart failure [188], and many other genes are potential candidates for treatment of heart failure. Genetic modification of Ca^{2+}-handling factors could be a viable treatment approach, because myocardial contractility is dependent on ventricular Ca^{2+} handling, and factors that are involved in myocyte apoptosis, survival, and proliferation could also be targeted to preserve cardiac function. In addition to treating acute damage, cardiac function could be preserved by reducing chronic remodeling after infarction; heme oxygenase-1 markedly reduced fibrosis and ventricular remodeling in an animal model of myocardial infarction [189]. However, despite the wide range of potential approaches and genetic targets, clinical investigations of gene therapy for treatment of heart failure have yet to be completed.

As the genetic factors influencing arrhythmia are identified, it may be possible to generate impulses in bradyarrhythmic individuals, negatively modulate activity in tachyarrhythmic patients [190], or suppress malignant arrhythmias like Brugada syndrome with long-term gene therapy. Reliable genetic modulation could replace rhythmic therapy devices and reduce the treatment costs associated with pacemakers and defibrillators. To date, these approaches have not been attempted in the clinical setting.

46.6 SAFETY

Results from numerous studies indicate that the transfer of proteins and genes to the human system is safe and feasible [8–10,34]; however, long-term safety data from ongoing large-scale studies on cardiovascular gene therapy are not available. For this reason, a number of safety concerns are yet to be addressed, including the potential for angiogenesis-triggered malignancies, the impact of angiogenesis on physiologic or pathologic processes, and the toxic effects of specific growth factors.

Because anti-angiogenic therapies are proven effective for treating tumors, there is concern that pro-angiogenic growth factors may promote tumor development. To date, preclinical and clinical experiences with different growth factors have not identified an increased risk for malignancies, although long-term follow-up periods are rare. Two of the 88 patients at our institution who received VEGF gene transfer therapy for critical limb ischemia have been diagnosed with cancer [191], and in the VIVA trial, tumor incidence was higher in the placebo group than in VEGF-treated patients [103]. Speculation about the risk for angiomata has been common, but this complication has never been reported in clinical studies nor in animal studies that involved transducing myoblasts or applying supraphysiologic DNA doses via plasmid or adenoviral vectors [192,193]. Consensus opinion suggests that the risk of malignancy is primarily determined by the age of the patients who are eligible for gene therapy.

Because high VEGF levels in ocular fluid levels have been observed in patients with active proliferative diabetic retinopathy and endangered vision, there were theoretical concerns that VEGF gene therapy would increase the risk of proliferative and/or hemorrhagic retinopathy in these patients [194]. This potential problem has not been observed in clinical studies or in more than 100 patients treated at our institution, one-third of whom had diabetes or remote retinopathy. Treatment with naked plasmid VEGF1 or VEGF2 DNA did not affect visual acuity or eye fundus (as assessed by serial funduscopic examinations) over a 4-year follow-up period.

The potential of VEGF and other pro-angiogenic factors to promote atherosclerotic plaque growth or destabilization has been debated [122,195–197]. This concern is based, in part, on the angiogenic potential of FGF2 and VEGF on vasa vasorum [198] and on a study by Moulton et al., who demonstrated that plaque area and intimal neovascularization were significantly reduced after treatment with angiogenesis inhibitors (e.g., endostatin, TNP-470) in hypercholesterolemic and apolipoprotein E-deficient mice [199]. Results from both preclinical and clinical studies indicate that angiogenic cytokines are not likely to accelerate atherosclerosis [160–163,200]. Instead, VEGF administration significantly reduced intimal thickening by promoting re-endothelialization, which refutes speculation that VEGF therapy may hasten the progression of atherosclerosis.

When gene therapy is administered via a viral vector, the vector itself could lead to side effects in the patient. Earlier generations of Ad vectors were shown to induce cell-cycle dysregulation [201], and Ad-vector-related death has been discussed in the past [202], although one fatality appears to have been caused by overdosing an immunocompromised patient [203]. Adenoviral vectors are used extensively in cardiovascular trials, and apart from occasional inflammation related to vector persistence [204], they have a very good safety profile. Nevertheless, continued awareness and further development of optimized vectors are warranted.

Angiogenic factors are accompanied by their own potential toxicities. Hypotension is associated with FGF2 [205] and may be more severe with recombinant VEGF proteins, especially when administered systemically or at higher doses [206,207]. Mechanistically, VEGF could induce hypotension by upregulating nitric oxide synthesis and subsequent nitric oxide–dependent arteriolar vasodilation. Hypotension has never been confirmed with VEGF gene transfer in animal or human studies [208,209], but the hypotensive effect has been dose-limiting in phase 1 trials that use both FGF2 and VEGF. Exposure to FGF2 has also been associated with renal insufficiency due to membranous nephropathy accompanied by proteinuria [210]; protein deposition in the glomerular membrane is the most likely underlying cause. Co-application of PDGFC or PDGFD and FGF may lead to tissue fibrosis and heart failure [211].

In transgenic mice, VEGF overexpression induced lethal vascular leakage, but the leakage was attenuated by co-overexpression of both VEGF and Ang1 [212]. Other reports have linked transient tissue edema to VEGFA and VEGFD as well as FGF4 gene transfer [85,96,213,214]; fluid accumulation was dose-dependent, correlated with capillary size, and coincided with maximal perfusion increase 5 to 6 days after transfer. Clinically, patients treated with VEGF for critical limb ischemia have experienced local, pedal edema that responded well to diuretics [147]; however, edema in limb muscle and (especially) in the myocardium can be hazardous, so strategies must be employed to minimize this risk [108]. When administered in moderate doses, long-lasting VEGF stabilizes growing vessels and appears to reduce, but not eliminate, edema [215]. Corticosteroids can counteract VEGF-mediated vascular permeability but also inhibit angiogenesis (Korpisalo, unpublished data) [10].

Overall, adverse effects in phase 1 trials of cardiovascular gene therapy were rare and mild; however, long-term safety data are lacking, so clinicians who perform this experimental therapy must remain aware of potential risks. The recent death of a young woman who received gene therapy for rheumatoid arthritis emphasizes the need for continued vigilance [216], although her death has not been attributed to gene therapy [217].

46.7 FUTURE PERSPECTIVES

New studies designed to advance the field of gene therapy continue to be initiated. In addition to refining techniques for therapeutic angiogenesis, which is currently the most advanced application of gene therapy, strategies to prevent postinterventional vascular remodeling and bypass-graft failure, stabilize vulnerable plaques and aneurysms, manage heart failure, hypertension, hyperlipidemia, and thrombotic states, and modulate rhythmic activity are being pursued. The identification of new genetic targets will proceed in parallel.

As the characterization of individual gene therapies becomes more complete, researchers can logically speculate about the effectiveness of treatment that combines two or more gene therapies. Preclinical investigations designed to identify the potential complementary or synergistic effects achieved with combinations of gene therapies are under way. The effectiveness of this strategy will be determined by the same variables that influence single-gene therapy, including (but not limited to) the model species, the delivery vector, the organ and disease treated, and the genes delivered.

Although cell therapy has produced promising results in small clinical studies of cardiovascular disease, methodological uncertainties, technical difficulties, and an inadequate understanding of the therapeutic mechanisms have limited its application [218,219]. Because gene and cell therapies appear to have mechanistic commonalities, it may be possible to increase the effectiveness of cell therapy by genetically modifying stem cells to improve the survival, differentiation, and functional integration of both transplanted and endogenous cells. Combined gene-cell therapy may also produce equivalent or greater benefit with a smaller dose than either individual therapy, thereby improving patient safety. This concept is supported by evidence that the paracrine effects of bone marrow cells can be augmented by overexpression of trophic factors in mesenchymal stem cells [220], and results from our laboratory indicate that a combination of VEGF2 gene transfer and EPC administration reduced cell apoptosis (through Akt activation) and infarct size, enhanced capillary development, and improved functional recovery after acute myocardial infarction in rats [221]. Shh and SDF-1α are likely candidates for future studies that combine gene and cell therapy because they have been shown to regulate the recruitment and incorporation of bone marrow–derived EPCs.

Gene and cell therapy can be combined in three ways: (1) gene therapy supplemented with drug-induced stem or progenitor cell mobilization, (2) combined administration of both gene therapy and ex vivo expanded cells, and (3) administration of genetically modified cells during cell therapy. In a large-animal study performed in our laboratory, cytokine-induced bone marrow cell mobilization combined with local VEGF gene transfer boosted myocardial recovery after infarction by enhancing bone marrow cell recruitment, retention, and incorporation; increased myocardial function and perfusion were accompanied by more dense capillary development [222]. However, intramyocardial injection of VEGF$_{165}$ plasmid combined with granulocyte-colony stimulating factor induced cell mobilization did not improve myocardial performance in patients with severe chronic ischemic heart disease in one report [223]. This latter result is surprising, and additional trials are needed to assess the potential of VEGF gene therapy combined with drug-induced cell mobilization for the treatment of ischemic heart disease. The ischemic blood supply may not be sufficient to ensure the survival of incorporated cells, thus, in an ongoing study at our institution, gene transfer is performed before cell transplantation, thereby initiating angiogenesis in the cardiac tissue before cell mobilization and homing (ClinicalTrials.gov, NCT00279539). Combined gene and cell therapy also produced unexpected results in a sheep model of myocardial infarction; simultaneous injection of VEGF protein and myoblast cells was no more effective than treatment with VEGF protein alone [224], which might be explained by rapid degradation of the VEGF protein.

Genetic modification of cells before administration may enable enhancement of cell survival and proliferation after incorporation [225,226], thereby improving the effectiveness of cell therapy. Genetic modifications can also change intercellular communication and electrical properties [227,228] and could potentially be used to treat or prevent arrhythmia or to eliminate arrhythmic risks that may be associated with certain cell lines.

46.8 CONCLUSIONS

The safety and feasibility of gene therapy have been demonstrated repeatedly in well-designed, randomized, controlled studies with large numbers of patients, and there is convincing clinical evidence to support the efficacy of VEGF gene transfer for therapeutic angiogenesis. However, conflicting efficacy results from recent phase 2 and 3 studies make the utility of gene therapy for cardiovascular disease somewhat uncertain. The effectiveness of gene therapy is influenced by a large number of complex factors, so identifying methods to optimize gene therapy in cardiovascular medicine will require detailed analyses of the clinical data. For studies that combine the delivery of recombinant protein or gene therapy with conventional revascularization procedures, it may be difficult to determine the relative contributions of the angiogenic agent and the surgical procedure.

Vector-delivered gene therapy appears to be safer and more effective than the administration of recombinant proteins. The disappointing results observed with intravascular administration of recombinant protein likely arise, at least in part, because the concentration of protein in the target organ fails to reach therapeutic levels. Methodological improvement of gene therapy requires large-scale, double-blind, randomized, controlled trials of adequate detection power to enable optimization of vector choice and delivery techniques. Because clinical choices are initially extrapolated from preclinical studies, the potential implications of species-specific variation must be considered. Development of the catheter-based NOGA delivery system has made gene delivery more attractive for myocardial applications because the therapy can now be administered without open-chest surgery or thoracotomy.

Gene therapy is most advanced for applications that involve therapeutic angiogenesis. Until recently, this application of gene therapy for treatment of myocardial and limb ischemia was restricted to highly symptomatic patients for whom conventional treatments had failed or were poorly suited. Recently, gene therapy for therapeutic angiogenesis has been made available to patients who are symptomatic but more stable, such as CAD patients with CCS class II/III angina. With continued advancement, cardiovascular gene therapy could become a routine treatment tool. The development of gene therapy for other indications has lagged, but numerous targets for genetic modification have been identified, so the expanded use of gene therapy seems almost inevitable. As more genetic modifications are characterized, combinations of factors will be investigated in an attempt to identify complementary or synergistic actions between them, and recent progress in cell therapy, administered both alone and in combination with gene therapy, provides more options for enhancing patient response and improving patient safety.

ACKNOWLEDGMENTS

We thank W. Kevin Meisner, PhD, ELS, for editorial support and Mickey Neely for administrative support. This work was supported in part by NIH grants HL-53354, HL-57516, HL-77428, HL-63414, HL-80137, PO1HL-66957. Jörn Tongers was supported by the German Heart Foundation and Solvay Pharmaceuticals, Jerome Roncalli by the French Federation of Cardiology.

REFERENCES

1. Ross, R., Atherosclerosis—an inflammatory disease, *N Engl J Med* 340(2), 115–126, 1999.
2. Libby, P. and Theroux, P., Pathophysiology of coronary artery disease, *Circulation* 111(25), 3481–3488, 2005.
3. Murray, C. J. and Lopez, A. D., Alternative projections of mortality and disability by cause 1990–2020: Global Burden of Disease Study, *Lancet* 349(9064), 1498–1504, 1997.
4. *2007 Heart and Stroke Statistical Update.* American Heart Association and American Stroke Association, Dallas, Texas, 75231-4596, 2007.
5. Hirsch, A. T., Criqui, M. H., Treat-Jacobson, D., Regensteiner, J. G., Creager, M. A., Olin, J. W., Krook, S. H., Hunninghake, D. B., Comerota, A. J., Walsh, M. E., McDermott, M. M., and Hiatt, W. R., Peripheral arterial disease detection, awareness, and treatment in primary care, *Jama* 286(11), 1317–1324, 2001.
6. Folkman, J., Tumor angiogenesis: Therapeutic implications, *N Engl J Med* 285, 1182–1186, 1971.
7. Yanagisawa-Miwa, A., Uchida, Y., Nakamura, F., Tomaru, T., Kido, H., Kamijo, T., Sugimoto, T., Kaji, K., Utsuyama, M., Kurashima, C., and Ito, H., Salvage of infarcted myocardium by angiogenic action of basic fibroblast growth factor, *Science* 257(5075), 1401–1403, 1992.
8. Yla-Herttuala, S. and Alitalo, K., Gene transfer as a tool to induce therapeutic vascular growth, *Nat Med* 9(6), 694–701, 2003.
9. Yla-Herttuala, S. and Martin, J. F., Cardiovascular gene therapy, *Lancet* 355(9199), 213–222, 2000.
10. Rissanen, T. T. and Yla-Herttuala, S., Current status of cardiovascular gene therapy, *Mol Ther* 15(7), 1233–1247, 2007.
11. Carmeliet, P., Mechanisms of angiogenesis and arteriogenesis, *Nat Med* 6(4), 389–395, 2000.
12. Arras, M., Ito, W. D., Scholz, D., Winkler, B., Schaper, J., and Schaper, W., Monocyte activation in angiogenesis and collateral growth in the rabbit hindlimb, *J Clin Invest* 101, 40–50, 1998.
13. Risau, W., Mechanisms of angiogenesis, *Nature* 386(6626), 671–674, 1997.
14. Risau, W., Differentiation of endothelium, *Faseb J* 9(10), 926–933, 1995.
15. Asahara, T., Murohara, T., Sullivan, A., Silver, M., van der Zee, R., Li, T., Witzenbichler, B., Schatteman, G., and Isner, J. M., Isolation of putative progenitor endothelial cells for angiogenesis, *Science* 275(5302), 964–967, 1997.
16. Asahara, T., Masuda, H., Takahashi, T., Kalka, C., Pastore, C., Silver, M., Kearne, M., Magner, M., and Isner, J. M., Bone marrow origin of endothelial progenitor cells responsible for postnatal vasculogenesis in physiological and pathological neovascularization, *Circ Res* 85(3), 221–228, 1999.
17. Takahashi, T., Kalka, C., Masuda, H., Chen, D., Silver, M., Kearney, M., Magner, M., Isner, J. M., and Asahara, T., Ischemia- and cytokine-induced mobilization of bone marrow-derived endothelial progenitor cells for neovascularization, *Nat Med* 5(4), 434–438, 1999.
18. Losordo, D. W., Schatz, R. A., White, C. J., Udelson, J. E., Veereshwarayya, V., Durgin, M., Poh, K. K., Weinstein, R., Kearney, M., Chaudhry, M., Burg, A., Eaton, L., Heyd, L., Thorne, T., Shturman, L., Hoffmeister, P., Story, K., Zak, V., Dowling, D., Traverse, J. H., Olson, R. E., Flanagan, J., Sodano, D., Murayama, T., Kawamoto, A., Kusano, K. F., Wollins, J., Welt, F., Shah, P., Soukas, P., Asahara, T., and Henry, T. D., Intramyocardial transplantation of autologous CD34 + stem cells for intractable angina: A phase I/IIa double-blind, randomized controlled trial, *Circulation* 115(25), 3165–3172, 2007.
19. Tateishi-Yuyama, E., Matsubara, H., Murohara, T., Ikeda, U., Shintani, S., Masaki, H., Amano, K., Kishimoto, Y., Yoshimoto, K., Akashi, H., Shimada, K., Iwasaka, T., and Imaizumi, T., Therapeutic angiogenesis for patients with limb ischaemia by autologous transplantation of bone-marrow cells: A pilot study and a randomised controlled trial, *Lancet* 360(9331), 427–435, 2002.
20. Ferrara, N., Gerber, H. P., and LeCouter, J., The biology of VEGF and its receptors, *Nat Med* 9(6), 669–676, 2003.
21. Ferrara, N., Vascular endothelial growth factor: Basic science and clinical progress, *Endocr Rev* 25(4), 581–611, 2004.
22. Carmeliet, P., Ferreira, V., Breier, G., Pollefeyt, S., Kieckens, L., Gertsenstein, M., Fahrig, M., Vandenhoeck, A., Harpal, K., Eberhardt, C., Declercq, C., Pawling, J., Moons, L., Collen, D., Risau, W., and Nagy, A., Abnormal blood vessel development and lethality in embryos lacking a single VEGF allele, *Nature* 380(6573), 435–439, 1996.
23. Ferrara, N., Carver-Moore, K., Chen, H., Dowd, M., Lu, L., O'Shea, K. S., Powell-Braxton, L., Hillan, K. J., and Moore, M. W., Heterozygous embryonic lethality induced by targeted inactivation of the VEGF gene, *Nature* 380(6573), 439–442, 1996.
24. Shalaby, F., Ho, J., Stanford, W. L., Fischer, K. D., Schuh, A. C., Schwartz, L., Bernstein, A., and Rossant, J., A requirement for Flk1 in primitive and definitive hematopoiesis and vasculogenesis, *Cell* 89(6), 981–990, 1997.
25. Kusano, K. F., Pola, R., Murayama, T., Curry, C., Kawamoto, A., Iwakura, A., Shintani, S., Ii, M., Asai, J., Tkebuchava, T., Thorne, T., Takenaka, H., Aikawa, R., Goukassian, D., von Samson, P., Hamada, H., Yoon, Y. S., Silver, M., Eaton, E., Ma, H., Heyd, L., Kearney, M., Munger, W., Porter, J. A., Kishore, R., and Losordo, D. W., Sonic hedgehog myocardial gene therapy: Tissue repair through transient reconstitution of embryonic signaling, *Nat Med* 11(11), 1197–1204, 2005.

26. Yamaguchi, J., Kusano, K. F., Masuo, O., Kawamoto, A., Silver, M., Murasawa, S., Bosch-Marce, M., Masuda, H., Losordo, D. W., Isner, J. M., and Asahara, T., Stromal cell-derived factor-1 effects on ex vivo expanded endothelial progenitor cell recruitment for ischemic neovascularization, *Circulation* 107(9), 1322–1328, 2003.

27. Askari, A. T., Unzek, S., Popovic, Z. B., Goldman, C. K., Forudi, F., Kiedrowski, M., Rovner, A., Ellis, S. G., Thomas, J. D., DiCorleto, P. E., Topol, E. J., and Penn, M. S., Effect of stromal-cell-derived factor 1 on stem-cell homing and tissue regeneration in ischaemic cardiomyopathy, *Lancet* 362(9385), 697–703, 2003.

28. Abbott, J. D., Huang, Y., Liu, D., Hickey, R., Krause, D. S., and Giordano, F. J., Stromal cell-derived factor-1alpha plays a critical role in stem cell recruitment to the heart after myocardial infarction but is not sufficient to induce homing in the absence of injury, *Circulation* 110(21), 3300–3305, 2004.

29. Pola, R., Ling, L. E., Silver, M., Corbley, M. J., Kearney, M., Blake Pepinsky, R., Shapiro, R., Taylor, F. R., Baker, D. P., Asahara, T., and Isner, J. M., The morphogen Sonic hedgehog is an indirect angiogenic agent upregulating two families of angiogenic growth factors, *Nat Med* 7(6), 706–711, 2001.

30. Gerber, H. P., Hillan, K. J., Ryan, A. M., Kowalski, J., Keller, G. A., Rangell, L., Wright, B. D., Radtke, F., Aguet, M., and Ferrara, N., VEGF is required for growth and survival in neonatal mice, *Development* 126(6), 1149–1159, 1999.

31. Maharaj, A. S., Saint-Geniez, M., Maldonado, A. E., and D'Amore, P. A., Vascular endothelial growth factor localization in the adult, *Am J Pathol* 168(2), 639–648, 2006.

32. Rissanen, T. T., Vajanto, I., Hiltunen, M. O., Rutanen, J., Kettunen, M. I., Niemi, M., Leppanen, P., Turunen, M. P., Markkanen, J. E., Arve, K., Alhava, E., Kauppinen, R. A., and Yla-Herttuala, S., Expression of vascular endothelial growth factor and vascular endothelial growth factor receptor-2 (KDR/Flk-1) in ischemic skeletal muscle and its regeneration, *Am J Pathol* 160(4), 1393–1403, 2002.

33. Takeshita, S., Weir, L., Chen, D., Zheng, L. P., Riessen, R., Bauters, C., Symes, J. F., Ferrara, N., and Isner, J. M., Therapeutic angiogenesis following arterial gene transfer of vascular endothelial growth factor in a rabbit model of hindlimb ischemia, *Biochem Biophys Res Commun* 227, 628–635, 1996.

34. Yla-Herttuala, S., Rissanen, T. T., Vajanto, I., and Hartikainen, J., Vascular endothelial growth factors: Biology and current status of clinical applications in cardiovascular medicine, *J Am Coll Cardiol* 49(10), 1015–1026, 2007.

35. Soker, S., Fidder, H., Neufeld, G., and Klagsbrun, M., Characterization of novel vascular endothelial growth factor (VEGF) receptors on tumor cells that bind VEGF165 via its exon 7-encoded domain, *J Biol Chem* 271(10), 5761–5767, 1996.

36. Brogi, E., Schatteman, G., Wu, T., Kim, E. A., Varticovski, L., Keyt, B., and Isner, J. M., Hypoxia-induced paracrine regulation of VEGF receptor expression, *J Clin Invest* 97, 469–476, 1996.

37. Gerber, H.-P., McMurtrey, A., Kowalski, J., Yan, M., Keyt, B. A., Dixit, V., and Ferrara, N., Vascular endothelial growth factor regulates endothelial cell survival through the phosphatidylinositol 3′-kinase/Akt signal transduction pathway. Requirement for Flk-1/KDR activation, *J Biol Chem* 273, 30336–30343, 1998.

38. Yuan, F., Chen, Y., Dellian, M., Safabakhsh, N., Ferrara, N., and Jain, R. K., Time-dependent vascular regression and permeability changes in established human tumor xenografts induced by an anti-vascular endothelial growth factor/vascular permeability factor antibody, *Proc Natl Acad Sci USA* 93(25), 14765–14770, 1996.

39. Fong, G. H., Rossant, J., Gertsenstein, M., and Breitman, M. L., Role of flt-1 receptor tyrosine kinase in regulating the assembly of vascular endothelium, *Nature* 376, 66–70, 1995.

40. Carmeliet, P. and Collen, D., Molecular analysis of blood vessel formation and disease, *Am J Physiol* 273, H2091–H2104, 1997.

41. Shalaby, F., Rossant, J., Yamaguchi, T. P., Gertsenstein, M., Wu, X.-F., Breitman, M. L., and Schuh, A. C., Failure of blood-island formation and vasculogenesis in Flk-1 deficient mice *Nature* 376, 62–66, 1995.

42. Jeltsch, M., Kaipainen, A., Joukov, V., Meng, X., Lakso, M., Rauvala, H., Swartz, M., Fukumura, D., Jain, R. K., and Alitalo, K., Hyperplasia of lymphatic vessels in VEGF-C transgenic mice, *Science* 276, 1423–1425, 1997.

43. Leung, D. W., Cachianes, G., Kuang, W. J., Goeddel, D. V., and Ferrara, N., Vascular endothelial growth factor is a secreted angiogenic mitogen, *Science* 246(4935), 1306–1309, 1989.

44. Conn, G., Soderman, D., Schaeffer, M.-T., Wile, M., Hatcher, V. B., and Thomas, K. A., Purification of glycoprotein vascular endothelial cell mitogen from a rat glioma cell line, *Proc Natl Acad Sci USA* 87, 1323–1327, 1990.

45. Ferrara, N. and Henzel, W. J., Pituitary follicular cells secrete a novel heparin-binding growth factor specific for vascular endothelial cells, *Biochem Biophys Res Commun* 161(2), 851–858, 1989.

46. Kalka, C., Masuda, H., Takahashi, T., Gordon, R., Tepper, O., Gravereaux, E., Pieczek, A., Iwaguro, H., Hayashi, S. I., Isner, J. M., and Asahara, T., Vascular endothelial growth factor(165) gene transfer augments circulating endothelial progenitor cells in human subjects, *Circ Res* 86(12), 1198–1202, 2000.

47. Kalka, C., Tehrani, H., Laudenberg, B., Vale, P. R., Isner, J. M., Asahara, T., and Symes, J. F., Mobilization of endothelial progenitor cells following gene therapy with VEGF$_{165}$ in patients with inoperable coronary disease, *Ann Thoracic Surg* 70, 829–834, 2000.

48. Asahara, T., Takahashi, T., Masuda, H., Kalka, C., Chen, D., Iwaguro, H., Inai, Y., Silver, M., and Isner, J. M., VEGF contributes to postnatal neovascularization by mobilizing bone marrow-derived endothelial progenitor cells, *EMBO J* 18(14), 3964–3472, 1999.

49. Miller, D. L., Ortega, S., Bashayan, O., Basch, R., and Basilico, C., Compensation by fibroblast growth factor 1 (FGF1) does not account for the mild phenotypic defects observed in FGF2 null mice, *Mol Cell Biol* 20(6), 2260–2268, 2000.

50. Ueno, H., Li, J. J., Masuda, S., Qi, Z., Yamamoto, H., and Takeshita, A., Adenovirus-mediated expression of the secreted form of basic fibroblast growth factor (FGF-2) induces cellular proliferation and angiogenesis in vivo, *Arterioscler Thromb Vasc Biol* 17(11), 2453–2460, 1997.

51. Coussens, L. M., Raymond, W. W., Bergers, G., Laig-Webster, M., Behrendtsen, O., Werb, Z., Caughey, G. H., and Hanahan, D., Inflammatory mast cells up-regulate angiogenesis during squamous epithelial carcinogenesis, *Genes Dev* 13(11), 1382–1397, 1999.

52. Li, J., Shworak, N. W., and Simons, M., Increased responsiveness of hypoxic endothelial cells to FGF2 is mediated by HIF-1alpha-dependent regulation of enzymes involved in synthesis of heparan sulfate FGF2-binding sites, *J Cell Sci* 115 (Pt 9), 1951–1959, 2002.

53. Calvani, M., Rapisarda, A., Uranchimeg, B., Shoemaker, R. H., and Melillo, G., Hypoxic induction of an HIF-1alpha-dependent bFGF autocrine loop drives angiogenesis in human endothelial cells, *Blood* 107(7), 2705–2712, 2006.

54. Giordano, F. J., Ping, P., McKirnan, M. D., Nozaki, S., DeMaria, A. N., Dillmann, W. H., Mathieu-Costello, O., and Hammond, H. K., Intracoronary gene transfer of fibroblast growth factor-5 increases blood flow and contractile function in an ischemic region of the heart, *Nat Med* 2(5), 534–539, 1996.

55. McKirnan, M. D., Guo, X., Waldman, L. K., Dalton, N., Lai, N. C., Gao, M. H., Roth, D. A., and Hammond, H. K., Intracoronary gene transfer of fibroblast growth factor-4 increases regional contractile function and responsiveness to adrenergic stimulation in heart failure, *Cardiac Vascular Regen* 1, 11–21, 2000.

56. Tabata, H., Silver, M., and Isner, J. M., Arterial gene transfer of acidic fibroblast growth factor for therapeutic angiogenesis in vivo: Critical role of secretion signal in use of naked DNA, *Cardiovasc Res* 35, 470–479, 1997.

57. Safran, M. and Kaelin, W. G., Jr., HIF hydroxylation and the mammalian oxygen-sensing pathway, *J Clin Invest* 111(6), 779–783, 2003.

58. Ceradini, D. J., Kulkarni, A. R., Callaghan, M. J., Tepper, O. M., Bastidas, N., Kleinman, M. E., Capla, J. M., Galiano, R. D., Levine, J. P., and Gurtner, G. C., Progenitor cell trafficking is regulated by hypoxic gradients through HIF-1 induction of SDF-1, *Nat Med* 10(8), 858–864, 2004.

59. Carmeliet, P., Dor, Y., Herbert, J. M., Fukumura, D., Brusselmans, K., Dewerchin, M., Neeman, M., Bono, F., Abramovitch, R., Maxwell, P., Koch, C. J., Ratcliffe, P., Moons, L., Jain, R. K., Collen, D., and Keshert, E., Role of HIF-1alpha in hypoxia-mediated apoptosis, cell proliferation and tumour angiogenesis, *Nature* 394(6692), 485–490, 1998.

60. Lindahl, P., Johansson, B. R., Leveen, P., and Betsholtz, C., Pericyte loss and microaneurysm formation in PDGF-B-deficient mice, *Science* 277(5323), 242–245, 1997.

61. Richardson, T. P., Peters, M. C., Ennett, A. B., and Mooney, D. J., Polymeric system for dual growth factor delivery, *Nat Biotechnol* 19 (11), 1029–1034, 2001.

62. Cao, R., Brakenhielm, E., Pawliuk, R., Wariaro, D., Post, M. J., Wahlberg, E., Leboulch, P., and Cao, Y., Angiogenic synergism, vascular stability and improvement of hind-limb ischemia by a combination of PDGF-BB and FGF-2, *Nat Med* 9(5), 604–613, 2003.

63. Li, X., Tjwa, M., Moons, L., Fons, P., Noel, A., Ny, A., Zhou, J. M., Lennartsson, J., Li, H., Luttun, A., Ponten, A., Devy, L., Bouche, A., Oh, H., Manderveld, A., Blacher, S., Communi, D., Savi, P., Bono, F., Dewerchin, M., Foidart, J. M., Autiero, M., Herbert, J. M., Collen, D., Heldin, C. H., Eriksson, U., and Carmeliet, P., Revascularization of ischemic tissues by PDGF-CC via effects on endothelial cells and their progenitors, *J Clin Invest* 115(1), 118–127, 2005.

64. Aoki, M., Morishita, R., Taniyama, Y., Kida, I., Moriguchi, A., Matsumoto, K., Nakamura, T., Kaneda, Y., Higaki, J., and Ogihara, T., Angiogenesis induced by hepatocyte growth factor in non-infarcted myocardium and infarcted myocardium: Upregulation of essential transcription factor for angiogenesis, ets, *Gene Ther* 7(5), 417–427, 2000.

65. Lobov, I. B., Brooks, P. C., and Lang, R. A., Angiopoietin-2 displays VEGF-dependent modulation of capillary structure and endothelial cell survival in vivo, *Proc Natl Acad Sci USA* 99(17), 11205–11210, 2002.

66. Thurston, G., Rudge, J. S., Ioffe, E., Zhou, H., Ross, L., Croll, S. D., Glazer, N., Holash, J., McDonald, D. M., and Yancopoulos, G. D., Angiopoietin-1 protects the adult vasculature against plasma leakage, *Nat Med* 6(4), 460–463, 2000.

67. Tashiro, K., Tada, H., Heilker, R., Shirozu, M., Nakano, T., and Honjo, T., Signal sequence trap: A cloning strategy for secreted proteins and type I membrane proteins, *Science* 261(5121), 600–603, 1993.

68. Shirozu, M., Nakano, T., Inazawa, J., Tashiro, K., Tada, H., Shinohara, T., and Honjo, T., Structure and chromosomal localization of the human stromal cell-derived factor 1 (SDF1) gene, *Genomics* 28(3), 495–500, 1995.

69. Balabanian, K., Lagane, B., Infantino, S., Chow, K. Y., Harriague, J., Moepps, B., Arenzana-Seisdedos, F., Thelen, M., and Bachelerie, F., The chemokine SDF-1/CXCL12 binds to and signals through the orphan receptor RDC1 in T lymphocytes, *J Biol Chem* 280(42), 35760–35766, 2005.

70. Tachibana, K., Hirota, S., Iizasa, H., Yoshida, H., Kawabata, K., Kataoka, Y., Kitamura, Y., Matsushima, K., Yoshida, N., Nishikawa, S., Kishimoto, T., and Nagasawa, T., The chemokine receptor CXCR4 is essential for vascularization of the gastrointestinal tract, *Nature* 393(6685), 591–594, 1998.

71. Orimo, A., Gupta, P. B., Sgroi, D. C., Arenzana-Seisdedos, F., Delaunay, T., Naeem, R., Carey, V. J., Richardson, A. L., and Weinberg, R. A., Stromal fibroblasts present in invasive human breast carcinomas promote tumor growth and angiogenesis through elevated SDF-1/CXCL12 secretion, *Cell* 121(3), 335–348, 2005.

72. Grunewald, M., Avraham, I., Dor, Y., Bachar-Lustig, E., Itin, A., Yung, S., Chimenti, S., Landsman, L., Abramovitch, R., and Keshet, E., VEGF-induced adult neovascularization: Recruitment, retention, and role of accessory cells, *Cell* 124(1), 175–189, 2006.

73. Nusslein-Volhard, C. and Wieschaus, E., Mutations affecting segment number and polarity in *Drosophila*, *Nature* 287, 795–801, 1980.

74. Brand-Saberi, B., Wilting, J., Ebensperger, C., and Christ, B., The formation of somite compartments in the avian embryo, *Int J Dev Biol* 40 (1), 411–420, 1996.

75. Johnson, R. L. and Tabin, C. J., Molecular models for vertebrate limb development, *Cell* 90(6), 979–990, 1997.

76. Roelink, H., Augsburger, A., Heemskerk, J., Korzh, V., Norlin, S., Ruiz i Altaba, A., Tanabe, Y., Placzek, M., Edlund, T., Jessell, T. M., and Dodd, J., Floor plate and motor neuron induction by vhh-1, a vertebrate homolog of hedgehog expressed by the notochord, *Cell* 76(4), 761–775, 1994.

77. Asai, J., Takenaka, H., Kusano, K. F., Ii, M., Luedemann, C., Curry, C., Eaton, E., Iwakura, A., Tsutsumi, Y., Hamada, H., Kishimoto, S., Thorne, T., Kishore, R., and Losordo, D. W., Topical sonic hedgehog gene therapy accelerates wound healing in diabetes by enhancing endothelial progenitor cell-mediated microvascular remodeling, *Circulation* 113(20), 2413–2424, 2006.

78. Lazarous, D. F., Shou, M., Stiber, J. A., Dadhania, D. M., Thirumurti, V., Hodge, E., and Unger, E. F., Pharmacodynamics of basic fibroblast growth factor: Route of administration determines myocardial and systemic distribution, *Cardiovasc Res* 36(1), 78–85, 1997.

79. Lazarous, D. F., Shou, M., Stiber, J. A., Hodge, E., Thirumurti, V., Goncalves, L., and Unger, E. F., Adenoviral-mediated gene transfer induces sustained pericardial VEGF expression in dogs: Effect on myocardial angiogenesis, *Cardiovasc Res* 44(2), 294–302, 1999.

80. Hedman, M., Hartikainen, J., Syvanne, M., Stjernvall, J., Hedman, A., Kivela, A., Vanninen, E., Mussalo, H., Kauppila, E., Simula, S., Narvanen, O., Rantala, A., Peuhkurinen, K., Nieminen, M. S., Laakso, M., and Yla-Herttuala, S., Safety and feasibility of catheter-based local intracoronary vascular endothelial growth factor gene transfer in the prevention of postan-

gioplasty and in-stent restenosis and in the treatment of chronic myocardial ischemia: Phase II results of the Kuopio Angiogenesis Trial (KAT), *Circulation* 107(21), 2677–2683, 2003.

81. Kastrup, J., Jorgensen, E., Ruck, A., Tagil, K., Glogar, D., Ruzyllo, W., Botker, H. E., Dudek, D., Drvota, V., Hesse, B., Thuesen, L., Blomberg, P., Gyongyosi, M., and Sylven, C., Direct intramyocardial plasmid vascular endothelial growth factor-A165 gene therapy in patients with stable severe angina pectoris A randomized double-blind placebo-controlled study: The Euroinject One trial, *J Am Coll Cardiol* 45(7), 982–988, 2005.

82. Wright, M. J., Wightman, L. M., Lilley, C., de Alwis, M., Hart, S. L., Miller, A., Coffin, R. S., Thrasher, A., Latchman, D. S., and Marber, M. S., In vivo myocardial gene transfer: Optimization, evaluation and direct comparison of gene transfer vectors, *Basic Res Cardiol* 96(3), 227–236, 2001.

83. Mathiesen, I., Electropermeabilization of skeletal muscle enhances gene transfer in vivo, *Gene Ther* 6(4), 508–514, 1999.

84. Yamashita, Y., Shimada, M., Tachibana, K., Harimoto, N., Tsujita, E., Shirabe, K., Miyazaki, J., and Sugimachi, K., In vivo gene transfer into muscle via electro-sonoporation, *Hum Gene Ther* 13(17), 2079–2084, 2002.

85. Rutanen, J., Rissanen, T. T., Markkanen, J. E., Gruchala, M., Silvennoinen, P., Kivela, A., Hedman, A., Hedman, M., Heikura, T., Orden, M. R., Stacker, S. A., Achen, M. G., Hartikainen, J., and Yla-Herttuala, S., Adenoviral catheter-mediated intramyocardial gene transfer using the mature form of vascular endothelial growth factor-D induces transmural angiogenesis in porcine heart, *Circulation* 109(8), 1029–1035, 2004.

86. Hao, X., Mansson-Broberg, A., Grinnemo, K. H., Siddiqui, A. J., Dellgren, G., Brodin, L. A., and Sylven, C., Myocardial angiogenesis after plasmid or adenoviral VEGF-A(165) gene transfer in rat myocardial infarction model, *Cardiovasc Res* 73(3), 481–487, 2007.

87. Wright, M. J., Rosenthal, E., Stewart, L., Wightman, L. M., Miller, A. D., Latchman, D. S., and Marber, M. S., beta-Galactosidase staining following intracoronary infusion of cationic liposomes in the in vivo rabbit heart is produced by microinfarction rather than effective gene transfer: A cautionary tale, *Gene Ther* 5(3), 301–308, 1998.

88. McMahon, J. M., Wells, K. E., Bamfo, J. E., Cartwright, M. A., and Wells, D. J., Inflammatory responses following direct injection of plasmid DNA into skeletal muscle, *Gene Ther* 5(9), 1283–1290, 1998.

89. Nalbantoglu, J., Pari, G., Karpati, G., and Holland, P. C., Expression of the primary coxsackie and adenovirus receptor is downregulated during skeletal muscle maturation and limits the efficacy of adenovirus-mediated gene delivery to muscle cells, *Hum Gene Ther* 10(6), 1009–1019, 1999.

90. Makinen, K., Manninen, H., Hedman, M., Matsi, P., Mussalo, H., Alhava, E., and Yla-Herttuala, S., Increased vascularity detected by digital subtraction angiography after VEGF gene transfer to human lower limb artery: A randomized, placebo-controlled, double-blinded phase II study, *Mol Ther* 6(1), 127–133, 2002.

91. Dishart, K. L., Denby, L., George, S. J., Nicklin, S. A., Yendluri, S., Tuerk, M. J., Kelley, M. P., Donahue, B. A., Newby, A. C., Harding, T., and Baker, A. H., Third-generation lentivirus vectors efficiently transduce and phenotypically modify vascular cells: Implications for gene therapy, *J Mol Cell Cardiol* 35(7), 739–748, 2003.

92. Cefai, D., Simeoni, E., Ludunge, K. M., Driscoll, R., von Segesser, L. K., Kappenberger, L., and Vassalli, G., Multiply attenuated, self-inactivating lentiviral vectors efficiently transduce human coronary artery cells in vitro and rat arteries in vivo, *J Mol Cell Cardiol* 38(2), 333–344, 2005.

93. Ozawa, C. R., Banfi, A., Glazer, N. L., Thurston, G., Springer, M. L., Kraft, P. E., McDonald, D. M., and Blau, H. M., Microenvironmental VEGF concentration, not total dose, determines a threshold between normal and aberrant angiogenesis, *J Clin Invest* 113(4), 516–527, 2004.

94. Tsurumi, Y., Takeshita, S., Chen, D., Kearney, M., Rossow, S. T., Passeri, J., Horowitz, J. R., Symes, J. F., and Isner, J. M., Direct intramuscular gene transfer of naked DNA encoding vascular endothelial growth factor augments collateral development and tissue perfusion, *Circulation* 94(12), 3281–3290, 1996.

95. Takeshita, S., Isshiki, T., and Sato, T., Increased expression of direct gene transfer into skeletal muscles observed after acute ischemic injury in rats, *Lab Invest* 74(6), 1061–1065, 1996.

96. Rissanen, T. T., Markkanen, J. E., Arve, K., Rutanen, J., Kettunen, M. I., Vajanto, I., Jauhiainen, S., Cashion, L., Gruchala, M., Narvanen, O., Taipale, P., Kauppinen, R. A., Rubanyi, G. M., and Yla-Herttuala, S., Fibroblast growth factor 4 induces vascular permeability, angiogenesis and arteriogenesis in a rabbit hindlimb ischemia model, *Faseb J* 17(1), 100–102, 2003.

97. Wright, M. J., Wightman, L. M., Latchman, D. S., and Marber, M. S., In vivo myocardial gene transfer: Optimization and evaluation of intracoronary gene delivery in vivo, *Gene Ther* 8(24), 1833–1839, 2001.

98. Walter, D. H., Cejna, M., Diaz-Sandoval, L., Willis, S., Kirkwood, L., Stratford, P. W., Tietz, A. B., Kirchmair, R., Silver, M., Curry, C., Wecker, A., Yoon, Y. S., Heidenreich, R., Hanley, A., Kearney, M., Tio, F. O., Kuenzler, P., Isner, J. M., and Losordo, D. W., Local gene transfer of phVEGF-2 plasmid by gene-eluting stents: An alternative strategy for inhibition of restenosis, *Circulation* 110(1), 36–45, 2004.

99. Vincent, K. A., Jiang, C., Boltje, I., and Kelly, R. A., Gene therapy progress and prospects: Therapeutic angiogenesis for ischemic cardiovascular disease, *Gene Ther* 14(10), 781–789, 2007.

100. Hendel, R. C., Henry, T. D., Rocha-Singh, K., Isner, J. M., Kereiakes, D. J., Giordano, F. J., Simons, M., and Bonow, R. O., Effect of intracoronary recombinant human vascular endothelial growth factor on myocardial perfusion: Evidence for a dose-dependent effect, *Circulation* 101(2), 118–121, 2000.

101. Henry, T. D. and Abraham, J. A., Review of preclinical and clinical results with vascular endothelial growth factors for therapeutic angiogenesis, *Curr Interv Cardiol Rep* 2(3), 228–241, 2000.

102. Henry, T. D., Rocha-Singh, K., Isner, J. M., Kereiakes, D. J., Giordano, F. J., Simons, M., Losordo, D. W., Hendel, R. C., Bonow, R. O., Eppler, S. M., Zioncheck, T. F., Holmgren, E. B., and McCluskey, E. R., Intracoronary administration of recombinant human vascular endothelial growth factor to patients with coronary artery disease, *Am Heart J* 142(5), 872–880, 2001.

103. Henry, T. D., Annex, B. H., McKendall, G. R., Azrin, M. A., Lopez, J. J., Giordano, F. J., Shah, P. K., Willerson, J. T., Benza, R. L., Berman, D. S., Gibson, C. M., Bajamonde, A., Rundle, A. C., Fine, J., and McCluskey, E. R., The VIVA trial: Vascular endothelial growth factor in ischemia for vascular angiogenesis, *Circulation* 107(10), 1359–1365, 2003.

104. Banai, S., Jaklitsch, M. T., Shou, M., Lazarous, D. F., Scheinowitz, M., Biro, S., Epstein, S. E., and Unger, E. F., Angiogenic-induced enhancement of collateral blood flow to ischemic myocardium by vascular endothelial growth factor in dogs, *Circulation* 89(5), 2183–2189, 1994.

105. Lopez, J. J., Laham, R. J., Stamler, A., Pearlman, J. D., Bunting, S., Kaplan, A., Carrozza, J. P., Sellke, F. W., and Simons, M., VEGF administration in chronic myocardial ischemia in pigs, *Cardiovasc Res* 40(2), 272–281, 1998.

106. Harada, K., Friedman, M., Lopez, J. J., Wang, S. Y., Li, J., Prasad, P. V., Pearlman, J. D., Edelman, E. R., Sellke, F. W., and Simons, M., Vascular endothelial growth factor administration in chronic myocardial ischemia, *Am J Physiol* 270 (5 Pt 2), H1791–H1802, 1996.

107. Hughes, C. G., Biswas, S. S., Yin, B., Baklanov, D. V., DeGrado, T. R., Coleman, R. E., Donovan, C. L., Lowe, J. E., Landolfo, K. P., and Annex, B. H., Intramyocardial but not intravenous vascular endothelial growth factor improves regional perfusion in hibernating porcine myocardium, *Circulation* 100, I-476, 1999.

108. Lee, R. J., Springer, M. L., Blanco-Bose, W. E., Shaw, R., Ursell, P. C., and Blau, H. M., VEGF gene delivery to myocardium: Deleterious effects of unregulated expression, *Circulation* 102(8), 898–901, 2000.

109. Mack, C. A., Patel, S. R., Schwarz, E. A., Zanzonico, P., Hahn, R. T., Ilercil, A., Devereux, R. B., Goldsmith, S. J., Christian, T. F., Sanborn, T. A., Kovesdi, I., Hackett, N., Isom, O. W., Crystal, R. G., and Rosengart, T. K., Biologic bypass with the use of adenovirus-mediated gene transfer of the complementary deoxyribonucleic acid for vascular endothelial growth factor 121 improves myocardial perfusion and function in the ischemic porcine heart, *J Thorac Cardiovasc Surg* 115(1), 168–176; discussion 176–177, 1998.

110. Esakof, D. D., Maysky, M., Losordo, D. W., Vale, P. R., Lathi, K., Pastore, J. O., Symes, J. F., and Isner, J. M., Intraoperative multiplane transesophageal echocardiograpy for guiding direct myocardial gene transfer of vascular endothelial growth factor in patients with refractory angina pectoris, *Hum Gene Ther* 10, 2315–2323, 1999.

111. Losordo, D. W., Vale, P. R., Symes, J. F., Dunnington, C. H., Esakof, D. D., Maysky, M., Ashare, A. B., Lathi, K., and Isner, J. M., Gene therapy for myocardial angiogenesis: Initial clinical results with direct myocardial injection of phVEGF165 as sole therapy for myocardial ischemia, *Circulation* 98(25), 2800–2804, 1998.

112. Symes, J. F., Losordo, D. W., Vale, P. R., Lathi, K. G., Esakof, D. D., Mayskiy, M., and Isner, J. M., Gene therapy with vascular endothelial growth factor for inoperable coronary artery disease, *Ann Thorac Surg* 68(3), 830–836; discussion 836–837, 1999.

113. Vale, P. R., Losordo, D. W., Milliken, C. E., Maysky, M., Esakof, D. D., Symes, J. F., and Isner, J. M., Left ventricular electromechanical mapping to assess efficacy of phVEGF(165) gene transfer for therapeutic angiogenesis in chronic myocardial ischemia, *Circulation* 102(9), 965–974, 2000.

114. Fortuin, F. D., Vale, P., Losordo, D. W., Symes, J., DeLaria, G. A., Tyner, J. J., Schaer, G. L., March, R., Snell, R. J., Henry, T. D., Van Camp, J., Lopez, J. J., Richenbacher, W., Isner, J. M., and Schatz, R. A., One-year follow-up of direct myocardial gene transfer of vascular endothelial growth factor-2 using naked plasmid deoxyribonucleic acid by way of thoracotomy in no-option patients, *Am J Cardiol* 92(4), 436–439, 2003.

115. Stewart, J. D., A phase 2 randomized, multicenter, 26-week study to assess the efficacy and safety of BIOBYPASS (adgfVEGF121.10) delivered through maximally invasive surgery versus maximal medical treatment in patients with severe angina, advanced coronary artery disease and no options for revascularization, *Circulation* 106, 2986-a, 2002.

116. Kornowski, R., Leon, M. B., Fuchs, S., Vodovotz, Y., Flynn, M. A., Gordon, D. A., Pierre, A., Kovesdi, I., Keiser, J. A., and Epstein, S. E., Electromagnetic guidance for catheter-based transendocardial injection: A platform for intramyocardial angiogenesis therapy. Results in normal and ischemic porcine models, *J Am Coll Cardiol* 35(4), 1031–1039, 2000.

117. Vale, P. R., Losordo, D. W., Tkebuchava, T., Chen, D., Milliken, C. E., and Isner, J. M., Catheter-based myocardial gene transfer utilizing nonfluoroscopic electromechanical left ventricular mapping, *J Am Coll Cardiol* 34(1), 246–254, 1999.

118. Losordo, D. W., Vale, P. R., Hendel, R. C., Milliken, C. E., Fortuin, F. D., Cummings, N., Schatz, R. A., Asahara, T., Isner, J. M., and Kuntz, R. E., Phase 1/2 placebo-controlled, double-blind, dose-escalating trial of myocardial vascular endothelial growth factor 2 gene transfer by catheter delivery in patients with chronic myocardial ischemia, *Circulation* 105(17), 2012–2018, 2002.

119. Unger, E. F., Banai, S., Shou, M., Lazarous, D. F., Jaklitsch, M. T., Scheinowitz, M., Klingbeil, C., and Epstein, S. E., Basic fibroblast growth factor enhances myocardial collateral flow in a canine model, *Am J Physiol* 266, H1588–H1595, 1994.

120. Lazarous, D. F., Scheinowtiz, M., Shou, M., Hodge, E., Rajanayagam, S., Hunsberger, S., Robison, W. G., Jr., Stiber, J. A., Correa, R., Epstein, S. E., and Unger, E. F., Effects of chronic systemic administration of basic fibroblast growth factor on collateral development in the canine heart, *Circulation* 91, 145–153, 1995.

121. Rajanayagam, M. A., Shou, M., Thirumurti, V., Lazarous, D. F., Quyyumi, A. A., Goncalves, L., Stiber, J., Epstein, S. E., and Unger, E. F., Intracoronary basic fibroblast growth factor enhances myocardial collateral perfusion in dogs, *J Am Coll Cardiol* 35, 519–526, 2000.

122. Lazarous, D. F., Shou, M., Scheinowitz, M., Hodge, E., Thirumurti, V., Kitsiou, A. N., Stiber, J. A., Lobo, A. D., Husnberger, S., Guetta, E., Epstein, S. E., and Unger, E. F., Comparative effects of basic fibroblast growth factor and vascular endothelial growth factor on coronary collateral development and arterial response to injury, *Circulation* 94, 1074–1082, 1996.

123. Schumacher, B., Pecher, P., vonSpecht, B. U., and Stegmann, T., Induction of neoangiogenesis in ischemic myocardium by human growth factors: First clinical results of a new treatment of coronary heart disease, *Circulation* 97, 645–650, 1998.

124. Schumacher, B., Stegmann, T., and Pecher, P., The stimulation of neoangiogenesis in the ischemic human heart by the growth factor FGF: First clinical results, *J Cardiovasc Surg* 39, 783–789, 1998.

125. Stegmann, T. J., Hoppert, T., Schlurmann, W., and Gemeinhardt, S., First angiogenic treatment of coronary heart disease by FGF-1: Long-term results after 3 years, *Cardiac Vascular Regen* 1, 5–10, 2000.

126. Sellke, F. W., Laham, R. J., Edelman, E. R., Pearlman, J. D., and Simons, M., Therapeutic angiogenesis with basic fibroblast growth factor: Technique and early results, *Ann Thoracic Surg* 65, 1540–1544, 1998.

127. Laham, R. J., Sellke, F. W., Edelman, E. R., Pearlman, J. D., Ware, J. A., Brown, D. L., Gold, J. P., and Simons, M., Local perivascular delivery of basic fibroblast growth factor in patients undergoing coronary bypass surgery: Results of a phase I randomized, double-blind, placebo-controlled trial, *Circulation* 100(18), 1865–1871, 1999.

128. Udelson, J. E., Dilsizian, V., Laham, R. J., Chronos, N., Vansant, J., Blais, M., Galt, J. R., Pike, M., Yoshizawa, C., and Simons, M., Therapeutic angiogenesis with recombinant fibroblast growth factor-2 improves stress and rest myocardial perfusion abnormalities in patients with severe symptomatic chronic coronary artery disease, *Circulation* 102, 1605–1610, 2000.

129. Unger, E. F., Goncalves, L., Epstein, S. E., Chew, E. Y., Trapnell, C. B., and Cannon, R. O., III, Effects of a single intracoronary injection of basic fibroblast growth factor in stable angina pectoris, *Am J Cardiol* 85(12), 1414–1419, 2000.

130. Laham, R. J., Chronos, N. A., Pike, M., Leimbach, M. E., Udelson, J. E., Pearlman, J. D., Pettigrew, R. I., Whitehouse, M. J., Yoshizawa, C., and Simons, M., Intracoronary basic fibroblast growth factor (FGF-2) in patients with severe ischemic heart disease: Results of a phase 1 open-label dose escalation study, *J Am Coll Cardiol* 36, 2132–2139, 2000.

131. Kleiman, N. S. and Califf, R. M., Results from late-breaking clinical trials sessions at ACCIS 2000 and ACC 2000. American College of Cardiology, *J Am Coll Cardiol* 36(1), 310–325, 2000.

132. Grines, C. L., Watkins, M. W., Helmer, G., Penny, W., Brinker, J., Marmur, J. D., West, A., Rade, J. J., Marrott, P., Hammond, H. K., and Engler, R. L., Angiogenic Gene Therapy (AGENT) trial in patients with stable angina pectoris, *Circulation* 105(11), 1291–1297, 2002.

133. Grines, C. L., Watkins, M. W., Mahmarian, J. J., Iskandrian, A. E., Rade, J. J., Marrott, P., Pratt, C., and Kleiman, N., A randomized, double-blind, placebo-controlled trial of Ad5FGF-4 gene therapy and its effect on myocardial perfusion in patients with stable angina, *J Am Coll Cardiol* 42(8), 1339–1347, 2003.

134. Henry, T. D., Grines, C. L., Watkins, M. W., Dib, N., Barbeau, G., Moreadith, R., Andrasfay, T., and Engler, R. L., Effects of Ad5FGF-4 in patients with angina: An analysis of pooled data from the AGENT-3 and AGENT-4 trials, *J Am Coll Cardiol* 50(11), 1038–1046, 2007.

135. Treat-Jacobson, D., Halverson, S. L., Ratchford, A., Regensteiner, J. G., Lindquist, R., and Hirsch, A. T., A patient-derived perspective of health related quality of life with peripheral arterial disease, *J Nursing Scholar* 34, 55–60, 2002.

136. Eneroth, M. and Persson, B. M., Amputation for occlusive arterial disease. A multicenter study of 177 amputees, *Int Orthopaed* 16, 382–387, 1992.

137. Campbell, W. B., Johnston, J. A., Kernick, V. F., and Rutter, E. A., Lower limb amputation: Striking the balance, *Ann Royal Coll Surg Engl* 76, 205–209, 1994.

138. Dawson, I., Keller, B. P., Brand, R., Pesch-Batenburg, J., and Hajo van Bockel, J., Late outcomes of limb loss after failed infrainguinal bypass, *J Vascul Surg* 21, 613–622, 1995.

139. Takeshita, S., Zheng, L. P., Brogi, E., Kearney, M., Pu, L. Q., Bunting, S., Ferrara, N., Symes, J. F., and Isner, J. M., Therapeutic angiogenesis. A single intraarterial bolus of vascular endothelial growth factor augments revascularization in a rabbit ischemic hind limb model, *J Clin Invest* 93(2), 662–670, 1994.

140. Connolly, D. T., Hewelman, D. M., Nelson, R., Olander, J. V., Eppley, B. L., Delfino, J. J., Siegel, R. N., Leimgruber, R. S., and Feder, J., Tumor vascular permeability factor stimulates endothelial cell growth and angiogenesis, *J Clin Investig* 84, 1470–1478, 1989.

141. Rivard, A., Silver, M., Chen, D., Kearney, M., Magner, M., Annex, B., Peters, K., and Isner, J. M., Rescue of diabetes-related impairment of angiogenesis by intramuscular gene therapy with adeno-VEGF, *Am J Pathol* 154(2), 355–363, 1999.

142. Isner, J. M. and Feldman, L., Gene therapy for arterial disease, *Lancet* 344, 1653–1654, 1994.

143. Isner, J. M., Pieczek, A., Schainfeld, R., Blair, R., Haley, L., Asahara, T., Rosenfield, K., Razvi, S., Walsh, K., and Symes, J. F., Clinical evidence of angiogenesis after arterial gene transfer of phVEGF165 in patient with ischaemic limb, *Lancet* 348(9024), 370–374, 1996.

144. Baumgartner, I., Pieczek, A., Manor, O., Blair, R., Kearney, M., Walsh, K., and Isner, J. M., Constitutive expression of phVEGF165 after intramuscular gene transfer promotes collateral vessel development in patients with critical limb ischemia, *Circulation* 97(12), 1114–1123, 1998.

145. Rauh, G., Gravereaux, E., Pieczek, A., Curry, C., Schainfeld, R., and Isner, J. M., Assessment of safety and efficiency of intramuscular gene therapy with VEGF-2 in patient with critical limb ischemia, *Circulation* 100, I-770, 1999.

146. Rauh, G., Gravereaux, E. C., Pieczek, A. M., Radley, S., Schainfeld, R. M., and Isner, J. M., Age <50 years and rest pain predict positive clinical outcome after intramuscular gene transfer of phVEGF$_{165}$ in patients with critical limb ischemia, *Circulation* 100, I-319, 1999.

147. Baumgartner, I., Rauh, G., Pieczek, A., Wuensch, D., Magner, M., Kearney, M., Schainfeld, R., and Isner, J. M., Lower-extremity edema associated with gene transfer of naked DNA encoding vascular endothelial growth factor, *Ann Intern Med* 132(11), 880–884, 2000.

148. Kusumanto, Y. H., van Weel, V., Mulder, N. H., Smit, A. J., van den Dungen, J. J., Hooymans, J. M., Sluiter, W. J., Tio, R. A., Quax, P. H., Gans, R. O., Dullaart, R. P., and Hospers, G. A., Treatment with intramuscular vascular endothelial growth factor gene compared with placebo for patients with diabetes mellitus and critical limb ischemia: A double-blind randomized trial, *Hum Gene Ther* 17(6), 683–691, 2006.

149. Isner, J. M., Baumgartner, I., Rauh, G., Schainfeld, R., Blair, R., Manor, O., Razvi, S., and Symes, J. F., Treatment of thromboangiitis obliterans (Buerger's disease) by intramuscular gene transfer of vascular endothelial growth factor: Preliminary clinical results, *J Vasc Surg* 28(6), 964–973; discussion 73–75, 1998.

150. Rajagopalan, S., Shah, M., Luciano, A., Crystal, R., and Nabel, E. G., Adenovirus-mediated gene transfer of VEGF(121) improves lower-extremity endothelial function and flow reserve, *Circulation* 104(7), 753–755, 2001.

151. Rajagopalan, S., Mohler, E. R., 3rd, Lederman, R. J., Mendelsohn, F. O., Saucedo, J. F., Goldman, C. K., Blebea, J., Macko, J., Kessler, P. D., Rasmussen, H. S., and Annex, B. H., Regional angiogenesis with vascular endothelial growth factor in peripheral arterial disease: A phase II randomized, double-blind, controlled study of adenoviral delivery of vascular endothelial growth factor 121 in patients with disabling intermittent claudication, *Circulation* 108(16), 1933–1938, 2003.

152. Baffour, R., Berman, J., Garb, J. L., Rhee, S. W., Kaufman, J., and Friedmann, P., Enhanced angiogenesis and growth of collaterals by in vivo administration of recombinant basic fibroblast growth factor in a rabbit model of acute lower limb ischemia: Dose–response effect of basic fibroblast growth factor, *J Vasc Surg* 16, 181–191, 1992.

153. Yang, H. T., Deschenes, M. R., Ogilvie, R. W., and Terjung, R. L., Basic fibroblast growth factor increases collateral blood flow in rats with femoral arterial ligation, *Circ Res* 79, 62–69, 1996.

154. Chlegoun, J. O., Martins, R. N., Mitchell, C. A., and Chirila, T. V., Basic FGF enhances the development of collateral circulation after acute arterial occlusion, *Biochem Biophys Res Commun* 185, 510–516, 1992.

155. Lazarous, D. F., Unger, E. F., Epstein, S. E., Stine, A., Arevalo, J. L., Chew, E. Y., and Quyyumi, A. A., Basic fibroblast growth factor in patients with intermittent claudication: Results of a phase I trial, *J Am Coll Cardiol* 36(4), 1239–1244, 2000.

156. Lederman, R. J., Mendelsohn, F. O., Anderson, R. D., Saucedo, J. F., Tenaglia, A. N., Hermiller, J. B., Hillegass, W. B., Rocha-Singh, K., Moon, T. E., Whitehouse, M. J., and Annex, B. H., Therapeutic angiogenesis with recombinant fibroblast growth factor-2 for intermittent claudication (the TRAFFIC study): A randomised trial, *Lancet* 359(9323), 2053–2058, 2002.

157. Comerota, A. J., Throm, R. C., Miller, K. A., Henry, T., Chronos, N., Laird, J., Sequeira, R., Kent, C. K., Bacchetta, M., Goldman, C., Salenius, J. P., Schmieder, F. A., and Pilsudski, R., Naked plasmid DNA encoding fibroblast growth factor type 1 for the treatment of end-stage unreconstructible lower extremity ischemia: Preliminary results of a phase I trial, *J Vasc Surg* 35(5), 930–936, 2002.

158. Powell, R. J., Dormandy, J., Simons, M., Morishita, R., and Annex, B. H., Therapeutic angiogenesis for critical limb ischemia: Design of the hepatocyte growth factor therapeutic angiogenesis clinical trial, *Vasc Med* 9(3), 193–198, 2004.

159. Kutryk, M. J., Foley, D. P., van den Brand, M., Hamburger, J. N., van der Giessen, W. J., deFeyter, P. J., Bruining, N., Sabate, M., and Serruys, P. W., Local intracoronary administration of antisense oligonucleotide against c-myc for the prevention of in-stent restenosis: Results of the randomized investigation by the Thoraxcenter of antisense DNA using local delivery and IVUS after coronary stenting (ITALICS) trial, *J Am Coll Cardiol* 39(2), 281–287, 2002.

160. Asahara, T., Bauters, C., Pastore, C., Kearney, M., Rossow, S., Bunting, S., Ferrara, N., Symes, J. F., and Isner, J. M., Local delivery of vascular endothelial growth factor accelerates reendothelialization and attenuates intimal hyperplasia in balloon-injured rat carotid artery, *Circulation* 91(11), 2793–2801, 1995.

161. Asahara, T., Chen, D., Tsurumi, Y., Kearney, M., Rossow, S., Passeri, J., Symes, J., and Isner, J., Accelerated restitution of endothelial integrity and endothelium-dependent function following phVEGF$_{165}$ gene transfer, *Circulation* 94, 3291–3302, 1996.

162. Van Belle, E., Tio, F., Couffinhal, T., Maillard, L., Passeri, J., and Isner, J. M., Stent endothelialization: Time course, impact of local catheter delivery, feasibility of recombinant protein administration, and response to cytokine expedition, *Circulation* 94, I-259, 1996.

163. Van Belle, E., Tio, F. O., Chen, D., Maillard, L., Chen, D., Kearney, M., and Isner, J. M., Passivation of metallic stents after arterial gene transfer of phVEGF165 inhibits thrombus formation and intimal thickening, *J Am Coll Cardiol* 29(6), 1371–1379, 1997.

164. Hiltunen, M. O., Laitinen, M., Turunen, M. P., Jeltsch, M., Hartikainen, J., Rissanen, T. T., Laukkanen, J., Niemi, M., Kossila, M., Hakkinen, T. P., Kivela, A., Enholm, B., Mansukoski, H., Turunen, A. M., Alitalo, K., and Yla-Herttuala, S., Intravascular adenovirus-mediated VEGF-C gene transfer reduces neointima formation in balloon-denuded rabbit aorta, *Circulation* 102(18), 2262–2268, 2000.

165. Leppanen, O., Rutanen, J., Hiltunen, M. O., Rissanen, T. T., Turunen, M. P., Sjoblom, T., Bruggen, J., Backstrom, G., Carlsson, M., Buchdunger, E., Bergqvist, D., Alitalo, K., Heldin, C. H., Ostman, A., and Yla-Herttuala, S., Oral imatinib mesylate (STI571/gleevec) improves the efficacy of local intravascular vascular endothelial growth factor-C gene transfer in reducing neointimal growth in hypercholesterolemic rabbits, *Circulation* 109(9), 1140–1146, 2004.

166. Khurana, R., Zhuang, Z., Bhardwaj, S., Murakami, M., De Muinck, E., Yla-Herttuala, S., Ferrara, N., Martin, J. F., Zachary, I., and Simons, M., Angiogenesis-dependent and independent phases of intimal hyperplasia, *Circulation* 110 (16), 2436–2443, 2004.

167. Vale, P. R., Wuensch, D. I., Rauh, G. F., Rosenfield, K., Schainfeld, R. M., and Isner, J. M., Arterial gene therapy for inhibiting restenosis in patients with claudication undergoing superficial femoral artery angioplasty, *Circulation* 98, I-66, 1998.

168. Janssens, S., Flaherty, D., Nong, Z., Varenne, O., van Pelt, N., Haustermans, C., Zoldhelyi, P., Gerard, R., and Collen, D., Human endothelial nitric oxide synthase gene transfer inhibits vascular smooth muscle cell proliferation and neointima formation after balloon injury in rats, *Circulation* 97(13), 1274–1281, 1998.

169. Numaguchi, Y., Naruse, K., Harada, M., Osanai, H., Mokuno, S., Murase, K., Matsui, H., Toki, Y., Ito, T., Okumura, K., and Hayakawa, T., Prostacyclin synthase gene transfer accelerates reendothelialization and inhibits neointimal formation in rat carotid arteries after balloon injury, *Arterioscler Thromb Vasc Biol* 19(3), 727–733, 1999.

170. Chiu-Pinheiro, C. K., O'Brien, T., Katusic, Z. S., Bonilla, L. F., Hamner, C. E., and Schaff, H. V., Gene transfer to coronary artery bypass conduits, *Ann Thorac Surg* 74(4), 1161–1166; discussion 1166, 2002.

171. Turunen, P., Puhakka, H. L., Heikura, T., Romppanen, E., Inkala, M., Leppanen, O., and Yla-Herttuala, S., Extracellular superoxide dismutase with vaccinia virus anti-inflammatory protein 35K or tissue inhibitor of metalloproteinase-1: Combination gene therapy in the treatment of vein graft stenosis in rabbits, *Hum Gene Ther* 17(4), 405–414, 2006.

172. Luo, Z., Akita, G. Y., Date, T., Treleaven, C., Vincent, K. A., Woodcock, D., Cheng, S. H., Gregory, R. J., and Jiang, C., Adenovirus-mediated expression of beta-adrenergic receptor kinase C-terminus reduces intimal hyperplasia and luminal stenosis of arteriovenous polytetrafluoroethylene grafts in pigs, *Circulation* 111(13), 1679–1684, 2005.

173. Ratner, M., Genentech discloses safety concerns over Avastin, *Nat Biotechnol* 22(10), 1198, 2004.

174. Lemstrom, K. B., Krebs, R., Nykanen, A. I., Tikkanen, J. M., Sihvola, R. K., Aaltola, E. M., Hayry, P. J., Wood, J., Alitalo, K., Yla-Herttuala, S., and Koskinen, P. K., Vascular endothelial growth factor enhances cardiac allograft arteriosclerosis, *Circulation* 105(21), 2524–2530, 2002.

175. Kibbe, M. R., Tzeng, E., Gleixner, S. L., Watkins, S. C., Kovesdi, I., Lizonova, A., Makaroun, M. S., Billiar, T. R., and Rhee, R. Y., Adenovirus-mediated gene transfer of human inducible nitric oxide synthase in porcine vein grafts inhibits intimal hyperplasia, *J Vasc Surg* 34(1), 156–165, 2001.

176. Nykanen, A. I., Krebs, R., Saaristo, A., Turunen, P., Alitalo, K., Yla Herttuala, S., Koskinen, P. K., and Lemstrom, K. B., Angiopoietin-1 protects against the development of cardiac allograft arteriosclerosis, *Circulation* 107(9), 1308–1314, 2003.

177. Mann, M. J., Whittemore, A. D., Donaldson, M. C., Belkin, M., Conte, M. S., Polak, J. F., Orav, E. J., Ehsan, A., Dell'Acqua, G., and Dzau, V. J., *Ex-vivo* gene therapy of human vascular bypass grafts with E2F decoy: The PREVENT single-centre, randomised, controlled trial, *Lancet* 354(9189), 1493–1498, 1999.

178. Alexander, J. H., Hafley, G., Harrington, R. A., Peterson, E. D., Ferguson, T. B., Jr., Lorenz, T. J., Goyal, A., Gibson, M., Mack, M. J., Gennevois, D., Califf, R. M., and Kouchoukos, N. T., Efficacy and safety of edifoligide, an E2F transcription factor decoy, for prevention of vein graft failure following coronary artery bypass graft surgery: PREVENT IV: A randomized controlled trial, *Jama* 294(19), 2446–2454, 2005.

179. Conte, M. S., Bandyk, D. F., Clowes, A. W., Moneta, G. L., Seely, L., Lorenz, T. J., Namini, H., Hamdan, A. D., Roddy, S. P., Belkin, M., Berceli, S. A., DeMasi, R. J., Samson, R. H., and Berman, S. S., Results of PREVENT III: A multicenter, randomized trial of edifoligide for the prevention of vein graft failure in lower extremity bypass surgery, *J Vasc Surg* 43(4), 742–751; discussion 751, 2006.

180. Grossman, M., Rader, D. J., Muller, D. W., Kolansky, D. M., Kozarsky, K., Clark, B. J., 3rd, Stein, E. A., Lupien, P. J., Brewer, H. B., Jr., Raper, S. E., et al., A pilot study of *ex vivo* gene therapy for homozygous familial hypercholesterolaemia, *Nat Med* 1(11), 1148–1154, 1995.

181. Cichon, G., Willnow, T., Herwig, S., Uckert, W., Loser, P., Schmidt, H. H., Benhidjeb, T., Schlag, P. M., Schnieders, F., Niedzielska, D., and Heeren, J., Non-physiological overexpression of the low density lipoprotein receptor (LDLr) gene in the liver induces pathological intracellular lipid and cholesterol storage, *J Gene Med* 6(2), 166–175, 2004.

182. Kastelein, J. J., Wedel, M. K., Baker, B. F., Su, J., Bradley, J. D., Yu, R. Z., Chuang, E., Graham, M. J., and Crooke, R. M., Potent reduction of apolipoprotein B and low-density lipoprotein cholesterol by short-term administration of an antisense inhibitor of apolipoprotein B, *Circulation* 114(16), 1729–1735, 2006.

183. Wang, H., Katovich, M. J., Gelband, C. H., Reaves, P. Y., Phillips, M. I., and Raizada, M. K., Sustained inhibition of angiotensin I-converting enzyme (ACE) expression and long-term antihypertensive action by virally mediated delivery of ACE antisense cDNA, *Circ Res* 85(7), 614–622, 1999.

184. Schillinger, K. J., Tsai, S. Y., Taffet, G. E., Reddy, A. K., Marian, A. J., Entman, M. L., Oka, K., Chan, L., and O'Malley, B. W., Regulatable atrial natriuretic peptide gene therapy for hypertension, *Proc Natl Acad Sci USA* 102(39), 13789–13794, 2005.

185. Ohno, N., Itoh, H., Ikeda, T., Ueyama, K., Yamahara, K., Doi, K., Yamashita, J., Inoue, M., Masatsugu, K., Sawada, N., Fukunaga, Y., Sakaguchi, S., Sone, M., Yurugi, T., Kook, H., Komeda, M., and Nakao, K., Accelerated reendothelialization with suppressed thrombogenic property and neointimal hyperplasia of rabbit jugular vein grafts by adenovirus-mediated gene transfer of C-type natriuretic peptide, *Circulation* 105(14), 1623–1626, 2002.

186. Furukoji, E., Matsumoto, M., Yamashita, A., Yagi, H., Sakurai, Y., Marutsuka, K., Hatakeyama, K., Morishita, K., Fujimura, Y., Tamura, S., and Asada, Y., Adenovirus-mediated transfer of human placental ectonucleoside triphosphate diphosphohydrolase to vascular smooth muscle cells suppresses platelet aggregation in vitro and arterial thrombus formation in vivo, *Circulation* 111(6), 808–815, 2005.

187. Gheorghiade, M. and Bonow, R. O., Chronic heart failure in the United States: A manifestation of coronary artery disease, *Circulation* 97(3), 282–289, 1998.

188. Leotta, E., Patejunas, G., Murphy, G., Szokol, J., McGregor, L., Carbray, J., Hamawy, A., Winchester, D., Hackett, N., Crystal, R., and Rosengart, T., Gene therapy with adenovirus-mediated myocardial transfer of vascular endothelial growth factor 121 improves cardiac performance in a pacing model of congestive heart failure, *J Thorac Cardiovasc Surg* 123(6), 1101–1113, 2002.

189. Liu, X., Pachori, A. S., Ward, C. A., Davis, J. P., Gnecchi, M., Kong, D., Zhang, L., Murduck, J., Yet, S. F., Perrella, M. A., Pratt, R. E., Dzau, V. J., and Melo, L. G., Heme oxygenase-1 (HO-1) inhibits postmyocardial infarct remodeling and restores ventricular function, *Faseb J* 20(2), 207–216, 2006.

190. Rosen, M. R., Brink, P. R., Cohen, I. S., and Robinson, R. B., Genes, stem cells and biological pacemakers, *Cardiovasc Res* 64(1), 12–23, 2004.

191. Isner, J. M., Vale, P. R., Symes, J. F., and Losordo, D. W., Assessment of risks associated with cardiovascular gene therapy in human subjects, *Circ Res*, 2001.

192. Schwartz, E. R., Speakman, M. T., Patterson, M., Hale, S. S., Isner, J. M., Kedes, L. H., and Kloner, R. A., Evaluation of the effects of intramyocardial injection of DNA expressing vascular endothelial growth factor (VEGF) in a myocardial infarction model in the rat—angiogenesis and angioma formation, *J Am Coll Cardiol* 35,1323–1330, 2000.

193. Springer, M. L., Chen, A. S., Kraft, P. E., Bednarski, M., and Blau, H. M., VEGF gene delivery to muscle: Potential role for vasculogenesis in adults, *Mol Cell* 2(5), 549–558, 1998.

194. Aiello, L. P., Avery, R. L., Arrigg, P. G., Keyt, B. A., Jampel, H. D., Shah, S. T., Pasquale, L. R., Thieme, H., Iwamoto, M. A., Park, J. E., Nguyen, H. V., Aiello, L. M., Ferrara, N., and King, G. L., Vascular endothelial growth factor in ocular fluid of patients with diabetic retinopathy and other retinal disorders, *N Engl J Med* 331(22), 1480–1487, 1994.

195. Inoue, M., Itoh, H., Ueda, M., Naruko, T., Kojima, A., Komatsu, R., Doi, K., Ogawa, Y., Tamura, N., Takaya, K., Igaki, T., Yamashita, J., Chun, T. H., Masatsugu, K., Becker, A. E., and Nakao, K., Vascular endothelial growth factor (VEGF) expression in human coronary atherosclerotic lesions: Possible pathophysiological significance of VEGF in progression of atherosclerosis, *Circulation* 98(20), 2108–2116, 1998.

196. Celletti, F. L., Hilfiker, P. R., Ghafouri, P., and Dake, M. D., Effect of human recombinant vascular endothelial growth factor165 on progression of atherosclerotic plaque, *J Am Coll Cardiol* 37(8), 2126–2130, 2001.

197. Celletti, F. L., Waugh, J. M., Amabile, P. G., Brendolan, A., Hilfiker, P. R., and Dake, M. D., Vascular endothelial growth factor enhances atherosclerotic plaque progression, *Nat Med* 7(4), 425–429, 2001.

198. Nabel, E. G., Yang, Z. Y., Plautz, G., Forough, R., Zhan, X., Haudenschild, C. C., Maciag, T., and Nabel, G. J., Recombinant fibroblast growth factor-1 promotes intimal hyperplasia and angiogenesis in arteries in vivo, *Nature* 362(6423), 844–846, 1993.

199. Moulton, K. S., Heller, E., Konerding, M. A., Flynn, E., Palinski, W., and Folkman, J., Angiogenesis inhibitors endostatin or TNP-470 reduce intimal neovascularization and plaque growth in apolipoprotein E-deficient mice, *Circulation* 99(13), 1726–1732, 1999.

200. Laitinen, M., Hartikainen, J., Hiltunen, M. O., Eranen, J., Kiviniemi, M., Narvanen, O., Makinen, K., Manninen, H., Syvanne, M., Martin, J. F., Laakso, M., and Yla-Herttuala, S., Catheter-mediated vascular endothelial growth factor gene transfer to human coronary arteries after angioplasty, *Hum Gene Ther* 11(2), 263–270, 2000.

201. Wersto, R. P., Rosenthal, E. R., Seth, P. K., Eissa, N. T., and Donahue, R. E., Recombinant, replication-defective adenovirus gene transfer vectors induce cell cycle dysregulation and inappropriate expression of cyclin proteins, *J Virol* 72(12), 9491–9502, 1998.

202. Hollon, T., Researchers and regulators reflect on first gene therapy death, *Nat Med* 6(1), 6, 2000.

203. Raper, S. E., Chirmule, N., Lee, F. S., Wivel, N. A., Bagg, A., Gao, G. P., Wilson, J. M., and Batshaw, M. L., Fatal systemic inflammatory response syndrome in a ornithine transcarbamylase deficient patient following adenoviral gene transfer, *Mol Genet Metab* 80(1–2), 148–158, 2003.

204. Dewey, R. A., Morrissey, G., Cowsill, C. M., Stone, D., Bolognani, F., Dodd, N. J., Southgate, T. D., Klatzmann, D., Lassmann, H., Castro, M. G., and Lowenstein, P. R., Chronic brain inflammation and persistent herpes simplex virus 1 thymidine kinase expression in survivors of syngeneic glioma treated by adenovirus-mediated gene therapy: Implications for clinical trials, *Nat Med* 5(11), 1256–1263, 1999.

205. Cuevas, P., Carceller, F., Ortega, S., Zazo, M., Nieto, I., and Gimenez-Gallego, G., Hypotensive activity of fibroblast growth factor, *Science* 254(5035), 1208–1210, 1991.

206. Hariawala, M. D., Horowitz, J. R., Esakof, D., Sheriff, D. D., Walter, D. H., Keyt, B., Isner, J. M., and Symes, J. F., VEGF improves myocardial blood flow but produces EDRF-mediated hypotension in porcine hearts, *J Surg Res* 63(1), 77–82, 1996.

207. Horowitz, J. R., Rivard, A., van der Zee, R., Hariawala, M., Sheriff, D. D., Esakof, D. D., Chaudhry, G. M., Symes, J. F., and Isner, J. M., Vascular endothelial growth factor/vascular permeability factor produces nitric oxide-dependent hypotension. Evidence for a maintenance role in quiescent adult endothelium, *Arterioscler Thromb Vasc Biol* 17(11), 2793–2800, 1997.

208. van der Zee, R., Murohara, T., Luo, Z., Zollmann, F., Passeri, J., Lekutat, C., and Isner, J. M., Vascular endothelial growth factor/vascular permeability factor augments nitric oxide release from quiescent rabbit and human vascular endothelium, *Circulation* 95(4), 1030–1037, 1997.

209. Murohara, T., Asahara, T., Silver, M., Bauters, C., Masuda, H., Kalka, C., Kearney, M., Chen, D., Symes, J. F., Fishman, M. C., Huang, P. L., and Isner, J. M., Nitric oxide synthase modulates angiogenesis in response to tissue ischemia, *J Clin Invest* 101(11), 2567–2578, 1998.

210. Mazue, G., Bertolero, F., Jacob, C., Sarmientos, P., and Roncucci, R., Preclinical and clinical studies with recombinant human basic fibroblast growth factor, *Ann N Y Acad Sci* 638,329–340, 1991.

211. Ponten, A., Li, X., Thoren, P., Aase, K., Sjoblom, T., Ostman, A., and Eriksson, U., Transgenic overexpression of platelet-derived growth factor-C in the mouse heart induces cardiac fibrosis, hypertrophy, and dilated cardiomyopathy, *Am J Pathol* 163(2), 673–682, 2003.

212. Thurston, G., Suri, C., Smith, K., McClain, J., Sato, T. N., Yancopoulos, G. D., and McDonald, D. M., Leakage-resistant blood vessels in mice transgenically overexpressing angiopoietin-1, *Science* 286(5449), 2511–2514, 1999.

213. Rissanen, T. T., Korpisalo, P., Markkanen, J. E., Liimatainen, T., Orden, M. R., Kholova, I., de Goede, A., Heikura, T., Grohn, O. H., and Yla-Herttuala, S., Blood flow remodels growing vasculature during vascular endothelial growth factor gene therapy and determines between capillary arterialization and sprouting angiogenesis, *Circulation* 112(25), 3937–3946, 2005.

214. Rissanen, T. T., Markkanen, J. E., Gruhala, M., Heikura, T., Puranen, A., Kettunen, M. I., Kholova, I., Kauppinen, R. A., Achen, M. G., Stacker, S. A., Alitalo, K., and Yla-Herttuala, S., VEGF-D is the strongest angiogenic and lymphangiogenic effector among VEGFs delivered into skeletal muscle via adenoviruses, *Circ Res* 92(10), 1098–1106, 2003.

215. Dor, Y., Djonov, V., Abramovitch, R., Itin, A., Fishman, G. I., Carmeliet, P., Goelman, G., and Keshet, E., Conditional switching of VEGF provides new insights into adult neovascularization and pro-angiogenic therapy, *EMBO J* 21(8), 1939–1947, 2002.

216. Hughes, V., Therapy on trial, *Nat Med* 13(9), 1008–1009, 2007.

217. Wadman, M., Gene therapy might not have caused patient's death, *Nature* 449(7160), 270, 2007.

218. Abdel-Latif, A., Bolli, R., Tleyjeh, I. M., Montori, V. M., Perin, E. C., Hornung, C. A., Zuba-Surma, E. K., Al-Mallah, M., and Dawn, B., Adult bone marrow-derived cells for cardiac repair: A systematic review and meta-analysis, *Arch Intern Med* 167(10), 989–997, 2007.

219. Wollert, K. C. and Drexler, H., Clinical applications of stem cells for the heart, *Circ Res* 96(2), 151–163, 2005.

220. Kurozumi, K., Nakamura, K., Tamiya, T., Kawano, Y., Ishii, K., Kobune, M., Hirai, S., Uchida, H., Sasaki, K., Ito, Y., Kato, K., Honmou, O., Houkin, K., Date, I., and Hamada, H., Mesenchymal stem cells that produce neurotrophic factors reduce ischemic damage in the rat middle cerebral artery occlusion model, *Mol Ther* 11(1), 96–104, 2005.

221. Shintani, S., Kusano, K., Ii, M., Iwakura, A., Heyd, L., Curry, C., Wecker, A., Gavin, M., Ma, H., Kearney, M., Silver, M., Thorne, T., Murohara, T., and Losordo, D. W., Synergistic effect of combined intramyocardial CD34 + cells and VEGF2 gene therapy after MI, *Nat Clin Pract Cardiovasc Med* 3 Suppl 1, S123–S128, 2006.

222. Kawamoto, A., Murayama, T., Kusano, K., Ii, M., Tkebuchava, T., Shintani, S., Iwakura, A., Johnson, I., von Samson, P., Hanley, A., Gavin, M., Curry, C., Silver, M., Ma, H., Kearney, M., and Losordo, D. W., Synergistic effect of bone marrow mobilization and vascular endothelial growth factor-2 gene therapy in myocardial ischemia, *Circulation* 110(11), 1398–1405, 2004.

223. Ripa, R. S., Wang, Y., Jorgensen, E., Johnsen, H. E., Hesse, B., and Kastrup, J., Intramyocardial injection of vascular endothelial growth factor-A165 plasmid followed by granulocyte-colony stimulating factor to induce angiogenesis in patients with severe chronic ischaemic heart disease, *Eur Heart J* 27 (15), 1785–1792, 2006.

224. Chachques, J. C., Duarte, F., Cattadori, B., Shafy, A., Lila, N., Chatellier, G., Fabiani, J. N., and Carpentier, A. F., Angiogenic growth factors and/or cellular therapy for myocardial regeneration: A comparative study, *J Thorac Cardiovasc Surg* 128(2), 245–253, 2004.

225. Stevens, K. R., Rolle, M. W., Minami, E., Ueno, S., Nourse, M. B., Virag, J. I., Reinecke, H., and Murry, C. E., Chemical dimerization of fibroblast growth factor receptor-1 induces myoblast proliferation, increases intracardiac graft size, and reduces ventricular dilation in infarcted hearts, *Hum Gene Ther* 18(5), 401–412, 2007.

226. Kutschka, I., Kofidis, T., Chen, I. Y., von Degenfeld, G., Zwierzchoniewska, M., Hoyt, G., Arai, T., Lebl, D. R., Hendry, S. L., Sheikh, A. Y., Cooke, D. T., Connolly, A., Blau, H. M., Gambhir, S. S., and Robbins, R. C., Adenoviral human BCL-2 transgene expression attenuates early donor cell death after cardiomyoblast transplantation into ischemic rat hearts, *Circulation* 114(1 Suppl), I174–I180, 2006.

227. Kizana, E., Chang, C. Y., Cingolani, E., Ramirez-Correa, G. A., Sekar, R. B., Abraham, M. R., Ginn, S. L., Tung, L., Alexander, I. E., and Marban, E., Gene transfer of connexin43 mutants attenuates coupling in cardiomyocytes: Novel basis for modulation of cardiac conduction by gene therapy, *Circ Res* 100(11), 1597–1604, 2007.

228. Potapova, I., Plotnikov, A., Lu, Z., Danilo, P. J., Valiunas, V., Qu, J., Doronin, S., Zuckerman, J., Shlapakova, I. N., Gao, J., Pan, Z., Herron, A. J., Robinson, R. B., Brink, P. R., Rosen, M. R., and Cohen, I. S., Human mesenchymal stem cells as a gene delivery system to create cardiac pacemakers, *Circ Res* 94(7), 952–959, 2004.

229. Stewart, D. J., Hilton, J. D., Arnold, J. M., Gregoire, J., Rivard, A., Archer, S. L., Charbonneau, F., Cohen, E., Curtis, M., Buller, C. E., Mendelsohn, F. O., Dib, N., Page, P., Ducas, J., Plante, S., Sullivan, J., Macko, J., Rasmussen, C., Kessler, P. D., and Rasmussen, H. S., Angiogenic gene therapy in patients with nonrevascularizable ischemic heart disease: A phase 2 randomized, controlled trial of AdVEGF(121) (AdVEGF121) versus maximum medical treatment. *Gene Ther.* 13(21): 1503–1511, 2006.

47 Gene Therapy for the Treatment of Cancer: From Laboratory to Bedside

Gianpietro Dotti, Barbara Savoldo, Fatma Okur, Raphaël F. Rousseau, and Malcolm K. Brenner

CONTENTS

47.1 STRATEGIES OF GENE TRANSFER FOR THE TREATMENT OF CANCER

Since the 1970s, a multidisciplinary approach combining surgery, chemotherapy, and radiation has lead to a dramatic improvement in survival for patients affected by malignant diseases. Cellular therapies, such as stem cell transplantation, have also made a significant contribution. Nonetheless, many patients are still resistant to standard therapies, which also have high and often unacceptable acute and chronic organ toxicity and an increased risk for secondary malignancies. Therefore, new strategies are needed to improve overall survival and decrease treatment-associated morbidity.

Gene transfer may be a component of these new approaches against cancer and may be used in a number of different ways:

1. *Modification of tumor cells*: By repairing genetic defects believed to be responsible for tumoral proliferation, for example, by restoring genes controlling cellular division or that induce programmed cell death (apoptosis).
2. *Sensitization of normal tissues or tumor cells in order to modify the therapeutic index of cytotoxic drugs*: Genes encoding enzymes that can transform a nontoxic prodrug into an active drug can be introduced into cancer cells, or we can modify normal tissues with genes that can protect them against cytotoxic agents.
3. *Modulation of tumor invasiveness*: By delivering genes that can inhibit the growth of new blood vessels to impede nutrient supply to the tumor cells (inhibition of neoangiogenesis).

4. *Enhancement of the antitumor immune response*: Either by inducing the recognition of tumor cells by the host's immune system or by enhancing the cytotoxic function of immune effector cells.

5. *Gene marking*: Even though gene marking is not a therapeutic intervention per se, this approach has helped investigators to understand the behavior and outcome of transduced cells once returned to the patient. This strategy is discussed elsewhere in this book (see Chapter 44).

47.2 MODIFICATION OF TUMOR CELLS

Progress made in identifying and understanding molecular aberrations has allowed much better characterization of the steps leading to cancer. Many of these aberrations alter key regulatory, survival, and differentiation processes in the normal cell cycle. Other aberrations lead to the production of abnormal fusion products, with subsequent gain or loss of critical functions in cell life and death.

Currently, the most widely used strategy is to insert a complementary DNA (cDNA) that encodes a functional protein to compensate for loss of function mutations. For example, p53 gene mutations generating inactive forms of the protein have been identified in numerous tumors and likely play a crucial role in the abnormal cell proliferation associated with malignant diseases. Transfer of wild-type p53 in p53-mutated or -deficient tumors reduces uncontrolled cell growth [1] and can induce apoptosis [2], and adenoviral [3], retroviral [4], or nonviral [5] p53 gene transfer have been explored in preclinical models [6], and subsequent clinical trials. Despite low rates of gene transfer, apparent clinical benefits are observed following p53 gene transfer in patients with ovarian cancer [7], head and neck tumors [8,9], and nonsmall cell lung carcinomas [10]. Indeed an adenoviral p53 formulation has been licensed as a drug in China and details are described elsewhere in this book (see Chapter 38). Likewise, alterations of p16 have been described in squamous head and neck carcinomas, and gene replacement approaches similar to those developed for p53 mutations are being explored [11].

In other tumors, the transformation event is caused by oncogenic fusion transcripts [12] or aberrant expression of antiapoptotic genes [13] determined by specific chromosomal translocations or by aberrant activation of oncogenes such as genes of the *ras* family (H-*ras*, N-*ras*, and K-*ras*) by a simple point mutation [14]. Gene transfer can help reverse these gain of function mutations by knocking-down oncogene activity. Knock-down of the mRNA from a mutated K-ras gene can be achieved by antisense RNA, thereby reducing production of the altered protein and delaying tumor cell growth *in vitro* and *in vivo* [15,16]. *Fos* oncogene has been similarly targeted in a murine model of mammary tumor [17]. Several other knock-down approaches using ribozymes [18] (see Chapter 18), antisense RNA [19] (see Chapter 19), or intracellular antibodies [20] have all shown promising results, but are not yet evaluated clinically. The recent discovery that double-stranded small interfering RNAs (siRNAs) can knockdown the expression

of a target gene very efficiently and with high specificity [21,22] has increased interest in this general approach. Systemic delivery or gene transfer of siRNAs can specifically downregulate oncogenes in tumor cells [23] or inhibitory molecules [24] or death signals in cells of the immune system, thereby improving the activity of the latter [25].

Numerous difficulties must be overcome before tumor correction strategies can be successful in a clinical trial. For example, the corrective gene must enter an extremely high proportion of malignant cells, unless there is a potent "bystander" effect on nontransduced tumor (postulated but unproven for p53 gene therapy). Secondly, targeting metastatic disease will usually be necessary and will require adequate distribution by blood stream and perhaps by replication competence of the therapeutic vector. Thirdly, correction of a single defect may be inadequate to eradicate the tumor cells, leaving instead a collection of "$n-1$ cells" (where n is the number of mutations required for malignancy to occur) capable of undergoing another mutation to restart the malignant process.

Another possible complication has recently come to light. Investigators have identified a subset of tumor cells that possess self-renewal capability and can generate more differentiated tumor cells. These cells have been dubbed "cancer stem cells" and have been isolated from a variety of cancers of the blood, breast, central nervous system, pancreas, skin, head and neck, colon, and prostate [26]. This concept may have profound implications for the design of approaches that specifically target these cells and thereby permanently eradicate the tumor.

Overall, exploitation of tumor correction strategies will require significant improvements in vector efficiency and targeting, and until these come to pass, the development of novel rationally targeted small molecules will likely dominate this "corrective" approach.

47.3 SENSITIZATION OF NORMAL TISSUES OR TUMOR CELLS

47.3.1 PRODRUG-METABOLIZING ENZYME

Introduction of a gene encoding an enzyme that metabolizes an otherwise inert molecule into a cytotoxic agent has frequently been used in tumor gene therapy. Although the herpes simplex virus thymidine kinase (HSV-Tk)-ganciclovir system has been the most widely applied, there are more than 20 such prodrug-metabolizing enzyme (PDME) systems currently in various stages of development or clinical trials [27,28]. For almost all of these, the concept is that the gene encoding the PDME is expressed in the cancer cell and metabolizes a small molecule to an active moiety, which then kills the tumor cell directly. The molecule may also diffuse either through intercellular gap junctions or in the extracellular space and destroy adjacent tumor cells. In this way, transduction of even a small proportion of tumor cells can produce a large "bystander" effect on adjacent tumor tissue. As discussed later in this section, bystander effects may also result from recruitment of components of the immune system, which are triggered by the products of dead and dying cells. Irrespective of the mechanisms, the bystander effect compensates for the low efficiency of transduction achieved by currently available vectors and may help to destroy a large tumor burden.

Brain tumors were an attractive initial target for PDME gene therapy. Because the tumors seldom metastasize, the goal of the therapy is the local eradication of the tumor. Hence, the major limitation of PDME, that it requires local inoculation of a tumor with the vector encoding the gene, does not represent a major disadvantage. In the mid-1990s, phaseI/II studies in patients with malignant glioma used intratumoral injection of retroviral vector-producing packaging cell lines (VPC) generating particles encoding HSV-Tk. These injections were followed by systemic administration of gancyclovir [29–31]. Because of initial promising results, a phase III study was conducted in patients with previously untreated glioblastoma [32]. In this study, 248 patients were randomized to conventional treatment (usually resection and radiotherapy) or conventional therapy plus adjuvant injection of VPC-HSV-Tk [32]. Unfortunately, this study failed to show improved progression or disease free survival in the patients receiving the study drug and currently this approach has been discontinued for this disease.

PDME does, however, continue to be studied in other cancers. Retinoblastoma is a localized tumor that is conventionally treated by enucleation or chemoradiotherapy [33]. Enucleation is obviously disabling and deforming, and if the tumor is bilateral it leads to blindness. The alternatives to chemotherapy and radiotherapy are less mutilating but are both associated with secondary malignancies. Hurwitz and colleagues enrolled eight patients with bilateral retinoblastoma in a phase I clinical trial [33]. These patients received bilateral intratumoral injection of adenovirus (Ad) type 5 encoding the thymidine kinase gene followed by administration of gancyclovir. This dose escalation study showed mild ocular toxicity at 10^{11} vp. and resolution of the vitreous tumor seeds in seven patients who received 10^{10} vp. One of these patients remained tumor-free 38 months after therapy and a second, modified study is planned.

The accessibility of locally recurrent prostate cancer means that these patients can receive multiple intratumoral injections of TK-encoding adenovectors [34–37]. Several studies have independently confirmed the safety of the procedure and suggested efficacy by showing prolongation of the median serum prostate-specific antigen (PSA) doubling time. The combination of this suicide gene approach with delivery of transgenes encoding immunomodulatory molecules such as IL-12 may further improve benefit by favoring an active antitumor immune response following the tumor destruction mediated by the PDME.

Other suicide gene therapies are being evaluated. Among the most developed of them is the cytosine deaminase system, which converts fluorocytosine to fluorourosil [38]. There are, however, concerns that this suicide system may be less potent than the Tk-ganciclovir prodrug system. Other molecules that metabolize drugs or trigger apoptotic pathways within tumor cells are also being considered. Perhaps the most important future trend is to attempt to enhance the bystander effect. At present, this is mediated predominantly by transfer of the small molecule cytotoxic drug from cell to cell. However, it is apparent that at least part of the bystander effect is dependent on an immune response generated to the lysed tumor. Hence, the bystander effect in immunocompromised animals has been observed to be substantially less than in those with intact immune systems. Investigators are now attempting to combine PDME genes with sequences encoding a variety of immunostimulatory molecules (see Section 47.5.5.1), including, but not limited to interleukin 2 (IL-2), IL-12, and granulocyte-macrophage colony-stimulating factor (GM-CSF) [39–41]. Data from these studies are yet to be evaluated. Efforts are also being made to generate vectors, which can selectively divide in malignant cells (conditionally replication-competent vectors; see Chapters 3 and 38) and may therefore spread their encoded PDME genes throughout the tumor bed [42].

PDME has also effectively been used as a means of controlling T cell therapies. For example, graft versus host disease (GvHD) may occur when donor T cells are given to patients after allogeneic stem cell transplantation in an effort to treat tumor relapse (graft versus tumor effect) or posttransplant infections. Several groups have infused donor T cells transduced with the HSV-Tk gene and reported successful abrogation of GvHD after treatment with ganciclovir [43,44]. However, the main limitation of this strategy is the high immunogenicity of the viral-derived TK protein that can eliminate transgenic T cells. The substitution of human-derived proteins should reduce the immunogenicity of the transgene. For example, the Fas receptor [45] is one of the most potent inducer of apoptosis. Fas has been modified to incorporate a specific endodomain that induces its dimerization and activation after exposure to a specific chemical inducer of dimerization (CID), analogs of which have been safely tested in a phase I study [46]. Recently, our group extended this concept by constructing a suicide gene based on an inducible caspase-9 molecule [47]. This gene has advantages over Fas attributable to higher potency, since recruitment of few molecules is sufficient to induce apoptosis. We are currently implementing a phase I clinical trial to test the safety and efficacy of this approach in patients receiving haploidentical transplantation [48].

As T cell therapies for cancer become more widespread, these suicide mechanisms will become extremely important in ensuring that the regimens are acceptably safe.

47.3.2 CYTOTOXIC DRUG-RESISTANCE GENE TRANSFER

The concept of dose intensification has long been current in modern oncology. In other words, it is believed that giving more of a cytotoxic drug over a longer period will cure a higher proportion of patients. Although there are many obvious exceptions to this rule, for many pediatric malignancies in particular, it is clear that failure to tolerate chemotherapy in the intended doses correlates well with an increased risk of relapse. For that reason, there is an interest in using genes, which will protect normal tissues from cytotoxic drugs while leaving malignant cells vulnerable to destruction. By increasing the therapeutic index in this way, it is hoped that more drug can be administered and a higher percentage of patients cured.

There are many different candidate drug-resistance genes that can be transferred, but perhaps the most widely studied is the human multidrug resistance-1 (MDR-1) gene. The gene product acts as a drug efflux pump and prevents accumulation of toxic small molecules, including a range of cytotoxic drugs such as mitoxantrone and daunorubicin. The primary toxicity of many of these cytotoxic drugs is on hematopoietic progenitor cells. Retroviral-mediated gene transfer of drug-resistance genes into hematopoietic stem cells has until recently been difficult to accomplish. The incorporation of fibronectin together with hematopoietic growth factors into the transduction regimen, together with repeated cycles of gene transfer, has allowed a significant proportion of hematopoietic cells to be protected with expression levels adequate to reduce the sensitivity of these stem cells to chemotherapeutic agents [49]. Other molecules extensively validated in preclinical studies include dihydrofolate reductase (DHFR), which confers resistance to antifolates, and O6-methylguanine-DNA-methyltransferase (MGMT), which confers resistance to the cytotoxic effects of alkylating agents.

Clinical trials have been reported using HSC-expressing MDR1 [49–53]. Although appealing in concept, several problems have progressively reduced enthusiasm for this general approach, including the limited levels of transgene expression and hence drug protection *in vivo*, and toxicities to other unprotected organ systems once drug dosages are escalated. Moreover, the development of myeloproliferative/leukemic disorders in some animal studies in which MDR transporter proteins are expressed in HSC has raised concerns about the oncogenic potential of these transgenes [54]. Despite these limitations, a number of additional clinical studies are planned. These studies hope to benefit from the higher levels of gene transfer to human HSC that can now be obtained using lentiviral vectors, which may also have lower oncogenic genotoxicity compared to oncoretroviral vectors [55].

47.4 ANTIANGIOGENESIS GENE THERAPY

Because angiogenesis is a prerequisite for the development of metastatic disease for solid tumors, and probably also for leukemias and lymphomas, an attack on newly formed blood vessels may help to impede the spread of disease. Recombinant endostatin is currently the most clinically studied inhibitor of angiogenesis. Although administration of the recombinant protein has proven safe, the clinical benefits were minimal [56,57]. One explanation for this failure is that sustained effective concentrations of endostatin might not be achieved because of the short half-life of the (recombinant) protein [58]. By contrast, intramuscular injection of an adeno-associated Ad encoding for the endostatin gene

[59] makes product continuously, and the more favorable pharmacokinetics may be a reason to reevaluate the contribution of endostatin to tumor control.

A number of other different large and small molecule inhibitors of vasculogenesis are currently under study and some of them are suitable for a gene therapy approach [60]. Much remains to be learned about the most appropriate route and cell delivery of angiogenesis inhibition, but as with any protein-based therapeutic, gene transfer should allow a continual delivery of the drug rather than the peak-and-trough concentrations that result from most forms of injection, and may thereby produce a more sustained and effective response.

47.5 GENE MODIFICATION OF THE IMMUNE RESPONSE

The specificity of the immune system affords many opportunities for targeted cancer therapy. Identification of antigens expressed on tumor cells (Table 47.1) and the improvements made in gene transfer techniques, together with a better understanding of the molecular and cellular mechanisms involved in the immune response against cancer, have given investigators tools to manipulate the immune system to induce an efficient immune response in the tumor-bearing host.

47.5.1 TUMOR ESCAPE

Even though many *in vitro* cancer gene and immunotherapy studies have been published since the early 1980s, and despite impressive improvements in gene transfer technology and techniques for immune manipulation, clinical efficacy is still poor. Interest now focuses on a better understanding of the mechanisms by which tumors can escape immunosurveillance *in vivo* (Table 47.2), perhaps the greatest impediment to successful application of immunotherapy strategies.

47.5.2 ANTIGEN PRESENTATION DEFECTS

An appropriate pattern of T helper and T cytotoxic cell stimulation has been proved to be essential in mounting an efficient immune response. Many tumors prevent the recruitment of T cell immune response by downregulation of major histocompatibility (MHC) molecules or of costimulatory and adhesion molecules.

Several immunotherapy strategies have been conceived to overcome these defects. Introduction of MHC molecules into tumors was one of the first strategies adopted clinically in the 1990s. Earlier murine models had shown that an increased expression of MHC class I molecules decreased tumorigenicity due to enhanced antigen presentation to CD8 + cytotoxic T lymphocytes (CTLs) [61]. Tumor immunogenicity has also been increased by gene transfer of both allogeneic MHC class I and II molecules [62], thus demonstrating the relevance of both CD8 cytotoxic and CD4 helper T cells in enhancing systemic immunity against cancer. Increased expression of MHC molecules can also be obtained indirectly by transducing

TABLE 47.1

Tumor Antigens Recognized by T Lymphocytes

Antigen	Primary Tumor[a]	Normal Tissues
Differentiation antigens		
Tyrosinase	Melanoma	Melanocytes
ACE	Colon, digestive tumors	Colon
Immunoglobulin (idiotype)	B lymphomas	B lymphocytes
gp100	Melanoma	Melanocytes, retina[b]
Melan A/Mart 1	Melanoma	Melanocytes, retina[b]
gp75/TRP-1	Melanoma	
PSA	Prostate	Prostate
Self-antigens expressed at low levels on normal tissues		
MAGE-1	Melanoma	Testis,[b] trophoblast
MAGE-3	Melanoma	Testis[b]
BAGE	Melanoma	Testis,[b] trophoblast
GAGE 1,2	Melanoma	Testis,[b] trophoblast
RAGE-1	Renal carcinoma	Retina[b]
NY-ESO-1	Melanoma, breast carcinoma, lung	Testis[b]
MUC-1	Breast	Lactating breast
Oncopeptides overexpressed in tumors		
HER2/neu	Breast carcinoma, ovarian carcinoma	
P53 (wild-type)	Squamous cell carcinoma	
CDK4	Melanoma	
B-catenin	Melanoma	
CASP-8	Squamous cell carcinoma	
P53 (mutated)	Many	
Ras	Colon, lung, pancreas	
MUM-1	Melanoma	
bcr/abl	Chronic myeloid leukemia	

[a] Antigen expressed in more than 25% of cases.

[b] These tissues do not express MHC class-1; therefore, antigen may not be presented.

tumor cells with cytokines able to induce MHC molecule upregulation on the cell surface. Examples of cytokines with this property are interferon-γ (IFN-γ) [63], IL-4, and tumor necrosis factor-α (TNF-α) [64].

Once the T cell receptor (TCR) has specifically interacted with the epitope, costimulatory signals induce a T cell response and prevent anergy [65]. A number of such signals have been identified of which the B7 family members, such as B7.1 (CD80) and B7.2 (CD86), are among the best known. These molecules are expressed on the antigen-presenting cell (APC) surface and bind to their cognate receptor, CD28, on the responding T cell [66]. Other costimulatory molecules, such as intercellular adhesion molecules (ICAM) and leukocyte function-associated antigens, are also important. The absence of costimulatory molecules on the tumor cell surface is one of the explanations for the unresponsiveness of CD8 + T cells in the tumor-bearing host. This important role of CD80/86-CD28 interaction in T cell activation made B7 genes an appealing target for gene transfer into tumor cells [67].

CD40 ligand (CD40L) also represents a possible candidate for gene transfer because it can enhance antigen presentation by tumor cells and induce maturation of professional APCs [68]. This molecule (normally expressed only by activated CD4 + T lymphocytes) interacts with the specific receptor CD40 and increases antigen presentation by malignant cells by upregulating costimulatory molecules (e.g., B7–1 and B7–2), adhesion molecules (e.g., ICAM-1 and ICAM-3), and MHC molecules. CD40L also increases antigen uptake by dendritic cells (DCs) and allows them to bypass CD4 + T helper cells in recruiting specific CTL [69]. Hence, CD40L gene transfer has been widely used in multiple preclinical and several clinical studies (described later in this chapter) with promising results [70–76].

Failure to take up tumor antigens or to present them adequately on APCs (mainly DCs) is commonly observed in tumors. Many immunotherapy strategies therefore aim to increase the efficiency of antigen presentation. In this context, GM-CSF has been widely used for transducing

TABLE 47.2

Mechanisms by Which Tumor Can Escape Immune Surveillance

Defects in antigen presentation

Adhesion deficiency

MHC molecules/pathway defects

Defects in antigen processing/transport

Defects in costimulatory pathways

Antigenic variants

Decoys

Microenvironment abnormalities

Inhibitory cytokines/ligand

Growth/survival factors, angiogenesis

Immune sanctuaries (eye, testis, central nervous system)

Establishment of latency (virus-induced tumors such as EBV, HPV, and HTLV)

Recruitment of T cells with inhibitory function (Treg cells)

T cell defects

Absence/deletion of specific T cell precursors

Anergy

Downregulation of TcR ζ chain

Mutations in signaling pathways

Deletion/defect in helper T cell

Defect in establishment of T cell memory

T cell inhibitors

---

autologous or allogeneic cancer cells and proved to be one of the most potent molecules *in vivo*, mainly acting as a critical factor for promoting maturation of DCs at the site of vaccine injection and thereby enhancing tumor antigen uptake and presentation [77].

47.5.3 T Cell Defects

T lymphocytes in the tumor-bearing host may be decreased in number or fail to home to tumor infiltrates. They may also be anergic and show increased apoptosis, impaired cytokine secretion, inhibition of proliferation, decreased cytotoxic capacity, defects in helper function, and absence of memory T cells. In part, these diverse deficits are attributable to an increased production of immunosuppressive factors expressed by the tumor itself or by regulatory T cells (Treg). Transforming growth factor-β (TGF-β) and IL-10 [78] and proapototic molecules such as Fas-ligand (Fas-L) that inhibit T cell functions or contribute to an early death of cytotoxic T cells that express high levels of Fas receptor [25,79,80] are commonly implicated.

Multiple strategies have been used to overcome the above T cell defects. Most commonly, tumor cells are transduced with genes encoding cytokines able to recreate an optimal microenvironment for T cells recruitment, activation, and expansion. Several cytokines can be used alone or in various combinations to enhance antitumor T cell-mediated immunity. Recruitment of cytotoxic CD8 + T cells able to

recognize tumor-specific antigens is observed with IL-4 [81], whereas CD4+ T cell recruitment is favored by TNF [82] and IL-7 [83].

Systemic administration of IL-2 increases the *in vivo* expansion of adoptively transferred antitumor CTLs [84]. However, the systemic toxicity [84] and the expansion of unwanted cells, including Treg [85], limit the clinical value of this strategy. Transgenic expression of IL-2 and IL-15 in tumor-specific CTLs increased the expansion of these cells in the tumor microenvironment [86–88] and enhanced antitumor activity [88]. Moreover, the coexpression of an efficient suicide gene allowed efficient elimination of transgenic CTLs, thereby increasing the safety and feasibility of this approach for clinical application [89,90].

Expression of class I antigens is enhanced by IFN-γ [91], which in turn increases tumor immunogenicity. A combination of these effects may also be obtained. For example, IL-2 can recruit both cytotoxic T and natural killer (NK) cells directly [92] and can also induce release of the secondary cytokine IFN-γ [93]. NK cells may also be important in generating an effective antitumor immune response [94]. These cells exert cytotoxic activity through the granzyme–perforin system and release inflammatory cytokines (including IFN-γ, IL-5, TNF-α, and GM-CSF). NK cells, unlike T lymphocytes, recognize and destroy target cells that lack MHC antigens, so they can be effective against tumors with downregulated surface expression of these molecules. IL-2 can also induce lymphokine-activated killer cells from both

T and NK cells [92], which have enhanced cytolytic activity against a broad range of tumor cells and act in an MHC-unrestricted manner.

The cytokine IL-12 also increases antitumor activity, by recruitment of cytotoxic T cells and NK cells [95]. Unfortunately, early clinical trials of recombinant IL-12 in cancer were rapidly discontinued due to lethal toxicity [96]. Transfer of the IL-12 gene to a restricted cell population, such as the tumor cells themselves, may retain the benefits of IL-12 in the local tumor environment whilst reducing its adverse systemic consequences by limiting peak levels [97]. Similarly, transgenic production of IL-12 by DCs increases the potency of antitumor vaccines [98]. Finally, in experimental systems, tumor-specific CTLs that are genetically manipulated to express IL-12 are resistant to the inhibitory effects of the Th2 tumor environment [99].

Costimulatory surface molecules such as B7.1 or CD40L may serve as accessory signals in the T cell activation process and prevent/overcome the T cell anergy induced by tumors [68,69]. Finally, chemokines (such as the T cell specific lymphotactin [LTN]) [100] may be used to attract immune-effector cells to the vaccination site, thereby enhancing the probability of adequate immune activation.

47.5.4 Microenvironment Abnormalities

Tumors can secrete substances able to induce immunosuppression. The best characterized of these are TGF-β, IL-10, vascular endothelial growth factor (VEGF), and Fas-L. TGF-β affects CTL function, inhibiting production of Th1 cytokines (especially IL-12), downregulating surface expression of IL-2 receptors on T cells, inhibiting antigen presentation on MHC class II molecules, and decreasing surface expression of costimulatory and adhesion molecules [78]. IL-10 shares many immune-inhibitory properties with TGF-β, reducing Th1 cytokine synthesis and making tumor cells insensitive to CTL-mediated lysis. IL-10 antisense gene transduction can restore immunogenicity when tumor cells produce high amounts of IL-10 [101]. VEGF serves to promote tumor angiogenesis and inhibit DC differentiation. Antisense gene transduction or gene therapy with soluble inhibitory receptors represents a promising strategy to block these VEGF effects [102]. The Fas/Fas-L pathway is crucial for the T cell contraction that follows elimination of infectious agents [103,104]. Many tumors can aberrantly express Fas-L and create an unfavorable environment for tumor specific Fas positive CTLs, inducing premature apoptosis [80]. As we describe in the following sections, T lymphocytes can be genetically modified to overcome some of these tumor immune escape mechanisms.

47.5.5 Clinical Applications of Immunogene Therapies

Gene transfer can be used in several ways to manipulate the immune system (Table 47.3). Passive immunotherapy consists of the adoptive transfer either of specific antitumor immune effectors, gene modified or not, or of specific antibodies (generated *ex vivo*) to the cancer patient. The use of cancer vaccines in active immunotherapy aims to induce efficient and long-lasting immune responses by direct stimulation

---

TABLE 47.3
Strategies in Cancer Immunotherapy

Passive Immunotherapy "Adoptive Immunotherapy"	Active Immunotherapy "Cancer Vaccines"
Transfer of effector cells	*In vivo immunization with*
• Sensitized *ex vivo*	• Purified antigen
Pulsed APCs[a]	• Immunodominant peptide
Transduced effectors	• "Naked" DNA encoding tumor antigen
Restricted/nonrestricted Epitopes	• Recombinant viruses encoding tumor antigen
• Expanded *ex vivo*	• Whole tumor cells
Bulk	Used as APCs
Cloned population	Modified to attract host APCs (cross-priming)
Transfer of specific monoclonal antibodies	*Adjuvant therapies*
Induce ADCC[b]	IL-2, IFN-γ: systemic administration
Coupled with toxin or radioisotope	IL-2, GM-CSF,[c] LTN[d]
	IL-12: in situ administration

[a] APC, antigen-presenting cell.
[b] ADCC, antibody-dependent cell-dependent cytotoxicity.
[c] GM-CSF, granulocyte-macrophage colony-stimulating factor.
[d] LTN, lymphotactin.

(with gene-modified cells or cell components) of the patient immune system.

47.5.5.1 Gene-Modified Autologous and Allogeneic Tumor Cells

Several clinical applications in humans have been reported using manipulated autologous cancer cells [105]. When transduced autologous tumor cell lines cannot be obtained (because the tumor is not accessible, the tumor cells do not grow *ex vivo*, or gene delivery is difficult), an immunogenic allogeneic tumor cell line can be a valid alternative. This approach unfortunately has several limitations: (1) The tumor antigens present on the autologous tumor population may be absent in the tumor cell line; (2) the antigen may be presented on a mismatched MHC molecule and, in the absence of cross-priming of host APCs/lymphocytes, may fail to be recognized by host T cells; (3) the tumor antigen may not contain peptides capable of being presented by host APC so an immunogenic allogeneic tumor in one individual may be nonimmunogenic in a second patient with a different human leucocyte antigen (HLA) type.

Among the many immunomodulatory gene products tested to date, vaccination with irradiated cancer cells engineered to secrete GM-CSF induced the most efficient, specific, and long-lasting immunity in several murine tumor models [77]. This approach, often referred as "GVAX cancer immunotherapy," has been translated in numerous phase I/II clinical trials for therapy of several types of malignancies, including renal cell carcinoma [106,107], metastatic melanoma [108,109], prostate cancer [110], and nonsmall cell lung cancer [111]. In early studies, tumor cells were collected from the patients and engineered to produce GM-CSF using retroviral vectors. However, the transduction system was rapidly switched to Ads, as these vectors can transduce both dividing and nondividing tumor cells and guarantee higher expression of the transgene. The number of patients enrolled in each phase I/II studies ranged from 10 to 100 and single or multiple doses (from two to eight) of the vaccine were administered to the patients. Overall, treatments were well-tolerated with the most common reported toxicities being inflammation at the injection site and modest flu-like symptoms. In the great majority of the studies, the immunologic response to the vaccine was evaluated as delayed-type hypersensitivity (DTH) skin responses against irradiated autologous cancer cells and in some cases against specific tumor-associated peptides. The proportion of patients with immune responses ranged from 20% to 80% depending on the tumor type, cell dose, and number of vaccinations. The clinical activity of this approach was modest, with mixed tumor responses or tumor regression for more than 3 years in just 5% to 8% of patients.

In addition to its limited efficacy, the above approach is hindered by the complexity of the vaccine preparation. For this reason, an alternative GVAX immunotherapy has been developed in which unmanipulated autologous tumor cells are coinjected with an allogeneic bystander GM-CSF producer cell line that functions as "universal vaccine" [112]. The cell line chosen (K562) lacks HLA class I or II antigens, so reduces the probability of inducing an allogeneic T cell-mediated response. In addition, the cell line produces large amounts of GM-CSF, so that few modified cells need to be given together with autologous unmodified tumor cells. This greatly increases the feasibility of the approach for clinical application. Two clinical trials evaluated this approach in patients with multiple myeloma [113] and nonsmall cell lung cancer [114]. Although these trials confirmed feasibility and showed the induction of immune responses, the clinical impact has remained marginal.

Other clinical trials directly use GM-CSF secreting allogeneic tumor cells alone. This approach is the simplest to scale-up clinically, since it uses an "off-the-shelf" product that can be manufactured, certified, and distributed as a drug. The rationale is that common tumor antigens will be cross-presented after uptake of tumor apoptotic bodies by local professional APCs. The most advanced clinical application of allogeneic GVAX is in patients with prostate cancer using two GM-CSF modified prostate cancer cell lines. Approximately 200 patients with recurrent prostate cancer [115] received the vaccines, which were well tolerated. Activity was demonstrated by reduction of PSA levels and by production of antibodies binding the allogeneic vaccine. Allogeneic vaccination increased the survival of metastatic patients to 20 months. Based on these studies, two phase III clinical trials including almost 600 patients have been initiated [115].

IL-2 has also emerged as a potential immunomodulatory molecule alone or in combination with other cytokines or chemokines, such as LTN. Clinical trials conducted by our group at St. Jude Children's Research Hospital and Baylor College of Medicine (BCM) used neuroblastoma cells transduced with retroviral vectors so that they expressed the IL-2 gene. Both autologous and allogeneic studies were instituted. In the autologous trial, patients received up to eight injections of their own tumor cells subcutaneously. More than half the patients produced specific antibody and a specific cytotoxic T cell response directed against the autologous neuroblasts. Of 10 patients, five had clinical tumor responses including one complete and one very good partial response [116]. In the study using allogeneic vaccine, however, the immunizing cell line induced no evident specific immunity and only one patient showed a partial response [116]. Of note, in both studies a significant number of children showed good tumor responses on subsequent treatment with low-dose oral etoposide. This interaction between genetic immunotherapy and low-dose chemotherapy has subsequently been observed in a number of tumor vaccine studies [117,118], and likely represents a genuine interactivity between these treatment modalities that may usefully be exploited for therapeutic benefit in the future.

A subsequent clinical study in neuroblastoma was based on animal data showing that the combination of LTN a T cell chemokine, and IL-2, the T cell growth factor, accelerated, and augmented the immune response to a tumor cell line [100]. Patients with relapsed or refractory neuroblastoma received either an autologous or allogeneic vaccine engineered to release IL-2 and LTN. In the allogeneic group, it was possible to observe specific antitumor immune responses to the immunizing cell line, and four patients entered complete remission, which was durable in three [119]. In the autologous study, the clinical results did not appear to be measurably superior to IL-2 alone [120]. Hence, in the allogeneic setting, the combination of two agents acting at different phases of the immune response may be superior to a single agent. Confirmation will be sought in larger studies.

Tumor vaccines with engineered autologous or allogeneic tumor cells are also used in patients with acute leukemia [121]. In one study, acute myelogenous leukemia (AML) blast cells are collected at diagnosis and then patients are vaccinated with tumor cells mixed with the K562 cell line producing GM-CSF. These vaccines can be given before and after autologous stem cell transplantation, with the intent of priming the antileukemic response during immune reconstitution after stem cell transplant, followed by additional vaccination to boost the immune response. Interim analysis suggested improved clearance of minimal residual disease in vaccinated patients [121].

The CD40/CD40L pathway plays a crucial role in enhancing the APC capacity of CD40+ malignant B lymphocytes [70,71,73] and is a key molecule for activation of professional APCs. We and others are producing leukemia-vaccines engineered to express CD40L alone or in combination with IL-2. We completed a phase I study in patients with acute lymphoblastic leukemia using autologous tumor cells expressing CD40L/IL-2 [75]. Ten patients received up to six vaccine injections without adverse events, and immunological evidence of antileukemia T-cell response was detected in more than half of the patients, even when the vaccines were given in the immediate posttransplant period [75]. The great majority of patients were in clinical remission at the time of vaccination, so the clinical impact of this adjuvant treatment is unknown. The manufacturing complexities of these autologous vaccines are considerable so feasibility of large-scale studies using this approach is questionable [122]. Once again, allogeneic gene modified tumor cells may be preferable, and will be evaluated.

Chronic lymphocytic leukemia (B-CLL) is an indolent leukemia, with plentiful availability of tumor cells, and may be a better target for autologous vaccination than an acute leukemic illness. We therefore used CD40L/IL-2 gene transfer into autologous cells, to examine the immunological and clinical responses to subcutaneous injections [74,76]. Vaccination induced a specific immunologic antitumor response *in vivo* as measured by specific IFNγ release by T cells in response to autologous tumor cells and also by the production of antibodies-binding tumor cells. The clinical benefit of this approach consisted in transient reduction of the lymphadenopathy and stabilization of the disease in some patients. Of note, the immunologic effects of the vaccine were sustained during the vaccination period (from two to eight vaccinations), but ceased rapidly after the suspension of the vaccine, suggesting that prolonged vaccination schedule is likely required to produce a

prolonged immune response and to generate clinical benefits [74]. In another study, B-CLL patients received autologous tumor vaccines expressing murine CD40L: this vaccine was administered i.v. [76]. A proportion of treated patients had reduced tumor burden and disease stabilization, but it is likely that these antitumor effects were mediated by direct apoptosis of the endogenous B-CLL cells due to the engagement of CD40L [123] rather than to indirect induction of an anti-B-CLL immune response.

In summary, genetic modification of tumor cells appears safe and is capable of generating specific humoral and cellular antitumor cytotoxic responses. There have been at least some tumor regressions and the approach is now being evaluated in a wider range of tumors and in a larger number of patients.

47.5.5.2 Cancer Therapy with Gene-Modified T Cells

Several studies have suggested the feasibility and apparent clinical efficacy of adoptive transfer of CTLs directed at tumor antigens [84,124,125]. By using gene-marked cells in these studies, it has not only been possible to determine the survival and homing of the infused T cells, but also to determine if they mediate adverse effects such as GvHD [124,126, 127]. Clinical studies and their results are described elsewhere in this book (see Chapter 44), and include Epstein–Barr virus (EBV)-associated posttransplant lymphoproliferative disorder (PTLD) [128–132], Hodgkin's disease (HD) [133,134], and nasopharyngeal carcinoma (NPC) [135–137].

Protecting T cells against tumor-induced downregulation. Clinical studies using CTL against EBV-related malignancies including PTLD, HD, and nasopharyngeal cancer NPC are promising and there have been complete and sustained tumor responses, particularly in patients with PTLD. Nonetheless, a significant fraction of patients with HD or NPC have incomplete responses after T cell therapy. This may be due to a lack of specificity of the EBV-specific CTL for the immunosubdominant LMP1 and LMP2 antigens that are all present on the HD/NPC tumor cells. In addition, these tumors secrete immunosuppressive cytokines and chemokines, which affect CTL function [138].

Gene transfer can be used to overcome both types of problems. By using DCs transduced with adenoviral vectors encoding either LMP2 or a mutated LMP1, it is possible to generate CTL that have high cytolytic activity *in vitro* to LMP2- or LMP1-positive targets when compared with conventional EBV-CTL [139,140]. We recently completed a clinical trial in which DCs were genetically modified to overexpress the weak tumor associated antigen LMP-2. Coculture of these gene-modified DC with patient T cells produced a high proportion of CTLs with specificity for the LMP-2 antigen [141]. Following adoptive transfer, these LMP-2-enriched CTL lines were safe, and 9/10 patients treated in remission of high-risk disease remained in remission, while 5/6 with active relapsed disease had a tumor response, which was complete in four and sustained >9 months [141].

Although LMP2- or LMP1-specific CTL may be more effective, there is a concern that they will remain vulnerable to the immunosuppressive factors secreted by the Hodgkin's Reed–Sternberg cell. The cytokine that has the most devastating effects on CTL proliferation and function is TGF-β [138]. This cytokine is secreted by a wide variety of tumors and allows the tumor to escape the immune response [138]. To overcome this capacity to inhibit the EBV-CTL, our group transduced CTL from patients with relapsed EBV-positive HD with a retrovirus vector expressing a dominant-negative TGF-β type II receptor (DNR). This prevents formation of the functional trimeric receptor. Cytotoxicity, proliferation, and cytokine release assays showed that exogenous TGF-β had minimal inhibitory effects on DNR-transduced CTLs [142]. This approach will be tested in a phase I clinical trial that has been recently approved by Food and Drug Administration.

Hodgkin's lymphoma uses multiple mechanisms to escape the immune recognition. Hence a combination of strategies will likely be required to overcome the tumor resistance. For example, Reed-Stenberg cells not only have abnormally high expression of Fas-L [143], which induces premature apoptosis of CTLs, they also secrete the chemokine Thymus- and activation-regulated chemokine (TARC) [144], which attracts T lymphocytes with Th2 properties [138] and Treg [145]. Transfer of a specific siRNA that knocks-down the Fas receptor in CTLs makes them resistant to Fas-L-mediated apoptosis [25], while transgenic IL-12 makes them resistant to the hostile Th2 environment [99]. Moreover, enhanced CTL proliferation in the tumor microenvironment through the autocrine production of cytokine (IL-2 or IL-15) further augments their antitumor effect [86–88], suggesting that transfer of these cytokine genes may overcome multiple mechanisms of resistance.

αβTCR transfer for tumor therapy. The generation of CTLs recognizing a broad spectrum of viral-derived-tumor associated antigens is feasible and can be scaled up to obtain sufficient cell numbers for adoptive T cell transfer in patients. In contrast, the reactivation and expansion *ex vivo* of CTLs specific for epitopes derived from the great majority of tumor-associated antigens, including cancer testis antigens and epitopes derived from fusion oncogenic proteins, is suboptimal for translation in clinical trials.

We can overcome this limitation by transferring the TCR genes (αβTCR) that recognize a specific tumor-associated epitope [146–148]. The proof-of-principle for this technology has been recently demonstrated by Rosenberg et al. in patients with metastatic melanoma. In this trial, polyclonal T cells were genetically modified to express an αβTCR specific for MART-1. The adoptive transfer of these cells induced objective tumor regression in 20% of the patients [149]. Currently, this technology has several limitations: (1) MHC restriction of transgenic αβTCR means the approach is feasible only in patients with a specific HLA type; (2) tumor cells that downregulate MHC will escape lysis; (3) cross-pairing of the transgenic αβTCR with the native αβTCR produces chimeric T cells with unknown specificity; (4) escape mutants rapidly emerge since the T cells target just one antigenic epitope.

Several strategies have been proposed to overcome these limitations. For example, αβTCR crosspairing can be minimized by modifications in αβTCR construction [150]. In addition, the selection of tumor-antigens crucial for tumor survival, and the combination of adoptive immunotherapy with other conventional treatments, may limit the possibility of tumor escape [151].

Chimeric T cells for tumor therapy. T lymphocytes can also be modified to express chimeric antigen receptors (CARs) derived from antibodies. If these CAR are specific for tumor antigens expressed on the cell surface, they may have considerable therapeutic potential [152–154]. CARs allow us to extend the recognition specificity of T lymphocytes beyond classical T cell epitopes by transducing cells with genes that encode the variable domain of an antigen-specific monoclonal antibody (MAb) single-chain fragment (ScFv) joined to a cytoplasmic signaling domain suce ζ- or γ-chain of the CD3αβTCR complex. This strategy can therefore be applied to every malignancy that expresses a tumor-associated antigen for which an MAb exists. CAR-transduced T cells potentially have numerous advantages over immunotherapies based on monoclonal antibodies or T lymphocytes alone: (i) Because there is no need to select and expand tumor-specific antigens from scanty precursors, large populations of antigen-redirected T lymphocytes can be obtained in a matter of weeks. (ii) Since CARs are MHC unrestricted, tumor escape by downregulation of HLA class I molecules or defects in antigen processing are bypassed. (iii) Because both CD4+ and CD8+ T cells can express the same CAR, the full network of T cell function is directed against tumor cells [155]. (iv) The presence of CAR-mediated effector function may be more likely to produce tumor cell lysis than humoral immune responses alone since the perforin/granzyme killing mechanism may be effective against cells that are relatively resistant to antibody and complement, whereas cytokine secretion upon T cell activation by tumor antigen recruits additional components of the immune system, amplifying the antitumor immune response. (v) Unlike intact antibodies, T cells can migrate through microvascular walls, and extravasate and penetrate the core of solid tumors to exert their cytolytic activity. (vi) Finally, a single T lymphocyte can sequentially kill a multiplicity of target cells.

Human T lymphocytes genetically engineered to express these CAR genes have exhibited specific lysis via the perforin/granzyme pathways, as well as cytokine secretion upon exposure to tumor cells expressing the cognate target antigen [156]. Examples of clinical applications to tumor treatment are given in Table 47.4 [155–160].

Although adoptively transferred chimeric receptor-transduced cells are protective in murine tumor models [155,156,161–163], few clinical trials have yet been reported. In human immunodeficiency virus (HIV) patients, T cells

TABLE 47.4

Examples of Different Targets of Chimeric TCR Constructed with scFvs

Target	Comment
GD$_2$	GD$_2$ expressed on melanomas and neuroblastomas
GD$_3$	GU$_3$ expressed on melanoma cells
HMW-MAA	High-molecular-weight melanoma-associated antigen
CII	Directed against type II collagen—use T cells as gene carriers in rheumatoid arthritis
EGP40	Colon cancer-associated antigen
EGP-2	Derived from MAb GA733.2, which binds the epithelial glycoprotein 2 protein, which is overexpressed on a variety of human carcinomas
CD30	Malignant cell population in Hodgkin's lymphoma expresses high amounts of this cell surface antigen
CEA	Anti-CEA chimeric TCR with scFv derived from the MAb BW431/26
ErbB2	ErbB2 is a type I growth factor receptor overexpressed in a high percentage of human adenocarcinomas
FR	Folate receptor is expressed on most ovarian carcinomas and some types of brain tumors
Neu/HER2	Human adenocarcinoma-associated growth factor receptor
κ-light chain of immunoglobulins	Directed against mature B-cell derived malignancies

were transduced to express a CAR binding the HIVgp120 envelop protein in order to kill HIV-infected cells [164]. CAR-T cells homed to the HIV infected tissues after adoptive transfer, but the antiviral benefits of this treatment are limited [164]. Recently, phase I studies have been conducted in cancer patients with neuroblastoma [165] and ovarian cancer [166], showing the feasibility of the approach, but also the lack of objective antitumor effects. These trials also suggested that the limited effects may be attributable to lack of persistence of the transgenic T cells.

Costimulatory signals are crucial for T cell function, since αβTCR stimulation in absence of costimulation induces T cell anergy. To enhance function and persistence, the second generation of CARs contains costimulatory endodomains (CD28, OX40, or 4–1BB), to ensure the transgenic T cells are fully activated after their encounter with their specific target [155,167–170]. In preclinical studies, T cells expressing CARs encoding costimulatory endodomains exhibit potent antitumor activity and secrete significant amounts of IL-2, enhancing their persistence *in vivo* [155,171]. Clinical studies have begun.

An alternative strategy is to incorporate the CAR in CTLs with well-defined specificity in their native antigen receptor [163,172,173]. These (usually) viral-specific CTLs can be repeatedly boosted and activated *in vivo* by the engagement of their native αβTCR with viral epitopes on professional APCs. By providing all necessary costimulatory signals, this strategy should maintain the pool of transgenic

CTLs, allowing the CAR to redirect the activated cells to tumor cells (Figure 47.1). This principle has been tested in preclinical studies [163,172,173]. A current clinical trial at BCM gives patients with relapsed neuroblastoma both polyclonal T cells and EBV-specific CTLs that are genetically modified to express two distinguishable but functionally identical CARs that target the G(D2a) antigen. This trial will show whether costimulation of G(D2) + EBV-specific CTL by latently infected B-lymphocytes in EBV-seropositive patients can increase their persistence compared to polyclonal G(D2) + T cells. If this hypothesis is confirmed, the potential use of an "EBV-based vaccine" to boost *in vivo* the CTLs may further enhance their persistence. Other investigators have evaluated similar approaches using CMV-specific [174] or allo-specific CTL [175].

47.5.6 CLINICAL VACCINES USING DENDRITIC CELLS

To overcome the antigen presentation defects of tumor cells, DCs can themselves be manipulated *ex vivo* and used as cancer vaccines, mainly acting by priming naive T cells [176]. Several clinically applicable methods are now available to isolate and expand DCs from peripheral blood and bone marrow. Once isolated, DCs can be induced to present tumor antigens by several strategies, including feeding with tumor cell lysates and apoptotic bodies or by using tumor-derived RNA or making DC–tumor cell hybrids. Specific tumor

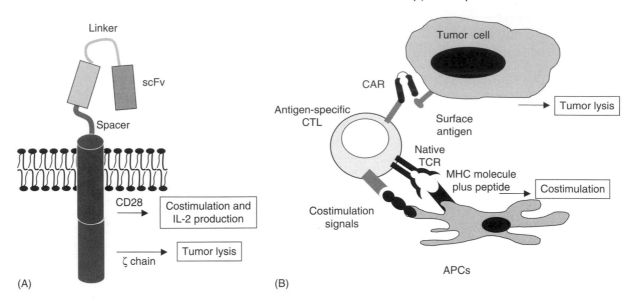

FIGURE 47.1 Strategies to express functional CARs in T cells. (A) Polyclonal T cells modified to express CAR incorporating a costimulatory endodomain (CD28) to provide full activation after engagement with the antigen. (B) CARs engrafted on antigen specific CTLs. CAR provides the antitumor effects and the native αβTCR provides appropriate costimulation by engagement with specific antigens expressed on professional APCs.

antigen gene transfer into DC using viral or nonviral vectors is also possible [177].

Several clinical studies using active immunization with DCs have been completed in patients with metastatic melanoma, metastatic renal cell carcinoma, and B cell lymphoma. Most of them have used antigen-pulsed DC [178,179], but some investigators have reported induction of immune responses using gene-modified DC [180].

In the past few years, genetic modification of DC has been directed not only to load these cells with specific antigens but also to manipulate specific pathways that can increase their effects after vaccination. In particular the down regulation of inhibitory molecules such as SOCS-1 [24] or the prolonged activation of the CD40 pathway [181] significantly improved the antitumor effects of DC vaccine in mouse models. Both these strategies are now approaching clinical validation.

47.5.7 NUCLEIC ACID VACCINES

Nucleic acid vaccines induce an immune response targeted against a protein expressed *in vivo* subsequent to the administration of its encoding DNA or RNA. Investigators demonstrated that the intradermal and intramuscular injections of polynucleotide products generate long-lasting T cell and humoral immunity [182].

Nucleotide-based vaccines have several advantages over proteins and peptides. They provide prolonged antigen expression that can continuously stimulate the immune system, probably through an intracellular antigenic reservoir, resistant to antibody-mediated clearance. This may favor induction of immune memory even in the absence of booster immunization. Codelivery with plasmids encoding

cytokines or costimulatory molecules can further enhance the immune response [183]. Moreover, nucleotide vaccination leads to antigen processing through both the endogenous and exogenous pathways, so that specific CTL and helper T cells can be recruited.

In a murine model of B cell lymphoma, mice immunized with DNA constructs encoding the idiotype had specific antibody responses and were protected against tumor challenge. Use of DNA encoding a fusion protein of idiotype and GM-CSF greatly improved vaccine efficacy [178]. Improvement of vaccine efficacy was also observed when the idiotypic DNA was fused to the C fragment of tetanus toxoid [184]. These preliminary results indicate that DNA may be a simple and efficacious system of inducing immune responses against a weak tumor antigen, provided that additional stimuli are included with the DNA (e.g., GM-CSF or tetanus toxoid).

47.5.8 FUTURE PERSPECTIVES FOR GENETIC IMMUNOTHERAPY IN CANCERS

Tumors possess multiple and powerful strategies to escape immune surveillance and can develop new immune evasion mechanisms during disease progression. Many of these stratagems have been elucidated, and counterattack can now be planned thanks to major improvements in protein engineering, gene transfer, and cell therapy technologies. Why then have few major clinical success yet been reported? Two centuries past, in 1893, William Coley reported regression of tissue sarcoma in 10% of patients after they had been infected with live or heat-inactivated bacteria. Since then, clinical response rates after immunotherapy have not convincingly changed. Hence, our nightmare is that this is as good as it ever gets. But the reality is that, despite lack of concrete clinical progress, our increasing knowledge and skills offer a real promise of improved results in the near future. It is also clear that immunotherapy by itself is unlikely to be sufficient to eradicate tumors because human malignancies are too heterogeneous in terms of the antigens they express and their susceptibility to immune-mediated killing. Only integration with standard treatments will likely be successful. Active immunization will be used as adjuvant therapy to eradicate minimal residual disease in patients previously debulked of their tumor by surgery, chemoradiotherapy, or stem cell transplantation. Patients receiving allogeneic stem cell transplant represent an obvious paradigm. After transplantation, when a new immune system is reconstituting, cancer vaccines could induce the formation of potent and long-lasting immunity. Specific cytotoxic lymphocytes from the donor could also be expanded *ex vivo* in the presence of host malignant cells and administered after transplant should malignancy recur.

Although no "breakthrough" clinical success has been reported, a better understanding of immune evasion strategies and the availability of improved technologies of immune manipulation have opened the way for real immunothérapies of cancer that should ultimately deliver on the dreams of the nineteenth and twentieth centuries.

47.6 CONCLUSION

We have far to go before gene therapy of malignancy can truly be said to have made a major impact on these diseases. Nonetheless, over the past decade, new techniques including gene transfer and immunotherapy have produced unequivocal tumor responses even in advanced disease. Improvement of gene transfer for the treatment of cancer certainly relies on four major steps: (1) simplification of gene transfer protocols, still too complex to implement in a clinical environment; (2) controllable, tissue-specific regulation of transgene expression; (3) progress in the understanding of carcinogenesis mechanisms to improve therapeutic strategies; (4) improvement in the methodology of clinical trials, including optimal choice of the patient population, and monitoring of tumor and immune responses, within the tight frame of regulatory and cost-related issues. As we continue to make incremental advances in the application of these approaches, we can expect to see gene therapy increasingly supplement probably long before they can eventually supplant conventional cancer therapeutics.

REFERENCES

1. Kokunai T, Kawamura A, and Tamaki N. Induction of differentiation by wild-type p53 gene in a human glioma cell line. *J Neurooncol*. 1997;32:125–133.
2. Wang J, Bucana CD, Roth JA, and Zhang WW. Apoptosis induced in human osteosarcoma cells is one of the mechanisms for the cytocidal effect of Ad5CMV-p53. *Cancer Gene Ther*. 1995;2:9–17.
3. Gomez-Manzano C, Fueyo J, Kyritsis AP et al. Adenovirus-mediated transfer of the p53 gene produces rapid and generalized death of human glioma cells via apoptosis. *Cancer Res*. 1996;56:694–699.
4. Cai DW, Mukhopadhyay T, Liu Y, Fujiwara T, and Roth JA. Stable expression of the wild-type p53 gene in human lung cancer cells after retrovirus-mediated gene transfer. *Hum Gene Ther*. 1993;4:617–624.
5. Habib NA, Ding SF, el Masry R et al. Preliminary report: The short-term effects of direct p53 DNA injection in primary hepatocellular carcinomas. *Cancer Detect Prev*. 1996;20:103–107.
6. Nguyen DM, Wiehle SA, Koch PE et al. Delivery of the p53 tumor suppressor gene into lung cancer cells by an adenovirus/DNA complex. *Cancer Gene Ther*. 1997;4:191–198.
7. Buller RE, Runnebaum IB, Karlan BY et al. A phase I/II trial of rAd/p53 (SCH 58500) gene replacement in recurrent ovarian cancer. *Cancer Gene Ther*. 2002;9:553–566.
8. Lamont JP, Nemunaitis J, Kuhn JA, Landers SA, and McCarty TM. A prospective phase II trial of ONYX-015 adenovirus and chemotherapy in recurrent squamous cell carcinoma of the head and neck (the Baylor experience). *Ann Surg Oncol*. 2000;7:588–592.
9. Clayman GL, El Naggar AK, Lippman SM et al. Adenovirus-mediated p53 gene transfer in patients with advanced recurrent head and neck squamous cell carcinoma. *J Clin Oncol*. 1998;16:2221–2232.
10. Fujiwara T, Tanaka N, Kanazawa S et al. Multicenter phase I study of repeated intratumoral delivery of adenoviral p53 in patients with advanced non-small cell lung cancer. *J Clin Oncol*. 2006;24:1689–1699.
11. Rocco JW, Li D, Liggett WH, Jr. et al. p16INK4A adenovirus-mediated gene therapy for human head and neck squamous cell cancer. *Clin Cancer Res*. 1998;4:1697–1704.
12. Gilliland DG. Molecular genetics of human leukemias: New insights into therapy. *Semin Hematol*. 2002;39:6–11.
13. Thomadaki H and Scorilas A. BCL2 family of apoptosis-related genes: Functions and clinical implications in cancer. *Crit Rev Clin Lab Sci*. 2006;43:1–67.
14. Marshall MS. The effector interactions of p21ras. *Trends Biochem Sci*. 1993;18:250–254.
15. Mukhopadhyay T, Tainsky M, Cavender AC, and Roth JA. Specific inhibition of K-ras expression and tumorigenicity of lung cancer cells by antisense RNA. *Cancer Res*. 1991; 51:1744–1748.
16. Georges RN, Mukhopadhyay T, Zhang Y, Yen N, and Roth JA. Prevention of orthotopic human lung cancer growth by intratracheal instillation of a retroviral antisense K-ras construct. *Cancer Res*. 1993;53:1743–1746.
17. Arteaga CL and Holt JT. Tissue-targeted antisense c-fos retroviral vector inhibits established breast cancer xenografts in nude mice. *Cancer Res*. 1996;56:1098–1103.
18. Kashani-Sabet M and Scanlon KJ. Application of ribozymes to cancer gene therapy. *Cancer Gene Ther*. 1995;2:213–223.
19. Komata T, Kondo Y, Koga S et al. Combination therapy of malignant glioma cells with 2–5A-antisense telomerase RNA and recombinant adenovirus p53. *Gene Ther*. 2000;7:2071–2079.
20. Richardson JH and Marasco WA. Intracellular antibodies: Development and therapeutic potential. *Trends Biotechnol*. 1995;13:306–310.
21. Brummelkamp TR, Bernards R, and Agami R. A system for stable expression of short interfering RNAs in mammalian cells. *Science*. 2002;296:550–553.

22. Schomber T, Kalberer CP, Wodnar-Filipowicz A, and Skoda RC. Gene silencing by lentivirus-mediated delivery of siRNA in human CD34 + cells. *Blood.* 2004;103:4511–4513.

23. Landen CN, Jr., Chavez-Reyes A, Bucana C et al. Therapeutic EphA2 gene targeting in vivo using neutral liposomal small interfering RNA delivery. *Cancer Res.* 2005;65:6910–6918.

24. Shen L, Evel-Kabler K, Strube R, and Chen SY. Silencing of SOCS1 enhances antigen presentation by dendritic cells and antigen-specific anti-tumor immunity. *Nat Biotechnol.* 2004;22:1546–1553.

25. Dotti G, Savoldo B, Pule M et al. Human cytotoxic T lymphocytes with reduced sensitivity to Fas-induced apoptosis. *Blood.* 2005;105:4677–4684.

26. Al Hajj M. Cancer stem cells and oncology therapeutics. *Curr Opin Oncol.* 2007;19:61–64.

27. Beltinger C, Fulda S, Kammertoens T et al. Herpes simplex virus thymidine kinase/ganciclovir-induced apoptosis involves ligand-independent death receptor aggregation and activation of caspases. *Proc Natl Acad Sci USA.* 1999;96:8699–8704.

28. Beltinger C, Uckert W, and Debatin KM. Suicide gene therapy for pediatric tumors. *J Mol Med.* 2001;78:598–612.

29. Packer RJ, Raffel C, Villablanca JG et al. Treatment of progressive or recurrent pediatric malignant supratentorial brain tumors with herpes simplex virus thymidine kinase gene vector-producer cells followed by intravenous ganciclovir administration. *J Neurosurg.* 2000;92:249–254.

30. Ram Z, Culver KW, Oshiro EM et al. Therapy of malignant brain tumors by intratumoral implantation of retroviral vector-producing cells. *Nat Med.* 1997;3:1354–1361.

31. Bansal K and Engelhard HH. Gene therapy for brain tumors. *Curr Oncol Rep.* 2000;2:463–472.

32. Rainov NG. A phase III clinical evaluation of herpes simplex virus type 1 thymidine kinase and ganciclovir gene therapy as an adjuvant to surgical resection and radiation in adults with previously untreated glioblastoma multiforme. *Hum Gene Ther.* 2000;11:2389–2401.

33. Chevez-Barrios P, Chintagumpala M, Mieler W et al. Response of retinoblastoma with vitreous tumor seeding to adenovirus-mediated delivery of thymidine kinase followed by ganciclovir. *J Clin Oncol.* 2005;23:7927–7935.

34. Nasu Y, Kusaka N, Saika T, Tsushima T, and Kumon H. Suicide gene therapy for urogenital cancer: Current outcome and prospects. *Mol Urol.* 2000;4:67–71.

35. Herman JR, Adler HL, Aguilar-Cordova E et al. In situ gene therapy for adenocarcinoma of the prostate: A phase I clinical trial. *Hum Gene Ther.* 1999;10:1239–1249.

36. Miles BJ, Shalev M, Aguilar-Cordova E et al. Prostate-specific antigen response and systemic T cell activation after in situ gene therapy in prostate cancer patients failing radiotherapy. *Hum Gene Ther.* 2001;12:1955–1967.

37. Nasu Y, Saika T, Ebara S et al. Suicide gene therapy with adenoviral delivery of HSV-tK gene for patients with local recurrence of prostate cancer after hormonal therapy. *Mol Ther.* 2007;15:834–840.

38. Mullen CA, Kilstrup M, and Blaese RM. Transfer of the bacterial gene for cytosine deaminase to mammalian cells confers lethal sensitivity to 5-fluorocytosine: A negative selection system. *Proc Natl Acad Sci USA.* 1992;89:33–37.

39. Castleden SA, Chong H, Garcia-Ribas I et al. A family of bicistronic vectors to enhance both local and systemic antitumor effects of HSVtk or cytokine expression in a murine melanoma model. *Hum Gene Ther.* 1997;8:2087–2102.

40. Palu G, Cavaggioni A, Calvi P et al. Gene therapy of glioblastoma multiforme via combined expression of suicide and cytokine genes: A pilot study in humans. *Gene Ther.* 1999;6:330–337.

41. Jones RK, Pope IM, Kinsella AR, Watson AJ, and Christmas SE. Combined suicide and granulocyte-macrophage colony-stimulating factor gene therapy induces complete tumor regression and generates antitumor immunity. *Cancer Gene Ther.* 2000;7:1519–1528.

42. Avvakumov N and Mymryk JS. New tools for the construction of replication-competent adenoviral vectors with altered E1A regulation. *J Virol Methods.* 2002;103:41–49.

43. Bonini C, Ferrari G, Verzeletti S et al. HSV-TK gene transfer into donor lymphocytes for control of allogeneic graft-versus-leukemia. *Science.* 1997;276:1719–1724.

44. Ciceri F, Bonini C, Gallo-Stampino C, and Bordignon C. Modulation of GvHD by suicide-gene transduced donor T lymphocytes: Clinical applications in mismatched transplantation. *Cytotherapy.* 2005;7:144–149.

45. Thomis DC, Marktel S, Bonini C et al. A Fas-based suicide switch in human T cells for the treatment of graft-versus-host disease. *Blood.* 2001;97:1249–1257.

46. Iuliucci JD, Oliver SD, Morley S et al. Intravenous safety and pharmacokinetics of a novel dimerizer drug, AP1903, in healthy volunteers. *J Clin Pharmacol.* 2001;41:870–879.

47. Straathof KC, Pule MA, Yotnda P et al. An inducible caspase 9 safety switch for T-cell therapy. *Blood.* 2005;105:4247–4254.

48. Tey SK, Dotti G, Rooney CM, Heslop HE, and Brenner MK. Inducible caspase 9 suicide gene to improve the safety of allodepleted T cells after haploidentical stem cell transplantation. *Biol Blood Marrow Transplant.* 2007;13:913–924.

49. Moscow JA, Huang H, Carter C et al. Engraftment of MDR1 and NeoR gene-transduced hematopoietic cells after breast cancer chemotherapy. *Blood.* 1999;94:52–61.

50. Hanania EG, Giles RE, Kavanagh J et al. Results of MDR-1 vector modification trial indicate that granulocyte/macrophage colony-forming unit cells do not contribute to posttransplant hematopoietic recovery following intensive systemic therapy. *Proc Natl Acad Sci USA.* 1996;93:15346–15351.

51. Hesdorffer C, Ayello J, Ward M et al. Phase I trial of retroviral-mediated transfer of the human MDR1 gene as marrow chemoprotection in patients undergoing high-dose chemotherapy and autologous stem-cell transplantation. *J Clin Oncol.* 1998;16:165–172.

52. Cowan KH, Moscow JA, Huang H et al. Paclitaxel chemotherapy after autologous stem-cell transplantation and engraftment of hematopoietic cells transduced with a retrovirus containing the multidrug resistance complementary DNA (MDR1) in metastatic breast cancer patients. *Clin Cancer Res.* 1999;5:1619–1628.

53. Abonour R, Williams DA, Einhorn L et al. Efficient retrovirus-mediated transfer of the multidrug resistance 1 gene into autologous human long-term repopulating hematopoietic stem cells. *Nat Med.* 2000;6:652–658.

54. Modlich U, Kustikova OS, Schmidt M et al. Leukemias following retroviral transfer of multidrug resistance 1 (MDR1) are driven by combinatorial insertional mutagenesis. *Blood.* 2005;105:4235–4246.

55. Montini E, Cesana D, Schmidt M et al. Hematopoietic stem cell gene transfer in a tumor-prone mouse model uncovers low genotoxicity of lentiviral vector integration. *Nat Biotechnol.* 2006;24:687–696.

56. Eder JP, Jr., Supko JG, Clark JW et al. Phase I clinical trial of recombinant human endostatin administered as a short intravenous infusion repeated daily. *J Clin Oncol.* 2002;20:3772–3784.

57. Thomas JP, Arzoomanian RZ, Alberti D et al. Phase I pharmacokinetic and pharmacodynamic study of recombinant human endostatin in patients with advanced solid tumors. *J Clin Oncol.* 2003;21:223–231.

58. Clamp AR and Jayson GC. The clinical potential of antiangiogenic fragments of extracellular matrix proteins. *Br J Cancer.* 2005;93:967–972.

59. Tjin Tham Sjin RM, Naspinski J, Birsner AE et al. Endostatin therapy reveals a U-shaped curve for antitumor activity. *Cancer Gene Ther.* 2006;13:619–627.

60. Feldman AL, Alexander HR, Hewitt SM et al. Effect of retroviral endostatin gene transfer on subcutaneous and intraperitoneal growth of murine tumors. *J Natl Cancer Inst.* 2001;93:1014–1020.

61. Wallich R, Bulbuc N, Hammerling GJ et al. Abrogation of metastatic properties of tumour cells by de novo expression of H-2K antigens following H-2 gene transfection. *Nature.* 1985;315:301–305.

62. Plautz GE, Yang ZY, Wu BY et al. Immunotherapy of malignancy by in vivo gene transfer into tumors. *Proc Natl Acad Sci USA.* 1993;90:4645–4649.

63. Nemunaitis J, Bohart C, Fong T et al. Phase I trial of retroviral vector-mediated interferon (IFN)-gamma gene transfer into autologous tumor cells in patients with metastatic melanoma. *Cancer Gene Ther.* 1998;5:292–300.

64. Hallermalm K, Seki K, Wei C et al. Tumor necrosis factor-alpha induces coordinated changes in major histocompatibility class I presentation pathway, resulting in increased stability of class I complexes at the cell surface. *Blood.* 2001;98:1108–1115.

65. Gimmi CD, Freeman GJ, Gribben JG, Gray G, and Nadler LM. Human T-cell clonal anergy is induced by antigen presentation in the absence of B7 costimulation. *Proc Natl Acad Sci USA.* 1993;90:6586–6590.

66. Schwartz RH. Costimulation of T lymphocytes: The role of CD28, CTLA-4, and B7/BB1 in interleukin-2 production and immunotherapy. *Cell.* 1992;71:1065–1068.

67. Chen L, Ashe S, Brady WA et al. Costimulation of antitumor immunity by the B7 counterreceptor for the T lymphocyte molecules CD28 and CTLA-4. *Cell.* 1992;71:1093–1102.

68. Van Kooten C and Banchereau J. CD40-CD40 ligand: A multifunctional receptor-ligand pair. *Adv Immunol.* 1996;61:1–77.

69. Banchereau J and Steinman RM. Dendritic cells and the control of immunity. *Nature.* 1998;392:245–252.

70. Dilloo D, Brown M, Roskrow M et al. CD40 ligand induces an antileukemia immune response in vivo. *Blood.* 1997;90:1927–1933.

71. Dotti G, Savoldo B, Takahashi S et al. Adenovector-induced expression of human-CD40-ligand (hCD40L) by multiple myeloma cells. A model for immunotherapy. *Exp Hematol.* 2001;29:952–961.

72. Dotti G, Savoldo B, Yotnda P, Rill D, and Brenner MK. Transgenic expression of CD40 ligand produces an in vivo antitumor immune response against both CD40(+) and CD40(–) plasmacytoma cells. *Blood.* 2002;100:200–207.

73. Biagi E, Yvon E, Dotti G et al. Bystander transfer of functional human CD40 ligand from gene-modified fibroblasts to B-chronic lymphocytic leukemia cells. *Hum Gene Ther.* 2003;14:545–559.

74. Biagi E, Rousseau R, Yvon E et al. Responses to human CD40 ligand/human interleukin-2 autologous cell vaccine in patients with B-cell chronic lymphocytic leukemia. *Clin Cancer Res.* 2005;11:6916–6923.

75. Rousseau RF, Biagi E, Dutour A et al. Immunotherapy of high-risk acute leukemia with a recipient (autologous) vaccine expressing transgenic human CD40L and IL-2 after chemotherapy and allogeneic stem cell transplantation. *Blood.* 2006;107:1332–1341.

76. Wierda WG, Cantwell MJ, Woods SJ et al. CD40-ligand (CD154) gene therapy for chronic lymphocytic leukemia. *Blood.* 2000;96:2917–2924.

77. Borrello I and Pardoll D. GM-CSF-based cellular vaccines: A review of the clinical experience. *Cytokine Growth Factor Rev.* 2002;13:185–193.

78. Igney FH and Krammer PH. Immune escape of tumors: Apoptosis resistance and tumor counterattack. *J Leukoc Biol.* 2002;71:907–920.

79. Shiraki K, Tsuji N, Shioda T, Isselbacher KJ, and Takahashi H. Expression of Fas ligand in liver metastases of human colonic adenocarcinomas. *Proc Natl Acad Sci USA.* 1997; 94:6420–6425.

80. O'Connell J, O'Sullivan GC, Collins JK, and Shanahan F. The Fas counterattack: Fas-mediated T cell killing by colon cancer cells expressing Fas ligand. *J Exp Med.* 1996;184:1075–1082.

81. Giezeman-Smits KM, Okada H, Brissette-Storkus CS et al. Cytokine gene therapy of gliomas: Induction of reactive CD4 + T cells by interleukin-4-transfected 9L gliosarcoma is essential for protective immunity. *Cancer Res.* 2000;60:2449–2457.

82. Ascher AL, Mule JJ, Kasid A et al. Murine tumor cells transduced with the gene for tumor necrosis factor-alpha. Evidence for paracrine immune effects of tumor necrosis factor against tumors. *J Immunol.* 1991;146:3227–3234.

83. Hock H, Dorsch M, Diamantstein T, and Blankenstein T. Interleukin 7 induces CD4 + T cell-dependent tumor rejection. *J Exp Med.* 1991;174:1291–1298.

84. Dudley ME and Rosenberg SA. Adoptive-cell-transfer therapy for the treatment of patients with cancer. *Nat Rev Cancer.* 2003;3:666–675.

85. Ahmadzadeh M and Rosenberg SA. IL-2 administration increases CD4 + CD25(hi) Foxp3 + regulatory T cells in cancer patients. *Blood.* 2006;107:2409–2414.

86. Liu K and Rosenberg SA. Transduction of an IL-2 gene into human melanoma-reactive lymphocytes results in their continued growth in the absence of exogenous IL-2 and maintenance of specific antitumor activity. *J Immunol.* 2001;167:6356–6365.

87. Hsu C, Hughes MS, Zheng Z et al. Primary human T lymphocytes engineered with a codon-optimized IL-15 gene resist cytokine withdrawal-induced apoptosis and persist long-term in the absence of exogenous cytokine. *J Immunol.* 2005;175:7226–7234.

88. Quintarelli C, Vera JF, Savoldo B et al. Co-expression of cytokine and suicide genes to enhance the activity and safety of tumor-specific cytotoxic T lymphocytes. *Blood.* 2007;110:2793–2802.

89. Hsu C, Jones SA, Cohen CJ et al. Cytokine-independent growth and clonal expansion of a primary human CD8 + T-cell clone following retroviral transduction with the IL-15 gene. *Blood.* 2007;109:5168–5177.

90. Quintarelli C, Vera JF, Savoldo B et al. Co-expression of cytokine and suicide genes to enhance the activity and safety of tumor specific cytotoxic T lymphocytes. *Blood.* 2007;110:2793–2802.

91. Sigal RK, Lieberman MD, Reynolds JV et al. Low-dose interferon gamma renders neuroblastoma more susceptible to interleukin-2 immunotherapy. *J Pediatr Surg.* 1991;26:389–395.

92. Richards JM. Therapeutic uses of interleukin-2 and lymphokine-activated killer (LAK) cells. *Blood Rev.* 1989;3:110–119.

93. Bowman L, Grossmann M, Rill D et al. IL-2 adenovector-transduced autologous tumor cells induce antitumor immune responses in patients with neuroblastoma. *Blood.* 1998;92:1941–1949.

94. Brittenden J, Heys SD, Ross J, and Eremin O. Natural killer cells and cancer. *Cancer.* 1996;77:1226–1243.

95. Colombo MP and Trinchieri G. Interleukin-12 in anti-tumor immunity and immunotherapy. *Cytokine Growth Factor Rev.* 2002;13:155–168.

96. Leonard JP, Sherman ML, Fisher GL et al. Effects of single-dose interleukin-12 exposure on interleukin-12-associated toxicity and interferon-gamma production. *Blood.* 1997;90:2541–2548.

97. Sangro B, Melero I, Qian C, and Prieto J. Gene therapy of cancer based on interleukin 12. *Curr Gene Ther.* 2005;5:573–581.

98. Melero I, Duarte M, Ruiz J et al. Intratumoral injection of bone-marrow derived dendritic cells engineered to produce interleukin-12 induces complete regression of established murine transplantable colon adenocarcinomas. *Gene Ther.* 1999;6:1779–1784.

99. Wagner HJ, Bollard CM, Vigouroux S et al. A strategy for treatment of Epstein–Barr virus-positive Hodgkin's disease by targeting interleukin 12 to the tumor environment using tumor antigen-specific T cells. *Cancer Gene Ther.* 2004;11:81–91.

100. Dilloo D, Bacon K, Holden W et al. Combined chemokine and cytokine gene transfer enhances antitumor immunity. *Nat Med.* 1996;2:1090–1095.

101. Qin Z, Noffz G, Mohaupt M, and Blankenstein T. Interleukin-10 prevents dendritic cell accumulation and vaccination with granulocyte-macrophage colony-stimulating factor gene-modified tumor cells. *J Immunol.* 1997;159:770–776.

102. Davidoff AM, Nathwani AC, Spurbeck WW et al. rAAV-mediated long-term liver-generated expression of an angiogenesis inhibitor can restrict renal tumor growth in mice. *Cancer Res.* 2002;62:3077–3083.

103. Miyawaki T, Uehara T, Nibu R et al. Differential expression of apoptosis-related Fas antigen on lymphocyte subpopulations in human peripheral blood. *J Immunol.* 1992;149:3753–3758.

104. Bonfoco E, Stuart PM, Brunner T et al. Inducible nonlymphoid expression of Fas ligand is responsible for superantigen-induced peripheral deletion of T cells. *Immunity.* 1998;9:711–720.

105. Biagi E, Rousseau RF, Yvon E et al. Cancer vaccines: Dream, reality, or nightmare? *Clin Exp Med.* 2002;2:109–118.

106. Tani K, Azuma M, Nakazaki Y et al. Phase I study of autologous tumor vaccines transduced with the GM-CSF gene in four patients with stage IV renal cell cancer in Japan: Clinical and immunological findings. *Mol Ther.* 2004;10:799–816.

107. Simons JW, Jaffee EM, Weber CE et al. Bioactivity of autologous irradiated renal cell carcinoma vaccines generated by ex vivo granulocyte-macrophage colony-stimulating factor gene transfer. *Cancer Res.* 1997;57:1537–1546.

108. Luiten RM, Kueter EW, Mooi W et al. Immunogenicity, including vitiligo, and feasibility of vaccination with autologous GM-CSF-transduced tumor cells in metastatic melanoma patients. *J Clin Oncol.* 2005;23:8978–8991.

109. Soiffer R, Hodi FS, Haluska F et al. Vaccination with irradiated, autologous melanoma cells engineered to secrete granulocyte-macrophagecolony-stimulating factor by adenoviral-mediated gene transfer augments antitumor immunity in patients with metastatic melanoma. *J Clin Oncol.* 2003;21:3343–3350.

110. Simons JW, Mikhak B, Chang JF et al. Induction of immunity to prostate cancer antigens: Results of a clinical trial of vaccination with irradiated autologous prostate tumor cells engineered to secrete granulocyte-macrophage colony-stimulating factor using ex vivo gene transfer. *Cancer Res.* 1999;59:5160–5168.

111. Salgia R, Lynch T, Skarin A et al. Vaccination with irradiated autologous tumor cells engineered to secrete granulocyte-macrophage colony-stimulating factor augments antitumor immunity in some patients with metastatic non-small-cell lung carcinoma. *J Clin Oncol.* 2003;21:624–630.

112. Borrello I, Sotomayor EM, Cooke S, and Levitsky HI. A universal granulocyte-macrophage colony-stimulating factor-producing bystander cell line for use in the formulation of autologous tumor cell-based vaccines. *Hum Gene Ther.* 1999;10:1983–1991.

113. Borrello I, Biedrzycki B, and Sheetsd N. Autologous tumor combined with a GM-CSF-secreting cell line vaccine (GVAX@) following autologous stem cell transplant (ASCT) in multiple myeloma [abstract]. *Blood.* 2007;104:129a.

114. Nemunaitis J, Jahan T, Ross H et al. Phase 1/2 trial of autologous tumor mixed with an allogeneic GVAX vaccine in advanced-stage non-small-cell lung cancer. *Cancer Gene Ther.* 2006;13:555–562.

115. Hege KM, Jooss K, and Pardoll D. GM-CSF gene-modifed cancer cell immunotherapies: Of mice and men. *Int Rev Immunol.* 2006;25:321–352.

116. Bowman LC, Grossmann M, Rill D et al. Interleukin-2 gene-modified allogeneic tumor cells for treatment of relapsed neuroblastoma. *Hum Gene Ther.* 1998;9:1303–1311.

117. Ratto GB, Cafferata MA, Scolaro T et al. Phase II study of combined immunotherapy, chemotherapy, and radiotherapy in the postoperative treatment of advanced non-small-cell lung cancer. *J Immunother.* 2000;23:161–167.

118. Klingebiel T, Bader P, Bares R et al. Treatment of neuroblastoma stage 4 with 131I-meta-iodo-benzylguanidine, high-dose chemotherapy and immunotherapy. A pilot study. *Eur J Cancer.* 1998;34:1398–1402.

119. Rousseau RF, Haight AE, Hirschmann-Jax C et al. Local and systemic effects of an allogeneic tumor cell vaccine combining transgenic human lymphotactin with interleukin-2 in patients with advanced or refractory neuroblastoma. *Blood.* 2003;101:1718–1726.

120. Russell HV, Strother D, Mei Z et al. Phase I trial of vaccination with autologous neuroblastoma tumor cells genetically modified to secrete IL-2 and lymphotactin. *J Immunother.* 2007;30:227–233.

121. DeAngelo D, Alyea E, and Borrello I. Posttransplant immunotherapy with a GM-CSF-based tumor vaccine (GVAX@) following autologous stem cell transplant (ASCT) in acute myeloid leukemia [abstract]. *Blood.* 2007;104:129a.

122. Haining WN, Cardoso AA, Keczkemethy HL et al. Failure to define window of time for autologous tumor vaccination in patients with newly diagnosed or relapsed acute lymphoblastic leukemia. *Exp Hematol.* 2005;33:286–294.

123. Chu P, Deforce D, Pedersen IM et al. Latent sensitivity to Fas-mediated apoptosis after CD40 ligation may explain activity of CD154 gene therapy in chronic lymphocytic leukemia. *Proc Natl Acad Sci USA.* 2002;99:3854–3859.

124. Rooney CM, Smith CA, Ng CY et al. Use of gene-modified virus-specific T lymphocytes to control Epstein–Barr-virus-related lymphoproliferation. *Lancet.* 1995;345:9–13.

125. Dudley ME, Wunderlich JR, Robbins PF et al. Cancer regression and autoimmunity in patients after clonal repopulation with antitumor lymphocytes. *Science.* 2002;298:850–854.

126. Rosenberg SA, Aebersold P, Cornetta K et al. Gene transfer into humans—Immunotherapy of patients with advanced melanoma, using tumor-infiltrating lymphocytes modified by retroviral gene transduction. *N Engl J Med.* 1990;323:570–578.

127. Heslop HE, Ng CY, Li C et al. Long-term restoration of immunity against Epstein–Barr virus infection by adoptive transfer of gene-modified virus-specific T lymphocytes. *Nat Med.* 1996;2:551–555.

128. Rooney CM, Smith CA, Ng CY et al. Infusion of cytotoxic T cells for the prevention and treatment of Epstein–Barr virus-induced lymphoma in allogeneic transplant recipients. *Blood.* 1998;92:1549–1555.

129. Savoldo B, Goss JA, Hammer MM et al. Treatment of solid organ transplant recipients with autologous Epstein–Barr virus-specific cytotoxic T lymphocytes (CTLs). *Blood.* 2006;108:2942–2949.

130. Comoli P, Labirio M, Basso S et al. Infusion of autologous Epstein–Barr virus (EBV)-specific cytotoxic T cells for prevention of EBV-related lymphoproliferative disorder in solid organ transplant recipients with evidence of active virus replication. *Blood.* 2002;99:2592–2598.

131. Gustafsson A, Levitsky V, Zou JZ et al. Epstein–Barr virus (EBV) load in bone marrow transplant recipients at risk to develop posttransplant lymphoproliferative disease: Prophylactic infusion of EBV-specific cytotoxic T cells. *Blood.* 2000;95:807–814.

132. Dotti G, Heslop H, and Rooney C. Epstein–Barr virus cell based therapy. In: Robertson ES, ed. *Epstein Barr Virus.* Academic Press, London; 2005:669–690.

133. Roskrow MA, Suzuki N, Gan Y et al. Epstein–Barr virus (EBV)-specific cytotoxic T lymphocytes for the treatment of patients with EBV-positive relapsed Hodgkin's disease. *Blood.* 1998;91:2925–2934.

134. Bollard CM, Aguilar L, Straathof KC et al. Cytotoxic T lymphocyte therapy for Epstein–Barr virus+ Hodgkin's disease. *J Exp Med.* 2004;200:1623–1633.

135. Chua D, Huang J, Zheng B et al. Adoptive transfer of autologous Epstein–Barr virus-specific cytotoxic T cells for nasopharyngeal carcinoma. *Int J Cancer.* 2001;94:73–80.

136. Straathof KC, Bollard CM, Popat U et al. Treatment of nasopharyngeal carcinoma with Epstein–Barr virus–specific T lymphocytes. *Blood.* 2005;105:1898–1904.

137. Comoli P, Pedrazzoli P, Maccario R et al. Cell therapy of stage IV nasopharyngeal carcinoma with autologous Epstein–Barr virus-targeted cytotoxic T lymphocytes. *J Clin Oncol.* 2005;23:8942–8949.

138. Poppema S, Potters M, Visser L, and van den Berg AM. Immune escape mechanisms in Hodgkin's disease. *Ann Oncol.* 1998;9 Suppl 5:S21–S24.

139. Gahn B, Siller-Lopez F, Pirooz AD et al. Adenoviral gene transfer into dendritic cells efficiently amplifies the immune response to LMP2A antigen: A potential treatment strategy for Epstein–Barr virus—Positive Hodgkin's lymphoma. *Int J Cancer.* 2001;93:706–713.

140. Gottschalk S, Edwards OL, Sili U et al. Generating CTLs against the subdominant Epstein–Barr virus LMP1 antigen for the adoptive immunotherapy of EBV-associated malignancies. *Blood.* 2003;101:1905–1912.

141. Bollard CM, Gottschalk S, Leen AM et al. Complete responses of relapsed lymphoma following genetic modification of tumor-antigen presenting cells and T-lymphocyte transfer. *Blood.* 2007;110:2838–2845.

142. Bollard CM, Rossig C, Calonge MJ et al. Adapting a transforming growth factor beta-related tumor protection strategy to enhance antitumor immunity. *Blood.* 2002;99:3179–3187.

143. Verbeke CS, Wenthe U, Grobholz R, and Zentgraf H. Fas ligand expression in Hodgkin lymphoma. *Am J Surg Pathol.* 2001;25:388–394.

144. van den BA, Visser L, and Poppema S. High expression of the CC chemokine TARC in Reed-Sternberg cells. A possible explanation for the characteristic T-cell infiltrate in Hodgkin's lymphoma. *Am J Pathol.* 1999;154:1685–1691.

145. Marshall NA, Christie LE, Munro LR et al. Immunosuppressive regulatory T cells are abundant in the reactive lymphocytes of Hodgkin lymphoma. *Blood.* 2004;103:1755–1762.

146. Stanislawski T, Voss RH, Lotz C et al. Circumventing tolerance to a human MDM2-derived tumor antigen by TCR gene transfer. *Nat Immunol.* 2001;2:962–970.

147. Schumacher TN. T-cell-receptor gene therapy. *Nat Rev Immunol.* 2002;2:512–519.

148. Xue SA, Gao L, Hart D et al. Elimination of human leukemia cells in NOD/SCID mice by WT1-TCR gene-transduced human T cells. *Blood.* 2005;106:3062–3067.

149. Morgan RA, Dudley ME, Wunderlich JR et al. Cancer regression in patients after transfer of genetically engineered lymphocytes. *Science.* 2006;314:126–129.

150. Kuball J, Dossett ML, Wolfl M et al. Facilitating matched pairing and expression of TCR chains introduced into human T cells. *Blood.* 2007;109:2331–2338.

151. Dudley ME, Wunderlich JR, Yang JC et al. Adoptive cell transfer therapy following non-myeloablative but lymphodepleting chemotherapy for the treatment of patients with refractory metastatic melanoma. *J Clin Oncol.* 2005;23:2346–2357.

152. Hedrick SM. Chimeric T cell receptor-immunoglobulin molecules: Function and applications. *Int Rev Immunol.* 1993;10:279–290.

153. Pule M, Finney H, and Lawson A. Artificial T-cell receptors. *Cytotherapy.* 2003;5:211–226.

154. Dotti G and Heslop HE. Current status of genetic modification of T cells for cancer treatment. *Cytotherapy.* 2005;7:262–272.

155. Vera J, Savoldo B, Vigouroux S et al. T-lymphocytes redirected against the kappa light chain of human immunoglobulin efficiently kill mature B-lymphocyte derived malignant cells. *Blood.* 2006;108:3890–3897.

156. Hwu P, Yang JC, Cowherd R et al. In vivo antitumor activity of T cells redirected with chimeric antibody/T-cell receptor genes. *Cancer Res.* 1995;55:3369–3373.

157. Cooper LJ, Topp MS, Serrano LM et al. T-cell clones can be rendered specific for CD19: Toward the selective augmentation of the graft-versus-B-lineage leukemia effect. *Blood.* 2003;101:1637–1644.

158. Hombach A, Heuser C, Sircar R et al. Characterization of a chimeric T-cell receptor with specificity for the Hodgkin's lymphoma-associated CD30 antigen. *J Immunother.* 1999;22:473–480.

159. Rossig C, Bollard CM, Nuchtern JG, Merchant DA, and Brenner MK. Targeting of G(D2)-positive tumor cells by human T lymphocytes engineered to express chimeric T-cell receptor genes. *Int J Cancer.* 2001;94:228–236.

160. Jensen MC, Cooper LJ, Wu AM, Forman SJ, and Raubitschek A. Engineered CD20-specific primary human cytotoxic T lymphocytes for targeting B-cell malignancy. *Cytotherapy.* 2003;5:131–138.

161. Riviere I, Sadelain M, and Brentjens RJ. Novel strategies for cancer therapy: The potential of genetically modified T lymphocytes. *Curr Hematol Rep.* 2004;3:290–297.

162. Brentjens RJ, Latouche JB, Santos E et al. Eradication of systemic B-cell tumors by genetically targeted human T lymphocytes co-stimulated by CD80 and interleukin-15. *Nat Med*. 2003;9:279–286.

163. Savoldo B, Rooney CM, Di Stasi A et al. Epstein–Barr virus specific cytotoxic T lymphocytes expressing the anti-CD30{zeta} artificial chimeric T-cell receptor for immunotherapy of Hodgkin disease. *Blood*. 2007;110:2620–2630.

164. Mitsuyasu RT, Anton PA, Deeks SG et al. Prolonged survival and tissue trafficking following adoptive transfer of CD4zeta gene-modified autologous CD4(+) and CD8(+) T cells in human immunodeficiency virus-infected subjects. *Blood*. 2000;96:785–793.

165. Park JR, Digiusto DL, Slovak M et al. Adoptive transfer of chimeric antigen receptor re-directed cytolytic T lymphocyte clones in patients with neuroblastoma. *Mol Ther*. 2007;15:825–833.

166. Kershaw MH, Westwood JA, Parker LL et al. A phase I study on adoptive immunotherapy using gene-modified T cells for ovarian cancer. *Clin Cancer Res*. 2006;12:6106–6115.

167. Krause A, Guo HF, Latouche JB et al. Antigen-dependent CD28 signaling selectively enhances survival and proliferation in genetically modified activated human primary T lymphocytes. *J Exp Med*. 1998;188:619–626.

168. Finney HM, Lawson AD, Bebbington CR, and Weir AN. Chimeric receptors providing both primary and costimulatory signaling in T cells from a single gene product. *J Immunol*. 1998;161:2791–2797.

169. Pule MA, Straathof KC, Dotti G et al. A chimeric T cell antigen receptor that augments cytokine release and supports clonal expansion of primary human T cells. *Mol Ther*. 2005;12:933–941.

170. Imai C, Mihara K, Andreansky M et al. Chimeric receptors with 4–1BB signaling capacity provoke potent cytotoxicity against acute lymphoblastic leukemia. *Leukemia*. 2004;18:676–684.

171. Kowolik CM, Topp MS, Gonzalez S et al. CD28 costimulation provided through a CD19-specific chimeric antigen receptor enhances in vivo persistence and antitumor efficacy of adoptively transferred T cells. *Cancer Res*. 2006;66:10995–11004.

172. Rossig C, Bollard CM, Nuchtern JG, Rooney CM, and Brenner MK. Epstein–Barr virus-specific human T lymphocytes expressing antitumor chimeric T-cell receptors: Potential for improved immunotherapy. *Blood*. 2002;99:2009–2016.

173. Cooper LJ, Al Kadhimi Z, Serrano LM et al. Enhanced antilymphoma efficacy of CD19-redirected influenza MP1-specific CTLs by cotransfer of T cells modified to present influenza MP1. *Blood*. 2005;105:1622–1631.

174. Heemskerk MH, Hoogeboom M, de Paus RA et al. Redirection of antileukemic reactivity of peripheral T lymphocytes using gene transfer of minor histocompatibility antigen HA-2-specific T-cell receptor complexes expressing a conserved alpha joining region. *Blood*. 2003;102:3530–3540.

175. Kershaw MH, Westwood JA, and Hwu P. Dual-specific T cells combine proliferation and antitumor activity. *Nat Biotechnol*. 2002;20:1221–1227.

176. Fong L and Engleman EG. Dendritic cells in cancer immunotherapy. *Annu Rev Immunol*. 2000;18:245–273.

177. Reinhard G, Marten A, Kiske SM et al. Generation of dendritic cell-based vaccines for cancer therapy. *Br J Cancer*. 2002;86:1529–1533.

178. Hsu FJ, Benike C, Fagnoni F et al. Vaccination of patients with B-cell lymphoma using autologous antigen-pulsed dendritic cells. *Nat Med*. 1996;2:52–58.

179. Bendandi M, Gocke CD, Kobrin CB et al. Complete molecular remissions induced by patient-specific vaccination plus granulocyte-monocyte colony-stimulating factor against lymphoma. *Nat Med*. 1999;5:1171–1177.

180. Kirk CJ and Mule JJ. Gene-modified dendritic cells for use in tumor vaccines. *Hum Gene Ther*. 2000;11:797–806.

181. Hanks BA, Jiang J, Singh RA et al. Re-engineered CD40 receptor enables potent pharmacological activation of dendritic-cell cancer vaccines in vivo. *Nat Med*. 2005;11:130–137.

182. Yankauckas MA, Morrow JE, Parker SE et al. Long-term anti-nucleoprotein cellular and humoral immunity is induced by intramuscular injection of plasmid DNA containing NP gene. *DNA Cell Biol*. 1993;12:771–776.

183. Conry RM, Widera G, LoBuglio AF et al. Selected strategies to augment polynucleotide immunization. *Gene Ther*. 1996;3:67–74.

184. King CA, Spellerberg MB, Zhu D et al. DNA vaccines with single-chain Fv fused to fragment C of tetanus toxin induce protective immunity against lymphoma and myeloma. *Nat Med*. 1998;4:1281–1286.

48 p53 Tumor Suppressor Opens Gateways for Cancer Therapy

Mu-Shui Dai, Jayme R. Gallegos, and Hua Lu

CONTENTS

48.1 INTRODUCTION

The tumor suppressor p53 is one of the most important and intensively studied molecules in biomedical research. Since its discovery 28 years ago, more than 43,000 articles have been published about p53. These studies cover nearly all aspects of biomedical research, encompassing biochemistry, biophysics, molecular biology, cellular biology, genetics, pharmacology, toxicology, metabolism, immunology, bioinformatics, as well as clinical research. Tremendous effort has been spent elucidating the mechanisms underlying p53's tumor suppressive function and how it is regulated. Still, there is much ground to cover before p53 and its signaling pathways are fully understood. However, this fact does not prevent the application of our current knowledge to the development of strategies for treating cancer patients using the p53 pathway as a therapeutic target. Indeed, a number of strategies, such as introduction of functional wild-type p53 into cancer cells and inhibition of MDM2-mediated p53 suppression, have been investigated in recent years. In this chapter, we will review p53's properties and functions, as well as its regulation in response to diverse cellular stressors. We will also briefly describe recent progress in the development of anticancer therapies that target the MDM2-p53 feedback loop. p53 gene delivery-based gene therapy will be discussed in Chapter 49.

The p53 protein is a stress-activated transcription factor; therefore activated p53 can either induce or repress the transcription of many target genes. The proteins encoded by these target genes are involved in the regulation of multiple biological functions, including cell cycle, apoptosis, cell senescence, differentiation, angiogenesis, cell migration, and DNA repair [1]. Diverse stressors, including DNA damage, oncogene activation, hypoxia/anoxia, ribonucleotide depletion, and loss of support/survival signals, stabilize the p53 protein and enhance its activity [2]. The importance of p53 in tumor suppression is highlighted by the fact that more than half of all types of human tumors harbor mutations or deletions in the p53 gene, and the remainder often have impaired function of the p53 pathway through the involvement of direct or indirect p53 regulators [3–6]. Germ-line mutations of p53 have been identified in individuals with the cancer-prone Li-Fraumeni syndrome [7,8]. Similar to human cancers, mice homozygous for inactivated p53 are highly susceptible to spontaneous tumorigenesis [9], and transgenic mice expressing hot-spot gain-of-function p53 mutations develop tumors in various tissues [10,11]. These studies establish p53 as a principal "guardian of the genome" and demonstrate that p53 plays an essential role in protecting the organism from tumorigenesis.

The p53 protein possesses the typical structural domains of a transcription factor, as well as several unique domains. These features include the DNA-binding domain, the transactivation domain, the oligomerization domain, the basic regulatory region, and the proline-rich domain. These features of the p53 protein allow for the dynamic regulation of p53's stability and activity in response to various external and internal cellular stressors. The central DNA-binding domain mediates sequence-specific binding to chromatin [12–14]. The majority of p53 gene mutations, which are found in human cancers occur in this domain, emphasizing the importance of this region for p53's function [15]. These mutations alter the conformation of p53 and affect the folding of the DNA-binding domain, therefore disrupting the capacity of p53 to bind to its DNA target, rendering it inactive. This domain has also been shown to interact with the ASPP (Ankyrin repeat, SH3 domain, and proline-rich domain containing) family proteins ASPP1 and ASPP2, allowing p53 to preferably activate transcription of proapoptotic genes such as *Bax* and *PIG3* [16].

The N-terminal, bipartite acidic transactivation domain makes contacts with basal transcription factors and coactivators, thus initiating transcriptional activation of target genes [17,18]. The C-terminal oligomerization domain allows p53 to form a tetramer and is required for its transcriptional activity [12]. The basic regulatory region at the extreme C-terminus is thought to regulate the sequence-specific binding activity of the central core DNA-binding domain and contributes to p53's ability to recognize several forms of DNA that resemble structures caused by DNA-damaging agents [12,19–24]. Finally, the p53 N-terminal proline-rich domain, containing five copies of the sequence PXXP, has been shown to be important for the p53-induced apoptotic response to DNA-damaging agents [25–30].

48.2 BIOLOGICAL FUNCTIONS OF p53

Upon activation, p53 binds to its cognate DNA response elements (p53RE) in the genome and activates or represses the transcription of genes residing in the vicinity of these binding sites. There are over 4000 putative p53-binding sites existing in the human genome [31]. More than 150 p53 target genes have been described and many more will be revealed with the development of advanced molecular technology. The proteins encoded by these genes contribute to diverse biological functions of p53 in multiple ways, including inducing cell cycle arrest, apoptosis, senescence, and angiogenesis [1]. In addition, p53 may facilitate DNA repair directly or indirectly through the induction of genes associated with DNA repair. These cellular responses to p53 activation can be variable and highly dependent on both cell type and the nature of the sustained damage.

Proper cell cycle checkpoints ensure that genomic integrity is maintained throughout cell division. p53 plays a role in both the G1 and the G2 checkpoints of the cell cycle, in part by induction of its target genes *p21^WAF1/CIP1* (p21 will be used hereafter), *14-3-3-σ*, and *GADD45*. The p21 protein inhibits cyclin D-dependent kinases (CDK). CDKs phosphorylate Rb, thereby causing the dissociation of Rb from E2F, allowing E2F to activate the expression of proteins important for the progression of the cell cycle [32]. As a result, p21 maintains the Rb–E2F complex and indirectly inhibits E2F activity, preventing the G1-S transition [15,33]. The 14-3-3-σ and GADD45 proteins inhibit cyclin B-CDC2 kinase activity, which is essential for the G2-M transition. In response to DNA damage, 14-3-3σ binds to phosphorylated Cdc25, a tyrosine protein phosphatase for CDC2, and sequesters Cdc25 in the cytoplasm where it cannot activate CDC2. Then the GADD45 protein dissociates CDC2 from cyclin B, blocking the G2-M phase transition [34,35]. Thus, p53 also mediates the G2 cell cycle arrest [36,37].

Depending on the type and duration of the stress and the cellular growth conditions, p53 can activate different subsets of target genes with proapoptotic activity [1,38]. These genes encode the cell membrane proteins Fas/CD95, KILLER/DR5, and PERP [39–43], the cytoplasmic proteins PIDD and PIG (p53-inducible gene family), and mitochondrial proteins, such as BAX, NOXA [44], PUMA [45], p53AIP1 [46], BID, and others. These proteins trigger the death-receptor-mediated [47,48] and mitochondrial-mediated apoptotic pathways [49,50]. In addition, p53 can interact directly with antiapoptotic proteins, such as Bcl-XL and Bcl-2, to exert its apoptogenic function in the mitochondria, independent of its transcription activity [51–53]. Also, activation of autophagy by the p53-induced protein DRAM has been described as an important contribution to the apoptotic response [54]. Therefore, primarily by inducing cell cycle arrest or apoptosis, p53 provides a crucial surveillance mechanism for allowing cells to either recover from stress or to be eliminated from the replicative pool, thus preventing growing cells from undergoing malignant transformation.

In addition, p53 plays an important role in maintenance of genomic stability by mediating DNA repair [55–57]. It has been shown that p53 is involved in various types of DNA repair, including nucleotide excision repair (NER), base excision repair (BER), nonhomologous end-joining (NHEJ) and homologous recombination (HR) [58–62]. For example, p53-dependent transcriptional activity is important for regulation of NER by p53 [61]. p53 binds to the NER-associated helicases XPB and XPD and modulates their activities [63,64]. It also regulates the expression of the DDB2 and XPC [65–67], and serves as a chromatin accessibility factor for NER of DNA damage [68]. Further, p53 also binds to RAD51 and RAD54, major components of the HR machinery, and controls the level of HR [69,70]. Therefore, p53 regulates DNA repair as well as the DNA damage response.

In addition to its role in gene maintenance, p53 stimulates the expression of genes important for suppression of blood vessel formation (angiogenesis). Angiogenesis is critical for tumor progression [71]. At least three mechanisms account for this inhibitory effect of p53 on angiogenesis: Interference with the central regulators of hypoxia that mediate angiogenesis, inhibition of the production of proangiogenic factors, and direct increase of the production of endogenous angiogenesis inhibitors. These mechanisms license p53 to shut down the angiogenic potential of cancer cells and prevent tumor growth, progression, and metastasis [71]. Recently, p53 has been shown to inhibit hypoxia-inducible factor-1 (HIF-1) activity; HIF-1 induces angiogenic factors in response to hypoxia and impairs cardiac angiogenesis in response to pressure overload. As a consequence, p53 prevents the development of cardiac hypertrophy and induces systolic dysfunction in response to sustained pressure overload, therefore fulfilling a crucial function in the transition from cardiac hypertrophy to heart failure [72].

Of further interest, p53 also can activate the transcription of some noncoding RNAs, resulting in cell growth inhibition and apoptosis. For example, p53 induces the expression of miRNA-34a, which also contributes to p53-mediated cell cycle arrest and apoptosis [73–76]. Moreover, p53 represses RNA polymerase (Pol I)-mediated transcription of precursor rRNAs and Pol III-mediated transcription of tRNAs and 5S rRNA, leading to inhibition of ribosomal biogenesis [77].

p53 has been shown to repress the Pol II-mediated transcription of U1 snRNA [78] and Pol III-mediated transcription of U6 snRNA [79,80]. Therefore, there are many layers to p53's role in cell growth.

The tumor suppressive function of p53 is validated concretely by several *in vivo* mouse models. It is firmly established that p53 knockout mice die within 10 months due to a variety of spontaneous tumors [9]. Remarkably, restoration of endogenous p53 expression in p53-deficient tumors leads to complete regression of these tumors due to cell cycle arrest, apoptosis, senescence, and initiation of an innate immune response [81,82]. These studies place important emphasis on the fact that, although cancer arises from a combination of mutations in oncogenes and tumor suppressor genes, p53 deficiency is required for maintenance of aggressive tumors. Also, these *in vivo* studies provide an incredibly solid foundation for cancer therapeutic strategies aimed at reintroduction of p53's function.

48.3 MDM2: A FEEDBACK INHIBITOR OF p53

The ability of p53 to induce apoptosis or cell cycle arrest can be detrimental to normal cell growth if left uncontrolled. Therefore, it is essential for a cell to tightly control p53 activity during normal development and cell growth. Under physiological conditions, p53 is maintained at an extremely low and inert level with a half-life of approximately 30 min. This rapid turnover of p53 is due to its ubiquitylation-mediated proteasomal degradation. Although a number of ubiquitin ligases, such as Pirh2, COP1, and alternative reading frame (ARF)-BP1, have been shown to ubiquitylate p53 [83], the central and most extensively studied ubiquitin ligase is the oncoprotein MDM2. The *mdm2* gene was originally identified on a mouse double minute chromosome in the 3T3DM cell line [84]. It can immortalize and, in cooperation with Ras, transform rat embryonic fibroblasts [85]. Consistent with this study, overexpression or gene amplification of *mdm2* has been shown in a variety of human tumors, particularly in soft tissue sarcomas, carcinomas, leukemias, lymphomas, breast and lung cancers [86–91]. More recent data have shown that a naturally occurring polymorphism (SNP309) within the *mdm2* promoter leads to an increase in *mdm2* mRNA and protein in human populations [92], which may be related to higher incidence of cancers.

MDM2 is a nuclear phosphoprotein, which possesses several important functional domains, including the p53-binding domain, a central acidic region with a C4 zinc finger, and a C-terminal RING domain, which confers MDM2's E3 ligase activity. MDM2's N-terminal p53-interacting domain mediates MDM2's binding to the N-terminal transcriptional activation domain of p53, thus interfering with p53's ability to interact with the transcription machinery [93,94]. The central acidic domain of MDM2 is pivotal for MDM2-mediated p53 degradation, but not p53's ubiquitylation [95,96]. A number of proteins, such as the ribosomal proteins L5 and L23, and ARF, bind to this domain, leading to inhibition of MDM2-mediated p53 degradation. The C-terminal side of the acidic domain contains a C4 zinc finger domain, which has recently been shown to mediate the binding of MDM2 to ribosomal protein L11. Mutation of residue Cys 305 to either Phe or Ser resulted in the loss of L11 binding to MDM2 and stabilization of p53, indicating this region may also play an important role in controlling p53 degradation [97]. The C-terminal RING finger domain is required for the E3 ligase activity of MDM2 [98]. MDM2 also contains a nuclear localization signal (NLS) and a nuclear export signal (NES), which mediates the shuttling of MDM2 between the cytoplasm and the nucleus and also provides a mechanism to regulate p53's activity [99,100]. Further, within the RING domain, amino acids 464–471 can function as a nucleolar localization signal (NoLS) [101]. All of these domains in MDM2 are crucial for regulating p53's stability and activity.

MDM2 inhibits p53's function through several mechanisms. MDM2 binds p53 specifically, linking their N-terminal domains. This binding conceals the N-terminal transcription activation domain of p53 at its target promoters, preventing the interaction of p53 with the basal transcription machinery. Also, by occupying at p53's target promoters with p53, MDM2 can also interact with histones and promote mono-ubiquitylation of histone H2B in the vicinity of a p53-binding site [102–105]. These actions lead to the inhibition of p53's transcriptional activity [94,106]. In addition, this binding initiates the ubiquitylation of p53 at several C-terminal lysine residues, catalyzed by the C-terminal RING-finger domain of MDM2; this ubiquitylation results in p53's degradation by the 26S proteasome [98,107]. MDM2 was recently found to differentially catalyze monoubiquitylation and polyubiquitylation of p53 in a dosage-dependent manner [108]. As a consequence, low levels of MDM2 activity induce mono-ubiquitylation and nuclear export of p53, whereas high levels promote polyubiquitylation and nuclear degradation of p53. It seems likely that these distinct mechanisms are employed under different physiological settings. For example, MDM2-mediated polyubiquitylation and nuclear degradation may play a critical role in suppressing p53's function during the later stages of a DNA damage response, or when MDM2 is malignantly overexpressed [109,110]. On the other hand, MDM2-mediated monoubiquitylation and subsequent cytoplasmic translocation of p53 may represent an important means of p53 regulation in unstressed cell, where MDM2 is maintained at low levels [111–114]. Moreover, MDM2 was also reported to promote NEDD8 conjugation of p53. The C-terminal glycine residue of the ubiquitin-like protein NEDD8 can be covalently linked to Lys 370, 372, or 373 of p53. This modification inhibits p53's transcriptional activity without affecting p53's protein stability [115]. The lysine residues modified by neddylation are three of the six lysines also targeted by ubiquitylation. Whether neddylation augments ubiquitylation is not yet clear. Interestingly, the *mdm2* gene is a downstream target gene of p53 [116,117], thus forming a negative feedback loop [118,119]. Indeed, genetic disruption of *p53* rescues the lethal phenotype of *mdm2* knockout mice [120,121], firmly validating that MDM2 is a critical inhibitor of p53.

48.4 p53 STRESS RESPONSE

To activate p53, cells must overcome the MDM2-p53 negative feedback circuit. Multiple pathways can lead to activation of p53 in response to a wide variety of cellular stressors, including DNA damage, oncogenic stress, ribosomal stress, and others, such as those induced by hypoxia, reactive oxygen species, telomere erosion, and the loss of survival signals [122,123]. All of these stressors lead to disruption of the negative control of p53 imposed by MDM2 through shared or distinct pathways or cellular components.

DNA damage triggers an Ataxia telangiectasia mutated kinase (ATM) or ataxia telangiectasia RAD3-related kinase (ATR) kinase-dependent phosphorylation cascade and results in p53 activation. In response to ionizing radiation (IR), p53 is phosphorylated at Ser 15 by ATM kinase [124–127] and at Ser 20 by Chk2, which is phosphorylated by ATM [128–130]. In response to UV damage, p53 is phosphorylated at Ser 15 by ATR kinase [131,132] and at Ser 20 by Chk1, which is phosphorylated and activated by ATR [133]. Although phosphorylation of Ser 15 and Ser 20 did not diminish the binding of an N-terminal p53 peptide to MDM2, subsequent phosphorylation of Thr 18 drastically reduced p53-MDM2 binding [134]. Since phosphorylation of Thr 18 requires prior phosphorylation on Ser 20, DNA damage-induced phosphorylation of p53 at the N-terminal residues within the MDM2 binding region impairs the binding of MDM2 to p53 and blocks its inhibitory effect on p53. Similar to p53, phosphorylation of MDM2 also plays a role in p53's activation during a DNA damage response. Most MDM2 phosphorylation sites are clustered within MDM2's N-terminal p53-binding domain and the central acidic domain. For example, MDM2 is phosphorylated by DNA-PK at Ser 17. This phosphorylation might play a role in blocking the MDM2—p53 interaction [135]. ATM phosphorylates MDM2 at Ser 395 and impairs MDM2's ability to promote p53 degradation, possibly through phosphorylation-dependent inhibition of p53's nuclear export by MDM2 [136–138]. In addition to the regulation of p53's stability upon DNA damage, phosphorylation also regulates the recruitment of transcriptional coactivators such as p300/CBP to p53, thus enhancing p53's transcriptional activity [139]. Taken together, DNA damage triggers the activation of p53 through phosphorylation of both p53 and MDM2, impairing MDM2's ability to bind to p53, therefore relieving its inhibitory effect on p53.

The MDM2–p53 feedback loop is also subjected to regulation through protein–protein interaction. One critical player of this regulatory mechanism is ARF (p14ARF in human, p19ARF in mouse) that is encoded by the INK4a locus and translated in an ARF, when compared to the reading frame for the CDK inhibitor p16 [140]. ARF activates p53 in response to aberrant growth and proliferation signals, such as those induced by the overexpression of the oncogenes Ras [141], c-Myc [142], E2F [143], E1A [144], or β-catenin [145]. It binds to the central acidic domain of MDM2 and directly inhibits MDM2 ubiquitin ligase activity, both *in vitro* and in cells [146], thus leading to stabilization and activation of p53

[146–150]. Because of this function, ARF also acts as an important tumor suppressor [137,151,152].

Another group of proteins, which activate p53 through direct interaction with MDM2 and suppression of MDM2's activity, are ribosomal proteins. Recently, at least four ribosomal proteins, including L11, L5, L23, and S7 [83,153–158], have been shown to interact with MDM2 in response to ribosomal stress caused by perturbation of ribosomal biogenesis.

The ribosome is a fine-tuned cellular machine that translates cellular mRNA through a static, higher-ordered cellular process, into proteins [159,160]. To produce a ribosome, eukaryotic cells must assemble about 79 ribosomal proteins with four different ribosomal RNA (rRNA) species (28S, 18S, 5.8S, and 5S) into ribosomal subunits in the nucleolus [161,162]. Notably, all three RNA polymerases (I, II and III) are involved in this process and are coordinated to ensure the high efficiency and accuracy of ribosome production. Together, these complex processes are called ribosomal biogenesis and of fundamental importance for normal cell growth and proliferation. Therefore, it is also perfectly coupled with cell growth and proliferation. Illustrating this point are studies showing that interference with ribosome production severely retards animal growth and development, at both the cellular level and the organism level.

Since ribosomal biogenesis occurs primarily in the nucleolus and many external and internal stimuli lead to the disruption of the nucleolus, it is understandable that perturbation of the nucleolus or nucleolar protein production would be linked to p53 activity along with other types of stress [163]. This specific type of stress is often referred to ribosomal (or nucleolar) stress, and can be triggered by actinomycin D or 5-fluorouracil (5-FU) treatment [164–166], serum starvation [167], the expression of dominant-negative Bop1 [168], or the genetic disruption of ribosomal protein S6 and TIF-IA [169,170]. In response to ribosomal stress, free L5, L11, L23, and S7 may be released to the nucleus or the cytoplasm where they bind to MDM2 and inhibit MDM2-mediated p53 suppression [83,153–158]. These studies suggest that p53-dependent cell cycle checkpoints monitor the malfunction ribosomal biogenesis. Interestingly, like ARF, these individual ribosomal proteins are small basic proteins. They also bring up several important questions. Why do so many basic nucleolar proteins bind to and inhibit MDM2's function? Do these nucleolar proteins collaborate to produce an optimal stress response? Would they play a role in response to different nucleolar stressors? Finally, how might the regulation of ribosomal proteins play a role in preventing tumorigenesis? All of these questions and others remain.

48.5 OTHER REGULATORS OF THE MDM2–p53 FEEDBACK LOOP

Besides the aforementioned proteins, the MDM2–p53 feedback loop is also subjected to regulation by many other proteins. The transcriptional coactivators p300 and CBP appear to exert a dual function on this loop [171]. p300/CBP acetylates p53 and stimulates its activity. This acetylation can be inhibited by MDM2 [172,173]. Additionally, p300/CBP interacts with MDM2 in nuclear body-like structures, where MDM2 might be protected from proteasomal degradation [174] and cooperates with MDM2 to degrade p53 [171,175,176]. Consistently, MDM2 mutants lacking the p300/CBP-binding domain within MDM2's central acidic domain failed to degrade p53, but still promoted monoubiquitylation of p53 [177,178]. More recently, p300/CBP was shown to act as an E4 enzyme to assist MDM2 in polyubiquitylation of p53 [179]. It is yet unclear what physiological conditions may cause p300 to regulate this portion of the feedback loop and if the overall outcome of this stimulus would be the positive or negative regulation of p53.

Another key regulator of MDM2 is its homolog, MDMX, which assists MDM2 in downregulating p53's function [180]. MDMX shares significant homology with MDM2 in its N-terminal p53-binding domain and its C-terminal RING-finger domain [181]. Like MDM2, MDMX binds p53 and inhibits its function [182–184]. As in the case of MDM2, genetically targeting the p53 gene also rescues the lethal phenotype of *mdmx* knockout mice, suggesting that MDMX is critical for MDM2-p53 feedback regulation as well [185–187]. Increased expression of MDMX is frequently observed in human tumors [188–190]. However, unlike MDM2, the expression of MDMX is not regulated by p53 [180], and MDMX alone does not ubiquitylate p53 [182,186,191,192]. Also distinct from MDM2, MDMX appears to reside mostly in the cytoplasm [186,193], but can be recruited to the nucleus by MDM2 [186,194]. The nuclear import of MDMX is also induced by DNA damage signals, such as γ irradiation [195]. In the absence of MDMX, MDM2 is relatively ineffective at downregulating p53 because of its extremely short half-life. MDMX aids MDM2 through an interaction between MDM2 and MDMX's RING-finger domains. This interaction sufficiently stabilizes MDM2 and enables it to degrade p53 at its optimal turnover rate [194].

MDMX is also degraded by MDM2 [196,197]. Moreover, ARF prevents MDM2 from degrading p53 and shifts MDM2 activity to degrade MDMX instead [197]. Therefore, MDM2 and MDMX may have different roles in inhibiting p53. MDMX is thought to enhance MDM2-mediated p53 ubiquitylation and degradation [198,199], consequently repressing p53's function. Interestingly, in response to ionizing or UV irradiation, MDMX is phosphorylated at Ser 376 by ChK2 or ChK1 and this phosphorylation leads to the interaction of MDMX with 14-3-3 proteins. As a result, MDMX loses its ability to suppress p53, thus leading to p53 activation [200,201]. Therefore, to activate p53, stress signals must turn on cellular mechanisms that surmount the negative control by MDM2 and MDMX.

Finally, the MDM2–p53 feedback loop is regulated by deubiquitylation. Herpes virus-associated ubiquitin-specific protease (HAUSP), an ubiquitin hydrolase, was shown to be a direct antagonist of MDM2 activity and acts by specifically deubiquitylating p53 after stimulation by DNA damage, thus protecting p53 from MDM2-mediated degradation [202]. However, HAUSP was also shown to bind and to deubiquitylate MDM2 and MDMX, thus stabilizing both proteins [203,204]. This effect appears to be more dominant, as knockdown or knockout of HAUSP activates p53 function [203]. In contrast to HAUSP, another deubiquitylation enzyme called USP2a has recently been shown to specifically bind to and deubiquitylate MDM2, but not p53, thus enhancing MDM2-mediated p53 degradation. Consistently, reduction of USP2a levels destabilizes MDM2 and causes the accumulation and activation of p53 [205]. These studies suggest that deubiquitylation also regulates the MDM2-p53 feedback loop. Whether these deubiquitylases play a role in tumorigenesis would be an interesting and critical question for future investigation.

The above-discussed and other p53 regulators not discussed are listed in Table 48.1, highlighting the extreme complexity of p53 regulation in cells.

TABLE 48.1
Upstream Regulators of p53

Protein	Type of Molecule	Role	References
(A) Enzymatic activators			
E4F1	Atypical ubiquitin ligase	Ubiqutylation	[224]
p300/CBP	Acetyltransferase	Acetylation	[171,179,225]
PCAF	Acetyltransferase	Acetylation	[102,226,227]
PML/p300	Tumor suppressor/acetyltransferase complex	Transcription	[228]
Set7/9	Lysine methyltransferase	Methylation	[269]
NQO1	NADH oxidioreductase	20S proteasome associated factor	[230]
Pin 1	Prolyl isomerase	Phosphorylation alteration/enhancement	[231,232]
p38	Ser/Thr kinase	Phosphorylation	[233–235]
ATM/ATR	Ser/Thr kinases	Phosphorylation	[4,236]
CK1	Ser/Thr kinase	Phosphorylation	[134,237]
Chk ½	Ser/Thr kinases	Phosphorylation	[238]

(continued)

TABLE 48.1 (continued)
Upstream Regulators of p53

Protein	Type of Molecule	Role	References
DNAPK	Ser/Thr kinase	Phosphorylation	[239,240]
ERK	Ser/Thr kinase	Phosphorylation	[235,241,242]
MAPK	Ser/Thr kinase	Phosphorylation	[242,243]
JNK	Ser/Thr kinase	Phosphorylation	[244–246]
Daxx/Axin/HIPK2	Ser/Thr kinase complex	Phosphorylation (UV response)	[247]
FACT (SSRP1/SPT 16)/CK2	Ser/Thr kinase/cofactor complex	Phosphorylation	[248,249]
c-Abl	Tyr kinase	p53 binding and Phosphorylation of MDM2	[250,251]
(B) Enzymatic Repressors			
HDAC	Deacetylase	Deacetylation	[252–254]
Sir2α	Deacetylase	Deacetylation	[255]
FBX011	NEDD ligase	Neddylation	[256]
Set8/PR-Set7	Lysine methyltransferase	Methylation	[257]
Smyd2	Lysine methyltransferase	Methylation	[258]
Pias (1, xβ, y)	SUMO ligase	Sumoylation	[259,260]
Sumo 1	SUMO ligase	Sumoylation	[261]
ArfBP1 (HECTH9/MULE)	Ubiquitin ligase	Ubiquitylation	[262]
Carps	Ubiquitin ligase	Ubiquitylation	[263,264]
CHIP	Ubiquitin ligase	Ubiquitylation	[263,265]
E6AP	Ubiquitin ligase	Ubiquitylation	[266]
Mdm2	Ubiquitin ligase	Ubiquitylation/Neddylation	[98,107]
PIRH2	Ubiquitin ligase	Ubiquitylation	[267]
WWP1	Ubiquitin ligase	Ubiquitylation	[268,269]
Daxx/HAUSP/MDM2/MDMX	Ubiquitin ligase complex	Ubiquitylation	[270]
LAMA/EC5S/VHL	Ubiquitin ligase complex	Ubiquitylation	[271]
YY1/MDM2	Ubiquitin ligase complex	Ubiquitylation	[272,273]

Protein	Type of Molecule	Role	p53's Fate	References
(C) Nonenzymatic Interactors				
ASPP1/2	Binding protein	Cell cycle/apoptosis	Activation	[16,274,275]
VHL	Binding protein	Hypoxia/tumor suppressor	Activation	[276]
Topors	RING family zinc finger Protein	Binding protein	Activation	[277]
WRN	Helicase	Binding protein	Activation	[152,278]
Ribosomal proteins (L5, L11, L23, S7)	Ribosomal subunits	Binding proteins to MDM2	Activation	[83,153–158]
Sp1	Transcription factor	Transcription	Activation	[279]
p14/p19Arf	Tumor suppressor	Cell Cycle/MDM2 inhibitor	Activation	[280,281]
iASPP	Binding protein	Cell cycle/apoptosis	Inactivation	[275,282]
Hsp 90	Chaper one	Conformation	Inactivation	[283]
Jab-1	Shuttling factor	Cell cycle	Inactivation	[284,285]

48.6 STRATEGIES FOR TARGETING p53 IN CANCER THERAPY

The understanding of p53's biological function and its regulation provides a basis for targeting p53 for anticancer drug development. Over the past decade, a number of attempts have been made to develop drugs that either rescue p53's activity by overexpressing its wild-type form in cancers, or enhance p53's activity by interfering with the MDM2–p53 interaction or MDM2's ubiquitin ligase activity (Table 48.2).

Some of the approaches currently explored to activate or rescue the wild-type function of p53 are the small molecules cp-31398, PRIMA-1, and MIRA-1, and the recombinant adenoviral

p53, known as Gendicine. CP-31398, PRIMA-1, and MIRA-1 were developed as chaperone molecules to aid in refolding of mutant p53 in cancer tissue so that it can assume a proper wild-type conformation. CP-31398 had the disadvantage in that it could only chaperone the newly translated p53 protein. However, recent tests with PRIMA-1 and MIRA-1 are very promising and demonstrate that these compounds can not only chaperone the folding of the newly produced p53 but also refold the mutant p53 already present in the cells [206–208]. The description of Gendicine, a recombinant adenovirus encoding the human *p53* tumor suppressor gene (rAd-p53), and its clinical studies are discussed by Dr. Zhaohui Peng and his

TABLE 48.2

Chemothereputic Agents Targeting p53 Pathways

Compound	Form	Pathway Target	Trials	References
Chalcone derivatives	Flavonoid intermediate	Disruption of p53/MDM2 interaction—activity questionable	Cells, animals	[208,220]
Chlorofusin	Fungal metabolite	Disruption of p53/MDM2 interation—activity questionable	Cells, animals	[208,219]
CP-31398	Small molecule	Reactivation of endogenous mutant p53—only newly synthesized	Mouse	[206,208]
Gendicine	Recombinant adenovirus	Direct expression of wild-type p53	Progress to clinical trials	See chapter
HLI98	Small molecule	MDM2 E3 ligase inhibition	Cells	[215]
MDM2 silencing	Oligonuclotides	MDM2 downregulation	Cells, mouse, human	[209–211,213]
MIRA-1	Small molecule	Reactivation of endogenous mutant p53	Mouse	[206–208]
Nutlin	Small molecule	MDM2/binding E3 ligase inhibition	Cells, mouse	[208,222]
PRIMA-1	Small molecule	Reactivation of endogenous mutant p53	Mouse	[206–208]
RITA	Small molecule	Disruption of p53/MDM2 interaction	Cells, mouse	[208,286]

colleagues in Chapter 49, and represents a paradigm for clinical application p53 as an anticancer agent.

Since aberrant overexpression of MDM2 occurs in subset of tumors with wild-type p53, it is also necessary to overcome MDM2's inhibition of p53 by downregulating its expression, either by directly inhibiting its ubiquitin ligase activity or compromising its interaction with p53 to restore p53 function in some tumors. Over past years, several strategies that target MDM2 for inhibition have been explored: (1) Inhibition of MDM2 expression by antisense oligonucleotides has been shown to activate p53 in various wild-type p53-containing tumor cell lines and has antitumor activity in xenograft tumor models in nude mice [209–211]. These antisense oligonucleotides synergistically enhance the antitumor effect of chemotherapeutics and radiation therapy [212–214]. Interestingly, the antisense MDM2 inhibitors also have antitumor activities in human cancers with p53 deficiency, reflecting their inhibitory effect on p53-independent function of MDM2 [210,213].

As noted above, MDM2 is the central negative regulator of p53, acting as an ubiquitin ligase to target p53 for proteasome-mediated degradation. Thus inhibition of MDM2's E3 ligase activity would stabilize p53 for activation. Recently, small molecule inhibitors have been identified to possess such an inhibitory effect on MDM2. One of such compounds, named HLI98, inhibits MDM2-mediated p53 ubiquitylation and induces p53-dependent apoptosis in cancer cells [215]. The major drawback for this class of compounds is their low selectivity and potency. To screen more selective small molecules for the desired specificity would increase the feasibility of using them in cancer therapy.

Finally, a potential way to activate p53 is through inhibition of the MDM–p53 binding. MDM2 contains a well-defined, relatively deep hydrophobic pocket in its the N- terminus (residues 25–109) where the transactivational domain of p53 binds, thereby concealing p53 from interacting with the transcriptional machinery [216]. The minimal MDM2-binding site on p53 was subsequently mapped to residues 18–26 [93,217,218]. This pocket is filled by three primary side chains (Phe 19,

Trp 23, and Leu 26) from the helical region of the p53 peptide [216,217]. Therefore, it is possible to design small molecules to mimic p53's binding to MDM2. A number of such molecules have been investigated, including chalcone derivatives, cholorofusin, nutlin, and reactivation of p53 and induction of tumor cell apoptosis (RITA). Chalcone derivatives are present in many antioxidant-rich foods and are intermediates in the production of flavanoids. They were the first found inhibitors of the MDM–p53 interaction, as was cholorofusin, a fungal metabolite [208, 219,220]. However, their activity and cell and animal models are currently unconfirmed. The small molecule inhibitors nutlin and RITA are potent and selective MDM2 antagonists, which bind to MDM2, blocking its suppression of p53 [221,222] *in vitro* and *in vivo* tumor models and are promising for future study [223].

In summary, p53 and the MDM2–p53 feedback loop are highly relevant to cancer formation and progression. Hence, using p53 as an anticancer gene therapy or targeting this loop for anticancer therapy presents a very promising approach. Other alternative strategies could be designed by either disrupting MDMX–p53 binding or screening compounds that target the central domain of MDM2 or MDMX, thus inhibiting their ability to inactivate p53. Although we have a long path to march in order to develop strategies stemmed from these concepts for effective cancer treatment, such a triumphant day is within reach, given that tremendous effort will continuingly be spent expanding upon the wealth of knowledge already established in this exciting and advancing arena.

REFERENCES

1. Oren, M. Decision making by p53: Life, death and cancer. *Cell Death Differ* 10, 431–442 (2003).
2. Vousden, K.H. and Lu, X. Live or let die: The cell's response to p53. *Nat Rev Cancer* 2, 594–604 (2002).
3. Soussi, T., Dehouche, K., and Beroud, C. p53 website and analysis of p53 gene mutations in human cancer: Forging a link between epidemiology and carcinogenesis. *Hum Mutat* 15, 105–113 (2000).

4. Vogelstein, B., Lane, D., and Levine, A.J. Surfing the p53 network. *Nature* 408, 307–310 (2000).

5. Levine, A.J., Momand, J., and Finlay, C.A. The p53 tumour suppressor gene. *Nature* 351, 453–456 (1991).

6. Hollstein, M., Sidransky, D., Vogelstein, B., and Harris, C.C. p53 mutations in human cancers. *Science* 253, 49–53 (1991).

7. Srivastava, S., Zou, Z.Q., Pirollo, K., Blattner, W., and Chang, E.H. Germ-line transmission of a mutated p53 gene in a cancer-prone family with Li-Fraumeni syndrome. *Nature* 348, 747–749 (1990).

8. Malkin, D. et al. Germ line p53 mutations in a familial syndrome of breast cancer, sarcomas, and other neoplasms. *Science* 250, 1233–1238 (1990).

9. Donehower, L.A. et al. Mice deficient for p53 are developmentally normal but susceptible to spontaneous tumours. *Nature* 356, 215–221 (1992).

10. Lang, G.A. et al. Gain of function of a p53 hot spot mutation in a mouse model of Li-Fraumeni syndrome. *Cell* 119, 861–872 (2004).

11. Olive, K.P. et al. Mutant p53 gain of function in two mouse models of Li-Fraumeni syndrome. *Cell* 119, 847–860 (2004).

12. Wang, Y. et al. p53 domains: Identification and characterization of two autonomous DNA-binding regions. *Genes Dev* 7, 2575–2586 (1993).

13. Pavletich, N.P., Chambers, K.A., and Pabo, C.O. The DNA-binding domain of p53 contains the four conserved regions and the major mutation hot spots. *Genes Dev* 7, 2556–2564 (1993).

14. Bargonetti, J., Manfredi, J.J., Chen, X., Marshak, D.R., and Prives, C. A proteolytic fragment from the central region of p53 has marked sequence-specific DNA-binding activity when generated from wild-type but not from oncogenic mutant p53 protein. *Genes Dev* 7, 2565–2574 (1993).

15. Ko, L.J. and Prives, C. p53: Puzzle and paradigm. *Genes Dev* 10, 1054–1072 (1996).

16. Bergamaschi, D. et al. ASPP1 and ASPP2: Common activators of p53 family members. *Molecular Cell Biol* 24, 1341–1350 (2004).

17. Fields, S. and Jang, S.K. Presence of a potent transcription activating sequence in the p53 protein. *Science* 249, 1046–1049 (1990).

18. Raycroft, L., Wu, H.Y., and Lozano, G. Transcriptional activation by wild-type but not transforming mutants of the p53 anti-oncogene. *Science* 249, 1049–1051 (1990).

19. Liu, Y., Lagowski, J.P., Vanderbeek, G.E., and Kulesz-Martin, M.F. Facilitated search for specific genomic targets by p53 C-terminal basic DNA binding domain. *Cancer Biol Ther* 3 (2004).

20. Reed, M. et al. The C-terminal domain of p53 recognizes DNA damaged by ionizing radiation. *Proc Natl Acad Sci USA* 92, 9455–9459 (1995).

21. Lee, S., Elenbaas, B., Levine, A., and Griffith, J. p53 and its 14kDa C-terminal domain recognize primary DNA damage in the form of insertion/deletion mismatches. *Cell* 81, 1013–1020 (1995).

22. Jayaraman, J. and Prives, C. Activation of p53 sequence-specific DNA binding by short single strands of DNA requires the p53 C-terminus. *Cell* 81, 1021–1029 (1995).

23. Hupp, T.R., Sparks, A., and Lane, D.P. Small peptides activate the latent sequence-specific DNA binding function of p53. *Cell* 83, 237–245 (1995).

24. Bayle, J.H., Elenbaas, B., and Levine, A.J. The carboxyl-terminal domain of the p53 protein regulates sequence-specific DNA binding through its nonspecific nucleic acid-binding activity. *Proc Natl Acad Sci USA* 92, 5729–5733 (1995).

25. Baptiste, N., Friedlander, P., Chen, X., and Prives, C. The proline-rich domain of p53 is required for cooperation with anti-neoplastic agents to promote apoptosis of tumor cells. *Oncogene* 21, 9–21 (2002).

26. Walker, K.K. and Levine, A.J. Identification of a novel p53 functional domain that is necessary for efficient growth suppression. *Proc Natl Acad Sci USA* 93, 15335–15340 (1996).

27. Sakamuro, D., Sabbatini, P., White, E., and Prendergast, G.C. The polyproline region of p53 is required to activate apoptosis but not growth arrest. *Oncogene* 15, 887–898 (1997).

28. Venot, C. et al. The requirement for the p53 proline-rich functional domain for mediation of apoptosis is correlated with specific PIG3 gene transactivation and with transcriptional repression. *The Embo J* 17, 4668–4679 (1998).

29. Zhu, J., Jiang, J., Zhou, W., Zhu, K., and Chen, X. Differential regulation of cellular target genes by p53 devoid of the PXXP motifs with impaired apoptotic activity. *Oncogene* 18, 2149–2155 (1999).

30. Zhu, J., Zhang, S., Jiang, J., and Chen, X. Definition of the p53 functional domains necessary for inducing apoptosis. *J Biol Chem* 275, 39927–39934 (2000).

31. Wang, L. et al. Analyses of p53 target genes in the human genome by bioinformatic and microarray approaches. *J Biol Chem* 276, 43604–43610 (2001).

32. el-Deiry, W.S. p21/p53, cellular growth control and genomic integrity. *Curr Top Microbiol Immunol* 227, 121–137 (1998).

33. Gottlieb, T.M. and Oren, M. p53 in growth control and neoplasia. *Biochim Biophys Acta* 1287, 77–102 (1996).

34. Zhan, Q. et al. Association with Cdc2 and inhibition of Cdc2/Cyclin B1 kinase activity by the p53-regulated protein Gadd45. *Oncogene* 18, 2892–2900 (1999).

35. Jin, S. et al. GADD45-induced cell cycle G2-M arrest associates with altered subcellular distribution of cyclin B1 and is independent of p38 kinase activity. *Oncogene* 21, 8696–8704 (2002).

36. Hermeking, H. et al. 14–3–3 sigma is a p53-regulated inhibitor of G2/M progression. *Mol Cell* 1, 3–11 (1997).

37. Taylor, W.R. and Stark, G.R. Regulation of the G2/M transition by p53. *Oncogene* 20, 1803–1815 (2001).

38. Benchimol, S. p53-dependent pathways of apoptosis. *Cell Death Differ* 8, 1049–1051 (2001).

39. Wu, G.S. et al. KILLER/DR5 is a DNA damage-inducible p53-regulated death receptor gene. *Nat Genet* 17, 141–143 (1997).

40. Wu, G.S., Burns, T.F., Zhan, Y., Alnemri, E.S., and El-Deiry, W.S. Molecular cloning and functional analysis of the mouse homologue of the KILLER/DR5 tumor necrosis factor-related apoptosis-inducing ligand (TRAIL) death receptor. *Cancer Res* 59, 2770–2775 (1999).

41. Muller, M., Scaffidi, C.A., Galle, P.R., Stremmel, W., and Krammer, P.H. The role of p53 and the CD95 (APO-1/Fas) death system in chemotherapy-induced apoptosis. *Eur Cytokine Netw* 9, 685–686 (1998).

42. Attardi, L.D. et al. PERP, an apoptosis-associated target of p53, is a novel member of the PMP-22/gas3 family. *Genes Dev* 14, 704–718 (2000).

43. Owen-Schaub, L.B. et al. Wild-type human p53 and a temperature-sensitive mutant induce Fas/APO-1 expression. *Molec Cell Biol* 15, 3032–3040 (1995).

44. Oda, E. et al. Noxa, a BH3-only member of the Bcl-2 family and candidate mediator of p53-induced apoptosis. *Science* 288, 1053–1058 (2000).

45. Nakano, K. and Vousden, K.H. PUMA, a novel proapoptotic gene, is induced by p53. *Mol Cell* 7, 683–694 (2001).

46. Oda, K. et al. p53AIP1, a potential mediator of p53-dependent apoptosis, and its regulation by Ser-46-phosphorylated p53. *Cell* 102, 849–862 (2000).

47. Takimoto, R. and El-Deiry, W.S. Wild-type p53 transactivates the KILLER/DR5 gene through an intronic sequence-specific DNA-binding site. *Oncogene* 19, 1735–1743 (2000).

48. Wu, G.S., Kim, K., and el-Deiry, W.S. KILLER/DR5, a novel DNA-damage inducible death receptor gene, links the p53-tumor suppressor to caspase activation and apoptotic death. *Adv Exp Med Biol* 465, 143–151 (2000).

49. Lin, Y., Ma, W., and Benchimol, S. Pidd, a new death-domain-containing protein, is induced by p53 and promotes apoptosis. *Nat Genet* 26, 122–127 (2000).

50. Polyak, K., Xia, Y., Zweier, J.L., Kinzler, K.W., and Vogelstein, B. A model for p53-induced apoptosis. *Nature* 389, 300–305 (1997).

51. Dumont, P., Leu, J.I., Della Pietra, A.C., 3rd, George, D.L., and Murphy, M. The codon 72 polymorphic variants of p53 have markedly different apoptotic potential. *Nat Genet* 33, 357–365 (2003).

52. Mihara, M. et al. p53 has a direct apoptogenic role at the mitochondria. *Molec Cell* 11, 577–590 (2003).

53. Chipuk, J.E. et al. Direct activation of Bax by p53 mediates mitochondrial membrane permeabilization and apoptosis. *Science* 303, 1010–1014 (2004).

54. Crighton, D. et al. DRAM, a p53-induced modulator of autophagy, is critical for apoptosis. *Cell* 126, 121–134 (2006).

55. Liu, M.T. et al. Epstein–Barr virus latent membrane protein 1 represses p53-mediated DNA repair and transcriptional activity. *Oncogene* 24, 2635–2646 (2005).

56. Nowak, M.A. et al. The role of chromosomal instability in tumor initiation. *Proc Natl Acad Sci USA* 99, 16226–16231 (2002).

57. Avkin, S. et al. p53 and p21 regulate error-prone DNA repair to yield a lower mutation load. *Mol Cell* 22, 407–413 (2006).

58. Sengupta, S. and Harris, C.C. p53: Traffic cop at the crossroads of DNA repair and recombination. *Nat Rev Mol Cell Biol* 6, 44–55 (2005).

59. Smith, M.L. and Seo, Y.R. p53 regulation of DNA excision repair pathways. *Mutagenesis* 17, 149–156 (2002).

60. Bertrand, P., Saintigny, Y., and Lopez, B.S. p53's double life: Transactivation-independent repression of homologous recombination. *Trends Genet* 20, 235–243 (2004).

61. Adimoolam, S. and Ford, J.M. p53 and regulation of DNA damage recognition during nucleotide excision repair. *DNA Repair (Amst)* 2, 947–954 (2003).

62. Zurer, I. et al. The role of p53 in base excision repair following genotoxic stress. *Carcinogenesis* 25, 11–19 (2004).

63. Leveillard, T. et al. Functional interactions between p53 and the TFIIH complex are affected by tumour-associated mutations. *Embo J* 15, 1615–1624 (1996).

64. Wang, X.W. et al. p53 modulation of TFIIH-associated nucleotide excision repair activity. *Nat Genet* 10, 188–195 (1995).

65. Rubbi, C.P. and Milner, J. p53 is a chromatin accessibility factor for nucleotide excision repair of DNA damage. *Embo J* 22, 975–986 (2003).

66. Hwang, B.J., Ford, J.M., Hanawalt, P.C., and Chu, G. Expression of the p48 xeroderma pigmentosum gene is p53-dependent and is involved in global genomic repair. *Proc Natl Acad Sci USA* 96, 424–428 (1999).

67. Adimoolam, S. and Ford, J.M. p53 and DNA damage-inducible expression of the xeroderma pigmentosum group C gene. *Proc Natl Acad Sci USA* 99, 12985–12990 (2002).

68. Wang, Q.E. et al. Tumor suppressor p53 dependent recruitment of nucleotide excision repair factors XPC and TFIIH to DNA damage. *DNA Repair (Amst)* 2, 483–499 (2003).

69. Linke, S.P. et al. p53 interacts with hRAD51 and hRAD54, and directly modulates homologous recombination. *Cancer Res* 63, 2596–2605 (2003).

70. Sengupta, S. et al. BLM helicase-dependent transport of p53 to sites of stalled DNA replication forks modulates homologous recombination. *Embo J* 22, 1210–1222 (2003).

71. Teodoro, J.G., Evans, S.K., and Green, M.R. Inhibition of tumor angiogenesis by p53: A new role for the guardian of the genome. *J Mol Med* 85, 1175–1186 (2007).

72. Sano, M. et al. p53-induced inhibition of Hif-1 causes cardiac dysfunction during pressure overload. *Nature* 446, 444–448 (2007).

73. Chang, T.C. et al. Transactivation of miR-34a by p53 broadly influences gene expression and promotes apoptosis. *Mol Cell* 26, 745–752 (2007).

74. He, L. et al. A microRNA component of the p53 tumour suppressor network. *Nature* 447, 1130–1134 (2007).

75. Raver-Shapira, N. et al. Transcriptional activation of miR-34a contributes to p53-mediated apoptosis. *Mol Cell* 26, 731–743 (2007).

76. Tarasov, V. et al. Differential regulation of microRNAs by p53 revealed by massively parallel sequencing: miR-34a is a p53 target that induces apoptosis and G1-arrest. *Cell Cycle* 6, 1586–1593 (2007).

77. White, R.J. RNA polymerases I and III, growth control and cancer. *Nat Rev Mol Cell Biol* 6, 69–78 (2005).

78. Gridasova, A.A. and Henry, R.W. The p53 tumor suppressor protein represses human snRNA gene transcription by RNA polymerases II and III independently of sequence-specific DNA binding. *Mol Cell Biol* 25, 3247–3260 (2005).

79. Cairns, C.A. and White, R.J. p53 is a general repressor of RNA polymerase III transcription. *Embo J* 17, 3112–3123 (1998).

80. Chesnokov, I., Chu, W.M., Botchan, M.R., and Schmid, C.W. p53 inhibits RNA polymerase III-directed transcription in a promoter-dependent manner. *Mol Cell Biol* 16, 7084–7088 (1996).

81. Ventura, A. et al. Restoration of p53 function leads to tumour regression in vivo. *Nature* 445, 661–665 (2007).

82. Xue, W. et al. Senescence and tumour clearance is triggered by p53 restoration in murine liver carcinomas. *Nature* 445, 656–660 (2007).

83. Dai, M.S., Jin, Y., Gallegos, J.R., and Lu, H. Balance of Yin and Yang: Ubiquitylation-mediated regulation of p53 and c-Myc. *Neoplasia* 8, 630–644 (2006).

84. Cahilly-Snyder, L., Yang-Feng, T., Francke, U., and George, D.L. Molecular analysis and chromosomal mapping of amplified genes isolated from a transformed mouse 3T3 cell line. *Somat Cell Mol Genet* 13, 235–244 (1987).

85. Finlay, C.A. The mdm-2 oncogene can overcome wild-type p53 suppression of transformed cell growth. *Mol Cell Biol* 13, 301–306 (1993).

86. Dworakowska, D. et al. MDM2 gene amplification: A new independent factor of adverse prognosis in non-small cell lung cancer (NSCLC). *Lung Cancer* 43, 285–295 (2004).

87. Bueso-Ramos, C.E. et al. The human MDM-2 oncogene is overexpressed in leukemias. *Blood* 82, 2617–2623 (1993).

88. Cordon-Cardo, C. et al. Molecular abnormalities of mdm2 and p53 genes in adult soft tissue sarcomas. *Cancer Res* 54, 794–799 (1994).

89. Deb, S.P. Cell cycle regulatory functions of the human oncoprotein MDM2. *Mol Cancer Res* 1, 1009–1016 (2003).

90. Momand, J., Jung, D., Wilczynski, S., and Niland, J. The MDM2 gene amplification database. *Nucleic Acids Res* 26, 3453–3459 (1998).

91. Watanabe, T., Ichikawa, A., Saito, H., and Hotta, T. Overexpression of the MDM2 oncogene in leukemia and lymphoma. *Leuk Lymphoma* 21, 391–397, color plates XVI following 395 (1996).

92. Bond, G.L. et al. A single nucleotide polymorphism in the MDM2 promoter attenuates the p53 tumor suppressor pathway and accelerates tumor formation in humans. *Cell* 119, 591–602 (2004).

93. Chen, J., Marechal, V., and Levine, A.J. Mapping of the p53 and mdm-2 interaction domains. *Mol Cell Biol* 13, 4107–4114 (1993).

94. Oliner, J.D. et al. Oncoprotein MDM2 conceals the activation domain of tumour suppressor p53. *Nature* 362, 857–860 (1993).

95. Kawai, H., Wiederschain, D., and Yuan, Z.M. Critical contribution of the MDM2 acidic domain to p53 ubiquitination. *Mol Cell Biol* 23, 4939–4947 (2003).

96. Meulmeester, E. et al. Critical role for a central part of Mdm2 in the ubiquitylation of p53. *Mol Cellular Biol* 23, 4929–4938 (2003).

97. Lindstrom, M.S., Deisenroth, C., and Zhang, Y. Putting a finger on growth surveillance: Insight into MDM2 zinc finger-ribosomal protein interactions. *Cell Cycle* 6, 434–437 (2007).

98. Fang, S., Jensen, J.P., Ludwig, R.L., Vousden, K.H., and Weissman, A.M. Mdm2 is a RING finger-dependent ubiquitin protein ligase for itself and p53. *J Biol Chem* 275, 8945–8951 (2000).

99. Freedman, D.A. and Levine, A.J. Nuclear export is required for degradation of endogenous p53 by MDM2 and human papillomavirus E6. *Mol Cell Biol* 18, 7288–7293 (1998).

100. Roth, J., Dobbelstein, M., Freedman, D.A., Shenk, T., and Levine, A.J. Nucleo-cytoplasmic shuttling of the hdm2 oncoprotein regulates the levels of the p53 protein via a pathway used by the human immunodeficiency virus rev protein. *Embo J* 17, 554–564 (1998).

101. Lohrum, M.A., Ashcroft, M., Kubbutat, M.H., and Vousden, K.H. Contribution of two independent MDM2-binding domains in p14(ARF) to p53 stabilization. *Curr Biol* 10, 539–542 (2000).

102. Jin, Y., Zeng, S.X., Dai, M.S., Yang, X.J., and Lu, H. MDM2 inhibits PCAF (p300/CREB-binding protein-associated factor)-mediated p53 acetylation. *J Biol Chem* 277, 30838–30843 (2002).

103. Lu, H. and Levine, A.J. Human TAFII31 protein is a transcriptional coactivator of the p53 protein. *Proc Natl Acad Sci USA* 92, 5154–5158 (1995).

104. Lu, H., Lin, J., Chen, J., and Levine, A.J. The regulation of p53-mediated transcription and the roles of hTAFII31 and mdm-2. *Harvey Lectures* 90, 81–93 (1994).

105. Minsky, N. and Oren, M. The RING domain of Mdm2 mediates histone ubiquitylation and transcriptional repression. *Mol Cell* 16, 631–639 (2004).

106. Momand, J., Zambetti, G.P., Olson, D.C., George, D., and Levine, A.J. The mdm-2 oncogene product forms a complex with the p53 protein and inhibits p53-mediated transactivation. *Cell* 69, 1237–1245 (1992).

107. Honda, R., Tanaka, H., and Yasuda, H. Oncoprotein MDM2 is a ubiquitin ligase E3 for tumor suppressor p53. *FEBS Lett* 420, 25–27 (1997).

108. Li, M. et al. Mono- versus polyubiquitination: Differential control of p53 fate by Mdm2. *Science* 302, 1972–1975 (2003).

109. Xirodimas, D.P., Stephen, C.W., and Lane, D.P. Cocompartmentalization of p53 and Mdm2 is a major determinant for Mdm2-mediated degradation of p53. *Exp Cell Res* 270, 66–77 (2001).

110. Shirangi, T.R., Zaika, A., and Moll, U.M. Nuclear degradation of p53 occurs during down-regulation of the p53 response after DNA damage. *Faseb J* 16, 420–422 (2002).

111. Stommel, J.M. et al. A leucine-rich nuclear export signal in the p53 tetramerization domain: Regulation of subcellular localization and p53 activity by NES masking. *Embo J* 18, 1660–1672 (1999).

112. Boyd, S.D., Tsai, K.Y., and Jacks, T. An intact HDM2 RING-finger domain is required for nuclear exclusion of p53. *Nat Cell Biol* 2, 563–568 (2000).

113. Freedman, D.A. and Levine, A.J. Regulation of the p53 protein by the MDM2 oncoprotein—thirty-eighth G.H.A. Clowes Memorial Award Lecture. *Cancer Res* 59, 1–7 (1999).

114. Geyer, R.K., Yu, Z.K., and Maki, C.G. The MDM2 RING-finger domain is required to promote p53 nuclear export. *Nat Cell Biol* 2, 569–573 (2000).

115. Xirodimas, D.P., Saville, M.K., Bourdon, J.C., Hay, R.T., and Lane, D.P. Mdm2-mediated NEDD8 conjugation of p53 inhibits its transcriptional activity. *Cell* 118, 83–97 (2004).

116. Perry, M.E., Piette, J., Zawadzki, J.A., Harvey, D., and Levine, A.J. The mdm-2 gene is induced in response to UV light in a p53-dependent manner. *Proc Natl Acad Sci USA* 90, 11623–11627 (1993).

117. Barak, Y., Juven, T., Haffner, R., and Oren, M. mdm2 expression is induced by wild type p53 activity. *Embo J* 12, 461–468 (1993).

118. Picksley, S.M. and Lane, D.P. The p53–mdm2 autoregulatory feedback loop: A paradigm for the regulation of growth control by p53? *Bioessays* 15, 689–690 (1993).

119. Wu, X., Bayle, J.H., Olson, D., and Levine, A.J. The p53-mdm-2 autoregulatory feedback loop. *Genes Dev* 7, 1126–1132 (1993).

120. Jones, S.N., Roe, A.E., Donehower, L.A., and Bradley, A. Rescue of embryonic lethality in Mdm2-deficient mice by absence of p53. *Nature* 378, 206–208 (1995).

121. Montes de Oca Luna, R., Wagner, D.S., and Lozano, G. Rescue of early embryonic lethality in mdm2-deficient mice by deletion of p53. *Nature* 378, 203–206 (1995).

122. Giaccia, A.J. and Kastan, M.B. The complexity of p53 modulation: Emerging patterns from divergent signals. *Genes Dev* 12, 2973–2983 (1998).

123. Oren, M. and Rotter, V. Introduction: p53—the first twenty years. *Cell Mol Life Sci* 55, 9–11 (1999).

124. Nakagawa, K., Taya, Y., Tamai, K., and Yamaizumi, M. Requirement of ATM in phosphorylation of the human p53 protein at serine 15 following DNA double-strand breaks. *Mol Cell Biol* 19, 2828–2834 (1999).

125. Siliciano, J.D. et al. DNA damage induces phosphorylation of the amino terminus of p53. *Genes Dev* 11, 3471–3481 (1997).

126. Banin, S. et al. Enhanced phosphorylation of p53 by ATM in response to DNA damage. *Science* 281, 1674–1677 (1998).

127. Canman, C.E. et al. Activation of the ATM kinase by ionizing radiation and phosphorylation of p53. *Science* 281, 1677–1679 (1998).

128. Shieh, S.Y., Ahn, J., Tamai, K., Taya, Y., and Prives, C. The human homologs of checkpoint kinases Chk1 and Cds1 (Chk2) phosphorylate p53 at multiple DNA damage-inducible sites. *Genes Dev* 14, 289–300 (2000).

129. Chehab, N.H., Malikzay, A., Appel, M., and Halazonetis, T.D. Chk2/hCds1 functions as a DNA damage checkpoint in G(1) by stabilizing p53. *Genes Dev* 14, 278–288 (2000).

130. Hirao, A. et al. DNA damage-induced activation of p53 by the checkpoint kinase Chk2. *Science* 287, 1824–1827 (2000).

131. Kapoor, M., Hamm, R., Yan, W., Taya, Y., and Lozano, G. Cooperative phosphorylation at multiple sites is required to activate p53 in response to UV radiation. *Oncogene* 19, 358–364 (2000).

132. Tibbetts, R.S. et al. A role for ATR in the DNA damage-induced phosphorylation of p53. *Genes Dev* 13, 152–157 (1999).

133. Zhao, H. and Piwnica-Worms, H. ATR-mediated checkpoint pathways regulate phosphorylation and activation of human Chk1. *Mol Cell Biol* 21, 4129–4139 (2001).

134. Sakaguchi, K. et al. Damage-mediated phosphorylation of human p53 threonine 18 through a cascade mediated by a casein 1-like kinase. Effect on Mdm2 binding. *J Biol Chem* 275, 9278–9283 (2000).

135. Mayo, L.D., Turchi, J.J., and Berberich, S.J. Mdm-2 phosphorylation by DNA-dependent protein kinase prevents interaction with p53. *Cancer Res* 57, 5013–5016 (1997).

136. de Toledo, S.M., Azzam, E.I., Dahlberg, W.K., Gooding, T.B., and Little, J.B. ATM complexes with HDM2 and promotes its rapid phosphorylation in a p53-independent manner in normal and tumor human cells exposed to ionizing radiation. *Oncogene* 19, 6185–6193 (2000).

137. Khosravi, R. et al. Rapid ATM-dependent phosphorylation of MDM2 precedes p53 accumulation in response to DNA damage. *Proc Natl Acad Sci USA* 96, 14973–14977 (1999).

138. Maya, R. et al. ATM-dependent phosphorylation of Mdm2 on serine 395: Role in p53 activation by DNA damage. *Genes Dev* 15, 1067–1077 (2001).

139. Buschmann, T., Adler, V., Matusevich, E., Fuchs, S.Y., and Ronai, Z. p53 phosphorylation and association with murine double minute 2, c-Jun NH2-terminal kinase, p14ARF, and p300/CBP during the cell cycle and after exposure to ultraviolet irradiation. *Cancer Res* 60, 896–900 (2000).

140. Zhang, Y. and Xiong, Y. Control of p53 ubiquitination and nuclear export by MDM2 and ARF. *Cell Growth Differ* 12, 175–186 (2001).

141. Palmero, I., Pantoja, C., and Serrano, M. p19ARF links the tumour suppressor p53 to Ras. *Nature* 395, 125–126 (1998).

142. Zindy, F. et al. Myc signaling via the ARF tumor suppressor regulates p53-dependent apoptosis and immortalization. *Genes Dev* 12, 2424–2433 (1998).

143. Bates, S. et al. p14ARF links the tumour suppressors RB and p53. *Nature* 395, 124–125 (1998).

144. de Stanchina, E. et al. E1A signaling to p53 involves the p19(ARF) tumor suppressor. *Genes Dev* 12, 2434–2442 (1998).

145. Damalas, A., Kahan, S., Shtutman, M., Ben-Ze'ev, A., and Oren, M. Deregulated beta-catenin induces a p53- and ARF-dependent growth arrest and cooperates with Ras in transformation. *Embo J* 20, 4912–4922 (2001).

146. Honda, R. and Yasuda, H. Association of p19(ARF) with Mdm2 inhibits ubiquitin ligase activity of Mdm2 for tumor suppressor p53. *Embo J* 18, 22–27 (1999).

147. Zhang, Y., Xiong, Y., and Yarbrough, W.G. ARF promotes MDM2 degradation and stabilizes p53: ARF-INK4a locus deletion impairs both the Rb and p53 tumor suppression pathways. *Cell* 92, 725–734 (1998).

148. Tao, W. and Levine, A.J. P19(ARF) stabilizes p53 by blocking nucleo-cytoplasmic shuttling of Mdm2. *Proc Natl Acad Sci USA* 96, 6937–6941 (1999).

149. Llanos, S., Clark, P.A., Rowe, J., and Peters, G. Stabilization of p53 by p14ARF without relocation of MDM2 to the nucleolus. *Nature Cell Biol* 3, 445–452 (2001).

150. Midgley, C.A. et al. An N-terminal p14ARF peptide blocks Mdm2-dependent ubiquitination in vitro and can activate p53 in vivo. *Oncogene* 19, 2312–2323 (2000).

151. Ashcroft, M., Kubbutat, M.H., and Vousden, K.H. Regulation of p53 function and stability by phosphorylation. *Mol Cell Biol* 19, 1751–1758 (1999).

152. Blattner, C., Tobiasch, E., Litfen, M., Rahmsdorf, H.J., and Herrlich, P. DNA damage induced p53 stabilization: No indication for an involvement of p53 phosphorylation. *Oncogene* 18, 1723–1732 (1999).

153. Dai, M.S. et al. Ribosomal protein L23 activates p53 by inhibiting MDM2 function in response to ribosomal perturbation but not to translation inhibition. *Mol Cell Biol* 24, 7654–7668 (2004).

154. Dai, M.S. and Lu, H. Inhibition of MDM2-mediated p53 ubiquitination and degradation by ribosomal protein L5. *J Biol Chem* 279, 44475–44482 (2004).

155. Chen, D., Zhang, Z., Li, M., Wang, W., Li, Y., Rayburn, E.R., Hill, D.L., Wang, H., and Zhang, R. Ribosomal protein S7 as a novel modulator of p53-MDM2 interaction: Binding to MDM2, stabilization of p53 protein, and activation of p53 function. *Oncogene* 26, 5029–5037 (2007).

156. Lohrum, M.A., Ludwig, R.L., Kubbutat, M.H., Hanlon, M., and Vousden, K.H. Regulation of HDM2 activity by the ribosomal protein L11. *Cancer Cell* 3, 577–587 (2003).

157. Zhang, Y. et al. Ribosomal protein L11 negatively regulates oncoprotein MDM2 and mediates a p53-dependent ribosomal-stress checkpoint pathway. *Mol Cell Biol* 23, 8902–8912 (2003).

158. Jin, A., Itahana, K., O'Keefe, K., and Zhang, Y. Inhibition of HDM2 and activation of p53 by ribosomal protein L23. *Mol Cell Biol* 24, 7669–7680 (2004).

159. Rudra, D. and Warner, J.R. What better measure than ribosome synthesis? *Genes Dev* 18, 2431–2436 (2004).

160. Ruggero, D. and Pandolfi, P.P. Does the ribosome translate cancer? *Nat Rev Cancer* 3, 179–192 (2003).

161. Warner, J.R. The economics of ribosome biosynthesis in yeast. *Trends Biochem Sci* 24, 437–440 (1999).

162. Hannan, K.M., Hannan, R.D., and Rothblum, L.I. Transcription by RNA polymerase I. *Front Biosci* 3, d376–d398 (1998).

163. Rubbi, C.P. and Milner, J. Disruption of the nucleolus mediates stabilization of p53 in response to DNA damage and other stresses. *Embo J* 22, 6068–6077 (2003).

164. Gilkes, D.M., Chen, L., and Chen, J. MDMX regulation of p53 response to ribosomal stress. *Embo J* 25, 5614–5625 (2006).

165. Xiao, S.W., L.C., Sun, Y., Su, X., Li, D.M., Xu, G., Zhu, G.Y., XU, B., and Zhang, S.W. Clinical effectiveness of recombinant adenovirus-p53 combined with hyperthermia in advanced soft tissue sarcoma (a report of 13 cases). Collection of 10th National Conference of Hyperthermia (2007).

166. Ashcroft, M., Taya, Y., and Vousden, K.H. Stress signals utilize multiple pathways to stabilize p53. *Mol Cell Biol* 20, 3224–3233 (2000).

167. Bhat, K.P., Itahana, K., Jin, A., and Zhang, Y. Essential role of ribosomal protein L11 in mediating growth inhibition-induced p53 activation. *Embo J* 23, 2402–2412 (2004).

168. Pestov, D.G., Strezoska, Z., and Lau, L.F. Evidence of p53-dependent cross-talk between ribosome biogenesis and the cell cycle: Effects of nucleolar protein Bop1 on G(1)/S transition. *Mol Cell Biol* 21, 4246–4255 (2001).

169. Yuan, X. et al. Genetic inactivation of the transcription factor TIF-IA leads to nucleolar disruption, cell cycle arrest, and p53-mediated apoptosis. *Mol Cell* 19, 77–87 (2005).

170. Panic, L. et al. Ribosomal protein S6 gene haploinsufficiency is associated with activation of a p53-dependent checkpoint during gastrulation. *Mol Cell Biol* 26, 8880–8891 (2006).

171. Kawai, H., Nie, L., Wiederschain, D., and Yuan, Z.M. Dual role of p300 in the regulation of p53 stability. *J Biol Chem* 276, 45928–45932 (2001).

172. Ito, A. et al. p300/CBP-mediated p53 acetylation is commonly induced by p53-activating agents and inhibited by MDM2. *Embo J* 20, 1331–1340 (2001).

173. Kobet, E., Zeng, X., Zhu, Y., Keller, D., and Lu, H. MDM2 inhibits p300-mediated p53 acetylation and activation by forming a ternary complex with the two proteins. *Proc Natl Acad Sci USA* 97, 12547–12552 (2000).

174. Zeng, S.X., Jin, Y., Kuninger, D.T., Rotwein, P., and Lu, H. The acetylase activity of p300 is dispensable for MDM2 stabilization. *J Biol Chem* 278, 7453–7458 (2003).

175. Thomas, A. and White, E. Suppression of the p300-dependent mdm2 negative-feedback loop induces the p53 apoptotic function. *Genes Dev* 12, 1975–1985 (1998).

176. Grossman, S.R. et al. p300/MDM2 complexes participate in MDM2-mediated p53 degradation. *Mol Cell* 2, 405–415 (1998).

177. Argentini, M., Barboule, N., and Wasylyk, B. The contribution of the acidic domain of MDM2 to p53 and MDM2 stability. *Oncogene* 20, 1267–1275 (2001).

178. Zhu, Q., Yao, J., Wani, G., Wani, M.A., and Wani, A.A. Mdm2 mutant defective in binding p300 promotes ubiquitination but not degradation of p53: Evidence for the role of p300 in integrating ubiquitination and proteolysis. *J Biol Chem* 276, 29695–29701 (2001).

179. Grossman, S.R. et al. Polyubiquitination of p53 by a ubiquitin ligase activity of p300. *Science* 300, 342–344 (2003).

180. Shvarts, A. et al. MDMX: A novel p53-binding protein with some functional properties of MDM2. *Embo J* 15, 5349–5357 (1996).

181. Sharp, D.A., Kratowicz, S.A., Sank, M.J., and George, D.L. Stabilization of the MDM2 oncoprotein by interaction with the structurally related MDMX protein. *J Biol Chem* 274, 38189–38196 (1999).

182. Jackson, M.W. and Berberich, S.J. MdmX protects p53 from Mdm2-mediated degradation. *Mol Cell Biol* 20, 1001–1007 (2000).

183. Rallapalli, R., Strachan, G., Tuan, R.S., and Hall, D.J. Identification of a domain within MDMX-S that is responsible for its high affinity interaction with p53 and high-level expression in mammalian cells. *J Cell Biochem* 89, 563–575 (2003).

184. Marine, J.C. and Jochemsen, A.G. Mdmx as an essential regulator of p53 activity. *Biochem Biophys Res Commun* 331, 750–760 (2005).

185. Parant, J. et al. Rescue of embryonic lethality in Mdm4-null mice by loss of Trp53 suggests a nonoverlapping pathway with MDM2 to regulate p53. *Nat Genet* 29, 92–95 (2001).

186. Migliorini, D. et al. Mdm4 (Mdmx) regulates p53-induced growth arrest and neuronal cell death during early embryonic mouse development. *Mol Cell Biol* 22, 5527–5538 (2002).

187. Finch, R.A. et al. mdmx is a negative regulator of p53 activity in vivo. *Cancer Res* 62, 3221–3225 (2002).

188. Riemenschneider, M.J., Knobbe, C.B., and Reifenberger, G. Refined mapping of 1q32 amplicons in malignant gliomas confirms MDM4 as the main amplification target. *Int J Cancer* 104, 752–757 (2003).

189. Riemenschneider, M.J. et al. Amplification and overexpression of the MDM4 (MDMX) gene from 1q32 in a subset of malignant gliomas without TP53 mutation or MDM2 amplification. *Cancer Res* 59, 6091–6096 (1999).

190. Ramos, Y.F. et al. Aberrant expression of HDMX proteins in tumor cells correlates with wild-type p53. *Cancer Res* 61, 1839–1842 (2001).

191. Stad, R. et al. Mdmx stabilizes p53 and Mdm2 via two distinct mechanisms. *Embo Rep* 2, 1029–1034 (2001).

192. Stad, R. et al. Hdmx stabilizes Mdm2 and p53. *J Biol Chem* 275, 28039–28044 (2000).

193. Rallapalli, R., Strachan, G., Cho, B., Mercer, W.E., and Hall, D.J. A novel MDMX transcript expressed in a variety of transformed cell lines encodes a truncated protein with potent p53 repressive activity. *J Biol Chem* 274, 8299–8308 (1999).

194. Gu, J. et al. Mutual dependence of MDM2 and MDMX in their functional inactivation of p53. *J Biol Chem* 277, 19251–19254 (2002).

195. Li, C., Chen, L., and Chen, J. DNA damage induces MDMX nuclear translocation by p53-dependent and -independent mechanisms. *Mol Cell Biol* 22, 7562–7571 (2002).

196. de Graaf, P. et al. Hdmx protein stability is regulated by the ubiquitin ligase activity of Mdm2. *J Biol Chem* 278, 38315–38324 (2003).

197. Pan, Y. and Chen, J. MDM2 promotes ubiquitination and degradation of MDMX. *Mol Cell Biol* 23, 5113–5121 (2003).

198. Linares, L.K., Hengstermann, A., Ciechanover, A., Muller, S., and Scheffner, M. HdmX stimulates Hdm2-mediated ubiquitination and degradation of p53. *Proc Natl Acad Sci USA* 100, 12009–12014 (2003).

199. Ghosh, M., Huang, K., and Berberich, S.J. Overexpression of Mdm2 and MdmX fusion proteins alters p53 mediated transactivation, ubiquitination, and degradation. *Biochemistry* 42, 2291–2299 (2003).

200. Jin, Y. et al. 14–3–3gamma binds to MDMX that is phosphorylated by UV-activated Chk1, resulting in p53 activation. *Embo J* 25, 1207–1218 (2006).

201. LeBron, C., Chen, L., Gilkes, D.M., and Chen, J. Regulation of MDMX nuclear import and degradation by Chk2 and 14–3–3. *Embo J* 25, 1196–1206 (2006).

202. Li, M. et al. Deubiquitination of p53 by HAUSP is an important pathway for p53 stabilization. *Nature* 416, 648–653 (2002).

203. Li, M., Brooks, C.L., Kon, N., and Gu, W. A dynamic role of HAUSP in the p53-Mdm2 pathway. *Mol Cell* 13, 879–886 (2004).

204. Meulmeester, E. et al. Loss of HAUSP-mediated deubiquitination contributes to DNA damage-induced destabilization of Hdmx and Hdm2. *Mol Cell* 18, 565–576 (2005).

205. Stevenson, L.F. et al. The deubiquitinating enzyme USP2a regulates the p53 pathway by targeting Mdm2. *Embo J* 26, 976–986 (2007).

206. Bykov, V.J. et al. Restoration of the tumor suppressor function to mutant p53 by a low-molecular-weight compound. *Nat Med* 8, 282–288 (2002).

207. Bykov, V.J. et al. PRIMA-1(MET) synergizes with cisplatin to induce tumor cell apoptosis. *Oncogene* 24, 3484–3491 (2005).

208. Levesque, A.A. and Eastman, A. p53-based cancer therapies: Is defective p53 the Achilles heel of the tumor? *Carcinogenesis* 28, 13–20 (2007).

209. Chen, L. et al. Ubiquitous induction of p53 in tumor cells by antisense inhibition of MDM2 expression. *Mol Med* 5, 21–34 (1999).

210. Wang, H. et al. Anti-tumor efficacy of a novel antisense anti-MDM2 mixed-backbone oligonucleotide in human colon cancer models: p53-dependent and p53-independent mechanisms. *Mol Med* 8, 185–199 (2002).

211. Wang, H. et al. MDM2 oncogene as a target for cancer therapy: An antisense approach. *Int J Oncol* 15, 653–660 (1999).

212. Bianco, R., Ciardiello, F., and Tortora, G. Chemosensitization by antisense oligonucleotides targeting MDM2. *Current Cancer Drug targets* 5, 51–56 (2005).

213. Zhang, R., Wang, H., and Agrawal, S. Novel antisense anti-MDM2 mixed-backbone oligonucleotides: Proof of principle, in vitro and in vivo activities, and mechanisms. *Current Cancer Drug Targets* 5, 43–49 (2005).

214. Zhang, Z. et al. Radiosensitization by antisense anti-MDM2 mixed-backbone oligonucleotide in in vitro and in vivo human cancer models. *Clin Cancer Res* 10, 1263–1273 (2004).

215. Yang, Y. et al. Small molecule inhibitors of HDM2 ubiquitin ligase activity stabilize and activate p53 in cells. *Cancer Cell* 7, 547–559 (2005).

216. Kussie, P.H. et al. Structure of the MDM2 oncoprotein bound to the p53 tumor suppressor transactivation domain. *Science* 274, 948–953 (1996).

217. Bottger, A. et al. Molecular characterization of the hdm2-p53 interaction. *J Mol Biol* 269, 744–756 (1997).

218. Bottger, V. et al. Identification of novel mdm2 binding peptides by phage display. *Oncogene* 13, 2141–2147 (1996).

219. Duncan, S.J. et al. Isolation and structure elucidation of chlorofusin, a novel p53-MDM2 antagonist from a Fusarium sp. *J Am Chem Soc* 123, 554–560 (2001).

220. Stoll, R. et al. Chalcone derivatives antagonize interactions between the human oncoprotein MDM2 and p53. *Biochemistry* 40, 336–344 (2001).

221. Efeyan, A. et al. Induction of p53-dependent senescence by the MDM2 antagonist nutlin-3a in mouse cells of fibroblast origin. *Cancer Res* 67, 7350–7357 (2007).

222. Vassilev, L.T. et al. In vivo activation of the p53 pathway by small-molecule antagonists of MDM2. *Science* 303, 844–848 (2004).

223. Tovar, C. et al. Small-molecule MDM2 antagonists reveal aberrant p53 signaling in cancer: Implications for therapy. *Proc Natl Acad Sci USA* 103, 1888–1893 (2006).

224. Le Cam, L. et al. E4F1 is an atypical ubiquitin ligase that modulates p53 effector functions independently of degradation. *Cell* 127, 775–788 (2006).

225. Lill, N.L., Grossman, S.R., Ginsberg, D., DeCaprio, J., and Livingston, D.M. Binding and modulation of p53 by p300/CBP coactivators. *Nature* 387, 823–827 (1997).

226. Jin, Y., Zeng, S.X., Lee, H., and Lu, H. MDM2 mediates p300/CREB-binding protein-associated factor ubiquitination and degradation. *J Biol Chem* 279, 20035–20043 (2004).

227. Liu, L. et al. p53 sites acetylated in vitro by PCAF and p300 are acetylated in vivo in response to DNA damage. *Mol Cell Biol* 19, 1202–1209 (1999).

228. Pearson, M. et al. PML regulates p53 acetylation and premature senescence induced by oncogenic Ras. *Nature* 406, 207–210 (2000).

229. Chuikov, S. et al. Regulation of p53 activity through lysine methylation. *Nature* 432, 353–360 (2004).

230. Asher, G. and Shaul, Y. p53 proteasomal degradation: Poly-ubiquitination is not the whole story. *Cell Cycle* 4, 1015–1018 (2005).

231. Zacchi, P. et al. The prolyl isomerase Pin1 reveals a mechanism to control p53 functions after genotoxic insults. *Nature* 419, 853–857 (2002).

232. Zheng, H. et al. The prolyl isomerase Pin1 is a regulator of p53 in genotoxic response. *Nature* 419, 849–853 (2002).

233. Bulavin, D.V. et al. Phosphorylation of human p53 by p38 kinase coordinates N-terminal phosphorylation and apoptosis in response to UV radiation. *Embo J* 18, 6845–6854 (1999).

234. Keller, D. et al. The p38MAPK inhibitor SB203580 alleviates ultraviolet-induced phosphorylation at serine 389 but not serine 15 and activation of p53. *Biochem Biophys Res Commun* 261, 464–471 (1999).

235. She, Q.B., Chen, N., and Dong, Z. ERKs and p38 kinase phosphorylate p53 protein at serine 15 in response to UV radiation. *J Biol Chem* 275, 20444–20449 (2000).

236. Efeyan, A. and Serrano, M. p53: Guardian of the genome and policeman of the oncogenes. *Cell Cycle* 6, 1006–1010 (2007).

237. Dumaz, N., Milne, D.M., and Meek, D.W. Protein kinase CK1 is a p53-threonine 18 kinase which requires prior phosphorylation of serine 15. *FEBS Lett* 463, 312–316 (1999).

238. Dasika, G.K. et al. DNA damage-induced cell cycle checkpoints and DNA strand break repair in development and tumorigenesis. *Oncogene* 18, 7883–7899 (1999).

239. Lees-Miller, S.P., Sakaguchi, K., Ullrich, S.J., Appella, E., and Anderson, C.W. Human DNA-activated protein kinase phosphorylates serines 15 and 37 in the amino-terminal transactivation domain of human p53. *Mol Cell Biol* 12, 5041–5049 (1992).

240. Wang, Y. and Eckhart, W. Phosphorylation sites in the amino-terminal region of mouse p53. *Proc Natl Acad Sci USA* 89, 4231–4235 (1992).

241. Persons, D.L., Yazlovitskaya, E.M., and Pelling, J.C. Effect of extracellular signal-regulated kinase on p53 accumulation in response to cisplatin. *J Biol Chem* 275, 35778–35785 (2000).

242. Wu, G.S. The functional interactions between the p53 and MAPK signaling pathways. *Cancer Biol Ther* 3, 156–161 (2004).

243. Hildesheim, J. et al. Gadd45a regulates matrix metalloproteinases by suppressing DeltaNp63alpha and beta-catenin via p38 MAP kinase and APC complex activation. *Oncogene* 23, 1829–1837 (2004).

244. Fuchs, S.Y. et al. JNK targets p53 ubiquitination and degradation in nonstressed cells. *Genes Dev* 12, 2658–2663 (1998).

245. Hu, M.C., Qiu, W.R., and Wang, Y.P. JNK1, JNK2 and JNK3 are p53 N-terminal serine 34 kinases. *Oncogene* 15, 2277–2287 (1997).

246. Milne, D.M., Campbell, L.E., Campbell, D.G., and Meek, D.W. p53 is phosphorylated in vitro and in vivo by an ultraviolet radiation-induced protein kinase characteristic of the c-Jun kinase, JNK1. *J Biol Chem* 270, 5511–5518 (1995).

247. Li, Q. et al. Daxx cooperates with the Axin/HIPK2/p53 complex to induce cell death. *Cancer Res* 67, 66–74 (2007).

248. Keller, D.M. and Lu, H. p53 serine 392 phosphorylation increases after UV through induction of the assembly of the CK2.hSPT16.SSRP1 complex. *J Biol Chem* 277, 50206–50213 (2002).

249. Keller, D.M. et al. A DNA damage-induced p53 serine 392 kinase complex contains CK2, hSpt16, and SSRP1. *Mol Cell* 7, 283–292 (2001).

250. Jing, Y. et al. c-Abl tyrosine kinase activates p21 transcription via interaction with p53. *J Biochem* 141, 621–626 (2007).

251. Kharbanda, S., Yuan, Z.M., Weichselbaum, R., and Kufe, D. Determination of cell fate by c-Abl activation in the response to DNA damage. *Oncogene* 17, 3309–3318 (1998).

252. Harris, S.L. and Levine, A.J. The p53 pathway: Positive and negative feedback loops. *Oncogene* 24, 2899–2908 (2005).

253. Juan, L.J. et al. Histone deacetylases specifically downregulate p53-dependent gene activation. *J Biol Chem* 275, 20436–20443 (2000).

254. Luo, J., Su, F., Chen, D., Shiloh, A., and Gu, W. Deacetylation of p53 modulates its effect on cell growth and apoptosis. *Nature* 408, 377–381 (2000).

255. Luo, J. et al. Negative control of p53 by Sir2alpha promotes cell survival under stress. *Cell* 107, 137–148 (2001).

256. Abida, W.M., Nikolaev, A., Zhao, W., Zhang, W., and Gu, W. FBXO11 promotes the Neddylation of p53 and inhibits its transcriptional activity. *J Biol Chem* 282, 1797–1804 (2007).

257. Shi, X. et al. Modulation of p53 function by SET8-mediated methylation at lysine 382. *Mol Cell* 27, 636–646 (2007).

258. Huang, J. et al. Repression of p53 activity by Smyd2-mediated methylation. *Nature* 444, 629–632 (2006).

259. Kahyo, T., Nishida, T., and Yasuda, H. Involvement of PIAS1 in the sumoylation of tumor suppressor p53. *Mol Cell* 8, 713–718 (2001).

260. Schmidt, D. and Muller, S. Members of the PIAS family act as SUMO ligases for c-Jun and p53 and repress p53 activity. *Proc Natl Acad Sci USA* 99, 2872–2877 (2002).

261. Rodriguez, M.S. et al. SUMO-1 modification activates the transcriptional response of p53. *Embo J* 18, 6455–6461 (1999).

262. Chen, D. et al. ARF-BP1/Mule is a critical mediator of the ARF tumor suppressor. *Cell* 121, 1071–1083 (2005).

263. McDonald, E.R., 3rd and El-Deiry, W.S. Suppression of caspase-8- and -10-associated RING proteins results in sensitization to death ligands and inhibition of tumor cell growth. *Proc Natl Acad Sci USA* 101, 6170–6175 (2004).

264. Yang, W. et al. CARPs are ubiquitin ligases that promote MDM2-independent p53 and phospho-p53ser20 degradation. *J Biol Chem* 282, 3273–3281 (2007).

265. Esser, C., Scheffner, M., and Hohfeld, J. The chaperone-associated ubiquitin ligase CHIP is able to target p53 for proteasomal degradation. *J Biol Chem* 280, 27443–27448 (2005).

266. Longworth, M.S. and Laimins, L.A. Pathogenesis of human papillomaviruses in differentiating epithelia. *Microbiol Mol Biol Rev* 68, 362–372 (2004).

267. Leng, R.P. et al. Pirh2, a p53-induced ubiquitin-protein ligase, promotes p53 degradation. *Cell* 112, 779–791 (2003).

268. Chen, C. and Matesic, L.E. The Nedd4-like family of E3 ubiquitin ligases and cancer. *Cancer Metastasis Rev* (2007).

269. Laine, A. and Ronai, Z. Regulation of p53 localization and transcription by the HECT domain E3 ligase WWP1. *Oncogene* 26, 1477–1483 (2007).

270. Tang, J. et al. Critical role for Daxx in regulating Mdm2. *Nat Cell Biol* 8, 855–862 (2006).

271. Cai, Q.L., Knight, J.S., Verma, S.C., Zald, P., and Robertson, E.S. EC5S ubiquitin complex is recruited by KSHV latent antigen LANA for degradation of the VHL and p53 tumor suppressors. *PLoS Pathogens* 2, e116 (2006).

272. Gronroos, E., Terentiev, A.A., Punga, T., and Ericsson, J. YY1 inhibits the activation of the p53 tumor suppressor in response to genotoxic stress. *Proc Natl Acad Sci USA* 101, 12165–12170 (2004).

273. Sui, G. et al. Yin Yang 1 is a negative regulator of p53. *Cell* 117, 859–872 (2004).

274. Samuels-Lev, Y. et al. ASPP proteins specifically stimulate the apoptotic function of p53. *Mol Cell* 8, 781–794 (2001).

275. Sullivan, A. and Lu, X. ASPP: A new family of oncogenes and tumour suppressor genes. *Br J Cancer* 96, 196–200 (2007).

276. Roe, J.S. and Youn, H.D. The positive regulation of p53 by the tumor suppressor VHL. *Cell Cycle* 5, 2054–2056 (2006).

277. Lin, L. et al. Topors, a p53 and topoisomerase I-binding RING finger protein, is a coactivator of p53 in growth suppression induced by DNA damage. *Oncogene* 24, 3385–3396 (2005).

278. Blattner, C., Hay, T., Meek, D.W., and Lane, D.P. Hypophosphorylation of Mdm2 augments p53 stability. *Mol Cell Biol* 22, 6170–6182 (2002).

279. Gualberto, A. and Baldwin, A.S., Jr. p53 and Sp1 interact and cooperate in the tumor necrosis factor-induced transcriptional activation of the HIV-1 long terminal repeat. *J Biol Chem* 270, 19680–19683 (1995).

280. Kamijo, T. et al. Functional and physical interactions of the ARF tumor suppressor with p53 and Mdm2. *Proc Natl Acad Sci USA* 95, 8292–8297 (1998).

281. Sherr, C.J. and Weber, J.D. The ARF/p53 pathway. *Curr Opin Genet Dev* 10, 94–99 (2000).

282. Bergamaschi, D. et al. iASPP oncoprotein is a key inhibitor of p53 conserved from worm to human. *Nat Genet* 33, 162–167 (2003).

283. Tsutsumi, S. and Neckers, L. Extracellular heat shock protein 90: A role for a molecular chaper one in cell motility and cancer metastasis. *Cancer Sci* 98, 1536–1539 (2007).

284. Lee, E.W., Oh, W., and Song, J. Jab1 as a mediator of nuclear export and cytoplasmic degradation of p53. *Mol Cells* 22, 133–140 (2006).

285. Oh, W. et al. Jab1 induces the cytoplasmic localization and degradation of p53 in coordination with Hdm2. *J Biol Chem* 281, 17457–17465 (2006).

286. Issaeva, N. et al. Small molecule RITA binds to p53, blocks p53-HDM-2 interaction and activates p53 function in tumors. *Nat Med* 10, 1321–1328 (2004).

49 Recombinant Adenoviral-p53 Agent for Treatment of Cancers

Zhaohui Peng, Qing Yu, and Jingya Zhu

CONTENTS

49.1 INTRODUCTION

Over the past decade, gene therapy has been increasingly applied in clinical studies as a new mode of medical intervention. According to data published by the *Journal of Gene Medicine* (http://www.wiley.co.uk/genetherapy/clinical/), there were a total of 1309 approved clinical trials for gene therapy across the world at the end of July of 2007. Among these clinical trials, 66.5% were for the treatment of cancer.

Of these cancer gene therapy trials, 62 used recombinant adenovirus encoding the p53 tumor suppressor (rAd-p53). More than 20 kinds of cancer indications have been treated with the rAd-p53 agent, such as head and neck squamous cell carcinoma (HNSCC), lung cancer, breast cancer, and liver cancers. A number of clinical treatment regimens have been evaluated, including administration of rAd-p53 agent alone, or in combination with conventional therapies, such as radiotherapy, chemotherapy, and surgery.

Encouraging clinical responses have been reported by a number of study groups. Lang et al. (2003) reported the results of a phase I clinical trial study in which rAd-p53 was administered to 15 patients with recurrent glioma. rAd-p53 was injected intratumorally at doses between 3×10^{10} and 3×10^{12} viral particles (VP). Three days after rAd-p53 injection the tumor was resected and more rAd-p53 was injected into the tumor bed. Of the 15 patients treated, one survived more than 3 years without evidence of recurrence, four patients experienced no recurrence for more than 6 months after treatment, and two of these four patients survived for more than 1 year. In a multicenter phase II trial involving 25 patients with nonsmall cell lung cancer, 7.5×10^{12} VP of rAd-p53 was injected intratumorally in combination with cisplatin and vinorelbine. No significant difference in tumor response was observed from rAd-p53-injected lesions and noninjected lesions (52% versus 48%, respectively), yet the rAd-p53-treated lesions appeared to be smaller than the nontreated controls (Schuler et al., 2001). In a separate clinical study of 24 patients with nonsmall cell lung cancer (Nemunaisitis et al., 2000), rAd-p53 was injected intratumorally at doses between 1×10^6 and 1×10^{11} plaque-forming units (PFU)/injection in combination with cisplatin. Seventeen patients achieved stable disease, two patients achieved partial response (PR), and four patients had progressive disease. Intratumoral injection of rAd-p53 in combination with cisplatin was well tolerated and there was evidence of clinical efficacy. rAd-p53 was also used for nonsmall cell lung cancer treatment in combination with radiation therapy (Swisher et al., 2003). The rAd-p53 dose ranged from 3×10^{11} to 3×10^{12} VP/injection. Radiation was given concurrently over 6 weeks to a total of 60 Gy. Of the 19 patients treated, 1 showed complete regression (CR) (5%), 11 demonstrated partial regression (PR) (58%), 3 showed stable disease (16%), 2 showed progressive disease (11%), and 2 were unable to be evaluated (11%). In a bladder carcinoma trial, 12 patients received either intratumoral or intravascular injections of rAdp53 at doses of 7.5×10^{11} to 7.5×10^{13} VP (Kuball et al., 2002). Higher transduction efficiency was observed when using the intravascular delivery method. Seven of the 11 patients who could be evaluated had evidence of p53 expression by reverse transcription-polymerase chain reaction (RT-PCR) analysis. Nine of the 12 patients were alive at a median follow-up of 30 months. Pagliaro et al. (2003) reported a similar phase I study in which rAd-p53 was instilled intravesicularly at 1×10^{12} VP/dose in 13 patients with locally advanced transitional cell carcinoma of the bladder. A preliminary antitumorigenic effect was observed at a treatment dose of 1×10^{12} VP on days 1 and 4. Clayman et al. (1998) reported the results of a clinical study in which rAd-p53 was applied as a single agent to treat advanced recurrent HNSCC. Thirty-three patients received an intratumoral injection of rAd-p53 at doses up to 1×10^{11} PFU/injection. Of the 17 patients who could be evaluated, 2 patients showed objective tumor regression of greater than 50%, 6 patients presented stable disease for up to 3.5 months, and 9 patients showed progressive disease. Cristofanilli et al. (2006) reported a prospective, open-label, phase II trial. Thirteen patients with locally advanced breast cancer (LABC) were treated with six 3-week cycles of primary systemic therapy (PST), which consisted of intratumoral injections of Ad5CMV-p53 for 2 consecutive days, plus docetaxel and doxorubicin followed by surgery. p53 status was determined at baseline and was assessed immediately after the first injection (up to 48 h). Clinical response was assessed by clinical and radiologic methods. Serial biopsies showed an increase in *p53* messenger RNA (mRNA) and *p21$^{WAF1/Cip1}$* mRNA. All 12 patients who could be evaluated achieved an objective clinical response. The surgical specimens revealed scattered tumor cells with extensive leukocytes infiltration into the tumor (predominantly T-lymphocytes). At a median follow-up of 37 months (range, 30–41 months), four patients (30%) developed systemic recurrence and two patients died. The estimated breast cancer-specific survival rate at 3 years was 84% (95% confidence interval, 65.7%–100%). There was no increase in systemic toxicity. Cristofanilli et al. (2003) reported another clinical study with Adp53 combined with chemotherapy in the treatment of LABC. Twelve patients were enrolled. Patients received up to six cycles of the combination of Adp53, docetaxel, and doxorubicin. A total of nine patients had completed induction chemotherapy, seven patients had surgery. Clinical response (CR + PR): 8 (90%); partial regression (PR) 1 (10%); stable disease (SD) or progress disease (PD) 0 (0%). A phase I, pharmacokinetic, and pharmacodynamic study of intravenously administered adp53 in patients with advanced cancer was studied by another group (Tolcher et al., 2006). Seventeen patients were treated with escalating doses of Adp53, ranging from 3×10^{10} to 3×10^{12} VP, intravenously over 30 min on days 1, 2, and 3, every 28 days. No severe toxicity was observed during the treatment operation. Adenovirus was observed in the circulation 24 h after administration and the p53 transgene was detected in tumor tissue distant from the site of administration. Thus it was feasible to systemically deliver Adp53 for cancer treatment.

In all the reported clinical studies, the most common side effects were pain at the injection site, fatigue, and development of self-limited fever. Overall, rAd-p53 treatment was well-tolerated by patients without serious side effects. To commercialize Adp53 for cancer treatment, Introgen Therapeutics (Austin, TX) is currently pursuing a number of late-phase clinical studies using rAd-p53 for the treatment of cancers, of which treatment for HNSCC is in phase III clinical study. There has been an active clinical gene therapy program using rAd-p53 for the treatment of HNSCC in China since 1998. After extensive multiyear and multicenter clinical studies, a recombinant human adenoviral–p53 injection (trademarked as Gendicine) (Figure 49.1), developed by Shenzhen SiBiono GeneTech (SiBiono; Shenzhen, China), was approved by the State Food and Drug Administration of China (SFDA) on October 16, 2003 for the treatment of HNSCC. The product was formally launched in the market in April 2004. Gendicine became the world's first gene therapy product approved by a government agency for the treatment of cancer indications and set a new milestone in the history of gene therapy.

This chapter presents a general description of the clinical research and applications of Gendicine and its antitumorigenic

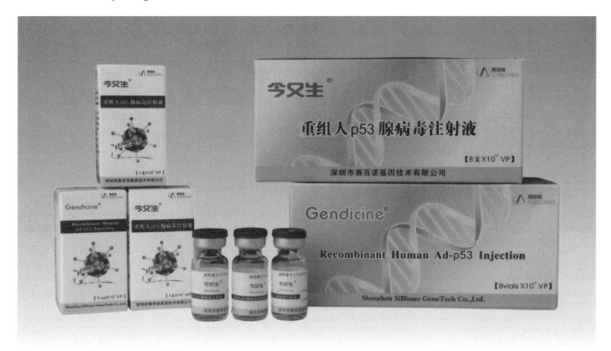

Gendicine (recombinant human Ad-p53 injection)

1×10^{12} viral particles/vial

Made by SiBiono GeneTech Co., Ltd. China

FIGURE 49.1 Package of Gendicine.

mechanisms. The p53 tumor suppressor gene and its biological function and regulation are also briefly outlined, while the detailed p53 regulation and the recent progress in p53 research are reviewed by Hua Lu and his colleagues in a separate chapter (Chapter 48). In addition, an overview about the manufacturing process, quality control, and quality assurance of Gendicine is provided.

49.2 GENDICINE AND ITS ANTITUMORIGENIC MECHANISMS

Gendicine is a recombinant human serotype 5 adenovirus in which the E1 region is replaced by a human wild-type p53 expression cassette. The *p53* gene is driven by a Rous sarcoma virus (RSV) promoter with a bovine growth hormone (BGH) poly (A) tail. The recombinant adenovirus (rAd) is produced in the SBN-Cel producer cells grown in a bioreactor. The SBN-Cel producer cell is a propriety cell line subcloned from the human embryonic kidney (HEK) 293 cells. Virus produced from the bioreactor is further processed and chromatographically purified to produce the recombinant human Ad-p53 injection product. After more than two decades of study, the *p53* gene is now widely regarded as the "guardian of the genome." It has been estimated that at least half of all human malignancies are related to a mutation of the *p53* gene (Shiraishi et al., 2004). p53's ever-increasing research knowledge base facilitated the application of its pathways to the clinic, of which

p53 gene therapy is a very promising treatment. It is expected that the clinical application will, in turn, help us to a greater understanding of p53's function.

The active pharmaceutical ingredient in Gendicine is the recombinant adenoviral vector encoding the *p53* gene. The adenoviral vector functions as a delivery vehicle for the therapeutic *p53* gene. After Gendicine administration, the adenoviral particle infects the targeted tumor cells and delivers the adenoviral genome, carrying the therapeutic *p53* gene, to the cytoplasm and the nucleus for transcription and translation. p53 then exerts its antitumorigenic activities by one or more of the following mechanisms:

- Simultaneously triggering apoptotic pathways in tumor cells by a transcription-dependent mechanism in the nucleus (Muller et al., 1998; Bouvard et al., 2000; Matsuda et al., 2002; Taha et al., 2004), and by a transcription-independent mechanism in the mitochondria (Chipuk et al., 2004; Leu et al., 2004) and Golgi apparatus (Bennett et al., 1998; Ding et al., 2000)
- Activation of immune response factors such as natural killer (NK) cells (Yen et al., 2000; Cerwenka and Lanier, 2003; Rosenblum et al., 2004) to exert "bystander effects"
- Inhibition of DNA repair and antiapoptotic functions in tumor cells (Sah et al., 2003)
- Downregulation of the expression of (1) multidrug resistance genes (Krishna and Mayer, 2000) to

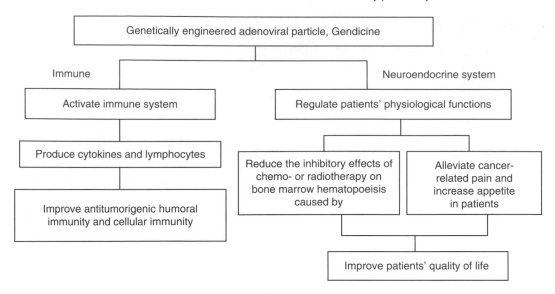

FIGURE 49.2   Proposed therapeutic mechanisms of Gendicine adenovirus particle.

revert the resistance of tumor cells against radio- and chemotherapies, (2) the vascular endothelial growth factor (VEGF) gene (Dameron et al., 1994; Pal et al., 2001) to block the blood supply to the tumor tissue, and (3) matrix metalloproteinase (MMP) (Toschi et al., 2000; Ala-aho et al., 2002; Sun et al., 2004) to suppress tumor cell adhesion, infiltration, and metastasis

- Blockage of the transcription of survival signals in tumor cells (Singh et al., 2002; Yin et al., 2003; Rother et al., 2004) to inhibit the growth of tumor cells in any stage of the cell cycle
- Limitation of the uptake of glucose (Schwartzenberg-Bar-Yoseph et al., 2004) and the production of ATP (Brasseur et al., 1997; Iiizumi et al., 2002) in tumor cells

After intratumoral injection of Gendicine, we observed marked lymphocytic infiltration and obvious inhibition of VEGF activity in biopsies of tumor lesions of patients enrolled in clinical trials of Gendicine.

We believe, in addition to the therapeutic *p53* gene, that the adenoviral particle component of Gendicine also plays an important role in Gendicine's therapeutic effect. The adenoviral delivery vehicle is well known in triggering a strong immune response in patients. It may also induce various hormones and cytokines, thereby affecting the endocrine and immune systems in the body. The triggering of these systems may increase humoral immunity, cellular immunity, and stimulate NK cells to target tumor cells more effectively. Evidence is found for these increased immune activities in the development of grade I/II self-limited fevers in approximately 32% of treated patients. Fever is usually regarded as a side effect in clinical practice; however, in the case of Gendicine treatment, it may reflect the effectiveness of Gendicine in mobilizing the body's immune system. An enhanced immune response in advanced or terminal stage cancer patients is likely to be beneficial for tumor containment.

In clinical studies, we also observed that Gendicine reduced the side effects caused by conventional chemo- and radiation therapies. A significant observation was that some patients showed improved appetite and general health status approximately 2 days after receiving Gendicine treatment. This result is a positive clinical development for cancer patients who suffer from severe side effects caused by radio- and chemotherapy. The mechanism leading to the improvement is not yet known. We suspect it is related to the rAd particle component of Gendicine, as shown in Figure 49.2. Further study is under way to better understand the clinical implications of these results.

49.3 SAFETY AND EFFICACY OF GENDICINE

49.3.1 Safety

Gendicine is an active rAd genetically engineered to express the human *p53* gene. The *p53* gene exists ubiquitously in normal cells and is one of the most widely studied tumor suppressor genes in the human body. Because of its importance in maintaining genome stability, the *p53* gene has been dubbed the "guardian of the genome." The adenoviral vector is a replication-incompetent serotype 5 adenovirus (Ad5) with a deletion in the E1 region, which limits the virus so that it is capable of only one infection cycle. Furthermore, the adenoviral genome does not integrate into the host's genomic DNA. Ad5 has one of the weakest pathogenicity levels in the adenovirus family. Wild-type Ad5 infection generally results in mild upper respiratory disease and fever. Although one fatality has been reported in a study of using adenoviral vector to treat a patient with ornithine transcarbamylase (OTC) deficiency syndrome, the death was later found to be caused by the patient's overwhelming immune reaction against a high systemic dose of adenoviral vector (Chen et al., 2003; Han et al., 2003; Zhang et al., 2003a,b). No fatalities have been reported in other cancer gene therapy trials using

TABLE 49.1

Comparison of Clinical Responses of Patients Receiving GTRT and RT

Groups	\multicolumn{5}{c}{4 Weeks}					\multicolumn{5}{c}{8 Weeks}					\multicolumn{5}{c}{12 Weeks (Confirmation)}				
	N	CR	PR	SD	PD	N	CR	PR	SD	PD	N	CR	PR	SD	PD
GTRT	63	5	41	17	0	62	28	29	5	0	56	36	16	4	0
		8%	65%	27%	0%		45%	47%	8%	0%		64%	29%	7%	0%
RT	72	0	29	41	2	71	6	44	20	1	63	12	38	13	0
		0%	40%	57%	3%		8%	62%	28%	0%		19%	60%	21%	0%

adenoviral vectors. Data from our own clinical studies demonstrated that Gendicine is safe for clinical use.

In a phase I clinical study (Han et al., 2003) from 1998 to May 2000, 12 patients with advanced laryngeal cancer were given various doses of Gendicine before and after surgery. The study results showed that administration of Gendicine did not change the healing process of the surgical wound. No other side effects were observed besides self-limited grade I or II (below 40°C) fever. Results from extensive multicenter controlled and randomized clinical studies involving 135 patients (Zhang et al., 2003a) confirmed that grade I/II self-limited fever was the most commonly observed side effect, occurring in approximately 50%–60% of Gendicine treated patients (Table 49.1). In a few rare cases, patient fever reached as high as 40°C. Development of fever was observed as soon as approximately 3h after injection, lasted about 4h, and then disappeared spontaneously. On some occasions, the fever lasted more than 10h. A common antipyretic was used to effectively control the patient's temperature. It was also observed that Gendicine in combination with radiotherapy (GTRT) did not exacerbate side effects resulting from radiotherapy.

To date, Gendicine has been administered to more than 6000 patients with a variety of cancers from around 40 countries with distinct ethnic backgrounds. The route of administration included intratumoral injection (Figure 49.3), intrapleural and intraperitoneal infusion, intravenous injection, intervention administration, and endotracheal and

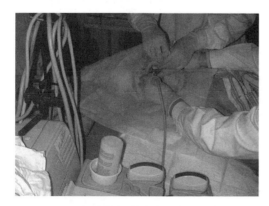
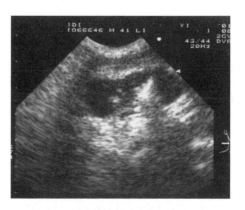

Intratumoral injection of Gendicine

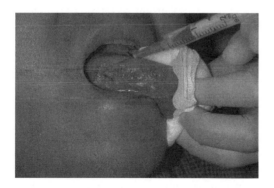

(A)   Injection of Gendicine under ultrasonic guidance   (B)   Showing a strong echo after Gendicine injection

FIGURE 49.3 (See color insert following blank page 206.) Injection of Gendicine.

intravesicular instillation. Gendicine has been used in combination with other therapies including radiotherapy, chemotherapy, hyperthermia, surgery and biological therapies. Deterioration of the side effects associated with the traditional therapies was not observed in any of the combination treatments. GTRT or chemotherapy did not exacerbate any side effects.

Based on our clinical study experience, the recommended dose of Gendicine treatment is one vial (1×10^{12} VP) per injection, per week. An average treatment course involves eight injections for a total of 8 weeks. The maximum accumulative use of Gendicine in a single case was 85 doses (85×10^{12} VP) within a successive 40 weeks. No unexpected side effect was observed in the patients. Our extensive clinical experience showed that Gendicine is safe and well tolerated by treated patients.

49.3.2 THERAPEUTIC EFFECTS

The therapeutic *p53* gene carried by Gendicine exists ubiquitously in normal cells. The rAd can infect, with varying efficiency, almost all human cells, including dividing cells and resting cells. Those unique biological properties make Gendicine a wide-spectrum antitumorigenic agent. Its efficacy in treating cancer has been demonstrated in our clinical studies and the continued patient use after marketing approval. Furthermore, the clinical efficacy of Gendicine appears to be independent of the endogenous *p53* gene status of tumor cells. The published data on Gendicine clinical studies are presented below.

49.3.2.1 Advanced Head and Neck Squamous Cell Carcinoma

49.3.2.1.1   Gendicine in Combination with Surgery for Treatment of Laryngeal Carcinoma
From 1998 to 1999, Gendicine was used in a phase I study for the treatment of 12 patients (Han et al., 2003) with laryngeal carcinoma with an average clinical symptom for 41 months. Seven of the 12 patients did not receive any treatment prior to Gendicine administration and 5 of the 12 patients had a history of one or multiple tumor recurrences. One of the patients had received six laser surgeries because of frequent relapses averaging about 9 months. The patients were divided into three groups receiving escalating doses of Gendicine. Intratumoral injection was administered at doses of 1×10^{10}, 1×10^{11}, and 1×10^{12} VP every the other day for a total of 10 injections. According to 6 years of follow-up data, 11 of the 12 patients were still alive, and 9 remained tumor-free. The 6-year overall survival (OS) rate is 91.7%. This result compared very favorably to the data from the American Joint Committee on Cancers (2002) which showed that the 5-year survival rate for treatment of laryngeal carcinoma by conventional therapies was 68.6%–71.1% at stage I, 55.5%–59.9% at stage II, and 46.04%–50.1% at stage III, respectively. Therefore, Gendicine can effectively prevent tumor recurrence and prolong the lifespan of patients suffering from laryngeal carcinoma.

49.3.2.1.2   Gendicine in Combination with Radiation Therapy for Treatment of Head and Neck Squamous Cell Carcinoma
Significant synergetic effects have been demonstrated using GTRT, chemotherapy, surgery, and hyperthermia for treatment of cancers (Chen et al., 2003; Zhang et al., 2003 a,b). A multicenter, concurrently controlled, randomized clinical trial using Gendicine on 135 patients with HNSCC was conducted. Of the enrolled patients, 77% had late stage III to IV cancer and either failed radio-/chemotherapies or were not eligible for surgery. A majority (85%) of the patients had nasopharyngeal cancer. The patients were divided randomly into two groups: one group received GTRT and the other group received radiotherapy (RT) alone. A conventional or three-dimensional conformal radiotherapy was used at doses of 70 Gy/35 f/7–8 w for the RT treatment. In the GTRT group, Gendicine was given at a dose of 1×10^{12} VP three days prior to the radiotherapy per week for a total of 8 weeks. Radiotherapy in the GTRT group was kept the same as that used in the RT group. The objective tumor response was evaluated using computed tomography (CT) or magnetic resonance imaging (MRI), according to the World Health Organization (WHO) response criteria. Usually, it is somewhat misleading to use the total response rate in judging clinical efficacy because of the frequently observed high instance of relapse for patients with PR. Instead, we believe that it is more appropriate to use the CR rate to evaluate true clinical efficacy. The short-term efficacy data, shown in Table 49.1 below, showed that the response rate in the GTRT group was 93%, of which 64% was CR and 29% was PR. The response rate in the RT group was 79% of the patients, of which 19% was CR and 60% was PR. There is a significant difference ($P < 0.01$) between the two groups in either the CR rate or the PR rate. The CR rate in the GTRT group was threefold higher than in the RT group. Furthermore, it is very encouraging to note that 4 weeks after treatment, 65% of patients exhibited PR in the GTRT group whereas 40% of patients in the RT group showed PR and 57% showed SD (stable disease). GTRT showed obvious synergistic effects for cancer treatment (Figure 49.4).

Sun et al. (2005) reported the clinical outcome of combination of Gendicine with radiotherapy for treatment of refractory or recurrent cancers of the head and neck. Twenty-five patients were enrolled, 6 at stage III and 19 at stage IV (five of them with distant metastasis). Twelve patients had squamous cell carcinoma (SCC) (58%) and 13 had nonsquamous cell cancers (non-SCC) including 2 patients with adenocarcinoma, 3 with adenoid cystic carcinoma, 6 with sarcoma, 1 with melanoma, and 1 with basaloma. Gendicine was administered either by direct intratumoral injection or by ultrasound-guided percutaneous intratumoral injection at a dose of $1–2 \times 10^{12}$ VP once a week. The total doses of Gendicine ranged from 3×10^{12} to 20×10^{12} VP (median of 8×10^{12} VP). The radiotherapy dose for patients receiving Gendicine ranged from 40 to 76 Gy (median of 70 Gy). The mean follow-up for surviving patients was 15 months (ranged from 3 to 52 months). The median survival was 8 months. The 1- and 2-year survival

CT image   Nasopharyngeal photography

Before Gendicine treatment

NPC at stage T2N2M0, III, tumor size was 23.5 cm². Fully occupying naso-pharyngeal cavity.

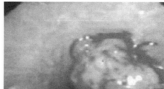
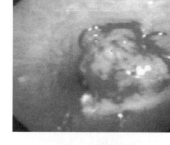
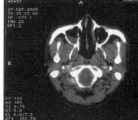

After Gendicine treatment

40 Gy radiation therapy in combination with six injections of Gendicine. Tumor showing almost CR.

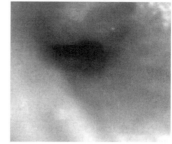

FIGURE 49.4 Efficacy of Gendicine in combination with radiation therapy for treatment of advanced nasopharyngeal carcinoma.

rates were 52.4% and 26.3%, respectively, with no significant differences between SCC and non-SCC. The 1-year tumor-free survival rate was 33.3% for all patients, 16.7% for those with SCC, and 55.5% for those with non-SCC. A higher local control rate was noted for patients with non-squamous tumors ($p = 0.0594$). The tumor response was evaluated at the end of the eighth week after beginning gene therapy. CR and PR were seen in 2 (8.0%) and 14 patients (56.0%), respectively. Nine patients (36.0%) had stable response. Patients with recurrent tumors had a poor tumor response rate ($p = 0.0060$). No significant difference between the tumor response rate and the histological type was noted.

49.3.2.1.3 Five-year Follow-Up for Patients with Nasopharyngeal Carcinoma Using Gendicine in Combination with Radiation Therapy

According to the 5-year follow-up results for the 49 patients (24 cases in the GTRT Group, 25 cases in the RT-only group) with stage II/III nasopharyngeal carcinoma (NPC) presented by the principal investigator (PI), Dr. Zhang Shanwen of Beijing Tumor Hospital, at the 10th National Conference of Tumor Hyperthermia, Gendicine showed noticeable improvement of the local-regional tumor control and prolonged the survival time for patients with NPC. The 5-year local regional failure of GTRT and RT groups is 4.4% and 47.6%, respectively. The local regional control rate of the GTRT group is 10 times higher than that of the RT group. The 5-year overall and tumor-free survival rates are 65.4% and 64.3% for the GTRT group versus 51.9% and 35.6% for the RT group, respectively. The 5-year overall and tumor-free survival rates of

the GTRT group are 13.5% and 36% higher than those of the RT group, respectively (Tables 49.2 through 49.4). The 5-year median time to recurrence is 43 months for the GTRT group and 38 months for the RT group. A new multicenter, concurrently controlled, randomized phase IV clinical study on 300 patients with NPC has been initiated and is in progress.

49.3.2.1.4 Gendicine in Combination with Chemotherapy for Treatment of Advanced Head and Neck Carcinoma

Li and Huang (2006) reported a clinical and biomolecular combination therapy study using subselective intra-arterial Gendicine infusion with induction chemotherapy for locally advanced head and neck carcinoma (Table 49.5). Eighty-eight patients with locoregionally advanced HNSCC or adenoid cystic carcinoma of head and neck (ACCHN) were randomly assigned

TABLE 49.2

Local-Regional Recurrence and Metastasis Rate for NPC Patients

Year	Local-Regional Recurrence Rate (%)		Metastasis Rate (%)	
	GTRT	**RT**	**GTRT**	**RT**
1	0	7.41	7.69	3.7
2	0	19.02	11.54	20.06
3	4.35	23.52	15.75	29.31
4	4.35	23.52	24.18	29.31
5	4.35	47.56	33.90	48.59

TABLE 49.3

5-Year Follow-Up Survival Time of the Patients with NPC

Group	Median Survival Time (months)	5-Year OS Rate (%)	Median Disease-Free Survival Time (months)	5-Year Disease-Free Survival Rate (%)
GTRT	39	65.4	34	64.3
RT	34	51.9	28	35.6

TABLE 49.4

5-Year Follow-Up Survival Rate of the Patients with NPC

Year	Tumor-Free Survival Rate (%) GTRT	RT	OS Rate (%) GTRT	RT
1	92.31	92.31	92.31	92.59
2	88.46	65.38	88.46	66.67
3	80.59	57.69	76.92	51.85
4	72.53	57.69	69.23	51.85
5	63.23	26.97	65.38	51.85

TABLE 49.5

Clinical Response of Patients in the Three Groups

	Primary Tumor (%)	Cervical Metastases (%)	Distant Metastases (%)
Group I	30 (100)	19 (100)	4 (100)
CR	11 (37)	0 (0)	0 (0)
PR	13 (43)	11 (58)	3 (75)
SD or PD	6 (20)	8 (42)	1 (25)
Group II	28 (100)	17 (100)	3 (100)
CR	5 (18)	0 (0)	0 (0)
PR	13 (46)	6 (35)	2 (67)
SD or PD	10 (36)	11 (65)	1 (33)
Group III	30 (100)	18 (100)	4 (100)
CR	5 (17)	0 (0)	0 (0)
PR	13 (43)	6 (33)	2 (50)
SD or PD	12 (40)	12 (67)	2 (50)

to three groups. The groups were (I) Gendicine in combination with chemotherapeutic agents ($n = 30$), (II) Gendicine alone ($n = 28$), and (III) chemotherapy alone ($n = 30$). Patients in groups I and II received 10 cycles of Gendicine infusion within 6 weeks. The selected dose of Gendicine was delivered in 20 min using 10 mL syringes once every day for 4 days as a treatment cycle. The Gendicine doses ranged from 1×10^{12} to 2×10^{12} VP for patients with unilateral and bilateral catheters, respectively. The Gendicine doses remained the same throughout the study. Patients were closely monitored for 4 h after each Gendicine administration. Patients in group I received two periods of combination chemotherapy, which was similar to those in group III. The first combination chemotherapy period began 2 days after the first Gendicine infusion. Between the two treatment periods there was a 7-day recovery break.

The chemotherapeutic agents used for treatment of squamous cell carcinoma were Carboplatin (CP), Bleomycin (BLM), and Methotrexate (MTX). The chemotherapeutic agents used for treatment of adenocarcinoma (including adenoid cystic carcinoma and mucoepidermoid carcinoma) were CP, 5-Fluorouracil (5-FU), and cyclophosphamide (CTX). The primary tumor lesion of 60 patients responded to the therapy (CR + PR), of which 21 patients showed CR and 39 patients showed PR. The CR rate (37%) of group I was significantly higher than those of group II and III (18% and 17%, respectively) ($p < 0.05$). No significant difference in therapeutic outcome was found between group II and III ($p > 0.05$). The primary tumor lesions of 28 patients did not respond to the treatment (SD or PD), of which 6 (20%) were in group I, 10 (36%) were in group II, and 12 (40%) were in group III.

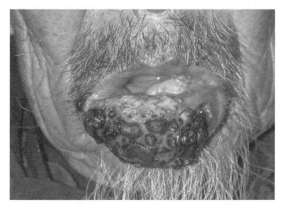 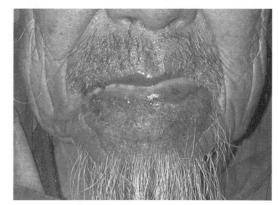

Before Gendicine administration   2 Months after Gendicine treatment

Male, 67 years, suffering from SCC of lower lip, T4N2aM0, was enrolled into the group
and treated with Gendicine in combination with chemotherapy (CP, BLM, and MTX).
Ten injections of Gendicine (4×10^{12} VP/injection, one injection every 5 days)
and two courses of chemotherapy. The tumor showed complete clinical recovery.

FIGURE 49.5 (See color insert following blank page 206.) Efficacy of Gendicine in combination with chemotherapy for treatment of advanced squamous cell carcinoma.

The rate of the nonresponders (SD and PD) in group I was significantly lower than those in group II and III ($p < 0.05$). Although we found promising therapeutic responses on cervical nodal metastasis and distant metastasis of patients in group I, statistical analysis was not performed because of the limited number of patients (Figure 49.5).

The CR rate of group I was statistically higher than that of the other two groups. Intraarterial infusion of Gendicine resulted in successful transfer of the exogenous wild-type *p53* gene and subsequent expression of functional p53 protein. The expressed p53 protein activated Bcl-2, inactivated Bax, and ultimately induced tumor cell apoptosis. As a result, Gendicine in combination with chemotherapy produced an additive therapeutic effect with no apparent added side effects.

49.3.2.1.5 Advanced Liver Cancer

Guan et al. (2005a), at West China Hospital, Sichuan University, China, reported the results of a study using Gendicine in combination with hepatic arterial chemoembolization (TACE) on patients with advanced hepatic cell carcinoma (HCC). A total of 150 patients were enrolled. Sixty-eight patients were treated using Gendicine-TACE (GT-TACE) combination by intratumoral injection of Gendicine under CT guidance or transcatheter perfusion at $1–4 \times 10^{12}$ VP per dose, one dose/week for 4 weeks. Gendicine was administered 2–5 days after a TACE

was carried out. Eighty-two patients were treated by TACE alone as the control group. For both the treatment and control groups, TACE was performed with 5-fluorouracil (5-Fu) 1 g, hydroxycamptothecin (HCPT), Adriamycin (ADM) 40 mg, and arterial embolization with iodized oil (10–30 mL).

Patient response rates were 67.6% and 51.2% in the GT-TACE group and the TACE alone group, respectively. The difference is statistically significant ($P < 0.05$) (Table 49.6). Compared with patients in the control group, patients in the GT-TACE group experienced better pain relief and improvement of Karnofsky performance status. It is important to point out that 1-year survival rate was 76.5% (52/68) in the GT-TACE group and only 23.2% (19/82) in the control group. The difference is statistically significant ($P < 0.01$) (Table 49.7). The results showed that Gendicine in combination with TACE is effective for treatment of HCC patients.

Zhu Zhibing (2004) at the Second People's Hospital of Shenzhen reported a clinical study on 38 patients with refractory, inoperable advanced HCC. Thirty of the 38 patients received Gendicine treatment alone and the other eight patients were treated with Gendicine in combination with hyperthermia. All patients received the same Gendicine treatment at a dose of $1–2 \times 10^{12}$ VP/wk (injection) for a total of 4 weeks, by intralesional injection via percutaneous hepatic paracentesis or hepatic arterial infusion. The preliminary

TABLE 49.6

Short-Term Efficacy of Gendicine in Combination with TACE for Treatment of HCC Patients

Groups	Case No.	CR	PR	NC	PD	CR + PR (%)
Gendicine–TACE	68	0	45	15	7	67.6
TACE	82	0	42	27	13	51.2

TABLE 49.7

Survival Rate of HCC Patients after Gendicine in Combination with TACE Treatment

	3 Months	6 Months	12 Months
Gendicine-TACE	89.71%	76.13%	43.30%
TACE	68.15%	36.98%	24.02%

TABLE 49.8

Survival Time of the Patients with Advanced HCC Receiving Gendicine Treatment

Group	Case Numbers	Average Survival Time (days)	Range of Survival Time (days)
Gendicine + chemotherapy	14	238.1 ± 119.9	14 to 405
Chemotherapy	16	80.7 ± 35.9	18 to 167

TABLE 49.9

Karnofsky Scores of the Patients with Advanced HCC Receiving Gendicine Treatment

Group	Before Treatment	After Treatment	P Value
Gendicine + chemotherapy	54.3 ± 12.2	67.9 ± 23.3	<0.05
Chemotherapy	60.0 ± 12.6	50.6 ± 22.4	>0.05

clinical data for the 30 patients receiving Gendicine alone showed 2/30 of PR, 24/30 of SD, and 4/30 of PD. Twenty-six of the 30 patients showed an improvement of Karnofsky performance status from 55.79 ± 11.30 to 61.05 ± 21.64. All the patients tolerated the Gendicine treatment well.

Chen Shixi (2007) at Cancer Hospital of Jiangsu Province reported a preliminary clinical study of Gendicine on 30 patients with advanced hepatocellular carcinoma (HCC). The 30 patients were divided into a treatment group (14 cases) and a control group (16 cases). For the treatment group, a dose of 10^{12} VP of Gendicine in combination with 20 mg of HCPT were infused into the hepatic artery once a week, for a total of 3 weeks as a treatment course. The 14 patients in the treatment group each received one to four courses of treatment. Only 20 mg of HCPT was used for the control group. Patients in the treatment group survived between 14 to 405 days with an average of 238.1 days. Patients in the control survived between 18 to 167 days with an average of 80.7 days. There was a significant difference in the average survival between these two groups ($P < 0.05$) (Table 49.8). The KPS score of the treatment group was also statistically higher than that of the control group (Table 49.9).

Zhang et al. (2004, 2007) at the Friendship Hospital of Dalian in China reported the preliminary clinical observation on interventional Gendicine injection guided by ultrasonography (US) combined with sequential chemoembolization for cancerous embolization in portal vein, for treatment of HCC. 21G needle was inserted into portal vein tumor thrombi (PVTT) under US guide and Gendicine was injected slowly in the first day. On the third day, Chemoembolization will be finished in the same way. Results showed that in 16 cases of primary hepatocellular carcinoma (PHC) with PVTT treated by sequential therapy of Gendicine and Chemoembolization, the mass of tumor thrombus was decreased in all patients after two to three sequential therapies. Blood flow can be seen in the tumor thrombus under US. The clinical symptoms were partially relieved. There was no severe side effect related to therapy observed in all patients. It indicated that this measure is an effective, safe, and easy procedure.

49.3.2.1.6 Advanced Lung Cancer

Fifteen patients with stage IIIb–IV lung cancer were treated with Gendicine by intratumoral injection at a dose of 1×10^{12} VP (Weng et al., 2004). Gendicine was injected by percutaneous lung paracentesis under CT guidance once a week for 4 consecutive weeks as a treatment course. Clinical response was evaluated by CT image, tumor biopsy and a 2-month follow-up. The results were PR in 5 of the 15 (33.3%) patients, SD in 7 of the 15 (46.7%) patients, and PD in 3 of the 15 (20%) patients.

Tumor biopsy for 6 of the 12 patients having PR and SD revealed obvious tumor tissue necrosis and reduction in tumor cell number.

Fifteen patients with advanced lung cancer were treated with Gendicine in combination with conventional chemotherapy in Sichuan University (Guan et al., 2005b). Gendicine were infused into the bronchial artery via a catheter with a dose of $1–4 \times 10^{12}$ VP each week for 4 weeks. Two to five days after each Gendicine infusion, the patient was treated with cisplatin, 5-Fu, or etoposide (DDP 100 mg, 5-FU 1.0 g, and VP-16 100 mg). Of all the 15 patients, the total response rate (CR + PR) was 46.7% (seven out of 15). Tumor lesion disappearance was seen in one case, significant tumor regression was seen in six cases, mediastinal lymph nodes shrinkage was seen in three cases, and regression of thoracic fluid was seen in one case. Fourteen out of 15 (93.3%) of the cases had alleviation of clinical symptoms. Results from this preliminary study showed that Gendicine in combination with BAI is safe and effective in lung cancer treatment. Eight of the 13 patients showed positive response after a 1–4 month follow-up, including 1 CR, 6 PR, and one patient with thoracic liquid reduction. Eleven of the 13 patients showed alleviation of clinical symptoms such as cough, thoracodynia, hemoptysis, and dyspnea.

49.3.2.1.7 Advanced Soft Tissue Sarcoma

Clinical application of Gendicine in combination with hyperthermia and radiotherapy for treatment of advanced soft tissue sarcoma was studied by Xiao et al. (2007) (Figure 49.6). Thirteen patients with advanced soft tissue sarcoma were enrolled. All patients received intratumoral injection of Gendicine at 1×10^{12} VP once a week for a total of 8 weeks. Two days after Gendicine injection, all patients received additional hyperthermia treatment, once or twice a week for a total of 9 weeks. Twelve patients were concurrently treated with radiation therapy with the conventional fractionation of 2 Gy/f, five fractions a week to a total dose of 16~70 Gy in 8~35 f spanning 2~8 w. The average total radiation was 56.3 Gy. One patient was concurrently treated with DDP-based chemotherapy, once a month with a total two cycles. Patients were monitored for adverse events and tumor response. The response rate was assessed by CT analysis 2 months after treatment. Among the 13 patients CR was observed in two cases (15.4%), PR in four cases (30.8%), and SD in seven cases (53.8%). The 1-year, 2-year, and 3-year survival rates were 75%, 50%, and 33.3%, respectively. Intratumoral injection of Gendicine in combination with hyperthermia was found to be safe and effective for treatment of advanced soft tissue sarcoma.

49.3.2.1.8 Pancreatic Cancer

Chen Jie et al. (2007) at Department of Gastroenterology, Chang Hai Hospital of the Second Military Medical University, in Shanghai, treated eight patients with unresectable pancreatic cancer by intratumoral injection of Gendicine

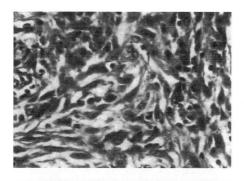

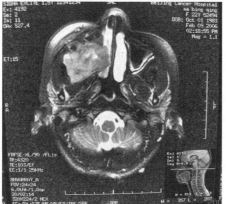

Before Gendicine treatment

Eiomyosarcoma

13 Months after 6 injections of Gendicine in combination with a 50 Gy radiation therapy

Tumor showing CR

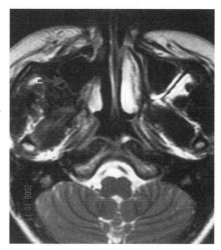

After Gendicine treatment

FIGURE 49.6 (See color insert following blank page 206.)   Efficacy of Gendicine in combination with radiation therapy for treatment of eiomyosarcoma.

with a paracentetic needle guided by endoscopic ultrasound (EUS) once a week for 4–5 weeks. Gimcetabin chemotherapy was used at the same time. The average treatment time was about 10 min. After treatment, one patient complained of a burning sensation in the stomach and a flushed face, and two patients felt febrile (37.5%). All these symptoms subsided spontaneously. One patient died of cancer during the course of the treatment. Significant relief of the pain caused by the cancer was observed in all the remaining seven patients. One patient experienced a complete relief from the pain and gained 2 kg weight in 2 months. Image scans showed stable tumor lesions in all patients. EUS-guided intratumoral injection of Gendicine appeared to be a safe and reliable approach for treatment of unresectable pancreatic cancer.

49.3.2.1.9 Esophageal carcinoma

Lu et al. (2007) at Xinxiang Medical University in China, observed intratumoral injection of Gendicine by endoscopy in combination with radiotherapy for treatment of 15 cases with esophageal carcinoma. A prospective randomized control method was adopted. Thirty patients with unresectable esophageal cancer were randomly divided into two groups. The 15 cases used in GTRT group, or radiotherapy alone. Gendicine was intratumorally injected once a week for a total of 6 weeks and radiotherapy was given 3 days after the first injection of Gendicine. Then the patients were followed up every 2 months. In combination group, 5/15 patients reached CR (33.3%) and 8/15 patients PR (53.3%). In radiation therapy group, 2/15 patients CR (13.3%) and 8/15 patients PR (53.3%). Compared to the efficacy rate between the two groups, $P > 0.05$ and CR was 2.5 times higher in combination group than in radiation therapy group. Eight of 30 patients presented with self-limited fever and 10/30 patients felt pain in the injection region; no other side effects were noted.

49.3.2.1.10 Other Advanced Solid Tumors

Ding et al. (2005) at the Cancer Center of the Sun Yat-Sen University reported the use of Gendicine for treatment of 24 patients with advanced solid tumors (comprising 13 distinct tumor types). All the patients failed conventional therapies, such as chemotherapy and radiation therapy. The treatment regimen was 1×10^{12} VP per week for a total of 4 weeks. Administration routes included intratumoral injection, intrabronchial spray, intraperitoneal injection, artery infusion, and intravenous injection. Gendicine in combination with a variety of therapies was used, including 18 cases with chemotherapy, two cases with radiation therapy, one case with radio- and chemotherapies, one case with abdominal thermotherapy and Gefitinib, one case with immunotherapy, and one case with Gendicine treatment alone. One patient withdrew from the study due to early disease progression. Twenty patients received four injections of Gendicine, two patients received eight injections of Gendicine, and one patient received 20 injections of Gendicine. Results from the 21 of valuable cases showed PR in 5/21 (24%) cases, SD in 5/21 (24%) cases, and PD in 11 cases. Clinical responses were observed in 10/21 (48%) patients. A common side effect was grade I–II self-limited fever, with two patients experiencing grade III fever. Other rare side effects were pain at the injection site, shivering, and muscle soreness.

Zhang et al. (2003a) at the Department of Radiation Therapy of the Beijing Cancer Hospital reported a clinical study of Gendicine in combination with thermotherapy in seven patients with advanced malignant tumors. Gendicine was injected intratumorally at 1×10^{12} VP per week for a total of 4 weeks. On day 3 after Gendicine administration, hyperthermia was applied to advanced malignant tumors using either a 915 MHz microwave to warm tumor tissue up to 43°C–44°C or heating by a 40 MHz to increase the temperature to 42°C–43°C. The data showed one CR, two PR, and four SD at the end of eight doses of Gendicine treatment. Biopsy for CR and PR cases revealed tumor necrosis (Figure 49.7).

Zhang et al. (2005) reported a study on 21 patients with advanced malignant tumors receiving Gendicine treatment. As a treatment course, patient received weekly Gendicine injection for a total of 4 weeks. All patients were assessed for response and toxicity by clinical observation, CT or MRI and short-time follow-up. Significant reduction in tumor volume ($t = 2.04$, $P < 0.05$) and the Karnofsky score ($t = 2.66$, $P < 0.05$) (Tables 49.10 and 49.11) improved after Gendicine treatment.

Another study of Gendicine on malignant tumor patients was carried out by Qi et al. (2005) at Beijng Tongren Hospital. Twenty-three patients with advanced cancers were treated with Gendicine by intravenous drip and intratumoral injection. Ten patients were treated with Gendicine alone. Three of the 10 patients exhibited PR, 4 of the 10 patients had stable disease at the last time of follow-up, and 3 of the 10 patients showed progressive disease. Total clinical response was 70%. The other 13 patients were treated by Gendicine in combination with chemotherapy. Of the 13 patients, 6 showed PR (6 of 13 patients, 46.2%), 4 showed SD (4 of 13 patients, 30.8%), 3 showed PD (3 of 13 patients, 23.1%). On average, the Karnofsky Performance Score increased over 10 for both groups. Total clinical response was 76.9%.

49.3.2.1.11 Cancerous Ascites

A significant effect was reported using Gendicine to treat cancerous ascites (Zhu et al., 2005). Thirteen patients with advanced cancers (eight patients with gastric carcinomas, four with colon carcinoma, and one with carcinoma of the gall bladder) and having large amount of ascites were treated with Gendicine via peritoneal paracentesis and intraperitoneal infusion at a dose of $1–2 \times 10^{12}$ VP per week for a total of 4 weeks. Six of the 13 patients also had jaundice. Clinical response was evaluated through measurement of abdominal girth, CT, or MRI and 1 month of follow-up. After 3 weeks of treatment, 7 of the 13 patients showed significant reduction in ascites build-up, alleviation of disease symptoms, such as abdominal distention and shortness of breath, and improvement in Karnofsky performance status ($P < 0.05$). Except for the development of self-limited fever, no other side effects were observed. These results suggest that Gendicine can be used to treat patients with cancerous ascites for relief of disease symptoms and improvement of the quality of life.

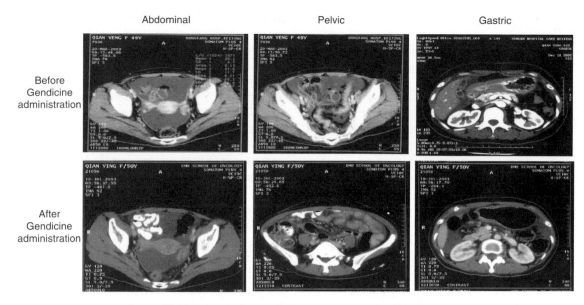

Patient: Y. Qian, female 49 years old, case # 9529300, Beijing Tumor Hospital.
Advanced gastric cancer (Borrmann IV) with less differentiated gastric adenocarcinoma
showing huge gastric antral tumor with all around adhesion and extensive metastatic
nodes in abdominal and pelvic cavities, and thickened peritoneum and omentum, as well
as large amounts of ascitic fluid in abdominal cavity. As the tumor was inoperable,
patient received conventional chemotherapy for 3 months; however, it was ineffective
and did not stop the tumor progression.
Patient was perfused abdominally with Gendicine, 1×10^{12} VP in 1000 ml physiological
saline twice a week, in combination with DDP 40 mg or 5-Fu 1000 mg. After 3 weeks of
treatment, her ascites disappeared and the severe side effects caused by chemotherapy
were relieved. After Gendicine treatment for 12 weeks, CT imaging showed that
abnormally thick gastric antrum with all around adhesion were obviously reduced, and so
did the thickened peritoneum and omentum. Tumor recurrence was shown
negative by CT imaging after 12-month follow-up.

(A)

(B)

FIGURE 49.7 (A) Efficacy of Gendicine in combination with chemotherapy for treatment of advanced gastric carcinoma. (B) Comparison of biopsy and blood vessel growth in gastric carcinoma before and after Gendicine administration.

TABLE 49.10
Tumor Size and Patients' KPS Score before and after Treatment

	n	Tumor Size	KPS Score
Before treatment	21	17.21 ± 8.77	43.33 ± 11.97
After treatment	21	11.72 ± 8.66	57.14 ± 20.53

Liu et al. (2006) at Xijing Hospital, Xi'an, Shanxi, treated 32 patients with malignant mesothelioma and alimentary tract carcinoma, including esophageal carcinoma, gastric cancer, pancreatic cancer, colon cancer, diagnosed by pathological evidence or image and serum tumor markers. Following tumor effusions, Gendicine was administered via peritoneal perfusion in 100 mL normal saline. After 72 h, the patients received radiofrequency hyperthermia. The treatment was performed with a frequency of one cycle per week. The volume of the effusions was examined at the beginning and 2 weeks later. There were two cases of CR, three PR, and three patients with no discernable effect (NE) in eight patients with malignant mesothelioma. The clinical response rate was 62%. There were 5 CR (1 case of esophageal carcinoma, 2 gastric cancer, 1 colon cancer), 7 PR (1 esophageal carcinoma, 4 gastric cancer, 2 colon cancer), and 12 NE in the 24 malignant alimentary tract cancer patients. The overall clinical response rate was 50%.

Another study of the use of Gendicine in combination with chemotherapy for treatment of malignant body cavity effusion was reported by Cao et al. (2005), from Hebei Provincial Qinhuangdao Seaport Hospital. Fifty patients with malignant body cavity effusion were randomly divided into two groups. The treatment group were given intracavitary administration of Gendicine 1×10^{12} VP after puncture drainage, which was followed 48 h later by intracavitary administration of 60 mg/m^2 cisplatin once a week for 3–4 weeks. The control group was given the same intracavitary therapy as the treatment group but without Gendicine. The total response rates of the treatment group and the control group were 85.7% and 51.7% ($P < 0.05$), respectively. The common side effect was self-limited fever.

Results from the above extensive clinical application of Gendicine show that Gendicine is a wide-spectra anticancer agent with minimal side effects. Gendicine can be used in monotherapy as well as in combination with traditional cancer treatment approaches such as chemotherapy, radiotherapy, and surgery, etc. In addition to achieve synergistic effects with the traditional therapies, Gendicine treatment can also alleviate the common side effects caused by radiotherapy and chemotherapy and achieves a better improvement in the patients' quality of life.

49.4 MANUFACTURE OF GENDICINE

The following is a general description of the large-scale production technologies that are used for the manufacture of Gendicine.

49.4.1 LARGE-SCALE CELL CULTURE TECHNOLOGY

Large-scale culture of the production cells is a critical step in the production of Gendicine. The technology has undergone significant development in the past decade, from the first-generation adherent cell culture using roller bottles, CellCube bioreactor (Corning Life Sciences, Acton, MA), and packed bed Celligen plus bioreactor (New Brunswick Scientific, Edison, NJ), to the second-generation suspension, serum-free culture in large-scale bioreactors. Gendicine is produced using a patented large-scale bioreactor system. The patented producer cell line, SBN-Cel, is a subclone derived from the HEK293 cell line through a repeated cell clone selection processes. The SBN-Cel clone showed stronger attachment to culture surface, faster growth rate (doubling time approximately 18 h) and better viral productivity than the parental HEK293 cells.

49.4.2 CHROMATOGRAPHY PURIFICATION TECHNOLOGY

For large-scale production of Gendicine, a downstream processing and automated chromatography purification process have been developed and optimized. Downstream processing

TABLE 49.11
Clinical Response after Treatment

Diagnosis	n	Clinical Responses				Effective Rate (CR + PR) (%)
		CR	PR	NC	PD	
NPC	6	0	5	1	0	83.33
NSCLC	5	0	1	4	0	20
HCC	4	0	0	2	2	0
Metastasis cancer	6	0	2	3	1	33.33
Total	21	0	8	10	3	38.10

includes tangential flow filtration for harvest clarification and tangential flow ultrafiltration for concentration and diafiltration. The concentrate is further treated with Benzonase to breakdown large cellular DNA. The material is further clarified and chromatographically purified using an automated chromatography system (FPLC). After a single-step column purification, more than 98% purity can be achieved for the Gendicine final product. Approximately 4×10^{15} VP purified final product can be produced from a single batch 14 L bioreactor run. Both master/working cell banks and master/working virus banks (MVB/WVB), which are critical raw materials for the commercial production of Gendicine, have been established and qualified.

49.5 QUALITY CONTROL AND QUALITY ASSURANCE OF GENDICINE

In response to the rapid development in the gene therapy field and to promote gene therapy research and eventual commercialization in China, The SFDA published "Points to Consider for Human Gene Therapy and Product Quality Control" in 2003 (SFDA 2003a,b; SiBiono 2004). This document was published in the *Biopharm International* journal in May 2004. This guideline is one of the most comprehensive documents published by a government agency for gene therapy product quality control. It specifies that extensive quality control testing should be performed during the manufacture of gene therapy products. Product release testing should be performed

on crude harvest, bulk product (drug substance), and the final drug product. However, depending on the nature of the manufacturing process, some testing should be performed on in-process samples if the excipients included in the final product formulation interfere with the testing.

Table 49.12 outlines the SFDA specified 12 quality control tests, and the 21 test methods and specifications performed on in-process and final product samples during the manufacture of Gendicine. The tests can be divided into the following three groups.

49.5.1 GENERAL TESTS REQUIRED FOR BIOPHARMACEUTICALS

This group includes physical characteristics, chemistry characteristics, purity and detection of residual impurities, sterility testing, mycoplasma testing, general safety testing, and bacterial endotoxin testing. These tests are generally applied to all biopharmaceuticals.

49.5.2 BASIC TESTS

This group includes identity and efficacy tests. The purpose of these tests is to demonstrate the presence of the correct therapeutic gene, gene expression, and biological activity of the expressed gene product. Restriction mapping and PCR analysis are used to identify the presence of the correct therapeutic gene. Efficacy testing comprises a

TABLE 49.12
Quality Control Methods and Specifications for Gendicine

No	Assays		Method	Specification
1	Physical characteristics	Appearance	Examine under light	Opalescent
		Recoverable volume	Capacity assay	≥ 1.5 mL/dose
2	Chemical characteristics		pH	8.0–8.5
3	Identity	RE mapping	Restriction mapping	Given segment
		p53 gene	PCR analysis	396 bp
4	Purity		HPLC	$\geq 95.0\%$
5	Viral titer	VP	A260 assay	$\geq 1.0 \times 10^{12}$ VP/dose
		Infectious Unit	TCID50	IU
		IU/VP		$\geq 3.3\%$
6	Gene Efficacy	Gene expression	Western blot	Positive
		Bioactivity	Saos-2 cell bioassay	Positive
7	RCA		A549 cell bioassay	$\leq 1 RCA/3 \times 10^{10}$ VP
8	AAV Detection		PCR	Negative
9	Residuals	BSA	Cell agglutination	≤ 50.0 ng/dose
		HCP	ELISA	≤ 100 ng/dose
		Cell DNA	Southern blot	≤ 10 ng/dose
		Benzonase	ELISA	≤ 1 ng/dose
10	Microbe	Sterility testing	Membrane filtration	Negative
		Mycoplasma testing	DNA fluorescent	Negative
11	General safety		Guinea pig and suckling mice	Safe
12	Endotoxin		LAL	≤ 10.0 EU/ dose

measurement of gene expression and biological activities of the expressed gene product. Gene expression is measured by Western blot or enzyme-linked immunosorbent assay (ELISA) of cells infected with the rAd product (e.g., expression of p53 protein in H1299 cells). Bioactivity of the expressed gene product is assayed on a specific cell infected with the rAd product (e.g., induction of cell apoptosis by p53 on SAOS-2 cells).

49.5.3 Unique Tests

These tests include virus particle titer determination, infectivity, IU/VP ratio, and detection of replication-competent adenovirus (RCA).

49.5.3.1 Virus Particle Titer Determination

This procedure is generally done using the A260 UV absorption method. In the presence of sodium dodecyl sulfate (SDS), one absorption unit at 260 nm equals to 1.1×10^{12} VP/ml.

49.5.3.2 Infectivity

Infectivity is measured using a $TCID_{50}$ method by serial dilution. Infectivity titer is calculated using the formula: $T = 10^{1+d(S-0.5)}$ (IU/100 μl), wherein d is dilution log S is the sum of infection rate from the highest dilution.

49.5.3.3 IU/VP Ratio

According to the SFDA guidelines, the specific activity (IU/VP ratio) of clinical grade rAd needs to be at least 3.3%. Gendicine generally has an IU/VP ratio of 4.0% with an infectious titer of $4–5 \times 10^{10}$ IU/ml, exceeding the guideline requirement.

49.5.3.4 Purity Test

Although purity determination is not unique to rAd product, the use of A260/280 as a purity indication for rAd is unique and should be in the range of 1.2 to 1.3. High-performance liquid chromatography (HPLC) is also used for purity determination. According to HPLC analysis, Gendicine generally has a purity of greater than 98%, exceeding the SFDA guideline specification.

49.5.3.5 Detection of RCA

Determination of RCA level is an important safety criterion for rAd product. RCA can arise by homologous recombination between adenovirus vector and the host cell genome during the adenovirus production process. Amplification in A549 cells is generally used to detect the presence of RCA. The RCA level in Gendicine meets the less than $1RCA/3 \times 10^{10}$ VP, as specified in the SFDA guideline document.

In addition to performing all the necessary quality control tests, Gendicine is produced following strict good manufacturing practices (GMP) regulations, as outlined in ICH Q7A guidelines. SiBiono has instituted an independent quality assurance function to ensure the consistent production of high-quality Gendicine product.

In the seven years since the submission of phase I clinical trial application, in March 1998 to March 2004, when the GMP production certificate was issued, Gendicine underwent five stringent reviews and approval steps. These steps included application for initiation of a phase I clinical trial, application for initiation of phase II/III clinical trials, submission of a new drug license application, submission of a production license application, and submission of a GMP certification application. Strict review by an expert advisory team is conducted for each phase of the clinical trial application. Clinical trials can only begin after the sponsor satisfactorily addresses any questions the reviewers have and the application is approved.

After the approval of the new drug license, the Guangdong Food and Drug Administration inspected SiBiono's facility and submitted three batches of sealed samples to the National Institute for the Control of Pharmaceutical and Biological Products (NICPBP) for lot release testing. The SFDA granted the production license only after three lots of NICPBP successfully passed all the tests. China's Drug Law specifies that new drug has to obtain production license first before the sponsor can apply for GMP certificate for its facility. Prior to market launch, the sponsor has to apply for pricing approval from the State Price Bureau. There are two stages of GMP certificate approval. The first GMP certificate is valid for 1 year. A second GMP certificate application needs to be resubmitted in the second year. After review and approval, the second GMP certificate is valid for 5 years. Gendicine's production facility and the more than 1100 SOPs used for the production of Gendicine all passed the second GMP certification.

49.6 SUMMARY AND PROSPECTS

Gendicine, the first gene therapy product approved by a government agency, has been shown to be safe and effective for treatment of a variety of cancers. Since its market introduction in China in April 2004, more than 6000 cancer patients of diverse ethnic background have received Gendicine treatment. Although effective as a monotherapy, Gendicine showed synergistic therapeutic effects when used in combination with other traditional cancer treatment methods such as surgery, chemo- and radiotherapies. More importantly, Gendicine also alleviated the deleterious side effects commonly associated with both chemo- and radiotherapies.

As the first gene therapy product for a revolutionary new way of cancer treatment, significant basic research and clinical studies are required to understand the complex biological and therapeutic mechanisms of Gendicine's action. Dr. Lu and his colleagues provide a comprehensive review of the biological mechanisms of the roles of the p53 tumor suppressor gene in Chapter 48. Working closely with clinical collaborators, our scientists at SiBiono have begun to elucidate the therapeutic mechanisms for Gendicine. We believe data from the additional postapproval clinical studies that are currently underway at Sibiono to broadening the approved clinical indications for Gendicine will further our understanding of the therapeutic mechanism and provide insight for the development of the next generation of cancer gene therapy products.

49.7 POINTS TO CONSIDER FOR HUMAN GENE THERAPY AND PRODUCT QUALITY CONTROL*

49.7.1 INTRODUCTION

Gene therapy is a medical intervention based on the modification of the genetic material of living cells. Currently, gene therapy is restricted in application to somatic cells. Based on transferring methods, gene therapy can be classified into two categories: *ex vivo* and *in vivo*. *Ex vivo* gene therapy refers to cells being modified *ex vivo* for subsequent administration to humans, while *in vivo* refers to cells being altered *in vivo* by giving gene therapy directly to the subject. The products of *ex vivo* gene therapy are cells that are modified and are intended to be administered to the patient. *Ex vivo* gene therapy is expected to be performed in well-established medical care establishments with specially trained medical professionals and GMP facilities. The products of *in vivo* gene therapy are recombinant DNA or RNA in the form of naked DNA, DNA complex, or viral vectors that are manipulated by genetic technologies. Both *ex vivo* and *in vivo* gene therapy products are subject to the regulations in this guidance. Because of the complexities of the different modalities of gene therapy, it is not possible to generalize a common guidance that is suitable for all kinds of products. However, the following basic principles should be followed when sponsoring the development of a gene therapy product:

1. Safety and efficacy of the product should be ensured. A comprehensive assessment of the benefit and risk of the product should be conducted.
2. New and innovative ideas should be promoted when sponsoring a gene therapy product development. Considering the uniqueness of gene therapy relative to traditional chemically synthesized and genetically engineered protein medicines, there will be certain flexibilities for the regulation of novel gene therapy products. Gene therapy as a form of medical intervention is still in its early phase of development. SFDA expects the sponsors to not just follow this guidance but also conduct rigorous scientific study to ensure the development of a safe and efficacious gene therapy product.

---

* This document was worked out and translated into English mainly by Dr. Zhaohui Peng, SiBiono GeneTech Co., Ltd. Shenzhen 518057, China, and officially issued by SFDA on March 20, 2003. Up to date, it is a systemic gene therapy guideline.

49.7.2 APPLICATION CONTENT AND PRODUCT QUALITY CONTROL

The following should be included in the application proposal.

49.7.2.1 Construction of the DNA Expression Cassette and the Gene Delivery System

49.7.2.1.1 Therapeutic Target Gene
A detailed description of the clonal origin of the therapeutic target gene, including a patent search on the gene, should be provided. Method of gene cloning and sequence identity should be provided.

49.7.2.1.2 Vector
Information supplied should include restriction mapping and gene-bank data for the vector. Known regulatory elements such as promoters, enhancers, and PolyA should be identified. If there is change in the vector backbone gene structure (such as deletion, mutation, or insertion), the DNA sequence data should be provided. For a new viral vector, it is necessary to provide information on the material, method of construction, and testing of the new vector.

For nonviral gene delivery systems, plasmid is needed to express the target gene in human cells. In addition to naked DNA, another component is generally used to form complex with the DNA. This guidance does not cover oligoribonucleotide (such as antisense RNA, ribozyme, and siRNA) products.

49.7.2.1.3 DNA Expression Cassette
A detailed description of the cloning procedure, the methods and materials used, and DNA sequence data should be given. Known regulatory elements such as promoters, enhancers, and PolyA should be identified. Restriction mapping of the gene expression cassette and the kinetics of gene expression should be provided.

49.7.2.1.4 Construction of the Gene Delivery System (Including Viral and Nonviral Gene Delivery Systems)
49.7.2.1.4.1 Viral Gene Delivery System Including Adenoviral Vector, Retroviral Vector, and Adeno-Associated Viral (AAV) Vector
A thorough description of the clonal origin of the viral vector should be provided. The methods and materials used for the construction of the viral vector should be included. Testing methods and results should also be provided. General testing should include structural analysis (e.g., restriction mapping and PCR), complete sequencing of the viral genome ($\leq 40\,kb$), gene expression and bioactivity analysis, sodium dodecyl sulfate-polyacrylamide gel electrophoresis (SDS-PAGE), DNA sequencing of the gene expression cassette, Western blot analysis of the expressed protein, transduction efficiency analysis, negative-staining transmission electron microscopy of the purified viral vector, replication-competent virus detection, testing for adventitious agents, and analysis for residual process contaminants.

49.7.2.1.4.2 Nonviral Gene Delivery System

Nonviral gene delivery systems encompass naked DNA, mammalian cell carrier systems, and other carrier systems such as liposome, polypeptide, and gold particles. The nature and characteristics of the nonviral delivery system should be described adequately. To prevent allergic reaction to penicillin in some patient populations, it is recommended that a kanamycin- or neomycin-resistant gene be used as the drug selection gene.

For the physical delivery system, a detailed description of the delivery method, procedure, efficiency of gene delivery and expression, gene stability after delivery, and bioactivity should be included. Evidence of absence of gene arrangement and mutation should be provided. Detailed testing results should be given including plasmid restriction mapping, PCR analysis, DNA sequencing of the gene expression cassette, SDS-PAGE, and Western blot analysis of the expressed protein.

49.7.2.2 Generation and Characterization of Cell Banks and Engineered Bacteria Banks

A three-tier bank system should be established. These typically include a primary seed bank, a master cell bank, and a working cell bank. For engineered bacteria, these typically include a primary bacteria bank, a master bacteria bank, and a working bacteria bank.

49.7.2.2.1 Cell Bank

Primary seed bank. The origin, cell passage history, and cell characterization, as well as cell culture and banking procedures, should be described.

Master cell bank. The master cell bank should be derived from one or more ampoules of the primary seed bank by serial subculture to a specific passage number. The passage number should be identified.

Working cell bank. The working cell bank is derived from one or more ampoules of the master cell bank by serial subculture to a specific passage number. The passage number should be identified.

Characterization of cell banks. The master cell bank or working cell bank should be characterized according to guidelines in "Guidance for application of clinical trials for human gene therapy." Additional testing should include:

1. Testing of susceptibility of virus infection and production
2. Testing the status and stability of the transfected gene and level of gene expression
3. Testing for generation of replication competent viruses

If a helper virus is needed in the testing, the origin, methods of preparation and separation, and virus passage history should be provided.

49.7.2.2.2 Bacteria Bank

A master bacteria bank and a working bacteria bank should be established and tested according to the specifications of the "Chinese Pharmacopoeia."

1. Testing of the master bacteria bank
   (a) Uniformity and the identity of bacteria host strain: Testing should include the origin, genotype, and phenotype of the bacteria strain. The genotype should be tested by random amplified polymorphic DNA (RAPD). The phenotype should be tested for the specific marker or antibiotic resistant gene marker of the engineered plasmids.
   (b) Purity: The bacteria should be shown to be free of other adventitious agents, such as fungi and bacteria.
   (c) Stability: The ratio of transformed bacteria, propagation condition, copy numbers, level of gene expression, and passage number should be qualified.
   (d) Sequencing of gene expression cassette: The inserted gene of interest and associated regulatory elements should be sequenced.
2. Working bacteria bank. Testing of the working bacteria bank should be performed in accordance with the requirements for the master bacteria bank mentioned above with the exception of the sequencing of the gene expression cassette.

49.7.2.3 Manufacture of Gene Therapy Product

49.7.2.3.1 General Requirements

1. Description of manufacturing facility and environment GMP regulations should be followed in the manufacturing of gene therapy products. This rule applies to both the *ex vivo* cell products and recombinant viral products. Recombinant viral products intended for preclinical and clinical studies should be manufactured using a validatable production process.
2. A detailed description of the manufacturing process, raw materials, and the components used should be provided.
3. Manufacturing process controls should be provided. Critical process parameters should be identified, controlled, and recorded.
4. Batch production records and testing records of one lot of product should be provided.
5. One lot of product should be tested by institutes that are designated by SFDA for lot release testing. The testing report should be provided.

49.7.2.3.2 Recombinant Viral Vector as Gene Therapy Product

An MVB and a WVB should be established and tested. The WVB should be derived from the MVB. A detailed description of the origin, construction, cloning, passage, and storage of the virus bank should be provided. Quality control testing procedures outlined in this document should be followed. Production should be initiated from producer cells derived from working cell bank. Nonqualified cells should not be used for production purposes. For a viral vector product, the producer cell infection should be carried using a virus from the WVB. Infection should not be performed using a nonqualified virus. Production can be performed using large-scale attachment-dependent cell or suspension cell culture. Virus banks can be prepared directly from a cell culture without further purification. Buffer formulation for virus banks should be qualified to protect virus infectivity during storage.

49.7.2.3.3 Nonviral Plasmid DNA Complex as Gene Therapy Product

A detailed description of the following aspects should be provided:

1. Manufacturing process and quality control of the recombinant plasmid DNA
2. Origin, characteristics, and method of preparation of the liposome if used
3. Origin, characteristics, and method of preparation of the polypeptide if used
4. Origin, characteristics, and method of preparation of other components in the product complex
5. Method of preparation and quality control of the final product complex

For gene therapy products in the form of gene gun and other physical gene delivery approaches, detailed description of the following items should be provided:

(a) Origin and preparation of the therapeutic gene
(b) Gene expression
(c) Propagation of the naked DNA
(d) DNA purification
(e) Equipment characteristics
(f) Method of gene delivery

49.7.2.3.4 Genetically Modified Somatic Cells as Products

This category encompasses *ex vivo* gene therapy products. Manufacture of this category of products should follow guidelines in "Guidelines for Study of Somatic Cell Therapy and Quality Control." Detailed description of the process for cell propagation, media used for cell culture, method of cell collection, method of *ex vivo* cell transduction and selection, and formulation used for washing and storage of the transduced cells, as well as testing performed on the transduced cells should be provided.

49.7.2.4 Quality Control

Sufficient quality control testing should be performed during the manufacture of gene therapy products. Product release testing should be performed on the final drug product. However, depending on the nature of the manufacturing process, some testing should be performed on process samples if the excipients included in the final product formulation interfere with the testing.

49.7.2.4.1 Recombinant Viral Vector as Gene Therapy Product

In reference to the domestic and international viral vector development status, testing of a recombinant adenovirus (rAd) product is used as an example of the quality control testing for viral vector products. The testing can be used as a reference for other viral vector products.

49.7.2.4.1.1 Quality Control for Rad Gene Therapy Product

1. Crude harvest
   (a) Sterility tests to be performed in accordance with "SFDA Guidance for Biological Products, subpart—Sterility Testing" 2000 edition.
   (b) Measurement of virus particles
   (c) Titration of infectious virus titer
   (d) Mycoplasma testing
2. Bulk product (drug substance)
   (a) Sterility test
   (b) Measurement of virus particles
   (c) Endotoxin testing
3. Final product (drug product)
   (a) Particulates
   (b) Physical appearance
   (c) Recoverable volume
   (d) pH
   (e) Identity testing by restriction mapping analysis of the gene expression cassette
   (f) Purity: A260/280 nm or HPLC
   (g) Virus particle concentration: Spectrophotometric OD260 method
   (h) Potency: Testing level of gene expression and bioactivity testing
   (i) RCA. A549 cell-culture detection method (no more than 1 RCA in 3×10^{10} VP)
   (j) AAV. Testing of the presence of AAV using PCR
   (k) Testing of residuals. Residuals, such as host cell DNA, host cell protein, bovine serum albumin for process using fetal bovine serum, and other residuals specific to the production process should be tested and quantified
   (l) Sterility testing
   (m) General safety testing
   (n) Endotoxin testing
   (o) Infectivity testing (such as infectious titer by TCID50)
   (p) Stability study: Stability of three batches of final product during real-time storage, freeze, and thaw; and accelerated stability study should be provided

49.7.2.4.2 Quality Control for ex Vivo Gene Therapy Using Retrovirus Gene Therapy Product

Manufacture of gene therapy products for *ex vivo* transduction of somatic cells should follow guidelines in "Guidelines for Study of Somatic Cell Therapy and Quality Control." Testing for replication-competent retroviruses (RCR) should be emphasized. Testing should be performed for the master cell bank, the working cell bank, end-of-production cells, and final virus products. Five percent of the total supernatant should be tested for RCR. One percent of total pooled end-of-production cells or 10^8 cells, whichever is less, should be cocultured with a permissive cell line (such as *Mus dunni* cells) for a minimum of five passages in order to amplify any potential RCR present. Then, the amplified material is detected in a combination of two appropriate indicator cell assays, such as PG4 S+L- cells, maker rescue, or RT/PCR method. All assays should include positive (such as supernatant from COS4070A cells) and negative controls to assess specificity, sensitivity, and reproducibility of the methods used.

49.7.2.4.2.1 Retroviral Vector Producer Cells as Product
Testing should include:
- (a) Morphology, karyology, surface marker and homogeneity of the cells
- (b) Cell viability and concentration
- (c) Stability, copy number, and level of expression of the inserted gene (particularly, describe the risk of insertional mutagenesis in target cells and method of evaluation)

49.7.2.4.2.2 Quality Control of Thawed Cells
1. These tests should be performed for thawed cells:
   - (a) Percentage of cell recovery and viability after thaw
   - (b) Transduction efficiency
2. These tests should be performed for supernatant from thawed cells:
   - (a) Sterility testing
   - (b) Endotoxin testing
   - (c) General safety testing
   - (d) Transduction efficiency
   - (e) Mycoplasma testing
3. Retroviral vector supernatant as product

The following testing should be performed:
   - (a) Virus titer
   - (b) Restriction mapping
   - (c) Transduction efficiency
   - (d) RCR testing (it is critically important to test the level of RCRs in the product considering the nature of chromosomal integration of retrovirus and associated potential genetic toxicity)
   - (e) Testing for sterility and endotoxin
   - (f) Measurement of residual bovine serum albumin (BSA)
   - (g) Mycoplasma testing

Note: For cell products, the expiration is very short. In those cases, the cell product can be released pending results from testing that requires a longer time. If testing results are positive, the cell product should be immediately recalled and investigation should be initiated for the positive results.

49.7.2.4.3. Nonviral DNA Vectors as Gene Therapy Product
The following testing should be performed:

1. Purity: Testing for percentage of supercoiled DNA, residual bacterial RNA and DNA, and residual bacterial host protein

2. DNA restriction mapping
3. Origin and characteristics of liposome, polypeptide and gold particles used (due to the inherent variation in the production of liposome and polypeptide, measures should be implemented to ensure the consistency of the production process)
4. Sterility testing
5. Endotoxin testing
6. Allergy testing
7. General safety testing
8. Bioactivity testing
9. Testing for process residuals

49.7.2.5 Evaluation of Efficacy for Gene Therapy Product

49.7.2.5.1 In Vitro Testing
Data should be provided to show therapeutic effect when the exogenous gene is delivered into the target cells. For *ex vivo* cell therapy, data should be provided to show the gene expression level in the transduced cells.

49.7.2.5.2 In Vivo Testing
Animal study data should be provided to show the gene expression level in the target tissue cells, biodistribution of the product and gene expression, and therapeutic effect. The selection of the animal model should be justified. The route of administration should simulate that of clinical situation. If limited by the animal model, and the clinical route of administration cannot be used, justification should be provided for selection of other routes of administration. If efficacy cannot be produced in animal model studies, surrogate indicators for efficacy should be provided.

49.7.2.6 Safety Evaluation

In addition to the above quality control requirements, the following product safety information should also be provided.

49.7.2.6.1 Overall Safety Evaluation
For *ex vivo* somatic cell therapy products, the following information should also be provided:

1. For growth-factor-dependent cells, cells cannot be used if loss of growth-factor-dependence is observed during cell expansion.
2. For allogeneic somatic cell therapy products, immunological safety data should be provided.
3. For xenogeneic somatic cell therapy products, the survival of the xenogeneic cells and *in vivo* safety data should be provided.
4. Tumorigenicity testing should be performed for all somatic cell therapy products. For cancer vaccine products, procedures taken to attenuate the growth potential of cells derived from the tumor tissue should be provided. The tumorigenicity testing should include testing in nude mice and colony formation in soft agar.
5. Testing for cell homogeneity: For cancer vaccine products, measures should be provided detailing the

separation of nontumor cells from tumor cells. If nontumor cells are selected from tumor cells (e.g., lymphocytes), measures used to eliminate tumor cells should be provided.

6. Testing for replication competent viruses—See related parts of these guidelines.
7. Additives. In addition to the additives used for cell culture and product storage, additives are used in some *ex vivo* or *in vivo* gene therapy products. The safety profile of those additives should be provided from animal studies.

49.7.2.6.2 Genetics Study

For *in vivo* gene therapy products, the biodistribution of the viral and nonviral DNA products and expression of the recombinant DNA in target and nontarget tissue should be provided. For *ex vivo* gene therapy products, the biodistribution and gene expression of the transduced cells should be provided. Risks for stem cell transduction should be carefully evaluated.

49.7.2.6.3 Toxicology Study

Toxicity evaluation is an important part of the safety testing.

1. For *in vivo* gene therapy products, a detailed description of the toxicology study should be given. In addition to the study of potential toxicity of the therapeutic gene product, a safety evaluation of the delivery system should be performed.

   The toxicology study should include an acute toxicity study (maximum tolerated dose) and a chronic toxicity study. A relatively wide range of doses should be tested, such that a reasonable dose–response curve for toxicity can be obtained. The dose range should include at least a dose equivalent to that to be used in a clinical trial and higher doses.

   The routes of product administration should simulate that to be used in clinical trial study. Justification should be provided if any other route of administration is used. Product doses based on either body weight or total body surface area are preferred to facilitate comparisons across species. In addition to observations for standard toxicities, genetic toxicity specific to gene therapy products should also be included.

2. For autologous cancer vaccine, lymphocyte, or macrophage *ex vivo* gene therapy products, toxicity study in an animal model is not required. However, if specific additives (e.g., gelatin, suture, or special catheters) are included in the *ex vivo* gene therapy product, or the gene therapy product is delivered to critical organs (e.g., heart, brain, or liver), a toxicity study in an animal model is required. If cell growth factors are supplemented in the *ex vivo* cell expansion process, a safety study is required on the growth factors. For subcutaneous and intramuscular

administration, information on any injection site reaction should be provided.

49.7.2.6.4 Immunology Studies

The immunological response, including anaphylaxis, repellency, and autoimmune reaction, should be evaluated for both viral and nonviral gene therapy products after administration. A strategy to deal with potential immunological reactions should be established. For gene therapy products that are used to elicit immunological responses, such as cancer vaccines, cytokines, or genetically modified immune cells, changes to the immunological system and their potential side effects should be described. A strategy for monitoring and resolving the potential side effects should be established. Refer to "Guidelines for Study of Somatic Cell Therapy and Quality Control" for details.

49.7.2.6.5 Tumorigenicity Studies

See related section in this guidance. (Author's note: For viral gene therapy product derived from recombinant retrovirus and AAV, tumorigenisis caused from insertional mutagenesis should be carefully studied.)

49.7.2.7 Clinical Trial of Gene Therapy Products

Clinical trials of gene therapy products are different from those with other biotechnology products. First, gene therapy products tend to be very complex. For example, *ex vivo* gene therapy products require a medical specialist to administer the product to the targeted tissues. In some cases, surgery is required in order to administer the product. Secondly, gene therapy product development is still in its early phase with possible unknown risks. Intensive patient monitoring should be implemented during clinical trial. The following information should be included when clinical trial plans are submitted.

1. Sponsor GMP compliance certificate
2. Clinical study site and biography of the principal investigator
3. Route of administration, dose, time and period of treatment (if surgery is required for product administration, detailed description of the surgery procedure is required)
4. General clinic status index and laboratory test
5. Signed consent forms of patients and family members
6. Molecular biology analysis of target and nontarget tissues
7. Recording and reporting of side effects
8. Follow-up strategy and plan
9. Medical intervention plan should be provided to deal with unexpected immune reactions

49.7.2.8 Ethics Studies

Special attention should be paid to medical ethics during clinical trials of gene therapy products. Details can be found in SFDA's GCP (Good Clinical Practice) regulations. The study plan and potential risks associated with the clinical study should be clearly communicated to the patient and family members. Patients have the right to choose medical treatment options and terminate participation in the gene therapy clinical trial study. The patient's medical history should be kept private. The patient cannot be enrolled in the study until the patient or family member signs the study consent form.

ACKNOWLEDGMENTS

We would like to thank Drs. Hua Lu, Mu-Shui Dai, and Ms. Jayme Gallegos at the Department of Biochemistry and Molecular Biology, School of Medicine, Indiana University, Indianapolis and Dr. Huijun Zou Ring, Director of Pharmacogenetics, DNA Direct Inc., San Francisco, California for active discussion and proofreading.

REFERENCES

Ala-aho R, Grenman R, Seth P, and Kahari VM. 2002. Adenoviral delivery of p53 gene suppresses expression of collagenase-3 (MMP-13) in squamous carcinoma cells. *Oncogene* 21(8): 1187–1195.

Bennett M, Macdonald K, Chan SW, Luzio JP, Simari R, and Weissberg P. 1998. Cell surface trafficking of Fas: A rapid mechanism of p53-mediated apoptosis. *Science* 282(5387): 290–293.

Brasseur G, Tron P, Dujardin G, Slonimski PP, and Brivet-Chevillotte P. 1997. The nuclear ABC1 gene is essential for the correct conformation and functioning of the cytochrome bc1 complex and the neighbouring complexes II and IV in the mitochondrial respiratory chain. *Eur J Biochem* 246: 103–111.

Cao XJ, Zhang Y, Li HL, Wang HP, Wang ZY, and Wei L. 2005. Treatment of malignant body cavity effusion with recombinant human p53 adenovirus injection combining chemotherapy observation of clinical effects. *China Pharmacy* 16(23).

Cerwenka A and Lanier LL. 2003. NKG2D ligands: Unconventional MHC class I-like molecules exploited by viruses and cancer. *Tissue Antigens* 61(5): 335–343.

Chen CB, Pan JJ, and Xu LY. 2003. Recombinant adenovirus p53 agent injection combince with radiotherapy in treatment of nasopharyngeal carcinoma: A phase II clinical trial. *Natl Med J China* 83(23): 2033–2035.

Chen J, Jin ZD, Li ZS, Jiang YP, Zhan XB, and Wang LW. 2007. Clinical observation on short-term outcome of intratumoral injection of recombinant human adenovirus p53 for treatment of unresectable pancreatic cancer. *Chin J Pancreatol* 7(2): 75–77.

Chen SX, Chen J, Xu WD, Yin GW, and Xi W. 2007. A preliminary clinical study on p53 gene in the therapy of advanced hepatocellullar carcinoma. *J Intervent Radiol* 16(2): 127–129.

Chipuk JE, Kuwana T, Bouchier-Hayes L, Droin NM, Newmeyer DD, Schuler M, and Green DR. 2004. Direct activation of Bax by p53 mediates mitochondrial membrane permeabilization and apoptosis. *Science* 303(5660): 1010–1014.

Clayman GL, Naggar AK, Lippman SM, Henderson YC, Frederick M, Merritt JA, Zumatein LA, Timmons TM, Liu TJ, Ginsberg L, Roth JA, Hong WK, Bruso P, and Goepfert H. 1998. Adenovirus-mediated p53 gene transfer in patients with advanced recurrent head and neck squamous cell carcinoma. *J Clin Oncol* 16(6): 2221–2232.

Cristofanilli, M, Krishnamurthy S, Gurra L, Bisotooni K, Arun B, Walters R, Booser D, Coffee K, Valero V, and Hortobagyi GN. 2003. Ad5CMV-p53 combined with docetaxel (T) and doxorubicin (D) as induction chemotherapy (IC) for patients with locally advanced breast cancer (LABC): Preliminary report of safety and efficacy. *Am Soc Clin Oncol* 22: 241.

Cristofanilli M, Krishnamurthy S, Guerra L, Broglio K, Arun B, Booser DJ, Menander K, Van Wart Hood J, Valero V, and Hortobagyi GN. 2006. A nonreplicating adenoviral vector that contains the wild-type p53 transgene combined with chemotherapy for primary breast cancer: Safety, efficacy, and biologic activity of a novel gene-therapy approach. *Cancer* 107(5): 935–944.

Dameron KM, Volpert OV, Tainsky MA, and Bouck N. 1994. Control of angiogenesis in fibroblasts by p53 regulation of thrombospondin-1. *Science* 265: 1582–1584.

Ding HF, Lin YL, Mcgill G, Juo P, Zhu H, Blenis J, Yuan JY, and Fisher DE. 2000. Essential role for caspase-8 in transcription-independent apoptosis triggered by p53. *J Biol Chem* 275: 8905–8911.

Ding Y, Zhang XS, Peng RQ, Zhang R, Zhang NH, Li ZM, Liu JY, Ma J, Cheng X, Su YS, and Zeng YX. 2005. Safety and primary efficacy of recombinant human adenovirus-p53 injection on advanced solid tumor. *Chin J Clin Pharmacol Ther* 10(9): 1025–1029.

Greene FL, Page DL, Fleming ID, et al. 2002. *Cancer Staging Manual*, 6 Ed. American Joint Committee on Cancer, Springer-Verlag, New York.

Guan YS, Sun L, Zhou XP, Li X, He Q, and Liu Y. 2005. Combination therapy with recombinant adenovirus-p53 injection (rAd-p53) via transcatheter hepatic arterial chemoembolization for advanced hepatic carcinoma. *World Chin J Digestol* 13(1): 125–127.

Guan YS, Liu Y, He Q, Yang L, Li X, and Sun L. 2005. p53 gene (Gendicine) therapy combining with bronchial artery infusion for treatment of lung cancer, short-time follow-up in 15 cases. *Chin J Intervent Imaging Ther* 2(6): 405–408.

Han DM, Huang ZG, Zhang W, Yu ZK, Wang Q, Ni X, Chen XH, Pan JH, and Wang H. 2003. Effectiveness of recombinant adenovirus p53 injection on laryngeal cancer: Phase I clinical trial and follow up. *Natl Med J China* 83(23): 2029–2032.

Iiizumi M, Arakawa H, Mori T, Ando A, and Nakamura Y. 2002. Isolation of a novel gene, CABC1, encoding a mitochondrial protein that is highly homologous to yeast activity of bc1 complex. *Cancer Res* 62: 1246–1250.

Krishna R and Mayer LD. 2000. Multidrug resistance (MDR) in cancer. Mechanisms, reversal using modulators of MDR and the role of MDR modulators in influencing the pharmacokinetics of anticancer drugs. *Eur J Pharm Sci* 11(4): 265–283.

Kuball J, Wen SF, Leissner J, Atkins D, Meinhardt P, Quijano E, Engler H, Hutchins B, Maneval DC, Grace MJ, Fritz MA, Storkel S, Thuroff JW, Huber C, and Schuler M. 2002. Successful adenovirus-mediated wild-type p53 gene transfer in patients with bladder cancer by intravesical vector instillation. *J Clin Oncol* 20: 957–965.

Lang FF, Bruner JM, Fuller GN, Aldape K, Prados MD, Chang S, Berger MS, McDemott MW, Kunwar SM, Junck LR, Chandler W, Zwiebel JA, Kaplan RS, and Yung WK. 2003. Phase I trial of adenovirus-mediated p53 gene therapy for recurrent glioma: Biological and clinical results. *J Clin Oncol* 21: 2508–2518.

Leu J, Dumont P, Hafey M, Murphy ME, and George DL. 2004. Mitochondrial p53 activates Bak and causes disruption of a Bak-Mcl1 complex. *Nat Cell Biol* 6(5): 443–450.

Li LJ and Huang YD. 2006. Combination therapy of subselective intraarterial rAd-p53 infusion with induction chemotherapy for locally advanced head and neck carcinoma: A clinical and bimolecular trial. *J Mol Ther* in press.

Liu DH, Liu WC, Fanl I, Sheng R, Xue Y, Wang W, Wang J, and Huang Y. 2006. Clinical effect of p53 gene—Gendicine in treatment of malignant effusions. *Chin J Can Prev Treat* 13(14): 1108–1109.

Lu P, et al. Observation and reflection of rAd-p53 gene injected intratumorally by endoscopy combined radiotherapy in the treatment of 15 cases with esophageal carcinoma. 15th ESCGT coference, Rotterdam, the Netherlands, October 27–30, 2007.

Matsuda K, Yoshida K, Taya Y, Nakamura K, Nakamura Y, and Araka WH. 2002. p53AIP1 regulates the mitochondrial apoptotic pathway. *Cancer Res* 62: 2883–2889.

Muller M, Wilder S, Bannasch D, Israeli D, Lehlbach K, Weber ML, Friedman SL, Galle PR, Stremmel W, Oren M, and Krammer PH. 1998. p53 activates the CD95 (APO-1/Fas) gene in response to DNA damage by anticancer drugs. *J Exp Med* 188: 2033–2045.

Nemunaisitis J, Swisher SG, Timmons T, Connors D, Mack M, Doerksen L, Weill D, Wait J, Lawtence DD, Kemp BL, Fossella F, Glisson BS, Hong WK, Khuri FR, Kurie JM, Lee JJ, Lee JS, Nguyen DM, Nesbitt JC, Perez-Soler R, Pisters KM, Putmam JB, Richli WR, Shin DM, Walsh GL, Merritt J, and Roth J. 2000. Adenovirus-mediated p53 gene transfer in sequence with cisplatin to tumors of patients with non-small-cell lung cancer. *J Clin Oncol* 18(3): 609–622.

Pagliaro LC, Keyhani A, Williams D, Woods D, Liu B, Perrotte P, Station JW, Merritt JA, Grossman HB, and Dinney CP. 2003. Repeated intravesical instillations of an adenoviral vector in patients with locally advanced bladder cancer: A phase I study of p53 gene therapy. *J Clin Oncol* 21: 2247–2253.

Pal S, Datta K, and Mukhopadhyay D. 2001. Central role of p53 on regulation of vascular permeability factor/vascular endothelial growth factor (VPF/VEGF) expression in mammary carcinoma. *Cancer Res*, 61: 6952–6957.

Qi XD, Hang DM, Ni X, Yang ZB, Niu Q, Li ZX, and Jiang NJ. 2005. The clinical effection of recombinant human Adp53 injection—Gendicine in malignant tumor patients. Poster at 2005 ISCGT China conference.

Rosenblum MD, Olasz E, Woodliff JE, Johnson BD, Konkol MC, Gerber KA, Orentas RJ, Sandford G, and Truitt RL. 2004. CD200 is a novel p53-target gene involved in apoptosis-associated immune tolerance. *Blood* 103(7): 2691–2698.

Rother K, Jlhne C, Spiesbach K, Haugwitz U, Tschop K, Wasner M, Klein-Hitpass L, Moroy T, Mossner J, and Engeland K. 2004. Identification of Tcf-4 as a transcriptional target of p53 signalling. *Oncogene* 23(19): 3376–3384.

Sah NK, Munshi A, Nishikawa T, Mukhopadhyay T, Roth JA, and Meyn RE. 2003. Adenovirus-mediated wild-type p53 radiosensitizes human tumor cells by suppressing DNA repair capacity. *Mol Cancer Ther* 2(11): 1223–1231.

Schuler M, Herrmann R, De Greve JL, Stewart AK, Gatzemier U, Stewart DJ, Laufman L, Gralla R, Kuball J, Buhl R, Heussel CP, Kommoss F, Perruchoud AP, Shepherd FA, Fritz MA, Horowitz JA, Huber C, and Rochlitz C. 2001. Adenovirus-mediated wild-type p53 gene transfer in patients receiving chemotherapy for advanced non-small-cell lung cancer: Results of a multi-center phase II study. *J Clin Oncol* 19: 1750–1758.

Schwartzenberg-Bar-Yoseph F, Armoni M, and Karnieli E. 2004. The tumor suppressor p53 down-regulates glucose transporters GLUT1 and GLUT4 gene expression. *Cancer Res*, 64: 2627–2633.

SFDA. 2003a. The State Biological Products Standardization Commission of the People's Republic of China. Guidance for Human Gene Therapy Research and its Products Quality Control. Beijing 2–6.

SFDA. 2003b. Points to Consider for Human Gene Therapy and Product Quality Control.

Shiraishi K, Kato S, Han SY, Han SY, Liu W, Otsuka K, Sakayori M, Ishida T, Takeda M, Kanamara R, Ohuchi N, and Ishioka C. 2004. Isolation of temperature-sensitive p53 mutations from a comprehensive missense mutation library. *J Biol Chem* 279(1): 348–355.

SiBiono 2004. Points to Consider for Human Gene Therapy and Product Quality Control. The document by SiBiono GeneTech Co., Ltd. and national institute for the control of pharmaceutical and biological products. *BioPharm Intl* 17(5): 73–76.

Singh B, Reddy PG, Goberdhan A, Walsh C, Dao S, Ngai I, Chou TC, O-Charoenrat P, Levine AJ, Rao PH, and Stoffel A. 2002. p53 regulates cell survival by inhibiting PIK3CA in squamous cell carcinomas. *Genes Dev* 16(8): 984–993.

Sun Y, Zeng XR, Wenger L, Firestein GS, and Cheung HS. 2004. p53 down-regulates matrix metalloproteinase-1 by targeting the communications between AP-1 and the basal transcription complex. *J Cell Biochem* 92(2): 258–269.

Sun Y, Zhang SW, Xiao SW, Zhu GY, and Xu B. 2005. Clinical outcome of combination of adenoviral mediated p53 gene with radiotherapy for advanced cancers in head and neck. Collection of 2005 ISCGT China Conference. C8–C20.

Swisher SG, Roth JA, Komaki R, Gu J, Lee JJ, Hick M, Ro JY, Hong WK, Merritt JA, Ahrar K, Atkinson NE, Correa AM, Dolormente M, Dreiling L, El-Naggar AK, Fossella F, Francisco R, Glisson B, Grammer S, Herbst R, Huaringa A, Kemp B, Khuri FR, Kurie JM, Liao Z, McDonnell TJ, Morice R, Morello F, Munden R, Papadimitrakopolou V, Pisters KM, Putnam JB Jr, Sarabia AJ, Shelton T, Stevens C, Shin DM, Smythe WR, Vaporciyan AA, Walsh GL, and Yin M. 2003. Induction of p53-regulated genes and tumor regression in lung cancer patients after intratumoral delivery of adenoviral p53 (INGN 201) and radiation therapy. *Clin Cancer Res* 9(1): 93–101.

Taha TA, Osta W, Kozhaya L, Birlawski J, Johnson KR, Gillanders WE, Dbaibo GS, Hannun YA, and Oheid LM. 2004. Down-regulation of sphingosine kinase-1 by DNA damage: Dependence on proteases and p53. *J Biol Chem* 279(19): 20546–20554.

Tolcher AW, Hao D, de Bono J, Miller A, Patnaik A, Hammond LA, Smetzer L, Van Wart Hood J, Merritt J, Rowinsky EK, Takimoto C, Von Hoff D, and Eckhardt SG. 2006. Phase I, pharmacokinetic, and pharmacodynamic study of intravenously administered Ad5CMV-p53, an adenoviral vector containing the wild-type p53 gene, in patients with advanced cancer. *J Clin Oncol* 24: 2052–2058.

Toschi E, Rota R, Antonini A, Melillo G, and Capogrossi MC. 2000. Wild-Type p53 gene transfer inhibits invasion and reduces matrix metalloproteinase-2 levels in p53-mutated human melanoma cells. *J Invest Dermatol* 114: 1188–1194.

Weng Z, Qin TL, Tan SY, Liu JL, Sui J, and Zhu ZB. 2004. Clinical observation of advanced pulmonary carcinoma injected with rAd p53 carcinoma injected with rAd p53. *Shenzhen J Integ Trad Chin West Med* 14(4): 206–210.

Xiao SW, Liu CQ, Sun Y, Su X, Li DM, Xu G, Zhu GY, Xu B, and Zhang SW. 2007. Clinical effectiveness of recombinant adenovirus-p53 combined with hyperthermia in advanced soft tissue sarcoma (a report of 13 cases). Collection of 10th National conference of hyperthermia.

Yen N, Ioannides CG, Xu K, Swisher SG, Lawrence DD, Kemp BL, El-Naggar AK, Cristiano RJ, Fang B, Glisson BS, Hong WK, Khuri FR, Kurie JM, Lee JJ, Lee JS, Merritt JA, Mukhopadhyay T, Nesbitt JC, Nguyen D, Perez-Soler R, Pisters KM, Putnam JB Jr, Schrump DS, Shin DM, Walsh GL, and Roth JA. 2000. Cellular and humoral immune responses to adenovirus and p53 protein antigens in patients following intratumoral injection of an adenovirus vector expressing wild-type. P53 (Ad-p53). *Cancer Gene Ther* 7(4): 530–536.

Yin Y, Liu YX, Jin YJ, Hall EJ, and Barrett JC. 2003. PAC1 phosphatase is a transcription target of p53 in signalling apoptosis and growth suppression. *Nature* 422: 527–531.

Zhang SW, Xiao SW, and Lv YY. 2003a. Thermosensitized effects of adenovirus-mediated p53(Ad-p53): Preclinical study and a phase II clinical trial in China. *Jpn J Hyperthermic Oncol* 19(3): 141–149.

Zhang SW, Xiao SW, Liu CQ, Sun Y, Su X, Li DM, Xu G, Cai Y, Zhu GY, Xu B, and Lv YY. 2003b. Treatment of head and neck squamous cell carcinoma by recombinant adenovirus-p53 combined with radiotherapy. *Natl Med J China* 83(23): 2023–2028.

Zhang XZ, Zhang WM, Yu Q, Chen ZJ, and Peng ZH. 2004. Progress in cancer therapy of rAd/p53 *Oncology Prog* 2(Suppl): 56–63.

Zhang QH, Wang D, Peng N, Xie JY, Yang ZX, Xiao SQ, and Wang G. 2005. Study of treatment of malignant tumour in patients with rAd-p53 Poster at 2005 ISCGT China conference.

Zhang YW et al. 2007. Preliminary clinical observation on interventional p53 gene injection guided by ultrasonographic (US) combined with sequential chemoembolization for cancerous embolization in portal vein. 15th ESCGT Conference, Rotterdam, the Netherlands, October 27–30.

Zhu ZB, Liu JL, Sui J, Weng Z, Tan SY, Qin TL, and Li MS. 2004. Study of treatment of advanced hepatobiliary carcinoma in patients with rAd-p53. *Chin J Intern Med* 3(11): 11–14.

Zhu ZB, Shui J, Liu JL, Tan SY, Weng Z, and Li MS. 2005. Study of treatment of carcinous ascites in Patients with rAd-p53. *Chin J Composite Clin Hyg* 7(2): 22–24.

50 Gene Marking Studies of Hematopoietic Cells

Catherine M. Bollard, Siok K. Tey, Malcolm K. Brenner, and Helen E. Heslop

CONTENTS

50.1 INTRODUCTION

The transfer of genetic material into a target cells offers a potentially powerful strategy in the treatment of a wide variety of diseases. While gene marking does not confer therapeutic benefit, much of the biological information gained from these trials has been proving to be valuable for improving the outcome of therapies such as stem cell transplantation [1,2]. Gene marking can be used to track the *in vivo* behavior of almost any tissue [3], and the success of the strategy depends on tracking the marked cell not only for its entire lifespan, but also for the lifespan of all its progeny. The vectors used must therefore integrate in the host cell DNA and efficiently replicate with the cell. For this reason, the majority of marker studies performed to date have used murine retroviral vectors. Although these agents are able to integrate stably, they have many limitations. In particular, they are only able to integrate into dividing cells and have low transduction efficiency.

The aims of this chapter are threefold: (1) to identify the reasons why gene marking studies have been performed and why this research was important to the ever-broadening field of gene transfer, (2) to discuss the vectors and marker genes employed and the factors that influence gene transfer efficiency and gene expression, and (3) to review the concerns about the safety issues surrounding gene transfer.

50.2 GENE MARKING OF HEMATOPOIETIC PROGENITOR CELLS

The principle of gene marking is the transfer of a unique DNA sequence (e.g., a nonhuman gene) into a host cell (e.g., T-cell, hematopoietic stem cell [HSC], etc.) allowing the gene or the gene product to be easily detected, thereby serving as a marker for these labeled cells. Detection of the marker gene may be phenotypic or genotypic. For example, the frequently used bacterial neomycin resistance (*neo*) marker gene encodes the enzyme neomycin phosphotransferase, which can inactivate neomycin or its analogs such as G418. When cells are expressing the *neo* gene, they become resistant to G418 in culture medium and can be detected by the ability to grow in G418. The transgene can also be readily detected in these cells using the polymerase chain reaction (PCR) technique [2].

50.2.1 GENE MARKING IN AUTOLOGOUS STEM CELL TRANSPLANTATION

In the first gene marking trials involving HSCs, the question posed was whether relapse following autologous stem cell transplant was due to a contribution from contaminating malignant cells in the stem cell harvest [4]. In order to answer this question, a portion of the HSC product was marked at the time of harvest with a murine retroviral vector encoding the neomycin resistance gene. If the patient subsequently relapsed, it was possible to detect whether the marker gene was present in the malignant cells. Since 1991, studies were initiated using this approach in a variety of malignancies treated by autologous HSC transplantation [4–13]. As shown in Table 50.1, the underlying malignancies treated with marked autologous HSCs include acute myeloid leukemia (AML), chronic myeloid leukemia (CML), acute lymphoblastic leukemia (ALL), neuroblastoma, lymphoma, multiple myeloma, and breast cancer.

The initial studies evaluating this approach took place at St. Jude Children's Research Hospital (SJCRH) in pediatric

TABLE 50.1
Clinical Gene Marking Studies in Autologous HSC Transplantation

Institution	Patient Population (Number)	Target Cell and Transduction Method	Findings	Reference
Single marking studies				
St. Jude Children's Research Hospital	Pediatric AML (12); neuroblastoma (8). Extended study included 25 and 12 patients, respectively	Marrow, 6 h culture with vector	Marking detected in marrow progenitor and peripheral blood (PB) in 15 of 18 patients at 1 month	[2,14,16, 20,21]
			Marking detected in neutrophils, T cells, and B cells	
			Extended follow-up detected marked cells in blood and marrow in 32 of 37 patients at 2 years with persistence for at least 9 years	
			Marked tumor cells detectable in two of two AML and three of three neuroblastoma relapses in initial report. Follow-up detected marked tumor cells in three of four AML and four of five neuroblastoma relapses	
M.D. Anderson	Adult CML (4)	CD34-enriched marrow. 6 h culture with vector	Marked normal cells and leukemia cells for up to 280 days	[17]
Indiana University	Adult AML (4); ALL (1)	Marrow, 4 h culture with vector	Minimal detection of marked cells. No marked tumor cells in two relapses	[18]
University of Toronto	Adults. Multiple myeloma (15)	Stroma-based long term (21 days) marrow culture	Marked normal cells detected for up to 2 years	[35]
University Hospital, Freiburg, Germany	Adult. CML (3)	CD34-enriched PB. Cultured with Flt-3 ligand, SCF, IL-3, ±IL-6, ±G-CSF (5-day culture)	Marked normal cells detected in all patients early on (day 11) but no significant detection after 9 months	[69]
Karolinska Institute	Adult. Multiple myeloma (6)	CD34 selected marrow or PB. Transduced in presence of SCF, IL-3, IL-6, bFGF (24 h culture)	Marked normal marrow cells detected up to 5 years	[23]
			No marked myeloma cells in three cases of disease progression	

TABLE 50.1 (continued)
Clinical Gene Marking Studies in Autologous HSC Transplantation

Institution	Patient Population (Number)	Target Cell and Transduction Method	Findings	Reference
Double marking studies				
St. Jude Children's Research Hospital	Pediatric. AML (12)	Marrow purged with 4HC versus IL-2, 6 h culture with vector	Marked cells from 4HC arm persisted for >3 years. Marked cells from IL-2 treated arm detected transiently only No marked tumor cells in 3 relapses	[20,21,25]
National Institutes of Health	Adult. Multiple myeloma (6); breast cancer (5)	CD34 selected PB versus marrow, 72 h culture with SCF, IL-3, ± IL-6	Marked cells persisted for >18 months with contribution from both PB and marrow No marked tumor cells in three breast cancer relapses	[19]
National Institutes of Health	Adult. Multiple myeloma (3); breast cancer (2)	CD34 selected PB versus marrow, 6 h culture with vector. No growth factors	Marked cells detected in four of five patients early posttransplant (two from PB source only, one marrow only, one both) No marking detectable at 1 year	[22]
National Institutes of Health	Adult. Multiple myeloma (2); breast cancer (2)	CD34 selected PB versus marrow, 72 h culture on autologous stroma layer	Marked cells detected early posttransplant in one of four patients from marrow only. No marking detectable at 1 year No marked tumor cells in one breast cancer relapse	[22]
M.D. Anderson	Adult. Follicular lymphoma (3)	CD34 selected PB versus marrow, 4 to 6 h culture	Marked cells detected from both PB and marrow up to 9 months No marked tumor cells in one relapse	[24]

AML, acute myeloid leukemia; bFGF, basic fibroblast Growth Factor; G-CSF, granulocyte colony stimulating factor; 4HC, 4-hydroperoxycyclophosphamide; IL-2, interleukin-2; IL-3, interleukin-3; IL-6, interleukin-6; PB, peripheral blood; SCF, stem cell factor.

patients receiving autologous bone marrow transplantation (BMT) as part of therapy for AML or neuroblastoma [4,6,13]. In both studies, one-third of the marrow was marked in a 6 h transduction protocol in the absence of growth factors. Four of 12 AML patients who received marked marrow relapsed. In three of the four patients, detection of both the transferred marker and a tumor-specific marker in the same cells at the time of relapse provided unequivocal evidence that the residual malignant cells in the marrow were a source of leukemic recurrence [14]. For example, in one patient the malignant blasts had the unique phenotypic features of CD34 and CD56 coexpression and a complex t (1:8:21) translocation, resulting in an AML1/ETO fusion transcript that could be identified by PCR. Malignant cells sorted by this phenotype contained both the AML1/ETO fusion transcript and the transferred neomycin gene [14]. In the neuroblastoma study [6], five of the nine patients relapsed, and gene marked neuroblastoma cells, coexpressing the neuroblastoma-specific antigen GD2 together with the transferred marker gene, were detected in four cases [15,16]. In one of these patients, marked neuroblastoma cells were detected in an extramedullary relapse in the liver [16]. Similarly, marked malignant cells coexpressing the t(9;22) and *neo* transcripts were found in two patients with CML at relapse [17]. These data show that in these three malignancies, marrow harvested in apparent clinical remission may contain residual tumorigenic cells that can contribute to a subsequent relapse. In contrast, marked cells have not been found in adult patients relapsing after autografts for ALL [18]. However, for logistic reasons, only around 10% of the marrow was marked in this study and gene transfer to normal cells was very low. In a study at the NIH, autologous transplantation for breast cancer used CD34 selected HSC from blood and marrow. In the two patients who have relapsed the marker gene was not found [19]. This may reflect a lower efficiency of the marking of breast cancer cells or the CD34+ selection procedure used may have reduced contamination with malignant cells. Alternatively, relapse in breast cancer may occur predominantly from residual disease in the patient.

As only modest levels of gene transfer into malignant progenitor cells is obtained, a definitive conclusion about the contribution of marrow-based disease to recurrence can only be drawn if a relapse is marked. Therefore, although the presence of gene-marked cells within relapse samples is proof that reinfused cells contribute to relapse, their absence is less informative. Studies where all relapses were negative for gene-marked cells include three children with relapsed AML after transplantation with purged marrow [20,21], two adults with relapsed AML after BMT [18], and four relapsed breast cancer [19,22], three progressive myeloma [23], and one relapsed follicular lymphoma [24] after CD34 selected stem cell transplantation. It is possible that these findings reflect true differences in tumor biology or stem cell product; however, other factors, in particular tumor marking efficiency and chance effect inherent to small studies, cannot be ruled out. For example, unmarked relapses might mean that marrow does not significantly contribute to recurrence or may mean that relapse is generated by only a few marrow-derived malignant cells that have escaped being marked because of the inefficiency of the transduction process.

50.3 USE OF MULTIPLE VECTORS

50.3.1 Double Gene Marking to Monitor Purging

The overall implication of the above studies was that residual malignant cells can contribute to relapse and the outcome of HSC rescue may be improved by strategies targeted at residual malignant cells [21]. One means of achieving this is purging of the transplanted marrow and gene-marking techniques are useful in the evaluation of the purging technologies. In second-generation studies, two genetically distinct marker genes were used to compare either marrow purging verses no purging, or two different purging techniques [12,25,26]. Two closely related vectors, G1Na and LNL6, were used. These vectors differ in their 3′ noncoding sequences so primers can be designed allowing the vectors to be discriminated by virtue of the differing fragment sizes they produce after PCR amplification. In the AML study [25], 15 patients were enrolled between September 1993 and November 1996. At least 10^8 marrow cells/kg were frozen without manipulation as a safety backup. The remaining marrow mononuclear cells were divided in half and randomized to marking with G1Na or LNL6. The aliquots were then randomly assigned to the two purging techniques being evaluated. Initially the pharmacologic "gold standard" hydroperoxycyclophosphamide (4HC) was compared with an immunologic purge by culture with IL-2 [27]. Later in the study, CD15 antibodies (instead of IL-2) were used in conjunction with 4HC [28]. At the time of transplant, both aliquots were reinfused. If the patients relapsed, the premise was that detection of either marker would reveal which of the compared purging techniques had failed.

Over the 3-year period, 15 patients received gene marked marrow. In five patients, aliquots of marrow were purged with 4HC and IL-2 and in three patients the marrow was purged with 4HC and CD15. PCR studies on PB granulocytes and mononuclear cells consistently showed a stronger signal from the 4HC-purged fraction than that from the IL-2-purged fraction, regardless of which vector was used for marking each aliquot. These observations suggested that the 4HC-purged fraction was making a greater contribution to hematopoietic reconstitution than the IL-2-purged fraction thus resulting in the substitution of CD15 antibodies for IL-2. Seven patients received marrow purged with 4HC alone. This was due to either insufficient number of cells harvested or unavailability of one clinical grade retroviral vector. In all, three of the fifteen patients relapsed. Two patients relapsed early at 2 and 3 months and were noninformative, as marked malignant cells were not

detected. A third patient relapsed at 20 months and his blasts were also negative [20]. The *neo* gene was detected in normal hematopoietic and immune system cells at a lower level than in the studies using unpurged marrow.

50.4 GENE TRANSFER TO NORMAL CELLS

These marking studies also provided information on the transfer of marker genes to normal hematopoietic cells and showed that marrow autografts contribute to long-term hematopoietic reconstitution after transplant [2,21] (Table 50.1). Long-term transfer up to 9 years has been seen in the mature progeny of marrow precursor cells, including PB T and B cells and neutrophils [20]. It was also detected in lymphoblastoid cell lines and cytotoxic T cell (CTL) lines derived from these patients. However, the level of gene transfer varied and was highest in marrow clonogenic hematopoietic progenitors, where an average of 6% of myeloid colonies were G418-resistant at 3 months. The levels of transfer into marrow progenitors are higher than predicted from animal models and may be attributed to the fact that marrow was harvested during regeneration after intensive chemotherapy, when a higher than normal proportion of stem cells are in cycle [21]. In the mature PB cells, the level of gene transfer was, however, some fivefold lower than in the marrow progenitors with levels of 0%–1% seen [29].

Useful information was gained from these initial studies leading to the development of more complex gene marking protocols (Table 50.1). These include the following. (i) Double gene marking to allow simultaneous study of two distinctly treated cell populations, for example, of unmanipulated HSCs versus stem cells treated with growth-stimulatory agents. (ii) The use of more purified populations of early progenitor cells to allow the "true" pluripotent tem cell to be identified phenotypically and increase the efficiency of gene transfer by more precise tailoring of transduction conditions.

50.4.1 Double Gene Marking to Compare Long-Term Reconstitution from Different Populations of Hematopoietic Progenitor Cells

No *in vitro* assay can yet assess the capacity of a cell population to produce short- and long-term repopulation of humans. The use of double gene marking with distinguishable vectors in a single patient potentially allows (i) comparison of these properties between different sources of putative stem cells, such as PB and marrow, (ii) determination of the function of stem cell subpopulations, for example CD34$^+$ CD38$^+$ versus CD34$^+$ CD38$^-$ progenitor cells, (iii) the consequences of *ex vivo* manipulation, such as culture of putative stem cells on stromal support or with cytokines. By using the two distinguishable retroviral markers, it is possible to compare quantitatively *in vivo* and within each patient the short- and long-term reconstituting capacity of different populations of HSC.

One aim of clinical stem cell transplantation is to minimize the period of marrow aplasia posttransplant by increasing the number of progenitor cells infused. This approach is only of value if progenitor cell manipulation does not induce loss of self-renewal capacity when the treated cells are reinfused. One strategy is to expand the progenitor cells at least 50-fold *ex vivo* using stimulatory cytokines [19,30] or stromal [20,31] components. There are, however, concerns that stimulatory cytokines will not only induce primitive cells to expand but will also cause them to differentiate and lose their capacity for self-renewal. In other words, the cytokine stimulation *ex vivo* may produce a faster initial engraftment but at the cost of later graft failure. Gene marking can be used to not only address concerns about the effects of growth-stimulating agents on pluripotent progenitor cells and long-term engraftment, but also determine if *ex vivo* stimulation increases gene transfer efficiency.

The NIH study by Dunbar et al. [19] compared the reconstitution of PB and bone marrow in patients receiving autologous HSC rescue for myeloma or breast cancer. They differentially marked mobilized PB and marrow CD34+ cells using two distinguishable *neo* containing retroviral vectors and infused both populations of stem cells into the same patient. An average of 21% of colony forming units (CFUs) were transduced using a 3-day culture in the presence of SCF, IL-3, and IL-6. The level of gene transfer into committed progenitor cells posttransduction was as high as 50% using this approach. However the engraftment of the marked cells was not accelerated compared to the earlier SJCRH studies [2,14,25,32] in which growth factors were not used. In addition, although vector sequences were found in myeloid and lymphoid cells at 0.01%–0.001% for up to 18 months, these levels were lower than the earlier SJCRH studies [33,34]. These findings may reflect the different patient population both by virtue of age (adult versus pediatric) and disease (breast cancer/myeloma versus leukemia/neuroblastoma). Alternatively, this may indicate that that culture with growth factors commits transduced cells to differentiation, so that high-level, long-term engraftment is not obtained.

Although the study by Dunbar et al. showed low levels of gene transfer detected *in vivo*, the marker gene derived from PB and marrow HSC demonstrated that each can contribute to long-term recovery [19]. It is of interest that the marked-PB stem cells (PBSCs) contributed to hematopoiesis earlier and for a longer period compared to the gene marked-marrow cells. It is possible that this phenomenon may be related to physiological differences in these two populations of HSCs. This includes the ability of each HSC to contribute to hematopoiesis, the cell cycle status of the HSC or the susceptibility of the HSC to retroviral transduction [31].

Another promising method by which progenitor cells may be expanded and their transducibility increased is the initiation of long-term cultures or the addition of stromal support components to culture. One group [11,35], established marrow in long-term cultures where an autologous stroma developed over the course of 3-week culture. The cultures were transduced weekly. Initially a mean of 16% of gene marked pro-

genitors were seen in the early posttransplant period. This level dropped to 3% by 2 years posttransplant and only 0.01% of the circulating blood cells were marked.

Although the use of double gene markers was hampered by the complexity of the studies, the approach was a powerful way of investigating the biology of different stem cell populations and of determining the consequences of stem cell manipulation.

50.4.2 CD34+ SELECTED HSC TRANSPLANTATION

In the context of gene marking, the justification for CD34+ selection is that it reduces the required volume of vector supernatant. It may also serve as a means of marrow purging in patients in whom the malignant cells are CD34− or co-express CD34+ and one or more lineage (lin) commitment antigens. The marking approach can be used to discover the relative ability of distinct populations of CD34+ cells to produce long-lived, multilineage reconstitution following reinfusion. The strategy can also be used to determine if lin-positive and -negative CD34 subsets make a different contribution to short and long-term reconstitution. The potential disadvantage of selecting CD34+ cells or their subsets is that immune reconstitution may be delayed because mature lymphocytes are not transferred with the graft. This delay may then increase the risks of subsequent neoplastic change if undetected replication-competent retrovirus has contaminated the vector as demonstrated in an animal study [36].

The advantages of CD34 selection may therefore be counterbalanced, at least in part, by a decrease in the margin of safety for marker studies. Nonetheless, many of the later marker studies (as shown in Table 50.1) used CD34+ cells as the vector target [12,19,33].

50.5 ROLE OF MARKER STUDIES TO VALIDATE EFFORTS TO INCREASE THE LEVEL OF GENE TRANSFER TO THE HSC

50.5.1 ROLE OF CELL CYCLE

Attempts to induce HSCs to cycle generally use various cytokines/growth factor combinations. However, it has been demonstrated that culture of human CD34+ CD38− cells in serum free media with IL-3, IL-6 SCF, Flt-3 ligand, and thrombopoietin or megakaryocyte-derived growth factor (MDGF) for 9 days can increase both the number of HSCs cycling and the length of time they can be maintained *in vitro* [37–39]. HSCs may also be induced to cycle *in vivo* as a normal physiologic response to hematopoietic cytoablation or cytokine mobilization. In the SJCRH studies in pediatric patients [2,14,25,32], a 6 h incubation with retroviral supernatant was employed for transduction. The success of the transduction efficiency and long-term persistence of marker gene expression in these studies compared to similar studies in adults [19,35] may, however, be attributed to the fact that HSCs from children have an endogenously higher proliferation index. There is also evidence that treatment of animals with a combination of granulocyte-stimulating factor (G-CSF) and SCF mobilizes marrow progenitor cells that are more primitive and more readily transduced than after treatment with G-CSF alone [40].

50.5.2 Role of Stromal Support Elements

Some groups add stromal supports such as a recombinant fragment of fibronectin to bring the retroviral vector and target cell into close apposition thereby increasing the efficiency of progenitor cell marking [41,42]. Several studies have tried to address the extent of long-term engraftment of retrovirally transduced PB progenitor cells using recombinant fibronectin [43–45]. One study [45] assessed the effects of recombinant fibronectin on marrow repopulating cells and on the efficiency of gene transfer using the multidrug resistance (MDR1) gene or *neo* gene to ascertain the fate of the treated cells. In this study, half of the enriched CD34+ cells were transduced with MDR1 and the other half with *neo*. There was low level gene marking of granulocytes (0.01%–1%) by PCR analysis up to 6 months posttransplant. In all six patients, there was a higher level of engraftment of MDR1-containing cells relative to *neo*, suggesting that MDR1 overexpression was beneficial to the engraftment potential of hematopoietic cells.

One study [42] reports the highest level of long-term engraftment of retrovirally transduced PBSCs. The 11 enrolled patients were all adults (17–51 years) undergoing tandem autologous PBSC transplants for germ cell tumors. In this study, MDR1 was used as a marker gene and in the hope that it may render the HSC resistant to oral etoposide [46], thereby allowing for further dose-intensive therapy without the delay of prolonged cytopenia posttransplant. The PBSCs were mobilized using G-CSF. The initial harvest was unmanipulated and used for the initial transplant. The second collection was CD34+ selected, stimulated *ex vivo* with cytokines (SCF, IL-6, MGDF, G-CSF), and cultured with MDR-1 retroviral vector on plates coated with recombinant fibronectin (CH-296) then cryopreserved on day 5. The median gene transfer efficiency of all colonies immediately after transduction was 14% (4%–52%). At 12 months postinfusion of the MDR1-PBSCs, four of the seven evaluable patients had detectable levels (5%–15%) of transgene-containing colonies in their bone marrow samples. This study also showed that exposure of CD34+ cells to CH-296 did not adversely affect either engraftment kinetics or long-term hematopoietic function.

These studies therefore indicate the feasibility of using MDR gene transfer as a means of enriching marrow for MDR-transduced cells. It remains unclear whether or not the approach will be valuable for protecting and selecting other cell types following chemotherapy.

50.6 GENE MARKING STUDIES OF T CELLS

50.6.1 Gene Marking of Tumor-Infiltrating T Lymphocytes

The first gene marking protocol was conducted to determine the fate of lymphoid cells that infiltrate solid tumors [47]. These tumor-infiltrating lymphocytes (TILs) were obtained from tumor biopsies, expanded *ex vivo*, and then re-infused. It was therefore important to determine if these infused TILs were tumor-specific and ascertain their distribution and persistence at tumor sites after infusion. Initially, studies to assess the ability of TILs to traffic to and remain at tumor sites postinfusion used TILs, which were radioactively labeled with [111]In [48,49]. However, this approach requires a high labeling to enhance the detection level and duration, resulting in potentially toxic radiation exposure to cells from the gamma radiation emitted by the isotope. In addition, the half-life of ^{111}In is only 2.8 days and it is therefore not suitable for monitoring TILs over extended periods of time [50].

It was apparent that a need existed for an alternative labeling technique. The genetic marking of the TIL *ex vivo* using a *neo* encoding retroviral vector helped answer several questions about TILs in cancer therapy. After *neo*-marked TILs were infused in to the patient and tumor sites biopsied, it was shown that large numbers of gene-marked TILs could be safely returned to patients, and that the marker could be transiently detected in the PB by PCR analysis using *neo*-specific primers [51,52]. Analysis of PB and tumor deposits for presence of the marker gene suggested that TIL could persist for up to 2 months [51]. Studies in one patient showed presence of the marker gene in biopsy tissue from a tumor deposit [53]. In a French study marked TIL were detected for up to 260 days and TIL were also detected in 4/8 tumor biopsies after therapy [54]. A subsequent double-marking study compared the survival and tracking of PB lymphocytes (PBL) and TIL by marking with distinguishable retroviral vectors [55]. Both marked PBL and TIL could be detected in PB for 4 months and no selective homing of TILs to tumor sites compared to unmanipulated PBL was seen [55]. None of these studies therefore provided support for the theory that TILs are capable of selective homing.

50.6.2 Gene Marking to Track Adoptively Transferred CTLs After Allogeneic BMT

Several studies have shown the feasibility and apparent clinical efficacy of adoptive transfer of CTLs directed at viral or tumor antigens [54–63]. The transfer of a marker gene has allowed monitoring of adoptive transfer approaches to determine the survival of infused T cells. In addition, the T cells can be tracked to learn if they can home to sites of disease and if they mediate adverse effects such as graft versus host disease (GVHD). The clinical results of gene marking approaches using T cells are summarized in Table 50.2.

Matched unrelated-donor or mismatched family-donor BMT results in a high risk of GVHD due to the greater genetic disparity between the donor and the recipient. An effective means of reducing this risk is to T cell deplete the donor marrow *ex vivo*. The disadvantage of this approach is that recipients of the T-cell-depleted bone marrow have delayed immune recovery and an increased incidence of viral infections. One such infection is Epstein–Barr virus lymphoproliferative disease (EBV-LPD), which is due to a proliferation of

TABLE 50.2

Clinical Gene Marking Studies on T Cells

Institution	Disease (Number of Patients)	Target Cell Type	Findings	Reference
National Cancer Institute	Melanoma (5)	TILs	Marked cells detected in blood and tumor sites for up to 2 months	[47]
Centre Leon Berard, Lyon, France	Melanoma or renal cell carcinoma (5)	TILs	Marked cells detected in blood for up to 260 days. Marked cells detected in four of eight tumor deposits	[54]
University of California Los Angeles	Melanoma (6) or renal cell carcinoma (3)	Double marking study using TILs versus PBL	Marked cells from both components detectable in blood, tumor deposits, and normal peripheral tissues (muscle, fat, skin) for up to 99 days. No evidence for preferential trafficking of TILs to tumor deposits	[55]
St. Jude Children's Research Hospital	Prophylaxis or treatment of EBV-associated posttransplant lymphoproliferative disorder in T-cell depleted allogeneic BMT (39)	EBV-specific CTLs	Marked cells detected in unmanipulated PB for up to 18 weeks using semiquantitative PCR, and up to 6.7 years using real-time PCR. Remained detectable following *ex vivo* antigenic stimulation for up to 10 years. Initial expansion *in vivo* and long term antigen responsiveness *in vivo* and *ex vivo*. Accumulation of marked cells in tumor sites	[57,59,60]
Center for Cell and Gene Therapy, Houston	EBV-positive Hodgkin's Lymphoma (7)	EBV-specific CTLs	Marked cells detected in PB for 3–12 months. Accumulation of marked cells in tumor sites	[61,65]
Fred Hutchinson Cancer Research Center	HIV-positive patients (9)	HIV gag-specific CD8 + CTL clones modified with HyTK (6 patients) or neo(3 patients)	T cell mediated eradication of cells modified with HyTK Neo-marked cells detected in blood for 3–4 weeks. Accumulation of neo-marked cells adjacent to HIV-infected cells in lymph nodes	[56,62]

CTL, cytotoxic T lymphocytes; EBV, Epstein–Barr virus; HIV, human immunodeficiency virus; HyTK, hygromycin phosphotransferase and herpes simplex virus thymidine kinase; PCR, polymerase chain reaction.

EBV-infected B cells that are highly immunogenic and are not eliminated in the immunocompromised host. This complication occurs in 5%–30% of patients receiving T-depleted marrows from mismatch family or unrelated donors. Donor-derived EBV-specific T cells were generated by culturing donor T cells with donor-derived EBV-infected lymphoblastoid cell lines [57,59,60,64]. To determine whether these cells persisted and whether they caused adverse effects they were marked with the *neo* gene before administration. Twenty-six patients received *neo*-marked CTLs. Infusion of virus-specific CTLs produced a virus-specific immune response as documented by a fall in EBV DNA levels [57]. The gene-marking component of this study allowed the demonstration of the persistence of the *neo* gene (using sensitive real-time

quantitative PCR) in the PB in more than 75% of patients, for up to 6.7 years, with the median being 227 days [60]. Three patients who were treated for clinically evident EBV-LPD attained prolonged remission after CTL infusion and in situ hybridization as well as semiquantitative PCR showed that the gene marked CTL had selectively accumulated at disease sites [59].

50.6.3 GENE MARKING OF EBV-SPECIFIC CTLs FOR RELAPSED EBV-POSITIVE HODGKIN DISEASE

Marking of autologous EBV-specific polyclonal CTLs has also been used in a clinical study for patients with relapsed EBV genome positive Hodgkin disease [61]. Seven patients with multiply relapsed Hodgkin disease have received gene-marked EBV-specific CTL in a phase I dose-escalation study. Gene-marked CTLs were found in PB up to 12 months following infusion. One patient had erosion of tumor through the L upper lobe bronchus and died 2 months after CTL infusion. In situ PCR revealed gene-marked CTLs within part of the mediastinal tumor but not at the site of L upper lobe bronchus erosion. In another patient, EBV-specific CTLs were localized to a malignant pleural effusion 3 weeks after CTL infusion as assayed by gene marking. The conclusions from this study were that marked CTLs expanded *in vivo* and were found to persist for up to 12 months as tracked by RT-PCR analysis [65].

50.6.4 GENE MARKING OF T CELLS IN HIV THERAPY

To study the survival of normal T cells in patients with HIV, PBLs from unaffected syngeneic twins were transduced with a *neo* vector and transferred to their HIV-infected sibling [63]. Marked CD4$^+$ T cells persisted in the circulation for 4–18 weeks after transfer in all patients and at 6 months, marked cells were found in lymphoid tissues [63]. In subsequent studies, autologous HIV gag-specific CD8$^+$ CTL clones genetically marked with the LN retrovirus were infused [62]. The infused CTLs were seen to accumulate adjacent to HIV-infected lymph nodes and transiently reduced the levels of circulating productively infected CD4$^+$ T cells. This decline was transient, likely due to a lack of CD4$^+$ help. However the study did provide some rationale for pursuing this approach in conjunction with strategies to circumvent the requirement for CD4 cells.

50.7 GENE MARKING OF MESENCHYMAL CELLS

Over the last few years, there has been increasing interest in the use of mesenchymal cells to treat patients with genetic disorders affecting mesenchymal tissues, including bone, cartilage, and muscle. To evaluate this therapeutic option, Horwitz et al. infused gene-marked, donor marrow-derived mesenchymal cells to treat six children who had undergone standard BMT for severe osteogenesis imperfecta [66]. The marking component of this study showed that the mesenchymal cells could engraft in bone, skin, and marrow stroma [66]. Of interest the marker gene could not be detected in one patient likely due to an immune response directed at the neo gene [66].

50.8 SAFETY CONCERNS

Safety is clearly an important consideration in gene marking studies. Since 1991, more than 170 patients have been treated with gene-marked HSCs or T cells, and none have had discernible deleterious effects. In particular, there has been no evidence of selective or autonomous expansion of marked cells on follow-up extending beyond 10 years [20,59,67]. In contrast to this impeccable safety record, four children with

X-linked severe combined immunodeficiency (SCID-X1) treated with autologous CD34$^+$ selected stem cells transduced with common γ-chain subsequently developed T cell lymphoproliferation as a result of insertional mutagenesis to the LMO2 proto-oncogene [68]. The superior safety profile of marking studies can likely be explained by the genetic modification of a smaller number of cells, each with fewer integrants, and by the absence of a transgene with growth promoting activities. The target cell type is also likely to be important. To date, insertional mutagenesis has not been reported in T cell gene transfer. Nonetheless, the fact that a previously hypothetical risk of retroviral vectors had now become reality changed many people's perception of the safety of gene marking. Since gene marking had no direct therapeutic benefit to the patients receiving the modified cells, any risk of oncogenesis was now considered unacceptable; however, the probability of such an event is low. We have to accept that for the foreseeable future, gene therapy of all types will be held to a much higher level of accountability and safety than any other therapeutic modality. It is troubling to realize that few of our most successful treatment options for life threatening conditions—from organ transplantation to cancer chemotherapy—would have been adopted had they too been held to "gene therapy standards of safety."

Paradoxically, it was further investigation into the retroviral associated leukemias in the affected patients that reanimated the field. These studies focused on analyzing the sites of viral integration in the host genome, and greatly improved on the methodologies available for this purpose. Although their primary intent was to establish how the leukemias had originated, the availability of more robust techniques for integrant site analyses provided additional sophistication to marking studies and greatly increased the information they could provide.

50.9 INTEGRANT ANALYSES

50.9.1 INTEGRATION SITE ANALYSIS TO ANALYZE CLONAL FATE

Measures of the presence and overall level of marker gene positivity in a cell population are highly informative, but determining the sites in the genome at which the provirus integrates provides a further level of sophistication to the analyses. Every retroviral integration event results in a host integrant junction that is unique to the transduced cell and its progeny. Integration sites as clonality markers enable one to not only enumerate clones but also to track the fate of individual clones *in vivo* over time. It is therefore possible to find out, for example, the minimum number of transduced cells that have contributed to hematopoietic reconstitution [69] or tumor relapse [16], the kinetics of individual clonal contribution [69], whether clonal dominance evolves *in vivo* [70,71], and even the true pluripotency of transduced cells [71,72].

The earliest available techniques for integration analysis were insensitive and had poor resolution. Amplification of the host integrant junction where the sequence at one end is unknown requires an adaptation of PCR amplification techniques. An early example of clonal analysis in humans was the study of gene marked neuroblastoma relapse using inverted PCR [7]. In this example, the host integrant junction was amplified by PCR prior to Southern blotting. PCR amplification was made possible by inverting the sequence [73]. In each of two neuroblastoma samples studied, two integration sites were detected. Assuming that approximately 1% of reinfused tumor cells were gene marked [32], and that there was one integrant per transduced cell, it could be estimated that a minimum of 200 tumor cells had contributed to relapse [16].

Techniques for host integrant junction analysis have become much more sensitive and are now capable of analyzing complex polyclonal populations. Broadly speaking, these techniques are based on ligation-mediated PCR (LM-PCR) in which a known sequence is ligated to the unknown flanking end to allow PCR amplification. Linear amplification-mediated PCR (LAM-PCR) has been most widely used [71,74–76]. This involves repeated cycles of linear PCR using a biotinylated primer, enrichment of resulting fragments on streptavidin magnetic beads followed by second strand synthesis with random hexamer priming. The resulting double-stranded DNA is digested and ligated to a complementary ligation cassette, thereby giving rise to DNA fragments with known sequences on both ends: The vector sequence in one and the ligation cassette sequence in the other. Nested exponential PCR is then performed and the resulting bands can be isolated and sequenced [71,75,76]. This method has a sensitivity of 1/100 to 1/1000 [76] and could detect more than 50 integration sites at a time [75].

LAM-PCR has provided important insight into the true pluripotency of gene-modified progenitor cells. In the SCID-X1 trial, clonality analysis has detected common integration sites shared by T cells, B cells, neutrophils, and monocytes, thereby demonstrating the transduction of pluripotent common progenitor cells [77]. Clonal analysis of patients from the adenosine deaminase (ADA)–SCID trial found that in one patient, the majority of T cells, as well as some myeloid cells, shared a common integration site. This clonal dominance remained stable without any evidence of progression to leukemia in over 8 years of observation [78]. LAM-PCR has also been used to characterize the kinetics of PB repopulation following autologous PBSC transplantation in adults with CML [74]. Depending on the level of gene-marking, between three and 59 different insertion sites could be identified in each patient *in vivo*, with different clones appearing at different time points [74].

50.9.2 Mapping Integration Sites

With the completion of the human genome-sequencing project, the precise sites of retroviral integration can be readily mapped. This allows the identification of any integration "hotspots" and integration sites that have the potential for causing insertional mutagenesis [79]. In the CML study, a total of 69 unique sites were identified in three patients using LAM-PCR [74]. Fifty-one could be mapped using BLAST search tool, of which around 80% were found to be within or

near genes. Mapping of integration sites in T cell therapy has been largely been performed on patients who received HSV-TK-modified donor T cells [80,81]. In one study, 300 integration sites from the infusion product or patient PB T cells were detected, sequenced, and mapped [80]. Of these, 52% were within transcribed region of genes and 35% were within 30 kb of genes. An analysis of 40 clones with integration within or near genes found that 20% of integration could affect expression of nearby genes. Reassuringly, there were no discernable functional differences between transduced and nontransduced T cells and no evidence of clonal selection at up to 9 years.

50.9.3 Each Patient Can Become His or Her Own Control by Incorporating Marker Sequences into Therapeutic Transgenes

The added information obtained from integrant analyses was insufficient to overcome the concerns investigators had developed about pure marking studies. Fortunately, it quickly became apparent that it was possible to incorporate marker oligonucleotide sequences in therapeutic transgenes that could then be tracked *in vivo*. In this way, a single patient could receive two or more genetically distinguishable populations of cells all expressing the same therapeutic transgene. Hence, it would still be possible, in a single patient, to compare the consequences of expressing the transgenes in different or differently treated cell populations. As alluded to earlier, the enormous heterogeneity of human responses makes such internally controlled experiments extremely powerful even with only small numbers of patients. One example is the study of gene modified T cells for the treatment of cancer.

50.10 IMPROVING MARKER STUDIES

50.10.1 Marker Gene

Most of the clinical studies reported to date have used *neo* as the reporter gene. This marker has the advantage that its safety has been widely studied in many different animal models and it can be detected both phenotypically and genotypically. In gene marking, the concern is that the activity of neomycin phosphotransferase (the product of the *neo* gene) is also likely to phosphorylate cellular proteins and may thereby modify the growth or differentiation of cells expressing the gene. *In vitro*, this effect has been observed in the HL60 cell line [82]. There is evidence that a similar phenomenon occurs *in vivo*. There is the almost uniform observation that the proportion of cells positive for the *neo* marker gene in circulation is 0.5–1 log below the number of cells that are positive as determined by *in vitro* colony assays in the presence of the neomycin analog G418 [15]. It may be that *neo* expression retards progenitor growth and differentiation *in vivo* thus rapidly diminishing the numbers of *neo* progeny detectable. It is also possible that the expression of the transgene (i.e., *neo*) by the transduced cell results in an immune response that eliminates the transduced population. For some gene products, such as the hygromycin-thymidine kinase (Hy-TK) fusion protein, the potency of this immune response is such that the response is a rapid elimination of large numbers of transduced cells in less than 48 h [56]. It is as yet unclear if so potent an immune response regularly occurs after *neo* transduction, since neo-positive cells can be detected for up to 9 years after infusion and neo-marked T cells can be readily expanded *in vivo* by appropriate antigenic stimulation [59,60]. If the discrepancy between levels of marking in CFU and mature cells is a consequence of the immunogenicity of the marker signal, then the interpretation of certain marker studies may be rendered more difficult.

Because of the limitations of the *neo* gene as a marker, several alternatives have been proposed. Amongst the most widely used in studies are green fluorescent protein (GFP) and cell surface markers such as the truncated (low-affinity) receptor for nerve growth factor (dLNGFR) [83–85]. Unfortunately, many of these have the same limitations as neo and there is concern that some of these proteins (e.g., GFP) may prove to be significantly immunogenic *in vivo* [86]. Moreover, aberrant expression of cell-surface molecules may lead to unwanted cell trafficking or harmful intercellular contacts, even if the cell–surface molecule has been modified to preclude intracellular signaling. One such concern is highlighted in a murine model using the same truncated nerve growth factor receptor (dLNGFR) marker gene that had been used in clinical studies [85]. This gene was introduced by a replication-incompetent murine retroviral vector into bone marrow cells, which were then transplanted into irradiated mice. Marrow was then harvested after 28 weeks, pooled, and transplanted into secondary irradiated mice. These recipients subsequently developed hematopoietic disorders, carrying the same leukemic clone with a single vector copy integrated into the transcription factor gene Evi1 [87], and it was suggested that the capacity of dLNGFR to produce aberrant signal transduction contributed to the oncogenic events.

Because of the above concerns that marker gene may be immunogenic, may affect cell growth and may contribute to oncogenicity, the most ideal marker may in fact be a nonexpressed sequence [66].

50.10.2 USE OF ALTERNATIVE VECTORS

For gene marking studies, in which stable transduction is required and in which all daughter cells should contain equal amounts of the marker gene, integrating vectors are required. To date, this has meant using Moloney murine leukemia virus (MoMuLV) [21,26,88]. The main limitation of the Moloney-based vectors is that they can only effectively transduce actively dividing cells, a major limitation when human HSCs are the targets because too few of these cells are in cycle at any one time.

Because of these limitations, human and feline lentiviral vectors have attracted increasing attention as gene delivery systems [89,90]. Lentiviral vectors have been reported to readily transduce hemopoietic progenitor cells [91]. Because they form a more stable preintegration complex than MoMuLV, they are also able to infect quiescent subsets of primitive hemopoietic stem cells, such as the CD34+ CD38− or the CD38− lin negative population, where they persist and integrate once these cells enter cycle [92]. HIV can also efficiently infect terminally differentiated cells such as neurons [93] and in both types of cell, high levels of gene expression have been reported.

50.11 FUTURE DEVELOPMENTS

Noninvasive tracking of gene-marked cells using imaging techniques is being actively developed. Until recently, imaging has been limited to bioluminescence technology suitable only for small animals. However, human imaging is now possible through the use of radionuclide imagine (positron emission tomography [PET] or single photon emission tomography [SPECT]) or magnetic resonance imaging (MRI) [94]. Several approaches have been taken but in general, they involve the selective accumulation of labeled-substance by gene-modified cells. One example is the phosphorylation of uracil or guanosine derivatives (^{124}I-or ^{131}I-FIAU or [17] F-FHBG, respectively) by HSV-TK, thereby preventing their efflux from cells. Another is the active transport of iodide by human sodium iodide symporter (hNIS). In addition to cell tracking, both systems have potential for therapeutic application: HSV-TK transduced cells can be eliminated by ganciclovir and radioactive iodide can be used to ablate hNIS-expressing cells. At present, human imaging is restricted to gene-marked tumor cells [94,95]. However, FIAU imaging of EBV-specific T cells has been shown to be feasible in mouse models [96] and will likely translate into human application.

Increasing interest in cell therapies using stem cells derived from, or targeted to, nonhematopoietic tissues, will doubtless widen the range of cells to which the above marker techniques can be applied. The use of hematopoietic or cardiac stem cells to treat heart failure, for example [97,98], is one important area in which cell tracking and fate determination would greatly simplify assessment of the contribution of the infused stem cell product to the cardiac tissues, and would allow comparison of multiple strategies in a single patient. The use of mesenchymal cells to treat tissue injury or prevent GVHD is another example in which the approach would help resolve many of the current controversies [99,100].

50.12 CONCLUSION

Gene marking is a useful scientific tool in the study of HSC transplantation and adoptive cellular therapy. It has also guided us as to appropriate techniques for *ex vivo* manipulation to augment gene transfer. The large number of patients and the long follow-up available for many of them form a useful source of reference for therapeutic gene transfer.

The full potential of gene marking studies is yet to be realized. While it is hoped that this valuable approach will not be curtailed by inappropriately perceived risks, it seems likely that the future of gene marking lies in incorporating multiple distinguishable marker sequences within transgenes that offer some therapeutic potential, thereby evolving from purely marking studies to a combination of marker/therapy.

REFERENCES

1. Miller AR, Skotzko MJ, Rhoades K et al. Simultaneous use of two retroviral vectors in human gene marking trials: Feasibility and potential applications. *Hum. Gene Ther.* 1992;3:619–624.

2. Brenner MK, Rill DR, Holladay MS et al. Gene marking to determine whether autologous marrow infusion restores long-term haemopoiesis in cancer patients. *Lancet* 1993; 342:1134–1137.

3. Rosenberg SA, Blaese RM, and Anderson WF. The N2-TIL human gene transfer clinical protocol. *Hum. Gene Ther.* 1990;1:73–92.

4. Brenner M, Mirro J, Jr., Hurwitz C et al. Autologous bone marrow transplant for children with AML in first complete remission: Use of marker genes to investigate the biology of marrow reconstitution and the mechanism of relapse. *Hum. Gene Ther.* 1991;2:137–159.

5. Deisseroth AB, Kantarjian H, Talpaz M et al. Autologous bone marrow transplantation for CML in which retroviral markers are used to discriminate between relapse which arises from systemic disease remaining after preparative therapy versus relapse due to residual leukemia cells in autologous marrow: A pilot trial. *Hum. Gene Ther.* 1991;2:359–376.

6. Santana VM, Brenner MK, Ihle J et al. A phase I trial of high-dose carboplatin and etoposide with autologous marrow support for treatment of stage D neuroblastoma in first remission: Use of marker genes to investigate the biology of marrow reconstitution and the mechanism of relapse. *Hum. Gene Ther.* 1991;3:257–272.

7. Cornetta K, Tricot G, Broun ER et al. Retroviral-mediated gene transfer of bone marrow cells during autologous bone marrow transplantation for acute leukemia. *Hum. Gene Ther.* 1992;3:305–318.

8. Dunbar CE, Nienhuis AW, Stewart FM et al. Genetic marking with retroviral vectors to study the feasibility of stem cell gene transfer and the biology of hemopoietic reconstitution after autologous transplantation in multiple myeloma, chronic myelogenous leukemia, or metastatic breast cancer. *Hum. Gene Ther.* 1993;4:205–222.

9. Bjorkstrand B, Gahrton G, Sirac Dilber M et al. Retroviral-mediated gene transfer of CD34-enriched bone marrow and peripheral blood cells during autologous stem cell transplantation for multiple myeloma. *Hum. Gene Ther.* 1994;5:1279–1286.

10. Schuening F, Miller AD, Torok-Storb B et al. Study on contribution of genetically marked peripheral blood repopulating cells to hematopoietic reconstitution after transplantation. *Hum. Gene Ther.* 1994;5:1523–1534.

11. Stewart AK, Dube ID, Kamel-Reid S, and Keating A. A Phase 1 study of autologous bone marrow transplantation with stem cell gene marking in multiple myeloma. *Hum. Gene Ther.* 1995;6:107–110.

12. Gahrton G, Bjorkstrand B, Dilber MS et al. Gene marking and gene therapy in multiple myeloma. *Adv. Exp. Med. Biol.* 1998;451:493–497.

13. Santana VM, Brenner MK, Ihle J et al. A phase I trial of high-dose carboplatin and etoposide with autologous marrow support for treatment of relapse/refractory neuroblastoma without apparent bone marrow involvement: Use of marker genes to investigate the biology of marrow reconstitution and the mechanism of relapse. *Hum. Gene Ther.* 1991;2:273–286.

14. Brenner MK, Rill DR, Moen RC et al. Gene-marking to trace origin of relapse after autologous bone marrow transplantation. *Lancet* 1993;341:85–86.

15. Brenner MK, Heslop HE, Rill D et al. *Gene Transfer and Bone Marrow Transplantation.* Cold Spring Harbor Symposia on Quantititive Biology. Cold Springer Harbor, NY: Cold Spring Harbor Press; 1994:691–697.

16. Rill DR, Santana VM, Roberts WM et al. Direct demonstration that autologous bone marrow transplantation for solid tumors can return a multiplicity of tumorigenic cells. *Blood* 1994;84:380–383.

17. Deisseroth AB, Zu Z, Claxton D et al. Genetic marking shows that Ph+ cells present in autologous transplants of chronic myelogenous leukemia (CML) contribute to relapse after autologous bone marrow in CML. *Blood* 1994;83:3068–3076.

18. Cornetta K, Srour EF, Moore A et al. Retroviral gene transfer in autologous bone marrow transplantation for adult acute leukemia. *Hum. Gene Ther.* 1996;7:1323–1329.

19. Dunbar CE, Cottler-Fox M, O'Shaunessy JA et al. Retrovirally marked CD34-enriched peripheral blood and marrow cells contribute to long term engraftment after autologous transplantation. *Blood* 1995;85:3048–3057.

20. Rill DR, Sycamore DL, Smith SS et al. Long term in vivo fate of human hemopoietic cells transduced by moloney-based retroviral vectors [abstract]. *Blood* 2000;96:844a.

21. Bollard CM, Heslop HE, and Brenner MK. Gene-marking studies of hematopoietic cells. *Int. J. Hematol.* 2001;73:14–22.

22. Emmons RV, Doren S, Zujewski J et al. Retroviral gene transduction of adult peripheral blood or marrow-derived CD34+ cells for six hours without growth factors or on autologous stroma does not improve marking efficiency assessed in vivo. *Blood* 1997;89:4040–4046.

23. Alici E, Bjorkstrand B, Treschow A et al. Long-term follow-up of gene-marked CD34+ cells after autologous stem cell transplantation for multiple myeloma. *Cancer Gene Ther.* 2007;14:227–232.

24. Bachier CR, Giles RE, Ellerson D et al. Hematopoietic retroviral gene marking in patients with follicular non-Hodgkin's lymphoma. *Leuk. Lymphoma* 1999;32:279–288.

25. Brenner MK, Krance R, Heslop HE et al. Assessment of the efficacy of purging by using gene marked autologous marrow transplantation for children with AML in first complete remission. *Hum. Gene Ther.* 1994;5:481–499.

26. Brenner MK. Gene transfer and the treatment of haematological malignancy. *J. Intern. Med.* 2001;249:345–358.

27. Klingemann HG, Eaves CJ, Barnett MJ et al. Transplantation of patients with high risk acute myeloid leukemia in first remission with autologous marrow cultured in interleukin-2 followed by interleukin-2 administration. *Bone Marrow Transpl.* 1994;14:389–396.

28. Selvaggi KJ, Wilson JW, Mills LE et al. Improved outcome for high-risk acute myeloid leukemia patients using autologous bone marrow transplantation and monoclonal antibody-purged bone marrow. *Blood* 1994;83:1698–1705.

29. Heslop HE, Rill DR, Horwitz EM et al. Gene marking to assess tumor contamination in stem cell grafts for acute myeloid leukemia. In: Dicke KA, Keating A, eds. *Autologous Blood and Marrow Transplantation.* Charlottesville, VA: Carden Jennings; 1999:513–520.

30. Brugger W, Heimfeld S, Berenson RJ, Mertelsmann R, and Kanz L. Reconstitution of hematopoiesis after high dose chemotherapy by autologous progenitor cells generated ex vivo. *N. Engl. J. Med.* 1995;333:283–287.

31. Bienzle D, Abrams-Ogg ACG, Kruth SA et al. Gene transfer into hematopoietic stem cells: Long-term maintenance of in vitro activated progenitors without marrow ablation. *Proc. Natl. Acad. Sci. USA* 1994;91:350–354.

32. Rill DR, Moen RC, Buschle M et al. An approach for the analysis of relapse and marrow reconstitution after autologous marrow transplantation using retrovirus-mediated gene transfer. *Blood* 1992;79:2694–2700.

33. Dunbar CE, Bodine DM, Sorrentino B et al. Gene transfer into hematopoietic cells: Implications for cancer therapy. *Ann. NY Acad. Sci.* 1994;716:216–224.

34. Bodine DM, Moritz T, Donahue RE et al. Long-term in vivo expression of a murine adenosine deaminase gene in rhesus monkey hematopoietic cells of multiple lineages after retroviral mediated gene transfer into CD34+ bone marrow cells. *Blood* 1993;82:1975–1980.

35. Stewart AK, Sutherland DR, Nanji S et al. Engraftment of gene-marked hematopoietic progenitors in myeloma patients after transplant of autologous long-term marrow cultures [in process citation]. *Hum. Gene Ther.* 1999;10:1953–1964.

36. Donahue RE, Kessler SW, Bodine D et al. Helper virus induced T cell lymphoma in nonhuman primates after retroviral mediated gene transfer. *J. Exp. Med.* 1992;176:1125–1135.

37. Kobayashi M, Laver JH, Lyman SD et al. Thrombopoietin, steel factor and the ligand for flt3/flk2 interact to stimulate the proliferation of human hematopoietic progenitors in culture. *Int. J. Hematol.* 1997;66:423–434.

38. Piacibello W, Sanavio F, Garetto L et al. Extensive amplification and self-renewal of human primitive hematopoietic stem cells from cord blood. *Blood* 1997;89:2644–2653.

39. Ku H, Yonemura Y, Kaushansky K, and Ogawa M. Thrombopoietin, the ligand for the Mpl receptor, synergizes with steel factor and other early acting cytokines in supporting proliferation of primitive hematopoietic progenitors of mice. *Blood* 1996;87:4544–4551.

40. Bodine DM, Seidel NE, and Orlic D. Bone marrow collected 14 days after in vivo administration of granulocyte colony-stimulating factor and stem cell factor to mice has 10-fold more repopulating ability than untreated bone marrow. *Blood* 1996;88:89–97.

41. Hanenberg H, Xiao XL, Dilloo D et al. Co-localization of retrovirus and target cells on specific fibronectin adhesion domains for increased genetic transduction of mammalian cells. *Nat. Med.* 1996;2:876–882.

42. Abonour R, Williams DA, Einhorn L et al. Efficient retrovirus-mediated transfer of the multidrug resistance 1 gene into autologous human long-term repopulating hematopoietic stem cells [in process citation]. *Nat. Med.* 2000;6:652–658.

43. Hesdorffer C, Ayello J, Ward M et al. Phase I trial of retroviral-mediated transfer of the human MDR1 gene as marrow chemoprotection in patients undergoing high-dose chemotherapy and autologous stem-cell transplantation. *J. Clin. Oncol.* 1998;16:165–172.

44. Hanania EG, Giles RE, Kavanagh J et al. Results of MDR-1 vector modification trial indicate that granulocyte/macrophage colony-forming unit cells do not contribute to posttransplant hematopoietic recovery following intensive systemic therapy. *Proc. Natl. Acad. Sci. USA* 1996;93:15346–15351.

45. Moscow JA, Huang H, Carter C et al. Engraftment of MDR1 and NeoR gene-transduced hematopoietic cells after breast cancer chemotherapy. *Blood* 1999;94:52–61.

46. Ward M, Richardson C, Pioli P et al. Transfer and expression of the human multiple drug resistance gene in human CD34+ cells. *Blood* 1994;84:1408–1414.

47. Rosenberg SA, Aebersold P, Cornetta K et al. Gene transfer into humans—Immunotherapy of patients with advanced melanoma, using tumor-infiltrating lymphocytes modified by retroviral gene transduction. *N. Engl. J. Med.* 1990;323:570–578.

48. Fisher B, Packard BS, Read EJ et al. Tumor localization of adoptively transferred indium-111 labeled tumor infiltrating lymphocytes in patients with metastatic melanoma. *J. Clin. Oncol.* 1989;7:250–261.

49. Griffith KD, Read EJ, Carrasquillo JA et al. In vivo distribution of adoptively transferred indium-111-labeled tumor infiltrating lymphocytes and peripheral blood lymphocytes in patients with metastatic melanoma. *J. Natl. Cancer Inst.* 1989;81:1709–1717.

50. Cai Q, Rubin JT, and Lotze MT. Genetically marking human cells: Results of the first clinical gene transfer studies. *Cancer Gene Ther.* 1995;2:125–136.

51. Aebersold P, Kasid A, and Rosenberg SA. Selection of gene-marked tumor infiltrating lymphocytes from post-treatment biopsies: A case study. *Hum. Gene Ther.* 1990;1:373–384.

52. Morgan RA, Cornetta K, and Anderson WF. Applications of the polymerase chain reaction in retroviral mediated gene transfer and the analysis of gene marked human TIL cells. *Hum. Gene Ther.* 1990;1:135–150.

53. Hwu P and Rosenberg SA. The use of gene-modified tumor infiltrating lymphocytes for cancer therapy. *Ann. NY Acad. Sci.* 1994;716:188–199.

54. Merrouche Y, Negrier S, Bain C et al. Clinical application of retroviral gene transfer in oncology: Results of a French study with tumor-infiltrating lymphocytes transduced with the gene of resistance to neomycin. *J. Clin. Oncol.* 1995;13:410–418.

55. Economou JS, Belldegrun AS, Glaspy J et al. In vivo trafficking of adoptively transferred Interleukin-2 expanded tumor-infiltrating lymphocytes and peripheral blood lymphocytes: Results of a double gene marking trial. *J. Clin. Invest.* 1996;97:515–521.

56. Riddell SR, Elliot M, Lewinsohn DA et al. T-cell mediated rejection of gene-modified HIV-specific cytotoxic T lymphocytes in HIV-infected patients. *Nat. Med.* 1996;2:216–223.

57. Rooney CM, Smith CA, Ng C et al. Use of gene-modified virus-specific T lymphocytes to control Epstein–Barr virus-related lymphoproliferation. *Lancet* 1995;345:9–13.

58. Walter EA, Greenberg PD, Gilbert MJ et al. Reconstitution of cellular immunity against cytomegalovirus in recipients of allogeneic bone marrow by transfer of T-cell clones from the donor. *N. Engl. J. Med.* 1995;333:1038–1044.

59. Heslop HE, Ng CYC, Li C et al. Long-term restoration of immunity against Epstein–Barr virus infection by adoptive transfer of gene-modified virus-specific T lymphocytes. *Nat. Med.* 1996;2:551–555.

60. Rooney CM, Smith CA, Ng CYC et al. Infusion of cytotoxic T cells for the prevention and treatment of Epstein–Barr virus-induced lymphoma in allogeneic transplant recipients. *Blood* 1998;92:1549–1555.

61. Roskrow MA, Suzuki N, Gan Y-J et al. EBV-specific cytotoxic T lymphocytes for the treatment of patients with EBV positive relapsed Hodgkin's disease. *Blood* 1998;91:2925–2934.

62. Brodie SJ, Lewinsohn DA, Patterson BK et al. In vivo migration and function of transferred HIV-1-specific cytotoxic T cells [see comments]. *Nat. Med.* 1999;5:34–41.

63. Walker RE, Carter CS, Muul L et al. Peripheral expansion of pre-existing mature T cells is an important means of CD4+ T-cell regeneration HIV-infected adults. *Nat. Med.* 1998;4:852–856.

64. Smith CA, Ng CYC, Heslop HE et al. Production of genetically modified EBV-specific cytotoxic T cells for adoptive transfer to patients at high risk of EBV-associated lymphoproliferative disease. *J. Hematother.* 1995;4:73–79.

65. Bollard CM, Aguilar L, Straathof KC et al. Cytotoxic T lymphocyte therapy for Epstein–Barr Virus + Hodgkin's disease. *J. Exp. Med.* 2004;200:1623–1633.

66. Horwitz EM, Gordon PL, Koo WK et al. Isolated allogeneic bone marrow-derived mesenchymal cells engraft and stimulate growth in children with osteogenesis imperfecta: Implications for cell therapy of bone. *Proc. Natl. Acad. Sci. USA* 2002;99:8932–8937.

67. Brenner MK and Heslop HE. Is retroviral gene marking too dangerous to use? *Cytotherapy* 2003;5:190–193.

68. Hacein-Bey-Abina S, von Kalle C, Schmidt M et al. LMO2-associated clonal T cell proliferation in two patients after gene therapy for SCID-X1. *Science* 2003;302:415–419.

69. Ott MG, Schmidt M, Schwarzwaelder K et al. Correction of X-linked chronic granulomatous disease by gene therapy, augmented by insertional activation of MDS1-EVI1, PRDM16 or SETBP1. *Nat. Med.* 2006;12:401–409.

70. Kustikova O, Fehse B, Modlich U et al. Clonal dominance of hematopoietic stem cells triggered by retroviral gene marking. *Science* 2005;308:1171–1174.

71. Schmidt M, Carbonaro DA, Speckmann C et al. Clonality analysis after retroviral-mediated gene transfer to CD34+ cells from the cord blood of ADA-deficient SCID neonates. *Nat. Med.* 2003;9:463–468.

72. Schmidt M, Hacein-Bey-Abina S, Wissler M et al. Clonal evidence for the transduction of CD34+ cells with lymphomyeloid differentiation potential and self-renewal capacity in the SCID-X1 gene therapy trial. *Blood* 2005;105:2699–2706.

73. Triglia T, Peterson MG, and Kemp DJ. A procedure for in vitro amplification of DNA segments that lie outside the boundaries of known sequences. *Nucleic Acids Res.* 1988;16:8186.

74. Glimm H, Schmidt M, Fischer M et al. Efficient marking of human cells with rapid but transient repopulating activity in autografted recipients. *Blood* 2005;106:893–898.

75. Schmidt M, Zickler P, Hoffmann G et al. Polyclonal long-term repopulating stem cell clones in a primate model. *Blood* 2002;100:2737–2743.

76. Schmidt M, Hoffmann G, Wissler M et al. Detection and direct genomic sequencing of multiple rare unknown flanking DNA in highly complex samples. *Hum. Gene Ther.* 2001;12:743–749.

77. Shwartzwaelder K, Howe SJ, Schmidt M et al: Gammaretrovirus-mediated correction of SCID-X1 is associated with skewed vector integration site distribution in vivo. *J. Clin. Invest.* 2007;117:2241–2249.

78. Woods NB, Bottero V, Schmidt M, von KC, and Verma IM. Gene therapy: Therapeutic gene causing lymphoma. *Nature* 2006;440:1123.

79. Deichmann A, Hacien-Bey-Abina S, Schmidt M et al: Vector integration is non-random, clustered and influences the fate of lymphopoiesis in SCID-X1 gene therapy. *J. Clin. Invest.* 2007;117:2225–2232.

80. Recchia A, Bonini C, Magnani Z et al. Retroviral vector integration deregulates gene expression but has no consequence on the biology and function of transplanted T cells. *Proc. Natl. Acad. Sci. USA* 2006;103:1457–1462.

81. Giordano FA, Fehse B, Hotz-Wagenblatt A et al. Retroviral vector insertions in T-lymphocytes used for suicide gene therapy occur in gene groups with specific molecular functions. *Bone Marrow Transplant.* 2006;38:229–235.

82. von Melchner H and Housman DE. The expression of neomycin phosphotransferase in human promyelocytic leukemia cells (HL60) delays their differentiation. *Oncogene* 1987;2:137–140.

83. Pawliuk R, Kay R, Lansdorp P, and Humphries RK. Selection of retrovirally transduced hematopoietic progenitor cells using CD24 as a marker of gene transfer. *Blood* 1994;84:2868–2877.

84. Misteli T and Spector DL. Applications of the green fluorescent protein in cell biology and biotechnology. *Nat. Biotechnol.* 1998;15:961–964.

85. Bonini C, Ferrari G, Verzeletti S et al. HSV-TK gene transfer into donor lymphocytes for control of allogeneic graft versus leukemia. *Science* 1997;276:1719–1724.

86. Stripecke R, Carmen VM, Skelton D et al. Immune response to green fluorescent protein: Implications for gene therapy. *Gene Ther.* 1999;6:1305–1312.

87. Li Z, Dullmann J, Schiedlmeier B et al. Murine leukemia induced by retroviral gene marking. *Science* 2002;296:497.

88. Wivel NA and Wilson JM. Methods of gene delivery. *Hematol. Oncol. Clin. North Am.* 1998;12:483–501.

89. Amado RG and Chen IS. Lentiviral vectors—the promise of gene therapy within reach? *Science* 1999;285:674–676.

90. Buchschacher GL, Jr. and Wong-Staal F. Development of lentiviral vectors for gene therapy for human diseases. *Blood* 2000;95:2499–2504.

91. Sutton RE, Wu HT, Rigg R, Bohnlein E, and Brown PO. Human immunodeficiency virus type 1 vectors efficiently transduce human hematopoietic stem cells. *J. Virol.* 1998;72:5781–5788.

92. Case SS, Price MA, Jordan CT et al. Stable transduction of quiescent CD34(+)CD38(−) human hematopoietic cells by HIV-1-based lentiviral vectors. *Proc. Natl. Acad. Sci. USA* 1999;96:2988–2993.

93. Naldini L, Blomer U, Gallay P et al. In vivo gene delivery and stable transduction of nondividing cells by a lentiviral vector. *Science* 1996;272:263–267.

94. Penuelas I, Haberkorn U, Yaghoubi S, and Gambhir SS. Gene therapy imaging in patients for oncological applications. *Eur. J. Nucl. Med. Mol. Imaging* 2005;32 Suppl. 2: S384–S403.

95. Penuelas I, Mazzolini G, Boan JF et al. Positron emission tomography imaging of adenoviral-mediated transgene expression in liver cancer patients. *Gastroenterology* 2005; 128:1787–1795.

96. Koehne G, Doubrovin M, Doubrovina E et al. Serial in vivo imaging of the targeted migration of human HSV-TK-transduced antigen-specific lymphocytes. *Nat. Biotechnol.* 2003; 21:405–413.

97. Mathur A and Martin JF. Stem cells and repair of the heart. *Lancet* 2004;364:183–192.

98. Oettgen P, Boyle AJ, Schulman SP, and Harc JM. Cardiac stem cell therapy. Need for optimization of efficacy and safety monitoring. *Circulation* 2006;114:353–358.

99. Bengel FM, Schachinger V, and Dimmeler S. Cell-based therapies and imaging in cardiology. *Eur. J. Nucl. Med. Mol. Imaging* 2005;32 Suppl 2:S404–S416.

100. Le Blanc K and Ringden O. Immunobiology of human mesenchymal stem cells and future use in hematopoietic stem cell transplantation. *Biol. Blood Marrow Transplant.* 2005;11: 321–334.

51 Regulatory Aspects of Gene Therapy

Stephanie L. Simek and Lilia Bi

CONTENTS

51.1 OVERSIGHT OF GENE THERAPY IN THE UNITED STATES

51.1.1 FDA Regulatory Authority

For over the past 100 years, the American public has been concerned about the purity and quality of their food and medicines going as far back as 1848 when the state of California passed a pure food and drink law. By the end of the nineteenth century scientific discoveries were taking place that were being applied to the prevention and treatment of dangerous diseases. Live vaccines were being used for smallpox and rabies, killed-vaccines for cholera, typhoid and plague, and antitoxins for diphtheria and tetanus. During this time, laws regulating biologics were being enacted in Europe and Russia, whereas in the United States even though there was a concern regarding the safety of biologics there was an eagerness to use these products without the implementation of regulatory safeguards. In 1901, 13 children in St. Louis died of tetanus after

receiving diphtheria antitoxin from a horse. This tragedy lead in 1902 to Congress making appropriations to establish food standards and in the same year, the Biologics Control Act was passed to license and regulate interstate sale of serum, vaccines, and other biologics used to prevent or treat disease in humans. The 1902 Act established federal inspection of licensed facilities, prohibition of false labeling, and the concept of a dating period during which a biological product would be medically used. In 1906, the Congress passed the Food and Drug Act and prohibited interstate commerce of misbranded and adulterated foods, drinks, and drugs. The basis of the 1906 law rested on the regulation of product labeling and not on premarket approval. In 1938, the Federal Food Drug and Cosmetic (FD&C) Act was enacted, making the 1906 Food and Drug Act obsolete. The FD&C Act extended controls to cosmetics and therapeutic devices and also required predistribution clearance of new drugs based on safety. In addition, this act authorized standards of identity and quality for foods and drugs as well as authorizing factory inspections.

In 1944, the Public Health Service Act (PHSA) was established and consolidated the major rule making authority for biological products under Sections 351 and 352. The PHSA requires that the product and the establishment where the product is manufactured meet standards to ensure continued safety, purity, and potency of the biologic.

Today the primary mission of the Food and Drug Administration (FDA) is to protect the public health by assuring the safety, efficacy, and security of human and veterinary drugs, biological products, medical devices, our nation's food supply, cosmetics, and products that emit radiation. The FDA is also responsible for advancing the public health by helping speed innovations that make medicines and foods more effective, safer, and more affordable and helping the public get the accurate, science-based information they need to use medicines and foods to improve their health. This is accomplished by upholding established regulatory principles, such as quality control (QC), sound scientific rationale, and risk–benefit assessment. During the regulatory process, it is important that these principles be applied in a way that will encourage early product development while ensuring patient protection. This is particularly challenging for the field of gene therapy, where some of the risks are still unknown while potential benefits to patients may be great. To apply these principles, the Center for Biologics Evaluation and Research (CBER) takes a comprehensive approach to the regulation of biological products. This approach involves scientific review of submissions, a strong research program to support the regulatory process, surveillance, and policy development based on sound science to ensure that the manufacture of biological products and the conduct of clinical trials are performed in accordance with the regulations and statutes set forth to safeguard subjects enrolled in clinical trials.

CBER is one of six centers comprising the FDA and is responsible for the regulation of biological products. The authority to regulate biologics is promulgated by both the FD&C Act and the PHSA. These acts outline binding procedures for the agency and the sponsor. The FD&C Act provides the legal interpretation that a "biologic product" is also a "drug," and Section 351 of the PHSA makes provisions for the regulation of biological products through licensure of the product and the establishment in which the product is manufactured. Under the PHSA a biologic is defined as "a virus, therapeutic serum, toxin, antitoxin, vaccine, blood, blood component or derivative, allergenic product, analogous product, or arsphenamine or derivative of arsphenamine (or any other trivalent organic arsenic compound), applicable to the prevention, treatment or cure of a disease or condition of human beings."

Regulations pertaining to conduct of clinical investigations using biological products are outlined in title 21 of the Code of Federal Regulations (CFR). Regulations are interpretations of the laws and are binding like laws. The regulations covered under 21 CFR 312 specify requirements necessary for submission of Investigational New Drug (IND) applications, while standards for licensure of a biologic product are described under 21 CFR 601 and 610. FDA's regulatory

authority in the somatic cell and gene therapy area was established in the Federal Register notice of October 14, 1993 entitled "Application of Current Statutory Authorities to Human Somatic Cell Therapy Products and Gene Therapy Products." This document establishes that somatic cell and gene therapy products are biological products and as such are subject to the licensing provisions of the PHSA.

Besides the statutes and regulations by which the FDA governs its day-to-day interactions with industry and academia, the agency also issues guidance documents, which are not binding but rather describe CBER's policy and regulatory approach to specific product areas. Examples of documents that present recommendations and give relevant guidance to somatic cell and gene therapy products are provided in Section 51.5.

51.1.2 STATUS OF GENE THERAPY IN THE UNITED STATES

As provided in "Guidance for Industry: Gene Therapy Clinical Trials—Observing Subjects for Delayed Adverse Events," a gene therapy product is defined as "all products that mediate their effects by transcription and/or translation of transferred genetic material and/or by integrating into the host genome and that are administered as nucleic acids, viruses, or genetically engineered microorganisms. The products may be used to modify cells *in vivo* or transferred to cells *ex vivo* prior to administration to the recipient." Examples, which fall under this definition, include peripheral blood stem cells modified with a viral vector, intramuscular injection of a plasmid DNA vector, and direct injection of a viral vector. Since submission of the first human gene therapy IND in 1989, 552 human gene therapy INDs have been submitted to CBER for review (through June 2007). The review of gene therapy INDs is coordinated through the Office of Cellular, Tissue, and Gene Therapies (OCTGT) within CBER.

Currently there are 314 active gene therapy INDs at CBER (Figure 51.1). While more than half of the INDs submitted in

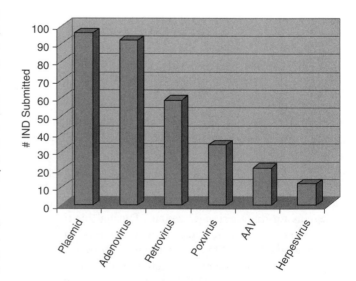

FIGURE 51.1 Schematic of currently active INDs showing the types and number of vectors being used in U.S. gene therapy clinical trials.

the early 1990s involved *ex vivo* modification of cells using gene therapy vectors, since 1995 the number of INDs involving direct administration of gene therapy vectors has been increasing. Currently, almost 60% of all INDs use direct vector administration. Overall, almost 95% of INDs submitted have involved expression or replacement of a gene with therapeutic intent and comprise indications as broad as cancer, human immunodeficiency virus (HIV), cystic fibrosis, hemophilia, peripheral and arterial vascular disease, arthritis, Parkinson's disease as well as many rare diseases (i.e., x-linked SCID, Fanconi's anemia). In approximately 30% of the INDs, nonviral plasmid vectors were used for gene delivery. A variety of viral vector systems are currently used in clinical studies, such as retroviral, adenoviral, adeno-associated virus (AAV), vaccinia, and herpes viral vectors. As a general rule, most of these vector systems consist of replication defective virus containing a therapeutic transgene. Although not considered a gene therapy vector system in the strictest sense, the investigational use of oncolytic viruses for certain cancer indications has been increasing over the past few years. This class of viruses usually does not contain a therapeutic gene but rather is a replication competent virus (RCV), which functions by producing lysis of tumor cells when administered directly into the tumor. Although oncolytic viruses are not considered to be gene therapy vectors, their oversight and regulation follows a similar pathway as traditional gene therapy products. Bacterial vectors are also being studied as a vector system for gene delivery. Most of these bacterial vectors are attenuated to reduce virulence or are inactivated by heat or other means.

For the area of gene therapy, CBER's regulatory approach has included an element of public process. The goal of this public interaction has been to increase the community's understanding of the CBER review process and requirements, to allow deliberation of ethical and social issues that surround the area of gene therapy, to receive input into CBER gene therapy policy development, and to provide accurate information to the public about the progress of gene therapy clinical trials. CBER's approach has included convening its Cellular, Tissue and Gene Therapies, Advisory Committee (CTGTAC), formerly known as the Biologic Response Modifiers Advisory Committee (BRMAC) to discuss relevant issues regarding the safe use of gene therapy products and conduct of clinical trials. In addition, FDA routinely takes part in forums with industry, trade groups, academia, and the public in order to foster public understanding in the area of gene therapy, which is essential for continued progress of the field. CBER considers this interaction with the scientific community important when developing policy and guidances on gene therapy. Another large part of this process has been facilitated by the National Institutes of Health (NIH), Office of Biotechnology Activities (OBA), through quarterly meetings of the Recombinant DNA Advisory Committee (RAC).

51.1.3 Oversight of FDA/RAC

Unlike other areas under clinical investigation which are regulated by the FDA, gene therapy clinical investigations are subject to the oversight of two federal agencies within the Department of Health and Human Services, the FDA and the NIH, OBA. FDA and NIH have complimentary roles with respect to oversight of human gene therapy clinical trials. FDA's primary role is to ensure that manufacturers of gene therapy products produce quality products that have been demonstrated to be safe for administration into subjects. NIH's primary role is to evaluate the quality of the science involved in gene therapy research, fund scientists who perform gene therapy clinical studies, and review and evaluate the composition of Institutional Biosafety Committees (IBC). Oversight of human gene therapy clinical trials at the NIH involves a public process of review and discussion, conducted by the RAC that ensures public awareness of clinical trial registration and follow-up. The RAC meets quarterly for public discussion of proposals which are deemed novel. In contrast, FDA review of gene therapy INDs is confidential and conducted by agency reviewers on an ongoing basis.

Both the NIH and FDA deliberate preclinical and clinical issues; however, the RAC's responsibilities extend beyond safety and efficacy to the consideration of the ethical, legal, and social implications of such research. The FDA also provides careful and thorough review of product manufacturing related to product safety, purity, potency, and identity. Currently, information regarding human gene therapy clinical trials must be submitted to both the FDA and the NIH.

51.1.3.1 History of RAC

The RAC was established in 1975 as a result of public concern over the potential risks of the new field of recombinant DNA (rDNA) research. Scientists worldwide had voluntarily halted their research and met in Asilomar, California, to debate the future of the use of rDNA technology. The RAC evolved from these debates and met for the first time just after the Asilomar meeting. The RAC mission was to advise the NIH Director and to review in public each experiment involving recombinant DNA research. Subsequently NIH established the OBA to provide administrative support to the RAC. Over the first few meetings, the RAC set minimum standards for biological and physical containment of rDNA molecules. This was accomplished through public debate and with input from scientists and lay representatives, including ethicists and economists. In 1976, as a result of these discussions, the "NIH Guidelines for Research Involving Recombinant DNA Molecules" (NIH Guidelines) were published in the Federal Register. The NIH Guidelines provided for submission and review of rDNA experiments by the RAC and also provided for the element of public debate of rDNA research. The NIH Guidelines are not regulations, but establish their authority through the NIH funding process. Investigators receiving NIH funding for Research Involving Recombinant DNA Molecules, or who are affiliated with an institution that has NIH funding for recombinant DNA research, must comply with the NIH Guidelines. In addition, they ask for voluntary compliance by non-NIH funded investigators. This process has provided a precedent for the public discussion and consideration of gene therapy clinical trials conducted today.

In 1982, in response to the report of the President's Commission, entitled, "Splicing Life: Social and Ethical Issues of Genetic Engineering with Human Beings," the Human Gene Therapy Subcommittee (HGTS) to the RAC was established to review the application of rDNA technology to human gene therapies. The first human gene therapy protocol was approved after public discussion by this committee and separately by the FDA in 1990. Over time, as the public concern over rDNA experiments subsided and the interest in gene therapy increased, the HGTS merged with the full RAC and the combined group discussed and approved each gene therapy protocol prior to its initiation. In 1994, an accelerated review process was adopted for certain categories of clinical trials that had been routinely reviewed by the RAC and determined not to represent significant risk to human health and the environment. Under this mechanism, such protocols were subject only to written review by several RAC members, and OBA approval, outside of the quarterly meetings.

The RAC is currently composed of 21 members from the disciplines of science, medicine, law, ethics, and members of patient and other lay communities. Members of the RAC meet quarterly for the public review and discussion of novel gene therapy protocols. However, they no longer approve or disapprove protocols. Currently, their function is purely to provide a platform for public discussion of novel issues involved in gene therapy clinical trials. In addition, the RAC sponsors the Gene Therapy Policy Conferences (GTPC),

which provides a mechanism for in-depth discussion of relevant gene therapy issues. For this forum, a panel of experts is convened with the goal of reaching consensus and developing guidance in a particular area. GTPC discussions have focused on topics such as genetic enhancement, inadvertent germline transmission, use of lentiviral vectors, prenatal gene therapy, and immunogenicity of AAV vectors. In 2002, OBA formed an RAC Informed Consent Working Group composed of members of the RAC, outside experts, and representatives of the FDA and the Office for Human Research Protections (OHRP) to assist the in the development of the "NIH Guidance on Informed Consent for Gene Transfer Research." This is a Web-based (http://www4.od.nih.gov/oba/rac/ic/index.html) guidance document that is intended to serve as a resource and learning tool for individuals involved in gene transfer studies and others with an interest in this field.

51.1.3.2 Dual Submissions

As illustrated in Figure 51.2, there is a parallel path of submissions required prior to initiation of a gene therapy clinical trial. Sponsors of a gene therapy clinical trial must submit an IND application to CBER, FDA for review under a 30-day review cycle and may not proceed until the IND is found to be acceptable. In addition, the FDA requires that a sponsor receives the approval of the Institutional Review Boards (IRBs) and IBCs affiliated with the institutions at which the trial will be conducted. Concurrently or prior to submission of an IND to the FDA, investigators conducting the clinical

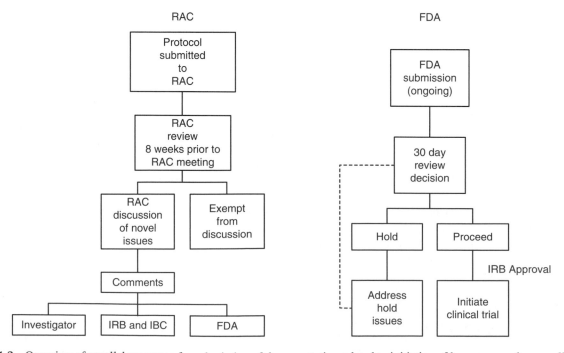

FIGURE 51.2 Overview of parallel processes for submission of documentation related to initiation of human gene therapy clinical trials to the RAC and the FDA.

trial should also submit their protocol and information specified in Appendix M of the NIH Guidelines to NIH, OBA for review by the RAC. Appendix M is a section included in the NIH Guideline that describes points to consider in the design and submission of human gene therapy trials, including the standards and procedures to which investigators need to adhere. Based on the changes made in May 2002 to the NIH Guidelines for Research Involving Recombinant DNA Molecules, all investigators/sponsors who receive or who are sponsored by an entity that receives NIH support for recombinant DNA research must submit their clinical protocol to the RAC for review. Failure to follow the NIH Guidelines can result in withdrawal of NIH funding to the investigator and any supporting institutions. Although an investigator may submit the protocol to the FDA and IRB prior to submittal to the RAC, review by the RAC is required before final IBC approval, so the committee can be informed of the RAC recommendations before making its final determination. The OBA submission must be received no less than 8 weeks prior to the next scheduled RAC meeting. RAC members will review the submitted Appendix M and will notify the sponsor within 15 days of receiving the submission the outcome of the initial review and whether the protocol has been selected for public RAC review. A gene therapy clinical trial will be judged as exempt from, or in need of, full RAC/public discussion based on the following factors: Novelty of the vector or gene delivery system, special disease concerns, unique applications of the gene therapy research, or important social or ethical issues raised by the proposed research. Recommendations and comments resulting from discussion at the RAC meetings are forwarded to the investigator, the IRB, IBC, and the FDA.

Due to timing and regulatory requirements, there are situations where the FDA decision to allow a clinical trial to proceed must be made before public discussion can occur. In this situation, the investigator/sponsor is reminded by the FDA of the need to comply with NIH guidelines regarding DNA recombinant research and that enrollment of subjects into the clinical trial may not begin until review and, if selected, public discussion of the protocol by the RAC has taken place. In order to avoid this potential conflict, sponsors are encouraged to submit to NIH and if selected go through public discussion prior to submitting their IND to FDA. In general, public discussion of gene therapy clinical trials has been highly beneficial since it allows for consideration of societal and ethical issues surrounding the field of gene therapy and ensures the continued public acceptance and progression of the field.

51.2 IND PROCESS

As mandated under Section 505 of the FD&C Act and Section 351 of the PHSA, it is illegal to sell or distribute into interstate commerce any biologic unless it is licensed or under an IND Exemption (21 CFR 312.2). Submission of an IND to provide for this exemption allows clinical investigation of the product to proceed in order to determine safety, dosage, and effectiveness. This investigational process is usually divided into three phases, with each phase providing the next step in support of product licensure.

51.2.1 DEFINITION OF IND PHASES

As illustrated in Figure 51.3, product development begins with the pre-IND stage and progresses through the investigational or IND stage (phases I–III) where data are obtained to support product licensure. This is followed by the postlicensing stage (phase IV), during which postmarketing studies are often performed. An IND may be submitted for one or more phases of an investigation, although in general the phases of a clinical study are conducted sequentially. The FDA's primary objective when reviewing all phases of an IND is to assure the safety and rights of the patient, and later in phases II and III, to help assure that the validity and quality of the scientific data used to evaluate the product are adequate to assess product safety and efficacy.

A phase I trial includes the initial introduction of an IND into humans. The primary focus of a phase I study is to monitor product safety in a specific patient population although it should be noted that assessment of product safety remains a primary issue throughout product development. Phase I studies should be designed to determine the metabolic and pharmacologic actions of the drug in humans, the side effects associated with increasing doses and, if possible, to gain early evidence of product effectiveness. During the phase I study, the investigator should focus on obtaining sufficient information about the drug's pharmacokinetics and pharmacological effects that would permit for the design of well-controlled, scientifically valid, phase II studies. In a phase I study, the product may also be assessed for structure–activity relationships as well as the mechanism of action in humans. The total number of subjects included in a phase I gene therapy study varies but is typically between 10 and 40.

Phase II of an IND should include controlled clinical studies conducted to evaluate the effectiveness of the product for a particular indication in patients with the disease or condition under study. These studies should be designed to determine

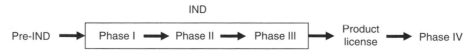

FIGURE 51.3 Phases of product development and approval.

common short-term side effects and risks associated with the biologic product. Phase II studies are typically well-controlled, closely monitored, and conducted in a relatively small number of patients, usually involving no more than several hundred patients.

Phase III studies are expanded controlled and uncontrolled trials that are performed after preliminary data for the effectiveness of the product have been obtained. These studies are intended to gather the additional information about product effectiveness and safety that is needed to evaluate the overall benefit–risk relationship of the product and to provide an adequate basis for physician labeling. Phase III studies usually include from several hundred to several thousand patients depending on the clinical indication and patient population.

Clinical evaluation of a biological product rarely ends with issuance of a biologics license. Phase IV or the postmarketing stage refers to the ongoing period of development after the product is licensed. Examples of postmarketing studies include clinical studies to extend claims or usage for the addition of a new patient population or indication, studies to demonstrate product comparability after manufacturing changes, and studies to validate surrogate clinical endpoints that are required in cases of expedited review and accelerated approval.

51.2.2   Pre-IND Phase

An IND application and the application process itself can be a challenge for novice sponsors. Therefore, before IND submission, CBER encourages an early interaction, in the form of a pre-IND meeting, to discuss preclinical animal testing, product development, and clinical trial plans. While the central focus of the pre-IND meeting is to define what is needed to support the IND submission, another important goal is to create a dialogue between CBER and the sponsor/investigator, which can be maintained throughout the process of product development. The pre-IND meeting is arranged at the request of the sponsor; however, before CBER can grant a pre-IND meeting, the sponsor will need to prepare a meeting package, which needs to be submitted at least 30 days prior to the meeting. Although the investigational plan does not have to be in its final form, the sponsor should be prepared to briefly describe all aspects of the proposed clinical study in the meeting package. This should encompass a description of the biological product, product manufacturing, and testing schemes, established preclinical data, preclinical studies performed and any plans for additional preclinical studies, the proposed clinical protocol and the general proof-of-concept behind the proposal. Most importantly, the pre-IND meeting should be used as a mechanism for focusing on unresolved issues relating to preclinical studies, clinical studies or product development. It is recommended that questions and objectives be submitted with the pre-IND package addressing any concerns or potential problems with the investigational plan that would require guidance or discussion with the agency before submission of the IND. Past experience has proven that early

identification and resolution of these issues will ultimately enhance and accelerate the product development process.

51.2.3   IND Phase

Before submitting an IND, the sponsor must have generated enough preclinical data to ensure the safety of the proposed clinical trial. Preclinical data can be generated using either *in vivo* animal studies or *in vitro* studies, or both, to assess the product's activity, efficacy, pharmacology, pharmacokinetics, and toxicity. The preclinical data should be adequate to support the proposed clinical trial with recommendation of initial safe dose and dose escalation scheme in humans. The schedule or duration of treatment and route of administration should mimic those planned for the clinical trial. There should be sufficient safety data available to determine endpoints for monitoring in the clinic.

The general requirements for safety evaluation for gene therapy products are similar to other biologics such as the pharmacologic profiles; "proof-of-concept" studies; dose–response relationship, and toxicology profile. In addition, issues specific for gene therapy that can be addressed with preclinical studies include biodistribution or trafficking of vectors, level and persistence of gene expression, and germline alteration. From the Chemistry, Manufacturing, and Controls (CMC) point of view, the sponsor should also have a well-developed and controlled product manufacturing scheme and have collected data regarding product characterization and consistency to support the manufacturing process. This data should also be used to support proposed specifications for product QC and release.

Once these issues have been addressed, the next step is the preparation of the IND submission. For assistance in preparation of an adequate and complete IND, the sponsor should contact the Office of Communication, Training and Manufacturers Assistance (OCTMA) to request an IND submission package. The package contains forms to be submitted, copies of the IND regulations, informed consent, and IRB regulations, information pertaining to good laboratory practices/good manufacturing practices (GLP/GMP), and the essentials required for conducting adequate and well-controlled clinical trials. An IND submission package can also be obtained at the FDA Web site:www.fda.gov/cber/ind/ind.htm

51.2.3.1   IND Submission

The content and format of an IND submission is specified in 21 CFR under part 312.23. This part lists, in order, the items that a sponsor (person who takes responsibility for, and conducts a clinical investigation) should submit in the IND (Figure 51.4). The IND should be submitted to CBER in triplicate and upon receipt the sponsor of the IND will be issued an acknowledgement letter containing the date of receipt, the assigned IND number and a reminder of their responsibility for submission to NIH, OBA according to Appendix M of the NIH Guidelines. The IND receipt date begins the official

Cover sheet—Form FDA 1572

Table of contents

Introductory statement and general investigational plan

Investigator's brochure
    Required if product is supplied to clinical investigators other than sponsor

Chemistry, manufacture, and control information
    Description of composition, manufacture, and purification of the investigational drug

    Substance and drug product

    Description of placebo

    Labeling

    Environmental analysis requirements

Pharmacology and toxicology information

IRB Approved consent form

Previous human experience with the investigational drug

Additional information

FIGURE 51.4 Content and format of the IND submission as specified in 21CFR 312.23.

review clock and the IND review will take place over the next 30 calendar days. INDs automatically become effective 30 days after receipt unless FDA notifies the sponsor that the IND is subject to clinical hold.

In CBER, where human gene therapy INDs are reviewed, the IND review team is composed of a product reviewer, a pharmacology/toxicology reviewer, a clinical reviewer, and a regulatory project manager who handles administrative aspects of the IND review. The product reviewer is responsible for coordinating the review team and ensuring consistency within a product area. Consult reviewers are used on an ad hoc basis, depending on the product and its application, and could come from other offices in CBER or other centers within the agency, depending on the required expertise.

At any point in the review process, a reviewer may call the sponsor to request additional information or to discuss deficiencies in the IND. After each reviewer has completed their review, the team meets to discuss the file and make a final decision on the status of the application and whether the clinical trial may proceed or will be placed on clinical hold. The IND may also be discussed at office level meetings for further input. The review decision is communicated to the sponsor by phone within the 30-day period. The reviewing office also issues a letter giving the details of hold issues (if any), review comments or requests for further information. If the file is placed on clinical hold the letter is issued 30 days from the decision date.

The criteria for placing INDs on clinical hold are addressed in 21 CFR 312. Phase I INDs may be placed on clinical hold, as covered under 21 CFR 312.42(b), if: human subjects are exposed to unreasonable and significant risk of illness or injury; the IND does not contain sufficient information to allow adequate assessment of the risk; the information in the investigators brochure is misleading, erroneous, or materially incomplete; or the clinical investigators are not qualified to conduct the study. In addition, for investigational studies of drugs intended to treat a life-threatening disease or condition that affects both men and women, phase I INDs may be placed on clinical hold if men or women of reproductive potential are excluded from the clinical trial because of reproductive risk, or the potential of reproductive risk, due to the reproductive or developmental toxicity associated with the drug. Phase II and phase III INDs may be put on clinical hold if any of the conditions cited above apply or if the protocol design is deficient to meet the objectives of the proposal.

In order to proceed with the clinical study the sponsor must correct the deficiencies identified during the review and submit the additional information or data in an amendment to the IND. Once the amendment containing the complete response to clinical hold has been submitted and reviewed (within 30 days of receipt date) by the FDA, the sponsor will be notified as to whether the hold issues have been adequately addressed and if the clinical trial may proceed. The review decision is communicated to the sponsor by phone within the 30-day period and a letter is issued.

51.2.3.1.1 Annual Reports

An annual report describing the progress of the investigation should be submitted to the IND within 60 days of the anniversary date that the IND went into effect. 21 CFR part 312.33 should be consulted for specific details, but in general the annual report should provide information on the status of each study in progress or completed during the previous year. To this effect the annual report should include an update on the following: the title of studies, the number of subjects enrolled and their status, serious adverse experiences observed, summary of available study results if the study is completed, information relevant to understanding of the drug's action, results of additional preclinical studies performed during the year, manufacturing changes, and an investigational plan for the coming year. The annual report should also provide results of product characterization and lot release testing for all lots of product produced during the year under report. For human gene therapy trials, reports should be submitted annually that include additional information relevant to gene therapy vectors such as assessment of evidence of gene expression, biological activity, immune response, evidence of gonadal distribution, results from long-term patient follow-up, and status of requests for autopsy. The update on information requested in the March 6, 2000 gene therapy letter (posted on the CBER Web site: www.fda.gov/cber/ind/ind.htm) may be provided in the annual report or be provided in a separate amendment.

51.2.3.2 Master File Submission

Another mechanism available for submission of information to the FDA is the Master File (MF). The procedures for submitting an MF are outlined in part 314.420 of 21 CFR. In contrast to the IND, which contains manufacturing, preclinical, and clinical information, the MF could contain product manufacturing or facilities information only. Submission of an MF allows the MF holder to incorporate the information by reference when the MF holder submits an IND. The MF holder may also authorize other sponsors to rely on the information to support an IND submission without disclosing the information to the IND sponsors. There are four types of MF submissions, of which Type II is the one most commonly used for biological product development. Information submitted in a Type II MF can include: information on drug substance or drug product specifically regarding product manufacturing and purification schemes, standard operating procedures (SOPs), lot release protocols, test methods and release acceptance criteria, descriptions of tissue culture media components, and other proprietary information needed to support an IND application. A Type V MF, the second most commonly used, can include detailed information about facilities, processes, or articles used in the manufacturing, processing, packaging, and storing of gene therapy products. The IND sponsor should obtain a letter from the MF holder granting written permission for a cross-reference, and a copy of the cross-reference letter should be submitted to the IND and should identify by name, reference number, volume, and page number the exact information that each IND sponsor is authorized to incorporate. A copy of this cross reference letter should also be submitted to the MF. For example, in the case where retroviral vector supernatant is provided to an IND holder by a manufacturer and the method of manufacture is proprietary, the manufacturer

could submit an MF documenting retroviral vector production and testing in support of the IND for review by CBER staff. As product lots are produced, lot release data should be submitted to the product MF as well as to the cross referenced IND for CBER review.

MFs are neither approved nor disapproved and are often reviewed only in the context of an IND application. CBER will comment on the contents of the MF and ask for clarifying information in order to more adequately review an IND, which cross-references the MF. Importantly, the MF has to contain complete information to support the decision to allow the associated IND to proceed. If there are safety concerns raised because of insufficient information provided in the MF, which is cross-referenced by an IND, then the IND may be put on clinical hold until all identified concerns are addressed by the MF sponsor.

51.3 REVIEW CONSIDERATIONS: EARLY PHASES

51.3.1 PHASED-IN APPROACH

Product characterization and QC of gene therapy products include issue and concerns common to most biological products, such as demonstration of safety, development of methods for assessment of potency, determination of identity and purity, as well as product stability. In addition, development of specifications for each of these parameters is an important part of product development and characterization. Specifications are the quality standards (i.e., tests, analytical procedures, and acceptance criteria) that confirm the quality of products and other materials used in the production of a product. Acceptance criteria are the numerical limits, ranges, or other criteria for the tests described. FDA believes that certain release specifications, such as those related to product safety, should be in place prior to initiating phase I clinical studies. As product development proceeds, additional specifications for product quality and manufacturing consistency should be developed and implemented. FDA recommends that proposed analytical procedures and release acceptance criteria for the final product be based on scientific data and manufacturing experience obtained during development of the product. During phase I, analytical procedures are usually based on the CFR methods or alternative methods, if appropriate. If an alternative to the CFR method is used, it is recommended that sponsors initiate validation of the alternative method by phase III. For phase I trials, release acceptance criteria are generally based on data obtained from lots used in preclinical studies. During phase II trials, these release acceptance criteria should be refined and tightened based on data generated during the phase I study.

Another aspect common among biologics is control and regulation of not only the product, but also each step of the manufacturing process. Adherence to this approach, which is encompassed under current good manufacturing practices (cGMPs) provides for quality and safety throughout the process and will lead to consistency of product lots.

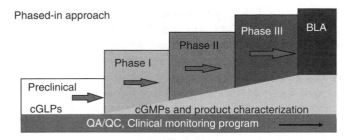

FIGURE 51.5 Phased-in approach.

In the area of gene therapy, a flexible approach to regulatory requirements has been attempted in order to find a balance between ensuring patient safety and fostering development of the field. In order to facilitate this process a phased-in approach to product characterization and compliance with cGMP has been adopted. This approach involves a progressive scale of requirements for product characterization and compliance with cGMPs, which increase as the study moves from phase I toward phase III (Figure 51.5). The first step in this approach, which occurs during phase I of the clinical trial but is required throughout all of product development, is the demonstration of product safety. As the clinical trial precedes so should the characterization of the product, such that by phase III, the product is fully characterized with regards to safety, purity, identity, and potency. In addition, by phase III, the product should be produced under full compliance with cGMP regulations although complete validation of all methods and processes are not required until licensure. Requirements for licensure of a biological product, product standards, and cGMPs are specified in Parts 610, 210, and 211 of Section 21 of the CFR. The phased-in approach applies to each aspect of gene therapy product manufacture, including vector development, establishment and characterization of cell banks, product manufacture, and characterization, establishment of release acceptance criteria for the final product, and control of reagents and components used during manufacture (ancillary products). It is equally important that a QC program, which is separate from manufacturing, be in place at the earliest phase of product manufacture. The QC unit should be responsible for ensuring the quality of the product and for product release for clinical use. The role of the QC unit is described in Part 211 of Section 21 of the CFR. Control of each of these aspects is important to assure product consistency and safety and is the focus of the CMC review.

51.3.2 CMC Review

Each step of product development, characterization, manufacture, and control should be described in the IND under the Chemistry Manufacturing and Controls section. Sufficient information is required in this section to assure the proper identification, quality, and safety of the investigational drug. For phase I trials, data must be submitted to establish product safety. In addition, information regarding product characterization should include a description of the product's physical, chemical, or biological characteristics, the method of manufacture, analytical methods used during manufacture and testing, and initial product specifications. The emphasis of the review of an IND submission is on data; therefore, in all cases data, as opposed to conceptual information, should be provided in support of the IND. Valid scientific principles should be applied throughout product development with regard to product safety, characterization, and QC of the manufacturing process. To ensure product quality, there should also be proper testing in place for all the components and materials used during product manufacture. During the review process, a case-by-case approach is applied in order to ensure that requirements and recommendations are satisfied in a manner appropriate to a particular product and manufacturing method. An example of some of the information that is considered during the review includes production and purification of the gene therapy vector, preparation of *ex vivo* gene-modified cells, and final formulation of the product. This information will allow the FDA reviewer to assess the identity, quality, purity, and potency of the product.

An early step in the development of a gene therapy product is construction of the initial gene therapy vector. The information supplied to the IND regarding the vector construct should include a description of the history and detailed derivation of the gene therapy vector, a description of vector components and their sources, including vector constructs used during generation of the final vector and a vector diagram. The vector diagram should include relevant restriction sites, gene insert, and regulatory elements, such as a promoter, enhancer, and polyadenylation signal. Currently the FDA requires that all gene therapy vectors less than 40 kb be fully sequenced prior to initiation of a phase I clinical trial. A fully annotated DNA sequence and sequence analysis report should be submitted, identifying all open reading frames (expected and unexpected) and genes encoded in the vector. The report should provide an evaluation of the significance of all discrepancies between the expected sequence found in a relevant current database and the investigational vector, and evaluation and significance of any unexpected sequence elements, including open reading frames. For vectors over 40 kb, any testing performed by restriction endonuclease analysis and sequence analysis of the gene insert, flanking regions (at least 500 base pairs either side) and any regions of the vector, which have been modified, should be submitted.

The next component important for the manufacture of a gene therapy product is a cell or cell bank system. Cells can be a component used during production of the vector product or *ex vivo* genetically modified and administered as final product. These cells can be either autologous or allogeneic. Information that should be submitted in the CMC section regarding the cells includes the cell source, collection method, whether a mobilization protocol is used, and donor screening and testing (if applicable). Requirements for screening and testing donors of human cells and tissues are described in 21 CFR Part 1271. If autologous cells are used, then it should be determined whether the donor tests positive for specific pathogens such as HIV or cytomegalovirus (CMV). It is important that any culture methods used to propagate the cells do not cause an increase in the level of pathogen. If allogeneic cells are used these cells should be screened and tested for adventitious agents such as HIV-1, HIV-2, hepatitis B virus (HBV, surface and core antigen), hepatitis C (HCV), human T-lymphotropic virus types 1 and 2 (HTLV-1, HTLV-2), CJD (screening only), and CMV. Further information concerning donor screening and testing can be found in "Guidance for Industry: Eligibility Determination for Donors of Human Cells, Tissues, and Cellular and Tissue-Based Products (HCT/Ps) 2007."

If a cell bank system is utilized to assure consistency of production then the next step in product development is usually the establishment of a master cell bank (MCB). MCB characterization includes testing to establish the properties and stability of the cells and is performed on a single-time basis. For the IND submission, information on the history of the cell line, as well as culture and storage conditions of the MCB should be included. Testing should be performed to establish identity, which includes cellular phenotype and genotype, cellular isoenzyme expression, and stability. If the cell line has been engineered to express exogenous genes, then tests should be performed to establish the presence and stability of the transgene. Safety testing should be performed to demonstrate the MCB is sterile and free of mycoplasma, endotoxin, and adventitious agents (through *in vitro* and *in vivo* assays). Depending on the origin, MCBs should be tested for species-specific pathogens. For example, if the MCB is of human origin then testing for human pathogens such as CMV, HIV-1 and 2, HTLV-1 and 2, EBV, B19, AAV, HBV, and HCV should be performed. Analytical methods used for testing should be summarized and initial release acceptance criteria for qualification of the MCB established. The MCB is usually the first tier of a two-tier cell bank system; the second is the working cell bank (WCB), which is derived from one or more vials of the MCB. The amount of information needed for characterization of the WCB is generally less extensive and includes testing for sterility, mycoplasma, and adventitious agents by *in vitro* assay and identity. Additional information on cell banking procedures and cell bank characterization can be found in the "Points to Consider in the Characterization of Cell Lines to Produce Biologicals 1993."

Depending on the gene therapy vector, the next step in gene therapy product development may be the establishment of a master viral bank (MVB). MVBs are generated by infecting or transducing the specific vector into cells grown from one or more vials of banked cells. As with MCBs, a second-tier or MWVB can also be generated. Requirements for MCB/WCB testing and characterization would also apply to testing of the MVB/MWVB, with the addition of tests for the presence of RCV related to the vector.

All procedures used during production and purification of the gene therapy vector product should be included in the CMC section of the IND submission. Safety testing and characterization of gene therapy vectors should be performed on bulk production lots, before and after purification steps and final product after formulation and fill, as appropriate for the particular assay. Safety testing should include tests for sterility, mycoplasma, endotoxin, and RCV, as well as tests for adventitious virus by *in vitro* assay. Testing for general safety is required for licensure of all gene therapy vector products unless the product is exempt under 21 CFR 610.11(g). Product identity for a gene therapy vector product may be established through the physical or chemical characteristics of the product using *in vitro*, *in vivo*, or when appropriate, molecular methods. If the final product is an *ex vivo* gene modified cell product, then the final cell product also needs to be tested for identity. Although it is recommended to initiate identity testing early in product development it is not absolutely required at phase I. Purity of a gene therapy vector is typically established through determination of levels of residual materials such as cellular DNA, RNA, and protein, and ancillary products. With regard to product potency, for early phase trials the level of gene expression is acceptable as a measure of potency. However, before initiation of a phase III clinical trial, a potency assay will need to be developed and in place that measures the relative biological activity of the product. As with testing of the MCB and MVB, analytical methods used should be summarized in the IND and initial release acceptance criteria for qualification of the gene therapy vector product established. This summary is often presented in the form of a Certificate of Analysis.

To date approximately 40% of gene therapy clinical trials have involved the genetic modification of autologous or allogeneic cells *ex vivo*. This has been accomplished using viral and nonviral vectors. Information regarding the process of *ex vivo* modification of cells should be documented in the IND submission as follows: a description of the source of cells, results of donor screening (if applicable), method of cell collection and processing, culture conditions, and the procedure for *ex vivo* modification of cells. Safety testing and characterization of the *ex vivo* gene modified cells should include assessment of sterility, mycoplasma, endotoxin, identity, potency, and freedom from RCV. Analysis of phenotypic markers as well as confirmation of the integrity of the genetic insert may be used to confirm cell identity. Care should also be taken to assure patient specificity for autologous products using proper labeling and tracking systems. Additional parameters to be assessed include transduction efficiency, longevity of gene expression, and cell viability. If the cells are irradiated or frozen and thawed prior to use, then the effect of these parameters on cell viability will need to be assessed. In addition, the vector used for *ex vivo* modification should be tested as described above before use in the *ex vivo* modification of the cell product. Analytical methods used should be summarized in the IND and initial release acceptance criteria for qualification of the *ex vivo* modified cells established.

For gene therapy vector products and *ex vivo* gene modified cell products, a stability program that assesses product safety, activity, and integrity should be established during early phases of investigation. The objectives of stability testing during early phases are to establish that the product is stable for the duration of the clinical trial and to collect information needed to develop a final formulation and dating period. The submission to the IND should include a brief description of the proposed stability study, the test methods to be used to monitor stability, and preliminary data if available.

The phased-in approach provides flexibility to product manufacture, characterization and testing during early phases of product development. However, testing and assay qualification should be performed in early phases to support data collection, development of release acceptance criteria, and compliance with cGMP in preparation for phase III studies.

Early in product development there should be adequate controls in place for monitoring the manufacturing process. It is recommended that prior to the production of clinical product; there should be the establishment and implementation of written procedures that are well-defined. These should include the responsibilities and procedures applicable to the QC unit, such as responsibility for reagent quality, review, and approval of production procedures, testing procedures, and release acceptance criteria, and responsibility for releasing or rejecting clinical lots based on review of production records. There should be a summary of the QC plan in place to prevent, detect, and correct deficiencies that may compromise product integrity or function or that may lead to the possible transmission of adventitious agents.

51.3.3 REAGENTS AND MATERIALS

The manufacturing process for gene therapy products entails the use of reagents and source materials of different quality, variability, and risk for the introduction of adventitious agents. To ensure that patients receive a safe, consistent, and potent product, it is important to have a qualification program in place for reagents and source materials used in manufacturing. Reagents and materials are those components that are essential for cellular growth, differentiation, selection, purification, or other critical manufacturing steps, but that are not intended to be part of the final product. Examples include fetal bovine serum, trypsin, growth factors, cytokines, monoclonal antibodies, cell separation devices, medium and medium components. These reagents can affect the safety, potency, and purity of the final product, especially with regard to the introduction of adventitious agents.

Ideally, licensed or clinical grade reagents should be used for preparation of gene therapy products; however, these are often not readily available. Recommendations for the use of reagents that are not clinical grade during early phases of product development involve establishing a qualification program and release acceptance criteria for each reagent. The qualification program should include adequate characterization, including safety testing, functional analysis, and a demonstration of purity. The purity of a given reagent need not be 100%, but the purity profile between different lots of the reagent should remain consistent. The extent of reagent testing required will depend on the point at which the reagent is used in the manufacturing process and the biological system used for production of the reagent. For example, in the case of

a growth factor or cytokine, if a murine monoclonal antibody is used during production or purification then the reagent will have to be tested for adventitious agents and murine retrovirus. If there are known toxicities associated with the reagent, testing for residual levels of the reagent in the final product preparation should be performed. It should be noted that the use of reagents produced under full cGMP is recommended for phase III trials. If human albumin is used, there should be a procedure in place to ensure that no recalled lots were used during manufacture or preparation of the product. If using human AB serum, it is important to ensure serum is obtained from an approved blood bank and meets all blood donor criteria. For all other reagents that are human-derived information should be provided to ascertain whether it is a licensed product, clinical, or research grade, and provide a Certificate of Analysis or information regarding testing of the donor or reagent. If porcine products are used then a Certificate of Analysis or other documentation that the products are free of porcine parvovirus should be included in the IND. If a reagent is derived from bovine material, it should be sourced from countries free of BSE. The following information should be included in the IND submission: the identity of the bovine material, the source of the material, information on the location where the herd was born, raised, and slaughtered, and any other information relevant to the likelihood that the animal may have ingested animal feed prohibited under 21 CFR 589.2000. Bovine material may be introduced at different points in production of a reagent, and the information described above should be provided for all bovine materials used. For more information, see "Proposed Rule: Use of Materials Derived from Cattle in Medical Products Intended for Use in Humans and Drugs Intended for Use in Ruminants," Federal Register 72(8) 1581–1619, January 12, 2007. In addition, a Certificate of Analysis should be provided in the IND to document that bovine materials are free of bovine viruses listed in 9 CFR 113.53.

51.3.4 GENE THERAPY–SPECIFIC SAFETY ISSUES

As the development of gene therapy vectors for therapeutic gene delivery broadens and moves forward, the possibility of approved therapeutic products becomes more of a reality. However, the use of these therapeutic products in clinical trials and regulatory approval requires a good safety profile. The use of gene therapy vectors raises certain safety issues that are specific to each vector type. Concerns specific to the safe use of gene therapy vectors include the generation of RCV, vector-associated toxicity (as opposed to toxicity related to the specific transgene), viral shedding, and the risk of inadvertent modification of a recipient's germline. Because of their potential effect on the patient, the public health and future generations, these issues need to be addressed during development of a gene therapy product and should be documented in the IND. Since RCV can arise from recombination events during manufacture, tests designed to detect RCV should be performed at multiple stages of manufacture and for each final lot.

Testing for replication competent retrovirus (RCR) in the production of gammaretroviral and lentiviral vectors needs to be performed at multiple points in production including MVB, WVB, vector supernatants, end of production cells, and *ex vivo* modified cells. A permissive cell line such as *Mus dunni* should be used to test cells producing vector containing the amphotropic murine leukemia virus envelope. If an ecotropic packaging cell line is utilized, then an ecotropic gammaretroviral assay, for the detection of low-level viral contamination, will need to be performed.

Retroviral vector supernatant also must be tested by amplification on a permissive cell line such as *Mus dunni*, followed by detection in an appropriate indicator cell assay (PG-4 S + L–). Based on research reports, it has been demonstrated that detection of RCR/RCL in *ex vivo* modified cells requires cells to be cultured for at least 4 days posttransduction. Therefore, *ex vivo* gene modified cells that are cultured for 4 or more days would require RCR testing as a product release assay. If *ex vivo* gene modified cells are cultured for less than 4 days, then archiving of cells is recommended in place of active RCR testing. Further information regarding RCR testing is contained in the CBER guidance "Guidance for Industry: Supplemental Guidance on Testing for Replication Competent Retrovirus in Retroviral Vector Based Gene Therapy Products and During Follow-up of Patients in Clinical Trials Using Retroviral Vectors, 11/28/2006." In studies using adenoviral vectors, MVBs and WVBs, as well as the final vector product, must be tested for replication competent adenovirus (RCA). The FDA recommends that the level of RCA in the final product be less than 1 in 3×10^{10} viral particles. Adenoviral vectors pose a specific safety risk since toxicity has been observed when these vectors are administered at high doses. Research data suggest that a portion of the observed toxicity is due to the vector particle itself and is not related to the actual infection of the cells by the vector. As a result, CBER requires that adenoviral vector doses be based on the actual particle count as opposed to the infectious titer, which is commonly the unit for quantifying viruses. In addition, the agency asks that all sponsors of INDs using adenoviral vectors report the ratio of viral particle to infectious unit for each clinical lot of vector. The current recommended limit for this ratio is set at ≤30.

After the death in 1999 of a subject enrolled in an adenoviral gene therapy clinical trial the NIH, OBA established the adenoviral Safety and Toxicity Working Group (AdSAT), whose mission was to conduct an in-depth review and evaluation of safety and toxicity data generated from adenovirus gene therapy clinical trials. One outcome of this working group was a recommendation that a qualitative and quantitative adenovirus standard be developed. In response to this recommendation, the FDA, Williamsburg BioProcessing Foundation, and members from industry and academia formed a working group to develop and characterize an adenoviral reference material (ARM). It is FDA's recommendation that the ARM be used by manufacturers of adenoviral vectors to standardize viral particle and titer measurements. In addition, use of the ARM will provide a means of analyzing the safety and efficacy of adenoviral vectors produced by different manufactures and allow for comparison between different clinical trials.

Safety testing goes beyond testing of the product, and should include monitoring programs for all subjects administered gene therapy vectors. When replication-competent viral vectors are used in a clinical trial, it is important to demonstrate that there is no shedding of virus to health care professionals or family members. One way to ensure that replication-competent virus are not shed to third-party individuals is to monitor patients for viral shedding by analyzing samples such as urine or throat swabs. Montoring should be continued until the presence of infectious virus is no longer detectable in patients' samples.

For *in vivo* administered gene therapy vectors, data must be submitted demonstrating the extent to which a vector is able to disseminate out of the injection site and distribute to the gonads. This data will provide the information required to assess the risk of inadvertent gene transfer to germ cells, which may result in genetic changes in subsequent progeny. In general, these data should be obtained during the course of product development and provided to the agency for review. However, in cases where a novel vector, route of administration, or vector delivery system is proposed, preclinical studies may be required before initiation of the phase I study. Biodistribution data may be obtained either from studies in animals, analysis of clinical samples or from a combination of preclinical and clinical sample analyses. Clinical data demonstrating vector biodistribution should be derived from peripheral blood cells and semen samples taken during the treatment and follow-up periods for the clinical trial and from gonadal tissues obtained when possible at autopsy from consenting patients. The agency should be updated of the status of these studies at least at the time of each annual report in order to guide further product development and track any potential adverse events. A statement explaining the risk for genetic alteration of sperm or eggs and possible outcome for a fetus and future child, which could occur as a result of study participation, should be included in the informed consent document. This statement should clearly explain the status of biodistribution studies to assess gonadal distribution for each particular clinical trial and the fact that the likelihood of an adverse outcome is currently unknown. In addition, the consent form should advise study participants to practice birth control for a suitable period of time.

A primary safety concern when using gammaretrovial or lentiviral vectors to transduce human hematopoietic stem cells/progenitor cells is the risk of oncogenesis. This risk can occur by: (1) unintended generation of replication competent retrovirus/lentivirus (RCR/RCL) leading to neoplasia due to RCR/RCL infection of the target or nontarget cells *in vivo* or (2) vector integration into chromosomal DNA leading to neoplasia due to insertional mutagenesis. These risks should be addressed in nonclinical and clinical studies. This theoretical risk has recently become an actual risk by the report of four children, enrolled in a French trial for treatment of X-SCID, who developed leukemia after administration of a gammaretrovial vector.

The FDA has developed a long-term follow-up program that will aid in the monitoring of all subjects receiving gene therapy products for unanticipated adverse events that manifest long after administration of the gene therapy product and after the observation period for acute toxicity has ended. FDA issued a guidance for industry "Gene Therapy Clinical Trials—Observing Participants for Delayed Adverse Events" 2007, which provides recommendations regarding the design of studies to include the collection of data on delayed adverse events in participants who have been exposed to investigational gene therapy products. The guidance provides recommendations on the following: (1) methods to assess the risk of gene therapy–related delayed adverse events following exposure to gene therapy products; (2) guidance for determining the likelihood that long-term follow-up observations on study participants will provide scientifically meaningful information; and (3) specific advice regarding the duration and design of long-term follow-up observations.

Factors likely to increase the risk of delayed adverse events following exposure to gene therapy products include persistence of the viral vector, integration of genetic material into the host genome, prolonged expression of the transgene, and altered expression of the host's genes. Persistence of the viral vector, sometimes associated with latency, could permit continued expression of the gene or delayed effects of the viral infection. Integration of genetic material from a viral vector into the host cell genomic DNA raises the risk of malignant transformation. Prolonged expression of the transgene may also be associated with long-term risks resulting from unregulated cell growth and malignant transformation, autoimmune-like reaction to self antigens, and unpredictable adverse events. Altered expression of the host genes could also result in unpredictable and undesirable biologic events. One way to determine persistence of a gene therapy vector is to conduct a biodistrubution study in an appropriate animal model over multiple time points. FDA would consider vector persistence to be any vector that can be detected throughout all of the time points tested in the study without any downward trend over several of the measured time points.

51.4 REVIEW CONSIDERATIONS: LATER PHASES

The previous sections have given an overview of the review process, and CBER's expectations for product development as the clinical investigation proceeds through early phases. To better understand CBER's requirements for product licensure and establishment standards, a sponsor should be familiar with practices that are governed by law and specified in Parts 600, 601, 606, 610, 210, and 211 of Section 21 of the CFR. Part 600, 601, 606, and 610 provide information pertaining to product release requirements and general provisions for licensure and in some cases will provide information on specific assay methodology and Parts 210 and 211 explain the cGMPs.

Although the regulations by their nature prescribe requirements, there is a level of flexibility built into the regulations that allow for modification of required test methods or manufacturing processes. The provision for test method modification is found under 21 CFR 610.9, entitled "Equivalent Methods and Processes." This provision is especially useful for gene therapy products, where conditions of the manufacturing process and many times the product itself make it difficult to perform standard assays. To apply this provision, a sponsor should provide supporting evidence for why a specific method is not ideal for the gene therapy application and present data, with appropriate controls, to demonstrate that the modified method will provide assurances of the safety, purity, potency, and effectiveness of the product equal to or greater than the assurance provided by the method or process specified in the general standards or additional standards (21CFR 610.9). Data to support the modification can be accumulated during the early IND phases and then validated in support of its use as an established method prior to licensure.

For a product to meet the requirements for licensure it must be fully characterized prior to submission of the license application with regard to safety, purity, potency, and identity. Each of these aspects will be discussed in the context of requirements for initiation of a phase III trial and product licensure.

Demonstration of product safety requires the implementation of specific tests, which measure sterility, mycoplasma, endotoxin, and general safety of the product. Product safety also includes demonstration that the product is free from adventitious virus, which to support the phase III study and product licensure includes *in vitro* and *in vivo* adventitious virus testing of the MCB, MVB, and final product, and when appropriate testing for RCV. Testing for general safety is not as rigidly adhered to during phases I and II studies but by phase III, the required standards and methods specified in 21 CFR 610.11 should be in place. An exception from testing for general safety is made for cellular therapy products [21CFR 610.11(g)(1)] and a sponsor may request for exception from testing for general safety for certain types of gene therapy products, such as plasmid DNA products, adenovirus vector products or *ex vivo* genetically modified cells [21CFR 610.11(g)(2)] if the purity of these products can be demonstrated through extensive testing and characterization.

Tests for potency of a gene therapy product should consist of either *in vitro* or *in vivo* tests, or both, which have been specifically designed for each product [21 CFR 610.10]. Tests for potency should indicate the specific ability or capacity of the product to effect a given result. Acceptability of potency assays will be determined on a case-by-case basis. However, based on current regulatory statutes, they must meet certain criteria such as: indicate biological activity(s) specific to the product; results must be available for lot release; provide quantitative readout; meet predefined acceptance and/or rejection criteria; include appropriate reference material and/or controls; be validated for licensure; measure activity of all active components; and be stability indicating. A suitable potency assay should be one that is quantitative in nature and measures an appropriate biological activity. If development of a quantitative biological assay is not possible, then a quantitative physical assay, which correlates with and is used in conjunction with a qualitative biological assay, can be used.

Product identity can be demonstrated through the use of an assay or assays that are specific for each product in a manner that will adequately identify it as the designated product and distinguish it from any other product being processed in the same facility. At phase I, a test for identity is not required but it should be in place by the start of phase II, so that data can be collected, specifications determined, and the assay validated by licensure.

Product purity is defined as freedom from extraneous material except that unavoidable in the manufacturing process. Testing for purity of a gene therapy product could involve assays for residual protein, DNA, RNA, solvents used during production and purification, or ancillary products used during manufacture such as cytokines, antibodies, or serum. As with all of the previously mentioned testing, a quantitative assay for purity should be in place by the start of phase III, with development of the assay being initiated much earlier in the investigational trial.

To complete product characterization in support of phase III and product licensure, a stability testing program that includes, but is not limited to, assays that measure product integrity, potency, sterility and, in the case of an *ex vivo* gene modified product, viability is needed. A stability protocol for study of the bulk and final drug products, as well as all MCBs and MVBs, should be defined so that stability data generated during phase III will be appropriate to support licensure. In general, the stability program should be initiated at phase I so that by phase II, the objective of obtaining real-time data to support stability of the investigational formulation can be met. For phase III, data collected should be used to support the proposed expiration-dating period of the final drug product, as well as the container and closure system. The determination of which tests should be included in the stability program will be product-specific. In addition, stability testing of any product intermediates should be in place to support the validation of the duration and conditions of storage of the bulk product.

51.4.1 Current Good Manufacturing Practices

cGMP is defined as a set of current, scientifically sound methods, practices or principles that are implemented and documented during product development and production to ensure consistent manufacture of safe, pure, and potent products (§501(a)(2)(B)). Some major elements of cGMP include detailed record keeping, development of written procedures or SOPs, institution of QC program and analytical assays, and equipment and process validation. cGMP also requires a program be in place for the certification and training of personnel and for environmental monitoring. The agency believes that applying QC principles to the production of investigational products (i.e., interpreting and implementing cGMPs consistent with good scientific methodology) will facilitate the initiation of investigational trials in humans and protect trial subjects. cGMPs play an important role in control and regulation of not only the product but all steps of the manufacturing

process. Adherence to cGMPs provides for quality and safety throughout the process and will lead to reproducible and consistent performance of product lots. It should be noted that cGMPs apply to the manufacturing process as well as to the facilities. As product development proceeds toward phase III so should the validation of the conditions under which the product is manufactured, controlled, and characterized. Full adherence to cGMP regulations is expected to develop as the clinical trial advances. For example, process validation, methods validation, in-process testing, and establishment of specifications should be incorporated as part of a phased-in approach to fulfilling regulatory requirements.

51.4.2 Product Comparability

Changes in the manufacturing process, equipment, or facilities often occur during product development and can result in change in the biological product itself. A manufacturer must fully describe any change to a biological product in the IND or license application, regardless of whether the change occurs prior to or after product approval. The manufacturing change should be assessed and the resulting product compared to the existing product to assure that the change does not alter the safety, purity, potency, or integrity of the final product. Determinations of product comparability may be based on a combination of *in vitro* or *in vivo* studies ranging from chemical, physical, and biological assays, assessment of pharmacokinetics or pharmacodynamics and toxicity in animals to clinical testing. The type of study required would depend on the extent of the change and the phase of clinical development in which the change occurs. Product comparability should be demonstrated through side-by-side analyses of the product lots manufactured under the established procedures and qualification lots of the product manufactured by the new procedure. If a sponsor can demonstrate comparability with nonclinical data, additional clinical safety or efficacy trials with the new product generally will not be needed. FDA will determine if comparability data are sufficient to demonstrate that additional clinical studies are unnecessary. Example of changes that would require a comparability study includes any change in the manufacturing scheme or site, changes to the MCB, modification of the vector product, a change in fermentation, isolation or purification, change in storage container, or product formulation. The final section of this chapter contains references that provide additional guidance on the demonstration of product comparability.

Gene therapy is continuing to evolve with the promise of new products on the horizon for treatment of diseases such as cancer, hemophilia, and many rare diseases. CBER is aware of this promise as well as the public's concern about the safety of human gene therapy studies. CBER is working with NIH, academia, and industry to help ensure that gene therapy products are safe while promoting the further development of this promising area of clinical studies. In particular CBER continues to work with the American Society of Gene Therapy (ASGT), individual sponsors, and the scientific community when we develop new policy and guidance.

51.5 GUIDANCE DOCUMENTS AND OTHER REFERENCES

CBER Documents Relevant to Gene therapy.

All current and past regulatory documents can be obtained from the CBER Web site: CBER_info@CBER.FDA.Gov. or www.fda.gov/cber/publications.

Application of Current Statutory Authorities to Cell and Gene therapy Products, Federal Register/ Vol. 58, No. 1977 /Oct. 14, 1993.

Guidance for Human Somatic Cell Therapy and Gene therapy. CBER, FDA, March 1998.

Points to Consider in the Characterization of Cell Lines to Produce Biologicals, CBER, FDA, 1993.

Guidance for Industry Submitting Type V Drug Master Files to the Center for Biologics Evaluation and Research, CBER, August 2001.

Draft Guidance for FDA Review Staff and Sponsors: Content and Review of Chemistry, Manufacturing, and Control (CMC) Information for Human Gene Therapy Investigational New Drug Applications (INDs), CBER, November 2004.

Points to Consider in the Production and Testing of New Drugs and Biologicals Produced by Recombinant DNA Technology, CBER, FDA, 1985 and Supplement: Nucleic Acid Characterization and Genetic Stability, CBER FDA, 1992.

Points to Consider in the Manufacture and Testing of Monoclonal Antibody Products for Human Use, CBER, FDA, (1997).

FDA Guidance Concerning Demonstration of Comparability of Human Biological Product, Including Therapeutic Biotechnology-derived Products, CBER, FDA, 1996.

ICH Guideline Q5C. Quality of Biotechnological Products: Stability Testing of Biotechnological/Biological Products, November 1995.

Revisions to the General Safety Requirements for Biological Products; Companion Document to Direct Final Rule, Federal Register/ Vol. 63, No. 75/ April 20, 1998.

Draft Guidance for Industry: INDs for Phase 2 and 3 Studies of Drugs, Including Specified Therapeutic Biotechnology-Derived Products, Chemistry Manufacturing and Controls Content and Format April 20, 1999.

Guidelines for Research Involving Recombinant DNA Molecules (NIH Guidelines), April 1998.

Guideline on Validation of the Limulus Amebocyte Lysate test as an End-Product Endotoxin Test for Human and Animal Parenteral Drugs, Biological Products and Medical Devices, 1987.

International Conference on Harmonization; Guidance on Viral Safety Evaluation of Biotechnology Products Derived From Cell Lines of Human or Animal Origin, Federal Register, Sept. 24, 1998, Vol. 63, Number 185.

ICH document Q5D, Derivation and Characterization of Cell Substrates Used for Production of Biotechnological/Biological Products.

ICH Harmonized Tripartite Guidelines—"Impurities: Guidance for residual Solvents" and "Impurities in New Drug Substances" and "Impurities in New Drug Products."

ICH Q5C: Quality of Biotechnological Products: Stability Testing of Biotechnological/Biological Products (November 1995).

Guidelines for Research Involving Recombinant DNA Molecules (NIH Guidelines), April 1998.

Final Rule: Eligibility Determination for Donors of Human Cells, Tissues, and Cellular and Tissue-Based Products, (69 FR 29786, May 25, 2004).

Guidance for Industry: Eligibility Determination for Donors of Human Cells, Tissues, and Cellular and Tissue-Based Products (HCT/Ps), dated February 2007.

Revisions to the General Safety Requirements for Biological Products, 68 FR 10157, March 4, 2003.

Proposed Rule: Use of Materials Derived from Cattle in Medical Products Intended for Use in Humans and Drugs Intended for Use in Ruminants, Federal Register 72(8) 1581–1619, January 12, 2007.

Guidance for Industry: Supplemental Guidance on Testing for Replication Competent Retrovirus in Retroviral Vector Based Gene Therapy Products and During Follow-up of Patients in Clinical Trials Using Retroviral Vectors—11/28/2006.

Guidance for Industry: Gene therapy Clinical Trials—Observing Subjects for Delayed Adverse Events, 11/28/2006.

CBER SOPP 8101.1: Scheduling and Conduct of Regulatory Review Meetings with Sponsors and Applicants, May 18, 2007.

Guidance for Industry: INDs—Approaches to Complying with CGMP during Phase 1, January 2006.

ACKNOWLEDGMENTS

We would like to thank Drs Kimberly Benton, Daniel Takefman, and Celia Witten from OCTGT, CBER for critical reading, advice, and editorial suggestions.

ACRONYMS

ARM	adenoviral reference material
CBER	Center for Biologics Evaluation and Research
CFR	Code of Federal Regulations
cGMP	Current good manufacturing practices
CMC	Chemistry Manufacturing and Controls
FDA	Food and Drug Administration
FD&C Act	Food Drug and Cosmetic Act
GLP	good laboratory practices
GTPC	Gene Therapy Policy Conferences
IBC	Institutional Biosafety Committee
IRB	Institutional Review Board
kb	kilobases
MCB	master cell bank
MF	master file
MVB	master viral bank
NIH	National Institutes of Health
OCTMA	Office of Communication Training and Manufacture Assistance
OBA	Office of Biotechnology Activities
PHSA	Public Health Service Act
PTC	points to consider
RAC	Recombinant DNA Advisory Committee
RCA	replication competent adenovirus
RCV	replication competent virus
RCR	replication competent retrovirus
SOP	standard operating procedures
WCB	working cell bank

Index

A